TABLE OF ATOMIC WEIGHTS AND NUMBERS

Based on the 1985 Report of the Commission on Atomic Weights of the International Union of Pure and Applied Chemistry and for the elements as they exist naturally on earth. Scaled to the relative atomic mass of carbon-12. The estimated uncertainties in values, between ± 1 and ± 9 units in the last digit of an atomic weight, are in parentheses after the atomic weight. (From *Pure and Applied Chemistry*, Vol. 58 (1986), pp. 1677–1692. Copyright © 1986 IUPAC)

Element	Symbol	Atomic Number	Atomic Weight		Element	Symbol	Atomic Number	Atomic Weight	
Actinium	Ac	89	227.0278	(L)	Neodymium	Nd	60	144.24(3)	(g)
Aluminum	Al	13	26.981539(5)		Neon	Ne	10	20.1797(6)	(g, m)
Americium	Am	95	243.0614	(L)	Neptunium	Np	93	237.0482	(L)
Antimony	Sb	51	121.75(3)		Nickel	Ni	28	58.69(1)	
Argon	Ar	18	39.948(1)	(g, r)	Niobium	Nb	41	92.90638(2)	
Arsenic	As	33	74.92159(2)		Nitrogen	N	7	14.00674(7)	(g, r)
Astatine	At	85	209.9871	(L)	Nobelium	No	102	259.1009	(L)
Barium	Ba	56	137.327(7)		Osmium	Os	76	190.2(1)	(g)
Berkelium	Bk	97	247.0703	(L)	Oxygen	O	8	15.9994(3)	(g, r)
Beryllium	Be	4	9.012182(3)		Palladium	Pd	46	106.42(1)	(g)
Bismuth	Bi	83	208.98037(3)		Phosphorus	P	15	30.973762(4)	
Boron	B	5	10.811(5)	(g, m, r)	Platinum	Pt	78	195.08(3)	
Bromine	Br	35	79.904(1)		Plutonium	Pu	94	244.0642	(L)
Cadmium	Cd	48	112.411(8)	(g)	Polonium	Po	84	208.9824	(L)
Calcium	Ca	20	40.078(4)	(g)	Potassium	K	19	39.0983(1)	
Californium	Cf	98	242.0587	(L)	Praseodymium	Pr	59	140.90765(3)	
Carbon	C	6	12.011(1)	(r)	Promethium	Pm	61	144.9127	(L)
Cerium	Ce	58	140.115(4)	(g)	Protactinium	Pa	91	231.03588(2)	(Z)
Cesium	Cs	55	132.90543(5)		Radium	Ra	88	226.0254	(L)
Chlorine	Cl	17	35.4527(9)		Radon	Rn	86	222.0176	(L)
Chromium	Cr	24	51.9961(6)		Rhenium	Re	75	186.207(1)	
Cobalt	Co	27	58.93320(1)		Rhodium	Rh	45	102.90550(3)	
Copper	Cu	29	63.546(3)	(r)	Rubidium	Rb	37	85.4678(3)	(g)
Curium	Cm	96	247.0703	(L)	Ruthenium	Ru	44	101.07(2)	(g)
Dysprosium	Dy	66	162.50(3)	(g)	Samarium	Sm	62	150.36(3)	(g)
Einsteinium	Es	99	252.083	(L)	Scandium	Sc	21	44.955910(9)	
Erbium	Er	68	167.26(3)	(g)	Selenium	Se	34	78.96(3)	
Europium	Eu	63	151.965(9)	(g)	Silicon	Si	14	28.0855(3)	
Fermium	Fm	100	257.0951	(L)	Silver	Ag	47	107.8682(2)	(g)
Fluorine	F	9	18.9984032(9)		Sodium	Na	11	22.989768(6)	
Francium	Fr	87	223.0197	(L)	Strontium	Sr	38	87.62(1)	(g, r)
Gadolinium	Gd	64	157.25(3)	(g)	Sulfur	S	16	32.066(6)	(r)
Gallium	Ga	31	69.723(4)		Tantalum	Ta	73	180.9479(1)	
Germanium	Ge	32	72.61(2)		Technetium	Tc	43	98.9072	(L)
Gold	Au	79	196.96654(3)		Tellurium	Te	52	127.60(3)	(g)
Hafnium	Hf	72	178.49(2)		Terbium	Tb	65	158.92534(3)	
Helium	He	2	4.002602(2)	(g, r)	Thallium	Tl	81	204.3833(2)	
Holmium	Ho	67	164.93032(3)		Thorium	Th	90	232.0381(1)	(g, r, Z)
Hydrogen	H	1	1.00794(7)	(g, m, r)	Thulium	Tm	69	168.93421(3)	
Indium	In	49	114.82(1)		Tin	Sn	50	118.710(7)	(g)
Iodine	I	53	126.90447(3)		Titanium	Ti	22	47.88(3)	
Iridium	Ir	77	192.22(3)		Tungsten	W	74	183.85(3)	
Iron	Fe	26	55.847(3)		(Unnilhexium)	(Unh)	106	263.118	(L, n)
Krypton	Kr	36	83.80(1)	(g, m)	(Unnilpentium)	(Unp)	105	262.114	(L, n)
Lanthanum	La	57	138.9055(2)	(g)	(Unnilquadium)	(Unq)	104	266.11	(L, n)
Lawrencium	Lr	103	260.105	(L)	(Unnilseptium)	(Uns)	107	262.12	(L, n)
Lead	Pb	82	207.2(1)	(g, r)	Uranium	U	92	238.0289(1)	(g, m, Z)
Lithium	Li	3	6.941(2)	(g, m, r)	Vanadium	V	23	50.9415(1)	
Lutetium	Lu	71	174.967(1)	(g)	Xenon	Xe	54	131.29(2)	(g, m)
Magnesium	Mg	12	24.3050(6)		Ytterbium	Yb	70	173.04(3)	(g)
Manganese	Mn	25	54.93805(1)		Yttrium	Y	39	88.90585(2)	
Mendelevium	Md	101	258.10	(L)	Zinc	Zn	30	65.39(2)	
Mercury	Hg	80	200.59(3)		Zirconium	Zr	40	91.224(2)	(g)
Molybdenum	Mo	42	95.94(1)						

(g) Geologically exceptional specimens of this element are known that have different isotopic compositions. For such samples, the atomic weight given here may not apply as precisely as indicated.

(L) This atomic weight is for the relative mass of the isotope of longest half-life. The element has no stable isotopes.

(m) Modified isotopic compositions can occur in commercially available materials that have been processed in undisclosed ways, and the atomic weight given here might be quite different for such samples.

(n) Name and symbol are assigned according to systematic rules developed by the IUPAC.

(r) Ranges in isotopic compositions of normal samples obtained on earth do not permit a more precise atomic weight for this element, but the tabulated value should apply to any normal sample of the element.

(Z) Despite having no stable isotopes, the terrestrial compositions of samples of the long-lived isotopes allow a meaningful atomic weight.

FUNDAMENTALS

OF

CHEMISTRY

WITH QUALITATIVE ANALYSIS

FUNDAMENTALS
OF
CHEMISTRY

WITH QUALITATIVE ANALYSIS

Third Edition

James E. Brady
St. John's University, New York

John R. Holum
Augsburg College, Minnesota

WILEY

John Wiley & Sons

New York   Chichester   Brisbane   Toronto   Singapore

Book production supervised by
    Lucille Buonocore and Dawn Reitz.
Cover and text designed by Kevin Murphy.
Cover art by Roy Wiemann.
Photo researched by Safra Nimrod.
Illustrations by John Balbalis.
Manuscript editor was Josephine Della Peruta,
    under the supervision of Deborah Herbert.

ISBN 0-471-84491-8

Printed in the United States of America

10 9 8 7 6 5 4 3 2 1

Preface

Since the first edition of this text was published, the emphasis in teaching the first year course in chemistry for science majors has shifted from a predominantly "principles" approach to one that places more emphasis on descriptive chemistry—chemical reactions and properties. This transformation blends well with the basic philosophy that we brought to this book from the very beginning, that chemistry is fundamentally an experimental science, and that observed facts must form the foundation for theory and explanations. Therefore, when we approached this revision we welcomed the opportunity to incorporate more of the flavor of descriptive chemistry while continuing to maintain a solid treatment of chemical principles. At the same time, we wished to preserve the character of the previous edition, which has a relaxed, nonthreatening writing style, pedagogical aids designed to assist students in acquiring problem-solving skills, and many examples that illustrate how chemistry relates to the world in which we live.

Features in Organization and Content

In realizing our goals, we have combined our own thoughts with input from users of previous editions as well as suggestions by reviewers. The result has been some major changes in the flow of topics. For example, those familiar with the previous edition will note that much of the material about the periodic table that had appeared in Chapter 5 now occurs much earlier in Chapter 2. In addition, we have changed the sequence of topics dealing with thermodynamics, kinetics, and equilibrium. The first chapter on equilibrium now occurs at a location that places it approximately at the beginning of the second semester; thermodynamics logically precedes the chapter on electrochemistry; and kinetics is placed after thermodynamics and immediately before chapters dealing with the chemistry of the elements.

Throughout the revision special care was taken to present accurate, up-to-date treatments of important and current topics. Examples are discussions of the bonding in the water molecule (page 282) and of sulfide precipitations (page 650). New to this edition is an Appendix of tables of data that will be helpful to the instructor in preparing additional exercises and exam questions.

Although there have been organizational changes in the text, it continues to provide a logical development of concepts. In a broad sense, the text can be divided into a number of major units as follows:

Introduction to the Science and Structure of Chemistry   (Chapters 1 and 2). Students are introduced to the concept of chemical change, the tools of science and measurement, the periodic table and some properties of the elements and their compounds, and the basics of chemical nomenclature. Chapter 1 has been extensively revised to introduce students to chemical reactions and chemical properties more quickly. Much of former Chapter 5 is now incorporated in Chapter 2. Together, these chapters give students a core of chemical facts that serve as a foundation for future chapters.

Quantitative Aspects of Chemistry   (Chapters 3 to 5). These chapters include the stoichiometry of compounds and chemical reactions as well as thermochemistry. Additional material has been included on the stoichiometry of reactions in solution in Chapter 4, which should aid in integrating the text with the laboratory. In thermochemistry, the emphasis is now almost entirely on joules as an energy unit.

Atomic Structure and Bonding   (Chapters 6 to 8). In this unit we begin by examining the electronic structures of atoms and some atomic properties that vary systematically in the periodic table. The basic features of ionic and covalent bonding are now presented together in one chapter along with the shapes of molecules and their influence on molecular polarity. A thermochemical discussion of bond energies complements treatments of valence bond and molecular orbital theories.

Properties of Pure Substances   (Chapters 9 and 10). The emphasis is on the physical properties of gases, liquids, and solids, and the influence of intermolecular forces in determining these properties. The nature of intermolecular forces is now discussed early in Chapter 10, which makes subsequent discussions of the properties of liquids and solids more meaningful.

Chemical Reactions in Mixtures   (Chapters 11 and 12). Acid-base and metathesis reactions are discussed first along with the Brønsted and Lewis approaches, followed by an entire chapter devoted to redox reactions. A new section on periodic trends in acid-base strengths has been added, as well as a discussion of metal complexes as an illustration of Lewis acid-base chemistry. Solution stoichiometry and titrations are treated; and normality and equivalents are discussed in separate sections for those who may choose to omit these topics.

Physical Properties of Mixtures   (Chapter 13). This chapter is devoted to the physical properties of colloids and solutions. The discussion of the dissolution process has been completely rewritten and examines the interplay between intermolecular forces and the tendency toward randomness.

Equilibrium   (Chapters 14 and 15). The phenomenon of chemical equilibrium is examined first for gaseous systems and then for heterogeneous reactions, including solubility equilibria. The effect of complex ion formation on equilibrium is discussed. The unit concludes with acid-base equilibria. The section on hydrolysis now focuses on the acid-base properties of ions.

Factors that Control the Outcomes of Reactions   (Chapters 16 to 18). Having studied *what* happens in reactions, we now explore the reasons *why*. The interplay between energy and entropy in determining the spontaneity of events is studied, followed by the relationship between thermodynamics and equilibrium. Electrochemistry, in its ability to predict the outcome of reactions, can be seen as a practical extension of thermodynamics. Rates of reactions determine whether we ever get to see reactions that are thermodynamically possible.

Inorganic Chemicals and Reactions   (Chapters 19 to 22). The focus is on important chemicals and reactions. These chapters have been enlarged considerably in this edition, with an emphasis on applying chemical principles learned in earlier chapters. The material is presented so that instructors can easily choose to cover topics they consider to be most important to their students.

Nuclear Chemistry and the Chemistry of Carbon (Chapters 23 and 24). The origin of nuclear energy and stability is explored along with applications of nuclear phenomena to chemistry and an up-to-date discussion of nuclear energy production. Organic chemistry and biochemistry serve as a bridge to the second-year course in chemistry.

Although the order of topics presented here follows a logical development of the course material, we recognize that other equally valid organizations are possible, and the chapters have been carefully written to permit alternative orders of coverage. Some are suggested in the Instructor's Guide.

Descriptive Inorganic Chemistry

A question that many teachers have been asking themselves is "How shall I bring more descriptive chemistry into my course?" There is probably no single best answer to this, so in organizing this book our aim has been to provide the maximum degree of flexibility and effectiveness. First, we have sought to make descriptive chemistry interesting by illustrating many important reactions with really useful photographs. Second, we interweave a large amount of descriptive chemistry in discussions related to principles.

For example, Chapter 1 has been reorganized to provide an early introduction to chemical reactions and chemical properties as well as an introduction to the tools of measurement. Chapter 2 incorporates discussions of the basic structure of matter with the periodic table and properties of the elements and some of their compounds. And in Chapter 11 we discuss metal complexes to illustrate Lewis acid-base chemistry. Third, we have included 16 interchapter units called *Chemicals in Use* that focus on important chemicals and chemical processes. In making this revision, we have added some, deleted others, and revised many. Each now includes a set of questions, but their basic goal is the same: to provide interesting glimpses of how chemistry impacts on our lives. Fourth, new to this edition are three "interchapters" called *Qual Topics* that each illustrate how chemical principles discussed in the preceding chapter are brought to bear on the separation and identification of cations in solution. They illustrate practical applications of principles and so are valuable even for students who will not be preforming qualitative analyses in the laboratory. They will be especially useful, however, for courses that place a heavy emphasis on qualitative analysis in their lab components, and they supplement material discussed in the "Qual version" of this text. Finally, we have expanded considerably Chapters 19 through 22, which are devoted entirely to the chemistry of the nonmetals and metals. By having this flexible approach available, teachers can choose a presentation best suited to their course.

Features in Problem Solving

In this edition we continue our commitment toward developing problem-solving skills through a planned program of review and reinforcement. Within chapters there are numerous worked examples, many of which have been rewritten to improve clarity. Each begins with a brief statement identifying the type of problem being solved. Worked examples are usually followed by Practice Exercises that encourage students to test their understanding. At the ends of chapters there are extensive sets of Review Exercises, sorted by category. Answers to all the Practice Exercises and to a representative sampling of the Review Exercises are given in Appendix C. Finally, there are additional sets of exercises called Tests of Facts and Concepts that occur after groups of related chapters. These exercises review concepts covered in sets of two or more chapters,

and their answers are only to be found in the Instructor's Guide.

Features That Enhance Student Interest

A goal of this text throughout all its editions has been to make chemistry as interesting as possible. We continue to be careful to maintain a conversational writing style and we use common chemicals whenever possible as examples in both theoretical discussions and in problems and exercises. To help students appreciate the importance of chemistry to other disciplines and to their lives in general, we have opened each chapter with a photograph that, in most instances shows how the chapter contents relate to some familiar aspect of the world around us. We also continue to employ numerous Special Topics placed strategically within chapters throughout the text. As noted earlier, the interchapter Chemicals in Use focus on common and important chemicals and have been deliberately written in a light and interesting style.

Features in Learning Aids

Each chapter incorporates a variety of learning aids to improve student understanding and performance. A chapter begins with a brief table of contents so students can see at a glance what topics will be covered. Each section then begins with a brief title followed by a statement that summarizes the key lesson to be learned. This gives students a brief preview of the section content and serves as an aid later during review. Throughout the text there are frequent margin comments, figures, and photographs that supplement discussions. New terms are presented in blue boldface type when they are introduced and defined, and important equations and definitions are boxed in red to set them apart. All of the boldface terms are also located in a Glossary at the end of the book where each term is accompanied by a reference to one or more principal sections where the term is discussed. (A glossary by chapter is also provided in the Study Guide.) Another review aid is the chapter summaries that provide an overview of important concepts.

James E. Brady

John R. Holum

Acknowledgments

We begin by paying special tribute to our wives, Mary Holum and June Brady, and to our children Liz, Ann, and Kathryn Holum and to Mark and Karen Brady for their constant support and encouragement. Our appreciation is also extended to our colleagues Earl Alton of Augsburg College and to Ernest Birnbaum, Neil Jespersen, Eugene Holleran, and William Pasfield of St. John's University for their helpful discussions and stimulating ideas. We also recognize the support of our colleagues at the administrative level — Dean Ryan LaHurd and President Charles S. Anderson of Augsburg College and Rev. Joseph Breen, C. M. of St. John's University. It is with particular pleasure that we thank the staff at Wiley for their careful work, encouragement, and sense of humor, particularly our editor Dennis Sawicki, our picture editor Safra Nimrod, our designer Kevin Murphy, our illustrator John Balbalis, our production supervisors Lucille Buonocore and Dawn Reitz and their delightful boss Linda Indig. We also extend our most sincere appreciation to Dr. Donald W. Murphy of AT & T Bell Laboratories for providing us with a photograph of a magnet levitated over one of the newly discovered superconducting ceramics, to Dr. Harry Allcock of Penn State University for a photograph of a polyphosphazene polymer, to Sandra Olmstead for her valuable suggestions passed along during her help with the proofreading, and to the colleagues listed below whose, careful reviews, helpful suggestions, and thoughtful criticisms of the manuscript have been of such great value to us in developing this book. And finally, we are especially grateful to Dr. Jo Beran of Texas A & I University for the use of some of the specifics of his qual analysis procedures in Chapter 25 of the Qual Version of this text.

Joe F. Allen
Clemson University

Roger D. Barry
Northern Michigan University

Jo A. Beran
Texas A & I University

Charles Chakoumakos
University of Maine at Farmington

Geoffrey Davies
Northeastern University

Michael I. Davis
University of Texas at El Paso

John M. DeKorte
Northern Arizona University

Gordon J. Ewing
New Mexico State University

Roy G. Garvey
North Dakota State University

Stephen J. Hawkes
Oregon State University

Barton Houseman
Goucher College

Paul Hunter
Michigan State University

Jan G. Jaworski
Miami University, Ohio

Henry C. Kelly
Texas Christian University

Russell D. Larsen
Texas Tech University

Edmund Leddy Jr
Miami-Dade Community College

Marvin F. Lofquist
Ferris State College

R. Kent Murmann
University of Missouri — Columbia

Sandra Olmstead
Augsburg College

Harry E. Pence
State University of New York, College at Oneonta

William T. Scroggins
El Camino College

Jacob A. Seaton
Stephen F. Austin State University

Morris D. Taylor
Eastern Kentucky University

Richard S. Treptow
Chicago State University

J. E. B.

J. R. H.

Supplementary Material for Students and Teachers

A complete package of supplements has been assembled to aid the teacher in presenting the course and to help students accomplish problem solving and other study assignments.

Study Guide for Fundamentals of Chemistry Third edition by James E. Brady and John R. Holum. This softcover book has been carefully structured to assist students in mastering the important subjects in the text. For each section there is a brief statement of objectives, followed by a brief review that highlights major topics, sometimes with additional worked-out examples. Most sections have a brief "Self-Test," with answers provided, that supplements the text exercises. Each chapter concludes with a glossary of key terms introduced in the chapter.

Laboratory Manual for Fundamentals of Chemistry Third edition by Jo A. Beran. This manual includes a thorough techniques section covering 19 basic lab techniques, photographs of important manipulations, and 46 carefully planned experiments geared to topic development in the main text.

Solutions Manual for Fundamentals of Chemistry Third edition by Ernest R. Birnbaum and Paul Gaus. This softcover supplement provides detailed solutions to all of the numerical problems and answers to all the nonnumerical questions that appear in Practice Exercises and end-of-chapter Review Exercises.

Teachers Manual for Fundamentals of Chemistry This manual, available only to instructors, contains suggestions for alternate topic sequences, detailed chapter objectives and chapter rationales, answers to the *Tests of Facts and Concepts* questions, and assistance with qualitative analysis.

Teachers Manual for the Laboratory Manual This supplement for instructors provides outlines of every experiment, lab suggestions and cautions, expected experimental data, chemicals and equipment to be used, and details on reagent solution preparation.

Test Bank/Microtest Available in both hard copy and software (Apple®, Macintosh®, and IBM® compatible) versions, this carefully prepared test resource contains 1800 questions in all.

Transparencies Instructors who adopt this book may obtain from Wiley, without charge, a set of one hundred $8\frac{1}{2}$ in. $\times$ 11 in. transparencies that duplicate key illustrations from the text.

Video Package Developed at the University of Illinois, and available from Wiley, this package features a range of 2 to 3 minute video lecture demonstrations showing chemical reactions and applications and single topic sequences on subjects such as thermodynamics.

J. E. B.

J. R. H.

Contents

FUNDAMENTALS
OF
CHEMISTRY
WITH QUALITATIVE ANALYSIS

"Geese at Daybreak" The sight of geese silhouetted against the morning sun is likely to lift the spirits of anyone. But to scientists, this scene has a richer meaning. They also see the wonders of the physics of flight, the biology of feathered creatures, and the remarkable profusion of chemicals in nature. Chemicals, of course, are what this book is about, and we find them everywhere — in the air and clouds, and in geese that fill the morning air. Chemicals are what make us "tick," and understanding something of the way they influence our lives is a goal worth keeping in mind as you begin this course. It can give you, too, a better appreciation of the world around us.

THE SCIENCE OF CHEMISTRY

1.1 WHY STUDY CHEMISTRY

Chemistry is needed in most programs in the sciences because it describes and explains the composition and behavior of so many different substances.

It is difficult today to imagine our world without the many conveniences, and even necessities, that owe their existence to science. We take for granted such things as telephones, radio and TV, electronic calculators, plastics and synthetic fibers that make up our clothes, fresh fruits and vegetables out of season, and pharmaceuticals for the prevention and cure of disease. All these and more exist now because of the extraordinary curiosity that human beings have about the world around them. The exercise of this curiosity and the study of nature, of course, are what science is all about, and in this book we focus on one particular branch of science called chemistry.

In chemistry we study the compositions of things to find out what they are made of. We seek to learn how their composition affects their characteristics and behavior, so that perhaps we can make new materials with better properties that satisfy particular needs. To do this we have to learn how substances undergo changes in composition and properties—changes that we call "chemical reactions." And beyond all this, we seek to understand better the underlying workings of nature.

The success of chemistry in achieving these goals is all around us and is perhaps best illustrated by the profusion of various plastics that serve so many functions today. Some are hard and impact resistant, and can be used to make football helmets. Others are soft and pliable, such as vinyl plastics used in place of leather in so many products. And still others are soft clear films that we use as plastic food wrappings. These materials were not discovered purely by accident, but evolved from the deliberate efforts of chemists to develop new and better materials.

In your chemistry class you are likely to find students planning to major in a variety of disciplines. This is because so many sciences draw on various chemical principles and concepts. Engineers must know the chemical behavior of the materials they use. Biologists must know a lot of chemistry in order to understand such basic processes as metabolism and energy conversion in organisms. People in the field of medical technology must understand the chemical basis of the tests and analyses they perform. Pharmacists must understand the chemistry of drug reactions and interactions, and nutritionists must have some understanding of the way the body functions chemically, so they can provide the proper nutrients.

The list of specialties that require a knowledge of chemistry could go on for several pages, and no doubt you would find your particular field of specialization there. But the fact that you plan to go into one of these fields need not be your only reason for studying chemistry. Throughout this book you will learn how chemistry touches your personal life in many, many ways. Thus, studying chemistry can help you understand better the world in which you must function day after day.

1.2 CHEMICAL REACTIONS AND CHEMICAL CHANGE

The central focus of chemistry is on the way chemicals interact with each other to form new materials with new properties.

Often people think of chemicals as things only found in laboratories. Actually, the only place we don't find them is in a vacuum. They are everywhere else. All the foods we eat, the clothing we wear, the books we read, and literally everything we come in contact with are composed of chemicals.

One of the marvelous things about chemicals is that they often undergo chemical reactions or chemical changes, during which new substances with new characteristics and compositions are formed. As you might expect, chemical reactions are at the heart of chemistry, and we will see many examples of them throughout this book. Since this may be your first exposure to the study of chemical reactions, however, let's begin with an example that demonstrates in a rather startling way just how dramatic these changes can be. To do this we will look at the reaction that takes place between a substance called sodium and another substance called chlorine.

First, let's examine the characteristics of the starting materials, sodium and chlorine. Sodium is a metal, and has many of the characteristics that are usually associated with metals. For example, it is shiny and conducts electricity well. It is a very soft metal and is easily cut with a knife, as shown in Figure 1.1. Notice the bright shiny metallic surface of the freshly cut metal. Also notice that the outside of the bar of sodium is coated with a white crust. This was formed by the rapid chemical reaction of sodium with oxygen and moisture in the air. The crust consists of the products of these reactions and, as you can see, it has an appearance that is quite different from that of sodium.

Sodium is a rather dangerous chemical to work with. Its reaction with liquid water is particularly violent, producing much heat and liberating a flammable gas called hydrogen. During this reaction a substance called sodium hydroxide is also formed. Sodium

Scientists hold many national and international meetings each year to communicate their findings to other scientists working in the same or related fields.

If you've had a prior course in chemistry, much of what is discussed in this first chapter may be familiar. Nevertheless, be sure you review it thoroughly, because much of the material discussed later draws on these introductory topics.

FIGURE 1.1 Sodium is a soft metal that is easily cut with a knife. The typical metallic luster of the freshly exposed surface will quickly fade as the sodium reacts with oxygen and moisture in the air.

hydroxide is commonly known as lye, which is very corrosive toward flesh. As a result, contact of sodium with your skin can cause severe chemical burns.

Chlorine has many characteristics that differ greatly from those of sodium. It is a pale, yellow-green gas, as shown in Figure 1.2. If you have ever smelled liquid laundry bleach such as Clorox®, you have smelled chlorine, which escapes from the bleach in small amounts. In concentrated form chlorine is especially dangerous to inhale, causing severe lung damage that can easily lead to death. In fact, chlorine was used as a war gas during World War I.

When metallic sodium and gaseous chlorine are allowed to come into contact with each other they react violently, with the liberation of a great deal of heat, as shown in Figure 1.3. The substance formed in this reaction is a white powder, quite different in appearance from either sodium or chlorine. Its chemical name is sodium chloride, but you are more likely to recognize it by its common name—table salt.

If you think about this reaction for a moment, perhaps it will strike you as really quite amazing, and even a bit like magic. Here we have two chemicals, sodium and chlorine, whose ingestion can produce severe medical problems or even death. Yet, when they react with each other they produce a substance that our bodies cannot do without! Such startling events are what make chemistry fascinating, and as you study this course perhaps you will also discover for yourself some of that same sense of magic that has caught the imagination of others and produced generations of chemists.

FIGURE 1.2 Chlorine is a pale yellow-green gas.

FIGURE 1.3 A small piece of sodium is melted in a metal spoon before being thrust into the flask containing the chlorine. The sodium burns rapidly in the chlorine atmosphere as sodium chloride is formed. (The sodium used in this reaction had been stored in kerosene to protect it from oxygen and moisture. Some black smoke, which is caused by the reaction of chlorine with small amounts of kerosene still clinging to the sodium, can also be seen in the flask.

1.3 MATTER AND ENERGY

Matter has mass, occupies space, and is able to do work when it possesses energy.

The description of the reaction between sodium and chlorine in the previous section illustrates the two principal realms of interest of scientists who study chemistry. Sodium, chlorine, and sodium chloride are examples of *matter,* and the reaction of sodium and chlorine releases *energy.*

Matter is anything that has *mass* and occupies space. It is the stuff our universe is made of, and includes not only the chemicals just mentioned, but everything else that is tangible. Air, water, rock, and even people are composed of matter.

In our definition of matter we have used the term mass. Mass and weight are often used interchangeably, although the two terms refer to slightly different things. Mass refers to the amount of matter in an object; weight refers to the force with which the object is attracted by gravity. For example, a golf ball consists of a certain amount of matter, and its mass is the same regardless of its location. However, on the moon the golf ball would weigh only $\frac{1}{6}$ as much as on earth because the gravitational attraction of the moon is only $\frac{1}{6}$ that of earth. Because mass does not vary from place to place, we use mass rather than weight when we specify the amount of matter in an object. The way we measure mass and the units we use to express it will be discussed in Section 1.10.

Closely related to matter is energy. Energy is usually defined as the capacity to do work; it is something matter possesses if the matter is able, in some way, to accomplish the movement of objects. For example, you know that you possess energy because you are able to do the work of moving your limbs against the opposing forces on them caused by gravity. A moving freight train has energy as evidenced by its ability to do work on (crush) a car stalled in its path. We know oil and coal possess energy because they can be burned and the heat released can be harnessed to run machinery.

In chemistry we are concerned with both matter and energy, because the chemicals we work with are forms of matter and when they react with each other they gain or lose energy. We've seen this in the reaction of sodium with chlorine, and we know it to be true for the reaction of coal with oxygen in the air. To really understand chemistry, therefore, we must learn more about matter as well as energy, and we will begin by devoting the remainder of this section to a discussion of energy.

Nearly everyone is familiar with the word *energy,* yet few people really understand what it is. This is because energy is so different from matter. We can't touch it, see it, or smell it; all we can do is observe its effects on matter. Therefore, let's begin our discussion by learning *how* matter can "possess energy."

Gasoline

Pizza

Camera

Some examples of matter.

Kinetic Energy

Basically, there are two kinds of energy that matter can have. One is kinetic energy and the other is potential energy. Kinetic energy is the energy an object has because of its motion, and it depends on both the object's mass and its speed. For instance, you know that a Cadillac hit by a Volkswagen at 20 miles per hour will suffer much less damage than a Cadillac hit by a more massive freight train moving at the same speed. Because it has more mass than a Volkswagen, the freight train has more kinetic energy. You also know that a Volkswagen at 80 miles per hour will do more damage than one traveling at only 20 miles per hour. Kinetic energy increases with an object's speed.

There is a simple equation[1] that relates the amount of kinetic energy (KE) an object has to the object's mass and speed:

---

[1] Many equations will be numbered to make it easy to refer to them later on in the book or in class discussions. However, the really important equations, such as this one, will be put in "boxes." Be sure you learn the boxed equations.

$$KE = \tfrac{1}{2}mv^2 \qquad\qquad (1.1)$$

where m is the mass and v is the velocity (speed).

Potential Energy

Kinetic energy is easy for most people to understand because they have seen its effects firsthand and because kinetic energy can be calculated by a simple equation. Potential energy, however, is more abstract. Potential energy is stored energy. For example, when you wind an alarm clock, the energy that you expend winding it is stored in a spring; you have given the spring potential energy. This potential energy is gradually changed back to kinetic energy as the timepiece operates and the hour and minute hands are set into motion.

The foods that we eat also possess potential energy, but of a somewhat different kind that is sometimes called chemical energy. This is the energy that is stored in chemicals, and it can be released during chemical reactions. For example, the foods we eat possess potential energy that is released by the process we call metabolism. This energy ultimately appears as movements of muscles and as body heat. Sodium and chlorine also possess potential energy. When they react they release energy and form sodium chloride, which has less potential energy. The difference between the potential energy of the sodium chloride and the potential energy of the starting materials is the energy that appears in the form of the heat given off in this reaction.

Unlike kinetic energy, there is no single, simple equation that can always be used to calculate the amount of potential energy that an object has.

How We Observe Energy

Energy makes its presence known in a variety of different ways. It can appear entirely as *mechanical energy*, that is, as energy of motion or kinetic energy. We observe this energy when it is transferred to other objects and makes them move. We also observe energy when it is transferred as electrical energy, light energy, or sound energy. Eventually, however, all energy ends up as heat.

The concept of heat immediately brings to mind the notion of temperature. It is important, however, to remember that heat and temperature are quite different. Heat is a form of energy. Temperature is a measure of the *intensity* of heat, and heat always flows without outside help from warm objects to cool ones. A hot cup of coffee, for example, becomes cool because heat flows from the hotter coffee into the cooler surroundings.

Both fires could be at the same temperature, but the forest fire has more heat in it than the flame of the match.

1.4 PROPERTIES OF MATTER

Samples of matter are identified by their characteristics.

Now that you have some notion of what energy is, let's return to our discussion of matter. We begin by presenting a simple problem. Suppose you were searching through a pile of books for your chemistry book. What is it about the book that would help you find it? No doubt you would look for such things as size, color, and the printing on the cover. These are characteristics that books have that help you distinguish among them. Similarly, in chemistry we use the characteristics, or properties, of matter to distinguish one kind from another. To help us organize our thinking, we classify the properties of matter into different types.

Physical Properties

Physical properties are characteristics that can be specified without reference to other substances. For example, at room temperature in an open container, water exists as a liquid. This *existence as a liquid* is a physical property that we observe for all samples of pure water under these conditions. Other examples of physical properties are mass and volume. If we wish to specify the amount of matter in something, we give its mass; if we wish to specify the amount of space something occupies, we give its volume. Notice that both mass and volume are properties that depend on the size of the sample. A large sample of water has both a larger mass and a larger volume than a small sample. Properties such as these that depend on the size of the sample are called extensive properties.

Extensive properties are not particularly useful for identifying substances. More useful are properties that do not depend on the size of the sample. These are called intensive properties, and some examples are melting point, color, and electrical conductivity. All samples of pure metallic copper, for example, melt at the same temperature, are reddish in color when polished, and are good electrical conductors.

The melting point is simply the temperature at which the substance melts.

As we said, we use physical properties to identify substances. For another example, suppose you were given two polished, metallic objects and were told that one was iron and one was gold. Would you be able to tell which was which? Of course you would. You know that gold is a bright yellow metal and that iron has a much darker, silvery color.[2] You also know that iron is attracted by a magnet, and you might know that gold is not. Thus, magnetism is another physical property — an *intensive property* — that we could use to tell these two metals apart.

Chemical Properties

A chemical property of a substance is any chemical reaction that the substance can undergo. If an iron nail, for example, is exposed to oxygen and moisture, it undergoes a chemical reaction in which another substance called iron oxide — rust — forms. This ability of iron to react with oxygen and water to form rust is one chemical property of iron, and we could use this fact to distinguish iron from gold, which is unaffected by oxygen and moisture. Here we see another way to tell these two metals apart — one based on chemical properties. Notice that when we *observe* a chemical property, the substance invariably changes to something else with a different composition. When we observe a physical property, no such alteration occurs. This is the chief distinction between a chemical and a physical property.

---

[2] Color can often be a useful property for identification, but there are instances where you can be fooled, particularly when particle size is very small. For example, you know that silver is a white metal with a high luster; however, in a very finely divided state, as in the image on black-and-white photographic film or paper, metallic silver appears black.

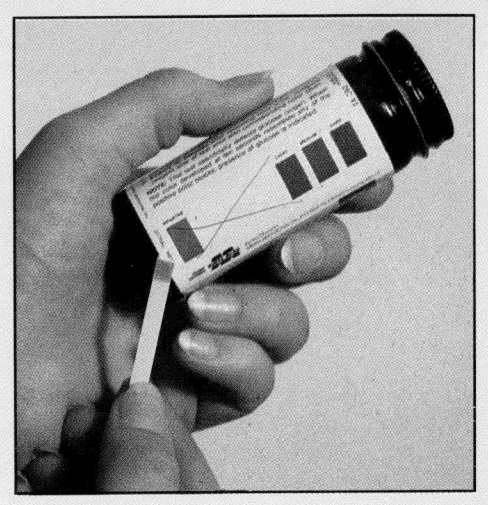
1.5 ELEMENTS, COMPOUNDS, AND MIXTURES

In chemistry we classify matter into three broad categories: elements, compounds, and mixtures.

Just a casual glance around your room suggests that matter occurs in many different forms. For example, plastics, paper, wood, steel, and paint are just some of the things that are likely within your view. If we choose to study these materials, where do we begin? Scientists, whenever faced with such variety in nature, search for ways to organize and classify what they see. This has led to the three general classes of matter described in the title of this section — elements, compounds, and mixtures — listed in order of their degree of complexity. Let's see what "degree of complexity" means.

Elements

If a chemical reaction changes one substance into two or more others, a decomposition has occurred. For example, if we pass electricity through a water solution, two gases form — hydrogen and oxygen. In this example, we have decomposed water into two new substances. Scientists have carried out innumerable such decompositions, and some of the new substances have defied all efforts at further decomposition — at least into anything that can be stored and looked at. These substances that are stable to further decomposition are called elements. Elements are the simplest, least complex forms of matter that we encounter in the laboratory or anywhere else, and they cannot be decomposed into simpler substances by chemical reactions. Some examples already mentioned are sodium, chlorine, copper, hydrogen, oxygen, iron, and gold. Other familiar examples include silver, nickel, lead, chromium, zinc, sulfur, and neon.

Chemical reactions are unable to decompose hydrogen and oxygen into simpler substances.

So far, scientists have discovered 90 elements in nature and they have made 18 more, for a total of 108. Their names are given inside the front cover of this book. The list may seem long, but many elements are rare, and we will be most interested in only a relatively small number. But these 108 elements, by themselves as well as in chemically combined forms, account for all matter in all its enormous variety everywhere in the known universe. This recognition that, out of the seemingly infinite variety of things, there are only just over a hundred elementary substances has greatly simplified the chemist's goal to understand the workings of nature.

Compounds

By means of chemical reactions, elements combine in various proportions to give all the more complex substances in nature. Hydrogen and oxygen combine to form water; sodium and chlorine combine to form sodium chloride (salt). Water and salt are examples of compounds, and both are more complex than their elements because they can be decomposed to elements. A compound is a substance formed from two or more elements in which the elements are always combined in the same fixed (i.e., constant) proportions by mass. For example, in any sample of pure water there is always eight times as much oxygen (by mass) as hydrogen. Similarly, in any sample of pure sodium chloride, the mass of chlorine is always 1.542 times the mass of sodium.

In reading this "chemist's description" of the composition of a compound, there is a subtle but very important point to be understood. Consider, for example, the compound sodium chloride. We say that this compound is composed of sodium and chlorine. However, it is very important to understand that in the compound these two elements are not present in the same form as we find them in their pure states. As we noted earlier, the pure element sodium is a metal, but salt certainly has no metallic characteristics. Similarly, pure chlorine is a pale green gas, but salt is not green and there is nothing gaseous about it. When these or other elements react to form a compound, their individual properties are lost and in their place we find the properties of the compound.

Mixtures containing sugar can have variable compositions.

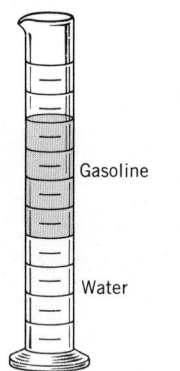

FIGURE 1.4 A two-phase heterogeneous mixture.

Mixtures

Elements and compounds are examples of what chemists call pure substances.[3] The composition of a pure substance is always the same, regardless of its source. All samples of table salt, for example, contain sodium and chlorine combined in the same proportions by mass, and all samples of water contain the same proportions by mass of hydrogen and oxygen. Pure substances are rare, however. Usually we encounter mixtures of compounds or elements. Unlike elements and compounds, mixtures can have variable compositions. Sugar and water are examples of compounds because each consists of elements chemically combined in *definite* proportions. But, when we dissolve sugar in water, we can vary the proportions almost as much as we wish. (Some people like their coffee sweeter than others!)

Mixtures are either homogeneous or heterogeneous. A homogeneous mixture has the same properties throughout the sample. Examples are thoroughly stirred mixtures of salt in water or sugar in water. We call a homogeneous mixture a solution (solutions need not be liquids, just homogeneous).

A heterogeneous mixture consists of two or more regions called phases that differ in properties. A mixture of gasoline and water, for example, is a two-phase mixture in which the gasoline floats on the water as a separate layer (Figure 1.4). The phases in a mixture don't have to be chemically different substances like gasoline and water, however. A mixture of ice and liquid water, for example, is a two-phase mixture in which each phase has the same chemical composition; yet another property — solid versus liquid — is not the same throughout the mixture.

---

[3] We have used the term *substance* rather loosely until now. Strictly speaking, *substance* really means *pure substance*. Each unique chemical element and compound is a substance; a mixture consists of two or more substances.

(a) (b)

FIGURE 1.5 *(a)* Samples of powdered sulfur and powdered iron. *(b)* A mixture of sulfur and iron is made by stirring the two powders together.

FIGURE 1.6 Formation of the mixture is a physical change; it hasn't changed the iron and sulfur into a compound of these two elements. The mixture can be separated by pulling the iron out with a magnet.

One of the important ways that mixtures differ from compounds is in the changes that occur when they form. Consider, for example, iron and sulfur — two elements — pictured in Figure 1.5. We can make a mixture of iron and sulfur simply by dumping them together and stirring them. Both elements retain their original properties. The process we use to create this mixture is called a physical change, rather than a chemical change, because no new chemical substances form. To separate this mixture, we would similarly use just physical changes. For example, we could remove the iron by stirring the mixture with a magnet — a physical operation. The iron pieces would stick to the magnet as we pull it out, leaving the sulfur (Figure 1.6). The mixture could also be separated by treating it with a liquid called carbon disulfide, which dissolves the sulfur but does not dissolve the iron. Filtering the sulfur solution from the solid iron, followed by evaporation of the liquid carbon disulfide from the sulfur solution, gives the original components — iron and sulfur — separated from each other.

As we have noted earlier, formation of a compound from its elements involves a chemical reaction. Iron and sulfur, for example, form a compound often called "fool's gold" because of its appearance (Figure 1.7). In the compound, the elements no longer have the same properties they had before they were combined. Just as the formaton of a compound involves a chemical reacton, so also does decomposition. The decomposition of fool's gold into iron and sulfur is a chemical reaction.

The relationship among elements, compounds, and mixtures is shown in Figure 1.8.

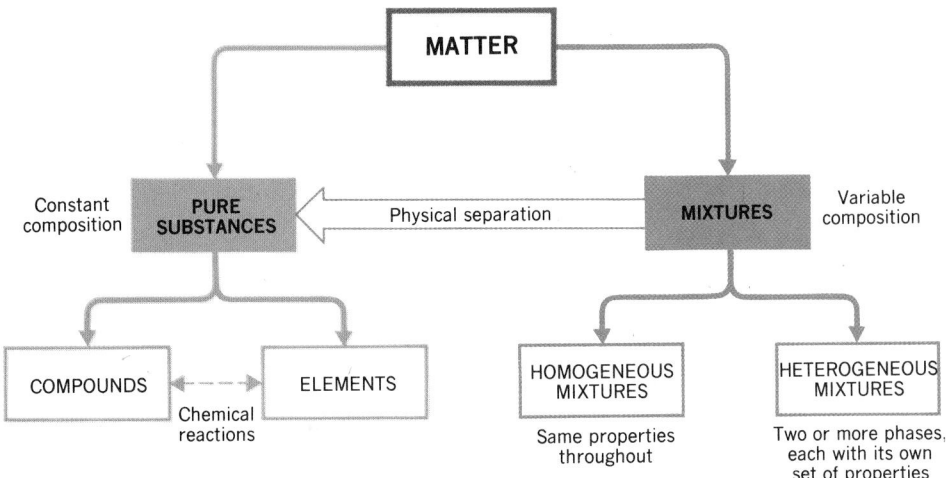

FIGURE 1.8 Classification of matter.

FIGURE 1.7 Iron pyrite. Its color accounts for its nickname, "fool's gold."

1.6 CHEMICAL SYMBOLS, FORMULAS, AND EQUATIONS

A chemical symbol is a shorthand notation for the name of an element; a chemical formula gives the composition of a compound; a chemical equation gives a "before-and-after" picture of a chemical reaction.

Chemical Symbols

As you are probably aware, in chemistry we use a system of symbols to write the formulas of compounds and to describe chemical reactions by chemical equations. In this system, each of the chemical elements is assigned its own unique symbol. In most cases, these are formed from one or two letters of the English name for the element. For instance, the symbol for carbon is C, for bromine it is Br, and for barium it is Ba. In some instances, the symbols are derived from the Latin names which were given to those elements long ago. Table 1.1 contains a list of elements whose symbols come to us in this way.[4]

Regardless of the origin of the symbol, the first letter is always capitalized and the second letter is always written lowercase. For example, the symbol for copper is Cu, not CU. Be very careful to follow this rule so you can avoid confusion between such symbols as Co (cobalt) and CO (carbon monoxide).

Latin was the universal language of science in the early days of chemistry.

Chemical Formulas

Chemical formulas convey several kinds of information. In one sense the formula is a shorthand way of writing the name for a compound, in the same way that chemical symbols are shorthand notations for the names of elements. The most important characteristic of a formula, however, is that it specifies the composition of a chemical substance.

The smallest bits of an element that retain their chemical identity are atoms. Scientists have reached this conclusion based on many observations and measurements, which we will discuss in Chapter 2. When atoms combine chemically to form more complex substances, they come together in definite proportions. Often, the atoms join to produce larger stable particles called molecules. Sometimes the combining atoms acquire electrical charges and cling together because of the attraction between opposite charges.

Regardless of how the atoms of the various elements are held to each other in a compound, each is identified by its chemical symbol in the compound's formula. In addition, subscripts are used in the formula to tell us how many atoms of each of the elements are present. For example, the iron oxide in rust has the formula Fe_2O_3, which tells us that the compound is composed of iron and oxygen. It also tells us that in this

Later in the book, when we wish to refer to individual atoms or molecules in general, we will simply use the term particle.

TABLE 1.1 Elements That Have Symbols Derived from Their Latin Names

Element	Symbol	Latin Name
Sodium	Na	Natrium
Potassium	K	Kalium
Iron	Fe	Ferrum
Copper	Cu	Cuprum
Silver	Ag	Argentum
Gold	Au	Aurum
Mercury	Hg	Hydrargyrum
Antimony	Sb	Stibium
Tin	Sn	Stannum
Lead	Pb	Plumbum

[4] The symbol for tungsten is W, from the name *wolfram*. This is the only element whose symbol is neither related to its English name nor derived from its Latin name.

compound there are two atoms of iron for every three atoms of oxygen. In many formulas we find atoms written without subscripts; when no subscript is written, we assume it to be 1. For instance, the compound water has the formula H_2O (which you probably already know). This formula tells us that there are *two* H atoms for every *one* O atom. Similarly, the formula for chloroform is $CHCl_3$, which indicates that one atom of carbon, one atom of hydrogen, and three atoms of chlorine have combined.

For more complicated compounds, we sometimes find formulas containing parentheses. An example is the formula for urea, $CO(NH_2)_2$. This formula tells us that the atoms within the parentheses, NH_2, are repeated twice. The formula for urea could be writen as CON_2H_4, but there are good reasons for writing certain formulas with parentheses, as we will see later.

Chemical formulas are used to describe not only compounds, but also some of the elements themselves, which normally occur as simple molecules. Examples are oxygen and nitrogen, the major gases found in the atmosphere. When these elements are not combined with other elements in compounds, we always find them as molecules having the formulas O_2 and N_2.

Other elements that normally occur as simple two-atom molecules are H_2, F_2, Cl_2, Br_2 and I_2.

Hydrates   Certain compounds form crystals that contain water molecules. An example is ordinary plaster — the material often used to coat the interior walls of buildings. Plaster consists of crystals of calcium sulfate, $CaSO_4$, that contain two molecules of water for each $CaSO_4$. These water molecules are not held very tightly and can be driven off by heating the crystals. The dried crystals absorb water again if exposed to moisture, and the amount of water absorbed always gives crystals in which the H_2O to $CaSO_4$ ratio is 2 to 1. Compounds such as this, whose crystals contain water molecules in fixed ratios, are quite common and are called hydrates. The formula for this hydrate of calcium sulfate is written $CaSO_4 \cdot 2H_2O$ to show that there are two molecules of water per $CaSO_4$. The raised dot is intended to suggest that the water molecules are not bound too tightly in the crystal and can be removed.

Sometimes the dehydration (removal of water) of hydrate crystals produces changes in color. An example is copper sulfate, which is sometimes used as an agricultural fungicide. Copper sulfate forms blue crystals with the formula $CuSO_4 \cdot 5H_2O$ in which there are five water molecules for each $CuSO_4$. When these blue crystals are heated strongly, the water is driven off and the solid that remains, now nearly pure $CuSO_4$, is almost white (Figure 1.9). If left exposed to the air, the $CuSO_4$ will absorb moisture and form blue $CuSO_4 \cdot 5H_2O$ again.

(a)   (b)

FIGURE 1.9 *(a)* Blue crystals of copper sulfate, $CuSO_4 \cdot 5H_2O$. *(b)* When heated, these crystals lose water. The nearly white solid that remains is pure $CuSO_4$.

EXAMPLE 1.1   COUNTING THE ATOMS IN FORMULAS

Problem: How many atoms of each element are expressed in the formula for sodium bicarbonate (baking soda), $NaHCO_3$?

Solution: We simply look at the subscripts following the symbols for the elements. Remember that the subscript 1 is implied when no subscript is written. This formula tells us that the smallest unit of the compound contains

1	Na	1	C
1	H	3	O

EXAMPLE 1.2   COUNTING THE ATOMS IN FORMULAS

Problem: How many atoms of each element are represented in the formula for calcium bicarbonate, $Ca(HCO_3)_2$?

Solution: The subscript 2 outside the parentheses tells us that every atom within appears twice. Within the parentheses are one H, one C, and three O, so in the formula for the compound there is a total of two H, two C, and six O. Obviously, there is also one Ca in the formula. To summarize, then, the formula has

1	Ca	2	C
2	H	6	O

EXAMPLE 1.3   COUNTING THE ATOMS IN FORMULAS

Problem: How many atoms of each element are represented in the formula for cobalt chloride, $CoCl_2 \cdot 6H_2O$?

Solution: In the formula $CoCl_2$, we count one Co and two Cl. In *each* H_2O there are two H and one O, so in six H_2O there are $6 \times 2 = 12$ H and $6 \times 1 = 6$ O. Therefore, the formula of the hydrate has

1	Co	12	H
2	Cl	6	O

PRACTICE EXERCISE 1   How many atoms of each element are expressed by the formulas (a) $NiCl_2$, (b) $FeSO_4$, (c) $Ca_3(PO_4)_2$, (d) $CuSO_4 \cdot 5H_2O$?

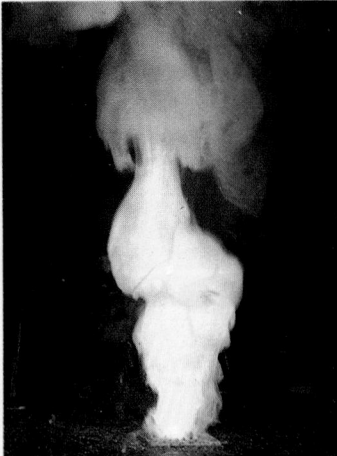

FIGURE 1.10 A mixture of zinc (Zn) and sulfur (S) reacts violently when ignited and forms the compound zinc sulfide, ZnS.

Chemical Equations

A chemical equation describes what happens when a chemical reaction occurs. It uses chemical formulas to provide us with a before-and-after picture of the chemical substances involved. Consider, for example, the reaction that occurs when a mixture of powdered zinc (Zn) and sulfur (S) is ignited. The reaction, shown in Figure 1.10, is quite violent and produces a substance called zinc sulfide, which has the formula ZnS. This reaction is represented by the chemical equation

$$Zn + S \longrightarrow ZnS$$

The two substances Zn and S, which appear before the arrow, are called the reactants. These are the substances that exist before the reaction occurs. After the arrowhead is ZnS, which is the product of the reaction. In this example, only one substance is formed in the reaction, so there is only one product. As we will see, however, in most chemical reactions there is more than one product. The products are the substances that are formed and that exist after the reaction is over. The arrow means "reacts to yield." Thus, this equation tells us that *zinc and sulfur react to yield zinc sulfide.*

Coefficients

Many reactions are more complex than the one we've just examined. An example is the burning of butane, C_4H_{10}, the fluid in disposable cigarette lighters (Figure 1.11). The equation for this chemical reaction is

$$2C_4H_{10} + 13O_2 \longrightarrow 8CO_2 + 10H_2O$$

In this equation there are numbers, called coefficients, in front of the formulas. These are present to balance the equation. An equation is balanced if there is the same number of atoms of each element indicated on both sides of the arrow. The 2 before the C_4H_{10} tells us that two molecules of butane react. This involves a total of 8 carbon atoms and 20 hydrogen atoms. On the right we find 8 molecules of CO_2, which contains a total of 8 carbon atoms. Similarly, 10 water molecules contain 20 hydrogen atoms. Finally, we can count 26 oxygen atoms on both sides of the equation. We will learn how to balance equations in Chapter 4.

In a chemical equation it is sometimes necessary to specify the physical states of the reactants and products, that is, whether they are solids, liquids, or gases. This is done by writing s, ℓ, or g in parentheses after the chemical formulas. For example, the equation for the combustion of the carbon in a charcoal briquet can be written

$$C(s) + O_2(g) \longrightarrow CO_2(g)$$

At times we will also find it useful to indicate that a particular substance is dissolved in water. We do this by writing aq, meaning *aqueous solution*, in parentheses after the formula. For instance, the reaction between stomach acid (HCl) and the active ingredient in Tums®, $CaCO_3$, is

$$2HCl(aq) + CaCO_3(s) \longrightarrow CaCl_2(aq) + H_2O(\ell) + CO_2(g)$$

FIGURE 1.11 Combustion of butane.

s = solid
ℓ = liquid
g = gas

Aqueous means "containing water." → Dissolved in water

EXAMPLE 1.4 READING CHEMICAL EQUATIONS

Problem: The chemical equation for the combustion of acetylene, C_2H_2, is

$$2C_2H_2 + 5O_2 \longrightarrow 4CO_2 + 2H_2O$$

How many atoms of oxygen are shown on each side of this equation?

Solution: On the left we see $5O_2$, which represents 10 oxygen atoms (5 molecules, each with 2 oxygens). On the right we have $4CO_2$, which contains 8 oxygens (4×2), plus $2H_2O$, which contains 2 oxygens (2×1). The total on the right is $8 + 2 = 10$ oxygen atoms.

PRACTICE EXERCISE 2 How many atoms of each element appear on each side of the equation

$$Mg(OH)_2 + 2HCl \longrightarrow MgCl_2 + 2H_2O$$

PRACTICE EXERCISE 3 Rewrite the equation in Practice Exercise 2 to show that $Mg(OH)_2$ is a solid, HCl and $MgCl_2$ are dissolved in water, and H_2O is a liquid.

1.7 HOW WE STUDY CHEMISTRY—THE SCIENTIFIC METHOD

Careful observations and the collection of facts form the foundation of explanations that are broadened into scientific theories.

In the preceding sections we have given you a brief introductory glimpse of the science of chemistry, and just like the other natural sciences, it is not a static subject. Chemistry is

FIGURE 1.12 A scientist working in a modern chemical research laboratory.

Webster's defines *empirical* as "pertaining to, or founded upon, experiment or experience."

constantly changing as new discoveries are made by scientists who work in university and industrial laboratories. Before we proceed further with our description of this subject, it is worthwhile examining how science advances and the kinds of "tools" that scientists need to make progress.

Scientists are really very much like other people. They experience the same emotions, desires, fears, and prejudices that afflict all human beings. Perhaps the one thing that makes them a little different from others, however, is that they have learned to be very aware of the physical and chemical changes that take place around them and they have learned to analyze what they see in an orderly, logical way.

Empirical Facts and Scientific Laws

Scientists study nature by methods that are not very different from the way small children explore the world around them. When a child touches a hot stove and gets burned, the idea is planted that there is some relationship between the stove and pain. If brave enough, the child perhaps will test his or her idea by touching the stove again. The logic we apply to our work as scientists follows a sequence of steps that is only slightly more formalized than that used by the child testing the stove. The sequence normally starts when some particular observation fires our curiosity and raises some questions about the behavior of nature. Usually, we begin our search for answers by studying the work of other scientists in areas that are related to the one of interest to us. As our knowledge and understanding grow, we plan experiments that permit us to make observations. Most often this is done in a laboratory (Figure 1.12). We use laboratories because they allow us to control our experiments, so that the observations we make are reproducible, that is, we can be sure that if we again do the same things in the same way, we will get the same results. From our observations, we learn certain empirical facts—so named because we learn them by observing the behavior of some physical, chemical, or biological system. For example, suppose we wished to study the behavior of gases, such as the air we breathe. We would soon discover that the volume of a gas depends on a number of factors, including the weight of the gas, its temperature, and its pressure. The existence of these factors and the various relationships among them represent our empirical facts.

The empirical facts we gather from our observations are referred to as data, and we carefully record them. Then, looking at our data, we may be able to summarize some of the information we have discovered. Such summaries of data are called generalizations. For instance, one generalization we would probably make from our observation of gases is that when the temperature of a gas rises, the gas tends to expand and occupy a larger volume. If we were to repeat our experiments many times with numerous different gases, we would find that this generalization is uniformly applicable to them all. Such a broad generalization, based on the results of many experiments, is called a law.

Hypotheses and Theories

As useful as they may be in summarizing the results of experiments, laws can only state what happens. They do not explain *why* substances behave the way they do. Human beings are curious creatures, though, and we seek explanations. Therefore, based on the results of our experiments, we formulate a hypothesis—a tentative explanation of why nature behaves in certain ways. We use a hypothesis to make predictions of new behavior and then design experiments to test our predictions. If the results of our experiments prove that our hypothesis was incorrect, we *must* discard it and seek a new one. However, if the explanation survives repeated testing, it is gradually accepted and achieves the status of a theory. A theory is a tested explanation of why nature behaves as it does. Because theories are usually very broad and have many subtle implications, it is virtually impossible to perform every conceivable test that might disprove a theory. Therefore, we can rarely prove that a theory is correct. Generally, all we can do is find additional experimental support for it.

Scientific Method

The sequence of steps just described — observation, formulation of a theory, and testing of a theory by additional observation — is called the scientific method. Despite its name, this method isn't used only by those who label themselves scientists. The child touching the hot stove also learns by using these steps, as does an automobile mechanic who listens to your engine and makes various tests to determine why it doesn't run well. The mechanic replaces parts and then starts the engine again to see if the problem has been found. Doctors certainly use the scientific method all the time in diagnosing illness. Temperature, pulse, blood pressure, and "where it hurts" are all pieces of data from which a physician forms a hypothesis about what is causing your pain. Treatment is given based on this hypothesis, and your progress is observed to test it. In short, we all use the scientific method, usually more by instinct than by design.

The scientific method is merely a formal description of the way humans approach many of the experiences of life.

From the preceding discussion you may get the impression that scientific progress always proceeds in a planned and orderly fashion. This isn't true — luck sometimes plays an important role. For example, in 1828 Frederick Wöhler, a German chemist, was heating a substance called ammonium cyanate in an attempt to add support to one of his theories. His experiment, however, produced an unexpected substance that he analyzed and found to be urea (a constituent of urine). This was the first time anyone had ever made a substance produced only by living creatures from a chemical not having a life origin. The fact that this could be done was an important discovery. Nevertheless, if it had not been for Wöhler's application of the scientific method to his unexpected results, the significance of his experiment may have gone unnoticed.

As a final note, it is significant that the most spectacular and dramatic changes in science occur when major theories are proven incorrect. This happens only rarely, but when it does occur, scientists are set scrambling and exciting new frontiers are opened.

1.8 MEASUREMENT AND THE METRIC SYSTEM

Progress in science requires the making of measurements and a generally accepted system of units.

The application of the scientific method to chemistry involves more than observing the colors of chemicals or the evolution or absorption of heat during a reaction. These are qualitative observations; they do not involve numerical information. However interesting, they are relatively of little usefulness in chemistry, or in any other science. Qualitative observations simply do not provide enough information. In order to make progress, every science must resort to quantitative observations, and these involve making measurements.

The process of measurement is a common and necessary activity in all our lives. We measure the speed of our car to avoid getting a traffic ticket. We buy much of our food according to how much it weighs. Then we get on a scale to see whether we can afford to eat the amount we buy!

A necessary aspect of all measurements is a uniform and generally accepted system of units. In the United States we normally use the English system of units for everyday purposes. We measure distance in units of inches, feet, or miles; volume in ounces, quarts, or gallons; and weight in ounces, pounds, or tons. Sometimes it is necessary to convert from one unit to another, and no doubt you've become adept at performing simple conversions such as 6 ft = 72 in. However, in the English system there is no common, simple relationship between the various units. If, for some reason, you wanted to know how many inches there are in one mile, you would probably use the following:

$$12 \text{ in.} = 1 \text{ ft}$$
$$5280 \text{ ft} = 1 \text{ mi}$$

FIGURE 1.13 Metric units are becoming a common sight on many consumer products and highways.

Computing the answer requires a calculator or time-consuming arithmetic with pencil and paper. Because conversions such as these can be so tedious, the English system of units is not well suited to scientific work.

In the sciences, and in all of the other industrialized countries of the world, another set of units is used. Originally called the metric system, this set of units uses a decimal system in which conversion from one size unit to another involves simply moving the decimal point to the left or right an appropriate number of spaces. This is a clear advantage over our English system and accounts for its widespread use. In fact, the United States is in the process of a voluntary conversion to the metric system, and many metric units have become commonplace on consumer products (Figure 1.13). Soft drinks and alcoholic beverages, for example, are now routinely packaged in containers having volumes measured in liters or milliliters. Road signs often display distances in both miles and kilometers, and packaged goods on grocery shelves give their contents in grams as well as ounces or pounds.

Only four countries do not use the metric system: the United States, Burma, South Yemen, and Brunei.

The Metric Conversion Act (1975) calls for a board to coordinate the voluntary switch to metric units in the United States.

1.9 THE INTERNATIONAL SYSTEM OF UNITS

The International System of Units is the modern version of the metric system.

The metric system has gradually evolved since it was first formulated by the French Academy of Sciences during the 1790s. In 1960, the General Conference on Weights and Measures adopted a modified version of the metric system called the International System of Units, abbreviated SI (from the French name, *Le Système International d'Unités*).

There are differences between some of the units of the old metric system and those of the SI. For the most part, scientists have accepted the new SI units readily but, as happens when any change is made, there are a few holdouts. In addition, much of the existing scientific literature was written before the SI was adopted. It is therefore a good idea to be familiar with the units of both systems. Although the primary emphasis here will usually be on the SI, the important differences will be pointed out to you as we encounter them.

SI Base Units

The SI begins with the set of base units given in Table 1.2. Not all of these units will have meaning for you now, but we will encounter most of them later in this book and their meanings will be made clear when they are needed. The size of each base unit has been very precisely defined. For example, the international standard for the SI unit of mass, the kilogram, is a carefully preserved platinum–iridium alloy block stored at the International Bureau of Weights and Measures in France. This metal block is *defined* as having a mass of *exactly* one kilogram. Most countries keep in some central location a carefully

TABLE 1.2 The SI Base Units

Quantity	Unit	Symbol
Length	Meter	m
Mass	Kilogram	kg
Time	Second	s
Electric current	Ampere	A
Temperature	Kelvin	K
Amount of substance	Mole	mol
Luminous intensity	Candela	cd

calibrated replica of the international standard kilogram. For example, the National Bureau of Standards in Washington, D.C. has its own standard kilogram (Figure 1.14) which as near as possible is a duplicate of the one in France. This U.S. kilogram serves indirectly as the calibrating standard for all "weights" used for scales and balances in this country. Thus, the masses you measure on the balances in your general chemistry lab can be traced to the U.S. standard, and therefore ultimately to the international standard.

Objects created by human hands, of course, can be destroyed or lost and therefore are not the most desirable choices for standards. In the SI, only the kilogram is defined by such an object. The other base units are established in terms of reproducible physical phenomena. For example, in October 1983 the General Conference of Weights and Measures defined the meter as exactly the distance traveled by light in a vacuum in 1/299,792,458th of a second. Because light and a vacuum are available to any scientist, everyone has access to this standard.

FIGURE 1.14 The standard kilogram belonging to the National Bureau of Standard in Washington, D.C. It consists of a platinum-iridium alloy, and its mass has been very accurately and precisely compared to the international standard kilogram which is kept at the International Bureau of Weights and Measures in France.

Derived Units

The base units, which form the foundation of the SI, are used to define additional derived units. For example, we know that to calculate the area of a rectangular room, we multiply its length by its width. Similarly, the *unit* for area is derived by multiplying the *unit* for length by the *unit* for width.

$$\text{length} \times \text{width} = \text{area}$$
$$(\text{meter}) \times (\text{meter}) = (\text{meter})^2$$
$$\text{m} \times \text{m} = \text{m}^2$$

The SI derived unit for area is therefore m^2 (meter squared, or square meter). In obtaining this unit, we employ a very important concept that we will use repeatedly throughout this book when we perform calculations. *Units undergo the same kinds of mathematical operations that numbers do.* Thus $m \times m = m^2$ just as $3 \times 3 = 3^2$. We will say more about how we use this concept in Section 1.13.

In the English system we measure area in square feet (ft^2) or square yards (yd^2). Carpeting, for example, is priced by the square yard.

PRACTICE EXERCISE 4   What would be the SI derived unit for (a) volume (of a room, for example) and (b) speed? (*Hint:* In your car you measure speed in miles/hour.)

Decimal Multipliers

When making measurements, we sometimes find that the basic units are awkward. For instance, the meter (approximately 39 in.) is not convenient for expressing the dimensions of very small objects such as bacteria. Nor is it convenient for expressing very large distances such as those between planets or stars. In the SI, we can modify the basic units so that they more closely suit our needs by using decimal multipliers. These factors and the prefixes that we use to identify them are given in Table 1.3. Those that we will encounter most frequently are given in boldface. Be sure to learn them.

Larger and smaller metric units are created by multiplying the basic units by appropriate decimal factors.

TABLE 1.3 Decimal Multipliers That Serve as SI Unit Prefixes

Prefix	Pronunciation	Symbol	Multiplication Factor[a]
exa	*Texaco*	E	$1,000,000,000,000,000,000 = 10^{18}$
peta	*petal*	P	$1,000,000,000,000,000 = 10^{15}$
tera	*terrace*	T	$1,000,000,000,000 = 10^{12}$
giga	*gig* as in *gig*gle; *a* as in *above*	g	$1,000,000,000 = 10^9$
mega	*mega*phone	M	$1,000,000 = 10^6$
kilo	kill-o (*o* as in *over*)	k	$1000 = 10^3$
hecto	heck-to (*to* as in *total*)	h	$100 = 10^2$
deka	deck-a (as in *above*)	da	$10 = 10^1$
deci	*deci*mal	d	$0.1 = 10^{-1}$
centi	*centi*grade	c	$0.01 = 10^{-2}$
milli	*milli*tant	m	$0.001 = 10^{-3}$
micro	*micro*phone	μ	$0.000\,001 = 10^{-6}$
nano	nan-oh (*a* as in *nanny* goat)	n	$0.000\,000\,001 = 10^{-9}$
pico	peek-oh	p	$0.000\,000\,000\,001 = 10^{-12}$
femto	fem-toe (*fem* as in *feminine*)	f	$0.000\,000\,000\,000\,001 = 10^{-15}$
atto	*ato*mize	a	$0.000\,000\,000\,000\,000\,001 = 10^{-18}$

[a] Exponential forms such as 10^6 or 10^{-15} will be explained in Section 1.12.

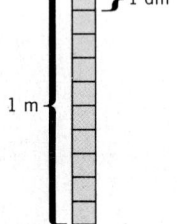

When the name of a unit is preceded by one of these prefixes, the size of the unit is modified by that decimal multiplier. For instance, the prefix *kilo* implies 1000. Therefore, a kilometer is a unit of length equal to 1000 meters. Similarly, the prefix *deci* signifies $\frac{1}{10}$th. A decimeter, then, is $\frac{1}{10}$th of a meter (1 decimeter = 0.1 meter). Notice also that each prefix has its own abbreviation which is added to the abbreviation of the unit that it modifies. Thus, 1 km = 1000 m and 1 dm = 0.1 m.

PRACTICE EXERCISE 5   Write out the full names of these abbreviations: (a) nm, (b) cm, (c) km, (d) pm, (e) mm.

1.10 LABORATORY MEASUREMENTS

The most common measurements you will make in the laboratory are those of mass, volume, temperature, and length.

You will routinely carry out many measurements in the course of your scientific studies. Among the most common measurements you will make are those of length, volume, mass, and temperature. Probably, you already have a good idea of how much a foot, a quart, or a pound is. The SI units, however, may be another story. Table 1.4 gives some of

TABLE 1.4 Some Useful Conversions

	English to Metric	Metric to English
Length	1 in. = 2.54 cm 1 yd = 0.9144 m 1 mile = 1.609 km	1 m = 39.37 in. 1 km = 0.6215 mile
Mass	1 lb = 453.6 g 1 oz = 28.35 g	1 kg = 2.205 lb
Volume	1 gal = 3.786 L 1 qt = 946.4 mL 1 oz (liquid) = 29.6 mL	1 L = 1.057 qt

the common conversions between the English system and the SI.[5] You should become familiar with the SI units and begin to develop a feel for their size.

Length

The SI base unit for length is the meter (m). Unfortunately, the meter is awkward to use for most laboratory purposes. More convenient units of length are the centimeter (cm) and the millimeter (mm). They are related to the meter in the following way.

$$1 \text{ cm} = 0.01 \text{ m}$$
$$1 \text{ mm} = 0.001 \text{ m}$$

although it is often more useful to remember the relationship

$$1 \text{ m} = 100 \text{ cm} = 1000 \text{ mm}$$

Many 12-in. rulers have one side that is marked in centimeters and millimeters (Figure 1.15). Note that there are 10 mm in each centimeter. Also, notice that the line marking the fifth millimeter division is made slightly longer to locate the midpoint between the larger centimeter lines. This is a common practice when dividing a scale into 10 parts.

It is useful to remember that 1 in. equals approximately 2.5 cm (1 in. = 2.54 cm, exactly).

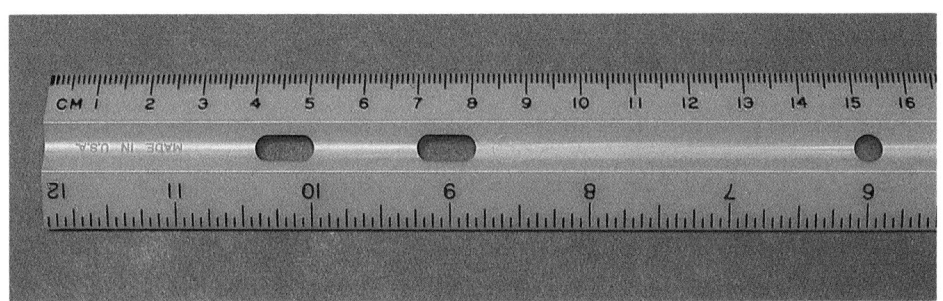

FIGURE 1.15 Centimeters and millimeters are conveniently sized units for most laboratory measurements of length. Here is a common ruler that is marked in both English units and metric units.

Volume

To calculate the volume of a room you would multiply its length by its width, and then multiply that answer by the height of the room. If each of these dimensions were measured in meters, the unit for volume would be the cubic meter (m³). This is somewhat more than a cubic yard—obviously, much too large to use for laboratory purposes.

Cubic meter (m³) is the answer to Practice Exercise 4a.

The traditional unit of volume used for the measurement of liquids in the metric system is the liter (L). In SI terms, a liter is defined as one cubic decimeter.

$$1 \text{ L} = 1 \text{ dm}^3$$

However, the liter is even too large for most laboratory purposes, and the glassware we use is usually marked in milliliters (mL, or ml).[6]

$$1 \text{ liter} = 1000 \text{ milliliters}$$
$$1 \text{ L} = 1000 \text{ mL}$$

---

[5] Originally, these conversions were established by measurement. For example, if a metric ruler is used to measure the length of an inch, it is found that 1 in. equals 2.54 cm. Later, to avoid confusion about the accuracy of such measurements, it was agreed that these relationships would be taken to be exact. For instance, 1 in. is now defined as *exactly* 2.54 cm. Exact conversions also exist for the other quantities in Table 1.3, but for simplicity many have been rounded off. For example, 1 lb = 453.59237 g, exactly.

[6] The use of the abbreviations L for liter and mL for milliliter is rather recent. Confusion between the printed letter l and the number 1 prompted the change from l for liter and ml for milliliter. You can expect to encounter the abbreviation ml quite frequently in other books and on the markings of some laboratory glassware.

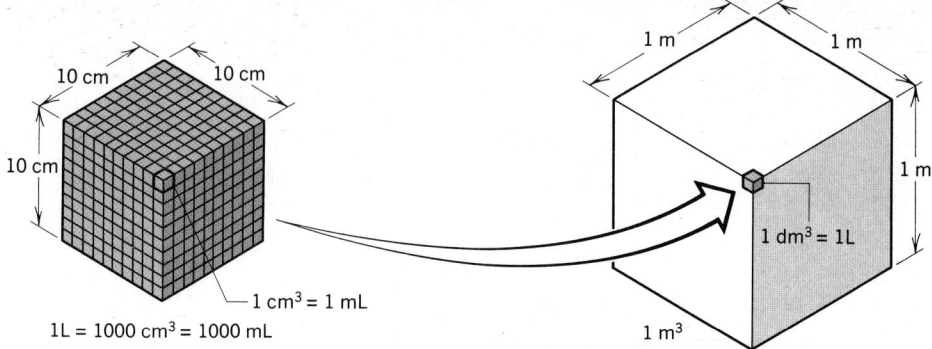

Since 1 dm = 10 cm, 1 dm³ = (10 cm)³ = 1000 cm³. Therefore, 1 mL is exactly 1 cm³ (cubic centimeter). The cm³ is sometimes abbreviated cc, although the SI frowns on this symbol.

$$1\ dm^3 = 1\ L$$
$$1\ cm^3 = 1\ mL$$

Some typical laboratory glassware used for volume measurements is shown in Figure 1.16.

FIGURE 1.16 Some apparatus used to measure volumes of liquids in the laboratory.

Mass

In the SI, the base unit for mass is the kilogram (kg). However, the gram (g), which is $\frac{1}{1000}$ of a kilogram (0.001 kg), is a more convenient unit for most laboratory measurements.

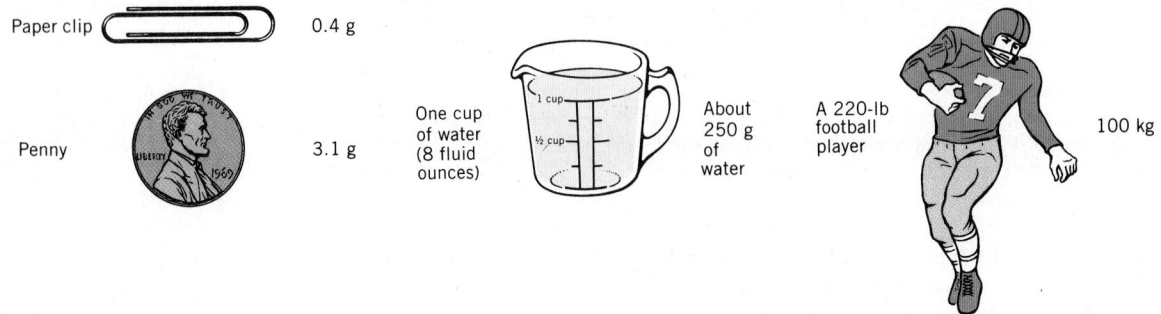

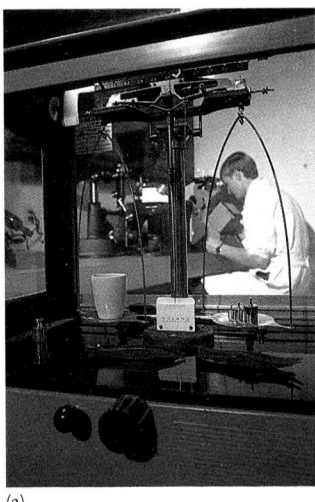
(a)

(b)

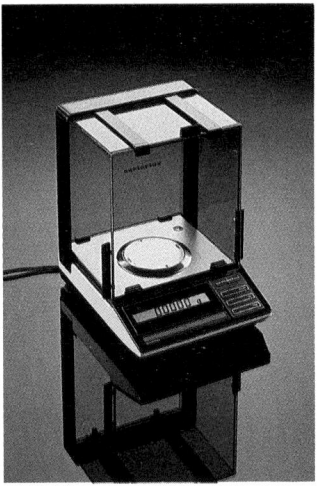
(c)

FIGURE 1.17 Typical laboratory balances. *(a)* A traditional two-pan balance. *(b)* A modern top-loading balance. *(c)* A modern analytical balance capable of measurements to the nearest 0.0001 g.

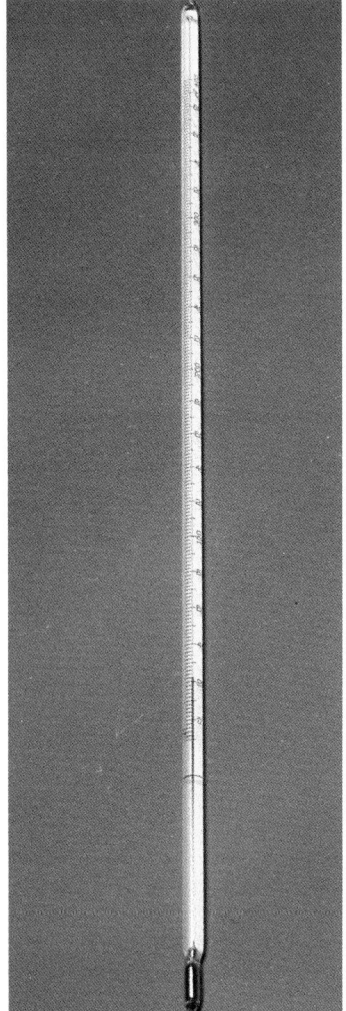

FIGURE 1.18 A typical laboratory thermometer.

Biologists who study microorganisms carry out many of their experiments at 37 °C because that is normal human body temperature.

The mass of a sample is determined by a procedure called weighing, even though we do not really measure the sample's weight as we would that of a piece of fish or some vegetables on a spring scale in a grocery store or supermarket. Instead, we *compare* the sample's weight (the force of its attraction by gravity) with the weights of standard masses. The apparatus that we use is called a balance. Examples of typical laboratory balances are shown in Figure 1.17. To find the mass of our sample, we place it on one pan and the standard masses on the other. When the weight of the sample and the weights of the standards are in balance (when they match), their masses are the same.

Notice that two of the balances in Figure 1.17 have only one pan. This is true of most modern balances. The standard masses supplied with these balances by the manufacturer are located within the balance case, where they are protected from dust (and the probing fingers of curious people). In some balances, such as the one shown in Figure 1.17*b*, there are no standard masses at all. These electronic balances work on an altogether different principle, which is too complex to discuss here.

Temperature

Temperature is something with which we are all familiar, and *hot, warm, cool,* and *cold* are terms we all use quite frequently. To actually measure the temperature, we use a thermometer—a long glass tube with a very thin bore connected to a reservoir containing a liquid, usually mercury (Figure 1.18). As the temperature rises, the liquid in the reservoir expands and its length in the column increases.

Thermometers are graduated, or marked, in degrees according to one of two temperature scales. Both scales use the temperature at which water freezes[7] and the temperature at which it boils as reference points. On the Fahrenheit scale, water freezes at 32 °F and boils at 212 °F. This is probably the scale with which you are most familiar. In recent years, however, you have probably noticed an increased use of the Celsius scale (formerly called the centigrade scale), especially in weather forecasts. This is the scale we employ in the sciences. On the Celsius scale, water freezes at 0 °C and boils at 100 °C (Figure 1.19 on page 24).

As you can see, on the Celsius scale there are 100 degree units between the freezing point of water and its boiling point, while on the Fahrenheit scale there are 180 degree units between these reference points. Each Celsius degree, therefore, is nearly twice as

---

[7] Water freezes and ice melts at the *same* temperature. A mixture of ice and water has a temperature of 32 °F or 0 °C. If heat is removed, some water freezes and more ice is formed. If heat is added, some ice melts and more water is formed. The temperature, however, remains constant.

FIGURE 1.19 Comparison of the Celsius and Fahrenheit temperature scales.

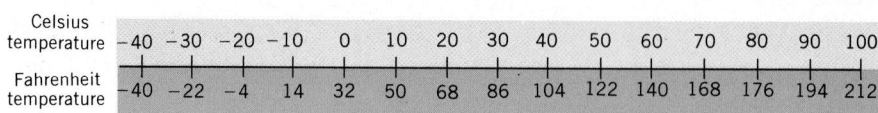

large as a Fahrenheit degree (actually, 5 Celsius degrees are equal to 9 Fahrenheit degrees). To convert between temperatures in °C, which we can represent as t_C, and temperatures in °F, t_F, we use the following relationship.

$$t_F = \frac{9}{5} t_C + 32 \qquad (1.2)$$

EXAMPLE 1.5   CONVERTING FROM °C TO °F

Problem: Thermal pollution is a serious problem near power plants. For example, trout will die if the water temperature rises above approximately 25°C. What is this temperature in °F?

Solution: We can use Equation 1.2.

$$t_F = \frac{9}{5} t_C + 32$$

Substituting, we get

$$t_F = \frac{9}{5} (25) + 32$$
$$= 45 + 32$$
$$= 77$$

Trout die at temperatures above 77 °F.

EXAMPLE 1.6   CONVERTING FROM °F TO °C

Problem: A student, writing a letter to a friend in Europe, wanted to impress on him how warm the recent weather has been. That day the temperature had climbed to 95 °F. What Celsius temperature should the student have reported to his friend?

Solution: To convert to °C, we simply substitute the Fahrenheit temperature into Equation 1.2 and solve for t_C. Thus,

$$t_F = \frac{9}{5} t_C + 32$$

$$95 = \frac{9}{5} t_C + 32$$

Rearranging, we get

$$t_C = \frac{5}{9} (95 - 32)$$

$$= \frac{5}{9} (63)$$

$$= 35$$

The temperature had been 35 °C that day.

PRACTICE EXERCISE 6   What Celsius temperature corresponds to a room temperature of 86 °F? What Fahrenheit temperature corresponds to −17.8 °C?

PRACTICE EXERCISE 7   Hypothermia is a condition caused by the loss of deep-body heat. When body temperature falls below 90 °F, unconsciousness can occur. Death from heart failure can occur when the body temperature drops below about 85 °F. What are these temperatures in °C?

The SI uses the Kelvin temperature scale, which, not surprisingly, has a degree unit called the **kelvin (K).** Notice that the symbol for this unit is K, not °K. Kelvin temperatures must be used in many equations in which the temperature enters directly into the calculations. We will come across this situation many times throughout the book.

Note that the name of the unit (the kelvin) begins with a lowercase letter. The name of the temperature scale (the Kelvin scale) and the symbol for the kelvin (K) use a capital letter K.

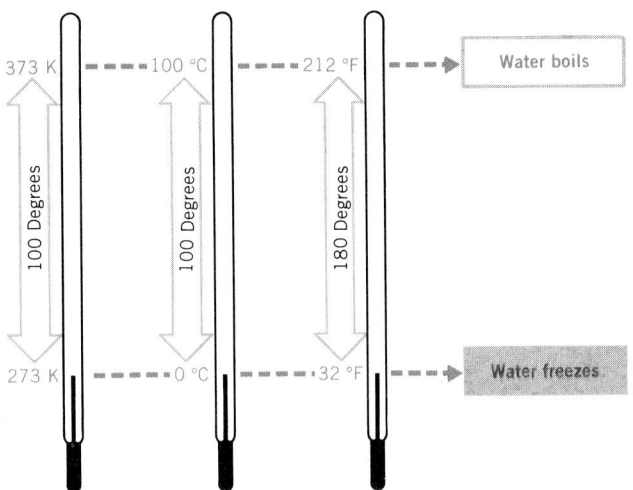

FIGURE 1.20 Comparison among Kelvin, Celsius, and Fahrenheit scales.

Figure 1.20 shows how the Kelvin, the Celsius, and the Fahrenheit temperature scales relate to each other. You can see that the kelvin is exactly the same size as the Celsius degree. The only difference between these two scales is the zero point. The zero point on the Kelvin scale—called **absolute zero**—corresponds to the lowest temperature that is possible. It is 273.15 degree units lower than the zero point on the Celsius scale. This means that 0 K equals −273.15 °C and 0 °C equals 273.15 K. Thermometers are never marked with the Kelvin scale; so if we need to express temperature in kelvins, we have to do some arithmetic. The relationship between kelvins and degrees Celsius is

$$T_K = t_C + 273.15$$

In our calculations, we'll nearly always round this to the nearest degree and use the equation

$$T_K = t_C + 273 \tag{1.3}$$

We will use a capital T for the Kelvin temperature, and a lowercase t as in t_C for the Celsius temperature.

EXAMPLE 1.7   CONVERTING BETWEEN °C AND KELVINS

Problem: Liquid oxygen (sometimes abbreviated LOX) is used in liquid-fuel rockets. Its boiling point is −183 °C. What is this temperature in kelvins?

Solution: Using Equation 1.3

$$T_K = -183 + 273$$
$$= 90$$

Liquid oxygen boils at 90 K.

PRACTICE EXERCISE 8   A substance is heated from 300 K to 315 K. What is the change in temperature expressed in °C?

1.11   SIGNIFICANT FIGURES

The number of significant figures indicates the precision or reliability of a measurement.

When a scientist writes down a number that has been obtained from a measurement, two kinds of information are given at the same time. One of these, of course, is the magnitude of the measurement. *The other is the extent of its reliability.*

Let's look at an example. Suppose you hired two different people to measure the width of your room. The first person reports that the room is 11.2 ft wide. The second tells you that he measured the width at the same place and obtained 11.13 ft. Obviously the room can have only one true width at any given place. What, then, do these two numbers tell us?

The first number, 11.2 ft, implies that the width of the room was measured with a tape measure that required the person using it to estimate the tenths place. Figure 1.21*a* illustrates how a tape measure such as this would be marked. Because the tenths place must be estimated, different people measuring the width of the room might report distances that differ by ±0.1 ft; that is, we might expect another person's estimate to be as much as 0.1 ft larger or 0.1 ft smaller than the original estimate of 11.2 ft. Therefore, we must view the reported value of 11.2 ft as uncertain by ±0.1 ft.

The second measurement, 11.13 ft, suggests that the tape measure used to otain it was more finely divided, as shown in Figure 1.21*b*. The markings on this scale allow the user to be certain of the tenths place, but require that the hundredths place be estimated. Therefore, we can expect the second measurement to be uncertain by about ±0.01 ft.

FIGURE 1.21 *(a)* Measurement of length with a tape measure marked only every 1 ft. An estimate must be made of the tenths place. *(b)* A portion of the scale of another tape measure that is marked every 0.1 ft.

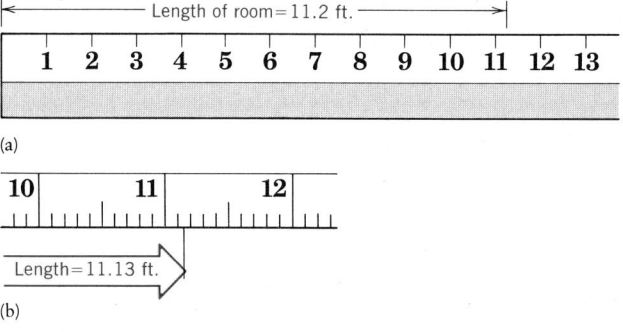

(a)

(b)

We would certainly expect the second measurement to be more reliable than the first because it has a smaller amount of uncertainty. We are told this by the way the two measurements are recorded. *The reliability of a piece of data is indicated by the number of digits used to represent it.* Digits that result from measurement such that only the digit farthest to the right is not known with certainty are called significant figures. The *number* of significant figures in a number is equal to the number of digits known for sure *plus* one that is uncertain. The first measurement, 11.2 ft, has only three significant figures. The second, more reliable measurement, 11.13 ft, has four significant figures.

Accuracy and Precision

These are frequently used terms with which you should be familiar. Accuracy refers to freedom from error, mistake, or misfunction of the measuring instrument. The closer a measurement corresponds to the actual, true value, the more accurate it is. Therefore, accurate measurements depend on careful calibration to be sure our thermometers, rulers, graduated cylinders, and other instruments provide us with correct values. Accuracy also depends on the skill of the person using them. This is a skill that is acquired through practice, which is a reason science courses usually have a lab component.

Precision refers to how *reproducible* measurements are. The first value for the width of the room mentioned earlier (11.2 ft) is reproducible to the nearest 0.1 ft. By this we mean that we can be reasonably sure that anyone else measuring the width of the room with this same tape measure will obtain a value that differs from 11.2 ft by no more than 0.1 ft. The second value (11.13 ft) is reproducible to the nearest 0.01 ft and it is considered to be more precise. In general, the more significant figures there are in a reported value, the greater its precision.

We usually assume that a very precise measurement is also highly accurate. We can be painfully wrong, however, if our instruments are improperly calibrated. For example, a poorly marked ruler might measure an 11.00 ft distance to be 10.45 ft. Even though we may be able to reproduce the measurement to the nearest 0.01 ft, the value is incorrect by a much larger margin. This is why the calibration of measuring instruments is so important.

How accurate would measurements be with this ruler?

A measurement made with great precision won't be highly accurate if the measuring device is improperly calibrated.

Combining Numbers in Calculations

When we perform several different measurements, the results usually must be combined through arithmetic calculations to give some desired final answer. For instance, calculating the area of a rectangular room requires two measurements—the length and the width—that are multiplied together to give the answer we want. If one of these measurements is very precise but the other is not, we can't expect too much precision in the calculated area. To get an idea of how reliable the calculated area is, we need to have a way of being sure that the answer reflects the precision of the original measurements. To make sure this happens, we follow certain rules to avoid answers that may contain too many or two few significant figures.

For multiplication and division, the number of significant figures in the answer should not be greater than the number of significant figures in the least precise factor. Let's look at a typical problem involving some measured numbers.

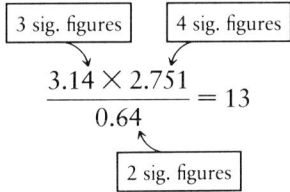

The answer to this problem that is displayed on a calculator[8] is 13.497093. However, the rule says that the answer should have only as many significant figures as the least precise factor. Since the least precise factor, 0.64, has only two significant figures, the answer should have only two. This means that we must round off the calculator answer to 13.

*For addition or subtraction, the answer should have the same number of decimal places as the quantity having the **least** number of decimal places.* As an example, consider the following problem involving the addition of measured values.

$$
\begin{array}{r}
3.247 \\
41.36 \\
+\ \underline{125.2} \\
169.8
\end{array}
$$
     ⟵ (This number has only 1 decimal place.)

     ⟵ (Answer has been rounded to 1 decimal place.)

Most calculations in this text will be carried to three significant figures.

---

[8] Calculators usually give answers with too many figures. An exception is when the answer has zeros at the right that are significant digits. An answer of 1.200 would be displayed simply as 1.2; if the zeros belong in the answer, don't forget to write them down.

The reason for this rule is fairly obvious. The digits listed beneath the 7 in 3.247 are unspecified. We know nothing about them. Therefore, when 7 is added to these unknown digits, the result is completely uncertain and we are not justified in writing the 7 in the answer. The same argument applies to the hundredths column. Beneath the 4 and 6 is another unknown digit. When this column is added the answer will be unknown. We are not justified in writing a digit in the answer in the hundredths place.

To obtain the sum in this problem we can round off the terms that are being added.

$$3.2 + 41.4 + 125.2 = 169.8$$

Many people prefer to add the original numbers and then round off the answer. If we enter the original numbers into a calculator, we obtain the sum 169.807. Rounding to the first decimal place again gives 169.8.

Not all the numbers we use come from measurements. Sometimes in performing calculations, we use numbers that come from a direct count of objects or that result from definitions. Such numbers are called exact numbers and are considered to possess an infinite number of significant figures. An example is the number of inches in one foot. By definition, there are exactly 12 in. in 1 ft, no more or less. When we write 1 ft = 12 in., there is absolutely no uncertainty in either quantity, so we can express them to any desired number of digits with complete confidence. Therefore, when using exact numbers in calculations, we can forget them as far as significant figures are concerned. We determine the number of significant figures in the answer in the usual way, but we take into account only those numbers that arise from measurement.

EXAMPLE 1.8   CALCULATIONS USING EXACT NUMBERS

Problem: A student measured her height at 64.25 in. What is her height measured in feet?

Solution: We know that to obtain feet from inches we must divide the number of inches by 12. The arithmetic involved is 64.25 ÷ 12. We can consider the 12 to contain as many significant figures as we wish. The answer, then, should contain 4 significant figures because that is the number of significant figures in 64.25

$$\frac{64.25 \text{ in.}}{12 \text{ in./ft}} = 5.354 \text{ ft}$$

PRACTICE EXERCISE 9   Perform the following calculations, making sure that each answer has the proper number of significant figures. Assume that *each* number is the result of a measurement: (a) 4.8 × 392; (b) 7.255 ÷ 81.334; (c) 0.2983 + 1.52; (d) 14.5403 − 0.022.

PRACTICE EXERCISE 10   Calculate the height of the student in Example 1.8 in units of (a) yards and (b) miles. Be sure to express the answer to the correct number of significant figures.

1.12 SCIENTIFIC NOTATION

Very large and very small numbers are easier to work with when they are put into exponential forms.

Some numbers that we deal with — such as the number of red blood cells in the body — are very large. Others — such as the mass of a single blood cell — are very tiny. To avoid the problems of doing arithmetic with such cumbersome quantities, it is often convenient to express them in scientific notation (often called exponential notation). In this system numbers are written as the product of two factors. The first is a decimal number that usually ranges between 1 and 10, and the second is 10 raised to an appropriate power. For example, 500 can be written as

$$500 = 5 \times 100 = 5 \times 10 \times 10$$
$$= 5 \times 10^2$$

Notice that the exponent on the 10 is equal to the number of places that the decimal has been moved to the left to obtain the 5. For instance, we can write 6253 as

$$6253 = 6.253 \times 10^3$$

decimal is moved 3 places to the left

$6253 = 6.253 \times 1000$
$= 6.253 \times 10^3$

When writing small numbers in scientific notation, we must move the decimal point to the *right* to give a number between 1 and 10. The exponent on the 10 is therefore negative. For example,

$$0.00063 = 6.3 \times 10^{-4}$$

decimal is moved 4 places to the right

$0.00063 = 6.3 \times 0.0001$
$= 6.3 \times 10^{-4}$

EXAMPLE 1.9   WRITING NUMBERS IN SCIENTIFIC NOTATION

Problem: There are approximately 25,000,000,000,000 red blood cells in the average human.[9] Write this number in scientific notation.

Solution: To write this number in scientific notation, we must move the decimal point 13 places to the left to locate it between the 2 and the 5. Therefore, we write the number as 2.5×10^{13}.

EXAMPLE 1.10   CONVERTING FROM SCIENTIFIC NOTATION TO DECIMAL NOTATION

Problem: A typical bacteria cell has a length of about 5×10^{-6} m. Write this number in standard decimal notation.

$10^{-6} = 0.000001$ *Solution:*   The exponential part of this number, 10^{-6}, tells us we must move the decimal 6 places to the
$5 \times 10^{-6} = 5 \times 0.000001$ left to write the number in the usual decimal form. The answer is 0.000005 m.
$= 0.000005$

PRACTICE EXERCISE 11   Write these numbers in scientific notation: (a) 23,000; (b) 21,700,000; (c) 0.0015; (d) 0.000027.

PRACTICE EXERCISE 12   Write these numbers in standard notation: (a) 2.7×10^3; (b) 3.5×10^{28}; (c) 2×10^{-12}.

Besides providing a compact way of writing very large or very small numbers, scientific notation allows us to avoid the confusion that can result when zeros may be significant figures in a number. Suppose, for instance, that you are told that the value of a measurement is 200 cm. Just how precise is this measurement? Is it 200 ± 100 cm, 200 ± 10 cm, or 200 ± 1 cm? When a number such as 200 is written in standard notation, there is no indication of its precision. However, by presenting the measurement in

[9] Outside the United States, a comma is sometimes used as a decimal marker (i.e., decimal point). So that we may avoid confusion in international dealings, numbers are written using thin spaces instead of commas to separate groups of digits: for example, 25 000 000 000 000. Small numbers such as 0.00063 are written 0.000 63.

scientific notation, we can use the decimal portion to indicate the correct number of significant figures.

$$
\begin{aligned}
200 \text{ cm} &= 2 \times 10^2 \text{ cm} &\quad \text{(1 significant figure)} \\
&= 2.0 \times 10^2 \text{ cm} &\quad \text{(2 significant figures)} \\
&= 2.00 \times 10^2 \text{ cm} &\quad \text{(3 significant figures)}
\end{aligned}
$$

Throughout the remainder of this book we will assume that values such as 200 cm, 760 mm, or 450 mL have three significant figures unless we specify otherwise. Only if we wish to indicate fewer than three significant figures will we write such numbers expressed in scientific notation. This conforms to the practice used by most scientists.

It would be a good idea for you to review the procedures for handling arithmetic computations that involve numbers in exponential form, as provided in Appendix A. *You should do this even if you have an electronic calculator that can deal with numbers expressed in scientific notation.*

1.13 THE FACTOR-LABEL METHOD

Units, like numbers, can be multiplied or divided (canceled).

You are undoubtedly accustomed to performing simple conversions such as changing feet to inches or pounds to ounces. Probably, you view such problem solving as routine because you are very familiar with the units; you've developed a "feel" for them. In chemistry, however, some of the units will be new to you, and setting up a problem to obtain the correct answer may be a challenge. The factor-label method is a system that we use to help us to perform the proper arithmetic. We use it to make sure we have set up a problem correctly. The method consists of using *conversion factors* to change the units given to us in the data to the units asked for in the answer.

given quantity × conversion factor(s) = desired quantity

A conversion factor is a fraction obtained from a *valid relationship or equality between units*. For example, suppose we wish to convert 3.00 in. to centimeters. The relationship between inches and centimeters is

1.00 in. = 2.54 cm

If we divide both sides of this equation by 1.00 in., we obtain a conversion factor.

$$
\frac{1.00 \text{ in.}}{1.00 \text{ in.}} = \frac{2.54 \text{ cm}}{1.00 \text{ in.}} = 1
$$

Notice that we have canceled the units inches from the numerator and denominator. *Units behave just as numbers do in mathematical operations.* This fact is a key part of the factor-label method. Also notice that the conversion factor 2.54 cm/1.00 in. is equal to 1.00 in./1.00 in., which is the same as 1. This means that if we multiply some quantity by 2.54 cm/1.00 in., it is effectively the same as multiplying by 1, in the sense that the absolute magnitude of the given quantity isn't changed; we only change its *units* of measurement. Now let's see how we use this conversion factor by converting 3.00 in. to centimeters.

$$
3.00 \text{ in.} \times \frac{2.54 \text{ cm}}{1.00 \text{ in.}} = 7.62 \text{ cm}
$$

$$
\binom{\text{given}}{\text{quantity}} \times \binom{\text{conversion}}{\text{factor}} = \binom{\text{desired}}{\text{quantity}}
$$

Again, we cancel the unit inches. The only units left are centimeters, which are the units we are looking for.

To solve a problem, you first have to collect all the information you need. The factor-label method will then help you set up the arithmetic correctly.

To construct a valid conversion factor, the relationship between the units must be true. For example, the statement

3 ft = 41 in.

is false. It could not possibly be used to construct a useful conversion factor.

We can also construct a conversion factor, 1.00 in./2.54 cm, from the relationship between inches and centimeters. What would have happened if we had used this factor in our calculation?

$$3.00 \text{ in.} \times \frac{1.00 \text{ in.}}{2.54 \text{ cm}} = 1.18 \text{ in.}^2/\text{cm}$$

In this case, inches do not cancel! We get units of in.²/cm because in. × in. = in.² Even though our calculator is very good at arithmetic, the quantity 1.18 is not the number of centimeters in 3.00 in. *The factor-label method lets us know the answer is wrong because the units are wrong!*

We will use the factor-label method extensively throughout this book to aid us in setting up problems. Below are several examples that further illustrate how the method is used. Even though this method may seem strange to you now, you should practice using it, because it is a powerful tool that can help you solve problems correctly.

These units are nonsense! Who ever heard of square inches per centimeter?

Most chemists and physicists use the factor-label method. There must be a good reason why.

EXAMPLE 1.11   CALCULATION USING THE FACTOR-LABEL METHOD

Problem: Convert 3.25 m to millimeters.

Solution: The problem can be stated in mathematical form as

$$3.25 \text{ m} = ? \text{ mm}$$

To solve the problem we must have a relationship between meters and millimeters. Using the information in Table 1.3, we can write

$$1 \text{ mm} = 10^{-3} \text{ m}$$

We could certainly use this to construct our conversion factor, but most people find it easier to simply remember the number of millimeters in 1 meter,

$$1000 \text{ mm} = 1 \text{ m}$$

Of course, both expressions are equivalent. Let's use the second to construct the two possible conversion factors relating meters and millimeters.

$$\frac{1 \text{ m}}{1000 \text{ mm}} \quad \text{and} \quad \frac{1000 \text{ mm}}{1 \text{ m}}$$

Which one shall we use? We know that we must eliminate meters, and this requires using the second factor so that meters cancel.

$$3.25 \text{ m} \times \frac{1000 \text{ mm}}{1 \text{ m}} = 3250 \text{ mm}$$

EXAMPLE 1.12   CALCULATION USING THE FACTOR-LABEL METHOD

Problem: A liter, which is slightly more than a quart, is defined as 1 cubic decimeter (1 dm³). How many liters are there in 1 cubic meter (1 m³)?

Solution: This problem isn't difficult if we solve it step-by-step. Let's begin by stating it in mathematical form, as we did in Example 1.11.

$$1 \text{ m}^3 = ? \text{ L}$$

We are given the relationship between cubic decimeters and liters,

$$1 \text{ L} = 1 \text{ dm}^3$$

If we can change cubic meters to cubic decimeters, we can use this relationship to finally get liters. But we don't find in our tables a relationship between cubic units; all we can get is

$$1 \text{ dm} = 10^{-1} \text{ m}$$

which we can also state as

$$10 \text{ dm} = 1 \text{ m}$$

Since we know that units undergo the same kind of mathematical operations as numbers, all we need to do to obtain the relationship between the cubic units is to cube each side. Thus,

$$\begin{aligned} 1 \text{ m}^3 &= 1 \text{ m} \times 1 \text{ m} \times 1 \text{ m} \\ &= 10 \text{ dm} \times 10 \text{ dm} \times 10 \text{ dm} \\ &= 1000 \text{ dm}^3 \end{aligned}$$

Now we can solve the problem. First, we use the factor $\dfrac{1000 \text{ dm}^3}{1 \text{ m}^3}$, which cancels m³.

$$1 \text{ m}^3 \times \frac{1000 \text{ dm}^3}{1 \text{ m}^3} = 1000 \text{ dm}^3$$

Then we use the factor $\dfrac{1 \text{ L}}{1 \text{ dm}^3}$, which cancels dm³.

$$1000 \text{ dm}^3 \times \frac{1 \text{ L}}{1 \text{ dm}^3} = 1000 \text{ L}$$

Thus, 1m³ = 1000 L.

Usually, when a problem involves the use of two or more conversion factors, they can be "strung together" to avoid having to compute intermediate results. For example, this problem would be set up as follows.

$$1 \text{ m}^3 \times \frac{1000 \text{ dm}^3}{1 \text{ m}^3} \times \frac{1 \text{ L}}{1 \text{ dm}^3} = 1000 \text{ L}$$

The first factor had to have m³ in the denominator, the second then had to have dm³ in the denominator so that the units of the answer could be liters.

EXAMPLE 1.13   CALCULATION USING THE FACTOR-LABEL METHOD

Problem: In 1975 the world record for the long jump was 29.21 ft. Use the factor-label method to convert this distance to meters.

Solution: One set of relationships that we can use to construct the necessary conversion factors is

$$\begin{aligned} 1 \text{ ft} &= 12 \text{ in.} \\ 1 \text{ in.} &= 2.54 \text{ cm} \\ 100 \text{ cm} &= 1 \text{ m} \end{aligned}$$

The given value of 29.21 ft is converted to meters by setting up successive conversion factors in such a way that all the units except meters will cancel.

Keep an eye on the units that you're trying to get rid of. Set up the units to cancel, and the arithmetic will take care of itself.

$$29.21 \text{ ft} \times \frac{12 \text{ in.}}{1 \text{ ft}} \times \frac{2.54 \text{ cm}}{1 \text{ in.}} \times \frac{1 \text{ m}}{100 \text{ cm}} = 8.903 \text{ m}$$

ft to in.
ft to cm
ft to m

Notice that if we were to stop after the first conversion factor, the units of our answer would be inches. After the second factor the units would be centimeters. After the third factor we can stop, because the units are finally meters — the units we want. Also notice that the answer has been rounded to four significant figures because that is how many there were in the measured distance.

This is not the only way to solve the problem. Other sets of conversion factors could have been chosen. For example, we could also have used

$$3 \text{ ft} = 1 \text{ yd}$$
$$1 \text{ yd} = 0.9144 \text{ m}$$

Then the problem would be set up as

$$29.21 \cancel{\text{ft}} \times \frac{1 \cancel{\text{yd}}}{3 \cancel{\text{ft}}} \times \frac{0.9144 \text{ m}}{1 \cancel{\text{yd}}} = 8.903 \text{ m}$$

Many problems that you meet, just like this one, have more than one path to the answer. There isn't any *one* correct way to set up the solution. The important thing is for you to be able to reason your way through a problem and find some set of relationships that can take you from the given information to the answer. The key is to keep in mind the units that must be gotten rid of by cancellation.

EXAMPLE 1.14   CALCULATION USING THE FACTOR-LABEL METHOD

Problem: Show that the proper units are obtained in converting 77 °F to °C using Equation 1.2.

Solution: Equation 1.2 was presented earlier without units because at that time you were not ready to see how units are used in problem solving. In this equation, the conversion is based on the relative sizes of the Fahrenheit and Celsius degrees: nine degree units on the Fahrenheit scale corresponds to exactly five degree units on the Celsius scale. This can be expressed in equation form as

$$9 \text{ °F} = 5 \text{ °C}$$

Equation 1.2

$$t_F = \frac{9}{5} t_C + 32$$

Whenever you add or subtract numbers, their units must be identical.

The 32 in Equation 1.2 arises because 0 °C corresponds to 32 °F. Therefore, the units in Equation 1.2 are

$$t_F = \frac{9 \text{ °F}}{5 \text{ °C}} t_C + 32 \text{ °F}$$

Substituting 77 °F for t_F and rearranging, we get

$$t_C = \frac{5 \text{ °C}}{9 \text{ °F}} (77 \text{ °F} - 32 \text{ °F})$$

$$= \frac{5 \text{ °C}}{9 \cancel{\text{°F}}} (45 \cancel{\text{°F}})$$

$$= 25 \text{ °C}$$

We see that the units of t_C are °C, which is correct.

PRACTICE EXERCISE 13   Use the factor-label method to perform the following conversions: (a) 3.00 yd to inches; (b) 1.25 km to centimeters; (c) 3.27 mm to feet; (d) 20.2 miles/gal to liters/kilometer.

PRACTICE EXERCISE 14   Show that Equation 1.2 gives the correct units in converting 30 °C to °F.

1.14 DENSITY AND SPECIFIC GRAVITY

Laboratory measurements provide data for the calculation of density—a useful intensive physical property.

Earlier we stressed the importance of quantitative observations in the development of chemistry, and in the preceding several sections you have learned about some of the "tools" that we need to make measurements and to use them in calculations. These include the system of units that we use in expressing measured quantities, the methods we use to assure the proper number of significant figures in values obtained from calculations, and the technique of using units to guide us to the proper arithmetic in problem solving. In this section we examine a physical property called density, which illustrates how these various tools are used.

In Section 1.4 you learned that extensive properties (those that depend on the size of a sample) are not particularly valuable for purposes of identification. Mass and volume are two examples. If you had a sample of a silvery metal, you could not use either its mass or volume alone to determine what the metal is. This is because each of these quantities depends on how big the sample is.

One of the interesting things about extensive properties is that if you take their ratios, the resulting quantity is usually independent of sample size. In effect, the sample size cancels out, and the calculated quantity becomes an intensive property. One useful property obtained this way is density, which is defined as the ratio of an object's mass to its volume. Using the symbols d for density, m for mass, and V for volume, we can express this mathematically as

$$d = \frac{m}{V} \qquad (1.4)$$

Notice that in determining an object's density we must make two measurements: one of them is the object's mass and the other is its volume.

Densities of Some Common Substances in g/cm³ at Room Temperature

Water	1.00
Aluminum	2.70
Iron	7.86
Silver	10.5
Gold	19.3
Glass	2.2
Air	0.0012

EXAMPLE 1.15   CALCULATING DENSITY

Problem: A student measured the volume of an iron nail to be 0.880 cm³ (or 0.880 mL). She found that its mass was 6.92 g. What is the density of iron?

Solution: To calculate density we have to take the ratio of mass to volume.

$$\text{density} = \frac{6.92 \text{ g}}{0.880 \text{ cm}^3}$$

$$= 7.86 \text{ g/cm}^3$$

This could also be written as

$$\text{density} = 7.86 \text{ g/mL}$$

because 1 cm³ = 1 mL.

EXAMPLE 1.16   SIGNIFICANT FIGURES IN DENSITY CALCULATIONS

Problem: A student measured the volume of a sample of isopropyl alcohol (rubbing alcohol) using a graduated cylinder and recorded a value of 14.2 mL. The mass of the liquid was then measured and found to be 11.090 g.
(a) What is the density of the liquid, calculated to the correct number of significant figures?
(b) To improve the precision of the density value, which measurement must be improved?

Solution: (a) As before, we have to take the ratio of mass to volume:

$$\text{density} = \frac{11.090 \text{ g}}{14.2 \text{ mL}}$$

$$= 0.781 \text{ g/mL}$$

A calculator actually gives a value of 0.780985915, but we have rounded the answer to three significant figures, which is the number in the least precise measurement (the volume).

(b) A density value with "improved precision" would be one with a larger number of significant figures. In this particular calculation, it is the number of significant figures in the volume that limits the number of significant figures in the calculated density. If we were to measure the volume more precisely (that is, if we were able to obtain more significant figures in the volume measurement), we would then be justified in reporting more significant figures in the density. Thus, it is the volume measurement that must be improved. Because a graduated cylinder can only be read to the nearest 0.1 mL (at best), a different measuring device would have to be used.

Making a more precise measurement of the mass will not improve the precision of the density value in this particular example.

The density of iron calculated in Example 1.15 tells us how much of this kind of matter is packed into each cubic centimeter (or milliliter). It gives us iron's *mass per unit volume* —in this case, in units of *grams per milliliter*. Remember that in such written or verbal expressions, the word *per* means "divided by." (For example, speed in miles per hour means miles/hour.)

Each pure substance has its own characteristic density. Gold, for example, is much more dense than iron. Each cubic centimeter of gold has a mass of 19.3 g; its density is 19.3 g/cm³. By comparison, the density of water is 1.00 g/cm³, and the density of air at room temperature is only about 0.0012 g/cm³.

There is more mass in 1 cm³ of gold than there is in 1 cm³ of iron.

Most substances expand slightly as their temperature is increased. This is what makes a thermometer work. The same amount of matter occupies a larger volume at a higher temperature, so the amount of mass packed into each cubic centimeter decreases as the temperature rises. This means that the density decreases with increasing temperature, and a density value reported in a reference table is normally accompanied by the temperature at which the density was measured. Actually, for solids and liquids the size of the change of density with temperature is quite small and can be ignored in most cases. Only when very precise measurements are involved is it really necessary to be concerned about it. Density is also useful because it provides a conversion factor between mass and volume, as illustrated in the following examples.

Density of Water as a Function of Temperature

Temperature (°C)	Density (g/cm³)
10	0.999700
15	0.999099
20	0.998203
25	0.997044
30	0.995646

EXAMPLE 1.17   CALCULATIONS USING DENSITY

Problem: A sample of vegetable oil had a density of 0.916 g/mL. Calculate the mass of 0.500 L of this oil.

Solution: The density tells us the mass of each milliliter of the oil. Let's express this as an equality:

$$1.00 \text{ mL oil} = 0.916 \text{ g oil}$$

This relationship can be used to write two conversion factors.

$$\frac{0.916 \text{ g oil}}{1.00 \text{ mL oil}} \quad \text{and} \quad \frac{1.00 \text{ mL oil}}{0.916 \text{ g oil}}$$

Notice that the first is simply the density written with the number 1.00 before the units mL, and the second is just the inverse of the density. The question now is which of these shall we use in our calculation.

In using the factor-label method, we choose conversion factors that get rid of undesired units and give us the units we want. The starting quantity has units of volume (0.500 L, or 500 mL). To get rid of these units and obtain mass units, we must use the first conversion factor. Since the volume unit in the density is mL, we will work with the value 500 mL for the volume of the oil.

$$500 \text{ mL oil} \times \left(\frac{0.916 \text{ g oil}}{1.00 \text{ mL oil}}\right) = 458 \text{ g oil}$$

The sample of oil has a mass of 458 g (slightly more than one pound).

EXAMPLE 1.18   CALCULATIONS USING DENSITY

Problem: The density of a typical piece of Spanish mahogany is 0.86 g/cm³. What is the volume of a piece of mahogany weighing 75 g?

Solution: The density tells us that 1 cm³ = 0.86 g for this wood. We can use this relationship to construct a conversion factor.

$$75 \text{ g mahogany} \times \frac{1 \text{ cm}^3 \text{ mahogany}}{0.86 \text{ g mahogany}} = 87 \text{ cm}^3 \text{ mahogany}$$

PRACTICE EXERCISE 15   A bar of aluminum has a volume of 1.45 mL. Its mass is 3.92 g. What is its density?

PRACTICE EXERCISE 16   What volume would 2.86 g of silver occupy? The density of silver is 10.5 g/cm³. What is the mass of 16.3 cm³ of silver?

Specific Gravity

The numerical value for the density of a substance depends on the units used for mass and volume. For example, if we express mass in grams and volume in milliliters, the density of water is 1.00 g/mL. However, if mass is given in pounds and volume in gallons, the density of water is 8.34 lb/gal; and if mass is in pounds and volume is in cubic feet, water's density is 62.4 lb/ft³. In a similar way, the densities of other substances also differ for different units. One way to avoid having to tabulate densities in all sorts of different units is to tabulate specific gravities instead. The specific gravity of a substance is defined as the ratio of the density of the substance to the density of water.

Although the density of water varies slightly with temperature, a value of 1.00 g/cm³ can be used if the water is near room temperature and only three significant figures are needed.

$$\text{sp. gr.} = \frac{d_{\text{substance}}}{d_{\text{water}}} \qquad (1.5)$$

The specific gravity tells us how much denser than water a substance is — if its specific gravity is 2.00, then it is twice as dense as water; if its specific gravity is 0.50, then it is only half as dense as water. This means that if we know the density of water in a particular set of units, we can multiply it by a substance's specific gravity to obtain the substance's density in these same units.

$$d_{\text{substance}} = \text{sp. gr.}_{\text{substance}} \times d_{\text{water}} \qquad (1.6)$$

Rearranging Equation 1.5 gives Equation 1.6.

Let's look at a simple example.

EXAMPLE 1.19   USING SPECIFIC GRAVITY

Problem: Methanol, a liquid fuel that can be made from coal, has a specific gravity of 0.792. Calculate its density in units of g/mL, lb/gal, and lb/ft³.

Solution: All we need to solve the problem is the density of water in the desired units. In the discussion above, we saw that

$$d_{\text{water}} = 1.00 \text{ g/mL}$$
$$= 8.34 \text{ lb/gal}$$
$$= 62.4 \text{ lb/ft}^3$$

To obtain the density of methanol in g/mL we use Equation 1.6.

$$d_{methanol} = 0.792 \times 1.00 \text{ g/mL}$$
$$= 0.792 \text{ g/mL}$$

Similarly, for the other units,

$$d_{methanol} = 0.792 \times 8.34 \text{ lb/gal}$$
$$= 6.61 \text{ lb/gal}$$

and

$$d_{methanol} = 0.792 \times 62.4 \text{ lb/ft}^3$$
$$= 49.4 \text{ lb/ft}^3$$

Notice that when density is expressed in g/mL, its numerical value is the same as the specific gravity.

PRACTICE EXERCISE 17   The density of aluminum is 2.70 g/mL. What is its specific gravity? What is aluminum's density expressed in lb/ft³?

PRACTICE EXERCISE 18   Ethyl acetate is a clear, colorless solvent having a fruity odor that is used in the manufacture of plastics. Its specific gravity is 0.902. What is its density in g/mL and in lb/gal?

SUMMARY

Chemical Reactions.   When substances undergo a chemical reaction (or chemical change) they change into new substances. This is accompanied by changes in the characteristics of the materials involved. Before the reaction begins we observe the characteristics of the starting materials; after the reaction is over we observe the properties of the new substances that are formed. Often these changes in characteristics are very pronounced.

Matter and Energy.   Matter is anything that has mass and occupies space. Mass is proportional to the amount of matter in a substance, and differs from weight, which is determined by how mass is attracted to the earth by gravity. Energy is the capacity to do work. An object can have energy as kinetic energy (caused by motion) or as potential energy (stored energy). Eventually, all forms of energy are converted to heat. Heat and temperature are different: heat is energy; temperature is a measure of the intensity of heat.

Properties of Matter.   When studying matter we are concerned with its physical properties and chemical properties. Physical properties can be described without reference to other substances. Chemical properties relate to how substances change into other substances. Properties are most useful if they are independent of sample size.

Elements, Compounds, and Mixtures.   An element is a substance that cannot be decomposed into something simpler by a chemical reaction. Elements combine in fixed proportions to form compounds. Elements and compounds are pure substances that may be combined in varying proportions to give mixtures. A one-phase mixture is a solution and is homogeneous. If a mixture consists of two or more phases, it is heterogeneous. Formation or separation of a mixture occurs by physical changes because the chemical properties of the components do not change. Formation or decomposition of a compound takes place by a chemical change.

Symbols, Formulas, and Equations.   Each element has been assigned an internationally agreed upon symbol. These symbols are used to write formulas for chemical compounds in which the symbol stands for an atom of the element. Subscripts are used to specify how many atoms of each kind are present. Some compounds form crystals, called hydrates, that contain molecules of water in definite proportions.

Chemical equations present before-and-after descriptions of chemical reactions. When balanced, the equation contains coefficients that make the numbers of atoms of each kind the same among the reactants and products.

Scientific Method.   Observations are made and the empirical data that are collected from many experiments are often summarized in scientific laws. Hypotheses and theories that explain the data are tested by other experiments. Although the general pattern in the development of science is the cycle of observation–explanation–observation–explanation . . . , many discoveries are made accidentally by people trained to be observant.

Measurement and Units.   Measurement is necessary for the progress of science. A modified metric system called the SI has been adopted. This system defines a set of base units (Table 1.2) that are used to derive units for other quantities, such as area or volume. The size of units can be modified with decimal multipliers (Table 1.3), which make them more closely suit our needs in particular circumstances.

Laboratory Measurements.   In the laboratory we routinely measure mass, volume, temperature, and length. Convenient units for these purposes are grams, milliliters, degrees Celsius or kelvins, and centimeters or millimeters.

Significant Figures.   The precision of a measured quantity is expressed by the number of significant figures that it contains. This is the number of digits known with certainty plus the one that possesses some uncertainty. A measured value is precise if it contains many significant figures, and is therefore highly reproducible. A measurement is accurate if its value lies very close to the true value. When measurements are combined by multiplication or division, the answer should not contain more significant figures than the least precise factor. When addition or subtraction is used, the answer is rounded to the same number of decimal places as the quantity having the fewest number of decimal places. Exact numbers do not enter into determining the number of significant figures in a calculated quantity because they have no

error. Scientific notation is useful for writing large or small numbers in compact form and for expressing unambiguously the number of significant figures in a number.

Factor-Label Method.   The factor-label method is based on the ability of units to undergo the same mathematical operations as numbers. Conversion factors are constructed from valid relationships between units. Unit cancellation serves as a guide to the use of conversion factors and aids us in correctly setting up the arithmetic for a problem.

Density and Specific Gravity.   Density is a useful intensive property that is defined as the ratio of a sample's mass to its volume. Density is a conversion factor relating mass to volume. Specific gravity is the ratio of a sample's density to that of water. The numerical values of specific gravity and density are the same if density is expressed in the units g/mL.

REVIEW EXERCISES

Answers to questions whose numbers are printed in color are given in Appendix C.
Difficult questions are marked with asterisks.

General

1.1   You are probably not a chemistry major. After some thought, give two reasons why a course in chemistry will be beneficial to *you* in the pursuit of your particular major.

1.2   With what is the science of chemistry concerned?

1.3   After you read this question, look around and then list 10 items you can see that are made of synthetic materials not found in nature.

1.4   In answering Review Exercise 1.3, what have you learned about the contribution of chemistry to modern civilization.

Chemical Reactions

1.5   What is a chemical reaction?

1.6   Lye is a common name for a substance called sodium hydroxide. Muriatic acid is the common name for another substance whose chemical name is hydrochloric acid. Either of these chemicals can cause severe burns if left in contact with the skin, but when water solutions of them are mixed together in just the right proportions, the resulting solution contains only sodium chloride. From this description, how do you know that a chemical reaction occurs between sodium hydroxide and hydrochloric acid?

1.7   If you swallow a water solution of baking soda, a gas (carbon dioxide) forms in your stomach, which causes you to burp. Has a chemical reaction occurred? Explain your answer.

Matter and Energy

1.8   Define *matter* and *energy*. Which of the following are examples of matter? (a) air, (b) a pencil, (c) a cheese sandwich, (d) a squirrel, (e) your mother.

1.9   How do mass and weight differ?

1.10   Define (a) *kinetic energy* and (b) *potential energy*. What two things determine how much kinetic energy an object possesses?

1.11   State the equation used to calculate an object's kinetic energy. Define the symbols used in the equation.

1.12   How do *temperature* and *heat* differ?

1.13   When gasoline burns, it reacts with oxygen in the air and forms carbon dioxide and water vapor. How does the potential energy of the gasoline and oxygen compare with the potential energy of the carbon dioxide and water vapor?

1.14   What is *mechanical energy*?

Properties of Matter

1.15   What is a *physical property*? What is a *chemical property*? What is the chief distinction between physical and chemical properties?

1.16   Define *intensive property* and *extensive property*.

1.17   Choose an object on your desk and list as many of its properties as you can. Which are physical properties and which are chemical properties? Which properties are independent of the size of the object?

1.18   Suppose you were told that behind a screen there were two samples of liquid, one of them water and the other gasoline. You are told that Sample 1 occupies 3 fluid ounces and Sample 2 occupies 7 fluid ounces.
    (a) What kind of property (intensive/extensive) is volume?
    (b) Can you use the information given to you in this question to determine which sample is water and which is gasoline?

1.19 Name two intensive properties that you *could* use to distinguish between water and gasoline. Give one chemical property you could use.

Elements, Compounds, and Mixtures

1.20 Define (a) *element*, (b) *compound*, (c) *mixture*, (d) *homogeneous*, (e) *heterogenous*, (f) *phase*, (g) *solution*, and (h) *physical change*.

1.21 What kind of change is needed to separate a compound into its elements?

1.22 If seawater is boiled, the water evaporates and the steam can be condensed to give pure water. Left behind is solid salt. Are the changes described here chemical or physical?

Chemical Symbols

1.23 What is the chemical symbol for each of the following elements? (a) chlorine, (b) sulfur, (c) iron, (d) silver, (e) sodium, (f) phosphorus, (g) iodine, (h) copper, (i) mercury, (j) calcium.

1.24 What is the name of each of the following elements? (a) K, (b) Zn, (c) Si, (d) Sn, (e) Mn, (f) Mg, (g) Ni, (h) Al, (i) C, (j) N.

Chemical Formulas

1.25 What is the difference between an atom and a molecule?

1.26 How many atoms of hydrogen are represented in each of the following formulas? (a) $KHCO_3$, (b) H_2SO_4, (c) C_3H_8, (d) $HC_2H_3O_2$, (e) $(NH_4)_2SO_4$, (f) $(CH_3)_3COH$.

1.27 How many atoms of each kind are represented in the following formulas? (a) Na_3PO_4, (b) $Ca(H_2PO_4)_2$, (c) C_4H_{10}, (d) $Fe_3(AsO_4)_2$, (e) $Cu(NO_3)_2$, (f) $MgSO_4 \cdot 7H_2O$

1.28 Asbestos, a known cancer-causing agent, has as a typical formula, $Ca_3Mg_5(Si_4O_{11})_2(OH)_2$. How many atoms of each element are given in this formula?

1.29 What elements would you expect to find in (a) zinc sulfide, (b) magnesium nitride, (c) calcium phosphide, and (d) carbon tetrachloride?

1.30 Write the formulas and names of the elements that exist in nature as molecules that are each composed of two atoms.

Chemical Equations

1.31 The combustion of a thin wire of magnesium metal (Mg) in an atmosphere of pure oxygen produces the brilliant light of a flashbulb. After the reaction, a thin film of magnesium oxide is seen on the inside of the bulb. The equation for the reaction is

$$2Mg + O_2 \longrightarrow 2MgO$$

(a) State in words how this equation is read.
(b) Give the formula(s) of the reactants.
(c) Give the formula(s) of the products.

1.32 What are the numbers called that are written in front of the formulas in a balanced chemical equation?

1.33 Consider the balanced equation

$$2Fe(NO_3)_3 + 3Na_2CO_3 \longrightarrow Fe_2(CO_3)_3 + 6NaNO_3$$

(a) How many atoms of Na are on each side of the equation?
(b) How many atoms of C are on each side of the equation?
(c) How many atoms of O are on each side of the equation?

1.34 Consider the balanced equation for the combustion of octane, a component of gasoline:

$$2C_8H_{18} + 25O_2 \longrightarrow 16CO_2 + 18H_2O$$

(a) How many atoms of C are on each side of the equation?
(b) How many atoms of H are on each side of the equation?
(c) How many atoms of O are on each side of the equation?

1.35 Rewrite the chemical equation in the preceding question so that it specifies gasoline and water as liquids and oxygen and carbon dioxide (CO_2) as gases.

Scientific Method

1.36 What steps are involved in the scientific method?

1.37 What is the function of a laboratory?

1.38 Define (a) *data*, (b) *hypothesis*, (c) *law*, and (d) *theory*.

Metric and SI Units

1.39 What advantages does the metric system of units have over the English system?

1.40 Several commercial products that use metric units are pictured in Figure 1.13. Can you find any others among the items with which you come in contact each day?

1.41 What does the abbreviation *SI* stand for?

1.42 Which SI base unit is defined in terms of a physical object?

1.43 On page 7 we saw that we could calculate kinetic energy by using the equation $KE = \frac{1}{2}mv^2$. The SI unit for mass is the kilogram (kg), and the derived unit for velocity or speed is meter/second (m/s). What is the derived unit for energy?

1.44 What is the meaning of these prefixes? (a) *centi*, (b) *milli*, (c) *kilo*, (d) *micro*, (e) *nano*, (f) *pico* (g) *mega*.

1.45 What abbreviation is used for each of the prefixes named in Review Exercise 1.44?

1.46 What units are most useful in the laboratory for measuring (a) length, (b) volume, and (c) mass?

1.47 How do mass and weight differ?

1.48 How is mass measured? What is the difference between a balance and a scale?

1.49 What number should replace the question mark in each of the following?
(a) 1 cm = ? m   (c) 1 m = ? pm   (e) 1 g = ? kg
(b) 1 km = ? m   (d) 1 dm = ? m   (f) 1 mg = ? g

1.50 What numbers should replace the question marks below?
(a) 1 nm = ? m   (d) 1 Mg = ? g
(b) 1 μg = ? g   (e) 1 mg = ? g
(c) 1 kg = ? g   (f) 1 dg = ? g

Temperature

1.51 What reference do we use in calibrating the scale of a thermometer? What temperature on the Celsius scale do we assign to each of these reference points?

1.52 Perform the following conversions.
  (a) 24 °C to °F   (d) 50 °F to °C
  (b) 10 °C to °F   (e) 30 °C to K
  (c) 41 °F to °C   (f) −10 °C to K

1.53 Perform the following conversions.
  (a) 85 °F to °C   (d) 215 K to °C
  (b) −5 °F to °C   (e) 315 K to °C
  (c) −40 °C to °F   (f) 25 °C to K

1.54 A clinical thermometer registers a patient's temperature to be 37.13 °C. What is this temperature in °F?

1.55 The coldest permanently inhabited place on earth is the Siberian village of Oymyakon in the USSR. In 1964 the temperature reached a shivering −96 °F! What is this temperature in °C?

1.56 Helium has the lowest boiling point of any liquid. It boils at 4 K. What is its boiling point in °C and °F?

1.57 When an object is heated to high temperature, it glows and gives off light. The color balance of this light depends on the temperature of the glowing object. Photographic lighting is described, in terms of its color balance, as a temperature in kelvins. For example, a certain electronic flash gives a color balance (called color temperature) rated at 5800 K. What is this temperature expressed in °C?

Significant Figures

1.58 Define *significant figures*.

1.59 What is *accuracy*? What is *precision*?

1.60 How many significant figures do the following measured quantities have?
  (a) 2.75 cm   (d) 0.0021 kg
  (b) 39.24 mm   (e) 0.0006080 m
  (c) 12.0 g   (f) 0.002 mL

1.61 How many significant figures do the following measured quantities have?
  (a) 0.240 g   (c) 0.0008 kg   (e) 1.00005 L
  (b) 11.303 m   (d) 615.0 mg   (f) 3.505 mm

1.62 Perform the following arithmetic and round off the answers to the correct number of significant figures. Assume that all of the numbers were obtained by measurement.
  (a) 0.022×315
  (b) $83.25 - 0.1075$
  (c) $(84.4 \times 0.02)/(31.22 \times 9.8)$
  (d) $(33.4 + 112.7 + 0.002)/(6.488)$
  (e) $(315.44 - 208.1) \times 8.8175$

1.63 Perform the following arithmetic and round off the answers to the correct number of significant figures. Assume that all of the numbers were obtained by measurement.
  (a) $3.58/1.739$
  (b) $4.02 + 0.001$
  (c) $(22.4 \times 8.3)/(1.142 \times 0.002)$
  (d) $(1.345 + 0.022)/(13.36 \times 8.4115)$
  (e) $(74.335 - 74.332)/(4.75 \times 1.114)$

Scientific Notation

1.64 Express the following numbers in scientific notation *without using a calculator*. Assume three significant figures in each number.
  (a) 245   (d) 45,000,000
  (b) 31,000   (e) 0.0000000400
  (c) 0.00287   (f) 324,000

1.65 Express the following numbers in scientific notation *without using a calculator*. Assume, in this problem, that only the nonzero digits are significant figures.
  (a) 3389   (c) 81,300,000   (e) 2.33
  (b) 0.000025   (d) 0.0225   (f) 18,300

1.66 Write the following numbers in standard, nonexponential form.
  (a) 2.1×10^3 (d) 4.6×10^{-10}
  (b) 3.35×10^{-4} (e) 34.6×10^{-2}
  (c) 3.8×10^6 (f) 8.5×10^4

1.67 Write the following numbers in standard, nonexponential form:
  (a) 4.27×10^{-4} (d) 2.85×10^{-3}
  (b) 7.11×10^7 (e) 5.0000×10^4
  (c) 33.5×10^{-6} (f) 17.2×10^5

1.68 Without using a calculator, perform the following arithmetic.
  (a) $\dfrac{(1.0 \times 10^7) \times (4.0 \times 10^5)}{(2.0 \times 10^8)}$
  (b) $\dfrac{(4.0 \times 10^{-5}) \times (6.0 \times 10^{10})}{(3.0 \times 10^{-2})}$
  (c) $\dfrac{(5.0 \times 10^{-4}) \times (2.0 \times 10^{-6})}{(1.0 \times 10^{-12})}$
  (d) $(3.0 \times 10^4) + (2.1 \times 10^5)$
  (e) $(8.0 \times 10^{12})/(2.0 \times 10^{-3})^2$

1.69 Without using a calculator, perform the following arithmetic.
  (a) $\dfrac{(8.0 \times 10^6)}{(2.0 \times 10^5) \times (1.0 \times 10^3)}$
  (b) $\dfrac{(1.6 \times 10^{15}) \times (1.0 \times 10^{-5})}{(8.0 \times 10^4)}$
  (c) $\dfrac{(4.5 \times 10^{28})}{(3.0 \times 10^{-6})^2}$
  (d) $(1.4 \times 10^5) - (3.0 \times 10^4)$
  (e) $(3.3 \times 10^{-4}) + (2.52 \times 10^{-2})$

Factor-Label Method

1.70 What is a conversion factor? We *cannot* use the equation 1 yd = 2 ft to construct a proper conversion factor relating yards to feet. Why? Can we construct a conversion factor relating centimeters to meters from the equation 1 cm = 1000 m?

1.71 In 1 hour there are 3600 seconds. By what conversion factor would you multiply 250 seconds to convert it to hours? By what conversion factor would you multiply 3.84 hours to convert it to seconds?

1.72 If you were to convert the measured length 4.165 ft to yards by multiplying by the conversion factor (1 yd/3 ft),

is just the opposite of the one written above for the dehydration of gypsum.

$$(CaSO_4)_2 \cdot H_2O + 3H_2O \longrightarrow 2CaSO_4 \cdot 2H_2O$$

This reaction, like so many others that you will encounter in your study of chemistry, produces heat that is released into the surroundings, which is why the plaster becomes warm as it hardens. Another interesting thing is that as the "dihydrate" (the hydrate with two waters) is formed, the crystals expand somewhat and become interlocked. This explains why plaster casts retain all the details of the molds that they fill—the plaster expands to fill all the tiny crevices. The interlocking of the expanding crystals also explains why you can start with powdered plaster of Paris and finish with a solid. The interlocking of the tiny crystals holds them together tightly in a solid mass.

Partial dehydration of the dihydrate gives plaster of Paris, but complete dehydration is also possible if the gypsum is heated strongly at a high temperature. The product of this heating—pure $CaSO_4$—is used by industry in making plasterboard, which has almost completely replaced wet plaster in coating interior walls and ceilings in buildings constructed today. If you live in a modern home or apartment building, your walls and ceilings are almost surely covered with plasterboard. The completely dehydrated gypsum is also used to make an especially tough plaster finish called Keen's cement, which is applied to walls of hospitals, stores, and railroad stations. Dehydrated gypsum is even used as a filler in products such as candy and paint.

So far we've described the uses of dehydrated forms of gypsum. The mineral itself also has some important uses, as well as some not-too-beneficial effects on the environment. On the positive side is the fine-grained crystalline form of the mineral called alabaster. It is an exceptionally soft stone, easily carved by sculptors. When highly polished, it takes on a beautiful appearance as we can see in Figure 1b. More commonly, gypsum has a chalky appearance when dug from the earth. One of its uses in this form is as a fertilizer where it provides a source of calcium for plants as it gradually dissolves. Other uses are in portland cement, to keep it from hardening too quickly, and as a pigment in some paints.

We don't normally think of plaster as being particularly soluble in water. If you place a piece of white plaster in a glass of water, nothing much appears to happen. Nevertheless, if you were to analyze the water afterward, you would find that a very small amount of the solid had dissolved. This slight ability of gypsum to dissolve in water is responsible for the appearance of small amounts of calcium in municipal water supplies in parts of the country where there are gypsum deposits underground. Rain water seeping through the gypsum dissolves some of it, and this $CaSO_4$ collects in the ground water, later to be brought to the surface by wells. The problem with this dissolved calcium sulfate (more specifically, with the calcium in the

FIGURE 1b The translucent nature of polished alabaster gives this carved figure an inner glow.

water) is that it interferes with the action of soap by forming an insoluble solid with it. Water that displays this problem is called *hard water*. Exactly what the chemical reason for its "hardness" and how the water is "softened" chemically are discussed in Special Topic 11.1 on page 395. If you've had an earlier course in chemistry, you might find that interesting reading now. Otherwise, keep our current discussion in mind when you come to it in Chapter 11.

Questions

1. How many atoms of each element are represented in the formula for gypsum?
2. Write the chemical equation for the dehydration of gypsum to give plaster of Paris.
3. What chemical reaction occurs when plaster of Paris is mixed with water?
4. Why does plaster of Paris form a solid mass when the moist paste sets?
5. What is Keen's cement?
6. What is the name of the stone used by sculptors that is composed of gypsum?
7. Why is it difficult to use soap effectively in well water taken from regions that contain underground deposits of gypsum?

Hydrogen and oxygen, two gaseous elements, form when electricity is passed through certain aqueous solutions, as seen in this close up. Together with roughly 100 other elements, they are assigned particular places in one of chemistry's most useful tools, the periodic table, the major topic of this chapter.

ATOMS, ELEMENTS, AND THE PERIODIC TABLE

2.1 LAWS OF CHEMICAL COMBINATION

In a compound, the elements always are combined in a constant ratio by mass.

In this chapter our goal is to continue to expand your knowledge of chemical and physical properties, which we began in Chapter 1 with an illustration of the reaction between sodium and chlorine. We start here with a discussion of the basic building blocks of all matter—atoms and the subatomic particles of which they are composed. With this knowledge we can then begin to explore the kinds of substances formed by atoms when they combine with each other. You will also learn how chemical information is organized so that chemical and physical properties can be understood.

In modern science, we have come to take for granted the existence of atoms. Even in nontechnical circles we routinely hear talk of *atomic energy* and *nuclear energy*. However, human knowledge of the existence of atoms is relatively recent, and science had not progressed very far before firm experimental evidence of atoms was uncovered.

FIGURE 2.1 Cutting a piece of gold successively into equal parts gives smaller and smaller pieces, but each is still gold. (Eventually, a magic knife would be needed to work on pieces invisible to the eye or under a microscope.) In time, however, a piece of gold would be reached that couldn't be cut to give smaller pieces of gold; in fact, it couldn't be cut at all. This is an atom of gold.

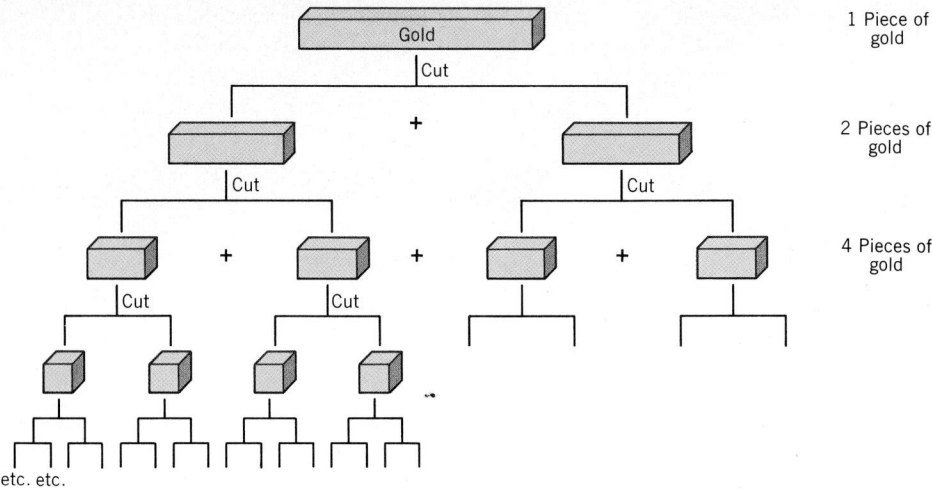

The concept of atoms actually began nearly 2500 years ago. In that time, some Greek philosophers, for reasons of their own, wondered if matter could be cut into an infinite number of pieces. For example, it is easy to tell that if a piece of, say, gold is cut in two, the smaller pieces are still gold (Figure 2.1). If each of these is cut in two again, the four still-smaller pieces are also samples of gold. But how far could this be carried, even in the imagination? The Greek thinkers reasoned that there had to be a limit — that there must exist a sample of gold that, if cut in two, would not give a new (and smaller) piece of gold. In fact, they reasoned that the smallest sample of gold can't be cut at all. The Greek word for "not cut" is *atomos*, from which we get our word *atom*.

The concept of atoms remained a philosophical belief, having limited scientific usefulness, until the discovery of two quantitative laws of chemical combination — the law of conservation of mass and the law of definite proportions. Evidence for these laws came from the work of many scientists in the eighteenth and early nineteenth centuries. Their discoveries supplied powerful clues to the fundamental nature of all matter, because they provided scientists with some of the earliest and most convincing evidence that elements consist of atoms.

To be *conserved* means to be kept from being lost.

Law of Conservation of Mass.   No detectable gain or loss in mass occurs in chemical reactions. Mass is conserved.

Law of Definite Proportions.   In a given chemical compound, the elements are always combined in the same proportion by mass.

The law of conservation of mass means that if we place chemical reactants in a sealed vessel that permits no matter to escape, the mass of the vessel and its contents after a reaction will be identical to its mass before. Although this may seem quite obvious to us now, it wasn't quite so clear in the early history of modern chemistry. When chemical changes were being studied back in the seventeenth and eighteenth centuries, often one or more of the chemicals involved were gases that entered or escaped unnoticed, so consistent patterns in the ways that the masses changed were hard to find. Not until

At one time gases weren't even recognized as substances.

scientists made sure that *all* reactants and *all* products were included when masses were measured could the law of conservation of mass rest on solid evidence.

The law of definite proportions can be illustrated by a beautiful, naturally occurring mineral that geologists have called iron pyrite and that gold seekers have ruefully dubbed "fool's gold." (Figure 1.7 on page 11 shows a sample of iron pyrite. Although it has the beautiful appearance of gold, it sells for about $5/lb.) Let us see how it illustrates the law of definite proportions.

If we were to take samples of iron pyrite, each with a mass of 1.000 g, and decompose them into their elements, we would always obtain 0.4655 g of iron and 0.5345 g of sulfur. The ratio of sulfur to iron is invariably

$$\frac{0.5345 \text{ g sulfur}}{0.4655 \text{ g iron}} = \frac{1.148 \text{ g of sulfur}}{1.000 \text{ g of iron}}$$

Provided that samples of iron pyrite are pure and are not contaminated by other minerals or by soil or sand, they all contain sulfur and iron combined in the same ratio by mass, namely, 1.148 g of sulfur to 1.000 g of iron. The size of the pyrite sample doesn't matter. Thus, a 20.000-g sample of pure iron pyrite contains 10.690 g of sulfur combined with 9.310 g of iron. The ratio of 10.690 g of sulfur to 9.310 g of iron can be reduced to 1.148 g sulfur/1.000 g iron, which is the same as for the smaller sample. Regardless of the source or the size of the sample, the elements sulfur and iron are always combined in pure iron pyrite in the same proportion by mass. Figure 2.2 illustrates these relationships.

$$\frac{10.690}{9.310} = 1.148$$

What is true about pure iron pyrite is true about all other compounds — they obey the law of definite proportions. For example, no matter where in the world a sample of pure water is obtained, its elements — hydrogen and oxygen — are always combined in the definite ratio of 7.937 g of oxygen to 1.000 g of hydrogen. Table salt, to take another example, is a chemical combination of the elements sodium and chlorine in an invariable ratio of 1.542 g of chlorine to 1.000 g of sodium. Common aspirin is made of three elements always in a proportion of 13.406 g of carbon to 1.000 g of hydrogen to 7.937 g of oxygen.

The consistency summarized by the law of definite proportions has proved to be so complete that this law is now used to *define* what is a chemical compound and what is not. A chemical compound is any substance consisting of two or more chemically combined elements present in a definite proportion by mass, as we learned in Chapter 1.

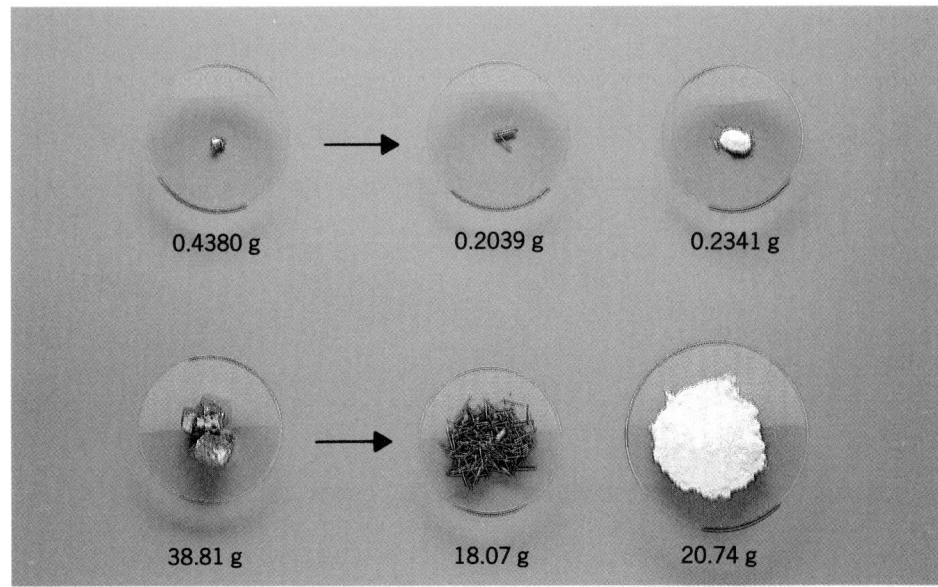

| 0.4380 g | 0.2039 g | 0.2341 g |

| 38.81 g | 18.07 g | 20.74 g |

FIGURE 2.2 Two samples of iron pyrite, FeS_2. The first sample has a mass of 0.4380 g and when decomposed yields 0.2039 g of iron and 0.2341 g of sulfur. The second sample weighs 38.81 g and when decomposed yields 18.07 g of iron and 20.74 g of sulfur. In each sample, the elements are combined in the same ratio by mass:

0.2341 g S/0.2039 g Fe =
20.74 g S/18.07 g Fe =
1.148 g Fe/1.000 g S

Like charges repel!

2.2 DALTON'S ATOMIC THEORY

Dalton proposed that each element consists of indestructible particles called atoms that are identical in mass.

The law of definite proportions begs a question also raised by the law of conservation of mass: "What must be true about the nature of matter, given the truth of this law?" In other words, what is matter made of?

At the beginning of the nineteenth century, John Dalton (1766–1844), an English scientist, used the Greek concept of atoms to make sense out of the emerging law of definite proportions. Dalton, in fact, believed that this law and the law of conservation of mass *compelled* a belief in atoms as the real and ultimate particles of which all elements and compounds are made.

Dalton reasoned that if atoms really exist, they must have certain properties to account for the two laws of chemical combination. He described such properties, and the list constitutes what we now call Dalton's atomic theory.

Dalton's Atomic Theory

1. Matter consists of definite particles called atoms.
2. Atoms are indestructible. In chemical reactions, the atoms rearrange but they do not themselves break apart.
3. The atoms of one particular element are all identical in mass (and other properties).
4. The atoms of different elements differ in mass (and other properties).
5. When atoms of different elements combine to form compounds, new and more complex particles form. However, their constituent atoms are always present in a definite numerical ratio.

First, let's see how this theory explains the law of conservation of mass. According to the theory, atoms are indestructible and each has its own characteristic mass. Further, a chemical reaction simply involves the reordering of atoms from one combination to another. If there are a certain number of atoms present at the beginning of a reaction, and no atoms are allowed to enter or escape, then the same atoms must also be present after the reaction is over. Since each atom has its own characteristic mass, which does not change during a reaction, the total mass *must* remain constant. This, of course, is the law of conservation of mass.

The law of definite proportions is equally easy to explain. One of the key parts of the theory is that every atom of a given element has the same mass, which differs from the mass of the atoms of every other element. The theory also states that in a given compound the atoms of the various elements are always combined in a definite numerical ratio. Suppose we let the symbol A stand for one element and B for another. Also suppose that these elements form a compound in which there is one atom of A for each atom of B. In other words, the formula of the compound is AB. Let's also suppose that an atom of B has twice the mass of an atom of A. This means that, regardless of the size of the sample of the compound, every time we find an atom of A with a certain mass we will also find an atom of B with twice as much mass. In this compound, the ratio of the mass of B to the mass of A must always be 2 to 1. This condition is exactly what the law of definite proportions requires—in any sample of the compound, the elements A and B are always present in the same proportion by mass.

Law of Multiple Proportions

One powerful piece of evidence for Dalton's theory came when Dalton and other scientists studied elements that could combine to give two (or more) compounds. Iron and sulfur, for example, form more than one compound. If we assume simple ratios, a few of the many theoretically possible combinations of iron and sulfur atoms would be as follows.

				Atom Ratio
Fe + S	⟶	Fe S [FeS]		1 Fe to 1 S
Fe + S + S	⟶	S Fe S [FeS$_2$]		1 Fe to 2 S
Fe + Fe + S	⟶	Fe S Fe [Fe$_2$S]		2 Fe to 1 S
Fe + Fe + S + S + S	⟶	S Fe S Fe S [Fe$_2$S$_3$]		2 Fe to 3 S

Fe is an iron atom and S is a sulfur atom.

Suppose that FeS and FeS$_2$ exist. If Dalton is right, then to make FeS$_2$ from a *fixed* mass of iron — say, 1.000 g of iron — should require exactly twice the mass of sulfur needed to make FeS from 1.000 g of iron. This is because, for a given amount of iron, there are twice as many sulfur atoms in FeS$_2$ as in FeS. In other words, the ratio of the *mass* of sulfur in FeS$_2$ to its mass in FeS should be 2 to 1 — a ratio of simple whole numbers, provided each compound contains the same mass of iron. And this result has been repeatedly observed. Here are the mass ratios in two known compounds of iron and sulfur.

Mineral	Mass Ratio of the Elements	
Pyrite (FeS$_2$)	1.000 g iron to 1.148 g sulfur	Mass ratio of 2 to 1 for sulfur
Troilite (FeS)	1.000 g iron to 0.574 g sulfur	

Pyrite and troilite are common, not official, names.

Compare the masses of sulfur that combine in different ways with 1.000 g of iron.

$$\frac{1.148 \text{ g sulfur}}{0.574 \text{ g sulfur}} = \frac{2}{1}$$

It would be extremely hard to imagine any other view of matter besides Dalton's atomic theory that could account for a simple, whole-number ratio like this.

Many other examples of elements forming two or more compounds are known. One example involves the elements tin and oxygen. In one compound, these two combine in a ratio of 1.000 g of oxygen to 7.420 g of tin. In another compound, the mass ratio is 1.000 g of oxygen to 3.710 g of tin. Now compare the two masses of tin that combine with 1.000 g of oxygen.

"Tin cans" are actually made of steel with a thin tin coating that resists corrosion.

$$\frac{7.420 \text{ g tin}}{3.710 \text{ g tin}} = \frac{2}{1}$$

The ratio is one of simple whole numbers, 2 to 1. Out of these and many other examples came the third great law of chemical combinations, the law of multiple proportions.

Law of Multiple Proportions.   Whenever two elements form more than one compound, the different masses of one element that combine with the same mass of the other element are in the ratio of small whole numbers.

Some Problems with Dalton's Theory

Dalton turned out to be wrong in some ways, but luckily the errors did not affect the basic postulate that atoms do exist. One error was the idea that atoms are indestructible.

In this century, scientists have invented "atom-smashing" machines that develop enough energy to split atoms into a number of fragments, with such names as electrons, protons, and neutrons. Dalton's theory worked because in all *chemical* changes atoms do not fragment into such smaller particles, at least not in any way that observably affects mass relationships. In many chemical reactions atoms do pass some of their tiniest, least massive particles — electrons — back and forth, but as we will see, the total number of electrons is conserved. Therefore, the total mass of a compound is unchanged compared to the sum of the masses of the elements used to make it.

Isotopes

Another incorrect postulate in Dalton's atomic theory was the idea that all of the atoms of an element have identical masses. Actually, most elements are uniformly intermingled mixtures of two or more unique substances called the isotopes of the element. All of the isotopes of the given element have very nearly the same *chemical* properties, but their atoms have slightly different masses. An iron nail, for example, is made up of a mixture of four isotopes. One contributes 91.66% of the atoms in any iron sample. If we arbitrarily assign to the atoms of this particular isotope a mass of 1.000000 mass units, then the relationships by mass of the four isotopes of iron are as follows.

Isotope Number	Percentage of All Atoms	Mass Units per Atom
1	5.82	0.964382
2	91.66	1.000000
3	2.19	1.017887
4	0.33	1.035727

The existence of isotopes did not affect the development of Dalton's theory for two important reasons. First, regardless of where on the earth or in the atmosphere an element happens to be found, the relative proportions of its different isotopes are essentially constant. As a result, every sample of a particular element has the same isotopic composition, and the *average* mass per atom is the same from sample to sample. Second, all the isotopes of a given element have virtually identical *chemical* properties — all give the same kinds of chemical reactions. For example, any one of the four isotopes of iron can combine with sulfur to give either FeS_2 or FeS. As a result, even though the existence of isotopes makes Dalton's third postulate untrue, an element behaves *chemically* as if it were true.

2.3 WHAT ATOMS ARE MADE OF

The chief particles that make up atoms are electrons, protons, and neutrons.

To understand why the isotopes of an element consist of atoms of different masses but are chemically alike, we have to look briefly at the composition of an atom. We have mentioned that contrary to Dalton's original hypothesis, atoms are not the smallest bits of matter. Atoms are themselves composed of still smaller pieces that are called *subatomic particles*. At this point we will not go into the experiments that disclosed what these subatomic particles are like — this will be done in Chapter 6 — but the results of these experiments will be useful to know now.

There are three principal subatomic particles: protons, neutrons, and electrons. The protons and neutrons are found together in a very tiny, extremely dense core particle called a nucleus located in the center of the atom. Because they are found in nuclei, protons and neutrons are sometimes called nucleons. Electrons, each of which has a

Protons are in *all* nuclei. Except for ordinary hydrogen, all nuclei also contain neutrons.

TABLE 2.1 Properties of Subatomic Particles

Particle	Mass (g)	Mass (amu)	Electrical Charge	Symbol
Electron	$9.1093897 \times 10^{-28}$	0.0005485712	1 $-$	$_{-1}^{0}e$
Proton	1.672649×10^{-24}	1.007271605	1 $+$	$_{1}^{1}H$, $_{1}^{1}p$
Neutron	1.674954×10^{-24}	1.008665	0	$_{0}^{1}n$

much smaller mass than a proton or neutron, occupy the volume near the nucleus. Just how the electrons are distributed throughout this volume is an exceedingly important matter, because the electron distribution determines the chemical properties of the atom. We will study the distribution of electrons in atoms in detail in Chapter 6. In summary, then, this is the basic picture that we have of the structure of an atom: an atom consists of a dense nucleus surrounded by electrons.

Two important properties of subatomic particles are their relative masses and the electrical charges they carry. Table 2.1 gives their masses in grams in the second column. Because the numbers are so exceptionally small, they have been recalculated on the basis of another unit of mass, called the atomic mass unit, or amu, where

$$1 \text{ amu} = 1.6605402 \times 10^{-24} \text{ g}$$

The third column of Table 2.1 gives the masses of the subatomic particles in amu. Notice that even on the amu scale the mass of an electron is much smaller than that of either nucleon—only about 1/1830 that of a proton or neutron.

In the SI, the symbol for atomic mass unit is u. We will use the older symbol amu because it is easier to remember.

Electrical Properties of Subatomic Particles

If you have ever felt a jolt or seen a spark when you touch a metal object or a person after walking across certain rugs when the air is very dry, you are familiar with an electrical charge. The spark occurs when an electrical charge moves through the air from one place to another. There are two kinds of charge, and you have no doubt experienced both. Have you ever tried to throw away a small piece of paper or plastic, only to have it stick stubbornly to your fingers? Very annoying! This happens because you are carrying one kind of charge and the plastic or paper has picked up the opposite kind. *Opposite charges attract.* This is a very important rule in chemistry and physics.

Perhaps you have seen your hair stand on end after blow drying. If you're ever caught outside on a hill or mountain during an electrical disturbance, you'll see the same thing. (You will also be in deep trouble if you don't get down!) The individual hairs behave as if they repel each other, because each hair is carrying the same kind of electrical charge. *Like charges repel.* This is another, equally important rule. We will use these two rules of electrical behavior to explain an astonishing amount of chemistry.

Attraction and repulsion are opposites, and one way to signify opposites is to assign one a positive sign and the other a negative sign. Two of the three subatomic particles are electrically charged. The proton carries one unit of positive charge, 1 $+$. The electron has one unit of the opposite charge, one negative charge, or 1 $-$. Because the neutron is uncharged or neutral (hence its name, neutron), we can say that its charge is zero (0). Since all of the atom's protons are in its nucleus, the nucleus is a positively charged particle.

If free to move, protons repel each other—*like charges repel.* (In a nucleus, the protons are not free to move; what holds them is called a nuclear force, and it involves some of the other subatomic particles that we will not study.) Similarly, if free to move, electrons repel each other. However, protons and electrons attract each other—*unlike charges attract.* This keeps the atom intact.

Like charges repel!

Atomic Numbers and Mass Numbers

The atoms of all isotopes of the same element have an identical number of protons. Atoms found in different elements have different numbers of protons. Thus, each element has associated with it a unique number, called its atomic number, that equals the number of protons in the nuclei of any of its atoms.

$$\text{atomic number} = \text{number of protons}$$

What makes isotopes of the same element different are the numbers of neutrons in their nuclei. The atoms of each isotope of a given element have an identical number of protons but a number of neutrons that is unique to the isotope. The numerical sum of the protons and neutrons in the atoms of a particular isotope is called the mass number of the isotope.

The mass number is just the number of nucleons, and it has no units.

$$\text{mass number} = \text{number of protons} + \text{number of neutrons}$$

Thus, every isotope is fully defined by two numbers, its atomic number and its mass number. Sometimes these numbers are added to the left of the chemical symbol of an element as a superscript and a subscript, respectively. The isotope of uranium used in nuclear reactors, for example, can be symbolized as follows.

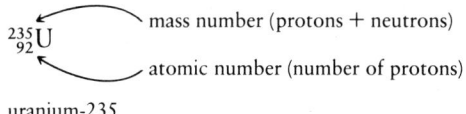

$$^{235}_{92}\text{U} \quad \text{mass number (protons + neutrons)}$$
$$\text{atomic number (number of protons)}$$

uranium-235

The name of this isotope is, as indicated, uranium-235. Each atom contains 92 protons and $(235 - 92) = 143$ neutrons. In writing the symbol for the isotope, the atomic number is often omitted because it is really redundant. Every atom of uranium has 92 protons and an atomic number of 92, and every atom that has an atomic number of 92 must be uranium. It would not be unusual, therefore, for this uranium-235 isotope to be represented simply as ^{235}U.

PRACTICE EXERCISE 1   Write the symbol for the isotope of plutonium (Pu) with 146 neutrons. The atomic number of plutonium is 94.

PRACTICE EXERCISE 2   How many neutrons are in each atom of $^{4}_{2}\text{He}$?

Atoms are electrically neutral particles, and the only way they can be neutral is to have exactly equal numbers of protons and electrons. Under this condition, the total positive charge (from the protons) equals and therefore cancels the total negative charge (from the electrons).

$$\text{atomic number} = \text{number of protons} = \text{number of electrons}$$

The algebraic sum of 92+ and 92− is zero.

For example, an atom of $^{235}_{92}\text{U}$ has 92 protons and 92 electrons, so it is electrically neutral.

We can now replace Dalton's definition of an atom with a better definition. We define an atom as an electrically neutral particle having *one* nucleus and a number of electrons equaling the atomic number.

2.4 ATOMIC WEIGHTS

An atomic weight is the average relative mass of an element's naturally occurring atoms on a scale that uses atoms of carbon-12 as the reference.

One of the most useful and far-reaching aspects of Dalton's atomic theory was the notion that the atoms of each element have a characteristic mass. It opened the door to the determination of chemical formulas and ultimately to the most useful device chemists have for organizing chemical information—the modern periodic table, a copy of which is located on the inside front cover to this book.

Let's consider a practical lab problem to see why the concept of characteristic atomic masses is useful. Suppose we wished to attempt to make a sample of pyrite (fool's gold), which has the formula FeS_2. We can tell from the formula that the ratio by *atoms*, iron to sulfur, must be 1 to 2. The practical question is, "How can we measure the right amounts of iron and sulfur in the lab in order to be sure we have the proper ratio by atoms?"

Since we can't pick up and count atoms like so many cups and saucers, we have to develop an operation that we can do in the lab to get whatever ratio by atoms we require. What solves this problem is the existence of *two* ratios that can be known. One is the measured mass ratio of the elements in a compound whose formula is already known, and from this we can obtain the relative masses of the atoms of the different elements. The second is the ratio by atoms of the elements in the desired compound, and this we get from the chemical formula. As we will see, this means that we can use mass ratios, which we can work with experimentally, to get atom ratios.

Let's see how we can get the relative masses of the atoms in a compound. We found out earlier that the mass ratio of the elements in troilite, FeS, is 1.000 g of iron to 0.574 g of sulfur. We also know that the atom ratio in this compound is 1 atom of iron to 1 atom of sulfur. The only way that both ratios could be true is if an iron atom is heavier than a sulfur atom, and that it must be heavier by a mass ratio of 1.000/0.574, which can also be expressed as 1.742/1.000 (1.000/0.574 = 1.742/1.000). In other words, an atom of iron must be 1.742 times as heavy as an atom of sulfur. This is the only way that the total mass of iron could be 1.742 times larger than the total mass of sulfur, given that there are equal numbers of iron and sulfur atoms in the sample.

How does this help us in our search for a way of obtaining the right amounts of iron and sulfur to make FeS_2? Once we know the relative masses of the atoms of iron and sulfur, we can obtain any atom ratio we want in the lab just by taking the right mass ratio. If we want an iron-to-sulfur atom ratio of 1 to 1, then we take 1.742 g of iron for every 1.000 g of sulfur. To make FeS_2, we want an iron-to-sulfur atom ratio of 1 to 2, so we need twice as much sulfur and take 1.742 g of iron for every $2 \times (1.000 \text{ g}) = 2.000$ g of sulfur.

The Carbon-12 Atomic Mass Scale

The preceding discussion reveals why it is useful to know the relative masses of atoms of the elements. By knowing these relative masses we are able to select amounts of the elements (in grams) needed to give any desired atom ratio. This fact is *so* useful that in the years following Dalton's presentation of his atomic theory, chemists worked hard to determine a complete set of relative masses for all the elements that were known. These are called atomic masses or atomic weights. *Atomic weight* is the term that has been used most widely by scientists in the United States, and for that reason it is the term that we will use most of the time. However, you should not feel uncomfortable with either *atomic mass* or *atomic weight*.

In establishing a table of atomic weights, it is necessary to have a reference against which to compare the relative masses. Currently, the agreed-upon reference is the most abundant isotope of carbon, which is carbon-12 or $^{12}_{6}C$. By definition, an atom of this isotope is defined as having a mass of *exactly* 12.000 amu. In other words, an amu (atomic mass unit) is defined as $\frac{1}{12}$th of the mass of one atom of carbon-12.

The definition of the size of the atomic mass unit is really quite arbitrary. It could just as easily have been selected to be $\frac{1}{24}$th of the mass of a carbon atom, or $\frac{1}{10}$th of the mass of an iron atom, or any other value. Why $\frac{1}{12}$th of the mass of a ^{12}C atom? First, carbon is a very common element, available to any scientist. Second, and most important, by choosing the amu of this size, the atomic masses of nearly all the other elements are almost whole numbers, with the lightest atom having a mass of approximately 1. (The lightest atom of all, an atom of the most abundant isotope of hydrogen, has a mass of 1.007825 amu when carbon-12 is assigned a mass of exactly 12 amu.)

Chemists, of course, do not generally work with samples that consist of just one

In modern terms, the atomic weight of an element is the average mass of the element's atoms (as they occur in nature) relative to an atom of carbon-12, which is assigned a mass of 12 units.

At the time the atomic weight scale was first developed, the mass in grams of 1 amu was unknown. The value given on page 51 was determined experimentally at a later date.

isotope, but with whatever *mixture* of isotopes comes with a given element as it occurs naturally. For nearly all elements, as we said, the isotopic composition is a constant regardless of where in nature a particular element is obtained. Because of this isotopic constancy, we can speak of an *average atom* of an element — average in terms of mass — taking into account the relative numbers of each isotope in a sample of the element. For example, naturally occurring hydrogen is a mixture of two isotopes in the relative proportions given in the margin. The "average atom" of the element hydrogen, as it occurs in nature, has a mass that is 0.0839 that of a carbon-12 atom. Since 0.0839×12.000 amu $= 1.0079$ amu, the average atomic weight of hydrogen is 1.0079 amu. Notice that this average value is just a little larger than the atomic mass of ^1H because naturally occurring hydrogen also contains a little ^2H.

The atomic weights of the rest of the elements are obtained in the same way. The average atom of naturally occurring iron, for example, is 4.6359 times heavier than an atom of ^{12}C, so the average atomic weight of iron is 4.6359×12.0000 amu $= 55.847$ amu. Similarly, the average atom of naturally occurring sulfur has a mass that is 2.672 times that of ^{12}C, so the average atomic weight of sulfur is 2.672×12.000 amu $= 32.06$ amu. A complete table of atomic weights is given inside the front cover of this book. You will no doubt find many occasions to refer to it in later chapters.

Hydrogen Isotope	Percentage Abundance
^1H	99.985
^2H	0.015

The amu is sometimes called the **dalton.**

1 amu = 1 dalton

PRACTICE EXERCISE 3   Aluminum atoms have a mass that is 2.24845 times that of an atom of ^{12}C. What is the atomic weight of aluminum?

PRACTICE EXERCISE 4   How much heavier than an atom of ^{12}C is the average atom of naturally occurring copper? Refer to the table inside the front cover of the book for the necessary data.

PRACTICE EXERCISE 5   Suppose the atomic mass unit were defined as $\frac{1}{20}$th of the mass of a carbon-12 atom (i.e., suppose one atom of carbon-12 were assigned a mass of 20.000 amu). What would be the atomic weights of hydrogen, iron, and sulfur?

2.5 WHAT COMPOUNDS ARE MADE OF

Chemical compounds are made up of either neutral molecules or ions of opposite charge.

In discussing the three laws of chemical combination, we learned about whole-number ratios of atoms in compounds. This language could leave the impression that *intact* atoms occur in compounds, and this is not exactly true. For example, the atoms of hydrogen in water no longer have the same properties as free atoms of hydrogen. Nor does the atom of oxygen in H_2O behave in the same way as a free, uncombined atom of oxygen. When atoms combine with other atoms, they lose their individual properties and the properties that we observe are those of the new substance that has been formed. In the new substance, the atoms are held to each other by special forces of attraction called chemical bonds. The ways in which these forces develop are discussed in detail in Chapters 7 and 8. In this section we simply introduce the kinds of particles that atoms change to when chemical bonds form. For now, that is all you will have to be concerned with.

Molecular Compounds

In many compounds, atoms are linked together in very tiny, discrete, electrically neutral particles called molecules. Some contain as few as two atoms, and are referred to as *diatomic* molecules. An example is the compound carbon monoxide, CO, a poisonous gas that is one of the substances in the exhaust of automobile engines. Even some elements can occur as diatomic molecules instead of as individual atoms. As noted in Chapter 1 (page 13), the air we breathe contains oxygen and nitrogen in the form of the diatomic molecules O_2 and N_2. Some other elements also normally occur in nature as diatomic molecules and are included in the list in the margin. (Be sure you learn this list; you will encounter these elements frequently during this course.) Most molecules are more complex than these, however, and contain more atoms. Molecules of ordinary table sugar, for example, have the formula $C_{12}H_{22}O_{11}$, and some molecules that occur in plastics and in living organisms contain millions of atoms.

The bonds that hold atoms to each other in molecules are electrical in nature and arise from the sharing of electrons between one atom and another. We will discuss bonding in detail later. What is important to know about molecules now is that the group of atoms that make up a molecule move about together and behave as a single particle, just as the various parts that make up a bicycle move about as one unit. Compounds that exist as molecules are called molecular compounds, and the formulas that describe the compositions of molecules are called molecular formulas.

The other elements with diatomic molecules are

Hydrogen	H_2
Fluorine	F_2
Chlorine	Cl_2
Bromine	Br_2
Iodine	I_2

Ions and Ionic Compounds

Under appropriate conditions, the atoms of some elements can accept one or more additional electrons, and atoms of certain other elements are able to give up one or more electrons. A chlorine atom, for example, can accept one electron, and a sodium atom can lose one. In fact, this is exactly what happens when we let chlorine come in contact with metallic sodium. In the violent reaction that produces sodium chloride, each sodium atom gives up one electron to a chlorine atom, and new particles form. The changes that occur can be illustrated in equation form if we take the liberty of treating an electron as if it were a chemical substance and give it a symbol, e^-. (To focus attention on just one feature—electron tranfer—we will also treat chlorine as consisting of atoms, not diatomic molecules.)

$$Na \longrightarrow Na^+ + e^-$$
$$Cl + e^- \longrightarrow Cl^-$$

The electron released by the sodium atom in the first equation is the electron accepted by the chlorine atom in the second. Actually, the reaction of sodium with chlorine doesn't occur in two separate steps. The transfer of electrons takes place directly from sodium atoms to chlorine atoms. The change is broken down here into two equations to focus attention on what happens to each atom. *These transfers create particles that are no longer referred to as atoms.* Each *atom* of sodium (atomic number 11) has 11 electrons, so it is neutral. When it loses one electron, only 10 remain, so now there is a net charge of $1+$ on the surviving particle. This is why it isn't correct to call this new particle an *atom* of sodium. Instead, we call it an *ion,* a sodium ion, and we write its symbol as Na^+. An ion *is an electrically charged particle having roughly the size of an atom or molecule.* The neutral chlorine atom, by accepting an electron, also loses neutrality and becomes an ion. Because it has one extra electron, it has a charge of $1-$. It is called a chloride ion, and its symbol is Cl^-. Thus, ions are either positively or negatively charged, and compounds that are composed of ions are called ionic compounds.

Any chemical substance that we can hold in our hands or store in a bottle is electrically neutral. Therefore, if it is composed of ions, both positive and negative ions must be present. In addition, these ions must be present in a ratio that guarantees that the algebraic sum of the positive and negative charges is zero. The only ratio in which sodium and chloride ions can assemble, therefore, is 1 to 1, one Na^+ to one Cl^-. In this ratio (and only in this ratio) the net charge is zero because $1+$ cancels $1-$. This explains why the formula of sodium chloride is $NaCl$ rather than something like $NaCl_2$ or Na_2Cl. We will use this requirement of electrical neutrality later in this chapter to help write the formulas of other ionic compounds.

Na = sodium *atom*
Na^+ = sodium *ion*

Notice that the charges on the ions are omitted in writing the formula.

Particles, in this context, are microscopic — in fact, supermicroscopic; substances are macroscopic.

Before continuing, let's review certain key terms. Elements and compounds are names for different kinds of *substances*. Atoms, ions, and molecules are names for different kinds of *particles* of which substances are composed. We can see and manipulate samples of substances, but no one has ever seen a naked atom, ion, or molecule. Even the most powerful microscopes have produced only blurry dots caused by individual atoms, as seen in Figure 2.3.

2.6 THE FIRST PERIODIC TABLE

The large volume of information gathered by scientists can only be understood if it is organized.

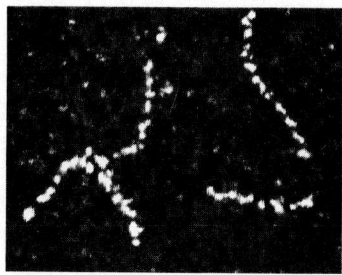

FIGURE 2.3 The white spots in this photograph were caused by individual atoms of the element thorium.

So far in this book we have mentioned a variety of different kinds of substances. We discussed elements and compounds, and among compounds we noted that some are ionic and some are molecular. Among elements such as sodium and chlorine, we mentioned metallic and nonmetallic properties. If we were to continue on this way, without attempting to build our subject around some central organizing structure, it would not be long before we became buried beneath a mountain of information in the form of seemingly unconnected facts. To make some sense out of it all, and to use the information effectively, we must in some way organize it. In this way, important similarities, differences, and trends can be seen. Finding these relationships can help us understand why certain elements are metals and others are nonmetals, and why certain elements often form ionic compounds while others seldom do.

The periodic table is often called the periodic chart.

The need for organization was recognized by many early chemists, and led directly to the development of a special way of arranging the chemical elements — the modern periodic table, located on the inside front cover of this book. By the middle of the nineteenth century, many of the quantitative laws of chemistry had been discovered, which allowed scientists to gather a wealth of empirical facts about the chemical and physical properties of the elements. Just knowing facts, however, is not very satisfying to scientists; they want to know *why* the facts they discover are true. Dalton's atomic theory, developed in the early 1800s, had helped provide some early answers, but many questions remained. Being able to answer them depended on finding some order in the jumble of chemical information.

There were numerous attempts to discover relationships among the chemical and physical properties of the elements. A number of different sequences of elements were tried in search of some sort of order or pattern. A few of these arrangements came quite close, at least in some respects, to our current table, but either they were flawed in some way or they were presented to the scientific community in a manner that did not lead to their acceptance.

The periodic table we use today is based primarily on the efforts of a Russian chemist, Dmitri Ivanovich Mendeleev (1834–1907), and a German physicist, Julius Lothar Meyer (1830–1895). Working independently, these scientists developed similar periodic tables only a few months apart in 1869. Mendeleev is usually given the credit, however, because he had the good fortune to publish first.

Mendeleev was preparing a chemistry textbook for his students at the University of St. Petersburg. He found that when he arranged the elements in order of increasing atomic weight, similar chemical properties occurred over and over again at regular intervals. For instance, the elements lithium (Li), sodium (Na), potassium (K), rubidium (Rb), and cesium (Cs) have similar chemical properties. Each of these forms a water-soluble chlorine compound with the general formula MCl: LiCl, NaCl, KCl, RbCl, and CsCl. Moreover, the elements that immediately follow each of these elements also constitute a set with similar chemical properties. Beryllium (Be) follows lithium, magnesium (Mg) follows sodium, calcium (Ca) follows potassium, strontium (Sr) follows rubidium, and barium (Ba) follows cesium. All of these elements form a water-soluble chlorine com-

pound with the general formula MCl_2: $BeCl_2$, $MgCl_2$, $CaCl_2$, $SrCl_2$, and $BaCl_2$. Based on extensive observations of this type, Mendeleev devised the original form of the periodic law. It stated that *the chemical and physical properties of the elements vary in a periodic way with their atomic weights.* Mendeleev used this law to construct his periodic table, which is illustrated in Figure 2.4.

Periodic refers to the recurrence of properties at regular intervals.

    The elements in Mendeleev's table are arranged in rows, called periods, in order of increasing atomic weight. When the rows are broken at the right places and stacked, the elements fall naturally into columns, called groups, so that elements in any given column have similar chemical properties. Mendeleev's genius rested on his placing elements with similar properties in the same group even though this left occasional gaps in the table. For

FIGURE 2.4 Mendeleev's periodic table roughly as it appeared in 1871. The numbers next to the symbols are atomic weights.

	Group I	Group II	Group III	Group IV	Group V	Group VI	Group VII	Group VIII
1	H 1							
2	Li 7	Be 9.4	B 11	C 12	N 14	O 16	F 19	
3	Na 23	Mg 24	Al 27.3	Si 28	P 31	S 32	Cl 35.5	
4	K 39	Ca 40	— 44	Ti 48	V 51	Cr 52	Mn 55	Fe 56, Co 59 Ni 59, Cu 63
5	(Cu 63)	Zn 65	— 68	— 72	As 75	Se 78	Br 80	
6	Rb 85	Sr 87	?Yt 88	Zr 90	Nb 94	Mo 96	— 100	Ru 104, Rh 104 Pd 105, Ag 108
7	(Ag 108)	Cd 112	In 113	Sn 118	Sb 122	Te 125	I 127	
8	Cs 133	Ba 137	?Di 138	?Ce 140	—	—	—	— —
9	—	—	—	—	—	—	—	
10	—	—	?Er 178	?La 180	Ta 182	W 184	—	Os 195, Ir 197 Pt 198, Au 199
11	(Au 199)	Hg 200	Tl 204	Pb 207	Bi 208	—		
12	—	—	—	Th 231	—	U 240	—	— — — —

TABLE 2.2 Comparison of Some Predicted Properties of Eka-silicon and Observed Properties of Germanium

Property	Observed for Silicon (Si)	Predicted for Eka-Silicon (Es)	Observed for Tin (Sn)	Found for Germanium (Ge)
Atomic weight	28	72	118	72.59
Melting point (°C)	1410	high	232	947
Density (g/cm³)	2.33	5.5	7.28	5.35
Formula of oxide	SiO_2	EsO_2	SnO_2	GeO_2
Density of oxide (g/cm³)	2.66	4.7	6.95	4.23
Formula of chloride	$SiCl_4$	$EsCl_4$	$SnCl_2$	$GeCl_4$
Boiling point of chloride (°C)	57.6	100	114	84

example, he placed arsenic (As) in Group V under phosphorus because its chemical properties were similar to those of phosphorus, even though this left gaps in Groups III and IV. Mendeleev reasoned, correctly, that the elements that belonged in these gaps had simply not yet been discovered. Based on the location of these gaps, however, Mendeleev was able to predict with remarkable accuracy the properties of these yet-to-be-found substances. In fact, his predictions helped serve as a guide in the search for these missing elements. Table 2.2 compares Mendeleev's predicted properties of "eka-silicon" and some of its compounds with the actual properties measured for the element germanium.

Two elements, tellurium (Te) and iodine (I), caused Mendeleev some problems. According to the best estimates at that time, the atomic weight of tellurium was greater than that of iodine. Yet if these elements were placed in the table according to their atomic weights, they would not fall into the proper groups required by their properties. Therefore, Mendeleev switched their order and in so doing violated his own periodic law. (Actually, he believed that his law was correct and that the atomic weight of tellurium had been incorrectly measured, but it wasn't true.)

The table that Mendeleev developed is in many ways similar to the one we use today. One of the main differences, though, is that Mendeleev's table lacks the elements helium (He) through radon (Rn). In Mendeleev's time, none of these elements had yet been found because they are relatively rare and because they have virtually no tendency to undergo chemical reactions. When they finally were discovered, beginning in 1894, another problem arose. Two more elements, argon (Ar) and potassium (K), did not fall into the groups required by their properties if they were placed in the table in the order required by their atomic weights. Another switch was necessary. Now there were two exceptions to Mendeleev's periodic law. Apparently, then, atomic weight was not the true basis for the periodic repetition of the properties of the elements. To determine what the true basis was, however, scientists had to await the discoveries of the atomic nucleus, the proton, and atomic numbers.

2.7 THE MODERN PERIODIC TABLE

In our modern periodic table, the elements are arranged in the order of their increasing atomic number.

In Section 2.3, you learned that the nucleus of an atom contains protons and neutrons and that it is surrounded by the atom's electrons. The number of protons in the nucleus, you recall, is the atom's atomic number. One of the gratifying dividends of the discovery of atomic numbers was that the elements in Mendeleev's table turned out to be arranged in precisely the order of increasing atomic number. In other words, if we take atomic numbers as the basis for arranging the elements in sequence, no annoying switches are

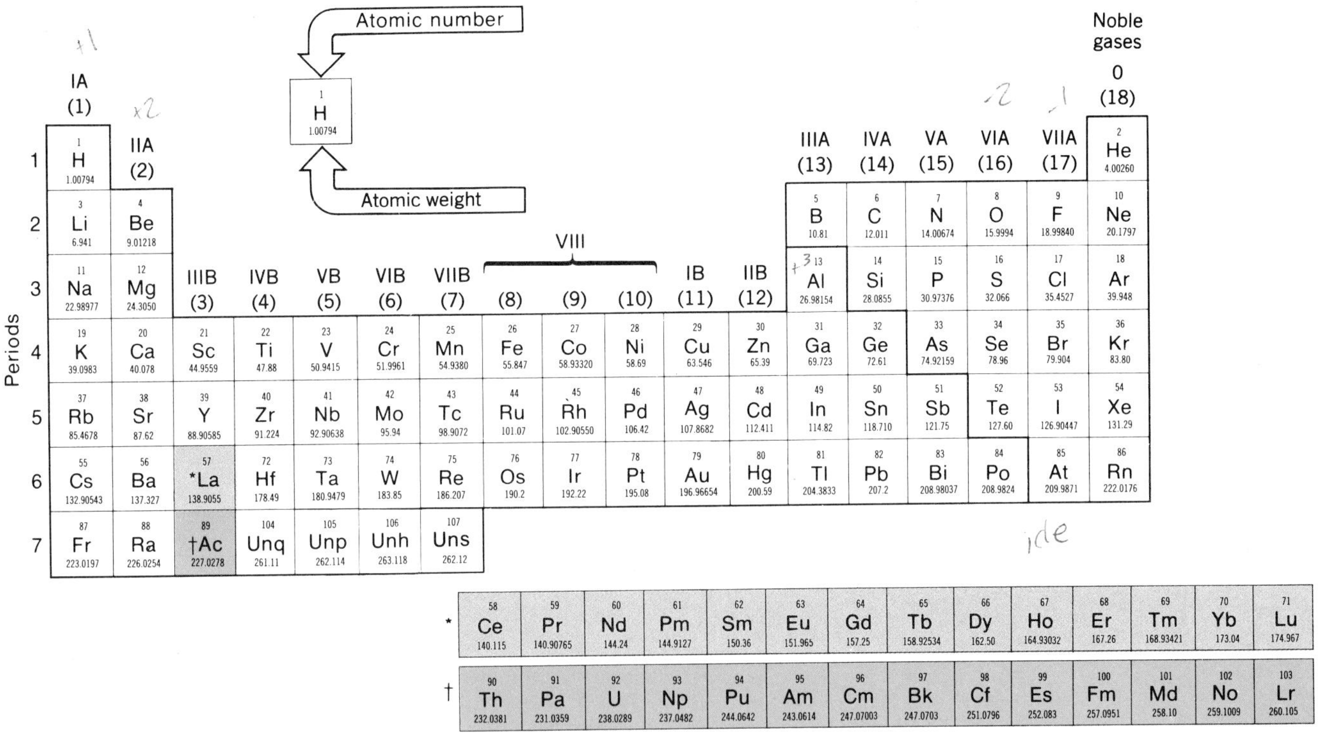

FIGURE 2.5 The modern periodic table.

required and the elements Te and I or Ar and K are no longer a problem. This leads to our present statement of the periodic law.

The Periodic Law.   The chemical and physical properties of the elements vary in a periodic way with their atomic *numbers*.

The fact that it is the atomic number — the number of protons in the nucleus of an atom — that determines the order of elements in the table is very significant. We will see later that this has very important implications with regard to the relationship between the number of electrons in an atom and the atom's chemical properties.

Our modern periodic table, which appears on the inside cover of the book, is also reproduced in Figure 2.5 We will refer to the table frequently, so it is important for you to become familiar with it and with some of the terminology applied to it.

Terminology Associated with the Periodic Table

As in Mendeleev's table, the elements are arranged in rows called periods, but here they are arranged in order of increasing atomic number. For identification purposes the periods are numbered. We will find these numbers useful later on. Below the main body of the table are two long rows of 14 elements each that actually belong in the main body of the table following La ($Z = 57$) and Ac ($Z = 89$), as shown in Figure 2.6. They are almost always placed below the table simply to conserve space. Fully spread out, the table requires too much room to be conveniently printed on one page. Notice that in the fully extended form of the table, with all the elements arranged in their proper locations, there is a great deal of empty space. An important requirement of a detailed atomic theory, which we will get to in Chapter 6, is that it must explain not only the repetition of properties, but also why the empty space appears.

We use the symbol Z to stand for atomic number.

1 H																	2 He
3 Li	4 Be											5 B	6 C	7 N	8 O	9 F	10 Ne
11 Na	12 Mg											13 Al	14 Si	15 P	16 S	17 Cl	18 Ar
19 K	20 Ca	21 Sc	22 Ti	23 V	24 Cr	25 Mn	26 Fe	27 Co	28 Ni	29 Cu	30 Zn	31 Ga	32 Ge	33 As	34 Se	35 Br	36 Kr
37 Rb	38 Sr	39 Y	40 Zr	41 Nb	42 Mo	43 Tc	44 Ru	45 Rh	46 Pd	47 Ag	48 Cd	49 In	50 Sn	51 Sb	52 Te	53 I	54 Xe
55 Cs	56 Ba	57 La	72 Hf	73 Ta	74 W	75 Re	76 Os	77 Ir	78 Pt	79 Au	80 Hg	81 Tl	82 Pb	83 Bi	84 Po	85 At	86 Rn
87 Fr	88 Ra	89 Ac	104 Unq	105 Unp	106 Unh	107 Uns											

58 Ce	59 Pr	60 Nd	61 Pm	62 Sm	63 Eu	64 Gd	65 Tb	66 Dy	67 Ho	68 Er	69 Tm	70 Yb	71 Lu
90 Th	91 Pa	92 U	93 Np	94 Pu	95 Am	96 Cm	97 Bk	98 Cf	99 Es	100 Fm	101 Md	102 No	103 Lr

FIGURE 2.6 Extended form of the periodic table. The two long rows are placed in their proper places in the table.

International Union of Pure and Applied Chemistry is generally abbreviated IUPAC.

If you continue in chemistry, you will certainly need to know the customary U.S. system because so many references use it.

The IUPAC scheme also recognizes the groupings called the lanthanide and actinide series.

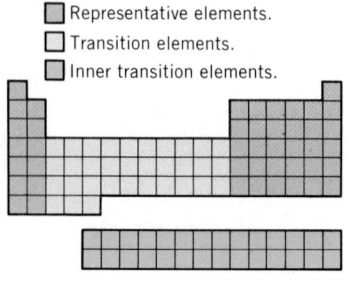

☐ Representative elements.
☐ Transition elements.
☐ Inner transition elements.

Again, as in Mendeleev's table, the vertical columns are called groups. However, there is not uniform agreement on how they should be numbered. In the past, the groups were labeled with Roman numerals and divided into A-groups and B-groups, separated by the three short columns headed by Fe, Co, and Ni (Group VIII), as indicated in Figure 2.5. This corresponds quite closely to Mendeleev's original designations, and is preferred by many chemists in the United States. However, another version of the table, popular in Europe, has the first seven groups from left to right labeled A, followed by the three short columns headed by Fe, Co, and Ni, and then the next seven groups labeled B. In an attempt to standardize the table, the International Union of Pure and Applied Chemistry, an international body of scientists responsible for setting standards in chemistry, has officially adopted a third system in which the groups are simply numbered sequentially from left to right using Arabic numerals. Thus, Group IA in the old system is Group 1 in the IUPAC table, and Group VIIA in the old system is Group 17 in the IUPAC table. In Figure 2.5 and on the inside front cover of the book, we have used both the older U.S. labels as well as those preferred by the IUPAC. Because of the lack of uniform agreement among chemists on how the groups should be specified, and because many chemists still prefer the more traditional A-group/B-group designations in Figure 2.5, we will use just the latter when we wish to specify a particular group.

As we have already noted, the elements in a given group bear similarities to each other. Because of such similarities, groups are frequently referred to as families of elements. The elements in the longer columns — the A-groups and Group 0 — are known as the representative elements. Those that fall into the B-groups in the center of the table are called transition elements. The two long rows of elements below the table are called the inner transition elements, and each row is named after the element that it follows in the main body of the table. Thus, elements 58–71 are called the lanthanide elements, and elements 90–103 are called the actinide elements.

Some of the groups have acquired common names. The Group IA elements, for example, are metals. They form compounds with oxygen that dissolve in water to give solutions that are strongly alkaline, or caustic. As a result, they are called the alkali metals, or simply the alkalis. The Group IIA elements are also metals. Their oxygen compounds are alkaline, too, but many compounds of the Group IIA elements are unable to dissolve in water and are found in deposits in the ground. Because of their properties and where they are found, the Group IIA elements became known as the alkaline earth metals.

On the right side of the table, in Group 0, are the noble gases. They used to be called the inert gases until it was discovered that the heavier members of the group show a small degree of reactivity. The term *noble* is used when we wish to suggest a very limited degree of chemical reactivity. Gold, for instance, is often referred to as a noble metal because so few chemicals are capable of reacting with it.

Finally, the elements of Group VIIA are called the halogens, derived from the Greek words meaning sea or salt. Chlorine (Cl), for example, is found in familiar table salt, a compound that accounts in large measure for the salty taste of seawater.

PRACTICE EXERCISE 6   Circle the correct choices.
(a) Representative elements are: K, Cr, Pr, Ar, Al.
(b) A halogen is: Na, Fe, O, Cl, Cu.
(c) An alkaline earth metal is: Rb, Ba, La, As, Kr.
(d) A noble gas is: H, Ne, F, S, N.
(e) An alkali metal is: Zn, Ag, Br, Ca, Li.
(f) An inner transition element is: Ce, Pb, Ru, Xe, Mg.

2.8 METALS, NONMETALS, AND METALLOIDS

Elements are classified as metals, nonmetals, or metalloids according to such properties as electrical conductivity and luster.

The periodic table is as important to chemists and chemistry students as good maps are to travelers. The table organizes all sorts of chemical and physical information about the elements and their compounds. It allows us to study systematically the way properties vary with an element's position within the table and, in turn, makes the similarities and differences among the elements easier to understand and remember.

Even a casual inspection of samples of the elements reveals that some are familiar metals and that others, equally familiar, are not metals. Most of us are already familiar with metals such as lead, iron, or gold and nonmetals such as oxygen or nitrogen. A closer look at the nonmetallic elements, though, reveals that some of them, silicon and arsenic to name two, have properties that lie between those of true metals and true nonmetals. These elements are called metalloids. Division of the elements into these three categories—metals, nonmetals, and metalloids—is not an even one, however (see Figure 2.7). Most of the elements are metals, slightly over a dozen are nonmetals, and only a handful are metalloids.

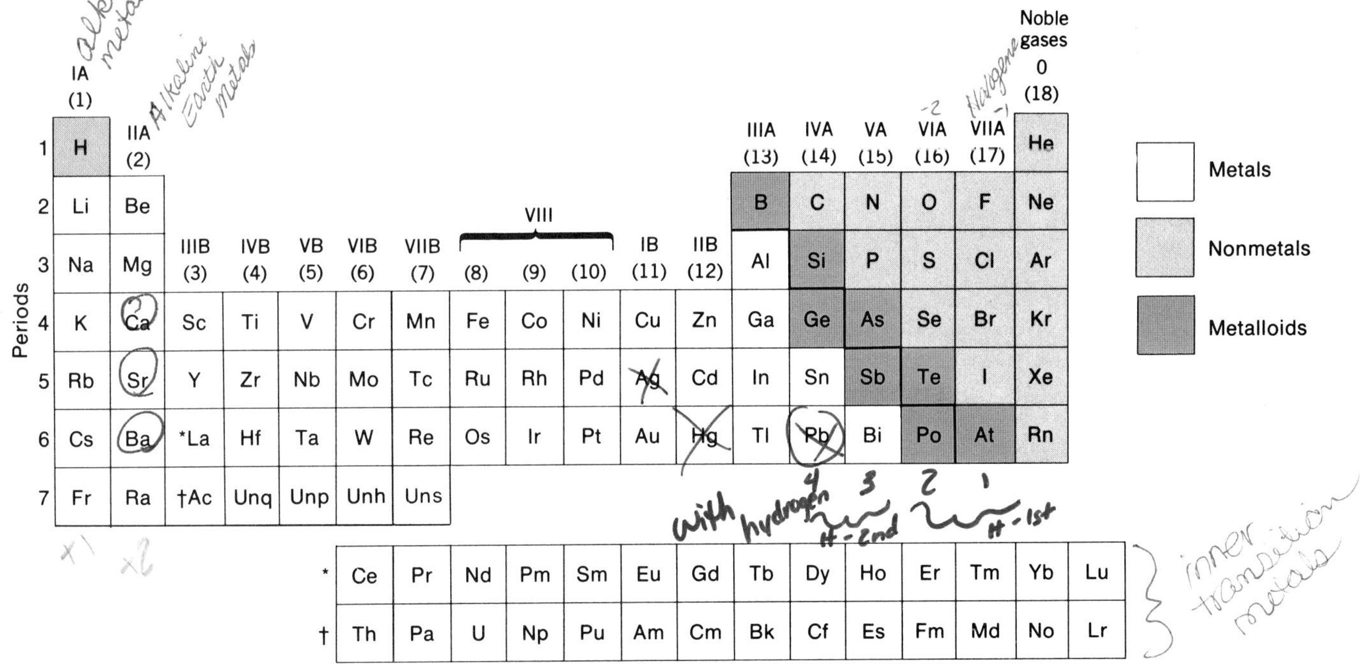

FIGURE 2.7 Distribution of metals, nonmetals, and metalloids in the periodic table.

This shine is so unique that it's called a *metallic luster*.

You probably know a metal when you see one. Metals tend to have a shine or luster that is easily recognized. For example, the silvery sheen of a freshly exposed surface of sodium in Figure 1.1 (page 5) would most likely lead you to identify this element as a metal even if you had never seen or heard of it before. We also know that metals conduct electricity. Few of us would poke an iron nail we were holding in our hand into an electrical outlet. In addition, we know that metals conduct heat very well. On a cool day, metals always feel colder to the touch than do neighboring nonmetallic objects because metals conduct heat away from your hand very rapidly. Nonmetals seem less cold because they can't conduct heat away as quickly and therefore their surfaces warm up faster.

Other properties that metals possess, to varying degrees, are malleability—the ability to be hammered or rolled into thin sheets—and ductility—the ability to be drawn into wire. For example, the production of sheet steel (Figure 2.8) for automobiles and household appliances depends on the malleability of iron and steel, and the manufacture of electrical wire (Figure 2.9) is based on the ductility of copper.

Thin lead sheets are used for sound-deadening because the soft, easily deformed lead absorbs the sound vibrations.

Another important physical property that we usually think of when metals are mentioned is hardness. Some, such as chromium or iron, are indeed quite hard; but others, like copper and lead, are rather soft. The alkali metals are so soft that they can be cut with a knife, but they are also so chemically reactive that we rarely get to see them as free elements.

All of the metallic elements, except mercury, are solids at room temperature. Mercury's low freezing point (−39 °C) and fairly high boiling point (357 °C) make it useful as a fluid in thermometers. Most of the other metals have much higher melting points, and some are used primarily because of this. Tungsten, for example, has the highest melting point of any metal (3400 °C, or 6150 °F), which explains its use as a filament in electric light bulbs. Cobalt and chromium are combined in alloys (mixtures of metals) such as "stellite" to make high-speed, high-temperature cutting tools for industry. This alloy retains its hardness even at high temperatures, and the cutting tools therefore remain sharp.

A bead of the liquid metal mercury on a porcelain surface.

The chemical properties of metals vary tremendously. Some, such as gold and platinum, are very unreactive toward almost all chemical agents. This property, plus their

FIGURE 2.8 The malleability of iron is demonstrated in the production of sheet steel. Here we see a slab of red-hot steel as it comes from between rollers that flatten the steel plate. Repeated rolling of the steel eventually produces the kind of sheet steel used in automobiles and home appliances.

FIGURE 2.9 The ductility of copper allows it to be drawn into wire. Here copper wire passes through one die after another as it is drawn into thinner and thinner wire.

natural beauty and their rarity, makes them highly prized for use in jewelry. Other metals, however, are so reactive that few people except chemistry students ever have an opportunity to see them. For instance, in Chapter 1 you learned that the metal sodium reacts very quickly with oxygen or moisture in the air, and its bright metallic surface tarnishes almost immediately. On the other hand, compounds of sodium are quite stable and very common. Examples are table salt ($NaCl$), baking soda ($NaHCO_3$), lye ($NaOH$), and bleach ($NaOCl$). We will have more to say about the chemical properties of metals in Section 2.9.

Nonmetals

We see many objects each day that are clearly not metals. Some examples are plastics, wood, concrete, and glass. These aren't elements, though. Most often, we encounter the nonmetallic elements in the form of compounds or mixtures of compounds. There are, however, some nonmetals that are very important to us in their elemental forms. The air we breathe, for instance, contains mostly nitrogen, N_2, and oxygen, O_2. Both are gaseous, colorless, and odorless nonmetals. Since we can't see, taste, or smell them, however, it's difficult to experience their existence. (Although if you step into an atmosphere without oxygen, your body will very quickly tell you that something is missing!) Probably the most commonly *observed* nonmetallic element is carbon. We find it as the graphite in pencils, as coal, and as the charcoal used for barbecues. It also occurs in the more valuable form of diamonds. Although diamond and graphite differ in appearance, each is a form of elemental carbon.

Photographs of some of the nonmetallic elements appear in Figure 2.10. Their properties are almost completely opposite those of metals. Each of these elements lacks the characteristic appearance of a metal. They are poor conductors of heat and, with the exception of the graphite form of carbon, are also poor conductors of electricity. The electrical conductivity of graphite appears to be an accident of molecular structure, since the structures of metals and graphite are completely different.

Many of the nonmetals are solids at room temperature and atmospheric pressure, while many others are gases. All of the Group 0 elements are gases in which the particles consist of single atoms. The other gaseous elements — hydrogen, oxygen, nitrogen,

The Group 0 elements are said to be *monatomic*.

FIGURE 2.10 Some nonmetallic elements. (a) Diamonds of gem quality. (b) Powdered graphite. (c) Powdered sulfur. (d) Red phosphorus. This is one of several forms of phosphorus and is used on the striking surfaces of matchbooks. (e) Liquid bromine. Note the deep red color of the liquid and how the liquid vaporizes. (f) Solid iodine. Note the violet color of the vapor. (g) Gaseous chlorine.

(a)

(b)

(c)

(d)

(e)

(f)

(g)

fluorine, and chlorine—are composed of diatomic molecules. Their formulas are H_2, O_2, N_2, F_2, and Cl_2. As we learned in Section 2.5, a molecule is diatomic if it is composed of two atoms. Bromine and iodine are also diatomic, but bromine is a liquid and iodine is a solid at room temperature.

The nonmetallic elements lack the malleability and ductility of metals. A lump of sulfur crumbles when hammered and breaks apart when pulled on. Diamond cutters rely on the brittle nature of carbon when they split a gem-quality stone by carefully striking a quick blow with a sharp blade.

As with metals, nonmetals exhibit a broad range of chemical reactivity. Fluorine, for instance, is extremely reactive. It reacts readily with almost all the other elements. At the other extreme is helium, the gas used to inflate children's balloons and the Goodyear blimp. This element does not react with anything. Chemists find helium useful when they want to provide a totally inert (unreactive) atmosphere inside some apparatus.

Mercury and bromine are the only two liquid elements at room temperature and pressure.

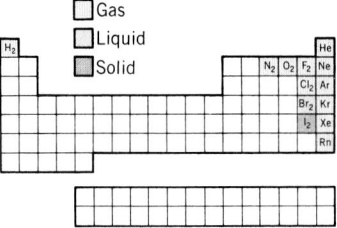

Monatomic and diatomic nonmetals.

Metalloids

The properties of metalloids lie between those of metals and those of nonmetals. This shouldn't surprise us since the metalloids are located between the metals and the nonmetals in the periodic table. In most respects, metalloids behave as nonmetals, both chemically and physically. However, in their most important physical property, electrical conductivity, they somewhat resemble metals. Metalloids tend to be semiconductors—they conduct electricity, but not nearly so well as metals. This property, particularly as found in silicon and germanium, is responsible for the remarkable progress made during the last decade in the field of solid-state electronics. The newest hi-fi stereo systems, television receivers, and CB radios rely heavily on transistors made from semiconductors. Perhaps the most amazing advance of all has been the fantastic reduction semiconductors have allowed in the size of electronic components. To it, we owe the development of small and versatile hand-held calculators and microcomputers. The heart of these devices is a microcircuit printed on a tiny silicon chip (Figure 2.11).

Trends In the Periodic Table

The occurrence of the metalloids between the metals and the nonmetals is our first example of trends in properties within the periodic table. We will see frequently that as we move from position to position across a period or down a group, chemical and physical properties change in a more or less regular fashion. There are few abrupt changes

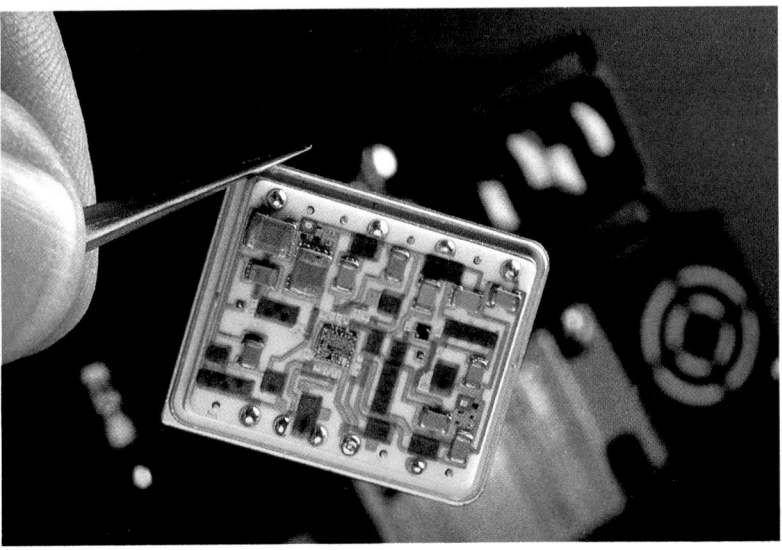

FIGURE 2.11 The heart of this tiny electronic device is the microcircuit in the center, which is set into the surface of a tiny silicon chip. Silicon's semiconductor properties make possible microelectronic devices such as this.

in the characteristics of the elements as we scan across a period or down a group. The location of the metalloids can be seen, then, as an example of the gradual transition between metallic and nonmetallic properties. From left to right across period 3, we go from aluminum, an element that has every appearance of a metal; to silicon, which is a semiconductor; to phosphorus, which is an element with clearly nonmetallic properties. A similar gradual change is seen going down Group IVA. Carbon is certainly a nonmetal, silicon and germanium are metalloids, and tin and lead are metals. Trends such as these are useful to spot, because they help us remember properties.

2.9 SOME CHEMICAL PROPERTIES OF METALS

Metals react with nonmetals to form ionic compounds.

A property possessed by nearly every element is the ability to combine with other elements to form compounds of varying degrees of complexity. However, not all combinations of elements appear possible. For example, sodium reacts vigorously with chlorine to form sodium chloride (NaCl), but no compound is formed between sodium and iron. When we examine chemical properties of the elements, though, we find that certain generalizations are possible. It is worth looking at these now for a number of reasons. First, they provide a general framework within which we can study chemical behavior in greater detail later. Second, they illustrate how the periodic table can make it easier to remember chemical properties. And finally, they show us that to understand the chemical behavior of the elements, we need a theoretical model that explains what actually occurs between atoms when they react.

An important property of metals is their ability to combine chemically with nonmetals, and among metal–nonmetal compounds are found certain similarities and differences that helped Mendeleev construct his table. Examples are the formulas of compounds formed by metals with the halogens of Group VIIA and with oxygen. Some of these are shown in Table 2.3.

Earlier in our discussion of Mendeleev's periodic table, we noted similarities among the formulas of the chlorine compounds of the alkali metals of Group IA—all form chlorides with the general formula MCl, where M stands for the symbol of a Group IA metal. Similar compounds are formed by the alkali metals with the other halogens. Thus, sodium forms the compounds NaF, NaCl, NaBr, and NaI. The other alkali metals form compounds with similar formulas, and altogether they might be generalized by the formula MX, where M stands for the symbol of an alkali metal and X stands for the symbol of a halogen.

Hydrogen is a special case. It is not a metal like the other Group IA elements and does not react by losing electrons.

TABLE 2.3 Formulas of Halogen and Oxygen Compounds of the Metals of Group IA and Group IIA

Group IA	Fluorine	Chlorine	Bromine	Iodine	Oxygen
Lithium	LiF	LiCl	LiBr	LiI	Li_2O
Sodium	NaF	NaCl	NaBr	NaI	Na_2O
Potassium	KF	KCl	KBr	KI	K_2O
Rubidium	RbF	RbCl	RbBr	RbI	Rb_2O
Cesium	CsF	CsCl	CsBr	CsI	Cs_2O
Group IIA					
Beryllium	BeF_2	$BeCl_2$	$BeBr_2$	BeI_2	BeO
Magnesium	MgF_2	$MgCl_2$	$MgBr_2$	MgI_2	MgO
Calcium	CaF_2	$CaCl_2$	$CaBr_2$	CaI_2	CaO
Strontium	SrF_2	$SrCl_2$	$SrBr_2$	SrI_2	SrO
Barium	BaF_2	$BaCl_2$	$BaBr_2$	BaI_2	BaO

TABLE 2.4 Some Ions Formed from Metals (*M*) and Nonmetals (*X*)

M^+	M^{2+}	M^{3+}	X^{2-}	X^-
Group IA	**Group IIA**	**Group IIIA**	**Group VIA**	**Group VIIA**
Li^+	Be^{2+}		O^{2-}	F^-
Na^+	Mg^{2+}	Al^{3+}	S^{2-}	Cl^-
K^+	Ca^{2+}		Se^{2-}	Br^-
Rb^+	Sr^{2+}		Te^{2-}	I^-
Cs^+	Ba^{2+}			

Atoms of Group IIA metals each combine with two halogen atoms. Thus, we find compounds such as MgF_2, $MgCl_2$, $MgBr_2$, and MgI_2. Altogether these Group IIA–Group VIIA compounds can be represented by the general formula MX_2. (In this case, *M* stands for the symbol of a Group IIA metal.)

Metal oxides — compounds of metals with oxygen — have long been important substances because oxygen itself is such a readily available chemical and because O_2, the molecular form of oxygen found in the air, reacts with nearly all metals. Here again we find similarities in chemical formulas that are particularly easy to see for the Group IA and IIA metals. The alkali metals all form oxides with the general formula M_2O (for example, Li_2O and Na_2O), and the alkaline earth metals all form oxides with the general formula MO (for example, MgO and CaO).

A key aspect of the reactions of metals with nonmetals, which explains formulas like the ones we've just discussed, is the transfer of one or more electrons from an atom of the metal to an atom of the nonmetal. As you learned in Section 2.5, this changes atoms into ions. The metal atom, by losing one or more electrons, becomes a positively charged ion. The nonmetal atom, by acquiring one or more electrons, becomes a negatively charged ion. In referring to these particles we will frequently call a positively charged ion a **cation** (pronounced *CAT-ion*) and a negatively charged ion an **anion** (pronounced *AN-ion*).[1]

We will have to wait until a later chapter to study the reasons why certain atoms gain or lose one electron each, while other atoms gain or lose two or more electrons. For now, however, we can use the periodic table as the basis for some generalizations that can help us remember the kinds of ions formed by many of the representative elements. For example, except for hydrogen, the neutral atoms of each of the Group IA elements always lose one electron when they react, thereby becoming ions with a charge of 1+. Similarly, atoms of the Group IIA elements always lose two electrons when they react; so these elements always form ions with a charge of 2+. In Group IIIA, the only important positive ion we need consider now is that of aluminum, Al^{3+}; an aluminum atom loses three electrons when it reacts.

All these ions are listed in Table 2.4. Notice that the number of positive charges on each of these cations is the same as the group number when we use the traditional numbering of groups in the periodic table. Thus, sodium is in Group IA and forms an ion with a 1+ charge, barium (Ba) is in Group IIA and forms an ion with a 2+ charge, and aluminum is in Group IIIA and forms an ion with a 3+ charge. Although this generalization doesn't work for all the metallic elements (it doesn't work for the transition elements, for instance), it does help us remember what happens to the metallic elements of Groups IA and IIA and aluminum when they react.

Among the nonmetals on the right side of the periodic table we also find some useful generalizations. For example, when they combine with metals, the halogens (Group VIIA) form ions with a 1 − charge and the nonmetals in Group VIA form ions

At a later time we will see that some of the alkali metals form other kinds of compounds with oxygen.

[1] The names *cation* and *anion* come from the way the ions behave when electrically charged metal plates called electrodes are dipped into a solution that contains them. We will discuss this in detail in Chapter 17.

with a $2-$ charge. Notice that the number of negative charges on the ion is equal to the number of spaces to the right that we have to move to get to a noble gas.

Formulas of Ionic Compounds

A substance is electrically neutral if the total positive charge equals the total negative charge.

Because all chemical compounds are electrically neutral, the ions in an ionic compound always occur in a ratio such that the total positive charge is equal to the total negative charge. As we noted in Section 2.5, this is why the formula for sodium chloride is NaCl; the 1-to-1 ratio of Na^+ to Cl^- gives electrical neutrality. In calcium chloride, however, it takes two Cl^- to balance the charge of a single Ca^{2+}, so the formula is given as $CaCl_2$.

In ionic compounds, such as sodium chloride and calcium chloride, molecules do not exist. Recall that in a molecule, atoms are linked together to form discrete particles that stay intact as the molecule moves about. In a water molecule, for example, it is safe to say that two hydrogen atoms "belong" to each oxygen atom in a particle having the formula H_2O. However, in NaCl it is impossible to say that a particular Na^+ ion belongs to a particular Cl^- ion (see Figure 2.12). The ions in a crystal of NaCl are simply stacked in the most efficient way, so that positive ions and negative ions can be as close to each other as possible. In this way, the attractions between oppositely charged ions, which are responsible for holding the compound together, can be as strong as possible. Because molecules don't exist in ionic compounds, the subscripts in their formulas are always chosen to specify the smallest whole-number ratio of the ions. This is why the formula of sodium chloride is given as NaCl rather than Na_2Cl_2 and Na_3Cl_3. Although the smallest unit of an ionic compound can't be called a molecule, the idea of "smallest unit" is still quite often useful. Therefore, we take the smallest unit of an ionic compound to be whatever is represented in its formula and call this unit a formula unit. Thus, one formula unit of NaCl consists of one Na^+ and one Cl^-, whereas one formula unit of $CaCl_2$ consists of one Ca^{2+} and two Cl^-. (In a larger sense, we can use the term *formula unit* to refer to whatever is represented by the formula. Sometimes the formula specifies a set of ions, as in NaCl; sometimes it is a molecule, as in O_2 or H_2O; sometimes it can be just an ion, as in Cl^- or Ca^{2+}; and sometimes it might be just an atom, as in Na.)

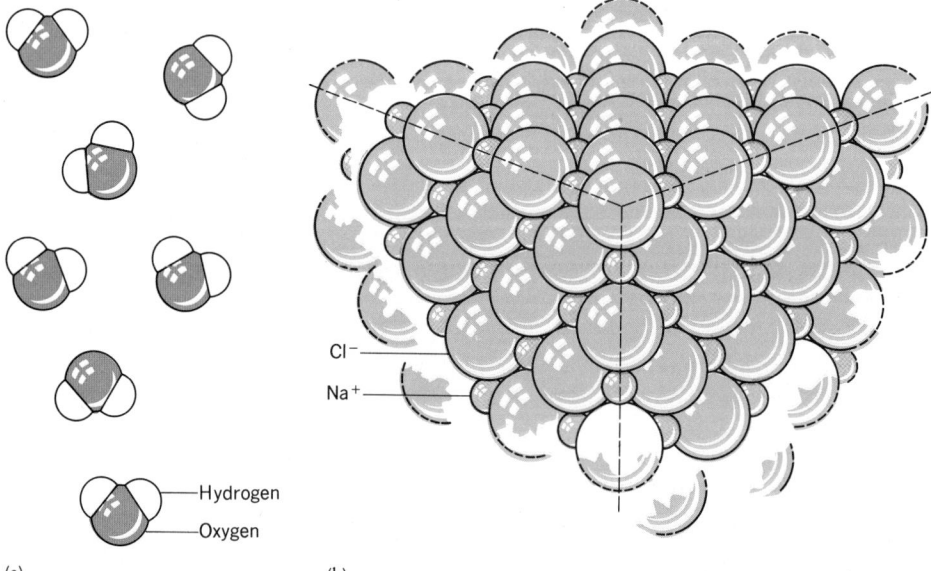

FIGURE 2.12 *(a)* In water there are discrete molecules that each consist of one atom of oxygen and two atoms of hydrogen. Each particle has the formula H_2O. *(b)* In sodium chloride, ions are packed in the most efficient way. Each Na^+ is surrounded by six Cl^-, and each Cl^- is surrounded by six Na^+. Because individual molecules do not exist, we simply specify the ratio of ions as NaCl.

Cl⁻

Na⁺

Hydrogen

Oxygen

(a)   (b)

The arguments presented in the preceding two paragraphs form the basis for the rules that we use in writing formulas of ionic compounds. There are only three to remember:

Rules for Writing Formulas of Ionic Compounds

1. The positive ion is given first in the formula. This isn't required by nature, but it is a custom we always follow.
2. The subscripts in the formula must produce an electrically neutral formula unit. (Nature does require electrical neutrality.)
3. The subscripts should be the smallest set of whole numbers possible.

Example 2.1 illustrates how we use these rules.

EXAMPLE 2.1   WRITING FORMULAS FOR IONIC COMPOUNDS

Problem: Write the formulas for the ionic compounds formed from (a) Al and Cl, (b) Al and O, and (c) Ba and S.

Solution: In each case, the ions must be combined in a ratio that produces an electrically neutral formula unit with the smallest set of whole number subscripts.

(a) For these elements the ions are Al^{3+} and Cl^-. We can obtain a neutral formula unit by combining one Al^{3+} and three Cl^-. (The charge on Cl is 1−; the 1 is understood.)

$$1(3+) + 3(1-) = 0$$

The formula is $AlCl_3$.

(b) For these elements the ions are Al^{3+} and O^{2-}. In the formula that we seek, there must be the same number of positive charges as negative charges. This number must be a whole number multiple of both 3 and 2. The smallest number that has both 3 and 2 as factors is 6, so there must be two Al^{3+} and three O^{2-} in the formula

$$\begin{aligned} 2(3+) &= 6+ \\ 3(2-) &= \underline{6-} \\ \text{sum} &\quad 0 \end{aligned}$$

The formula is Al_2O_3.

(c) The ions here are Ba^{2+} and S^{2-}. Since the charges are equal but opposite, their ratio in the compound is 1 to 1. The formula is BaS.

PRACTICE EXERCISE 7   Write formulas for ionic compounds formed from (a) Na and F, (b) Na and O, (c) Mg and F, and (d) Al and S.

There is another rather simple way to obtain the formulas of the compounds in Example 2.1. The procedure is to make the subscript for one ion equal to the *number* of charges on the other. For example, for Al^{3+} and Cl^{1-}, we can write

which gives

$$Al_1Cl_3 \quad \text{or simply} \quad AlCl_3$$

For the ions of Al^{3+} and O^{2-} we write

This gives

$$Al_2O_3$$

TABLE 2.5 Ions of Some Transition and Post-transition Metals

Transition Metals

Chromium	Cr^{2+}, Cr^{3+}
Manganese	Mn^{2+}, Mn^{3+}
Iron	Fe^{2+}, Fe^{3+}
Cobalt	Co^{2+}, Co^{3+}
Nickel	Ni^{2+}
Copper	Cu^{+}, Cu^{2+}
Zinc	Zn^{2+}
Silver	Ag^{+}
Cadmium	Cd^{2+}
Gold	Au^{3+}
Mercury[a]	Hg_2^{2+}, Hg^{2+}

Post-transition Metals

Tin	Sn^{2+}, Sn^{4+}
Lead	Pb^{2+}, Pb^{4+}
Bismuth	Bi^{3+}

[a] Hg_2^{2+} is actually two Hg^+ stuck together.

Fe_2O_3 is the iron oxide in rust.

If you follow this method, you can be fooled if you are not careful. For example, if you apply this method to the compound formed from the ions Ba^{2+} and S^{2-}, it gives the formula Ba_2S_2. However, by convention we always choose the smallest whole-number ratio of ions (rule 3 above). Notice in Ba_2S_2 that both subscripts are divisible by 2. To obtain the correct formula, we reduce the subscripts to the smallest set of whole numbers, which gives BaS. Exercising appropriate care, you might go back and try this method on Practice Exercise 7.

Many of our most important chemicals are ionic compounds. We have mentioned NaCl, common table salt, and $CaCl_2$, a substance used to melt ice on walkways in the winter and to keep dust down on dirt roads in the summer. Other examples are sodium fluoride, NaF, used by dentists to give fluoride treatments to teeth, and calcium oxide, CaO, an important ingredient in cement.

Transition Metals and Post-transition Metals

The transition elements are located in the center of the periodic table, from Group IIIB on the left to Group IIB on the right. All of them lie to the left of the metalloids, and they all are metals. Included here are some of our most familiar metals, including iron, chromium, copper, silver, and gold.

Most of the transition metals are much less reactive than the metals of Groups IA and IIA, but when they react they also transfer electrons to nonmetal atoms to form ionic compounds. However, the charges on the ions of the transition metals do not follow as straightforward a pattern as do those of the alkali and alkaline earth metals. One of the characteristic features of the transition metals is the ability of many of them to form more than one positive ion. Iron, for example, can form two different ions, Fe^{2+} and Fe^{3+}. This means that iron can form more than one compound with a given nonmetal. For example, with chloride ion, Cl^-, iron forms two compounds, with the formulas $FeCl_2$ and $FeCl_3$. With oxygen, we find the compounds FeO and Fe_2O_3. As usual, we see that the formulas contain the ions in a ratio that gives electrical neutrality. Some of the most common ions of the transition metals are given in Table 2.5.

PRACTICE EXERCISE 8   Write formulas for the chlorides and oxides formed by (a) chromium and (b) copper.

The prefix *post* means "after."

The post-transition metals are those metals that occur in the periodic table immediately following a row of transition metals. The two most common and important ones are tin (Sn) and lead (Pb). These post-transition metals are quite different from the metals that precede the transition metals. One of the most significant differences is their ability to form two different ions, and therefore two different compounds with a given nonmetal. For example, tin forms two oxides, SnO and SnO_2. Lead also forms two oxides that have similar formulas (PbO and PbO_2). The ions that these metals form are also included in Table 2.5.

Compounds Containing Ions Composed of More Than One Element

A compound is *diatomic* if it is composed of molecules that contain only two atoms. It is a *binary compound* if it contains two different elements, regardless of the number of each. BrCl is a binary compound and it is also diatomic; CH_4 is a binary compound, but it isn't diatomic.

The metal compounds that we have discussed so far have been binary compounds — compounds formed between *two different* elements. There are many other ionic compounds that contain more than two elements. These substances usually contain polyatomic ions, which are ions that are themselves composed of two or more atoms linked by the same kinds of bonds that hold molecules together. Polyatomic ions differ from molecules, however, in that they contain either too many or too few electrons to make them electrically neutral. Table 2.6 lists some important polyatomic ions. The formulas of ionic compounds formed from them are determined in the same way as are those of binary ionic compounds — the ratio of the ions must be such that the formula unit is electrically neutral, and the smallest set of whole-number subscripts is used.

TABLE 2.6 Some Polyatomic Ions

Ion	Name (Alternate Name in Parentheses)
NH_4^+	Ammonium ion
H_3O^+	Hydronium ion[a]
OH^-	Hydroxide ion
CN^-	Cyanide ion
NO_2^-	Nitrite ion
NO_3^-	Nitrate ion
ClO^-	Hypochlorite ion
ClO_2^-	Chlorite ion
ClO_3^-	Chlorate ion
ClO_4^-	Perchlorate ion
MnO_4^-	Permanganate ion
$C_2H_3O_2^-$	Acetate ion
CO_3^{2-}	Carbonate ion
HCO_3^-	Hydrogen carbonate ion (bicarbonate ion)[b]
SO_3^{2-}	Sulfite ion
SO_4^{2-}	Sulfate ion
HSO_4^-	Hydrogen sulfate ion (bisulfate ion)
CrO_4^{2-}	Chromate ion
$Cr_2O_7^{2-}$	Dichromate ion
PO_4^{3-}	Phosphate ion (orthophosphate ion)
HPO_4^{2-}	Monohydrogen phosphate ion
$H_2PO_4^-$	Dihydrogen phosphate ion

[a] You will only encounter this ion in aqueous solutions.

[b] Although "hydrogen carbonate ion" is formally correct, "bicarbonate ion" is what you will see and hear the most. We'll use "bicarbonate" too.

EXAMPLE 2.2 WRITING FORMULAS CONTAINING POLYATOMIC IONS

Problem: Write the formula of the ionic compound formed from Ca^{2+} and PO_4^{3-}.

Solution: As before, we write the positive ion first and then exchange subscripts and numbers of charges.

$$Ca^{2+} \qquad PO_4^{3-}$$

The formula is written with parentheses to show the number of PO_4^{3-} ions.

$$Ca_3(PO_4)_2$$

PRACTICE EXERCISE 9 Write formulas of ionic compounds formed from (a) Na^+ and CO_3^{2-}, (b) NH_4^+ and SO_4^{2-}, (c) potassium ion and acetate ion, (d) strontium ion and nitrate ion, and (e) Fe^{3+} and acetate ion.

Polyatomic ions are found in a great number of very important compounds. Some common ones are $CaSO_4$ (in plaster of Paris), $NaHCO_3$ (baking soda), $NaOCl$ (liquid household bleach), $NaNO_2$ (sodium nitrite, a meat preservative), $MgSO_4$ (Epsom salts), and $NH_4H_2PO_4$ (ammonium dihydrogen phosphate, a fertilizer).

2.10 SOME CHEMICAL PROPERTIES OF NONMETALS

Nonmetals form ionic compounds with metals and form molecular compounds with each other.

Of all the elements, only the noble gases (He through Rn) are content to exist simply as individual atoms. In the free state, when they are not combined with other different elements, the rest of the nonmetallic elements exist in molecular forms. As you've already learned, atoms such as oxygen, nitrogen, or chlorine combine with other like atoms to form diatomic molecules such as O_2, N_2, and Cl_2, which are gases. Except for iodine, I_2, those that are solids have rather complex structures, which we won't discuss now.

Although relatively few in number, the nonmetallic elements are found in more compounds than are the metals. As you learned in the last section, they combine with metals to form ionic compounds. However, nonmetals also combine with other nonmetals and with the metalloids to form a wide variety of nonionic, molecular substances.

Nonmetal–Nonmetal Compounds

The formulas of compounds formed between metals and nonmetals are relatively simple, being determined only by the requirement of electrical neutrality (i.e., the balancing of the charges of the ions). Compounds between nonmetals, however, are held together by entirely different kinds of chemical bonds, which arise from the mutual sharing of electrons instead of from electron transfer. We will discuss these bonds, which are called *covalent bonds,* at considerable length in Chapters 7 and 8. All we need to know now is that they differ from ionic bonds, and that they enable nonmetals to combine in a variety of ways, giving formulas of varying degrees of complexity. This complexity reaches a maximum with compounds in which carbon is combined with a handful of other elements such as hydrogen, oxygen, and nitrogen. There are so many of these compounds, in fact, that their study encompasses the chemical specialties called organic chemistry and biochemistry.

Hydrogen Compounds   At this early stage we can only begin to look for signs of order among the many nonmetal–nonmetal compounds, so we will restrict our discussion to just the simple compounds that the nonmetals form with hydrogen and oxygen. The easiest place to start is with the simple hydrides of the nonmetals, that is, their compounds with hydrogen. The formulas of these compounds are given in Table 2.7.[2]

Notice that within a given group, the formulas are similar. Also notice that for a given nonmetal, the number of hydrogens in the compound is the same as the number of electrical charges that the simple ion of the nonmetal has in its compounds with metals. Thus, oxygen forms the ion O^{2-} when it combines with metals, and in water the oxygen is combined with *two* hydrogen atoms. This should not be taken to mean that the nonmetals exist as ions in these compounds. Indeed, this is not the case at all. But it will help you to remember what the formulas are.

Many of the nonmetals form more than one compound with hydrogen. Carbon, for example, forms thousands of them. Petroleum is a complex mixture of these compounds, which are called hydrocarbons. We will leave further discussion of the hydrogen compounds of the nonmetals until later in the book.

Oxygen Compounds   Earlier we mentioned that the formulas of the oxides of the metals, especially those of Group IA and IIA, are characteristic of the group. For instance, all the Group IIA metals form oxides with the general formula MO. Among the nonmetal oxides we find greater complexity because of the different kinds of chemical

Remember: H_2, N_2, O_2, F_2, and Cl_2 are gases; Br_2 is a liquid; I_2 is a solid.

Molecular Formulas of the Nonmetals
Group Number

IVA	VA	VIA	VIIA
C[a]	N_2	O_2	F_2
	P_4	S_8	Cl_2
	As_4	Se_8	Br_2
			I_2

[a] Carbon forms crystals of graphite and diamond that contain enormous numbers of atoms linked in either a two- or three-dimensional interlocking network.

TABLE 2.7 Simple Hydrogen Compounds of the Nonmetallic Elements

Period	Group			
	IVA	VA	VIA	VIIA
2	CH_4	NH_3	H_2O	HF
3	SiH_4	PH_3	H_2S	HCl
4	GeH_4	AsH_3	H_2Se	HBr
5		SbH_3	H_2Te	HI

[2] This table shows how the formulas are normally written. The order in which the hydrogens appear in the formula is not of concern to us now. Instead, we are interested in the *number* of hydrogens that combine with a given nonmetal.

bonds that are involved. Sulfur, for example, forms two oxides, SO_2 and SO_3, both of which are industrially important and both of which are also serious air pollutants. Nitrogen oxides are even more numerous: NO, NO_2, N_2O, N_2O_3, N_2O_4, and N_2O_5. Despite this complexity, however, there is still some order to be found, as we can see in Table 2.8 where we have included formulas of some of the oxides of the nonmetals and metalloids. Patterns such as those shown in the table also helped Mendeleev fix the locations of the nonmetallic elements in their respective groups in the periodic table.

2.11 PROPERTIES OF IONIC AND MOLECULAR COMPOUNDS

Properties such as high melting point, brittleness, and the electrical behavior of ionic compounds are explained by the strong attractions between oppositely charged ions.

The properties of ionic compounds reflect the way ions interact with each other. The attractive force between ions of opposite charge is very large, as is the repelling force between ions of like charge. Therefore, in an ionic compound its ions arrange themselves so that the attractions between unlike-charged ions are at a maximum and the repulsions between like-charged ions are at a minimum.

In a crystal at room temperature, the ions just rattle about within a cage of other ions. They are held too tightly to move away. When heat is added to raise the temperature of the crystal, neighboring ions bounce off each other more violently. Eventually a temperature is reached at which the violent motions overcome the attractions between the ions and the crystal collapses. The compound melts, but because the net attractions are so large, the required temperature is very high. All ionic compounds, therefore, are solids at room temperature and tend to have high melting points.

You have probably never seen an ionic substance melt. Most of them melt well above room temperature. For example, ordinary table salt, NaCl, melts at 801 °C. Some ionic substances melt only at extremely high temperatures. For instance, aluminum oxide, Al_2O_3, melts at about 2000 °C, and for this reason it is used in special bricks that line the inside walls of furnaces.

Another property of ionic compounds is that their solids are generally hard and brittle. Molecular substances, such as paraffin wax or car wax, tend to be soft and easily crushed, but a crystal of rock salt is much harder. When struck by a hammer, however, the salt crystal shatters. The slight movement of a layer of ions within an ionic crystal suddenly places ions of the *same* charge next to one another, and for that instant there are large repulsive forces that split the solid, as illustrated in Figure 2.13.

TABLE 2.8 Simplest Formulas of Some Oxides of the Nonmetallic Elements

Group IVA	Group VA	Group VIA
Carbon	*Nitrogen*	· · ·
CO_2	N_2O_3	
	N_2O_5	
Silicon	*Phosphorus*	*Sulfur*
SiO_2	P_2O_3	SO_2
	P_2O_5	SO_3
Germanium	*Arsenic*	*Selenium*
GeO_2	As_2O_3	SeO_2
	As_2O_5	SeO_3
	Antimony	*Tellurium*
	Sb_2O_3	TeO_2
	Sb_2O_5	TeO_3

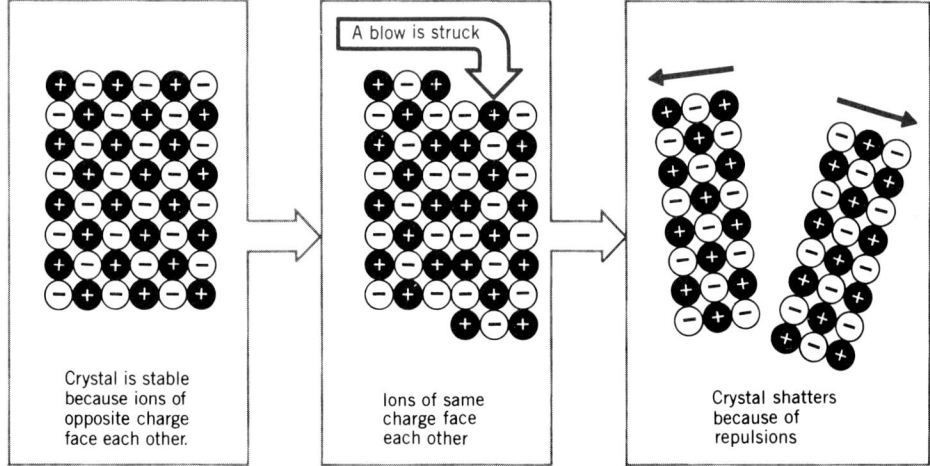

FIGURE 2.13 An ionic crystal shatters when struck because ions of like charge repel and force the crystal apart.

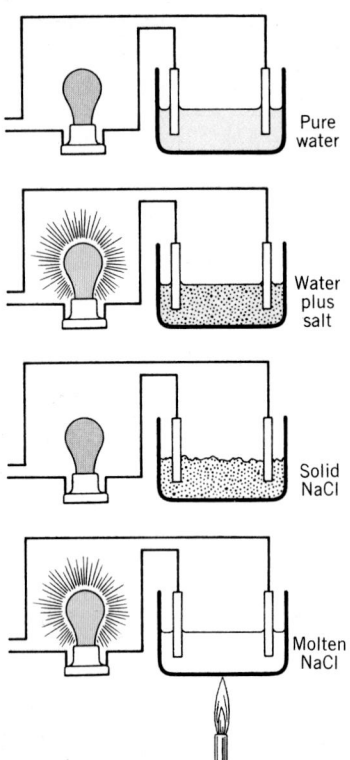

Pure water

Water plus salt

Solid NaCl

Molten NaCl

FIGURE 2.14 An apparatus to test for electrical conductivity. The electrodes are dipped into the substance to be tested. If the light bulb glows, the sample is an electrical conductor. Here we see that neither pure water nor solid sodium chloride conduct. Salt does conduct, however, if it is melted or dissolved in water.

In the solid state, ionic compounds do not conduct electricity. This is because electrical conductivity requires the movement of electrical charges, and in the solid the attractive forces prevent the movement of ions through the crystal. When the solid is melted, however, the ions become free to move about and the liquid conducts electricity well.

This can be easily demonstrated experimentally with the help of the apparatus depicted in Figure 2.14, which consists of a pair of metal electrodes that can be dipped into a container holding a substance whose electrical conductivity we wish to test. One of these electrodes is wired to an electric light bulb that will glow if electricity is able to pass between the two electrodes. When this apparatus is used to test the electrical conductivity of solid salt crystals, the bulb fails to light. However, if a flame is applied, the bulb lights brightly as soon as the salt melts.

The apparatus in Figure 2.14 can also be used to show that water solutions of ionic compounds conduct electricity. If the electrodes are dipped into pure distilled water, the bulb remains dark — pure water is not a conductor of electricity. But when a pinch of salt is added and the mixture is stirred, the bulb glows brightly, which shows that the solution does conduct. The reason it conducts is that the ions become separated when the salt dissolves, and they are therefore free to move about. This freedom of movement of the charged ions permits the conduction of electricity[3] and has a profound influence on the reactions of ionic compounds, as we will see shortly.

The properties of substances whose atoms are attached to each other by covalent bonds (electron sharing) usually differ markedly from those of ionic compounds. In these substances we find particles that are molecules. Within the individual molecules the atoms are held to each other very strongly, but between neighboring molecules the attractions are very weak. These weak attractions are responsible for many of the properties of molecular substances, just as the strong attractions between ions are responsible for many of the properties of ionic compounds. For example, molecular compounds such as water and candle wax tend to have low melting points. The molecules of these substances don't have to bounce around inside their crystals as violently as do the ions in an ionic crystal to overcome the attractive forces and become a liquid. Crystals of molecular compounds are also usually soft, because the molecules easily slide past each other.

Molecular substances differ from ionic compounds in their electrical characteristics, too. Molecules are uncharged particles, so they do not conduct electricity in the solid state or when melted. Most molecular substances also will not conduct electricity when dissolved in water. For example, if the electrodes of the conductivity apparatus shown in Figure 2.14 are dipped into a solution of sugar in water, the light bulb won't glow. Sugar molecules carry no electrical charge, so there are no electrical charges in the solution to provide conduction. As we will see in the next section, however, there are certain kinds of molecules that react with water to give ions, and their solutions do conduct electricity.

2.12 ACIDS AND BASES

Among the most important compounds containing metals and nonmetals are those that furnish hydrogen ions — acids — and those that give hydroxide ions — bases.

A method that we often use to establish some order among chemical facts is to categorize reactions and the chemicals that participate in them into certain classes. One important class of chemical reaction takes place between substances that we call acids and substances that we call bases — acids always react with bases in a predictable way.

---

[3] This kind of conduction takes place by a different means than the conduction in metals, and is described in more detail in Chapter 17.

Some of our most familiar chemicals as well as many of our most important laboratory agents are acids or bases. The vinegar in a salad dressing, the sour juices of a lemon, and the liquid in the battery of an automobile are similar in at least one respect — they are all acids. The white crystals of lye in drain cleaners, creamy milk of magnesia, and solutions of ammonia are all bases. Many acids and bases are also found within our bodies — who hasn't heard of stomach acid?

Acids are substances that release hydrogen ions, H^+, when placed in water. Actually these H^+ ions become attached to water molecules and travel about as ions with the formula H_3O^+ (they are called hydronium ions). Nevertheless, the active ingredient is the H^+. Bases are substances that give hydroxide ions, OH^-, when placed in water.

Think of H_3O^+ as a carrier for H^+ in water.

For all acids, the hydrogen atoms that are available as H^+ are usually written first in the formula. Thus, HNO_3 is an acid that can furnish one H^+, H_2CO_3 and H_2SO_3 are acids that can furnish two H^+, and H_3PO_4 is an acid that can supply three H^+. Later we will encounter many acids that also contain one or more hydrogens that are not able to be released as H^+. An example is acetic acid, $HC_2H_3O_2$, the substance that gives vinegar its sour taste. This acid can only furnish one H^+ — the one written first in its formula. In water, acetic acid can separate into H^+ and $C_2H_3O_2^-$ ions.

The hydrogen ion, H^+, consists of just a bare hydrogen nucleus, which is a proton. Acids that can supply just one H^+ are therefore often referred to as monoprotic acids. Examples are HCl, HNO_3, and $HC_2H_3O_2$. Acids that can supply more than one H^+ are called polyprotic acids. We could also be more specific and refer to H_2CO_3 and H_2SO_3 as diprotic acids and to H_3PO_4 as a triprotic acid. We will encounter these terms again in later chapters.

Some common acids.

Some common bases.

Neutralization

A property of acids and bases that we discussed earlier is their reaction with each other, which destroys their acidic and basic characteristics. For example, if hydrochloric acid and lye are mixed in just the right proportions, a solution containing nothing but ordinary table salt is formed. Hydrochloric acid, HCl, separates into ions in water. One of these ions is H^+ and the other is Cl^-. Lye is sodium hydroxide, NaOH, which separates into Na^+ and OH^- ions in water. When the two solutions are mixed, the H^+ and OH^- ions come in contact and react to form a molecule of water, H_2O. Since the H^+ and OH^- disappear, the solution is no longer acidic or basic. The only ions left are Na^+ and Cl^-, the ions found in table salt.

The reaction of an acid and a base is called neutralization because the H^+ and OH^- cancel or neutralize each other. We have already discussed the reaction between hydrochloric acid and sodium hydroxide.

$$HCl + NaOH \longrightarrow H_2O + NaCl$$

In general, the reaction of an acid with a base produces water and an ionic compound, in this case, sodium chloride, or salt. Because this is such a general reaction, the term *salt* has taken on a broader meaning than just NaCl. We will use the word salt to mean *any* ionic compound that doesn't contain OH^- or O^{2-}. Our general statement about the reaction of acids with bases can therefore be stated as follows: *whenever any acid reacts with any base in water, one product is water and the other is a salt.* Another example is the reaction of nitric acid, HNO_3, with NaOH.

If an ionic compound contains OH^- or O^{2-} it is called a base.

$$HNO_3 + NaOH \longrightarrow H_2O + NaNO_3$$

Sodium nitrate, $NaNO_3$, is ionic and is also called a salt. Perhaps you have recognized the NO_3^- ion as one of the polyatomic ions listed in Table 2.6. Maybe you have also noticed that acid–base neutralization reactions provide another way to prepare salts besides the direct combination of the elements described earlier.

Acid Salts

The reaction of a monoprotic acid such as HCl with a base such as NaOH can only lead to complete neutralization and the formation of a salt such as NaCl. With polyprotic acids, however, the product of neutralization depends on the proportions of acid and base used in the reaction. This is because polyprotic acids can be partially neutralized to give anions that still contain acidic hydrogens. For example, if sulfuric acid is combined with sodium hydroxide in a ratio of one formula unit of acid to two formula units of base, then complete neutralization does take place.

An acidic hydrogen is one that can be released as H^+.

Many acid salts have useful applications. As its active ingredient, this familiar product contains sodium hydrogen sulfate (sodium bisulfate), which the manufacturer calls "sodium acid sulfate."

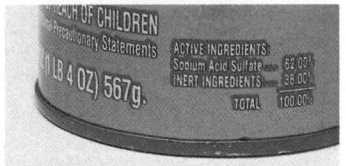

$$H_2SO_4 + 2NaOH \longrightarrow 2H_2O + Na_2SO_4$$

However, if the acid and base are combined in a one-to-one ratio, only half of the available hydrogens are neutralized.

$$H_2SO_4 + NaOH \longrightarrow H_2O + NaHSO_4$$

The salt $NaHSO_4$, which can be isolated as crystals by evaporating the reaction mixture, is referred to as an acid salt because it contains the anion HSO_4^-, which is still capable of furnishing additional H^+. In fact, solutions of $NaHSO_4$ are acidic.

There are many examples of acid salts. One of the most common is sodium bicarbonate, $NaHCO_3$, which we can consider to be the product of the partial neutralization of carbonic acid, H_2CO_3. Sodium bicarbonate is better known as baking soda.

Acid–Base Properties of Metal and Nonmetal Oxides

The tendency for elements to form acids or bases depends on their location in the periodic table, and this in turn helps us in the classification of elements. One of the controlling factors is the way an element's oxide reacts with water. In general, *when metal oxides react with water they form bases, and when nonmetal oxides react with water they form acids.*

Typical examples of the formation of bases are the reactions of sodium oxide, Na_2O, and calcium oxide, CaO, with water.

$$Na_2O + H_2O \longrightarrow 2NaOH \qquad \text{sodium hydroxide}$$
$$CaO + H_2O \longrightarrow Ca(OH)_2 \qquad \text{calcium hydroxide}$$

Metal oxides are called **basic anhydrides**. *Anhydride* means "without water."

Calcium oxide is an important ingredient in cement. When water is added to the cement the reaction above is one of many that occurs.

Typical examples of the formation of acids are the reactions of CO_2 and SO_2 with water.

$$CO_2 + H_2O \longrightarrow H_2CO_3 \qquad \text{carbonic acid}$$
$$SO_2 + H_2O \longrightarrow H_2SO_3 \qquad \text{sulfurous acid}$$

Nonmetal oxides are called **acidic anhydrides**.

Carbonic acid and sulfurous acid are too unstable to be isolated as pure compounds, but water solutions of them are quite common. Atmospheric CO_2 dissolved in groundwater exists partly as CO_2 molecules and partly as carbonic acid. It slowly dissolves limestone and is responsible for the large limestone caves found in various locations. Pollution of the air by SO_2 makes rain slightly acidic. This has caused a great deal of damage to marble statues in many parts of the world (marble is a form of limestone, $CaCO_3$).

2.13 NAMING CHEMICAL COMPOUNDS

The naming of chemical compounds follows a systematic procedure, internationally agreed upon by chemists, that allows only one formula to be associated with a name.

In conversation, chemists rarely use formulas to describe compounds. Instead, names are used. For example, you already know that water is the name for the compound having the formula H_2O and that sodium chloride is the name of NaCl.

At one time there was no uniform procedure for assigning names to compounds, and those who discovered compounds used whatever method they wished. Without some sort of system, however, the problem of remembering names for the rapidly increasing number of compounds soon became impossible. The search for a solution led chemists around the world to agree on a systematic method for naming substances. As a result, we are able to write the correct formula for any given compound given the correct name, and vice versa.

In this section we discuss the nomenclature (naming) of simple inorganic compounds. In general, these are substances that would *not* be considered to be derived from hydrocarbons such as methane (CH_4), ethane (C_2H_6), and other carbon–hydrogen compounds. The hydrocarbons and compounds that can be thought of as coming from them are called organic compounds (although they need not occur in living systems). We will have more to say about naming them in Chapter 24.

Binary Compounds Containing a Metal and a Nonmetal

For salts of a metal and a nonmetal, the name of the metal is given first, followed by the name of the negative ion formed from the nonmetal. The latter is formed from the stem of the nonmetal name plus the suffix *-ide*. An example is our old friend, NaCl — sodium chlor*ide*. Table 2.9 lists some common monatomic negative ions and their names. Other examples of compounds formed from them are

CaO   calcium oxide
ZnS   zinc sulfide
Mg_3N_2 magnesium nitride

The *-ide* suffix is usually used only for monatomic ions, although there are two common exceptions — *hydroxide* (OH^-) and *cyanide* (CN^-).

Many metals, particularly transition metals and post-transition metals, are able to form more than one ion. Iron, a typical example, forms ions with either a 2+ or a 3+

TABLE 2.9 Monatomic Negative Ions

H^-	hydride	N^{3-}	nitride	O^{2-}	oxide	F^-	fluoride
C^{4-}	carbide	P^{3-}	phosphide	S^{2-}	sulfide	Cl^-	chloride
Si^{4-}	silicide	As^{3-}	arsenide	Se^{2-}	selenide	Br^-	bromide
				Te^{2-}	telluride	I^-	iodide

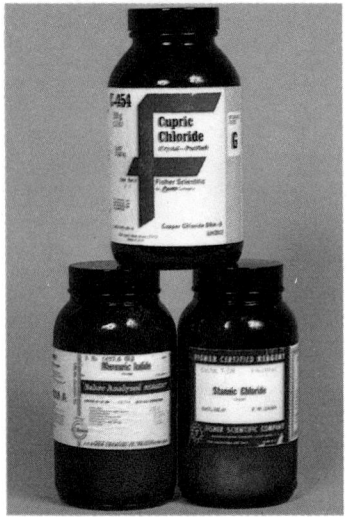

The older system of nomenclature is still used on the labels of many laboratory chemicals.

charge (Fe^{2+} or Fe^{3+}). Salts containing these different iron ions have different formulas, so in their names it is necessary to specify which iron ion is present. There are two ways of doing this. In the old system, the suffix *-ous* is used to specify the ion with the lower charge and the suffix *-ic* is used to specify the ion with the higher charge. With this method, we use the Latin stem for elements having symbols not derived from their English names.

Fe^{2+}	ferrous ion	$FeCl_2$	ferrous chloride
Fe^{3+}	ferric ion	$FeCl_3$	ferric chloride
Cu^+	cuprous ion	$CuCl$	cuprous chloride
Cu^{2+}	cupric ion	$CuCl_2$	cupric chloride

A list that contains additional examples is given in Table 2.10. Notice that mercury is an exception — we use the English stem when naming its ions.

TABLE 2.10 Metals That Form More Than One Ion

Cr^{2+}	chromous	Mn^{2+}	manganous	Fe^{2+}	ferrous	Cu^+	cuprous
Cr^{3+}	chromic	Mn^{3+}	manganic	Fe^{3+}	ferric	Cu^{2+}	cupric
Hg_2^{2+}	mercurous[a]	Sn^{2+}	stannous	Pb^{2+}	plumbous	Co^{2+}	cobaltous
Hg^{2+}	mercuric	Sn^{4+}	stannic	Pb^{4+}	plumbic	Co^{3+}	cobaltic

[a] Notice that Hg_2^{2+} contains two joined Hg^+ ions.

The currently preferred method for naming ions of metals that can have more than one charge in compounds is called the Stock system. Here we use the English name followed, without a space, by the numerical value of the charge written as a Roman numeral in parentheses.

Fe^{2+}	iron(II)	$FeCl_2$	iron(II) chloride
Fe^{3+}	iron(III)	$FeCl_3$	iron(III) chloride
Cr^{2+}	chromium(II)	CrS	chromium(II) sulfide
Cr^{3+}	chromium(III)	Cr_2S_3	chromium(III) sulfide

Copper(I) sulfate is Cu_2SO_4.
Copper(II) sulfate is $CuSO_4$.

Remember that the Roman numeral is the positive charge on the metal ion; it is *not* necessarily a subscript in the formula. You must figure out the formula from the ion charges, as discussed in Section 2.9. Even though the Stock system is now preferred, chemical companies still label bottles of chemicals using the old system. These old names also appear in the older scientific literature, which still holds much excellent data. Unfortunately, this means that you must know both systems.

PRACTICE EXERCISE 10   Name the compounds K_2S, Mg_3P_2, $NiCl_2$, and Fe_2O_3 using the Stock system where appropriate.

PRACTICE EXERCISE 11   Write formulas for (a) aluminum sulfide, (b) strontium fluoride, (c) titanium(IV) oxide, and (d) chromous bromide.

Binary Compounds between Two Nonmetals

In naming binary compounds containing two nonmetals, we usually use a method that specifies the actual numbers of atoms in a molecule. This system makes use of the following Greek prefixes.

NO - Nitric oxid

mono- = 1(often omitted)	hexa- = 6	
di- = 2	hepta- = 7	} use with non-metals with oxygen
tri- = 3	octa- = 8	
tetra- = 4	nona- = 9	
penta- = 5	deca- = 10	

NH₃ - Amonia

For example, NO_2 is nitrogen *di*oxide and N_2O_4 is *di*nitrogen *tetr*oxide (we drop the *a* for ease of pronunciation).

Some other examples are

HCl	hydrogen chloride (*mono-* omitted)	$AsCl_3$	arsenic trichloride
CO	carbon monoxide	SF_6	sulfur hexafluoride

PRACTICE EXERCISE 12   Name the following compounds using Greek prefixes when needed: PCl_3, SO_2, Cl_2O_7.

Binary Acids and Their Salts

The binary compounds of hydrogen with many of the nonmetals are acidic, and in their water solutions they are referred to as binary acids. Some examples are HCl and H_2S. In naming these substances as acids, we add the prefix *hydro-* and the suffix *-ic* to the stem of the nonmetal name, followed by the word *acid*. For example, water solutions of hydrogen chloride and hydrogen sulfide are named as follows:

HCl	*hydro*chlor*ic acid*
H_2S	*hydro*sulfur*ic acid*

Salts formed by neutralizing these acids contain the negative ion of the nonmetal and always end in *-ide*.

PRACTICE EXERCISE 13   Name the water solutions of the following acids: HF, HBr.

Oxoacids and Their Salts

Acids that contain hydrogen, oxygen, plus another element are called oxoacids. Examples are H_2SO_4 and HNO_3. These acids do not take the prefix *-hydro*. Many nonmetals form two or more oxoacids that differ in the number of oxygen atoms in their formulas, and they are named according to which one has the larger or smaller number of oxygens. The acid with the larger number of oxygens takes the suffix *-ic* and the one with the fewer number of oxygens takes the suffix *-ous*.

H_2SO_4	sulfur*ic acid*	HNO_3	nitr*ic acid*
H_2SO_3	sulfur*ous acid*	HNO_2	nitr*ous acid*

The halogens can occur in as many as four different oxoacids. The oxoacid with the most oxygens has the prefix *per-*, and the one with the least has the prefix *hypo-*.

HClO	*hypo*chlor*ous* acid (usually written HOCl)	$HClO_3$	chlor*ic* acid
$HClO_2$	chlor*ous* acid	$HClO_4$	*per*chlor*ic* acid

The neutralization of oxoacids produces negative polyatomic ions. There is a very simple relationship between the name of the polyatomic ion and its parent acid.

(1) -*ic* acids give -*ate* ions: HNO_3 (nit*ric* acid) NO_3^- (nit*rate* ion)
(2) -*ous* acids give -*ite* ions: H_2SO_3 (sulfur*ous* acid) SO_3^{2-} (sulf*ite* ion)

In naming polyatomic anions, the prefixes *per* and *hypo* carry over from the name of the parent acid. Thus perchloric acid, $HClO_4$, gives perchlorate ion, ClO_4^-, and hypochlorous acid, $HClO$, gives hypochlorite ion, ClO^-.

PRACTICE EXERCISE 14   The formula for arsenic acid is H_3AsO_4. What is the name of the salt Na_3AsO_4?

Acid Salts

In the previous section it was noted that polyprotic acids may be partially neutralized to give ions that still contain hydrogen. For example, H_2SO_4 can be partially neutralized to give the HSO_4^- ion. In naming acid salts formed by these kinds of ions, we specify the number of hydrogens that can still be neutralized if the salt were to be treated with additional base.

$NaHSO_4$ sodium hydrogen sulfate
NaH_2PO_4 sodium dihydrogen phosphate

For acid salts of diprotic acids, the prefix *bi-* is still often used.

$NaHCO_3$ sodium *bi*carbonate

Notice that the prefix *bi-* does *not* mean "two"; it means that there is an acidic hydrogen in the compound.

PRACTICE EXERCISE 15   What is the formula for sodium bisulfite? $Na\,HSO_3$

Common Names

Not every compound is named according to the systematic procedure outlined above. Many familiar substances were discovered long before a systematic method for naming them had been developed and they acquired common names that are so well known that no attempt has been made to rename them. For example, following the scheme described above we might expect that H_2O would have the name dihydrogen oxide. Although this isn't wrong, the common name *water* is so well known that it is always used. Another example of a compound known by its common name is ammonia, NH_3, whose odor you have no doubt experienced while using household ammonia solutions or the window cleaner Windex®.

Another class of compounds for which common names are often used is very complex substances. A common example is sucrose, which is the chemical name for table sugar, $C_{12}H_{22}O_{11}$. The structure of this compound is pretty complex, and its name assigned following the systematic method is equally complex. It is much easier to say the simple name sucrose, and be understood, than to struggle with the cumbersome systematic name for this compound.

SUMMARY

Laws of Chemical Combination.   When accurate masses of all the reactants and products in a reaction are measured and compared, no observable changes in mass accompany chemical reactions (the law of conservation of mass). The mass ratios of the elements in any compound are constant regardless of the source of the compound or how it is prepared (the law of definite proportions). Whenever two elements form more than one compound, then the different masses of one element that combine with a fixed mass of the other are in a ratio of small whole numbers (the law of multiple proportions).

Dalton's Atomic Theory.   The three laws of chemical combination make most sense if we assume that matter consists of atoms that can't break up during chemical reactions and that have masses that don't alter throughout these changes. Most elements consist of a small number of isotopes whose masses differ slightly. However, all of the isotopes of an element have very nearly identical chemical properties and the percentages of the various isotopes that make up an element are generally so constant throughout the world that we can say that the average mass of their atoms is a constant.

Atoms, Ions, and Molecules.   Atoms consist of electrons, protons, and neutrons. The protons (with a charge of $1+$) and the neutrons (0 charge) make up the atomic nucleus, and the number of protons is called the atomic number of the element. Each element has a different atomic number. The electrons (with a charge of $1-$) are found outside the nucleus, but their number equals the atomic number in a neutral atom. Isotopes of an element have identical atomic numbers but different numbers of neutrons. An element's atomic weight is the relative mass of its atoms on a scale in which atoms of carbon-12 have a mass of exactly 12.000 amu.

When atoms chemically combine to form compounds, the reactions produce either ions of opposite charge or neutral molecules, depending on the elements involved. An ionic compound always contains ions of both positive and negative charge. The ions always occur in a ratio that ensures that the net electrical charge on the formula unit is zero. When molecules form, the atoms are linked by the sharing of electrons. Molecules carry no electrical charge.

The Periodic Table.   The search for similarities and differences among the properties of the elements led Mendeleev to discover that when the elements are placed in (approximate) order of increasing atomic weight, similar properties recur at regular, repeating intervals. In the modern periodic table the elements are arranged in rows, called periods, in order of increasing atomic number. The rows are stacked so that elements in the columns, called groups or families, have similar chemical and physical properties. The A-group elements (IUPAC Groups 1, 2, and 13–18) are called representative elements; the B-group elements (IUPAC Groups 3–12) are called transition elements. The two long rows of inner transition elements located below the main body of the table consist of the lanthanides, which follow La ($Z = 57$), and the actinides, which follow Ac ($Z = 89$). Certain groups are given

family names: for instance, Group IA (Group 1), the alkali metals (the alkalis); Group IIA (Group 2), the alkaline earth metals; Group VIIA (Group 17), the halogens; Group 0 (Group 18), the noble gases.

Metals, Nonmetals, and Metalloids.   Most elements are metals; they occupy the lower left-hand region of the periodic table (to the left of a line drawn approximately from boron, B, to astatine, At). Nonmetals are found in the upper right-hand region of the table. Metalloids occupy a narrow band between the metals and nonmetals.

Metals have a characteristic luster, tend to be ductile and malleable, and conduct electricity. They react with nonmetals to form ionic compounds called salts. Nonmetals tend to be brittle, lack "metallic" luster, and are nonconductors of electricity. Many nonmetals are gases. Besides combining with metals, nonmetals combine with each other to form molecules without undergoing electron transfer. Bromine (a nonmetal) and mercury (a metal) are the two elements that are liquids at ordinary room temperature. Metalloids have properties intermediate between those of metals and nonmetals.

When ionic binary compounds are formed, electrons are transferred from a metal to a nonmetal. The metal atom becomes a positive ion; the nonmetal atom becomes a negative ion. The formulas of ionic compounds are controlled by the requirement that the compound must be electrically neutral. Many ionic compounds also contain polyatomic ions—ions that are composed of two or more atoms. Ionic compounds tend to be brittle, high-melting, nonconducting solids. When melted or dissolved in water, however, they do conduct electricity. Most molecular compounds tend to be soft and low-melting.

Acids and Bases.   Acids are substances that react with water to give hydronium ion, H_3O^+, which we usually abbreviate simply as H^+. Bases in water give hydroxide ion, OH^-. Neutralization of an acid by a base gives a salt plus water. Partial neutralization of polyprotic acids gives acid salts. Metal oxides that react with water form bases and are called basic anhydrides; nonmetal oxides that react with water give acids and are called acidic anhydrides.

Naming Compounds.   International agreement between chemists provides a system that allows us to write a single formula from a compound's name. For salts, the Stock system is preferred, but the older system must also be learned. Compounds between nonmetals use Greek prefixes to specify number. Binary acids are *hydro . . . ic acids* and give *-ide* anions. Oxoacids and their anions are related: *-ic acid* produces *-ate* ion; *-ous acid* produces *-ite* ion. Acid salts use the prefix *bi-*, or contain the name, hydrogen.

REVIEW EXERCISES

Answers to questions whose numbers are printed in color are given in Appendix C.
Difficult questions are marked with asterisks.

Laws of Chemical Combination and Dalton's Theory

2.1   Name and state the three laws of chemical combination.

2.2   In your own words, give the five postulates of Dalton's atomic theory.

2.3   Which postulate of Dalton's theory is based on the law of conservation of mass?

2.4   Which postulate of Dalton's theory is based on the law of definite proportions?

2.5   Which postulates of Dalton's theory are not strictly correct, and why didn't these errors affect the apparent validity of the theory?

2.6   In your own words, describe how Dalton's theory explains the law of conservation of mass and the law of definite proportions.

2.7   Which of the laws of chemical combination is used to define the term *compound?*

2.8   Laughing gas is a compound formed from nitrogen and oxygen in which there are 1.75 g of nitrogen to 1.00 g of oxygen. Below are given the compositions of several nitrogen–oxygen compounds. Which of these is laughing gas?
(a) 6.35 g nitrogen, 7.26 g oxygen
(b) 4.63 g nitrogen, 10.58 g oxygen
(c) 8.84 g nitrogen, 5.05 g oxygen
(d) 9.62 g nitrogen, 16.5 g oxygen
(e) 14.3 g nitrogen, 40.9 g oxygen

2.9   A compound of nitrogen and oxygen has the formula NO. In this compound there are 1.143 g of oxygen for each 1.000 g of nitrogen. A different compound of nitrogen and oxygen has the formula NO_2. How many grams of oxygen would be combined with each 1.000 g of nitrogen in NO_2?

2.10   Tin forms two compounds with chlorine, $SnCl_2$ and $SnCl_4$.
(a) When combined with the same mass of tin, what would be the ratio of the weights of chlorine in the two compounds?
(b) In the compound $SnCl_2$, 0.597 g of chlorine is combined with each 1.000 g of tin. In $SnCl_4$, how many grams of chlorine would be combined with 1.000 g of tin?

Atoms and Isotopes

2.11   What are the names, symbols, electrical charges, and masses (expressed in amu) of the three subatomic particles introduced in this chapter?

2.12   Where is nearly all of the mass of an atom located? Explain your answer in terms of what contributes to this mass.

2.13   What does *amu* stand for? What symbol is given to this quantity in the SI?

2.14   What are the names of the nucleons we have studied?

2.15   Define the terms *atomic number* and *mass number.*

2.16   Which is better related to the chemistry of an element, its mass number or its atomic number? Give a brief explanation. (Remember that there is much more to be said about this matter than is presented in this chapter.)

2.17   How are isotopes of the same element alike? How do they differ?

2.18   What is one property that atoms of two *different* elements might possibly have in common?

2.19   Consider the symbol $^a_bX^c_d$, where X stands for the chemical symbol for an element. What information is given in locations (a) *a*, (b) *b*, (c) *c*, and (d) *d*?

2.20   Write the symbols of the isotopes that contain the following. (Use the table of atomic weights and numbers printed inside the front cover for additional information, as needed.)
(a) An isotope of iodine whose atoms have 78 neutrons.
(b) An isotope of strontium whose atoms have 52 neutrons.
(c) An isotope of cesium whose atoms have 82 neutrons.
(d) An isotope of fluorine whose atoms have 9 neutrons.

2.21   Give the numbers of neutrons, protons, and electrons in the atoms of each of the following isotopes. (Use the table of atomic weights and numbers printed inside the front cover for additional information, as needed.)
(a) radium-226   (c) $^{206}_{82}Pb$
(b) carbon-14   (d) $^{23}_{11}Na$

2.22   Give the numbers of electrons, protons, and neutrons in the atoms of each of the following isotopes. (The table of atomic weights and numbers printed inside the front cover may have to be consulted for some needed information.)
(a) cesium-137   (c) $^{238}_{92}U$
(b) iodine-131   (d) $^{197}_{79}Au$

Atomic Weights

2.23   Write the symbol for the isotope that forms the basis of the atomic weight scale. What is the mass of this atom expressed in atomic mass units?

2.24   The actual mass of the amu is $1.6605665 \times 10^{-24}$ g. Using this value, calculate the mass in grams of one atom of carbon-12.

*2.25   In the compound CH_4, 0.33597 g of hydrogen is combined with 1.000 g of carbon-12. Use this information to calculate the atomic weight of the element hydrogen.

2.26   If an atom of carbon-12 had been assigned a relative mass of 24.0000 amu, what would be the atomic weight of hydrogen relative to this mass?

Molecules, Ions, and Compounds

2.27   Consider the sodium atom and the sodium ion.
(a) Write the chemical symbol of each.
(b) Do these particles have the same number of nuclei?
(c) Do they have the same number of protons?
(d) Could they have different numbers of neutrons?
(e) Do they have the same number of electrons?

2.28   If an atom gains an electron to become an ion, what kind of electrical charge does the ion have?

2.29   How many electrons, protons, and neutrons are in each of the following particles? (a) $^{35}_{17}Cl^-$, (b) $^{56}_{26}Fe^{3+}$, (c) $^{64}_{29}Cu^{2+}$, (d) $^{39}_{19}K^+$.

2.30   Describe what kind of event must occur if the atoms of two different elements are to react to form (a) an ionic compound or (b) a molecular compound.

The Periodic Table

2.31 What observation led Mendeleev to develop his periodic table?

2.32 On the basis of their positions in the periodic table, why is it not surprising that strontium-90, a dangerous radioactive isotope of strontium, replaces calcium in newly formed bones?

2.33 Why would you reasonably expect cadmium to be a contaminant in zinc but not in silver?

2.34 In the refining of copper, sizable amounts of silver and gold are recovered. Why is this not surprising?

2.35 Below are some data for the elements sulfur and tellurium.

	Sulfur	Tellurium
Atomic mass	32.06	127.60
Melting point (°C)	112.8	449.5
Boiling point (°C)	445	990
Formula of oxide	SO_2	TeO_2
Melting point of oxide (°C)	−72.7	733
Density (g/cm³)	2.07	6.25

Had the element selenium not been known in the time of Mendeleev, he would have called it eka-sulfur. Estimate its properties as an average of those of sulfur and tellurium.

2.36 What was the original form of the periodic law? Try to state it in your own words.

2.37 What is a "period"? What is a "group"?

2.38 Why were there gaps in Mendeleev's periodic table?

2.39 In the text, we identified two places in the periodic table where the atomic weight order was reversed. Using the table on the inside front cover, locate two other places where this occurs.

2.40 State the modern form of the periodic law, using your own words.

2.41 Make a rough sketch of the periodic table and mark off those areas where you would find (a) the representative elements, (b) the transition elements, and (c) the inner transition elements.

2.42 What group numbers are used to designate the representative elements following (a) the traditional U.S. system for designating groups and (b) the IUPAC system.

2.43 Supply the IUPAC group numbers that correspond to the following customary U.S. designations: (a) Group IA, (b) Group VIIA, (c) Group IIIB, (d) Group IB, (e) Group IVA.

2.44 Based on discussions in this chapter, explain why it is unlikely that scientists will discover a new element, never before observed, having an atomic weight of approximately 73.

2.45 Which is an alkali metal? Ca, Cu, In, Li, S.

2.46 Which is a halogen? Ce, Hg, Si, O, I.

2.47 Which is a transition element? Pb, W, Ca, Cs, P.

2.48 Which is a noble gas? Xe, Se, H, Sr, Zr.

2.49 Which is a lanthanide element? Th, Sm, Ba, F, Sb.

2.50 Which is an actinide element? Ho, Mn, Pu, At, Na.

2.51 Which is an alkaline earth metal? Mg, Fe, K, Cl, Ni.

Physical Properties of Metals, Nonmetals, and Metalloids

2.52 Give five physical properties that we usually observe for metals.

2.53 Why is mercury used in thermometers? Why is tungsten used in light bulbs?

2.54 What property of metals allows them to be drawn into wire?

2.55 Gold can be hammered into sheets so thin that some light can pass through them. What property of gold allows such thin sheets to be made?

2.56 Only two metals are colored (the rest are "white," like iron or lead). You have surely seen both of them. Which metals are they?

2.57 Which nonmetals occur as monatomic gases (gases whose particles consist of single atoms)?

2.58 Which nonmetals occur in nature as diatomic molecules? Which are gases?

2.59 Which two elements exist as liquids at room temperature and pressure?

2.60 Which physical property of metalloids distinguishes them from metals and nonmetals?

2.61 Sketch the shape of the periodic table and mark off those areas where we find (a) metals, (b) nonmetals, and (c) metalloids.

2.62 Which metals can you think of that are commonly used to make jewelry? Why isn't iron used to make jewelry?

2.63 With what kind of elements do metals react?

2.64 What is an ionic compound? What holds an ionic compound together? Can we identify individual molecules in an ionic compound?

2.65 What is the difference between a binary compound and one that is diatomic? Give examples that illustrate this difference.

2.66 What is a post-transition metal?

2.67 Define *cation* and *anion*.

Formulas, Ions, and Ionic Compounds

2.68 Referring to the periodic table, but without looking at Table 2.4, give the symbols for ions of (a) K, (b) Br, (c) Mg, (d) S, and (e) Al.

2.69 Use the periodic table, but not Table 2.4, to write the symbols for ions of (a) barium, (b) oxygen, (c) fluorine, (d) strontium, and (e) rubidium.

2.70 What rules do we apply when we write formulas for ionic compounds?

2.71 Write formulas for ionic compounds formed between (a) Na and Br, (b) K and I, (c) Ba and O, (d) Mg and Br, and (e) Ba and F.

2.72 Which of the following formulas are incorrect? (a) NaO_2, (b) RbCl, (c) K_2S, (d) Al_2Cl_3, (e) MgO_2.

*2.73 From what you have learned in Section 2.9, write correct balanced equations for the reactions between (a) calcium and chlorine, (b) magnesium and oxygen, (c) aluminum and oxygen, and (d) sodium and sulfur.

2.74 What are the formulas for (a) cyanide ion, (b) ammonium ion, (c) nitrate ion, (d) sulfite ion, and (e) chlorate ion?

2.75 What are the formulas for (a) hypochlorite ion, (b) bisulfate ion, (c) phosphate ion, (d) dihydrogen phosphate ion, and (e) permanganate ion?

2.76 What are the names of the following ions? (a) $Cr_2O_7{}^{2-}$, (b) OH^-, (c) $C_2H_3O_2{}^-$, (d) $CO_3{}^{2-}$, (e) $ClO_4{}^-$.

2.77 Write formulas for the ionic compounds formed from (a) K^+ and nitrate ion, (b) Ca^{2+} and acetate ion, (c) ammonium ion and Cl^-, (d) Fe^{3+} and carbonate ion, and (e) Mg^{2+} and phosphate ion.

2.78 Write formulas for the ionic compounds formed from (a) Zn^{2+} and hydroxide ion, (b) Ag^+ and chromate ion, (c) Ba^{2+} and sulfite ion, (d) Rb^+ and sulfate ion, and (e) Li^+ and bicarbonate ion.

2.79 Write the symbol for the ion formed when titanium (Ti) loses four electrons.

2.80 Write formulas for two compounds formed between O^{2-} and (a) lead, (b) tin, (c) manganese, (d) iron, and (e) copper.

2.81 Write formulas for the ionic compounds formed from Cl^- and (a) cadmium ion, (b) silver ion, (c) zinc ion, and (d) nickel ion.

Nonmetal–Nonmetal Compounds

2.82 With what kinds of elements do nonmetals combine?

2.83 In what major way do compounds formed between two nonmetals differ from those formed between a metal and a nonmetal?

2.84 Why are nonmetals found in more compounds than are metals, even though there are fewer nonmetals than metals?

2.85 Which are the only elements that exist as free, individual atoms when not chemically combined with other elements?

2.86 Without referring to Table 2.7, but using the periodic table, write chemical formulas for the simplest hydrogen compounds of (a) carbon, (b) nitrogen, (c) tellurium, and (d) iodine.

2.87 Astatine forms a compound with hydrogen. Predict its chemical formula.

2.88 Under appropriate conditions, tin can be made to form a simple molecular compound with hydrogen. What would you predict its formula to be?

2.89 Based on what you've learned in this section, what are likely formulas of the two oxides of bismuth?

Properties of Ionic and Molecular Compounds

2.90 What are typical physical properties of ionic compounds?

2.91 Why does molten KCl conduct electricity even though solid KCl does not?

2.92 What happens when an ionic compound dissolves in water that allows its solution to conduct electricity?

2.93 In what basic way do the chemical bonds in molecular substances differ from those in salts such as NaCl?

2.94 Why are ionic compounds so brittle?

2.95 Why do most ionic compounds have high melting points?

2.96 Naphthalene (moth flakes) consists of soft crystals that melt at a relatively low temperature of 80 °C. Does this substance have characteristics of an ionic or a molecular compound?

2.97 Compare the properties of ionic and molecular compounds.

Acids and Bases

2.98 Define *acid, base, neutralization,* and *salt.*

2.99 What is the name of the ion H_3O^+?

2.100 Why is H_3O^+ often written simply as H^+?

2.101 Milk of magnesia contains $Mg(OH)_2$ and stomach acid contains HCl. Write the chemical reaction that occurs when you swallow milk of magnesia.

2.102 Many common antacids contain aluminum hydroxide, $Al(OH)_3$. Write an equation for the neutralization of the HCl of stomach acid by $Al(OH)_3$.

*2.103 Complete and balance the following equations for complete neutralization.
(a) $HNO_2 + KOH \longrightarrow$
(b) $HCl + Ca(OH)_2 \longrightarrow$
(c) $H_2SO_4 + NaOH \longrightarrow$
(d) $HClO_4 + Al(OH)_3 \longrightarrow$
(e) $H_3PO_4 + Ba(OH)_2 \longrightarrow$

2.104 Define *monoprotic acid, polyprotic acid,* and *diprotic acid.*

2.105 What is an acidic anhydride? What is a basic anhydride?

2.106 Write the formulas for all the acid salts that could be formed from the reaction of NaOH with the triprotic acid H_3PO_4.

2.107 An oxide of a certain element produced an acidic solution when dissolved in water. The oxide has the empirical formula X_2O_3, where X represents the symbol for the element. Is the element X a metal or a nonmetal? In which group in the periodic table does it belong?

Naming Compounds

2.108 Name the following.
(a) CaS   (c) $AlBr_3$ (e) Na_3P
(b) NaF   (d) Mg_2C (f) Li_3N

2.109 Name the following, using both the old nomenclature system and the Stock system.
(a) $CrCl_3$ (c) CuO   (e) SnO_2
(b) Mn_2O_3 (d) Hg_2Cl_2 (f) PbS

2.110 Name the following.
(a) SiO_2 (c) XeF_4 (e) P_4O_{10}
(b) ClF_3 (d) S_2Cl_2 (f) N_2O_5

2.111 The periodate ion has the formula IO_4^-. What would be the name for the acid HIO_4?

2.112 Name the following. If necessary, refer to Table 2.6 on page 71.
(a) $NaNO_2$ (c) $KMnO_4$ (e) $BaSO_4$
(b) K_3PO_4 (d) $NH_4C_2H_3O_2$ (f) $Fe_2(CO_3)_3$

2.113 What would be the formula for chromic acid?

2.114 Write formulas for the following.
(a) sodium monohydrogen phosphate
(b) lithium selenide
(c) sodium hydride
(d) chromic acetate
(e) nickel(II) cyanide
(f) iron(III) oxide
(g) stannic sulfide
(h) antimony pentafluoride
(i) dialuminum hexachloride
(j) tetraarsenic decaoxide

(k) magnesium hydroxide
(l) cupric bisulfate

2.115 Write formulas for the following.
(a) ammonium sulfide
(b) chromium(III) sulfate
(c) molybdenum(IV) sulfide
(d) tin(IV) chloride
(e) iron(III) oxide
(f) calcium bromate
(g) mercury(II) acetate
(h) barium bisulfite
(i) silicon tetrafluoride
(j) boron trichloride
(k) stannous sulfide
(l) calcium phosphide

2.116 Name the following oxoacids and give the names of the salts formed from them by neutralization with NaOH: (a) $HOCl$, (b) HIO_2, (c) $HBrO_3$, (d) $HClO_4$.

CHEMICALS IN USE

SODIUM CHLORIDE

2

Almost everybody in the world is familiar with sodium chloride, NaCl, and English-speaking people call it *salt*. It is as vital to our lives as air, water, and food. The migrations and patterns of settlement of whole tribes and peoples in ancient times were governed as much by the availability of salt as that of anything else. Even in early U.S. history, according to the historian Frederick Jackson Turner, little westward migration could occur until sources of salt were found beyond the Allegheny mountains in what are now western parts of Virginia, West Virginia, New York, and Kentucky.

Our Need for Salt

Every cell of the human body contains the ions of salt, Na^+ and Cl^-. Altogether, the average adult human has about 160 to 175 g of salt, enough to fill six one-ounce salt shakers. To compensate for small daily losses by perspiration and excretion in the urine, we each need about 200 mg a day — roughly one-tenth of a teaspoonful.

The juices of all meats contain enough salt so that people whose diets include meat can satisfy all of their daily needs without supplementary salt. The need for extra salt arose when ancient tribes discovered how to raise crops and relied more and more on them for their diets. Vegetables and cereal grains generally cannot alone replace the normal, daily salt losses of the body.

Another extremely important use of salt that ancient peoples discovered was that meat does not spoil when it is coated heavily with crystalline salt or soaked in concentrated salt brine. Decay-causing microorganisms cannot live in a high concentration of salt. Until modern refrigeration came into widespread use, meatpackers literally packed meat in salt. Without salting, only deep freezing can indefinitely preserve meat. Fishermen who go to sea for weeks to harvest fish carry tons of salt to preserve the catch, unless they are equipped with deep-freezing systems. No wonder that sailors have long been called "old salts."

So essential has salt been both to the body and for food preservation that in some salt-poor regions it has enjoyed the status of money itself. When Marco Polo visited Tibet on his way to China in the thirteenth century, he found coins made of rock salt bearing the seal of the ruler.

Ancient Rome had to use armed troops to guard the salt supplies obtained from salt-making operations near Ostia, Italy, southwest of Rome. The soldiers who protected and transported the salt along the Via Salarium (the Salt Road) were originally paid in salt. In time, they were given money with which to buy the salt, and this payment was called

FIGURE 2a Most of the world's supply of salt for human use is obtained by the evaporation of water from seawater spread out in large salt pans, such as those, on the left in this photograph, taken at a salt-making operation in Brazil. The windmill provides the energy for pumping the salt water.

FIGURE 2b The red color of certain microorganisms that can live in very salty water give these salt pans in California the appearance of blood.

their *salarium,* the Latin root of the English word, *salary.* Herein, too, is the origin of the expression "worth your salt."

Manufacture of Salt

The salt available to us in grocery stores has a purity that we take for granted, but only in certain deposits of solid salt found beneath the earth's surface is the purity truly high. Thus, people described as "the salt of the earth" are those with high moral purity and integrity. Quite often, the salt that can be obtained from solid deposits contains claylike solids plus ions of calcium, magnesium, and sulfate. The listeners to the Sermon on the Mount were unbewildered by a reference to "salt that has lost its taste" and that is good only to be "thrown out and trodden under foot" (Matthew 5:13). The salt sold to poor people in that time was often of poor quality because it was contaminated by so much earthy matter. The "salt that has lost its taste" was the residue after the sodium chloride had been leached away.

Salt isn't actually "made" in the way that aspirin is made from substances that aren't aspirin. The "manufacture" of salt refers instead to harvesting it from its natural sources and purifying it. Since the most ancient of times, people near the ocean have let sunlight evaporate seawater to obtain salt. The salt content of the oceans isn't high — 3.3% on the average (meaning 3.3 g NaCl per 100 g of seawater). Despite this, seawater is still the principal source of salt for most countries. Taken together, the world's oceans have enough salt — over 500,000 cubic miles of it — to cover the entire area of the United States to a depth of nearly one and a half miles.

To obtain salt from the oceans, seawater is allowed to flow into a salt pan, a low-lying area of several acres surrounded by dikes. As sunlight works on the still water, the undissolved matter (sand and clay) settles. As more and more water evaporates, the least soluble substances (the sulfates and carbonates of calcium) begin to deposit from the solution. Now the concentrated brine is allowed to flow successively into crystallizing pans where further evaporation leaves "sea salt" of varying grades, the highest being 96% NaCl. Figure 2a depicts various stages of these operations.

In the Middle East and many other regions, salt is obtained from the natural brines of salt marshes and swamps. These brines are frequently home to the few microorganisms that can thrive in saltwater, and they include some that lend a striking red color to the salt pan. (See Figure 2b.) Water that is both red and salty — characteristics of blood — might be confused as blood. Evidently, the army of Moab made this mistake (2 Kings 3:22–23) when, looking at the sun glinting red off the salt pans near the camp of the combined armies of Judah, Israel, and Edom, they thought they had slaughtered each other. These salt pans were probably those of Sodom, a city whose name may be from the Hebrew words for "field" *(sade)* and "red" *(adom).*

Rock salt is the other principal source of salt. It can be obtained by conventional methods of underground mining, or by injecting water into the deposits, pumping out the brine, and then processing the brine in various ways.

Industrial Uses of Salt

About 75% of the annual U.S. salt production, which is about 43 million tons/year, is used to make other chemicals. The largest single uses are for the manufacture of chlorine, Cl_2, hydrochloric acid, and sodium hydroxide, NaOH. (We will return to this in *Chemicals in Use 3.*) The manufacture of plastics such as Saran Wrap® and PVC (polyvinyl chloride used in making clear bottles), drugs such as aspirin, many solvents such as carbon tetrachloride, antifreeze compounds, and a very large number of other commercially important substances requires the use of chlorine. For these purposes and as well as to furnish municipal water supply systems with chlorine for killing disease-causing organisms, nearly 11 million tons of chlorine are made from salt each year in the United States. Thus salt not only supports life at its most physical level, it supports the standards of living of all industrialized countries.

Questions

1. What are the chief sources of sodium chloride?
2. In what ways was salt vital to ancient peoples?
3. What is the average percentage concentration of salt in the ocean?
4. Which three chemicals of industrial importance are manufactured from salt?

Buying jelly beans of different flavors in equal *numbers* is easy. We just have them weighed in 1 to 1 ratios by mass because the beans all have about the same mass. Obtaining atoms, ions, or molecules in equal numbers is almost as easy. We weigh them in ratios according to their formula weights, as we'll study in this chapter.

MASS RELATIONSHIPS FOR PURE SUBSTANCES

CHEMICAL CALCULATIONS

FORMULA WEIGHTS AND MOLECULAR WEIGHTS

THE MOLE

PERCENTAGE COMPOSITION

EMPIRICAL FORMULAS

MOLECULAR FORMULAS

3.1 CHEMICAL CALCULATIONS

The ratios by mass among the chemicals in a reaction are constant regardless of the scale at which the reaction is carried out in the lab.

Although the idea of atoms was conceived long ago, scientists found it of little value until John Dalton showed how it could be used to explain observable and *measurable* facts. This was really the key to Dalton's success — the ability to explain the quantitative laws of chemical combination (the laws of conservation of mass and definite proportions) known at that time. But without these quantitative laws, there would have been nothing to explain, and scientific knowledge would still be a hodgepodge of unexplained facts.

The importance of quantitative laboratory measurements is probably even greater today than in Dalton's time. In this chapter and the next we explore what some of these measurements are and how they relate to chemical composition and reactions. You will see that they allow us to determine the formulas of compounds and calculate the amounts of chemicals that a chemist (or chemistry student) might bring together in a particular experiment.

A term chemists often use when they refer to the masses of chemicals actually involved in an experiment is the *scale of the reaction.* In a small-scale reaction, the chemicals might be in milligram-to-gram quantities. A large-scale reaction, on the other hand, might involve hundreds or thousands of grams or even tons of reactants. The laws of conservation of mass and of definite proportions ensure, however, that the *relative* masses of reacting chemicals are identical for a given reaction regardless of the scale. These laws, familiar to Dalton, are still used by chemists and chemical engineers to scale the amounts of reactants up or down using simple proportions or their equivalents in the factor-label method. As our introduction to chemical calculations, we will work a simple example that shows this.

EXAMPLE 3.1   USING THE LAW OF DEFINITE PROPORTIONS IN A CHEMICAL CALCULATION

Problem:

Calcium chloride removes water from the air by reacting with it to form a hydrate. Compounds that do this are said to be **deliquescent.**

Calcium chloride is a compound whose crystals are able to draw large amounts of moisture from the air, so it is often used to dehumidify damp basements and other humid places. In this compound, composed of the elements calcium and chlorine, it is always found that for each 1.000 g of calcium (Ca) there are 1.769 g of chlorine (Cl). How many grams of chlorine would be required to make calcium chloride on a scale large enough to use 14.25 g of calcium?

Solution:

This question provides our first opportunity to apply the factor-label method to a problem in chemistry. As you learned in the first chapter, this method uses the cancellation of units as a guide in setting up the arithmetic. To apply the method, we search for valid relationships between quantities, and then we use these relationships to construct conversion factors that are then used to arrive at the answer.

Before we attempt to perform any calculations, we have to understand *how* to solve the problem. First, what does this question have to do with the law of definite proportions? As you know, the law states that in all samples of *any* given compound the elements are *always* present in the same ratio by mass. If you look at the data in the problem, you see that the mass of chlorine in calcium chloride is 1.769 times as large as the mass of calcium. (We are told that 1.769 g of Cl combines with 1.000 g of Ca.) Therefore, if the mass of Ca in another sample is 14.25 g, the mass of chlorine must again be 1.769 times as large, and 1.769×14.25 g $= 25.21$ g. Thus, 25.21 g of Cl will combine with 14.25 g of Ca.

We've gotten an answer to the problem by one line of reasoning, and certainly a valid way. The factor-label method, however, gives us another line of reasoning that is usually faster while being equally as sure a way of finding the same answer.

Step 1. We first have to express the problem in terms of the factor-label method. In this approach, we need to find a conversion factor that lets us change grams of calcium to grams of chlorine.

$$\text{(grams calcium)} \times \text{(conversion factor)} = \text{(grams chlorine)}$$

or

$$\text{(14.25 g Ca)} \times \text{(conversion factor)} = \text{g Cl}$$

Step 2. We examine the statement of the problem for information that allows us to construct an appropriate conversion factor. The given data define a relationship between the masses of calcium and chlorine in this compound. This allows us to write an "equation" that relates these masses in this reaction:

$$1.769 \text{ g Cl} = 1.000 \text{ g Ca}$$

We have used an equal sign here without intending that this sign mean that 1.769 g of Cl is *identical* to 1.000 g of Ca. There is, unfortunately, no universally accepted symbol for the phrase "is chemically equivalent to," which is closer to the meaning of the equals sign here. And there really is no need for one, because the context will always make clear what is meant. Most chemists use an equal sign, therefore, in more than one way. Here, when we state that 1.769 g Cl = 1.000 g Ca in calcium chloride, we obviously do not mean that the two chemicals are identical. Instead, we mean simply that these are masses of the two elements that would exactly combine to form calcium chloride. We are saying that 1.769 g of Cl is equivalent to 1.000 g of Ca when they combine to form calcium chloride. Our equation, 1.769 g Cl = 1.000 g Ca, is the law of definite proportions for calcium chloride.

Once we have the relationship between the two masses of chemicals in calcium chloride, we can construct two conversion factors:

> Sometimes it helps to read the divisor line as *per;* thus, we can say for the first factor "1.769 g Cl *per* 1.000 g Ca."

$$\frac{1.769 \text{ g Cl}}{1.000 \text{ g Ca}} \quad \text{and} \quad \frac{1.000 \text{ g Ca}}{1.769 \text{ g Cl}}$$

Step 3. We choose the conversion factor that will allow us to cancel the units "g Ca."

$$14.25 \text{ g Ca} \times \frac{1.769 \text{ g Cl}}{1.000 \text{ g Ca}} = 25.21 \text{ g Cl}$$

Thus, 14.25 g of Ca requires 25.21 g of Cl in making calcium chloride on this scale.

In solving chemistry problems by the factor-label method, we always consider the same three steps introduced in Example 3.1. First, we express the problem in terms of the factor-label method. Second, we look for relationships between *units* involved in the problem. Third, we choose conversion factors that allow us to cancel unwanted units and obtain the units of the answer.

To minimize the danger that the factor-label method might become just an unthinking tool, we should ask after applying this method, "Does the size of the answer make sense?" In Example 3.1, it does, because the mass of Cl in calcium chloride must *always* be greater than the mass of Ca; the initial data — the law of definite proportions about calcium chloride — told us this. Thus this reasoning also reassures us that we did not take the wrong conversion factor.

We will use the factor-label method in future problems, and if you practice the same approach in working homework problems you will discover what a powerful tool it is for problem solving.

PRACTICE EXERCISE 1   In a sample of sodium chloride (table salt) there is 2.059 g of chlorine (Cl) combined with each 1.335 g of sodium (Na). In another sample of sodium chloride, there is 2.366 g of Cl. How many grams of Na are in this sample?

In principle, all chemical calculations involving combinations of chemicals can be solved by the strategies just studied. All we need are the proportions by mass of each element in each compound. The most common way to obtain such data for a known compound is to use the formula of the compound, a table of atomic weights, and the concept of a formula weight that we will next study.

3.2 FORMULA WEIGHTS AND MOLECULAR WEIGHTS

The sum of the atomic weights of the atoms in a chemical formula is the formula weight of the compound.

In Chapter 2 we introduced the concept of average atomic weight. Oxygen, for example, has an atomic weight of 16.0 amu and hydrogen has an atomic weight of 1.0 amu (in both cases rounded to the nearest tenth). When these atoms combine to form a molecule of water, the weight of this molecule (H_2O) is just the weight of two hydrogen atoms plus the weight of one oxygen atom. This must be true, because atoms do not change mass

> The law of conservation of mass ensures that a formula unit has a mass equal to the sum of the masses of the atoms in it.

when they combine in chemical reactions. The mass of the average water molecule must therefore be 18.0 amu (2 × 1.0 amu + 16.0 amu = 18.0 amu). This value, 18.0 amu, is called the molecular weight of water.

As you learned in the last chapter, there are many substances that do not exist as molecules. Salt (sodium chloride) is an example. For this substance we just specify the ratio of the atoms combined and we call the formula (NaCl, in this case) a *formula unit*. Formula unit is really a general term, and we can use it for H_2O as well as for NaCl. Similarly, we can consider the term formula weight to be a generalization of *molecular weight*. It is simply the sum of the masses of the atoms expressed in the chemical formula. The term *formula weight* can even be applied to elements. For example, the formula of sodium is Na and one formula unit of Na consists of one atom so its mass (that is, its formula weight) is 22.98977 amu.

The best way to learn how to calculate formula weights is by working some examples. We will introduce here an operating procedure that will be standard for this book. Round atomic weights to three significant figures *before* using them in calculations, unless more significant figures are suggested by the data in the problem.

EXAMPLE 3.2   CALCULATING A FORMULA WEIGHT

Problem: Calculate the formula weight of aspirin, $C_9H_8O_4$.

Solution: First assemble the atomic weights, rounding their values (taken from the table inside the front cover) to three significant figures. For any atom that appears more than once in the formula, we multiply its rounded atomic weight by the subscript of its symbol and then add the numbers.

$$C_9H_8O_4 = \quad 9C \quad + \quad 8H \quad + \quad 4O$$
$$= (9 \times 12.0) + (8 \times 1.01) + (4 \times 16.0)$$
$$= \quad 108 \quad + \quad 8.08 \quad + \quad 64.0$$
$$= 180, \text{ the formula weight of aspirin}$$
$$\text{(correctly rounded)}$$

For formulas that have parentheses, remember that a subscript outside a parenthesis is a multiplier for all atoms inside. For example, in the formula $Ca(C_2H_3O_2)_2$ there are 1 calcium, 4 carbon, 6 hydrogen, and 4 oxygen atoms.

EXAMPLE 3.3   CALCULATING A FORMULA WEIGHT

Problem: Find the formula weight of calcium acetate, $Ca(C_2H_3O_2)_2$, the stiffening agent in "canned heat." Round atomic weights to three significant figures.

Solution:

Sterno® is one brand of canned heat. The actual fuel is alcohol.

$$Ca(C_2H_3O_2)_2 = \quad Ca \quad + \quad 4C \quad + \quad 6H \quad + \quad 4O$$
$$= (1 \times 40.1) + (4 \times 12.0) + (6 \times 1.01) + (4 \times 16.0)$$
$$= 158.2, \text{ the formula weight of calcium acetate}$$

PRACTICE EXERCISE 2   Calculate the formula weight of each substance.
(a) Sodium chloride, NaCl.
(b) Sucrose (table sugar), $C_{12}H_{22}O_{11}$.
(c) Calcium propionate, $Ca(C_3H_5O_2)_2$, a food additive.
(d) Ferrous ammonium sulfate, $(NH_4)_2Fe(SO_4)_2$.

3.3   THE MOLE

The lab-sized unit of a substance is its formula weight taken in grams, a quantity called one mole of the substance.

Formula weights and atomic weights are useful quantities to have because with them we can count out formula units in whatever ratio we wish. We introduced you to this idea in Section 2.4, but it is so important that it is worth reviewing here. Let us start by taking our thinking to the level of individual atoms and molecules.

Suppose we wished to make molecules of carbon monoxide, CO, in such a way that there would be no carbon or oxygen atoms left over to contaminate the product. Obviously, this would require that we take equal numbers of carbon atoms and oxygen atoms. If we took 10 atoms of C and 10 atoms of O, for example, we could combine them to form 10 molecules of CO. To scale this up, we could take 10,000 atoms of C and 10,000 atoms of O. Once again, we would have just the right number of atoms of each kind to make CO molecules with nothing left over.

The problem with this approach, of course, is that atoms and molecules are too tiny to count individually. We are able to get around this obstacle, however, precisely because each element has its own characteristic atomic weight and each formula has its own formula weight. We know, for example that an atom of oxygen weighs 1.33 times as much as an atom of carbon, because this is exactly the ratio of their atomic weights:

$$\frac{16.0 \text{ amu (for one atom of O)}}{12.0 \text{ amu (for one atom of C)}} = \frac{1.33}{1}$$

If we take samples of oxygen and carbon in this mass ratio of 1.33 to 1, we *must* obain equal numbers of their atoms. For instance, suppose we actually had a balance capable of measuring atomic mass units (amu) directly. Then, if we took 16.0 amu of oxygen and 12.0 amu of carbon—a mass ratio of 1.33 amu to 1 amu—we would have exactly one atom of each and a 1 to 1 ratio *by atoms*.

In the lab our balances measure in grams, so we might weigh out 16.0 g of oxygen and 12.0 g of carbon. Once again we have a mass ratio of 1.33 to 1, so we *must* have a 1 to 1 ratio by atoms. Of course, this time we have an enormous number of atoms in each sample, but the important fact is they are in just the right ratio to give us carbon monoxide molecules with no atoms of C or O left over.

When we weigh out a sample of an element such that its mass in grams is numerically equal to the element's atomic weight, we always obtain the same number of atoms no matter which element we choose. Thus 12.0 g of carbon has the same number of atoms as 16.0 g of oxygen, or 32.1 g of sulfur, or 55.8 g of iron. (See Figure 3.1.)

It costs time and money to purify products contaminated by leftover reactants (or anything else).

Element	Atomic Weight
C	12.0
O	16.0
S	32.1
Fe	55.8

1 mol sulfur
1 mol iron
1 mol copper
1 mol mercury

FIGURE 3.1 Each quantity of these elements contains the same number of atoms—Avogardo's number.

FIGURE 3.2 One mole of four different compounds. Each contains the same number of formula units, although the number of atoms is different from sample to sample. This is because the number of atoms per formula unit is different for each compound.

What is especially valuable about this relationship between ratios by atomic masses and ratios by atoms is that it also extends to compounds. The formula weight of H_2O is 18.0 amu. If you apply the same reasoning as in the preceding paragraphs, you will see that when we take 18.0 g of H_2O, the sample contains the same number of molecules of water as there are atoms in 12.0 g of carbon. Similarly, the formula weight of NaCl is 58.5 amu, and a 58.5-g sample of NaCl contains the same number of formula units of NaCl as there are atoms in 12.0 g of carbon or molecules in 18.0 g of water.

Na	23.0
Cl	35.8
NaCl	58.5

We have been relating other elements to carbon in this discussion for a reason. The carbon-12 isotope, which makes up 98.89% of naturally occurring carbon, is the reference used by the SI in its definition of the base unit for amount of chemical substance, the mole, abbreviated, mol. One mole of a chemical substance is the amount of it that contains as many elementary entities (atoms, molecules, or formula units) as there are atoms in 12 g (exactly) of carbon-12. Thus the amount of NaCl that equals one mole is 58.5 g, because this much NaCl has the same number of formula units as there are atoms in 12.0 g of carbon-12. The amount of H_2O that equals one mole is 18.0 g, because this mass has as many molecules of H_2O as there are atoms in 12.0 g of carbon-12. Figure 3.2 shows one-mole samples of several compounds.

The mole concept is one of the most important in all of chemistry. Learning to use it in chemical reasoning and for chemical calculations is one of the most critically important aspects of this course. When you are able to grasp the mole concept, much of the rest of chemistry comes quite easily.

Avogadro's Number

More precisely, 1 mol has 6.022137×10^{23} formula units.

The number of formula units in one mole is 6.02×10^{23} and is called Avogadro's number in honor of Italian scientist Amadeo Avogadro (1776–1856). It is a pure number with a special name, like many other numbers. For example,

$$2 = pair$$
$$12 = dozen$$
$$144 = gross$$
$$6.02 \times 10^{23} = Avogadro's\ number$$

You might be wondering why chemists would let such a complicated number become associated with a base unit in the SI. Actually, it was unavoidable once the mole was defined as the number of atoms in exactly 12 g of carbon-12. Measurement of this number (by methods we are not yet ready to discuss) gives Avogadro's number, and from it we can calculate the size of the amu. Here are the relationships we need:

$$12 \text{ amu} = 1 \text{ atom C}$$
$$1 \text{ mol C} = 6.02 \times 10^{23} \text{ atom C}$$
$$1 \text{ mol C} = 12 \text{ g C}$$

By the factor-label method,

$$1 \text{ amu} \times \frac{1 \text{ atom C}}{12 \text{ amu}} \times \frac{1 \text{ mol C}}{6.02 \times 10^{23} \text{ atom C}} \times \frac{12 \text{ g}}{1 \text{ mol C}} = 1.66 \times 10^{-24} \text{ g}$$

This is the value we saw earlier in Chapter 2 for the mass of 1 amu. We can also use it to show that the mass of a mole of any other substance is numerically equal to its formula weight in grams. For example, the atomic weight of sodium is 23.0, so 1 atom of Na has a mass of 23.0 amu. Let's now calculate the number of grams of sodium that contains Avogadro's number of sodium atoms (i.e., the mass of one mole of sodium atoms).

$$6.02 \times 10^{23} \text{ Na atoms} \times \frac{23.0 \text{ amu}}{1 \text{ Na atom}} \times \frac{1.66 \times 10^{-24} \text{ g}}{1 \text{ amu}} = 23.0 \text{ g Na}$$

Thus the logic of Avogadro's number is clear. Avogadro's number of formula units of a substance provides a mass in grams numerically equal to something familiar about the substance, its formula weight.

> Avogadro's number equals the number of formula units per mole.

The mass in grams of a substance that equals one mole is often called its molar mass, and the units are g/mol. Since the formula weight of aspirin, for example, is 180, its molar mass is 180 g/mol. In other words, 1 mol aspirin = 180 g aspirin. This information lets us construct the following conversion factors for aspirin:

$$\frac{180 \text{ g aspirin}}{1 \text{ mol aspirin}} \quad \text{and} \quad \frac{1 \text{ mol aspirin}}{180 \text{ g aspirin}}$$

We will repeatedly use molar masses to construct such conversion factors. Before showing an example, let's make sure we understand Avogadro's number by working a problem.

EXAMPLE 3.4   UNDERSTANDING AVOGADRO'S NUMBER

Problem: How many carbon atoms are in 6.00 g of C?

Solution: Work from the basic meaning of Avogadro's number, the number of formula units per mole of substance. Since carbon's atomic weight is 12.0, one mole of C consists at the same time of 12.0 g of C and 6.02×10^{23} atoms of C. These facts give us a choice of two conversion factors:

$$\frac{6.02 \times 10^{23} \text{ atoms C}}{12.0 \text{ g C}} \quad \text{and} \quad \frac{12.0 \text{ g C}}{6.02 \times 10^{23} \text{ atoms C}}$$

This answer makes sense because 6 g is half of 12 g (a mole of C) so our answer is half of Avogadro's number of C atoms.

If we multiply what was given, the 6.00 g of carbon, by the first conversion factor, the units "g C" will cancel and our answer will be in atoms of carbon:

$$6.00 \text{ g C} \times \frac{6.02 \times 10^{23} \text{ atoms C}}{12.0 \text{ g C}} = 3.01 \times 10^{23} \text{ atoms C}$$

PRACTICE EXERCISE 3   How many molecules of water, H_2O, are in 250 g of water, roughly one cup?

Using the Mole Concept

Since the mass of 1 mol of a substance is calculated from its formula weight, it is vitally important when using the mole concept to correctly specify the formula actually meant. It is not enough, for example, to say "1 mol of oxygen." We have to specify the formula we mean — O (atomic oxygen) or O_2 (ordinary oxygen, the form present in air) — before we can make our intentions clear. One mole of O consists of Avogadro's number of *atoms* (and has a mass of 16.0 g). But one mole of O_2 consists of Avogadro's number of *molecules,* each molecule being made of *two* oxygen atoms. And one mole of O_2 molecules has a total mass of 2×16.0 g $= 32.0$ g. Ozone, O_3, is another form in which oxygen atoms can be combined, and one mole of ozone has Avogadro's number of O_3 molecules, each made of *three oxygen atoms.* So one mole of O_3 has a total mass of 3×16.0 g $= 48.0$ g.

One of the advantages of the mole concept is that it lets us think about formulas on two levels at the same time. One level is that of atoms or other formula units (molecules or sets of ions) and the other level is that of practical, lab-sized quantities. When we think about H_2O at the first level, we can easily tell that a dozen of its molecules are made from two dozen atoms of H and one dozen of O. However, if we switch to the more practical, lab-sized level, it is just as easy to think about one *mole* of H_2O and to view this quantity as consisting of two moles of H and one mole of O.

H_2O consists of	2H	+	O
1 molecule H_2O	2 atoms H		1 atom O
1 dozen H_2O molecules	2 dozen H atoms		1 dozen O atoms
6.02×10^{23} H_2O molecules	12.04×10^{23} H atoms		6.02×10^{23} O atoms
1 mol H_2O molecules	2 mol H atoms		1 mol O atoms
18.0 g H_2O	2.0 g H		16.0 g O

The symbol *mol* stands for both the singular and the plural.

The numbers in all but the last row are in the same ratio regardless of the scale, whether we deal with single particles or with moles of them. After planning an experiment at the mole level, it is easy to convert numbers of moles into corresponding masses of chemicals to meet any desired needs.

EXAMPLE 3.5   USING THE MOLE CONCEPT

Problem?   How many moles of sulfur atoms must be combined with 2.0 mol of iron atoms to give iron pyrite, FeS_2?

Solution: Let's begin by expressing the problem in factor-label terms:

$$2.0 \text{ mol Fe} = ? \text{ mol S}$$

This equation asks us to find out how many moles of S atoms "equal" 2.0 mol of Fe atoms when they combine to form FeS_2. (As we've said before, the equals sign in this context doesn't mean that iron and sulfur literally equal each other. It means that an amount of one is equivalent to some amount of the other in a chemical sense.)

To solve this problem, as expressed by the above "equation," we have to find a relationship between the amount of iron and the amount of sulfur that combine to form the compound, and then use that relationship to construct an appropriate conversion factor.

The formula of the compound gives us the atom ratio. FeS_2 means that the atom ratio of iron to sulfur is 1 to 2. The mole concept allows us to scale this up to lab-sized quantities, because moles of atoms combine in the same ratio as the individual atoms themselves.

Therefore, we can say that in FeS_2 the ratio is 1 mol of Fe to 2 mol of S, so we can say that in this compound,

$$1 \text{ mol Fe} = 2 \text{ mol S}$$

This actually gives us two conversion factors:

$\dfrac{1 \text{ mol Fe}}{2 \text{ mol S}}$ means $\dfrac{1 \text{ mol Fe atoms}}{2 \text{ mol S atoms}}$

$$\frac{1 \text{ mol Fe}}{2 \text{ mol S}} \quad \text{and} \quad \frac{2 \text{ mol S}}{1 \text{ mol Fe}}$$

We were given 2.0 mol of Fe, so we want to use the second of these factors to make the unit "mol Fe" cancel. This converts the quantity of iron into the quantity of sulfur that is chemically equivalent to it in FeS_2.

$$2.0 \text{ mol Fe} \times \frac{2 \text{ mol S}}{1 \text{ mol Fe}} = 4.0 \text{ mol S}$$

Thus, 4.0 mol S can combine with 2.0 mol Fe.

EXAMPLE 3.6 USING THE MOLE CONCEPT

Problem: Methane (natural gas) has the formula CH_4. If a sample of methane contains 0.30 mol C, how many moles of H are present?

Solution: Because the formula says that 1 atom of carbon combines with 4 atoms of hydrogen, we automatically know that 1 mole of C combines with 4 mol of H. We therefore have two ratios to choose from as conversion factors.

$$\frac{1 \text{ mol C}}{4 \text{ mol H}} \quad \text{and} \quad \frac{4 \text{ mol H}}{1 \text{ mol C}}$$

To get "mol C" to cancel and leave our answer in "mol H" we have to use the second factor.

$$0.30 \text{ mol C} \times \frac{4 \text{ mol H}}{1 \text{ mol C}} = 1.2 \text{ mol H}$$

PRACTICE EXERCISE 4   How many moles of Fe combine with 0.22 mol of O to give rust, Fe_2O_3?

PRACTICE EXERCISE 5   In 0.50 mol of P_4O_{10} how many moles of P and how many moles of O are present?

Converting between Grams and Moles

In laboratory work, there are two important kinds of calculations involving moles. One is a moles-to-grams conversion and the other is the opposite, a grams-to-moles conversion. We need the first to be able to shift from theory to practice and the second to check practice against theory. "Theory" deals with numbers of *moles* implied by a formula or an equation, and "practice" deals with the numbers of *grams* we have to weigh out or that are produced by some experiment. Always keep in mind that laboratory balances read in *grams*, not moles, so in practical work we always have to translate moles into grams. Both kinds of conversions work from the same idea — the definition of the mole as a molar mass. Let us first see how a moles-to-grams conversion is done.

EXAMPLE 3.7 CALCULATING GRAMS FROM MOLES

Problem: Calcium arsenate, $Ca_3(AsO_4)_2$, is a poison sometimes used to kill insects on plants. What is the mass of 0.586 mol of calcium arsenate?

Content

Solution:

Ca: $3 \times 40.1 = 120$
As: $2 \times 74.9 = 150$
O: $8 \times 16.0 = \underline{128}$
     398

In just about any kind of problem involving moles, the first step should be to find the formula weights of the chemicals involved. For calcium arsenate, this is 398, so the molar mass of calcium arsenate is 398 g/mol. From this we can construct two conversion factors. *There is probably no more important use of a formula weight than to give us two conversion factors.* In our example, they are

$$\frac{398 \text{ g } Ca_3(AsO_4)_2}{1 \text{ mol } Ca_3(AsO_4)_2} \quad \text{and} \quad \frac{1 \text{ mol } Ca_3(AsO_4)_2}{398 \text{ g } Ca_3(AsO_4)_2}$$

The problem calls for an answer in units of *mass,* so we have to use the first factor to cancel "mol $Ca_3(AsO_4)_2$."

$$0.586 \text{ mol } Ca_3(AsO_4)_2 \times \frac{398 \text{ g } Ca_3(AsO_4)_2}{1 \text{ mol } Ca_3(AsO_4)_2} = 233 \text{ g } Ca_3(AsO_4)_2$$

The mass of calcium arsenate (233 g) found in Example 3.7, which is the same as 0.586 mol of calcium arsenate, contains exactly the same number of formula units as there are in 0.586 mol of water, 0.586 mol of glucose, 0.586 mol of iron, or 0.586 mol of hydrogen.

PRACTICE EXERCISE 6   Calculate the number of grams in 0.586 mol of each of the following substances.
(a) Water, H_2O.
(b) Glucose, $C_6H_{12}O_6$, a sugar in grape juice and honey.
(c) Iron, Fe.
(d) Methane, CH_4.

EXAMPLE 3.8   CALCULATING MOLES FROM GRAMS

Problem: Sodium bicarbonate, $NaHCO_3$, is one ingredient of baking powder. In one experiment in a series of tests of different ratios of ingredients, a scientist used 21.0 g $NaHCO_3$. How many moles of $NaHCO_3$ were in this sample?

Solution: First find the formula weight of $NaHCO_3$. It's 84.0 , which gives us the choice of these two conversion factors.

Don't confuse baking *powder* with baking *soda*. The latter is sodium bicarbonate, $NaHCO_3$; the former is a special mixture of compounds.

$$\frac{84.0 \text{ g } NaHCO_3}{1 \text{ mol } NaHCO_3} \quad \text{and} \quad \frac{1 \text{ mol } NaHCO_3}{84.0 \text{ g } NaHCO_3}$$

To change grams to moles we have to pick the second factor, which lets us cancel grams.

$$21.0 \text{ g } NaHCO_3 \times \frac{1 \text{ mol } NaHCO_3}{84.0 \text{ g } NaHCO_3} = 0.250 \text{ mol } NaHCO_3$$

PRACTICE EXERCISE 7   Calculate the number of moles of each substance in 100.0 g of each of the following samples:
(a) Ammonia, NH_3.   (c) Gold, Au.
(b) Cholesterol, $C_{27}H_{46}O$.   (d) Ethyl alcohol, C_2H_6O.

PRACTICE EXERCISE 8   Why does 100.0 g of ammonia, NH_3, have so many more moles than 100.0 g of cholesterol, $C_{27}H_{46}O$?

PRACTICE EXERCISE 9   If 40.0 g of an element contains 12.04×10^{23} atoms, what is its atomic weight?

3.4 PERCENTAGE COMPOSITION

The percentage by weight of an element in a compound is the same as the number of grams of this element present in 100 grams of the compound.

One of the important experimental tasks in chemical research is to determine the chemical formula of a substance. Such work begins with qualitative analysis, a set of procedures by which all of the elements making up the substance are identified. This is followed by quantitative analysis in which the mass of each element in a sample of the substance is determined. As we will see in the next section, the data obtained from quantitative analyses allow us to calculate the subscripts in chemical formulas.

With a formula, we can find a formula weight, which is essential to all mole calculations.

Percentages by weight are the usual forms for describing the relative masses of the elements in a compound, and a list of these percentages is called the compound's percentage composition. The percentage by weight[1] of an element in a compound is equivalent to the number of grams of the element present in 100 g of the compound.

These percentages are usually precise to four significant figures because the experimental data are this precise.

EXAMPLE 3.9   CALCULATING A PERCENTAGE COMPOSITION FROM MASS DATA

Problem: A sample of a liquid with a mass of 8.657 g was decomposed into its elements and gave 5.217 g of carbon, 0.9620 g of hydrogen, and 2.478 g of oxygen. What is the percentage composition of this compound?

Solution: Let's give the label A to this liquid. We were given the grams of each element in 8.657 g of A, not in 100.00 g, so we just do a straightforward percentage calculation.

Recall that a percentage is calculated by

$$\% = \frac{part}{whole} \times 100$$

$$\frac{5.217 \text{ g C}}{8.657 \text{ g } A} \times 100 = 60.26\% \text{ C}$$

$$\frac{0.9620 \text{ g H}}{8.657 \text{ g } A} \times 100 = 11.11\% \text{ H}$$

$$\frac{2.478 \text{ g O}}{8.657 \text{ g } A} \times 100 = \underline{28.62\% \text{ O}}$$
$$99.99\%$$

Allowing for slight errors caused by rounding or in getting the data, the percentages must add up to 100.00%. The above results tell us that in 100.00 g of this liquid there are 60.26 g of carbon, 11.11 g of hydrogen, and 28.62 g of oxygen.

PRACTICE EXERCISE 10   A quantitative analysis of 0.4620 g of an unknown solid found 0.1945 g of carbon, 0.02977 g of hydrogen, and 0.2377 g of oxygen. Calculate the percentage composition.

Another kind of calculation is to find the theoretical percentage composition from a formula. This is necessary, for example, when two or more formulas are possible for the same elements, and we want to decide which formula is correct. By comparing what has been experimentally found to be the percentage composition with what is calculated for each candidate, a choice among the candidates can be made. For example, suppose a compound was found to contain calcium and oxygen in percentages of 71.47% calcium and 28.53% oxygen. Do these data match CaO, CaO_2, or another formula? We will work this as an example.

---

[1] Percentage by *weight* is the most widely used expression rather that the more correct term, percentage by *mass*, and we will follow this common usage.

EXAMPLE 3.10   CALCULATING PERCENTAGE COMPOSITIONS FROM FORMULAS

Problem: What is the percentage composition of CaO?

Solution: First calculate the formula weight of CaO; do this to *four* significant figures (to match the number in the given percentages, above).

CaO is commonly called "quick lime."

$$Ca + O = CaO$$
$$40.08 + 16.00 = 56.08$$

Remember that these data mean the following:

56.08 g CaO contains 40.08 g Ca
56.08 g CaO contains 16.00 g O

We have to find what percentage of CaO is Ca and is O.

For Ca

$$\frac{40.08 \text{ g Ca}}{56.08 \text{ g CaO}} \times 100 = 71.47\% \text{ Ca in CaO}$$

For O

$$\frac{16.00 \text{ g O}}{56.08 \text{ g CaO}} \times 100 = \underline{28.53\%} \text{ O in CaO}$$
$$100.00\%$$

Notice that the theoretical percentages do add up to 100.00%, as they must, and that they agree with the observed data if the formula is CaO.

PRACTICE EXERCISE 11

What is the percentage by weight of each element in butane, C_4H_{10}, a liquid fuel used in cigarette lighters? (Remember that the experimental data can be obtained with a precision of four significant figures, so round atomic weights in problems such as this to four, not three, significant figures.)

This margin of error is small enough to permit a distinction between two similar formulas in nearly all cases.

In actual experimental work the calculated and the experimental percentages almost never agree exactly. There are just too many opportunities for small errors to creep in — errors caused by the limitations of equipment as well as human errors. Consequently, we have to have some understanding about what is "good enough." Although there is not unanimous agreement on this point, most chemists probably would say that for each element a theoretical and an observed percentage should be within 0.30% of each other in order to conclude that a particular formula is correct. For example, if the percentage of carbon in a compound is calculated to be 47.46%, then the value obtained using quantitative analysis should be in the range of 47.16% to 47.76%.

PRACTICE EXERCISE 12

Chromium carbonyl has been tested as a gasoline additive to improve engine performance. By quantitative analysis, its percentage composition was found to be 23.75% Cr, 32.69% C, and 43.55% O. Which formula do these data fit, $Cr(CO)_2$ or $Cr(CO)_6$?

3.5 EMPIRICAL FORMULAS

The empirical formula of a compound uses the lowest possible whole-number subscripts to describe the compound's composition.

When the quantitative analysis of a compound gives us the masses of each element in the *same-sized* sample, we can convert these masses into proportions by moles by a grams-to-moles calculation. With atomic particles, of course, proportions by moles are numerically the same as proportions by atoms, which are just the numbers we need to construct a chemical formula. The formula obtained this way is called an *empirical formula*. An empirical formula is one that uses as subscripts the smallest whole numbers that describe the ratios of atoms in a compound. We will work an example to show how we can use the mass data for a sample to find an empirical formula.

Empirical is from a Greek term meaning "experienced," suggesting an *experimental* result.

EXAMPLE 3.11 CALCULATING AN EMPIRICAL FORMULA FROM DATA OBTAINED BY QUANTITATIVE ANALYSIS

Problem: A sample of an unknown compound with a mass of 2.571 g was found to contain 1.102 g of carbon and 1.469 g of oxygen. What is its empirical formula?

Solution: First, notice that the data are precise to *four* significant figures. When you see this, remember to round atomic weights to four significant figures too. Now convert 1.102 g of C into moles of C and convert 1.469 g of O into moles of O, using their respective atomic weights in the usual way.

$$1.102 \text{ g C} \times \frac{1 \text{ mol C}}{12.01 \text{ g C}} = 0.09176 \text{ mol C}$$

$$1.469 \text{ g O} \times \frac{1 \text{ mol O}}{16.00 \text{ g O}} = 0.09181 \text{ mol O}$$

We could now write the formula $C_{0.09176}O_{0.09181}$; it does specify the relative numbers of moles of each element in the compound. But we want a formula that also expresses the relative numbers of *atoms,* and for that reason we need whole numbers as subscripts. To change these subscripts into a set of whole numbers *that are in the same ratio,* we have to divide each of them by some common divisor. From the set of subscripts calculated thus far, the easiest common divisor is the smaller number in the set. *This is always the way to begin the search for whole-number subscripts—we pick the smallest number of the set as the divisor.* Here, we divide both numbers by 0.09176.

$$C_{\frac{0.09176}{0.09176}} O_{\frac{0.09181}{0.09176}} = C_{1.000}O_{1.001}$$

The resulting subscripts are acceptably close to whole numbers; the formula is CO, carbon monoxide.

As a rule, if the calculated subscript expressed to the proper number of significant figures differs from a whole number by only several units in the last place, we can safely round to the whole number.

Sometimes our strategy of using as our divisor the smallest number of the appropriate set does not seem to work. Sometimes division of the initial set of subscripts by the smallest one does not give simple whole numbers. In the next worked example, we will see how to handle such a situation.

EXAMPLE 3.12 CALCULATING AN EMPIRICAL FORMULA FROM DATA OBTAINED BY QUANTITATIVE ANALYSIS

Problem: When a sample with a mass of 2.448 g of a compound present in liquefied petroleum gas was analyzed, it was found to contain 2.003 g of carbon and 0.4448 g of hydrogen. What is its empirical formula?

Solution: Remembering to work with atomic weights to four significant figures, we first convert the masses into their respective moles.

$$2.003 \ \cancel{g \ C} \times \frac{1 \ \text{mol C}}{12.11 \ \cancel{g \ C}} = 0.1655 \ \text{mol C}$$

$$0.4448 \ \cancel{g \ H} \times \frac{1 \ \text{mol H}}{1.008 \ \cancel{g \ H}} = 0.4441 \ \text{mol H}$$

These results let us write the following formula:

$$C_{0.1655}H_{0.4441}$$

Our first effort to change the ratio of 0.1655 to 0.4441 into whole numbers is to divide both by the smaller, 0.1655.

$$C_{\frac{0.1655}{0.1655}} H_{\frac{0.4441}{0.1655}} = C_{1.000}H_{2.667}$$

The *mole* ratio in this substance is 1 to 2.667, but we can't use these numbers to stand for the numbers of individual atoms in the *atom* ratio, as given in a formula, because it isn't possible to have a fraction of an atom.

This time the procedure seemingly failed. The subscript for H, 2.667, is much too far from a whole number to round off. In a *mole* sense, the ratio of 1 to 2.667 is correct; we just have not found a way to state it in whole numbers. Let's multiply each subscript by a whole number, 2.

$$C_{1 \times 2}H_{2.667 \times 2} = C_2 H_{5.334}$$

This didn't work either; 5.334 is obviously so far from a whole number that we cannot round it off. Let's try using 3 instead of 2 on the earlier ratio of 1 to 2.667.

$$C_{1 \times 3}H_{2.667 \times 3} = C_3 H_{8.001}$$

Now we can safely round. The empirical formula of the compound is C_3H_8. This is propane, a common home heating and cooking fuel.

Let us review the approach demonstrated in Example 3.12. We first converted the mass data into mole data. Then we divided each of the mole numbers by the smallest to make at least one number in the mole ratio equal to 1. When the other numbers were not close enough to whole numbers, we multiplied the numbers in the mole ratio, including the one already equal to 1, by 2. When this did not work, we tried a factor of 3 (and so on when necessary) until all of the mole numbers were whole numbers.

PRACTICE EXERCISE 13   A 5.438-g sample of ludlamite, a grayish-blue mineral sometimes used in ceramics, was found to contain 2.549 g of iron, 1.947 g of oxygen, and 0.9424 g of phosphorus. What is its empirical formula?

Empirical Formulas from Percentage Composition

Quite often it is not possible to obtain by quantitative analysis the masses of *all* of the elements in a compound from the *same* initial sample. When this is true, two or more analyses are performed on different samples, and the experimental results are reported as percentages. In these situations, we have to be able to calculate an empirical formula from a complete percentage composition.

When we start with percentage composition data, the "trick" is to imagine that we have a 100-g sample of the compound. If we do this, then the percentages become the masses of the elements in this sample. For example, SO_2 is 50.0% S and 50.0% O, by weight. If we had a 100-g sample of SO_2, it would consist of 50.0 g of S and 50.0 g of O. Once we have grams, we are back into the by-now-familiar (we hope) grams-to-moles business. Therefore, the strategy is to change the percentages to grams, convert the grams to moles, and then reduce these numbers to simple whole numbers in the usual way. Let's work an example.

EXAMPLE 3.13 CALCULATING EMPIRICAL FORMULAS FROM PERCENTAGE COMPOSITION

Problem: Barium carbonate, a white powder used in paints, enamels, and ceramics, has the following composition: Ba, 69.58%; C, 6.090%; O, 24.32%. What is its empirical formula?

Solution: If we had a 100.00-g sample of this compound, it would contain 69.58 g of Ba, 6.090 g of C, and 24.32 g of O. So we convert the grams (given by the percentages) to moles.

$$\text{Ba:} \quad 69.58 \text{ g Ba} \times \frac{1 \text{ mol Ba}}{137.3 \text{ g Ba}} = 0.5068 \text{ mol Ba}$$

$$\text{C:} \quad 6.090 \text{ g C} \times \frac{1 \text{ mol C}}{12.01 \text{ g C}} = 0.5071 \text{ mol C}$$

$$\text{O:} \quad 24.32 \text{ g O} \times \frac{1 \text{ mol O}}{16.00 \text{ g O}} = 1.520 \text{ mol O}$$

Our preliminary empirical formula is then

$$Ba_{0.5068}C_{0.5071}O_{1.520}$$

We next divide each subscript by the smallest, 0.5068:

$$Ba_{\frac{0.5068}{0.5068}} C_{\frac{0.5071}{0.5068}} O_{\frac{1.520}{0.5068}} = Ba_1 C_{1.001} O_{2.999}$$

The result is acceptably close to $BaCO_3$, which is the empirical formula of barium carbonate.

PRACTICE EXERCISE 14 Calomel is the common name of a white powder once used in the treatment of syphilis. Its composition is 84.98% mercury and 15.02% chlorine. What is its empirical formula?

Empirical Formulas from Combustion Analyses

We next want to correct an impression that may have emerged from the way that the problems have been devised and solved thus far. These problems might have suggested that compounds are always broken down completely to their elements when they are analyzed. Actually, this is seldom done. Instead, analysts change the compound being analyzed into other *compounds*. In choosing the reactions to accomplish this, the goal is to separate the elements in the original sample from each other, and then to capture these elements in compounds whose formulas are known. If care is taken so that nothing is lost during the reactions, it is not a difficult task to calculate, from the masses of the new compounds obtained, the masses of the elements originally present in the sample being analyzed. Of course, once these masses are known, we can calculate the empirical formula. We can best explain how this works by means of an example.

We will work an example using a compound consisting entirely of carbon, hydrogen, and oxygen. Such compounds burn completely when ignited in pure oxygen — the reaction is called combustion — and the sole products are carbon dioxide and water. In other words, all of the carbon atoms in the original compound end up in the form of CO_2 molecules, and all of the hydrogen atoms emerge in molecules of H_2O. These two products of the combustion are separated, collected, and weighed (Figure 3.3). Of course, each mole of CO_2 produced means one mole of carbon in the original sample, so we can do a moles-to-moles and then a moles-to-grams conversion to find the grams of

FIGURE 3.3 A combustion "train" used to analyze carbon and hydrogen in organic compounds. The sample rests in a platinum "boat," and it is strongly heated as a stream of pure, dry oxygen flows over it. The sample burns to carbon dioxide and water (with the CuO ensuring complete oxidation). The water vapor is absorbed by $MgClO_4$ (magnesium perchlorate), which has no effect on the carbon dioxide. The latter is absorbed by NaOH (sodium hydroxide). The changes in the weights of these two tubes, which can be disconnected, are equal to the weights of H_2O and CO_2 formed in the combustion and can be used to calculate the percentages of carbon and hydrogen in the sample.

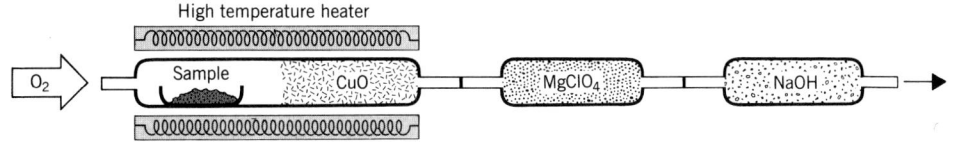

Part of the oxygen in the CO_2 and H_2O comes from the compound being burned and part from the O_2 used for combustion.

carbon in the original sample. And each mole of water obtained means two moles of hydrogen *atoms*, so we can then calculate the grams of hydrogen in the initial sample. We cannot avoid these calculations of the *grams* of C and H. We cannot, in other words, just use the moles of these elements directly because we have no direct way of knowing the moles of oxygen in the sample. So we add the grams of C and H and subtract this sum from the mass of the sample to find the grams of the remaining element, oxygen. After this, we can follow the steps outlined in Examples 3.11 and 3.12.[2]

EXAMPLE 3.14   CALCULATING EMPIRICAL FORMULAS FROM MASS DATA OBTAINED BY COMBUSTION ANALYSIS

Problem: A sample of a liquid consisting of only C, H, and O and having a mass of 0.5438 g was burned in pure oxygen and 1.039 g of CO_2 and 0.6369 g of H_2O were obtained. What is the empirical formula of the compound?

Solution: We first do grams-to-mole calculations to compute the moles of the elements C and H from the masses of the products, CO_2 and H_2O.

$$1.039 \text{ g CO}_2 \times \frac{1 \text{ mol CO}_2}{44.01 \text{ g CO}_2} \times \frac{1 \text{ mol C}}{1 \text{ mol CO}_2} = 0.02361 \text{ mol C}$$

$$0.6369 \text{ g H}_2\text{O} \times \frac{1 \text{ mol H}_2\text{O}}{18.02 \text{ g H}_2\text{O}} \times \frac{2 \text{ mol H}}{1 \text{ mol H}_2\text{O}} = 0.07073 \text{ mol H}$$

We now can find the grams of C and H.

$$0.02361 \text{ mol C} \times \frac{12.01 \text{ g C}}{1 \text{ mol C}} = 0.2836 \text{ g C}$$

$$0.07073 \text{ mol H} \times \frac{1.008 \text{ g H}}{1 \text{ mol H}} = 0.07130 \text{ g H}$$

Sum of the masses, C + H = 0.3549 g

The mass of the oxygen in the sample is then found by taking the difference:

$$\begin{aligned} \text{mass of original sample} &= 0.5438 \text{ g} \\ -(\text{mass of C + H}) &= \underline{-0.3549 \text{ g}} \\ \text{mass of oxygen in sample} &= 0.1889 \text{ g} \end{aligned}$$

Now we can compute the moles of oxygen and thus complete our list of the numbers of moles of the elements present in the sample.

$$0.1889 \text{ g O} \times \frac{1 \text{ mol O}}{16.00 \text{ g O}} = 0.01181 \text{ mol O}$$

Our first formula, then, is

$$C_{0.02361}H_{0.07073}O_{0.01181}$$

Dividing each by the smallest, 0.01181, gives

$$C_{\frac{0.02361}{0.01181}} H_{\frac{0.07073}{0.01181}} O_{\frac{0.01181}{0.01181}} = C_{1.999}H_{5.989}O_{1.000}$$

The results are acceptably close to C_2H_6O, the answer. (The substance is ethyl alcohol.)

PRACTICE EXERCISE 15   A compound known to contain only barium, sulfur, and oxygen was analyzed by first heating a sample with a mass of 0.8778 g at a very high temperature in air. All of the

---

[2] Another strategy would be to calculate the *percentages* of C and H and subtract their sum from 100% to find the percentage of oxygen. Then we could calculate the empirical formula from percentage data as outlined in Example 3.13.

barium atoms end up in barium oxide, BaO, a solid. All of the sulfur atoms emerge in molecules of sulfur trioxide, SO_3, a gas that can easily be trapped and weighed. The weights of the products were 0.5771 g of BaO and 0.3012 g of SO_3. Calculate the empirical formula of this compound of barium. (Arrange the symbols of the elements in the order Ba, S, and O in the formula.) What is the name of the compound?

3.6 MOLECULAR FORMULAS

A molecular formula gives the actual composition of one molecule.

Since ionic substances do not consist of discrete, individual molecules, the empirical formula is the only kind we can write for them. This formula is completely adequate, of course, because all we need — in fact, all we can get — is a description of the ratios of the ions. Molecular compounds, however, consist of individual molecules, and their formulas ought to tell us as much as can be known about the complete composition of one molecule. For example, the common bleaching agent hydrogen peroxide has an empirical formula of HO, meaning a 1-to-1 ratio of H and O atoms. Each molecule of this compound, however, has two H atoms and two O atoms. The ratio of 2 to 2 is equivalent to the ratio of 1 to 1, so the empirical formula, HO, is correct. But it would be more informative with this molecular compound to write its formula as H_2O_2. A formula that gives the composition of one molecule is called a molecular formula.

Experimentally, a quantitative analysis of hydrogen peroxide for its composition cannot alone tell between the formulas HO and H_2O_2. Both correspond to identical percentage compositions. H_2O_2, however, has exactly twice the formula weight of HO. Thus we can use a measurement of the formula weight of hydrogen peroxide to help us decide if the subscripts in its molecular formula should be some simple multiple of those of its empirical formula. There are many ways to determine formula weights. (We will study one in Section 9.8 and another in Section 13.6.) For the moment, however, we will only study how to use a formula weight and an empirical formula to write a molecular formula.

EXAMPLE 3.15 DETERMINING A MOLECULAR FORMULA FROM AN EMPIRICAL FORMULA USING THE FORMULA WEIGHT

Problem: Styrene, the raw material for polystyrene foam plastics, has the empirical formula of CH. Its formula weight is 104. What is its molecular formula?

Solution: First we compute the formula weight of the given empirical formula. For CH, this comes to $12.0 + 1.01 = 13.0$. The molar mass of CH is 13.0 g/mol. This is certainly far from 104, so we next ask, How far? How much larger is a molar mass of 104 g/mol than one of 13.0 g/mol? To find out, we divide the larger by the smaller.

$$\frac{104 \text{ g/mol}}{13.0 \text{ g/mol}} = 8.00$$

Formula weights obtained by most methods aren't good to more than three significant figures.

Thus 104 is 8 times larger than 13.0, so the correct molecular formula of styrene must have subscripts 8 times those in CH. Styrene, therefore, is C_8H_8.

PRACTICE EXERCISE 16 Calomel has a formula weight of 427. We found its empirical formula in Practice Exercise 14. What is its molecular formula?

PRACTICE EXERCISE 17 A compound with 87.42% N and 12.58% H has a formula weight of 32.1. What is its molecular formula?

In general, chemists use molecular formulas, not empirical formulas, when working with molecular compounds. Sometimes the two types are the same, as in H_2O. But hydrogen peroxide is always given as H_2O_2, never as HO. Similarly, styrene (Example 3.15) is never represented as CH. Its molecular formula, C_8H_8, is used instead.

SUMMARY

Mole Concept.   The sum of the atomic weights of all of the atoms appearing in a chemical formula gives the formula weight. One mole of any substance is an amount with the same number of formula units as there are atoms of carbon in 12 g of carbon-12. This quantity contains Avogadro's number of formula units (6.02×10^{23}). The formula weight taken in grams is called the molar mass of a substance.

Chemical Formulas.   The actual composition of a molecule is given by its molecular formula. An empirical formula gives the ratio of atoms, but in the smallest whole numbers. The empirical formula is generally the *only* one we write for ionic compounds. In the case of a molecular compound, to see if its empirical formula is also its molecular formula, the formula weight has to be known and compared with the calculated values of the candidate formulas. Empirical formulas can be calculated from the masses of the elements obtained by quantitative analysis of a known sample of the compound or from percentage compositions. The percentage of an element in a compound is the same as the number of grams of it in 100.00 grams of the compound.

REVIEW EXERCISES

Answers to questions whose numbers are printed in color are given in Appendix C.
Difficult questions are marked with asterisks.

Chemical Calculations — Avogadro's Number and Moles

3.1   Aluminum carbide was prepared from the elements in an experiment in which 6.26 g of carbon combined with 18.74 g of aluminum. In a second experiment, 8.240 g of carbon was used. How many grams of aluminum were needed in the second experiment?

3.2   A compound of silicon and chlorine has been used by naval vessels to generate smoke screens. In one experiment, 1.652 g of silicon was combined with 8.348 g of chlorine to make this compound. If another experiment was devised to use 10.00 g of silicon, how many grams of chlorine would be needed?

3.3   To what kinds of substances does the term *formula weight* apply?

3.4   What are the formula weights of the following compounds?
(a) Sodium carbonate, Na_2CO_3.
(b) Ammonium tetraborate, $(NH_4)_2B_4O_7$ (a fire-retardant for wood and textiles).
(c) Calcium cyclamate, $Ca(C_6H_{12}NSO_3)_2$ (a sweetening agent).

3.5   Calculate the formula weights of the following.
(a) Calcium nitrate, $Ca(NO_3)_2$ (used in matches, fire-crackers, and explosives).
(b) Magnesium phosphate, $Mg_3(PO_4)_2$ (used in antacids).
(c) Ascorbic acid, $H_2C_6H_6O_6$ (vitamin C).

3.6   Which has the larger mass, a molecule of water or a mole of water?

3.7   In your own words, what is the difference between a molecule and a mole?

3.8   What is the relationship between formula weight and molar mass?

3.9   What are the units of molar mass?

3.10   How many atoms of potassium are in a sample having a mass of 39.1 g? (The atomic weight of potassium is 39.1.) What is the name of this number?

3.11   A sample of magnesium contains 12.04×10^{23} atoms.
(a) How many moles does this sample contain?
(b) What is the mass of this sample, in grams?

3.12   What is the mass of 0.100 mol of each of the substances given in Review Exercise 3.4?

3.13   What is the mass of 0.250 mol of each of the substances given in Review Exercise 3.5?

3.14   How many moles of water are in 9.00 g of water, H_2O?

3.15   How many moles of sodium nitrate are in 1.70 g of sodium nitrate, $NaNO_3$, a substance used in fertilizers and to make gunpowder.

3.16   Ammonium sulfate, $(NH_4)_2SO_4$, is a fertilizer used to supply both nitrogen and sulfur. How many grams of ammonium sulfate are in 35.8 mol of $(NH_4)_2SO_4$?

3.17   A 0.500-mol sample of table sugar, $C_{12}H_{22}O_{11}$, weighs how many grams?

3.18   A solution of zinc chloride, $ZnCl_2$, in water is used to soak the ends of wooden fenceposts to preserve them from rotting while they are stuck in the ground. One ratio used is 840 g $ZnCl_2$ to 4 L water. How many moles of $ZnCl_2$ are in 840 g of $ZnCl_2$?

3.19   In the early 1970s, thallium sulfate, Tl_2SO_4, a powerful poison, was illegally used in poison baits to control predators such as coyotes on western rangeland. Hundreds of bald and golden eagles died after taking the baits. A 1.00-kg can of Tl_2SO_4 contains how many moles of this compound?

3.20 Borazon, one crystalline form of boron nitride, BN, is very likely the hardest of all substances. If one sample contains 3.02×10^{23} atoms of boron, how many atoms and how many grams of nitrogen are also in this sample?

3.21 When water is dropped onto calcium carbide, CaC_2, acetylene (C_2H_2) forms, which can be ignited to give light. Calcium carbide has been used to make signal flares for shipboard use. If one sample of CaC_2 contains 12.04×10^{23} atoms of carbon, how many atoms and how many grams of calcium are also in the sample?

3.22 If iodine is not in a person's diet, a thyroid condition called goiter develops. Iodized salt is all that it takes to prevent this disfiguring condition. Calcium iodate, $Ca(IO_3)_2$, is added to table salt to make iodized salt. How many atoms of iodine are in 0.500 mol of $Ca(IO_3)_2$? How many grams of calcium iodate are needed to supply this much iodine?

3.23 Ammonium carbonate, $(NH_4)_2CO_3$, is present in a 1-to-1 ratio by mass with powdered soap in "ammoniated washing powder." How many atoms of nitrogen are in 0.750 mol of ammonium carbonate? How many grams of this compound supply this much nitrogen?

3.24 Ammonium nitrate, NH_4NO_3, is used as a fertilizer and to manufacture explosives. How many atoms of nitrogen are in 0.665 mol of this substance? How many grams of ammonium nitrate supply this much nitrogen?

3.25 "Diammonium phosphate," $(NH_4)_2HPO_4$, made from phosphate rock and ammonia, is widely used in fertilizers as a source of both nitrogen and phosphorus. How many grams are in 6.26 mol of this substance?

3.26 Calcium dihydrogen phosphate, $Ca(H_2PO_4)_2$, is a component of both "superphosphate" and "triple superphosphate" fertilizers. How many grams are in 4.34 mol of this substance?

3.27 A 100-lb sack of "diammonium phosphate" (Review Exercise 3.25) contains how many moles of this substance (calculated to three significant figures)?

3.28 A hopper containing 525 lb of calcium dihydrogen phosphate (Review Exercise 3.26) has how many moles of this compound?

3.29 Sodium perborate, $NaBO_3$, is present in "oxygen bleach"; it acts by releasing oxygen, which has bleaching ability. How many grams of sodium perborate are in 4.65 mol of $NaBO_3$?

3.30 Barium sulfate, $BaSO_4$, is given to patients as a thick slurry in flavored water before X rays are taken of the intestinal tract. The barium blocks X rays, and the tract therefore casts a shadow that is seen on the X-ray film. How many grams are in 0.568 mol of barium sulfate?

3.31 One recipe for making soap calls for mixing 13 oz of sodium hydroxide ("lye"), NaOH, dissolved in 5.0 cups of water, with 6.0 lb of tallow. Assume that the formula of tallow is $C_{57}H_{110}O_6$. The mole ratio theoretically should be 3.0 mol of NaOH to 1.0 mol of tallow. Calculate the actual mole ratio of NaOH to tallow represented by this recipe. (If excess NaOH remains in the soap, the soap will be very harsh on skin. If excess tallow is present, the soap will be greasy and not a good cleaner.)

3.32 One recipe for hydroponic plant food calls for 1.50 oz of KNO_3, 1.00 oz of $CaSO_4$, 0.750 oz of $MgSO_4$, 0.500 oz of $CaHPO_4$, and 0.250 oz of $(NH_4)_2SO_4$—all dissolved in 10.0 gal of water. How many grams and how many moles of each of these solutes are represented by these quantities? (Hydroponics is a method of raising plants through the use of a balanced nutrient medium, with or without any mechanical support of soil, sand, or gravel.)

3.33 In the mid-1980s, U.S. production of ammonia, NH_3, an important fertilizer and raw material for the chemical industry, was about 32 billion (32×10^9) pounds per year. How many moles per year is this?

3.34 In the mid-1980s, about 40 million (40×10^6) tons of sulfuric acid were made in the United States each year. How many moles is this? (1 ton = 2000 lb.)

Percentage Composition

3.35 Calculate the percentage of nitrogen in the two important nitrogen fertilizers ammonia, NH_3, and urea, $CO(NH_2)_2$.

3.36 Calculate the percentages of nitrogen and phosphorus in each of the following "ammonium phosphate" fertilizers: $NH_4H_2PO_4$ and $(NH_4)_2HPO_4$.

3.37 Phencyclidine ("angel dust") is $C_{17}H_{25}N$. A sample suspected of being this dangerous drug was found to have a percentage composition of 83.71% C, 10.42% H, and 5.61% N. Do these data acceptably match the theoretical data for phencyclidine? (Calculate percentages to four significant figures.)

3.38 The drug known as LSD has the formula $C_{20}H_{25}N_3O$. One suspected sample contained 74.07% C, 7.95% H, and 9.99% N. Are these data consistent for LSD within the allowed limits of error and rounding? (Calculate the theoretical percentages to four significant figures.)

Empirical and Molecular Formulas

3.39 To determine the empirical formula of a new substance, what information must be obtained?

3.40 What is the difference between an empirical formula and a molecular formula?

3.41 The molecular formulas of some substances are as follows. Write their empirical formulas.
(a) Acetylene, C_2H_2 (used in oxyacetylene torches).
(b) Glucose, $C_6H_{12}O_6$ (the chief sugar in blood).
(c) Octane, C_8H_{18} (a component of gasoline).
(d) Ammonium nitrate, NH_4NO_3 (a fertilizer component).

3.42 What information must be obtained to calculate a molecular formula from an empirical formula?

3.43 Describe in your own words how we can use the percentage composition of a substance to figure out an empirical formula. (How do we interpret percentage data?)

3.44 Explain why calculations based on percentage composition give only empirical formulas, not molecular formulas.

3.45 A radioactive form of sodium pertechnetate is used as a brain-scanning agent in medical diagnosis. An analysis of a 0.9872-g sample found 0.1220 g of sodium and 0.5255 g of technetium. The remainder was oxygen. Calculate the empirical formula of sodium pertechnetate. (Use the value of 98.907 as the atomic weight of Tc and arrange the atomic symbols in the formula in the order NaTcO.)

3.46 Potassium persulfate (Anthion®) is used in photography to remove the last traces of hypo from photographic papers and plates. A 0.8162-g sample was found to contain 0.2361 g of potassium and 0.1936 g of sulfur; the rest was oxygen. The formula weight of this compound was measured to be 270. What are the empirical and molecular formulas of potassium persulfate? (Arrange the atomic symbols in the formulas in the order KSO.)

3.47 Adenosine triphosphate (ATP) is an important substance in all living cells. A sample with a mass of 0.8138 g was analyzed and found to contain 0.1927 g of carbon, 0.02590 g of hydrogen, 0.1124 g of nitrogen, and 0.1491 g of phosphorus. The remainder was oxygen. Its formula weight was determined to be 507. Calculate the empirical and molecular formulas of adenosine triphosphate. (Arrange the atomic symbols in alphabetical order in the formulas.)

3.48 Realgar (re-AL-gar) is a deep red pigment used in painting. A 0.6817-g sample was found to contain 0.4774 g of arsenic; the remainder was sulfur. The formula weight of realgar was found to be 428. What are the empirical and molecular formulas of this pigment? (Arrange the symbols in the order AsS.)

3.49 Isobutylene is a raw material for making synthetic rubber. A sample with a mass of 0.6481 g was found to contain 0.5555 g of carbon; the rest was hydrogen. Its formula weight was determined to be 57. What are the empirical and molecular formulas of isobutylene? (Place the atomic symbols in the formulas in the order CH.)

3.50 Cyanuric acid is used for such different purposes as making synthetic sponges and killing weeds. A sample with a mass of 0.5627 g was found to contain 0.1570 g of carbon, 0.01317 g of hydrogen, and 0.1832 g of nitrogen, with the balance being oxygen. Its formula weight was found to be 129. Calculate the empirical and molecular formulas of cyanuric acid, arranging the atomic symbols in alphabetical order.

3.51 Hypophosphoric acid is one of several acids containing both oxygen and phosphorus. Its formula weight is 162. A sample with a mass of 0.8821 g was found to contain 0.0220 g of hydrogen and 0.3374 g of phosphorus, with the remainder being oxygen. Calculate its empirical and molecular formulas. (Arrange the atomic symbols in the formulas in the order HPO.)

3.52 The chief compound in the mineral celestine consists of strontium, sulfur, and oxygen. The percentage composition is 47.70% Sr and 17.46% S; the remainder is oxygen. Its formula weight is 184. What are the empirical and molecular formulas of this compound? (Arrange the atomic symbols in the formulas in the order Sr, S, and O.)

3.53 One compound of mercury with a formula weight of 519 contains 77.26% Hg, 9.25% C, and 1.17% H, and the remainder is oxygen. Calculate its empirical and molecular formulas, arranging the atomic symbols in the order Hg, C, H, and O.

3.54 One of the most deadly poisons, strychnine, has a formula weight of 334 and the composition 75.42% C, 6.63% H, 8.38% N; the rest is oxygen. Calculate the empirical and molecular formulas of strychnine, arranging the atomic symbols in alphabetical order.

3.55 Lactic acid, the substance that causes sour milk to taste as it does, has a formula weight of 90. It consists of 40.00% C and 6.71% H, with the rest being oxygen. Calculate the empirical and molecular formulas of lactic acid, arranging the atomic symbols in alphabetical order.

*3.56 Sodium reacts with oxygen to form sodium peroxide.
(a) What is its empirical formula if it can be made by combining 0.4681 g of sodium with 0.3258 g of oxygen?
(b) On this basis, how many grams of sodium can combine with 1.000 g of oxygen?
(c) Another known oxide of sodium is called sodium oxide. Using only your answer to part b, calculate some possible values for the grams of sodium combined with 1.000 g of oxygen in sodium oxide.
(d) A sample of sodium oxide was found to contain 1.145 g of sodium and 0.3983 g of oxygen. How many grams of sodium can combine with 1.000 g of oxygen to make sodium oxide?
(e) Comparing the answers of parts c and d, is the law of multiple proportions illustrated by these two compounds? What is the ratio of the masses of sodium in the two oxides that combine with 1.000 g of oxygen?
(f) Calculate the empirical formula of the second oxide.

*3.57 Potassium reacts with oxygen to form a superoxide.
(a) What is the empirical formula of this compound if it can be made by combining 0.5634 g of potassium with 0.4611 g of oxygen?
(b) How many grams of oxygen can combine with 1.000 g of potassium to make potassium superoxide?
(c) Another known oxide of potassium is called potassium oxide. Using only your answer to part b, calculate some possible values for the grams of oxygen that are present with 1.000 g of potassium in this oxide.
(d) Potassium oxide contains its elements in a ratio of 0.8298 g of potassium to 0.1698 g of oxygen. What is the empirical formula of potassium oxide?
(e) How many grams of oxygen can combine with 1.000 g of potassium to form potassium oxide?
(f) In what ratio do the answers to parts b and e stand? What law of chemical combination is illustrated by these examples?

3.58 During early research on a sex hormone, thousands of liters of urine collected from pregnant mares were processed to isolate less than a gram of the hormone. It was found to consist entirely of carbon, hydrogen, and oxygen. When a sample of this hormone with a mass of 6.853 mg was burned, 20.08 mg CO_2 and 5.023 mg of H_2O were obtained. The formula weight of the substance was found to be 270. What is its molecular formula?

3.59 Vitamin D activity is shown by several compounds consisting wholly of carbon, hydrogen, and oxygen. When a sample of one of them with a mass of 5.676 mg was burned in oxygen, 17.536 mg of CO_2 and 5.850 mg of H_2O were obtained. The formula weight was found to be 385. What is the molecular formula of this compound?

3.60 Pictured below is the balance arm of a simple weighing balance. Each vertical line represents a 1-g unit of mass. Three drawings just like the first are also shown, except that the numbers are omitted. The vertical lines still stand for units of 1 g, however. The total capacity of the balance is 100 g.

(a) Using drawing *a*, mark the balance for its potential use in measuring water (H_2O) in units of *moles* of water. Place small marks on the balance arm corresponding to 1.0 mol, 1.5 mol, 2.0 mol, 2.5 mol, and so on up in 0.5-mol increments as far as you can go in the drawing.

This scale could now be used to obtain samples of water directly by moles, thus avoiding the need to make a moles-to-gram calculation whenever water is used as a chemical.

(b) Mark drawing *b* for use in measuring moles of methyl alcohol (CH_4O) directly. Do this as instructed in part a.

(c) Mark drawing *c* for measuring moles of aluminum (Al) in 0.5-mol units, as instructed in part a.

(d) If the research of a particular chemist routinely required the use of 500 chemicals, each with a different formula weight, how many mole-reading balances would be needed? If you were providing the money for this research, which approach would you have the chemist use—grams-reading or moles-reading balances?

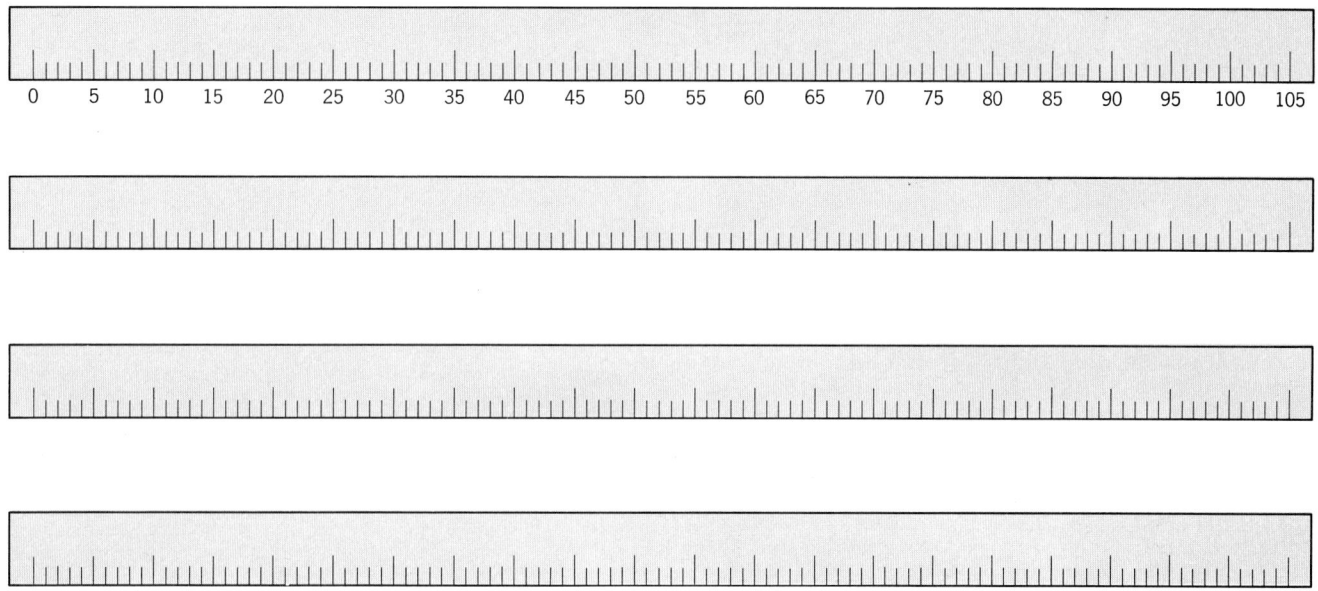

CHEMICALS IN USE
SODIUM HYDROXIDE
AND HYDROCHLORIC ACID

3

Sodium hydroxide (NaOH) and hydrochloric acid (HCl) are among the most commonly used laboratory chemicals. We will survey some of their properties here and point out how to handle them carefully. First, however, we will see how both are made from the same raw materials, water and ordinary salt (NaCl), the subject of *Chemicals in Use 2.*

Manufacture of NaOH and HCl from NaCl

When a direct current of electricity is passed through a solution of sodium chloride in water in suitably designed equipment, sodium hydroxide, hydrogen, and chlorine are produced. (Details of this operation are described in Chapter 17.)

$$2NaCl + 2H_2O \xrightarrow{\text{DC electricity}} 2NaOH + Cl_2 + H_2$$

The NaOH is still in solution, but the water can be evaporated to give solid sodium hydroxide.

The two gases produced by this reaction, H_2 and Cl_2, can be mixed in a separate apparatus where they react to give gaseous hydrogen chloride.

$$H_2 + Cl_2 \longrightarrow 2HCl$$

When gaseous HCl is passed into water, it dissolves and the resulting solution, not gaseous HCl, is what is called hydrochloric acid.

Thus, from inexpensive, abundant raw materials — salt and water — plus the energy of electricity, chemical companies manufacture sodium hydroxide and hydrochloric acid. Both rank among the top 50 chemicals manufactured per year in the United States. Sodium hydroxide, commonly called lye, is used to make glass, soap, and other laundry products and many chemicals for the pharmaceutical and baking industries. One important use of hydrochloric acid is to remove metal oxide coatings on metals, such as iron, that would interfere with the further processing of the metals. The operation is called "pickling" the metal.

Handling Solid Sodium Hydroxide

Most chemists see sodium hydroxide in the form of hard, white chips or pellets kept in airtight bottles. Whether as a solid or as a solution, sodium hydroxide slowly draws two substances from air, water and carbon dioxide. Enough water can be taken up by solid NaOH to form a very

FIGURE 3a Pellets from a freshly opened bottle of sodium hydroxide (left) gradually draw moisture from the air and develop coatings of very concentrated sodium hydroxide solution (right).

concentrated solution of NaOH that coats the material (Figure 3a). The CO_2 in air reacts with NaOH to form sodium bicarbonate, $NaHCO_3$.

$$NaOH + CO_2 \longrightarrow NaHCO_3$$

Even the best grades of commercial NaOH are 2 to 3% contaminated.

When dispensing sodium hydroxide, always wear safety glasses and be aware of the nearest eye wash fountain. Dusts of NaOH can come from the solid form, which might drift into the eyes. Always weigh solid NaOH into a glass container, never onto paper.

Handling Concentrated Hydrochloric Acid

Most laboratories purchase hydrochloric acid in its most concentrated form, as 36% HCl. This means that each 100 g of the solution contains 36 g of HCl. The vapor above concentrated HCl has an extremely sharp odor. Never inhale it. It can very seriously damage the soft tissues in your lungs. As a general rule — when checking the odor of a chemical, start with the bottle cap and approach it very cautiously.

If you spill some concentrated hydrochloric acid on your skin, rinse it off as quickly as possible. It's not quite as dangerous to the skin as other concentrated acids (such as sulfuric and nitric acids), but it can cause a burn. The eyewash fountain must be used promptly if any acid gets into your eyes. Then see a physician.

Concentrated HCl is seldom set out for use in beginning chemistry labs, not because some of its uses are dangerous but more because it is just too concentrated. It contains 11 to 12 mol of HCl per liter of solution, and few laboratory applications call for concentrations greater than 1 mol/L.

Some Chemical Properties of Hydrochloric Acid

HCl dissolves many metal oxides. In the "pickling" of iron products, concentrated hydrochloric acid reacts with iron oxides. For example,

$$6HCl + Fe_2O_3 \longrightarrow 2FeCl_3 + 3H_2O$$

The $FeCl_3$ stays in the solution and can be rinsed away.

Hydrochloric acid also dissolves many metals, including iron, but metals vary widely in how fast this kind of reaction occurs. (Iron, for example, reacts much more slowly with the acid than does its oxide coating, so pickling the iron results in no important loss of the metal.) Aluminum and zinc, on the other hand, react quite rapidly.

$$Zn + 2HCl \longrightarrow ZnCl_2 + H_2$$
$$2Al + 6HCl \longrightarrow 2AlCl_3 + 3H_2$$

The metals dissolve as their compounds with chlorine form, and the hydrogen gas escapes. When lead is subjected to concentrated hydrochloric acid, the $PbCl_2$ that initially forms does not dissolve in the solution. Instead, it adheres to the surface of the lead quite strongly and thus protects the lead from further reaction.

In later chapters we will learn when to expect differences in reactivities of metals with acids and how to predict the solubilities of various kinds of compounds. These are very practical matters because the environment offers many opportunities for metals to be in contact with an acidic medium, whether it's acid rain, the natural acids in certain foods and beverages, the acids produced by the decay of various plants, or acid spills from batteries or other causes.

Hydrochloric acid reacts with metal carbonates and bicarbonates, whether the latter are soluble or insoluble in water, to liberate CO_2. Sodium carbonate reacts, for example, as follows.

$$2HCl + Na_2CO_3 \longrightarrow 2NaCl + CO_2 + H_2O$$

Sodium bicarbonate reacts in a very similar way to give the same products. Only the proportions are different.

$$HCl + NaHCO_3 \longrightarrow NaCl + CO_2 + H_2O$$

Since these reactions destroy the acid, it's perfectly proper to say that they also represent the *neutralization* of an acid. Solid sodium carbonate or bicarbonate may be sprinkled onto an acid spill as a first step in a cleanup. You can tell when you've used enough because the fizzing caused by escaping CO_2 stops.

Some Other Chemical Properties of Sodium Hydroxide

We have already seen how sodium hydroxide reacts with carbon dioxide. Sodium hydroxide also neutralizes hydrochloric acid (or any acid), as we learned in Chapter 2.

$$NaOH + HCl \longrightarrow NaCl + H_2O$$

This is a very rapid reaction, and you can see that in an overall sense it is just the opposite from the electricity-driven, two-step conversion of NaCl and water into NaOH and HCl that we studied earlier. The reaction generates a great deal of heat per mole of each substance. This is one reason why concentrated solutions of acids and bases are not often used.

Questions

1. Write the equation for the reaction of (a) NaOH with HCl, (b) HCl with $NaHCO_3$, and (c) HCl with Na_2CO_3.

2. Potassium hydroxide (KOH) has the same kinds of reactions as sodium hydroxide. Write the equation for its reaction with HCl.

3. Why must supplies of NaOH be kept in well-capped bottles?

4. Name some uses of NaOH and HCl in industry.

All chemical reactions, including those that occur as food is prepared, follow the same rule — reactants combine in definite proportions by moles. Over the centuries, for example, recipes have evolved that correspond to the best proportions for pasta, which a chef is preparing here. We will study the fundamental laws governing the best proportions of reactants for all chemical reactions in this chapter.

MASS RELATIONSHIPS
IN CHEMICAL REACTIONS

STOICHIOMETRY
BALANCING EQUATIONS
STOICHIOMETRIC CALCULATIONS
PERCENTAGE YIELD AND LIMITING REACTANTS

WHEN REACTANTS ARE IN SOLUTION
MOLAR CONCENTRATIONS
MAKING DILUTE SOLUTIONS FROM CONCENTRATED SOLUTIONS

4.1 STOICHIOMETRY

Stoichiometry is the quantitative description of the proportions by moles of the substances in a chemical reaction.

When doing experiments, it is unscientific and often dangerous to approach reactions haphazardly. Therefore, when we go into the laboratory to work with reactions, we should almost always use a balanced equation. This is true whether the equation is for an actual reaction known to occur or for one that we predict. A balanced equation helps us because it tells us the reaction's stoichiometry.

The stoichiometry of a reaction is the description of the relative quantities *by moles* of the reactants and products as they are given to us by the coefficients of the balanced equation. In a sense, stoichiometry is the molar bookkeeping of chemistry, and the books have to balance. Quantitative data about a chemical reaction are available only from its stoichiometry. This applies not only to the study of chemistry itself, but also to the uses of chemistry in such different fields as agriculture, clinical analysis, the study of drug reactions, food chemistry, inhalation therapy, nutrition, crime lab analysis, geochemistry — the list could go on and on. The analysis of a reaction's stoichiometry

Stoke-ee-ah-meh-tree, from the Greek *stoicheion* ("element") and *metron* ("measure").

begins with writing and balancing its equation, because this is the work that gives the coefficients — the numbers that disclose the proportions by moles.

4.2 BALANCING EQUATIONS

In a balanced equation, all atoms that appear among the reactants must appear somewhere among the products.

In Section 1.6, we first learned the terms *reactant, product,* and *coefficient,* and some general ideas about balanced chemical equations. By a balanced equation we mean one in which the number of the atoms of each element appearing among the reactants equals their number among the products. *Balanced* also means, therefore, that the sum of the masses of the reactants equals the sum of the masses of the products.

A balanced equation is a shorthand but quantitative description of a chemical reaction. For example, we might say, "Hydrogen reacts with oxygen to give water." This does not alone reveal the reacting proportions. To show this we convert the statement into a balanced equation. We first set down the chemical symbols for the reactants, H_2 and O_2, separated by a plus sign to stand for "reacts with." Then we write an arrow that points to the chemical symbol for the product, H_2O. Read the arrow as "react to yield." Thus, "Hydrogen and oxygen react to yield water:"

$$H_2 + O_2 \longrightarrow H_2O \qquad \text{(unbalanced)}$$

The reaction cannot happen as written, because one of the two atoms of oxygen on the left does not show up in the product. To bring this equation into balance, we find the coefficients for each symbol that ensure that all atoms of each element on one side of the arrow show up on the other side too. If we write "2" next to H_2O, then we can show two atoms of oxygen on both left and right sides:

$$H_2 + O_2 \longrightarrow 2H_2O \qquad \text{(unbalanced)}$$

Since "$2H_2O$" also means $2 \times 2 = 4$ atoms of hydrogen, we now have four atoms of hydrogen on the right but only two on the left. To make the numbers of hydrogen atoms equal on both sides, we write a coefficient of "2" for H_2 on the left side.

$$2H_2 + O_2 \longrightarrow 2H_2O \qquad \text{(balanced)}$$

WARNING   Do not change subscripts within a formula to get an equation to balance.

For example, writing H_2O_2 instead of $2H_2O$ in the first effort to get our equation to balance could never work, because the formula H_2O_2 stands for hydrogen peroxide, a substance entirely different from H_2O. Once we knew we had to write the formula of water and wrote it correctly as H_2O, the only way we could have doubled the oxygen atoms to the right of the arrow would have been to write $2H_2O$. (Of course, this also doubled the hydrogen atoms on the right, so we weren't quite through.) Let's try this in another example.

Margin notes

The equality of these sums is required by the law of conservation of mass, and in chemistry this also means a conservation of atoms.

Remember, these two elements occur as diatomic molecules.

Coefficients are multipliers for every subscript within a formula.

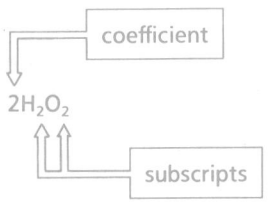

H_2O_2 is a bleach.

EXAMPLE 4.1   WRITING A BALANCED EQUATION

Problem: Write the balanced equation for the formation of table salt, NaCl (sodium chloride), from sodium (Na) and gaseous chlorine (Cl_2).

Solution:

Here we must use the fact that the element chlorine occurs as diatomic molecules, not as individual atoms.

First, write the correct formulas in the style for an equation.

$$Na + Cl_2 \longrightarrow NaCl \quad \text{(unbalanced)}$$

The subscripts show two chlorine atoms on the left, but only one on the right. This hints at the best single piece of advice in balancing an equation—let the subscripts of one substance suggest the coefficients of another. For example, the subscript of 2 in Cl_2 suggests that 2 could be the coefficient for NaCl. Try it.

$$Na + Cl_2 \longrightarrow 2NaCl \quad \text{(unbalanced)}$$

You can see how easily this brought the chlorine atoms into balance, but now there are two sodium atoms on the right and only one on the left. Therefore, write a coefficient of 2 for Na.

$$2Na + Cl_2 \longrightarrow 2NaCl \quad \text{(balanced)}$$

Now two sodium atoms are on the left and two are on the right, and two chlorine atoms are on the left and two are on the right. The equation is balanced.

Example 4.1 illustrates a strategy for judging what coefficients to pick—letting subscripts suggest coefficients. For example, in the presence of electrical sparks or high-energy ultraviolet light, ordinary molecular oxygen (O_2) changes to another form, ozone (O_3). To symbolize this change we begin with the formulas.

Ozone is dangerous to health even in low concentrations in air.

$$O_2 \longrightarrow O_3 \quad \text{(unbalanced)}$$

The subscript of O_3 suggests a coefficient for O_2, just as the subscript in O_2 suggests a coefficient for O_3, because $2 \times 3 = 3 \times 2$, a balance. Therefore we can write

$$3O_2 \longrightarrow 2O_3$$

Now there are six oxygen atoms on the left (3×2) and six on the right (2×3). Let's try this strategy in a similar example.

EXAMPLE 4.2 BALANCING A CHEMICAL EQUATION BY USING SUBSCRIPTS TO SUGGEST COEFFICIENTS

Problem: Although bright and shiny, aluminum objects are covered with a tight, invisible coating of aluminum oxide (Al_2O_3) that forms when freshly exposed aluminum (Al) reacts with oxygen. Write the balanced equation for this reaction.

Solution: First, the chemical symbols are set up in the pattern of an equation.

$$Al + O_2 \longrightarrow Al_2O_3 \quad \text{(unbalanced)}$$

Aluminum oxide is found in nature in bauxite, an ore from which metallic aluminum is isolated.

It does not matter too much where we start, so let us start with the imbalance in oxygen atoms. The subscript 3 in Al_2O_3 suggests that 3 should be the coefficient for O_2, just as the 2 in O_2 suggests that 2 should be the coefficient for Al_2O_3.

$$Al + 3O_2 \longrightarrow 2Al_2O_3 \quad \text{(unbalanced)}$$

Of course, doubling Al_2O_3 doubles the aluminum atoms on the right, from 2 to 4. But this is easily solved by writing a coefficient of 4 in front of Al.

$$4Al + 3O_2 \longrightarrow 2Al_2O_3 \quad \text{(balanced)}$$

PRACTICE EXERCISE 1 Balance these equations by letting subscripts suggest coefficients.
(a) $P + O_2 \longrightarrow P_4O_{10}$ (This is the reaction that occurs when some white flares are used.)

(b) $N_2 + H_2 \longrightarrow NH_3$ (This is the equation for the industrial synthesis of ammonia, a valuable chemical and fertilizer.)

Many reactions involve chemicals whose formulas include polyatomic ions, which sometimes are in parentheses. *If such ions do not change, treat them as units when balancing the equation.*

EXAMPLE 4.3   BALANCING EQUATIONS WHEN FORMULAS INCLUDE POLYATOMIC IONS

Problem: Water that contains dissolved calcium compounds is called "hard" water, meaning that it is hard to use soap in it. One way to "soften" such water is to add sodium carbonate, Na_2CO_3. Sodium carbonate removes the calcium ions by forming calcium carbonate, $CaCO_3$, which is insoluble in water. Balance the equation that illustrates this reaction, in which we somewhat arbitrarily let the calcium ion be in the form of calcium nitrate, $Ca(NO_3)_2$.

Ca^{2+} ions form a scum with negative ions from soap.

$$Ca(NO_3)_2(aq) + Na_2CO_3(aq) \longrightarrow CaCO_3(s) + NaNO_3(aq) \quad \text{(unbalanced)}$$

Remember,

Solution: Neither the nitrate ion (NO_3^-) nor the carbonate ion (CO_3^{2-}) changes from one side to the other in this equation. Therefore, treat them as units. There are two nitrate ions on the left but only one on the right, so put a 2 in front of $NaNO_3$.

(aq) = aqueous solution
(s) = solid

$$Ca(NO_3)_2 + Na_2CO_3 \longrightarrow CaCO_3 + 2NaNO_3 \quad \text{(balanced)}$$

That brings everything into balance. As a check, notice that there is 1 Ca on each side, $2NO_3$ on each side, 2Na on each side, and $1CO_3$ on each side.

PRACTICE EXERCISE 2   Balance the following equations.
(a) $Mg + O_2 \longrightarrow MgO$
(b) $CH_4 + Cl_2 \longrightarrow CCl_4 + HCl$
(c) $NO + O_2 \longrightarrow NO_2$
(d) $NaOH + H_2SO_4 \longrightarrow Na_2SO_4 + H_2O$
(e) $CH_4 + O_2 \longrightarrow CO_2 + H_2O$
(f) $C_2H_6 + O_2 \longrightarrow CO_2 + H_2O$
(g) $Al(OH)_3 + H_2SO_4 \longrightarrow Al_2(SO_4)_3 + H_2O$

4.3   STOICHIOMETRIC CALCULATIONS

Coefficients give relative quantities of reactants and products by moles as well as by molecules.

As we balance an equation, our thinking is in terms of the small particles represented by the formulas. For example, in the reaction between hydrogen and oxygen that gives water, the equation that we wrote

$$2H_2 + O_2 \longrightarrow 2H_2O$$

can mean

$$2 \text{ molecules } H_2 + 1 \text{ molecule } O_2 \longrightarrow 2 \text{ molecules } H_2O$$

These coefficients must be treated as *exact* numbers because atoms cannot be broken here.

However, when we use a balanced equation to plan how much of each reactant to use in an actual experiment, we have to shift our thinking to huge collections of molecules — to moles. We saw in Chapter 3 that it is impossible to count out molecules. We can make this shift from molecules to moles by taking advantage of a simple rule from mathematics — multiplying a set of numbers (such as coefficients) by any constant number does not alter the *ratios* among the members of the set. If the original numbers are, say, 1, 2, and 3, and we multiply each by 2 to get the new set 2, 4, and 6, the *ratios* within the set are unchanged. If we select Avogadro's number (6.02×10^{23}) as the multiplier, we can get lab-sized units of each chemical.

$$2 \times (6.02 \times 10^{23} \text{ molecules}) H_2 + 1 \times (6.02 \times 10^{23} \text{ molecules}) O_2 \longrightarrow$$
$$2 \times (6.02 \times 10^{23} \text{ molecules}) H_2O$$

The essential $2:1:2$ ratio has not been changed by this multiplication. But the *scale* of the reaction has shifted to the mole level, because 6.02×10^{23} formula units of any substance is 1 mol of it. Now we can interpret the coefficients as follows.

$$2 \text{ mol } H_2 + 1 \text{ mol } O_2 \longrightarrow 2 \text{ mol } H_2O$$

The ratio of *moles* of molecules is identical to the ratio of *molecules* — it has to be, since equal numbers of moles have equal numbers of molecules.

The ratio of the coefficients for any given chemical reaction is set by nature; we cannot change this ratio. The decision left for us is the *scale* of the reaction — how much do we want to use or make? The number of options is infinite. We could have

$$0.02 \text{ mol } H_2 + 0.01 \text{ mol } O_2 \longrightarrow 0.02 \text{ mol } H_2O$$

or

$$1.36 \text{ mol } H_2 + 0.68 \text{ mol } O_2 \longrightarrow 1.36 \text{ mol } H_2O$$

or

$$88 \text{ mol } H_2 + 44 \text{ mol } O_2 \longrightarrow 88 \text{ mol } H_2O$$

In every case, the relative mole quantities of H_2 to O_2 to H_2O are $2:1:2$. We could say that 2 mol H_2, 1 mol O_2, and 2 mol H_2O are *equivalent* to each other *in this reaction*. This does not mean that one chemical can actually substitute for any of the other chemicals. It means that a specific mole quantity of one substance *requires* the presence of a specific mole quantity of each of the other substances in accordance with the ratios of coefficients.

Figure 4.1 on page 118 shows five different scales for the reaction of iron with sulfur to make iron sulfide, FeS. Notice that the *mole* ratios are the same regardless of the scale.

One kind of practical problem involving a balanced equation is the calculation of the mole quantities of all of the substances when given a certain number of moles of one of them. In the following example we will see how to use the coefficients (as moles) to set up conversion factors in working these problems.

Because one mole of any substance has *exactly* the same number of formula units as one mole of any other substance, we can use the coefficients as exact numbers when interpreting them on a mole basis.

EXAMPLE 4.4 WORKING WITH MOLE RELATIONSHIPS

Problem: Two atoms of sulfur react with three molecules of oxygen to form two molecules of sulfur trioxide, an air pollutant.

$$2S + 3O_2 \longrightarrow 2SO_3$$

How many *moles* of sulfur react in this way with 9 mol of O_2?

FIGURE 4.1 Atom/molecule ratios, mole ratios, and the *scale* of a reaction. Iron and sulfur, when heated, combine as follows: $Fe + S \rightarrow FeS$. Sulfur atoms have smaller masses than iron atoms—by a factor of 32.1 to 55.8, the ratio of their atomic weights. The piles of sulfur therefore have smaller masses than those of iron, but the *atoms* in each pile are always in the 1 to 1 ratio that nature requires for this reaction. It's the *scale* of the reaction that increases from top to bottom.

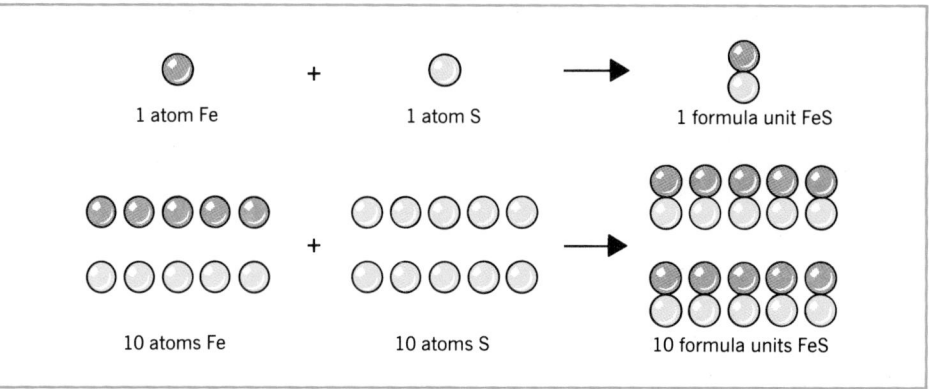

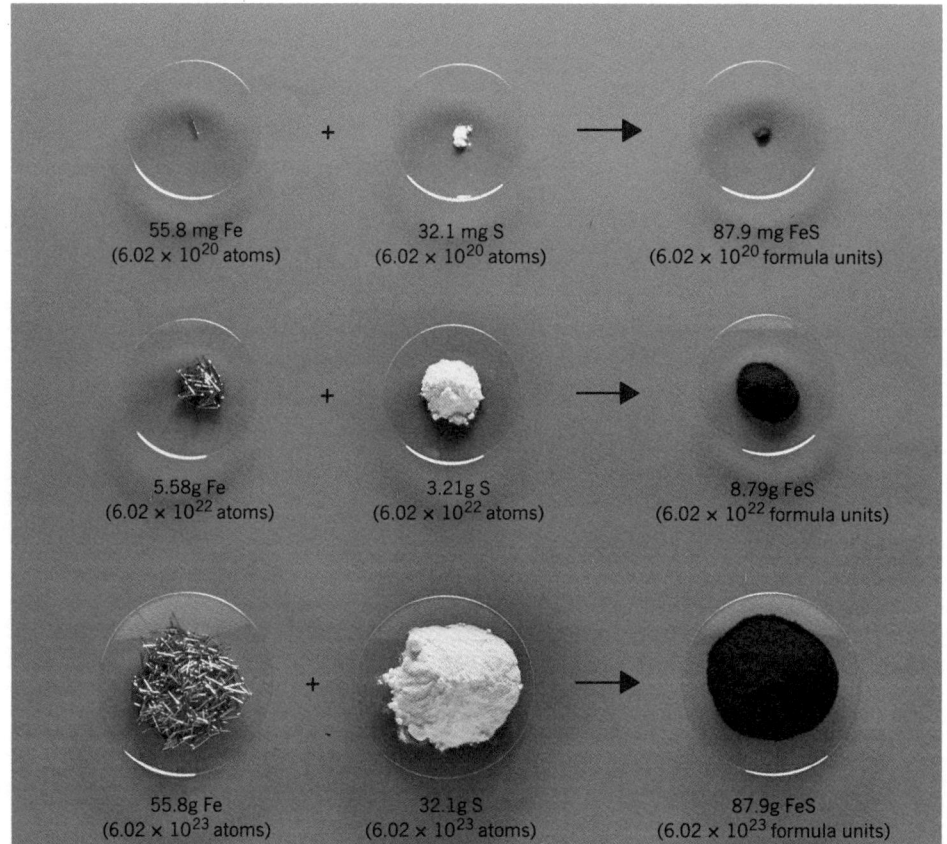

Solution: Our goal here is to learn how to work stoichiometry problems using the factor-label method. Therefore, even if the answer seems obvious to you, carefully go through the solution, because in most such problems the numbers aren't as simple as those given here.

First, where do we begin? Since we are going to have to do some calculations, let's start by restating what is asked in mathematical form.

$$9 \text{ mol } O_2 = ? \text{ mol } S$$

To use the factor-label method, we need a conversion factor that relates these quantities. Then we will be able to express the solution as

$$9 \text{ mol } O_2 \times (\text{conversion factor}) = ? \text{ mol } S$$

The critically important next step is to realize that it is the coefficients in the equation that give us the relationship between the amounts of sulfur and oxygen on a mole basis. These coefficients are 2 and 3 — 2 atoms of sulfur to 3 molecules of oxygen, which translates to 2 mol S to 3 mol O_2. This relationship can be stated by either of two conversion factors.

Never omit any of the units. Here, they must include both "mol" and the chemical symbol, S or O_2, as the case may be.

$$\frac{2 \text{ mol S}}{3 \text{ mol } O_2} \quad \text{or} \quad \frac{3 \text{ mol } O_2}{2 \text{ mol S}}$$

In choosing which conversion factor to use, we just make sure that the units that we want to get rid of, "mol O_2", appears in the denominator. Therefore, we multiply "9 mol O_2" by the first conversion factor:

$$9 \text{ mol } O_2 \times \frac{2 \text{ mol S}}{3 \text{ mol } O_2} = 6 \text{ mol S}$$

We have just found that 6 mol of S is needed to react with 9 mol of O_2.

PRACTICE EXERCISE 3   How many moles of iron, Fe, can be made from Fe_2O_3 by the use of 18 mol of carbon monoxide, CO, in the following reaction?

$$Fe_2O_3 + 3CO \longrightarrow 2Fe + 3CO_2$$

PRACTICE EXERCISE 4   How many moles of H_2O are produced when 6 mol of O_2 is consumed in burning methyl alcohol, CH_3OH, according to the following equation?

$$2CH_3OH + 3O_2 \longrightarrow 2CO_2 + 4H_2O$$

PRACTICE EXERCISE 5   Solutions of iron(III) chloride, $FeCl_3$, are used in photoengraving and to make ink. This compound can be made by the following reaction.

$$2Fe + 3Cl_2 \longrightarrow 2FeCl_3$$

(a) Set up the three possible pairs of conversion factors based on the coefficients of this equation. One pair involves Fe and Cl_2; another, Fe and $FeCl_3$; and the third, Cl_2 and $FeCl_3$. Select from these six conversion factors to solve the next problems.
(b) How many moles of $FeCl_3$ form from 24 mol of Cl_2?
(c) How many moles of Fe are needed to combine with 24 mol of Cl_2 by this reaction?
(d) If 0.500 mol of Fe is to be used by this reaction, how many moles of Cl_2 are needed, and how many moles of $FeCl_3$ form?

    The reaction in Practice Exercise 5 was described and problems related to it were solved in terms of moles. However, laboratory weighing balances are marked not in moles but in grams. Therefore, a reaction given in moles sooner or later has to be worked out in grams before it can be carried out. The only way available for relating *grams* of one compound to *grams* of another in a particular reaction is to go through their respective moles during the calculations.

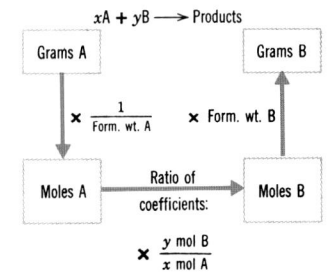

EXAMPLE 4.5   WORKING MASS–MOLE PROBLEMS

Problem: Aluminum oxide (Al_2O_3), a buffing powder, is to be made by combining 5.00 g of aluminum with oxygen (O_2). How much oxygen is needed in moles? In grams?

Solution: We outlined the basic approach to this kind of problem in the preceding paragraph. How-ever, rather than just marching through the various steps in the calculation, let's examine the reasoning involved, so you can see how the units suggest what information is needed and how to set up the arithmetic. As in the preceding example, let's begin by restating the question in mathematical form.

$$5.00 \text{ g Al} = ? \text{ mol } O_2$$

Once we've obtained an answer to this question, we can convert moles of oxygen to grams, which will then complete the problem.

As in Example 4.4, the essential chemical knowledge required here is realizing that *any time we have a problem dealing with the stoichiometry of a chemical reaction, we should always begin with a balanced equation.* We looked at this particular equation in Example 4.2.

$$4Al + 3O_2 \longrightarrow 2Al_2O_3$$

As before, the coefficients will give us the relationship between *moles* of the reactants. For aluminum and oxygen *in this reaction* we have two conversion factors:

$$\frac{4 \text{ mol Al}}{3 \text{ mol } O_2} \quad \text{and} \quad \frac{3 \text{ mol } O_2}{4 \text{ mol Al}}$$

The second will let us convert from moles of aluminum to moles of oxygen, but we are given grams of aluminum in the problem. Therefore, we must first change grams of aluminum to moles of aluminum. For this we need the atomic weight of Al, from which we get

$$1 \text{ mol Al} = 27.0 \text{ g Al}$$

Now we can calculate the number of moles of aluminum in 5.00 g.

$$5.00 \text{ g Al} \times \frac{1 \text{ mol Al}}{27.0 \text{ g Al}} = 0.185 \text{ mol Al}$$

The next step is to compute the moles of O_2 needed. Here we use the conversion factor constructed from the coefficients in the equation.

$$0.185 \text{ mol Al} \times \frac{3 \text{ mol } O_2}{4 \text{ mol Al}} = 0.139 \text{ mol } O_2$$

This is the answer at the mole level: 0.139 mol of O_2 is needed to combine with 0.185 mol of Al to make Al_2O_3 by our equation. This problem also calls for the grams of O_2, and we know from the formula weight of O_2 that there is 32.0 g O_2/mol O_2. Therefore,

$$0.139 \text{ mol } O_2 \times \frac{32.0 \text{ g } O_2}{1 \text{ mol } O_2} = 4.45 \text{ g } O_2$$

This is the second answer: it takes 4.45 g of O_2 (0.139 mol) to combine with 5.00 g of Al (0.185 mol) to make Al_2O_3.

PRACTICE EXERCISE 6   Calculate how many grams of iron can be made from 16.5 g of Fe_2O_3 by the following equation.

$$Fe_2O_3 + 3H_2 \longrightarrow 2Fe + 3H_2O$$

PRACTICE EXERCISE 7   Calculate how many grams of $K_2Cr_2O_7$ are needed to make 35.8 g of I_2 according to the following equation.

$$K_2Cr_2O_7 + 6NaI + 7H_2SO_4 \longrightarrow Cr_2(SO_4)_3 + 3I_2 + 7H_2O + 3Na_2SO_4 + K_2SO_4$$

PRACTICE EXERCISE 8   Iodine chloride, ICl, can be made by the following reaction between iodine, I_2, potassium iodate, KIO_3, and hydrochloric acid.

$$2I_2 + KIO_3 + 6HCl \longrightarrow 5ICl + KCl + 3H_2O$$

Calculate how many grams of iodine are needed to prepare 28.6 g of ICl by this reaction.

EXAMPLE 4.6   WORKING MASS–MOLE PROBLEMS

Problem: During its combustion, ethane (C_2H_6) combines with oxygen (O_2) to give carbon dioxide and water. A sample of ethane was burned completely and the water that formed had a mass of 1.61 g. How much ethane, in moles and in grams, was in the sample?

Solution: First, set up the equation using the formulas given.

$$C_2H_6 + O_2 \longrightarrow CO_2 + H_2O$$

Then balance it.

$$2C_2H_6 + 7O_2 \longrightarrow 4CO_2 + 6H_2O$$

Then compute the needed formula weights.

$$\begin{array}{ll} H_2O & 18.0 \\ C_2H_6 & 30.0 \end{array}$$

To calculate how many grams of one substance correspond to a certain number of grams of another, we have to switch over to moles. Otherwise we couldn't use the equation's coefficients. Therefore, we change the given quantity, which is 1.61 g of H_2O, to moles using the formula weight of H_2O.

$$1.61 \text{ g } H_2O \times \frac{1 \text{ mol } H_2O}{18.0 \text{ g } H_2O} = 0.0894 \text{ mol } H_2O$$

Now we must calculate how many moles of ethane are needed to make 0.0894 mol H_2O by this reaction. From the coefficients of the equation we can set up two possible conversion factors.

Coefficients are never in grams but in moles.

$$\frac{6 \text{ mol } H_2O}{2 \text{ mol } C_2H_6} \quad \text{and} \quad \frac{2 \text{ mol } C_2H_6}{6 \text{ mol } H_2O}$$

We pick the second because when we multiply it by 0.0894 mol H_2O to find moles of C_2H_6, the units cancel properly.

$$0.0894 \text{ mol } H_2O \times \frac{2 \text{ mol } C_2H_6}{6 \text{ mol } H_2O} = 0.0298 \text{ mol } C_2H_6$$

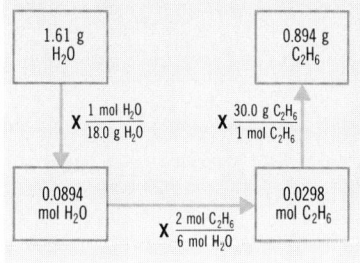

This is the answer at the mole level: 0.0298 mol of C_2H_6 must be burned to give 0.0894 mol (1.61 g) of H_2O. The question also calls for the answer in grams of C_2H_6, so we next use the formula weight of C_2H_6 to calculate the number of grams.

$$0.0298 \text{ mol } C_2H_6 \times \frac{30.0 \text{ g } C_2H_6}{1 \text{ mol } C_2H_6} = 0.894 \text{ g } C_2H_6$$

This is the other answer: 0.894 g of C_2H_6 combines with oxygen to make 1.61 g of H_2O.

PRACTICE EXERCISE 9   The nitrite ion (NO_2^-) in potassium nitrite is changed to the nitrate ion by the action of potassium permanganate ($KMnO_4$) in sulfuric acid solution.

$$5KNO_2 + 2KMnO_4 + 3H_2SO_4 \longrightarrow 5KNO_3 + 2MnSO_4 + K_2SO_4 + 3H_2O$$

How many moles and how many grams of $KMnO_4$ are needed to carry out this reaction on 11.4 g of KNO_2?

PRACTICE EXERCISE 10   The chief component of glass is silica for which the formula SiO_2 can be used. Silica is dissolved by hydrofluoric acid, HF, according to the following reaction that produces silicon tetrafluoride, SiF_4, a gas at room temperature.

$$SiO_2 + 4HF \longrightarrow SiF_4 + 2H_2O$$

How many grams and how many moles of SiF_4 can be produced from 63.4 g of HF?

PRACTICE EXERCISE 11   Copper(I) iodide, CuI, is not stable enough to last long in storage, so it is generally made just prior to use. It can be prepared from copper sulfate and hydriodic acid by the following reaction.

$$2CuSO_4 + 4HI \longrightarrow 2CuI + 2H_2SO_4 + I_2$$

If 10.4 g of $CuSO_4$ is used, calculate the number of grams of HI needed and the number of grams of each of the products that are produced. Show that the mass data are in accordance with the law of conservation of mass in chemical reactions.

EXAMPLE 4.7   WORKING MOLE PROBLEMS BY ACCUMULATING CONVERSION FACTORS

Problem: How might conversion factors be accumulated to shorten the solution to a problem such as that solved in Example 4.6?

Solution: Go back to Example 4.6 and review the three *separate* calculations. In the first, the result was 0.0894 mol H_2O. In the second, this result was converted to 0.0298 mol C_2H_6. The final answer came from the third conversion—of 0.0298 mol C_2H_6 to 0.894 g C_2H_6. You can save time by accumulating the conversion factors as follows.

$$1.61 \text{ g } H_2O \times \frac{1 \text{ mol } H_2O}{18.0 \text{ g } H_2O} \times \frac{2 \text{ mol } C_2H_6}{6 \text{ mol } H_2O} \times \frac{30.0 \text{ g } C_2H_6}{1 \text{ mol } C_2H_6} = 0.894 \text{ g } C_2H_6$$

$$\downarrow$$
0.0894 mol H_2O

$$\downarrow$$
0.0298 mol C_2H_6

$$\downarrow$$
0.894 g C_2H_6

Notice that all units cancel except the unit for the final answer. If these cancellations had not led to this result, then we would have known with certainty that the solution had not been set up properly. The first stage, of course, changes g H_2O into mol H_2O. Then mol H_2O is converted to the equivalent amount of C_2H_6—mol C_2H_6. Finally, there is a shift back to the gram level when mol C_2H_6 is converted into g C_2H_6. Through the method of accumulating conversion factors, the operations are rolled into one long setup. Then the arithmetic is done. In brief, we start with a mass of one substance, move to the mole level, stay there to find moles of any other substance in the reaction, and then move back to the gram level of that other compound. *The calculation pathway from grams of one substance to grams of anything else in the reaction always goes through the mole relationship in the balanced equation.* If the only information available is the balanced equation, then there is no way to avoid using the coefficients in going from grams of one component to grams of another.

PRACTICE EXERCISE 12   Use the strategy of Example 4.7 to work Practice Exercise 9 again.

PRACTICE EXERCISE 13   Work Practice Exercise 10 again using the strategy of Example 4.7.

PRACTICE EXERCISE 14   Rework Practice Exercises 6 and 7 by the method of accumulating conversion factors.

PRACTICE EXERCISE 15   Use accumulation of conversion factors to work Practice Exercise 8.

4.4 PERCENTAGE YIELD AND LIMITING REACTANTS

If one reactant is entirely used up before any of the others, then that reactant limits the maximum yield of the product.

Sometimes in experimental work one of the two or more reactants is deliberately used in a mole quantity that exceeds the requirements set by the reaction's stoichiometry. This might be done to help ensure that *all* of another (perhaps expensive) reactant is fully used. At other times this strategy is used to make a reaction occur faster. For example, a large excess of oxygen makes things burn more rapidly.

Whatever the reason for the use of an excess of a reactant, we cannot make *more* product from the excess, because the yield — the amount of product that forms — is limited by the reactant that is completely used. Once this reactant is gone, the reaction has to stop, and no additional products can form. The reactant that determines how much product can form is called the limiting reactant. The theoretical yield of a product is the maximum yield obtainable as calculated on the basis of the amount of the limiting reactant used.

EXAMPLE 4.8 CALCULATING LIMITING REACTANTS AND THEORETICAL YIELDS

Problem: Methane, CH_4, burns in oxygen to give carbon dioxide and water according to the following equation.

$$CH_4 + 2O_2 \longrightarrow CO_2 + 2H_2O$$

In one experiment, a mixture of 0.250 mol of methane was burned in 1.25 mol of oxygen in a sealed steel vessel. Find the limiting reactant, if any, and calculate the theoretical yield (in moles) of water.

Solution: In a limiting reactant problem, we could restate the question in two ways. For example, we might ask how many moles of CH_4 could 1.25 mol of O_2 actually handle?

$$1.25 \text{ mol } O_2 = ? \text{ mol } CH_4$$

Or we could ask how many moles of O_2 would 0.250 mol of CH_4 require for this reaction?

$$0.250 \text{ mol } CH_4 = ? \text{ mol } O_2$$

Either way will work because the question we want answered — What is the limiting reactant? — will emerge when we compare the answer obtained by either approach to the original amounts taken for the experiment. So let's work at the first approach: How many moles of CH_4 can 1.25 mol of O_2 handle? To do this we have to find a conversion factor for the following transformation.

$$1.25 \text{ mol } O_2 \times (\text{conversion factor}) = ? \text{ mol } CH_4$$

The coefficients of the chemical equation gives us the following factors.

$$\frac{2 \text{ mol } O_2}{1 \text{ mol } CH_4} \quad \text{and} \quad \frac{1 \text{ mol } CH_4}{2 \text{ mol } O_2}$$

We have to use the second of these to find the moles of CH_4.

$$1.25 \text{ mol } O_2 \times \frac{1 \text{ mol } CH_4}{2 \text{ mol } O_2} = 0.625 \text{ mol } CH_4$$

Since 1.25 mol O_2 can handle 0.625 mol of CH_4, and we were given only 0.250 mol of CH_4, we clearly have more than enough O_2. Oxygen, in other words, is not the limiting reactant. Methane must be. But let's double check. Let us use the first conversion factor on 0.250 mol of CH_4.

$$0.250 \text{ mol } CH_4 \times \frac{2 \text{ mol } O_2}{1 \text{ mol } CH_4} = 0.500 \text{ mol } O_2$$

Since 0.250 mol of CH_4 needs only 0.500 mol of O_2 and we were given 1.25 mol of O_2, we obviously have more than enough O_2, so we are certain that CH_4 is the limiting reactant. This answers the first question.

For the second question, the maximum yield of either product is tied to the moles of limiting reactant taken. The question about the yield of water can be restated as follows, now that we know that CH_4 is the limiting reactant.

$$0.250 \text{ mol } CH_4 = ? \text{ mol } H_2O$$

In other words, we need a conversion factor for the following:

$$0.250 \text{ mol } CH_4 \times (\text{conversion factor}) = ? \text{ mol } H_2O$$

Once again, the equation supplies our need, because its coefficients let us write the following conversion factors.

$$\frac{1 \text{ mol } CH_4}{2 \text{ mol } H_2O} \quad \text{and} \quad \frac{2 \text{ mol } H_2O}{1 \text{ mol } CH_4}$$

To get the units in the answer right, we have to use the second.

$$0.250 \text{ mol } CH_4 \times \frac{2 \text{ mol } H_2O}{1 \text{ mol } CH_4} = 0.500 \text{ mol } H_2O$$

This tells us that when any excess of oxygen reacts with 0.250 mol of CH_4, the theoretical yield of H_2O is 0.500 mol. The mixture of products, of course, is contaminated by the unreacted oxygen, but using an excess of oxygen ensured that no carbon monoxide, CO, was produced, as often happens when something burns in a limited supply of oxygen.

PRACTICE EXERCISE 16   When it is heated in the presence of carbon monoxide, iron(III) oxide is converted to iron according to the following reaction.

$$Fe_2O_3 + 3CO \longrightarrow 2Fe + 3CO_2$$

In one experiment 0.300 mol of Fe_2O_3 was heated with 1.20 mol of CO. Which was the limiting reactant, and what was the theoretical yield of Fe in moles?

EXAMPLE 4.9   CALCULATING LIMITING REACTANTS AND THEORETICAL YIELDS

Problem: Chloroform, $CHCl_3$, reacts with chlorine, Cl_2, to form carbon tetrachloride, CCl_4, and hydrogen chloride, HCl. In one experiment the reactants were initially present in a ratio of 1 to 1 *by mass*; specifically, 25.0 g of $CHCl_3$ was mixed with 25.0 g of Cl_2. Which is the limiting reactant? What is the maximum yield of CCl_4 in moles and in grams?

Solution: As usual we have to start with a balanced equation, which we can construct with the information given.

$$CHCl_3 + Cl_2 \longrightarrow CCl_4 + HCl$$

We will need formula weights, so we might as well calculate them next:

$CHCl_3$	119
Cl_2	71.0
CCl_4	154

As always in stoichiometric problems, get the calculations to the mole level.

Remember that we have to compare the *mole* ratios required by the coefficients of the equation with the *mole* ratios actually taken of the reactants. We therefore have to convert the grams of reactants to moles, which we will do next.

$CHCl_3$: $25.0 \text{ g } CHCl_3 \times \dfrac{1 \text{ mol } CHCl_3}{119 \text{ g } CHCl_3} = 0.210 \text{ mol } CHCl_3$ taken

Cl_2: $25.0 \text{ g } Cl_2 \times \dfrac{1 \text{ mol } Cl_2}{71.0 \text{ g } Cl_2} = 0.352 \text{ mol of } Cl_2$ taken

Now comes the key step in determining which is the limiting reactant — comparison of the ratio of the moles of reactants taken with the ratio of the moles required by the coefficients of the equation. These are 1 to 1, 1 mol of $CHCl_3$ to 1 mol of Cl_2. Hence, 0.210 mol of $CHCl_3$ needs only 0.210 mol of Cl_2, *which is less than the 0.352 mol of Cl_2 actually taken*. There is plenty of chlorine. The limiting reactant, then, is $CHCl_3$. When it is gone, no more products can form, regardless of how much chlorine is still present.

To solve the last part of the problem, the theoretical yield of CCl_4, we note in the equation that there is a 1-to-1 ratio between the limiting reactant, $CHCl_3$, and CCl_4. The theoretical yield is always based on the limiting reactant, so 0.210 mol of $CHCl_3$ can be changed into a maximum of 0.210 mol of CCl_4, the theoretical yield (in moles). This translates into grams in the usual way:

$$0.210 \text{ mol } CCl_4 \times \frac{154 \text{ g } CCl_4}{1 \text{ mol } CCl_4} = 32.3 \text{ g } CCl_4$$

Thus, the theoretical yield of CCl_4 in this experiment is 32.3 g (0.210 mol).

EXAMPLE 4.10   CALCULATING LIMITING REACTANTS AND THEORETICAL YIELDS

Problem: Aluminum chloride, Al_2Cl_6, can be made by the reaction of aluminum with chlorine according to the following equation.

$$2Al + 3Cl_2 \longrightarrow Al_2Cl_6$$

What is the limiting reactant if 20.0 g of Al and 30.0 g of Cl_2 are used, and how much Al_2Cl_6 can theoretically form?

Solution: **Step 1.** Use the balanced equation given above.

Step 2. Formula weights: Al, 27.0; Cl_2, 71.0; Al_2Cl_6, 267.

Step 3. Convert what is given into moles:

Aluminum: $20.0 \text{ g Al} \times \dfrac{1 \text{ mol Al}}{27.0 \text{ g Al}} = 0.741 \text{ mol Al}$

Chlorine: $30.0 \text{ g } Cl_2 \times \dfrac{1 \text{ mol } Cl_2}{71.0 \text{ g } Cl_2} = 0.423 \text{ mol } Cl_2$

Step 4. Compare these calculated moles with the coefficients of the equation, which tell us that 2 mol of Al is needed for 3 mol of Cl_2. Let's see how much Cl_2 is actually needed for 0.741 mol of Al:

$$0.741 \text{ mol Al} \times \frac{3 \text{ mol } Cl_2}{2 \text{ mol Al}} = 1.11 \text{ mol } Cl_2 \text{ (needed)}$$

In other words, 0.741 mol of Al requires 1.11 mol of Cl_2 to react completely. Obviously, the 0.423 mol of Cl_2 actually supplied falls far short of this. Therefore, chlorine is the limiting reactant. Let's check this by finding out how much aluminum is needed for 0.423 mol of Cl_2.

$$0.423 \text{ mol } Cl_2 \times \frac{2 \text{ mol Al}}{3 \text{ mol } Cl_2} = 0.282 \text{ mol Al (needed)}$$

In other words, the chlorine we have available needs only 0.282 mol of Al — far less than the 0.741 mol of Al actually supplied — so this calculation verifies that Al is in excess, and that Cl_2 is the limiting reactant.

Step 5. Calculate the theoretical yield. Since chlorine is the limiting reactant, calculate the yield of Al_2Cl_6 using the moles of chlorine actually consumed, 0.423 mol Cl_2. The balanced equation tells us that 1 mol of Al_2Cl_6 can be produced from 3 mol of Cl_2; therefore,

$$0.423 \text{ mol } Cl_2 \times \frac{1 \text{ mol } Al_2Cl_6}{3 \text{ mol } Cl_2} = 0.141 \text{ mol } Al_2Cl_6$$

Next, convert this answer to its equivalent in grams.

The formula weight of Al_2Cl_6 is 267, so this appears in the conversion factor.

$$0.141 \; \text{mol } Al_2Cl_6 \times \frac{267 \text{ g } Al_2Cl_6}{1 \text{ mol } Al_2Cl_6} = 37.6 \text{ g } Al_2Cl_6$$

(theoretical yield)

The theoretical yield of Al_2Cl_6 is 37.6 g.

PRACTICE EXERCISE 17   What is the limiting reactant when 10.0 g of sodium and 20.0 g of chlorine, Cl_2, are mixed to produce sodium chloride, NaCl?

Side Reactions, By-products, and Percentage Yields

Maybe Murphy was a chemist.

Many reactions, particularly those of carbon compounds, do not occur exactly according to balanced equations. Methane, for example, reacts with chlorine to give chiefly chloromethane and hydrogen chloride. The balanced equation is as follows.

$$\underset{\text{methane}}{CH_4} + Cl_2 \longrightarrow \underset{\text{chloromethane}}{CH_3Cl} + HCl$$

CH_2Cl_2 = dichloromethane
$CHCl_3$ = trichloromethane
CCl_4 = tetrachloromethane
(carbon tetrachloride)

However, it is always found that some dichloromethane also forms, along with traces of trichloro- and tetrachloromethane. The reason is that one of the initial products, chloromethane, can also react with chlorine. This gives some dichloromethane (as well as more hydrogen chloride), and dichloromethane can itself react with any still-unchanged chlorine, and so forth. (Write the equations for these reactions.) In other words, this reaction is not stoichiometrically neat and clean. The reactions that produce the extra products are called side reactions, and the extra products are called by-products (as opposed to the main product (chloromethane in our example).

$$\% \text{ yield} = \frac{\text{actual}}{\text{predicted}} \times 100$$

Side reactions cut down on how much main product forms, so they reduce the percentage yield. The percentage yield of a product is the ratio, calculated as a percentage, of the actual amount of product to the amount predicted by the reaction's stoichiometry. For example, if 0.50 mol of CH_4 is used as the limiting reactant, but only 0.40 mol of CH_3Cl (instead of 0.50 mol) is isolated, then the percentage yield is

$$\% \text{ yield} = \left(\frac{0.40 \; \text{mol } CH_3Cl}{0.50 \; \text{mol } CH_3Cl} \right) \times 100 = 80\%$$

EXAMPLE 4.11   CALCULATING PERCENTAGE YIELD

Problem: A chemist new to the behavior of chlorine toward compounds containing carbon and hydrogen tried to make dichloromethane (CH_2Cl_2) by mixing 0.250 mol of chloromethane (CH_3Cl) and 0.250 mol of chlorine (Cl_2), expecting the following reaction to proceed cleanly, with no side reactions.

$$CH_3Cl + Cl_2 \longrightarrow CH_2Cl_2 + HCl$$

Inevitably, some chloroform ($CHCl_3$) and carbon tetrachloride (CCl_4) formed and some chloromethane remained unchanged. When the mixture of products was separated, a yield of 12.8 g of dichloromethane was obtained. Calculate the percentage yield of dichloromethane.

Solution: First we calculate the theoretical yield. Since all coefficients are 1, then 0.250 mol of CH_3Cl ought to yield 0.250 mol of CH_2Cl_2 if no side reactions occur. So we change 0.250 mol CH_2Cl_2 to grams, using the formula weight of CH_2Cl_2, 85.0,

$$0.250 \; \cancel{\text{mol } CH_2Cl_2} \times \frac{85.0 \text{ g } CH_2Cl_2}{1 \; \cancel{\text{mol } CH_2Cl_2}} = 21.3 \text{ g } CH_2Cl_2$$

The theoretical yield of CH_2Cl_2 is 21.3 g. Next, calculate the percentage yield. Only 12.8 g of CH_2Cl_2 was isolated. Therefore, the percentage yield is

$$\frac{12.8 \; \cancel{\text{g } CH_2Cl_2}}{21.3 \; \cancel{\text{g } CH_2Cl_2}} \times 100 = 60.1\%$$

The percentage yield of CH_2Cl_2 from CH_3Cl was 60.1%.

PRACTICE EXERCISE 18   One of the steps in one industrial synthesis of sulfuric acid (H_2SO_4) from sulfur is the conversion of sulfur dioxide into sulfur trioxide by the reaction

$$2SO_2 + O_2 \longrightarrow 2SO_3$$

In one "run," 1.75 kg of SO_2 was used and 1.72 kg of SO_3 was isolated from the mixture of products. What was the percentage yield of SO_3?

4.5 WHEN REACTANTS ARE IN SOLUTION

Whenever possible, reactions are carried out with all of the reactants in the same fluid phase.

Generally speaking, the particles of reacting substances, be they atoms, ions, or molecules, can react with each other only if they collide. For such collisions to occur, the particles obviously have to be free to move around. They can do this only in the liquid and gaseous states, so when working with solids we normally try to find a liquid that can dissolve the solids and form a solution. Before going on we first have to learn some of the common terms used to describe solutions.

A solution is a uniform mixture of particles of atomic, ionic, or molecular size. A minimum of two substances are present. One is called the solvent and all the others are called the solutes. The solvent is the medium into which the other substances are mixed or dissolved. A solute is any substance dissolved by the solvent. A solvent can be a solid, a liquid, or a gas, but we will confine our attention in this section to aqueous solutions — those for which the solvent is water. In an aqueous solution of sugar, the solute is sugar and the solvent is water. Club soda is a solution of a gas, carbon dioxide (the solute), in water (the solvent). Antifreeze is mostly a solution of the liquid ethylene glycol (the solute) in water. (If we mix two liquids, which one is called the solute and which one is called the solvent aren't important. Usually, the liquid present in the greater amounts is arbitrarily called the solvent, although water, if present in any amount, is often given this honor.)

We will discuss the properties of solutions in much more depth in Chapter 13.

More Vocabulary for Solutions

Several terms are used to describe solutions. A dilute solution is one in which the ratio of solute to solvent is very small, for example, a few crystals of sugar in a glass of water. In a concentrated solution the ratio of solute to solvent is large. Syrup, for example, is a concentrated solution of sugar in water. A saturated solution is one in which no more solute can dissolve at a particular temperature. If the solute is a solid, any excess of it in a saturated solution just sits on the bottom of the container. An unsaturated solution is one in which the ratio of solute to solvent is lower than that of the corresponding saturated solution. If more solute is added to an unsaturated solution, at least some of it will dissolve.

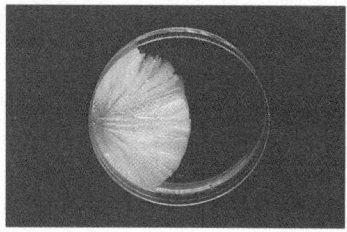
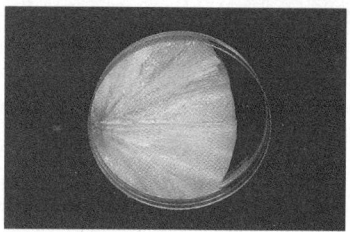
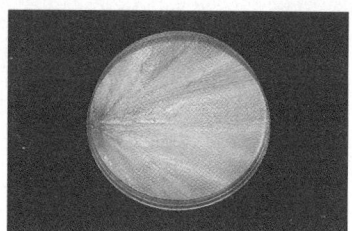

FIGURE 4.2 Addition of a small seed crystal of sodium acetate to a supersaturated solution of this salt causes crystallization to occur rapidly. Crystallization continues until the remaining liquid is just saturated.

For nearly all solids, the warmer the solvent the more concentrated the solution can be made.

Probably the most frequently used units of solubility are g solute/100 g solvent.

It isn't easy, but sometimes a supersaturated solution can be made. This is an unstable system in which the ratio of dissolved solute to solvent is higher than that of a saturated solution. We can sometimes make a supersaturated solution by carefully cooling a saturated solution. Usually, the cooler a solution is, the lower the maximum possible ratio of solute to solvent is. If we are very careful—using dust- and lint-free solvent and avoiding any shaking or stirring of the solution—the solute that should come out of solution as we cool it stays dissolved. However, if we tap the container, scratch its inner wall with a stirring rod, or add a "seed" crystal of the pure solute, the excess solute usually separates from the solution immediately. This event can be very dramatic and pretty to observe (see Figure 4.2). The formation of a solid from a solution is called precipitation, and the solid itself is called the precipitate.

The amount of solute needed to make a saturated solution in a given quantity of solvent at a specific temperature is called the solubility of the solute in the given solvent. Table 4.1 gives some examples that show how widely solubilities can vary. Notice that a saturated solution can still be relatively dilute. For example, only a very small amount of lead sulfate is dissolved in 100 g of water in a saturated solution.

TABLE 4.1 Solubilities of Some Compounds in Water

Substance	Formula	Solubility (g/100 g water)[a]
Ammonium chloride	NH_4Cl	29.7 (0 °C)
Boric acid	H_3BO_3	6.35 (30 °C)
Calcium chloride	$CaCl_2$	74.5 (20 °C)
Copper sulfide	CuS	3.3×10^{-5} (18 °C)
Lead sulfate	$PbSO_4$	4.3×10^{-3} (25 °C)
Sodium hydroxide	NaOH	42 (0 °C)
		347 (100 °C)

[a] At the temperature given in parentheses.

4.6 MOLAR CONCENTRATIONS

The unit of moles per liter is one of the most useful units for describing concentrations.

Sometimes a particular reactant is available from the stockroom only as a solution. In fact, it often is very convenient to have a way to get the amount of *solute* we need simply by pouring out a certain volume of the solution. The question now is, "How do we calculate the volume of a *solution* that can deliver a specific amount of its solute?"

This problem is solved if we know the concentration of the solution. The concentration of a solution is the ratio of solute to some given unit of solution. The units can be anything we wish, but for stoichiometric calculations, the best units are the moles of solute per liter of solution. The special name for this ratio is molar concentration, or molarity, abbreviated M.

Notice that it's per liter *of solution,* not per liter of solvent.

The standard abbreviation of "solution" is *soln.*

$$M = \frac{\text{mol solute}}{\text{L soln}} = \frac{\text{mol solute}}{1000 \text{ mL soln}}$$

(a)   (b)   (c)   (d)   (e)

FIGURE 4.3 Preparation of a solution having some particular molar concentration. The flask is a volumetric flask, and it has an etched line on its neck that marks a known volume. (*a*) An accurately weighed sample of the solute is put into the flask. (*b*) Some distilled (or deionized) water is added. (*c*) The contents of the flask are swirled to bring the solute into solution. (*d*) More water is added until the surface of the solution reaches the etched line. (*e*) The flask is stoppered and then shaken to make the solution uniform.

Thus a bottle with the label "0.10 *M* NaCl" contains a solution of sodium chloride with a concentration of 0.10 mol of sodium chloride per liter of solution (or per 1000 mL of solution). Figure 4.3 shows how a volumetric flask is used to make a solution of known molarity.

EXAMPLE 4.12   USING MOLAR CONCENTRATIONS IN MOLE PROBLEMS

Problem: A student needs 0.250 mol of NaCl and all that is available is a solution labeled "0.400 *M* NaCl." What volume of the solution should be used? Give the answer in milliliters.

Solution: The best approach is to use the information on the label of the bottle, the molar concentration, to construct ratios that can be used in a factor-label approach. The label tells us we could use either of the following two ratios, in which we use "1000 mL" in place of "1 L."

Treat 1 or 1000 here as exact numbers.

$$\frac{0.400 \text{ mol NaCl}}{1000 \text{ mL NaCl soln}} \quad \text{or} \quad \frac{1000 \text{ mL NaCl soln}}{0.400 \text{ mol NaCl}}$$

Now we apply the factor-label method. Since we want our answer in a unit of volume, we must use the second ratio.

$$0.250 \text{ mol NaCl} \times \frac{1000 \text{ mL NaCl soln}}{0.400 \text{ mol NaCl}} = 625 \text{ mL NaCl soln}$$

Thus, 625 mL of 0.400 *M* NaCl contains 0.250 mol of NaCl.

PRACTICE EXERCISE 19   A glucose solution with a molar concentration of 0.200 *M* is available. What volume of this solution must be measured to obtain 0.00100 mol of glucose? Give your answer in milliliters.

Another common problem involving molar concentrations is to calculate the grams of solute needed to make a given volume of solution having a specified molarity. We will see again in the next example how we can "translate the label"—use the value of the molarity—to set up two possible conversion factors to solve such a problem.

EXAMPLE 4.13   PREPARING SOLUTIONS WITH SPECIFIC MOLAR CONCENTRATIONS

Problem: How can 500 mL of 0.150 M Na_2CO_3 solution be prepared?

Solution: Although the problem is stated in a way that it normally arises in the lab, what we really need to know is how many *grams* of Na_2CO_3 (sodium carbonate) are in 500 mL of 0.150 M Na_2CO_3 solution. Although the label reads in M (mol/L), the balance reads in grams, and to make the solution we first have to weigh out grams. Of course, before we can calculate grams we have to calculate moles. In this example, how many *moles* of Na_2CO_3 are present in 500 ml of 0.150 M Na_2CO_3 solution? The key step is to "translate the label," which means to use the concentration stated on the bottle's label to set up conversion factors. (And it is still a good idea to use 1000 mL for 1 L whenever the desired volume is in milliliters instead of liters.) For example, "0.150 M Na_2CO_3" lets us write

$$\frac{0.150 \text{ mol } Na_2CO_3}{1000 \text{ mL } Na_2CO_3 \text{ soln}} \quad \text{and} \quad \frac{1000 \text{ mL } Na_2CO_3 \text{ soln}}{0.150 \text{ mol } Na_2CO_3}$$

To find how many *moles* of Na_2CO_3 we need for 500 mL of Na_2CO_3 solution, we use the factor-label method.

$$500 \text{ mL } Na_2CO_3 \text{ soln} \times \frac{0.150 \text{ mol } Na_2CO_3}{1000 \text{ mL } Na_2CO_3 \text{ soln}} = 0.0750 \text{ mol } Na_2CO_3$$

Convert moles of Na_2CO_3 to grams using the formula weight of 106.

$$0.0750 \text{ mol } Na_2CO_3 \times \frac{106 \text{ g } Na_2CO_3}{1 \text{ mol } Na_2CO_3} = 7.95 \text{ g } Na_2CO_3$$

The answer, then, is that we need 7.95 g Na_2CO_3 to make 500 mL of 0.150 M Na_2CO_3. We do not add the solute to 500 mL of pure water, however, because this could give a final volume different than 500 mL. The amount of solute in each milliliter therefore would not be precisely known. Instead, we dissolve the 7.95 g of Na_2CO_3 in a small amount of water and then dilute it to exactly 500 mL.

PRACTICE EXERCISE 20   How can we prepare 250 mL of 0.200 M $NaHCO_3$?

Special pieces of glassware called volumetric flasks are available for preparing solutions of known concentrations. One was pictured in Figure 4.3. A line etched on the narrow neck of a volumetric flask shows where the liquid level should reach for the flask to contain exactly the volume stated on its side. Since liquids expand or contract slightly with changes in temperature, the given capacity of the flask is correct only for the temperature value etched on the flask. In the most careful work, the capacity of a flask is checked against precise standards. (Any laboratory operation in which a stated capacity or a stated mass is checked against precisely known standards is called a calibration.)

PRACTICE EXERCISE 21   How can you prepare 250 mL of 0.0500M glucose, $C_6H_{12}O_6$?

Stoichiometry of Reactions in Solutions

Now that we know what *molarity* means and how to make solutions of known molar concentrations, let's learn how to do mole calculations when solutions, rather than the pure substances, are used.

EXAMPLE 4.14   STOICHIOMETRIC CALCULATIONS WHEN REACTANTS ARE IN SOLUTION

Problem: What volume of 0.556 *M* HCl has enough hydrochloric acid to combine exactly with 25.4 mL of aqueous sodium hydroxide with a concentration of 0.458 *M*? The equation for the reaction is

$$HCl(aq) + NaOH(aq) \longrightarrow NaCl(aq) + H_2O$$

Solution: This kind of a problem is nothing more than a variation of the problem of matching the given moles of one chemical to the moles of another according to the ratio of their coefficients in the balanced equation. What is different is that the chemicals are provided already in solution, and after finding the *moles* needed, we then have to calculate what *volume* contains these moles. So we first find the moles of NaOH in 25.4 mL of 0.458 *M* NaOH. The molarity of the solution gives us these conversion factors.

$$\frac{0.458 \text{ mol NaOH}}{1000 \text{ mL NaOH soln}} \quad \text{and} \quad \frac{1000 \text{ mL NaOH soln}}{0.458 \text{ mol NaOH}}$$

To see how much NaOH is in 25.4 mL of NaOH soln, we use the first.

$$25.4 \text{ mL NaOH soln} \times \frac{0.458 \text{ mol NaOH}}{1000 \text{ mL NaOH soln}} = 0.0116 \text{ mol NaOH}$$

Now we use a ratio of coefficients from the balanced equation to calculate how many moles of HCl are equivalent to 0.0116 mol NaOH.

$$0.0116 \text{ mol NaOH} \times \frac{1 \text{ mol HCl}}{1 \text{ mol NaOH}} = 0.0116 \text{ mol HCl}$$

the ratio of the coefficients from the equation

But this 0.0116 mol HCl comes only in a solution with a concentration of 0.556 *M*. This molarity gives us the following conversion factors.

$$\frac{0.556 \text{ mol HCl}}{1000 \text{ mL HCl soln}} \quad \text{and} \quad \frac{1000 \text{ mL HCl soln}}{0.556 \text{ mol HCl}}$$

Therefore to find the volume of HCl that holds 0.0116 mol of HCl, we do the following.

$$0.0116 \text{ mol HCl} \times \frac{1000 \text{ mL HCl soln}}{0.556 \text{ mol HCl}} = 20.9 \text{ mL HCl soln}$$

We could also have solved this problem by accumulating conversion factors before doing the calculation.

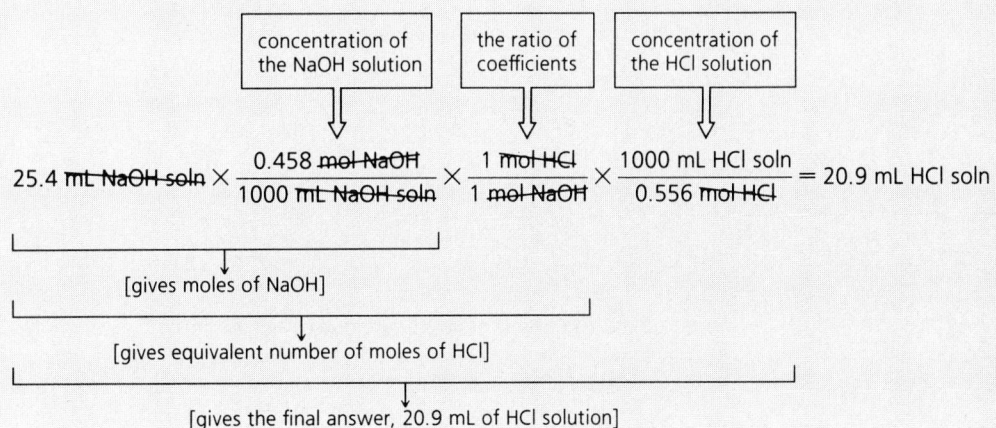

EXAMPLE 4.15   STOICHIOMETRIC CALCULATIONS WHEN REACTANTS ARE IN SOLUTION

Problem: How many milliliters of 0.114 M H_2SO_4 solution provide the sulfuric acid required to react with the sodium hydroxide in 32.2 mL of 0.122 M NaOH according to the following equation?

$$H_2SO_4(aq) + 2NaOH(aq) \longrightarrow Na_2SO_4(aq) + 2H_2O$$

Solution: The molarity of the NaOH solution gives us these conversion factors.

$$\frac{0.122 \text{ mol NaOH}}{1000 \text{ mL NaOH soln}} \quad \text{and} \quad \frac{1000 \text{ mL NaOH soln}}{0.122 \text{ mol NaOH}}$$

Therefore in 32.2 mL of 0.122 M NaOH there is

$$32.2 \text{ mL NaOH soln} \times \frac{0.122 \text{ mol NaOH}}{1000 \text{ mL NaOH soln}} = 0.00393 \text{ mol NaOH}$$

How many moles of sulfuric acid react with and are chemically equivalent to this much NaOH *in the reaction given?* To find out, we multiply the moles of NaOH by a conversion factor based on the coefficients in the equation.

$$0.00393 \text{ mol NaOH} \times \frac{1 \text{ mol } H_2SO_4}{2 \text{ mol NaOH}} = 0.00197 \text{ mol } H_2SO_4$$

Now we calculate what volume of $H_2SO_4(aq)$ contains 0.00197 mol of H_2SO_4; and we use the given concentration of the $H_2SO_4(aq)$ to set up the correct conversion factor. A concentration of 0.114 M H_2SO_4 means either of two conversion factors.

$$\frac{0.114 \text{ mol } H_2SO_4}{1000 \text{ mL } H_2SO_4 \text{ soln}} \quad \text{or} \quad \frac{1000 \text{ mL } H_2SO_4 \text{ soln}}{0.114 \text{ mol } H_2SO_4}$$

Therefore,

$$0.00197 \text{ mol } H_2SO_4 \times \frac{1000 \text{ mL } H_2SO_4 \text{ soln}}{0.114 \text{ mol } H_2SO_4} = 17.3 \text{ mL } H_2SO_4 \text{ soln}$$

Thus, 17.3 mL of 0.114 M H_2SO_4 can react with 32.2 mL of 0.122 M NaOH according to the equation given. This problem could also be worked by accumulating conversion factors.

$$32.2 \text{ mL NaOH soln} \times \frac{0.122 \text{ mol NaOH}}{1000 \text{ mL NaOH soln}} \times \frac{1 \text{ mol } H_2SO_4}{2 \text{ mol NaOH}} \times \frac{1000 \text{ mL } H_2SO_4 \text{ soln}}{0.114 \text{ mol } H_2SO_4} = 17.2 \text{ mL } H_2SO_4 \text{ soln}$$

The answer here isn't exactly the same as the answer obtained by solving the problem stepwise, because several roundings can sometimes lead to small discrepancies. Don't worry about this.

PRACTICE EXERCISE 22   What volume of 0.337 M KOH provides enough solute to combine with the sulfuric acid in 18.6 mL of 0.156 M H_2SO_4? The reaction is

$$2KOH(aq) + H_2SO_4(aq) \longrightarrow K_2SO_4(aq) + 2H_2O$$

The reactions used in Examples 4.14 and 4.15 and in Practice Exercise 22 are examples of a very important general reaction introduced in Section 2.12 — the neutralization of an acid by a base. Any reaction that destroys either an acid or a base is called a neutralization. In the next exercise, for example, sodium carbonate is used to destroy or neutralize an acid. One product, carbon dioxide, is a gas that fizzes out of the solution.

PRACTICE EXERCISE 23   Hydrochloric acid reacts with sodium carbonate as follows.

(g) = gas

$$2HCl(aq) + Na_2CO_3(aq) \longrightarrow 2NaCl(aq) + CO_2(g) + H_2O$$

What volume of 0.224 M HCl is neutralized by the Na_2CO_3 in 24.2 mL of 0.284 M Na_2CO_3?

4.7   MAKING DILUTE SOLUTIONS FROM CONCENTRATED SOLUTIONS

When a specific amount of concentrated solution is diluted, the original amount of solute becomes part of a larger volume.

If solvent is added to a solution, the solute is spread out through a larger volume and the number of moles per unit volume (the number of moles per liter, for example) decreases. We say the solution has been diluted. Sometimes such dilution is a natural part of carrying out an experiment. For example, if solutions of two different solutes are mixed, the total volume increases and becomes occupied by both solutes. As a result, the concentration of each solute is diminished. At other times, dilution is a deliberate process. For instance, stock solutions of common chemicals are often prepared in large volumes of preset concentrations. Thus, the chief supply of hydrochloric acid in your stockroom may be 1.00 M HCl solution. Your experiment, however, may call for a less concentrated solution, so the more concentrated solution must be diluted before you can use it.

The first step in carrying out a dilution is to calculate how much of the more concentrated solution is needed to make a given volume of the less concentrated solution. Then the necessary amount of the concentrated solution is measured and additional solvent is added to give the desired final volume. For routine work, ordinary graduated cylinders can be used for these measurements, but for highly precise work accurately calibrated glassware is used, as shown in Figure 4.4. (CAUTION: When highly concentrated acids such as sulfuric acid are diluted, the mixing *must* begin by adding the acid *to water;* never add water to a concentrated acid. Otherwise, so much heat may be generated so quickly that the solution will boil and spatter you with acid!)

Many common acids are sold and shipped in very concentrated solutions that are then diluted.

FIGURE 4.4 Preparing solutions by dilutions. (*a*) The calculated volume of the more concentrated solution is withdrawn from the stock solution using a pipet. (*b*) It is allowed to flow into a volumetric flask of the correct capacity. (*c*) and (*d*) Water (or other solvent) is added as the contents are shaken until the final volume reaches the etched line. Then the solution is transferred to a dry reagent bottle for storage.

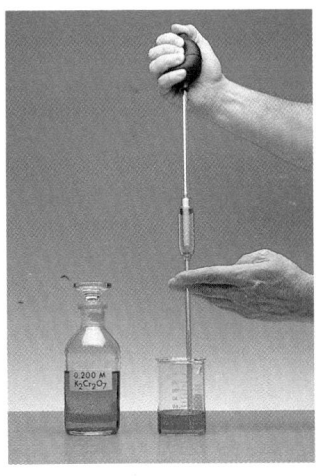
(a)

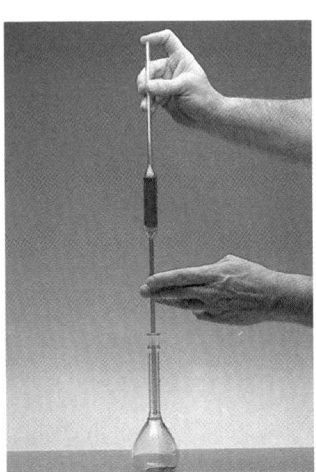
(b)

(c)

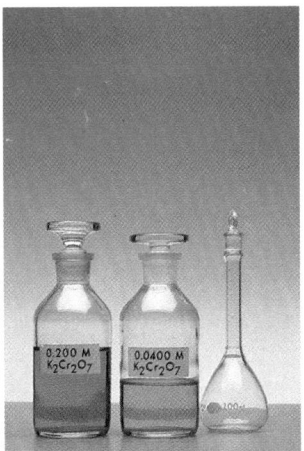
(d)

Dilution Calculations

All calculations having to do with dilution are based on the same simple fact: As a solution is diluted, the number of moles of solute doesn't change; the solute is simply spread through a larger volume. In the previous section, you learned that you can calculate the moles of solute in a solution by multiplying the solution's volume (in liters) by its molarity.

$$\text{volume (L)} \times \text{molarity} = \text{moles of solute}$$

$$\text{liters} \times \frac{\text{moles}}{\text{liter}} = \text{moles of solute}$$

Since the number of moles of solute doesn't change during a dilution, then it must be true that the product of the initial volume and molarity (before dilution) is equal to the product of the final volume and molarity (after dilution). If we let V_i and M_i stand for the initial volume and molarity, and V_f and M_f stand for the final volume and molarity, we can write a simple equation that applies to dilution problems:

$$V_i \times M_i = V_f \times M_f \tag{4.1}$$

One of the interesting things about this equation is that the units for volume are the same on both sides of this equation, so they always cancel. This means that the volume unit doesn't have to be in liters, which is often an awkward unit for ordinary laboratory work. The milliliter, or any other volume unit that pleases us, can be used, as long as the same unit is used on both sides of the equation.

EXAMPLE 4.16   DOING THE CALCULATIONS FOR MAKING DILUTIONS

Problem: How can we prepare 100 mL of 0.0400 M $K_2Cr_2O_7$ from 0.200 M $K_2Cr_2O_7$?

Solution: We first assemble the data.

$$V_i = ? \qquad V_f = 100 \text{ mL}$$
$$M_i = 0.200\ M \qquad M_f = 0.0400\ M$$

Next, we use Equation 4.1.

$$V_i \times 0.200\ M = 100 \text{ mL} \times 0.0400\ M$$
$$V_i = \frac{100 \text{ mL} \times 0.0400\ M}{0.200\ M}$$
$$V_i = 20.0 \text{ mL}$$

This answer means that we would place 20.0 mL of 0.200 M $K_2Cr_2O_7$ into a 100-mL volumetric flask and then add water until the final volume is 100 mL. (See Figure 4.4.)

PRACTICE EXERCISE 24   Describe how to make 500 mL of 0.20 M NaOH from 0.50 M NaOH.

Many times in the lab we do not need the kind of precision made possible by using pipets and volumetric flasks when we make a dilute solution from one more concentrated. We can get along with simpler equipment by taking advantage of an important fact. *When water is added to a dilute aqueous solution — 1 M or less — the volumes are additive,* at least to one and often two significant figures in the final concentration. When such precision is acceptable, the calculation for a dilution problem involves finding the volume of water to be added to the concentrated solution. We will work an example to illustrate how to do this.

EXAMPLE 4.17   DOING CALCULATIONS FOR MAKING DILUTIONS

Problem: How many milliliters of water would have to be added to 100 mL of 0.40 *M* HCl to give a solution with a concentration of 0.10 *M*?

Solution: First let's assemble the data:

$$V_i = 100 \text{ mL} \qquad V_f = ?$$
$$M_i = 0.40 \text{ } M \qquad M_f = 0.10 \text{ } M$$

Now we can use Equation 4.1.

$$100 \text{ mL} \times 0.40 \text{ } M = V_f \times 0.10 \text{ } M$$
$$V_f = 400 \text{ mL}$$

Now comes the tricky part of the solution. The *final* volume must be 400 mL, but we are starting with 100 mL. We must therefore add an additional 300 mL of water.

PRACTICE EXERCISE 25   How many milliliters of water would have to be added to 300 mL of 0.5 *M* NaOH to give a solution with a concentration of 0.2 *M*?

SUMMARY

Stoichiometry.   Because equal numbers of moles contain equal numbers of molecules, the coefficients in a balanced equation provide the key to calculating the quantities of chemicals involved in a reaction. At one level, these numbers give ratios in terms of molecules (or other kinds of formula units). For laboratory work, they also give the same ratios in terms of moles. When balancing an equation, only the coefficients can be adjusted; the subscripts within the formulas cannot be changed.

Mole Relationships.   If x is the coefficient of substance A and y is the coefficient of substance B in some equation, for example,

$$xA + mW + \cdots \longrightarrow yB + nZ + \cdots$$

then the relationship of A to B in terms of moles is given by either of two conversion factors.

$$\frac{x \text{ mol } A}{y \text{ mol } B} \quad \text{or} \quad \frac{y \text{ mol } B}{x \text{ mol } A}$$

Regardless of the units in which substance A is taken — grams or volume of some solution — we can find the quantity of substance B that is related to A by this equation only by working the problem at the mole level. Any other units must first be converted to the number of moles. See the figure at the right.

Yields of Products.   When one of two or more reactants is taken in a mole quantity less than the stoichiometry requires, it is a limiting reactant and the calculation of yield is based on this chemical. Side reactions or the failure of some of a reactant to react are reasons why the actual yield can be less than theoretical.

Stoichiometry When Reactants Are in Solution.   Solutions provide a fluid medium that allows particles to collide and react.

Several common qualitative terms are used to describe concentrations of solutions (the ratio of solute to solution volume) — *dilute, concentrated, saturated, unsaturated, supersaturated* — but for calculating the mole quantities of reactants, the *molar concentration* is the best description. Molar concentration describes the moles of solute per liter of solution. In working mole problems that involve molar concentrations, always write out the units of molarity (including the formula of the solute). This is what "translate the label" means. The general equation for calculating how to make a dilute solution from a concentrated solution is

$$V_i \times M_i = V_f \times M_f$$

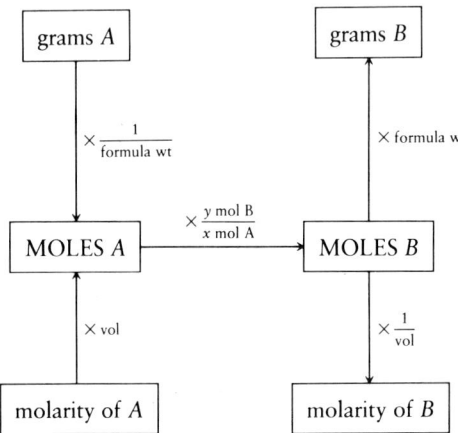

REVIEW EXERCISES

Answers to questions whose numbers are printed in color are given in Appendix C.
Difficult questions are marked with asterisks.

Balanced Equations

4.1   Why is a balanced equation important in experimental work?

4.2   To balance an equation, which numbers do we change, the subscripts of the formulas or their coefficients?

4.3   When sulfur trioxide (SO_3), which is present in smoggy air in trace concentrations, reacts with water, sulfuric acid (H_2SO_4), a very corrosive acid, forms as the only product. Write a balanced equation for this reaction and describe its stoichiometry in words.

4.4   Write the equation that expresses in acceptable chemical shorthand the information given in the statement, "Iron can be made to react with molecular oxygen to give iron oxide having the formula Fe_2O_3."

4.5   Balance the following equations.
(a) $Ca(OH)_2 + HCl \longrightarrow CaCl_2 + H_2O$
(b) $AgNO_3 + CaCl_2 \longrightarrow Ca(NO_3)_2 + AgCl$
(c) $Fe_2O_3 + C \longrightarrow Fe + CO_2$
(d) $NaHCO_3 + H_2SO_4 \longrightarrow Na_2SO_4 + H_2O + CO_2$
(e) $C_4H_{10} + O_2 \longrightarrow CO_2 + H_2O$

4.6   Balance the following equations.
(a) $SO_2 + O_2 \longrightarrow SO_3$
(b) $P_4O_{10} + H_2O \longrightarrow H_3PO_4$
(c) $Pb(NO_3)_2 + Na_2SO_4 \longrightarrow PbSO_4 + NaNO_3$
(d) $Fe_2O_3 + H_2 \longrightarrow Fe + H_2O$
(e) $Al + H_2SO_4 \longrightarrow Al_2(SO_4)_3 + H_2$

4.7   Balance the following equations.
(a) $Mg(OH)_2 + HBr \longrightarrow MgBr_2 + H_2O$
(b) $HCl + Ca(OH)_2 \longrightarrow CaCl_2 + H_2O$
(c) $Al_2O_3 + H_2SO_4 \longrightarrow Al_2(SO_4)_3 + H_2O$
(d) $KHCO_3 + H_3PO_4 \longrightarrow K_2HPO_4 + H_2O + CO_2$
(e) $C_9H_{20} + O_2 \longrightarrow CO_2 + H_2O$

4.8   Balance the following equations.
(a) $CaO + HNO_3 \longrightarrow Ca(NO_3)_2 + H_2O$
(b) $Na_2CO_3 + Mg(NO_3)_2 \longrightarrow MgCO_3 + NaNO_3$
(c) $(NH_4)_3PO_4 + NaOH \longrightarrow Na_3PO_4 + NH_3 + H_2O$
(d) $LiHCO_3 + H_2SO_4 \longrightarrow Li_2SO_4 + H_2O + CO_2$
(e) $C_4H_{10}O + O_2 \longrightarrow CO_2 + H_2O$

Stoichiometric Calculations

4.9   What important fact or principle makes it possible for us to think of coefficients either as representing the numbers of individual molecules (or atoms or ions, as the case may be) of substances or as representing the numbers of *moles* of substances?

4.10   If an experimenter carelessly thinks that the coefficients of a balanced equation give proportions by masses, and carries out an experiment on that basis, there will usually be an insufficient amount of one reactant. What in general terms is this reactant called?

4.11   If two substances react completely on both a 1-to-1 ratio by mass and a 1-to-1 ratio by moles, what must be true about these substances?

4.12   A mixture of 0.1 mol of magnesium and 0.1 mol of chlorine reacted completely to form magnesium chloride according to the equation $Mg + Cl_2 \longrightarrow MgCl_2$.
(a) What information describes the *stoichiometry* of this reaction?
(b) What information gives the *scale* of the reaction?

4.13   A chemist describes a particular experiment in this way: "0.0400 mol of H_2O_2 decomposed into 0.0400 mol of H_2O and 0.0200 mol O_2." Express the chemistry of this reaction by a conventional equation.

4.14   The octane present in gasoline burns according to the following equation.

$$2C_8H_{18} + 25O_2 \longrightarrow 16CO_2 + 18H_2O$$
octane

(a) How many moles of O_2 are needed to react fully with 4 mol of octane?
(b) How many moles of CO_2 can form from 1 mol of octane?
(c) How many moles of water are produced by the combustion of 6 mol of octane?
(d) If this reaction is to be used to synthesize 8 mol of CO_2, how many moles of oxygen are needed? How many moles of octane?

4.15   The alcohol in "gasohol" burns according to the following equation.

$$C_2H_6O + 3O_2 \longrightarrow 2CO_2 + 3H_2O$$

(a) If 25 mol of ethyl alcohol burns this way, how many moles of oxygen are needed?
(b) If 30 mol of oxygen is consumed by this reaction, how many moles of alcohol are used up? How many moles of carbon dioxide are formed?
(c) In one test, 23 mol of carbon dioxide was produced by this reaction. How many moles of oxygen were consumed?
(d) In another test, 41 mol of water is collected from this reaction. How many moles of alcohol had been burned? How many moles of oxygen were used up? How many moles of CO_2 also formed?

4.16   The combustion of a sample of butane, C_4H_{10} (lighter fluid), produced 2.46 g water.

$$2C_4H_{10} + 13O_2 \longrightarrow 8CO_2 + 10H_2O$$

(a) How many moles of water formed?
(b) How many moles of butane burned?
(c) How many grams of butane burned?
(d) How much oxygen was used up in moles? In grams?

4.17 One way to change iron ore, Fe_2O_3, into metallic iron is to heat it together with hydrogen.

$$Fe_2O_3 + 3H_2 \longrightarrow 2Fe + 3H_2O$$

(a) How many moles of iron are made from 25 mol of Fe_2O_3?

(b) How many moles of hydrogen are needed to make 30 mol of Fe?

(c) If 120 mol of H_2O forms, how many grams of Fe_2O_3 were used up?

4.18 Terephthalic acid, an important raw material for making Dacron, a synthetic fiber, is made from *para*-xylene by the following reaction.

$$\underset{\substack{para- \\ xylene}}{C_8H_{10}} + 3O_2 \xrightarrow{\text{(special conditions)}} \underset{\substack{\text{terephthalic} \\ \text{acid}}}{C_8H_6O_4} + 2H_2O$$

How much terephthalic acid could be made from 154 g of *para*-xylene in moles? In grams?

4.19 Adipic acid, a raw material for nylon, is made industrially by the oxidation of cyclohexane.

$$5O_2 + \underset{\text{cyclohexane}}{2C_6H_{12}} \xrightarrow{\text{(special conditions)}} \underset{\text{adipic acid}}{2C_6H_{10}O_4} + 2H_2O$$

(a) How many moles of oxygen would be needed to make 40.0 mol of adipic acid by this reaction?

(b) If 164 g of cyclohexane is used, what is the theoretical yield of adipic acid in moles? In grams?

4.20 The Solvay process is used to make sodium carbonate, Na_2CO_3, a chemical that ranked 11th among all chemicals in annual production in 1986. The process begins with the passing of ammonia and carbon dioxide through a solution of sodium chloride. This makes sodium bicarbonate and ammonium chloride:

$$H_2O + NaCl + NH_3 + CO_2 \longrightarrow NH_4Cl + NaHCO_3$$

In the next step, sodium bicarbonate is heated to make sodium carbonate.

$$2NaHCO_3 \longrightarrow Na_2CO_3 + CO_2 + H_2O$$

(a) How many moles of sodium carbonate could, in theory, be made from 100 mol of NaCl?

(b) What is the theoretical yield of sodium carbonate from 546 g of NaCl, in moles? In grams?

4.21 Rock phosphate, a mineral so important as a source of fertilizer for the world's food supply that it's called white gold, is a mixture of calcium phosphate, $Ca_3(PO_4)_2$, calcium hydroxide, $Ca(OH)_2$, and calcium fluoride, CaF_2. Some references give $Ca_3(PO_4)_2 \cdot Ca(OH)_2$ as a sufficiently accurate formula. Where calcium fluoride is prominent, the formula is given as $Ca_3(PO_4)_2 \cdot CaF_2$. [The centered dot separating $Ca_3(PO_4)_2$ from CaF_2 is a standard technique used in writing the formula of a complex substance when somewhat distinct chemicals are bound in a stoichiometric ratio.] Ordinary "superphosphate" fertilizer is made from phosphate rock by the action of sulfuric acid.

$$3H_2SO_4 + Ca_3(PO_4)_2 \cdot Ca(OH)_2 \longrightarrow$$
$$\underset{\text{phosphate rock}}{}$$
$$Ca(H_2PO_4)_2 + 3CaSO_4 + 2H_2O$$

(a) How much sulfuric acid (in moles and in metric tons) is needed to convert 125 metric tons of phosphate rock (1 metric ton $= 10^6$ g) into "superphosphate" by this equation?

(b) The mixture of calcium dihydrogen phosphate and calcium sulfate produced by this equation (plus some water) is marketed as the "superphosphate." The 125-metric-ton batch (part a) would produce how much calcium dihydrogen phosphate (in metric tons) and how much calcium sulfate (in the same unit)?

4.22 The chief process for converting iron ore, Fe_2O_3, into iron involves the combined action of coal and oxygen. Partial combustion of the carbon in coal gives carbon monoxide.

$$2C + O_2 \longrightarrow 2CO$$

In a series of steps, CO acts on Fe_2O_3, with the overall result

$$Fe_2O_3 + 3CO \longrightarrow 2Fe + 3CO_2$$

(a) In an experiment done on a very small scale to test the efficiency of new approaches, a sample of 324 g of Fe_2O_3 was converted to iron. How much iron, in theory, could form from this sample, in moles? In grams?

(b) The actual yield of iron in this test was 198 g. What was the percentage yield?

4.23 The Synthane process, developed by the U.S. Bureau of Mines for coal gasification, converts the carbon in coal into methane, CH_4, by first heating the pulverized coal with steam and oxygen and then heating the resulting mixture of gases with carbon monoxide and hydrogen. A simple statement of the overall process is

$$C + 2H_2 \longrightarrow CH_4$$

A feed of 13,800 g of coal yields 580 g of methane.

(a) If the coal were pure carbon, what would be the percentage yield?

(b) Recalculate the percentage conversion if subbituminous coal with 44.0% (w/w) carbon is used. [Note: "44.0% (w/w)" means 44.0 g of carbon per 100 g of coal; "w/w" refers to a weight-to-weight percentage.)

4.24 One way to make phosphoric acid, H_3PO_4, a chemical needed to make "triple superphosphate" fertilizer from phosphate rock, is to boil a solution of tetraphosphorus decaoxide, P_4O_{10}, in water.

$$P_4O_{10} + 6H_2O \longrightarrow 4H_3PO_4$$

(a) In a small-scale laboratory experiment, 88.6 g of P_4O_{10} was changed to phosphoric acid in this way. How much phosphoric acid formed in moles? In grams?

(b) If the experiment were done so that the final volume was 500 mL, what was the molarity of the phosphoric acid solution produced?

(c) In 1986, 18.41×10^9 lb of phosphoric acid was made in the United States. If all of it had been made by this reaction (which is untrue), how much tetraphosphorus decaoxide would have been needed, in moles? In metric tons?

4.25 When fuels containing sulfur are burned, the gases leaving the furnace for the smokestack contain SO_2 from the reaction

$$S + O_2 \longrightarrow SO_2$$

Unless emissions of SO_2 are controlled, bodies of water downwind will slowly become more acidic. One technique for removing SO_2 from smokestack gases is to let the gases interact with wet limestone, $CaCO_3$. The net effect is

$$SO_2 + CaCO_3 + \text{other chemicals} \longrightarrow$$
$$CaSO_3 + \text{other products}$$

Roughly 400 million metric tons (400×10^6 metric tons) of coal is burned each year to make electricity in the United States. If the average sulfur content is 2.5 g S/100 g coal, then
(a) How many metric tons of sulfur are in 400 million metric tons of coal?
(b) How much sulfur dioxide can form from this sulfur in moles? In metric tons? (1 metric ton = 10^6 g)
(c) How many metric tons of $CaSO_3$ could be made?
(d) This much $CaSO_3$ represents a solid-waste disposal problem. To get an idea of its magnitude, if all of it were loaded into railroad coal cars, each carrying 100 metric tons of $CaSO_3$, how many cars would be needed to haul it away?

4.26 When we convert such naturally occurring sulfide ores as galena (lead sulfide, PbS) or covellite (copper sulfide, CuS) to their respective metals, the first step is heating them in air. The process is called roasting the ore, and it changes the sulfides to their corresponding oxides—PbO or CuO in our examples. If we let M represent Pb or Cu, we can represent the roasting of both ores as follows.

$$2MS + 3O_2 \longrightarrow 2MO + 2SO_2$$

In one test operation, 543 g of a sulfide—either PbS or CuS—gave 452 g of the corresponding oxide. Assuming that this was the theoretical yield, which sulfide had been used? Show your calculations.

4.27 Although mercury is a metal, at room temperature it is a liquid, and it has a density of 13.6 g/mL. How many kilograms of mercuric sulfide, HgS, theoretically are needed to prepare one *commercial flask*. This standard unit contains 2.55 L of mercury. The reaction used to prepare this sample is as follows.

$$4HgS + 4CaO \longrightarrow 4Hg + 3CaS + CaSO_4$$

4.28 An unknown compound consisting only of carbon and hydrogen was burned in pure oxygen to give carbon dioxide, CO_2, and water H_2O. No side reactions took place. The

carbon dioxide was quantitatively trapped by the following reaction.

$$CO_2(g) + CaO(s) \longrightarrow CaCO_3(s)$$

When a sample of the unknown compound with a mass of 5.334 mg was processed by these steps, there was obtained 37.39 mg of $CaCO_3$ and 7.570 mg of water. What was the empirical formula of the unknown compound? (Use atomic weights to four significant figures.) Its formula weight was found to be 114. What was its molecular formula?

Limiting Reactants and Percentage Yields

4.29 Powdered aluminum and iron(III) oxide react with a large evolution of heat (per mole) according to the following equation.

$$2Al + Fe_2O_3 \longrightarrow Al_2O_3 + 2Fe$$

In one experiment, 3.00 mol of Al was mixed with 1.25 mol of Fe_2O_3.
(a) Which reactant, if either, was the limiting reactant?
(b) Calculate the theoretical yield (in moles) of iron.

4.30 Phosphorus pentachloride reacts with water to give phosphoric acid and hydrogen chloride according to the following equation.

$$PCl_5 + 4H_2O \longrightarrow H_3PO_4 + 5HCl$$

In one experiment, 0.600 mol of PCl_5 was slowly added to 4.80 mol of water.
(a) Which reactant, if either, was the limiting reactant?
(b) Calculate the theoretical yields (in moles) of H_3PO_4 and HCl.

4.31 Zinc and sulfur react to form zinc sulfide according to the equation

$$Zn + S \longrightarrow ZnS$$

If 25.0 g of zinc and 30.0 g of sulfur are mixed,
(a) Which chemical is the limiting reactant?
(b) How many grams of ZnS will be formed?
(c) How many grams of the excess reactant will remain after the reaction is over?

4.32 Silver nitrate, $AgNO_3$, reacts with ferric chloride, $FeCl_3$, to give silver chloride, AgCl, and ferric nitrate, $Fe(NO_3)_3$. In a particular experiment, it was planned to mix a solution containing 25.0 g of $AgNO_3$ with another solution containing 45.0 g of $FeCl_3$.
(a) Write the chemical equation for the reaction.
(b) Which reactant is the limiting reactant?
(c) What is the maximum number of moles of AgCl that could be obtained from this mixture?
(d) What is the maximum number of grams of AgCl that could be obtained?
(e) How many grams of the reactant in excess will remain after the reaction is over?

4.33 In Exercise 4.25 it was stated that solid calcium carbonate, $CaCO_3$, is able to remove sulfur dioxide from waste gases by the reaction

$$CaCO_3 + SO_2 + \text{other reactants} \longrightarrow$$
$$CaSO_3 + \text{other products}$$

In a particular experiment, 255 g of $CaCO_3$ was exposed to 135 g of SO_2 in the presence of an excess amount of the other chemicals required for the reaction.
(a) What was the theoretical yield of $CaSO_3$?
(b) If only 198 g of $CaSO_3$ was isolated from the products, what was the percentage yield of $CaSO_3$ in this experiment?

*4.34 A research supervisor told a chemist to make 100 g of chlorobenzene from the reaction of benzene with chlorine and to expect a yield no higher than 65%. What is the minimum quantity of benzene that can give 100 g of chlorobenzene if the yield is 65%? The equation for the reaction is

$$C_6H_6 + Cl_2 \xrightarrow{\text{(special conditions)}} C_6H_5Cl + HCl$$
benzene   chlorobenzene

*4.35 Certain salts of benzoic acid have been used as food additives for decades. The potassium salt of benzoic acid, potassium benzoate, can be made by the action of potassium permanganate on toluene.

$$C_7H_8 + 2KMnO_4 \longrightarrow KC_7H_5O_2 + 2MnO_2 + KOH + H_2O$$
toluene   potassium benzoate

If the yield of potassium benzoate cannot realistically be expected to be more than 68%, what is the minimum number of grams of toluene needed to achieve this yield while producing 10.0 g of $KC_7H_5O_2$?

Solutions

4.36 Can a saturated solution be dilute? Use data in Table 4.1 to cite an example.

4.37 Can a concentrated solution be unsaturated? Using Table 4.1, supply an illustration.

4.38 In what way is the concept of *concentration* different from that of *solubility*?

4.39 A sample of lead sulfate ($PbSO_4$) with a mass of 1.25 g was shaken for an hour with 20 g of water at 25 °C.
(a) What is the final concentration of lead sulfate in g/100 g H_2O?
(b) Is this solution unsaturated or saturated? (Use data in Table 4.1.)
(c) Is the solution concentrated or dilute?
(d) What would happen if you added 1.00 g more of lead sulfate?

4.40 A solution was prepared by mixing 100 mL of methyl alcohol with 100 mL of water. Which is the solvent and which is the solute, or doesn't it matter? Explain. How would most people answer a question such as this if 1 mL of water were added to 100 mL of methyl alcohol?

Molarity

4.41 Write out in your own words all the information given on the following label found on a reagent bottle: "0.500 *M* NaCl."

4.42 Does the label "0.500 *M* NaCl" tell how much solution is in the bottle?

4.43 Does the label "0.500 *M* NaCl" tell how much water was used to prepare, say, 1 L of the solution?

4.44 Why is it unimportant from the standpoint of mole problems to know exactly how much *solvent* was used to prepare a given quantity of a solution having some specific molar concentration?

4.45 What is the difference between "1 mol NaCl" and "1 *M* NaCl"?

4.46 What is the difference between a molecule, a mole, and a molar concentration?

4.47 Suppose you prepare a 0.200 *M* salt solution in a 250-mL volumetric flask and then accidentally spill some of it. What happens to the concentration of the solution in the bottle?

4.48 Chemical handbooks have extensive tables giving to several significant figures the density of pure water at each degree Celsius between its freezing and boiling points. Well-equipped laboratories have constant-temperature baths, which are large vats of water that can be kept at a given temperature to within at least 0.1 degree. (Some can provide much more precise control.) Given enough time, anything put into a constant-temperature bath comes to the bath temperature. These labs, of course, have high-precision analytical balances. Describe in your own words how you could use the facilities of such a laboratory to check the calibration of a volumetric flask with the legend "100 mL at 20 °C" etched on its side.

4.49 How many millimoles make one mole?

4.50 Prove that the numerical value of a concentration given in moles per liter is the same as if it were stated in millimoles per milliliter.

4.51 Calculate the number of grams of each solute that has to be taken to make each of the following solutions.
(a) 250 mL of 0.100 *M* NaCl
(b) 100 mL of 0.440 *M* $C_6H_{12}O_6$ (glucose)
(c) 500 mL of 0.500 *M* H_2SO_4

4.52 How much solute, in grams, is needed to make each of the following solutions?
(a) 250 mL of 0.00100 *M* Na_2SO_4
(b) 100 mL of 0.250 *M* Na_2CO_3
(c) 500 mL of 0.400 *M* NaOH

4.53 What is the molarity of pure water?

4.54 What is the molarity of pure sulfuric acid? Its density is 1.94 g/mL.

4.55 Concentrated ammonia contains 26 g NH_3 per 100 mL of solution. What is its molarity?

4.56 What is the molarity of a solution of sodium chloride that contains 26.00 g NaCl/100.0 g solution? Its density is 1.199 g/mL.

4.57 What is the molarity of a solution of acetic acid, $HC_2H_3O_2$, in water (vinegar) that contains 5.50 g acetic acid per 100.0 g water? The density of this solution is 1.0077 g/mL.

Reactions in Solution

4.58 For an experiment, a student needed freshly made calcium carbonate, which can be made by mixing solutions of calcium chloride and potassium carbonate.

$$CaCl_2(aq) + K_2CO_3(aq) \longrightarrow CaCO_3(s) + 2KCl(aq)$$

The reactants were available as "0.500 M $CaCl_2$" and "0.750 M K_2CO_3."
(a) If the yield is 100%, what volumes of these solutions (in milliliters) should be mixed to produce 12.0 g of $CaCO_3$?
(b) If the yield is only 92%, what volumes have to be mixed to get 12.0 g of $CaCO_3$?

4.59 If some sulfuric acid is spilled on a lab bench, a safe way to neutralize it is to sprinkle solid sodium bicarbonate on it. The following reaction destroys the acid.

$$2NaHCO_3 + H_2SO_4 \longrightarrow Na_2SO_4 + 2CO_2 + 2H_2O$$

Bicarbonate is added and stirred into the acid with a glass rod until the CO_2 stops fizzing out.
(a) If 40 mL of 6.0 M H_2SO_4 is spilled, will 25 g of $NaHCO_3$ be enough?
(b) If not, what percentage of the acid will be neutralized?

4.60 A 0.321-g sample of sodium carbonate, which was contaminated by sodium chloride, was dissolved in water, and the resulting solution required 35.4 mL of 0.144 M HCl to react completely with the sodium carbonate it contained. The reaction is

$$2HCl(aq) + Na_2CO_3(aq) \longrightarrow 2NaCl(aq) + CO_2(g) + H_2O$$

(The impurity in the sample does not interfere with this analysis.) How much Na_2CO_3 was present in the sample in grams? What was the percentage purity of Na_2CO_3 in the sample?

4.61 What is the molar concentration of a sulfuric acid solution if 15.46 mL is neutralized by 33.48 mL of 0.1048 M NaOH as follows.

$$2NaOH(aq) + H_2SO_4(aq) \longrightarrow Na_2SO_4(aq) + 2H_2O$$

*4.62 Sodium sulfate, Na_2SO_4, reacts in an aqueous solution with lead nitrate, $Pb(NO_3)_2$, as follows.

$$Na_2SO_4(aq) + Pb(NO_3)_2(aq) \longrightarrow PbSO_4(s) + 2NaNO_3(aq)$$

The lead sulfate that forms is essentially insoluble in water, as data in Table 4.1 show, while sodium nitrate is very soluble and remains in solution. In one analysis, 35.3 mL of a solution of sodium sulfate reacted exactly with 32.5 mL of a lead nitrate solution. The lead sulfate that precipitated was collected, dried, and weighed, and it was found to have a mass of 1.13 g. There were no side reactions. What were the molar concentrations of the original solutions of sodium sulfate and lead nitrate?

*4.63 In aqueous solution, magnesium chloride, $MgCl_2$, reacts cleanly without side reactions with silver nitrate, $AgNO_3$, as follows.

$$MgCl_2(aq) + 2AgNO_3(aq) \longrightarrow 2AgCl(s) + Mg(NO_3)_2(aq)$$

The newly formed silver chloride, AgCl, is very insoluble in water, but the other product, $Mg(NO_3)_2$, remains in solu-

tion. In one experiment, the solute in 25.8 mL of silver nitrate solution reacted exactly with the solute in 19.5 mL of magnesium chloride solution. The newly formed silver chloride was collected, dried, and weighed, and it was found to have a mass of 0.696 g. What were the molar concentrations of the solutions of the original reactants?

4.64 Magnesium, calcium, and zinc all react with hydrochloric acid as follows (where M represents any of these metallic elements).

$$M(s) + 2HCl(aq) \longrightarrow MCl_2(aq) + H_2(g)$$

After a piece of one of these metals had reacted completely with the acid in 55.8 mL of 1.24 M HCl, the resulting solution was evaporated to dryness. The residue of MCl_2 had a mass of 4.72 g. Which metal had been used?

4.65 An unknown solution was prepared by dissolving 2.833 g of either $CaCl_2$ or $FeCl_3$ in water to make 250.0 mL of solution. When 14.35 mL of this solution was mixed with 0.1184 M $AgNO_3$, the reaction was complete when 24.76 mL of the $AgNO_3$ solution was added. What was the solute? The possible reactions are as follows.

$$CaCl_2(aq) + 2AgNO_3(aq) \longrightarrow 2AgCl(s) + Ca(NO_3)_2(aq)$$
$$FeCl_3(aq) + 3AgNO_3(aq) \longrightarrow 3AgCl(s) + Fe(NO_3)_3(aq)$$

4.66 The chemical process for making photographic film involves the chemical conversion of metallic silver to silver nitrate by the following reaction.

$$3Ag + 4HNO_3 \longrightarrow 3AgNO_3 + NO + 2H_2O$$

In one operation a block of silver with the dimensions $40.0 \times 40.0 \times 100.0$ cm was dissolved in concentrated nitric acid. (The density of silver is 10.5 g/cm³.)
(a) What is the theoretical yield of silver nitrate in kilograms?
(b) How many liters of 16.0 M HNO_3 are needed, in theory?

4.67 Concentrated nitric acid is 16 M HNO_3. How many milliliters would you need to prepare 100 mL of 0.50 M HNO_3?

4.68 Concentrated acetic acid is 17.4 M $HC_2H_3O_2$. How many milliliters would you need to prepare 1.00 L of 1.00 M acetic acid?

4.69 (a) How many milliliters of water would have to be added to 150 mL of 0.450 M KCl solution to give a solution whose concentration is 0.100 M KCl? (Assume volumes are additive.)
(b) If 60.0 mL of water was added to 80.0 mL of 0.500 M Na_2CO_3, what would be the final molar concentration of the solute?

4.70 How many milliliters of water would have to be added to 250 mL of 0.20 M HCl to make the final concentration 0.10 M?

4.71 If 150 mL of water is added to 100 mL of 0.15 M Na_2CO_3, what is the final molar concentration of Na_2CO_3?

*4.72 A solution with a concentration of 10 g NaOH per liter (of solution) was diluted by the addition of 500 mL of water. The final concentration was found to be 0.15 M NaOH. What was the volume of the initial solution in milliliters?

TEST OF FACTS AND CONCEPTS: CHAPTERS 1–4

A number of fundamental concepts and problem-solving skills that were studied in the preceding chapters will carry forward into the rest of this book. It would be a good idea to pause here and see how well you have grasped the concepts, how familiar you are with important terms, and how able you are at working chemical problems. Don't be discouraged if some of the problems turn out to be difficult. Most students of chemistry have experienced the frustration of initially not remembering what had only one or two weeks earlier seemed clear and straightforward. But if such frustrations take you back to a review, you will also experience what so many others have — what was once learned well (but since forgotten) comes back very quickly.

Some of the problems require data or other information found in tables in this book, including those inside the covers. Freely use these tables as needed.

1. A rectangular box was found to be 24.6 cm wide, 0.35140 m high, and 7,424 mm deep.
   (a) How many significant figures are in each measurement?
   (b) Calculate the volume of the box in units of cm^3. Be sure to express your answer to the correct number of significant figures.
   (c) Use the answer in part b to calculate the volume of the box in cubic feet.
   (d) Suppose the box was solid and composed entirely of zinc, which has a specific gravity of 7.140. What would be the mass of the box in kilograms?

2. The atoms of an isotope of plutonium, Pu, each contain 94 protons, 110 neutrons, and 94 electrons. Write a symbol for this element that incorporates its mass number and atomic number.

3. An atom of an isotope of nickel has a mass number of 60. How many protons, neutrons, and electrons are in this atom?

4. A solution was found to contain particles consisting of 12 neutrons, 10 electrons, and 11 protons. Write the chemical symbol for this particle, consulting the periodic table as needed.

5. Classify each either as a *substance* or as a *particle* of which a substance can consist.
   (a) Ion   (e) Compound
   (b) Mixture   (f) Molecule
   (c) Isotope   (g) Element
   (d) Atom   (h) Nucleus

6. Is it possible to have a visible sample of each? If not, explain.
   (a) An isotope of iron
   (b) An atom of iron
   (c) A molecule of water
   (d) A mole of water
   (e) An ion of sodium
   (f) A formula unit of sodium chloride

7. Make a sketch of the general shape of the modern periodic table and mark off those areas where we find the metals, metalloids, and nonmetals.

8. Which of the following elements would most likely be found together in nature: Ca, Hf, Sn, Cu, Zr?

9. Match an element on the left with a description on the right.
   Calcium   Halogen
   Iron   Noble gas
   Helium   Alkali metal
   Gadolinium   Alkaline earth metal
   Iodine   Transition metal
   Sodium   Inner transition metal

10. Define *ductile* and *malleable*.

11. Which metal is a liquid at room temperature? Which metal has the highest melting point?

12. What is the most important property that distinguishes a metalloid from a metal or a nonmetal?

13. Give the symbols of the post-transition metals.

14. Give chemical formulas for the following.
    (a) potassium nitrate
    (b) calcium carbonate
    (c) cobalt(II) phosphate
    (d) lithium hydrogen sulfate
    (e) magnesium sulfite
    (f) iron(III) bromide
    (g) magnesium nitride
    (h) aluminum selenide
    (i) cupric perchlorate
    (j) disodium hydrogen phosphate
    (k) nitrous acid
    (l) hypoiodous acid
    (m) barium bisulfite
    (n) bromine pentafluoride
    (o) dinitrogen pentaoxide
    (p) strontium acetate
    (q) ammonium dichromate
    (r) copper(I) sulfide

15. Give chemical names for the following.
    - (a) $NaClO_3$
    - (b) HIO_3
    - (c) $Ca(H_2PO_4)_2$
    - (d) $NaMnO_4$
    - (e) AlP
    - (f) ICl_3
    - (g) PCl_3
    - (h) $HC_2H_3O_2$
    - (i) K_2CrO_4
    - (j) $HOCl$
    - (k) $Ca(CN)_2$
    - (l) $MnCl_2$
    - (m) $NaNO_2$
    - (n) $Fe(HSO_4)_3$

16. Write a chemical equation for the complete neutralization of H_3PO_4 by NaOH.

17. A certain compound is hard, brittle, has a high melting point, and conducts electricity when melted but not when solid. What kind of compound is this?

18. Why do we always write empirical formulas for ionic compounds?

19. Which ion exists in abundance in all solutions of strong acids?

20. Which ion makes a solution basic?

21. Define *monoprotic acid, diprotic acid,* and *polyprotic acid.* What is the general definition of a *salt*?

22. Which of the following oxides are acidic and which are basic: P_4O_6, Na_2O, SeO_3, CaO, PbO, SO_2?

23. Which of the following are binary substances: Al_2O_3, Cl_2, MgO, NO_2, $HClO_4$?

24. Write the formulas of any acid salts formed by the reaction of the following acids with potassium hydroxide.
    - (a) sulfurous acid
    - (b) nitric acid
    - (c) hypochlorous acid
    - (d) phosphoric acid
    - (e) carbonic acid

25. The formula weight of a substance is 60.2. Therefore, what is the mass in grams of one of its molecules?

26. A sample of a compound with a mass of 204 g consists of 1.00×10^{23} molecules. What is its formula weight?

27. Calculate the formula weight of $Fe_4[Fe(CN)_6]_3$.

28. How many grams of copper(II) nitrate trihydrate, $Cu(NO_3)_2 \cdot 3H_2O$, are present in 0.118 mol?

29. How many moles of nickel(II) iodide hexahydrate, $NiI_2 \cdot 6H_2O$, are in a sample of 15.7 g of this compound?

30. A sample of 0.5866 g of nicotine was analyzed and found to consist of 0.4343 g C, 0.05103 g H, and 0.1013 g N. Calculate the percentage composition of nicotine.

31. A compound of potassium had the following percentage composition: K, 37.56%; H, 1.940%; P, 29.79%. The rest was oxygen. Calculate the empirical formula of this compound (arranging the atomic symbols in the order K H P O.)

32. How many molecules of ethyl alcohol, C_2H_5OH, are in 1.00 oz? The density of ethyl alcohol is 0.798 g/mL (1 oz = 29.6 mL).

33. What volume in liters is occupied by a sample of ethylene glycol, $C_2H_6O_2$, that consists of 5.00×10^{24} molecules. The density of ethylene glycol is 1.11 g/mL.

34. If 2.56 g of chlorine, Cl_2, are to be used to prepare dichlorine heptoxide, Cl_2O_7, how many moles and how many grams of oxygen, O_2, are needed?

35. Balance the following equations.
    - (a) $Fe_2O_3 + HNO_3 \longrightarrow Fe(NO_3)_3 + H_2O$
    - (b) $C_{21}H_{30}O_2 + O_2 \longrightarrow CO_2 + H_2O$

36. How many moles of nitric acid, HNO_3, are needed to react with 2.56 mol of Cu in the following reaction?

$$3Cu + 8HNO_3 \longrightarrow 3Cu(NO_3)_2 + 2NO + 4H_2O$$

37. Under the right conditions, ammonia can be converted to nitric oxide, NO, by the following reaction.

$$4NH_3 + 5O_2 \longrightarrow 4NO + 6H_2O$$

How many moles and how many grams of oxygen are needed to react with 56.8 g of ammonia by this reaction?

38. To neutralize the acid in 10.0 mL of 18.0 M H_2SO_4 that was accidentally spilled on a laboratory bench top, solid sodium bicarbonate was used. The container of sodium bicarbonate was known to weigh 155.0 g before this use and out of curiosity its mass was measured as 144.5 g afterwards. The reaction that neutralizes sulfuric acid this way is as follows.

$$H_2SO_4 + 2NaHCO_3 \longrightarrow Na_2SO_4 + 2CO_2 + 2H_2O$$

Was sufficient sodium bicarbonate used? Calculate the limiting reactant and the maximum yield in grams of sodium sulfate.

39. How many mL of concentrated sulfuric acid (18.0 M) are needed to prepare 125 mL of 0.144 M H_2SO_4?

40. The density of concentrated phosphoric acid solution is 1.689 g solution/mL solution at 20 °C. It contains 144 g H_3PO_4 per 1.00×10^2 mL of solution.
    - (a) Calculate the molar concentration of this solution.
    - (b) Calculate the grams of this solution required to hold 50.0 g H_3PO_4.

41. A mixture consists of lithium carbonate (Li_2CO_3) and potassium carbonate (K_2CO_3). These react with hydrochloric acid as follows.

$$Li_2CO_3(s) + 2HCl(aq) \longrightarrow 2LiCl(aq) + H_2O + CO_2(g)$$
$$K_2CO_3(s) + 2HCl(aq) \longrightarrow 2KCl(aq) + H_2O + CO_2(g)$$

When 4.43 g of this mixture was analyzed, it consumed 53.2 mL of 1.48 M HCl. Calculate the number of grams of each carbonate and their percentages.

42. Dolomite is a mineral consisting of calcium carbonate ($CaCO_3$) and magnesium carbonate ($MgCO_3$). When dolomite is strongly heated, its carbonates decompose to their oxides (CaO and MgO) and carbon dioxide is expelled.
    - (a) Write the separate equations for these decompositions of calcium carbonate and magnesium carbonate.
    - (b) When a dolomite sample with a mass of 5.78 g was heated strongly, the residue had a mass of 3.02 g. Calculate the masses in grams and the percentages of calcium carbonate and magnesium carbonate in this sample of dolomite.

43. Adipic acid, $C_6H_{10}O_4$, is a raw material for making nylon, and it can be prepared in the laboratory by the following reaction between cyclohexene, C_6H_{10}, and sodium dichromate, $Na_2Cr_2O_7$ in sulfuric acid.

$$3C_6H_{10}(\ell) + 4Na_2Cr_2O_7(aq) + 16H_2SO_4(aq) \longrightarrow$$
$$3C_6H_{10}O_4(s) + 4Cr_2(SO_4)_3(aq) + 4Na_2SO_4(aq) + 16H_2O$$

There are side reactions. These plus losses of product during its purification reduce the overall yield. A typical yield of purified adipic acid is 68.6%.
(a) To prepare 12.5 g of adipic acid in 68.6% yield requires how many grams of cyclohexene?
(b) The only available supply of sodium dichromate is its dihydrate, $Na_2Cr_2O_7 \cdot 2H_2O$. (Since the reaction occurs in an aqueous medium, the water in the dihydrate causes no problems, but it does contribute to the mass of what is taken of this reactant.) How many grams of this dihydrate are also required in the preparation of 12.5 g of adipic acid in a yield of 68.6%?

44. One of the ores of iron is hematite, Fe_2O_3, mixed with other rock. One sample of this ore is 31.4% hematite. How many tons of this ore are needed to make 1.00 ton of iron if the percentage recovery of iron from the ore is 91.5% (1 ton = 2000 lb)?

45. Gold occurs in the ocean in a range of concentration of 0.1 to 2 mg of gold per ton of seawater. Near one coastal city the gold concentration of the ocean is 1.5 mg/ton.
(a) How many tons of seawater have to be processed to obtain 1.0 troy ounce of gold if the recovery is 65% successful? (The troy ounce, 31.1 g, is the standard "ounce" in the gold trade.)
(b) If gold can be sold for 455 dollars per troy ounce, what is the breakeven point in the dollar-cost per ton of processed seawater for extracting gold from the ocean at this location?

46. One way to prepare iodine is to mix sodium iodate, $NaIO_3$, with hydriodic acid, HI. The following reaction occurs.

$$NaIO_3 + 6HI \longrightarrow 3I_2 + NaI + 3H_2O$$

Calculate the number of moles and the number of grams of iodine that can be made this way from 16.4 g of $NaIO_3$.

47. C.I. Pigment Yellow 45 ("sideran yellow") is a pigment used in ceramics, glass, and enamel. When analyzed, a 2.164 g sample of this substance was found to contain 0.5259 g of Fe and 0.7345 g of Cr. The remainder was oxygen. Calculate the empirical formula of this pigment.

48. When 6.584 g of one of the hydrates of sodium sulfate (Na_2SO_4) was heated so as to drive off all of its water of hydration, the residue of anhydrous sodium sulfate had a mass of 2.889 g. What is the formula of the hydrate?

49. A white solid was known to be the anhydrous form of either sodium carbonate (Na_2CO_3) or sodium bicarbonate ($NaHCO_3$). Either reacts with hydrochloric acid (HCl) to give sodium chloride, water, and carbon dioxide, but the mole proportions are not the same.
(a) Write the balanced equation for each reaction.
(b) It was found that 0.5128 g of the solid reacted with 47.80 mL of 0.2024 M HCl, and that the addition of more acid caused the formation of no more carbon dioxide. Perform the calculations that establish which substance the unknown solid was.

As a passenger during takeoff in a jet airliner, like this Boeing 737, you may harbor a few worries. Will traffic controllers control? Will wing flaps flap? Will the pilot pilot? But have you ever worried about the chemical reactions in the great jet engines, reactions that produce the hot expanding and thrusting gases? The complete reliability of the thermochemistry of reactions is studied in this chapter.

ENERGY CHANGES
IN CHEMICAL REACTIONS

5.1 ENERGY—SOURCES AND UNITS

All forms of energy can be converted quantitatively into heat.

In the last few chapters we have studied how to write formulas, balance equations, and do stoichiometric calculations. Until now, however, we have paid little attention to another aspect of reactions—the energy changes that accompany them. These are important, because most of the energy that civilization uses is derived from chemical reactions, especially those of oxygen with various fuels. In addition, the study of energy changes provides much useful scientific information. For example, in Chapter 6 we will see that energy changes that occur when atoms emit light provide clues to the basic structures of atoms. And energy changes of chemical reactions directly reflect changes in the bonds that bind atoms together in compounds.

What Is Energy?

Let us begin our study by reviewing what we mean by energy. We learned in Chapter 1 that energy is not a physical thing but an ability—the ability to do work. *Work,* as understood by scientists, means more than towing barges and lifting bales. The concept goes beyond physical labor to embrace a number of actions, but they all come down to the pushing or pulling of something against an opposing force. For example, when hot gases expand in the cylinder of a gasoline engine, they push back the piston and ultimately move the car. This moving of an object is called mechanical work.

Hot expanding gases can also be made to push on the blades of gas turbine engines.

Everybody agrees that a battery has energy. It also has the ability to do work, and it delivers its energy by pushing electrons through a wire. We call this pushing of electrons *electrical work*. The current might run a small motor (and so be changed into mechanical work); or it might operate a small pocket calculator; or it might go into a flashlight's bulb and so be changed into both *light* and *heat*, two other forms of energy.

In Chapter 1 we also learned that physical objects, be they cars or chemicals, can have energy in two ways. One is kinetic energy — energy of motion — the mechanical energy that we described in the last paragraph. It can be calculated from the object's mass (m) and its velocity (v) by the equation

$$KE = \tfrac{1}{2}mv^2$$

The other kind of energy is potential energy, which we defined earlier as stored energy. Springs store energy when they are compressed, and the energy that is stored is supplied by the person who does the compressing. Lifting your book off your desk also puts stored energy into the book. You have to expend some of your own energy to increase the distance between the earth and the book, because the book and the earth are attracted to each other. This energy can be recovered by dropping the book on your hand — which, if the transfer occurs suddenly, your hand might not appreciate.

These last two examples illustrate how energy is stored. *Whenever objects that attract each other are pulled apart, their potential energy (stored energy) increases.* This is how energy is stored by the book when it is raised from the desk. *Whenever objects that feel a repelling force are pushed together, the potential energy also increases.* This is how a spring stores energy as it is compressed.

It is very important for you to learn how potential energy is related to attractions and repulsions, because changes in potential energy are extremely important in chemistry. For example, the parts of an atom — electrons and nuclei — experience an attraction for each other, and as the distances between these particles vary in an atom, so do their potential energies. And these energies, we will see, have a profound influence on the structures of atoms, and therefore on all of the chemical properties of the elements.

Chemical Energy

The term *chemical energy* was first introduced in Chapter 1. Chemical energy is the special name often given to the form of potential energy that arises from the forces of attraction that bind atoms together in compounds. These forces of attraction are called *chemical bonds,* and we will discuss how and why they are formed at a later time. What is important now is the idea that when chemicals react to form new substances, atoms are exchanged as old bonds break and new ones form. This process changes the potential energies of the atoms. Sometimes a reaction's products have more potential energy than its reactants; in other reactions the products have less potential energy than the reactants. These changes in potential energy as the reactants change to products are observed by us either as energy being absorbed by the reacting mixture or as energy being given off.

Much if not all of the energy released or absorbed usually appears as heat, so what we *observe* are temperature changes.

Changes in chemical (potential) energy occur in nearly every chemical reaction, and it is our goal in this chapter to study how these energy changes are measured and, to some degree, what they tell us about the substances that react. Before getting to this subject, however, it is worthwhile to take a brief look at the sources of energy that fuel our civilization, because in most cases this energy comes from chemical reactions that we harness for our benefit.

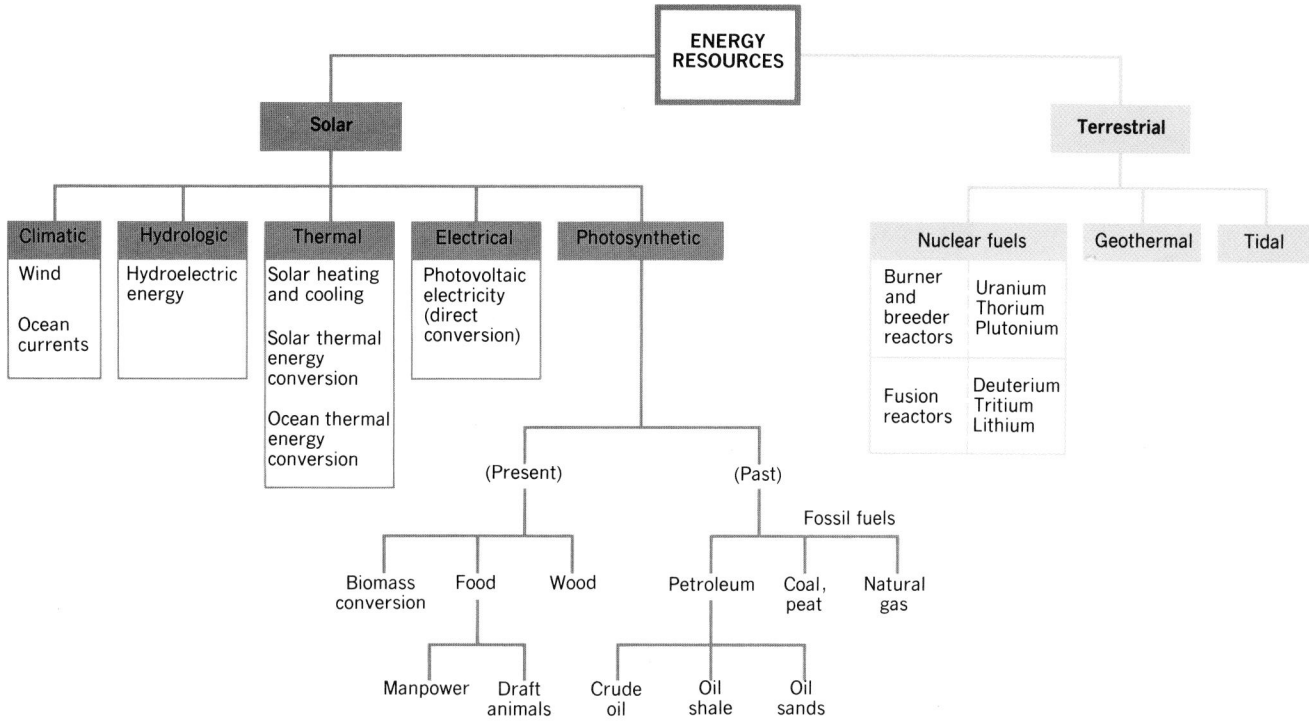

FIGURE 5.1 The principal sources of energy on our planet.

Sources of Energy

Figure 5.1 outlines our principal sources of energy. An energy source—which we naturally define according to *human* needs—is any natural change in our environment that we can use to make it easier to do our tasks, or any material that we can get to undergo such a change. The wind, for example, is a natural *change*. It blows where it wills, but clever mariners long ago discovered that if they put sails into the wind, they could rest at the oars. We can expect to see wind energy tapped more and more, particularly by small systems.

Coal, another energy source, is a natural *material* whose chemical energy operated most of the world's trains until the middle of this century. Coal is still our principal source of energy for generating electrical energy at large, central power stations. (See Figure 5.2.)

Sunlight is another important energy source. Each year, the solar energy received and absorbed by the earth is equivalent to roughly 46 times the initial world reserves of minable coal, or to about 425 times the initial world reserves of crude oil. Even if the earth's human population quadrupled and the rate at which each person, on the average, consumed energy were to multiply by a factor of 200, the resulting rate of consuming energy would still be less than 1% of that absorbed annually from the sun. Special Topic 5.1 discusses some of the technologies currently under investigation by which we might be able someday to channel more solar energy into meeting human needs and wants.

A miniscule 0.04% of the incoming solar energy is used by green plants and oceanic phytoplankton to make complex chemical compounds that are vital to all living things. This process in plants is called photosynthesis, and it uses solar energy trapped by the green pigment chlorophyll to change simple, low-energy molecules—carbon dioxide and water—into high-energy molecules that the plants need. We can represent the overall change as follows.

$$\text{Solar energy} + CO_2 + H_2O \longrightarrow (CH_2O) + O_2$$

Wind power is proportional to the *cube* of the wind velocity.

FIGURE 5.2 Unit coal trains—trains devoted exclusively to the shipment of coal—move around the clock from western coal fields to central station power plants near metropolitan areas. Each of these cars holds more than 100 tons of coal.

About 200 billion tons of carbon (as CO_2) are fixed annually by photosynthesis.

Each year about 400 billion tons of oxygen are released annually.

Oil shale and oil sand in various deposits in the world hold huge quantities of energy, too.

The formula (CH_2O) represents only the fundamental unit in one of the products, a carbohydrate. Glucose, for example, is $C_6H_{12}O_6$ [which superficially looks like $(CH_2O)_6$]. The other product is oxygen, and photosynthesis is the principal means whereby the world's supply of oxygen is constantly regenerated. Glucose is an important source of chemical energy for both plants and animals, but since animals cannot do photosynthesis they (and we) have to rely on plants for the nutritional fruits of solar energy.

Thanks to biological and geological changes that occurred long ago, the remains of dead plants changed into what we now call the fossil fuels. These are principally coal, petroleum, and natural gas whose energy came originally from the sun. We value them not only as sources of energy but also for the chemicals obtained or made from them. Virtually all pharmaceuticals, pesticides, dyes, synthetic fibers, plastics, and coatings (paints) are now made from one or another of the fossil fuels.

If we reflect on the involvement of sunlight in our lives, we soon realize that the sun is directly or indirectly the source of nearly all forms and quantities of usable energy on earth—particularly the chemical energy in fossil fuels, wood, and the foods we eat. Even wind and water power are ultimately possible because of the sun's energy, as explained in Special Topic 5.1.

Units of Energy

To measure the amounts of energy involved in chemical reactions, we must have a unit of energy. Over many decades a variety of units have emerged, some more useful than

FIGURE 5.4 The black, rectangular solar panels on the arms of Skylab 4 constantly face the sun and recharge the batteries of this NASA satellite.

conversion, involves the use of plant wastes to make fuel. *Biomass* means any combustible or fermentable materials made by plants, including agricultural wastes such as straw, stalks, and manure, as well as wastepaper, wood, and surplus grains. Perhaps you or a friend have used gasohol, a blend of ethyl alcohol and gasoline, in a car. The alcohol can be made from any wastes or surplus products that contain carbohydrates, such as sugars or starch, and many large-scale plants are now operating. Methane, which is the chief constituent in natural gas, can be made from manure, and some farmers have small-scale operations that make methane for use in warming poultry houses and barns. Municipal trash includes great quantities of biomass and other burnable materials—paper, cardboard, tires, and plastics, for example. Many cities reduce trash-disposal problems and generate municipal energy by burning their trash in special generators.

Direct photovoltaic energy conversion, the last of the listed solar technologies, is the use of sunlight to make electricity directly rather than by means of a steam-powered generator run on solar-generated heat. (See Figure 5.4.) When sunlight falls on certain materials, such as properly treated silicon, it generates a flow of electricity, and much research is presently going into the development of silicon-based solar cells.

At the moment the solar technologies are generally more expensive than traditional sources of fuels—at least when it comes to very large-scale applications. Advances in engineering and the development of materials are certain to bring these costs down.

surface area of the earth, so they absorb huge quantities of solar energy. However, the efficiency of the natural mixing of ocean water is not great, and the surface water is warmer than that hundreds of feet below. Anytime you have a difference in temperature, you can arrange for a flow of energy, and this means you can devise a generator to make electricity. Such generators could be anchored offshore from coastal cities and electricity could be piped ashore by underwater cables.

The fifth-listed solar energy technology, biomass

others in chemistry. The SI energy unit is derived from the SI base units that we discussed in Chapter 1. This unit is called the *joule*, and it is based on the definition of kinetic energy:

$$KE = \tfrac{1}{2}mv^2$$

One joule, abbreviated J, is the kinetic energy possessed by an object with a mass of 2 kg moving with a velocity of 1 m/s.

The joule is named after J. P. Joule (1818–1889), an important English physicist.

$$1 \text{ J} = \frac{1}{2} (2 \text{ kg}) (1 \text{ m/s})^2$$

or,

$$1 \text{ J} = 1 \text{ kg m}^2/\text{s}^2$$

If you dropped 4.4 lb of butter on your foot from a height of 4 in., you would deliver about 1 J of energy to your foot. This is actually a very small amount of energy, especially when we consider chemicals reacting on a mole scale. We will more often encounter the larger unit, the kilojoule, kJ.

$$1 \text{ kJ} = 1000 \text{ J}$$

The Special Place of Heat

One of the important facts about our world is that all forms of energy can be converted quantitatively into heat. For example, the mechanical energy of a moving car is converted entirely into heat by the frictional action of the brakes, and the brake shoes and drums can become very hot, indeed. When a current of electrons is directed into a poor conductor, something with high resistance such as the heating element in a toaster, the electrical energy changes entirely into heat. The full convertibility of energy in other forms into heat gives us a way to measure the other kinds of energy, and the measurement of heat is relatively simple.

As we will see very soon, one way to measure heat is to absorb it into water and observe the rise in temperature. The traditional unit of energy is based on this technique. This energy unit is called the calorie, abbreviated cal, and was originally defined as the amount of heat that can raise the temperature of one gram of water by one degree Celsius. Since the development of the SI, however, the calorie has been redefined in terms of the joule. This relationship, which you should learn, is

This is now the official definition of the calorie.

$$1 \text{ cal} = 4.184 \text{ J}$$

Like the joule, the calorie is a very small amount of energy, so the larger unit, the kilocalorie, kcal, is often used.

The Calorie in nutrition is really the kilocalorie.

$$1 \text{ kcal} = 1000 \text{ cal}$$

Often it is necessary to convert between kilocalories and kilojoules, and the relationship is simple:

$$1 \text{ kcal} = 4.184 \text{ kJ}$$

Special Topic 5.2 describes other energy units that are common in various fields.

5.2 SPECIFIC HEAT AND HEAT CAPACITY

When we can measure the change in temperature of a known mass of water, we can calculate how much energy caused this change.

The physical properties of a substance that concern its ability to absorb heat without changing chemically are called its thermal properties. Three examples are *heat capacity*, *molar heat capacity*, and *specific heat capacity*, which is usually just called *specific heat*.

Specific Heat and Molar Heat Capacity

The specific heat of any substance is the energy needed to raise the temperature of one gram of it by one degree Celsius. It can be calculated by the equation

$$\text{specific heat} = \frac{(\text{energy absorbed})}{(\text{mass of sample in g}) \times (\text{temperature change in } ^\circ C)} \tag{5.1}$$

"Specific" in *specific* heat, *specific* gravity, and other quantities always signifies some kind of ratio of quantities.

If we use energy units of joules, the SI unit of energy, the energy units of specific heat are

$$\text{specific heat} = \frac{J}{g \, ^\circ C}$$

If we use the older unit of calorie, the units of specific heat are

$$\text{specific heat} = \frac{cal}{g \, ^\circ C}$$

Since the Celsius degree and the kelvin are identical in size, the units could be (and often are) expressed as cal/g K or J/g K.

Often we find it necessary to use specific heat data to calculate amounts of energy absorbed when a sample of a substance changes temperature. In these instances, we rearrange Equation 5.1 to solve for the energy change.

$$(\text{energy change}) = (\text{specific heat}) \times (\text{mass of sample}) \times (\text{temperature change}) \tag{5.2}$$

$$J \quad = \quad \frac{J}{g \, ^\circ C} \quad \times \quad (g) \quad \times \quad ^\circ C$$

$$cal \quad = \quad \frac{cal}{g \, ^\circ C} \quad \times \quad (g) \quad \times \quad ^\circ C$$

Many students find Equation 5.2 easier to remember than Equation 5.1.

The specific heat of any substance varies with the temperature, but not by much. The specific heat of water, for example, is exactly 1 cal/g $^\circ$C for the one degree between 14.5 $^\circ$C and 15.5 $^\circ$C. (The calorie was originally *defined* this way.) At 25 $^\circ$C, the specific heat of water is 0.99895 cal/g $^\circ$C (4.1796 J/g $^\circ$C), and to two significant figures, the specific heat of liquid water is 1.0 cal/g $^\circ$C (4.2 J/g $^\circ$C) over the whole range between its freezing and boiling points. Table 5.1 gives the specific heats for many common substances at room temperature.

TABLE 5.1 **Specific Heats**

Substance	Specific Heat (25 °C)	
	J/g °C	**cal/g °C**
Carbon (graphite)	0.711	0.170
Copper	0.387	0.0924
Ethyl alcohol	2.45	0.586
Gold	0.129	0.0308
Granite	0.803	0.192
Iron	0.4498	0.1075
Lead	0.128	0.0305
Olive oil	2.0	0.47
Silver	0.235	0.0562
Water, liquid	4.1796	0.99895

EXAMPLE 5.1   CALCULATING WITH SPECIFIC HEATS

Problem: If a gold ring with a mass of 5.5 g changes temperature from 25.0 °C to 28.0 °C, how much energy (in joules) has it absorbed?

Solution: To calculate the amount of energy absorbed, we need three quantities, the specific heat of gold, the mass of the sample, and the *change* in temperature. In Table 5.1 we find that the specific heat of gold is 0.129 J/g °C at 25 °C. We are also given the mass of the sample, 5.5 g. The temperature goes from 25.0 °C to 28.0 °C, so the change in temperature is 3.0 °C. Substituting these values into Equation 5.2 gives us

$$\text{(heat absorbed)} = \frac{0.129 \text{ J}}{\cancel{g} \, \cancel{°C}} \times (5.5 \, \cancel{g}) \times (3.0 \, \cancel{°C})$$

$$= 2.1 \text{ J (rounded from 2.1285 J)}$$

Thus, only 2.1 J raises the temperature of 5.5 g of gold by 3.0 °C.

PRACTICE EXERCISE 1   The temperature of a sample of 250 g of water is changed from 25.0 °C to 30.0 °C. How much energy was transferred into the water to cause this change? Calculate your answer in joules and in calories.

We have defined thermal properties in terms of the heat that a substance must *absorb* for a given *increase* in temperature. The other side of the coin is that specific heat represents the heat a substance must *lose* if its temperature is to *decrease*. Specific heat, thus, tells us not only how relatively hard it is to increase the temperature of a specific amount of substance but also how relatively difficult it is to cool it.

As we said, *specific heat* is the shortened name for *specific heat capacity*, the heat that can be absorbed by specifically one gram per degree change in temperature. Sometimes it is more useful to work with a mole quantity instead of the gram. The molar heat capacity is the energy needed to raise the temperature of one mole of a substance by one degree Celsius. In terms of the joule,

$$\frac{J}{g \, °C} \times \frac{g}{mol} = \frac{J}{mol \, °C}$$

specific   molar   molar heat
heat   mass   capacity

$$\text{molar heat capacity} = \frac{J}{mol \, °C} \qquad (5.3)$$

Using the calorie,

$$\text{molar heat capacity} = \frac{cal}{mol \, °C} \qquad (5.4)$$

Since the molar mass of water is 18.0 g/mol, the molar heat capacity of water is 18.0 times its specific heat, namely 75.3 J/mol °C or 18.0 cal/mol °C.

Heat Capacity

The quantitative measure of the overall effect of heat on the temperature of some given object — a beaker of an aqueous solution, a person, or a lake, for example — is called the object's heat capacity. The heat capacity is the amount of heat that must move into or out of the object to change its temperature by one degree Celsius. The usual units of heat capacity are joules per degree Celsius or calories per degree Celsius.

The units cal/K or J/K could be used instead of cal/°C and J/°C without affecting the numerical values.

$$\text{heat capacity} = \frac{J}{°C} \qquad (5.5)$$

$$\text{heat capacity} = \frac{cal}{°C} \qquad (5.6)$$

(Sometimes larger units of heat such as the kilocalorie or the kilojoule are used to describe an object's heat capacity.)

Specific heat is heat capacity *per gram.* Molar heat capacity is heat capacity *per mole*

It is important that you distinguish carefully between *heat capacity* and *specific heat. Heat capacity* is an extensive property; it refers to a thermal property of a whole object, just as *mass* (another extensive property) applies to an entire sample or a whole object. *Specific heat*, on the other hand, is an intensive property; it refers to the heat capacity *per unit of mass*. Thus to calculate the heat capacity of some specific object, we only have to multiply the specific heat by the mass, as the following analysis shows, using one particular set of units.

$$\text{heat capacity} = \text{specific heat} \times \text{mass}$$

$$\frac{J}{°C} = \frac{J}{g \ °C} \times g \qquad\qquad \frac{cal}{°C} = \frac{cal}{g \ °C} \times g$$

EXAMPLE 5.2 CALCULATING A HEAT CAPACITY

Problem: How many kilojoules can 250 g of water absorb for each degree Celsius change in the temperature range near room temperature? In other words, what is the heat capacity of 250 g of water?

Solution: From Table 5.1 we see that the specific heat of water for our temperature range is (rounded) 4.18 J/g °C. Therefore,

$$\text{heat capacity} = \text{specific heat} \times \text{mass}$$

$$= 4.18 \ \frac{J}{g \ °C} \times 250 \ g$$

$$= 1.05 \times 10^3 \ J/°C \ (\text{rounded from } 1.045)$$

$$= 1.05 \ kJ/°C$$

The heat capacity of 250 g of water is 1.05 kJ/°C

EXAMPLE 5.3 CALCULATING THE ENERGY BUDGET FOR A TEMPERATURE CHANGE

Problem: Suppose the 250-g sample of water in the previous example underwent a change in temperature from 25.0 °C to 26.4 °C. How much heat caused this change? (Give your answer in kilojoules.)

Solution: From the previous example, the heat capacity of this water sample is 1.05 kJ/°C. The temperature change is (26.4 − 25.0) °C = 1.4 °C. The heat capacity of the water sample supplies two conversion factors.

An energy *budget* is the energy to be lost or gained by a system in a given change.

$$\frac{1.05 \ kJ}{1 \ °C} \qquad \text{and} \qquad \frac{1 \ °C}{1.05 \ kJ}$$

We therefore multiply the given, 1.4 °C, by the first factor.

1.05 kJ/°C means $\frac{1.05 \ kJ}{1 \ °C}$

$$1.4 \ °C \times \frac{1.05 \ kJ}{1 \ °C} = 1.5 \ kJ \ (\text{rounded from } 1.47)$$

The temperature change of 1.4 °C required 1.5 kJ of energy for this sample of water.

The preceding example served the important purpose of showing how we can measure energy using thermometer readings and previously determined values of heat capacity. This is the basis for experimental work in thermochemistry, which is the subject of Section 5.3.

PRACTICE EXERCISE 2 The temperature of 335 g of water changed from 24.5 °C to 26.4 °C. How much heat did this sample absorb? Calculate your answer in kilojoules and kilocalories. (Hint: you can either calculate the heat capacity of the water sample first and use the result to find the answer, or you can set up a chain calculation.)

The Unusual Thermal Properties of Water

Notice in Table 5.1 the unusually high specific heat of water. It is higher than the values for almost all known materials, for example, roughly 10 times that of iron and nearly 33 times that of gold. In other words, the heat that raises the temperature of 1 g of water 1 °C raises the temperature of 1 g of iron by 10 °C. Or, we could compare iron and water in terms of cooling. If we remove a calorie of heat from a gram of iron, its temperature falls by 10 °C. But the same loss of heat from a gram of water makes its temperature fall by only 1 °C. You can see that water's high specific heat means that it is not possible to make the temperature of water swing widely by trifling exchanges of heat.

Ecologists point out that the high specific heat of water is one of nature's "tremendous trifles." It helps to make it much easier for the human body to maintain a steady temperature of 37 °C, for example, and an even body temperature is vital to survival. Even a few degrees of change is deadly. Since the adult body is about 60% water by mass, it has a high overall heat capacity. This means that the body can exchange considerable energy with the environment but experience only a very small effect on temperature. The human body, in other words, has a large thermal "cushion," giving it time to adjust to large and sudden changes in outside temperature with little fluctuation of its core temperature. Hypothermia, discussed in Special Topic 5.3, is an emergency brought on when the body cannot prevent its core temperature from decreasing.

An infant's body is about 80% water.

Cities by oceans or very large bodies of water (such as one of the Great Lakes) have relatively mild climates compared to cities 50 or so miles inland because of such a "cushion."

5.3 THERMOCHEMISTRY: ENERGY CHANGES IN CHEMICAL REACTIONS

For most reactions, we are interested in the heat absorbed or evolved when the chemical change occurs under conditions of constant pressure.

Earlier we mentioned that nearly every chemical reaction occurs with either the absorption or release of energy, and learning about the amounts of energy involved is another important aspect of investigating stoichiometry. It is so important, in fact, that the study of these energy changes is thought of as a separate subject and is given the name *thermochemistry*. Specifically, thermochemistry deals with transfers of energy between reacting chemicals and the world surrounding them. We will begin our study of thermochemistry by defining some terms.

Systems, Surroundings, and Boundaries

The word **system** refers to that particular part of the universe we wish to study. The system might be the chemicals reacting in a beaker or the chemicals in a battery reacting to give electricity, or the system could be a living cell.

The word surroundings refers to whatever is entirely outside the defined system, everything in the universe except the system itself.

A boundary, real or imaginary, separates the system from its surroundings. When the system is a solution in a beaker, for example, the boundary exists wherever the solution contacts the beaker or the air above it. Here we see a clearly defined, real boundary. Sometimes scientists have to think of an imaginary boundary. For example, suppose we were considering exchanges of energy between the earth and the rest of the universe. Where does our system—the earth and its atmosphere—end and the surroundings begin? The atmosphere just gets thinner and thinner as we get farther from the earth, so we have to define some arbitrary boundary at a particular altitude. One choice, for example, might be the imaginary "boundary" that separates the top of the stratosphere from the bottom of the mesosphere at an altitude of about 50 km (about 31 miles).

If the boundary can prevent any transfer of heat between the system and the surroundings, we say that the system is *insulated* from its surroundings. Styrofoam makes a good insulating boundary for keeping a cup of coffee hot, for example, but no material is a perfect insulator.

Another term that we will use frequently is the state of a system. Each system has a *state* defined by listing its temperature, pressure, volume, and composition (including concentration terms). We say that a system undergoes a change of state whenever any change occurs in one or more of the variables that define the system.

Notice carefully that the term *state* in thermodynamics has a broader meaning than it has in such expressions as "liquid state" or "solid state."

Heats of Reaction

In thermochemistry we are concerned with the exchange of energy between a chemical system and its surroundings. Sometimes chemical changes are able to bring energy into the system. These are called endothermic changes. An example is the charging of a battery, in which energy from an external source becomes stored in the battery in the form of chemical (potential) energy. Photosynthesis is also endothermic as far as the plant is concerned, and the needed energy is imported from the sun. When endothermic changes occur by themselves within an uninsulated system, we can often notice a cooling effect in the surroundings. This is what we sense, for instance, when we use an "instant cold" compress, which can be purchased in most pharmacies.

Many chemical reactions are able to release energy to the surroundings. Such changes are described as exothermic. A typical example is the combustion of gasoline. Heat transfers away from the system (if uninsulated) and into the surroundings, where the temperature increases.

The form of the energy absorbed or released during a change can vary. It sometimes appears as light or as electrical work, but often it occurs only as heat. Earlier we learned, however, that all forms of energy can be converted entirely to heat. And in the previous section we learned how we can use temperature readings and heat capacity data to measure this heat. This is why we generally describe the energy changes in a chemical reaction in terms of amounts of heat. When the entire energy change of a reaction involves heat, the amount of heat is called the heat of reaction and is usually represented by the symbol q.

To show the *direction* with which energy transfers, we give it either a positive or a negative sign. If heat leaves the system, q has a minus sign (and you know that a minus sign often is used to indicate the loss of something).

Containers holding endothermic changes sometimes become noticeably cooler.

Test tubes or flasks holding exothermic reactions nearly always become warm to the touch.

exo means "out"
endo means "in"
therm means "heat"

Exothermic changes $q \, (-)$

If the heat enters the system, q gets a positive sign.

$$\text{Endothermic changes} \qquad q\,(+)$$

The actual amount of heat of reaction for a given change in a system depends to some extent on the conditions under which we carry out the reaction. It depends, for example, on the physical states of the reactants and products; it depends on the initial temperature of the system; and it depends somewhat on whether the volume of the system or its pressure is held constant or is permitted to change.

The pressure of the atmosphere is caused by the gravitational attraction of the earth for the surrounding air.

To simplify matters just a little, specialists in thermochemistry noted long ago that most of the reactions of importance to us take place in open containers where the system is under a constant pressure — the pressure of the surrounding atmosphere. This is true of reactions in open beakers and true even for the biochemical reactions within us, because we are also under atmospheric pressure. In this chapter we will be concerned mostly with heats of reaction under constant pressure. Since such heat values aren't always the same as when a system reacts under changing pressures we need special terms to use for the different circumstances.

Enthalpy

Another name for enthalpy is *heat content*.

The first of these terms is for the total value of the energy of a system when it is at constant pressure; this is called the enthalpy of the system, symbolized by the letter H. When a system reacts at constant pressure and absorbs or evolves energy, we say that it experiences an enthalpy change, represented as ΔH (read "delta H") and defined by the equation

$$\Delta H = H_{final} - H_{initial} \qquad (5.7)$$

Notice:

   H = hydrogen atom
   H = enthalpy

H_{final} is the enthalpy of the system in its final state and $H_{initial}$ is the enthalpy of the system in its initial state.

For a chemical reaction, the initial state refers to the reactants and the final state to the products, so for a chemical reaction this equation can be rewritten as follows.

$$\Delta H = H_{products} - H_{reactants} \qquad (5.8)$$

Unless we specify otherwise, whenever we write ΔH, we mean ΔH for the *system*, not the surroundings.

(From time to time we will encounter other quantities that are defined as a difference between two absolute values. In each situation, *we will subtract the value for the initial state from the value for the final state.* Simply put, this always means "products minus reactants.")

After having gone to all this effort to give a formal definition of ΔH, it is perhaps a bit disappointing to learn that we cannot actually calculate it from measured values of H_{final} and $H_{initial}$. This is because the total enthalpy of the system depends on its *total* kinetic energy plus its *total* potential energy, and these values can never be determined. We can see why by a little reflection about our universe.

Suppose, for example, that we are interested in the total enthalpy possessed by the reactants in a chemical reaction about to happen in a beaker resting on the laboratory bench. The system might *appear* to be standing still, and thus it might seem to have no kinetic energy. However, it is on the surface of a possibly shifting continental plate that is on a rotating planet that moves about the sun, which itself moves through a galaxy that moves within the universe. We cannot know all of these velocities, and so the system's total kinetic energy cannot be measured or even calculated. Likewise, we cannot know its total potential energy. We have no way to measure, for example, the potential energies involving the attractions caused by gravity between the beaker and all of the other bodies in the universe. If we cannot measure H, then what good is it?

Our formal definition of H actually serves only one purpose — to define the meaning of positive and negative values of ΔH. We said that when a system absorbs energy from its surroundings, the change is endothermic. This means that the system's final enthalpy, whatever its absolute value is, has to be larger than its initial enthalpy.

$$\text{In endothermic changes:} \qquad H_{final} > H_{initial}.$$

Therefore, if we could calculate ΔH from H_{final} and $H_{initial}$, we would subtract a smaller value from a larger one, so the difference (ΔH) would have a positive sign.

In endothermic changes: ΔH is positive.

On the other hand, if a change is exothermic, the system loses energy and its final enthalpy would have to be less than its initial enthalpy.

$$\text{In exothermic changes:} \qquad H_{final} < H_{initial}.$$

Now, to calculate ΔH from $H_{initial}$ and H_{final} (if we could) we would subtract a larger number from one that is smaller, so the value of ΔH for all exothermic changes is negative.

In exothermic changes: ΔH is negative.

Although we cannot measure the total enthalpy, H, the good news is that we really do not need to know its value. We care only about what our system could do for us (or to us!) right here at a particular place on this planet. For example, when we want to know the yield of energy from burning gasoline, we really do not care what its total enthalpy is in either the initial or the final state. All we care about is by how much the enthalpy *changes*, because it is only this enthalpy change that is available to us. In other words, we really have no need for the absolute values of H_{final} and $H_{initial}$. All we need is the difference, ΔH, and this is something we *can* measure, and we will soon see how.

Enthalpy Change and Heat of Reaction at Constant Pressure

The enthalpy change for a reaction can make itself known in the surroundings either as work or as heat energy or as some of each. If the reaction is occurring in a battery, for example, at least part of the enthalpy change can appear as electrical work. In other circumstances, all of the enthalpy change of the reaction may involve only heat. We might, for example, let the chemical reaction of the battery happen with no connection to electrical circuits, in which case all of the enthalpy would appear as heat and none as work. When *all* of the enthalpy change appears as heat, we have a way to measure the enthalpy change for a reaction, because in this circumstance ΔH *is equal to the heat of reaction at constant pressure.*

$$\Delta H = q \qquad \text{(at constant pressure)}$$

We can now see how the sign of q, defined earlier, is actually determined by the sign of ΔH. ΔH is negative and so q is negative for exothermic changes. ΔH is positive and so q is positive for endothermic changes.

Sometimes you'll see the symbol q given as q_p, where the subscript is meant to designate constant pressure.

Energy Conservation

In the discussion thus far, one important truth about our world has been strongly implied but never stated in so many words — the amount of energy that leaves a system is exactly the same as the amount that goes into the surroundings. *No energy is lost.* It just transfers from one place to another, and some of it might change from one form to another. The formal statement of this truth is called the law of conservation of energy.

> Law of Conservation of Energy.   The energy of the universe is constant; it can be neither created nor destroyed, but only transferred and transformed.

State Functions

A state function is any physical property whose value does not depend on the system's history. Some examples are pressure, volume, and temperature. For example, a system's temperature at a particular moment does not depend on what its temperature was the day before, nor does it depend on *how* the system reached its current temperature. If the system's temperature is now 25 °C, this is all we have to know about its temperature. We do not have to specify how it got there. Also, if the temperature were to rise to 35 °C, the change in temperature, Δt, is simply the difference between the final and the initial temperatures.

$$\Delta t = t_{final} - t_{initial}$$

We do not have to know what caused the temperature to change to calculate this difference. All that we need are the initial and final values. This independence from the *method* by which a change occurs is an especially important property of state functions, and being able to recognize when some function or property is a state function simplifies many useful calculations, as we will often see.

Enthalpy as a State Function

Enthalpy is a particularly important state function. The enthalpy of a system in a given state cannot depend on how the system happened to arrive in this state. Even though we cannot measure a system's total kinetic and potential energies, these can have only specific values for a given system in any given state. This must therefore mean that when a system changes from one state to another, the *change* in enthalpy depends only on the initial and final states and is independent of how the change takes place. Since the value of ΔH does not depend on how the change occurs, an enthalpy change is also a state function. This is very useful to know, because when we measure the heat of a reaction we do not have to worry about *how* the reaction is occurring, but only that it is. To determine ΔH, we only have to be sure of our initial and final states and then measure the total amount of heat absorbed or evolved as the system changes between these states. Let us now see how such measurements can be made.

5.4 MEASURING HEATS OF REACTION: CALORIMETRY

The change in temperature caused by a reaction, combined with the values of the specific heat and the mass of the reacting system, makes it possible to determine the heat of reaction.

Earlier we discussed how heat energy can be measured, that is, by observing how the temperature of a known mass of water (or other substance) changes when heat is either added or removed. This is basically how most heats of reaction are determined. The reaction is carried out in some insulated container, where the heat absorbed or evolved by the reaction causes the temperature of the contents to change. This temperature change is measured and the amount of heat that caused the change is calculated by multiplying the temperature change by the heat capacity of the system.

The apparatus used to measure the temperature change for a reacting system is called a calorimeter (that is, a *calorie meter*). The science of using such a device and the data obtained with it is called calorimetry. The design of a calorimeter is not standard; it can vary according to the precision desired. One very simple design used in many general chemistry labs is the "coffee cup" calorimeter, which usually consists of two nested styrofoam coffee cups. Styrofoam, we mentioned earlier, is a very good insulator.

When a reaction occurs at constant pressure inside a Styrofoam coffee-cup calorimeter, the enthalpy change involves heat, and little heat is lost to the lab (or gained from it). If the reaction evolves heat, for example, very nearly all of it stays inside the calorimeter where it causes an easily measured temperature rise. Using the heat capacity of the calorimeter (predetermined), the amount of heat absorbed or evolved by the reaction is calculated. This is illustrated in Example 5.4.

EXAMPLE 5.4   DETERMINING THE HEAT OF REACTION

Problem: The reaction of an acid such as HCl with a base such as NaOH in water involves the exothermic reaction

$$HCl(aq) + NaOH(aq) \longrightarrow NaCl(aq) + H_2O$$

$$1.00\ M\ HCl = \frac{1.00\ mol\ HCl}{1000\ mL\ HCl\ soln}$$

In one experiment, a student placed 50.0 mL of 1.00 M HCl in a coffee-cup calorimeter and carefully measured its temperature to be 25.5 °C. To this was added 50.0 mL of 1.00 M NaOH solution whose temperature was also 25.5 °C. The mixture was quickly stirred, and the student noticed that the temperature of the mixture rose to 32.4 °C. What was the heat of reaction in calories and in joules? (How many calories were released by the reaction of the acid and the base? How many joules evolved in this reaction?) For simplicity, assume that the specific heat of all of the solutions was 1.00 cal/g °C, and that the density of each solution was 1.00 g/mL.

Solution: We can calculate the number of calories evolved from the specific heat if we know (1) the mass of the solution and (2) the temperature change because, by Equation 5.2,

$$\text{heat} = (\text{specific heat}) \times (\text{mass}) \times (\text{temperature change})$$

To calculate the mass and the temperature change, we imagine that the solutions mix first, and then the reaction occurs. This is okay, because the energy change is a *state function.* It isn't important *how* things happen—all we need is the initial state (the separate solutions) and the final state (the combined solutions after the reaction is over). Whether the chemicals react as they are mixed or after they have mixed does not matter.

The mass of the system is found by using the density (remembering that we have to add two volumes):

$$(50.0\ mL + 50.0\ mL) \times 1.00\ \frac{g}{mL} = 100\ g$$

The temperature change is simply $(t_{final} - t_{initial})$

$$32.4\ °C - 25.5\ °C = 6.9\ °C$$

Therefore, the heat evolved is

$$\text{heat} = \frac{1.00\ cal}{g \cdot °C} \times 100\ g \times 6.9\ °C$$

$$\text{heat} = 6.9 \times 10^2\ cal$$

This is actually the heat *gained* by the solution, and therefore it is the energy *lost* by the reacting chemicals. In this sense, the reaction is exothermic, and q is negative.

$$q = -6.9 \times 10^2\ cal$$

Finally, to convert to joules, we use the relationship, 1 cal = 4.184 J.

$$q = -6.9 \times 10^2\ cal \times \frac{4.184\ J}{1\ cal}$$

$$q = -2.9 \times 10^3\ J$$

The calculation that we performed in Example 5.4 gives us the heat of reaction for the specific amounts of chemicals used in that particular experiment. If more acid and base had been used, more heat would have evolved; if less of the reactants were used, then less heat would have evolved. You can see that if we want to compare the heats of reaction for different reactions, we have to decide on a particular quantity. Usually the heat of reaction *per mole* is used for such comparisons. For instance, in the reaction of HCl with NaOH, we would compute how much heat is given off if one mole of acid reacts with one mole of base. In Example 5.4, the mole quantities of HCl and NaOH can be calculated from the volumes of their solutions and their concentrations. For the acid, the moles of HCl taken is

$$50.0 \text{ mL HCl} \times \frac{1.00 \text{ mol HCl}}{1000 \text{ mL HCl}} = 0.0500 \text{ mol HCl}$$

The same quantity of base, 0.0500 mol NaOH, was used. To calculate the energy evolved *per mole* of acid or base, we simply divide the number of calories (or joules) by the number of moles of the selected reactant.

$$q = \frac{-6.9 \times 10^2 \text{ cal}}{0.0500 \text{ mol}} = -14 \times 10^3 \text{ cal/mol} \ (-14 \text{ kcal/mol})$$

$$q = \frac{-2.9 \times 10^3 \text{ J}}{0.0500 \text{ mol}} = -5.8 \times 10^4 \text{ J/mol} \ (-58 \text{ kJ/mol})$$

The final units are cal/mol or J/mol, but since the numbers are so large we have also expressed them, as is common practice, in kcal/mol and kJ/mol.

In the previous section we saw that the heat of reaction measured under conditions of constant pressure corresponds to the enthalpy change for the reaction,

$$\Delta H_{\text{reaction}} = q \quad \text{(at constant pressure)}$$

Therefore, for the neutralization of an acid with a base, the enthalpy change, often called the *enthalpy of reaction*, is $\Delta H = -14$ kcal/mol or $\Delta H = -58$ kJ/mol.

Remember:

10^3 cal = 1 kcal
10^3 J = 1 kJ

PRACTICE EXERCISE 3   In Chapter 2 we mentioned that when sulfuric acid dissolves in water, a great deal of heat is given off. The enthalpy change for this process is called the *enthalpy of solution*. To measure it, 175 g of water was placed in a coffee-cup calorimeter and chilled to 10.0 °C. Then 4.90 g of pure sulfuric acid (H_2SO_4), also at 10.0 °C, was added, and the mixture was quickly stirred with a thermometer. The temperature rose rapidly to 14.9 °C. Assume that the value of the specific heat of the solution is 4.184 J/g °C, and that all of the heat evolved is absorbed by the solution. Calculate q for the formation of this solution, and calculate the enthalpy of solution in units of kilojoules per mole of H_2SO_4. (Hint: Remember to use the *total* mass of the solution.)

The Bomb Calorimeter

A type of calorimeter often used in precise measurements of heats of reaction is called the *bomb calorimeter* (Figure 5.5). It is used to measure energy changes for reactions that will not happen until they are deliberately initiated, for example, combustions which must be ignited. The reactants are put into the "bomb," which is then sealed and immersed in a large, well-insulated vat of water. When the reaction is set off, any heat that is liberated is absorbed by the bomb, the water, and any piece of the equipment sticking into the water, and the temperature of the entire contents of the vat rises. The stirrer ensures that any heat released becomes uniformly distributed before the final

It's called a "bomb" because early models reminded the scientists involved of a bomb (and sometimes the device exploded when the reaction was initiated).

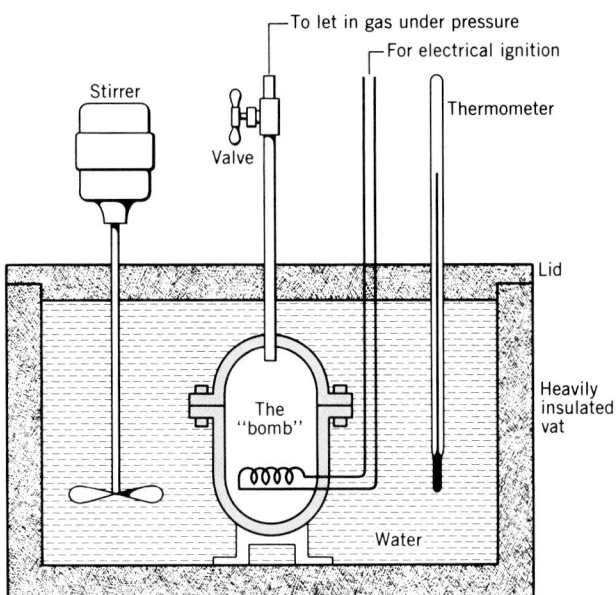

FIGURE 5.5 A bomb calorimeter. The water bath is usually equipped with devices for adding or removing heat from the water to keep the temperature constant.

temperature is read. From the temperature change and the heat capacity of the calorimeter (water plus everything in the water), the heat liberated is calculated.

When the reaction is combustion, the enthalpy change is called the **heat of combustion,** and its units are usually kilojoules per mole or kilocalories per mole. As we will see soon, heats of combustion provide data from which other heats of reaction can be calculated. The next example shows how bomb calorimetry data can be used to calculate a heat of combustion.

EXAMPLE 5.5   DETERMINING A HEAT OF COMBUSTION

Problem: A sample of sucrose (table sugar) with a mass of 1.32 g is burned in a bomb calorimeter. The heat capacity of this calorimeter had been previously found to be 9.43 kJ/°C. The temperature changed from 25.00 °C to 27.31 °C. Calculate the heat of combustion of sucrose in units of kilojoules per mole. The formula of sucrose is $C_{12}H_{22}O_{11}$.

Solution: The change in temperature is 2.31 °C. For each degree increase, the reaction has evolved 9.43 kJ, as we know from the heat capacity. Therefore the total heat evolved is

$$2.31 \; \text{°C} \times 9.43 \; \frac{kJ}{1 \; \text{°C}} = 21.8 \; kJ$$

This heat was produced by the combustion of 1.32 g sucrose; this corresponds in moles to

$$1.32 \; \text{g sucrose} \times \frac{1 \; mol \; sucrose}{342.3 \; \text{g sucrose}} = 3.86 \times 10^{-3} \; mol \; sucrose$$

Therefore, the heat evolved per mole of sucrose is

$$\frac{21.8 \; kJ}{3.86 \times 10^{-3} \; mol} = 5.65 \times 10^{3} \; kJ/mol$$

Since combustion is exothermic, this should be given a minus sign and reported as -5.65×10^{3} kJ/mol.

PRACTICE EXERCISE 4   Gram for gram, fats in food have much more chemical energy than sugar. One component of fat is stearic acid, $C_{18}H_{36}O_2$. When a sample of 1.02 g of stearic acid was burned
completely in a bomb calorimeter, the temperature of the calorimeter rose by 4.26 °C.
The heat capacity of the calorimeter (the same one used in Example 5.5) was 9.43 kJ/°C.
Calculate the molar heat of combustion of stearic acid in kilojoules per mole.

Heats of Reaction at Constant Pressure versus at Constant Volume

Strictly speaking, the value we obtained for the heat of combustion in Example 5.5
should not really be called an enthalpy change. Enthalpy changes correspond to heats of
reaction measured at constant pressure, but in a bomb calorimeter it is the volume that
stays the same, not necessarily the pressure. Under these conditions, any gases that are
formed are confined and the pressure builds up. For reasons that we will not go into at
this time, it happens that the heat of reaction measured when the reaction is carried out
at constant volume is not quite the same as the heat of reaction measured under
conditions of constant pressure. The difference is small, however, even when large
amounts of gases are produced or consumed and the pressure changes are very large. For
example the combustion of one mole of octane (a component of gasoline) at 100 °C
liberates 1220 kcal (5104 kJ) at constant pressure. The identical reaction carried out in a
bomb calorimeter at the same temperature liberates at constant volume about 2.6 kcal/
mol more heat, or 1223 kcal/mol (rounded). The difference amounts to about 0.2%.

   The reason why the heats of reaction at constant pressure and at constant volume
differ will be discussed in Chapter 16. From a practical standpoint, they are so nearly
alike that we are only concerned about the difference when very precise calculations are
involved on systems in which large amounts of gases are formed or consumed. To avoid
any problems, however, we will deal with constant pressure conditions for the remainder
of this chapter so we can limit our discussions to enthalpy changes.

5.5 STANDARD HEATS OF REACTION

**Heats of various reactions can be compared if they are measured under the same
standard conditions of temperature and pressure.**

The value of ΔH for a reaction depends not only on the specific reaction but also on the
temperature and pressure. This is because these conditions are needed to define the state
of a system, and ΔH is a state function. We cannot directly compare enthalpy changes for
different reactions, therefore, unless their values of ΔH are measured under the same
conditions of temperature and pressure.

Standard Conditions for Enthalpy Changes

Scientists have agreed to a somewhat arbitrary set of reference conditions of temperature
and pressure for comparing enthalpy data. These conditions have been chosen to make
them easy to establish during an experiment. Thus the standard reference temperature is
25 °C, which is just slightly above normal room temperature. We can easily keep a
reaction at this temperature by immersing the reaction vessel in a vat of water kept at
25 °C with a thermostatically controlled heater.

   The reference pressure is approximately the pressure exerted by our atmosphere.
This pressure, of course, varies somewhat from day to day according to the weather as
well as from place to place. You have no doubt heard of high-pressure and low-pressure
air masses while listening to weather forecasts. The standard pressure, therefore, must be

carefully defined. We will discuss pressure in detail in Chapter 9, but for now we can use both the name and a reasonably acceptable definition of the standard pressure. It is called one standard atmosphere; its symbol is atm, and we can consider it to be approximately the average pressure exerted by the earth's atmosphere at sea level.

Standard Heat of Reaction

When the enthalpy change of a reaction is determined with all reactants and products at 1 atm and some given temperature, and when the scale of the reaction is in the *moles* specified by the coefficients of the equation, then ΔH is the standard enthalpy change or the standard heat of reaction. To show that a pressure of 1 atm is used, the symbol ΔH is given a superscript, $\circ$, to make the symbol ΔH°. Values of ΔH° usually correspond to an initial and final temperature of 25 °C, unless specified otherwise.

The units of ΔH° are normally kilojoules or kilocalories. For example, a reaction between gaseous nitrogen and hydrogen produces gaseous ammonia according to the equation

$$N_2(g) + 3H_2(g) \longrightarrow 2NH_3(g)$$

When specifically 1.000 mol of N_2 and 3.000 mol of H_2 react to form 2.000 mol of NH_3 at 25 °C and 1 atm, the reaction releases 92.38 kJ. Hence, for the reaction as given by the above equation, $\Delta H^\circ = -92.38$ kJ.

Thermochemical Equations

Often it is useful to make the enthalpy change of a reaction part of its equation. When we do this, we have to be very careful about the coefficients, and we must indicate the physical states of all the reactants and products. The reaction between gaseous nitrogen and hydrogen to form gaseous ammonia, for example, releases 92.38 kJ if 2.000 mol of NH_3 forms.

But if we were to make twice as much, or 4.000 mol, of NH_3 from 2.000 mol of N_2 and 6.000 mol of H_2, then twice as much heat (184.8 kJ) would be released. On the other hand, if only 0.5000 mol of N_2 and 1.500 mol of H_2 were to react to form only 1.000 mol of NH_3, then only half as much heat (46.19 kJ) would be released.

An equation that includes its value of ΔH° is called a thermochemical equation. The following three thermochemical equations for the formation of ammonia, for example, give the quantitative data described in the preceding paragraph and correctly specify the physical states of all substances.

$$N_2(g) + 3H_2(g) \longrightarrow 2NH_3(g) \qquad \Delta H^\circ = -92.38 \text{ kJ}$$
$$2N_2(g) + 6H_2(g) \longrightarrow 4NH_3(g) \qquad \Delta H^\circ = -184.8 \text{ kJ}$$
$$\tfrac{1}{2}N_2(g) + \tfrac{3}{2}H_2(g) \longrightarrow NH_3(g) \qquad \Delta H^\circ = -46.19 \text{ kJ}$$

When we read a thermochemical equation, we always interpret the coefficients as *moles*. This is why we can use fractional coefficients in such equations, where normally we avoid them (because we cannot have fractions of molecules). In thermochemical equations, however, fractions are allowed, because we can have fractions of moles.

PRACTICE EXERCISE 5   The reaction of 2.000 mol of gaseous hydrogen with 1.000 mol of gaseous oxygen to form 2.000 mol of liquid water releases 517.8 kJ, provided that all reactants and products are brought to 25 °C and 1 atm. Write a thermochemical equation for the formation of 1.000 mol of liquid water.

A perpetual motion machine is one that creates more energy than it uses and so could run forever (or at least until key parts wear out). The U.S. Patent Office became so tired of receiving patent applications for these impossibilities that it took the step of insisting that all applications be accompanied by *working* models. No applications have come in since.

Here's how such a machine would work in principle. Suppose that it costs only 50 kcal to break CO_2 back to C and O_2. If we get 94 kcal/mol by burning carbon to carbon dioxide but only 50 kcal/mol to recycle the system back to C + O_2, our profit is

44 kcal/mol. Over and over, we could let the elements burn, get 94 kcal/mol, save 50 kcal/mol for recycling, and get 44 kcal/mol free each time. We could use this energy, say, to swat flies, bake cookies, repair wornout parts of the machine, or run something else while we took it easy. Very nice—but impossible. We don't know *why* it's impossible, except that this is the way of nature. All we have done is *discover* that it is impossible, and out of this discovery—often repeated by would-be inventors of perpetual motion machines—came the law of conservation of energy.

Thermochemical Equations for Experimentally Difficult Reactions

Once we have the thermochemical equation for a particular reaction, we automatically have all we need to write the thermochemical equation of the reverse reaction. The thermochemical equation for the combustion of carbon to give carbon dioxide, for example, is

$$C(s) + O_2(g) \longrightarrow CO_2(g) \qquad \Delta H° = -393.5 \text{ kJ}$$

The reverse reaction would be the decomposition of carbon dioxide to carbon and oxygen. Although this reaction is very hard to carry out experimentally, we can still know what its value of $\Delta H°$ would have to be from the law of conservation of energy. This law requires that $\Delta H°$ for the decomposition must be $+393.5$ kJ. If the values of ΔH for forward and reverse reactions were not equal but opposite in sign, then perpetual motion machines would be possible (which they aren't), as Special Topic 5.4 discusses. So we can write the thermochemical equation for the decomposition of carbon dioxide to carbon and oxygen as follows.

$$CO_2(g) \longrightarrow C(s) + O_2(g) \qquad \Delta H° = +393.5 \text{ kJ}$$

Thus, if we have $\Delta H°$ for a given reaction, then $\Delta H°$ for the reverse reaction has the same numerical value, but its algebraic sign is reversed. This extremely useful fact makes data available that would otherwise be impossible to measure.

5.6 HESS'S LAW OF HEAT SUMMATION

When thermochemical equations are added to give some new equation, their values of $\Delta H°$ are also added to give the $\Delta H°$ of the new equation.

When a real sample of carbon burns, all sorts of processes occur, but the overall ΔH depends only on the initial and final states.

Earlier we noted that the enthalpy change for a reaction is a state function. Its value is determined only by the enthalpies of the initial and final states of the chemical system, and not by the path taken by the reactants as they form the products. To appreciate the significance of this, let us consider again the combustion of carbon to give carbon dioxide. There are two ways or two paths by which this net reaction can occur. One is the complete combustion in one step, and its thermochemical equation is

$$C(s) + O_2(g) \longrightarrow CO_2(g) \qquad \Delta H° = -393.5 \text{ kJ} \ (-94.05 \text{ kcal})$$

The second path to carbon dioxide from its elements involves two steps. The first is the combination of carbon with just enough oxygen to form carbon monoxide. Then, in the second step, this CO is burned with additional oxygen to produce CO_2. Both steps are exothermic, and their thermochemical equations are

$$C(s) + \tfrac{1}{2}O_2(g) \longrightarrow CO(g) \qquad \Delta H° = -110.5 \text{ kJ} \ (-26.42 \text{ kcal})$$
$$CO(g) + \tfrac{1}{2}O_2(g) \longrightarrow CO_2(g) \qquad \Delta H° = -283.0 \text{ kJ} \ (-67.63 \text{ kcal})$$

Notice that if we add the amount of heat liberated in the first step to the amount released in the second, the total is the same as the heat given off by the one-step reaction that we described first.

$$(-110.5 \text{ kJ}) + (-283.0 \text{ kJ}) = -393.5 \text{ kJ}$$

Or, in kilocalories,

$$(-26.42 \text{ kcal}) + (-67.63 \text{ kcal}) = -94.05 \text{ kcal}$$

Enthalpy Diagrams

The enthalpy relationships involved in adding thermochemical equations are most easily visualized by means of an enthalpy diagram, such as that given in Figure 5.6. Horizontal lines in such a diagram correspond to different absolute values of enthalpy, H. A horizontal line drawn higher in the diagram represents a larger value of H than a line drawn lower, which corresponds to a smaller value of H. *Changes* in enthalpy, ΔH, are represented by the vertical distances between these lines. For an exothermic reaction, the reactants are always on a higher line than the products, as seen by the location of $C(s) + O_2(g)$ in Figure 5.6, because in exothermic changes the reactants have more chemical energy (and more enthalpy, therefore) than the products. Think of a horizontal line as representing the *sum* of the enthalpies of all of the substances on the line (and in the physical states specified).

To the left in Figure 5.6 we see an arrow pointing down that connects the reactants, $C(s) + O_2(g)$, with the final product, $CO_2(g)$. On the right the overall change stops at the intermediate products, $CO(g) + \tfrac{1}{2}O_2(g)$ — those made (or left over) in the first step of the two-step path to the final product. The total decrease in energy, however, is the same regardless of which path is taken, so the total energy evolved in the two-step path has to be the same as in the one-step reaction.

Hess's Law

Scientists have observed many other examples illustrating how enthalpy changes can be added when thermochemical equations are added. These experiences led to the formulation of Hess's law of heat summation, named in honor of an early student of thermochemistry.

> **Law of Heat Summation (Hess's Law).** For any reaction that can be written in steps, the standard heat of reaction is the same as the sum of the standard heats of reaction for the steps.

One of the most useful applications of Hess's law is calculation of the value of $\Delta H°$ for a reaction whose $\Delta H°$ is unknown or cannot be measured. Hess's law says that we can add thermochemical equations, including their values of $\Delta H°$, to obtain some desired thermochemical equation and its $\Delta H°$. Applying this concept often involves manipulating thermochemical equations in ways that are already familiar. The technique is best illustrated by examples, and we begin with a familiar system — the reaction of carbon with oxygen.

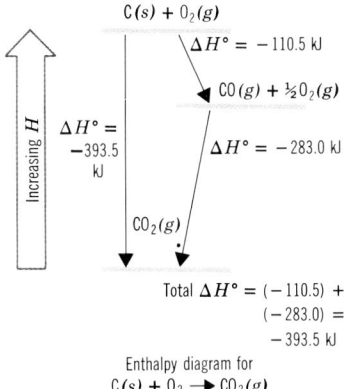

Total $\Delta H° = (-110.5) +$
$(-283.0) =$
-393.5 kJ

Enthalpy diagram for
$C(s) + O_2 \rightarrow CO_2(g)$

FIGURE 5.6 An enthalpy diagram for the formation of carbon dioxide gas from its elements. Whether it is carried out in one direct step or two steps, the overall enthalpy change is the same.

An exothermic change always has a *downward*-pointing arrow in an enthalpy diagram.

EXAMPLE 5.6   MANIPULATING THERMOCHEMICAL EQUATIONS

Problem: Consider the following thermochemical equations:

$$C(s) + \tfrac{1}{2}O_2(g) \longrightarrow CO(g) \qquad \Delta H° = -110.5 \text{ kJ}$$
$$CO(g) + \tfrac{1}{2}O_2(g) \longrightarrow CO_2(g) \qquad \Delta H° = -283.0 \text{ kJ}$$

Use them to find $\Delta H°$ in kilojoules for the reaction

$$C(s) + O_2(g) \longrightarrow CO_2(g)$$

Solution: This is a particularly simple problem, but it illustrates a few important points about all such problems. Let's add the two given thermochemical equations:

$$C(s) + \tfrac{1}{2}O_2(g) \longrightarrow CO(g) \qquad\qquad \Delta H° = -110.5 \text{ kJ}$$
$$\underline{CO(g) + \tfrac{1}{2}O_2(g) \longrightarrow CO_2(g) \qquad\qquad \Delta H° = -283.0 \text{ kJ}}$$

$$C(g) + \tfrac{1}{2}O_2(g) + \tfrac{1}{2}O_2(g) + CO(g) \longrightarrow CO(g) + CO_2(g)$$

This resulting equation can be simplified in two ways. First, $CO(g)$ appears twice but on opposite sides of the arrow; so we can cancel both formulas, just as we can cancel terms found on two sides of a mathematical equation. We can always cancel identical formulas on opposites of the arrow in this manner, *provided that the chemicals are in identical physical states* (for example, if both are gases). The second simplification is to combine the two $\tfrac{1}{2}O_2(g)$ terms. We can add chemical formulas like this *provided that they are in the same physical state*. This gives one $O_2(g)$ on the left. These simplifications give us the desired "target" equation:

$$C(s) + O_2(g) \longrightarrow CO_2(g)$$

Hess's law says that if we can add two (or more) equations to obtain a target equation, we simply add their $\Delta H°$ values to find the $\Delta H°$ for the target equation. Restating the solution to the problem then gives

$$C(s) + \tfrac{1}{2}O_2(g) \longrightarrow CO(g) \qquad \Delta H° = -110.5 \text{ kJ}$$
$$\underline{CO(g) + \tfrac{1}{2}O_2(g) \longrightarrow CO_2(g) \qquad \Delta H° = -283.0 \text{ kJ}}$$

$$C(g) + O_2(g) \longrightarrow CO_2(g) \qquad \Delta H° = -393.5 \text{ kJ}$$

EXAMPLE 5.7   MANIPULATING THERMOCHEMICAL EQUATIONS

Problem: Carbon monoxide is often used in metallurgy to remove oxygen from metal oxides and thereby give the free metal. The thermochemical equation for the reaction of CO with iron(III) oxide, Fe_2O_3, is

This reaction also occurs in making iron from iron ore.

$$Fe_2O_3(s) + 3CO(g) \longrightarrow 2Fe(s) + 3CO_2(g) \qquad \Delta H° = -6.39 \text{ kcal}$$

Use this equation and the equation for the combustion of CO

$$CO(g) + \tfrac{1}{2}O_2(g) \longrightarrow CO_2(g) \qquad \Delta H° = -67.63 \text{ kcal}$$

to calculate the value of $\Delta H°$ in kilocalories for the reaction

$$2Fe(s) + \tfrac{3}{2}O_2(g) \longrightarrow Fe_2O_3(s)$$

Solution: As in the previous example, we must combine the given equations in such a way that we obtain the final equation. Then we just add the corresponding values of $\Delta H°$ to obtain the $\Delta H°$ of the target equation. This time, however, simple addition of the given equations doesn't give the equation we want. How, then, do we proceed?

We first have to manipulate the given equations so that when we add them we will get the target equation. During this manipulation, we must keep an eye on where we are going. By using the following reasoning, we can manipulate the given equations to achieve our goal, namely, the adjusted equations and their sum, given below.

Step 1. The target equation must have 2Fe on the *left,* but the first equation above has 2Fe to the right of the arrow. To move it to the left, we must reverse the entire equation, remembering also to reverse the sign of $\Delta H°$. "Keeping an eye out," we also are pleased to note that this manipulation places Fe_2O_3 to the right of the arrow, which is where it has to be after we add our adjusted equations.

Step 2. There must be $\frac{3}{2}O_2$ on the left, and we must be able to cancel *three* CO and *three* CO_2 when the equations are added. If we multiply by 3 the second of the equations given above, we will obtain the necessary coefficients. We must also multiply the value of $\Delta H°$ of this equation by 3, because three times as much chemicals are now involved in the reaction.

Here are the adjusted equations followed by their sum:

$$2Fe(s) + 3CO_2(g) \longrightarrow Fe_2O_3(s) + 3CO(g) \qquad \Delta H° = +6.39 \text{ kcal}$$
$$3CO(g) + \tfrac{3}{2}O_2(g) \longrightarrow 3CO_2(g) \qquad \begin{array}{l} \Delta H° = 3(-67.63 \text{ kcal}) \\ = -202.9 \text{ kcal} \end{array}$$

$$\overline{2Fe(s) + \tfrac{3}{2}O_2(g) \longrightarrow Fe_2O_3(s) \qquad \Delta H° = -196.5 \text{ kcal}}$$

Notice once again that when we reverse a thermochemical equation, we must reverse the sign of $\Delta H°$, and when we multiply the coefficients of an equation by some factor, we must multiply the value of $\Delta H°$ by the same factor.

PRACTICE EXERCISE 6

Ethanol is also called ethyl alcohol or grain alcohol.

Ethanol, C_2H_5OH, is made industrially by the reaction of water with ethylene, C_2H_4. Calculate the value of $\Delta H°$ for the reaction

$$C_2H_4(g) + H_2O(\ell) \longrightarrow C_2H_5OH(\ell)$$

given the following thermochemical equations:

$$C_2H_4(g) + 3O_2(g) \longrightarrow 2CO_2(g) + 2H_2O(\ell) \qquad \Delta H° = -1411.1 \text{ kJ}$$
$$C_2H_5OH(\ell) + 3O_2(g) \longrightarrow 2CO_2(g) + 3H_2O(\ell) \qquad \Delta H° = -1367.1 \text{ kJ}$$

5.7 STANDARD HEATS OF FORMATION AND HESS'S LAW

The $\Delta H°$ for a reaction can be obtained by subtracting the sum of the standard heats of formation of the reactants from the sum of the standard heats of formation of the products.

In the last section we saw that the value of $\Delta H°$ for a reaction is independent of the path that the reaction takes to convert reactants into products. We can measure the $\Delta H°$ for a one-step path, or we can add up the values of $\Delta H°$ for a sequence of reactions that lead ultimately to the same final state of the system. Either way, we obtain the same value for $\Delta H°$. This fact has given us a simple way to compute a value of $\Delta H°$ for a reaction from thermochemical data available in published tables. These tabulated data are the enthalpy changes that accompany the formation of compounds from their elements. To use such data, we figure out an alternative path for an overall reaction — a path that involves first the decomposition of the reactants into their elements, then the combination of these elements into the compounds found among the products. This is illustrated in a general way in Figure 5.7 on page 168 for a system in which the reactants and the products each have lower enthalpies than their elements.

The net enthalpy change for the system depicted by Figure 5.7 is the algebraic sum of the $\Delta H°$ for the decomposition of the reactants and the $\Delta H°$ for the formation of the products.

$$\Delta H° = (\Delta H° \text{ for decomposition of reactants}) + (\Delta H° \text{ for formation of products}) \quad (5.9)$$

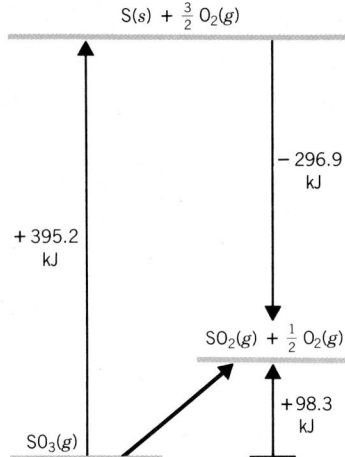

FIGURE 5.7 Enthalpy diagram showing how to use standard heats of formation to calculate the standard enthalpy of a reaction. The reaction used here to illustrate the method is

$$SO_3(g) \longrightarrow SO_2(g) + \tfrac{1}{2}O_2(g)$$

(The heavy arrow in the drawing signifies this change.) Since the value of ΔH_f° for $SO_3(g)$ is -395.2 kJ/mol (from Table 5.2), it costs $+395.2$ kJ to decompose one mole of SO_3 into its elements (shown on the upper enthalpy line). When one mole of sulfur and one of oxygen combine to give sulfur dioxide, the reaction liberates the ΔH_f° for one mole of $SO_2(g)$, -296.9 kJ. (We show $\tfrac{1}{2}$ mole of oxygen riding along from the upper line to the enthalpy line for SO_2. It's the other product of the overall reaction; remember that elements in their standard states have zero values of enthalpies of formation.) The difference between the enthalpy line for SO_3 and the line for $(SO_2 + \tfrac{1}{2}O_2)$ is $+98.3$ kJ. This is the value of ΔH° that we sought for the given reaction.

ΔH_f° is just a special kind of ΔH°.

Remember, ΔH°, unlike ΔH_f°, is for whatever mole quantities are given by the equation's coefficients. ΔH_f° is always for just one mole of the product being formed.

Now let us see how we can obtain a value for the first term after the equals sign in Equation 5.9. The value of ΔH° for the *decomposition* of the reactants has to be numerically equal but opposite in sign to the value of ΔH° for the *formation* of the reactants.

(ΔH° for decomposition of reactants) $= -(\Delta H^\circ$ for formation of reactants)

Substituting this into Equation 5.9 gives us

$$\Delta H^\circ = -(\Delta H^\circ \text{ for formation of reactants}) + (\Delta H^\circ \text{ for formation of products})$$

and rearranging terms gives

$$\Delta H^\circ = (\Delta H^\circ \text{ for formation of products}) - (\Delta H^\circ \text{ for formation of reactants}) \quad (5.10)$$

This important equation tells us that tabulated values of ΔH° for the formation of reactants and products can be used to compute ΔH° for the reaction. This is particularly useful when it is difficult or impossible to determine ΔH° directly. As we will soon see, we can do this calculation without having to manipulate thermochemical equations.

Standard Heats of Formation

With the preceding paragraphs as background, let us define in a precise way what we mean by "ΔH° of formation." We must begin, however, by defining the term *standard state*. Any substance in its most stable physical form (gas, liquid, or solid) at 25 °C and under a pressure of 1 atm is said to be in its standard state. The element oxygen, for example, is in its standard state when it exists as *molecules* of O_2 — not as atoms and not as molecules of ozone (O_3) — at 25 °C and 1 atm. The element carbon is in its standard state when it exists as graphite — not as diamond — at 25 °C and 1 atm. (Diamond is also a form of carbon, but is actually slightly less stable than graphite.)

The quantities that we will use to compute values of ΔH° for reactions are called standard enthalpies of formation or standard heats of formation. The standard heat of formation of a compound, ΔH_f° is the amount of heat absorbed or evolved when *one mole of the compound is formed from its elements in their standard states*. Thus, the thermochemical equation for the formation of one mole of liquid water from oxygen and hydrogen in their standard states is

$$H_2(g) + \tfrac{1}{2}O_2(g) \longrightarrow H_2O(\ell) \qquad \begin{array}{l} \Delta H_f^\circ = -285.9 \text{ kJ/mol} \\ \text{or } \Delta H_f^\circ = -68.32 \text{ kcal/mol} \end{array}$$

The standard enthalpy change for this reaction, that is, the enthalpy change at 25 °C and 1 atm, is called the *standard heat of formation of liquid water*.

The subscript "f" in the symbol ΔH_f° always means that we are dealing with a reaction in which *one mole* of a compound is made *from its elements*. This is a point that often causes confusion among students. Each of the following equations, for example, involves the formation of $CO_2(g)$.

$$C(s) + O_2(g) \longrightarrow CO_2(g)$$
$$CO(g) + \tfrac{1}{2}O_2(g) \longrightarrow CO_2(g)$$
$$2C(s) + 2O_2(g) \longrightarrow 2CO_2(g)$$

However, only the first involves *just* the elements as reactants and the formation of just *one mole* of CO_2. In the second equation, one of the reactants is a compound, carbon monoxide, and in the third, two moles of CO_2 are formed. Only the first equation, therefore, has a standard enthalpy change that we identify as ΔH_f°. The standard enthalpy change for the second reaction would be written simply as ΔH°.

Values of standard heats of formation for a variety of substances are given in Table 5.2. Notice that the values given for the elements are zero. The ΔH_f° for any element in

TABLE 5.2 Standard Enthalpies of Formation of Typical Substances

Substance	ΔH_f° kJ/mol	ΔH_f° kcal/mol	Substance	ΔH_f° kJ/mol	ΔH_f° kcal/mol
Ag(s)	0.00	0.00	$H_2O_2(\ell)$	−187.6	−44.84
AgCl(s)	−127.0	−30.36	HCl(g)	−92.30	−22.06
Al(s)	0.00	0.00	HI(g)	26.6	6.35
$Al_2O_3(s)$	−1669.8	−399.09	$HNO_3(\ell)$	−173.2	−41.40
C(s, graphite)	0.00	0.00	$H_2SO_4(\ell)$	−811.32	−193.91
CO(g)	−110.5	−26.42	$HC_2H_3O_2(\ell)$	−487.0	−116.4
$CO_2(g)$	−393.5	−94.05	Hg(ℓ)	0.00	0.00
$CH_4(g)$	−74.848	−17.889	Hg(g)	60.84	14.54
$CH_3Cl(g)$	−82.0	−19.6	$I_2(s)$	0.00	0.00
$CH_3I(g)$	14.2	3.40	K(s)	0.00	0.00
$CH_3OH(\ell)$	−238.6	−57.02	KCl(s)	−435.89	−104.18
$CO(NH_2)_2(s)$ (urea)	−333.19	−79.634	$K_2SO_4(s)$	−1433.7	−342.66
$CO(NH_2)_2(aq)$	−319.2	−76.30	$N_2(g)$	0.00	0.00
$C_2H_2(g)$	226.75	54.194	$NH_3(g)$	−46.19	−11.08
$C_2H_4(g)$	52.284	12.496	$NH_4Cl(s)$	−315.4	−75.38
$C_2H_6(g)$	−84.667	−20.236	NO(g)	90.37	21.60
$C_2H_5OH(\ell)$	−277.63	−66.356	$NO_2(g)$	33.8	8.09
Ca(s)	0.00	0.00	$N_2O(g)$	81.57	19.49
$CaCO_3(s)$	−1207	−288.5	$N_2O_4(g)$	9.67	2.31
$CaCl_2(s)$	−795.0	−190.0	Na(s)	0.00	0.00
CaO(s)	−635.5	−151.9	$NaHCO_3(s)$	−947.7	−226.5
$Ca(OH)_2(s)$	−986.59	−235.80	$Na_2CO_3(s)$	−1131	−270.3
$CaSO_4(s)$	−1432.7	−342.42	NaCl(s)	−411.0	−98.23
$CaSO_4 \cdot \frac{1}{2}H_2O(s)$	−1575.2	−376.47	NaOH(s)	−426.8	−102.0
$CaSO_4 \cdot 2H_2O(s)$	−2021.1	−483.06	$Na_2SO_4(s)$	−1384.5	−330.90
$Cl_2(g)$	0.00	0.00	$O_2(g)$	0.00	0.00
Fe(s)	0.00	0.00	Pb(s)	0.00	0.00
$Fe_2O_3(s)$	−822.2	−196.5	PbO(s)	−219.2	−52.40
$H_2(g)$	0.00	0.00	S(s)	0.00	0.00
$H_2O(g)$	−241.8	−57.80	$SO_2(g)$	−296.9	−70.96
$H_2O(\ell)$	−285.9	−68.32	$SO_3(g)$	−395.2	−94.45

A more complete table of standard enthalpies of formation is located in Appendix D at the back of the book.

its standard state is zero. This makes sense, if you think about it — there is no enthalpy change if you "form an element in its standard state from itself."

Using Hess's Law with Standard Heats of Formation

Now we are ready to apply Hess's law using ΔH_f° values. We simply apply the idea that we developed earlier in this section — the ΔH° for any reaction must be the difference between the total enthalpies of formation of the products and those of the reactants.

Any reaction can be generalized by the equation

$$aA + bB + \cdots \longrightarrow nN + mM + \cdots$$

where a, b, n, etc., are the coefficients of substances A, B, N, etc. We can find the value of the reaction's ΔH° by the following equation, called the Hess's law equation.

$$\Delta H^\circ = [n\Delta H_f^\circ(N) + m\Delta H_f^\circ(M) + \cdots] - [a\Delta H_f^\circ(A) + b\Delta H_f^\circ(B) + \cdots]$$

What the Hess's law equation amounts to is nothing more than the following:

$$\Delta H° = [\text{sum of } \Delta H° \text{ of products}] - [\text{sum of } \Delta H° \text{ of reactants}] \quad (5.11)$$

The following examples show how Equation 5.11 can be used.

EXAMPLE 5.8   USING $\Delta H_f°$ IN HESS'S LAW CALCULATIONS

Problem: Some chefs keep baking soda, $NaHCO_3$, handy to put out grease fires. When thrown on the fire, baking soda partly smothers the fire, and the heat decomposes it to give CO_2, which further smothers the flame. The equation for the decomposition of $NaHCO_3$ is

$$2NaHCO_3(s) \longrightarrow Na_2CO_3(s) + H_2O(g) + CO_2(g)$$

Use the data in Table 5.2 to calculate the $\Delta H°$ for this reaction in kilojoules.

Solution: We calculate the values of $\Delta H°$ for the products, taken as a set, and the values of $\Delta H°$ for the reactants, also taken as a set. Then we subtract the latter from the former to calculate $\Delta H°$. We compute values of $\Delta H°$ for the products using $\Delta H_f°$ data from a table of such data and we use the coefficients of the chemical equation as the moles of products formed. We compute values of $\Delta H°$ for the reactants in the same way. Thus,

$$\Delta H° = [1 \text{ mol } Na_2CO_3 \times \Delta H_f°(Na_2CO_3) + 1 \text{ mol } H_2O \times \Delta H_f°(H_2O) + 1 \text{ mol } CO_2 \times \Delta H_f°(CO_2)]$$
$$- [2 \text{ mol } NaHCO_3 \times \Delta H_f°(NaHCO_3)]$$

Looking up the values of $\Delta H_f°$ in Table 5.2 and substituting give

$$\Delta H° = \left[1 \text{ mol}\left(-1131 \frac{kJ}{mol}\right) + 1 \text{ mol}\left(-241.8 \frac{kJ}{mol}\right) + 1 \text{ mol}\left(-393.5 \frac{kJ}{mol}\right)\right]$$
$$- \left[2 \text{ mol}\left(-947.7 \frac{kJ}{mol}\right)\right]$$
$$= (-1766 \text{ kJ}) - (-1895 \text{ kJ})$$
$$= +129 \text{ kJ}$$

Thus, under standard conditions, the reaction is endothermic by 129 kJ.

EXAMPLE 5.9   CALCULATING WITH HESS'S LAW EQUATION

Problem: What is the $\Delta H°$ in kilocalories for the combustion of 1 mol of ethanol, $C_2H_5OH(\ell)$, to form gaseous carbon dioxide and gaseous water?

Solution: We first have to write the balanced equation:

$$C_2H_5OH(\ell) + 3O_2(g) \longrightarrow 2CO_2(g) + 3H_2O(g)$$

The form that the Hess law equation takes for this reaction is as follows.

$$\Delta H° = [2 \text{ mol } CO_2 \times \Delta H_f°(CO_2) + 3 \text{ mol } H_2O \times \Delta H_f°(H_2O)]$$
$$- [1 \text{ mol } C_2H_5OH \times \Delta H_f°(C_2H_5OH) + 3 \text{ mol } O_2 \times \Delta H_f°(O_2)]$$

$$= \left[2 \text{ mol}\left(-94.05 \frac{kcal}{mol}\right) + 3 \text{ mol}\left(-57.80 \frac{kcal}{mol}\right)\right]$$

Remember, $\Delta H_f°$ is zero for all elements in their most stable forms.

$$- \left[1 \text{ mol}\left(-66.356 \frac{kcal}{mol}\right) + 3 \text{ mol}\left(0 \frac{kcal}{mol}\right)\right]$$
$$= -295.1 \text{ kcal}$$

This is the heat of combustion of 1.000 mol of ethanol under standard conditions. Notice that $\Delta H_f°$ for oxygen is zero, as we have already explained.

When you make these calculations, be sure to be very careful about the physical state of the compounds in Table 5.2. There are two values of ΔH_f° for water, one for gaseous water and one for liquid water. We had to pick the value for gaseous water in Examples 5.8 and 5.9 because that is what the equations specified.

PRACTICE EXERCISE 7   Calculate ΔH° in kilojoules for the following reactions.
(a) $2NO(g) + O_2(g) \longrightarrow 2NO_2(g)$
(b) $NaOH(s) + HCl(g) \longrightarrow NaCl(s) + H_2O(g)$

Values of ΔH_f° from Standard Heats of Combustion

The previous two examples illustrated how useful it is to have values of ΔH_f°. The experimental determination of these values by the *direct* combination of the elements, however, is often either difficult or impossible. For example, to measure directly the heat of formation of sucrose, $C_{12}H_{22}O_{11}$, we would have to carry out the following reaction:

$$12C(s) + 11H_2(g) + \tfrac{11}{2}O_2(g) \longrightarrow C_{12}H_{22}O_{11}(s)$$

But no one has ever been able to figure out how to make this reaction occur directly under any conditions, so there is no direct way to measure the ΔH_f° of sucrose. How, then, can we obtain values of ΔH_f° for compounds such as sucrose?

If the compound in question can be burned — which is usually far easier to do than make it from its elements — then we have a source of energy data that we can use to calculate its ΔH_f°. This is because the products of combustion are nearly always compounds whose values of ΔH_f° are known or can be measured by direct means. For instance, the combustion of sucrose in an atmosphere of pure oxygen proceeds by the following equation.

Measuring heats of combustion with a bomb calorimeter was described on page 160.

$$C_{12}H_{22}O_{11}(s) + 12O_2(g) \longrightarrow 12CO_2(g) + 11H_2O(\ell)$$

If the standard enthalpy change for this reaction can be measured, and if we can look up the values of ΔH_f° for three of the four chemicals in the equation (H_2O, CO_2, and O_2), then we can use the Hess law equation to find the ΔH_f° of the remaining substance, sucrose.

The enthalpy change for the combustion of *one* mole of a compound under standard conditions is called the standard heat of combustion, and its symbol is $\Delta H_{combustion}^\circ$. Its value for sucrose is -5640.9 kJ/mol. Let us now see how we can use a standard heat of combustion to calculate ΔH_f° for a compound that is impossible to make from its elements.

$\Delta H_{combustion}^\circ$ is also a special kind of ΔH°.

EXAMPLE 5.10   USING $\Delta H_{combustion}^\circ$ TO CALCULATE ΔH_f°

Problem: One of the "building blocks" for proteins such as those in muscles and sinews is an amino acid called glycine, $C_2H_5NO_2$. The equation for its combustion is

$$4C_2H_5NO_2(s) + 9O_2(g) \longrightarrow 8CO_2(g) + 10H_2O(\ell) + 2N_2(g)$$

The value of its $\Delta H_{combustion}^\circ$ is -973.49 kJ/mol. (Remember this means kilojoules *per mole* of glycine, not for four moles.) Using this information and values of ΔH_f° calculate the ΔH_f° for glycine.

Solution: For this problem, Hess's law equation (Equation 5.11) becomes

$$\Delta H^\circ = [8 \text{ mol} \times \Delta H_f^\circ(CO_2) + 10 \text{ mol} \times \Delta H_f^\circ(H_2O) + 2 \text{ mol} \times \Delta H_f^\circ(N_2)]$$
$$- [4 \text{ mol} \times \Delta H_f^\circ(C_2H_5NO_2) + 9 \text{ mol} \times \Delta H_f^\circ(O_2)]$$

The first term, ΔH°, is obtained from the standard heat of combustion of glycine. Since the chemical equation for this reaction is for the combustion of *four* moles of glycine, we have to multiply $\Delta H_{combustion}^\circ$ by four.

$$\Delta H^\circ = 4 \text{ mol} \times \left(-973.49 \frac{kJ}{mol}\right) = -3894.0 \text{ kJ}$$

From Table 5.2 we find the following values for ΔH_f°: $CO_2(g)$, -393.5 kJ/mol: $H_2O(\ell)$, -285.9 kJ/mol. The values for both nitrogen and oxygen, elements, are zero. When we use these data, we get

$$-3894.0 \text{ kJ} = \left[8 \text{ mol} \times \left(-393.5 \frac{kJ}{mol}\right) + 10 \text{ mol} \times \left(-285.9 \frac{kJ}{mol}\right) + 0\right]$$
$$- [4 \text{ mol} \times \Delta H_f^\circ(C_2H_5NO_2) + 0]$$

Therefore, by rearranging,

$$4 \text{ mol} \times \Delta H_f^\circ(C_2H_5NO_2) = +3894.0 \text{ kJ} - 3148 \text{ kJ} - 2859 \text{ kJ}$$
$$= -2113 \text{ kJ}$$

And

$$\Delta H_f^\circ(C_2H_5NO_2) = \frac{-2113 \text{ kJ}}{4 \text{ mol}}$$
$$= -528.3 \text{ kJ/mol}$$

Thus, the standard heat of formation of glycine is -528.3 kJ/mol, and we have seen how we can determine this quantity without making glycine directly from its elements.

PRACTICE EXERCISE 8   Write the thermochemical equation for the combustion of one mole of sucrose from data given earlier in this section, and calculate the standard heat of formation of this compound.

SUMMARY

Energy Sources and Units.   Because electrical attractions and repulsions occur within atoms, molecules, and ions, substances have potential energy, often called chemical energy, to varying degrees. As the particles of substances rearrange during chemical reactions, some of their potential energy might be converted into heat that evolves; such reactions are exothermic. In other reactions, called endothermic reactions, the particles absorb heat from their surroundings, and this heat becomes part of the potential energy of the products. The fossil fuels are huge reservoirs of chemical energy, which ultimately came from the sun as photosynthesis occurred in ancient plants.

The SI unit for energy is based on the definition of kinetic energy and is called the joule. An older and still much used unit of energy is the calorie, which is related to the joule by the equation 1 cal = 4.184 J (exactly).

Specific Heat and Heat Capacity.   Two thermal properties that all substances have are specific heat and heat capacity. Heat ca-

pacity, like mass, is a function of the size of the sample. Hence, it is more significant to define a quantity for each substance that gives us the heat capacity per gram (or per mole). Heat capacity is the number of joules (or kilocalories) needed to change the temperature of the entire sample by one degree Celsius. The specific heat is this heat capacity per gram. If calculated on a per mole basis, instead, the quantity is called the molar heat capacity. Water has an unusually high specific heat.

Thermochemistry.   The heat of a reaction, q, can be calculated from the temperature change caused when known quantities of reactants undergo the reaction in a system of known heat capacity. The system might give off heat to its surroundings (raising their temperature) or it might take heat from the surroundings (cooling them). The law of conservation of energy ensures that the energy that leaves a system is identical in magnitude to the energy that enters its surroundings. The value of q depends not just on the particular reaction but also on its scale, the temperature of the system, the pressure, and other variables that define the state of the system.

When the system is under constant pressure, the heat of the reaction is called its enthalpy change, ΔH. (If the reaction is conducted under conditions of constant volume, the heat of the reaction might be slightly different from that obtained at constant pressure.) Exothermic reactions have negative values of ΔH; endothermic changes have positive values. (Only enthalpy *changes* can be determined, not absolute values of enthalpy.)

An enthalpy change is a state function, since its value depends only on the initial and final states of the system and not on the path taken between these states.

The law of conservation of energy ensures that an enthalpy change for a forward reaction has the same magnitude as but an opposite sign to the enthalpy change for the reverse reaction (assuming that the conditions are otherwise constant).

Enthalpy changes can be determined by calorimetry. A "coffee-cup" system works well when low precision is acceptable. A bomb calorimeter can be used for reactions that don't happen until initiated, for example, combustions.

Standard Heats of Reaction.   When enthalpy changes for different reactions are compared, the reactions should be studied under the same conditions of temperature and pressure. In thermochemistry, standard conditions are 25 °C and 1 atm of pressure. An enthalpy change measured under these conditions is called the standard heat of reaction. Its value is a function of the amounts of substances involved in the reaction. Its symbol is $\Delta H°$, where the degree sign (°) signifies that standard conditions are involved. If the moles of reactants are doubled, the standard heat of the reaction is twice as large. (The units for $\Delta H°$ are joules or kilojoules, calories or kilocalories.)

If the enthalpy change is for the combustion of *one* mole of a compound, all substances involved in their standard states, then it is called the standard heat of combustion, $\Delta H°_{combustion}$, with units of energy per mole, for example, kilojoules per mole.

If the enthalpy change is for the formation of *one* mole of a compound under standard conditions from its elements *in their standard states,* it is called the standard heat of formation of the compound, $\Delta H°_f$, with units of energy per mole of compound formed (for example, kJ/mol or kcal/mol).

A balanced chemical equation that includes the enthalpy change, whatever its type, and that specifies the physical states of the substances is called a thermochemical equation. Like mathematical equations, they can be added, reversed (reversing also the sign of the enthalpy change), multiplied by a constant multiplier (doing the same to the enthalpy change), and similarly manipulated. (If formulas are canceled or added, they must be of substances in identical physical states.)

Hess's Law.   The law of heat summation is possible because ΔH is a state function and does not depend on the kinds or numbers of steps between the initial and final states. The $\Delta H°$ of a reaction can be calculated as the sum of the $\Delta H°$ values of the products minus the sum of the $\Delta H°$ values of the reactants, where the coefficients in the equation and values of $\Delta H°_f$ from tables are used to calculate values of $\Delta H°$ for products and reactants. Otherwise, values of $\Delta H°$ can be determined by the manipulation of any combination of thermochemical equations that can be made to add up to the final net equation.

When standard heats of formation cannot be measured directly, they can sometimes be calculated from standard heats of other reactions, for example, standard heats of combustion, using the Hess law equation and standard heats of formation of all of the reactants and products except the one whose value of $\Delta H°_f$ is the target.

REVIEW EXERCISES

Answers to questions whose numbers are printed in color are given in Appendix C.
Difficult questions are marked with asterisks.

Energy Sources and Units

5.1 What is meant by the term *chemical energy?*

5.2 What is meant by an *energy source?*

5.3 What are the names of the principal fossil fuels, and what does *fossil* signify in this term?

5.4 Write an equation that describes the overall effect of photosynthesis.

5.5 How is the joule defined?

5.6 How was the calorie originally defined?

5.7 How are the calorie and the joule related? (Give an equation.)

5.8 Use an equation to show the relationship between the kilocalorie and the kilojoule.

5.9 In 458 kcal there are how many joules? How many kilojoules?

5.10 How many kilocalories are in 5225 J?

5.11 The British thermal unit, or BTU, is a unit of energy much more commonly used to describe fuels than is either the kilocalorie or the joule. One BTU is the energy that raises the temperature of 1 lb of water by 1 °F.

$$1 \text{ J} = 9.48 \times 10^{-4} \text{ BTU}$$

One calorie equals how many BTU?

5.12 For convenience in working with huge quantities of energy, the unit *quad* is used. In the early 1980s, for example, the United States consumed roughly 80 quads of energy per year.

$$1 \text{ quad} = 1 \times 10^{15} \text{ BTU}$$

Using information and results from Review Exercise 5.11, calculate how many kilocalories are in 1 quad. (Calculate to two significant figures.)

5.13 A car weighing 4.0×10^{13} lb is traveling at 55 miles/hr. How much kinetic energy does it have, in kilojoules? Use conversion factors found in various tables in this book.

5.14 A baseball weighing 5.0 oz traveling at 66 miles/hr has how much kinetic energy, in kilojoules? Use conversion factors found in various tables in this book.

Specific Heat and Heat Capacity

5.15 Which kind of substance needs more energy to undergo a rise of 5 degress in temperature—something with a *high* specific heat or something with a *low* specific heat? Explain.

5.16 Which kind of substance experiences the greater rise in temperature when it absorbs 10 kcal—something with a *high* specific heat or something with a *low* specific heat?

5.17 If specific heat were defined in units of kcal/kg K, would the numerical values in Table 5.1 change? If so, in what way? Explain.

5.18 Explain how water's high specific heat helps guard against swings in the body's core temperature as the outside temperature fluctuates.

5.19 How much heat in kilocalories has to be removed from 225 g of water to lower its temperature from 25.0 °C to 10.0 °C? (This would be like cooling a glass of lemonade.)

5.20 To bring 1.0 kg of water from 25 °C to 99 °C takes how much heat input, in kilocalories? This would be like making four cups of coffee.

*5.21 Fat tissue is 85% fat and 15% water. The complete breakdown of the fat itself converts it to CO_2 and H_2O, and releases about 9.0 kcal/g (of the fat in the fat tissue).
 (a) How many kilocalories are released by a loss of 1.0 lb of fat *tissue* in a weight-reduction program?
 (b) A person running at 8.0 miles/hr expends about 5.0×10^2 kcal/hr of extra energy. How far does a person have to run to "burn off" 1.0 lb of fat *tissue* by this means alone?

*5.22 A well-nourished person adds about 0.50 lb of fat tissue for each 3.5×10^2 kcal of food energy taken in over and above that needed. Suppose you decided to reduce your weight simply by omitting butter but keeping every other aspect of your diet and your activities the same. How many days would be needed to lose 1.0 lb of fat *tissue* by this strategy alone? The caloric content of butter is 9.0 kcal/g. Suppose that you have been eating 0.25 lb of butter a day.

5.23 Calculate the molar heat capacity of iron in J/mol °C. Its specific heat is 0.4498 J/g °C.

5.24 What is the molar heat capacity of ethyl alcohol, C_2H_5OH, in unts of J/mol °C, if its specific heat is 0.586 cal/g °C?

Enthalpy

5.25 What do we mean by *system?*

5.26 In the bomb calorimeter of Figure 5.5, what specifically is included in the "system?"

5.27 What is meant by *surroundings?* In the bomb calorimeter of Figure 5.5, is the insulation in the walls of the vat part of the system or part of the surroundings?

5.28 What is another term for the *enthalpy* of a system? What is meant by H_{final} and $H_{initial}$ for a system undergoing a change?

5.29 What is the equation that defines ΔH in terms of $H_{reactants}$ and $H_{products}$?

5.30 What is the law of conservation of energy?

5.31 If the enthalpy of a system increases by, say, 100 kJ, what must be true about the enthalpy of the surroundings? Explain.

5.32 What is the sign of ΔH for an exothermic change?

5.33 When ΔH is described as a *state function,* what does this mean? Is the temperature of a system a state function?

Calorimetry

5.34 The science of calorimetry involves what kind of study?

5.35 What is a bomb calorimeter and for what kinds of reactions is it used?

5.36 If a value of heat capacity (for which the energy part of the unit is the calorie) is multiplied by a value of a temperature change, what units are possible for the result? What additional multiplication has to be done to obtain the same units if a value of specific heat were used instead of heat capacity?

5.37 Why isn't the heat of a reaction measured in a bomb calorimeter called an enthalpy change?

5.38 A vat of 5.45 kg of water underwent a drop in temperature from 60.30 °C to 57.60 °C. How much energy in kilocalories and in kilojoules left the water? Use a value of 1.00 cal/g °C for the specific heat of water in this range of temperatures.

5.39 Nitric acid, HNO_3, reacts with potassium hydroxide as follows: $HNO_3(aq) + KOH(aq) \longrightarrow KNO_3(aq) + H_2O(\ell)$. In one experiment a student placed 55.0 mL of 1.3 M HNO_3 in a coffee-cup calorimeter, noted that the temperature was 23.5 °C, and added 55.0 mL of 1.3 M KOH also at 23.5 °C. After quickly stirring the mixture with a thermometer, its temperature was seen to rise to 31.8 °C. Calculate the heat of this reaction in calories and joules. Assume that the specific heats of all solutions were 1.00 cal/g °C and that the densities of all solutions were 1.00 g/mL. Calculate the heat of reaction per mole of acid using units of kilocalories per mole.

5.40 When a reaction was carried out in a bomb calorimeter, 16.44 kJ of energy was released. The initial temperature of the calorimeter was 23.518 °C and the heat capacity of the system was 18.56 kJ/°C. Calculate the final temperature of the calorimeter.

5.41 A reaction in a bomb calorimeter liberated 1.05×10^4 J and caused the temperature in the calorimeter to rise to 26.13 °C. The heat capacity of the system was 26.6 kJ/°C. What was the initial temperature in the calorimeter?

*5.42 A dilute solution of hydrochloric acid with a mass of 610.29 g and containing 0.33183 mol of HCl was exactly neutralized in a calorimeter by the sodium hydroxide in 615.31 g of a comparably dilute solution. The temperature rose from 16.784 °C to 20.610 °C. The specific heat of the hydrochloric acid solution was 4.031 J/g °C, and that of the sodium hydroxide solution was 4.046 J/g °C. The heat capacity of the calorimeter was 77.99 J/°C. Use these data to calculate a value for the enthalpy of neutralization that proceeds according to the following equation.

$$HCl(aq) + NaOH(aq) \longrightarrow NaCl(aq) + H_2O(\ell)$$

Assume that the individual solutions that were mixed together made independent contributions to the total heat capacity of the system after these solutions were mixed.

*5.43 A dilute solution of hydrochloric acid with a mass of 610.28 g, a specific heat of 4.031 J/g °C, and containing 0.33143 mol of HCl was exactly neutralized by the potassium hydroxide in 619.69 g of KOH solution having a specific heat of 4.003 J/g °C. The heat capacity of the calorimeter was 77.99 J/°C. As a result of the neutralization, the temperature rose from 15.533 °C to 19.410 °C. Use these data to calculate the enthalpy change for the reaction. Its equation is as follows.

$$HCl(aq) + KOH(aq) \longrightarrow KCl(aq) + H_2O(\ell)$$

Assume that the individual solutions make independent contributions to the total heat capacity of the system when they are mixed.

Standard Heats of Reaction

5.44 Why do standard reference values for temperature and pressure have to be selected when we consider and compare heats of reaction for various reactions?

5.45 Under what conditions can the value of ΔH for a reaction be symbolized by $\Delta H°$?

5.46 What distinguishes a thermochemical equation from an ordinary chemical equation?

5.47 If a fraction such as $\frac{1}{2}$ appears as a coefficient in a thermochemical equation, what unit does it signify?

5.48 Aluminum and iron(III) oxide, Fe_2O_3, react and form aluminum oxide, Al_2O_3, and iron. For each mole of aluminum used, 426.9 kJ of energy is released under standard conditions. Write the thermochemical equation that shows the consumption of 4 mol of Al. (All the substances are solids.)

5.49 The value of $\Delta H°$ for the reaction (the combustion of benzene, C_6H_6)

$$2C_6H_6(\ell) + 15O_2(g) \longrightarrow 12CO_2(g) + 6H_2O(\ell)$$

is 6542 kJ when measured under standard conditions. What is $\Delta H°$ for the combustion of 1.500 mol of benzene?

5.50 The following equation represents the dehydration of calcium hydroxide to make quick lime, CaO, a substance present in cement.

$$Ca(OH)_2(s) \longrightarrow CaO(s) + H_2O(\ell) \qquad \Delta H° = +65.3 \text{ kJ}$$

One of the reactions when cement is mixed with water is the reverse of this reaction. Write the thermochemical equation for the reaction of 10 mol of quicklime with water under standard conditions.

Hess's Law and Thermochemical Equations

5.51 Construct an enthalpy diagram that shows the enthalpy changes for a one-step conversion of germanium, Ge(s), into its dioxide, $GeO_2(s)$, and a two-step conversion—first to the monoxide, GeO(s), and then the oxidation of the monoxide to the dioxide. The relevant thermochemical equations are as follows.

$$Ge(s) + \tfrac{1}{2}O_2(g) \longrightarrow GeO(s) \qquad \Delta H° = -255 \text{ kJ}$$
$$Ge(s) + O_2(g) \longrightarrow GeO_2(s) \qquad \Delta H° = -534.7 \text{ kJ}$$

Using this diagram, determine the value of $\Delta H°$ for the reaction

$$GeO(s) + \tfrac{1}{2}O_2(g) \longrightarrow GeO_2(s)$$

5.52 Construct an enthalpy diagram for the formation of $NO_2(g)$ from its elements in one step. Do the same for a two-step process, the first step being

$$\tfrac{1}{2}N_2(g) + \tfrac{1}{2}O_2(g) \longrightarrow NO(g) \qquad \Delta H° = +90.37 \text{ kJ}$$

and the second being

$$NO(g) + \tfrac{1}{2}O_2(g) \longrightarrow NO_2(g) \qquad \Delta H° = ?$$

The equation for the one-step formation of NO_2 is

$$\tfrac{1}{2}N_2(g) + O_2(g) \longrightarrow NO_2(g) \qquad \Delta H° = +33.8 \text{ kJ}$$

Be careful to notice if a value of $\Delta H°$ is positive or negative, because in an enthalpy diagram arrows must point upward for endothermic steps and downward for exothermic steps. Using this diagram, determine the value of $\Delta H°$ for the second step of the two-step process.

5.53 What is Hess's law of constant heat summation?

5.54 What important feature of enthalpy makes Hess's law possible?

5.55 Show how the equations

$$N_2O_4(g) \longrightarrow 2NO_2(g) \qquad \Delta H° = +57.93 \text{ kJ}$$
$$2NO(g) + O_2(g) \longrightarrow 2NO_2(g) \qquad \Delta H° = -113.14 \text{ kJ}$$

can be manipulated to give the value of $\Delta H°$ for the following change.

$$2NO(g) + O_2(g) \longrightarrow N_2O_4(g)$$

5.56 Show how two of the thermochemical equations of Review Exercise 5.52 can be manipulated to give the complete thermochemical equation for the conversion of 1 mol of NO(g) into $NO_2(g)$.

5.57 We can generate hydrogen chloride by heating a mixture of sulfuric acid and potassium chloride according to the reaction

$$2KCl(s) + H_2SO_4(\ell) \longrightarrow 2HCl(g) + K_2SO_4(s)$$

Calculate $\Delta H°$ for this reaction from the following thermochemical equations. Give your answer in kilojoules.

$$HCl(g) + KOH(s) \longrightarrow KCl(s) + H_2O(\ell)$$
$$\Delta H° = -203.6 \text{ kJ}$$
$$H_2SO_4(\ell) + 2KOH(s) \longrightarrow K_2SO_4(s) + 2H_2O(\ell)$$
$$\Delta H° = -342.4 \text{ kJ}$$

5.58 Calculate $\Delta H°$ for the following reaction, which describes the preparation of an unstable acid, HNO_2, nitrous acid.

$$HCl(g) + NaNO_2(s) \longrightarrow HNO_2(\ell) + NaCl(s)$$

Use the following thermochemical equations. Calculate the answer in kilojoules.

$$2NaCl(s) + H_2O(\ell) \longrightarrow 2HCl(g) + Na_2O(s)$$
$$\Delta H° = +507.31 \text{ kJ}$$
$$NO(g) + NO_2(g) + Na_2O(s) \longrightarrow 2NaNO_2(s)$$
$$\Delta H° = -427.14 \text{ kJ}$$
$$NO(g) + NO_2(g) \longrightarrow N_2O(g) + O_2(g)$$
$$\Delta H° = -42.68 \text{ kJ}$$
$$2HNO_2(\ell) \longrightarrow N_2O(g) + O_2(g) + H_2O(\ell)$$
$$\Delta H° = +34.35 \text{ kJ}$$

5.59 Barium oxide, BaO, can be used to neutralize pure sulfuric acid, H_2SO_4. The equation is

$$BaO(s) + H_2SO_4(\ell) \longrightarrow BaSO_4(s) + H_2O(\ell)$$

What is the standard enthalpy change of this reaction? The following thermochemical equations can be used.

$$SO_3(g) + H_2O(\ell) \longrightarrow H_2SO_4(\ell) \quad \Delta H° = -78.2 \text{ kJ}$$
$$BaO(s) + SO_3(g) \longrightarrow BaSO_4(s) \quad \Delta H° = -213 \text{ kJ}$$

5.60 Exothermic reactions occur spontaneously more frequently than do endothermic changes. In which direction will the following system most likely react, left to right or right to left? (In other words, calculate $\Delta H°$ for the following reaction first; then answer the question.)

$$2Ag(s) + Zn(NO_3)_2(aq) \longrightarrow Zn(s) + 2AgNO_3(aq)$$

The following thermochemical equations can be used.

$$Cu(NO_3)_2(aq) + Zn(s) \longrightarrow Zn(NO_3)_2(aq) + Cu(s)$$
$$\Delta H° = -61.7 \text{ kcal}$$
$$2AgNO_3(aq) + Cu(s) \longrightarrow Cu(NO_3)_2(aq) + 2Ag(s)$$
$$\Delta H° = -25.3 \text{ kcal}$$

5.61 Copper metal can be obtained by roasting copper oxide, CuO, with carbon monoxide. What is the value of $\Delta H°$ in kilojoules for this reaction, which goes according to the equation

$$CuO(s) + CO(g) \longrightarrow Cu(s) + CO_2(g)$$

The following thermochemical data are known.

$$2CO(g) + O_2(g) \longrightarrow 2CO_2(g) \quad \Delta H° = -566.1 \text{ kJ}$$
$$2Cu(s) + O_2(g) \longrightarrow 2CuO(s) \quad \Delta H° = -310.5 \text{ kJ}$$

5.62 An aqueous solution of calcium hydroxide, $Ca(OH)_2$, is called limewater. It neutralizes hydrochloric acid, HCl(aq), as follows.

$$Ca(OH)_2(aq) + 2HCl(aq) \longrightarrow CaCl_2(aq) + 2H_2O(\ell)$$

Calculate $\Delta H°$ in kilojoules for this reaction. The following thermochemical equations can be used as needed.

$$CaO(s) + 2HCl(aq) \longrightarrow CaCl_2(aq) + H_2O(\ell)$$
$$\Delta H° = -186 \text{ kJ}$$
$$CaO(s) + H_2O(\ell) \longrightarrow Ca(OH)_2(s) \quad \Delta H° = -62.3 \text{ kJ}$$
$$Ca(OH)_2(s) \xrightarrow{water} Ca(OH)_2(aq) \quad \Delta H° = -12.6 \text{ kJ}$$

5.63 What is the value for $\Delta H°$ of neutralization of lithium hydroxide, LiOH(aq), by hydrochloric acid, HCl(aq), by the following equation?

$$LiOH(aq) + HCl(aq) \longrightarrow LiCl(aq) + H_2O(\ell)$$

The following thermochemical equations can be used in addition to any prepared, as needed, with the help of data in Table 5.2.

$$Li(s) + \tfrac{1}{2}O_2(g) + \tfrac{1}{2}H_2(g) \longrightarrow LiOH(s) \quad \Delta H° = -487.0 \text{ kJ}$$
$$2Li(s) + Cl_2(g) \longrightarrow 2LiCl(s) \quad \Delta H° = -815.0 \text{ kJ}$$
$$LiOH(s) \xrightarrow{in\ water} LiOH(aq) \quad \Delta H° = -19.2 \text{ kJ}$$
$$HCl(g) \xrightarrow{in\ water} HCl(aq) \quad \Delta H° = -77.0 \text{ kJ}$$
$$LiCl(s) \xrightarrow{in\ water} LiCl(aq) \quad \Delta H° = -36.0 \text{ kJ}$$

5.64 The standard heat of formation of gaseous hydrogen bromide was first evaluated by means of the standard enthalpy values measured for the following reactions. The last three are standard heats of solution.

$$Cl_2(g) + 2KBr(aq) \longrightarrow Br_2(aq) + 2KCl(aq)$$
$$\Delta H° = -96.2 \text{ kJ}$$
$$H_2(g) + Cl_2(g) \longrightarrow 2HCl(g) \quad \Delta H° = -184 \text{ kJ}$$
$$HCl(aq) + KOH(aq) \longrightarrow KCl(aq) + H_2O(\ell)$$
$$\Delta H° = -57.3 \text{ kJ}$$
$$HBr(aq) + KOH(aq) \longrightarrow KBr(aq) + H_2O(\ell)$$
$$\Delta H° = -57.3 \text{ kJ}$$
$$HCl(g) \longrightarrow HCl(aq) \quad \Delta H° = -77.0 \text{ kJ}$$
$$Br_2(g) \longrightarrow Br_2(aq) \quad \Delta H° = -4.2 \text{ kJ}$$
$$HBr(g) \longrightarrow HBr(aq) \quad \Delta H° = -79.9 \text{ kJ}$$

Write the thermochemical equation for the formation of 1 mol of HBr(g) from its elements, including its value for $\Delta H_f°$.

Hess's Law and Standard Heats of Formation

5.65 Which of the following equations has a $\Delta H°$ that would properly be labeled as a $\Delta H_f°$?
(a) $CaCO_3(s) \longrightarrow CaO(s) + CO_2(g)$
(b) $Ca(s) + \tfrac{1}{2}O_2(g) \longrightarrow CaO(s)$
(c) $2Fe(s) + O_2(g) \longrightarrow 2FeO(s)$
(d) $SO_2(g) + \tfrac{1}{2}O_2(g) \longrightarrow SO_3(g)$

5.66 Write the thermochemical equations that would be used in connection with values of $\Delta H_f°$ for each of the following compounds. Obtain values of $\Delta H_f°$ (in kJ/mol) from Table 5.2.
(a) Acetic acid, $HC_2H_3O_2(\ell)$

(b) Sodium bicarbonate, $NaHCO_3(s)$
(c) Gypsum, $CaSO_4 \cdot 2H_2O(s)$
(d) Plaster of paris, $CaSO_4 \cdot \frac{1}{2}H_2O(s)$
(e) Methyl alcohol, $CH_3OH(\ell)$

5.67 Using data in Table 5.2, calculate $\Delta H°$ in kilojoules for the following reactions.
(a) $2H_2O_2(\ell) \longrightarrow 2H_2O(\ell) + O_2(g)$
(b) $HCl(g) + NaOH(s) \longrightarrow NaCl(s) + H_2O(\ell)$
(c) $CH_4(g) + Cl_2(g) \longrightarrow CH_3Cl(g) + HCl(g)$
(d) $2NH_3(g) + CO_2(g) \longrightarrow \underset{\text{urea}}{CO(NH_2)_2(s)} + H_2O(\ell)$

5.68 Sulfur trioxide, a trace air pollutant, reacts with water to produce corrosive sulfuric acid. What is $\Delta H°$ for the following reaction?

$$SO_3(g) + H_2O(\ell) \longrightarrow H_2SO_4(\ell)$$

Use $\Delta H_f°$ data from Table 5.2, and give your answer in kilojoules.

5.69 Nitrogen dioxide, an air pollutant, combines with water to make nitric acid (HNO_3), a corrosive acid, and nitrogen monoxide (NO). What is $\Delta H°$ for the following reaction in kilojoules?

$$3NO_2(g) + H_2O(\ell) \longrightarrow 2HNO_3(\ell) + NO(g)$$

5.70 Iron oxides can be changed into iron metal by the action of very hot carbon. What is $\Delta H°$ for the following reaction in kilojoules?

$$Fe_2O_3(s) + 3C(s) \longrightarrow 2Fe(s) + 3CO(g)$$

5.71 We can remove carbon dioxide from air by passing the air through solid granules of sodium hydroxide. What is $\Delta H°$ for the following reaction in kilocalories?

$$CO_2(g) + NaOH(s) \longrightarrow NaHCO_3(s)$$

5.72 Write thermochemical equations for the combustion of one mole of each of the following compounds. The products in each case are gaseous carbon dioxide and liquid water. (Standard conditions are assumed.) The numbers given in parentheses after each formula are the standard heats of combustion.
(a) Acetylene, $C_2H_2(g)$ (−1299.6 kJ/mol)
(b) Methyl alcohol, $CH_3OH(\ell)$ (−726.51 kJ/mol)
(c) Diethyl ether, $C_4H_{10}O(\ell)$ (−2751.1 kJ/mol)
(d) Toluene, $C_7H_8(\ell)$ (−3909 kJ/mol)

5.73 From $\Delta H_f°$ data for the formation of $CO_2(g)$ and $H_2O(\ell)$, and the thermochemical equation for the combustion of glucose,

$$C_6H_{12}O_6(s) + 6O_2(g) \longrightarrow 6CO_2(g) + 6H_2O(g)$$
$$\Delta H_{\text{combustion}}° = -2.82 \times 10^3 \text{ kJ/mol}$$

estimate $\Delta H_f°$ for glucose, in kilojoules per mole.

5.74 Palmitic acid, $C_{16}H_{32}O_2$, is typical of the materials available from fats and oils insofar as chemical energy is concerned.
(a) Write the thermochemical equation of the complete combustion of one mole of this compound, a solid, assuming that gaseous carbon dioxide and liquid water form. The standard heat of combustion of palmitic acid is −2380 kcal/mol.
(b) Using the molar heat of combustion of palmitic acid and data in Table 5.2, estimate the value of $\Delta H_f°$ for palmitic acid.

5.75 Using data available in this chapter, calculate the heat of combustion of ethylene, C_2H_4, in kJ/mol. The equation is

$$C_2H_4(g) + 3O_2(g) \longrightarrow 2CO_2(g) + 2H_2O(\ell)$$

5.76 Phosphorus burns in air to give tetraphosphorus decaoxide.

$$4P(s) + 5O_2(g) \longrightarrow P_4O_{10}(s) \qquad \Delta H° = -3062 \text{ kJ}$$

The product combines with water to give phosphoric acid.

$$P_4O_{10}(s) + 6H_2O(\ell) \longrightarrow 4H_3PO_4(\ell)$$
$$\Delta H° = -257.2 \text{ kJ}$$

Using these equations and such others from the chapter as needed, write the thermochemical equation for the formation of 1 mol of $H_3PO_4(\ell)$ from the elements, including the value for $\Delta H_f°$.

CHEMICALS IN USE
SODIUM CARBONATE

4

In ancient Egypt, soda ash (the common name for sodium carbonate, Na_2CO_3) was regarded as sacred and was reserved largely for ritual purifications and for preparing mummies for burial. Today it is one of the important heavy chemicals of industrialized nations—"heavy" because of the large tonnages used. In the late 1980s, sodium carbonate ranked among the top 15 chemicals produced in the United States—over 8.6 million tons each year. Approximately 50% is used in making glass, and the rest is consumed in the manufacture of other chemicals, soaps, detergents, and cleansers (Figure 4a).

The term *soda ash* has not always signified pure sodium carbonate, as it does today. "Soda" is the English equivalent of *natron,* a term that goes back to the Greek *nitron.* The Latin name *natrium,* meaning sodium (Na), evolved from this, but ancient nitron was a mixture of the carbonates of sodium and potassium. The exact composition varied with the source.

When plants are dried and burned, the sodium and potassium they contain are left behind, giving ashes rich in sodium and potassium carbonates—Na_2CO_3 and K_2CO_3. These compounds are water soluble and can be leached from the ashes with boiling water. Most land plants use little sodium, so the extracts of their ashes contain mostly potassium carbonate, K_2CO_3, which became known as "potash." Seaweed and certain salt marsh plants give ashes rich in sodium carbonate, and Na_2CO_3 became "soda ash." Until the late 1700s, plant ashes were the most common sources of potassium and sodium carbonates.

Extracts of plant ashes—aqueous solutions of sodium and potassium carbonates—are chemically basic, which means that they neutralize acids. The term alkaline, which we take to mean basic, comes from the Arabic word *al-qali* ("from plant ash"). The cleansing abilities of alkaline solutions were known from the most ancient of times. No one knows who did it, but someone back then, no doubt accidentally, spilled some fat or oil into such an extract and probably didn't bother to throw it away for several days. Given enough time, a mixture such as this changes; the fat or oil appears to dissolve, and the product is a cleansing agent superior to soda ash itself. Soap had, in fact, been discovered.

By the ninth century, the manufacture of soap was a common cottage activity throughout Europe. It was not "wasted" on personal hygiene but was used instead to remove grease and other impurities from wool and other fibers. By the early eighteenth century, the demands of the textile industry for cleansing alkalies and soaps far exceeded the supply, and many families who lived by the sea earned livelihoods from "kelping"—harvesting and burning kelp and extracting the ashes to make soda ash and soap.

The Industrial Synthesis of Sodium Carbonate

Industrial and commercial demands have always had a way of inspiring inventors. As a further inducement to the development of a cheap source of sodium carbonate, the French Academy of Sciences in 1775 offered a prize of 12,000 francs to anyone who developed a commercially successful method for making soda ash from common salt. Nicolas Leblanc (1742–1806) succeeded where several had failed, but he had the misfortune to live at the time of the French Revolution. His patron, the Duke of Orleans, was executed at the guillotine, and the soda ash factory was confiscated and dismantled. Leblanc's patent was nullified;

FIGURE 4a A commercial cleanser that uses sodium carbonate.

he was forced to reveal the secrets of his method; he never received the prize money; and he was mocked and hounded into taking his own life. (In belated recognition of his contribution, the French erected a statue in his honor 80 years later.)

The Leblanc process begins by mixing salt and sulfuric acid in cast iron pans. Heating drives off both the hydrogen chloride that forms and the water present in the sulfuric acid.

$$NaCl + H_2SO_4 \xrightarrow{heat} NaHSO_4 + HCl(g)$$

The residue in the pan is now mostly a mixture of unreacted sodium chloride and newly formed sodium hydrogen sulfate. This mixture is transferred to a roasting oven and heated to make sodium sulfate as well as more hydrogen chloride.

$$NaHSO_4 + NaCl \xrightarrow{heat} Na_2SO_4 + HCl(g)$$

The sodium sulfate next is mixed with limestone ($CaCO_3$) and coal and heated to a high temperature. The changes that now happen are complex. The most simple expression of the net result is the equation

$$Na_2SO_4 + CaCO_3 + 2C \xrightarrow{heat} Na_2CO_3 + CaS + 2CO_2(g)$$

The solid mixture of sodium carbonate and calcium sulfide still contains unchanged coal, so it became known as "black ash." This black ash is pulverized and leached with warm water, which quickly dissolves the sodium carbonate. When the aqueous extract is filtered and evaporated, the residue is the desired sodium carbonate. The Leblanc process dominated the soda ash industry for most of the nineteenth century, and according to historians of technology it marked for industry an important transition from craft knowledge to chemical knowledge.

The Leblanc process suffered from a number of problems. It generated an air pollutant, hydrogen chloride, which had a devastating effect on the surrounding environment. It left a water-insoluble residue for which no use could be found; it consumed considerable energy; and using coal, it was also a dirty procedure. All these drawbacks disappeared with the development throughout the last half of the nineteenth century of another way to make sodium carbonate. The newer method, which entirely displaced the Leblanc process, is called the Solvay process, in honor of Ernest and Alfred Solvay of Belgium, who made the important contributions.

The details of the Solvay process are discussed more fully in Chapter 21, but the principal chemical reaction takes advantage of the low solubility of $NaHCO_3$ in cold water.

$$NaCl(aq) + NH_3(aq) + CO_2(aq) + H_2O \xrightarrow{cold} NaHCO_3(s) + NH_4Cl(aq)$$

The NaCl, of course, is salt and the CO_2 is obtained by decomposing limestone, $CaCO_3$. After the sodium bicar-

FIGURE 4b The mining of trona from natural deposits at Green River, Wyoming.

bonate is filtered from the mixture, it is dried and heated, which causes it to partially decompose and give off carbon dioxide and water. The product left behind is the desired sodium carbonate.

$$2NaHCO_3(s) \xrightarrow{heat} Na_2CO_3(s) + CO_2(g) + H_2O(g)$$

Although the Solvay process is used to manufacture much of the Na_2CO_3 made in other countries, its use in the United States has practically ceased because of the availability of low-cost natural sodium carbonate from brines in California and from deposits in Wyoming of an ore called *trona* — a mixture of sodium carbonate and sodium bicarbonate with a composition that can be approximated by the formula $Na_2CO_3 \cdot NaHCO_3 \cdot 2H_2O$ (Figure 4b).

Questions

1. Write the chemical formulas for soda ash and potash.
2. What is the chief use of Na_2CO_3 in industry.
3. What prompted the large demand for sodium carbonate during the seventeenth and early eighteenth centuries?
4. Write chemical equations for the reactions that take place during the Leblanc process.
5. Why is HCl gas such a hazardous air pollutant?
6. Write chemical equations for the chief reactions in the Solvay process.
7. Sodium carbonate was used to neutralize 20,000 gallons of HNO_3 that spilled in a railroad yard in Denver, Colorado, in 1983.
   (a) Write the equation for the reaction of nitric acid with sodium carbonate, assuming the products are sodium nitrate, carbon dioxide, and water.
   (b) Suppose the nitric acid solution contained 101 g of HNO_3 per 100 mL and that the solid that was used was the monohydrate, $Na_2CO_3 \cdot H_2O$ (which is much cheaper than the anhydrous compound). How many tons (1 ton = 2000 lb) of the solid were needed to neutralize the acid spill?

The extreme precision of a laser beam is used here to deposit patterns of materials on metal surfaces. The pure colors of laser light come from electronically excited atoms, and the study of similar light emitted by atoms provided the clues that led to our present detailed knowledge of the structures of atoms.

THE ELECTRONIC STRUCTURES OF ATOMS

6.1 THE DISCOVERY OF SUBATOMIC PARTICLES

Experiments by physicists on the electrical conduction of gases at low pressures
opened the door to the subatomic world of electrons, protons, and neutrons.

It would seem, at first glance, that the complexity of nature is so immense that attempting to understand chemical and physical behavior is a hopeless task. Why, for example, does metallic sodium react violently with water, whereas gold is resistant to attack by even concentrated acids? Why does water form a solid (ice) that is less dense than the liquid, thereby allowing the solid to float in the liquid? Why are heavy metals such as lead and mercury poisonous? And why are some reactions endothermic and others exothermic? There's no end to such questions, and in a beginning course like this we can only hope to start to find some answers.

In Chapter 2, you saw that some of the complexity of chemistry can be simplified by organizing the elements into the periodic table. This shows why, for example, strontium-90 collects in the bones of growing children along with calcium — both elements are in the same group in the periodic table and should be expected to have similar chemical properties. With a little thought, however, you will see there is an even more profound question here: What is there about calcium atoms and strontium atoms that causes them to behave in similar ways? Being able to answer this question would enable us to understand *why* calcium and strontium happen to fall in the same periodic group. Similarly, we would like to know why some groups contain only three elements, while others are longer and contain five or six.

To answer some of these basic "why" questions of chemical behavior, even in an elementary way, we must once again look at the structure of the atom. But this time we must look even deeper. We begin by briefly examining how our knowledge of basic subatomic structure came about, that is, how we came to know that the atom is composed of electrons and nuclei, and how the nuclear particles, the protons and neutrons were discovered.

The Discovery of the Electron

The first hint that matter was electrical in nature came in 1834 when Michael Faraday, a British physicist, found that chemical changes occur when an electrical current is passed through certain chemical solutions. Later in the nineteenth century, some studies of the effects of electricity on matter were done with *gas discharge tubes* (Figure 6.1). These are glass tubes that contain a gas at a low pressure. Each is fitted with a pair of metal wires called *electrodes,* one at each end of the tube, that can be connected to a source of electricity. When this is done, electricity flows through the tube and the gas within glows. (Modern neon signs work this way.)

This flow of electricity is called an electric discharge, which is how these tubes got their names.

The physicists who first studied this phenomenon did not know what caused the tube to glow, but tests soon revealed that something was moving from the negative electrode, which they called the cathode, to the positive electrode, which they called the anode. The physicists called these emissions *rays,* and because the rays came from the cathode, they called them cathode rays.

Simple experiments soon showed that the cathode rays were electrically charged particles, rather than just light waves. For example, cathode rays could make a paddle wheel turn when it was placed in their way inside the tube. When a metal plate outside the tube was given a positive charge, the cathode rays bent toward it, which meant that the particles in the cathode rays were negatively charged. Remember, "unlike charges attract."

The early experiments with gas discharge tubes gave only qualitative information. In 1897 the British physicist J. J. Thomson constructed a special gas discharge tube to make quantitative measurements of the properties of cathode rays. In some ways, the apparatus he used was similar to a television picture tube, as Figure 6.2 shows.

During a typical experiment, a high voltage was passed between the cathode and the anode of the tube. Because the anode had a hole in it, some of the cathode ray particles passed through and traveled on to strike the face of the tube at point 1, where they produced a bright spot on a phosphor (a chemical that glows when struck by a beam of

(a)

(b)

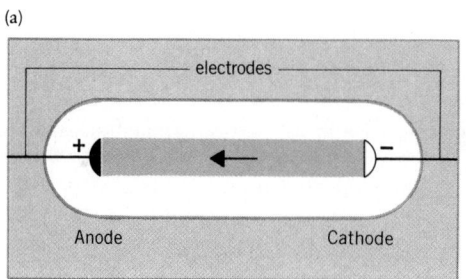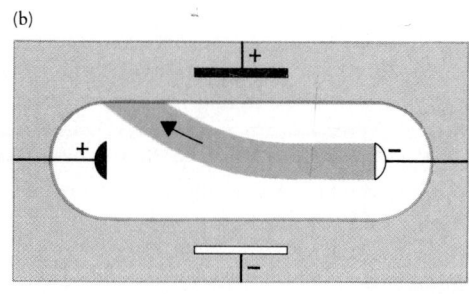

FIGURE 6.1 *(a)* A gas discharge tube. *(b)* Deflection of a cathode ray toward a positively charged plate.

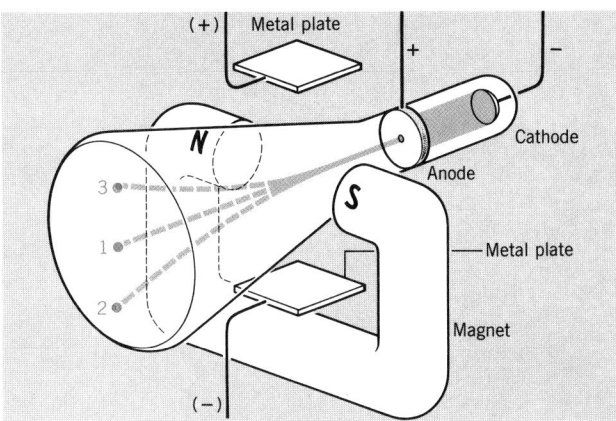

FIGURE 6.2 Thomson's cathode ray tube, which was used to measure the charge-to-mass ratio for the electron.

cathode rays) that coated the tube's inner surface. Thomson also fitted the neck of the tube with a magnet and a pair of metal plates, which could be given electrical charges. He found that if the metal plates were left uncharged while the cathode rays passed between the poles of the magnet, the path of the negatively charged particles was curved downward to point 2 by the magnetic field. On the other hand, when the magnet was removed and the metal plates were given opposite electrical charges, as shown in Figure 6.2, the cathode ray particles were deflected upward (toward the positive plate) to point 3.

Thomson found that by carefully controlling the charge on the plates when the plates and the magnet were both around the tube, he could make the cathode rays strike the tube at point 1 again. In other words, he was able to cancel the effect of the magnetic field by applying an electric field that tended to bend the path of the cathode rays in the opposite direction.

By measuring the strengths of the magnetic and electric fields needed to balance each other, Thomson was able to calculate the first bit of quantitative information about a cathode ray particle — the ratio of its charge to its mass (often expressed as e/m, where e stands for the charge and m stands for the mass). The charge-to-mass ratio has a value of -1.76×10^8 coulombs/gram, where the coulomb (C) is the SI unit of electrical charge, and the negative sign reflects the negative charge on the particle. Notice that he wasn't able to measure just the charge or just the mass; instead, his calculations only gave the value of their ratio.

One of the most important discoveries that Thomson made in his experiments was that he always obtained the same value for the charge-to-mass ratio, regardless of the materials used to construct the tube or the nature of the residual gas in the tube through which the cathode rays passed. This suggested that cathode ray particles are found in everything, and that they are basic, fundamental particles of matter. They are, in fact, the electrons that we discussed earlier.

Today's TV tubes work on this principle. Different phosphors give different colored lights when cathode rays strike them.

A modern television picture tube.

Determination of the Charge and the Mass of the Electron

In 1909 a researcher at the University of Chicago, Robert Millikan, designed a clever experiment that enabled him to measure the electron's charge. This experiment is illustrated in Figure 6.3. During an experiment he would spray a fine mist of oil droplets above a pair of parallel metal plates, the top one of which had a small hole in it. As the oil drops settled, some would pass through this hole into the space between the plates, where he would irradiate them briefly with X rays. The X rays knocked electrons off

SPECIAL TOPIC 6.1 *THE MASS SPECTROMETER AND ATOMIC WEIGHTS*

When an electron is knocked off a molecule in a gas, as in Millikan's oil drop experiment, the particle that remains behind carries a positive charge—it's a positive ion. Unlike that of the electron, the ratio of the charge-to-mass for positive ions is not the same for all substances. Some molecules have large masses and give heavy ions, while some have small masses and give light ions. The mass (and therefore the charge-to-mass ratio) depends on the chemical nature of the gas.

The device that is used to study the positive ions produced from gas molecules is called a mass spectrometer (illustrated in the figure at the right). In a mass spectrometer, the positive ions are created by passing an electrical discharge through a sample of the particular gas being studied. As the positive ions are formed, they are attracted toward a negatively charged electrode that has a small hole in its center. Some of the positive ions pass through this hole and travel onward through a tube that passes between the poles of a powerful magnet.

One of the properties of charged particles, both positive and negative, is that their paths become curved as they pass through a magnetic field. This is exactly what happens to the positive ions in the mass spectrometer as they pass between the poles of the magnet. However, the extent to which their paths are bent depends on the masses of the ions. This is because the path of a heavy ion, like that of a speeding cement truck, is difficult to change, but the path of a light ion, like the path of a lightweight auto, is influenced more easily. As a result, heavy ions emerge from between the magnet's poles along different lines than the lighter ions. In effect, an entering beam containing ions of different mass is sorted by the magnet into a number of beams, each containing ions of

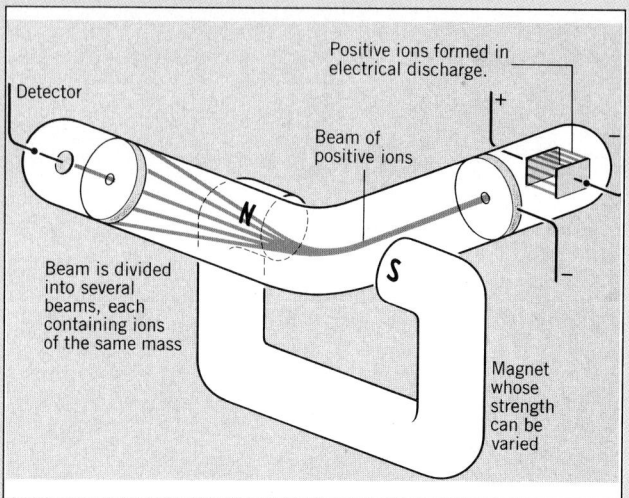

the same mass. This spreading out of the ion beam thus produces an array of different beams called a *mass spectrum.*

In practice, the strength of the magnetic field is gradually changed, which sweeps the beams of ions across a detector located at the end of the tube. As a beam of ions strikes the detector, its intensity is measured and the masses of the particles in the beam are computed based on the strength of the magnetic field and the geometry of the apparatus.

Among the benefits derived from measurements using the mass spectrometer are very accurate isotopic masses and relative isotopic abundances. These have served as the basis for the very precise values of the atomic masses that you find in the periodic table.

molecules in the air, and the electrons became attached to the oil drops, which thereby were given an electrical charge. By observing the rate of fall of the charged drops both when the metal plates were electrically charged and when they were not, Millikan was able to calculate the amount of charge carried by each drop. When he examined his

FIGURE 6.3 Millikan's famous oil drop experiment. Electrons, which are ejected from air molecules by the X rays, are picked up by very small drops of oil falling through the tiny hole in the upper metal plate. By observing the rate of fall of the charged oil drops, with and without electrical charges on the metal plates, Millikan was able to calculate the charge carried by an electron.

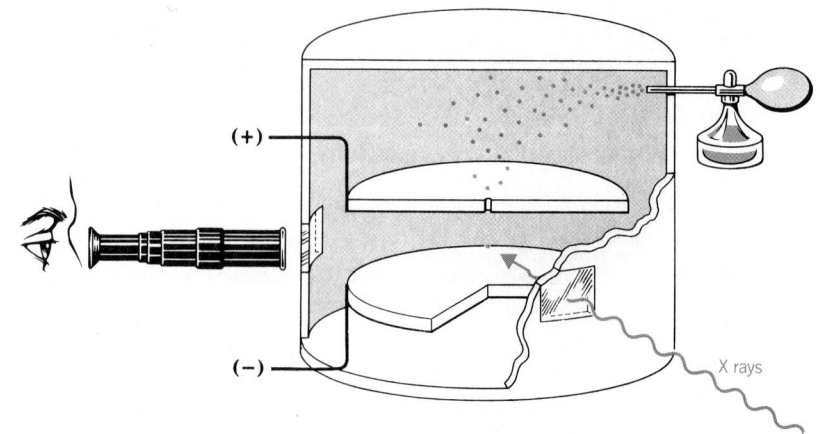

results, he found that all the values he obtained were whole-number multiples of -1.60×10^{-19} C. He reasoned that since a drop could only pick up whole numbers of electrons, this value must be the charge carried by each individual electron.

When atoms lose or gain electrons, as they do when they become ions, their electrical charges always change by multiples of 1.60×10^{-19} C. Therefore, when we specify an ion's charge, we are really giving it in units of this size. For example, an ion with a charge of $1+$ has an actual charge, expressed in SI units, of $+1.60 \times 10^{-19}$ C, and an ion with a charge of $2-$ has an actual charge of -3.20×10^{-19} C. These actual charges are rarely needed, however. Normally we only need to know the number of charges and their sign.

Once Millikan had measured the electron's charge, its mass could then be calculated from Thomson's charge-to-mass ratio.

$$e/m = -1.76 \times 10^8 \text{ C/g}$$

Solving for the mass, m, and substituting -1.60×10^{-19} C for e gives

$$m = \frac{-1.60 \times 10^{-19} \text{ C}}{-1.76 \times 10^8 \text{ C/g}}$$
$$= 9.09 \times 10^{-28} \text{ g}$$

More precise measurements have since been made, and the mass of the electron is currently reported to be $9.1093897 \times 10^{-28}$ g.

The Discovery of the Proton

After the discovery of the electron, gas discharge tubes were modified for additional experiments. In one series, the cathode was shaped like a disk with a hole in its center, and the tube's inner surface *behind* the cathode was coated with a phosphor (Figure 6.4). In the new experiments, the phosphor was placed where the electron beam could not touch it, yet it still glowed. Obviously, something was moving through the hole in the cathode, in a direction opposite that of the cathode rays. It didn't take long to find out that a stream of particles was moving through the hole in the cathode and that these particles had a positive charge. The beam, for example, could be deflected toward a negatively charged plate positioned outside the tube. The mechanism for making these positive particles, as the experimenters reasoned things out, was probably collisions between electrons in the cathode rays and atoms of the residual gas in the tube. Such a collision knocked off an electron from the atom, leaving behind a particle with a positive charge.

Unlike cathode rays, the masses of the positive particles varied according to the gas present in the tube. The lightest positive particles (hydrogen ions, H^+) were observed when the tube had hydrogen gas in it, yet these particles were still about 1800 times as heavy as an electron. When other gases were used, their masses always seemed to be whole-number multiples of the mass observed for hydrogen atoms. This suggested the

The positively charged particles in atoms contribute much more to the mass than do the electrons.

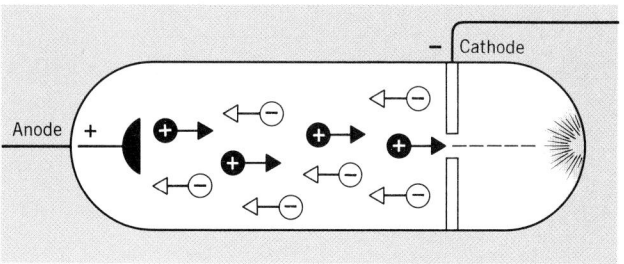

FIGURE 6.4 Positively charged particles are made when cathode rays (electrons) strike atoms of residual gas. They are attracted to the cathode, and some sail through the hole to strike the phosphor and generate a flash of light.

possibility that clusters of the positively charged particles made from hydrogen atoms made up the positively charged particles of other gases. In other words, since the lightest of all positive particles came from hydrogen, it seemed likely that all gases—and all matter, in fact—were made of combinations of the particles in hydrogen. The hydrogen atom, minus an electron, thus seemed to be a fundamental particle in all matter, and this particle was therefore named the proton, after the Greek *proteios,* meaning "of first importance."

The Discovery of the Atomic Nucleus

The effect of the cathode ray (an electron beam) on matter that led to the discovery of the proton suggested many similar experiments. The subatomic "bullets" did not all have to be made by gas discharge tubes, either. Elements had been discovered that spontaneously sent out showers of subatomic particles in a phenomenon called *radioactivity.* Some radioactive elements send out electrons, but others send out much larger particles having masses four times those of the proton and bearing two positive charges. These were called alpha particles.

Early in this century, Hans Geiger and Ernest Marsden, working under Ernest Rutherford at Great Britain's Manchester University, studied what happened when alpha rays hit thin metal foils. Most of the alpha particles sailed right on through as if the foils were virtually empty space (Figure 6.5). A significant number of alpha particles,

FIGURE 6.5 Alpha particles are scattered in all directions by a thin metal foil. Some hit something very massive head-on and are deflected backward. Many sail through. Some, making near misses with the massive "cores" (nuclei), are still deflected, because alpha particles have the same kind of charge (+) as these cores. "Like charges repel."

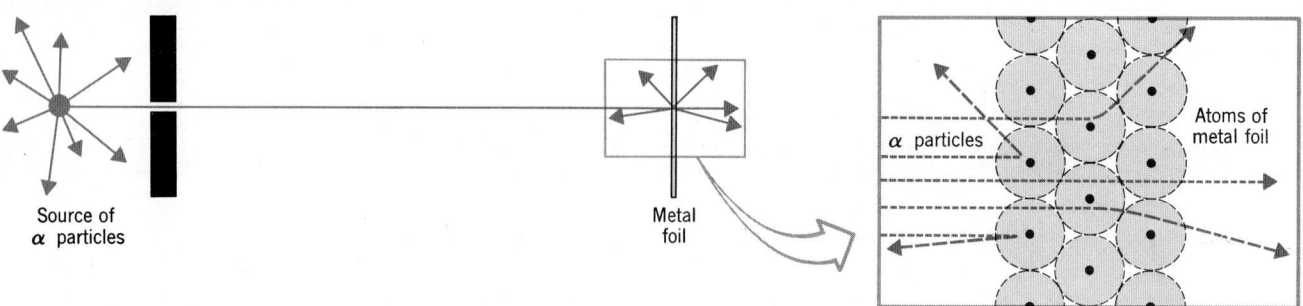

Source of α particles

Metal foil

α particles

Atoms of metal foil

TABLE 6.1 Properties of Subatomic Particles

Particle	Mass (amu)	Mass (g)	Electrical Charge
Electron	0.0005485712	$9.1093897 \times 10^{-28}$	$1- (-1.60 \times 10^{-19}$ C)
Proton	1.00727605	$1.6726231 \times 10^{-24}$	$1+ (+1.60 \times 10^{-19}$ C)
Neutron	1.008665	1.674954×10^{-24}	0

however, were deflected at very large angles. Some were even deflected backward, as if they had hit stone walls. Rutherford was so astounded that he compared the effect to that of firing a 15-in. artillery shell at a piece of tissue paper and having it come back and hit the gunner. He reasoned that only something extraordinarily massive, compared with the alpha particle, could cause such an occurrence. From studying the angles of deflection of the particles, Rutherford determined that whatever it was in the foil that was so massive had to be positively charged. However, since most of the alpha particles went straight through, he further reasoned that the metal atoms in the foils must be mostly empty space. Rutherford's conclusion was that virtually all of the mass of an atom must be concentrated in a particle having a very small volume located in the center of the atom. He called this massive particle the atom's nucleus.

The Discovery of the Neutron

From the way alpha particles were scattered by a metal foil, Rutherford and his students were able to estimate the number of positive charges on the nucleus of an atom of the metal. This had to be equal to the number of protons in the nucleus, of course. But when they computed the nuclear mass based on this number of protons, the value always fell short of the actual mass. In fact, Rutherford found that only about half of the nuclear mass could be accounted for by protons. This led him to suggest that there were other particles in the nucleus that had a mass close to or equal to that of a proton, but with no electrical charge. This suggestion initiated a search that finally ended in 1932 with the discovery of the neutron by Sir James Chadwick, a British physicist. Table 6.1 summarizes the properties found for the three principal subatomic particles.

Chadwick received the 1935 Nobel Prize in physics for his discovery of the neutron.

6.2 ELECTROMAGNETIC RADIATION AND ATOMIC SPECTRA; ENERGY LEVELS IN ATOMS

The light given off by atoms that have absorbed energy shows that the electrons in an atom can possess only certain specific amounts of energy.

The picture of the atom that emerged from the work of Rutherford and his students — a small dense nucleus containing protons and neutrons surrounded by electrons in the remaining volume of the atom — raises important chemical questions. Although the nucleus determines the mass of the atom and the number of electrons needed to give the atom electrical neutrality, the nucleus does not play a direct part in chemical reactions. When two or more atoms join together to form a compound, the nuclei of the atoms stay relatively far apart. Only the atoms' outer reaches — the regions inhabited by electrons — come in close contact. The chemical properties of the elements, then, must be determined by the electrons of the various atoms, and the similarities and differences in these properties must have something to do with the way these electrons are distributed around the particular nuclei.

How electrons are distributed about the nucleus is called the atom's electronic structure. The basic clue to the electronic structures of the various elements comes from the study of the light emitted when atoms of the elements are excited, or energized. To learn about this, however, we must first learn a little about light itself.

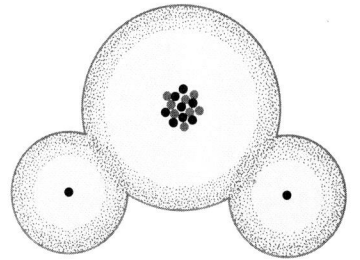

Atoms not drawn to scale, as they are joined in water, H_2O. Nuclei stay far apart and only the outer parts of the atoms touch.

Electromagnetic Energy

Thus far in our discussion of energy we have concentrated largely on one form of it — heat energy. In Chapter 5, we saw that many reactions give off energy, whereas other reactions need a continuous supply of energy to continue to occur. Another important kind of energy in chemistry is called electromagnetic energy, popularly known as light energy. Many chemical systems emit visible light as they react. An example is when anything burns.

Electromagnetic energy is energy carried through space or matter by means of wavelike oscillations. These waves are like water waves in that something goes up and down, but, unlike water waves, what oscillates is not matter. The oscillations are systematic fluctuations in the intensities of very tiny electrical and magnetic forces. The space in which these oscillations occur is called the electromagnetic field. Both electrical and magnetic forces are present in an electromagnetic wave, and each changes rhythmically with time. In other words, the value of each force goes through a maximum, then to zero and down to a minimum (negative) value, and back up again through zero to the maximum value. Figure 6.6 shows this for the electrical component. The magnetic component would be depicted in a plane at right angles to the plane carrying the electrical component. Each oscillation is called one *cycle*. The successive series of these oscillations occurring through space from the origin of the light is called electromagnetic radiation (popularly, a light wave). The number of cycles per second is called the frequency of the electromagnetic radiation, and its symbol is v (a Greek letter pronounced "new").

FIGURE 6.6 The electrical force associated with electromagnetic radiation fluctuates rhythmically. *(a)* Two cycles of fluctuation are shown; therefore, the frequency is 2 Hz. *(b)* An electromagnetic radiation frozen in time. This curve shows how the electrical force varies along the direction of travel. The distance between two maximum values is the wavelength of the electromagnetic radiation.

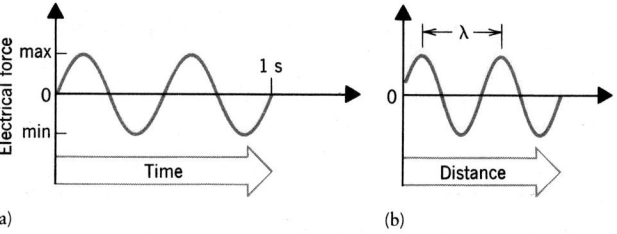

The concept of frequency extends beyond just electromagnetic radiation to other events that recur at regular intervals. For example, you go to school 5 days *per week*, or you pay your bills once *each month*. Each of these statements describes how frequently an event occurs, and they have in common the notion of *per unit of time*, or 1/time. In the SI, the unit of time is the second, so frequency is given the unit "per second," which is 1/second, or (second)$^{-1}$. This unit is given the special name hertz (Hz).

The SI symbol for the *second* is s.

$$s^{-1} = 1/s$$

$$1 \text{ Hz} = 1 \text{ s}^{-1}$$

As the radiation moves away from its source, the maximum values of the electrical and magnetic forces are regularly spaced in the field. The distance separating maximum values is called the radiation's wavelength, symbolized by λ (another Greek letter, *lambda*). See Figure 6.6*b*. Because wavelength is a distance, it has distance units (for example, meters).

If we multiply the wavelength by frequency, the result is the velocity of the wave. We can see this if we analyze the units.

For any wave, the product of its wavelength and its frequency equals the velocity of the wave.

$$\text{meters} \times \frac{1}{\text{second}} = \frac{\text{meters}}{\text{second}} = \text{velocity}$$

(In SI units)   m $\times$ 1/s   =   m/s   =   m s^{-1}

The velocity of electromagnetic radiation in a vacuum is a constant, 3.00×10^8 m/s (or m s^{-1}) and is commonly called the *speed of light*. This is an important physical constant

When an element is bombarded with a high-energy electron beam, the element emits X rays, a form of high-energy, penetrating, electromagnetic radiation. This is how the X rays are generated when you have your teeth or chest x-rayed. What is particularly interesting is that each element emits its own characteristic X-ray spectrum. Only certain X-ray frequencies are given off.

In 1912, Henry G. J. Moseley was studying the relationship between these X-ray frequencies and an element's location in the periodic table. No correlation was found between atomic weight and X-ray frequency, but Moseley found that he was able to relate an element's X-ray frequency to a characteristic integer which was the same as the element's position number in the periodic table. This integer is the element's atomic number. Moseley's discovery allowed scientists to experimentally measure the atomic numbers of new elements and confirm their correct positions in the periodic table.

Moseley died in 1915 at the age of 27 during the

Henry G. J. Moseley.

British invasion of Gallipoli in Turkey. His death led to the British government's policy of assigning noncombat duties to its scientists during World War II.

that has implications far beyond our present discussion, and is given the symbol *c*.

$$c = 3.00 \times 10^8 \text{ m s}^{-1}$$

From the preceding discussion we obtain a very important relationship that allows us to convert between λ and v.

$$\lambda \times v = c = 3.00 \times 10^8 \text{ m s}^{-1} \tag{6.1}$$

Electromagnetic radiation comes in a large range of frequencies called the electromagnetic spectrum, shown in Figure 6.7. Each portion of the spectrum has a popular name. For example, radio waves are electromagnetic radiations having very low frequencies (and therefore very long wavelengths). Microwaves, which also have low frequencies, are absorbed by molecules in food, and the energy that these molecules take on raises their temperature. As a result, the food cooks. (Keep your hands and eyes out of the way of microwave radiation; it can cook you, too.) Infrared radiation consists of the range of frequencies that can make molecules of most substances vibrate internally. Each substance absorbs a uniquely different set of infrared frequencies. A plot of the frequencies absorbed versus the intensities of absorption is called an infrared spectrum. It can be used to identify a compound, because each infrared spectrum is as unique as a set of fingerprints. (See Figure 6.8.) Many substances absorb visible and ultraviolet radiations in unique ways, too, and they have visible and ultraviolet spectra. Gamma rays are at the

Remember that there is an inverse relationship between wavelength and frequency. The lower the frequency, the longer the wavelength.

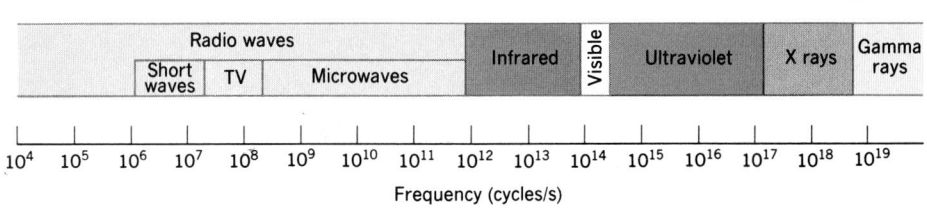

FIGURE 6.7 The electromagnetic spectrum.

FIGURE 6.8 Infrared spectrum of methyl alcohol, wood alcohol (courtesy Sadtler Research Laboratories, Inc., Philadelphia, Pa.).

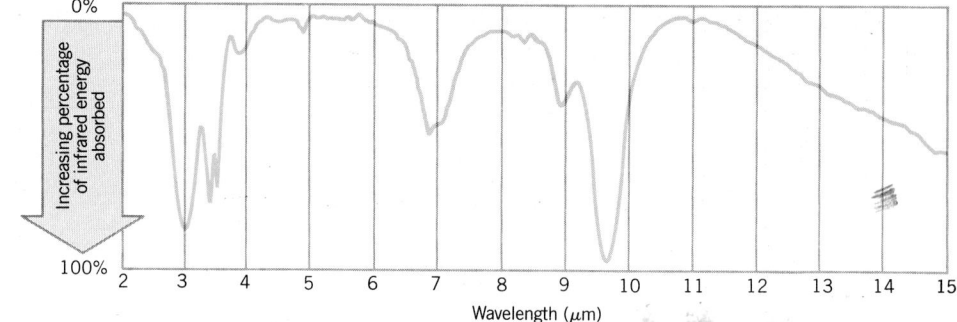

high-frequency end of the electromagnetic spectrum. They are produced by certain elements that are radioactive. X rays are very much like gamma rays, but they are usually made by special equipment. Both X rays and gamma rays penetrate living things easily.

FIGURE 6.9 *(a)* A diagram showing how white light is refracted by a glass prism, which spreads out the colors of the visible spectrum. *(b)* In this color photograph we see the continuous rainbow of colors formed from white light.

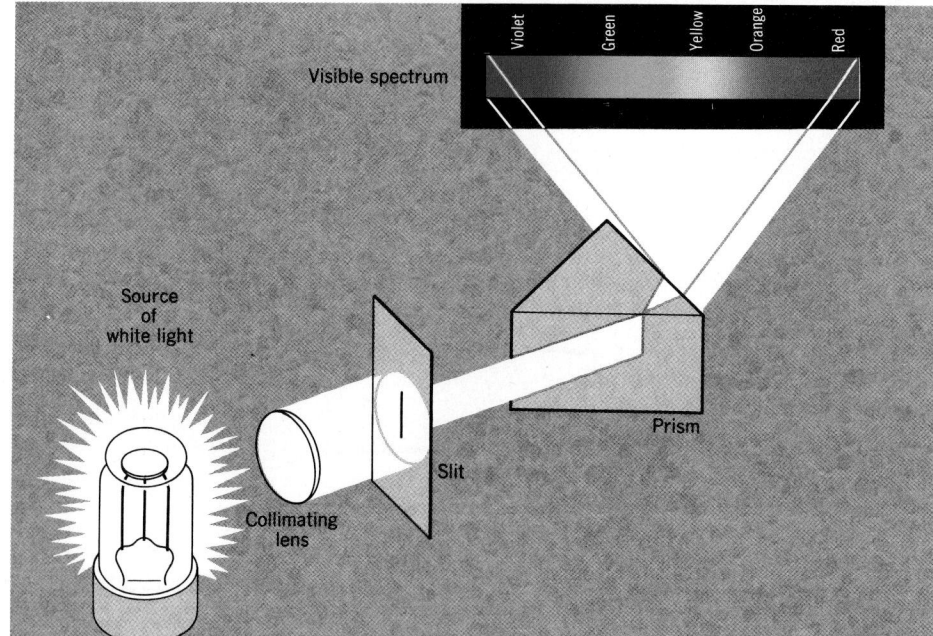

Almost all the time, you are bombarded with electromagnetic radiations from all these portions of the electromagnetic spectrum. Radio and TV signals pass through you; you feel infrared radiation when you sense the warmth of a radiator; X rays and gamma rays fall on you from space; and light from a lamp reflects from the page you're reading into your eyes. Of all these radiations, your eyes are able to sense only a very narrow band of wavelengths ranging from about 400 to 700 nm (corresponding to frequencies of 7.5×10^{14} to 4.3×10^{14} Hz). This band is called the visible spectrum and consists of all the colors you can see, from red through orange, yellow, green, blue, indigo, and violet. White light is composed of all these colors in roughly equal amounts, and can be separated into them by focusing a beam of white light through a prism, which spreads the various wavelengths apart. This is shown diagrammatically in Figure 6.9a. A photograph showing the production of a visible spectrum is shown in Figure 6.9b.

EXAMPLE 6.1   CALCULATIONS INVOLVING WAVELENGTH AND FREQUENCY

Problem: Yellow light has a wavelength of 600 nm. What is its frequency in hertz?

Solution: To convert between wavelength and frequency we must use Equation 6.1.

$$\lambda \times \nu = c$$

Solving for the frequency, ν, gives

$$\nu = \frac{c}{\lambda}$$

Next, we substitute the values for c (3.00×10^8 m s^{-1}) and for the wavelength. But we have to be careful here. To make the units cancel correctly, we have to express the wavelength in meters. The unit for wavelength given in the problem is *nm*, which means *nanometer*. In Chapter 1 you learned that the prefix *nano-* means "$\times 10^{-9}$", so we can write the wavelength as 600×10^{-9} m. Substituting this value, we get

$$\nu = \frac{3.00 \times 10^8 \ \cancel{m} \ s^{-1}}{600 \times 10^{-9} \ \cancel{m}}$$

$$= 5.00 \times 10^{14} \ s^{-1}$$

The unit s^{-1} is a hertz (Hz), so the frequency is

$$\nu = 5.00 \times 10^{14} \ Hz$$

EXAMPLE 6.2   CALCULATIONS INVOLVING WAVELENGTH AND FREQUENCY

Problem: Radio station WGBB on Long Island, New York, broadcasts at a frequency of 1240 kHz. What is the wavelength of these radio waves, expressed in meters?

Solution: Once again, we use the equation

$$\lambda \times \nu = c$$
$$= 3.00 \times 10^8 \ m \ s^{-1}$$

Recalling that the prefix k in kHz stands for "$\times 10^3$," and that 1 Hz = 1 s^{-1}, we can write the frequency, ν, as $\nu = 1240 \times 10^3$ s^{-1}. Now we solve for the wavelength and substitute.

$$\lambda = \frac{3.00 \times 10^8 \ m \ s^{-1}}{\nu}$$

$$= \frac{3.00 \times 10^8 \ m \ \cancel{s^{-1}}}{1240 \times 10^3 \ \cancel{s^{-1}}} = 242 \ m$$

These radio waves are longer than the length of two football fields.

This is the wavelength in meters.

PRACTICE EXERCISE 1   Green light has a wavelength of 550 nm. What is its frequency in hertz?

PRACTICE EXERCISE 2   A certain FM radio station broadcasts at a frequency of 93.5 MHz (megahertz). What is the wavelength of these radio waves, expressed in meters?

The Energy of a Light Wave

In 1900, Max Planck (1858–1947), a German physicist, launched one of the greatest upheavals in the history of science when he proposed that electromagnetic radiation is emitted only in tiny packets of energy later called photons. Each photon pulses with a frequency, v, and each travels with the speed of light. Planck proposed and Albert Einstein (1879–1955) confirmed that *the energy of a radiation is proportional to its frequency,* not to its intensity or brightness as had been believed up to that time.

$$\text{energy of a photon} = E = hv \qquad (6.2)$$

where h is a proportionality constant now called Planck's constant. The energy of one photon is called one quantum of energy.

The value of Planck's constant is $h = 6.63 \times 10^{-34}$ J s. It has units of energy (joules) multiplied by time (seconds).

Planck's and Einstein's discovery was really quite surprising. If a particular event requiring energy, such as photosynthesis in green plants, is initiated by the absorption of light, it is the frequency of the light that is important, not its intensity or brightness. An analogy would be a group of pole-vaulters trying to get over a wall. If they have long enough poles, each can clear the wall. If the poles are too short, however, they can batter the wall with as much intensity as they want, but they'll never clear the top. Not even doubling or tripling the number of vaulters with short poles will get anyone across.

Atomic Spectra

The spectrum described in Figure 6.9 is called a continuous spectrum because it contains light of all colors. It is formed when the light from the sun, or any other object that's been heated to very high temperatures (such as the filament in an electric light bulb), is split by a prism and displayed on a screen. A rainbow after a summer shower is a continuous spectrum that most people have seen. In this case, the colors contained in sunlight are spread out by tiny water droplets in the air.

A somewhat different kind of spectrum is produced if we examine light that is given off by a gas such as hydrogen when an electric discharge passes through it. This discharge

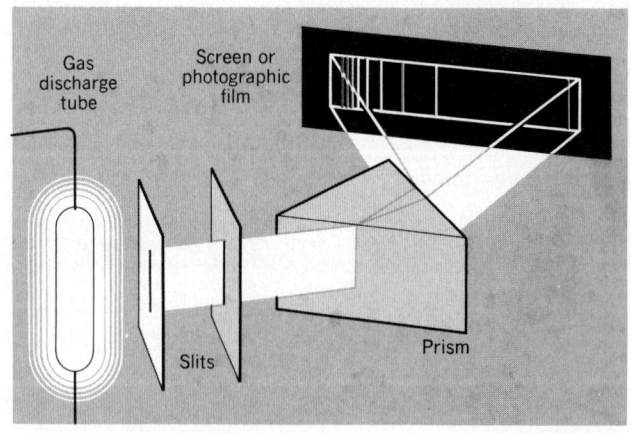

FIGURE 6.10 Production of a line spectrum. The light emitted by excited atoms is formed into a narrow beam and passed through a prism. This light beam is divided into relatively few narrow beams with frequencies that are characteristic of the particular element that is emitting the light.

SPECIAL TOPIC 6.4 PHOTOELECTRICITY AND ITS APPLICATIONS

One of the earliest clues to the relationship between the frequency of light and its energy was the discovery of the photoelectric effect. In the latter part of the nineteenth century, it was found that certain metals acquired a positive charge when they were illuminated by light. Apparently, light is capable of kicking electrons out of the surface of the metal.

When this phenomenon was studied in detail, it was discovered that electrons could only be made to leave a metal's surface if the frequency of the incident radiation was above some certain value, which was named the threshold frequency. This threshold frequency differs for different metals, depending on how tightly the metal atom holds onto electrons. Above the threshold frequency, the kinetic energy of the emitted electron increases with increasing frequency, but its kinetic energy does not depend on the intensity of the light. In fact, if the frequency of the light is below the minimum frequency, no electrons are observed at all, no matter how bright the light is. To physicists of that time, this was very perplexing. The explanation of the phenomenon was finally given by Albert Einstein in the form of a very simple equation.

$$KE = h\nu - w$$

where KE is the kinetic energy of the electron that is emitted, $h\nu$ is the energy of the photon of frequency ν, and w is the minimum energy needed to eject the electron from the metal's surface. Stated another way, part of the energy of the photon is needed just to get the electron off the surface of the metal. This amount is w. Any energy left over ($h\nu - w$) appears as the electron's kinetic energy.

Besides its important theoretical implications, the photoelectric effect has received a large number of practical applications. For example, automatic "electric eye" door openers use this phenomenon by sensing the interruption of a light beam caused by the person wishing to use the door. The phenomenon is also responsible for photoconduction by certain substances that are used in light meters in cameras and other devices. The production of sound in motion pictures makes use of a strip along the edge of the film (called the sound track) that causes the light passing through it to fluctuate in intensity according to the frequency of the sound that's been recorded. A photocell converts this light to a varying electric current that is amplified and played through speakers in the theater. Even the sensitivity of photographic film to light is related to the release of photoelectrons within tiny grains of silver bromide that are suspended in a coating on the surface of the film.

is an electric current that *excites*, or energizes, the atoms of the gas, and they emit this energy in the form of light as they return to a lower energy state. When a narrow beam of this light is passed through a prism, as shown in Figure 6.10, a continuous spectrum is *not* produced. Instead, only a few colors are observed, displayed as a series of individual lines. This series of lines is called the element's atomic spectrum. Figure 6.11 shows the

Atoms of an element can also be excited by adding them to the flame of a Bunsen burner.

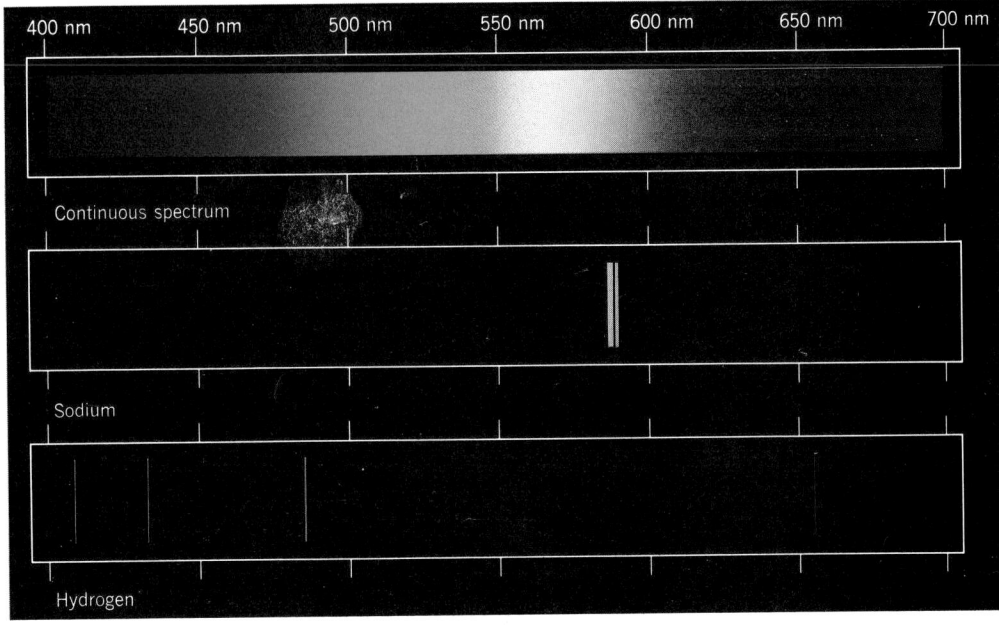

FIGURE 6.11 (Top) The continuous visible spectrum produced by an incandescent lamp. (Center) The atomic spectrum (line spectrum) produced by sodium. (Bottom) The atomic spectrum (line spectrum) produced by hydrogen.

Atomic spectra like those shown in Figure 6.11 are also called *emission spectra* or *line spectra*

$E = hv$
$h = 6.63 \times 10^{-34}$ J s
$E = (6.63 \times 10^{-34}$ J s$)$
    $\times (4.57 \times 10^{14}$ s$^{-1})$
    $= 3.03 \times 10^{-19}$ J

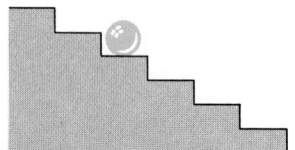

FIGURE 6.12 Ball on a staircase. The ball can have only certain amounts of potential energy when at rest.

The potential energy of the ball at rest is quantized.

visible portions of the atomic spectra of two common elements, sodium and hydrogen, and how they compare with a continuous spectrum. Notice that the spectra of these elements are quite different. In fact, each element has its own unique atomic spectrum that is as characteristic as a fingerprint.

The Significance of Atomic Spectra

Earlier we saw that there is a simple relationship between the frequency of light and its energy, $E = hv$. The fact that excited atoms emit light of only certain characteristic frequencies tells us that only certain characteristic energy changes take place within the atom. For instance, in the spectrum of hydrogen there is a red line (see Figure 6.11) that has a wavelength of 656 nm and a frequency of 4.57×10^{14} Hz. The energy of each photon of this light is 3.03×10^{-19} J. Therefore, when a hydrogen atom emits light of this frequency, the energy of the atom decreases by 3.03×10^{-19} J. What is very special here is that whenever a hydrogen atom emits red light, its frequency is 4.57×10^{14} Hz and the energy change within the atom is *always exactly* 3.03×10^{-19} J, never more and never less. Atomic spectra, then, tell us that *when an excited atom loses energy, not just any arbitrary amount is lost.* Only certain specific energy changes can occur, which is why only certain specific frequencies of light are emitted. This is the only way atomic spectra can be explained.

How is it that atoms of a given element always undergo exactly the same specific energy changes? The answer seems to be that in an atom an electron can have only certain definite amounts of energy and no others. In the words of science, we say that the electron is restricted to certain energy levels. We also say that the energy of the electron is quantized, meaning once again that the electron's energy in a particular atom can have only certain values and no others.

The energy of an electron in an atom might be compared to the potential energy of a ball on a staircase (see Figure 6.12). The ball can only come to rest on a step, and on each step it will have some specific amount of potential energy. If the ball is raised to a higher step, its potential energy will be increased. When it drops to a lower step, its potential energy decreases. But each time the ball stops, it stops on one of the steps, never in between. Thus, the ball at rest can only have certain specific amounts of potential energy, which are determined by the energy levels of the various steps of the staircase. So it is with an electron in an atom. The electron can only have energies corresponding to the set of electron energy levels in the atom. When the atom is supplied with energy, as in a gas discharge tube, an electron is raised from a low-energy level to a higher one. When the electron drops back, energy equal to the difference between the two levels is released and emitted as a photon. Because only certain energy jumps can occur, only certain frequencies can appear in the spectrum.

The existence of specific energy levels in atoms, as implied by atomic spectra, forms the foundation of all theories about electronic structure. Any model of the atom that attempts to describe the positions or motions of electrons must also account for atomic spectra.

6.3 THE BOHR MODEL OF THE ATOM

The first model of the atom to meet with some success imagined the electron to be revolving about the nucleus in orbits of fixed energy.

The discovery that the energy of electrons is quantized led to attempts to develop theoretical models of the way electrons behave in atoms. The goals were to explain how electrons move, where they are located, and how they change energy to give off photons of light. Physicists were faced with a problem, however. None of the physical laws that seemed to govern the motion of large objects, like baseballs or people, were able to account for the strange behavior of electrons.

In 1913 Niels Bohr (1885–1962), a Danish physicist, proposed a theoretical model for the hydrogen atom. He chose hydrogen because its atoms are the simplest, having only one electron about the nucleus, and because it produces the simplest spectrum with the fewest lines. In his model, Bohr imagined the electron to move around the nucleus following fixed paths, or orbits, much as a planet moves around the sun. His model also restricted the sizes of the orbits and the energy that the electron could have in a given orbit. The equation Bohr derived for the energy of the electron includes a number of physical constants such as the mass of the electron, its charge, and Planck's constant. It also contains an integer, *n*, that Bohr called a quantum number. Each of the orbits can be specified by its value of *n*. If all the constants are combined, Bohr's equation is

$$E = \frac{-k}{n^2} \tag{6.3}$$

where *E* is the energy of the electron and *k* is the combined constant (its value is 2.18×10^{-18} J). The allowed values of *n* range from 1 to ∞, with all integers permitted (i.e., *n* could equal 1, 2, 3, 4, . . . , ∞). Therefore, the energy of the electron in any particular orbit could be calculated.

Because of the negative sign in Equation 6.3, the lowest (most negative) energy value occurs when *n* = 1, which corresponds to the *first Bohr orbit*. This lowest energy state is called the ground state. According to Bohr's theory also, this orbit brings the electron closest to the nucleus.

When the hydrogen atom absorbs energy, as it does in a discharge tube, the electron is raised from the orbit having *n* = 1 to a higher orbit — to *n* = 2 or *n* = 3 or even higher. Then, when the electron drops back to a lower orbit, energy is emitted in the form of light (see Figure 6.13). Since the energy of the electron in a given orbit is fixed, a drop from one particular orbit to another, say from *n* = 2 to *n* = 1, always releases the same amount of energy, and the frequency of the light emitted because of this change in energy is always precisely the same.

Niels Bohr won the 1922 Nobel Prize in physics for his work on atomic structure.

It's really rather amazing that an equation as simple as this, involving only a constant and a set of integers, can be used to calculate the energies that an electron can have in the hydrogen atom.

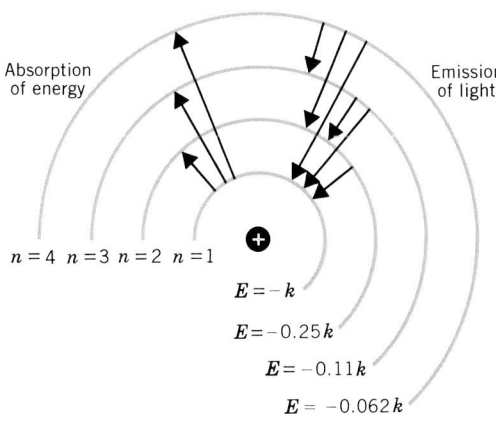

FIGURE 6.13 Absorption of energy and emission of light by the hydrogen atom. When the atom absorbs energy, the electron is raised to a higher energy level. When the electron falls to a lower energy level, light of a particular energy and frequency is emitted.

Bohr's model of the atom was both a success and a failure. It successfully predicted the frequencies of the lines in the hydrogen spectrum, so there seemed to be some validity to the theory. Nevertheless, the model was a total failure for atoms with more than one electron. Still, though the theory met with only limited success, introduction of the ideas of quantum numbers and fixed energy levels was an important step forward.

6.4 THE WAVE NATURE OF MATTER

Very small particles show properties that we identify with waves.

Bohr's efforts to develop a theory of electronic structure were doomed from the very beginning because the classical laws of physics — those known in his day — simply do not apply to particles as tiny as the electron. Since all of the objects that had been studied by scientists until that time were large and massive in comparison with the electron, no one had yet detected the limits of classical physics. Classical physics fails for atomic particles because matter is not really as our physical senses perceive it. Under appropriate circumstances, small bits of matter, such as an electron, behave not like solid particles, but instead like waves. This idea was first proposed in 1924 by a young French graduate student, Louis de Broglie.

In Section 6.2 we saw that light waves are characterized by their wavelengths and their frequencies. The same is true of matter waves. De Broglie suggested that the wavelength of a matter wave, λ, is given by the equation

De Broglie was awarded a Nobel Prize in 1929.

$$\lambda = \frac{h}{mv} \qquad (6.4)$$

where h is Planck's constant, m is the particle's mass, and v is its velocity.

The concept of a particle of matter behaving as a wave rather than as a solid object may at first seem difficult to comprehend. This book certainly seems solid enough, and if you drop it on your toe, it surely doesn't seem to be a wave, at least not as we generally think of waves in the ocean. The reason for the book's apparent solidity is that in de Broglie's equation (Equation 6.4) the mass appears in the denominator. This means that heavy objects have extremely short wavelengths. The peaks of the matter waves for heavy objects are so close together that the wave properties go unnoticed and can't even be measured experimentally. But tiny particles with very low masses have much longer

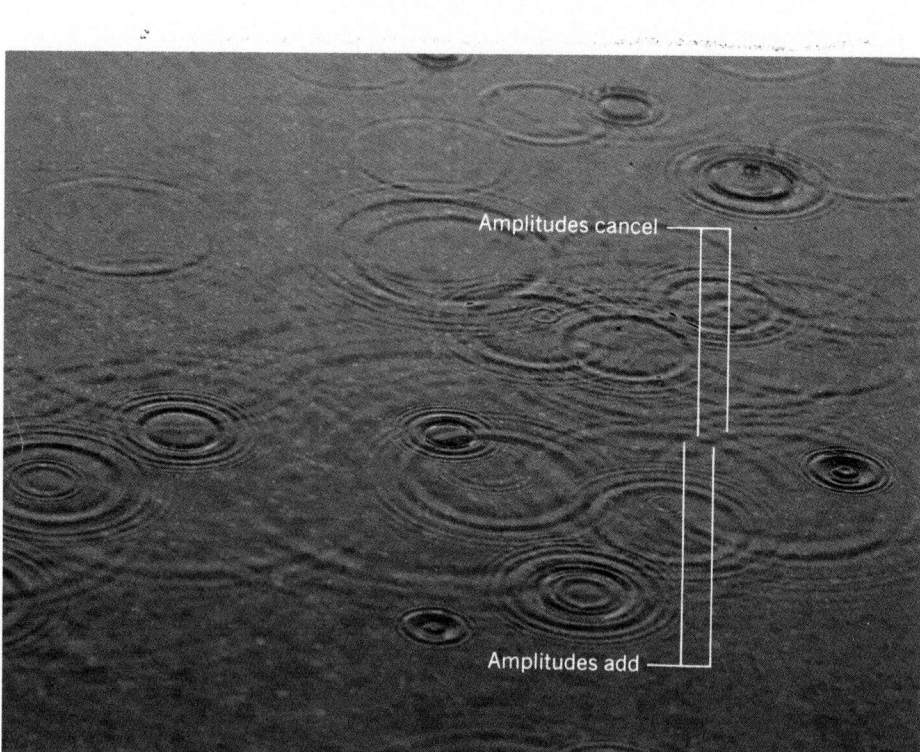

FIGURE 6.14 Diffraction of water waves produced by raindrops falling on a pond. As the waves cross, the amplitudes increase where the waves are in phase and cancel where they are out of phase.

Amplitudes cancel

Amplitudes add

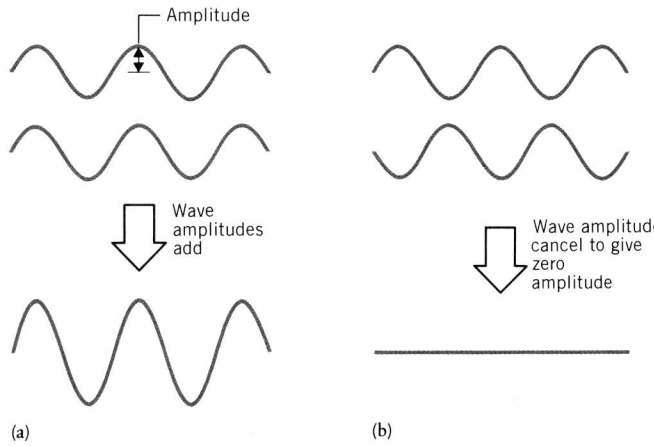

FIGURE 6.15 *(a)* Waves in phase—constructive interference. *(b)* Waves out of phase—destructive interference.

wavelengths; therefore, their wave properties become an important part of their overall behavior.

By now you may have begun to wonder if there is in fact any way to prove that matter has wave properties. Actually, these properties are easily shown by a common phenomenon that you have often observed. For example, when raindrops fall on a quiet pond, ripples spread out from where the drops strike the water, as shown in Figure 6.14. When two sets of ripples cross, there are places where the waves are *in phase*—the peak of one wave coincides with the peak of the other. At these points, the height of the water is equal to the sum of the heights of the two crossing waves. At other places the crossing waves cancel each other because the waves are *out of phase.* This reinforcement and cancellation of wave intensities, referred to, respectively, as constructive and destructive interference, is the phenomenon called diffraction. It is examined more closely in Figure 6.15.

Diffraction occurs with water waves, light waves, and matter waves. For light waves it can be shown in an experiment such as that illustrated in Figure 6.16. The two beams of light formed by the closely spaced pair of slits interfere with each other. On the screen, we see bands of light where the light waves reinforce each other and bands of darkness where the waves cancel. The display of light and dark bands is called a diffraction pattern.

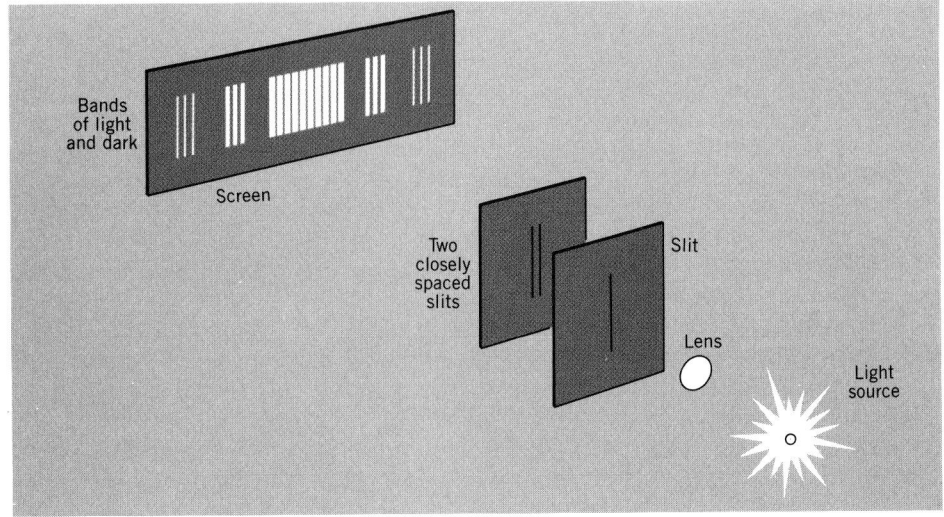

FIGURE 6.16 Production of a diffraction pattern.

SPECIAL TOPIC 6.5 *THE ELECTRON MICROSCOPE*

The usefulness of a microscope in studying small specimens is limited by its ability to distinguish between closely spaced objects. We call this ability the resolving power of the microscope. Through optics, it is possible to increase the magnification and thereby increase the resolving power, but only within limits. These limits depend on the wavelength of the light that is used. Objects with diameters less than the wavelength of the light cannot be seen in detail. Since visible light has wavelengths of the order of about 200 to 800 nm, objects smaller than this can't be examined with a microscope that uses visible light.

The electron microscope uses electron waves to "see" very small objects. De Broglie's equation,

$\lambda = h/mv$, suggests that if an electron, proton, or neutron has a very high velocity, its wavelength will be very small. In the electron microscope, electrons are accelerated to high speeds across high-voltage electrodes. This gives electron waves with typical wavelengths of about 0.006 to 0.001 nm that strike the sample and are then focused magnetically onto a fluorescent screen where they form a visible image. Because of other difficulties, the actual resolving power is considerably less than this—generally on the order of 6 to 1 nm. Some high resolution electron microscopes, however, are able to reveal the shadows of individual atoms in very thin specimens through which the electron beam passes.

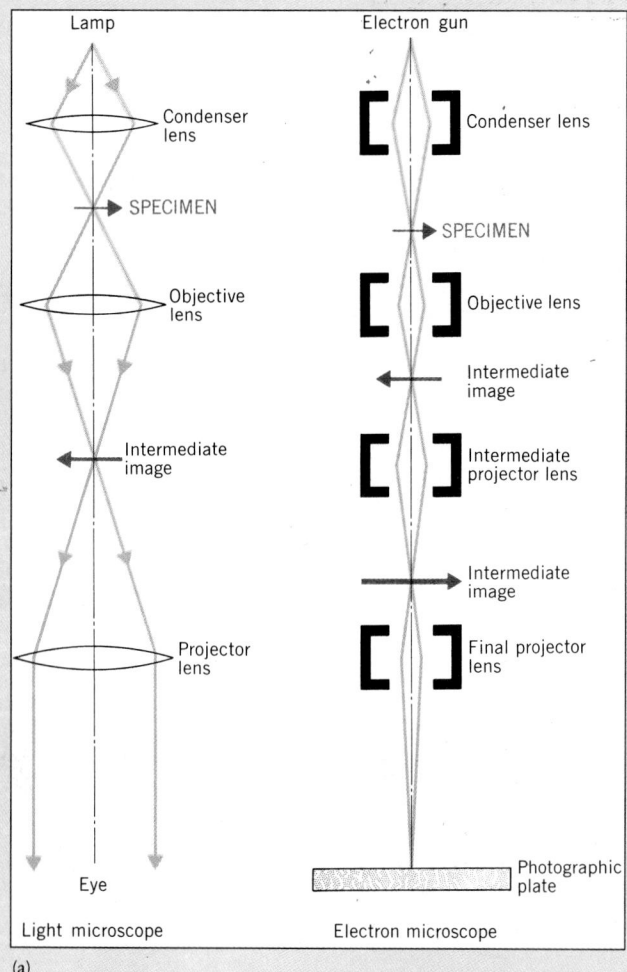

(a)

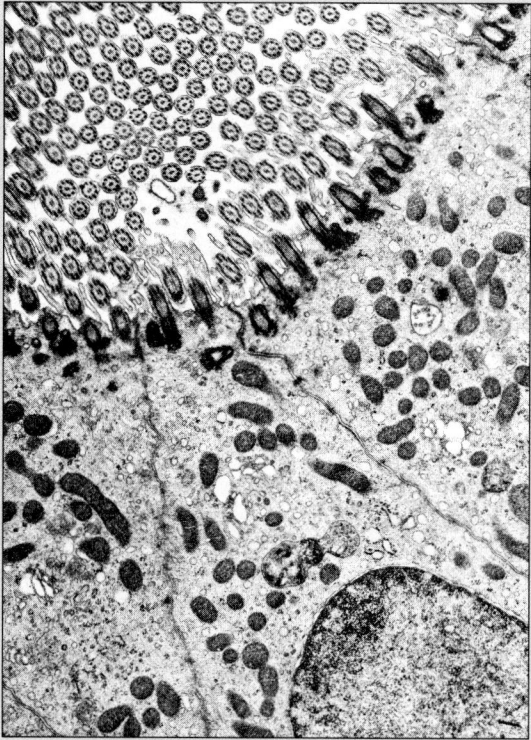

(b)

(a) Schematics of both the light microscope and the electron microscope (from A. G. Marshall, *Biophysical Chemistry*, 1978, John Wiley and Sons, New York, used by permission). *(b)* Electron micrograph of the retina of a rabbit, magnification 8,400 ✕.

Diffraction is a phenomenon that can only be explained as a property of waves. We have seen how it can be demonstrated with water waves and light waves. Experiments can also be done to show that electrons, protons, and neutrons experience diffraction—in fact, that is the principle on which the electron microscope is based (Special Topic 6.5). These experiments prove that these small particles also have wave properties.

6.5 ELECTRON WAVES IN ATOMS

Electron waves in atoms can be identified by three quantum numbers.

Current theories of electronic structure are based on the wave properties of the electron. In fact, the theory has been given the name wave mechanics. It is also called quantum mechanics because the theory predicts quantized energy levels. In 1926 Erwin Schrödinger (1887–1961), an Austrian physicist, became the first scientist to successfully apply the concept of the wave nature of matter to an explanation of electronic structure. His work, for which he won a Nobel Prize in 1933, and the theory that has developed from it are highly mathematical. Fortunately, we need only a qualitative understanding of electronic structure, and the main points of the theory can be understood without all the math.

First, however, we must learn a little more about waves. There are basically two kinds of waves, traveling waves and standing waves. On a lake or ocean the wind produces waves whose crests and troughs move across the water's surface, as shown in Figure 6.17. The water moves up and down while the crests and troughs travel horizontally in the direction of the wind. These are examples of traveling waves.

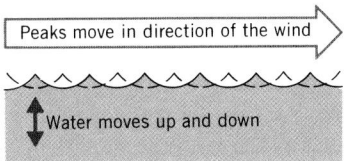

Peaks move in direction of the wind

Water moves up and down

FIGURE 6.17 Traveling waves.

A more important kind of wave for us is the standing wave. An example is the vibrating string of a guitar. When the string is plucked, its center vibrates up and down while the ends, of course, remain fixed. The crest, or point of maximum amplitude, of the wave occurs at one position. At the ends of the string are points of zero amplitude, called nodes, and their positions are also fixed. A standing wave, then, is one in which the crests and nodes do not change position.

The wavelength of this wave is actually twice the length of the string.

As you know, many notes can be played on a guitar string by shortening its effective length with a finger placed at frets along the neck of the instrument. But even without shortening the string, we can play a variety of notes. For instance, if the string is touched momentarily at its midpoint at the same time as it is plucked, the string vibrates as shown in Figure 6.18 and produces a tone that is an octave higher. The wave that produces this higher tone has a wavelength exactly half of that formed when the untouched string is plucked. In Figure 6.18 we see that other wavelengths are possible, too — each gives a different note.

Notes played this way are called harmonics.

If you examine Figure 6.18, you will see that there are some restrictions on the possible wavelengths. Not just any wavelength is possible, because the nodes at either end of the string are in fixed positions. The *only* waves that can occur are those for which a half wavelength can be repeated *exactly* a whole number of times. Expressed another

FIGURE 6.18 Standing waves on a guitar string.

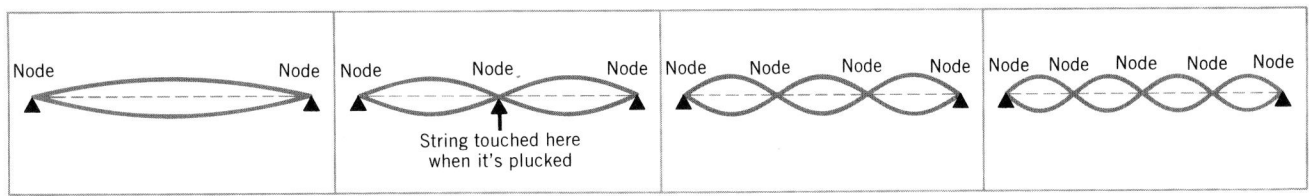

Node · Node
Node · Node · Node
Node · Node · Node · Node
Node · Node · Node · Node · Node

String touched here when it's plucked

way, the length of the string is a whole-number multiple of half wavelengths. In a mathematical form we could write this as

$$L = n\left(\frac{\lambda}{2}\right)$$

where L is the length of the string, λ is a wavelength (therefore, $\lambda/2$ is half the wavelength), and n is an integer. We see that whole numbers (similar to quantum numbers) appear quite naturally in determining which vibrations can occur.

Knowing something about standing waves, we can now look at matter waves, concentrating on electron waves in particular. To study this topic, you may find it best to read through the entire discussion rather quickly to get an overview of the subject, and then read it again more slowly. All of the information fits together in the end, somewhat like the pieces of a puzzle.

The theory of quantum mechanics tells us that in the atom, electron waves are standing waves. Similar to guitar strings, electrons can have many different waveforms or wave patterns. Each of these waveforms, which are called orbitals, has a characteristic energy. Not all of the energies are different, but most are. *Energy changes within an atom are simply the result of an electron changing from a wave pattern with one energy to a wave pattern with a different energy*.

We will be interested in two properties of orbitals — their energies and their shapes. Their energies are important because we normally find atoms in their most stable states, referred to as their ground states, which occur when the electrons assume waveforms having the lowest possible energies. The shapes of the wave patterns (i.e., where their amplitudes are large and where they are small) are important because the theory tells us that the amplitude of a wave at any particular location is related to the likelihood of finding the electron there. This will be important when we study how and why atoms form chemical bonds to each other.

In much the same way that the characteristics of a wave on a guitar string can be related to an integer, wave mechanics tells us that the electron waves (orbitals) can be characterized by a set of *three* integer quantum numbers, n, ℓ, and m_ℓ. In discussing the energies of the orbitals, it is usually most convenient to sort the orbitals into groups according to these quantum numbers.

The Principal Quantum Number, *n*

The quantum number n is called the principal quantum number, and all orbitals that have the same value of n are said to be in the same shell. The values of n can range from $n = 1$ to $n = \infty$. The shell with $n = 1$ is called the *first shell*, the shell with $n = 2$ is the *second shell*, and so forth. (The term *shell* comes from an early notion that atoms could be thought of as similar to onions, with the electrons being arranged in layers around the nucleus.) The various shells are also often identified by letters, beginning (for no significant reason) with K for the first shell ($n = 1$).

n	1	2	3	4	. . .
shell	K	L	M	N	. . .

The principal quantum number serves to determine the size of the electron wave — how far the wave effectively extends from the nucleus. In effect, it is related to how far from the nucleus we are likely to find the electron — the higher the value of n, the larger is the electron's average distance from the nucleus. This quantum number is also related to the energy of the orbital. As n increases, the energies of the orbitals also increase.

Bohr's theory took into account only the principal quantum number n. His theory worked fine for hydrogen because it just happens to be the one element in which all orbitals having the same value of n also have the same energy. Bohr's theory failed for atoms other than hydrogen, however, because orbitals with the same value of n can have different energies when the atom has more than one electron.

Electron waves are described by the term *orbital* to differentiate them from the notion of *orbits*, which was part of the Bohr model of the atom.

"Most stable" almost always means "lowest energy."

Bohr was lucky — he chose hydrogen to construct his theory.

The Secondary Quantum Number, ℓ

The secondary quantum number, ℓ, divides the shells into smaller groups of orbitals called subshells. The value of n determines the possible values of ℓ. For a given n, ℓ may range from $\ell = 0$ to $\ell = n - 1$. Thus when $n = 1$, ℓ can have only one value, 0. This means that when $n = 1$, there is only one subshell (the shell and subshell are really identical). When $n = 2$, ℓ can have values of 0 or 1. (The maximum value of $\ell = n - 1 = 2 - 1 = 1$.) This means that when $n = 2$, there are two subshells. One has $n = 2$ and $\ell = 0$, and the other has $n = 2$ and $\ell = 1$. Table 6.2 summarizes the relationship between n and the possible values of ℓ.

Subshells could be identified by their value of ℓ. However, to avoid confusing the numerical values of n with those of ℓ, a letter code is normally used to specify the value of ℓ.

value of ℓ	0	1	2	3	4	5	. . .
letter designation	s	p	d	f	g	h	. . .

To designate a particular subshell, we write the value of its principal qua... number followed by the letter code for the subshell. For example, the subshell with $n = 2$ and $\ell = 1$ is the 2p subshell; the subshell with $n = 4$ and $\ell = 0$ is the 4s subshell. Notice that because of the relationship between n and ℓ, every shell has an s subshell (1s, 2s, 3s, etc.); all the shells except the first have a p subshell (2p, 3p, 4p, etc.); and all but the first and second shells have a d subshell (3d, 4d, etc.); and so forth.

ℓ is also called the azimuthal quantum number.

The number of subshells in a given shell is equal to n for that shell.

TABLE 6.2 Relationship between n and ℓ

Value of n	Values of ℓ
1	0
2	0, 1
3	0, 1, 2
4	0, 1, 2, 3
5	0, 1, 2, 3, 4
. . .	. . .

PRACTICE EXERCISE 3   What subshells would be found in the shells with $n = 3$ and $n = 4$?

Whereas the principal quantum number primarily describes the energy and size of an orbital, the secondary quantum number determines the shape of the orbital, which we will examine more closely later. Except for hydrogen, the subshells within a given shell differ slightly in energy, with the energy of the subshell increasing with increasing ℓ. This means that within a given shell, the s subshell is lowest in energy, p is the next lowest, followed by d, then f, and so on. For example,

$$4s < 4p < 4d < 4f$$
increasing energy →

The Magnetic Quantum Number, m_ℓ

The third quantum number, m_ℓ, is known as the magnetic quantum number. It splits the subshells into individual orbitals. This quantum number describes how an orbital is oriented in space relative to other orbitals. As with ℓ, there are restrictions as to the possible values of m_ℓ; they can range from $-\ell$ to $+\ell$. When $\ell = 0$, m_ℓ can only have the value 0 because $+0$ and -0 are the same. An s subshell, then, has but a single orbital. When $\ell = 1$, the possible values of m_ℓ are -1, 0, and $+1$. A p subshell therefore has three orbitals — one with $\ell = 1$ and $m_\ell = 1$, another with $\ell = 1$ and $m_\ell = 0$, and a third with $\ell = 1$ and $m_\ell = -1$. Following similar reasoning, we find that a d subshell has five orbitals and an f subshell has seven orbitals. The numbers of orbitals in the subshells are easy to remember because they follow a simple arithmetic progression.

s	p	d	f
1	3	5	7

m_ℓ is used to explain additional lines that appear in the spectra of atoms when they emit light while in a magnetic field. That's how it got its name.

PRACTICE EXERCISE 4   How many orbitals are there in a g subshell?

TABLE 6.3 Summary of Relationships Among n, ℓ, and m_ℓ

Value of n	Value of ℓ	Values of m_ℓ	Subshell	Number of Orbitals
1	0	0	1s	1
2	0	0	2s	1
	1	−1, 0, +1	2p	3
3	0	0	3s	1
	1	−1, 0, +1	3p	3
	2	−2, −1, 0, +1, +2	3d	5
4	0	0	4s	1
	1	−1, 0, +1	4p	3
	2	−2, −1, 0, +1, +2	4d	5
	3	−3, −2, −1, 0, +1, +2, +3	4f	7

Summary

We are finally ready, now, to look at the whole picture. The relationships among all three quantum numbers are summarized in Table 6.3. In addition, the relative energies of the subshells in an atom containing two or more electrons are depicted in Figure 6.19. Several important features should be noticed. First, note that each orbital on this energy diagram is indicated by a separate circle — one for an s subshell, three for a p subshell, and so forth. Second, notice that all the orbitals of a given subshell have the *same* energy. Third, note that, in going upward on the energy scale, the spacing between successive shells decreases as the number of subshells increases. This leads to the overlapping of shells having different values of n. For instance, the 4s subshell is lower in energy than the 3d subshell, 5s is lower than 4d, and 6s is lower than 5d. In addition, the 4f subshell is below the 5d subshell and 5f is below 6d.

We will see shortly that Figure 6.19 is very useful for predicting the electronic structures of atoms. Before discussing this, however, we must examine another very important property of the electron, called spin.

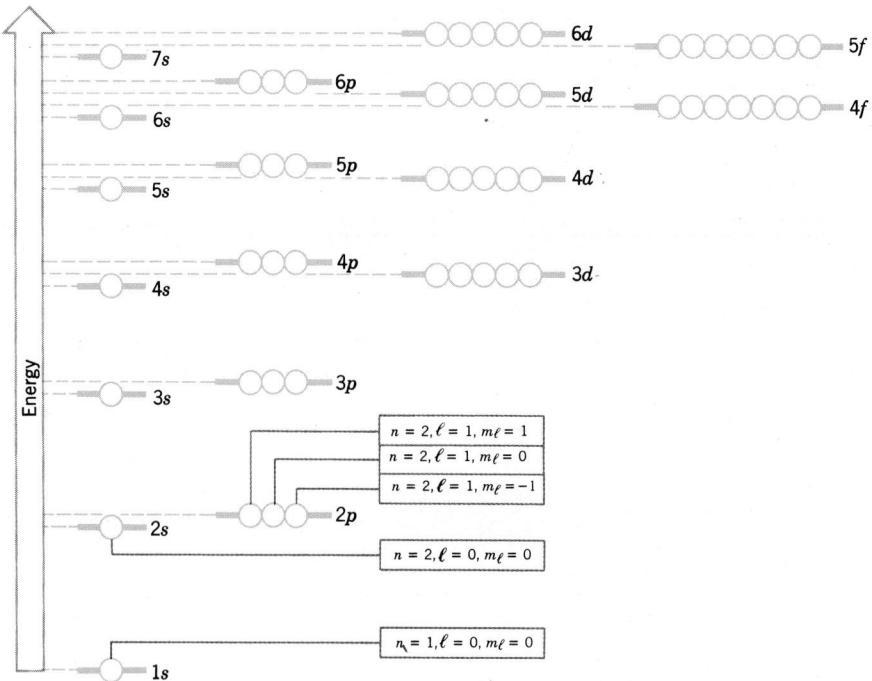

FIGURE 6.19 Approximate energy level diagram for atoms with two or more electrons. The quantum numbers associated with the orbitals in the first two shells are also shown.

6.6 ELECTRON SPIN

The magnetic properties of the electron, which can be explained by the rotation of electrical charge, limit the number of electrons per orbital to two.

Earlier it was stated that an atom is in its most stable state (its ground state) when its electrons have wave patterns with the lowest possible energies. This occurs when the electrons "occupy" the lowest energy orbitals that are available. But what determines how the electrons "fill" these orbitals? Fortunately, there are some simple rules that can help. These govern both the maximum number of electrons that can be in a particular orbital and how orbitals having the same energy can be filled. One important factor that influences the distribution of electrons is the phenomenon known as electron spin.

The concept of electron spin is based on the fact that electrons behave as tiny magnets. This can be explained by imagining that an electron spins about its axis, much like a toy top. The revolving electrical charge of the electron creates its own magnetic field. The same effect is used to make electric motors work. The passage of electrical charge through the curved windings of an electric motor sets up magnetic forces that push and pull, thereby causing the rotor within the device to turn.

Electron spin gives us a fourth quantum number for the electron, called the spin quantum number, m_s. Again like the toy top, the electron can spin in either of two directions. Therefore, the spin quantum number can take on either of two possible values; they are $m_s = +\frac{1}{2}$ or $m_s = -\frac{1}{2}$. The actual values of m_s and the reason that they are not integers aren't very important, but the fact that there are *only* two values is very significant.

In 1925 an Austrian physicist, Wolfgang Pauli (1900–1958), expressed the importance of electron spin in determining electronic structure. The Pauli exclusion principle states that no two electrons in the same atom may have identical values for all four quantum numbers. To understand what this means, suppose that two electrons occupy the 1s orbital of an atom. Each electron would have $n = 1$, $\ell = 0$, and $m_\ell = 0$. Since these three quantum numbers are the same for both electrons, the exclusion principle requires that their fourth quantum numbers (their spin quantum numbers) be different; one electron would have to have $m_s = +\frac{1}{2}$ and the other, $m_s = -\frac{1}{2}$. No more than two electrons can occupy the 1s orbital simultaneously because there are only two possible values of m_s. Thus the Pauli exclusion principle is really telling us that the maximum number of electrons in any orbital is two, and that when two electrons are in the same orbital, they must have opposite spins.

The limit of two electrons per orbital also limits the maximum electron populations of the shells and subshells. For the subshells we have

Electrons moving in curved paths through wires in the large electro-magnet suspended from this crane produce a magnetic field that is strong enough to pick up sizable portions of scrap steel.

Pauli received the 1945 Nobel Prize in physics for his discovery of the exclusion principle.

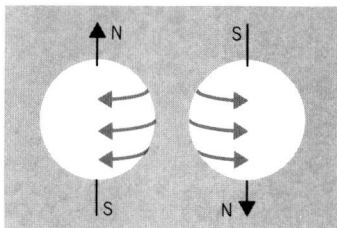

The electron can spin in either of two directions.

Subshell	Number of Orbitals	Maximum Number of Electrons
s	1	2
p	3	6
d	5	10
f	7	14

The maximum electron population per shell is shown below.

Shell	Subshells	Maximum Shell Population[a]
1	1s	2
2	2s 2p	8 (2 + 6)
3	3s 3p 3d	18 (2 + 6 + 10)
4	4s 4p 4d 4f	32 (2 + 6 + 10 + 14)

[a] In general, the maximum electron population of a shell is $2n^2$.

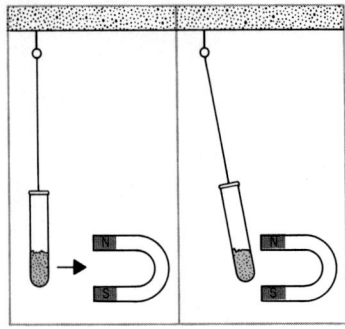

A paramagnetic substance is attracted to a magnetic field.

Diamagnetic substances are actually weakly repelled by a magnetic field.

We have seen that two electrons occupying the same orbital must have different values of m_s. When this occurs, we say that the spins of the electrons are *paired,* or simply that the electrons are *paired.* Such pairing leads to the cancellation of the electrons' magnetic effects because the north pole of one electron magnet is opposite the south pole of the other. Atoms having more electrons that spin in one direction than in the other are said to contain *unpaired* electrons. For these atoms, the magnetic effects do not cancel and the atoms themselves become tiny magnets which can be attracted to an external magnetic field. This weak attraction to a magnet of a substance containing unpaired electrons is called *paramagnetism.* Substances in which all the electrons are paired are not attracted to a magnet and are said to be *diamagnetic.* The measurement of paramagnetism or diamagnetism, then, provides experimental verification of the presence or absence of unpaired electrons. In addition, the quantitative measurement of the strength of the attraction of a paramagnetic substance toward a magnetic field permits the calculation of the number of unpaired electrons in its atoms, molecules, or ions.

6.7 ELECTRON CONFIGURATIONS

The electron configuration of an element is obtained by placing electrons into the lowest available orbitals, while heeding the Pauli exclusion principle and spreading the electrons as much as possible among orbitals of identical energy.

The Pauli exclusion principle and the energy level diagram in Figure 6.19 allow us to predict which orbitals in a particular atom will be populated by electrons and the number of electrons that will be found in each. This arrangement of electrons is called the atom's electronic structure or electron configuration. Knowing how to predict electron configurations is important because the arrangement of electrons controls an atom's chemical properties. Let's look at specific examples to see how all this works.

We will begin with the hydrogen atom, whose atomic number, Z, is 1. A neutral hydrogen atom has one electron. In its ground state this electron will occupy the lowest energy orbital that's available, which is the 1s orbital. To indicate the population of a subshell, we use a superscript with the subshell designation. Thus the electron configuration of hydrogen is written as

$$H \quad 1s^1$$

Another way of expressing electron configurations that we will sometimes find useful is the orbital diagram. In it, each orbital is represented by a circle and arrows are used to indicate the individual electrons, heads up for spin in one direction and heads down for spin in the other. The orbital diagram for hydrogen is simply

$$H \quad \textcircled{\uparrow}$$
$$1s$$

Next, let's look at helium, for which Z = 2. This atom has two electrons, both of which are permitted to occupy the 1s orbital. The electron configuration of helium can therefore be written as

$$He \quad 1s^2 \quad \text{or} \quad He \quad \textcircled{\uparrow\downarrow}$$
$$1s$$

Notice that the orbital diagram clearly indicates that the electrons in the 1s orbital are paired.

We can proceed in the same fashion to predict successfully the electron configurations of most of the elements in the periodic table. For example, the next elements in the table are lithium, Li (Z = 3), and beryllium, Be (Z = 4), which have three and four electrons, respectively. After the $1s$ subshell is filled with two electrons, the next lowest energy orbital is the $2s$. Therefore, we can represent the electronic structures of lithium and beryllium as

Li $1s^22s^1$ or   Li   (↿⇂)   (↑)

Be $1s^22s^2$ or   Be   (↿⇂)   (↿⇂)

                                        $1s$ $2s$

Following beryllium, we have boron, B ($Z = 5$). Referring to Figure 6.19, we see that the first four electrons of this atom complete the $1s$ and $2s$ subshells, so the fifth electron must be placed into the $2p$ subshell.

B $1s^22s^22p^1$

In the orbital diagram for boron, the fifth electron can be put into any one of the $2p$ orbitals—which one doesn't matter because they are all of equal energy.

B   (↿⇂)   (↿⇂)   (↑)(◯)(◯)

      $1s$ $2s$ $2p$

Notice, however, that when we give this orbital diagram we show *all* of the orbitals of the $2p$ subshell even though two of them are empty.

Next we come to carbon, which has six electrons. As before, the first four electrons complete the $1s$ and $2s$ orbitals. The remaining two electrons go in the $2p$ subshell to give

C $1s^22s^22p^2$

Now, however, to give the orbital diagram, we have to make a decision as to where to put the two p electrons. (At this point you may have an unprintable suggestion! But try to bear up. It's really not all that bad.) To make this decision, we use Hund's rule. This rule states that *when electrons are placed in a set of orbitals of equal energy, they are spread out as much as possible to give as few paired electrons as possible*. Both theory and experiment have shown that following this rule leads to the electron arrangement with the lowest energy. For carbon, it means that the two p electrons are in separate orbitals and their spins are in the same direction.

C   (↿⇂)   (↿⇂)   (↑)(↑)(◯)

      $1s$ $2s$ $2p$

Using the Pauli exclusion principle and Hund's rule, we can now complete the orbital diagrams for the rest of the elements of the second period.

                            $1s$ $2s$ $2p$

N $1s^22s^22p^3$ (↿⇂)   (↿⇂)   (↑)(↑)(↑)

O $1s^22s^22p^4$ (↿⇂)   (↿⇂)   (↿⇂)(↑)(↑)

F $1s^22s^22p^5$ (↿⇂)   (↿⇂)   (↿⇂)(↿⇂)(↑)

Ne $1s^22s^22p^6$ (↿⇂)   (↿⇂)   (↿⇂)(↿⇂)(↿⇂)

We could, of course, continue to predict electron configurations using Figure 6.19 as a guide and following our filling rules. But how could *you* remember the sequence of energy levels if Figure 6.19 were not available for use? One device that can help is shown in Figure 6.20. The procedure is to follow the diagonal arrows beginning at the bottom. This tells us that the subshells should be filled in the following order: $1s$, $2s$, $2p$, $3s$, $3p$, $4s$, $3d$, $4p$, $5s$, $4d$, $5p$, $6s$, $4f$, $5d$, $6p$, $7s$, $5f$, $6d$, $7p$, etc.

It doesn't matter which two orbitals are shown as occupied. Any of these are okay for carbon.

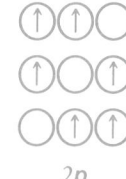

$2p$

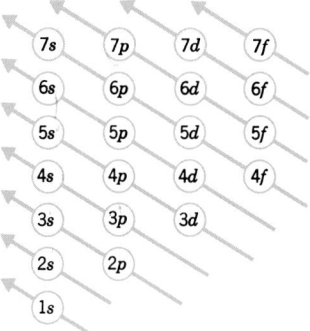

FIGURE 6.20 A way to remember the filling order of subshells. Write the subshell designations as shown and follow the diagonal arrows, starting at the bottom.

EXAMPLE 6.3   WRITING ELECTRON CONFIGURATIONS

Problem: Write the electron configuration for manganese, Mn ($Z = 25$).

Solution: First, we determine how many electrons are in a manganese atom. Since Mn has $Z = 25$, there are 25 electrons. Following Figure 6.20, we begin with the $1s$ subshell and fill the rest of the subshells until we run out of electrons. Remember the maximum subshell populations: $s = 2$, $p = 6$, $d = 10$, and $f = 14$. This gives

$$\text{Mn} \quad 1s^2 2s^2 2p^6 3s^2 3p^6 4s^2 3d^5$$

Some people prefer to write all subshells of a given shell together.

$$\text{Mn} \quad 1s^2 2s^2 2p^6 3s^2 3p^6 3d^5 4s^2$$

The orbital diagram for manganese would be

Notice that the manganese atom has five unpaired electrons.

EXAMPLE 6.4   WRITING ELECTRON CONFIGURATIONS

Problem: Predict the electron configuration of bismuth, Bi ($Z = 83$).

Solution: Once again we must follow the filling sequence of Figure 6.20. We get

$$\text{Bi} \quad 1s^2 2s^2 2p^6 3s^2 3p^6 4s^2 3d^{10} 4p^6 5s^2 4d^{10} 5p^6 6s^2 4f^{14} 5d^{10} 6p^3$$

If we group subshells of the same shell together,

$$\text{Bi} \quad 1s^2 2s^2 2p^6 3s^2 3p^6 3d^{10} 4s^2 4p^6 4d^{10} 4f^{14} 5s^2 5p^6 5d^{10} 6s^2 6p^3$$

PRACTICE EXERCISE 5   Predict the electron configurations for (a) Mg, (b) Ge, (c) Cd, and (d) Gd. Group subshells of the same shell together.

PRACTICE EXERCISE 6   Write orbital diagrams for (a) Na, (b) S, and (c) V.

Abbreviated Electron Configurations

When considering the chemical properties of elements, we are rarely interested in electrons that are buried deep within the atom. Our attention is usually focused only on the electron configuration of the electrons in the outer shells of the atom. This is because the outer electrons are the ones that are involved in chemical reactions. The inner electrons, called the core electrons, of one atom are not exposed to the electrons of other atoms when chemical bonds are formed. To direct attention to these outer electrons, we often write electron configurations in an abbreviated, or shorthand, form. Consider, for instance, the elements in period 3 of the periodic table, taking sodium and magnesium as representative examples. The electron configurations of these elements are

$$\text{Na} \quad 1s^2 2s^2 2p^6 3s^1$$
$$\text{Mg} \quad 1s^2 2s^2 2p^6 3s^2$$

The outer electrons are in the third shell; the core ($1s^2 2s^2 2p^6$) is identical for both. This core configuration is the same as that of the noble gas neon. To write the shorthand configuration for an element we indicate what the core is by placing in brackets the symbol of the noble gas whose electron configuration is the same as the core configuration. This is followed by the configuration of the outer electrons for the particular element. The noble gas used is almost always the one that occurs at the end of the period preceding the period containing the element whose configuration we wish to represent. Thus, for sodium and magnesium we would write

$$\text{Na} \quad [\text{Ne}]3s^1$$
$$\text{Mg} \quad [\text{Ne}]3s^2$$

EXAMPLE 6.5 WRITING SHORTHAND ELECTRON CONFIGURATIONS

Problem: What is the shorthand electron configuration of manganese?

Solution: Manganese is in period 4 and has 25 electrons (because $Z = 25$). The preceding noble gas is argon, Ar, which has an atomic number of 18. This means that an argon atom has 18 electrons. When we write the shorthand configuration, the first 18 electrons are represented by placing the symbol Ar in brackets. From Figure 6.20, this corresponds to completed 1s, 2s, 2p, 3s, and 3p subshells. The remaining seven electrons are distributed as $4s^2 3d^5$. Therefore, the shorthand electron configuration for Mn is

$$\text{Mn} \quad [\text{Ar}]4s^2 3d^5$$

Placing the electrons that are in the highest shell farthest to the right gives

$$\text{Mn} \quad [\text{Ar}]3d^5 4s^2$$

PRACTICE EXERCISE 7   Placing the electrons that are in the highest shell farthest to the right, write shorthand configurations for (a) P and (b) Sn.

6.8 SOME UNEXPECTED ELECTRON CONFIGURATIONS

Filled and half-filled subshells have extra stability that sometimes affects electron configurations.

The rules that you've learned for predicting electron configurations work most of the time—but not always. Appendix B gives the electron configurations of all of the elements as determined experimentally. Close examination reveals that there are quite a few exceptions to the rules. Fortunately, most of these exceptions are of little consequence to us because the elements involved are relatively rare and their chemistry will not be important to us in this course. Some of the exceptions are important, though, because they occur with common elements, notably, chromium and copper.

Following the rules, we would expect the configurations to be

$$\text{Cr} \quad [\text{Ar}]3d^4 4s^2$$
$$\text{Cu} \quad [\text{Ar}]3d^9 4s^2$$

However, the actual electron configurations, determined experimentally, are

$$\text{Cr} \quad [\text{Ar}]3d^5 4s^1$$
$$\text{Cu} \quad [\text{Ar}]3d^{10} 4s^1$$

The corresponding orbital diagrams are

Cr [Ar] ↑ ↑ ↑ ↑ ↑   ↑

Cu [Ar] ↑↓ ↑↓ ↑↓ ↑↓ ↑↓   ↑

          3d 4s

Notice that for chromium, an electron is "borrowed" from the 4s subshell to give a 3d subshell that is exactly half-filled. For copper the 4s electron is borrowed to give a completely filled 3d subshell. A similar thing happens with silver and gold, which have filled 4d and 5d subshells, respectively.

$$\text{Ag} \quad [\text{Kr}]4d^{10}5s^1$$
$$\text{Au} \quad [\text{Xe}]4f^{14}5d^{10}6s^1$$

Apparently, half-filled and filled subshells (particularly the latter) have some special stability that makes such borrowing energetically favorable. This subtle, but nevertheless important, phenomenon affects not only the ground state configurations of atoms but also the relative stabilities of some of the ions formed by the transition elements.

6.9 ELECTRON CONFIGURATIONS AND THE PERIODIC TABLE

The similarities among elements within groups and the structure of the periodic table can be explained by electron configurations.

In Chapter 2 we saw that when the periodic table was constructed, atoms with similar chemical properties were arranged in vertical columns called groups. The basic structure and shape of the periodic table that result form one of the strongest empirical supports for the quantum theory that we have been using to predict electron configurations.

Consider, for example, the way the table is laid out (Figure 6.21). On the left there are *two* columns of elements, on the right there is a block of *six* columns, in the center there is a block of *ten* columns of elements, and below the table there are two rows consisting of *fourteen* elements each. These numbers—2, 6, 10, and 14—are *precisely* the numbers of electrons that the quantum theory tells us can occupy the s, p, d, and f subshells, respectively!

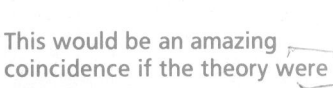
This would be an amazing coincidence if the theory were wrong.

FIGURE 6.21 Overall structure of the periodic table.

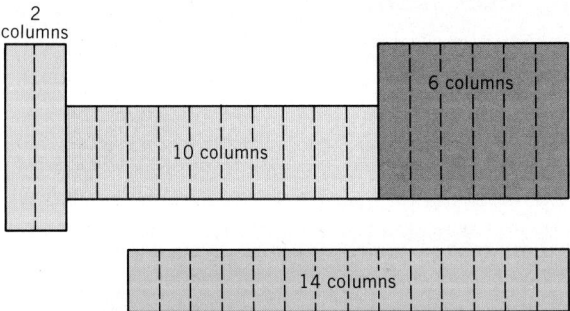

The similarities in group properties can also be explained. Because only the outer parts of atoms touch during chemical reactions, we might expect to find similarities in the outer-shell electron configurations of elements that have similar chemical properties —and we do. Let's look at the alkali metals, Group IA. Going by our rules, we get the following configurations.

Li	$1s^2 2s^1$
Na	$1s^2 2s^2 2p^6 3s^1$
K	$1s^2 2s^2 2p^6 3s^2 3p^6 4s^1$
Rb	$1s^2 2s^2 2p^6 3s^2 3p^6 3d^{10} 4s^2 4p^6 5s^1$
Cs	$1s^2 2s^2 2p^6 3s^2 3p^6 3d^{10} 4s^2 4p^6 4d^{10} 5s^2 5p^6 6s^1$

Each of these elements has one outer electron which is in an s subshell. We know that when they react, the alkali metals lose one electron to form ions with a charge of 1+. For each, then, the electron that is lost is this outer s electron, and the electron configuration of the ion that is formed is the same as that of the preceding noble gas.

Li^+	$1s^2$	He	$1s^2$
Na^+	$1s^2 2s^2 2p^6$	Ne	$1s^2 2s^2 2p^6$
K^+	$1s^2 2s^2 2p^6 3s^2 3p^6$	Ar	$1s^2 2s^2 2p^6 3s^2 3p^6$

etc.

Valence Shell Electron Configurations

For the representative elements (those in the longer columns), the only electrons that are normally important in controlling chemical properties are the ones in the outermost shell — that is, the occupied shell having the highest value of n. This outer shell is known as the valence shell, and the electrons in it are called valence electrons. (The term *valence* comes from the study of chemical bonding and relates to the combining capacity of an element, but that's not important here.) Even with the transition elements, we need only concern ourselves with the outermost shell and the d subshell just below. For example, iron has the configuration

$$Fe \quad 1s^2 2s^2 2p^6 3s^2 3p^6 3d^6 4s^2$$

Only the $4s$ and $3d$ electrons play a role in the chemistry of iron. In general, the electrons below the outer s and d subshells of a transition element — the *core* electrons — are relatively unimportant. In every case these core electrons have the electron configuration of a noble gas.

For the representative elements there is a very simple way to determine the electron configuration of the valence shell using the periodic table. Consider the following two examples.

$$C \quad 1s^2 2s^2 2p^2$$
$$S \quad 1s^2 2s^2 2p^6 3s^2 3p^4$$

The valence shell is shown in color. Now, notice that carbon is in period 2 of the table. This number, 2, is the same as the value of n for carbon's valence shell. For sulfur, in period 3, the valence shell is the third shell, for which $n = 3$.

Next, notice that to get to carbon by moving from left to right in period 2, we first have to pass through the group of two columns. This corresponds to filling the $2s$ subshell. Then we have to go two spaces into the group of six columns to finally get to carbon. These two spaces are the number of electrons that go into the $2p$ subshell. For sulfur, in period 3, we pass through the group of two columns (hence $3s^2$) and then go four spaces into the group of six columns (hence $3p^4$).

We see that the periodic table serves as a useful guide in writing electron configurations. The value of n for the outer shell of elements in any given period is equal to the period number. Then, we simply move from left to right across the period, placing electrons into an s subshell as we cross through the block of two columns, and into a p subshell as we cross the block of six columns. This is summarized in Figure 6.22.

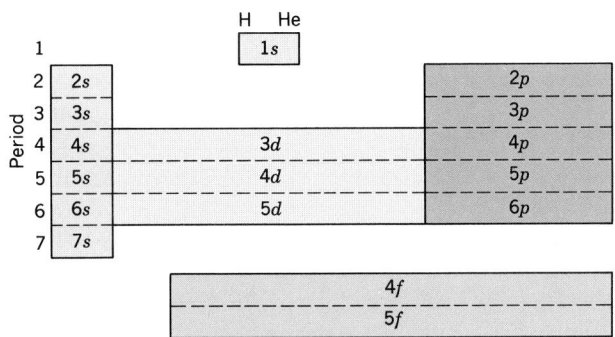

FIGURE 6.22 Subshells that become filled as we cross periods.

PRACTICE EXERCISE 8   Give the electron configurations of the valence shells of (a) N, (b) Si, and (c) Sr.

Earlier in this section we saw that the electrons that determine the chemical properties of iron, a transition element, are in the $4s$ and $3d$ subshells. To obtain the subshell population of a transition element we can also use its location in the table as a guide. To get to iron in period 4, we first pass through the set of two columns ($4s^2$) and then enter the block of 10 columns. Here we fill a d subshell whose value of n is equal to the period number minus one (Figure 6.22). To reach iron, then, we move six spaces into this $(4 - 1)d$ or $3d$ subshell; thus the configuration of this subshell is $3d^6$.

EXAMPLE 6.6   WRITING ELECTRON CONFIGURATIONS FROM THE PERIODIC TABLE

Problem: What is the electron configuration of zirconium, Zr ($Z = 40$)?

Solution: Zirconium is in period 5. The preceding noble gas is Kr. To reach Zr we pass across the first two columns ($5s^2$) and then through two spaces in the center. The last two electrons enter the $(5 - 1)d$ or $4d$ subshell. Thus, for Zr we get

$$\text{Zr} \quad [\text{Kr}]4d^25s^2$$

PRACTICE EXERCISE 9   Write the shorthand electron configurations for (a) Ni and (b) Ru.

EXAMPLE 6.7   WRITING VALENCE SHELL ELECTRON CONFIGURATIONS

Problem: Predict the electron configuration of the valence shell of arsenic ($Z = 33$).

Solution: Arsenic is a period 4 element. The preceding noble gas is Ar. To reach As, we cross the s block to get $4s^2$, the d block to get $(4 - 1)d^{10}$ or $3d^{10}$, and then move three spaces into the p block to get $4p^3$. This gives

$$\text{As} \quad [\text{Ar}]3d^{10}4s^24p^3$$

Among the representative elements, completed subshells below the outer shell are unimportant in determining chemical properties. For elements such as arsenic we are only interested in the valence shell. The valence shell is the one with highest n, in this case, 4, so the configuration of the valence shell is $4s^24p^3$.

PRACTICE EXERCISE 10   What are the electron configurations for the valence shells of (a) Se, (b) Sn, and (c) I?

6.10 WHERE THE ELECTRON SPENDS ITS TIME

The s, p, d, and f orbitals have shapes and directional orientations that describe where and how electrons are most probably distributed.

The same theory that tells us of the energies of atomic orbitals also describes the shapes of electron waves. How do we know these shapes are correct? We don't for sure, but many of the predictions that have been made using the theory seem to be borne out by experiments. This gives the theoretical explanations strength and support. For example, the fact that the results of wave mechanics account for the shape of the periodic table so very well gives the theory a good deal of credibility.

The difficulty in describing where electrons are in an atom stems from the basic problem that we face when we attempt to picture a particle as a wave. There is nothing in our worldly experience that is comparable. The way we get around this perplexing

conceptual problem, so that we can still think of the electron as a particle in the usual sense, is to speak in terms of the statistical probability of the electron being found at a particular place.

Describing the electron's position in terms of statistical probability is based on more than simple convenience. The German physicist Werner Heisenberg showed mathematically that there are limits to our ability to measure both a particle's velocity and its position at the same instant. This was Heisenberg's famous uncertainty principle. The theoretical limitations on measuring speed and position are not significant for large objects. However, for small particles such as the electron, these limitations prevent us from ever knowing or predicting exactly where an electron will be at a particular instant. So we speak of probabilities instead.

Wave mechanics views the probability of finding an electron at a given point in space as equal to the square of the amplitude of the electron wave at that point. It seems quite reasonable to relate probability to amplitude, or intensity, because where a wave is intense its presence is strongly felt. The amplitude is squared because, mathematically, the amplitude can be either positive or negative, but probability only makes sense if it is positive. Squaring the amplitude assures us that the probabilities will be positive. We need not be very concerned about this point, however.

The concept of electron probability leads to two very important and frequently used ideas. One is that an electron behaves as if it were spread out around the nucleus in a sort of electron cloud. Figure 6.23 illustrates the way the probability of finding the electron varies for a 1s orbital. In those places where there are large numbers of dots, the amplitude of the wave is large and the probability of finding the electron is large.

The other important idea is electron density, which relates to how much of the electron's charge is packed into a given volume. Because of its wave nature, the electron (and its charge) is spread out around the nucleus. In regions of high probability there is a high concentration of electrical charge and the electron density is large; in regions of low probability, the electron density is small. In looking at the way the electron density distributes itself in atomic orbitals, we are interested in three things — the *shape* of the orbital, its *size,* and its *orientation* in space relative to other orbitals.

Electron density doesn't end abruptly at some particular distance from the nucleus. It gradually fades away. To define the size and shape of an orbital, it is useful to picture some imaginary surface enclosing, say, 90% of the electron density of the orbital, and on which the probability of finding the electron is everywhere the same. For the 1s orbital in Figure 6.23, we would find that if we go out a given distance from the nucleus in *any* direction, the probability of observing the electron would be the same. This means that all the points of equal probability would lie on the surface of a sphere, so we can say that the shape of the orbital is spherical. In fact, all *s* orbitals are spherical. As suggested earlier, their size increases with increasing *n*. This is illustrated in Figure 6.24. Notice that

Heisenberg won the 1932 Nobel Prize in physics for his work.

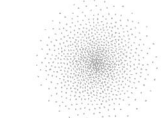

FIGURE 6.23 Electron probability distribution for a 1s electron.

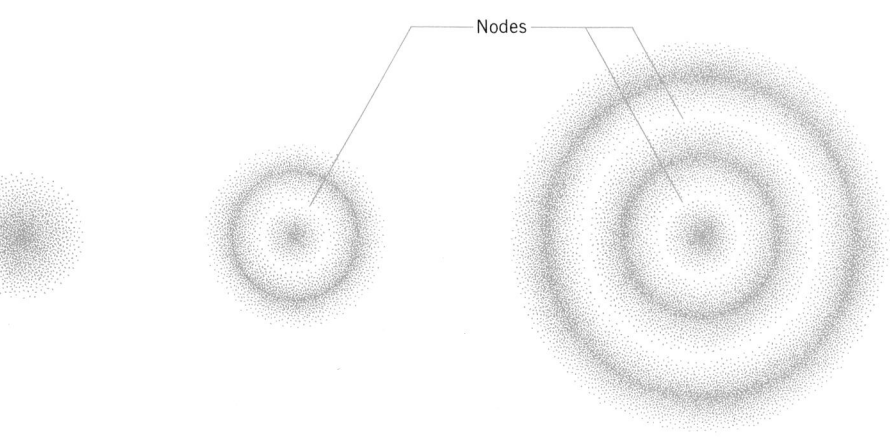

Nodes

1s   2s   3s

FIGURE 6.24 Size variation among *s* orbitals. Orbitals become larger with increasing *n*. The diagrams represent cross sections of the spherical electron density patterns.

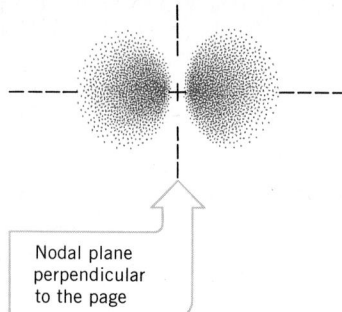

Nodal plane
perpendicular
to the page

FIGURE 6.25 The probability
distribution in a *p* orbital.

The *f* orbitals are even more
complex than the *d* orbitals, but
we will have no need to discuss
their shapes.

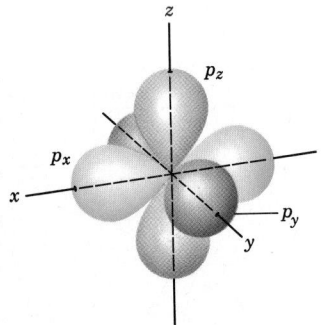

FIGURE 6.26 The orientations of
the three *p* orbitals in a *p* subshell.

FIGURE 6.27 The shapes and
directional properties of the five *d*
orbitals of a *d* subshell.

beginning with the 2*s* orbital, there are certain places where the electron density drops to zero. These are the nodes of the electron wave. It is interesting that electron waves have nodes just like the waves on a guitar string. For electron waves, however, the nodes consist of imaginary *surfaces* on which the electron density is zero.

The *p* orbitals are quite different from *s* orbitals, as shown in Figure 6.25. Notice that the electron density is equally distributed in two regions on opposite sides of the nucleus. Figure 6.25 shows the *two* lobes of *one* 2*p* orbital. The electron density of the lobes is concentrated about an imaginary line that passes through the nucleus. Between the two lobes, there is a nodal plane — an imaginary flat surface on which every point has an electron density of zero.

A *p* subshell consists of three orbitals whose directions lie at 90° to each other along the axes of an imaginary *xyz* coordinate system (Figure 6.26). For convenience in referring to them, the orbitals are often labeled according to the axis along which they lie. The *p* orbital concentrated along the *x* axis is labeled p_x, and so forth. As with *s* orbitals, the sizes of the *p* orbitals increase with increasing *n*; as *n* increases they extend farther from the nucleus.

The shapes of the *d* orbitals, illustrated in Figure 6.27, are a bit more complex than are those of the *p* orbitals. Because of this, and because there are five orbitals in a *d* subshell, we haven't attempted to draw all of them at the same time on the same set of coordinate axes. Notice that four of the five *d* orbitals have the same shape and consist of four lobes of electron density. These orbitals differ only in their orientations around the nucleus (their labels come from the mathematics of wave mechanics). The fifth *d* orbital, labeled d_{z^2}, has two lobes that point in opposite directions along the *z* axis plus a doughnut-shaped ring of electron density around the center that lies in the *x*–*y* plane. We will see that the *d* orbitals are important in the formation of chemical bonds in certain molecules, and that their shapes and orientations are important in understanding the properties of the transition metals, which will be discussed in Chapter 22.

6.11 SIZES OF ATOMS AND IONS

Atomic and ionic size is determined by the balance between the attractions the electrons feel for the nucleus and the repulsions they feel for each other.

In this and the following two sections, we will take a brief look at some of the properties of atoms that are related to their electron configurations and that are responsible for many of the chemical characteristics of the various elements. The first of these properties is size, both of atoms and of ions that are formed from them.

The very nature of the wave concept of the electronic structure of the atom makes it difficult to define exactly what we mean by the "size" of an atom or ion. As we've seen, the electron cloud doesn't simply stop at some particular distance from the nucleus; instead it gradually fades away. Nevertheless, atoms and ions do behave in many ways as though they have characteristic sizes. For example, in a whole host of hydrocarbons—

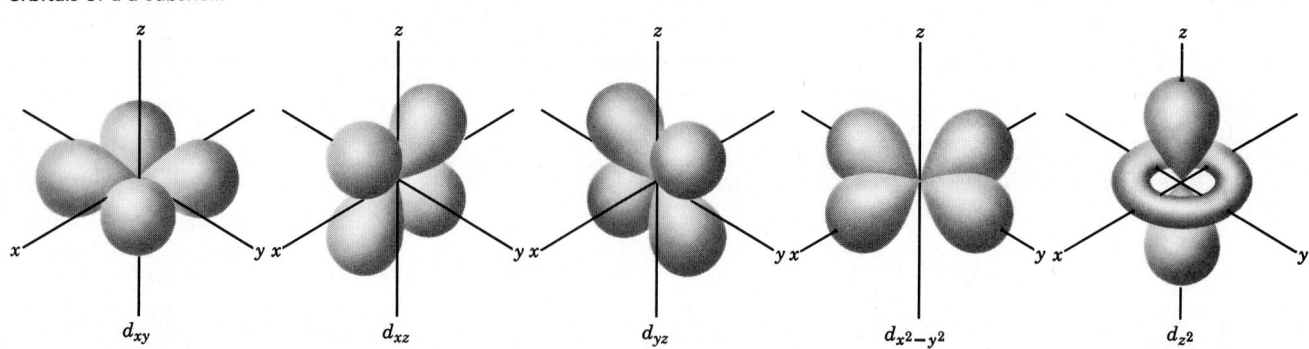

d_{xy} d_{xz} d_{yz} $d_{x^2-y^2}$ d_{z^2}

ranging from methane (CH_4, natural gas) to octane (C_8H_{18}, in gasoline) to many others —the distance between the nuclei of carbon and hydrogen atoms is virtually the same. This would suggest that carbon and hydrogen have the same relative sizes in each of these compounds.

The C—H distance in most hydrocarbons is about 110 pm.

It is difficult to appreciate how small atoms really are. They are incredibly tiny. We've already seen how little they weigh in Chapter 3. In size, atoms range from about 1.4×10^{-10} to 5.7×10^{-10} m in diameter. Their radii, which is the usual way that size is specified, range from about 7.0×10^{-11} to 2.9×10^{-10} m. These are difficult numbers to comprehend. A million carbon atoms placed side by side in a line would extend a little less than 0.2 mm, or about the diameter of the period at the end of this sentence.

The sizes of atoms and ions are rarely expressed in meters because the numbers are so cumbersome. Instead, a unit is chosen that makes the values easier to comprehend. A unit that scientists have traditionally used is called the **angstrom** (symbolized Å), which is defined as

$$1 \text{ Å} = 1 \times 10^{-10} \text{ m}$$

However, the angstrom is not an SI unit, and in many current scientific journals, atomic dimensions are given in picometers, or sometimes in nanometers (1 pm = 10^{-12} m and 1 nm = 10^{-9} m). In this book, we will express atomic dimensions in picometers, but because much of the earlier scientific literature has these quantities in angstroms, you may someday find it useful to remember the conversions:

$$1 \text{ Å} = 100 \text{ pm}$$
$$1 \text{ Å} = 0.1 \text{ nm}$$

The actual determination of atomic size is a difficult and complex task. We will touch on this subject briefly in Chapter 10. For our purposes, though, it isn't really necessary to know how atomic radii are measured or even what the actual radii of atoms and ions are. However, it is useful to know something about the relative sizes of different atoms and how the sizes of ions compare with their neutral atoms.

Atomic size varies in a more or less systematic way in the periodic table, as can be seen in Figure 6.28.

Large atoms are found in the lower left of the periodic table, and small atoms are found in the upper right.

> Atoms become larger from top to bottom in a group; they become smaller from left to right in a period.

It is worthwhile examining the reasons for these variations because the same things that influence atomic size also affect other atomic properties. There are two principal factors to consider. One is the value of the principal quantum number of the valence electrons, and the other is the strengths of the attractions felt by the valence electrons toward the center of the atom.

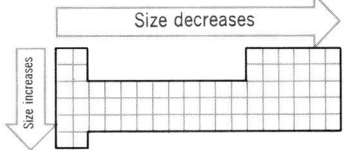

Going from top to bottom within a group in the periodic table, the outer electron configurations of the atoms remain the same except for the increase in the value of n. For instance, the valence shell configuration for lithium is $2s^1$; for sodium, it is $3s^1$; for potassium, it is $4s^1$; and so forth for the rest of the elements in Group IA. Earlier we saw that the size of a given orbital, such as an s orbital, increases as the value of its principal quantum number increases. Therefore, the orbital containing the valence electrons becomes larger as we descend a group, and the atoms grow in size. The same argument applies whether the valence shell orbitals are s or p.

Moving from left to right across a period, electrons are added to the same shell. The orbitals holding the valence electrons all have the same value of n, so changes in the principal quantum number certainly can't be responsible for the size changes that occur. In this case we have to examine the strengths of the attractions that the valence electrons feel in the direction of the nucleus — the larger these attractions, the more the valence electrons are pulled toward the center of the atom and the smaller the size of the atom.

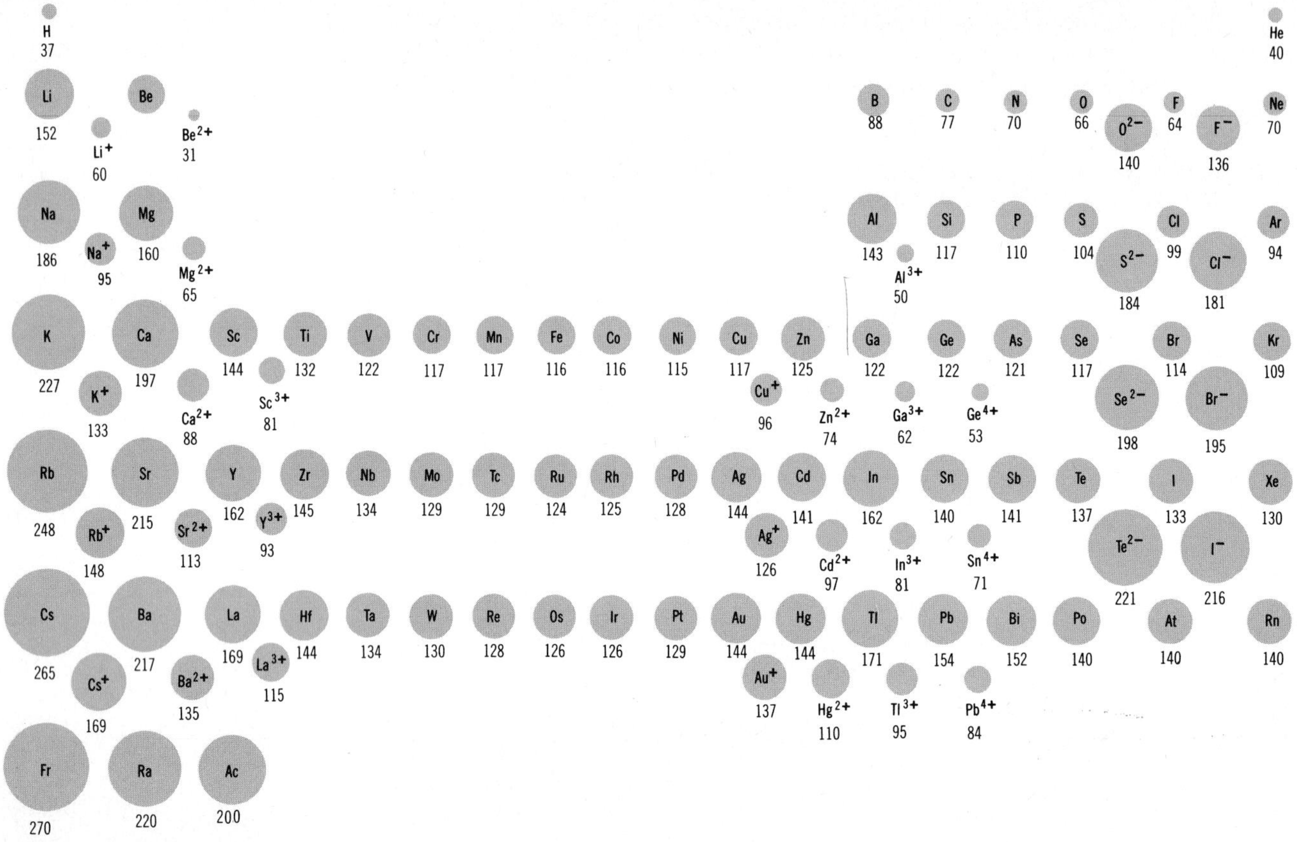

H 37   He 40

Li 152, Li$^+$ 60   Be, Be^{2+} 31   B 88, C 77, N 70, O 66, O^{2-} 140, F 64, F$^-$ 136, Ne 70

Na 186, Na$^+$ 95, Mg 160, Mg^{2+} 65   Al 143, Al^{3+} 50, Si 117, P 110, S 104, S^{2-} 184, Cl 99, Cl$^-$ 181, Ar 94

K 227, K$^+$ 133, Ca 197, Ca^{2+} 88, Sc 144, Sc^{3+} 81, Ti 132, V 122, Cr 117, Mn 117, Fe 116, Co 116, Ni 115, Cu 117, Cu$^+$ 96, Zn 125, Zn^{2+} 74, Ga 122, Ga^{3+} 62, Ge 122, Ge^{4+} 53, As 121, Se 117, Se^{2-} 198, Br 114, Br$^-$ 195, Kr 109

Rb 248, Rb$^+$ 148, Sr 215, Sr^{2+} 113, Y 162, Y^{3+} 93, Zr 145, Nb 134, Mo 129, Tc 129, Ru 124, Rh 125, Pd 128, Ag 144, Ag$^+$ 126, Cd 141, Cd^{2+} 97, In 162, In^{3+} 81, Sn 140, Sn^{4+} 71, Sb 141, Te 137, Te^{2-} 221, I 133, I$^-$ 216, Xe 130

Cs 265, Cs$^+$ 169, Ba 217, Ba^{2+} 135, La 169, La^{3+} 115, Hf 144, Ta 134, W 130, Re 128, Os 126, Ir 126, Pt 129, Au 144, Au$^+$ 137, Hg 144, Hg^{2+} 110, Tl 171, Tl^{3+} 95, Pb 154, Pb^{4+} 84, Bi 152, Po 140, At 140, Rn 140

Fr 270, Ra 220, Ac 200

FIGURE 6.28 Variations in atomic and ionic radii in the periodic table. Values are in picometers.

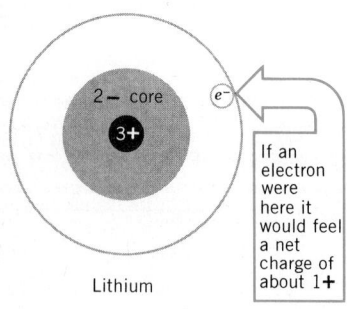

Lithium

2 – core
3+
If an electron were here it would feel a net charge of about 1+

An electron spends very little time between the nucleus and another electron in the same shell, so it shields that other electron poorly.

The principal factor controlling these attractions is the way electrons in inner shells help offset, or partially neutralize, the positive charge of the nucleus, thereby exposing the outer electrons to only a fraction of the full nuclear charge. To understand this, consider the element lithium, which has the electron configuration $1s^2 2s^1$. The electrons beneath the valence shell—those in the $1s^2$ core—are tightly packed around the nucleus and for the most part lie between the nucleus and the electron in the outer shell. This core has a charge of 2– and it surrounds a nucleus that has a charge of 3+. The outer $2s$ electron, when "looking toward" the center of the atom, "sees" the 3+ charge of the nucleus reduced to only about 1+ because of the intervening 2– charge of the core. In other words, the 2– charge of the core effectively neutralizes two of the positive charges of the nucleus, so the net charge that the outer electron feels (we call it the effective nuclear charge) is only about 1+. This is illustrated rather simplistically in the figure in the margin.

Although electrons in inner shells shield the electrons in higher shells quite effectively from the nuclear charge, electrons in the *same* shell are much less effective at shielding each other. For example, in the element beryllium ($1s^2 2s^2$) each of the electrons in the outer $2s$ orbital is shielded quite well from the nuclear charge by the inner $1s^2$ core, but one $2s$ electron doesn't shield the other $2s$ electron very well at all. This is because electrons in the same shell are at about the same average distance from the nucleus, and in attempting to stay away from each other they only spend a very small amount of time one below the other, which is what's needed to provide shielding. Since electrons in the same shell hardly shield each other at all from the nuclear charge, the effective nuclear charge is determined primarily by the difference between the charge on the nucleus and the charge on the core.

With this as background, we are now ready to understand why size decreases from left to right across a period. As we move to the right, it is the outer shell that increases in population; the inner core remains the same. For example, from lithium to fluorine the number of electrons in the second shell increases from 1 to 7, but the core continues to be $1s^2$. However, while the outer shell is becoming more populated, the charge on the nucleus is also increasing (e.g., from 3+ for Li to 9+ for F), and the *difference* between the nuclear charge and the charge on the core also increases. In other words, as we go across the period, the effective nuclear charge felt by the valence electrons increases. This causes these valence electrons to be drawn inward, and thereby causes the sizes of the atoms to decrease.

Across a row of transition elements or inner transition elements, the size variations are less pronounced than they are among the representative elements. This is because the outer shell configuration remains essentially the same while an inner shell is filled. From atomic numbers 21 to 30, for example, the outer electrons occupy the $4s$ subshell while the $3d$ subshell is gradually completed. The amount of shielding provided by the addition of electrons to this inner $3d$ level is greater than the amount of shielding that would occur if the electrons were added to the outer shell, so the effective nuclear charge felt by the outer electrons increases more gradually. As a result, the decrease in size with increasing atomic number is also more gradual.

Sizes of Ions

Figure 6.28 also shows how the ions of many of the elements compare with the neutral atoms in size. As you can see, when atoms gain or lose electrons to form ions, rather significant size changes take place. The reasons for these changes are easy to understand and remember. *Adding electrons creates an ion that is larger than the neutral atom; removing electrons produces an ion that is smaller than the neutral atom.*

When electrons are added, the mutual repulsions between them increase, and this causes them to push apart and occupy a larger volume. Therefore, *negative ions are always larger than the atoms from which they are formed.*

By similar reasoning, we should expect that removing an electron from the valence shell would decrease electron–electron repulsions and thereby allow the remaining electrons to come closer together and be pulled closer to the nucleus. For example, the radius of Fe^{2+} is 76 pm, whereas that of Fe^{3+} is 64 pm. When elements belonging to the representative elements form positive ions, the entire valence shell is generally emptied, as we will see in the next chapter. Removal of this outer shell exposes the inner core of electrons, which naturally has a smaller volume than the neutral atom. Therefore, *positive ions are always smaller than the atoms from which they are formed.*

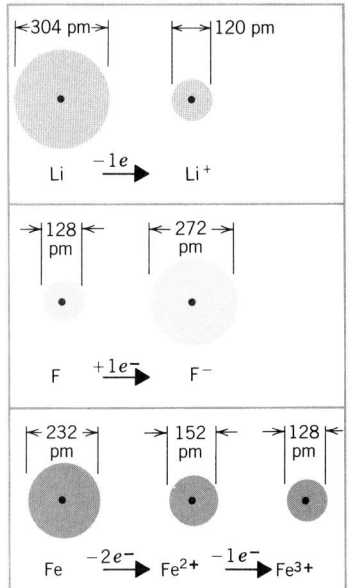

PRACTICE EXERCISE 11   Use the periodic table to choose the largest atom or ion in each set.
(a) Ge, Te, Se, Sn   (c) Fe, Fe^{2+}, Fe^{3+}
(b) C, F, Br, Ga   (d) O, O^{2-}, S, S^{2-}

6.12 IONIZATION ENERGY

The energy required to strip an electron from an isolated atom is called the ionization energy.

The ionization energy (abbreviated IE) is the energy needed to remove an electron from an isolated, gaseous atom. For an element X, it is the energy associated with the change

$$X(g) \longrightarrow X^+(g) + e^-$$

In effect, the ionization energy is a measure of how tightly the electrons are held by the atom. Usually it is expressed in units of kilojoules per mole (kJ/mol), which is really the energy needed to remove one electron from one mole of atoms. This, of course, amounts

Ionization energy is also called ionization potential.

TABLE 6.4 Successive Ionization Energies in kJ/mol for Hydrogen through Magnesium

	1	2	3	4	5	6	7	8
H	1312							
He	2372	5250						
Li	520	7297	11,810					
Be	899	1757	14,845	21,000				
B	800	2426	3659	25,020	32,820			
C	1086	2352	4619	6221	37,820	47,260		
N	1402	2855	4576	7473	9442	53,250	64,340	
O	1314	3388	5296	7467	10,987	13,320	71,320	84,070
F	1680	3375	6045	8408	11,020	15,160	17,860	92,010
Ne	2080	3963	6130	9361	12,180	15,240	—	—
Na	496	4563	6913	9541	13,350	16,600	20,113	25,666
Mg	737	1450	7731	10,545	13,627	17,995	21,700	25,662

to interpreting the coefficients in the equation above on a mole basis, just as you learned to do in the last chapter.

Table 6.4 gives the ionization energies of the first 12 elements. As you can see, atoms with more than one electron have more than one ionization energy. These correspond to the stepwise removal of electrons, one after the other. Lithium, for example, has three ionization energies because it has three electrons. To remove the outer 2s electrons from one mole of isolated lithium atoms to give one mole of gaseous lithium ions, Li^+, requires 520 kJ; so the first IE of lithium is 520 kJ/mol. The second IE of lithium is 7297 kJ/mol, and the third IE is 11,810 kJ/mol. In general, successive ionization energies always increase because each subsequent electron is being pulled away from an increasingly more positive ion.

The magnitudes of the ionizations energies of atoms are really quite enormous. For example, compared with most elements, lithium holds its valence electron rather loosely. Yet, the amount of energy needed to remove just one electron from one mole of lithium atoms (6.9 g Li) could raise the temperature of two liters of water (slightly more than a half-gallon) by 62 °C or 112 °F. Such huge energy requirements indicate how very tightly atoms hold on to their electrons.

Within the periodic table there are trends in the way IE varies that are useful to know and to which we will refer in later discussions. We can see these by examining a graph of first ionization energy versus atomic number, shown in Figure 6.29.

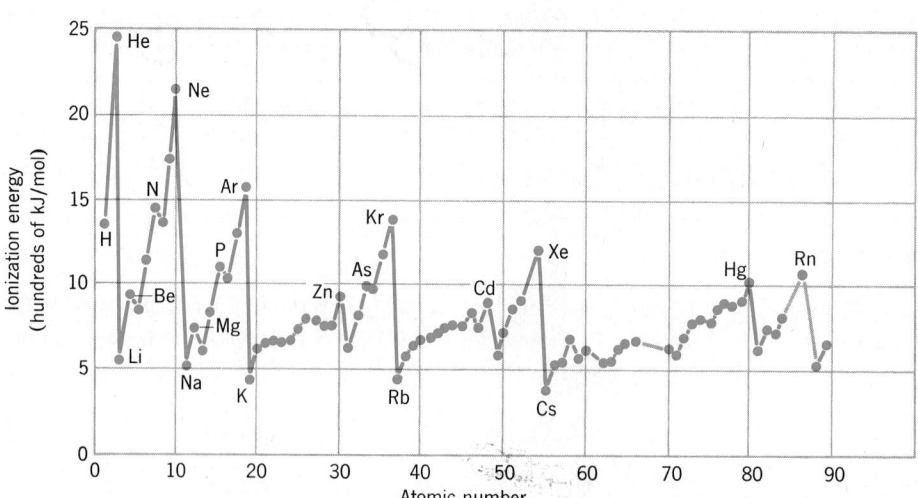

FIGURE 6.29 The variation of first ionization energy with atomic number.

Ionization energy increases from bottom to top within a group, and increases from left to right within a period.

The trend within a group can be seen easily by observing how the ionization energies vary for either the alkali metals (Li through Cs) or the noble gases (He through Rn). The change in IE from Li through Ne or from Na through Ar illustrates the trend across a period. These periodic trends are summarized in Figure 6.30. It is often helpful to remember that the trends in IE are just the opposite of the trends in atomic size within the periodic table.

The same factors that affect atomic size also affect ionization energy. As the value of n increases going down a group, the orbitals become larger and the outer electrons are farther from the nucleus. Electrons farther from the nucleus are bound less tightly, so IE decreases from top to bottom. Of course, this is just the same as saying that it increases from bottom to top.

As you can see, the horizontal variation of IE is somewhat irregular. For example, in period 2 the IE increases from Li to Be, but then decreases from Be to B. Another reversal occurs between nitrogen and oxygen. These irregularities can be explained, but we will not attempt to do so here. All you're expected to know is that there is a gradual overall increase in IE as we move from left to right across a period. The reason for this overall trend is the gradual increase in effective nuclear charge felt by the valence electrons. As we've seen, this draws the valence electrons closer to the nucleus, thereby leading to a decrease in atomic size as we move from left to right. But the increasing effective nuclear charge also causes the valence electrons to be held more tightly, and that makes it more difficult to remove them. As a result, the IE undergoes a gradual increase from left to right.

The results of these trends place those elements with the largest IE in the upper right-hand corner of the periodic table. It is very difficult to cause these atoms to lose electrons. In the lower left-hand corner of the table are the elements that have loosely held valence electrons. These elements form positive ions relatively easily, as we learned in Chapter 2.

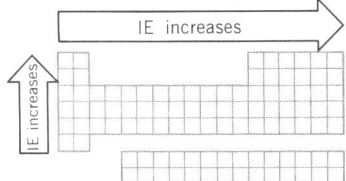

FIGURE 6.30 Variation of ionization energy within the periodic table.

Stability of the Noble Gas Configuration

If we examine Table 6.4, we find that for a given element there is a gradual increase in IE until all of the electrons in its valence shell have been removed. Then there is a very much larger increase in IE as the core is broken into. These huge increases in IE occur for core electron configurations that are the same as the electron configuration of one of the noble gases. For example, for Be the large jump in IE occurs after the pair of 2s valence electrons have been removed. The inner core of a Be atom consists of $1s^2$, which is the same as the electron configuration of helium, and it is from this "noble gas electron configuration" that the third electron of Be is removed.

The data in Table 6.4 suggest that although it may be moderately difficult to empty the valence shell of an atom, it is extremely difficult to break into the noble gas configuration of the core electrons. In effect, we are saying that this noble gas core is very stable, and it is one of the factors that explains the very low degree of reactivity of the noble gas elements themselves. The electron configurations of these elements are so stable that they resist attempts to cause them to change—in other words, these elements resist chemical change and are very unreactive. We will say more about this topic in the next chapter.

PRACTICE EXERCISE 12   Use the periodic table to choose the atom with the largest IE: (a) Na, Sr, Be, Rb; (b) B, Al, C, Si.

6.13 ELECTRON AFFINITY

Energy is usually released when an electron is added to an isolated neutral atom.

In the last section we saw that energy is needed, and therefore work must be done, to remove an electron from an atom. Each electron in an atom is attracted to the nucleus, and this attraction must be overcome if an electron is to be removed. If an extra electron is added to an atom to create a negative ion, this electron will also come under the influence of the nucleus, and work will have to be done to remove it to make the atom neutral once again. The energy needed to remove an electron from a negative ion is equal in magnitude to the energy given off when the electron is added to the atom to create the ion. This energy, which is associated with the change

$$X(g) + e^- \longrightarrow X^-(g)$$

is called the electron affinity (abbreviated EA).

For nearly all the elements, the addition of one electron is exothermic, and therefore the EA is given as a negative value. However, when a second electron must be added, as in the formation of the oxide ion, O^{2-}, work must be done to force the electron into an already negative ion. This change is therefore endothermic and the EA is given as a positive value. Notice that the sign convention used here agrees with the one that we followed in determining the sign of ΔH in Chapter 5. Electron affinities of some common elements are given in Table 6.5.

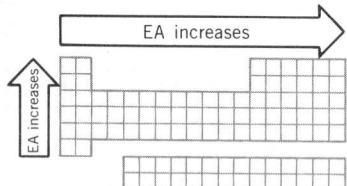

EA increases

EA increases

FIGURE 6.31 Variation of electron affinity within the periodic table.

TABLE 6.5 Electron Affinities for Some Elements

Element	EA (kJ/mol)[a]	Process
Fluorine	−344	$F + e^- \longrightarrow F^-$
Chlorine	−349	$Cl + e^- \longrightarrow Cl^-$
Bromine	−325	$Br + e^- \longrightarrow Br^-$
Oxygen	−142	$O + e^- \longrightarrow O^-$
	+844	$O^- + e^- \longrightarrow O^{2-}$
Hydrogen	−72	$H + e^- \longrightarrow H^-$
Sodium	−50	$Na + e^- \longrightarrow Na^-$

[a] A negative sign means energy is evolved when an electron is added.

Trends in EA parallel the trends in IE.

Trends in EA are very similar to those for IE. EA generally increases from left to right in a period and increases from bottom to top in a group (see Figure 6.31). This shouldn't be surprising, because a valence shell that loses electrons easily will have little attraction for additional electrons. On the other hand, a valence shell that holds its electrons tightly will also tend to hold an additional electron tightly.

SUMMARY

Subatomic Particles.   Experiments with gas discharge tubes led to the discovery of the electron and the proton. Thomson measured the charge-to-mass ratio of cathode ray particles and found them to be fundamental particles present in all matter. Millikan measured the charge on the electron. Rutherford proposed the nuclear model of the atom to explain the scattering of alpha particles by thin metal foils. His measurements led to the search for the electrically neutral nuclear particle called the neutron. Neutrons and protons have masses of about 1 amu, and are about 1800 times heavier than an electron.

Electromagnetic Energy.   Electromagnetic energy, or light energy, travels through space at a constant speed of 3.00×10^8 m/s in the form of waves. The wavelength, λ, and frequency, v, of the wave are related by the equation $\lambda v = c$, where c is the speed of light. The SI unit for frequency is the hertz (Hz). Light also behaves as if it consists of packets of energy called photons. The energy delivered by a photon is proportional to the frequency of the light, and is given by the equation $E = hv$, where h is Planck's constant. White light is composed of many frequencies and can be split into a continuous spectrum. Visible light represents only a small portion of the entire electromagnetic spectrum, which also includes X rays, ultraviolet light, infrared, microwaves, and radio and TV waves.

Atomic Spectra.   The occurrence of line spectra tells us that atoms can emit energy only in discrete amounts and suggests that the energy of the electron is quantized, that is, the electron is restricted to definite energy levels in an atom. Neils Bohr recognized this and, although his theory was later shown to be incorrect, he was the first to introduce the idea of quantum numbers.

Matter Waves.   The wave behavior of electrons and other tiny particles, which can be demonstrated by diffraction experiments, was suggested by de Broglie. Schrödinger applied this to the atom and launched the theory we call wave mechanics or quantum mechanics. This theory tells us that electron waves in atoms are standing waves whose crests and nodes are stationary. Each standing wave, or orbital, is characterized by three quantum numbers, n, ℓ, and m_ℓ. Shells are designated by n, subshells by ℓ, and orbitals within subshells by m_ℓ.

Electron Configurations.   The Pauli exclusion principle, based on the concept of electron spin, limits orbitals to a maximum population of two electrons with paired spins. The electron configuration of an element in its ground state is obtained by filling orbitals beginning with the $1s$ subshell and following the Pauli exclusion principle and Hund's rule (electrons spread out as much as possible in orbitals of equal energy). Sometimes we represent electron configurations using orbital diagrams. Unex-

pected configurations occur for chromium and copper because of the extra stability of half-filled and filled subshells.

Periodic Table.   We can divide the periodic table into sets of columns that can help us to write electron configurations. The valence shell configuration can be obtained from an element's period number and the sets of columns that must be crossed to reach the element in the table.

Orbital Shapes.   All s orbitals are spherical; each p orbital consists of two lobes with a nodal plane between them. A p subshell has three p orbitals whose axes are mutually perpendicular and point along the x, y, and z axes of an imaginary coordinate system centered at the nucleus. In each orbital the electron is conveniently viewed as a sort of cloud with a varying electron density. Four of the five d orbitals in a d subshell have the same shape, with four lobes of electron density each. The fifth has two lobes of electron density pointing in opposite direction along the z axis and a ring of electron density in the x–y plane.

Atomic Properties.   Atomic radii decrease from left to right and bottom to top in the table. Negative ions are larger than the atoms from which they are formed; positive ions are smaller than the atoms from which they are formed. Ionization energy (IE) is the energy needed to remove an electron from a gaseous atom. Electron affinity (EA) is energy released when an electron is added to an atom. IE and EA increase from left to right and from bottom to top in the periodic table.

REVIEW EXERCISES

Answers to questions whose numbers are printed in color are given in Appendix C. Difficult questions are marked with asterisks.

Discovery of Subatomic Particles

6.1   Describe the contributions of the following scientists.
    (a) Thomson   (d) Moseley
    (b) Millikan   (e) Chadwick
    (c) Rutherford

6.2   What are cathode rays? How do the masses of cathode ray particles compare with the masses of positive particles produced in a gas discharge tube?

6.3   Calculate the value of the charge-to-mass ratio for a proton.

6.4   An alpha particle is actually the nucleus of a helium atom, and carries a 2+ charge. Its charge-to-mass ratio is 4.82×10^4 C/g. What is the mass of an alpha particle in grams?

6.5   What would be the charge-to-mass ratio of a neutron?

Electromagnetic Radiation

6.6   In general terms, why do we call light "electromagnetic energy"?

6.7   What is meant by the "frequency of light"? What symbol is used for it, and what is the SI unit (and symbol) for frequency?

6.8   What do we mean by the "wavelength" of light? What symbol is used for it?

6.9   Sketch a picture of a wave and label its wavelength and its amplitude.

6.10   What names do we give to various ranges of frequencies in the electromagnetic spectrum, beginning from the lowest-frequency side?

6.11   What is meant by "visible spectrum"?

6.12   Give the equation that relates the wavelength and frequency of a light wave.

6.13   What is the frequency in hertz of blue light having a wavelength of 425 nm?

6.14   A certain substance strongly absorbs infrared light having a wavelength of $6.50\ \mu$m. What is the frequency in hertz of this light?

6.15   Ozone protects the earth's inhabitants from the harmful effects of ultraviolet light arriving from the sun. This shielding is a maximum for UV light having a wavelength of 295 nm. What is the frequency in hertz of this light?

6.16   Radar signals are electromagnetic radiations in the microwave region of the spectrum. A typical radar signal has a wavelength of 3.19 cm. What is its frequency in hertz?

6.17 In New York City, radio station WCBS broadcasts its FM signal at a frequency of 101.1 megahertz (MHz). What is the wavelength of this signal in meters?

6.18 Sodium vapor lamps are often used in residential street lighting. They give off a yellow light having a frequency of 5.09×10^{14} Hz. What is the wavelength of this light in nanometers?

6.19 How is the frequency of a particular type of radiation related to the energy associated with it? (Give an equation, defining all symbols.)

6.20 Show that $E = hc/\lambda$.

6.21 Examine each of the following pairs and state which of the two has the higher *energy*.
(a) Microwaves and infrared.
(b) Visible light and infrared.
(c) Ultraviolet light and X rays.
(d) Visible light and ultraviolet light.

6.22 What is a photon?

6.23 What is a quantum of energy?

6.24 Calculate the energy in joules of a photon of red light having a frequency of 4.0×10^{14} Hz.

6.25 Calculate the energy in joules of a photon of green light having a wavelength of 550 nm.

6.26 Microwaves are used to heat food in microwave ovens. The microwave radiation is absorbed by moisture in the food. This heats the water, and as the water becomes hot, so does the food. How many photons having a wavelength of 3.00 mm would have to be absorbed by 1.00 g of water to raise its temperature by 1.00 °C?

Atomic Spectra

6.27 What is an atomic spectrum? How does it differ from a continuous spectrum?

6.28 What fundamental fact is implied by the existence of atomic spectra?

Bohr Atom

6.29 Describe Niels Bohr's model of the structure of the hydrogen atom.

6.30 In qualitative terms, how did Bohr's model account for the atomic spectrum of hydrogen?

6.31 What is the term used to describe the lowest-energy state of an atom?

6.32 In what way was Bohr's theory both a success and a failure?

Wave Nature of Matter

6.33 How does the behavior of very small particles differ from that of the larger, more massive objects that we meet in everyday life? Why don't we notice this same behavior for the larger, more massive objects?

6.34 Describe the phenomenon called diffraction. How can this be used to demonstrate that de Broglie's theory was correct?

6.35 What is the difference between a *traveling wave* and a *standing wave*?

Electron Waves in Atoms

6.36 What are the names used to refer to the theories that apply the matter–wave concept to electrons in atoms?

6.37 What is the term used to describe a particular waveform of a standing wave for an electron?

6.38 What are the two properties of orbitals in which we are most interested? Why?

Quantum Numbers

6.39 What are the allowed values of the principal quantum number?

6.40 What is the value for n for (a) the K shell and (b) the M shell?

6.41 What is the letter code for a subshell with (a) $\ell = 1$, (b) $\ell = 3$, and (c) $\ell = 5$?

6.42 Give the values of n and ℓ for the following subshells: (a) 2s, (b) 3d, (c) 5f.

6.43 For the shell with $n = 4$, what are the possible values of ℓ?

6.44 Why does every shell contain an s subshell?

6.45 What are the possible values of m_ℓ for a subshell with (a) $\ell = 1$ and (b) $\ell = 3$?

6.46 How many orbitals are found in (a) an s subshell, (b) a p subshell, (c) a d subshell, and (d) an f subshell?

6.47 How many orbitals are there in an h subshell ($\ell = 5$)?

6.48 Give the complete set of quantum numbers for all of the electrons that could populate the 2p subshell of an atom.

6.49 What is the value of n for the valence shells of (a) Sn, (b) K, (c) Br, and (d) Bi.

Electron Spin

6.50 What physical property of electrons leads us to propose that they spin like a toy top?

6.51 What is the name of the magnetic property exhibited by atoms that contain unpaired electrons?

6.52 What are the possible values of the spin quantum number?

Electron Configuration of Atoms

6.53 What do we mean by the term *electronic structure*?

6.54 Within any given shell, how do the energies of the s, p, d, and f subshells compare?

6.55 What fact about the energies of subshells was responsible for the apparent success of Bohr's theory about electronic structure?

6.56 How do the energies of the orbitals belonging to a given subshell compare?

6.57 What is the Pauli exclusion principle? What effect does it have on the populating of orbitals by electrons?

6.58 Give the electron configurations of the elements in period 2 of the periodic table.

6.59 Predict the electron configurations of (a) S, (b) K, (c) Ti, and (d) Sn.

6.60 Predict the electron configurations of (a) As, (b) Cl, (c) Fe, and (d) Si.

6.61 Give the correct electron configurations of (a) Cr and (b) Cu.

6.62 Give orbital diagrams for (a) Mg and (b) Ti.

6.63 Give orbital diagrams for (a) As and (b) Ni.

6.64 How many unpaired electrons would be found in the ground state of (a) Mg, (b) P, and (c) V?

6.65 Write the shorthand electron configurations for (a) Ni, (b) Cs, (c) Ge, and (d) Br.

6.66 Write the shorthand electron configurations for (a) Al, (b) Se, (c) Ba, and (d) Sb.

6.67 How are the electron configurations of the elements in a given group similar?

6.68 Define the terms *valence shell* and *valence electrons*.

6.69 Give the configuration of the valence shell for (a) Na, (b) Al, (c) Ge, and (d) P.

6.70 Give the configuration of the valence shell for (a) Mg, (b) Br, (c) Ga, and (d) Pb.

6.71 How many unpaired electrons are expected to be found in the ground states of the following atoms: (a) Mn, (b) As, (c) S, (d) Sr, (e) Ar?

6.72 Which of the following atoms in their ground states are expected to be diamagnetic: (a) Ba, (b) Se, (c) Zn, (d) Si?

Shapes of Electron Orbitals

6.73 In what general terms do we describe an electron's location in an atom?

6.74 Sketch the approximate shapes of *s* and *p* orbitals.

6.75 How does the size of a given type of orbital vary with *n*?

6.76 How are the *p* orbitals of a given *p* subshell oriented relative to each other?

6.77 What is a *nodal plane*?

Atomic and Ionic Size

6.78 What is the meaning of *effective nuclear charge*? How does the effective nuclear charge felt by the outer electrons vary going down a group? How does it change as we go from left to right across a period?

6.79 Choose the larger atom in each pair: (a) Na or Si; (b) P or Sb.

6.80 Choose the larger atom in each pair: (a) Al or Cl; (b) Al or In.

6.81 Choose the largest atom among the following: Ge, As, Sn, Sb.

6.82 In what region of the periodic table are the largest atoms found? Where are the smallest atoms found?

6.83 Place the following in order of increasing size: N^{3-}, Mg^{2+}, Na^+, F^-, O^{2-}, Ne. (*Hint:* Count the number of electrons and the nuclear charge in each of them.)

6.84 Why are the size changes among the transition elements more gradual than those among the representative elements?

6.85 Choose the larger particle in each pair: (a) Na or Na^+; (b) Co^{3+} or Co^{2+}; (c) Cl or Cl^-.

Ionization Energy

6.86 Define *ionization energy*.

6.87 Choose the atom with the higher ionization energy in each pair: (a) B or C; (b) O or S; (c) Cl or As.

6.88 Why is the noble gas configuration so stable?

6.89 Explain *why* ionization energy increases from left to right in a period and decreases from top to bottom in a group.

6.90 Why is an atom's second ionization energy always larger than its first ionization energy?

Electron Affinity

6.91 Define *electron affinity*.

6.92 Choose the atom with the more exothermic EA in each pair: (a) Cl or Br; (b) Se or Br; (c) Si or Ga.

6.93 Why is the second electron affinity of an atom ($M^- + e^- \rightarrow M^{2-}$) always endothermic?

CHEMICALS IN USE

SILICON

Silicon is certainly one of the more common elements. In fact, its abundance is exceeded only by that of oxygen. The earth's crust is composed of about 27.7% silicon by weight, or about 21.2% if atoms are counted. Silicon occurs in many different minerals, most of which are silicate rocks such as those described in Unit 6 of this *Chemicals in Use* series.

Preparation and Purification of Silicon

The raw material from which silicon is extracted is quartz sand, which is largely composed of silicon dioxide, SiO_2. The silicon is recovered from it by reduction with carbon.

$$SiO_2(s) + 2C(s) \longrightarrow Si(s) + 2CO_2(g)$$

To get the reaction to go, the reaction mixture is heated in an *electric furnace*—a furnace in which the contents are heated to very high temperatures by passing a heavy electric current through them.

Silicon, shown in Figure 5a, is a dark solid with a high melting point (1410 °C). It has somewhat of a metallic appearance, although its electrical conductivity is much less than that of a typical metal. In fact, the semiconductor properties of silicon are what make this element useful in the electronics industry.

The silicon used in electronic devices must be of extremely high purity, which requires that the product that comes from the electric furnace be purified. First, the silicon is converted to $SiCl_4$ by reaction with chlorine.

$$Si + 2Cl_2 \longrightarrow SiCl_4$$

Silicon tetrachloride is a liquid at room temperature and has a low boiling point (59 °C). Therefore, the mixture is simply heated, which vaporizes the $SiCl_4$ and leaves the impurities behind. The silicon is then recovered by allowing the $SiCl_4$ to react with H_2.

$$SiCl_4 + 2H_2 \longrightarrow Si + 4HCl$$

Although the silicon made by this method is quite pure, it is still not pure enough to be used to make electronic devices, such as the "chips" in microcomputers and electronic calculators. The final purification is done by *zone refining* (Figure 5b). A bar of silicon supported at both ends is circled by a heating element that can melt a very thin wafer of the material. After this molten band of silicon has formed, the heating element is moved slowly from one end of the bar to the other. As it moves, the molten zone follows; silicon melts at the forward edge of the zone and other silicon crystallizes at the trailing edge. While this is happening, the impurities collect in the melted region, and the silicon that solidifies behind it is very, very pure. When the heater finally reaches the end of the bar, it is turned off and brought back to the beginning, and the process is repeated. After several passes, the silicon has become ex-

FIGURE 5a On the left, a sample of crystalline elemental silicon. On the right, silicon wafers used to manufacture electronic devices for computers and electronic calculators.

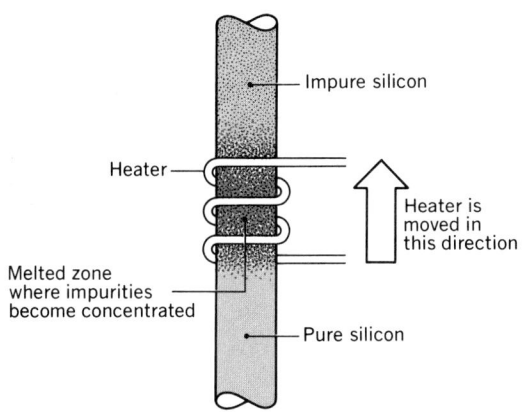

FIGURE 5b Zone refining. (Adapted from J. E. Brady and G. E. Humiston, *General Chemistry: Principles and Structure*, 3rd ed., John Wiley & Sons, 1982.)

tremely pure, except for a small portion of the bar at one end where the impurities have accumulated. This end is cut off and discarded, and the remaining bar contains silicon with impurity levels ranging from a few parts per million down to as little as a few parts per billion.

Semiconductor Properties of Silicon

In a solid, the atoms are pressed closely together. The atomic orbitals in the outer shells are forced to overlap, and the individual orbitals of all the atoms merge to give a number of *energy bands* that extend throughout the solid. Each band is composed of an enormous number of very closely spaced energy levels.

Certain energy bands are given special names. For example, the band composed of the atoms' valence shells is called the **valence band.** In sodium, for instance, the outer 3s orbitals of the sodium atoms merge to give a half-filled valence band — half-filled because the 3s orbital of each sodium atom is only half-filled. Because the valence band of sodium is only partially filled, electrons can jump from atom to atom with ease, and sodium is a good conductor of electricity.

A band of energy levels that extends through the solid and is either partially occupied or completely empty is able to transport electrons and is called a **conduction band.** In sodium, the conduction band and the valence band are the same, but this is not always the case. In solid magnesium, for example, the valence shell of each atom has two electrons in its 3s orbital, so the 3s band is filled. However, the energy difference between the 3s and 3p subshells in magnesium is small enough that in the solid the empty 3p band overlaps the 3s band (Figure 5c). Electrons can easily move between the two bands, so the filled 3s band supplies electrons that can be transported by the empty conduction band. As a result, magnesium is also a good electrical conductor.

In a nonmetallic solid, there is a large energy gap between the filled valence band and the conduction band (Figure 5c). For example, in diamond (an elemental form of carbon), each atom is bonded to four others by a sharing of electrons in such a way that the valence shells of the carbon atoms are completed. The valence band formed from the 2s and 2p orbitals of the carbon atoms is full, and the nearest conduction band is formed by orbitals in the next higher shell (3s and 3p). The difference in energy between the second shell and third shell is large, so there is a large energy gap between the filled valence band and the empty conduction band. Because the conduction band is empty, and because it cannot be easily populated by electrons from the filled valence band, diamond and other nonmetallic solids are poor conductors of electricity.

Silicon and other metalloids owe their unusual electrical properties to a relatively small energy gap between the filled valence band and the empty conduction band (Figure 5c). Silicon has the same structure as diamond, so its valence band is filled and at a lower energy than its empty conduction band. Because the conduction band is not normally occupied by any of silicon's electrons, silicon is not a good electrical conductor; that is, it doesn't display metallic conduction. However, the band gap is small enough in silicon that thermal energy (heat energy) possessed by the solid at room temperature is able to raise some of the electrons from the valence band to the conduction band where they are able to move through the solid. These are the electrons that are responsible for silicon's semiconductivity.

Questions

1. Write the balanced chemical equation for the reaction used to prepare elemental silicon from silicon dioxide.

2. Write balanced chemical equations for the two steps in the chemical purification of silicon.

3. What is the name given to the energy band in a solid that is formed from the atom's valence orbitals?

4. What is the name given to an energy band in a solid that is empty or only partially filled with electrons?

5. In your own words, explain why nonmetallic elements such as diamond and phosphorus are nonconductors of electricity (insulators).

6. Why are metalloids, such as silicon and germanium, semiconductors of electricity?

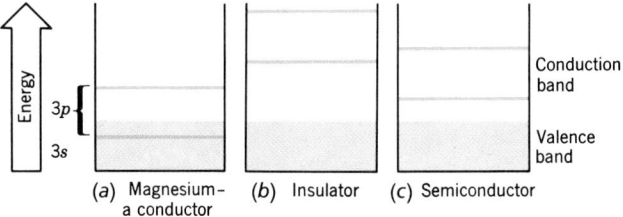

FIGURE 5c Energies of the valence band and conduction band in three kinds of solids. *(a)* Magnesium — a conductor. *(b)* An insulator. *(c)* A semiconductor.

Two beautiful copper-containing minerals. In the center is green *malachite*, $Cu_2(CO_3)(OH)_2$.
Surrounding it is the rather rare mineral *azurite*, $Cu_3(CO_3)_2(OH)_2$, named for its azure-blue color.

CHEMICAL BONDS BETWEEN ATOMS

7.1 THE FORMATION OF IONIC COMPOUNDS

When elements with low ionization energies react with elements with high electron affinities, ionic compounds form.

In the last chapter, we told you that the distributions of electrons in the outer parts of atoms are responsible for chemical properties. When atoms that react with each other touch, changes occur among the electrons in these regions. These changes give rise to chemical bonds—the forces of attraction that bind atoms to each other in compounds.

Understanding the nature and origin of chemical bonds is an important part of understanding chemistry, because changes in these bonding forces constitute the underlying basis for all chemical reactions. Old bonds break and new ones form when chemicals react. For instance, when we digest meat, the bonds that hold the amino acids together in proteins are broken and the amino acid molecules are set free. Then processes in our cells recombine them to form the specific proteins that we need for survival. Both the complex structures and the functions of starch, proteins, and DNA are determined not just by the atoms present, but also by the chemical bonds that bind the atoms to each other within their molecules.

You learned in Chapter 2 that there are two principal classes of bonding forces. One occurs in molecules and involves the sharing of electrons; it is called *covalent bonding*. The other involves the transfer of electrons from atoms of one element to atoms of another; this produces ions and is called *ionic bonding*. We will study both of them, but ionic bonding is simpler to understand, so we will discuss it first.

Ionic Bonds

As you know, ionic compounds are formed when metals react with nonmetals. The example we used earlier was sodium chloride—table salt. You saw that each sodium atom loses one electron to form a sodium ion, Na^+, and each chlorine atom gains one electron and becomes a chloride ion, Cl^-. Once formed, these ions cling together because their opposite charges attract one another. This attraction between positive and negative ions in an ionic compound is what we call the ionic bond.

The reason for the attraction between Na^+ and Cl^- ions—the fact that opposite charges attract—is not difficult to understand. But *why* are electrons transferred between these and other atoms? *Why* does sodium form Na^+ and not Na^- or Na^{2+}? And *why* does chlorine form Cl^- instead of Cl^+ or Cl^{2-}? To answer these questions we have to consider a number of factors that, taken together, determine whether an ionic compound can form and what its formula will be. These factors affect the numbers of electrons lost or gained, and they are all related to the total energy of the system of reactants and products. *For any stable compound to form from its elements, there must be a net lowering of the energy.* In other words, energy must be released.

Three major contributing factors affect the energy in the formation of an ionic compound. One is the removal of electrons from the atoms that become cations (e.g., sodium). Formation of a cation requires an input of energy—the ionization energy. A second factor is the energy change that accompanies the addition of one or more electrons to the atoms that become anions (e.g., chlorine). This energy is the electron affinity. As we saw earlier, the formation of an anion is usually exothermic—energy is released.

The ionization energy (IE) and electron affinity (EA) are energies associated with the changes of isolated gaseous atoms to isolated gaseous ions. A crystal of salt, however, does not consist of isolated ions (see Figure 7.1). You've already learned that the ions in salt are packed tightly together in a regular pattern. This pattern is referred to as a *lattice*, and it has a lower energy than the isolated ions. To understand this, let's imagine that we wished to vaporize a crystal of salt, so that the ions become spread far apart. In the salt crystal the attractive forces outweigh those of repulsion, so to accomplish this vaporization we have to pull the ions apart. This requires work, so vaporizing the crystal increases the ions' potential energy and is endothermic. The reverse process—the imaginary process that forms the lattice from isolated ions—must therefore lead to a lowering of the potential energy of the system and be exothermic. The amount that the energy of the system is lowered because of the mutual attractions of its ions is called the lattice energy.

The lattice energy is the major stabilizing factor for ionic compounds. In almost every case, the energy input required by the ionization energy is larger than the energy recovered by the electron affinity, so the IE and EA combined have a net energy-raising effect. If it were not for the large energy-lowering effect of the lattice energy, formation of ionic compounds would be endothermic and they simply wouldn't be formed. In a stable ionic compound, therefore, the loss of the lattice energy produces an energy-lowering effect large enough to offset the net energy-raising effect of the IE and EA.

Now we can explore what happens to atoms when they react. Right from the beginning, we can see why metals tend to form positive ions and nonmetals tend to form negative ions. At the left of the periodic table are the metals—elements with small IE and EA. Relatively little energy is needed to remove electrons from them to produce positive

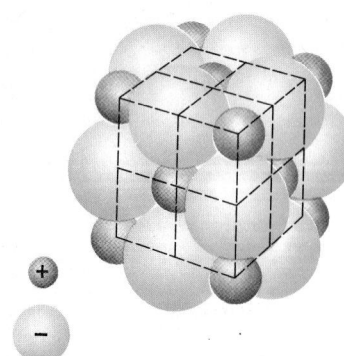

FIGURE 7.1 Packing of ions in NaCl.

The lattice energy would be released if gaseous ions condense to form the solid ionic compound.

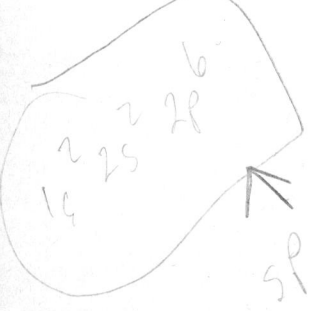

ions. At the upper right of the periodic table are the nonmetals — elements with large IE and EA. It is very difficult to remove electrons from these elements, but sizable amounts of energy are released when they gain electrons. On an energy basis, it is least "expensive" to form a cation from a metal and an anion from a nonmetal, so it is relatively easy for the energy-lowering effect of the lattice energy to exceed the net energy-raising effect of the IE and EA. Therefore, metals combine with nonmetals to form ionic compounds simply because ionic bonding is favored over other types whenever atoms of small IE combine with atoms of large EA.

Formation of Ions by the Representative Elements

Now let's look at how the electronic structures of the elements affect the kinds of ions that they form. We will begin by examining what happens when sodium loses an electron. The electronic structure of Na is

$$\text{Na} \qquad 1s^2 2s^2 2p^6 3s^1$$

The electron that is lost is the one least tightly held. For sodium that is the single outer $3s$ electron. The electronic structure of the Na^+ ion, then, is

$$\text{Na}^+ \qquad 1s^2 2s^2 2p^6$$

For sodium:

    1st IE = 496 kJ/mol
    2nd IE = 4563 kJ/mol

The removal of the first electron from Na does not require much energy because the first IE of Na is small. Therefore, an input of energy equal to the first IE can be easily recovered by the exothermic lattice energy of ionic compounds containing the Na^+ ion. However, removal of a second electron from sodium is very difficult — the second IE of Na is enormous. The amount of energy that must be invested to create a Na^{2+} ion is therefore much greater than the amount of energy that can be recovered by the lattice energy, so overall the formation of a compound containing Na^{2+} is very energetically unfavorable. This is why we never observe compounds that contain this ion, and why sodium stops losing electrons once it has achieved a noble gas configuration.

A somewhat similar situation exists for other metals, too. Consider calcium, for example. We know that his metal forms ions with a 2+ charge. This means that when it reacts, a calcium atom loses its two outermost electrons.

$$\begin{aligned} \text{Ca} &\qquad 1s^2 2s^2 2p^6 3s^2 3p^6 4s^2 \\ \text{Ca}^{2+} &\qquad 1s^2 2s^2 2p^6 3s^2 3p^6 \end{aligned}$$

For calcium:

    1st IE = 590 kJ/mol
    2nd IE = 1140 kJ/mol
    3rd IE = 4940 kJ/mol

The two $4s$ electrons of Ca are not held too tightly, so the total amount of energy that must be invested to remove them (the sum of the first and second IE) can be recovered easily by the large lattice energy of a Ca^{2+} compound. However, the removal of yet another electron from calcium to form Ca^{3+} requires breaking into the noble gas core. As in the case of sodium, a tremendous amount of energy is needed to accomplish this — much more than would be regained by the lattice energy of a Ca^{3+} compound. Therefore, a calcium atom loses just two electrons when it reacts.

In the case of sodium and calcium, we find that the stability of the noble gas core that lies below the outer shell of electrons of these metals effectively limits the number of electrons that they lose, and that the ions that are formed have a noble gas electron configuration. A similar configuration also tends to be the fate of nonmetals when they form anions.

Chlorine and oxygen are typical nonmetals that form anions when they react with metals such as sodium or calcium. When a chlorine atom reacts, it gains one electron. For chlorine we have

$$\text{Cl} \qquad 1s^2 2s^2 2p^6 3s^2 3p^5$$

and when an electron is gained, its configuration becomes

$$\text{Cl}^- \qquad 1s^2 2s^2 2p^6 3s^2 3p^6$$

At this point, electron gain ceases, because if another electron were to be added, it would have to enter an orbital in the next higher shell.

With oxygen, a similar situation exists. The formation of the oxide ion, O^{2-}, gives oxygen a noble gas configuration without much difficulty,

$$O\ (1s^22s^22p^4) + 2e^- \longrightarrow O^{2-}\ (1s^22s^22p^6)$$

and the large lattice energies of metal oxides leads to stable compounds. However, we never observe the formation of O^{3-} because, once again, the last electron would have to enter an orbital in the next higher shell, and this is very energetically unfavorable.

What we see here is that energy factors cause many atoms to form ions that have a noble gas electron configuration. In turn, this leads to the useful generalization that _when they form ions, atoms of most of the representative elements tend to gain or lose electrons until they have obtained a configuration that is the same as that of the nearest noble gas._

Ions of Transition Metals

As you learned in Chapter 2, the transition elements are metals and form cations when they react. However, the situation here is a bit more complicated than for the representative elements. Many transition elements are able to form more than one cation because they have a partially filled d subshell that is just slightly lower in energy than the outer s subshell. When a transition metal forms a positive ion, it always loses electrons from its outer s subshell first. Once these are gone, any further electron loss takes place from the partially filled d subshell. Iron is a typical example. Its electron configuration is

$$Fe \quad [Ar]3d^64s^2$$

Remember, the electrons that an atom loses first *always* come from the shell with largest n.

When iron reacts, it loses its $4s$ electrons fairly easily to give Fe^{2+}. But because the $3d$ subshell is close in energy to the $4s$, it is not very difficult to remove still another electron to give Fe^{3+}.

$$Fe^{3+} \quad [Ar]3d^5$$

Because so many of the transition elements are able to form ions in a similar way, the ability to form more than one positive ion is usually cited as one of the characteristic properties of the transition elements. Frequently, one of the ions formed has a 2+ charge, which arises from the loss of the two outer s electrons. Ions with larger positive charges result when additional d electrons are lost. Unfortunately, it is not easy to predict exactly which ions can form for a given transition metal, nor is it simple to predict their relative stabilities. As your study of chemistry progresses, you will gradually learn which ions are formed by the common metals. In fact, you began this in Chapter 2.

EXAMPLE 7.1   ELECTRON CONFIGURATIONS OF IONS

Problem: How does the electron configuration of the valence shell of nitrogen change when it forms the N^{3-} ion?

Solution: The electron configuration for nitrogen is

$$N \quad [He]2s^22p^3$$

To form N^{3-}, three electrons must be gained. These would enter the $2p$ subshell because it is the lowest available energy level. The configuration for the ion would be

$$N^{3-} \quad [He]2s^22p^6$$

PRACTICE EXERCISE 1   How does the valence shell electron configuration of a sulfur atom change when it forms the S^{2-} ion? How does the electron configuration of magnesium change when it forms Mg^{2+}?

7.2 ELECTRON BOOKKEEPING: LEWIS SYMBOLS

Lewis symbols are used to show the valence electrons of an atom or ion.

In the last section, you saw how the valence shells of atoms change when electrons are transferred during the formation of ions. We will soon see that some atoms share their valence electrons with each other when they form covalent bonds. In these discussions of bonding it is useful to be able to keep track of valence electrons. To help us do this, we use a simple bookkeeping device called Lewis symbols, named after their inventor, the famous American chemist, G. N. Lewis (1875–1946).

To draw the Lewis symbol for an element, we write its chemical symbol surrounded by a number of dots (or other similar symbol), which represent the atom's valence electrons. For example, the element lithium, which has one valence electron in its $2s$ subshell, has the Lewis symbol

Li·

In fact, each element in Group IA has a similar Lewis symbol, because each has only one valence electron. The Lewis symbols for all of the Group IA elements are

Li·   Na·   K·   Rb·   Cs·

The Lewis symbols for the eight A-group elements of period 2 are[1]

Group	IA	IIA	IIIA	IVA	VA	VIA	VIIA	0
Symbol	Li·	·Be·	·B·	·C·	·N:	·O:	·F:	:Ne:

The elements in each group below those given have Lewis symbols identical to the respective period 2 element except, of course, for the chemical symbol of the element. Notice that when an atom has more than four valence electrons, the additional electrons are shown to be paired with others. Also notice that *for the representative elements, the group number is equal to the number of valence electrons* when the U.S. convention for numbering groups in the periodic table is followed.

Sometimes Lewis symbols are called dot symbols.

We will generally follow the U.S. convention for this reason.

EXAMPLE 7.2   WRITING LEWIS SYMBOLS

Problem: What is the Lewis symbol for arsenic, As?

Solution: Arsenic is in Group VA and therefore has five valence electrons. The first four are placed about the symbol for the arsenic atom as

·As·

The fifth electron is paired with one of the first four to give

·As:

Note: Equally valid would be

·As·   or   :As·   or   ·As·

---

[1] For beryllium, boron, and carbon, the number of unpaired electrons in the Lewis symbol doesn't agree with the number predicted from the atom's electron configuration. Boron, for example, has two electrons paired in its $2s$ orbital and a third electron in one of its $2p$ orbitals; therefore, there is actually only one unpaired electron in a boron atom. The Lewis symbols are drawn as shown, however, because when beryllium, boron, and carbon form bonds, they *behave* as if they have two, three, and four unpaired electrons, respectively.

PRACTICE EXERCISE 2   Write Lewis symbols for (a) Se, (b) I, and (c) Ca.

Although we will use Lewis symbols mostly to follow the fate of valence electrons in covalent bonds, they can also be used to describe what happens during the formation of ions. For example, when a sodium atom reacts with a chlorine atom, the electron transfer can be depicted as

$$Na \overset{\frown}{\cdot} + \cdot \overset{..}{\underset{..}{Cl}}: \longrightarrow Na^+ + \left[: \overset{..}{\underset{..}{Cl}} : \right]^-$$

The valence shell of the sodium atom is emptied, so no dots remain. The outer shell of chlorine, which formerly had seven electrons, gains one electron to give a total of eight. The brackets are drawn around the chloride ion to show that the electrons represented by the dots are its exclusive property.

We can diagram a similar reaction between calcium and chlorine atoms.

$$: \overset{..}{\underset{..}{Cl}} \cdot \overset{\frown}{} Ca \overset{\frown}{} \cdot \overset{..}{\underset{..}{Cl}} : \longrightarrow Ca^{2+} + 2 \left[: \overset{..}{\underset{..}{Cl}} : \right]^-$$

EXAMPLE 7.3   USING LEWIS SYMBOLS

Problem: Using Lewis symbols, diagram the reaction that occurs between sodium and oxygen atoms to give Na^+ and O^{2-} ions.

Solution: First we draw the Lewis symbols for Na and O.

$$Na \cdot \qquad \cdot \overset{..}{\underset{..}{O}} :$$

It takes two electrons to complete the octet around oxygen. Each Na can supply only one, so we need two Na atoms. Therefore,

$$Na \overset{\frown}{\cdot} + \cdot \overset{..}{\underset{\cdot}{O}} : + \overset{\frown}{} Na \longrightarrow 2Na^+ + \left[: \overset{..}{\underset{..}{O}} : \right]^{2-}$$

Don't forget to put the brackets around the oxide ion.

PRACTICE EXERCISE 3   Diagram the reaction between magnesium and oxygen atoms to give Mg^{2+} and O^{2-} ions.

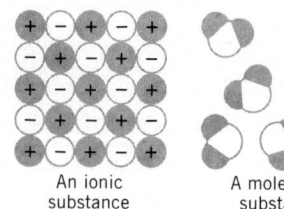

An ionic substance   A molecular substance

As a rough rule, ionic bonding occurs between metals and nonmetals, and covalent bonding occurs when nonmetals combine with nonmetals.

7.3 ELECTRON SHARING

The sharing of electron pairs provides an alternative to ion formation when atoms form bonds.

Most of the substances we encounter in our daily lives are not ionic. Rather than existing as collections of electrically charged particles (ions), they occur as electrically neutral combinations of atoms called molecules. Water, as you already know, consists of molecules made from two hydrogen atoms and one oxygen atom, and the formula for one particle of this compound is H_2O. Most substances consist of much larger molecules. For example, the formula for table sugar is $C_{12}H_{22}O_{11}$, and gasoline is composed of hydrocarbon molecules such as octane, C_8H_{18}.

Earlier we saw that for ionic bonding to occur, the energy-lowering effect of the lattice energy must be greater than the combined energy-raising effects of the ionization

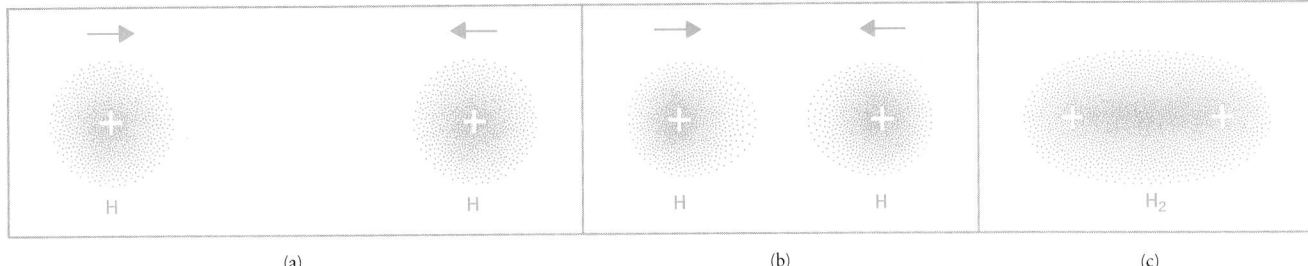

(a)   (b)   (c)

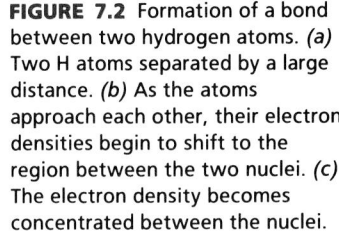

FIGURE 7.2 Formation of a bond between two hydrogen atoms. *(a)* Two H atoms separated by a large distance. *(b)* As the atoms approach each other, their electron densities begin to shift to the region between the two nuclei. *(c)* The electron density becomes concentrated between the nuclei.

energy (IE) and electron affinity (EA). Many times this is not possible, particularly when the IE of all the atoms involved is large, as happens when nonmetals combine with other nonmetals. In such instances, nature uses a different way to lower the energy — electron sharing.

Let's look at what happens when two hydrogen atoms join together to form the H_2 molecule (Figure 7.2). As the two atoms approach each other, the electron of each atom begins to feel the attraction of both nuclei. This causes the electron density around each nucleus to shift toward the region between the two atoms. Therefore, as the distance between the nuclei decreases, there is an increase in the probability of finding either electron near either nucleus. In effect, then, each of the hydrogen atoms in this H_2 molecule now has a share of two electrons.

When the electron density shifts to the region between the two hydrogen atoms, it attracts each of the nuclei and pulls them together. Being of the same charge, however, the two nuclei also repel each other, as do the two electrons. In the molecule that forms, therefore, the atoms are held at a distance at which these attractions and repulsions are balanced. Overall, the nuclei are kept from separating, and the net force of attraction produced by the sharing of the pair of electrons is called a covalent bond.

Every covalent bond is characterized by two quantities, the average distance between the nuclei held together by the bond, and the energy needed to separate the two atoms to produce neutral atoms again. In the hydrogen molecule, the attractive forces pull the nuclei to a distance of 75 pm, and this distance is called the bond length or bond distance. Because a covalent bond holds atoms together, work must be done (energy must be supplied) to separate them. When the bond is *formed*, an equivalent amount of energy is released. The amount of energy released when the bond is formed (or the amount of energy needed to "break" the bond) is called the bond energy.

Formation of a covalent bond releases the bond energy, which means that as the bond forms, the energy of the atoms decreases. Figure 7.3 shows how the energy changes when two hydrogen atoms form H_2. We see that the minimum energy occurs at a bond

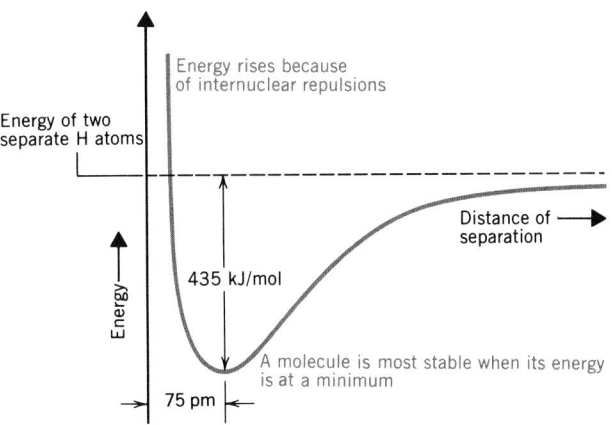

Energy rises because of internuclear repulsions

Energy of two separate H atoms

Distance of separation

435 kJ/mol

Energy

A molecule is most stable when its energy is at a minimum

75 pm

FIGURE 7.3 Energy of two hydrogen atoms as they approach each other.

The formation of the covalent bond has lowered the energy.

distance of 75 pm, and that 1 mol of hydrogen molecules is more stable than 2 mol of hydrogen atoms by 435 kJ. In other words, the bond energy of H_2 is 435 kJ/mol.

Before joining, each of the separate hydrogen atoms has one electron in its $1s$ orbital. When these electrons are shared, the $1s$ orbital of each atom has, in a sense, become filled. The electrons have also become paired, as required by the Pauli exclusion principle — that is, m_s, is $+\frac{1}{2}$ for one of the electrons and $-\frac{1}{2}$ for the other. In general, we almost always find that the electrons involved become paired when atoms form covalent bonds. In fact, a covalent bond is sometimes referred to as an electron pair bond.

Recall that m_s is the spin quantum number.

To keep track of the electrons in covalent bonds, we often use Lewis symbols in much the same way as we use them for ionic bonds. For the covalent bond, the electrons that are shared are shown as a pair of dots between the symbols of the bonded atoms. The formation of H_2 from hydrogen atoms, for example, can be depicted as

$$H \cdot + H \cdot \longrightarrow H : H$$

Formulas constructed from the Lewis symbols for atoms are called Lewis formulas or Lewis structures.

Usually, however, the electron pair in the covalent bond is represented as a dash. For example, the hydrogen molecule is given as

$$H - H$$

This kind of representation of a molecule is called a structural formula; it shows which atoms are present in a molecule *and* how they are attached to each other.

7.4   COVALENT BONDING AND THE OCTET RULE

The tendency of atoms to acquire a noble gas electron configuration can often be used to predict the number of covalent bonds that an atom will form.

We have seen that an electron configuration corresponding to that of a noble gas is very stable. When ions form, electrons tend to be gained or lost until a noble gas configuration is achieved. The stability of the noble gas configuration also influences the number of electrons an atom tends to acquire by sharing — it often controls the number of covalent bonds that an atom forms.

Hydrogen, with just one electron in its $1s$ orbital, can achieve a noble gas configuration (that of helium) by obtaining a share of one electron from another atom. The Lewis structure of H_2 implies that both atoms have access to both electrons in the bond.

$$\boxed{H : H}$$

(Two electrons can be counted around each of the H atoms.)

Hydrogen atoms only form one covalent bond.

Since hydrogen obtains a stable valence shell configuration when it shares just one pair of electrons with another atom, hydrogen atoms only form one covalent bond.

The Octet Rule

The valence shells of the noble gases other than helium all contain eight electrons, and the tendency of many atoms to acquire such an outer shell electron configuration forms the basis of the octet rule: *When atoms react, they tend to achieve an outer shell having eight electrons.* Many of the representative elements (sodium and chlorine, for instance) follow this rule when they form ions. In this case, they achieve an octet by the gain or loss of electrons. When atoms other than hydrogen form covalent bonds, however, an octet is achieved by acquiring electrons through sharing.

The octet rule is only applied to the representative elements. In general, it doesn't work for the transition elements.

Often, the octet rule can be used to explain the number of covalent bonds an atom forms. This number normally equals the number of electrons the atom needs to have a total of eight electrons (an octet) in its outer shell. For example, the halogens (Group VIIA) all have seven valence electrons. The Lewis symbol for a typical member of this group, chlorine, is

$$\cdot \ddot{\underset{\displaystyle ..}{Cl}} :$$

We can see that only one electron is needed to complete an octet. Of course, chlorine can actually gain this electron and become a chloride ion. This is what it does when it forms ionic compounds such as sodium chloride (NaCl). But when chlorine combines with another nonmetal, the transfer of electrons is not energetically favorable. Therefore, in forming such compounds as HCl or Cl_2, chlorine gets the one electron it needs by forming a covalent bond.

$$H \cdot \; \overset{\frown}{+} \; \cdot \ddot{\underset{..}{Cl}} : \; \longrightarrow \; H : \ddot{\underset{..}{Cl}} :$$

$$: \ddot{\underset{..}{Cl}} \cdot \; \overset{\frown}{+} \; \cdot \ddot{\underset{..}{Cl}} : \; \longrightarrow \; : \ddot{\underset{..}{Cl}} : \ddot{\underset{..}{Cl}} :$$

The arrows here just show which electrons become shared.

The HCl and Cl_2 molecules can also be represented using the "dash" bond.

$$H - \ddot{\underset{..}{Cl}} : \quad \text{and} \quad : \ddot{\underset{..}{Cl}} - \ddot{\underset{..}{Cl}} :$$

There are many nonmetals that form more than one covalent bond. For example, the three most important elements in biochemical systems are carbon, nitrogen, and oxygen.

$$\cdot \overset{\displaystyle .}{\underset{\displaystyle .}{C}} \cdot \quad \cdot \overset{\displaystyle .}{\underset{\displaystyle ..}{N}} \cdot \quad \cdot \overset{\displaystyle .}{\underset{\displaystyle ..}{O}} :$$

The simplest compounds that these elements form with hydrogen are methane, ammonia, and water, respectively. Since hydrogen forms only one covalent bond, the formulas for the molecules of these compounds are easy to predict using the octet rule. Carbon has four electrons, so it needs to get a share of four more electrons to achieve an octet. This means that it forms covalent bonds with four hydrogen atoms. Using the same reasoning, we can see that nitrogen forms covalent bonds with three hydrogen atoms, and oxygen, with two.

In most of the compounds in which they occur, carbon forms four covalent bonds, nitrogen forms three, and oxygen forms two.

$$
\begin{array}{ccc}
H & H & H \\
\ddot{} & \ddot{} & \ddot{} \\
H : C : H & H : N : H & H : \ddot{\underset{..}{O}} : \\
H & \underset{..}{} & \\
H & & \\
\text{or} & \text{or} & \text{or} \\
\end{array}
$$

$$
\begin{array}{ccc}
H & H & H \\
| & | & | \\
H - C - H & H - N - H & H - \ddot{\underset{..}{O}} : \\
| & & \\
H & & \\
\text{methane} & \text{ammonia} & \text{water}
\end{array}
$$

These are further examples of structural formulas — formulas that show how the atoms in a molecule are attached to each other. We will use structural formulas many times throughout this book.

EXAMPLE 7.4   USING THE OCTET RULE TO PREDICT STRUCTURAL FORMULAS

Problem: Use the octet rule to predict the formula of the compound formed from hydrogen and sulfur.

Solution: First we write Lewis symbols for each atom.

$$\text{H}\cdot \qquad \cdot\ddot{\underset{\cdot\cdot}{\text{S}}}:$$

(Sulfur has six electrons because it is in Group VIA.)
        Sulfur needs two electrons to complete its octet, so it bonds to two hydrogen atoms.

$$\text{H}:\ddot{\underset{\cdot\cdot}{\text{S}}}: \qquad \text{or} \qquad \text{H}-\ddot{\underset{\cdot\cdot}{\text{S}}}:$$

Thus, the molecule has the formula H_2S. This, by the way, is the substance that gives rotten eggs their foul odor.

PRACTICE EXERCISE 4   Use the octet rule to predict the formulas of compounds formed from (a) P and H and (b) S and F.

Multiple Bonds

The bond produced by the sharing of one pair of electrons between two atoms is called a single bond. So far, we have discussed only molecules that contain single bonds.

Each molecule to the right has only single bonds.

$$:\ddot{\underset{\cdot\cdot}{\text{Cl}}}-\ddot{\underset{\cdot\cdot}{\text{Cl}}}: \qquad \text{H}-\text{H} \qquad \text{H}-\overset{\overset{\displaystyle\text{H}}{|}}{\underset{\underset{\displaystyle\text{H}}{|}}{\text{C}}}-\text{H}$$

There are, however, many molecules in which more than one pair of electrons are shared between two atoms. For example, we know that nitrogen, the most abundant gas in the atmosphere, occurs in the form of diatomic molecules: that is, each of its molecules, N_2, is composed of two nitrogen atoms. As we've just seen in the previous section, the Lewis symbol for nitrogen is

$$\cdot\dot{\underset{\cdot\cdot}{\text{N}}}\cdot$$

and each nitrogen atom needs three electrons to complete its octet. When the N_2 molecule is formed, each of the nitrogen atoms shares three electrons with the other.

$$:\overset{\cdot}{\text{N}}\cdot \longleftrightarrow \cdot\overset{\cdot}{\text{N}}:$$

The result is called a triple bond. Notice that in the Lewis formula for the molecule, the three shared pairs of electrons are placed between the two atoms. We count all of these electrons as though they belong to both of the atoms. Each nitrogen therefore has an octet.

8 electrons ────┐   ┌──── 8 electrons

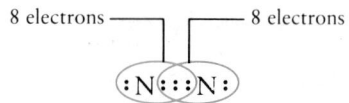

The triple bond is usually represented by three dashes, so the bonding in the N_2 molecule is normally shown as

$$:N \equiv N:$$

Double bonds also occur in molecules. A prime example is the CO_2 molecule, which contains a carbon atom that is bonded covalently to two separate oxygen atoms. We can diagram the formation of the bonds in CO_2 as follows.

The central carbon atom shares two of its electrons with each of the oxygen atoms, and each oxygen shares two electrons with carbon. The result is the formation of two <u>double bonds</u>. Once again, if we circle the valence shell electrons that "belong" to each atom, we can see that each has an octet.

The structural formula for CO_2, using dashes, is

$$:O = C = O:$$

7.5 WHEN THE OCTET RULE FAILS

The atoms in some molecules cannot obey the octet rule because there are either two few or too many electrons.

Sometimes it is just impossible to write a Lewis structure in which all of the atoms in a molecule obey the octet rule. This happens most often when an atom forms more than four bonds. Examples are PCl_5 and SF_6, in which there are five P—Cl bonds and six S—F bonds, respectively. Since each covalent bond requires the sharing of a pair of electrons, phosphorus and sulfur must exceed eight electrons in their outer shells. The Lewis formulas of these two molecules are shown below. In the next section we will discuss methods that you can learn to help you write structures such as these.

SF_6 is used as a gaseous insulator in high-voltage electrical equipment.

| 8 electrons around each Cl | 8 electrons around each F |
| 10 electrons around P | 12 electrons around S |

If an atom forms more than four bonds, it must end up with more than four electron pairs in its valence shell.

Elements in period 2 such as carbon or nitrogen, never exceed an octet simply because their valence shells, having $n = 2$, can hold a maximum of only 8 electrons. Elements in periods below period 2, however, sometimes do exceed an octet, because their valence shells can hold more than 8 electrons. For example, the valence shell for elements in period 3, for which $n = 3$, can hold a maximum of 18 electrons, and the valence shell for period 4 elements can hold as many as 32 electrons.

In some molecules (but not many), an atom has less than an octet. The most common examples are compounds of beryllium and boron.

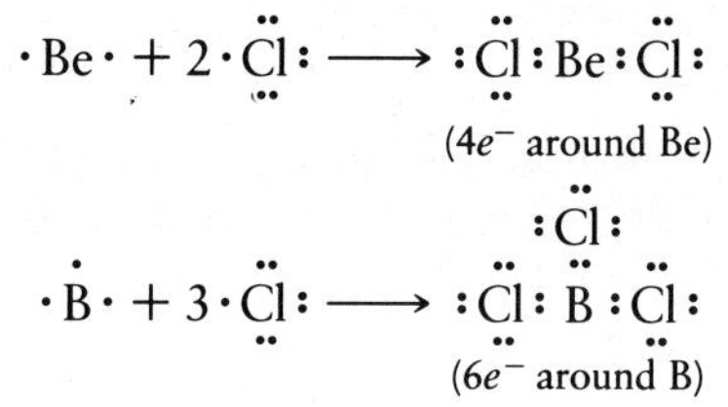

$$\cdot Be \cdot + 2 \cdot \overset{..}{\underset{..}{Cl}}: \longrightarrow :\overset{..}{\underset{..}{Cl}}:Be:\overset{..}{\underset{..}{Cl}}:$$

(4e⁻ around Be)

$$\cdot B \cdot + 3 \cdot \overset{.}{\underset{..}{Cl}}: \longrightarrow \begin{matrix} :\overset{..}{\underset{..}{Cl}}: \\ :\overset{..}{\underset{..}{Cl}}: B :\overset{..}{\underset{..}{Cl}}: \end{matrix}$$

(6e⁻ around B)

The method that we will discuss in the next section doesn't work for these molecules; they must simply be learned as exceptions.

7.6 DRAWING LEWIS STRUCTURES

In drawing Lewis structures we try, whenever possible, to have all atoms obey the octet rule.

Polyatomic ions are held together by covalent bonds, too.

Being able to draw Lewis structures for molecules or polyatomic ions is important, because they often form the basis for much chemical reasoning. Also, we will soon see that quite accurate predictions can be made concerning the shapes of molecules if we know their Lewis structures.

The method of writing Lewis structures can be broken down into a number of steps. The first is to decide which atoms are bonded to each other, so that we know where to put the dots. This is not always a simple matter. Many times the formula suggests the way the atoms are arranged because the central atom is usually written first. Examples are CO₂ and ClO₄⁻, which have the *skeletal structures* (i.e., arrangements of atoms),

O C O   and   O Cl O
                        O
                        O

Sometimes, obtaining the skeletal structure is not quite so simple. How, for example, would we predict that the skeletal structure of nitric acid, HNO₃, is

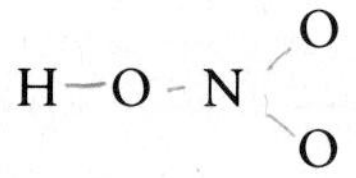

H—O - N ⟨ O   O   (correct)

rather than some other structure such as one of the following?

H
O N O   or   H O O N O   (incorrect)
O

The answer is not obvious. Nitric acid belongs to that group of substances called oxoacids (Section 2.13). It happens that the hydrogen atoms that can be released from these molecules are always bonded to oxygen, which is in turn bonded to the other nonmetal atom. Therefore, recognizing HNO₃ as the formula of an oxoacid allows us to predict that the three oxygen atoms are bonded to the nitrogen, and the hydrogen is bonded to one of the oxygens.

EXAMPLE 7.5   WRITING SKELETAL STRUCTURES

Problem: What is the probable skeletal structure of sulfuric acid, H_2SO_4?

Solution: From our discussion of HNO_3, we should expect that the four oxygens are bonded to the sulfur and that the two hydrogens are bonded to two of these oxygens. This would give

$$O$$
$$H \quad O \quad S \quad O \quad H$$
$$O$$

It is not important to which oxygens we attach the hydrogens.

PRACTICE EXERCISE 5   Predict reasonable skeletal structures for SO_2, NO_3^-, $HClO_3$, and H_3PO_4.

There are times when no reasonable basis can be found for choosing a particular skeletal structure. If you must make a guess, choose the most symmetrical arrangement of atoms.

Once you've decided on the skeletal structure, the next step in writing the Lewis structure is to count all of the *valence electrons* to find out how many dots must appear in the final formula. Using the periodic table, locate the groups in which the elements in the formula occur to determine the number of valence electrons contributed by each atom. If the structure you wish to draw is that of an ion, add one additional valence electron for each negative charge or remove a valence electron for each positive charge.

EXAMPLE 7.6   COUNTING VALENCE ELECTRONS

Problem: How many dots, representing electrons, must appear in the Lewis structures of SO_3, NO_3^-, and NH_4^+?

Solution:

Notice that when the customary U.S. numbering of groups in the periodic table is used, the group number is equal to the number of valence electrons for the representative elements.

SO_3	sulfur (Group VIA) contributes six electrons	$6 \times 1 = 6$
	oxygen (Group VIA) contributes six electrons each	$6 \times 3 = 18$
		Total $24e^-$
NO_3^-	nitrogen (Group VA) contributes five electrons	$5 \times 1 = 5$
	oxygen (Group VIA) contributes six electrons each	$6 \times 3 = 18$
	add another electron for the $1-$ charge	1
		Total $24e^-$
NH_4^+	nitrogen (Group VA) — five electrons	$5 \times 1 = 5$
	hydrogen (Group IA) — one electron each	$1 \times 4 = 4$
	subtract one electron for the $1+$ charge	-1
		Total $8e^-$

PRACTICE EXERCISE 6   How many dots should appear in the Lewis structures of SO_2, PO_4^{3-}, and NO^+?

Once we know the total number of valence electrons, we place them in the skeletal formula, always in groups of two. Start by placing a pair in each bond. Then, if possible, complete the octets of the atoms attached to the central atom. Finally, check to see if the central atom has an octet. (Remember, however, that the maximum number of electrons in the valence shell of hydrogen is two.)

EXAMPLE 7.7   WRITING LEWIS STRUCTURES

Problem: Write the Lewis structure for the SO_4^{2-} ion.

Solution: We would expect the skeletal structure to be

$$
\begin{array}{c}
\text{O} \\
\text{O} \quad \text{S} \quad \text{O} \\
\text{O}
\end{array}
$$

The total number of electrons in the formula is 32 ($6e^-$ from the sulfur, plus $6e^-$ from each oxygen, plus $2e^-$ for the 2− charge). We begin to distribute the electrons in the skeletal structure by placing a pair in each bond. This gives

$$
\begin{array}{c}
\text{O} \\
\text{O} : \text{S} : \text{O} \\
\text{O}
\end{array}
$$

This leaves $32 - 8 = 24e^-$. We use these to complete the octets around the oxygens.

$$
\begin{array}{c}
: \ddot{\text{O}} : \\
: \ddot{\text{O}} : \text{S} : \ddot{\text{O}} : \\
: \ddot{\text{O}} :
\end{array}
$$

The final structure has 32 dots, and each atom obeys the octet rule. Since we are dealing with an ion, we should indicate its charge. We do this as follows.

Remember, a dash stands for two electrons.

$$
\left[\begin{array}{c}
: \ddot{\text{O}} : \\
: \ddot{\text{O}} : \text{S} : \ddot{\text{O}} : \\
: \ddot{\text{O}} :
\end{array} \right]^{2-}
\quad \text{or} \quad
\left[\begin{array}{c}
: \ddot{\text{O}} : \\
| \\
: \ddot{\text{O}} - \text{S} - \ddot{\text{O}} : \\
| \\
: \ddot{\text{O}} :
\end{array} \right]^{2-}
$$

EXAMPLE 7.8   WRITING LEWIS STRUCTURES

Problem: What is the Lewis structure for the ClO_2^- ion?

Solution: The skeletal structure would be

$$\text{O} \quad \text{Cl} \quad \text{O}$$

In this ion there are $7 + 12 + 1 = 20$ electrons. First, we put a pair in each bond.

$$\text{O} : \text{Cl} : \text{O}$$

Next, we complete the octets of the oxygens. This uses another $12e^-$.

$$: \ddot{\text{O}} : \text{Cl} : \ddot{\text{O}} :$$

There are four electrons left, which we place on the Cl.

$$: \ddot{\text{O}} : \ddot{\text{Cl}} : \ddot{\text{O}} :$$

All atoms have an octet. To indicate the charge, we write the structure of the ion as

$$
\left[: \ddot{\text{O}} : \ddot{\text{Cl}} : \ddot{\text{O}} : \right]^-
\quad \text{or} \quad
\left[: \ddot{\text{O}} - \ddot{\text{Cl}} - \ddot{\text{O}} : \right]^-
$$

EXAMPLE 7.9 WRITING LEWIS STRUCTURES

Problem: One of the oxoacids of chlorine is chloric acid, $HClO_3$. Give a reasonable Lewis structure for this molecule.

Solution: First, we choose a reasonable skeletal structure. Since the substance is an oxoacid, we expect the hydrogen to be bonded to an oxygen, which in turn would be bonded to the chlorine. The other two oxygens would also be bonded to the chlorine. This gives

$$\begin{array}{c} O \\ H \quad O \quad Cl \quad O \end{array}$$

The total number of electrons is 26 ($1e^-$ from H, $6e^-$ from each O, and $7e^-$ from Cl). We begin to distribute them in the structure by placing a pair in each bond.

$$\begin{array}{c} O \\ \ddot{} \\ H:O:Cl:O \end{array}$$

In Section 7.4 you learned that hydrogen only forms one covalent bond.

This still leaves $26 - 8 = 18e^-$. No additional electrons are needed around the H, because $2e^-$ are all that can occupy its valence shell. Therefore, we next complete the octets of the oxygens.

$$\begin{array}{c} :\ddot{O}: \\ H:\ddot{O}:Cl:\ddot{O}: \end{array}$$

We have now used a total of $24e^-$, so there are two electrons left. These are placed on the Cl to give

$$\begin{array}{c} :\ddot{O}: \\ H:\ddot{O}:\ddot{Cl}:\ddot{O}: \end{array}$$

which we can also write as

$$\begin{array}{c} :\ddot{O}: \\ | \\ H-\ddot{O}-Cl-\ddot{O}: \end{array}$$

The chlorine and the three oxygens have octets, and the hydrogen is completed with $2e^-$, so we are finished.

Sometimes, following the rules we've been using, you will find that there are either too few electrons to complete the octets of all the atoms, or there are electrons left over after all the octets have been filled. We have two rules to handle these problems:

1. When there are not enough electrons to give every atom an octet, multiple bonds must be created. (But remember that Be and B are exceptions.)
2. When electrons are left over, they are always placed on the central atom, in pairs.

The next three examples illustrate how this works.

EXAMPLE 7.10 WRITING LEWIS STRUCTURES

Problem: Write the Lewis structure for the air pollutant SO_3 (sulfur trioxide).

Solution: The skeletal structure is

Simply for convenience, we've arranged the oxygens symmetrically around the sulfur.

$$\begin{array}{c} O \\ O \quad S \\ O \end{array}$$

There are 24 valence electrons to be distributed in it. First, we place two electrons in each bond.

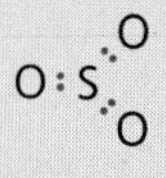

$$O:S:O$$ with O above and O below

There are $24 - 6 = 18e^-$ left. Next, we complete the octets of the oxygens.

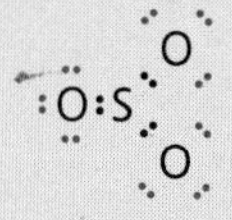

We're not finished yet, however, because there is not an octet around sulfur. We cannot simply add more dots because the total must be 24. The procedure that we follow, therefore, is to move a pair of electrons that we have shown to belong solely to an oxygen into a sulfur–oxygen bond so that they can also be counted as belonging to sulfur. In other words, we place a double bond between sulfur and one of the oxygens. It doesn't matter which oxygen we choose for this honor.

gives

We can also draw this structure as

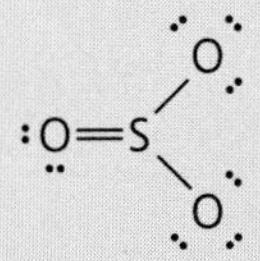

Notice that each atom has an octet.

EXAMPLE 7.11   WRITING LEWIS STRUCTURES

Problem: What is the Lewis structure for the carbon monoxide molecule, CO?

Solution: The skeletal structure is simply

$$C \quad O$$

There is a total of $4 + 6 = 10$ valence electrons. We begin with a pair in the CO bond.

$$C:O$$

Now we try to complete the octets with the remaining 8 electrons.

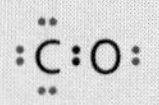

$$:\overset{..}{\underset{..}{C}}:\overset{..}{\underset{..}{O}}:$$

Obviously there aren't enough, which means that carbon monoxide must have a multiple bond of some sort. Therefore, we must shift enough electrons around so that oxygen also has an octet. Since the oxygen atom needs 4 more electrons, we must move two more pairs into the bond, which gives us a triple bond.

gives $:C:::O:$

Using dashes, the formula is

$$:C \equiv O:$$

EXAMPLE 7.12 WRITING LEWIS STRUCTURES

Problem: What is the Lewis structure for SF_4?

Solution: The skeletal structure is

$$\begin{array}{ccc} & F & \\ F & S & F \\ & F & \end{array}$$

and there must be $6 + 28 = 34$ electrons in the structure. First, we place 2 electrons in each bond, and then we complete the octets about the fluorine atoms. This uses 32 electrons.

$$\begin{array}{ccc} & :\overset{\cdot\cdot}{\underset{\cdot\cdot}{F}}: & \\ :\overset{\cdot\cdot}{\underset{\cdot\cdot}{F}}: & \overset{\cdot\cdot}{\underset{\cdot\cdot}{S}}: & :\overset{\cdot\cdot}{\underset{\cdot\cdot}{F}}: \\ & :\overset{\cdot\cdot}{\underset{\cdot\cdot}{F}}: & \end{array}$$

There are still 2 electrons left, even though all the atoms have an octet. Remember that if there are electrons left over after all of the octets have been completed, *they are all placed on the central atom in pairs.* This gives us

We've redrawn the molecule to make room for the extra electrons on sulfur.

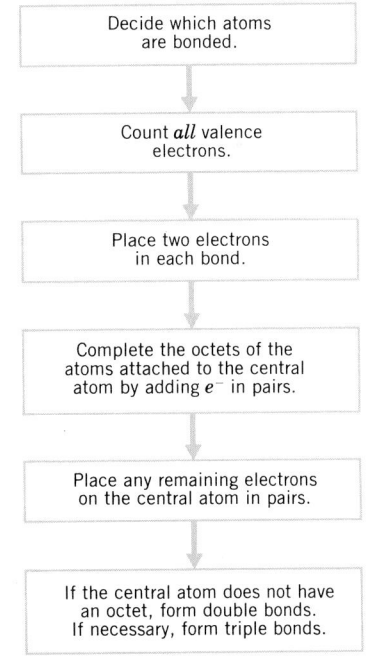

We see that, in the SF_4 molecule, sulfur violates the octet rule.

Figure 7.4 reviews the steps that we follow when we draw Lewis structures. Practice Exercise 7 will give you some practice in applying them.

FIGURE 7.4 Summary of steps in the writing of Lewis structures.

Decide which atoms are bonded.

↓

Count *all* valence electrons.

↓

Place two electrons in each bond.

↓

Complete the octets of the atoms attached to the central atom by adding e^- in pairs.

↓

Place any remaining electrons on the central atom in pairs.

↓

If the central atom does not have an octet, form double bonds. If necessary, form triple bonds.

PRACTICE EXERCISE 7   Draw Lewis structures for OF_2, NH_4^+, SO_2, NO_3^-, ClF_3, and $HClO_4$.

7.7 (RESONANCE:) WHEN LEWIS STRUCTURES FAIL

The bonding in some molecules and ions cannot be adequately described by a single Lewis structure.

Lewis structures are useful for keeping tabs on valence electrons, and most of the time the bonding pictures that they give seem reasonable with respect to properties that we can check experimentally. Two such properties are bond length, the distance between the nuclei of the bonded atoms, and bond energy, the energy required to separate the bonded atoms to give neutral particles. For example, as we mentioned in Section 7.3, measurement has shown the H_2 molecule to have a bond length of 75 pm (Figure 7.5) and a bond energy of 435 kJ/mol, which means that it takes 435 kJ to break the bonds of 1 mol of H_2 molecules to give 2 mol of hydrogen atoms.

Bond length and bond energy depend on the amount of electron density in the bond between two atoms. The higher the electron density, the more tightly the nuclei are held and the closer they are drawn together. We therefore should expect to find a double bond to be shorter and stronger than a single bond, because the electron density produced by two pairs of electrons is greater than that produced by one pair. Similarly, a triple bond should be shorter and stronger than a double bond. Such expectations are borne out by the data in Table 7.1, which gives typical bond lengths and bond energies for single, double, and triple carbon–carbon bonds.

Bond length = 75 pm

FIGURE 7.5 The hydrogen molecule.

TABLE 7.1 Average Bond Lengths and Bond Energies Measured for Carbon–Carbon Bonds

Bond	Bond Length (pm)	Bond Energy (kJ/mol)
C—C	154	348
C=C	134	615
C≡C	120	812

There are some molecules and ions for which we cannot write Lewis structures that agree with experimental measurements of bond length and bond energy. One example is the formate ion, CHO_2^-, which is produced by neutralizing formic acid, $HCHO_2$ (the substance that causes the stinging sensation in bites from fire ants). The skeletal structure for this ion is

Formic acid is an example of an organic acid. It has the structure

$$H \quad C \quad \begin{matrix} O \\ \\ O \end{matrix}$$

and, following the usual steps, we would write its Lewis structure as

$$\left[H-C \underset{\ddot{\underset{\cdot}{O}}\cdot}{\overset{\ddot{O}\cdot}{\diagup}} \right]$$

This structure suggests that one carbon–oxygen bond should be longer than the other, but experiment shows that they are identical. In fact, the C—O bond lengths are about halfway between that expected for a single bond and that expected for a double bond. The Lewis structure doesn't match the experimental evidence, and there's no way to write one that does. It would require showing all of the electrons in pairs and, at the same time, showing 1.5 pairs of electrons per bond.

The way we get around problems like that of the formate ion is through the use of a concept called resonance. We view the actual structure of the molecule or ion, which we cannot draw satisfactorily, as a composite, or average, of a number of Lewis structures that we can draw. For example, for formate we write

$$\left[\begin{array}{c} \ddot{O}: \\ \| \\ H-C-\ddot{O}: \end{array} \right]^{-} \longleftrightarrow \left[\begin{array}{c} :\ddot{O}: \\ | \\ H-C=\ddot{O}: \end{array} \right]^{-}$$

No atoms have been moved; the electrons have just been redistributed.

where we have simply shifted electrons around in going from one structure to another. The bond between the carbon and a particular oxygen is depicted as a single bond in one structure and as a double bond in the other. The average of these is 1.5 bonds — halfway between a single and a double bond — which is in agreement with experimental findings. These two Lewis structures are called resonance structures or contributing structures, and the actual structure of the ion is said to be a resonance hybrid of the two resonance structures that we have drawn. The double-ended arrow is drawn to show that the true hybrid structure is a composite of both resonance structures.

The term *resonance* is often misleading to the beginning student. The word itself suggests that the actual structure flip-flops back and forth between the two structures shown. This is *not* the case! A mule, which is the hybrid offspring of a donkey and a horse, isn't a donkey one minute and a horse the next! Although it may have characteristics of both parents, a mule is a mule. A resonance hybrid has characteristics of its "parents," but it never has the exact structure of any of them.

There is a simple way to determine when resonance should be applied to Lewis structures. If you find that you must move electrons to create a double bond while following the procedure developed in the previous section, the number of resonance structures is equal to the number of equivalent choices for the location of the double bond. For example, in drawing the Lewis structure for SO_3 following our procedure, we reach the stage

$$\begin{array}{c} :\ddot{O}: \\ \cdot S \cdot \\ :\ddot{O} \quad \ddot{O}: \end{array}$$

The three oxygens in SO_3 are said to be *equivalent:* that is, they are all alike in their chemical environment. The only thing that each oxygen is bonded to is the sulfur atom.

A double bond must be created to give sulfur an octet. Since it can be placed in any one of three positions, there are *three* resonance structures for this molecule.

$$\begin{array}{ccc} \ddot{O} \diagdown_S \diagup \ddot{O} & \ddot{O} \diagdown_S \diagup \ddot{O} & \ddot{O} \diagdown_S \diagup \ddot{O} \\ | & | & \| \\ :\ddot{O}: & :\ddot{O}: & :\ddot{O}: \end{array}$$

Notice that each structure is the same except for the location of the double bond. All atoms are in identical relative locations.

EXAMPLE 7.13 DRAWING RESONANCE STRUCTURES

Problem: Draw resonance structures for SO_2.

Solution: The SO_2 molecule has $6 + 12 = 18$ electrons, and the expected skeletal structure is

O S O

Placing electrons according to our procedure gives

$$:\overset{..}{O}:\overset{..}{S}:\overset{..}{O}:$$

To give sulfur an octet, we must move a pair of electrons to make a sulfur–oxygen double bond. There are two choices as to where this double bond can be placed, so there must be two resonance structures.

$$:\overset{..}{O}=S-\overset{..}{\underset{..}{O}}: \longleftrightarrow :\overset{..}{\underset{..}{O}}-S=\overset{..}{O}:$$

The real SO_2 molecule is a hybrid of these, and both S—O bonds have identical properties.

PRACTICE EXERCISE 8   Draw the resonance structures for the nitrate ion, NO_3^-.

7.8 COORDINATE COVALENT BONDING

Sometimes one atom supplies both of the electrons that are shared in a covalent bond.

When a hydrogen atom forms a covalent bond with a chlorine atom, the pair of electrons that they share between them is made up of one electron from each of the two atoms.

$$H\cdot + \cdot\overset{..}{\underset{..}{Cl}}: \longrightarrow H:\overset{..}{\underset{..}{Cl}}:$$

As we saw in Section 7.4, similar bonds are formed between a pair of chlorine atoms and when hydrogen forms bonds to carbon, nitrogen, and oxygen. Even in the N_2 molecule, the triple bond is formed from three electrons from each of the two nitrogen atoms. Sometimes, however, bonds are formed in which both electrons of the shared pair come from the same atom. Let's look at some examples.

When ammonia, NH_3, is placed into an acidic solution, it picks up a hydrogen ion, H^+, and becomes NH_4^+. Let's keep tabs on all of the electrons when this happens.

All electrons are alike, of course. The dots and crosses are used just so we can see where the electrons came from.

$$\begin{array}{c} H \\ \overset{\bullet\times}{} \\ H\overset{\times}{\bullet}N\overset{\bullet}{\underset{\bullet\times}{}}: + H^+ \longrightarrow \\ H \end{array} \left[\begin{array}{c} H \\ \overset{\bullet\times}{} \\ H\overset{\times}{\bullet}N\overset{\bullet}{\underset{\bullet\times}{}}H \\ H \end{array} \right]^+$$

The H^+ ion has a vacant valence shell that can accommodate two electrons. When the H^+ is bonded to the nitrogen of NH_3, the nitrogen donates both of the electrons to the bond. This type of bond, in which both electrons of the shared pair come from one of the two atoms, is called a coordinate covalent bond. Even though we make a distinction as to the origin of the electrons, once the bond is formed, it is really the same as any other covalent bond. We can't tell where the electrons in the bond came from after the bond has been formed. In NH_4^+, for instance, all four of the N—H bonds are equivalent once they have been formed, and no distinction is made between them. We usually write the structure of NH_4^+ as

$$\left[\begin{array}{c} H \\ | \\ H-N-H \\ | \\ H \end{array} \right]^+$$

Another example of a coordinate covalent bond occurs when a molecule having an incomplete valence shell reacts with a molecule having valence shell electrons that aren't being used in bonding.

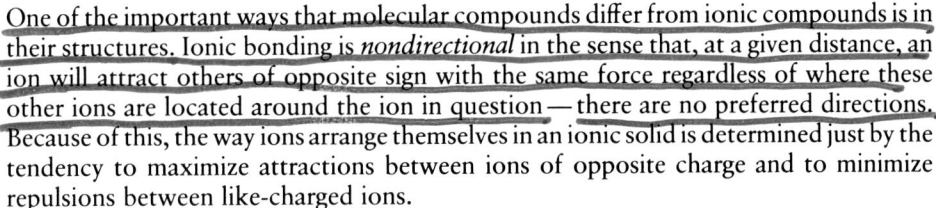

Compounds like BCl_3NH_3, which are formed by simply joining two smaller molecules, are called **addition compounds.**

An arrow sometimes is used to represent the donated pair of electrons in the coordinate covalent bond.

7.9 SHAPES OF MOLECULES: VSEPR THEORY

The three-dimensional shapes of molecules can be predicted if we assume that electron pairs in the valence shells of atoms stay as far apart as possible.

One of the important ways that molecular compounds differ from ionic compounds is in their structures. Ionic bonding is *nondirectional* in the sense that, at a given distance, an ion will attract others of opposite sign with the same force regardless of where these other ions are located around the ion in question—there are no preferred directions. Because of this, the way ions arrange themselves in an ionic solid is determined just by the tendency to maximize attractions between ions of opposite charge and to minimize repulsions between like-charged ions.

In molecular compounds, quite a different situation exists. Covalent bonds are highly *directional*. That is, for a given central atom in a molecule there are preferred orientations for the atoms attached to it, because covalent bonds are not formed with equal ease in all directions. As a result, in polyatomic molecules the atoms remain in the same relative orientations, regardless of whether the substance is a solid, liquid, or gas, and we can say that the molecules have a definite structure or shape.

The shapes of molecules are very important because many of their physical and chemical properties depend upon the three-dimensional arrangements of their atoms. For example, the functioning of enzymes, which are substances that control how fast biochemical reactions occur, requires that there be a very precise fit between one molecule and another. Even slight alterations in molecular geometry can destroy this fit and deactivate the enzyme, which in turn prevents the biochemical reaction involved from occurring. Nerve poisions work this way.

The best theoretical explanations of molecular shapes are based on quantum mechanics and the orbital pictures of electrons that we discussed in Chapter 6. We will have more to say about these later. There is a very simple theory, however, that is remarkably effective in predicting the shapes of molecules formed by the representative elements. It is called the valence shell electron pair repulsion theory (VSEPR theory, for short). Despite its long name, the principle behind the theory is very simple. The theory is based on the idea that valence shell electron pairs, being negatively charged, stay as far apart from each other as possible so that the repulsions between them are at a minimum. Let's look at an example.

Consider the $BeCl_2$ molecule. We've seen that its Lewis structure is

$$:\!\overset{..}{\underset{..}{Cl}}\!:Be:\!\overset{..}{\underset{..}{Cl}}\!:$$

But how are these atoms arranged? Is $BeCl_2$ linear or is it nonlinear, that is, do the atoms lie in a straight line, or do they form some angles less than 180°?

According to VSEPR theory, we can predict the shape of a molecule by looking at the electron pairs in the valence shell of the central atom. For $BeCl_2$, there are two pairs of electrons around the beryllium atom. The question is, How can they locate themselves to be as far apart as possible? The answer, of course, is that minimum repulsions will occur when the electron pairs are on opposite sides of the nucleus. We can represent this as

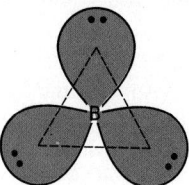

to suggest the approximate locations of the electron clouds of the valence shell electron pairs. In order for the electrons to be *in* the Be—Cl bonds, the Cl atoms must be placed where the electrons are; the result is that we predict that a $BeCl_2$ molecule should be linear.

$$Cl—Be—Cl$$

Let's look at another example, the BCl_3 molcule. Its Lewis structure is

$$
\begin{array}{c}
:\!\overset{..}{Cl}\!: \\[2pt]
:\!\overset{..}{\underset{..}{Cl}}\!:B:\!\overset{..}{\underset{..}{Cl}}\!:
\end{array}
$$

Here, the central atom has three electron pairs. What arrangement will lead to minimum repulsions? As you may have guessed, the electron pairs will be as far apart as possible when they are located at the corners of a triangle with the boron in the center.

When we attach the Cl atoms we obtain a triangular molecule.

$$
\begin{array}{c}
Cl \\
| \\
B \\
Cl^{\diagup}\ ^{\diagdown}Cl
\end{array}
$$

Actually, we say the shape is planar triangular, because all four atoms lie in the same plane. Experimentally, it's been proven that this is indeed the shape of BCl_3.

Number of Electron Pairs	Shape	Example

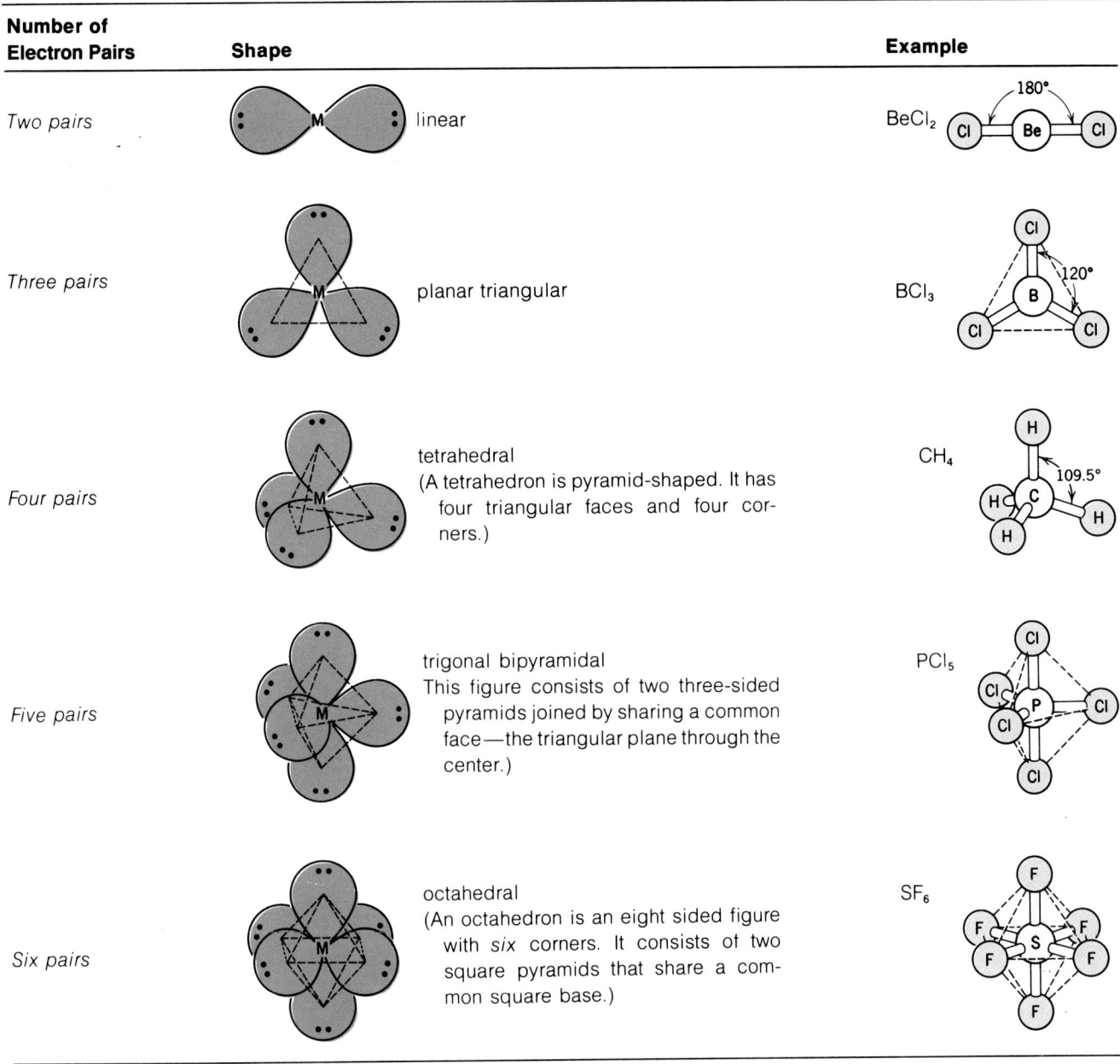

Two pairs	linear	$BeCl_2$
Three pairs	planar triangular	BCl_3
Four pairs	tetrahedral (A tetrahedron is pyramid-shaped. It has four triangular faces and four corners.)	CH_4
Five pairs	trigonal bipyramidal This figure consists of two three-sided pyramids joined by sharing a common face—the triangular plane through the center.)	PCl_5
Six pairs	octahedral (An octahedron is an eight sided figure with *six* corners. It consists of two square pyramids that share a common square base.)	SF_6

FIGURE 7.6 Shapes expected for different numbers of electron pairs.

Figure 7.6 shows the shapes expected for different numbers of electron pairs around the central atom. Some of these geometric figures may be new to you, so you should try especially hard to visualize them.

EXAMPLE 7.14 PREDICTING MOLECULAR SHAPES

Problem: Carbon tetrachloride was once used as a cleaning fluid, until it was discovered that it causes liver damage if absorbed by the body. What is the shape of the CCl_4 molecule?

Solution: To predict the molecular shape, we need the Lewis structure of the molecule.

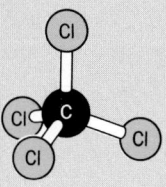

There are four electron pairs in the valence shell of carbon. These would have a minimum repulsion when arranged tetrahedrally, so the molecule would be tetrahedral.

PRACTICE EXERCISE 9   What would be the shape of $SbCl_5$?

Molecular Shapes When Some Electron Pairs Are Not in Bonds

In the molecules that we have considered so far, all the electron pairs of the central atom have been in bonds. This isn't always the case, though. Some molecules have a central atom with one or more pairs of electrons that are not shared with another atom. Still, because they are in the valence shell, these unshared electron pairs—also called lone pairs—affect the geometry of the molecule. An example is $SnCl_2$.

$$:\overset{..}{\underset{..}{Cl}}:Sn:\overset{..}{\underset{..}{Cl}}:$$

There are three pairs of electrons around the tin atom—the two in the bonds plus the lone pair. As in BCl_3, the mutual repulsions of the three pairs will place them at the corners of a triangle.

Adding on the two chlorine atoms gives

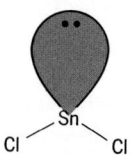

We can't describe this molecule as triangular, even though that is how the electron pairs are arranged. *Molecular shape describes the arrangement of atoms, not the arrangement of electron pairs.* Therefore, we describe the shape of the $SnCl_2$ molecule as being nonlinear or bent.

Notice that when there are three electron pairs around the central atom, two different molecular shapes are possible. If all three electron pairs are in bonds, as in BCl_3, a molecule with a planar triangular shape is formed. If one of the electron pairs is a lone pair, as in $SnCl_2$, the arrangement of the atoms in the molecule is said to be nonlinear. The predicted shapes of both, however, are derived by first noting the triangular arrangement of electron pairs around the central atom and *then* adding the necessary number of atoms.

The most common type of molecule or ion has four electron pairs — an octet — in the valence shell of the central atom. When these electron pairs are used to form four bonds, as in methane (CH_4), the resulting molecule is tetrahedral, as we've seen. There are many examples, however, where some of the pairs are lone pairs. For example,

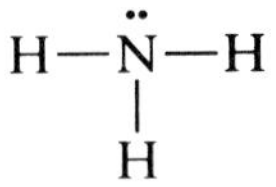

$$H - \overset{\cdot\cdot}{N} - H \qquad \text{1 lone pair}$$ 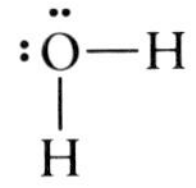
$$\overset{|}{H}$$

$$:\overset{\cdot\cdot}{O} - H \qquad \text{2 lone pairs}$$
$$\overset{|}{H}$$

Figure 7.7 shows how the lone pairs affect the shapes of molecules of this type.

Number of Pairs in Bonds	Number of Lone Pairs	Structure	
4	0	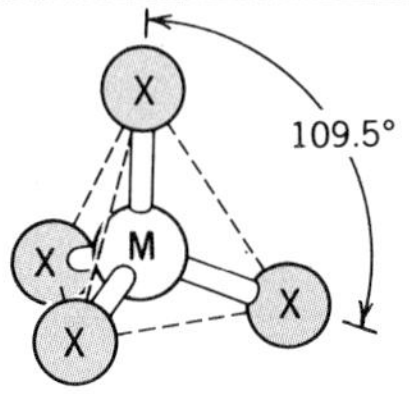	Tetrahedral (Example, CH_4) All bond angles are 109.5°.
3	1	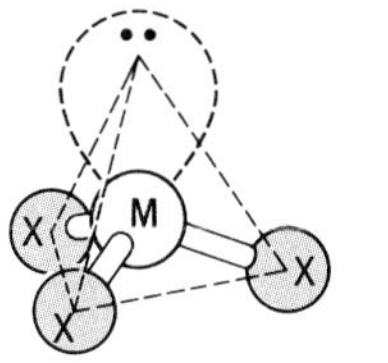	Trigonal pyramidal (Pyramid-shaped) (Example, NH_3)
2	2	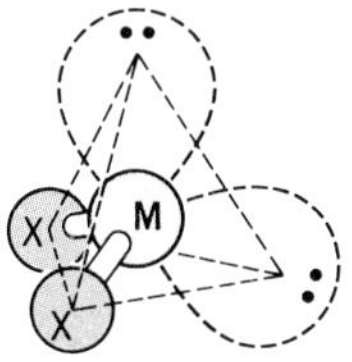	Nonlinear, bent (Example, H_2O)

FIGURE 7.7 Molecular shapes with four electron pairs around the central atom.

EXAMPLE 7.15   PREDICTING THE SHAPES OF MOLECULES AND IONS

Problem: Do we expect the ClO_2^- ion to be linear or nonlinear?

Solution: To apply VSEPR theory, we first need the Lewis structure for the ClO_2^- ion. Following our procedure, we obtain

$$\left[:\ddot{O}—\ddot{Cl}—\ddot{O}: \right]^-$$

Counting electron pairs, we see that there are four around the chlorine. Four electron pairs (according to the theory) are always arranged tetrahedrally. For the electron pairs, this gives

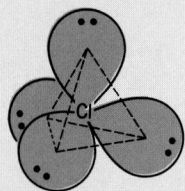

Now we add the two oxygens (it doesn't matter where in the tetrahedron).

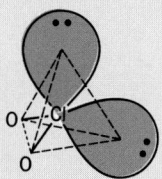

We see that the O—Cl—O angle is less than 180°, and the ion is therefore nonlinear.

The geometric figure is called a *trigonal bipyramid;* the molecule is described as having a *trigonal bipyramidal* structure.

When five electron pairs are present around the central atom, they are directed toward the corners of a trigonal bipyramid. Molecules such as PCl_5 have this geometry, as we saw in Figure 7.6. In the trigonal bipyramid, not all atoms have the same nearest-neighbor environments. Those at the top and bottom each have three neighbors at angles of 90° and one at 180°. Those in the triangular plane through the center of the molecule have two neighbors at 90° and two at 120°. When lone pairs occur in this structure, they always are located in the triangular plane. Figure 7.8 shows the kinds of geometries that we find for different numbers of lone pairs.

EXAMPLE 7.16   PREDICTING THE SHAPES OF MOLECULES AND IONS

Problem: Xenon is one of the noble gases, and is generally quite unreactive. In fact, it was long believed that all the noble gases were totally unable to form compounds. It came as quite a surprise, therefore, when it was discovered that some compounds could be made. One of these is xenon difluoride, XeF_2. What would we expect the geometry of XeF_2 to be?

Solution: To apply VSEPR theory, we need a Lewis structure. The outer shell of xenon, of course, has a noble gas structure, which contains 8 electrons. Each fluorine has 7 valence electrons, so the total number of electrons that must be placed in the structure is 22. These are distributed following our usual procedure, taking Xe as the central atom. The result is

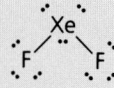

Next we count electron pairs around xenon — there are five of them. When there are five electron pairs, they are arranged in a trigonal bipyramid. (Continued on next page, bottom.)

FIGURE 7.8 Molecular shapes with five electron pairs around the central atom.

Number of Pairs in Bonds	Number of Lone Pairs	Structure	
5	0		Trigonal bipyramidal (Example, PCl_5)
4	1		Unsymmetrical tetrahedron (Example, SF_4)
3	2		T-Shaped (Example, ClF_3)
2	3		Linear (Example, I_3^-)

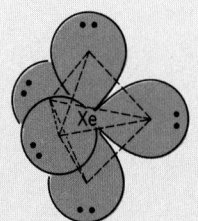

Now we must add the fluorine atoms. In a trigonal bipyramid, the lone pairs *always* occur in the triangular plane through the center, so the fluorines go on the top and bottom. This gives

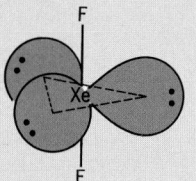

The three atoms, F—Xe—F, are arranged in straight line — the molecule is linear.

Finally, we come to those molecules or ions that have six electron pairs around the central atom. When all are in bonds, as in SF_6, the molecule is octahedral. When one lone pair is present, the molecule or ion has the shape of a square pyramid, and when two lone pairs are present the molecule or ion has a square planar structure. These shapes are shown in Figure 7.9.

No common molecule or ion with six electron pairs around the central atom has more than two of them as lone pairs.

FIGURE 7.9 Molecular shapes with six electron pairs around the central atom.

Number of Pairs in Bonds	Number of Lone Pairs	Structure	
6	0		Octahedral (Example, SF_6) All bond angles are 90°.
5	1		Square pyramidal (Example, BrF_5)
4	2		Square planar (Example, XeF_4)

EXAMPLE 7.17 PREDICTING THE SHAPES OF MOLECULES AND IONS

Problem: What is the probable geometry of the $XeOF_4$ molecule, which contains an oxygen atom and four fluorine atoms each bonded to xenon?

Solution: First we draw the Lewis structure.

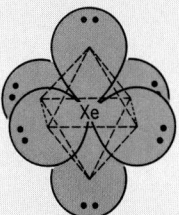

There are six electron pairs around xenon, and they must be arranged octahedrally.

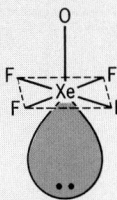

Next we attach the oxygen and four fluorines. The most symmetrical structure is

which we would describe as a square pyramid. Although the oxygen might, in principle, be placed in one of the positions at the corners of the square base (in place of one of the fluorines), the structure shown happens to be the actual structure of the molecule.

Shapes of Molecules and Ions with Double or Triple Bonds

Our discussion to this point has dealt only with the shapes of molecules or ions having single bonds. Luckily, the presence of double or triple bonds does not complicate matters at all. In a double bond, both electron pairs must stay together between the two atoms; they can't wander off to different locations in the valence shell. This is also true for the three pairs of electrons in a triple bond. *For the purposes of predicting molecular geometry, then, we can treat double and triple bonds just as we do single bonds.* For example, the Lewis formula for CO_2 is

$$\ddot{O}=C=\ddot{O}$$

The carbon atom has no lone pairs. Therefore, the two groups of electron pairs that make up the double bonds are located on opposite sides of the nucleus and a linear molecule is formed. Similarly, we would predict the following shapes for SO_2 and SO_3.

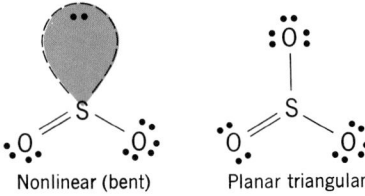

Nonlinear (bent)   Planar triangular

We only need to consider one of the resonance structures to reach a conclusion about molecular structure.

EXAMPLE 7.18   PREDICTING THE SHAPES OF MOLECULES AND IONS

Problem: The Lewis structure for the very poisonous gas hydrogen cyanide, HCN, is

$$H—C≡N:$$

Is HCN a linear or nonlinear molecule?

Solution: The triple bond behaves like a single bond for the purposes of predicting overall geometry. Therefore, the triple and single bonds would locate themselves 180° apart, and HCN would be linear.

PRACTICE EXERCISE 10   Predict the geometry of ClO_3^-, XeO_4, and OF_2.

PRACTICE EXERCISE 11   Predict the geometry of CO_3^{2-}.

7.10 POLAR BONDS AND ELECTRONEGATIVITY

The relative abilities of bonded atoms to attract electrons determines whether or not the electrons in the bond are shared equally.

When two identical atoms form a covalent bond, as in H_2 or Cl_2, each has an equal share of the electron pair in the bond. The electron density at both ends of the bond is the same, because the electrons are equally attracted to both nuclei. However, when different kinds of atoms combine, as in HCl, the attractions usually are not equal. Generally, one of the nuclei attracts the electrons more strongly than the other.

The effect of unequal attractions for the bonding electrons is an unbalanced distribution of electron density within the bond. For example, it has been found that a chlorine atom attracts electrons more strongly than does a hydrogen atom. In the HCl molecule, therefore, the electron cloud is pulled more tightly around the Cl, and that end of the molecule experiences a slight buildup of negative charge. The electron density that shifts toward the chlorine is removed from the hydrogen, which causes the hydrogen end to acquire a slight positive charge.

In HCl, electron transfer is incomplete. The electrons are still shared, but unequally. The charges on either end of the molecule are less than full 1 + and 1 − charges — they are partial charges, normally indicated by the lowercase Greek letter delta, δ (see Figure 7.10). (In HCl, for example, the hydrogen carries a partial charge of +0.17 and the

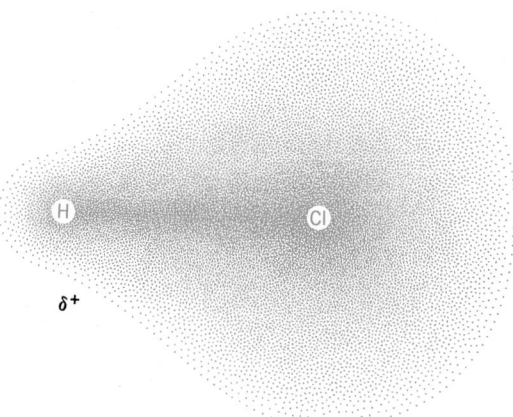

δ^+

δ^-

FIGURE 7.10 Unsymmetrical distribution of electrons in the HCl bond.

chlorine a partial charge of -0.17.) Partial charges can also be indicated on Lewis structures. For example,

$$H-\overset{..}{\underset{..}{Cl}}\colon$$
$$\delta+ \quad \delta-$$

A bond that carries partial positive and negative charges on opposite ends is called a polar bond, or a polar covalent bond. The term *polar* comes from the notion of poles of opposite charge at either end of the bond. Because there are two poles of charge involved, the bond is said to be a dipole.

The polar bond in HCl causes the molecule as a whole to have opposite charges on either end, so we say that HCl is a polar molecule. The HCl molecule as a whole is also a dipole, and the extent of its polarity is expressed quantitatively through its dipole moment, which is found by multiplying the amount of charge on either end by the distance between the charges.

The degree to which a covalent bond is polar depends on the relative abilities of the bonded atoms to attract electrons. If one of the atoms has a strong attraction for electrons and the other attracts electrons weakly, then the bond will be very polar because the electron density will be shifted to a large extent toward the atom that attracts electrons the most. On the other hand, if the differences in the bonded atoms are small, then the electrons will be shared almost equally.

The term that we use to describe the relative attraction of an atom for the electrons in a bond is called the electronegativity of the atom. In HCl, for example, chlorine is more electronegative than hydrogen. The electron pair of the covalent bond spends more of its time around the more electronegative atom, which is why that end of the bond acquires a partial negative charge.

The concept of electronegativity has been put on a quantitative basis — numerical values have been assigned for each element, as shown in Figure 7.11. This information is useful because the *difference* in electronegativity provides an estimate of the degree of polarity of a bond. In addition, the relative magnitudes of the electronegativities indicate which end of the bond carries the negative charge. For instance, fluorine is more electronegative than chlorine. Therefore, we would expect an HF molecule to be more polar than an HCl molecule. In addition, hydrogen is less electronegative than either

Experimental measurement of dipole moments is one way to check our theoretical models.

Linus Pauling, winner of two Nobel Prizes, was the first scientist to set up a table of electronegativity values.

FIGURE 7.11 Table of electronegativities.

H 2.1																	
Li 1.0	Be 1.5											B 2.0	C 2.5	N 3.1	O 3.5	F 4.1
Na 1.0	Mg 1.3											Al 1.5	Si 1.8	P 2.1	S 2.4	Cl 2.9
K 0.9	Ca 1.1	Sc 1.2	Ti 1.3	V 1.5	Cr 1.6	Mn 1.6	Fe 1.7	Co 1.7	Ni 1.8	Cu 1.8	Zn 1.7	Ga 1.8	Ge 2.0	As 2.2	Se 2.5	Br 2.8
Rb 0.9	Sr 1.0	Y 1.1	Zr 1.2	Nb 1.3	Mo 1.3	Tc 1.4	Ru 1.4	Rh 1.5	Pd 1.4	Ag 1.4	Cd 1.5	In 1.5	Sn 1.7	Sb 1.8	Te 2.0	I 2.2
Cs 0.9	Ba 0.9	La 1.1	Hf 1.2	Ta 1.4	W 1.4	Re 1.5	Os 1.5	Ir 1.6	Pt 1.5	Au 1.4	Hg 1.5	Tl 1.5	Pb 1.6	Bi 1.7	Po 1.8	At 2.0
Fr 0.9	Ra 0.9	Ac 1.0	Lanthanides: 1.0–1.2													
			Actinides: 1.0–1.2													

fluorine or chlorine, so in both of these molecules the hydrogen bears the positive charge.

$$H\!-\!\ddot{\underset{..}{F}}\colon \qquad H\!-\!\ddot{\underset{..}{Cl}}\colon$$
$$\delta + \quad \delta - \qquad \delta + \quad \delta -$$

PRACTICE EXERCISE 12   In each of the following bonds, choose the atom that carries the partial negative charge: (a) P—Br, (b) Si—Cl, (c) S—Cl.

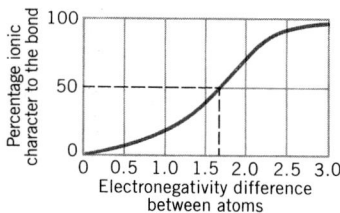

FIGURE 7.12 Variation of percentage ionic character with electronegativity difference.

The concept of electronegativity shows us that there is no sharp dividing line between ionic and covalent bonding. Ionic bonding and nonpolar covalent bonding simply represent the two extremes. Ionic bonding occurs when the difference in electronegativity between two atoms is very large; the more electronegative atom acquires essentially complete control of the bonding electrons. In a nonpolar covalent bond, there is no difference in electronegativity, so the pair of bonding electrons is shared equally.

$$Cs^{+}\left[\ddot{\underset{..}{\textrm{:}F\textrm{:}}}\right]^{-} \qquad\qquad \ddot{\underset{..}{\textrm{:}F}}\overset{..}{\underset{..}{\textrm{ }}}\ddot{\underset{..}{F\textrm{:}}}$$

"bonding pair" held   bonding pair
exclusively by fluorine   shared equally

The degree of polarity, which we might think of as the amount of ionic character of the bond, varies in a continuous way with changes in the electronegativity difference (Figure 7.12). The bond becomes more than 50% ionic when the electronegativity difference exceeds 1.7.

Within the periodic table, electronegativity varies in a more or less systematic way, and the trends follow those for the ionization energy (IE) discussed in Chapter 6. Thus, atoms with large ionization energies also have large electronegativities. This really shouldn't be surprising. An atom that has a small IE will give away an electron more easily than an atom with a large IE, just as an atom with a small electronegativity will lose its share of an electron pair more readily than an atom with a large electronegativity.

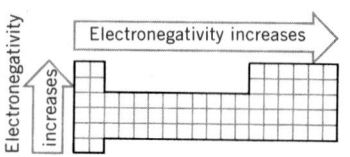

Metals have low electronegativities; nonmetals have high electronegativities.

As you know, sodium (Group IA) combines with chlorine (upper right-hand corner of the table) to give ionic NaCl.

The figure at the left illustrates the trends in electronegativity within the periodic table — electronegativity increases from bottom to top in a group, and form left to right in a period. Elements in the same region of the table (for example, the nonmetals) have similar electronegativities, which means that if they form bonds with each other, the electronegativity differences will be small and the bonds will be mostly covalent. On the other hand, if elements from widely separated regions of the table react, large electronegativity differences will occur and the bonds will be predominantly ionic. This is what occurs, for example, when an element from Group IA or Group IIA reacts with an element from the upper right-hand corner of the periodic table.

7.11 MOLECULAR SHAPES AND THE POLARITY OF MOLECULES

Molecular shape is a determining factor in controlling whether a molecule with polar bonds is a polar molecule.

One of the main reasons we are concerned about the polarity of molecules is because many physical properties, such as melting point and boiling point, are affected by it. This is because polar molecules attract each other — the positive end of one polar molecule

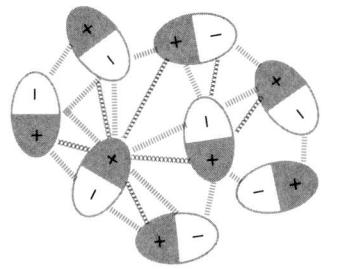

|||||||| Attractions
xxxxxxx Repulsions

FIGURE 7.13 Attractions between dipoles. Molecules tend to orient themselves so that the positive end of one dipole is near the negative end of another.

attracts the negative end of another (Figure 7.13). The strength of the attraction depends both on the amount of charge on either end of the molecule and on the distance between the charges — in other words, it depends on the molecule's dipole moment.

The dipole moment of a molecule is a property that can be determined experimentally, and when this is done, an interesting observation is made. There are many molecules that have no dipole moment even though they contain bonds that are polar. Stated differently, they are nonpolar molecules, even though they have polar bonds. The reason for this can be seen if we examine the key role that molecular structure plays in determining molecular polarity.

Let's begin with the H—Cl molecule, which has only two atoms and therefore only one bond. As you've learned, this bond is polar, and opposite ends of the bond carry partial charges of opposite sign. Because there are only two atoms in the molecule, which are located at the ends of the bond, the molecule as a whole has ends with equal but opposite charges. A molecule with equal but opposite charges on opposite ends is polar, so HCl is a polar molecule. In fact, any molecule composed of just two atoms that differ in electronegativity must be polar.

For molecules that contain more than two atoms, we have to consider the combined effects of all the polar bonds. Sometimes, when all the atoms attached to the central atom are the same, the effects of the individual polar bonds cancel and the molecule as a whole is nonpolar. Some examples are shown in Figure 7.14. In this figure the dipoles associated with the bonds themselves — the bond dipoles — are shown as arrows crossed at one end, ↦. The arrow points in the direction of the negative end of the bond dipole and the crossed end is the positive end.

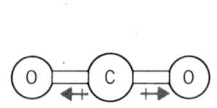

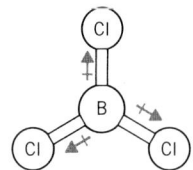

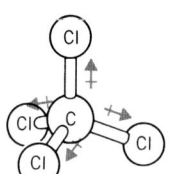

FIGURE 7.14 In symmetric molecules the bond dipoles cancel to give nonpolar molecules.

In CO_2 both bonds are identical, so each bond dipole is of the same magnitude. Because CO_2 is a linear molecule, these bond dipoles point in opposite directions and work against each other. The net result is that their effects cancel, and CO_2 is a nonpolar molecule. Although it isn't as easy to visualize, the same thing also happens in BCl_3 and CCl_4. In each of these molecules the effects of one of the bond dipoles is cancelled by the effects of the others.

Perhaps you've noticed that the structures of the molecules in Figure 7.14 correspond to three of the basic geometries that we used in the VSEPR theory to derive the shapes of molecules. Molecules with the remaining two structures — trigonal bipyramidal and octahedral — also are nonpolar if all the atoms attached to the central atom are

Bond dipoles can be treated as vectors, and the polarity of a molecule is predicted by taking the vector sum of the bond dipoles. If you've studied vectors before, this may help your understanding.

FIGURE 7.15 *(a)* A molecule with the general formula AX_5 and a trigonal bipyramidal structure. The set of three bond dipoles in the triangular plane in the center (in blue) cancel, as do the linear set of dipoles (red). The molecule is nonpolar. *(b)* An octahedral molecule AX_6 consists of three sets of linear pairs of bond dipoles. Concellation occurs for each set, so the molecule is nonpolar.

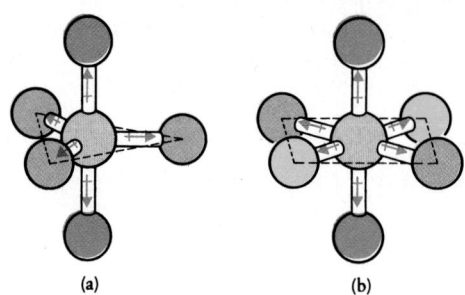

(a)   (b)

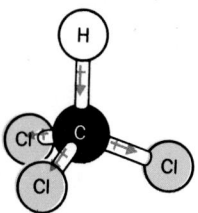

FIGURE 7.16 Bond dipoles in the chloroform molecule, $CHCl_3$. The small bond dipole of the C—H bond actually adds to the effects of the C—Cl bond dipoles. This causes $CHCl_3$ to be a polar molecule.

Because C is slightly more electronegative than H, the C—H bond dipole points toward the carbon and adds to the effect of the C—Cl bond dipoles.

the same. Examples are shown in Figure 7.15. The trigonal bipyramidal structure can be viewed as a planar triangular set of atoms (shown in blue) plus a pair of atoms arranged linearly (shown in red). All the bond dipoles in the planar triangle cancel, as do the two dipoles of the bonds arranged linearly, so the molecule is nonpolar. Similarly, we can look at the octahedral molecule as consisting of three sets of linear arrangements of bond dipoles. Cancellation of bond dipoles occurs in each set, so the octahedral molecule is also nonpolar.

If all the atoms attached to the central atom are not the same, or if there are lone pairs in the valence shell of the central atom, the molecule is usually polar. For example, in $CHCl_3$, one of the atoms in the tetrahedral structure is different from the others. The C—H bond is less polar than the C—Cl bonds, and the bond dipoles do not cancel (Figure 7.16).

Two familiar molecules that have lone pairs in the valence shell of the central atom are shown in Figure 7.17. Here the bond dipoles are oriented in such a way that their effects do not cancel. In water, for example, each bond dipole points partially in the same direction, toward the oxygen atom, and because of this the bond dipoles partially add to give a net dipole moment for the molecule. The same thing happens in ammonia where three bond dipoles point partially in the same direction and add to give a polar NH_3 molecule.

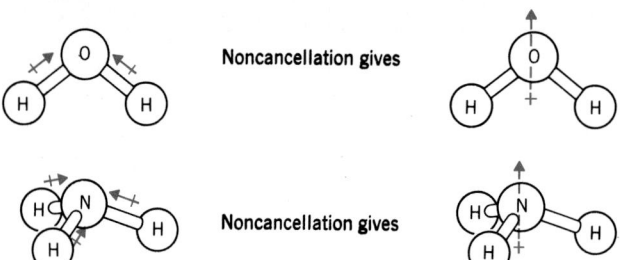

Noncancellation gives

Noncancellation gives

FIGURE 7.17 When lone pairs occur on the central atom, the bond dipoles usually do not cancel, and polar molecules result.

Not every structure that contains lone pairs on the central atom produces polar molecules. The following are two exceptions.

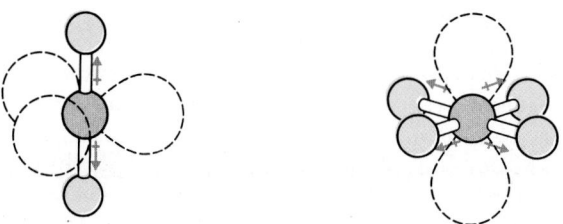

In the first case, we have a pair of bond dipoles arranged linearly, just as in CO_2. In the second, the bonded atoms lie at the corners of a square, which can be viewed as two sets of linear arrangements of bond dipoles. If the atoms attached to the central atom are the same, cancellation of bond dipoles is bound to occur and produce nonpolar molecules. This means that molecules such as linear XeF_2 and square planar XeF_4 are nonpolar.

Based on the preceding discussions, let's see how we can use VSEPR theory to determine whether molecules are expected to be polar or nonpolar.

EXAMPLE 7.19 PREDICTING MOLECULAR POLARITY

Problem: Would we expect the molecule PCl_3 to be polar or nonpolar?

Solution: The electronegativities of the atoms (P = 2.1, Cl = 2.9) tell us that the individual P—Cl bonds will be polar. Therefore, if we can predict the structure of the molecule, we should be able to predict whether or not the molecule is polar. First we draw the Lewis structure following our usual procedure.

There are four electron pairs around the phosphorus, so according to VSEPR theory, they should be arranged tetrahedrally. This means that the PCl_3 molecule should have a trigonal pyramidal shape, as shown below. Because of the structure, the bond dipoles do not cancel, and we expect that the molecule will be polar.

EXAMPLE 7.20 PREDICTING MOLECULAR POLARITY

Problem: Would you expect the molecule SO_3 to be polar or nonpolar?

Solution: Oxygen is more electronegative than sulfur, so we expect the S—O bonds to be polar. Let's look at the molecular structure next to see if the bond dipoles cancel. The Lewis structure is

This is one of three resonance structures. All the S—O bonds are equivalent.

VSEPR theory tells us that the molecule should have a planar triangular shape. We saw in Figure 7.14 that such a molecule is nonpolar when all the attached atoms are the same, because the bond dipoles cancel.

EXAMPLE 7.21 PREDICTING MOLECULAR POLARITY

Problem: Would you expect the molecule HCN to be polar or nonpolar?

Solution: Once again we have polar bonds because carbon is slightly more electronegative than hydrogen and nitrogen is slightly more electronegative than carbon. The Lewis structure of HCN is

$$H—C\equiv N\!:$$

VSEPR theory predicts a linear shape, but the two bond dipoles do not cancel. One reason is that they are not of equal magnitude, which we know because the difference in electronegativity between C and H is 0.4, whereas the difference in electronegativity between C and N is 0.6. The other reason is because both bond dipoles point in the same direction — from the atom of low electronegativity to the one of high electronegativity. This is illustrated below.

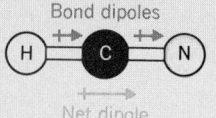

Bond dipoles

Net dipole

Notice that the bond dipoles add to give a net dipole moment for the molecule.

PRACTICE EXERCISE 13   Which of the following molecules would be expected to be polar? (a) SF_6, (b) SO_2, (c) BrCl, (d) AsH_3, (e) CF_2Cl_2.

SUMMARY

Ionic Bonding.   The forces of attraction between positive and negative ions in an ionic compound are called ionic bonds. The formation of ionic compounds by electron transfer is favored when atoms of low ionization energy react with atoms of high electron affinity, that is, when metals combine with nonmetals. The chief stabilizing influence in the formation of ionic compounds is the lattice energy — the energy needed to separate the ions completely, or the energy lowering that occurs when gaseous ions are brought together to form the crystal of the ionic compound. When atoms of the representative elements form ions, they usually gain or lose enough electrons to achieve a noble gas electron configuration.

Covalent Bonding.   Electron sharing between atoms occurs when electron transfer is energetically too "expensive." Shared electrons attract the positive nuclei and this leads to a lowering of the energy of the atoms as a covalent bond forms. Electrons generally become paired when they are shared. An atom tends to share enough electrons to complete its valence shell. Except for hydrogen, the valence shell usually holds eight electrons, which forms the basis of the octet rule. The octet rule states that atoms, particularly those of the representative elements, tend to acquire eight electrons in their outer shells when they form bonds. Single, double, and triple bonds involve atomic nuclei sharing one, two, and three pairs of electrons, respectively. Some molecules don't obey the octet rule. Compounds of B and Be often have less than an octet around the central atom. Atoms of the elements of period 2 cannot have more than an octet because their outer shells can hold only eight electrons. Elements in period 3 and below can exceed an octet if they form more than four bonds.

Bond energy (the energy needed to separate the bonded atoms) and bond length (the distance between the nuclei of the atoms connected by the bond) are two experimentally measurable quantities that can be related to the number of pairs of electrons in the bond. For bonds between the same atoms, bond energy increases and bond length decreases as we go from single to double to triple bonds.

Lewis Symbols and Lewis Structures.   Lewis symbols are a bookkeeping device used to keep track of valence electrons during bond formation. The Lewis symbol of an element consists of the element's chemical symbol surrounded by a number of dots equal to the number of valence electrons. The Lewis structure for a molecule or polyatomic ion uses pairs of dots between chemical symbols to represent shared pairs of electrons. Electron pairs in covalent bonds usually are represented by dashes — one dash equals two electrons. The following procedure is used to draw the Lewis structure: (1) decide on the skeletal structure (remember that the central atom is usually first in the formula); (2) count all the valence electrons, taking into account the charge, if any; (3) place a pair of electrons in each bond; (4) complete the octets of atoms other than the central atom (but remember that hydrogen can only have two electrons); (5) place any left over electrons on the central atom in pairs; (6) if the central atom still lacks an octet, move electron pairs to make double or triple bonds.

Resonance.   Two or more atoms in a molecule or polyatomic ion are chemically equivalent if they are attached to the same kinds of atoms or groups of atoms. Bonds to chemically equivalent atoms must be the same — they must have the same bond length and the same bond energy, which means they must involve the sharing of the same number of electron pairs. Sometimes the Lewis structures we draw suggest that the bonds to chemically equivalent atoms are not the same. Typically, this occurs when it is necessary to form multiple bonds during the construction of a Lewis structure. When alternatives exist for the location of this multiple bond among two or more equivalent atoms, then each possible Lewis structure is actually a resonance structure or contributing structure, and we draw them all. In drawing resonance structures, the relative locations of the nuclei must be identical in all. Remember that none of the resonance structures corresponds to a real molecule, but their composite — the resonance hybrid — does approximate the actual structure of the molecule or ion.

Coordinate Covalent Bonding. For bookkeeping purposes, we sometimes single out a covalent bond whose electron pair originated from one of the two bonded atoms. An arrow is sometimes used to indicate the donated pair of electrons. Once formed, a coordinate covalent bond is no different than any other covalent bond.

VSEPR Theory. Molecular geometry can be predicted by assuming that electron pairs in the valence shell of an atom stay as far apart as possible. Figures 7.6 through 7.9, along with the Lewis structure, usually give the correct structure of the molecule or polyatomic ion.

Polar Molecules and Electronegativity. Electronegativity is the attraction an atom has for electrons in a bond. When two atoms

of different electronegativity form a covalent bond, the electrons in the bond are shared unequally and the bond is polar; it has opposite partial charges on opposite ends. A nonpolar bond is formed when the two atoms have the same electronegativity, and an ionic bond is formed when the difference in electronegativity is very large. A covalent bond is more than 50% ionic if the electronegativity difference is larger than 1.7. A diatomic molecule is polar if its bond is polar. The extent of polarity is determined by the dipole moment of the molecule — the magnitude of the partial charge on either end multiplied by the distance between the partial charges. In polyatomic molecules, bond dipoles can cancel to give nonpolar molecules if (a) all the atoms attached to the central atom are alike, and (b) if the molecule has one of the structures in Figure 7.6. When lone pairs are in the valence shell of the central atom, the molecule is usually polar. Two exceptions are shown on page 258.

REVIEW EXERCISES

Answers to questions whose numbers are printed in color are given in Appendix C.

Ionic Bonding

7.1 What must be true about the total energy of a collection of atoms for a stable compound to be formed from them?

7.2 What is an *ionic bond?*

7.3 How is the tendency to form ionic bonds related to the IE and EA of the atoms involved?

7.4 What is the lattice energy? In what ways does it contribute to the stability of ionic compounds?

7.5 Explain what happens to the electron configurations of Mg and Br when they react to form the compound $MgBr_2$.

7.6 Describe what happens to the electron configuration of a nitrogen atom when it forms the nitride ion, N^{3-}.

7.7 Magnesium forms the ion Mg^{2+}, but not the ion Mg^{3+}. Why?

7.8 Why doesn't chlorine form the ion Cl^{2-}?

7.9 Why do many of the transition elements in period 4 form ions with a 2+ charge?

Lewis Symbols

7.10 Write Lewis symbols for the following atoms: (a) Si, (b) Sb, (c) Ba, (d) Al.

7.11 Use Lewis symbols to diagram the reactions between (a) Ca and Br, (b) Al and O, and (c) K and S.

Electron Sharing

7.12 In terms of energy, why doesn't ionic bonding occur when two nonmetals react with each other?

7.13 Describe what happens to the electron density around two hydrogen atoms as they come together to form an H_2 molecule.

7.14 What holds the two nuclei together in a covalent bond?

7.15 What happens to the energy of two hydrogen atoms as they approach each other?

7.16 What factors determine the bond distance in a covalent bond?

7.17 Why is a covalent bond also often called an electron pair bond?

7.18 Is bond formation endothermic or exothermic?

7.19 How much energy, in joules, is required to break the bond in one hydrogen molecule? The bond energy of H_2 is 435 kJ/mol.

7.20 How many grams of water could have their temperature raised from 25 °C (room temperature) to 100 °C (the boiling point of water) by the amount of energy released in the formation of 1 mol of H_2 from hydrogen atoms?

Covalent Bonding and the Octet Rule

7.21 Use Lewis structures to diagram the formation of (a) Br_2, (b) H_2O, and (c) NH_3 from neutral atoms.

7.22 What is the *octet rule?* What is responsible for it?

7.23 Use the octet rule to predict the formula of the simplest compound formed from hydrogen and (a) Se, (b) As, and (c) Si. (Remember, however, that the valence shell of hydrogen can hold only two electrons.)

7.24 What would be the formula for the simplest compound formed from (a) P and Cl, (b) C and F, and (c) Br and Cl?

7.25 Why do period-2 elements never form more than four covalent bonds? Why are period-3 elements able to exceed an octet when they form bonds?

7.26 Define (a) *single bond,* (b) *double bond,* and (c) *triple bond.*

7.27 What is a structural formula?

7.28 Hydrogen cyanide has the Lewis structure H—C≡N:. Draw circles enclosing electrons to show that carbon and nitrogen obey the octet rule.

7.29 Why doesn't hydrogen obey the octet rule? How many covalent bonds does a hydrogen atom form?

Failure of the Octet Rule

7.30 How many electrons are in the valence shells of (a) Be in $BeCl_2$, (b) B in BCl_3, and (c) H in H_2O?

7.31 What is the minimum number of electrons that would be expected to be in the valence shell of As in $AsCl_5$?

Drawing Lewis Structures

7.32 What would be the skeletal structures for (a) $SiCl_4$, (b) PF_3, (c) PH_3, and (d) SCl_2?

7.33 How many dots must appear in the Lewis structures of the molecules in Review Exercise 7.32?

7.34 Give Lewis structures for the molecules in Review Exercise 7.32.

7.35 Give Lewis structures for (a) CS_2, (b) CN^-, (c) SeO_3, and (d) SeO_2.

7.36 Give Lewis structures for (a) HNO_2, (b) $HClO_2$, and (c) H_2SeO_3.

7.37 Give Lewis structures for (a) NO^+, (b) NO_2^-, (c) $SbCl_6^-$, and (d) IO_3^-. Which contain multiple bonds?

7.38 Give Lewis structures for (a) TeF_4, (b) ClF_5, (c) XeF_2, and (d) XeF_4.

7.39 Give Lewis structures for (a) ClO_4^-, (b) AsH_3, (c) PCl_4^+, and (d) PCl_6^-.

7.40 Draw the Lewis structure for formaldehyde, CH_2O. The central atom in the molecule is carbon, which is attached to two hydrogens and an oxygen.

7.41 Draw Lewis structures for (a) $GeCl_4$, (b) CO_3^{2-}, (c) PO_4^{3-}, and (d) O_2^{2-}.

Bond Length and Bond Energy

7.42 Define *bond length* and *bond energy*.

7.43 How should the N—O bond lengths compare in the ions NO_3^- and NO_2^-?

7.44 The energy required to break the H—Cl bond to give H^+ and Cl^- ions would not be called the H—Cl bond energy. Why?

7.45 How are bond energy and bond length related to the number of pairs of electrons shared between atoms in a bond?

Resonance

7.46 Draw the resonance structures for CO_3^{2-}.

7.47 Arrange the following in order of increasing C—O bond length: CO, CO_3^{2-}, CO_2.

7.48 How many resonance structures could be written for the N_2O_4 molecule? Its skeletal structure is

$$
\begin{array}{ccc}
O & & O \\
 & N \quad N & \\
O & & O
\end{array}
$$

7.49 The Lewis structure of CO_2 was given as $:\ddot{O}{=}C{=}\ddot{O}:$; but two other resonance structures can also be drawn for it. What are they?

7.50 What is a resonance hybrid? How does it differ from the resonance structures drawn for a molecule?

7.51 Write Lewis structures for the nitrate ion, NO_3^-, and for nitric acid, HNO_3. How many equivalent oxygens are there in each structure? Compare the number of resonance structures for each.

Coordinate Covalent Bonds

7.52 What is a coordinate covalent bond?

7.53 Once formed, how (if at all) does a coordinate covalent bond differ from an ordinary covalent bond?

7.54 BCl_3 has an incomplete valence shell. Show how it could form a coordinate covalent bond with a water molecule.

7.55 Use Lewis structures to show that the hydronium ion, H_3O^+, can be considered to be formed by the creation of a coordinate covalent bond between H_2O and H^+.

7.56 What is an *addition compound*?

VSEPR Theory

7.57 What arrangements of electron pairs are expected when the valence shell of the central atom contains (a) three pairs, (b) six pairs, (c) four pairs, or (d) five pairs of electrons?

7.58 Practice drawing the tetrahedron, the trigonal bipyramid, and the octahedron.

7.59 What molecular shapes are expected when the central atom has in its valence shell (a) three bonding electron pairs and one lone pair, (b) four bonding electron pairs and two lone pairs, and (c) two bonding electron pairs and one lone pair?

7.60 Predict the shapes of (a) FCl_2^+, (b) AsF_5, (c) AsF_3, and (d) SeO_2.

7.61 Predict the shapes of (a) TeF_4, (b) $SbCl_6^-$, (c) NO_2^-, and (d) PO_4^{3-}.

7.62 Predict the shapes of (a) IO_4^-, (b) ICl_4^-, (c) TeF_6, and (d) ICl_2^-.

7.63 Predict the shapes of (a) CS_2, (b) BrF_4^-, (c) ICl_3, and (d) SeO_3.

7.64 Preduct the shapes of (a) SbH_3, (b) PCl_4^+, (c) ClO_3^-, and (d) SiO_4^{4-}.

7.65 Acetylene, a gas used in welding torches, has the Lewis structure H—C≡C—H. What would you expect the H—C—C bond angle to be in this molecule?

7.66 Ethylene, a gas used to ripen tomatoes artificially, has the Lewis structure

$$
\begin{array}{cc}
H & H \\
| & | \\
H{-}C & {=}C{-}H
\end{array}
$$

What would you expect the H—C—H and H—C—C bond angles to be in this molecule? (*Caution:* Don't be fooled by the way the structure is drawn here.)

7.67 Formaldehyde has the Lewis structure

$$
\begin{array}{c}
H \\
| \\
H{-}C{=}\ddot{O}:
\end{array}
$$

What would you predict its shape to be?

Polar Bonds and Electronegativity

7.68 What is a *polar covalent bond*? Define *dipole moment*.

7.69 Define *electronegativity*.

7.70 Which element has the highest electronegativity? Which is the second most electronegative element?

7.71 Which elements are assigned electronegativities of zero?

7.72 Using Figure 7.11, choose the atom in each of the following bonds that carries the partial positive charge: (a) N—S, (b) Si—I, (c) N—Br.

7.73 Which of the bonds in Review Exercise 7.72 is the most polar?

7.74 Which of the following bonds is most polar? In each bond, choose the atom that carries the partial negative charge: (a) Hg—I, (b) P—I, (c) Si—F, (d) Mg—N.

7.75 Are any of the bonds in Review Exercise 7.74 more ionic than covalent, based on electronegativity differences?

7.76 If an element has a low electronegativity, is it likely to be a metal or a nonmetal?

7.77 Without referring to the table of electronegativity values, use the periodic table to choose the element in each set that has the highest electronegativity: (a) Si, As, Ge, P (b) P, Mg, Ba, Sb (c) B, F, Te, P

7.78 HF is a polar molecule. Draw its Lewis structure and indicate the partial charges in the molecule.

Predicting Molecular Polarity

7.79 Why is it useful to know the polarities of molecules?

7.80 How do we indicate a bond dipole when we draw the structure of a molecule?

7.81 What condition must be met if a molecule having polar bonds is to be nonpolar?

7.82 Use a drawing to show why the SO_2 molecule is polar.

7.83 Which of the following molecules would be expected to be polar? (a) HBr, (b) $POCl_3$, (c) CH_2O, (d) $SnCl_4$, (e) $SbCl_5$.

7.84 Which of the following molecules would be expected to be polar? (a) PBr_3, (b) SO_3, (c) $AsCl_3$, (d) ClF_3, (e) BCl_3.

CHEMICALS IN USE
SILICATE MINERALS AND SOIL

Oxygen and silicon alone make up slightly over 80% of the atoms in the earth's crust, and they contribute about 75% to the crustal mass. These atoms occur in the form of a family of minerals called the silicates. (A *mineral* is a specific solid in the earth's crust that has a more or less definite stoichiometry, like an individual compound. *Rocks* are complex mixtures of minerals. An *ore* is a rock or a mineral that can be processed at a profit.)

Quartz, which is virtually pure silicon dioxide and which forms beautiful crystals, is probably the most familiar member of the silicate family. Other silicates are various feldspars, micas, and hornblendes.

Silicon–Oxygen Tetrahedra

The bonds that hold these minerals together include ionic bonds and covalent bonds. In all of them, the dominant structural feature is the silicon–oxygen tetrahedron (Figure 6a). Each oxygen in one tetrahedron extends one bond to its central silicon atom and another bond to a silicon atom in an adjacent tetrahedron. It isn't possible to tell where one tetrahedron ends and the next one starts. An analogy is a neatly stacked pile of oranges (oxygen atoms) at a supermarket. Each regularly spaced, unfilled cavity between oranges could hold a small cherry (a silicon atom). In other words, the silicon–oxygen tetrahedra are interconnected in a vast network of bonds that gives the mineral unusual strength. There are several ways in which the tetrahedra can be organized, as seen in Figure 6b.

Ions in Silicate Minerals

The arrays of silicon–oxygen tetrahedra in silicate minerals are never so perfectly organized that no crevices and cavities are present. These have an important part in the value of weathered silicates in soil, because the cavities can hold a number of chemical species that are needed for plant growth — metal ions, hydroxide ions, hydrogen ions, and even water molecules (as water of hydration). In semiarid regions where the rainfall is sparse, the ability of soil to trap and hold water in this way makes it possible to raise certain crops by a technique known as dryland farming.

The presence of species other than silicon and oxygen interspersed in the silicates also makes it difficult to write

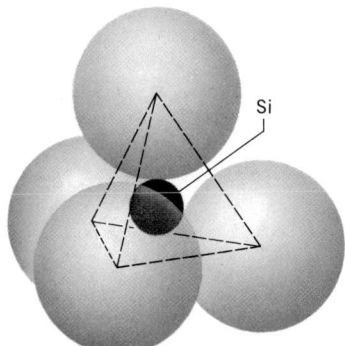

FIGURE 6a The silicon–oxygen tetrahedron.

exact formulas for a particular mineral. Quartz has the tightest packing of the tetrahedra, and virtually no other ions or molecules are present. Consequently, quartz crystals are able to take very regular forms of great beauty. In the mineral mica, the tetrahedra are organized in sheets, so this mineral is easily cleaved into flakes and leaflets.

Soil Formation

The silicates are found mostly in igneous rocks, those that once were hot, thick fluids like lava. Such rocks are changed to soil by weathering caused by both chemical and physical agents and by the introduction of the remains of plants and animals. The bulk of useful soil consists of silicate minerals, and the quality of such soil depends heavily on the trace metal elements that are also present.

When weathering occurs, a regrouping of molecule-sized fragments takes place among the minerals present in the rock. Thus, the silicate-based clays, so important in fertile soil, form by regroupings that bring aluminum ions, water molecules, and hydroxy groups into the systems. Figure 6c shows a model of the mineral kaolinite, just one of several varieties of clay. The layer-cake arrangement of atoms in kaolinite extensively uses entrapped water molecules to make the layers stick together. At the surfaces of the particles, there are many electron-rich sites that help the soil hold more water molecules as well as metal ions needed for plant growth. The aluminum ions in clay help to hold such anions as nitrate, sulfate, phosphate, and nitrite ions, which are also needed by plants.

Individual	Single chain	Double chain	Sheet	3-dimensional
Olivine	Pyroxene augite	Amphibole hornblende	Biotite (mica)	Quartz

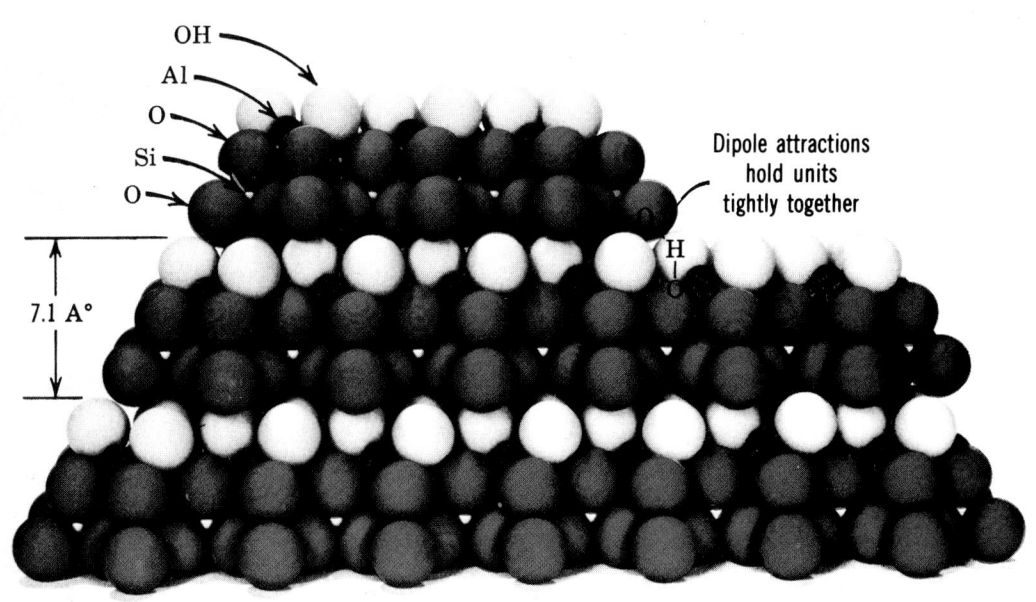

FIGURE 6b The most common arrangements of silicon–oxygen tetrahedra in the silicate minerals and their relationship to weathering. (From H. D. Foth, *Fundamentals of Soil Science,* 6th ed., 1978. John Wiley and Sons, Inc., New York. Used by permission.)

$\longrightarrow$ Increased weathering resistance $\longrightarrow$

Oxygen-silicon ratio				
4	3	2.7	2.5	2
Approximate Empirical Formula				
SiO_4	SiO_3	Si_4O_{11}	Si_2O_5	SiO_2

FIGURE 6c The structure of kaolinite. (From H. D. Foth, *Fundamentals of Soil Science,* 6th ed., 1978. John Wiley and Sons, Inc., New York. Used by permission.)

OH
Al
O
Si
O

Dipole attractions hold units tightly together

H

7.1 A°

Questions

1. Which term describes the kind of chemical bonding in silica — ionic, molecular, covalent (network), or metallic?

2. Describe the geometric arrangement of oxygen atoms around a silicon atom in silica? Does it fit the VSEPR theory?

3. How do species besides O and Si become incorporated into silicates?

4. Describe in general terms how the chemical weathering of silicates to fertile soil occurs.

5. How is water held within otherwise dry-appearing silicate-based soil?

Unknowingly, a spider depends on the strengths of covalent bonds that bind atoms together in the fibrous protein that makes up its silk web. In this chapter we will explore further the nature of covalent bonds and the theories that have evolved to explain them.

MODERN THEORIES OF COVALENT BONDING

THE NEED FOR MORE-ADVANCED THEORIES
VALENCE BOND THEORY
HYBRID ORBITALS
COORDINATE COVALENT BONDS AND HYBRID ORBITALS

DOUBLE AND TRIPLE BONDS
BOND ENERGIES AND THEIR MEASUREMENT
MOLECULAR ORBITAL THEORY
DELOCALIZED MOLECULAR ORBITALS

8.1 THE NEED FOR MORE-ADVANCED THEORIES

Because Lewis structures cannot explain why or how covalent bonds are formed, we must look to other theories based on wave mechanics.

So far, we have taken a very simple view of covalent bonding based on the octet rule and the use of Lewis structures for electron bookkeeping. Lewis structures, however, tell us nothing about *why* covalent bonds are formed or *how* electrons manage to be shared between atoms. Nor does the VSEPR theory, as useful and accurate as it generally is at predicting molecular geometry, explain *how* the electron pairs in the valence shell of an atom manage to avoid each other. Thus, as helpful as these simple models are, we must look beyond them if we wish to understand more fully the covalent bond and the factors that determine molecular geometry.

Modern theories of bonding are based on the principles of wave mechanics. This is the theory, you recall, that gives us the electron configurations of atoms and the description of the shapes of atomic orbitals. When wave mechanics is applied to molecules, it considers how the orbitals of the atoms that come together to form a covalent bond interact with each other, and in so doing it attempts to explain in detail how atoms share electrons.

There are fundamentally two theories of covalent bonding that have evolved, and in many ways, they complement one another. They are called the valence bond theory (or VB theory, for short) and the molecular orbital theory (MO theory). They differ principally in the way they construct a theoretical model of the bonding in a molecule. For example, the valence bond theory imagines individual atoms, each with its own orbitals and electrons, coming together and forming the covalent bonds of the molecule. On the other hand, the molecular orbital theory doesn't concern itself with how the molecule is formed. It just views a molecule as a collection of positively charged nuclei surrounded in some way by electrons that occupy a set of molecular orbitals, in much the same way that the electrons in an atom occupy atomic orbitals. (In a sense, this theory would look at an atom as if it were a special case—a molecule having only one positive center, instead of many.)

In their simplest forms, the VB and MO theories appear to be rather different. However, it has been found that both theories can be extended and refined to give the same results. This is as it should be, of course, since they both attempt to explain the same set of facts—the structures and shapes of molecules, and the strengths of chemical bonds. We will examine both theories in an elementary way, but the greater emphasis will be on the VB theory because it is easier to comprehend.

8.2 VALENCE BOND THEORY

The overlap of atomic orbitals provides a way for a pair of electrons to be shared between two nuclei.

In Section 7.3, we described the formation of the covalent bond in the hydrogen molecule, H_2. We saw that as two hydrogen atoms approach each other, the electron density shifts toward the region between the two nuclei. In the molecule, both electrons are able to move around both nuclei. Now, let's look at the formation of this covalent bond in terms of orbitals and electrons. In an isolated hydrogen atom, there is one electron in the atom's $1s$ atomic orbital. When a hydrogen molecule is formed, the atomic orbitals of two atoms merge so as to allow the electrons to move back and forth between the two nuclei. Valence bond theory gives us a way of describing how this merging of orbitals occurs.

According to VB theory, a bond between two atoms is formed when a pair of electrons is shared by two overlapping atomic orbitals, one orbital from each of the atoms joined by the bond. By overlap of orbitals we simply mean that portions of two atomic orbitals from different atoms share the same space. An important part of the theory is that only *one* pair of electrons, with their spins paired, can be shared by two overlapping orbitals. This electron pair becomes concentrated in the region of overlap and helps "cement" the nuclei together, so the strength of the covalent bond—and the extent to which the energy of the atoms is lowered—is determined in part of the extent to which the orbitals overlap. Because of this, atoms tend to position themselves so that the maximum amount of orbital overlap occurs. As we will see, this behavior is one of the major factors that control the shapes of molecules.

The formation of a hydrogen molecule according to VB theory is shown in Figure 8.1. As the two atoms approach each other, their $1s$ orbitals overlap, thereby giving the

Remember, *two electrons with paired spins* can be shared between two overlapping orbitals.

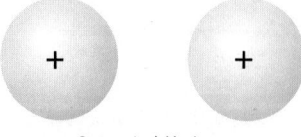 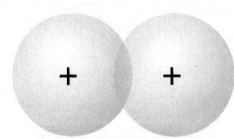 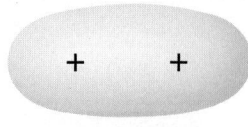

FIGURE 8.1 The formation of the hydrogen molecule according to valence bond theory.

Separated H atoms   Overlapping of orbitals   Covalent bond in H_2

H—H bond. The description of the bond in this molecule provided by VB theory is essentially the same as that discussed in Section 7.3.

Next, let's look at a molecule that is just a bit more complex then H_2—hydrogen fluoride, HF. Following the rules in Section 7.6, we would write its Lewis structure as

$$H\!:\!\ddot{\underset{\cdot\cdot}{F}}\!:$$

and we could diagram the formation of the bond as

$$H\cdot + \cdot\ddot{\underset{\cdot\cdot}{F}}\!: \longrightarrow H\!:\!\ddot{\underset{\cdot\cdot}{F}}\!:$$

The H—F bond is formed by the pairing of electrons—one from hydrogen and one from fluorine. To explain this according to VB theory, we must have two half-filled orbitals—one from each atom—that can be joined by overlap. (They must be half-filled, because we can't place more than two electrons into the bond.) To see clearly what must happen, it is best to look at the orbital diagrams of the valence shells of hydrogen and fluorine.

H ⟨↑⟩
1s

F ⟨↑↓⟩ ⟨↑↓⟩⟨↑↓⟩⟨↓⟩
2s   2p

The requirements for bond formation are met by overlapping the half-filled 1s orbital of hydrogen with the half-filled 2p orbital of fluorine; there are then two orbitals plus two electrons whose spins can adjust so they are paired. The formation of the bond is illustrated in Figure 8.2.

FIGURE 8.2 The formation of the hydrogen fluoride molecule according to valence bond theory. Only one of fluorine's 2p orbitals is shown.

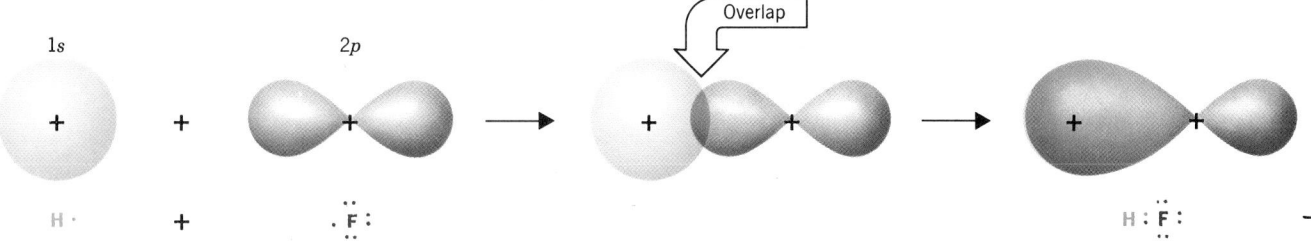

The overlap of orbitals provides a means for sharing electrons, thereby allowing each atom to complete its valence shell. It is sometimes convenient to indicate this using orbital diagrams. For example, the diagram below shows how the fluorine atom completes its 2p subshell by acquiring a share of an electron from hydrogen.

F (in HF) ⟨↑↓⟩ ⟨↑↓⟩⟨↑↓⟩⟨↑↓⟩ (colored arrow is the H electron)
2s   2p

Since the Lewis and VB descriptions of the formation of the H—F bond both account for the completion of the atoms' valence shells, a Lewis structure can be viewed, in a very qualitative sense, as a shorthand notation for the valence bond description of the molecule.

Now, let's turn our attention to a more complicated molecule, hydrogen sulfide, H_2S. This is a nonlinear molecule, and experiment has shown that the H—S—H bond

H_2S is the compound that gives rotten eggs their foul odor.

angle, the angle formed between the two H—S bonds, is about 92°.

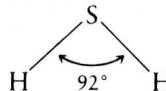

The orbital diagram for sulfur's valence shell is

S   ⬆⬇   ⬆⬇ ⬆ ⬆
    3s   3p

Sulfur has two p orbitals that each contain only one electron. Each of these can overlap with the 1s orbital of a hydrogen atom, as shown in Figure 8.3. This overlap completes the valence shell of sulfur because each hydrogen provides one electron.

S (in H$_2$S)   ⬆⬇   ⬆⬇ ⬆⬇ ⬆⬇   (colored arrows are H electrons)
              3s   3p

Two orbitals from different atoms never overlap with opposite ends of the same p orbital simultaneously.

In Figure 8.3, notice that when the 1s orbital of each hydrogen atom overlaps with a p orbital of sulfur, the best overlap occurs when the hydrogen atoms are located along the imaginary y and z axes drawn through the center of the sulfur atom. The angle formed between these two H—S bonds, the predicted bond angle, is 90°. This is very close to the actual bond angle of 92° found by experiment. Thus, the VB theory requirement for maximum overlap quite nicely explains the geometry of the hydrogen sulfide molecule.

FIGURE 8.3 Bonding in H$_2$S. The hydrogen 1s orbitals must locate themselves so that they can overlap with the two partially filled 3p orbitals of sulfur.

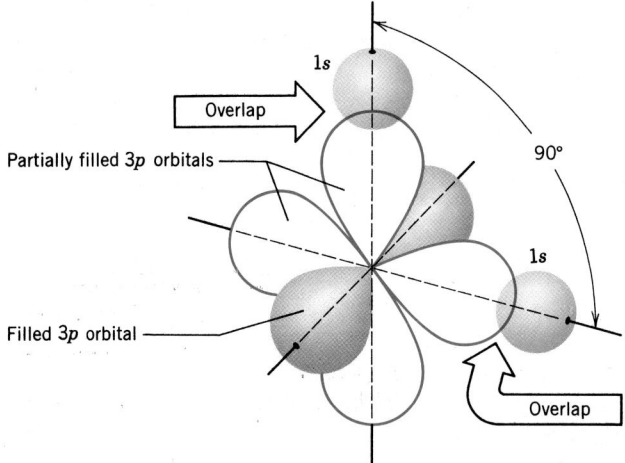

FIGURE 8.4 The fluorine molecule in valence bond theory. The two completely filled p orbitals on each fluorine atom are omitted, for clarity.

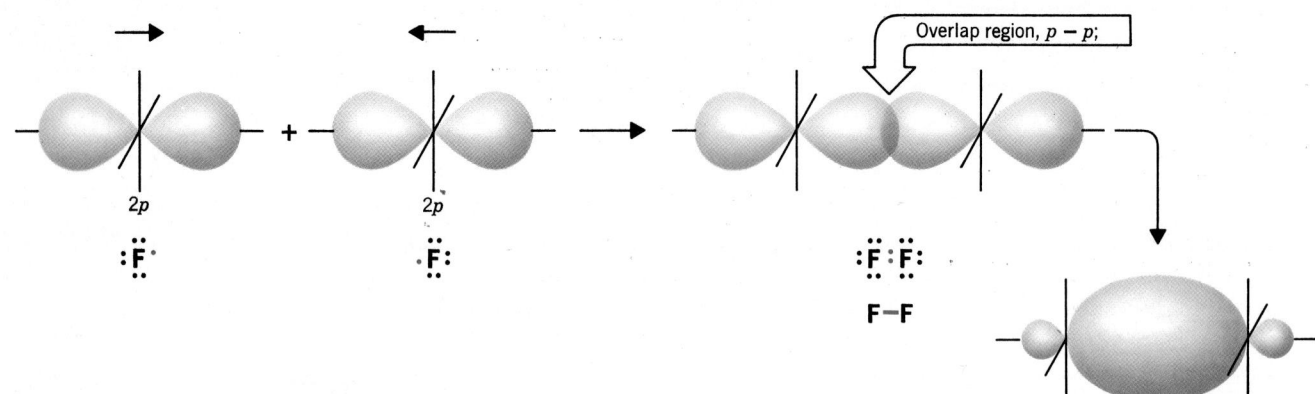

The overlap of p orbitals with each other is also possible. For example, we can consider the bonding in the fluorine molecule, F_2, to occur by the overlap of two $2p$ orbitals, as shown in Figure 8.4. The formation of the other diatomic molecules of the halogens, all of which are held together by single bonds, could be similarly described.

PRACTICE EXERCISE 1   Use the principles of VB theory to explain the bonding in HCl.

PRACTICE EXERCISE 2   The phosphine molecule, PH_3, has a pyramidal shape with H—P—H bond angles of 93.7°. Explain the bonding in PH_3 using VB theory.

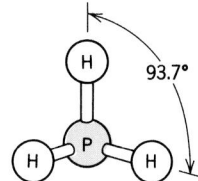

8.3 HYBRID ORBITALS

The mixing of atomic orbitals gives new orbitals that are able to form stronger bonds.

The approach that we have taken so far has worked well with some simple molecules. Their shapes are explained very nicely by the overlap of simple atomic orbitals. It does not take long, however, to find molecules with shapes and bond angles that fail to fit the model that we have developed. For example, methane, CH_4, has a shape that the VSEPR theory predicts (correctly) to be tetrahedral. The H—C—H bond angles in this molecule are 109.5°. No simple atomic orbitals are oriented at this angle with respect to each other. Therefore, before we can explain the bonds in more complicated molecules such as CH_4, we must look at how atomic orbitals of the same atom can interact with each other when bonds are formed.

Theoreticians who study wave mechanics have found that when atoms form bonds, their simple s, p, and d orbitals often *mix* to form new atomic orbitals, called hybrid atomic orbitals. These new orbitals have new shapes and new directional properties. The reason for this mixing can be seen if we look at their shapes.

One kind of hybrid atomic orbital is formed by mixing an s orbital and a p orbital. This creates *two* new orbitals called sp hybrid orbitals (the sp is used to designate the kinds of orbitals from which the hybrid was formed). Their shapes and directional properties are illustrated in Figure 8.5. Notice that each of the hybrid orbitals has the same shape—each has one large lobe and another much smaller lobe. The large lobe extends farther from the nucleus than either the s or p orbital from which the hybrid orbital was formed. This allows the hybrid orbital to overlap more effectively with an orbital on another atom when a bond is formed. In general, the greater the overlap of two orbitals, the stronger the bond. Therefore, hybrid orbitals allow atoms to form stronger, more stable bonds than would be possible if just simple atomic orbitals were used.

Another point to notice in Figure 8.5 is that the large lobes of the two sp hybrid orbitals point in opposite directions—that is, they are 180° apart. If bonds are formed by overlap of these hybrids with orbitals of other atoms, the other atoms will occupy positions on opposite sides of this central atom. Let's look at a specific example, the linear beryllium hydride molecule, BeH_2, as it would be formed in the gas phase.[1]

At a given internuclear distance, the greater "reach" of a hybrid orbital gives better overlap than either an s or p orbital.

The mathematics of wave mechanics predicts this 180° angle between sp hybrid orbitals.

[1] In the solid state, BeH_2 has a complex structure not consisting of simple BeH_2 molecules.

FIGURE 8.5 *sp* Hybridization.

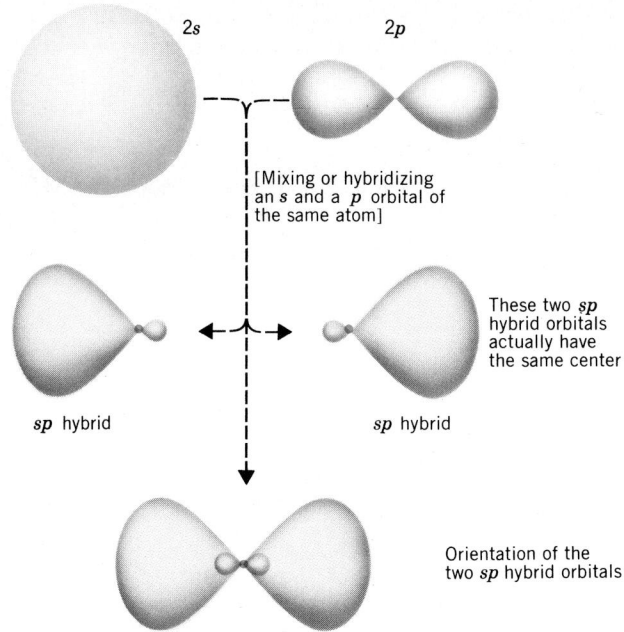

The orbital diagram for the valence shell of beryllium is

Note that the 2s orbital is filled and the three 2p orbitals are empty. For bonds to form at a 180° angle between beryllium and the two hydrogen atoms, two conditions must be met: (1) the two orbitals that beryllium uses to form the Be—H bonds must be aligned oppositely, at 180°, and (2) each of the beryllium orbitals must contain only one electron. The reason for the first requirement is obvious—the overlap of Be and H orbitals must give the correct experimentally measured bond angle. The reason for the second is that each bond must contain two electrons. Since a hydrogen 1s orbital supplies one electron, the beryllium orbital with which it overlaps must also contain one electron. A filled Be orbital won't do, because overlap of a half-filled hydrogen 1s with a filled orbital on Be would produce a "bond" with three electrons in it—a situation that's forbidden by VB theory. An empty Be orbital won't do either, because overlap with a hydrogen 1s orbital would give a bond with only one electron in it. Although such a bond isn't forbidden, it would be much weaker than a bond with two electrons, so electron pair bonds are definitely preferred. The net effect of all this is that when the Be—H bonds form, the electrons of the beryllium atom become unpaired, and the resulting half-filled s and p atomic orbitals become hybridized.

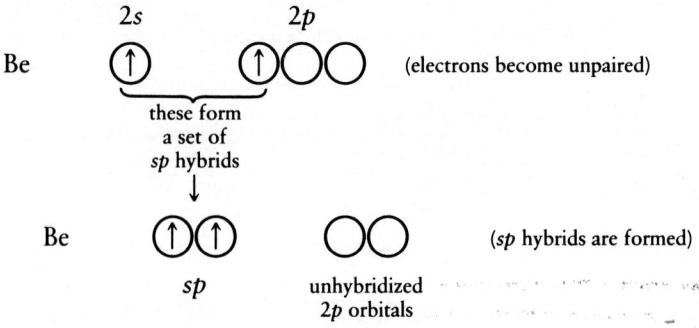

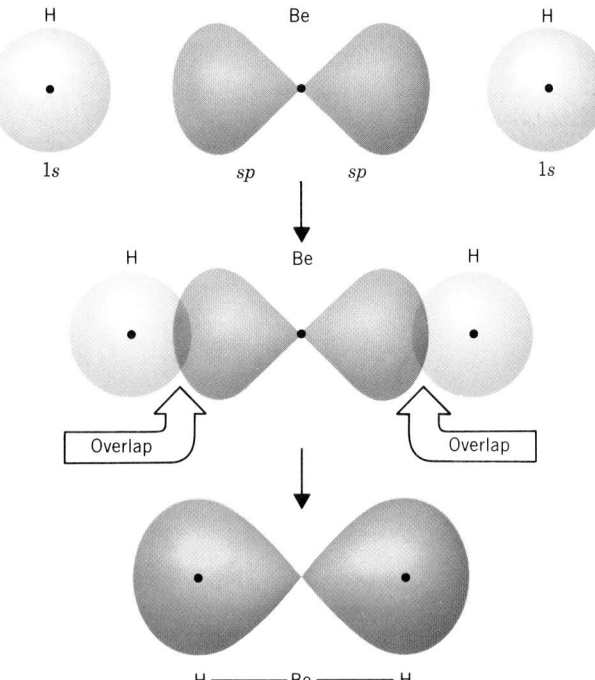

FIGURE 8.6 Bonding in BeH₂ according to valence bond theory. Only the larger lobe of each *sp* hybrid orbital is shown.

Now the 1s orbitals of the hydrogen atoms can overlap with the *sp* hybrids of beryllium as shown in Figure 8.6. Because the two *sp* hybrid orbitals of beryllium are identical in shape, the two Be—H bonds are identical except for the directions in which they point — we say that the bonds are *equivalent*. Since the bonds point in opposite directions, the molecule should be linear. The orbital diagram for beryllium in this molecule is

Be (in BeH₂)   ⬆⬇ ⬆⬇   ◯ ◯   (colored arrows are H electrons)
    sp   unhybridized
                2*p* orbitals

These *sp* hybrid orbitals are said to be *equivalent* because they have the same size, shape, and energy.

This explanation of the bonding in BeH₂ may seem to be a sort of theoretical "magic show" — a bit like pulling a rabbit out of a hat. It really isn't. A beryllium atom *does* indeed have a $1s^2 2s^2$ configuration, and the Be *does* require two singly occupied orbitals with which the 1s orbitals of hydrogen can overlap. The unpairing of electrons in the second shell would place one electron in the 2s orbital and one in a 2p orbital. It can be shown mathematically that the s and p orbitals combine by simple addition or subtraction of their electron waves to give the new hybrid orbitals. From a theoretical point of view, all the pieces of the puzzle fit together very well. If this does seem like magic, it is only because the theory must be treated very qualitatively in a course at this level. We use pictures because the actual math is too complex to deal with here.

If we draw the Lewis structure for BeH₂, H:Be:H, and apply VSEPR theory, we predict that this molecule should be linear, which is the same conclusion we came to using VB theory and the concept of hybrid atomic orbitals. This is no accident. As we will see, VB theory generally does a very good job of explaining the structures predicted by VSEPR theory, not only for linear molecules, but for molecules with other kinds of hybrid orbitals and shapes as well.

Other Hybrid Orbitals

Hybrid atomic orbitals can be formed by mixing more than just two simple atomic orbitals. If an s orbital and two p orbitals combine, *three* hybrid orbitals, each similar in shape to the *sp* hybrids, are formed. They are called *sp²* hybrid orbitals, the superscript 2

The number of hybrid orbitals of a given kind is always equal to the number of atomic orbitals that are mixed.

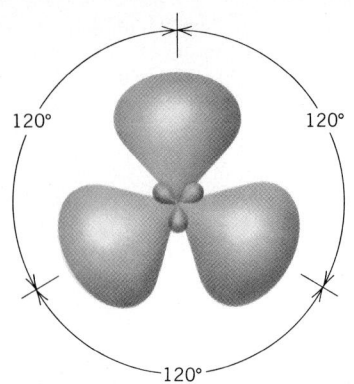

FIGURE 8.7 Directional properties of sp^2 hybrid orbitals. The axes of the orbitals are in the same plane separated by angles of 120°.

specifying the *number* of p orbitals taking part in the formation of the hybrids. (A superscript of 1 is implied for the contribution of the s orbital.) Also, notice that the number of hybrids in the set is equal to the number of simple atomic orbitals from which the hybrids are formed. The directional properties of sp^2 hybrid orbitals are illustrated in Figure 8.7. All three large lobes, which have the same shape, are in the same plane and point toward the corners of a triangle.

According to VB theory, boron trichloride, BCl_3, is a molecule in which the central boron atom uses sp^2 hybrids for bonding. A boron atom has the valence shell configuration

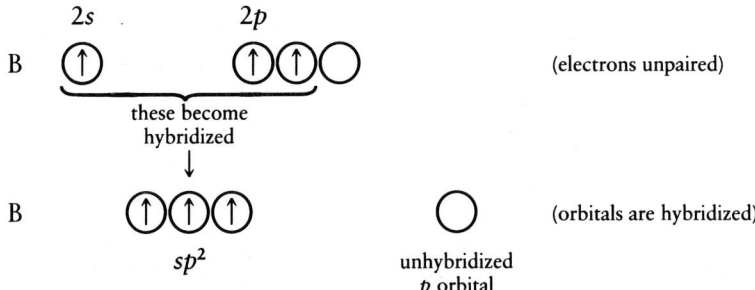

To form three bonds, boron must have three half-filled orbitals, so its $2s$ electrons must become unpaired. The resulting half-filled s and p orbitals then become hybridized.

A chlorine atom has the valence shell configuration

The half-filled $3p$ orbital of each chlorine atom overlaps with one of the sp^2 hybrids of boron to give

These bonds are illustrated in Figure 8.8. Notice that all three bonds appear to be alike—they are equivalent. This agrees with experiment, which shows that all three B—Cl bonds have the same length and the same bond energy. The geometry of the molecule is also explained. Because the sp^2 hybrids point toward the corners of a triangle and because the chlorine atoms must be located so that their p orbitals can overlap with the hybrids, the BCl_3 molecule has a *planar triangular shape*. Once again, this is the same shape that we would predict using VSEPR theory and the Lewis structure of BCl_3.

VB theory, like VSEPR theory, must explain the experimental fact that BCl_3 is a planar triangular molecule.

Table 8.1 lists some important types of hybrid orbitals. The directional properties of the various hybrids are shown in Figure 8.9. Let's look at an example showing how this information can be used to explain bonding and geometry.

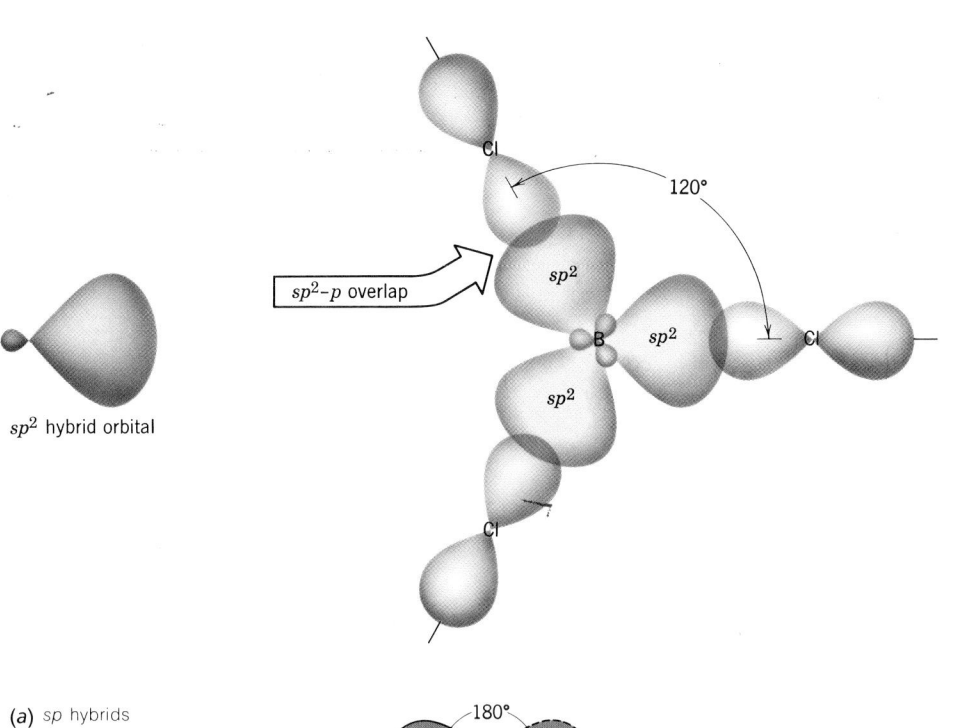

*sp*² hybrid orbital

*sp*²–*p* overlap

120°

*sp*²

*sp*²

*sp*²

B

Cl

Cl

Cl

FIGURE 8.8 Bonding in BCl₃. The four atoms as well as the bond axes all lie in the same plane.

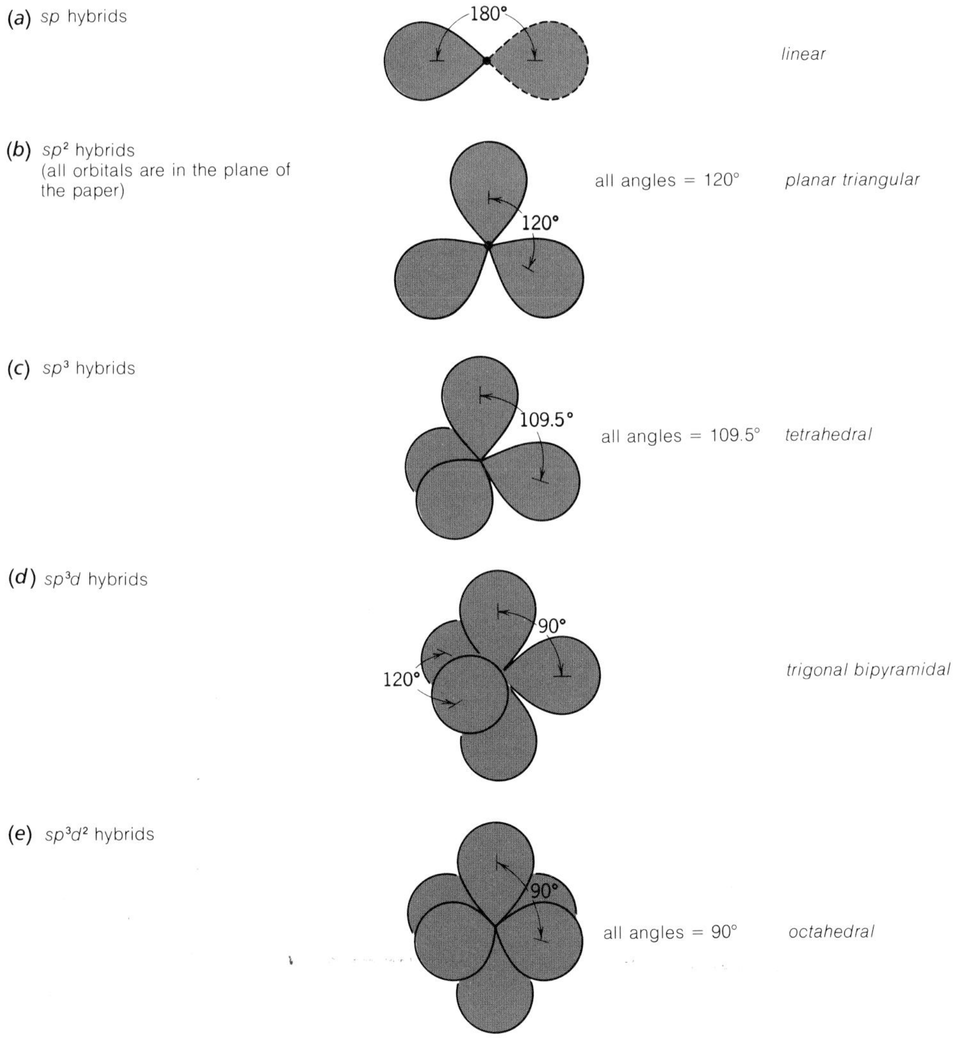

(*a*) *sp* hybrids

180°

linear

(*b*) *sp*² hybrids
(all orbitals are in the plane of
the paper)

all angles = 120°   *planar triangular*

120°

(*c*) *sp*³ hybrids

109.5°

all angles = 109.5°   *tetrahedral*

(*d*) *sp*³*d* hybrids

90°

120°

trigonal bipyramidal

(*e*) *sp*³*d*² hybrids

90°

all angles = 90°   *octahedral*

FIGURE 8.9 The orientations of hybrid orbitals in space.

TABLE 8.1 Hybrid Orbitals

Hybrid	Orbitals Mixed	Orientation in Space
sp	$s + p$	Linear
sp^2	$s + p + p$	Planar triangular
sp^3	$s + p + p + p$	Tetrahedral
sp^3d	$s + p + p + p + d$	Trigonal bipyramidal
sp^3d^2	$s + p + p + p + d + d$	Octahedral

EXAMPLE 8.1   EXPLAINING BONDING WITH HYBRID ORBITALS

Problem: What kinds of hybrid orbitals would be found in methane, CH_4? What is the expected geometry of this molecule?

Solution: We begin by looking at the valence shell of carbon.

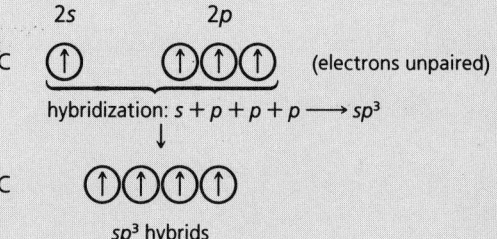

To form four C—H bonds, carbon needs four half-filled orbitals. The $2s$ electrons must therefore become unpaired, which places one of them in the unoccupied p orbital. Then the orbitals needed for bonding are hybridized.

Finally we form the four bonds.

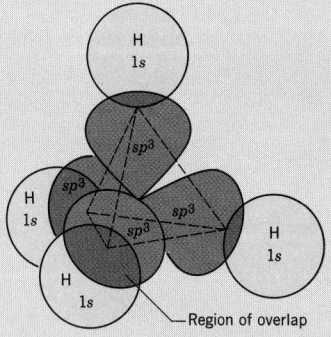

Since sp^3 hybrids point to the corners of a tetrahedron, the CH_4 molecule should be tetrahedral. Methane is, in fact, tetrahedral, and all four of its C—H bonds are equivalent.

In methane, carbon forms four single bonds with hydrogen atoms by using sp^3 hybrid orbitals. Carbon uses these same kinds of orbitals in *all* of its compounds in which it is bonded to four other atoms by single bonds. This makes the tetrahedral orientation of atoms around carbon one of the primary structural features of organic compounds, and organic chemists routinely think in terms of "tetrahedral carbon," shown in Figure 8.10.

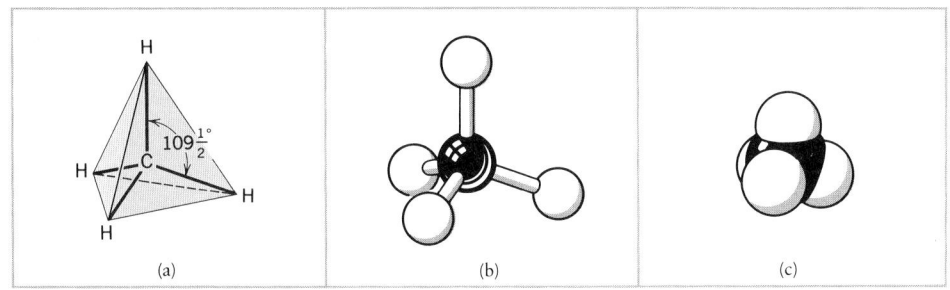

(a) (b) (c)

FIGURE 8.10 The "tetrahedral carbon." *(a)* The heavy lines are the axes of the bonds. *(b)* A ball-and-stick model of the CH₄ molecule. *(c)* A scale model of CH₄ that indicates the relative volumes occupied by the electron clouds.

In many compounds, carbon atoms are bonded to other carbon atoms. The simplest such compound that contains only single bonds is called ethane, C_2H_6.

$$
\begin{array}{ccc}
& H & H \\
& | & | \\
H- & C- & C-H \\
& | & | \\
& H & H
\end{array}
$$

In this molecule, the carbons are bonded together by the overlap of sp^3 hybrid orbitals (Figure 8.11). One of the most important characteristics of this bond is that the overlap of the orbitals in the C—C bond is hardly affected at all if one portion of the molecule rotates relative to the other about the bond axis. Such rotation, therefore, is said to occur freely. This free rotation permits different possible relative orientations of the atoms in the molecule. These different relative orientations are called conformations. With complex molecules, the number of possible conformations is enormous. For example, Figure

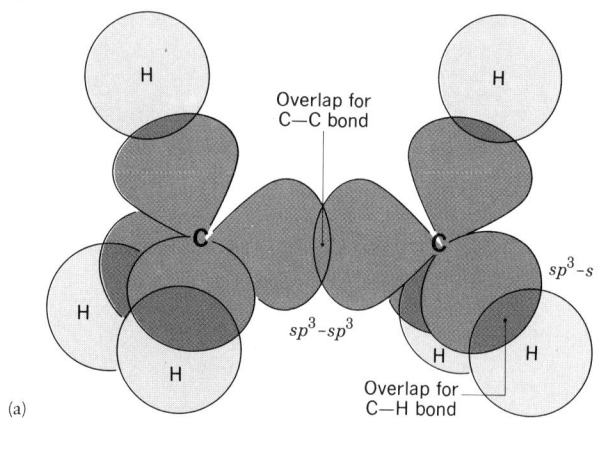

Overlap for C—C bond

sp^3-s

sp^3-sp^3

Overlap for C—H bond

(a)

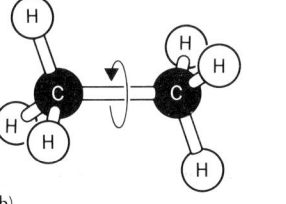

(b)

FIGURE 8.11 The bonds in the ethane molecule. *(a)* Overlap of orbitals. *(b)* The degree of overlap of the sp^3 orbitals in the carbon–carbon bond is not appreciably affected by the rotation of the two

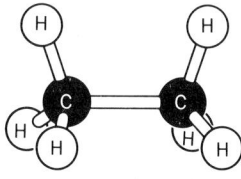

groups relative to each other about that bond.

8.12 illustrates just three of the many possible conformations of pentane, C_5H_{12}, one of the low-molecular-weight organic compounds found in gasoline.

$$H-\overset{\overset{\displaystyle H}{|}}{\underset{\underset{\displaystyle H}{|}}{C}}-\overset{\overset{\displaystyle H}{|}}{\underset{\underset{\displaystyle H}{|}}{C}}-\overset{\overset{\displaystyle H}{|}}{\underset{\underset{\displaystyle H}{|}}{C}}-\overset{\overset{\displaystyle H}{|}}{\underset{\underset{\displaystyle H}{|}}{C}}-\overset{\overset{\displaystyle H}{|}}{\underset{\underset{\displaystyle H}{|}}{C}}-H$$ (pentane)

FIGURE 8.12 Just three of the innumerable conformations of the carbon chain in pentane, C_5H_{12} (hydrogen atoms not shown). Free rotation about single bonds makes these different conformations possible.

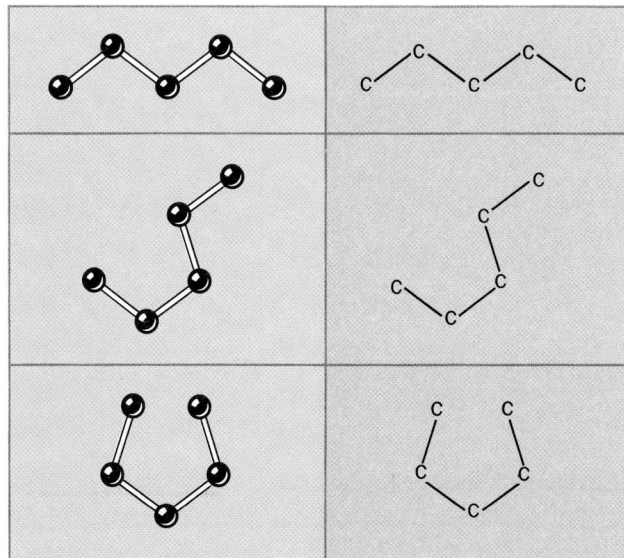

Hybridization When the Central Atom Has More Than an Octet

Earlier we saw that certain molecules have atoms that must violate the octet rule because they form more than four bonds. In these cases, the atom must reach beyond its s and p valence shell orbitals to form sufficient half-filled orbitals for bonding. This is because the s and p orbitals provide a maximum of only four orbitals. But where can the atom find additional orbitals? The answer can be understood best through an example.

EXAMPLE 8.2   EXPLAINING BONDING WITH HYBRID ORBITALS

Problem: What kind of hybrid orbitals would be used by sulfur in sulfur hexafluoride? What would be the geometry of SF_6?

Solution: Once again we look at the configuration of the valence shell of the central atom.

To form six bonds, all of sulfur's valence electrons must be used. But how can six half-filled orbitals be formed? The answer is that we must reach into sulfur's unoccupied $3d$ subshell. Every atom has a $3d$ subshell, whether or not it is occupied. A free sulfur atom has an unoccupied $3d$ subshell that is fairly close in energy to its $3s$ and $3p$ subshells. When necessary (as in the formation of SF_6), sulfur can utilize its $3d$ orbitals to form hybrids.
Let's rewrite the orbital diagram to show the vacant $3d$ subshell.

There is no $2d$ subshell—that is why period-2 elements never exceed an octet.

Unpairing all of the electrons to give six half-filled orbitals, followed by hybridization, gives six equivalent sp^3d^2 hybrid orbitals.

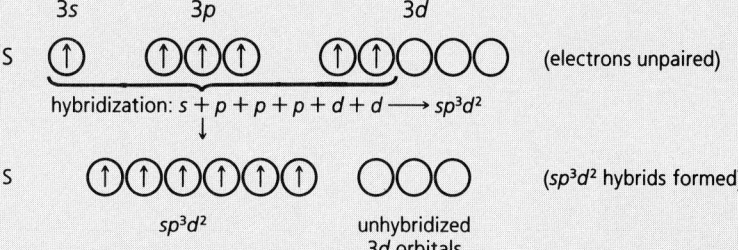

Finally, the six S—F bonds are formed. The half-filled $2p$ orbitals of the six fluorine atoms overlap with the sp^3d^2 hybrids.

From Table 8.1 and Figure 8.9, we see that sp^3d^2 hybrids point toward the corners of an octahedron, and this is where the fluorine atoms must be located to give the best orbital overlap. This model of the bonding in SF_6, therefore, suggests that the molecule is octahedral, which it is.

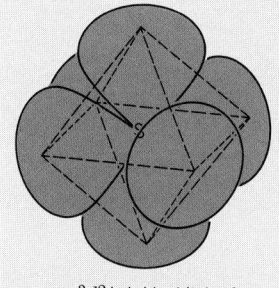

sp^3d^2 hybrid orbitals of sulfur in SF_6

Using VSEPR Theory to Predict Hybridization

We have seen that if we know the kind of hybrid orbitals that the central atom uses in a molecule, we can predict the geometry of the molecule. But this can also be viewed from the other direction. *If we know the geometry of a molecule, we can predict the kind of hybrid orbitals that are used.* Since VSEPR theory works so well in predicting geometry, we can use it to help us obtain VB descriptions of bonding. For example, the Lewis structures of CH_4 and SF_6 are

In CH_4, there are four electron pairs around carbon. The VSEPR model tells us that they should be arranged tetrahedrally. The only hybrid orbitals that are tetrahedral are sp^3 hybrids, and we have seen that they explain the structure of this molecule well. Similarly, VSEPR theory tells us that the six electron pairs around sulfur should be arranged octahedrally. In Table 8.1, the only octahedrally oriented hybrids are sp^3d^2 — the sulfur in the SF_6 molecule must use these hybrids.

PRACTICE EXERCISE 3   What kind of hybrid orbitals would be used by the central atom (underlined) in (a) $\underline{Si}H_4$ and (b) $\underline{P}Cl_5$.

Hybridization in Molecules That Have Lone Pairs of Electrons

Methane, CH_4, is a tetrahedral molecule with sp^3 hybridization of the orbitals of carbon and H—C—H bond angles that are each equal to 109.5°. In ammonia, NH_3, the H—N—H bond angles are 107°, and in water the H—O—H bond angle is 104.5°. Both NH_3 and H_2O have H—X—H bond angles that are close to the bond angles expected for a molecule whose central atom has sp^3 hybrids. The use of sp^3 hybrids by oxygen and nitrogen, therefore, is often used to explain the geometry of H_2O and NH_3.

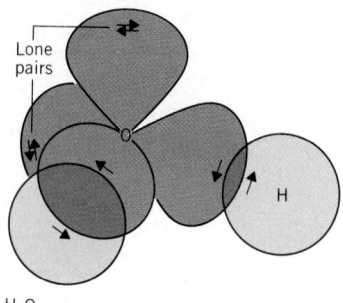

H_2O

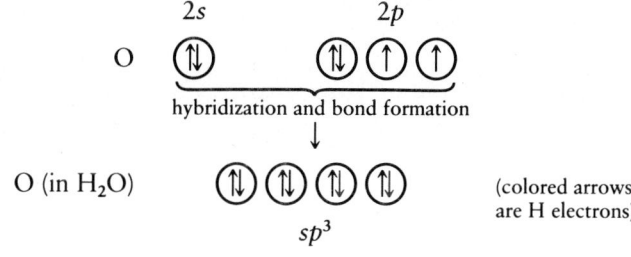

O (in H_2O)   (colored arrows are H electrons)

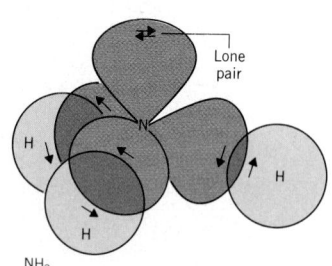

NH_3

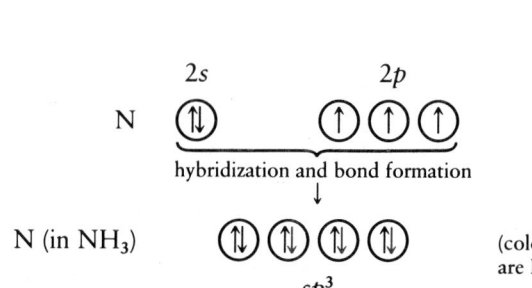

N (in NH_3)   (colored arrows are H electrons)

According to these descriptions, not all of the hybrid orbitals of the central atom must be used for bonding. Lone pairs (unshared pairs) of electrons can be accommodated in them too. In fact, there is evidence to suggest that the lone pair on the nitrogen in ammonia does in fact reside in an sp^3 hybrid orbital. (The case is less convincing for water, however, as noted in Special Topic 8.1.)

EXAMPLE 8.3   EXPLAINING BONDING WITH HYBRID ORBITALS

Problem: What kind of hybrid orbitals would sulfur have in sulfur tetrafluoride, SF_4?

Solution: Let's begin by constructing the Lewis structure for the molecule. Following the rules given in Section 7.6, we obtain

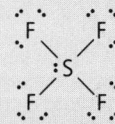

VSEPR theory predicts that the electron pairs around the sulfur should be arranged in a trigonal bipyramidal fashion, and the only hybrids that match this are sp^3d. To see how they are formed, we look at the valence shell of sulfur, including its vacant $3d$ subshell.

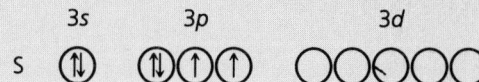

To form the four bonds to fluorine there must be four half-filled orbitals, so we unpair two electrons.

Next, we form the hybrids. In doing this, we use *all* the valence shell orbitals that have electrons in them.

Over the years there's been quite a bit of controversy about the nature of the bonding in the water molecule. Specifically, the question has been: What kind of orbitals does oxygen really use in forming its bonds to the hydrogen atoms?

In our discussion of this molecule, we've given what has become a standard explanation for the H—O—H bond angle—the use by oxygen of sp^3 hybrid orbitals to overlap with the $1s$ orbitals of the hydrogen atoms. Based on bond angle data alone, this is plausible because the angles between these kinds of hybrids is quite close to the bond angle in H_2O. It also fits with the general rule of thumb that we've followed in the remainder of the discussion of VB theory—that atoms generally use hybrid orbitals to form bonds to neighboring atoms. By following this rule, we obtain reasonable explanations of molecular structure. Nevertheless, as useful as this rule may be for other molecules, there is considerable evidence that it doesn't really work well for water.

In our discussion of hybridization, we've ignored the fact that forming hybrid orbitals requires energy, which is usually more than paid back by the formation of strong covalent bonds. In oxygen and other period 2 elements, however, it is especially "expensive" to form the hybrids, because the energy of the $2s$ orbital is so much lower than the energy of the $2p$ orbitals. This makes it less attractive for these elements to use hybrid orbitals if unhybridized orbitals can also do the job, and this does appear to be the case with water, based on detailed calculations and some experimental

evidence that are beyond the scope of this discussion. But, if oxygen uses its pure $2p$ orbitals to form the O—H bonds, how come the bond angle isn't close to 90°, as it is in H_2S (page 270) where sulfur uses its $3p$ orbitals to form the S—H bonds?

To explain the spreading of the H—O—H angle in water from 90° to 104°, we have to consider the small size of the oxygen atom. As you can see in the figure below, if the bond angle were 90°, the two hydrogen atoms would interpenetrate each other, which would cause their completed valence shells to overlap. This can't happen, however, because two pairs of electrons can never occupy the same space, so the angle spreads to relieve this problem. In H_2S, the bond angle doesn't have to increase appreciably from 90° because the sulfur atom is larger and the hydrogens don't interfere with each other.

The point of this discussion is that often there are alternative theoretical explanations of the same phenomenon (in this instance, the bonding in water). Some are simple and others are more complex. Which one we use depends on how precise we must be. The VSEPR theory, for example, is very simple, but it doesn't attempt to explain bonding in terms of the atomic orbitals involved. The model we've provided in this chapter using hybrid orbitals also works well most of the time, so we find it useful. But even this model is not completely accurate all the time, so when it really matters, many factors have to be weighed in deciding what is as close to the "truth" as possible.

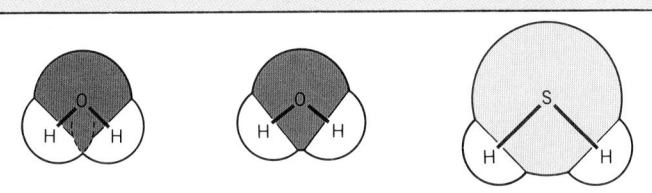

(left) If the bond angle in H_2O were 90°, the H atoms would overlap severely. (center) At a bond angle of 104°, the hydrogens don't interfere with each other very much. (right) Because S is larger than O, the bond angle in H_2S can be much closer to 90°.

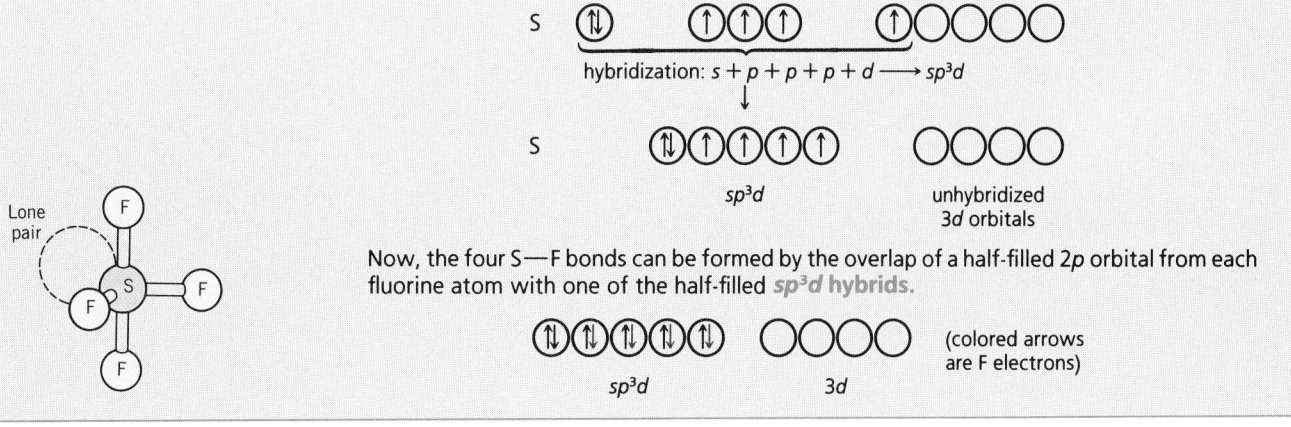

Now, the four S—F bonds can be formed by the overlap of a half-filled $2p$ orbital from each fluorine atom with one of the half-filled *sp^3d hybrids*.

PRACTICE EXERCISE 4   What kinds of hybrid orbitals would the central atom (underlined) use in (a) $\underline{P}Cl_3$ and (b) $\underline{Cl}F_3$.

8.4 COORDINATE COVALENT BONDS AND HYBRID ORBITALS

A filled orbital and an empty orbital can overlap to form a coordinate covalent bond.

In Section 7.8 we defined a coordinate covalent bond as a bond in which both of the shared electrons are provided by just one of the joined atoms. For example, boron trifluoride, BF_3, can combine with an additional fluoride ion to form the tetrafluoroborate ion, BF_4^-, according to the equation

$$BF_3 + F^- \longrightarrow BF_4^-$$
tetrafluoroborate ion

Using x's for boron's electrons and dots for fluorine's electrons, this reaction can be shown as

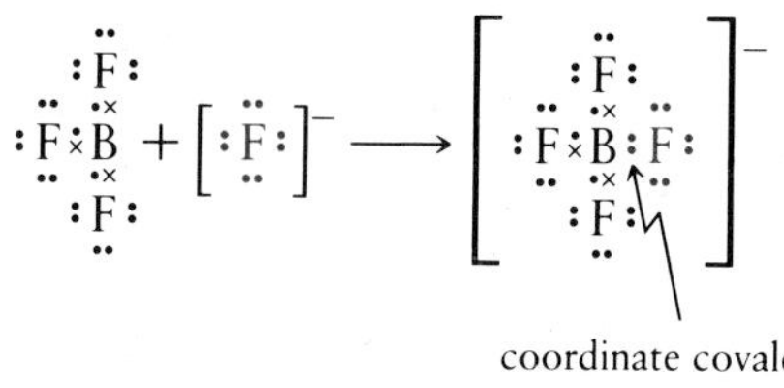

coordinate covalent bond

As we mentioned previously, the coordinate covalent bond is really no different from any other covalent bond once it has been formed. The distinction between them is made *only* for bookkeeping purposes. One place where such bookkeeping is useful is in keeping track of the orbitals and electrons used when atoms bond together.

The VB theory requirements for bond formation — two overlapping orbitals sharing two electrons — can be satisfied in two ways. One, as we have seen, is the overlapping of two half-filled orbitals. This gives us an "ordinary" covalent bond. The other is the overlapping of one filled orbital with one empty orbital. The shared pair of electrons is donated by the atom with the filled orbital and a coordinate covalent bond is formed.

The use of X's and dots does not mean that some electrons are different than others. We are just observing the origin of the electrons in the bond.

EXAMPLE 8.4   EXPLAINING COORDINATE COVALENT BONDS

Problem: Explain the bonding and predict the geometry for BF_4^-.

Solution: The orbital diagram for boron is

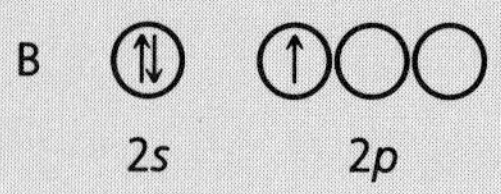

To form four bonds, we need four hybrid orbitals. These can be sp^3 hybrids formed from the $2s$ and $2p$ subshells. Boron has enough electrons to half-fill only three of the hybrids. This gives

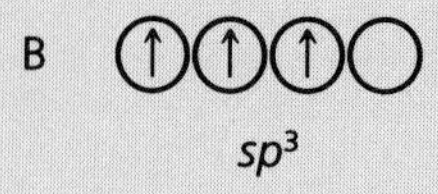

Boron forms three ordinary covalent bonds with fluorine atoms plus one coordinate covalent bond with a fluoride ion.

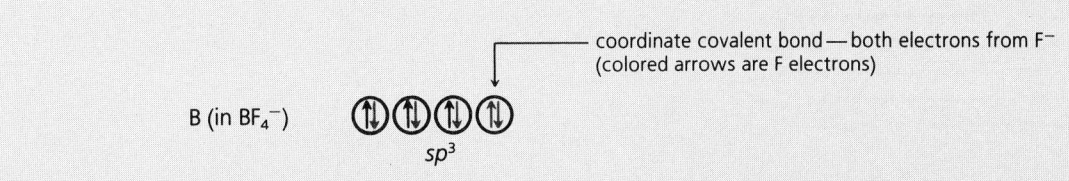

coordinate covalent bond—both electrons from F⁻
(colored arrows are F electrons)

B (in BF₄⁻)

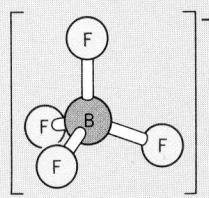

sp^3

According to VSEPR theory, we would expect that BF₃ would be planar triangular and BF₄⁻ would be tetrahedral.

Finally, we should expect the BF₄⁻ ion to have a tetrahedral shape, because that is the geometry of sp^3 hybrids.

PRACTICE EXERCISE 5   What hybrid orbitals are used by phosphorus in PCl₆⁻? Draw the orbital diagram for phosphorus in PCl₆⁻. What is the geometry of PCl₆⁻?

8.5 DOUBLE AND TRIPLE BONDS

The "sideways" overlap of p orbitals allows more than one pair of electrons to be shared between two atoms.

The types of overlap of orbitals that we have described so far produce bonds in which the electron density is concentrated most heavily between the nuclei of the two atoms along an imaginary line that joins their centers. Any bond of this kind, whether formed from the overlap of s orbitals, p orbitals, or hybrid orbitals (Figure 8.13), is called a sigma bond or σ bond (σ is the Greek letter *sigma*).

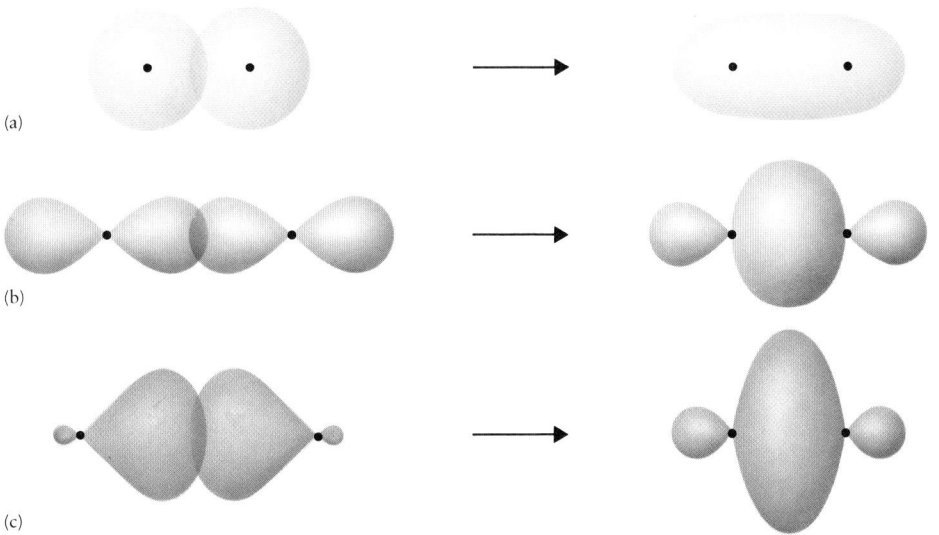

FIGURE 8.13 Sigma bonds. (a) From the overlap of s orbitals. (b) From the end-to-end overlap of p orbitals. (c) From the overlap of hybrid orbitals.

There is another way that p orbitals can overlap, which produces a bond in which the electron density is concentrated in *two* separate regions that lie on opposite sides of the imaginary line joining the two nuclei. This bond, called a pi bond, or π bond, is shown

FIGURE 8.14 Formation of a π bond. Two p orbitals overlap sideways instead of end to end. The electron density is concentrated above and below the bond axis.

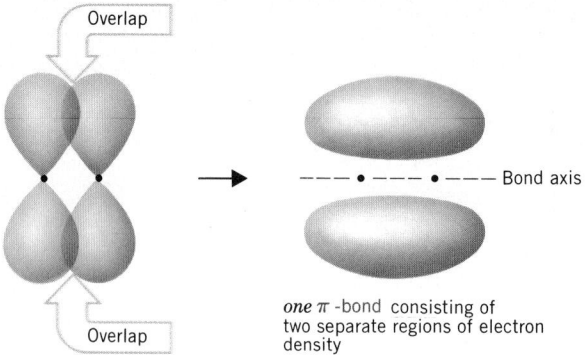

Overlap

Overlap

Bond axis

one π -bond consisting of two separate regions of electron density

in Figure 8.14. Notice that a π bond, like a p orbital, consists of two parts. Each region of electron density makes up only half of a π bond; it takes *both* of them to make up *one π bond*.

The formation of a π bond allows two atoms to share more than one pair of electrons between them. We can use ethene (also called ethylene), C_2H_4, as an example. The Lewis structure for this molecule contains a double bond.

Polyethylene, a common plastic, is made from C_2H_4.

$$\begin{array}{ccc} H & & H \\ \diagdown & & \diagup \\ & C = C & \\ \diagup & & \diagdown \\ H & & H \end{array}$$

We can explain the bonding in this molecule in the following way. First, the skeletal structure of the molecule — the molecular framework — is assembled by connecting the atoms together with σ bonds that are formed by the overlap of hybrid orbitals. The number of hybrid orbitals needed by each atom depends on the number of σ bonds it must form. In the case of ethene, each carbon atom is bonded to three other atoms — two hydrogens and the other carbon. This means that each carbon atom needs *three* hybrid orbitals for the σ bonds. These can be formed from the 2s orbital and two of the 2p orbitals.

Generally, the number of hybrid orbitals that an atom needs is equal to the sum of the number of *atoms* to which it is bonded plus the number of *lone pairs* that it has.

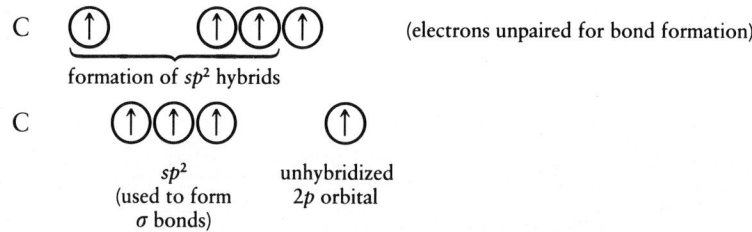

C ⊕   ⊕⊕⊕   (electrons unpaired for bond formation)

formation of sp^2 hybrids

C ⊕⊕⊕   ⊕

sp^2 (used to form σ bonds)   unhybridized 2p orbital

Notice that the carbon atom has an unpaired electron in an unhybridized p orbital. This p orbital is oriented perpendicular to the triangular plane of the sp^2 hybrid orbitals, as shown in Figure 8.15.

The C—H σ bonds in C_2H_4 are formed by the overlap of the sp^2 orbitals of carbon with the 1s orbitals of hydrogen. The C—C σ bond is formed by overlap of one sp^2 orbital from each of the two carbon atoms (also shown in Figure 8.15). Finally, the remaining unhybridized p orbitals, one from each carbon atom, also overlap to produce a π bond.

In C₂H₄, the measured H—C—H bond angle is 118°, and the measured H—C—C bond angle is 121°.

In C_2H_4, the predicted bond angles are all 120°, which is the angle between the sp^2 hybrid orbitals. The actual bond angles are, in fact, very close to this. Our description of the double bond, which requires the sideways overlap of the unhybridized p orbitals,

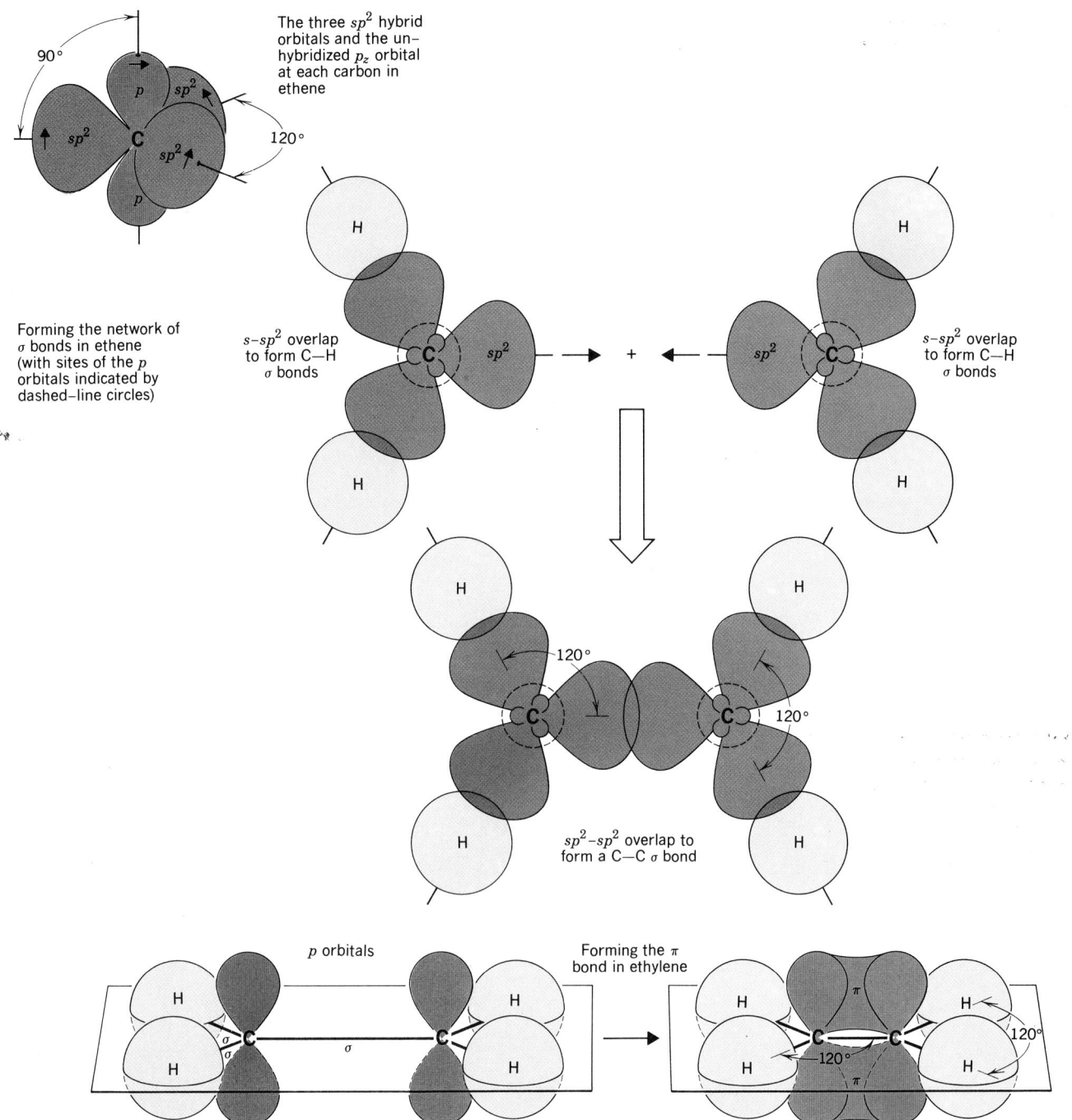

The three sp^2 hybrid orbitals and the un-hybridized p_z orbital at each carbon in ethene

Forming the network of σ bonds in ethene (with sites of the p orbitals indicated by dashed-line circles)

$s-sp^2$ overlap to form C—H σ bonds

$s-sp^2$ overlap to form C—H σ bonds

sp^2-sp^2 overlap to form a C—C σ bond

p orbitals

Forming the π bond in ethylene

FIGURE 8.15 The carbon–carbon double bond. (Adapted from J. R. Holum, *Organic and Biological Chemistry*, 2nd edition, 1986, John Wiley & Sons, New York.)

also predicts that all six atoms should be in the same plane, which they are. In addition, notice that the electron pairs in the σ and π bonds successfully avoid each other, which minimizes their mutual repulsions. The σ-bond electrons are concentrated along the bond axis, and the π-bond electrons are concentrated in spaces above and below the bond axis.

Restricted rotation about
double bonds has important
consequences in organic
chemistry.

One of the most important properties of double bonds is that rotation of one portion of a molecule relative to the rest about the axis of the double bond occurs only with great difficulty. In Figure 8.16 we see that as one group:

$$=C\diagup^{\Large H}_{\diagdown H}$$

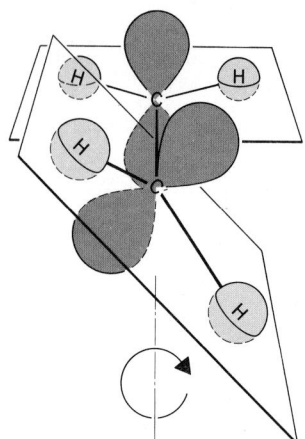

is rotated relative to the other group about the carbon–carbon bond, the unhybridized p orbitals are no longer oriented so that they can overlap. This destroys the π bond. In effect, rotation about a double bond involves bond breaking. In general, breaking a bond requires a large amount of energy, more than is normally available to the molecules at room temperature, so rotation around the double bond axis usually doesn't take place.

In almost every instance, a double bond consists of a σ bond and a π bond. Another example is the compound methanal, also called formaldehyde (the substance used as a preservative for biological specimens and as an embalming fluid). The Lewis structure of this compound is

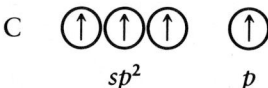

$$\diagup^{\Large H}_{\diagdown H}C=\ddot{O}{\,\cdot\cdot}$$

FIGURE 8.16 Restricted rotation about a double bond. Rotation of the CH$_2$ closest to us relative to the one at the rear causes the unhybridized p orbitals to become misaligned, thereby destroying the π bond.

As with ethene, the carbon forms sp^2 hybrids, leaving an unpaired electron in an unhybridized p orbital.

C   (↑)(↑)(↑)   (↑)
       sp^2 p

The oxygen can also form sp^2 hybrids with the placing of electron pairs in two of them and an unpaired electron in the third. This means that the remaining unhybridized p orbital also has an unpaired electron.

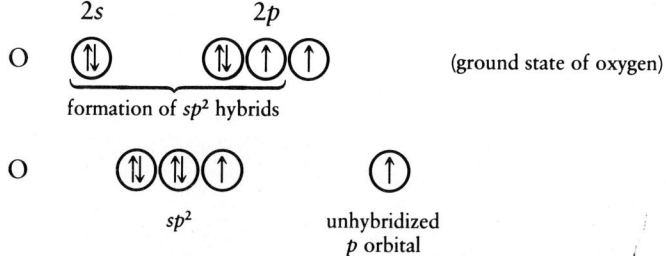

$$2s \qquad\qquad 2p$$

O   (↑↓)   (↑↓)(↑)(↑)   (ground state of oxygen)

formation of sp^2 hybrids

O   (↑↓)(↑↓)(↑)   (↑)
              sp^2 unhybridized
                                  p orbital

Figure 8.17 shows how the carbon, hydrogen, and oxygen atoms come together to form the molecule. As before, the basic framework of the molecule is formed by the σ bonds. These determine the molecular shape. The carbon–oxygen double bond also contains a π bond from the overlap of the unhybridized p orbitals.

Now let's look at a molecule containing a triple bond. An example is ethyne, C$_2$H$_2$, which is probably better known as acetylene (a gas used as a fuel for welding torches). This molecule has the Lewis structure

$$H{-}C{\equiv}C{-}H$$

In acetylene, each carbon needs two hybrid orbitals to form two σ bonds—one to a hydrogen atom and one to the other carbon atom. These can be provided by mixing the

Forming the
network of
σ bonds
in methanal

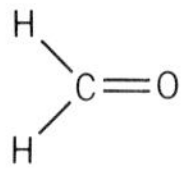

1s

sp^2

s–sp^2 overlap
to form C—H
σ bonds

sp^2

sp^2

C

sp^2

Carbon

+

sp^2

sp^2

O

Outline of
unhybridized
p_z orbital

sp^2

Oxygen

H
      C=O
H

Methanal

H

sp^2–sp^2 overlap to
form the C—O σ bond

Unshared pairs
of electrons
on oxygen in
sp^2 orbitals

H

Forming the
π bond in
methanal

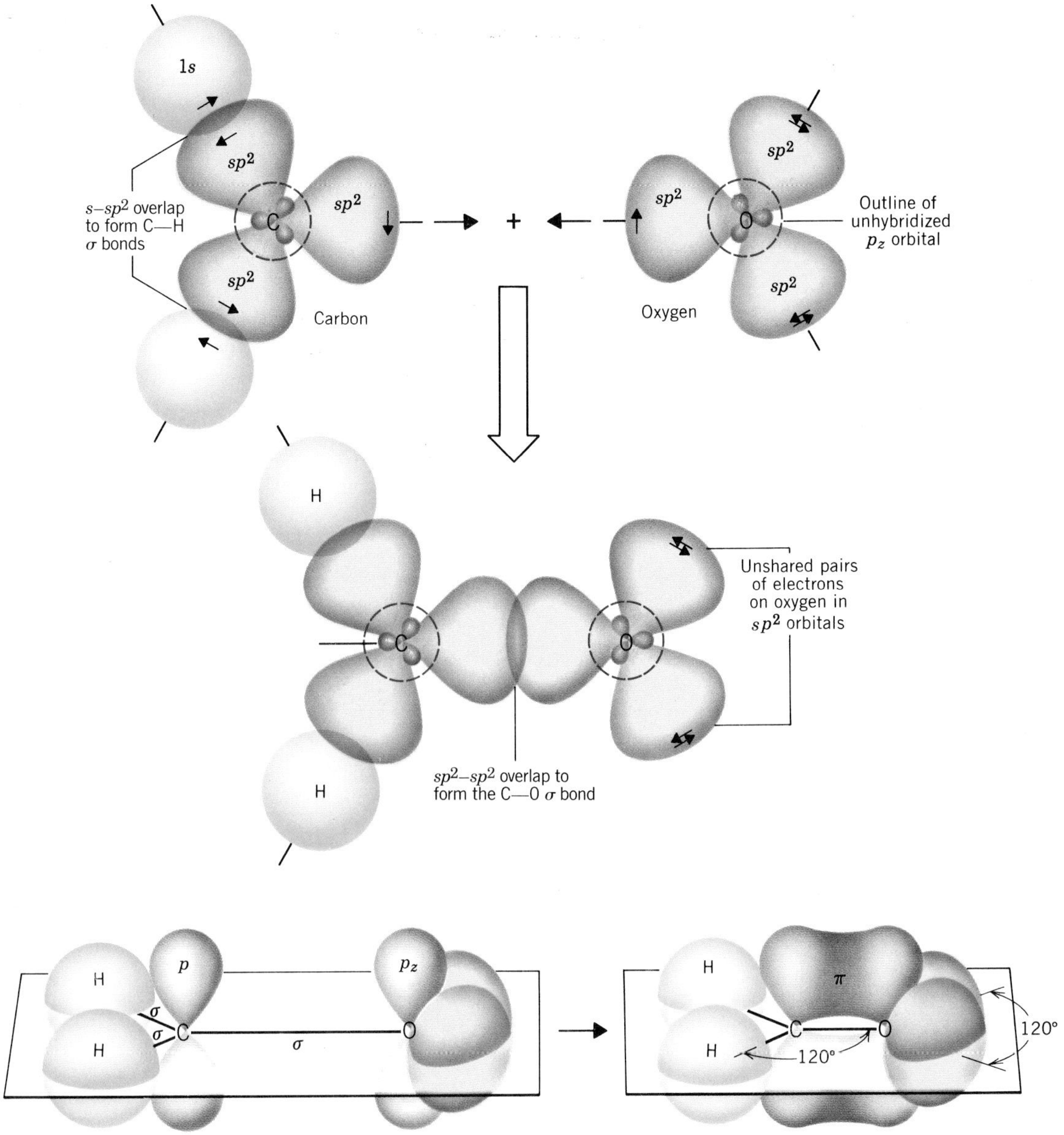

H

p

σ
σ
C

H

p_z

σ

O

H

π

C

120°

H

120°

O

120°

FIGURE 8.17 The carbon–oxygen
double bond in methanal (formaldehyde).

2s and one of the 2p orbitals to form *sp* hybrids. To help us visualize the bonding, we will
imagine that there is an *xyz* coordinate system centered at each carbon atom and that it is
the $2p_x$ orbital that becomes mixed in the hybrid orbitals.

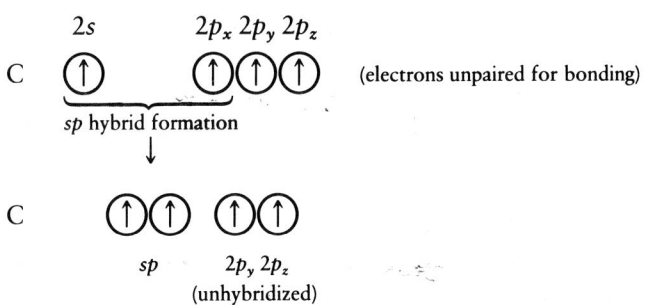

2s $2p_x$ $2p_y$ $2p_z$

C   ↑   ↑ ↑ ↑   (electrons unpaired for bonding)

sp hybrid formation

C   ↑ ↑   ↑ ↑

sp $2p_y$ $2p_z$
        (unhybridized)

We label the orbitals p_x, p_y, and
p_z just for convenience—they
are really all equivalent.

FIGURE 8.18 The carbon–carbon triple bond in ethyne (acetylene).

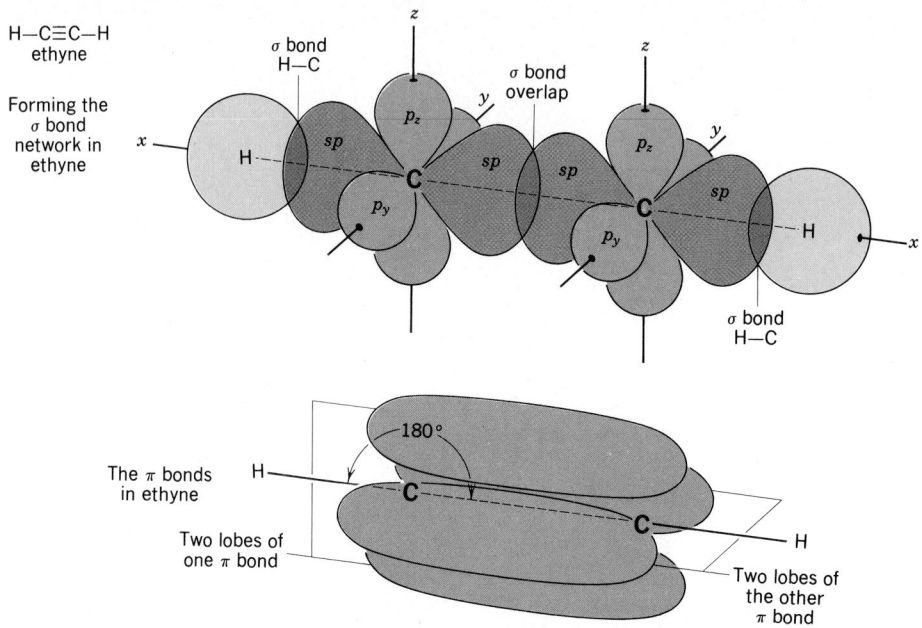

H—C≡C—H
ethyne

Forming the σ bond network in ethyne

The π bonds in ethyne

Two lobes of one π bond

Figure 8.18 shows how the molecule is formed. The sp orbitals point in opposite directions and are used to form the σ bonds. The unhybridized $2p_y$ and $2p_z$ orbitals are perpendicular to the C—C bond axis and overlap sideways to form *two* separate π bonds, which surround the C—C σ bond. Notice that we now have three pairs of electrons in three bonds — one σ bond and two π bonds — whose electron densities are concentrated in different places. The three electron pairs therefore manage to avoid each other as much as possible. Also notice that the use of sp hybrid orbitals for the σ bonds allows us to predict that the molecule will be linear, so all four atoms should lie on the same straight line. This is, in fact, the structure that has been found for acetylene experimentally.

FIGURE 8.19 The triple bond in nitrogen, N_2.

:N≡N:

Nitrogen
The σ bond in N_2 is made by the overlap of two sp hybrid orbitals. Each nitrogen has an unshared pair of electrons in its other sp orbital.

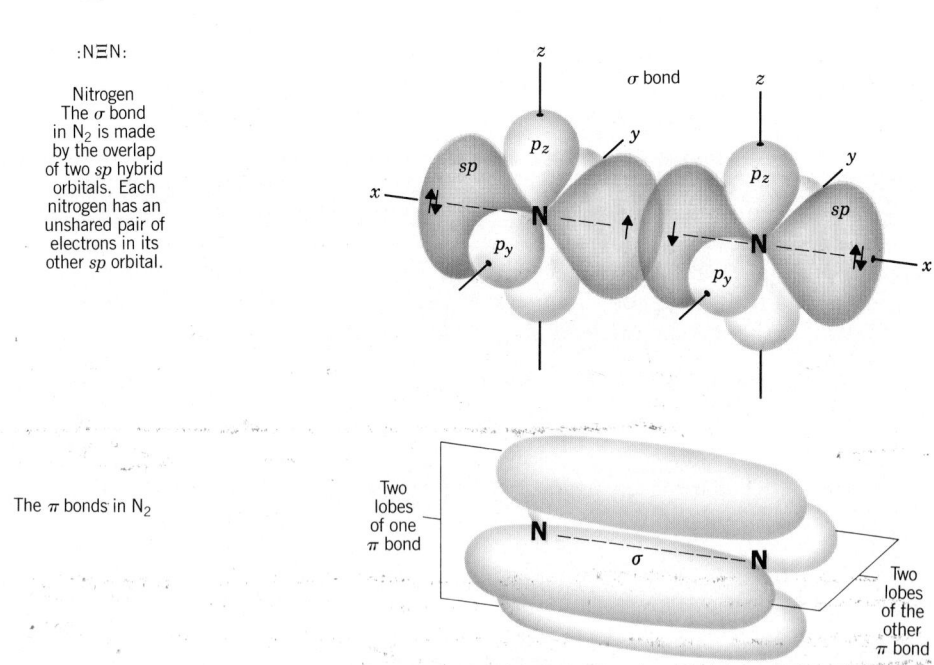

The π bonds in N_2

Similar descriptions can be used to explain the bonding in other molecules that have triple bonds. Figure 8.19, for example, shows how the nitrogen molecule, N_2, is formed. In it, too, the triple bond is composed of one σ bond and two π bonds.

8.6 BOND ENERGIES AND THEIR MEASUREMENT

Heats of formation provide data from which bond energies can be computed.

We have seen that the valence bond theory allows explanation of the formation of covalent bonds. By appropriate combinations and overlaps of orbitals, atoms can form single, double, or triple bonds, which adequately explains the structures of molecules. In another sense, the structures themselves serve as a check on the theory; if the theory didn't account for the structures correctly, it would be discarded. Structures, however, are not the only check on bonding theories. There are other bond properties that are related to the amount of electron density in a bond that can tell us whether or not a theory is on the right track. One such property, which has a profound influence on the chemical properties of substances, is the bond energy.

The concept of bond energy was introduced in Chapter 7 in our discussion of the failure of Lewis structures to adequately describe the bonding in certain molecules, and the need for the concept of resonance. The bond energy — the amount of energy needed to break a chemical bond to give electrically neutral fragments — is a very useful quantity to know in the study of chemical properties. Bond energies tell us a lot about a molecule. For example, they can tell us whether there are single, double, or triple bonds in the molecule. As we saw in the last chapter, the bond energy increases as the number of bonds between two atoms increases from one to two to three.

Bond energies are also important in determining the chemical properties of a substance. During chemical reactions, bonds within the reactants are broken and new ones are formed as the products appear. The first step — bond breaking — is one of the factors that controls the reactivity of substances. Elemental nitrogen, for example, has a very low degree of reactivity, which is generally attributed to the very strong triple bond in N_2. Any reactions that involve the breaking of this bond in one step simply do not occur. When N_2 does react, it is by a stepwise involvement of its three bonds, one at a time.

Bond energy is also called bond dissociation energy.

Bond strength relates to bond energy. Strong bonds have large bond energies.

The formation of the nitrogen oxides from the elements is endothermic.

Measurement of Bond Energies

The bond energies of simple molecules such as H_2, O_2, and Cl_2 are usually measured spectroscopically. That is, the light that is emitted by the molecules when they are energized (excited) by a flame or an electric arc is analyzed, and the amount of energy needed to break the bond is computed.

For more complex molecules, thermochemical data can be used to calculate bond energies in a Hess's law kind of calculation. We will use the standard heat of formation of methane, CH_4, to illustrate how this is accomplished. However, before we can attempt such a calculation, we must take a moment to define a new thermochemical quantity that we will call the atomization energy, symbolized ΔH_{atom}. This is the amount of energy that is needed to rupture all the chemical bonds in one mole of gaseous molecules to give gaseous atoms as products. For example, the change corresponding to the atomization of methane is

$$CH_4(g) \longrightarrow C(g) + 4H(g)$$

and the enthalpy change for the process is ΔH_{atom}. For this particular molecule, ΔH_{atom} corresponds to the total amount of energy needed to break all the C—H bonds in one mole of CH_4; therefore, division of ΔH_{atom} by 4 would give the average C—H bond energy in methane, expressed in kJ/mol (or kcal/mol).

The emission spectrum of a molecule gives information about the energy levels in the molecule, just as the emission spectrum of an atom gives information about atomic energy levels.

The kind of bond breaking described here divides the electrons of the bond between the two atoms. It could be symbolized as

$$A\!:\!B \longrightarrow A\cdot + \cdot B$$

Atomization energies are always endothermic. The formation of a molecule from its gaseous elements is always exothermic.

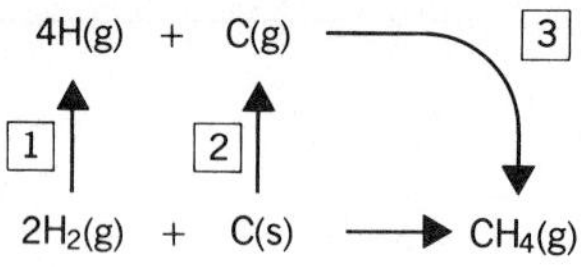

FIGURE 8.20 Two paths for the formation of methane from its elements in their standard states. Steps 1, 2, and 3 of the upper path involve the formation of gaseous atoms of the elements and the formation of the bonds in CH_4.

When the equations are added, $C(g)$ and $4H(g)$ cancel.

TABLE 8.2 Standard Heats of Formation of Gaseous Atoms from the Elements

Atom	ΔH_f° kJ/mol	kcal/mol
C	715.0	170.9
N	472.7	113.0
O	249.2	59.56
H	218.0	52.10
F	78.91	18.86
Cl	121.0	28.92
Br	111.9	26.74
S	274.7	65.65

TABLE 8.3 Average Bond Energies in Polyatomic Molecules

Bond	Bond Energy kJ/mol	kcal/mol
C—H	413	98.8
C—O	351	84.0
C=O	715	171
C—C	348	83.1
C=C	615	147
C≡C	812	194
C—N	292	69.7
C=N	615	147
C≡N	891	213
C—F	439	105
C—Cl	328	78.5
H—O	464	111
H—N	391	93.4

Figure 8.20 shows how we can use the standard heat of formation, ΔH_f°, to calculate the atomization energy. Across the bottom we have the chemical equation for the formation of CH_4 from its elements. The enthalpy change for this reaction, of course, is ΔH_f°. In this figure we also can see an alternative three-step path that leads to $CH_4(g)$. One step is the breaking of H—H bonds in the H_2 molecules to give gaseous hydrogen atoms, another is the vaporization of carbon to give gaseous carbon atoms, and the third is the combination of the gaseous atoms to form CH_4 molecules. These changes are labeled 1, 2, and 3 in the figure.

Since H is a state function, the net enthalpy change from one state to another is the same regardless of the path that we follow. This means that the sum of the enthalpy changes along the upper path must be the same as the enthalpy change along the lower path, ΔH_f°. Perhaps this can be more easily seen in Hess's-law terms if we write the changes along the upper path in the form of thermochemical equations.

Steps 1 and 2 have enthalpy changes that are called *standard heats of formation of gaseous atoms*. Values for these quantities have been measured for many of the elements, and some of them are given in Table 8.2. Step 3 is the opposite of atomization, and its enthalpy change will therefore be the negative of ΔH_{atom}. (recall that if we reverse a reaction, we change the sign of its ΔH).

(Step 1)	$2H_2(g) \longrightarrow 4H(g)$	$\Delta H_1^\circ = 4\Delta H_f^\circ[H(g)]$
(Step 2)	$C(s) \longrightarrow C(g)$	$\Delta H_2^\circ = \Delta H_f^\circ[C(g)]$
(Step 3)	$4H(g) + C(g) \longrightarrow CH_4(g)$	$\Delta H_3^\circ = -\Delta H_{atom}$
	$2H_2(g) + C(s) \longrightarrow CH_4(g)$	$\Delta H^\circ = \Delta H_f^\circ[CH_4(g)]$

Notice that by adding the first three equations, we get the equation for the formation of CH_4 from its elements in their standard states. This means that adding the ΔH° values of the first three equations should give ΔH_f° for CH_4.

$$\Delta H_1^\circ + \Delta H_2^\circ + \Delta H_3^\circ = \Delta H_f^\circ[CH_4(g)]$$

Let's substitute for $\Delta H_1^\circ, \Delta H_2^\circ,$ and ΔH_3°, and then solve for ΔH_{atom}. First, we substitute for the ΔH° quantities.

$$4\Delta H_f^\circ[H(g)] + \Delta H_f^\circ[C(g)] + (-\Delta H_{atom}) = \Delta H_f^\circ[CH_4(g)]$$

Next, we solve for $(-\Delta H_{atom})$

$$-\Delta H_{atom} = \Delta H_f^\circ[CH_4(g)] - 4\Delta H_f^\circ[H(g)] - \Delta H_f^\circ[C(g)]$$

Changing signs and rearranging the right side of the equation just a bit gives

$$\Delta H_{atom} = 4\Delta H_f^\circ[H(g)] + \Delta H_f^\circ[C(g)] - \Delta H_f^\circ[CH_4(g)]$$

Now all we need are values for the ΔH_f°'s on the right side. From Table 8.2 we obtain $\Delta H_f^\circ[H(g)]$ and $\Delta H_f^\circ[C(g)]$, and the value of $\Delta H_f^\circ[CH_4(g)]$ is obtained from Table 5.2.

$$\Delta H_f^\circ[H(g)] = 218.0 \text{ kJ/mol}$$
$$\Delta H_f^\circ[C(g)] = 715.0 \text{ kJ/mol}$$
$$\Delta H_f^\circ[CH_4(g)] = -74.8 \text{ kJ/mol}$$

Substituting these values gives

$$\Delta H_{atom} = 1662 \text{ kJ/mol (rounded)}$$

and division by 4 gives an estimate of the average C—H bond energy in this molecule.

$$\text{Bond energy} = \frac{1662 \text{ kJ/mol}}{4}$$
$$= 416 \text{ kJ/mol of C—H bonds}$$

This value is quite close to the one in Table 8.3, which is an average of the C—H bond energies in many different compounds. The other bond energies in Table 8.3 are also based on thermochemical data and were obtained by similar calculations.

Uses of Bond Energies

An amazing thing about covalent bond energies is that they are very nearly the same in many different compounds. This suggests, for example, that a C—H bond is very nearly the same in CH_4 as it is in a large number of other compounds that contain this kind of bond. That alone is very useful information, because it explains why molecules having similar bonds behave in similar ways. If the properties of a particular kind of bond varied much from compound to compound, this would not be true.

Because the bond energy doesn't vary much from compound to compound, we can use tabulated bond energies to estimate the heats of formation of substances. This is illustrated in Example 8.5.

EXAMPLE 8.5 USING TABULATED BOND ENERGIES

Problem: Use the bond energies in Table 8.3 to estimate the heat of formation of methanol vapor, $CH_3OH(g)$. The structural formula for methanol is

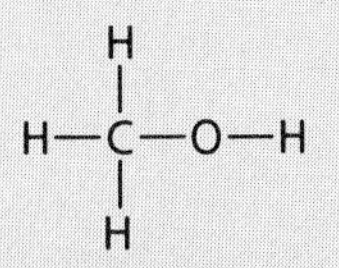

$$H-\underset{\underset{H}{|}}{\overset{\overset{H}{|}}{C}}-O-H$$

Solution: We solve this problem in much the same way that we calculated the bond energy of methane in the discussion above. We set up two paths from the elements to the compound, as shown in Figure 8.21. The lower path has an enthalpy change corresponding to $\Delta H_f^\circ[CH_3OH(g)]$, while the upper path takes us to the gaseous elements and then through the energy released when the bonds in the molecule are formed. This latter energy can be computed from the bond energies in Table 8.3. As before, the sum of the energy changes along the upper path must be the same as the energy change along the lower path, and this permits us to compute $\Delta H_f^\circ[CH_3OH(g)]$.

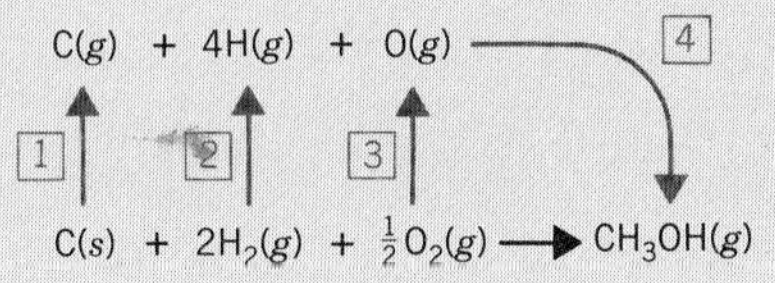

FIGURE 8.21 The formation of methanol vapor from its elements, following alternative paths.

$$C(g) + 4H(g) + O(g) \xrightarrow{\boxed{4}}$$
$$\boxed{1}\uparrow \quad \boxed{2}\uparrow \quad \boxed{3}\uparrow \qquad \downarrow$$
$$C(s) + 2H_2(g) + \tfrac{1}{2}O_2(g) \longrightarrow CH_3OH(g)$$

Steps 1, 2, and 3 involve the formation of the gaseous atoms from the elements, and their enthalpy changes are gotten from Table 8.2.

$$\Delta H_1^\circ = \Delta H_f^\circ[C(g)] = 1\ mol \times 715.0\ kJ/mol = 715.0\ kJ$$
$$\Delta H_2^\circ = 4\Delta H_f^\circ[H(g)] = 4\ mol \times (218.0\ kJ/mol) = 872.0\ kJ$$
$$\Delta H_3^\circ = \Delta H_f^\circ[O(g)] = 1\ mol \times (249.2\ kJ/mol) = 249.2\ kJ$$

Adding these values gives a total energy input for the first three steps of 1836.2 kJ. In other words, the net ΔH° for the first three steps is +1836.2 kJ.

The formation of the CH_3OH molecule from the gaseous atoms is exothermic—energy is always released when atoms become joined by a covalent bond. In this molecule we can count three C—H bonds, one C—O bond, and one O—H bond. Their formation releases energy equal to their bond energies, which we obtain from Table 8.3.

Bond	Energy (kJ)
3(C—H)	3 × (413 kJ/mol) = 1239
C—O	351
O—H	464

Adding these together gives a total of 2054 kJ. ΔH° for this step is therefore −2054 kJ.

(because it is exothermic). Now we can compute the total enthalpy change for the upper path.

$$\Delta H° = (+1836 \text{ kJ}) + (-2054 \text{ kJ})$$
$$\Delta H° = -218 \text{ kJ}$$

The value just calculated must be equal to the $\Delta H°$ for the lower path in Figure 8.21, which is $\Delta H_f°$ for $CH_3OH(g)$. Experimentally, it has been found that $\Delta H_f°$ for this molecule (in the vapor state) is -201 kJ/mol. At first glance, the agreement doesn't seem very good, but on a relative basis the calculated value (-218 kJ) differs from the experimental one by only about 8%.

PRACTICE EXERCISE 6   Using the data in Table 8.3, calculate the atomization energy of formic acid, $HCHO_2$, which has the structural formula

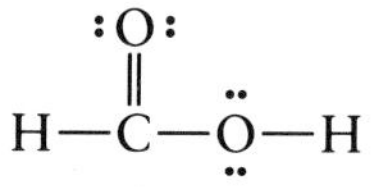

$$
\begin{array}{c}
:\!O\!: \\
\|\\
H\!-\!C\!-\!\ddot{O}\!-\!H \\
\cdot\cdot
\end{array}
$$

PRACTICE EXERCISE 7   Use the data in Tables 8.2 and 8.3 to calculate the value of $\Delta H_f°$ for $CH_3Cl(g)$. Compare your calculated value to the experimentally measured $\Delta H_f°$ given in Table 5.2.

Comparisons between measured and calculated bond energies have sometimes helped chemists understand unusual properties of some substances. Consider, for example, the case of benzene, C_6H_6. This molecule has its six carbon atoms arranged in a hexagonal ring, and bonded to each carbon is one of the hydrogen atoms. One of the Lewis structures that we can draw for the molecule is

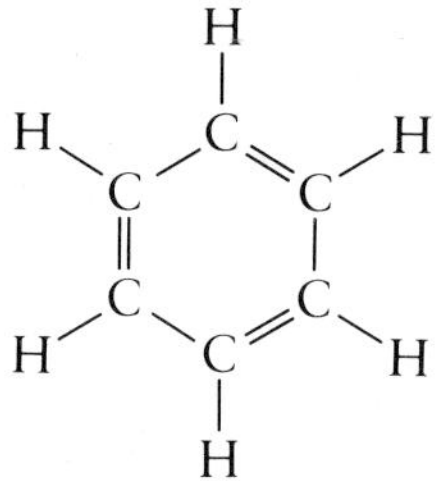

Benzene, however, doesn't behave chemically as if this is its structure. It doesn't undergo reactions that are typical of hydrocarbons with carbon–carbon double bonds, and the ring structure has a great tendency to remain intact during chemical reactions. This Lewis structure is inadequate in other ways, too. In Chapter 7 you learned that the bond length is related to the number of electron pairs that are shared between two bonded atoms. On that basis we would anticipate that some of the carbon–carbon bonds in benzene should be longer than others. Actually, they are all the same, and to account for this we say that the actual structure of benzene is a resonance hybrid. The resonance structures that are drawn to represent the hybrid are

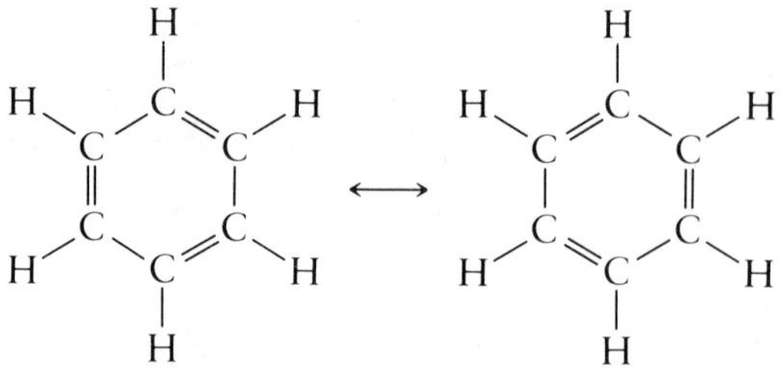

If we used either one of these Lewis structures as if it really existed in an attempt to calculate the heat of formation of C_6H_6, just as in Example 8.5, the value we would obtain is about $+230$ kJ/mol. However, the experimentally measured ΔH_f° for C_6H_6 is only about $+84$ kJ/mol. The difference between the calculated and experimental values is much too large to be just experimental error, and it means that the actual molecule is more stable than we computed it to be by about 146 kJ/mol.

In general, it is found that for molecules for which resonance is used to explain equality of bond lengths—as in benzene here and in molecules such as SO_2 and SO_3, which were discussed in Chapter 7—the actual ΔH_f° is more exothermic (less endothermic) than the ΔH_f° calculated using tabulated bond energies. In other words, resonance leads to an additional stabilization of the molecule, and the energy difference between the calculated and actual ΔH_f° is called the resonance energy or stabilization energy. We will discuss the bonding in these kinds of molecules in more detail later in this chapter.

8.7   MOLECULAR ORBITAL THEORY

The overlap of atomic orbitals produces two kinds of "molecular orbitals"— bonding orbitals, which help stabilize a molecule, and antibonding orbitals, which tend to destabilize a molecule.

Molecular orbital theory takes the view that a molecule is really not too much different from an atom. They are different, of course, because a molecule has several positive nuclei instead of only one. They are similar, however, in that both an atom and a molecule have energy levels that correspond to various orbitals that can be populated with electrons. In atoms, these orbitals are called atomic orbitals; in molecules, they are called molecular orbitals. (We will frequently call them MOs.)

In general, the actual shapes and energies of the molecular orbitals of a molecule cannot be determined exactly. Nevertheless, theoreticians have found that reasonably good estimates of their shapes and energies can be obtained by combining the electron waves corresponding to the atomic orbitals of the atoms that make up the molecule. These waves interact by constructive or destructive interference just like other waves that we've seen. Their intensities are either added or subtracted when the atomic orbitals overlap. The way this occurs can be seen if we look at the overlap of a pair of $1s$ orbitals from two atoms in a molecule like H_2 (Figure 8.22). The *two* $1s$ orbitals combine when the molecule is formed to give *two* MOs. In one, the electron density from the two orbitals is added together between the nuclei. This gives a buildup of negative charge and helps hold the nuclei near each other. Such an MO is said to be a bonding molecular orbital. In the other MO, cancellation of the electron waves reduces the electron density between the nuclei. The absence of much negative charge between the nuclei allows them to repel each other strongly, so this MO is called an antibonding molecular orbital. Antibonding MOs tend to destabilize a molecule when occupied by electrons.

The number of MOs formed is always equal to the number of atomic orbitals that are combined.

FIGURE 8.22 Interaction of $1s$ atomic orbitals to produce bonding and antibonding molecular orbitals. These are σ-type orbitals because the electron density is symmetrical around the imaginary line that passes through both nuclei.

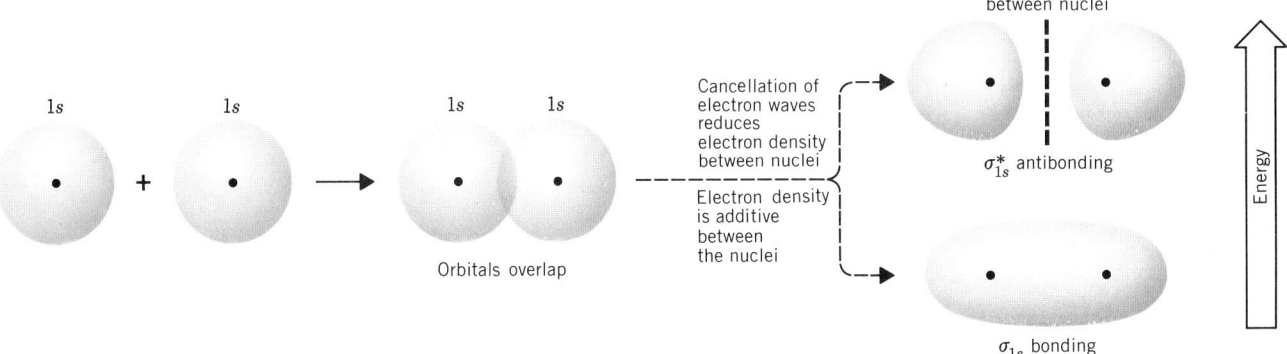

Both bonding and antibonding MOs formed by the overlap of *s* orbitals have their maximum electron density on an imaginary line that passes through the two nuclei. Earlier, we called bonds that have this property sigma bonds. Molecular orbitals like this are also designated as sigma (σ), an asterisk is used to indicate the MOs that are antibonding, and a subscript is written to report which atomic orbitals make up the MO. For example, the bonding MO formed by the overlap of $1s$ orbitals is symbolized as σ_{1s} and the antibonding MO is written as σ_{1s}^*.

Bonding MOs are lower in energy than antibonding MOs formed from the same atomic orbitals. This is also depicted in Figure 8.22. When electrons populate molecular orbitals, they fill the lower-energy, bonding MOs first. The rules that apply to filling MOs are the same as those for filling atomic orbitals: *electrons spread out over orbitals of equal energy and two electrons can only occupy the same orbital if their spins are paired.*

Why Some Molecules Exist and Others Do Not

Let's see how molecular orbital theory can be used to account for the existence of certain molecules, as well as the nonexistence of others. Figure 8.23 is an MO energy level diagram for H_2. The energies of the separate $1s$ orbitals are indicated at the left and right; those of the molecular orbitals are shown in the center. The H_2 molecule has two electrons, and both can be placed in the σ_{1s} orbital. The shape of this bonding orbital, shown in Figure 8.22, should be familiar. It's the same as the shape of the electron cloud that we described using the valence bond theory.

FIGURE 8.23 Molecular orbital energy-level diagram for H_2.

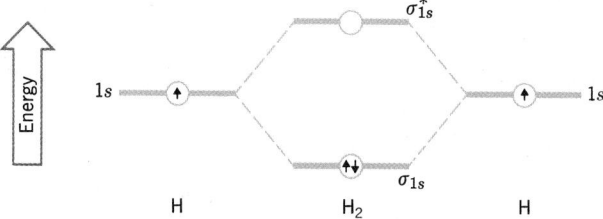

Next, let's consider what happens when two helium atoms come together. Why can't a stable molecule of He_2 be formed? Figure 8.24 is the energy diagram for He_2. Notice that both bonding and antibonding orbitals are filled. In situations such as this there is a net destabilization because the antibonding MO is raised in energy, relative to the orbitals of the separated atoms, more than the bonding MO is lowered. Thus, the total energy of He_2 is greater than that of two separate He atoms, so the "molecule" immediately comes apart. In general, the effects of *antibonding electrons* (those in antibonding MOs) cancel the effects of an equal number of bonding electrons, and molecules with equal numbers of bonding and antibonding electrons are unstable. If we remove an antibonding electron from He_2 to give He_2^+, there would be a net excess of bonding electrons. The existence of the He_2^+ ion should therefore be possible. Actually, He_2^+ can be observed when an electric discharge is passed through a helium-filled tube.

He_2^+ has two bonding electrons and one antibonding electron.

FIGURE 8.24 Molecular orbital energy-level diagram for He_2.

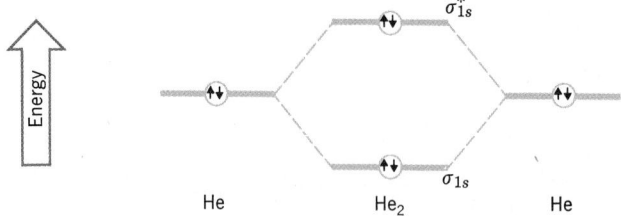

Bond Orders

According to the classical description, covalent bonds are formed by the sharing of *pairs* of electrons — one pair gives a single bond, two pairs give a double bond, and three pairs give a triple bond. To translate the MO description into these terms, we can compute the bond order as

$$\text{bond order} = \frac{(\text{number of bonding electrons}) - (\text{number of antibonding electrons})}{2}$$

For the H_2 molecule, we have

$$\text{bond order} = \frac{2-0}{2} = 1$$

A bond order of 1 corresponds to a single bond. For He_2 we have

$$\text{bond order} = \frac{2-2}{2} = 0$$

A bond order of zero means no bond exists. The He_2 molecule is unable to exist; however, the He_2^+ ion is able to exist, and its calculated bond order is $\frac{1}{2}$.

The bond order is the number of *pairs* of electrons shared between the two atoms. Note that it doesn't have to be a whole number.

Bonding in Diatomic Molecules of Period 2

The outer shell of a period-2 element consists of $2s$ and $2p$ subshells. When atoms of this period bond to each other, the atomic orbitals of these subshells interact strongly to produce molecular orbitals. The $2s$ orbitals, for example, overlap to form σ_{2s} and σ_{2s}^* molecular orbitals having essentially the same shapes as the σ_{1s} and σ_{1s}^* molecular orbitals, respectively. Figure 8.25 shows the shapes of the bonding and antibonding MOs pro-

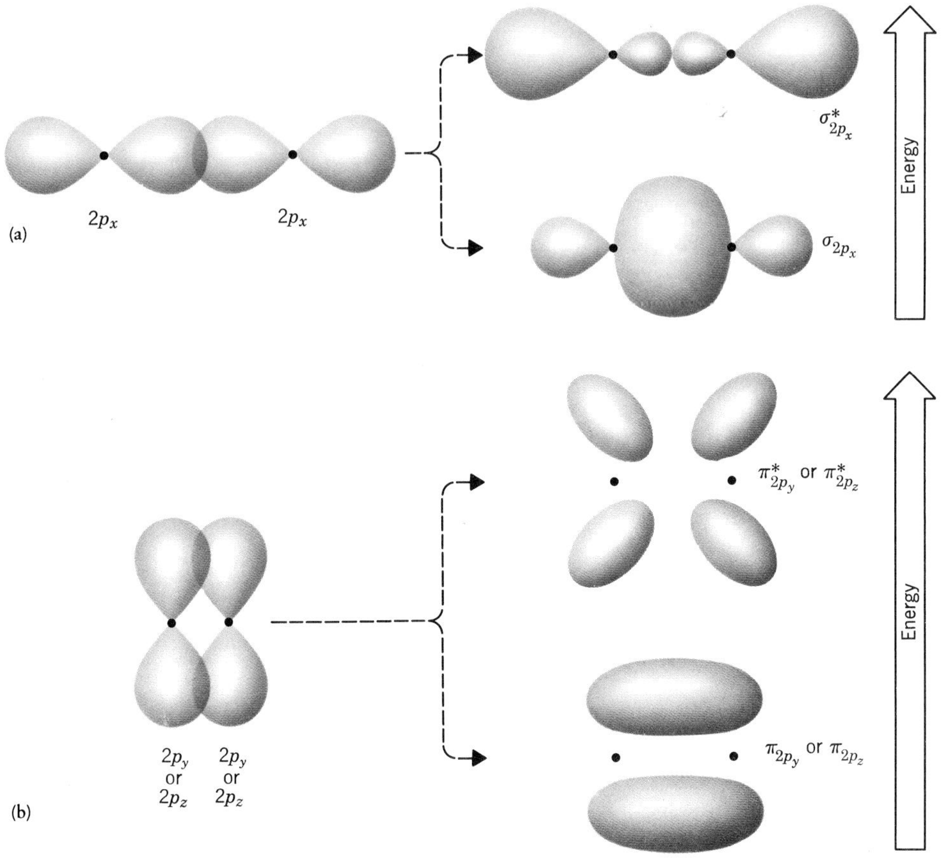

FIGURE 8.25 Production of molecular orbitals by the overlap of p orbitals. *(a)* Two p_x orbitals that point at each other give bonding and antibonding σ-type MOs. *(b)* Perpendicular to the $2p_x$ orbitals are $2p_y$ and $2p_z$ orbitals that overlap to give two sets of bonding and antibonding π-type MOs.

duced when the $2p$ orbitals overlap. If we label those that point toward each other as $2p_x$, a set of bonding and antibonding MOs are formed that we can label as σ_{2p_x} and $\sigma^*_{2p_x}$. The $2p_y$ and $2p_z$ orbitals, which are perpendicular to the $2p_x$ orbitals, then overlap sideways to give π-type molecular orbitals. They are labeled π_{2p_y} and $\pi^*_{2p_y}$, and π_{2p_z} and $\pi^*_{2p_z}$.

The approximate relative energies of the MOs formed from the second shell orbitals are shown in Figure 8.26. Using this energy diagram, we can predict the electronic structures of diatomic molecules of period 2. These *MO electron configurations* are obtained using the same rules that are applied to the filling of atomic orbitals in atoms.

1. No more than two electrons, with spins paired, can occupy any orbital.
2. Electrons fill the lowest-energy orbitals that are available.
3. Electrons spread out as much as possible, with spins unpaired, over orbitals that have the same energy.

FIGURE 8.26 Approximate relative energies of molecular orbitals in second period diatomic molecules.

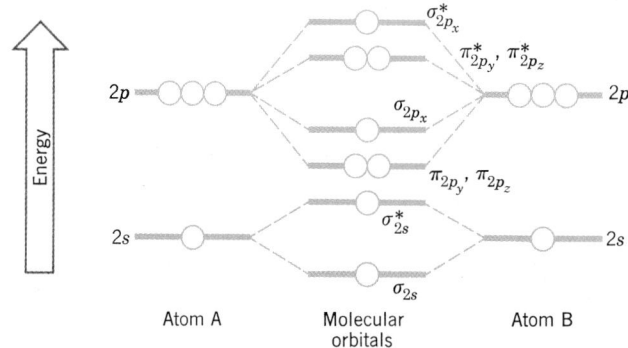

Applying these rules to the valence electrons of period-2 atoms gives the MO electron configurations shown in Table 8.4. Let's see how well MO theory performs by examining data that are available for these molecules.

TABLE 8.4 Molecular Orbital Populations and Bond Orders for Period-2 Diatomic Molecules[a]

		Li_2	Be_2	B_2	C_2	N_2	O_2	F_2	Ne_2
	$\sigma^*_{2p_x}$	◯	◯	◯	◯	◯	◯	◯	⬆⬇
	$\pi^*_{2p_y},\pi^*_{2p_z}$	◯◯	◯◯	◯◯	◯◯	◯◯	⬆ ⬆	⬆⬇ ⬆⬇	⬆⬇ ⬆⬇
	σ_{2p_x}	◯	◯	◯	◯	⬆⬇	⬆⬇	⬆⬇	⬆⬇
Energy	π_{2p_y},π_{2p_z}	◯◯	◯◯	⬆ ⬆	⬆⬇ ⬆⬇	⬆⬇ ⬆⬇	⬆⬇ ⬆⬇	⬆⬇ ⬆⬇	⬆⬇ ⬆⬇
	σ^*_{2s}	◯	⬆⬇	⬆⬇	⬆⬇	⬆⬇	⬆⬇	⬆⬇	⬆⬇
	σ_{2s}	⬆⬇	⬆⬇	⬆⬇	⬆⬇	⬆⬇	⬆⬇	⬆⬇	⬆⬇
Number of bonding electrons		2	2	4	6	8	8	8	8
Number of antibonding electrons		0	2	2	2	2	4	6	8
Bond order		1	0	1	2	3	2	1	0
Bond energy (kJ/mol)		110	—	300	612	953	501	129	—
Bond length (pm)		267	—	158	124	109	121	144	—

[a] σ_{2p_x} is thought to be lower in energy than π_{2p_y} and π_{2p_z} in O_2 and F_2, but this does not affect our conclusions about the net number of bonds in these molecules or the number of unpaired electrons that they have. Therefore, we have used the same energy diagram for all period-2 diatomic molecules to minimize confusion.

According to Table 8.4, MO theory predicts that molecules of Be_2 and Ne_2 should not exist at all because they have bond orders of zero. In beryllium vapor and in gaseous neon, no evidence of Be_2 or Ne_2 has ever been found. MO theory also predicts that diatomic molecules of the other period-2 elements should exist because they all have bond orders greater than zero. These molecules have, in fact, been observed. Although lithium, boron, and carbon are complex solids under ordinary conditions, they can be vaporized, and, in the vapor, molecules of Li_2, B_2, and C_2 can be detected. Nitrogen, oxygen, and fluorine, as you know, are gaseous elements that exist as N_2, O_2, and F_2.

In Table 8.4, we also see that the predicted bond order increases from boron to carbon to nitrogen and then decreases from nitrogen to oxygen to fluorine. As the bond order increases, the net number of bonding electrons increases, so more electron density is concentrated between the nuclei. This greater concentration of negative charge should bind the nuclei more tightly and therefore give a stronger bond. The attraction between the increased electron density and the positive nuclei should also draw the nuclei closer to the center of the bond, thereby decreasing the bond length. The *experimentally measured* bond energies and bond lengths given in Table 8.4 follow these predictions quite nicely.

Molecular orbital theory is particularly successful in explaining the electronic structure of the oxygen molecule. Experiments have shown that O_2 is paramagnetic; it contains two unpaired electrons. In addition, the bond length in O_2 is about what is expected for an oxygen–oxygen double bond. These data are not explained by valence bond theory. For example, if we write a Lewis structure for O_2 that shows a double bond and also obeys the octet rule, all the electrons appear in pairs.

Molecular oxygen is attracted weakly by a magnet.

:Ö::Ö: (not acceptable based on experimental evidence, because all electrons are paired)

On the other hand, if we indicate the unpaired electrons, the structure shows only a single bond and doesn't obey the octet rule.

:Ö:Ö: (not acceptable based on experimental evidence, because of the O—O single bond)

With MO theory, we don't have any of these difficulties. According to this theory, the two electrons that are placed in the π^* orbitals of O_2 will spread out over these orbitals with their spins unpaired because these orbitals have the same energy. The electrons in the two antibonding π^* orbitals cancel the effects of two electrons in the two bonding π orbitals, so the net bond order is 2 and the bond is a double bond.

PRACTICE EXERCISE 8   The MO energy level diagram for the nitric oxide molecule, NO, is essentially the same as that shown in Table 8.4 for the period-2 diatomic molecules. Indicate which MOs are populated in NO and calculate the bond order for the molecule.

8.8 DELOCALIZED MOLECULAR ORBITALS

When the unhybridized p orbitals on three or more atoms are arranged so that they form a continuous overlapping sequence, the shared pi electrons can spread over more than two atoms.

One of the least satisfying aspects of the way valence bond theory explains chemical bonding is the need to write resonance structures for certain molecules and ions. In

Chapter 7 we discussed this in terms of Lewis structures, which we have since learned are really equivalent to shorthand notations for valence bond descriptions of molecules. For example, let's look at the formate ion, CHO_2^-, which we described in Chapter 7 as a resonance hybrid of the following two structures.

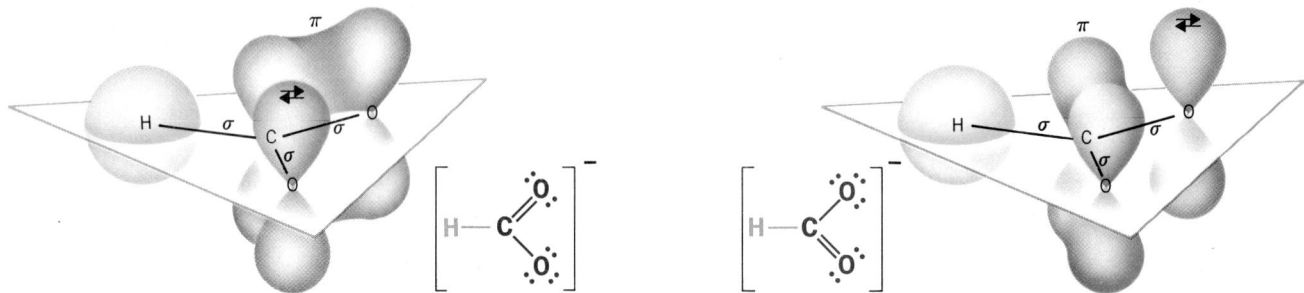

FIGURE 8.27 Valence bond descriptions of the resonance structures of the formate ion, CHO_2^-. Only the p orbitals of oxygen that can form π bonds to carbon are shown. The ion is planar because carbon uses sp^2 orbitals to form the σ-bond framework.

In terms of the overlap of orbitals, VB theory would picture the formate ion as shown in Figure 8.27. The σ-bond framework is formed by overlap of the sp^2 hybrid orbitals of carbon with the $1s$ orbital of hydrogen and the p orbitals of the oxygen atoms. As you can see in Figure 8.27, the π bond can be formed with either oxygen atom, which is how we obtain the two resonance structures.

Molecular orbital theory avoids the problem of resonance by recognizing that electron pairs can sometimes be shared among overlapping orbitals from three or more atoms. In the CHO_2^- ion, it allows all *three p* orbitals to overlap so as to form one large π-type molecular orbital that spreads over all three nuclei as shown in Figure 8.28. Since the π electrons in this MO are not required to stay "localized" between just two nuclei, we say that the bond is delocalized. Delocalized π-type molecular orbitals permit a single description of the electronic structure of molecules and ions.

FIGURE 8.28 Molecular orbital theory allows the electron pair in the π-type bond to be spread out, or delocalized, over all three atoms of the CO_2^- unit in CHO_2^-.

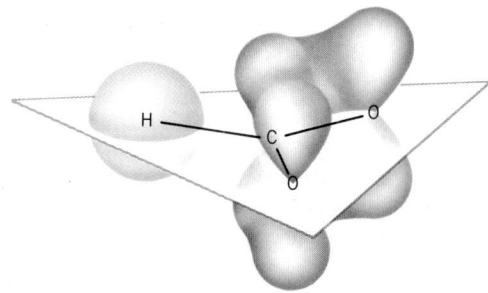

A **localized bond** is one in which the electrons spend all their time shared between just two atoms. The electrons in a delocalized bond become shared among more than two atoms.

The delocalized nature of π bonds that extend over three or more nuclei can be indicated with dotted lines rather than dashes when Lewis-type structural formulas are drawn. The formate ion, for example, can be drawn as

$$\left[H-C \overset{O}{\underset{O}{<}} \right]^-$$

Delocalized π bonds are quite common in many kinds of molecules and ions. Some other examples are

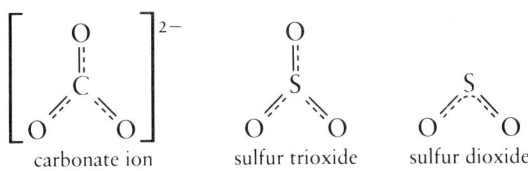

carbonate ion   sulfur trioxide   sulfur dioxide

The bonding in each of these ions or molecules is explained by just one structure in the MO theory.

An important example from the realm of organic chemistry is benzene, C_6H_6. As you learned in Section 8.6, this molecule consists of six carbon atoms arranged in a ring, with a hydrogen atom attached to each carbon (see Figure 8.29). The sigma bond framework of benzene requires that the carbon atoms have sp^2 hybrid orbitals because each carbon forms three σ bonds. This leaves each carbon atom with a half-filled unhybridized p orbital perpendicular to the plane of the ring. These p orbitals overlap to give a delocalized π-electron cloud that looks something like two doughnuts with the sigma bond framework sandwiched between them.

Benzene is an important industrial solvent, but it is also quite toxic.

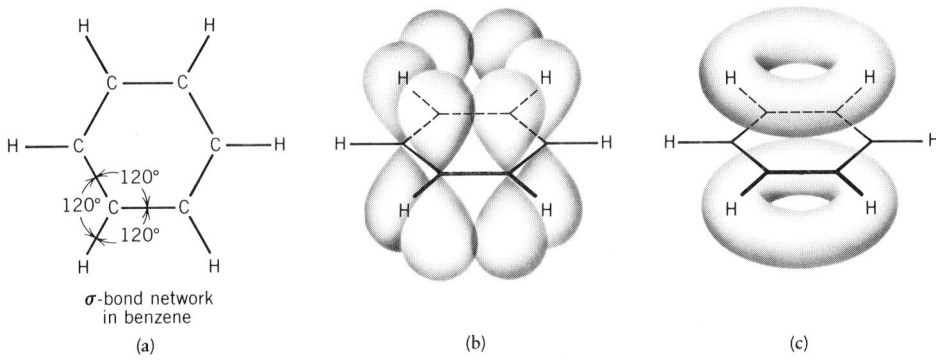

σ-bond network in benzene

(a)   (b)   (c)

FIGURE 8.29 Benzene. *(a)* The σ-bond framework. All atoms lie in the same plane. *(b)* The unhybridized p orbitals at each carbon prior to side-to-side overlap. *(c)* The double doughnut-shaped space for the π electrons.

In our previous discussion of benzene, we described the molecule as a hybrid of two resonance structures.

two resonance structures for benzene

For simplicity, these structures are usually drawn as hexagons. It is assumed that at each corner there is a carbon that is also bonded to a hydrogen.

To indicate the delocalized π-electron cloud, the dashes representing the alternating double bonds are replaced by a circle.

Functionally, the terms *resonance energy* and *delocalization energy* are the same; they just come from different approaches to bonding theory.

One of the special characteristics of delocalized bonds, like those found in CO_3^{2-}, SO_3, SO_2, and C_6H_6, is that they make a molecule or ion more stable than it would be if it had localized bonds. In Section 8.6, for example, the standard heat of formation of benzene was discussed, and we saw that the actual structure of this molecule is considerably more stable than the one with localized single and double bonds. This extra degree of stability, you may recall, is described by the resonance energy, or stabilization energy. In the molecular orbital theory, we no longer speak of resonance — instead, we refer to the electrons as being delocalized. The extra stability that is associated with this delocalization is therefore described, in the language of MO theory, as the delocalization energy.

SUMMARY

Valence Bond (VB) Theory. This is one of two wave mechanics-based theories of chemical bonding that are discussed in this chapter. According to VB theory, a covalent bond is formed between two atoms when an atomic orbital on one atom overlaps with an atomic orbital on the other and a pair of electrons with paired spins is shared between the overlapping orbitals. In general, the better the overlap of the orbitals, the stronger the bond. A given atomic orbital can only overlap with one other on a different atom, so a given atomic orbital can only form one bond with one other atom. Hybrid atomic orbitals, formed by mixing pure s, p, and/or d orbitals, overlap better with other orbitals than the pure atomic orbitals from which they are formed, so bonds formed by hybrid orbitals are stronger than those formed by ordinary atomic orbitals. Sigma bonds (σ bonds) are formed by the following kinds of orbital overlap: $s-s$, $s-p$, end-to-end $p-p$, and overlap of hybrid orbitals. Sigma bonds allow free rotation around the bond axis. The side-by-side overlap of p orbitals produces a pi bond (π bond). Pi bonds do not permit free rotation around the bond axis because doing so would destroy the bond. In other words, rotation about the axis of a π bond involves bond breaking. In complex molecules, the basic molecular framework is built with σ bonds. A double bond consists of one σ bond and one π bond. A triple bond consists of one σ bond and two π bonds.

Bond Energies. The bond energy is the energy needed to break a covalent bond to produce electrically neutral particles. The sum of all the bond energies of a molcule is the atomization energy, ΔH_{atom}. Using two separate paths from the elements to a compound, we can calculate bond energies from ΔH_f° for the compound, or we can estimate ΔH_f° for a compound using tabulated bond energies. In these kinds of calculations, we use tabulated ΔH_f° values for the gaseous atoms of the elements. Molecules whose structures are described by resonance are more stable than any of their individual resonance structures by an amount called the resonance energy.

Molecular Orbital (MO) Theory. This theory begins with the supposition that molecules, like atoms, have to be treated as collections of nuclei and electrons, and that the electrons are in molecular orbitals of different energies. Molecular orbitals can spread over two or more nuclei, and can be considered to be formed by the constructive and destructive interference of the electron waves corresponding to the atomic orbitals of the atoms in the molecule. Bonding MOs concentrate electron density between nuclei; antibonding MOs remove electron density from between nuclei. The rules for the filling of MOs are the same as those for atomic orbitals. The ability of MO theory to describe delocalized orbitals avoids the need for resonance theory. Delocalization of bonds leads to a lowering of the energy and, therefore, to more stable molecular structures.

REVIEW EXERCISES

Answers to questions whose numbers are printed in color are given in Appendix C.

Modern Bonding Theories

8.1   What is the theoretical basis of both valence bond (VB) theory and molecular orbital (MO) theory?

8.2   What shortcomings of Lewis structures and VSEPR theory do VB and MO theories attempt to overcome?

8.3   What is the main difference in the way VB and MO theories view the bonds in a molecule?

Valence Bond Theory

8.4   What is meant by orbital overlap?

8.5   How does VB theory explain electron sharing between atoms in a covalent bond?

8.6   Use drawings to describe how the covalent bond in H_2 is formed.

8.7   Use drawings to describe how VB theory would explain the formation of the H—Br bond in hydrogen bromide.

8.8 Hydrogen selenide is one of nature's most foul-smelling substances. Molecules of H₂Se have H—Se—H bond angles very close to 90°. How would VB theory explain the bonding in H₂Se?

8.9 Use drawings to show how VB theory explains the bonding in the F₂ molecule.

Hybrid Orbitals

8.10 What term is used to describe the mixing of atomic orbitals of the same atom?

8.11 Why do atoms usually prefer to use hybrid orbitals for bonding?

8.12 Sketch figures that illustrate the directional properties of the following hybrid orbitals: (a) *sp*, (b) *sp²*, (c) *sp³*, (d) *sp³d*, (e) *sp³d²*.

8.13 Why do period-2 elements never use *sp³d* or *sp³d²* hybrid orbitals for bond formation?

8.14 What relationship is there, if any, between Lewis structures and valence bond descriptions of molecules?

8.15 Use orbital diagrams to explain how the BeCl₂ molecule is formed. What kind of hybrid orbitals does beryllium use in BeCl₂?

8.16 Draw Lewis structures for the following and use the geometry predicted by VSEPR theory for the electron pairs to determine what kind of hybrid orbitals the central atom uses in bond formation: (a) ClO₃⁻, (b) SO₃, (c) OF₂, (d) SbCl₆⁻, (e) BrCl₃, (f) XeF₄.

8.17 What hybrid orbitals are used by tin in SnCl₆²⁻? Draw the orbital diagram for Sn in SnCl₆²⁻. What is the geometry of SnCl₆²⁻?

8.18 Use orbital diagrams to describe the bonding in (a) SnCl₄ and (b) SbCl₅.

8.19 Use VSEPR theory to help you describe the bonding in the following molecules according to VB theory: (a) AsCl₃, (b) ClF₃, (c) SbCl₅, (d) SeCl₂.

8.20 Using orbital diagrams, describe how *sp³* hybridization occurs in each atom: (a) carbon, (b) nitrogen, (c) oxygen.

8.21 Sketch the way the orbitals overlap to form the bonds in each substance: (a) CH₄, (b) NH₃, (c) H₂O.

8.22 We explained the bond angles of 107° in NH₃ by using *sp³* hybridization of the central nitrogen. Had the original *unhybridized p* orbitals of the nitrogen been used to overlap with 1*s* orbitals of each hydrogen, what would have been the H—N—H bond angles? Why?

8.23 Cyclopropane is a triangular molecule with C—C—C bond angles of 60°. Explain why the σ bonds joining carbon atoms in cyclopropane are weaker than the carbon–carbon bonds in the noncyclic propane.

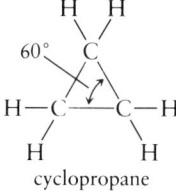
cyclopropane   propane

8.24 What facts about boron strongly suggested the need to consider *sp²* hybridization of its second shell atomic orbitals?

8.25 Using sketches of orbitals and orbital diagrams, describe *sp²* hybridization at (a) boron and (b) carbon.

8.26 Phosphorus trifluoride, PF₃, has F—P—F bond angles of 97.8°.
 (a) How would VB theory use hybrid orbitals to explain these data?
 (b) How would VB theory use unhybridized orbitals to account for these data?
 (c) Do either of these models work very well?

8.27 If the central oxygen in the water molecule did not use *sp³* hybridized orbitals (or orbitals of any other kind of hybridization), what would be the bond angle in H₂O (assuming no angle-spreading force)?

Coordinate Covalent Bonds and VB Theory

8.28 The ammonia molecule, NH₃, can combine with a hydrogen ion, H⁺ (which has an empty 1*s* orbital), to form the ammonium ion, NH₄⁺. (This is how ammonia can neutralize acid and therefore function as a base.) Sketch the geometry of the ammonium ion, indicating the bond angles.

8.29 Use orbital diagrams to show the formation of a coordinate covalent bond in H₃O⁺ when H₂O reacts with H⁺.

Multiple Bonds and Hybrid Orbitals

8.30 What are the characteristics of σ and π bonds?

8.31 Why is free rotation permitted about a σ-bond axis but not about a π-bond axis?

8.32 Using sketches, describe the bonds and bond angles in ethene, C₂H₄.

8.33 Sketch the way the bonds form in ethyne, C₂H₂.

8.34 The six-membered ring of carbons can hold a double bond but not a triple bond. Explain.

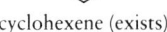
cyclohexene (exists)   cyclohexyne (unknown)

8.35 A nitrogen atom can also undergo *sp²* hybridization when, for example, it becomes part of a carbon–nitrogen double bond, as in H₂C=NH.
 (a) Using a sketch, show the electron configuration of *sp²* hybridized nitrogen just before the overlapping occurs to make this double bond.
 (b) Using sketches (and the analogy to the double bond in C₂H₄), describe the two bonds of the carbon–nitrogen double bond.
 (c) Describe the geometry of H₂C=NH (using a sketch that shows all expected bond angles).

8.36 A nitrogen atom, somewhat like carbon, can undergo *sp* hybridization and then become joined to carbon by a triple bond $-C\equiv N:$. This triple bond consists of one σ bond and two π bonds.

(a) Write the orbital diagram for sp hybridized nitrogen as it would look before any bonds form.

(b) Using the carbon–carbon triple bond as the analogy, and drawing pictures to show which atomic orbitals overlap with which, show how the three bonds of the triple bond in $-C\equiv N:$ form.

(c) Again using sketches, describe all the bonds in hydrogen cyanide, $H-C\equiv N:$.

(d) What is the likeliest $H-C-N$ bond angle in HCN?

8.37 How does VB theory treat the benzene molecule? (Draw sketches describing the orbital overlaps and the bond angles.)

8.38 Tetrachloroethylene, a common dry-cleaning solvent, has the formula C_2Cl_4. Its structure is

Use VB theory to describe the bonding in this molecule.

8.39 Phosgene, $COCl_2$, was used as a war gas during World War I. It reacts with moisture in the lungs of its victims to form CO_2 and gaseous HCl, which cause the lungs to fill with fluid. Phosgene is a simple molecule having the structure

Describe the bonding in this molecule, using VB theory.

Bond Energies

8.40 Define *bond energy* and *atomization energy*. What is *bond dissociation energy*?

8.41 Of what use are bond energies in the study of chemistry?

8.42 What data in Section 8.6 can be used to calculate the bond energy of H_2? What is the value of the bond energy in this molecule?

8.43 Calculate the bond energy in the nitrogen molecule and the oxygen molecule.

8.44 Use the data in Table 8.3 to compute the approximate atomization energy of NH_3.

8.45 Approximately how much energy would be released during the formation of the bonds in one mole of acetone molecules? Acetone, the solvent usually found in nail polish remover, has the structural formula

8.46 The standard heat of formation of ethylene, $C_2H_4(g)$, is $\Delta H_f^\circ = +52.284$ kJ/mol. Calculate the $C=C$ bond energy in this molecule.

8.47 The standard heat of formation of ethanol vapor, $C_2H_5OH(g)$ is $\Delta H_f^\circ = -235.3$ kJ/mol. Use the data in Table 8.2 and the average bond energies for $C-C$, $C-H$, and $O-H$ bonds to estimate the $C-O$ bond energy in this molecule. The structure of the molecule is

8.48 Carbon disulfide, CS_2, has the Lewis structure $:S=C=S:$ and for $CS_2(g)$, $\Delta H_f^\circ = +115.3$ kJ/mol. Use the data in Table 8.2 to calculate the average $C=S$ bond energy in this molecule.

8.49 For $SF_6(g)$, $\Delta H_f^\circ = -1096$ kJ/mol. Use the data in Table 8.2 to calculate the average $S-F$ bond energy in SF_6.

8.50 Gaseous hydrogen sulfide, H_2S, has $\Delta H_f^\circ = -20.15$ kJ/mol. Use data in Table 8.2 to calculate the average $S-H$ bond energy in this molecule.

8.51 Use the results of Review Exercise 8.49 and the data in Table 8.2 to calculate the standard heat of formation of $SF_4(g)$. The measured value of ΔH_f° for $SF_4(g)$ is -171.7 kcal/mol. What is the percentage difference between your calculated value of ΔH_f° and the experimentally determined value?

8.52 For the substance $SO_2F_2(g)$, $\Delta H_f^\circ = -858$ kJ/mol. The structure of the SO_2F_2 molecule is

Use the value of the $S-F$ bond energy calculated in Review Exercise 8.49 and the data in Table 8.2 to determine the average $S=O$ bond energy in SO_2F_2.

8.53 Use the data in Tables 8.2 and 8.3 to estimate the standard heat of formation of acetylene, $H-C\equiv C-H$, in the gaseous state.

8.54 What would be the approximate heat of formation of CCl_4 vapor at 25°C and 1 atm?

8.55 Which substance should have the more exothermic heat of formation, CF_4 or CCl_4?

8.56 Would you expect the value of ΔH_f° for SO_3, computed from tabulated bond energies, to be very close the experimentally measured value of ΔH_f°? Justify your answer.

8.57 The benzene molecule is considerably more stable than we would have expected it to be based on calculations using tabulated bond energies. What is this extra stabilization energy called in VB theory?

Molecular Orbital Theory

8.58 Why is the higher-energy MO in H_2 called an antibonding orbital?

8.59 Using a sketch, describe the MOs of H_2 and their relationship to their parent atomic orbitals.

8.60 Explain why He$_2$ does not exist but H$_2$ does.

8.61 How does MO theory account for the paramagnetism of O$_2$?

8.62 On the basis of MO theory, explain why Li$_2$ molecules can exist but Be$_2$ molecules cannot.

8.63 Use the MO energy diagram to predict which in each pair has the greater bond energy: (a) O$_2$ or O$_2^+$, (b) O$_2$ or O$_2^-$, (c) N$_2$ or N$_2^+$.

8.64 What are the bond orders in (a) O$_2^+$, (b) O$_2^-$, and (c) C$_2^+$?

8.65 What relationship have we found between bond order and bond energy?

8.66 Sketch the shapes of π_{2p_y} and $\pi_{2p_y}^*$ MOs.

8.67 What is a delocalized MO?

8.68 Use a Lewis-type structure to indicate delocalized bonding in the nitrate ion.

8.69 There exists a hydrocarbon called butadiene, which has the molecular formula C$_4$H$_6$ and the structure

The C$=$C bond lengths are 134 pm (about what is expected for a carbon–carbon double bond), but the C$-$C bond length in this molecule is 147 pm, which is shorter than a normal C$-$C single bond. The molecule is planar (i.e., all the atoms lie in the same plane).

(a) What kind of hybrid orbitals do the carbon atoms use in this molecule to form the carbon–carbon bonds?

(b) Between which pairs of carbon atoms do we expect to find sideways overlap of p orbitals (i.e., π-type p-p overlap)?

(c) On the basis of your answer to part b, do you expect to find localized or delocalized π bonding in the carbon chain in this molecule?

(d) Based on your answer to part c, explain why the center carbon–carbon bond is shorter than a carbon–carbon single bond?

8.70 What problem encountered by VB theory does MO theory avoid by delocalized bonding?

8.71 Draw the representation of the benzene molecule that indicates its delocalized π system.

8.72 What effect does delocalization have on the stability of the electronic structure of a molecule?

8.73 What is delocalization energy? How is it related to resonance energy?

TEST OF FACTS AND CONCEPTS: CHAPTERS 5-8

In Chapters 5 through 8 we have studied interrelationships among energy, reactions, and the structure of matter. The energy budgets of chemical reactions inform us about the relative energy contents or enthalpies of substances. The existence of electronic energy levels within atoms underlies all modern theories of atomic and molecular structure. We have seen how the periodic table helps us organize our knowledge not just of the properties of the elements but also of their structures. The following questions will help you test your mastery of these essential aspects of chemistry before we apply them extensively in the remaining chapters.

1. If we think of the system as consisting of 225 g of pure water at 25.00 °C and 1 atm of pressure, what value do we use for the heat capacity of this system? What value do we use for the specific heat of the water? (For the energy part of the unit, use the calorie.)

2. In what specific ways do a standard enthalpy of reaction and a standard enthalpy of formation differ?

3. Why was the concept of a *standard state* introduced into chemistry?

4. Under what conditions is an element in its standard state? Which form of oxygen can exist in the standard state of oxygen?

5. The specific heat of helium is 1.24 cal/g °C and of nitrogen is 0.249 cal/g °C. How many joules can one mole of each gas absorb when its temperature increases 1.00 °C?

6. A calorimeter vat in which a stirring motor, a thermometer, and a "bomb" are immersed absorbed the heat released by the combustion of 0.514 g of benzoic acid. The thermochemical equation for its combustion is as follows.

$$2C_7H_6O_2(s) + 15O_2(g) \longrightarrow 14CO_2(g) + 6H_2O(\ell)$$
$$\Delta H° = -6452 \text{ J}$$

The temperature of the calorimeter rose from 24.112 °C to 24.866 °C. What is the heat capacity of this calorimeter in J/°C?

7. A sample of 10.1 g of ammonium nitrate, NH_4NO_3, was dissolved in 125 g of water in a coffee cup calorimeter. The temperature changed from 24.5 °C to 18.8 °C. Calculate the heat of solution of ammonium nitrate in kJ/mol. Assume that the energy exchange involves only the solution and that the specific heat of the solution is 4.18 J/g°C.

8. When 0.6484 g of cetyl palmitate, $C_{32}H_{64}O_2$ (a fruit wax), was burned in a bomb calorimeter with a heat capacity of 11.99 kJ/°C, the temperature of the calorimeter rose from 24.518 °C to 26.746 °C. Calculate the molar heat of combustion of cetyl palmitate in kJ/mol.

9. The value of ΔH for some chemical reaction is called a *state function*. What does this mean?

10. The combustion of methane (the chief component of natural gas) follows the equation

$$CH_4(g) + 2O_2(g) \longrightarrow CO_2(g) + 2H_2O(g)$$

$\Delta H°$ for this reaction is -802.3 kJ. How many grams of methane must be burned to provide enough heat to raise the temperature of 250 mL of water from 25.0 °C to 50.0 °C?

11. Label the following thermal properties as *intensive* or *extensive* properties.
    (a) specific heat   (b) heat capacity   (c) $\Delta H_f°$
    (d) $\Delta H°$ (e) molar heat capacity

12. The thermochemical equation for the combustion reaction of half a mole of carbon monoxide is as follows.

$$\tfrac{1}{2}CO(g) + \tfrac{1}{4}O_2(g) \longrightarrow \tfrac{1}{2}CO_2(g) \quad \Delta H° = -141.482 \text{ kJ}$$

Write the thermochemical equation for
    (a) The combustion of 2 mol of $CO(g)$.
    (b) The decomposition of 1 mol of $CO_2(g)$ to $O_2(g)$ and $CO(g)$.

13. The standard heat of combustion of eicosane, $C_{20}H_{42}(s)$, a typical component of candle wax, is 1.332×10^4 kJ/mol, when it burns in pure oxygen, and the products are cooled to 25 °C. The only products are $CO_2(g)$ and $H_2O(\ell)$. Calculate the value of the standard heat of formation of eicosane (in kJ/mol) and write the corresponding thermochemical equation.

14. Using data in Table 5.2 calculate values for the standard heats of reaction (in kilojoules) for the following reactions.
    (a) $H_2SO_4(\ell) \longrightarrow SO_3(g) + H_2O(\ell)$
    (b) $C_2H_6(g) \longrightarrow C_2H_4(g) + H_2(g)$

15. Calculate the standard heat of formation of calcium carbide, $CaC_2(s)$, in kJ/mol using the following thermochemical equations.

$$Ca(s) + 2H_2O(\ell) \longrightarrow Ca(OH)_2(s) + H_2(g)$$
$$\Delta H° = -414.79 \text{ kJ}$$
$$2C(s) + O_2(g) \longrightarrow 2CO(g) \quad \Delta H° = -221.0 \text{ kJ}$$
$$CaO(s) + H_2O(\ell) \longrightarrow Ca(OH)_2(s) \quad \Delta H° = -65.19 \text{ kJ}$$
$$2H_2(g) + O_2(g) \longrightarrow 2H_2O(\ell) \quad \Delta H° = -571.8 \text{ kJ}$$
$$CaO(s) + 3C(s) \longrightarrow CaC_2(s) + CO(g)$$
$$\Delta H° = +462.3 \text{ kJ}$$

16. What are the three principle particles that make up the atom? On the atomic mass scale, what are their approximate masses? What are their electrical charges?

17. What are the possible quantum numbers of the electrons in

the *valence shells* of (a) sulfur, (b) strontium, (c) lead, (d) bromine, (e) boron?

18. If a given shell has $n = 4$, which kinds of subshells (s, p, etc.) does it have? What is the maximum number of electrons that could populate this shell?

19. A beam of green light has a wavelength of 500 nm. What is the frequency of the light? What is the energy, in joules, of one photon of this light? What is the energy, in joules, of one *mole* of photons of this light? Would blue light have more or less energy per photon than this light?

20. Arrange the following kinds of electromagnetic radiation in order of increasing frequency: X rays, blue light, radio waves, gamma rays, microwaves, red light, infrared light, ultraviolet light.

21. What is a continuous spectrum? How does it differ from an atomic spectrum?

22. What experimental evidence is there that matter has wavelike properties?

23. Use the periodic table to predict the electron configurations of (a) tin, (b) germanium, (c) silicon, (d) lead, and (e) nickel.

24. Give the electron configurations of the ions (a) Pb^{2+}, (b) Pb^{4+}, (c) S^{2-}, (d) Fe^{3+}, and (e) Zn^{2+}.

25. What causes an atom, molecule, or ion to be paramagnetic? Which of the ions in the preceding question are paramagnetic? What term describes the magnetic properties of the others?

26. Give the shorthand electron configurations of (a) Ni, (b) Cr, (c) Sr, (d) Sb, and (e) Po.

27. Define *ionization energy* and *electron affinity*. In terms of these properties, which kinds of elements tend to react to form ionic compounds?

28. In general, the second ionization energy of an atom is larger than the first, the third is larger than the second, and so on. Why?

29. Sketch the shapes of (a) an s orbital, (b) a p orbital, (c) the $3d_{xz}$ orbital.

30. What is meant by the term *electron density*?

31. Give orbital diagrams for the valence shells of selenium and thallium.

32. Which ion would be larger: (a) Fe^{2+} or Fe^{3+}, (b) O^- or O^{2-}?

33. Which of the following pairs of elements would be expected to form ionic compounds: (a) Br and F, (b) H and P, (c) Ca and F?

34. Use Lewis symbols to diagram the reaction of calcium with sulfur to form CaS.

35. Draw Lewis structures for (a) SbH_3, (b) IF_3, (c) $HClO_2$, (d) C_2^{2-}, (e) AsF_5, (f) O_2^{2-}, (g) HCO_3^-, (h) TeF_6, (i) HNO_3.

36. Use the VSEPR theory to predict the shapes of (a) $SbCl_3$, (b) IF_5, (c) AsH_3, (d) BrF_2^-, (e) OF_2.

37. What kinds of hybrid orbitals are used by the central atom in each of the species in the preceding question?

38. Referring to your answers to questions 35 and 36, which of the following molecules would be nonpolar: SbH_3, IF_3, AsF_5, $SbCl_3$, or OF_2?

39. Oxalate ion has the following arrangement of atoms.

$$\begin{array}{ccc} O & & O \\ & C\ \ C & \\ O & & O \end{array}$$

Draw all of its resonance structures.

40. What is meant by the term *overlap of orbitals*?

41. What are sigma bonds? What are pi bonds? How are sigma and pi bonds used to explain the formation of double and triple bonds?

42. A certain element X was found to form three compounds with chlorine having the formulas XCl_2, XCl_4, and XCl_6. One of its oxides has the formula XO_3, and X reacts with sodium to form the compound Na_2X.
    (a) Is X a metal or a nonmetal?
    (b) In which group in the periodic table is X located?
    (c) In which periods in the periodic table could X possibly be located?
    (d) Draw Lewis structures for XCl_2, XCl_4, XCl_6, and XO_3. Which has multiple bonding?
    (e) What would be the molecular structures of XCl_2, XCl_4, XCl_6, and XO_3? Which are polar molecules?
    (f) The element X also forms the oxide XO_2. Draw its Lewis structure.
    (g) How would the bond lengths in XO_3, compare to those in XO_2?
    (h) What kinds of hybrid orbitals would X use for bonding in XCl_4 and XCl_6?
    (i) If X were to form a compound with aluminum, what would be its formula?
    (j) Which compound of X would have the more ionic bonds, Na_2X or MgX?
    (k) If X were in period 5, what would be the electron configuration of its valence shell?

43. Glycine, one of the important amino acids, has the structure

$$H-\underset{\underset{H}{|}}{\overset{\overset{H}{|}}{N}}-\underset{\underset{H}{|}}{\overset{}{C}}-\overset{\overset{O}{\|}}{C}-O-H$$

Calculate the atomization energy of this molecule from the data in Table 8.3.

44. Use bond energies in Table 8.3 to calculate the approximate energy that would be absorbed or given off in the formation of 25.0 g of C_2H_6 by the reaction (in the gas phase)

$$H-C\equiv C-H + 2H_2 \longrightarrow H-\underset{\underset{H}{|}}{\overset{\overset{H}{|}}{C}}-\underset{\underset{H}{|}}{\overset{\overset{H}{|}}{C}}-H$$

45. What are bonding and antibonding molecular orbitals? How do they differ?

46. Describe how molecular orbital theory explains the bonding in the oxygen molecule.

47. What is a delocalized orbital? How does molecular orbital theory avoid the concept of resonance?

When liquid vaporizes in a confined space, the pressure of its vapor can become enough to move mountains. Seen here is a towering fountain of molten rock spewed skyward by gas pressures deep within the Mauan Ulu volcano in Hawaii. We'll study the relationships among the pressure, temperature, and volume of a gas in this chapter.

PROPERTIES OF GASES

9.1 QUALITATIVE FACTS ABOUT GASES

The four important measurable properties of a gas are volume, temperature, pressure, and mass.

Thus far we have concentrated on the kinds of matter — elements and compounds — on the kinds of particles that make up matter — atoms, ions, and molecules — and on both the quantitative and periodic relationships among them. With this chapter, we start a systematic study of the *states* of matter — solids, liquids, and gases.

At a still more basic level, the kinds of particles are the electron, proton, and neutron.

We take up gases first, partly because their physical properties are the easiest to understand and partly because important features about gases help to explain some of the properties of liquids and solids. Moreover, gases are important substances in their own right. One of them, oxygen, is essential minute by minute to our lives. And minute by minute we have to remove another gas, carbon dioxide, a waste product of our body's use of oxygen. Another gas, ammonia, is one of the world's most important agricultural fertilizers, and it is made from two gases, hydrogen and nitrogen. Nitrogen makes up

Liquid ammonia is injected directly into soil by special machinery.

80% (v/v) means 80% by volume.

nearly 80% (v/v) of air. The noble gas argon, at low pressure, is the gas inside most light bulbs, and "neon" lights are named after another noble gas. You can see that several gases are familiar substances.

Properties Common to All Gases

One of the remarkable facts about gases is that despite wide differences in chemical properties, they all more or less obey the same set of physical laws — the gas laws. In fact, a question we want to raise is why there are such gas laws for gases but not comparable "liquid laws" or "solid laws" for the other states of matter.

Although you may not know the formal wording of the gas laws, you no doubt are familiar with some of the ways gases behave, physically. Tire pumps, tire punctures, and balloons give most people some experience with gas pressure. And the "feel" of a balloon gives us the sense that gas pressure acts equally in all directions on its container. The warning on all aerosol cans, "Do Not Incinerate," suggests that an increase in temperature on a closed can will make the internal pressure increase perhaps enough to blow the can apart. To study such matters as how pressure and temperature affect gases we need to learn how four physical properties of gases are related — volume, temperature, pressure, and amount (in grams or moles).

The volume of a gas is always the same as the volume of its container, because gases spread out to fill whatever space they are given. The symbol for volume is V, and the units used are usually liters (L) or milliliters (mL). (The SI derived unit of volume, the cubic meter, is rarely used in chemistry.)

1 m³ = 1000 L (about 260 gallons)

The symbol for temperature is T, and a gas temperature is expressed in kelvins (K). (We will see why later.) When the temperature is given in any other unit, it always has to be changed to kelvins in any calculations involving gases.

The amount of matter in a gas sample is usually expressed in moles (n) when we know its chemical identity. Otherwise, grams or milligrams can be used. Actually, in most experiments the amount of the gas sample remains constant because the containers are closed and nothing escapes. Most of the gas laws, therefore, concern interrelationships among volume, temperature, and pressure for a fixed amount of the gas. In the next section, we will study the meaning of pressure and the units chemists most use for it.

With the force distributed over a large area, the net pressure is small (top). When the force acts on a small area, the net pressure is large (bottom).

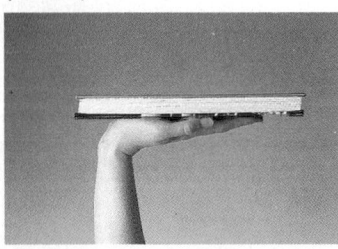

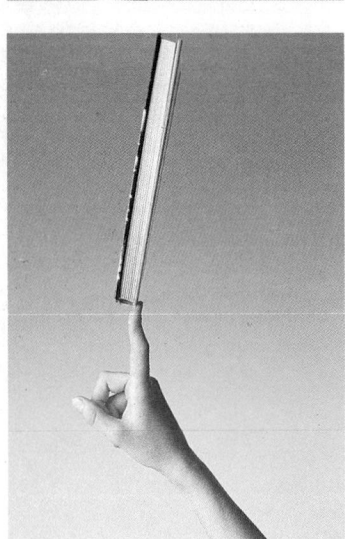

9.2 PRESSURE

In chemistry, pressure — force per unit area — is usually expressed in units of the torr or the atmosphere.

Pressure is created by applying a force to some surface, so to understand pressure we have to be clear about *force*.

A force is whatever causes something to change its motion or direction, to make it start moving, go faster or slower, veer to one side, come, go, rise, fall, or crack apart. A force is also whatever can act to counterbalance any of these changes. When *nothing* is happening, you can be sure that two or more forces are exactly balanced.

Force per Unit Area — Pressure

Pressure is force *per unit area*, calculated by dividing the force by the area on which the force acts.

$$\left(\text{pressure} = \frac{\text{force}}{\text{area}} \right)$$

This distinction between force and pressure is important. You no doubt have experienced the comfort of distributing a constant force over a broader area many times. A number of books (creating a force) in a backpack causes less shoulder pain when the pack has wide straps than if it has straps made of twine. The force (the weight) is the same in both cases, but the pressure on any small part of your shoulder is less with the wide straps. They distribute the force over a larger area, so the ratio of force to area (pressure) is everywhere less. We will do a calculation to dramatize this important point.

EXAMPLE 9.1   CALCULATING PRESSURE

Problem: Assume that this book has a weight of 4.25 lb and that your palm has an area of 19.5 in². Compare the pressures the book can exert when it rests cover-side-down on your palm to when it is balanced on an edge on your index finger. Assume the affected area of your finger measures 0.500×0.0625 in. ($\frac{1}{2} \times \frac{1}{16}$ in.). Calculate to three significant figures.

Solution: In the first situation,

$$\text{pressure} = \frac{\text{force}}{\text{area}} = \frac{4.25 \text{ lb}}{19.5 \text{ in.}^2}$$

$$= 0.218 \text{ lb/in.}^2$$

In the second experience, the area is 0.500×0.0625 in. $= 0.0313$ in.². Therefore,

$$\text{pressure} = \frac{\text{force}}{\text{area}} = \frac{4.25 \text{ lb}}{0.0313 \text{ in.}^2}$$

$$= 136 \text{ lb/in.}^2$$

Over 600 times the pressure results from moving the constant force from the large area to the smaller area.

PRACTICE EXERCISE 1   Butter is often sold in a 1.00-lb container that measures $5.00 \times 4.72 \times 1.22$ in. What pressure does this much butter cause when its container is lying with its largest area down? What is the pressure when it is lying with its long and narrow side down? (Do the calculations to three significant figures.)

Atmospheric Pressure and the Barometer

The earth exerts a gravitational force of attraction on anything with mass, ranging from a leaf to the moon to the air itself. The earth's gravitational force pulling on the air creates an opposing force, that of the air pushing on the earth. This opposing force per unit area is atmospheric pressure.

The presence of an atmospheric pressure can be demonstrated by a very simple experiment, shown in Figure 9.1, in which we see the effects of pumping the air from a steel can. Before the air is removed, the walls of the can feel the atmospheric pressure identically inside and out. When air is pumped out, however, only the air on the outside presses on the walls, and the can collapses.

(a)

(b)

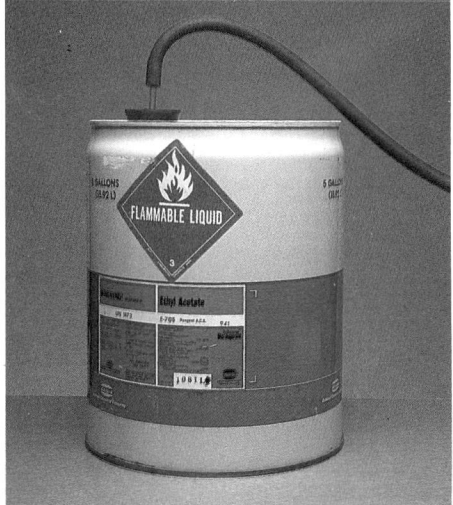
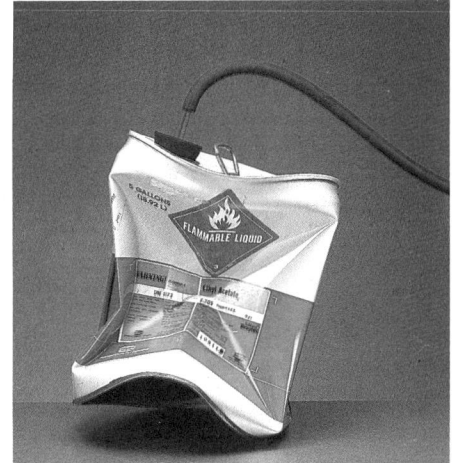

FIGURE 9.1 The effect of an unbalanced pressure. *(a)* The pressure inside this can is the same as outside — the pressures are balanced. *(b)* When a vacuum pump reduces the pressure inside the can, the unbalanced outside pressure quickly and violently makes the can collapse.

Evangelista Torricelli

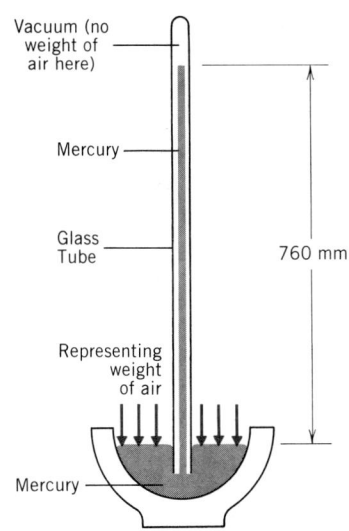

Vacuum (no weight of air here)

Mercury

Glass Tube

760 mm

Representing weight of air

Mercury

FIGURE 9.2 The mercurial or Torricelli barometer.

The simplest device for measuring the atmospheric pressure is called a Torricelli barometer (Figure 9.2). It consists of a long narrow tube (80 cm or more in length) that is sealed at one end, filled with mercury, and then inverted into a dish of mercury. When the tube is inverted, some mercury runs out, but not all. The atmosphere, pushing on the surface of the mercury in the dish, attempts to force the liquid up the tube. Meanwhile, the downward pressure caused by the weight of the mercury in the tube attempts to force liquid out. When these two pressures become equal, the mercury level stops falling.

The height of the mercury column, measured from the surface of the mercury in the dish, is directly proportional to the atmospheric pressure. On days when the atmospheric pressure is high, mercury is forced into the tube and the height of the column rises. When the atmospheric pressure drops (because of an approaching storm, for example), some mercury flows out of the tube and the height of the column drops. Generally most people live where the height of the mercury column fluctuates around 730 to 760 mm.

Units of Pressure

Because any metal, including mercury, expands or contracts as the temperature increases or decreases, the height of the mercury column in the barometer varies a little with temperature. It also varies with altitude. On the top of the earth's highest peak, Mount Everest, for example, the rare atmosphere can hold the mercury to a height of only about 250 mm. At sea level and at ordinary temperatures, however, the height of the mercury column fluctuates around a value of 760 mm — some days it is a little higher and some a little lower. Scientists long ago decided to call a pressure capable of supporting a column of mercury 760 mm high, measured at 0 °C, one standard atmosphere, symbolized atm.

For ordinary laboratory work in chemistry and biology, in which we normally measure changes or differences in pressure, the atmosphere is an awkward-sized unit; it is too large. A more convenient unit is the *millimeter of mercury, or mm Hg* for short. This is the pressure exerted by a column of mercury 1 mm high at 0 °C. It has the special name torr (after the inventor of the barometer).

$$1 \text{ torr} = 1 \text{ mm Hg} \tag{9.1}$$

This also means, of course, that

$$1 \text{ atm} = 760 \text{ torr} \tag{9.2}$$

In the SI, the unit of pressure is called the pascal, symbolized Pa. The pascal is

Mercury, a shiny, metallic element (page 62), is a liquid above −39 °C. Handle it with care. Its vapor is poisonous.

1 Pa = 1 N/m², where N is the symbol for the *newton,* the SI unit of force.

related to the atmosphere as follows. In fact, this is now the modern definition of the standard atmosphere.

$$1 \text{ atm} = 101{,}325 \text{ Pa (exactly)}$$

Although we will not use the pascal as a unit, it may see increasing use as time goes on. (A major reference in clinical chemistry, for example, has urged clinical chemists to report all pressures of respiratory gases—oxygen and carbon dioxide—in kilopascals.) The torr and occasionally the atm are the units that we will use throughout this book.

EXAMPLE 9.2 CONVERTING AMONG UNITS OF PRESSURE

Problem: Calculate how many pascals are in 1 torr (to six significant figures).

Solution: We know two relationships from which to construct conversion factors.

$$1 \text{ atm} = 760 \text{ torr (exactly)}$$
$$1 \text{ atm} = 101{,}325 \text{ Pa (exactly)}$$

Therefore we can set up the following sequence.

$$1 \text{ torr} \times \frac{1 \text{ atm}}{760 \text{ torr}} \times \frac{101{,}325 \text{ Pa}}{1 \text{ atm}} = 133.322 \text{ Pa}$$

Thus, 1 torr = 133.322 Pa (You can see from this that the pascal is a very small amount of pressure—much smaller than the torr. This is why the *kilopascal*, kPa, is usually used by chemists when they want to use the SI. 1 kPa = 10^3 Pa.)

PRACTICE EXERCISE 2   Suppose one day you find the pressure in the chemistry laboratory to be 740 torr. What is this pressure in atmospheres, in kilopascals, and in millimeters of Hg?

Manometers

The value to a chemist of the small-sized torr (or mm Hg) as a pressure unit can be seen if we examine how pressures of gases are measured in the laboratory where we study samples of gases that we have trapped in some vessel. We might have obtained a gas as the product of some reaction, for example, and trapped it for further study. The device we use to measure pressures of gases in such situations is called a manometer. Two types are illustrated in Figure 9.3.

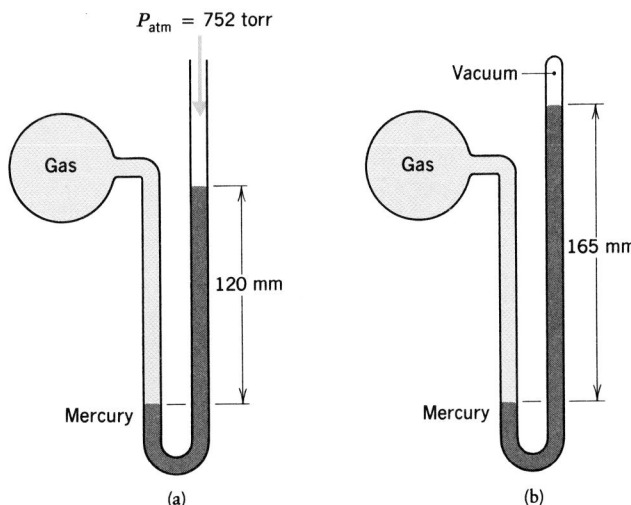

$P_{\text{atm}} = 752 \text{ torr}$

Gas

120 mm

Mercury

(a)

Vacuum

Gas

165 mm

Mercury

(b)

FIGURE 9.3 *(a)* An open-end manometer. *(b)* A closed-end manometer.

Some advantages of mercury over other liquids are its low reactivity, low melting point, and high density (which allows short tubes).

The first is called an open-end manometer and consists of a U-shaped tube partly filled with a liquid, usually mercury. One arm of the U is open to the atmosphere where atmospheric pressure bears on the mercury surface. The other end is exposed to the pressure of the entrapped gas. In Figure 9.3a it is clear that the entrapped gas exerts a larger pressure than the atmosphere—that is why the mercury is forced up in the arm open to the air. The difference in the height of the mercury in the two arms of this manometer is proportional to the difference in pressure between the two sides. In Figure 9.3a this difference is 120 mm Hg, which corresponds to a pressure difference of 120 torr. The pressure of the trapped gas is therefore 120 torr larger than the atmospheric pressure.

$$P_{gas} = P_{atm} + 120 \text{ torr}$$

The atmospheric pressure at the time of the experiment, measured separately with a barometer, is given in Figure 9.3a as 752 torr, so the pressure of the gas is

$$P_{gas} = 752 \text{ torr} + 120 \text{ torr}$$
$$= 872 \text{ torr}$$

EXAMPLE 9.3   USING AN OPEN-END MANOMETER

Problem: At the end of a chemical reaction, begun at 745 torr and carried out in the large bulb of the open-end manometer of Figure 9.3a, the mercury level in the arm exposed to the air stood 26 mm *lower* than the mercury level open to the bulb. What was the final pressure inside the bulb? (The atmospheric pressure that day, measured with a barometer, was 745 torr.)

Solution: The difference in height, 26 mm, corresponds to a pressure difference of 26 torr. But the pressure inside the bulb must be *less* than the atmospheric pressure, because the pressure of the atmosphere has succeeded in forcing mercury out of the open arm and into the arm next to the bulb. We therefore have to *subtract* the pressure difference from the atmospheric pressure.

$$P_{gas} = P_{atm} - 26 \text{ torr}$$

Since $P_{atm} = 745$ torr, we have

$$P_{gas} = 745 \text{ torr} - 26 \text{ torr}$$
$$= 719 \text{ torr}$$

The chemical reaction thus caused the pressure to decrease from 745 to 719 torr. (Perhaps some gas originally present was a reactant, and was not replaced by another gas.)

PRACTICE EXERCISE 3   When a reaction was carried out within the bulb of an open-end manometer, the height of the mercury column in the arm open to the air became 66 mm higher than the mercury level in the other arm. What was the pressure within the bulb? The atmospheric pressure was 752 torr.

PRACTICE EXERCISE 4   Convert the pressures in torr given by the answers to Example 9.3 and Practice Exercise 3 into atmospheres.

If you worked Practice Exercise 4, you may guess why chemists prefer the torr to the atmosphere. It is just easier to read and pronounce numbers greater than 1 than decimals less than 1.

In Figure 9.3b we see the second type of manometer. It is called a closed-end manometer for obvious reasons. It is prepared essentially just as one would start making a Torricelli barometer—a tube is completely filled with mercury. Only now the lower end is bent around into a U-shape, and then connected to a bulb where a gas could be trapped. The space inside the sealed end above the mercury level is a vacuum, just as in a barometer, so the difference in height of the two mercury levels—165 mm in Figure 9.3b—corresponds directly to the actual pressure inside the bulb, 165 torr. Thus a closed-end manometer gives us pressure directly without the need to do a calculation involving the atmospheric pressure.

9.3 THE PRESSURE–VOLUME RELATIONSHIP

At constant temperature, the volume of a fixed mass of gas is inversely proportional to the pressure acting on it.

Now that we have the physical units to describe gases, we can begin a study of the ways that gases respond to changes in pressure, volume, and temperature. How, for example, does the volume of a gas change if you change the pressure on it? Over 300 years ago, the English scientist Robert Boyle (1627–1691) used a simple J-tube such as pictured in Figure 9.4 to get data concerning this question.

Robert Boyle

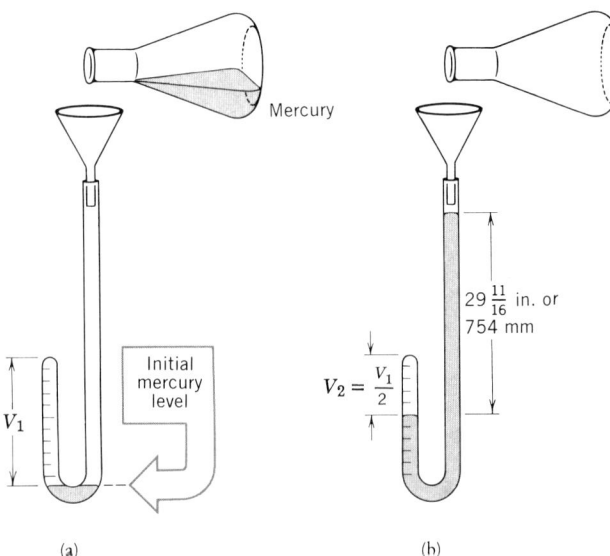

(a)   (h)

FIGURE 9.4 The J-tube experiment and the effect of pressure on volume. Boyle used a J-tube with a much longer open end to test the effects of still higher pressures.

Boyle sealed the short end of the J-tube and poured mercury into its open end until it just barely trapped the air in the closed, short end. See Figure 9.4a. Since he did not add enough mercury to squeeze this air, it was now at atmospheric pressure. But he now had a fixed mass of air. Then he added more mercury (Figure 9.4b) until the volume of the entrapped gas was cut exactly in half. Boyle then measured the difference in the heights of the two columns of mercury, and found it to be (in today's SI terms) 754 mm. This corresponds very closely to 1 atm (760 torr). This was the extra pressure on the entrapped gas, which initially had a pressure of 1 atm, so the total pressure on this gas was now 2 atm.

By doubling the pressure, Boyle had reduced the volume by one-half. During the measurements, no air escaped—the mass of the gas was fixed—and the temperature did not change.

Boyle obtained much more data than these, of course, and the curve in Figure 9.5 shows how volume varies inversely with pressure, provided that no gas escapes and the temperature is constant. Other scientists found similar data with other gases, using the same conditions. This reciprocal relationship between pressure and volume is now expressed as the pressure–volume law or Boyle's law.

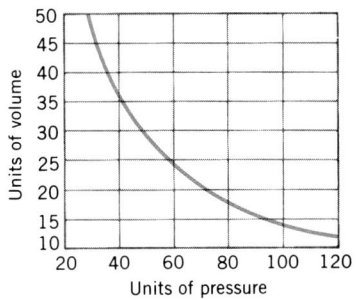

FIGURE 9.5 The relationship of volume to pressure—Boyle's law. The unit of pressure here is the same unit that Boyle used, *inches of mercury*. Notice that the volume is reduced by half when the gas pressure goes from 30 to 60 units and then by still under half as the pressure increases from 60 to 120 units.

> **Pressure–Volume Law (Boyle's Law).**   The volume of a given sample of gas held at constant temperature is inversely proportional to the applied pressure.
>
> $$V \propto \frac{1}{P} \qquad (T \text{ and mass are constant}) \qquad (9.3)$$

We can remove the proportionality sign, $\propto$, by introducing a proportionality constant, C.

$$V = \left(\frac{1}{P}\right)(C) \qquad (9.4)$$

Let us take a look at how we use the relationship in Equation 9.4 to predict what happens to a gas if we change its volume or pressure, while keeping its temperature constant. We will use the symbols P_1 and V_1 to stand for the initial values of pressure and volume — the values before the change. By rearranging Equation 9.4, we get

$$P_1V_1 = C \qquad (C \text{ is the proportionality constant}) \qquad (9.5)$$

But after any change in P or V (at constant temperature on the same sample) the constant C must still be the same. Otherwise we could not call it a constant. Therefore, at new values, which we will call P_2 and V_2, we can write

$$P_2V_2 = C \qquad (9.6)$$

The constant C in Equation 9.6 is the same as the one in Equation 9.5.

Since C is the same in both Equations 9.5 and 9.6, we can combine them and thus have an expression for Boyle's law that avoids the constant, C.

$$P_1V_1 = P_2V_2 \qquad (9.7)$$

When we know any three of the four values in Equation 9.7, we can calculate the fourth, which typically will be either a final pressure, P_2, or a final volume, V_2. With a little mental rearranging of equation 9.7, in other words, we can find a new pressure, P_2, by multiplying the old pressure, P_1, by a ratio of volumes, V_1/V_2.

As long as V_1 and V_2 are in the same units they can be in any units, e.g., mL, L, etc.

$$P_2 = P_1 \times \left(\frac{V_1}{V_2}\right) \qquad (9.8)$$

a ratio of volumes

Similarly, to find a new volume, we can multiply the old volume by a ratio of pressures.

As long as P_1 and P_2 are in the same units they can be in any units, e.g., torr, atm, Pa, etc.

$$V_2 = V_1 \times \left(\frac{P_1}{P_2}\right) \qquad (9.9)$$

a ratio of pressures

EXAMPLE 9.4   CALCULATING WITH THE PRESSURE–VOLUME LAW

Problem: A sample of oxygen with a volume of 500 mL at a pressure of 760 torr is to be compressed to a volume of 450 mL. What pressure is needed to do this if the temperature is kept constant?

Solution: You will always find it easier to work any gas law problem if you assemble all pertinent data first.

$$V_1 = 500 \text{ mL} \qquad P_1 = 760 \text{ torr}$$
$$V_2 = 450 \text{ mL} \qquad P_2 = ?$$

Now comes the thinking step. We ask, "What *direction* must the pressure change?" By *direction,* we mean must the pressure increase or decrease? The pressure–volume law, and even our own experiences with tire pumps, tells us that pressure and volume change in opposite directions. Since in this problem the volume is to decrease, the pressure *must* increase. What we need is a pressure-*raising* ratio of volumes to multiply by the initial pressure. The two possible ratios of volumes are

$$\frac{500 \text{ mL}}{450 \text{ mL}} \quad \text{and} \quad \frac{450 \text{ mL}}{500 \text{ mL}}$$

Only the first ratio can make P_1 *increase,* because it is larger than 1. The other ratio is less than 1. Therefore,

We could also have calculated P_2 by using Equation 9.7 directly.

$$P_2 = 760 \text{ torr} \times \frac{500 \text{ mL}}{450 \text{ mL}}$$
$$= 844 \text{ torr (rounded from 844.444)}$$

PRACTICE EXERCISE 5   A sample of nitrogen has a volume of 880 mL and a pressure of 740 torr. If the volume is changed to 440 mL at the same temperature, what will be the pressure?

PRACTICE EXERCISE 6   If 200 mL of helium at 760 torr is given a pressure of 800 torr, what volume will the sample have, assuming no change in temperature?

Ideal Gas

No real gas obeys the pressure–volume law *exactly* over a wide range of conditions. Most gases, however, come quite close, particularly when their pressures are low (near or less than 1 atm) and the constant temperatures selected are relatively high. Under these conditions, a gas is not close to condensing to a liquid. Only when a gas is under conditons that favor its changing to a liquid — high pressures and low temperatures — does it poorly fit the Boyle's law equation.

The hypothetical gas that would obey Boyle's law exactly can still be imagined, however, and such a gas is called an ideal gas. We say that real gases approach ideal gas behavior as the pressure is lowered and the temperature is raised.

An ideal gas would fit all of the gas laws exactly.

9.4 PARTIAL PRESSURE RELATIONSHIPS

Each gas in a mixture contributes its own partial pressure to the total pressure as if it acted independently of all the others.

In many applications of the pressure–volume law — studies of respiration, for example — we have to work with a mixture of gases, not a single pure substance. How does this complication affect the study of gases? John Dalton's study of this matter led to the law of partial pressures, sometimes called Dalton's law of partial pressures.

Law of Partial Pressures (Dalton's Law). The total pressure exerted by a mixture of nonreacting gases is the sum of their individual partial pressures.

$$P_{\text{total}} = P_a + P_b + P_c + \cdots \tag{9.10}$$

The partial pressure of a gas in a mixture of gases is the pressure that it would exert on the container if it were present all alone in that same volume (and temperature). The subscript letters in Equation 9.10 identify individual gases, except that in real situations the formulas of the gases are used instead. The symbol for the partial pressure of oxygen in air, for example, would be P_{O_2} (or, sometimes, PO_2).

EXAMPLE 9.5 CALCULATING PARTIAL PRESSURES

Problem: Imagine that we have three cylinders of the same volume, each fitted with a pressure gauge, and all empty (i.e., each contains a vacuum). We put 4.49 g of nitrogen in the first cylinder, and its pressure reads 300 torr. We put 3.42 g of oxygen in the second container, and its pressure reads 200 torr. What will be the final pressure in the third cylinder if we put 4.49 g of nitrogen and 3.42 g of oxygen into it? (Assume a constant temperature throughout and that no chemical reaction occurs.)

Solution: The law of partial pressures tells us that each gas acts as if it were alone in the third cylinder. Therefore,

$$P_{total} = P_{N_2} + P_{O_2}$$
$$= 300 \text{ torr} + 200 \text{ torr} = 500 \text{ torr}$$

Thus to get both gases into the same container, a total pressure of 500 torr is necessary.

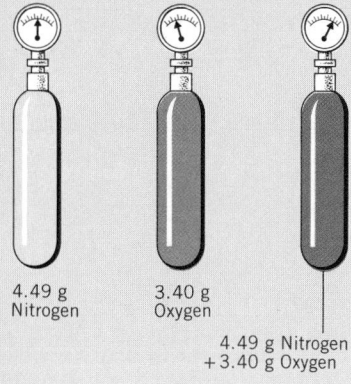

4.49 g
Nitrogen

3.40 g
Oxygen

4.49 g Nitrogen
+ 3.40 g Oxygen

PRACTICE EXERCISE 7   At sea level and 0 °C, the partial pressure of nitrogen in clean dry air is 601 torr when the total pressure is 760 torr. If the only other constituent is oxygen, what is its partial pressure under these conditions?

Collecting a Gas over Water

When a gas is prepared in the laboratory, it is normally collected over water using the kind of apparatus illustrated in Figure 9.6. (If the gas reacts with water, then another

FIGURE 9.6 Collecting a gas over water. As the gas bubbles through the water, water vapor goes into the gas, so the total pressure inside the bottle includes the partial pressure of the water vapor at the temperature of the water.

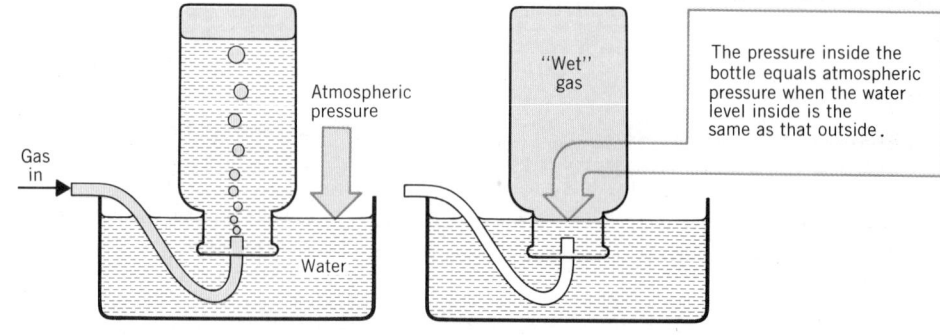

Atmospheric
pressure

Gas
in

Water

"Wet"
gas

The pressure inside the bottle equals atmospheric pressure when the water level inside is the same as that outside.

TABLE 9.1 Vapor Pressure of Water at Various Temperatures

Temperature (°C)	Vapor Pressure (torr)	Temperature (°C)	Vapor Pressure (torr)
0	4.579	50	92.51
5	6.543	55	118.0
10	9.209	60	149.4
15	12.79	65	187.5
20	17.54	70	233.7
25	23.76	75	289.1
30	31.82	80	355.1
35	42.18	85	433.6
37[a]	47.07	90	527.8
40	55.32	95	633.9
45	71.88	100	760.0

[a] Body temperature.

A more extensive table of the vapor pressure of water at various temperatures is in Appendix D.

approach is necessary.) A gas collected over water contains water vapor — as much water vapor as the gas can hold at the temperature used. Actually, a space above any liquid always contains some of the vapor of the liquid.

The vapor from a liquid exerts a pressure, just like any other gas, and this pressure is called the vapor pressure of the substance. For a given substance, its value depends *only* on the temperature. Table 9.1, for example, gives the vapor pressure of water at different temperatures.

Even the mercury in a barometer has a tiny vapor pressure — 0.0011 torr at 20 °C.

When water vapor is present in a gas we treat it like any other gas in partial pressure calculations. One common situation that nicely illustrates this involves the air we inhale and exhale. Table 9.2 gives the values for the partial pressures of the four chief gases in this air. Notice how much the partial pressure of water vapor changes as air is inhaled into the moist alveoli of the lungs, where water vapor is picked up. The temperature there — body temperature — is 37 °C, and Table 9.1 tells us that the vapor pressure of water at 37 °C is 47 torr. This is therefore the partial pressure of water vapor both in the air of the alveoli and in exhaled air. Notice also in Table 9.2 how the partial pressures of the other gases adjust as oxygen is removed from air in the lungs and waste carbon dioxide is added to the air. Essentially no nitrogen is removed or added, but since the total pressure in the lungs is the same as outside, the partial pressure of nitrogen must adjust. The total pressure is 760 torr everywhere, as the sums of the partial pressures in Table 9.2 also show.

The alveoli are the lung's 300 million thin-walled air sacs enmeshed in beds of fine blood capillaries, the terminals of the successively branching tubes that make up the lungs.

When we collect a gas over water we usually want to know how much *dry* gas we obtained, and we can use data for the wet gas to find out. From Dalton's law, we can write

$$P_{total} = P_{dry\ gas} + P_{water}$$

TABLE 9.2 The Changing Composition of Air during Respiration

Gas	Partial Pressure (torr)		
	Inhaled Air	Exhaled Air	Alveolar Air[a]
Nitrogen	594.70	569	570
Oxygen	160.00	116	103
Carbon dioxide	0.30	28	40
Water vapor	5.00[b]	47	47
Total	760.00	760	760

[a] Alveolar air is air within the alveoli.

[b] The average value representing air of about 20% relative humidity, a familiar weather report term that we will not further define.

By rearranging, we find the gas's "dry pressure."

$$P_{\text{dry gas}} = P_{\text{total}} - P_{\text{water}} \tag{9.11}$$

We look up in a table the value of P_{water} at the temperature of the water used (e.g., Table 9.1). And we read P_{total} from a barometer in the laboratory. Now we are back to a Boyle's law calculation to find out what volume of dry gas we would have obtained at this temperature. And we can calculate this volume at any pressure we choose, such as 760 torr, the standard atmosphere, instead of the pressure that day in the lab. We will work a problem to show how this is done.

EXAMPLE 9.6   CORRECTING FOR THE WATER VAPOR PRESENT WHEN A GAS IS COLLECTED OVER WATER

Problem: Suppose we collect oxygen over water at 20 °C in the apparatus of Figure 9.6. To be sure that the pressure inside the collecting bottle is actually the atmospheric pressure (the value of P_{total} we will use), all we have to do is make sure that the water level is the same inside and outside the bottle. Having done this, we mark the bottle so we can later measure the volume of the entrapped gas by filling the bottle with water to that mark and then measuring the water volume. In this way we find that the volume of the collected, wet gas is 310 mL, and we find that the atmospheric pressure is 738 torr.

Calculate the partial pressure of the oxygen in the bottle, and find what volume the *dry* oxygen would occupy if it were at 760 torr (and 20 °C). Calculate to three significant figures.

Solution: First we find the partial pressure of the water vapor in the bottle. At 20 °C, this is 17.5 torr (Table 9.1), so we can use this value in Equation 9.11.

$$\begin{aligned} P_{O_2} &= P_{\text{total}} - P_{\text{water}} \\ &= 738 \text{ torr} - 17.5 \text{ torr} \\ &= 721 \text{ torr (rounded)} \end{aligned}$$

This is the pressure the oxygen would exert in this same volume if the oxygen were completely dry. We already know the volume — 310 mL — so we have both an initial pressure and an initial volume. We were asked to find the final volume at a final pressure of 760 torr, so all we have to do now is a simple Boyle's law calculation.

The final pressure is greater than the initial pressure, so we have to multiply the initial volume by a volume-decreasing ratio of the two pressures — 721 torr/760 torr.

$$V_2 = 310 \text{ mL} \times \frac{721 \text{ torr}}{760 \text{ torr}} = 294 \text{ mL}$$

We could also have calculated V_2 by plugging the data into Equation 9.7.

The answer means that if we could take the water vapor out of the sample of wet oxygen, the dry oxygen would occupy a volume of 294 mL at 760 torr pressure.

PRACTICE EXERCISE 8   Suppose you prepared a sample of nitrogen and collected it over water at 15 °C at a total pressure of 745 torr. This sample occupied 310 mL. Find the partial pressure of the nitrogen in this sample, and calculate its dry volume at 760 torr.

9.5 THE TEMPERATURE–VOLUME RELATIONSHIP

Hot-air ballooning was just becoming a sensation in Charles's time, but the heat had to be generated by burning straw!

At constant pressure, the volume of a fixed mass of gas is directly proportional to the Kelvin temperature.

In the pressure–volume law, the temperature was kept constant. Suppose that the pressure on a gas is kept constant instead, and the temperature is varied. How does the volume of a fixed mass of gas change? Jacques Alexander Cèsar Charles (1746–1823), a

FIGURE 9.7 Hot air balloons. Propane gas burners just inside the lower throat of a hot-air balloon heat the air in the balloon. The air expands in volume, and because the balloon has an open end, some air spills out. This reduces the *density* of the remaining air, and the balloon rises in the more dense outside air.

French physicist, became interested in this during the time in French history when groups were competing to develop balloons as flying machines. And if you have ever seen splendidly colored, hot-air balloons in flight (Figure 9.7), you can appreciate how exciting the first flights must have been.

Figure 9.8 shows how we could maintain a gas at constant pressure while studying the effects of temperature on volume (without letting any gas escape, unlike in hot-air balloons). The large bulb holding the gas sample is connected by a flexible U-shaped tube—a bent glass tube and a flexible plastic or rubber tube—to a leveling bulb, which

To keep details from obscuring principles, Figure 9.8 excludes gadgets by which gas samples can be admitted into the large bulb.

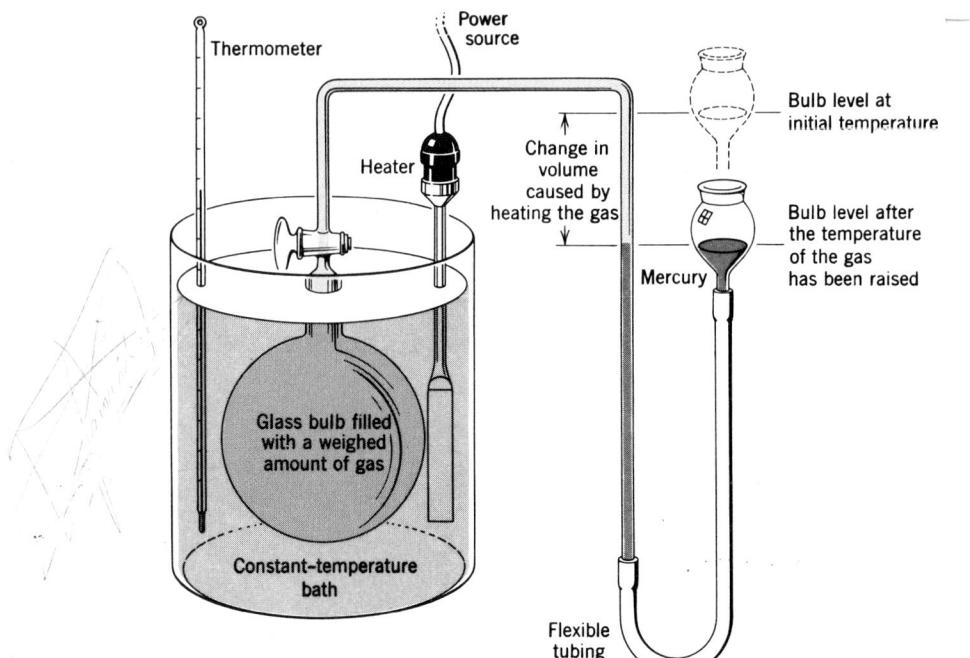

FIGURE 9.8 Charles' law measurements. Shown here is an apparatus for studying how the volume of a gas changes with temperature while the pressure is held constant.

is open to the air. The leveling bulb contains enough mercury to allow the mercury level to rise in the flexible U-tube. As long as the mercury levels in the leveling bulb and in the other part of the U-tube are identical, the pressure inside the gas bulb must be identical with the atmospheric pressure in the lab. If the pressure in the gas bulb were to increase, it would force mercury out of the U-tube and into the leveling bulb. But now the experimenter would lower this bulb until the mercury levels were again the same. This restores the pressure in the gas bulb to atmospheric pressure, but the gas now occupies more volume, as shown in Figure 9.8. You can see that by raising and lowering the mercury bulb we could keep the pressure in the gas bulb a constant. Then, either with a heating unit or a refrigerating unit (not seen in the figure), we would change the temperature of the gas and measure how this affects the volume.

The bath in Figure 9.8 is usually stirred to keep its temperature uniform.

Figure 9.9 plots some of the temperature–volume data for different masses of the same gas under constant pressure. This gas happens to liquefy at −100 °C. Each plot is a straight line, however, so it is easy to extrapolate (that is, reasonably extend) the line to lower temperatures (dashed lines). Such extrapolations are just reasonable predictions of how the gas would behave if it never changed to a liquid.

FIGURE 9.9 Charles' law plots. Each line shows how the gas volume changes with temperature for a different mass of the same gas.

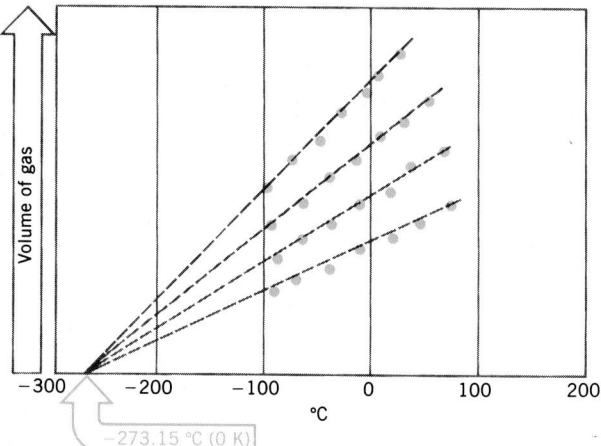

All of the solid lines in Figure 9.9 show that the volume of each fixed mass of gas is directly proportional to temperature (at constant pressure). But notice that all the lines point, when extrapolated, to −273 °C. They all reach this temperature when the corresponding gas volumes become zero. But for a real gas, a zero volume is, of course, impossible. This is how scientists recognized that −273 °C must represent the coldest attainable temperature. To imagine a lower temperature would imply that a gas could have a negative volume. So −273 °C (actually, after refinements, −273.15 °C) came to be called absolute zero and was designated zero on the Kelvin scale of temperatures.

Other gases behave in the same way Charles found, so we now have another gas law, the temperature–volume law, often called Charles' law.

> Temperature–Volume Law (Charles' Law).   The volume of a given sample of gas held at a constant pressure is directly proportional to the Kelvin temperature.
>
> $$V \propto T \qquad (P \text{ and mass are constant}) \qquad (9.12)$$

Charles' law can be stated in this very simple way only by using the *absolute* temperature scale, so we use kelvins in all calculations involving this and other gas laws.

We can write proportionality 9.12 as an equation if we introduce a proportionality constant, which we will designate as C' (where the prime, ', says that this is not the same constant used for Boyle's law).

$$V = C'T \qquad (9.13)$$

Rearranging,

$$\frac{V}{T} = C' \qquad (9.14)$$

We can use subscripts as we did with the pressure–volume law to restate the relationship as follows, and thus make an evaluation of C' unnecessary.

$$\frac{V_1}{T_1} = \frac{V_2}{T_2} \qquad (9.15)$$

This equation must be true because each side equals the same constant, C' (as long as the amount and the pressure of the gas remain constant).

Since we normally want to find either V_2 or T_2 after some change, we can rearrange Equation 9.15 in one of two ways. In the first, the final volume is seen as the product of the initial volume and a ratio of Kelvin temperatures.

$$V_2 = V_1 \left(\frac{T_2}{T_1}\right) \qquad (9.16)$$

a ratio of Kelvin temperatures

Both T_1 and T_2 must not only have the same unit, but this unit *must* be the kelvin.

In the other form, we see that a final temperature is the product of the initial temperature and a ratio of volumes.

$$T_2 = T_1 \left(\frac{V_2}{V_1}\right) \qquad (9.17)$$

a ratio of volumes

The temperatures must always be in Kelvins, but the volumes can be in any unit provided the same unit for volume appears everywhere.

EXAMPLE 9.7 CALCULATING WITH THE TEMPERATURE–VOLUME LAW

Problem: Anesthetic gas is normally given to a patient when the room temperature is 20.0 °C and the patient's body temperature is 37.0 °C. What would this temperature change do to 1600 mL of gas if the pressure and mass stay constant? (Calculate to three significant figures.)

Solution: As with pressure–volume calculations, we first collect the data in one place and *change all Celsius temperatures to kelvins.*

Nowadays, anesthetic gases are nearly always nonflammable, like N_2O (nitrous oxide) and unlike ether.

$$V_1 = 1600 \text{ mL} \qquad T_1 = 293 \text{ K} (20.0 + 273)$$
$$V_2 = ? \qquad T_2 = 310 \text{ K} (37.0 + 273)$$

In this problem we see that the amount of gas and the pressure remain constant, so we know we are dealing with the temperature-volume law. The final volume, therefore, is just the initial volume multiplied by a ratio of temperatures. Let's use our reasoning approach once again to decide whether this temperature ratio is (293 K/310 K) or (310 K/293 K). To do this we examine how the temperature is changing and then decide how this will affect the volume. In this case, we see that the temperature is increasing. We know now that warming a gas at constant pressure causes it to expand, so the final volume *must* be larger than the initial volume. Therefore, we must multiply the initial volume by a volume-raising ratio of temperatures, which means that the temperature ratio must be larger than 1 and that the larger temperature must be in the numerator. Thus,

We could also calculate V_2 by solving Equation 9.15 to give

$$V_2 = V_1\left(\frac{T_2}{T_1}\right)$$

$$V_2 = 1600 \text{ mL} \times \frac{310 \text{ K}}{293 \text{ K}}$$

$$= 1.69 \times 10^3 \text{ mL (rounded from 1692.8328)}$$

Thus the volume increases by about 90 mL, and this change is less than 6%. Still, it is a factor that anesthesiologists and inhalation therapists have to consider.

PRACTICE EXERCISE 9    To make 300 mL of oxygen at 20.0 °C change its volume to 250 mL, what must be done to the sample if its pressure and mass are to be held constant?

9.6 THE COMBINED GAS LAW

The pressure–volume law and the temperature–volume law are special cases of a more general law, the combined gas law.

In many experiments with gases, keeping either the temperature or the pressure of a gas constant is not attempted. Doing either is not always necessary because of a more general law, the *combined gas law*, that relates the temperature, pressure, and volume of a fixed mass of gas.

> Combined Gas Law.   The volume of a given sample of gas is proportional to the ratio of its Kelvin temperature and its pressure.
>
> $$V \propto \frac{T}{P} \quad \text{(at constant mass)}$$

By introducing a proportionality constant, C'', we restate this law in the form of the following equation.

$$V = \left(\frac{T}{P}\right)C'' \tag{9.18}$$

Rearranging this equation, we have

$$\frac{PV}{T} = C'' \quad \text{(at constant mass)} \tag{9.19}$$

By using subscripts in the usual way, we can restate the relationship of Equation 9.19 in a way that relieves us from the need to evaluate C''.

$$\frac{P_1V_1}{T_1} = \frac{P_2V_2}{T_2} \quad \text{(at constant mass)} \tag{9.20}$$

This equation is true because both sides equal the same constant, C''. As always, the temperature must be in kelvins, but pressure and volume can be in any units if used consistently.

An ideal gas would obey this law exactly.

Equation 9.19 is the combined gas law in mathematical form, and Equation 9.20 is the form most useful for calculations. This law, to be valid, must contain each of the other gas laws as special cases, as we illustrate in the next example.

EXAMPLE 9.8   RELATING THE COMBINED GAS LAW TO THE OTHER GAS LAWS

Problem: Prove that the equation for the combined gas law (Equation 9.20) reduces to the equation for the pressure–volume law under the right conditions.

Solution: The "right conditions" for the pressure–volume law are constant mass and temperature. This means that $T_1 = T_2$, so the two can be cancelled in Equation 9.20.

$$\frac{P_1V_1}{\cancel{T_1}} = \frac{P_2V_2}{\cancel{T_2}}$$

Therefore, $P_1V_1 = P_2V_2$, which is one way to state the pressure–volume law (Equation 9.7).

PRACTICE EXERCISE 10   Prove that the combined gas law (Equation 9.20) reduces to the temperature–volume law (Equation 9.15) under the right conditions.

Pressure–Temperature Law

Equation 9.20 contains another gas law, the pressure–temperature law discovered by Jacques Charles and Joseph Gay-Lussac (1778–1850), a French scientist. Using Example 9.8 as a guide, work Practice Exercise 11 to discover how the pressure of a gas changes with the Kelvin temperature when the mass and the *volume* of the sample are fixed.

PRACTICE EXERCISE 11   To what does Equation 9.20 reduce if the gas volume is constant? You will then have a mathematical expression for the pressure–temperature law. Write a statement of this law similar to the statements for the other gas laws. (Prepare both a written statement and an equation.)

PRACTICE EXERCISE 12   A steel cylinder contains neon at a pressure of 760 torr and a temperature of 25 °C. If the cylinder survives a fire in the building where the temperature reaches 800 °C, what would be the internal pressure in this cylinder (assuming no gas escapes and the change in volume of the cylinder caused by metal expansion is negligible)? Express the result in torr and atmospheres (to three significant figures).

EXAMPLE 9.9   USING THE COMBINED GAS LAW

Problem: If a sample of argon, the gas in electric light bulbs, is at a pressure of 760 torr when the volume is 100 mL and the temperature is 35.0 °C, what must its temperature be if its pressure becomes 720 torr and its volume 200 mL? (Calculate to three significant figures.)

Solution: As always, we first assemble the data (changing °C to K).

$$V_1 = 100 \text{ mL} \qquad P_1 = 760 \text{ torr} \qquad T_1 = 308 \text{ K } (35.0 + 273)$$
$$V_2 = 200 \text{ mL} \qquad P_2 = 720 \text{ torr} \qquad T_2 = ?$$

Equation 9.20 can be re-arranged to give:

$$T_2 = T_1\left(\frac{V_2}{V_1}\right)\left(\frac{P_2}{P_1}\right)$$

If we solve Equation 9.20 for T_2, we find that to obtain our answer we multiply T_1 by both a ratio of volumes and a ratio of pressures (see the marginal note). Of course, we could plug our numbers directly into this equation and get the answer, but we will follow the reasoning approach again because it reinforces your knowledge of the behavior of gases and it can also save you some grief if you have a tendency to make mistakes in algebra.

To obtain the volume and pressure ratios, we consider each change separately. First, let's examine the change in volume. Notice that the volume is increasing from 100 mL to 200 mL. How must the temperature change to cause this volume change? We know gases expand when warmed, so this volume change must be associated with an increase in the temperature. The volume ratio, therefore, has to be larger than 1, so it must be

$$\text{volume ratio} = \frac{200 \text{ mL}}{100 \text{ mL}}$$

When we examine the pressure, we see that it is dropping from 760 torr to 720 torr. How must the temperature change to cause this pressure change? We know that warming a confined gas increases its pressure, so cooling it must decrease the pressure. Therefore, the pressure change in this problem must be associated with a temperature *decrease;* the pressure ratio must be smaller than 1.

$$\text{pressure ratio} = \frac{720 \text{ torr}}{760 \text{ torr}}$$

Notice that in deciding what these ratios must be, we have obtained apparently conflicting signals about how the temperature is changing. The volume change suggests a temperature increase while the pressure change suggests a temperature decrease. There is really nothing wrong with this. We could just as easily have carried out the overall change in two steps, first increasing the temperature at constant pressure (the initial pressure) to cause the volume change, and then lowering the temperature at constant volume (the final volume) to cause the pressure to change to its final volume. The actual temperature change is just the net result of both operations, and is obtained by applying both of these ratios to the initial temperature as follows.

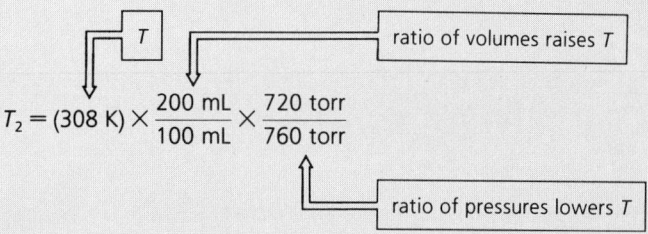

$$T_2 = (308 \text{ K}) \times \frac{200 \text{ mL}}{100 \text{ mL}} \times \frac{720 \text{ torr}}{760 \text{ torr}}$$

$$= 584 \text{ K (rounded from 583.579), or 311 °C (584 − 273)}$$

Thus the net effect is a large increase in temperature. The small temperature-decreasing effect of the pressure change is overwhelmed by the large effect of the volume change, which required the increase in temperature.

PRACTICE EXERCISE 13   Rearrange Equation 9.20 for the general gas law to give equations that express the following relationships in terms of P, V, and T, but use appropriate subscripts for initial and final conditions.
(a) $P_2 = P_1 \times$ (ratio of volumes) $\times$ (ratio of kelvins)
(b) $V_2 = V_1 \times$ (ratio of pressures) $\times$ (ratio of kelvins)
(c) $T_2 = T_1 \times$ (ratio of volumes) $\times$ (ratio of pressures). This is what we did in Example 9.9.

PRACTICE EXERCISE 14   What will be the final pressure of a sample of nitrogen with a volume of 950 m³ at 745 torr and 25.0 °C if it is heated to 60.0 °C and given a final volume of 1150 m³? (Calculate to three significant figures.)

9.7   THE IDEAL GAS LAW

For an ideal gas, the ratio of *PV* to *nT*, *PV/nT*, is a constant called the universal gas constant and symbolized as *R*.

John Dalton exploited mass relationships in chemical reactions to support his atomic theory. Other chemists studied volume relationships instead, looking particularly at reactions where all reactants and products are gases. Gay-Lussac, for example, observed some whole-number ratios for gas volumes that excited the interest of Avogadro.

Gay-Lussac's Law of Combining Volumes

Many reactions occur in which gases constitute both reactants and products. Hydrogen gas reacts with chlorine gas to give gaseous hydrogen chloride. Hydrogen combines with oxygen to give water, which is a gas above 100 °C. Hydrogen also can be made to react with nitrogen to give gaseous ammonia. Gay-Lussac studied reactions and observed what we now call Gay-Lussac's Law of Combining Volumes.

> Gay-Lussac's Law of Combining Volumes. When measured at the same pressures and temperatures, the volumes of gaseous reactants and gaseous products are in ratios of simple, whole numbers.

Some examples that illustrate this law are as follows, where the relative volumes of substances are given beneath their names.

hydrogen + chlorine ⟶ hydrogen chloride
[1 vol]   [1 vol]   [2 vol]

hydrogen + oxygen ⟶   water vapor
[2 vol]   [1 vol]   [2 vol]

hydrogen + nitrogen ⟶   ammonia
[3 vol]   [1 vol]   [2 vol]

Avogadro's Principle

Notice that the sums of the volumes on either side of the arrow are not necessarily equal. But neither are the sums of the coefficients on either side of the arrow in an equation. What intrigued Avogadro about the volume data was that the volumes were in simple whole-number ratios, just like coefficients in a balanced equation. It's as if the numbers representing the volume ratios were the same as the numbers representing the ratios *by molecules*. If this were really so—if the numbers can stand for both gas volumes and gas molecules—then something must be true about gases, a fact we now call Avogadro's Principle.

> Avogadro's Principle. When measured at the same temperature and pressure, equal volumes of gases contain equal numbers of molecules or, equivalently, equal numbers of moles.
>
> $V \propto n$ (at constant T and P, where n = number of moles)

Under Avogadro's principle we can explain Gay-Lussac's observations in terms of balanced chemical equations as follows, where the volume data and the mole data are expressed in the same numbers.[1]

---

[1] We note here that the combining volume data were used to show that hydrogen, chlorine, and oxygen must consist not of atoms but of molecules, specifically *di*atomic molecules at least. If we assume, for example, that hydrogen chloride is correctly formulated as molecules of HCl (not H_2Cl_2, or H_3Cl_3, or higher), then the only way *two* volumes of hydrogen chloride could arise from just *one* volume of hydrogen and *one* of chlorine is if each particle of hydrogen and chlorine consisted of two atoms of hydrogen and two of chlorine, respectively. Otherwise, if hydrogen consisted of single atoms, H, and chlorine consisted of single atoms, Cl, then 1 vol of H and 1 vol of Cl would be able to give just 1 vol of HCl, not 2. (If hydrogen chloride were actually H_2Cl_2, then hydrogen would have to be H_4 and chlorine, Cl_4, and so on.)

$$H_2(g) \quad + \quad Cl_2(g) \quad \longrightarrow \quad 2HCl(g)$$

hydrogen + chlorine $\longrightarrow$ hydrogen chloride

[1 vol H_2]   [1 vol H_2]   [2 vol HCl]

[1 mol H_2]   [1 mol Cl_2]   [2 mol HCl]

$$2H_2(g) \quad + \quad O_2(g) \quad \longrightarrow \quad 2H_2O(g)$$

hydrogen + oxygen $\longrightarrow$ water vapor

[2 vol H_2]   [1 vol O_2]   [2 vol H_2O]

[2 mol H_2]   [1 mol O_2]   [2 mol H_2O]

$$3H_2(g) \quad + \quad N_2(g) \quad \longrightarrow \quad 2NH_3(g)$$

hydrogen + nitrogen $\longrightarrow$ ammonia

[3 vol H_2]   [1 vol N_2]   [2 vol NH_3]

[3 mol H_2]   [1 mol N_2]   [2 mol NH_3]

Standard Molar Volume

On the basis of Avogadro's principle, the volume occupied by one mole of any gas — its molar volume — must be the same *for all gases* at the same temperature and pressure. For comparing experimental molar volumes of real gases, scientists have agreed to use 1 atm (760 torr) and 273.15 K (0 °C) as the standard conditions of temperature and pressure, or STP for short. Under STP, the volume of one mole of a gas is 22.4 L, or very close to this, as the experimental data in Table 9.3 show. This volume of 22.4 L at STP is taken to be the volume of 1 mol of an *ideal gas* and is called the standard molar volume.

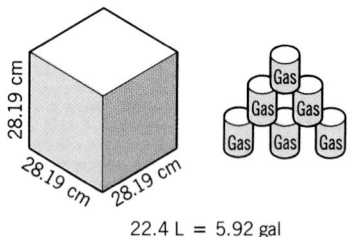

28.19 cm

28.19 cm   28.19 cm

22.4 L = 5.92 gal

1.00 mol of ideal gas = 22.4 L of gas   (at STP)

TABLE 9.3 Molar Volumes of Some Gases at STP

Gas	Formula	Molar Volume (L)	Molar Mass (g)
Helium	He	22.398	4.00
Argon	Ar	22.401	39.9
Hydrogen	H_2	22.410	2.02
Nitrogen	N_2	22.413	28.0
Oxygen	O_2	22.414	32.0
Carbon dioxide	CO_2	22.414	44.0

Ideal Gas Law

In the last section we learned that the combined gas law is given by Equation 9.19, $PV/T = C''$. The constant C'' is actually a constant only in one special circumstance, when the number of *moles* of gas in the sample is fixed. If we change the moles, we also get a new value for C''. Avogadro's law lets us factor the moles of gas into the combined gas law and thus make an even more general gas law.

Suppose that we double the number of moles, n, at constant P and T. The only way we can do this is to double the volume of the gas, so the combined gas law can hold only if the value of C'' also doubles. In other words, the value of the constant C'' must be directly proportional to the number of moles in the sample. Hence, we can write

$$C'' \propto n$$

or

$$C'' = n \times \text{(a new constant)}$$

This new constant of proportionality is given the symbol R and called the universal gas constant. R is the same for all gases. We can now write the combined gas law in a still more general form, one called the ideal gas law (or, in some references, the universal gas law): $PV/T = nR$. This is usually rearranged and written as follows.

$$PV = nRT \qquad\qquad (9.21)$$

The value of R depends on the units we pick for P and V. (T is always in kelvins.) For one mole of an ideal gas, for which $V = 22.4$ L, $T = 273$ K, and $P = 1.00$ atm, we can find R as follows.

$$R = \frac{PV}{nT} = \frac{(1.00\ \text{atm})(22.4\ \text{L})}{(1.00\ \text{mol})(273\ \text{K})} = 0.0821\ \text{L atm mol}^{-1}\ \text{K}^{-1}$$

EXAMPLE 9.10 CALCULATING WITH THE UNIVERSAL GAS CONSTANT

Problem: A sample of oxygen at 21.0 °C and 740 torr weighs 16.0 g. What volume does it occupy?

Solution: We must first convert the given data into the units used in the universal gas constant, $0.0821\ \text{L atm mol}^{-1}\text{K}^{-1}$. Thus the temperature is $273 + 20.0 = 293$ K. The moles of oxygen are found as follows, since the molecular weight of O_2 is 32.0.

$$16.0\ \text{g } O_2 \times \frac{1\ \text{mol } O_2}{32.0\ \text{g } O_2} = 0.500\ \text{mol } O_2$$

The pressure in atmospheres, since 1 atm = 760 torr, is

$$740\ \text{torr} \times \frac{1\ \text{atm}}{760\ \text{torr}} = 0.974\ \text{atm}$$

Let's now collect the data.

$$n = 0.500\ \text{mol} \qquad\qquad T = 293\ \text{K}$$
$$R = 0.0821\ \text{L atm mol}^{-1}\ \text{K}^{-1} \qquad P = 0.974\ \text{atm}$$

We can now use the ideal gas law, Equation 9.21, to find V, cancelling units where possible.

$$V = \frac{nRT}{P} = \frac{(0.500\ \text{mol})(0.0821\ \text{L atm mol}^{-1}\text{K}^{-1})(293\ \text{K})}{(0.974\ \text{atm})}$$
$$= 12.3\ \text{L (rounded from 12.3487)}$$

PRACTICE EXERCISE 15 What volume in liters does a sample of methane, CH_4, weighing 10.2 g occupy at 25.0 °C and 755 torr?

Determining Formula Weights of Gases

The remarkable fact about gases is that when we know any three of the four variables for any sample of any gas — n, P, V, and T — we can calculate the fourth. (This works best, of course, under conditions when the real gas acts most nearly like an ideal gas — at higher temperatures and lower pressures, when it is not close to changing to a liquid.) Thus if we know P, V, and T for a sample for which we know the *mass* but not n itself, we can calculate n using the equation for the ideal gas law. Then by taking the ratio of grams to moles, we can calculate the formula weight of the gas. This is all a formula weight is, the ratio of grams to moles. We will work an example.

When any three of the four variables are fixed, the fourth can have only one value.

EXAMPLE 9.11   CALCULATING THE FORMULA WEIGHT OF A GAS

Problem: A liquid can be decomposed by electricity into two gases. In one experiment, one of the gases was collected. The sample had a mass of 1.090 g, a volume of 850 mL, a pressure of 746 torr, and a temperature of 25 °C. Calculate its formula weight.

Solution: To calculate a formula weight, we need to know the number of moles of the substance in a sample of known mass. Then we divide grams by moles to give the number of grams per mole, which numerically equals the formula weight.

In this problem we are given the mass of gas, so we have part of the necessary information given us directly. We have to calculate the other part — the number of moles in the sample. This we can obtain using the ideal gas law and the values of P, V, and T — the data given in the problem.

$$V = 0.850 \text{ L (since 850 mL} = 0.850 \text{ L)}$$

$$P = 0.982 \text{ atm} \left(\text{since } 746 \text{ torr} \times \frac{1 \text{ atm}}{760 \text{ torr}} = 0.982 \text{ atm} \right)$$

$$T = 298 \text{ K } (25.0 + 273)$$

Now we use the ideal gas law equation to find n, the number of moles.

$$(0.982 \text{ atm}) \times (0.850 \text{ L}) = (n) \times (0.0821 \text{ L atm mol}^{-1} \text{ K}^{-1}) \times (298 \text{ K})$$
$$n = 0.0341 \text{ mol (rounded)}$$

The ratio of grams to moles, the formula weight, is then found by

$$\frac{1.090 \text{ g}}{0.0341 \text{ mol}} = 32.0 \text{ g/mol (rounded)}$$

Thus the formula weight of the gas is 32.0. (The gas is oxygen.)

PRACTICE EXERCISE 16   A sample of the other gas obtained by the electrical decomposition of the liquid of Example 9.11 has a volume of 817 mL at 745 torr and 25 °C and a mass of 0.0682 g. What is its formula weight? What is its likely formula?

PRACTICE EXERCISE 17   How many milliliters of nitrogen, N_2, would have to be collected at 744 torr and 28 °C to have a sample containing 0.015 mol of N_2?

9.8 STOICHIOMETRY OF REACTIONS BETWEEN GASES

The ideal gas law equation lets us work with experimental variables other than mass in stoichiometric calculations for reactions between gases.

Vacuum Lines

Major laboratories retain full-time glassblowers to help put together vacuum lines and make other special glassware.

Scientists who study reactions between gases have developed some very sophisticated equipment to keep gases from escaping during a reaction. The central piece of apparatus in experiments with gases is a *vacuum line* connected to a vacuum pump and fitted with as many valves, reacting chambers, connecting tubes, and pressure gauges as desired. See Figure 9.10. Most gases can be changed to liquids (or solids) when their containers are immersed in liquid nitrogen (−196 °C) or liquid helium (−269 °C), so labs with vacuum lines generally have large, wide-mouthed vacuum bottles holding supplies of such super-cold liquids.

To transfer a gas from one part of the line to another, a liquefied (or solidifed) sample is simply allowed to warm up and develop a pressure. The gas then flows through

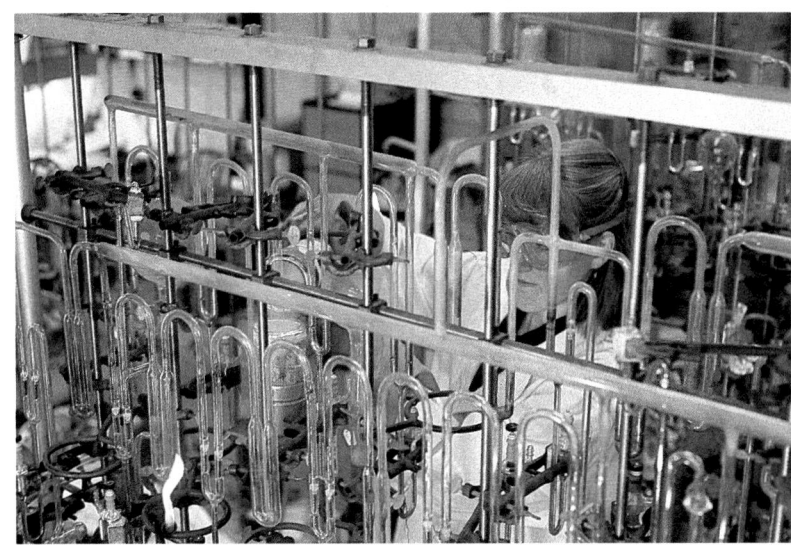

FIGURE 9.10 Vacuum line.

a valve or a stopcock to another containment chamber kept at a lower pressure and usually also chilled to trap the gas in its liquid (or solid) state. The volumes of these containment chambers are known (probably measured as the vacuum line was built); the pressure of a gas sample in such a space can be easily measured; and the temperature can also be found. In other words, it is not hard to determine P, V, and T for any gas sample in a particular part of a vacuum line.

Here is where the ideal gas law now comes in handy. We can use it to calculate the moles of the gas, n, without having to measure its mass. Of course, the mass can be separately measured by closing off a particular chamber, removing it from the vacuum line, and weighing it both empty and with the gas sample in it. Sometimes this has to be done, but the ideal gas law often lets the experimenter avoid the whole operation.

Mole Problems and Gases

Up to now we have learned how to solve mole problems given the masses of chemicals, or volumes of solutions of known molarities. Now we will look at an example in which gases are the reactants and we start with data on P, V, and T, rather than masses or moles.

EXAMPLE 9.12   WORKING MOLE PROBLEMS INVOLVING GASES

Problem: The industrial synthesis of ammonia (NH_3) by the Haber process involves the following reaction.

$$3H_2(g) + N_2(g) \longrightarrow 2NH_3(g)$$

If 725 L of nitrogen, initially at 740 torr and 25.0 °C (298 K), is to be used to make ammonia, how much hydrogen is needed in moles, in grams, and in liters (at 25.0 °C and 760 torr)?

Solution: Instead of being given a certain mass, we start with a certain volume of a reactant. So instead of going from the grams to the mole level, we have to go from the gas volume to the mole level to answer the question. We use, therefore, the ideal gas law equation ($PV = nRT$) to go from volume to moles, remembering to transform any given data into the units needed for R (L atm mol^{-1} K^{-1}). The given volume is already in liters; the temperature is provided in kelvins. We change 740 torr to atmospheres by

$$740 \text{ torr} \times \frac{1 \text{ atm}}{760 \text{ torr}} = 0.974 \text{ atm}$$

Now we can solve for n, the number of moles of nitrogen.

$$n = \frac{PV}{RT} = \frac{(0.974\ \text{atm})(725\ \text{L})}{(0.0821\ \text{L atm mol}^{-1}\text{K}^{-1})(298\ \text{K})}$$
$$= 28.9\ \text{mol N}_2$$

From here on, it's a routine mole problem. Using the coefficients of the balanced equation, we devise a conversion factor to find the moles of H_2 required for 28.9 mol of N_2 in this reaction.

$$28.9\ \text{mol N}_2 \times \frac{3\ \text{mol H}_2}{1\ \text{mol N}_2} = 86.7\ \text{mol H}_2$$

Thus 86.7 mol of H_2 is needed to react with the given volume of 725 L of nitrogen. We now find the grams of H_2 in 86.7 mol of H_2, using the formula weight of H_2, 2.02.

$$86.7\ \text{mol H}_2 \times \frac{2.02\ \text{g H}_2}{1\ \text{mol H}_2} = 175\ \text{g H}_2$$

Finally, we reuse the ideal gas law to calculate the volume of hydrogen at 760 torr (1.00 atm) and 25.0 °C (298 K) from the moles of hydrogen.

$$V = \frac{nRT}{P} = \frac{(86.7\ \text{mol})(0.0821\ \text{L atm mol}^{-1}\text{K}^{-1})(298\ \text{K})}{1.00\ \text{atm}}$$
$$= 2.12 \times 10^3\ \text{L of H}_2$$

PRACTICE EXERCISE 18   How many moles and grams of ammonia form in the Experiment of Example 9.12? How many liters of NH_3 (at STP) does this much ammonia represent?

9.9 LAW OF EFFUSION

Less dense gases effuse more rapidly than more dense gases.

If you have ever had a pet dog or cat that has lost a "stare-down" with a skunk, you know about the fact if not the name of diffusion. Diffusion is the spontaneous intermingling of one substance with another, such as the spreading of the fragrance of irate skunk throughout the room air. When the diffusing substances have the same final concentration everywhere in the chamber, the process is over. Another way to describe the end result when gases are the diffusing fluids is to say that the partial pressures of the gases become everywhere identical in all parts of the chamber. When a fragrance is more or less concentrated in one corner of a room, its partial pressure is higher there than later, after diffusion is over.

A useful term for discussing a situation in which the value of some physical quantity such as concentration or partial pressure changes from one place to another is gradient. Just after you add a spoon of sugar to hot coffee and before you stir it, a concentration gradient for sugar forms in the cup. Just after the unhappy pet enters the room seeking comfort, there is a partial pressure gradient for the alien odor. The point that we want to make here, one of the great facts about natural processes, is that "nature abhors gradients." Given the chance, gradients in nature tend to disappear, some rapidly and some only over eons of time, until there is as much of an evenness as possible.

Effusion is a process similar in nature to diffusion. The effusion of a gas is its movement through an extremely tiny opening into a region of lower pressure. See Figure 9.11. The natural impulse for effusion to occur under the kind of arrangement pictured in this figure is essentially the same as for diffusion — "nature abhors gradients." This

There is an altitude gradient at every hill. Road engineers call it a *grade*.

FIGURE 9.11 Effusion.

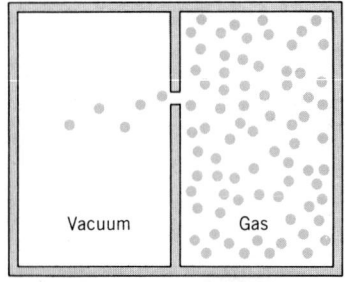

Vacuum   Gas

statement, however, speaks only to the *direction* of a natural change, not to how rapidly the change occurs.

An English scientist, Thomas Graham (1805–1869), studied the rates at which various gases effuse, and he found that the more dense the gas is the slower it effuses. The exact relationship between rate and gas density, *d*, is called the law of gas effusion, or sometimes Graham's law.

Law of Gas Effusion (Graham's Law).   The rate of effusion of a gas is inversely proportional to the square root of its density when the pressure and temperature of the gas are constant.

$$\text{effusion rate} \propto \frac{1}{\sqrt{d}} \qquad \text{(constant } P \text{ and } T\text{)} \qquad (9.22)$$

The proportionality of expression 9.22 can be changed to an equation by using a constant C^*.

$$(\text{effusion rate}) \times \sqrt{d} = C^* \qquad \text{(constant } P \text{ and } T\text{)} \qquad (9.23)$$

What is noteworthy about the constant, C^*, is that it is virtually the same for all gases. This important fact allows us to predict the relative rates of effusion of two gases under the same conditions simply by comparing their densities. Let us let subscripts A and B represent gases A and B. Knowing that C^* is the same for both A and B, we can use Equation 9.23 to write the following relationship.

$$(\text{effusion rate})_A \times \sqrt{d_A} = (\text{effusion rate})_B \times \sqrt{d_B} \qquad (9.24)$$

Gases with low densities effuse faster than those with high densities under the same conditions.

By rearranging Equation 9.24 we can see how the ratio of effusion rates of two gases relates to the ratio of their densities.

$$\frac{(\text{effusion rate})_A}{(\text{effusion rate})_B} = \sqrt{\frac{d_B}{d_A}} \qquad (9.25)$$

EXAMPLE 9.13   USING THE LAW OF GAS EFFUSION

Problem: A store receives a shipment of defective balloons. Each has a tiny pinhole of the same size. If one balloon is filled with helium and another is filled with air to the same volume and pressure, which balloon will deflate faster and how much faster? The density of helium at room temperature is 0.00016 g/mL and that of air is 0.0012 g/mL.

Solution: Equation 9.25 is the one to use here.

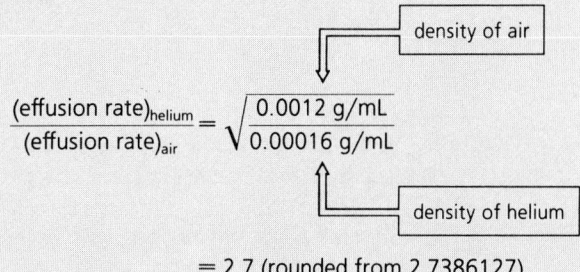

$$\frac{(\text{effusion rate})_{\text{helium}}}{(\text{effusion rate})_{\text{air}}} = \sqrt{\frac{0.0012 \text{ g/mL}}{0.00016 \text{ g/mL}}}$$

$$= 2.7 \text{ (rounded from 2.7386127)}$$

This answer means that helium effuses 2.7 times as fast as air through the same-size pinhole. In fact, helium even effuses through the taut rubber material of a defect-free child's balloon. You may well remember from your own childhood the disappointment of finding that your helium-filled balloon went absolutely limp overnight.

(Left) Two newly filled balloons —air-filled on the left and helium-filled on the right. *(Right)* Same balloons, one day later. Helium has escaped by effusion from its balloon.

PRACTICE EXERCISE 19   Natural gas pipelines cannot be kept totally free of developing microscopic leaks. Natural gas is methane, which has a density at room temperature and 760 torr of 0.654 g/L. Under these conditions the density of hydrogen is 0.0818 g/L. This difference in density, in the light of Graham's law, suggests a potential problem in using natural gas pipelines for sending hydrogen if we should ever adopt hydrogen as an alternative fuel to methane. What is this problem? Hint: Show how the rate of effusion of hydrogen compares with that of methane.

Graham's Law and Formula Weights

Sometimes data on gas densities are not available or are hard to look up. When this happens, we can still use Graham's law, because the density of a gas is directly proportional to its formula weight. We can show this easily by going back to the basic relationship among moles, mass, and formula weight and by substituting this into the ideal gas law. The n or number of moles in the ideal gas law, $PV = nRT$, can be calculated by dividing the mass of a sample by its formula weight (which we will symbolize by f.wt.) (Remember that a formula weight is a ratio of grams to moles, or g/mol.)

$$\text{No. of moles} = \frac{\text{No. of grams}}{\text{formula weight}} = \frac{g}{g/mol} = \frac{g}{\text{f.wt.}}$$

Therefore, we can substitute the expression g/f.wt. for n into the ideal gas law:

$$PV = nRT$$

So

$$PV = \frac{g}{\text{f.wt.}} \times RT$$

By a little clever rearranging we can "solve" this equation for f.wt.

$$\text{f.wt.} = \frac{g}{V} \times \frac{RT}{P}$$

Notice, now, that g/V is really an expression for mass per unit volume, which is density, d. Therefore, we can rewrite this equation as follows:

$$\text{f.wt.} = d \times \frac{RT}{P} \tag{9.26}$$

Since the ratio RT/P is a constant when the experiment is run at fixed values for T and P, we can see that Equation 9.26 tells us what we set out to show — the density of a gas is proportional to its formula weight. In other words,

$$\text{gas density} = d = \text{f.wt.} \times (\text{a constant}) \qquad (\text{at fixed } T \text{ and } P)$$

With this shown, we can now rewrite Equation 9.25 into the most useful variation for calculating relative rates of effusion with Graham's law, a variation in which we can avoid the constant of proportionality.

$$\frac{(\text{effusion rate})_A}{(\text{effusion rate})_B} = \sqrt{\frac{d_B}{d_A}} = \sqrt{\frac{\text{f.wt.}(B)}{\text{f.wt.}(A)}} \qquad (9.27)$$

Gases with low formula weights effuse faster than those with high formula weights under the same conditions.

Now let's work an example to show how Equation 9.27 can be used.

EXAMPLE 9.14   USING GRAHAM'S LAW

Problem: Under the same conditions of temperature and pressure, does hydrogen iodide or ammonia effuse faster? Calculate the relative rates at which they effuse.

Solution: The formula weight of ammonia, NH_3, is 17.0 and that of hydrogen iodide, is 128, so ammonia effuses faster. To find out by how much, we apply Equation 9.27 as follows:

$$\frac{(\text{rate for } NH_3)}{(\text{rate for HI})} = \sqrt{\frac{128}{17.0}}$$
$$= 2.74$$

Thus ammonia effuses almost three times as fast as hydrogen iodide.

PRACTICE EXERCISE 20   Solar energy may some day be used to split water into hydrogen and oxygen and the hydrogen then used as a fuel. The different rates of effusion of hydrogen and oxygen from a mixture of the two and through a very tiny hole might be the basis of separating them. Which gas effuses more rapidly, and by what relative amount?

9.10 THE KINETIC THEORY OF GASES

The gas laws can be explained by assuming that gases consist of many small particles moving randomly according to the laws of physics.

If you have ever taken something apart to see how it works—a clock, a radio, a car engine—you'll be able to appreciate the impulses in scientists back in the nineteenth century who asked, "How do gases work?" These people were aware of the very predictable behavior of gases, and they knew the gas laws. So they wondered what had to be true about a gas to explain its conformity to the gas laws. The kinetic theory of gases was the answer. This theory begins with a set of postulates that suppose what gases must be like, and it assumes that the laws of physics and statistics hold just as much for the particles of a gas as for any other objects.

Postulates of the Kinetic Theory of Gases

1. A gas consists of an extremely large number of very tiny particles that are in constant, random motion.
2. The gas particles themselves occupy a net volume so small in relation to the volume of their container that their contribution to the total volume can be ignored.

The kinetic theory begins with this *model* of an ideal gas.

In fact, the particles are assumed to be so small that they have no dimensions at all.

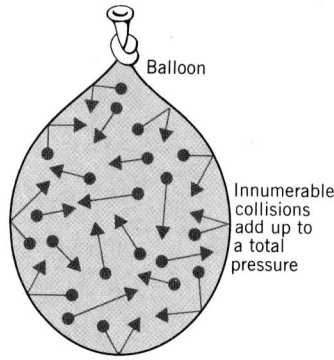

Balloon

Innumerable
collisions
add up to
a total
pressure

The molecules in air (25 °C)
have an average velocity of
about 400 km/s (900 miles per
hour)!

The kinetic energy at the peak
of the curve is called the *most
probable value* of the kinetic
energy.

In Section 16.2, we will see that
the product, *PV*, does have the
dimensions of energy.

3. The particles often collide in perfectly elastic collisions with themselves and with the walls of the container, and they move in straight lines between collisions. (This would be true only if the particles exert no forces of attraction or repulsion on each other.)

Physics, the study of matter, energy, and physical changes, has a branch called mechanics that deals with the laws of behavior of moving objects. (Billiard players master the applications of these laws, if not their mathematical forms.) In effect, the kinetic theory treats gases as supersmall, constantly moving billiard balls that bounce off each other and the walls of their container. They are so small, that their individual volumes can be ignored. Using mechanics and statistics, physicists were able to show that gas pressure is the net effect of innumerable collisions made by gas particles on the walls. They calculated the relationship between pressure and volume at constant temperature, and the result agreed beautifully with the pressure–volume law observed by Boyle.

Kinetic Theory and Gas Temperature

The biggest triumph of the kinetic theory came with its explanation of gas temperature. Each gas particle has a particular mass and velocity. At any given instant, some particles are momentarily stopped and have zero velocity. Others, after being struck particularly hard, acquire a very high velocity. Since collisions are random, the velocity of any one particle changes constantly. But a gas sample has a huge number of particles, so we could envision an average velocity.

Any object with both mass and velocity has energy of motion called kinetic energy (KE), as we learned in Section 1.2. $KE = \frac{1}{2}(mv^2)$, where m = mass and v = velocity. If we can speak of an average velocity there must therefore also be an average kinetic energy. *Individual* particles can have momentary energies ranging from zero to high values, but overall there is an average kinetic energy. Figure 9.12 is a plot of the way the various values of kinetic energy are distributed among the particles. The peak of the curve represents the most frequently occurring value of kinetic energy. The *average* kinetic energy lies at a slightly higher value because the curve is not symmetrical.

When physicists used the postulates of the kinetic theory to find an equation for pressure, they found that the product of pressure and volume, *PV*, is proportional to the average kinetic energy of the particles.

$$PV \propto \text{average KE} \qquad (9.28)$$

But we know from a source entirely different from mechanics and statistics—the equation for the ideal gas law—that the value of *PV* is proportional to the Kelvin temperature.

$$PV \propto T \qquad (9.29)$$

Equations 9.28 and 9.29 tell us that *PV* is proportional both to *T* and to the average KE. *The Kelvin temperature of a gas must, therefore, be directly proportional to the average*

FIGURE 9.12 The distribution of kinetic energies among gas particles. The curve with the higher maximum shows how at room temperature (300 K) the relative number of particles having a particular value of kinetic energy changes with changes in the kinetic energy. The flatter curve shows what happens to the plot when the temperature is made higher.

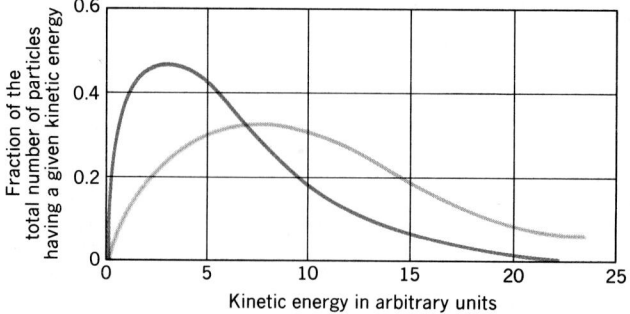

kinetic energy of its particles.

$$T \propto \text{average KE} \tag{9.30}$$

This extremely important relationship between gas temperature and the average kinetic energy of gas particles gives us a most illuminating glimpse into the realm of atoms and molecules. We can now understand, for example, what happens to gas particles when a gas is heated. Because the absorbed heat energy goes to increasing their average kinetic energy, it must therefore go to increasing their velocities. No change in the mass of a gas particle can occur, so the only way the quantity $\frac{1}{2}(mv^2)$, the KE, can change is by a change in v, velocity. Thus, heating a gas makes its particles move faster, on the average.

The effect of temperature on the kinetic energy distribution in a gas can also be seen in Figure 9.12. Notice that at the higher temperature the fractions of molecules with low kinetic energies decrease and the fractions with large kinetic energies increase. This causes the average kinetic energy to shift to a higher value at the higher temperature. The kinetic energy distributions illustrated in Figure 9.12 are extremely important because, as it turns out, they apply not only to particles in gases but also to the particles in a liquid or a solid, as we will see in the next chapter.

9.11   KINETIC THEORY AND THE GAS LAWS

The kinetic theory explains each gas law.

The kinetic theory is the theory about gases, and if it is any good it should explain the facts about gases, the gas laws. Let's see how it does this.

Pressure–Temperature Law

The kinetic theory tells us that an increase in gas temperature increases the average velocity of gas particles. At higher velocities, the particles must strike the container's walls more frequently and with greater force. But at constant volume, the *area* being struck is still the same, so the force per unit area—the pressure—must therefore increase (Figure 9.13). And this is how gas pressure is proportional to gas temperature (under constant volume and mass)—the pressure–temperature law.

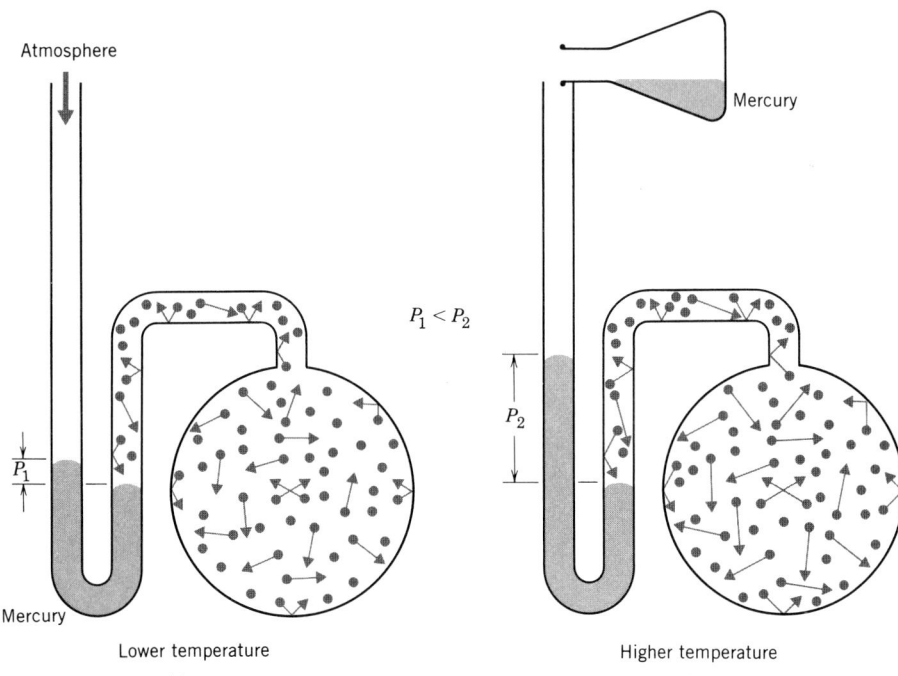

Atmosphere

Mercury

$P_1 < P_2$

P_2

P_1

Mercury

Lower temperature

(a)

Higher temperature

(b)

FIGURE 9.13 The kinetic theory and the pressure–temperature law. *(a)* Here the gas exerts only a small pressure, P_1, over and above the atmospheric pressure. *(b)* Raising the gas temperature raises the pressure to P_2, the extra mercury being added to keep the gas from expanding its volume. At the higher temperature the gas particles move with greater average velocity (as suggested by the longer motion arrows), so normally they would take up more space. The extra mercury is a measure of the increased pressure needed to prevent this.

Temperature – Volume Law (Charles' Law)

Suppose one "wall" of the gas container is a mercury level that can move in or out so that the pressure inside can always equal the pressure outside. When an increase in temperature acts to increase the inside pressure — explained just above in terms of the kinetic theory — the mercury wall moves out instead and so gives the gas more volume (Figure 9.14). And this is how gas volume is proportional to gas temperature (under constant pressure and mass) — Charles's law.

FIGURE 9.14 The kinetic theory and the temperature – volume law. The pressure is the same in both (a) and (b), as indicated by the mercury levels. However, at the lower temperature (a), the gas takes up less volume. At the higher temperature (b), the gas particles have a higher average energy and velocity. To hold the pressure constant (a condition of Charles' law), the gas has to be given more room by letting some mercury flow out of the tube and back into the reservoir.

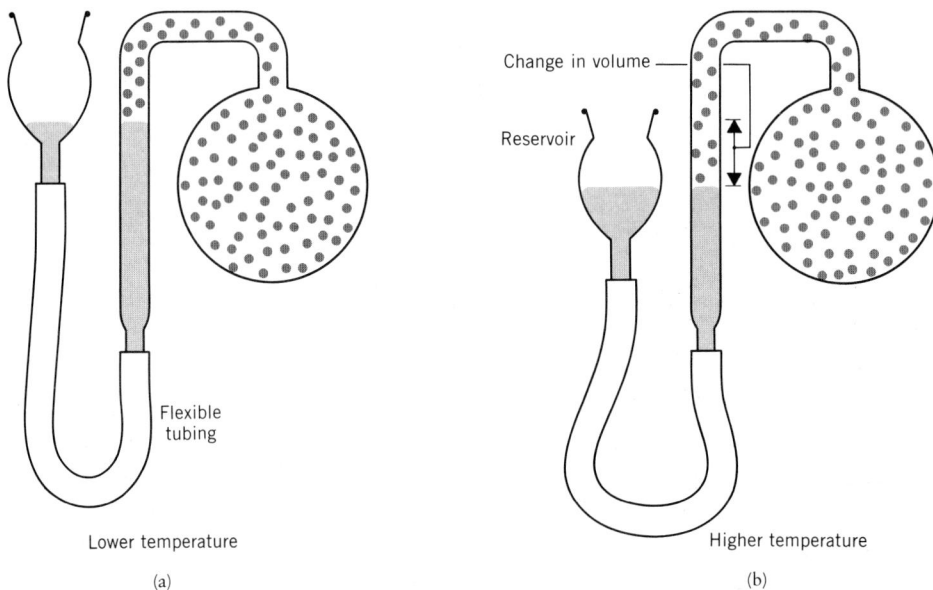

Pressure – Volume Law (Boyle's Law)

Suppose that one wall of the gas container is a movable piston that we can pull out (or push in) and thereby change the gas volume. If we make the gas volume larger without changing its temperature, the gas thins out and there are fewer gas particles per unit volume. The particles still have the same average kinetic energy (since the temperature is the same), but fewer strike any unit area of the walls per second. The pressure, therefore, must decrease. If we made the volume smaller at constant temperature (and constant

FIGURE 9.15 The kinetic theory and the pressure – volume law. When the gas volume is made smaller in going from (a) to (b) the frequency of the collisions per unit area of the container's walls increases. Therefore, the pressure increases.

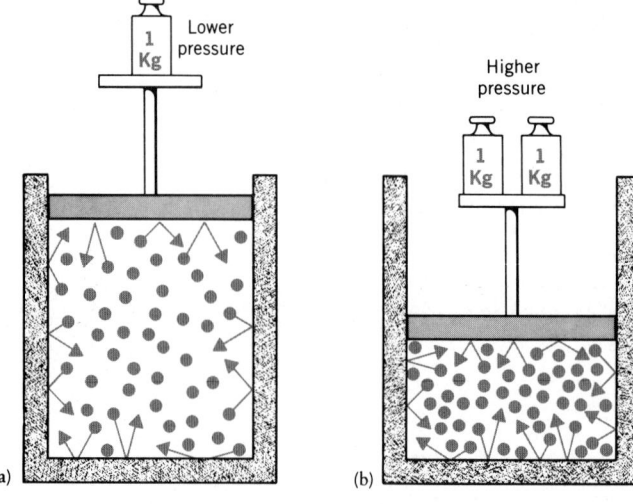

average kinetic energy), the particles would strike a unit wall area more frequently and create a higher pressure (Figure 9.15). And this, in a qualitative sense, is how the kinetic theory explains Boyle's law — gas volume and pressure are inversely proportional (under constant temperature and mass).

Dalton's Law of Partial Pressures

The law of partial pressures is actually evidence for the postulate in the kinetic theory that gas particles neither attract nor repel each other — that they act independently. Only if the particles of each gas in a mixture of gases act independently can the partial pressures add up in a simple way to give the total pressure (Figure 9.16).

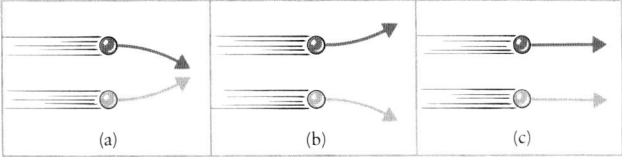

(a)   (b)   (c)

FIGURE 9.16 The kinetic theory and Dalton's law of partial pressures. *(a)* If gas particles attracted each other, the total pressure would be less than the sum of the partial pressures, because they couldn't hit the container's walls with as much force. *(b)* If gas particles repelled each other, the total pressure would be greater than the sum of the partial pressures, because the repulsions would give extra force to each strike at the walls. *(c)* By neither attracting nor repelling each other, the gas particles behave independently and their partial pressures add up to the total pressure (Dalton's law).

Law of Effusion (Graham's Law)

The conditions of Graham's Law are that the rates of effusion of two gases with different formula weights must be compared at the same pressure and temperature. When two gases are at the same temperature, their particles must have the same average kinetic energy. This conclusion, you will recall, was deduced from the postulates of the kinetic theory using the laws of physics and statistics. Remember that $KE = \frac{1}{2}(mv^2)$. When two gases have different formula weights, the value of m in this equation is different for the two. So if we change m in going from one gas to the other, the only way KE can still be a constant is for the average v to be different in each gas. When m is small, as for a low-formula-weight gas, v (or rather, v^2) must be larger to compensate. In other words, the gas with the lower formula weight has particles moving around with a greater velocity than the gas with the higher formula weight. And this is how the lower-formula-weight gas effuses more rapidly than the other (Figure 9.17). Because of the velocity-*squared* term in the equation for kinetic energy, a *square-root* term must occur in the equation that compares rates of effusion (Equation 9.27).

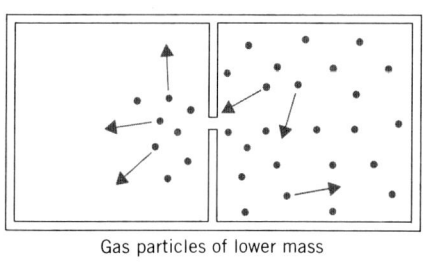

Gas particles of lower mass

(a)

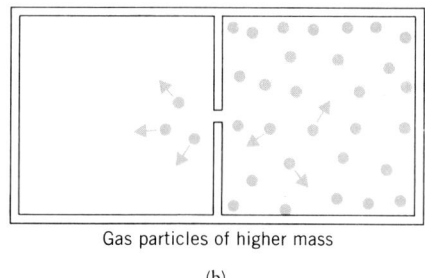

Gas particles of higher mass

(b)

FIGURE 9.17 The kinetic theory and the law of effusion (Graham's law). The chambers have the same volumes, and the pinholes are the same size and large enough for the larger particles in *(b)* to get through. The values of *P* and *T* are the same in both parts, so the average kinetic energy of the small particles in *(a)* is the same as that of the larger particles in *(b)*. In *(a)*, however, the particles must have higher average *velocities* to compensate for their lower mass in the equation for kinetic energy, $KE = \frac{1}{2}mv^2$. Therefore, the smaller particles in *(a)* effuse more rapidly than the larger particles in *(b)*.

Kinetic Theory and the Ideal Gas Law

The equation for the ideal gas law, $PV = nRT$, nowhere identifies a particular gas. It's true for all gases — a most remarkable fact about our natural world. The kinetic theory accommodates this fact as follows. We will first rearrange the ideal gas law equation.

$$P = \frac{n}{V}RT$$

This form of the ideal gas law tells us that the pressure of any gas is directly proportional to *its molar concentration* — n/V is moles per liter — when the temperature is constant.

The question, then, is why should gases at the same molar concentration and the same temperature have the same pressure *regardless of the gas?*

Different gases at the same temperature consist of particles with the same average kinetic energy, according to the kinetic theory. It can be shown (and we will not go into the details) that moving particles with the same kinetic energy exert the same individual force per hit as they strike a given wall area. Of course, if we multiply the force per hit by the number of particles hitting the given area, we get the total force on this area, or force per unit area — pressure. But the total number of particles hitting a given area is proportional not to the identities of the particles but to their concentration — their moles per liter, to pick the most useful units. Different gases at identical molar concentrations and temperatures, therefore, ought to exert identical pressures — and they do.

The Ease of Compressibility of Gases

The chief reason why we have gas laws for gases but not comparable laws for liquids or solids is that gases are mostly empty space. Gas molecules do not touch each other, except when they collide. There is no other (or nearly no other) interaction between them, so the chemical identity of the gas does not matter. When the pressure on a gas is increased, the molecules are denied some of their "empty space" as the volume shrinks in accordance with Boyle's law. Thus gases are compressible. Liquids and solids, with little if any "empty spaces," cannot respond like gases to compression.

Kinetic Theory and Absolute Zero

Why should there be a floor on how cold things can get? Why should there be an *absolute* zero of coldness? If we think about the relationship of temperature to average kinetic energy, as given by the kinetic theory, $T \propto$ average KE, we can see that if the average KE falls to zero, the temperature must become zero, also. But KE $= \frac{1}{2}(mv)^2$, and we know that m cannot become zero. So the only way KE can become zero is if v goes to zero. A gas particle cannot move any slower than it does at a dead standstill, so if the gas molecules stop moving entirely, the gas is as cold as anything can get — absolute zero.

9.12 REAL GASES — DEVIATIONS FROM THE IDEAL GAS LAW

The atoms or molecules of real gases have real volumes and do attract each other slightly, so real gases cannot obey the laws of an ideal gas perfectly.

A real gas deviates from ideal behavior in two important ways. First, the model of an ideal gas, as postulated by the kinetic theory, assumes that gas molecules individually have no volume, but of course they do. In the aggregate, they take up part of the total space, a portion called the *excluded volume*. It is because the excluded volume at ordinary pressures is such a tiny fraction of the total volume — the value of V in the gas laws — that real gases fit the gas laws so well. Because of the excluded volume, however, we cannot indefinitely reduce a gas volume by half each time we double the pressure (Boyle's law). This is why we have said that real gases behave most like an ideal gas when the pressure is relatively low.

The second reason for deviations from ideal gas behavior is that the particles of a real gas do attract each other, unlike the assumption in the model of an ideal gas. Let us now study one of the ways by which scientists have modified the gas law equations to give better fits to data for real gases.

Van der Waals' Equation

Van der Waals won the 1910 Nobel prize in physics.

J. D. van der Waals (1837–1923), a Dutch scientist, was one of many scientists who tried to modify the gas law equations to get a better fit to the data. He first corrected for the

TABLE 9.4 Van Der Waals Constants

Substance	a (L² atm/mol²)	b (L/mol)
Noble Gases		
Helium, He	0.03412	0.02370
Neon, Ne	0.2107	0.01709
Argon, Ar	1.345	0.03219
Krypton, Kr	2.318	0.03978
Xenon, Xe	4.194	0.05105
Other Gases		
Hydrogen, H_2	0.02444	0.02661
Oxygen, O_2	1.360	0.03183
Nitrogen, N_2	1.390	0.03913
Methane, CH_4	2.253	0.04278
Carbon dioxide, CO_2	3.592	0.04267
Ammonia, NH_3	4.170	0.03707
Water, H_2O	5.464	0.03049
Ethyl alcohol, C_2H_5OH	12.02	0.08407

excluded volume. He reasoned that the excluded volume should be subtracted from the measured volume of the real gas, V_r, before the latter is used in a gas law calculation. This correction would give the volume, V_i, for the gas if it were actually ideal. Letting n be the number of moles of gas in a sample and b stand for the actual correction—the *excluded volume per mole*—the ideal gas volume is then found by the following equation.

$$V_i = V_r - nb \qquad (9.31)$$

The measured volume, V_r, is greater than the ideal volume.

As we would expect, gases with large molecules have larger values of b than those with small molecules, as seen in the data of Table 9.4.

Van der Waals next reasoned that the measured pressure of the real gas, P_r, under the influences of attractions between molecules, must be less than the pressure, P_i, if the gas were ideal. We can understand this by reflecting on what the forces of attraction must do to the paths taken by moving gas molecules. As illustrated in Figure 9.18a, the particles of an ideal gas move in straight lines between collisions—a postulate of the kinetic theory. But if gas molecules attract each other in a real gas, when they get close they must change directions and move in curved paths as they pass each other, as illustrated in Figure 9.18b. This would make real gas molecules travel longer distances to reach the walls, so they would take more time and would therefore hit any unit area of

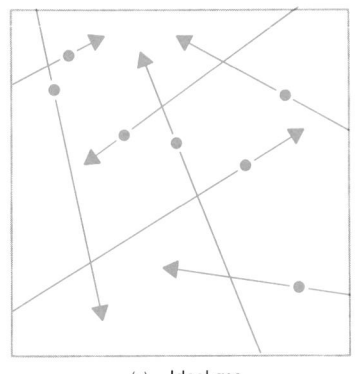

(a)   Ideal gas

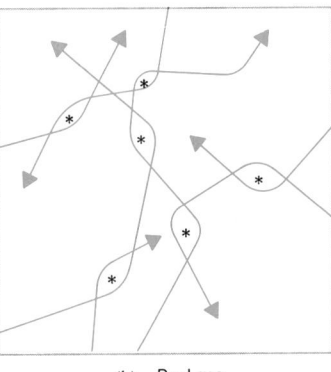

(b)   Real gas

FIGURE 9.18 The effect of forces of attraction between the molecules of a real gas. *(a)* In an ideal gas, the molecules are postulated to travel in straight lines. *(b)* In a real gas, slight attractions between molecules make the paths curve around the points (marked by asterisks) where one molecule passes close to another.

the walls less frequently than they would if ideal. The result of a decreased collision frequency in a real gas would mean a lower pressure. To correct for this, van der Waals added a term to the measured pressure, P_r, as follows.

$$P_i = P_r + \frac{n^2a}{V_r^2} \qquad (9.32)$$

In this equation, n is the number of moles in the gas sample, V_r, is the measured volume of the real gas, and a is the specific correction term. The quantity a is proportional to the strengths of the attractive forces between the gas molecules. You can see from Equation 9.32 that the larger the value of a — meaning the stronger the attractive forces — the greater must be the total correction to the measured pressure. But of course the stronger the attractive forces are, the greater will be the deviations in the paths of the moving particles. As a result, the collisions with the walls will occur even less frequently. So with strong attractive forces, implied by large values of a, large corrections to the measured pressure are needed. Table 9.4 includes values of a for several gases.

In Equation 9.32, we can also see that because of the V_r^2 term in the denominator, the pressure correction is only significant when the volume is small — in other words, when the molecules are packed tightly together. This makes sense, because only under these conditions will there be many close passes between neighbors. When the volume is large (i.e., when the pressure is low) most of the molecules in the gas will follow nearly straight paths and the pressure of the gas will deviate only slightly from that expected for an ideal gas. We see, once again, that a real gas should behave most like an ideal gas when its pressure is low.

If we now substitute the expressions for the volume and the pressure of the gas, as corrected, into the equation for the ideal gas law, we obtain what is called van der Waals' equation of state for a real gas. (We drop the subscripts, because the values of P, V, and T in this equation must refer to the *measured* or real values; they are the only values accessible by direct experiment.)

$$\left(P + \frac{n^2a}{V^2}\right)(V - nb) = nRT \qquad (9.33)$$

The constants a and b are called the van der Waals constants. They are determined by making careful measurements of P, V, and T for the real gas in question under a number of different sets of conditions. Then trial calculations are made to figure out what values of the van der Waals constants give the best match between the observed data and van der Waals' equation of state for a real gas. The usefulness of a and b lies in the information they imply about molecular size and attractions between molecules.

SUMMARY

Gas Properties.   To describe the physical state of a gas we use as many of the following measurable properties as we need.

Property	Symbol	Usual Units
Pressure	P	torr, atm
Volume	V	mL, L
Temperature	T	kelvins (always)
Amount of gas	n	mol
Partial pressure	P_a	torr (a stands for the formula of the gas)
Density	$d = m/V$	g/L, mg/L

Gas Laws.   An ideal gas is a hypothetical gas that obeys the gas laws exactly over all ranges of pressure and temperature. Real gases exhibit ideal gas behavior most nearly at low pressures and high temperatures, that is, under conditions remote from those that liquefy a gas.

Pressure–Volume Law (Boyle's Law).   Volume varies inversely with pressure at constant temperature and mass. $V \propto 1/P$.

Law of Partial Pressures (Dalton's Law).   The total pressure of a mixture of gases is the sum of the partial pressures of the individual gases. $P_{\text{total}} = P_a + P_b + P_c + \cdots$.

Temperature–Volume Law (Charles' Law).   Volume varies directly with Kelvin temperature at constant pressure and mass. $V \propto T$.

Temperature–Pressure Law.   Pressure varies directly with Kelvin temperature at constant volume and mass. $P \propto T$.

Gay-Lussac's Law of Combining Volumes.   The volumes of gaseous reactants and products are in ratios of simple, whole numbers when compared at identical temperatures and pressures.

Avogadro's Principle.   Equal volumes of gases contain equal numbers of moles when compared at the same temperature and pressure.

Combined Gas Law. PV divided by T for a given gas sample is a constant.

Ideal Gas Law. $PV = nRT$.

Law of Effusion (Graham's Law).   The rate of effusion of a gas varies inversely with the square root of its density (or the square root of its formula weight) at constant pressure and temperature.

Kinetic Theory of Gases.   An ideal gas consists of innumerable very hard particles, with individual volumes so small they can be ignored, between which no forces of attraction or repulsion act. They are in constant, chaotic, random motion. When the laws of physics and statistics are applied to this model, and the results compared with the ideal gas law, the Kelvin temperature of a gas is proportional to the average kinetic energy of the gas particles. Pressure is the result of forces of collisions of the particles with the container's walls.

Real Gases.   Because individual gas particles do have real volumes and because small forces of attraction do exist between them, real gases do not exactly obey the gas laws. The van der Waals equation of state for a real gas makes corrections for the excluded volume of a gas and for the attractive forces.

REVIEW EXERCISES

Answers to questions whose numbers are printed in color are given in Appendix C. Difficult questions are marked with asterisks.

Concept of Pressure

9.1   What is the difference between force and pressure?

9.2   What causes the atmosphere to exert a force on the earth?

9.3   Why is the high density of mercury an advantage in a Torricelli barometer?

9.4   Mercury has a low vapor pressure. What advantage does this give to the use of mercury in a Torricelli barometer?

9.5   At 20 °C the density of mercury is 13.5 g/mL and that of water is 1.00 g/mL. Also at 20 °C, the vapor pressure of mercury is 0.0012 torr and that of water is 18 torr. Give and explain two reasons why water would be an inconvenient fluid to use in a Torricelli barometer.

9.6   How are torr and mm Hg related?

9.7   How are mm Hg and atm related?

9.8   If a force of 200 lb acts on an area of 1 ft², what is the pressure in pounds per square inch?

9.9   Which exerts the higher pressure, a force of 100 lb acting on 25 in.² or a force of 25 lb acting on 5 in.²?

9.10   A scientist observed that the atmospheric pressure in the laboratory was 744 mm Hg. What was the pressure in torr?

9.11   One day in the lab the atmospheric pressure was 755 torr. What was it in mm Hg?

9.12   A scientist observed that the atmospheric pressure was 750 torr. What fraction of one standard atmosphere is that?

9.13   What would be the pressure in torr if it is 0.850 atm?

9.14   A pressure of 0.445 atm would be how many torr?

9.15   A pressure of 755 torr would be how many kilopascals?

9.16   What are the partial pressures in kilopascals of nitrogen and oxygen in inhaled air? (Use the data in Table 9.2.)

9.17   In kilopascals, what are the partial pressures of nitrogen and oxygen in exhaled air? (Use the data in Table 9.2.)

*9.18   A suction pump works by creating a vacuum into which some fluid will rise. Suppose that you are working for a construction company and someone tells you to take a suction pump and remove the water from a pit. The water level is more than 35 ft below the place where you must put the pump. Can any suction pump do the job? (The density of water is 1.00 g/mL, and the density of mercury is 13.6 g/mL.)

*9.19   One of the oldest units for atmospheric pressure is pounds per square inch (lb/in.²). Find out what a standard atmosphere is in these units. Give the answer in three significant figures. *Hint:* Use the results of Review Exercise 9.18. Calculate the mass in pounds of a uniform column of water 33.9 ft high having an area of 1.00 in.² at its base. (The density of water is 1.00 g/mL; 1 mL = 1 cm³; 1 lb = 454 g; 1 in. = 2.54 cm.)

Specific Gas Laws

9.20   State the pressure–volume law both in words and in the form of an equation.

9.21   Give the temperature–volume law both in words and in the form of an equation.

9.22   What is the law of partial pressures? State it in words and in the form of an equation.

9.23   In what way does the law of partial pressures provide evidence for a postulate of the kinetic theory?

9.24 Of the four important variables in the study of the physical properties of gases, which are assumed to be held at constant values in each of these laws?
(a) Boyle's law
(b) Charles' law
(c) law of partial pressures
(d) pressure–temperature law
(e) Graham's law of effusion
(f) Avogadro's law

9.25 To compress nitrogen at 760 torr from 750 mL to 500 mL, what must the new pressure be if the temperature is kept constant?

9.26 If oxygen at 950 torr is allowed to expand at constant temperature until its pressure is 760 torr, how much larger will the volume become?

9.27 Helium in a 100-mL container at a pressure of 500 torr is transferred to a container with a volume of 250 mL. What is the new pressure (a) if no change in temperature occurs? (b) If its temperature changes from 20 to 15 °C?

9.28 What will have to happen to the temperature of a sample of methane if 1000 mL at 740 torr and 25.0 °C is given a pressure of 814 torr and a volume of 900 mL?

9.29 A sample of nitrogen at 760 torr with a volume of 100 mL is carefully compressed at constant temperature in successive changes of pressure, equaling 20 torr at a time, until the final pressure is 1000 torr. Calculate each new volume and prepare a plot of P versus V, showing P on the horizontal axis.

9.30 A sample of helium is confined in a heavy-walled container with a volume of 450 mL fitted with a pressure gauge. The initial pressure is 760 torr. The helium is then heated in increments of 20 K starting at 293 K, until the final temperature is 393 K. Prepare a plot of P versus T, using the horizontal axis for T.

9.31 When nitrogen is prepared and collected over water at 30 °C and a total pressure of 738 torr, what is its partial pressure in torr?

9.32 If you were to prepare oxygen and collect it over water at a temperature of 10 °C and a total pressure of 751 torr, what would be its partial pressure in torr?

9.33 A sample of carbon monoxide was prepared and collected over water at a temperature of 20 °C and a total pressure of 749 torr. It occupied a volume of 275 mL. Calculate the partial pressure of this gas in the sample (in torr) and its dry volume (in mL) under a pressure of 760 torr.

9.34 A sample of hydrogen was prepared and collected over water at 25 °C and a total pressure of 736 torr. It occupied a volume of 295 mL. Calculate its partial pressure (in torr) and what its dry volume would be (in mL) under a pressure of 760 torr.

9.35 What volume of "wet" methane would you have to collect at 20.0 °C and 740 torr to be sure the sample contained 240 mL of dry methane at the same pressure?

9.36 What volume of "wet" oxygen would you have to collect if you needed the equivalent of 260 mL of dry oxygen at 760 torr and the atmospheric pressure in the lab that day

was 746 torr? The oxygen is to be collected over water at a temperature of 15.0 °C.

9.37 Under conditions in which the density of carbon dioxide is 1.96 g/L and that of nitrogen is 1.25 g/L, which gas will effuse more rapidly? What will be the ratio of the rates of effusion of nitrogen to carbon dioxide?

9.38 Uranium hexafluoride is a white solid that readily passes directly into the vapor state. At room temperature (20.0 °C), its vapor pressure is 120 torr. A trace amount (about 0.7%) of the uranium in this compound is of the type that can be used in nuclear power plants or atomic bombs. It's called uranium-235. Essentially all of the rest, called uranium-238, is useless for direct application in power plants or bombs. In fact, its presence interferes. During World War II a massive government effort was made to separate the two kinds of uranium. Gas effusion was used because the density of the hexafluoride made from uranium-235 is 2.2920 g/L at 120 torr and 20.0 °C while under these same conditions the density of the hexafluoride made from uranium-238 is 2.3119 g/L. (In the gaseous states, these compounds behave remarkably like ideal gases.) Calculate how much more rapidly the lower-density compound will effuse compared to the higher-density compound. (The very small difference that you will find was actually enough, but repeated effusions were necessary.)

Combined and Ideal Gas Laws

9.39 Calculate the value of the universal gas constant if the standard molar volume is in units of milliliters, and STP are expressed as 760 torr and 273 K.

9.40 What is the difference between a molecule, a mole, and a molar volume of a gas (e.g., oxygen)?

9.41 At STP, how many molecules of hydrogen are in 22.4 L?

9.42 Show how the ideal gas law reduces to each of the simpler laws under the appropriate conditions.
(a) Boyle's law   (c) pressure–temperature law
(b) Charles' law   (d) Avogadro's law

9.43 A sample of helium at a pressure of 745 torr and in a volume of 2.55 L was heated from 25.0 to 75.0 °C. The volume of the container expanded to 2.75 L. What was the final pressure (in torr) of the helium?

9.44 When a sample of neon with a volume of 665 mL and a pressure of 1.00 atm was heated from 15.0 to 65.0 °C, its volume became 695 mL. What was its final pressure (in atm)?

9.45 What must be the new volume of a sample of oxygen (in L) if 2.75 L at 742 torr and 25.0 °C is heated to 37.0 °C under conditions that let the pressure change to 760 torr?

9.46 When 275 mL of nitrogen at 742 torr and 15.0 °C was warmed to 30.0 °C, the pressure became 760 torr. What was the final volume (in mL)?

9.47 A sample of krypton with a volume of 6.25 L, a pressure of 765 torr, and a temperature of 20.0 °C expanded to a volume of 9.55 L and a pressure of 375 torr. What was its final temperature (in °C)?

9.48 A sample of Freon, a refrigerating gas, in a volume of 445 mL, at a pressure of 1.50 atm, and at a temperature of 25.0 °C was compressed into a volume of 225 mL with a pressure of 2.00 atm. To what temperature (in °C) did it have to change?

9.49 After a sample of argon with a volume of 525 mL was heated from 21.0 to 85.0 °C, its volume changed to 585 mL and its pressure became 795 torr. What must have been its initial pressure in torr?

9.50 After a sample of nitrogen with a volume of 675 mL and a pressure of 1.00 atm was compressed to a volume of 340 mL and a pressure of 2.00 atm, its temperature was 27.0 °C. What must have been its starting temperature (in °C)?

*9.51 Fuel is ignited in a diesel engine when it is injected into hot compressed air. The compression is what heats the air. In a typical high-speed diesel engine the chamber in the cylinder has a diameter of 10.8 cm and a length of 13.3 cm. [Find the volume by the equation, volume = 3.14 × (radius)^2 × length.] On compression, the length of the chamber is *reduced* by 12.7 cm (a "5-in. stroke"). The compression of the air changes its pressure from 1.0 to 34.0 atm. The initial temperature of the air is 363 K. What will be its final temperature (in K and °C) just before the fuel injection, as a result of the compression? (Calculate to three significant figures.)

*9.52 Suppose early one cool morning (61.0 °F) you take your bike for a long bike ride. You check the tire pressure and find that it's 50.0 lb/in^2. (That means that the air pressure in the tire is actually 50.0 lb/in^2 + 14.7 lb/in.^{2} = 64.7 lb/in^2. Tire gauges tell us how much *over* the atmospheric pressure the measured pressure is, and we may take atmospheric pressure to be 14.7 lb/in.^{2}.) By late afternoon the temperature has climbed to 98.5 °F. This rise in temperature plus the heat created by road friction makes the air temperature inside the tire 105 °F. What will the pressure gauge read now, assuming that the volume of air in the tire and the atmospheric pressure have not changed?

9.53 What volume (in L) does 1.00 mol of nitrogen occupy at 25.0 °C and 760 torr?

9.54 A sample of 1.00 mol of oxygen at 50.0 °C and 740 torr occupies what volume (in L)?

9.55 A sample of 4.25 mol of hydrogen at 20.0 °C occupies a volume of 25.0 L. Under what pressure (in atm) is this sample?

9.56 If a steel cylinder with a volume of 1.50 L contains 10.0 mol of oxygen, under what pressure (in atm) is the oxygen if the temperature is 27.0 °C?

9.57 A steel cylinder containing 83.4 mol of helium with a volume of 10.0 L is under what pressure (in atm) at room temperature (25 °C)?

9.58 When the pressure in a certain gas cylinder with a volume of 4.50 L reaches 500 atm, the cylinder is likely to explode. If this cylinder contains 40.0 mol of argon at 25.0 °C, is it on the verge of exploding? Calculate the pressure (in atm).

9.59 A steel cylinder of nitrogen has a volume of 25.0 L at 24.0 °C and a nitrogen pressure of 150 atm. How many moles of nitrogen does this cylinder hold?

9.60 After the contents of a steel cylinder of pressurized helium with a volume of 18.4 L have been depleted until the pressure is 1.00 atm, the cylinder is called "empty." How many moles of helium remain in the cylinder if the temperature is 25.0 °C?

9.61 When a steel cylinder of oxygen with a volume of 14.5 L was used to supply oxygen to an oxyacetylene torch, the pressure in the cylinder changed from 208 to 201 atm. The temperature of the oxygen was 25.0 °C at both times the pressures were read. How many moles and how many grams of oxygen had been used up?

9.62 When a small cylinder of hydrogen with a volume of 855 mL and a pressure of 115 atm at 26.0 °C was used to supply hydrogen for a reaction in another vessel, the pressure in the hydrogen cylinder after the experiment was half as much (still at 26.0 °C). How many moles and how many grams of hydrogen were taken?

9.63 What is the density (in g/L) of each of the following gases at STP? (Carry the calculations to three significant figures.) (a) CH_4 (methane), (b) O_2, (c) H_2.

9.64 Calculate the density (in mg/mL) of each of the following gases at STP. (Do the calculations to three significant figures.) (a) He, (b) N_2, (c) C_2H_6 (ethane).

9.65 What density (in g/L) does oxygen have at 25.0 °C and 745 torr?

9.66 At 752 torr and 20.0 °C, what is the density of argon (in g/L)?

9.67 At 22.0 °C and a pressure of 755 torr, a gas was found to have a density of 1.14 g/L. Calculate its formula weight.

9.68 A gas was found to have a density of 1.76 mg/mL at 24.0 °C and a pressure of 741 torr. What was its formula weight?

9.69 Working at a vacuum line, a chemist isolated a gas in a weighing bulb with a volume of 255 mL, at a temperature of 25.0 °C, and under a pressure in the bulb of 10.0 torr. The gas weighed 12.1 mg. What was the formula weight of this gas?

9.70 A chemist isolated 6.3 mg of one of the many boron hydrides in a weighing bulb of a vacuum line. The volume of this bulb was 385 mL. At 25.0 °C, the pressure in the bulb was 11 torr.
(a) What was the formula weight of this hydride?
(b) What was the molecular formula, BH_3, B_2H_6, or B_4H_{10}?

9.71 How many liters of chlorine at STP are needed to react with 5.00 L of H_2 (also at STP) in the following reaction?

$$H_2(g) + Cl_2(g) \longrightarrow 2HCl(g)$$

9.72 In the Haber process for the synthesis of ammonia

$$N_2(g) + 3H_2(g) \longrightarrow 2NH_3(g)$$

how many liters of N_2 are needed to react completely with 45.0 L of H_2 if both gases are at STP?

9.73 In the first step of the Ostwald process for making nitric acid, ammonia reacts with oxygen at 650 °C and 1 atm. The following reaction occurs.

$$4NH_3(g) + 5O_2(g) \longrightarrow 4NO(g) + 6H_2O(g)$$

How many liters of oxygen at 650 °C and 1 atm are needed to react with 48 L of NH_3 also at 650 °C and 1 atm?

9.74 Ethylene, C_2H_4, reacts with hydrogen under pressure to give ethane, C_2H_6, according to the following equation.

$$C_2H_4(g) + H_2(g) \longrightarrow C_2H_6(g)$$

A sample of 14.0 g of ethylene requires how many liters of hydrogen at STP for this reaction?

9.75 A common laboratory preparation of hydrogen on a small scale uses the reaction of zinc with hydrochloric acid that produces zinc chloride and hydrogen.
  (a) Write a balanced equation for this reaction.
  (b) If 10.0 L of hydrogen at 760 torr and 25.0 °C is wanted, how much zinc is needed, in theory. Give your answer in moles and in grams.
  (c) How many moles of HCl are needed for part b?
  (d) If the hydrochloric acid is available as 6.00 *M* HCl, how many milliliters of this solution are sufficient?

9.76 In an experiment designed to prepare a small amount of hydrogen by the reaction of zinc and hydrochloric acid (Review Exercise 9.75), a student was limited to using a gas-collecting bottle with a maximum capacity of 325 mL. The method involved collecting the hydrogen over water. What are the minimum grams of zinc and milliliters of 8.00 *M* HCl needed to produce the *wet* hydrogen that can exactly fit this collecting bottle at a pressure of 745 torr and a temperature of 20 °C?

9.77 Ammonia is converted to ammonium sulfate, an important fertilizer, by the reaction

$$2NH_3 + H_2SO_4 \longrightarrow (NH_4)_2SO_4$$

  (a) If 225 kg of ammonium sulfate is to be made in one batch, how many liters of ammonia at STP are needed?
  (b) How many moles of H_2SO_4 are required?
  (c) If the H_2SO_4 is in the form of a 6.00 *M* solution, what volume of this solution, in liters, is needed?

9.78 Carbon dioxide can be made in the lab by the reaction of hydrochloric acid with calcium carbonate.

$$CaCO_3(s) + 2HCl(aq) \longrightarrow CaCl_2(aq) + H_2O + CO_2(g)$$

How many grams of calcium carbonate and how many milliliters of 6.00 *M* HCl are needed to prepare 475 mL of dry CO_2 if it is to be collected at 20.0 °C and 755 torr?

9.79 One of the concerns about the world's dwindling supply of natural gas (methane, CH_4, mostly) is the need for ammonia, a fertilizer that is essential if enough food is to be raised to meet world needs. The overall equation for one process for making ammonia is described by the following equa-

tion, in which N_4O is used as an approximate formula for air, one of the raw materials.

$$7CH_4 + 10H_2O + 4N_4O \longrightarrow 16NH_3 + 7CO_2$$

Methane is sold in units of *tcf*, where 1 *tcf* = 1 thousand cubic feet = 10^3 ft^3 = 28.3×10^3 L at STP.
  (a) What volume of ammonia in liters at STP can be made by this process from 1.00 *tcf* of methane (also at STP)?
  (b) How many moles of ammonia does this represent?
  (c) One thousand cubic feet of methane therefore represents how many kilograms of ammonia?

9.80 One industrial synthesis of acetylene, a gas used as a raw material for making countless synthetic drugs, dyes, and plastics, is the addition of water to calcium carbide.

$$\underset{\substack{\text{calcium} \\ \text{carbide}}}{CaC_2} + 2H_2O \longrightarrow Ca(OH)_2 + \underset{\text{acetylene}}{C_2H_2}$$

  (a) In a small-scale test to improve efficiency, 100 g of CaC_2 is converted to acetylene. What is the theoretical yield of acetylene in moles? In liters (at STP)?
  (b) To make 1.00×10^6 L of acetylene (at STP) by this method requires how much calcium carbide in moles? In kilograms?

9.81 A sample of an unknown gas having a mass of 3.620 g was made to decompose into 2.172 g of oxygen and 1.448 g of sulfur. Prior to the decomposition, this sample occupied a volume of 1120 mL when its pressure was 750 torr and its temperature 25.0 °C.
  (a) What is the percentage composition of the elements in this gas?
  (b) What is the empirical formula of the gas?
  (c) What is its molecular formula?

9.82 A sample of an unknown gas with a mass of 1.620 g occupied a volume of 941 mL when its pressure was 748 torr and its temperature was 20.00 °C. When made to decompose into its elements, 1.389 g of carbon and 0.2314 g of hydrogen were obtained.
  (a) What is the percentage composition of the elements in this gas?
  (b) Determine its empirical formula.
  (c) Determine its molecular formula.

9.83 A sample of a new antimalarial drug with a mass of 0.2394 g was made to undergo a series of reactions that changed all of the nitrogen in the compound into N_2. This gas, when collected over water at 23.80 °C at a total pressure of 746.0 torr, had a volume of 18.90 mL. The vapor pressure of water at 23.80 °C is 22.110 torr.
  (a) Calculate the percentage of nitrogen in the sample.
  (b) When a sample of this drug with a mass of 6.478 mg was burned in pure oxygen, 17.57 mg of CO_2 and 4.319 mg of H_2O were obtained. What are the percentages of carbon and hydrogen in this compound? Assuming that any undetermined element is oxygen, write an empirical formula for the compound.
  (c) The formula weight of the compound was found to be 324. What is its molecular formula?

*9.84 In one analytical procedure for determining the percentage of nitrogen in unknown compounds, a weighed sample is made to decompose to nitrogen gas, which is collected over water at known temperatures and pressures. The volume of the nitrogen is then translated into grams and then into percentages.

(a) Show that the following equation can be used to calculate the percentage of nitrogen in a sample having a mass of W grams when the nitrogen gas with a volume of V mL is collected over water at t_c °C, the total pressure in torr is P, and the vapor pressure of water is P_{H_2O}.

$$\text{Percentage N} = 0.04489 \times \frac{V(P - P_{H_2O})}{W(273 + t_c)}$$

(b) Use this equation to calculate the percentage of nitrogen in the sample described in Review Exercise 9.83.

Kinetic Theory of Gases

9.85 What model of a gas is proposed by the kinetic theory?

9.86 To what properties of an ideal gas is its temperature proportional?
(a) According to the kinetic theory.
(b) According to the combined gas law.

9.87 Gases, if left to themselves, must migrate from a region of high pressure to a region of low pressure. In what specific way can the kinetic theory be used to explain this direction of migration?

9.88 What aspects of the kinetic theory of gases guarantee that two gases given the freedom to mix will always mix entirely?

9.89 If the molecules of a gas at constant volume are somehow given a lower average kinetic energy, what physical properties of the gas will change (and in what direction)?

9.90 Explain *how* heating makes a gas expand at constant pressure. (Describe a mechanical model that connects the rising temperature to the response of the gas — expansion.)

9.91 Explain *how* heating makes a confined gas have a higher pressure at constant volume.

9.92 According to van der Waals, what postulates of the kinetic theory aren't strictly true?

9.93 What equation does the van der Waals equation become more and more like when the volume of a gas becomes larger and larger without any change in temperature or the number of moles?

9.94 What does a large value for the van der Waals constant b indicate is probably true about the molecules of the gas?

9.95 A small value for the van der Waals constant a suggests something about the molecules of a gas. What?

CHEMICALS IN USE
AIR AS A NATURAL RESOURCE

When an air sample is cleaned of its trace amounts of dusts and water vapor, what remains is called *pure dry air*. It is a mixture of several gases, each present in a remarkably constant percentage by volume (Figure 7a). A percentage by volume tells us the number of liters of a component in 100 liters of a gas mixture.

Pure dry air sampled anywhere — over the equator or the polar regions, at any altitude up to as high as 60 to 75 miles — has a constant composition of 78.084% N_2 and 20.948% O_2. These add up to 99.032%, and the noble gas argon makes up nearly all of the remainder. The rest of the noble gases are also present in air, but at much lower concentrations. We will survey here some of the most important uses of the nitrogen and oxygen in air. Superconductivity, which uses liquid nitrogen, is discussed in Special Topic 19.1.

Air as a Chemical Resource for the Biosphere

The plants and animals of the earth's biosphere — the sum total of all regions where living things occur — depend absolutely on air for both nitrogen and oxygen. The need for oxygen to support the reactions of living systems is well known. Molecular oxygen is also needed to support the processes of combustion and decay. If dead plant and animal matter did not decay, the earth would be littered by the remains to depths difficult to imagine, assuming that living things could continue to grow.

Unlike the widespread need for molecular oxygen, only a few microorganisms use molecular nitrogen from air. But on these few depends much of the success of all other living systems. Higher organisms use the compounds

made from ammonia by certain single-celled species. These live cooperatively — biologists call it symbiosis — with the root systems of legumes like alfalfa, clover, peas, and beans. The process is called *nitrogen fixation*. Other single-celled organisms in soil convert ammonia to nitrite and nitrate ions. And plants use these simple nitrogen substances to make the building blocks of all proteins, the amino acids.

Amino acids are made of atoms of nitrogen, carbon, hydrogen, and oxygen. They share the features shown in the general structure, and they differ in the group of atoms (G) attached at one carbon atom. Thus

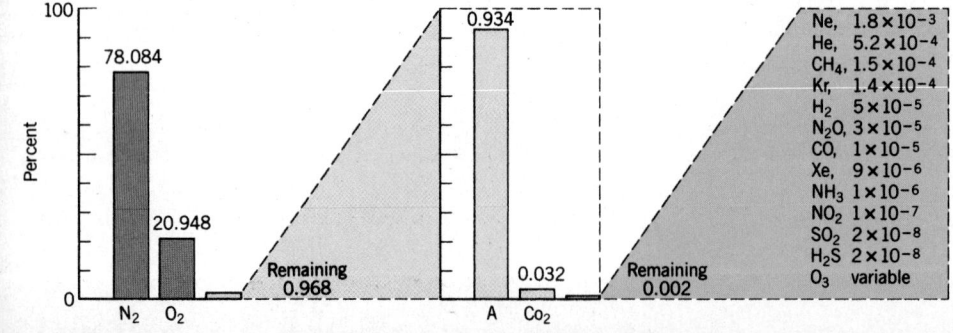

amino acid — general structure   alanine — a typical amino acid

in alanine, $G = CH_3$, and alanine is just one of about 20 of the amino acids used by plants and animals. (Sulfur atoms occur in one amino acid, where $G = CH_2-S-H$.)

Animals are unable to make all of the amino acids they need, so they rely on plants to make them. When plant proteins are digested, their amino acids are released to be used by the eater. The exclusively plant-eating animals get their amino acids directly, and meat-eating animals get them from other animals that feed on plants.

When animals and plants die, microorganisms reconvert their proteins into simpler nitrogen compounds. Some nitrogen atoms get into nitrite and nitrate ions and some into atmospheric nitrogen. Thus a cycle exists in nature —

Ne,	1.8×10^{-3}
He,	5.2×10^{-4}
CH_4,	1.5×10^{-4}
Kr,	1.4×10^{-4}
H_2	5×10^{-5}
N_2O,	3×10^{-5}
CO,	1×10^{-5}
Xe,	9×10^{-6}
NH_3	1×10^{-6}
NO_2	1×10^{-7}
SO_2	2×10^{-8}
H_2S	2×10^{-8}
O_3	variable

FIGURE 7a The components of pure dry air. The numbers are percentages by volume.

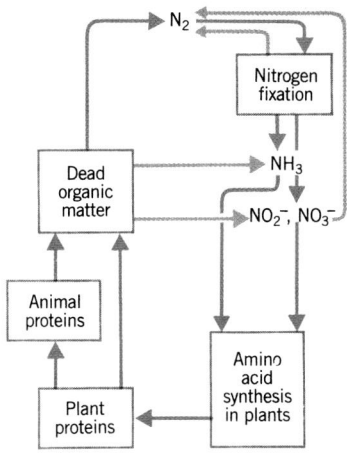

FIGURE 7b The nitrogen cycle.

the nitrogen cycle (Figure 7b)—in which nitrogen atoms move between air, earth, and living systems.

Air as a Resource for Industry and Agriculture

Most people are now familiar with the use of liquid oxygen (LOX) in rocket engines. Anyone who has been around hospitals knows of the uses of oxygen or oxygen-enriched air in medicine. Still another use of oxygen is in one step in making steel. A blast of oxygen sent through molten iron helps to separate impurities that would make the steel too brittle, as we will study in Chapter 21.

In most agriculture, the natural nitrogen cycle does not replenish soil nitrogen at a high enough rate. So artificial nitrogen fertilizers are used—chiefly ammonia, urea, and ammonium nitrate. All are made from the nitrogen and oxygen in air, and the needed hydrogen comes from methane. Synthetic ammonia is made by an industrial operation called the Haber–Bosch process. Its overall reaction is

$$N_2 + 3H_2 \longrightarrow 2NH_3$$

It does not happen except at elevated pressures and temperatures under which the reactant gases are made to flow through special metal screens on whose surfaces the reaction can go much more readily than in their absence. The metal used is an example of a *catalyst,* a substance that is itself not permanently changed by a reaction but helps to make the reaction occur either under milder conditions or at a faster rate under tougher conditions. (In Chapter 18 we will say much more about catalysts.) Without the Haber–Bosch catalyst, the ammonia synthesis almost certainly could not be made to work economically.

The high costs of achieving the elevated temperatures and pressures required by the Haber–Bosch process have stimulated considerable research to find a way to "fix" atmospheric nitrogen under the same conditions used by legumes and their symbiotic microorganisms. World agriculture would be able to feed people at lower costs if such research were successful. Particular attention is being paid

to the catalysts present in the living systems that fix nitrogen. They are fragile compounds that include both iron and molybdenum, and they are quickly inactivated by exposure to oxygen.

Some industrial ammonia is converted into urea, which is a solid, by the following overall reaction with carbon dioxide.

$$2NH_3 + CO_2 \longrightarrow NH_2-\overset{\overset{\displaystyle O}{\|}}{C}-NH_2 + H_2O$$
$$\text{urea}$$

Soil bacteria and water work together to break urea back down to ammonia (and CO_2)—the reverse of this reaction—so urea is thought of as a solid and hence easier-to-handle form of ammonia.

Another good solid nitrogen fertilizer is ammonium nitrate, NH_4NO_3, made from ammonia and nitric acid.

$$NH_3 + HNO_3 \longrightarrow NH_4NO_3$$

Essentially all nitric acid is made today from ammonia by a method called the *Ostwald process.* Its several steps (to be studied in Chapter 19) add up to the following equation. (Notice that oxygen, obtained from air, is also a reactant.)

$$NH_3 + 2O_2 \xrightarrow{\text{(several steps)}} HNO_3 + H_2O$$

Most nitric acid production goes to ammonium nitrate, which also happens to be an ingredient in several explosives. The manufacture of other explosives, like nitroglycerin and TNT, also requires nitric acid. Peacetime uses of explosives include the excavation of tunnels, mining, and other earth-moving operations that would be next to impossible by manual labor, even if all the laborers were slaves. Nitric acid is also needed to make nylon and some other nitrogen-containing synthetic fibers and plastics.

You can see that both the oxygen and the nitrogen in air are essential to the manufacture of a number of commodities deemed important not just to modern agriculture but also to modern civilization.

Questions

1. Because the percentages of nitrogen and oxygen in air are so constant throughout time and place, and because they add up nearly to 100%, it might be argued that air is a (slightly impure) *compound* instead of a mixture of nitrogen and oxygen. Using the percentages given for these two gases, calculate the best empirical formula for this "compound."

2. What process inserts the nitrogen from air into the nitrogen cycle, and, in broad terms, what is this cycle?

3. Write the overall equations for the Haber–Bosch process and the Ostwald process. What are the names of three nitrogen fertilizers?

4. The biosphere alone uses oxygen in three important processes. What are they?

5. Why do you suppose that liquid oxygen instead of the less expensive liquid air is used in rocketry?

Farmers in some southern states protect their crops from unseasonable frost damage by spraying the vegetation with water during the hours of subfreezing temperatures. Even though ice forms, as long as liquid water is present the system's temperature cannot fall below the water-ice equilibrium temperature of 0 °C. The temperature inside the sprayed plants, therefore, cannot drop below freezing and the crop survives.

LIQUIDS, SOLIDS, AND CHANGES OF STATE

10.1 PHYSICAL PROPERTIES VERSUS CHEMICAL PROPERTIES

Physical properties of substances are as important as chemical properties.

Although much of the study of chemistry is devoted to chemical properties and reactions, the physical properties of substances often concern us more on a day-to-day basis. For example, farmers worry about whether expected precipitation will come in liquid or solid form. Rain (a liquid) is usually welcomed; hail (a solid) is almost universally dreaded because of the damage it can inflict on crops. The fact that water expands when it solidifies may bring disaster (or at least an expensive repair bill) to the automobile owner who has neglected to put antifreeze in the car's radiator if the outside temperature drops very far below freezing. During the summer, the high rate of evaporation of paint solvent can make painting in the hot sun difficult. And in Denver, the traditional "three-minute egg" isn't nearly as well cooked as most people like, so it's boiled a little longer.

These are just a few examples that show how the physical properties of substances and the transformations among the three states of matter influence our lives. One of the goals of chemistry has been to understand what determines these properties. From this understanding has come a more complete knowledge of the structure of matter and an ability (at least to some degree) to design chemicals that possess the kinds of properties we want. Examples include the synthetic fibers from which many of our modern fabrics are woven — rayon, nylon, and polyesters. These are substances not found in nature and without which life would be much less interesting.

We began our study of the physical properties of substances in the last chapter where we discussed the properties of gases. In this chapter we will go on to investigate the properties of the other two states of matter — liquids and solids — as well as what happens when substances change from one state to another.

Many common substances, such as synthetic fibers, were developed simply because they have desirable physical properties as well as chemical properties.

10.2 WHY GASES DIFFER FROM LIQUIDS AND SOLIDS

Physical properties depend on forces of attraction between molecules, which are strong in liquids and solids and very weak in gases.

When we studied the gas laws in Chapter 9, one comforting fact was that we didn't have to worry about the chemical composition of a gas. For the most part, the gas laws work quite well regardless of whether we are studying H_2, CO_2, or any other substance, provided it's a gas under the specified conditions of temperature and pressure. A useful consequence of this, for example, is that we can use the ideal gas law to calculate molecular weights of gases. Such calculations rely on the fact that one mole of *any* gas, regardless of its composition, occupies the same volume at a given temperature and pressure. But why is this so?

Actually, it isn't. In our discussion of real gases, we discovered that the "ideal" gas laws are not obeyed exactly by real substances. For example, for any real gaseous substance at a given temperature and pressure there is a small difference between the actual volume and the volume calculated from the ideal gas law. Two reasons were given: the gas molecules themselves do occupy a very small part of the total volume of the gas, and there are weak attractive forces between the molecules. But just the existence of *gas laws,* which we are able to use at all with reasonable success, depends on the fact that (1) the volume of the molecules is nearly insignificant compared to the total volume occupied by the gas and (2) the attractive forces between the gas molecules are very weak, and are thus also nearly insignificant. We don't find similar "liquid laws" or "solid laws" because in liquids and solids these two factors can't be ignored; they play a very important role in determining the properties of these "condensed" states of matter.

In liquids and solids, the molecules are packed together very tightly and there is little empty space between them. For a liquid or a solid, then, most of the volume is taken up by the molecules themselves, and there is little empty space into which other molecules can be squeezed. These states of matter therefore lack the ability to be compressed into smaller volumes as easily as gases. This is one of the most obvious differences between liquids and gases and between solids and gases.

Most of the physical properties of gases, liquids, and solids are actually controlled by the strengths of intermolecular attractions — the attractive forces that exist between neighboring particles. In liquids and solids, these forces are much stronger than in gases. But why?

Anyone who has ever played with magnets has discovered an important property of attractive forces between things: Whether these forces are magnetic or electrical, their strengths decrease quite rapidly as the distance between the attracting particles increases. For example, when two magnets are placed near each other, the attraction between them can be quite strong, but if they are far apart hardly any attraction is felt at all. This same phenomenon is what is primarily responsible for that fact that gases have properties that depend very little on chemical composition, whereas the properties of liquids and solids are influenced by chemical composition to a very large extent.

In a gas the molecules are far apart. At these large distances, the attractive forces are very weak, so the differences caused by dissimilarities in chemical makeup are very small and are hardly noticeable at all. As a result, all gases seem to be pretty much alike. However, in a liquid or solid, where the molecules are very close to each other, these attractive forces become quite strong. Differences between the intermolecular attractions caused by variations in chemical composition become magnified, and this causes unlike substances to behave quite differently.

10.3 INTERMOLECULAR ATTRACTIONS

Dipole–dipole forces, London forces, and hydrogen bonding are the principal kinds of intermolecular attractions.

From the preceding discussion, it should be obvious that to understand and explain the properties of liquids and solids, we need to know about the kinds and relative strengths of intermolecular attractions. Before we proceed further, therefore, let's consider what they are and how their relative strengths vary.

First, it is important to realize that the attractions *between* molecules are always much weaker than the attractions *within* molecules. For example, in a molecule of HCl, the hydrogen atom and chlorine atom are held to each other very tightly by a covalent bond. Neighboring HCl molecules are attracted to each other by much weaker forces. Therefore, when a particular chlorine atom moves, the hydrogen atom bonded to it is forced to follow along, and the HCl molecule remains intact as it moves about, as illustrated in Figure 10.1.

The strengths of the attractions within molecules (the chemical bonds) determine chemical properties. The attractions *between* molecules determine the physical properties of substances.

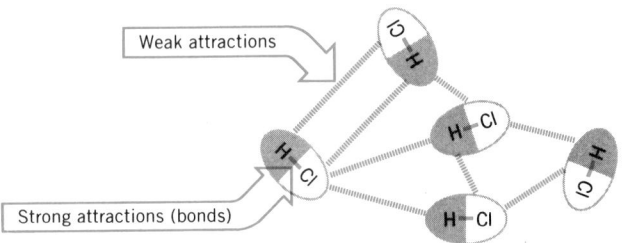

FIGURE 10.1 Strong attractions exist between H and Cl atoms within HCl molecules. Weaker attractions exist between neighboring HCl molecules.

Dipole–Dipole Attractions

In Chapter 7 we saw that HCl is an example of a polar molecule. Molecules like this have a partial positive charge at one end and a partial negative charge at the other. They tend to line up so that the positive end of one dipole is near the negative end of another. Thermal energy, however, causes the molecules to jiggle about, so the alignment isn't perfect. Nevertheless, there is still a net attraction between the polar molecules, as shown in Figure 10.2. We call these attractive forces dipole–dipole attractions. Generally, they are only about 1% as strong as a covalent bond, and their strengths decrease quite rapidly as the distance between molecules increases.

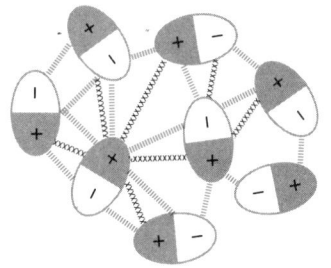

Attractions (⸺) are greater than repulsions (✕✕✕), so the molecules feel a net attraction to each other.

FIGURE 10.2 Attractions between polar molecules occur because the molecules tend to align themselves so that opposite charges are near each other.

London Forces

It is fairly easy to understand why there are attractions between polar molecules such as HCl or SO_2. But even between the particles of nonpolar substances, such as the atoms of the noble gases and nonpolar molecules such as Cl_2 and CH_4, there are attractive forces. For example, these nonpolar substances can be condensed to liquids and even solids if cooled to low enough temperatures, so there must be attractions between their particles that cause them to cling together in the liquid or solid state. These attractions between nonpolar particles are considerably weaker than those between polar molecules, however. Evidence for this comes from comparing properties that depend on the strengths of intermolecular attractions. As you will learn later in this chapter, one such property is boiling point — the higher the boiling point, the stronger are the attractions between molecules in the liquid.

Table 10.1 gives data for two polar substances, HCl and H_2S, and two nonpolar substances, F_2 and Ar. All four are somewhat similar in that they have similar formula weights and the same number of electrons. Notice that the two nonpolar substances have very similar boiling points. Notice also that the two polar compounds have similar boiling points. However, the nonpolar substances have boiling points that are at least 100 degrees lower than those of the polar substances, which tells us that the attractions in the nonpolar liquids are considerably weaker than those in the polar liquids.

TABLE 10.1 Boiling Points of Some Polar and Nonpolar Substances

Substance	Boiling Point (°C)	Formula Weight	Number of Electrons
Polar			
HCl	−84.9	36	18
H_2S	−60.7	34	18
Nonpolar			
F_2	−188.1	38	18
Ar	−185.7	40	18

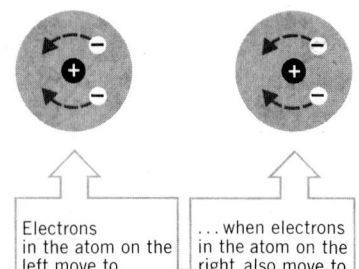

Electrons in the atom on the left move to the left...   ...when electrons in the atom on the right also move to the left.

FIGURE 10.3 Instantaneous or "frozen" views of the electron density in two neighboring particles.

In 1930, Fritz London, a German physicist, offered a simple explanation for the weak attractions between nonpolar particles. In any atom or molecule the electrons are constantly moving, and if we could examine this motion in two neighboring particles, we would find that the movement of electrons in one particle influences the movement of electrons in the other. This is because electrons repel each other and tend to stay as far apart as they can. Therefore, as an electron of one particle gets near the other particle, electrons on the second particle are pushed away. This happens continually as the electrons move around, so to some extent, the electron density in both particles flickers back and forth in a synchronous fashion. This is illustrated in Figure 10.3, which depicts

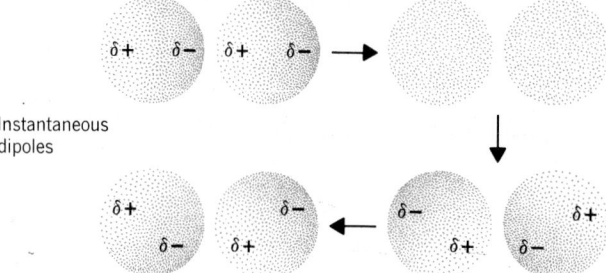

Instantaneous dipoles

a series of instantaneous "frozen" views of the electron density. Notice that at any given moment the electron density of a particle can be unsymmetrical, with more negative charge on one side than on the other. For that particular instant, the particle becomes a dipole, and we call it a momentary dipole or instantaneous dipole.

As the instantaneous dipole forms in one particle, it pushes electron density around in its neighbor, which causes the electron density there to become unsymmetrical, too. As a result, this second particle also becomes a dipole, and we call it an induced dipole because it is caused by, or *induced* by, the formation of the first dipole. Because of the way the dipoles are formed, they always have the positive end of one near the negative end of the other, so there is a dipole–dipole attraction between the two particles. It is a very short-lived attraction, however, because the electrons keep moving, which causes the dipoles to vanish as quickly as they were formed. Then, perhaps a moment later, the dipoles will reappear in a different orientation and there will be another brief dipole–dipole attraction. In this way the short-lived dipoles cause momentary tugs between the particles. When averaged over a period of time, the overall attraction tends to be quite weak because the attractive forces are only "turned on" part of the time.

The momentary dipole–dipole attractions that we've just discussed are called *instantaneous dipole-induced dipole attractions*, or London forces, to distinguish them from the kind of dipole–dipole attractions that exist continually in substances such as HCl.

London forces are the only kind of attraction present between nonpolar molecules. They also are present between polar molecules (in addition to the regular dipole–dipole attractions), and they even occur between ions. However, because London forces are weak, their effects contribute relatively little to the overall attractions between ions.

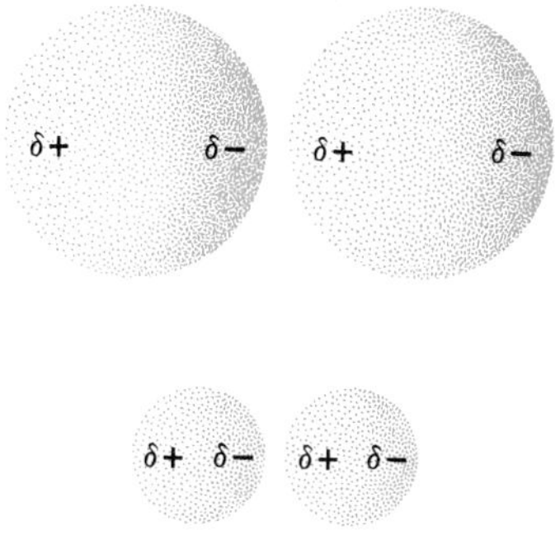

The strengths of London forces depend on several factors. One of them is the size of the electron cloud of the particle. In general, when the electron cloud is large, the outer electrons are not held very tightly by the nucleus (or nuclei, if the particle is a molecule). This makes the electron cloud "mushy" and rather easily deformed (Figure 10.4), which is precisely the condition that most favors the formation of an instantaneous dipole or the creation of an induced dipole. Particles with large electron clouds therefore form short-lived dipoles more easily and experience stronger London forces than do similar particles with small electron clouds. The effects of size can be seen if we compare the boiling points of the halogens or the noble gases (Table 10.2). Those that are large have higher boiling points (and therefore stronger intermolecular attractions) than those that are small.

FIGURE 10.4 A large electron cloud is easily deformed; a small electron cloud is not.

TABLE 10.2 Boiling Points of the Halogens and Noble Gases

Group VIIA	Boiling Point (°C)	Group 0	Boiling Point (°C)
F_2	-188.1	He	-268.6
Cl_2	-34.6	Ne	-245.9
Br_2	58.8	Ar	-185.7
I_2	184.4	Kr	-152.3
		Xe	-107.1
		Rn	-61.8

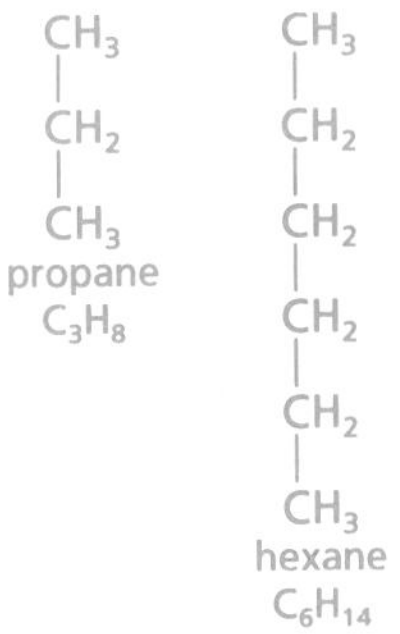

A second factor that affects the strength of London forces is the number of atoms in a molecule. For molecules containing the same elements, the London forces increase with the number of atoms. The hydrocarbons (see Table 10.3) are a good example. They are composed of chains of carbon atoms with hydrogen atoms bonded to the carbon atoms all along the chain. If we compare two hydrocarbon molecules of different chain length — for example, C_3H_8 and C_6H_{14} — the one having the longer chain also has the higher boiling point. This means that the molecule with the longer chain length experiences the stronger attractive forces. This is so because it has more places along its length

CH_3
|
CH_2
|
CH_3
propane
C_3H_8

CH_3
|
CH_2
|
CH_2
|
CH_2
|
CH_2
|
CH_3
hexane
C_6H_{14}

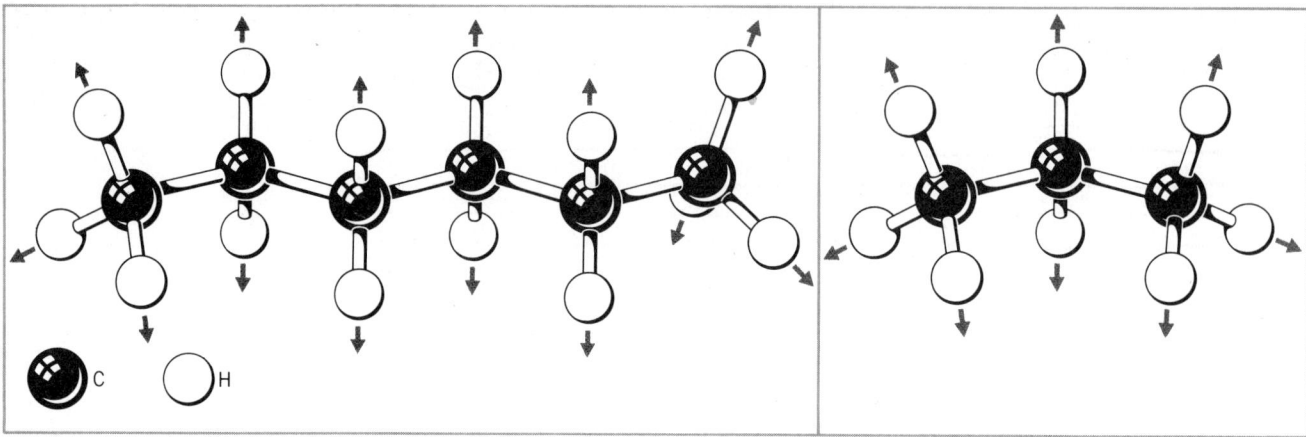

FIGURE 10.5 The C_6H_{14} molecule has more places along its length where it can be held than does the shorter C_3H_8 molecule.

where it can be attracted to other molecules, as illustrated in Figure 10.5. Even if the strength of attraction at each location is about the same, the *total* attraction experienced by the longer C_6H_{14} molecule is greater than that felt by the shorter C_3H_8 molecule. In Table 10.3 you can see that some of the hydrocarbons have boiling points that are quite high, which shows that the cumulative effects of London forces can result in very strong attractions.

TABLE 10.3 Boiling Points of Some Hydrocarbons[a]

Molecular Formula	Boiling Point (°C at 1 atm)
CH_4	−161.5
C_2H_6	−88.6
C_3H_8	−42.1
C_4H_{10}	−0.5
C_5H_{12}	36.1
C_6H_{14}	68.7
.	.
.	.
.	.
$C_{10}H_{22}$	174.1
.	.
.	.
.	.
$C_{22}H_{46}$	327

[a] The molecules of each hydrocarbon in this table have carbon chains of the type C—C—C—C— etc., that is, one carbon follows another.

Hydrogen Bonds

The last kind of attractive force that we will discuss is a special case of dipole–dipole attraction. When hydrogen is covalently bonded to a very small, highly electronegative atom such as fluorine, oxygen, or nitrogen, unusually strong dipole–dipole attractions are often observed. There are two reasons for this. First, because of the large electronegativity difference, the F—H, O—H, or N—H bonds are very polar. The ends of these dipoles carry a substantial fraction of one charge. Second, because of the small size of the atoms involved, the charge on the end of a dipole is also highly concentrated. This makes

Hydrogen bonding between water molecules in ice causes the molecules to be farther apart in the solid than in the liquid. This makes ice less dense than liquid water, which is why icebergs like this one float.

it particularly effective at attracting the end of opposite charge on a neighboring dipole. These two factors combine to produce attractions, called hydrogen bonds, that are about five times stronger than normal dipole–dipole attractions. The strength of hydrogen bonds causes molecules that experience them to have some very unusual and special properties. Many molecules in biological systems contain N—H and O—H bonds, and hydrogen bonding in these substances is one of the principal factors that determine their overall structures and shapes, as we will see in Chapter 24. In biochemistry and molecular biology, the *shape* of a molecule is just as important as its structure, as we suggested in our study of VSEPR theory. Even slight changes in structure can cause changes in molecular shape, which can make a substance—a protein or an enzyme, for example, or a gene—unfit for any biological use and possibly even lethally dangerous to an organism. Any factor that affects molecular shape, such as the hydrogen bond, is thus extremely important in any study of life at the molecular level.

10.4 SOME GENERAL PROPERTIES OF LIQUIDS AND SOLIDS

Some physical properties depend primarily on how closely packed the molecules are, and others depend primarily on the strengths of the intermolecular attractions.

We begin our study of liquids and solids by examining some of their general properties. Two of these, their *resistance to being compressed* and their *relative rates of diffusion*, depend mostly on how tightly packed the molecules are. Other properties, such as *retention of volume or shape, surface tension* and a solid's or liquid's *tendency to evaporate,* depend much more on the strengths of the intermolecular attractive forces.

Properties That Depend Primarily on Tightness of Packing

Compressibility   In Section 10.2, we noted that in a liquid or solid most of the space is taken up by the molecules, and that there is very little empty space into which to crowd other molecules. As a result, we cannot compress liquids or solids to a smaller volume by applying pressure, so we say that these states of matter are virtually incompressible. This is a property that we often make use of. For example, when you "step on the brakes" of a car, you count on the incompressibility of the brake fluid to transmit the pressure you apply with your foot through the narrow tubes of the brake lines to the brake shoes on the wheels. The brake shoes press against surfaces on the wheels and cause the car to

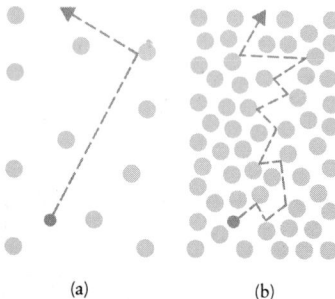

Hydraulic machinery, such as this, uses the incompressibility of liquids to transmit forces that accomplish work.

(a)   (b)

FIGURE 10.6 *(a)* Diffusion in a gas is rapid because relatively few collisions occur between widely spaced molecules. *(b)* Diffusion in a liquid is slow because of many collisions between closely spaced particles.

FIGURE 10.7 A glass carefully filled with water above the rim.

For liquids having strong intermolecular attractions, this "skin" is strong and the surface tension is high.

stop. If some air should get into the brake lines, you're in big trouble, because when you apply pressure to the brakes it simply compresses the air and doesn't force the brake shoes to stop the car. A similar application of the incompressibility of liquids is the foundation of the engineering science of hydraulics, which uses fluids to transmit forces that lift or move heavy objects.

Diffusion   This is another property that varies with the state of matter. It occurs rapidly in gases, slowly in liquids, and hardly at all in solids. For example, the scent of cologne on someone stepping into an elevator is quickly sensed by all present because the odor spreads rapidly through the air. On the other hand, a spoonful of sugar added to a cup of coffee doesn't produce much immediate sweetening unless the coffee is stirred. But if left long enough, the sugar will gradually mix and the coffee will become uniformly sweetened, even without being stirred. However, if a piece of copper and a piece of steel are clamped together, they can be left for years with no noticeable mixing of the two metals.

In gases, molecules can diffuse rapidly because they travel relatively long distances between collisions — their movement from one place to another is not greatly hindered, as illustrated in Figure 10.6. In liquids, however, a given molecule suffers many collisions as it moves about, so it takes longer to move from place to place and diffusion is much slower. Diffusion in solids is almost nonexistent at room temperature because the particles of a solid are held tightly in place. At high temperatures, though, the particles of a solid sometimes have enough kinetic energy to jiggle their way past each other, and diffusion can occur slowly. Such high-temperature solid-state diffusion is used in the production of electronic devices such as transistors. It allows for small, carefully controlled amounts of some impurity — arsenic, for example — to be added to a semiconductor such as silicon or germanium. This process, called "doping," allows the conductivity of these materials to be modified in desirable ways.

Properties That Depend Primarily on the Strengths of Intermolecular Attractions

Retention of Volume and Shape   These are obvious properties. You know from firsthand experience that a solid such as ice keeps both its shape and volume when transferred from one container to another, but the same amount of liquid water just keeps the same volume; the liquid conforms to the shape of the container. The reason for this behavior is not very difficult to understand.

Gases expand spontaneously to fill whatever volume is available to them because the attractive forces are so weak. In liquids and solids, however, the attractions are much stronger — so strong, in fact, that they are able to hold the particles closely together and thereby prevent liquids and solids from expanding. As a result, liquids and solids keep the same volume regardless of the size of their container. In a solid, the attractions are even stronger than in a liquid and hold the particles more or less rigidly in place. This keeps a solid from changing its shape when moved from one container to another.

Surface Tension   A characteristic phenomenon of the liquid state is surface tension. Among other things, it allows us to fill a water glass above the rim (Figure 10.7), something that most of us have done at one time or another. The phenomenon is caused by the difference between the attractions felt by molecules at a liquid's surface and those felt by molecules within the liquid. These attractions are illustrated in Figure 10.8. A molecule within the liquid is surrounded by others on all sides. A molecule at the surface, however, has neighbors beside and below it, but none above. Thus the attractions it feels are not uniform; instead they occur mainly in the direction of the bulk of the liquid. As a result, the entire surface of the liquid experiences a pull toward the center, which makes it behave something like a "skin."

The only way to increase the surface area of the liquid—and by so doing make the "skin" larger—is to pull molecules to the surface from the interior. This requires energy (work) because molecules within the liquid experience a greater total number of attractions than those at the surface. The surface tension of a liquid is related to the energy needed to expand the liquid's surface area—the more energy needed, the larger the surface tension. This also means that the surface tension is related to the strengths of the attractive forces in the liquid. When these attractions are strong, a lot of energy must be expended to overcome them and it is difficult to move a molecule to the surface, so the surface tension is large. On the other hand, when the intermolecular attractive forces are relatively weak, the surface tension is small.

Understanding how surface tension affects the properties of liquids requires that you understand how energy (especially potential energy) is related to stability. In general, a system becomes more stable when its potential energy decreases. That's why a yardstick standing on end falls over; by doing so its potential energy decreases and it reaches a lower-energy, more stable position. Now, because energy is needed to expand the surface area of a liquid, energy is released if this surface area becomes smaller. Any liquid will therefore seek to minimize its surface area because that *lowers* the potential energy. For a given volume, the shape having the smallest surface area is a sphere, so liquids spontaneously tend to assume spherical shapes if they can. This is why rain drops are spherical.

The tendency to minimize surface area also explains why a glass can be filled above its rim with water. The water piles up, trying to assume a spherical shape. Working against this is gravity, which tends to pull the water down. If too much water is added to the glass, the effects of gravity finally win and the "skin" breaks—the water overflows. Surface tension also causes moist grains of sand or soil to stick together. Farmers often squeeze a handful of soil to judge its moisture content. If it's dry, the soil will crumble, but if it is moist, the soil will stick together and form a ball. Similarly, children quickly learn that sand under water will slip through their fingers. But if the sand is merely damp, the small grains cling to each other, and it is possible to build a sand castle. These things are possible because the water between the particles of sand or soil attempts to minimize its surface area and, by so doing, draws the grains of soil or sand together.

A property that we associate with liquids, especially water, is their ability to wet things. Wetting is the spreading of a liquid across a surface. Water, for example, will wet clean glass, such as the windshield of a car, by forming a thin film over the surface of the glass. Water won't wet a greasy windshield, however. Instead, it forms tiny, and very annoying, beads of water.

For wetting to occur, there must be attractive forces between the liquid and the surface that are of about the same strength as the attractions between molecules of the liquid. When a liquid has a low surface tension (and therefore weak intermolecular attractions), it easily wets solid surfaces. Gasoline is an example. It is composed of nonpolar hydrocarbon molecules that only attract each other by London forces. The weak attractions within the liquid are readily overcome by attractions to almost any surface, so the liquid easily spreads to a thin film. If you've ever spilled a little gasoline, you experienced firsthand this tendency of the liquid to spread over surfaces.

Water is an example of a liquid with a large surface tension. In liquid water, the attractive forces are strong hydrogen bonds. Water wets glass because the surface of the solid contains lots of oxygen atoms, so the water molecules can hydrogen-bond to the glass nearly as well as they can hydrogen-bond to each other. Therefore, part of the energy needed to expand the water's surface area is recovered by the formation of hydrogen bonds to the glass surface. When the glass is coated by a film of oil or grease, however, conditions are quite different. Oil and grease molecules are usually nonpolar, so their attractions to other molecules occur through London forces. These are weak forces compared with hydrogen bonds, so the attractions within liquid water are much stronger than the attractions between water molecules and the greasy surface. These

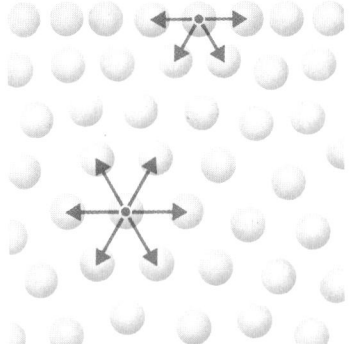

Molecules within the bulk of the liquid have more neighbors and, therefore, experience more attractions than those at the surface.

FIGURE 10.8 Molecules at the surface of a liquid are attracted toward the interior. Those within the liquid are uniformly attracted in all directions.

Liquids tend to form spherical drops because a sphere has the smallest surface area for a given volume.

The surface tension of water is roughly two to three times larger than the surface tension of any common organic solvent. The strong intermolecular hydrogen bonding between water molecules is responsible.

(a) Water wets a clean glass surface. *(b)* If the surface is greasy, the water doesn't wet it. The water resists spreading and forms a bead instead.

(a)   (b)

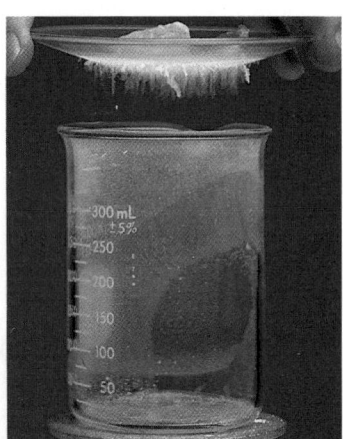

Naphthalene sublimes when heated, and the vapor condenses to give beautiful flaky crystals on the underside of a watch glass cooled with ice.

water–surface attractions can't overcome the water's surface tension, so the water doesn't spread out; it forms beads of water instead. This is one of the reasons detergents are used for such chores as doing laundry or washing floors. The detergents contain chemicals called surfactants that drastically lower the surface tension of water. This makes the water "wetter," which allows the detergent solution to spread more easily across the surface to be cleaned.

Evaporation   Finally, we come to one of the most important properties of liquids and solids — their tendency to evaporate, which involves a change of state from liquid to gas or from solid to gas. Everyone has seen liquids evaporate. This is what happens to water when streets, wet from a rain shower, gradually dry. Solids can also change to a vapor by evaporation. An example is solid carbon dioxide, commonly called dry ice. It is "dry" ice because it doesn't melt; instead, at atmospheric pressure it just changes directly to gaseous CO_2. Perhaps an even more common example is naphthalene, the chemical name for the substance in some brands of moth flakes. Have you ever noticed that they gradually disappear over a period of time. Naphthalene also evaporates without melting. This direct conversion of solid to vapor is called by a special name — sublimation.

To understand evaporation, we have to examine the motions of molecules in liquids and solids. Although the molecules in these states are held near each other by intermolecular attractions, they are not motionless. The molecules bounce around against their neighbors, and just as molecules in a gas have various kinetic energies, so do those in a liquid or a solid. In fact, at a given temperature, there is exactly the same distribution of kinetic energies in a liquid or a solid as there is in a gas; Figure 9.12 on page 334 applies to liquids and solids as well as gases. This means that some molecules have low kinetic energies and move slowly, while others with higher kinetic energies move faster. A small fraction of the molecules have very large kinetic energies and therefore very high velocities. If one of these high-speed molecules reaches the surface, and if it is moving fast enough, it may be able to escape the attractions of its neighbors and enter the vapor. When this happens, we say the molecule has left by evaporation.

One of the things we notice about the evaporation of a liquid is that it produces a cooling effect. Have you ever come out of the water after swimming and been chilled by a breeze? Evaporation of water from your body produced this effect. In fact, our bodies use the evaporation of water to maintain a constant body temperature. During warm weather or vigorous exercise, we perspire, and the evaporation of the perspiration cools the skin. You've probably also used the cooling effect caused by evaporation by blowing gently on the surface of a hot bowl of soup or a cup of coffee. The stream of air stirs the liquid, bringing more hot liquid to the surface to be cooled as it evaporates.

We can easily understand why liquids become cool during evaporation by examining Figure 10.9, which illustrates the kinetic energy distribution in a liquid at a given temperature. Marked along the horizontal axis is the minimum kinetic energy that a molecule needs to escape the attractions of other molecules in the liquid. Only molecules with kinetic energies equal to or greater than this minimum can leave the liquid. Others

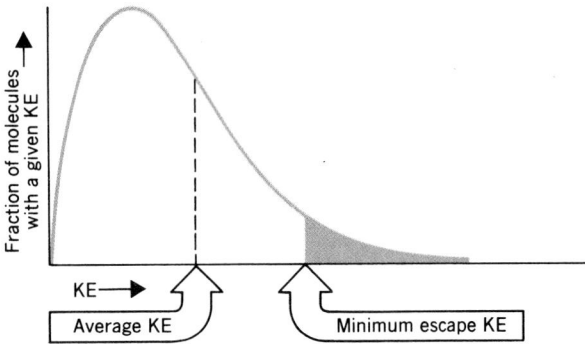

FIGURE 10.9 The minimum kinetic energy needed by a molecule to escape from the surface of the liquid is much larger than the average kinetic energy. The shaded area represents the total fraction of molecules that have the "escape KE."

with less kinetic energy may begin to leave, but before they can escape they slow to a stop and then fall back. The situation is somewhat like attempting to launch a space ship using a cannon. If the velocity of the projectile is not very great, it will not rise very far before it slows to a halt and falls to earth. However, if it is going fast enough (if it has its "escape velocity"), it can overcome earth's gravity and leave.

In Figure 10.9, notice that the minimum kinetic energy needed to escape is much larger than the average kinetic energy. This means that the molecules that evaporate are those with large kinetic energies, and it also means that the average kinetic energy of those that remain decreases. You might think of this as being similar to removing all the tall people from a classroom. When this is done, the average height of those who are left decreases.

In Chapter 9 you learned that the temperature is directly proportional to the average kinetic energy. Since evaporation lowers the average kinetic energy of the molecules that remain in the liquid, the temperature of the liquid must also drop — in other words, evaporation causes the liquid that remains behind to become cool.

Factors That Control the Rate of Evaporation

One of the important things we are going to be concerned about later in this chapter is the *rate of evaporation* of a liquid — in other words, how fast the liquid evaporates. There are several factors that control this. You are probably already aware of one of them — the surface area of the liquid. Since evaporation occurs from the liquid's surface, and not from within, it makes sense that as the surface area is increased, more molecules are able to escape from the liquid and the liquid evaporates more quickly. For liquids having the same surface area, the rate of evaporation depends on two factors: (a) the temperature, and (b) the strengths of intermolecular attractions. Let's examine each of them separately.

Figure 10.10 illustrates the kinetic energy distributions for the same liquid at two temperatures. There are two important features of this figure. First, we see that the same

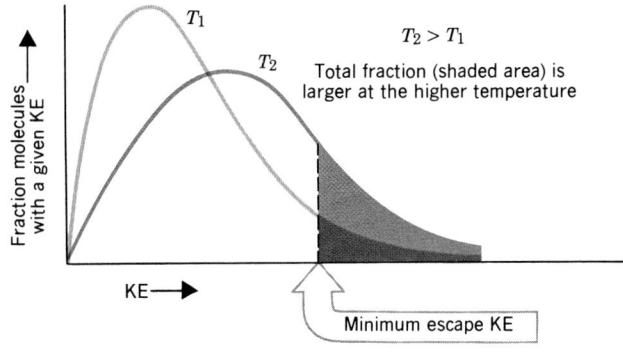

FIGURE 10.10 Increasing the temperature increases the total fraction of molecules with enough kinetic energy to escape from the liquid.

minimum kinetic energy is needed to escape at both temperatures. This is determined by the kinds of attractive forces between the molecules, and is independent of temperature. Second, the shaded area of the curve represents the *total* fraction of molecules having kinetic energies equal to or greater than the minimum. At the higher temperature, this total fraction is larger, which means that at the higher temperature a greater total fraction has the ability to evaporate. As you might expect, when more molecules have the energy to evaporate, more do, and *the rate of evaporation of a given liquid is greater at a higher temperature.* This should be no surprise; you already know that hot water evaporates faster than cold water.

The effect of intermolecular attractions on evaporation rate can be seen by studying Figure 10.11. Here we see kinetic energy distributions for two *different* liquids, both at the same temperature. For convenience, we have labeled the liquids A and B. In liquid A, the attractive forces are weak; they might be of the London type, for example. As we see, the minimum kinetic energy needed by these molecules to escape is not very large because the molecules are not attracted very strongly to others in the liquid. In liquid B, the attractive forces are much stronger — they might be hydrogen bonds, for instance. Because of these stronger attractions, molecules of B are held more tightly by the liquid and must have a larger kinetic energy to evaporate. We can see quite clearly from the figure that the fraction of molecules having enough energy to evaporate is greater for A than for B, which means that A will evaporate faster than B. In general, then, *the stronger the intermolecular attractive forces, the slower is the rate of evaporation at a given temperature.* You are probably also aware of this phenomenon. Gasoline, whose molecules experience weak London forces of attraction, evaporates much faster than water, whose molecules feel the effects of much stronger hydrogen bonds.

FIGURE 10.11 Kinetic energy distribution in two liquids, *A* and *B*, at the same temperature.

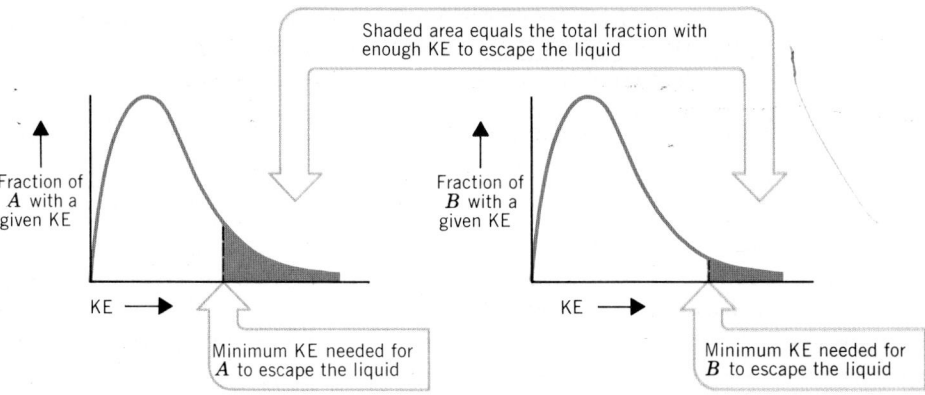

Shaded area equals the total fraction with enough KE to escape the liquid

Fraction of *A* with a given KE

KE ⟶

Minimum KE needed for *A* to escape the liquid

Fraction of *B* with a given KE

KE ⟶

Minimum KE needed for *B* to escape the liquid

10.5 CHANGES OF STATE AND DYNAMIC EQUILIBRIUM

When two opposing changes occur at equal rates, so that their effects cancel, a state of dynamic equilibrium exists.

A change of state is also called a phase change.

A change of state occurs when a substance is transformed from one physical state to another. The evaporation of a liquid and the sublimation of a solid, described in the preceding section, are two examples. Others are the melting of a solid such as ice and the freezing of a liquid such as water.

One of the interesting things about changes of state is that at any particular temperature they always tend toward a condition called dynamic equilibrium. The concept of dynamic equilibrium is one of the most important principles in all of chemistry, because such a condition is the fate (or at least the "goal") of all chemical and physical changes. To understand the concept, let's examine in some detail what happens when a liquid evaporates.

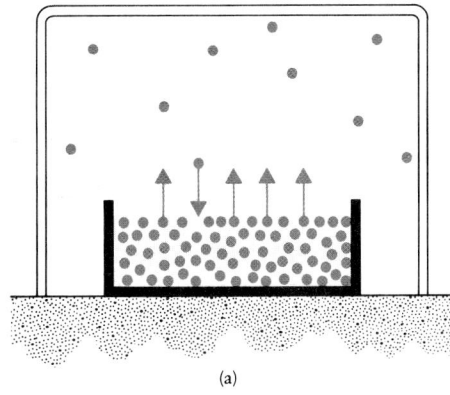 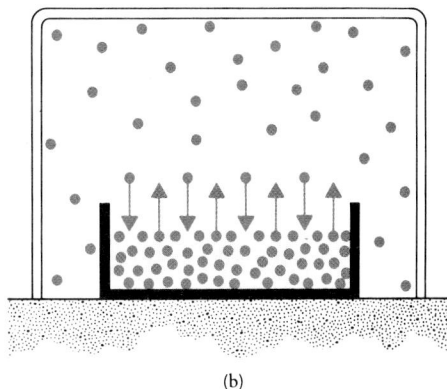

FIGURE 10.12 (a) A liquid begins
to evaporate into a closed container. (b) Equilibrium is reached
when the rate of evaporation
equals the rate of condensation.

(a)   (b)

You already know what happens if the evaporation occurs from an open container
—the volume of the liquid will gradually decrease until all of it is gone. This is because
the molecules that leave the liquid simply wander away into the atmosphere and are lost
from the liquid forever. However, if the container holding the liquid is sealed, the
molecules cannot wander very far (Figure 10.12a). They collect as a vapor over the liquid.
As they fly around in the vapor, the molecules collide with each other, with the walls of
the container, and with the surface of the liquid itself. Those that strike the liquid's
surface tend to stick because their kinetic energy becomes scattered among the surface
molecules. It is somewhat like throwing a ping-pong ball into a large box of ping-pong
balls. The incoming ball knocks others around, and by giving them kinetic energy it loses
some of its own. Because its kinetic energy has been reduced, there is a high probability
that the incoming ball won't bounce out.

As you might expect, the rate at which molecules collide with a liquid's surface, and
thereby return to the liquid in the process called condensation, depends on the number
of molecules in the vapor. When there are only a few molecules in the vapor, the number
of collisions per second with the liquid's surface will be small; when there are many
molecules in the vapor, there will be many such collisions each second.

With this as background, let's consider what happens when a liquid is added to a
container in which none of its molecules are in the vapor. At first, we only have
evaporation because there is nothing to condense; the rate of condensation is zero. As
time passes and evaporation continues, more and more molecules collect in the vapor
phase and there is a rise in the number of collisions per second that these vapor molecules
make with the surface of the liquid. This means that the rate of condensation increases,
and it continues to do so until the rate at which molecules leave the liquid is the same as
the rate at which they return (Figure 10.12b). From that moment on, the number of
molecules in the vapor will remain the same, because over a given period of time the
number that enter the vapor is the same as the number that leave. At this point we have
reached a situation in which two opposing effects—evaporation and condensation—
are occurring at equal rates and their effects cancel; we have reached a condition called
dynamic equilibrium. It is an *equilibrium* because there is no apparent change in the
system of liquid and vapor; the number of vapor molecules remains constant as does the
number of liquid molecules. It is a *dynamic* equilibrium because activity has not ceased.
Molecules continue to evaporate and condense; they just do so at equal rates.

Similar equilibria are reached in melting and sublimation, too. When a solid melts at
a temperature called its melting point, there is a dynamic equilibrium between molecules
in the solid and those in the liquid. If no heat is added or removed from the solid-liquid
mixture, molecules leave the solid and enter the liquid at the same rate as molecules leave
the liquid and become part of the solid. There is an equilibrium between the solid and the
liquid. For sublimation, the situation is exactly the same as in the evaporation of a liquid
into a sealed container. After a few moments, the rate at which molecules leave the solid

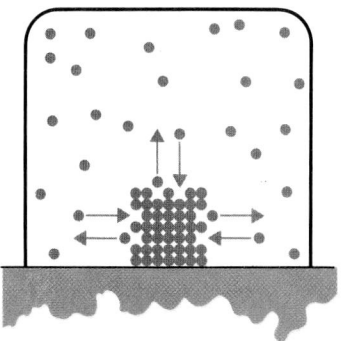

A solid–vapor equilibrium.
Molecules evaporate and condense
on the crystal at equal rates when
equilibrium is reached.

and enter the vapor is the same as the rate at which molecules leave the vapor and become attached to the solid; a solid–vapor equilibrium is established. All these various equilibria have very profound effects on the properties of solids and liquids, as we shall soon see.

10.6 ENERGY CHANGES DURING CHANGES OF STATE

The energy change that accompanies the change from one state to another gives a measure of the differences in the strengths of the intermolecular attractions in the two states.

Another important feature of changes of state is that they are accompanied by energy changes. Events such as the melting of a solid, the evaporation of a liquid, and the sublimation of a solid are all endothermic—they absorb heat. You have witnessed examples of this yourself. You have put ice cubes into a glass of soda to keep it cool; the reason is because ice absorbs heat as it melts. Similarly, the evaporation of perspiration cools your body because water absorbs heat from the body when it evaporates. And during the summer, ice cream trucks carry dry ice because as the solid CO_2 sublimes, it absorbs heat and keeps the ice cream from melting.

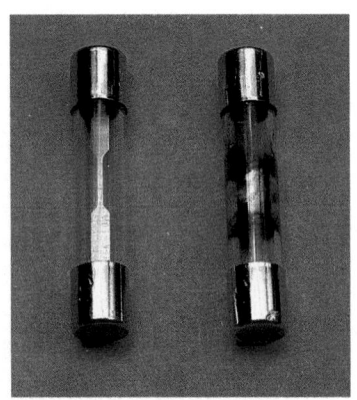

Fusion means melting. The thin metal band in an electrical fuse protects a circuit by melting if too much current is passed through it. On the right is a fuse that has done its job.

When particles that attract each other are pulled apart, their potential energy increases. The amount of energy corresponding to this increase is absorbed from the surroundings.

In Chapter 5 you learned that exchanges of heat between a system and its surroundings that take place at constant pressure are called enthalpy changes. Therefore, the heat absorbed or evolved in a change of state at constant pressure is also an enthalpy change and has associated with it a value of ΔH. Usually, these enthalpy changes are expressed on a "per mole" basis and are given special names that identify the kind of change involved. For example, the molar heat of fusion, ΔH_{fusion}, is the heat absorbed by one mole of a solid when it melts to give a liquid at the same temperature and pressure. (Fusion is a word that means "melting.") Similarly, the molar heat of vaporization, $\Delta H_{vaporization}$, is the heat absorbed when one mole of a liquid is changed to one mole of a vapor at a constant temperature and pressure, and the molar heat of sublimation, $\Delta H_{sublimation}$, is the heat absorbed by one mole of a solid when it sublimes to give one mole of vapor, once again at a constant temperature and pressure.

The values of ΔH for fusion, vaporization, and sublimation are all positive because the phase change in each case is accompanied by a net increase in energy (in other words, $H_{final} > H_{initial}$). This is because the molecules must overcome to one degree or another the attractive forces of their neighbors, and that gives rise to an increase in their potential energies. For example, when a solid melts, the particles must overcome the attractions that tend to hold them in place, and overcoming attractions always involve increasing the potential energy. If the temperature is to remain constant during the phase change (i.e., if the average kinetic energy is to stay the same), the energy that becomes stored as potential energy must be absorbed from outside the solid, so a cooling effect is observed in the surroundings. A similar cooling effect is observed when a liquid evaporates because the intermolecular attractions between the molecules of the liquid must be overcome, and this also requires an input of energy from outside if the temperature is to stay constant. And it is not difficult to see that when a solid sublimes, the molecules must overcome the attractive forces in the solid as they move into the vapor, so sublimation is also endothermic.

Energy Changes and Intermolecular Attractions

The magnitude of ΔH for a change of state is determined both by the strengths of the intermolecular attractions and by the amount that these forces change during the change of state. For example, when a solid substance melts, the particles undergo relatively small changes in distance, and there is only a small change in the potential energy. On the other hand, when the same amount of the same liquid evaporates, there are much larger changes in intermolecular distances and a much larger increase in potential energy. Because of this, for a given substance the value of $\Delta H_{vaporization}$ is much larger than the

TABLE 10.4 Some Typical Heats of Vaporization

Substance	$\Delta H_{vaporization}$ kJ/mol	kcal/mol	Type of Attractive Force
H_2O	+40.6	+9.70	Hydrogen bonding
NH_3	+21.7	+5.19	Hydrogen bonding
HCl	+15.6	+3.72	Dipole–dipole
SO_2	+24.3	+5.81	Dipole–dipole
F_2	+5.9	+1.4	London
Cl_2	+10.0	+2.39	London
Br_2	+15.0	+3.58	London
I_2	+22.0	+5.26	London
CH_4	+8.16	+1.95	London
C_2H_6	+15.1	+3.61	London
C_3H_8	+16.9	+4.03	London
C_6H_{14}	+30.1	+7.20	London

value of ΔH_{fusion}. For the same substance, the value of $\Delta H_{sublimation}$ is even larger than $\Delta H_{vaporization}$ because the attractive forces in the solid are larger than those in the liquid. As you would expect, overcoming these stronger forces requires more energy.

When a liquid evaporates, or when a solid sublimes, the particles go from a situation where the attractive forces are very strong to one in which the attractive forces are so small that they can almost be ignored. Therefore, values of $\Delta H_{vaporization}$ and $\Delta H_{sublimation}$ give us directly the energy needed to separate molecules from each other. We can examine such values to obtain reliable comparisons of the strengths of intermolecular attractions. Let's look at some values of $\Delta H_{vaporization}$ (Table 10.4) to see how they agree with what we have learned about the kinds and strengths of intermolecular attractions.

First, notice that the heats of vaporization of water and ammonia are very large, which is just what we would expect for hydrogen-bonded substances. By comparison, a nonpolar substance of similar molecular weight, CH_4, has a very small heat of vaporization. Note also that polar substances such as HCl and SO_2 have fairly large heats of vaporization compared with nonpolar substances with the same number of electrons. For example, HCl and F_2 both have 18 electrons and similar formula weights, but the heat of vaporization of polar HCl is much larger than that of nonpolar F_2. We can make a similar comparison between SO_2 (32 electrons) and Cl_2 (34 electrons). Once again the polar substance (SO_2) has a larger heat of vaporization than the nonpolar one (Cl_2).

Heats of vaporization also reflect the factors that control the strengths of London forces. For example, the data in Table 10.4 show the effect of chain length on the intermolecular attractions between hydrocarbons; as the chain length increases from one carbon in CH_4 to six carbons in C_6H_{14}, the heat of vaporization also increases. This is in agreement with an increase in the total strengths of the London forces. In our discussion of London forces, we also mentioned that their strengths increase as the electron clouds of the particles become larger, and we used the boiling points of the halogens to illustrate this. The heats of vaporization of the halogens in Table 10.4 are in agreement with this, too.

The stronger the attractions, the more the potential energy will increase when the molecules become separated, and the larger will be the value of ΔH.

The Lewis structure of SO_2 tells us that the molecule is non-linear, and this means that it must be polar.

10.7 VAPOR PRESSURES

Molecules that evaporate from a liquid into a closed container exert a pressure that remains constant after the rates of evaporation and condensation become equal.

When a liquid evaporates, the molecules that enter the vapor behave just like the molecules of any other gas. One form of this behavior is that they exert a pressure, called

FIGURE 10.13 Measuring the vapor pressure of a liquid.

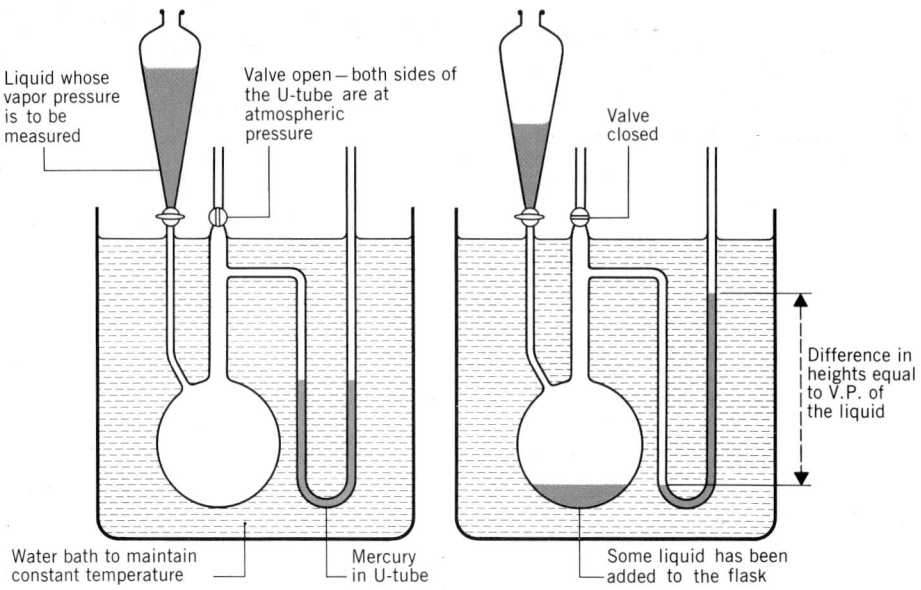

Liquid whose vapor pressure is to be measured

Valve open—both sides of the U-tube are at atmospheric pressure

Valve closed

Difference in heights equal to V.P. of the liquid

Water bath to maintain constant temperature

Mercury in U-tube

Some liquid has been added to the flask

In chemistry, when we use the word *equilibrium,* we really mean *dynamic equilibrium.*

the vapor pressure. From the very moment a liquid begins to evaporate into the vapor space above it, there is a vapor pressure. If the evaporation is taking place inside a sealed container, this pressure grows until finally equilibrium is reached. Once the rates of evaporation and condensation become equal, the number of molecules in the vapor remains constant, and the vapor exerts a constant pressure. This final pressure is called the equilibrium vapor pressure of the liquid. In general, when we refer to the vapor pressure, we really mean the *equilibrium* vapor pressure.

We can measure the vapor pressure of a liquid with an open-end manometer such as in the apparatus shown in Figure 10.13. Initially, both sides of the apparatus are open to the atmosphere, so that the pressure on both sides is the same. A small amount of liquid is added to the flask from the funnel and the left side is then sealed immediately. As some of the liquid evaporates, the pressure in the flask rises, forcing the fluid in the left side of the U-shaped tube downward. The increase in pressure, as measured by the difference in the levels in the manometer, is equal to the pressure exerted by the vapor—the vapor pressure of the liquid.

Humidity is a measure of how nearly saturated the air is with moisture. At 100% humidity, the partial pressure of water vapor is equal to the equilibrium vapor pressure of water at the temperature of the air.

An important fact about the vapor pressure is that *its magnitude doesn't depend on the amount of liquid in the flask, the volume of the flask, or the surface area of the liquid, just as long as some liquid remains when equilibrium is reached.* Increasing the amount of liquid increases the liquid's surface area, and that increases the rate of evaporation. But the larger surface area also presents a larger target for returning molecules, so the rate of condensation also increases. The net result is that the rates of evaporation and condensation are affected equally, so no net change in vapor pressure is produced.

To understand why the vapor pressure doesn't depend on the size of the container, imagine a liquid and its vapor contained in a cylinder with a movable piston. Withdrawing the piston increases the volume of the vapor and momentarily lowers the pressure. This means that fewer molecules of the vapor strike a given surface area of the liquid each instant, and the rate of condensation will have decreased. The rate of evaporation hasn't changed, however, so more of the substance is evaporating than condensing. This condition will continue until there are enough molecules in the vapor to make the condensation rate again equal to the rate of evaporation, and at this point the vapor pressure will have returned to its original value. By similar reasoning, we can expect that decreasing a container's volume should increase the condensation rate without changing the rate of evaporation. Condensation, therefore, occurs faster than evaporation until equilibrium is finally restored and the vapor pressure drops to its original value again. In summary, then, changing the volume of a vessel in which there is a liquid–vapor equilibrium produces a momentary change in pressure, but when equilibrium is restored, the vapor pressure returns to its original value.

PRACTICE EXERCISE 1   Suppose a liquid is in equilibrium with its vapor in a piston–cylinder apparatus like that described in the preceding paragraph. If the piston is withdrawn a short way and the system is allowed to return to equilibrium, what will have happened to the *total number* of molecules in both the liquid and the vapor?

Factors That Affect the Vapor Pressure

In our discussion of the vapor pressure so far, we have found that ultimately it is determined by the rate of evaporation per unit surface area of a liquid. When the volume of the container changes, the rate of evaporation is unaffected. Instead, the change in volume alters the rate of condensation, and the number of molecules in the vapor then increases or decreases just enough to make the two rates the same. The only way the vapor pressure can be changed is to change the rate of evaporation.

Earlier we saw that two things affect the rate of evaporation from a given surface area of a liquid. One is the temperature and the other is the strengths of the intermolecular attractions. As the temperature increases, so does the rate of evaporation and so does the vapor pressure. This is shown in Figure 10.14, which is a graph of vapor pressure versus temperature for several liquids.

You also saw that as the strengths of the intermolecular attractions become larger, the rate of evaporation decreases. This means that the vapor pressure should also decrease, which it does. Figure 10.14 also illustrates the effects of intermolecular attractions. Ether, $(C_2H_5)_2O$, is a nonpolar substance whose molecules are attracted to each other by relatively weak London forces. At any given temperature, ether has a higher vapor pressure than water, whose molecules are attracted to each other by much stronger hydrogen bonds. We also see that at a given temperature acetic acid has a lower vapor pressure than water, and that propylene glycol, $C_3H_6(OH)_2$, has an even lower vapor pressure than acetic acid. This tells us that the intermolecular attractions are stronger in acetic acid than in water, and those in propylene glycol are stronger than the attractions in acetic acid. Notice that we can use vapor pressures as an indication of the relative strengths of the attractive forces in liquids.

The equilibrium vapor pressure is determined just by the rate of evaporation per unit area of the liquid's surface.

QUESTION

Substances with high vapor pressures, such as gasoline, evaporate quickly from open containers.

A liquid with a high vapor pressure is said to be volatile.

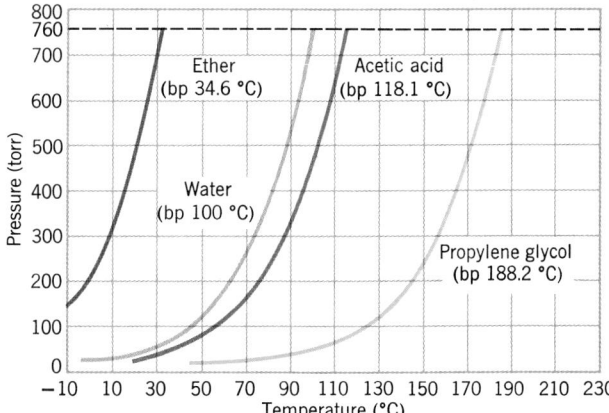

FIGURE 10.14 Variation of vapor pressure with temperature for some common liquids.

Anyone who has watched much television has certainly seen advertisements that praise the virtues of freeze-dried instant coffee. Campers routinely carry freeze-dried foods because they are lightweight. In addition, bacteria cannot grow and reproduce in the complete absence of moisture, so freeze-dried foods need no refrigeration. Therefore, freeze-drying is clearly a useful way of preserving foods. It is accomplished by first freezing the food (or brewed coffee) and then placing it in a chamber that is connected to high-capacity vacuum pumps. These pumps lower the pressure in the chamber below the vapor pressure of ice, which causes the ice crystals to sublime. Drying foods in this way has one important advantage over other methods—the delicate molecules responsible for the flavor of foods are not destroyed, as they would be if the food were heated. Freeze-dried foods may also be easily reconstituted simply by adding water.

Freeze-drying is also used by biologists to preserve tissue cultures and bacteria. Removing moisture from tissue and storing it at low temperatures allows the cells to survive in what amounts to a state of suspended animation for periods of at least several years.

It's common to find freeze-dried foods among the supplies carried by campers.

Vapor Pressures of Solids

Solids have vapor pressures just as liquids do. In a crystal, the particles are not stationary. They vibrate back and forth about their equilibrium positions. At a given temperature there is a distribution of kinetic energies, so some particles vibrate slowly while others vibrate with a great deal of energy. Some particles at the surface have high enough energies to break away from their neighbors and enter the vapor state. When particles in the vapor collide with the crystal, they can be recaptured, so condensation can occur too. Eventually, the concentration of particles in the vapor reaches a point where the rate of sublimation from the solid is the same as the rate of condensation, and a dynamic equilibrium is established. The pressure of the vapor that is in equilibrium with the solid is called the equilibrium vapor pressure of the solid. As with liquids, the equilibrium vapor pressure is usually referred to simply as the vapor pressure. Like that of a liquid, the vapor pressure of a solid is determined by the strengths of the attractive forces between its particles and by the temperature.

10.8 DYNAMIC EQUILIBRIUM AND LE CHÂTELIER'S PRINCIPLE

Disturbing a system at equilibrium causes it to change in a way that counteracts the disturbing influence and brings the system to equilibrium again.

In the last section, you learned that when the temperature of a liquid is raised, its vapor pressure increases. We could have reached this conclusion by analyzing what happens to the system in the following way. Initially, the liquid is in equilibrium with its vapor, which exerts a certain pressure. When the temperature is increased, equilibrium no longer exists because evaporation occurs more rapidly than condensation. Eventually, as the concentration of molecules in the vapor increases, the system reaches a new equilibrium in which there is more vapor. This greater amount of vapor exerts a larger pressure.

What happens to the vapor pressure of a liquid when the temperature is raised is an example of a general phenomenon. Whenever a system at equilibrium is subjected to a disturbance that upsets the equilibrium, the system changes in a way that will return it to equilibrium again. For a liquid–vapor equilibrium, such a disturbance is a change of temperature, as we saw in the preceding paragraph.

We will deal with many kinds of equilibria, both chemical and physical, from time to time. It would be very time consuming and sometimes very difficult to carry out a detailed analysis each time we wish to know the effects of some disturbance on the equilibrium system. Fortunately, there is a relatively simple and fast method of predicting the effect of a disturbance. It is based on a principle proposed in 1888 by a brilliant French chemist, Henry Le Châtelier (1850–1936).

> **Le Châtelier's Principle**   When a system in equilibrium is subjected to a disturbance that upsets the equilibrium, the system responds in a direction that tends to counteract the disturbance and restore equilibrium in the system.

Let's see how we can apply Le Châtelier's principle to a liquid–vapor equilibrium that is subjected to a temperature increase. To do this we have to ask ourselves, how do we go about increasing the temperature of a system? The answer, of course, is that we add heat to it. When the temperature is increased, it is the addition of heat that is really the disturbing influence.

If we write the liquid–vapor equilibrium in the form of a chemical equation, we can include the energy change as follows.

$$\text{heat} + \text{liquid} \rightleftharpoons \text{vapor} \qquad (10.1)$$

This tells us that heat is absorbed by the liquid when it changes to the vapor and that heat is released when the vapor condenses to a liquid.

Dynamic equilibrium is indicated in a chemical equation by a pair of arrows pointing in opposite directions, $\rightleftharpoons$, which implies opposing changes happening at equal rates.

When heat is added to a liquid–vapor system that is at equilibrium, Le Châtelier's principle tells us that the system will try to adjust in a way that counteracts the disturbance. The system will attempt to change in a way that absorbs some of the heat that is added. This can happen if some liquid evaporates, because vaporization is an endothermic change. When liquid evaporates, the amount of vapor increases and the pressure rises. Thus, we have reached the correct conclusion in a very simple way.

We often use the term position of equilibrium to refer to the relative amounts of the substances on both sides of the double arrows in an equilibrium equation such as Equation 10.1. Then, we think of how a disturbance affects the position of equilibrium. For example, increasing the temperature increases the amount of vapor and decreases the amount of liquid, and we say that the position of equilibrium has shifted — in this case, it has shifted in the direction of the vapor, or it has shifted to the right. In using Le Châtelier's principle, it is convenient to think of a disturbance as "shifting the position of equilibrium" in one direction or another in the equilibrium equation.

A rise in temperature moves an equilibrium in the direction of an endothermic change.

Now, let's use Le Châtelier's principle to analyze what happens when the temperature of a liquid–vapor equilibrium is lowered. To lower the temperature, heat must be removed. The system responds by undergoing a change in a direction that tends to replace the lost heat — the position of equilibrium shifts to the left because as some vapor condenses, some heat is evolved. At the new position of equilibrium there is more liquid and less vapor. Since there is less vapor, the pressure is lower. Once again, we come to the proper conclusion — that the vapor pressure is lowered by lowering the temperature.

A decrease in temperature favors a net change that is exothermic.

PRACTICE EXERCISE 2   Use Le Châtelier's principle to predict how a temperature increase will affect the vapor pressure of a solid. (*Hint:* solid + heat $\rightleftharpoons$ vapor.)

10.9 WHAT HAPPENS WHEN LIQUIDS BOIL

The pressure of the vapor in the bubbles within a boiling liquid is equal to the pressure of the atmosphere.

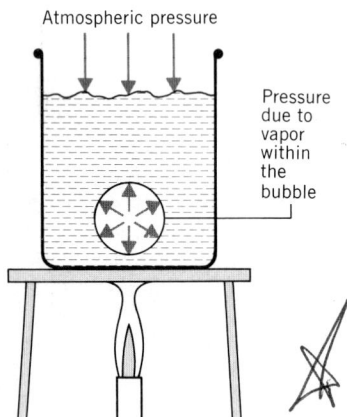

We know that the water in this pot of vegetables is boiling because we can see bubbles of steam rising to the surface.

FIGURE 10.15 The pressure of the vapor within a bubble in a boiling liquid pushes the liquid aside against the opposing pressure of the atmosphere.

On the top of Mt. Everest, water boils at only 69 °C.

If you were asked to check whether a pot of water were boiling, what would you look for? The answer, of course, is *bubbles.* When a liquid boils, large bubbles usually form at many places on the inner surface of the container and rise to the top. If you were to place a thermometer into the boiling water to measure its temperature, you would find it remains constant, regardless of how you adjust the flame under the pot. A hotter flame just makes the water bubble faster, but it doesn't raise the temperature. *Any liquid remains at a constant temperature while it is boiling.* This temperature is called the boiling point.

If you measure the boiling point of water in Philadelphia, New York, or any place else that is nearly at sea level, your thermometer will read 100 °C or very close to it. However, if you try this experiment in Denver, Colorado, you will find that the water boils at about 95 °C. Denver is a mile above sea level, and the atmospheric pressure there is lower, so we find that the boiling point depends on the atmospheric pressure.

These observations raise some interesting questions. Why do liquids boil? And why does the boiling point depend on the pressure of the atmosphere? The answers become apparent when we discover that inside the bubbles of a boiling liquid is *the liquid's vapor.* When water boils, the bubbles contain water vapor (steam); when alcohol boils, the bubbles contain alcohol vapor. As a bubble grows, liquid evaporates into it, and the pressure of the vapor pushes the liquid aside, as shown in Figure 10.15. Opposing this, however, is the pressure of the atmosphere pushing down on the top of the liquid in the container. The atmospheric pressure attempts to collapse the bubble. The only way the bubble can exist and grow is if the pressure of the vapor within it is equal to the pressure exerted by the atmosphere. In other words, there cannot be bubbles of vapor until the temperature of the liquid rises to a point at which its vapor pressure equals the atmospheric pressure. Thus, in scientific terms, the boiling point is defined as the temperature at which the vapor pressure of the liquid is equal to the prevailing atmospheric pressure.

Now we can easily understand why water boils at a lower temperature in Denver than it does in New York City. Because the atmospheric pressure is lower in Denver, the water there doesn't have to be heated to as high a temperature to make its vapor pressure equal to the atmospheric pressure. This phenomenon, besides being interesting, can also influence the way we do certain things. The lower temperature of boiling water at places with high altitudes, like Denver, makes it necessary to cook foods longer. In fact, on the top of a very high mountain, water boils at such a low temperature that foods won't cook in it at all. At the other extreme, a pressure cooker is a device that raises the pressure over the boiling water and thereby raises the boiling point. This makes foods cook more quickly.

To make it possible to compare the boiling points of different liquids, chemists have chosen, somewhat arbitrarily, to record boiling points measured at a reference pressure of 1 atm. The boiling point of a liquid at 1 atm is called its normal boiling point. If a boiling point is reported without also mentioning the pressure at which it was measured, we assume it to be the normal boiling point. Notice that in Figure 10.14 on page 365, we can find the normal boiling points for ether, water, acetic acid, and propylene glycol by observing the temperatures at which their vapor pressure curves cross the 1-atm pressure line.

Earlier we mentioned that the boiling point is a property whose value depends on the strengths of the intermolecular attractions in a liquid. This is now easy to understand. When the attractive forces are strong, the liquid has a low vapor pressure at a given temperature, so it must be heated to a high temperature to bring its vapor pressure up to atmospheric pressure. High boiling points are therefore a result of strong intermolecular attractions.

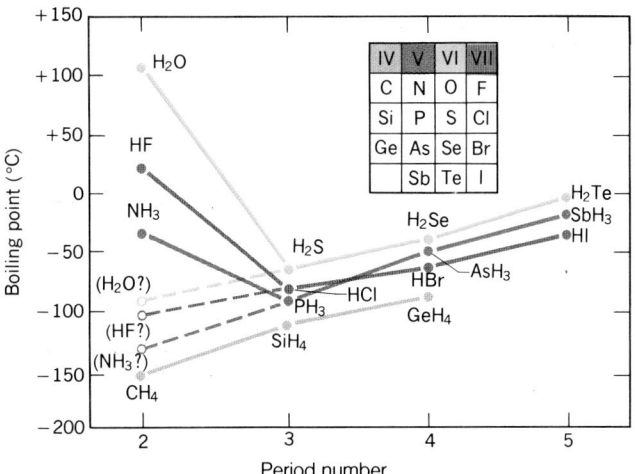

FIGURE 10.16 Boiling points of the hydrogen compounds of elements of Groups IVA, VA, VIA, and VIIA of the periodic table. The actual boiling points of HF, H_2O, and NH_3 are higher than expected, based on the trends shown by the other hydrogen compounds, because of hydrogen bonding.

The effects of intermolecular attractions on boiling point are easily seen by examining Figure 10.16, which is a graph of the boiling points versus period number for the hydrogen compounds of the elements in Groups IVA, VA, VIA, and VIIA. Notice, first, that there is a gradual increase in boiling point for the hydrogen compounds of the Group IVA elements (CH_4 through GeH_4). These substances are composed of nonpolar tetrahedral molecules, and experience relatively weak London forces. The boiling points increase from CH_4 to GeH_4 simply because the molecules become larger.

When we look at the hydrogen compounds of the other nonmetals, we find the same trend from period 3 through period 5. Thus, for the series PH_3, AsH_3, SbH_3, there is a gradual increase in boiling point, corresponding again to the increasing strengths of London forces. Similar increases occur for H_2S, H_2Se, and H_2Te and for HCl, HBr, and HI. Significantly, however, the period 2 members of each of these series — NH_3, H_2O, and HF — have much higher boiling points than might otherwise be expected. The reason is that each is involved in hydrogen bonding, which is a much stronger attraction than London forces.

One of the most interesting and far-reaching consequences of hydrogen bonding is that it causes water to be a liquid, rather than a gas, at room temperature. If it were not for hydrogen bonding, water would have a boiling point somewhere near -80 °C and could not exist as a liquid except at still lower temperatures. At such low temperatures it is unlikely that life as we know it could have developed.

PRACTICE EXERCISE 3   The atmospheric pressure at the top of Mt. McKinley in Alaska, 3.85 miles above sea level, is 330 torr. Use Figure 10.15 to estimate the boiling point of water at the top of this mountain.

10.10 CRYSTALLINE SOLIDS

In a crystal the atoms, molecules, or ions are arranged in a highly regular repeating pattern.

Figure 10.17 on page 370 is a photograph of crystals of one of our most familiar chemicals, sodium chloride — ordinary table salt. Notice that each particle is very nearly a perfect little cube. You might think that the manufacturers went to a lot of trouble and expense to make such uniformly shaped crystals. Actually, they could hardly avoid it. Whenever a solution of NaCl is evaporated, the crystals that form have edges that intersect at 90° angles. Cubes, then, are the norm, not the exception for NaCl.

FIGURE 10.17 Crystals of table salt. The size of the tiny cubic sodium chloride crystals can be seen in comparison with a penny.

The regular, symmetrical shapes of snowflakes are caused by the highly organized packing of water molecules within crystals of ice.

When most substances freeze, or when they separate out as a solid from a solution that is being evaporated, they normally form crystals that have highly regular features. For example, the crystals have flat surfaces that meet at angles which are characteristic for a given substance. The regularity of these surface features reflects a high degree of order among the particles that lie within the crystal. This is true whether the particles are atoms, molecules, or ions.

The particles in crystals are arranged in patterns that repeat over and over again in all directions. The overall pattern that results is called a crystal lattice. Its high degree of regularity is the principal feature that makes solids different from liquids—a liquid lacks this long-range repetition of structure because the particles in a liquid are jumbled and disorganized as they move about.

Because there are millions of chemical compounds, it might seem that an enormous number of different kinds of lattices are possible. If this were true, studying solids would be hopelessly complex. Fortunately, however, the number of *kinds* of lattices that are mathematically possible is quite limited. This fact has allowed for a great deal of progress in understanding solid structures.

To describe the structure of a crystal it is convenient to view it as being composed of a huge number of simple, basic units called unit cells. By repeating this simple structural unit up and down, back and forth, in all directions, we can build the entire lattice. This is illustrated in Figure 10.18 for the simplest and most symmetrical of all unit cells, called

FIGURE 10.18 *(a)* A simple cubic unit cell, showing the locations of the lattice positions. *(b)* A simple cubic unit cell with atoms having their nuclei at the corners. *(c)* A portion of a simple cubic lattice built by stacking simple cubic unit cells.

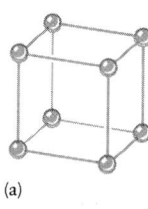

(a)

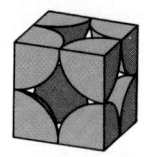

(b)

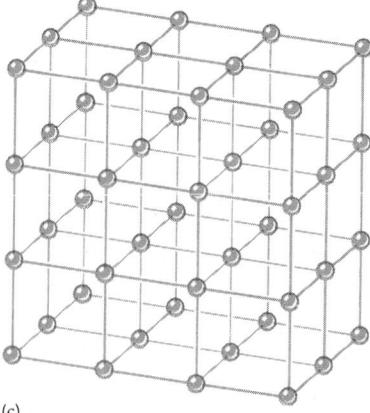

(c)

the simple cubic. This unit cell is a cube having atoms (or molecules or ions) at each of its eight corners. Stacking these unit cells gives a simple cubic lattice.

Two other cubic unit cells are also possible: face-centered cubic and body-centered cubic. The face-centered cubic (fcc) unit cell has identical particles at each of the corners plus another in the center of each face, as shown in Figure 10.19. Many common metals — copper, silver, gold, aluminum, and lead, for example — form crystals that have face-centered cubic lattices. Each of these metals has the same *kind* of lattice, but the sizes of their unit cells differ because the sizes of the atoms differ (Figure 10.20).

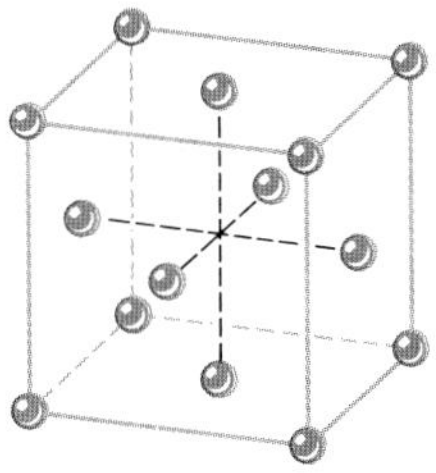

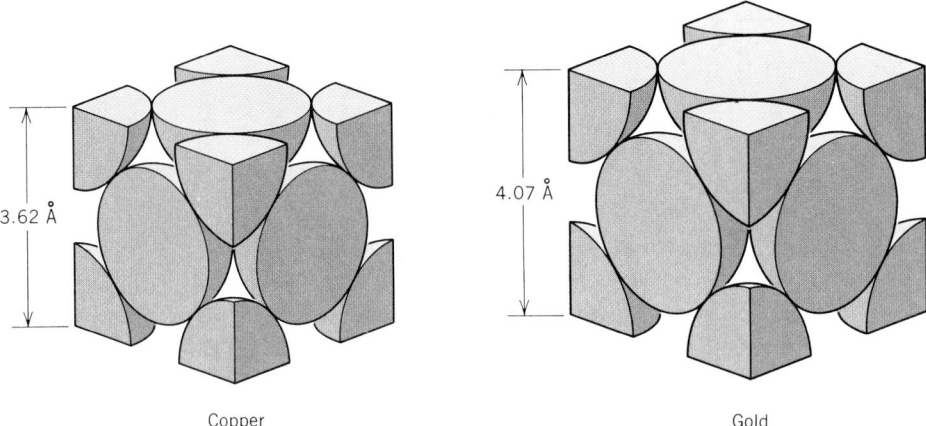

3.62 Å

4.07 Å

Copper

Gold

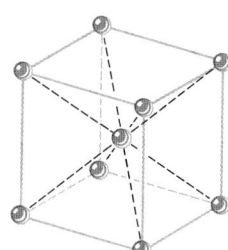

FIGURE 10.19 A face-centered cubic unit cell.

FIGURE 10.20 Two similar face-centered cubic unit cells. The atoms are arranged in the same way, but their unit cells have edges of different lengths because the atoms are of different sizes.

The body-centered cubic (bcc) unit cell has identical particles at each corner plus one in the center of the cell, as illustrated in Figure 10.21. The body-centered cubic lattice is also common among a number of metals — examples are chromium, iron, and platinum. Again, these are substances with the same *kind* of lattice, but the dimensions of the lattices reflect the sizes of the particular atoms.

Not all unit cells are cubic. Some have edges of different lengths or edges that intersect at angles other than 90°. Although you should realize that these other unit cells and the lattices they form exist, we will limit the remainder of our discussion to cubic lattices.

We have seen that a number of metals have fcc or bcc lattices. The same is true for many compounds. Figure 10.22, for example, is a cutaway view of a portion of a sodium chloride crystal. The gray particles represent Na⁺ ions. Notice that they are located at the lattice positions that correspond to a face-centered cubic unit cell. The Cl⁻ ions (green) fill the spaces between the Na⁺ ions. Sodium chloride is said to have a face-centered

FIGURE 10.21 A body-centered cubic unit cell.

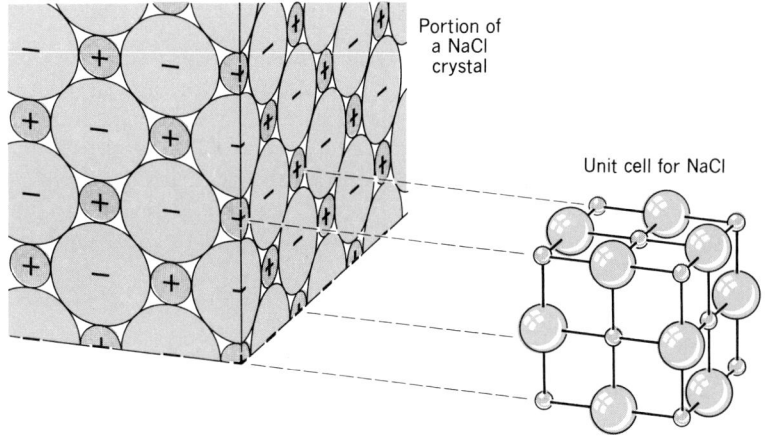

Portion of a NaCl crystal

Unit cell for NaCl

FIGURE 10.22 A face-centered cubic unit cell can be seen in the structure of NaCl.

cubic lattice, and the cubic shape of this lattice is the reason that NaCl crystals take on a cubic shape when they form.

Many of the alkali halides (Group IA – VIIA compounds), such as NaCl and KCl, crystallize with an fcc lattice. Since sodium chloride and potassium chloride both have the same kind of lattice, Figure 10.22 also could be used to describe the unit cell of KCl. The sizes of their unit cells are different, however, because K^+ is a larger ion than Na^+.

A major point to be learned from the preceding discussion is that a single lattice type can be used to describe the structures of many different crystals. For this reason, a handful of different lattice types is all we need to describe the crystal structure of every possible chemical element or compound.

FIGURE 10.23 X rays emitted from atoms are in phase in some directions and out of phase in other directions.

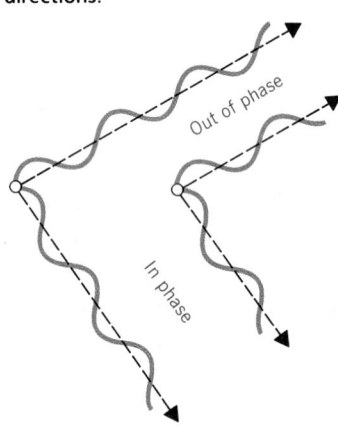
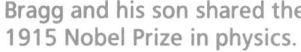

10.11   X-RAY DIFFRACTION

Measuring the angles at which X rays of known wavelength are scattered by a crystal allows calculation of the distances between planes of atoms in the solid.

When atoms are bathed in X rays, they absorb some of the radiation and then emit it again in all directions. In effect, each atom becomes a tiny X-ray source. If we look at radiation from two such atoms (Figure 10.23), we find that the X rays emitted are in phase in some directions and out of phase in others. In Chapter 6 we saw that such constructive and destructive interference produces a diffraction pattern.

In a crystal, there are enormous numbers of atoms evenly spaced throughout the lattice. When the crystal is bathed in X rays, intense beams of diffracted X rays caused by constructive interference appear in certain specific directions, while in other directions no X rays appear because of destructive interference. The diffraction pattern thus produced can be detected using photographic film, as is illustrated in Figure 10.24. (The film is darkened only where the X rays strike.)

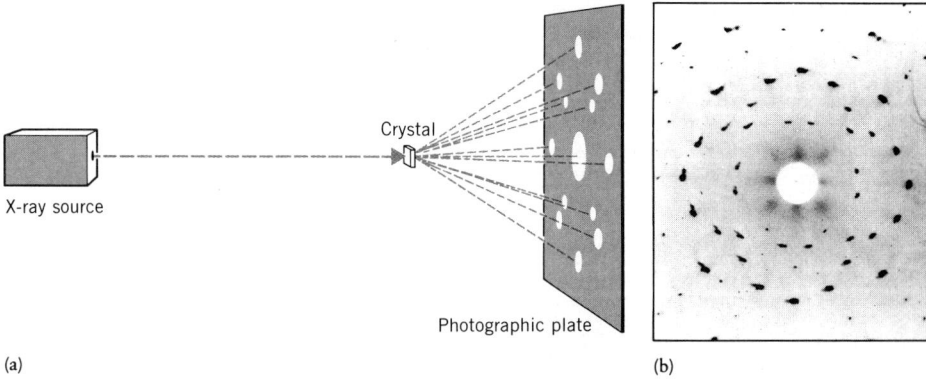

Crystal

X-ray source

Photographic plate

(a)

(b)

FIGURE 10.24 X-ray diffraction. *(a)* The production of an X-ray diffraction pattern. *(b)* An X-ray diffraction pattern produced by sodium chloride.

Bragg and his son shared the 1915 Nobel Prize in physics.

In 1913, the British physicist William Henry Bragg and his son William Lawrence Bragg discovered that just a few variables control the appearance of such an X-ray diffraction pattern. These are shown in Figure 10.25, which illustrates the proper conditions necessary to obtain constructive interference of the X rays from successive layers of atoms (planes of atoms) in a crystal. A beam of X rays having a wavelength λ strikes the layers at an angle θ. Constructive interference causes an intense diffracted beam to emerge at the same angle θ. The Braggs derived an equation relating λ, θ, and the distance between the planes of atoms, d,

$$n\lambda = 2d \sin \theta \qquad (10.2)$$

where n is an integer (i.e., n can equal 1 or 2 or 3, etc.). This equation, called the Bragg equation, is the basic tool used by scientists in their study of solid structures. Let's briefly examine how they use it.

Biochemists have found that X-ray diffraction is extremely useful for studying the structures of large molecules. The most famous example of this use was the experimental determination of the molecular structure of DNA. DNA is found in the nuclei of cells and serves to carry an organism's genetic information. In 1953, using X-ray diffraction photographs of DNA

fibers obtained by Rosalind Franklin and Maurice Wilkins, James Watson and Francis Crick came to the conclusion that the DNA structure consists of the now-famous double helix. We will examine this important structure in more detail in Chapter 24. Watson, Crick, and Wilkins shared the 1962 Nobel Prize in physiology and medicine for their discovery.

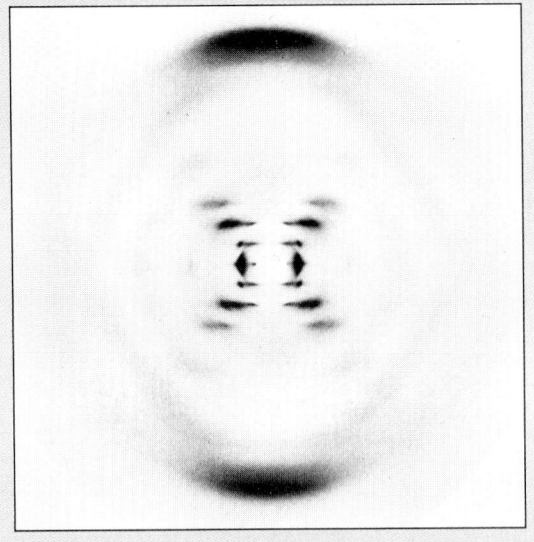

(a) X-ray diffraction photograph of DNA. (b) DNA double helix—a model of a short section of a molecule showing 10 amine pairs. The white rings are deoxyribose units and the red sections are phosphate units of the main chains. Projecting from these chains are the paired amines: guanine (gray), cytosine (violet), adenine (light green), and thymine (dark green). Within the amine pairs are atoms of oxygen (red) and nitrogen (blue).

(a)

(b)

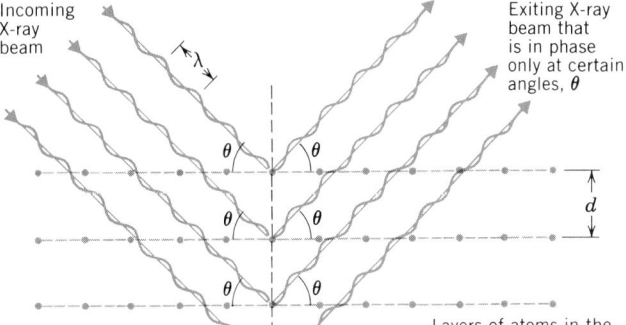

FIGURE 10.25 Diffraction of X rays from successive layers of atoms in a crystal.

In any crystal, many different sets of planes can be passed through the atoms. Figure 10.26 illustrates this idea in two dimensions for a simple pattern of points. When a crystal produces an X-ray diffraction pattern, many spots are observed because of diffraction from the many sets of planes. The physical geometry of the apparatus that is used to record the diffraction pattern allows the measurement of the angles at which the diffracted beams emerge from each distinct set of planes. Knowing the wavelength of the X rays, the values of n (which can be determined), and the measured angles, θ, the distances between planes of atoms, d, can be computed. The next step is to use the calculated interplanar distances to work backward to deduce where the atoms in the crystal must be located so that layers of atoms are indeed separated by these distances. If this sounds like a difficult task, it is! Some sophisticated mathematics as well as computers are needed to

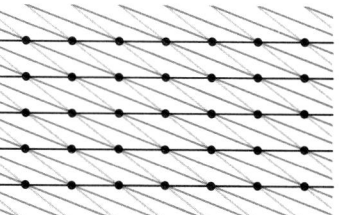

FIGURE 10.26 A two-dimensional pattern of points with many possible sets of parallel lines. In a crystal there are many sets of parallel planes.

accomplish it. The efforts, however, are well rewarded because the calculations give the locations of atoms within the unit cell and the distances between them. This information, plus a lot of chemical "common sense," is used by chemists to arrive at the shapes and sizes of the molecules in the crystal.

10.12   PHYSICAL PROPERTIES AND CRYSTAL TYPES

The physical properties of crystals are determined by the kinds of particles they are composed of and the attractive forces between them.

You know from personal experience that solids exhibit a wide range of physical properties. Some, such as diamond, are very hard; others, such as naphthalene (moth flakes) or ice, are soft, by comparison, and are easily crushed. Some solids, such as salt crystals or iron, have high melting points, whereas others, such as candle wax, melt at low temperatures. Some conduct electricity well, but others are nonconducting.

Physical properties such as these depend on the kinds and strengths of the attractive forces that hold the particles together in the solid. Even though we can't make exact predictions about such properties, some generalizations do exist. In making these generalizations, we can divide crystals into several types according to the kinds of particles located at sites in the lattice and the kinds of attractions that exist between the particles.

We have already discussed some of the properties of ionic crystals in Chapter 2. We saw that they are hard, have high melting points, and are brittle. When they melt, the resulting liquids conduct electricity well. These properties reflect the strong attractive forces between ions of opposite charge as well as the repulsions that occur when ions of like charge are placed near each other.

Molecular crystals are solids in which the lattice sites are occupied either by atoms —as in solid argon or krypton—or by molecules—as in solid CO_2, SO_2, or H_2O. Such solids tend to be soft and have low melting points because the particles in the solid experience relatively weak intermolecular attractions. The crystals are soft because little effort is needed to separate the particles or to move them past each other. The solids melt at low temperatures because the particles need little kinetic energy to break away from the solid. If the crystals contain only individual atoms, as in solid argon or krypton, or if they are composed of nonpolar molecules, as in naphthalene, the only attractions between the particles are relatively weak London forces. In crystals containing polar molecules, such as sulfur dioxide, the major forces that hold the particles together are dipole–dipole attractions. In crystals such as water, the primary forces of attraction are due to hydrogen bonding.

Covalent crystals are solids in which lattice positions are occupied by atoms that are covalently bonded to other atoms at neighboring lattice sites. The result is a crystal that is essentially one gigantic molecule. These solids are sometimes called *network solids* because of the interlocking network of covalent bonds extending throughout the crystal in all directions. A typical example is the diamond, the structure of which is illustrated in Figure 10.27. Covalent crystals tend to be hard and to have very high melting points because of the strong attractions between covalently bonded atoms. Other examples of covalent crystals are quartz (SiO_2—typical grains of sand) and silicon carbide (SiC—a common abrasive used in sandpaper).

Metallic crystals have properties that are quite different from those of the other three types of crystals that we've discussed. They conduct heat and electricity well, and they have the luster that we characteristically associate with metals. A number of different models have been developed to describe metals. The simplest one views the crystal as having positive ions at the lattice positions which are surrounded by electrons in a cloud that spreads throughout the entire solid, as illustrated in Figure 10.28. The electrons in this cloud belong to no single positive ion, but rather to the crystal as a whole. Because the electrons aren't localized on any one atom, they are free to move easily, which

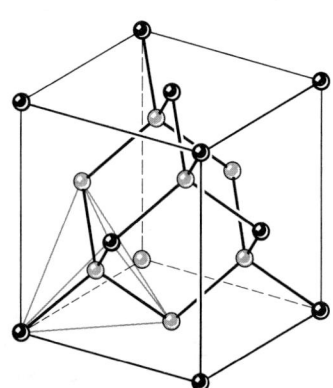

FIGURE 10.27 The structure of diamond. Notice that each carbon atom is covalently bonded to four others at the corners of a tetrahedron. This structure extends throughout an entire diamond crystal. (In diamond, of course, all the atoms are identical. They are different shades of gray here to make it easier to visualize the structure.)

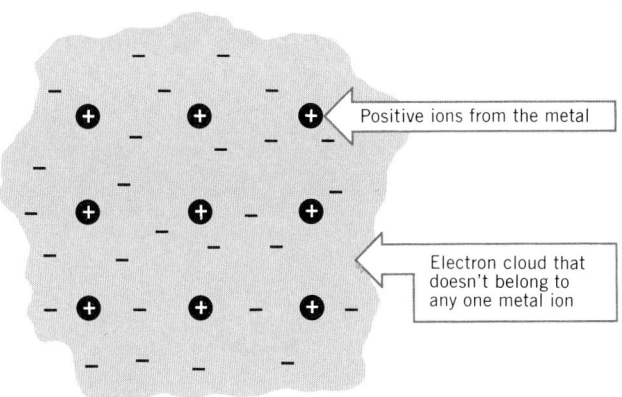

FIGURE 10.28 The "electron sea" model of a metallic crystal.

accounts for the electrical conductivity of metals. By their movement, the electrons can also transmit kinetic energy rapidly through the solid, so metals are also good conductors of heat. This model explains the luster of metals, too. When light shines on the metal, the loosely held electrons vibrate easily and readily reemit the light with essentially the same frequency and intensity.

It is not possible to make many simple generalizations about the melting points of metals. We've seen before that some, such as tungsten, have very high melting points, whereas others, such as mercury, have quite low melting points. To some degree, the melting point depends on the charge of the positive ions in the metallic crystals. The atoms of the Group IA metals tend to exist as cations with a 1 + charge, and they are only weakly attracted to the "electron sea" that surrounds them. Atoms of the Group IIA metals, however, each lose two electrons to the metallic lattice and form ions with a 2 + charge. These more highly charged ions are attracted more strongly to the surrounding electron sea, so the Group IIA metals have higher melting points than their neighbors in Group IA. For example, magnesium melts at 650 °C, whereas sodium melts at only 98 °C. Those metals with very high melting points, such as tungsten, must have *very* strong attractions between their atoms, which suggests that there probably is some covalent bonding between them as well.

The different ways of classifying crystals and a summary of their general properties are given in Table 10.5.

TABLE 10.5 Types of Crystals

Crystal Type	Particles Occupying Lattice Sites	Type of Attractive Force	Typical Examples	Typical Properties
Ionic	Positive and negative ions	Attractions between ions of opposite charge	NaCl $CaCl_2$ $NaNO_3$	Hard; high melting points; nonconductors of electricity as solids, but conductors when melted
Molecular	Atoms or molecules	Dipole–dipole attractions London forces Hydrogen bonding	HCl, SO_2 N_2, Ar, CH_4 H_2O	Soft; low melting points; nonconductors of electricity in both solid and liquid
Covalent (network)	Atoms	Covalent bonds between atoms	Diamond SiC (silicon carbide) SiO_2 (sand, quartz)	Very hard; very high melting points; nonconductors of electricity

(Continued)

TABLE 10.5 **Types of Crystals** (Continued)

Crystal Type	Particles Occupying Lattice Sites	Type of Attractive Force	Typical Examples	Typical Properties
Metallic	Positive ions	Attractions between positive ions and an electron cloud that extends throughout the crystal	Cu Ag Fe Na Hg	Range from very hard to soft; melting points range from high to low; conduct electricity well in both solid and liquid; have characteristic luster

EXAMPLE 10.1   IDENTIFYING CRYSTAL TYPES

Problem: The metal osmium, Os, forms an oxide with the formula OsO_4. The soft crystals of OsO_4 melt at 40 °C, and the resulting liquid does not conduct electricity. In what form does OsO_4 probably exist in the solid?

Solution: The characteristics of the OsO_4 crystals — softness and low melting point — suggest that solid OsO_4 exists as molecular crystals that contain *molecules* of OsO_4. This is further supported by the fact that liquid OsO_4 does not conduct electricity, which is evidence for the lack of ions in the liquid.

PRACTICE EXERCISE 4   Boron nitride, which has the empirical formula BN, melts under pressure at 3000 °C and is as hard as a diamond. What is the probable solid type for this compound?

PRACTICE EXERCISE 5   Crystals of elemental sulfur are easily crushed and melt at 113 °C to give a clear yellow liquid that doesn't conduct electricity. What is the probable crystal type for solid sulfur?

10.13   NONCRYSTALLINE SOLIDS

Some liquids cease to flow as they are cooled and never become crystalline when they solidify.

If a cubic salt crystal is broken, the pieces still have flat faces that intersect at 90° angles. On the other hand, if you shatter a piece of glass, the pieces often have surfaces that are not flat. Instead, they tend to be smooth and curved. This behavior illustrates a major difference between crystalline solids, like NaCl, and noncrystalline solids, or amorphous solids, such as glass.

The word *amorphous* is derived from the Greek word *amorphos,* which means "without form." Amorphous solids do not have long-range repetitive internal structures such as those found in crystals. In some ways their structures, being jumbled, are more like liquids than solids. Examples of amorphous solids are ordinary glass and many plastics. In fact, *glass* is sometimes used as a general term to refer to any amorphous solid.

Substances that form amorphous solids usually consist of long, chainlike molecules that are intertwined in the liquid state somewhat like long strands of cooked spaghetti. To form a crystal from the melted material, these long molecules would have to become untangled and line up in specific patterns. But as the liquid is cooled, the molecules move more slowly. Unless the liquid is cooled extremely slowly, the molecular motion decreases too rapidly for the untangling to take place, and the substance solidifies with the molecules still intertwined.

Pieces of broken glass have sharp edges, but their surfaces are not flat planes.

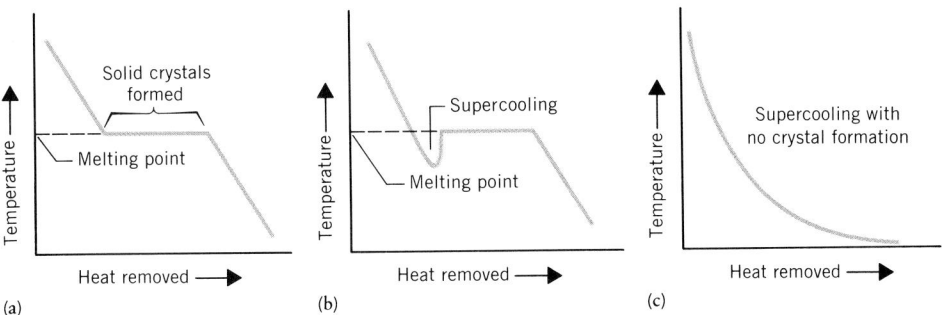

Compared with substances that produce crystalline solids, those that form amorphous solids behave quite oddly when they are cooled. Those that form crystals solidify at a constant temperature, as shown in Figure 10.29*a*. As the liquid is cooled, it eventually reaches the substance's freezing point, and crystals begin to form. Even though more heat continues to be removed, the temperature remains constant until all of the liquid has frozen. Only then does the temperature of the solid begin to drop.

Sometimes a liquid can actually be cooled below its freezing point. As the temperature approaches the freezing point, particles of the liquid may not be aligned in just the proper way for them to form a crystal. Thus, the temperature may continue to fall below the freezing point until, by chance, the particles in some small portion of the liquid suddenly find themselves properly arranged for a small crystal to form. This crystal grows rapidly, and the temperature rises again to the freezing point. The temperature then stays constant until all the liquid has frozen, as shown in Figure 10.29*b*. While the liquid has a temperature below its freezing point, it is said to be supercooled.

With substances that form amorphous solids, the solidification of the melted material into highly ordered crystals never occurs because the molecules can't become untangled before they are frozen in place at a low temperature. Sometimes amorphous solids are therefore described as supercooled liquids. This term connotes the kind of structural disorder found in liquids. It also suggests that the material's constituent molecules retain some residual ability at least to flex their chains if not to diffuse throughout the material and, over a long period of time, achieve an improved degree of crystalline orderliness. But the rate at which such change occurs is typically so small that under ordinary conditions it cannot be observed and the material is fully rigid. Glass, for example, which is a typical amorphous solid, over a very long time will develop regions of higher and higher order, as revealed by X ray diffraction patterns of old glass. Figure 10.29*c* shows a typical cooling curve for an amorphous solid. No sharp breaks occur in the curve as the material is cooled until it becomes rigid.

Amorphous solids also soften gradually when they are heated. This is the reason you can heat glass tubing in a flame to soften it so that you can bend it. By contrast, if you warm an ice cube (crystalline water), it won't become soft gradually—at 0 °C it will suddenly melt and drip all over you!

Some scientists prefer to reserve the term *solid* for crystalline substances, so they refer to glass as a "supercooled liquid."

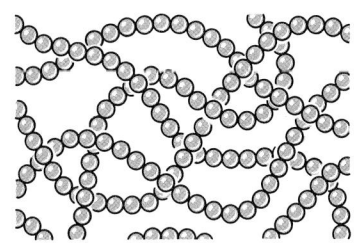

In an amorphous solid, long molecules are tangled and disorganized.

10.14 PHASE DIAGRAMS

Graphical representations help us view the pressure–temperature relationships among the various phases of a substance.

Sometimes it is useful to know whether a substance will be a liquid, a solid, or a gas at a particular temperature and pressure. A simple way of determining this is to use a phase diagram—a graphical representation of the pressure–temperature relationships that apply to the equilibria between the phases of the substance.

FIGURE 10.30 The phase diagram for water, distorted to emphasize certain features. Usually a phase diagram is drawn properly to scale so that its temperature and pressure scales may be read accurately.

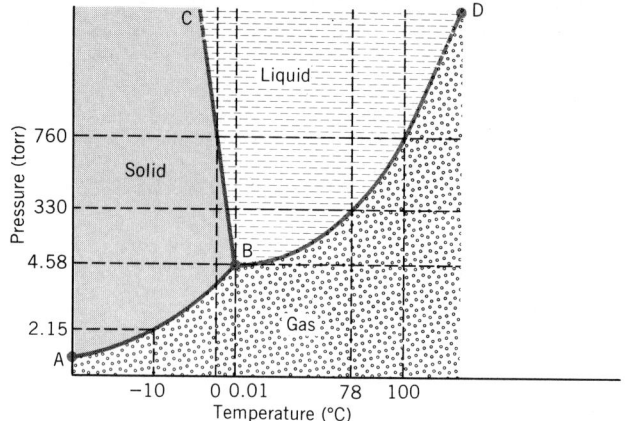

Figure 10.30 is the phase diagram for water. As we will see, it is not really as complicated as it appears at first glance. On it, there are three lines that intersect at a common point. These lines give temperatures and pressures at which equilibria between phases can exist. For example, line *AB* is the vapor pressure curve for the solid (ice). Every point on this line gives a temperature and a pressure at which ice and its vapor are in equilibrium. For instance, at −10 °C ice has a vapor pressure of 2.15 torr. This kind of information would be very useful to know if you wanted to design a system for making freeze-dried coffee. It tells you that ice will sublime at −10 °C if a vacuum pump can reduce the pressure of water vapor above the ice to less than 2.15 torr.

Line *BD* is the vapor pressure curve for liquid water. It gives the temperatures and pressures at which the liquid and vapor are in equilibrium. Notice that when the temperature is 100 °C, the vapor pressure is 760 torr. Therefore, this diagram also tells us that water will boil at 100 °C when the pressure is 1 atm (760 torr), because that is the temperature at which the vapor pressure equals 1 atm. At the top of Mt. McKinley, in Alaska, the atmospheric pressure is only about 330 torr. The phase diagram tells us that the boiling point of water there will be 78 °C.

The melting point and boiling point can be read directly from the phase diagram.

The solid–vapor equilibrium line, *AB*, and the liquid–vapor line, *BD*, intersect at a common point, *B*. Because this point is on both lines, there is equilibrium between all three phases at the same time.

$$\text{liquid} \rightleftharpoons \text{solid}$$
vapor

The temperature and pressure at which this occurs is called the triple point. For water the triple point occurs at 0.01 °C and 4.58 torr. Every known chemical substance except helium (Section 19.2) has its own characteristic triple point, which is controlled by the balance of intermolecular forces in the solid, liquid, and vapor.

In the SI, the triple point of water is used to define the Kelvin temperature of 273.16 K.

Line *BC*, which extends upward from the triple point, is the solid–liquid equilibrium line or *melting point line*. It gives temperatures and pressures at which solid–liquid equilibria occur. At the triple point, melting of ice occurs at +0.01 °C; at 760 torr, melting occurs at 0 °C. Thus, increasing the pressure on ice lowers its melting point.

Ice floats because it is less dense than liquid water.

The decrease in the melting point of ice that occurs when there is an increase in pressure can be predicted using Le Châtelier's principle and the knowledge that liquid water is more dense than ice. (More water molecules are packed into a given volume of liquid than in the same volume of solid water.) Let's consider the equilibrium

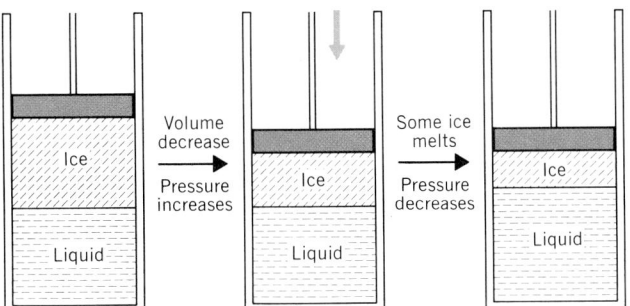

FIGURE 10.31 The effect of pressure on the equilibrium, H₂O (solid) ⇌ H₂O (liquid).

$$H_2O(s) \rightleftharpoons H_2O(\ell)$$

set up in an apparatus like that shown in Figure 10.31. If the piston is pushed in slightly, the pressure rises. According to Le Châtelier's principle, the system should respond in a way that will reduce the pressure. The only way that this can happen is for some of the ice to melt so that the ice–liquid mixture won't require as much space. Then the molecules won't push as hard against each other, and the pressure will drop. Thus, a pressure increase favors a volume decrease.

Now, suppose we have ice at a pressure just below the solid–liquid line, *BC*. If we raise the pressure to a point just above the line, the ice will melt and become a liquid. This will happen even if we don't change the temperature. By changing the pressure *at constant temperature,* we change the system from a solid whose melting point is above that temperature to a liquid whose melting point is below that temperature. This can only be true if the melting point becomes lower as the pressure is raised.

Water is very unusual. Almost all other substances have melting points that increase with pressure. Consider, for example, the phase diagram for carbon dioxide shown in Figure 10.32. For CO₂ the solid–liquid line slants to the right. Also notice that solid carbon dioxide has a vapor pressure of 1 atm at −78 °C. This is the temperature of dry ice, which sublimes at atmospheric pressure.

Besides specifying phase equilibria, the three intersecting lines on a phase diagram serve another important purpose—they define regions of temperature and pressure at which only a single phase can exist. For example, between lines *BC* and *BD* in Figure 10.30 are temperatures and pressures at which water exists as a liquid. At 760 torr, water is a liquid anywhere between 0 °C and 100 °C. For instance, we are told by the diagram that we can't have ice with a temperature of 25 °C if the pressure is 760 torr. (Of course, you already knew this from firsthand experience; ice never has a temperature of 25 °C.) The diagram also says that we can't have water vapor with a pressure of 760 torr when the temperature is 25 °C. (You've already seen that the temperature has to rise to 100 °C for the vapor pressure to reach 760 torr). Instead, we are told by the phase diagram that the *only* phase for pure water at 25 °C and 1 atm is the liquid. Below 0 °C at 760 torr, water is a solid; above 100 °C at 760 torr, water is a vapor. On the phase diagram for water, the phases that can exist in the different temperature–pressure regions are marked.

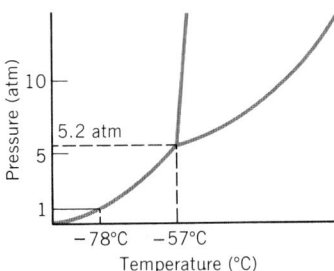

FIGURE 10.32 The phase diagram for CO₂.

EXAMPLE 10.2 USING A PHASE DIAGRAM

Problem: What phase would we expect for water at 0 °C and 4.58 torr?

The fact that the evaporation of a liquid is endothermic and the condensation of a gas is exothermic forms the basis for the operation of refrigerators and air conditioners. These devices use a gas that has a high critical temperature—ammonia or a halogenated methane such as CCl_2F_2 (Freon)—to pump heat from one place to another.

In a refrigerator, a compressor squeezes the gas into a small volume, which causes the pressure to increase. At the same time, the work done compressing the gas becomes stored as heat—the gas becomes warm. The warm gas is then circulated through cooling coils that are usually located on the back of the refrigerator. (Did you ever notice that these coils are warm?) When the gas cools under pressure, it liquefies when its temperature drops below its critical temperature, and this releases the substance's heat of vaporization, which is also given off to the surroundings. After being cooled, the pressurized liquid moves back into the refrigerator where it passes through a nozzle into a region of low pressure. At this low pressure, the liquid evaporates, and heat equal to the heat of vaporization is absorbed, which cools the inside of the refrigerator. Then the gas moves back to the compressor, and the cycle is repeated.

Through condensation and evaporation, the compressor uses the gas as the fluid in a heat pump that absorbs heat from inside the refrigerator and dumps it outside. An air conditioner works in essentially the same way, except that a room becomes the equivalent of the inside of a regrigerator, and the heat pump dumps the heat into the air outside.

Solution: First we find 0 °C on the temperature axis. Then we move upward until we intersect a line corresponding to 4.58 torr. This intersection occurs in the "solid" region of the diagram. At 0 °C and 4.58 torr, then, water exists as a solid.

EXAMPLE 10.3 USING A PHASE DIAGRAM

Problem: What phase changes occur if the pressure on water at 0 °C is gradually increased from 2.15 torr to 800 torr?

Solution: At 0 °C and 2.15 torr, water exists as a gas. As we move upward along the 0 °C line, we first encounter the solid–vapor line. As we go above this line, the water will freeze and become ice. Continuing the climb at 0 °C, we next encounter the solid–liquid line at 760 torr. Above 760 torr the solid will melt. At 800 torr and 0 °C, the water will be liquid.

PRACTICE EXERCISE 6   What phase changes will occur if water at −20 °C and 2.15 torr is heated to 50 °C under constant pressure?

PRACTICE EXERCISE 7   What phase will water be in if it is at a pressure of 330 torr and a temperature of 50 °C?

There is one final aspect of the phase diagram that has to be mentioned. The vapor pressure line for the liquid, which begins at point B, terminates at point D, which is known as the critical point. The temperature at D is called the critical temperature, T_c, and the pressure at D is called the critical pressure, P_c. Above the critical temperature, a

FIGURE 10.33 What happens when a liquid is heated in a sealed container. (a) Below the critical temperature. (b) Above the critical temperature.

More dense liquid at the bottom can be detected by the interface between the phases

Densities of "liquid" and "vapor" have become the same — there is only one phase

(a)

(b)

distinct liquid phase can't exist, regardless of the pressure. Figure 10.33 illustrates what happens to a substance as it approaches its critical point.

In Figure 10.33a, we see a liquid in a container with some vapor above it. We can distinguish between the two phases because they have different densities, which causes them to bend light differently. This allows us to see the interface, or surface, between the more dense liquid and the less dense vapor.

If this liquid is now heated, two things happen. First, more liquid evaporates. This causes an increase in the number of molecules per cubic centimeter of vapor which, in turn, causes the density of the vapor to increase. Second, the liquid expands (just like mercury does in a thermometer). This means that a given mass of liquid occupies more volume, so its density decreases. As the temperature of the liquid and vapor continue to rise, the vapor density and the liquid density approach each other. Eventually these densities become equal, and there no longer is any distinction between the liquid and the vapor—everything is the same (Figure 10.33b). The temperature at which this occurs is the critical temperature. At this temperature the pressure of the vapor is called the critical pressure. A substance with a temperature above its critical temperature and a density near its liquid density is described as a supercritical fluid.

The values of the critical temperature and critical pressure are unique for every chemical substance. As with the other physical properties that we've discussed, the values T_c and P_c are controlled by the intermolecular attractions. Liquids with strong intermolecular attractions, such as water, tend to have high critical temperatures. Substances with weak intermolecular attractions tend to have low critical temperatures.

This interface between liquid and gas is called the **meniscus**.

Supercritical CO_2 has solvent properties that allow it to dissolve and remove caffeine from coffee beans. Some producers of decaffeinated coffee have recently switched from using potentially toxic solvents to the use of supercritical CO_2, because of the nontoxic nature of any traces of CO_2 that might remain in their decaffeinated product.

SUMMARY

Physical Properties: Gases, Liquids, and Solids. Although the chemical properties of substances are very important, physical properties often have a greater effect on our daily lives. Most physical properties depend primarily on intermolecular attractions. In gases, these attractions are weak because the molecules are so far apart. They are much stronger in liquids and solids, whose particles are packed together tightly.

Intermolecular Attractions. Polar molecules attract each other primarily by dipole–dipole attractions, which arise because the positive end of one dipole attracts the negative end of another. Nonpolar molecules are attracted to each other by London forces, which are instantaneous dipole-induced dipole attractions. London forces are present between all particles, including atoms, polar and nonpolar molecules, and ions. London forces increase with increasing size of the particle's electron cloud; they also increase with increasing chain length among molecules such as the hydrocarbons. Hydrogen bonding, a special case of dipole–dipole attractions, occurs between molecules in which hydrogen is covalently bonded to a small, very electronegative atom—principally, nitrogen, oxygen, or fluorine.

General Properties of Liquids and Solids. Properties that depend mostly on closeness of packing of particles are incompressibility and diffusion. Diffusion is slow in liquids and almost nonexistent in solids at room temperature. Properties that depend mostly on intermolecular attractions are retention of volume and shape, surface tension, and ease of evaporation. Both solids and liquids retain volume; solids retain shape when transferred from one vessel to another. Surface tension is related to the energy needed to expand a liquid's surface area, and it causes liquids to form spherical drops. Liquids can wet a surface if its molecules are attracted to the surface about as strongly as they are attracted to each other. The most important property of liquids and solids is evaporation, which is endothermic, and produces a cooling effect. The rate of evaporation increases with increasing surface area, increasing temperature, and decreasing intermolecular attractions. Evaporation of a solid is called sublimation.

Changes of State. Changes from one physical state to another, such as melting, vaporization, or sublimation, can occur as dynamic equilibria. In a dynamic equilibrium, opposite processes occur at equal rates, so there is no apparent change in the composition of the system. For liquids and solids, equilibria are established when vaporization occurs in a sealed container. A solid is in equilibrium with its liquid at the melting point. For melting, vaporization of a liquid, and sublimation of a solid, the corresponding enthalpy changes are ΔH_{fusion}, $\Delta H_{vaporization}$, and $\Delta H_{sublimation}$. Their values are positive. $\Delta H_{vaporization}$ and $\Delta H_{sublimation}$ are measures of the strengths of the intermolecular attractive forces.

Vapor Pressure. When the rates of evaporation and condensation of a liquid are equal, the vapor exerts a pressure called the equilibrium vapor pressure (or more commonly, just the vapor pressure). The vapor pressure is controlled by the rate of evaporation *per unit surface area*. When the intermolecular attractive forces are large, the rate of evaporation is small and the vapor pressure is small. Vapor pressure increases with increasing temperature because the rate of evaporation increases as the temperature rises. The vapor pressure is independent of the *total* surface area of the liquid because as the surface area increases, the rates of evaporation and condensation are both increased equally; there is no change in the number of molecules in the vapor, so there is no change in the vapor pressure.

Le Châtelier's Principle. Le Châtelier's principle allows us to predict how a system in dynamic equilibrium will respond to a disturbance that upsets the equilibrium. The system undergoes a change that counters the disturbance and this returns the system to equilibrium, if it can. By this principle, we find that raising the temperature favors an endothermic change.

Boiling Point. A substance boils when its vapor pressure equals the prevailing atmospheric pressure. The normal boiling point of a liquid is the temperature at which its vapor pressure equals 1 atm. Substances with high boiling points have strong intermolecular attractions.

Crystalline Solids. Crystalline solids have very regular features that are determined by the highly ordered arrangements of particles within their lattices. The simplest portion of the lattice is the unit cell. Three cubic unit cells are possible—simple cubic, face-centered cubic, and body-centered cubic. Many different substances can have the same kind of lattice. Information about crystal structures is obtained from the crystal's X-ray diffraction pattern. Distances between planes of atoms can be calculated by the Bragg equation, $n\lambda = 2d \sin \theta$.

Crystal Types. Crystals can be divided into four general types: ionic, molecular, covalent, and metallic. Their properties depend on the kinds of particles within the lattice and on the attractions between the particles, as summarized in Table 10.5. A noncrystalline or amorphous solid is formed when a liquid is cooled without crystallization occurring. Such a solid has no sharply defined melting point and is called a supercooled liquid.

Phase Diagrams. Temperatures and pressures at which equilibrium can exist between phases are given graphically in a phase diagram. The three equilibrium lines intersect at the triple point. The liquid–vapor line terminates at the critical point. At the critical temperature, a liquid has a vapor pressure equal to its critical pressure. Above the critical temperature a single phase exists. The equilibrium lines also divide the phase diagram into temperature–pressure regions in which a substance can exist in just a single phase. According to Le Châtelier's principle, raising the pressure favors a change that leads to a smaller volume.

REVIEW EXERCISES

Answers to questions whose numbers are printed in color are given in Appendix C.

Comparing the States of Matter

10.1   List four items in your room whose physical properties account for their use in a particular application.

10.2   Under what conditions would we expect gases to obey the gas laws best?

10.3   Why is the behavior of gases affected very little by their chemical compositions?

10.4   Why are the intermolecular attractive forces stronger in liquids and solids than they are in gases?

Intermolecular Attractions

10.5   Describe dipole–dipole attractions.

10.6   What are London forces? How are they affected by molecular size?

10.7   What are hydrogen bonds?

10.8   Which nonmetals, besides hydrogen, are most often involved in hydrogen bonding?

10.9   Which is expected to have the higher boiling point, C_8H_{18} or C_4H_{10}? Explain your choice.

10.10   Ethanol and dimethyl ether have the same molecular formula, C_2H_6O. Ethanol boils at 78.4 °C, whereas dimethyl ether boils at −23.7 °C. Their structural formulas are

$$CH_3CH_2OH \qquad CH_3OCH_3$$
$$\text{ethanol} \qquad\qquad \text{dimethyl ether}$$

Explain why the boiling point of the ether is so much lower than the boiling point of ethanol.

10.11   How do the strengths of covalent bonds and dipole–dipole attractions compare? How do the strengths of ordinary dipole–dipole attractions compare with the strengths of hydrogen bonds?

10.12   Explain why London forces are called instantaneous dipole-induced dipole forces.

10.13   What kinds of intermolecular attractive forces (dipole–dipole, London, hydrogen bonding) are present in the following substances?
(a) HF   (c) PCl_3 (e) SO_2
(b) CS_2 (d) SF_6

General Properties of Liquids and Solids

10.14   Name two physical properties of liquids and solids that are controlled primarily by how tightly packed the particles are. Name three that are controlled mostly by the strengths of the intermolecular attractions.

10.15 Why does diffusion occur more slowly in liquids than in gases? Why does diffusion occur extremely slowly in solids?

10.16 Compare how shape and volume change when (a) a gas, (b) a liquid, and (c) a solid is transferred from one container to another.

10.17 Why are liquids and solids so difficult to compress?

10.18 On the basis of kinetic theory, would you expect the rate of diffusion in a liquid to increase or decrease as the temperature is increased? Explain your answer.

10.19 What is surface tension? Why do molecules at the surface of a liquid behave differently from those within the interior?

10.20 What kinds of observable effects are produced by the surface tension of a liquid?

10.21 What relationship is there between surface tension and the intermolecular attractions in the liquid?

10.22 Which liquid is expected to have the larger surface tension at a given temperature, CCl_4 or H_2O? Explain your answer.

10.23 What is *wetting*? What is a *surfactant*? How does it function?

10.24 Polyethylene plastic consists of long chains of carbon atoms, each of which is also bonded to hydrogens as shown below:

$$\cdots\ -\!\!\begin{array}{c}H\\|\\C\\|\\H\end{array}\!\!-\!\!\begin{array}{c}H\\|\\C\\|\\H\end{array}\!\!-\!\!\begin{array}{c}H\\|\\C\\|\\H\end{array}\!\!-\!\!\begin{array}{c}H\\|\\C\\|\\H\end{array}\!\!-\!\!\begin{array}{c}H\\|\\C\\|\\H\end{array}\!\!-\!\!\begin{array}{c}H\\|\\C\\|\\H\end{array}\!\!-\!\!\begin{array}{c}H\\|\\C\\|\\H\end{array}\!\!-\!\!\begin{array}{c}H\\|\\C\\|\\H\end{array}\!\!-\!\!\begin{array}{c}H\\|\\C\\|\\H\end{array}\!\!-\ \cdots$$

Water forms beads when placed on a polyethylene surface. Why?

10.25 The structural formula for glycerin is

$$H\!-\!\begin{array}{c}H\\|\\C\\|\\OH\end{array}\!\!-\!\!\begin{array}{c}H\\|\\C\\|\\OH\end{array}\!\!-\!\!\begin{array}{c}H\\|\\C\\|\\OH\end{array}\!\!-\!H$$

Would you expect this liquid to wet a glass surface? Explain your answer.

10.26 Why does evaporation lower the temperature of a liquid?

10.27 On the basis of the distribution of kinetic energies of the molecules of a liquid, explain why increasing the liquid's temperature increases the rate of evaporation.

10.28 How is the rate of evaporation of a liquid affected by increasing the surface area of the liquid? How is the rate of evaporation affected by the strengths of intermolecular attractive forces?

10.29 During the cold winter months, snow often gradually disappears without melting. How is this possible? What is the process called?

10.30 How is freeze-drying accomplished? What are its advantages?

10.31 Why do puddles of water evaporate more quickly in the summer than in the winter?

Changes of State and Equilibrium

10.32 What is a "change of state"?

10.33 Why do molecules of a vapor that collide with the surface of a liquid tend to be captured by the liquid?

10.34 Under what conditions is an equilibrium established in the evaporation of a liquid? Why do we call it a *dynamic equilibrium*?

10.35 Under what conditions is a dynamic equilibrium established between the liquid and solid forms of a substance?

10.36 What is the temperature called at which there is an equilibrium between a liquid and a solid?

Energy Changes That Accompany Changes of State

10.37 What is the name given to the enthalpy change involved in (a) the conversion of one mole of liquid to one mole of vapor, (b) the conversion of one mole of solid to one mole of vapor, and (c) the conversion of one mole of solid to one mole of liquid?

10.38 The molar heat of vaporization of water at 25 °C is 43.9 kJ/mol. How much energy would be required to vaporize 1.00 L (1.00 kg) of water?

10.39 The molar heat of vaporization of acetone, C_3H_6O, is 30.3 kJ/mol at its boiling point. How much heat would be liberated by the condensation of 1.00 g of acetone?

10.40 Why is $\Delta H_{vaporization}$ larger than ΔH_{fusion}? How does $\Delta H_{sublimation}$ compare with $\Delta H_{vaporization}$?

10.41 Would the "heat of condensation," $\Delta H_{condensation}$, be exothermic or endothermic?

10.42 Rain forms when water vapor condenses in clouds. Explain why this is the source of energy for producing winds in storms.

10.43 Ethanol (grain alcohol) has a molar heat of vaporization of 39.3 kJ/mol. Ethyl acetate, a common solvent, has a molar heat of vaporization of 32.5 kJ/mol. Which of these substances has the larger intermolecular attractions?

10.44 Acetic acid has a heat of fusion of 10.8 kJ/mol and a heat of vaporization of 24.3 kJ/mol.

$$HC_2H_3O_2(s) \longrightarrow HC_2H_3O_2(\ell)$$
$$\Delta H_{fusion} = 10.8 \text{ kJ/mol}$$
$$HC_2H_3O_2(\ell) \longrightarrow HC_2H_3O_2(g)$$
$$\Delta H_{vaporization} = 24.3 \text{ kJ/mol}$$

Estimate the value for the heat of sublimation of acetic acid, in kilojoules per mole.

10.45 Suppose 50.0 g of water at 80 °C is added to 100 g of ice at 0 °C. The molar heat of fusion of water is 6.01 kJ/mol, and the specific heat of water is 4.18 J/g °C. Based on these data, (a) what will be the final temperature of the mixture and (b) how many grams of ice will melt?

10.46 A burn caused by steam is much more serious than one caused by the same amount of boiling water. Why?

10.47 Arrange the following substances in order of their increasing values of $\Delta H_{vaporization}$: (a) HF, (b) CH_4, (c) CF_4, (d) HCl.

Vapor Pressure

10.48 Define *equilibrium vapor pressure*. Why do we call the equilibrium involved a *dynamic equilibrium*?

10.49 Explain why changing the volume of a container in which there is a liquid–vapor equilibrium has no effect on the vapor pressure.

10.50 What effect does increasing the temperature have on the vapor pressure of a liquid?

10.51 Below are the vapor pressures of some relatively common chemicals measured at 20 °C. Arrange these substances in order of increasing intermolecular attractive forces.

Benzene, C_6H_6	80 torr
Acetic acid, $HC_2H_3O_2$	11.7 torr
Acetone, C_3H_6O	184.8 torr
Diethyl ether, $C_4H_{10}O$	442.2 torr
Water	17.5 torr

10.52 Why does humid air at 30 °C contain more moisture per cubic meter than the same volume of humid air at 25 °C?

10.53 Why does rain often form when humid air is forced to rise over a mountain range?

10.54 Why does moisture condense on the outside of a cool glass of soda in the summertime?

10.55 Why do we feel more uncomfortable in humid air at 90 °F than in dry air at 90 °F?

10.56 Why does the air in heated buildings in the winter have such a low humidity?

10.57 Why doesn't increasing the surface area of a liquid cause an increase in the equilibrium vapor pressure?

10.58 When warm moist air sweeps in from the ocean and rises over a mountain range, it expands and cools. Explain how this cooling is related to the attractive forces between gas molecules. Why does this cause rain to form? When the air drops down the far side of the mountain range, its pressure rises as it is compressed. Explain why this causes the air temperature to rise. How does the humidity of this air compare with the air that originally came in off the ocean? Now, explain why the coast of California is lush farmland, whereas the plains on the eastern side of the Rocky Mountains are arid and dry.

Le Châtelier's Principle

10.59 State Le Châtelier's principle in your own words.

10.60 What do we mean by the "position of equilibrium"?

10.61 Use Le Châtelier's principle to predict the effect of adding heat in the equilibrium: solid + heat $\rightleftharpoons$ liquid.

Boiling Points of Liquids

10.62 Define *boiling point* and *normal boiling point*.

10.63 Why does the boiling point vary with atmospheric pressure?

10.64 Mt. Kilimanjaro in Tanzania is the tallest peak in Africa (19,340 ft). The normal barometric pressure at the top of this mountain is about 345 torr. At what Celsius temperature would water be expected to boil there? (See Figure 10.14.)

10.65 Why is the boiling point a useful property with which to identify liquids?

10.66 The boiling points of some common substances are given here. Arrange these substances in order of increasing strengths of intermolecular attractions.

Ethanol C_2H_5OH	78.4 °C
Ethylene glycol, $C_2H_4(OH)_2$	197.2 °C
Water	100 °C
Diethyl ether, $C_4H_{10}O$	34.5 °C

10.67 Explain how a pressure cooker works.

10.68 The radiator cap of an automobile engine is designed to maintain a pressure of approximately 15 lb/in.^{2} above normal atmospheric pressure. How does this help prevent the engine from "boiling-over" in warm weather?

10.69 Butane, C_4H_{10}, has a boiling point of -0.5 °C (which is 31 °F). Despite this, liquid butane can be seen sloshing about inside a typical butane lighter, even at room temperature. Why isn't the butane boiling inside the lighter at room temperature?

10.70 Which would be expected to have a higher molar heat of vaporization, water (bp 100 °C) or alcohol (bp 78.4 °C)?

10.71 Why does H_2S have a lower boiling point than H_2Se? Why does H_2O have a much higher boiling point than H_2S?

10.72 Explain why HF has a lower boiling point than H_2O, even though HF is more polar than H_2O and forms stronger hydrogen bonds.

Crystalline Solids and X-Ray Diffraction

10.73 What surface features do crystals have that suggest a high degree of order among the particles within them?

10.74 What is a crystal lattice? What is a unit cell? What relationship is there between a unit cell and a crystal lattice?

10.75 Describe simple cubic, face-centered cubic, and body-centered cubic unit cells.

10.76 Make a sketch of a layer of sodium and chloride ions in a NaCl crystal. Indicate how the ions are arranged in a face-centered cubic pattern.

10.77 How do the crystal structures of copper and gold differ? In what way are they similar? Based on the location of the elements in the periodic table, what kind of crystal structure would you expect for silver?

10.78 Only 14 different kinds of crystal lattices are possible. How can this be true, considering the fact that there are millions of different chemical compounds that are able to form crystals?

10.79 Write the Bragg equation and define the symbols.

10.80 Explain, qualitatively, how an X-ray diffraction pattern of a crystal and the Bragg equation provide information that allows chemists to figure out the structures of molecules.

10.81 How many atoms are contained within a simple cubic unit cell? (*Hint:* To answer the question, add up the *parts* of atoms shown in Figure 10.18b.)

10.82 How many copper atoms are within the face-centered cubic unit cell of copper. (*Hint:* See Figure 10.20 and add up all the *parts* of atoms in the fcc unit cell.)

10.83 Copper crystallizes with a face-centered cubic unit cell. The length of the edge of a unit cell is 362 pm. Sketch the face of a unit cell, showing the nuclei of the copper atoms at the lattice points. The atoms are in contact along the diagonal from one corner to another. The length of this diagonal is four times the radius of a copper atom. What is the atomic radius of copper?

10.84 Silver forms face-centered cubic crystals. The atomic radius of a silver atom is 144 pm. Draw the face of a unit cell with the nuclei of the silver atoms at the lattice points. The atoms are in contact along the diagonal. Calculate the length of an edge of this unit cell.

Crystal Types

10.85 What kinds of particles are located at the lattice sites in a metallic crystal?

10.86 What kinds of attractive forces exist between particles in (a) molecular crystals, (b) ionic crystals, and (c) covalent crystals?

10.87 Why are covalent crystals sometimes called network solids?

10.88 Tin(IV) chloride, $SnCl_4$, has soft crystals with a melting point of -30.2 °C. The liquid is nonconducting. What type of crystal is formed by $SnCl_4$?

10.89 Magnesium chloride is ionic. What general physical properties are expected of its crystals?

10.90 Elemental boron is a semiconductor, is very hard, and has a melting point of about 2250 °C. What type of crystal is formed by boron?

10.91 Gallium crystals are shiny and conduct electricity. Gallium melts at 29.8 °C. What type of crystal is formed by gallium?

10.92 Titanium(IV) bromide forms soft orange-yellow crystals that melt at 39 °C to give a liquid that doesn't conduct electricity. The liquid boils at 230 °C. What type of crystals does $TiBr_4$ form?

10.93 Columbium is another name for one of the elements. This element is shiny, soft, and ductile. It melts at 2468 °C and conducts electricity. What kind of solid does columbium form?

10.94 Elemental phosphorus consists of soft white waxy crystals that are easily crushed and melt at 44 °C. The solid does not conduct electricity. What type of crystal does phosphorus form?

Amorphous Solids

10.95 What does the word *amorphous* mean?

10.96 What is an amorphous solid? What happens when a substance that forms an amorphous solid is cooled from the liquid state to the solid state?

10.97 What is supercooling?

10.98 Compare what happens when glass is heated in a flame with what occurs when a crystal, such as ice, is heated.

Phase Diagrams

10.99 Sketch the phase diagram for a substance that has a triple point at -10.0 °C and 0.25 atm, melts at -8.0 °C at 1 atm, and has a normal boiling point of 80 °C.

10.100 Based on the phase diagram of Review Exercise 10.99, below what pressure will the substance undergo sublimation?

10.101 Based on the phase diagram in Review Exercise 10.99, how does the density of the liquid compare with the density of the solid?

10.102 For most substances, the solid is more dense than the liquid. Use Le Châtelier's principle to explain why the melting point of such substances should *increase* with increasing pressure.

10.103 Define *critical temperature* and *critical pressure*.

10.104 What phases are in equilibrium at the triple point?

10.105 Why doesn't CO_2 have a normal boiling point?

10.106 For CO_2, what phase(s) should exist at (a) -60 °C and 6 atm, (b) -60 °C and 2 atm, (c) -40 °C and 10 atm, and (d) -57 °C and 5.2 atm?

10.107 Describe what happens when CO_2 at 2 atm is warmed from -80 °C to 0 °C at constant pressure.

10.108 Describe what happens when the pressure on CO_2 is gradually raised from 1 atm to 20 atm at a constant temperature of -56.8 °C. What happens if the pressure is increased from 1 atm to 20 atm at a constant temperature of -58 °C?

10.109 Looking at the phase diagram for CO_2, how can we tell that solid CO_2 is more dense than liquid CO_2?

10.110 If a substance exists as a liquid at room temperature and atmospheric pressure, what can we say about its triple point?

10.111 At room temperature, hydrogen can be compressed to very high pressures without liquefying. On the other hand, butane becomes a liquid at high pressure (at room temperature). What does this tell us about the critical points of hydrogen and butane?

CHEMICALS IN USE

GLASS

A *glass,* as we learned in chapter 10, is any amorphous solid, one having no regularly repeating order to its constituent particles. Most chemists use the term this way, but in the public's mind, "glass" is understood in narrower terms. The American Society for Testing and Materials (ASTM), an organization that sets widely accepted standards for analytical procedures and for quality control, also prefers to limit the term. To the ASTM, *glass* is an inorganic material made by the thermal fusion of solids and cooled without crystallizing until it is rigid.

One substance that can be melted and then cooled in this manner is quartz sand, which is essentially all silicon dioxide — silica. The trouble is that it melts at a very high temperature — above 1700 °C — and becomes rigid so quickly when cooled that it is very hard to work into desired shapes.

Soda – Lime – Silica Glass

When sodium carbonate — soda ash — is mixed with silica sand, the melting point is lowered from a temperature of over 1700 °C (for pure silica) to the range of 700 – 850 °C. In other words, sodium carbonate acts as a fluxing agent, something that can help prepare a flux or a molten mass. When a molten mixture of silica and sodium carbonate cools, a glass forms, but it is relatively soluble in water. It is called "water glass," and it is used, for example, to coat eggs to keep air from entering them and causing spoilage.

If limestone — calcium carbonate — is added along with the sodium carbonate to the silica sand, the melting-point lowering is still quite acceptable, but the glass that forms is much less soluble in water. These facts were known to Egyptian glass workers at least 2000 years ago when this, the most common kind of glass and called soda – lime – silica glass, began its continuous history of manufacture. If too much sodium carbonate is used, the resistance to water drops; if too much lime is used, the glass tends to "devitrify," meaning that crystalline regions develop. A typical batch is made up of 75% silica (of as uniform grain size as possible), 10% limestone, and 15% sodium carbonate.

Sometimes some of the limestone is replaced by magnesium carbonate and aluminum oxide. Since nearly all silica contains traces of iron compounds, the glass product is green in color. This kind of green glass is still the cheapest glass made. In time, glass workers found that by adding traces of oxides such as those of cobalt, arsenic, and selenium, the green color could be mostly neutralized.

Although the ingredients for soda – lime – silica glass are limestone and sodium carbonate along with silica, the high temperatures involved decompose the carbonates to oxides.

$$CaCO_3(s) \longrightarrow CaO(s) + CO_2(g)$$
$$\text{limestone} \qquad \text{quick lime}$$
$$Na_2CO_3(s) \longrightarrow Na_2O + CO_2$$

Thus, the melting of the initial batch generates considerable gas, and the mixture has to be maintained in the molten state for a few days to allow time for this gas to escape from the very thick, viscous fluid.

The final glass can be given a variety of bright colors. The addition of finely divided copper selenide (CuSe), cadmium selenide (CdSe), or metallic gold produces a ruby red glass. When copper or cobalt compounds are dissolved in the melt, the glass is blue. Chromium compounds are used to make bright green glass. (See Figure 8a.)

Soda – lime – silica glass is the "soft glass" familiar to chemistry students for making glass stirring rods and bent glass items fashioned from glass tubing. It is easily fire polished using a Bunsen burner flame because its sharp edges soften readily. (The familiar yellow color that this softening glass imparts to the flame is caused by the sodium ions in the glass.)

Borosilicate Glass

One of the problems with soda – lime – silica glass is that when heated it expands, and it contracts unevenly as it cools unevenly. This results in stress points in the cooled glass which make it break or shatter very readily. In other words, it is difficult to work with this kind of glass in situations involving wide changes in temperature. Table 8a shows the relative thermal expansions of various kinds of glasses, and soda – lime – silica glass has nearly the poorest showing with fifteen times the thermal expandability of pure silica glass.

If much of the sodium carbonate is replaced by boron oxide, B_2O_3, and some of the limestone is replaced by

FIGURE 8a The addition of traces of transition metal oxides to molten glass makes possible the striking colors seen in this stained glass window at the cathedral of Chartres, France.

aluminum oxide, Al_2O_3, the resulting glass is much more thermally stable. It is called borosilicate glass, and the most common brand is Pyrex®. Although it softens at a higher temperature — typically requiring oxygen–acetylene or oxygen–hydrogen flames — hot pieces can be fused together and annealed to give seals that withstand mechanical shocks very well. "Annealing" means cooling the softened material slowly and uniformly and thus minimizing the development of stress points. Borosilicate glass is the material used to make nearly all laboratory glassware (Figure 8b) and glass kitchen ware. When glass of even lower thermal expandability is needed, then pure silica glass or a near relative, Vycor®, is used (Table 8a).

Lead Crystal Glass

When lead oxide, PbO, is used as the fluxing agent instead of sodium carbonate, the resulting glass has a high refractive index (it can "bend" light rays well). When cut or

TABLE 8a Thermal Expandabilities of Various Glasses

Type of Glass	Relative Expandability
Silica glass	1.0
Vycor®	1.4
Borosilicate glass	6.1
Soda–lime–silica glass	15
Lead crystal glass	16

polished to angular designs, it sparkles with a brilliance that has attracted glass artisans for over 300 years.

Safety Glass

It is probably safe to say that the automobile industry would not have been possible without the use of a glass that could not shatter and throw shards when struck. There are two general ways to make such a glass. One is to seal a thin layer of a clear plastic between two sheets of glass. When this laminated glass breaks, its fragments are held by the plastic. If several alternating layers of glass and plastic are laminated together to a thickness of 1.5 to 2 in., the product is bulletproof glass.

Safety glass can also be made by a special process of heat treatment. The glass sheet is heated until its temperature is just below the softening point and then a cold blast of air is spread over its surfaces. The glass nearest these surfaces becomes subject to great compressional strains that aid the glass in resisting forces of bending and twisting. When it receives a sharp, glass-breaking blow, it instantly develops cracks over its entire expanse. The glass particles are small and relatively harmless, and do not fly about. (Laminated glass in the same circumstances develops cracks that radiate from the point of impact.)

FIGURE 8b Laboratory beakers are made by machines from borosilicate glass.

Questions

1. How does the ASTM define glass?
2. Why is pure quartz seldom used to make ordinary glass items?
3. Sodium carbonate is called a fluxing agent for silica. What does this mean?
4. Why is calcium carbonate a part of the recipe for glass?
5. SiO_2 is an acidic anhydride. What basic anhydrides form during the manufacture of soda–lime–silica glass? (Give their names and formulas.)
6. What property of borosilicate glass suits it better to laboratory glassware than soda–lime–silica glass?
7. What is the fluxing agent (name and formula) in lead crystal glass, and what makes this glass attractive to glass artisans?

Nitric acid pours from a tank car in a major acid spill in Denver, Colorado in April of 1983. Already the acid is attacking metals as indicated by the yellowish-brown gas you can see rising from the scene. The acid was neutralized by sodium carbonate, as shown in the newspaper photo on page 404. We continue our study of acids, bases, and neutralization in this chapter.

ACIDS, BASES, AND IONIC REACTIONS

ELECTROLYTES
ACIDS AND BASES AS ELECTROLYTES
IONIC REACTIONS IN AQUEOUS SOLUTIONS
PREDICTING WHEN METATHESIS REACTIONS WILL OCCUR
STOICHIOMETRY OF IONIC REACTIONS: ACID–BASE
    TITRATIONS

EQUIVALENT WEIGHTS AND NORMALITY
BRØNSTED CONCEPT OF ACIDS AND BASES
PERIODIC TRENDS IN THE STRENGTHS OF ACIDS
LEWIS CONCEPT OF ACIDS AND BASES
COMPLEX IONS

11.1 ELECTROLYTES

Solutes that yield electrically conducting solutions give ions that are free to roam about in the solution.

In Chapter 2, you were introduced to some of the chemical and physical properties of the elements and their compounds. Since then, you've had the opportunity to learn about the structures of atoms and molecules, and why the elements form certain kinds of compounds. (For example, nonmetals combine with each other to form covalently bonded molecules, but they combine with metals to form predominantly ionic compounds.) With this as background, we now return to discussions of chemical reactions in this chapter and the next. We begin here by studying how and why reactions between ions occur in aqueous solution. An important class of substances that participate in such reactions is made up of acids and bases, and later in this chapter we will see that the acid–base phenomenon extends far beyond the limits of water as a solvent.

You learned in Section 2.11 that aqueous solutions of ionic compounds such as NaCl or $CaCl_2$ are able to conduct electricity. Not all chemical compounds behave this way, however. For example, if sugar is dissolved in water we find that the solution is not electrically conducting. It is really not difficult to understand why some solutes produce solutions that conduct and others do not. We mentioned that for electrical conduction to take place, there must be electrical charges that are able to move. When an ionic compound is dissolved in water, the ions that are packed tightly together in the solid become separated—we say the compound dissociates. As the ions enter the solution they become surrounded by water molecules, a process called hydration, and we say the ions have become hydrated in the solution. These hydrated ions are able to wander about in the solution and behave more or less independently, and their ability to move is what accounts for the electrical conductivity of the solution. However, when a molecular substance such as sugar dissolves in water, the molecules become dispersed throughout the solution, but they stay intact. There are no charged particles in the solution, so the solution is a nonconductor.

The margin note:

For an ionic compound, ions are present both in the solid and in aqueous solutions.

The ability to produce a conducting solution is a special property of a solute—a property that is given a special name. Any solute that dissolves in water to give a solution that contains ions, and which therefore conducts electricity, is called an electrolyte. On the other hand, a solute that remains undissociated in solution is called a nonelectrolyte.

Dissociation Reactions

We will wait until Chapter 13 to discuss the details of the physical changes that take place between the solute and the solvent when an ionic compound dissolves in water. For now, we can simply describe the overall change that occurs by a chemical equation as follows, using NaCl as a familiar example.

$$NaCl(s) \longrightarrow Na^+(aq) + Cl^-(aq)$$

The *aq*, which stands for aqueous, in parentheses after the formulas of the ions means that the ions are dissolved in water and are hydrated, and by writing the formulas of the ions separately we mean that in the solution the ions are essentially independent of each other. A similar equation that represents what occurs when calcium chloride, $CaCl_2$, dissolves in water is

$$CaCl_2(s) \longrightarrow Ca^{2+}(aq) + 2Cl^-(aq)$$

which shows that two Cl^- ions are present in the solution for each Ca^{2+}. Once again we have used *s* and *aq* in parentheses to point out the changes that take place.

Salts that contain polyatomic ions also dissociate in aqueous solution, and it is important to remember that in the solution the polyatomic ions remain intact. For instance, when the salt sodium sulfate, Na_2SO_4, dissolves in water, the solution contains sodium ions and sulfate ions.

The atoms in a polyatomic ion, you recall, are held to each other by covalent bonds.

$$Na_2SO_4(s) \longrightarrow 2Na^+(aq) + SO_4^{2-}(aq)$$

You can see that it is especially important that you know both the formulas and charges of the polyatomic ions to write equations such as these properly. If necessary, you should refer to Table 2.6 for review.

Often, in writing chemical equations for dissociation reactions or for other reactions involving ions in solution, the symbols (*s*) and (*aq*) following the formulas are omitted. When we're discussing reactions in an aqueous solution, they are "understood." Therefore, you should not be fooled when you see an equation such as

$$CaCl_2 \longrightarrow Ca^{2+} + 2Cl^-$$

Unless something is said to the contrary, it means

$$CaCl_2(s) \longrightarrow Ca^{2+}(aq) + 2Cl^-(aq)$$

PRACTICE EXERCISE 1   Write chemical equations that show what happens when these solid ionic compounds dissolve in water: (a) $MgCl_2$, (b) $Al(NO_3)_3$, (c) Na_2CO_3, (d) $(NH_4)_2SO_4$.

11.2 ACIDS AND BASES AS ELECTROLYTES

Molecular substances that cause a solution to become acidic or basic are electrolytes, too.

Svante Arrhenius almost failed his doctoral examination at the university in Uppsala, Sweden, in 1884 when he proposed that ions form directly when salts dissolve in water. It was revolutionary at that time to think that free sodium ions or chloride ions could exist in a solution. His theory of electrolytes was kept alive because it provided logical explanations not just for one observation or one phenomenon, but for several. The theory accounts for the electrical conduction of salt solutions, and we will see in Chapter 13 that it accounts for some of the unusual physical properties of salt solutions.

Ionic compounds are not the only substances that produce solutions that conduct electricity. In Chapter 2, you learned that acids, such as HCl, are substances that react with water to give ions — one of them being the hydronium ion, H_3O^+. In its pure state, hydrogen chloride is molecular, and at room temperature it is a gas. The hydrogen and chlorine atoms of each molecule are joined by a covalent bond. But when HCl is dissolved in water, we've seen that it reacts to give H_3O^+ and Cl^-.

$$HCl(g) + H_2O \longrightarrow H_3O^+(aq) + Cl^-(aq)$$

Other acids, such as HNO_3 and H_2SO_4, are also molecular in their pure states, and it is their reaction with water that produces ions. Acids, therefore, are exceptions to the rule that molecular substances are nonelectrolytes.

Recognition that acids are substances that release H^+ ions when they are dissolved in water, and that the reaction of an acid and a base really involves the combination of "hydrogen ion" with hydroxide ion to form water, led Arrhenius to one of the earliest definitions of acids and bases. According to Arrhenius, but restated in modern terms, an *acid* is a substance that produces H_3O^+ in water and a *base* is a substance that gives OH^- in water. This, of course, is the same definition that we gave in Chapter 2. In this chapter we will see how this definition is too narrow.

Ionization Reactions

When sodium chloride dissolves in water, the ions that exist in the crystals become separated. We used the term *dissociation* to describe this change because "to dissociate" means "to separate." However, when hydrogen chloride dissolves in water, a chemical change takes place. Each HCl molecule gives up an H^+ ion to a water molecule, which leaves a chloride ion behind. We can diagram the change using Lewis structures as follows.

Since ions now exist where none did before, the change is called an ionization reaction, and it produces a solution that conducts electricity. For this reason, HCl and other acids are classified as electrolytes, too.

Arrhenius was one of Sweden's most famous chemists. He was awarded the Nobel Prize in chemistry in 1903 for his theory of electrolytes.

A theory is only as good as its explanations and predictions.

If gaseous HCl is cooled to about −85 °C, it condenses to a liquid that doesn't conduct electricity. No ions are present in pure liquid HCl.

The Arrhenius concept only recognizes acids and bases in aqueous solutions. Later we will see that this restriction is removed by broader acid–base definitions.

Since ionic compounds already contain ions, we don't describe their dissolving in water as ionization — we call it *dissociation,* which means breaking apart.

Svante Arrhenius, Swedish physicist and chemist.

Strong and Weak Electrolytes

Sodium chloride and hydrogen chloride are both examples of strong electrolytes—electrolytes that break up essentially 100% into ions in water. In other words, there are no molecules of either solute detectable in their aqueous solutions — they exist entirely as ions.

All ionic compounds are strong electrolytes, even those of very limited solubility. For example, when solid $CaCO_3$ (chalk) is placed in water, hardly any dissolves; the solubility of this compound is very low. But the small amount that does dissolve is fully dissociated into Ca^{2+} ions and CO_3^{2-} ions.

Also included among the strong electrolytes are the metal hydroxides that provide hydroxide ions in water. These compounds are bases, and chemists refer to any base that is a strong electrolyte as a strong base. *Strong* carries the identical meaning in both terms — that the substance (electrolyte or base) exhibits a high percentage dissociation in water. Sodium hydroxide and all the other Group IA hydroxides are strong bases. The hydroxides of magnesium, calcium, strontium, and barium (Group IIA metals) are also strong bases despite the fact that they are not very soluble in water. But whatever does dissolve separates 100% into metal ions and hydroxide ions.

What we have said about bases extends to acids. Hydrogen chloride, as we said, ionizes 100% in water, and this solution is called hydrochloric acid. Since hydrochloric acid is a strong electrolyte, it is also a strong acid — an acid whose percentage ionization is 100%. Nitric acid, HNO_3, is another example of a strong acid. Its molecules ionize completely in water according to the equation

$$HNO_3(aq) + H_2O \longrightarrow H_3O^+(aq) + NO_3^-(aq)$$

$HC_2H_3O_2$

Acetic acid—only the H in red is acidic

Not all molecular acids form aqueous solutions that conduct electricity well. Acetic acid, responsible for the sour taste of vinegar, is a common example. (As we mentioned in Section 2.12, we frequently write its formula as $HC_2H_3O_2$ to indicate that only the first hydrogen in the formula can separate as H^+; the other three hydrogens are too strongly held to one of the carbon atoms.) When the electrodes of a conductivity apparatus (Figure 2.14, page 74) are dipped into a 1 M solution of acetic acid, the light bulb glows dimly, but when they are dipped into a 1 M solution of NaCl or HCl, the bulb burns brightly. This means that the acetic acid solution is not as good a conductor as either of the other two solutions. In aqueous acetic acid, only a small fraction (about 0.5%) of the acetic acid molecules have reacted with water to form ions. We can write the ionization reaction as follows.

$$HC_2H_3O_2(aq) + H_2O \xrightarrow{\text{small percentage}} H_3O^+(aq) + C_2H_3O_2^-(aq)$$

Most acids are weak; only a few are strong.

The remaining 99.5% of the acetic acid molecules exist in the solution entirely un-ionized. Because of its low percentage ionization, acetic acid is classified as a weak electrolyte. For the same reason, acetic acid is also a weak acid. Other weak acids include carbonic acid, H_2CO_3, and nitrous acid, HNO_2.

Certain bases are also weak electrolytes with a low percentage ionization. An example is aqueous ammonia. Pure ammonia, NH_3, is a gas at room temperature, but it is quite soluble in water. Solutions of ammonia are commonly used around the home as cleansing and water-softening agents, where they are simply called "ammonia," although *aqueous ammonia* would be more proper. When the electrodes of a conductivity apparatus are dipped into a 1 M solution of aqueous ammonia, the light bulb glows dimly, indicating that aqueous ammonia is a poor conductor and a weak electrolyte. That it conducts electricity at all is because a small percentage of ammonia molecules have reacted with water to give ammonium ions and hydroxide ions.

$NH_3(g)$ = gaseous ammonia
$NH_3(aq)$ = aqueous ammonia

$$NH_3(aq) + H_2O \xrightarrow{\text{small percentage}} NH_4^+(aq) + OH^-(aq)$$

In 1 *M* $NH_3(aq)$, less than 1% of the NH_3 molecules have reacted as shown, so the concentration of ions, the carriers of electricity, is low. Because only a small percentage of ammonia molecules are able to generate hydroxide ions in water, aqueous ammonia is also classified as a weak base.

From these examples, two simple generalizations follow.

1. Weak acids or bases are weak electrolytes.
2. Strong acids or bases are strong electrolytes.

Equilibrium

Perhaps after reading the discussion above you are wondering, "If some of the acetic acid molecules or ammonia molecules in a solution can react with water, why can't the others?" Actually, they all have the potential to react. What we are observing here is a chemical example of a dynamic equilibrium—a concept introduced in Chapter 10 during our discussion of the vapor pressure of a liquid.

Consider what happens when pure acetic acid is added to pure water. In the solution, molecules of acetic acid frequently and continually collide with molecules of water. Because of the chemical nature of this particular solute, however, there is only a small probability that during one of these collisions an acetic acid molecule will actually react with a water molecule to form ions. What this means is that the rate at which ions form is quite small compared with the rate of collision.

The probability that ions will form in a collision varies from one solute to another.

Once hydronium ions and acetate ions, $C_2H_3O_2^-$, have formed, they wander about in the solution and, on occasion, a cation encounters an anion. But because of their chemical nature, when H_3O^+ and $C_2H_3O_2^-$ collide there is a *high* probability that they will react to form an $HC_2H_3O_2$ molecule. This reaction, which we call the *reverse reaction,* can also be expressed as an equation.

Remember, a *cation* is a positive ion and an *anion* is a negative ion.

$$H_3O^+(aq) + C_2H_3O_2^-(aq) \longrightarrow H_2O + HC_2H_3O_2(aq)$$

Let us go over the dynamics of this solution again. Just after adding pure $HC_2H_3O_2$ to pure water, the concentrations of the ions, H_3O^+ and $C_2H_3O_2^-$, build up because of the ionization reaction, which we can call the *forward reaction.* But as their concentrations increase, the ions naturally encounter each other more frequently. So the reverse reaction occurs faster and faster. Eventually the concentrations of the ions become large enough *to make the rate of the reverse reaction equal the rate of the forward reaction.* At this point, ions disappear in the solution as fast as they form, so their concentrations remain constant from this time on. The entire chemical system of H_2O, $HC_2H_3O_2$, H_3O^+, and $C_2H_3O_2^-$ has reached a state of balance—a state of dynamic equilibrium. Recall that we use double arrows ($\rightleftharpoons$) to indicate that a system has reached equilibrium. For the ionization of $HC_2H_3O_2$, we write the chemical equilibrium as follows.

$$HC_2H_3O_2(aq) + H_2O \rightleftharpoons H_3O^+(aq) + C_2H_3O_2^-(aq)$$

The reaction read from left to right is the *forward reaction,* and the one read in the opposite direction is the *reverse reaction.*

In discussing an equilibrium such as this, we will often talk about the *extent of completion* of a forward or reverse reaction. To call acetic acid a *weak* acid, for example, is just another way of saying that the forward reaction in this equilibrium is far from completion.

Water itself is a very weak electrolyte. When two H_2O molecules collide, there is a small chance that an H^+ will be transferred from one molecule to another. But this is just a forward reaction, which we can represent as follows.

$$H_2O + H_2O \longrightarrow H_3O^+(aq) + OH^-(aq)$$

Notice that it produces hydronium ions and hydroxide ions in equal numbers, so the solution is neither acidic nor basic.

The reverse of this reaction has a very strong tendency to occur any time H_3O^+ and OH^- encounter each other. You have already learned that when an acid neutralizes a base, hydronium ions and hydroxide ions react to form water. Whenever a H_3O^+ ion encounters a OH^- ion, a H^+ is transferred to the OH^- and two water molecules are created once again. These two opposing reactions give the following equilibrium.

$$H_2O + H_2O \rightleftharpoons H_3O^+(aq) + OH^-(aq)$$

The extent to which the forward reaction proceeds toward completion is very small. In pure water, only about $2 \times 10^{-7}\%$ of the water is present in ionized form. Nevertheless, we will see in Chapter 15 that it is a very important equilibrium to consider in all solutions of acids and bases.

For a weak electrolyte, only a very small percentage of the solute is actually ionized at any instant when equilibrium is reached. The reason is that the reverse reaction has a high probability of occurring when a pair of the ions meet. This causes the reverse reaction to occur rapidly, which thereby prevents the ion concentrations from becoming too high. For a strong electrolyte, however, there is a high probability that the molecules will react to form ions, but an extremely low probability that the ions will react when they meet. In a solution of HCl, for example, the Cl^- ion has very little tendency to acquire an H^+ from an H_3O^+ when these ions bump into each other. As a result, the reverse reaction has virtually no tendency to occur, so in a very brief time all of the HCl is converted to ions—it becomes 100% ionized.

Since the reverse reaction can't be detected in a solution of a strong electrolyte, we omit the reverse arrow in the chemical equation—in other words, we don't write the equation for the dissociation of a strong electrolyte as an equilibrium.

Only those molecular compounds that are acids or bases ionize in water. Other molecular compounds are nonelectrolytes.

PRACTICE EXERCISE 2

formic acid, $HCHO_2$

Formic acid, $HCHO_2$, is the substance that is responsible for the painful bites of fire ants. It is a weak electrolyte and reacts with water in the same manner as acetic acid. Write a chemical equation that illustrates the equilibrium for this reaction.

PRACTICE EXERCISE 3

Write an equation that represents the ionization of ammonia in water as an equilibrium.

11.3 IONIC REACTIONS IN AQUEOUS SOLUTIONS

When solutions of electrolytes react, it is usually the ions themselves that are involved in the reaction.

Compounds that are electrolytes undergo many reactions with each other, and in virtually every case the reactions take place in aqueous solutions. The reason solutions are used to carry out reactions was mentioned in Chapter 4—in a solution the particles that make up the solute are dispersed, so they can intermingle freely. Now we've learned that ionic compounds and certain molecular compounds exist in aqueous solutions in the form of separated ions. When solutions of such solutes react, it is the ions themselves that participate in the reaction. It's no surprise, then, that we call these changes ionic reactions.

There are many commonplace examples of ionic reactions. For instance, in many communities the water contains dissolved ions that interfere with the action of soap. It is called "hard water" because these ions—usually Ca^{2+}, Mg^{2+}, Fe^{2+}, and Fe^{3+}—react with ions in the soap to form a scum, as discussed in Special Topic 11.1, so it is "hard" to get a lather using this water. Alleviating the problem is called softening the water, and homemakers sometimes add washing soda, which is $Na_2CO_3 \cdot 10H_2O$, to accomplish the job. The washing soda reacts with the "hardness ions" and thereby prevents them from interfering with the action of the soap. Softening the water is also a reaction between ionic compounds in aqueous solution. And in the last section we noted that the neutral-

Precipitation reactions occur around us all the time and we hardly ever take notice — until they cause a problem. One common problem is caused by **hard water,** groundwater that contains the "hardness ions," Ca^{2+}, Mg^{2+}, Fe^{2+}, or Fe^{3+}, in concentrations high enough to form precipitates with ordinary soap. Soap normally consists of the sodium salts of long-chain organic acids derived from animal fats or oils. An example in sodium stearate, $NaC_{17}H_{35}CO_2$. The negative ion of the soap forms an insoluble "scum" with hardness ions, which reduces the effectiveness of the soap for removing dirt and grease.

Hardness ions can be removed from water in a number of ways. One way is to add hydrated sodium carbonate, $Na_2CO_3 \cdot 10H_2O$, often called washing soda, to the water. The carbonate ion forms insoluble precipitates with the hardness ions; an example is $CaCO_3$.

$$Ca^{2+}(aq) + CO_3^{2-}(aq) \longrightarrow CaCO_3(s)$$

Once precipitated, the hardness ions are not available to interfere with the soap.

Another problem with hard water that contains bicarbonate ion is the precipitation of insoluble carbonates on the inner walls of hot water pipes. When heated, solutions containing bicarbonate ion lose CO_2.

$$2HCO_3^-(aq) \longrightarrow H_2O + CO_2(g) + CO_3^{2-}(aq)$$

Gases become less soluble as the temperaure is raised. As CO_2 is driven from the solution, the HCO_3^- is

Boiler scale on the inside of this water pipe has nearly sealed it shut.

gradually converted to CO_3^{2-}, which is able to precipitate the hardness ions. This precipitate, which sticks to the inner walls of pipes and hot water boilers, is called boiler scale. In locations that have high concentrations of Ca^{2+} and HCO_3^- in the water supply, boiler scale is a very serious problem, as illustrated in the accompanying photograph.

---

ization reaction of an acid and a base involves the reaction of hydronium ions and hydroxide ions in water.

Equations for Ionic Reactions

In writing a chemical equation to describe an ionic reaction, we have several options. Consider the reaction that takes place, for example, when 100 mL of a 1.0 M solution of cadmium nitrate, $Cd(NO_3)_2$, is added to 100 mL of a 1.0 M solution of sodium sulfide, Na_2S. As shown in Figure 11.1, a bright orange-yellow solid precipitates as the two solutions are mixed. Analysis of the solid would reveal that its composition is CdS, and if we filter the solid from the mixture (Figure 11.2) and evaporate the filtrate (the solution

Recall that a precipitate is a solid that forms in a solution.

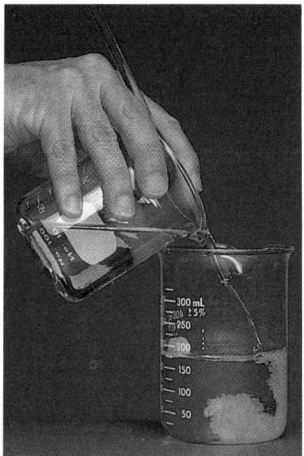 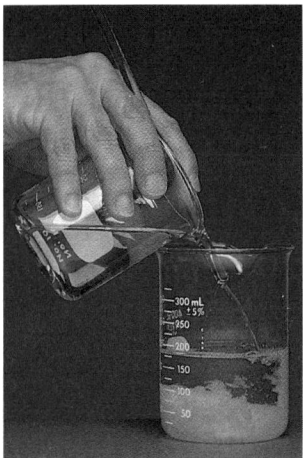 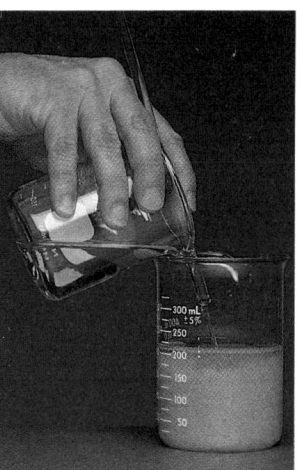

FIGURE 11.1 The addition of 1.0 M $Cd(NO_3)_2$ solution to a 1.0 M Na_2S solution gives a bright orange-yellow precipitate of cadmium sulfide, CdS.

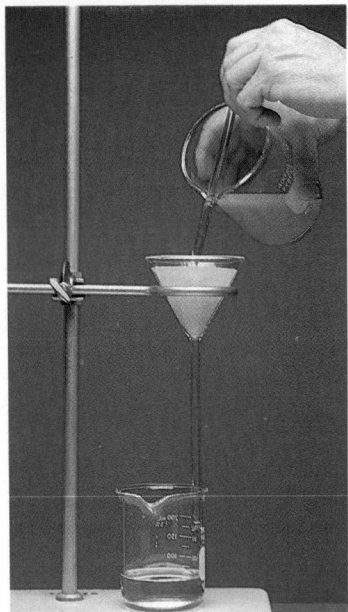

FIGURE 11.2 Filtration is a procedure that separates a solid precipitate from a solution.

that passes through the filter paper), crystals of sodium nitrate are left behind. From this information, we can write the chemical equation

$$Cd(NO_3)_2(aq) + Na_2S(aq) \longrightarrow CdS(s) + 2NaNO_3(aq)$$

This kind of reaction, in which ions change partners, is called metathesis (pronounced *me-TATH-e-sis*). Another name that's sometimes used is double replacement. In this example, cadmium ions replace sodium ions in one compound and sodium ions replace cadmium ions in the other.

The equation we have just written is often called a molecular equation because the formulas of the reactants and products are written as if they were molecules, that is, no indication has been made that the soluble compounds are dissociated into ions. (Thus we say "molecular" equation even though none of these substances consists of molecules. All consist of ions. "Molecular" in this context simply means that the *complete formulas of all reactants and products are shown.*)

Ionic Equations

A more accurate description of the reaction as it actually takes place is provided by an ionic equation. In an ionic equation the formulas of any strong electrolytes are written in dissociated form to show that the solute exists in the solution in the form of separated ions. On the other hand, the formulas of solids and weak electrolytes are written in molecular form. For a solid, this shows that the ions are not free to move away from each other, but instead are forced to stay together within the crystals of the solid. For a weak electrolyte, writing the molecular formula shows that this is the predominant form of the solute in the solution; most of the solute exists as molecules and only a very small fraction is present in the form of ions. The ionic equation for the reaction that we've been discussing is

$$Cd^{2+}(aq) + 2NO_3^-(aq) + 2Na^+(aq) + S^{2-}(aq) \longrightarrow CdS(s) + 2Na^+(aq) + 2NO_3^-(aq)$$

To obtain the ionic equation we have divided each of the soluble, ionic compounds into the ions that are released when the salts dissolve, but we have left intact the formula of the solid.

Figure 11.3 illustrates what happens when the $Cd(NO_3)_2$ and Na_2S solutions are

FIGURE 11.3 The reaction of $Cd(NO_3)_2$ with Na_2S. (*a*) These ionic compounds exist as separated ions in their aqueous solutions. (*b*) When the two solutions are combined, the cadmium ions, Cd^{2+}, and sulfide ions, S^{2-}, collect together and form an insoluble precipitate of cadmium sulfide, CdS. However, the sodium ions, Na^+, and the nitrate ions, NO_3^-, remain free.

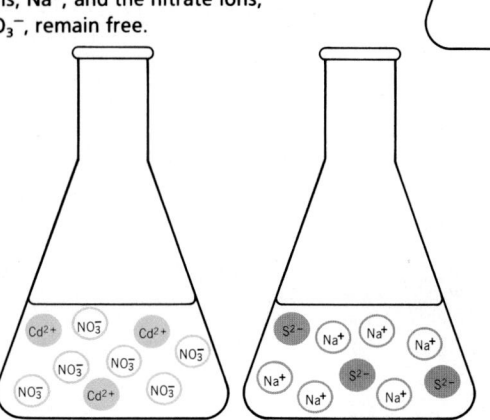

(a)

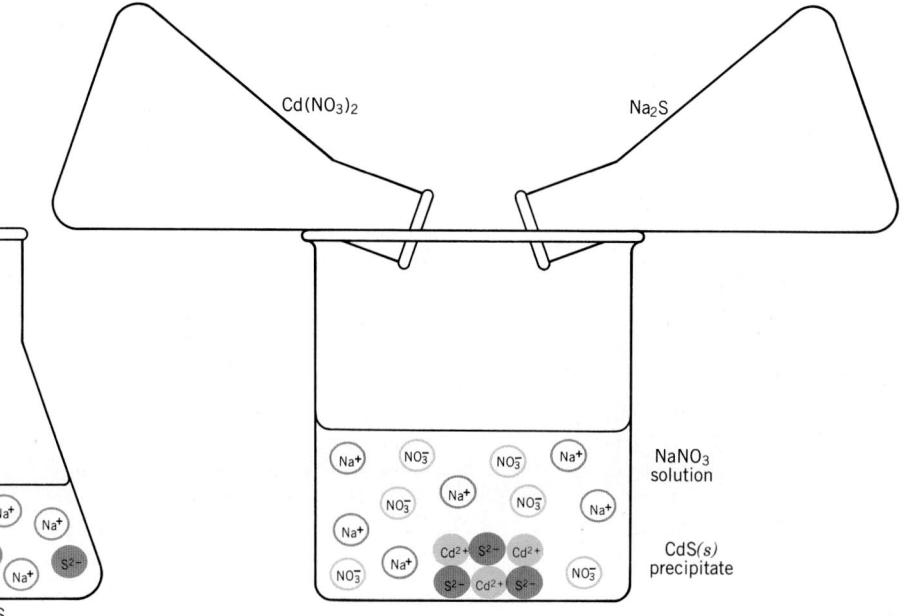

(b)

mixed, and the ionic equation lets us appreciate why the reaction takes place. Cadmium sulfide has a very low solubility in water. If some is placed in a beaker full of water, so little dissolves that nothing appears to happen at all. A trace does dissolve, but when the solution becomes saturated, the concentrations of Cd^{2+} and S^{2-} are very, very small (about 3×10^{-14} mol/L). When we combine 1 M solutions of $Cd(NO_3)_2$ and Na_2S, therefore, the concentrations of cadmium ion and sulfide ion are initially much, much higher than required to give a saturated solution of CdS. If nothing happened after the solutions were mixed, the resulting solution would be supersaturated in CdS. But a supersaturated solution is an unstable system, so cadmium sulfide precipitates, and it continues to do so until the solution reaches its true saturation level.

An ionic equation, like any other chemical equation, should be balanced, and this one is. It satisfies, in fact, the *two* criteria for a balanced ionic equation.

If mixing two solutions produces ion concentrations larger than would be found in a saturated solution of a particular salt, then that salt will tend to precipitate.

---

1. Material balance.   All atoms on one side of the arrow must appear somewhere on the other side.

2. Electrical balance.   The net charge on the left equals the net charge on the right. (The overall electrical neutrality of the solution requires an electrical balance.)

Net Ionic Equations

Notice in the ionic equation above that two of the ions appear in exactly the same way on both sides of the arrow. They are the nitrate ion, $NO_3^-(aq)$, and the sodium ion, $Na^+(aq)$. What this tells us is that during the reaction these ions are left unaffected—they are present before the reaction starts and they are present after the reaction is complete. They don't actually participate in the reaction at all. The real reaction is one between $Cd^{2+}(aq)$ and $S^{2-}(aq)$ to form the solid, CdS(s). Ions that do not participate in a reaction, but are present in the reaction mixture, are frequently called spectator ions—in a sense they must "stand around and watch the reaction take place" without getting involved themselves.

A third way to describe an ionic reaction is to ignore the spectator ions and focus attention on the species that are actually involved. For the reaction we have been studying, if we cross out the spectator ions,

$$Cd^{2+}(aq) + \cancel{2NO_3^-(aq)} + \cancel{2Na^+(aq)} + S^{2-}(aq) \longrightarrow CdS(s) + \cancel{2Na^+(aq)} + \cancel{2NO_3^-(aq)}$$

the equation that remains is

$$Cd^{2+}(aq) + S^{2-}(aq) \longrightarrow CdS(s)$$

An ionic equation from which we have removed the spectator ions is called a net ionic equation. Its real usefulness is that it allows us to *generalize;* that is, it allows us to extend what we have learned about one reaction to other similar reactions. In this case, it calls to our attention the fact that any soluble compound that provides Cd^{2+} will react in solution with any soluble compound that provides S^{2-} to form a precipitate of cadmium sulfide. This is very helpful; like any generalization, it greatly reduces the amount of facts about the chemical properties of individual compounds that have to be learned. If we have a solution containing, for example, $CdCl_2$ and another containing K_2S, we now can confidently know that a precipitate of CdS will form if the solutions are mixed. This is because the solution of $CdCl_2$ contains Cd^{2+} ions and the solution of K_2S provides S^{2-} ions, and these ions—regardless of their origins—always form an insoluble precipitate of CdS in water.

$CdCl_2$ and K_2S are soluble salts.

EXAMPLE 11.1   WRITING EQUATIONS FOR IONIC REACTIONS

Problem: When solutions of sodium sulfate, Na_2SO_4, and barium nitrate, $Ba(NO_3)_2$, are mixed, a precipitate of barium sulfate is formed. The other product is sodium nitrate, which is soluble in water. Write molecular, ionic, and net ionic equations for the reaction.

Solution: We begin by constructing the molecular equation using entire "molecular" formulas for the reactants and products. Remember that at this stage we do not attempt to balance it — writing an equation and balancing it should always be approached as a two-step process.

$$Na_2SO_4(aq) + Ba(NO_3)_2(aq) \longrightarrow BaSO_4(s) + NaNO_3(aq)$$

Notice that we've used aq and s to indicate which are soluble and which are not. Now we balance the equation. This is easily done here by placing a coefficient of 2 in front of $NaNO_3$.

$$Na_2SO_4(aq) + Ba(NO_3)_2(aq) \longrightarrow BaSO_4(s) + 2NaNO_3(aq)$$

This is the complete, balanced molecular equation. To construct the ionic equation from it, we have to write any soluble ionic compounds in dissociated form. The formula of the solid is left in molecular form. This gives

Notice that *one* $Ba(NO_3)_2$ gives *one* Ba^{2+} and *two* NO_3^-. Also notice that *two* $NaNO_3$ give *two* Na^+ and *two* NO_3^-.

$$2Na^+(aq) + SO_4^{2-}(aq) + Ba^{2+}(aq) + 2NO_3^-(aq) \longrightarrow BaSO_4(s) + 2Na^+(aq) + 2NO_3^-(aq)$$

Next, we obtain the net ionic equation by crossing out any ion that appears the same on both sides of the ionic equation. Thus,

$$\cancel{2Na^+}(aq) + SO_4^{2-}(aq) + Ba^{2+}(aq) + \cancel{2NO_3^-}(aq) \longrightarrow BaSO_4(s) + \cancel{2Na^+}(aq) + \cancel{2NO_3^-}(aq)$$

gives

$$SO_4^{2-}(aq) + Ba^{2+}(aq) \longrightarrow BaSO_4(s)$$

This is the net ionic equation. Notice that it satisfies both criteria for a balanced equation. It is balanced both materially and electrically.

PRACTICE EXERCISE 4   Milk of magnesia is a suspension of solid magnesium hydroxide, $Mg(OH)_2$, in water. This solid can be made by adding a solution of sodium hydroxide, NaOH, to a solution of magnesium chloride, $MgCl_2$, which causes $Mg(OH)_2$ to precipitate and leaves sodium chloride in solution. Write molecular, ionic, and net ionic equations for this reaction.

Perhaps at this point you are wondering which is *the* correct way to write the equation for a metathesis reaction. The answer is that no one of these is more correct than another, nor is one better than another. Each has its own usefulness. For example, we need the *molecular equation* for planning an experiment, because it identifies the actual reactants and lets us work out the amounts to use. We need the *ionic equation* to really see what's happening in the solution when the reaction is taking place. And we need the *net ionic equation* when we want to generalize the reaction — to appreciate that more than one set of reactants can lead to the same net reaction. The net ionic equation focuses our attention most sharply on the chemical change that is actually taking place in a solution when ionic compounds react.

11.4   PREDICTING WHEN METATHESIS REACTIONS WILL OCCUR

If a precipitate, a weak electrolyte, or a gas can form, a metathesis reaction can take place.

In the preceding section we saw that a net reaction occurs between salts when a precipitate forms. However, the formation of a precipitate is only one of several "driving forces" for change in metathesis reactions. Two others are the formation of a weak electrolyte and the formation of a gas. In this section we will learn how to use these to predict the outcome of metathesis reactions.

Reactions in Which a Precipitate Is Formed

If we had a lot of information about the solubilities of compounds, we could predict whether or not a precipitation reaction would take place when solutions of various compounds are mixed. For example, knowing that calcium carbonate is insoluble allows us to predict that if a solution is prepared that contains both Ca^{2+} and CO_3^{2-} ions, a precipitate of $CaCO_3$ will be formed. And if we knew that $CaCl_2$, Na_2CO_3, and $NaCl$ are all soluble, then we would have all the information that we need to predict the reaction between calcium chloride and sodium carbonate.

Actually, the terms *soluble* and *insoluble* do not have precise, quantitative meanings, because there is no sharp dividing line between compounds that are soluble and those that are not. Normally, if a precipitate of a particular compound is formed even when dilute solutions of its constituent ions are mixed, then that compound is considered to be insoluble. For now, we can give you some general solubility rules that apply nearly all the time, and which you can use to predict the outcome of metathesis reactions. We will return to the question of solubility again in Chapter 14, at which time we will examine the topic in more quantitative terms.

Soil scientists consider salts that are less soluble than $CaSO_4 \cdot 2H_2O$ (gypsum) as "insoluble," and others as soluble.

Solubility Rules

Soluble Compounds

1. All compounds of the alkali metals (Group IA) are soluble.
2. All salts containing NH_4^+, NO_3^-, ClO_4^-, ClO_3^-, and $C_2H_3O_2^-$ are soluble.
3. All chlorides (Cl^-), bromides (Br^-), and iodides (I^-) are soluble, *except* those of Ag^+, Pb^{2+}, and Hg_2^{2+} (note the subscript "2").
4. All sulfates are soluble, *except* those of Pb^{2+}, Ca^{2+}, Sr^{2+}, Hg_2^{2+}, and Ba^{2+}.

Insoluble Compounds

5. All hydroxides (OH^-) and metal oxides (containing O^{2-}) are insoluble, *except* those of Group IA and Ca^{2+}, Sr^{2+}, and Ba^{2+}. When metal oxides do dissolve, they give hydroxides (their solutions do not contain O^{2-} ions). For example,

$$Na_2O(s) + H_2O \longrightarrow 2NaOH(aq)$$

6. All compounds that contain PO_4^{3-}, CO_3^{2-}, SO_3^{2-}, and S^{2-} are insoluble, *except* those of Group IA and NH_4^+.

"Insoluble" actually means "very sparingly soluble"—only trace amounts dissolve. Nothing is 100% insoluble in water.

Before we look at how these rules are used to predict metathesis reactions in Examples 11.2 and 11.3, let's see what they tell us. Rule 1, for example, states that all compounds of the alkali metals are soluble. This means that you can expect that all salts containing Na^+ or K^+, or any of the Group IA metal ions, regardless of the anion, are soluble. If one of the reactants in a metathesis were Na_3PO_4, you would know that it is soluble, and therefore you would write it in dissociated form when you write the ionic equation. Similarly, Rule 6 states, in part, that all sulfite salts (SO_3^{2-}) are *insoluble* except those of the alkali metals and ammonium ion. If one of the products in a metathesis reaction were $FeSO_3$, you would expect it to be insoluble, and therefore you would write its formula in undissociated form when you construct the ionic equation.

EXAMPLE 11.2   PREDICTING METATHESIS REACTIONS

Problem: Write molecular, ionic, and net ionic equations for the reaction between $Pb(NO_3)_2$ and $Fe_2(SO_4)_3$.

Solution: We begin with the molecular equation. Our reactants are

Write equations in two steps:

$$Pb(NO_3)_2 + Fe_2(SO_4)_3$$

1. Write correct formulas.
2. Balance the equation.

To write the products, we interchange NO_3^- and SO_4^{2-}. However, we must be very careful to write the correct formulas of the products.[1] From the formulas of the reactants, we know that the positive ions here are Pb^{2+} and Fe^{3+}. The correct formulas of the products, then, are $PbSO_4$ and $Fe(NO_3)_3$.

At this point, the unbalanced equation is

$$Pb(NO_3)_2 + Fe_2(SO_4)_3 \longrightarrow PbSO_4 + Fe(NO_3)_3$$

When it is balanced, we obtain the *molecular equation.*

$$3Pb(NO_3)_2 + Fe_2(SO_4)_3 \longrightarrow 3PbSO_4 + 2Fe(NO_3)_3$$

Next, we expand this to give the ionic equation in which soluble compounds are written in dissociated (separated) form as ions, and insoluble compounds are written in "molecular" form. To do this, we must examine the formulas using our solubility rules to determine which compounds are soluble.

$Pb(NO_3)_2$	soluble	(rule 2)
$Fe_2(SO_4)_3$	soluble	(rule 4)
$PbSO_4$	insoluble	(rule 4)
$Fe(NO_3)_3$	soluble	(rule 2)

Now we can write the ionic equation,

$$3Pb^{2+}(aq) + 6NO_3^-(aq) + 2Fe^{3+}(aq) + 3SO_4^{2-}(aq) \longrightarrow 3PbSO_4(s) + 2Fe^{3+}(aq) + 6NO_3^-(aq)$$

The net ionic equation is obtained by canceling spectator ions. This leaves

$$3Pb^{2+}(aq) + 3SO_4^{2-}(aq) \longrightarrow 3PbSO_4(s)$$

Finally, we can reduce the coefficients to

$$Pb^{2+}(aq) + SO_4^{2-}(aq) \longrightarrow PbSO_4(s)$$

EXAMPLE 11.3   PREDICTING METATHESIS REACTIONS

Problem: What reaction (if any) occurs between KNO_3 and NH_4Cl?

Solution: First we write the molecular equation.

$$KNO_3 + NH_4Cl \longrightarrow KCl + NH_4NO_3$$

Next we determine solubilities. The rules tell us that all four of these compounds are soluble. When we expand this, we obtain the ionic equation.

$$K^+(aq) + NO_3^-(aq) + NH_4^+(aq) + Cl^-(aq) \longrightarrow K^+(aq) + Cl^-(aq) + NH_4^+(aq) + NO_3^-(aq)$$

If we cancel spectator ions, nothing is left! This tells us that there is *no* net reaction. If we mix KNO_3 and NH_4Cl solutions, all we obtain is a mixture of the ions. No reaction occurs.

PRACTICE EXERCISE 5   Predict what reaction (if any) would occur upon mixing solutions of (a) $AgNO_3$ and NH_4Cl, (b) Na_2S and $Pb(C_2H_3O_2)_2$, and (c) $BaCl_2$ and NH_4NO_3.

---

[1] Some students might be tempted (without thinking) to write $Pb(SO_4)_2$ and $Fe_2(NO_3)_3$, or even $Pb(SO_4)_3$ and $Fe_2(NO_3)_2$. This is a common error. Always be careful to figure out the charges of the ions that must be combined in the formula. Then take the ions in a ratio that gives a neutral formula unit.

Reactions in Which a Weak Electrolyte Is Formed

When we can predict an event that removes ions from solution, as we have just seen, we can predict a chemical reaction. It is not always necessary that an insoluble compound form, however, to have a reaction that takes ions out of solution. The formation of a soluble compound will serve just as well, provided that it is a sufficiently weak electrolyte—like water or some very weak acid. Let us look first at the formation of water as a "driving force" for chemical reactions.

In Section 2.12 we learned that in any acid–base neutralization in aqueous solution, the products are a salt and water. If we analyze such a reaction in terms of its net ionic equation, we find that the reaction is actually one between hydronium ion and hydroxide ion. Consider, for example, the reaction of hydrochloric acid with sodium hydroxide. If we write its equation in the following way, it certainly has the appearance of a metathesis reaction although no precipitate forms.

$$HCl(aq) + NaOH(aq) \longrightarrow HOH + NaCl(aq)$$

HOH is water, of course.

Let us now convert this into the net ionic equation. The first step, recall, is to divide the strong electrolytes into the ions they provide in the solution. We must keep in mind that water is a very weak electrolyte, so we have to leave its formula intact. For simplicity, we also can begin to use H^+ for the hydrogen ion in water. The ionic equation is therefore

$$H^+(aq) + Cl^-(aq) + Na^+(aq) + OH^-(aq) \longrightarrow H_2O + Na^+(aq) + Cl^-(aq)$$

We realize, of course, that the more accurate description of the reaction uses H_3O^+ instead of H^+ to represent the hydrogen ion. In this form, the ionic equation becomes

$$H_3O^+(aq) + Cl^-(aq) + Na^+(aq) + OH^-(aq) \longrightarrow 2H_2O + Na^+(aq) + Cl^-(aq)$$

If we now cancel the spectator ions, we obtain

$$H^+(aq) + OH^-(aq) \longrightarrow H_2O$$

or

$$H_3O^+(aq) + OH^-(aq) \longrightarrow 2H_2O$$

These are the two ways to write the net ionic equation for the neutralization of any strong acid by any strong base in aqueous solution. We see that the neutralization reaction occurs because the formation of a weak electrolyte—water—removes ions from the solution. Water is such a weak electrolyte that its formation is a very powerful driving force for chemical reactions. We will now look at three other general situations that illustrate this important fact.

Reactions in which water can be formed tend to be driven to completion, *even when one of the reactants is insoluble.* For example, insoluble hydroxides readily neutralize and dissolve in acids because the neutralization reaction forms water as one of the products. This is illustrated by the reaction of HCl, the acid in your stomach, with insoluble $Mg(OH)_2$, the white solid in the milky aqueous slurry called milk of magnesia, commonly sold as a nonprescription stomach antacid.

Molecular Equation

$$Mg(OH)_2(s) + 2HCl(aq) \longrightarrow MgCl_2(aq) + 2H_2O$$

Ionic Equation

$$Mg(OH)_2(s) + 2H^+(aq) + 2Cl^-(aq) \longrightarrow Mg^{2+}(aq) + 2Cl^-(aq) + 2H_2O$$

Milk of magnesia is a home remedy for acid indigestion.

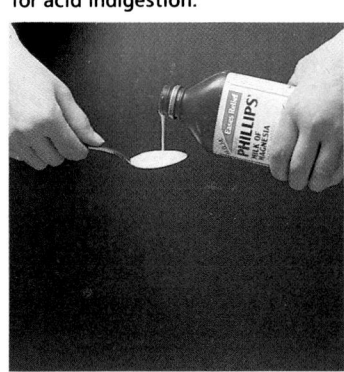

Net Ionic Equation

$$Mg(OH)_2(s) + 2H^+(aq) \longrightarrow Mg^{2+}(aq) + 2H_2O$$

These equations show how milk of magnesia acts to neutralize stomach acid. If you used a strong, *soluble* base, like NaOH, to neutralize excess stomach acid, you would permanently scar the tissues in your mouth and esophagus before the base got to the stomach. (This could even kill you.) Hydroxide ions *in solution* attack tissue vigorously because in solution they are free to move and make such attacks. But the hydroxide ions in the water-insoluble $Mg(OH)_2$ of milk of magnesia are not free to move like this, so they get to the stomach without attacking anything on the way. In the stomach, hydrogen ions of stomach acid *move to them* where they wait in the microcrystals of $Mg(OH)_2$. (Of course, Mg^{2+} ions are also released from the solid state, and if you over-use milk of magnesia you can load your body with harmful levels of magnesium ion.)

Insoluble metal oxides also react with acids, because an oxide ion in the solid can combine with two H^+ ions to give water. For instance, rust, which is an iron oxide with the formula Fe_2O_3, can be removed from steel by treating the metal with hydrochloric acid, HCl. In fact, this is one of the more important uses of this acid in industry. The equation for the reaction can be represented as

Removing rust from steel by dissolving it in acid is called pickling *the steel.*

Molecular Equation

$$Fe_2O_3(s) + 6HCl(aq) \longrightarrow 2FeCl_3(aq) + 3H_2O$$

Ionic Equation

$$Fe_2O_3(s) + 6H^+(aq) + 6Cl^-(aq) \longrightarrow 2Fe^{3+}(aq) + 6Cl^-(aq) + 3H_2O$$

Net Ionic Equation

$$Fe_2O_3(s) + 6H^+(aq) \longrightarrow 2Fe^{3+}(aq) + 3H_2O$$

Another example of how the formation of water is a driving force for a chemical reaction involves weak acids, like acetic acid ($HC_2H_3O_2$) or carbonic acid (H_2CO_3). Although only small percentages of the molecules of weak acids ionize in solution to give H^+ ions (and some anion), virtually 100% of weak acid molecules will give up H^+ ions to OH^- when a metal hydroxide is added. Acetic acid, $HC_2H_3O_2$, for example, reacts quantitatively (completely) with sodium hydroxide as follows.

Molecular Equation

$$HC_2H_3O_2(aq) + NaOH(aq) \longrightarrow NaC_2H_3O_2(aq) + H_2O$$

Ionic Equation

$$HC_2H_3O_2(aq) + Na^+(aq) + OH^-(aq) \longrightarrow Na^+(aq) + C_2H_3O_2^-(aq) + H_2O$$

Net Ionic Equation

$$HC_2H_3O_2(aq) + OH^-(aq) \longrightarrow C_2H_3O_2^-(aq) + H_2O$$

PRACTICE EXERCISE 6   Write molecular, ionic, and net ionic equations for the following reactions.
(a) $Ni(OH)_2 + HClO_4$
(b) $Al_2O_3 + H_2SO_4$ (assume complete neutralization of the sulfuric acid)
(c) $H_2CO_3 + KOH$ (assume complete neutralization of the carbonic acid)

Neutralization reactions that form water are not the only ones that are driven by the formation of a weak electrolyte. For example, if we add hydrochloric acid to a solution containing sodium acetate, $NaC_2H_3O_2$, the following reaction occurs.

$$NaC_2H_3O_2(aq) + HCl(aq) \longrightarrow NaCl(aq) + HC_2H_3O_2(aq)$$

Once again, let us analyze it by writing a net ionic equation. We begin by constructing the ionic equation. $NaC_2H_3O_2$, HCl, and NaCl are all strong electrolytes, so we write them in dissociated forms. However, $HC_2H_3O_2$ is a weak acid, and therefore it is a weak electrolyte. This means that we must use its molecular formula. The ionic equation, then, is

$$Na^+(aq) + C_2H_3O_2^-(aq) + H^+(aq) + Cl^-(aq) \longrightarrow Na^+(aq) + Cl^-(aq) + HC_2H_3O_2(aq)$$

Next, we cancel the spectator ions, which gives the net ionic equation.

$$C_2H_3O_2^-(aq) + H^+(aq) \longrightarrow HC_2H_3O_2(aq)$$

Thus, we see that a net reaction occurs because the formation of a weak electrolyte — this time, a weak acid — removes ions from the solution.

PRACTICE EXERCISE 7   Write the molecular, ionic, and net ionic equations for the reaction, if any, between hydrochloric acid and sodium fluoride, NaF. (If a metathesis reaction occurs, the formulas of the products would be NaCl and HF.)

PRACTICE EXERCISE 8   Acid–base neutralizations can produce precipitates, too. Write molecular, ionic, and net ionic equations for the reaction between $Ba(OH)_2$ and H_2SO_4.

Reactions in Which a Gas Is Formed

Sometimes a product of a metathesis reaction is a substance that normally is a gas at room temperature. An example is hydrogen sulfide, H_2S, the compound that gives rotten eggs their foul odor. It is a weak electrolyte and is formed when a strong acid such as HCl is added to a metal sulfide such as sodium sulfide, Na_2S. Hydrogen sulfide has a low solubility in water, so it tends to bubble out of the solution as it forms and escape as a gas. Once it has escaped, there is no possible way for it to participate in any sort of reverse reaction, so its departure drives the reaction to completion. The molecular, ionic, and net ionic equations for the reaction are

Molecular Equation

$$2HCl(aq) + Na_2S(aq) \longrightarrow 2NaCl(aq) + H_2S(g)$$

Ionic Equation

$$2H^+(aq) + 2Cl^-(aq) + 2Na^+(aq) + S^{2-}(aq) \longrightarrow 2Na^+(aq) + 2Cl^-(aq) + H_2S(g)$$

Net Ionic Equation

$$2H^+(aq) + S^{2-}(aq) \longrightarrow H_2S(g)$$

In Section 2.12 it was mentioned that carbonic acid and sulfurous acid are too unstable to be isolated in pure form. Because of this instability, when they form in appreciable amounts as the products of metathesis reactions, they decompose into their respective anhydrides (CO_2 and SO_2) and most of these molecules leave the solutions as gases.

$$H_2CO_3(aq) \longrightarrow H_2O + CO_2(g)$$
$$H_2SO_3(aq) \longrightarrow H_2O + SO_2(g)$$

NaHCO₃ is soluble in water (all
Na⁺ salts are soluble).

FIGURE 11.4 The reaction of
sodium bicarbonate with hydrochloric acid. The bubbles contain the
gas carbon dioxide.

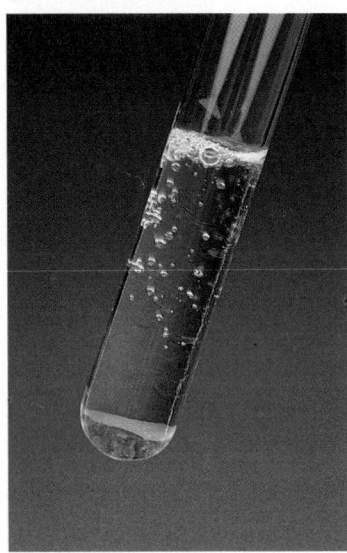

The carbonate ion is the acid
neutralizer in the Na₂CO₃ being
pumped by a snow-blower onto the
acid spill pictured in the chapter-
opening photograph of this chapter.

As an example, let's study the reaction between hydrochloric acid and sodium bicarbonate ($NaHCO_3$), which is illustrated in Figure 11.4. As the sodium bicarbonate solution is added to the hydrochloric acid solution, bubbles of the gas carbon dioxide are released. This is the same reaction that occurs when you take sodium bicarbonate to soothe an upset stomach. Stomach acid is HCl and its reaction with the $NaHCO_3$ produces CO_2 (burp!).

In this reaction, hydronium ions from the HCl solution react with bicarbonate ions to give molecules of the weak electrolyte carbonic acid, H_2CO_3. The equation for the reaction is

$$H_3O^+(aq) + HCO_3^-(aq) \longrightarrow H_2CO_3(aq) + H_2O$$

However, as we've said, carbonic acid molecules are not very stable and they easily and quickly decompose into carbon dioxide and water. This is what happens as the H_2CO_3 concentration begins to build up, and soon CO_2 bubbles from the solution.

$$H_2CO_3(aq) \longrightarrow H_2O + CO_2(g)$$

The net overall change that occurs is therefore

$$H_3O^+(aq) + HCO_3^-(aq) \longrightarrow 2H_2O + CO_2(g)$$

or, using H^+ as an abbreviation for H_3O^+,

$$H^+(aq) + HCO_3^-(aq) \longrightarrow H_2O + CO_2(g)$$

Carbonic and sulfurous acids represent just two compounds that decompose into gaseous products when they are formed in metathesis reactions. A more complete list, showing the decomposition reactions, is given in Table 11.1. You should learn these reactions so that you can use them in writing ionic and net ionic equations.

TABLE 11.1 Gases Formed in Metathesis Reactions

Compound	How It Is Formed
H_2S	Net reaction: $2H^+ + S^{2-} \longrightarrow H_2S(g)$
CO_2	$2H^+ + CO_3^{2-} \longrightarrow H_2CO_3$ or $H^+ + HCO_3^- \longrightarrow H_2CO_3$ Then: $H_2CO_3 \longrightarrow H_2O + CO_2(g)$ Net reactions: $2H^+ + CO_3^{2-} \longrightarrow H_2O + CO_2(g)$ $H^+ + HCO_3^- \longrightarrow H_2O + CO_2(g)$
SO_2	$2H^+ + SO_3^{2-} \longrightarrow H_2SO_3$ or $H^+ + HSO_3^- \longrightarrow H_2SO_3$ Then: $H_2SO_3 \longrightarrow H_2O + SO_2(g)$ Net reactions: $2H^+ + SO_3^{2-} \longrightarrow H_2O + SO_2(g)$ $H^+ + HSO_3^- \longrightarrow H_2O + SO_2(g)$
NH_3	$NH_4^+ + OH^- \longrightarrow NH_3(g) + H_2O$ (In writing a metathesis reaction, you may find your-self writing "NH_4OH." This is really a solution of NH_3 in water. NH_4OH doesn't actually exist.)

EXAMPLE 11.4 PREDICTING METATHESIS REACTIONS

Problem: Write molecular, ionic, and net ionic equations for the reaction of $NaHSO_3$ with HCl.

Solution: We begin by writing the molecular equation. Taking it to be a typical double replacement reaction, we interchange cations between the two compounds. This gives

$$NaHSO_3 + HCl \longrightarrow NaCl + H_2SO_3$$

But we now know that H_2SO_3 decomposes into H_2O and $SO_2(g)$, so we rewrite the equation as

$$NaHSO_3 + HCl \longrightarrow NaCl + H_2O + SO_2(g)$$

This is the molecular equation. To obtain the ionic equation we divide the electrolytes into their ions. This gives

$$Na^+(aq) + HSO_3^-(aq) + H^+(aq) + Cl^-(aq) \longrightarrow Na^+(aq) + Cl^-(aq) + H_2O + SO_2(g)$$

Finally, we cancel spectator ions to get the net ionic equation.

$$HSO_3^-(aq) + H^+(aq) \longrightarrow H_2O + SO_2(g)$$

PRACTICE EXERCISE 9 Write molecular, ionic, and net ionic equations for the reaction of $(NH_4)_2SO_4$ with (a) NaOH and (b) $Ba(OH)_2$.

Writing Molecular Equations from Net Ionic Equations

Once we know the factors that control whether or not a metathesis reaction will occur, we can work backward to decide which chemicals to work with in the laboratory to get the reaction to take place. Usually, this requires a more thorough knowledge of solubility rules and the reactions that produce gases than does writing a net ionic equation from a molecular equation. This is because there are often many alternate choices, so we have to sort through various possibilities to pick a reasonable one. Since more than one reaction will usually work, a question that asks you to write a molecular equation from a net ionic equation generally has more than one answer. This can be best appreciated by an example.

EXAMPLE 11.5 WRITING MOLECULAR EQUATIONS FROM NET IONIC EQUATIONS

Problem: Silver ion, Ag^+, reacts with chloride ion to form silver chloride, AgCl.

$$Ag^+(aq) + Cl^-(aq) \longrightarrow AgCl(s)$$

Write a molecular equation for this reaction.

Solution: Solving this problem involves choosing the right set of reactants. First, let's consider what conditions they must meet. Both the silver ions and chloride ions are in solution, so the compounds that we choose for them must be soluble in water. In addition, the compound formed from the anion of the silver salt and the cation of the chloride salt must also be water-soluble so these ions will be spectator ions and therefore cancel when the net ionic equation is formed. There are many possible choices here. We can be sure our conditions are met if we choose an anion for the silver salt that produces only soluble compounds—this way we know the silver compound will be soluble, as will the other compound formed in the metathesis reaction. One anion that meets this requirement is nitrate, NO_3^-. Another is the acetate ion, $C_2H_3O_2^-$. For no special reason, let's choose $AgNO_3$ as one of the reactants.

For a given net ionic equation, there are usually many different molecular equations from which to choose.

Once we've chosen nitrate as the anion of the silver salt, it doesn't much matter what we choose as the chloride salt, as long as it is soluble. Remembering the solubility rules, we realize that the only cations that we have to avoid are Pb^{2+} and Hg_2^{2+}, because their chlorides are insoluble. Therefore, for no special reason again, let's choose Ca^{2+} as the cation of the chloride salt. Our chosen reactants, then, are $AgNO_3$ and $CaCl_2$. Now we can write the molecular equation in the usual way, being sure to write correct formulas for all the salts.

$$AgNO_3 + CaCl_2 \longrightarrow AgCl + Ca(NO_3)_2$$

Finally, we balance the equation

$$2AgNO_3 + CaCl_2 \longrightarrow 2AgCl + Ca(NO_3)_2$$

This is the completed molecular equation—one of many that we could use in the laboratory.

11.5   STOICHIOMETRY OF IONIC REACTIONS: ACID–BASE TITRATIONS

Ionic reactions have many uses in the area of chemical analysis.

There are many ways in which ionic reactions are used, both in and outside the laboratory. Sometimes they are used to synthesize useful industrial and consumer chemicals. For instance, one of the critical steps in the manufacture of photographic film is the formation of silver bromide by a precipitation reaction. This AgBr is incorporated into the emulsion of the film, and is what makes the film sensitive to light. In the laboratory, one of the principal uses of ionic reactions is in chemical analysis.

Quantitative chemical analysis might be thought of as applied stoichiometry. Here we find that ionic reactions are used to separate ionic components of a mixture, to isolate specific ions in compounds of known composition where their amounts can be measured, and even in the direct measurement of the concentrations of substances in solutions.

Before we see how ionic reactions are applied, it would be worthwhile to take another look at some aspects of stoichiometry, to observe how they apply to these kinds of reactions. An important tool for dealing with reactions in solution is the concentration unit, molarity—the number of moles of solute per liter of solution. Let's look at an example that reviews how we use molarity in stoichiometric calculations.

EXAMPLE 11.6   WORKING PROBLEMS IN SOLUTION STOICHIOMETRY

Problem: How many milliliters of 0.100 M $AgNO_3$ solution are needed to react completely with 25.0 mL of 0.400 M $CaCl_2$ solution?

Solution: The first step in any stoichiometry problem is to establish the chemical equation for the reaction. Remembering the solubility rules, we can write the following balanced molecular equation.

$$2AgNO_3(aq) + CaCl_2(aq) \longrightarrow 2AgCl(s) + Ca(NO_3)_2(aq)$$

We see that we anticipate a precipitate of silver chloride.

Now we plan the strategy for solving the problem. In Chapter 4 we saw that we can use the molarity and volume to calculate the number of moles of solute in a sample of a solution. Therefore, from the data given, we can calculate the number of moles of $CaCl_2$. Next, we use the coefficients of the balanced equation to calculate the number of moles of $AgNO_3$ that are needed. Finally, we compute the volume of the 0.100 M $AgNO_3$ solution needed.

First, the number of moles of $CaCl_2$. Recall that we use molarity as a conversion factor relating moles and volume. Thus 0.400 M $CaCl_2$ means

$$\frac{0.400\ \text{mol}\ CaCl_2}{1.00\ \text{L}} \qquad \frac{0.400\ \text{mol}\ CaCl_2}{1000\ \text{mL}}$$

Therefore,

$$25.0\ \text{mL} \times \left(\frac{0.400\ \text{mol}\ CaCl_2}{1000\ \text{mL}}\right) = 0.0100\ \text{mol}\ CaCl_2$$

Next, we use the coefficients in the balanced equation to determine how many moles of silver nitrate are needed.

$$0.0100\ \text{mol}\ CaCl_2 \times \frac{2\ \text{mol}\ AgNO_3}{1\ \text{mol}\ CaCl_2} = 0.0200\ \text{mol}\ AgNO_3$$

And finally, we use the concentration of the $AgNO_3$ solution as a conversion factor to calculate the volume of this solution that is needed to provide 0.0200 mol $AgNO_3$.

$$0.0200\ \text{mol}\ AgNO_3 \times \frac{1000\ \text{mL}}{0.100\ \text{mol}\ AgNO_3} = 200\ \text{mL}$$

Our calculations tell us that we must use 200 mL of the $AgNO_3$ solution.

PRACTICE EXERCISE 10   (a) How many moles of HCl are in 50.0 mL of 0.600 M HCl solution? (b) How many milliliters of 0.600 M HCl solution are needed to supply 4.00×10^{-2} mol HCl?

PRACTICE EXERCISE 11   How many milliliters of 0.500 M KOH are needed to react completely with 60.0 mL of 0.250 M $FeCl_2$ solution to precipitate $Fe(OH)_2$?

Now let's look at a problem that illustrates some of the complexities that can arise in ionic reactions. The following example illustrates how we can use ionic and net ionic equations in working limiting reactant problems.

EXAMPLE 11.7   STOICHIOMETRY OF IONIC REACTIONS

Problems: Suppose that 40.0 mL of 0.200 M NaOH solution is added to 25.0 mL of 0.300 M $MgCl_2$ solution. What weight of $Mg(OH)_2$ will be formed, and what will be the concentrations of the ions in the solution after reaction is complete?

Solution: Let's begin by writing balanced molecular, ionic, and net ionic equations for the reaction. The molecular equation is

$$MgCl_2(aq) + 2NaOH(aq) \longrightarrow Mg(OH)_2(s) + 2NaCl(aq)$$

The ionic equation is

$$Mg^{2+}(aq) + 2Cl^-(aq) + 2Na^+(aq) + 2OH^-(aq) \longrightarrow Mg(OH)_2(s) + 2Na^+(aq) + 2Cl^-(aq)$$

and canceling spectator ions gives the net ionic equation,

$$Mg^{2+}(aq) + 2OH^-(aq) \longrightarrow Mg(OH)_2(s)$$

Actually, the reaction occurs as the solutions are mixed. But for purposes of calculation, we may pretend that they can be mixed first and then allowed to react.

To solve a problem of this kind, the best approach is to begin by tabulating the number of moles of each ion in the solution *after* mixing, but *before* reaction. First, let's determine the number of moles of $MgCl_2$ and NaOH.

$$40.0\ \text{mL} \times \left(\frac{0.200\ \text{mol}\ NaOH}{1000\ \text{mL}}\right) = 8.00 \times 10^{-3}\ \text{mol}\ NaOH$$

$$25.0\ \text{mL} \times \left(\frac{0.300\ \text{mol}\ MgCl_2}{1000\ \text{mL}}\right) = 7.50 \times 10^{-3}\ \text{mol}\ MgCl_2$$

From this information, we can obtain the number of moles of each ion. In doing this, notice we take into account that 1 mol of $MgCl_2$ gives 2 mol Cl^-.

Moles of Ions before Reaction

Mg^{2+}	7.50×10^{-3} mol
Cl^-	15.0×10^{-3} mol
Na^+	8.00×10^{-3} mol
OH^-	8.00×10^{-3} mol

Now we refer to the net ionic equation, where we see that only Mg^{2+} and OH^- react, and that each mole of Mg^{2+} requires two moles of OH^-. This means that 7.50×10^{-3} mol Mg^{2+} would require 15.0×10^{-3} mol OH^-. This is more OH^- than we have. Insufficient OH^- is available, so OH^- must be the limiting reactant; that is, all the OH^- will be used, and it will react with

$$8.00 \times 10^{-3} \text{ mol OH}^- \times \left(\frac{1 \text{ mol Mg}^{2+}}{2 \text{ mol OH}^-} \right) = 4.00 \times 10^{-3} \text{ mol Mg}^{2+}$$

Now we can tabulate the number of moles of each ion in the mixture after the reaction is complete. For Mg^{2+}, the amount remaining equals the initial amount minus the amount that reacted. The numbers of moles of Cl^- and Na^+ are the same as before, because they are spectator ions; nothing happens to them. Finally, there is no OH^- in the solution, because it has all reacted. (Actually, this is only an approximation that will be considered in more detail in Chapter 14.)

There is really a very small amount of OH^- left in the solution, but in computations such as these we can ignore it and assume that the reaction proceeds 100% to completion.

Moles of Ions after Reaction

Mg^{2+}	$(7.50 \times 10^{-3}) - (4.00 \times 10^{-3}) = 3.50 \times 10^{-3}$ mol
Cl^-	15.0×10^{-3} mol
Na^+	8.00×10^{-3} mol
OH^-	0.00 mol

The question asks for the *concentrations* of the ions in the reaction mixture, so we must now divide each of these numbers of moles by the total volume of the final solution (40.0 mL + 25.0 mL = 65.0 mL), expressed in liters (0.0650 L). For example, for Mg^{2+}, its concentration is

$$\frac{3.50 \times 10^{-3} \text{ mol}}{0.0650 \text{ L}} = 0.0538 \text{ mol/L} = 0.0538 \ M$$

Performing similar calculations for the other ions gives

Concentrations of Ions after Reaction

Mg^{2+}	$0.0538 \ M$
Cl^-	$0.231 \ M$
Na^+	$0.123 \ M$
OH^-	$0.0 \ M$

Now we can turn our attention to the amount of $Mg(OH)_2$ that is formed. If 4.00×10^{-3} mol Mg^{2+} reacts, then 4.00×10^{-3} mol $Mg(OH)_2$ must be formed. The formula weight of $Mg(OH)_2$ is 58.3, so the weight of $Mg(OH)_2$ formed is

$$4.00 \times 10^{-3} \text{ mol Mg(OH)}_2 \times \left(\frac{58.3 \text{ g Mg(OH)}_2}{1 \text{ mol Mg(OH)}_2} \right) = 0.233 \text{ g Mg(OH)}_2$$

PRACTICE EXERCISE 12   A solution is labeled $0.230 \ M$ $Fe_2(SO_4)_3$. How many moles of Fe^{3+} and SO_4^{2-} are in 75.0 mL of this solution?

PRACTICE EXERCISE 13   How many moles of $BaSO_4$ will form if 20.0 mL of $0.600 \ M$ $BaCl_2$ is mixed with 30.0 mL of $0.500 \ M$ $MgSO_4$? What will the concentrations of each ion be in the final reaction mixture?

The previous example is about as complex as calculations ever become when we deal with ionic reactions. We are ready now to view an example of how these reactions are used in chemical analysis. A principle that is often used is the isolation of one component of a compound or mixture in some other compound whose formula we know. Example 11.8 illustrates a typical approach to the problem.

EXAMPLE 11.8   CHEMICAL ANALYSIS USING IONIC REACTIONS

Problem: A certain insecticide was known to contain carbon, hydrogen, and chlorine. Reactions were carried out on a 1.000-g sample of the compound that converted all the chlorine to chloride ion present in an aqueous solution. This solution was treated with an excess amount of $AgNO_3$ solution and the AgCl precipitate that was formed was collected and weighed. Its mass was 2.022 g. What was the percentage by weight of Cl in the original insecticide sample? (Notice that the data permit 4 significant figures in the calculated answer.)

Solution: The solution of the problem is quite simple. The strategy is to determine the weight of Cl in 2.022 g of AgCl. We then assume that all this chlorine was in the original 1.000-g sample, and calculate the percentage Cl.

Good experimental technique is needed to ensure that no Cl is lost during the reactions that ultimately give the AgCl.

The formula weight of AgCl is 143.3, and the atomic weight of Cl is 35.45. In one mole of AgCl there must be one mole of Cl, so in 143.3 g AgCl there must be 35.45 g Cl. With this information, we can now calculate the weight of Cl in 2.022 g AgCl.

$$2.022 \; \cancel{g\,AgCl} \times \left(\frac{35.45 \; g \; Cl}{143.3 \; \cancel{g\,AgCl}} \right) = 0.5002 \; g \; Cl$$

The percentage Cl in the sample can be calculated as

$$\%Cl = \frac{\text{weight of Cl}}{\text{weight of sample}} \times 100$$

$$= \frac{0.5002 \; g \; Cl}{1.000 \; g \; sample} \times 100$$

$$= 50.02\% \; Cl$$

PRACTICE EXERCISE 14   A sample of a mixture containing $CaCl_2$ and $MgCl_2$ weighed 2.000 g. The sample was dissolved in water and H_2SO_4 was added until precipitation of $CaSO_4$ was complete. The $CaSO_4$ was filtered, dried completely, and weighed. A total of 0.736 g of $CaSO_4$ was obtained.
(a) How many moles of Ca^{2+} were in the $CaSO_4$?
(b) How many moles of Ca^{2+} were in the original 2.000-g sample?
(c) How many moles of $CaCl_2$ were in the 2.000-g sample?
(d) How many grams of $CaCl_2$ were in the 2.000-g sample?
(e) What was the percentage of $CaCl_2$ in the mixture?

Titrations

Titration is an important laboratory procedure that is used in performing chemical analyses. The apparatus that is employed is shown in Figure 11.5. The long tube is called a buret, and it is marked for volumes, usually in increments of 0.10 mL. In a typical analysis, a solution containing an unknown amount of the substance to be analyzed is placed in the receiving flask. Precisely measured volumes of a solution of known concentration—a standard solution—are then added from the buret. This addition is

FIGURE 11.5 Phenolphthalein is often used as an indicator in the titration of a strong acid with a strong base.

continued until some visual effect, such as a color change, signals that the two reactants have been combined in just the right ratio to give complete reaction. The valve at the bottom of the buret is called a stopcock, and it permits the analyst to control the amount of titrant (the solution in the buret) that is delivered to the receiving flask. This permits the addition of the titrant to be stopped once the reaction is complete.

The titration procedure is applicable to many different kinds of reactions. In this chapter we focus our attention on acid–base titrations in which the reaction is neutralization, but we will see other applications of the procedure in Chapter 12. When an acid–base titration is performed, the analyst adds a drop or two of an indicator solution to the solution in the receiving flask prior to the start of the titration. An acid–base indicator is a dye that has one color in an acidic solution and a different color in a basic solution. Two common examples are litmus,[2] which is red in acid and blue in base, and phenolphthalein, which is colorless in acid and pink in base. (The theory of acid–base indicators is discussed in Chapter 15.) In performing the titration, the analyst looks for a change in color that signals that the solution has changed from acidic to basic (or basic to acidic, depending on the nature of the reactants in the buret and flask). Usually this color change is very abrupt—as the end of the reaction is approached, the change in color normally takes place with the addition of only one drop of the titrant. When the color change finally occurs, delivery of the titrant is stopped—the end point has been reached—and the total volume of the titrant that was added to the receiving flask is recorded.

With this as background, now, we can examine some sample calculations.

EXAMPLE 11.9   USING TITRATION DATA TO CALCULATE MOLARITIES

Problem: A student prepared a solution of hydrochloric acid that was approximately 0.1 M and wished to determine its precise concentration. A 25.00-mL portion of the HCl solution was transferred to a flask, and after a few drops of indicator were added, the HCl solution was titrated with 0.0775 M NaOH solution. The titration required exactly 37.46 mL of the standard NaOH solution. What was the exact molarity of the HCl solution?

Solution: The chemical equation for the neutralization reaction is

$$HCl(aq) + NaOH(aq) \longrightarrow NaCl(aq) + H_2O$$

From the molarity and volume of the NaOH solution, we calculate the number of moles of NaOH consumed in the titration.

$$37.46 \text{ mL} \times \left(\frac{0.0775 \text{ mol NaOH}}{1000 \text{ mL}} \right) = 2.90 \times 10^{-3} \text{ mol NaOH}$$

The coefficients in the equation tell us that NaOH and HCl react in a one-to-one mole ratio, so in this titration, 2.90×10^{-3} mol HCl was in the flask. To calculate the molarity of the HCl, we simply take the ratio of the number of moles of HCl that reacted (2.90×10^{-3} mol HCl) to the volume (in liters) of the HCl solution that was used (0.02500 L).

$$\text{molarity of HCl soln} = \frac{2.90 \times 10^{-3} \text{ mol HCl}}{0.02500 \text{ L}}$$

$$= 0.116 \text{ mol HCl/L}$$

The molarity of the hydrochloric acid was 0.116 M.

---

[2] Litmus paper, commonly found among the items in a locker in the general chemistry lab, consists of strips of absorbent paper that have been soaked in a solution of litmus and then dried. Red litmus paper is used to test if a solution is basic; if it is basic, a drop of the solution will turn the dye blue. To test if the solution is acidic, blue litmus paper is used. A drop of the solution will turn blue litmus paper pink if the solution is acidic.

EXAMPLE 11.10   USING ACID–BASE TITRATIONS IN CHEMICAL ANALYSIS

Problem: A solution of ammonia in water was analyzed by titrating the ammonia with hydrochloric acid. The net ionic reaction is

$$NH_3(aq) + H_3O^+(aq) \longrightarrow NH_4^+(aq) + H_2O$$

In the analysis, a 5.00-g sample of the ammonia solution was placed in a flask and titrated with 1.00 M HCl, using an appropriate indicator. The titration required 29.86 mL of the HCl solution. What is the percentage by weight of NH_3 in the ammonia solution?

Solution: First we calculate the number of moles of HCl used in the titration.

$$29.86 \text{ mL} \times \left(\frac{1.00 \text{ mol HCl}}{1000 \text{ mL}} \right) = 0.0299 \text{ mol HCl}$$

Since HCl is a monoprotic acid, 0.0299 mol HCl must supply 0.0299 mol H_3O^+, and according to the coefficients in the equation above, this must react with 0.0299 mol NH_3.

$$0.0299 \text{ mol HCl} \times \left(\frac{1 \text{ mol } H_3O^+}{1 \text{ mol HCl}} \right) \times \left(\frac{1 \text{ mol } NH_3}{1 \text{ mol } H_3O^+} \right) = 0.0299 \text{ mol } NH_3$$

This is the number of moles of NH_3 in the 5.00-g sample. The mass of this NH_3 is

$$0.0299 \text{ mol } NH_3 \times \left(\frac{17.0 \text{ g } NH_3}{1.000 \text{ mol } NH_3} \right) = 0.508 \text{ g } NH_3$$

The percentage of NH_3 in the sample is

$$\%NH_3 = \frac{\text{weight of } NH_3}{\text{weight of sample}} \times 100$$

$$= \frac{0.508 \text{ g}}{5.00 \text{ g}} \times 100$$

$$= 10.2\%$$

PRACTICE EXERCISE 15   In a titration, a sample of H_2SO_4 solution having a volume of 15.00 mL required 36.42 mL of 0.147 M NaOH solution for complete neutralization. What is the molarity of the H_2SO_4 solution?

PRACTICE EXERCISE 16   "Stomach acid" is hydrochloric acid. A sample of gastric juices having a volume of 5.00 mL required 11.00 mL of 0.0100 M KOH solution for neutralization in a titration. What was the molar concentration of HCl in this fluid? If we assume a density of 1.00 g/mL for the fluid, what was the percentage by weight of HCl?

11.6 EQUIVALENT WEIGHTS AND NORMALITY

For certain kinds of reactions, it is possible to define chemical quantities that always combine in a one-to-one ratio, regardless of the coefficients in the balanced equation.

In problems dealing with the stoichiometry of chemical reactions, the first step has always been to write a balanced chemical equation. This is because the coefficients of the equation determine the ratios by moles in which the reactants combine and the products are formed. For most reactions, there is no sure way of knowing the stoichiometric relationships except by obtaining a balanced equation. However, two important kinds of reactions obey certain rules that allow us to bypass the chemical equation in many cases. One is acid–base reactions, which we discuss here, and the other kind is called oxidation–reduction, which will be discussed in Chapter 12.

In an acid–base neutralization reaction, hydrogen ions are supplied by the acid and hydroxide ions are supplied by the base. Regardless of the acid or base, however, the net ionic equation for the neutralization reaction is always exactly the same.

$$H_3O^+(aq) + OH^-(aq) \longrightarrow 2H_2O$$

or, written more simply by using H^+ as an abbreviation for the hydronium ion,

$$H^+(aq) + OH^-(aq) \longrightarrow H_2O$$

Now, suppose that 36 g of some acid was able to furnish *exactly* one mole of H^+, and 40 g of some base was able to furnish *exactly* one mole of OH^-. Because one mole of H^+ reacts with one mole of OH^-, as specified in the equation for the neutralization, we know that the 36 g of the acid is just enough to react with the 40 g of base. Notice that we can make this statement without knowing what the acid or base is; we only have to know how much of the acid gives 1 mol H^+ and how much of the base gives 1 mol OH^-.

Equivalents

The kind of reasoning that we have just gone through forms the basis of the definition of another kind of chemical quantity called the equivalent, abbreviated eq. The exact definition depends on the kind of reaction, acid–base or oxidation–reduction. But the equivalent in both of these reactions is always defined in such a way that equivalents of reactants *always* react in a one-to-one ratio. This is the key concept to remember! If there are two reactants, A and B,

---

1 eq of reactant A reacts with exactly 1 eq of reactant B

Well, how shall we define equivalents so they obey the requirement in the statement above? For acid and bases, the equivalent is defined as follows.

One equivalent of an acid is the amount of the acid that is able to furnish one mole of hydrogen ion.

One equivalent of a base is the amount of the base that is able to furnish one mole of hydroxide ion.

From the chemical equation for the neutralization reaction, we see that these definitions assure a one-to-one combining ratio for equivalents of acids and bases.

Let's look now at how this applies to some real chemicals and how the equivalent is related to the more familiar chemical unit, the mole. Consider the acids HCl and H_2SO_4. One mole of HCl is enough acid to supply one mole of H^+, so *one* mole of HCl must be the same as *one* equivalent of HCl.

$$1 \text{ mol } HCl = 1 \text{ eq } HCl$$

On the other hand, *one* mole of H_2SO_4 is enough acid to supply *two* moles of H^+, provided the H_2SO_4 is completely neutralized. For complete neutralization, then, *one* mole of H_2SO_4 must be equal to *two* equivalents of acid.

$$1 \text{ mol } H_2SO_4 = 2 \text{ eq } H_2SO_4$$
(for complete reaction)

Thus, there is a simple relationship between moles and equivalents for acids. *The number of equivalents in one mole of an acid is equal to the number of hydrogens that are neutralized when one molecule of the acid reacts.*

EXAMPLE 11.11 WORKING WITH EQUIVALENTS

Problem: How many equivalents of acid are in 1 mol of H_3PO_4 (a) if the acid is completely neutralized to PO_4^{3-}, and (b) if the acid is only partially neutralized to $H_2PO_4^-$?

Solution: (a) If the H_3PO_4 reacts completely with a base, all three hydrogens will be neutralized. Therefore, there are three equivalents of acid per mole of H_3PO_4.

$$1 \text{ mol } H_3PO_4 = 3 \text{ eq } H_3PO_4$$

(b) When H_3PO_4 is neutralized to give the ion $H_2PO_4^-$, only one hydrogen is neutralized per molecule of the acid, so there is only one equivalent per mole of the acid in this case.

$$1 \text{ mol } H_3PO_4 = 1 \text{ eq } H_3PO_4$$

PRACTICE EXERCISE 17 If oxalic acid, $H_2C_2O_4$, is neutralized completely, how many equivalents are in 1.00 mol of oxalic acid? If it is only partially neutralized to give the acid salt $NaHC_2O_4$, how many equivalents are in one mole of the oxalic acid?

Determining the number of equivalents per mole of base is also very simple. Consider the two bases NaOH and $Ba(OH)_2$. One mole of NaOH is enough base to supply one mole of OH^-, so an equivalent of NaOH must be the same as one mole of NaOH.

$$1 \text{ mol NaOH} = 1 \text{ eq NaOH}$$

One mole of $Ba(OH)_2$ is able to supply two moles of OH^-, so one mole of $Ba(OH)_2$ must be two equivalents of base.

In the reactions we will consider, bases will be completely neutralized.

$$1 \text{ mol } Ba(OH)_2 = 2 \text{ eq } Ba(OH)_2$$

Thus, *for a metal hydroxide, the number of equivalents in one mole of the base is equal to the number of hydroxides in one formula unit of the base.*

PRACTICE EXERCISE 18 How many equivalents of base are in 1.00 mol of $Al(OH)_3$?

Equivalent Weights

In working problems using equivalents, it is often useful to know the mass of an equivalent of each of the reactants. For example, in the beginning of this section we described a case in which 36 g of an acid gives 1 mol of H^+ and 40 g of a base gives 1 mol of OH^-. These quantities correspond to one equivalent each of acid and base. Knowing how much an equivalent of each of them weighs establishes a mass relationship between the two reactants that can be used in stoichiometric calculations. Thus, if we had a sample of 20 g of the base (0.5 eq of base), we know that we would only need 18 g of the acid (0.5 eq of acid). *Knowing the weights of equivalents of reactants establishes a mass relationship between them that can be used in calculations.*

The weight in grams of one equivalent is called the equivalent weight. Calculating its value for a particular compound is most easily done using the formula weight and a knowledge of the number of equivalents per mole.

EXAMPLE 11.12   CALCULATING EQUIVALENT WEIGHTS

Problem: Calculate the equivalent weight of sulfuric acid, assuming that it will be completely neutralized when it reacts.

Solution: The formula weight of sulfuric acid, H_2SO_4, is 98.0. Therefore, 1 mol of H_2SO_4 weighs 98.0 g. For complete neutralization,

$$1 \text{ mol } H_2SO_4 = 2 \text{ eq } H_2SO_4$$

Therefore,

$$1 \text{ eq } H_2SO_4 = 0.5 \text{ mol } H_2SO_4$$

Since 0.5 mol of H_2SO_4 weighs 49.0 g,

$$1 \text{ eq } H_2SO_4 = 49.0 \text{ g } H_2SO_4$$

We could also work this problem using the factor-label method. We wish to know the number of grams corresponding to one equivalent, so we can state the problem

$$1 \text{ eq} = ? \text{ g}$$

$$\text{Equiv. wt.} = \frac{\text{formula wt.}}{\text{no. eq per mol}}$$

To solve this, we multiply 1 eq by conversion factors to obtain the units grams.

$$1 \text{ eq } H_2SO_4 \times \left(\frac{1 \text{ mol } H_2SO_4}{2 \text{ eq } H_2SO_4} \right) \times \left(\frac{98.0 \text{ g } H_2SO_4}{1 \text{ mol } H_2SO_4} \right) = 49.0 \text{ g } H_2SO_4$$

PRACTICE EXERCISE 19   Calculate the equivalent weight of H_3PO_4, assuming (a) complete neutralization, and (b) partial neutralization to give the HPO_4^{2-} ion.

Let's look now at how we can use equivalent weights in solving problems in stoichiometry.

EXAMPLE 11.13   USING EQUIVALENT WEIGHTS IN SOLVING PROBLEMS

Problem: How many grams of $Ca(OH)_2$ would be needed to completely neutralize 42.6 g of H_3PO_4?

Solution: To solve the problem using equivalents, we first compute the equivalent weights of each reactant. For $Ca(OH)_2$, there are two equivalents per mole.

$$1 \text{ mol } Ca(OH)_2 = 2 \text{ eq } Ca(OH)_2$$

and 1 mol of $Ca(OH)_2$ weighs 74.1 g. The weight of 1 eq is therefore

$$1 \text{ eq } Ca(OH)_2 \times \left(\frac{1 \text{ mol } Ca(OH)_2}{2 \text{ eq } Ca(OH)_2} \right) \times \left(\frac{74.1 \text{ g } Ca(OH)_2}{1 \text{ mol } Ca(OH)_2} \right) = 37.1 \text{ g } Ca(OH)_2$$

For complete neutralization, 1 mol of H_3PO_4 equals 3 eq.

$$1 \text{ mol } H_3PO_4 = 3 \text{ eq } H_3PO_4$$

and 1 mol of H_3PO_4 weighs 98.0 g. Therefore,

$$1 \text{ eq } H_3PO_4 \times \left(\frac{1 \text{ mol } H_3PO_4}{3 \text{ eq } H_3PO_4} \right) \times \left(\frac{98.0 \text{ g } H_3PO_4}{1 \text{ mol } H_3PO_4} \right) = 32.7 \text{ g } H_3PO_4$$

Since 1 eq $Ca(OH)_2$ reacts with 1 eq H_3PO_4, then 37.1 g $Ca(OH)_2$ reacts with 32.7 g H_3PO_4. We can use this weight relationship to construct a conversion factor for solving the problem.

$$42.6 \text{ g } H_3PO_4 \times \left(\frac{37.1 \text{ g } Ca(OH)_2}{32.7 \text{ g } H_3PO_4} \right) = 48.3 \text{ g } Ca(OH)_2$$

The amount of $Ca(OH)_2$ needed is therefore 48.3 g.

PRACTICE EXERCISE 20   Using equivalent weights, determine the number of grams of H_2SO_4 needed to react completely with 14.0 g KOH.

Normality

In our previous dealings with the stoichiometry of reactions in solution, we found that molarity was a convenient unit of concentration to use because it relates moles of solute to the volume of solution. This relationship permits us to dispense moles of solute by simply measuring volumes of solution—a process that is easily accomplished in the laboratory. Now we have learned about another chemical unit that is similar to the mole—the equivalent. In working problems with equivalents when the reactants are dissolved in solution, it is convenient to define a quantity similar to molarity, but in terms of equivalents instead of moles. This quantity is called the normality. The normality of a solution is a concentration unit that expresses the number of equivalents of solute per liter of solution.

$$\text{normality} = \frac{\text{equivalents of solute}}{1 \text{ liter of solution}}$$

Thus, a solution that contains one equivalent of solute per liter of solution has a concentration that we specify as "one normal," abbreviated 1 N. Like the molarity, it can be considered to be a conversion factor that relates equivalents and volume. For example, a solution labeled 2.00 N H_2SO_4 contains 2.00 eq of H_2SO_4 per liter (1000 mL), and provides the two conversion factors

$$\frac{2.00 \text{ eq } H_2SO_4}{1000 \text{ mL soln}} \qquad \frac{1000 \text{ mL soln}}{2.00 \text{ eq } H_2SO_4}$$

Often, in working problems, it is helpful to consider the relationship between normality and molarity. For example, consider their definitions.

$$\text{molarity} = \frac{\text{moles of solute}}{1.00 \text{ L soln}} \qquad \text{normality} = \frac{\text{equivalents of solute}}{1.00 \text{ L soln}}$$

Notice that only the numerators in these fractions are different. Normality has equivalents in the numerator, whereas molarity has moles in the numerator. In the previous discussion we saw that there is a simple relationship between moles and equivalents— the number of equivalents of a reactant is either equal to the number of moles or is a multiple of the number of moles. Similarly, the normality of a solution is either equal to its molarity or is a multiple of the molarity. Let's look at a specific example.

EXAMPLE 11.14   CONVERTING MOLARITY TO NORMALITY

Problem: A solution of phosphoric acid is labeled 0.100 M. It is to be used in a reaction in which the acid will be completely neutralized. What is the normality of the solution?

Solution: First, let's translate the label for molarity to see what it means.

$$0.100 \text{ } M \text{ } H_3PO_4 = \frac{0.100 \text{ mol } H_3PO_4}{1.00 \text{ L soln}}$$

The solution contains 0.100 mol H_3PO_4 in one liter. If the phosphoric acid is to be completely neutralized, there are three equivalents per mole. In other words, the number of equivalents is three times the number of moles.

$$1 \text{ mol } H_3PO_4 = 3 \text{ eq } H_3PO_4$$

Normality = molarity × no. eq per mol

This means that 0.100 mol of the acid is 0.300 eq of the acid, so the solution has a concentration that we can express as

$$\frac{0.300 \text{ eq } H_3PO_4}{1.00 \text{ L soln}}$$

Therefore, the normality of the H_3PO_4 is 0.300 N.

The lesson to be learned from Example 11.14 is that we can obtain the normality of a solution from its molarity if we know the number of equivalents of solute per mole of solute. This number, multiplied by the molarity, gives the normality.

PRACTICE EXERCISE 21   Calculate (a) the normality of a 0.10 M $Ba(OH)_2$ solution and (b) the molarity of a 0.60 N H_2SO_4 solution, both of which are used in reactions in which they are completely neutralized.

Titrations Using Normality as a Concentration Unit

One of the benefits of defining equivalents and normality is that it makes calculations for titrations very simple. Consider the units that we obtain if we multiply the normality of a solution by the volume that is used in a particular reaction.

$$\underbrace{\frac{\text{eq solute}}{\text{L soln}}}_{N} \times \underbrace{\text{L soln}}_{V} = \text{eq solute}$$

Normality (N) times volume (V) gives equivalents. In any reaction, the number of equivalents of one reactant that are consumed is always exactly equal to the number of equivalents of the other reactant that are consumed — that's the way equivalents are defined. This permits us to write the following relationship that applies to any reaction between two substances, say, A and B, in which the concentrations are expressed in normality.

$$N_A V_A = N_B V_B \qquad (11.1)$$

CAUTION: This equation only works using normality. If you use molarity instead, you may get wrong answers.

EXAMPLE 11.15   TITRATIONS USING NORMALITY

Problem: A solution of sulfuric acid was prepared having a concentration of approximately 0.2 N. It was standardized by titrating 25.00 mL of it with 0.150 N KOH. The titration required 34.60 mL of the base. What is the actual normality of the acid?

Solution: This is really a very simple problem. Because the concentrations are expressed in normality, we don't need a balanced equation to solve it. All we do is substitute values into Equation 11.1. Taking the H_2SO_4 to be reactant A, and KOH to be reactant B, we can solve for the normality of A to obtain

$$N_A = \frac{N_B V_B}{V_A}$$

Substituting values gives

$$N_A = \frac{(0.150\ N) \times (34.60\ \text{mL})}{(25.00\ \text{mL})}$$

$$= 0.208\ N$$

The normality of the H_2SO_4 solution is 0.208 N.

PRACTICE EXERCISE 22   How many milliliters of 0.200 N NaOH are needed to react with 45.0 mL of 0.150 N H_2SO_4?

PRACTICE EXERCISE 23   How many equivalents of HNO_3 are in 35.0 mL of 0.600 N HNO_3 solution?

PRACTICE EXERCISE 24   A 1.000-g sample of an "unknown" acid was dissolved in water, and in a titration required 31.6 mL of 0.200 N NaOH solution for complete neutralization.
(a) How many equivalents of NaOH were used?
(b) How many equivalents of acid reacted?
(c) What is the equivalent weight of the acid?

11.7 BRØNSTED CONCEPT OF ACIDS AND BASES

Acids are proton donors and bases are proton acceptors.

All of our discussions of acids and bases until this time have been centered around their reactions in aqueous solutions, where acids give hydronium ions and bases give hydroxide ions. This is derived from the way Arrhenius defined acids and bases — in terms of the ions that they produce in water, and the reaction of an acid with a base in water is really the reaction between H_3O^+ and OH^- to give the solvent, H_2O. But this is a very restrictive definition. There are many reactions that occur in solvents other than water, and even reactions that occur in the absence of any solvent whatsoever, that have all the trappings of acid–base reactions even though they do not involve H_3O^+ and OH^-.

In water, the reaction between ammonia (a base) and hydrochloric acid gives the salt ammonium chloride. According to the Arrhenius approach, the reaction would be considered to occur between OH^- formed in the ionization of the weak base NH_3,

$$NH_3(aq) + H_2O \rightleftharpoons NH_4^+(aq) + OH^-(aq)$$

and H_3O^+ formed by the ionization of HCl in water

$$HCl(aq) + H_2O \longrightarrow H_3O^+(aq) + Cl^-(aq)$$

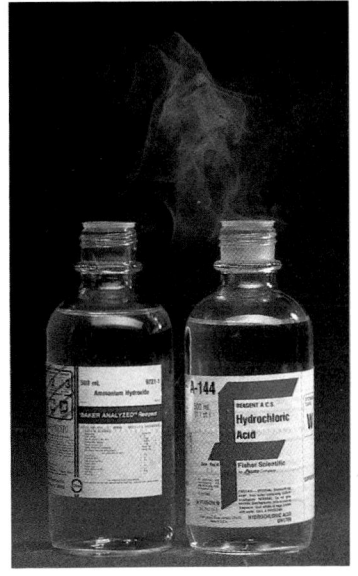

FIGURE 11.6 The reaction of gaseous HCl and gaseous NH_3 to form NH_4Cl. The HCl and NH_3 are escaping from their concentrated aqueous solutions and are forming microcrystals of NH_4Cl, which cause the cloud.

However, the reaction between HCl and NH_3 does not require the presence of water. In fact, no solvent is needed at all. In the photograph in Figure 11.6 we see a white cloud of NH_4Cl formed by the reaction between gaseous ammonia and gaseous hydrogen chloride.

$$NH_3(g) + HCl(g) \longrightarrow NH_4Cl(s)$$

In both aqueous and gas-phase reactions, the same reactants are involved (NH_3 and HCl) and the same ultimate product is formed (NH_4Cl). It certainly seems sensible that if one is an acid–base reaction, we ought to consider the other to be one too. We can't do this using the Arrhenius definition, however, because the gaseous reaction doesn't involve H_3O^+ and OH^-.

The limitations of the Arrhenius definition of acids and bases are removed by a broader definition that was proposed by the Danish scientist Johannes Brønsted (1879 – 1947) in 1923.[3] Brønsted recognized that in many reactions, both in and out of water, an H^+ is transferred from one atom to another. this is what happens, for example, when gaseous ammonia and hydrogen chloride react.

These similarities led Brønsted to define acids and bases in terms of the exchange of an H^+ ion between two particles. An H^+ ion, of course, is simply the nucleus of a hydrogen atom, and this nucleus consists of just one particle, the proton. Another name for the hydrogen ion, therefore, is the proton. This is what the hydrogen ion is called in Brønsted's definition of acids and bases.

> Brønsted's Definition of Acids and Bases
> An acid is a proton donor.
> A base is a proton acceptor.

According to this definition, even in the gaseous state HCl is an acid in its reaction with NH_3 because in the reaction it donates a proton to the NH_3 molecule. Similarly, ammonia is a base because it accepts a proton from the HCl molecule.

Conjugate Acids and Bases

Under the Brønsted definition, it is convenient to consider any acid–base reaction as an equilibrium, with both a forward and reverse reaction. For example, when hydrogen cyanide, HCN (an infamous, deadly poison), is dissolved in water, the solution is somewhat acidic because HCN is a weak acid. We can write the equation for the ionization as an equilibrium.

$$HCN(aq) + H_2O \rightleftharpoons H_3O^+(aq) + CN^-(aq)$$

As you learned earlier, in the equilibrium both the forward and reverse reactions are occurring at equal rates.

If you examine the forward reaction, you will see that the HCN donates its proton to the water molecule. This is how it becomes a cyanide ion. Therefore, in the forward reaction the HCN is behaving as a Brønsted acid. Since water is accepting a proton from the HCN, water is behaving as a Brønsted base.

Now let's look at the reverse reaction—the one read from right to left. In this reaction, it is the H_3O^+ that is behaving as a Brønsted acid by donating a proton to the CN^- ion, and the CN^- ion is behaving as a Brønsted base by accepting the proton.

The reaction between HCN and H_2O is typical of proton transfer reactions in general. In them we can always identify *two* acids (e.g., HCN and H_3O^+) and *two* bases (e.g., H_2O and CN^-). Notice, too, that the acid on the right (H_3O^+) is formed from the base on the left (H_2O), and the base on the right (CN^-) is formed from the acid on the left

[3] Thomas Lowry (1874–1936), a British scientist working independently of Brønsted, developed many of the same ideas but did not carry them as far. Some references call this view of acids and bases the Brønsted–Lowry theory.

(HCN). Two substances, such as H_3O^+ and H_2O, that differ from each other by just *one* proton are referred to as a conjugate acid–base pair. Hydronium ion and water are a conjugate acid–base pair, and we say that hydronium ion is the conjugate acid of water, and that water is the conjugate base of hydronium ion. In this reaction, the other conjugate acid–base pair is HCN and CN^-; the conjugate acid of CN^- is HCN and the conjugate base of HCN is CN^-. One of the common ways of designating the members of the conjugate acid–base pairs in an equation is to connect them by a line, as follows.

conjugate pair

$$HCN(aq) + H_2O \rightleftharpoons H_3O^+(aq) + CN^-(aq)$$
acid   base   acid   base

conjugate pair

EXAMPLE 11.16 WRITING THE FORMULA OF A CONJUGATE BASE

Problem: What is the conjugate base of nitric acid, HNO_3?

Solution: A conjugate base can always be found by removing one H^+ from a given acid. Removing one H^+ (both the atom and the charge) from HNO_3 leaves NO_3^-. The nitrate ion, NO_3^-, is the conjugate base of HNO_3.

EXAMPLE 11.17 WRITING THE FORMULA OF A CONJUGATE BASE

Problem: The hydrogen sulfate ion, HSO_4^-, has a conjugate base. Write its formula.

Solution: Deleting an H^+ from HSO_4^- leaves SO_4^{2-}. [Notice that the charge goes from 1− to 2− because $(1-) - (1+) = (2-)$.]

EXAMPLE 11.18 WRITING THE FORMULA OF A CONJUGATE ACID

Problem: What is the formula of the conjugate acid of OH^-?

Solution: To find a conjugate acid we add one H^+ to the formula of the given base. Adding H^+ to OH^- gives H_2O. Water is the conjugate acid of the hydroxide ion.

EXAMPLE 11.19 WRITING THE FORMULA OF A CONJUGATE ACID

Problem: The phosphate ion, PO_4^{3-}, is a rather strong Brønsted base. What is the formula of its conjugate acid?

Solution: Adding H^+ to the formula PO_4^{3-} gives HPO_4^{2-}. The conjugate acid of PO_4^{3-} is HPO_4^{2-}. Always be sure that the electrical charge is correctly shown.

PRACTICE EXERCISE 25   Write the formula of the conjugate base for each of the following Brønsted acids.
(a) H_2O (c) HNO_2 (e) $H_2PO_4^-$ (g) H_2
(b) HI   (d) H_3PO_4 (f) HPO_4^{2-} (h) NH_4^+

PRACTICE EXERCISE 26   Write the formula of the conjugate acid for each of the following Brønsted bases.
(a) HO_2^- (c) PO_4^{3-} (e) NH_2^- (g) $H_2PO_4^-$
(b) SO_4^{2-} (d) $C_2H_3O_2^-$ (f) NH_3 (h) HPO_4^{2-}

EXAMPLE 11.20   IDENTIFYING BRØNSTED ACIDS AND BASES

Problem: Sodium hydrogen sulfate is used in the manufacture of certain kinds of cement and to clean oxide coatings from metals. Its anion, HSO_4^-, reacts as follows with the phosphate anion, PO_4^{3-}.

$$HSO_4^-(aq) + PO_4^{3-}(aq) \longrightarrow SO_4^{2-}(aq) + HPO_4^{2-}(aq)$$

Identify the two Brønsted acids and the two Brønsted bases in this reaction.

Solution: We know that there must be an acid and a base on each side of the arrow, and the acid on one side is related to the base on the other. The members of a conjugate acid–base pair must be alike except in one way — they differ by a single hydrogen ion. If you examine the formulas in the equation, you will see that there are two of them containing "PO_4." They have to belong to the same conjugate pair, and the one with the greater number of hydrogens, HPO_4^{2-}, must be the acid, while the other, PO_4^{3-}, must be the base. Therefore, one acid–base pair is HPO_4^{2-} and PO_4^{3-}. The other two ions (HSO_4^- and SO_4^{2-}) belong to the second acid–base pair; HSO_4^- is the acid and SO_4^{2-} is the base.

conjugate pair

$$HSO_4^-(aq) + PO_4^{3-}(aq) \longrightarrow SO_4^{2-}(aq) + HPO_4^{2-}(aq)$$

conjugate pair

EXAMPLE 11.21   IDENTIFYING CONJUGATE ACID–BASE PAIRS

Problem: Identify the conjugate acid–base pairs in the following reaction.

$$C_2H_3O_2^-(aq) + H_3O^+(aq) \rightleftharpoons HC_2H_3O_2(aq) + H_2O$$

Solution: To pick out the acid–base pairs, we look for formulas that differ only by one H^+. It's easy to see that one pair must be $C_2H_3O_2^-$ and $HC_2H_3O_2$, and the other pair must be H_3O^+ and H_2O. In each pair, the acid is the substance with one more hydrogen, so the two acids are H_3O^+ and $HC_2H_3O_2$, and the two bases are $C_2H_3O_2^-$ and H_2O.

conjugate pair

$$C_2H_3O_2^-(aq) + H_3O^+(aq) \rightleftharpoons HC_2H_3O_2(aq) + H_2O$$

conjugate pair

PRACTICE EXERCISE 27   When aqueous solutions of sodium bicarbonate and sodium dihydrogen phosphate are mixed in the proper mole ratios of solutes, the following reaction occurs.

$$HCO_3^-(aq) + H_2PO_4^-(aq) \longrightarrow H_2CO_3(aq) + HPO_4^{2-}(aq)$$

Identify the two Brønsted acids and the two Brønsted bases in this reaction. Rewrite the equation, showing the two conjugate acid–base pairs.

PRACTICE EXERCISE 28   If some of the strong cleaning agent trisodium phosphate were mixed with household vinegar, which contains acetic acid, the following equilibrium would be established. (The products are favored.) Identify the pairs of conjugate acids and bases.

$$PO_4^{3-}(aq) + HC_2H_3O_2(aq) \rightleftharpoons HPO_4^{2-}(aq) + C_2H_3O_2^-(aq)$$

Amphoteric Substances

Some molecules or ions are able to function as either an acid or a base, depending on the kind of substance with which they react. The most common example is water. When gaseous hydrogen chloride is dissolved in water, it reacts as follows.

$$HCl(g) + H_2O \longrightarrow H_3O^+(aq) + Cl^-(aq)$$
$$\underset{\text{acid}}{} \quad \underset{\text{base}}{}$$

In this reaction, the H_2O is behaving as a base because it is accepting a proton from the HCl molecule. On the other hand, when water reacts with ammonia,

$$NH_3(aq) + H_2O \rightleftharpoons NH_4^+(aq) + OH^-(aq)$$
$$\underset{\text{base}}{} \quad \underset{\text{acid}}{}$$

it is serving as an acid by donating a proton to the NH_3 molecule.

Substances that can be either acids or bases depending on the other substance present are said to be amphoteric (from *amphoteros* — Greek for "partly one and partly the other"). Another term that is sometimes used to stress that it is the proton donating/accepting ability that is of central concern is amphiprotic.

As suggested above, amphoteric substances need not be molecules. Some ions also have this property, especially the anions found in acid salts. For example, the bicarbonate ion found in baking soda can either donate a proton to a base, or it can accept a proton from an acid. Toward hydroxide ion, therefore, HCO_3^- is an acid.

$$HCO_3^-(aq) + OH^-(aq) \longrightarrow CO_3^{2-}(aq) + H_2O$$

Toward hydronium ion, HCO_3^- is a base.

$$HCO_3^-(aq) + H_3O^+(aq) \longrightarrow H_2CO_3(aq) + H_2O$$

PRACTICE EXERCISE 29 Illustrate, using chemical equations, how the anion in Na_2HPO_4 is amphoteric.

Relative Strengths of Conjugate Acids and Bases

Let's consider again the equilibrium that is present when a weak acid such as HCN is dissolved in water.

$$HCN(aq) + H_2O \rightleftharpoons H_3O^+(aq) + CN^-(aq)$$

In this reaction, only a very small percentage of the HCN is ionized in the solution — most of the HCN is present as molecules. In other words, the position of equilibrium lies to the left. The location of the position of equilibrium in a reaction such as this tells us something about the relative strengths of the two Brønsted acids in the reaction, as well as the relative strengths of the two Brønsted bases.

In the reaction above, the two acids are HCN and H_3O^+. Each is competing in an attempt to donate its proton to a base. The fact that nearly all the protons end up in the HCN molecules, and only relatively few spend their time in the H_3O^+ ions, means that the hydronium ion must be a better proton donor than hydrogen cyanide.

By a similar argument we can conclude that CN^- is a stronger base than H_2O. Both of these bases are competing for the available protons. At equilibrium, most of the protons are found in HCN molecules rather than in H_3O^+ ions, which means that the cyanide ions must be more effective at winning protons than water molecules. Of course, this is the same as saying that cyanide ion is a stronger base than water.

Relative acidities of some acids. The weaker the acid is, the lower it is on this list. (Only a few weak acids are given.)

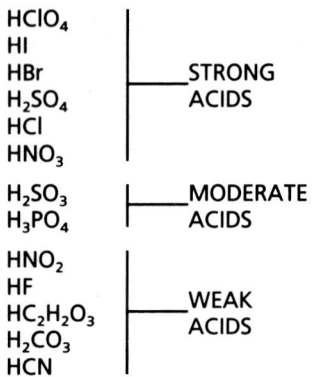

In an equilibrium between Brønsted acids and bases, the position of equilibrium always favors the weaker species. In the example above, for instance, the position of equilibrium favors HCN and H_2O — the weaker of the two acids and bases. Stated another way, *stronger acids and bases tend to react to produce their weaker conjugates.*

There is another useful generalization that we can make with respect to the relative strengths of the members of conjugate acid–base pairs. *The stronger a Brønsted acid is, the weaker is its conjugate base.* For example, HCl is a very strong Brønsted acid. We know this because in water it is 100% ionized — it is strong enough to give away all of its protons to water molecules.

$$HCl(aq) + H_2O \xrightarrow{100\%} H_3O^+(aq) + Cl^-(aq)$$

At the same time, we can also see that the chloride ion (the conjugate base of HCl) is a very weak Brønsted base. Even in the presence of H_3O^+, the Cl^- ion isn't able to capture any protons.

An extension of the generalization above is that *weak Brønsted acids have very strong conjugate bases.* As an example, let's compare the ions OH^- and O^{2-}. In this pair, the hydroxide ion is the conjugate acid and the oxide ion is the conjugate base. The hydroxide ion is a very weak acid. In fact, you've known it only as a base so far, because of its strong tendency to react with H_3O^+. The conjugate base of OH^- is the oxide ion, and it is a very strong base. In water, oxide ion reacts completely.

$$O^{2-}(aq) + H_2O \xrightarrow{100\%} OH^-(aq) + OH^-(aq)$$

By now you may have realized that there must be variations in weakness among acids or bases described here simply as weak. Indeed there are, and we will return to this topic in Chapter 15 where we develop a quantitative analysis of relative weakness.

11.8 PERIODIC TRENDS IN THE STRENGTHS OF ACIDS

The strengths of acids vary systematically with the location of the principal element in the periodic table.

One of the chief goals of our study throughout this book is to acquaint you with the periodic table and to give you a feel for the way chemists use the table to correlate and remember chemical facts. One kind of chemical fact that you learned in earlier sections is that some acids are strong and some are weak, meaning that some react completely with water to form ions while others react incompletely.

In this section we will see that the periodic table can be used to help us remember the strong and weak acids, and that it can help us predict the relative strengths of acids. We will also see how the concept of electronegativity helps us understand the relative strengths of acids — why, for example, some acids, such as H_2SO_4, are strong but very similar acids, such as H_2SO_3, are much weaker.

Oxoacids

Strong Oxoacids

HNO_3	$HClO_4$
H_2SO_4	$HClO_3$

Recall that acids made of hydrogen, oxygen, and some other element are called oxoacids. They include many common and important acids, and Table 11.2 gives a list of some. Those that are *strong* are marked by asterisks. If you have not already done so, you should memorize the names and formulas of these strong acids, since there are relatively few of them. Once you know which are strong, you will know that any others are weak.

The ionizable hydrogens of oxoacids are attached to oxygen atoms and these, in turn, are held by some other atom — usually that of a nonmetal or a metalloid — found at or near the center of the molecule. Oxoacids, in other words, have one or more O—H

TABLE 11.2 Some Oxoacids of the Nonmetals and Metalloids

Group IVA	Group VA		Group VIA		Group VIIA	
H_2CO_3 Carbonic acid	*HNO_3	Nitric acid[a]	—		HFO	Hypofluorous acid
	HNO_2	Nitrous acid				
	H_3PO_4	Phosphoric acid	*H_2SO_4	Sulfuric acid	*$HClO_4$	Perchloric acid
	H_3PO_3	Phosphorous acid[b]	H_2SO_3	Sulfurous acid	*$HClO_3$	Chloric acid
					$HClO_2$	Chlorous acid
					HClO	Hypochlorous acid
	H_3AsO_4	Arsenic acid	H_2SeO_4	Selenic acid	$HBrO_4$	Perbromic acid
	H_3AsO_3	Arsenous acid	H_2SeO_3	Selenous acid	$HBrO_3$	Bromic acid
					HIO_4 (H_5IO_6)	Periodic acid[c]
					HIO_3	Iodic acid

[a] Strong acids are marked with asterisks.
[b] This acid only gives two H^+ per molecule, not three as its formula might suggest.
[c] H_5IO_6 is formed from $HIO_4 + 2H_2O$.

groups attached to a central atom. For example, here are the Lewis structures of two oxoacids of Group VIA elements. The central atom in sulfuric acid is S, and in selenic acid it is Se.

$$H-O-\underset{\underset{O}{|}}{\overset{\overset{O}{|}}{S}}-O-H \qquad H-O-\underset{\underset{O}{|}}{\overset{\overset{O}{|}}{Se}}-O-H$$

$$\begin{array}{cc} H_2SO_4 & H_2SeO_4 \\ \text{sulfuric acid} & \text{selenic acid} \end{array}$$

Sulfuric acid is a stronger acid than selenic acid, and the question is "why?" But we might first ask a prior question, "why should either be acidic at all?"

To answer the last question, we know that to be acidic means that H^+ can transfer to H_2O in an ionization. This H atom in the oxoacids is bound to an atom of the second most electronegative element, oxygen, so the O—H bond is a very polar bond. Even in the un-ionized molecular acid, the H atom already is part way toward having a positive charge. It is part way, in other words, toward separating as H^+.

$$\overset{\delta^-}{O}-\overset{\delta^+}{H}$$
polarity of O—H
bond in oxoacids

Thus, one reason why oxoacids with O—H groups are acidic in the first place is because the O—H bond is already partly ionic.

The more polar the O—H bond is, the greater is δ^+ on H, so the more readily can this H transfer as H^+ to H_2O, and the stronger the acid will be. Here is where the central atom exerts an additional influence on acid strength. When comparing oxoacids of similar structures, the size of the δ^+ on H in the O—H unit varies in a systematic way with the location of the central atom in the periodic table.

> When the central atoms are in the same group and they hold the same number of oxygen atoms, the acid strength increases from bottom to top within the group.

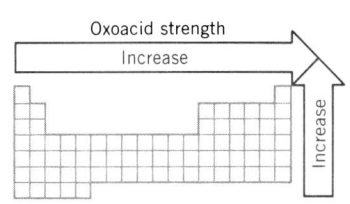

The reason for this is now easy to understand. Central atoms of the same group have electronegativities that increase from bottom to top. And if we consider the three-atom sequence, X—O—H, we find that as X becomes more electronegative, it draws electron density toward itself and away from the oxygen, which makes the oxygen even more able to draw electron density away from the hydrogen. An increase in the electronegativity of X, therefore, increases the shift of electron density down the entire X—O—H chain toward the X and away from the H and leaves a larger δ^+ on the H. And the larger the δ^+ on H, the more able is the H to transfer as H^+.

We can now understand why H_2SO_4 is a stronger acid than H_2SeO_4. Their central atoms are in the same group, with sulfur being nearer the top. Electronegativities increase from bottom to top in a group, so we know that sulfur is more electronegative than selenium. An S atom, therefore, can cause a larger δ^+ on the H of an adjoining O—H group than can an Se atom. So sulfuric acid is a stronger acid than selenic acid.

The same trend holds among the Group VA elements. Phosphoric acid, H_3PO_4, is a stronger acid than arsenic acid, H_3AsO_4, because P is more electronegative than As, which is below P in Group VA.

Another generalization about the relative strengths of oxoacids also stems from the relative electronegativities of central atoms.

> For oxoacids with the same number of oxygens on the central atom, acid strength increases from left to right in a period.

Electronegativities, of course, also increase from left to right within a period. Phosphorus, for example, is just to the left of sulfur in the third period, and is less electronegative than sulfur. The P atom in H_3PO_4, therefore, cannot contribute to the size of δ^+ on the H of an O—H group as much as S, so H_3PO_4 is a weaker acid than H_2SO_4.

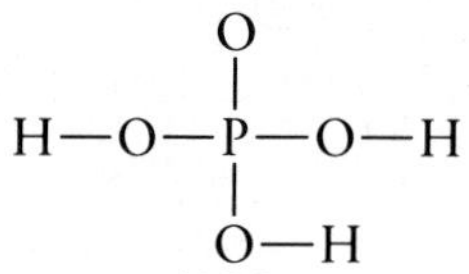

<pre>
        O   O
        | |
H—O—P—O—H   H—O—S—O—H
        | |
       O—H   O
      H₃PO₄   H₂SO₄
  phosphoric acid   sulfuric acid
</pre>

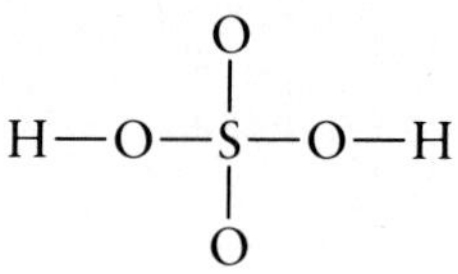

Still another factor that contributes to the relative acidities of oxoacids is their number of *lone oxygens,* those bonded to the central atom but not also bonded to any H atoms. As you can see in their structures above, two such lone oxygens occur in H_2SO_4 but only one is present in H_3PO_4. Following ionization, three lone oxygens are in the HSO_4^- anion but only two are in the $H_2PO_4^-$ anion. The lone oxygens in such anions, not the oxygens still bonded to hydrogens, bear the major burden of carrying the negative charge. The negative charge actually spreads uniformly to these lone oxygens. But by such spreading of a single charge, each of the three lone oxygens in HSO_4^- carries less of the anion's negative charge than is carried by any of the two lone oxygens in $H_2PO_4^-$. Less negative charge means, of course, less ability to attract something positively charged. As a result, any given lone oxygen in the HSO_4^- ion is less able than any lone oxygen in the $H_2PO_4^-$ ion to recapture H^+ from H_3O^+ in the solution. In other words, HSO_4^- cannot become H_2SO_4 again as readily as $H_2PO_4^-$ can become H_3PO_4. So if we compare solutions of H_2SO_4 and H_3PO_4 of equal concentrations, there will be a larger percentage of HSO_4^- ions than of $H_2PO_4^-$ ions. Sulfuric acid, in other words, will be more fully ionized. Of course, this is how we define acid strength—in terms of the percentage of the acid molecules that become ionized in solution—so we say that sulfuric acid is a stronger acid than phosphoric acid.

PRACTICE EXERCISE 30 Predict the stronger acid in each pair: (a) $HClO_4$ or $HBrO_4$, (b) H_3AsO_4 or H_2SeO_4.

In the two boxed generalizations made earlier, we stated an important condition — the number of oxygens held by the central atom must be the same in the formulas of the acids being compared. This is because acid strength also varies with the number of such oxygens, when we are comparing acids with the same central atom.

> For a given central atom, the acid strength of an oxoacid increases with the number of oxygens it holds.

Sulfuric acid (H_2SO_4), for example, is a stronger acid than sulfurous acid (H_2SO_3). Nitric acid (HNO_3) is a stronger acid than nitrous acid (HNO_2).

The extra oxygen on S in sulfuric acid contributes to an additional withdrawal of electron density from each O—H bond and thus makes each O—H bond more polar in this acid than in sulfurous acid. So the size of δ^+ on H in sulfuric acid is greater than it is in sulfurous acid, making sulfuric acid the stronger acid of the two. The same argument explains why nitric acid is a stronger acid than nitrous acid. Also at work here is the effect of the number of lone oxygens. In the HSO_4^- ion there are three lone oxygens and in HSO_3^- there are two. The greater number of lone oxygens in the HSO_4^- ion leads to a greater spreading out of the negative charge than occurs in the HSO_3^- ion. Consequently, HSO_4^- has a lesser tendency than HSO_3^- to recapture H^+. So in solutions of equal concentrations of sulfuric acid and sulfurous acid, the percentage ionization in H_2SO_4 is higher, and sulfuric acid is the stronger acid. The same factors help to make nitric acid a stronger acid than nitrous acid.

PRACTICE EXERCISE 31 Predict the order of acid strength for the four oxoacids of chlorine in Table 11.2.

Acids That Do Not Contain Oxygen

The binary compounds between hydrogen and many nonmetals are also acidic; they are called binary acids. HCl is an example already mentioned. Table 11.3 lists the binary hydrogen compounds that are acids in water. Those that are strong are once again

Strong Binary Acids
HCl HBr HI

TABLE 11.3 Binary Compounds of Hydrogen and Nonmetals That Are Acidic[a]

Group VI	Group VII
(H_2O)	HF Hydrofluoric acid
H_2S Hydrosulfuric acid	*HCl Hydrochloric acid
H_2Se Hydroselenic acid	*HBr Hydrobromic acid
H_2Te Hydrotelluric acid	*HI Hydriodic acid

[a] The names apply to the solutions of these compounds in water. Strong acids are marked with asterisks.

marked by asterisks. Be sure to add them to your memorized list of strong acids. Water, a very weak acid as we have mentioned, is included. Its reaction with itself, however, produces H_3O^+ and OH^- ions in equal numbers, so pure water is neutral. Even though the ionization of water occurs to a very small extent, the resulting system is extremely important. We will have much to say about it in Chapter 15.

The relative strengths of the binary acids can be correlated with the periodic table in two ways.

> The strengths of the binary acids increase from left to right in a period.

These horizontal trends in acid strength are also consistent with trends in electronegativity. As we go from S to Cl or from O to F, the electronegativity increases, so the H—X bond becomes more polar, making δ^+ on H greater. Hence the H separates more easily as H^+, and the acid is stronger. So HCl is a stronger acid than H_2S and HF is a stronger acid than H_2O.

The final correlation will seem at first to violate what we have been emphasizing — the importance of electronegativity.

Binary acid strength
Increase

Increase

> Within a group, the strengths of binary acids increase from top to bottom.

Among the binary acids of the halogens, for example, the following is the order of relative acidity.

$$HF < HCl < HBr < HI$$

HF is the *weakest* acid in the series, and HI is the strongest. How can this be, since F is the *most* electronegative of the halogens, and I is the least electronegative? Shouldn't the H—F bond be more polar than that H—I bond? In fact it is. Therefore, why isn't H—F the most acidic of these binary acids of the halogens? Evidently, in our theory about acid strengths, relative electronegativities must be just one factor.

Another factor concerns the strength of the bond from the nonmetal atom to H. The farther that the electron pair of this bond is from the nucleus of the nonmetal atom, the weaker is this bond. In the oxoacids that we studied, the bond was always between the same two elements, H and O. In the binary acids, this is not true. Instead, as we proceed down a group in the periodic table, the nonmetal atom holding H becomes progressively larger. With its increasing size, its ability to hold the H atom becomes progressively weaker. As we descend the halogen group, this loss in H—X bond strength tends to make the acids stronger. It more than compensates for the lessening size of $\delta +$ on H as X becomes less electronegative, which tends to make acids weaker. The net effect of these two opposing factors is an increase in the strengths of the binary acids of the halogens as we descend the group.

In the series of the binary acids of Group VIA elements we see the identical trend. Their formulas are of the general type H_2X, and the farther X is from the top of the group the stronger H_2X is as an acid.

PRACTICE EXERCISE 32   Using only the periodic table, and without looking at Table 11.3, choose the stronger acid of each pair: (a) H_2Se or HBr, (b) H_2Se or H_2Te, (c) CH_3OH or CH_3SH.

11.9 LEWIS CONCEPT OF ACIDS AND BASES

G. N. Lewis defined an acid as an electron pair acceptor and a base as an electron pair donor.

In the Brønsted view of acids and bases, the key reaction is simply the transfer of a proton from the acid to the base. Looking at acids and bases in this way frees us from the need to have water present in an acid–base reaction, but we are still restricted to reactants that donate or accept protons. There are other reactions, however, with the "look and feel" of acid–base reactions, but which do not involve proton transfer. For example, sulfur dioxide, which is the anhydride of sulfurous acid, combines with calcium oxide, which is the anhydride of calcium hydroxide, to give calcium sulfite ($CaSO_3$).

$$CaO(s) + SO_2(g) \longrightarrow CaSO_3(s)$$

The product of this reaction is the same as the salt formed when sulfurous acid and calcium hydroxide react in water.

$$H_2SO_3(aq) + Ca(OH)_2(aq) \longrightarrow CaSO_3(s) + 2H_2O$$

In considering reactions such as these, G. N. Lewis saw a way to broaden the concept of acids and bases beyond the Brønsted view. Lewis acids and bases are defined as follows.

Lewis Definitions of Acids and Bases

1. An acid is any ionic or molecular species that can accept a pair of electrons in the formation of a coordinate covalent bond.
2. A base is any ionic or molecular species that can donate a pair of electrons in the formation of a coordinate covalent bond.
3. Neutralization is the formation of a coordinate covalent bond between the donor (base) and the acceptor (acid).

Keep in mind that the notion of a coordinate covalent bond just helps us keep track of where the electrons in the bond came from. Once formed, the bond is just like any other covalent bond.

The reaction between ammonia and boron trichloride, which was described in Chapter 7, is a classic example of a Lewis acid–base reaction.

The ammonia molecule has an unshared pair of electrons and is a Lewis base. The boron atom in BCl_3 has only six electrons in its valence shell and seeks two more to achieve an octet. By accepting the electrons from the ammonia molecule, BCl_3 is a Lewis acid. As we learned in Section 7.8, the coordinate covalent bond that is formed between the boron and the nitrogen can be represented by an arrow.

The Lewis concept includes both the Arrhenius and Brønsted acids and bases as special cases. When hydronium ion reacts with hydroxide ion, for example, a proton is transferred. In the Lewis view, this proton is an acid that leaves one particle and becomes joined to another. When it becomes attached to the hydroxide ion, a Lewis acid–base "neutralization" takes place.

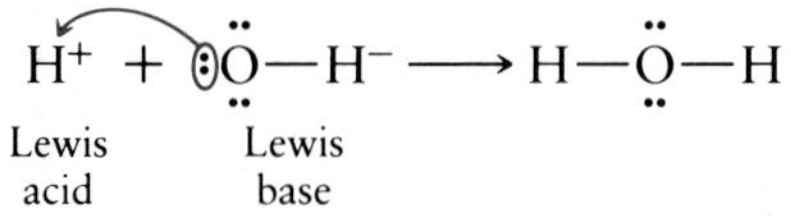

$$H^+ + \overset{..}{O}-H^- \longrightarrow H-\overset{..}{\underset{..}{O}}-H$$

Lewis   Lewis
acid   base

Earlier in this section we commented that the Lewis approach frees us from proton transfer as a criterion for an acid–base reaction. We can analyze the reaction between sulfur dioxide and oxide ion (present in CaO) in the following way.

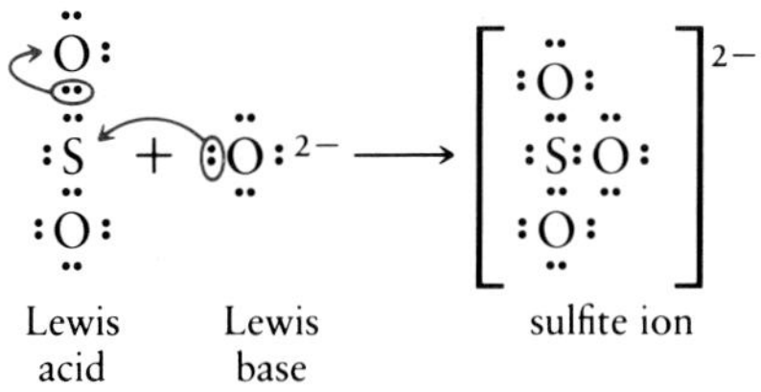

$$\underset{\substack{Lewis\\acid}}{\overset{\displaystyle :\overset{..}{O}:}{:S} :\overset{..}{O}:} \; + \; \underset{\substack{Lewis\\base}}{:\overset{..}{O}:^{2-}} \longrightarrow \left[\begin{array}{c} :\overset{..}{O}: \\ :S:\overset{..}{O}: \\ :\overset{..}{O}: \end{array} \right]^{2-}$$

sulfite ion

Notice that the oxide ion supplies the electron pair for the covalent bond and that the electrons shift about in the SO_2 molecule to make room for the additional bond. Here is another example that involves the movement of some atoms as well.

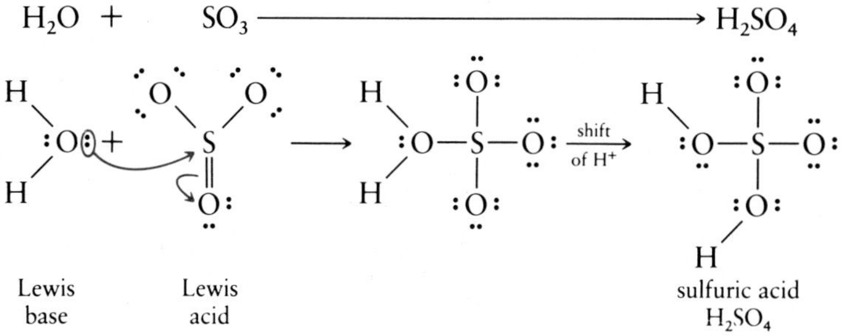

$$H_2O \; + \; SO_3 \longrightarrow H_2SO_4$$

Lewis   Lewis   sulfuric acid
base   acid H_2SO_4

PRACTICE EXERCISE 33

This behavior of OH^- toward CO_2 means that a standard solution of NaOH or KOH exposed to the air, which contains CO_2, will not long remain at the normality given on the label of the bottle.

When carbon dioxide is bubbled into aqueous sodium hydroxide, the gas is instantly trapped as the bicarbonate ion. (In fact, this reaction makes it hard to store standard NaOH solutions in contact with air, in which traces of CO_2 are always present.) The reaction can be written as follows.

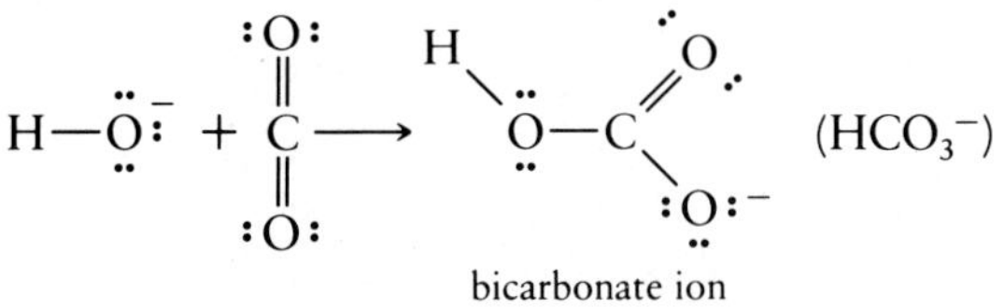

bicarbonate ion

Which is the Lewis acid and which is the Lewis base?

11.10 COMPLEX IONS

Many metal ions combine with molecules or negative ions by Lewis acid–base reactions to form complex ions.

In our previous discussions of metal-containing compounds, we left you with the impression that the only kinds of bonds in which metals are ever involved are ionic bonds.

For some metals, such as the alkali metals of Group IA, this is essentially true. But for many others, especially the transition metals and the post-transition metals, it is not. This is because the ions of many of these metals are able to behave as Lewis acids, and by participating in Lewis acid–base reactions they become covalently bonded to other atoms. Copper(II) ion is a typical example.

In water solutions of copper(II) salts, such as $CuSO_4$ or $Cu(NO_3)_2$, the copper is not present as simple Cu^{2+} ions. Instead, each Cu^{2+} ion becomes bonded to four water molecules, which gives a pale blue ion with the formula $Cu(H_2O)_4{}^{2+}$. We call it a complex ion because it is composed of a number of simpler species (i.e., it is *complex*, not simple). The chemical equation for the formation of the $Cu(H_2O)_4{}^{2+}$ ion is

The origin of the colors of complexes will be discussed in Chapter 22.

$$Cu^{2+} + 4H_2O \longrightarrow Cu(H_2O)_4{}^{2+}$$

which can be diagrammed using Lewis structures as follows:

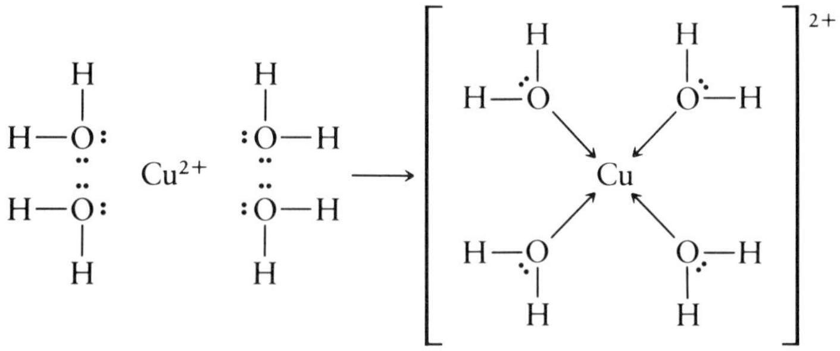

As you can see, in this analysis the Cu^{2+} ion accepts pairs of electrons from the water molecules, so Cu^{2+} is a Lewis acid and the water molecules are each Lewis bases.

The study of the properties, reactions, structures, and bonding in complex ions such as $Cu(H_2O)_4{}^{2+}$ has become an important specialty within chemistry, and as such it has developed its own terminology. For example, a Lewis base that attaches itself to a metal ion is called a ligand, from the Latin word *ligare*, meaning "to bind." As we will see, ligands can be neutral molecules with unshared pairs of electrons (like the water molecule) or they can be anions. The atom in the ligand that actually provides the electron pair is called the donor atom, and the metal ion is the acceptor. The result of combining a metal ion with one or more ligands is a complex ion, or simply just a complex. Using just the word "complex" avoids problems when the particle formed is electrically neutral, as sometimes happens. Compounds that contain complex ions are generally referred to as coordination compounds because the bonds in a complex ion can be viewed as coordinate covalent bonds.

Ligands

As we've noted, ligands may be either anions or neutral molecules. In either case, they are Lewis bases and, therefore, contain at least one atom with one or more lone pairs of electrons.

Anions that serve as ligands include many simple monatomic ions such as the halides (F^-, Cl^-, Br^-, I^-) and sulfide ion (S^{2-}). Common polyatomic anions that function as ligands are nitrite ion ($NO_2{}^-$), cyanide ion (CN^-), hydroxide ion (OH^-), thiocyanate ion (SCN^-), and thiosulfate ion ($S_2O_3{}^{2-}$). This is really only a small sampling, not a complete list.

The most common neutral molecule that serves as a ligand is water, and most of the reactions of metal ions in aqueous solutions are actually reactions of their complex ions—ions in which the metal is attached to some number of water molecules. This number isn't always the same, however. For example, copper(II) forms the complex ion

FIGURE 11.7 (*a*) Aqueous CuSO₄ is a blue solution. (*b*) Aqueous CuSO₄ to which ammonia has been added is deep blue in color.

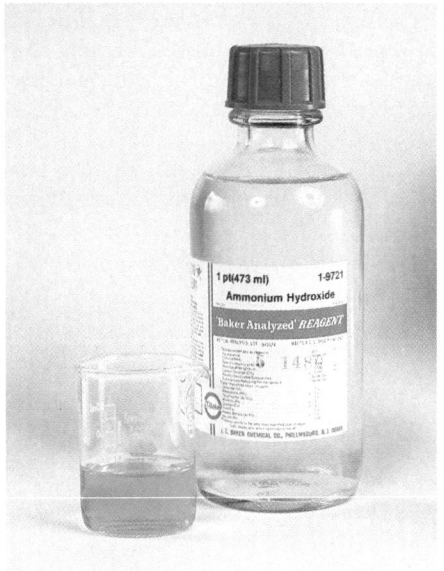

(a)

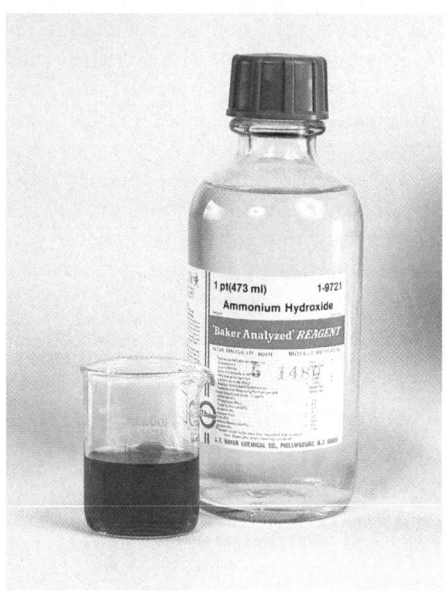

(b)

Cu(H₂O)₄²⁺ (as we've noted earlier). but cobalt(II) combines with water molecules to form Co(H₂O)₆²⁺. Another common neutral ligand is ammonia, NH₃, which has one lone pair of electrons on the nitrogen atom. If ammonia is added to an aqueous solution containing the Cu(H₂O)₄²⁺ ion, for example, the color deepens dramatically as ammonia molecules displace water molecules (see Figure 11.7).

$$Cu(H_2O)_4^{2+}(aq) + 4NH_3(aq) \longrightarrow Cu(NH_3)_4^{2+}(aq) + 4H_2O$$
$$\text{(pale blue)} \qquad\qquad\qquad\qquad \text{(deep blue)}$$

From the Latin dentatus *("toothed").*

Each of the ligands that we have discussed so far is able to use just one atom to attach itself to a metal ion. Such ligands are called monodentate ligands, indicating that they have only "one tooth" with which to "bite" the metal ion. There are also many ligands that have two or more donor atoms, and collectively they are referred to as polydentate ligands. Among the most common polydentate ligands are those with two donor atoms. These are called bidentate ligands, and when they form complexes, both donor atoms become attached to the same metal ion. Examples of bidentate ligands are oxalate ion and ethylenediamine.

$$\begin{array}{c} :O::O: \\ \| \| \\ ^-:\ddot{O}-C-C-\ddot{O}:^- \\ \phantom{..:\ddot{O}}\text{oxalate ion} \end{array} \qquad \ddot{N}H_2-CH_2-CH_2-\ddot{N}H_2$$
$$\text{ethylenediamine}$$

When these ligands become attached to a metal ion, ring structures are formed as shown.

an oxalate complex   an ethylenediamine complex

Such structures are important in "complex ion chemistry," and we will study them in greater detail in Chapter 22.

One of the most common polydentate ligands is a compound called ethylenediaminetetraacetic acid, mercifully abbreviated EDTA. The H atoms attached to the oxygen atoms are easily removed as protons, which gives an anion with a charge of 4 −. The structures of EDTA and its anion, EDTA⁴⁻, are as follows, with the donor atoms printed in red.

Pronounced ethyl-ene-di-amine-tetra-acetic acid.

$$\text{EDTA}$$

$$\text{EDTA}^{4-}$$

As you can see, the EDTA⁴⁻ ion has six donor atoms, and this permits it to wrap itself around a metal ion and form very stable complexes.

EDTA is a particularly important ligand. It is relatively nontoxic, which allows it to be used in trace amounts in foods to retard spoilage. If you look at the labels on bottles of salad dressings, for example, you will find that one of the ingredients often is $CaNa_2EDTA$. The EDTA⁴⁻ available from this salt forms soluble complex ions with any traces of metal ions that might otherwise promote reactions of the salad oils with oxygen, and thereby lead to spoilage. (The calcium salt is used because the EDTA⁴⁻ ion would otherwise extract Ca^{2+} ions from bones, and that would be harmful.) Many shampoos contain Na_4EDTA to soften water. EDTA⁴⁻ binds to Ca^{2+}, Mg^{2+}, and Fe^{3+} ions, which removes them from the water and prevents them from interfering with the action of soaps in the shampoo. EDTA is also sometimes added in small amounts to whole blood to prevent clotting. It ties up calcium ions, which the clotting process requires. EDTA has even been used in poison treatment because it can help remove poisonous heavy metal ions such as Pb^{2+} from the body when they have been accidentally ingested.

Writing Formulas for Complexes

When we write the formula for a complex, we follow two rules:

1. The symbol for the metal ion is always given first, followed by the ligands.

2. The charge on the complex is the algebraic sum of the charge on the metal ion and the charges on the ligands.

For example, the formula of the complex ion of Cu^{2+} and water, which we mentioned earlier, was written $Cu(H_2O)_4{}^{2+}$ with the copper first followed by the ligands. The charge on the complex is 2 + because the copper ion has a charge of 2 + and the water molecules are neutral. Copper(II) ion also forms a complex ion with four cyanide ions (CN^-). In this

case the metal ion contributes two positive charges and the four ligands contribute a total of four negative charges (one for each cyanide ion). The algebraic sum is therefore $(2+) + (4-) = 2-$, so the complex ion has a charge of $2-$. The formula of the ion is then $Cu(CN)_4^{2-}$.

Often, the formula for a complex ion is enclosed within parentheses, with the charge on the ion shown outside. Thus, the two complexes just mentioned would be written as $[Cu(H_2O)_4]^{2+}$ and $[Cu(CN)_4]^{2-}$. The purpose of this is to emphasize that the ligands are attached to the metal ion and are not free to roam about. Brackets are needed particularly when a complex ion contains more than one kind of ligand or when it is isolated in a salt. For example, one of the many complex ions formed by chromium(III) ion contains as ligands five water molecules and one chloride ion. To indicate that all of them are attached to the Cr^{3+} ion, we enclose the formula within brackets and write it as $[Cr(H_2O)_5Cl]^{2+}$. This ion can also be isolated as a solid chloride salt, and the formula for the solid is written $[Cr(H_2O)_5Cl]Cl_2$, which clearly shows that five water molecules and a chloride ion are bonded to the chromium, and the other two chloride ions are present to provide electrical neutrality for the salt. Also, notice that the rule for writing the formula for the salt conforms to previous practice; we write the cation first, followed by the anion. In this case, the cation is the complex ion $[Cr(H_2O)_5Cl]^{2+}$.

In Chapter 1 you learned about hydrates, and one of the compounds described was copper sulfate, which forms beautiful blue crystals with the formula $CuSO_4 \cdot 5H_2O$. It was much too early then to make the distinction, but the formula should have been given as $[Cu(H_2O)_4]SO_4 \cdot H_2O$ to show that four of the five water molecules are held in the crystal as part of the complex ion $[Cu(H_2O)_4]^{2+}$. The fifth water molecule is held in the crystal by being hydrogen bonded to the sulfate ion. Many hydrates of metal salts actually contain complex ions of the metals in which water is the ligand. Cobalt salts such as cobalt(II) chloride, for example, crystallize from aqueous solutions as hexahydrates (meaning they containing six water molecules per formula unit of the salt). The compound $CoCl_2 \cdot 6H_2O$ (Figure 11.8) actually is $[Co(H_2O)_6]Cl_2$, and contains the pink complex $[Co(H_2O)_6]^{2+}$. This ion also gives solutions of cobalt(II) salts a pink color as can also be seen in Figure 11.8. Although most hydrates of metal salts contain complex ions, the distinction is seldom made, and it's acceptable to write the formula for these hydrates and any others in the usual fashion.

Figure 11.8 Cobalt chloride taken from a reagent bottle has the formula $CoCl_2 \cdot 6H_2O$. It actually contains the complex ion, $Co(H_2O)_6^{2+}$, which imparts a rose color to both the crystals and solutions of the compound.

EXAMPLE 11.22   WRITING THE FORMULA FOR A COMPLEX ION

Problem: Write the formula for the complex ion formed by the metal ion Cr^{3+} and six NO_2^- ions as ligands. Decide whether the complex could be isolated as a chloride salt or a potassium salt, and write the formula for the appropriate salt.

Solution: First, let's calculate the charge on the complex. Six NO_2^- ions contribute a total charge of $6-$; the metal contributes a charge of $3+$. The algebraic sum is $(6-) + (3+) = 3-$. The formula is therefore

$$[Cr(NO_2)_6]^{3-}$$

Since the complex is an anion, it requires a cation to form a neutral salt, so the complex would be isolated as a potassium salt, not as a chloride salt. For the salt to be electrically neutral, there must be three K^+ ions for each anionic complex. The salt is therefore

$$K_3[Cr(NO_2)_6]$$

PRACTICE EXERCISE 34   Write the formula of the complex ion formed by Ag^+ and two thiosulfate ions, $S_2O_3^{2-}$. If we were able to isolate this complex ion as its ammonium salt, what would be the formula for the salt?

PRACTICE EXERCISE 35   Aluminum chloride crystallizes from aqueous solutions as a hexahydrate. Write the formula for the salt and suggest a formula for the complex ion formed by aluminum ion and water.

Why Complexes Are Important

As you may have already guessed, the number of complexes is huge. They occur widely in nature, and we encounter them every day. We already mentioned the uses of EDTA in household products and as an additive in food products. This barely scratches the surface, however. Most of the trace metals that our bodies require are involved as complex ions. In hemoglobin, for example, which is the carrier of oxygen in the blood, it is an Fe^{2+} ion held in a complex ion that actually binds the oxygen. A similar complex of Co^{2+} is part of vitamin B_{12}. Chlorophyll, the green pigment in leaves that absorbs solar energy for photosynthesis, has a Mg^{2+} ion held in a complex. Scientists have even found that some complexes help them fight disease. A very simple complex containing platinum(II) ion, chloride ion, and ammonia molecules has the formula $[Pt(NH_3)_2Cl_2]$. One particular form of this complex that goes by the name cisplatin is used as an anticancer drug. Thus, no matter where we look in the plant or animal kingdoms we are almost certain to find complexes.

Many enzymes require metal ions in acceptor roles.

SUMMARY

Electrolytes.   Substances that dissociate or ionize in water to produce cations and anions are electrolytes; those that do not are called nonelectrolytes. When the ions enter the solution they become hydrated. Electrolytes include salts and metal hydroxides as well as molecular acids and bases that ionize by reaction with water. Remember that the atoms of polyatomic ions remain together as these ions enter a solution.

Acids and Bases as Electrolytes.   The modern version of the Arrhenius definition of acids and bases is that an acid is a substance that produces H_3O^+ ions when dissolved in water, and a base produces OH^- ions when dissolved in water. Strong acids and bases are strong electrolytes; weak acids and bases are weak electrolytes. In a solution of a weak electrolyte there is a chemical equilibrium between the nonionized molecules of the solute and the ions formed by the reaction of the solute with water.

Ionic Reactions.   Reactions between electrolytes usually involve the ions themselves. Equations for these reactions can be written in three different ways. In molecular equations, complete formulas for all reactants and products are used. In an ionic equation, soluble strong electrolytes are written in dissociated (ionized) form; "molecular" formulas are used for solids and weak electrolytes. A net ionic equation is obtained by eliminating spectator ions from the ionic equation.

Metathesis Reactions.   Metathesis or double replacement reactions occur when a solid, a weak electrolyte, or a gas is formed among the products. Metal oxides react with acids because of the formation of the very weak electrolyte, water. You should learn the solubility rules, and remember that all salts are strong electrolytes. Remember that all strong acids and bases are strong electrolytes, too. Everything else is a weak electrolyte or a nonelectrolyte. Be sure to learn the reactions that produce gases in metathesis reactions, which are found in Table 11.1.

Solution Stoichiometry and Titrations.   Molarity is one of the convenient concentration units for quantitative work with ionic reactions. In limiting reactant problems, calculate the moles of each ion in the reaction mixture before reaction occurs; determine the limiting reactant; then calculate the moles of ions in the reaction mixture after the reaction. In computing final ion concentrations, remember to use the final total volume of the solution. Titration is a laboratory technique that is used to make quantitative measurements of amounts of solutions needed to obtain complete reaction. In an acid–base titration, the endpoint is normally detected visually using an acid–base indicator.

Equivalents.   Equivalents are defined so they always react in a one-to-one ratio. An equivalent of an acid supplies one mole of H^+, and an equivalent of a base supplies one mole of OH^- or is by some other species able to neutralize one mole of H^+ ion. The number of equivalents in one mole of an acid is equal to the number of hydrogens of the acid that are neutralized. Normally,

the number of equivalents in one mole of base is equal to the number of hydroxides in one formula unit of the base.

Equivalent Weights and Normality.   The equivalent weight is the mass of one equivalent. Equivalent weights are used to establish mass relationships that can be used in stoichiometric calculations. Normality is the number of equivalents of solute per liter of solution. The product of normality and volume (in liters) gives equivalents. For reactions between solutions, the equation that applies is

$$N_A V_A = N_B V_B$$

Brønsted Acids and Bases.   A Brønsted acid is a proton donor; a Brønsted base is a proton acceptor. In a Brønsted acid–base reaction, there are two conjugate acid–base pairs. The benefits of the Brønsted definition include the freedom from being forced to consider only acid–base reactions in water. A substance that can be either an acid or a base, depending on the nature of the other reactant, is amphoteric (amphiprotic).

Oxoacids, which contain oxygen atoms in addition to hydrogen and another element, increase in strength as the number of oxygen atoms increases. Oxoacids having the same number of oxygens increase in strength from bottom to top within a group and from left to right across a period. Binary acids contain only hydrogen and another nonmetal. Their strength increases from top to bottom within a group and left to right across a period.

Lewis Acids and Bases.   Under the Lewis definition, an acid accepts a pair of electrons in the formation of a covalent bond, and a base donates an electron pair in the formation of a covalent bond. Lewis acid–base reactions do not require the presence of a solvent, or even a proton. Arrhenius and Brønsted acid–base reactions are special cases of Lewis acid–base reactions.

Complex Ions.   Coordination compounds contain complex ions formed from a metal ion and a number of ligands. Ligands are Lewis bases and may be monodentate, bidentate, or, in general, polydentate, depending on the number of donor atoms that they contain. Water is the most common monodentate ligand. A common bidentate ligand is ethylenediamine (en) and a common polydentate ligand is ethylenediaminetetraacetic acid (EDTA), which has six donor atoms.

REVIEW EXERCISES

Answers to questions whose numbers are printed in color are given in Appendix C. Difficult questions are marked with asterisks.

Electrolytes

11.1   What is an electrolyte? What is a nonelectrolyte?

11.2   Define *dissociation* as it applies to ionic compounds that dissolve in water. What does *hydration* mean?

11.3   Write equations for the dissociation of the following ionic compounds in water: (a) LiCl, (b) $BaCl_2$, (c) $Al(C_2H_3O_2)_3$, (d) $(NH_4)_2CO_3$, (e) $FeCl_3$.

11.4   Write equations for the dissociation of the following ionic compounds in water: (a) $CuSO_4$, (b) $Al_2(SO_4)_3$, (c) $CrCl_3$, (d) $(NH_4)_2HPO_4$, (e) $KMnO_4$.

Acids and Bases As Electrolytes

11.5   How did Arrhenius define an acid and a base?

11.6 What is the difference between dissociation and ionization as applied to the discussion of electrolytes?

11.7 Pure $HClO_4$ is molecular. Write an equation for its reaction with water to give H_3O^+ and ClO_4^- ions.

11.8 What is the difference between a strong electrolyte and a weak electrolyte?

11.9 If a substance is a weak electrolyte, what does this mean in terms of the tendency of the ions to react to reform the molecular compound? How does this compare with strong electrolytes?

11.10 Nitrous acid, HNO_2, is a weak acid. Write an equation showing its reaction with water.

11.11 Hydrazine is a toxic substance that can be formed when household ammonia is mixed with a bleach such as Clorox. Its formula is N_2H_4, and it is a weak base. Write a chemical equation showing its reaction with water.

11.12 Why don't we use double arrows in the equation for the reaction of a strong acid with water?

11.13 $HClO_3$ is a strong acid. Write an equation for its reaction with water.

11.14 When diprotic acids, such as H_2CO_3, react with water to release their H^+, they do so in two steps, each of which is a dynamic equilibrium. Write chemical equations that illustrate this for H_2CO_3.

11.15 Phosphoric acid is a weak acid that undergoes ionization in three steps. Write chemical equations for each of these reactions.

Ionic Reactions

11.16 Define *ionic reaction*. What is another name for *metathesis reaction*? What is a *precipitate*? What are *spectator ions*?

11.17 What are the differences among molecular, ionic, and net ionic equations?

11.18 What *two* conditions must be fulfilled by a balanced ionic equation?

11.19 Write ionic and net ionic equations for these reactions.
(a) $(NH_4)_2CO_3(aq) + MgCl_2(aq) \longrightarrow$
$2NH_4Cl(aq) + MgCO_3(s)$
(b) $CuCl_2(aq) + 2NaOH(aq) \longrightarrow$
$Cu(OH)_2(s) + 2NaCl(aq)$
(c) $3FeSO_4(aq) + 2Na_3PO_4(aq) \longrightarrow$
$Fe_3(PO_4)_2(s) + 3Na_2SO_4(aq)$
(d) $2AgC_2H_3O_2(aq) + NiCl_2(aq) \longrightarrow$
$2AgCl(s) + Ni(C_2H_3O_2)_2(aq)$

11.20 Write *balanced* ionic and net ionic equations for these reactions.
(a) $CuSO_4(aq) + BaCl_2(aq) \longrightarrow CuCl_2(aq) + BaSO_4(s)$
(b) $Fe(NO_3)_3(aq) + LiOH(aq) \longrightarrow$
$LiNO_3(aq) + Fe(OH)_3(s)$
(c) $Na_3PO_4(aq) + CaCl_2(aq) \longrightarrow$
$Ca_3(PO_4)_2(s) + NaCl(aq)$
(d) $Na_2S(aq) + AgC_2H_3O_2(aq) \longrightarrow$
$NaC_2H_3O_2(aq) + Ag_2S(s)$

11.21 The following equation *is not* balanced. What is wrong with it?
$$NO_2 + H_2O \longrightarrow NO_3^- + 2H^+$$

11.22 Aqueous solutions of sodium sulfide, Na_2S, and copper nitrate, $Cu(NO_3)_2$, are mixed. A precipitate of copper sulfide, CuS, forms at once. Left behind is a solution of sodium nitrate, $NaNO_3$. Write the net ionic equation for this reaction.

11.23 Silver bromide is "insoluble." What does this mean about the concentrations of Ag^+ and Br^- in a saturated solution of $AgBr$? Explain why a precipitate of $AgBr$ forms when solutions of the soluble salts $AgNO_3$ and $NaBr$ are mixed.

11.24 Silver bromide (mentioned in the preceding question) is the chief light-sensitive substance used in the manufacture of photographic film. It can be prepared by mixing solutions of $AgNO_3$ and $NaBr$. Taking into account that $NaNO_3$ is soluble, write molecular, ionic, and net ionic equations for this reaction.

11.25 Trisodium phosphate, Na_3PO_4, is a useful cleaning agent, but it must be handled with care because its solutions are quite caustic. If a solution of Na_3PO_4 is added to one containing a calcium salt such as $CaCl_2$, a precipitate of calcium phosphate is formed. Write molecular, ionic, and net ionic equations for the reaction.

Predicting Double Replacement Reactions

11.26 Study the solubility rules on page 399; then choose the compounds that are soluble in water.
(a) $Ca(NO_3)_2$ (c) $Ni(OH)_2$ (e) $BaSO_4$
(b) $FeCl_2$ (d) $AgNO_3$ (f) $CuCO_3$

11.27 After studying the solubility rules on page 399, pick the compounds that are insoluble in water.
(a) $AgCl$ (d) $Ca_3(PO_4)_2$
(b) $Cr_2(SO_4)_3$ (e) $Al(C_2H_3O_2)_3$
(c) $(NH_4)_2CO_3$ (f) ZnO

11.28 Complete and balance the following reactions, and then write the ionic and net ionic equations. If all ions cancel, indicate that no reaction (N.R.) takes place.
(a) $Na_2SO_3 + Ba(NO_3)_2 \longrightarrow$
(b) $K_2S + ZnCl_2 \longrightarrow$
(c) $NH_4Br + Pb(C_2H_3O_2)_2 \longrightarrow$
(d) $NH_4ClO_4 + Cu(NO_3)_2 \longrightarrow$

11.29 Complete and balance the following reactions, and then write the ionic and net ionic equations. If all ions cancel, indicate that no reaction (N.R.) takes place.
(a) $(NH_4)_2S + NaCl \longrightarrow$
(b) $Cr_2(SO_4)_3 + K_2CO_3 \longrightarrow$
(c) $Sr(OH)_2 + MgCl_2 \longrightarrow$
(d) $Ba(NO_3)_2 + Na_2SO_3 \longrightarrow$

11.30 Complete and balance the molecular, ionic, and net ionic equations for the following reactions.
(a) $HNO_3 + Cr(OH)_3 \longrightarrow$
(b) $HClO_4 + NaOH \longrightarrow$
(c) $Cu(OH)_2 + HC_2H_3O_2 \longrightarrow$
(d) $ZnO + HBr \longrightarrow$

11.31 Complete and balance molecular, ionic, and net ionic equations for the following reactions.
(a) $NaHSO_3 + HBr \longrightarrow$
(b) $(NH_4)_2CO_3 + NaOH \longrightarrow$
(c) $(NH_4)_2CO_3 + Ba(OH)_2 \longrightarrow$
(d) $FeS + HCl \longrightarrow$

11.32 How would the electrical conductivity of a solution of $Ba(OH)_2$ change as a solution of H_2SO_4 is added slowly to it? Use a net ionic equation to justify your answer.

11.33 Washing soda is $Na_2CO_3 \cdot 10H_2O$. Explain, using chemical equations, how this substance is able to remove Ca^{2+} ions from "hard water."

11.34 If a solution of trisodium phosphate, Na_3PO_4, is poured into seawater, precipitates of calcium phosphate and magnesum phosphate are formed. (Magnesium and calcium ions are among the principal ions found in seawater.) Write net ionic equations for these reactions.

11.35 Choose reactants that would yield the following net ionic equations. Write molecular equations for each.
(a) $HCO_3^- + H^+ \longrightarrow H_2O + CO_2(g)$
(b) $Fe^{2+} + 2OH^- \longrightarrow Fe(OH)_2(s)$
(c) $Ba^{2+} + SO_3^{2-} \longrightarrow BaSO_3(s)$
(d) $2Ag^+ + S^{2-} \longrightarrow Ag_2S(s)$
(e) $ZnO(s) + 2H^+ \longrightarrow Zn^{2+} + H_2O$

11.36 Suppose that you wished to prepare copper(II) carbonate by a precipitation reaction involving Cu^{2+} and CO_3^{2-}. Which of the following pairs of reactants could you use as solutes?
(a) $Cu(OH)_2 + Na_2CO_3$ (d) $CuCl_2 + K_2CO_3$
(b) $CuSO_4 + (NH_4)_2CO_3$ (e) $CuS + NiCO_3$
(c) $Cu(NO_3)_2 + CaCO_3$

11.37 Explain why the following reactions take place.
(a) $CrCl_3 + 3NaOH \longrightarrow Cr(OH)_3 + 3NaCl$
(b) $ZnO + 2HBr \longrightarrow ZnBr_2 + H_2O$
(c) $MnCO_3 + H_2SO_4 \longrightarrow MnSO_4 + H_2O + CO_2$
(d) $Na_2C_2O_4 + 2HNO_3 \longrightarrow 2NaNO_3 + H_2C_2O_4$

Solution Stoichiometry

11.38 Calculate the number of moles of each of the ions in the following solutions.
(a) 25.0 mL of 1.30 M KOH
(b) 37.5 mL of 0.50 M $CaCl_2$
(c) 20.0 mL of 0.50 M $(NH_4)_2CO_3$
(d) 35.0 mL of 0.40 M $Al_2(SO_4)_3$

11.39 How many milliliters of 0.30 M $NiCl_2$ solution are needed to react completely with 25.0 mL of 0.10 M Na_2CO_3 solution? How many grams of $NiCO_3$ will be formed?

11.40 Epsom salts is $MgSO_4 \cdot 7H_2O$. How many grams of this compound are needed to react completely with 100.0 mL of 0.100 M $BaCl_2$ solution?

11.41 How many milliliters of 0.200 M NaOH are needed to completely neutralize 15.0 mL of 0.300 M H_3PO_4?

11.42 How many grams of baking soda, $NaHCO_3$, are needed to react with 150 mL of stomach acid having an HCl concentration of 0.050 M?

*__**11.43**__ In Review Exercise 11.42, how many milliliters of wet CO_2 gas would be formed at a temperature of 37 °C (body temperature) and a pressure of 1.00 atm?

11.44 How many milliliters of 0.400 M $CaCl_2$ would be needed to react completely with 35.0 mL of 0.600 M $AgNO_3$ solution? The equation for the reaction is

$$2AgNO_3(aq) + CaCl_2(aq) \longrightarrow Ca(NO_3)_2(aq) + 2AgCl(s)$$

11.45 How many milliliters of ammonium sulfate solution having a concentration of 0.300 M are needed to react completely with 60.0 mL of 1.00 M NaOH solution? The net ionic equation for the reaction is

$$NH_4^+(aq) + OH^-(aq) \longrightarrow NH_3(g) + H_2O$$

11.46 Suppose that 5.00 g of solid Fe_2O_3 is added to 20.0 mL of 0.500 M HCl solution. What will the concentration of the Fe^{3+} be when all the HCl has reacted? What weight of Fe_2O_3 will not have reacted?

11.47 Suppose that 30.0 mL of 0.400 M NaCl is added to 30.0 mL of 0.300 M $AgNO_3$.
(a) How many moles of AgCl would precipitate?
(b) What would be the concentrations of each of the ions in the reaction mixture after the reaction?

11.48 A mixture is prepared by adding 20.0 mL of 0.200 M Na_3PO_4 to 30.0 mL of 0.150 M $Ca(NO_3)_2$.
(a) What weight of $Ca_3(PO_4)_2$ will be formed?
(b) What will be the concentrations of each of the ions in the mixture after reaction?

11.49 Solid metal carbonates react with acids to liberate CO_2. How many milliliters of 0.500 M H_2SO_4 are needed to dissolve 0.500 g of copper(II) carbonate?

11.50 10.0 mL of 0.100 M HCl is added to 60.0 mL of water that contains 2.00×10^{-3} mol of solid $Mg(OH)_2$. After the reaction is over, how many moles of $Mg(OH)_2$ remained undissolved? What are the concentrations of the ions in the final solution?

Titrations and Chemical Analyses

11.51 Describe the following.
(a) buret   (d) endpoint
(b) titrant   (e) stopcock
(c) indicator   (f) standard solution

11.52 In a titration, 24.00 mL of 0.100 M NaOH was needed to react with 20.00 mL of HCl solution. What is the molarity of the acid?

11.53 A 10.0-mL sample of vinegar, containing acetic acid, $HC_2H_3O_2$, was titrated using 0.500 M NaOH solution. The titration required 13.40 mL of the base.
(a) What was the molar concentration of acetic acid in the vinegar?

(b) Assuming the density of the vinegar to be 1.0 g/mL, what was the percent (by weight) of acetic acid in the vinegar?

11.54 Lactic acid, $HC_3H_5O_3$, is a monoprotic acid that is formed when milk sours. A 20.0-mL sample of a solution of lactic acid required 18.35 mL of 0.160 M NaOH to reach an end point in a titration. How many moles of lactic acid were in the sample?

11.55 Magnesium sulfate forms a hydrate known as Epsom salts. A student dissolved 1.24 g of this hydrate in water and added a $BaCl_2$ solution until precipitation of $BaSO_4$ was complete. The precipitate was filtered, dried, and found to weigh 1.174 g.
(a) How many moles of $BaSO_4$ were formed?
(b) How many moles of $MgSO_4$ were in the original 1.24-g sample?
(c) How many grams of H_2O were in the original 1.24-g sample?
(d) Determine the formula of Epsom salts?

11.56 A sample of iron chloride weighing 0.300 g was dissolved in water and the solution was treated with $AgNO_3$ solution to precipitate the chloride as AgCl. After precipitation was complete, the AgCl was filtered, dried, and found to weigh 0.678 g.
(a) How many grams of Cl were in the iron chloride sample?
(b) Determine the empirical formula of the iron chloride.

11.57 A 1.500-g sample of a mixture of limestone, $CaCO_3$, and rock was pulverized and then treated with 50.0 mL of 0.200 M HCl. The mixture was warmed to expel the last traces of CO_2 and the unreacted HCl was then titrated with 0.500 M NaOH. The volume of base required was 34.60 mL.
(a) How many moles of NaOH were used in the titration?
(b) How many moles of HCl remained after reaction with the $CaCO_3$?
(c) How many moles of $CaCO_3$ had reacted?
(d) What was the percentage by weight of $CaCO_3$ in the original sample?

11.58 Aspirin is a monoprotic acid called acetylsalicylic acid. Its formula is $HC_9H_7O_4$. A certain pain reliever was analyzed for aspirin by dissolving 0.250 g of it in water and titrating it with 0.0300 M KOH solution. The titration required 29.40 mL of base. What is the percentage by weight of aspirin in the drug?

Equivalents and Equivalent Weights

11.59 If two substances, A and B, react with each other, by what ratio do their equivalents react? How is the equivalent defined for acids and bases?

11.60 How many equivalents are in one mole of the following acids and bases?
(a) HBr

(b) $Ca(OH)_2$
(c) H_3AsO_4 (if completely neutralized)
(d) LiOH

11.61 How many equivalents are in one mole of the following acids and bases?
(a) H_2SO_4 (if neutralized to HSO_4^-)
(b) $Cr(OH)_3$
(c) $H_2C_6H_6O_6$ (if completely neutralized)
(d) $Fe(OH)_2$

11.62 How many equivalents of $Ba(OH)_2$ are needed to react with (a) 0.250 eq H_2SO_4, (b) 0.150 eq H_3PO_4, (c) 0.440 eq HCl, and (d) 0.350 eq barbituric acid.

11.63 What is the equivalent weight of H_3PO_4 if (a) it is completely neutralized? (b) It is neutralized to the $H_2PO_4^-$ ion?

11.64 How many equivalents are in 15.0 g of the following.
(a) NaOH
(b) H_3AsO_4 (if neutralized completely)
(c) H_2S (if neutralized completely)
(d) $Ca(OH)_2$

11.65 How many grams of H_3PO_4 are completely neutralized by 45.0 g of $Ba(OH)_2$? Work the problem using equivalents.

Normality

11.66 Define *normality*. Compare its definition with that of *molarity*.

11.67 A solution of H_2SO_4 is labeled 0.500 M. It is to be used in a reaction in which it will be completely neutralized. What is the normality of the solution?

11.68 What would be the normality of a 0.200 M H_3PO_4 solution if it were to be used in a reaction in which the H_3PO_4 is (a) Completely neutralized? (b) Neutralized to $H_2PO_4^-$ ion? (c) Neutralized to HPO_4^{2-} ion?

11.69 How many grams of $Ba(OH)_2$ are needed to prepare 250 mL of a 0.0100 N solution of this base?

11.70 How many equivalents are in 35.0 mL of 0.500 N $H_2C_2O_4$?

11.71 A volume of 35.40 mL of 0.140 N NaOH is needed to neutralize completely 40.20 mL of H_2SO_4 solution.
(a) What is the normality of the H_2SO_4 solution?
(b) What is the molarity of the H_2SO_4 solution?

11.72 Benzoic acid was determined to have a molecular weight of 122. A 0.100-g sample of it was titrated with 0.100 N NaOH solution. 8.20 mL of the base was required for complete neutralization.
(a) What is the equivalent weight of the benzoic acid?
(b) Is benzoic acid a mono-, di-, or triprotic acid?

11.73 Citric acid is found to have a molecular weight of 192. A sample of citric acid weighing 0.200 g required 31.25 mL of 0.100 N NaOH for complete neutralization.
(a) What is the weight of one equivalent of citric acid?
(b) How many hydrogen ions can one citric acid molecule furnish in reaction with a base?

Brønsted Acids and Bases

11.74 How is a Brønsted acid defined? How is a Brønsted base defined?

11.75 Give the conjugate acids of the following.
(a) NH_3 (d) CO_3^{2-}
(b) N_2H_4 (e) NH_2^-
(c) C_5H_5N

11.76 Give the conjugate bases of the following.
(a) NH_3 (d) H_5IO_6
(b) HCO_3^- (e) NH_4^+
(c) HCN

11.77 Identify the conjugate acid–base pairs in the following reactions.
(a) $HNO_3 + N_2H_4 \rightleftharpoons NO_3^- + N_2H_5^+$
(b) $NH_3 + N_2H_5^+ \rightleftharpoons NH_4^+ + N_2H_4$
(c) $H_2PO_4^- + CO_3^{2-} \rightleftharpoons HPO_4^{2-} + HCO_3^-$
(d) $HIO_3 + HC_2O_4^- \rightleftharpoons IO_3^- + H_2C_2O_4$

11.78 Identify the conjugate acid–base pairs in the following reactions.
(a) $HSO_4^- + SO_3^{2-} \rightleftharpoons HSO_3^- + SO_4^{2-}$
(b) $S^{2-} + H_2O \rightleftharpoons HS^- + OH^-$
(c) $CN^- + H_3O^+ \rightleftharpoons HCN + H_2O$
(d) $H_2Se + H_2O \rightleftharpoons HSe^- + H_3O^+$

11.79 What is meant by the term *amphoteric?* Give two chemical equations that illustrate the amphoteric nature of water.

11.80 The position of equilibrium in the equation below lies far to the left. Identify the conjugate acid–base pairs. Which of the two acids is stronger?

$$HOCl(aq) + H_2O \rightleftharpoons H_3O^+(aq) + OCl^-(aq)$$

11.81 Consider the following: CO_3^{2-} is a weaker base than hydroxide ion, and HCO_3^- is a stronger acid than water. In the equation below, would the position of equilibrium lie to the left or to the right? Justify your answer.

$$CO_3^{2-}(aq) + H_2O \rightleftharpoons HCO_3^-(aq) + OH^-(aq)$$

11.82 Acetic acid, $HC_2H_3O_2$, is a weaker acid than nitrous acid, HNO_2. How do the strengths of the bases $C_2H_3O_2^-$ and NO_2^- compare?

11.83 Nitric acid, HNO_3, is a very strong acid. It is 100% ionized in water. In the reaction below, would the position of equilibrium lie to the left or to the right?

$$NO_3^-(aq) + H_2O \rightleftharpoons HNO_3(aq) + OH^-(aq)$$

Trends in Acid–Base Strength

11.84 Choose the stronger acid: (a) H_2S or H_2Se; (b) H_2Te or HI; (c) HIO_3 or HIO_4; (d) H_2SeO_4 or $HClO_4$.

11.85 Choose the stronger acid: (a) HBr or HCl; (b) H_2O or HF; (c) H_3AsO_3 or H_3AsO_4; (d) H_2S or HBr.

11.86 Suppose that a new element were discovered. Based on the discussions in this chapter, what properties (both physical and chemical) might be used to classify the element as a metal or a nonmetal?

11.87 Astatine, atomic number 85, is radioactive and does not occur in appreciable amounts in nature. On the basis of what you have learned in his chapter, answer the following.
(a) How would the acidity of HAt compare to HI?
(b) What might be a formula for an oxoacid of At?

11.88 Explain why nitric acid is a stronger acid than nitrous acid.

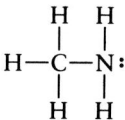

11.89 Explain why H_2S is a stronger acid than H_2O.

11.90 Explain why $HClO_4$ is a stronger acid than H_2SeO_4.

Lewis Acids and Bases

11.91 What are the Lewis definitions of acids and bases?

11.92 Explain why the addition of a proton to a water molecule to give H_3O^+ is a Lewis acid–base neutralization reaction.

11.93 Methylamine has the formula CH_3NH_2 and the structure

Use Lewis structures to illustrate the reaction of methylamine with boron trifluoride, BF_3.

11.94 Aluminum chloride, $AlCl_3$, forms molecules with itself with the formula Al_2Cl_6. Its structure is

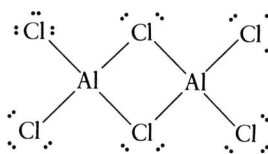

Use Lewis structures to show how the reaction $2AlCl_3 \rightarrow Al_2Cl_6$ is a Lewis acid–base reaction.

11.95 Use Lewis structures to show the Lewis acid–base reaction between SO_2 and H_2O to give H_2SO_3. Identify the Lewis acid and the Lewis base in the reaction.

11.96 Explain why the oxide ion, O^{2-}, can function as a Lewis base but not as a Lewis acid.

Complex Ions

11.97 The formation of the complex ion $[Cu(H_2O)_4]^{2+}$ is described as a Lewis acid–base reaction. Explain.
(a) What is the formula of the Lewis acid and the Lewis base in this reaction?

(b) What is the formula of the ligand?

(c) What is the name of the species that provides the *donor atom?*

(d) What atom is the donor atom and why is it so designated?

(e) What is the name of the species that is the *acceptor?*

11.98 To be a ligand, a species should also be a Lewis base. Explain.

11.99 Why are substances that contain complex ions often called *coordination compounds?*

11.100 Give the names of two species we mentioned that are electrically neutral, monodentate ligands.

11.101 Give the formulas of four species that have 1− charges and are monatomic, monodentate ligands.

11.102 Use Lewis structures to diagram the formation of $Cu(NH_3)_4^{2+}$ and $CuCl_4^{2-}$ ions from their respective components.

11.103 What must be true about the structure of a ligand classified as *bidentate?*

11.104 How many donor atoms does $EDTA^{4-}$ have?

11.105 Explain how a salt of $EDTA^{4-}$ can retard the spoilage of salad dressing.

11.106 How does a salt of $EDTA^{4-}$ in shampoo make the shampoo work better in hard water?

11.107 The cobalt(III) ion, Co^{3+}, forms a 1:1 complex with $EDTA^{4-}$. What is the net charge, if any, on this complex, and what would be a suitable formula for it (using the symbol EDTA)?

11.108 The iron(III) ion forms a complex with six cyanide ions, called the ferricyanide ion. What is the net charge on this complex ion, and what is its formula?

11.109 The silver ion forms a complex ion with two ammonia molecules. What is the formula of this ion? Can this complex ion exist as a salt with the sodium ion or with the chloride ion? Write the formula of the possible salt. (Use brackets and parentheses correctly.)

11.110 In which complex, $[Cu(H_2O)_4]^{2+}$ or $[Cu(NH_3)_4]^{2+}$, are the bonds from the acceptor to the donor atoms stronger? What evidence can you cite for this? Describe what one does in the lab and what one sees during a test for aqueous copper(II) ion that involves these species.

CHEMICALS IN USE
LIMESTONE, CHEMICAL EROSION, AND LIMESTONE CAVERNS

9

Of all of the types of stone used in constructing buildings, limestone and the closely related dolomite are employed in greater masses than any other in the United States. Limestone consists mostly of calcium carbonate, $CaCO_3$, made of the calcium ion and the carbonate ion. The magnesium ion can replace the calcium ion, and when its relative abundance is about the same as the calcium ion the mineral is called dolomite and has the formula $MgCa(CO_3)_2$.

Formation of Limestone

Limestone originated largely by the chemical activities of living marine organisms like snails, clams, corals, and algae. These use carbon dioxide and calcium ions from their surrounding water to build calcium carbonate into shells and other hardened portions of their systems. When these organisms die and decay, this calcium carbonate remains as a deposit.

Limestone Caverns and Chemical Erosion

Most people have seen pictures or have at least heard of the Carlsbad Caverns (New Mexico), Mammoth Cave (Kentucky), the Wind Caves (South Dakota), and Luray Cavern (Virginia). These occur in limestone formations that underwent an interesting kind of chemical erosion.

Chemical erosion of limestone occurs when it is in contact with water containing acids, whether they are organic acids from the decay of plants, or the acids in "acid rain," or carbonic acid, H_2CO_3. Carbonic acid forms when carbon dioxide, produced in the environment by decay and combustion, dissolves in water. The maximum concentration of CO_2 is 0.033 M at 1 atm and 25 °C, but within capillary cracks of a limestone formation there is a "bottling effect," and the pressure is slightly higher than 1 atm. This effect has a major function in the chemical erosion that leads to limestone caverns. At a higher pressure, more CO_2 is dissolved in water, more H_2CO_3 forms from it, and more H^+ appears in solution by the ionization of H_2CO_3. The extra H^+ ions attack the surrounding limestone. Let us study this in more detail using equilibrium expressions and Le Châtelier's principle (Section 10.8).

At a higher pressure, a gas is more soluble in water. No doubt you have guessed this from the behavior of CO_2 in carbonated beverages. In the unopened bottle, the pressure is higher than the atmosphere's. This keeps CO_2 in solution. As soon as you open the bottle, the pressure suddenly becomes the same as the atmosphere's, and some CO_2 fizzes out. We can use the following equilibrium to discuss this in terms of Le Châtelier's principle.

$$CO_2(g) + \text{pressure} \xrightleftharpoons{\text{water}} CO_2(aq) \qquad (1)$$

An increase in pressure would put a stress on this equilibrium. To relieve the stress, in accordance with Le Châtelier's principle, the equilibrium would shift to the right because this shift would cause the number of gas molecules to decrease and that would lower the pressure. Such a shift, of course, would put more CO_2 into the dissolved state. If the pressure were next reduced, then the equilibrium would shift to the left, and CO_2 would come out of the solution.

A small fraction of CO_2 molecules dissolved in water participate in the following equilibrium to give a small concentration of carbonic acid, H_2CO_3.

$$CO_2(aq) + H_2O \rightleftharpoons H_2CO_3(aq) \qquad (2)$$

If, because of higher pressure, there were more CO_2 in solution—more $CO_2(aq)$—then this equilibrium would shift to the right to use up some of this extra CO_2, and more $H_2CO_3(aq)$ would form.

Carbonic acid is a weak acid, so a small percentage of its molecules ionize according to the following equilibrium.

$$H_2CO_3(aq) \rightleftharpoons H^+(aq) + HCO_3^-(aq) \qquad (3)$$

At 1 atm and 25 °C, the concentration of H^+ is about 2×10^{-5} M. This number, however, would be greater if the pressure were higher because the concentration of H_2CO_3 would be greater. The higher pressure, as we explained, shifts equilibria (1) and (2) to the right, which makes more H_2CO_3, and more of this shifts equilibrium (3) to the right. We now have explained why a higher pressure makes the carbonated water more acidic.

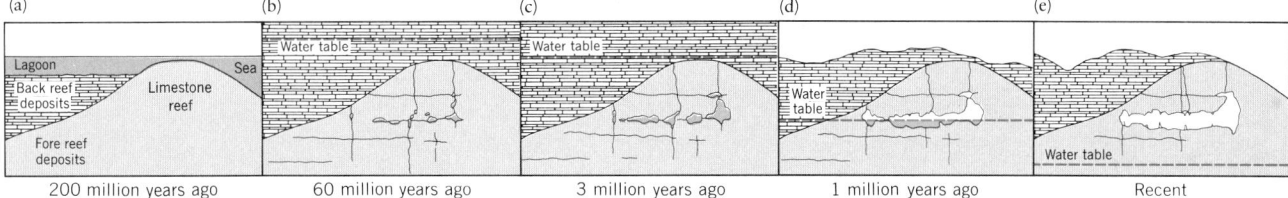

(a)	(b)	(c)	(d)	(e)
200 million years ago	60 million years ago	3 million years ago	1 million years ago	Recent

FIGURE 9a The sequence of events that led to the formation of the Carlsbad Caverns of New Mexico. (a) A limestone reef formed in an ancient sea. (b) The sea level dropped, but the water table remained above the reef. Water rich in CO_2 infiltrated the limestone. (c) The water-filled spaces grew in size. (d) The water table dropped as geological events ocurred. (e) As CO_2-rich water percolated through the fissures and as the water table continued to drop, stalagtites and stalagmites grew. (Courtesy of John Barnell, N.M.)

It is the hydrogen ion that attacks limestone; specifically it attacks the carbonate ions present in this rock and changes them to bicarbonate ions, HCO_3^-. Thus the higher the acidity, the more the following equilibrium shifts to the right to give species that are much more soluble in water than $CaCO_3$.

$$CaCO_3(s) + H^+(aq) \rightleftharpoons Ca^{2+}(aq) + HCO_3^-(aq) \quad (4)$$

It is this change of carbonate ions to the more soluble bicarbonate ions that allows limestone to undergo chemical erosion. And because of the "bottling effect" that leads to a higher concentration of H^+ in the water in fine cracks, the limestone surrounding such cracks dissolves more rapidly than limestone elsewhere. These relatively simple chemical properties account for the hollowing of many of the great limestone caverns of the world as well as for the formation of stalagtites, stalagmites, columns, walls, travertine terraces, drip curtains, and other beautiful features that fascinate tourists and geologists alike.

The sequence that probably happened to create limestone caverns, at least those that formed below the water table, is as follows. See Figure 9a. (The water table is the imaginary surface that exists beneath the ground below which the ground is saturated with water.) First, water rich

in carbon dioxide invades fine cracks and fissures. Then these openings enlarge as the limestone slowly dissolves. The enlarging spaces remain filled with water that, being below ground and "bottled," is still rich in carbon dioxide. Slowly these eroding actions create large spaces sometimes of immense size.

Eventually, much of the water itself finds outlets, the caves and caverns drain, the underground "bottles" of space become uncorked, so to speak. However, many fissures in the walls of the caverns remain, and water still rich in carbon dioxide trickles through them carrying away dissolved calcium ions and bicarbonate ions.

When these seepages leave the fissures, the pressure drops slightly and the water loses some of its dissolved carbon dioxide. This loss of CO_2 leads to leftward shifts of equilibria (1)–(4). When (4) shifts to the left, insoluble $CaCO_3$ precipitates. In other words, after the water leaves a crack in the cavern wall, the $CaCO_3$ comes out of solution. When this happens *before* a drop lets go from a cavern ceiling, a spikelike formation, called a stalagtite ("it hangs tight"), grows just a little more downward. Otherwise, the calcium carbonate leaves the solution when the drop hits the cave floor and helps to build up a heavier growth called a stalagmite. If the water drips from a long, meandering joint crack, the result is a drip curtain. Sometimes stalagmites and stalagtites grow enough toward each other to join, giving a column. Columns can even join to make walls. Some of these features of limestone caverns are shown in Figure 9b.

Questions

1. What is the formula and name of the chief substance in limestone and how does limestone differ from dolomite?

2. Write the equilibrium expressions for the following systems.
   (a) The formation of carbonic acid from dissolved carbon dioxide.
   (b) The ionization of carbonic acid.
   (c) The action of hydrogen ion on calcium carbonate as it occurs within a fissure of a limestone cavern.
   (d) The dissolving of carbon dioxide under pressure.

3. Marble is a special form of calcium carbonate. Marble statues corrode when in contact with rainwater rich in dissolved acids. Write an equation that explains how this happens.

4. A drop in pressure shifts several equilibria involved with the formation of stalagtites. Writing the changes that occur *as forward reactions only*, not the equilibrium expressions, explain how stalagtites form.

FIGURE 9b Stalagtites and stalagmites are prominent features of most large limestone caverns.

Metal plates used by the U. S. Government's Bureau of Engraving and Printing to print one-dollar bills. The images on these plates were etched into the metal by using acid. When acids dissolve metals, the reaction changes the metal atoms to metal ions—a process that involves the transfer of electrons from one substance to another. Electron transfer reactions are discussed in this chapter.

OXIDATION–REDUCTION REACTIONS

12.1 OXIDATION–REDUCTION REACTIONS

Many chemical reactions involve the transfer of electron density from one atom to another.

In our attempt to understand chemistry, one of the things we do is classify chemical reactions into various categories. We did this in Chapter 11, where we examined reactions that we grouped under headings such as *neutralization* and *metathesis*. By doing this we are able to see patterns in some reactions that permit us to anticipate what will happen in others. For instance, finding that acids and bases react to form water and a salt lets us predict that HNO_2 will react with NaOH to form H_2O and $NaNO_2$.

    With the knowledge that you've gathered from previous chapters about the different kinds of bonds that atoms form with each other, we now turn our attention to another very important class of chemical reactions — those that involve the transfer of

$HNO_2 + NaOH$
acid   base

$H_2O + NaNO_2$
water   salt

electron density from one atom to another. Two examples are the reaction of sodium with chlorine and the reaction of hydrogen with oxygen.

$$2Na + Cl_2 \longrightarrow 2NaCl$$
$$2H_2 + O_2 \longrightarrow 2H_2O$$

At first glance, these reactions appear to be very different from each other. In NaCl, ions have been formed, and that has certainly involved the transfer of electron density; an electron is transferred completely from a sodium to a chlorine atom as the Na^+ and Cl^- ions are created. But how about the reaction that produces water? Here we have the formation of a molecule held together by covalent bonds — bonds in which electrons are shared. How does this reaction involve a transfer of electron density? To answer this question, we have to look closely at the bonds in both the reactants and the products.

Hydrogen and oxygen molecules are nonpolar. This is because both atoms in an H_2 or O_2 molecule are the same, so the electronegativity difference between them is zero. In a nonpolar molecule, the electron pair in the bond is shared equally, and neither atom carries a partial charge. Stated another way, each atom in an H_2 or O_2 molecule is electrically neutral. Now let's look at the product, water. The electronegativities of hydrogen and oxygen are quite different, oxygen being more electronegative than hydrogen. This means that the O—H bonds are quite polar, with the hydrogen carrying a substantial positive partial charge and the oxygen carrying a substantial negative partial charge.

Now we can look at what happens to the electron density around an atom during the reaction. A hydrogen atom begins with a zero partial charge in H_2 and finishes with a partial positive charge in H_2O. Similarly, oxygen begins with a zero partial charge in O_2 and finishes with a partial negative charge in H_2O. Thus, during the reaction there is a shift of electron density from a hydrogen atom to an oxygen atom, and it is in this sense that the reaction of H_2 with O_2 is similar to the reaction of Na with Cl_2.

Many chemical reactions involve (or at least *appear* to involve) a shift of electron density from one atom to another. Collectively, such reactions are called oxidation–reduction reactions, or simply redox reactions. The term oxidation refers to the loss of electrons by one reactant, and reduction refers to the gain of electrons by another. For example, the reaction between sodium and chlorine involves a loss of electrons by sodium (oxidation of sodium) and a gain of electrons by chlorine (reduction of chlorine).

$$Na \longrightarrow Na^+ + e^- \qquad \text{(oxidation)}$$
$$Cl_2 + 2e^- \longrightarrow 2Cl^- \qquad \text{(reduction)}$$

We say that sodium is oxidized and chlorine is reduced.

Oxidation and reduction *always* occur together. No substance is ever oxidized unless something else is reduced. Otherwise, electrons would appear as a product of the reaction, and this is never observed. During a redox reaction, then, some substance *must* accept the electrons that another substance loses. This electron-accepting substance is called the oxidizing agent because it helps something else to be oxidized. The substance that supplies the electrons is called the reducing agent because it helps something else to be reduced. Sodium is a reducing agent, for example, when it supplies electrons to chlorine. In the process, sodium is oxidized. Chlorine is an oxidizing agent when it accepts electrons from the sodium, and when that happens, chlorine is reduced to chloride ion. One way to remember this is

The substance that is oxidized is the reducing agent.
The substance that is reduced is the oxidizing agent.

Electronegativity is a measure of the relative attraction an atom has for electrons in a bond.

The *oxidizing agent* causes oxidation to occur by accepting electrons. The *reducing agent* causes reduction to occur by supplying electrons.

Redox reactions are very common. Whenever you use a battery, a redox reaction occurs. The battery is constructed so that the electron transfer must take place through a wire. On its way, we use the energy provided by the electron transfer to light a flashlight or to power a pocket calculator. The metabolism of foods, which supplies our bodies with energy, also occurs by a series of redox reactions that use oxygen to convert carbohydrates and fats to carbon dioxide and water. Ordinary household bleach works by oxidizing substances that stain fabrics, making them easier to remove from the fabric or rendering them colorless.

Liquid bleach contains hypochlorite ion, OCl⁻, as the oxidizing agent.

EXAMPLE 12.1 IDENTIFYING OXIDATION AND REDUCTION AND OXIDIZING AND REDUCING AGENTS

Problem: Calcium and oxygen react to form calcium oxide, CaO, an ionic compound.

$$2Ca + O_2 \longrightarrow 2CaO$$

Which element is oxidized and which is reduced? What are the oxidizing and reducing agents?

Solution: When calcium reacts with oxygen, it forms Ca^{2+} ions, which means that the calcium atoms must *lose* electrons.

$$Ca \longrightarrow Ca^{2+} + 2e^-$$

This is oxidation; calcium is oxidized. Since calcium is oxidized, it must be the reducing agent.
   When oxygen reacts with calcium, it forms O^{2-} ions, which means that the oxygen atoms must *gain* electrons.

$$O_2 + 4e^- \longrightarrow 2O^{2-}$$

This is reduction; oxygen is reduced. This means that oxygen is the oxidizing agent.

PRACTICE EXERCISE 1 Identify the substances oxidized and reduced and the oxidizing and reducing agents in the reaction of calcium and chlorine to form calcium chloride, $CaCl_2$, an ionic compound.

12.2 OXIDATION NUMBERS

An oxidation number is the charge an atom in a compound would have if the electron pairs in the bonds belonged entirely to the more electronegative atoms.

Because redox reactions are so important, chemists have developed a special bookkeeping system to follow what happens when they occur. The system employs a device called oxidation numbers. Roughly speaking, an oxidation number is the charge that an atom would have if the electrons in each bond belonged entirely to the more electronegative element. For example, in assigning oxidation numbers to hydrogen and chlorine in the polar H—Cl molecule, we imagine that both electrons of the bond are in the sole possession of the more electronegative chlorine atom. If this were the case, then the charge on hydrogen would be 1+ and the charge on chlorine would be 1−. Therefore, in HCl

To be sure not to confuse oxidation numbers with actual electrical charges, we will specify the sign before the number when writing oxidation numbers, and after the number when writing electrical charges. Thus +1 is an oxidation number, and 1+ stands for a single positive electrical charge.

H   oxidation number $= +1$
Cl   oxidation number $= -1$

For covalently bonded molecules such as HCl, of course, we know that the atoms never carry more than partial positive or negative charges. Nevertheless, oxidation numbers are assigned as if each compound were ionic. It is important to remember, therefore, that the oxidation numbers assigned to atoms in compounds *do not* have to correspond to the actual charges on the atoms—sometimes they do, but often they do not.

A term that is frequently used interchangeably with oxidation number is oxidation state. In the HCl molecule, hydrogen has an oxidation number of $+1$ and is said to be in the $+1$ oxidation state. Similarly, the chlorine in HCl is said to be in the -1 oxidation state. We can use our new terms to put the concept of a redox reaction on a quantitative and therefore more useful basis.

> A redox reaction is a chemical reaction in which changes in oxidation numbers occur.

We will find it easy to follow electron transfers in redox reactions by taking note of the changes in oxidation numbers. To do this, however, we have to be able to figure out the oxidation numbers of atoms in a compound simply and quickly.

Fortunately, the assignment of oxidation numbers to atoms does not require continual reference to a table of electronegativity values. There are some simple rules and generalizations that make the task relatively easy.

First, we know that any free atom of an element carries no electrical charge. Therefore, we assign it an oxidation number of zero. When atoms of the same element combine to produce more complex molecular forms, such as O_2, N_2, P_4, or S_8, the bonds between the atoms are nonpolar. The atoms in any one of these molecules carry no partial charges because their electronegativities are identical—thus, their oxidation numbers are also zero. This gives us our first rule.

> Any element, when not combined with atoms of a different element, has an oxidation number of zero.

When an ionic compound is formed from two elements, complete electron transfer occurs. The atom with lower electronegativity loses one or more electrons to the atom with the higher electronegativity. Consider sodium chloride, for example. The electrons in the Na—Cl "bond" are claimed exclusively by the chloride ion.

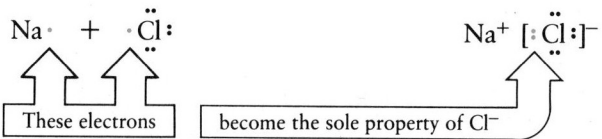

The sodium and chloride ions are therefore assigned oxidation numbers equal to their charges.

$$Na^+ \quad \text{oxidation number} = +1$$
$$Cl^- \quad \text{oxidation number} = -1$$

In general,

> Any simple monatomic ion (one-atom ion) has an oxidation number equal to its charge.

In any compound (either ionic or covalent), the algebraic sum of all of the electrical charges is equal to zero. The formula of the compound always reflects this because it is written without a charge (e.g., NaCl, HCl, or $CaCl_2$). To conform with the electrical neutrality of a compound, the sum of the oxidation numbers of its atoms must always equal zero. Thus, for NaCl

$$
\begin{array}{lll}
Na^+ & \text{oxidation number} = +1 \\
Cl^- & \text{oxidation number} = -1 \\
\hline
NaCl & & \text{sum} = 0
\end{array}
$$

and for $CaCl_2$

$$
\begin{array}{lll}
Ca^{2+} & \text{oxidation number} = +2 \\
Cl^- & \text{oxidation number} = -1 \\
Cl^- & \text{oxidation number} = -1 \\
\hline
CaCl_2 & & \text{sum} = 0
\end{array}
$$

We can extend this concept to polyatomic ions, such as NO_3^- or SO_4^{2-}, and require that the sum of the oxidation numbers of their atoms add up to the *charge on the ion*. In general, then,

> The sum of the oxidation numbers of all of the atoms in a formula must equal the charge written for the formula.

This summation rule is particularly useful when the oxidation numbers are known for all but one of the atoms in a formula. To make use of this rule, however, we must rely on some elements that always (or at least *almost* always) have the same oxidation numbers in their compounds. The periodic table and what we have learned earlier can help us remember them. For example, we have seen that each of the alkali metals in Group IA forms an ion with a 1+ charge and each element in Group IIA forms an ion with a 2+ charge. Similarly, aluminum, the only element that is important to us in Group IIIA, forms an ion with a 3+ charge. These are the only ions formed by these elements, so we can state the following:

> In compounds, the oxidation number of any Group IA metal is always +1, the oxidation number of any Group IIA element is always +2, and the oxidation number of aluminum (in Group IIIA) is always +3.

On the right side of the periodic table are the nonmetals. You should recall that when they are found in *binary compounds with metals,* the nonmetals occur as ions with a negative charge that is equal to the number of electrons that their atoms require to achieve an octet. For example, when the halogens in Group VIIA form binary compounds with metals, such as $FeCl_2$ or $CrBr_3$, they occur as negative ions with a charge equal to 1−. In compounds of this type, each of the halogens has an oxidation number of −1. Similarly, if we refer to Table 2.4 on page 67, we see that any Group VIA nonmetal will have an oxidation number of −2 in binary compounds with metals, and any nonmetal in Group VA will have an oxidation number of −3.

Remember, a binary compound consists of two different elements.

> In binary compounds with metals, the oxidation number of a nonmetal is equal to the charge of its simple monatomic anion.

When nonmetals are combined with other nonmetals in molecules or polyatomic ions such as NO_3^- or SO_4^{2-}, fewer generalizations are possible about their oxidation numbers. Only fluorine, hydrogen, and oxygen have relatively unchanging oxidation numbers from one compound to another.

Because fluorine is the most electronegative of all the elements, in its compounds it is *always* assigned an oxidation number of -1. In other words, for the purposes of assigning oxidation numbers, we imagine that fluorine has total possession of the one electron that it acquires a share of when it forms a bond with another atom.

Hydrogen is generally assigned an oxidation number of $+1$ in its compounds. The only exceptions are when hydrogen occurs in a binary compound with a metal, such as LiH, where it has an oxidation number of -1. Notice that because Li only forms a $1+$ ion and therefore has an oxidation number of $+1$, hydrogen must be assigned an oxidation number of -1 to satisfy the summation rule. Thus, the exceptions for hydrogen are easy to recognize.

Oxygen, the second most electronegative element, nearly always occurs with an oxidation number of -2 in its compounds. Exceptions are rare and only occur when the rules for fluorine and hydrogen would be violated. Thus, in the compound OF_2, oxygen must be assigned an oxidation number of $+2$ because *fluorine is always* -1 *in its compounds* and because the sum of the oxidation numbers must be zero.

For OF_2

$$
\begin{array}{lr}
\text{F} & 2 \text{ atoms} \times (-1) = -2 \\
\text{O} & \underline{1 \text{ atom} \times (+2) = +2} \\
& \text{sum} = 0
\end{array}
$$

Similarly, in hydrogen peroxide, H_2O_2, oxygen is assigned an oxidation number of -1, which we get by *not* violating the rule for hydrogen.

For H_2O_2

$$
\begin{array}{lr}
\text{H} & 2 \text{ atoms} \times (+1) = +2 \\
\text{O} & \underline{2 \text{ atoms} \times (-1) = -2} \\
& \text{sum} = 0
\end{array}
$$

The rules for fluorine, hydrogen, and oxygen can be summarized as follows:

In compounds,
  F is always -1
  O is almost always -2
  H is almost always $+1$

Now let's look at some examples that illustrate how we assign oxidation numbers to atoms in compounds and polyatomic ions.

EXAMPLE 12.2   ASSIGNING OXIDATION NUMBERS

Problem: Molybdenum disulfide, MoS_2, has a structure that allows it to behave as a dry lubricant, much like graphite. What are the oxidation numbers of the atoms in MoS_2?

Solution: This is a binary compound of a metal (molybdenum) with a Group VIA nonmetal (sulfur). Therefore, we take the oxidation number of sulfur to be -2. Then we can use the sum rule to find the oxidation number of molybdenum. Since MoS_2 is electrically neutral, the sum of the oxidation numbers must be zero.

$$
\begin{array}{ll}
S & 2 \text{ atoms} \times (-2) = -4 \\
Mo & \underline{1 \text{ atom} \times (x) = \quad x} \\
& \text{sum} = \quad 0
\end{array}
$$

The value of x must be $+4$ for the sum to be 0. Therefore,

$$
\begin{array}{l}
S = -2 \\
Mo = +4
\end{array}
$$

EXAMPLE 12.3 ASSIGNING OXIDATION NUMBERS

Problem: Liquid laundry bleach contains hypochlorite ion, OCl^-, as its active ingredient. Assign oxidation numbers to the atoms in OCl^-.

Solution: Both oxygen and chlorine are nonmetals. In this case, it is oxygen whose oxidation number is assigned first—we take it to be -2. We can then use the summation rule to find the oxidation number of Cl.

$$
\begin{array}{lll}
O & 1 \text{ atom} \times (-2) = -2 & \text{(O is almost always } -2) \\
Cl & \underline{1 \text{ atom} \times (x) = \quad x} \\
& \text{sum} = -1
\end{array}
$$

The value of x must be $+1$ for the sum to be -1. Therefore,

$$
\begin{array}{l}
O = -2 \\
Cl = +1
\end{array}
$$

EXAMPLE 12.4 ASSIGNING OXIDATION NUMBERS

Problem: A chemical used to remove unexposed silver compounds from photographic film is sodium thiosulfate, $Na_2S_2O_3$. What are the oxidation numbers of the atoms in this compound?

Solution: Once again we rely on the sum rule.

$$
\begin{array}{lll}
Na & 2 \text{ atoms} \times (+1) = +2 & \text{(Na is in group IA)} \\
S & 2 \text{ atoms} \times (x) = \quad 2x \\
O & \underline{3 \text{ atoms} \times (-2) = -6} & \text{(O is almost always } -2) \\
& \text{sum} = \quad 0 & \text{(summation rule)}
\end{array}
$$

We can solve for x algebraically.

$$
\begin{array}{l}
(+2) + (2x) + (-6) = 0 \\
\quad\quad\quad\quad 2x = +4 \\
\quad\quad\quad\quad\ x = +2
\end{array}
$$

The oxidation number of sulfur is $+2$. Therefore,

$$
\begin{array}{l}
Na = +1 \\
O = -2 \\
S = +2
\end{array}
$$

It is possible for oxidation numbers to have fractional values. How this happens is illustrated in the next example.

EXAMPLE 12.5 ASSIGNING OXIDATION NUMBERS

Problem: Acetone, C_3H_6O, is a common solvent in fingernail polish remover. What is the average oxidation number of carbon in this compound?

Solution: Again we apply the summation rule.

$$
\begin{array}{lll}
\text{C} & 3 \text{ atoms} \times (x) = 3x & \\
\text{H} & 6 \text{ atoms} \times (+1) = +6 & \text{(H is almost always } +1) \\
\text{O} & \underline{1 \text{ atom} \times (-2) = -2} & \text{(O is almost always } -2) \\
 & \text{sum} = 0 & \text{(summation rule)} \\
 & 3x = -4 & \\
 & x = -\tfrac{4}{3} & \\
\end{array}
$$

The oxidation number of carbon is $-\tfrac{4}{3}$.[1]

The charge on the polyatomic ion is the sum of the oxidation numbers of its atoms.

Many compounds contain familiar polyatomic ions such as ammonium (NH_4^+), sulfate (SO_4^{2-}), nitrate (NO_3^-), and acetate ($C_2H_3O_2^-$). The charges on these ions can be considered to represent the net oxidation number of the atoms that make up the ion. We can use this sometimes to help assign oxidation numbers, as shown in Example 12.6.

EXAMPLE 12.6   ASSIGNING OXIDATION NUMBERS

Problem: What is the oxidation number of chromium in the compound, $Cr(NO_3)_3$?

Solution: The nitrate ion has a charge of $1-$, which we take to be its net oxidation number. Then we apply the summation rule.

$$
\begin{array}{lll}
\text{Cr} & 1 \text{ atom} \times (x) = x & \\
NO_3^- & \underline{3 \text{ ions} \times (-1) = -3} & \\
 & \text{sum} = 0 & \\
\end{array}
$$

The oxidation number of chromium must be $+3$.

PRACTICE EXERCISE 2   Assign oxidation numbers to each atom in $NiCl_2$, Mg_2TiO_4, $K_2Cr_2O_7$, SO_4^{2-}.

PRACTICE EXERCISE 3   What is the average oxidation number of carbon in isopropyl alcohol (rubbing alcohol), C_3H_8O?

Oxidation numbers have several uses. One of them is in analyzing redox reactions to identify the substances that are oxidized and reduced.

> Oxidation leads to an increase in oxidation number.
> Reduction leads to a decrease in oxidation number.

Let's see how this applies to the reaction of hydrogen with chlorine. To avoid ever confusing oxidation numbers with actual electrical charges, we write oxidation numbers directly above the symbols of the elements in the formula.

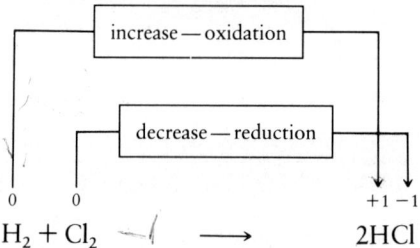

----

[1] What we have calculated is the average oxidation number of carbon. There are ways of assigning oxidation numbers to the individual carbon atoms in the molecule. However, they require a knowledge of the structure of the molecule and therefore we won't explore them any further.

The changes in oxidation number tell us that hydrogen is oxidized and chlorine is reduced.

12.3   USING OXIDATION NUMBERS TO BALANCE EQUATIONS

In redox equations, the total increase in oxidation number must equal the total decrease.

A useful feature of oxidation numbers is that they provide a rather easy means for balancing redox equations. We have discussed balancing equations before, but these were relatively simple equations, with small coefficients, that could be balanced by inspection. Redox reactions, however, are sometimes complex, as illustrated by the equation for the oxidation of isopropyl alcohol to acetone.

$$3C_3H_8O + 2CrO_3 + 3H_2SO_4 \longrightarrow Cr_2(SO_4)_3 + 3C_3H_6O + 6H_2O$$
$$\text{isopropyl alcohol} \qquad\qquad\qquad\qquad\qquad\qquad \text{acetone}$$

Equations such as this can be balanced by requiring that the total increase in oxidation number be equal to the total decrease. The rationale for this is easily seen if we consider a reaction in which electrons are clearly transferred from one atom to another — the reaction between sodium and chlorine. In this reaction, each sodium atom loses one electron as it forms a Na^+ ion, and its oxidation number increases by one unit (from 0 to +1). At the same time, each chlorine atom gains one electron as it forms a Cl^- ion, and its oxidation number decreases by one unit (from 0 to −1). To maintain electrical neutrality, so that electrons are neither a reactant nor a product in the overall reaction, it is necessary that the number of electrons gained be equal to the number lost and that the total increase in oxidation number be equal to the total decrease.

To illustrate how we use oxidation numbers to balance equations, we will begin with the very simple one for the reaction of sodium with chlorine. (Actually, it is simpler to balance the equation by inspection, but this time we will do it the hard way.) To help us follow what happens, we will use the symbol Δ (delta) to stand for a change in oxidation number.

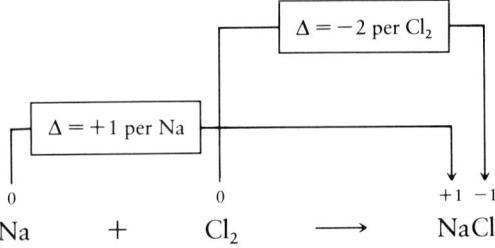

This technique for balancing redox equations is called the **oxidation number-change method.**

The first step is to assign oxidation numbers to each atom in the equation, and we've done this just as before. Next, we analyze how oxidation numbers change. Here we see that each Cl has its oxidation number decreased by 1 (Cl goes from 0 to −1). Since a Cl_2 molecule contains two Cl atoms, the total change when one Cl_2 reacts is −2, where the negative sign for the change means a *decrease*. For the sodium, there is an increase in oxidation number of 1, so the change when one sodium reacts is +1. Next comes the critical step — we *force* the total increase in oxidation number to be equal to the total decrease. To do this we double the number of Na that are reacting by doubling the coefficient of Na.

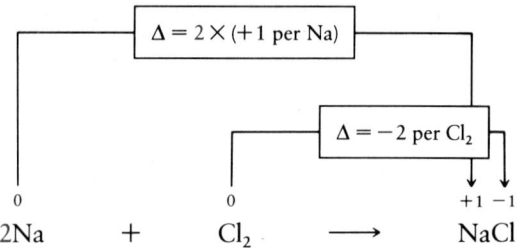

$$2Na \quad + \quad Cl_2 \quad \longrightarrow \quad NaCl$$

Now the total increase in oxidation number ($+2$) equals the total decrease in oxidation number (-2). Finally, we place a 2 in front of NaCl to balance the Cl and Na on the left.

$$2Na + Cl_2 \longrightarrow 2NaCl$$

Now let's see how this method works with a more complex equation such as the oxidation of isopropyl alcohol to acetone described above. As we balance the equation we will divide the procedure into a series of steps that can be followed to balance any redox equation.

Step 1. *Write the correct formulas for all the reactants and products.*

$$C_3H_8O + CrO_3 + H_2SO_4 \longrightarrow Cr_2(SO_4)_3 + C_3H_6O + H_2O$$

Step 2. *Assign oxidation numbers to the atoms in the equation.*

$$\overset{-2\,+1\,-2}{C_3H_8O} + \overset{+6\,-2}{CrO_3} + \overset{+1\,\,-2}{H_2\overbrace{SO_4}} \longrightarrow \overset{+3\,\,-2}{Cr_2\overbrace{(SO_4)}_3} + \overset{-\frac{4}{3}\,+1\,-2}{C_3H_6O}$$

Because SO_4^{2-} appears the same on both sides of the equation, we have used its net oxidation number.

This tells us that C is oxidized and Cr is reduced; so C_3H_8O is the reducing agent and CrO_3 is the oxidizing agent.

Step 3. *Identify which atoms change oxidation number.*

Inspection reveals that carbon and chromium change. Carbon goes from -2 to $-\frac{4}{3}$ and chromium goes from $+6$ to $+3$. We will now focus on the formulas containing these atoms.

Step 4. *Make the number of atoms that change oxidation number the same on both sides by inserting temporary coefficients.*

Both formulas containing carbon have three carbon atoms; the temporary coefficients of C_3H_8O and C_3H_6O are each 1. $Cr_2(SO_4)_3$ contains two Cr while CrO_3 contains one. We place a temporary coefficient of 2 in front of CrO_3. Now there are the same numbers of C and Cr on both sides.

$$\overset{-2}{1C_3H_8O} + \overset{+6}{2CrO_3} + H_2SO_4 \longrightarrow \overset{+3}{1Cr_2(SO_4)_3} + \overset{-\frac{4}{3}}{1C_3H_6O} + H_2O$$

Step 5. *Compute the **total** change in oxidation number for the oxidation and reduction that occur.*

For the elements that change, we calculate the *total* oxidation number that they contribute to each formula. In $1C_3H_8O$ there are 3 carbons, each contributing -2; the total is -6. In $2CrO_3$ there are 2 chromiums, each contributing $+6$; the total is $+12$, and so forth. Then we note how the total oxidation number for each element changes.

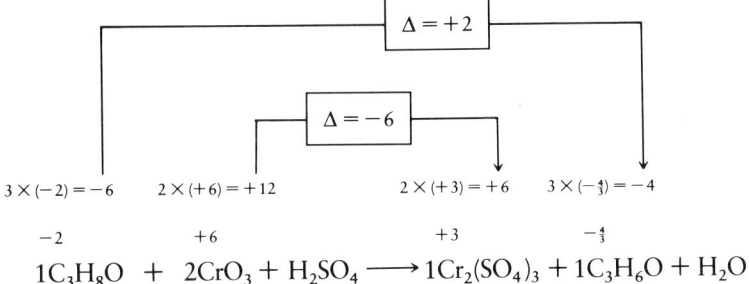

$$3 \times (-2) = -6 \qquad 2 \times (+6) = +12 \qquad 2 \times (+3) = +6 \qquad 3 \times (-\tfrac{4}{3}) = -4$$

$$\overset{-2}{} \qquad \overset{+6}{} \qquad \overset{+3}{} \qquad \overset{-\tfrac{4}{3}}{}$$

$$1C_3H_8O \; + \; 2CrO_3 + H_2SO_4 \longrightarrow 1Cr_2(SO_4)_3 + 1C_3H_6O + H_2O$$

For carbon, the total change is $+2$ (going from -6 to -4 involves an *increase* of 2 units). For Cr the total change is -6 (going from $+12$ to $+6$ involves a *decrease* of 6 units).

Step 6. *Make the total increase in oxidation number equal the total decrease by multiplication using appropriate factors.*

$$\text{For carbon} \qquad \Delta = (+2) \times 3 = +6$$
$$\text{For chromium} \qquad \Delta = (-6) \times 1 = -6$$

The factors 3 and 1 are used to multiply the temporary coefficients.

$$\underbrace{3C_3H_8O}_{3 \times 1} + \underbrace{2CrO_3}_{1 \times 2} + H_2SO_4 \longrightarrow \underbrace{1Cr_2(SO_4)_3}_{1 \times 1} + \underbrace{3C_3H_6O}_{3 \times 1} + H_2O$$

Step 7. *Balance the remainder by inspection.*

We have now found the correct coefficients for all but two formulas. There are $3SO_4$ on the right, so we need a coefficient of 3 in front of H_2SO_4. Next, we look at the hydrogens. There are 30 H on the left. On the right there are 18 H in the acetone, so we must have 12 more in H_2O. Therefore, the coefficient of H_2O must be 6. This gives (finally!)

$$3C_3H_8O + 2CrO_3 + 3H_2SO_4 \longrightarrow Cr_2(SO_4)_3 + 3C_3H_6O + 6H_2O$$

The procedure may seem long, but after just a bit of practice you will find that it's not difficult.

PRACTICE EXERCISE 4   Balance the following redox equations.
(a) $KCl + MnO_2 + H_2SO_4 \longrightarrow K_2SO_4 + MnSO_4 + Cl_2 + H_2O$
(b) $KMnO_4 + FeSO_4 + H_2SO_4 \longrightarrow K_2SO_4 + MnSO_4 + Fe_2(SO_4)_3 + H_2O$
(c) $Zn + HNO_3 \longrightarrow Zn(NO_3)_2 + NH_4NO_3 + H_2O$
This last equation has some HNO_3 that changes oxidation number and some that does not. It may help to write the formula for HNO_3 twice so that you can deal with the two types separately.

$$Zn + HNO_3 + HNO_3 \longrightarrow Zn(NO_3)_2 + NH_2NO_3 + H_2O$$

12.4 BALANCING REDOX EQUATIONS BY THE ION-ELECTRON METHOD

Dividing a reaction into two parts that are balanced separately provides a simple method for obtaining a net ionic equation for a redox reaction.

Many redox reactions in aqueous solution involve ions — they are ionic reactions. An example is the reaction of laundry bleach with substances in the wash water. The active ingredient in the bleach, you may recall from Example 12.3, is hypochlorite ion, OCl^-, and it is this ion that is the oxidizing agent in these reactions. Just as we found it helpful to analyze metathesis reactions by writing ionic and net ionic equations, so too it is helpful to have such equations in studying redox reactions. Balancing net ionic equations for redox reactions is especially easy if we follow a procedure called the ion-electron method.

In the ion-electron method, we divide the oxidation and reduction processes into separate equations called half-reactions that are balanced separately. Then we combine the balanced half-reactions to obtain the fully balanced net ionic equation. The procedure is very systematic, and once you learn to use it, balancing redox equations will be simple.

> Both $FeCl_3$ and $SnCl_2$ are soluble salts.

To illustrate how we apply the method, let's balance the net ionic equation for the reaction of iron(III) chloride, $FeCl_3$, with tin(II) chloride, $SnCl_2$, which reduces the iron to Fe^{2+} and oxidizes the tin to Sn^{4+}. In the reaction the chloride ion is unaffected — it's a spectator ion.

We begin by writing a skeleton equation, which shows only the ions (or sometimes molecules, too) involved in the reaction. The reactants are Fe^{3+} and Sn^{2+}, and the products are Fe^{2+} and Sn^{4+}. The skeleton equation is therefore

$$Fe^{3+} + Sn^{2+} \longrightarrow Fe^{2+} + Sn^{4+}$$

> Each half-reaction must have the same elements on both sides. If Sn is on one side, it must also be on the other side. Keep this in mind when dividing a skeleton equation into half-reactions.

The next step is to divide the equation into two half-reactions — in this case, one for tin and one for iron. We choose one of the reactants, let's say Sn^{2+}, and write it at the left of an arrow. On the right, we write what Sn^{2+} changes to, which is Sn^{4+}. This gives us the beginnings of one half-reaction. (We still have to balance it.) For the second half-reaction, we write the other reactant, Fe^{3+}, on the left and the other product, Fe^{2+}, on the right.

$$Sn^{2+} \longrightarrow Sn^{4+}$$
$$Fe^{3+} \longrightarrow Fe^{2+}$$

> Electrons *must* be added to whichever side of the half-reaction is more positive (or less negative).

Now we balance them so that each one obeys both criteria for a balanced ionic equation — both atoms and charge have to balance. Obviously, the atoms balance in each equation. There is only one atom on each side in each half-reaction. The charge, however, is not balanced. To bring the charge into balance, we add electrons. For the first half-reaction, 2 electrons are added to the right, so the net charge on both sides will be 2+. In the second half-reaction, one electron is added to the left; that makes the net charge on both sides equal to 2+.

$$Sn^{2+} \longrightarrow Sn^{4+} + 2e^-$$
$$Fe^{3+} + e^- \longrightarrow Fe^{2+}$$

In any redox reaction, the number of electrons lost is always equal to the number gained. We used this in the last section to balance redox equations, and we use it here, too. To make the electron gain equal to the electron loss, the second reaction has to occur twice, each time the first reaction occurs once. We indicate this by multiplying *each* of the coefficients in the second equation by 2.

$$Sn^{2+} \longrightarrow Sn^{4+} + 2e^-$$
$$2(Fe^{3+} + e^- \longrightarrow Fe^{2+})$$

In an oxidation, the electrons appear as a product in the half-reaction, and in reduction they appear as a reactant. That's the way we identify which is oxidation and which is reduction when we balance an equation this way.

This gives

$$Sn^{2+} \longrightarrow Sn^{4+} + 2e^-$$
$$2Fe^{3+} + 2e^- \longrightarrow 2Fe^{2+}$$

Finally, we recombine the balanced half-reactions by adding them together. The result is

$$Sn^{2+} + 2Fe^{3+} + 2e^- \longrightarrow Sn^{4+} + 2Fe^{2+} + 2e^-$$

Canceling the 2 electrons from both sides gives the final balanced equation.

$$Sn^{2+} + 2Fe^{3+} \longrightarrow Sn^{4+} + 2Fe^{2+}$$

Did you notice that in the ion-electron method, we don't use oxidation numbers?

Redox Reactions Involving H^+ or OH^-

In many oxidation-reduction reactions in aqueous solutions, H^+ or OH^- ions play an important role, as do water molecules. For example, if $K_2Cr_2O_7$ and $FeSO_4$ solutions are mixed, it is observed that the acidity of the solution decreases as dichromate ion, $Cr_2O_7^{2-}$, oxidizes iron(II). This is because the reaction uses up H^+ as a reactant and it produces H_2O as a product. In other reactions, OH^- is consumed, while in still others, H_2O is a reactant. Another fact is that in many cases the products (or even the reactants) of a redox reaction will differ depending on the acidity of the solution. For example, in an acidic solution $Cr_2O_7^{2-}$ oxidizes Fe^{2+} to give Cr^{3+} and Fe^{3+} as products. On the other hand, if the reaction is carried out in basic solution, the reactants are written as CrO_4^{2-} and $Fe(OH)_2$. This is because $Cr_2O_7^{2-}$ changes to CrO_4^{2-} in basic solution and because $Fe(OH)_2$ is insoluble. The products of the reaction are CrO_2^- and $Fe(OH)_3$.

Because H^+ or OH^- can be consumed or produced by a redox reaction, and because the products can change during the reaction if the solution changes from acidic to basic (or vice versa), redox reactions are generally carried out in solutions containing a substantial excess of either acid or base. Therefore, before we can apply the ion-electron method, we have to know whether the reaction occurs in an acidic or a basic solution. (This information will always be given to you in this book.)

Balancing Redox Equations for Acidic Solutions

As we just learned, $Cr_2O_7^{2-}$ reacts with Fe^{2+} in an acidic solution to give Cr^{3+} and Fe^{3+} as products. This information gives the skeleton equation,

$$Cr_2O_7^{2-} + Fe^{2+} \longrightarrow Cr^{3+} + Fe^{3+}$$

We can then use the following steps to find the balanced net ionic equation.

Step 1. *Divide the skeleton equation into half-reactions.*

We create the beginnings of two half-reactions. Except for hydrogen and oxygen, the same elements have to appear on both sides of a given half-reaction.

$$Cr_2O_7^{2-} \longrightarrow Cr^{3+}$$
$$Fe^{2+} \longrightarrow Fe^{3+}$$

As we balance the equation, the ion-electron method will tell us how H^+ and H_2O are involved in the reaction. We don't need to know this information in advance.

Step 2. *Balance atoms other than H and O.*

There are two Cr atoms on the left and only one on the right in the first half-reaction, so a 2 is placed in front of Cr^{3+}. The second half-rection is already balanced in terms of atoms, so nothing need be done to it.

$$Cr_2O_7{}^{2-} \longrightarrow 2Cr^{3+}$$
$$Fe^{2+} \longrightarrow Fe^{3+}$$

We use H_2O, not O or O_2, to balance oxygen atoms, because H_2O is actually present in the solution.

Step 3. *Balance oxygen atoms by adding H_2O to the side that needs O.*

There are seven oxygen atoms on the left of the first half-reaction. These are balanced by adding $7H_2O$ (which contains 7 oxygen atoms) to the right. (We have just discovered that water is a product in this particular reaction.)

$$Cr_2O_7{}^{2-} \longrightarrow 2Cr^{3+} + 7H_2O$$
$$Fe^{2+} \longrightarrow Fe^{3+}$$

By balancing oxygen with H_2O, we have created an imbalance in H. We correct this by adding H^+.

Step 4. *Balance hydrogen by adding H^+ to the side that needs H.*

The first half-reaction has 14H on the right; we add $14H^+$ to the left. When you do this step (or others) *be careful to write the charges on the ions.* If they are omitted, you will not obtain a balanced equation in the end.

$$14H^+ + Cr_2O_7{}^{2-} \longrightarrow 2Cr^{3+} + 7H_2O$$
$$Fe^{2+} \longrightarrow Fe^{3+}$$

Now each half-reaction is balanced in terms of atoms. Next we will balance the charge.

Step 5. *Balance the charge by adding electrons.*

First compute the net electrical charge on each side. For the first half-reaction we have

$$\underbrace{14H^+ + Cr_2O_7{}^{2-}}_{\text{net charge} = (14+) + (2-) = 12+} \longrightarrow \underbrace{2Cr^{3+} + 7H_2O}_{\text{net charge} = 2(3+) = 6+}$$

The difference between the net charges on each side is equal to the number of electrons that must be added to the most positive (or least negative) side. In this half-reaction, $(12+) - (6+) = 6+$. This means that we have to add $6e^-$ to the left side.

$$6e^- + 14H^+ + Cr_2O_7{}^{2-} \longrightarrow 2Cr^{3+} + 7H_2O$$

Check the net charge after adding e^- — it must be the same on both sides.

This half-reaction is now complete — it is balanced in terms of both atoms and charge.

The second half-reaction ($Fe^{2+} \rightarrow Fe^{3+}$) is easy to balance. We simply add one electron to the right.

$$Fe^{2+} \longrightarrow Fe^{3+} + e^-$$

Now this half-reaction is completed too.

Step 6. *Make the electrons gained equal to the electrons lost and then add the two half-reactions.*

At this point we have the two half-reactions,

$$6e^- + 14H^+ + Cr_2O_7{}^{2-} \longrightarrow 2Cr^{3+} + 7H_2O$$
$$Fe^{2+} \longrightarrow Fe^{3+} + e^-$$

Six electrons are gained in the first half-reaction but only one electron is lost in the second. Therefore, the second half-reaction must be multiplied by 6 before it can be added to the first.

$$6e^- + 14H^+ + Cr_2O_7^{2-} \longrightarrow 2Cr^{3+} + 7H_2O$$
$$6(Fe^{2+} \longrightarrow Fe^{3+} + e^-)$$
$$\text{(Sum)} \ 6e^- + 14H^+ + Cr_2O_7^{2-} + 6Fe^{2+} \longrightarrow 2Cr^{3+} + 7H_2O + 6Fe^{3+} + 6e^-$$

Step 7. *Cancel anything that is the same on both sides.*

This is the final step (at last!). Six electrons are canceled from each side to give the final equation.

$$14H^+ + Cr_2O_7^{2-} + 6Fe^{2+} \longrightarrow 2Cr^{3+} + 7H_2O + 6Fe^{3+}$$

Notice that *both* charge and atoms balance.

In some reactions you may have H_2O or H^+ on both sides—for example, $6H_2O$ on the left and $2H_2O$ on the right. Cancel as many as you can. For example,

$$\ldots + 6H_2O \longrightarrow \ldots + 2H_2O$$

gives

$$\ldots + 4H_2O \longrightarrow \ldots$$

The following is a summary of the steps used for balancing redox reactions that occur in an acidic solution.

Ion-Electron Method—Acidic Solution
Step 1. Divide equation into half-reactions.
Step 2. Balance atoms other than H and O.
Step 3. Balance O by adding H_2O.
Step 4. Balance H by adding H^+.
Step 5. Balance net charge by adding e^-.
Step 6. Make e^- gain and loss equal; then add half-reactions.
Step 7. Cancel anything that's the same on both sides.

If you don't skip any steps, you will always obtain a properly balanced equation.

At this point you should be aware that the ion-election method gives a balanced redox equation without the use of oxidation numbers. This is an important advantage of the method. In fact, when you use the ion-electron method, you should be careful to avoid oxidation numbers altogether.

EXAMPLE 12.7 **USING THE ION-ELECTRON METHOD**

Problem: Balance the following equation. The reaction occurs in an acidic solution.
$$MnO_4^- + H_2SO_3 \longrightarrow SO_4^{2-} + Mn^{2+}$$

Solution: **Step 1**
$$MnO_4^- \longrightarrow Mn^{2+}$$
$$H_2SO_3 \longrightarrow SO_4^{2-}$$

Step 2. This step is not required for this particular reaction because all atoms other than H and O are in balance.

Step 3
$$MnO_4^- \longrightarrow Mn^{2+} + 4H_2O$$
$$H_2O + H_2SO_3 \longrightarrow SO_4^{2-}$$

Step 4
$$8H^+ + MnO_4^- \longrightarrow Mn^{2+} + 4H_2O$$
$$H_2O + H_2SO_3 \longrightarrow SO_4^{2-} + 4H^+$$

Step 5
$$5e^- + 8H^+ + MnO_4^- \longrightarrow Mn^{2+} + 4H_2O$$
$$H_2O + H_2SO_3 \longrightarrow SO_4^{2-} + 4H^+ + 2e^-$$

Step 6
$$2(5e^- + 8H^+ + MnO_4^- \longrightarrow Mn^{2+} + 4H_2O)$$
$$5(\quad H_2O + H_2SO_3 \longrightarrow SO_4^{2-} + 4H^+ + 2e^-)$$

$$10e^- + 16H^+ + 2MnO_4^- + 5H_2O + 5H_2SO_3 \longrightarrow 2Mn^{2+} + 8H_2O + 5SO_4^{2-} + 20H^+ + 10e^-$$

Step 7. Cancel $10e^-$, $16H^+$, and $5H_2O$ from each side. The final equation is

$$2MnO_4^- + 5H_2SO_3 \longrightarrow 2Mn^{2+} + 3H_2O + 5SO_4^{2-} + 4H^+$$

Notice that the equation has the same number of atoms on each side and the same net charge on each side. It is therefore a balanced equation.

PRACTICE EXERCISE 5   What is the balanced equation for the following reaction in acidic solution?

$$Cu + NO_3^- \longrightarrow Cu^{2+} + NO$$

Balancing Redox Equations for Basic Solutions

In basic solutions the concentration of H^+ is very small; the dominant species are H_2O and OH^-. Strictly speaking, these should be used to balance the half-reactions. However, the simplest way to obtain a balanced equation for a redox reaction that occurs in a basic solution is to first pretend that the solution is acidic. We can begin to balance the equation using the seven steps we just learned. Then we can use a simple three-step procedure to convert the equation to the correct form for a basic solution.

Suppose, for example, that we wanted to balance the following equation for a basic solution.

$$SO_3^{2-} + MnO_4^- \longrightarrow SO_4^{2-} + MnO_2$$

Following steps 1 through 7 for acidic solutions gives us

$$2H^+ + 3SO_3^{2-} + 2MnO_4^- \longrightarrow 3SO_4^{2-} + 2MnO_2 + H_2O$$

Conversion of this equation to one appropriate for a basic solution takes advantage of the fact that H^+ and OH^- react in a 1-to-1 ratio to give H_2O. The procedure involves the following three additional steps.

Step 8. *Add the same number of OH^- as there are H^+ to **both** sides of the equation.*

The equation for acidic solution has $2H^+$ on the left, so we add $2OH^-$ to each side. This gives

Remember to add the OH^- to *both* sides; otherwise you will *upset* the balance of atoms and charge.

$$2OH^- + 2H^+ + 3SO_3^{2-} + 2MnO_4^- \longrightarrow 3SO_4^{2-} + 2MnO_2 + H_2O + 2OH^-$$

Step 9. *Combine OH^- and H^+ to form H_2O.*

The left side has $2OH^-$ and $2H^+$. We combine them to give $2H_2O$.

$$2H_2O + 3SO_3^{2-} + 2MnO_4^- \longrightarrow 3SO_4^{2-} + 2MnO_2 + H_2O + 2OH^-$$

Step 10. *Cancel any H_2O that you can.*

In this equation, one H_2O can be canceled from both sides. The final equation, balanced for a basic solution, is

$$H_2O + 3SO_3^{2-} + 2MnO_4^- \longrightarrow 3SO_4^{2-} + 2MnO_2 + 2OH^-$$

PRACTICE EXERCISE 6   Balance the following equation for a basic solution.

$$MnO_4^- + C_2O_4^{2-} \longrightarrow MnO_2 + CO_3^{2-}$$

12.5 REACTIONS OF METALS WITH ACIDS

Many metals react with acids to liberate hydrogen.

Now that you've learned what oxidation and reduction mean, and how to keep track of electrons when they are transferred, we will turn our attention in this and the next few sections to some especially important kinds of redox reactions. One of these is the reaction of a metal with an acid.

When a nonchemist hears the word *acid,* an image of some horrible, corrosive liquid often comes to mind. Such visions are not without some justification, because acids are quite reactive chemicals. However, they're not nearly as dangerous as some people think, provided care is taken in handling them. One of the properties that is probably responsible, at least in part, for the bad reputation that acids seem to have is the tendency of acids to react with certain metals. Battery acid (which is sulfuric acid, H_2SO_4) will begin to dissolve unprotected iron or steel parts of an automobile if it's spilled and not flushed away with water. A similar reaction occurs even faster, accompanied by vigorous bubbling, if the acid is spilled on the zinc surface of a galvanized garbage pail.

In general, the reaction of an acid with a metal is a redox reaction in which the metal is oxidized and the acid is reduced. But in these reactions, the part of the acid that is reduced depends on the composition of the acid itself as well as on the metal.

When sulfuric acid reacts with the zinc coating on a galvanized garbage pail or a galvanized nail, the reaction is

$$Zn(s) + H_2SO_4(aq) \longrightarrow ZnSO_4(aq) + H_2(g)$$

The bubbling that's observed is caused by the release of the hydrogen gas. If we assign oxidation numbers, we can analyze the oxidation–reduction changes that occur. Following the rules developed in the Section 12.2, we have

$$
\begin{array}{c}
\Delta = +2 \text{ per } H_2 \\
\Delta = -2 \text{ per } Zn \\
\underset{0}{Zn}(s) + \underset{+1\ +6\ -2}{H_2SO_4}(aq) \longrightarrow \underset{+2\ +6\ -2}{ZnSO_4}(aq) + \underset{0}{H_2}(g)
\end{array}
$$

The oxidation number of zinc increases from 0 to $+2$, which means that the zinc is oxidized, and the oxidation number of hydrogen decreases from $+1$ to 0, which means that the hydrogen of the acid is reduced.

An even clearer picture of what occurs is seen if we write the net ionic equation. Sulfuric acid, you recall, is a strong electrolyte, so in an aqueous solution it exists in ionic form. Similarly, zinc sulfate is a soluble salt, and it is also a strong electrolyte. When we write the soluble strong electrolytes in ionic form and then cancel spectator ions, we obtain the net ionic equation

$$Zn(s) + 2H^+(aq) \longrightarrow Zn^{2+}(aq) + H_2(g)$$

It's natural to think, "If an acid can do that to a metal, imagine what it will do to me!"

A galvanized (zinc-coated) iron nail dipped into a dilute solution of sulfuric acid. The acid dissolves the zinc and liberates hydrogen gas, which bubbles to the surface.

Remember, oxidizing agents are reduced and reducing agents are oxidized.

FIGURE 12.1 An iron nail reacts with a solution of hydrochloric acid. As in the reaction of the zinc coating on a galvanized nail, the reaction of the iron reduces the hydrogen ions to elemental hydrogen, H_2.

In many cases,

metal + acid $\longrightarrow$ H_2 + salt

Assigning oxidation numbers reveals once again that zinc is oxidized, and we see that it is the H^+ of the acid (actually present in solution as H_3O^+) that is reduced.

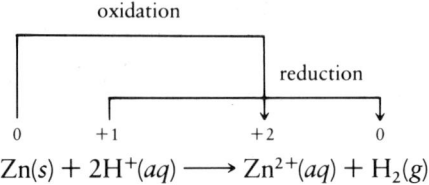

$$Zn(s) + 2H^+(aq) \longrightarrow Zn^{2+}(aq) + H_2(g)$$

Stated another way, the H^+ of the acid is serving as the oxidizing agent.

Many metals react with acids just as zinc does—by being oxidized by hydrogen ions. In these reactions a metal salt and gaseous hydrogen are the products. Another example is shown in Figure 12.1, where we see an iron nail dissolving in hydrochloric acid. The bubbles are hydrogen gas, and if the solution is evaporated after the nail has dissolved, the salt $FeCl_2$ can be recovered. From this information we can now write the molecular equation for the reaction as

$$Fe(s) + 2HCl(aq) \longrightarrow FeCl_2(aq) + H_2(g)$$

Hydrochloric acid is a strong acid, so it is fully ionized, and $FeCl_2$ is a soluble salt, so it is completely dissociated in solution. Therefore, following the rules discussed in Chapter 11, the ionic and net ionic equations are

$$Fe(s) + 2H^+(aq) + 2Cl^-(aq) \longrightarrow Fe^{2+}(aq) + 2Cl^-(aq) + H_2(g)$$

and

$$Fe(s) + 2H^+(aq) \longrightarrow Fe^{2+}(aq) + H_2(g)$$

PRACTICE EXERCISE 7   Write balanced molecular, ionic, and net ionic equations for the reaction of hydrochloric acid with (a) magnesium and (b) aluminum. (Both are oxidized by hydrogen ions.)

The ease with which metals lose electrons varies considerably from one metal to another. Some metals are easily oxidized and others are not. For many, such as iron and zinc, hydrogen ion is a sufficiently strong oxidizing agent to do the job, so when they are placed in a solution of an acid they are oxidized while the hydrogen ions are reduced. Such metals are said to be more *active* than hydrogen (H_2), and they dissolve in acids like HCl and H_2SO_4 to give hydrogen gas and a metal salt that contains the anion of the acid. For other metals, however, hydrogen ions are not powerful enough to cause their oxidation. An example is the common metal, copper. Copper is significantly more difficult to oxidize than either zinc or iron, and H^+ simply is not able to react with it. If a copper penny is dropped into sulfuric acid or hydrochloric acid, it just sits there. No reaction occurs. Copper is an example of a metal that is less active than H_2.

Oxidizing Power of Acids

A solution of hydrochloric acid contains H_3O^+ ions (which we abbreviate as H^+) and Cl^- ions, and we've seen that the hydrogen ion in these solutions is able to serve as an oxidizing agent by being reduced to H_2. The other ion in the solution, Cl^-, has no tendency at all to serve as an oxidizing agent, because it would have to become a Cl^{2-} ion if it were to gain an electron. From our earlier discussions about bonding, this is just not feasible—too much energy would be required. In a solution of HCl, therefore, the only oxidizing agent is H^+.

In an aqueous solution of sulfuric acid, we find a similar situation. In water this acid ionizes to give hydrogen ions and sulfate ions. Although the sulfate ion can be reduced (to SO_3^{2-}, for example), hydrogen ion is more easily reduced. Therefore, when we add a metal such as zinc, it is the H^+, rather than the SO_4^{2-}, that removes the electrons from the Zn.

Compared with many other chemicals that we will study, the hydrogen ion in water is really a rather poor oxidizing agent, so solutions of HCl and H_2SO_4 have rather poor oxidizing abilities. For this reason, they are often called nonoxidizing acids, even though they are able to oxidize certain metals. Actually, when we use this label we are implying that the *anion* of the acid is a weaker oxidizing agent than H^+.

From the preceding discussion, you've probably guessed that there are also acids that we call oxidizing acids whose anions are stronger oxidizing agents than H^+. An example is nitric acid, HNO_3. When dissolved in water, this compound ionizes to give H^+ and NO_3^- ions. However, in this case the nitrate ion is a more powerful oxidizing agent than the hydrogen ion. In the competition for electrons, therefore, it is the nitrate ion that generally wins, and when nitric acid reacts with a metal, it is the nitrate ion that is reduced.

Because the NO_3^- ion is a stronger oxidizing agent than H^+, it is able to oxidize metals that H^+ cannot. For example, if a copper penny is dropped into concentrated nitric acid, it reacts violently. The reaction is shown in Figure 12.2. The reddish brown gas is nitrogen dioxide, NO_2, which is formed by the reduction of the NO_3^- ion. The molecular equation for the reaction is

$$4HNO_3(aq) + Cu(s) \longrightarrow Cu(NO_3)_2(aq) + 2NO_2(g) + 2H_2O$$

If we change this to a net ionic equation and then assign oxidation numbers, we can see easily which is the oxidizing agent and which is the reducing agent. Nitrate ion is reduced, so it is the oxidizing agent; copper is oxidized, so it is the reducing agent.

$$\overset{\Delta = +1}{\overbrace{\underset{0}{Cu(s)} + 2\underset{+5}{NO_3^-}(aq) + 4H^+(aq) \longrightarrow \underset{+2}{Cu^{2+}}(aq) + 2\underset{+4}{NO_2}(g) + 2H_2O}}$$
$$\Delta = -2$$

Notice that in this reaction, no hydrogen gas is formed. The H^+ ions simply become part of water molecules.

When oxidizing acids react with metals, it is often difficult to predict the products.

The strongest oxidizing agent in a solution of a "nonoxidizing" acid is H^+.

Notice that even though it is the nitrogen who's oxidation number is decreasing, we identify the *substance* that contains this nitrogen, the NO_3^- ion, as the oxidizing agent.

FIGURE 12.2 A copper penny reacts violently with concentrated nitric acid, as this sequence of photographs shows. The dark red-brown vapors are nitrogen dioxide, the same gas that gives smog its characteristic color.

Reduction of the nitrate ion, for example, can produce all sorts of oxidation states for the nitrogen, depending on the reducing power of the metal and the concentration of the acid. When concentrated nitric acid reacts with a metal, it often produces NO_2 as the reduction product. When dilute nitric acid is used to dissolve a metal, the product is often nitric oxide, NO. This is what occurs with copper, for example. The molecular and net ionic equations for these reactions are

Concentrated HNO₃

$$Cu(s) + 4HNO_3(aq) \longrightarrow Cu(NO_3)_2(aq) + 2NO_2(g) + 2H_2O$$

$$Cu(s) + 4H^+(aq) + 2NO_3^-(aq) \longrightarrow Cu^{2+}(aq) + 2NO_2(g) + 2H_2O$$

Dilute HNO₃

$$3Cu(s) + 8HNO_3(aq) \longrightarrow 3Cu(NO_3)_2(aq) + 2NO(g) + 4H_2O$$

$$3Cu(s) + 8H^+(aq) + 2NO_3^-(aq) \longrightarrow 3Cu^{2+}(aq) + 2NO(g) + 4H_2O$$

If very dilute nitric acid is allowed to react with a metal that is a very good reducing agent, such as zinc, the nitrogen can be reduced all the way down to the -3 oxidation state it has in NH_4^+ (or NH_3). The net ionic equation is

$$4Zn(s) + 10H^+(aq) + NO_3^-(aq) \longrightarrow 4Zn^{2+}(aq) + NH_4^+(aq) + 3H_2O$$

Nitric acid is quite a powerful oxidizing acid. All metals except the very unreactive ones, such as platinum and gold, are attacked by it. It also does a good job of oxidizing organic substances, so it is wise to be especially careful when working with this acid in the laboratory. If you should spill any on your skin, wash it off immediately and seek the help of your lab teacher.

Nitric acid causes severe skin burns, so be careful when you work with it in the laboratory.

12.6 DISPLACEMENT OF ONE METAL BY ANOTHER FROM COMPOUNDS

A metal can reduce the cation of another less active metal.

When we observe hydrogen being evolved in the reaction of a metal with an acid, we are witnessing a special case of a much more general phenomenon — one element displacing (pushing out) another element from a compound by means of a redox reaction. In the case of a metal–acid reaction, it is the metal that displaces hydrogen from the acid, changing $2H^+$ to H_2. This happens, you recall, if the metal is "more active" than H_2.

Another reaction of this same general type occurs when one metal displaces another metal from its compounds, and is illustrated by the experiment shown in Figure 12.3.

FIGURE 12.3 The reaction of zinc with copper ion. (left) A piece of shiny zinc next to a beaker containing a copper sulfate solution. (center) When the zinc is placed in the solution, copper ions are reduced to the free metal while the zinc dissolves. (right) After a while the zinc becomes coated with a red-brown layer of copper. Notice that the solution is a lighter blue than before.

Here we see a brightly polished strip of metallic zinc that is dipped into a solution of copper sulfate. After being in the solution for a while, a reddish brown deposit of metallic copper has formed on the zinc, and if the solution were analyzed, we would find that it now contains zinc ions, as well as some remaining unreacted copper ions.

Copper ions as $[Cu(H_2O)_4]^{2+}$ give the solution its blue color.

The results of this experiment can be summarized by the chemical equation

$$Zn(s) + CuSO_4(aq) \longrightarrow Cu(s) + ZnSO_4(aq)$$

The redox changes become quite clear if we write the net ionic equation for the reaction. Both copper sulfate and zinc sulfate are soluble salts, so they are completely dissociated. This gives the ionic equation

$$Zn(s) + Cu^{2+}(aq) + SO_4^{2-}(aq) \longrightarrow Cu(s) + Zn^{2+}(aq) + SO_4^{2-}(aq)$$

Sulfate ion is a spectator ion, so it cancels. The result is

$$Zn(s) + Cu^{2+}(aq) \longrightarrow Cu(s) + Zn^{2+}(aq)$$

Looking at the equation, we see that zinc is oxidized and copper ion is reduced. Stated another way, the zinc has reduced the copper ion to metallic copper, and copper ion has oxidized metallic zinc to zinc ion. In the process, zinc ions have taken the place of the copper ions in the compound and in the solution.

The reaction of zinc with copper ion is quite similar to the reaction of zinc with sulfuric acid. The more "active" zinc replaces another less "active" element in a compound. Furthermore, it is easy to show that the reverse reaction doesn't occur. In Figure 12.4 we see what happens if a brightly polished piece of copper is dipped into a solution of zinc sulfate. No matter how long the copper is in the solution, its surface remains untarnished. This means that the reaction of copper with zinc sulfate doesn't occur.

$$Cu(s) + ZnSO_4(aq) \longrightarrow \text{no reaction}$$

In other words, the less "active" copper is unable to displace the more "active" zinc from the solution.

The Activity Series

Throughout the discussion in the preceding paragraph we have used the word *active* to mean "easily oxidized." In other words, an element that is more easily oxidized will displace one that is less easily oxidized from its compounds. The relative ease of oxidation of two metals can be established in experiments just as simple as the ones pictured in Figures 12.3 and 12.4. After such comparisons are made for many pairs, the metals can be

FIGURE 12.4 Although metallic zinc will displace copper from a solution containing Cu^{2+} ion, metallic copper will not displace Zn^{2+} from its solutions. Here we see that the copper bar is unaffected by being dipped into a solution of zinc sulfate.

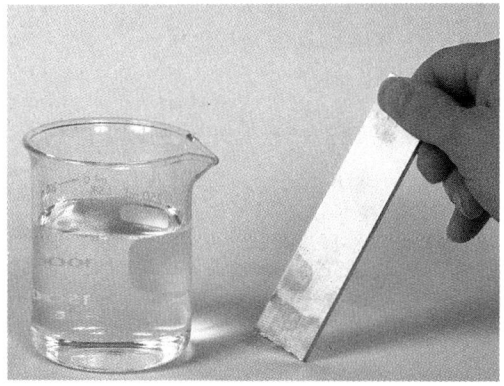

TABLE 12.1 Activity Series for Hydrogen and Some Typical Metals

	Element	Oxidation Product
LEAST ACTIVE	Gold	Au^{3+}
	Mercury	Hg^{2+}
	Silver	Ag^+
	Copper	Cu^{2+}
	HYDROGEN	H^+
	Lead	Pb^{2+}
	Tin	Sn^{2+}
	Cobalt	Co^{2+}
	Cadmium	Cd^{2+}
	Iron	Fe^{2+}
	Chromium	Cr^{3+}
	Zinc	Zn^{2+}
	Manganese	Mn^{2+}
	Aluminum	Al^{3+}
	Magnesium	Mg^{2+}
	Sodium	Na^+
	Calcium	Ca^{2+}
	Strontium	Sr^{2+}
	Barium	Ba^{2+}
	Potassium	K^+
MOST ACTIVE	Rubidium	Rb^+
	Cesium	Cs^+

INCREASING EASE OF OXIDATION

H^+ will react with any metal below it in Table 12.1.

Metallic sodium reacts violently with water. Reduction of water molecules liberates hydrogen. The heat of the reaction ignites the sodium metal, which can be seen burning and sending sparks from the surface of the water.

arranged in order of their ease of oxidation to give what is often called an **activity series** like the one shown in Table 12.1. According to the way the metals have been arranged in this table, those at the bottom are more easily oxidized (more active) than those at the top. This means that a given element will be displaced from its compounds by any metal below it in the table.

Notice that we have included hydrogen in the activity series. This is because certain metals can displace hydrogen from solutions containing H^+, just as they can displace other metals from solutions of their ions. From the position of hydrogen in the table, we can identify the metals that are able to produce H_2 by such a displacement reaction. In other words, we can use the table to identify which elements are able to be oxidized by H^+. These are the metals that are capable of reacting with the nonoxidizing acids discussed in the last section. On the other hand, metals that are above hydrogen in the table will not react with acids having H^+ as the strongest oxidizing agent.

Metals at the very bottom of the table are extremely strong reducing agents. They are so reactive, in fact, that they are able to reduce the hydrogen in water. Sodium, for example, reacts vigorously with water to liberate H_2. The reaction is

$$2Na(s) + 2H_2O \longrightarrow H_2(g) + 2NaOH(aq)$$

Having the activity series in Table 12.1 at our disposal easily permits us to make predictions of the outcome of metal-displacement redox reactions, as illustrated in Example 12.8.

EXAMPLE 12.8   USING THE ACTIVITY SERIES TO PREDICT REDOX REACTIONS

Problem: What will happen if an iron nail is dipped into a solution containing copper sulfate? If a reaction occurs, write its chemical equation.

In constructing an equation, always be sure the formulas of reactants and products are correct.

Solution: According to the activity series in Table 12.1, iron is more easily oxidized than copper, so the iron will displace the copper from the copper sulfate. To write a chemical equation for the reaction, we have to know the final oxidation state of the iron. In the table, this is indicated as $+2$, so the equation for the reaction is

$$Fe(s) + CuSO_4(aq) \longrightarrow Cu(s) + FeSO_4(aq)$$

Notice that we have been careful to write the correct formula for the iron-containing product. The equation is balanced as it stands, so we have answered the question.

EXAMPLE 12.9 USING THE ACTIVITY SERIES TO PREDICT REDOX REACTIONS

Problem: What will happen if an iron nail is dipped into a solution of aluminum sulfate?

Solution: Scanning the activity series, we see that aluminum is more easily oxidized than iron. This means that aluminum would displace iron from an iron compound. But is also means that iron *cannot* displace aluminum from its compounds. Therefore, we conclude that no reaction will be observed.

$$Fe(s) + Al_2(SO_4)_3(aq) \longrightarrow \text{no reaction}$$

PRACTICE EXERCISE 8 Write a chemical equation for the reaction that will occur, if any, when (a) magnesium metal is added to a solution of copper chloride and (b) silver metal is added to a solution of aluminum sulfate. If no reaction will occur, write "no reaction" in place of the products.

12.7 PERIODIC TRENDS IN THE REACTIVITY OF METALS

Variations in the ease of oxidation of metals follow trends in their ionization energies.

If you were an engineer designing parts for some piece of machinery, you would consider several factors in choosing the right metal for the job. Certainly, some of them would be physical properties — strength and hardness, for example. However, you would also be concerned about the chemical properties of the metal. What is the tendency of the metal to react with air and moisture? Will the metal react with any chemicals that come into contact with it while the machine is operating? These are chemical questions whose answers would be very important to the lifespan of your machine.

When we raise questions about the reactivity of a metal, we are concerned with how easily the metal is oxidized. This is because in nearly every compound containing a metal, the metal exists in a positive oxidation state. Therefore, when a free metal reacts to form a compound, it is oxidized. If we speak of the tendency of a metal to be oxidized, we are talking about the tendency of a metal to react with oxidizing agents, and this tendency is often referred to as the metal's "reactivity." For example, a metal like sodium, which is very easily oxidized, is said to be very reactive, whereas a metal like platinum, which is very difficult to oxidize, is said to be unreactive.

In general, reactivity refers to the tendency of a substance to react. For metals, the kind of reaction is oxidation.

There are several ways to compare how easily metals are oxidized. In the last section, we saw that by comparing the abilities of metals to displace each other from compounds we are able to establish their relative ease of oxidation. This was the basis for the construction of the activity series shown in Table 12.1.

Another way of obtaining this information is to study the reactions of metals with acids. We have already seen that some metals react with acids, while others do not. Certainly, those that are oxidized by H^+ are more reactive than those that aren't. And even among metals that are oxidized by hydrogen ions, the speed with which the reaction occurs generally parallels the ease with which the metal loses electrons. For example, in

If a metal is above hydrogen in Table 12.1, it will not react with nonoxidizing acids.

FIGURE 12.5 The relative ease of oxidation of metals is reflected in their rates of reaction with hydrogen ions of an acid. The products are hydrogen gas and the metal ion in solution. All three test tubes contain HCl*(aq)* at the same concentration. The first also contains pieces of iron, the second, pieces of zinc, and the third, pieces of magnesium. Among these three metals, the ease of oxidation increases from iron to magnesium.

Figure 12.5, we see samples of iron, zinc, and magnesium reacting with solutions of hydrochloric acid. In each case the HCl concentration is the same, but we see that the magnesium reacts more rapidly than zinc, which reacts more rapidly than iron. You can see that the order of reactivity in Table 12.1 is the same; magnesium is more easily oxidized than zinc, which is more easily oxidized than iron.

As useful as the activity series is in predicting the outcome of certain redox reactions, it is difficult to remember in detail. Furthermore, in many instances it is sufficient just to know approximately where an element stands in relation to others in a broad range of reactivity. This is where the periodic table can be especially useful to us once again, because there are trends and variations in reactivity within the periodic table that are rather simple to identify and to remember.

Figure 12.6 illustrates how the ease of oxidation (reactivity) of metals varies in the periodic table. In general, these trends roughly parallel the variations in ionization energy. You might expect this, since the ionization energy is a measure of how difficult it is to remove an electron from an atom. The relationship between reactivity and ionization energy (IE) is only an approximate one, however, because the IE refers to isolated gaseous atoms, and when we consider ease of oxidation, we are concerned with removal of electrons from the metal in its usual solid form as well as hydration of the cation that is formed.

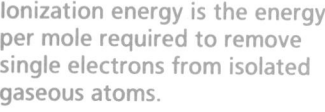

Ionization energy is the energy per mole required to remove single electrons from isolated gaseous atoms.

Examining Figure 12.6, we see that the metals that are most easily oxidized are found at the far left. For example, the metals in Group IA are so easily oxidized that all of them react with water to liberate hydrogen. It is precisely because they are so reactive that we rarely encounter them as free metals. They have no useful applications that require exposure to the atmosphere because of their reactivity toward moisture and oxygen. The same is true of the heavier metals in Group IIA, calcium through barium. These elements also react with water to liberate hydrogen.

You should study this figure so that you know where the very reactive elements are found and where the very unreactive ones are found.

In Section 6.12, you learned how the ionization energies of the elements vary within

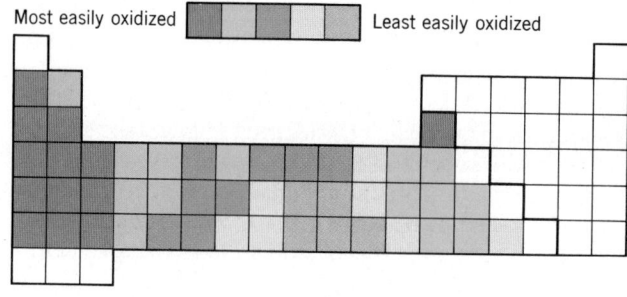

FIGURE 12.6 The variation in the ease of oxidation of metals with position in the periodic table.

the periodic table. Recall that the ionization energy increases as we move from left to right in a period. This places metals with the lowest ionization energies at the left of the table — in other words, in Groups IA and IIA. It should be no surprise, therefore, that these elements are very easy to oxidize. You also learned that the ionization energy decreases as we go down in a group. This explains why the heavier elements in Group IIA are especially easy to oxidize compared with those at the top of the group.

In Figure 12.6 we can also locate the metals that are the most difficult to oxidize. They occur for the most part among the heavier transition elements in the center of the periodic table. Here we find the very unreactive elements platinum and gold — metals used to make fine jewelry. Their bright luster and lack of any tendency to corrode combine to make them particularly attractive for this purpose. This same lack of reactivity also is responsible for their industrial uses. Gold, for example, is used to coat the electrical contacts in low-voltage circuits found in microcomputers, because even small amounts of corrosion on more reactive metals would be sufficient to impede the flow of electricity so much as to make the devices unreliable.

12.8 MOLECULAR OXYGEN AS AN OXIDIZING AGENT

Oxygen reacts with many compounds and elements to form oxides.

Humans long ago discovered that certain substances were able to be burned, and they used this knowledge to heat their homes and cook their food. To this day, civilization depends primarily on the combustion of fuels of various kinds to provide the energy necessary to run machinery and generate electricity, as well as for cooking and heating. In fact, almost everyone is well aware that the fate of our economy hangs on the continued availability of relatively inexpensive fuels that can give us this energy.

In 1789, a French chemist named Antoine Lavoisier discovered that combustion involves the reaction of oxygen in the air with chemicals in the fuel. In fact, the term *oxidation* was invented to describe reactions in which one of the reactants was elemental oxygen. Oxidation meant the reaction of a substance with O_2. It was only much later that chemists finally realized that the reaction of oxygen with various substances was a special case of a much broader class of chemical reactions involving changes in oxidation numbers. This realization led eventually to the extension of the meaning of the term oxidation to cover all these similar reactions.

Oxygen is a plentiful chemical — it's in the air and available to anyone, chemist or not, who wants to use it. Furthermore, it is a very reactive oxidizing agent, so it is no surprise that the reactions of O_2 have been well studied. When substances combine with oxygen, the reaction products are generally oxides — molecular oxides when the oxygen is combined with nonmetals and ionic oxides when the oxygen is combined with metals. In this section we will look briefly at some typical reactions of oxygen, both with compounds and with elements, to learn more about the kinds of substances that are formed.

Combustion of Organic Compounds

If circumstances made it necessary for you to build a fire to keep warm, you no doubt would look around for twigs, logs, or other pieces of wood to use as fuel. You know that wood burns. Experience has also taught you that certain other substances burn. Examples are natural gas and gasoline. In fact, if you drive a car, it is probably powered by the combustion of gasoline. These fuels are examples of substances or mixtures of substances that chemists call organic compounds — compounds whose structures are determined primarily by the linking together of carbon atoms by covalent bonds. When organic compounds burn, the products of the reactions are usually easily predicted.

Fuels such as natural gas, gasoline, kerosene, heating oil, and diesel fuel are examples of hydrocarbons—compounds containing only the elements carbon and hydrogen. Natural gas is composed principally of the compound methane, CH_4. Gasoline is a mixture of hydrocarbons, the most familiar of which is octane, C_8H_{18}. Kerosene, heating oil, and diesel fuel are mixtures of hydrocarbons in which the molecules contain even more atoms of carbon and hydrogen.

When hydrocarbons burn in a plentiful supply of oxygen, the products of combustion are always carbon dioxide and water. Thus, methane and octane combine with oxygen according to the equations

Hydrocarbon + O_2 $\longrightarrow$
          CO_2 + H_2O
(not balanced, of course)

$$CH_4 + 2O_2 \longrightarrow CO_2 + 2H_2O$$
$$2C_8H_{18} + 25O_2 \longrightarrow 16CO_2 + 18H_2O$$

Many people don't realize that water is one of the products of the combustion of hydrocarbons, even though they have seen evidence for it. If you live in the northern United States, you surely have seen clouds of condensed water vapor coming from the exhaust pipes of automobiles on cold winter days. Even in warmer climates, you may have noticed that shortly after you first start a car, water drops fall from the exhaust pipe. This is water that has been formed during the combustion of the gasoline. It condenses from the engine exhaust gases when they come in contact with the cool metal of the exhaust pipe.

PRACTICE EXERCISE 9   Write a balanced equation for the combustion of butane, C_4H_{10}, in an abundant supply of oxygen. Butane is the fuel used in disposable cigarette lighters.

When there is less than an abundant supply of oxygen during the combustion of a hydrocarbon, not all of the carbon is converted to carbon dioxide. Instead, some of it forms carbon monoxide. This is one of the pollution problems caused by gasoline engines, as you know. If the oxygen supply is very limited, only the hydrogen is converted to the oxide (water). For example, the combustion of methane in a very limited oxygen supply follows the equation

$$CH_4 + O_2 \longrightarrow C + 2H_2O$$

This finely divided form of carbon is also called *carbon black*.

The carbon formed in this reaction is very finely divided. It would be called soot by almost anyone observing the reaction. Nevertheless, it has considerable commercial value when collected and marketed under the name lampblack. This sooty form of carbon is used to manufacture inks and much of it is used in the production of rubber tires, where it serves as a binder and a filler.

Earlier we mentioned various fuels you might choose if you were to build a fire. One of them was wood. The chief combustible ingredient in wood is cellulose—a fibrous material that gives plants their structural strength. Cellulose is composed of the elements carbon, hydrogen, and oxygen. Each cellulose molecule consists of many small, identical groups of atoms that are linked together to form a very long molecule, although the length of each molecule may be different. For this reason we don't even attempt to specify the molecular formula; we use the empirical formula instead. The empirical formula for cellulose is $C_6H_{10}O_5$, which is the formula of the small units that make up the large cellulose molecule. When cellulose burns, the products are also carbon dioxide and water. The only difference between its reaction and the reaction of a hydrocarbon with oxygen is that some of the oxygen in the products comes from the cellulose.

The formula for cellulose can be expressed as $(C_6H_{10}O_5)_n$, which indicates that the molecule contains the $C_6H_{10}O_5$ unit repeated some large number n times.

$$C_6H_{10}O_5 + 6O_2 \longrightarrow 6CO_2 + 5H_2O$$

The combustion of other compounds containing these same three elements produces the same products and follows similar equations.

PRACTICE EXERCISE 10   Ethanol, C_2H_5OH, has been mentioned as an alternative to gasoline as an automotive fuel. In fact, it is mixed with gasoline and the mixture is sold under the name gasohol. Write a chemical equation for the combustion of ethanol.

Reactions of Metals with Oxygen

Combustion is normally taken to mean a rapid reaction of a substance with oxygen in which both heat and light are given off. Although the reactions of the organic compounds discussed above fit this description, we don't often think of metals as undergoing this kind of reaction. But have you ever fired a flash bulb to take a photograph? The source of light is the reaction of the metal magnesium with oxygen. A close look at a fresh flashbulb reveals the fine web of thin magnesium wire within the glass envelope. This wire is surrounded by an atmosphere of oxygen, a clear colorless gas. When the flashbulb is used, a small electrical current is passed through the thin wire, causing it to become hot enough to ignite, and it burns rapidly in the oxygen atmosphere. The equation for the reaction is

$$2Mg + O_2 \longrightarrow 2MgO$$

Most metals react directly with oxygen, although not as spectacularly, and usually we refer to the reaction as corrosion or tarnishing because the oxidation products dull the shiny metal surface. For example, a common metal found around the home is aluminum. It has all sorts of applications, from aluminum foil to aluminum window frames. Aluminum is a rather easily oxidized metal, as can be seen from its position in the activity series, and a freshly exposed surface of the metal very quickly reacts with oxygen and becomes coated with a thin film of aluminum oxide, Al_2O_3. Fortunately, this oxide coating adheres very tightly to the surface of the metal and makes it very difficult for additional oxygen to combine with the aluminum. Therefore, further oxidation of the aluminum occurs very slowly.

We're not as lucky with iron and steel as with aluminum. Although less reactive than aluminum, iron is oxidized fairly easily, especially in the presence of moisture. As you know, under these conditions the iron corrodes—it rusts. Rust is a form of iron(III) oxide, Fe_2O_3, that also contains an appreciable amount of absorbed water.

Although the rusting of iron is a slow reaction, the combination of iron with oxygen can be speeded up if the metal is heated to a very high temperature. This is illustrated in the photograph in Figure 12.7.

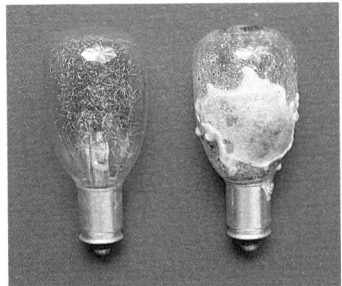

Fine magnesium wire in an atmosphere of oxygen fills the flashbulb at the left. After being used (right), the interior of the bulb is coated with a white film of magnesium oxide.

Rust is said to be a hydrous oxide. The rusting of iron is described in more detail in Special Topic 22.1.

PRACTICE EXERCISE 11   The oxide of iron formed in the reaction shown in Figure 12.7 is Fe_2O_3. Write an equation for the reaction.

FIGURE 12.7 Steel is cut by a stream of pure oxygen whose reaction with the red-hot metal produces enough heat to melt the steel and send a shower of burning steel sparks flying.

Reactions of Nonmetals with Oxygen

Nitrogen is one nonmetal that doesn't react readily with oxygen, especially at normal temperatures. But nitrogen oxides formed at high temperatures in automobile engines are a serious source of air pollution.

Most nonmetals combine as readily with oxygen as do the metals. Usually, these reactions occur rapidly enough to be described as combustion. The most important reaction of this type is the combustion of carbon as a source of heat. Without much effort, you would probably call to mind coal and charcoal as common carbon fuels. Both are used in large amounts—to generate electricity and to broil hamburgers. If plenty of oxygen is available, the combustion product is CO_2, but when the supply of oxygen is limited, CO can be formed as well. This is one reason why manufacturers that package charcoal briquettes print a warning on the bag that the charcoal shouldn't be used indoors for cooking or heating unless very good ventilation is available.

There is strong evidence that SO_2 released by coal-burning power plants in the Midwest is responsible for acid rain in the northeastern United States and Canada.

Another nonmetal that readily burns in oxygen is sulfur. In the manufacture of sulfuric acid, the first step is the combustion of sulfur to produce sulfur dioxide. Sulfur dioxide is also formed when sulfur compounds burn, and the presence of both sulfur and sulfur compounds as impurities in coal and petroleum is a major source of air pollution. Combustion of these fuels releases their sulfur content in the form of SO_2, which enters the atmosphere, where it drifts on the wind until it finally dissolves in rain water. Then, as a dilute solution of sulfurous acid, it falls to earth as one component of acid rain. The acid rain problem is discussed in greater detail in Section 20.2.

PRACTICE EXERCISE 12

Elemental phosphorus can exist in a white, waxy form that consists of molecules having the formula P_4. When phosphorus burns in an abundant supply of oxygen, it forms an oxide composed of molecules of P_4O_{10}. Write an equation for the reaction.

12.9 REDOX REACTIONS IN THE LABORATORY

Redox reactions play a useful role in many analytical procedures in the laboratory.

In the last several sections we've described various types of oxidation–reduction reactions, and as you've seen, many of them find a place in our day-to-day lives. Redox reactions, of course, are also important in the chemistry laboratory. Many reactions that are useful in chemical analyses are of this type, as are many of the reactions that are carried out in the preparation of new compounds. Redox reactions are therefore of value to "synthetic chemists"—chemists who synthesize new substances.

Common Oxidizing and Reducing Agents

When a chemist plans a reaction, a natural question is, "What chemicals should be used?" If a redox reaction is planned, there are many different compounds that might be chosen as oxidizing or reducing agents, but certain compounds are chosen more frequently than others simply because they are readily available, relatively inexpensive, and able to do the job (oxidation or reduction) most of the time.

Oxidizing Agents   Three of the most common oxidizing agents found in the laboratory are the permanganate ion, MnO_4^-, the dichromate ion, $Cr_2O_7^{2-}$, and the chromate ion, CrO_4^{2-}. These are usually found on the stockroom shelf as sodium or potassium salts. For example, permanganate ion is normally obtained from the salt potassium permanganate, $KMnO_4$, and if you wish to use chromate or dichromate ion as an oxidizing agent, you would probably find them as either sodium or potassium salts—for example, Na_2CrO_4 and $Na_2Cr_2O_7$.

All three of the anions mentioned above are very powerful oxidizing agents, and care should be exercised in handling them to avoid their contact with organic materials. They readily supply oxygen in the formation of the normal oxidation products of organic substances, so they have the potential to cause fires.

Permanganate ion, chromate ion, and dichromate ion have characteristic colors. Permanganate ion is deep purple in aqueous solution, dichromate ion is reddish orange, and chromate ion is yellow. Solutions containing these ions are shown in Figure 12.8.

When an oxidizing agent reacts, it is reduced, and if the oxidizing agent is permanganate ion, the products of the reduction reaction depend on the acidity of the solution in which the reaction occurs. In an acidic solution, the manganese in the permanganate ion is usually reduced to the manganese(II) ion, Mn^{2+}. The half-reaction is

$$MnO_4^-(aq) + 8H^+(aq) + 5e^- \longrightarrow Mn^{2+}(aq) + 4H_2O$$

The Roman numerial in parentheses following the name of the element is the element's oxidation number.

However, if the solution is neutral, or even slightly basic, the manganese is reduced to only the +4 oxidation state, where it forms the compound MnO_2. Manganese(IV) oxide, or manganese dioxide as it is better known, is insoluble in water, so when it is formed it makes its appearance as a brownish precipitate.

$$MnO_4^-(aq) + 2H_2O + 3e^- \longrightarrow MnO_2(s) + 4OH^-(aq)$$

Chromate and dichromate ions have something in common—both contain chromium in the +6 oxidation state, and one can be changed into the other in aqueous solutions by the appropriate addition of either an acid or a base. For example, if an acid is added to a solution of chromate ion, the following reaction takes place.

$$2CrO_4^{2-}(aq) + 2H^+(aq) \longrightarrow Cr_2O_7^{2-}(aq) + H_2O$$

On the other hand, if a base is added to a solution of dichromate ion, the chromate ion is formed.

$$Cr_2O_7^{2-}(aq) + 2OH^-(aq) \longrightarrow 2CrO_4^{2-}(aq) + H_2O$$

FIGURE 12.8 Some polyatomic ions have characteristic colors that they impart to their compounds. On the left is a solution of potassium chromate containing the yellow CrO_4^{2-} ion. In the center is a solution of potassium dichromate containing the red-orange $Cr_2O_7^{2-}$ ion. On the right is a solution of potassium permanganate containing the violet MnO_4^- ion.

Thus, in acidic solutions it is the dichromate ion that carries chromium(VI), and in basic solutions it is the chromate ion. As in the case of the permanganate ion, the reduction products that form when these ions function as oxidizing agents depend on the acidity of the solution in which the redox reaction takes place. In acidic solutions, where the chromium(VI) is present in the form of dichromate ion, reduction usually produces chromium(III) ion.

$$Cr_2O_7^{2-}(aq) + 14H^+(aq) + 6e^- \longrightarrow 2Cr^{3+}(aq) + 7H_2O$$

In a mildly basic solution, where the chromium(VI) is present in chromate ion, reduction produces the insoluble chromium(III) hydroxide.

$$CrO_4^{2-}(aq) + 4H_2O \cdot \longrightarrow Cr(OH)_3(s) + 5OH^-(aq)$$

These chemical facts about MnO_4^-, CrO_4^{2-}, and $Cr_2O_7^{2-}$ should be remembered.

PRACTICE EXERCISE 13   Write balanced chemical equations for the oxidation of Fe^{2+} to Fe^{3+} in an acidic solution by (a) permanganate ion and (b) dichromate ion.

Reducing Agents   If you search the stockroom shelves for a suitable reducing agent to combine with some oxidizing agent in a reaction that you're planning, there are several common compounds that you might consider. One choice includes compounds that contain the sulfite ion, SO_3^{2-}, or the bisulfite ion, HSO_3^-. As you may recall, these anions are formed by the neutralization (either complete or partial) of sulfurous acid, H_2SO_3. In the lab, we usually find them as either sodium or potassium salts — for example, Na_2SO_3 or $NaHSO_3$.

The sulfite ion is a fairly potent reducing agent, and when it reacts with an oxidizing agent it is oxidized to sulfate ion.

$$SO_3^{2-}(aq) + H_2O \longrightarrow SO_4^{2-}(aq) + 2H^+(aq) + 2e^-$$

In Section 20.2, we'll have more to say about the nature of aqueous sulfurous acid.

Actually, if the solution is acidic the reactant is bisulfite ion, HSO_3^-, or sulfurous acid, H_2SO_3. This is because H_2SO_3 is a weak acid, and you learned in Chapter 11 that if a strong acid is added to the anion of a weak acid (that is, if a solution of the anion of the weak acid is acidified), a metathesis reaction will tend to occur in which the weak electrolyte (weak acid) is formed. Depending on how acidic a solution of sulfite ion is made, either one or two H^+ are added to the SO_3^{2-} ion, giving HSO_3^- or H_2SO_3.

$$SO_3^{2-}(aq) + H_3O^+(aq) \longrightarrow HSO_3^-(aq) + H_2O \quad \text{(in slightly acidic solution)}$$
$$HSO_3^-(aq) + H_3O^+(aq) \longrightarrow H_2SO_3(aq) + H_2O \quad \text{(in acidic solution)}$$

PRACTICE EXERCISE 14

Write half-reactions for the oxidation of (a) bisulfite ion and (b) sulfurous acid to sulfate ion in an acidic solution.

Another weak acid that serves as a common reducing agent for reactions in aqueous solutions is oxalic acid, $H_2C_2O_4$. We find it on the shelf either as the acid itself, or in the form of a salt such as $Na_2C_2O_4$. When the salt is dissolved in a solution that is acidic, the anion combines with H^+ to form the molecular acid, just as sulfite ion combines with H^+ to form H_2SO_3 in an acidic solution. Oxidation of oxalic acid in an acidic solution produces carbon dioxide.

$$H_2C_2O_4(aq) \longrightarrow 2CO_2(g) + 2H^+(aq) + 2e^-$$

Some other reducing agents that are useful in the laboratory include thiosulfate ion, $S_2O_3^{2-}$, and tin(II) ion. Thiosulfate ion is especially useful for trapping halogens, such as chlorine, that might otherwise be released into the air in the lab. Chlorine is itself a powerful oxidizing agent. It is a gas having an irritating odor and in even relatively small concentrations can cause lung damage if inhaled. Therefore, if you are using chlorine in a reaction, or if chlorine is being produced by a reaction, it is wise to prevent it from escaping into the air. Carrying out the reaction in an apparatus that directs any gases into a solution containing thiosulfate ion will remove chlorine from them by chemical reaction.

Household bleach solutions have an odor caused by small amounts of Cl_2 that are released from them.

$$4Cl_2(g) + S_2O_3^{2-}(aq) + 5H_2O \longrightarrow 2SO_4^{2-}(aq) + 10H^+(aq) + 8Cl^-(aq)$$

When tin(II) is used as a reducing agent, it is oxidized to tin(IV).

$$Sn^{2+}(aq) \longrightarrow Sn^{4+}(aq) + 2e^-$$

This reducing agent is often used in analytical applications to reduce a metal ion in solution to a lower oxidation state prior to its oxidation by another chemical in a titration. We will discuss these kinds of reactions very shortly. In the meantime, these chemical facts about SO_3^{2-}, $H_2C_2O_4$, $S_2O_3^{2-}$, and Sn^{2+} should be learned.

Stoichiometry of Redox Reactions

The quantitative relationships between chemical quantities in redox reactions are handled in exactly the same way as those described in Chapter 4. In general, to perform stoichiometric calculations we need a balanced chemical equation for the reaction, because the coefficients in the equation tell us the ratios by moles in which the reactants combine and in which the products are formed. Earlier in this chapter you learned how to balance redox equations, both by the oxidation number-change method and by the ion-electron method, so obtaining a balanced equation should no longer present any great difficulty to you, providing you know the reactants and the products. The discussion of common oxidizing agents and reducing agents at the beginning of this section can be of help in anticipating what the products are, as illustrated in Example 12.10.

EXAMPLE 12.10 WORKING STOICHIOMETRY PROBLEMS FOR REDOX REACTIONS

Problem: How many grams of potassium dichromate are required to oxidize the Fe^{2+} in 21.00 g of $FeSO_4$ to Fe^{3+} if the reaction is carried out in an acidic solution?

Solution: The first step is to write a balanced chemical equation. In this problem, it is probably easiest to work with the net ionic equation and use the ion-electron method to balance it. First, of course, we must have the reactants and products. When dissolved in water, $K_2Cr_2O_7$ dissociates into K^+ and $Cr_2O_7^{2-}$ ions, and $FeSO_4$ dissociates into Fe^{2+} and SO_4^{2-} ions. Therefore, the reactants are $Cr_2O_7^{2-}$ and Fe^{2+}. Earlier we learned that dichromate ion is reduced to chromium(III) ion in acidic solution, and the problem here tells us that the Fe^{2+} is being oxidized to Fe^{3+}. The products of the reaction are therefore Cr^{3+} and Fe^{3+}, and the skeleton equation that we have to balance is

$$Cr_2O_7^{2-} + Fe^{2+} \longrightarrow Cr^{3+} + Fe^{3+}$$

Applying the ion-electron method in the usual way gives the balanced equation

$$Cr_2O_7^{2-} + 6Fe^{2+} + 14H^+ \longrightarrow 6Fe^{3+} + 2Cr^{3+} + 7H_2O$$

Looking at the coefficients in this equation, we can easily see the mole relationship between the reactants.

$$1 \text{ mol } K_2Cr_2O_7 = 6 \text{ mol } FeSO_4$$

Now all we need to solve the problem are the formula weights of $K_2Cr_2O_7$ and $FeSO_4$. These are 294.2 and 151.9, respectively. Applying the factor-label method that you learned in Chapter 4 gives

A solution of $Na_2Cr_2O_7$ being added to an acidic solution of Fe^{2+}.

$$21.00 \text{ g FeSO}_4 \times \frac{1 \text{ mol FeSO}_4}{151.9 \text{ g FeSO}_4} \times \frac{1 \text{ mol } K_2Cr_2O_7}{6 \text{ mol FeSO}_4} \times \frac{294.2 \text{ g } K_2Cr_2O_7}{1 \text{ mol } K_2Cr_2O_7} = 6.779 \text{ g } K_2Cr_2O_7$$

The reaction of 21.00 g $FeSO_4$ requires 6.779 g $K_2Cr_2O_7$.

PRACTICE EXERCISE 15   Potassium peroxydisulfate, $K_2S_2O_8$, is a powerful oxidizing agent that can oxidize Cr^{3+} to $Cr_2O_7^{2-}$ in acid solution. In the reaction, the peroxydisulfate ion, $S_2O_8^{2-}$, is reduced to sulfate ion. Write a balanced net ionic equation for the reaction, and calculate the number of grams of $K_2S_2O_8$ required to oxidize the chromium in 34.0 g of $Cr(NO_3)_3$.

The redox reactions that we have been discussing in this chapter take place, for the most part, in aqueous solutions. It is quite common in dealing with the stoichiometry of these reactions to work with solutions of either known or unknown concentrations. In Chapter 4 you learned that the concentration unit most suited to problems in stoichiometry is molarity. In working problems involving redox reactions, you will find molarity to be a handy concentration unit too, as shown in Example 12.11.

EXAMPLE 12.11 SOLUTION STOICHIOMETRY IN REDOX REACTIONS

Problem: How many milliliters of 0.125 M $K_2Cr_2O_7$ are required to react completely with 25.0 mL of 0.250 M $FeSO_4$ solution if the solution is acidic?

Solution: The chemical equation developed in Example 12.10 provides the necessary mole relationship.

$$1 \text{ mol } K_2Cr_2O_7 = 6 \text{ mol } FeSO_4$$

To solve the problem, we first determine the number of moles of $FeSO_4$ that are available. To do this we translate the concentration into a conversion factor. Recall that two conversion factors can be written — we must simply choose the correct one.

$$\frac{0.250 \text{ mol } FeSO_4}{1000 \text{ mL}} \qquad \frac{1000 \text{ mL}}{0.250 \text{ mol } FeSO_4}$$

Using the first factor gives us moles of $FeSO_4$.

$$25.0 \text{ mL} \times \frac{0.250 \text{ mol } FeSO_4}{1000 \text{ mL}} = 0.00625 \text{ mol } FeSO_4$$

Next, we determine the number of moles of $K_2Cr_2O_7$ that must react. This is given by

$$0.00625 \text{ mol } FeSO_4 \times \frac{1 \text{ mol } K_2Cr_2O_7}{6 \text{ mol } FeSO_4} = 0.00104 \text{ mol } K_2Cr_2O_7$$

Finally, we translate the concentration of the $K_2Cr_2O_7$ into a conversion factor to calculate the volume of that solution required in the reaction.

$$\frac{0.125 \text{ mol } K_2Cr_2O_7}{1000 \text{ mL}} \qquad \frac{1000 \text{ mL}}{0.125 \text{ mol } K_2Cr_2O_7}$$

Using the second conversion factor gives

$$0.00104 \text{ mol } K_2Cr_2O_7 \times \frac{1000 \text{ mL}}{0.125 \text{ mol } K_2Cr_2O_7} = 8.32 \text{ mL}$$

Thus, 8.32 mL of the $K_2Cr_2O_7$ solution is required. Of course, all of these computations could be performed at once by stringing together the conversion factors.

$$25.0 \text{ mL} \times \frac{0.250 \text{ mol } FeSO_4}{1000 \text{ mL}} \times \frac{1 \text{ mol } K_2Cr_2O_7}{6 \text{ mol } FeSO_4} \times \frac{1000 \text{ mL}}{0.125 \text{ mol } K_2Cr_2O_7} = 8.33 \text{ mL}$$

The small difference between the two answers is caused by rounding off in the step-by-step solution.

PRACTICE EXERCISE 16   How many milliliters of 0.200 *M* $FeSO_4$ are needed to reduce 50.0 mL of 0.400 *M* $Na_2Cr_2O_7$ in an acidic solution? The products of the reaction are Cr^{3+} and Fe^{3+}.

Redox Titrations

A chemical analysis tells us quantitatively what the composition of a sample is. It gives us the identity of the substances that make up the sample and it gives us their proportions, usually by mass. These kinds of data are very important, both to chemists and to others who depend on a knowledge of the composition of substances to make decisions. For example, a mining engineer surely must know the percentage of iron in a potential iron ore if a decision must be made about mining the ore. If the percentage of iron in the ore is too low, mining it would be unprofitable and could spell economic disaster for the mining corporation. For chemists, the analysis of new compounds is an essential first step in determining empirical formulas, and a knowledge of the formulas of compounds is necessary to understand why they behave as they do.

Because so many different substances undergo redox reactions, it is natural that some of these reactions are useful for performing chemical analyses. Often, the reactions that are chosen for analytical purposes are those that can be carried out in solution by a titration. In this procedure, you recall, a solution of one reactant is added to a solution of

A blood analysis is important to a physician when he or she treats a patient.

A solution of $KMnO_4$ being added to a stirred acidic solution containing Fe^{2+}. The reaction oxidizes the pale blue-green Fe^{2+} to Fe^{3+} while the MnO_4^- is reduced to the almost colorless Mn^{2+} ion. The purple color of the permanganate will continue to be destroyed until all of the Fe^{2+} has reacted. Only then will the iron-containing solution take on a pink or purple color. This ability of MnO_4^- to signal the completion of the reaction makes it especially useful in redox titrations.

the other from a buret — a long tube that is accurately calibrated to measure volumes of solutions that are dispensed from it.

One of the most useful reactants for redox titrations is potassium permanganate, especially if the reaction can be carried out in an acidic solution. Earlier in this section you learned that the permanganate ion is a powerful oxidizing agent; therefore, it will oxidize most substances that are capable of being oxidized. That's one reason why it is used. Especially important, though, is the fact that the MnO_4^- ion has a deep purple color and its reduction product in acidic solution is the almost colorless Mn^{2+} ion. As a solution of $KMnO_4$ is added from a buret to a solution of a reducing agent, the chemical reaction that occurs forms a nearly colorless product. Therefore, as the $KMnO_4$ solution is added to the solution being titrated, the purple color continues to be destroyed as long as there is any reducing agent left. However, after the last trace of the reducing agent has been consumed, the MnO_4^- ion in the next drop of titrant has nothing to react with, so it colors the solution pink. This signals the end of the titration. In this way, permanganate serves as its own indicator in redox titrations. Example 12.12 illustrates a typical analysis using $KMnO_4$ in a redox titration.

In dilute solutions, the MnO_4^- ion appears pink.

EXAMPLE 12.12   REDOX TITRATIONS IN CHEMICAL ANALYSIS

Problem: All the iron in a 2.000-g sample of an iron ore was dissolved in an acidic solution and converted to Fe^{2+}, which was then titrated with 0.1000 M $KMnO_4$ solution. In the titration the iron was oxidized to Fe^{3+}. The titration required 27.45 mL of the $KMnO_4$ solution.
(a) How many grams of iron were in the sample?
(b) What was the percentage iron in the sample?
(c) If the iron was present in the sample as Fe_2O_3, what was the percentage by weight of Fe_2O_3 in the sample?

Solution: To solve the problem, we must first have a balanced chemical equation. The reactants are Fe^{2+} and MnO_4^-, and the products are Fe^{3+} and Mn^{2+}. The skeleton equation is therefore

$$Fe^{2+} + MnO_4^- \longrightarrow Fe^{3+} + Mn^{2+}$$

Balancing it by the ion-electron method for acidic solutions gives

$$5Fe^{2+} + MnO_4^- + 8H^+ \longrightarrow 5Fe^{3+} + Mn^{2+} + 4H_2O$$

Now, before we begin the calculations, let's analyze the problem to see where we are going. From the volume of the $KMnO_4$ solution and its concentration we can determine the number of moles of titrant used. The coefficients of the equation then allow us to compute the number of moles of iron(II) that reacted. Since all the iron had been previously changed to iron(II), this is the number of moles of iron in the sample. Converting moles of iron to grams of iron gives the mass of iron in the sample, from which the percentage of iron can be calculated easily. Let's tackle the first two parts of the problem now, and then return to the last part afterward.

Recall that the reactant that is added from the buret is called the *titrant.*

The number of moles of $KMnO_4$ used is calculated from the volume of the solution used and its concentration.

$$27.45 \ \cancel{mL} \times \frac{0.1000 \ mol \ KMnO_4}{1000 \ \cancel{mL}} = 0.002745 \ mol \ KMnO_4$$

The chemical equation tells us that five moles of iron react for each mole of permanganate consumed. The number of moles of iron that reacted is therefore

$$0.002745 \ \cancel{mol \ KMnO_4} \times \frac{5 \ mol \ Fe}{1 \ \cancel{mol \ KMnO_4}} = 0.01372 \ mol \ Fe$$

and the mass of iron is

$$0.01372 \ \cancel{mol \ Fe} \times \frac{55.85 \ g \ Fe}{1 \ \cancel{mol \ Fe}} = 0.7651 \ g \ Fe$$

This is the answer to part (a) of the problem. Next, we calculate the percentage iron in the sample. This is equal to the mass of iron divided by the mass of the sample, all multiplied by 100.

$$\% \ Fe = \frac{mass \ of \ Fe}{mass \ of \ sample} \times 100$$

Substituting gives

$$\% \ Fe = \frac{0.7651 \ g}{2.000 \ g} \times 100 = 38.26\% \ Fe$$

Now we can work on the last part of the question. Earlier in the problem we determined the number of moles of iron that reacted—0.01372 mol Fe. How many moles of Fe_2O_3 would have contained this number of moles of iron? That's the critical question we have to answer. From the chemical formula, we can state the following ratios.

$$\frac{1 \ mol \ Fe_2O_3}{2 \ mol \ Fe} \qquad \frac{2 \ mol \ Fe}{1 \ mol \ Fe_2O_3}$$

The first provides the conversion factor that we need.

$$0.01372 \ \cancel{mol \ Fe} \times \frac{1 \ mol \ Fe_2O_3}{2 \ \cancel{mol \ Fe}} = 0.006860 \ mol \ Fe_2O_3$$

This is the number of moles of Fe_2O_3 in the sample. The formula weight of Fe_2O_3 is 159.7, so the weight of this compound in the sample was

$$0.006860 \ \cancel{mol \ Fe_2O_3} \times \frac{159.7 \ g \ Fe_2O_3}{1 \ \cancel{mol \ Fe_2O_3}} = 1.096 \ g \ Fe_2O_3$$

Finally, the percentage Fe_2O_3 is

$$\% \ Fe_2O_3 = \frac{1.096 \ g \ Fe_2O_3}{2.000 \ g \ sample} \times 100 = 54.80\% \ Fe_2O_3$$

Thus the ore sample was 54.80% (w/w) Fe_2O_3.

PRACTICE EXERCISE 17   A sample of a tin ore weighing 0.3000 g was dissolved in an acid solution and all the tin in the sample was changed to tin(II). The solution was titrated with 8.08 mL of 0.0500 M $KMnO_4$ solution, which oxidized the tin(II) to tin(IV).
(a) What is the balanced equation for the reaction in the titration?
(b) How many grams of tin were in the sample?
(c) What was the percentage by weight of tin in the sample?
(d) If the tin in the sample had been present in the compound SnO_2, what would have been the percentage by weight of SnO_2 in the sample?

Titrations of Iodine with Thiosulfate

Another useful pair of reactants in redox titrations is iodine, I_2, and thiosulfate ion, $S_2O_3^{2-}$. The reaction that occurs between them is

$$I_2(s) + 2S_2O_3^{2-}(aq) \longrightarrow 2I^-(aq) + S_4O_6^{2-}(aq)$$

In the reaction, elemental iodine is reduced to iodide ion as the thiosulfate ion is oxidized to the tetrathionate ion, $S_4O_6^{2-}$. There are problems, though, with the reactants as written. To be able to be used in a redox titration, both reactants must be in solution so the reaction can occur quickly. Only in this way can the endpoint in the titration be easily detected. The difficulty here is that molecular iodine is insoluble in water. Fortunately, the problem is overcome rather easily because iodine combines readily with iodide ion to form the triiodide ion, I_3^-.

$$I_2(s) + I^-(aq) \longrightarrow I_3^-(aq)$$

This holds the iodine in solution, and when the thiosulfate ion is added the reaction that occurs is

$$I_3^-(aq) + 2S_2O_3^{2-}(aq) \longrightarrow 3I^-(aq) + S_4O_6^{2-}(aq) \qquad (12.1)$$

Even though the reactant is I_3^-, the "active ingredient" is I_2.

There are two factors that make this reaction especially valuable in titrations. One of them is the ease with which the endpoint can be detected. In any solution containing triiodide ion, there is actually an equilibrium between I^- and I_2, so there is a very small amount of molecular iodine in the solution. Iodine has the peculiar property of being able to be adsorbed on the surface of starch molecules, and when adsorbed, it produces a very deep blue color that can be seen easily even when the amount of iodine adsorbed is very small. This phenomenon is used in the following way. During the titration a suspension of starch is added to the reaction mixture, causing it to become deeply colored. This color persists while the thiosulfate solution is added from the buret until finally enough thiosulfate has been added to consume the last traces of the iodine. At this point, all the iodine has been used up, so none is left to form the blue color with the starch, and the blue color disappears. Thus, the sudden disappearance of the blue color marks the end of the titration.

$I_2 + I^- \rightleftharpoons I_3^-$

The endpoint in this titration is very easy to detect.

The second factor that makes the reactants useful in redox titrations is the ease with which I_2 can be formed from iodide ion by reaction with other substances. For example, iodide ion is a sufficiently strong reducing agent to cause the reduction of copper(II) ion in solution to the +1 oxidation state. In the process, the iodide ion itself is oxidized to molecular iodine. The actual reaction that takes place in the presence of excess I^- ion is

$$2Cu^{2+}(aq) + 5I^-(aq) \longrightarrow 2CuI(s) + I_3^-(aq) \qquad (12.2)$$

The I_2 that is formed is incorporated into the I_3^-.

This reaction is *quantitative* — that is, it proceeds far enough to completion, without any side reactions, to produce one mole of I_3^- for each two moles of Cu^{2+}. This means that if we carry out this reaction between copper(II) ion and iodide ion, and then titrate the I_3^- with a standard solution of thiosulfate, we can determine the amount of I_3^- in the solution and relate this *directly* to the amount of copper(II) that had reacted to form the I_3^-. The usefulness of this in the analysis of a sample is illustrated in Example 12.13.

EXAMPLE 12.13   TITRATIONS USING REDOX REACTIONS

Problem: A sample weighing 0.2000 g was known to contain $CuSO_4$. It was dissolved in water and the copper in the solution was allowed to react with a solution containing excess iodide ion.

The I_3^- that was formed was titrated with 0.01000 M $Na_2S_2O_3$ solution using starch as an indicator. The titration required 35.65 mL of the thiosulfate solution. What was the percentage by weight of $CuSO_4$ in the original 0.2000-g sample?

Solution: Solving this problem is really quite simple, provided that we plan our approach to it. We begin with the quantities that are known — the concentration of the thiosulfate solution and the volume of it required in the titration. From the volume of the $Na_2S_2O_3$ solution and its molarity we can calculate the number of moles of thiosulfate needed to react with the I_3^-. From the coefficients of Equation 12.1 above, we can then calculate the number of moles of I_3^- that reacted. Next, we use the coefficients of Equation 12.2 to calculate the number of moles of Cu^{2+} used to form the I_3^-. Since there is one mole of copper per mole of $CuSO_4$, this is also the number of moles of $CuSO_4$ in the original sample. Then we can calculate the grams of $CuSO_4$ and finally the percentage of $CuSO_4$ in the sample.

Now that we've chosen our strategy, we can begin doing the computations. First, we translate the label for the concentration of the thiosulfate solution. A concentration of 0.01000 M gives us the conversion factors

$$\frac{0.01000 \text{ mol } S_2O_3^{2-}}{1000 \text{ mL}} \qquad \frac{1000 \text{ mL}}{0.01000 \text{ mol } S_2O_3^{2-}}$$

Using the first of them and the volume of the thiosulfate solution that was consumed in the titration, we calculate the number of moles of $S_2O_3^{2-}$ that reacted.

$$35.65 \text{ mL } S_2O_3^{2-} \times \frac{0.01000 \text{ mol } S_2O_3^{2-}}{1000 \text{ mL}} = 3.565 \times 10^{-4} \text{ mol } S_2O_3^{2-}$$

In Equation 12.1 we see that two moles of $S_2O_3^{2-}$ react with one mole of I_3^-, so the number of moles of I_3^- that reacted was

$$3.565 \times 10^{-4} \text{ mol } S_2O_3^{2-} \times \frac{1 \text{ mol } I_3^-}{2 \text{ mol } S_2O_3^{2-}} = 1.782 \times 10^{-4} \text{ mol } I_3^-$$

Next, the coefficients of Equation 12.2 give the relationship

$$1 \text{ mol } I_3^- = 2 \text{ mol } Cu^{2+}$$

The number of moles of copper(II) ion that reacted was therefore

$$1.782 \times 10^{-4} \text{ mol } I_3^- \times \frac{2 \text{ mol } Cu^{2+}}{1 \text{ mol } I_3^-} = 3.564 \times 10^{-4} \text{ mol } Cu^{2+}$$

This is also the number of moles of $CuSO_4$ in the original sample. The formula weight of $CuSO_4$ is 159.6, so the mass of $CuSO_4$ in the sample was

$$3.564 \times 10^{-4} \text{ mol } CuSO_4 \times \frac{159.6 \text{ g } CuSO_4}{1 \text{ mol } CuSO_4} = 0.05688 \text{ g } CuSO_4$$

Finally, the percentage of $CuSO_4$ in the sample is given by the relationship

$$\% \text{ } CuSO_4 = \frac{\text{mass of } CuSO_4}{\text{mass of sample}} \times 100$$

Substituting values gives

$$\% \text{ } CuSO_4 = \frac{0.05688 \text{ g } CuSO_4}{0.2000 \text{ g sample}} \times 100 = 28.44\% \text{ } CuSO_4$$

The percentage of $CuSO_4$ in the sample was 28.44%.

PRACTICE EXERCISE 18 A 1.00-mL sample of household laundry bleach, which contains sodium hypochlorite, $NaOCl$, was added to a solution containing excess iodide ion. The hypochlorite ion reacts with iodide ion as follows.

$$2H^+ + OCl^- + 3I^- \longrightarrow I_3^- + Cl^- + H_2O$$

After the reaction, the I_3^- was titrated with a 0.0500 M solution of $Na_2S_2O_3$, using starch as an indicator. The titration required 28.20 mL of the thiosulfate solution.
(a) How many moles of I_3^- reacted with the thiosulfate solution in the titration?
(b) How many moles of OCl^- reacted to form the I_3^-?
(c) What was the molar concentration of OCl^- in the laundry bleach?

12.10   EQUIVALENT WEIGHTS AND NORMALITY IN REDOX REACTIONS

For redox reactions, the definition of the equivalent is based on the requirement that the total number of electrons lost by one substance must equal the total number of electrons gained by another.

In the last chapter the concept of equivalents was introduced as a tool for dealing with the stoichiometry of acid–base reactions, and we mentioned at that time that the concept can also be applied to redox reactions. The chief benefit achieved by working with equivalents instead of moles is that equivalents of reactants *always* react in a one-to-one ratio. In a redox reaction, this means that the number of equivalents of oxidizing agent consumed is equal to the number of equivalents of reducing agent that react.

> (number of equivalents of oxidizing agent) =
> (number of equivalents of reducing agent)

All we have to do to use this relationship is to choose an appropriate definition of the equivalent.

You learned that in any redox reaction, the number of electrons lost by one substance is always equal to the number of electrons gained by another. In fact, this was the basis for our methods for balancing equations for redox reactions. It also provides us with a means of defining equivalents of oxidizing and reducing agents so that the one-to-one ratio of equivalents always holds.

> For any substance involved in a redox reaction, an equivalent is defined as the amount of the substance that either gains or loses one mole of electrons.

Calculating the Number of Equivalents per Mole

Consider the reaction of potassium permanganate as an oxidizing agent. When the reaction occurs in an acidic solution, we've seen that the permanganate ion is reduced to the manganese(II) ion.

$$MnO_4^-(aq) + 8H^+(aq) + 5e^- \longrightarrow Mn^{2+}(aq) + 4H_2O$$

In the reaction, each mole of MnO_4^- gains five moles of electrons, so it must correspond to five equivalents. Therefore, we can say that for reactions in acidic solution, one mole of potassium permanganate equals five equivalents of potassium permanganate.

$$1 \text{ mol } KMnO_4 = 5 \text{ eq } KMnO_4$$

When permanganate ion is reduced in a neutral or slightly basic solution, we've seen that the manganese is reduced to MnO_2.

$$MnO_4^-(aq) + 2H_2O + 3e^- \longrightarrow MnO_2(s) + 4OH^-(aq)$$

Three moles of electrons are gained by each mole of permanganate, so there are three equivalents per mole of $KMnO_4$ for this reduction.

$$1 \text{ mol } KMnO_4 = 3 \text{ eq } KMnO_4$$

These examples illustrate that the number of equivalents per mole depends on the number of electrons transferred during the redox reaction. To determine the number of equivalents per mole, we must be able to write a balanced half-reaction for the change, which is equivalent to saying that the oxidation states of the reactants and the products must both be known.

EXAMPLE 12.14   WORKING WITH EQUIVALENTS

Problem: How many equivalents of $Na_2Cr_2O_7$ are in one mole of this compound if, during a redox reaction, the chromium is reduced to the $+3$ oxidation state?

Solution: In $Na_2Cr_2O_7$, the oxidation state of the chromium is $+6$. Therefore, each chromium changes oxidation state by three units when the dichromate ion is reduced. Since the formula contains two chromium atoms, each formula unit of $Na_2Cr_2O_7$ must undergo a change in oxidation number of six units. This means, of course, that one mole of sodium dichromate gains six moles of electrons when it is reduced, so there are six equivalents for each mole of $Na_2Cr_2O_7$.

Notice that we must consider both the change in oxidation number and the number of atoms that change in the reactant.

$$1 \text{ mol } Na_2Cr_2O_7 = 6 \text{ eq } Na_2Cr_2O_7$$

PRACTICE EXERCISE 19   How many equivalents are there in one mole of oxalic acid, $H_2C_2O_4$, when it is oxidized to CO_2?

Equivalent Weights

The equivalent weight, you recall, is simply the weight of one equivalent. Calculating its value for a particular substance requires the formula weight of the compound and a knowledge of the number of equivalents per mole. Remember that the number of equivalents per mole varies with different conditions, as we have just seen; so the equivalent weight is also a quantity that varies with conditions.

EXAMPLE 12.15   CALCULATING EQUIVALENT WEIGHTS FOR REDOX REAGENTS

Problem: Calculate the equivalent weight of $Na_2Cr_2O_7$ if it is to be reduced to Cr^{3+} in a redox reaction.

Solution: The two pieces of information that we need are the formula weight of $Na_2Cr_2O_7$,

$$1 \text{ mol } Na_2Cr_2O_7 = 262.0 \text{ g } Na_2Cr_2O_7$$

and the number of equivalents per mole, which was computed in Example 12.14,

$$1 \text{ mol } Na_2Cr_2O_7 = 6 \text{ eq } Na_2Cr_2O_7$$

Using these in factor-label form is simple if we state the problem as

$$1 \text{ eq } Na_2Cr_2O_7 = ? \text{ g } Na_2Cr_2O_7$$

$$1 \text{ eq } Na_2Cr_2O_7 \times \frac{1 \text{ mol } Na_2Cr_2O_7}{6 \text{ eq } Na_2Cr_2O_7} \times \frac{262.0 \text{ g } Na_2Cr_2O_7}{1 \text{ mol } Na_2Cr_2O_7} = 43.7 \text{ g } Na_2Cr_2O_7$$

The equivalent weight of $Na_2Cr_2O_7$ is therefore 43.7 g.

PRACTICE EXERCISE 20   Calculate the equivalent weight of Na_2CrO_4 if it is to be reduced to Cr^{3+} when it reacts as an oxidizing agent.

Using Equivalents to Solve Problems in Stoichiometry

Solving problems with equivalents saves us from having to write a balanced chemical equation. Although this is not much of a benefit in acid–base reactions, it can be a help in working with redox reactions, because their equations are usually more complex. We can see this in Example 12.16.

EXAMPLE 12.16   USING EQUIVALENT WEIGHTS IN SOLVING PROBLEMS

Problem: How many grams of $KMnO_4$ must be used to react completely with 15.00 g $SnCl_2$ in an acidic solution? In the reaction, the manganese in the $KMnO_4$ is reduced to Mn^{2+} and the tin(II) is oxidized to Sn^{4+}.

Solution: First, we calculate the equivalent weights of each of the reactants. For the $KMnO_4$, the manganese changes oxidation number from $+7$ to $+2$, which corresponds to a gain of 5 electrons per $KMnO_4$. Therefore,

$$1 \text{ mol } KMnO_4 = 5 \text{ eq } KMnO_4$$

The formula weight of $KMnO_4$ is 158.0, so one mole of the compound weighs 158.0 g. The weight of one equivalent is

$$1 \text{ eq } KMnO_4 \times \frac{1 \text{ mol } KMnO_4}{5 \text{ eq } KMnO_4} \times \frac{158.0 \text{ g } KMnO_4}{1 \text{ mol } KMnO_4} = 31.60 \text{ g } KMnO_4$$

Therefore,

$$1 \text{ eq } KMnO_4 = 31.60 \text{ g } KMnO_4$$

For the tin, the change in oxidation number is 2; therefore

$$1 \text{ mol } SnCl_2 = 2 \text{ eq } SnCl_2$$

A mole of $SnCl_2$ weighs 189.6 g, so

$$1 \text{ eq } SnCl_2 \times \frac{1 \text{ mol } SnCl_2}{2 \text{ eq } SnCl_2} \times \frac{189.6 \text{ g } SnCl_2}{1 \text{ mol } SnCl_2} = 94.80 \text{ g } SnCl_2$$

Therefore,

$$1 \text{ eq } SnCl_2 = 94.80 \text{ g } SnCl_2$$

When $KMnO_4$ and $SnCl_2$ react, we know that one equivalent of $KMnO_4$ *must* react with exactly one equivalent of $SnCl_2$—equivalents are defined to give this one-to-one relationship. Therefore,

$$1 \text{ eq } KMnO_4 = 1 \text{ eq } SnCl_2$$

or

$$31.60 \text{ g } KMnO_4 = 94.80 \text{ g } SnCl_2$$

Now we can use this relationship to make a conversion factor to solve the problem.

$$15.00 \text{ g } SnCl_2 \times \frac{31.60 \text{ g } KMnO_4}{94.80 \text{ g } SnCl_2} = 5.000 \text{ g } KMnO_4$$

The mass of $KMnO_4$ required in the reaction is 5.000 g.

PRACTICE EXERCISE 21   How many grams of Na_2SO_3 are required to react with 12.0 g Na_2CrO_4 if the sulfite ion, SO_3^{2-}, is oxidized to sulfate, and the chromate ion, CrO_4^{2-}, is reduced to $Cr(OH)_3$?

Normality

The concentration unit, normality, has the same kinds of applications in problems dealing with redox reactions as in problems involving acids and bases. Normality, you recall, is the number of equivalents of solute per liter of solution. It is easy to obtain the normality from the molarity if you know the number of equivalents per mole — simply multiply the molarity (moles per liter) by the number of equivalents per mole. For example, suppose that you had a solution of sodium dichromate labeled 0.100 M $Na_2Cr_2O_7$. If this solution were to be used in a reaction in which the dichromate ion is reduced to Cr^{3+}, there are 6 eq of $Na_2Cr_2O_7$ per mole — we saw this in Example 12.14. We can calculate the normality of the solution as

$$\text{normality} = \frac{0.100 \text{ mol } Na_2Cr_2O_7}{1.00 \text{ L soln}} \times \frac{6 \text{ eq } Na_2Cr_2O_7}{1 \text{ mol } Na_2Cr_2O_7} = \frac{0.600 \text{ eq } Na_2Cr_2O_7}{1.00 \text{ L soln}}$$

$$= 0.600 \text{ N } Na_2Cr_2O_7$$

A solution is labeled 0.200 M $KMnO_4$. It is to be used in a reaction in which the manganese is to be reduced to Mn^{2+}.
(a) How many equivalents are there per mole of $KMnO_4$?
(b) What is the normality of this solution?
(c) Could we use this same normality value if the permanganate were being reduced to MnO_2?

This only works for normality! You *cannot*, in general, use a similar equation in which the concentrations are expressed in molarity. This is because moles do not always combine in a 1-to-1 ratio.

Normality is a useful concentration unit to use in titrations, and the calculations are the same for redox reactions as for acid–base reactions. Since the number of equivalents must be the same in two solutions that react with each other completely, we can use the simple equation (derived in Section 11.6)

$$N_A V_A = N_B V_B \tag{12.3}$$

The product $N_A V_A$ gives the number of equivalents of substance A in its solution, and the product $N_B V_B$ gives the number of equivalents of substance B in its solution.

EXAMPLE 12.17   WORKING TITRATION PROBLEMS USING NORMALITY

Problem: A solution of $KMnO_4$ has been prepared, but its concentration is not known accurately. To measure its normality, a 25.00-mL sample of a 0.200 N solution of $H_2C_2O_4$ was made strongly acidic (to ensure reduction of MnO_4^- to Mn^{2+}) and titrated. The reaction required 21.65 mL of the $KMnO_4$ solution. What is the normality of the $KMnO_4$ solution?

Solution: Because the concentrations are expressed in normality, this is a very simple problem. We don't even need a balanced chemical equation to solve it. All we have to do is substitute values into Equation 12.3. Taking the $KMnO_4$ to be reactant B and $H_2C_2O_4$ to be reactant A, we can solve for the normality of B to obtain

$$N_B = \frac{N_A V_A}{V_B}$$

Substituting values gives

$$N_B = \frac{(0.200 \text{ N})(25.00 \text{ mL})}{(21.65 \text{ mL})}$$

$$= 0.231 \text{ N}$$

The normality of the $KMnO_4$ solution is 0.231 N.

PRACTICE EXERCISE 23   A 25.00-mL sample of a solution of Na_2SO_3 was acidified and titrated with the $KMnO_4$ solution described in Example 12.17. A volume of 31.05 mL of the $KMnO_4$ solution was required for complete reaction. Calculate the normality of the Na_2SO_3 solution.

PRACTICE EXERCISE 24   If 21.68 mL of the $KMnO_4$ solution in Example 12.17 was needed to titrate a sample containing tin(II) chloride, in which the Sn^{2+} was oxidized to the +4 oxidation state, how many moles of tin were in the sample? (*Hint:* How many equivalents of tin are there per mole of tin in this reaction?)

PRACTICE EXERCISE 25   This exercise brings together many of the concepts in this chapter to show you how they can be combined in a typical chemical analysis using redox reactions: A 10.0-g sample of liquid bleach (an aqueous solution of NaOCl) was added to 50.0 mL of a 0.400 N Na_2SO_3 solution. This caused the OCl^- to be reduced to Cl^-. After the reaction was over, some SO_3^{2-} still remained in the solution, and it was titrated with 0.216 N $KMnO_4$ solution. The titration required 27.30 mL of the $KMnO_4$.
(a) How many equivalents of $KMnO_4$ were used in the titration?
(b) How many equivalents of Na_2SO_3 reacted with the $KMnO_4$?
(c) How many equivalents of Na_2SO_3 were in the original 50.0 mL?
(d) How many equivalents of OCl^- reacted?
(e) How many moles of NaOCl were in the sample of bleach?
(f) What was the percentage by weight of NaOCl in the bleach solution?

SUMMARY

Oxidation–Reduction. Oxidation is the loss of electrons; reduction is the gain of electrons. Both always occur together in redox reactions. The substance oxidized is the reducing agent; the substance reduced is the oxidizing agent. Oxidation numbers are a device that we use to keep tabs on electrons when they are transferred. They are assigned according to the rules on pages 446–448. An increase in oxidation number is oxidation; a decrease in oxidation number is reduction.

Balancing Redox Equations. In a balanced redox equation, the number of electrons gained by one substance is always equal to the number lost by another substance. Oxidation numbers are used in the oxidation number-change method to establish the coefficients of those substances being oxidized and reduced; the remainder of the equation is balanced by inspection. The ion-electron method divides a skeleton net ionic equation into two half-reactions that are balanced separately before being recombined to give the final balanced net ionic equation for the reaction. For reactions in basic solution, the equation is balanced as if it occurred in an acidic solution, and then the balanced equation is converted to its proper form for basic solution by adding an appropriate number of OH^-. Remember, don't use oxidation numbers when balancing equations by the ion-electron method!

Metal–Acid Reactions. In nonoxidizing acids, the strongest oxidizing agent is H^+. The reaction of a metal with a nonoxidizing acid gives hydrogen and a salt of the acid. Only metals more active than hydrogen react this way. These are metals that are located below hydrogen in the activity series shown in Table 12.1. Oxidizing acids such as HNO_3 contain an anion that is a stronger oxidizing agent than H^+, and are able to oxidize many metals that nonoxidizing acids cannot.

Metal-Displacement Reactions. If one metal is more easily oxidized than another, it can displace the other metal from its compounds by a redox reaction. Any metal in the activity series of Table 12.1 will displace others above it in the table from their compounds. Within the periodic table, the most reactive metals (those that are most easily oxidized) are located at the left in Groups IA and IIA. The least reactive metals are located in the center of the periodic table among the heavier transition metals.

Oxidations by Molecular Oxygen. Combustion is the reaction of a substance with oxygen, accompanied by the evolution of heat and light. Combustion of a hydrocarbon in the presence of excess oxygen gives CO_2 and H_2O. If the supply of oxygen is limited, some CO is formed, and in a very limited supply of oxygen only the hydrogen combines to form H_2O—the carbon appears in a very finely divided, elemental form. If the organic compound contains carbon, hydrogen, and oxygen, combustion still gives the same products—CO_2 and H_2O. Metals also combine with oxygen, but only sometimes is the reaction violent enough to be considered combustion. The products are ionic metal oxides. Most nonmetals also burn in oxygen to give molecular oxides.

Redox Reaction in the Laboratory.   Common oxidizing agents in the lab are MnO_4^-, $Cr_2O_7^{2-}$, and CrO_4^{2-}. Chromate ion can be converted to dichromate ion by adding acid, and dichromate ion can be changed to chromate ion by adding base. In acidic solution, the MnO_4^- ion is reduced to Mn^{2+}, but in neutral or slightly basic solution it is reduced to insoluble MnO_2. In an acidic solution, the $Cr_2O_7^{2-}$ ion is usually reduced to Cr^{3+}, while in slightly basic solution, the CrO_4^{2-} ion is reduced to insoluble $Cr(OH)_3$.

Common reducing agents include sulfite ion (which is usually oxidized to sulfate ion), oxalic acid (which is oxidized to carbon dioxide), thiosulfate ion (which is usually oxidized to sulfate ion), and tin(II) ion [which is oxidized to tin(IV)].

Working stoichiometry problems with redox reactions follows the same procedures as developed in Chapters 4 and 11. The balanced equation gives coefficients that relate moles of substances involved in the reaction. Molarity is a convenient unit of concentration to use in dealing with problems in solution stoichiometry. Potassium permanganate is an especially useful titrant in redox titrations because it serves as its own indicator. Iodine (triiodide) titrations using thiosulfate ion as the titrant and starch as an indicator are also very useful.

Equivalents and Normality.   Stoichiometry problems involving redox reactions can be solved without a balanced chemical equation by using equivalents. One equivalent of a substance in a redox reaction either gains or loses one mole of electrons. Equivalents *always* combine in a one-to-one ratio. The equivalent weight is the mass in grams of one equivalent. Normality is the number of equivalents of solute per liter of solution. For the titration of two substances A and B, the following equation applies.

$$N_A V_A = N_B V_B$$

REVIEW EXERCISES

Answers to questions whose numbers are printed in color are given in Appendix C. Difficult questions are marked with asterisks.

Oxidation–Reduction

12.1   Define *oxidation* and *reduction*.

12.2   Analyze what happens to the electron density in the atoms during the reaction $H_2 + Cl_2 \rightarrow 2HCl$.

12.3   Analyze what happens to the electron density in the atoms during the reaction $Mg + Cl_2 \rightarrow MgCl_2$.

12.4   Why are the reactions in Review Exercises 12.2 and 12.3 classified as redox reactions?

12.5   In the reaction $H_2 + Cl_2 \rightarrow 2HCl$, which substance is the oxidizing agent and which is the reducing agent? Which substance is oxidized and which is reduced?

12.6   Why must both oxidation and reduction occur simultaneously during a redox reaction?

Oxidation Numbers

12.7   Roughly speaking, what is the definition of an oxidation number?

12.8   In the compound As_4O_6, arsenic has an oxidation number of $+3$. What is the oxidation state of the arsenic in this compound?

12.9   Assign oxidation numbers to the atoms indicated by boldface type: (a) $\mathbf{S}^{2-}$, (b) $\mathbf{S}O_2$, (c) $\mathbf{P}_4$, (d) $\mathbf{P}H_3$.

12.10   Assign oxidation numbers to the atoms indicated by boldface type: (a) $\mathbf{Cl}O_4^-$, (b) $\mathbf{Cr}Cl_3$, (c) $\mathbf{Sn}S_2$, (d) $\mathbf{Au}(NO_3)_3$.

12.11   Assign oxidation numbers to all of the atoms in the following compounds: (a) Na_2HPO_4, (b) $BaMnO_4$, (c) $Na_2S_4O_6$, (d) ClF_3.

12.12   Assign oxidation numbers to all the atoms in these ions: (a) NO_3^-, (b) SO_3^{2-}, (c) NO^+, (d) $Cr_2O_7^{2-}$.

12.13   Assign oxidation numbers to nitrogen in the following.
(a) NO (d) N_2O_5 (g) NH_2OH
(b) NO_2 (e) N_2H_4 (h) NH_3
(c) N_2O_3 (f) N_2 (i) NaN_3

12.14   Is the following a redox reaction? Explain.

$$2NO_2 \longrightarrow N_2O_4$$

12.15   Assign oxidation numbers to each atom in the following: (a) $NaOCl$, (b) $NaClO_2$, (c) $NaClO_3$, (d) $NaClO_4$.

12.16   Assign oxidation numbers to the elements in the following. (a) $Ca(VO_3)_2$, (b) $SnCl_4$, (c) MnO_4^{2-}, (d) MnO_2.

12.17   Assign oxidation numbers to the elements in the following. (a) PbS, (b) $TiCl_4$, (c) $Sr(IO_3)_2$, (d) Cr_2S_3.

12.18   What is average oxidation number of carbon in (a) C_2H_5OH (grain alcohol), (b) $C_{12}H_{22}O_{11}$ (sucrose — table sugar), (c) $CaCO_3$ (limestone), (d) $NaHCO_3$ (baking soda)?

12.19   Based on trends in electronegativity in the periodic table, what oxidation numbers would you assign to the atoms in (a) $BrCl$ and (b) SCl_2?

12.20   If the oxidation number of nitrogen in a certain molecule changes from $+3$ to -2 during a reaction, is the nitrogen oxidized or reduced? How many electrons are gained (or lost) by each nitrogen atom?

Balancing Redox Equations Using Oxidation Numbers

12.21   Balance the following redox equations, using oxidation numbers.
(a) $HNO_3 + H_3AsO_3 \longrightarrow H_3AsO_4 + NO + H_2O$
(b) $NaI + HOCl \longrightarrow NaIO_3 + HCl$
(c) $KMnO_4 + H_2C_2O_4 + H_2SO_4 \longrightarrow$
$\qquad CO_2 + K_2SO_4 + MnSO_4 + H_2O$
(d) $H_2SO_4 + Al \longrightarrow Al_2(SO_4)_3 + SO_2 + H_2O$

12.22 Balance the following redox equations, using oxidation numbers.
(a) $K_2Cr_2O_7 + HCl \longrightarrow KCl + CrCl_3 + Cl_2 + H_2O$
(b) $NaIO_3 + NaI + HCl \longrightarrow NaCl + I_2 + H_2O$
(c) $Cu + HNO_3 \longrightarrow Cu(NO_3)_2 + NO + H_2O$
(d) $Cu + HNO_3 \longrightarrow Cu(NO_3)_2 + NO_2 + H_2O$

12.23 Balance the following redox equations, using oxidation numbers.
(a) $Cu + H_2SO_4 \longrightarrow CuSO_4 + H_2O + SO_2$
(b) $SO_2 + HNO_3 + H_2O \longrightarrow H_2SO_4 + NO$
(c) $Zn + H_2SO_4 \longrightarrow ZnSO_4 + H_2S + H_2O$
(d) $I_2 + HNO_3 \longrightarrow HIO_3 + NO_2 + H_2O$

12.24 Balance the following equation. Note that the same substance is both oxidized and reduced.

$$I_2 + NaOH \longrightarrow NaI + NaIO_3 + H_2O$$

Ion-Electron Method

12.25 Balance the following half-reactions occurring in an acidic solution. Indicate whether each is an oxidation or a reduction.
(a) $BiO_3^- \longrightarrow Bi^{3+}$ (c) $Pb^{2+} \longrightarrow PbO_2$
(b) $NO_3^- \longrightarrow NH_4^+$ (d) $Cl_2 \longrightarrow ClO_3^-$

12.26 Balance the following half-reactions occurring in a basic solution. Indicate whether each is an oxidation or a reduction.
(a) $Fe \longrightarrow Fe(OH)_2$
(b) $SO_2Cl_2 \longrightarrow SO_3^{2-} + Cl^-$
(c) $Mn(OH)_2 \longrightarrow MnO_4^{2-}$
(d) $H_4IO_6^- \longrightarrow I_2$

12.27 Balance these equations for reactions occurring in an acidic solution.
(a) $S_2O_3^{2-} + OCl^- \longrightarrow Cl^- + S_4O_6^{2-}$
(b) $NO_3^- + Cu \longrightarrow NO_2 + Cu^{2+}$
(c) $IO_3^- + AsO_3^{3-} \longrightarrow I^- + AsO_4^{3-}$
(d) $SO_4^{2-} + Zn \longrightarrow Zn^{2+} + SO_2$
(e) $NO_3^- + Zn \longrightarrow NH_4^+ + Zn^{2+}$
(f) $Cr^{3+} + BiO_3^- \longrightarrow Cr_2O_7^{2-} + Bi^{3+}$
(g) $I_2 + OCl^- \longrightarrow IO_3^- + Cl^-$
(h) $Mn^{2+} + BiO_3^- \longrightarrow MnO_4^- + Bi^{3+}$
(i) $H_3AsO_3 + Cr_2O_7^{2-} \longrightarrow H_3AsO_4 + Cr^{3+}$
(j) $I^- + HSO_4^- \longrightarrow I_2 + SO_2$

12.28 Balance these equations for reactions occurring in an acidic solution.
(a) $Sn + NO_3^- \longrightarrow SnO_2 + NO$
(b) $PbO_2 + Cl^- \longrightarrow PbCl_2 + Cl_2$
(c) $Ag + NO_3^- \longrightarrow NO_2 + Ag^+$
(d) $Fe^{3+} + NH_3OH^+ \longrightarrow Fe^{2+} + N_2O$
(e) $HNO_2 + I^- \longrightarrow I_2 + NO$
(f) $C_2O_4^{2-} + HNO_2 \longrightarrow CO_2 + NO$
(g) $HNO_2 + MnO_4^- \longrightarrow Mn^{2+} + NO_3^-$
(h) $H_3PO_2 + Cr_2O_7^{2-} \longrightarrow H_3PO_4 + Cr^{3+}$
(i) $VO_2^+ + Sn^{2+} \longrightarrow VO^{2+} + Sn^{4+}$
(j) $XeF_2 + Cl^- \longrightarrow Xe + F^- + Cl_2$

12.29 Balance equations for these reactions occurring in a basic solution.
(a) $CrO_4^{2-} + S^{2-} \longrightarrow S + CrO_2^-$
(b) $MnO_4^- + C_2O_4^{2-} \longrightarrow CO_2 + MnO_2$

(c) $ClO_3^- + N_2H_4 \longrightarrow NO + Cl^-$
(d) $NiO_2 + Mn(OH)_2 \longrightarrow Mn_2O_3 + Ni(OH)_2$
(e) $SO_3^{2-} + MnO_4^- \longrightarrow SO_4^{2-} + MnO_2$
(f) $CrO_2^- + S_2O_8^{2-} \longrightarrow CrO_4^{2-} + SO_4^{2-}$
(g) $SO_3^{2-} + CrO_4^{2-} \longrightarrow SO_4^{2-} + CrO_2^-$
(h) $O_2 + N_2H_4 \longrightarrow H_2O_2 + N_2$
(i) $Fe(OH)_2 + O_2 \longrightarrow Fe(OH)_3 + OH^-$
(j) $Au + CN^- + O_2 \longrightarrow Au(CN)_4^- + OH^-$

*12.30 Balance the following equations by the ion-electron method.
(a) $NBr_3 \longrightarrow N_2 + Br^- + HOBr$ (basic solution)
(b) $ClNO_2 \longrightarrow NO_3^- + Cl^-$ (acidic solution)
(c) $Cl_2 \longrightarrow Cl^- + ClO_3^-$ (basic solution)
(d) $H_2SeO_3 + H_2S \longrightarrow S + Se$ (acidic solution)
(e) $MnO_2 + SO_3^{2-} \longrightarrow Mn^{2+} + S_2O_6^{2-}$ (acidic solution)
(f) $BrO_3F \rightleftharpoons BrO_4^- + F^-$ (basic solution)
(g) $XeO_3 + I^- \longrightarrow Xe + I_2$ (acidic solution)
(h) $HXeO_4^- \longrightarrow XeO_6^{4-} + Xe + O_2$ (basic solution)
(i) $(CN)_2 \longrightarrow CN^- + OCN^-$ (basic solution)

Reactions of Metals with Acids

12.31 Write balanced ionic and net ionic equations for the reactions of the following metals with hydrochloric acid to give hydrogen plus the metal ion in solution.
(a) Manganese (gives Mn^{2+})
(b) Cadmium (gives Cd^{2+})
(c) Tin (gives Sn^{2+})
(d) Nickel (gives Ni^{2+})
(e) Chromium (gives Cr^{3+})

12.32 Write balanced ionic and net ionic chemical equations for the reaction of each metal in Review Exercise 12.31 with dilute sulfuric acid.

12.33 What is a nonoxidizing acid? Give two examples. What is the oxidizing agent in a nonoxidizing acid?

12.34 Suggest chemical equations for the oxidation of metallic silver to silver(I) with (a) dilute HNO_3 and (b) concentrated HNO_3.

12.35 What is the strongest oxidizing agent in solutions of nitric acid?

12.36 When hot and concentrated, sulfuric acid is a fairly strong oxidizing agent. Write a balanced molecular equation for the oxidation of metallic copper to copper(II) ion by hot concentrated H_2SO_4, in which the sulfur is reduced to SO_2.

Displacement Reactions and the Activity Series

12.37 The following chemical reactions are observed to occur in aqueous solution.
(a) $2Al + 3Cu^{2+} \longrightarrow 2Al^{3+} + 3Cu$
(b) $2Al + 3Fe^{2+} \longrightarrow 3Fe + 2Al^{3+}$
(c) $Pb^{2+} + Fe \longrightarrow Pb + Fe^{2+}$
(d) $Fe + Cu^{2+} \longrightarrow Fe^{2+} + Cu$
(e) $2Al + 3Pb^{2+} \longrightarrow 3Pb + 2Al^{3+}$
(f) $Pb + Cu^{2+} \longrightarrow Pb^{2+} + Cu$
Arrange the metals Al, Pb, Fe, and Cu in order of increasing ease of oxidation.

12.38 In the preceding question, were *all* the experiments described actually necessary to establish the order?

12.39 According to the activity series in Table 12.1, which of the following metals react with nonoxidizing acids? (a) Silver, (b) gold, (c) zinc, (d) magnesium.

12.40 In each pair below, choose the metal that would be expected to react most rapidly with a nonoxidizing acid such as HCl.
(a) Aluminum or iron
(b) Zinc or nickel
(c) Cadmium or magnesium

12.41 Use Table 12.1 to predict the outcome of the following reactions. If no reaction occurs, write N.R. If a reaction occurs, write a balanced equation for it.
(a) $Fe + Mg^{2+} \longrightarrow$
(b) $Cr + Pb^{2+} \longrightarrow$
(c) $Ag^+ + Fe \longrightarrow$
(d) $Ag + Au^{3+} \longrightarrow$

12.42 Use Table 12.1 to predict the outcome of the following displacement reactions. If no reaction occurs, write N.R. If a reaction occurs, write a balanced chemical equation for it.
(a) $Mg + Fe^{2+} \longrightarrow$
(b) $Au + Ag^+ \longrightarrow$
(c) $Cd + Zn^{2+} \longrightarrow$
(d) $Mn + Co^{2+} \longrightarrow$
(e) $Cr + Sn^{2+} \longrightarrow$

12.43 Use Table 12.1 to predict whether the following displacement reactions should occur. If no reaction occurs, write N.R. If a reaction does occur, write a balanced chemical equation for it.
(a) $Zn + Sn^{2+} \longrightarrow$
(b) $Cr + H^+ \longrightarrow$
(c) $Pb + Cd^{2+} \longrightarrow$
(d) $Mn + Pb^{2+} \longrightarrow$
(e) $Zn + Co^{2+} \longrightarrow$

12.44 Which metals in Table 12.1 will *not* react with nonoxidizing acids?

12.45 Which metals in Table 12.1 will react with water? Write chemical equations for each of these reactions.

Trends in Reactivity of Metals

12.46 When we say that aluminum is more reactive than iron, which property of these elements are we concerned with?

12.47 In what groups in the periodic table are the most reactive metals found? Where do we find the least reactive metals?

12.48 How is the ionization energy of a metal related to its reactivity?

12.49 Arrange the following metals in their approximate order of reactivity (most reactive first, least reactive last) based on their locations in the periodic table: (a) iridium, (b) silver, (c) calcium, (d) iron.

Oxidation by O_2

12.50 Define *combustion*.

12.51 Why is "loss of electrons" described as oxidation?

12.52 Are covalent oxides formed by metals or by nonmetals? Explain your answer based on periodic trends in electronegativity, which you learned in Chapter 7.

12.53 Write balanced chemical equations for the complete combustion (in the presence of excess oxygen) of the following:
(a) C_6H_6 (benzene, an important industrial chemical and solvent)
(b) C_3H_8 (propane—a gaseous fuel used in many stoves)
(c) $C_{21}H_{44}$ (a component of paraffin wax)
(d) $C_{12}H_{26}$ (a component of kerosene)
(e) $C_{18}H_{38}$ (a component of diesel fuel)

12.54 Methanol, CH_3OH, has been suggested as an alternative to gasoline as an automotive fuel. Write a balanced chemical equation for its complete combustion.

12.55 Metabolism of carbohydrates such as glucose, $C_6H_{12}O_6$, produces the same products as complete combustion. Write a chemical equation representing the metabolism (combustion) of glucose.

12.56 Sucrose, $C_{12}H_{22}O_{11}$, is ordinary table sugar. Write a balanced chemical equation representing the metabolism of sucrose. (See Review Exercise 12.55.)

12.57 The standard heat of formation of sucrose, $C_{12}H_{22}O_{11}$, is -2230 kJ/mol. Use data in Table 5.2 (page 169) to compute the amount of energy (in kJ) released by metabolizing 1 oz (28.4 g) of sucrose.

12.58 For ethanol, C_2H_5OH, which is mixed with gasoline to make the fuel gasohol, $\Delta H_f^\circ = -277.63$ kJ/mol. Calculate the number of kilojoules released by burning completely 1 gal (6.56 kg) of ethanol. Use data in Table 5.2 to help in the computation.

12.59 What products are produced in the combustion of $C_{10}H_{22}$ (a) if there is an excess of oxygen available? (b) If there is a slightly limited oxygen supply? (c) If there is a very limited supply of oxygen?

12.60 If one of the impurities in diesel fuel has the formula C_2H_6S, what products will be formed when it burns? Write a balanced chemical equation for the reaction.

12.61 Burning ammonia in an atmosphere of oxygen produces stable N_2 molecules as one of the products. What is the other product? Write the balanced equation.

12.62 Write chemical equations for the reaction of oxygen with (a) zinc, (b) aluminum, (c) magnesium, (d) iron, and (e) calcium.

Common Chemicals for Redox Reactions

12.63 Give the names and chemical formulas for (a) two substances that are common oxidizing agents, and (b) two substances that are common reducing agents.

12.64 Write chemical equations for the conversion of (a) $Cr_2O_7^{2-}$ to CrO_4^{2-} and (b) CrO_4^{2-} to $Cr_2O_7^{2-}$.

12.65 What manganese-containing product is formed when MnO_4^- ion is reduced (a) in an acidic solution and (b) in a basic solution?

12.66 What chromium-containing product is formed when (a) $Cr_2O_7{}^{2-}$ is reduced in an acidic solution and (b) $CrO_4{}^{2-}$ is reduced in a slightly basic solution?

12.67 What is the usual product formed when $SO_3{}^{2-}$ is oxidized?

12.68 Write a balanced net ionic equation for the reaction of $MnO_4{}^-$ with $HSO_3{}^-$ ion in an acidic solution.

12.69 Write a balanced net ionic equation for the reaction of OCl^- with $S_2O_3{}^{2-}$. The OCl^- is reduced to Cl^-.

12.70 Write a balanced net ionic equation for the oxidation of $H_2C_2O_4$ by $Cr_2O_7{}^{2-}$ ion in an acidic solution.

12.71 Iodine oxidizes thiosulfate ion to $S_4O_6{}^{2-}$ ion. Write a net ionic equation for this reaction.

12.72 Write a net ionic equation for the reaction of permanganate ion with tin(II) ion in an acidic solution.

Stoichiometry of Redox Reactions

12.73 Lead(IV) oxide reacts with hydrochloric acid to give chlorine. The equation for the reaction is

$$PbO_2 + 4Cl^- + 4H^+ \longrightarrow PbCl_2 + 2H_2O + Cl_2$$

How many grams of chlorine are formed from 15.0 g of PbO_2?

12.74 How many grams of magnesium are needed to displace all the silver from 25.0 g of silver nitrate?

12.75 Manganese(II) ion is oxidized to permanganate ion by bismuthate ion, $BiO_3{}^-$, in an acidic solution. In the reaction, $BiO_3{}^-$ is reduced to Bi^{3+}.
(a) Write a balanced net ionic equation for the reaction.
(b) How many grams of $NaBiO_3$ are needed to oxidize the manganese in 15.0 g of $Mn(NO_3)_2$?

12.76 Sodium iodate reacts with sodium sulfite according to the equation

$$NaIO_3 + 3Na_2SO_3 \longrightarrow 3Na_2SO_4 + NaI$$

(a) In this reaction, which substance is the oxidizing agent?
(b) How many grams of $NaIO_3$ are needed to react with 5.00 g of Na_2SO_3?

12.77 A sample of a chromium-containing alloy weighing 3.000 g was dissolved in acid, and all the chromium in the sample was oxidized to $CrO_4{}^{2-}$. It was then found that 3.09 g of Na_2SO_3 was required to reduce the $CrO_4{}^{2-}$ to $Cr(OH)_3$ in a basic solution, with the $SO_3{}^{2-}$ being oxidized to $SO_4{}^{2-}$.
(a) Write a balanced equation for the reaction of $CrO_4{}^{2-}$ with $SO_3{}^{2-}$.
(b) How many moles of $CrO_4{}^{2-}$ reacted with Na_2SO_3?
(c) How many grams of Cr were in the alloy sample?
(d) What is the weight percentage of Cr in the alloy?

12.78 Solder is an alloy containing the metals tin and lead. A particular sample of this alloy weighing 1.50 g was dissolved in acid. All the tin was then converted to the +2 oxidation state. Next, it was found that 0.368 g of $Na_2Cr_2O_7$ was required to oxidize the Sn^{2+} to Sn^{4+} in an

acidic solution. In the reaction the chromium was reduced to Cr^{3+} ion.
(a) Write a balanced net ionic equation for the reaction between the Sn^{2+} and $Cr_2O_7{}^{2-}$ in an acidic solution.
(b) Calculate the number of grams of tin that were in the sample of solder.
(c) What was the percentage by weight of tin in the solder?

Redox Reactions in Solution

12.79 How many milliliters of 0.150 M $KMnO_4$ solution are needed to react completely with 35.0 mL of 0.150 M $SnCl_2$ solution, given the chemical equation for the reaction.

$$16H^+ + 2MnO_4{}^- + 5Sn^{2+} \longrightarrow 5Sn^{4+} + 2Mn^{2+} + 8H_2O$$

12.80 In an acidic solution, $HSO_3{}^-$ ion reacts with $ClO_3{}^-$ ion to give $SO_4{}^{2-}$ ion and Cl^- ion.
(a) Write a balanced net ionic equation for the reaction.
(b) How many milliliters of 0.100 M $NaClO_3$ solution are needed to react completely with 25.0 mL of 0.400 M $NaHSO_3$ solution?

12.81 Hydrogen peroxide (H_2O_2) can be purchased in drug stores for use as an antiseptic. A sample of such a solution weighing 1.000 g was acidified with H_2SO_4 and titrated with a 0.2000 M solution of $KMnO_4$ solution. The net ionic equation for the reaction is

$$6H^+ + 5H_2O_2 + 2MnO_4{}^- \longrightarrow 5O_2 + 2Mn^{2+} + 8H_2O$$

The titration required 17.60 mL of $KMnO_4$ solution.
(a) How many grams of H_2O_2 reacted?
(b) What is the percentage by weight of the H_2O_2 in the original antiseptic solution?

12.82 Sodium nitrite, $NaNO_2$, is used as a perservative in meat products such as frankfurters and bologna. In an acidic solution, nitrite ion is converted to nitrous acid, HNO_2, which reacts with permanganate ion according to the equation

$$H^+ + 5HNO_2 + 2MnO_4{}^- \longrightarrow$$
$$5NO_3{}^- + 2Mn^{2+} + 3H_2O$$

A 1.000-g sample of a water-soluble solid containing $NaNO_2$ was dissolved in dilute H_2SO_4 and titrated with 0.01000 M $KMnO_4$ solution. The titration required 11.60 mL of the $KMnO_4$ solution. What was the percentage of $NaNO_2$ in the original 1.000-g sample?

12.83 Both calcium chloride, $CaCl_2$, and sodium chloride are used to melt ice and snow on roads in the winter. A certain company was marketing a mixture of these two compounds for this purpose. A chemist, wishing to analyze the mixture, dissolved 2.651 g of it in water and precipitated the calcium by adding sodium oxalate, $Na_2C_2O_4$.

$$Ca^{2+} + C_2O_4{}^{2-} \longrightarrow CaC_2O_4(s)$$

The calcium oxalate was then carefully filtered from the solution, dissolved in sulfuric acid, and titrated with

0.1000 M $KMnO_4$ solution. The reaction that occurred was

$$6H^+ + 5H_2C_2O_4 + 2MnO_4^- \longrightarrow$$
$$10CO_2 + 2Mn^{2+} + 8H_2O$$

The titration required 23.88 mL of the $KMnO_4$ solution.
(a) How many moles of $C_2O_4^{2-}$ were present in the CaC_2O_4 precipitate?
(b) How many grams of $CaCl_2$ were in the original 2.651-g sample?
(c) What was the percentage of $CaCl_2$ in the sample?

12.84 Another way to analyze a sample for nitrite ion (see Review Exercise 12.82) is to acidify a solution containing NO_2^- and then allow the HNO_2 that is formed to react with iodide ion in the presence of excess I^-. The reaction is

$$2HNO_2 + 2H^+ + 3I^- \longrightarrow 2NO + 2H_2O + I_3^-$$

Then the I_3^- is titrated with $Na_2S_2O_3$ solution using starch as an indicator.

$$I_3^- + 2S_2O_3^{2-} \longrightarrow 3I^- + S_4O_6^{2-}$$

In a typical analysis, a 0.9470-g sample that was known to contain $NaNO_2$ was treated as described above. The titration required 29.00 mL of 0.3000 M $Na_2S_2O_3$ solution.
(a) How many moles of I_3^- had been produced in the first reaction?
(b) How many moles of NO_2^- had been in the original 0.9470-g sample?
(c) What was the percentage of $NaNO_2$ in the original sample?

Equivalents and Equivalent Weights

12.85 How many equivalents are in one mole of the following?
(a) $KClO_3$, if the chlorine is reduced to Cl_2
(b) $KClO_3$, if the chlorine is reduced to Cl^-
(c) NiO_2, if the nickel is reduced to Ni^{2+}
(d) Cr_2O_3, if the chromium is oxidized to CrO_4^{2-}

12.86 How many equivalents are in one mole of the following?
(a) $NaOCl$, if the hypochlorite ion is oxidized to ClO_3^-
(b) H_3AsO_3, if it is oxidized to H_3AsO_4
(c) I_2, it if is oxidized to IO_3^-
(d) $NaClO_4$, if the perchlorate ion is reduced to Cl^-

12.87 If CrO_4^{2-} ion is reduced to CrO_2^- ion in a basic solution, what is the equivalent weight of Na_2CrO_4?

12.88 If Mn_2O_3 is oxidized to MnO_4^-, what is the equivalent weight of Mn_2O_3?

12.89 What is the mass in grams of 0.250 eq of SO_2 if it is oxidized to SO_4^{2-}?

12.90 What is the mass in grams of 1.45 eq of HNO_3 if it is reduced to NO?

12.91 How many grams of SO_2 will be oxidized to SO_4^{2-} by 35.0 g $KMnO_4$ if the permanganate ion is reduced to Mn^{2+}? Work the problem using equivalents.

12.92 How many grams of $NaIO_3$ will be reduced to I_2 by reaction with 0.250 g of $NaNO_2$ if the NO_2^- is oxidized to NO_3^-? Solve the problem using equivalents.

12.93 How many grams of $Na_2S_2O_3$ are needed to react with the $NaOCl$ in 100 mL of household bleach, which has a concentration of 0.70 M $NaOCl$? In the reaction, $S_2O_3^{2-}$ is oxidized to SO_4^{2-}, and OCl^- is reduced to Cl^-.

Normality

12.94 What is the normality of a 0.200 M solution of $NaIO_3$ if in a reaction the IO_3^- is to be reduced to I_2?

12.95 A solution is labeled 0.500 N $NaIO_3$ for use in a reaction in which the IO_3^- is to be reduced to I_2. How would this same solution be relabeled if it were to be used in a reaction in which the IO_3^- is reduced to I^-?

12.96 How many grams of $Ba(OH)_2$ are needed to prepare 100.0 mL of 0.0100 N solution of this base?

12.97 How many grams of $K_2Cr_2O_7$ are needed to prepare 250 mL of 0.100 N solution if the $Cr_2O_7^{2-}$ will be reduced to Cr^{3+} in a reaction?

12.98 How many equivalents are in 25.0 mL of 0.100 N K_2CrO_4?

12.99 How many equivalents are in 35.0 mL of 0.500 N $H_2C_2O_4$?

12.100 A 0.1000 N solution of $H_2C_2O_4$ was used to determine the exact concentration of a freshly prepared solution of $KMnO_4$. A volume of 25.00 mL of the 0.1000 N $H_2C_2O_4$ solution required 22.45 mL of the $KMnO_4$ solution for complete reaction. What is the normality of the $KMnO_4$ solution?

12.101 A mixture of $NaCl$ and $NaNO_2$ was to be used in processing meat to make bologna. Before being used, it was analyzed as follows. A 1.000-g portion of the mixture was dissolved in water and acidified. The HNO_2 that was formed was titrated with 0.200 N $K_2Cr_2O_7$, which oxidized the HNO_2 to NO_3^-. The titration required 21.74 mL of the $K_2Cr_2O_7$ solution.
(a) How many equivalents of HNO_2 reacted in the titration?
(b) How many moles of $NaNO_2$ were in the 1.000-g sample?
(c) What was the percentage of $NaNO_2$ in the sample?

12.102 A 0.5000-g sample of an alloy containing manganese was dissolved in acid and the resulting solution was treated with $NaBiO_3$, which oxidized the manganese to MnO_4^-. The permanganate ion was titrated with 0.2000 N $H_2C_2O_4$, which required 28.80 mL for complete reaction. In the reaction, the MnO_4^- was reduced to Mn^{2+}.
(a) How many equivalents of MnO_4^- reacted in the titration?
(b) How many moles of MnO_4^- reacted?
(c) How many grams of manganese were in the alloy?
(d) What was the percentage by mass of manganese in the alloy?

12.103 All the copper in a sample of a copper ore weighing 0.8470 g was dissolved to give a solution containing Cu^{2+}. To this was added a solution containing an excess amount of KI. The reaction that occurred was

$$2Cu^{2+} + 5I^- \longrightarrow 2CuI + I_3^-$$

The I_3^- was titrated with 0.1000 N $Na_2S_2O_3$, which caused the I_3^- to be reduced to I^-. The titration required 18.90 mL of the $Na_2S_2O_3$ solution.

(a) How many equivalents of I_3^- reacted in the titration?
(b) How many moles of I_3^- reacted in the titration?
(c) How many grams of copper were in the ore sample?
(d) What was the percentage of copper in the ore sample?

CHEMICALS IN USE

PHOTOGRAPHY

Compounds of silver are not very stable, and are rather easily decomposed. This is also true of compounds of some other metals—gold, for example—but what makes silver special is that the decomposition of its compounds is promoted by the absorption of light. Despite many attempts, no one has yet found any compounds that quite rival those of silver for light sensitivity, and nearly all the film available today uses silver compounds as the light-sensing medium.

Photographic Film and Paper

Photographic film and paper are similar in some respects. Both have a light-sensitive coating called an *emulsion* spread on a supporting base. In film, this base is a flexible clear plastic; in photographic paper the base is either a white paper or, in modern times, a white plastic.

The emulsion consists of gelatin and one or more of the silver halides (AgCl, AgBr, and AgI). The light sensitivity increases from AgCl to AgI, and most negative film uses either AgBr or a mixture of AgBr and AgI. In color film, there are several emulsion layers separated by filter layers to enable the recording of color information. In photographic paper, the emulsion usually contains a mixture of AgBr and AgCl.

Preparation of the photographic emulsion involves precipitation of the silver halides in a metathesis reaction of the type that we discussed in Chapter 11. A solution of $AgNO_3$ is added slowly to a solution that contains KBr (perhaps with a small percentage of KI) and a small amount of gelatin. The net ionic equation is

$$Ag^+(aq) + Br^-(aq) \longrightarrow AgBr(s)$$

The gelatin serves several functions, including keeping the silver halide crystals suspended when the emulsion is spread on the film or paper base.

Formation of the Photographic Image

When you snap a photograph, light falls very briefly on the photographic film. In this brief instant, the silver halide crystals undergo a very subtle change. It is believed that the light absorbed by a crystal causes some electrons to be set free, and these wander through the crystal and become trapped at certain active sites located at the surface. Nearby silver ions then diffuse to these sites and become reduced to silver atoms by the trapped electrons. These silver atoms form the beginnings of the image that will eventually appear on the film. However, at this early stage, no image is yet visible. To the eye, the film would appear unchanged. The still-invisible impression made by the light is called the "latent image," the term *latent* meaning hidden or dormant.

After you've finished a roll of film, your next step is to bring it to a store to be "developed." The processing of the film involves several chemical reactions, all of which actually take place within the emulsion layer itself. If the film is of the black-and-white variety, the processing is quite simple. The first step is to treat the film with a solution called the "developer," which contains one or more mild reducing agents. This causes the silver bromide to be reduced to metallic silver, but the reduction occurs most rapidly at those sites where silver atoms of the latent image reside. In a sense, the silver atoms of the latent image serve as nuclei upon which additional silver atoms can be deposited as they are formed in the reduction reaction. As a result, small grains of silver grow only where the film has been exposed to light. The silver particles are so tiny they appear black, and the image is called a *negative* (Figure 10a) because those areas that were bright in the original scene show up black on the film.

FIGURE 10a A negative and the positive print made from it.

Photographic developers are normally quite basic. They have to be for the special reducing agents to function. Therefore, after the development process is complete and the developer solution has been poured off, the film is treated with a dilute acetic acid solution called a "stop bath." The acid neutralizes the small amount of base in the residual developer absorbed in the emulsion of the film, and that halts the development process.

If you were to look at the film after development, you would see the negative image surrounded by milky-white undeveloped silver bromide. This must be removed from the film because it would gradually turn dark on exposure to light. The unreacted AgBr is removed by a solution called the "fixer." The fixer contains sodium or ammonium thiosulfate, $Na_2S_2O_3$ or $(NH_4)_2S_2O_3$. The thiosulfate ion reacts with the AgBr and causes it to dissolve by forming a water-soluble complex ion, $Ag(S_2O_3)_2{}^{3-}$.

$$AgBr(s) + 2S_2O_3{}^{2-}(aq) \longrightarrow Ag(S_2O_3)_2{}^{3-}(aq) + Br^-(aq)$$

After the excess AgBr has been removed from the emulsion, the fixer is poured off and the film is washed thoroughly with water and then dried.

Production of a black-and-white print follows essentially the same sequence as the production of the negative. Light is projected through the negative onto a piece of photographic paper, forming a latent image. The paper is then treated with a developer–stop bath–fixer sequence to produce the final picture. This time a positive image is formed because the paper is not exposed under the dark areas of the negative but is exposed through the clear areas. In this way, dark regions of the negative appear bright in the print, and light areas appear dark, just like in the original scene.

Color Photographs

If you look very closely at a color television screen, you will see that the image is made up of tiny dots of three colors. They are red, green, and blue, and they constitute what are called the primary colors. By combining light of just these three colors in the proper proportions we can reproduce any color in the rainbow, and that is exactly what the color TV does. Color film produces its hues by the same principle, with the primary colors being produced by dyes in the film's emulsion.

There are two types of color film that you have probably encountered. Reversal film is the kind used to make color slides. The second type, which is the one we will discuss in some detail below, works much like black and white film; the scene is produced on the film as a color negative, which is then printed on special "color paper" to give a positive color print. Both kinds of film are similar in one respect; they contain three emulsion layers. Each is affected by just one of the primary colors. As light from the subject passes through the film, the color image is sepa-

A color photograph and its color negative. In the negative, the colors are reversed; the green leaves appear red, the yellow flower petals appear blue, and the red center of the flower appears green.

rated into three images, each being the contribution of just one of the primary colors.

Each light-sensitive layer in a typical color negative film contains a mix of silver halides as the light-sensing medium. Also incorporated into each emulsion layer are portions of dye molecules that are called *dye couplers*. When the film is processed, a special color developer is used. This developer becomes oxidized as it reduces silver ion in places that had been exposed to light. The oxidation products, which are produced in proportion to the amount of silver reduced, combine with the dye couplers in the emulsion layers to produce images made up of dye molecules. Then the silver itself (which is not needed anymore) is removed by oxidizing it to Ag^+ using a mild oxidizing agent such as $Fe(CN)_6{}^{3-}(aq)$, and unexposed silver halides are removed by dissolving them in a thiosulfate fixing bath. This leaves the color negative with its image made up of only dye molecules (Figure 10b).

If you examine a color negative (which is somewhat difficult because of its overall orange tint, necessary to improve color quality in the finished print), you will see that it is a "negative" in two ways: light areas in the subject appear dark, and the colors of the subject are reversed. Thus, a blue dye covers areas in the subject that are yellow, a red dye covers subject areas that are green, and a green dye covers subject areas that are red. These dyes serve as filters that block light of their opposite colors when the negative is printed; blue blocks yellow, red blocks green, and green blocks red. This causes a second reversal in colors when the print is made, and thereby gives the correct colors in the finished photograph (Figure 10b).

Questions

1. Give the chemical names and formulas of the principal light-sensitive chemicals used in photographic emulsions.
2. Write a net ionic equation that shows how one of the chemicals in the answer to Question 1 is prepared.
3. What chemical action does the developer cause in the processing of photographic film and paper?
4. Write a chemical equation that illustrates the function of the sodium thiosulfate in the fixer solution used in the final step in the processing of black and white film.
5. What are the three primary colors from which a color image is formed on color film?

The solubilities of air's nitrogen and oxygen in blood increase with pressure, as this diver well knows. If she ascends too rapidly, the dissolved gases that leave her blood at lesser pressures will form microbubbles that cannot get through small capillaries, and the painful, sometimes fatal, condition called the "bends" will occur. The solubilities of all kinds of solutes in various solvents are studied in his chapter.

SOLUTIONS AND COLLOIDS

KINDS OF MIXTURES
WHY SOLUTIONS FORM
HEATS OF SOLUTION
EFFECT OF TEMPERATURE ON SOLUBILITY
EFFECT OF PRESSURE ON THE SOLUBILITIES OF GASES
CONCENTRATION EXPRESSIONS

LOWERING OF THE VAPOR PRESSURE
ELEVATION OF THE BOILING POINT
DEPRESSION OF THE FREEZING POINT
DIALYSIS AND OSMOSIS
COLLIGATIVE PROPERTIES OF SOLUTIONS OF ELECTROLYTES

13.1 KINDS OF MIXTURES

Solutions, colloidal dispersions, and suspensions are three important kinds of mixtures.

Of the three fundamental kinds of substances introduced in Chapter 1 — elements, compounds, and mixtures — our daily experiences with matter are nearly always with mixtures. Sugar, salt, and possibly a few medications are pure compounds; the water we drink is usually a very dilute solution of traces of salts, particularly if it is "hard water." As for elements, the metals in coins and jewelry are almost never single, metallic elements but mixtures of two or more metals. Aluminum foil is as pure a metal as most people ever see.

Type of Matter	Example
Elements	107
Compounds	A few million (and growing)
Mixtures	Probably infinite

Mixtures, as you learned, do not obey the law of definite proportions. Since their components can be mixed together (a *physical,* not a chemical, change) in almost any proportion, the number of possible mixtures in the universe is infinite. Taking just sugar, $C_{12}H_{22}O_{11}$, and water, we know we can make a mixture of these so concentrated that it oozes like cold syrup or so dilute that the sweetness of the sugar can hardly be detected.

TABLE 13.1 Solutions, Colloidal Dispersions, and Suspensions Compared

Particle Sizes Become Progressively Larger		
Solutions	**Colloidal Dispersions**	**Suspensions**
All particles are on the order of atoms, ions, or small molecules (0.1 – 1 nm)	Particles of at least one component are large clusters of atoms, ions, or small molecules (1 – 1000 nm)	Particles of at least one component can be seen with a low-power microscope (1000 nm)
Most stable to gravity	Less stable to gravity	Unstable to gravity
Homogeneous	Homogeneous but borderline	Inhomogeneous
Transparent (but often colored)	Often translucent or opaque; can be transparent	Not transparent
No Tyndall effect	Tyndall effect	Not applicable (suspensions cannot be transparent)
No Brownian movement	Brownian movement	Particles separate
Not separable by filtration	Not separable by filtration	Separable by filtration

The *chemical* properties of such solutions can usually be figured out by a knowledge of the individual chemical properties of their components. The physical properties, however, are another matter, and this is why we have this chapter.

The simplest types of mixtures are suspensions, colloidal dispersions, and solutions, which are compared and contrasted in Table 13.1. Solutions are completely homogeneous, which means that no matter how small the sample taken, the composition is everywhere the same. Suspensions, on the other hand, are not considered to be homogeneous, although if they are stirred vigorously they approach it. Finely ground clay in water or fine dust blowing in the wind are examples. Colloidal dispersions lie between these extremes. Many appear to be homogeneous and are stable on standing, like the milk you buy or mayonnaise. Others, like a misty fog, do not stay homogeneous indefinitely.

The underlying factor that chiefly distinguishes among solutions, colloidal dispersions, and suspensions is the relative sizes of the particles in the mixture. Only in

FIGURE 13.1 The centrifuge. (*a*) Cutaway sketch of a typical inclined-tube laboratory centrifuge. (*b*) Common inclined-tube laboratory centrifuge. (*c*) An ultracentrifuge with rotor speeds of 20,000 to 60,000 revolutions per minute. The chamber can be cooled or warmed and air friction can be reduced by evacuating the chamber.

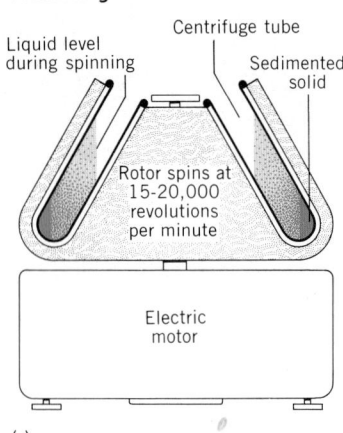

Liquid level during spinning
Centrifuge tube
Sedimented solid
Rotor spins at 15-20,000 revolutions per minute
Electric motor

(a)

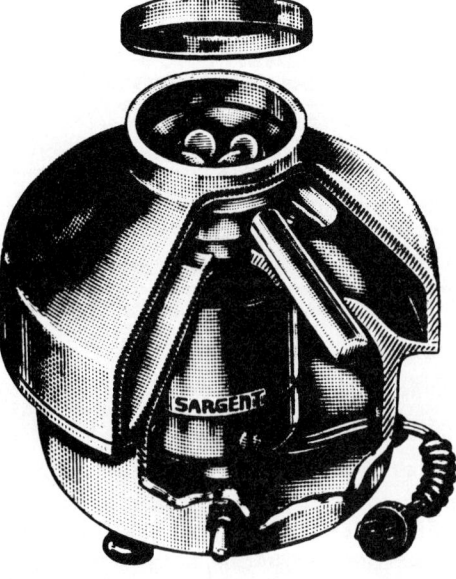

(b)

(c)

solutions are the intermixed particles at the atomic, ionic, or molecular level of size. In suspensions, the suspended particles are so large that if the mixture is a fluid it must be stirred continuously or its components will separate fairly quickly. In colloidal dispersions, the dispersed particles are intermediate in size between those of a solution and a suspensions. Let us now take a closer look at each type.

Suspensions

A suspension is a mixture in which the particles of one (or more) of the substances have at least one dimension *larger* than 1000 nanometers (1000 nm or 1 μm). Finely divided clay in water is an example, but unless it is stirred or shaken the larger particles soon settle out.

Suspensions of solids in fluids can nearly always be separated by filtration using ordinary filter paper or by the use of a centrifuge. In a centrifuge (Figure 13.1), a sample is whirled at high speed, which creates a strong gravitation-like force sufficient to pull the suspended matter down through the fluid.

Colloidal Dispersions

A colloidal dispersion — often simply called a colloid — is a mixture in which the dispersed particles have at least one dimension in the range of 1 to 1000 nm. Table 13.2 gives examples of colloidal dispersions. Colloidally dispersed particles are generally too small to be trapped by ordinary filter paper. Sulfur, for example, sometimes forms as a colloidal dispersion. This solid element is not soluble in water, but when it forms in water from dissolved reactants, crystals large enough to be separated by filtration often fail to form. Instead, the sulfur emerges as a colloidal dispersion with a very pale yellowish color.

The tendency of colloidal dispersions in a fluid state not to separate is aided by the buffeting that the dispersed particles experience from the constantly moving molecules of the "solvent." This erratic movement of colloidally dispersed particles is called Brownian movement after an English botanist, Robert Brown (1773–1858), who first noticed it when he used a microscope to study pollen grains in water. Particles in suspension, of course, experience the same buffeting, but they are too large to be kept in suspension without outside help — the extra buffeting caused, for example, by stirring or wind action.

Colloidal dispersions that do eventually separate are those in which the dispersed particles, over time, grow too large. You can, for example, make a colloidal dispersion of olive oil in vinegar — a salad dressing — by vigorously shaking the mixture. This breaks at least some of the oil into microdroplets in the right range of size for colloidal dispersions.

"Colloidal," from the Greek *kolla*, "glue"; old-fashioned glues formed colloidal dispersions in water.

Brown's discovery (1827) was early and very dramatic evidence for the existence of atoms and molecules and their kinetic motions.

TABLE 13.2 Colloidal Systems

Type	Dispersed Phase[a]	Dispersing Medium[b]	Common Example
Foam	Gas	Liquid	Soap suds, whipped cream
Solid foam	Gas	Solid	Pumice, marshmallow
Liquid aerosol	Liquid	Gas	Mist, fog, clouds, liquid air pollutants
Emulsion	Liquid	Liquid	Cream, mayonnaise, milk
Solid emulsion	Liquid	Solid	Butter, cheese
Smoke	Solid	Gas	Dust and particulates in smog
Sol	Solid	Liquid	Starch in water, jellies,[c] paints
Solid sol	Solid	Solid	Alloys, pearls, opals

[a] The colloidal particles constitute the *dispersed phase*.

[b] The continuous matter into which the colloidal particles are dispersed is called the *dispersing medium*.

[c] Sols that become semisolid or semirigid, like gelatin desserts and fruit jellies, are called *gels*.

Stability here means the resistance to the separation of the colloidal particles from the dispersion.

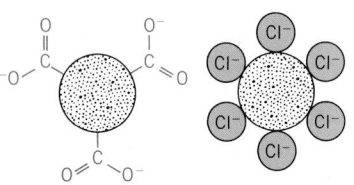

FIGURE 13.2 Colloidal particles often bear electrical charges that stabilize the dispersion. On the left is a particle whose extremely large molecules carry negatively charged groups. On the right, the colloidal particle has attracted chloride ions to itself. In either case, these colloidal particles repel each other and cannot join together.

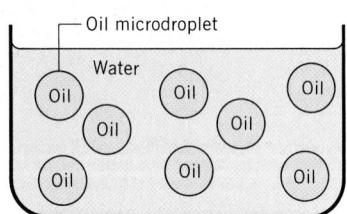

Oil in water emulsion

FIGURE 13.3 An oil-in-water emulsion.

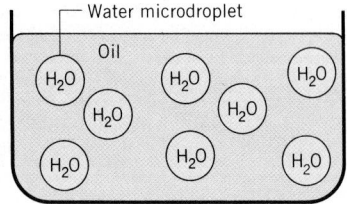

Water in oil emulsion

FIGURE 13.4 A water-in-oil emulsion.

But, as you know, the microdroplets eventually coalesce and grow in mass until gravity wins over Brownian buffeting.

Evidently, to prepare a *stable* colloidal dispersion, we have to do two things — make the dispersed particles initially small enough and then keep them from joining together or coalescing. The dispersed particles will not coalesce if they carry the same kind of electrical charge — all positive or all negative. (Ions of opposite charge would be present out in the solvent to keep the system electrically neutral.) Some of the most stable dispersions occur when colloidal particles have preferentially attracted to their surfaces ions of just one kind of charge from some dissolved salt. The dispersed particles of most sols — colloidal dispersions of solids in a fluid — are examples. Alternatively, stable dispersions might form when extremely large, like-charged ions, such as those of proteins, are involved. See Figure 13.2.

Emulsions — dispersions of one liquid in another — often are relatively stable colloidal dispersions, provided a third component called an emulsifying agent is also present. Mayonnaise, for example, is an oil-in-water emulsion (Figure 13.3) of an edible oil in a dilute solution of an edible organic acid. The emulsifying agent is the protein in egg yolk. Molecules of this protein form a "skin" around each microdroplet of the oil, which keeps the microdroplets from coalescing. (And if you've ever made mayonnaise, you know how vigorously the mixture must be whipped as it is made. This makes the oil droplets very small, and ensures that each gets its coating of the emulsifying agent.) In homogenized milk, the microparticles of butterfat are coated with the milk protein, casein. Water-in-oil emulsions are also possible (Figure 13.4), and margarine is an example; a soybean product is the emulsifying agent.

A useful colloid that you might employ in the lab is starch in water, particularly if you use the I_3^-/I_2 system described in the previous chapter. Colloidal starch, as we said, gives an intensely purple-black color with I_2, so it is a common test for iodine. Starch molecules are themselves very large, and it does not take many to make a colloid-sized particle. Usually, colloidal starch mixtures have a milky look or they can even be opaque, because colloidal particles are large enough to reflect and scatter visible light. But they can be prepared so dilute that they appear clear and transparent, like water. Even such dilute dispersions, however, can scatter light.

As seen in Figure 13.5, when a beam of light is focused on a dilute and otherwise clear-looking dispersion of starch in water, the path of the light beam can be seen from the side. This light-scattering by colloidal dispersions is called the Tyndall effect, after John Tyndall (1820–1893), a British scientist. Solutes in solutions, however, involve species — ordinary ions and molecules — too small to scatter light, so solutions do not give the Tyndall effect.

The Tyndall effect is quite common. Whenever you see sunlight "streaming"

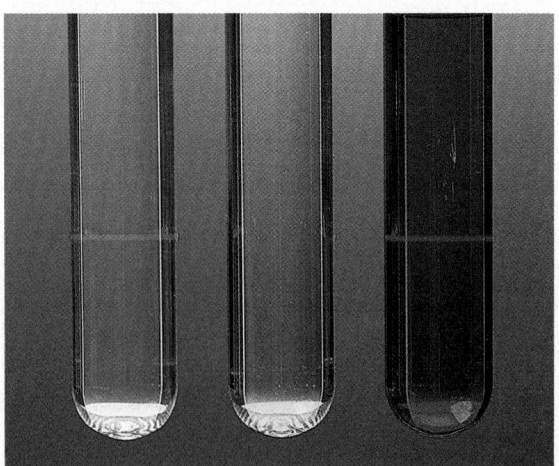

FIGURE 13.5 The Tyndall effect. A pencil-line thin red laser beam passes through the liquid in three test tubes. The first contain a colloidal dispersion of starch, the second a solution of sodium chromate, and the third a colloidal dispersion of Fe_2O_3. All three appear transparent, and in the absence of the Tyndall effect we might think they are all solutions. However, the Tyndall effect reveals that the first and third are colloids, not true solutions.

When we look at a part of the sky well away from the sun on a reasonably clear day, we see the sky as blue. Why isn't it black?

No part of our sky is totally free from colloid-sized particles, and these scatter sunlight back toward our eyes. Then why isn't the sky white, like the sun? Why is it blue?

And why is the western sky after sunset often a beautiful orange or red?

The answers lie in the fact that not all frequencies of sunlight are scattered by colloidal particles with the same intensity. White light from the sun is a mixture of the frequencies of light from all colors of the spectrum —from the lowest visible frequency, red, through orange, yellow, green, blue, and indigo, to the highest frequency, violet. And the intensity of *scattered* light varies with the fourth power of the frequency. Thus, blue to violet frequencies are the most intensely scattered. So when the midday sunlight streams across the sky, these frequencies are preferentially scattered toward our vision, and we see a blue sky. When we're enjoying the midday sun, those far to the east or the west of us are seeing either a sunset or a sunrise. And as we see the scattered blue light, they see more of

Light-scattering colors the sky at sunset.

what isn't scattered so well—the oranges and reds.

When the sky has a lot of colloidal particles — maybe from a forest fire or from heavy smog—so much blue light is scattered as the setting sunlight streams toward us that the light that reaches us is left mostly vivid in reds and oranges. Thin clouds can cause their own variations, as you can see in the accompanying photograph.

through openings in a forest canopy or between clouds, you are observing the Tyndall effect as light is scattered by colloidally dispersed particles of smoke or microdroplets of liquids exuded by trees. Not all wavelengths of visible light are equally well scattered by colloidal dispersions. As a result, when the sun is near the horizon at sunrise or sunset and the air is polluted by smoke, brilliant sky colors are seen, as further discussed in Special Topic 13.1.

Solutions

A solution is a homogeneous mixture in which all particles have the sizes of atoms, small molecules, and small ions—average diameters in the range 0.05 to 0.25 nm, sometimes up to 1 nm. Table 13.3 gives several examples of the kinds of solutions possible. Solutions do not exhibit either the Tyndall effect or Brownian movement.

TABLE 13.3 Solutions

Gaseous Solutions	
Gas in a gas	Air
Liquid in a gas	(If droplets are present—a colloidal system)
Solid in a gas	(If particles are present—a colloidal system)
Liquid Solutions	
Gas in a liquid	Carbonated beverages (CO_2 in water)
Liquid in a liquid	Vinegar (acetic acid in water), gasoline
Solid in a liquid	Sugar in water, seawater
Solid Solutions	
Gas in a solid	Alloy of hydrogen in palladium[a]
Liquid in a solid	Benzene in rubber (rubber cement)
Solid in a solid	Carbon in iron (steel)[a]

[a] Some doubt exists that these are a true solutions.

If small samples were taken from different parts of the same solution, they would have identical densities. The force of gravity is not strong enough to pull heavier solute particles downward through the lighter solvent molecules. There is too much kinetic motion occurring in the solution to allow such to happen, but the ultracentrifuge can change this.

Cesium chloride, $CsCl$ (formula weight 168.4), is very soluble even in cold water (162 g/100 g H_2O at 0.7 °C), and its solutions are not thick and viscous. Cesium ions are heavy enough that when a cesium chloride solution is spun at superhigh rotational speeds in an ultracentrifuge, they migrate toward the bottom of the tube. Of course, they have to drag along chloride ions so that the solution is everywhere neutral. The result is a gradual change in the concentration of $CsCl$ from top to bottom in the tube. And this causes a density gradient in the tube, with the density increasing steadily from top to bottom, from barely over 1 g/mL to nearly 2 g/mL.

Suppose now we had included in the tube a mixture of substances made of large molecules of varying densities. Objects placed into a liquid tend to settle to whichever point in the liquid has a density that matches their own. (If that point is never reached, then the object settles to the bottom. Submarine operators, for example, adjust the quantities of seawater in the craft's holding tanks to finetune the density of the submarine to match a particular ocean depth.) Getting back to the cesium chloride solution, assume that we have included in it a mixture of nucleic acids, for example, the bearers of genes, and extraordinarily large molecules.

As the tube is whirled in an ultracentrifuge, and the $CsCl$ density gradient is created, the nucleic acid molecules sort themselves out according to their own masses at different places in the tube. It might take hours or even days for it to work, but it can be done. If the centrifuge tube is made of plastic, the tube and centrifuged contents could be frozen, and then cut up into zones, each zone with a nucleic acid of a particular formula weight. Thus, the mixture of nucleic acids is separated.

In Section 4.5 we introduced most of the important terms needed to talk about solutions — solute, solvent, solubility, dilute, concentrated, unsaturated, saturated, and supersaturated, for example. You should review these before you go on, if you are not sure of them.

You have often observed in the lab that solutes cannot be removed from solutions by filtration, and that the force of gravity does not cause solutes to separate either. Ultracentrifuges, however, can create forces that are much more powerful than ordinary gravity, and they are used in an interesting application described in Special Topic 13.2.

13.2 WHY SOLUTIONS FORM

The combined influence of nature's tendency toward randomness and the effects of intermolecular attractions determines whether substances are soluble in each other.

Something can dissolve in a solvent either by a chemical reaction or by a physical mixing. An example of the former occurs if we dissolve sodium oxide, Na_2O, in water. By letting the solution evaporate at or near room temperature, a white residue remains that is sodium hydroxide, $NaOH$, not sodium oxide. The following chemical reaction has occurred.

Na_2O gives this reaction even with the moisture in air, so bottles of Na_2O must be tightly capped.

$$Na_2O(s) + H_2O \longrightarrow 2NaOH(aq)$$

If we dissolve sodium chloride in water, however, the solid remaining after we let the solution evaporate is sodium chloride. Although we can represent the formation of the solution in the form of an equation,

$$NaCl(s) \longrightarrow Na^+(aq) + Cl^-(aq)$$

our ability to recover the original solute by the physical process of evaporating the water lets us classify the formation of the solution as a physical, not a chemical change. The process is simply the dissociation of the preexisting ions of the crystal into the solution. Only solutions formed by such physical changes will concern us in this section.

The formation of a solution by the physical mixing of its components is such a common event that we take it almost for granted. By experience, we have come to expect certain substances to dissolve in certain solvents and not in others; that is, we have developed some intuitive feelings about solubility. Experience has shown us, for example, to expect that motor oil will not dissolve in water, yet anyone who has operated an outboard motor or a moped knows that oil dissolves in gasoline, because the fuel mixture has to consist of such a solution. We also know that sugar dissolves in water, but we had better keep sugar out of our gas tank because its insolubility in gasoline means it will clog the fuel lines of our car.

When we examine the kinds of substances that dissolve in different kinds of solvents, we discover an important principle often quoted as the like dissolves like rule —polar and ionic compounds tend to be soluble only in polar solvents, whereas nonpolar compounds tend to be soluble only in nonpolar solvents. Thus polar sugar molecules and ionic salt crystals are soluble in polar water, and the nonpolar molecules in oil are soluble in gasoline, a solvent composed of nonpolar molecules. Our goal in this section is to understand why, and to do this we are going to take a detailed look at the *process* whereby solutions form.

Water is the most polar of all common solvents.

A Tendency toward Randomness — The Driving Force for Dissolving

In Chapter 9 you learned that gases mix spontaneously with each other. This is illustrated in Figure 13.6. We start with a container holding two gases, separated from each other by a removable panel. When the panel is slid away, the gas molecules begin to mingle and in a very short time they have formed a uniform mixture — a solution. The gas molecules do this without any outside help, and once the solution has formed we can wait forever without ever seeing the molecules separate from one another and return to their original, unmixed state. You have witnessed such solutions form often — the way the odor of a perfume or a bouquet of flowers soon seems to fill an entire room, for example.

In observing the mixing of gases, we are witnessing one of nature's strongest driving forces for change: a tendency toward an increase in disorder or randomness — toward mixing things up. Its importance extends well beyond the process of forming a solution, and we will discuss it in a more formal way in Chapter 16. But for now, let us see how it is related to the formation of a solution.

At the instant the panel is slid away in Figure 13.6, we have two gases in the same container, but they are separated and at opposite ends. This represents considerable order, like lines of boys and girls before a sixth grade dance. Just think about how unlikely it is that the molecules will stay this way, flying about without ever intermingling. Not too probable, is it? The relatively ordered distribution of the gas molecules in the container suddenly becomes highly improbable just as soon as the panel is removed. A much more probable distribution is one in which the molecules have mixed. In other words, a disordered mixed-up distribution of gas molecules is more likely to occur than the ordered arrangement we started with, and nature simply allows the molecules to mingle to form this more disordered state.

Nature's drive toward disorder is a powerful one. You can appreciate this if you consider for a moment how hard it is to keep your room neat. Somehow, without any apparent effort, your room becomes a mess, but to keep it neat and clean requires constant attention.

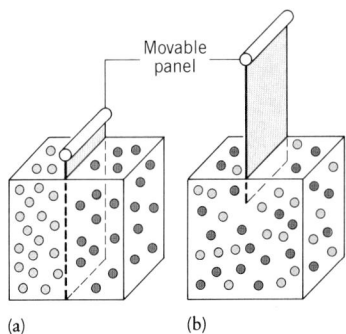

FIGURE 13.6 When two gases initially in separate compartments (*a*) suddenly find themselves in the same container, they mix spontaneously (*b*).

Solutions in Liquids — The Importance of Attractive Forces

In understanding why gaseous solutions form, the only factor we have to consider is the increase in disorder that accompanies the mixing. The molecules in a gas are so far apart and attractive forces between them are so weak that these forces play a negligible role.

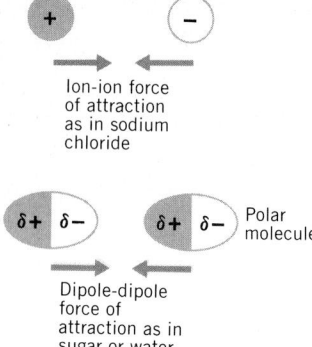

Ion-ion force
of attraction
as in sodium
chloride

Polar
molecule

Dipole-dipole
force of
attraction as in
sugar or water

Two liquids are miscible if they
are completely soluble in all
proportions. The opposite of
miscible is immiscible.

ethyl alcohol

benzene

The drive toward disorder
doesn't win in benzene–water
combinations.

When solids or liquids dissolve in liquid solvents, however, attractive forces are important. As you have already learned, in solids and liquids the attractive forces between their particles—molecules or ions—are strong because opposite full or partial charges are present and the particles are so close together. These attractive forces act as brakes on the separation and intermingling of particles, events that must happen if solutions are to form. Success in getting a liquid solution to form, therefore, depends on the strengths of the attractive forces in *both solvent and solute*. These strengths ultimately determine whether nature's natural drive toward disorder can or cannot win out. To see this, let us examine some typical solvent–solute combinations.

Solutions of Liquids in Liquids

If we add ethyl alcohol (the kind of alcohol in alcoholic beverages) to water, it dissolves completely in any proportion we want. We say that water and ethyl alcohol are completely miscible. Benzene, C_6H_6, on the other hand, is a liquid that is virtually insoluble in water. To understand why, let us take a close look at what happens in each case when the liquids are combined.

Ethyl alcohol has a molecular structure that includes an O—H group (a group, incidentally, with *no* tendency to separate as an anion from the rest of the molecule). Because of the O—H group, ethyl alcohol molecules can interact with water molecules by hydrogen bonding. Water molecules are able to hydrogen bond to it, and it can form hydrogen bonds to water molecules, as shown in Figure 13.7. As a result, water molecules attract alcohol molecules almost as strongly as they attract each other. So the forces that must be overcome when water molecules are pushed apart to make room for alcohol molecules are compensated by similar attractions to the incoming alcohol molecules. The attractions between solvent and solute make it easy, in a sense, for the alcohol molecules to move into the water and form the solution. Similarly, attractions between molecules of water and alcohol make it easy for water molecules to move into the alcohol, so these two liquids dissolve in each other. Because forces of attraction are thus accommodated as this solution forms, nature's tendency toward the greater disorder of a solution can work its way.

FIGURE 13.7 Ethyl alcohol molecules, C_2H_5—O—H, experience hydrogen bonding ($\cdots$) between themselves in pure alcohol. Hydrogen bonds also occur in pure water. When these two liquids form a solution, hydrogen bonds can easily form between molecules of water and those of alcohol. Thus, attractive forces between molecules in the pure liquids are replaced by similar forces in the solution, and the solution easily forms.

Quite a different situation occurs if we try to dissolve benzene in water. Benzene molecules have no O—H groups and are otherwise nonpolar. Between them are just relatively weak London forces of attraction. They are not attracted to water molecules very strongly at all, so they are not "helped along" by attractions to the water molecules that must be pushed aside to make room for them. The water molecules, in other words, stick together too tightly to be pushed aside by the benzene molecules. Even though the tendency toward the greater disorder of a solution is present, the attractive forces between the water molecules are just too strong to be overcome by benzene molecules. So benzene is insoluble in water.

It is also easy to see why a solution of water in benzene cannot be made. Just suppose for a moment that we did manage to disperse water molecules in this liquid. As they move about, they would occasionally encounter each other. Because H_2O molecules attract each other so much more strongly than they attract benzene molecules, they would stick together at each such encounter. This would continue to happen until all the water was in a separate phase. A solution of water in benzene would thus not be stable, and it would not spontaneously form.

Although benzene cannot dissolve water, it does dissolve nonpolar liquids such as carbon tetrachloride, CCl$_4$, quite well. In this case, CCl$_4$ molecules have about as weak forces of attraction between them as do benzene molecules. So CCl$_4$ molecules can easily leave their own kind and with little effort push aside and mingle with the benzene molecules. Nature's drive toward disorder easily overcomes what little resistance there is, and the solution readily forms.

In summary, we see that when the strengths of intermolecular attractions are similar in solute and solvent, as in alcohol–water mixtures and in benzene–carbon tetrachloride mixtures, mutual solubility is possible. Hence, the expression "like dissolves like" summarizes many situations. When the strengths of intermolecular attractions are quite different, on the other hand, mutual insolubility is the result.

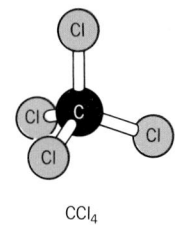

CCl$_4$

The distinction between solvent and solute is unimportant when both are liquids.

Solutions of Solids in Liquids

The situation here is only slightly different from that in solutions of liquids in liquids. Let us first examine what happens when a salt, such as NaCl, dissolves in water.

Figure 13.8 shows a section of a crystal of NaCl in contact with water. Where the solvent contacts the crystal, the dipoles of the water molecules orient themselves so that

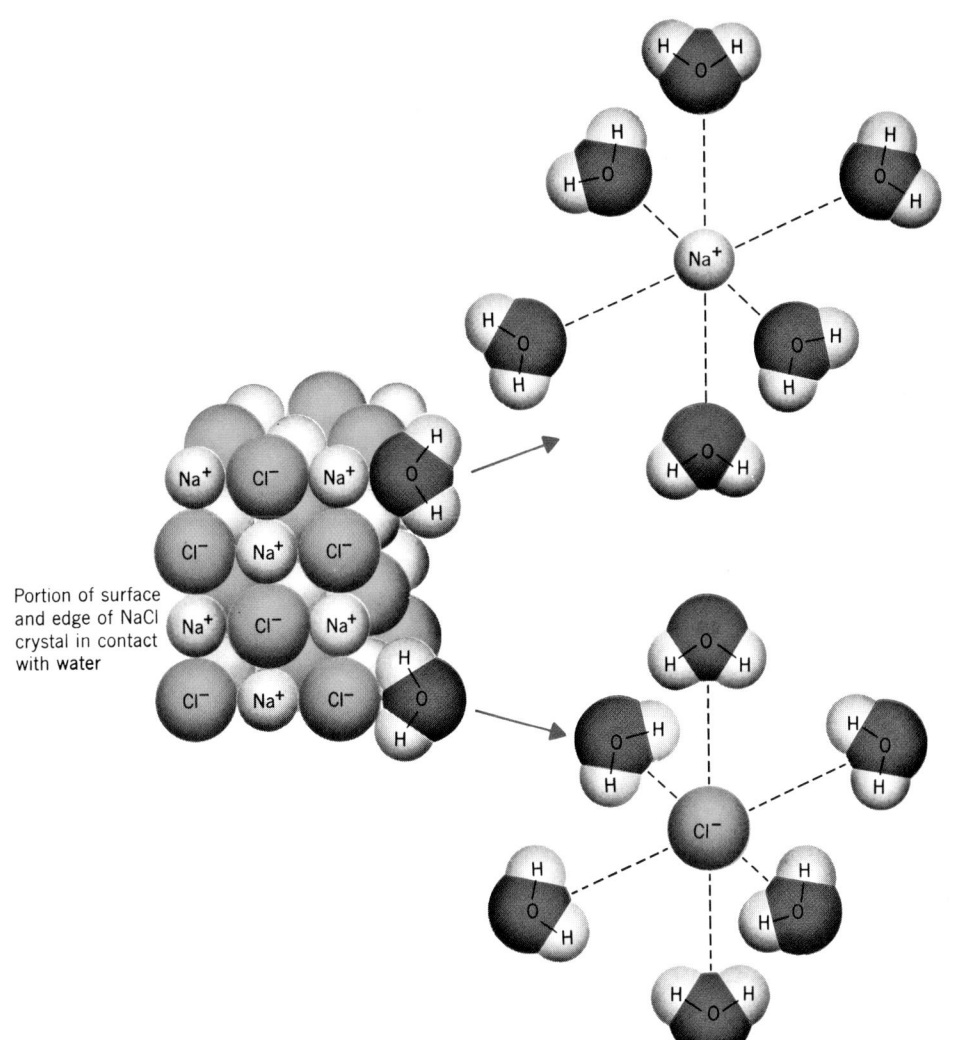

Portion of surface and edge of NaCl crystal in contact with water

FIGURE 13.8 Hydration of ions. Hydration involves a complex redirection of forces of attraction and repulsion. Before this solution forms, water molecules are attracted only to each other; and Na$^+$ and Cl$^-$ ions have only each other in the crystal to be attracted to. In the solution, the ions have water molecules to take the places of their oppositely charged counterparts; and water molecules find ions more attractive than even other water molecules.

their negative ends point toward positive Na^+ ions and their positive ends point at negative Cl^- ions. In other words, *ion–dipole* attractions occur that tend to pull ions from the crystal. Notice that an ion at the corner of a crystal is not held nearly as well in the solid as an ion elsewhere — it has only three nearby neighbors of opposite charge, as compared to six such ions within the crystal. So a corner ion is particularly vulnerable. Ions along crystal edges, held by just four oppositely charged neighbors, are also vulnerable. Through collisions with the crystal and ion–dipole attractions, water molecules dislodge these ions, leaving new corners and edges to be attacked.

Once ions are free, they become completely surrounded by water dipoles. We say that the ion has become hydrated. Similar events occur with other solvents and other solutes, and in general, when a solute particle becomes surrounded by solvent molecules we say it has become solvated. Hydration is just a special case of solvation.

Ionic compounds are able to dissolve in water because the attractions between water dipoles and the ions are able to overcome the attractions of the ions for each other in the crystal. And this allows nature's tendency for disorder to work its dissolving ways.

Similar events also explain why solids composed of polar molecules, like those of sugar, dissolve in water. As shown in Figure 13.9, attractions between the solvent dipoles and the solute dipoles are able to dislodge molecules from the crystal and bring them into solution. Again we see that "like dissolves like" summarizes experience.

The same kind of reasoning explains why nonpolar solids such as wax are soluble in nonpolar solvents such as benzene. In wax, the nonpolar hydrocarbon molecules attract each other very weakly. They easily slip away from the crystal even though the attractions between the solvent (benzene) and the solute (wax) are weak themselves.

When attractive forces within solute and solvent are vastly different, we once again cannot make the two form a solution. Thus, ionic solids or very polar molecular solids (such as sugar) are insoluble in nonpolar solvents such as benzene. Benzene molecules are unable to attract ions or very polar molecules with enough force to overcome the much stronger attractions within ionic or polar solids. Once again we have a situation in which nature's drive toward disorder is stymied by interionic or intermolecular forces.

FIGURE 13.9 Solvation of polar molecules. The molecules of polar molecular compounds can trade forces of attraction to each other (in the crystal) for forces of attraction to the molecules of a polar solvent (in the solution).

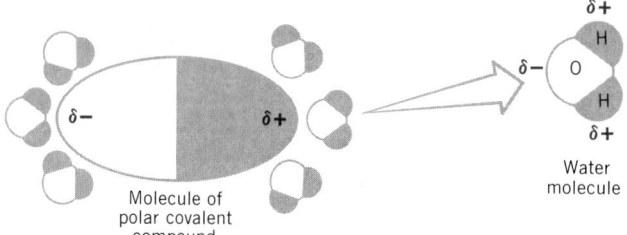

How Soaps and Detergents Work

The "like dissolves like" rule-of-thumb helps explain why hot soapy water easily removes oily and greasy matter from fabrics or skin but hot water alone does not. All detergents have a common feature whether they are ionic or molecular. Their structures include a long, nonpolar, hydrocarbon "tail" holding a very polar or ionic "head." A typical system in ordinary soap, for example, has the following anion. (The charge-balancing cation is Na^+ in most soaps.)

$$CH_3CH_2CH_2CH_2CH_2CH_2CH_2CH_2CH_2CH_2CH_2CH_2CH_2CH_2CH_2CH_2CH_2\overset{\overset{\displaystyle O}{\|}}{C}-O^-$$

hydrocarbon tail   anionic head

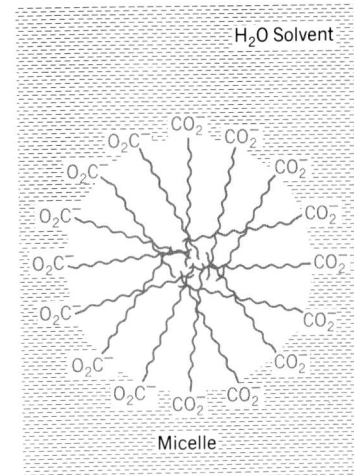

H₂O Solvent / Micelle

FIGURE 13.10 The formation of micelles in soap is driven by the water-avoiding properties of the hydrophobic tails of soap units and the water-attracting properties of their hydrophilic heads. (From J. E. Brady and G. E. Humiston, *General Chemistry Principles and Structure,* 4th ed. Copyright © 1986, John Wiley & Sons, New York. Used by permission.)

The tail of this species is like all hydrocarbons — nonpolar. And it would have a strong tendency to dissolve in oil or grease but not in water. Its natural tendency is to avoid water; it is hydrophobic (water-fearing). It is dragged into solution in hot water only because of the ionic head, which can be hydrated. (And this rather dramatically speaks to the significance of hydration as a special kind of driving force.) The ionic head is hydrophilic (water-loving). To minimize contacts between the hydrophobic tails and water and to maximize contacts between the hydrophilic heads and water, the soap's anions gather into colloidal-sized groups called soap micelles, illustrated in Figure 13.10.

When soap micelles encounter a greasy film, their anions find something else that can accommodate hydrophobic tails. As illustrated in Figure 13.11, the tails of innumerable anions work their way into the film. But they cannot drag the hydrophilic heads entirely in. The heads stay out in the water. With a little agitating help from the washing machine, the greasy film breaks up into countless tiny droplets *each pincushioned by projecting negatively charged heads.* Since the droplets, now colloidally dispersed, all have the same kind of charge, they repel each other and cannot again coalesce.

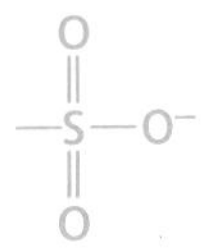

For all practical purposes, the greasy film has been dissolved, and we send it on its way to feed bacteria at wastewater treatment plants or in the ground. (The bacteria also eat the soap. In the early days of synthetic detergents, some products could not be eaten by bacteria. In some localities, enough detergent built up in the water supply that water taps foamed. And giant rafts of suds drifted on rivers into lakes.)

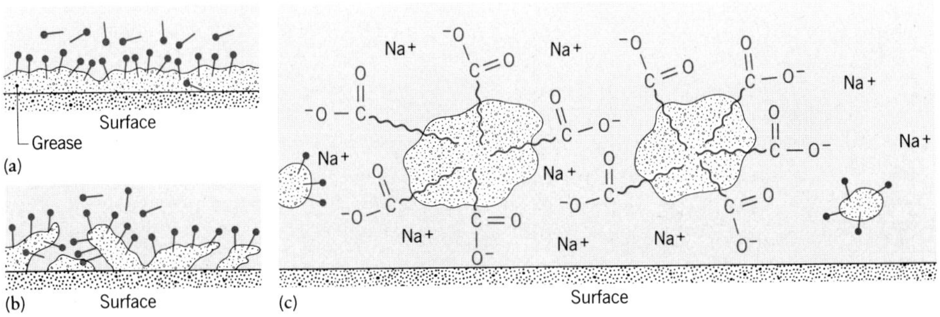

Hydrocarbon units are nonpolar because C—H and C—C bonds are nonpolar.

In synthetic detergents, the hydrophilic group is not

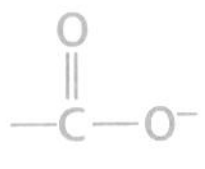

$$\underset{\displaystyle \overset{O}{\|}}{-C-O^-}$$

Instead, it often is

$$\underset{\displaystyle \underset{O}{\|}}{\overset{\displaystyle \overset{O}{\|}}{-S-O^-}}$$

FIGURE 13.11 How soap loosens grease that might be coating a cloth surface. (*a*) The hydrophobic tails of soap anions work their way into the grease layer. The hydrophilic heads stay exposed to water. (*b*) The grease layer breaks up. (*c*) Microdroplets of grease form, each pincushioned and made like-charged by soap anions. These are now easily poured out with the wash water. (From J. R. Holum, *Fundamentals of General, Organic, and Biological Chemistry,* 3rd ed. Copyright © 1986, John Wiley & Sons, New York. Used by permission.)

13.3 HEATS OF SOLUTION

The heat of solution depends on the relative strengths of the intermolecular attractions in the solute, in the solvent, and in the solution.

In our previous discussion of why some solutions form and others do not, we relied heavily on the notion of intermolecular (or interionic) attractions. When changes occur in these attractive forces as the solution forms, an observable and measurable consequence is an energy change. The energy exchanged between the system and its surroundings when one mole of a solute dissolves in a solvent at constant pressure to make a dilute solution is called the *molar enthalpy of solution,* or usually just the heat of solution, ΔH_{soln}.

The reason for an exchange of energy is that formation of a solution involves overcoming attractive forces in the solute and solvent and establishing new ones in the solution. It costs energy to separate the particles of solute and solvent and make them spread apart to make room for each other. Such an energy cost increases the potential energy of the system and is therefore endothermic. But once the spread-apart particles come back together as a solution, new attractive forces are created and there is a lowering of the system's potential energy — an exothermic step. The heat of solution, ΔH_{soln}, is simply the net result of these two opposing energy contributions. To look at this more closely, let us follow the energy changes that occur when a solid dissolves in a liquid.

$$\Delta H_{soln} = H_{soln} - H_{components}$$

The magnitude of ΔH_{soln} depends somewhat on the final concentration of the solution being made.

Lattice Energies and Solvation Energies

In Chapter 5 you learned that we could use Hess's law to calculate heats of reaction from the heats of formation of the reactants and the products. This works because an enthalpy change is a *state function* — its magnitude does not depend on the path used to go from one state to another. The same property helps us analyze heats of solution and explain why some solutions form endothermically and others exothermically.

We will pick a two-step path that takes us from the solid solute and liquid solvent to the solution. See the enthalpy diagram of Figure 13.12. The first step is to separate the solid into its individual particles — in effect, we imagine that the solid is vaporized. The second step brings these gaseous solute particles into the solvent where they become solvated.

As you can see from Figure 13.12, the first step is endothermic and it increases the potential energy of the system. The particles in the solid attract each other, so energy

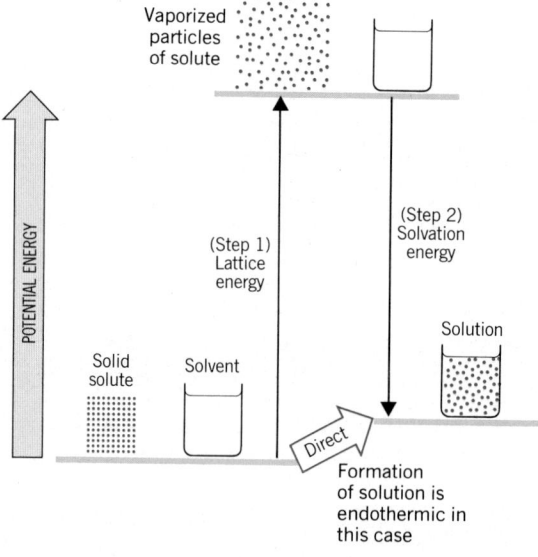

FIGURE 13.12 Enthalpy diagram for a solid dissolving in a liquid. In the real world, the solution is formed directly as indicated by the red arrow. We can analyze the energy change by imagining the two separate steps, because enthalpy changes are functions of state and are independent of path. The energy change along the direct path is the algebraic sum of Step 1 and Step 2.

TABLE 13.4 Lattice Energies, Hydration Energies, and Heats of Solution for Some Alkali Metal Halides

Compound	Lattice Energy (kJ/mol)[a]	Hydration Energy (kJ/mol)	ΔH_{soln}[b] Calculated ΔH_{soln} (kJ/mol)	ΔH_{soln}[b] Measured ΔH_{soln} (kJ/mol)
LiCl	+833	−883	−50	−37.0
NaCl	+766	−770	−4	+3.9
KCl	+690	−686	+4	+17.2
LiBr	+787	−854	−67	−49.0
NaBr	+728	−741	−13	−0.602
KBr	+665	−657	+8	+19.9

[a] These values are the true lattice energies given opposite signs. True lattice energies have negative values, since they relate to the exothermic formation of crystalline lattices from their constituent, gaseous ions (or molecules).

[b] Heats of solution refer to the preparation of extremely dilute solutions.

must be supplied to separate them. This energy is called the *lattice energy*.[1] In the second step, gaseous solute particles enter the solvent and are solvated. As they do, the potential energy of the system decreases, because solute and solvent particles attract each other. This step is therefore exothermic, and its energy is called the solvation energy, which is the general term. The special name hydration energy is used for aqueous solutions.

The heat of solution—the net energy change—is the difference between the energy required for step 1 and the energy released for step 2. It is easy to see now why some heats of solution reflect overall endothermic processes and others exothermic. When the energy required for step 1 exceeds the energy released in step 2, the formation of the solution is overall endothermic. But when more energy is released in step 2 than is needed for step 1, the solution forms exothermically.

Although this analysis seems reasonable, is there any way to check if it actually works? Let us test it for some solutions of salts in water. We can do this because lattice energies and hydration energies have been estimated for salts, and Table 13.4 contains some of the data. Here we also have experimental heats of solution to compare with those we can calculate from lattice energies and hydration energies. Figures 13.13 and

When attractive forces increase, potential energy decreases; when they decrease, potential energy increases.

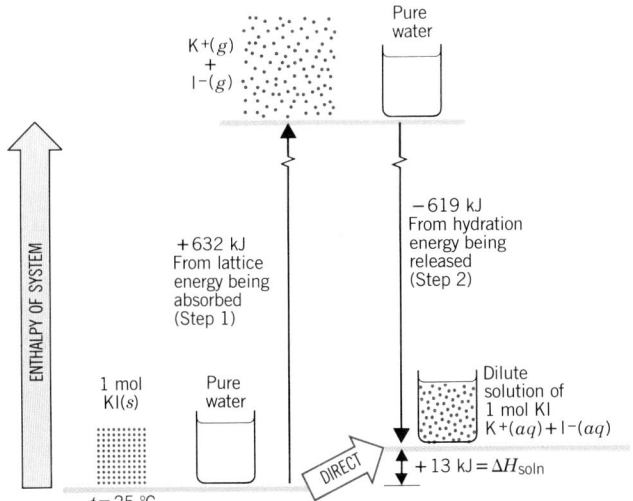

FIGURE 13.13 The formation of aqueous potassium iodide.

Step 1:	$KI(s) \longrightarrow K^+(g) + I^-(g)$	$\Delta H = +632$ kJ
Step 2:	$K^+(g) + I^-(g) \longrightarrow K^+(aq) + I^-(aq)$	$\Delta H = -619$ kJ
Net:	$KI(s) \longrightarrow K^+(aq) + I^-(aq)$	$\Delta H_{soln} = +13$ kJ

[1] The concept of lattice energy applies to any kind of crystalline solid—ionic or molecular. In Chapter 7, you may recall, we discussed lattice energies as contributing to the stabilities of ionic compounds.

FIGURE 13.14 The formation of aqueous sodium bromide.

Step 1: $NaBr(s) \longrightarrow Na^+(g) + Br^-(g)$ $\Delta H = +728$ kJ
Step 2: $Na^+(g) + Br^-(g) \longrightarrow Na^+(aq) + Br^-(aq)$ $\Delta H = -741$ kJ
Net: $NaBr(s) \longrightarrow Na^+(aq) + Br^-(aq)$ $\Delta H_{soln} = -13$ kJ

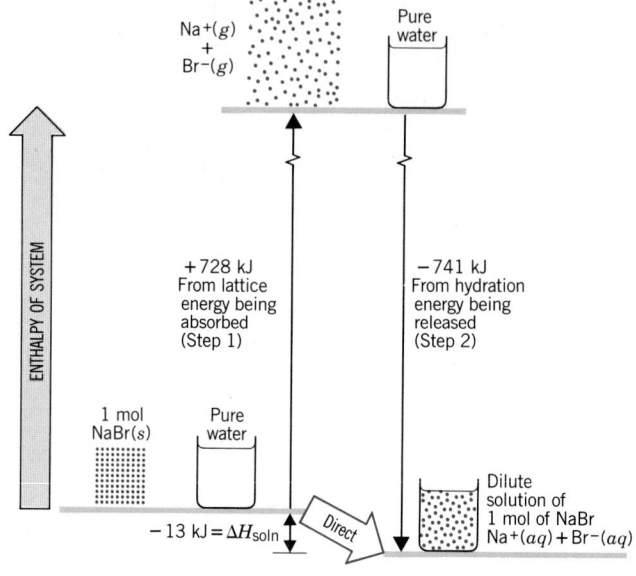

Small percentage errors in very large numbers can cause huge percentage changes in the *differences* between such numbers (as you can discover by working Review Exercise 13.30).

13.14 provide enthalpy diagrams showing how the heats of solution can be calculated for the formation of aqueous solutions of two salts, KI and NaBr.

The agreement between calculated and measured values in Table 13.4 is not particularly impressive in an absolute sense. This is partly because the lattice energies and hydration energies are not precisely known and partly because the model that we have used in our analysis is just too simple to give good agreement. But notice that when "theory" predicts relatively large heats of solution, the experimental values are also relatively large, and that both values have the same sign. Notice also that the changes in values show the same trends when we compare the three chloride salts — LiCl, NaCl, and KCl — or the three bromide salts — LiBr, NaBr, and KBr.

Solutions of Liquids in Liquids

As we move to consider heats of solutions when liquids dissolve in liquids, we will use a three-step process to go from the initial to the final state. See Figure 13.15. First, we imagine that the molecules of one liquid are moved apart just far enough to make room for molecules of the other liquid. (We will designate one liquid as the solvent.) Since we have to overcome forces of attraction, this increases the system's potential energy and so the step is endothermic. The second step is like the first, only it is done to the other liquid — the solute — whose molecules move apart slightly to make room for those of the solvent. On an enthalpy diagram (Figure 13.16) we have climbed two energy steps and have both the solvent and the solute in their slightly "expanded" conditions. The third step lets nature's drive for randomness take its course, as the molecules of expanded solvent and solute come together and intermingle. Since the molecules of these two now experience mutual forces of attraction, the system's potential energy decreases, and step 3 is exothermic. The value of ΔH_{soln} will, again, be the net energy of these steps.

The enthalpy diagram in Figure 13.16 shows the case when the sum of the energy costs for steps 1 and 2 is equal to the energy released in step 3, so the overall ΔH_{soln} is zero. This is very nearly the case when we make a solution of benzene and carbon tetrachloride. Attractive forces between molecules of benzene are almost exactly the same as those between molecules of CCl_4, or between molecules of benzene and those of CCl_4. If all such intermolecular forces were identical, the net ΔH_{soln} would be exactly zero, and the resulting solution would be called an ideal solution. Carbon disulfide, CS_2,

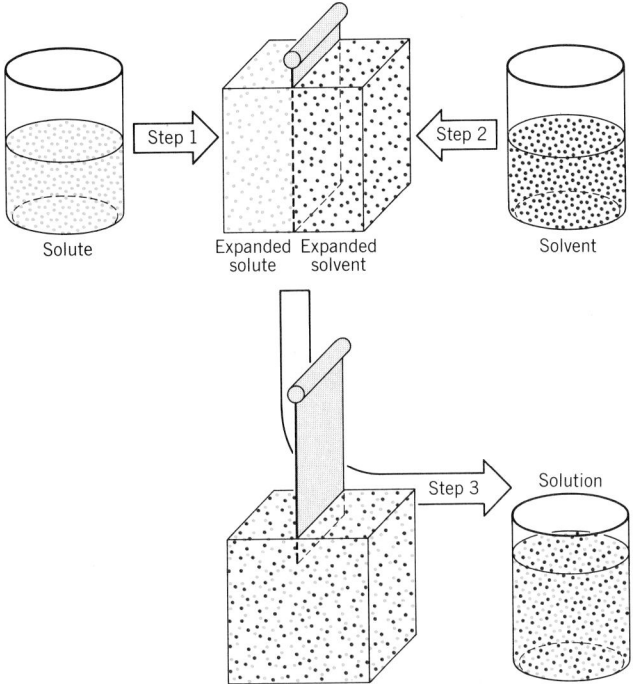

FIGURE 13.15 Hypothetical steps in the analysis of the enthalpy change for the formation of a solution of two liquids.

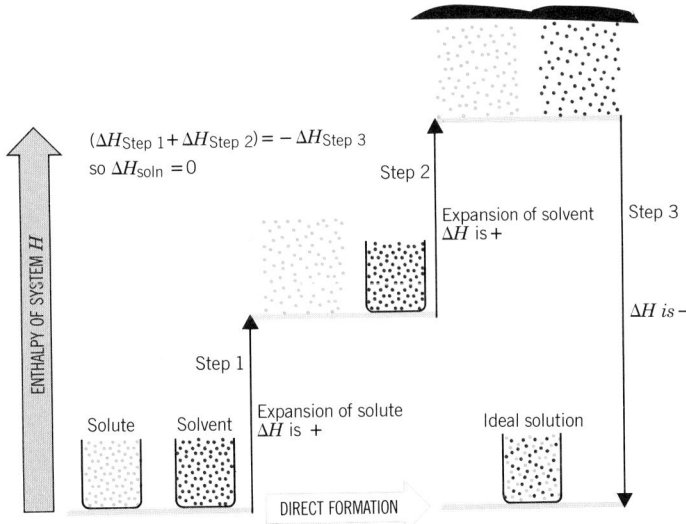

$$(\Delta H_{\text{Step 1}} + \Delta H_{\text{Step 2}}) = -\Delta H_{\text{Step 3}}$$
so $\Delta H_{\text{soln}} = 0$

Step 2

Expansion of solvent
ΔH is +

Step 3

ΔH is −

Step 1

Expansion of solute
ΔH is +

Solute   Solvent

Ideal solution

DIRECT FORMATION

ENTHALPY OF SYSTEM H

FIGURE 13.16 Enthalpy changes in the formation of an ideal solution. The direct path represents how a solution actually forms, but the net result is the same when we imagine the steps shown here.

hydrogen
bonding in
acetone–water
solutions

and benzene constitute another pair of nonpolar liquids that form an almost ideal solution.

Acetone and water form a solution exothermically. With these, the third step releases more energy than the sum of the first two (Figure 13.17*a*) chiefly because

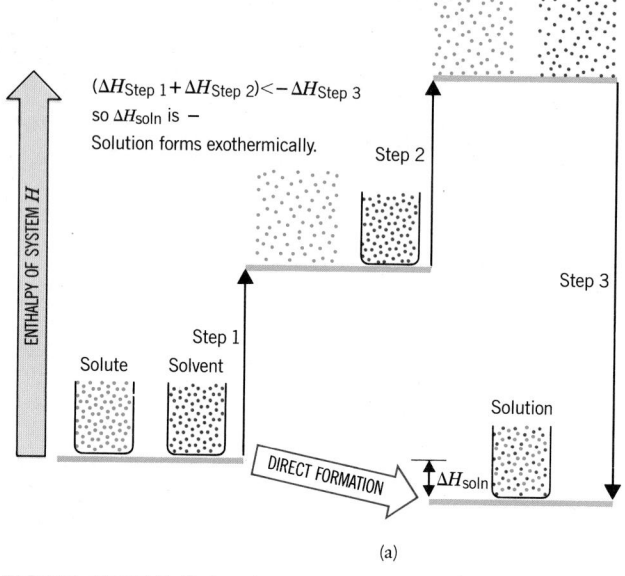

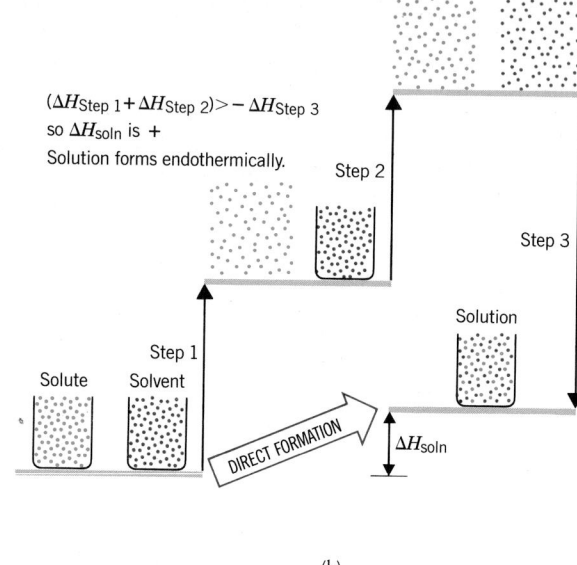

FIGURE 13.17 Enthalpy changes when a real solution forms. (*a*) The process is exothermic. (*b*) The process is endothermic.

$CH_3CH_2CH_2CH_2CH_2CH_3$
hexane

molecules of water and acetone attract each other more strongly than acetone molecules attract each other.

Ethyl alcohol and hexane form a solution endothermically (Figure 13.17*b*). Now the third step cannot provide enough of a release of energy to compensate for the costs of steps 1 and 2. The problem is chiefly that ethyl alcohol molecules attract each other more strongly than they can attract hexane molecules. Hexane molecules cannot push their way in and among those of ethyl alcohol without breaking up some of the hydrogen bonding. This absorbs energy, and the solution becomes cool as it forms.

Solutions of Gases in Liquids

When a gas is the solute, it is already expanded, so there is really no energy cost associated with "expanding the solute." The solvent, however, must be expanded slightly to accommodate the molecules of the gas, and this does require a small energy input — small because the attractive forces do not change very much when the solvent molecules are moved apart just a little bit. When the gas molecules finally fill the spaces made for them by the expanding solvent, there is a significant lowering of the potential energy, because the gas molecules change from a condition in which the attractive forces are almost zero to a condition in which there are significant forces of attraction. The net energy change associated with the two steps of expanding the solvent slightly and then filling the spaces with solute molecules is simply the solvation energy, and we see that for gases this is all we really need to contend with. And because solvation is exothermic, all gases dissolve in liquids exothermically.

13.4 EFFECT OF TEMPERATURE ON SOLUBILITY

Le Châtelier's principle and heats of solution help us predict how the solubility of a compound will change with temperature.

As defined in Section 4.5, *solubility* refers to the mass of solute that forms a saturated solution with a given mass of solvent at a specified temperature. The units most commonly used are grams of solute per 100 g of solvent. Since solubility refers to a *saturated* solution, if we try to make more solute dissolve, any extra solute we add to the system will just sit there as a separate phase. There will be a coming and going of solute particles between the dissolved and undissolved phases, of course, because in a saturated solution we have equilibrium between them. We can write this equilibrium as follows.

$$(solute)_{undissolved} \rightleftharpoons (solute)_{dissolved}$$
[solute contacts   [solute is in a
the saturated   saturated
solution]   solution]

Figure 13.18 illustrates this system. As long as we keep the temperature constant, we cannot shift this equilibrium. (If a gas is the solute, then we have to specify a constant pressure as well. For the moment, we will discuss just solutions involving solids and liquids.) We can shift this equilibrium by subjecting it to a specific stress—changing its temperature. Recall that according to Le Châtelier's principle, an equilibrium will shift in whichever direction most directly absorbs a stress.

Suppose that making more solute dissolve into an already saturated solution absorbs heat. We could then rewrite the equilibrium as follows.

$$(solute)_{undissolved} + heat \rightleftharpoons (solute)_{dissolved}$$

You can see that if we raise the temperature of the saturated solution, we can make more solute dissolve. Raising the temperature adds heat—the stress—so the equilibrium shifts to the right to absorb the stress and more solute dissolves. How much more dissolves depends on the individual solutes and solvents, and there are wide variations as seen in Figure 13.19 where the changing solubilities of several solids in water are plotted. The solubility of ammonium nitrate, for example, increases very rapidly with temperature, but that of sodium chloride increases very little between 0 and 100 °C.

One salt in Figure 13.19, cerium(III) sulfate, becomes less soluble with increasing temperature. For this compound, heat must be *removed* from a saturated solution to make more solute dissolve. Now the equilibrium in a saturated solution must be written with heat as a *product*.

$$(solute)_{undissolved} \rightleftharpoons (solute)_{dissolved} + heat$$

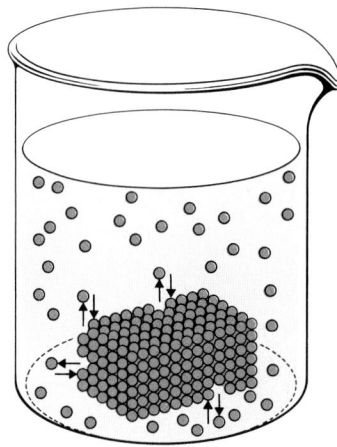

FIGURE 13.18 In a saturated solution dynamic equilibrium exists between the undissolved solute and the solute in the solution.

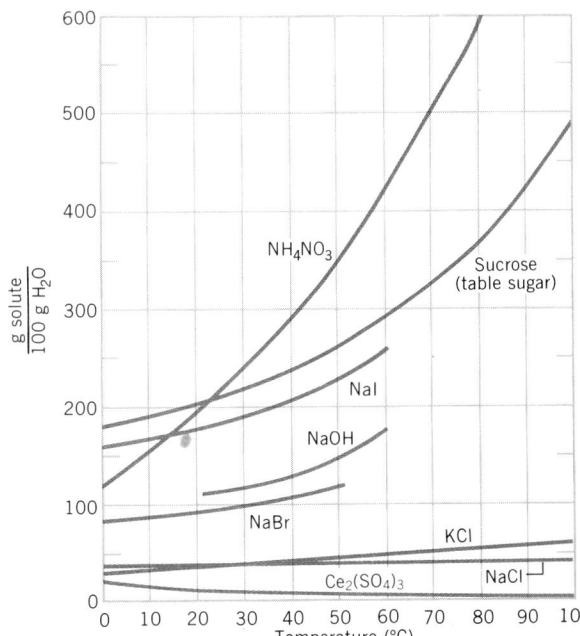

FIGURE 13.19 Solubility in water versus temperature for several substances.

If we *raised* the temperature on this system, the equilibrium could absorb the stress—the added heat—only by shifting to the left. So some of the dissolved solute would come out of solution. Few systems involving liquid solvents and solid or liquid solutes are like this.

Gases dissolve exothermically in liquids at all concentrations, so the associated equilibrium is

$$(gas)_{undissolved} \rightleftharpoons (gas)_{dissolved} + heat$$

When we heat a solution of a gas in a liquid, this equilibrium must shift to the left to absorb the stress, so the solubilities of gases in liquids always decrease with temperature. Table 13.5 gives data on the solubilities of several common gases in water at different temperatures that illustrate this generalization.

TABLE 13.5 Solubilities of Common Gases in Water[a]

Gas	0 °C	20 °C	50 °C	100 °C
Nitrogen, N_2	0.0029	0.0019	0.0012	0
Oxygen, O_2	0.0069	0.0043	0.0027	0
Carbon dioxide, CO_2	0.335	0.169	0.076	0
Sulfur dioxide, SO_2	22.8	10.6	4.3	1.8 (at 90 °C)
Ammonia, NH_3	89.9	51.8	28.4	7.4 (at 96 °C)

[a] Solubilities are in grams of solute per 100 g of water.

Fractional Crystallization

Fractional crystallization is a procedure for the purification of a solid—for separating it into fractions, one which is the solid we want purified and the other(s), the impurities.

Let us assume that we have a solid contaminated with another. Also suppose that our desired solid is very soluble in a hot solvent but much less soluble at a low temperature. We can then recrystallize the impure sample by dissolving it in enough *boiling* solvent to make a saturated solution *at the boiling point* of the solution. When this solution is cooled, nearly always, the solute we want begins to come out of solution—to crystallize. Given the demands of crystal lattices for uniformity, the ions or molecules of only one species are permitted to crystallize in a given lattice. So the ions or molecules of the impurities generally cannot get mixed into the crystals of the desired compound as it precipitates. The impurities remain in solution, if it all works, even at the lowest temperature to which the solution is cooled—usually that of an ice-water slush. The desired solid is now at least much purer than before. We can collect its crystals by filtration (and maybe even rinse them with a little of the ice-cold pure solvent we used). We have lost some of the desired compound, because the filtrate is still saturated in it. But we have not lost much if we have picked the solvent well and chilled it to a low temperature. In any event, the purpose is *purity* in this operation.

13.5 EFFECT OF PRESSURE ON THE SOLUBILITIES OF GASES

When the temperature is constant, the solubility of a gas in a liquid is directly proportional to the partial pressure of the gas on the solution.

Club soda is a solution of carbon dioxide in water. It is bottled under pressure so that more CO_2 will be dissolved, because the solubility of any gas in any liquid is increased by pressure. When the bottle of club soda is opened, the pressure drops back to atmospheric pressure, so carbon dioxide fizzes out. In a short time, the soda is "flat," because little CO_2 remains in it. Soft drinks, beer, and champagne go flat for the same reason

Solids needed for precise physical measurements or for medical tests are often recrystallized several times.

If the solid melts at a temperature below that of the hot solution, it will separate as an oil as the solution starts to cool, and no purification occurs.

When the cork is eased out far enough, the CO_2 dissolved under pressure in the champagne can give the final pop.

When people work in a space where the air pressure is much above normal, they have to be careful to return slowly to the atmosphere. Otherwise, they face the danger of the "bends"—severe pains in joints and muscles, fainting, possible deafness, paralysis, and death. Workers building deep tunnels, where higher-than-normal air pressure is maintained to keep water out, are at risk. So are deep-sea divers.

The bends can develop because the solubilities of both nitrogen and oxygen in blood are higher under higher pressure, as Henry's law says. Once the blood is enriched in these gases it must not be allowed to lose

them suddenly. If microbubbles of nitrogen or oxygen appear at blood capillaries, they will block the flow of blood. Such a loss is particularly painful at joints, and any reduction in blood flow to the brain can be extremely serious.

For each atmosphere of pressure above the normal, about 20 minutes of careful decompression is usually recommended. This allows time for the respiratory system to gather and expel excess nitrogen, which can't leave any other way. Excess oxygen can be used up by normal metabolism.

when opened. (They go flat a lot faster if the beverages are also warm, because gases are less soluble in liquids at higher temperatures—as we learned in the previous section.)

Air is not very soluble in water under ordinary pressures, but at elevated pressures both the oxygen and nitrogen in air become increasingly soluble, as the plots in Figure 13.20 show. Special Topic 13.3 describes the dangerous implications of these changes to those who must work under higher-than-normal air pressures.

It is easy to understand why the solubility of any gas in a liquid should increase with pressure. The relevant equilibrium is

$$\text{gas} + \text{solvent} \rightleftharpoons \text{solution}$$

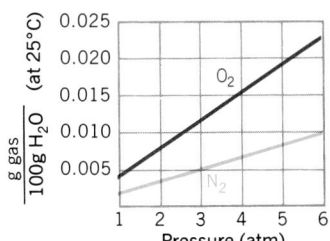

FIGURE 13.20 Solubility in water versus pressure for two gases.

Suppose we've established this equilibrium and then increase the pressure—by reducing the volume available to the gas, for example. Le Châtelier's principle says that the system will attempt to change in a way to counteract this "stress" and bring the pressure down somewhat. This can happen if some molecules leave the gas phase and enter the solution. Then there will be fewer molecules in the gas phase, and these fewer molecules exert less pressure. Thus increasing the pressure of the gas causes the equilibrium to be shifted to the right and causes the solubility of the gas to increase. Conversely, if we reduce the pressure on the system, the equilibrium must shift to the left and some dissolved gas will leave the solution.

If we let our imaginations take us down to the molecular level of events, we can see how the effect of pressure on gas solubility works. Imagine a closed container partly filled with a solution of some gas in a liquid. One wall of the container is movable, as indicated in Figure 13.21. We start on the left with equilibrium; gas molecules come and go at equal rates between the dissolved and undissolved states. The rate at which they enter the solution is proportional to the frequency with which they collide with the surface of the solution. This frequency must increase as we increase the pressure of the

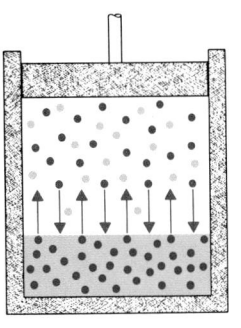
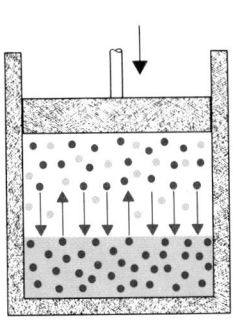
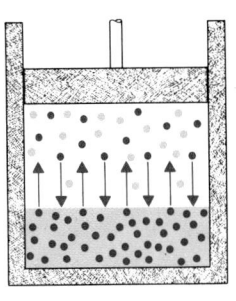

(a)   (b)   (c)

FIGURE 13.21 How pressure increases the solubility of a gas in a liquid. (*a*) At some specific pressure, equilibrium exists between the vapor phase and the solution. (*b*) An increase in pressure puts stress on the equilibrium. (*c*) More gas dissolves and equilibrium is restored.

gas (Figure 13.21*b*) because the gas molecules are squeezed closer together. More of them can now collide with a given area of the solution in each second of time. Since liquids and liquid solutions are incompressible, the increased pressure alone has no effect on the frequency with which gas molecules can leave the solution. Thus the increased pressure has increased the rate of the forward reaction in our equilibrium over the rate of the reverse reaction.

As more gas dissolves, the forward rate slows down. During this time, the reverse rate increases, because at higher and higher concentrations of the gas, there are simply more gas molecules in solution at each unit area of the surface. Their frequency of escape is proportional to this concentration. But the forward reaction dominates until enough additional gas has dissolved to cause the opposing rates to equalize. We again have equilibrium (Figure 13.21*c*), but it is a new equilibrium because more gas molecules are in solution.

For gases that do not react with the solvent, there is a simple relationship between gas pressure and gas solubility — the pressure–solubility law, often called Henry's law.

William Henry (1775–1836) was an English chemist and physician.

> **Pressure — Solubility Law (Henry's Law).**   The concentration of a gas in a liquid at any given temperature is directly proportional to the partial pressure of the gas on the solution.
>
> $$C_g = k_g P_g$$

Henry's law constants can be found in certain handbooks.

In the equation, C_g = the concentration of the gas, k_g = the proportionality constant, and P_g = the partial pressure of the gas above the solution. (Constant temperature is assumed.) The equation is true only at relatively low concentrations and pressures and for gases that do not react with the solvent.

In a calculation using Henry's law we are usually given data on an initial concentration at some initial pressure and asked to find what the concentration of the gas will be at some new pressure. We can do this without first finding the proportionality constant because we know it will have the same value at both pressures. (Otherwise, we would not have a *law*.) Thus, using subscripts 1 and 2 in the usual way (and omitting *g* as a subscript — it's understood), we know that

$$\frac{C_1}{P_1} = k$$

and

$$\frac{C_2}{P_2} = k$$

So, an alternate expression of Henry's law is

$$\frac{C_1}{P_1} = \frac{C_2}{P_2}$$

EXAMPLE 13.1   CALCULATING WITH THE PRESSURE–SOLUBILITY LAW (HENRY'S LAW)

Problem: At 20 °C the solubility of N_2 in water is 0.0150 g/L when the partial pressure of nitrogen is 580 torr. What will be the solubility of N_2 in water at 20 °C when its partial pressure is 800 torr? (After studying this example, read about the "bends" in Special Topic 13.3.)

Solution: Let us gather the data first.

$$C_1 = 0.0150 \text{ g/L} \qquad C_2 = ?$$
$$P_1 = 580 \text{ torr} \qquad P_2 = 800 \text{ torr}$$

Using the alternate equation for Henry's law, we have

$$\frac{0.0150 \text{ g/L}}{580 \text{ torr}} = \frac{C_2}{800 \text{ torr}}$$

Thus,

$$C_2 = \frac{(800 \text{ torr}) \times (0.0150 \text{ g/L})}{(580 \text{ torr})}$$

$$= 0.0207 \text{ g/L}$$

The solubility under the higher pressure is 0.0207 g/L.

PRACTICE EXERCISE 1   How many grams of nitrogen and oxygen are dissolved in 100 g of water at 20 °C when the water is saturated with air? Above this solution, at a total pressure of 760 torr, the air is saturated with water vapor and the partial pressure of nitrogen is 586 torr and the partial pressure of oxygen is 156 torr. The solubility of oxygen in water at a total pressure of 760 torr is 0.00430 g O_2/100 g H_2O. The solubility of nitrogen in water at a total pressure of 760 torr is 0.00190 g N_2/100 g H_2O.

Solubilities of Gases That Are Strongly Hydrated

Some of the gases in Table 13.5—sulfur dioxide, ammonia, and, to a lesser extent, carbon dioxide—are far more soluble than oxygen or nitrogen. Part of the reason is that molecules of these gases have polar bonds, and water molecules are attracted to the sites of partial charges at their ends. In other words, hydrogen bonding occurs to hold solute molecules in solution. Ammonia molecules, in addition, not only can accept hydrogen bonds from water (O—H··N) but also can donate them through their N—H bonds (N—H··O), as illustrated in Figure 13.22.

These more soluble gases also react to some extent with water as the following chemical equilibria form.

$$CO_2(aq) + H_2O \rightleftharpoons H_2CO_3(aq) \rightleftharpoons H^+(aq) + HCO_3^-(aq)$$
$$SO_2(aq) + H_2O \rightleftharpoons H^+(aq) + HSO_3^-(aq)$$
$$NH_3(aq) + H_2O \rightleftharpoons NH_4^+(aq) + OH^-(aq)$$

These reactions help retain the gases in solutions.

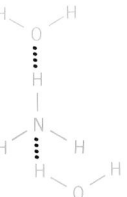

FIGURE 13.22 An ammonia molecule in an aqueous solution can donate a hydrogen bond to a water molecule (top) and accept another (beneath). (The dotted lines represent hydrogen bonds.)

13.6 CONCENTRATION EXPRESSIONS

Mole fractions, mole percents, and molal concentrations are ways to indicate the ratio of solute particles to solvent particles.

Having completed our investigation of how solutions form, how concentrated they can be, and what factors affect concentration, we can turn to physical properties displayed more or less by all solutions. These properties include changes in the vapor pressure of a liquid when a solute is present, depression of the freezing point of a solvent by a solute, or elevation of its boiling point. And they also include phenomena such as osmosis and dialysis. These properties vary in magnitude with the relative proportions *by particles*— meaning, by *moles*—of solute and solvent. When we know the *molar* concentration of a solution, however, we do not have this information.

Information concerning a solution's molar concentration is needed when we deal with the stoichiometry of chemical reactions in solution. But molarity gives the ratio of solute to volume of *solution*, not volume of solvent. In this section, therefore, we will study other concentration expressions that give us, directly or indirectly, ratios *by moles* of all of the components of a solution, including the solvent.

When we study how a solute affects a melting point or a boiling point, we obviously intend to operate at temperatures different from room temperature, at which a solution is usually prepared. This poses another problem with molar concentrations; their values are sensitive to the temperature of the solution. Solutions expand somewhat in volume when heated and contract when cooled, but the number of moles of solute stays the same. So the number of moles of solute per unit volume (1 L) of solution — the molarity — changes with temperature. If, however, we prepare a solution according to specified *mole* ratios of all components, *including the solvent*, such ratios could not vary with temperature. Mole ratios, after all, are ultimately ratios by particles, and the *numbers* of particles are unaffected by temperature.

Mole Fractions and Mole Percents

One way to describe the relative numbers of molecules of all components of a solution is by their mole fractions. The mole fraction of any component in a mixture is the ratio of the number of moles of it to the total number of moles of all components present. Expressed mathematically,

$$X_A = \frac{n_A}{n_A + n_B + n_C + \cdots + n_X} \qquad (13.1)$$

where X_A is the mole fraction of component A, and n_A, n_B, n_C, ..., n_X, are the numbers of moles of each component, A, B, C, ..., X, respectively. The sum of all mole fractions for a mixture must always equal 1.

EXAMPLE 13.2   CALCULATING MOLE FRACTIONS

Problem: What are the mole fractions of each component in a solution that consists of 1.00 mol of C_2H_5OH, 0.500 mol of CH_3OH, and 6.00 mol of H_2O?

Solution: We have to apply Equation 13.1 to each component, in turn.

C_2H_5OH

$$X_{C_2H_5OH} = \frac{1.00}{1.00 + 0.500 + 6.00} = 0.133$$

CH_3OH

$$X_{CH_3OH} = \frac{0.500}{1.00 + 0.500 + 6.00} = 0.0667$$

H_2O

$$X_{H_2O} = \frac{6.00}{1.00 + 0.500 + 6.00} = 0.800$$

$$\text{total} = 1.000$$

The units — moles — were not shown because they would all cancel. Mole fractions have no units.

Even though mole fractions have no formal units, always remember that they stand for the ratio of moles of a component to total moles in the mixture.

Sometimes, we multiply a mole fraction by 100 and call the result the mole percent. Thus the mixture in Example 13.2 had 13.3 mole percent of ethyl alcohol, 6.70 mole percent methyl alcohol, and 80.0 mole percent of water—all adding up to 100%.

PRACTICE EXERCISE 2   What are the mole fractions and mole percents of the components in an antifreeze solution made by mixing 500 g of methyl alcohol (CH_3OH) with 500 g of water?

PRACTICE EXERCISE 3   What are the mole fractions and mole percents of the solute and solvent in a solution made by dissolving 43.88 g of NaCl in 1000 g of water?

Mole Fractions and Mole Percents for Gas Mixtures from Partial Pressures

With gas mixtures, we can find mole fractions simply from partial pressure data. Let us see how.

When we know the partial pressure, P_A, for any one gas, A, in a gas mixture with a total volume V at a temperature T, we can use the ideal gas law equation, $PV = nRT$, to calculate the number of moles of A present.

$$n_A = \frac{P_A V}{RT}$$

Using different letters to identify individual gases in the usual way, we can use this expression for each component in Equation 13.1 for gas mixtures. All gases in the mixture, of course, have the same values of V, R, and T. Thus, starting with Equation 13.1

$$X_A = \frac{n_A}{n_A + n_B + n_C + \cdots + n_X}$$

we substitute expressions for n for the individual gases to give us

$$X_A = \frac{P_A V / RT}{P_A V / RT + P_B V / RT + P_C V / RT + \cdots + P_X V / RT}$$

$$= \frac{(V/RT)}{(V/RT)(P_A + P_B + P_C + \cdots + P_X)}$$

We can factor out V/RT.

or

$$X_A = \frac{P_A}{P_A + P_B + P_C + \cdots + P_X}$$

The denominator in the last equation is the sum of the partial pressures of all the gases in the mixture. But according to Dalton's law of partial pressures, this sum equals the total pressure of a gas mixture. Therefore this equation simplifies as follows to give a way to show that the mole fraction of any gas in a mixture is simply the ratio of its partial pressure to the total pressure.

$$\text{mole fraction of gas } A = X_A = \frac{P_A}{P_{total}} \qquad (13.2)$$

EXAMPLE 13.3 CALCULATING THE MOLE FRACTION OF A GAS FROM PARTIAL PRESSURES

Problem: What are the mole fractions and mole percents of nitrogen and oxygen in air when the partial pressures are 160 torr for oxygen and 600 torr for nitrogen? (Assume no other gases are present.)

Solution: Let us use Equation 13.2 to find the mole fraction of N_2 first.

$$X_{N_2} = \frac{P_{N_2}}{P_{total}} = \frac{600 \text{ torr}}{P_{total}}$$

But the total pressure is the sum of the partial pressures, so

$$X_{N_2} = \frac{600 \text{ torr}}{600 \text{ torr} + 160 \text{ torr}}$$

$$= 0.789, \text{ or } 78.9 \text{ mole percent of } N_2$$

Now we do the same for oxygen.

$$X_{O_2} = \frac{160 \text{ torr}}{600 \text{ torr} + 160 \text{ torr}}$$

$$= 0.211, \text{ or } 21.1 \text{ mole percent of } O_2$$

You can easily see that the two mole percents add up to 100%.

PRACTICE EXERCISE 4   What is the mole fraction and the mole percent of oxygen in exhaled air if its partial pressure is 116 torr and the total pressure is 760 torr?

Molal Concentration

Notice that it's per kilogram of *solvent,* not of solution.

The number of moles of solute per kilogram of *solvent* is called the molal concentration or the molality of a solution. The usual symbol for molality is m.

$$\text{molality} = m = \frac{\text{mol of solute}}{\text{kg of solvent}}$$

If, for example, we dissolve 0.5000 mol of sugar in 1000 g (1.000 kg) of water, we have made a 0.5000 molal solution of sugar. We would not even have to use a volumetric flask, because we weigh the solvent.

The molality of a solution is an indirect expression of the ratio of moles of solute to moles of solvent, because the mass of the solvent is known. If needed, we could convert the 1 kg of solvent into moles of solvent. The fact that such a calculation is seldom necessary does not affect this interpretation.

To contrast molality with molarity, suppose we put the same quantity of sugar — 0.5000 mol — into a one liter volumetric flask and added water to the 1-liter mark, making sure everything dissolved. The calibration mark on a volumetric flask correlates with the flask's specified volume *only at a given temperature,* which is etched on the flask. If our solution were at this temperature, then we would have a 0.5000 molar solution. *But we would not know the mass of water used,* and it would be very unlikely that this mass would be 1.000 kg. Unless we actually measured the mass of water used to make the 0.5000 *molar* solution, or knew the density of the solution, we would have no way to calculate its *molality.*

It is very important that you learn the distinction between *molarity* and *molality.* Molarity is useful in determining the moles of solute obtained when we pour a specified volume of solution. Molality is useful in describing the ratio of moles of solute to moles of solvent. The value of a concentration given in units of molality does not change with temperature, but that of molarity does. Let us now work a problem that illustrates molality.

EXAMPLE 13.4 CALCULATING MOLAL CONCENTRATION

Problem: An experiment calls for an aqueous 0.150 *m* solution of sodium chloride. To prepare a solution with this concentration, how many grams of NaCl would have to be dissolved in 500 g of water?

Solution: As with almost all problems involving concentrations, our first step is to use the given concentration to prepare a conversion factor. Thus, "0.150 *m* NaCl" gives us the following two ratios, where we substitute 1000 g for 1 kg.

$$\frac{0.150 \text{ mol NaCl}}{1000 \text{ g H}_2\text{O}} \quad \text{and} \quad \frac{1000 \text{ g H}_2\text{O}}{0.150 \text{ mol NaCl}}$$

To calculate the moles of NaCl we need for 500 g of H_2O, we use the first ratio, because then the units will cancel properly.

$$500 \text{ g H}_2\text{O} \times \frac{0.150 \text{ mol NaCl}}{1000 \text{ g H}_2\text{O}} = 0.0750 \text{ mol NaCl}$$

This gives us the *moles* of NaCl needed. We next convert 0.0750 mol of NaCl to grams of NaCl. (The formula weight of NaCl is 58.5, which means, of course, 58.5 g NaCl/mol NaCl.)

$$0.0750 \text{ mol NaCl} \times \frac{58.5 \text{ g NaCl}}{1 \text{ mol NaCl}} = 4.39 \text{ g NaCl}$$

Thus, when 4.39 g of NaCl is dissolved in 500 g of H_2O, the concentration is 0.150 *m* NaCl.
With a little practice, you will be able to set up a string of conversion factors and do the calculation at the end. For example,

$$500 \text{ g H}_2\text{O} \times \frac{0.150 \text{ mol NaCl}}{1000 \text{ g H}_2\text{O}} \times \frac{58.5 \text{ g NaCl}}{1 \text{ mol NaCl}} = 4.39 \text{ g NaCl}$$

PRACTICE EXERCISE 5 As we will soon study, water freezes at a lower temperature when it contains solutes. To study the effect of methyl alcohol on the freezing point of water, we might begin by preparing a series of solutions of known molalities. Calculate the number of grams of methyl alcohol (CH_3OH) needed to prepare a 0.250 *m* solution, using 2000 g of water.

PRACTICE EXERCISE 6 If you prepare a solution by dissolving 4.00 g of NaOH in 250 g of water, what is the molality of the solution? (*Hint:* Molality, remember, is the ratio of moles of solute to *kilograms* of solvent, so be sure to convert grams of water to kilograms.)

Percentage Concentrations

Before we use mole fractions or molalities in a study of certain physical properties of solutions, we will complete the list of the chief ways by which chemists describe concentrations.[2] The rest of the list are various kinds of *percent concentrations*.

In most chemical laboratories, a *percent concentration* means percent by weight or weight/weight percent. The symbol for it is usually % (w/w). This is the number of grams of solute per 100 grams of *solution* (not solvent, *solution*). A solution labeled "0.9% (w/w) NaCl," for example, is one in which the ratio of solute to solution is 0.9 g of NaCl to 100 g of solution.

[2] We will now yield to the widespread practice of using "percent" when "percentage" is correct.

EXAMPLE 13.5   USING WEIGHT/WEIGHT PERCENTS

Problem: How many grams of a 4.00% (w/w) solution of NaCl is needed to obtain 0.500 g of NaCl?

Solution: The given concentration gives us the following conversion factors.

$$\frac{4.00 \text{ g NaCl}}{100 \text{ g solution}} \quad \text{and} \quad \frac{100 \text{ g solution}}{4.00 \text{ g NaCl}}$$

We want 0.500 g of NaCl from this solution, so we use the second conversion factor.

$$0.500 \text{ g NaCl} \times \frac{100 \text{ g solution}}{4.00 \text{ g NaCl}} = 12.5 \text{ g of 4.00\% (w/w) solution}$$

Thus, if we take 12.5 g of the 4.00% (w/w) NaCl solution, we will also be taking 0.500 g of NaCl.

PRACTICE EXERCISE 7   Hydrochloric acid can be purchased from chemical supply houses as a solution that is 37% (w/w) HCl. What mass of this solution contains 7.5 g of HCl?

EXAMPLE 13.6   PREPARING WEIGHT/WEIGHT PERCENT SOLUTIONS

Problem: "White vinegar" can be made by preparing a 5.0% (w/w) solution of acetic acid in water. How would you make 500 g of such a solution?

Solution: The mass of acetic acid first has to be calculated. Then it has to be combined with enough water to make the total solution have a mass of 500 g. To find the mass of solute, we use the given concentration to give us our choice of conversion factors. Thus, 5.0% (w/w) acetic acid means

$$\frac{5.0 \text{ g acetic acid}}{100 \text{ g solution}} \quad \text{and} \quad \frac{100 \text{ g solution}}{5.0 \text{ g acetic acid}}$$

We have to use the first factor to get the units to work out.

More than two-significant-figure precision is not often sought when solutions are made up to a percent concentration.

$$500 \text{ g solution} \times \frac{5.0 \text{ g acetic acid}}{100 \text{ g solution}} = 25 \text{ g acetic acid}$$

Thus, if we dissolve 25 g of acetic acid in enough water to make the final mass equal to 500 g, we would have a 5.0% (w/w) solution of acetic acid—the same as white vinegar. This, of course, would require 475 g of water (500 g − 25 g).

Because we know that the density of water at room temperature is (to two significant figures) 1.0 g/mL, we can easily see that the 475 g of water needed in this example is the same as 475 mL of water. When greater precision is not needed, a chemist will generally make this mental calculation and then mix 25 g of acetic acid with 475 mL of water to make the 5.0% (w/w) acetic acid solution. It is just easier to measure a large volume than a large mass because large graduated cylinders are readily available.

PRACTICE EXERCISE 8   One common laboratory reagent is aqueous 10% (w/w) NaOH. How could 750 g of this solution be prepared without weighing a huge mass of solvent?

When both solute and solvent are liquids, or both are gases, then it can be convenient to use *percent by volume*—symbolized as % (v/v)—which is the number of *volumes* of one component in 100 volumes of the entire solution. (Notice again that the emphasis is on the volume of the final *solution*.) The unit of volume can be any unit, provided that the same unit is used for both solute and solution. (True percentages, of course, have no formal units because they cancel in the final calculation, but we still have to remember how each kind of percent concentration is defined.)

EXAMPLE 13.7 USING VOLUME/VOLUME PERCENTS

Problem: A 40% (v/v) solution of ethylene glycol in water gives an antifreeze that will protect a vehicle's cooling system to −24 °C (−12 °F). What volume of ethylene glycol has to be used to make 5 quarts of this solution?

Solution: First, we use the given concentration to construct conversion factors. Thus, "40% (v/v) ethylene glycol" means

$$\frac{40 \text{ vol ethylene glycol}}{100 \text{ vol solution}} \quad \text{and} \quad \frac{100 \text{ vol solution}}{40 \text{ vol ethylene glycol}}$$

We have to use the first factor to get the answer in units of ethylene glycol. And we substitute "quarts" for "vol" now.

$$5 \text{ quarts solution} \times \frac{40 \text{ quarts ethylene glycol}}{100 \text{ quarts solution}} = 2 \text{ quarts ethylene glycol}$$

Thus if we dissolve 2 quarts of ethylene glycol in enough water to make the final volume of the solution 5 quarts, the antifreeze will be 40% (v/v) ethylene glycol. (Incidentally, we could not assume that all we had to do was add 2 quarts of ethylene glycol to 3 quarts of water to make 5 quarts of solution. When two liquids mix, the final volume is sometimes less than the sum of their initial volumes, and sometimes more. Sometimes it is the same, but this cannot be assumed. It depends on how individual molecules find room for each other in the solution.)

PRACTICE EXERCISE 9

A 35% (v/v) solution of wood alcohol (methyl alcohol) in water gives substantially the same antifreeze protection as 40% (v/v) ethylene glycol (referring to Example 13.7). How would you prepare 20 quarts of this solution? Such a solution, incidentally, eventually loses its methyl alcohol, because this solute (unlike ethylene glycol) readily evaporates. With its loss, antifreeze protection is lost.

Methyl alcohol boils at 64.7 °C.
Ethylene glycol boils at 197.6 °C.

Some fields of science employ still other kinds of percent concentrations, but they are not true percents (the units do not cancel) and they are not preferred by chemists. In case you might run into them we have described a few in Special Topic 13.4.

Converting Among Concentration Units

Many made-up solutions are available from chemical supply houses and stockrooms, as you have probably already seen in the lab. Sometimes in working with a solution whose molarity is known we run into a situation where we would like to know its molality or its percent (w/w) concentration, too. Is there any way we can calculate one kind of concentration from another? We will work some examples to show how.

The only additional information we need for some calculations are formula weights and densities, which are often given in tables in reference handbooks for chemistry. These conversions are easy, *provided you know the definitions of each kind of concentration* and can prepare conversion factors from given concentrations. In fact, working through these examples and doing practice exercises is a superb way to get a final and lasting grasp of all the concentration expressions.

EXAMPLE 13.8 CONVERTING FROM WEIGHT PERCENT TO MOLALITY

Problem: What is the molality of a 10.0% (w/w) NaCl solution?

Solution: We are given a ratio of 10.0 g of NaCl to 100 g of *solution*. This is what the percent (w/w) means. But the units of molality are

$$\frac{\text{mol of solute}}{\text{kg of solvent}}$$

What we need to do, then, is use the units of the percent concentration to work our way to the units of molality. The steps are

Step 1. Calculate the moles of NaCl in 10.0 g of NaCl.

Step 2. Calculate the kilograms of H_2O in 100 g of this *solution*. (Remember it is grams of *solution* with percent concentrations, but kilograms (or 1000 g) of *solvent* with molality.

Step 3. Calculate the ratio of these, because molality is the ratio of moles of solute to kilograms of solvent.

We find the moles of NaCl in 10.0 g of NaCl in the usual way, using the molar mass of NaCl, 58.5 g NaCl/mol NaCl.

$$10.0 \text{ g NaCl} \times \frac{1 \text{ mol NaCl}}{58.5 \text{ g NaCl}} = 0.171 \text{ mol NaCl}$$

Next, step 2, we note that 100 g of this *solution*, which has 10.0 g of solute, must therefore have 90.0 g of solvent (100 g − 10 g). So the kilograms of solvent are found by

$$90.0 \text{ g H}_2\text{O} \times \frac{1 \text{ kg H}_2\text{O}}{1000 \text{ g H}_2\text{O}} = 0.0900 \text{ kg H}_2\text{O}$$

Remember, both numbers in 1 kg = 1000 g are defined as exact.

Finally, step 3, we calculate *m*, the ratio of moles of solute to kilograms of solvent.

$$\frac{0.171 \text{ mol NaCl}}{0.0900 \text{ kg H}_2\text{O}} = 1.90 \text{ } m$$

Thus, 10.0% (w/w) NaCl is also 1.90 *m* NaCl.

EXAMPLE 13.9 CONVERTING FROM WEIGHT PERCENT TO MOLE FRACTIONS

Problem: What are the mole fractions and mole percents of the components in 10.0% (w/w) NaCl, the same solution we used in the previous example?

Solution: Our goal now is to move the units from those of percent (w/w),

$$\frac{\text{g solute}}{100 \text{ g solution}}$$

to those implied by a mole fraction,

$$\frac{\text{mol solute}}{\text{total moles}}$$

The amount of solute from the previous example is 0.171 mol of NaCl. We also found in that example that there is 90.0 g of water in 100 g of the solution. And the moles of water in 90.0 g of H_2O (at 18.0 g H_2O/mol H_2O) is

$$90.0 \text{ g } H_2O \times \frac{1 \text{ mol } H_2O}{18.0 \text{ g } H_2O} = 5.00 \text{ mol } H_2O$$

So the total number of moles of both solute and solvent in 10.0% (w/w) NaCl is

$$0.171 \text{ mol} + 5.00 \text{ mol} = 5.17 \text{ mol (correctly rounded)}$$

Now we can calculate mole fractions of solute and solvent.

For NaCl

$$X_{NaCl} = \frac{0.171 \text{ mol}}{5.17 \text{ mol}} = 0.0331 \text{ (or 3.31 mole percent)}$$

For H_2O

$$X_{H_2O} = \frac{5.00 \text{ mol}}{5.17 \text{ mol}} = 0.967 \text{ (or 96.7 mole percent)}$$

For a two-component system like this, we would not need this last calculation. The mole fractions must add up to 1 (or the percents, to 100%), so we could have found the mole fraction of water simply by subtracting the mole fraction of NaCl from 1.00. (Or we could have calculated the mole percent of water by subtracting the mole percent of NaCl from 100.)

PRACTICE EXERCISE 10   A certain sample of concentrated hydrochloric acid is 37.0% (w/w) HCl. (a) Calculate the molality of this solution. (b) What are the mole percents of HCl and H_2O in the solution?

EXAMPLE 13.10   CALCULATING PERCENTS BY WEIGHT FROM MOLE FRACTIONS

Problem: So-called "100 proof" alcohol has the following mole fractions of components: for water, 0.765, and for ethyl alcohol, 0.235. Calculate the percents (w/w) of water and ethyl alcohol in 100 proof ethyl alcohol. The formula weights are: H_2O, 18.0; and ethyl alcohol, C_2H_5OH, 46.1.

Solution: As in all of these kinds of calculations, we work from the meanings of the given concentration units toward the final units. To work toward percents by weight, we have to work toward a ratio calculated from *grams* of components. Let us start with water. When the mole fraction of water is 0.765, we know that there are 0.765 mol of H_2O in a total of 1.00 mol of all components. So, in terms of grams of water, at 18.0 g H_2O/mol H_2O, we have

$$0.765 \text{ mol } H_2O \times \frac{18.0 \text{ g } H_2O}{1 \text{ mol } H_2O} = 13.8 \text{ g } H_2O$$

Next, we take ethyl alcohol. When its mole fraction is 0.235, we know that there is 0.235 mol of C_2H_5OH in 1.00 mol of all components. So, in terms of grams of C_2H_5OH, at 46.1 g C_2H_5OH/mol C_2H_5OH, we have

$$0.235 \text{ mol } C_2H_5OH \times \frac{46.1 \text{ g } C_2H_5OH}{1 \text{ mol } C_2H_5OH} = 10.8 \text{ g } C_2H_5OH$$

Thus, in "1.00 mol" of this ethyl alcohol solution there are 13.8 g of H_2O and 10.8 g of C_2H_5OH. To convert to percents by weight, we need the total mass of the solution, which is 13.8 g + 10.8 g = 24.6 g. For the 13.8 g of water, its percentage of 24.6 g is

$$\frac{13.8 \text{ g}}{24.6 \text{ g}} \times 100 = 56.1\% \text{ (w/w)}$$

And for the 10.8 g of ethyl alcohol, its percentage of 24.6 g is

$$\frac{10.8 \text{ g}}{24.6 \text{ g}} \times 100 = 43.9\% \text{ (w/w)}$$

(Note that the percents add up to 100%, as they must.) Thus, 100 proof alcohol is 56.1% (w/w) water and 43.9% (w/w) ethyl alcohol.

PRACTICE EXERCISE 11   In a two-component solution of benzene in carbon tetrachloride, the mole fraction of benzene is 0.450. What is the percent (w/w) of benzene in this solution?

EXAMPLE 13.11   CONVERTING MOLALITY INTO MOLE FRACTIONS

Problem: Ammonium nitrate is sometimes a component in nitrogen fertilizers. Suppose that an aqueous solution of NH_4NO_3 has a concentration of 2.48 m. What are the mole fractions of NH_4NO_3 and water in this solution?

Solution: The given molal concentration means that we can choose from the following conversion factors, as needed.

$$\frac{2.48 \text{ mol } NH_4NO_3}{1 \text{ kg water}} \quad \text{and} \quad \frac{1 \text{ kg water}}{2.48 \text{ mol } NH_4NO_3}$$

We have to work our way toward a ratio of moles of solute to total number of *moles,* so we need to find the moles of water in 1 kg (and remember that we treat this value as exact). Using 1000 g = 1 kg (both exact numbers) and knowing that the formula weight of water is 18.0, we have

$$1 \text{ kg } H_2O \times \frac{1000 \text{ g}}{1 \text{ kg}} \times \frac{1 \text{ mol } H_2O}{18.0 \text{ g } H_2O} = 55.6 \text{ mol } H_2O$$

So the total moles, water + NH_4NO_3, is

$$2.48 \text{ mol} + 55.6 \text{ mol} = 58.1 \text{ mol (correctly rounded)}$$

Now we can calculate the mole fraction of NH_4NO_3.

$$X_{NH_4NO_3} = \frac{2.48 \text{ mol}}{58.1 \text{ mol}} = 0.0427$$

The mole fraction of water is simply

$$X_{H_2O} = 1.0000 - 0.0427$$
$$= 0.9573$$

PRACTICE EXERCISE 12   A solution of $CaCl_2$ has a concentration of 4.57 m. What are the mole fractions and mole percents of $CaCl_2$ and H_2O in this solution?

When we have to do conversions that involve molar concentrations, then we need data on densities as well as formula weights, as we will see in the next example.

EXAMPLE 13.12   CONVERTING A PERCENT BY WEIGHT INTO A MOLAR CONCENTRATION

Problem: A certain supply of concentrated hydrochloric acid is 36.0% (w/w) in HCl. The density of the solution is 1.19 g/mL. Calculate the molar concentration of HCl in this solution.

Solution: Let us first make sure we know our starting and ending units. We are to work our way from a percent by weight with units of

$$\frac{\text{g HCl}}{100 \text{ g solution}}$$

to molarity, with units of

$$\frac{\text{mol HCl}}{1000 \text{ mL solution}}$$

The need for the density stems from the need to get *g solution* into *mL solution.* So let us exploit the given density first, because it relates grams and milliliters. Its value tells us that in this solution we can use the following conversion factors.

We treat the 1 in 1 mL as exact.

$$\frac{1.19 \text{ g solution}}{1 \text{ mL solution}} \quad \text{and} \quad \frac{1 \text{ mL solution}}{1.19 \text{ g solution}}$$

Since we have to deal with 1000 mL of solution as we work from grams of solution toward the molar concentration, we now calculate how many grams of this solution are in 1000 mL.

$$1000 \text{ mL solution} \times \frac{1.19 \text{ g solution}}{1 \text{ mL solution}} = 1.19 \times 10^3 \text{ g solution}$$

36% HCl means 36.0% (w/w), or 36.0 g HCl/100 g HCl solution.

But this mass of solution is 36.0% (w/w) in HCl; 36.0% of this mass is pure HCl. So the quantity of pure HCl in 1.19×10^3 g of solution is found by

$$1.19 \times 10^3 \text{ g solution} \times \frac{36.0 \text{ g HCl}}{100 \text{ g solution}} = 428 \text{ g HCl}$$

Thus, there is 428 g of HCl in 1.19×10^3 g of solution. But, remember that this mass of solution has a volume of 1000 mL or 1 L. So we are nearly home — to the moles per liter or molarity. We just need to convert 428 g of HCl to moles of HCl. The formula weight of HCl is 36.5, so we have

$$428 \text{ g HCl} \times \frac{1 \text{ mol HCl}}{36.5 \text{ g HCl}} = 11.7 \text{ mol HCl}$$

We made it! There are 11.7 mol of HCl in 1 L of 36% (w/w) HCl, so the molarity of this solution is 11.7 *M*.

PRACTICE EXERCISE 13 Hydrobromic acid can be purchased as 40.0% (w/w) HBr. The density of this solution is 1.38 g/mL. What is the molar concentration of HBr in this solution?

EXAMPLE 13.13 CONVERTING FROM MOLARITY TO PERCENT BY WEIGHT

Problem: Perchloric acid, $HClO_4$ (formula weight, 100.5), can be purchased at a concentration of 9.20 *M*. The density of this solution is 1.54 g/mL. What is the percent by weight of perchloric acid in this solution?

Solution: When the molarity is 9.20 *M*, we have the following ratio.

$$\frac{9.20 \text{ mol HClO}_4}{1000 \text{ mL solution}}$$

What we want is the ratio of *grams* of $HClO_4$ to *grams* of solution, taken as a percent. This is what a percent by weight means. So, in 9.20 mol of $HClO_4$, we have

$$9.20 \text{ mol HClO}_4 \times \frac{100.5 \text{ g HClO}_4}{1 \text{ mol HClO}_4} = 925 \text{ g HClO}_4$$

And in 1000 mL solution we have the following mass of solution.

$$1000 \text{ mL solution} \times \frac{1.54 \text{ g solution}}{1 \text{ mL solution}} = 1.54 \times 10^3 \text{ g solution}$$

So the ratio of grams of solute to grams of solution, taken as a percent, is

$$\frac{925 \text{ g (HClO}_4)}{1.54 \times 10^3 \text{ g (solution)}} \times 100 = 60.0\% \text{ (w/w)}$$

Thus, 9.20 M HClO$_4$ is 60.0% (w/w) HClO$_4$.

PRACTICE EXERCISE 14   Hydriodic acid can be purchased as 5.51 M HI. The density of this solution is 1.50 g/mL. What is the percent concentration (w/w) of HI in this solution?

13.7 LOWERING OF THE VAPOR PRESSURE

The vapor pressure of any component in a mixture is lowered by the presence of the other components.

After the Greek *kolligativ*, depending on number and not on nature.

The physical properties of solutions to be studied in this and succeeding sections are called colligative properties, because they depend mostly on the relative populations of particles in mixtures, not on their chemical identities. In this section, we will look at the effect of a solute on the vapor pressure of a solvent in a liquid solution. The effect depends on whether the solute is volatile or nonvolatile, so we will study these separately.

Volatile means "can evaporate," has a low boiling point.
Nonvolatile means "cannot evaporate," has a very high boiling point.

How a Nonvolatile Solute Affects the Vapor Pressure of a Solution

We'll study solutions of ions in Section 13.11.

All liquid solutions of nonvolatile solutes have lower vapor pressures than their pure solvents. The extent of the lowering of the vapor pressure of the solvent depends on whether the solute is molecular (and does not ionize) or is ionic. We will consider nonvolatile molecular solutes, like sugar, first, because there is a particularly simple law that correlates the solution's vapor pressure with the mole fraction of the solvent. This is the vapor pressure–concentration law, or Raoult's law, and the best statement of it is just its equation.

François Marie Raoult (1830–1901) was a French scientist.

> **Vapor Pressure–Concentration Law (Raoult's Law)**
>
> $$P_{\text{solution}} = X_{\text{solvent}} \times P^\circ_{\text{solvent}} \qquad (13.3)$$

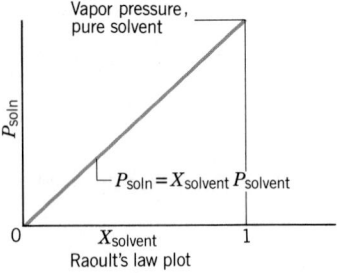

Raoult's law plot

P_{solution} is the vapor pressure of the solution, P°_{solvent} is the vapor pressure of the pure solvent, and X_{solvent} is the mole fraction of the solvent in this solution. Thus, the vapor pressure of the solution is a fraction — the mole fraction, X_{solvent} — of the pure solvent's vapor pressure (at the same temperature). A plot of P_{solution} versus X_{solvent} for a solution that obeyed Raoult's law at all concentrations would be linear, as shown in the margin.

The reason for Raoult's law is easy to understand. The equilibrium vapor pressure of a liquid reflects how far the following equilibrium favors the vapor.

$$\text{solvent molecules in solution} \rightleftharpoons \text{solvent molecules in vapor state}$$

When the rate of evaporation — the forward change — is relatively large, the vapor pressure is relatively high. When we add nonvolatile solute molecules to the solvent, they

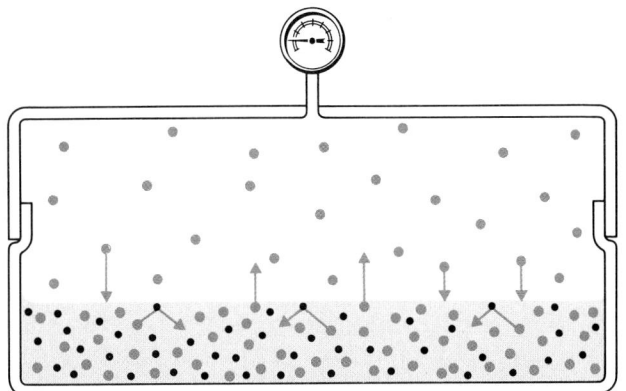

FIGURE 13.23 Lowering of vapor pressure by a nonvolatile solute. Because particles of the solute are present, the opportunities for escape of the solvent particles are reduced. Therefore, the vapor pressure of this solution is less than the vapor pressure of the pure solvent (at the same temperature).

take up spaces at the surface of the solution and leave the solvent's molecules with a fraction of their former surface population. Thus, the solute molecules, which cannot evaporate, interfere with the escape of solvent molecules into the vapor state. They reduce the rate of the forward change. But they do not interfere with the return of solvent molecules to the solution. See Figure 13.23. The equilibrium thus shifts to the left and the vapor pressure decreases until the forward and reverse changes are again equal.

The interference caused by the solute molecules and therefore the extent to which the vapor pressure is *lowered* depends on the fraction of molecules at the surface which are the solute. If 25% of these molecules are solute, then the vapor pressure is lowered by 25%. Another way of looking at this, however, is that 75% of the molecules at the surface are those of the *solvent*, and the vapor pressure, having been lowered by 25%, is now 75% of its value for the pure solvent. This is exactly what Raoult's law says, because if 75% of the molecules are those of the solvent, then the mole fraction of the solvent must be 0.75, and 0.75 multiplied by the vapor pressure of pure solvent gives a value that is 75% of P°_{solvent}.

EXAMPLE 13.14   RAOULT'S LAW CALCULATION FOR NONVOLATILE SOLUTES

Problem: Carbon tetrachloride has a vapor pressure of 100 torr at 23 °C. This solvent can dissolve candle wax, which is essentially nonvolatile. Although candle wax is a mixture, we can take its molecular formula to be $C_{22}H_{46}$ (formula weight, 310). What will be the vapor pressure at 23 °C of a solution prepared by dissolving 10.0 g of wax in 40.0 g of CCl_4? The formula weight of CCl_4 is 154.

Solution: We have to find the mole fraction of the solvent before we can use Raoult's law to answer this question. So we have to calculate the moles of solute and solvent first.

For CCl_4

$$40.0 \text{ g } CCl_4 \times \frac{1 \text{ mol } CCl_4}{154 \text{ g } CCl_4} = 0.260 \text{ mol } CCl_4$$

For $C_{22}H_{46}$

$$10.0 \text{ g } C_{22}H_{46} \times \frac{1 \text{ mol } C_{22}H_{46}}{310 \text{ g } C_{22}H_{46}} = 0.0323 \text{ mol } C_{22}H_{46}$$

The total number of moles is the sum: 0.292 mol. Now, we can find the mole fraction of CCl_4.

$$X_{CCl_4} = \frac{0.260 \text{ mol}}{0.292 \text{ mol}} = 0.890$$

The mole fraction of CCl_4, the solvent, is 0.890. The vapor pressure of the solution will be this fraction of the vapor pressure of pure CCl_4 (100 torr) as calculated by Raoult's law.

$$P_{solution} = 0.890 \times 100 \text{ torr}$$
$$= 89.0 \text{ torr}$$

The presence of the wax has lowered the vapor pressure of CCl_4 from 100 to 89.0 torr.

PRACTICE EXERCISE 15   Dibutyl phthalate, $C_{16}H_{22}O_4$ (formula weight 278), is an oil sometimes worked into plastic articles to give them greater softness. It has a negligible vapor pressure at room temperature. (It has to be heated to 148 °C before its vapor pressure is even 1 torr.) What is the vapor pressure, at 20 °C, of a solution of 20.0 g of dibutyl phthalate in 50.0 g of octane, C_8H_{18} (formula weight, 114)? The vapor pressure of pure octane at 20 °C is 10.5 torr.

The Effect of a Volatile Solute on the Vapor Pressure of a Solution

When two (or more) components of a liquid solution can evaporate, the vapor contains molecules of each. Each volatile component contributes its own partial pressure to the total pressure and this partial pressure is directly proportional to the component's mole fractions in the solution. This is reasonable because the rate of evaporation of each compound has to be lowered by molecules of the others in the same way we described just above for nonvolatile solutes. So, by Dalton's law, the total vapor pressure will be the sum of the partial pressures. To calculate these partial pressures, we use Raoult's law for each component. For component A, present in a mole fraction X_A, its partial pressure, P_A, is found as a fraction of its vapor pressure when pure, P_A°; thus,

$$P_A = X_A P_A^\circ$$

And the partial pressure of component B would be

$$P_B = X_B P_B^\circ$$

Similar calculations would be done on other components, but we will stick to two-component systems.

Remember, P_A and P_B here are the *partial* pressures as calculated by Raoult's law.

The total pressure of the solution of liquids A and B is then

$$P_{total} = P_A + P_B$$

EXAMPLE 13.15   APPLYING RAOULT'S LAW TO SOLUTIONS OF VOLATILE SOLUTES

Problem: Acetone is a solvent for both water and molecular liquids that do not dissolve in water, like benzene. At 20 °C acetone has a vapor pressure of 162 torr. The vapor pressure of water at 20 °C is 17.5 torr. What is the vapor pressure of a solution of acetone and water with 50.0 mole percent of each? (We assume that Raoult's law applies.)

Solution: To find P_{total} we need to calculate the individual partial pressures and then add them.

$$P_{acetone} = 162 \text{ torr} \times 0.500 = 81.0 \text{ torr}$$
$$P_{water} = 17.5 \text{ torr} \times 0.500 = 8.75 \text{ torr}$$
$$P_{total} = 89.8 \text{ torr}$$

Thus the vapor pressure of the solution is much higher than that of pure water but much less than that of pure acetone.

PRACTICE EXERCISE 16   At 20 °C, the vapor pressure of cyclohexane, a hydrocarbon solvent, is 66.9 torr and that of toluene (another solvent) is 21.1 torr. What is the vapor pressure of a solution of these two at 20 °C when each is present at a mole fraction of 0.500?

Figure 13.24 shows how the total vapor pressure of a two-component solution of volatile liquids is related to composition and partial pressures according to Raoult's law. Each of the steeper lower lines shows how the *partial* vapor pressure of one component changes as its own mole fraction changes. Any point on the top line represents the total pressure at a particular pair of values of mole fractions. (Try this out yourself with a millimeter ruler. The distance from a point on the baseline to one of the steeper lines plus the distance *from the same base point* to the other steeper line gives the distance from this base point to the top line.)

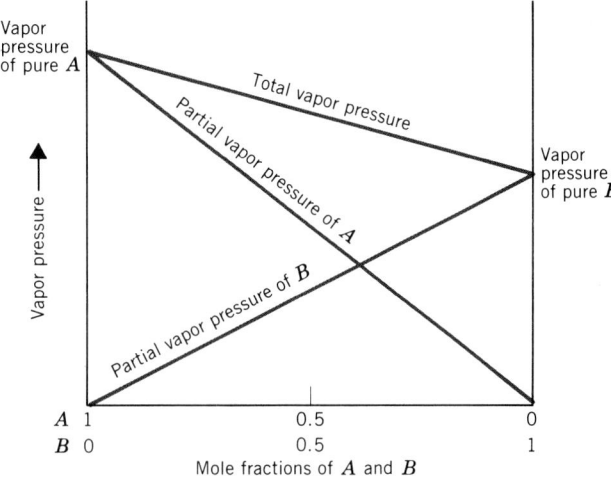

FIGURE 13.24 The vapor pressure of an ideal, two-component solution of volatile compounds.

Ideal Solutions and Deviations from Raoult's Law

As we described in Section 13.3, an *ideal solution* is one for which $\Delta H_{solution}$ is zero, and therefore one in which forces of attraction between both like molecules and unlike molecules are identical. Only an ideal solution would obey Raoult's law exactly. Only an ideal solution would give plots like those in Figure 13.24, which we just discussed.

Not very many real, two-component solutions come close to being an ideal solution. Figure 13.25 shows two of the typical ways in which real two-component solutions

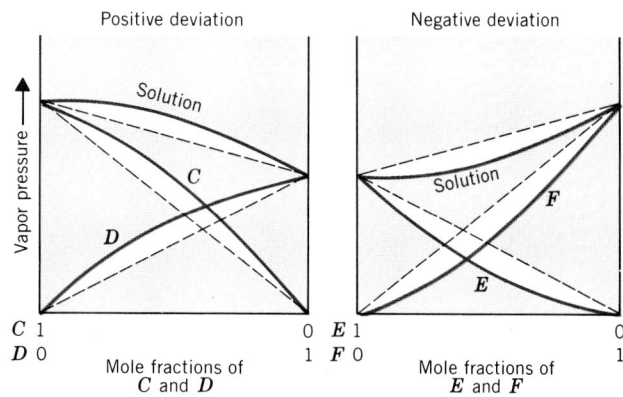

FIGURE 13.25 Typical deviations from ideal behavior of the total vapor pressures of real, two-component solutions of volatile substances.

behave. The individual partial pressures are not *linear* with composition as Raoult's law demands. So the sums of these partial pressures are seldom linear either. The sums given by the upper, total-pressure plots show either upward or *positive deviations* or they give downward or *negative deviations* from Raoult's law behavior. What causes this?

Let us identify the molecules of a two-component solution as *A* and *B*. Solutions with negative deviations generally form exothermically. In them, the attractive forces between *A* and *B* are stronger than the sum of the attractive forces between *A* and *A* and between *B* and *B*, as we analyzed on page 508. The extra attractive force between *A* and *B*, therefore, is an extra interference with the escape of either *A* or *B* into the vapor phase. So the partial pressure plots (the lower, steeper plots) for *A* and *B* sag, which makes the plot for total pressure sag, too. Thus, solutions that form exothermically tend to show negative deviations.

Solutions with positive deviations from Raoult's law generally form endothermically. In them the attractive forces between *A* and *B* are *less* than the sum of the attractive forces from *A* to *A* and from *B* to *B*. In the solution of *A* and *B*, it is as if *A* and *B* molecules are not held back as much by their own kind. They are released to evaporate at faster rates. This causes their individual *partial* pressure plots to curve upward. So the plot of total pressure must curve upward, too. Thus, solutions that form endothermically tend to show positive deviations. Figure 13.26 summarizes the relationships between Raoult's law deviations and the relative attractive forces — *A* to *A*, *B* to *B*, and *A* to *B* — in the solute, the solvent, and the solution.

Acetone – water solutions show negative deviations.

Ethyl alcohol – hexane solutions show positive deviations. (Hexane is a nonpolar component of gasoline.)

FIGURE 13.26 Deviations from Raoult's law and attractions between solute and solvent in a liquid-liquid two-component solution. Letting A represent one component and B another component, we can represent the attractive forces between molecules within pure A as "A to A", and within pure B as "B to B". In the solution, these forces can still operate, but we have to add the intermolecular attractive forces of A to B. The net effect of these forces on ΔH_{soln} and Raoult's law plots (vapor pressure versus composition) are as follows.

Relative Attractive Forces	ΔH_{soln}	Raoult's law plots
[A to A, B to B] > A to B	positive (endothermic)	positive deviations
[A to A, B to B] = A to B (ideal solution)	zero	obeyed perfectly
[A to A, B to B] < A to B	negative (exothermic)	negative deviations

13.8 ELEVATION OF THE BOILING POINT

The boiling point of a solution of a nonvolatile solute is higher than the boiling point of the solvent.

When the only volatile component of a solution is the solvent, then the vapor pressure of the *solution* is less than the vapor pressure that the solvent has at any given temperature. See Figure 13.27. At the temperature at which the pure solvent normally boils, the vapor pressure of the *solution* is still not equal to the atmospheric pressure. So to make the vapor pressure of the solution come up to atmospheric pressure, we have to increase the temperature of the solution further. The presence of a nonvolatile solute thus elevates the boiling point of the solution.

The most common application of this property of solutions occurs in the use of permanent-type antifreezes. These protect the liquid in a vehicle's cooling system not just from freezing but also from boiling over. These products are based on either

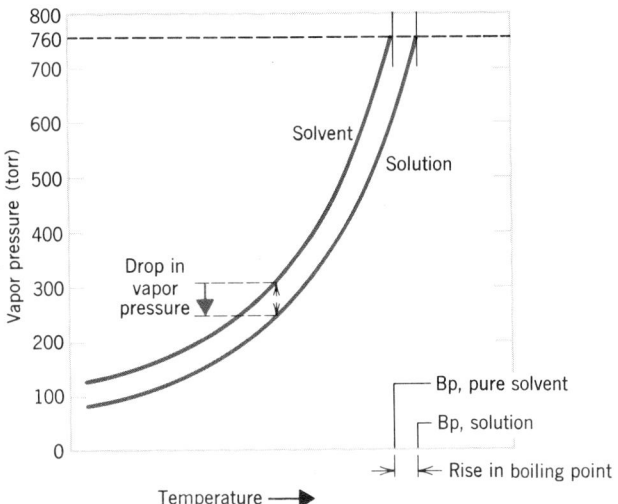

FIGURE 13.27 Boiling point elevation. Shown here are plots of vapor pressures versus temperatures for a solvent (upper curve) and for a solution of a nonvolatile solute in the same solvent (lower curve).

propylene glycol or ethylene glycol, which are high-boiling, nearly nonvolatile, water-soluble liquids, and they elevate the boiling point as well as lower the freezing point when dissolved in water.

$HOCH_2CH_2OH$
ethylene glycol
(bp 197.6 °C)

CH_3
|
$HOCH_2CH_2OH$
propylene glycol
(bp 188.2 °C)

Molal Boiling Point Elevation Constants

The number of degrees of elevation of a boiling point caused by a nonvolatile solute is directly proportional to its molal concentration, m. (We continue to deal only with molecular solutes. We will consider ionic solutes in Section 13.11.)

$$\Delta t \propto m$$

where $\Delta t = (bp_{solution} - bp_{solvent})$. With a proportionality constant, this becomes

$$\Delta t = k_b m \tag{13.4}$$

bp = boiling point

This equation works best for solutions with low solute concentrations. The constant, k_b, is called the *molal boiling point elevation constant*. Each solvent has its own value of k_b, and a few examples are given in Table 13.6. The units of k_b are those of $\Delta t/m$,

$$\frac{\Delta t}{m} = \frac{°C}{\text{mol-solute/kg-solvent}} = \frac{°C\ \text{kg-solvent}}{\text{mol-solute}}$$

TABLE 13.6 Molal Boiling Point Elevation and Freezing Point Depression Constants

Solvent	Bp (°C)	k_b	Mp (°C)	k_f
Water	100	0.51	0	1.86
Acetic acid	118.3	3.07	16.6	3.57
Benzene	80.2	2.53	5.45	5.07
Chloroform	61.2	3.63	—	—
Camphor	—	—	178.4	37.7
Cyclohexane	80.7	2.69	6.5	20.0

EXAMPLE 13.16 CALCULATING WITH MOLAL BOILING POINT ELEVATION CONSTANTS

Problem: At what temperature will a 0.15 m solution of sugar in water boil when the pressure is 1 atm?

Solution: We first calculate the *change*, Δt, in boiling point using the value of k_b for water from Table 13.6, which is 0.51 °C kg-solvent/mol-solute.

$$\Delta t = k_b \times m$$

$$= 0.51 \frac{\text{kg-solvent}}{\text{mol-solute}} \times 0.15 \frac{\text{mol-solute}}{\text{kg-solvent}}$$

$$= 0.077 \text{ °C}$$

Since pure water boils at 100.000 °C (at 1 atm), the solution boils at 100.000 °C + 0.077 °C = 100.077 °C. The elevation is not much, but it can be measured with good equipment.

PRACTICE EXERCISE 17   Estimate the boiling point of a permanent-type antifreeze that is a 16 m solution of ethylene glycol, a solution that corresponds roughly to about equal weights of ethylene glycol and water and is commonly used in radiators. (We say "estimate" because we have to assume that Equation 13.4 is satisfactory although the molality is very high.)

Formula Weights from Boiling Point Elevation Data

Equation 13.4 can be used to determine molal concentration if we measure Δt and know the value of k_b for the solvent. And once we have calculated m, we can work backward and calculate the formula weight of the solute, as we will show in the next example.

EXAMPLE 13.17   CALCULATING A FORMULA WEIGHT FROM BOILING POINT ELEVATION

Problem: A solution made by dissolving 10.00 g of an unidentified compound in 100 g of water boiled at 100.45 °C at 1 atm. What is the formula weight of the compound. (The solute was not an electrolyte.)

Solution: From Table 13.6, the K_b for water is 0.51 °C kg-solvent/mol-solute. The value of Δt is found by

$$\Delta t = 100.45 \text{ °C} - 100.00 \text{ °C}$$
$$= 0.45 \text{ °C}$$

Using Equation 13.4, we find m as

$$0.45 \text{ °C} = 0.51 \frac{\text{°C kg-solvent}}{\text{mol-solute}} \times m$$

or

$$m = \frac{0.45 \text{ °C}}{0.51 \frac{\text{°C kg-solvent}}{\text{mol-solute}}} = 0.88 \frac{\text{mol-solute}}{\text{kg-solvent}}$$

From this molal concentration we can calculate the number of *moles* of solute. The solution we know has 100 g or 0.100 kg of solvent, so it also contains

$$0.88 \frac{\text{mol-solute}}{\text{kg-solvent}} \times 0.100 \text{ kg-solvent} = 0.088 \text{ mol-solute}$$

From this calculation we now know that the 10.0 g of solute corresponds to 0.088 mol of solute, so the ratio of grams to moles — the formula weight — is found by

$$\frac{10.0 \text{ g solute}}{0.088 \text{ mol solute}} = 114 \text{ g/mol}$$

Giving due consideration to significant figures, the formula weight of the compound is between 110 and 120.

PRACTICE EXERCISE 18 A solution prepared by dissolving 11.0 g of a nonvolatile molecular solute in 100 g chloroform (bp 61.20 °C) was found to boil at 63.53 °C. What is the formula weight of the solute?

13.9 DEPRESSION OF THE FREEZING POINT

The freezing point of a solution is lower than the freezing point of the pure solvent.

A solution freezes at a temperature below the temperature at which the pure solvent freezes. We say that the solute causes a *freezing point depression*. Why this should be can be explained if we remember that *pure* crystals of one component form when something starts to crystallize. As we cool a solution, solvent molecules eventually lose enough average kinetic energy to enable them to settle into the lattices of crystals of pure solvent. Although these solvent crystals do not incorporate molecules of the solute, as the crystals grow, the solute molecules get in the way and interfere with the growth of solvent crystals. So, to compensate, we have to take more kinetic energy from the solvent—we have to cool the system. Thus, the solute depresses the freezing point.

Molal Freezing Point Depression Constants

The number of degrees, Δt, by which a solution's freezing point is depressed is proportional to the molality, m, at least at relatively low concentrations.

$$\Delta t \propto m$$

With a proportionality constant, the *freezing point depression constant, k_f*, we convert this to an equation.

$$\Delta t = k_f m \tag{13.5}$$

where $\Delta t = (\text{fp}_{\text{solvent}} - \text{fp}_{\text{solution}})$. The similarity to Equation 13.4 is obvious, and the units of k_f, °C kg-solvent/mol-solute, are the same as those of k_b. Each solvent has its own value of k_f, and several are given in Table 13.6.

fp = freezing point

EXAMPLE 13.18 CALCULATING WITH MOLAL FREEZING POINT DEPRESSION CONSTANTS

Problem: At what temperature will a 0.250 m solution of sugar in water freeze?

Solution: From Table 13.6, the k_f for water is 1.86 °C kg-solvent/mol-solute. So, using Equation 13.5,

$$\Delta t = 1.86 \frac{°C\ \text{kg-solvent}}{\text{mol-solute}} \times 0.250 \frac{\text{mol solute}}{\text{kg solvent}}$$

$$= 0.465\ °C$$

The freezing point of water, 0.000 °C, is lowered by 0.465 °C; so the *solution* itself freezes at −0.465 °C.

PRACTICE EXERCISE 19 Estimate the freezing point of the antifreeze solution in Practice Exercise 17. (The actual freezing point, incidentally, is quite close to the calculated value.)

The depression of a freezing point by a solute is the principle behind the use of salt to melt ice on city streets as well as the use of antifreeze in vehicle cooling systems.

Antifreezes contain corrosion inhibitors, but the Cl⁻ ion in salts accelerates the corrosion of metals by air and water.

Formula Weights from Freezing Point Depression Data

We will work an example to show how freezing point depression data can be used to determine formula weights.

EXAMPLE 13.19   CALCULATING A FORMULA WEIGHT FROM FREEZING POINT DEPRESSION

Problem: A solution made by dissolving 5.65 g of an unknown compound in 110 g of benzene froze at 4.39 °C. What is the formula weight of the solute?

Solution: From Table 13.6, for benzene

$$k_f = 5.07 \frac{°C \cdot kg\text{-benzene}}{mol\text{-solute}}$$

Pure benzene freezes at 5.45 °C, so the amount of freezing point depression is

$$\Delta t = 5.45\ °C - 4.39\ °C = 1.06\ °C$$

Using Equation 13.5, we calculate *m* as follows.

$$1.06\ °C = 5.07 \frac{°C \cdot kg\text{-benzene}}{mol\text{-solute}} \times m$$

$$m = \frac{1.06\ °C}{5.07 \dfrac{°C \cdot kg\text{-benzene}}{mol\text{-solute}}} = 0.209 \frac{mol\text{-solute}}{kg\text{-benzene}}$$

From this molal concentration we can calculate the number of *moles* of solute. The solution we know contains 110 g or 0.110 kg of benzene, so

$$0.209 \frac{mol\text{-solute}}{kg\text{-benzene}} \times 0.110\ kg\text{-benzene} = 0.0230\ mol\text{-solute}$$

With this calculation, we now know that the original 5.65 g of solute consists of 0.0230 *mol* of solute, so the ratio of grams to moles — the formula weight — is

$$\frac{5.65\ g\text{-solute}}{0.0230\ mol\text{-solute}} = 246\ g/mol$$

The formula weight of the solute is 246.

PRACTICE EXERCISE 20   A solution made by dissolving 3.46 g of an unknown compound in 85.0 g of benzene froze at 4.13 °C. What is the formula weight of the unknown?

PRACTICE EXERCISE 21   If 3.46 g of the unknown of Practice Exercise 20 were dissolved in 85.0 g of molten camphor, instead of benzene, at what temperature would the solution freeze? (Large freezing point depressions are often measured with greater precision than smaller values. Which solvent, then, benzene or camphor, offers better precision for the determination of a formula weight by this method?)

13.10 DIALYSIS AND OSMOSIS

When two solutions of different concentrations are separated by the right kind of membrane, their concentrations change in the direction of becoming equal.

In living things, membranes of various kinds keep mixtures organized and separated. These membranes are *semipermeable,* which means that they let water molecules as well as small ions and other small molecules pass through. But they do not allow the passage

Latin *permeare,* "to go through."

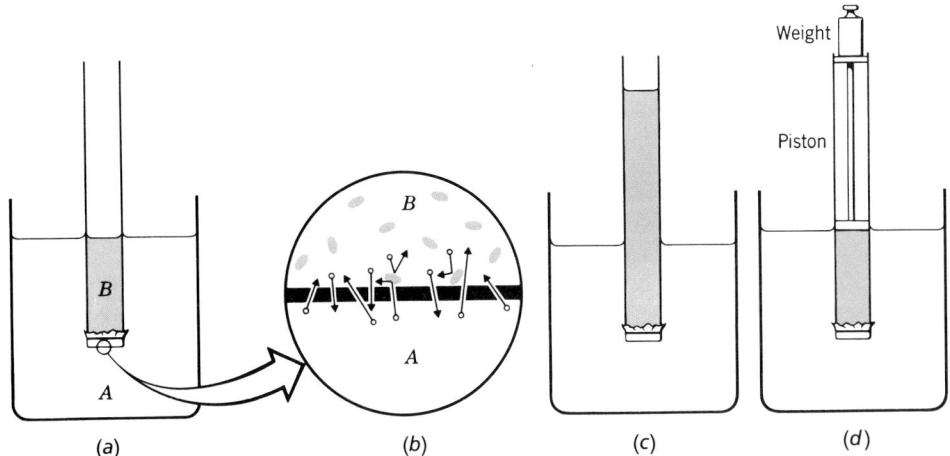

FIGURE 13.28 Osmosis and osmotic pressure. (*a*) Initial conditions. A solution, *B*, is separated from pure water, *A*, by an osmotic membrane, but no change has yet occurred. (*b*) Enlarged view at the membrane. Water molecules (small circles) are moving into *B* more easily and hence more rapidly than they are leaving *B* for *A*. The solute molecules (larger dots) interfere with the movement of water from *B* to *A*. (c) After a while the volume of fluid in the tube has increased visibly. Osmosis has occurred. (*d*) A back-pressure is needed to prevent osmosis. This is the osmotic pressure of the solution.

of very large molecules, like those of proteins. This phenomenon, the selective passage of small ions and molecules through a membrane, is called dialysis, and the membrane is a *dialyzing membrane.*

In the limiting case of dialysis, osmosis, only solvent molecules can get through the membrane, now called an *osmotic membrane.* Such membranes are rare, but they can be made. We will consider osmosis first.

One theory that explains osmosis is illustrated in Figure 13.28. An osmotic membrane separates two solutions from each other, one more concentrated in solute than the other. The solvent in each solution has its own "escaping tendency," which is just like the vapor pressure of a solution. The solution with the lower *solute* concentration has, of course, the higher mole fraction of *solvent.* Therefore, solvent molecules in the less concentrated solution (or the pure solvent) have a higher escaping tendency than those in the more concentrated solution. So solvent molecules escape through the membrane more frequently from the less concentrated side and pass into the more concentrated solution.

Some solvent molecules come the other way, of course, but not as frequently. The effect is a net shift of solvent through the membrane from the less concentrated side to the more concentrated side. This is osmosis, and it would continue, in principle, until the concentrations of the solutions became equal. Only then would the escaping tendencies of the solvent on both sides of the membrane became equal. In principle, of course, osmosis could not continue indefinitely if one liquid were pure solvent, instead of the more dilute solution. The more concentrated solution, no matter how much it became diluted, would always have a higher concentration of solute than the pure solvent. What sometimes stops osmosis, however, is the development of a back pressure.

Osmotic Pressure

Osmosis has all the appearance of there being something that creates a pressure that "forces" the solvent through the membrane. But there is no "some*thing,*" only unequal concentrations separated by an osmotic membrane plus nature's driving force to get everything as mixed up as possible. (And we can easily imagine that if we removed the membrane, the two solutions would intermingle and eventually become everywhere uniform in concentration — meaning fully mixed up.)

As illustrated in Figure 13.28c, as solvent moves into the more concentrated solution a column of solution rises. This creates a back pressure caused by the weight of the liquid in this column. We could actually prevent osmosis, and even force it to reverse, by putting an opposing pressure in place, as illustrated in Figure 13.28*d*. The exact back-pressure needed to prevent any osmotic flow *when one of the liquids is pure solvent* is called the osmotic pressure of the solution. Its symbol is Π.

The symbol Π is capital pi, the capital of π.

We are using the word *pressure* here in a somewhat different way than we have done before. Pure water alone, for example, does not "have" a special pressure called osmotic pressure. What it has is an escaping tendency that nature exploits as osmosis in a very special situation — when the water is separated from a solution by an osmotic membrane. Only because the escaping tendency of water from the solution is less than from the pure water can there be a net flow of water into the solution.

It might be easier to understand this with reference to Figure 13.29 where we have placed a beaker of water and a beaker of some solution in an enclosed space that is already saturated in water vapor. Water molecules escape from both beakers. They return to the beakers, too. At equilibrium over the beaker of water, the rates of escape and return would be equal, but there is now a complication. The rate of return of water molecules at the *solution* is greater than the rate of escape there. This is because solute molecules in the solution interfere with evaporation, as we studied in connection with the effect of a solute on vapor pressure. Thus, more water molecules enter the solution than leave, and this takes them from the vapor state. To replace them, and keep the vapor phase saturated in water, more water molecules evaporate from the pure water. The net effect is that water moves from the beaker of pure water into the vapor space and then down into the solution. In osmosis, instead of beakers separating the liquids, we have an osmotic membrane, but the same principles are at work.

It is not hard to imagine that the more concentrated is the solution in Figure 13.29, the more water will transfer to it. Thus the osmotic pressure of a solution is proportional to the relative concentrations of its molecules of solute and solvent. We have learned that *molality* serves us well to express such a ratio, but in any *dilute* aqueous solution the molality and molarity are virtually identical, and we will use *molarity* instead.

The osmotic pressure of a dilute aqueous solution is proportional to its molar concentration.

$$\Pi \propto M$$

Osmotic pressure is also proportional to the Kelvin temperature.

$$\Pi \propto T$$

Since Π is proportional both to M and T, it is proportional to the product of the two.

$$\Pi \propto MT$$

The proportionality constant turns out to be our old friend, the gas constant, R, so for a dilute solution we can write

$$\Pi = MRT \qquad (13.6)$$

Of course, $M = \text{mol/L} = n/V$, where $n = $ moles and $V = $ volume in liters. Substituting n/V for M in the equation for Π gives us an equation for osmotic pressure identical in form to the ideal gas law.

$$\Pi V = nRT$$

Osmotic pressures can be very high even in dilute solutions, as the next example shows.

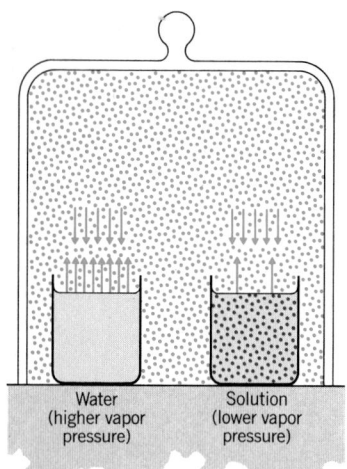

FIGURE 13.29 The vapor pressure of the pure solvent is greater than the vapor pressure of the solution made from this solvent. So liquid will transfer from the pure solvent to the solution.

Water (higher vapor pressure)   Solution (lower vapor pressure)

EXAMPLE 13.20   CALCULATING OSMOTIC PRESSURE

Problem: A very dilute solution, 0.0010 *M* sugar in water, is separated from pure water by an osmotic membrane. What osmotic pressure develops at 25 °C or 298 K? The gas constant $R = $ 0.0821 L atm/mol K.

Solution: All we have to do is apply Equation 13.6.

$$\Pi = MRT$$

$$= 0.0010 \; \frac{mol}{L} \times 0.0821 \; \frac{L \; atm}{mol \; K} \times 298 \; K$$

$$= 0.024 \; atm$$

The osmotic pressure is 0.024 atm. Let us see what this is in torr. Since 1 atm = 760 torr,

$$0.024 \; atm \times \frac{760 \; torr}{1 \; atm} = 18 \; torr$$

A pressure of 18 torr, from the previous example, corresponds to 18 mm Hg. Mercury has a density of 13.59 g/mL. Since the density of the sugar solution in this example is essentially the same as the density of water, we can convert mm Hg into the equivalent pressure expressed in inches of water. We will not show this calculation, but an osmotic pressure of 18 torr could support a column of water (or this dilute sugar solution) 10 in. high. A 0.1 M sugar solution, still relatively dilute, could support a column 100 times as high — about 1000 in., or about 83 ft! Thus osmosis is a part of the explanation for the rise of sap in trees.

When soil water is too concentrated in dissolved salts, the plant can't get the water it needs. In fact, the plant tends to *lose* water to the soil, as well as to evaporation, and it wilts.

PRACTICE EXERCISE 22

An instrument called an osmometer (Figure 13.30) measures osmotic pressure. And we can use a measured osmotic pressure to determine the molar concentration of a solution using Equation 13.6. What is the molar concentration of a solution with an osmotic pressure of 26.0 torr at 20 °C? Suppose 100 mL of this solution contains 0.122 g of solute. What is the molecular weight of the solute? Calculate the freezing point of this solution assuming that at this low concentration molality and molarity are equal. For substances with large molecular weights, which kind of measurement, osmotic pressure or freezing point depression, is more likely to provide more reliable values of molecular weights?

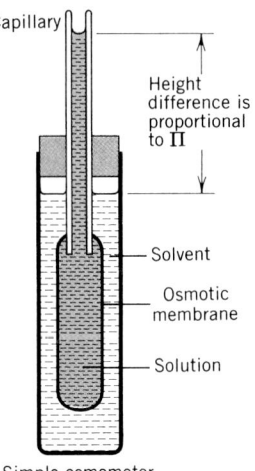

Capillary
Height difference is proportional to Π
Solvent
Osmotic membrane
Solution
Simple osmometer

FIGURE 13.30 Simple osmometer.

Dialysis

Although membranes in living systems are not true osmotic membranes, they do hold back large colloid-sized molecules and ions, such as those of proteins. When a colloidal dispersion is separated from water by a dialyzing membrane, therefore, the dispersion will receive water molecules from the other side faster than it lets them escape. This is just like osmosis, but it is now called dialysis because if other solutes consisting of small ions and molecules are also present, they will pass back and forth through the membrane. Only the presence of the larger particles evokes the osmosis-like behavior. In this situation, there is an associated *colloidal osmotic pressure*.

One practical use of dialysis is the separation of solutes in true solution from

Our kidneys are organs that cleanse the blood of many of the wastes of metabolism. If the kidneys quit working, these wastes will cause death.

The artificial kidney provides a way to take over a person's own kidneys should they develop problems. It works by dialysis, and the procedure is called **hemodialysis.** The blood is diverted from the body into a long coiled dialyzing membrane, often cellophane. A solution called the dialysate is constantly circulated around the outside of this tube.

The dialysate contains in solution all of the solutes that must remain in the blood at the concentrations they have in blood—NaCl, for example, plus traces of

other ionic and molecular compounds. The dialysate will not initially contain any of the wastes to be removed.

As the blood flows within the dialyzing membrane, its sodium ions move out into the dialysate. But they return at the same rate because the concentrations inside and out are the same. Impurities in the blood, such as urea, however, dialyze through the membrane faster than they can return. Thus, the blood experiences a loss of the impurity. Proteins, other large molecules, and blood cells cannot, of course, pass through the membrane, so they stay in the blood. Eventually, the blood reenters the body.

colloidal substances. Cellophane makes an acceptable dialyzing membrane. When, for example, an aqueous solution of sodium chloride that contains colloidally dispersed starch is placed into a cellophane bag and this is immersed in water, the ions of the salt dialyze out leaving just the starch in the bag. See Figure 13.31. Water moves in both directions, but more goes into the bag than leaves because of the colloidal osmotic pressure created by the starch.

If the liquid surrounding the cellophane bag also contained NaCl *at the same concentration* as the system inside, then sodium ions and chloride ions would dialyze through the cellophane at equal rates. No net removal of salt from the contents of the bag would occur. Hemodialysis is a special application of these principles that is used to cleanse the blood, as discussed in Special Topic 13.5.

FIGURE 13.31 Using dialysis to separate true solutes from colloidally dispersed particles. Solvent can come and go across the dialyzing membrane. So can the ions, but they are continuously swept away by the flowing water so the ions leave the bag much faster than they return. In time, only the colloidally dispersed starch will remain inside.

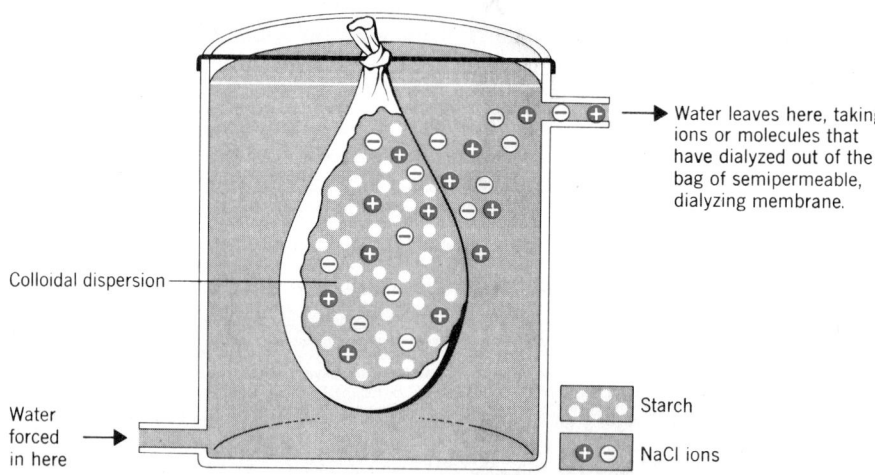

Water leaves here, taking ions or molecules that have dialyzed out of the bag of semipermeable, dialyzing membrane.

Colloidal dispersion

Water forced in here

Starch

NaCl ions

13.11 COLLIGATIVE PROPERTIES OF SOLUTIONS OF ELECTROLYTES

Mole for mole, electrolytes have two to three times the effect on colligative properties as molecular substances.

The k_f for water is 1.86 °C kg H₂O/mol solute.

A 1 *m* solution of NaCl, an ionic compound, freezes at −3.37 °C, instead of −1.86 °C, the expected freezing point of a 1 *m* molecular compound dissolved in water. This much greater depression of the freezing point by the salt — almost twice as much — is not hard to understand if we remember that colligative properties depend on the concentrations of *particles*, not their chemical identities. In 1 *m* NaCl, the solute has dissociated.

$$NaCl(s) \longrightarrow Na^+(aq) + Cl^-(aq)$$

So, 1 *m* NaCl actually has a 2 *m* concentration in dissolved solute particles. Theoretically, then, 1 *m* NaCl should freeze at $2 \times (-1.86 \,°C) = -3.72 \,°C$.

If we made up a solution of 1 *m* H_2SO_4, we would have to consider the following dissociation.

$$H_2SO_4(l) \longrightarrow 2H^+(aq) + SO_4^{2-}(aq)$$

Thus, 1 mol of H_2SO_4 can give a total of 3 mol of ions — 2 mol of H^+ and 1 mol of SO_4^{2-}. A 1 *m* solution of this acid has a calculated freezing point of $-5.58 \,°C$ $(3 \times -1.86 \,°C)$. It actually freezes at $-4.22 \,°C$.

EXAMPLE 13.21 PREDICTING A COLLIGATIVE PROPERTY OF AN ELECTROLYTE

Problem: Estimate the freezing point of 0.106 *m* aqueous $MgCl_2$, assuming that it ionizes completely.

Solution: When $MgCl_2$ dissolves in water it breaks up as follows.

$$MgCl_2(s) \longrightarrow Mg^{2+}(aq) + 2Cl^-(aq)$$

Since 1 mol of $MgCl_2$ gives 3 mol of ions, the true molality of what we have called 0.106 *m* $MgCl_2$ is three times as great.

$$\text{effective molality} = 3 \times (0.106) = 0.318 \frac{\text{mol-solute}}{\text{kg-solvent}}$$

Since, $\Delta t = k_f m$,

$$\Delta t = 1.86 \frac{°C \, \text{kg-solvent}}{\text{mol-solute}} \times 0.318 \frac{\text{mol-solute}}{\text{kg-solvent}}$$
$$= 0.519 \,°C$$

Thus the freezing point is depressed below 0.000 °C by 0.591 °C, so we calculate that this solution freezes at −0.591 °C. It actually freezes at −0.517 °C.

PRACTICE EXERCISE 23 Calculate the freezing point of aqueous 0.237 *m* LiCl on the assumption that it is 100% dissociated. Calculate what its freezing point would be if the percent dissociation were 0%.

Ions Interact with Each Other in Aqueous Solutions

Neither the 1 *m* NaCl nor the 1 *m* H_2SO_4 solution described earlier in this section freezes quite as low as calculated. In Example 13.21 we saw a similar discrepancy. What is wrong, of course, is not nature; it is our assumption of 100% separation into ions. The assumption is not quite correct. To be sure, the ions in all these aqueous solutions are hydrated, so the solvent helps to shield them from each other, but this shielding is not perfect. The ions are not really 100% free of each other so the solutes do not behave as if they were 100% dissociated. As solutions of electrolytes are made more and more dilute, however, the observed and calculated freezing points come closer and closer together. At greater dilutions, ions interact with each other less and less, so the solutes behave more and more as if they were 100% separated into their ions.

Jacobus Hendricus van't Hoff (1852–1911), a Dutch chemist, won the first Nobel Prize in chemistry (1901).

Chemists compare the degrees of dissociation of electrolytes at different dilutions by a quantity called the van't Hoff factor, i. It is the ratio of the observed freezing point depression to the value calculated on the assumption that the solute dissolves as un-ionized molecules.

$$i = \frac{(\Delta t_f)_{\text{measured}}}{(\Delta t_f)_{\text{calculated as a nonelectrolyte}}}$$

The van't Hoff factors for several electrolytes at different dilutions are given in Table 13.7. Notice that with decreasing concentration the experimental van't Hoff factors agree better and better with their corresponding hypothetical van't Hoff factors—calculated on the assumption of 100% dissociation. Thus, for NaCl, KCl, and $MgSO_4$, which break up into two ions, the hypothetical van't Hoff factor would be 2 if dissociation were 100%. For K_2SO_4, the theoretical value of i is 3.

The improvement in dissociation with greater dilution is not the same for all salts. In going from 0.1 to 0.001 m, the increase in dissociation of KCl, as measured by the change in i, is only about 7%. But for K_2SO_4, the improvement in dissociation for the same dilution is about 22%. The difference is in the anion, which has twice the charge in K_2SO_4 as in KCl. Potassium ions bear charges of $1+$; chloride ions, $1-$; and sulfate ions, $2-$. The attraction between $(1+)$ and $(2-)$ is stronger than between $(1+)$ and $(1-)$. Hence, letting $(1+)$ and $(2-)$ ions get farther apart by dilution has a greater effect on letting the ions act as independent species than giving $(1+)$ and $(1-)$ ions more room. When both cation and anion are doubly charged, the improvement in percent dissociation with dilution is even greater. Thus, there is an almost 50% increase in the value of i for $MgSO_4$ as we go from a 0.1 to a 0.001 m solution.

TABLE 13.7 Van't Hoff Factors versus Concentration

Salt	van't Hoff Factor			If 100% Dissociation Occurred
	Concentration (mol salt/kg water)			
	0.1	0.01	0.001	
NaCl	1.87	1.94	1.97	2.00
KCl	1.85	1.94	1.98	2.00
K_2SO_4	2.32	2.70	2.84	3.00
$MgSO_4$	1.21	1.53	1.82	2.00

The Percent Ionization of Weak Electrolytes Can Be Estimated from Freezing Point Depression Data

In 1.00 m aqueous acetic acid, the following equilibrium exists.

$$HC_2H_3O_2(aq) \rightleftharpoons H^+(aq) + C_2H_3O_2^-(aq)$$

The solution freezes at -1.90 °C, which is only a little lower that it would be expected to freeze (-1.86 °C) if no ionization occurred. So evidently some ionization, but not much, has happened. We can estimate the percent ionization by the following calculation.

We first use the data to calculate the apparent molality of the solution—the molality of all dissolved species, $HC_2H_3O_2 + C_2H_3O_2^- + H^+$.

$$\text{apparent molality} = \frac{\Delta t}{k_f} = \frac{1.90 \text{ °C}}{1.86 \dfrac{\text{°C kg-solvent}}{\text{mol-solute}}} = 1.02 \frac{\text{mol solute}}{\text{kg solvent}}$$

If there is 1.02 mol of all solute species in 1 kg of solvent we have 0.02 mol of solute species more than we started with, because we began with 1.00 mol of acetic acid. Since we get just one *extra* particle for each acetic acid molecule that ionizes, the extra 0.02 mol of particles must have come from the ionization of 0.02 mol of acetic acid. So the percent ionization is

$$\text{percent ionization} = \frac{\text{(mol of acid ionized)}}{\text{(mol of acid available)}} \times 100 = \frac{0.02}{1.00} \times 100 = 2\%$$

In other words, the percent ionization in 1.00 m acetic acid is estimated to be 2% by this procedure. (Electrical measurements offer better precision, and the percent ionization of acetic acid at this concentration is actually less than 1%.)

The Association of Solute Molecules Makes Them Less Effective in Producing Colligative Properties

Some solutes produce *weaker* colligative effects than their molal concentrations would lead us to predict. This is not often seen, but it points to another phenomenon that we should note — the phenomenon of association. This is the opposite of solute dissociation. Benzoic acid molecules in benzene, for example, associate as *dimers* held together by hydrogen bonds, indicated by the dotted lines in the following.

benzoic acid   benzoic acid dimer

C_6H_5— means

Because of this association, the depression of the freezing point of a 1 m solution of benzoic acid in benzene is almost one-half the calculated value. Because of formation of the dimer, the effective formula weight of benzoic acid is twice as much as normally calculated, which lessens the molal concentration of particles by half and this reduces the effect on the freezing point by half.

Di- signifies 2; so a dimer is the result of the combination of two single molecules.

SUMMARY

Mixtures.   Relative particle sizes and homogeneity distinguish among the three simplest kinds of mixtures — suspensions, colloidal dispersions, and solutions. Suspensions have particles with at least one dimension larger than 1000 nm, they are not homogeneous, and they settle out spontaneously in response to gravity. They can be separated by filtration.

Colloidal Dispersions.   In colloidal dispersions, the dispersed particles have one or more dimensions in the range of 1 to 1000 nm. They are considered to be homogeneous, but we're at a borderline. They separate more slowly than suspensions in response to gravity, but many can be kept indefinitely if the dispersed particles bear like charges or are coated with an emulsifying agent. Colloidal particles can scatter light (Tyndall effect), display the Brownian movement, cannot ordinarily be separated by filtration, and contribute to osmotic pressure.

Solutions.   In solutions, the dissolved particles have dimensions on the order of 0.05 to 0.25 nm (sometimes up to 1 nm). Solutions are fully homogeneous. They do not display the Tyndall effect or the Brownian movement, they cannot be separated by filtration, and they remain physically stable indefinitely.

When gaseous solutions form, nature's strong tendency toward randomness is a powerful driving force, and it encounters little if any resistance because of attractions between gas molecules.

When liquid solutions form by physical means, generally polar or ionic solutes dissolve well in polar solvents, like water. Nonpolar, molecular solutes dissolve well in nonpolar solvents. These observations are behind the "like dissolves like" rule. And they are explained in terms of relative forces of attraction between solute and solvent particles and how these forces affect nature's tendency toward greater randomness.

When the solvent is polar, the solute particles have to be either ions or polar molecules; only these can replace the attractive forces between solvent molecules with similar forces between solvent and solute particles. And only polar solvent molecules can break down attractive forces within the solute. Ion–dipole attractions and the solvation of dissolved species are major factors in forming solutions of ionic or polar solutes in something polar, like water. When both solute and solvent are nonpolar, nature's tendency toward more randomness works with little resistance from attractive forces.

The "like dissolves like" rule helps explain how detergents work. Detergent molecules have hydrophobic tails, which embed themselves in greasy films, and hydrophilic heads, which remain exposed to water. The grease breaks up into microdroplets with the hydrophilic parts of the detergents sticking from their surfaces, thus keeping the droplets from coalescing.

Heats of Solution.   The molar enthalpy of solution, the heat of solution, is the net of the energy cost to separate the molecules or ions of the solute from each other and the energy gain from the subsequent intermingling of these particles.

When a gas dissolves in a gas, the energy cost to separate the particles is virtually zero—they're already separated. And they remain separated in the solution, so the net heat of solution is very small if not zero.

Using a simple model with solid solutes dissolving in liquid solvents, the energy cost is the lattice energy, and the energy gain is the solvation energy (or, when water is the solvent, the hydration energy).

When liquids dissolve in liquids, if the energy cost equals the energy released, the resulting solution is an ideal solution. Usually, some net energy exchange occurs.

Gases dissolve in liquids exothermically, because there is usually some solvation energy (energy released), but there is no cost to separate gas molecules from each other.

Most solids dissolve into their saturated solutions when the temperature is raised, because for most solids this process is endothermic. When the process is exothermic, the effect of raising the temperature on a saturated solution is to drive some solute out of solution. Since all gases dissolve exothermically in liquid solvents, an increase in temperature expels gases from such solutions.

When a solid solute has a wide variation of solubility with temperature, it can be purified by fractional crystallization.

Pressure and Gas Solubility.   At pressures not too much different from atmospheric pressure, the solubility of a gas in a liquid is directly proportional to the partial pressure of the gas (at a given temperature)—Henry's law. A few gases, such as ammonia, carbon dioxide, and sulfur dioxide, are quite soluble in water because they actually react with water.

Concentration Expressions.   To study and use colligative properties, concentrations that express the ratios of solute particles (moles) to solvent particles (also moles, or their equivalent in fixed masses of solvent) are used. Mole fractions (or mole percents) and molalities answer this need. When a dilute solution is used, its molarity and molality are close enough to be used interchangeably, and molarity works more conveniently in studying osmotic pressures.

A mole fraction is the ratio of the moles of one component to the total number of moles of all components in the mixture. It can be expressed as a percent, too—the mole percent. When the mixture consists of gases, the mole fraction of any gas is the ratio of its partial pressure to the total pressure.

The molality of a solution, m, is the ratio of the moles of solute to the kilograms of *solvent* (not solution but solvent).

Percents by weight and percents by volume are concentrations used often for reagents when direct information about moles of solutes is unimportant.

Colligative Properties.   Colligative properties depend on the physical presence and concentrations of particles, not their chemical identities. The lowering of the vapor pressure, the depression of a freezing point, the elevation of a boiling point, and the osmotic pressure of a solution are proportional to the concentration of the solution.

According to Raoult's law, the vapor pressure of a solution of a nonvolatile (molecular) solute is the vapor pressure of the pure solvent times the mole fraction of this solvent. If the components of a solution are volatile liquids, then the Raoult's law calculations find the partial pressures of the vapors of the individual liquids. The sum of these partial pressures equals the total vapor pressure of the solution.

A solution that would obey Raoult's law exactly is an ideal solution. When the heat of solution is zero, the solution is ideal. Solutions of liquids that form exothermically usually display negative deviations from Raoult's law. Those that form endothermically have positive deviations.

In proportion to its molal concentration, a solute depresses the freezing point of a solution and elevates its boiling point, compared with the pure solvent. The proportionality constants are the freezing point depression constant and the boiling point elevation constant, and they differ from solvent to solvent. Freezing point and boiling point data from a solution made from known masses of both solute and solvent can be used to calculate the formula weights of the solutes.

When a solution is separated from pure solvent (or a less concentrated solution) by an osmotic membrane, a net flow of solvent occurs into the more concentrated solution. The backpressure that must be applied to prevent this osmosis is the solution's osmotic pressure, and it is proportional to the product of the Kelvin temperature and the molar concentration of the solution (when dilute). The proportionality constant is R, the ideal gas constant, so the equation relating these variables is $\Pi V = nRT$.

When the membrane separating two different solutions is a dialyzing membrane, it permits the passage not only of solvent molecules but also of small solute molecules or ions. Only very large molecules are denied passage, so only when they are present will a net flow—dialysis—occur across the membrane. But this can be used to remove small solute ions and molecules from colloidal dispersions.

Colligative Properties of Electrolytes.   Because an electrolyte releases more ions in solution than indicated by the molal (or molar) concentration, a solution of an electrolyte has more pronounced colligative properties than a solution of molecular compound at the same concentration. The dissociation of a strong electrolyte approaches 100% in very dilute solutions, particularly when both ions are singly charged. Weak electrolytes provide solutions more like those of nonelectrolytes.

The ratio of the value of a particular colligative property, like Δt for a freezing point depression, to the value expected under no dissociation is the van't Hoff factor, i. A solute whose formula unit breaks into two ions, like $NaCl$ or $MgSO_4$, would have a van't Hoff factor of 2 if it were 100% dissociated. Observed van't Hoff factors approach those corresponding to 100% dissociation only as the solutions are made more and more dilute.

REVIEW EXERCISES

Answers to questions whose numbers are printed in color are given in Appendix C.
Difficult questions are marked with asterisks.

Kinds of Mixtures

13.1   In terms of chemical composition, what are the three kinds of mixtures?

13.2   What feature about mixtures distinguishes them from elements and compounds?

13.3   What single feature most distinguishes suspensions, colloidal dispersions, and solutions from each other?

Colloidal Dispersions

13.4   List three factors that can contribute to the stability of a colloidal dispersion.

13.5   What general name is given to a colloidal dispersion of one liquid in another?

13.6   What is an emulsifying agent? Give an example.

13.7   What is the Tyndall effect and why do colloidal dispersions but not solutions show it?

13.8   What causes the Brownian movement in fluid colloidal dispersions?

13.9   What is a sol, and how can one often be stabilized?

13.10   What simple test could be used to tell if a clear, aqueous fluid contained colloidally dispersed particles?

Why Solutions Form

13.11   What does "like" refer to in the "like dissolves like" rule?

13.12   Iodine, I_2, is very slightly soluble in water but dissolves readily in carbon tetrachloride. Why won't it dissolve in water and what makes the formation of the carbon tetrachloride solution so easy?

13.13   Iodine dissolves far better in ethyl alcohol (to form "tincture of iodine") than in water. What does this tell us about molecules of ethyl alcohol, as compared with water molcules?

13.14   What is the "driving force" behind the spontaneous mixing of two gases and behind our never seeing them spontaneously become unmixed?

13.15   Acetone and water are *miscible* in all proportions. What does this mean?

13.16   Methyl alcohol, CH_3—O—H, and water are miscible in all proportions. Explain how the O—H unit in methyl alcohol contributes to this.

13.17   Gasoline and water are immiscible. What does this mean? And explain why, in terms of structural features of their molecules and forces of attraction between them.

13.18   Explain how ion–dipole forces bring KCl into solution in water.

13.19   What does "hydration" refer to, and what kinds of substances, electrolytes or nonelectrolytes, does it tend to help most to become soluble in water?

13.20   In terms of forces of attraction, explain why KCl does not dissolve in benzene.

13.21   What kinds of substances are more soluble in water, hydrophilic compounds or hydrophobic compounds?

13.22   Soap, as a solute in water, spontaneously forms micelles. What are they and what is the driving force for their formation?

13.23   How does a soap help dissolve an oily material that water alone will not dissolve?

13.24   Explain why KCl will not dissolve in CCl_4.

Heats of Solution

13.25   How do we define *heat of solution*?

13.26   No solution actually forms by the two-step process, one ending before the other starts, that we used as a model for the formation of a solution of some ionic compound in water. When we are interested in the molar enthalpy of solution, however, we can get away with this model. Explain. And what makes these particular two steps attractive for the purpose?

13.27   The ΔH_{soln} for a soluble compound is, say, $+26$ kJ/mol, and a relatively saturated solution is prepared in an insulated container (e.g., a coffee-cup calorimeter). How will the temperature of the system change — increase or decrease?

13.28   Referring to Review Exercise 13.27, which value for this compound would numerically be larger, its lattice energy or its hydration energy?

13.29   Consider the formation of aqueous potassium chloride.
(a) Write the thermochemical equations for the two steps in the formation of a solution of KCl in water. (We studied thermochemical equations in Section 5.5.) The lattice energy of KCl is -690 kJ/mol and the hydration energy is -686 kJ/mol.
(b) Write the sum of these two equations in the form of a thermochemical equation (showing also the net ΔH).

13.30   Referring to Table 13.4, suppose the lattice energy of KCl were just 2% greater than the value given. What then would be the calculated ΔH_{soln} (in kJ/mol)? How would this new value compare with that given in the table? (By what percent has the value changed?) [The point, of course, is that small percentage errors in two large numbers, such as lattice energy and hydration energy, can cause large percentage changes in the *differences* between them.]

13.31   The energies of the hydrations of individual ions add up to the hydration energy of a salt. Which would be expected to have the larger hydration energy, Al^{3+} or Li^+? Why?

13.32   Suggest a reason why the value of ΔH_{soln} for a gas such as CO_2 is negative.

13.33   The value of ΔH_{soln} for the formation of an acetone–water solution is negative. Explain this in general terms that discuss intermolecular forces of attraction.

13.34   The value of ΔH_{soln} for the formation of an ethyl alcohol–hexane solution is positive. Explain this in general terms that involve intermolecular forces of attraction.

13.35   How can Le Châtelier's principle be applied to explain how the addition of heat causes undissolved solute to go into solution (in the cases of most saturated solutions)?

Temperature and Solubility

13.36   If a saturated solution of NH_4NO_3 at 70 °C is cooled to 10 °C, how many grams of solute will separate if the quantity of the solvent is 100 g?

13.37   A hot concentrated solution in which equal masses of NaI and NaBr are dissolved is slowly cooled from 50 °C. Which salt precipitates first? (Use data in Figure 13.19.)

13.38   Compound A is soluble in both ethyl alcohol and water. In water, 35.2 g of A dissolves in 100 g of solvent at the boiling point, but at 5 °C, only 1.2 g of A dissolves. In ethyl alcohol, 36.8 g of A dissolves in 100 g of solvent at the boiling point, and at 5 °C, 15.2 g of A dissolves. You are given 30 g of A contaminated with 1 g of another compound, B. B is very soluble in both hot water and hot ethyl alcohol. At 5 °C, the solubility of B in water is 1.2 g/100 g H_2O and in ethyl alcohol it has a solubility at 5 °C of 3.6 g/100 g alcohol.
(a) Which solvent should be picked to purify A by fractional crystallization? Explain.
(b) How would you proceed in the lab to carry out this operation? (It should not be necessary to do any calculations to answer this question.)

Pressure and Solubility

13.39   What is Henry's law?

13.40   The solubility of methane, the chief component of Bunsen burner gas, in water at 20 °C and 1.0 atm pressure is 0.025 g/L. What will be its solubility at 1.4 atm and 20 °C?

*13.41   At 740 torr and 20 °C, nitrogen has a solubility in water of 0.018 g/L. At 620 torr and 20 °C, its solubility is 0.015 g/L. Does nitrogen obey the gas pressure—solubility law? (Do the necessary calculations to support your answer.)

13.42   What makes ammonia so much more soluble in water than nitrogen?

13.43   What makes carbon dioxide more soluble in water than oxygen?

13.44   Sulfur dioxide is an air pollutant wherever sulfur-containing fuels have been used. Rainfall that washes sulfur dioxide out of the air and onto the land is called acid rain. How does sulfur dioxide contribute to this?

13.45   Years ago it was common to call an aqueous solution of ammonia "ammonium hydroxide." (Some chemical supply houses still use this name.) Suggest a reason for this. Offer a reason why this name is considered a poor choice.

Expressions of Concentration

13.46   A solution is prepared by mixing 60 g of toluene (C_7H_8) and 60 g of chlorobenzene (C_6H_5Cl). What is the mole fraction and the mole percent of each component?

13.47   An antifreeze solution for an auto radiator is prepared by mixing 2.50 L of methyl alcohol (CH_3OH, density 0.780 g/mL) with 2.50 L of water (density 1.00 g/mL) What is the mole fraction and mole percent of each component?

13.48   A solution consisted of 33 mole percent methyl alcohol and 25 mole percent ethyl alcohol in water. What was the concentration of the water in mole percent?

13.49   What are the mole percents of the components of the air inside the lungs when they have the following partial pressures? For N_2, 570 torr; O_2, 103 torr; CO_2, 40 torr; and water vapor, 47 torr. (Calculate to two significant figures.)

13.50   Assuming no other components are present, (a) what are the mole percents of oxygen and nitrogen in the air at the top of Mt. Everest (elevation 8.8 km) on a day when $P_{N_2} = 197$ torr and $P_{O_2} = 53$ torr. (b) In air at sea level and 1 atm pressure, $P_{N_2} = 600$ torr and $P_{O_2} = 160$ torr. Again assuming no other components, what are the mole fractions of oxygen and nitrogen in this air? (c) Comparing the answers calculated for parts a and b, what has to be the reason why it is hard to breathe without supplemental oxygen at high altitudes? (Do all calculations to two significant figures.)

13.51   A solution is made by dissolving 18.0 g of glucose (formula weight 180) in 1.00 kg of water.
(a) What is its molal concentration?
(b) What is the mole fraction of glucose?
(c) What is the mole fraction of water?

13.52   If you dissolved 10.0 g NaCl in 1.00 kg of water, what would be its molal concentration? The volume of this solution is virtually identical to the original volume of the 1.00 kg of water. Therefore, what is the molar concentration of this solution? What would have to be true about any solvent for one of its dilute solutions to have essentially the same molar and molal concentrations?

13.53   A solution of methyl alcohol, CH_3OH, in water has a concentration of 1.50 m. Calculate the following.
(a) The mole fractions of methyl alcohol and water.
(b) The percent (w/w) of methyl alcohol.

13.54   An aqueous solution of Na_2SO_4 has a concentration of 0.370 m. Its density is 1.0436 g/mL. Calculate the following.
(a) The mole fractions of Na_2SO_4 and H_2O.
(b) The percent (w/w) concentration of Na_2SO_4.
(c) The molar concentration of Na_2SO_4.

13.55   In an aqueous solution of sulfuric acid, the concentration is 2.40 mole percent of acid. The density of the solution is

1.079 g/mL. Calculate the following.
(a) The molal concentration of H_2SO_4.
(b) The percent (w/w) of the acid.
(c) The molarity of the solution.

13.56 In an aqueous solution of KNO_3 (formula weight 101.1), the concentration is 3.28 mole percent of the salt. The density of the solution is 1.1039 g/mL. Calculate the following.
(a) The molal concentration of KNO_3.
(b) The percent (w/w) of KNO_3.
(c) The molarity of the solution.

13.57 A solution of NaCl in water has a concentration of 22.0% (w/w). Calculate the following.
(a) The molality of this solution.
(b) The mole fractions of NaCl and water.

13.58 A solution of HNO_3 in water has a concentration of 10.00% (w/w). Its density is 1.0543 g/mL. Calculate the following.
(a) The molality of this solution.
(b) The mole percent of HNO_3.
(c) The molarity of the HNO_3.

13.59 A solution of sodium nitrate has a concentration of 0.733 M. Its density is 1.0392 g/mL. Calculate the following.
(a) The molality of the solution.
(b) The mole percent of $NaNO_3$ in the solution.
(c) The percent (w/w) concentration of $NaNO_3$.

13.60 A solution of hydrochloric acid has a concentration of 0.275 M and a density of 1.0031 g/mL. Calculate the following.
(a) The molality of the solution? (How does it compare numerically with the given molarity?)
(b) The percent (w/w) of HCl.
(c) The mole percent of HCl in the solution.

13.61 A solution of ethyl alcohol, C_2H_5OH, in water has a concentration of 6.211 mol/L. At 20 °C, its density is 0.9539 g/mL. Calculate the following.
(a) The molality of the solution.
(b) The percent (w/w) concentration of the alcohol.
(c) The mole percent of the alcohol.
(d) The density of ethyl alcohol is 0.7893 g/mL and the density of water is 0.99823 g/mL at 20 °C. Calculate the percent (v/v) of ethyl alcohol in this solution.

13.62 A solution of NH_3 in water is at a concentration of 1.00% (w/w). Its density is 0.9938 g/mL. Calculate the following.
(a) The molarity of the solution.
(b) The molality of the solution.
(c) The mole percent of NH_3.

Dilutions Revisited

13.63 In Section 4.7, page 134, an equation (4.1) was developed that we can use to carry out the calculations for finding the volume of a concentrated solution of some molarity that, when diluted, would give a certain volume of a dilute solution. Very similar equations can be derived to use when concentrations are given in percent—either weight/weight or volume/volume. Derive these equations for the following. They will be used in the four review exercises that follow this one.
(a) Weight/weight percent concentrations.
(b) Volume/volume percent concentrations.

13.64 A stock solution is 10.0% (w/w) H_2SO_4. How would you prepare 250 g of 1.50% (w/w) H_2SO_4?

13.65 Aqueous silver nitrate is available as 1.00% (w/w) $AgNO_3$. How would you prepare 50.0 g of a 0.200% (w/w) $AgNO_3$ solution?

13.66 An antifreeze solution that gives protection to −12 °C is 25% (v/v) ethylene glycol in water. How could this solution be diluted to give 8 quarts of a 15% (v/v) solution (which would give protection to −5 °C)?

13.67 A student chemistry club at an Austrian university decided to make its own peppermint liquor for a party (probably to celebrate finishing the study of concentration expressions). The club added a few drops of peppermint flavor and some green food dye to properly diluted ethyl alcohol. The alcohol was available as 95.5% (v/v) ethyl alcohol in water. To prepare 2.00 L of 100 proof alcohol—50.0% (v/v)—how should one proceed?

Lowering of the Vapor Pressure

13.68 What specific fact about a physical property of a solution must be true to call it a colligative property?

13.69 What colligative properties of solutions were studied in this chapter?

13.70 What is Raoult's law for liquid solutions with nonvolatile solutes?

13.71 What causes a solution with a nonvolatile solute to have a lower vapor pressure than the solvent at the same temperature?

13.72 How is Raoult's law used when we want to calculate the total vapor pressure of a liquid solution made up of two *volatile* liquids?

13.73 What kinds of data would have to be obtained to find out if a binary solution of two miscible liquids is almost exactly an ideal solution?

13.74 At 25 °C, the vapor pressure of water is 23.8 torr. Assuming that the solution described in Review Exercise 13.51 is ideal, calculate its vapor pressure.

13.75 The vapor pressure of water at 20 °C is 17.5 torr. A 20% (w/w) solution of ethylene glycol in water is prepared. Assuming that the solute is nonvolatile, do a calculation to estimate the vapor pressure of the solution.

13.76 Pentane and heptane are two hydrocarbon liquids present in gasoline. At 20 °C, the vapor pressure of pentane is 420 torr and the vapor pressure of heptane is 36 torr. What is the total vapor pressure of a solution prepared with a concentration of 30.0 mole percent pentane in heptane? (Calculate to two significant figures.)

13.77 Benzene and toluene help give lead-free gasoline good engine performance. At 40 °C, the vapor pressure of benzene is 180 torr, and that of toluene is 60 torr. To prepare a solution of these that will have at 40 °C a total vapor pressure of 96 torr requires what mole percent concentrations of each? (Calculate to two significant figures.)

13.78 At 21.0 °C, a solution of 19.35 g of a nonvolatile, nonpolar compound in 34.88 g of ethyl bromide, C_2H_5Br, had a vapor pressure of 336.0 torr. The vapor pressure of pure ethyl bromide at this temperature is 400.0 torr. Calculate the following.
(a) The mole fractions of solute and solvent.
(b) The moles of solute present.
(c) The formula weight of the solute.

13.79 Aqueous solutions of ethyl alcohol show a *positive deviation* from the expected relationship between the vapor pressure of an ideal binary solution and its mole fraction composition. What does this mean? Devise a figure as part of your answer, and explain the positive deviation in terms of intermolecular attractions.

Freezing Point Depression and Boiling Point Elevation

13.80 When an aqueous solution of sodium chloride starts to freeze, why don't the ice crystals contain ions of the salt?

13.81 Explain why a nonvolatile solute dissolved in water makes the system have (a) a higher boiling point than water, and (b) a lower freezing point than water.

13.82 Ethylene glycol, $C_2H_6O_2$, is used in some antifreeze mixtures. Protection against freezing to as low as −40 °F is sought.
(a) How many moles of solute are needed per kilogram of water to ensure this protection?
(b) The density of ethylene glycol is 1.11 g/mL. To how many milliliters of solute does your answer to part a correspond?
(c) Look up and use conversion factors to calculate how many quarts of ethylene glycol should be mixed with each quart of water to get the desired protection.

13.83 Calculate what would be the boiling point of 2.00 m sugar in water. What would be its freezing point? (It's largely the sugar in ice cream that makes it hard to keep ice cream frozen hard.)

13.84 Glycerol, $C_3H_8O_3$ (formula weight 92), is essentially a nonvolatile liquid that is very soluble in water. A solution is made by dissolving 46.0 g of glycerol in 250 g of water. Calculate the following.
(a) The boiling point of the solution at 1 atm.
(b) Its freezing point.
(c) Its vapor pressure at 25 °C. (At this temperature, the vapor pressure of water is 23.8 torr.)

13.85 A solution of 12.00 g of an unknown molecular compound dissolved in 200.0 g of benzene froze at 3.45 °C. Calculate the formula weight of the unknown.

13.86 A solution of 14 g of an unknown (molecular) compound in 1.0 kg of benzene boiled at 81.7 °C. Calculate the formula weight of the unknown.

13.87 The action of hot concentrated nitric acid (dissolved in sulfuric acid) on benzene produces nitrobenzene, $C_6H_5NO_2$, as the chief product. A by-product is often obtained, which consists of 42.86% C, 2.40% H, and 16.67% N—all weight/weight percents, of course. The remainder is oxygen.
(a) Calculate the empirical formula of this by-product.
(b) The boiling point of a solution of 5.5 g of this compound in 45 g of benzene was 1.84 °C higher than that of benzene. Calculate a formula weight based on these data, and determine the molecular formula of the compound.

*13.88 An experiment calls for the use of the dichromate ion, $Cr_2O_7^{2-}$, as an oxidizing agent for isopropyl alcohol, C_3H_8O. The oxidation is done in acid, and the chief product is acetone, C_3H_6O.
(a) Write the balanced, net ionic equation for this oxidation.
(b) The dichromate ion is available when needed only as sodium dichromate dihydrate. In theory, how many grams of this substance are needed to oxidize 25.0 g of isopropyl alcohol according to the balanced equation?
(c) The amount of acetone actually isolated was 16.4 g. Calculate the percentage yield.
(d) The reaction produces a volatile by-product. When a sample with a mass of 8.654 mg was burned in oxygen, it was converted into 22.368 mg of carbon dioxide and 10.655 mg of water, the sole products. (Assume that any unaccounted-for element is oxygen.) Calculate the percentage composition of this by-product and determine its empirical formula.
(e) A solution prepared by dissolving 1.338 g of the by-product in 115.0 g of benzene had a freezing point of 4.87 °C. Calculate the formula weight of the by-product and write its molecular formula.

Dialysis and Osmosis

13.89 What is meant by *osmosis?*

13.90 What is the difference between osmosis and dialysis?

13.91 What is the difference between a dialyzing membrane and an osmotic membrane?

13.92 In osmosis, why *must* the net migration of solvent be from the side of the membrane less concentrated in solute to the side more concentrated in solute?

13.93 Two glucose solutions of unequal molarity are separated by an osmotic membrane. Which solution will *lose* water, the one with the higher or the lower molarity?

13.94 Two glucose solutions of unequal molarity are separated by a dialyzing membrane. Given the various driving forces in nature, what change or changes should be expected?

13.95 Which aqueous solution has the higher osmotic pressure, 10% (w/w) glucose, $C_6H_{12}O_6$, or 10% (w/w) sucrose, $C_{12}H_{22}O_{11}$? (Both are molecular compounds.)

13.96 What is the osmotic pressure in torr of a 0.010 M aqueous solution of a molecular compound at 25 °C?

13.97 (a) Show that the following equation is true.

$$\text{molar mass} = \frac{(\text{mass in g})RT}{\Pi V}$$

(b) An aqueous solution of a compound with a very high formula weight was prepared in a concentration of 2.0 g/L at 298 K. Its osmotic pressure was 0.021 torr. Calculate the formula weight of the compound.

13.98 When a solid is *associated* in a solution, what does this mean? And what difference does it make to expected colligative properties?

Colligative Properties of Electrolytes

13.99 Why are colligative properties of solutions of ionic compounds usually more pronounced that those of solutions of the same molalities of molecular compounds?

13.100 The vapor pressure of water at 20 °C is 17.5 torr. If the solute in a solution made from 10.0 g of NaCl in 1.00 kg of water is 100% dissociated (and is an ideal solution), what is the vapor pressure of this solution at 20 °C?

13.101 Which aqueous solution, if either, has the lower freezing point, 10% (w/w) NaCl or 10% (w/w) NaI?

13.102 Which aqueous solution, if either, has the higher boiling point, 0.50 m NaI or 0.50 m Na_2CO_3?

13.103 Which aqueous solution, if either, has the higher osmotic pressure, 1.5% (w/w) glucose ($C_6H_{12}O_6$, formula weight 180) or 1.5% (w/w) NaCl?

13.104 What is the van't Hoff factor? What is its calculated value for all molecular solutes? For all ionic solutes of the Na_2SO_4 type?

13.105 The van't Hoff factor for the solute in 0.100 m $NiSO_4$ is 1.19. What would this factor be if the solution behaved as if it were 100% dissociated?

13.106 The van't Hoff factor for the solute in 0.118 m LiCl is 1.89.
(a) Calculate the freezing point of this solution.
(b) This solution is roughly as dilute as the solution described in Review Exercise 13.105. Explain why the van't Hoff factor for LiCl is so much greater.

13.107 Consider an aqueous 1.00 m solution of Na_3PO_4.
(a) Calculate the boiling point of this solution on the assumption that it does not ionize at all in solution.
(b) Do the same calculation assuming that its van't Hoff factor reflects 100% dissociation into ions.
(c) The 1.00 m solution boils at 101.183 °C at 1 atm. Calculate the van't Hoff factor for the solute in this solution.

13.108 A 1.00 m aqueous solution of HF freezes at −1.91 °C. According to these data, what is the percent ionization of HF in this solution?

13.109 An aqueous solution of a weak electrolyte, HX, with a concentration of 0.125 m has a freezing point of −0.261 °C. What is the percent ionization of this compound (to two significant figures)?

TEST OF FACTS AND CONCEPTS: CHAPTERS 9–13

We pause again to provide an opportunity for you to test your grasp of concepts, your knowledge of scientific terms, and your skills at solving chemistry problems. If you can answer the following questions, you are in a good position to go on to the next chapters.

1. A 15.5-L sample of neon at 25.0 °C and a pressure of 748 torr is kept at 25.0 °C as it is allowed to expand to a final volume of 25.4 L. What is the final pressure?

2. An 8.95-L sample of nitrogen at 25.0 °C and 1.00 atm is compressed to a volume of 0.895 L and a pressure of 5.56 atm. What is its final temperature?

3. A mixture of propane and air explodes at 466 °C. If a 20.0-L sample of such a mixture originally at 25.0 °C and 1.00 atm is to be detonated by the heat that is generated by compression alone, what must be the final pressure if the final volume is to be 1.00 L?

4. A sample of oxygen-enriched air with a volume of 12.5 L at 25.0 °C and 1.00 atm consists of 45.0% (v/v) oxygen and 55.0% (v/v) nitrogen. What are the partial pressures of oxygen and nitrogen (in torr) in this sample after it has been warmed to a temperature of 37.0 °C and is still at a final volume of 12.5 L?

5. What is the formula weight of a gaseous element if 6.45 g occupies 1.92 L at 745 torr and 25.0 °C? Which element is it?

6. What is the formula weight of a gaseous element if at room temperature it effuses through a pinhole 2.16 times as rapidly as xenon? Which element is it?

7. Briefly and qualitatively explain how the model of an ideal gas, as described by the kinetic theory of gases, explains the following.
   (a) Boyle's law.   (d) The meaning of gas temperature.
   (b) Graham's law.   (e) Pressure-temperature law.
   (c) Charles' law.   (f) Absolute zero.

8. Which has a higher value of the van der Waals a constant, a gas whose molecules are polar or one whose molecules are nonpolar? Explain.

9. What is van der Waals b constant used to correct, and in what way is this correction accomplished?

10. Potassium hypobromite, KOBr, converts ammonia to nitrogen by the following reaction.

$$3KOBr + 2NH_3 \longrightarrow N_2 + 3KBr + 3H_2O$$

To prepare 475 mL of dry N_2, when measured at 24.0 °C and 738 torr, what is the minimum number of grams of KOBr required?

11. Hydrogen peroxide, H_2O_2, is decomposed by potassium permanganate according to the following reaction.

$$5H_2O_2 + 2KMnO_4 + 3H_2SO_4 \longrightarrow$$
$$5O_2 + 2MnSO_4 + K_2SO_4 + 8H_2O$$

What is the minimum number of milliliters of 0.125 M KMnO$_4$ required to prepare 375 mL of dry O_2 when the gas volume is measured at 22.0 °C and 738 torr?

12. One way to make chlorine is to let manganese dioxide, MnO_2, react with hydrochloric acid according to the following equation.

$$4HCl + MnO_2 \longrightarrow Cl_2 + MnCl_2 + 2H_2O$$

What is the minimum volume (in mL) of 6.44 M HCl needed to prepare 525 mL of dry chlorine when the gas is obtained at 24.0 °C and 742 torr?

13. A sample of 248 mL of wet nitrogen gas was collected over water at a gas pressure of 736 torr and a temperature of 21.0 °C. (The vapor pressure of water at 21.0 °C is 18.7 torr.) The nitrogen was produced by the reaction of sulfamic acid, HNH_2SO_3, with 425 mL of a solution of sodium nitrite according to the following equation.

$$NaNO_2 + HNH_2SO_3 \longrightarrow N_2 + NaHSO_4 + H_2O$$

Calculate what must have been the molar concentration of the sodium nitrite.

14. How many grams of 4.00% (w/w) solution of KOH in water are needed to neutralize completely the acid in 10.0 mL 0.256 M H_2SO_4?

15. Calculate the molar concentration of 15.00% (w/w) Na_2CO_3 solution at 20.0 °C where its density is 1.160 g/mL.

16. The solubility of *pure* oxygen in water in 20.0 °C and 760 torr is 4.30×10^{-2} g O_2/L H_2O. When air is in contact with water and the *air* pressure is 585 torr at 20 °C, how many grams of oxygen of the air dissolve in 1.00 L of water? The average concentration of oxygen in the air is 21.1% (vol/vol).

17. What two factors are principally responsible for the differences in the behavior of gases and liquids?

18. If the ideal gas law worked well for all substances at all temperatures and pressures, what volume would 1.00 mol of water vapor occupy at 25 °C and 1.00 atm? What volume does 1.00 mol of water actually occupy under these conditions?

19. Which properties of liquids and solids are controlled chiefly by the closeness of the packing of molecules in these states? Which properties are determined chiefly by the strengths of the intermolecular attractions?

20. Consider the molecule $POCl_3$, in which phosphorus is the central atom and is bonded to an oxygen atom and three chlorine atoms.
    (a) Draw the Lewis structure of $POCl_3$ and predict its geometry.

(b) Is the molecule polar or nonpolar? Explain.

(c) What kinds of attractive forces would be present between $POCl_3$ molecules in the liquid?

21. What kinds of attractive forces, including chemical bonds, would be present between the particles in
(a) $H_2O(\ell)$ (c) $CH_3OH(\ell)$ (e) $NaCl(s)$
(b) $CCl_4(\ell)$ (d) $BrCl(\ell)$ (f) $Na_2SO_4(s)$

22. What is a change of state? What terms are used to describe the energy changes associated with the change (a) solid → liquid, (b) solid → gas, and (c) liquid → gas?

23. What is a dynamic equilibrium? In terms of Le Châtelier's principle and the "equation"

$$\text{liquid} + \text{heat} \rightleftharpoons \text{vapor}$$

explain why raising the temperature of a liquid increases the liquid's equilibrium vapor pressure.

24. Can a solid have a vapor pressure? How would the vapor pressure of a solid vary with temperature?

25. Based on what you've learned in these chapters, explain
(a) Why a breeze cools you when you're perspiring.
(b) Why droplets of water form on the outside of a glass of cold soda on a warm, humid day.
(c) Why you feel more uncomfortable on a warm, humid day than on a warm, dry day.
(d) The origin of the energy in a violent thunderstorm.
(e) Why clouds form as warm, moist air flows over a mountain range.

26. How do the magnitudes of ΔH_{fusion}, $\Delta H_{\text{vaporization}}$, and $\Delta H_{\text{sublimation}}$ compare for a given substance?

27. Make sketches of (a) a face-centered cubic unit cell, (b) a body-centered cubic unit cell, and (c) a simple cubic unit cell. Which type of unit cell does NaCl have?

28. Trimethylamine $(CH_3)_3N$, is a substance responsible in part for the smell of fish. It has a boiling point of 3.5 °C and a molecular weight of 59.1. Dimethylamine, $(CH_3)_2NH$, has a similar odor and boils at a slightly higher temperature, 7 °C, even though it has a somewhat lower molecular weight (45.1). How can this be explained in terms of the kinds of attractive forces between their molecules?

29. Methanol, CH_3OH, commonly known as wood alcohol, has a boiling point of 64.7 °C. Methylamine, a fishy-smelling chemical found in herring brine, has a boiling point of −6.3 °C. Ethane, a hydrocarbon present in petroleum, has a boiling point of −88 °C.

```
      H                 H  H              H  H
      |                 |  |              |  |
  H—C—O—H          H—C—N—H          H—C—C—H
      |                 |                 |  |
      H                 N                 H  H
   methanol        methylamine          ethane
 b.p., 64.7 °C    b.p., −6.3 °C      b.p., −88 °C
```

Each has nearly the same molecular weight. Account for the large differences in their boiling points in terms of the attractive forces between their molecules.

30. Tin tetraiodide (stannic iodide) has the formula SnI_4. It forms soft, yellow to reddish crystals that melt at about 143 °C.

The crystals sublime at about 180 °C. What type of crystal does SnI_4 form? What kind of bonding occurs in SnI_4?

31. A certain compound has the formula MCl_2. Crystals of the compound melt at 772 °C and give a liquid that is electrically conducting. What kind of crystal does this compound form?

32. What general properties are expected of covalent crystals?

33. Silicon dioxide, SiO_2, forms very hard crystals that melt at 1610 °C to yield a liquid that does not conduct electricity. What crystal type does SiO_2 form?

34. Sketch the phase diagram for a substance that has a triple point at 25 °C and 100 torr, a normal boiling point of 150 °C, and a melting point at 1 atm of 27 °C. Is the solid more dense or less dense than the liquid? Where on the curve would the critical temperature and critical pressure be? What phase would exist at 30 °C and 10.0 torr?

35. What is the difference between a strong electrolyte and a weak electrolyte? Formic acid, $HCHO_2$, is a weak acid. Write a chemical equation showing its reaction with water.

36. Write an equation showing the reaction of water with itself to form ions.

37. Methylamine, CH_3NH_2, is a weak base. Write a chemical equation showing its reaction with water.

38. Write molecular, ionic, and net ionic equations for the reaction that occurs when a solution containing hydrochloric acid is added to a solution of the weak base, methylamine (CH_3NH_2).

39. Which is the stronger acid, H_3PO_3 or H_3PO_4? How can one tell without a table of weak and strong acids?

40. Which is the stronger acid, H_2S or H_2Te?

41. X, Y, and Z are all nonmetallic elements in the same period of the periodic table where they occur, left to right, in the order given. Which would be a stronger binary acid than the binary acid of Y, the binary acid of X or the binary acid of Z? Explain.

42. Which of the following salts would be classified as soluble?
(a) $Ca_3(PO_4)_2$ (f) $Au(ClO_4)_3$ (k) $ZnSO_4$
(b) $Ni(OH)_2$ (g) $Cu(C_2H_3O_2)_2$ (l) Na_2S
(c) $(NH_4)_2HPO_4$ (h) $AgBr$ (m) $CoCO_3$
(d) $SnCl_2$ (i) KOH (n) $BaSO_3$
(e) $Sr(NO_3)_2$ (j) Hg_2Cl_2 (o) MnS

43. What are the *two* criteria that must be met in order for an ionic equation to be balanced correctly?

44. Write molecular, ionic, and net ionic equations for any reactions that would occur between the following pairs of compounds. If no reaction, write "N.R."
(a) $CuCl_2(aq)$ and $(NH_4)_2CO_3(aq)$
(b) $HCl(aq)$ and $MgCO_3(s)$
(c) $ZnCl_2(aq)$ and $AgC_2H_3O_2(aq)$
(d) $HClO_4(aq)$ and $NaCHO_2(aq)$
(e) $MnO(s)$ and $H_2SO_4(aq)$
(f) $FeS(s)$ and $HCl(aq)$

45. How many milliliters of dry CO_2, measured at STP, could be evolved in the reaction between 20.0 mL of 0.100 M $NaHCO_3$ and 30.0 mL of 0.0800 M HCl?

46. How many milliliters of 0.200 M $BaCl_2$ must be added to

27.0 mL of 0.600 M Na_2SO_4 to give complete reaction between their solutes?

47. What weight of $Mg(OH)_2$ will be formed when 30.0 mL of 0.200 M $MgCl_2$ solution is mixed with 25.0 mL of 0.420 M NaOH solution? What will be the molar concentrations of the ions remaining in solution?

48. How many grams of CO_2 must be dissolved in 300 mL of 0.100 M Na_2CO_3 solution to change the solute into $NaHCO_3$?

49. A certain toilet cleaner uses $NaHSO_4$ as its active ingredient. In an analysis, 0.500 g of the cleaner was dissolved in 30.0 mL of distilled water and required 24.60 mL of 0.105 M NaOH for complete neutralization in a titration. The net ionic reaction was

$$HSO_4^- + OH^- \longrightarrow H_2O + SO_4^{2-}$$

What was the percentage by weight of $NaHSO_4$ in the cleaner?

50. A volume of 28.50 mL of a freshly prepared solution of KOH was required to titrate 50.00 mL of 0.0922 M HCl solution. What was the molarity of the KOH solution?

51. What is the equivalent weight of H_3AsO_4 if it is
   (a) Neutralized to give $H_2AsO_4^-$?
   (b) Neutralized to give $HAsO_4^{2-}$?
   (c) Neutralized to give AsO_4^{3-}?

52. How many equivalents are in 0.200 mol of
   (a) H_3PO_4, if it is neutralized to give HPO_4^{2-}?
   (b) H_2SO_4, if it is completely neutralized?

53. How many equivalents of HIO_3 are in 25.0 mL of 0.10 M HIO_3 if
   (a) HIO_3 is neutralized to give IO_3^-?
   (b) HIO_3 is reduced to give I_2?
   (c) HIO_3 is reduced to give I^-?

54. How many milliliters of Cl_2 gas, measured at 25 °C and 740 torr, are needed to react with 10.0 mL of 0.10 M NaI if the I^- is oxidized to IO_3^- and Cl_2 is reduced to Cl^-?

55. How many milliliters of 0.200 N $HClO_4$ solution are needed to react completely with 40.0 mL of 0.450 N $Ba(OH)_2$?

56. What are the conjugate acids of (a) HSO_3^-, (b) N_2H_4?

57. What are the conjugate bases of (a) HSO_3^-, (b) N_2H_4, (c) $C_5H_5NH^+$?

58. Identify the conjugate acid-base pairs in the reaction

$$CH_3NH_2 + NH_4^+ \rightleftharpoons CH_3NH_3^+ + NH_3$$

59. What is the definition of an amphoteric substance?

60. What is a Lewis acid? What is a Lewis base?

61. Use Lewis structures to diagram the reaction between the Lewis base OH^- and the Lewis acid SO_3.

62. Show how ethylenediamine and oxalate ion are able to function as bidentate ligands. What is the chelate effect?

63. How does EDTA aid in softening "hard water"?

64. Assign oxidation numbers to the atoms in the following formulas: (a) As_4, (b) $HClO_2$, (c) $MnCl_2$, (d) $V_2(SO_4)_3$.

65. Balance the following equations by the oxidation-number-change method.
   (a) $K_2Cr_2O_7 + HCl \longrightarrow KCl + Cl_2 + H_2O + CrCl_3$
   (b) $KOH + SO_2 + KMnO_4 \longrightarrow K_2SO_4 + MnO_2 + H_2O$

66. Balance the following equations by the ion-electron method for acidic slutions.
   (a) $Cr_2O_7^{2-} + Cl^- \longrightarrow Cr^{3+} + Cl_2$
   (b) $H_3AsO_3 + MnO_4^- \longrightarrow H_2AsO_4^- + Mn^{2+}$

67. Balance the following equations by the ion-electron method for basic solutions.
   (a) $I^- + CrO_4^{2-} \longrightarrow CrO_2^- + IO_3^-$
   (b) $SO_2 + MnO_4^- \longrightarrow MnO_2 + SO_4^{2-}$

68. In the previous two questions, identify the oxidizing agents and reducing agents.

69. Complete and balance the following equations if a reaction occurs.
   (a) $Sn(s) + HCl(aq) \longrightarrow$
   (b) $Cu(s) + HNO_3(concd) \longrightarrow$
   (c) $Zn(s) + Cu^{2+}(aq) \longrightarrow$
   (d) $Ag(s) + Cu^{2+}(aq) \longrightarrow$

70. Write a balanced chemical equation for the combustion of cetane, $C_{16}H_{34}$, a hydrocarbon present in diesel fuel, (a) in the presence of excess oxygen, (b) in a somewhat limited supply of oxygen, and (c) in a severely limited supply of oxygen.

71. Stearic acid, $C_{17}H_{35}CO_2H$, is derived from animal fat. Write a balanced chemical equation for the combustion of stearic acid in an abundant supply of oxygen.

72. Methanethiol, CH_3SH, is a foul-smelling gas produced in the intestinal tract by bacteria acting on albumin in the absence of air. Write a chemical equation for the combustion of CH_3SH in an excess supply of oxygen.

73. Why is $KMnO_4$ such a useful laboratory oxidizing agent? Write half-reactions for the reduction of MnO_4^- in (a) acidic solution and (b) basic solution.

74. Where in the periodic table are the very reactive metals located? Where are the least reactive ones located?

75. Write molecular equations for the reaction of O_2 with (a) magnesium, (b) aluminum, (c) phosphorus, (d) sulfur.

76. List three common laboratory oxidizing agents. List four common laboratory reducing agents.

77. Why is starch added in an iodine-thiosulfate titration?

78. *Bordeaux mixture* is traditionally prepared by mixing copper sulfate and calcium hydroxide in water. The resulting suspension of copper hydroxide is sprayed on trees and shrubs to fight fungus diseases. This fungicide is also available in commercial preparations. In an analysis of one such product, a sample weighing 0.238 g was dissolved in hydrochloric acid. Excess KI solution was then added and the iodine that was formed was titrated with 0.01669 M $Na_2S_2O_3$ solution using starch as an indicator. The titration required 28.62 mL of the thiosulfate solution. What was the percentage by weight of copper in the sample of Bordeaux mixture?

79. Compound A is a white solid with a high melting point.

When it melts, it conducts electricity. In which solvent is it likely to be more soluble, in water or in gasoline? Explain.

80. Compound *XY* is an ionic compound that dissociates as it dissolves in water. The lattice energy of *XY* is -600 kJ/mol. The hydration energy of its ions is -610 kJ/mol.
    (a) Write the thermochemical equations for the two steps in the formation of a solution of *XY* in water.
    (b) Write the sum of these two equations in the form of a thermochemical equation, showing the net ΔH.
    (c) Draw an enthalpy diagram for the formation of this solution.

81. A 0.270 *M* KOH solution has a density of 1.01 g/mL. Calculate the percent concentration (w/w) of KOH.

82. A 5.30 *M* solution of glycerol in water has a density of 1.11 g/mL. Calculate its percent concentration (w/w) of glycerol ($C_3H_8O_3$) and the mole fraction of glycerol present.

83. At 20 °C a 40.00% (v/v) solution of ethyl alcohol, C_2H_5OH, in water, has a density of 0.9369 g/mL. The density of pure ethyl alcohol at this temperature is 0.7907 g/mL and of water is 0.9982 g/mL.
    (a) Calculate the molar concentration and the molal concentration of C_2H_5OH in this solution.
    (b) Calculate the concentration of C_2H_5OH in this solution in mole fractions and mole percents.
    (c) The vapor pressure of ethyl alcohol at 20 °C is 41.0 torr and of water is 17.5 torr. If the 40.00% (v/v) solution were ideal, what would be the vapor pressure of each component over the solution?

84. Estimate the boiling point of 1.0 molal $Al(NO_3)_3$ if it breaks up entirely into Al^{3+} and NO_3^- ions in solution.

85. Squalene is an oil found chiefly in shark liver oil but also present in low concentrations in olive oil, wheat germ oil, and yeast. A qualitative analysis disclosed that its molecules consist entirely of carbon and hydrogen. When a sample of squalene with a mass of 0.5680 g was burned in pure oxygen, there was obtained 1.8260 g of carbon dioxide and 0.6230 g of water.
    (a) Calculate the empirical formula of squalene.
    (b) When 0.1268 g of squalene was dissolved in 10.50 g of molten camphor, the freezing point of this solution was 177.3 °C. (The melting point of pure camphor is 178.4 °C, and its molal freezing point depression constant is 37.5 $\frac{°C \text{ kg-camphor}}{\text{mol-solute}}$.) Calculate the formula weight of squalene and write its molecular formula.

The cliffs overlooking the sea at Dover, England, are composed of chalk, $CaCO_3$, formed from sea shells long ago when that part of Britain was submerged. Sea creatures such as clams and oysters extract calcium ions and carbonate ions from seawater and, at higher concentrations, cause them to precipitate as $CaCO_3$. Equilibria involving the solubilities of salts such as $CaCO_3$ are among the topics discussed in this chapter.

CHEMICAL EQUILIBRIUM

DYNAMIC EQUILIBRIUM IN CHEMICAL SYSTEMS
REACTION REVERSIBILITY
THE EQUILIBRIUM LAW FOR A REACTION
EQUILIBRIUM LAWS FOR GASEOUS REACTIONS
THE RELATIONSHIP BETWEEN K_p AND K_c
EQUILIBRIUM CONSTANTS: WHAT THEY TELL US

LE CHÂTELIER'S PRINCIPLE AND CHEMICAL EQUILIBRIA
EQUILIBRIUM CALCULATIONS
HETEROGENEOUS EQUILIBRIA
SOLUBILITY PRODUCT
COMPLEX ION EQUILIBRIA

14.1 DYNAMIC EQUILIBRIUM IN CHEMICAL SYSTEMS

When a chemical equilibrium is established, the concentrations of the reactants and products do not change with time because the forward and reverse reactions occur at equal rates.

You have already encountered the concept of a dynamic equilibrium several times earlier in this book. You saw that it can be used to explain the vapor pressures of liquids and solids, and its application to chemical systems first appeared in our discussion of weak electrolytes. For example, we described a number of acids as "weak acids" because when they are in an aqueous solution there is an equilibrium between the un-ionized molecules of the acid and the ions that are produced by their reaction with water. For instance, acetic acid rapidly establishes the equilibrium

$$HC_2H_3O_2(aq) + H_2O \rightleftharpoons H_3O^+(aq) + C_2H_3O_2^-(aq)$$

Recall that this equation describes a condition in which two opposing reactions occur at equal rates. Let's review again how this comes about. First, consider what happens to the forward reaction—the reaction of acetic acid molecules with water to form the ions. When the acetic acid and water are first mixed, many collisions occur each second between their molecules and this leads to the rapid formation of the ions. As a result, the concentrations of the ions increase rapidly and the concentration of the acetic acid decreases rapidly. However, as the acetic acid is used up, fewer collisions between the molecules are possible, so the rate of the forward reaction slows down.

Now consider what happens to the reverse reaction—the reaction of the "products" H_3O^+ and $C_2H_3O_2^-$ to form acetic acid and water molecules. Before the acetic acid begins to ionize, of course, there are no ions in the solution that can react, so there is no reverse reaction. But once the acetic acid begins to react, the concentrations of the ions build up, and their collisions with each other can lead to the formation of $HC_2H_3O_2$ and H_2O. As the ions become more concentrated because of the forward reaction, they collide more frequently, so this reverse reaction occurs faster.

What we see is that as time goes by, the forward reaction (from left to right) becomes slower and the reverse reaction (from right to left) becomes faster. Eventually, their rates become equal and the ions are forming at the same rate that they are disappearing. From that time on, there is no change in the *numbers* of ions or of acetic acid molecules in the solution; the concentrations remain constant and the system is in a state of dynamic equilibrium.

It is an *equilibrium* because the concentrations don't change. It is *dynamic* because the opposing reactions never cease.

The reaction of acetic acid with water is just one example of a general phenomenon. For almost any reaction, the concentrations change rapidly at first and then approach steady values as shown in Figure 14.1. In fact, if the system is not disturbed, these concentrations will never change. Keep in mind, however, that even after equilibrium has been reached, both the forward and reverse reactions continue to occur. This is what is implied by the double arrows, $\rightleftharpoons$, that we use to indicate equilibrium.

FIGURE 14.1 As a reaction proceeds, the concentrations of the reactants and products approach steady (constant) values.

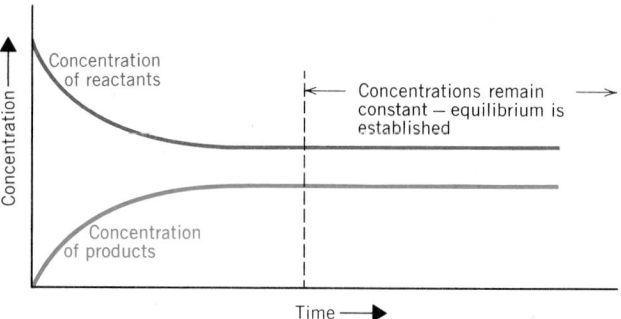

Almost all chemical systems eventually reach a state of dynamic equilibrium, although sometimes this equilibrium is extremely difficult (or even impossible) to detect. This is because in some reactions the amounts of either the reactants or products that are present at equilibrium are virtually zero. For instance, when a strong acid such as HCl is placed in water, it reacts so completely to produce H_3O^+ and Cl^- that there is no detectable amount of HCl molecules remaining—we say that the HCl is completely ionized. Similarly, in water vapor at room temperature, there are no detectable amounts of either H_2 or O_2 from the equilibrium

$$2H_2O(g) \rightleftharpoons 2H_2(g) + O_2(g)$$

Sometimes an equal sign (=) is used in place of the double arrows.

The water molecules are so stable that we can't detect whether any of them decompose. Even in cases such as these, however, it is often convenient to presume that an equilibrium does exist.

In this chapter we will study the equilibrium condition both qualitatively and quantitatively. Knowing the factors that affect the composition of an equilibrium system allows us to have some control over the outcome of a reaction. For example, the formation of the pollutant nitric oxide (NO) by reaction of N_2 and O_2 in a gasoline engine can be reduced, in a predictable way, by lowering the combustion temperature within the engine. In this chapter we will learn why. Because many biological substances are weak acids, knowledge of the principles of equilibrium helps biologists to understand how organisms control the acidity of fluids within them. We'll use the basic principles of equilibrium developed in this chapter to learn more about acids and bases in Chapter 15.

14.2 REACTION REVERSIBILITY

An equilibrium mixture will have the same composition whether approached from either pure reactants or pure products.

The ionization of acetic acid is an example of a reaction that is able to proceed in either direction. Acetic acid molecules are able to react with water molecules to form the ions, and the ions are able to react to form the molecules. Most chemical reactions are like this, and for these reactions the composition of the equilibrium mixture of reactants and products is found to be independent of whether we begin the reaction from the "reactant side" or the "product side." To illustrate this, let's consider some experiments that we might perform at 25 °C on the gaseous decomposition of N_2O_4 into NO_2,

$$N_2O_4(g) \rightleftharpoons 2NO_2(g)$$

Suppose we set up the two experiments shown in Figure 14.2. In the first 1-liter flask we place 0.0350 mol N_2O_4. Since no NO_2 is present, some of the N_2O_4 must decompose for the mixture to reach equilibrium, and we would find that at equilibrium the concentration of N_2O_4 has dropped to 0.0292 mol/L and the concentration of NO_2 has become 0.0116 mol/L. In the second 1-liter flask we place 0.0700 mol of NO_2 (*precisely* the amount of NO_2 that would form if 0.0350 mol of N_2O_4 decomposed completely). In this second flask there is no N_2O_4 present initially, so that enough NO_2 molecules must *combine*, following the reverse reaction, until there is sufficient N_2O_4 to give equilibrium. When we measure the concentrations at equilibrium in the second flask, we find, once again, 0.0292 mol/L of N_2O_4 and 0.0116 mol/L of NO_2.

0.0700 mol of NO_2 could be formed from 0.0350 mol of N_2O_4.

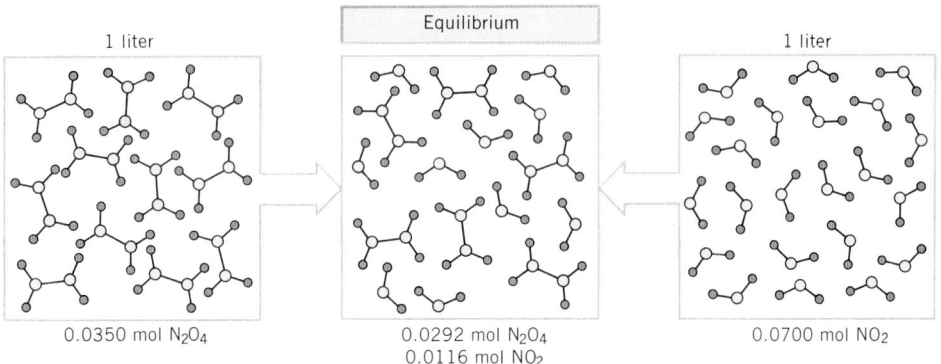

0.0350 mol N_2O_4 0.0292 mol N_2O_4 0.0116 mol NO_2 0.0700 mol NO_2

FIGURE 14.2 Reaction reversibility. The same equilibrium composition is reached from either the forward or reverse direction.

What we see here is that the identical composition of the system results whether we begin with pure NO_2 or pure N_2O_4, just as long as the total amount of nitrogen and oxygen to be divided between these two substances is the same. For a given overall composition we always reach the same equilibrium concentrations whether equilibrium is approached from the forward or reverse direction. It is in this sense that chemical reactions are said to be reversible.

14.3 THE EQUILIBRIUM LAW FOR A REACTION

At equilibrium the concentrations of the reactants and products, expressed in moles per liter, are interrelated in a predictable way by the equilibrium law.

For any chemical system at equilibrium there exists a simple relationship among the molar concentrations of the reactants and products. To show you this we will use the gaseous reaction of hydrogen and iodine to form the compound hydrogen iodide. The equation for the reaction is

$$H_2(g) + I_2(g) \rightleftharpoons 2HI(g)$$

Let's see what happens if we set up several experiments in which we start with different amounts of the reactants and/or product, as illustrated in Figure 14.3. As you can see, when equilibrium is reached the concentrations of H_2, I_2, and HI are different for each experiment. This isn't particularly surprising, but what is amazing is that the relationship among the concentrations is very simple and can actually be predicted from the balanced equation for the reaction.

For each experiment in Figure 14.3, if we square the molar concentration of HI at equilibrium and then divide this by the product of the equilibrium molar concentrations of H_2 and I_2, we always obtain the same numerical value. This is shown in Table 14.1 where we have used square brackets around formulas as symbols representing molar concentrations. Thus [HI] means the concentration of HI in units of mol/L, and [H_2] means the molar concentration of H_2. This is the standard way of symbolizing the molar concentration of a chemical, and you will see it used often throughout the rest of the book.

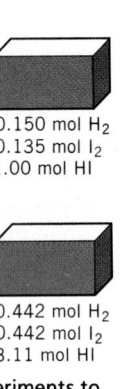

Initial amount / Measured composition at equilibrium

$H_2(g) + I_2(g) \rightleftharpoons 2HI(g)$

I   10 L
1.00 mol H_2 0.222 mol H_2
1.00 mol I_2 0.222 mol I_2
0.00 mol HI   1.56 mol HI

II   10 L
0.00 mol H_2 0.350 mol H_2
0.100 mol I_2 0.450 mol I_2
3.50 mol HI   2.80 mol HI

III 10 L
0.0150 mol H_2 0.150 mol H_2
0.00 mol I_2 0.135 mol I_2
1.27 mol HI   1.00 mol HI

IV   10 L
0.00 mol H_2 0.442 mol H_2
0.00 mol I_2 0.442 mol I_2
4.00 mol HI   3.11 mol HI

FIGURE 14.3 Four experiments to study the equilibrium among H_2, I_2, and HI at 440 °C.

TABLE 14.1 Equilibrium Concentrations and the Mass Action Expression

Experiment	Equilibrium Concentrations (mol/L)			$\dfrac{[HI]^2}{[H_2][I_2]}$
	[H_2]	**[I_2]**	**[HI]**	
I	0.0222	0.0222	0.156	$(0.156)^2/(0.0222)(0.0222) = 49.4$
II	0.0350	0.0450	0.280	$(0.280)^2/(0.0350)(0.0450) = 49.8$
III	0.0150	0.0135	0.100	$(0.100)^2/(0.0150)(0.0135) = 49.4$
IV	0.0442	0.0442	0.311	$(0.311)^2/(0.0442)(0.0442) = 49.5$
				Average $= 49.5$

The fraction used to calculate the values in the last column of Table 14.1,

$$\frac{[HI]^2}{[H_2][I_2]}$$

is called the mass action expression. The origin of this term isn't important; just consider it a name used to refer to this fraction. The numerical value of the mass action expression is called the reaction quotient, and is often symbolized by the letter **Q**. In Table 14.1, we see that when H_2, I_2, and HI are in dynamic equilibrium at 440 °C, the reaction quotient is equal to essentially a constant value of 49.5. In fact, if we repeated the experiments in Figure 14.3 over and over again starting with different amounts of H_2, I_2, and HI, we would always obtain the same reaction quotient, provided the systems had reached equilibrium and the temperature was 440 °C. For this reaction, therefore, we can write

$$\frac{[HI]^2}{[H_2][I_2]} = 49.5 \text{ (at 440 °C)} \tag{14.1}$$

This is the equilibrium law for the system. It tells us that for a reaction mixture of these three gases to be at equilibrium at 440 °C, the value of the mass action expression (the reaction quotient) must equal 49.5. If the reaction quotient has any other value, then the gases are not in equilibrium and they must react further. In effect, Equation 14.1 is a condition that must be fulfilled for equilibrium to exist, and the constant 49.5 which characterizes this equilibrium condition is called the equilibrium constant. In general, the equilibrium constant is symbolized by K_c (the subscript c because we write the mass action expression using molar concentrations). Thus, we could state the equilibrium law as follows:

$$\frac{[HI]^2}{[H_2][I_2]} = K_c = 49.5 \text{ (at 440 °C)} \tag{14.2}$$

As you've probably noticed, we have repeatedly noted the temperature when referring to the value of K_c for this reaction. This is because the value of the equilibrium constant changes when the temperature changes. In other words, if we had performed the experiments in Figure 14.3 at a temperature other than 440 °C, we would have obtained a different value for K_c.

Predicting the Equilibrium Law

An important fact about the mass action expression and the equilibrium law is that it can always be predicted from the balanced chemical equation for the reaction. For example, for the general chemical equation

$$dD + eE \rightleftharpoons fF + gG$$

where D, E, F, and G represent chemical formulas and d, e, f, and g are coefficients, the mass action expression is

$$\frac{[F]^f[G]^g}{[D]^d[E]^e}$$

The coefficients in the balanced equation are the same as the exponents in the mass action expression. The condition for equilibrium in this reaction would be

$$\frac{[F]^f[G]^g}{[D]^d[E]^e} = K_c$$

where the concentrations that satisfy the equation are equilibrium concentrations.

Notice that in writing the mass action expression the molar concentrations of the products are always placed in the numerator and those of the reactants appear in the denominator. Also note that after being raised to appropriate powers the concentrations are *multiplied,* not added. Using plus signs between concentration terms in the mass action expression is a common error that you should be sure to avoid.

Equation 14.1 is the *equilibrium condition* for the reaction of H_2, I_2, and HI at 440 °C.

In general, it is necessary to specify the temperature when giving a value of K_c, because K_c changes when the temperature changes.

For the reaction
$$CH_4(g) + H_2O(g) \rightleftharpoons CO(g) + 3H_2(g),$$
$K_c = 1.78 \times 10^{-3}$ at 800 °C
$K_c = 4.68 \times 10^{-2}$ at 1000 °C
$K_c = 5.67$ at 1500 °C

EXAMPLE 14.1 WRITING THE EQUILIBRIUM LAW

Problem: Write the equilibrium law for the reaction

$$N_2(g) + 3H_2(g) \rightleftharpoons 2NH_3(g)$$

This is the reaction used for the industrial preparation of ammonia.

Solution: The coefficients become exponents. We have to remember to place the products in the numerator and the reactants in the denominator. Also, we must remember that when there is more than one reactant or product, the concentrations are multiplied, *not* added.

$$\frac{[NH_3]^2}{[N_2]^1[H_2]^3} = K_c$$

PRACTICE EXERCISE 1   Write the equilibrium law for the following.

$$2H_2(g) + O_2(g) \rightleftharpoons 2H_2O(g)$$
$$NH_3(aq) + H_2O(\ell) \rightleftharpoons NH_4^+(aq) + OH^-(aq)$$

The rule that we always write the concentrations of the products in the numerator of the mass action expression and the concentrations of the reactants in the denominator hasn't been established by nature. It is simply a custom that chemists have agreed on. Certainly, if the mass action expression is equal to a constant,

$$\frac{[HI]^2}{[H_2][I_2]} = K_c$$

its reciprocal is also equal to a constant,

$$\frac{[H_2][I_2]}{[HI]^2} = \frac{1}{K_c} = K_c'$$

The reason that chemists have chosen to stick to a set pattern—always placing the concentrations of the products in the numerator—is to avoid having to specify mass action expressions along with tabulated values of equilibrium constants. If we have the chemical equation for the equilibrium, we can always construct the correct mass action expression from it. For example, suppose we're told that at a particular temperature $K_c = 10.0$ for the reaction

$$2NO_2(g) \rightleftharpoons N_2O_4(g)$$

From the chemical equation we can write the correct mass action expression and the correct equilibrium law.

$$K_c = \frac{[N_2O_4]}{[NO_2]^2} = 10.0$$

The balanced chemical equation contains all the information we need to write the equilibrium law.

14.4 EQUILIBRIUM LAWS FOR GASEOUS REACTIONS

For gaseous reactions, the equilibrium law can be written using either molar concentrations or partial pressures in the mass action expression.

When all the reactants and products are gases, we can formulate mass action expressions in terms of partial pressures instead of molar concentrations. This is possible because of a relationship between the molar concentration of any one gas in the mixture and its partial pressure. This relationship comes from the ideal gas law,

$$PV = nRT$$

where P stands for the partial pressure of the gas. Solving for P gives

$$P = \frac{n}{V} RT$$

The quantity n/V has units of mol/L, and is simply the molar concentration, M, so we can write

$$P = (\text{molar concentration}) \times RT \qquad (14.3)$$

Doubling the molar concentration in a gas without changing its volume or temperature doubles the pressure.

Thus the partial pressure of a gaseous reactant or product is directly proportional to its molar concentration at a given temperature.

The relationship expressed in Equation 14.3 is what lets us write the mass action expression for reactions between gases either in terms of molarities or in terms of partial pressures. Of course, when we make such a switch we can't expect the numerical values of the equilibrium constants to be the same, so we use two different symbols for K. When molar concentrations are used, we use the symbol K_c. When partial pressures are used, then K_p is the symbol. For example, the equilibrium law for the reaction of nitrogen and hydrogen to form ammonia

$$N_2(g) + 3H_2(g) \rightleftharpoons 2NH_3(g)$$

can be written in either of the following ways.

$$\frac{[NH_3(g)]^2}{[N_2(g)][H_2(g)]^3} = K_c$$

because *concentrations* are used in the mass action expression

$$\frac{p_{NH_3(g)}^2}{p_{N_2(g)} p_{H_2(g)}^3} = K_p$$

because partial *pressures* are used in the mass action expression

At equilibrium, the molar concentrations can be used to calculate K_c, while the partial pressures of the gases at equilibrium can be used to calculate K_p. We will discuss how to convert from K_p to K_c and vice versa in the next section.

EXAMPLE 14.2   WRITING THE EQUILIBRIUM LAW FOR GASEOUS REACTIONS

Problem: What is the expression for K_p for the reaction

$$N_2O_4(g) \rightleftharpoons 2NO_2(g)$$

Solution: For K_p we use partial pressures in the mass action expression. Therefore,

$$K_p = \frac{p_{NO_2}^2}{p_{N_2O_4}}$$

PRACTICE EXERCISE 2   Using partial pressures, write the equilibrium law for the reaction
$H_2(g) + I_2(g) \rightleftharpoons 2HI(g)$.

14.5 THE RELATIONSHIP BETWEEN K_p AND K_c

Converting between K_p and K_c involves using the coefficients in the balanced equation to find the difference between the total moles of products and the total moles of reactants.

For some reactions K_p is equal to K_c, but for many others the two have different values. It is therefore desirable to have a way to calculate one from the other. Converting between K_p and K_c takes advantage of the relationship between partial pressure and molarity described in the previous section.

Equation 14.3 can be used to change K_p to K_c by substituting

$$(\text{molar concentration}) \times RT$$

for the partial pressure of each gas in the expression for K_p. Similarly, K_c can be changed to K_p by solving Equation 14.3 for the molar concentrations, and then substituting the result, P/RT, into the appropriate expression for K_c. This sounds like a lot of work, and it is. Fortunately, there is a general equation that we can use to make these conversions.

$$K_p = K_c(RT)^{\Delta n_g} \tag{14.4}$$

In this equation, Δn_g is the change in the number of moles of gas in going from the reactants to the products.

$$\Delta n_g = (\text{moles of gaseous products}) - (\text{moles of gaseous reactants})$$

The numerical value of Δn_g is obtained from the coefficients of the balanced equation for the reaction. For example, the equation for the reaction of N_2 with H_2 to give ammonia,

$$N_2(g) + 3H_2(g) \rightleftharpoons 2NH_3(g) \tag{14.5}$$

Δn_g is calculated from the coefficients of the equation, taking them to stand for moles.

tells us that *two* moles of NH_3 are formed when *one* mole of N_2 and *three* moles of H_2 react. In other words, two moles of gaseous product are formed from a total of four moles of gaseous reactants. That's a decrease of two moles of gas, so Δn_g for this reaction equals -2.

For some reactions, the value of Δn_g is equal to zero. An example is the decomposition of HI. The equilibrium for this reaction is

$$2HI(g) \rightleftharpoons H_2(g) + I_2(g)$$

Notice that if we take the coefficients to mean moles, there are two moles of gas on each side of the equation. This means that $\Delta n_g = 0$. Since (RT) raised to the zero power is equal to 1, $K_p = K_c$.

EXAMPLE 14.3   CONVERTING K_c TO K_p

Problem: At 500 °C, the reaction between N_2 and H_2 to form ammonia has $K_c = 6.0 \times 10^{-2}$. What is the numerical value of K_p for this reaction?

Solution: The equation that we wish to use is

$$K_p = K_c(RT)^{\Delta n_g}$$

In the discussion above, we saw that $\Delta n_g = -2$ for this reaction. All we need now are appropriate values of R and T. The temperature, T, must be expressed in kelvins. (When used to stand for temperature, a capital letter T in a equation always means the absolute temperature.) Next we must choose the right value for R. Referring back to Equation 14.3, we see that partial pressures would be in atm and concentrations would be in mol/L. The only value of R that is consistent with these units is 0.0821 L atm mol^{-1} K^{-1}, and this is the *only* value of R that can be used in Equation 14.4. Assembling the data, then, we have

In general, equilibrium constants are written without units.

$$K_c = 6.0 \times 10^{-2}$$
$$\Delta n_g = -2$$
$$T = (500 + 273) = 773 \text{ K}$$
$$R = 0.0821 \text{ L atm mol}^{-1} \text{ K}^{-1}$$

Substituting these values into the equation for K_p gives

$$[63.5]^{-2} = \frac{1}{[63.5]^2}$$
$$= 2.48 \times 10^{-4}$$

$$K_p = (6.0 \times 10^{-2})[(0.0821) \times (773)]^{-2}$$
$$= (6.0 \times 10^{-2})[63.5]^{-2}$$
$$= (6.0 \times 10^{-2}) \times (2.48 \times 10^{-4})$$
$$= 1.5 \times 10^{-5}$$

Notice that in this case K_p has a numerical value quite different from that of K_c.

EXAMPLE 14.4   CALCULATING K_c FROM K_p

Problem: At 25 °C, K_p for the reaction

$$N_2O_4(g) \rightleftharpoons 2NO_2(g)$$

has a value of 0.140. Calculate the value of K_c.

Solution: Once again, the equation that we need is

$$K_p = K_c(RT)^{\Delta n_g}$$

To obtain Δn_g, we take the difference between the number of moles of gaseous products (2 mol NO$_2$) and the number of moles of gaseous reactants (1 mol N$_2$O$_4$). Thus, $\Delta n_g = 2 - 1 = +1$. Now let's tabulate the data.

$$K_p = 0.140$$
$$\Delta n_g = +1$$
$$T = (25 + 273) = 298 \text{ K}$$
$$R = 0.0821 \text{ L atm mol}^{-1} \text{ K}^{-1}$$

Substituting,

$$0.140 = K_c[(0.0821) \times (298)]^1$$
$$= K_c(24.5)$$

Solving for K_c gives

$$K_c = \frac{0.140}{24.5}$$
$$= 5.71 \times 10^{-3}$$

Once again, there is a substantial difference between the values of K_p and K_c.

PRACTICE EXERCISE 3   Methanol, CH$_3$OH, is a promising fuel that can be synthesized from carbon monoxide and hydrogen according to the reaction

$$CO(g) + 2H_2(g) \rightleftharpoons CH_3OH(g)$$

For this reaction at 200 °C, $K_p = 3.8 \times 10^{-2}$. What is the value of K_c at this temperature?

PRACTICE EXERCISE 4   Nitrous oxide, N$_2$O, is a gas used as an anesthetic — it is sometimes called "laughing gas." This compound has a strong tendency to decompose into nitrogen and oxygen following the equation

$$2N_2O(g) \rightleftharpoons 2N_2(g) + O_2(g)$$

but the reaction is so slow that the gas appears to be stable at room temperature (25 °C). The decomposition reaction has $K_c = 7.3 \times 10^{34}$. What is the value of K_p for this reaction at 25 °C?

14.6 EQUILIBRIUM CONSTANTS: WHAT THEY TELL US

The larger the value of K, the farther the reaction will have proceeded toward completion when equilibrium is reached.

A useful bonus of always writing the mass action expression with the product concentrations in the numerator is that the size of the equilibrium constant gives us a *direct* indication of how far the reaction has proceeded toward the formation of the products when equilibrium is reached. For example, the reaction

$$2H_2(g) + O_2(g) \rightleftharpoons 2H_2O(g)$$

has $K_c = 9.1 \times 10^{80}$ at 25 °C. This means that if there is an equilibrium between these reactants and product,

$$K_c = \frac{[H_2O]^2}{[H_2]^2[O_2]} = \frac{9.1 \times 10^{80}}{1}$$

By writing K_c as a fraction, $(9.1 \times 10^{80})/1$, we can see that the only way for the numerator to be so much larger than the denominator is for the concentration of H_2O to be enormous in comparison to the concentrations of H_2 and O_2. This means that at equilibrium most of the hydrogen and oxygen atoms are found in H_2O; very few are present as H_2 and O_2. Thus, the large value of K_c indicates that the reaction between H_2 and O_2 goes essentially to completion.

The reaction between N_2 and O_2 to give NO

$$N_2(g) + O_2(g) \rightleftharpoons 2NO(g)$$

has a very small equilibrium constant, $K_c = 4.8 \times 10^{-31}$ at 25 °C. The equilibrium law for this reaction is

$$\frac{[NO]^2}{[N_2][O_2]} = 4.8 \times 10^{-31}$$

Since $10^{-31} = 1/10^{31}$, we can write this as

$$\frac{[NO]^2}{[N_2][O_2]} = \frac{4.8}{10^{31}}$$

Actually, you would need about 200,000 L of water vapor at 25 °C just to find one molecule of O_2 and two molecules of H_2.

In air at 25 °C, the equilibrium concentration of NO should be about 10^{-17} mol/L. It is usually higher because NO is formed in various reactions, such as those responsible for air pollution caused by automobiles.

Here the denominator is huge compared with the numerator. The concentrations of N_2 and O_2 must therefore be very much larger than the concentration of NO. This means that in a mixture of N_2 and O_2 at this temperature, very little NO is formed. The reaction proceeds hardly at all toward completion before equilibrium is reached.

The relationship between the equilibrium constant and the extent to which the reaction proceeds toward the formation of products when equilibrium occurs can be summarized as follows.

K very large	Reaction proceeds far toward completion
$K \approx 1$	Concentrations of reactants and products are nearly the same at equilibrium
K very small	Hardly any products are formed

Note that we have omitted the subscript for K in this summary. The same qualitative predictions about the extent of reaction apply whether we use K_p or K_c.

One of the ways that we can use equilibrium constants is to compare the extents to which two or more reactions proceed to completion, as shown in Example 14.5. Care has to be exercised in making such comparisons, however, because unless the K's are greatly different, the comparison will be valid only if both reactions have the same number of reactant and product molecules appearing in their balanced chemical equations.

EXAMPLE 14.5   USING K_c TO ESTIMATE EXTENT OF REACTION

Problem: Which of the following two reactions would tend to proceed farthest to completion when they reach equilibrium?

(1) $2NO(g) + O_2(g) \rightleftharpoons 2NO_2(g)$ $K_c = 3.4 \times 10^{13}$
(2) $4NH_3(g) + 5O_2(g) \rightleftharpoons 4NO(g) + 6H_2O(g)$ $K_c = 5 \times 10^{198}$

Solution: We know that the larger the value of K, the farther the reaction proceeds toward completion. Since 5×10^{198} is much greater than 3.4×10^{13}, reaction (2) goes farther to completion than reaction (1).

PRACTICE EXERCISE 5   Which of the following reactions will tend to proceed farthest toward completion?

(1) $2HBr(g) \rightleftharpoons H_2(g) + Br_2(g)$ $K_c = 7.0 \times 10^{-20}$
(2) $Si(s) + O_2(g) \rightleftharpoons SiO_2(s)$ $K_c = 2 \times 10^{142}$
(3) $C_2H_4(g) + H_2(g) \rightleftharpoons C_2H_6(g)$ $K_c = 1.2 \times 10^{19}$

14.7 LE CHÂTELIER'S PRINCIPLE AND CHEMICAL EQUILIBRIA

If an equilibrium in a system is upset, the system will tend to react in a direction that will reestablish equilibrium.

Although it is possible to perform calculations that tell us what the composition of an equilibrium system is, many times we really don't need to know exactly what the equilibrium concentrations are. Instead, we may simply want to know what actions we should take to maximize or minimize the amount of a given product or reactant. For instance, if we were designing gasoline engines, one of the things we would like to know is what could be done to reduce nitrogen oxide pollutants. Or, if we were preparing ammonia, NH_3, by the reaction of N_2 and H_2, we might want to know how to increase the yield of NH_3.

Le Châtelier's principle, introduced in Chapter 10, provides us with the means for making qualitative predictions about changes in chemical equilibria. It does this in much the same way that it allows us to predict the effects of outside influences on equilibria that involve physical changes, such as liquid–vapor equilibria. Recall that Le Châtelier's principle states, in effect, that *if an outside influence upsets an equilibrium, the system responds in a direction that counteracts the disturbing influence such that equilibrium is reestablished.*

The following factors influence the relative amounts of products and reactants in a chemical system at equilibrium.

1. ***Adding or removing a reactant or product.*** If we add or remove a reactant or product, we will change one of the concentrations in the system. This means that the value of the reaction quotient will change; it will no longer equal K_c and the system will not be at equilibrium. For the system to return to equilibrium, the concentrations will have to change in some way so that the reaction quotient will equal K_c once again. This change is brought about by the chemical reaction proceeding either to the right or left. Le Châtelier's principle tells us which way the reaction goes. According to Le Châtelier's principle the reaction will proceed in a direction that will partially offset the change in concentration. For example, if we have the equilibrium

$$3H_2(g) + N_2(g) \rightleftharpoons 2NH_3(g)$$

and add some H_2, the system will respond by using up some of the H_2 by reaction with N_2. Therefore, some of the N_2 will also be consumed (because it reacts with the H_2) and some NH_3 will be formed. Thus, adding H_2 leads to the production of more NH_3 and we say that the equilibrium "shifts to the right." The equilibrium also shifts to the right if we remove some NH_3. In this case H_2 reacts with some N_2 to replace a portion of the NH_3 that has been removed.

Figure 14.4 illustrates how the concentrations change when we add H_2 or remove NH_3. At time t_1, we add some H_2 and its concentration suddenly increases. To restore equilibrium, some of the H_2 reacts with N_2 to form more NH_3. Therefore, the concentrations of H_2 and N_2 both drop and the concentration of NH_3 increases. The relative size of each change is determined by the stoichiometry of the reaction. As the balanced equation would predict, the H_2 concentration decreases three times as much as the N_2 concentration. Similarly, the NH_3 concentration increases by twice as much as the N_2 concentration decreases.

A similar analysis of the changes in concentrations can be made when NH_3 is removed at time t_2. The NH_3 concentration drops suddenly and then increases as some of it is replaced and the system approaches equilibrium again. The concentrations of H_2 and N_2 both decrease because they are the substances that react to replace part of the NH_3 that was removed.

An important point to notice in Figure 14.4 is that not all of the H_2 that we've added is removed when the equilibrium is reestablished. Neither is all the NH_3 replaced when the reaction adjusts to the final equilibrium. The concentra-

> The reaction shifts in a direction that will remove a substance that's been added or replace a substance that's been removed.

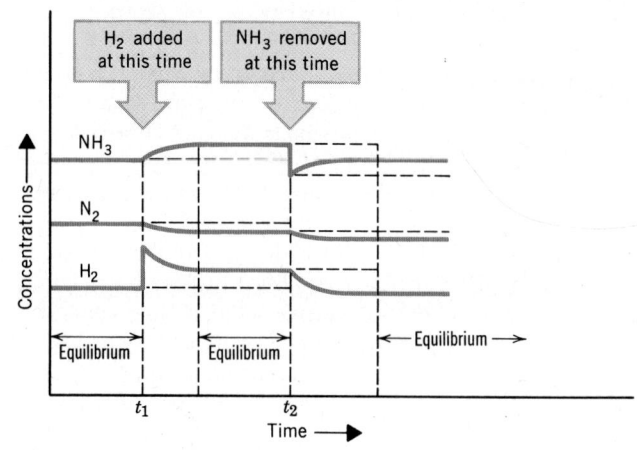

FIGURE 14.4 Effects of the addition of H_2 and removal of NH_3 on the equilibrium

$$N_2(g) + 3H_2(g) \rightleftharpoons 2NH_3(g)$$

When H_2 is added, some N_2 reacts and more NH_3 is formed as the system returns to equilibrium. When NH_3 is removed, more N_2 reacts with H_2 as the system again comes back to equilibrium.

tions in the new equilibrium systems have all changed, but in such a way that the mass action expression is still equal to the same value of K_c.

2. *Changing the volume in gaseous reactions.* Changing the volume of a system composed of gaseous reactants and products changes the pressure. For example, if we reduced the volume by half, we would expect the pressure to double. Is there any way that the system could oppose this change?

Again, let's consider the equilibrium

$$N_2(g) + 3H_2(g) \rightleftharpoons 2NH_3(g)$$

If the reaction proceeds to the right, two NH_3 molecules appear when four molecules (one N_2 and three H_2) disappear, and the number of molecules in the reaction vessel is reduced. You know by now that the fewer the number of molecules, the lower the pressure. Therefore, this equilibrium will respond to pressure increases (caused by volume decreases) by shifting to the right. The production of more NH_3 tends to lower the pressure and counter the effects of the volume decrease.

Decreasing the volume of a gaseous reaction mixture causes the reaction to decrease the number of molecules of gas, if it can.

Now let's look at the equilibrium

$$H_2(g) + I_2(g) \rightleftharpoons 2HI(g)$$

When this reaction occurs in either direction there is no change in the number of molecules of gas. This reaction, then, cannot respond to pressure changes, so changing the volume of the reaction vessel has virtually no effect on the equilibrium.

The simplest way to analyze the effects of a volume change on an equilibrium system is to count the number of molecules of *gaseous substances* on both sides of the equation. *An increase in pressure always drives the reaction in the direction of the fewest number of molecules of **gas**.*

As a final note here, pressure changes have essentially no effect on reactions involving only liquids or solids. Substances in these states are virtually incompressible, and reactions involving them have no way of counteracting pressure changes.

3. *Changes of temperature.* To understand how a chemical reaction at equilibrium responds to a change in temperature, we must know how the energy of the system changes when reaction occurs. For instance, the reaction to produce NH_3 from N_2 and H_2 is exothermic; heat is evolved when NH_3 molecules are formed ($\Delta H_f^\circ = -46.19$ kJ/mol from Table 4.2). If we include heat as a product in the equilibrium equation

You may recall that this same kind of analysis was used in Chapter 13 to predict how solubility changes with temperature.

$$N_2(g) + 3H_2(g) \rightleftharpoons 2NH_3(g) + \text{heat}$$

analyzing the effects of temperature changes becomes simple.

When we raise the temperature, we do so by adding heat from some external source, such as a Bunsen burner. The equilibrium above can counter the temperature increase by absorbing some of the added heat. It does this by proceeding from right to left because the decomposition of some NH_3 to give more N_2 and H_2 is endothermic and consumes some of the heat that we've added. Thus, raising the temperature shifts this equilibrium to the left.

Increasing the temperature shifts a reaction in a direction that produces an endothermic change.

An important point to note here is that when we change the temperature, the concentrations change even though the volume stays the same and no chemical substances have been added or removed. The system comes to a new equilibrium in which the value of the mass action expression has changed, which

As everyone knows, one of the most serious causes of air pollution is the automobile. All sorts of obnoxious chemicals are emitted in the exhaust, and there have been a variety of methods devised to control the amounts of these pollutants. For instance, most cars today are equipped with catalytic converters (see Chapter 18). These devices mix air with the exhaust gases and promote oxidation of unburned fuel and carbon monoxide to carbon dioxide. They also cause the decomposition of nitrogen oxides to elemental nitrogen and oxygen. Catalytic converters are expensive, however, and other methods to accomplish the same goals have also been studied.

When air is drawn into a car's engine, both N_2 and O_2 are present. During the combustion of the gasoline, oxygen reacts with the hydrocarbons in the fuel to produce CO_2, CO, and H_2O. However, N_2 and O_2 can also form NO.

$$N_2(g) + O_2(g) \rightleftharpoons 2NO(g)$$

At room temperature, K_c for this reaction is 4.8×10^{-31}. Its small value tells us that the equilibrium concentration of NO should be very small. Therefore, we don't find N_2 reacting with O_2 under ordinary conditions. The atmosphere, for instance, is quite stable.

The reaction of N_2 and O_2 to form NO is endothermic. Le Châtelier's principle tells us that at high temperatures, such as those found in the cylinders of a gasoline or diesel engine during combustion, this equilibrium should be shifted to the right. Therefore, at high temperatures some NO does form. Unfortunately, when the exhaust leaves the engine, it cools so rapidly that the NO can't decompose. The reaction rate at the lower temperature becomes too slow. The result is that some NO exists with the exhaust gases. Once in the atmosphere, the NO is oxidized to NO_2, which is responsible for the brownish haze often associated with severe air pollution. See Figure 19.14 on page 813.

One way to reduce the amount of NO pollution of the atmosphere is to reduce the amount of NO that's formed in automobile engines. Since the extent to which the reaction for the formation of NO proceeds to completion increases as the temperature is raised, it is clear that the amount of NO that is formed can be reduced by simply running the combustion reaction at a lower temperature. One method that has been used to accomplish this is to lower the compression ratio of the engine. This is the ratio of the volume of the cylinder when the piston is at the bottom of its stroke divided by the volume after the piston has compressed the air–fuel mixture. At high compression ratios, the air–fuel mixture is heated to a high temperature before it's ignited. After combustion, the gases are very hot, which favors the production of NO. Lowering the compression ratio lowers the maximum combustion temperature which decreases the tendency for NO to be formed. Unfortunately, lowering the compression ratio also lowers the efficiency of the engine, which makes for poorer fuel economy.

Another method for controlling NO emissions that has been experimented with involves mixing water with the air–fuel mixture. Some of the heat from the combustion is absorbed by the water vapor, so the mixture of exhaust gases doesn't get as hot as it would otherwise. At these lower temperatures, the concentration of the NO in the exhaust is drastically reduced.

The only factor that can change K_c or K_p for a given reaction is a change in temperature.

means that K has changed. In other words, the value of K changes when we change the temperature, and it isn't difficult to figure out the direction of the change. Let's look once again at the equilibrium law for the reaction.

$$\frac{[NH_3]^2}{[N_2][H_2]^3} = K_c$$

As we've noted, when we raise the temperature the concentration of NH_3 decreases while the concentrations of N_2 and H_2 increase. Therefore, the numerator of the mass action expression becomes smaller and the denominator becomes larger. This gives a smaller reaction quotient and therefore a smaller value of K_c. The conclusion, therefore, is that *when we raise the temperature of an exothermic reaction, the value of the equilibrium constant becomes smaller.* Of course, just the opposite occurs for an endothermic reaction.

K decreases with rising temperature for a reaction that is exothermic, when read from left to right.

4. *Effect of a catalyst.* There are certain substances that affect the speeds of some chemical reactions without actually being used up in the reaction. These substances are called *catalysts*, and we will discuss how they function in Chapter 18. For now, however, we just want to know if they can affect chemical equilibria, and the answer is that they cannot. The reason is that a catalyst affects both the forward and reverse reaction equally. Both are speeded up to the same degree, so

adding a catalyst to a system has no net affect on the system's equilibrium composition. The only effect a catalyst has on a chemical reaction is to bring it to equilibrium faster.

> Catalysts cause reactions to come to equilibrium more rapidly.

5. **Addition of an inert gas at constant volume.** A change in volume is not the only way to change the pressure in a vessel containing gaseous reactants and products at equilibrium. The pressure can also be changed by keeping the volume the same and adding another gas. If this gas is not able to react with any of the gases already present — that is, if the added gas is *inert* toward the gaseous reactants and products in equilibrium — then there will be no change in the position of equilibrium, even though the pressure has changed. The reason is easy to see. As long as the volume of the container doesn't change, and as long as the added gas isn't able to react, the *concentrations* of the reactants and products — the ratio of their moles to the volume of the container — will not change. This means that the values of the concentrations will continue to satisfy the equilibrium law, the reaction quotient will continue to equal K_c, and no changes in the amounts of the reactants or products will occur.

EXAMPLE 14.6   APPLICATION OF LE CHÂTELIER'S PRINCIPLE

Problem: The reaction $N_2O_4(g) \rightleftharpoons 2NO_2(g)$ is endothermic, with $\Delta H° = +56.9$ kJ. How will the amount of NO_2 at equilibrium be affected by (a) adding N_2O_4, (b) lowering the pressure by increasing the volume of the container, (c) raising the temperature, and (d) adding a catalyst to the system? Which of these changes will alter the value of K_c?

Solution: (a) Adding N_2O_4 will cause the equilibrium to shift to the right. *The amount of NO_2 will increase.*

(b) When the pressure in the system drops, the system responds by producing more molecules of gas, which will tend to raise the pressure and partially offset the change. Since more gas molecules are formed if some N_2O_4 decomposes, the *amount of NO_2 at equilibrium will increase.*

(c) Because the reaction is endothermic, we write the equation showing heat as a *reactant.*

$$\text{heat} + N_2O_4(g) \rightleftharpoons 2NO_2(g)$$

Since raising the temperature is accomplished by adding heat, the system will respond by absorbing heat. This means that the equilibrium will shift to the right. Therefore, when equilibrium is reestablished, *there will be more NO_2 present.*

(d) A catalyst has no effect on a chemical equilibrium. Catalysts affect only the speed of a reaction; they cause reactions to reach equilibrium more rapidly. Therefore, the amount of NO_2 at equilibrium will not be affected.

The only change that alters K_c is the temperature change. Raising the temperature (adding heat) will increase K_c for this endothermic reaction.

PRACTICE EXERCISE 6   Consider the equilibrium $PCl_3(g) + Cl_2(g) \rightleftharpoons PCl_5(g)$, for which $\Delta H° = -88$ kJ. How will the amount of Cl_2 at equilibrium be affected by (a) adding PCl_3, (b) adding PCl_5, (c) raising the temperature, and (d) decreasing the volume of the container? How (if at all) will each of these changes affect K_p for the reaction?

14.8   EQUILIBRIUM CALCULATIONS

The equilibrium law serves as the focal point in performing all equilibrium calculations.

Sometimes it is necessary to have more than merely a qualitative knowledge of equilibrium concentrations. For example, life forms are generally quite sensitive to the hydrogen ion concentration in their surroundings. If you wished to grow bacteria in a solution of a weak acid, it would be important to know (or to be able to calculate) the H^+ concentration in the solution.

Equilibrium calculations can be performed using either K_p or K_c for gaseous reactions, but for reactions in solution we must use K_c. Whether we deal with concentrations or partial pressures, however, the same basic principles apply. To keep things simple, we will restrict ourselves to calculations involving K_c and molar concentrations.

Overall, we can divide equilibrium calculations into two main categories:

1. Calculating equilibrium constants from known equilibrium concentrations.

2. Calculating one or more equilibrium concentrations using the known value of K_c.

Calculating K_c from Equilibrium Concentrations

One obvious way to determine the value of the equilibrium constant is to carry out the reaction and actually measure the concentrations of reactants and products after equilibrium has been reached. As an example, let's look again at the decomposition of N_2O_4.

$$N_2O_4(g) \rightleftharpoons 2NO_2(g)$$

In Section 14.2, we saw that if 0.0350 mol of N_2O_4 is placed into a 1-liter flask at 25 °C, the concentrations of N_2O_4 and NO_2 at equilibrium will be

$$[N_2O_4] = 0.0292 \text{ mol/L}$$
$$[NO_2] = 0.0116 \text{ mol/L}$$

To calculate K_c for this reaction, we substitute the equilibrium concentrations into the mass action expression.

$$\frac{[NO_2]^2}{[N_2O_4]} = K_c$$

$$\frac{(0.0116)^2}{(0.0292)} = K_c$$

Performing the arithmetic gives

$$K_c = 0.00461$$
$$= 4.61 \times 10^{-3}$$

Although calculating an equilibrium constant this way is easy, sometimes we have to do a little work to figure out what all the concentrations are, as shown in Example 14.7.

EXAMPLE 14.7   CALCULATING K_c FROM EQUILIBRIUM CONCENTRATIONS

Problem: At a certain temperature, a mixture of H_2 and I_2 was prepared by placing 0.100 mol of H_2 and 0.100 mol of I_2 into a 1.00-liter flask. After a period of time the equilibrium

$$H_2(g) + I_2(g) \rightleftharpoons 2HI(g)$$

was established. The purple color of the I_2 vapor was used to monitor the reaction, and from the decreased intensity of the purple color it was determined that, at equilibrium, the I_2 concentration had dropped to 0.020 mol/L. What is the value of K_c for this reaction at this temperature?

Solution: To calculate the value of K_c, we must substitute the equilibrium concentrations of H_2, I_2, and HI into the mass action expression, because for this reaction at equilibrium

$$\frac{[HI]^2}{[H_2][I_2]} = K_c$$

But what are the equilibrium concentrations? It appears that we've been given only the value of $[I_2]$. To obtain the others, we must examine the initial concentrations (the concentrations of reactants and products that were initially placed into the reaction flask) and figure out how they changed as the reaction came to equilibrium. (We will have to perform similar analyses in many of our equilibrium calculations in this chapter and the next.) To help organize the information and to approach the problem systematically, we will arrange the data in a table, using the chemical equation as a guide. The completed table for this problem is shown below.

	$H_2(g)$ +	$I_2(g) \rightleftharpoons$	$2HI(g)$
Initial concentrations (M)	0.100	0.100	0.000
Changes in concentration (M)	-0.080	-0.080	$+2(0.080)$
Equilibrium concentrations (M)	0.020	0.020	0.160

Let's see how the entries in the table were obtained.

Initial Concentrations. Under the formulas we've written the initial *molar* concentrations — that is, the initial concentrations expressed in moles per liter. In making these entries, you must always remember to calculate the ratio of moles to liters (unless molarity is given). The volume is 1.00 liter in this problem; therefore, the initial concentrations of H_2 and I_2 were 0.100 mol/1.00 L = 0.100 M. (Had the volume been 2.00 liters, the initial concentrations would have been 0.100 mol/2.00 L = 0.0500 M.) A value of 0.000 is placed under HI because none of this substance was originally placed in the reaction flask.

Changes in Concentration. Because the original mixture was not at equilibrium, a chemical reaction had to occur until equilibrium was reached. Since no HI was present initially, the reaction had to proceed to the right; there has to be some HI in the flask at equilibrium. As the reaction proceeded to the right, all of the concentrations changed. To obtain the entries in the "change" row we have to do some figuring. We know from the statement of the problem that the equilibrium concentration of I_2 is 0.020 M. Since it began as 0.100 M, its concentration must have *decreased* by 0.080 mol per liter (0.100 − 0.080 = 0.020). In the table we've entered the change with a *minus* sign to indicate a *decrease* in concentration.

Once we know how one of the concentrations changed, we can quickly figure how the others changed. The stoichiometry of the reaction controls this. Since H_2 and I_2 react in a one-to-one mole ratio, if 0.080 mol of I_2 reacted per liter, than 0.080 mol of H_2 must also have reacted per liter. The H_2 concentration must have decreased by the same amount as the I_2 concentration. Since two moles of HI are formed for each mole of I_2 that disappears, the HI concentration must have *increased* by *twice* the amount that the I_2 concentration decreased — thus the plus sign and the factor of 2.

Equilibrium Concentrations. By using the minus and plus signs in the "change" row to show decreases and increases in concentration, the equilibrium concentrations always can be calculated as

$$\left(\begin{array}{c}\text{equilibrium} \\ \text{concentration}\end{array}\right) = \left(\begin{array}{c}\text{initial} \\ \text{concentration}\end{array}\right) + \left(\begin{array}{c}\text{change in} \\ \text{concentration}\end{array}\right)$$

For example,

$$[H_2] = (0.100) + (-0.080) = 0.020 \text{ } M$$
$$[HI] = (0.000) + (+0.160) = 0.160 \text{ } M$$

Once values have been obtained for all of the equilibrium concentrations, they are substituted into the mass action expression to calculate K_c.

$$K_c = \frac{[HI]^2}{[H_2][I_2]}$$

$$= \frac{(0.160)^2}{(0.020)(0.020)}$$

$$K_c = 64$$

In any system, the initial concentrations are controlled by the person doing the experiment.

The changes in concentration are controlled by the stoichiometry of the reaction.

In the preceding example, there are some key points that apply in constructing the concentration table under the chemical equation. These apply not only to this problem, but to others that you will see that deal with equilibrium calculations.

1. You must remember that the only values that we may substitute into the mass action expression in the equilibrium law are *equilibrium concentrations* — the values that appear in the last row of our table.

By expressing concentrations in mol/L, we follow what happens to reactants and products in each liter of solution.

2. When we enter initial concentrations into the table, they should be in units of moles per liter (mol/L). The initial concentrations are those present in the reaction mixture when it's prepared; we imagine that no reaction occurs until everything is mixed.

3. The changes in concentration always occur in the *same ratio* as the coefficients in the balanced equation for the equilibrium. For example, if we were dealing with the equilibrium

$$3H_3(g) + N_2(g) \rightleftharpoons 2NH_3(g)$$

and found that the $N_2(g)$ concentration decreases by 0.10 M during the approach to equilibrium, the entries in the "change" row would be as follows:

$$
\begin{array}{ccc}
3H_2(g) & + & N_2(g) & \rightleftharpoons & 2NH_3(g) \\
\end{array}
$$

change in concentration: $-3 \times (0.10)$ $\quad$ $-1 \times (0.10)$ $\qquad$ $+2 \times (0.10)$

$$\downarrow \qquad\qquad \downarrow \qquad\qquad \downarrow$$

$$-0.30 \qquad\qquad -0.10 \qquad\qquad +0.20$$

If the initial concentration of some reactant is zero, its change must be positive (an increase) because the final concentration can't be negative.

4. In constructing the "change" row, be sure the reactant concentrations all change in the same direction, and that the product concentrations all change in the *opposite* direction. If the concentrations of the reactants decrease, *all* the entries for the reactants in the "change" row should have a minus sign, and *all* the entries for the products should be positive.

Keep these ideas in mind as we construct the concentration tables for other equilibrium problems.

PRACTICE EXERCISE 7 In a particular experiment, it was found that when $O_2(g)$ and $CO(g)$ were mixed and allowed to react according to the equation

$$2CO(g) + O_2(g) \rightleftharpoons 2CO_2(g)$$

the O_2 concentration had decreased by 0.030 mol/L when the reaction reached equilibrium. How did the concentrations of CO and CO_2 change?

PRACTICE EXERCISE 8 An equilibrium was established for the reaction

$$CO(g) + H_2O(g) \rightleftharpoons CO_2(g) + H_2(g)$$

at 500 °C. (This is an industrially important reaction for the preparation of hydrogen.) At equilibrium, the following concentrations were found in the reaction vessel: [CO] = 0.180 M, [H_2O] = 0.0411 M, [CO_2] = 0.150 M, and [H_2] = 0.200 M. What is the value of K_c for this reaction?

PRACTICE EXERCISE 9 A student placed 0.20 mol of $PCl_3(g)$ and 0.10 mol of $Cl_2(g)$ into a 1.00-liter flask at 250 °C. The reaction

$$PCl_3(g) + Cl_2(g) \rightleftharpoons PCl_5(g)$$

was allowed to come to equilibrium, at which time it was found that the flask contained 0.12 mol of PCl_3.

(a) What were the initial concentrations of the reactants and product?
(b) What were the changes in concentration?
(c) What were the equilibrium concentrations?
(d) What is the value of K_c for this reaction?

Calculating Equilibrium Concentrations Using K_c

The simplest calculation of this type occurs when all but one of the equilibrium concentrations are known, as illustrated in Example 14.8.

EXAMPLE 14.8 USING K_c TO CALCULATE CONCENTRATIONS AT EQUILIBRIUM

Problem: The reversible reaction

$$CH_4(g) + H_2O(g) \rightleftharpoons CO(g) + 3H_2(g)$$

has been suggested as a possible source of hydrogen. The hydrogen would be shipped to its destination as CH_4 through our natural gas pipelines. At 1500 °C, an equilibrium mixture of these gases was found to have the following concentrations: $[CO] = 0.300$ M, $[H_2] = 0.800$ M, and $[CH_4] = 0.400$ M. At 1500 °C, $K_c = 5.67$ for this reaction. What was the equilibrium concentration of $H_2O(g)$ in this mixture?

Solution: The first step, once we have the chemical equation for the equilibrium, is to write the equilibrium law for the reaction.

$$K_c = \frac{[CO(g)][H_2(g)]^3}{[CH_4(g)][H_2O(g)]}$$

The equilibrium constant and all of the equilibrium concentrations except that for H_2O are known. We simply substitute these values into the equilibrium law and solve for the unknown quantity.

$$5.67 = \frac{(0.300)(0.800)^3}{(0.400)[H_2O(g)]}$$

Multiplying both sides by $(0.400)[H_2O(g)]$ to clear fractions,

$$5.67(0.400)[H_2O(g)] = (0.300)(0.800)^3$$
$$2.27[H_2O(g)] = 0.154$$
$$[H_2O(g)] = \frac{0.154}{2.27}$$
$$[H_2O(g)] = 0.0678 \ M$$

PRACTICE EXERCISE 10 Ethyl acetate, $CH_3CO_2C_2H_5$, is an important solvent used in lacquers, adhesives, the manufacture of plastics, and even as a food flavoring. It is produced from acetic acid and ethanol by the reaction

$$CH_3CO_2H(\ell) + C_2H_5OH(\ell) \rightleftharpoons CH_3CO_2C_2H_5(\ell) + H_2O(\ell)$$

$$\underset{\text{acetic acid}}{} \quad \underset{\text{ethanol}}{}$$

At 25 °C, $K_c = 4.10$ for this reaction. In a reaction mixture, the following equilibrium concentrations were observed: $[CH_3CO_2H] = 0.210$ M, $[CH_3CO_2C_2H_5] = 0.910$ M, and $[H_2O] = 0.00850$ M. What was the concentration of C_2H_5OH in the mixture?

Calculating Equilibrium Concentrations Using K_c and Initial Concentrations

A more complex type of calculation involves the use of initial concentrations and K_c to compute equilibrium concentrations. Although some of these problems can be so complicated that a computer is needed to solve them, we can learn the general principles involved by working on simple calculations. Even these, however, require a little applied algebra. This is where the concentration table, built up as in Example 14.7, can be very helpful.

EXAMPLE 14.9   USING K_c TO CALCULATE EQUILIBRIUM CONCENTRATIONS

Problem: The reaction

$$CO(g) + H_2O(g) \rightleftharpoons CO_2(g) + H_2(g)$$

has $K_c = 4.06$ at 500 °C. If 0.100 mol of CO and 0.100 mol of $H_2O(g)$ are placed in a 1.00-liter reaction vessel at this temperature, what are the concentrations of the reactants and products when the system reaches equilibrium?

Solution: The key to solving this problem is that at equilibrium the mass action expression *must* equal K_c.

$$\frac{[CO_2][H_2]}{[CO][H_2O]} = 4.06$$

We must find values for the concentrations that satisfy this condition. Because we don't know what these concentrations are, we must represent them as unknowns, algebraically. This is where we use the concentration table. We need quantities to enter into the "initial concentrations," "changes in concentration," and "equilibrium concentrations" rows. Let's figure them out and then build the table.

Initial Concentrations. The initial concentrations of CO and H_2O were each 0.100 mol/1.00 L = 0.100 M. Since no CO_2 or H_2 was initially placed into the reaction vessel, their initial concentrations both were zero.

Changes in Concentration. Some CO_2 and H_2 had to form for the reaction to reach equilibrium. This also means that some CO and H_2O must have reacted. How much? If we knew the answer, we could calculate the equilibrium concentrations. The changes in concentration are our unknown quantities.

Let us allow x to be equal to the number of moles per liter of CO that reacted. The change in the concentration of CO would then be $-x$ (it is negative because the change *decreased* the CO concentration). Because CO and H_2O react one for one, the change in the H_2O concentration is also $-x$. Since one mole each of CO_2 and H_2 is formed from one mole of CO, the CO_2 and H_2 concentrations each increased by x (their changes are $+x$).

We could just as easily have chosen x to be the number of mol/L of H_2O that reacts or the mol/L of CO_2 or H_2 that is formed. There's nothing special about having chosen CO to define x.

Equilibrium Concentrations. We obtain the equilibrium concentrations as

$$\begin{pmatrix} \text{equilibrium} \\ \text{concentration} \end{pmatrix} = \begin{pmatrix} \text{initial} \\ \text{concentration} \end{pmatrix} + \begin{pmatrix} \text{change in} \\ \text{concentration} \end{pmatrix}$$

Now we can construct the concentration table.

The coefficients of x can be the same as the coefficients in the balanced chemical equation.

	CO(g) +	H₂O(g) ⇌	CO₂(g) +	H₂(g)
Initial concentrations (M)	0.100	0.100	0.0	0.0
Changes in concentration (M)	$-x$	$-x$	$+x$	$+x$
Equilibrium concentrations (M)	$0.100 - x$	$0.100 - x$	x	x

Note that the last line in this table merely tells us that the equilibrium CO and H_2O concentrations are equal to the number of moles per liter that were present initially minus the number of moles per liter that reacted. The equilibrium concentrations of CO_2 and H_2 are simply the number of moles per liter of each that formed, since no CO_2 or H_2 was present initially.

The equilibrium concentration values must satisfy the equation given by the equilibrium law.

Next we substitute the quantities from the "equilibrium concentration" row into the mass action expression and solve for x.

$$\frac{[CO_2][H_2]}{[CO][H_2O]} = 4.06$$

Substituting,

$$\frac{(x)(x)}{(0.100 - x)(0.100 - x)} = 4.06$$

which we can write as

$$\frac{x^2}{(0.100 - x)^2} = 4.06$$

In this problem we can solve the equation for x most easily by taking the square root of both sides.

$$\frac{x}{(0.100 - x)} = \sqrt{4.06} = 2.01$$

Clearing fractions gives

$$x = 2.01(0.100 - x)$$
$$x = 0.201 - 2.01x$$

Collecting terms in x gives

$$x + 2.01x = 0.201$$
$$3.01x = 0.201$$
$$x = \frac{0.201}{3.01}$$
$$= 0.0668$$

Now that we know the value of x, we can calculate the equilibrium concentrations from the last row of our table.

$$[CO] = 0.100 - x = 0.100 - 0.0668 = 0.033 \ M$$
$$[H_2O] = 0.100 - x = 0.100 - 0.0668 = 0.033 \ M$$
$$[CO_2] = x = 0.0668 \ M$$
$$[H_2] = x = 0.0668 \ M$$

In the preceding example, we were able to simplify the solution by taking the square root of both sides of the algebraic equation that we obtained when we substituted equilibrium concentrations into the mass action expression. You can't expect to do this always, but keep an eye open for ways to simplify the algebra. None of the problems that you will encounter in this book will require very extensive mathematical skills.

PRACTICE EXERCISE 11   During an experiment, 0.200 mol of H_2 and 0.200 mol of I_2 were placed into a 1.00-liter vessel. The reaction

$$H_2(g) + I_2(g) \rightleftharpoons 2HI(g)$$

was allowed to come to equilibrium. For this reaction, $K_c = 49.5$. What were the equilibrium concentrations of H_2, I_2, and HI? (*Hint:* In constructing the change column, remember that *two* moles of HI are produced for each mole of H_2 or I_2 that reacts. If the H_2 concentration decreases by x, the HI concentration increases by $2x$.)

Equilibrium Calculations When K_c Is Very Small

Many chemical reactions have equilibrium constants that are either very large or very small. For example, most of the weak acids that we find in biological systems have very small values for K_c. Only very tiny amounts of products form when these weak acids react with water.

When the K_c for a reaction is very small the equilibrium calculations can often be considerably simplified.

EXAMPLE 14.10   SIMPLIFYING EQUILIBRIUM CALCULATIONS FOR REACTIONS WITH SMALL K_c

Problem: Hydrogen, a potential fuel, is found in great abundance in water. Before the hydrogen can be used as a fuel, however, it must be separated from the oxygen; the water must be split into H_2 and O_2. One possibility is thermal decomposition, but this requires very high temperatures. Even at 1000 °C, $K_c = 7.3 \times 10^{-18}$ for the reaction

$$2H_2O(g) \rightleftharpoons 2H_2(g) + O_2(g)$$

If at 1000 °C the H_2O concentration in a reaction vessel is set initially at 0.100 M, what will the H_2 concentration be when the reaction reaches equilibrium?

Solution: Once again, we know that at equilibrium

$$\frac{[H_2]^2[O_2]}{[H_2O]^2} = 7.3 \times 10^{-18}$$

We now set up the concentration table.

Initial Concentrations. The initial concentration of H_2O is 0.100 M; those of H_2 and O_2 are both 0.0 M.

Changes in Concentration. It is best to define x in terms of a reactant or product whose coefficient is 1. We will let x be the number of moles/liter of O_2 that are formed, so the O_2 concentration will increase by x. The H_2 concentration will increase by $2x$ because two H_2 molecules are formed for each O_2 that is formed. The H_2O concentration will decrease by $2x$ because two H_2O molecules react to give each O_2 molecule. This gives us

	$2H_2O(g) \rightleftharpoons$	$2H_2(g) +$	$O_2(g)$
Initial concentrations (M)	0.100	0.0	0.0
Changes in concentration (M)	$-2x$	$+2x$	$+x$
Equilibrium concentration (M)	$0.100 - 2x$	$2x$	x

Once again, notice that we let the coefficients of x equal the coefficients in the balanced chemical equation.

When we substitute the equilibrium quantities into the mass action expression we get

$$\frac{(2x)^2 x}{(0.100 - 2x)^2} = 7.3 \times 10^{-18}$$

or

$(2x)^2 x = (4x^2)x = 4x^3$

$$\frac{4x^3}{(0.100 - 2x)^2} = 7.3 \times 10^{-18}$$

Even before we solve the problem, we know that hardly any H_2 and O_2 will be formed, because K_c is so small.

This is a cubic equation (one term involves x^3) and is rather difficult to solve unless we can simplify it. In this case we are able to do so because the very small value of K_c tells us that hardly any of the H_2O will decompose. Whatever the actual value of x, we know that it is going to be very small. This means that $2x$ will also be small, so when this tiny value is subtracted from 0.100, the result will still be very, very close to 0.100. We will make an assumption, then, that the denominator will be essentially unchanged from 0.100 by subtracting $2x$; that is, we will assume that $0.100 - 2x \approx 0.100$. This assumption greatly simplifies the math. We now have

$$\frac{4x^3}{(0.100)^2} = 7.3 \times 10^{-18}$$

$$4x^3 = 0.0100(7.3 \times 10^{-18}) = 7.3 \times 10^{-20}$$

$$x^3 = \frac{7.3 \times 10^{-20}}{4} = 1.8 \times 10^{-20}$$

$$x = \sqrt[3]{1.8 \times 10^{-20}}$$
$$= 2.6 \times 10^{-7}$$

The y^x key on your calculator can be used to obtain the cube root by calculating $(1.8 \times 10^{-20})^{0.333}$.

Notice that the value of x that we've obtained is indeed very small. If we double it and subtract the answer from 0.100, we still get 0.100 when we round to the third decimal place.

$$0.100 - 2x = 0.100 - 2(2.6 \times 10^{-7}) = 0.09999948 \xrightarrow[\text{to}]{\text{rounds}} 0.100$$

Always be sure to check your assumption when solving a problem of this kind.

This check verifies that our assumption was valid. Finally, we have to obtain the H_2 concentration. Our table gives

$$[H_2] = 2x$$

Therefore,

$$[H_2] = 2(2.6 \times 10^{-7}) = 5.2 \times 10^{-7} \, M$$

The simplifying assumption made in the preceding example is valid because a very small number is *subtracted* from a much larger one. We could also have neglected x (or $2x$) if it were a very small number that was being *added* to a much larger one. Remember that you can only neglect an x that's added or subtracted; you can never drop an x that occurs as a multiplying or dividing factor. Some examples are

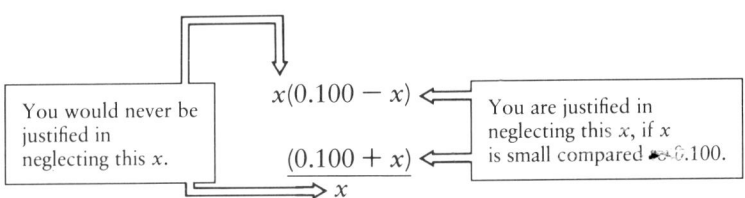

As a rule of thumb, you can expect that these simplifying assumptions will be valid if the concentration from which x is subtracted, or to which x is added, is at least 1000 times greater than K. For instance, in the preceding example, $2x$ was subtracted from 0.100. Since 0.100 is much larger than $1000 \times (7.3 \times 10^{-18})$, we expect the assumption $0.100 - 2x \approx 0.100$ to be valid. However, even when we *expect* the simplifying assumption to be valid, we should always check to see if it really is after we've finished the calculation, just as we did in Example 14.10. If the assumption proves invalid, then we have to find some other way to solve the algebra.

PRACTICE EXERCISE 12

In air at 25 °C and 1 atm, the N_2 concentration is 0.033 M and the O_2 concentration is 0.00810 M. The reaction

$$N_2(g) + O_2(g) \rightleftharpoons 2NO(g)$$

has $K_c = 4.8 \times 10^{-31}$ at 25 °C. Taking the N_2 and O_2 concentrations given above as initial values, calculate the equilibrium NO concentration that would exist in our atmosphere from this reaction at 25 °C.

14.9 HETEROGENEOUS EQUILIBRIA

The concentrations of reactants or products that are pure liquids or solids are constant and do not appear in the mass action expression.

In a *homogeneous reaction*—or in a *homogeneous equilibrium*—all of the reactants and products are in the same phase. Equilibria among gases are homogeneous because all gases mix freely with each other and a single phase exists. There are also many equilibria in which reactants and products are dissolved in the same liquid phase. Examples are the ionization equilibria of weak acids, which we'll look at closely in Chapter 15.

When more than one phase exists in a reaction mixture, we call it a *heterogeneous reaction*. Common examples are the burning of wood and the discharge of an automobile battery (where there's a reaction between the plates in the battery and the sulfuric acid solution in contact with them.) Another example is the thermal decomposition of sodium bicarbonate (baking soda) that occurs when it is sprinkled on a fire.

$$2NaHCO_3(s) \longrightarrow Na_2CO_3(s) + H_2O(g) + CO_2(g)$$

Many cooks keep a box of baking soda nearby because this reaction makes baking soda an excellent fire extinguisher for burning fats or oil. The fire is smothered by all the products.

Heterogeneous reactions are also able to reach equilibrium, just as homogeneous reactions can. If $NaHCO_3$ is placed in a sealed container so that no CO_2 or H_2O can escape, the gases and solids come to a *heterogeneous equilibrium*.

$$2NaHCO_3(s) \rightleftharpoons Na_2CO_3(s) + H_2O(g) + CO_2(g)$$

Following our usual procedure, we can write the equilibrium law for this reaction as

$$\frac{[Na_2CO_3(s)][H_2O(g)][CO_2(g)]}{[NaHCO_3(s)]^2} = K$$

However, the equilibrium law for reactions involving pure liquids and solids can be written in an even simpler form. This is because the concentration of a pure liquid or solid is unchangeable; that is, for any pure liquid or solid, the *ratio* of moles of substance to volume of substance is a constant. For example, if we had a one-mole crystal of $NaHCO_3$, it would occupy a volume of 38.9 cm³. Two moles of $NaHCO_3$ would occupy twice this volume, 77.8 cm³ (Figure 14.5), but the ratio of moles to volume (i.e., the molar concentration) remains the same. For $NaHCO_3$, the concentration of the substance in the solid is

$$1\ mol/0.0389\ L = 2\ mol/0.0778\ L = 25.7\ mol/L$$

FIGURE 14.5 The concentration of a substance in the solid state is a constant. Doubling the number of moles also doubles the volume, but the ratio of moles to volume remains the same.

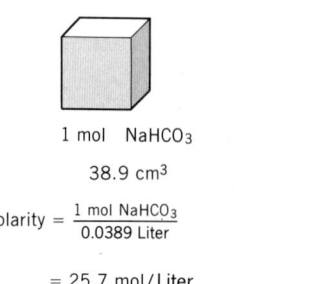

1 mol   NaHCO₃

38.9 cm³

$$Molarity = \frac{1\ mol\ NaHCO_3}{0.0389\ Liter}$$

$$= 25.7\ mol/Liter$$

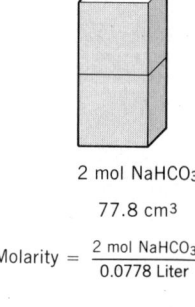

2 mol NaHCO₃

77.8 cm³

$$Molarity = \frac{2\ mol\ NaHCO_3}{0.0778\ Liter}$$

$$= 25.7\ mol/Liter$$

This is the concentration of $NaHCO_3$ in the solid, regardless of the size of the solid sample. In other words, the concentration of $NaHCO_3$ is constant, provided that some of it is present in the reaction mixture.

Similar reasoning shows that the concentration of Na_2CO_3 in pure solid Na_2CO_3 is a constant, too. This means that our equilibrium law now has three constants, K plus two of the concentration terms. It makes sense to combine all of the numerical constants together.

$$[H_2O(g)][CO_2(g)] = \frac{K[NaHCO_3(s)]^2}{[Na_2CO_3(s)]} = K_c$$

The equilibrium law for heterogeneous reactions is written without concentration terms for pure solids or liquids. Equilibrium constants that are given in tables represent all of the constants combined.

EXAMPLE 14.11   WRITING THE EQUILIBRIUM LAW FOR HETEROGENEOUS REACTIONS

Problem: The air pollutant sulfur dioxide can be removed from a gas mixture by passing the mixture over calcium oxide. The equation is

$$CaO(s) + SO_2(g) \rightleftharpoons CaSO_3(s)$$

Write the equilibrium law for this reaction.

Solution: The concentrations of the two solids, CaO and $CaSO_3$, are incorporated into the equilibrium constant for the reaction. The only concentration term that should appear in the mass action expression is that of SO_2. The equilibrium law is

$$\frac{1}{[SO_2(g)]} = K_c$$

PRACTICE EXERCISE 13   Write the equilibrium law for the following heterogeneous reactions.
(a) $2Hg(\ell) + Cl_2(g) \rightleftharpoons Hg_2Cl_2(s)$
(b) $NH_3(g) + HCl(g) \rightleftharpoons NH_4Cl(s)$

14.10 SOLUBILITY PRODUCT

A saturated solution of an "insoluble" salt is a heterogeneous equilibrium, for which the mass action expression is written as a product just of ion concentrations, each raised to an appropriate power.

The solubility rules given in Chapter 11 tell us that $PbCl_2$ and AgCl are both insoluble salts. However, if chloride ion is added slowly to a mixture containing both Pb^{2+} and Ag^+, nearly all of the silver is precipitated as AgCl before any $PbCl_2$ is formed. This is because silver chloride is much less soluble than lead chloride. To study such differences in solubility, we must examine solubility equilibria quantitatively. Let's begin by looking at the compound $CaCO_3$. When this substance is placed in water, a small amount dissolves and we obtain the equilibrium,

$$CaCO_3(s) \rightleftharpoons Ca^{2+}(aq) + CO_3^{2-}(aq)$$

As usual, we can write the equilibrium condition as

$$\frac{[Ca^{2+}(aq)][CO_3^{2-}(aq)]}{[CaCO_3(s)]} = K$$

This equilibrium, however, is a heterogeneous one. At the bottom of the solution rests pure solid $CaCO_3$. We saw earlier that the concentration of a substance within a pure solid is a constant. This means that the denominator of our mass action expression, $[CaCO_3(s)]$, is a constant that we can combine with K to obtain still another constant.

$$[Ca^{2+}(aq)][CO_3^{2-}(aq)] = K[CaCO_3(s)]$$
$$\text{another constant}$$

This new constant is called the solubility product constant, K_{sp}, because it is equal to a product of concentration terms. There's no denominator left in the mass action expression and the equilibrium condition is

$$[Ca^{2+}][CO_3^{2-}] = K_{sp}$$

The mass action expression for a solubility equilibrium is often called the ion product because it is a product of ion concentrations.

Many salts produce more than one of a given kind of ion when they dissociate. An example is silver chromate, Ag_2CrO_4. Each formula unit of Ag_2CrO_4 gives two silver ions and one chromate ion. Therefore, the equation representing the solubility equilibrium is

$$Ag_2CrO_4(s) \rightleftharpoons 2Ag^+(aq) + CrO_4^{2-}(aq)$$

As usual, the exponents on the concentration terms in the mass action expression are equal to the coefficients in the balanced equation. Therefore, the solubility product expression for Ag_2CrO_4 is

$$[Ag^+]^2[CrO_4^{2-}] = K_{sp}$$

PRACTICE EXERCISE 14

Write the K_{sp} expression for these equilibria.
(a) $BaCrO_4(s) \rightleftharpoons Ba^{2+}(aq) + CrO_4^{2-}(aq)$
(b) $Ag_3PO_4(s) \rightleftharpoons 3Ag^+(aq) + PO_4^{3-}(aq)$

Calculating K_{sp} from Solubility Data

One way of obtaining the value of K_{sp} for a salt is to measure its molar solubility in water — the number of moles of solute dissolved in one liter of its saturated solution. The salt is added to water and the mixture is stirred continuously for some period of time to ensure that equilibrium has been established. Then a portion of the solution is filtered to remove solids and analyzed to determine the amount of the salt that is in it. From data of this type we can calculate the molar solubility, from which the K_{sp} can be computed, as illustrated in Examples 14.12 through 14.14.

EXAMPLE 14.12   CALCULATING K_{sp} FROM SOLUBILITY DATA

Problem: In Section 11.5 it was mentioned that the light-sensitive ingredient in nearly all photographic film is AgBr. One liter of water is able to dissolve 7.1×10^{-7} mol of AgBr. What is K_{sp} for this salt?

Solution: In this problem we have been given the molar solubility of silver bromide — the number of moles of the solute that are needed to give one liter of a saturated solution. We can use this molar solubility, 7.1×10^{-7} mol/L, to construct the concentration table for the equilibrium. Our equation is

$$AgBr(s) \rightleftharpoons Ag^+(aq) + Br^-(aq)$$

If AgBr is placed in pure water, the initial concentrations of Ag^+ and Br^- are zero. The molar solubility tells us that 7.1×10^{-7} mol of AgBr dissolves per liter. From the stoichiometry of the equation, 7.1×10^{-7} mol of AgBr will give 7.1×10^{-7} mol of Ag^+ and 7.1×10^{-7} mol of Br^-. The concentrations of these ions therefore increase by this amount. We can now calculate the equilibrium concentrations from the initial concentrations and the changes.

	$AgBr(s) \rightleftharpoons Ag^+(aq)$	$+$	$Br^-(aq)$
Initial concentrations (M)		0.0	0.0
Changes in concentrations when AgBr dissolves (M)	No entry in this column	$+7.1 \times 10^{-7}$	$+7.1 \times 10^{-7}$
Equilibrium concentrations (M)		7.1×10^{-7}	7.1×10^{-7}

No entry appears under AgBr because its concentration term doesn't appear in the ion product.

To calculate K_{sp}, we substitute the equilibrium concentrations into the K_{sp} expression.

$$K_{sp} = [Ag^+][Br^-]$$
$$= (7.1 \times 10^{-7})(7.1 \times 10^{-7})$$
$$K_{sp} = 5.0 \times 10^{-13}$$

EXAMPLE 14.13 CALCULATING K_{sp} FROM SOLUBILITY DATA

Problem: The molar solubility of silver chromate, Ag_2CrO_4, in pure water is 6.7×10^{-5} mol/L. What is K_{sp} for Ag_2CrO_4?

Solution: The equilibrium equation is

$$Ag_2CrO_4(s) \rightleftharpoons 2Ag^+(aq) + CrO_4^{2-}(aq)$$

and the K_{sp} expression is

$$K_{sp} = [Ag^+]^2[CrO_4^{2-}]$$

Once again, we have none of the ions from the salt in the water initially, so their initial concentrations are equal to zero. When, in one liter, 6.7×10^{-5} mol of Ag_2CrO_4 dissolves, we must obtain 6.7×10^{-5} mol of CrO_4^{2-} and $2 \times (6.7 \times 10^{-5})$ mol of Ag^+. Knowing this, we can now construct the concentration table.

	$Ag_2CrO_4(s) \rightleftharpoons 2Ag^+(aq)$	$+$	$CrO_4^{2-}(aq)$
Initial concentrations (M)		0.0	0.0
Changes in concentrations when Ag_2CrO_4 dissolves		$+2 \times (6.7 \times 10^{-5})$ $= 1.3 \times 10^{-4}$	$+6.7 \times 10^{-5}$
Equilibrium concentrations		1.3×10^{-4}	6.7×10^{-5}

For each Ag_2CrO_4 that dissolves, we get two Ag^+ and one CrO_4^{2-}

Substituting the equilibrium concentrations into the mass action expression for K_{sp} gives

$$K_{sp} = (1.3 \times 10^{-4})^2(6.7 \times 10^{-5})$$
$$= 1.1 \times 10^{-12}$$

EXAMPLE 14.14 CALCULATING K_{sp} FROM SOLUBILITY DATA

Problem: The molar solubility of $PbCl_2$ in a 0.10 M NaCl solution is 1.7×10^{-3} mol/L. What is K_{sp} for $PbCl_2$?

Solution: Again, we begin by writing the equation for the equilibrium and the K_{sp} expression.

$$PbCl_2(s) \rightleftharpoons Pb^{2+}(aq) + 2Cl^-(aq)$$
$$K_{sp} = [Pb^{2+}][Cl^-]^2$$

As usual, we assume that any soluble salt is fully dissociated in water.

In this problem the $PbCl_2$ is being dissolved in a solution of NaCl, which already contains 0.10 M Cl^- (It also contains 0.10 M Na^+, but that's not important here because Na^+ doesn't affect the equilibrium and therefore doesn't appear in the K_{sp} expression.) The initial concentration of Pb^{2+} is zero because none was in the solution to begin with, but the initial concentration of Cl^- is 0.10 M. When the $PbCl_2$ dissolves in the NaCl solution, the Pb^{2+} concentration increases by 1.7×10^{-3} M and the Cl^- concentration increases by $2 \times (1.7 \times 10^{-3}$ $M)$. Now we can build our concentration table.

For each $PbCl_2$ that dissolves, we get one Pb^{2+} and two Cl^-.

	$PbCl_2(s) \rightleftharpoons Pb^{2+}(aq)$ +	$2Cl^-(aq)$
Initial concentrations (M)	0.0	0.10
Changes in concentrations when $PbCl_2$ dissolves (M)	$+1.7 \times 10^{-3}$	$+2 \times (1.7 \times 10^{-3})$ $= +3.4 \times 10^{-3}$
Equilibrium concentrations (M)	1.7×10^{-3}	$0.10 + 0.0034 = 0.10$ (to two significant figures)

Substituting the values for the equilibrium concentrations into the K_{sp} expression gives

$$K_{sp} = (1.7 \times 10^{-3})(0.10)^2$$
$$= 1.7 \times 10^{-5}$$

PRACTICE EXERCISE 15   One liter of water is able to dissolve 2.15×10^{-3} mol of PbF_2. What is K_{sp} for PbF_2?

PRACTICE EXERCISE 16   The molar solubility of $CoCO_3$ in a 0.10 M Na_2CO_3 solution is 1.0×10^{-9} mol/L. What is K_{sp} for $CoCO_3$?

PRACTICE EXERCISE 17   The molar solubility of PbF_2 in a 0.10 M $Pb(NO_3)_2$ solution is 3.1×10^{-4} mol/L. Calculate K_{sp} for PbF_2.

Calculating Molar Solubility from K_{sp}

Besides calculating K_{sp} from solubility information, we can also compute solubility if we know the value of K_{sp}. Some typical K_{sp} values are in Table 14.2. We will discuss the special situation with insoluble oxides and sulfides in Qual Topic II on page 650.

EXAMPLE 14.15   CALCULATING MOLAR SOLUBILITY FROM K_{sp}

Problem: The value of K_{sp} for AgCl is 1.8×10^{-10}. What is the molar solubility of AgCl in pure water?

Solution: The equation is

$$AgCl(s) \rightleftharpoons Ag^+(aq) + Cl^-(aq)$$

for which

$$K_{sp} = [Ag^+][Cl^-]$$

All of the AgCl that dissolves breaks down into Ag^+ and Cl^-.

Since no Ag^+ or Cl^- is present initially, their initial concentrations are zero. If we let x be the molar solubility (i.e., the number of moles of AgCl that dissolve per liter), then the Ag^+ and Cl^- concentrations will each increase by x.

	$AgCl(s) \rightleftharpoons Ag^+(aq) + Cl^-(aq)$	
Initial concentrations (M)	0.0	0.0
Changes in concentrations when AgCl dissolves (M)	$+x$	$+x$
Equilibrium concentrations (M)	x	x

TABLE 14.2 Solubility Product Constants[a]

Type	Salt	Ions of Salt	K_{sp}(25 °C)
Halides	$CaF_2 \rightleftharpoons Ca^{2+} + 2F^-$		3.9×10^{-11}
	$PbF_2 \rightleftharpoons Pb^{2+} + 2F^-$		3.6×10^{-8}
	$AgCl \rightleftharpoons Ag^+ + Cl^-$		1.8×10^{-10}
	$AgBr \rightleftharpoons Ag^+ + Br^-$		5.0×10^{-13}
	$AgI \rightleftharpoons Ag^+ + I^-$		8.3×10^{-17}
	$PbCl_2 \rightleftharpoons Pb^{2+} + 2Cl^-$		1.7×10^{-5}
	$PbBr_2 \rightleftharpoons Pb^{2+} + 2Br^-$		2.1×10^{-6}
	$PbI_2 \rightleftharpoons Pb^{2+} + 2I^-$		7.9×10^{-9}
Hydroxides	$Al(OH)_3 \rightleftharpoons Al^{3+} + 3OH^-$		3×10^{-34}(b)
	$Ca(OH)_2 \rightleftharpoons Ca^{2+} + 2OH^-$		6.5×10^{-6}
	$Fe(OH)_2 \rightleftharpoons Fe^{2+} + 2OH^-$		7.9×10^{-16}
	$Fe(OH)_3 \rightleftharpoons Fe^{3+} + 3OH^-$		1.6×10^{-39}
	$Mg(OH)_2 \rightleftharpoons Mg^{2+} + 2OH^-$		7.1×10^{-12}
	$Zn(OH)_2 \rightleftharpoons Zn^{2+} + 2OH^-$		3.0×10^{-16}(c)
Carbonates	$Ag_2CO_3 \rightleftharpoons 2Ag^+ + CO_3^{2-}$		8.1×10^{-12}
	$CaCO_3 \rightleftharpoons Ca^{2+} + CO_3^{2-}$		4.5×10^{-9}(d)
	$SrCO_3 \rightleftharpoons Sr^{2+} + CO_3^{2-}$		9.3×10^{-10}
	$BaCO_3 \rightleftharpoons Ba^{2+} + CO_3^{2-}$		5.0×10^{-9}
	$CoCO_3 \rightleftharpoons Co^{2+} + CO_3^{2-}$		1.0×10^{-10}
	$NiCO_3 \rightleftharpoons Ni^{2+} + CO_3^{2-}$		1.3×10^{-7}
	$ZnCO_3 \rightleftharpoons Zn^{2+} + CO_3^{2-}$		1.0×10^{-10}
Chromates	$Ag_2CrO_4 \rightleftharpoons 2Ag^{2+} + CrO_4^{2-}$		1.2×10^{-12}
	$PbCrO_4 \rightleftharpoons Pb^{2+} + CrO_4^{2-}$		1.8×10^{-14}(e)
Sulfates	$CaSO_4 \rightleftharpoons Ca^{2+} + SO_4^{2-}$		2.4×10^{-5}
	$SrSO_4 \rightleftharpoons Sr^{2+} + SO_4^{2-}$		3.2×10^{-7}
	$BaSO_4 \rightleftharpoons Ba^{2+} + SO_4^{2-}$		1.1×10^{-10}
	$PbSO_4 \rightleftharpoons Pb^{2+} + SO_4^{2-}$		6.3×10^{-7}

A more extensive table of K_{sp} values is given in Appendix D.

[a] Data for all salts (except lead chromate) are from A. E. Martell and R. M. Smith, *Critical Stability Constants*, Vol. 4. Plenum Press, New York, 1976. K_{sp} for $PbCrO_4$ is from R. C. Weast, editor, *Handbook of Chemistry and Physics*, 64th ed., 1984. CRC Press, Boca Raton, Fla.

[b] Alpha form.
[c] Amorphous form.
[d] Calcite form.
[e] At 18 °C.

Substituting,

$$K_{sp} = (x)(x) = 1.8 \times 10^{-10}$$
$$x^2 = 1.8 \times 10^{-10}$$
$$x = 1.3 \times 10^{-5}$$

The molar solubility of AgCl is 1.3×10^{-5} M in pure water.

EXAMPLE 14.16   CALCULATING SOLUBILITY FROM K_{sp}

Problem: Lead iodide, PbI_2, has $K_{sp} = 7.9 \times 10^{-9}$. What is the molar solubility of PbI_2 in water?

Solution: We have

$$PbI_2(s) \rightleftharpoons Pb^{2+}(aq) + 2I^-(aq)$$
$$K_{sp} = [Pb^{2+}][I^-]^2$$

The initial concentrations of Pb^{2+} and I^- are zero. If we let x be the molar solubility of PbI_2, then the Pb^{2+} concentration will increase by x and the I^- concentration will increase by $2x$.

	$PbI_2(s) \rightleftharpoons Pb^{2+}(aq) + 2I^-(aq)$	
Initial concentrations (M)	0.0	0.0
Changes in concentrations when PbI_2 dissolves (M)	$+x$	$+2x$
Equilibrium concentrations (M)	x	$2x$

The coefficients of x can be the same as the coefficients in the balanced chemical equation.

Substituting,

$$K_{sp} = (x)(2x)^2 = 7.9 \times 10^{-9}$$

Squaring $2x$ gives $4x^2$. Therefore,

$$4x^3 = 7.9 \times 10^{-9}$$
$$x^3 = 2.0 \times 10^{-9}$$
$$x = 1.3 \times 10^{-3}$$

The molar solubility of PbI_2 is 1.3×10^{-3} mol/L.

PRACTICE EXERCISE 18   What is the molar solubility of AgBr in water? (Obtain K_{sp} from Table 14.2.)

PRACTICE EXERCISE 19   What is the molar solubility of Ag_2CO_3 in water? (Obtain K_{sp} from Table 14.2.)

The Common Ion Effect

Suppose we establish the equilibrium

$$CaCO_3(s) \rightleftharpoons Ca^{2+}(aq) + CO_3^{2-}(aq)$$

and then add additional Ca^{2+} by dissolving some $CaCl_2$ in the solution. The resulting increase in the Ca^{2+} concentration will upset the equilibrium and, according to Le Châtelier's principle, the equilibrium will shift to the left to use up some of the Ca^{2+} that we added. This will cause some $CaCO_3$ to precipitate and less $CaCO_3$ will be left in solution when equilibrium is finally reestablished.

In this example, the Ca^{2+} in the final solution has two sources — the $CaCl_2$ that was added and the $CaCO_3$ that remains in the solution when equilibrium is reached again. Since calcium ion is common to both salts, it is called a **common ion.** The addition of the common ion lowers the solubility of the $CaCO_3$. The $CaCO_3$ is less soluble in the presence of $CaCl_2$ (or any other soluble calcium salt) than it is in pure water. This lowering of the solubility by the addition of a common ion is called the common ion effect.

Knowing about the common ion effect can be useful. If you wanted to remove Ca^{2+} from a solution (e.g., from hard water where the Ca^{2+} forms an insoluble precipitate with soap), you could add Na_2CO_3. By making sure that there is a relatively high concentration of CO_3^{2-} you could reduce the solubility of the $CaCO_3$ and thereby minimize the amount of free Ca^{2+} in the solution.

EXAMPLE 14.17 CALCULATION INVOLVING THE COMMON ION EFFECT

Problem: What is the molar solubility of PbI_2 in a 0.10 M NaI solution? For PbI_2, $K_{sp} = 7.9 \times 10^{-9}$.

Solution: The equilibrium is

$$PbI_2(s) \rightleftharpoons Pb^{2+}(aq) + 2I^-(aq)$$

and

$$K_{sp} = [Pb^{2+}][I^-]^2$$

The solution into which the PbI_2 is placed contains no Pb^{2+}, so the initial concentration of Pb^{2+} is zero. However, this solution does contain I^-. The salt, NaI, is fully dissociated in water, so the 0.10 M NaI solution contains 0.10 M Na^+ and 0.10 M I^-. Therefore, the initial concentration of I^- is 0.10 M. (As in Example 14.14, the solution also contains Na^+, but it isn't important because Na^+ is not involved in the equilibrium.) When x mol of PbI_2 now dissolves per liter, we will obtain x mol/L of Pb^{2+} and $2x$ mol/L of additional I^-. Setting up the concentration table,

	$PbI_2(s) \rightleftharpoons Pb^{2+}(aq) + 2I^-(aq)$	
Initial concentrations (M)	0.0	0.10
Changes in concentrations when PbI_2 dissolves (M)	$+x$	$+2x$
Equilibrium concentrations (M)	x	$0.10 + 2x$

Because K_{sp} is so small, we can reasonably expect that x, or $2x$, will be very small. In problems of this type it is safe to assume that $0.10 + 2x \approx 0.10$. (You may recall that this is the same type of simplification that we made earlier in Section 14.8). Substituting the final equilibrium quantities into the K_{sp} expression gives

$$K_{sp} = (x)(0.10)^2 = 7.9 \times 10^{-9}$$
$$x = \frac{7.9 \times 10^{-9}}{(0.10)^2}$$
$$x = [Pb^{2+}] = 7.9 \times 10^{-7}$$

Even though we felt confident our assumption was valid, we still had to check it at the end just to be sure.

Note that $2x$, which equals 1.6×10^{-6}, is much smaller than 0.10, just as we anticipated when we made our simplification. Also, note that in a 0.10 M NaI solution, the molar solubility of PbI_2 is 7.9×10^{-7} M, compared with 1.3×10^{-3} M in pure water (Example 14.16).

When problems like Example 14.17 are solved, the greatest number of errors are made when students double the initial concentration just because they see a coefficient of 2 before the I^- in the chemical equation for the equilibrium. In analyzing solubility problems, view the formation of the final solution as a two-step process. You begin with a solvent into which the "insoluble solid" will be placed. In some problems the solvent may be pure water, in which case the initial concentrations of the ions will be zero. In other problems, Example 14.17, for instance, the solvent will be a solution that contains a common ion. If this is so, decide what the concentration of the common ion is and enter this value in the "initial concentration" row. Then imagine that the solid is added

READ THIS PARAGRAPH!

to the solvent and a little of it dissolves. The amount that dissolves is what gives the values in the "change" row, and these entries must be in the same ratio as the coefficients in the balanced equation for the equilibrium. If you follow this two-step approach to setting up the problem, you will avoid many of the common errors that others make by rushing to an answer.

PRACTICE EXERCISE 20   What is the molar solubility of AgI in 0.20 M NaI solution? (Obtain K_{sp} from Table 14.2.)

PRACTICE EXERCISE 21   What is the molar solubility of $Fe(OH)_3$ in a solution with a hydroxide ion concentration of 0.050 M?

Predicting When Precipitation Occurs

Unsaturated, saturated, and supersaturated solutions were described in Section 4.5.

We can use K_{sp} to predict whether or not a precipitate will form in a given solution, provided we know the concentrations of the ions that appear in the K_{sp} expression. For example, we know that in a saturated solution of $CaCO_3$ the ion product, $[Ca^{2+}][CO_3^{2-}]$, is exactly equal to K_{sp}. Examining a portion of this solution shows that it is stable; no precipitate is forming in it. An unsaturated solution of $CaCO_3$ contains concentrations of Ca^{2+} and CO_3^{2-} that are lower than those in a saturated solution, so in the unsaturated solution the ion product, $[Ca^{2+}][CO_3^{2-}]$, is less than K_{sp}. No precipitate will form in an unsaturated solution because it is actually capable of dissolving more of the salt.

A supersaturated solution is one that contains more solute than is necessary for saturation. It is unstable and there is a tendency for the extra solute to precipitate. In a supersaturated solution of $CaCO_3$, the ion concentrations would be greater than in a saturated solution, and the ion product would be larger than K_{sp}. Comparison of K_{sp} with the ion product calculated for a given solution, therefore, serves to indicate whether or not a precipitate will be formed.

Precipitate will form	Ion product $>$ K_{sp} (supersaturated)
No precipitate will form	$\begin{cases} \text{Ion product} = K_{sp} \text{ (saturated)} \\ \text{Ion product} < K_{sp} \text{ (unsaturated)} \end{cases}$

EXAMPLE 14.18   PREDICTING WHETHER OR NOT A PRECIPITATE WILL FORM

Problem: A student wished to prepare 1.0 L of a solution containing 0.015 mol of NaCl and 0.15 mol of $Pb(NO_3)_2$. She was concerned that a precipitate of $PbCl_2$ might form. The K_{sp} of $PbCl_2$ is 1.7×10^{-5}. Can she expect to observe a precipitate of $PbCl_2$ in this mixture?

Solution: First, the ion product for $PbCl_2$ should be written

We can get the ion product by writing the equation for the dissociation of $PbCl_2$.

$PbCl_2(s) \rightleftharpoons Pb^{2+} + 2Cl^-$

$$[Pb^{2+}][Cl^-]^2$$

The solution that is being prepared will have the following concentrations.

$$[Pb^{2+}] = 0.15 \text{ mol}/1.0 \text{ L} = 0.15 \ M$$
$$[Cl^-] = 0.015 \text{ mol}/1.0 \text{ L} = 0.015 \ M$$

The value of the ion product is

$$[Pb^{2+}][Cl^-]^2 = (0.15)(0.015)^2$$
$$= 3.4 \times 10^{-5}$$

Since 3.4×10^{-5} is larger than K_{sp} (which equals 1.7×10^{-5}), a precipitate will form in the mixture.

PRACTICE EXERCISE 22   Will a precipitate of $CaSO_4$ form in a solution if the Ca^{2+} concentration is 0.0025 M and the SO_4^{2-} concentration is 0.030 M? For $CaSO_4$, $K_{sp} = 2.4 \times 10^{-5}$.

PRACTICE EXERCISE 23   Will a precipitate form in a solution containing 3.4×10^{-4} M CrO_4^{2-} and 4.8×10^{-5} M Ag^+? Use K_{sp} for Ag_2CrO_4 from Table 14.2.

EXAMPLE 14.19   PREDICTING WHETHER OR NOT A PRECIPITATE WILL FORM

Problem: If 50.0 mL of 0.0010 M $CaCl_2$ solution were added to 50.0 mL of 0.010 M Na_2SO_4 solution, would a precipitate of $CaSO_4$ be formed?

Solution: This problem is basically very similar to the previous one, except for one added bit of complication. Before we can compute the value of the ion product and compare it to K_{sp}, we must first take into account that the addition of one solution to another causes the ions in both of them to be diluted. Therefore, in this kind of problem, the first step is to consider dilution and calculate the concentrations in the combined mixture before any reaction takes place.

In dilution problems, a very simple equation applies.

$$M_iV_i = M_fV_f$$

where M_i and V_i are the initial molarity and volume, and M_f and V_f are the final molarity and volume. In this problem, the final volume is 100 mL, so the final concentrations are

$$[CaCl_2] = \frac{(50.0 \text{ mL})(0.0010 \text{ } M)}{100 \text{ mL}}$$
$$= 0.00050 \text{ } M$$
$$= 5.0 \times 10^{-4} \text{ } M$$

and

$$[Na_2SO_4] = \frac{(50.0 \text{ mL})(0.010 \text{ } M)}{100 \text{ mL}}$$
$$= 0.0050 \text{ } M$$
$$= 5.0 \times 10^{-3} \text{ } M$$

Since each $CaCl_2$ gives one Ca^{2+} ion and each Na_2SO_4 gives one SO_4^{2-} ion, the concentrations of the ions in the final mixture are

For $CaSO_4$, the dissociation reaction is

$$CaSO_4(s) \rightleftharpoons Ca^{2+} + SO_4^{2-}$$

Therefore, the ion product is $[Ca^{2+}][SO_4^{2-}]$

$$[Ca^{2+}] = 5.0 \times 10^{-4} \text{ } M$$
$$[SO_4^{2-}] = 5.0 \times 10^{-3} \text{ } M$$

These are the values that we now use to compute the ion product for $CaSO_4$.

$$[Ca^{2+}][SO_4^{2-}] = (5.0 \times 10^{-4})(5.0 \times 10^{-3})$$
$$= 2.5 \times 10^{-6}$$

In Table 14.2 the value of K_{sp} for $CaSO_4$ is given as 2.4×10^{-5}, so the ion product that we have calculated is smaller than K_{sp}, which means that no precipitate would form.

PRACTICE EXERCISE 24   Will a precipitate of $PbSO_4$ form if 100 mL of 1.0×10^{-3} M $Pb(NO_3)_2$ solution is added to 100 mL of 2.0×10^{-3} M $MgSO_4$ solution?

PRACTICE EXERCISE 25   Will a precipitate of $PbCl_2$ form if 50.0 mL of 0.10 M $Pb(NO_3)_2$ solution is added to 20.0 mL of 0.040 M NaCl solution?

14.11 COMPLEX ION EQUILIBRIA

In an aqueous solution, a complex exists in equilibrium with its components — a metal ion and the ligands.

Complex ions were discussed in Chapter 11, and as you may recall they are formed when Lewis bases (which we call *ligands*) attach themselves to metal ions in solution. An example is the blue $Cu(NH_3)_4{}^{2+}$ ion, which is formed by attaching four ammonia molecules to a Cu^{2+} ion. Complexes of this type are actually formed in a series of reactions in which one ligand after another becomes attached to the metal ion. Each of these reactions is an equilibrium, so the entire equilibrium system can be quite complicated. Fortunately, when the ligand concentration is large we can simplify the problem by just considering the overall reaction for the formation of the complex, and these will be the kinds of situations with which we shall work. Thus, the equilibrium for the formation of the copper – ammonia complex can be considered to be just

$$Cu^{2+}(aq) + 4NH_3(aq) \rightleftharpoons Cu(NH_3)_4{}^{2+}(aq)$$

Our principal goals here will be to study such equilibria and to learn how they can be used to influence the solubilities of salts.

Formation Constants of Complex Ions

Because the formation of a complex ion involves an equilibrium, there is an associated equilibrium law and an equilibrium constant, which is called the formation constant, K_{form}. For example, the equilibrium law for the formation of $Cu(NH_3)_4{}^{2+}$ is

$$\frac{[Cu(NH_3)_4{}^{2+}]}{[Cu^{2+}][NH_3]^4} = K_{form}$$

Sometimes this equilibrium constant is called a stability constant, because the larger its magnitude, the larger the equilibrium concentration of the complex, and therefore the more stable the complex must be. Some other complex ions and their formation constants are listed in Table 14.3. Notice that most of the metal ions that form complexes are those of transition elements. In fact, the tendency to form complex ions is often listed as one of the general properties of the transition metals.

In Chapter 11 you learned that the formulas of complex ions often are written using brackets; for example $[Cu(NH_3)_4]^{2+}$. Note that the charge appears outside the brackets. When brackets are used to mean molar concentration, they aren't used in writing the formula of the complex ion itself and the charge appears inside the brackets along with the rest of the formula of the complex.

Appendix D contains an even more extensive table of equilibrium constants for the formation of complexes.

TABLE 14.3 Formation Constants and Instability Constants for Some Complex Ions

Ligand	Equilibrium	K_{form}	K_{inst}
NH_3	$Ag^+ + 2NH_3 \rightleftharpoons Ag(NH_3)_2{}^+$	1.6×10^7	6.3×10^{-8}
	$Co^{2+} + 6NH_3 \rightleftharpoons Co(NH_3)_6{}^{2+}$	5.0×10^4	2.0×10^{-5}
	$Co^{3+} + 6NH_3 \rightleftharpoons Co(NH_3)_6{}^{3+}$	4.6×10^{33}	2.2×10^{-34}
	$Cu^{2+} + 4NH_3 \rightleftharpoons Cu(NH_3)_4{}^{2+}$	1.1×10^{13}	9.1×10^{-14}
	$Hg^{2+} + 4NH_3 \rightleftharpoons Hg(NH_3)_4{}^{2+}$	1.8×10^{19}	5.6×10^{-20}
F^-	$Al^{3+} + 6F^- \rightleftharpoons AlF_6{}^{3-}$	1×10^{20}	1×10^{-20}
	$Sn^{4+} + 6F^- \rightleftharpoons SnF_6{}^{2-}$	1×10^{25}	1×10^{-25}
Cl^-	$Hg^{2+} + 4Cl^- \rightleftharpoons HgCl_4{}^{2-}$	5.0×10^{15}	2.0×10^{-16}
Br^-	$Hg^{2+} + 4Br^- \rightleftharpoons HgBr_4{}^{2-}$	1.0×10^{21}	1.0×10^{-21}
I^-	$Hg^{2+} + 4I^- \rightleftharpoons HgI_4{}^{2-}$	1.9×10^{30}	5.3×10^{-31}
CN^-	$Fe^{2+} + 6CN^- \rightleftharpoons Fe(CN)_6{}^{4-}$	1.0×10^{24}	1.0×10^{-24}
	$Fe^{3+} + 6CN^- \rightleftharpoons Fe(CN)_6{}^{3-}$	1.0×10^{31}	1.0×10^{-31}

Data from L. G. Sillén and A. E. Martell, *Stability Constants of Metal–Ion Complexes*, The Chemical Society, London, Special Publication 17, 1964.

Instability Constants

In many chemical references the relative stabilities of complex ions are indicated in another way, and you should know about it. The inverses of formation constants are cited and are called instability constants, K_{inst}. In other words, instead of focusing attention on the *formation* of the complex ion, the equilibrium equation is written so as to show the *breakdown* of the ion as the left-to-right process. Using the complex between copper(II) ion and ammonia as our example again, we can rewrite the equilibrium as

$$Cu(NH_3)_4{}^{2+}(aq) \rightleftharpoons Cu^{2+}(aq) + 4NH_3(aq)$$

The instability constant, accordingly, is

$$K_{inst} = \frac{[Cu^{2+}][NH_3]^4}{[Cu(NH_3)_4{}^{2+}]} = \frac{1}{K_{form}}$$

K_{inst} is called an *instability* constant because the larger its value, the more *unstable* is the complex. This is also shown by the data in Table 14.3.

Effect of Complex Ion Formation on Solubility

The silver halides are notoriously insoluble compounds; the K_{sp} of AgBr, for example, is only 5.0×10^{-13} (at 25 °C). A saturated solution of this salt in water has only 7.1×10^{-7} mol/L of each of its ions, Ag^+ and Br^-. Suppose now that we have made a saturated solution of this salt, a solution in which undissolved AgBr(s) rests on the bottom of the flask so that there is a true equilibrium. Suppose further that we next add some aqueous ammonia to this system. Its molecules are strong ligands for silver ions, so they begin to form $Ag(NH_3)_2{}^+$ ions from the trace amounts of silver ions initially present in solution.

$$Ag^+(aq) + 2NH_3(aq) \rightleftharpoons Ag(NH_3)_2{}^+(aq)$$

In other words, the ammonia molecules upset the equilibrium present in the saturated solution.

$$AgBr(s) \rightleftharpoons Ag^+(aq) + Br^-(aq)$$

By pulling Ag^+ ions out of this equilibrium, the equilibrium must shift to replace them as best it can (Le Châtelier's principle). In other words, more AgBr must go into solution. This is a specific example of a general phenomenon:

> The solubility of a slightly soluble salt increases when one of its ions can be changed to a soluble complex ion.

Adding the ammonia introduces another equilibrium, and to analyze what happens, we can describe the net, overall effect as the sum of the two equilibria present.

New Equilibrium $Ag^+(aq) + 2NH_3(aq) \rightleftharpoons Ag(NH_3)_2{}^+(aq)$	$K_{form} = \dfrac{[Ag(NH_3)_2{}^+]}{[Ag^+][NH_3]^2}$
Original equilibrium $AgBr(s) \rightleftharpoons Ag^+(aq) + Br^-(aq)$	$K_{sp} = [Ag^+][Br^-]$
Sum of equilibria $AgBr(s) + 2NH_3(aq) \rightleftharpoons Ag(NH_3)_2{}^+(aq) + Br^-(aq)$	

The equilibrium constant for this overall equilibrium is defined in the usual way by an equation written so as not to include a term for the insoluble solid, AgBr(s), because it

Notice that Ag^+ cancels from both sides when the equations are added.

has a constant value anyway.

$$K_c = \frac{[Ag(NH_3)_2^+][Br^-]}{[NH_3]^2}$$

K_c for this expression can be obtained by multiplying K_{form} by K_{sp}, as we can show by making the appropriate substitutions and canceling.

If we add two equilibria to obtain a third one, we multiply their K_c's to obtain the new K_c.

$$K_c = K_{form} \times K_{sp} = \frac{[Ag(NH_3)_2^+]}{[Ag^+][NH_3]^2} \times [Ag^+][Br^-]$$

If we know the values of K_{form} and K_{sp} for this system, we can find K_c. Then if we know the molar concentration of the ligand, NH_3, we can calculate the maximum molar concentrations of $Ag(NH_3)_2^+$ and Br^- (the two are essentially equal because K_{form} is very large) that can be in solution. In other words, we can find the solubility of AgBr in a dilute ammonia solution. Let's see how this works.

EXAMPLE 14.20   CALCULATING THE SOLUBILITY OF A SLIGHTLY SOLUBLE SALT IN THE PRESENCE OF A LIGAND

Problem: How many moles of AgBr can dissolve in 1.0 L of 1.0 M NH_3?

Solution: We have the overall equilibrium.

$$AgBr(s) + 2NH_3(aq) \rightleftharpoons Ag(NH_3)_2^+(aq) + Br^-(aq)$$

We have the expression for K_c.

$$K_c = \frac{[Ag(NH_3)_2^+][Br^-]}{[NH_3]^2} = K_{form} \times K_{sp}$$

Since for AgBr, $K_{sp} = 5.0 \times 10^{-13}$, and for the complex ion, $K_{form} = 1.6 \times 10^7$, then

$$K_c = (1.6 \times 10^7)(5.0 \times 10^{-13})$$
$$= 8.0 \times 10^{-6}$$

With these preliminaries done, we can proceed to set up our usual concentration table, letting x stand for the number of moles of AgBr that dissolve in the 1.0 L of 0.10 M NH_3.

	AgBr(s) + 2NH$_3$(aq) $\rightleftharpoons$	Ag(NH$_3$)$_2^+$(aq) +	Br$^-$(aq)
Initial concentrations (*M*)	1.0	0.0	0.0
Changes in concentration caused by NH$_3$	$-2x$	$+x$	$+x$ (Note 1)
Equilibrium concentrations (*M*)	$(1.0 - 2x)$	x	x (Note 2)

Note 1. The 2 in $2x$ signifies that each mole of AgBr that dissolves consumes twice as many moles of NH_3.

Note 2. Letting the concentration of Br^- equal the concentration of $Ag(NH_3)_2^+$ is valid because and only because K_{form} is such a large number that essentially *all* Ag^+ ions that do dissolve out of the insoluble salt, AgBr, are changed to the complex ion. There are very few uncomplexed Ag^+ ions in the solution.

Using the values in the last row of the concentration table to substitute into the equation for K_c gives us

$$\frac{(x)(x)}{(1.0 - 2x)^2} = 8.0 \times 10^{-6}$$

To simplify this, we take the square root of both sides.

$$\frac{x}{(1.0 - 2x)} = 2.8 \times 10^{-3}$$

$x = 0.0028 - 2x(0.0028)$
$\quad = 0.0028 - 0.0056x$
$1.0056x = 0.0028$
$\quad\quad x = 0.0028$ (rounded)

Solving for x,

$$x = 2.8 \times 10^{-3}$$

In other words, 2.8×10^{-3} mol of AgBr dissolves in 1.0 L of 1.0 M NH$_3$. This isn't very much, of course, but in contrast, only 7.1×10^{-7} mol of AgBr dissolves in 1.0 L of pure water. Thus AgBr is nearly 4000 times more soluble in 1.0 M NH$_3$ than in pure water.

PRACTICE EXERCISE 26   Calculate the solubility of silver chloride in 0.10 M NH$_3$ and compare it with its solubility in pure water. For AgCl, $K_{sp} = 1.8 \times 10^{-10}$.

PRACTICE EXERCISE 27   How many moles of NH$_3$ have to be added to 1.0 L of water to dissolve 0.20 mol of AgCl? The complex ion Ag(NH$_3$)$_2{}^+$ forms. For AgCl, $K_{sp} = 1.8 \times 10^{-10}$.

SUMMARY

Dynamic Equilibrium.   When two opposing chemical reactions occur at equal rates, the system is in a state of dynamic equilibrium and the concentrations of reactants and products remain constant. For a given overall chemical composition, the amounts of reactants and products that are present at equilibrium are the same regardless of whether the equilibrium is approached from the direction of pure reactants, pure products, or any mixture of reactants and products.

Equilibrium Law.   The mass action expression is a fraction. The concentrations of the products, raised to powers equal to their coefficients in the chemical equation, are multiplied together in the numerator. The denominator is constructed in the same way from the concentrations of the reactants raised to powers equal to their coefficients. At equilibrium, the mass action expression is equal to the equilibrium constant, K_c. If partial pressures of gases are used in the mass action expression, K_p is obtained. The magnitude of the equilibrium constant is roughly proportional to the extent to which the reaction proceeds to completion when equilibrium is reached.

Relating K_p to K_c.   The values of K_p and K_c are only equal if the same number of moles of gas are represented on both sides of the chemical equation. When the number of moles of gas are different, K_p is related to K_c by the equation $K_p = K_c(RT)^{\Delta n_g}$. Remember to use $R = 0.0821$ L atm mol^{-1} K^{-1} and $T =$ absolute temperature. Also, be careful to calculate Δn_g as the difference between the number of moles of *gas* found among the products and the number of moles of *gas* found among the reactants.

Le Châtelier's Principle.   When an equilibrium is upset, a chemical change occurs in a direction that opposes the disturbing influence and brings the system to equilibrium again. Adding a reac-

tant or a product causes a reaction to occur that uses up part of what has been added. Removing a reactant or a product causes a reaction that replaces part of what has been removed. Increasing the pressure (by reducing the volume) drives a reaction in the direction of the fewest number of moles of gas. Pressure has virtually no effect on equilibria involving only solids and liquids. Raising the temperature causes an equilibrium to shift in an endothermic direction. The value of K increases with increasing temperature for reactions that are endothermic in the forward direction. A change in temperature is the *only* factor that changes K. Addition of a catalyst has no effect on an equilibrium.

Equilibrium Calculations.   The initial concentrations in a chemical system are determined by the person who combines the chemicals at the start of the reaction. The changes in concentration are determined by the stoichiometry of the reaction. Only equilibrium concentrations satisfy the equilibrium law. When these are used, the mass action expression is equal to K_c. When a change in concentration is expected to be very small compared to the initial concentration, the change may be neglected and the equation corresponding to the equilibrium law can be simplified. In general, this simplification is valid if the initial concentration is at least 1000 times larger than K.

Heterogeneous Equilibria and Solubility Product.   The mass action expression for a heterogeneous equilibrium never contains concentration terms for pure liquids or pure solids. When the heterogeneous equilibrium involves the solubility of a salt, the mass action expression is a product of ion concentrations raised to appropriate powers and the equilibrium constant, K_{sp}, is the solubility product constant. The molar solubility can be used to calculate K_{sp}, and K_{sp} can be used to calculate molar solubility. According to the common ion effect, a salt is less soluble in a solution containing one of its ions than it is in pure water. A solution containing a mixture of ions will produce a precipitate of a given salt only if the solution is supersaturated. Under these conditions the ion product is greater than the value of K_{sp} for the salt.

Complex Ion Equilibria.   Complex ions are formed when a number of molecules or anions become bound tightly to a metal ion. These species are formed especially easily by ions of the transition metals. The stabilities of complex ions vary widely, and they can be compared by means of formation constants (also called stability constants), K_{form}. Sometimes the reciprocals of formation constants are cited, and are called instability constants. The solubilities of "insoluble" salts are increased if the cation is able to form a complex ion with some other substance in the solution that is able to serve as a ligand. Some otherwise insoluble salts can be drawn completely into solution in this manner. We can use the value of K_{form} to calculate this effect on solubility, or we can find out how much of the ligand-contributing solute is needed to dissolve an "insoluble" salt.

REVIEW EXERCISES

Answers to questions whose numbers are printed in color are given in Appendix C. Difficult questions are marked with asterisks.

14.1   Why are chemical equilibria called *dynamic equilibria?*

14.2   Sketch a graph showing how the concentrations of the reactants and products of a typical chemical reaction vary with time during the course of the reaction. Assume no products are present at the start of the reaction.

14.3   What is meant when we say that chemical reactions are reversible?

Mass Action Expression, K_p and K_c

14.4   What relationship exists between the coefficients in a balanced chemical equation and the mass action expression for the reaction?

14.5   How is the mass action expression related to K_c?

14.6   In general, what units must the quantities in the mass action have if they are going to be related to K_c?

14.7   How is the term *reaction quotient* defined?

14.8   Under what conditions does the reaction quotient equal K_c?

14.9   What is an *equilibrium law?*

14.10   Write the equilibrium law for each of the following reactions in terms of molar concentrations:
(a) $2PCl_3(g) + O_2(g) \rightleftharpoons 2POCl_3(g)$
(b) $2SO_3(g) \rightleftharpoons 2SO_2(g) + O_2(g)$
(c) $N_2H_4(g) + 2O_2(g) \rightleftharpoons 2NO(g) + 2H_2O(g)$
(d) $N_2H_4(g) + 6H_2O_2(g) \rightleftharpoons 2NO_2(g) + 8H_2O(g)$

14.11   Write the equilibrium law for each of the following reactions in terms of molar concentrations:
(a) $H_2(g) + Cl_2(g) \rightleftharpoons 2HCl(g)$
(b) $\frac{1}{2}H_2(g) + \frac{1}{2}Cl_2(g) \rightleftharpoons HCl(g)$

How does the K_c for reaction (a) compare with the K_c for reaction (b)?

14.12   Write the equilibrium law for the reaction

$$2HCl(g) \rightleftharpoons H_2(g) + Cl_2(g)$$

How does K_c for this reaction compare with K_c for reaction (a) in Review Exercise 14.11?

14.13   Write the equilibrium law for the reactions in Review Exercise 14.11 in terms of partial pressures.

14.14   When a chemical equation and its equilibrium constant are given, why is it not necessary to also specify the form of the mass action expression?

14.15   At 225 °C, $K_p = 6.3 \times 10^{-3}$ for the reaction

$$CO(g) + 2H_2(g) \rightleftharpoons CH_3OH(g)$$

Would we expect this reaction to go nearly to completion?

14.16   Here are some reactions and their equilibrium constants.
(a) $2CH_4(g) \rightleftharpoons C_2H_6(g) + H_2(g)$ $K_c = 9.5 \times 10^{-13}$
(b) $CH_3OH(g) + H_2(g) \rightleftharpoons CH_4(g) + H_2O(g)$
      $K_c = 3.6 \times 10^{20}$
(c) $C_2H_4(g) + H_2O(g) \rightleftharpoons C_2H_5OH(g)$ $K_c = 8.2 \times 10^3$

Arrange these reactions in order of their increasing tendency to go toward completion.

Converting between K_p and K_c

14.17   State the equation relating K_p to K_c. In this equation, what is the definition of the term Δn_g?

14.18   Write equilibrium laws in terms of K_c and K_p for the reaction $2NO_2(g) \rightleftharpoons N_2O_4(g)$. Using units of mol/L for concentration and atm for partial pressures, show by means of unit cancellation that the only value of R that can be used in Equation 14.4 is $R = 0.0821$ L atm mol^{-1} K^{-1}.

14.19   For which of the following reactions would $K_p = K_c$?
(a) $N_2(g) + O_2(g) \rightleftharpoons 2NO(g)$
(b) $2H_2(g) + C_2H_2(g) \rightleftharpoons C_2H_6(g)$
(c) $2NO(g) + O_2(g) \rightleftharpoons 2NO_2(g)$
(d) $CO_2(g) + H_2(g) \rightleftharpoons CO(g) + H_2O(g)$

14.20   The reaction $CO(g) + 2H_2(g) \rightleftharpoons CH_3OH(g)$ has $K_p = 6.3 \times 10^{-3}$ at 225 °C. What is the value of K_c at this temperature?

14.21   One possible way of removing NO from the exhaust of a gasoline engine is to cause it to react with CO in the presence of a suitable catalyst.

$$2NO(g) + 2CO(g) \longrightarrow N_2(g) + 2CO_2(g)$$

At 300 °C, this reaction has $K_c = 2.2 \times 10^{59}$. What is K_p at 300 °C?

14.22 At 773 °C, the reaction $CO(g) + 2H_2(g) \rightleftharpoons CH_3OH(g)$ has $K_c = 0.40$. What is K_p at this temperature?

14.23 The reaction $COCl_2(g) \rightleftharpoons CO(g) + Cl_2(g)$ has $K_p = 4.6 \times 10^{-2}$ at 395 °C. What is K_c?

Le Châtelier's Principle

14.24 State Le Châtelier's principle *in your own words*.

14.25 How will the equilibrium

$$\text{heat} + CH_4(g) + 2H_2S(g) \rightleftharpoons CS_2(g) + 4H_2(g)$$

be affected by the following?
(a) The addition of $CH_4(g)$.
(b) The addition of $H_2(g)$.
(c) The removal of $CS_2(g)$.
(d) A decrease in the volume of the container.
(e) An increase in temperature.

14.26 The reaction $CO(g) + 2H_2(g) \rightleftharpoons CH_3OH(g)$ has $\Delta H° = -18$ kJ. How will the amount of CH_3OH present at equilibrium be affected by the following?
(a) Adding $CO(g)$.
(b) Removing $H_2(g)$.
(c) Decreasing the volume of the container.
(d) Adding a catalyst.
(e) Increasing the temperature.

14.27 In Review Exercises 14.25 and 14.26, which change(s) will alter K_c?

14.28 Consider the equilibrium $2NO(g) + Cl_2(g) \rightleftharpoons 2NOCl(g)$, for which $\Delta H° = -77.07$ kJ. How will the amount of Cl_2 at equilibrium be affected by the following?
(a) Removal of $NO(g)$.
(b) Addition of $NOCl(g)$.
(c) Raising the temperature.
(d) Decreasing the volume of the container.

14.29 Which of the equilibria in Review Exercise 14.19 will *not* be affected by a change in the volume of the container?

14.30 What would have to be true about $\Delta H°$ for a reaction to have an equilibrium constant that is the same at all temperatures?

Equilibrium Calculations

14.31 At 773 °C, a mixture of $CO(g)$, $H_2(g)$, and $CH_3OH(g)$ was allowed to come to equilibrium. The following equilibrium concentrations were then measured: $[CO] = 0.105$ M, $[H_2] = 0.250$ M, $[CH_3OH] = 0.00261$ M. Calculate K_c for the reaction $CO(g) + 2H_2(g) \rightleftharpoons CH_3OH(g)$.

14.32 At a certain temperature, $K_c = 0.18$ for the equilibrium

$$PCl_3(g) + Cl_2(g) \rightleftharpoons PCl_5(g)$$

Suppose a reaction vessel contained these three gases at the following concentrations: $[PCl_3] = 0.0520$ M, $[Cl_2] = 0.0140$ M, $[PCl_5] = 0.00600$ M.
(a) Is the system in a state of equilibrium?
(b) If not, which direction will the reaction have to proceed to get to equilibrium?

14.33 At 460 °C, the reaction

$$SO_2(g) + NO_2(g) \rightleftharpoons NO(g) + SO_3(g)$$

has $K_c = 85.0$. A reaction flask contains these gases at the following concentrations: $[SO_2] = 0.00150$ M, $[NO_2] = 0.00300$ M, $[NO] = 0.0100$ M, $[SO_3] = 0.0400$ M.
(a) Is the reaction at equilibrium?
(b) If not, which way will the reaction have to proceed to arrive at equilibrium?

14.34 The reaction $N_2O_4(g) \rightleftharpoons 2NO_2(g)$ has $K_p = 0.140$ atm^{-1} at 25 °C. In a reaction vessel containing these gases in equilibrium at this temperature, the partial pressure of N_2O_4 was 0.300 atm.
(a) What was the partial pressure of the NO_2 in the reaction mixture?
(b) What was the total pressure of the mixture of gases?

14.35 Ethylene, C_2H_4, and water react under appropriate conditions to give ethanol. The reaction is

$$C_2H_4(g) + H_2O(g) \rightleftharpoons C_2H_5OH(g)$$

An equilibrium mixture of these gases at a certain temperature had the following concentrations: $[C_2H_4] = 0.0222$ M, $[H_2O] = 0.0225$ M, $[C_2H_5OH] = 0.150$ M. What is the value of K_c?

14.36 At a certain temperature, the reaction $CO(g) + 2H_2(g) \rightleftharpoons CH_3OH(g)$ has $K_c = 0.500$. If a reaction mixture at equilibrium contains 0.210 M CO and 0.100 M H_2, what is the concentration of CH_3OH?

14.37 $K_c = 64$ for the reaction, $N_2(g) + 3H_2(g) \rightleftharpoons 2NH_3(g)$, at a certain temperature. Suppose it was found that an equilibrium mixture of these gases contained 0.280 M NH_3 and 0.00840 M N_2. What was the concentration of H_2 in the mixture?

14.38 At high temperature, 0.500 mol of HBr was placed in a 1.00-L container and allowed to decompose according to the reaction $2HBr(g) \rightleftharpoons H_2(g) + Br_2(g)$. At equilibrium the concentration of Br_2 was measured to be 0.130 M. What is K_c for this reaction at this temperature?

14.39 A 0.100-mol sample of formaldehyde vapor, CH_2O, was placed in a heated 1.00-L vessel and some of it decomposed. The reaction is

$$CH_2O(g) \rightleftharpoons H_2(g) + CO(g)$$

At equilibrium, the $CH_2O(g)$ concentration was 0.080 mol/L. Calculate the value of K_c for this reaction.

14.40 The equilibrium constant, K_c, for the reaction

$$SO_3(g) + NO(g) \rightleftharpoons NO_2(g) + SO_2(g)$$

was found to be 0.500 at a certain temperature. If 0.300 mol of SO_3 and 0.300 mol of NO were placed in a 2.00-L container and allowed to react, what would be the equilibrium concentration of each gas?

14.41 At a certain temperature the reaction

$$CO(g) + H_2O(g) \rightleftharpoons CO_2(g) + H_2(g)$$

has $K_c = 0.400$. Exactly 1.00 mol of each gas was placed in a 100-L vessel and the mixture was allowed to react. What was the equilibrium concentration of each gas?

14.42 The reaction $2HCl(g) \rightleftharpoons H_2(g) + Cl_2(g)$ has $K_c = 3.2 \times 10^{-34}$ at 25 °C. If a reaction vessel contains initially 2.00 mol/L of HCl and is allowed to come to equilibrium, what will be the concentrations of H_2 and Cl_2?

*14.43** The reaction $H_2(g) + Br_2(g) \rightleftharpoons 2HBr(g)$ has $K_c = 2.0 \times 10^9$ at 25 °C. If 0.100 mol of H_2 and 0.100 mol of Br_2 were placed in a 10.0-L container and allowed to react, what would all the equilibrium concentrations be at 25 °C?

*14.44** In Review Exercise 14.40 it was stated that at a certain temperature, $K_c = 0.500$ for the reaction

$$SO_3(g) + NO(g) \rightleftharpoons NO_2(g) + SO_2(g)$$

If 0.100 mol SO_3 and 0.200 mol NO were placed in a 2.0-L container and allowed to react, what would the NO_2 and SO_2 concentrations be at equilibrium?

Heterogeneous Equilibria

14.45 What is the difference between a heterogeneous and a homogeneous equilibrium?

14.46 Write the equilibrium law for each of the following heterogeneous reactions.
(a) $2C(s) + O_2(g) \rightleftharpoons 2CO(g)$
(b) $2NaHSO_3(s) \rightleftharpoons Na_2SO_3(s) + H_2O(g) + SO_2(g)$
(c) $2C(s) + 2H_2O(g) \rightleftharpoons CH_4(g) + CO_2(g)$
(d) $CaCO_3(s) + 2HF(g) \rightleftharpoons CaF_2(s) + H_2O(g) + CO_2(g)$

14.47 Why can we omit the concentrations of pure liquids and solids from the mass action expression of heterogeneous reactions?

14.48 The heterogeneous reaction $2HCl(g) + I_2(s) \rightleftharpoons 2HI(g) + Cl_2(g)$ has $K_c = 1.6 \times 10^{-34}$ at 25 °C. Suppose 1.00 mol of HCl and solid I_2 was placed in a 1.00-L container. What would be the equilibrium concentrations of HI and Cl_2 in the container?

14.49 Calculate the molar concentration of water in (a) 18.0 mL of H_2O, (b) 100.0 mL of H_2O, and (c) 1.00 L of H_2O. Assume that the density of water is 1.00 g/mL.

14.50 Refer to the latest edition of the *Handbook of Chemistry and Physics* to find the density of sodium chloride. Calculate the molar concentration of NaCl in pure solid NaCl.

Solubility Products

14.51 What is the difference between an *ion product* and an *ion product constant*?

14.52 The solubility product constant can be described as a *modified* equilibrium constant.
(a) Write the full equilibrium constant expression for the system

$$Ba_3(PO_4)_2(s) \rightleftharpoons 3Ba^{2+}(aq) + 2PO_4^{3-}(aq)$$

(b) How is this equilibrium constant expression modified and how is the modification justified?
(c) Write the solubility product constant expression for the equilibrium of part a.

14.53 Write the K_{sp} expression for these compounds.
(a) CaF_2 (c) $PbSO_4$ (e) PbI_2
(b) Ag_2CO_3 (d) $Fe(OH)_3$ (f) $Cu(OH)_2$

14.54 Write the K_{sp} expression for these compounds.
(a) AgI   (c) $PbCrO_4$ (e) $ZnCO_3$
(b) Ag_3PO_4 (d) $Al(OH)_3$ (f) $Zn(OH)_2$

14.55 What is the common ion effect? How does Le Châtelier's principle explain it?

14.56 With respect to K_{sp}, what conditions must be met if a precipitate is going to form in a solution?

14.57 Barium sulfate, $BaSO_4$, is so insoluble that it can be swallowed without significant danger, even though Ba^{2+} is toxic. At 25 °C, 1.00 L of water dissolves only 0.00245 g of $BaSO_4$.
(a) How many moles of $BaSO_4$ dissolve per liter?
(b) What are the molar concentrations of Ba^{2+} and SO_4^{2-} in a saturated $BaSO_4$ solution?
(c) Calculate K_{sp} for $BaSO_4$.

14.58 Magnesium hydroxide, $Mg(OH)_2$, found in milk of magnesia, has a solubility of 7.05×10^{-3} g/L at 25 °C.
(a) What is the solubility expressed in moles per liter?
(b) What are the Mg^{2+} and OH^- concentrations in moles per liter?
(c) Calculate K_{sp} for $Mg(OH)_2$.

14.59 At 25 °C the molar solubility of Ag_3PO_4 is 1.8×10^{-5} mol/L. Calculate K_{sp} for this salt.

14.60 The molar solubility of $Ba_3(PO_4)_2$ in water is 1.4×10^{-8} mol/L. Calculate K_{sp} for this salt.

14.61 It was found that the molar solubility of $BaSO_3$ in 0.10 M $BaCl_2$ solution is 8.0×10^{-6} M. What is the value of K_{sp} for $BaSO_3$?

14.62 A student prepared a saturated solution of $CaCrO_4$ and found that when 100 mL of this solution was evaporated, 0.416 g of $CaCrO_4$ was left behind. What is the value of K_{sp} for this salt?

14.63 Copper(I) chloride has $K_{sp} = 1.9 \times 10^{-7}$. Calculate the molar solubility of CuCl in (a) pure water, (b) 0.010 M HCl solution, (c) 0.10 M HCl solution, and (d) 0.10 M $CaCl_2$ solution.

14.64 Gold(III) chloride, $AuCl_3$, has $K_{sp} = 3.2 \times 10^{-25}$. Calculate the molar solubility of $AuCl_3$ in (a) pure water, (b) 0.020 M HCl solution, (c) 0.020 M $MgCl_2$ solution, and (d) 0.020 M $Au(NO_3)_3$ solution.

14.65 At 25 °C the value of K_{sp} for LiF is 1.7×10^{-3} and for BaF_2, 1.7×10^{-6}. In terms of moles per liter, which salt is the more soluble in water? Calculate the solubility of each in these units.

14.66 At 25 °C the value of K_{sp} for AgCN is 2.2×10^{-16} and for $Zn(CN)_2$, 3×10^{-16}. In terms of grams per 100 mL of solution, which salt is the more soluble in water? Calculate the solubility of each in terms of these units.

14.67 A salt whose formula is of the form MX has a value of K_{sp} equal to 2.0×10^{-10}. Another slightly soluble salt, MX_3, has to have what value of K_{sp} if the molar solubilities of the two salts are identical?

14.68 A salt having a formula of the type M_2X_3 has $K_{sp} = 1.0 \times 10^{-20}$. Another salt, M_2X, has to have what K_{sp} value if M_2X has twice the molar solubility of M_2X_3?

14.69   Calcium sulfate is found in plaster. At 25 °C the value of K_{sp} for $CaSO_4$ is 2.4×10^{-5}. What is the calculated solubility of $CaSO_4$ in water expressed in moles per liter?

14.70   Chalk is $CaCO_3$, and at 25 °C its $K_{sp} = 4.5 \times 10^{-9}$. What is the molar solubility of $CaCO_3$? How many grams of $CaCO_3$ dissolve in 100 mL of water?

14.71   Calculate the molar solubility of lead iodide, PbI_2, in water ($K_{sp} = 7.9 \times 10^{-9}$).

14.72   What is the molar solubility of Ag_2CO_3 in water (at 25 °C, $K_{sp} = 8.1 \times 10^{-12}$)?

14.73   What is the molar solubility of Ag_2CrO_4 in 0.10 M $AgNO_3$ at 25 °C? For Ag_2CrO_4 at 25 °C, $K_{sp} = 1.2 \times 10^{-12}$.

14.74   What is the molar solubility of $Mg(OH)_2$ in 0.10 M NaOH? (NaOH is a strong electrolyte and is completely dissociated into Na^+ and OH^- in aqueous solution.) For $Mg(OH)_2$, $K_{sp} = 7.1 \times 10^{-12}$.

14.75   Does a precipitate of $PbCl_2$ form when 0.0100 mol of $Pb(NO_3)_2$ and 0.0100 mol of NaCl are dissolved in 1.00 L of solution?

14.76   Silver acetate, $AgC_2H_3O_2$, has $K_{sp} = 4 \times 10^{-3}$. Does a precipitate form when 0.010 mol of $AgNO_3$ and 0.30 mol of $Ca(C_2H_3O_2)_2$ are dissolved in a total volume of 1.00 L of solution?

14.77   Does a precipitate of $PbBr_2$ form if 50.0 mL of 0.010 M $Pb(NO_3)_2$ is mixed with (a) 50.0 mL of 0.010 M KBr? (b) 50.0 mL of 0.10 M NaBr?

14.78   Would a precipitate of silver acetate form if 18.0 mL of 0.10 M $AgNO_3$ were added to 40.0 mL of 0.024 M $NaC_2H_3O_2$? For $AgC_2H_3O_2$, $K_{sp} = 4 \times 10^{-3}$.

*14.79   Both AgCl and AgI are "insoluble," but the solubility of AgI is much less than that of AgCl, as can be seen by their K_{sp} values. Suppose that a solution contained both Cl^- and I^- with concentrations as follows: $[Cl^-] = 0.10$ M, $[I^-] = 0.050$ M. If solid $AgNO_3$ were added to 1.00 L of this mixture (so that no appreciable change in volume occurred), what would the I^- concentration be when AgCl first begins to precipitate?

*14.80   Suppose that 50.0 mL of 0.10 M $AgNO_3$ were added to 50.0 mL of 0.050 M NaCl solution.
(a) What weight of AgCl would be formed?
(b) What would be the final concentrations of *all* the ions contributed by these salts?

*14.81   Calculate the molar solubility of $CaSO_4$ in 0.010 M $CaCl_2$ solution.

14.82   Suppose Na_2SO_4 is added gradually to 100 mL of a solution that contains both calcium ion (0.10 M) and strontium ion (0.10 M). What will the Sr^{2+} concentration be when $CaSO_4$ just begins to precipitate?

Complex Ion Equilibria

14.83   Using Le Châtelier's principle, explain how the addition of aqueous ammonia dissolves silver chloride. If HNO_3 is added after the AgCl has dissolved in the NH_3 solution, it causes AgCl to reprecipitate. Explain why.

14.84   For $PbCl_3^-$, $K_{form} = 2.5 \times 10^1$. If a solution containing this complex ion is diluted with water, $PbCl_2$ precipitates. Write the equations for the equilibria involved and use these equations together with LeChâtelier's principle to explain how this happens.

14.85   Write equilibrium expressions that correspond to K_{inst} for each of the following complex ions and write the chemical equations for K_{inst}. (a) $Co(NH_3)_6^{3+}$, (b) HgI_4^{2-}, (c) $Fe(CN)_6^{4-}$.

14.86   Write the equilibrium expressions that go with K_{form} for the following complex ions, and write the chemical equations for K_{form}: (a) $Hg(NH_3)_4^{2+}$, (b) SnF_6^{2-}, (c) $Fe(CN)_6^{3-}$.

14.87   The value of K_{inst} for $SnCl_4^{2-}$ is 5.6×10^{-2}.
(a) What is its value of K_{form}?
(b) Is this complex ion more or less stable than those in Table 14.3?

14.88   How many grams of solid NaCN have to be added to 1.0 L of water to dissolve 0.10 mol of $Fe(OH)_3$ in the form of $Fe(CN)_6^{3-}$? Use data, as needed, from Tables 14.2 and 14.3.

14.89   How many milliliters of ammonia gas at STP would have to be dissolved in 1.0 L of water to dissolve 1.0×10^{-4} mol of $CoCO_3$ in the form of $Co(NH_3)_6^{2+}$?

14.90   Silver ion forms a complex with thiosulfate ion, $S_2O_3^{2-}$, that has the formula $Ag(S_2O_3)_2^{3-}$. This complex has $K_{form} = 2.0 \times 10^{13}$. How many grams of AgBr ($K_{sp} = 5.0 \times 10^{-13}$) will dissolve in 500 mL of 2.0 M $Na_2S_2O_3$ solution?

QUAL TOPIC

QUALITATIVE ANALYSIS AND CHEMICAL EQUILIBRIA

One of the most intriguing problems in chemistry arises when a new substance is discovered. What is it made of? What elements are in it, and in what proportions? These are natural questions to a chemist, and finding the answers requires a chemical analysis.

If we only seek to discover what elements are contained in the unknown material, we perform a qualitative analysis — no numbers are involved. For example, a qualitative analysis of table salt reveals only that it is composed of the elements sodium and chlorine. On the other hand, if we are also interested in measuring the amounts of each component in the unknown, then we perform a quantitative analysis. (In fact, the term quantitative implies *numerical* measurement.) A quantitative analysis, perhaps performed by methods similar to those described earlier in Chapter 11, shows that table salt consists of 39.3% sodium and 60.7% chlorine, by weight.

You have already learned about some of the benefits of doing quantitative analyses. Information obtained from them can be used to calculate empirical formulas, for example. For a student of General Chemistry, the study of qualitative analysis is also valuable for several reasons. First, it gives you an opportunity to learn some chemical properties and reactions. Second, it reinforces chemical principles developed in earlier chapters. And third, if you have the opportunity to perform qualitative analyses in the laboratory, it teaches you new techniques and helps you refine them, and it gives you some of the sense of discovery that is such a rewarding part of science, as you search for clues that will help you identify what is in a sample of "unknown" composition.

This "Qual Topic" is the first of several special topics devoted to that part of inorganic qualitative analysis concerned with the detection and identification of common cations and anions. Our purpose here is to describe the chemical reactions and chemical principles that serve as the foundation for qualitative analysis. We begin first by studying the way qualitative analysis works; then we will examine how solubility and complex ion equilibria play their part.

How Qualitative Analysis Works

In qualitative analysis we make use of the specific chemical and physical properties of various cations and anions to devise tests that enable us to determine whether or not each of the ions is present. Usually these tests consist of chemical reactions whose outcomes permit us to draw specific conclusions. As an illustration, let's suppose we had a solution that we suspected to contain ammonium ion, NH_4^+. What chemical reaction (or perhaps, what series of chemical reactions) could we use to tell us if this ion is present?

In Section 11.4 you learned that one of the factors that leads to a net ionic reaction is the formation of a gas, and in Table 11.1 you saw that one such gas is ammonia, formed in the reaction of ammonium ion with hydroxide ion:

$$NH_4^+(aq) + OH^-(aq) \longrightarrow NH_3(g) + H_2O$$

Therefore, if we were to add some base (for instance, NaOH) to our solution, ammonia would be generated if the solution contained NH_4^+. The presence of the ammonia could be detected either by smell or by exposing moist red litmus paper to the vapors that escape from the solution (see Figure 1). In the latter case, we take advantage of the fact that ammonia is very soluble in water and that it is a weak base. The ammonia escaping from the solution will dissolve in the water that saturates the litmus paper and cause the red litmus to become blue.

Notice how the outcome of this test allows us to decide whether ammonium ion is present in the "unknown" solution. If ammonia is formed after NaOH is added to the solution, as determined by the odor of ammonia or by the effects of the vapors on the moist litmus paper, then the solution *must* have contained NH_4^+. If no ammonia is produced, then NH_4^+ must have been absent.

Many of the tests in qualitative analysis make use of metathesis reactions. Sometimes they rely on a gas being formed, as in the case of the test for ammonium ion, but usually they make use of precipitation reactions. Let's examine some of these and some of the kinds of problems that we encounter.

One of the properties of silver ion is that it forms a white insoluble precipitate with chloride ion.

$$Ag^+(aq) + Cl^-(aq) \longrightarrow \underset{\text{white}}{AgCl(s)}$$

Therefore, if a solution that contains chloride ion (for example, dilute HCl) is added to one that contains silver ion, we expect a white precipitate to form. Based on this alone, we might be tempted to conclude that to test for Ag^+ we simply add Cl^- and observe what happens. Actually, however, the life of an analytical chemist is not quite as simple as that. Can we really conclude that silver ion *must* be in an unknown solution if a white precipitate forms on the addition of chloride ion? The answer is no, because Ag^+ is not the only cation to form a white precipitate with Cl^-. Perhaps you recall the solubility rules on page 399 in which it was stated that all chlorides are soluble except those of Ag^+, Pb^{2+}, and Hg_2^{2+}. Each of these cations forms a white precipitate with Cl^- (AgCl, $PbCl_2$, and Hg_2Cl_2), so if a white precipitate does form, it merely suggests the presence of *at least* one of these cations. Determining which of them are actually present requires additional tests.

The discussion in the preceding paragraph illustrates one of the kinds of problems that influence the way a successful qualitative analysis is carried out. That is, it is not uncommon for two or more ions (e.g., Ag^+, Pb^{2+}, Hg_2^{2+}) to produce similar results when a testing reagent is added. A second obstacle occurs when the presence of one ion interferes with the test for another. For example, the ion Sb^{3+} forms an orange precipitate by reacting with hydrogen sulfide:

$$2Sb^{3+}(aq) + 3H_2S(aq) \longrightarrow \underset{\text{orange}}{Sb_2S_3(s)} + 6H^+(aq)$$

However, Pb^{2+} also forms a precipitate with sulfide ion (PbS), and its color is black. If both ions are in solution together when hydrogen sulfide is added, the black color of the PbS is so intense that it obscures the color of the Sb_2S_3, so the presence of antimony can't be detected.

To resolve such difficulties, the qualitative analysis of a sample is usually aproached stepwise. First, reactions are carried out to separate ions from each other, so they can't interfere with each other's tests; then other tests are performed that confirm either the presence or absence of each ion.

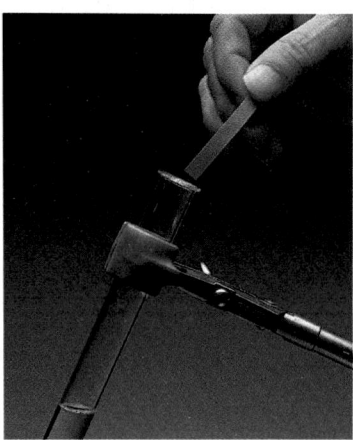

FIGURE 1 Addition of base (NaOH) to a solution that contains NH_4^+ liberates NH_3, which is detected by noting how it changes moist litmus paper from red to blue.

The standard qualitative analysis scheme for cations begins by separating the ions into groups (we will call them cation groups) based on the solubilities of their chlorides, sulfides, hydroxides, and carbonates. Each cation group, which now consists of a small number of ions, is then analyzed further. Usually, additional reactions are carried out that separate the ions in the group from each other; then confirmatory tests are performed to prove or disprove the presence of each.

The initial separation into cation groups is accomplished in the following order:

Cation Group 1. Here we have the ions Ag^+, Hg_2^{2+}, and Pb^{2+}. They are separated from other cations that might be present in a solution by precipitating them as their chlorides.

Cation Group 2. This group includes the ions Cu^{2+}, Hg^{2+}, Pb^{2+}, Sn^{4+}, Sn^{2+}, Bi^{3+}, Sb^{3+}, Cd^{2+}, and As^{3+} which are precipitated as sulfides from an acidic solution. Because of time constraints in laboratory courses, sometimes an abbreviated analytical scheme is chosen that limits this group to a smaller number of ions, for example, Cu^{2+}, Pb^{2+}, Sn^{4+}, Sn^{2+}, and Bi^{3+}. All the sulfides in this group are black except for CdS (yellow to orange), As_2S_3 (yellow), SnS_2 (yellow), and Sb_2S_3 (orange), as shown in Figure 2. If any of the first four ions are present, the intense black of their sulfides usually gives the cation group 2 precipitate a dark color.

FIGURE 2 From left to right, precipitates of CuS, CdS, As_2S_3, SnS_2, and Sb_2S_3.

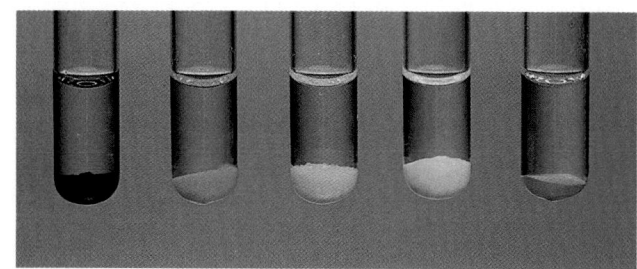

Cation Group 3. In this group we have the ions Co^{2+}, Ni^{2+}, Mn^{2+}, Fe^{2+}, Fe^{3+}, Al^{3+}, Cr^{3+}, and Zn^{2+}. Once again, a subset of these ions is sometimes chosen when lab time does not lend itself to the analysis of the entire group. Typically this might include the ions Ni^{2+}, Fe^{3+}, Fe^{2+}, Al^{3+}, Zn^{2+}, and Mn^{2+}. All of the ions of this group except Al^{3+} and

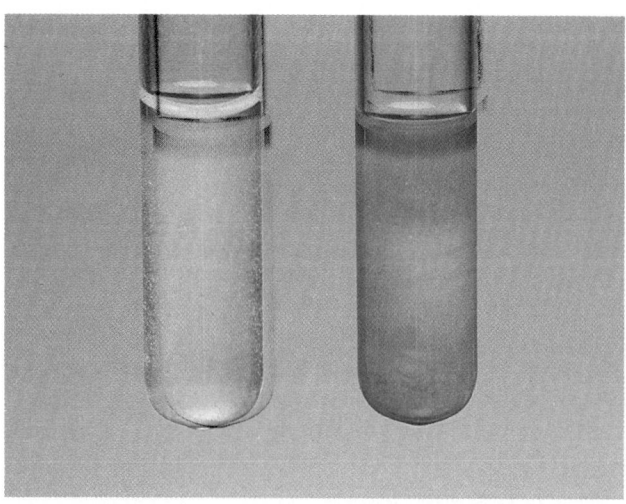

FIGURE 3 On the left, $Al(OH)_3$; on the right $Cr(OH)_3$. Actually, the formulas of these gelatinous precipitates are more complex than this, which we will discuss in Chapters 21 and 22.

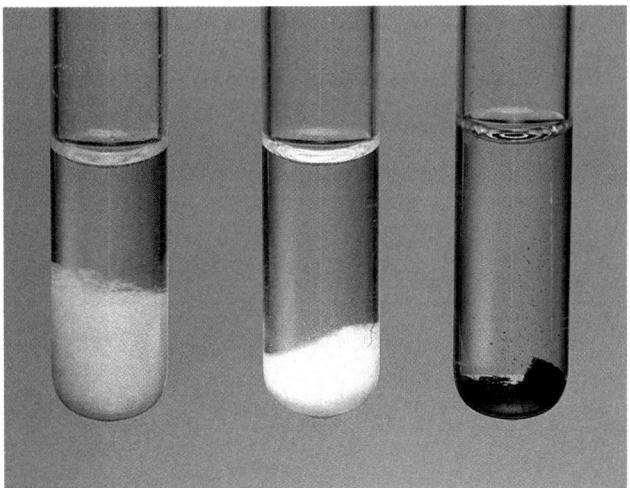

FIGURE 4 From left to right, MnS, ZnS, and FeS.

Cr^{3+} are precipitated as sulfides from a slightly basic solution. Under these conditions, aluminum and chromium are precipitated as hydroxides; their colors are shown in Figure 3. The color of the precipitate that forms when ions of this group are precipitated can also vary depending on the ions that are actually present. For example, MnS is pink, ZnS is white, FeS is black, and NiS is black (Figure 4). When one or more ions whose sulfides are black are in a mixture of these ions, the mixed precipitate that forms is dark in color and the individual colors of each solid can't be seen.

Cation Group 4.   This includes the ions Ba^{2+}, Sr^{2+}, and Ca^{2+} which are precipitated as their carbonates from a slightly basic solution. The precipitates are all white.

Cation Group 5.   This is often called the "soluble group" because the ions in it have survived chemical reagents that have precipitated the other cations. The group includes Mg^{2+}, NH_4^+, Na^+, and K^+.

Solubility and Complex Ion Equilibria

The separation of the ions into their respective groups, as well as many of the additional tests performed, provides an excellent opportunity to observe chemical equilibria at work and to demonstrate some of the principles developed earlier in Chapter 14. The separation and analysis of cation group I serves as a good example.

As noted above, cation group I consists of the ions Ag^+, Pb^{2+}, and Hg_2^{2+}, and we have already seen that they form white precipitates if Cl^- is added to a solution that contains them. This removes them from the solution, leaving behind any other cations (those of cation groups 2, 3, 4, and 5) that might be present. However, for the analysis of subsequent groups to be successful, these group 1 cations must be removed as completely as possible. Otherwise they interfere with tests that must be performed later. How is this accomplished?

Addition of HCl to the solution causes the insoluble chlorides to precipitate. To be certain that all the ions have been removed, the mixture is centrifuged, driving the precipitates to the bottom of the test tube and leaving a clear liquid above (see Figure 5). Another drop of HCl is added and if any unreacted Ag^+, Pb^{2+}, or Hg_2^{2+} is present, additional precipitate is formed. The mixture is centrifuged again, and the test for complete reaction is repeated until the addition of HCl causes no further precipitation. Completeness of precipitation is assured at this point because extra Cl^- (more than the

FIGURE 5 Testing for completeness of precipitation of AgCl. In this case, not all of the Ag^+ had been precipitated because the addition of more HCl solution gives more precipitate.

Too much chloride ion must be avoided, however, because it can cause AgCl to redissolve by forming the complex ion, $AgCl_2^-$.

$$AgCl(s) + Cl^-(aq) \rightleftharpoons AgCl_2^-(aq)$$

stoichiometric requirement) has been added. The extra chloride ion also reduces the solubility of the chlorides by the common ion effect.

After the group 1 chlorides have been precipitated as completely as possible, they are separated from the solution (which now contains the ions of subsequent groups) and analyzed further. Once again, chemical equilibria serve as the centerpiece.

The first step is the separation of Pb^{2+} from the other two cations. Here we make use of the fact that $PbCl_2$ is much more soluble than either of the other two chlorides. From their respective K_{sp} values,

$$\begin{array}{ll} PbCl_2 & K_{sp} = 1.7 \times 10^{-5} \\ AgCl & K_{sp} = 1.8 \times 10^{-10} \\ Hg_2Cl_2 & K_{sp} = 1.3 \times 10^{-18} \end{array}$$

the molar solubilities in water can be calculated to be as follows:

$$\begin{array}{ll} PbCl_2 & 0.016 \ M \\ AgCl & 0.000\ 013 \ M \\ Hg_2Cl_2 & 0.000\ 000\ 69 \ M \end{array}$$

At 100 °C, the solubility of the $PbCl_2$ is higher still (about 0.1 M), which is sufficient to cause any solid $PbCl_2$ to dissolve in boiling water. The solubilities of AgCl and Hg_2Cl_2 also increase, but not to such a significant degree, so the boiling water treatment effectively separates the lead from the silver and mercury.

After the hot water is separated from any remaining white precipitate, it must be tested to see if it actually contains any dissolved Pb^{2+}. Here we make use of the high degree of insolubility of lead chromate, $PbCrO_4$. A few drops of Na_2CrO_4 solution are added to the solution suspected to contain Pb^{2+}; the formation of a yellow precipitate confirms the presence of lead (Figure 6). On the other hand, if no yellow precipitate forms, then lead is absent.

If no precipitate remains after extraction with boiling water (i.e., if all the precipitate dissolves in boiling water), then Ag^+ and Hg_2^{2+} must have been absent from the unknown. Any precipitate that does remain at this point must be AgCl and/or Hg_2Cl_2, and the effect of complex ion formation on solubility is used next to separate the silver from the mercury. A set of equilibria similar to those discussed on page 585 is used. The procedure calls for the addition of aqueous ammonia, which causes the silver chloride to dissolve. The equilibria involved are

$$AgCl(s) \rightleftharpoons Ag^+(aq) + Cl^-(aq)$$

and

$$Ag^+(aq) + 2NH_3(aq) \rightleftharpoons Ag(NH_3)_2^+(aq)$$

As NH_3 is added, the second equilibrium is shifted to the right, causing the concentration of Ag^+ in the solution to drop. This, in turn, causes the first equilibrium to shift to the right and the silver chloride dissolves. The net chemical change is

$$AgCl(s) + 2NH_3(aq) \longrightarrow Ag(NH_3)_2^+(aq) + Cl^-(aq)$$

FIGURE 6 A yellow precipitate of $PbCrO_4$ is formed when a drop of Na_2CrO_4 solution is added to a solution that contains Pb^{2+}.

Addition of NH_3 to the AgCl and Hg_2Cl_2 mixture actually serves two purposes. First, it causes any AgCl in the precipitate to dissolve (so if all the precipitate dissolves in $NH_3(aq)$, we know that mercury(I) must have been absent from the unknown). Second, it causes any Hg_2Cl_2 in the mixture to undergo a redox reaction in which half of the mercury is oxidized while the remainder is reduced.

$$Hg_2Cl_2(s) + 2NH_3(aq) \longrightarrow Hg(\ell) + Hg(NH_2)Cl(s) + NH_4^+(aq) + Cl^-(aq)$$
white   black   white

The mixture of finely divided mercury (which is black) and the white $Hg(NH_2)Cl$ (called mercury amido chloride) generally appears dark gray or black. Therefore, if any remaining precipitate changes color from white to dark gray, the presence of mercury(I) is confirmed.

Now let's back up for a moment to just before aqueous ammonia is added to the precipitate that remains after extraction of $PbCl_2$ with boiling water. If a precipitate does remain at this point, it could be a mixture of AgCl and Hg_2Cl_2, or it could be either pure AgCl or pure Hg_2Cl_2. When aqueous ammonia is added, we know that the precipitate will dissolve completely if it consists of pure AgCl, and that it will turn black or dark gray if it contains any Hg_2Cl_2. But if Hg_2Cl_2 is present, it is usually hard to tell whether any AgCl is also present based on what happens when the $NH_3(aq)$ is added; it is difficult to tell whether part of the precipitate dissolves in the aqueous ammonia. Therefore, we must test the ammonia solution afterward to see whether or not it contains silver.

The final step in the analysis of cation group 1 is a confirmatory test for silver—a test of the ammonia solution that we suspect may contain $Ag(NH_3)_2^+$. This involves addition of HNO_3 until the solution is acidic. Here the H^+ of the acid combines with NH_3 to form NH_4^+.

$$NH_3(aq) + H^+(aq) \longrightarrow NH_4^+(aq)$$

If any silver is present, the removal of NH_3 from the solution causes the equilibrium

$$Ag^+(aq) + 2NH_3(aq) \rightleftharpoons Ag(NH_3)_2^+(aq)$$

to be shifted to the left. Since Cl^- is in the solution from the AgCl that had dissolved, the rising Ag^+ concentration causes AgCl to reprecipitate. The net chemical change is

$$Ag(NH_3)_2^+(aq) + 2H^+(aq) + 2Cl^-(aq) \longrightarrow AgCl(s) + 2NH_4^+(aq)$$

The confirmatory test for silver, therefore, is the appearance of a white precipitate when the ammonia-containing solution is acidified. If a white precipitate does form, silver is present in the unknown; if not, then silver is absent.

The separation and analysis of the ions in the other cation groups involve similar equilibria and reactions. We will discuss some of them too, but at a later time.

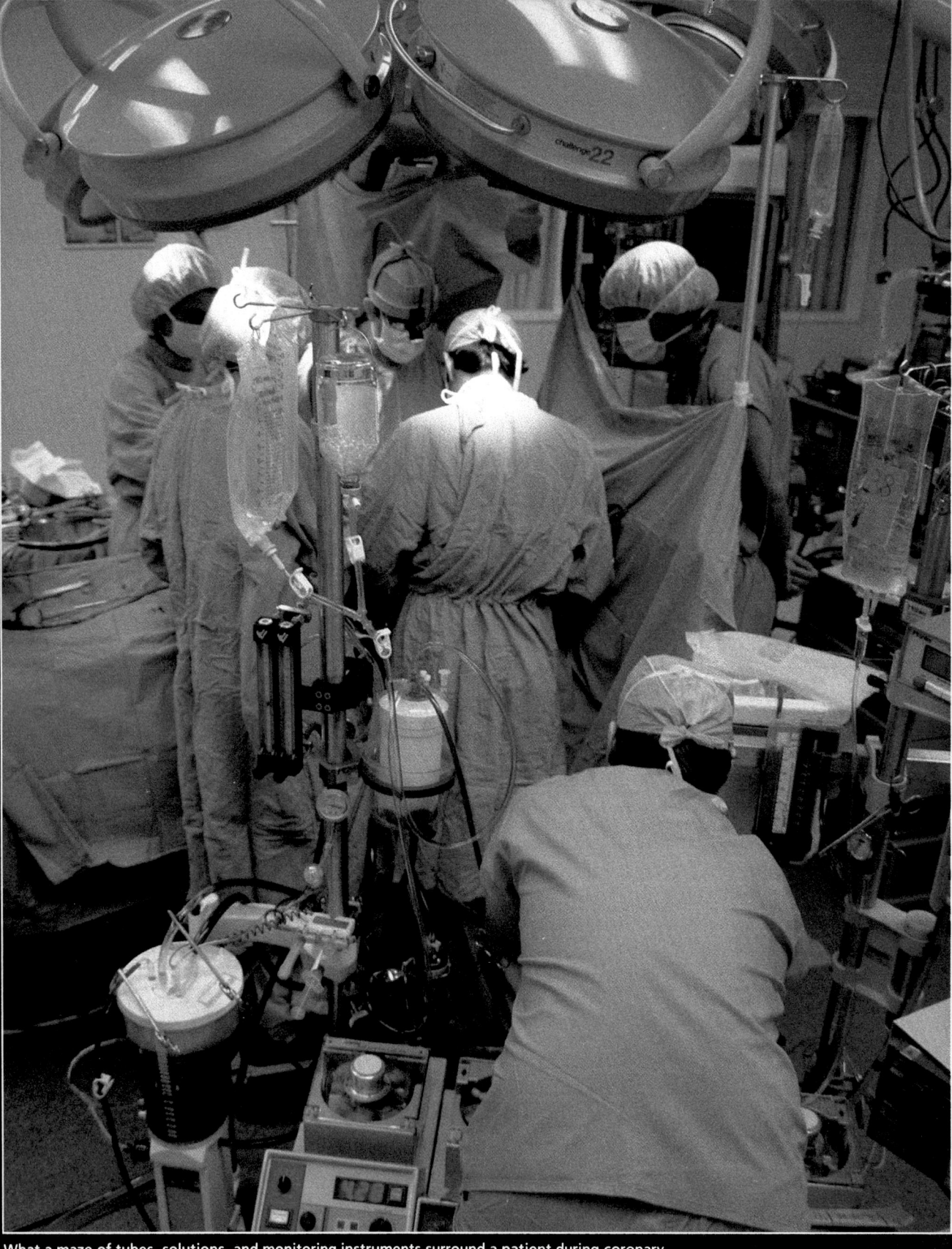

What a maze of tubes, solutions, and monitoring instruments surround a patient during coronary bypass surgery. One imperative is that the patient's blood not change in acidity. We continue our study of acids and bases in this chapter.

ACID – BASE EQUILIBRIA

15.1 THE ION-PRODUCT CONSTANT OF WATER

The product of the molar concentrations of hydrogen ion and hydroxide ion in water is a constant called the ion-product constant of water, K_w.

With the knowledge gained in the previous chapter about chemical equilibria, we can now develop quantitative ways to think about strong and weak acids and bases. These substances participate in dynamic equilibria in solution, so we should be able to use equilibrium constants to describe how weak or strong they are.

Because the most common solvent for acids and bases is water, we think of their relative strengths in relationship to this solvent and its own self-ionization. For water to serve as such a reference, we must put the equilibrium for this self-ionization on a quantitative basis.

In Section 11.3 we described this equilibrium as

$$H_2O + H_2O \rightleftharpoons H_3O^+(aq) + OH^-(aq)$$

The equilibrium constant for this system, including its numerical value at 25 °C, is

$$K_{eq} = \frac{[H_3O^+][OH^-]}{[H_2O]^2} = 3.2 \times 10^{-18} \text{ (25 °C)}$$

$$\frac{1000 \text{ g H}_2\text{O}}{L} \times \frac{1 \text{ mol H}_2\text{O}}{18.0 \text{ g H}_2\text{O}}$$
$$= 55.6 \text{ mol H}_2\text{O/L}$$

We can simplify this by noticing how little water's self-ionization affects the value of $[H_2O]$, initially 55.6 mol/L. At room temperature (25 °C), the molar concentrations of H_3O^+ and OH^- at equilibrium in water are each extremely small, only 1.0×10^{-7} mol/L. (They *must* equal each other because we get them in a one-to-one ratio.) Thus, water's self-ionization involves such a tiny loss of intact water molecules that it leaves the value of $[H_2O]$ essentially unchanged. We can therefore treat $[H_2O]$ as a constant, in other words. (The error in making this simplification is so extremely small — it calculates to be about 4×10^{-7}% — that we are certainly justified in proceeding this way.) This permits us to simplify the equation for K_{eq} by multiplying both sides by the constant value of $[H_2O]^2$. The result is a new constant, the product of one constant multiplied by another — $K_{eq} \times [H_2O]^2$ — and is called the ion-product constant of water, symbolized as K_w.

$$K_w = K_{eq} \times [H_2O]^2 = \frac{[H_3O^+][OH^-]}{[H_2O]^2} \times [H_2O]^2 = [H_3O^+][OH^-] \quad (15.1)$$

Before we push this relationship further, we want to take advantage of still another simplification.

In Chapter 11 we began to use H^+ to represent H_3O^+, so we will make this switch again (but always remember that H^+ means H_3O^+). Thus, the equilibrium in water can be represented as

$$H_2O \rightleftharpoons H^+(aq) + OH^-(aq)$$

With the use of H^+ for H_3O^+ in mind, we can rewrite the equation for K_w as follows.

$$K_w = [H^+][OH^-] \quad (15.2)$$

TABLE 15.1 K_w at Various Temperatures

Temperature (°C)	K_w
0	1.5×10^{-15}
10	3.0×10^{-15}
20	6.8×10^{-15}
25	1.0×10^{-14}
30	1.5×10^{-14}
40	3.0×10^{-14}
50	5.5×10^{-14}
60	9.5×10^{-14}

The value of K_w varies with temperature, as the data in Table 15.1 show. At 25 °C, $[H^+] = 1.0 \times 10^{-7}$ mol/L $= [OH^-]$, so at this temperature,

$$K_w = (1.0 \times 10^{-7}) \times (1.0 \times 10^{-7})$$

or

$$K_w = 1.0 \times 10^{-14} \text{ (25 °C)}$$

(From here on, unless we state otherwise, all calculations will assume that solutions are at 25 °C.)

Criteria for Acidic, Basic, and Neutral Solutions

A neutral solution (at any temperature) is *always* one in which the molar concentrations of H^+ and OH^- are equal. With this as a reference, we have the following criteria for acidic and basic solutions.

Acidic solution	$[H^+] > [OH^-]$
Basic solution	$[H^+] < [OH^-]$
Neutral solution	$[H^+] = [OH^-]$

EXAMPLE 15.1 CALCULATING [H⁺] FROM [OH⁻] OR [OH⁻] FROM [H⁺]

Problem: A sample of blood was found to have $[H^+] = 4.6 \times 10^{-8}$ M. Find the molar concentration of OH^-, and decide if the sample was acidic, basic, or neutral.

Solution: We know that $K_w = 1.0 \times 10^{-14} = [H^+][OH^-]$. So we put the given value of $[H^+]$ into this equation.

$$1.0 \times 10^{-14} = (4.6 \times 10^{-8})[OH^-]$$

Or,

$$[OH^-] = \frac{1.0 \times 10^{-14}}{4.6 \times 10^{-8}}$$
$$= 2.2 \times 10^{-7} \text{ mol/L}$$

Comparing $[H^+]$ with $[OH^-]$ shows that $[OH^-] > [H^+]$, so the blood is basic, although very slightly so (as it should be).

PRACTICE EXERCISE 1 An aqueous solution of baking soda, $NaHCO_3$, has a molar concentration of hydroxide ion of 7.8×10^{-6} mol/L. What is its molar concentration of hydrogen ion? Is the solution acidic, basic, or neutral?

15.2 THE pH CONCEPT

The value of $-\log[H^+]$ is called the pH of an aqueous solution.

In Example 15.1, we learned that the molar concentration of H^+ in blood is very low and that blood is slightly basic. The number 4.6×10^{-8}, however, is awkward to use, particularly when we want to compare the H^+ concentrations in two weakly basic solutions. For example, to compare a concentration of 4.6×10^{-8} mol H^+/L with a concentration of, say, 8.3×10^{-8} mol H^+/L, our eyes have to search two places. We have to compare the exponents on the 10; then after noting that they are the same, we have to compare 4.6 to 8.3. The pH concept was developed to help handle this problem, which occurs frequently because very low values of H^+ concentration often appear in discussions of weakly acidic or weakly basic solutions.

The pH of a solution is the negative of the logarithm of the molar concentration of hydrogen ion.

S. P. L. Sørenson (1868–1939) developed the pH concept in 1909.

$$pH = -\log[H^+] \qquad (15.3)$$

Since $[H^+] = 1.0 \times 10^{-7}$ mol/L in pure water (25 °C), the pH of pure water is 7.00, because

$$pH = -\log(1.0 \times 10^{-7})$$
$$= -(-7.00)$$
$$= 7.00 \text{ (see footnote 1)}$$

Notice that *common logs* (base 10 logs) are used in pH calculations. Be sure that you don't use the natural log function on your calculator by mistake.

[1] The rule for significant figures in logarithms is that *the number of decimal places in the logarithm of a number equals the number of significant figures in the number.* For example, 3.2×10^{-5} has just two significant figures, the two indicated by the "3.2" part of this number. (The 10^{-5} serves only to set off the decimal place since $3.2 \times 10^{-5} = 0.000032$.) The logarithm of this number, obtained with a pocket calculator, is -4.49485, but the digit to the *left* of the decimal point is actually derived from the exponential part of 3.2×10^{-5}. [$\log 3.2 \times 10^{-5} = \log(10^{-5}) + \log(3.2) = -5 + 0.50515 = -4.49485$.] Therefore, the number to the left of the decimal point in a logarithm cannot be counted as a significant figure. To show two significant figures in the logarithm, we have to round to two decimal places, to -4.49. In short, the number of significant figures in a logarithm is the number of digits to the *right* of the decimal point.

For some calculations, it is useful to have Equation 15.3 expressed in its equivalent exponential form. (And we now include the units, mol/L.)

$$[H^+] = 10^{-pH} \text{ mol/L} \tag{15.4}$$

Thus, we can also define the pH of a solution as the negative *power* (the *p* in pH) to which 10 must be raised to describe the molar concentration of H^+ (the "H" in pH).

Since the self-ionization of water gives hydroxide ions as well as hydrogen ions, there is an analogous pOH concept. The pOH of a solution is the negative of the logarithm of the molar concentration of OH^- ions.

$$pOH = -\log[OH^-]$$

In exponential form, this is the same as

$$[OH^-] = 10^{-pOH} \text{ mol/L} \tag{15.5}$$

In actual scientific work, pH values for solutions are normally used far more frequently than pOH values. Whenever we know one and need to find the other, we can use a very simple relationship among pH, pOH, and K_w. We already know that

$$[H^+][OH^-] = K_w = 1.0 \times 10^{-14}$$

By making substitutions using the relationships given by Equations 15.4 and 15.5 we can convert this equation into

$$(10^{-pH})(10^{-pOH}) = 1.0 \times 10^{-14}$$

Taking the logarithms of both sides gives us

$$(-pH) + (-pOH) = (-14.00)$$

Multiplying both sides by -1 gives us the following useful relationship between pH and pOH (at 25 °C).

$$pH + pOH = 14.00 \tag{15.6}$$

In other words, in an aqueous solution of *any* solute at 25 °C, the sum of the pH and the pOH is always 14.00.

Because pH occurs as a *negative* exponent in Equation 15.4, we have to be very careful to notice that *low* values of pH mean relatively *high* values of $[H^+]$, and *high* values of pH mean relatively *low* values of $[H^+]$. This reciprocal relationship must always be remembered in dealing with values of pH.

Since pure water at 25 °C has a pH of 7.00, we say that pure water *or any solution with a pH of 7.00* (at 25 °C) is neutral. Whenever the pH is less than 7.00, the solution is acidic, because then $[H^+] > [OH^-]$. When the pH of a solution is greater than 7.00, the solution is basic, because then $[H^+] < [OH^-]$.

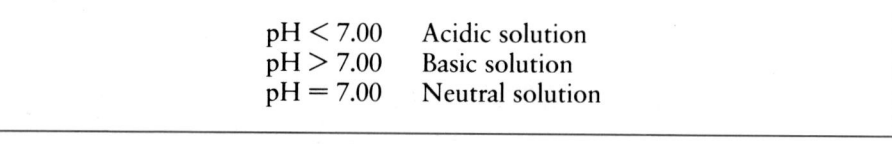

pH < 7.00	Acidic solution
pH > 7.00	Basic solution
pH = 7.00	Neutral solution

Table 15.2 gives the relationships among pH, $[H^+]$, $[OH^-]$, and pOH. Notice how the pH values become smaller as the concentration of hydrogen ion becomes larger. The pH for a number of common substances is given on a pH scale in Figure 15.1.

FIGURE 15.1 The pH scale.

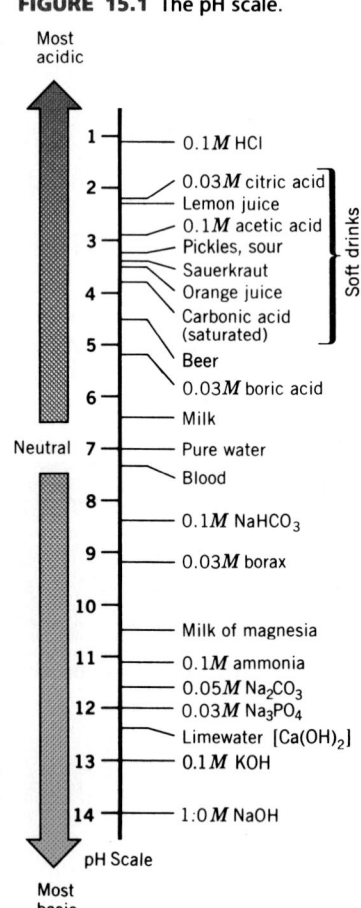

TABLE 15.2 pH, [H⁺], [OH⁻], and pOH[a]

pH	[H⁺]	[OH⁻]	pOH	
0	1	1×10^{-14}	14	
1	1×10^{-1}	1×10^{-13}	13	
2	1×10^{-2}	1×10^{-12}	12	
3	1×10^{-3}	1×10^{-11}	11	Acidic solutions
4	1×10^{-4}	1×10^{-10}	10	
5	1×10^{-5}	1×10^{-9}	9	
6	1×10^{-6}	1×10^{-8}	8	
7	1×10^{-7}	1×10^{-7}	7	Neutral solution
8	1×10^{-8}	1×10^{-6}	6	
9	1×10^{-9}	1×10^{-5}	5	
10	1×10^{-10}	1×10^{-4}	4	Basic solutions
11	1×10^{-11}	1×10^{-3}	3	
12	1×10^{-12}	1×10^{-2}	2	
13	1×10^{-13}	1×10^{-1}	1	
14	1×10^{-14}	1×10^{-0}	0	

[a] Concentrations are in mol/L at 25 °C.

Strong Acids and Bases and the pH Concept

We want to emphasize that the pH concept was devised to make it easier to handle a specific problem — how to express and to compare the concentrations of H⁺ ions when they have *very small* values involving awkward exponential terms. "Very small," in this case, means less than 1 mol H⁺/L. When the value of [H⁺] exceeds 1 mol/L, the pH works out to be a *negative* number. There is nothing wrong with this result, but we don't need it. We have no problem in comparing values of [H⁺] when they equal or exceed 1 mol/L, because the only way a solution can be this acidic is to have a *strong* acid present, one that ionizes virtually to 100%, as we learned in Section 11.2. The concentration of hydrogen ions in, say, 1.5 M HCl is for all practical purposes 1.5 mol H⁺/L, because each mole of HCl ionizes to give one mole of H⁺ (and one mole of Cl⁻, too, of course). In 1.2 M HBr, another strong acid, the value of [H⁺] is similarly 1.2 mol/L. Neither 1.8 nor 1.2 is an awkward, exponential number, and it is simple to compare them directly.

Strong acids, of course, can certainly be used in *dilute* solutions, and it is then again convenient to describe their acidities by the use of pH values. For example, a solution that is 0.0010 M HCl — a strong, monoprotic acid — has a pH of 3.00 since 0.0010 mol HCl/L contains 0.0010 mol H⁺/L and this is the same as 1.0×10^{-3} mol H⁺/L. When you compare the two expressions,

$$[H^+] = 1 \times 10^{-pH} \text{ mol/L}$$

which *defines* pH, and

$$[H^+] = 1.0 \times 10^{-3} \text{ mol/L}$$

which describes a particular solution, you can see by inspection that the pH of this solution must be 3.00. (Alas, life is seldom this easy, but we will soon see that more complicated pH problems are easily managed.)

What we have said about strong acids applies by analogy to strong bases such as NaOH or KOH. We can assume, for nearly all practical purposes, that these are 100% dissociated in water, so a solution that is, say, 1.1 M NaOH has a concentration of OH⁻ ions of 1.1 mol/L. For a concentration like this, greater than 1 mol OH⁻/L, we do not need a special number, like pOH, to help us describe it. When we work, however, with a *dilute* solution of a strong base — for example, 0.0010 M KOH — we can choose either a pOH or a pH description according to our needs. Thus, the pOH of 0.0010 M KOH is 3.00, because in the solution of this strong, fully dissociated base, $[OH^-] = 0.0010$ mol/L $= 1.0 \times 10^{-3}$ mol/L. When the pOH is 3.00, the pH is 11.00, because

$$14.00 - 3.00 = 11.00$$

pH Calculations

We will now work examples of the kinds of calculations involving pHs that you should learn how to do—calculating a pH from a value either of $[H^+]$ or of $[OH^-]$, and the opposite, finding the value of $[H^+]$ or $[OH^-]$ given a pH.

EXAMPLE 15.2   CALCULATING pH FROM $[H^+]$

Problem: Because rain washes pollutants out of the air, the lakes in many parts of the world are undergoing changes in pH. In a New England state, the water in one lake was found to have $[H^+] = 3.2 \times 10^{-5}$ mol/L. Calculate the pH and pOH of the lake's water and decide if it is acidic or basic.

Solution: Using Equation 15.3, which defines pH,

$$pH = -\log[H^+]$$
$$= -\log(3.2 \times 10^{-5})$$

Many lakes in southern Norway and Sweden no longer have fish because of the acid they have received by means of acid rain from the industrial complexes of the Midlands of England and the Ruhr Valley of Germany.

If you can enter numbers such as 3.2×10^{-5} into your pocket calculator, and if it has a way to find a value of a logarithm, the calculation is particularly easy. (If you do not have such a calculator, then you have to use a log table. Appendix A2 describes how to solve this kind of problem.) Using the calculator on this problem, we enter 3.2×10^{-5}, hit the $\boxed{\log}$ button, and obtain -4.49485. Hence,

$$pH = -(-4.49485)$$

After we change signs and round to two significant figures,

$$pH = 4.49$$

The pH is less than 7.00, so the lake's water is acidic—too acidic, in fact, for game fish to survive. The pOH of this lake water is

$$pOH = 14.00 - 4.49$$
$$= 9.51$$

EXAMPLE 15.3   CALCULATING pH FROM $[OH^-]$

Problem: What is the pH of a sodium hydroxide solution at 25 °C that has a concentration of 0.0026 mol NaOH/L?

Solution: The easiest way to find pH when working with a basic solution is to find pOH first, and then subtract it from 14.00. We know that in NaOH we have a strong base, so it is 100% dissociated in the solution. Therefore, since $[NaOH] = 0.0026$ mol/L, it is also true that $[OH^-] = 0.0026$ mol/L. Working with the definition of pOH,

$$pOH = -\log[OH^-]$$

we have,

$$pOH = -(\log 0.0026)$$
$$= -(-2.585026652) \text{ (unrounded)}$$
$$= 2.59 \text{ (correctly rounded)}$$

So,

$$pH = 14.00 - 2.59$$
$$= 11.41$$

PRACTICE EXERCISE 2   Calculate the values of both pH and pOH of the following solutions. (a) 0.020 M HCl. (b) 0.0050 M NaOH. (c) A blood specimen containing 7.2×10^{-8} mol H^+/L. Is the blood specimen slightly acidic or slightly basic? (d) 0.00035 M Ba(OH)$_2$, where this compound has to be considered to be 100% dissociated.

EXAMPLE 15.4   CALCULATING [H^+] FROM pH

Problem: "Calcareous soil" is soil rich in calcium carbonate (lime). The pH of such soil generally ranges from just over 7 to as high as 8.3. What value of [H^+] corresponds to a pH of 8.3? Is the soil slightly acidic or slightly basic?

Solution: To calculate [H^+] from pH, the alternative equation for pH, Equation 15.4, is used.

$$[H^+] = 10^{-pH} \text{ mol/L}$$

Since the pH is 8.3, we insert this value into the equation,

$$[H^+] = 10^{-8.3} \text{ mol/L}$$

If your pocket calculator has a key marked $\boxed{10^x}$, simply enter the pH *as a negative number,* -8.3 in our example, and hit this 10^x key. The answer, when correctly rounded, is

$$[H^+] = 5.0 \times 10^{-9} \text{ mol/L}$$

Since [H^+] $< 1 \times 10^{-7}$ mol/L, the soil is basic.

If you have a pocket calculator without the $\boxed{10^x}$ key, but it has the $\boxed{x^y}$ key, enter 10 as x and $-$pH as y. (Be sure to enter the pH as a *negative number.*) If your calculator has none of these functions, you will have to solve a pH-to-[H^+] calculation by the use of a log table. Appendix A.2 shows how.

PRACTICE EXERCISE 3   Find the values of [H^+], pOH, and [OH^-] that correspond to each of the following values of pH. (Remember that the answers should be rounded to two significant figures.)
(a) 2.90 (the approximate pH of lemon juice)
(b) 3.85 (the approximate pH of sauerkraut)
(c) 10.81 (the pH of milk of magnesia, a laxative)
(d) 4.11 (the pH of orange juice, on the average)
(e) 11.61 (the pH of dilute, household ammonia)

PRACTICE EXERCISE 4   Go back over the results of Practice Exercise 3 and tell if each solution is acidic or basic.

One of the deceptive aspects of the pH concept comes from its exponential definition. A change of just 1 pH unit means a tenfold change in [H^+]. When the pH is 0, then [H^+] = 1 mol/L, and you need only a liter of solution (about a quart) to have a mole of hydrogen ion present. At pH 1, only 1 pH unit more than 0, [H^+] = 10^{-1} mol/L, and you would need 10 L (about a medium-sized wastebasket) of solution to have 1 mol of hydrogen ion. At pH 5, you would need the largest railroad tankcar you have ever seen to hold the volume of solution containing 1 mol of H^+, because this ion is now at such a low concentration. If the pH of the water flowing over Niagara Falls were 10 (which it certainly is not), you would have to watch the falls about an hour to see 1 mol of H^+ go by. And at pH 14 — very alkaline — you would need a quarter of the volume of Lake Erie to find 1 mol of H^+.

Measuring pH

pH is measured either with a pH meter — the most accurate and precise method — or by touching an indicator paper with a drop of the solution and comparing the new color on the paper with a color code. As we learned in Section 11.5, dyes whose colors in aqueous

TABLE 15.3 Common Acid–Base Indicators

Name	Approximate pH Range	Color Change (lower to higher pH)
Methyl green	0.2–1.8	Yellow to blue
Thymol blue	1.2–2.8	Yellow to blue
Methyl orange	3.2–4.4	Red to yellow
Ethyl red	4.0–5.8	Colorless to red
Bromocresol purple	5.2–6.8	Yellow to purple
Bromothymol blue	6.0–7.6	Yellow to blue
Litmus	4.7–8.3	Red to blue
Cresol red	7.0–8.8	Yellow to red
Thymol blue	8.0–9.6	Yellow to blue
Phenolphthalein	8.2–10.0	Colorless to pink
Thymolphthalein	9.4–10.6	Colorless to blue
Alizarin yellow R	10.1–12.0	Yellow to red
Clayton yellow	12.2–13.2	Yellow to amber

solutions depend on the pH are called *acid–base indicators*. Table 15.3 gives several examples of indicator dyes. Litmus is available as "litmus paper," either red or blue. Below pH 4.7, litmus is red and above pH 8.3 it is blue. The transition for the color change occurs over a pH range of 4.7 to 8.3 with the center of this change at about 6.5, very nearly a neutral pH.

Some commercial test papers (e.g., Hydrion®) are impregnated with several dyes. The containers of these test papers carry the color code. (See Figure 15.2.) When we discuss acid–base titrations later in this chapter, we will see why it is useful to have indicators that change colors at widely different ranges of pH values.

FIGURE 15.2 The color of this Hydrion® test strip changed to purple when it was touched with a drop of the solution in the beaker. According to the color code, the pH of the solution is closer to 10 than to 8.

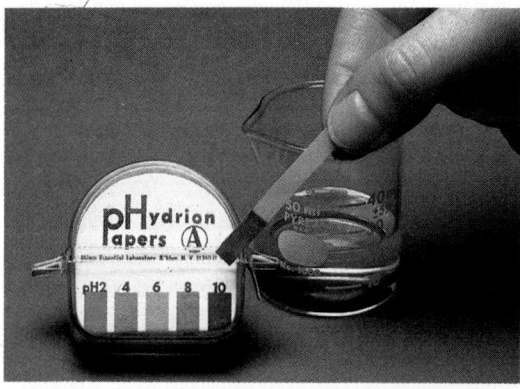

Values of pH can be determined to within ±0.01 pH unit with the better models of pH meters. (See Figure 15.3.) These instruments have to be used when solutions are

FIGURE 15.3 A pH meter.

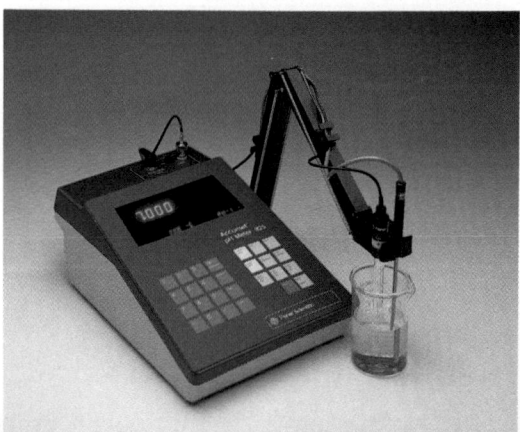

themselves highly colored. Another advantage of a pH meter is that it can be used to follow the change in the pH of a solution while some reaction is occurring.

15.3 ACID IONIZATION CONSTANTS

The relative strengths of acids are described quantitatively by acid ionization constants, K_a.

Many acids are weak electrolytes, and they ionize to an extent that is much less than 100%. We say that they are weak acids, but they are not equally so. Just how weak varies from acid to acid, so we could clearly use a quantitative method for comparing their relative strengths as acids.

K_a Values of Weak Acids

To make the discussion applicable to all weak acids, we will represent any such substance by the symbol HA, where A denotes the species that separates from H^+ when HA ionizes. It does not matter if HA is electrically neutral (e.g., acetic acid, $HC_2H_3O_2$), positively charged (e.g., the ammonium ion, NH_4^+), or negatively charged (e.g., the hydrogen sulfate ion, HSO_4^-). All these are Brønsted acids — donors of H^+. And if our acid happens to be diprotic or triprotic, we will consider the ionization of only one proton at a time.

$$\underset{\substack{\text{weak}\\\text{Brønsted}\\\text{acid}}}{HA} + H_2O \rightleftharpoons H_3O^+ + \underset{\substack{\text{conjugate}\\\text{base}}}{A^-}$$

Using the Brønsted acids already mentioned, examples of this equilibrium are

$$HC_2H_3O_2 + H_2O \rightleftharpoons H_3O^+ + C_2H_3O_2^-$$
$$NH_4^+ + H_2O \rightleftharpoons H_3O^+ + NH_3$$
$$HSO_4^- + H_2O \rightleftharpoons H_3O^+ + SO_4^{2-}$$

Ka my ass whole unit shit

The general form of the equilibrium law for these reactions is

$$K_c = \frac{[H_3O^+][A^-]}{[HA][H_2O]} \tag{15.7}$$

In dilute solutions, the *equilibrium value* of the moles of water per liter of *solution* is very nearly identical with its value in pure water. A reasonable approximation, therefore, is to take the value of $[H_2O]$ to be effectively a constant in Equation 15.7. This lets us simplify the equation by multiplying both sides by $[H_2O]$, as follows.

We can't emphasize it too much — equilibrium constant expressions involve molar concentrations of species *at equilibrium*.

$$K_c \times [H_2O] = \frac{[H_3O^+][A^-]}{[HA][H_2O]} \times [H_2O] = \text{a new constant}$$

The new constant is called the acid ionization constant, with the symbol K_a. Switching from H_3O^+ to H^+, the equation for K_a is

Some references call K_a the acid dissociation constant.

$$K_a = \frac{[H^+][A^-]}{[HA]} \qquad (\text{for } HA \rightleftharpoons H^+ + A^-) \tag{15.8}$$

TABLE 15.4 K_a Values for Acids[a]

Name	Formula	K_a (25 °C)
Perchloric acid	$HClO_4$	Large
Hydriodic acid	HI	Large
Hydrobromic acid	HBr	Large
Sulfuric acid	H_2SO_4	Large
Hydrochloric acid	HCl	Large
Nitric acid	HNO_3	Large
HYDRONIUM ION	H_3O^+	55
Sulfurous acid	H_2SO_3	1.2×10^{-2}
Hydrogen sulfate ion	HSO_4^-	1.0×10^{-2}
Phosphoric acid	H_3PO_4	7.1×10^{-3}
Citric acid[b]	$H_3C_6H_5O_7$	7.1×10^{-4}
Nitrous acid	HNO_2	7.1×10^{-4}
Hydrofluoric acid	HF	6.8×10^{-4}
Formic acid	HCO_2H	1.8×10^{-4}
Barbituric acid[c]	$HC_4H_3N_2O_3$	9.8×10^{-5}
Ascorbic acid[d]	$H_2C_6H_6O_6$	7.9×10^{-5}
Acetic acid	$HC_2H_3O_2$	1.8×10^{-5}
Carbonic acid	H_2CO_3	4.5×10^{-7}
Hydrogen sulfide (aq)	H_2S	9.5×10^{-8}
Hydrogen sulfite ion	HSO_3^-	6.6×10^{-8}
Dihydrogen phosphate ion	$H_2PO_4^-$	6.3×10^{-8}
Hypochlorous acid	$HOCl$	3.0×10^{-8}
Hydrogen cyanide (aq)	HCN	6.2×10^{-10}
Ammonium ion	NH_4^+	5.7×10^{-10}
Bicarbonate ion	HCO_3^-	4.7×10^{-11}
Monohydrogen phosphate ion	HPO_4^{2-}	4.5×10^{-13}
Hydrogen sulfide ion	SH^-	1×10^{-19}
WATER	H_2O	1.8×10^{-16}
Hydroxide ion	OH^-	(est) 1×10^{-36}

INCREASING ACID STRENGTH ↑

A more complete table of K_a values is located in Appendix D.

[a] Data for all acids, except citric, barbituric, and ascorbic acids as well as water and SH^-, are rounded to two significant figures from the values given in E. H. Martell and R. M. Smith, *Critical Stability Constants,* Plenum Press, New York, 1974. Data for the organic acids are from R. C. Weast, editor, *Handbook of Chemistry and Physics,* 64th ed., CRC Press, Boca Raton, Fla., 1984. For water, see R. Starkey, J. Norman, and M. Hintze, *J Chem. Ed.* **63** (1986), p. 473. For OH^-, H_2S, and SH^-, see R. J. Myers, *J. Chem. Ed.* **63** (1986), p. 687, and references cited therein.

[b] Citric acid is triprotic; K_a is for the first proton.

[c] Barbituric acid is monoprotic.

[d] Ascorbic acid (vitamin C) is diprotic; K_a is for the first proton.

The values of K_a for several acids are given in Table 15.4. The first seven, through H_3O^+, you will recognize as the common strong acids.

The first skill that must be learned in connection with K_a values is simply to write the equation for K_a, given a specific weak acid.

EXAMPLE 15.5   WRITING EQUATIONS FOR K_a FOR BRØNSTED ACIDS

Problem: What is the equation for K_a for acetic acid, $HC_2H_3O_2$, the acid present in wine gone sour?

Solution: First we have to write the equation for the equilibrium in aqueous acetic acid (letting H^+ represent H_3O^+).

$$HC_2H_3O_2(aq) \rightleftharpoons H^+(aq) + C_2H_3O_2^-(aq)$$

Next, we write the equation for K_a, remembering that the products appear in the numerator and the reactants in the denominator. The answer, therefore, is

$$K_a = \frac{[H^+][C_2H_3O_2^-]}{[HC_2H_3O_2]}$$

PRACTICE EXERCISE 5   Write the K_a equation for nitrous acid, HNO_2.

PRACTICE EXERCISE 6   Write the K_a equation for the monohydrogen phosphate ion, HPO_4^{2-}.

PRACTICE EXERCISE 7   If we treated H_2O as just another example of a weak acid, what would be the equation for its K_a? Similarly, if we treated H_3O^+ as just another Brønsted acid, what would be its K_a equation?

There are basically just two kinds of calculations involving K_a that we have to be concerned about.

1. Calculate the value of K_a for a specific acid from the pH of a solution for which we also know the *initial* concentration of the acid.

2. Calculate the equilibrium concentrations of H^+ and A^- from the *initial* concentration of a specific weak acid and its K_a value.

We will study situations in which a weak acid is the only solute in water. It is very important to remind ourselves that the value of the molar concentration of a weak acid is *not* the same as the value of $[H^+]$ in the solution. (That is why we call it a *weak* acid.) And if we want to know the value of $[H^+]$ in such a solution, we have to calculate it either from the solution's pH—which we can measure—or from the K_a of the acid and its molarity.

EXAMPLE 15.6   CALCULATING K_a FROM pH

Problem: Formic acid, $HCHO_2$, is a monoprotic acid. In a 0.100 M solution of formic acid, the pH is 2.38 at 25 °C. Calculate K_a for formic acid at this temperature.

Solution: In any problem like this we begin by writing the equation for the equilibrium so that we can set up the expression for K_a.

$$\underset{\text{formic acid}}{H-\overset{\overset{O}{\|}}{C}-O-H} = HCHO_2 \qquad\qquad HCHO_2 \rightleftharpoons H^+ + CHO_2^- \qquad K_a = \frac{[H^+][CHO_2^-]}{[HCHO_2]}$$

Remember, that the molar concentrations in the K_a expression all refer to *equilibrium* values. And we will note that whatever the equilibrium concentration of H^+ is, it is the same as that of CHO_2^-. But we have neither, only a pH. A simple calculation like that in Example 15.4, however, converts pH into $[H^+]$—into the equilibrium value of $[H^+]$.

$$\underset{\text{formate ion}}{H-\overset{\overset{O}{\|}}{C}-O^-} = CHO_2^-$$

$[H^+]$ means $[H^+]_{\text{at equilibrium}}$.
$[CHO_2^-]$ means
　　　　$[CHO_2^-]_{\text{at equilibrium}}$.
$[HCHO_2]$ means
　　　　$[HCHO_2]_{\text{at equilibrium}}$.

$$[H^+] = 10^{-pH} \text{ mol/L (not neglecting the units)}$$
$$= 10^{-2.38} \text{ mol/L}$$
$$= 4.2 \times 10^{-3} \text{ mol/L}$$
$$= 0.0042 \text{ mol/L (the form most useful next)}$$

Because $[CHO_2^-] = [H^+]$, we now know that $[CHO_2^-] = 0.0042$ mol/L. With these data we can set up a concentration table as we did in Chapter 14 in which we write the data beneath the formulas. (All concentrations in such a table, of course, are in moles per liter.)

	$HCHO_2$	$\rightleftharpoons$	H^+	+ CHO_2^-	
Initial concentrations	0.100		0	0	(Note 1)
Changes in concentrations caused by the ionization	−0.0042		+0.0042	+0.0042	(Note 2)
Final concentrations at equilibrium	(0.100 − 0.0042) = 0.096 (correctly rounded)		0.0042	0.0042	

Note 1. The *initial* concentration of the acid is what the label tells us, 0.100 *M*; it ignores whatever ionization occurs. However, by ignoring this we have to set the *initial* values of $[H^+]$ and $[CHO_2^-]$ equal to zero. (The ultratrace concentration of H^+ contributed by the self-ionization of water can be safely ignored. It is just too small, and we will ignore it in all calculations in this chapter.)

Note 2. These *changes* are the values we calculated from the pH. For every H^+ ion that forms by ionization, there is one less molecule of the initial acid. The minus sign in −0.0042 in the column for $HCHO_2$ is used because the initial concentration of this species is *decreased* by the ionization.

The last row of data gives us the equilibrium concentrations that we now use to calculate K_a.

$$K_a = \frac{(4.2 \times 10^{-3})(4.2 \times 10^{-3})}{0.096}$$
$$= 1.8 \times 10^{-4}$$

Formica is Latin for "ant." Formic acid is partly responsible for the pain of an ant sting.

Thus, the acid ionization constant for formic acid is 1.8×10^{-4}.

PRACTICE EXERCISE 8
Butyric acid is structurally similar to acetic acid—but what a difference in odor!

When butter turns rancid, its foul odor is mostly that of butyric acid, a weak acid. A 0.0100 *M* solution of butyric acid has a pH of 3.40 at 20 °C. Calculate the K_a of butyric acid at this temperature. (Use any symbol you wish for butyric acid and its conjugate base—for example, H*Bu* and *Bu*⁻.)

PRACTICE EXERCISE 9

At 60 °C, the pH of 0.0100 *M* butyric acid is 2.98. Calculate the K_a of butyric acid at this temperature.

EXAMPLE 15.7   CALCULATING THE VALUES OF $[H^+]$ AND pH FOR A SOLUTION OF A WEAK ACID FROM ITS K_a VALUE

Problem: The concentration of a sample of vinegar was found to be 0.75 *M* acetic acid, $HC_2H_3O_2$. Calculate the values of $[H^+]$ and pH of this sample.

Solution: We can use the same general approach given in Example 15.6—writing the chemical equilibrium and then preparing a concentration table.

	$HC_2H_3O_2$	$\rightleftharpoons$	H^+	+	$C_2H_3O_2^-$
Initial concentrations	0.75		0		0
Changes in concentrations caused by the ionization	−x		+x		+x (Note 1)
Final concentrations at equilibrium	(0.75 − x) ≈ 0.75 (Note 2)		x		x

In Example 15.7, the equilibrium concentration before use of the simplification described in Note 2 below the concentration table was given as

$$[HC_2H_3O_2] = (0.75 - x)$$

Instead of letting $(0.75 - x)$ equal 0.75, we will here use $(0.75 - x)$ in the equation for K_a and solve for x by the quadratic equation. This equation tells us that if $ax^2 + bx + c = 0$, then

$$x = \frac{-b \pm [b^2 - 4ac]^{1/2}}{2a}$$

The equation for K_a for acetic acid is

$$K_a = \frac{[H^+][C_2H_3O_2^-]}{[HC_2H_3O_2]} = \frac{(x)(x)}{(0.75 - x)} = 1.8 \times 10^{-5}$$

$$x^2 = (0.75 - x)(1.8 \times 10^{-5})$$
$$= 1.4 \times 10^{-5} - (1.8 \times 10^{-5})x$$

When we rearrange this into the form $ax^2 + bx + c = 0$, we get

$$x^2 + (1.8 \times 10^{-5})x - (1.4 \times 10^{-5}) = 0$$

Solving for the roots of x by the standard solution to a quadratic equation gives

$$x = \frac{-(1.8 \times 10^{-5}) \pm [(1.8 \times 10^{-5})^2 - 4(1)(-1.4 \times 10^{-5})]^{1/2}}{2(1)}$$

$$= \frac{-1.8 \times 10^{-5} \pm [3.2 \times 10^{-10} + 5.6 \times 10^{-5}]^{1/2}}{2}$$

$$= \frac{-1.8 \times 10^{-5} \pm [5.6 \times 10^{-5}]^{1/2}}{2}$$

$$= \frac{-1.8 \times 10^{-5} \pm 7.5 \times 10^{-3}}{2}$$

Because of the $\pm$ sign, one of the two roots of the quadratic would turn out to be negative. We ignore this root because it has no meaning. (We couldn't have a solution with a value of [H⁺] less than zero!) So we continue with the other root.

$$x = \frac{7.5 \times 10^{-3}}{2} \text{ (rounded from } 7.482 \times 10^{-3})$$

$$= 3.8 \times 10^{-3}$$

So, $[H^+] = 3.8 \times 10^{-3}$ mol/L.

In Example 15.7, after making the simplification that $(0.75 - x) = x$, we calculated that $[H^+] = 3.7 \times 10^{-3}$ mol/L. The difference in the results from the two approaches is only 3%—and much of this is error caused by rounding. Thus, the simplification is not only valid, but it saves an awful lot of work!

Note 1. We let $+x$ stand for the increase in the concentration of both H⁺ and $C_2H_3O_2^-$, and so the initial concentration of $HC_2H_3O_2$ is reduced by the same amount.

Note 2. This is an important simplification, one that makes the calculation much easier with almost no sacrifice in a precision. (We introduced this kind of simplification in Example 14.10 on page 572 for the same reason—to make the calculation much easier.) The problem we would have without doing this is as follows. If we use $(0.75 - x)$ for $[HC_2H_3O_2]$, we would soon have to solve for x in an equation with both an x^2 and an x term—a quadratic equation. In this example, the quadratic equation would be obtained by making the following substitutions from our concentration table, above. (The value of K_a is taken from Table 15.4.)

$$K_a = \frac{[H^+][C_2H_3O_2^-]}{[HC_2H_3O_2]} = \frac{(x)(x)}{(0.75 - x)} = 1.8 \times 10^{-5}$$

We would eventually have to solve for x in

$$x^2 + 1.8 \times 10^{-5}x - 1.35 \times 10^{-5} = 0$$

This can certainly be done, as demonstrated in Special Topic 15.1. (In Example 14.10 the problem was to solve a cubic equation, which is much harder.) But the question here (as in Example 14.10) is not if it can be done, but if it has to be done. Since K_a is very small, the calculated value of x is likely to be very small, too. If it is, then the value of $(0.75 - x)$ will likely be very nearly equal to 0.75, itself. Let us at least see if it works.

Letting $(0.75 - x) = 0.75$, and $x = [H^+] = [C_2H_3O_2^-]$,

$$K_a = \frac{[H^+][C_2H_3O_2^-]}{[HC_2H_3O_2]} = \frac{(x)(x)}{(0.75)} = 1.8 \times 10^{-5}$$

$$x^2 = (0.75)(1.8 \times 10^{-5})$$
$$= 1.35 \times 10^{-5}$$
$$x = 3.7 \times 10^{-3}$$

In other words, since $x = [H^+]$,

$$[H^+] = 3.7 \times 10^{-3} \text{ mol/L} = 0.0037 \text{ mol/L}$$

As you learned in Chapter 14, you should always check to see if this kind of approximation was valid. In this case, $(0.75 - 0.0037) = 0.7463$, which rounds properly to 0.75. The assumption was valid, the simplification worked and it certainly made the calculation a lot easier.

Having found that at equilibrium $[H^+] = 3.7 \times 10^{-3}$ mol/L, we next have to calculate the pH of this solution.

$$\begin{aligned} pH &= -\log[H^+] \\ &= -(\log 3.7 \times 10^{-3}) \\ &= -(-2.43) \\ pH &= 2.43 \text{ (the answer to the second part)} \end{aligned}$$

The simplifications that we can make in K_a calculations involving *weak* acids are so useful that we should review them. Whenever we want to calculate the values of $[H^+]$ and pH for a dilute solution of a weak acid, given K_a,

1. We can ignore the contribution to $[H^+]$ from the self-ionization of water.

2. We can drop the term, $[HA]_{\text{ionized}}$, in the expression for $[HA]_{\text{eq}}$:

$$\begin{aligned} [HA]_{\text{eq}} &= [HA]_{\text{init}} - [HA]_{\text{ionized}} \\ &= [HA]_{\text{init}} \end{aligned}$$

very small and dropped

All of the Practice Exercises and unstarred Review Exercises in this chapter have been selected in such a way that these assumptions work, but do not think that they are universally applicable. In extremely dilute solutions of weak acids, we cannot ignore the fact that the self-ionization of water (itself a weak, very weak electrolyte) can make a contribution to the value of $[H^+]$. On the other side, as the value of K_a gets closer to 10^{-3} and higher, we introduce larger and larger errors by letting $[HA]_{\text{init}}$ be equal to $[HA]_{\text{eq}}$. A general rule for evaluating how well the assumptions work is as follows.

If you find that the calculated value of x is smaller than one-tenth the value of $[HA]_{\text{initial}}$, then the assumptions work satisfactorily. Otherwise, if x is greater than one-tenth the value of $[HA]_{\text{initial}}$, then the quadratic solution described in Special Topic 15.1 should be used.

PRACTICE EXERCISE 10   Nicotinic acid, $HC_2H_4NO_2$, is a B vitamin. It is also a weak acid with $K_a = 1.4 \times 10^{-5}$. What is the $[H^+]$ and the pH of a 0.010 M solution?

Now that we have studied acid ionization constants, we have a better basis than percentage ionization for classifying acids as weak, moderate, or strong.

$K_a < 10^{-3}$	Weak acid
$K_a = 1$ to 10^{-3}	Moderate acid
$K_a > 1$	Strong acid

Percentage Ionization of Weak Acids

In discussing and comparing weak and strong acids, we have used the idea of percentage ionization without describing how such percentages can be obtained. The percentage ionization of something in solution is simply 100 times the ratio of the molecules that ionize to the number of original molecules. Thus, for a solution of a weak acid,

$$\text{percentage ionization} = \frac{\text{amount of HA ionized}}{\text{amount of HA initially available}} \times 100 \qquad (15.9)$$

We will work an example to show how to use this equation.

4.2×10^{-4}

EXAMPLE 15.8   CALCULATING A PERCENTAGE IONIZATION OF A WEAK ACID

Problem: What is the percentage ionization of acetic acid in 0.10 M $HC_2H_3O_2$?

Solution: We have been told the "amount of HA initially available," the denominator in Equation 15.9. What we must calculate is the "amount of HA ionized." So we start, as usual, with a table of concentrations.

	$HC_2H_3O_2$	$\rightleftharpoons$	H^+	+	$C_2H_3O_2^-$
Initial concentrations	0.10		0		0 (Note 1)
Changes in concentrations caused by the ionization	$-x$ (amount ionized		$+x$		$+x$ (Note 2)
Final concentrations at equilibrium	$(0.10 - x)$ ≈ 0.10		x		x

Note 1. Since no other species are present besides $HC_2H_3O_2$ to supply H^+ or $C_2H_3O_2^-$, the initial concentrations of these ions (before any ionization of $HC_2H_3O_2$) must be 0.
Note 2. We let $+x$ be the change in $[H^+]$ caused by the ionization of $HC_2H_3O_2$, the only source of H^+ in this solution. $HC_2H_3O_2$ is also the only source of $C_2H_3O_2^-$, so the same value, $+x$, is the change in $[C_2H_3O_2^-]$. Thus, the change in $[HC_2H_3O_2]$ must be $-x$, which is what we seek. It is the "amount of HA ionized," the numerator in Equation 15.9.
We solve for x in the usual way.

$$K_a = \frac{[H^+][C_2H_3O_2^-]}{[HC_2H_3O_2]} = \frac{(x)(x)}{(0.10)} = 1.8 \times 10^{-5}$$
$$x^2 = (0.10)(1.8 \times 10^{-5})$$
$$x = 1.3 \times 10^{-3}$$

In other words,

$$[H^+] = 1.3 \times 10^{-3} \text{ mol/L} = \text{the amount of acid ionized}$$

Now we use the definition of percentage ionization (Equation 15.9).

We can see by this calculation that when we talk about a *weak* acid, it's one with a low percentage ionization indeed.

$$\text{percentage ionization} = \frac{1.3 \times 10^{-3} \text{ mol/L}}{0.10 \text{ mol/L}} \times 100$$
$$= 1.3\%$$

In 0.10 M $HC_2H_3O_2$, the percentage ionization of acetic acid is 1.3%.

PRACTICE EXERCISE 11   Estimate the percentage ionization of acetic acid in solutions with concentrations of (a) 0.010 *M* and (b) 0.0010 *M*.

In Example 15.8, we found that in 0.1 M acetic acid the percentage ionization is 1.3%. If you worked Practice Exercise 11, you found that the *percentage* ionization increases as the solution of the weak acid becomes more dilute. In the more dilute solutions the opportunities for H^+ and A^- to recombine occur less frequently because they find each other less often. As a result, they make up a larger percentage of all solute particles. Even though there are fewer un-ionized acid molecules per liter, the percentage that are ionized has increased.

15.4 BASE IONIZATION CONSTANTS

Base ionization constants provide a way to compare the relative strengths of weak bases.

Strong bases, like sodium hydroxide, are strong electrolytes and fully dissociated in solution. Weak Brønsted bases, on the other hand, usually consist of molecules or ions that react with water, remove a proton from it, and generate a hydroxide ion. Both ammonia and the carbonate ion, for example, generate OH^- ions in water in the following equilibria.

$$NH_3(aq) + H_2O \rightleftharpoons NH_4^+(aq) + OH^-(aq)$$
$$CO_3^{2-}(aq) + H_2O \rightleftharpoons HCO_3^-(aq) + OH^-(aq)$$

Weak bases vary considerably in their abilities to accept protons from water molecules, and we compare these abilities with a special equilibrium constant. We will represent any base by the symbol B, regardless of its electrical charge, and write the equilibrium that B establishes in water as follows. (Unless stated otherwise, water will always be assumed to be the solvent in discussions of K_b.)

Common Brønsted bases are NH_3, HPO_4^{2-}, HCO_3^-, and OH^-.

$$B(aq) + H_2O \rightleftharpoons BH^+(aq) + OH^-(aq)$$

Brønsted base | conjugate acid

The base ionization constant for this equilibrium, K_b, is defined by the following equation.

$[H_2O]$ is incorporated into K_b just as it was into K_a.

$$K_b = \frac{[BH^+][OH^-]}{[B]} \qquad (15.10)$$

The values of K_b for several bases are given in Table 15.5. Always bear in mind that *the smaller the K_b, the weaker is the base.*

To use K_b values, we first have to be able to translate just the formula of some given base both into its chemical equilibrium in water and into the specific equation for its K_b. The next example shows how this is done.

EXAMPLE 15.9 WRITING EXPRESSIONS FOR K_b FOR BRØNSTED BASES

hydrazine

Problem: Hydrazine, N_2H_4, is a weak base, like ammonia. Write the equation for K_b for this base.

Solution: We first have to write the equation for the equilibrium that hydrazine establishes. To do this we write N_2H_4 and H_2O as the reactants, and we set down OH^- and the conjugate acid of N_2H_4 as the products. We figure out the formula of the conjugate acid of any base, you will

TABLE 15.5 K_b Values for Bases[a]

Name	Formula	K_b (25 °C)
Oxide ion	O^{2-}	1×10^{22}
Sulfide ion	S^{2-}	1×10^5
HYDROXIDE ION	OH^-	55
Phosphate ion	PO_4^{3-}	2.2×10^{-2}
Carbonate ion	CO_3^{2-}	2.1×10^{-4}
Ammonia	NH_3	1.8×10^{-5}
Cyanide ion	CN^-	1.6×10^{-5}
Hypochlorite ion	ClO^-	3.3×10^{-7}
Monohydrogen phosphate ion	HPO_4^{2-}	1.6×10^{-7}
Sulfite ion	SO_3^{2-}	1.5×10^{-7}
Hydrogen sulfide ion	SH^-	1.1×10^{-7}
Bicarbonate ion	HCO_3^-	2.6×10^{-8}
Acetate ion	$C_2H_2O_3^-$	5.7×10^{-10}
Ascorbate ion	$HC_6H_6O_6^-$	1.3×10^{-10}
Barbiturate ion	$C_4H_3N_2O_3^-$	1.0×10^{-10}
Formate ion	CHO_2^-	5.6×10^{-11}
Fluoride ion	F^-	1.5×10^{-11}
Citrate ion	$H_2C_6H_5O_7^-$	1.4×10^{-11}
Nitrite ion	NO_2^-	1.4×10^{-11}
Dihydrogen phosphate ion	$H_2PO_4^-$	1.4×10^{-12}
Sulfate ion	SO_4^{2-}	9.8×10^{-13}
Hydrogen sulfite ion	HSO_3^-	8.1×10^{-13}
WATER	H_2O	1.8×10^{-16}
Nitrate ion	NO_3^-	Very small
Chloride ion	Cl^-	Very small
Hydrogen sulfate ion	HSO_4^-	Very small
Bromide ion	Br^-	Very small
Iodide ion	I^-	Very small
Perchlorate ion	ClO_4^-	Very small

Additional values of K_b are located in Appendix D.

INCREASING BASE STRENGTH ↑

[a] K_b values (except for ammonia) were calculated from the K_a values obtained from the references cited for Table 15.4 and then rounded to two significant figures.

recall, by adding one H^+ to the formula of the base (remembering to adjust the charge correctly). So the conjugate acid of N_2H_4 is $N_2H_5^+$. Our equilibrium equation, then, is

Hydrazine is a rocket fuel.

$$N_2H_4(aq) + H_2O \rightleftharpoons N_2H_5^+(aq) + OH^-(aq)$$

Now we can write the equation for K_b, omitting H_2O and remembering that the products are always in the numerator. The answer, then, is

$$K_b = \frac{[N_2H_5^+][OH^-]}{[N_2H_4]}$$

PRACTICE EXERCISE 12

The proton-accepting site on aniline is the nitrogen atom, which has an unshared pair of electrons, like ammonia.

Write the equilibrium equations and the equations for K_b for each of the following Brønsted bases.
(a) CN^- (cyanide ion)
(b) $C_2H_3O_2^-$ (acetate ion)
(c) $C_6H_5NH_2$ (aniline)
(d) H_2O

Two kinds of calculations involve K_b.

1. The calculation of a value of K_b from $[OH^-]$ or pOH. (Or what amounts almost to the same thing, the calculation of K_b from pH, since $pOH = 14.00 - pH$.)

2. The calculation of equilibrium concentrations of $[H^+]$ or $[OH^-]$ from values of K_b and $[B]_{initial}$. (Or what is almost the same, the calculation of the pH of a solution of some weak base from K_b and $[B]_{initial}$, since we now can convert an equilibrium value of $[H^+]$ into the corresponding pH.)

EXAMPLE 15.10   CALCULATING K_b FROM pH

Problem: Methylamine, CH_3NH_2, is one of many substances that give herring brine its pungent odor. In 0.100 M CH_3NH_2, the pH is 11.80. What is the K_b of methylamine?

Solution: Doing this calculation is *very* similar to finding the K_a of a given weak acid, HA, from the pH at some value of $[HA]_{initial}$ (which we did in Example 15.6). The first step is to write the chemical equilibrium equation and, from it, the equation for K_b. The conjugate acid of CH_3NH_2, of course, is $CH_3NH_3^+$.

$$CH_3NH_2(aq) + H_2O \rightleftharpoons CH_3NH_3^+(aq) + OH^-(aq)$$

$$K_b = \frac{[CH_3NH_3^+][OH^-]}{[CH_3NH_2]}$$

At equilibrium, we know that $[OH^-] = [CH_3NH_3^+]$, since they are generated in equal numbers. We can find the value of $[OH^-]$ from the pOH of the solution, which we easily calculate from the pH as follows.

$$pH + pOH = 14.00$$

Since $pH = 11.80$,

$$pOH = 14.00 - 11.80$$
$$= 2.20$$

For the value of $[OH^-]$, we proceed from the pOH in the usual way.

$$[OH^-] = 10^{-pOH} \text{ mol/L}$$
$$= 10^{-2.20} \text{ mol/L}$$
$$= 0.0063 \text{ mol/L}$$

Now we can assemble a concentration table to help us figure out all the equilibrium concentrations needed to calculate K_b.

	$CH_3NH_2(aq)$ + H_2O	$\rightleftharpoons$ $CH_3NH_3^+(aq)$ +	$OH^-(aq)$
Initial concentrations	0.100	0	0
Changes in concentrations caused by the ionization	−0.0063	+0.0063	+0.0063
Final concentrations at equilibrium	(0.100 − 0.0063) = 0.094 (correctly rounded)	0.0063	0.0063

Notice that because we know both the initial concentration of CH_3NH_2 and the change in this value, we need make no assumptions in computing the final concentration.

Now we can calculate K_b.

$$K_b = \frac{(0.0063)(0.0063)}{(0.094)}$$
$$= 4.2 \times 10^{-4}$$

The base ionization constant of methylamine is 4.2×10^{-4}.

PRACTICE EXERCISE 13   Few substances are more effective in relieving intense pain than morphine. Morphine is an alkaloid—an alkali-like compound obtained from plants—and alkaloids are all weak bases. In 0.010 M morphine, the pH is 10.10. Calculate the K_b for morphine.

EXAMPLE 15.11   CALCULATING [OH⁻], pOH, AND pH FROM K_b AND $[B]_{initial}$

Problem: Calculate the values of $[OH^-]$, pOH, and pH in a 0.20 M solution of ammonia.

Solution: The best approach is just like that used to find $[H^+]$ from K_a in Example 15.7. We can make the same kinds of simplifying assumptions to make the calculations easier with no loss of precision. (Compare the following assumptions with those described on page 612 to see how similar they are.)

Assumption 1.   We ignore the contribution to the total $[OH^-]$ made by the self-ionization of water. It is just too small to matter.

Assumption 2. $[B]_{equilibrium} \approx [B]_{initial}$ for weak bases in moderately concentrated solutions.

Now we can prepare the concentration table.

	$NH_3(aq)$	$+ H_2O \rightleftharpoons$	$NH_4^+(aq)$	$+ OH^-(aq)$
Initial concentrations	0.20		0	0
Changes in concentrations caused by the ionization	$-x$		$+x$	$+x$
Final concentrations at equilibrium	$(0.20 - x)$ ≈ 0.20 (Note 1)		x	x

Note 1. Since x is expected to be small in relation to 0.20, the value of $(0.20 - x)$ will be nearly equal to 0.20, when correctly rounded (Assumption 2).

Now we can calculate the value of x from K_b, which is 1.8×10^{-5} (from Table 15.5).

$$K_b = \frac{[NH_4^+][OH^-]}{[NH_3]}$$

$$= \frac{(x)(x)}{0.20} = 1.8 \times 10^{-5}$$

$$x^2 = (0.20)(1.8 \times 10^{-5}) = 3.6 \times 10^{-6}$$
$$x = 0.0019$$

Checking our assumption, we see that the value of x is indeed much less than 0.20, so the approximation we made is acceptable. Since $x = [OH^-]$, we can now write

$$[OH^-] = 1.9 \times 10^{-3} \text{ mol/L}$$

Next, we have to calculate pOH. (And it is much easier to calculate the pH *after* we find pOH, since pH + pOH = 14.00.)

$$pOH = -\log[OH^-]$$
$$= -\log(1.9 \times 10^{-3})$$
$$= -(-2.72)$$
$$pOH = 2.72$$

Finally, we calculate the pH.

$$pH = 14.00 - pOH$$
$$= 14.00 - 2.72$$
$$pH = 11.28$$

Thus, in 0.20 M NH_3, $[OH^-] = 1.9 \times 10^{-3}$ mol/L, pOH = 2.72, and pH = 11.28.

PRACTICE EXERCISE 14 In Example 15.11, the pH of 0.20 M NH$_3$ was calculated. Suppose we dilute the solution tenfold to 0.020 M. Now what are the values of pH, pOH, and [OH$^-$]?

Concerning Aqueous Ammonia

The low value for K_b for ammonia means that the principal proton-accepting species — the chief base — in aqueous ammonia is molecular ammonia itself, NH$_3$, not OH$^-$. We have commented on this before, but now we see the quantitative data. Remember that no such compound as "ammonium hydroxide" exists, despite what you might see both in technical references and on bottle labels. "Ammonium hydroxide" must always be understood as meaning "aqueous ammonia."

15.5 pK_a AND pK_b

Alternatives to K_a and K_b for comparing the relative strengths of weak acids and bases are the negative logarithms of K_a and K_b, pK_a and pK_b.

We have seen that the values of K_a and K_b for weak acids and bases, like values of [H$^+$] for weakly acidic solutions, are small numbers usually expressed in exponential form. For the same reason that the pH concept was devised, we have analogous expressions for K_a and K_b, pK_a and pK_b, defined as follows.

$$pK_a = -\log K_a \qquad (15.11)$$
$$pK_b = -\log K_b \qquad (15.12)$$

Be sure to notice that pK_a and pK_b are defined as *negative* logarithms, as in the definition of pH. Therefore, the generalizations we want to carry forward from Equations 15.11 and 15.12 are the following.

The larger the pK_a, *the weaker is the acid.*
The larger the pK_b, *the weaker is the base.*

EXAMPLE 15.12   CALCULATING pK_a FROM K_a

Problem: The K_a for acetic acid is 1.8×10^{-5} (at 25 °C). What is pK_a at this temperature?

Solution: We simply plug the K_a into the equation that defines pK_a.

$$pK_a = -\log K_a$$
$$= -\log(1.8 \times 10^{-5})$$
$$pK_a = 4.74 \text{ (rounded as per footnote 1, page 601}$$

PRACTICE EXERCISE 15   What is the pK_a for nitrous acid, HNO$_2$? (Consult Table 15.4 for its K_a value.)

EXAMPLE 15.13   USING pK_a VALUES TO COMPARE STRENGTHS OF ACIDS

Problem: Hypoiodous acid, HIO, has a pK_a of 10.6. The pK_a of the closely related hypobromous acid, HBrO, is 8.64. Which is the weaker acid?

Solution: Remember the reciprocal relationship — the larger the pK_a, the weaker is the acid. So HIO is the weaker acid.

PRACTICE EXERCISE 16   The pK_a of hydrocyanic acid, HCN, is 9.2, and that of acetic acid if 4.76. Which is the stronger acid?

EXAMPLE 15.14   CALCULATING pK_b FROM K_b

Problem: The K_b of ammonia is 1.8×10^{-5} (at 25 °C). What is its pK_b?

Solution: This is no more difficult than plugging the K_b into the equation that defines pK_b.

$$pK_b = -\log K_b$$
$$= -\log(1.8 \times 10^{-5})$$
$$pK_b = 4.74 \text{ (rounded as per footnote 1)}$$

PRACTICE EXERCISE 17   The K_b for ethylamine, an ammonia-like compound, is 6.41×10^{-4}. Calculate its pK_b.

K_a and K_b for Conjugate Acid–Base Pairs

An important and very simple relationship exists between K_a and K_b for conjugate acid–base pairs.

$$K_a K_b = K_w \text{ (for a conjugate acid–base pair)} \qquad (15.13)$$

To prove Equation 15.13, we just substitute the expressions for K_a, K_b, and K_w into it and cancel what terms we can. Thus, we know that for the equilibrium in a weak acid we have

$$HA \rightleftharpoons H^+ + A^- \qquad K_a = \frac{[H^+][A^-]}{[HA]}$$

Similarly, for the conjugate base of HA, which here is A^-, we have

$$A^- + H_2O \rightleftharpoons HA + OH^- \qquad K_b = \frac{[HA][OH^-]}{[A^-]}$$

Now let us multiply the expressions for K_a and K_b and cancel what we can.

$$K_a \times K_b = \frac{[H^+]\cancel{[A^-]}}{\cancel{[HA]}} \times \frac{\cancel{[HA]}[OH^-]}{\cancel{[A^-]}}$$
$$= [H^+][OH^-]$$
$$= K_w \text{ (and this proves Equation 15.13)}$$

What makes Equation 15.13 particularly useful is that if we know K_b for a base, we can calculate K_a for its conjugate acid, since we can rearrange Equation 15.13 as follows.

$$K_a = \frac{K_w}{K_b} \qquad (15.14)$$

Likewise, if we know K_a for an acid, we can calculate K_b for its conjugate base, since we can also rearrange Equation 15.13 as follows.

$$K_b = \frac{K_w}{K_a} \qquad (15.15)$$

All the values of K_b in Table 15.5, for example, were calculated from the values of K_a in Table 15.4. In fact, many references furnish only a table of K_a values, expecting the reader to calculate K_b data as needed by Equation 15.13.

The relationship among K_a, K_b, and K_w leads to a simple relationship between pK_a and pK_b. If we take the logarithms of both sides of Equation 15.13, we obtain

$$\log(K_a \times K_b) = \log K_w$$

Or,

$$\log K_a + \log K_b = \log K_w$$

If we multiply both sides by -1, we get

$$(-\log K_a) + (-\log K_b) = (-\log K_w)$$

But this is the same as writing

$$pK_a + pK_b = -(\log 1.0 \times 10^{-14}) \text{ (at 25 °C)}$$

Since $\log 1.0 \times 10^{-14} = 14.00$, we have the following simple relationship between the pK_a and pK_b values for an acid and its conjugate base.

At 20 °C, $pK_w = 14.17$.
At 30 °C, $pK_w = 13.83$.

$$pK_a + pK_b = 14.00 \text{ (25 °C)} \qquad (15.16)$$

The negative logarithm of K_w has the symbol pK_w, so Equation 15.16 becomes the following general equation for use with a value of pK_w for any temperature.

$$pK_a + pK_b = pK_w$$

EXAMPLE 15.15 FINDING pK_a FROM pK_b OR pK_b FROM pK_a FOR CONJUGATE ACID–BASE PAIRS

Problem: The pK_a of acetic acid at 25 °C is 4.74. What is the pK_b of its conjugate base, the acetate ion, $C_2H_3O_2^-$? Write the equilibrium equation in which the conjugate base acts as a base in water, and write the expression for its K_b.

Solution:
$$pK_a + pK_b = 14.00$$
$$4.74 + pK_b = 14.00$$
$$pK_b = 14.00 - 4.74$$
$$= 9.26$$

The pK_b for the acetate ion is 9.26. The chemical equilibrium for which this value applies is one in which the acetate ion acts as a Brønsted base toward water:

$$C_2H_3O_2^-(aq) + H_2O \rightleftharpoons HC_2H_3O_2(aq) + OH^-(aq)$$

So the corresponding expression for K_b is

$$K_b = \frac{[HC_2H_3O_2][OH^-]}{[C_2H_3O_2^-]}$$

PRACTICE EXERCISE 18 Hydrocyanic acid, HCN, is a very weak acid with a pK_a of 9.2 at 25 °C. Calculate the pK_b of its conjugate base, CN^-. Write the equation for the chemical equilibrium in which CN^- acts as a Brønsted base in water, and then write the equation for K_b.

Reciprocal Relationships within Conjugate Acid–Base Pairs — A Revisit

If you worked Practice Exercise 18, you confirmed what was stated on page 422 — that the conjugate base of a weak acid is a fairly strong base. You found in this Exercise, for example, that for CN^-, $pK_b = 4.80$, making the cyanide ion almost as basic as ammonia ($pK_b = 4.74$). So the conjugate base, CN^-, of the weak acid, HCN, is a fairly strong base. We will restate the reciprocal relationships in general terms here.

> The conjugate base of a very weak acid is a relatively strong base.
> The conjugate acid of a very strong base is a relatively weak acid.

Or we could write the relationships in the following alternative forms, which suggest the same thing.

> The conjugate base of a strong acid is a weak base.
> The conjugate acid of a weak base is a strong acid.

Figure 15.4 displays these relationships graphically.

These relationships arise, of course, from Equations 15.14 and 15.15. In Equation 15.14, for example, we see that

$$K_a \propto \frac{1}{K_b}$$

where the proportionality constant is simply K_w.

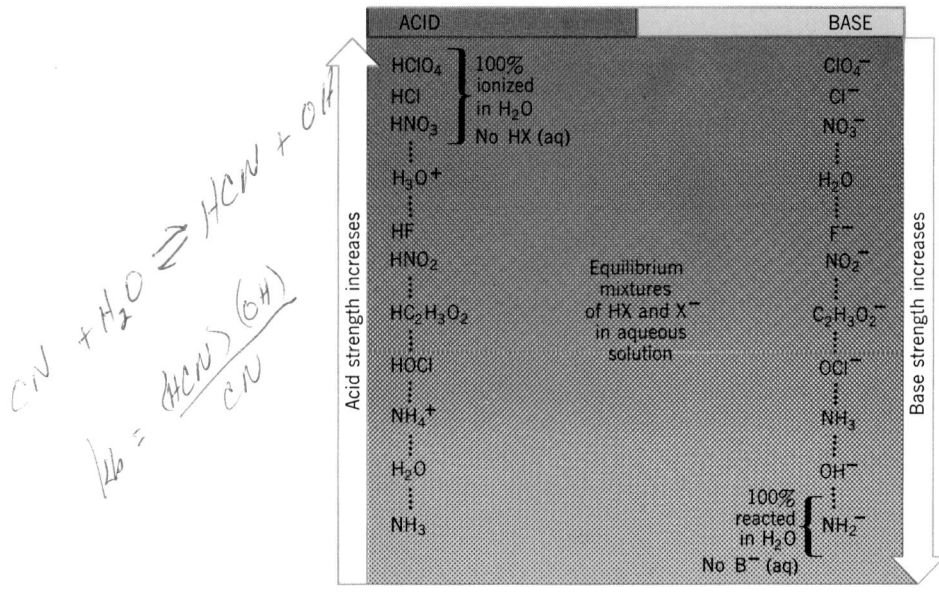

FIGURE 15.4 The relative strengths of conjugate acid–base pairs. The stronger the acid is, the weaker is its conjugate base. And the weaker the acid is, the stronger is its conjugate base. (From J. E. Brady and G. E. Humiston, *General Chemistry. Principles and Structure*, 4th ed. Copyright © 1986, John Wiley & Sons, New York. Used by permission.)

15.6 BUFFERS: THE CONTROL OF pH

When a solution is buffered, its solutes protect it against large changes in pH even when strong acids or bases are added.

We all know that acids cause things made of metals, like cars or battery terminals, to corrode faster. Chefs know that adding just a little lemon juice to milk makes it curdle. If the pH of your blood were to change from what it should be, 7.35, either to 7.00 or to 8.00, you would die. Thus, a change in the pH of the environment of some material or of a living system can cause chemical reactions, often unwanted. Are there any ways to protect systems against changes in pH?

Sometimes a chemist wants to study a reaction under conditions that would suppress any associated change in the pH of the reaction mixture. Are there ways to do this? In other experiments, the chemist might want to study a reaction at some predetermined pH and keep the pH constant. Can this be done?

By a careful choice of solutes, a chemist can ensure that a solution will not experience more than a trifling change in pH even if small amounts of a strong acid or a strong base are added. Such mixtures of solutes are called buffers, because they do just that — they buffer the system against a change in pH. The solution itself is said to be *buffered* or it is described as a *buffer solution*. There is almost no area of experimental work in chemistry, or in the applications of chemistry in molecular biology, microbiology, cell biology, or the sciences of foods, soils, nutrition, and clinical analysis in which the concept of buffers is not important. Happily, much of what we have studied in the previous sections of this chapter applies to buffers. Buffers introduce no new concepts, just a new application.

Components of Buffers

Usually a buffer consists of two solutes. One is a weak Brønsted acid, HA, and other is its conjugate base, A^-, *supplied by a soluble salt of the acid,* such as NaA. Thus a common buffer system consists of acetic acid plus sodium acetate, with the salt's acetate ion serving as the Brønsted base. Your blood uses carbonic acid (H_2CO_3), a weak diprotic acid, and the bicarbonate ion, its conjugate Brønsted base, to maintain a remarkably constant pH in the face of the body's steady production of organic acids by metabolism. (See Special Topic 15.2.) The weak acid can also be a cation, like NH_4^+, supplied by a salt such as NH_4Cl. In this case, the conjugate Brønsted base would be NH_3.

One important point about buffers is the distinction between keeping a solution at a particular pH and keeping it neutral — at a pH of 7. Although it is certainly possible to prepare a buffer to work at pH 7, buffers can be made that will work at any pH value throughout the pH scale. One topic we must study, therefore, is how to pick the weak acid and its salt — and their mole ratio — that would make a buffer good for some chosen pH.

Buffer systems do *not* protect a solution against a tide of strong acid or base, just relatively small amounts. So another topic we have to consider is the *capacity* of a buffer, the amount of strong acid or base it can absorb before the pH changes a lot.

For simplicity, we will first confine our discussion to the HA/A^- type of buffer system, like the acetic acid/sodium acetate buffer. The following reactions, which we've studied before, show *how* this system can buffer a solution.

Both key species in the buffer, HA and A^-, participate in the following equilibrium for the ionization of HA,

$$HA(aq) \rightleftharpoons H^+(aq) + A^-(aq)$$

If we add extra H^+ to the buffer, this equilibrium must shift to the left in accordance with Le Châtelier's principle. Thus the *net change* when the buffer neutralizes extra acid is

$$H^+(aq) + A^-(aq) \longrightarrow HA(aq)$$

If we add extra OH^- to the buffer, it will neutralize some H^+ in the equilibrium. This loss of H^+ forces some un-ionized HA to release more in accordance again with Le Châtelier's principle. The equation for the *net change* when OH^- is added to the buffer is

$$HA(aq) + OH^-(aq) \longrightarrow A^-(aq) + H_2O$$

Thus one member of a buffer team neutralizes any strong acid that might get into the solution and the other member neutralizes strong base.

A very simplified equation for one of the key equilibria involved with the transport and delivery of oxygen by the bloodstream is

$$HHb + O_2 \rightleftharpoons H^+ + HbO_2^-$$

where HHb represents the oxygen carrier, hemoglobin, and HbO_2^- is oxyhemoglobin. (Each hemoglobin molecule actually carries *four* O_2 molecules, but what we have written will serve the point of this special topic well enough.) The presence of H^+ in this equilibrium means that the entire transport of oxygen is sensitive to the pH of the bloodstream. Your system must control its blood pH or you will die.

The blood pH (as measured at *room* temperature) should be 7.35 or very, very close to this. Should the blood pH drift downward, say toward 7.20, the condition is called *acidosis* and emergency measures must be taken. Such varied conditions as uncontrolled diabetes, severe diarrhea, aspirin or narcotics overdose, emphysema, and starvation cause acidosis.

If the pH of the blood drifts upward toward, say, 7.5, the condition is called *alkalosis,* and this is equally life-threatening. Such diverse conditions as prolonged hysterics, continual vomiting, bicarbonate overdose,

kidney disease, and rapid breathing at a high altitude can bring on alkalosis.

The chief guardian of the pH of the blood is a buffer system—called the carbonate buffer—made up of bicarbonate ion and dissolved carbon dioxide plus carbonic acid. Dissolved CO_2 and H_2CO_3 are functionally equivalent as neutralizers of base, and we'll simplify the discussion by letting H_2CO_3 represent both. An enzyme, carbonic anhydrase, equilibrates $CO_2(aq)$ and $H_2CO_3(aq)$ exceedingly rapidly.

$$CO_2(aq) + H_2O \rightleftharpoons H_2CO_3(aq)$$

The carbonic acid can neutralize hydroxide ions which, if added, would increase the pH of the blood and cause alkalosis.

$$H_2CO_3(aq) + OH^-(aq) \longrightarrow HCO_3^-(aq) + H_2O$$

The bicarbonate ion can neutralize hydrogen ions which, if added, would cause a decrease in the pH of the blood and lead to acidosis.

$$HCO_3^-(aq) + H^+(aq) \longrightarrow H_2CO_3(aq)$$

Because both a base neutralizer and an acid neutralizer are present, the blood is buffered.

The pH of a Buffered Solution

We will assume that the buffer is made by dissolving some weak acid, HA, together with some of its salt, such as a sodium salt (NaA) in water. We generally want a fully soluble salt so that it is 100% ionized to provide what we are really after, the Brønsted base, A^-. The HA/A^- type of buffer system could involve any one of a number of weak acids of widely varying K_a values. So we cannot expect just any weak acid to work for the buffering of all ranges of pH. We therefore have to develop an equation with which we can determine at what pH a specific buffer system will work best.

Recall the equation *that defines* K_a for a weak acid, HA.

$$K_a = \frac{[H^+][A^-]}{[HA]}$$

Because we are after $[H^+]$ and then pH, let us rearrange this to give us an expression for $[H^+]$.

$$[H^+] = K_a \times \frac{[HA]}{[A^-]} \qquad (15.17)$$

All molar concentrations in Equation 15.17, remember, are *equilibrium* concentrations. When we deal with buffers, however, we can safely make some very important simplifications. We will show that we can use *initial* values of molarities of the weak acid, HA, and of the anion, A^- (as supplied by the salt) to be the same as the equilibrium values. Initial values are easy to use because we get them directly from the moles of solutes used to make the buffer solution.

The suppression of the ionization of H*A* by *A*⁻ is an example
of the common ion effect
(Section 14.10).

Because H*A* is a weak acid, little is ionized at equilibrium, even if it were the only solute. But when its anion, A^-, is also present (supplied by the salt) the ionization of H*A* is suppressed even more. The presence of A^- from the salt, in other words, acts as a stress on the following equilibrium and keeps it shifted to the left, in favor of un-ionized H*A*.

$$HA \rightleftharpoons H^+ + A^-$$

The result is that the value of $[HA]_{\text{equilibrium}}$ is essentially identical to that of $[HA]_{\text{initial}}$, the value we know about from preparing the solution. Thus our first simplification is

$$[HA]_{\text{equilibrium}} = [HA]_{\text{from initial concentration of acid}} = [\text{acid}]$$

Now let us see what we can do about $[A^-]_{\text{equilibrium}}$. Not much A^- is supplied by the ionization of the weak acid. ("Weak" implies this.) Nearly 100% of the A^- is supplied by the salt, instead, because the ionization of the weak acid is suppressed. Thus not much of anything changes the value $[A^-]$ *as initially supplied by the salt*. We conclude that the value of $[A^-]_{\text{equilibrium}}$ is essentially the same as $[A^-]_{\text{initial}}$. So our second simplification is

$$[A^-]_{\text{equilibrium}} = [A^-]_{\text{from initial concentration of salt}} = [\text{anion}]$$

These relationships define new terms, [acid] and [anion], which we can now substitute into Equation 15.17 for [H*A*] and [A^-], respectively. Always remember that [acid] refers to the *initial* concentration of the weak acid and [anion] refers to the *initial* concentration of the anion directly provided by the salt in the buffer system. So for a buffer system made of the H*A*/A^- pair, we have the following equation for [H⁺].

$$[H^+] = K_a \times \frac{[\text{acid}]}{[\text{anion}]} \tag{15.18}$$

Equation 15.18 is so similar to Equation 15.17 that most chemists just use Equation 15.17. This equation flows from the original definition of K_a (which must be remembered), and the substitutions allowed when working with buffers are then applied. Once these substitutions are made, it is easy to solve for [H⁺] (and then calculate the pH if desired).

There is a variation of Equation 15.18 involving pH and pK_a, instead of [H⁺] and K_a, that is popular among some scientists, particularly those in the life science, and we should tell you about it. If we take the logarithm of both sides of Equation 15.18, and then multiply every resulting term by -1, we get

$$-\log[H^+] = -\log K_a - \log \frac{[\text{acid}]}{[\text{anion}]}$$

Since we can recognize expressions for pH and pK_a in this, we can write

$$pH = pK_a - \log \frac{[\text{acid}]}{[\text{anion}]}$$

If we note that

$-\log a/b = +\log b/a$

$$-\log \frac{[\text{acid}]}{[\text{anion}]} = +\log \frac{[\text{anion}]}{[\text{acid}]}$$

then we can write

$$pH = pK_a + \log \frac{[\text{anion}]}{[\text{acid}]} \tag{15.19}$$

Equation 15.19 is sometimes called the *Henderson–Hasselbalch equation* by the scientists who use it.[2]

Before working some problems, let us review the implications of Equation 15.19 (or its predecessors). Two factors evidently govern the pH of a buffered solution. The first is the pK_a of the weak acid in the buffer pair. The second is the *ratio* of the initial molarities of this acid and the anion from the salt that provides the other buffer component. Notice particularly that if we prepare a buffer so as to make this ratio equal to 1, then the pH of the buffered solution equals the pK_a of the weak acid. Consequently, *what mostly determines where, on the pH scale, a buffer can work best is the pK_a of the acid.* So to prepare a specific buffer for use at a prechosen pH, we have to select the weak acid first.

Almost never can such an acid be found, however, whose pK_a exactly equals the pH we want. So we try for an acid whose pK_a is close to the pH we want — how close will be studied shortly. Then, by experimentally adjusting the ratio of [anion] to [acid], we can make a final adjustment. As you can see in Equation 15.19, the logarithm of this ratio adjusts the final pH.

The first calculation that we will study is how to find the pH of a given buffer solution.

When [anion] = [acid], then
$$\log \frac{[\text{anion}]}{[\text{acid}]} = \log 1 = 0$$

EXAMPLE 15.16 CALCULATING THE pH OF A BUFFERED SOLUTION

Problem: To study the effect of a weakly acidic medium on the rate of corrosion of a metal, a chemist prepared a buffer solution by making it 0.11 M $NaC_2H_3O_2$ (sodium acetate) and also 0.090 M $HC_2H_3O_2$ (acetic acid). What is the pH of this solution?

Solution: We note first, from Table 15.4, that K_a for acetic acid is 1.8×10^{-5}, and that from this we can calculate the pK_a to be 4.74. We will now show how to solve this problem in two ways — from the defining equation for K_a and then via the Henderson–Hasselbach equation.

From the Equation for K_a. The definition of K_a is

$$K_a = \frac{[H^+][A^-]}{[HA]}$$

Each formula unit of $NaC_2H_3O_2$ gives one anion, $C_2H_3O_2^-$, since sodium salts are fully dissociated in dilute solutions. When $[NaC_2H_3O_2] = 0.11\,M$, then $[C_2H_3O_2^-] = [\text{anion}] = 0.11\,M$.

Here is where we remember the crucial substitutions allowed when we are working with a buffer system — [anion] for [A^-] and [acid] for [HA]. Since [anion] = 0.11 mol/L, since [acid] = 0.090 mol/L, and since $K_a = 1.8 \times 10^{-5}$, we have

$$1.8 \times 10^{-5} = \frac{[H^+](0.11)}{(0.090)}$$

Solving this for [H^+], we have

$$[H^+] = \frac{(0.090)(1.8 \times 10^{-5})}{(0.11)}$$
$$= 1.5 \times 10^{-5}\ \text{mol/L}$$

[2] Be warned that you may sometimes see the Henderson–Hasselbalch equation written, instead, as

$$pH = pK_a + \log \frac{[\text{salt}]}{[\text{acid}]}$$

In other words, [salt] is used instead of [anion], as if the two were always identical in value. They are identical only when the cation of the salt is of the form M^+ (e.g., Na^+ or K^+) so that each formula unit of the salt furnishes only *one* anion. But when the cation is of the form M^{2+} (e.g., Ca^{2+}), then *two* anions are released by the dissociation of only one formula unit of the salt. With such salts the value of [anion] is twice the value of [salt]. It is safest to stick with the form of Equation 15.19.

Since pH $= -\log[H^+]$,

$$pH = -\log(1.5 \times 10^{-5})$$
$$= 4.82$$

It does not take long to whip all this out with a pocket calculator just from knowing the definitions of K_a and pH.

From the Henderson–Hasselbalch equation. Since [anion] $= 0.11$ mol/L and [acid] $= 0.090$ mol/L, and we know that $pK_a = 4.74$, we substitute these values into the Henderson–Hasselbalch equation as follows.

$$pH = 4.74 + \log \frac{(0.11)}{(0.090)}$$
$$= 4.74 + \log 1.2$$
$$= 4.74 + 0.079$$
$$pH = 4.82$$

Thus the buffer solution is at a pH of 4.82, and it will hold the pH at this value even if the reaction being studied generates a small amount of acid or base. (And it does not take long with a pocket calculator to do the calculation this way, either, but it does entail memorizing another long equation and remembering what is the numerator and what is the denominator in the logarithm term.)

PRACTICE EXERCISE 19   To compare the pH of the buffered solution of Example 15.16 with that of an unbuffered solution of 0.090 *M* acetic acid — the same as in the example — calculate the pH of 0.090 *M* acetic acid.

PRACTICE EXERCISE 20   How is the pH of a buffered solution related to the pK_a of the weak acid in the buffer if (a) the ratio of [anion] to [acid] is 10 to 1? (b) The ratio of [anion] to [acid] is 1 to 10?

PRACTICE EXERCISE 21   Calculate the pH of a buffered solution made up as 0.015 *M* sodium acetate and 0.10 *M* acetic acid.

Preparation of a Buffer

Let us return now to the selection of the solutes to prepare a buffer to work at a particular pH. The pK_a of the weak acid we pick should be within 1 unit of the desired pH. If we went outside this range, then we would have to make the [anion]/[acid] ratio either very large or very small, as suggested by Practice Exercise 20. Either choice might create problems involving solubilities or buffer capacity, which we will discuss shortly. Thus, once we have decided at what pH we want the solution buffered, we pick an acid whose value of pK_a fits the equation $pH = pK_a \pm 1$ or, rearranging terms, $pK_a = pH \pm 1$.

EXAMPLE 15.17   PREPARING A BUFFER SOLUTION

Problem: A solution buffered at pH 5.00 is needed in a chemistry experiment. Can we use acetic acid and sodium acetate to make it? If so, what ratio of acetate ion to acetic acid is needed?

Solution: We first check the pK_a of acetic acid to see if it is in the desired range. Since we want the pH to be 5.00, the pK_a of the selected acid should be 5.00 ± 1, meaning the range of 4.00 to 6.00. Since $K_a = 1.8 \times 10^{-5}$ for acetic acid, $pK_a = 4.74$. So the pK_a of acetic acid falls in the desired range, and acetic acid can be used together with the acetate ion to make the buffer. We will use Equation 15.19 to answer the second question: What ratio of solutes is needed?

$$pH = pK_a + \log \frac{[\text{anion}]}{[\text{acid}]}$$

$$5.00 = 4.74 + \log \frac{[\text{anion}]}{[\text{acid}]}$$

$$\log \frac{[\text{anion}]}{[\text{acid}]} = 5.00 - 4.74 = 0.26$$

$$\frac{[\text{anion}]}{[\text{acid}]} = 10^{0.26}$$

$$= 1.8$$

If $\log x = 0.26$, then $x = 10^{0.26}$.

The answer, which we can write as 1.8/1, is the ratio of molar *concentrations* of the anion and the acid. However, this ratio of concentrations is identical to the ratio of *moles* of anion to *moles* of acid, as we can see by this analysis of the units.

$$\frac{[\text{anion}]}{[\text{acid}]} = \frac{\dfrac{\text{mol anion}}{\text{t solution}}}{\dfrac{\text{mol acid}}{\text{t solution}}} = \frac{\text{mol anion}}{\text{mol acid}}$$

Thus, the answer that we calculated tells us that a solution containing a ratio of 1.8 mol of acetate ion to 1.0 mol of acetic acid is buffered at a pH of 5.00. To obtain this ratio, we could take 1.8 mol of $NaC_2H_3O_2$ to 1.0 mol of $HC_2H_3O_2$ or, if we wanted a buffer with a smaller capacity for handling the addition of extra acid or base, we could take 0.18 mol of $NaC_2H_3O_2$ to 0.10 mol of $HC_2H_3O_2$ in the same volume.

PRACTICE EXERCISE 22   A nutrition scientist needed an aqueous buffer for a pH of 3.90. Would formic acid and its salt, sodium formate, make a good pair for this purpose? If so, what mole ratio of the anion of this salt, HCO_2^-, to the acid, HCO_2H, is needed?

Buffer Capacity

In the previous example we hinted at the question of the capacity of a buffer — how much extra acid or base the solution can absorb before the buffer is essentially destroyed. It would always be possible to add so much strong acid to a buffered solution that all the base present would be neutralized. Or all of the buffer's acid could be destroyed by adding too much strong base. The *ratio* of anion to acid influences only the pH of the buffered solution, not its capacity. The buffer's *capacity* is determined by the sizes of the actual molarities of these components. So the chemist must decide before making the buffer solution what outer limits to the change in pH can be tolerated — 0.2 unit, 1.0 unit, or something else. Then, consistent with the necessary *ratio* of anion to acid, the chemist calculates the actual moles of each to dissolve in the solution.

A buffer of low capacity might be needed if the buffer's ions and molecules could, if too concentrated, interfere with some other use of the buffer solution.

The Effectiveness of a Buffer System

The question of buffer *capacity* is related to the question of buffer *effectiveness*. How well does a buffer system work? How well does it actually hold the pH? How well do we ask it to work? Suppose that 0.010 mol of a strong base such as NaOH was added to 1.0 L of the buffer solution described in Example 15.16. Without the buffer's presence, the concentration of OH^- ion would be 0.010 M. But the strong base is promptly neutralized by the weak acid of the buffer. Thus the addition of the strong base changes the concentration of the buffer's weak acid by neutralizing some of it.

$$\underset{\substack{\text{buffer}\\\text{component}}}{\text{H}A(aq)} + \underset{\substack{\text{added}\\\text{base}}}{\text{OH}^-(aq)} \longrightarrow \underset{\substack{\text{more}\\\text{forms}}}{A^-(aq)} + \text{H}_2\text{O}$$

For every mole of OH⁻ ion added we gain a mole of A^-, instead. And for every mole of OH⁻ added we lose a mole of HA. Let us now see how all this changes concentrations. In the 1.0 L of water, we started with 0.090 mol of HA. Of this, 0.010 mol is neutralized, which leaves (0.090 mol − 0.010 mol) = 0.080 mol of HA still in the 1.0 L. So the new moles/liter concentration of acid is 0.080 M. This neutralization, of course, makes the identical *change* in the amount of A^-. We started with 0.11 mol of A^- in 1.0 L, and we made 0.010 mol more of it for a total of 0.12 mol of A^-. This is in 1.0 L, as we said, so the value of [anion] is now 0.12 M. We will use Equation 15.19 to calculate the new pH.

$$\begin{aligned}
\text{pH} &= \text{p}K_a + \log\frac{\text{[anion]}}{\text{[acid]}}\\
&= 4.74 + \log\frac{(0.12)}{(0.080)}\\
&= 4.74 + \log 1.5 = 4.74 + 0.18\\
&= 4.92
\end{aligned}$$

Thus the addition of the strong base changed the pH from 4.82 to 4.92, only 0.10 unit. By comparison, had we added 0.010 mol of NaOH to 1.0 L of pure water, the pH would have changed from 7.00 to 12.00, a change of 5 pH units. We can see, on the one hand, that a buffer does not hold a pH *exactly* constant but, on the other hand, it does a very good job of limiting the change in pH to a very small amount.

The Ammonia/Ammonium Ion Buffer

Thus far, our attention has been on buffers that maintain pH values below 7, on the acid side. To hold a pH on the basic side, one commonly used buffer system involves a combination of ammonia and ammonium ion. An equation very similar to Equation 15.19, page 624 can be developed to calculate the pOH for such a system. Once the pOH is found, the pH can easily be calculated, since pH + pOH = 14.00. We begin with the equilibrium expression for the slight reaction of ammonia with water.

$$\text{NH}_3(aq) + \text{H}_2\text{O} \rightleftharpoons \text{NH}_4^+(aq) + \text{OH}^-(aq)$$

The equation for K_b is

$$K_b = \frac{[\text{NH}_4^+][\text{OH}^-]}{[\text{NH}_3]}$$

Taking the negative logarithms of both sides of this equation gives

$$-\log K_b = -\log[\text{OH}^-] - \log\frac{[\text{NH}_4^+]}{[\text{NH}_3]}$$

Our definitions of pK_b and pOH let us rewrite this equation as

$$\text{p}K_b = \text{pOH} - \log\frac{[\text{NH}_4^+]}{[\text{NH}_3]}$$

Of course, the quantities [NH$_4^+$] and [NH$_3$] mean the concentrations of these species *at equilibrium,* not the initial concentrations of ammonium ion and ammonia. However, the same kind of situation is present here as occurred during our development of Equation 15.19.

The cation, NH_4^+, is obtained from an ammonium salt, and nearly all these salts are strong electrolytes and 100% dissociated in a dilute solution. The very slight reaction of ammonia with water adds a negligible amount of NH_4^+ ions to that supplied by the ammonium salt, so we can substitute the *initial* value of the NH_4^+ concentration into our equation in place of $[NH_4^+]$. Similarly, the amount of NH_3 that actually changes into NH_4^+ as a result of our equilibrium is negligible, so we can substitute the *initial* concentration of ammonia into the equation in place of $[NH_3]$. Therefore, after rearranging terms, we obtain the following equation for use with any buffer like the ammonia/ammonium ion buffer, where a cation is the proton donor.

$$pOH = pK_b + \log \frac{[\text{cation}]}{[\text{base}]}$$

With the ammonia/ammonium ion buffer, $[\text{cation}]$ = the molar concentration of ammonium ions as supplied by the ammonium salt, and $[\text{base}]$ = the molar concentration of ammonia that is used to prepare the buffer; the value of pK_b for NH_3 is 4.74.

EXAMPLE 15.18   PREPARING AN AMMONIA/AMMONIUM ION BUFFER

Problem: A chemistry student needs 250 mL of a solution buffered at pH 9.00. How many grams of ammonium chloride have to be added to 250 mL of 0.200 M NH_3 to make such a buffer? (Assume that no change in volume occurs by the addition of this solid salt.)

Solution: To buffer at pH 9.00 means that the desired pOH is 5.00. Using this together with the value of $[\text{base}]$ given, 0.20 M, and the pK_b value of ammonia, 4.74, in the equation we developed,

$$pOH = pK_b + \log \frac{[\text{cation}]}{[\text{base}]}$$

gives us

$$5.00 = 4.74 + \log \frac{[\text{cation}]}{(0.20)}$$

Therefore,

$$\log \frac{[\text{cation}]}{(0.20)} = 0.26$$

$$\frac{[\text{cation}]}{(0.20)} = 10^{0.26} = 1.8$$

$$[\text{cation}] = 1.8 \times (0.20) = 0.36 \text{ mol } NH_4^+/L$$

Thus, the concentration of NH_4^+ in the buffer must be 0.36 M for the buffer to hold a pH of 9.00. Since we want 250 mL of solution and since NH_4^+ is to be supplied by NH_4Cl with a formula weight of 53.5, the mass of ammonium chloride that we need is found by

$$250 \text{ mL} \times \frac{0.36 \text{ mol salt}}{1000 \text{ mL}} \times \frac{53.5 \text{ g salt}}{1 \text{ mol salt}} = 4.8 \text{ g salt}$$

Thus, by dissolving 4.8 g of ammonium chloride in 250 mL of 0.20 M ammonia, the resulting solution is a buffer for pH 9.00.

PRACTICE EXERCISE 23   A buffer solution was prepared by dissolving 2.5 g of ammonium chloride in 125 mL of 0.24 M ammonia. At what pH will this solution serve as a buffer?

15.7 POLYPROTIC ACIDS

There is a separate acid ionization constant for the ionization of each hydrogen ion from a polyprotic acid.

Carbonic acid, H_2CO_3, is a diprotic acid, and there is a separate K_a for each step in its ionization. The first acid ionization constant, K_{a_1}, is for the following equilibrium.

$$H_2CO_3(aq) \rightleftharpoons H^+(aq) + HCO_3^-(aq)$$

$$K_{a_1} = \frac{[H^+][HCO_3^-]}{[H_2CO_3]} = 4.5 \times 10^{-7}$$

$$pK_{a_1} = 6.35$$

The second acid ionization constant of H_2CO_3, K_{a_2}, is for the following equilibrium.

$$HCO_3^-(aq) \rightleftharpoons H^+(aq) + CO_3^{2-}(aq)$$

$$K_{a_2} = \frac{[H^+][CO_3^{2-}]}{[HCO_3^-]} = 4.7 \times 10^{-11}$$

$$pK_{a_2} = 10.33$$

Notice that K_{a_1} is much greater than K_{a_2}. This is always true for diprotic acids because it is much easier for a neutral molecule, like H_2CO_3, to release a positively charged ion, H^+,

Additional K_a values of polyprotic acids are located in Appendix D.

TABLE 15.6 K_a and pK_a Values for Polyprotic Acids[a]

Acid	H$^+$	K_a	pK_a
Carbonic acid, H_2CO_3	1st	4.5×10^{-7}	6.35
	2nd	4.7×10^{-11}	10.33
Hydrogen sulfide, H_2S	1st	9.5×10^{-8}	7.02
	2nd	1×10^{-19}	19
Phosphoric acid, H_3PO_4	1st	7.1×10^{-3}	2.15
	2nd	6.3×10^{-8}	7.20
	3rd	4.5×10^{-13}	12.35
Sulfuric acid, H_2SO_4	1st	Large	
	2nd	1.0×10^{-2}	2.00
Selenic acid, H_2SeO_4	1st	Large	
	2nd	1.2×10^{-2}	1.92
Telluric acid, H_6TeO_6	1st	2×10^{-8}	7.7
	2nd	1×10^{-11}	11.0
	3rd	3×10^{-15}	14.5
Sulfurous acid, H_2SO_3	1st	1.2×10^{-2}	1.92
	2nd	6.6×10^{-8}	7.18
Selenous acid, H_2SeO_3	1st	3.5×10^{-3}	2.46
	2nd	5×10^{-8}	7.3
Tellurous acid, H_2TeO_3	1st	3×10^{-3}	2.5
	2nd	2×10^{-8}	7.7
Ascorbic acid, $H_2C_6H_6O_6$	1st	7.9×10^{-5}	4.10 (24 °C)
	2nd	1.6×10^{-12}	11.79 (16 °C)
Citric acid, $H_3C_6H_5O_7$	1st	7.1×10^{-4}	3.15 (18 °C)
	2nd	1.7×10^{-5}	4.77 (18 °C)
	3rd	6.4×10^{-6}	5.19 (18 °C)

[a] Data from references cited in Table 15.4 as well as from N. N. Greenwood and A. Earnshaw, *Chemistry of the Elements,* Pergamon Press, Oxford, 1984.

than it is for an already negatively charged ion, like HCO_3^-, to release H^+. Typically K_{a_1} is greater than K_{a_2} by a factor of between 10^4 and 10^5, as the data in Table 15.6 show. The second acid ionization constant for a triprotic acid is similarly much greater than the third, as the data for phosphoric acid, H_3PO_4, in Table 15.6 show.

The pH of Solutions of Polyprotic Acids

Because K_{a_1} is so much greater than K_{a_2} (or K_{a_3}), the pH of a dilute solution of a weak diprotic or triprotic acid can be calculated just from the value of K_{a_1}. The effect of second (or higher) ionizations on the value of $[H^+]_{equilibrium}$ are so small that after we round the answer to the correct number of significant figures, their contributions are entirely negligible. We will calculate $[H^+]_{equilibrium}$ for a hypothetical diprotic acid, H_2A, using only K_{a_1}. We will also show that the value of $[A^{2-}]_{equilibrium}$ numerically equals K_{a_2}, *provided that the diprotic acid is the only source of both H^+ and A^{2-}.*

EXAMPLE 15.19   CALCULATING $[H^+]$ AND $[A^{2-}]$ IN A SOLUTION OF A WEAK DIPROTIC ACID, H_2A

Problem: What are the values of $[H^+]$ and $[A^{2-}]$ at equilibrium in $0.100\ M\ H_2A$, when $K_{a_1} = 1.0 \times 10^{-5}$ and $K_{a_2} = 1.0 \times 10^{-9}$?

Solution: We will work with just the *first* ionization at the start, and we will prepare a concentration table in the usual way. As usual, x represents *changes* in concentrations caused by this ionization. This will take us two rows into the table.
   Then we will pause to enter a correction, y, to each value of x — row 3. The correction gives attention here to the second ionization, that of HA^-. The ionization of HA^- *decreases* the concentration of HA^- by y, and it *increases* the concentration of H^+ by the same amount, y. The adjusted concentrations are in the fourth row of the table.
   This may seem to be very complicated, but as you will see — the fifth row — there are some wonderful simplifications in store because K_{a_2} is so much smaller than K_{a_1}.

	H_2A	$\rightleftharpoons$ H^+	$+$ HA^-
Initial concentrations	0.100	0	0
Change in concentrations ignoring the second ionization	$-x$	$+x$	$+x$
Correction for second ionization	0 (Note 1)	$+y$	$-y$ (Note 2)
Final concentrations	$(0.100 - x)$	$(x + y)$	$(x - y)$ (Note 3)
Value after simplification	0.100	x	x

Note 1. There is no correction here because H_2A does not participate in the *second* ionization.
   Note 2. The second ionization *adds* something, y, to the concentration of H^+ as it simultaneously subtracts y from the concentration of HA^-.
   Note 3. Since $K_{a_2} \ll K_{a_1}$, the value of y is *very* small compared with that of x. So after we round correctly, both $(x - y)$ and $(x + y)$ reduce to x. Similarly, x is *very* small compared with 0.100, so $(0.100 - x)$ equals 0.100 after rounding.
   Now we use the values in the last row of the table together with K_{a_1} to find $[H^+]$ (which, of course, is x).

$$K_{a_1} = \frac{(x)(x)}{0.100} = 1.0 \times 10^{-5}$$

$$x^2 = (0.100)(1.0 \times 10^{-5}) = 1.0 \times 10^{-6}$$

$$x = 1.0 \times 10^{-3}$$

Note that x is indeed small compared to 0.100, which justifies one of our assumptions.

Therefore,

$$[H^+] = 1.0 \times 10^{-3} \text{ mol/L} \quad \text{(one answer)}$$

Now we will tackle the second part of the question — the equilibrium concentration of A^{2-}. We need the equation for K_{a_2} for this.

$$K_{a_2} = \frac{[H^+][A^{2-}]}{[HA^-]} \quad \text{(from } HA^- \rightleftharpoons H^+ + A^{2-}\text{)}$$

But both $[H^+]$ and $[HA^-]$ equal x from the concentration table, so

$$K_{a_2} = \frac{(x)[A^{2-}]}{(x)} = 1.0 \times 10^{-9}$$

Therefore,

We see that y, which equals $[A^{2-}]$, is in fact small compared to x, so $(x + y)$ and $(x - y)$ do round to just x.

$$[A^{2-}] = 1.0 \times 10^{-9} \text{ mol/L}$$

Notice that $[A^{2-}]$ equals K_{a_2}. This is true for the doubly negative ion produced in the ionization of any polyprotic acid provided, as we said, that no other species is in the solution that could furnish either H^+ or A^{2-} ions.

PRACTICE EXERCISE 24   The dihydrogen phosphate ion, $H_2PO_4^{2-}$, can be treated as a diprotic acid. (Its K_{a_1} would actually be the K_{a_2} value for H_3PO_4 in Table 15.6. And its K_{a_2} would be the K_{a_3} of phosphoric acid in this table.) Calculate $[H^+]$, pH, and $[PO_4^{3-}]$ in a 0.10 M solution of NaH_2PO_4, which, of course, supplies the dihydrogen phosphate ion.

15.8 HYDROLYSIS OF IONS

Many salts contain a cation or an anion that can react with water to make aqueous solutions of these salts either acidic or basic.

In many laboratory situations, an aqueous solution of some salt is prepared only to have the solution turn out to be either acidic or basic. If we were not aware that this could happen, we might unknowingly prepare a salt solution that could be more corrosive to metals than expected or that could contain a catalyst (for example, hydrogen ion) that might affect a reaction rate. Many salts involve an ion that reacts to some extent with water to upset its one-to-one mole ratio of $[H^+]$ to $[OH^-]$. The reactions of ions of salts with water is called the hydrolysis of ions, sometimes the *hydrolysis of salts*.

Hydrolysis means *reaction with water.*

When any ion hydrolyzes, it (or its hydrate) is merely reacting either as a Brønsted acid or a Brønsted base with water. All that we have learned, therefore, about calculating the pH values of solutions that contain such acids or bases applies to this section. The only different feature here is that the acid or base originates from a salt.

If just the anion of the salt hydrolyzes, it reacts as a base with water to accept a proton and generate OH^-. Letting A^- represent any anion, if it hydrolyzes it sets up the following equilibrium in which OH^- ion forms, making the solution basic.

$$A^-(aq) + H_2O \rightleftharpoons HA(aq) + OH^-(aq)$$

If just the cation of the salt hydrolyzes and it is a metal ion, the hydrolysis occurs in a somewhat complex manner. We will return to this phenomenon later in the section. If the cation is a nonmetal ion with protons to donate, like the ammonium ion, the effect of the cation is merely the effect of a weak Brønsted acid on pH.

If both the cation and the anion hydrolyze, the effect of the salt on the pH of its solution depends on whether the cation is better able to generate H_3O^+ than the anion can generate OH^-. These salts have to be examined on a case by case basis, and we will do little further with them.

We will first develop a simple strategy to predict, without the use of extensive tables, if one or both ions of a salt can hydrolyze and thus affect the pH of the solution. Then we will review pH calculations.

Predicting If The Anion of a Salt Hydrolyzes

To tell if the anion of a salt hydrolyzes is quite easy provided we remember which acids are strong, that is, which are 100% broken up into ions in water. The salt might be a chloride, like NaCl, for example, where the anion is Cl^-. This anion is a particularly weak Brønsted base. We need no table to tell us this because we know that its conjugate acid, HCl, is a strong acid. (The short list of strong acids should by now be learned.) A strong acid always has a weak conjugate base (Sections 11.7 and 15.5). Since Cl^- is a very weak base, it is unable to accept a proton from an acid, particularly not from water, so the chloride ion does not hydrolyze. It cannot generate OH^- as follows.

$$Cl^-(aq) + H_2O \xrightarrow{\quad\quad} HCl(aq) + OH^-(aq) \qquad \text{[does NOT occur]}$$

The anions of all strong monoprotic acids behave similarly. Thus, Cl^-, Br^-, I^-, NO_3^-, and ClO_4^- do not react with water and cannot themselves affect the pH of a solution. (We will study salts of polyprotic acids in the next section.)

The anions of all weak acids, on the other hand, are more or less basic, are able to accept protons from water, and so can hydrolyze. The acetate ion, $C_2H_3O_2^-$, for example, has acetic acid as its conjugate acid. This acid is not on the list of strong acids, so we know that it is a weak acid. Being a weak acid, its conjugate base is moderately strong. We can conclude, therefore, that the acetate ion will react with water or hydrolyze. The following equilibrium will form, which generates some hydroxide ion and makes the solution basic.

$$C_2H_3O_2^-(aq) + H_2O \rightleftharpoons HC_2H_3O_2(aq) + OH^-(aq)$$

To tell if an anion hydrolyzes, we thus follow a simple strategy. We first figure out what its conjugate acid is. Then we compare this acid with our list of strong acids. If it is on the list, its anion is not a base and does not hydrolyze. But if the conjugate acid is not on the list of strong acids, it must be a weak acid, so we then know that its anion is a base and does hydrolyze.

The common strong monoprotic acids are

HCl	HI	HNO₃
HBr	HClO₄	

> Anions whose conjugate acids are weak acids hydrolyze in water and make the solution basic. Anions of strong acids do not hydrolyze.

Predicting If the Cation of a Salt Hydrolyzes

Some but not all metal ions react with water and make a solution acidic. In general, the metal ions that hydrolyze are triply charged or doubly charged, have very small ionic radii, and so have high concentrations of positive charge. They strongly interact, therefore, with the δ^- charges on water molecules to form hydrates that are actually proton donors. The hexahydrate of aluminum, for example, is involved in the following equilibrium that makes the solution acidic.

$$Al(H_2O)_6^{3+}(aq) + H_2O \rightleftharpoons [Al(H_2O)_5(OH)]^{2+}(aq) + H_3O^+(aq)$$

The high positive-charge density on the metal ion draws considerable electron density from O—H bonds of the water molecules in the hydrate. This weakens an O—H bond and makes the O—H group a proton donor. We saw exactly the same effect in strong oxoacids (Section (11.8). The relatively high electronegativity of a central nonmetal atom in an acid like H_2SO_4 weakens a neighboring O—H bond and makes the system an acid.

None of the cations from Group IA have sufficiently high positive-charge densities, and none reacts with water to generate H_3O^+ ion. The same applies to all the metal ions from Group IIA, except Be^{2+}, which is a very tiny ion and so has a high enough positive-charge density to make it hydrolyze.

Common metal ions that do *not* hydrolyze are Li^+, Na^+, K^+, Mg^{2+}, Ca^{2+}, and Ba^{2+}.

> Metal ions from Group IA or IIA (except beryllium) do not hydrolyze. Other metal ions may hydrolyze and generate H^+ ions.

In some salts, the cation is not a metal ion. Salts of the ammonium ion are common examples. Here the cation is itself a weak Brønsted acid, so it is able to react to some extent with water to form hydronium ions and make the solution acidic.

$$NH_4^+(aq) + H_2O \rightleftharpoons NH_3(aq) + H_3O^+(aq)$$

> The ammonium ion of an ammonium salt hydrolyzes and tends to make an aqueous solution slightly acidic.

EXAMPLE 15.20   PREDICTING HOW A SALT AFFECTS THE pH OF ITS SOLUTION

Problem: Sodium phosphate, Na_3PO_4 (often called trisodium phosphate), is a common cleaning agent for walls and floors. How does it affect its solution? Does it make the solution acidic or basic?

Solution: Na_3PO_4 involves Na^+ and PO_4^{3-}. Na^+ does not hydrolyze, but PO_4^{3-} does. Its conjugate acid, HPO_4^{2-}, is not on our list of strong acids, so we can infer that it is a weak acid. We therefore expect PO_4^{3-} to be a relatively strong base and we expect it to hydrolyze as follows.

A solution of trisodium phosphate in water can be so alkaline that anyone using such a solution as a cleaning agent should wear rubber gloves.

$$PO_4^{3-}(aq) + H_2O \rightleftharpoons HPO_4^{2-}(aq) + OH^-(aq)$$

This reaction generates some OH^- ions, so the solution will be basic.

Notice that in Example 15.20 we did not need a table of Brønsted bases to predict that PO_4^{3-} would hydrolyze. We used our knowledge of just a few facts—which acids are strong acids in water and which cations do not hydrolyze—to figure out what we needed to know. We will work another example to practice using the list of strong aqueous acids to decide if a given salt can affect the pH of its solution.

EXAMPLE 15.21   PREDICTING IF A SALT AFFECTS THE ACIDITY OF ITS SOLUTION

Problem: Hydrazine, N_2H_4, is an ammonia-like base that reacts with HCl to form hydrazinium chloride, N_2H_5Cl. Does this salt make its aqueous solution acidic or basic?

Solution: Since it is described as ammonia-like, we can infer that hydrazine is a weak base. Therefore, its conjugate acid, the hydrazinium ion, $N_2H_5^+$, must be somewhat acidic, like the ammonium ion. It can be expected to react to some extent with water.

$$N_2H_5^+(aq) + H_2O \rightleftharpoons N_2H_4(aq) + H_3O^+(aq)$$

Because this reaction generates some hydrogen ions, the solution can be expected to be acidic.

PRACTICE EXERCISE 25   Determine without the use of tables if each ion can hydrolyze.
(a) CO_3^{2-} (c) S^{2-} (e) NO_2^-
(b) HPO_4^{2-} (d) Cu^{2+} (f) F^-

PRACTICE EXERCISE 26   How, if at all, does potassium acetate, $KC_2H_3O_2$, affect the acidity of its solution?

PRACTICE EXERCISE 27   How, if at all, does copper(II) nitrate, $Cu(NO_3)_2$, affect the acidity of its solution?

PRACTICE EXERCISE 28   Ammonium nitrate, NH_4NO_3, is a nitrogen fertilizer. What effect will the application of an aqueous solution of this fertilizer have on the pH of soil? Will it tend to raise the pH or lower it or leave it unchanged?

pH of Salt Solutions

In a salt where only the anion can hydrolyze, the calculation of the pH of its solution is identical to finding the pH of a solution of any Brønsted base, given its K_b and its molar concentration. We learned how in Example 15.11. We will work another example for review, because this kind of calculation is quite important.

EXAMPLE 15.22   CALCULATING THE pH OF A SALT SOLUTION

Problem: Potassium acetate hydrolyzes to give a basic solution, as predicted in Practice Exercise 26. What is the pH of a 0.15 M solution of potassium acetate? For $C_2H_3O_2^-$, $K_b = 5.7 \times 10^{-10}$.

Solution: The K^+ ion (a Group 1A cation) cannot affect the pH; only the anion, $C_2H_3O_2^-$, does. So the question is just like that in Example 15.11: What is the pH of a solution of a given Brønsted base? As in the earlier example, we begin with the chemical equilibrium and then assemble a concentration table.

	$C_2H_3O_2^-$	$+ H_2O \rightleftharpoons$	$HC_2H_3O_2$	$+ OH^-$
Initial concentrations	0.15		0	0
Changes in concentrations caused by the ionization	$-x$		$+x$	$+x$
Final concentrations at equilibrium	$(0.15 - x)$		x	x
	≈ 0.15 (Note 1)			

IMPORTANT

Note 1. This is the usual simplification based on the value of K_b and the expectation that x is too small to change 0.15 when the result is properly rounded.

Now we write the equation for K_b, make substitutions from the table into it, and calculate x.

$$K_b = \frac{[HC_2H_3O_2][OH^-]}{[C_2H_3O_2^-]} = 5.7 \times 10^{-10}$$

or

$$\frac{(x)(x)}{0.15} = 5.7 \times 10^{-10}$$

$$x^2 = (0.15)(5.7 \times 10^{-10})$$
$$x = 9.2 \times 10^{-6}$$

Notice that $x \lll 0.15$, so letting 0.15 stand for $(0.15 - x)$ works.

Thus the value of $[OH^-]$ is 9.2×10^{-6} mol/L. To find the pH of this solution, the easiest way to proceed is to use $[OH^-]$ to calculate pOH, and then calculate pH from the equation

$$pH + pOH = 14.00.$$

$$pOH = -\log(9.2 \times 10^{-6})$$
$$= 5.04$$

Therefore,

$$pH = 14.00 - 5.04$$
$$= 8.96$$

A 0.15 M solution of potassium acetate has a pH of 8.96 (and the solution is slightly basic).

For a salt solution where only the cation hydrolyzes, we have a way to calculate the pH only when the K_a of the cation is known. We have not studied such values for hydrated metal ions, and so will not consider them further. But we have studied situations involving weak Brønsted acids, like the ammonium ion. We will review an example.

EXAMPLE 15.23   CALCULATING THE pH OF AN AMMONIUM SALT SOLUTION

Problem: What is the pH of a 0.10 M solution of ammonium chloride? The K_a of NH_4^+ is 5.7×10^{-10}.

Solution: We know that the chloride ion in ammonium chloride does not hydrolyze, only the ammonium ion. So we next write the equilibrium equation for the hydrolysis of NH_4^+ and prepare the usual concentration table.

	H_2O +	NH_4^+	$\rightleftharpoons$	NH_3 +	H_3O^+
Initial concentrations		0.10		0	0
Changes in concentrations caused by the hydrolysis		$-x$		$+x$	$+x$
Final concentrations at equilibrium		$(0.10 - x)$ ≈ 0.10		x	x

The expression for K_a is

$$K_a = \frac{[NH_3][H_3O^+]}{[NH_4^+]} = 5.7 \times 10^{-10}$$

Therefore,

$$\frac{(x)(x)}{0.10} = 5.7 \times 10^{-10}$$
$$x^2 = 5.7 \times 10^{-11}$$
$$x = 7.5 \times 10^{-6}$$

As expected, x is small compared to 0.10 so letting $(0.10 - x)$ equal x works.

Therefore,

$$[H_3O^+] = 7.5 \times 10^{-6} \text{ mol/L}$$

Since

$$pH = -\log[H_3O^+]$$
$$pH = -\log(7.5 \times 10^{-6})$$
$$= 5.12$$

A 0.10 M solution of ammonium chloride has a pH of 5.12 (and so is slightly acidic).

PRACTICE EXERCISE 29   Calculate the pH of a 0.15 M solution of sodium lactate, $NaC_3H_5O_3$. Lactic acid, $HC_3H_5O_3$, the acid responsible for the sour taste of sour milk, is a monoprotic acid with $K_a = 1.4 \times 10^{-4}$.

15.9 HYDROLYSIS OF ANIONS OF POLYPROTIC ACIDS

When an anion of a polyprotic acid hydrolyzes, the pH is a function of the K_b of just this anion.

Sodium carbonate, Na_2CO_3, is the salt of an important, weak, diprotic acid, H_2CO_3. The salt's carbonate ion is a Brønsted base that hydrolyzes to give a basic solution involving the equilibrium

$$CO_3{}^{2-}(aq) + H_2O \rightleftharpoons HCO_3{}^-(aq) + OH^-(aq)$$

The equation for the base ionization constant, K_b, of $CO_3{}^{2-}$ is

$$K_b = \frac{[HCO_3{}^-][OH^-]}{[CO_3{}^{2-}]} = 2.1 \times 10^{-4} \qquad \text{(from Table 15.5)}$$

Notice in the chemical equilibrium that bicarbonate ion is another product, besides OH^-. *And $HCO_3{}^-$ is also a Brønsted base.* Therefore, we should also consider its own ability to hydrolyze and supply more OH^- ion. The hydrolysis of $HCO_3{}^-$ involves the equilibrium

$$HCO_3{}^-(aq) + H_2O \rightleftharpoons H_2CO_3(aq) + OH^-(aq)$$

The bicarbonate ion has its own K_b equation.

$$K_b = \frac{[H_2CO_3][OH^-]}{[HCO_3{}^-]} = 2.3 \times 10^{-8} \qquad \text{(from Table 15.5)}$$

The question now is, how should we calculate the pH of a solution in which OH^- is generated by the successive hydrolyses of a doubly (or triply) charged anion?

The answer is suggested when we compare the K_b values for successive hydrolyses, such as those just described. The K_b value of $HCO_3{}^-$, 2.3×10^{-8}, is much smaller than 2.1×10^{-4}, the K_b value of $CO_3{}^{2-}$, the anion directly furnished by the salt. $CO_3{}^{2-}$ is a stronger base than $HCO_3{}^-$ by roughly a factor of 10^4. This lets us ignore the hydrolysis of $HCO_3{}^-$. Its contribution to the total pool of OH^- will be just too small in relation to the OH^- supplied by the hydrolysis of its much stronger relative, $CO_3{}^{2-}$. So the answer to our question is that we can deal exclusively with the stronger base.

> This is the same approach that we used in calculating $[H^+]$ in solutions of polyprotic acids.

> When an anion of a polyprotic acid hydrolyzes, the pH is dominated by the hydrolysis of just this anion, and not other anions of the polyprotic acid that might also form in an equilibrium mixture.

Let's study an example of how this works.

EXAMPLE 15.24 CALCULATING THE pH OF A SOLUTION OF A SALT OF A WEAK DIPROTIC ACID

Problem: What is the pH of a 0.15 *M* solution of Na_2CO_3?

Solution: If we can ignore the hydrolysis of $HCO_3{}^-$, as we just discussed above, then the only relevant equation is

$$CO_3{}^{2-} + H_2O \rightleftharpoons HCO_3{}^- + OH^-$$

As we've done before, the K_b expression is

$$K_b = \frac{[HCO_3^-][OH^-]}{[CO_3^{2-}]} = 2.1 \times 10^{-4}$$

The concentration table is as follows.

	CO$_3^{2-}$	+	H$_2$O $\rightleftharpoons$ HCO$_3^-$	+	OH$^-$
Initial concentrations	0.15		0		0
Changes in concentrations caused by the ionization	$-x$		$+x$		$+x$
Final concentrations at equilibrium	$(0.15 - x)$		x		x

Let's first try a simplification that we have often used before, namely that $(0.15 - x) \approx 0.15$ as the value of $[CO_3^{2-}]$ in the equation for K_b.

$$\frac{[HCO_3^-][OH^-]}{[CO_3^{2-}]} = \frac{(x)(x)}{0.15} = 2.1 \times 10^{-4}$$

$$x^2 = (0.15)(2.1 \times 10^{-4})$$
$$= 3.2 \times 10^{-5}$$
$$x = 5.7 \times 10^{-3}$$

Notice that the value of x is, in fact, sufficiently smaller than 0.15 to let us use 0.15 as a good approximation of $(0.15 - x)$.

$$\text{Since } x = [OH^-] = 5.7 \times 10^{-3}$$
$$pOH = -\log(5.7 \times 10^{-3})$$
$$= 2.24$$

Hence,

$$pH = 14.00 - 2.24$$
$$= 11.76$$

Thus the pH of 0.15 M Na$_2$CO$_3$ is 11.76.

PRACTICE EXERCISE 30   What is the pH of a 0.20 M solution of Na$_2$SO$_3$ at 25 °C? For this diprotic acid, $K_{a_1} = 1.2 \times 10^{-2}$ and $K_{a_1} = 6.6 \times 10^{-8}$.

15.10 ACID–BASE TITRATIONS REVISITED

One goal of a well-run acid–base titration is to have the endpoint occur exactly at the equivalence point.

In Section 11.5, we studied the overall procedure for an acid–base titration, and we saw how titration data can be used to calculate the molar concentration of some "unknown." The goal during the actual procedure is to stop the addition of the titrant when stoichiometrically equivalent amounts of the two reactants have been combined. The point in the titration when this relationship is exactly true is called the equivalence point.

The equivalence point in an actual titration might not be identical with the endpoint. Ideally, it should be. But the endpoint occurs when the indicator changes color and the titration is stopped. Obviously, we want this to happen when the system is at the equivalence point. So the selection of a specific indicator requires some foresight. We will understand this better by studying how the pH of the solution being titrated changes with the addition of titrant.

Titration Curves

When the pH of a solution at different stages during a titration is plotted against the volume of standard solution added, we obtain a *titration curve*. The values of pH in these plots are calculated by the procedures studied in this chapter, so this discussion will serve as a review.

Titration of a Strong Acid by a Strong Base   Figure 15.5 gives the titration curve for the titration of a dilute solution of a typical strong, monoprotic acid, hydrochloric acid, with a dilute solution of a typical strong base, sodium hydroxide.

$$HCl(aq) + NaOH(aq) \longrightarrow NaCl(aq) + H_2O$$

At the start, before any NaOH has been added, the solution is simply 25.00 mL of 0.20 M HCl. Since HCl is a strong acid and is 100% ionized, we find the initial pH by a particularly simple calculation. Because of 100% ionization, $[H^+] = [HCl] = 0.20\ M$; so

$$\begin{aligned} pH &= -\log[H^+] \\ &= -\log(0.20) \\ &= 0.70 \quad \text{(initial pH)} \end{aligned}$$

We next calculate the pH at another point, after the addition of 10.00 mL of 0.20 M NaOH. We started with 0.0050 mol of HCl in the initial 25.00 mL of 0.20 M HCl, since

$$25.00\ \text{mL HCl solution} \times \frac{0.20\ \text{mol HCl}}{1000\ \text{mL HCl solution}} = 0.0050\ \text{mol HCl}$$

We are adding 10.00 mL of 0.20 M NaOH, and this quantity contains 0.0020 mol of NaOH, since

$$10.00\ \text{mL NaOH solution} \times \frac{0.20\ \text{mol NaOH}}{1000\ \text{mL NaOH solution}} = 0.0020\ \text{mol NaOH}$$

The coefficients of the neutralization equation tell us that 1 mol of HCl reacts with 1 mol of NaOH, so the addition of 0.0020 mol of NaOH neutralizes 0.0020 mol of HCl, leaving 0.0030 mol of HCl unneutralized. This is present in a solution that has grown in volume, of course. The new volume is (25.00 mL + 10.00 mL) = 35.00 mL, or 0.03500 L, so the concentration of HCl remaining is now

$$\frac{0.0030\ \text{mol HCl}}{0.03500\ \text{L}} = 0.086\ \text{mol HCl/L}$$

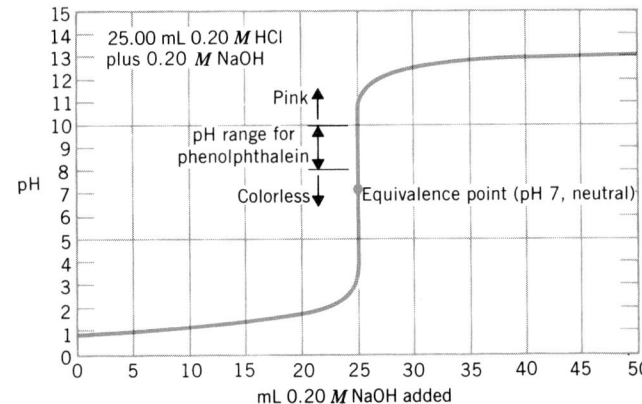

FIGURE 15.5 Titration curve for titrating a strong acid (0.20 M HCl) with a strong base (0.20 M NaOH).

In our solution now, because of 100% ionization, we have $[H^+] = [HCl] = 0.086$ mol/L. So the pH now is

$$pH = -\log 0.086$$
$$= 1.07$$

This is not a large increase from the initial pH of 0.70, and Figure 15.5 bears out comparable results of similar calculations. The pH rises slowly during the first part of the titration of the strong acid by the strong base. All that is happening is that the added base is changing some of the acid to a salt, NaCl, *neither of whose ions hydrolyzes*. And the concentration of remaining acid gradually becomes more and more dilute.

Next, let us move the titration to the point where 24.90 mL of NaOH has been added and the total volume has become 49.90 mL. This is just 0.10 mL short of the equivalence point. In effect, only 0.10 mL of the *original* acid solution remains to be neutralized. We find the moles of HCl remaining from the original acid solution as follows.

$$0.10 \ \cancel{\text{mL acid solution}} \times \frac{0.20 \text{ mol HCl}}{1000 \ \cancel{\text{mL acid solution}}} = 2.0 \times 10^{-5} \text{ mol HCl}$$

This quantity of HCl is now in 49.90 mL, or 0.04990 L, so the present concentration of HCl is

$$\frac{2 \times 10^{-5} \text{ mol HCl}}{0.04990 \text{ L}} = 4.0 \times 10^{-4} \text{ mol HCl/L}$$

This concentration of HCl is, of course, the same as $[H^+]$, so the pH of the solution is now

$$pH = -\log(4.0 \times 10^{-4})$$
$$= 3.40$$

We are only 0.10 mL of titrant away from the equivalence point and the pH is still a long way from 7.00.

By doing a similar set of calculations, we could show that when the volume of added NaOH has reached 24.99 mL — just 0.01 mL short of the equivalence point — the pH has still reached only 4.40. But with the final 0.01 mL of base, the pH jumps to 7.00, the pH of a dilute solution of NaCl.

By going beyond the equivalence point with the addition of 0.01 mL *more* base, the solution is made basic. It now contains 2.0×10^{-6} mol of NaOH in 50.01 mL for a concentration of 4.0×10^{-5} mol OH^-/L. (Remember, NaOH is a strong base and 100% dissociated.) The pOH for this solution ($-\log[OH^-]$) is 4.40, so the pH is (14.00 − 4.40) = 9.60. Thus, overshooting the equivalence point by only 0.01 mL makes the pH jump from 7.00 to 9.60. As we add more base, the pH continues to rise but also starts to level off. It is particularly striking that the pH changes by 5.20 units in going from 24.99 mL of added base to 25.01 mL. This volume change, only 0.02 mL, corresponds roughly to a third of a drop of titrant.

Just one drop delivers 0.05 to 0.06 mL.

Titration of a Weak Acid by a Strong Base   Figure 15.6 gives the titration curve for the titration of a typical weak acid (acetic acid, $HC_2H_3O_2$) with a typical strong base (NaOH).

$$NaOH(aq) + HC_2H_3O_2(aq) \longrightarrow NaC_2H_3O_2(aq) + H_2O$$

At the equivalence point, the solution contains sodium acetate, which has an anion but not a cation that hydrolyzes. We have already seen that the acetate ion hydrolyzes to give a slightly basic solution, so when the titration reaches the equivalence point, the solutions should be slightly basic. Let us look at some points on the titration curve.

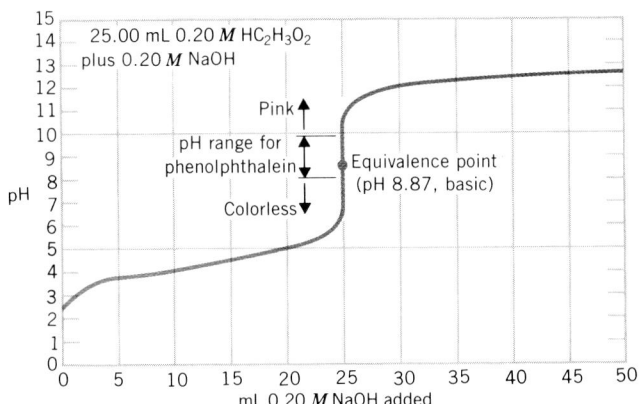

FIGURE 15.6 Titration curve for titrating a weak acid (0.20 *M* acetic acid) with a strong base (0.20 *M* NaOH).

Following the strategy of Example 15.7 (page 610), we can calculate that the pH of the untitrated solution of 0.20 *M* acetic acid is 2.72. The titration curve starts at this pH.

As we add NaOH, the solution will contain more and more sodium acetate and less and less acetic acid. The solution, in other words, has all the earmarks of a buffer—a weak acid plus the anion of the weak acid (supplied by the salt being formed by the neutralization). Therefore, we do buffer-type calculations to find the pH at intermediate points in the titration of a weak acid with a strong base. ✓

For example, when half of the stoichiometric quantity of NaOH has been added— 12.50 mL of 0.20 *M* NaOH—half of the acetic acid has been changed to sodium acetate. The solution is now equally concentrated in acetic acid and sodium acetate. It is just as if we had prepared a buffer consisting of equal concentrations of anion (acetate ion) and acid (acetic acid). We can calculate the pH of this solution just as we did in Example 15.16 (page 625). When [anion]/[acid] = 1/1, then, by Equation 15.19,

$$pH = pK_a + \log \frac{[anion]}{[acid]}$$
$$= 4.74 + \log 1$$
$$= 4.74$$

Moving on to the equivalence point in this titration, the pH of the solution at the equivalence point can be found by the same procedure we used in Example 15.22 (page 635) to calculate the pH of a solution of a salt that hydrolyzes. The moles of *salt* that have formed when the equivalence point is reached have to equal the moles of acid initially present in the 25.00 mL of original 0.20 *M* acetic acid. The original moles of acetic acid are found by

$$25.00 \text{ mL } HC_2H_3O_2 \times \frac{0.20 \text{ mol } HC_2H_3O_2}{1000 \text{ mL } HC_2H_3O_2} = 0.0050 \text{ mol } HC_2H_3O_2$$

So we have at the equivalence point 0.0050 mol of sodium acetate. This 0.0050 mol of salt is in 50.00 mL, or 0.05000 L, of solution, so its concentration is

$$\frac{0.0050 \text{ mol}}{0.05000 \text{ L}} = 1.0 \times 10^{-1} \text{ mol/L}$$

Since the salt, $NaC_2H_3O_2$, is 100% dissociated, we can use this concentration as the concentration of $C_2H_3O_2^-$ and calculate the pH by the same kind of calculation we did in Example 15.22. When we do this, we find that the pH of the solution at the equivalence point is 8.87, corresponding to a slightly alkaline solution. Now let us calculate the pH just 5.00 mL *before* the full 25.00 mL of NaOH solution has been added.

EXAMPLE 15.25   CALCULATING A pH AT A POINT ON A TITRATION CURVE

Problem: Calculate the pH of the resulting solution after 20.00 mL of 0.20 *M* NaOH has been added to 25.00 mL of 0.20 *M* HC$_2$H$_3$O$_2$.

Solution: The number of moles of acid in the original solution is found by the following calculation.

$$25.00 \text{ mL acid soln} \times \frac{0.20 \text{ mol HC}_2\text{H}_3\text{O}_2}{1000 \text{ mL acid soln}} = 0.0050 \text{ mol HC}_2\text{H}_3\text{O}_2$$

The number of moles of NaOH that is contributed when 20.00 mL of 0.20 NaOH solution is added is

$$20.00 \text{ mL NaOH soln} \times \frac{0.20 \text{ mol NaOH}}{1000 \text{ mL NaOH soln}} = 0.0040 \text{ mol NaOH}$$

This much base neutralizes the same amount of acid. The still unneutralized acid is, therefore, the difference,

$$0.0050 \text{ mol} - 0.0040 \text{ mol} = 0.0010 \text{ mol HC}_2\text{H}_3\text{O}_2$$

In other words, we have a solution with 0.0010 mol of a very slightly ionized acid (acetic acid). We also have in the same solution 0.0040 mol of fully dissociated salt (sodium acetate), so the quantity of acetate ion is also 0.0040 mol. The problem of finding the pH of this solution is just like the problem we worked in Example 15.16 for finding the pH of a buffer solution. Using Equation 15.19,

$$pH = pK_a + \log \frac{[\text{anion}]}{[\text{acid}]}$$

$$= 4.74 + \log \frac{(0.004)}{(0.001)}$$

$$= 4.74 + 0.60$$

$$= 5.34$$

Thus, the pH of the solution being titrated has become 5.34 after 20.00 mL has been added.

PRACTICE EXERCISE 31   Calculate the pH of the solution that results when 15.00 mL of 0.20 *M* NaOH is added to 25.00 mL of 0.20 *M* HC$_2$H$_3$O$_2$.

Titration of a Weak Base by a Strong Acid   We will not go through all of the calculations, but Figure 15.7 shows the titration curve for the titration of 0.20 *M* NH$_3$, a weak base, with 0.20 *M* HCl, a strong acid. Now the initial solution is that of a weak

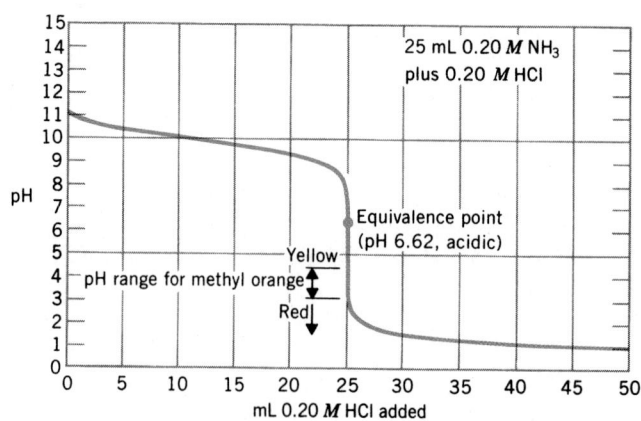

FIGURE 15.7 Titration curve for the titration of a weak base (0.20 *M* NH$_3$) with a strong acid (0.20 *M* HCl).

base, so we have to use the equation for the base ionization constant of NH_3 to find the pH before any acid has been added.

The solution at the equivalence point would contain only ammonium chloride, NH_4Cl. The cation, NH_4^+, of this salt hydrolyzes, as we learned on page 634, to give a slightly acidic solution. The pH at the equivalence point is found by using the K_a of NH_4^+ (5.7×10^{-10}, from Table 15.4) to calculate $[H^+]$ first, and then use this to calculate pH.

At points between the initial solution and the equivalence point, the calculations are those involving the ammonia/ammonium ion buffer. At points following the equivalence point, we are dealing essentially with a solution of the extra HCl being added, and we find pH values accordingly.

Selecting the Best Acid–Base Indicator

The *endpoint* is when we stop a titration. We want this to happen at the *equivalence point*. We want an indicator, therefore, that is in the middle of its color change *at the pH of the equivalence point*.

Each acid–base indicator changes color over a small *range* of pH, not sharply at one particular pH. These ranges were given in Table 15.3 on page 606 where we introduced indicators. Phenolphthalein, for example, changes a solution from colorless to pink as the solution's pH changes over a range of 8.2 to 10.0. See Figure 15.8. Phenolphthalein is therefore the perfect indicator for the titration of a weak acid by a strong base, where the pH of the equivalence point is on the basic side, as we saw in Figure 15.6. But it also works very well for the titration of a strong acid by a strong base, as we noted in Figure 15.5. Although the pH at the equivalence point, 7.00, is not yet in phenolphthalein's range, we saw that only a small fraction of a drop more titrant beyond the equivalence point raises the pH into this range.

FIGURE 15.8 Colors of some common acid-base indicators.

pH 8.2   pH 10.0
Phenolphthalein

pH 6.0   pH 7.6
Bromothymol blue

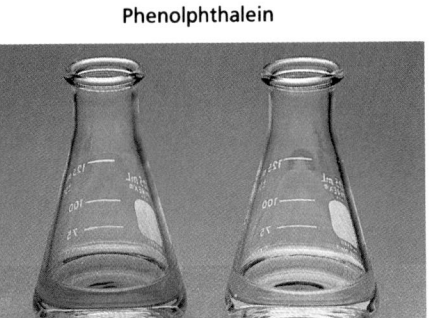

pH 3.2   pH 4.4
Methyl orange

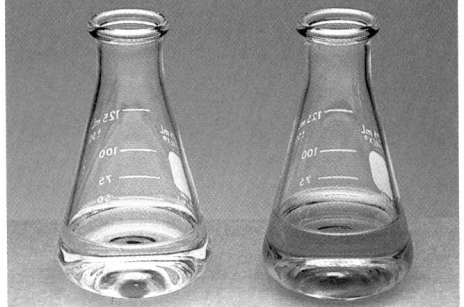

pH 9.4   pH 10.6
Thymolphthalein

For the titration of a weak base by a strong acid, indicators with color changes between pH 3 and 7 would work, as Figure 15.7 indicated. This figure showed that methyl orange (range 3.2–4.4) could be used, although its color change (red to yellow) is not easily noticed when the solutions are dilute.

One property an indicator should have is that its color change be as dramatic as possible. A major advantage of phenolphthalein, for example, is that its color change is very striking—colorless to pink. Another indicator, bromothymol blue, might appear to be a better indicator than phenolphthalein when the equivalence point is 7.00. A pH of 7.00 lies in the middle of bromothymol blue's range of 6.0 to 7.6 (Figure 15.8). But this indicator's color change—yellow to blue—is not as dramatic as that of phenolphthalein, particularly when concentrations are low. Cresol red, which changes color in the pH range 7.0 to 8.8 (Figure 15.8), also has the disadvantage of a less dramatic color change—yellow to red—than phenolphthalein.

Theory of Acid–Base Indicators   Most acid–base indicators are weak acids that are also dyes. We therefore represent an indicator with the formula H*In*. In its un-ionized state, H*In* has one color. Its conjugate base, *In*⁻, has a different color, the more strikingly different the better. In solution, the indicator is involved in the usual acid–base equilibrium:

$$\underset{\substack{\text{acid form}\\ \text{(one color)}}}{\text{H}In(aq)} \rightleftharpoons \text{H}^+(aq) + \underset{\substack{\text{base form}\\ \text{(another color)}}}{In^-(aq)}$$

The corresponding acid ionization constant, K_{In}, and pK_{In} are given by

$$K_{In} = \frac{[\text{H}^+][In^-]}{[\text{H}In]}$$

and

$$\text{p}K_{In} = -\log K_{In}$$

The equation for K_{In} simplifies when $[In^-] = [\text{H}In]$, which occurs when the titration has neutralized half of the H*In*. This midway point lies exactly in the middle of the indicator's color change, and this next calculation is intended to relate a property of the indicator, its pK_{In}, to the pH the solution will have at the equivalence point. Thus, when $[In^-] = [\text{H}In]$,

$$K_{In} = \frac{[\text{H}^+][\cancel{In^-}]}{[\cancel{\text{H}In}]}$$

or

$$K_{In} = [\text{H}^+]$$

Taking the negative logarithms of both sides gives us

$$-\log K_{In} = -\log[\text{H}^+]$$

or,

$$\text{p}K_{In} = \text{pH}_{\text{at equivalence point}}$$

This important result tells us that once we know the pH that the solution should have at the equivalence point of a titration, we also know the pK_{In} that the indicator should have. Selecting an indicator, therefore, amounts to finding one whose pK_{In} equals (or is as close as possible to) the pH at the equivalence point.

Still another factor involving indicators is that we want to use as little of one as possible. We just saw that they are weak acids, so they are also neutralized and they consume titrant. We have to accept some loss of titrant this way, but no more than

Suppose that the color of H*In* is red and the color of its anion, *In*⁻ is yellow. In terms of the eye's abilities, the ratio of [H*In*]/[*In*⁻] has to be roughly at least as large as 10 to 1 for the eye to see the color as definitely red. And this ratio has to be at least as small as 1 to 10 for the eye to see the color as definitely yellow. Therefore, the color transition that the eye actually sees at the endpoint of a titration happens as the ratio of [H*In*] to [*In*⁻] changes from 10/1 to 1/10.

We can calculate the relationship of pH to pK_a for the indicator at the start of this range as follows. We know that

$$K_{In} = \frac{[H^+][In^-]}{[HIn]}$$

Therefore, at the start of the range, we must have

$$K_{In} = [H^+]\frac{10}{1}$$

So

$$pK_{In} = pH - \log 10$$

Or, at the start of the transition we should have

$$pH = pK_{In} - 1$$

At the other end of the range, when [H*In*]/[*In*⁻] is 1/10, we have

$$K_{In} = [H^+]\frac{1}{10}$$

$$pK_{In} = pH - \log(10^{-1})$$

So at the end of the color transition, the pH should be

$$pH = pK_{In} + 1$$

Therefore, the *range* of pH over which a definite change in color occurs is

$$pH = pK_{In} \pm 1$$

A range of ±1 pH unit on either side of the value of pK_{In} corresponds to a span of 2 pH units over which the eye can definitely notice the color change.

absolutely necessary. So the best indicators are those with the most intense colors. Then, even ultratrace amounts can give striking colors.

The human eye's ability to discern color changes is also a factor. As discussed further in Special Topic 15.3, as much of a change as 2 pH units is needed for some indicators — more for litmus — before the eye sees the change in color. This is why tables of acid–base indicators, such as Table 15.3, provide approximate pH ranges for the color changes, instead of specific values of pK_{In}.

SUMMARY

The pH Concept and the Self-Ionization of Water. Water reacts with itself to produce trace (but equal) concentrations of H⁺ and OH⁻ ions whose values equal 1.0×10^{-7} mol/L at 25 °C. The product of their molar concentration is called the ion-product constant of water, K_w. Regardless of what solutes are present, this ion-product is a constant at a given temperature, and at 25 °C, $K_w = 1.0 \times 10^{-14}$.

To describe very low molar concentrations of H⁺ ions, chemists use the negative logarithm of [H⁺], called the pH of the solution, and generally it is a number between 0 and 14. A comparable expression, pOH, can be used to describe low OH⁻ ion concentrations: pOH = −log[OH⁻]. At 25 °C, pH + pOH = 14.00. A solution is acidic if the pH is less than 7.00 and basic if the pH is greater than 7.00. The smaller the pH, the more concentrated is the solution in hydrogen ions.

Acid and Base Ionization Constants. To describe the relative acidities of weak acids, their acid ionization constants, K_a, are used.

$$K_a = \frac{[H^+][\text{conjugate base of acid}]}{[\text{acid}]}$$

A similar equation is used to define the base ionization constant, K_b, for a weak base that reacts slightly as a proton acceptor in water.

$$K_b = \frac{[OH^-][\text{conjugate acid of the base}]}{[\text{base}]}$$

The smaller the values of K_a (or K_b), the weaker are the substances as Brønsted acids (or bases). To compare these relative strengths more easily, the negative logarithms of K_a and K_b, called pK_a and pK_b, respectively, are sometimes used. Because of the nature of negative logarithms, we have to remember that the *smaller* the pK_a, the *stronger* is the acid. Values of pK_a can be used to calculate the values of pK_b (at 25 °C) for the conjugate base of the acid, because

$$pK_a + pK_b = 14.00$$

The equation relating K_a and K_b for a conjugate acid–base pair is

$$K_a K_b = K_w$$

Buffers. A solution that contains a weak acid and one of its fully ionized salts (such as a sodium salt) is buffered at a pH that is generally less than 7. If the solute pair consists of ammonia and one of its salts with a strong, monoprotic acid, the solution is buffered at a pH greater than 7. The relationship among pH, pK_a, the molar concentration of the weak acid or [acid], and the molar concentration of the anion of the acid (as supplied by the salt) or [anion] is

$$pH = pK_a + \log \frac{[anion]}{[acid]}$$

(This equation is sometimes called the Henderson–Hasselbalch equation.) Generally, a buffer is most effective when pH = p$K_a \pm 1$. While the *ratio* [anion]/[acid] contributes to the value of the pH at which a buffer works, the individual concentrations of anion and acid give the buffer its *capacity* for taking on extra acid or base with little change in pH.

In a basic buffer made from a weak base and its salt, for example ammonia and an ammonium salt, the corresponding equation for pOH is

$$pOH = K_b + \log \frac{[cation]}{[base]}$$

where [cation] refers to the molar concentration of the conjugate acid of the base (for example, NH_4^+) as supplied by the *salt*, and [base] is the molar concentration of the weak base (e.g., NH_3).

Polyprotic Acids.   Polyprotic acids have K_a values for the ionization of each of its hydrogen ions. Successive values of K_a usually differ by a factor of 10^4 to 10^5. One consequence of this factor is that the pH of a dilute solution of a polyprotic acid can be calculated just from the value of K_{a_1}, provided that the acid is the only solute. When the acid is diprotic (e.g., H_2A, to use a general symbol), the value of K_{a_2} numerically equals the molar concentration of A^{2-}, again provided that the acid is the only solute present.

Hydrolysis of Ions.   When water can react with an ion of a dissolved salt, the balance between [H^+] and [OH^-] in the solution changes. In such reactions, called the hydrolysis of ions, the medium becomes acidic if the cation hydrolyzes. Metal ions with three or two positive charges generally can hydrolyze, but the metal ions of groups IA and IIA (except Be) do not hydrolyze. Cations, like NH_4^+, which are themselves proton donors, also hydrolyze to give slightly acidic solutions.

When the anion of a salt is that of a weak acid, it is a Brønsted base strong enough to hydrolyze. In this reaction, OH^- ion is generated, and the solution of the salt is made slightly basic.

When neither ion of a salt hydrolyzes, its aqueous solutions are neutral.

If both ions of a salt can hydrolyze, for example, if the salt comes from a weak acid and a weak base, the net effect has to be determined on a case by case basis.

The pH of solutions in which salts have hydrolyzed can be calculated using K_a or K_b values of the appropriate Brønsted acid or base present. If the anion of the salt is one from a polyprotic acid, where more than one equilibrium can contribute to the change in pH, the calculation can be simplified by assuming that all of the pH change is caused just by the anion released from the salt.

Acid–Base Titrations and Indicators.   In an acid–base titration, the pH at the equivalence point depends on how the ions of the salt made by the titration hydrolyze. Therefore, the indicator has to be picked so that its color change occurs as closely as possible to the pH that the solution will have at the equivalence point. The pK_{In} of the indicator ideally is within one unit of the pH at the equivalence point.

REVIEW EXERCISES

Answers to questions whose numbers are printed in color are given in Appendix C. Difficult questions are marked with asterisks.

Self-Ionization of Water and pH

15.1   How does the equation for the ion-product constant of water differ from that of the true equilibrium constant for the self-ionization of water?

15.2   How are acidic, basic, and neutral solutions in water defined (a) in terms of [H^+] and [OH^-]? (b) In terms of pH?

15.3   The value of K_w increases as the temperature of water is raised.
(a) What effect will increasing temperature have on the pH of pure water? (Will the pH increase, decrease, or stay the same?)
(b) As the temperature of pure water rises, will the water become more acidic, more basic, or remain neutral?
(c) Carefully explain your answers to both parts a and b.

15.4   At the temperature of the human body, 37 °C, the value of K_w is 2.4×10^{-14}. Calculate [H^+] and [OH^-], pH and pOH, and pK_w. What is the relation between pH, pOH, and pK_w at this temperature? Is water neutral at this temperature?

15.5   Deuterium oxide, D_2O, ionizes like water. At 20 °C its K_w or ion-product constant analogous to that of water is 8.9×10^{-16}. Calculate [D^+] and [OD^-] in deuterium oxide at 20 °C. Calculate also the pD, the pOD, and the pK_w.

15.6   Assuming 100% ionization of HCl in dilute solutions, what is the pH of 0.010 M HCl?

15.7   If nitric acid is 100% ionized in a 0.0050 M solution, what is the pH of this solution?

15.8   A sodium hydroxide solution is prepared by dissolving 6.0 g NaOH in 1.00 L of solution. Assuming that 100% dissociation occurs, what is the pOH and the pH of this solution?

15.9   A solution was made by dissolving 0.837 g Ba(OH)$_2$ in 100 mL final volume. If Ba(OH)$_2$ is fully broken up into its ions, what is the pOH and the pH of this solution?

15.10   A certain brand of beer had a hydrogen ion concentration equal to 1.9×10^{-5} mol/L. What is the pH of this beer?

15.11   A soft drink was put on the market with [H^+] = 1.4×10^{-5} mol/L. What is its pH?

Acid and Base Ionization Constants — K_a, K_b, pK_a, and pK_b

15.12 Benzoic acid, $C_6H_5CO_2H$, is an organic acid whose sodium salt, $C_6H_5CO_2Na$, has long been used as a safe food additive to protect beverages and many foods against harmful yeasts and bacteria. The acid is monoprotic. Write the equation for its K_a.

15.13 Write the equation for the equilibrium that the benzoate ion, $C_6H_5CO_2^-$ (Review Exercise 15.12), would produce in water as it functions as a Brønsted base. Then write the expression for the K_b of the conjugate base of benzoic acid.

15.14 The pK_a of HCN is 9.21 and that of HF is 3.17. Which is the stronger Brønsted *base*: CN^- or F^-?

15.15 What is the percent ionization in a 0.15 M solution of HF? What is the pH of this solution?

15.16 What is the percent ionization in 1.0 M acetic acid? What is the pH of this solution?

*15.17 The hydrogen sulfate ion, HSO_4^-, is a moderately strong Brønsted acid with a K_a of 1.0×10^{-2}.
   (a) Write the equilibrium expression for this acid.
   (b) What is the value of $[H^+]$ in 0.010 M HSO_4^- (furnished by the salt, $NaHSO_4$)? Do NOT make the simplifying assumption used in Example 15.7; solve the quadratic equation.
   (c) What is $[H^+]$ in 0.010 M HSO_4^-, calculated by using the simplifying assumption introduced in Example 15.7?
   (d) Using the results of part b and the equation that *defines* percentage ionization, calculate the percentage ionization of HSO_4^- into H^+ and SO_4^{2-} in 0.010 M HSO_4^-.
   (e) What percentage ionization is indicated by the result of part c?

15.18 Periodic acid, HIO_4, is an important oxidizing agent and a moderately strong acid. In a 0.10 M solution, $[H^+] = 3.8 \times 10^{-2}$ mol/L. Calculate the K_a and pK_a for periodic acid.

15.19 Chloroacetic acid, $HC_2H_2ClO_2$, is a stronger monoprotic acid than acetic acid. In a 0.10 M solution of this acid the pH was 1.96. Calculate the K_a and pK_a for chloroacetic acid.

15.20 *para*-Aminobenzoic acid, PABA, is a powerful sunscreening agent whose salts are used widely in sun tanning and screening lotions. The parent acid, which we may symbolize as H-*Paba*, is a weak acid with a pK_a of 4.92 (25 °C). What will be the $[H^+]$ and pH of a 0.030 M solution of this acid?

15.21 Barbituric acid, H-*Bar*, was discovered by Adolph von Baeyer (of Baeyer aspirin fame) and named after a friend, Barbara. It is the parent compound of widely used sleeping drugs, the barbiturates. Its pK_a is 4.01. What will be the $[H^+]$ and pH of a 0.050 M solution of H-*Bar*?.

15.22 Ethylamine, $CH_3CH_2NH_2$, has a strong, pungent odor similar to that of ammonia. Like ammonia, it is a Brønsted base. A 0.10 M solution has a pH of 11.86. Calculate the K_b and pK_b for the ethylamine, and find the pK_a for its conjugate acid, $CH_3CH_2NH_3^+$.

15.23 Hydrazine, N_2H_4, has been used as a rocket fuel. Like ammonia, it is a Brønsted base. A 0.15 M solution has a pH of 10.70. What is the K_b and pK_b for hydrazine and the pK_a of its conjugate acid?

15.24 Codeine, a cough suppressant extracted from crude opium, is a weak base with a pK_b of 5.79. What will be the pH of a 0.020 M solution of codeine? (Use *Cod* as a symbol for codeine.)

15.25 A hydroxy derivative of ammonia, hydroxylamine (NH_2OH), is a weak base with a pK_b of 8.04. What will be the pH of a 0.25 M solution? It reacts with water as follows: $NH_2OH + H_2O \rightleftharpoons NH_3OH^+ + OH^-$.

15.26 Many drugs that are natural Brønsted bases are put into aqueous solution as their much more soluble salts with strong acids. The powerful painkiller morphine, for example, is very slightly soluble in water, but morphine nitrate is quite soluble. We may represent morphine by the symbol *Mor* and its conjugate acid as H-*Mor*$^+$. The pK_b of morphine is 5.79. What is the pK_a of its conjugate acid, and what will be the calculated pH of a 0.20 M solution of H-*Mor*$^+$?

15.27 Quinine, an important drug in treating malaria, is a weak Brønsted base that we may represent as *Qu*. At 25 °C its pK_b is 5.48. To make it more soluble in water, it is put into a solution as its conjugate acid, which we may represent as H-*Qu*Cl. What is the calculated pH of a 0.15 M solution of H-*Qu*$^+$?

Buffers

15.28 Write ionic equations that illustrate how each pair of compounds can serve as a buffer pair.
   (a) H_2CO_3 and $NaHCO_3$ (the "carbonate" buffer in blood)
   (b) NaH_2PO_4 and Na_2HPO_4 (the "phosphate" buffer inside body cells)
   (c) NH_4Cl and NH_3

15.29 Show that Equation 15.20 can be rewritten as

$$pH = pK_a + \log \frac{[\text{base}]}{[\text{cation}]}$$

*15.30 Show that the following equation is equivalent to both Equation 15.19 and Equation 15.20 and could be used for buffers in general, whether they are to serve above or below a pH of 7.

$$pH = pK_a + \log \frac{[\text{conjugate base}]}{[\text{conjugate acid}]}$$

K_a is for the acid member of the conjugate acid/base pair.

15.31 Which buffer would be better able to hold a steady pH on the addition of strong acid, buffer 1 or buffer 2? Explain. *Buffer 1*: a solution containing 0.10 M NH_4Cl and 1 M NH_3. *Buffer 2*: a solution containing 1 M NH_4Cl and 0.10 M NH_3.

15.32 How many grams of sodium acetate, $NaC_2H_3O_2$, would have to be added to 1 L of 0.15 M acetic acid (pK_a = 4.74) to make the solution a buffer for pH 5.00?

15.33 How many grams of sodium formate, $NaCHO_2$, would have to be dissolved in 1.0 L of 0.12 M formic acid ($pK_a = 3.75$) to make the solution a buffer for pH 3.80?

15.34 What ratio of molar concentrations of NH_4Cl and NH_3 would buffer a solution at pH 9.25?

15.35 How many grams of ammonium chloride would have to be dissolved in 500 mL of 0.20 M NH_3 to prepare a solution buffered at pH 10.00?

*15.36 For an experiment involving what happens to the growth of a particular fungus in a slightly acidic medium, a biochemist needs 250 mL of an acetate buffer with a pH of 5.12. The buffer solution has to be able to hold the pH to within ± 0.10 pH unit of 5.12 even if 0.0100 mol of NaOH or 0.0100 mol of HCl enters the solution.
  (a) What is the minimum number of grams of acetic acid and of sodium acetate dihydrate that must be used to prepare this buffer? (Neglect the small volumes of NaOH and HCl involved.)
  (b) Describe this buffer by giving its molarity in acetic acid and its molarity in sodium acetate.
  (c) What is the pH of an unbuffered solution made by adding 0.010 mol of NaOH to 250 mL of pure water?
  (d) What is the pH of an unbuffered solution made by adding 0.010 mol of HCl to 250 mL of pure water?

*15.37 An analytical procedure for separating certain components of a dietary supplement requires that they be dissolved in a buffered solution having a pH of 3.80. This buffer has to be able to handle the introduction of 5.0×10^{-3} mol of either OH^- or H^+ without changing its pH by more than 0.050 unit.
  (a) Using data in Table 15.4, pick the best acid (and its sodium salt) that could be used for preparing 750 mL of this buffered solution.
  (b) Calculate the minimum number of grams of the pure acid and of the pure salt needed to make this buffered solution. (Ignore the small volumes involved with the added base or acid.)
  (c) Describe this buffer by giving its molarity in the selected acid and its molarity in the sodium salt of this acid.
  (d) If 5.0×10^{-3} mol of HCl were added to 750 mL of a solution of hydrochloric acid having a pH of 3.80, what would be the new pH?
  (e) If 5.0×10^{-3} mol of NaOH were added to 750 mL of a solution of hydrochloric acid with a pH of 3.80, what would be the new pH?

Polyprotic Acids

15.38 When sulfur dioxide, an air pollutant from the burning of sulfur-containing coal or oil, dissolves in water, some reacts to form sulfurous acid, usually represented as H_2SO_3.

$$H_2O + SO_2 \rightleftharpoons H_2SO_3$$

  (a) Use the data in Table 15.6 to decide whether this acid should be called a weak, moderate, or strong acid.
  (b) Write the expression for K_{a_1} and K_{a_2}.

  (c) Carbon dioxide, also produced by the burning of coal or oil, will similarly react with water to produce carbonic acid, H_2CO_3. Write the equilibria and K_a expressions of its K_{a_1} and K_{a_2}.
  (d) Which is the stronger acid: sulfurous or carbonic?

15.39 If water is treated as any other weak diprotic Brønsted acid, what is the equation expressing its K_a? How does this expression differ from K_w? Write the expression for K_{a_2} for water.

15.40 Phosphorous acid, H_3PO_3, is actually a diprotic acid; $K_{a_1} = 1.0 \times 10^{-2}$ and $K_{a_2} = 2.6 \times 10^{-7}$. What are the values of $[H^+]$, pH, and HPO_3^{2-} in a 1.0 M solution?

15.41 Tellurium, in the same family as sulfur, forms an acid analogous to sulfuric acid and called telluric acid. It exists, however, as H_6TeO_6 (which looks like $H_2TeO_4 + 2H_2O$). It is a diprotic acid. $K_{a_1} = 2.1 \times 10^{-8}$ and $K_{a_2} = 6.5 \times 10^{-12}$. Calculate $[H^+]$, pH, and $[H_4TeO_6^{2-}]$ in a 0.25 M solution of H_6TeO_6.

Hydrolysis of Ions

15.42 Aspirin is acetylsalicyclic acid, a monoprotic acid whose K_a value is 3.27×10^{-4}. Does a solution of the sodium salt of aspirin in water test acidic, basic, or neutral? Explain.

15.43 The K_b value of the oxalate ion, $C_2O_4^{2-}$, is 1.85×10^{-10}. Is a solution of $K_2C_2O_4$ acidic, basic, or neutral? Explain.

15.44 Consider the following compounds and suppose that 0.5 M solutions are prepared of each: NaI, KF, $(NH_4)_2SO_4$, KCN, $HC_2H_3O_2$, $CsNO_3$, and KBr. Write the formulas of those that have solutions that are (a) acidic, (b) basic, (c) neutral.

15.45 Calculate the pH of 0.20 M NaCN.

15.46 Calculate the pH of 0.10 M KNO_2.

15.47 Calculate the pH of 0.15 M Na_2SO_3.

15.48 Calculate the pH of 0.10 M K_2CO_3.

15.49 How many grams of $Na_2CO_3 \cdot 10H_2O$ have to be dissolved in 1.00 L of water at 25 °C to have a solution with a pH of 11.70?

15.50 Calculate the number of grams of NH_4Br that have to be dissolved in 1.00 L of water at 25 °C to have a solution with a pH of 5.15.

15.51 Calculate the number of grams of $Na_2SO_3 \cdot 7H_2O$ that you would have to dissolve in 0.50 L of water at 25 °C to have a solution with a pH of 10.00.

Acid–Base Titrations

15.52 What can make the titrated solution at the equivalence point in an acid–base titration have a pH not equal to 7.00? How does this possibility affect the choice of an indicator?

15.53 Explain why ethyl red is a better indicator than phenolphthalein in the titration of dilute ammonia by dilute hydrochloric acid.

15.54 What is a good indicator for titrating potassium hydroxide with hydrobromic acid? Explain.

15.55 In the titration of an acid with a base, what condition concerning the quantities of reactants ought to be true at the equivalence point?

15.56 When 50 mL of 0.10 M formic acid is titrated with 0.10 M sodium hydroxide, what is the pH at the equivalence point? (Be sure to take into account the change in volume during the titration.) What is a good indicator for this titration?

15.57 When 25 mL of 0.10 M aqueous ammonia is titrated with 0.10 M hydrobromic acid, what is the pH at the equivalence point? What is a good indicator?

*15.58 To obtain the data needed to plot a titration curve for the titration of a strong acid with a strong base, a chemist used 25.00 mL of 0.1000 N HCl. The normality of the base was also 0.1000 N, and this solution was added in small portions to the acid. Calculate the pH of the resulting solution after each of the following quantities of base had been added to the original solution (you must take into account the change in total volume).

(a) 0 mL	(d) 24.99 mL	(g) 25.10 mL
(b) 10.00 mL	(e) 25.00 mL	(h) 26.00 mL
(c) 24.90 mL	(f) 25.01 mL	(i) 50.00 mL

*15.59 Repeat Review Exercise 15.58 for the titration of 25.00 mL 0.1000 N acetic acid, $HC_2H_3O_2$, with 0.1000 N NaOH.

*15.60 Repeat Review Exercise 15.58 for the titration of 25.00 mL of 0.1000 N ammonia with 0.1000 N HCl.

QUAL TOPIC

II

SEPARATION OF IONS
BY SELECTIVE PRECIPITATION

In Qual Topic I we described the general approach to cation qualitative analysis. In that discussion we noted that the cations are grouped according to their ability to be separated from a mixture by various precipitating agents. Cation group 1, for instance, consists of the ions Ag^+, Pb^{2+}, and Hg_2^{2+}, which are removed from a mixture that might contain other cations by the addition of dilute HCl. In this case we make use of the low solubility of the chlorides of Ag^+, Pb^{2+}, and Hg_2^{2+}, and the much larger solubilities of the other metal chlorides. It is, in fact, the large differences in solubility that make it possible to selectively precipitate the cation group 1 ions as their chlorides.

Even when the salts of metal ions with a particular anion might be classified as "insoluble" according to solubility rules, the wide range of their solubility product constants (K_{sp} values) shows that their solubilities differ significantly. For example, both $Mg(OH)_2$ and $Al(OH)_3$ are insoluble according to the solubility rules on page 399. However, from their values of K_{sp} in Table 14.2 it can be calculated that in water the solubility of $Mg(OH)_2$ is 1.2×10^{-4} M, while the solubility of $Al(OH)_3$ is only 3×10^{-13} M. In other words, even though both of these metal hydroxides have low solubilities, aluminum hydroxide is less soluble in water than magnesium hydroxide by a factor of 4×10^8 (that's 400 million)! Because of this great difference in solubilities, it is possible to separate magnesium and aluminum ions from each other by carefully controlling the pH. In fact, this is precisely what is done in the qualitative analysis scheme when the ions of cation group 3 (which includes Al^{3+}) are separated from a mixture that could also contain Mg^{2+}. Ammonium chloride is added to the mixture, which is then made basic with aqueous ammonia. The buffering action of the NH_3/NH_4^+ mixture keeps the pH in a range where magnesium hydroxide is prevented from precipitating but aluminum hydroxide is not, so $Al(OH)_3$ precipitates with the other ions of cation group 3 and magnesium ion remains in the solution.

Precipitation of Metal Sulfides by Hydrogen Sulfide

In qualitative analysis, the most important example of selective precipitation is the separation of the cation group 2 sulfides from the metal ions in cation group 3, which also form "insoluble" sulfides. This, too, is accomplished by control of the pH, but the equilibria involved are a bit more complex than those involved in the separation of Al^{3+} from Mg^{2+}. They involve not only the solubility equilibria of the metal sulfides, but also equilibria of the diprotic acid, H_2S. To understand these equilibria, which is our next goal, it is best if we begin by examining again the reactions of another similar class of compounds — the metal oxides.

We have already seen that metal oxides such as Fe_2O_3 easily dissolve in dilute acid. For example,

$$Fe_2O_3(s) + 6H^+(aq) \longrightarrow 2Fe^{3+}(aq) + 3H_2O$$

Even when an oxide dissolves without the aid of an acid, it does so by chemical reaction with water, *not* a simple dissociation of ions that remain otherwise unchanged. For example, we learned that sodium oxide dissolves in water, but the solution does not contain the oxide ion, O^{2-}. Instead, the hydroxide ion forms. The equation is

$$Na_2O(s) + H_2O \longrightarrow 2NaOH(aq)$$

This actually involves the reaction of oxide ions with water as the crystals of Na_2O break up.

$$O^{2-}(s) + H_2O \longrightarrow 2OH^-(aq)$$

The oxide ion is simply too powerful a base to exist in water at any concentration worthy of experimental note. We can understand why from the extraordinarily high (estimated) value of K_b for O^{2-}, 1×10^{22} (Table 15.5). Thus there is no way to supply *oxide ions* to an aqueous solution to form an insoluble metal oxide directly. Oxide ions from any source react, instead, with water to generate the hydroxide ion. If an insoluble oxide instead of an insoluble hydroxide does precipitate from a solution, it forms because the specific metal ion is able to react with OH^-, extract O^{2-}, and leave H^+ or a neutralized form in the solution. The silver ion, for example, precipitates as brown silver oxide, Ag_2O, when OH^- is added to aqueous silver salts (Figure 1).

$$2Ag^+(aq) + 2OH^-(aq) \longrightarrow Ag_2O(s) + H_2O$$

When we drop down from oxygen to sulfur in the oxygen family, we find many similarities. But there are also some differences important enough for us to be able to sort cations according to their solubilities under just a few carefully specified circumstances. One is solubility in a solution saturated with hydrogen sulfide and at a pH of at least as low as 0.5 (equivalent, roughly, to 0.3 M HCl). To see how this can work, we will look more closely at hydrogen sulfide and the HS^- and S^{2-} ions.

One important similarity between oxides and sulfides (and, indirectly, between H_2O and H_2S) is that the sulfide ion, S^{2-}, like the oxide ion, does not exist in any ordinary aqueous solution. Thus, Na_2S, like Na_2O, dissolves in water by reacting with it, not by releasing an otherwise unchanged divalent anion, S^{2-}.

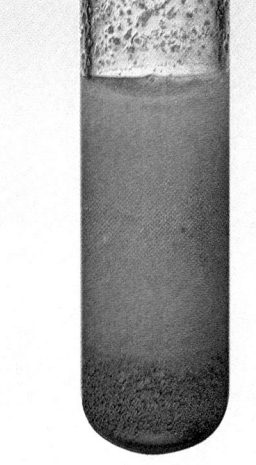

FIGURE 1 When NaOH is added to a solution that contains Ag^+, a brown precipitate of Ag_2O is formed.

$$Na_2S(s) + H_2O \longrightarrow 2Na^+(aq) + SH^-(aq) + OH^-(aq)$$

The sulfide ion, S^{2-}, has not been detected in an aqueous solution even in the presence of 8 M NaOH where one might think that the reaction

$$OH^- + SH^- \longrightarrow H_2O + S^{2-}$$

could generate some detectable S^{2-}. And an 8 M NaOH solution is at a concentration well outside the bounds of the "ordinary."

Just as some metal oxides can form by a reaction between a metal ion and hydroxide ion, many metal sulfides can also form when their metal ions are exposed to HS^- (which we earlier wrote as SH^- to emphasize its similarity to the OH^- ion). Many metal ions even react with H_2S to form sulfides directly. Thus, when hydrogen sulfide gas is passed into an aqueous solution of any one of a number of metal ions — Cu^{2+}, Pb^{2+}, and Ni^{2+}, for example — a sulfide precipitates. A typical reaction is

$$Cu^{2+}(aq) + H_2S(aq) \rightleftharpoons CuS(s) + 2H^+(aq) \qquad K = 1.7 \times 10^{15}$$

You can see from the value of the equilibrium constant that the forward reaction is essentially the only reaction. If we turn this equilibrium around, we would have something that looks very much like an equilibrium constant for defining a solubility product. Let's look at this possibility more closely.

K_{spa} Values for Metal Sulfides

Judging from the value of K for the reaction just described for Cu^{2+} with H_2S, CuS appears to be extremely insoluble in water. If we were to write the chemical equilibrium to describe this solubility in the traditional way, we would write

$$CuS(s) \rightleftharpoons Cu^{2+}(aq) + S^{2-}(aq)$$

But this assumes that S^{2-} exists in water, and it doesn't, as we discussed above. So we have to write its equilibrium instead as follows.

$$CuS(s) + H_2O \rightleftharpoons Cu^{2+}(aq) + HS^-(aq) + OH^-(aq)$$

The K_{sp} for CuS must then be expressed by the following equation.

$$K_{sp} = [Cu^{2+}][HS^-][OH^-]$$

The last column in Table 1 gives the K_{sp} values, defined in this new way, for several metal sulfides. A group of these metal sulfides are subclassified as *acid-insoluble sulfides* and another as *base-insoluble sulfides*. We will see why shortly. Notice particularly how much the K_{sp} values vary — from 3×10^{-53} to 3×10^{-14}, a spread of a factor of 10^{39}.

In qualitative analysis, the cation group 2 metals are separated from other cations that might also be present in a solution by precipitating them as their sulfides from an acidic solution that is saturated in H_2S. Under these conditions, the products in the solubility equilibria are different from HS^- and OH^-. In acid, HS^- and OH^- would be neutralized, leaving H_2S (and H_2O, which cancels out), so the CuS(s) solubility equilibrium in dilute acid is

$$CuS(s) + 2H^+(aq) \rightleftharpoons Cu^{2+}(aq) + H_2S(aq)$$

This changes the mass action expression for the solubility product equilibrium, which we will now call the acid solubility product. We also give it its own symbol for the equilibrium constant, K_{spa}, where the a in the subscript indicates that the medium is acidic.

$$K_{spa} = \frac{[Cu^{2+}][H_2S]}{[H^+]^2}$$

The first column in Table 1 gives several K_{spa} values for metal sulfides. Notice first that all K_{spa} values are about 10^{21} larger than the K_{sp} values. Metal sulfides are clearly vastly more soluble in dilute acid than in water. Yet several — the acid-insoluble sulfides — are still so insoluble that even the most soluble of these sulfides, SnS, barely dissolves, even in moderately concentrated acid.

As we mentioned earlier, the cation group 2 metals are separated from the other metals whose sulfides are insoluble in water by selectively precipitating them under controlled pH conditions. Let's work an example to show how we can calculate the conditions necessary to accomplish this, and at the same time see the reason for the classification of the metal sulfides as being "acid insoluble" or "base insoluble." Here is the problem:

Over what range of hydrogen ion concentrations (and pH) is it possible to separate Cu^{2+} from Ni^{2+} when both metal ions are present in a solution at a concentration of 0.010 M and the solution is saturated in H_2S (where its concentration is 0.1 M)?

Let's begin, as usual, by writing the chemical equilibria and their associated equilibrium expressions (the equations for their K_{spa}).

$$CuS(s) + 2H^+(aq) \rightleftharpoons Cu^{2+}(aq) + H_2S(aq) \qquad K_{spa} = \frac{[Cu^{2+}][H_2S]}{[H^+]^2}$$

TABLE 1 Metal Ions Separable by Selective Precipitation of Sulfides[a]

Metal Ion	Sulfide	K_{spa}	K_{sp}
Acid-Insoluble Sulfides			
Hg^{2+}	HgS (black)	2×10^{-32}	2×10^{-53}
Ag^+	Ag_2S	6×10^{-30}	6×10^{-51}
Cu^{2+}	CuS	6×10^{-16}	6×10^{-37}
Cd^{2+}	CdS	8×10^{-7}	8×10^{-28}
Pb^{2+}	PbS	3×10^{-7}	3×10^{-28}
Sn^{2+}	SnS	1×10^{-5}	1×10^{-26}
Base-Insoluble Sulfides (Acid-Soluble Sulfides)			
Zn^{2+}	α-ZnS	3×10^{-4}	3×10^{-25}
	β-ZnS	3×10^{-2}	3×10^{-23}
Co^{2+}	CoS	5×10^{-1}	5×10^{-22}
Ni^{2+}	NiS	4×10^{1}	4×10^{-20}
Fe^{2+}	FeS	6×10^{2}	6×10^{-19}
Mn^{2+}	MnS (pink)	3×10^{10}	3×10^{-11}
	(green)	3×10^{7}	3×10^{-14}

[a] Data are for 25 °C. See R. J. Myers, *J. Chem. Ed.*, vol. 63, 1986, p. 689.

$$NiS(s) + 2H^+(aq) \rightleftharpoons Ni^{2+}(aq) + H_2S(aq) \qquad K_{spa} = \frac{[Ni^{2+}][H_2S]}{[H^+]^2}$$

To separate the Cu^{2+} from the Ni^{2+}, we want one of them to precipitate as the sulfide while we hold the other in solution. By examining the K_{spa} values from Table 1 for NiS (4×10^1) and CuS (6×10^{-16}), we can see that NiS is much more soluble in an acidic solution than CuS. Therefore, we want to make the H^+ concentration large enough to prevent NiS from precipitating, and small enough that CuS does precipitate. The problem reduces to two questions: (1) "What hydrogen ion concentration would be needed to keep the Cu^{2+} in solution?" (This is just above the upper limit; we really want a lower H^+ concentration so CuS will precipitate.) (2) "What is the hydrogen ion concentration just before NiS precipitates?" (At any lower H^+ concentration, or higher pH, NiS will precipitate, so we want an H^+ concentration equal to or larger than this value.) Once we know these limits, we know that any hydrogen ion concentration in between these will do the job.

We will find the upper limit first. If CuS *doesn't* precipitate, the Cu^{2+} concentration will be 0.010M, so we substitute this value along with the H_2S concentration (0.1 M) into the expression for K_{spa}.

$$K_{spa} = 6 \times 10^{-16} = \frac{(0.010)(0.1)}{[H^+]^2}$$

Now we solve for $[H^+]$.

$$[H^+]^2 = \frac{(0.010)(0.1)}{6 \times 10^{-16}} = 1.7 \times 10^{12}$$

$$[H^+] = 1 \times 10^6 \ M \text{ (correctly rounded)}$$

If we could make the $[H^+] = 1 \times 10^6 \ M$, we could prevent CuS from forming. But wait a minute (as you're no doubt saying). It isn't possible to have a million moles of H^+ per liter. What this tells us is that no matter how acidic the solution is, we cannot prevent CuS from precipitating when we saturate the solution with H_2S, which is why CuS is classified as an "acid-insoluble sulfide." In other words, in this separation, we don't have to worry about the solution being so acidic that CuS is prevented from precipitating. But we do have to be concerned that the H^+ concentration is too low, however.

To obtain the lower limit, we calculate the $[H^+]$ required to give an equilibrium Ni^{2+} concentration equal to 0.010 M. If we keep the $[H^+]$ equal to or larger than this value, then NiS will be prevented from precipitating. The calculation is exactly like the one above. First, we substitute values into the K_{spa} expression.

$$K_{spa} = 4 \times 10^1 = \frac{(0.010)(0.1)}{[H^+]^2}$$

Once again, we solve for $[H^+]$.

$$[H^+]^2 = \frac{(0.010)(0.1)}{4 \times 10^1} = 2.5 \times 10^{-5}$$

$$[H^+] = 5 \times 10^{-3} \ M \text{ (correctly rounded)}$$

The pH is the negative logarithm of this, or

$$pH = 2.3$$

If we maintain the pH of the solution of 0.010 M Cu^{2+} and 0.010 M Ni^{2+} at 2.3 or lower (more acidic), as we make the solution saturated in H_2S, virtually all the Cu^{2+} will precipitate as CuS, but all the Ni^{2+} will stay in solution. Normally, we use a pH of about 0.5, corresponding to $[H^+] = 0.3 \ M$, which works quite well in separating the "acid-insoluble sulfides" from the "acid-soluble sulfides." We can also see why NiS is classified as a "base-insoluble sulfide"—if the solution is basic when it is made saturated in H_2S, the pH will surely be larger than 2.3 and NiS will precipitate. That's how Ni^{2+} and most of the other ions of cation group 3 are removed from a solution that might also contain cations from subsequent cation groups.

After a gathering, people leave an orderly arrangement of chairs and mingle in random groups. The unsynchronized movements of people that we see here are much like the random movements of molecules, a phenomenon that is responsible for the tendency toward randomness in all systems. This driving force for spontaneous change, which was first introduced to you in Chapter 13, is a topic examined more closely in the pages ahead.

THERMODYNAMICS: A GUIDE TO WHAT'S POSSIBLE

16.1 INTRODUCTION

Studying the energy flow accompanying chemical and physical changes tells us what is possible and what is not.

By now, we have studied quite a few kinds of chemical and physical changes. For example, we discussed the reactions of metals with acids, and one of our observations was that the hydronium ion in a solution of a strong acid like hydrochloric acid reacts with metals such as magnesium and zinc and oxidizes them to their metal ions. But hydronium ion does not oxidize copper or silver — we need an oxidizing agent stronger than H_3O^+ to attack these metals. Why? What is it that determines whether a given metal reacts with a particular oxidizing agent?

We have also discussed chemical equilibria. Perhaps you have wondered why some reactions have large equilibrium constants and proceed nearly to completion, while others hardly produce any products at all.

Questions such as "Can a reaction occur?" and "How far toward completion will it go?" are obviously of fundamental importance in chemistry. Being able to answer them, however, has value beyond just a better understanding of chemistry. The answers to such questions also go a long way toward helping us understand the world in which we live, because we are constantly concerned with whether certain changes are possible and what the results will be.

As you probably guessed from its title, this chapter deals with a subject called thermodynamics. Basically, thermodynamics is concerned with energy changes and the flow of energy from one substance to another. In fact, that's how the subject got its name; *thermo* implies heat and *dynamics* implies movement. Amazingly, a detailed study of energy transfers has led to an understanding of why certain changes are inevitable and why others are simply impossible. They let us understand why rivers flow downhill instead of up, and why stone walls crumble instead of being formed during the passage of time from heaps of sand.

Although you weren't aware of it, you were already introduced to some of the principles of thermodynamics in Chapter 5 when you learned to calculate enthalpy changes (heats of reaction). You should recall that enthalpy changes deal with the exchange of heat between chemical systems and their surroundings. Such transfers of heat, however, represent only one aspect of thermodynamics. Another is nature's drive toward disorder which we discussed briefly in Chapter 13. As you will see in the pages ahead, both energy and disorder are prime topics in thermodynamics.

16.2 ENERGY CHANGES IN CHEMICAL REACTIONS— A SECOND LOOK

The change in the total energy of the system is equal to the amount of heat absorbed or evolved by a reaction when it occurs without a change in volume.

Thermodynamics is a subject that is organized about a set of three fundamental, experimentally derived laws of nature. For ease of reference, they are identified by number — the First Law, the Second Law, and the Third Law. The First Law of Thermodynamics deals with exchanges of energy between a system and its surroundings. As we will see, it is basically a statement of the law of conservation of energy — a law, you recall, that serves as the foundation for Hess's law, which we used in our computations involving enthalpy changes in Chapter 5.

To understand the First Law of Thermodynamics, as well as the others, it is necessary to have a clear understanding of the energy changes that occur between a system and its surroundings. For this reason, we begin our study of thermodynamics by taking another look at thermochemistry. At the same time, we can clear up some questions raised in Chapter 5. There, you may recall, we mentioned that the energy given off or absorbed by a system when it undergoes a change depends on whether or not the pressure remains constant. In other words, the heat of reaction at constant volume often is not quite the same as the heat of reaction at constant pressure, ΔH. To understand why this is so, we must start again at the beginning.

Internal Energy

First, let's examine the total energy of a system. In thermodynamics, this total energy is called the system's internal energy. Regardless of the composition of a system, its internal energy is the sum of all the kinetic energies (KE) and potential energies (PE) of its particles. We will use the symbol E to stand for this total energy, so by definition

$$E_{system} = (KE)_{system} + (PE)_{system}$$

Despite the importance of this definition, we can never actually know the total energy of a system. As we commented in Chapter 5, the kinetic and potential energies of a system depend on velocities and attractions that we are unable to determine, so we can't measure E or calculate its value. This is no terrible crisis, however, because we aren't really concerned with what the total energy is; all we have to worry about is how the energy changes, because that is all that has any effect on our lives. This energy change is represented by the symbol ΔE and is defined by the equation

$$\Delta E = E_{final} - E_{initial}$$

or, for a chemical reaction,

$$\Delta E = E_{products} - E_{reactants}$$

Notice that we have followed the practice established in Chapter 5—"final minus initial" or "products minus reactants." This means that if a system absorbs energy from its surroundings during a change, its final energy is greater than its initial energy and ΔE is positive. This is what happens, for example, when the surroundings supply the energy needed to charge a battery. As the system (the battery) absorbs the energy, its energy increases and becomes available for later use.

ΔE is positive for an endothermic change.

Heat and Work

Next we must be concerned with *how* a system exchanges energy with its surroundings. The First Law considers two kinds of energy changes, heat and work, because these are the most common forms that the energy change takes. When a system undergoes a change, it may absorb heat or it may lose heat. Similarly, the system might do some work on the surroundings—for example, move a piston in an engine—and thereby lose some of its energy. Or the system might have some work done on it, as in the compression of a gas, and thereby increase its amount of stored energy. Usually the First Law is stated in the form of an equation.

$$\Delta E = q + w \qquad (16.1)$$

In this equation, q represents the amount of heat that is *absorbed by* the system, and w is the amount of work *done on* the system. In effect, the First Law states that the change in the internal energy of a system is the sum of the amount of energy it gains as heat plus the amount of energy it gains from having work done on it.

ΔE = (heat input) + (work input)

   In the preceding paragraph we have been very careful to define q and w precisely, because by doing so we establish the algebraic signs of these quantities. Thus, q is given a positive sign if heat flows into the system and a negative sign if heat flows out. (Note that this is consistent with the sign conventions we established in Chapter 5 for endothermic and exothermic changes.) Similarly, w is given a positive sign if the system gains energy by having work done on it, as in the winding of a spring in an alarm clock. On the other hand, if the system performs work, its energy reserves are depleted somewhat (i.e., the system loses energy) and w is given a negative sign. This is summarized as follows:

$q = (+)$	heat absorbed by system
$q = (-)$	heat released by system
$w = (+)$	work done on the system
$w = (-)$	work done by the system

ΔE Is a State Function, but q and w Are Not

An important fact about the internal energy E is that it is a state function—its value depends only on a system's particular set of conditions of temperature, pressure, and number of moles of each component. The magnitude of E does not depend on how the system happened to arrive in its current state. Because of this, the change in the internal energy, ΔE, that accompanies a change from one state to another depends only on what the two states are. ΔE does not depend on how the change occurs.

Although ΔE is independent of how a change takes place, the values of q and w are not. In other words, the path that's followed determines the form in which the energy appears. For example, consider the energy changes that occur when an automobile battery is discharged. Overall, this is an exothermic change because the internal energy of the battery decreases as its stored energy is gradually used up. The energy that is delivered, however, can appear partly as heat and partly as useful work, but exactly how much appears as heat and how much appears as work depend on how the battery is discharged. Nevertheless, the *sum* of the two, work plus heat, is the same under any circumstances because no matter how we carry out the change, the value of ΔE is always the same, provided that the initial and final states of the system are the same. It is in *this* sense that the First Law is a statement of the law of conservation of energy.

When we discharge a battery, our success in getting ΔE to emerge as much as possible in the form of useful work and as little as possible as wasted heat depends altogether on exactly *how* we let the exothermic change happen. Interestingly, and most significantly, the more slowly we let the battery discharge, the more is the available energy, ΔE, able to appear as work and the less able as heat. For example, if we simply short-circuit the battery by placing a heavy steel wrench across its two poles, the exothermic chemical reaction within the battery occurs rapidly and the wrench becomes very hot. (See Figure 16.1.) In this case, the ΔE appears *only* as heat. On the other hand, if we connect the battery to a tiny motor that draws current only slowly, we can run the motor for a long time and get as much of the value of ΔE in the form of work as possible with a low production of heat. Thus, even though ΔE is the same for both rapid and slow discharge, the magnitudes of q and w are not.

Heats of Reaction at Constant Volume

Most of the reactions that we carry out in the lab do not produce electricity, and therefore are not able to do electrical work. But there is another important kind of work

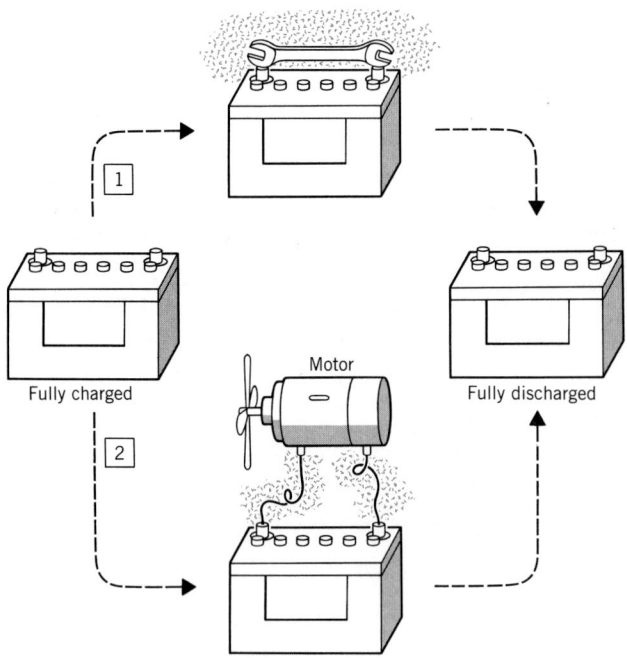

q and *w* are not state functions, but their sum (ΔE) is.

Be careful never to short the terminals of an automobile battery. The reaction can be quite violent!

FIGURE 16.1 The complete discharge of a battery along two different paths yields the same total amount of energy, ΔE. However, if the battery is simply shorted with a heavy wrench, as shown in Path 1, this energy appears entirely as heat. Path 2 gives part of the total energy as heat, but much of the energy appears as work done by the motor.

Fully charged   Fully discharged

Motor

in which these systems can be involved that is related to expansion or contraction under the influence of an outside pressure. For example, consider a gas contained in a cylinder fitted with a piston that is supported by the pressure of the atmosphere (Figure 16.2). When the pin is slid out, thereby allowing the piston to move, the gas will push the piston back against the opposing pressure of the atmosphere. In doing so, the gas will do some work. The amount of this work can be calculated by the simple equation

$$w = -P\Delta V \qquad (16.2)$$

where P is the *opposing pressure* against which the piston pushes and ΔV is the change in the volume of the system (the gas) during the expansion. The negative sign is used because the system is losing energy by doing the work on the surroundings.

Before continuing with our analysis, let's take a moment to see how pressure times a volume change equals work. We can do this by analyzing the units in relationship to the definition of work. Work is accomplished by moving an opposing force, F, through some distance or length, L. The amount of work accomplished is the product of this force and length.

$$\text{work} = F \times L$$

Now let's look at pressure times volume-change. Pressure is defined as force per unit area. But an area is simply length squared, or L^2. So we can write pressure as

$$P = \frac{F}{L^2}$$

Volume has dimensions of length cubed, or L^3, so pressure times the volume change works out to be

$$P\Delta V = \frac{F}{L^2} \times (L^3)$$
$$= F \times L$$

Thus, pressure times volume-change is really the same as force times distance, which is work. Actually, this should be no real surprise. We all know that the expanding gases in a cylinder of a car engine do work as they move a piston. And we know that we have to do work to push down on the handle of a tire pump. This kind of work is often called pressure–volume work, or P–V work.

If the only kind of work a system can do is this P–V work, then we can modify Equation 16.1 as follows.

$$\Delta E = q + w$$
$$= q + (-P\Delta V)$$
$$= q - P\Delta V$$

Suppose we now apply this equation to a chemical reaction occurring at constant volume in a bomb calorimeter (see Figure 5.5 on page 161). Because of the construction of the apparatus, the volume of the reaction mixture cannot change, so ΔV must be zero. This means, of course, that $P\Delta V$ also must be zero. Therefore, under conditions of constant volume,

$$\Delta E = q - 0$$

Thus, the heat of reaction measured in a bomb calorimeter is the heat of reaction at constant volume, often symbolized as q_v, and corresponds to ΔE for the reaction.

$$\Delta E = q_v$$

Enthalpy

Because most of the reactions that are of interest to us take place under conditions of constant pressure, rather than constant volume, thermodynamicists invented the quan-

If the gas is compressed by an external pressure P acting on the piston, work is *done on* the system, and w is positive.

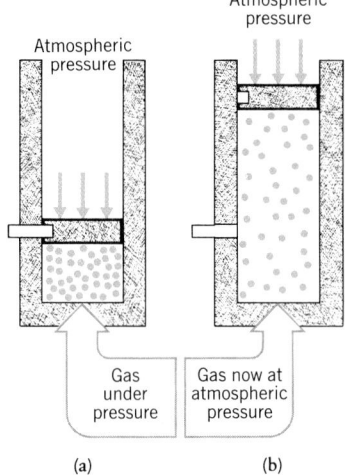

FIGURE 16.2 (*a*) A gas confined under pressure in a cylinder fitted with a piston that is held in place by a sliding pin. (*b*) When the piston is released, the gas inside the cylinder expands and pushes the piston upward against the opposing pressure of the atmosphere. As it does so, the gas does some work.

If there's no volume change, then $V_{final} = V_{initial}$ and ΔV must equal zero.

ΔE is the heat of reaction at constant volume.

tity that we call **enthalpy.** It is defined mathematically by the equation

$$H = E + PV$$

and if we consider only changes at constant pressure, but allow volume changes, then the change in enthalpy, ΔH, is

$$\Delta H = \Delta E + P\Delta V \qquad (16.3)$$

Let's see how Equation 16.3 applies to a chemical reaction taking place at constant pressure and during which the system expands against the opposing pressure of the atmosphere. As we saw in Equation 16.2, the work involved for the expanding system is

$$w = -P\Delta V$$

so, once again,

$$\Delta E = q + (-P\Delta V)$$
$$\Delta E = q - P\Delta V$$

Substituting this into Equation 16.3 gives

$$\Delta H = q - P\Delta V + P\Delta V$$

which reduces to

$$\Delta H = q_p$$

We use the subscript p to show that the heat involved this time is the heat of reaction at constant pressure. This equation simply says that the heat of reaction measured under conditions of constant pressure is equal to ΔH. As you should recall, this is exactly how we defined the enthalpy change in Chapter 5.

The Difference between ΔE and ΔH

Now let's look again at the equation $\Delta H = \Delta E + P\Delta V$, because it permits us to see why the heat of reaction at constant pressure (ΔH) differs from the heat of reaction at constant volume (ΔE). Rearranging this equation gives

$$\Delta H - \Delta E = P\Delta V$$

Thus ΔH and ΔE differ by $P\Delta V$. But what does this mean, physically?

Suppose that we have a reaction for which ΔE is negative (i.e., an exothermic reaction), and also suppose that during this reaction a gas is formed. Let's consider first what happens if we carry out this reaction in a bomb calorimeter, where the volume can't change. As the reaction occurs, a certain total amount of energy is given off. Since the volume is unable to change, ΔV must be zero, so the system can't do any "P–V" work. This means that all the energy must appear as heat, and the amount of heat that we measure is equal to ΔE.

The pressure inside the reaction vessel increases, but the gas can't do any work because the volume can't change.

Now, let's consider what happens if we carry out the reaction at a constant pressure. We could do this by letting the reaction take place inside a cylinder with a movable piston. Then, as the gas is formed, the piston can be pushed back against the opposing pressure of the atmosphere and the pressure within the system can remain constant. But, as the gas moves the piston it does some work, and the energy for this comes from the total energy given off by the reaction. Therefore, the amount of energy that's left over to be given off as the heat of reaction at constant pressure (ΔH) is equal to the total amount of energy released in the reaction (ΔE) *minus* the amount that was lost in pushing back the piston. Thus, the difference between ΔE and ΔH is the $P\Delta V$ work that the system does as it expands against the opposing pressure of the atmosphere.

The ΔV for reactions involving only solids and liquids are very tiny; ΔE and ΔH for these reactions are virtually identical in magnitude.

Whenever a reaction system expands under the constant pressure of the atmosphere, the measured ΔH is a little bit less than the ΔE for the reaction. This is simply a small penalty that we pay for the convenience of carrying out the reaction at constant pressure. On the other hand, if there is a decrease in the volume of the reacting system at constant pressure, then the value of ΔH is actually a bit larger than ΔE, and we get a bonus. In either case, though, the size of the $P\Delta V$ term is small for reactions at atmospheric pressure, and as we noted in Chapter 5, the differences between ΔE and ΔH are only important when very precise measurements are involved.

16.3 SPONTANEOUS CHANGE

Literally everything that happens begins with a spontaneous change somewhere.

Now that you've gotten to understand energy changes, we can turn our attention to one of the main goals of thermodynamics — finding relationships among the factors that control whether events are possible. Many such "possible" events are part of our lives all the time, and they occur without outside help. A book that slips from your hand will start to fall. The ice cubes in the cold drink you fix to quench your thirst on a warm day will gradually melt. Such events are examples of spontaneous changes — they occur by themselves without outside assistance. Once conditions are right for them to begin, they proceed on their own.

Some spontaneous changes occur very rapidly. An example is the set of biochemical reactions that take place when you accidentally touch something that is very hot — they cause you to jerk your hand away quickly. Other spontaneous events, such as the gradual erosion of a mountain, occur slowly and many years pass before a change is noticed. Still others occur at such an extremely slow rate under ordinary conditions that they appear not to be spontaneous at all. Gasoline – oxygen mixtures appear perfectly stable indefinitely at room temperature because under these conditions they react so very slowly; however, if heated, their rate of reaction increases and they can react explosively until all of one or the other is totally consumed.

Each day we also witness many events that are obviously not spontaneous. We many pass by a pile of bricks in the morning and later in the day find that they have become a brick wall. We know from experience that the wall didn't get there by itself. A pile of bricks becoming a brick wall is *not* spontaneous; it requires the intervention of a bricklayer. Similarly, the decomposition of water into hydrogen and oxygen is not spontaneous. We see water all the time and we know that it is stable. Nevertheless, we can cause water to decompose by passing an electric current through it in a process called *electrolysis*.

$$2H_2O(\ell) \xrightarrow{\text{electrolysis}} 2H_2(g) + O_2(g)$$

This process will continue, however, only as long as the electric current is maintained. As soon as the supply of electricity is cut off, the decomposition ceases. This example demonstrates the difference between spontaneous and nonspontaneous changes. Once a spontaneous event begins, it has a tendency to continue until it is finished. A nonspontaneous event, on the other hand, can continue only as long as it receives some sort of outside assistance.

Nonspontaneous changes have another common characteristic. They can occur only when some spontaneous change has occurred first. The bricklayer consumes food, and a series of spontaneous biochemical reactions then occur that supply the necessary muscle power to build the wall. Similarly, the nonspontaneous electrolysis of water requires some sort of spontaneous mechanical or chemical change to generate the needed electricity. In short, all nonspontaneous events occur at the expense of spontaneous ones. Everything that happens can be traced, either directly or indirectly, to some spontaneous change.

We take many spontaneous events for granted. What would you think if you dropped a book and it rose to the ceiling?

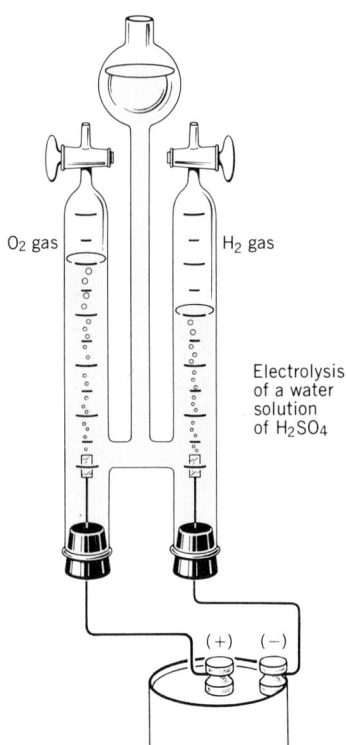

O₂ gas   H₂ gas

Electrolysis of a water solution of H_2SO_4

(+)   (−)

The driving of nonspontaneous reactions to completion by linking them to spontaneous ones is an important principle in biochemistry.

16.4 ENTHALPY CHANGES AND SPONTANEITY

Exothermic changes tend to be spontaneous.

Because spontaneous reactions are so important, it is necessary for us to understand the factors that favor spontaneity. We can begin by simply examining some everyday events such as those depicted in Figure 16.3. We know, for example, that skiers (and snowboarders) enjoy the downhill run because they don't have to work hard to make it happen; gravity and the tendency to slide to the bottom are on their side. We also know that water flows downhill and that heat is generated when gasoline burns. Each of these events is spontaneous. In fact, we expect them to happen and would be quite surprised if they didn't. Can we find some factor that all of them have in common?

If we study these events, we see that each of them leads to a lowering, or decrease, in the energy of the system. Both the snowboard and the water lose potential energy as they move from a higher to a lower altitude. Similarly, the chemical substances in the gasoline–oxygen mixture lose energy by evolving heat as they react to produce CO_2 and H_2O. Because these events are spontaneous, we conclude that *when a change lowers the energy of a system, it tends to occur spontaneously.* Since a change that lowers the energy of a system is said to be exothermic, we can state this factor another way — *exothermic changes have a tendency to proceed spontaneously.*

> Most, but not all, chemical reactions that are exothermic occur spontaneously.

In the preceding statement the word *tendency* should be emphasized. We will see that not every exothermic change is spontaneous, nor is every endothermic one nonspontaneous. The important point is that an energy decrease works as *one factor* in favor of spontaneity.

We now have to ask, how does this relate to chemical reactions? In Section 16.2 we saw that thermodynamics looks to ΔE and ΔH for the energy change in reacting systems. There we saw that it is ΔE that we observe when the reaction happens at constant volume, and it is ΔH that we measure when the reaction occurs at constant pressure. Since we are most interested in reactions at constant pressure, we will devote our attention to ΔH. This quantity, you recall, has a negative sign for an exothermic change, so we can say that reactions for which ΔH is negative *tend* to proceed spontaneously. The enthalpy change thus serves as *one* of the thermodynamic factors that influence whether or not a given process will occur by itself.

FIGURE 16.3 Three common spontaneous events—a person rides a ''snowboard'' downhill, water cascades over a waterfall, and fuel burns.

16.5 ENTROPY AND SPONTANEOUS CHANGE

An increase in randomness favors a spontaneous change.

In Chapter 13 we discussed the solution process, and you learned that one of the principal driving forces for the formation of a solution is the increase in disorder, or randomness, that occurs when particles of the solute mix with those of the solvent. In fact, this increase in disorder can be so important that it can outweigh energy effects. For example, the dissolving of NaI in water is endothermic (ΔH_{soln} is positive); nevertheless, NaI crystals will dissolve quite spontaneously in water, even though the energy effect would seemingly make the process nonspontaneous.

The dissolving of salts such as NaI in water is just one example of a change that occurs spontaneously even though it is endothermic. Other even more common examples are the melting of ice when the weather becomes warm and the evaporation of water from a puddle or a pond. Both of these changes are also endothermic and spontaneous, and they take place for the same reason that NaI dissolves in water. Each is accompanied by an increase in the randomness of the distribution of its particles. When ice melts, the water molecules leave the highly organized crystal state and become jumbled and disorganized in the liquid. When water evaporates, the molecules are no longer confined to the region of other water molecules; instead, they are free to roam about the entire atmosphere.

It is a universal phenomenon that something that brings about randomness is more likely to occur than something that brings about order. We can see the reason for this if we examine the close relationship between randomness and statistical probability. Suppose we try a simple experiment with a new deck of playing cards. When first unwrapped, they are separated according to suit and arranged numerically. Now, suppose we toss the deck in the air and let the cards fall to the floor. When we sweep them up and restack them, we will almost certainly find they have become disordered. (In fact, if they haven't become disordered, you would suspect some tampering with the deck!) We expect this disordering to occur because there are so many millions of ways for the cards to be disordered, but there is only one way for them to come together again in the original sequence. A disordered random sequence of cards is much more probable than the ordered one with which we began, and the spontaneous change from the ordered arrangement to the disordered one takes place simply because of the laws of chance.

The cards are highly ordered

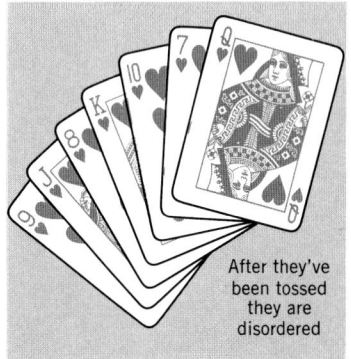

After they've been tossed they are disordered

Entropy

The same laws of chance that apply to the ordering of a deck of playing cards also apply to the ordering within distributions of particles in chemical and physical systems. Con-

FIGURE 16.4 (*a*) If water molecules are dropped into the container, they would not be expected to land in just the right places to give the highly ordered solid. If the way they land is determined purely by chance, the disordered liquid is more likely to occur. (*b*) Tossing bricks into the air is more likely to give a pile of bricks than a brick wall.

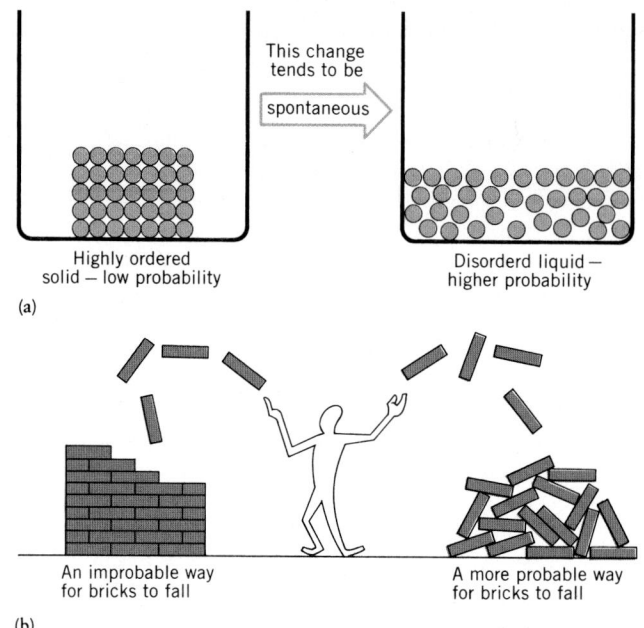

FIGURE 16.4 (*a*) If water molecules are dropped into the container, they would not be expected to land in just the right places to give the highly ordered solid. If the way they land is determined purely by chance, the disordered liquid is more likely to occur. (*b*) Tossing bricks into the air is more likely to give a pile of bricks than a brick wall.

sider, for example, the melting of an ice cube (Figure 16.4). Of all the possible ways of placing water molecules into a container, the highly ordered arrangement within the crystalline ice cube has a very low probability, while the disordered arrangement of the molecules found in the liquid is more probable. We can see this more clearly, perhaps, by analogy. Just imagine tossing bricks in the air and observing how they land. The chance that they will fall one on the other to form a brick wall is certainly very small; the brick wall is an improbable result. A much more likely arrangement is a jumbled, disordered pile. Thus, the collapse of a brick wall and the melting of an ice cube have something in common. In each case, the system passes spontaneously from a state of low probability to one of higher probability.

Because statistical probability is so important in determining the outcome of chemical and physical events, thermodynamics defines a quantity, called entropy (symbol S), that describes the degree of randomness of a system. The larger the value of the entropy, the larger is the degree of randomness of the system and, therefore, the larger is its statistical probability.

Like the enthalpy, entropy is a state function. It depends only on the state of the system, and therefore a change in entropy, ΔS, is independent of the path from start to finish. As with other thermodynamic quantities, ΔS is defined as "final minus initial" or "products minus reactants." Thus

$$\Delta S = S_{final} - S_{initial}$$

or, for chemical systems,

$$\Delta S = S_{products} - S_{reactants}$$

As you can see, when S_{final} is larger than $S_{initial}$ (or when $S_{products}$ is larger than $S_{reactants}$), the value of ΔS is positive. A positive value for ΔS means an increase in the randomness of the system during the change, and we have seen that this kind of change tends to be spontaneous. This leads to a general statement about entropy:

> Any event that is accompanied by an increase in the entropy of the system *tends* to occur spontaneously.

The greater the statistical probability, the greater the entropy.

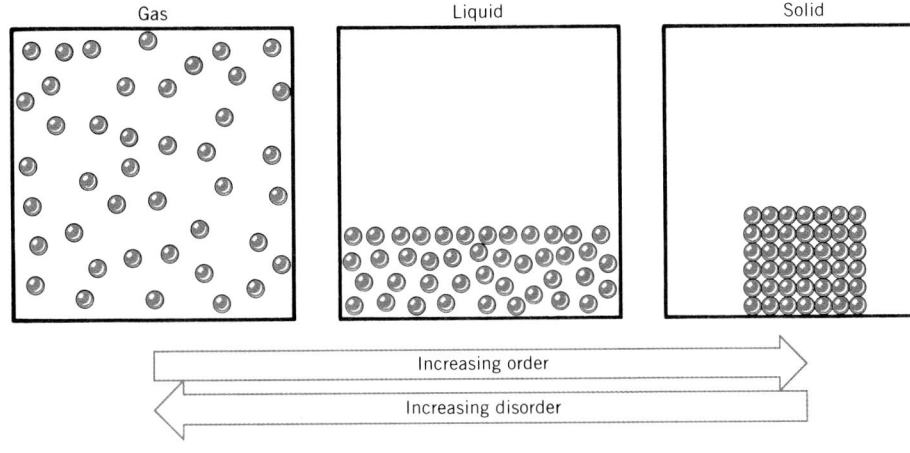

Gas   Liquid   Solid

Increasing order

Increasing disorder

FIGURE 16.5 Comparison of the entropies of the solid, liquid, and gaseous states of a substance. The crystalline solid is highly ordered and has a very low entropy. The liquid has a higher entropy because it is less ordered, but all the particles are still found at one end of the container. The gas has the highest entropy because the particles are randomly distributed throughout the entire container.

Predicting ΔS for Physical and Chemical Changes

Often (but not always), it is a relatively simple matter to predict whether a particular change will be accompanied by an increase or decrease in entropy. In doing this, one of the principal factors to consider is the physical states of the reactants and products. As we can see in Figure 16.5, a gas has a larger entropy than a liquid, which has a larger entropy than a solid. In fact, a gas has such a large entropy compared with a liquid or solid that changes that produce gases from liquids or solids are almost always accompanied by increases in entropy.

A major factor that affects the entropy of a system is the freedom of movement of its particles — the greater the freedom of movement, the larger the entropy. Notice that this gives us another way of arriving at the relative entropies of the solid, liquid, and gaseous states. In a solid this freedom of movement is at a minimum, and in a gas it is at a maximum, so a solid has the lowest entropy and a gas, the highest.

During chemical reactions, the freedom of movement of the atoms often changes because of changes in the complexity of the molecules. For example, consider the reaction

$$2NO_2(g) \longrightarrow N_2O_4(g)$$

Among the reactants are six atoms combined into two molecules of NO_2. Among the products, these same six atoms are confined to one molecule. In the reaction vessel, dividing the six atoms between two molecules allows them to spread out more and gives greater freedom of movement than when the atoms are combined into just one molecule. Therefore, we can conclude that as this reaction occurs, there is an entropy decrease.

EXAMPLE 16.1   PREDICTING THE SIGN OF THE ENTROPY CHANGE FOR A CHANGE OF STATE

Problem: Show that ΔS for the melting of ice is positive.

Solution: At a given temperature

$$S_{solid} < S_{liquid} < S_{gas}$$

Knowing this, it is now easy to predict the sign of the entropy change for the melting of ice. The change corresponds to

$$solid \longrightarrow liquid$$

Therefore

$$\Delta S = S_{liquid} - S_{solid}$$

Because S_{liquid} is larger than S_{solid}, ΔS must be a positive quantity.

EXAMPLE 16.2   PREDICTING THE SIGN OF THE ENTROPY CHANGE FOR A CHEMICAL REACTION

Problem: What is the expected algebraic sign of the entropy change for the reaction

$$2H_2(g) + O_2(g) \longrightarrow 2H_2O(\ell)$$

Solution: The entropy change is given by

$$\Delta S = S_{products} - S_{reactants}$$

and the sign of ΔS is determined by whichever is larger, $S_{products}$ or $S_{reactants}$. In this reaction, both reactants are gases, so the entropy of the reactants must be quite large. In addition, the system consists of only diatomic molecules. Now, consider the products. Here we have only liquid, so all the molecules are crowded into one end of the reaction vessel. This suggests that the entropy is low. Furthermore, now the system consists of triatomic molecules. Since there are fewer molecules, the atoms of the system have less freedom and have become more organized. This also suggests that the products have a lower entropy than the reactants. We can conclude, therefore, that for this reaction $S_{products} < S_{reactants}$. As a result, we predict that ΔS is negative.

PRACTICE EXERCISE 1   Predict the sign of the entropy change for (a) the condensation of steam to liquid water, and (b) the sublimation of a solid.

PRACTICE EXERCISE 2   Predict the sign of ΔS for the following reactions:
(a) $2SO_2(g) + O_2(g) \longrightarrow 2SO_3(g)$
(b) $2NaHCO_3(s) \longrightarrow Na_2CO_3(s) + CO_2(g) + H_2O(g)$

Based on the entropy change for the reaction in Example 16.2, we might be tempted to conclude that the reaction of hydrogen with oxygen is nonspontaneous. Of course, this just isn't true. These two chemicals react violently, as those who witnessed the Hindenburg disaster learned only too well. In this case, the large exothermic nature of the energy change for the reaction outweighs the negative entropy change. We see now the need for some way to analyze the relative importance of the energy and entropy changes in determining whether some chemical or physical change is spontaneous. This important topic is discussed in Section 16.7.

The fierce spontaneous reaction of hydrogen with oxygen in the air led to the fiery destruction of the Hindenburg, a German airship filled with hydrogen, in Lakehurst, New Jersey, in 1937. Thirty people died in the disaster.

Questions of energy resources and their exploitation, the decision of whether or not to concentrate on the use of nuclear fuels, the implications of careless pollution—all of these factors and more tie our social and economic well-being to that of our environment. The interaction between society and technology is obvious wherever we look, and technology draws on our finite world resources for raw materials, energy, and dumping places for waste. Our world is becoming too complex to allow careless or stupid decisions concerning how we manage our resources and economics because such decisions have very far-reaching effects. As we too often discover, these effects can be devastating. But how do we, as a society, make the right choices?

The answer may lie in the application of thermodynamic principles to our social and economic systems. In your study of chemistry, you have now encountered two instances where probabilities, rather than certainties, help us to understand nature. In Chapter 6 we used the probability concept to interpret the results of quantum mechanics; for instance, we speak of the probability of finding an electron at a certain location. In this chapter we have used probability to define

entropy and the tendency for systems to move in a direction of greater disorder. Both probability and the entropy concept are beginning to be seen as important principles in social systems as well as physical ones.

Professor Ilya Prigogine, the 1977 Nobel Prize winner in chemistry, has been working on mathematical models that establish parallels between the thermodynamic behavior of chemical or biological systems and the behavior of social systems. Economist Nicholas Georgescu-Roegen, professor emeritus at Vanderbilt University, has been studying the relationship between entropy and the economic process—how human activities hasten us toward ever greater entropy and disorder. His view is that we should "minimize future regrets" and maintain a balanced flow of energy by learning to use energy in amounts that we receive naturally.

The widespread interest in thermodynamic models of society stems from the hope that we can determine the effects of our decisions before we implement them and find out that we have made a serious mistake. As corny as it may sound, the future of the human race may hang in the balance.

The Second Law of Thermodynamics

One of the most far-reaching observations in science is incorporated into the Second Law of Thermodynamics, which states, in effect, that whenever a spontaneous event takes place in our universe, it is accompanied by an overall increase in entropy. Note that the increase in entropy that's referred to here is for the *universe*, not just the system. This means that a system's entropy can decrease, just as long as there is a larger increase in the entropy of the surroundings so that the *net* entropy change is positive. Because everything that happens relies on spontaneous changes of some sort, the entropy of the universe is constantly increasing.

In Section 16.7 we will see how the Second Law translates into concrete mathematical statements that we can use in analyzing the spontaneity of events. But for the moment, let's take a qualitative look at the "big picture" to see how the Second Law influences our daily activities. In particular, let's examine what the Second Law tells us about pollution and our efforts to control it. Can we avoid pollution of our environment entirely? Why are pollutants so easily spread and so difficult to eliminate? These are the kinds of questions we would like to answer.

Much of our time and effort is spent it tidying up our own home and neighborhood. We seem constantly involved in trying to decrease disorder in our lives. We clean up our desks and sweep the floor. We put out the garbage, mow the lawn, and rake leaves in the autumn. Overall our activities are spontaneous because the biochemical reactions driven by the foods we eat and digest are spontaneous and allow us to do these things. Since the total entropy of the universe must be increased by our spontaneous activities, the increased order that we create for ourselves has to be balanced by an even larger increase in disorder somewhere in our surroundings—the environment.

Pollution involves the scattering of undesirable substances through our environment and is accompanied by an increase in entropy. It is a direct result of our efforts to create an orderly world. For example, if we burn our leaves to get rid of them, the smoke

FIGURE 16.6 The release of hazardous materials into the environment can create great problems because the spread of pollutants is accompanied by large increases in entropy. Removing such materials from areas like that shown in this photo requires much effort and expense.

The major oil companies have spent large sums of money to remove gasoline that has leaked from many old steel storage tanks buried underground at service stations 20 or more years ago. This has been a special problem in areas that depend on ground water for drinking and cooking.

from our fire spreads through the neighborhood and disturbs our friends. When we expand our efforts beyond our home and attempt to provide order in our environment, our attempts are met by still more overall disorder. For instance, we use machinery to collect garbage so that we and our neighbors don't have to live in it. But the machinery, by burning fuel, generates air pollution that is even more difficult to clean up. It's a losing game. We are simply faced with the fact that we can never really eliminate pollution. All we can do is to try to keep it from places where it can do us great harm (Figure 16.6) and attempt to confine it to locations where it interferes with life as little as possible.

The entropy effect on pollution is important to understand, particularly when we consider the consequences of releasing harmful substances into the environment. Such materials include the highly toxic elements released in trace amounts when coal or nuclear fuels are used, as well as many toxic chemicals such as DDT and dioxin. When these chemicals are released, their spread is unavoidable because of the large entropy increase that occurs as they are scattered about. Eliminating them, once they have had an opportunity to disperse, is an almost impossible task because doing so requires an enormous expenditure of energy and *must* generate even more disorder around us. Only if the pollutant is extremely hazardous does it warrant the effort necessary to reduce its concentration in the environment, and even if this is done, we can never eliminate the pollutant entirely. We can only reduce our risk of injury. The surest way to overcome pollution, therefore, is not to create it in the first place. It is a trade-off between risk and benefit. For the benefits of activities that pollute (e.g., using fuels to make electricity), we try to reduce the risks as much as we judge possible and endure the rest.

16.6 THE THIRD LAW OF THERMODYNAMICS

The entropy of a pure crystal is zero at absolute zero.

The entropy of a substance (i.e., the extent of its thermodynamic disorder) varies with the temperature of the substance — the lower the temperature, the lower the entropy. For example, at a pressure of 1 atm and a temperature above 100 °C, water exists as a highly disordered gas with a very high entropy. If confined, the molecules of water vapor will be spread evenly throughout their container, and they will be in constant random motion. When the system is cooled, the water vapor eventually condenses to form a liquid. Although the molecules can still move somewhat freely, they are now confined to

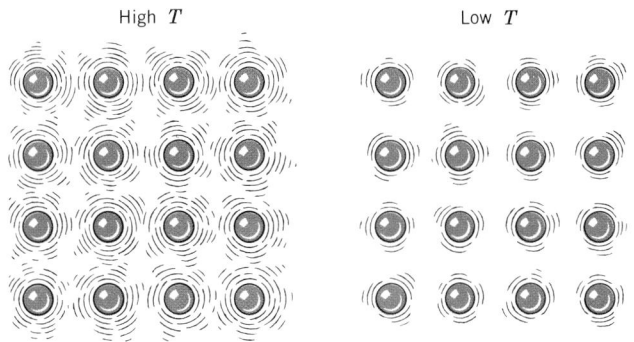

FIGURE 16.7 Greater disorder in a crystal at high temperature is caused by more violent molecular vibrations, which lead to a greater displacement of the atoms from their equilibrium lattice positions.

High *T*   Low *T*

the bottom of the container. Their distribution in the container is not as random as it was in the gas and the entropy of the liquid is lower. Further cooling decreases the entropy even more, and below 0 °C, the water molecules join together to form ice, a crystalline solid. The molecules are now in a highly ordered state, particularly in comparison to that of the gas, and the entropy of the system is very low.

Yet even in the crystalline form, the order isn't perfect and the water molecules still have some entropy. There is enough thermal energy left to cause them to vibrate or rotate within the general area of their lattice sites. Thus at any particular instant, we would find the molecules near, but probably not exactly at, their equilibrium lattice positions (Figure 16.7). If we cool the solid further, we decrease the thermal energy and the molecules spend less time away from their equilibrium lattice positions. The order of the crystal increases and the entropy decreases. Finally, at absolute zero, the ice will be in a state of perfect order and its entropy will be zero. This leads us to the statement of the Third Law of Thermodynamics. *At absolute zero, the entropy of a pure crystal is also zero.*

$$S = 0 \quad \text{at} \quad T = 0 \text{ K}$$

Because we know the point at which entropy has a value of zero, it is possible by measurement and calculation to determine the actual amount of entropy that a substance possesses at temperatures above 0 K. If the entropy of one mole of a substance is determined at a temperature of 298 K (25 °C) and a pressure of 1 atm, we call it the standard entropy, $S°$. Table 16.1 lists the standard entropies for a number of substances. Notice that entropy has the dimensions of energy/temperature (i.e., joules per kelvin or calories per kelvin). Once we know the entropies of a variety of substances we can calculate the standard entropy change, $\Delta S°$, for chemical reactions in much the same way as we calculated $\Delta H°$ in Chapter 5.

25 °C and 1 atm are the same standard conditions we used in our discussion of $\Delta H°$ in Chapter 5.

$$\Delta S° = (\text{sum of } S° \text{ of products}) - (\text{sum of } S° \text{ of reactants}) \qquad (16.4)$$

This is simply a Hess's law type of calculation. Note, however, that elements have nonzero $S°$ values and must be included in the bookkeeping.

If the reaction with which we are working corresponds to the formation of 1 mol of a compound from its elements, the $\Delta S°$ that we calculate can be referred to as the standard entropy of formation, $\Delta S_f°$. Values of $\Delta S_f°$ are not tabulated; if we need them, we must calculate them from values of $S°$.

EXAMPLE 16.3 CALCULATING $\Delta S°$ FROM STANDARD ENTROPIES

Problem: Urea (from urine) hydrolyzes slowly in the presence of water to produce ammonia and carbon dioxide.

$$CO(NH_2)_2(aq) + H_2O(\ell) \longrightarrow CO_2(g) + 2NH_3(g)$$
urea

TABLE 16.1 Standard Entropies of Some Typical Substances (25 °C, 1 atm)

Substance	Entropy, S°		Substance	Entropy, S°	
	J/mol K	cal/mol K		J/mol K	cal/mol K
$Ag(s)$	42.55	10.17	$H_2O(g)$	188.7	45.11
$AgCl(s)$	96.2	23.0	$H_2O(\ell)$	69.96	16.72
$Al(s)$	28.3	6.77	$HCl(g)$	186.7	44.62
$Al_2O_3(s)$	51.00	12.19	$HNO_3(\ell)$	155.6	37.19
$C(s, graphite)$	5.69	1.36	$H_2SO_4(\ell)$	157	37.5
$CO(g)$	197.9	47.30	$HC_2H_3O_2(\ell)$	160	38.2
$CO_2(g)$	213.6	51.06	$Hg(\ell)$	76.1	18.2
$CH_4(g)$	186.2	44.50	$Hg(g)$	175	41.8
$CH_3Cl(g)$	234.2	55.97	$K(s)$	64.18	15.34
$CH_3OH(\ell)$	126.8	30.3	$KCl(s)$	82.59	19.74
$CO(NH_2)_2(s)$	104.6	25.00	$K_2SO_4(s)$	176	42.0
$CO(NH_2)_2(aq)$	173.8	41.55	$N_2(g)$	191.5	45.77
$C_2H_2(g)$	200.8	48.00	$NH_3(g)$	192.5	46.01
$C_2H_4(g)$	219.8	52.54	$NH_4Cl(s)$	94.6	22.6
$C_2H_6(g)$	229.5	54.85	$NO(g)$	210.6	50.34
$C_2H_5OH(\ell)$	161	38.4	$NO_2(g)$	240.5	57.47
$Ca(s)$	154.8	36.99	$N_2O(g)$	220.0	52.58
$CaCO_3(s)$	92.9	22.2	$N_2O_4(g)$	304	72.7
$CaCl_2(s)$	114	27.2	$Na(s)$	51.0	12.2
$CaO(s)$	40	9.5	$Na_2CO_3(s)$	136	32.5
$Ca(OH)_2(s)$	76.1	18.2	$NaHCO_3(s)$	102	24.4
$CaSO_4(s)$	107	25.5	$NaCl(s)$	72.38	17.30
$CaSO_4 \cdot \frac{1}{2}H_2O(s)$	131	31.2	$NaOH(s)$	64.18	15.34
$CaSO_4 \cdot 2H_2(s)$	194.0	46.36	$Na_2SO_4(s)$	149.4	35.7
$Cl_2(g)$	223.0	53.29	$O_2(g)$	205.0	49.00
$Fe(s)$	27	6.5	$PbO(s)$	67.8	16.2
$Fe_2O_3(s)$	90.0	21.5	$S(s)$	31.9	7.62
$H_2(g)$	130.6	31.21	$SO_2(g)$	248.5	59.40
			$SO_3(g)$	256.2	61.24

A more complete table of thermodynamic data is located in Appendix D at the back of the book.

What is the standard entropy change, in J/K, for this reaction when 1 mol of urea reacts with water?

Solution: The standard entropies for the reactants and products are found in Table 16.1. Let's first collect the data and assemble them, as we've done in the table at the right

Substance	S° (J/mol K)
$CO(NH_2)_2(aq)$	173.8
$H_2O(\ell)$	69.96
$CO_2(g)$	213.6
$NH_3(g)$	192.5

Following Equation 16.4 we have

$$\Delta S° = [S°(CO_2) + 2S°(NH_3)] - [S°(CO(NH_2)_2) + S°(H_2O)]$$

$$= \left[1\ \text{mol} \times \left(\frac{213.6\ J}{\text{mol K}} \right) + 2\ \text{mol} \times \left(\frac{192.5\ J}{\text{mol K}} \right) \right]$$

$$- \left[1\ \text{mol} \times \left(\frac{173.8\ J}{\text{mol K}} \right) + 1\ \text{mol} \times \left(\frac{69.96\ J}{\text{mol K}} \right) \right]$$

$$= (598.6\ J/K) - (243.8\ J/K)$$
$$= 354.8\ J/K$$

Entropy is a state function, just like enthalpy. The value of ΔS doesn't depend on the "path" that is followed. In other words, we can proceed from one state to another in any way we like and ΔS will be the same. If we choose a path in which just a slight alteration in the system can change the direction of the process, we can measure ΔS directly. For example, at 25 °C an ice cube will melt and there is nothing we can do at that temperature to stop it. At 0 °C, however, it is simple to stop the melting process and reverse its direction. At 0 °C the melting of ice is said to be a reversible process. If we set up a change so that it is reversible, we can calculate the entropy change as $\Delta S = q/T$, where q is the heat added to the substance and T is the temperature at which the heat is added. We can understand this in the following way. In a system of molecules held together by attractive forces, disorder is introduced by adding heat, which makes the molecules move more violently. In a crystal, for instance,

the molecules vibrate about their equilibrium lattice positions, so that at any instant we really have a "not-quite-ordered" collection of particles. Adding heat increases molecular motion and leads to greater instantaneous disorder. In other words, if we could freeze the motion of the molecules, like a photographer's flash freezes action, we would see that after we have added some heat there is greater disorder and therefore greater entropy. This entropy increase is *directly proportional to the amount of heat added.* However, this added heat makes a more noticeable and significant difference if it is added at low temperature, where little disorder exists, rather than at high temperature where there is already substantial disorder. For a given quantity of heat the entropy change is therefore *inversely proportional to the temperature at which the heat is added.* Thus, $\Delta S = q/T$, and entropy has units of energy divided by temperature (e.g., J/K or cal/K).

PRACTICE EXERCISE 3 Calculate the standard entropy change in J/K for each of the following reactions.
(a) $CaO(s) + 2HCl(g) \longrightarrow CaCl_2(s) + H_2O(\ell)$
(b) $C_2H_4(g) + H_2(g) \longrightarrow C_2H_6(g)$

16.7 THE GIBBS FREE ENERGY

The Gibbs free energy gives us the net effect on spontaneity of the enthalpy and entropy changes.

We have now seen that two factors — enthalpy and entropy — determine whether or not a given physical or chemical event will be spontaneous. Sometimes these two factors work together. For example, when a stone wall crumbles, its enthalpy decreases and its entropy increases. Since a decrease in enthalpy and an increase in entropy both favor a spontaneous change, the two factors complement one another. In other situations, the effects of enthalpy and entropy are in opposition. Such is the case, as we have seen, in the melting of ice or the evaporation of water. The endothermic nature of these changes tends to make them nonspontaneous, while the increase in the randomness of the molecules tends to make them spontaneous. In the reaction of H_2 with O_2 to form H_2O, the enthalpy and entropy changes are also in opposition. In this case, the exothermic nature of the reaction is sufficient to overcome the negative value of ΔS and cause the reaction to be spontaneous.

It's easy to see why stone walls need almost constant attention.

When enthalpy and entropy oppose one another, their relative importance in determining spontaneity is far from obvious. In addition, temperature is a third factor that influences the direction of a spontaneous change. We have used the melting of ice or freezing of water to illustrate some points about enthalpy, entropy, and spontaneity. But we know that even a slight change in the temperature of an ice-water slush in equilibrium at 0 °C will cause the system to change to all liquid (if raised above 0 °C) or all solid (if lowered below 0 °C). Thus, we want to know how three factors — enthalpy, entropy, and temperature — interplay in determining spontaneity.

Specialists in thermodynamics, in studying the Second Law, defined a quantity called the Gibbs free energy, *G*, named to honor one of America's most important scientists, Josiah Willard Gibbs (1839–1903). (It's called *free energy* because it is related,

SPECIAL TOPIC 16.3 SYNTHETIC DIAMONDS

Ever since it was discovered that diamonds could be cut and polished to sparkle brightly in the light, they have captured the fascination of countless numbers of humans. They have caught the eyes of kings and queens, and have been celebrated in song. In the modern world, diamonds of lesser quality are valuable industrially because of their hardness. Diamond-studded cutting and grinding tools are used extensively by industries of all sorts. It's not difficult to imagine, therefore, the excitement of scientists when it was discovered in 1797 that diamonds are simply pure carbon. Could not the common graphite form of carbon be made into its more valuable cousin?

As you might expect, many attempted to make diamond from graphite. But try as they might, no one succeeded. In 1938, a careful thermodynamic analysis of the problem was performed, and the results of it are summarized in the accompanying figure. Here we have a graph of $\Delta G/T$ versus the absolute temperature for the process

$$C(s,\text{graphite}) \longrightarrow C(s,\text{diamond})$$

For this change to occur spontaneously, ΔG must be negative. Since T is always positive, then $\Delta G/T$ must be negative. On the graph, the regions where $\Delta G/T$ is negative are shaded.

Each line on the graph shows how $\Delta G/T$ varies with temperature for a given pressure. Thus, the line labeled 1 atm shows that at a pressure of 1 atm, $\Delta G/T$ becomes less positive as the temperature increases. However, notice that $\Delta G/T$ never becomes negative at 1 atm. Since ΔG (and therefore $\Delta G/T$) must be negative for the conversion to diamond to be spontaneous, it is thermodynamically impossible to convert graphite to diamond at atmospheric pressure, no matter how hard we try.

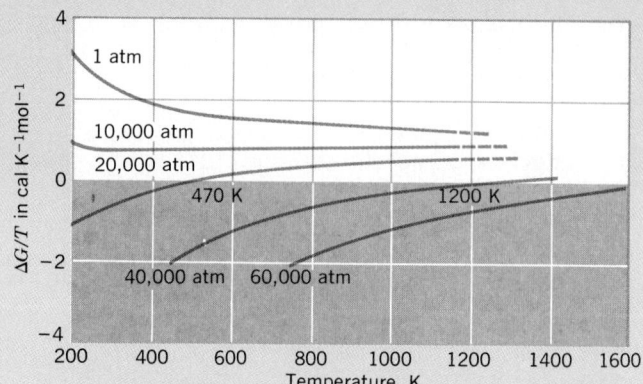

From *Chemical and Engineering News*, April 5, 1971, p. 51. Used by permission.

From the graph, we can also see that $\Delta G/T$ becomes less positive (more negative) with increasing pressure. At 20,000 atm, it is negative at temperatures below 470 K (approximately 200 °C), and at higher pressures $\Delta G/T$ becomes negative at still higher temperatures. From this graph, therefore, scientists were able to establish the conditions of temperature and pressure at which the conversion from graphite to diamond is at least *theoretically* possible. This was not the end of the problem, however, because suitable materials still had to be found that allowed the transition to occur at a reasonable speed. Success was elusive, but finally on February 16, 1955, the General Electric Company announced that their scientists had accomplished the task in their laboratories in Schenectady, New York. They had used pressures in excess of 100,000 atm and temperatures above 2800 °C. The diamonds were of industrial quality, and before long, General Electric was manufacturing over 1,000,000 carats of industrial diamonds annually.

The origin of *free* in *free energy* is discussed in Section 16.9.

as we will see later, to the maximum energy in a change that is "free" or "available" to do useful work.) The Gibbs free energy is defined as a function of enthalpy, H, entropy, S, and temperature, T, in the following way.

$$G = H - TS \qquad (16.5)$$

Because we cannot know H in an absolute sense, we cannot know G in an absolute sense either. But, as we've often commented, what affects us and what we need most to know about are *changes*, not absolute values. So we will consider a *change* in the Gibbs free energy, ΔG, and we will deal with systems at constant temperature and constant pressure. With these conditions in mind, Equation 16.5 becomes

$$\Delta G = \Delta H - T\Delta S \qquad (16.6)$$

Once again we have a state function, a quantity that is independent of the path of the event. This means that

$$\Delta G = G_{\text{final}} - G_{\text{initial}}$$

What is of particular importance to us is the fact that *a change can only be spontaneous if it is accompanied by a decrease in free energy*. In other words, for a change to be spontaneous, G_{final} must be less than $G_{initial}$ and ΔG must be negative. What does this mean in terms of the signs of ΔH and ΔS?

When a change is exothermic and is also accompanied by an increase in entropy, both factors favor spontaneity.

$$\Delta H \text{ is negative } (-)$$
$$\Delta S \text{ is positive } (+)$$
$$\Delta G = \Delta H - T\Delta S$$
$$= (-) - T(+)$$

In such a change, ΔG will be negative regardless of the value of the absolute temperature, T (which can only have positive values). Therefore, the change will occur spontaneously at *all* temperatures.

On the other hand, if a change is endothermic and is accompanied by a decrease in entropy, both factors work against spontaneity.

$$\Delta H \text{ is positive } (+)$$
$$\Delta S \text{ is negative } (-)$$
$$\Delta G = \Delta H - T\Delta S$$
$$= (+) - T(-)$$

In this case, ΔG will be positive at all temperatures and the change will always be nonspontaneous.

When ΔH and ΔS have the same sign, the temperature becomes critical in determining whether or not an event is spontaneous. If ΔH and ΔS are both positive,

$$\Delta G = (+) - T(+)$$

Only at relatively high temperatures will the value of $T\Delta S$ be larger than the value of ΔH so that their difference, ΔG, is negative. A familiar example is the melting of ice.

$$H_2O(s) \longrightarrow H_2O(\ell)$$

Here is a change that we know is endothermic and occurs with an increase in entropy. At temperatures above 0 °C (when the pressure is 1 atm), ice melts because the $T\Delta S$ term is bigger than the ΔH term. At lower temperatures, ice doesn't melt because the smaller value of T gives a smaller value for $T\Delta S$ and the difference, $\Delta H - T\Delta S$, is positive.

For similar reasons, when ΔH and ΔS are both negative, ΔG will be negative only at relatively low temperatures. The freezing of water is an example.

$$H_2O(\ell) \longrightarrow H_2O(s)$$

Energy is released as the solid is formed and the entropy decreases. You know, of course, that water freezes spontaneously at low temperatures, that is, below 0 °C.

Figure 16.8 summarizes the effects of the signs of ΔH and ΔS on ΔG, and hence on the spontaneity of a physical or chemical event.

Reactions that occur with a free energy decrease in the system are sometimes said to be **exergonic.** Those that occur with a free energy increase are sometimes said to be **endergonic.**

	ΔH	
	+	**−**
+	Spontaneous only at high T	Spontaneous at all T
−	Non-spontaneous at all T	Spontaneous only at low T

FIGURE 16.8 Summary of the effects of the signs of ΔH and ΔS on spontaneity.

16.8 STANDARD FREE ENERGIES

Standard free energies of formation can be used to obtain standard free energies of reaction.

When ΔG is determined at 25 °C (298 K) and 1 atm, we call it the standard free energy change, $\Delta G°$. There are a number of ways of obtaining $\Delta G°$ for a reaction. One of them is to compute $\Delta G°$ from $\Delta H°$ and $\Delta S°$.

$$\Delta G° = \Delta H° - (298 \text{ K}) \Delta S°$$

Experimental measurement of $\Delta G°$ is also possible, but we will discuss how this is done at a later time.

EXAMPLE 16.4   CALCULATING $\Delta G°$ FROM $\Delta H°$ AND $\Delta S°$

Problem: Compute $\Delta G°$ for the hydrolysis of urea, $CO(NH_2)_2$,

$$CO(NH_2)_2(aq) + H_2O(\ell) \longrightarrow CO_2(g) + 2NH_3(g)$$

Solution: We can calculate $\Delta H°$ from standard heats of formation found in Table 5.2. From Hess's law we can write

$$\Delta H° = [\Delta H_f°(CO_2) + 2\Delta H_f°(NH_3)] - [\Delta H_f°(CO(NH_2)_2) + \Delta H_f°(H_2O)]$$

$$= \left[1 \text{ mol} \times \left(\frac{-393.5 \text{ kJ}}{\text{mol}} \right) + 2 \text{ mol} \times \left(\frac{-46.19 \text{ kJ}}{\text{mol}} \right) \right]$$

$$- \left[1 \text{ mol} \times \left(\frac{-319.2 \text{ kJ}}{\text{mol}} \right) + 1 \text{ mol} \times \left(\frac{-285.9 \text{ kJ}}{\text{mol}} \right) \right]$$

$$= (-485.9 \text{ kJ}) - (-605.1 \text{ kJ})$$

$$= +119.2 \text{ kJ}$$

In Example 16.3, we calculated $\Delta S°$ for this reaction to be $+354.8$ J/K. Now, using the equation for the standard free energy change, we can compute $\Delta G°$. Note that we must be careful to express $\Delta H°$ and $\Delta S°$ in the same energy units.

$$\Delta G° = +119.2 \text{ kJ} - (298.2 \text{ K})(0.3548 \text{ kJ/K})$$

$$= +119.2 \text{ kJ} - 105.8 \text{ kJ} = +13.4 \text{ kJ}$$

To be exact,

$$t_K = t_C + 273.15$$

Note that we have expressed the temperature to four significant figures to conform with the number of significant figures in the data (25.0 + 273.2 = 298.2).

PRACTICE EXERCISE 4   Use the data in Tables 5.2 and 16.1 to calculate $\Delta G°$ for the formation of iron oxide (hematite), present in rust.

$$4Fe(s) + 3O_2(g) \longrightarrow 2Fe_2O_3(s)$$

In Section 5.7 we found it useful to tabulate standard heats of formation, $\Delta H_f°$, because we could use them following Hess's law to calculate $\Delta H°$ for many different reactions. **Standard free energies of formation, $\Delta G_f°$, can be used in similar calculations to obtain $\Delta G°$.**

$$\boxed{\Delta G° = (\text{sum of } \Delta G_f° \text{ of products}) - (\text{sum of } \Delta G_f° \text{ of reactants}) \quad (16.7)}$$

The $\Delta G_f°$ values for some typical substances are found in Table 16.2. Example 16.5 shows how we can use them to calculate $\Delta G°$ for a reaction.

TABLE 16.2 Standard Free Energies of Formation of Typical Substances

Substance	ΔG_f° kJ/mol	kcal/mol	Substance	ΔG_f° kJ/mol	kcal/mol
Ag(s)	0.00	0.00	$H_2O(g)$	−228.6	−54.64
AgCl(s)	−109.7	−26.22	$H_2O(\ell)$	−237.2	−56.69
Al(s)	0.00	0.00	HCl(g)	−95.27	−22.77
$Al_2O_3(s)$	−1576.4	−376.77	$HNO_3(\ell)$	−79.91	−19.10
C(s, graphite)	0.00	0.00	$H_2SO_4(\ell)$	−689.9	−164.9
CO(g)	−137.3	−32.81	$HC_2H_3O_2(\ell)$	−392.5	−93.8
$CO_2(g)$	−394.4	−94.26	$Hg(\ell)$	0.00	0.00
$CH_4(g)$	−50.79	−12.14	Hg(g)	+31.8	+7.59
$CH_3Cl(g)$	−58.6	−14.0	K(s)	0.00	0.00
$CH_3OH(\ell)$	−166.2	−39.73	KCl(s)	−408.3	−97.59
$CO(NH_2)_2(s)$	−197.2	−47.12	$K_2SO_4(s)$	−1316.4	−314.62
$CO(NH_2)_2(aq)$	−203.8	−48.72	$N_2(g)$	0.00	0.00
$C_2H_2(g)$	+209	+50.0	$NH_3(g)$	−16.7	−3.98
$C_2H_4(g)$	+68.12	+16.28	$NH_4Cl(s)$	−203.9	−48.73
$C_2H_6(g)$	−32.9	−7.86	NO(g)	+86.69	+20.72
$C_2H_5OH(\ell)$	−174.8	−41.77	$NO_2(g)$	+51.84	+12.39
$C_8H_{18}(\ell)$	+17.3	+4.14	$N_2O(g)$	+103.6	+24.76
Ca(s)	0.00	0.00	$N_2O_4(g)$	+98.28	+23.49
$CaCO_3(s)$	−1128.8	−269.78	Na(s)	0.00	0.00
$CaCl_2(s)$	−750.2	−179.3	$Na_2CO_3(s)$	−1048	−250.4
CaO(s)	−604.2	−144.4	$NaHCO_3(s)$	−851.9	−203.6
$Ca(OH)_2(s)$	−896.76	−214.33	NaCl(s)	−384.0	−91.79
$CaSO_4(s)$	−1320.3	−315.56	NaOH(s)	−382	−91.4
$CaSO_4 \cdot \frac{1}{2}H_2O(s)$	−1435.2	−343.02	$Na_2SO_4(s)$	−1266.8	−302.78
$CaSO_4 \cdot 2H_2O(s)$	−1795.7	−429.19	$O_2(g)$	0.00	0.00
$Cl_2(g)$	0.00	0.00	PbO(s)	−189.3	−45.25
Fe(s)	0.00	0.00	S(s)	0.00	0.00
$Fe_2O_3(s)$	−741.0	−177.1	$SO_2(g)$	−300.4	−71.79
$H_2(g)$	0.00	0.00	$SO_3(g)$	−370.4	−88.52

EXAMPLE 16.5   CALCULATING ΔG° FROM ΔG_f°

Problem: What is ΔG° for the combustion of ethyl alcohol (C_2H_5OH) to give $CO_2(g)$ and $H_2O(g)$?

Solution: First we need a balanced chemical equation.

$$C_2H_5OH(\ell) + 3O_2(g) \longrightarrow 2CO_2(g) + 3H_2O(g)$$

Using Equation 16.7,

$$\Delta G^\circ = [2\Delta G_f^\circ(CO_2) + 3\Delta G_f^\circ(H_2O)] - [\Delta G_f^\circ(C_2H_5OH) + 3\Delta G_f^\circ(O_2)]$$

As with ΔH_f°, $\Delta G_f^\circ = 0$ for an element in its normal state at 25 °C and 1 atm. Therefore,

$$\Delta G^\circ = [2 \text{ mol}(-394.4 \text{ kJ/mol}) + 3 \text{ mol}(-228.6 \text{ kJ/mol})]$$
$$-[1 \text{ mol}(-174.8 \text{ kJ/mol}) + 3 \text{ mol}(0 \text{ kJ/mol})]$$

$$\Delta G^\circ = (-1474.6 \text{ kJ}) - (-174.8 \text{ kJ})$$
$$= -1299.8 \text{ kJ}$$

PRACTICE EXERCISE 5   Calculate ΔG° in kJ for the following reactions, using the data in Table 16.2.
(a) $2NO(g) + O_2(g) \longrightarrow 2NO_2(g)$
(b) $Ca(OH)_2(s) + 2HCl(g) \longrightarrow CaCl_2(s) + 2H_2O(g)$

16.9 FREE ENERGY AND MAXIMUM WORK

ΔG is a measure of the maximum work that can be obtained from a reaction.

One of the most important uses to which we put spontaneous chemical reactions is the production of useful work. Fuels are burned in gasoline or diesel engines to power automobiles and heavy machinery. Chemical reactions in batteries start our autos and run all sorts of modern electronic gadgets, including your pocket calculator. And chemical reactions within our bodies pump blood, move muscles, transport nerve impulses, and do those other things that keep us alive and make everything else worthwhile.

When chemical reactions occur, however, their energy is not always harnessed to do work. For instance, if gasoline is burned in an open dish, the energy evolved is lost entirely as heat and no useful work is accomplished. Engineers, therefore, seek ways of capturing as much energy as possible in the form of work. One of their primary goals is to maximize the efficiency with which chemical energy is converted to work and to minimize the amount of energy transferred unproductively to the environment and lost as heat.

Scientists have discovered that the maximum conversion of chemical energy to work occurs if a reaction is carried out under conditions that are said to be reversible. A process is defined as reversible if its driving force is opposed by another force that is just a tiny bit weaker. An example of an almost-reversible process would be the use of a 12.00-V battery to charge an 11.99-V battery. The electricity from the 12.00-V battery is opposed by a "force" of 11.99 V.

Unfortunately, along with the "good news" that we get maximum work if we carry out a reaction reversibly, there is the "bad news" that a truly reversible process proceeds at an extremely slow speed. Even though we get the maximum amount of work, we get it so slowly that it is of no use to us. The goal, then, is to approach reversibility for maximum efficiency, but to carry out the reaction at a pace that will deliver work at acceptable rates.

The relationship of useful work to reversibility was illustrated earlier (Section 16.2) in our discussion of the discharge of an automobile battery. Recall that when the battery is shorted with the heavy wrench, no work is done and all the energy appears as heat. In this case there is nothing opposing the discharge, and it occurs in a most thermodynamically irreversible manner. However, when the current is passed through a small electric motor, the motor itself offers resistance to the passage of the electricity and the discharge takes place slowly. In this instance, the discharge occurs in a more nearly reversible manner because of the opposition provided by the motor, and a large amount of the available energy appears in the form of the work that is accomplished by the motor.

The preceding discussion leads naturally to the question, Is there a limit to the amount of the available energy in a reaction that can be harnessed as useful work? The answer to this question is to be found in the Gibbs free energy. The maximum amount of energy produced by a reaction that can be theoretically harnessed as work is equal to ΔG. This is the energy that need not be lost as heat and is therefore *free* to be used for work. Thus, by determining the value of ΔG, we can find out whether or not a given reaction will be an effective source of energy. Also, by comparing the actual amount of work derived from a given system with the ΔG values for the reactions involved, we can measure the efficiency of the system.

A reversible process requires an infinite number of tiny steps. This takes forever to accomplish.

EXAMPLE 16.6   CALCULATING MAXIMUM WORK

Problem: Calculate the maximum work available, expressed in kilojoules, from the oxidation of 1 mol of octane, $C_8H_{18}(\ell)$, by oxygen to give $CO_2(g)$ and $H_2O(\ell)$ at 25 °C and 1 atm.

Solution: First we need a chemical equation.

$$C_8H_{18}(\ell) + 12\tfrac{1}{2}O_2(g) \longrightarrow 8CO_2(g) + 9H_2O(\ell)$$

Since the free energy change is equal to the maximum work available, we must then calculate $\Delta G°$. (The reaction described here occurs under those conditions at which standard free energies are determined—how fortunate!)

$$\Delta G° = [8\Delta G_f°(CO_2) + 9\Delta G_f°(H_2O)] - [\Delta G_f°(C_8H_{18}) + 12.5\Delta G_f°(O_2)]$$

Referring to Table 16.2, and using values in kJ/mol,

$$\Delta G° = [8(-394.4) + 9(-237.2)] - [1(+17.3) + 12.5(0.0)]$$
$$\Delta G° = (-5290) - (+17.3)$$
$$= -5307 \text{ kJ}$$

Thus, at 25 °C and 1 atm, we can expect no more than 5307 kJ of work from the oxidation of one mole of C_8H_{18}.

PRACTICE EXERCISE 6   Calculate the maximum work that could be obtained at 25 °C and 1 atm from the oxidation of 1.00 mol of Al by O_2 to give Al_2O_3.

16.10 FREE ENERGY AND EQUILIBRIUM

Equilibrium exists when $\Delta G = 0$.

We have seen that when the value of ΔG for a given change is negative, the change will occur spontaneously. We have also seen that a change will be nonspontaneous when ΔG is positive. However, when ΔG is neither positive nor negative, the change will be neither spontaneous nor nonspontaneous—the system is in a state of equilibrium. This occurs when ΔG is equal to zero.

Let's again consider the freezing of water.

$$H_2O(\ell) \longrightarrow H_2O(s)$$

Because ΔS and ΔH for this "reaction" are both negative, ΔG is negative only at low temperatures (below 0 °C). Therefore, the reaction is spontaneous only at low temperatures. This agrees with our experience—we know that water freezes when it becomes very cold. At high temperatures, ΔG is positive, and the freezing is nonspontaneous. If ΔG is positive in the forward direction, it must be negative in the reverse direction. Therefore, at high temperatures (above 0 °C) it is melting that occurs spontaneously. Again this agrees with our experience.

Between high and low temperatures (at 0 °C), ΔG is equal to zero and an ice-water mixture exists in a condition of equilibrium. Neither freezing nor melting is spontaneous. As long as heat isn't added or removed from the system, the ice and liquid water can exist together indefinitely.

What does equilibrium mean as far as work is concerned? We have identified ΔG as a quantity that specifies the amount of work that is available from a system. Since at equilibrium, ΔG is zero, the amount of work available is zero also. Therefore, *when a system is at equilibrium, no work can be extracted from that system.* For an example, let's examine once more the common lead storage battery that we use to start our car.

When the battery is fully charged, there are virtually no products of the discharge reaction present. The chemical reactants, however, are present in large amounts. Therefore, the total free energy of the reactants far exceeds the total free energy of products and, since $\Delta G = G_{products} - G_{reactants}$, the ΔG of the system has a large negative value.

This means that a lot of work is available. As the battery discharges, the reactants are converted to products and $G_{products}$ gets larger while $G_{reactants}$ gets smaller; thus ΔG becomes less negative, and less work is available. Finally, the battery reaches equilibrium. The total free energies of the reactants and the products have become equal, and $\Delta G = 0$. No further work can be extracted; the battery is dead.

Equilibrium in Phase Changes

When we have equilibrium in any system, we know that $\Delta G = 0$. For a phase change such as $H_2O(\ell) \rightarrow H_2O(s)$, equilibrium can only exist at one particular temperature at atmospheric pressure. In this instance, that temperature is 0 °C. Above 0 °C, only liquid water can exist, and below 0 °C all the liquid will freeze to give ice. This gives us an interesting relationship between ΔH and ΔS for a phase change. Since $\Delta G = 0$,

$$\Delta G = 0 = \Delta H - T\Delta S$$

Therefore,

$$\Delta H = T\Delta S$$

and

$$\Delta S = \frac{\Delta H}{T} \tag{16.8}$$

Thus, if we know ΔH for the phase change and the temperature at which the two phases coexist, we can calculate ΔS for the phase change.

Another interesting relationship that we can obtain is

$$T = \frac{\Delta H}{\Delta S} \tag{16.9}$$

Thus, if we know ΔH and ΔS, we can calculate the temperature at which equilibrium will occur.

EXAMPLE 16.7   CALCULATING THE EQUILIBRIUM TEMPERATURE FOR A PHASE CHANGE

Problem: For the "reaction" $Br_2(\ell) \rightarrow Br_2(g)$, $\Delta H° = +31.0$ kJ/mol and $\Delta S° = 92.9$ J/mol K. Assuming that ΔH and ΔS are nearly temperature independent, calculate the approximate temperature at which $Br_2(\ell)$ will be in equilibrium with $Br_2(g)$ at 1 atm.

Solution: The temperature at which equilibrum exists is given by Equation 16.9,

$$T = \frac{\Delta H}{\Delta S}$$

If ΔH and ΔS do not depend on temperature,

$$T = \frac{\Delta H°}{\Delta S°}$$

Substituting the data given in the problem,

$$T = \frac{3.10 \times 10^4 \text{ J mol}^{-1}}{92.9 \text{ J mol}^{-1} \text{ K}^{-1}}$$

$$= 334 \text{ K}$$

The Celsius temperature is $334 - 273 = 61$ °C.

Notice that we were careful to express $\Delta H°$ in joules, not kilojoules, so that the units would cancel properly.

PRACTICE EXERCISE 7   The heat of vaporization of mercury is 14.5 kcal/mol. For Hg(ℓ), $S° = 18.2$ cal/mol K, and for Hg(g), $S° = 41.8$ cal/mol K. Calculate the temperature at which there is equilibrium between mercury vapor and liquid at a pressure of 1 atm.

16.11   PREDICTING THE OUTCOME OF CHEMICAL REACTIONS

$\Delta G°$ points the way in chemical reactions.

In a phase change such as $H_2O(\ell) \rightarrow H_2O(s)$, equilibrium can exist for a given pressure only at one particular temperature; for water at a pressure of 1 atm this temperature is 0 °C. At other temperatures, the "reaction" proceeds *entirely* to completion in one direction or another.

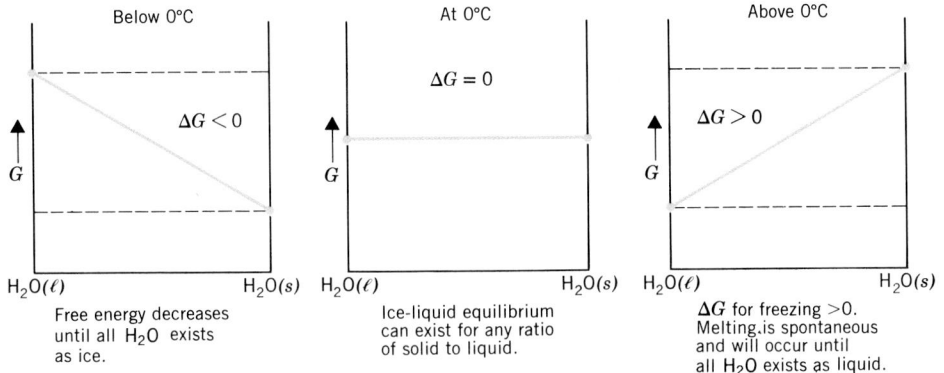

Below 0°C

$\Delta G < 0$

G

$H_2O(\ell)$ $H_2O(s)$

Free energy decreases until all H_2O exists as ice.

At 0°C

$\Delta G = 0$

G

$H_2O(\ell)$ $H_2O(s)$

Ice-liquid equilibrium can exist for any ratio of solid to liquid.

Above 0°C

$\Delta G > 0$

G

$H_2O(\ell)$ $H_2O(s)$

ΔG for freezing >0. Melting is spontaneous and will occur until all H_2O exists as liquid.

FIGURE 16.9 Free energy diagram for conversion of $H_2O(\ell)$ to $H_2O(s)$. At the left of each diagram, the system consists entirely of $H_2O(\ell)$. At the right is $H_2O(s)$. The horizontal axis represents the extent of conversion from $H_2O(\ell)$ to $H_2O(s)$.

Figure 16.9 illustrates how the free energy changes when $H_2O(\ell)$ becomes $H_2O(s)$ at different temperatures. Below 0 °C, the free energy decreases continually until *all* the liquid has frozen. Above 0 °C, the free energy decreases in the direction of $H_2O(s) \rightarrow H_2O(\ell)$ and continues to drop until all the solid has melted. Above or below 0 °C the system is unable to establish an equilibrium mixture of liquid and solid. However, at 0 °C there would be no change in free energy if either melting or freezing occurred; therefore, there is no driving force for either change. As long as the system is insulated from warmer or colder surroundings, any particular mixture of ice and water is stable and a state of equilibrium exists.

The free energy changes that occur in most chemical reactions are more complex than those in phase changes. Figure 16.10 shows the variation in free energy for the

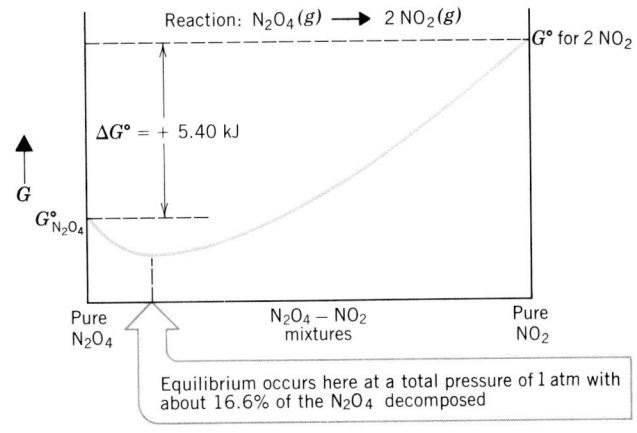

Reaction: $N_2O_4(g) \longrightarrow 2\,NO_2(g)$

$G°$ for 2 NO_2

$\Delta G° = +5.40$ kJ

G

$G°_{N_2O_4}$

Pure N_2O_4 $N_2O_4 - NO_2$ mixtures   Pure NO_2

Equilibrium occurs here at a total pressure of 1 atm with about 16.6% of the N_2O_4 decomposed

FIGURE 16.10 Free energy diagram for the decomposition of $N_2O_4(g)$. The minimum on the curve indicates the composition of the reaction mixture at equilibrium.

FIGURE 16.11 Free energy curve for a reaction having a negative $\Delta G°$. Since $G_B° < G_A°$, $\Delta G° < 0$.

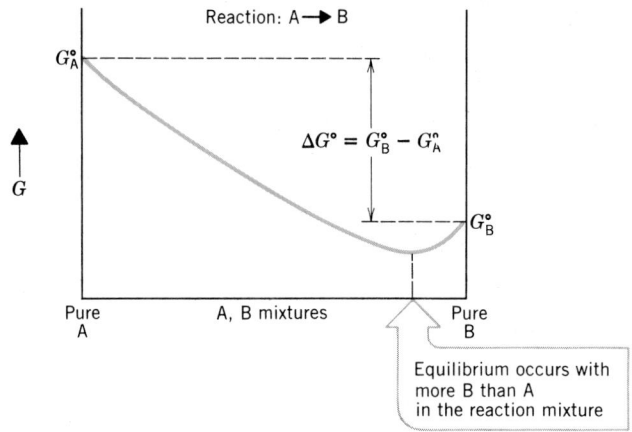

decomposition of N_2O_4, a rocket fuel, into NO_2, an air pollutant.

$$N_2O_4(g) \longrightarrow 2NO_2(g)$$

Notice that in going from reactant to product, the free energy of the reaction mixture drops *below* that of either pure N_2O_4 or pure NO_2.

Any system will spontaneously seek the lowest point on its free energy curve. If we begin with pure $N_2O_4(g)$, some $NO_2(g)$ will be formed, because proceeding in the direction of NO_2 leads to a lowering of the free energy. If we begin with pure $NO_2(g)$, a change also will occur. Going downhill on the free energy curve now takes place as the reverse reaction occurs [i.e., $2NO_2(g) \rightarrow N_2O_4(g)$]. Once the bottom of the "valley" is reached, the system has come to equilibrium. The composition of the mixture of N_2O_4 and NO_2 will remain constant because any change would require an uphill climb. Free energy increases are not spontaneous, so this doesn't happen.

Another thing to notice in Figure 16.10 is that some reaction takes place in the forward direction *even though $\Delta G°$ is positive*. For comparison, Figure 16.11 shows the shape of the free energy curve for a reaction having a negative $\Delta G°$. We see here that at equilibrium there has been a much greater conversion of reactants to products.

In general, the value of $\Delta G°$ for most reactions is much larger numerically than the $\Delta G°$ for the $N_2O_4 \rightarrow 2NO_2$ reaction. The extent to which a reaction proceeds is *very* sensitive to the size of $\Delta G°$. If the $\Delta G°$ value for a reaction is reasonably large—about 20 kJ or more—almost no *observable* reaction will occur when $\Delta G°$ is positive. However, the reaction will go almost to completion if $\Delta G°$ is both large and negative.[1] From a practical standpoint, then, the size and sign of $\Delta G°$ serve as an indicator of whether an observable spontaneous reaction will occur. In Section 16.12 we will see how $\Delta G°$ can be used to estimate quantitatively the extent to which a reaction will have proceeded toward completion by the time it reaches equilibrium.

EXAMPLE 16.8   USING $\Delta G°$ AS A PREDICTOR OF THE OUTCOME OF A REACTION

Problem: Would we expect to be able to observe the following reaction at 25 °C?

$$NH_4Cl(s) \longrightarrow NH_3(g) + HCl(g)$$

---

[1] To actually see a change take place, the speed of a spontaneous reaction must be reasonably fast. For example, the decomposition of the nitrogen oxides into N_2 and O_2 is thermodynamically spontaneous ($\Delta G°$ is negative) but their rates of decomposition are so slow that these substances appear to be stable and are obnoxious air pollutants.

Solution: First let's calculate $\Delta G°$ for the reaction using the data in Table 16.2. The procedure is the same as that discussed earlier.

$$\Delta G° = [\Delta G_f°(NH_3) + \Delta G_f°(HCl)] - [\Delta G_f°(NH_4Cl)]$$
$$= [(-16.7 \text{ kJ}) + (-95.27 \text{ kJ})] - (-203.9 \text{ kJ})$$
$$= +91.9 \text{ kJ}$$

Because $\Delta G°$ is large and positive, we shouldn't expect to observe the spontaneous formation of any products at this temperature.

PRACTICE EXERCISE 8   Use the data in Table 16.2 to determine whether the reaction $SO_2(g) + \frac{1}{2}O_2(g) \rightarrow SO_3(g)$ should "occur spontaneously" at 25 °C.

PRACTICE EXERCISE 9   Use the data in Table 16.2 to determine whether we should expect to see the formation of $CaCO_3(s)$ in the reaction $CaCl_2(s) + H_2O(g) + CO_2(g) \rightarrow CaCO_3(s) + 2HCl(g)$ at 25 °C.

Equilibrium and Temperature

So far, we have confined our discussion of the relationship of free energy and equilibrium to a special case, 25 °C. But what about other temperatures? Equilibria certainly can exist under conditions other than 25 °C, and it would be valuable to know how temperature affects the position of equilibrium.

At 25 °C, we've seen that the position of equilibrium is determined by the difference between the free energy of pure products and the free energy of pure reactants. This difference is given by $\Delta G°$, which is defined as

$$\Delta G° = G_{products}° - G_{reactants}°$$

At temperatures other than 25 °C, it is still the difference between the free energies of the products and reactants that determines the position of equilibrium, but we won't use the superscript, °, to denote this ΔG because we've reserved this superscript for standard conditions (25 °C and 1 atm). To avoid confusion, therefore, we will use the symbol $\Delta G'$ to stand for the equivalent of $\Delta G°$, but at a temperature other than 25 °C. Thus,

$$\Delta G' = G_{products}' - G_{reactants}'$$

where $G_{products}'$ and $G_{reactants}'$ are the total free energies of the pure products and pure reactants, respectively, at this other temperature.

Next, we must find a way to compute $\Delta G'$. Earlier we saw that $\Delta G°$ can be obtained from the equation

$$\Delta G° = \Delta H° - (298 \text{ K})\Delta S°$$

Of course, we could also write this as

$$\Delta G° = \Delta H° - T\Delta S°$$

and specify that the equation applies to the specific case where T is 298 K. However, if T has another value, we can express the equation as

$$\Delta G' = \Delta H' - T\Delta S'$$

Now we seem to be getting closer to our goal. If we can compute or estimate the values of $\Delta H'$ and $\Delta S'$ for a reaction, we have solved our problem.

The magnitudes of ΔH and ΔS are relatively insensitive to temperature changes. However, ΔG is very temperature sensitive because $\Delta G = \Delta H - T\Delta S$.

The size of $\Delta G'$ obviously depends very strongly on the temperature — the equation above has temperature as one of its variables. However, as mentioned in the last section, the magnitudes of the ΔH and ΔS for a reaction are relatively insensitive to temperature changes. As a result, we can use $\Delta H°$ and $\Delta S°$ as reasonable approximations[2] of $\Delta H'$ and $\Delta S'$. This allows us to rewrite the equation for $\Delta G'$ as

$$\Delta G' = \Delta H° - T\Delta S°$$

The following examples illustrate how this equation is useful.

EXAMPLE 16.9   CALCULATING $\Delta G'$ FOR A REACTION

Problem: Earlier we saw that the value of $\Delta G°$ for the reaction

$$N_2O_4(g) \longrightarrow 2NO_2(g)$$

has a value of $+5.40$ kJ. What is the value of $\Delta G'$ for this reaction at 100 °C?

Solution: To make use of the equation $\Delta G' = \Delta H° - T\Delta S°$, we need values of $\Delta H°$ and $\Delta S°$. The $\Delta H°$ for the reaction can be calculated from the data in Table 5.2 on p. 169. Here we find the following standard heats of formation.

$$N_2O_4(g) \quad \Delta H_f° = +9.67 \text{ kJ/mol}$$
$$NO_2(g) \quad \Delta H_f° = +33.8 \text{ kJ/mol}$$

We combine these by a Hess's law calculation to compute $\Delta H°$ for the reaction.

$$\Delta H° = [2\Delta H_f°(NO_2)] - [\Delta H_f°(N_2O_4)]$$
$$= \left[2 \text{ mol}\left(33.8 \frac{\text{kJ}}{\text{mol}}\right)\right] - \left[1 \text{ mol}\left(9.67 \frac{\text{kJ}}{\text{mol}}\right)\right]$$
$$= +57.9 \text{ kJ}$$

Next, we compute $\Delta S°$ for the reaction using data from Table 16.1.

$$N_2O_4(g) \quad S° = 304 \text{ J/mol K}$$
$$NO_2(g) \quad S° = 240.5 \text{ J/mol K}$$

This is also a Hess's law kind of calculation.

$$\Delta S° = \left[2 \text{ mol}\left(240.5 \frac{\text{J}}{\text{mol K}}\right)\right] - \left[1 \text{ mol}\left(304 \frac{\text{J}}{\text{mol K}}\right)\right]$$
$$= +177 \text{ J/K} = 0.177 \text{ kJ/K}$$

Now we can substitute into the equation for $\Delta G'$, using $T = 373$ K.

$$\Delta G' = (+57.9 \text{ kJ}) - (373 \text{ K})(0.177 \text{ kJ/K})$$
$$= -8.1 \text{ kJ}$$

Notice that at this higher temperature, the sign of $\Delta G'$ has become negative.

---

[2] Consider what happens to the enthalpy if we increase the temperature of a reaction. As the temperature rises, we increase the average kinetic energy of the particles of the system. This means that the enthalpy increases for both the reactants and the products. If we examine the initial and final states (i.e., the reactants and products), both at the *same* higher temperature, we find that each has had its enthalpy increased by nearly the same amount. The *difference* in their enthalpies, however, has not been affected very much. Thus, at a temperature other than 25 °C, the size of ΔH for a reaction is not much different than $\Delta H°$. As a reasonable approximation, therefore, we can assume that the $\Delta H'$ for a reaction is the same as $\Delta H°$. Following similar arguments, we could conclude that a reasonable approximation of $\Delta S'$ is $\Delta S°$. There are methods that can be used to compute fairly accurately $\Delta H'$ and $\Delta S'$ from $\Delta H°$ and $\Delta S°$, but they are beyond the scope of this book.

EXAMPLE 16.10 THE EFFECT OF TEMPERATURE ON EQUILIBRIUM

Problem: Using the results of Example 16.9, sketch a free energy diagram for the decomposition of N_2O_4 at 100 °C. For this reaction, how does the position of equilibrium change as the temperature is increased?

Solution: Since $\Delta G'$ is negative for the reaction at 100 °C, the G' of the reactants must be higher than the G' of the products. When we draw the free energy curve, the minimum therefore lies closer to the products than to the reactants, and the diagram has the following appearance.

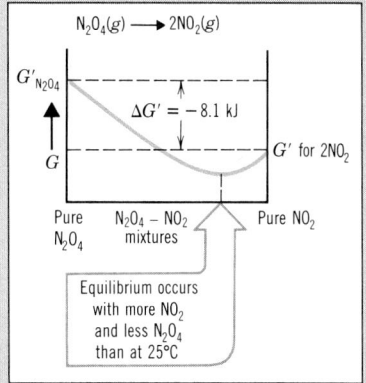

Comparing this diagram to the one in Figure 16.10, we see that at the higher temperature the minimum has moved toward the products. This means that at the higher temperature the reaction proceeds farther toward completion by the time equilibrium has been reached.

PRACTICE EXERCISE 10   Determine $\Delta G°$ for the reaction

$$2NaHCO_3(s) \longrightarrow Na_2CO_3(s) + CO_2(g) + H_2O(g)$$

using the data in Table 16.2. Then calculate $\Delta G'$ for the reaction at 200 °C using the data in Tables 5.2 and 16.1. How does the position of equilibrium change for this reaction as the temperature is increased?

16.12 CALCULATING EQUILIBRIUM CONSTANTS FROM THERMODYNAMIC DATA

The equilibrium constant for a reaction can be calculated from $\Delta G°$.

In the preceding section you saw that the position of equilibrium in a chemical reaction is determined by the value of $\Delta G°$ (or $\Delta G'$, if at a temperature other than 25 °C). You also learned in Chapter 15 that the larger the value of either K_c or K_p for a reaction, the farther the reaction will proceed toward completion by the time equilibrium is reached. Obviously, there is a relationship between $\Delta G°$ and the equilibrium constant. This relationship is simple enough, but it involves logarithms, as you can see by glancing at the equations below. If you still feel a bit uncertain about logarithms, even after using them in pH calculations, you are probably not alone; many of your classmates do, too. To help you, logarithms and their use are discussed in Appendix A, located at the back of the book.

Although we will not attempt to prove it, the following equation can be derived from the principles of thermodynamics:

If you are using a scientific calculator, you will probably find it best to work with natural logs.

$$\Delta G = -RT \ln K \qquad (16.10)$$

In this equation, R is the gas constant expressed in energy units ($R = 8.314$ J mol^{-1} K^{-1} or 1.987 cal mol^{-1} K^{-1}) and T is the absolute temperature in kelvins. The symbol "ln" in the last term means we take the *natural logarithm* of the equilibrium constant. Natural logarithms are logarithms to the base $e = 2.71828. \ldots$. In Equation 16.10, "ln K" means "To what power must we raise e to obtain the value of the equilibrium constant?"

Equation 16.10 can also be written using common logarithms.

The factor 2.303 converts common logs to natural logs.
$$2.303 \log x = \ln x$$

$$\Delta G^\circ = -2.303RT \log K \qquad (16.11)$$

Equations 16.10 and 16.11 are valuable because they allow us to relate ΔG° to K (sometimes called the thermodynamic equilibrium constant) quantitatively. For example, measurements of equilibrium constants allow us to calculate ΔG° and, therefore, equilibrium measurements serve as a source of thermodynamic data. We can also use these equations to calculate K from ΔG°. Therefore, if we have a way of obtaining ΔG° for a reaction, as in a Hess's law type of calculation, we can calculate K and predict quantitatively how far toward completion a reaction will go.

There is one complicating factor involving Equations 16.10 and 16.11 of which you should be aware. For reactions involving gases, the K that is calculated is K_p, with partial pressures expressed in atmospheres. For reactions involving liquid solutions, the calculated K is K_c.

EXAMPLE 16.11   CALCULATING ΔG° FROM K

Problem: The brownish haze associated with air pollution is caused by nitrogen dioxide, NO_2, a red-brown gas. Nitric oxide, NO, is oxidized to NO_2 by oxygen.

$$2NO(g) + O_2(g) \rightleftharpoons 2NO_2(g)$$

The value of K_p for this reaction is 1.7×10^{12} at 25.0 °C. What is ΔG° for the reaction, expressed in joules? In kilojoules?

Solution: If a scientific calculator is available, it is easiest to use Equation 16.10,

$$\Delta G^\circ = -RT \ln K_p$$

Once again, we have to be careful to calculate T to four significant figures.

Substituting $R = 8.314$ J/mol K (because the answer must be in joules), $T = 298.2$ K, and $K_p = 1.7 \times 10^{12}$, we find that

$$\Delta G^\circ = -(8.314 \times 298.2) \ln(1.7 \times 10^{12})$$

The natural log of $1.7 \times 10^{12} = 28.16$. Therefore,

$$\Delta G^\circ = -(8.314 \times 298.2 \times 28.16)$$
$$= -6.982 \times 10^4 \text{ J (to four significant figures)}$$

Expressed in kilojoules, $\Delta G^\circ = -69.82$ kJ

EXAMPLE 16.12   CALCULATING K FROM $\Delta G°$

Problem: Sulfur dioxide reacts with oxygen when it passes over the catalyst in automobile catalytic converters. The product is SO_3.

$$2SO_2(g) + O_2(g) \rightleftharpoons 2SO_3(g)$$

For this reaction, $\Delta G° = -33.4$ kcal at 25 °C. What is the value of K_p?

Solution: Once again we can use the equation

$$\Delta G° = -RT \ln K_p$$

This time we must use $R = 1.987$ cal/mol K and convert $\Delta G°$ from kilocalories to calories, so that calories will cancel.

$$\Delta G° = -33,400 \text{ cal}$$

Substituting values into the equation (taking $T = 298$ K),

$$-33,400 = -(1.987)(298) \ln K_p$$

Notice that we first solve for $\ln K_p$; then we take the antilogarithm to get K_p.

Next we multiply each side by -1 (to remove the negative signs) and then solve for the term $\ln K_p$.

$$\frac{33,400}{(1.987)(298)} = \ln K_p$$

This gives

$$\ln K_p = 56.4$$

Taking the antilogarithm,

$$K_p = e^{56.4}$$
$$= 3 \times 10^{24}$$

PRACTICE EXERCISE 11   The reaction $N_2(g) + 3H_2(g) \rightleftharpoons 2NH_3(g)$ has $K_p = 6.9 \times 10^5$ at 25 °C. Calculate $\Delta G°$ for this reaction in units of kilocalories.

PRACTICE EXERCISE 12   The reaction $H_2(g) + I_2(g) \rightleftharpoons 2HI(g)$ has $\Delta G° = +3.3$ kJ at 25 °C. What is the value of K_p at 25 °C?

Thermodynamic Equilibrium Constants at Temperatures Other Than 25 °C

In Section 16.11, we discussed the way temperature affects the position of equilibrium in a reaction. At 25 °C, the relative proportions of reactants and products at equilibrium are determined by $\Delta G°$, and we have just seen that we can calculate the value of the equilibrium constant at 25 °C from $\Delta G°$. As the temperature moves away from 25 °C, the position of equilibrium also changes because the value of ΔG changes — in Section 16.11 we used the symbol $\Delta G'$ to stand for the equivalent of $\Delta G°$, but at a temperature other than 25 °C. This $\Delta G'$ can be used to calculate K at temperatures other than 25 °C by the same methods that were used in Examples 16.11 and 16.12. The equation is

$$\Delta G' = -RT \ln K \qquad (16.12)$$

EXAMPLE 16.13   CALCULATING K AT TEMPERATURES OTHER THAN 25 °C

Problem: The decomposition of nitrous oxide, N_2O, has $K_p = 1.8 \times 10^{36}$ at 25 °C. The equation is

$$2N_2O(g) \rightleftharpoons 2N_2(g) + O_2(g)$$

For this reaction, $\Delta H° = -163$ kJ and $\Delta S° = +148$ J/K. What is the approximate value for K_p for this reaction at 40 °C?

Solution: The first step is to calculate the value of $\Delta G'$ so we can use it in Equation 16.12. Earlier you learned that ΔH and ΔS do not change very much with temperature, and that the value of $\Delta G'$ can be estimated by the equation

$$\Delta G' = \Delta H° - T\Delta S°$$

Substituting the values of $\Delta H°$ and $\Delta S°$ provided in the problem gives

$$\Delta G' = -163,000 \text{ J} - (313 \text{ K})(+148 \text{ J/K})$$

Notice that we've converted kilojoules to joules and used the Kelvin temperature (40 + 273 = 313 K). Performing the arithmetic gives

$$\Delta G' = -163,000 \text{ J} - (46,300 \text{ J})$$
$$= -209,000 \text{ J (rounded)}$$

The next step is to use this value of $\Delta G'$ to compute K_p with Equation 16.12. Substituting, with $R = 8.314$ J/mol K and $T = 313$ K, gives

$$-209,000 = -(8.314)(313) \ln K_p$$

As in Example 16.12, we multiply by -1 to get rid of the negative signs and then solve for $\ln K_p$.

$$\frac{209,000}{(8.314)(313)} = \ln K_p$$

This gives

$$\ln K_p = 80.3$$

Taking the antilog,

$$K_p = e^{80.3}$$
$$= 7 \times 10^{34}$$

Notice that at this higher temperature, N_2O is actually slightly more stable than at 25 °C, as reflected in the slightly smaller value for the equilibrium constant for its decomposition.

PRACTICE EXERCISE 13   The reaction $N_2(g) + 3H_2(g) \rightleftharpoons 2NH_3(g)$ has a standard heat of reaction of -92.4 kJ and a standard entropy of reaction equal to -198.3 J/K. Estimate the value of K_p for this reaction at 50 °C.

SUMMARY

First Law of Thermodynamics.   The change in the internal energy of a system, ΔE, equals the sum of the heat absorbed by the system, q, and the work done on the system, w. ΔE is a state function, but q and w are not. The values of q and w depend on how the change takes place. For pressure–volume work, $w = -P\Delta V$, where $\Delta V = V_{final} - V_{initial}$. The heat of reaction at constant volume, q_v, is equal to ΔE. ΔH differs from ΔE by the work expended in pushing back the atmosphere when the change occurs at constant atmospheric pressure. In general, the difference between ΔE and ΔH is quite small.

Spontaneity.   Spontaneous changes occur without outside assistance. Nonspontaneous changes require continuous help and can occur only if some other spontaneous event takes place first.

The energy change is one factor that influences spontaneity. Exothermic changes, with their negative values of ΔH, tend to proceed spontaneously. Events that occur with an increase in randomness also tend to occur by themselves. Randomness is associated with statistical probability—the more random a collection of molecules, the greater the probability of its occurrence. The thermodynamic quantity associated with randomness is entropy. An increase in entropy favors a spontaneous change. In general, gases have much higher entropies than liquids, which have somewhat higher entropies than solids. During a chemical reaction, the entropy tends to increase if the complexity of the molecules decreases (e.g., forming diatomic molecules from triatomic molecules).

Second Law of Thermodynamics. This law states that the entropy of the universe increases whenever a spontaneous change occurs. All of our activities, being traced ultimately to spontaneous events, increase the total entropy or disorder in the world. One of the noticeable effects of this is environmental pollution.

Third Law of Thermodynamics. The entropy of a pure crystalline substance is equal to zero at absolute zero (0 K). Because we know where the zero point is on the entropy scale, it is possible to measure absolute amounts of entropy possessed by substances. Standard entropies, $S°$, are calculated for 25 °C and 1 atm (Table 16.1) and can be used to calculate $\Delta S°$ for chemical reactions.

Gibbs Free Energy. The Gibbs free energy change, ΔG, allows us to determine the combined effects of enthalpy and entropy changes on the spontaneity of a chemical or physical change. A change is spontaneous only if the free energy of the system decreases (ΔG is negative). When ΔH and ΔS have the same algebraic sign, the temperature becomes the critical factor in determining spontaneity.

When ΔG is measured at 25 °C and 1 atm, it is the standard free energy change, $\Delta G°$. As with enthalpy changes, the standard free energy of formation, $\Delta G_f°$ (Table 16.2), can be used to obtain $\Delta G°$ for chemical reactions by a Hess's law type of calculation.

For any system, the value of ΔG is equal to the maximum amount of energy that can be obtained in the form of useful work. This maximum work can be obtained only if the change takes place reversibly. All real changes are irreversible and we always obtain less work than is theoretically available—the rest is lost as heat.

Free Energy and Equilibrium. When a system reaches equilibrium, $\Delta G = 0$ and no useful work can be obtained from it. At any particular pressure, an equilibrium between two phases (e.g., liquid ⇌ solid, or solid ⇌ vapor) can only occur at one temperature for a substance. The entropy change can be computed as $\Delta S = \Delta H/T$. The temperature at which the equilibrium occurs can be calculated from $T = \Delta H/\Delta S$.

In chemical reactions, a minimum on the free energy curve occurs partway between pure reactants and pure products. This minimum can be approached from either reactants or products, and the composition of the equilibrium mixture is determined by where the minimum lies along the reactant → product axis; when it lies close to the products, the proportion of product to reactants is large and the reaction goes far toward completion.

When a reaction has a value of $\Delta G°$ that is both large and negative, it will appear to occur spontaneously because a lot of products will be formed by the time equilibrium is reached. If $\Delta G°$ is large and positive, no *observable* reaction will occur because only tiny amounts of products will be formed. The sign and magnitude of $\Delta G°$ can therefore be used to predict the apparent spontaneity of a chemical reaction.

At a temperature other than 25 °C, the value of $\Delta G'$ can be calculated using $\Delta H°$ and $\Delta S°$ as approximations of $\Delta H'$ and $\Delta S'$. The equation is $\Delta G' = \Delta H° - T\Delta S°$. Computing $\Delta G'$ allows us to see how temperature affects the position of equilibrium in a chemical reaction.

Thermodynamic Equilibrium Constants. Equilibrium constants can be calculated from values of $\Delta G°$.

$$\Delta G° = -RT \ln K$$

or

$$\Delta G° = -2.303RT \log K$$

where $K = K_p$ for gaseous reactions and $K = K_c$ for reactions in solution. The value of K can be calculated for temperatures other than 25 °C by first calculating the appropriate value of $\Delta G'$.

REVIEW EXERCISES

Answers to questions whose numbers are printed in color are given in Appendix C. Difficult questions are marked by asterisks.

First Law of Thermodynamics

16.1 What concept is suggested by the term *thermodynamics?*

16.2 What kinds of energy contribute to the internal energy of a system? Why can't we measure or calculate a system's internal energy?

16.3 What is a state function?

16.4 How is a change in the internal energy defined in terms of the initial and final internal energies?

16.5 What is the algebraic sign of ΔE for an endothermic change? Why?

16.6 State the First Law of Thermodynamics in terms of an equation. Define the meaning of the symbols, including the significance of their algebraic signs.

16.7 A certain system absorbs 500 J of heat and has 200 J of work performed on it. What is the value of ΔE for the change? Is the overall change exothermic or endothermic?

16.8 The value of ΔE for a certain change is -1250 J. During the change, the system absorbs 603 J. How much work, expressed in joules, was accomplished by the system?

16.9 Which thermodynamic quantity corresponds to the heat

of reaction at constant volume? Which quantity corresponds to the heat of reaction at constant pressure?

16.10 How is the rate at which energy is withdrawn from a system related to the amount of that energy which can appear as useful work?

16.11 What are the units of $P\Delta V$ if pressure is expressed in pascals and volume is expressed in cubic meters?

16.12 If pressure is expressed in atmospheres and volume is expressed in liters, $P\Delta V$ has units of L atm (liters $\times$ atmospheres). In Chapter 9 you learned that 1 atm = 101,325 Pa, and in Chapter 1 you learned that 1 L = 1 dm³. Use this information to determine the number of joules corresponding to 1 L atm.

16.13 In Review Exercise 16.12 you found that 1 L atm = 101.325 J. Suppose that you were pumping an automobile tire with a hand pump that pushed 24.0 in.³ of air into the tire on each stroke, and that during one such stroke the opposing pressure in the tire was 30.0 lb/in.² *above* normal atmospheric pressure, 14.7 lb/in.². Calculate the number of joules of work accomplished during each stroke.

16.14 How are ΔE and ΔH related to each other?

16.15 If there is an increase in the number of moles of gas during an exothermic chemical reaction, which is larger, ΔE or ΔH? Why?

16.16 How much work is accomplished by a chemical reaction that occurs inside a bomb calorimeter?

*16.17 Consider the reaction between aqueous solutions of baking soda, $NaHCO_3$ and vinegar, $HC_2H_3O_2$.

$$NaHCO_3(aq) + HC_2H_3O_2(aq) \longrightarrow NaC_2H_3O_2(aq) + H_2O + CO_2(g)$$

If this reaction occurs at atmospheric pressure ($P = 1$ atm), how much work is done by the system in pushing back the atmosphere when 1.00 mol $NaHCO_3$ reacts at a temperature of 25 °C? (*Hint:* Review the gas laws.)

Spontaneous Change and Enthalphy

16.18 What is a "spontaneous change"?

16.19 List five changes that you have encountered recently that occurred spontaneously. List five changes that are nonspontaneous but that *you* have caused to occur.

16.20 Which of the items that you listed in Review Exercise 16.19 are exothermic and which are endothermic?

16.21 What role does the enthalpy change play in determining the spontaneity of an event?

16.22 Use the data from Table 5.2 to calculate $\Delta H°$ for the following reactions. On the basis of $\Delta H°$, which are favored to occur spontaneously?
(a) $CaO(s) + CO_2(g) \longrightarrow CaCO_3(s)$
(b) $C_2H_2(g) + 2H_2(g) \longrightarrow C_2H_6(g)$
(c) $3CaO(s) + 2Fe(s) \longrightarrow 3Ca(s) + Fe_2O_3(s)$
(d) $Ca(OH)_2(s) \longrightarrow CaO(s) + H_2O(\ell)$
(e) $2NaCl(s) + H_2SO_4(\ell) \longrightarrow Na_2SO_4(s) + 2HCl(g)$

16.23 Use the data from Table 5.2 to calculate $\Delta H°$ for the following reactions. On the basis of $\Delta H°$, which are vored to occur spontaneously?
(a) $2C_2H_2(g) + 5O_2(g) \longrightarrow 4CO_2(g) + 2H_2O(g)$
(b) $C_2H_2(g) + 5N_2O(g) \longrightarrow$
$$2CO_2(g) + H_2O(g) + 5N_2(g)$$
(c) $Fe_2O_3(s) + 2Al(s) \longrightarrow Al_2O_3(s) + 2Fe(s)$
(d) $NH_4Cl(s) \longrightarrow NH_3(g) + HCl(g)$
(e) $Ag(s) + KCl(s) \longrightarrow AgCl(s) + K(s)$

Entropy

16.24 When solid potassium iodide is dissolved in water, a cooling of the mixture occurs because the solution process is endothermic for these substances. Explain, in terms of what happens to the molecules and ions, why this mixing occurs spontaneously.

16.25 What is entropy?

16.26 Will the entropy change for each of the following be positive or negative?
(a) Moisture condenses on the outside of a cold glass.
(b) Raindrops form in a cloud.
(c) Gasoline vaporizes in the carburetor of an automobile engine.
(d) Air is pumped into a tire.
(e) Frost forms on the windshield of your car.
(f) Sugar dissolves in coffee.

16.27 On the basis of our definition of entropy, suggest why entropy is a state function.

16.28 State the Second Law of Thermodynamics.

16.29 In animated cartoons, visual effects are often created (for amusement) that show events that ordinarily don't occur in real life because they are accompanied by huge entropy decreases. Can you think of an example of this? Explain why there is an entropy decrease in your example.

16.30 How is entropy related to pollution? What are our chances of eliminating pollution from the environment?

16.31 When a coin is tossed, there is a 50–50 chance of it landing either heads or tails. Suppose that you tossed four coins. On the basis of the different possible outcomes, what is the probability of all four coins coming up heads? What is the probability of an even heads–tails distribution?

16.32 Suppose that you had two equal-volume containers sharing a common wall with a hole in it. Suppose there were four molecules in this system. What is the probability that all four would be in one container at the same time? What is the probability of finding an even distribution of molecules between the two containers? (*Hint:* Try Review Exercise 16.31 first.) What do the results suggest about why gases expand spontaneously?

16.33 Predict the algebraic sign of the entropy change for the following reactions.
(a) $PCl_3(g) + Cl_2(g) \longrightarrow PCl_5(g)$
(b) $SO_2(g) + CaO(s) \longrightarrow CaSO_3(s)$
(c) $CO_2(g) + H_2O(\ell) \longrightarrow H_2CO_3(aq)$
(d) $Ni(s) + 2HCl(aq) \longrightarrow H_2(g) + NiCl_2(aq)$

16.34 Predict the algebraic sign of the entropy change for the following reactions.
(a) $I_2(s) \longrightarrow I_2(g)$
(b) $Cl_2(g) + Br_2(g) \longrightarrow 2BrCl(g)$
(c) $NH_3(g) + HCl(g) \longrightarrow NH_4Cl(s)$
(d) $CaO(s) + H_2O(\ell) \longrightarrow Ca(OH)_2(s)$

Third Law of Thermodynamics

16.35 What is the Third Law of Thermodynamics?

16.36 Would you expect the entropy of an alloy (a solution of two metals) to be zero at 0 K? Explain your answer.

16.37 Why does entropy increase with increasing temperature?

16.38 Calculate $\Delta S°$ for the following reactions in J/K from the data in Table 16.1. On the basis of $\Delta S°$, which of these reactions are favored to occur spontaneously?
(a) $N_2(g) + 3H_2(g) \longrightarrow 2NH_3(g)$
(b) $CO(g) + 2H_2(g) \longrightarrow CH_3OH(\ell)$
(c) $2C_2H_6(g) + 7O_2(g) \longrightarrow 4CO_2(g) + 6H_2O(g)$
(d) $Ca(OH)_2(s) + H_2SO_4(\ell) \longrightarrow CaSO_4(s) + 2H_2O(\ell)$
(e) $S(s) + 2N_2O(g) \longrightarrow SO_2(g) + 2N_2(g)$

16.39 Calculate $\Delta S°$ for the following reactions in J/K, using the data in Table 16.1.
(a) $Ag(s) + \frac{1}{2}Cl_2(g) \longrightarrow AgCl(s)$
(b) $H_2(g) + \frac{1}{2}O_2(g) \longrightarrow H_2O(g)$
(c) $H_2(g) + \frac{1}{2}O_2(g) \longrightarrow H_2O(\ell)$
(d) $CaCO_3(s) + H_2SO_4(\ell) \longrightarrow$
$\qquad CaSO_4(s) + H_2O(g) + CO_2(g)$
(e) $NH_3(g) + HCl(g) \longrightarrow NH_4Cl(s)$

16.40 Calculate $\Delta S_f°$ for the following compounds in J/mol K.
(a) $C_2H_4(g)$ (d) $CaSO_4 \cdot 2H_2O(s)$
(b) $N_2O(g)$ (e) $HC_2H_3O_2(\ell)$
(c) $NaCl(s)$

16.41 Calculate $\Delta S_f°$ for the following compounds in J/mol K.
(a) $Al_2O_3(s)$ (d) $NH_4Cl(s)$
(b) $CaCO_3(s)$ (e) $CaSO_4 \cdot \frac{1}{2}H_2O(s)$
(c) $N_2O_4(g)$

16.42 Nitrogen dioxide, NO_2, an air pollutant, dissolves in rainwater to form a dilute solution of nitric acid. The equation for the reaction is

$$3NO_2(g) + H_2O(\ell) \longrightarrow 2HNO_3(\ell) + NO(g)$$

Calculate $\Delta S°$ for this reaction in J/K.

16.43 Good wine will turn to vinegar if it is left exposed to air because the alcohol is oxidized to acetic acid. The equation for the reaction is

$$C_2H_5OH(\ell) + O_2(g) \longrightarrow HC_2H_3O_2(\ell) + H_2O(\ell)$$

Calculate $\Delta S°$ for this reaction in J/K.

Gibbs Free Energy

16.44 What is the equation expressing the change in the Gibbs free energy for a reaction occurring at constant temperature and pressure?

16.45 In what circumstances will a change be spontaneous:
(a) At all temperatures?
(b) At low temperatures but not at high temperatures?
(c) At high temperatures but not at low temperatures?

16.46 In what circumstances will a change be nonspontaneous regardless of the temperature?

16.47 Phosgene, $COCl_2$, was used as a war gas during World War I. It reacts with the moisture in the lungs to produce HCl, which causes the lungs to fill with fluid, leading to the death of the victim. $COCl_2$ has a standard entropy, $S° = 284$ J/mol K and $\Delta H_f° = -223$ kJ/mol. Use this information and the data in Table 16.1 to calculate $\Delta G_f°$ for $COCl_2(g)$ in kJ/mol.

16.48 Aluminum oxidizes rather easily, but forms a thin protective coating of Al_2O_3 that prevents further oxidation of the aluminum beneath. Use the data for $\Delta H_f°$ (Table 5.2) and $S°$ to calculate $\Delta G_f°$ for $Al_2O_3(s)$ in kJ/mol.

16.49 Compute $\Delta G°$ in kJ for the following reactions, using the data in Table 16.2.
(a) $SO_3(g) + H_2O(\ell) \longrightarrow H_2SO_4(\ell)$
(b) $2NH_4Cl(s) + CaO(s) \longrightarrow$
$\qquad CaCl_2(s) + H_2O(\ell) + 2NH_3(g)$
(c) $CaSO_4(s) + 2HCl(g) \longrightarrow CaCl_2(s) + H_2SO_4(\ell)$
(d) $C_2H_4(g) + H_2O(g) \longrightarrow C_2H_5OH(\ell)$
(e) $Ca(s) + 2H_2SO_4(\ell) \longrightarrow$
$\qquad CaSO_4(s) + SO_2(g) + 2H_2O(\ell)$

16.50 Compute $\Delta G°$ in kJ for the following reactions, using the data in Table 16.2.
(a) $2HCl(g) + CaO(s) \longrightarrow CaCl_2(s) + H_2O(g)$
(b) $H_2SO_4(\ell) + 2NaCl(s) \longrightarrow 2HCl(g) + Na_2SO_4(s)$
(c) $3NO_2(g) + H_2O(\ell) \longrightarrow 2HNO_3(\ell) + NO(g)$
(d) $2AgCl(s) + Ca(s) \longrightarrow CaCl_2(s) + 2Ag(s)$
(e) $NH_3(g) + HCl(g) \longrightarrow NH_4Cl(s)$

16.51 Plaster of Paris, $CaSO_4 \cdot \frac{1}{2}H_2O(s)$, reacts with liquid water to form gypsum, $CaSO_4 \cdot 2H_2O(s)$. Write a chemical equation for the reaction and calculate $\Delta G°$ in kJ, using the data in Table 16.2.

16.52 When phosgene, the war gas described in Review Exercise 16.47, reacts with water vapor, the products are $CO_2(g)$ and $HCl(g)$. Write an equation for the reaction and compute $\Delta G°$ in kJ. $\Delta G_f°$ for $COCl_2$ is -210 kJ/mol.

16.53 Given the following,

$$4NO(g) \longrightarrow 2N_2O(g) + O_2(g) \qquad \Delta G° = -139.56 \text{ kJ}$$
$$2NO(g) + O_2(g) \longrightarrow 2NO_2(g) \qquad \Delta G° = -69.70 \text{ kJ}$$

calculate $\Delta G°$ for the reaction

$$2N_2O(g) + 3O_2(g) \longrightarrow 4NO_2(g)$$

16.54 Given these reactions and their $\Delta G°$ values,

$$COCl_2(g) + 4NH_3(g) \longrightarrow CO(NH_2)_2(s) + 2NH_4Cl(s)$$
$$\Delta G° = -79.36 \text{ kcal}$$
$$COCl_2(g) + H_2O(\ell) \longrightarrow CO_2(g) + 2HCl(g)$$
$$\Delta G° = -33.89 \text{ kcal}$$
$$NH_3(g) + HCl(g) \longrightarrow NH_4Cl(s) \qquad \Delta G° = -21.98 \text{ kcal}$$

calculate $\Delta G°$ for the reaction

$$CO(NH_2)_2(s) + H_2O(\ell) \longrightarrow CO_2(g) + 2NH_3(g)$$

16.55 Many biochemical reactions have positive values for $\Delta G°$ and seemingly should not be expected to be spontaneous. They occur, however, because they are chemically coupled with other reactions that have negative values of $\Delta G°$. An example is the set of reactions that forms the beginning part of the sequence of reactions involved in the metabolism of glucose, a sugar. Given these reactions and their corresponding $\Delta G°$ values,

glucose + phosphate $\longrightarrow$ glucose 6-phosphate + H_2O
$$\Delta G° = +13.13 \text{ kJ}$$
ATP + H_2O $\longrightarrow$ ADP + phosphate
$$\Delta G° = -32.22 \text{ kJ}$$

calculate $\Delta G°$ for the coupled reaction

glucose + ATP $\longrightarrow$ glucose 6-phosphate + ADP

Free Energy and Work

16.56 How is free energy related to useful work?

16.57 What is a reversible process?

16.58 When glucose is oxidized by the body to generate energy, this energy is stored in molecules of ATP (adenosine triphosphate). However, of the total energy released in the oxidation of glucose, only 38% actually becomes stored in the ATP. What happens to the rest of the energy?

16.59 Why are real, observable changes not considered to be reversible processes?

16.60 Gasohol is a mixture of gasoline and ethanol (grain alcohol), C_2H_5OH. Calculate the maximum work that could be obtained at 25 °C and 1 atm by burning 1 mol of C_2H_5OH.

$$C_2H_5OH(\ell) + 3O_2(g) \longrightarrow 2CO_2(g) + 3H_2O(g)$$

16.61 What is the maximum amount of useful work that could possibly be obtained at 25 °C and 1 atm from the combustion of 27.0 g of natural gas, $CH_4(g)$, to give $CO_2(g)$ and $H_2O(g)$?

16.62 Ethyl alcohol, C_2H_5OH, has been suggested as an alternative to gasoline as a fuel. In Example 16.5 we calculated $\Delta G°$ for combustion of 1 mol of C_2H_5OH; in Example 16.6 we calculated $\Delta G°$ for combustion of 1 mol of octane. Let's assume that gasoline has the same properties as octane (one of its major constituents). The density of C_2H_5OH is 0.7893 g/mL; the density of octane, C_8H_{18}, is 0.7025 g/mL. Calculate the maximum work that could be obtained by burning 1 gallon (3.78 liters) of both C_2H_5OH and C_8H_{18}. On a volume basis, which is a better fuel?

Free Energy and Equilibrium

16.63 In what way is free energy related to equilibrium?

16.64 Considering the fact that the formation of a bond between two atoms is exothermic and is accompanied by an entropy decrease, explain why all chemical compounds decompose into individual atoms if heated to a high enough temperature.

16.65 When a warm object is placed in contact with a cold one, they both gradually come to the same temprature. On a molecular level, explain how this is related to entropy and spontaneity.

16.66 Sketch the shape of the free energy curve for a chemical reaction that has a positive $\Delta G°$. Indicate the composition of the reaction mixture corresponding to equilibrum.

16.67 Sketch a graph to show how the free energy changes during a phase change such as melting.

16.68 Chloroform, formerly used as an anesthetic and now believed to be a carcinogen (cancer-causing agent), has a heat of vaporization $\Delta H_{vap} = 31.4$ kJ/mol. The change, $CHCl_3(\ell) \rightarrow CHCl_3(g)$ has $\Delta S° = 94.2$ J/mol K. At what temperature do we expect $CHCl_3$ to boil (i.e., at what temperature will liquid and vapor be in equilibrium at 1 atm pressure)?

16.69 For the melting of aluminum, $Al(s) \rightarrow Al(\ell)$, $\Delta H° = 10.0$ kJ/mol and $\Delta S° = 9.50$ J/mol K. Calculate the melting point of Al. (The actual melting point is 660°C.)

16.70 Isooctane, an important constituent of gasoline, has a boiling point of 99.3 °C and a heat of vaporization of 9.01 kcal/mol. What is ΔS (in cal/mol K) for the vaporization of 1 mol of isooctane?

16.71 Acetone (nail polish remover) has a boiling point of 56.2 °C. The change, $(CH_3)_2CO(\ell) \rightarrow (CH_3)_2CO(g)$, has $\Delta H° = 31.9$ kJ/mol. What is $\Delta S°$ for this change?

16.72 Determine whether the following reaction will be spontaneous.

$$C_2H_4(g) + 2HNO_3(\ell) \longrightarrow$$
$$HC_2H_3O_2(\ell) + H_2O(\ell) + NO(g) + NO_2(g)$$

16.73 Which of the following reactions (unbalanced) would be expected to be spontaneous at 25 °C and 1 atm?
(a) $PbO(s) + NH_3(g) \longrightarrow Pb(s) + N_2(g) + H_2O(g)$
(b) $NaOH(s) + HCl(g) \longrightarrow NaCl(s) + H_2O(\ell)$
(c) $Al_2O_3(s) + Fe(s) \longrightarrow Fe_2O_3(s) + Al(s)$
(d) $2CH_4(g) \longrightarrow C_2H_6(g) + H_2(g)$

16.74 Many reactions that have large, negative values of $\Delta G°$ are not actually observed to happen at 25 °C and 1 atm. Why?

Effect of Temperature on the Free Energy Change

16.75 Why is the value ΔG for a change so dependent on temperature?

16.76 What equation can be used to obtain an approximate value for $\Delta G'$ for a reaction at a temperature other than 25 °C?

16.77 Suppose a reaction has a negative $\Delta H°$ and a negative $\Delta S°$. Will more or less product be present at equilibrium as the temperature is raised?

16.78 Calculate the $\Delta G'$ in kJ for the following reactions at 100 °C, using data in Tables 5.2 and 16.1.

(a) $C_2H_4(g) + H_2(g) \longrightarrow C_2H_6(g)$

(b) $5SO_3(g) + 2NH_3(g) \longrightarrow$
$$2NO(g) + 5SO_2(g) + 3H_2O(g)$$

(c) $N_2O(g) + NO_2(g) \longrightarrow 3NO(g)$

Thermodynamic Equilibrium Constants

16.79 How is the equilibrium constant related to the standard free energy change for a reaction?

16.80 What is a *thermodynamic equilibrium constant*?

16.81 What is the value of $\Delta G°$ for a reaction having $K = 1$?

16.82 A potential reaction for conversion of coal to methane (natural gas) is

$$C(s) + 2H_2(g) \rightleftharpoons CH_4(g)$$

for which $\Delta G° = -50.79$ kJ. What is the value of K_p for this reaction at 25 °C? Does the value of K_p make studying this reaction as a method of natural gas production worthwhile?

16.83 One of the important reactions in living cells from which the organism draws energy is the reaction of adenosine triphosphate (ATP) with water to give adenosine diphosphate (ADP) plus free phosphate.

$$ATP + H_2O \rightleftharpoons ADP + phosphate$$

The value of $\Delta G'$ (the equivalent of $\Delta G°$, but at 37 °C) for this reaction is -8 kcal/mol. Calculate the equilibrium constant for the reaction.

16.84 What will be the value of the equilibrium constant for a reaction having $\Delta G° = 0$?

16.85 A certain reaction at 25 °C has $K_p = 10.0$. What is the value of $\Delta G°$ for this reaction?

16.86 Methanol, a potentially important fuel, can be made from CO and H_2. The reaction, $CO(g) + 2H_2(g) \rightleftharpoons CH_3OH(g)$, has $K_p = 6.25 \times 10^{-3}$ at 500 K. Calculate $\Delta G'$ for this reaction kJ.

16.87 Refer to the data in Table 16.2 to determine $\Delta G°$, in kJ, for the reaction $2NO(g) + 2CO(g) \rightleftharpoons N_2(g) + 2CO_2(g)$. What is the value of K_p for this reaction at 25 °C?

16.88 Use the data in Table 5.2 and Table 16.1 to calculate $\Delta G'$, in kJ, for the reaction $2NO(g) + 2CO(g) \rightleftharpoons N_2(g) + 2CO_2(g)$ at 500 °C. What is the value of K_p at this temperature?

CHEMICALS IN USE

11

AMMONIA

Throughout the nineteenth century a number of chemists sought a way to make ammonia directly from its elements. Most succeeded but with disappointingly low yields. They found that some kind of metal or metal oxide was needed to promote the reaction, that high pressures favored the conversion, and that high temperatures caused ammonia to break back down into its elements.

Fritz Haber (1868–1934), a German physical chemist, sorted out these factors and found the best set of conditions. These involved pressures and temperatures that taxed the properties of materials for making industrial equipment. But Karl Bosch, an engineer for the company that became interested in Haber's work (Badische Anilin and Soda Fabrik), solved these problems. Hence, the process is called the Haber–Bosch method. The first plant to use it went into production in 1913, and Haber won the 1918 Nobel Prize in chemistry for his research.

The formation of ammonia involves the following equilibrium:

$$N_2(g) + 3H_2(g) \rightleftharpoons 2NH_3(g) \qquad \Delta H^\circ = -92.2 \text{ kJ}$$

Since the substances are all gases, the coefficients give us not only mole ratios but also volume ratios (Avogadro's principle, Section 9.7). The reaction thus changes 4 volumes of reactants into 2 volumes of product. Hence, an increase in the pressure would favor the formation of ammonia. The equilibrium could shift to the right to absorb this volume-reducing stress in accordance with Le Châtelier's principle (which was known to Haber).

Since the forward reaction is exothermic, *heating* the mixture would shift the equilibrium to the left and lower the yield of ammonia. But many bonds have to be broken in order that others can form, so the mixture still must be heated even though this makes the position of the equilibrium less favorable.

Haber and his students made many measurements of the concentrations of ammonia present at equilibrium under several temperatures and pressures. Table 11a gives some of their results. You can see how the yield of ammonia increases with pressure (at a given temperature) and how it decreases with increasing temperature (at a given pressure), just as our analysis suggested.

The metal that promotes the reaction is an example of

a catalyst, a substance that in relatively small amounts can make a reaction occur under milder conditions or make it occur more rapidly under more severe conditions. No net, permanent change occurs to the catalyst. (We will have more to say about catalysts in Chapter 18.) The catalyst used in the Haber–Bosch process today consists of very small particles of iron doped with traces of alumina, silica, magnesium oxide, and potassium hydroxide. Without a catalyst, the rate of formation of ammonia is almost zero even at 500 °C, so we can see how vital the catalyst is. With it, most operations are run in the range of 380 °C to 450 °C at a pressure of about 200 atm. Figure 11a shows roughly the design of the reaction vessel in which the final step is carried out.

The raw materials that enter a Haber–Bosch plant are not the pure nitrogen and hydrogen of the simplified equation. Instead, the starting materials are water, air, and natural gas, which is almost entirely methane, CH_4. Water and methane supply the hydrogen, and air furnishes the nitrogen. Of course, air also supplies oxygen, but this is removed in a clever and resourceful way.

The methane and water are taken in sufficient proportions to supply enough hydrogen not only for the ammonia synthesis but also to react with all of the oxygen in the air. This regenerates some of the water, and it also supplies heat energy needed to run the whole operation. The carbon in methane ends up in carbon dioxide, and this is removed by its reaction with potassium carbonate.

Taking $[O_2 + 4N_2]$ to represent air in its almost exactly correct natural mole ratio of oxygen and nitrogen, we

TABLE 11a Mole Percent of Ammonia at Equilibrium

Temperature (°C)	Pressure (atm)						
	10	30	50	100	300	600	1000
200	51	68	74	82	90	95	98
300	15	30	39	52	71	84	93
400	3.9	10	15	25	47	65	80
500	1.2	3.5	5.6	11	26	42	57
600	0.49	1.4	2.3	4.5	14	23	31
700	0.23	0.68	1.1	2.2	7.3	13	13

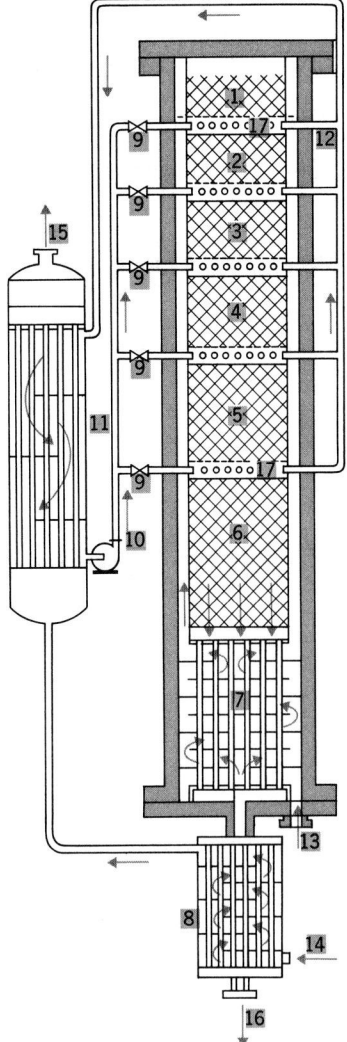

FIGURE 11a Schematic of an ammonia synthesizer employing the Haber–Bosch process. A mixture of nitrogen and hydrogen enters the system at 13 and then eventually flows through beds of the catalyst (1–6). The product exits at 16. The circulations indicated by the arrows involving pipes in 7–12, 14, 15, and 17 are flows of coolant to carry away the heat produced by this exothermic reaction. (From Supplement I, Volume III, Mellor's Comprehensive Treatise on Inorganic and Theoretical Chemistry, page 266, Longmans, Green and Co., London, 1864.)

can write the following equation for the entire overall operation.

$$7CH_4 + 2[O_2 + 4N_2] + 17H_2O + 7K_2CO_3 \longrightarrow 16NH_3 + 14KHCO_3$$

This equation represents the net effect of many steps. In the first, methane and steam are passed under high pressure (30 atm) and temperature (750 °C) over a nickel catalyst where the following reactions occur. Together they use all but about 9% of the CH_4. (Although these equations are written as equilibria, the subsequent reactions of the products help shift the equilibria to the right.)

$$CH_4 + H_2O \rightleftharpoons CO + 3H_2$$
$$CH_4 + 2H_2O \rightleftharpoons CO_2 + 4H_2$$

Now air is injected into the flowing gas stream, and some of the hydrogen just made reacts with its oxygen to leave water and nitrogen.

$$2H_2 + [O_2 + 4N_2] \rightleftharpoons 2H_2O + 4N_2$$

This raises the internal temperature to about 1100 °C, and some of the heat is used in two ways. It helps to make virtually all of the still unchanged CH_4 react with water, giving more hydrogen.

$$CH_4 + H_2O \rightleftharpoons CO + 3H_2$$

And some of the released heat is the energy that operates the gas compressors and that makes the high-temperature steam that is injected together with methane at the very start of the whole operation. This removal of heat cools the gas mixture (now chiefly CO, CO_2, H_2O, H_2, and N_2). This mixture is next channeled through a series of catalysts that remove the CO by the following reaction. More hydrogen is made.

$$CO + H_2O \rightleftharpoons CO_2 + H_2$$

Carbon monoxide must be removed because it inactivates ("poisons") the final Haber–Bosch iron catalyst. The small amount of CO not removed here is later almost 100% removed by conversion to methane over a nickel catalyst by the following reaction.

$$CO + 3H_2 \rightleftharpoons CH_4 + H_2O$$

The final mixture of gases that enters the last reaction vessel has only 1 to 2 μg of CO per liter.

Before the gas stream enters the last vessel, it is passed through a concentrated solution of potassium carbonate, which removes both water and carbon dioxide:

$$CO_2 + H_2O + K_2CO_3 \longrightarrow 2KHCO_3$$

(The potassium carbonate can be recovered by heating the potassium bicarbonate. Heat decomposes $KHCO_3$ by the reverse of this reaction.)

The gas stream now consists of hydrogen and nitrogen in the correct mole ratio for the final reaction that makes ammonia. The trace of argon from the original air is still present, but it causes no interference.

Teamwork among theoretical and experimental chemists and chemical engineers was responsible for the development of this process. Without the ammonia fertilizers made available by this operation, it is hard to imagine how agriculture could support the food needs of today's world population. The next major development in ammonia manufacture will no doubt be the industrial duplication of what nitrogen-fixing microorganisms have routinely done for eons of time, as mentioned in *Chemicals in Use 7*.

Questions

Where appropriate, include equations in your answers.

1. How is natural gas used in the industrial synthesis of NH_3?
2. What becomes of the carbon in CH_4 in the overall process?
3. How is O_2 removed from air in the Haber–Bosch operation?
4. Why does a high temperature lower the yield of ammonia?
5. Why does a high pressure raise the yield of ammonia?
6. Explain how carbon monoxide interferes with the Haber–Bosch reaction.

A researcher uses a portable computer to store data about the behavior of endangered species.
Devices such as portable computers, radios, and televisions, are possible because of our ability to

ELECTROCHEMISTRY

17.1 ELECTRICITY AND CHEMICAL CHANGE

Oxidation–reduction reactions can be brought about by electricity or used to produce electricity.

Oxidation and reduction, which we will consider here as the loss and gain of electrons, occur in many chemical systems. The rusting of iron, the photosynthesis that takes place in the leaves of trees and other green plants, and the conversion of foods to energy in the body are all examples of chemical changes that involve the transfer of electrons from one chemical species to another. When such reactions can be made to cause electrons to flow through a wire or when a flow of electrons makes a redox reaction happen, the processes are referred to as electrochemical changes. The study of these changes is called electrochemistry.

The applications of electrochemistry are widespread. Batteries, which produce electrical energy by means of chemical reactions, are used to power toys, flashlights, electronic calculators, pacemakers that maintain the rhythm of the heart, radios, tape recorders, and even some automobiles. In the laboratory, electrical measurements enable us to monitor chemical reactions of all sorts, even those in systems as tiny as a living cell. In industry, many important chemicals — including liquid bleach (sodium hypochlorite) and lye (sodium hydroxide, which is used to manufacture soap) — are manufactured by electrochemical reactions. In fact, if it were not for electrochemical reactions, the important structural metals aluminum and magnesium would be only laboratory curiosities. Most people would only see them in small amounts in museums.

In this chapter we will study the factors that affect the outcome of electrochemical changes. Besides the practical applications, fundamental information about chemical reactions is available from electrical measurements — for example, free energy changes and equilibrium constants. We will see, therefore, that electrochemistry is a very versatile tool for investigating chemical and biological systems.

17.2 ELECTROLYSIS

Electricity can provide the necessary energy to cause otherwise nonspontaneous reactions to occur.

When electricity is passed through a molten ionic compound or through a solution containing ions — an electrolyte — a chemical reaction called electrolysis occurs. A typical electrolysis apparatus, referred to as an electrolysis cell or electrolytic cell, is shown in Figure 17.1. This particular cell contains molten sodium chloride. (A substance undergoing electrolysis must be molten or in solution so that its ions can move freely and conduction can occur.) Inert electrodes — electrodes that won't react with the molten NaCl — are dipped into the cell and then connected to a source of direct current (DC) electricity.

When electricity starts to flow, chemical changes begin to take place. At the positive electrode, the anode, *oxidation* occurs as electrons are pulled from negatively charged chloride ions. The DC source pumps these electrons through the external electrical circuit to the negative electrode, the cathode. At the cathode, *reduction* takes place as the electrons are picked up by positively charged sodium ions.

Designating a particular electrode as the anode or cathode in an electrochemical cell depends on the type of chemical change — oxidation or reduction — that occurs there. In any electrochemical cell,

> *Anode* is the name of the electrode at which *oxidation* occurs.
> *Cathode* is the name of the electrode at which *reduction* occurs.

The chemical changes that take place at the electrodes can be expressed in the form of chemical equations.

$$Na^+(\ell) + e^- \longrightarrow Na(\ell) \qquad \text{(cathode)}$$
$$2Cl^-(\ell) \longrightarrow Cl_2(g) + 2e^- \qquad \text{(anode)}$$

You may recognize these as half-reactions of the type that we used in Chapter 12 to balance redox reactions by the ion-electron method. This is no accident. The half-reac-

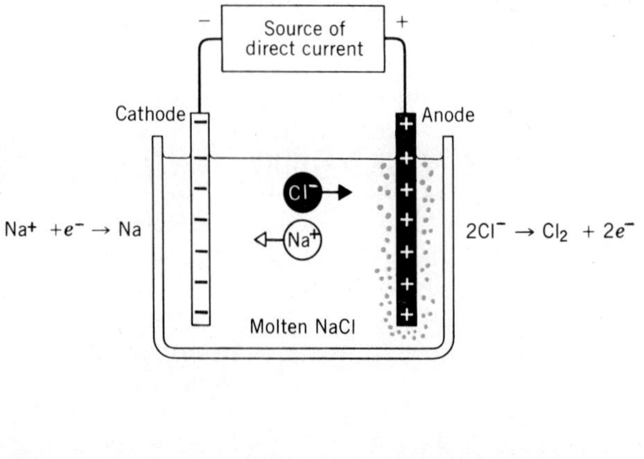

Sodium chloride melts at 801 °C.

When a direct current flows through a wire, electrons move in only one direction. In an alternating current, electrons are pulled and pushed back and forth through the wire. To study electrochemical changes, we must use direct current.

At the melting point of NaCl, metallic sodium is a liquid.

FIGURE 17.1 An electrolysis cell in which the passage of an electric current decomposes molten sodium chloride into metallic sodium and gaseous chlorine. Unless the products are kept apart, they react on contact to re-form NaCl.

tions that are generated by the ion-electron approach correspond to the chemical changes that occur at the electrodes during electrochemical changes. This is one of the bonuses that we get by learning the ion-electron method.

Conduction of Charge in Electrochemical Cells

When sodium chloride undergoes electrolysis (when it is electrolyzed), no electrons actually pass through the molten NaCl from one electrode to another. The electrical conduction here is quite different from that in a metal, where electrons carry the charge. In a molten salt such as sodium chloride, or in a solution of an electrolyte, it is the ions that move through the liquid that carry the charge. The transport of electrical charge by ions is called electrolytic conduction. In the case of molten NaCl, for example, negatively charged chloride ions gradually move toward the positive electrode, and positively charged sodium ions gradually move toward the negative electrode. Around each electrode, a layer of ions of opposite charge accumulates, and electrolysis is able to continue only because reactions of these ions deplete the layers and make room for more ions from the surrounding liquid. If the redox changes at the electrodes cease, the flow of electricity in the external circuit also stops.

Cell Reactions

To obtain the overall reaction that takes place in the electrolysis cell — we call it the cell reaction — we simply add the individual electrode reactions together. Before we can do this, however, we must make sure that the number of electrons gained is equal to the number lost. This is a requirement, you recall, that every redox reaction must obey, and the procedure that we use is the one that you learned in the ion-electron method. For the electrolysis of molten NaCl, we simply multiply the half-reaction for the reduction of sodium by 2, and then add the two half-reactions to obtain the net reaction.

$$\begin{array}{ll} 2Na^+(\ell) + 2e^- \longrightarrow 2Na(\ell) & \text{(cathode)} \\ 2Cl^-(\ell) \longrightarrow Cl_2(g) + 2e^- & \text{(anode)} \\ \hline 2Na^+(\ell) + 2Cl^-(\ell) + \cancel{2e^-} \longrightarrow 2Na(\ell) + Cl_2(g) + \cancel{2e^-} & \text{(cell reaction)} \end{array}$$

When electrons appear as a reactant, the process is reduction; when they appear as a product, it is oxidation.

As you know, table salt is quite stable. It doesn't normally decompose because the reaction of sodium and chlorine to form sodium chloride is highly spontaneous. We therefore often write the word *electrolysis* above the arrow in the equation to show that electricity is the driving force for this otherwise nonspontaneous reaction.

$$2Na^+(\ell) + 2Cl^-(\ell) \xrightarrow{\text{electrolysis}} 2Na(\ell) + Cl_2(g)$$

Electrolysis Reactions in Aqueous Solutions

When electrolysis is carried out in an aqueous solution, the net reaction is more difficult to predict because the oxidation and reduction of water can also occur. This happens, for example, when electrolysis is carried out in a solution of potassium nitrate. The products of the electrolysis are hydrogen and oxygen, as shown in Figure 17.2. At the

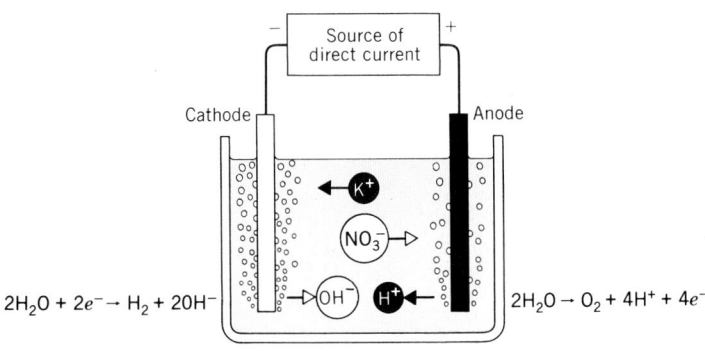

$2H_2O + 2e^- \rightarrow H_2 + 2OH^-$ $2H_2O \rightarrow O_2 + 4H^+ + 4e^-$

FIGURE 17.2 Electrolysis of a solution of potassium nitrate gives hydrogen gas and oxygen gas as products.

cathode, water is reduced.

$$2H_2O(\ell) + 2e^- \longrightarrow H_2(g) + 2OH^-(aq) \qquad \text{(cathode)}$$

At the anode, water is oxidized.

$$2H_2O(\ell) \longrightarrow O_2(g) + 4H^+(aq) + 4e^- \qquad \text{(anode)}$$

If acid–base indicators are placed in the solution, color changes confirm that the solution becomes basic around the cathode and acidic around the anode as the electrolysis proceeds (see Figure 17.3). In addition, the gases H_2 and O_2 can be collected.

The overall cell reaction can be obtained as before. Since the number of electrons lost has to equal the number gained, the cathode reaction must occur twice each time the anode reaction occurs once.

$$4H_2O(\ell) + 4e^- \longrightarrow 2H_2(g) + 4OH^-(aq)$$
$$2H_2O(\ell) \longrightarrow O_2(g) + 4H^+(aq) + 4e^-$$

After adding, we combine the coefficients for water and cancel the electrons from both sides. This gives

$$6H_2O(\ell) \longrightarrow 2H_2(g) + O_2(g) + 4H^+(aq) + 4OH^-(aq)$$

Notice that hydrogen ions and hydroxide ions are produced in equal numbers. If the solution is stirred, they combine to form water. The net change that takes place, then, is

$$2H_2O(\ell) \xrightarrow{\text{electrolysis}} 2H_2(g) + O_2(g)$$

FIGURE 17.3 Electrolysis of a solution of potassium nitrate, KNO_3, in the presence of indicators. The initial yellow color indicates that the solution is neutral (neither acidic nor basic). As the electrolysis proceeds, H^+ produced at the anode along with O_2 causes the solution there to become pink. At the cathode, H_2 is evolved and OH^- ions are formed, which turns the solution around that electrode a bluish violet. After the electrolysis is stopped and the solution is stirred, the color becomes yellow again as the H^+ and OH^- ions formed by the electrolysis neutralize each other.

At this point you may have begun to wonder whether the potassium nitrate serves any function. However, if electrolysis is attempted without the KNO_3 (i.e., using pure distilled water), nothing happens. There is no current flow, and no H_2 or O_2 is produced. Obviously, then, the potassium nitrate must have some function.

The presence of KNO_3 (or some other electrolyte) is necessary to maintain electrical neutrality in the vicinity of the electrodes. If the KNO_3 were not present and the electrolysis were to occur, the solution around the anode would become filled with H^+ ions, with no negative ions to counter the charge. Similarly, the solution surrounding the cathode would become filled with hydroxide ions with no nearby positive ions. Nature simply doesn't allow this to happen.

When KNO_3 is in the solution, K^+ ions move toward the cathode where they mingle with the OH^- ions as they are formed. The NO_3^- ions move toward the anode and mingle with the H^+ ions as they are produced there. In this way, at any moment, each small region of the solution contains the same number of positive and negative charges.

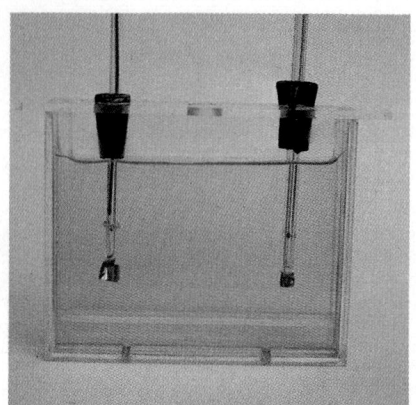

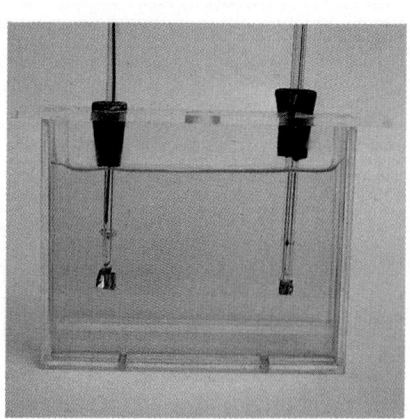

Predicting the outcome of electrolysis reactions in aqueous solutions can be tricky because the reactions at the electrode surfaces are complex, especially when they involve the evolution of hydrogen or oxygen. For example, oxidation of H_2O to give O_2 is thermodynamically easier (in terms of the sign and magnitude of $\Delta G°$) than the oxidation of Cl^- to give Cl_2. Therefore, you would expect that in the electrolysis of a solution of NaCl, where there are these competing reactions at the anode,

$$2H_2O \longrightarrow O_2(g) + 4H^+(aq) + 4e^-$$
$$2Cl^-(aq) \longrightarrow Cl_2(g) + 2e^-$$

oxygen should be evolved. However, because of the complexity of the electrode reactions, it is actually chlorine that is formed (provided the chloride ion concentration is reasonably high). Thus, even though evolution of O_2 is thermodynamically preferred, we don't see this happen because of other complicating factors.

Although it can be hard to anticipate beforehand what will happen in the electrolysis of aqueous solutions, we still can use what we learn experimentally about one electrolysis to predict what will happen in others. For instance, when a solution of $CuBr_2$ is electrolyzed, the cathode becomes coated with a reddish brown deposit of copper and the blue color of copper ion in the surrounding solution fades. At the same time, the solution around the anode acquires a reddish brown color. These observations tell us that copper ion is being reduced at the cathode and that bromide ion is being oxidized to bromine at the anode. The electrode reactions are

$$Cu^{2+}(aq) + 2e^- \longrightarrow Cu(s) \qquad \text{(cathode)}$$
$$2Br^-(aq) \longrightarrow Br_2(aq) + 2e^- \qquad \text{(anode)}$$

and the net cell reaction is

$$Cu^{2+}(aq) + 2Br^-(aq) \xrightarrow{\text{electrolysis}} Cu(s) + Br_2(aq)$$

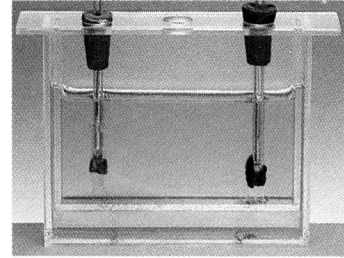

Electrolysis of a solution of copper(II) bromide. The blue color of the solution is due to copper ion (actually, $Cu(H_2O)_4{}^{2+}$). At the anode on the left, bromide ion is oxidized to bromine, which imparts an orange-yellow color to the solution around the electrode. At the cathode, copper is reduced. Here we see it as a black deposit building up on the electrode and flaking away to fall to the bottom of the apparatus.

The behavior of a particular ion in solution toward oxidation or reduction by electrolysis is the same regardless of the source of the ion. Therefore, once we know what happens in the electrolysis of $CuBr_2$, we can predict at least partially what will happen in the electrolysis of other salts that contain either of these ions. Solutions of $CuCl_2$, $CuSO_4$, $Cu(NO_3)_2$, and $Cu(C_2H_3O_2)_2$ all contain Cu^{2+} ion, and when they are electrolyzed we can expect that in each case Cu^{2+} will be reduced at the cathode. (Although we can't say at this point what will happen at the anode.) Similarly, solutions of KBr, NaBr, $CaBr_2$, and $FeBr_2$ all contain Br^-, so we can expect that at the anode bromide ion will be oxidized to give Br_2 when solutions of these salts are electrolyzed.

EXAMPLE 17.1 PREDICTING THE OUTCOME OF AN ELECTROLYSIS

Problem: From the results of the electrolysis of solutions of KNO_3 and of $CuBr_2$, predict the products that will appear if a solution of $Cu(NO_3)_2$ is electrolyzed. Give the net cell reaction.

Solution: A solution of $Cu(NO_3)_2$ contains Cu^{2+} and $NO_3{}^-$ ions. The Cu^{2+} will move toward the cathode and $NO_3{}^-$ will move toward the anode. The electrolyses described previously in this section tell us that Cu^{2+} is more easily reduced than water at the cathode and that water is more easily oxidized than $NO_3{}^-$ at the anode. The electrode reactions, therefore, will be

$$Cu^{2+}(aq) + 2e^- \longrightarrow Cu(s) \qquad \text{(cathode — reduction of } Cu^{2+})$$
$$2H_2O(\ell) \longrightarrow O_2(g) + 4H^+(aq) + 4e^- \qquad \text{(anode — oxidation of } H_2O)$$

To obtain the cell reaction, we must multiply the cathode reaction by 2 so that the numbers of electrons gained and lost are equal. This gives

$$2Cu^{2+}(aq) + 2H_2O(\ell) \longrightarrow 2Cu(s) + O_2(g) + 4H^+(aq)$$

PRACTICE EXERCISE 1   From the electrolysis reactions described in this section, predict the products that will form when a solution of KBr undergoes electrolysis. Write the equation for the net cell reaction.

17.3 STOICHIOMETRIC RELATIONSHIPS IN ELECTROLYSIS

Experimental measurements of electric current and time can be used to calculate the amount of chemical change during electrolysis.

Michael Faraday.

Much of the early research in electrochemistry was performed by a British scientist named Michael Faraday (1791–1867). It was he who coined the terms *anode, cathode, electrode, electrolyte,* and *electrolysis.* In about 1833, Faraday discovered that the amount of chemical change that occurs during electrolysis is directly proportional to the amount of electricity that is passed through an electrolysis cell. For example, the reduction of copper ion at a cathode is given by the equation

$$Cu^{2+}(aq) + 2e^- \longrightarrow Cu(s)$$

To deposit one mole of metallic copper requires two moles of electrons. To deposit two moles of copper requires four moles of electrons, and that takes twice as much electricity.

A unit normally used in electrochemistry to mean one mole of electrons is the faraday ($\mathscr{F}$), named in honor of Michael Faraday.

$$1 \ \mathscr{F} = 1 \ \text{mol} \ e^-$$

The half-reaction for an oxidation or reduction, therefore, relates the amount of chemical substance consumed or produced to the number of faradays that the electric current must supply.

There are electronic devices that measure amperes directly.

To use the faraday we must relate it to electrical measurements that can be made in the laboratory. The SI unit of electric current is the ampere (A) and the SI unit of charge is the coulomb (C). A coulomb is the amount of charge that passes by a given point in a wire when an electric current of one ampere flows for one second. This means that coulombs are the product of amperes of current multiplied by seconds. Thus

$$1 \ \text{coulomb} = 1 \ \text{ampere} \times 1 \ \text{second}$$

A · s means *amperes × seconds.* The dot means "multiplied by."

$$1 \ \text{C} = 1 \ \text{A} \cdot \text{s}$$

For example, if a current of 4 A flows for 10 s, 40 C pass by a given point in the wire.

$$(4 \ \text{A}) \times (10 \ \text{s}) = 40 \ \text{A} \cdot \text{s}$$
$$= 40 \ \text{C}$$

Experimentally, it has been determined that 1 faraday (1 mol of electrons) carries a charge of 96,485 C. We will use this number rounded to three significant figures.

$$1 \ \mathscr{F} = 96,500 \ \text{C} \qquad \text{(to three significant figures)}$$

We can also say that
$$1 \ \mathscr{F} = \frac{96,500 \ \text{C}}{1 \ \text{mol} \ e^-}$$

Now we have a way to relate laboratory measurements to the amount of chemical change that occurs during an electrolysis. Measuring current and time allows us to calculate the number of coulombs. From this we can get faradays, which we can then use to calculate the amount of chemical change produced.

EXAMPLE 17.2 QUANTITATIVE PROBLEMS ON ELECTROLYSIS

Problem: How many grams of copper are deposited on the cathode of an electrolytic cell if an electric current of 2.00 A is run through a solution of $CuSO_4$ for a period of 20.0 min?

Solution: First we convert minutes to seconds.

$$20.0 \text{ min} \times \left(\frac{60 \text{ s}}{1 \text{ min}} \right) = 1200 \text{ s}$$

Then we multiply the current by the time to obtain the number of coulombs $(1 \text{ A} \cdot \text{s} = 1 \text{ C})$.

$$(1200 \text{ s}) \times (2.00 \text{ A}) = 2400 \text{ A} \cdot \text{s}$$
$$= 2400 \text{ C}$$

Since $1 \, \mathscr{F} = 96{,}500 \text{ C}$,

$$2400 \text{ C} \times \left(\frac{1 \, \mathscr{F}}{96{,}500 \text{ C}} \right) = 0.0249 \, \mathscr{F}$$

Next we need the equation for the reduction of copper ion so that we can relate faradays to moles of copper. This equation is

$$Cu^{2+} + 2e^- \longrightarrow Cu$$

It provides us with the conversion factors

$$\frac{1 \text{ mol Cu}}{2 \, \mathscr{F}} \qquad \frac{2 \, \mathscr{F}}{1 \text{ mol Cu}}$$

Using the first of these along with the atomic weight of copper,

$$0.0249 \, \mathscr{F} \times \left(\frac{1 \text{ mol Cu}}{2 \, \mathscr{F}} \right) \times \left(\frac{63.5 \text{ g Cu}}{1 \text{ mol Cu}} \right) = 0.791 \text{ g Cu}$$

The electrolysis will deposit 0.791 g of copper on the cathode.

EXAMPLE 17.3 QUANTITATIVE PROBLEMS ON ELECTROLYSIS

Problem: Electroplating is an important application of electrolysis. How much time would it take in minutes to deposit 0.500 g of metallic nickel on a metal object using a current of 3.00 A? The nickel is reduced from the +2 oxidation state.

Solution: First, we need an equation for the reduction. Since the nickel is reduced to the free metal from the +2 state, we can write

$$Ni^{2+} + 2e^- \longrightarrow Ni$$

This gives the conversion factors

$$\frac{1 \text{ mol Ni}}{2 \, \mathscr{F}} \qquad \frac{2 \, \mathscr{F}}{1 \text{ mol Ni}}$$

We wish to deposit 0.500 g of Ni. This must be converted to moles.

$$0.500 \text{ g Ni} \times \left(\frac{1 \text{ mol Ni}}{58.7 \text{ g Ni}} \right) = 0.00852 \text{ mol Ni}$$

Next we determine the number of faradays required.

$$0.00852 \text{ mol Ni} \times \left(\frac{2 \, \mathscr{F}}{1 \text{ mol Ni}} \right) = 0.0170 \, \mathscr{F}$$

Then we calculate the number of coulombs.

$$0.0170 \, \mathscr{F} \times \left(\frac{96{,}500 \text{ C}}{1 \, \mathscr{F}} \right) = 1640 \text{ C} \qquad \text{(to three significant figures)}$$
$$= 1640 \text{ A} \cdot \text{s}$$

This tells us that the product of current multiplied by time equals 1640 A · s. The current is 3.00 A. Dividing 1640 A · s by 3.00 A gives the time required in seconds, which can then be converted to minutes.

$$\left(\frac{1640 \, A \cdot s}{3.00 \, A}\right) \times \left(\frac{1 \, min}{60 \, s}\right) = 9.11 \, min$$

The time required is 9.11 min.

EXAMPLE 17.4   QUANTITATIVE PROBLEMS ON ELECTROLYSIS

Problem: What current is needed to deposit 0.500 g of chromium metal from a solution of Cr^{3+} in a period of 1.00 hr?

Solution: This problem is quite similar to Example 17.3. First we write the equation for the reaction.

$$Cr^{3+} + 3e^- \longrightarrow Cr$$

This tells us that 3 $\mathscr{F}$ is needed for each mole of Cr produced. Now we can calculate the number of faradays required.

$$0.500 \, g \, Cr \times \left(\frac{1 \, mol \, Cr}{52.0 \, g \, Cr}\right) \times \left(\frac{3 \, \mathscr{F}}{1 \, mol \, Cr}\right) = 0.0288 \, \mathscr{F}$$

Then we convert faradays to coulombs.

$$0.0288 \, \mathscr{F} \times \frac{96,500 \, C}{1 \, \mathscr{F}} = 2780 \, C \qquad \text{(to three significant figures)}$$

$$= 2780 \, A \cdot s$$

Since we want the metal to be deposited in 1 hr (3600 s), the current is

$$\frac{2780 \, A \cdot s}{3600 \, s} = 0.772 \, A$$

The current required is 0.772 A.

PRACTICE EXERCISE 2   How many *moles* of hydroxide ion will be produced at the cathode during the electrolysis of water with a current of 4.00 A for a period of 200 s? The cathode reaction is $2e^- + 2H_2O \rightarrow H_2 + 2OH^-$.

PRACTICE EXERCISE 3   How long (in minutes) will it take a current of 10.0 A to deposit 3.00 g of gold from a solution of $AuCl_3$?

PRACTICE EXERCISE 4   What current must be supplied to deposit 3.00 g of gold from a solution of $AuCl_3$ in 20.0 min?

17.4   APPLICATIONS OF ELECTROLYSIS

Commercial applications of electrolysis include electroplating; the production of aluminum, magnesium, and sodium; the refining of copper; and the synthesis of sodium hydroxide and sodium hypochlorite.

Besides being a useful tool in the chemistry laboratory, electrolysis has many important industrial applications. In this section we will briefly examine the chemistry of electroplating and the production of some of our most common chemicals.

Electroplating

Electroplating — the application by electrolysis of a thin ornamental or protective coating of one metal over another — is a common technique for improving the appearance and durability of metal objects. For instance, a thin, shiny coating of metallic chromium is applied over steel automobile bumpers to make them attractive and to prevent rusting of the steel. Silver and gold plating is applied to jewelry made from less expensive metals, and silver plating is common on eating utensils (knives, forks, spoons, etc.). These thin metallic layers, generally 0.03 to 0.05 mm (0.001 to 0.002 inches) thick, are usually applied by electrolysis.

Figure 17.4 illustrates a typical apparatus used for plating silver. Silver ion in the solution is reduced at the cathode where it is deposited as metallic silver on the object to be plated. At the anode, silver from the metal bar is oxidized, replenishing the supply of the silver ion in the solution. As time passes, silver is gradually transferred from the bar at the anode onto the object at the cathode.

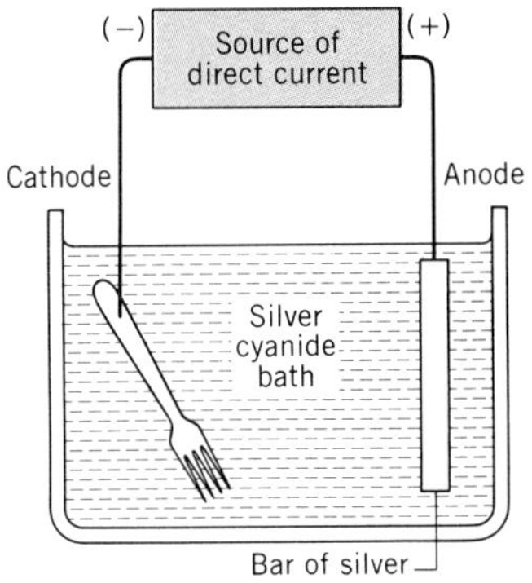

The exact composition of the electroplating bath varies, depending on the metal to be deposited, and can affect the appearance and durability of the finished surface. For example, silver deposited from a solution of silver nitrate ($AgNO_3$) does not stick to other metal surfaces very well. However, if it is deposited from a solution of silver cyanide containing $Ag(CN)_2^-$, the coating adheres well and is bright and shiny. Other metals that are electroplated from a cyanide bath are gold and cadmium. Nickel, which can also be applied as a protective coating, is plated from a nickel sulfate solution, and chromium is plated from a chromic acid (H_2CrO_4) solution.

FIGURE 17.4 Apparatus for electroplating silver.

Production of Aluminum

Until the latter part of the nineteenth century, aluminum was an uncommon metal— only the rich could afford aluminum products. A student at Oberlin College, 21-year-old Charles M. Hall, learned of this and began a series of experiments in an attempt to invent a cheap method of extracting the metal from its compounds. The difficulty that he faced was that aluminum is a very reactive element. It is difficult to produce as a free element by usual chemical reactions. Efforts to produce aluminum by electrolysis were unproductive because its anhydrous salts were difficult to prepare and its oxide, Al_2O_3, has such a high melting point (over 2000 °C) that no practical method of melting it could be found. In 1886 Hall discovered that Al_2O_3 dissolves in a mineral called cryolite, Na_3AlF_6, to give a conducting mixture with a relatively low melting point from which aluminum could be produced electrolytically.

A diagram of the apparatus used to produce aluminum is shown in Figure 17.5. Aluminum ore, called bauxite, contains Al_2O_3. The ore is purified and the Al_2O_3 is then added to the molten cryolite electrolyte in which it dissolves and dissociates. At the cathode, the aluminum ions are reduced to produce the free metal, which forms as a layer of molten aluminum below the less dense electrolyte. At the carbon anodes, oxide ion is oxidized to give free O_2.

$$Al^{3+} + 3e^- \longrightarrow Al(\ell) \quad \text{(cathode)}$$
$$2O^{2-} \longrightarrow O_2(g) + 4e^- \quad \text{(anode)}$$

The net cell reaction is

$$4Al^{3+} + 6O^{2-} \longrightarrow 4Al(\ell) + 3O_2(g)$$

Aluminum is used as a structural metal and in such products as aluminum foil, electrical wire, alloys, and kitchen utensils.

A large cell can produce as much as 900 lb of aluminum per day.

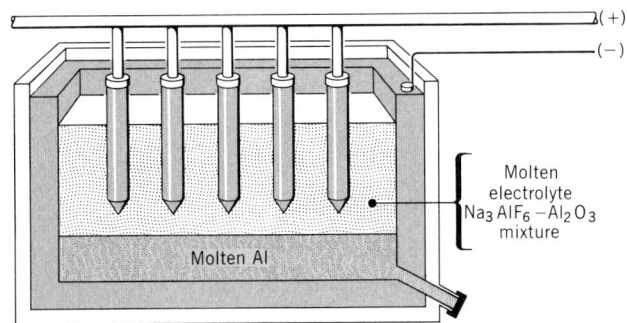

FIGURE 17.5 Diagram of the apparatus used to produce aluminum electrolytically by the Hall process.

A row of electrolytic cells fills this aluminum production plant in Ghana (a country located on the west coast of Africa). In the foreground, a crane is about to lift a large ladle filled with molten aluminum.

The oxygen produced at the anode attacks the carbon electrodes (producing CO_2), so the electrodes must be replaced frequently.

The production of aluminum consumes enormous amounts of electricity and is therefore very costly, not only in terms of dollars but also in terms of energy resources. For this reason, recycling of aluminum should receive higher priority as we seek to minimize our use of energy.

Production of Magnesium

Magnesium is a metal that has found a number of structural uses because of its light weight (low density). Around the home, for example, you may have a magnesium alloy ladder. Magnesium is also the wire inside flashbulbs. It produces a brilliant flash of light when it reacts with oxygen in the bulb to give MgO. The hydroxide of magnesium, $Mg(OH)_2$, is the creamy substance in milk of magnesia.

The major source of magnesium is seawater. On a mole basis, Mg^{2+} is the third most abundant ion in the ocean, exceeded only by sodium ion and chloride ion. To obtain magnesium, seawater is made basic, which precipitates $Mg(OH)_2$. This is separated by filtration and then dissolved in hydrochloric acid.

$$Mg(OH)_2 + 2HCl \longrightarrow MgCl_2 + 2H_2O$$

The solution of $MgCl_2$ is evaporated, and the resulting solid $MgCl_2$ is melted and electrolyzed. Free magnesium is deposited at the cathode and chlorine gas is produced at the anode.

$$MgCl_2(\ell) \xrightarrow{\text{electrolysis}} Mg(\ell) + Cl_2(g)$$

Production of Sodium

Sodium is prepared by the electrolysis of molten sodium chloride. The principles of this were discussed in Section 17.2. From a practical standpoint, the metallic sodium and the chlorine gas that are formed must be kept apart; otherwise they react violently to re-form NaCl. The Downs cell, illustrated in Figure 17.6, accomplishes this separation.

Both products of the electrolysis of molten NaCl are commercially important.

FIGURE 17.6 Cross section of the Downs cell used for the electrolysis of molten sodium chloride. The cathode is a circular ring that surrounds the anode. The electrodes are separated from each other by an iron screen. During the operation of the cell, molten sodium collects at the top of the cathode compartment, from which it is periodically drained. The chlorine gas bubbles out of the anode compartment and is collected.

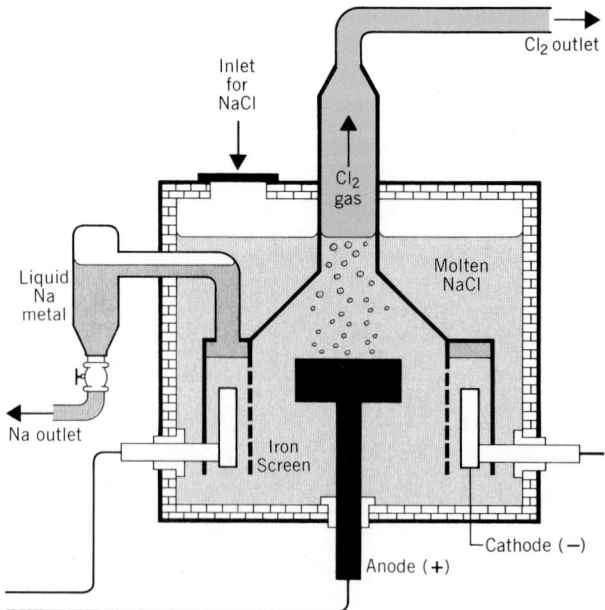

Chlorine is used to purify water and to manufacture many solvents as well as plastics such as polyvinyl chloride (PVC). Sodium is used in the manufacture of tetraethyllead (an additive for gasoline that is gradually being phased out), as a coolant in certain nuclear reactors, and in the production of sodium vapor lamps. Sodium lamps, which produce a bright yellow color, have the advantage of giving off most of their energy in a portion of the spectrum that humans can see. In terms of useful light output versus energy input, they are about 15 times more efficient than ordinary incandescent light bulbs and about three to four times more efficient than the bluish white mercury vapor lamps used for street lighting.

Refining of Copper

One of the most interesting and economically attractive applications of electrolysis is the purification or refining of metallic copper. When copper is first removed from its ore, it is about 99% pure. The impurities — mostly silver, gold, platinum, iron, and zinc — decrease the electrical conductivity of the copper significantly enough that it must be further refined before it can be used in electrical wire. Figure 17.7 illustrates how this is done.

The impure copper is used as the anode in an electrolysis cell that contains a solution of copper sulfate and sulfuric acid as the electrolyte. The cathode is a thin sheet of very pure copper. When the cell is operated at the correct voltage, only copper and impurities more easily oxidized than copper (iron and zinc) dissolve at the anode. The less active metals simply settle to the bottom of the container. At the cathode, copper ion is reduced, but the zinc ions and iron ions remain in solution because they are more difficult to reduce than copper. Gradually, the impure copper anode dissolves and the copper cathode grows. This copper is now about 99.95% pure. The sludge — called *anode mud* — that accumulates in the vessel is removed periodically and the value of the silver, gold, and platinum recovered from it virtually pays for the entire refining operation.

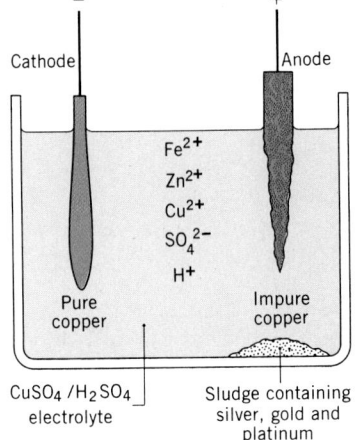

FIGURE 17.7 Purification of copper by electrolysis.

This is one of the chief sources of gold in the United States.

Anticipating a cut in electricity costs of about 40%, the town of Oyster Bay, New York, replaced all its incandescent street lights with efficient low-pressure sodium lamps. Residents now complain they can't tell the street lights from the yellow of a traffic light.

Copper cathodes, 99.96% pure, are pulled from the electrolytic refining tanks at Kennecott's Utah copper refinery. It takes about 28 days for the impure copper anodes to dissolve and deposit the pure metal on the cathodes.

FIGURE 17.8 Electrolysis of brine.

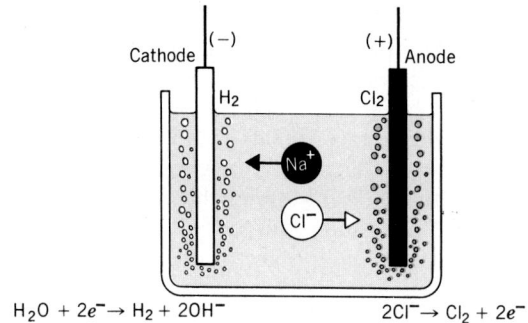

$$H_2O + 2e^- \rightarrow H_2 + 2OH^- \qquad\qquad\qquad 2Cl^- \rightarrow Cl_2 + 2e^-$$

Electrolysis of Brine

One of the most important commercial electrolysis reactions is the electrolysis of concentrated aqueous sodium chloride solutions called brine. An apparatus that could be used for this in the laboratory is shown in Figure 17.8. At the cathode, water is much more easily reduced than sodium ion, so H_2 is formed.

$$2e^- + 2H_2O \longrightarrow H_2(g) + 2OH^-(aq) \qquad \text{(cathode)}$$

As we noted earlier, even though water is more easily oxidized than chloride ion, complicating factors at the electrodes actually allow chloride ion to be oxidized instead. At the anode, therefore, we observe the formation of Cl_2.

$$2Cl^-(aq) \longrightarrow Cl_2(g) + 2e^- \qquad \text{(anode)}$$

The net cell reaction is therefore

$$2Cl^-(aq) + 2H_2O \longrightarrow H_2(g) + Cl_2(g) + 2OH^-(aq)$$

If we include the sodium ion that's in the solution, which is a spectator ion and not involved in the electrolysis directly, we can see why this is such an important reaction.

$$\underbrace{2Na^+(aq) + 2Cl^-(aq)}_{2NaCl(aq)} + 2H_2O \longrightarrow H_2(g) + Cl_2(g) + \underbrace{2Na^+(aq) + 2OH^-(aq)}_{2NaOH(aq)}$$

The electrolysis changes inexpensive salt to valuable chemicals: H_2, Cl_2, and NaOH. The H_2 is used to make chemicals, including hydrogenated vegetable oils; the Cl_2 is used to chlorinate drinking water and to make plastics, pesticides, and solvents. Sodium hydroxide, also known as lye or caustic soda, is one of industry's most important bases. Among its uses are the manufacture of soap and paper, the neutralization of acids in industrial reactions, and the purification of aluminum ores.

In the industrial electrolysis of brine to make NaOH, an apparatus as simple as that in Figure 17.8 can't be used. First, it is necessary to capture the H_2 and Cl_2 separately and to prevent them from mixing. Mixtures of H_2 and Cl_2 are explosive. Second, the NaOH from the reaction is contaminated with unreacted NaCl. And third, the NaOH would be contaminated by hypochlorite ion (OCl^-) formed by the reaction of Cl_2 with OH^-.

$$Cl_2 + 2OH^- \longrightarrow Cl^- + OCl^- + H_2O$$

Although this reaction is undesirable in the production of NaOH, it is used to make sodium hypochlorite solutions. The solution is stirred vigorously during the electrolysis reaction, so very little Cl_2 escapes. As a result, the solution of NaCl is changed gradually to a solution of NaOCl. When this solution is diluted, it is sold as liquid laundry bleach (e.g., Clorox®).

Most of the NaOH manufactured today is made in an apparatus called a diaphragm cell. The design varies somewhat, but Figure 17.9 illustrates its basic features. The cell consists of an iron wire mesh cathode that encloses a porous asbestos shell—the

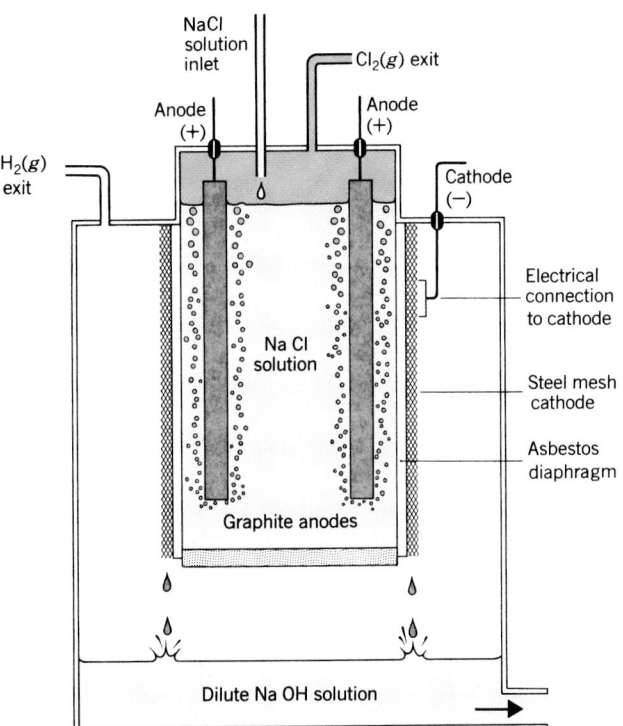

FIGURE 17.9 A diaphragm cell used in the production of NaOH by the electrolysis of aqueous NaCl. This is a cross section of a cylindrical cell in which the NaCl solution is surrounded by an asbestos diaphragm supported by an iron mesh cathode. (From J. E. Brady and G. E. Humiston, *General Chemistry: Principles and Structure,* 4th ed. Copyright © 1986, John Wiley & Sons, New York. Used by permission.)

diaphragm. The NaCl solution is added to the top of the cell and seeps slowly through the diaphragm. When it contacts the iron cathode, hydrogen is evolved and is pumped out of the surrounding space. The solution, now containing dilute NaOH, drips off the cell into the reservoir below. Meanwhile, within the cell, chlorine is generated at the anodes dipping into the NaCl solution. Because there is no OH⁻ in this solution, the Cl_2 can't react to form OCl⁻ ion and simply bubbles out of the solution and is captured.

The NaOH obtained from the diaphragm cell, although free of OCl⁻, is still contaminated by small amounts of NaCl. The production of pure NaOH is accomplished using a mercury cell, illustrated in Figure 17.10. In this apparatus, mercury serves as the

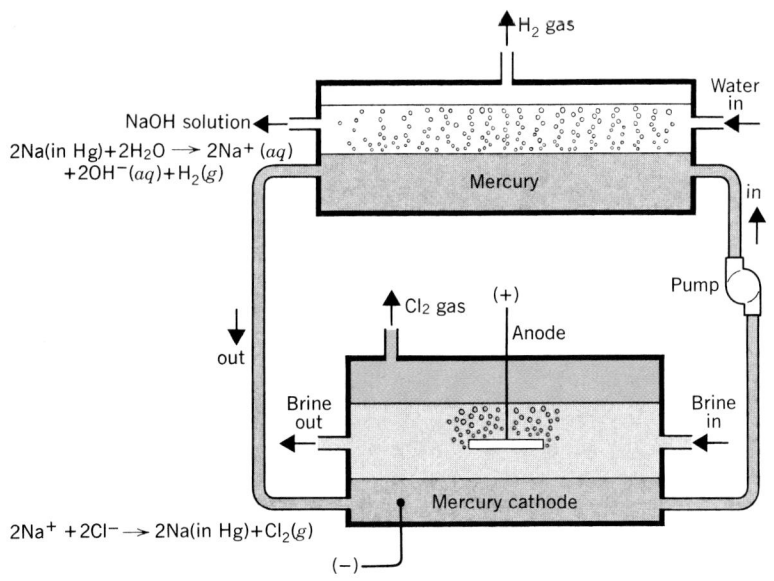

FIGURE 17.10 Electrolysis of aqueous sodium chloride (brine) using a mercury cell. At the anode, chloride ions are oxidized to chlorine, Cl_2. At the cathode, sodium ions are reduced to sodium atoms, which dissolve in the mercury. This mercury is pumped to a separate compartment, where it is exposed to water. As sodium atoms come to the surface of the mercury, they react with water to give H_2 and NaOH. (From J. E. Brady and G. E. Humiston, *General Chemistry: Principles and Structure,* 4th ed. Copyright © 1986, John Wiley & Sons, New York. Used by permission.)

cathode, and the cathode reaction actually involves the reduction of sodium ion to give sodium atoms, which dissolve in the liquid mercury. This mercury is pumped to another chamber where it is exposed to water that doesn't contain any NaCl. In this compartment, the sodium atoms react with water to give H_2 and NaOH. The net overall chemical change is still the conversion of NaCl into H_2, Cl_2, and NaOH, but the NaOH produced in the mercury cell isn't contaminated by NaCl. Although the use of the mercury cell gives pure NaOH, it poses an environmental hazard because of the potential for mercury pollution. The waste products discharged from plants using mercury cells must be carefully monitored to prevent this.

17.5 GALVANIC CELLS

Two half-reactions, one an oxidation and the other a reduction, can be set up so that the spontaneous electron transfer must occur through an external electrical circuit.

If you have silver fillings in your teeth, you may have experienced a strange and perhaps even unpleasant sensation while accidentally biting on a piece of aluminum foil. That sensation was caused by a very mild electric shock produced by a voltage difference between your metal fillings and the aluminum. In effect, you created a battery. Batteries not too much different from this serve to power all sorts of gadgets, as mentioned at the beginning of this chapter. The energy for this comes from spontaneous redox reactions in which the electron transfer is forced to take place through a wire. Cells that provide electricity in this way are called galvanic cells,[1] after Luigi Galvani (1737–1798), an Italian anatomist who discovered that electricity can cause the contraction of muscles.

If a shiny piece of metallic copper is placed into a solution of silver nitrate, a spontaneous reaction occurs. A grayish white deposit is formed on the copper, and the solution itself becomes pale blue. This is shown in Figure 17.11. The reaction that takes place is

$$2Ag^+(aq) + Cu(s) \longrightarrow Cu^{2+}(aq) + 2Ag(s)$$

No usable energy can be harnessed from this process, however, because the energy change that accompanies the reaction is lost as heat.

However, the same chemical reaction can occur and produce usable energy if the two half-reactions involved in this net reaction are made to occur in separate containers.

FIGURE 17.11 (Left) A coiled piece of copper wire next to a beaker containing a silver nitrate solution. (Center) When the copper wire is placed in the solution, copper dissolves, giving the solution its blue color, and metallic silver deposits as glittering crystals on the wire. (Right) After a while, much of the copper has dissolved and nearly all of the silver has deposited as the free metal.

[1] They are also often called voltaic cells, after Alessandro Volta (1745–1827), the inventor of the battery.

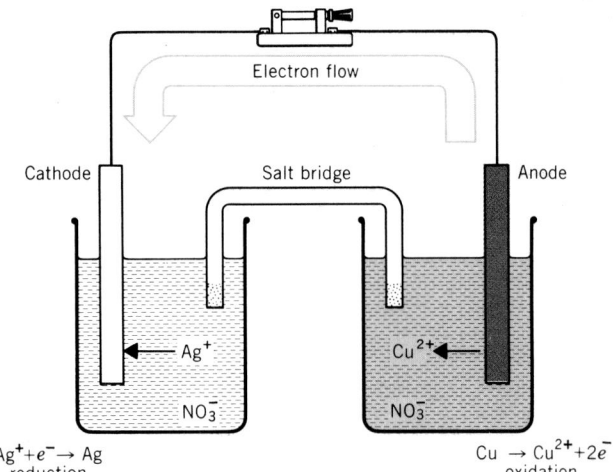

FIGURE 17.12 A galvanic cell.

An apparatus to accomplish this — a galvanic cell — is shown in Figure 17.12. On the left, a silver electrode is dipping into a solution of $AgNO_3$; on the right, a copper electrode is dipping into a $Cu(NO_3)_2$ solution. The two solutions are connected by a *salt bridge,* the function of which will be described shortly, and by an external electrical circuit. When the circuit is completed, reduction of Ag^+ to Ag occurs in the beaker on the left and oxidation of copper occurs in the beaker on the right. Because of the nature of the reactions taking place, we can identify the silver electrode as the cathode and the copper electrode as the anode.

$$Ag^+(aq) + e^- \longrightarrow Ag(s) \qquad \text{(reduction — cathode)}$$
$$Cu(s) \longrightarrow Cu^{2+}(aq) + 2e^- \qquad \text{(oxidation — anode)}$$

When these reactions take place, electrons left behind by oxidation of the copper travel through the external circuit as an electric current to the cathode where they are picked up by the silver ions, which are thereby reduced.

For a galvanic cell to work, the solutions in both compartments, or half-cells, must remain electrically neutral. However, when copper is oxidized, the solution surrounding the electrode becomes filled with Cu^{2+} ions. Similarly, when Ag^+ ions are reduced, NO_3^- ions are left behind in the solution. But these redox reactions can only take place if there are enough ions of opposite charge present to compensate for the ions that the reactions produce or remove. The salt bridge shown in Figure 17.12 serves to maintain this necessary electrical neutrality. A salt bridge is a tube filled with an electrolyte, commonly KNO_3 or KCl, and fitted with porous plugs at either end. During the cell reaction, either negative ions diffuse from the salt bridge into the copper half-cell or Cu^{2+} ions diffuse into the salt bridge to keep that half-cell electrically neutral. At the same time, the silver half-cell is kept electrically neutral by the diffusion of positive ions from the salt bridge or NO_3^- ions into the salt bridge. Without the salt bridge, then, no electrical current would be produced by the galvanic cell. Electrolytic contact — contact by means of a solution containing ions — must be maintained for the cell to function.

Charges on the Electrodes

In an electrolysis cell, the cathode carries a negative charge and the anode carries a positive charge. In a galvanic cell, these charges are reversed. At the anode of the cell in Figure 17.12, copper atoms leave the electrode and enter the solution as Cu^{2+} ions. The electrons that are left behind give the anode a negative charge. At the cathode, electrons

Oxidation *always* occurs at the anode, and reduction *always* occurs at the cathode.

are joining Ag^+ ions to produce neutral atoms, but the effect is the same as if Ag^+ ions became part of the electrode, so the cathode acquires a positive charge. The difference in charge between cathode and anode is what causes the electric current to flow when the circuit is complete. Remember, however, that it is the nature of the chemical change, not the electrical charge, that determines whether we label an electrode as a cathode or an anode.

Electrolytic Cell	Galvanic Cell
Cathode is negative (reduction).	Cathode is positive (reduction).
Anode is positive (oxidation).	Anode is negative (oxidation).

Even though the charges on the cathode and anode differ between electrolytic cells and galvanic cells, the ions in solution always move in the same direction. A cation is a positive ion that always moves away from the anode toward the cathode. In both types of cells, positive ions move toward the cathode. They are attracted there by the negative charge on the cathode in an electrolysis cell; they diffuse toward the cathode in our galvanic cell to balance the charge of negative ions left behind when the Ag^+ ions are reduced. Similarly, anions are negative ions that move away from the cathode and toward the anode. They are attracted to the positive anode in an electrolysis cell, and they diffuse toward the anode in our galvanic cell to balance the charge of the Cu^{2+} ions entering the solution.

EXAMPLE 17.5   DESCRIBING GALVANIC CELLS

Problem: The following spontaneous reaction occurs when metallic zinc is dipped into a solution of copper sulfate.

$$Zn(s) + Cu^{2+}(aq) \longrightarrow Zn^{2+}(aq) + Cu(s)$$

Describe a galvanic cell that could take advantage of this reaction. What are the half-cell reactions? Make a sketch of the cell and label the cathode and anode, the charges on each electrode, the direction of ion flow, and the direction of electron flow.

Solution: Two half-cells are needed. One must contain a zinc electrode that dips into a solution containing Zn^{2+} [e.g., $Zn(NO_3)_2$ or $ZnSO_4$]. The other must contain a copper electrode that dips into a solution containing Cu^{2+} [e.g., $Cu(NO_3)_2$ or $CuSO_4$].

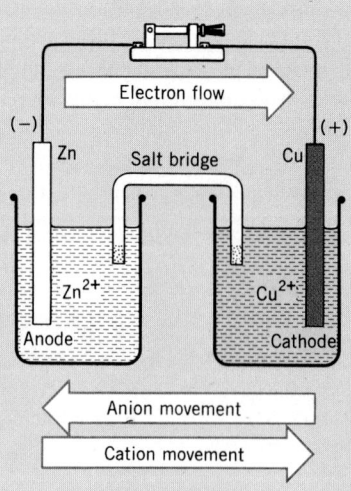

Dividing the overall reaction into half-reactions, we get

$$Zn \longrightarrow Zn^{2+} + 2e^- \quad \text{(oxidation)}$$
$$Cu^{2+} + 2e^- \longrightarrow Cu \quad \text{(reduction)}$$

Since zinc is oxidized, it is the anode and carries a negative charge; copper is the cathode and is positive. Anions move toward the anode and cations move toward the cathode. Electrons travel from the negative anode to the positive cathode via the external circuit.

PRACTICE EXERCISE 5   Sketch and label a galvanic cell that makes use of the following spontaneous redox reaction.

$$Mg(s) + Fe^{2+}(aq) \longrightarrow Mg^{2+}(aq) + Fe(s)$$

17.6 CELL POTENTIALS AND REDUCTION POTENTIALS

The voltage across the electrodes of a galvanic cell can be attributed to the difference in the tendencies of the two half-cells to undergo reduction.

Voltage or electromotive force (emf) can conveniently be thought of as the force with which an electric current is pushed through a wire. It is measured in terms of an electrical unit called the volt (V). Strictly speaking, voltage is a measure of the amount of energy that can be delivered by a coulomb of electrical charge as it passes through a circuit. A current flowing under an emf of 1 volt can deliver 1 joule of energy per coulomb.

$$1 V = 1 J/C \qquad (17.1)$$

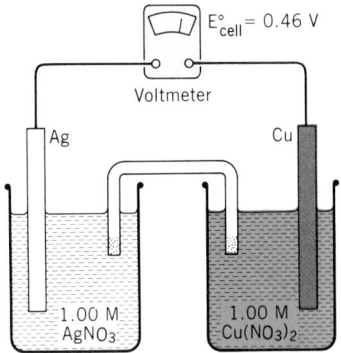

FIGURE 17.13 Cell designed to generate the standard cell potential.

If current is drawn from a cell, some of the cell's emf is lost overcoming its own internal resistance, and the measured voltage isn't E_{cell}.

We will find this definition important in Section 17.8.

The maximum emf of a galvanic cell is called the cell potential, E_{cell}. It is measured with a device that draws negligible current during the measurement. E_{cell} depends on the composition of the electrodes and the concentrations of the ions in each of the half-cells. For reference purposes, the standard cell potential, symbolized E_{cell}°, is the potential of the cell when all of the ion concentrations are 1.00 M, the temperature is 25 °C, and any gases that are involved in the cell reaction are at a pressure of 1 atm.

Cell potentials are rarely larger than a few volts. For example, the standard cell potential for the galvanic cell constructed from silver and copper electrodes shown in Figure 17.13 is only 0.46 V, and one cell in an automobile battery produces only about 2 V. Batteries that generate higher voltages contain a number of cells arranged in series so that their emfs are additive.

Reduction Potentials

As we will see, it is useful to imagine that the measured overall cell potential arises from a competition between the two half-cells for electrons. We think of each half-cell reaction as having a certain natural tendency to proceed as a reduction, the magnitude of which is expressed by its reduction potential (or standard reduction potential when the temperature is 25 °C, concentrations are 1 M, and the pressure is 1 atm). When two half-cells are connected, the one with the larger reduction potential—the one with the greater tendency to undergo reduction—acquires electrons from the half-cell with the lower reduction potential, which is therefore forced to undergo oxidation. The measured cell potential actually represents the magnitude of the *difference* between the reduction potential of one half-cell and the reduction potential of the other. In general,

A problem that has plagued humanity ever since the discovery of methods for obtaining iron and other metals from their ores has been corrosion—the reaction of a metal with substances in the environment. The rusting of iron, in particular, is a serious problem because iron and steel have so many uses. The process appears to require two important ingredients, oxygen and water, and involves electrochemical reactions. One of these reactions is the oxidation of iron.

$$Fe(s) \longrightarrow Fe^{2+}(aq) + 2e^-$$

The Fe^{2+} ions enter the water and move to the surface where they react with oxygen and water to form a hydrated form of ferric oxide, Fe_2O_3, which is rust. At the same time, hydrogen ions in the water are reduced.

$$H^+ + e^- \longrightarrow H$$

The hydrogen atoms either combine to form H_2 or react with oxygen to form H_2O. These reactions occur because hydrogen has a higher reduction potential than iron.

One way to prevent the rusting of iron is to coat it with another metal. This is done with "tin" cans, which are actually steel cans that have been coated with a thin layer of tin. However, if the layer of tin is scratched and the iron beneath is exposed, the corrosion is accelerated because iron has a lower reduction potential than tin—the iron becomes the anode in an electrochemical cell and is easily oxidized.

Another way to prevent corrosion is called *cathodic protection*. It involves placing the iron in contact with a metal that is more easily oxidized. This other metal is then the anode and tends to be oxidized. This keeps the iron itself from corroding. The photograph here shows how zinc wire was laid alongside the Alaskan pipeline, to which it is electrically connected. This wire protects the pipeline from

Zinc cable

A zinc wire is buried alongside the Alaskan pipeline. When electrically connected to the pipe, a galvanic cell is established with the zinc as the anode and the steel pipe as the cathode. Corrosion of the pipe is prevented because when oxidation does occur, it is the zinc that corrodes, not the steel.

corroding by serving as a *sacrificial anode*. Similar zinc anodes are placed on the exposed underwater metal surfaces of ships and boats to protect them from the corrosive effects of seawater. Galvanizing, the coating of iron with zinc, protects such common objects as chain-link fences and metal garbage pails.

$$E^\circ_{cell} = \begin{pmatrix} \text{standard reduction} \\ \text{potential of} \\ \text{substance reduced} \end{pmatrix} - \begin{pmatrix} \text{standard reduction} \\ \text{potential of} \\ \text{substance oxidized} \end{pmatrix} \qquad (17.2)$$

As an example, let's look at the silver–copper cell. From the cell reaction,

$$2Ag^+(aq) + Cu(s) \longrightarrow Cu^{2+}(aq) + 2Ag(s)$$

we can see that silver ion is reduced and copper is oxidized. If we compare the two possible reduction half-reactions,

$$Ag^+(aq) + e^- \longrightarrow Ag(s)$$
$$Cu^{2+}(aq) + 2e^- \longrightarrow Cu(s)$$

the one for Ag^+ *must* have a greater tendency to proceed than the one for Cu^{2+}, because it is the silver ion that is actually reduced. This means that the standard reduction potential of Ag⁺, $E°_{Ag^+}$, must be larger than the standard reduction potential of Cu^{2+}, $E°_{Cu^{2+}}$. If we knew the values of $E°_{Ag^+}$ and $E°_{Cu^{2+}}$, we could calculate $E°_{cell}$ with Equation 17.2 by subtracting the smaller reduction potential from the large one.

$$E°_{cell} = E°_{Ag^+} - E°_{Cu^{2+}}$$

Assigning Standard Reduction Potentials

Unfortunately there is no way to measure the standard reduction potential of an isolated half-cell. All we can measure is the difference that is produced when two half-cells are connected. Therefore, to assign values to the various standard reduction potentials, a reference electrode has been arbitrarily chosen and its standard reduction potential has been assigned a value of 0.00 V. This reference electrode is called the hydrogen electrode, illustrated in Figure 17.14. Gaseous hydrogen at a pressure of 1 atm is bubbled over a platinum electrode that is coated with very finely divided platinum, which causes the electrode reaction to occur rapidly. This electrode is surrounded by a solution whose temperature is 25 °C and in which the hydrogen ion concentration is 1.00 M. The half-cell reaction that occurs at the platinum surface is

$$2H^+(aq, 1.00\ M) + 2e^- \rightleftharpoons H_2(g,\ 1\ atm) \qquad E°_{H^+} = 0.00\ V \text{ at } 25\ °C$$

The double arrows indicate that the reaction is reversible—whether it occurs as oxidation or reduction depends on the reduction potential of the half-cell with which it is paired.

Figure 17.15 illustrates the hydrogen electrode connected to a copper half-cell to form a galvanic cell. To obtain the cell reaction, we have to know what is oxidized and what is reduced—we need to know which is the cathode and which is the anode. We can determine this by measuring the charges on the electrode, because we know that in a galvanic cell the cathode is the positive electrode and the anode is the negative electrode. When we use a voltmeter to measure the potential of the cell, we find that proper measurements are obtained only if the terminal labeled (+) is connected to the copper electrode and the terminal labeled (−) is connected to the hydrogen electrode. Thus, copper must be the cathode, and Cu^{2+} is reduced to Cu when the cell operates. Similarly, hydrogen must be the anode, and H_2 is oxidized to H^+. The half-reactions and cell reaction, therefore, are

$$
\begin{array}{ll}
Cu^{2+}(aq) + 2e^- \longrightarrow Cu(s) & \text{(cathode)} \\
H_2(g) \longrightarrow 2H^+(aq) + 2e^- & \text{(anode)} \\
\hline
Cu^{2+}(aq) + H_2(g) \longrightarrow Cu(s) + 2H^+(aq) & \text{(cell reaction)}
\end{array}
$$

Using Equation 17.2, we can express $E°_{cell}$ in terms of $E°_{Cu^{2+}}$ and $E°_{H^+}$.

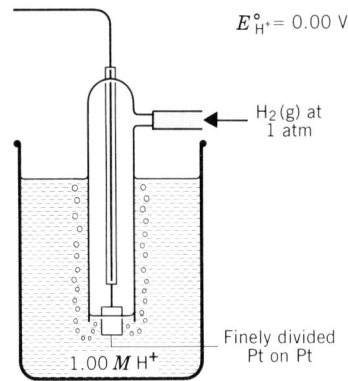

FIGURE 17.14 The hydrogen electrode. The half-reaction is $2H^+(aq) + 2e^- \rightleftharpoons H_2(g)$.

Remember, in a galvanic cell, the cathode is (+) and the anode is (−).

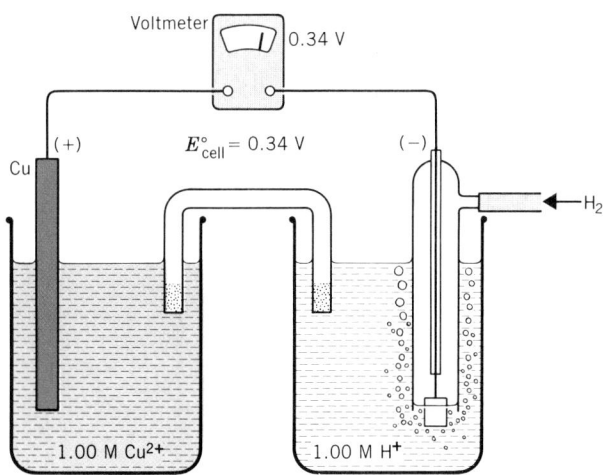

FIGURE 17.15 A galvanic cell composed of copper and hydrogen half-cells. The cell reaction is $Cu^{2+}(aq) + H_2(g) \rightarrow Cu(s) + 2H^+(aq)$.

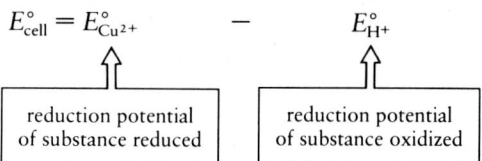

$$E°_{cell} = E°_{Cu^{2+}} \qquad - \qquad E°_{H^+}$$

reduction potential
of substance reduced

reduction potential
of substance oxidized

The measured standard cell potential is 0.34 V and $E°_{H^+}$ equals 0.00 V. Therefore,

$$0.34 \text{ V} = E°_{Cu^{2+}} - 0.00 \text{ V}$$

Relative to the hydrogen electrode, then, the standard reduction potential of Cu^{2+} is 0.34 V.

Now let's look at a galvanic cell set up between a zinc electrode and a hydrogen electrode, as shown in Figure 17.16. This time the voltmeter must have its positive terminal connected to the hydrogen electrode and its negative terminal connected to the zinc electrode—hydrogen is the cathode and zinc is the anode. This means that hydrogen ion is being reduced and zinc is being oxidized. The half-reactions and cell reaction are

$$
\begin{array}{ll}
2H^+(aq) + 2e^- \longrightarrow H_2(g) & \text{(cathode)} \\
Zn(s) \longrightarrow Zn^{2+}(aq) + 2e^- & \text{(anode)} \\
\hline
2H^+(aq) + Zn(s) \longrightarrow H_2(g) + Zn^{2+}(aq) & \text{(cell reaction)}
\end{array}
$$

From Equation 17.2, the standard cell potential is

$$E°_{cell} = E°_{H^+} - E°_{Zn^{2+}}$$

Substituting into this the measured standard cell potential of 0.76 V and $E°_{H^+} = 0.00$ V,

$$0.76 \text{ V} = 0.00 \text{ V} - E°_{Zn^{2+}}$$
$$E°_{Zn^{2+}} = -0.76 \text{ V}$$

The reduction potential of zinc is negative because the tendency of zinc ions to be reduced is less than that of H^+—that's why Zn is oxidized when it is paired with the hydrogen electrode.

The standard reduction potentials of many half-reactions can be compared to that for the hydrogen electrode in the manner described above. Table 17.1 lists values obtained for some typical half-reactions. They are arranged in decreasing order—the half-reactions at the top have the greatest tendency to occur as reduction, while those at the bottom have the least tendency to occur as reduction.

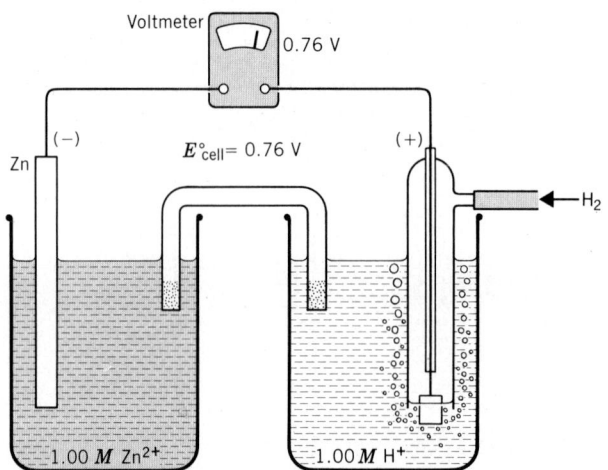

FIGURE 17.16 A galvanic cell composed of zinc and hydrogen half-cells. The cell reaction is $Zn(s) + 2H^+(aq) \rightarrow Zn^{2+}(aq) + H_2(g)$.

TABLE 17.1 Standard Reduction Potentials Measured at 25 °C

Half-Reaction	$E°$ (volts)
$F_2(g) + 2e^- \rightleftharpoons 2F^-(aq)$	+2.87
$PbO_2(s) + SO_4^{2-}(aq) + 4H^+(aq) + 2e^- \rightleftharpoons PbSO_4(s) + 2H_2O$	+1.69
$2HOCl(aq) + 2H^+(aq) + 2e^- \rightleftharpoons Cl_2(g) + 2H_2O$	+1.63
$MnO_4^-(aq) + 8H^+(aq) + 5e^- \rightleftharpoons Mn^{2+}(aq) + 4H_2O$	+1.49
$PbO_2(s) + 4H^+(aq) + 2e^- \rightleftharpoons Pb^{2+}(aq) + 2H_2O$	+1.46
$BrO_3^-(aq) + 6H^+(aq) + 6e^- \rightleftharpoons Br^-(aq) + 3H_2O$	+1.44
$Au^{3+}(aq) + 3e^- \rightleftharpoons Au(s)$	+1.42
$Cl_2(g) + 2e^- \rightleftharpoons 2Cl^-(aq)$	+1.36
$O_2(g) + 4H^+(aq) + 4e^- \rightleftharpoons 2H_2O$	+1.23
$Br_2(aq) + 2e^- \rightleftharpoons 2Br^-(aq)$	+1.07
$NO_3^-(aq) + 4H^+(aq) + 3e^- \rightleftharpoons NO(g) + 2H_2O$	+0.96
$Ag^+(aq) + e^- \rightleftharpoons Ag(s)$	+0.80
$Fe^{3+}(aq) + e^- \rightleftharpoons Fe^{2+}(aq)$	+0.77
$I_2(s) + 2e^- \rightleftharpoons 2I^-(aq)$	+0.54
$NiO_2(s) + 2H_2O + 2e^- \rightleftharpoons Ni(OH)_2(s) + 2OH^-(aq)$	+0.49
$Cu^{2+}(aq) + 2e^- \rightleftharpoons Cu(s)$	+0.34
$SO_4^{2-}(aq) + 4H^+(aq) + 2e^- \rightleftharpoons H_2SO_3(aq) + H_2O$	+0.17
$2H^+(aq) + 2e^- \rightleftharpoons H_2(g)$	0.00
$Sn^{2+}(aq) + 2e^- \rightleftharpoons Sn(s)$	−0.14
$Ni^{2+}(aq) + 2e^- \rightleftharpoons Ni(s)$	−0.25
$Co^{2+}(aq) + 2e^- \rightleftharpoons Co(s)$	−0.28
$PbSO_4(s) + 2e^- \rightleftharpoons Pb(s) + SO_4^{2-}(aq)$	−0.36
$Cd^{2+}(aq) + 2e^- \rightleftharpoons Cd(s)$	−0.40
$Fe^{2+}(aq) + 2e^- \rightleftharpoons Fe(s)$	−0.44
$Cr^{3+}(aq) + 3e^- \rightleftharpoons Cr(s)$	−0.74
$Zn^{2+}(aq) + 2e^- \rightleftharpoons Zn(s)$	−0.76
$2H_2O + 2e^- \rightleftharpoons H_2(g) + 2OH^-(aq)$	−0.83
$Al^{3+}(aq) + 3e^- \rightleftharpoons Al(s)$	−1.66
$Mg^{2+}(aq) + 2e^- \rightleftharpoons Mg(s)$	−2.37
$Na^+(aq) + e^- \rightleftharpoons Na(s)$	−2.71
$Ca^{2+}(aq) + 2e^- \rightleftharpoons Ca(s)$	−2.76
$K^+(aq) + e^- \rightleftharpoons K(s)$	−2.92
$Li^+(aq) + e^- \rightleftharpoons Li(s)$	−3.05

A more complete table of standard reduction potentials is located in Appendix D.

Since a substance that is reduced is an *oxidizing agent*, the substances located to the left of the double arrows are all oxidizing agents. The best oxidizing agents are those at the top of the table (e.g., F_2). When one half-reaction occurs as a reduction, another must occur as an oxidation. Substances that are oxidized are *reducing agents*. Reducing agents are located to the right of the double arrows and the best reducing agents are those found at the bottom of the table (e.g., Li).

EXAMPLE 17.6 CALCULATING HALF-CELL POTENTIALS

Problem: We mentioned earlier that the standard cell potential of the silver–copper galvanic cell has a value of 0.46 V. The cell reaction is

$$2Ag^+(aq) + Cu(s) \longrightarrow 2Ag(s) + Cu^{2+}(aq)$$

and we have seen that the reduction potential of Cu^{2+}, $E°_{Cu^{2+}}$, is 0.34 V. What is the value of $E°_{Ag^+}$, the reduction potential of Ag^+?

Solution: The substance that is reduced is Ag^+, which means that Cu is oxidized. According to Equation 17.2, then,

$$E°_{cell} = E°_{Ag^+} - E°_{Cu^{2+}}$$

Substituting the values for $E°_{cell}$ and $E°_{Cu^{2+}}$ gives

$$0.46\ V = E°_{Ag^+} - 0.34\ V$$
$$E°_{Ag^+} = 0.46\ V + 0.34\ V$$
$$= 0.80\ V$$

PRACTICE EXERCISE 6   The galvanic cell described in Practice Exercise 5 has a standard cell potential of 1.96 V. The standard reduction potential of Fe^{2+} corresponding to the half-reaction $Fe^{2+} + 2e^- \rightarrow Fe$ is -0.41 V. Calculate the standard reduction potential of magnesium. Check your answer by referring to Table 17.1.

17.7 USING STANDARD REDUCTION POTENTIALS

Remembering that the half-reaction with the more positive reduction potential occurs as a reduction allows us to predict the outcomes of redox reactions and calculate cell potentials.

The half-reactions and their standard reduction potentials found in Table 17.1 can be used in a number of ways to give information about galvanic cells. Equally useful is the fact that they can provide information about the outcome of redox reactions, whether they occur in a galvanic cell or not.

Example 17.7 shows how reduction potentials can be used to predict the standard cell potential and the overall cell reaction in a galvanic cell.

EXAMPLE 17.7   PREDICTING THE CELL REACTION AND CELL POTENTIAL

Problem: A galvanic cell was constructed using electrodes made of lead and lead dioxide (PbO_2) with sulfuric acid as the electrolyte. The half-reactions and their reduction potentials in this system are

$$PbO_2(s) + 4H^+(aq) + SO_4^{2-}(aq) + 2e^- \rightleftharpoons PbSO_4(s) + 2H_2O \qquad E°_{PbO_2} = 1.69\ V$$
$$PbSO_4(s) + 2e^- \rightleftharpoons Pb(s) + SO_4^{2-}(aq) \qquad E°_{PbSO_4} = -0.36\ V$$

What is the cell reaction and what is the standard potential of the cell?

Solution: In a competition for electrons, the half-reaction having the higher (more positive) reduction potential undergoes reduction. The half-reaction with the lower reduction potential is therefore forced to proceed as an oxidation. This means that here the first half-reaction given will occur as reduction and the second will be forced to reverse its direction. In the cell, the reactions are

$$PbO_2(s) + 4H^+(aq) + SO_4^{2-}(aq) + 2e^- \longrightarrow PbSO_4(s) + 2H_2O \qquad \text{(reduction)}$$
$$Pb(s) + SO_4^{2-}(aq) \longrightarrow PbSO_4(s) + 2e^- \qquad \text{(oxidation)}$$

Remember that half-reactions are combined following the same procedure as in the ion-electron method.

Adding the two half-reactions and canceling electrons gives the cell reaction,

$$PbO_2(s) + 4H^+(aq) + Pb(s) + 2SO_4^{2-}(aq) \longrightarrow 2PbSO_4(s) + 2H_2O$$

(This is the reaction that takes place in lead storage batteries used to start cars.) The cell potential can be obtained by using Equation 17.2.

$$E°_{cell} = (E°\text{ of substance reduced}) - (E°\text{ of substance oxidized})$$

Since the first half-reaction occurs as a reduction and the second as an oxidation,

$$E°_{cell} = E°_{PbO_2} - E°_{PbSO_4}$$
$$= (1.69\ V) - (-0.36\ V)$$
$$= 2.05\ V$$

EXAMPLE 17.8 PREDICTING THE CELL REACTION AND CELL POTENTIAL

Problem: What would be the cell reaction and the standard cell potential of a galvanic cell employing the following half-reactions?

$$Al^{3+}(aq) + 3e^- \rightleftharpoons Al(s) \qquad E°_{Al^{3+}} = -1.66\ V$$
$$Cu^{2+}(aq) + 2e^- \rightleftharpoons Cu(s) \qquad E°_{Cu^{2+}} = 0.34\ V$$

Which half-cell would be the anode?

Solution: Once again, the half-reaction with the more positive reduction potential occurs as a reduction; the other occurs as an oxidation. In this cell, then, Cu^{2+} is reduced and Al is oxidized. To obtain the cell reaction, we add the two half-reactions, remembering that the electrons must cancel.

$$3[Cu^{2+}(aq) + 2e^- \longrightarrow Cu(s)] \qquad \text{(reduction)}$$
$$2[Al(s) \longrightarrow Al^{3+}(aq) + 3e^-] \qquad \text{(oxidation)}$$
$$\overline{3Cu^{2+}(aq) + 2Al(s) \longrightarrow 3Cu(s) + 2Al^{3+}(aq)} \quad \text{(cell reaction)}$$

The anode in the cell is aluminum because that is where oxidation takes place.
To obtain the cell potential, we substitute into Equation 17.2.

$$E°_{cell} = E°_{Cu^{2+}} - E°_{Al^{3+}}$$
$$= (0.34\ V) - (-1.66\ V)$$
$$= 2.00\ V$$

Notice that we multiply the half-reactions by factors to make the electrons cancel, but *we do not multiply the reduction potentials by these factors.*[2] To obtain the cell potential, we simply subtract one reduction potential from the other.

PRACTICE EXERCISE 7 What is the overall cell reaction and the standard cell potential of a galvanic cell employing the following half-reactions?

$$NiO_2(s) + 2H_2O + 2e^- \rightleftharpoons Ni(OH)_2(s) + 2OH^-(aq) \qquad E°_{NiO_2} = 0.49\ V$$
$$Fe(OH)_2(s) + 2e^- \rightleftharpoons Fe(s) + 2OH^-(aq) \qquad E°_{Fe(OH)_2} = -0.88\ V$$

(These are the reactions in an Edison cell, a type of rechargeable storage battery.)

PRACTICE EXERCISE 8 What is the overall cell reaction and the standard cell potential of a galvanic cell employing the following half-reactions?

$$Cr^{3+}(aq) + 3e^- \rightleftharpoons Cr(s) \qquad E°_{Cr^{3+}} = -0.74\ V$$
$$MnO_4^-(aq) + 8H^+(aq) + 5e^- \rightleftharpoons Mn^{2+}(aq) + 4H_2O \qquad E°_{MnO_4^-} = +1.49\ V$$

Reduction potentials can also be used to predict the spontaneous reaction between the substances given in two half-reactions, even when these substances are not in a

[2] Reduction potentials are intensive quantities; they have the units volts, which are joules *per coulomb*. The same number of joules are available for each coulomb of charge regardless of the total number of electrons shown in the equation. Therefore, reduction potentials are never multiplied by factors before they are subtracted to give the cell potential.

galvanic cell. The procedure is very simple because we know that the half-reaction having the more positive reduction potential always undergoes reduction while the other is forced to undergo oxidation.

EXAMPLE 17.9   PREDICTING A SPONTANEOUS REACTION

Problem: What spontaneous reaction will occur if Cl_2 and Br_2 are added to a solution containing Cl^- and Br^-?

Solution: There are two possible reduction reactions.

$$Cl_2 + 2e^- \longrightarrow 2Cl^-$$
$$Br_2 + 2e^- \longrightarrow 2Br^-$$

Strictly speaking, the $E°$ values only tell us what to expect under standard conditions. However, only when $E°_{cell}$ is very small can changes in the concentrations change the direction of the spontaneous reaction.

Referring to Table 17.1, we find that Cl_2 has a more positive reduction potential (1.36 V) than does Br_2 (1.07 V). This means that Cl_2 will be reduced. When reaction occurs the half-reactions will be

$$Cl_2 + 2e^- \longrightarrow 2Cl^-$$
$$2Br^- \longrightarrow Br_2 + 2e^-$$

The net reaction will be

$$Cl_2 + 2Br^- \longrightarrow Br_2 + 2Cl^-$$

Experimentally, we find that chlorine does indeed oxidize bromide ion to bromine, and this fact is used in the synthesis of bromine.

When reduction potentials are listed in order of most positive to least positive (most negative) as they are in Table 17.1, the reactants and products of spontaneous redox reactions are easy to spot. If we choose any pair of half-reactions, the one higher up in the table has the more positive reduction potential and occurs as a reduction. The other half-reaction occurs as an oxidation. The reactants are found on the left side of the higher half-reaction and on the right side of the lower half-reaction.

EXAMPLE 17.10   PREDICTING A SPONTANEOUS REACTION

Problem: Referring to Table 17.1, predict the reaction that will occur when Ni and Fe are added to a solution that contains both Ni^{2+} and Fe^{2+}.

Solution: Ni^{2+} has a more positive (less negative) reduction potential than Fe^{2+}. Therefore, Ni^{2+} will be reduced and Fe (on the right of the lower reaction) will be oxidized.

$$Ni^{2+}(aq) + 2e^- \longrightarrow Ni(s) \qquad \text{(reduction)}$$
$$Fe(s) \longrightarrow Fe^{2+}(aq) + 2e^- \qquad \text{(oxidation)}$$
$$\overline{Ni^{2+}(aq) + Fe(s) \longrightarrow Ni(s) + Fe^{2+}(aq)} \qquad \text{(net reaction)}$$

PRACTICE EXERCISE 9   Use the positions of the half-reactions in Table 17.1 to predict the spontaneous reaction when Br^-, SO_4^{2-}, H_2SO_3, and Br_2 are mixed in an acidic solution.

Since we can predict the spontaneous redox reaction that will take place among a mixture of reactants, it should be possible to predict whether or not a particular reaction, as written, can occur spontaneously. This can be done by calculating the cell potential that corresponds to the reaction in question.

> For any functioning galvanic cell, the measured cell potential has a positive value.

Notice, for example, that to obtain the cell potential for a spontaneous reaction in our previous examples, we subtracted the reduction potentials in a way that gave a positive answer. Therefore, if we compute the cell potential for a particular reaction *based on the way the equation is written* and the potential comes out positive, we know the reaction is spontaneous. If the calculated cell potential comes out negative, however, the reaction is nonspontaneous. In fact, it is really spontaneous in the opposite direction.

EXAMPLE 17.11   DETERMINING WHETHER A REACTION IS SPONTANEOUS

Problem: Determine whether the following reactions are spontaneous as written. If they are not, give the reaction that is spontaneous.
(a) $Cu(s) + 2H^+(aq) \longrightarrow Cu^{2+}(aq) + H_2(g)$
(b) $3Cu(s) + 2NO_3^-(aq) + 8H^+(aq) \longrightarrow 3Cu^{2+}(aq) + 2NO(g) + 4H_2O$

Solution: (a) The half-reactions involved in this reaction are

$$Cu(s) \longrightarrow Cu^{2+}(aq) + 2e^-$$
$$2H^+(aq) + 2e^- \longrightarrow H_2(g)$$

The H^+ is reduced and Cu is oxidized. Therefore, from Equation 17.2,

$$E^\circ_{cell} = E^\circ_{H^+} - E^\circ_{Cu^{2+}}$$

Substituting values from Table 17.1,

$$E^\circ_{cell} = (0.00 \text{ V}) - (0.34 \text{ V})$$
$$= -0.34 \text{ V}$$

Now you know why copper doesn't dissolve in HCl—the only oxidizing agent is H^+.

The calculated cell potential is negative and therefore reaction (a) is *not* spontaneous in the forward direction. The spontaneous reaction is

$$H_2(g) + Cu^{2+}(aq) \longrightarrow Cu(s) + 2H^+(aq)$$

(b) The half-reactions involved are

$$Cu(s) \longrightarrow Cu^{2+}(aq) + 2e^-$$
$$NO_3^-(aq) + 4H^+(aq) + 3e^- \longrightarrow NO(g) + 2H_2O$$

The Cu is oxidized while the NO_3^- is reduced. According to Equation 17.2,

$$E^\circ_{cell} = E^\circ_{NO_3^-} - E^\circ_{Cu^{2+}}$$

Substituting values from Table 17.1 gives

$$E^\circ_{cell} = (0.96 \text{ V}) - (0.34 \text{ V})$$
$$= 0.62 \text{ V}$$

Copper will dissolve in HNO_3 because it contains NO_3^- as an oxidizing agent.

Since the calculated cell potential is positive, this reaction is spontaneous in the forward direction.

PRACTICE EXERCISE 10   Which of the following reactions occur spontaneously in the forward direction?
(a) $Br_2(aq) + Cl_2(g) + 2H_2O \longrightarrow 2Br^-(aq) + 2HOCl(aq) + 2H^+(aq)$
(b) $3Zn(s) + 2Cr^{3+}(aq) \longrightarrow 3Zn^{2+}(aq) + 2Cr(s)$

17.8 CELL POTENTIALS AND THERMODYNAMICS

Standard cell potentials can be used to calculate free energy changes and equilibrium constants.

The fact that cell potentials allow us to predict the spontaneity of redox reactions is no coincidence. There is a relationship between the cell potential and the free energy change for a reaction. In Chapter 16 we saw that ΔG for a reaction is a measure of the maximum useful work that can be obtained from a chemical reaction.

$$-\Delta G = \text{maximum work} \tag{17.3}$$

In an electrical system, work is supplied by the electric current that is pushed along by the potential of the cell. It can be calculated from the equation

$$\text{maximum work} = n\mathscr{F}E_{cell} \tag{17.4}$$

where n is the number of moles of electrons transferred, $\mathscr{F}$ is the faraday constant (96,500 coulombs per mole of electrons), and E_{cell} is the potential of the cell in volts. To see that Equation 17.4 gives *work* (which has the units of energy) we can analyze the units. In Equation 17.1 we saw that 1 V = 1 joule/coulomb. Therefore,

$$\text{maximum work} = (\text{mol } e^-) \times \left(\frac{\text{coulombs}}{\text{mol } e^-}\right) \times \left(\frac{J}{\text{coulomb}}\right) = \text{joule}$$

$$\underset{n}{\updownarrow} \qquad \underset{\mathscr{F}}{\updownarrow} \qquad \underset{E_{cell}}{\updownarrow}$$

Equating 17.3 and 17.4 gives us

$$\Delta G = -n\mathscr{F}E_{cell} \tag{17.5}$$

If we are dealing with the standard cell potential, we can calculate the standard free energy change.

$$\Delta G^\circ = -n\mathscr{F}E^\circ_{cell} \tag{17.6}$$

Remember, we can write

$$1\ \mathscr{F} = \frac{96{,}500\ \text{C}}{1\ \text{mol } e^-}$$

Note that when E_{cell} is positive, ΔG will be negative and the cell reaction will be spontaneous. We've used this idea earlier in predicting spontaneous cell reactions.

EXAMPLE 17.12 CALCULATING THE STANDARD FREE ENERGY CHANGE

Problem: Calculate ΔG° for the following reaction, given that its standard cell potential is 0.320 V at 25 °C.

$$NiO_2 + 2Cl^- + 4H^+ \longrightarrow Cl_2 + Ni^{2+} + 2H_2O$$

Solution: Since two Cl^- are oxidized to Cl_2, two electrons are transferred. Interpreting the coefficients as moles, 2 mol of electrons are transferred, so $n = 2$. Using Equation 17.3, we have

$$\Delta G^\circ = -(2\ \text{mol } e^-) \times \left(\frac{96{,}500\ \text{C}}{\text{mol } e^-}\right) \times \left(\frac{0.320\ \text{J}}{\text{C}}\right)$$

$$= -61{,}800\ \text{J} \quad \text{(rounded to three significant figures)}$$
$$= -61.8\ \text{kJ}$$

PRACTICE EXERCISE 11   Calculate ΔG° for the reaction that takes place in the galvanic cell described in Practice Exercise 8.

Determining Equilibrium Constants

One of the most useful applications of electrochemistry is the determination of equilibrium constants. In Chapter 16 we saw that ΔG° was related to the equilibrium constant (K_p for gaseous reactions and K_c for reactions in solution).

$$\Delta G° = -RT \ln K_c$$

Now we have seen that $\Delta G°$ is related to $E°_{cell}$.

$$\Delta G° = -n\mathscr{F}E°_{cell}$$

Therefore, $E°_{cell}$ and the equilibrium constant are also related. Equating the right sides of the two equations above and solving for $E°_{cell}$ gives

$$E°_{cell} = \frac{RT}{n\mathscr{F}} \ln K_c$$

For historical reasons, this equation is usually expressed in terms of common logs (base 10 logs), and for reactions at 25 °C, all the constants can be combined to give

$$E°_{cell} = \frac{0.0592}{n} \log K_c \qquad (17.7)$$

Notice that this equation uses common logarithms.

The constant 0.0592 has the units joules divided by coulombs, which is volts. As before, n is the number of moles of electrons that are transferred in the cell reaction as written.

EXAMPLE 17.13   CALCULATING EQUILIBRIUM CONSTANTS FROM $E°_{cell}$

Problem: Calculate K_c for the reaction in Example 17.12.

Solution: The reaction had $E°_{cell} = 0.320$ V and $n = 2$. Substituting these values into Equation 17.7 gives

$$0.320 \text{ V} = \frac{0.0592}{2} \log K_c$$

$$\frac{2(0.320)}{(0.0592)} = \log K_c$$

$$10.8 = \log K_c$$

Taking the antilog gives

$10^{10.8} = 6 \times 10^{10}$

$$K_c = 6 \times 10^{10}$$

PRACTICE EXERCISE 12   The calculated standard cell potential for the reaction

$$Cu^{2+}(aq) + 2Ag(s) \rightleftharpoons Cu(s) + 2Ag^+(aq)$$

is $E°_{cell} = -0.46$ V. Calculate K_c for this reaction. Would you expect a relatively large amount of products to form?

17.9 THE EFFECT OF CONCENTRATION ON CELL POTENTIALS

Altering the concentrations of the ions in the half-cells affects the cell potential in a way that depends on the logarithm of the mass action expression for the cell reaction.

When all of the ion concentrations in a cell are 1 M, the cell potential is equal to the standard potential. When the concentrations change, however, so does the potential.

For example, in an operating cell or battery, the potential gradually drops as the reactants are consumed. The cell approaches equilibrium, and when it gets there the potential has dropped to zero — the battery is dead.

The effect of concentration on the cell potential can be calculated from the Nernst equation,

$$E_{cell} = E_{cell}^\circ - \frac{2.303\,RT}{n\mathscr{F}} \log Q$$

where E_{cell} is the actual cell potential, E_{cell}° is the standard potential, R is the gas constant in energy units (joules), T is the absolute temperature, $\mathscr{F}$ is the faraday constant (96,500 C/mol e^-), n is the number of moles of electrons transferred in the cell reaction, and Q is the reaction quotient for the reaction. You may recognize the combination of constants that appear in front of the log term — you saw essentially these same constants earlier in our discussion of the determination of equilibrium constants from cell potentials. If we deal with reactions at 25 °C, all the constants can be combined to give the more easily remembered equation

$$E_{cell} = E_{cell}^\circ - \frac{0.0592}{n} \log Q$$

Walter Nernst (1864–1941) was a noted electrochemist and also a discoverer of the third law of thermodynamics.

Once again, note the use of common logarithms in this equation.

EXAMPLE 17.14 CALCULATING THE EFFECT OF CONCENTRATION ON E_{cell}

Problem: What is the potential of a copper–silver cell similar to that in Figure 17.12 if the Ag^+ concentration is 1.0×10^{-3} M and the Cu^{2+} concentration is 1.0×10^{-4} M? The standard potential of the cell is 0.46 V and the cell reaction is

$$2Ag^+(aq) + Cu(s) \longrightarrow 2Ag(s) + Cu^{2+}(aq)$$

Solution: First, let's write the correct form of the Nernst equation for this cell. Since two electrons are lost when one Cu atom is oxidized to give one Cu^{2+} ion, $n = 2$. The correct mass action expression for this reaction, as usual, has the product concentration in the numerator.

$$Q = \frac{[Cu^{2+}]}{[Ag^+]^2}$$

Notice that we have omitted the concentrations of the solids just as we did when we wrote similar expressions in Chapter 14. The Nernst equation for this system, then, is

$$E_{cell} = E_{cell}^\circ - \frac{0.0592\text{ V}}{2} \log \frac{[Cu^{2+}]}{[Ag^+]^2}$$

Substituting the values given in the problem, we have

$$E_{cell} = 0.46\text{ V} - \frac{0.0592\text{ V}}{2} \log \frac{1.0 \times 10^{-4}}{(1.0 \times 10^{-3})^2}$$

$$= 0.46\text{ V} - 0.0296\text{ V} \log (1.0 \times 10^2)$$

$$= 0.46\text{ V} - 0.0296\text{ V} (2.00)$$

$$= 0.46\text{ V} - 0.0592\text{ V}$$

$$= 0.40\text{ V}$$

The cell potential is 0.40 V.

EXAMPLE 17.15 CALCULATING THE EFFECT OF CONCENTRATION ON E_{cell}

Problem: A cell employs the following half-reactions.

$$Ni^{2+}(aq) + 2e^- \rightleftharpoons Ni(s) \qquad E^{\circ}_{Ni^{2+}} = -0.25 \text{ V}$$
$$Cr^{3+}(aq) + 3e^- \rightleftharpoons Cr(s) \qquad E^{\circ}_{Cr^{3+}} = -0.74 \text{ V}$$

Calculate the potential if $[Ni^{2+}] = 1.0 \times 10^{-4}$ and $[Cr^{3+}] = 2.0 \times 10^{-3} \ M$.

Solution: First we need the cell reaction. Nickel ion is reduced because it has the higher reduction potential. This means, of course, that chromium is oxidized. Making electron gain equal electron loss gives

$$\begin{array}{ll} 3[Ni^+(aq) + 2e^- \longrightarrow Ni(s)] & \text{(reduction)} \\ 2[Cr(s) \longrightarrow Cr^{3+}(aq) + 3e^-] & \text{(oxidation)} \\ \hline 3Ni^{2+}(aq) + 2Cr(s) \longrightarrow 3Ni(s) + 2Cr^{3+}(aq) & \text{(cell reaction)} \end{array}$$

Notice that six electrons are transferred overall, so $n = 6$. Now we can write the Nernst equation.

$$E_{cell} = E^{\circ}_{cell} - \frac{0.0592}{6} \log \frac{[Cr^{3+}]^2}{[Ni^{2+}]^3}$$

Next we need E°_{cell}. Since Ni^{2+} is reduced,

$$\begin{aligned} E^{\circ}_{cell} &= E^{\circ}_{Ni^{2+}} - E^{\circ}_{Cr^{3+}} \\ &= (-0.25 \text{ V}) - (-0.74 \text{ V}) \\ &= 0.49 \text{ V} \end{aligned}$$

Substituting values into the equation, we get

$$\begin{aligned} E_{cell} &= 0.49 \text{ V} - \frac{0.0592 \text{ V}}{6} \log \frac{(2.0 \times 10^{-3})^2}{(1.0 \times 10^{-4})^3} \\ &= 0.49 \text{ V} - 0.010 \text{ V} \log(4.0 \times 10^6) \\ &= 0.49 \text{ V} - 0.010 \text{ V} (6.60) \\ &= 0.49 \text{ V} - 0.066 \text{ V} \\ &= 0.42 \text{ V} \end{aligned}$$

PRACTICE EXERCISE 13 In a certain zinc–copper cell,

$$Zn(s) + Cu^{2+}(aq) \longrightarrow Zn^{2+}(aq) + Cu(s)$$

the ion concentrations are $[Cu^{2+}] = 0.0100 \ M$ and $[Zn^{2+}] = 1.0 \ M$. What is the cell potential? The standard cell potential is 1.10 V.

Determination of Concentrations from Cell Potentials

One of the principal uses of the relationship between concentration and cell potential is in the measurement of concentrations. Experimental determination of cell potentials combined with modern developments in electronics has provided a means of monitoring and analyzing the concentrations of all sorts of substances in solution, even some that are not themselves ionic and that are not involved directly in electrochemical changes. The basic concept behind all these approaches is illustrated in Example 17.16.

EXAMPLE 17.16 USING THE NERNST EQUATION TO DETERMINE CONCENTRATIONS

Problem: A chemist wanted to measure the concentration of Cu^{2+} in a large number of samples of water in which the copper ion concentration was expected to be quite small. The apparatus that was used consisted of a silver electrode dipping into a 1.00 M solution of $AgNO_3$. This half-cell was connected by a salt bridge to a second half-cell containing a copper electrode that was able to be dipped into each water sample, one after another. In the analysis of one of the samples, the cell potential was measured to be 0.62 V, with the copper serving as the anode. What was the concentration of copper ion in this particular sample of water?

Solution: The first step is to write the proper chemical equation, because we need it to compute E°_{cell} and to construct the mass action expression (reaction quotient) that will be used in the Nernst equation. Since copper is the anode, it is being oxidized. This also means that Ag^+ is being reduced. Therefore, the correct equation for the cell reaction is

$$Cu(s) + 2Ag^+(aq) \longrightarrow Cu^{2+}(aq) + 2Ag(s)$$

and the Nernst equation is

$$E_{cell} = E^\circ_{cell} - \frac{0.0592}{2} \log \frac{[Cu^{2+}]}{[Ag^+]^2}$$

The value of E°_{cell} can be obtained from the tabulated standard reduction potentials in Table 17.1. Following our usual procedure, recognizing that silver ion is reduced,

$$E^\circ_{cell} = E^\circ_{Ag^+} - E^\circ_{Cu^{2+}}$$
$$= 0.80 \text{ V} - 0.34 \text{ V}$$
$$= 0.46 \text{ V}$$

Now we can substitute values into the Nernst equation and solve for the concentration ratio in the mass action expression.

$$0.62 \text{ V} = 0.46 \text{ V} - \frac{0.0592}{2} \log \frac{[Cu^{2+}]}{[Ag^+]^2}$$

Solving for $\log ([Cu^{2+}/Ag^+]^2)$ gives

$$\log \frac{[Cu^{2+}]}{[Ag^+]^2} = -5.4$$

Taking the antilog gives us the value of the mass action expression.

$$\frac{[Cu^{2+}]}{[Ag^+]^2} = 4 \times 10^{-6}$$

Since we know that the concentration of Ag^+ is 1.00 M, we can now solve for the Cu^{2+} concentration.

$$\frac{[Cu^{2+}]}{(1.00)^2} = 4 \times 10^{-6}$$
$$[Cu^{2+}] = 4 \times 10^{-6} M$$

Notice that this is indeed a very small concentration, and it can be obtained very easily by simply measuring the potential generated by the electrochemical cell. Determining the concentrations in many samples is also very simple—just change the water sample and measure the potential again. The ease of such operations and the fact that they lend themselves well to automation and computer analysis make electrochemical analyses especially attractive to scientists.

PRACTICE EXERCISE 14 In the analysis of two other water samples by the procedure described in Example 17.16, cell potentials (E_{cell}) of 0.57 V and 0.82 V were obtained. Calculate the Cu^{2+} ion concentration in each of these samples.

PRACTICE EXERCISE 15 A galvanic cell was constructed by connecting a nickel electrode that was dipping into 1.20 M $NiSO_4$ solution to a chromium electrode that was dipping into a solution containing Cr^{3+} at an unknown concentration. The potential of the cell was measured to be 0.55 V, with the chromium serving as the anode. What was the concentration of Cr^{3+} in the solution of unknown concentration?

The Nernst equation tells us that the emf of a cell changes with the concentrations of the ions involved in the cell reaction. As noted in the conclusion of Example 17.16, one of the most useful results of this is that cell potential measurements provide a way to measure and monitor ion concentrations in aqueous solutions.

Scientists have developed a number of specialized electrodes that can be dipped into one solution after another and whose potential is affected in a reproducible way by the concentration of only one species in the solution. An example is the **glass electrode,** shown in the accompanying figure, which is used with a pH meter in the measurement of pH. The electrode is constructed from a hollow glass tube sealed with a special thin-walled glass membrane at the bottom. The tube is filled partway with a dilute solution of HCl, and dipping into this HCl solution is a silver wire coated with a layer of silver chloride. The potential of the electrode is controlled by the difference between the hydrogen ion concentrations inside and outside the thin glass membrane at the bottom. Since the H^+ concentration inside the electrode is constant, the electrode's potential varies only with the concentration of H^+ in the solution outside. In fact, this potential is proportional to the pH of the outside solution.

A glass electrode is always used with another reference electrode whose potential is a constant, and this gives a galvanic cell whose potential depends on the pH of the solution into which the electrodes are immersed. The measurement of the potential of this

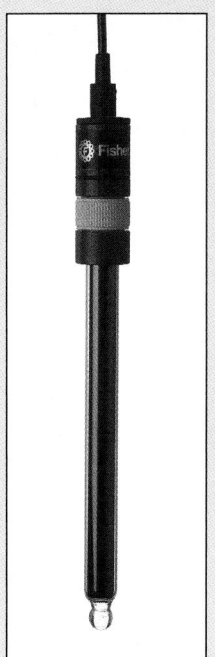

(a) Photograph of a typical glass electrode.

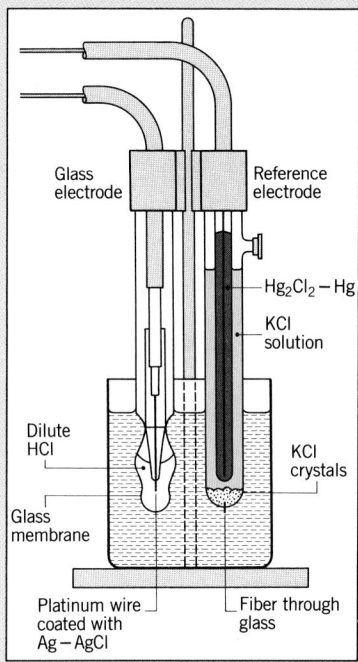

(b) Cutaway view of the construction of the electrode.

galvanic cell and the translation of the potential into pH are the job of a pH meter such as the one shown in Figure 15.3 on page 606.

17.10 PRACTICAL APPLICATIONS OF GALVANIC CELLS

Galvanic cells provide portable sources of electric power.

One of the principal uses of galvanic cells is the generation of portable electrical energy. In this section we will briefly examine some of the more common present-day uses of galvanic cells, popularly called batteries, and some promising future applications.

The Lead Storage Battery

The common lead storage battery used to start an automobile is composed of a number of galvanic cells, each having an emf of about 2 V (Example 17.7) connected in series so that their voltages will add. Most automobile batteries used today contain six such cells and give about 12 V, but 6-, 24-, and 32-V batteries are also available.

The construction of a typical lead storage battery is shown in Figure 17.17. The anode is composed of several plates of lead, and the cathode is composed of several plates of lead dioxide. The electrolyte is sulfuric acid. When the battery is discharging — for example, when it is delivering current to start a car — the electrode reactions are

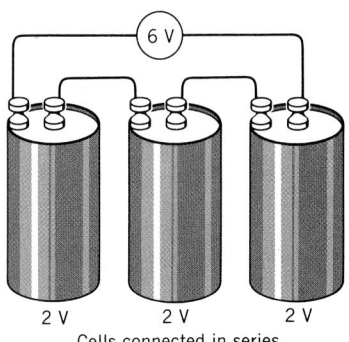

Cells connected in series

$$PbO_2(s) + 4H^+(aq) + SO_4^{2-}(aq) + 2e^- \longrightarrow PbSO_4(s) + 2H_2O \quad \text{(cathode)}$$
$$Pb(s) + SO_4^{2-}(aq) \longrightarrow PbSO_4(s) + 2e^- \quad \text{(anode)}$$

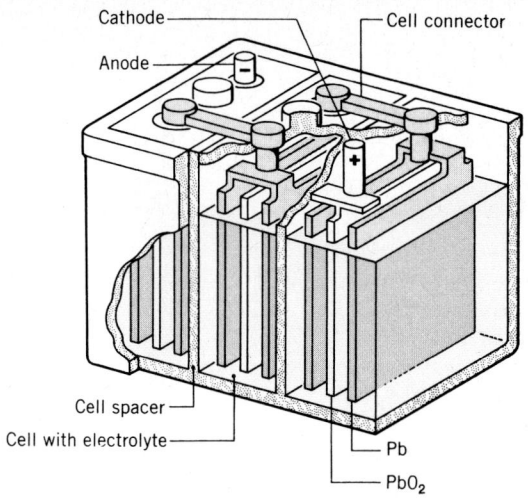

The net reaction taking place in each cell is

$$PbO_2(s) + Pb(s) + \underbrace{4H^+(aq) + 2SO_4{}^{2-}(aq)}_{2H_2SO_4} \longrightarrow 2PbSO_4(s) + 2H_2O$$

As the cell discharges, the sulfuric acid concentration decreases because it is used up in the cell reaction. This provides a convenient means of checking the state of a battery. Since the density of a sulfuric acid solution decreases as its concentration drops, the concentration can be determined very simply by measuring the density with a hydrometer (see Figure 17.18). A hydrometer consists of a rubber bulb that is used to draw the battery fluid into a glass tube containing a float. The depth to which the float sinks is inversely porportional to the density of the liquid — the deeper the float sinks, the lower is the density and the weaker is the charge on the battery. The narrow neck of the float is usually marked to indicate the state of charge of the battery.

The principal advantage of the lead storage battery is that the cell reactions that occur spontaneously during discharge can be reversed by the application of voltage from an external source. In other words, the battery can be recharged by electrolysis. The reaction is

$$2PbSO_4(s) + 2H_2O \xrightarrow{\text{electrolysis}} PbO_2(s) + Pb(s) + 4H^+(aq) + 2SO_4{}^{2-}(aq)$$

The major disadvantages of these batteries are that they are very heavy and that their corrosive sulfuric acid can spill out if one is upset.

The Zinc–Carbon Dry Cell

The ordinary, relatively inexpensive, 1.5-V **zinc–carbon dry cell** used to power flashlights, tape recorders, and the like is not really dry. A cutaway view of the internal construction of this type of cell is shown in Figure 17.19. The outer shell is made of zinc and serves as the anode — its exposed outer surface at the bottom is the negative end of the battery. The cathode — the positive terminal of the battery — consists of a carbon (graphite) rod surrounded by a moist paste of graphite powder, manganese dioxide, and ammonium chloride.

The anode reaction is simply the oxidation of the zinc.

$$Zn(s) \longrightarrow Zn^{2+}(aq) + 2e^- \qquad \text{(anode)}$$

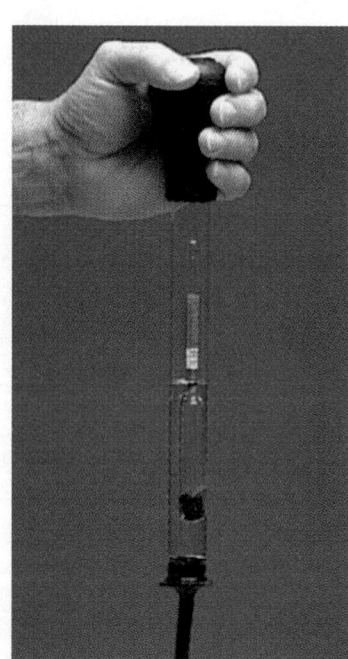

FIGURE 17.18 A battery hydrometer. Battery acid is drawn into the glass tube. The depth to which the float sinks is inversely proportional to the concentration of the acid and, therefore, to the state of charge of the battery.

The zinc carbon cell is also known as the Leclanché cell after its inventor, Georges Leclanché.

The cathode reaction is complex, and a mixture of products is formed. One of the major reactions is

$$2MnO_2(s) + 2NH_4^+(aq) + 2e^- \longrightarrow Mn_2O_3(s) + 2NH_3(aq) + H_2O \qquad \text{(cathode)}$$

The ammonia formed at the cathode reacts with some of the Zn^{2+} produced from the anode to form an ion having the formula $Zn(NH_3)_4^{2+}$. Because of the complexity of the cathode half-cell reaction, no simple overall cell reaction can be written.

The major advantages of the dry cell are its relatively low price and the fact that it normally works without leaking. One disadvantage is that the cell loses its ability to function rather rapidly under heavy current drain. This happens because the products of reaction can't diffuse away from the electrodes very easily. If unused for a while, however, these batteries can rejuvenate themselves somewhat. Another disadvantage is that they can't be recharged. Devices that claim to recharge zinc–carbon dry cell batteries simply drive the reaction products away from the electrodes, which allows them to function again. This doesn't work very many times, however. Before too long the zinc casing will develop holes and the battery must be discarded.

The common alkaline flashlight battery, or alkaline dry cell, also uses Zn and MnO_2 as reactants, but under basic conditions. The half-cell reactions are

$$Zn(s) + 2OH^-(aq) \longrightarrow ZnO(s) + H_2O + 2e^- \qquad \text{(anode)}$$
$$2MnO_2(s) + H_2O + 2e^- \longrightarrow Mn_2O_3(s) + 2OH^-(aq) \qquad \text{(cathode)}$$

and the voltage is about 1.54 V. It has a longer shelf-life and is able to deliver higher currents for longer periods than the less expensive zinc–carbon cell.

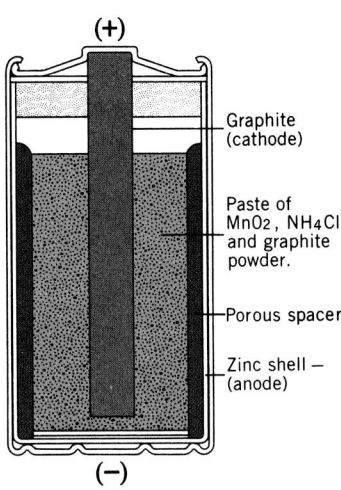

(+)

Graphite (cathode)

Paste of MnO_2, NH_4Cl and graphite powder.

Porous spacer

Zinc shell – (anode)

(−)

FIGURE 17.19 A cross section of a zinc–carbon dry cell.

The Nickel–Cadmium Storage Cell

The nickel–cadmium storage cell is the *nicad* battery that powers rechargeable electronic calculators, electric shavers, and power tools. It produces a potential of about 1.4 V, which is slightly lower than that of the zinc–carbon cell. The electrode reactions in the cell are

$$Cd(s) + 2OH^-(aq) \longrightarrow Cd(OH)_2(s) + 2e^- \qquad \text{(anode)}$$
$$2e^- + NiO_2(s) + 2H_2O \longrightarrow Ni(OH)_2(s) + 2OH^-(aq) \qquad \text{(cathode)}$$

The nickel–cadmium battery is rechargeable and can be sealed to prevent leakage, which is particularly important in electronic devices. For the storage of a given amount of electrical energy, however, the nicad battery is considerably more expensive than the lead storage battery.

Specialized Batteries

As everyone knows, the field of electronics has undergone fantastic changes in the past decade or so. Electronic calculators, digital watches, electronically controlled cameras that focus automatically and adjust themselves for the correct exposure, and even portable computers have been made possible by microelectronics centered around almost invisible circuits etched into the surfaces of silicon chips. This trend toward miniaturization has also been taken up by the makers of the batteries that power the electronics. Two of the common miniature batteries in use today are the "mercury battery" and the "silver oxide battery."

The Mercury Battery   This battery, one of the first small batteries to be developed for commercial use, first appeared in the early 1940s. It consists of a zinc anode and a mercury(II) oxide (HgO) cathode. The electrolyte is a concentrated solution of potas-

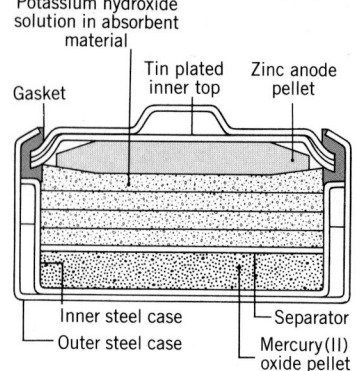

Potassium hydroxide solution in absorbent material

Gasket

Tin plated inner top

Zinc anode pellet

Inner steel case
Outer steel case

Separator
Mercury(II) oxide pellet

FIGURE 17.20 A mercury battery.

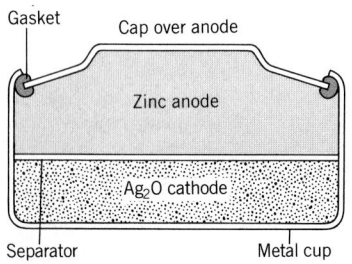

Gasket   Cap over anode

Zinc anode

Ag_2O cathode

Separator   Metal cup

FIGURE 17.21 A silver oxide battery.

If we could "burn" our petroleum in fuel cells to produce energy, fuel consumption could be halved.

sium hydroxide held in pads of absorbent material between the electrodes (Figure 17.20). The reactions that take place at the electrodes are relatively simple.

$$Zn(s) + 2OH^-(aq) \longrightarrow ZnO(s) + H_2O + 2e^- \quad \text{(anode)}$$
$$HgO(s) + H_2O + 2e^- \longrightarrow Hg(\ell) + 2OH^-(aq) \quad \text{(cathode)}$$

The net cell reaction is

$$Zn(s) + HgO(s) \longrightarrow ZnO(s) + Hg(\ell)$$

The mercury battery develops a potential of about 1.35 V. An especially useful feature of this cell is that its potential remains nearly constant throughout its life. By contrast, the potential of the ordinary dry cell drops off slowly but continuously as it is used.

The Silver Oxide Battery   This is a more recent entry into the world of miniature batteries and is now used in many wristwatches, calculators, and autoexposure cameras. The anode material in the cell (Figure 17.21) is zinc and the substance that reacts at the cathode is silver oxide, Ag_2O. The electrolyte is basic and the half-reactions that occur at the electrodes are

$$Zn(s) + 2OH^-(aq) \longrightarrow Zn(OH)_2(s) + 2e^- \quad \text{(anode)}$$
$$Ag_2O(s) + H_2O + 2e^- \longrightarrow 2Ag(s) + 2OH^-(aq) \quad \text{(cathode)}$$

The overall net cell reaction is

$$Zn(s) + Ag_2O(s) + H_2O \longrightarrow Zn(OH)_2(s) + 2Ag(s)$$

As you might expect, these cells are quite expensive because they contain silver. An advantage that they have over the mercury battery, however, is that they generate a potential of about 1.5 V.

Fuel Cells   The production of usable energy by the combustion of fuels is an extremely inefficient process. Modern electric power plants are able to harness only about 35 to 40% of the energy theoretically available from oil, coal, or natural gas. The gasoline or diesel engine has an efficiency of only about 25 to 30%. The rest of the energy is lost to the surroundings as heat. This is the reason that a car must have an effective cooling system.

Fuel cells are electrochemical cells that "burn" fuel under conditions that are much more nearly thermodynamically reversible than simple combustion. They therefore achieve much greater efficiencies—75% is quite feasible. Figure 17.22 illustrates a hydrogen–oxygen fuel cell. The electrolyte, a hot concentrated solution of potassium hydroxide in the center compartment, is in contact with two porous electrodes. Porous carbon and thin porous nickel electrodes have been used. Gaseous H_2 and O_2 under pressure are circulated so as to come in contact with the electrodes. At the cathode, oxygen is reduced.

$$O_2(g) + 2H_2O + 4e^- \longrightarrow 4OH^-(aq) \quad \text{(cathode)}$$

At the anode, hydrogen is oxidized to water.

$$H_2(g) + 2OH^-(aq) \longrightarrow 2H_2O + 2e^- \quad \text{(anode)}$$

Part of the water formed at the anode leaves as steam mixed with the circulating hydrogen gas. The net cell reaction, after making electron loss equal to electron gain, is

$$2H_2(g) + O_2(g) \longrightarrow 2H_2O$$

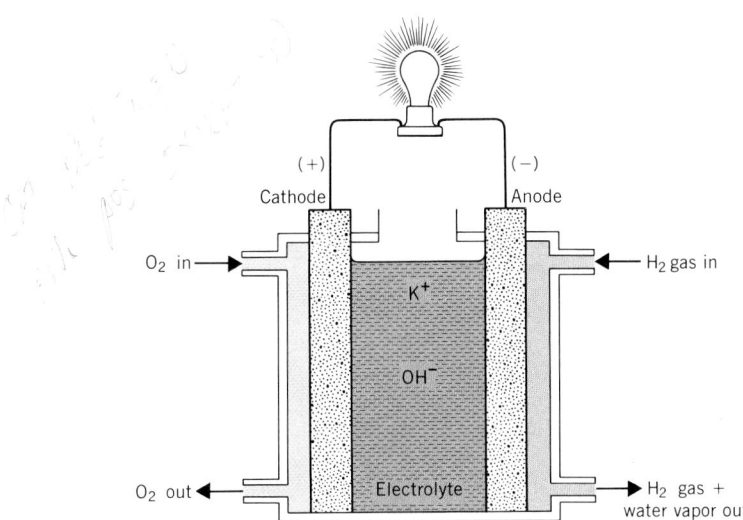

FIGURE 17.22 Hydrogen–oxygen fuel cell.

The advantages of the fuel cell are obvious. There is no electrode material to be replaced, as there is in an ordinary battery. The fuel can be fed in continuously to produce power. In fact, hydrogen–oxygen fuel cells have been used on spacecraft for just this reason. The disadvantages at the present time include their high cost and the rather large size that is necessary to produce useful quantities of electrical power.

In spacecraft, the product of the reaction in the hydrogen/oxygen fuel cell is used for drinking water.

SUMMARY

Electrolysis. In an electrolytic cell, a flow of electricity causes reduction at a negatively charged cathode and oxidation at a positively charged anode. Ion movement rather than electron transport occurs in the electrolyte. The electrode reactions are determined by which species is most easily reduced and which is most easily oxidized but in aqueous solutions complex surface effects at the electrodes can alter the natural order. In the electrolysis of water, an electrolyte must be present to maintain electrical neutrality at the electrodes.

Quantitative Aspects of Electrolysis. A faraday is 1 mol of electrons, and 96,500 C equals 1 faraday. The product of current (amperes) and time (seconds) gives coulombs. These relationships and the half-reactions occurring at anode or cathode permit us to relate the amount of chemical change to measurements of current and time.

Applications of Electrolysis. Electroplating, the production of aluminum and magnesium, the refining of copper, and the electrolysis of molten and aqueous sodium chloride are examples of practical applications of electrolysis.

Galvanic Cells. A spontaneous redox reaction, in which the individual half-reactions occur in separated half-cells, can force the electron transfer to occur through an external electrical circuit. Reduction occurs at the positively charged cathode and oxidation takes place at the negatively charged anode. The half-cells must be connected electrolytically to complete an electrical circuit. A salt bridge accomplishes this — it permits electrical neutrality to be maintained. The emf or voltage produced by a cell is equal to the standard cell potential when all ion concentrations are 1 M. The cell potential can be considered to be the difference between the reduction potentials of the half-cells. The half-cell with the higher reduction potential undergoes reduction and forces the other to undergo oxidation. Although the reduction potential of an isolated half-cell can't be measured, values are assigned by choosing the hydrogen electrode as a reference electrode having a reduction potential of 0.00 V. Species more easily reduced than H^+ have positive reduction potentials; those less easily reduced have negative reduction potentials. Reduction potentials can be used to predict the cell reaction and to calculate $E°_{cell}$. They can also be used to predict spontaneous redox reactions not occurring in galvanic cells and to predict whether or not a given reaction is spontaneous.

Thermodynamics and Cell Potentials. The values of $\Delta G°$ and K_c for a reaction can be calculated from $E°_{cell}$. The Nernst equation relates the cell potential to the standard cell potential and the mass action expression. It allows the cell potential to be calculated for ion concentrations other than 1 M.

Practical Galvanic Cells. The lead storage battery and the nickel–cadmium battery are rechargeable. The zinc–carbon cell —the common dry cell—is not. The common alkaline battery uses essentially the same reactions as the less expensive dry cell. The zinc–mercury(II) oxide cell and the zinc–silver oxide cell are commonly used in small electronic devices. Fuel cells consume fuel that can be fed continuously. For the production of usable energy, they offer much higher thermodynamic efficiencies than can be obtained by burning the fuel in conventional power plants or internal combustion engines.

REVIEW EXERCISES

Answers to questions whose numbers are printed in color are given in Appendix C.
Difficult questions are marked with asterisks.

Electrolysis

17.1 Define *oxidation* and *reduction*.

17.2 What is an electrochemical change?

17.3 How do we define anode and cathode?

17.4 What electrical charges do the anode and the cathode carry in an electrolytic cell?

17.5 How do electrolytic conduction and metallic conduction differ?

17.6 Why must electrolysis reactions occur in order for electrolytic conduction to continue?

17.7 What is the difference between a direct current and an alternating current?

17.8 What is meant by the term *cell reaction*?

17.9 What is the purpose of writing the word *electrolysis* over the arrow in a chemical equation?

17.10 Why must NaCl be melted before it is electrolyzed to give Na and Cl_2?

17.11 What does the term *inert electrode* mean?

17.12 Write the anode, cathode, and overall reactions for the electrolysis of molten NaCl.

17.13 Write half-reactions for the oxidation and the reduction of water.

17.14 What happens to the pH of the solution near the cathode and anode during the electrolysis of KNO_3?

17.15 What function does KNO_3 serve in the electrolysis of a KNO_3 solution?

17.16 Write the anode reaction for the electrolysis of an aqueous solution that contains (a) NO_3^-, (b) Br^-, and (c) NO_3^- and Br^-.

17.17 Write the cathode reaction for the electrolysis of an aqueous solution that contains (a) K^+, (b) Cu^{2+}, and (c) K^+ and Cu^{2+}.

17.18 What products would we expect at the electrodes if a solution containing both KBr and $Cu(NO_3)_2$ were electrolyzed?

Stoichiometric Relationships in Electrolysis

17.19 What is a faraday? What relationships relate faradays to current and time measurements?

17.20 How many coulombs are passed through an electrolysis cell by (a) a current of 4.00 A for 600 s? (b) A current of 10.0 A for 20 min? (c) A current of 1.50 A for 6.00 hr?

17.21 How many faradays correspond to the answers to each part of Review Exercise 17.20?

17.22 How many faradays are required to (a) reduce 0.20 mol Fe^{2+} to Fe? (b) Oxidize 0.70 mol Cl^- to Cl_2? (c) Reduce 1.50 mol Cr^{3+} to Cr? (d) Oxidize 1.0×10^{-2} mol Mn^{2+} to MnO_4^-?

17.23 How many faradays are required to (a) produce 5.00 g Mg from molten $MgCl_2$? (b) Form 41.0 g Cu from a $CuSO_4$ solution?

17.24 How many moles of Cr^{3+} would be reduced to Cr by the same amount of electricity that produces 10.0 g Ag from a $AgNO_3$ solution?

17.25 How many grams of $Fe(OH)_2$ are produced at an iron anode when a basic solution undergoes electrolysis at a current of 12.0 A for 10.0 min?

17.26 How many milliliters of gaseous H_2, measured at STP, would be produced at the cathode in the electrolysis of water with a current of 1.50 A for 5.00 min?

17.27 A solution of NaCl in water was electrolyzed with a current of 3.00 A for 10.0 min. How many milliliters of 0.100 M HCl would be required to neutralize the resulting solution?

17.28 How many hours would it take to produce 25.0 g of metallic chromium by the electrolytic reduction of Cr^{3+} with a current of 1.25 A?

17.29 How many hours would it take to generate 25.0 g lead from $PbSO_4$ during the charging of a storage battery using a current of 0.50 A? The half-reaction is $PbSO_4 + 2e^- \rightarrow Pb + SO_4^{2-}$.

17.30 How many amperes would be needed to produce 48.0 g of magnesium during the electrolysis of molten $MgCl_2$ in 1.00 hr?

17.31 A large electrolysis cell that produces metallic aluminum from Al_2O_3 by the Hall process is capable of yielding 900 lb (409 kg) of aluminum in 24 hr. What current is required?

17.32 How many grams of Cl_2 would be produced in the electrolysis of NaCl by a current of 2.50 A for 40.0 min?

*17.33 A solution containing vanadium (chemical symbol V) in an unknown oxidation state was electrolyzed with a current of 1.50 A for 30.0 min. It was found that 0.475 g of V was deposited on the cathode.
(a) How many equivalents of V were deposited?
(b) How many moles of V were deposited?
(c) What was the original oxidation state of the V ion?

Applications of Electrolysis

17.34 What is electroplating? Sketch an apparatus to electroplate silver.

17.35 Describe the Hall process for producing metallic aluminum. What electrochemical reaction is involved?

17.36 Describe how magnesium is recovered from seawater. What electrochemical reaction is involved?

17.37 How is metallic sodium produced? What are some uses of metallic sodium? Write equations for the anode and cathode reactions.

17.38 Describe the electrolytic refining of copper. What economic advantages offset the cost of electricity for this process? What chemical reactions occur at (a) the anode and (b) the cathode?

17.39 Describe the electrolysis of aqueous sodium chloride. How do the products of the electrolysis compare for stirred and unstirred reactions? Write chemical equations for the reactions that occur at the electrodes.

17.40 What are the advantages and disadvantages in the electrolysis of aqueous NaCl of (a) the diaphragm cell and (b) the mercury cell.

Galvanic Cells

17.41 What is a galvanic cell? What is a half-cell?

17.42 What is the function of a salt bridge?

17.43 In the copper–silver cell, why must the Cu^{2+} and Ag^+ solutions be kept in separate containers?

17.44 What processes take place at the anode and cathode in a galvanic cell? What electrical charge do the anode and cathode carry in a galvanic cell?

17.45 Explain how the movement of ions is the same in both galvanic and electrolytic cells.

17.46 When magnesium metal is placed into a solution of copper sulfate, the magnesium dissolves to give Mg^{2+} and copper metal is formed. Write a net ionic equation for this reaction. Describe how you could use the reaction in a galvanic cell. Which metal, copper or magnesium, is the cathode?

17.47 Aluminum will displace tin from solution according to the equation $2Al(s) + 3Sn^{2+}(aq) \rightarrow 2Al^{3+}(aq) + 3Sn(s)$. What would be the individual half-cell reactions if this were the cell reaction in a galvanic cell? Which metal would be the anode and which the cathode?

17.48 At first glance, the following equation may appear to be balanced: $MnO_4^- + Sn^{2+} \rightarrow SnO_2 + MnO_2$. What is wrong with it?

Cell Potentials and Reduction Potentials

17.49 What is the difference between a cell potential and a standard cell potential?

17.50 How are standard reduction potentials combined to give the standard cell potential for a spontaneous reaction?

17.51 What is emf? What are its units?

17.52 What ratio of units gives volts?

17.53 What are the units of amperes × volts × seconds?

17.54 Is it possible to measure the emf of an isolated half-cell? Explain your answer.

17.55 Describe the hydrogen electrode. What is the value of its standard reduction potential?

17.56 What do the positive and negative signs of reduction potentials tell us?

17.57 If $E°_{Cu^{2+}}$ had been chosen as a reference electrode with a potential of 0.00 V, what would the reduction potential of the hydrogen electrode be relative to it?

17.58 Using a voltmeter, how can you tell which is the anode and which is the cathode in a galvanic cell?

Using Standard Reduction Potentials

17.59 From the positions of the half-reactions in Table 17.1, determine whether the following reactions are spontaneous.
(a) $2Au^{3+} + 6I^- \longrightarrow 3I_2 + 2Au$
(b) $3Fe^{2+} + 2NO + 4H_2O \longrightarrow 3Fe + 2NO_3^- + 8H^+$
(c) $3Ca + 2Cr^{3+} \longrightarrow 2Cr + 3Ca^{2+}$

17.60 Use the data in Table 17.1 to determine which of the following reactions should occur spontaneously.
(a) $Br_2 + 2Cl^- \longrightarrow Cl_2 + 2Br^-$
(b) $Ni^{2+} + Fe \longrightarrow Fe^{2+} + Ni$
(c) $H_2SO_3 + H_2O + Br_2 \longrightarrow 4H^+ + SO_4^{2-} + 2Br^-$

17.61 Compare Table 12.1 with Table 17.1. What is the basis for the activity series for metals?

17.62 Use the data in Table 17.1 to calculate the standard cell potential for the reaction
$$NO_3^- + 4H^+ + 3Fe^{2+} \longrightarrow 3Fe^{3+} + NO + 2H_2O$$

17.63 Use the data in Table 17.1 to calculate the cell potential for the reaction
$$Cd^{2+} + Fe \longrightarrow Cd + Fe^{2+}$$

17.64 From the half-reactions below, determine the cell reaction and standard cell potential.
$$BrO_3^- + 6H^+ + 6e^- \rightleftharpoons Br^- + 3H_2O \quad E°_{BrO_3^-} = 1.44 \text{ V}$$
$$I_2 + 2e^- \rightleftharpoons 2I^- \quad E°_{I_2} = 0.54 \text{ V}$$

17.65 What is the standard cell potential and the net reaction in a galvanic cell that has the following half-reactions?
$$MnO_2 + 4H^+ + 2e^- \rightleftharpoons Mn^{2+} + 2H_2O$$
$$E°_{MnO_2} = 1.23 \text{ V}$$
$$PbCl_2 + 2e^- \rightleftharpoons Pb + 2Cl^-$$
$$E°_{PbCl_2} = -0.27 \text{ V}$$

17.66 What will be the spontaneous reaction among H_2SO_3, $S_2O_3^{2-}$, HOCl, and Cl_2? The half-reactions involved are
$$2H_2SO_3 + 2H^+ + 4e^- \rightleftharpoons S_2O_3^{2-} + 3H_2O$$
$$E°_{H_2SO_3} = 0.40 \text{ V}$$
$$2HOCl + 2H^+ + 2e^- \rightleftharpoons Cl_2 + 2H_2O$$
$$E°_{HOCl} = 1.63 \text{ V}$$

17.67 What will be the spontaneous reaction among Br_2, I_2, Br^-, and I^-? Use the data in Table 17.1.

17.68 Will the following reaction occur spontaneously?
$$SO_4^{2-} + 4H^+ + 2I^- \longrightarrow I_2 + H_2SO_3 + H_2O$$
Use the data in Table 17.1 to answer this question.

17.69 Use the data below to determine whether the reaction
$$S_2O_8^{2-} + Ni(OH)_2 + 2OH^- \longrightarrow 2SO_4^{2-} + NiO_2 + 2H_2O$$
will occur spontaneously.
$$NiO_2 + 2H_2O + 2e^- \rightleftharpoons Ni(OH)_2 + 2OH^-$$
$$E°_{NiO_2} = 0.49 \text{ V}$$
$$S_2O_8^{2-} + 2e^- \rightleftharpoons 2SO_4^{2-}$$
$$E°_{S_2O_8^{2-}} = 2.01 \text{ V}$$

Cell Potentials and Thermodynamics

17.70 Give the equation that relates the standard cell potential to the standard free energy change for a reaction.

17.71 What is the equation that relates the equilibrium constant to the standard cell potential?

17.72 What is the maximum amount of work, expressed in joules, that can be obtained from the discharge of a silver oxide battery (see p. 728) if 1.00 g of Ag_2O reacts? $E°$ for the cell reaction is 1.50 V.

*17.73 A watt is a unit of electrical power equal to one joule/second. (1 watt = 1 J/s.) How many hours can a calculator drawing 5×10^{-4} watt be operated by a mercury battery (see p. 728) having a cell potential equal to 1.34 V if a mass of 1.00 g of HgO is available as the cathode?

17.74 Given these half-reactions and their standard reduction potentials,

$$2ClO_3^- + 12H^+ + 10e^- \rightleftharpoons Cl_2 + 6H_2O$$
$$E°_{ClO_3^-} = 1.47 \text{ V}$$
$$S_2O_8^{2-} + 2e^- \rightleftharpoons 2SO_4^{2-}$$
$$E°_{S_2O_8^{2-}} = 2.01 \text{ V}$$

calculate (a) $E°_{cell}$, (b) $\Delta G°$ for the cell reaction, and (c) the value of K_c for the cell reaction.

17.75 Calculate $\Delta G°$ for the reaction

$$2MnO_4^- + 6H^+ + 5HCHO_2 \longrightarrow$$
$$2Mn^{2+} + 8H_2O + 5CO_2$$

for which $E°_{cell} = 1.69$ V.

17.76 Calculate $\Delta G°$ for the following reaction *as written*.

$$2Br^- + I_2 \longrightarrow 2I^- + Br_2$$

17.77 Calculate K_c for the system, $Ni^{2+} + Co \rightleftharpoons Ni + Co^{2+}$. Use the data in Table 17.1. Assume $T = 298$ K.

17.78 The system $2AgI + Sn \rightleftharpoons Sn^{2+} + 2Ag + 2I^-$ has a calculated $E°_{cell} = -0.015$ V. What is the value of K_c for this system?

17.79 Determine the K_c at 25 °C for the reaction

$$2H_2O + 2Cl_2 \rightleftharpoons 4H^+ + 4Cl^- + O_2$$

The Effect of Concentration on Cell Potentials

17.80 The principles of thermodynamics allow the following equation to be derived,

$$\Delta G = \Delta G° + RT \ln Q$$

where Q is the reaction quotient. Use this equation and what you learned in Section 17.8 to derive the Nernst equation.

17.81 The cell reaction during the discharge of a lead storage battery is

$$Pb(s) + PbO_2(s) + 4H^+(aq) + 2SO_4^{2-}(aq) \longrightarrow$$
$$2PbSO_4(s) + 2H_2O$$

The standard cell potential is 2.05 V. What is the correct form of the Nernst equation for this reaction?

17.82 The cell reaction

$$NiO_2(s) + 4H^+(aq) + 2Ag(s) \longrightarrow$$
$$Ni^{2+}(aq) + 2H_2O + 2Ag^+(aq)$$

has $E°_{cell} = 2.48$ V. What will be the cell potential at a pH of 6.00 when the concentrations of Ni^{2+} and Ag^+ are each 0.10 M?

17.83 $E°_{cell} = 0.135$ V for the reaction

$$3I_2(s) + 5Cr_2O_7^{2-}(aq) + 34H^+(aq) \longrightarrow$$
$$6IO_3^-(aq) + 10Cr^{3+}(aq) + 17H_2O$$

What is E_{cell} if $[Cr_2O_7^{2-}] = 0.10$ M, $[H^+] = 0.010$ M, $[IO_3^-] = 0.0010$ M, and $[Cr^{3+}] = 0.00010$ M?

*17.84 Suppose that a galvanic cell were set up having the net cell reaction

$$Zn(s) + 2Ag^+(aq) \longrightarrow Zn^{2+}(aq) + 2Ag(s)$$

The Ag^+ and Zn^{2+} concentrations in their respective half-cells initially are 1.00 M, and each half-cell contains 100 mL of electrolyte solution. If this cell delivers current at a constant rate of 0.10 A, what will the cell potential be after 10.00 hr?

17.85 A cell was set up having the following reaction.

$$Mg(s) + Cd^{2+}(aq) \longrightarrow Mg^{2+}(aq) + Cd(s) \quad E°_{cell} = 1.97 \text{ V}$$

The magnesium electrode was dipping into a 1.00 M solution of $MgSO_4$ and the cadmium was dipping into a solution of unknown Cd^{2+} concentration. The potential of the cell was measured to be 1.67 V. What was the unknown Cd^{2+} concentration?

17.86 A silver wire coated with AgCl is sensitive to the presence of chloride ion because of the half-cell reaction

$$AgCl(s) + e^- \rightleftharpoons Ag(s) + Cl^-(aq) \quad E°_{AgCl} = 0.22 \text{ V}$$

A student, wishing to measure the chloride ion concentration in a number of water samples, constructed a galvanic cell using the AgCl electrode as one half-cell and a copper wire dipping into 1.00 M $CuSO_4$ solution as the other half-cell. In one analysis, the potential of the cell was measured to be 0.09 V with the copper half-cell serving as the cathode. What was the chloride ion concentration in the water?

*17.87 A galvanic cell was set up having the following half-reactions.

$$Fe^{2+}(aq) + 2e^- \rightleftharpoons Fe(s) \quad E°_{Fe^{2+}} = -0.440 \text{ V}$$
$$Cu^{2+}(aq) + 2e^- \rightleftharpoons Cu(s) \quad E°_{Cu^{2+}} = +0.337 \text{ V}$$

The copper half-cell contained 100 mL of 1.00 M $CuSO_4$. The iron half-cell contained 50.0 mL of 0.100 M $FeSO_4$. To the iron half-cell was added 50.0 mL of 0.500 M NaOH solution. The mixture was stirred and the cell potential was measured to be 1.175 V. Calculate the value of K_{sp} for $Fe(OH)_2$.

*17.88 Suppose a galvanic cell was constructed using a Cu/Cu^{2+} half-cell (in which the molar concentration of Cu^{2+} was 1.00 M) and a hydrogen electrode having a partial pressure of H_2 equal to 1 atm. The hydrogen electrode dips into a solution of unknown hydrogen ion concentration, and the two half-cells are connected by a salt bridge. The precise value of $E^\circ_{Cu^{2+}}$ is $+0.337$ V.

(a) Derive an equation for the pH of the solution with the unknown hydrogen ion concentration, expressed in terms of E_{cell} and E°_{cell}.

(b) If the pH of the solution were 4.25, what would be the observed emf of the cell?

(c) If the emf of the cell were 0.660 V, what would be the pH of the solution?

Practical Galvanic Cells

17.89 What are the anode and cathode reactions during the discharge of a lead storage battery? How can a battery produce an emf of 12 V if the cell reaction has a standard potential of only 2 V?

17.90 What are the anode and cathode reactions during the charging of a lead storage battery?

17.91 What reactions occur in the ordinary dry cell?

17.92 What chemical reactions take place in an alkaline dry cell?

17.93 Give the half-cell reactions and the cell reaction in a nicad battery during discharge.

17.94 Describe the chemical reactions that take place in (a) a silver oxide battery and (b) a mercury battery.

17.95 What advantages do fuel cells offer over conventional means of obtaining electrical power by the combustion of fuels?

CHEMICALS IN USE
ION SELECTIVE ELECTRODES

12

The influence of electrochemistry is to be found almost anywhere we look. Electrolysis is used to manufacture a variety of chemicals that are indispensable in today's economy, and a growing assortment of galvanic cells run cameras and calculators and all sorts of other gadgets. There are other practical uses of electrochemistry that are less obvious, however, and many of them rely on the unique role that electrochemistry plays as an interface between chemical systems and electronic devices that display, record, and manipulate data. The key to using this interface has been the development of electrodes that are *selectively* sensitive to the kinds of chemicals of interest. These are called **ion selective electrodes** because of their ability to respond to certain specific ions while ignoring others, and by using them chemists have devised ways of measuring, monitoring, and even controlling the concentrations of a large variety of chemical species in solution.

In any galvanic cell, there are two electrodes. For the measurement of concentrations, one of them is a reference electrode whose reduction potential is constant. The other is called an indicating electrode; it is the ion selective electrode whose potential is sensitive to the concentration of the species of interest. Both electrodes are connected to some appropriate device that displays the emf of the cell or translates that potential into concentration information.

We introduced you to the concept of an ion selective electrode in Special Topic 17.2, where we discussed the measurement of pH by the use of the hydrogen glass electrode. This is one of a number of electrodes whose construction is based on the same basic design (Figure 12a). Inside is a reference solution, and dipping into it is the device's own internal reference electrode, which provides the electrical contact with the emf readout device. The design on the left is that of the hydrogen glass electrode described earlier. The thin glass membrane is ion sensitive and allows the electrode to detect a difference between the ion concentration inside and outside. By suitable changes in the composition of the glass membrane, this kind of electrode can be made selectively sensitive to the concentrations of a number of singly charged cations including H^+, Li^+, Na^+, K^+, Ag^+ and NH_4^+.

The glass electrode is called a solid-state membrane electrode because of the nature of the ion-sensitive material used in its construction. On the left in Figure 12a is illustrated the construction of a newer type based basically on the same principle. It uses an ionically conducting crystalline material such as LaF_3 or polycrystalline Ag_2S or AgCl. Although the two types appear different, they function in essentially the same way.

Another important class of ion selective electrode uses a different type of membrane material between the reference solution and the solution being tested. It is called a liquid ion-exchange membrane, and is composed of a water-insoluble organic liquid that's capable of transporting some specific ion (for example, Ca^{2+} or K^+) across the membrane boundary. This ability makes the electrode respond selectively to the particular ion being transported. The construction of a typical ion-exchange membrane electrode is shown in Figure 12b.

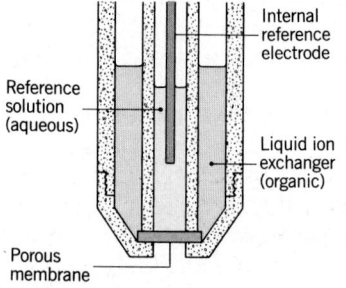

FIGURE 12b A liquid ion-exchange membrane electrode. The organic liquid ion exchanger seeps into the porous membrane and provides the ion selectivity of the electrode.

Practical Applications

Some of the most interesting applications of ion selective electrodes have been in the areas of medicine and biologi-

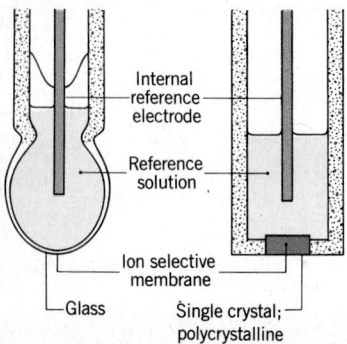

FIGURE 12a Ion selective electrodes with solid-state membranes.

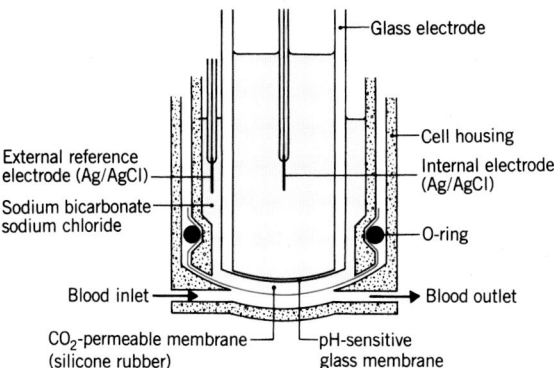

FIGURE 12c A schematic diagram of the CO_2 cell.

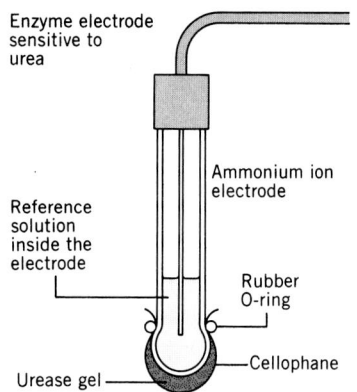

FIGURE 12d An enzyme electrode sensitive to urea.

cal research. For example, an important cell used to measure the carbon dioxide level in blood is based on the hydrogen glass electrode that we use to measure pH. A diagram of the cell is shown in Figure 12c. The key to the functioning of the cell is the thin silicone rubber membrane that separates the blood sample from a dilute $NaHCO_3$ solution in contact with the glass electrode. The rubber membrane is permeable to CO_2 which diffuses into the solution that contains the HCO_3^-. The CO_2 participates in the equilibrium

$$CO_2(aq) + H_2O \rightleftharpoons HCO_3^-(aq) + H^+(aq)$$

and thereby influences the pH of the solution, which is detected by the glass electrode. In an indirect way, then, the potential of the glass electrode reflects the CO_2 concentration just outside the glass membrane, and this in turn is a measure of the CO_2 concentration in the blood that's in contact with the silicone rubber membrane. In practice, the electrode is first calibrated with solutions of known CO_2 partial pressures and then used to measure the CO_2 content of a blood sample.

We mentioned earlier that by suitably modifying the glass membrane, electrodes can be made that respond to ions other than H^+. Sodium-type glass electrodes have been used in clinical laboratories to measure serum sodium levels. A clinical use of the second type of electrode in Figure 12a is in the measurement of chloride ion in sweat. Similarly, ion-exchange membrane electrodes have been used successfully to monitor serum levels of calcium ion and potassium ion.

In the laboratory there have been a number of interesting developments that have made biochemical reactions potentially easier to study. One very clever modification of the ammonium ion glass electrode, for example, is shown in Figure 12d. The glass membrane of the electrode is coated with a gel in which is embedded an enzyme such as urease — a biological catalyst that promotes the hydrolysis of one particular molecule, urea.

$$CO(NH_2)_2 + H_2O \xrightarrow{\text{urease}} CO_2 + 2NH_3$$

The gel–urease layer is held in place by a cellophane membrane through which the urea is able to diffuse. On contact with the enzyme, hydrolysis occurs and the ammonia that's formed then reacts with water in the equilibrium

$$NH_3(aq) + H_2O \rightleftharpoons NH_4^+(aq) + OH^-(aq)$$

The ammonium ion formed in this reaction is then detected by the electrode. Once again, we have an electrode that indirectly senses a particular chemical, and this time that chemical isn't even ionic. Other similar electrodes have been developed that are capable of monitoring the concentrations of ethanol, cholesterol and some other substances in body fluids. In fact, a glucose electrode is commercially available.

One of the most fascinating applications of ion selective electrodes has been the development of miniature devices that can be used to monitor ion concentrations in living cells. These tiny electrodes are made by drawing fine glass capillaries which are then filled at the tip with a suitable ion-exchange fluid. A reference solution is added, and finally an internal reference electrode is inserted. The finished electrode has a tip diameter of about 2×10^{-4} mm, which is small enough to be inserted into living cells without destroying them. With these electrodes, scientists have been able to measure ion concentration within cells, which is the kind of information needed to understand such things as the mechanical and electrical activities of nerves and muscles.

Questions

1. What equation forms the basis for measuring ion concentrations by the use of appropriately designed galvanic cells?

2. What ions can be measured by glass electrodes with suitable glass membranes?

3. What is an ion-exchange membrane electrode? How does it differ from a glass electrode?

4. In your own words, describe how the CO_2 electrode is able to detect varying CO_2 levels in the blood.

5. How does the enzyme substrate electrode work?

Insects are cold blooded creatures whose body temperatures follow the temperature of the surroundings. During cold weather, this bee's activities will slow because its body chemistry becomes slower at lower temperatures. Temperature is one of several factors that influence the speeds of chemical reactions, as we will see in this chapter.

RATES OF REACTION

18.1 SPEEDS AT WHICH REACTIONS OCCUR

By studying the speed of a chemical reaction we can learn details about how it occurs.

In Chapter 16 we studied the factors that determine whether or not reactions are possible, and in this discussion we saw that "spontaneity" is only one of the factors that determine whether or not we will observe a particular chemical change. Besides being possible, a reaction also must take place rapidly enough, and for this reason we now turn our attention to how speedily reactions occur.

If we consider chemical reactions that are a part of everyday life, we will notice that the amount of time it takes for different reactions to occur varies considerably. For example, when a mixture of gasoline and air is fed into the cylinders of an automobile engine and ignited, a very rapid (almost instantaneous) reaction occurs that we use to propel the car. Most reactions that we observe are much slower than this, however. The cooking of foods, for instance, involves chemical reactions and, as any hungry person waiting for a hamburger knows, these reactions take a while. Fortunately for us, so do the biochemical reactions involved in digestion or metabolism. If biochemical reactions were as rapid as the combustion of gasoline vapor, our lives would be over in an instant.

Some reactions going on around us are quite slow. A fresh coat of paint that contains linseed oil "dries" slowly, first by evaporation of the solvent, and then by gradual reaction of the linseed oil resins in the paint with oxygen. This may take weeks or even months. The gradual curing of cement and concrete is another example. Although cement solidifies relatively quickly, chemical reactions within the mixture continue for years as the cement becomes harder and harder.

In scientific terms, the speed at which a reaction occurs is called the rate of reaction. It tells us how quickly (or slowly) the reactants disappear and the products form. Such information can be quite valuable. For instance, studying the rates of reactions and the factors that control these rates allow the manufacturers of chemical products to improve productivity and hold down operating costs by adjusting the conditions of a reaction so that it takes place as efficiently as possible. Of more fundamental importance to us, however, is the fact that, by studying the rates of reactions, we can learn a great deal about the chemical steps that lead ultimately from the reactants to the products.

When most chemical reactions take place, the change described by the balanced chemical equation — the net overall change — almost always occurs by a series of simpler chemical reactions. Consider, for example, the combustion of propane, C_3H_8.

$$C_3H_8(g) + 5O_2(g) \longrightarrow 3CO_2(g) + 4H_2O(g)$$

This reaction simply cannot occur in a single, simultaneous collision that involves one propane molecule and five oxygen molecules. Anyone who has ever played pool or billiards knows how seldom three balls come together with only one "click." It is easy to understand, then, how unlikely is the simultaneous collision of six molecules in three-dimensional space. For this reaction to proceed rapidly, which indeed it does, some very much more probable events must be involved. The series of individual steps leading to the overall observed reaction is called the mechanism of the reaction. Information about reaction mechanisms is one of the dividends that the study of reaction rates can give.

Propane is the fuel used for cooking in many rural areas as well as in many campers and pleasure boats.

18.2 FACTORS THAT AFFECT REACTION RATES

The rate of a reaction depends both on the nature of the reactants and on the conditions under which the reaction occurs.

The rates of nearly all chemical reactions are controlled primarily by five factors: (1) the chemical nature of the reactants; (2) the ability of the reactants to come in contact with each other; (3) the concentrations of the reactants; (4) the temperature of the reacting system; and (5) the availability of agents called catalysts that affect the rate of the reaction but are not themselves consumed.

The Nature of the Reactants

Fundamental differences in chemical reactivity, which are controlled by the tendencies toward bond formation, are a major factor in determining the rate of a reaction. Some reactions are just naturally fast and others are naturally slow. For example, a freshly exposed surface of sodium tarnishes almost instantly if exposed to air and moisture because sodium loses electrons so easily. Iron also reacts with air and moisture — it forms rust — but the reaction is much slower under identical conditions because iron simply doesn't lose electrons as easily as sodium.

The Ability of the Reactants to Meet

Most reactions involve two or more reactants. Obviously, for the reaction to occur, the reactants must be able to come in contact with each other. This is one of the primary reasons that reactions are most often carried out in liquid solutions or in the gas phase. In these states, the reactants are able to intermingle on the molecular level, which allows their reacting particles — molecules or ions — to collide with each other easily.

Acids react with bicarbonates and carbonates to generate carbon dioxide and water. The net reactions are

$$H^+(aq) + HCO_3^-(aq) \longrightarrow H_2O + CO_2(g)$$
$$2H^+(aq) + CO_3^{2-}(aq) \longrightarrow H_2O + CO_2(g)$$

But for these reactions to occur, the H^+ must come into direct contact with the HCO_3^- or CO_3^{2-} ions. The makers of the well-known analgesic tablets Alka Seltzer® take full advantage of this fact. Among the ingredients in these tablets are sodium bicarbonate and citric acid. Crystals of these substances are able to exist in contact with each other without reacting because the hydrogen ions available from the citric acid cannot mingle with the bicarbonate ions in the sodium bicarbonate. Diffusion in solids is too slow for this to happen. But when the tablets are dropped into a glass of water *(plop, plop)*, the citric acid and sodium bicarbonate dissolve, thus allowing the mixing of the ions and the production of carbon dioxide *(fizz, fizz)*.

It is easy to find common examples of how the ability of the reactants to meet affects the speed of a reaction. For instance, liquid gasoline burns fairly rapidly, but exactly how fast it burns depends on how quickly air can come in contact with the surface of the burning liquid. However, if the gasoline is vaporized and mixed with air before it is ignited, the combustion reaction is virtually instantaneous—the mixture explodes. An explosion is simply an extremely rapid reaction that generates hot gases that expand very quickly.

When the reactants are present in different phases—when one is a gas and the other is a liquid or a solid, for example—the reaction that occurs is called a *heterogeneous reaction.* In these reactions, the reactants are able to come in contact with each other only where they meet at the interface between the two phases. The size of this area of contact determines the rate of the reaction. This area can be increased by decreasing the sizes of the particles of the reactants, as anyone who has ever tried to start a campfire knows. Lighting a large log with a match is very difficult. There is not enough contact between the wood and the oxygen in the air for combustion to occur easily. However, kindling—small twigs or slivers of wood—is relatively easy to light. Its greater surface area allows it to react more quickly. This is the reason that all good fire-builders always start with kindling.

If the particle sizes are extremely small, heterogeneous reactions can be explosive. Figure 18.1 shows the result of an explosion in a grain elevator that was used to store

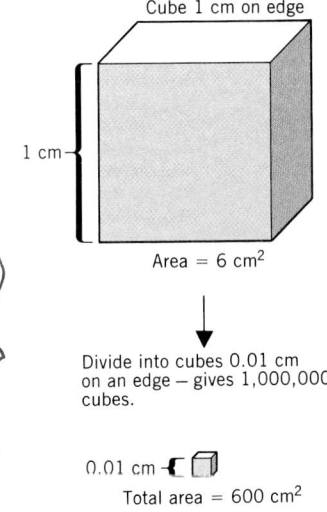

Cube 1 cm on edge

1 cm

Area = 6 cm²

Divide into cubes 0.01 cm on an edge—gives 1,000,000 cubes.

0.01 cm

Total area = 600 cm²

FIGURE 18.1 An explosion in a grain elevator in New Orleans, Louisiana, killed 35 people in December 1977. It occurred when a spark ignited very fine dust in the grain silos. Extremely rapid combustion of the dust produced the explosive effect.

wheat. The explosion occurred when fine particles of grain dust mixed with air were accidentally ignited by a chance spark.

When all the reactants are in the same phase, the reaction that occurs is called a homogeneous reaction. Examples are the combustion of gasoline vapor and other gaseous reactions as well as those reactions that occur in liquid solutions. Most of the topics in the rest of this chapter focus on homogeneous systems.

The Concentrations of the Reactants

The rates of both homogeneous and heterogeneous reactions are affected by the concentrations of the reactants. For example, wood burns relatively quickly in air but extremely rapidly in pure oxygen. The hazards of a pure oxygen atmosphere were tragically demonstrated in January 1967, when three astronauts died in a flash fire that swept through their oxygen-filled Apollo spacecraft during a training exercise at Cape Canaveral.

The Temperature of the System

Almost all chemical reactions occur faster at higher temperatures than they do at lower temperatures. If you have a Polaroid® camera, for instance, you may have noticed that the "instant picture" develops more quickly in warm weather than in cold weather. The reactions that develop the image occur faster at the higher temperature. You may have also noticed that insects begin to move more slowly in the autumn as the days become cooler. Insects are cold-blooded creatures whose body temperatures are determined by the temperatures of their surroundings. As they become cooler, their body chemistry slows down and they become sluggish.

The Presence of Catalysts

Catalysts are substances that increase the rates of chemical reactions without being used up. They affect our lives every moment — the enzymes that direct our body chemistry are catalysts. So are many of the different substances that the chemical industry uses to make gasoline, plastics, fertilizers, and myriad other products that have become virtual necessities in our lives. We will discuss how catalysts affect reaction rates later in the chapter.

18.3 MEASURING THE RATE OF REACTION

The rate of a reaction is obtained by measuring the concentration of a reactant or a product at regular time intervals during the reaction.

A rate is always expressed as a ratio in which units of time appear in the denominator. For example, if you have a job, you are probably paid at a certain rate — say, five dollars per hour. Since the word *per* can be translated to mean *divided by*, this pay rate can be written as a fraction.

$$\text{rate of pay} = \frac{5 \text{ dollars}}{1 \text{ hour}}$$

In general, $x^{-1} = 1/x$

Since the fraction $\dfrac{1}{1 \text{ hour}}$ can also be written as hour^{-1}, your pay rate can be rewritten as

$$\text{rate of pay} = 5 \text{ dollars hour}^{-1}$$

You might think of your rate of pay as a change in your wealth. While you are working, your wealth changes (increases) at a rate of 5 dollars/hour.

As a chemical reaction occurs, the concentrations of the reactants and products change with time. The concentrations of the reactants decrease as the reactants are used up, while the concentrations of the products increase as the products are formed. The rate of a chemical reaction, therefore, is expressed in terms of the rates at which the concentrations change. The unit of concentration that is generally used for this purpose is molar concentration, and the unit of time generally is the second (abbreviated s). This means that reaction rates are most frequently expressed with the units

$$\text{reaction rate} = \frac{\text{mol/L}}{\text{s}}$$

Since 1/L and 1/s can also be written as L^{-1} and s^{-1}, the units for the rate can be expressed as

$$\text{reaction rate} = \text{mol } L^{-1} s^{-1}$$

Thus, if the concentration of a certain reactant were changing by 0.5 mol/L each second, the rate of this reaction would be 0.5 mol L^{-1} s^{-1}.

For nearly all chemical reactions, the rate does not remain constant as the reaction progresses. As mentioned in Section 18.2, the rate usually depends on the reactant concentrations and these change as the reaction proceeds. Table 18.1, for example, contains data on the decomposition of hydrogen iodide at a temperature of 508 °C.

$$2HI(g) \longrightarrow H_2(g) + I_2(g)$$

These data, which show the way the HI concentration changes with time, are plotted in Figure 18.2. The vertical axis gives the molar concentration of HI.

In Figure 18.2, notice that the HI concentration drops fairly rapidly during the first 50 seconds of the reaction—the initial rate is relatively fast. Between 300 and 350 seconds, however, the concentration changes by only a small amount—the rate has slowed considerably. From this, we can see that the steepness of the concentration–time curve reflects the rate of the reaction—the steeper the curve, the higher the rate.

TABLE 18.1 Data, at 508 °C, for the Reaction $2HI(g) \rightarrow H_2(g) + I_2(g)$

Concentration of HI (mol/L)	Time (s)
0.100	0
0.0716	50
0.0558	100
0.0457	150
0.0387	200
0.0336	250
0.0296	300
0.0265	350

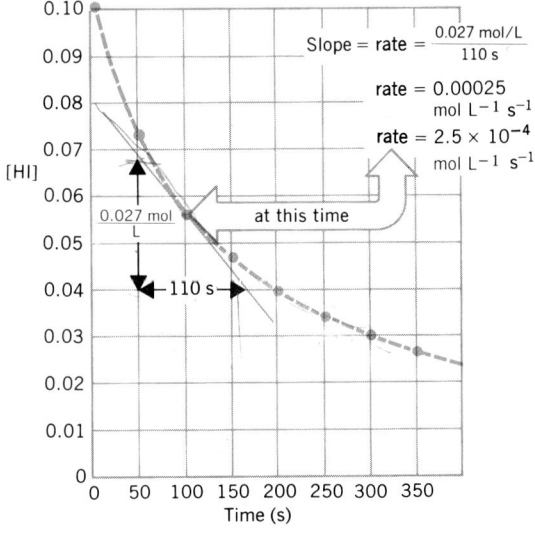

FIGURE 18.2 The change in the HI concentration with time for the reaction $2HI(g) \rightarrow H_2(g) + I_2(g)$ at 508 °C. The data that are plotted are in Table 18.1.

The exact rate at any particular time can be obtained by determining the slope of the line that is tangent to the concentration–time curve at that point. The slope, which is equal to the rate, is the ratio of the change in concentration to the change in time as read off the graph. In Figure 18.2, for example, the rate of the decomposition of hydrogen iodide is determined at a time 100 seconds from the start of the reaction. After the tangent to the curve is drawn, we measure a concentration change (0.027 mol/L) and a time change (110 s) from the graph. The ratio of these quantities,

$$\frac{0.027 \text{ mol/L}}{110 \text{ s}}$$

If necessary, see Appendix A for a discussion of slope.

gives the rate. We see that, at this point in the reaction, the HI concentration is decreasing at a rate of 2.5×10^{-4} mol L^{-1} s^{-1}. If you work Practice Exercise 1, you will see that the rate of the reaction later on is indeed lower.

PRACTICE EXERCISE 1   Use the graph in Figure 18.2 to estimate the reaction rate 250 seconds from the start of the reaction.

Although we usually measure the rate at which the reactants disappear, it is sometimes more practical to measure the rate at which the products are formed. For example, in the decomposition of HI, it is easiest to monitor the I$_2$ concentration because I$_2$ vapor is purple. Therefore, as the reaction proceeds, more and more purple vapor appears and the color becomes more intense. Instruments are available that can be used to relate the amount of light absorbed by the I$_2$ vapor to the concentration of I$_2$.

The rate at which the products form can be related to the rate at which the reactants disappear if the equation for the reaction is known. For the reaction

$$2HI(g) \longrightarrow H_2(g) + I_2(g)$$

the HI concentration decreases twice as fast as the I$_2$ concentration increases because 2 mol of HI disappear for each mole of I$_2$ that is formed. It doesn't really matter, then, which substance we use to follow a reaction because the rates at which all of the concentrations change are interrelated.

PRACTICE EXERCISE 2   In the reaction, $N_2O_4(g) \rightarrow 2NO_2(g)$, if the rate of formation of NO$_2$ were 0.010 mol L^{-1} s^{-1}, what would be the rate of decomposition for $N_2O_4(g)$?

18.4 CONCENTRATION AND RATE

An equation called the rate law expresses the way that the concentrations of the reactants affect the rate of a reaction.

We learned in the last section that the rates of most reactions change when the concentrations of the reactants change. The way that the reaction rate and the concentrations of the reactants are related can be expressed in the form of an equation. In general, the rate of a homogeneous reaction is found to be proportional to the product of the concentrations of the reactants, each raised to a power. For example, if we have a chemical equation such as

$$A + B \longrightarrow \text{products}$$

the rate of the reaction would be

$$\text{rate} \propto [A]^n[B]^m \tag{18.1}$$

The values of the exponents n and m must be determined by experiment—we will explore this further in a moment. The proportionality symbol, $\propto$, can be replaced by an equal sign if we introduce a proportionality constant, k, which is called the rate constant for the reaction. Equation 18.1 then takes the form

$$\text{rate} = k[A]^n[B]^m \tag{18.2}$$

This equation is called the rate law for the reaction. Once we have found values for k, n, and m, the rate law allows us to calculate the rate of the reaction if the concentrations are

known. For example, the reaction

$$H_2SeO_3 + 6I^- + 4H^+ \longrightarrow Se + 2I_3^- + 3H_2O \qquad (18.3)$$

would be expected to have a rate law of the form

$$\text{rate} = k[H_2SeO_3]^x[I^-]^y[H^+]^z$$

Experimentally, it has been found that for the initial rate of the reaction (that is, the rate when the reactants are first combined), $x = 1$, $y = 3$, and $z = 2$. The value of the rate constant k depends on the temperature (we will discuss why and how later in the chapter), and at 0 °C it has been determined that k for this reaction equals 5.0×10^5 L^5 mol^{-5} s^{-1}. Substituting these values gives the rate law

$$\text{rate} = (5.0 \times 10^5 \text{ L}^5 \text{ mol}^{-5} \text{ s}^{-1})[H_2SeO_3]^1[I^-]^3[H^+]^2 \qquad (18.4)$$

This equation allows us to calculate the rate of the reaction at 0 °C for any set of concentrations of H_2SeO_3, I^-, and H^+.

The value of k depends on the particular reaction being studied as well as the temperature at which the reaction occurs.

The units of the rate constant are generally such that the calculated rate will have the units mol L^{-1} s^{-1}.

EXAMPLE 18.1   CALCULATING REACTION RATE FROM THE RATE LAW

Problem: What is the rate of Reaction 18.3 at 0 °C when the reactant concentrations are $[H_2SeO_3] = 2.0 \times 10^{-2}$ M, $[I^-] = 2.0 \times 10^{-3}$ M, and $[H^+] = 1.0 \times 10^{-3}$ M?

Solution: We simply substitute the concentrations into the rate law, Equation 18.4. So that we can see how the units work out, we will write the rate constant's units in fraction form.

$$\text{rate} = \frac{5.0 \times 10^5 \text{ L}^5}{\text{mol}^5 \text{ s}} \times \left(\frac{2.0 \times 10^{-2} \text{ mol}}{\text{L}}\right) \times \left(\frac{2.0 \times 10^{-3} \text{ mol}}{\text{L}}\right)^3 \times \left(\frac{1.0 \times 10^{-3} \text{ mol}}{\text{L}}\right)^2$$

To perform the arithmetic, we first raise the concentrations *and* their units to the appropriate powers.

$$\text{rate} = \frac{5.0 \times 10^5 \text{ L}^5}{\text{mol}^5 \text{ s}} \times \left(\frac{2.0 \times 10^{-2} \text{ mol}}{\text{L}}\right) \times \left(\frac{8.0 \times 10^{-9} \text{ mol}^3}{\text{L}^3}\right) \times \left(\frac{1.0 \times 10^{-6} \text{ mol}^2}{\text{L}^2}\right) = \frac{8.0 \times 10^{-11} \text{ mol}}{\text{L s}}$$

The answer can also be written as

$$\text{rate} = 8.0 \times 10^{-11} \text{ mol L}^{-1} \text{ s}^{-1}$$

Notice that the answer has the usual units for reaction rate.

PRACTICE EXERCISE 3   The rate law for the decomposition of HI is rate $= k[HI]^2$. In Figure 18.2, the rate of the reaction was found to be 2.5×10^{-4} mol L^{-1} s^{-1} when the HI concentration was 0.0558 M. (a) What is the value of the rate constant for the reaction? (b) What are the units of k for this reaction?

In Chapter 14 you learned how to write the equilibrium law for a reaction, and you learned that the exponents on the concentration terms in the mass action expression are *always* equal to the coefficients in the balanced chemical equation. The situation with rate laws, however, is quite different. For example, take a look at Reaction 18.3 and its rate law, Equation 18.4. Notice that the exponents in the rate law appear to be unrelated to the coefficients in the overall chemical equation. This often is the case. *There is, in fact, no way to know for sure what the exponents in the rate law will be without doing experiments to determine them.*

The exponents in the rate law are not *always* different from the coefficients in the balanced chemical equation. Sometimes the coefficients and the exponents happen to be the same just by coincidence, as is the case with the rate law for the decomposition of HI.

$$2HI(g) \longrightarrow H_2(g) + I_2(g)$$

$$\text{rate} = k[HI]^2$$

There is no way of knowing this, however, without experimental data. You should never simply assume that they are the same — that's a trap into which many students fall.

The sum of the exponents in the rate law is referred to as the *order* of the reaction.[1] For instance, the decomposition of gaseous N_2O_5 into NO_2 and O_2 has the rate law

$$\text{rate} = k[N_2O_5]^1$$

This reaction has only one reactant. The exponent of the N_2O_5 concentration is 1 and therefore the reaction is said to be *first order*. The rate law for the decomposition of HI has an exponent of 2 for the HI concentration; it, then, is a *second-order* reaction.

When the rate law contains two or more reactants raised to powers, the order with respect to each reactant can be described. The rate law in Equation 18.4 tells us that Reaction 18.3 is *first order* with respect to H_2SeO_3, *third order* with respect to I^-, and *second order* with respect to H^+. The *overall order* of a reaction is the sum of the orders with respect to each reactant. Reaction 18.3 is a sixth-order reaction, overall.

The exponents in a rate law are usually whole numbers, but fractional and negative exponents are also found occasionally. A negative exponent means that the concentration term really belongs in the denominator and, as the concentration of the species increases, the rate of reaction decreases. Zero-order reactions also occur. These are reactions whose rate is independent of the concentration of the reactant.

2N$_2$O$_5$ $\longrightarrow$ 4NO$_2$ + O$_2$

When the exponent is equal to 1 it is usually omitted. Therefore, when no exponent is written, assume its value is 1.

$1 + 3 + 2 = 6$

PRACTICE EXERCISE 4

The reaction $BrO_3^- + 3SO_3^{2-} \rightarrow Br^- + 3SO_4^{2-}$ has the rate law rate = $k[BrO_3^-][SO_3^{2-}]$. What is the order of the reaction with respect to each reactant? What is the overall order of the reaction?

Determining Exponents of the Rate Law

We have stated several times in this chapter that the exponents of a rate law must be determined experimentally. *This is the only way for us to know for sure what these exponents are.*

To find the exponents of a rate law, we must study how changes in concentration affect the rate of the reaction. For example, suppose we have a reaction

$$A + B \longrightarrow \text{products}$$

for which the data in Table 18.2 have been obtained in a series of five experiments. The form of the rate law for the reaction would be

$$\text{rate} = k[A]^n[B]^m$$

For the first three sets of data, the concentration of B is constant. Changes in the rate are therefore due only to changes in the concentration of A. Now all we have to do is figure out what the order of the reaction must be with respect to A to give the observed changes.

Examining the data, we find that when $[A]$ is doubled, the rate doubles; when $[A]$ is tripled, the rate triples. What must the exponent be on the concentration of A for this to be true? The answer is that the exponent must be 1. We can see this if we try a few values for the concentration. When $[A] = 0.10$, $[A]^1 = 0.10$, and when $[A] = 0.20$, $[A]^1 = 0.20$.

TABLE 18.2 Concentration/Rate Data for the Reaction A + B → Products

Initial Concentrations		Initial Rate of Formation of Products
[A]	[B]	(mol L^{-1} s^{-1})
0.10	0.10	0.20
0.20	0.10	0.40
0.30	0.10	0.60
0.30	0.20	2.40
0.30	0.30	5.40

[1] The reason for describing the *order* of a reaction is that the mathematics involved in the treatment of the data are the same for all reactions having the same order. We will not go into this very deeply, but you should be familiar with the terminology used to describe the effects of concentration on the rates of reactions.

Thus, doubling the concentration of A doubles the value of $[A]^1$, and this will cause the rate to double.

In the final three sets of data, the concentration of B changes while the concentration of A is held constant. This time it is the concentration of B that affects the rate. Now we see that when $[B]$ is doubled, the rate increases by a factor of 4 (which equals 2^2), and when $[B]$ is tripled, the rate increases by a factor of 9 (which equals 3^2). The only way that B can affect the rate this way is if it concentration is squared in the rate law. This also can be shown using some trial values for $[B]$. If $[B] = 0.3$, then $[B]^2$ equals 0.09; if the concentration is doubled, so that $[B] = 0.6$, then $[B]^2$ equals 0.36. Note that 0.36 is *four* times as large as 0.09. Doubling the concentration therefore increases $[B]^2$ by a factor of 4, and that increases the rate by a factor of 4.

Having determined the exponents of the concentration terms, we now know that the rate law for this reaction must be

$$\text{rate} = k[A]^1[B]^2$$

To calculate the value of k, we need only substitute rate and concentration data into this rate law for any one of the sets of data.

$$k = \frac{\text{rate}}{[A][B]^2}$$

Using the data from the first set,

$$k = \frac{0.20 \text{ mol L}^{-1}\text{ s}^{-1}}{(0.10 \text{ mol L}^{-1})(0.10 \text{ mol L}^{-1})^2}$$

$$= \frac{0.20 \text{ mol L}^{-1}\text{ s}^{-1}}{0.0010 \text{ mol}^3 \text{ L}^{-3}}$$

Note that $\text{mol/mol}^3 = \text{mol}^{-2}$, and $L^{-1}/L^{-3} = L^2$. Thus, the value of k, with its correct units, is

$$k = 2.0 \times 10^2 \text{ L}^2 \text{ mol}^{-2} \text{ s}^{-1}$$

PRACTICE EXERCISE 5   Use the data from the other four experiments to calculate k for this reaction. What do you notice about the answers that you get?

The reasoning used to determine the order with respect to each reactant from experimental data is summarized in Table 18.3.

TABLE 18.3 Relation of the Order of a Reaction to Changes in Concentration and Rate

Factor by Which Concentration Is Changed	Factor by Which Rate Changes	Exponent on the Concentration Term Must Be:
2	⎰ Rate is ⎱	0
3	⎱ unchanged ⎰	0
4		0
2	$2 = 2^1$	1
3	$3 = 3^1$	1
4	$4 = 4^1$	1
2	$4 = 2^2$	2
3	$9 = 3^2$	2
4	$16 = 4^2$	2
2	$8 = 2^3$	3
3	$27 = 3^3$	3

EXAMPLE 18.2   DETERMINING THE EXPONENTS OF A RATE LAW

Problem: The following data were collected on the initial rates of decomposition of SO_2Cl_2 at a particular temperature.

$SO_2Cl_2(g) \longrightarrow SO_2(g) + Cl_2(g)$

SO_2Cl_2 is used in the manufacture of chlorophenol, an antiseptic.

Initial Concentration of SO_2Cl_2 (mol L^{-1})	Initial Reaction Rate (mol L^{-1} s^{-1})
0.100	2.2×10^{-6}
0.200	4.4×10^{-6}
0.300	6.6×10^{-6}

What is the rate law and what is the value of the rate constant?

Solution: We expect the rate law to be

$$\text{rate} = k[SO_2Cl_2]^n$$

The initial rate doubles (from 2.2×10^{-6} to 4.4×10^{-6}) when the initial concentration doubles; the rate triples when the concentration triples. The reaction must therefore be first order. Thus

$$\text{rate} = k[SO_2Cl_2]^1$$

Of course, we could get the same value using either of the other two sets of data.

The value of k, calculated from the first set of data, is

$$k = \frac{\text{rate}}{[SO_2Cl_2]}$$

$$= \frac{2.2 \times 10^{-6} \; \text{mol L}^{-1} \; s^{-1}}{0.10 \; \text{mol L}^{-1}}$$

$$= 2.2 \times 10^{-5} \; s^{-1}$$

EXAMPLE 18.3   DETERMINING THE EXPONENTS OF A RATE LAW

Problem: Isoprene, a monomer used to make natural rubber, forms a dimer called dipentene.

A polymer is a long chainlike molecule formed by linking many monomers. A dimer is formed from two monomers.

What is the rate law for this reaction, given the following data for the initial rates of formation of dipentene?

Initial Isoprene Concentration (mol L^{-1})	Initial Reaction Rate (mol L^{-1} s^{-1})
0.50	1.98
1.50	17.8

Solution: We expect the rate law to be of the form

$$\text{rate} = k[\text{isoprene}]^n$$

Comparing the two experiments, the isoprene concentration in the second is three times larger than it is in the first. The rate in the second experiment is 17.8/1.98 or 8.99 times larger than it is in the first experiment. The value 8.99 is very nearly 9, which is 3^2. Therefore, the value of n is 2 and the rate law is

$$\text{rate} = k[\text{isoprene}]^2$$

EXAMPLE 18.4 DETERMINING THE EXPONENTS OF A RATE LAW

Problem: The following data were measured for the reduction of nitric oxide with hydrogen.

$$2NO(g) + 2H_2(g) \longrightarrow N_2(g) + 2H_2O(g)$$

Initial Concentrations (mol L^{-1})		Initial Rate of Disappearance of NO (mol L^{-1} s^{-1})
[NO]	[H$_2$]	
0.10	0.10	1.23×10^{-3}
0.10	0.20	2.46×10^{-3}
0.20	0.10	4.92×10^{-3}

What is the rate law for this reaction?

Solution: We expect the rate law to be

$$\text{rate} = k[\text{NO}]^n[\text{H}_2]^m$$

First, we compare experiments 1 and 2. When the H$_2$ concentration is doubled, while [NO] remains constant, the rate is also doubled. The reaction, then, is first order with respect to H$_2$. Next, we compare experiments 1 and 3. In these, the NO concentration is doubled while the H$_2$ concentration remains constant. Doubling [NO] increases the rate by a factor of 4.92/1.23 or 4.00. Thus, the reaction must be second order with respect to NO. Now we can write the rate law.

$$\text{rate} = k[\text{NO}]^2[\text{H}_2]$$

PRACTICE EXERCISE 6   Ordinary sucrose (table sugar) reacts with water to produce two simpler sugars, glucose and fructose, that have the same molecular formulas.

$$\underset{\text{sucrose}}{C_{12}H_{22}O_{11}} + H_2O \longrightarrow \underset{\text{glucose}}{C_6H_{12}O_6} + \underset{\text{fructose}}{C_6H_{12}O_6}$$

In a particular series of experiments the following data were obtained.

Initial Sucrose Concentration (mol L^{-1})	Initial Rate of Reaction of Sucrose (mol L^{-1} s^{-1})
0.10	6.17×10^{-5}
0.20	1.23×10^{-4}
0.50	3.09×10^{-4}

What is the order of the reaction with respect to sucrose?

PRACTICE EXERCISE 7   A certain reaction that follows the equation $A + B \rightarrow C + D$ gave the following data.

[handwritten notes in margin: $r = k[A]^x[B]^y$; $k[A]^2[B]^-$]

Initial Concentrations (mol L^{-1})		Initial Rate of Reaction of A (mol L^{-1} s^{-1})
[A]	[B]	
0.40	0.30	1.0×10^{-4}
0.80	0.30	4.0×10^{-4}
0.80	0.60	1.6×10^{-3}

What is the rate law for the reaction? What is the value of the rate constant?

18.5 CONCENTRATION AND TIME

Mathematical expressions enable us to compute the concentration of a reactant remaining in a system at any particular time during the reaction.

The rate law tells us how the speed of a reaction varies with the concentrations of the reactants, and as we have seen, this rate changes during the reaction because the concentrations of the reactants decrease as the reaction proceeds. Often we wish to know more than just how fast a reaction is going, however. For instance, if we were manufacturing some chemical in a large reaction vessel, we might want to know what the concentrations of the reactants and products are at some specified time after the reaction has started, so we could decide whether it was time to harvest the products. And we might like to know how long it would take for the reactant concentrations to drop below some minimum optimum values, so we could replenish them. Obtaining information of this kind requires an expression of some sort that relates concentration to time.

The relationship between the concentration of a reactant and time can be obtained from the rate law of a reaction, but the derivation of the equation requires calculus, so we will only look at the results. Here it is useful to know the order of the reaction, because the mathematics is always the same for a reaction of a given order. Even so, for complex reactions these expressions become pretty complicated, and we will only consider very simple cases here, just to give you a taste of the subject.

First-Order Reactions

The relationship between concentration and time for a first-order reaction involves natural logarithms, which you learned about in Chapter 16. If necessary, you can review logarithms in Appendix A.

For a reaction involving a reactant A and having the rate law

$$\text{rate} = k[A]$$

it can be shown that the concentration of A is related to time by the equation

$\ln x$ is the natural logarithm of x; $\log x$ is the common, or base 10, logarithm of x.

$$\ln \frac{[A]_0}{[A]_t} = kt \qquad (18.5)$$

in which $[A]_t$ represents the concentration of A at some time t after the reaction has begun and $[A]_0$ is the concentration of A at the beginning of the reaction. This equation can also be expressed in terms of common logarithms by using the relationship $\ln x = 2.303 \log x$.

$$2.303 \log \frac{[A]_0}{[A]_t} = kt \qquad (18.6)$$

If you are using a scientific calculator for performing arithmetic, it has natural logarithm (ln) functions built in, so there is really no need to learn both equations. You can simply work problems using natural logarithms and Equation 18.5. However, if you are one of those brave souls who still use log tables, Equation 18.6 will be required. Let's look at an example that illustrates how these equations are used.

EXAMPLE 18.5   CALCULATING THE TIME REQUIRED FOR THE CONCENTRATION TO REACH A CERTAIN VALUE

Problem: Dinitrogen pentoxide, N_2O_5, is the anhydride of nitric acid, but it is not very stable. In the gas phase or in solution in a nonaqueous solvent, it decomposes by a first-order reaction into N_2O_4 and O_2. The rate law is

$$rate = k[N_2O_5]$$

At 45 °C, the rate constant for the reaction in carbon tetrachloride is $k = 6.22 \times 10^{-4}\,s^{-1}$. If the initial concentration of the N_2O_5 in the solution is 0.100 M, how long will it take for its concentration to drop to 0.0100 M?

Solution: To solve the problem, we just substitute values into Equation 18.5 and solve for t. Here are the data.

$$[N_2O_5]_0 = 0.100\ M$$
$$[N_2O_5]_t = 0.0100\ M$$
$$k = 6.22 \times 10^{-4}\,s^{-1}$$

Substituting,

$$\ln\left(\frac{0.100\ M}{0.0100\ M}\right) = (6.22 \times 10^{-4}\,s^{-1})\,t$$

$$\ln(10.0) = (6.22 \times 10^{-4}\,s^{-1})\,t$$

Taking the natural logarithm gives

$$2.30 = (6.22 \times 10^{-4}\,s^{-1})\,t$$

Solving for t,

$$t = \frac{2.30}{6.22 \times 10^{-4}\,s^{-1}}$$

$$= 3700\ s\ \text{(rounded to 3 significant figures)}$$

This is the time in seconds required for the concentration to drop to 0.0100 M. Expressed in minutes, the time is 61.7 min.

Let's use this same reaction, now, to calculate the amount of reactant that remains after some specified period of time. We could use Equation 18.5 once again, but if a scientific calculator is available, the simplest approach is to use an alternate form of Equation 18.5,

$$[A]_t = [A]_0\, e^{-kt} \qquad (18.7)$$

The use of this equation is illustrated in Example 18.6 below. If you don't have a scientific calculator and are using log tables to work problems such as these, you will find it best to work with Equation 18.6. The use of Equation 18.6 is shown in Example 18.7.

Examples 18.6 and 18.7 provide two approaches to solving the same problem. You shouldn't bother to learn both methods. Instead, pick the one that suits you best and stay with it.

Equation 18.5 can be derived by rearranging Equation 18.7 and then taking the natural logarithm of each side.

EXAMPLE 18.6   CALCULATING THE CONCENTRATION OF REACTANT AFTER A TIME t BY EQUATION 18.7

Problem: If the initial concentration of N_2O_5 in a carbon tetrachloride solution (at 45 °C) were 0.500 M, what would its concentration be after exactly one hour?

Solution: We must solve for $[N_2O_5]_t$. For this problem, Equation 18.7 takes the form

$$[N_2O_5]_t = [N_2O_5]_0 \, e^{-kt}$$

The value of $[N_2O_5]_0$ is 0.500 M, k is 6.22×10^{-4} s^{-1}, so we must express time in seconds, too, for the units to work out correctly. One hour is 3600 s (60 min × 60 s/min). Therefore, substituting gives

$$[N_2O_5]_t = (0.500 \ M) \, e^{-(6.22 \times 10^{-4} \, s^{-1})(3600 \, s)}$$
$$[N_2O_5]_t = (0.500 \ M) \, e^{-2.24}$$

Using a calculator to compute the value of $e^{-2.24}$ gives

$$[N_2O_5]_t = (0.500 \ M) \times (0.106)$$
$$= 0.0530 \ M$$

The answer here is quite sensitive to the numer of significant figures in the exponent on e and to the rounding off of this exponent.

One hour after the reaction has begun, the concentration of N_2O_5 will drop to 0.0530 M.

EXAMPLE 18.7   CALCULATING THE CONCENTRATION OF A REACTANT AFTER A TIME t BY EQUATION 18.6

Problem: The question here is the same as in Example 18.6. If the initial concentration of N_2O_5 in a carbon tetrachloride solution (at 45 °C) were 0.500 M, what would its concentration be after exactly one hour?

Solution: We must solve for $[N_2O_5]_t$, which is one of the terms in the concentration ratio in Equation 18.6. The best way to approach this problem is first to determine what the value of the ratio is and then to solve for the unknown concentration. If we let R stand for the concentration ratio

$$R = \frac{[N_2O_5]_0}{[N_2O_5]_t}$$

The approach to solving the problem in Example 18.7 can also be followed using Equation 18.5.

then Equation 18.6 becomes

$$2.303 \log R = kt$$

The value of k is 6.22×10^{-4} s^{-1} (from Example 18.5), and we must express time in seconds, too, for the units to work out correctly. One hour is 3600 seconds (60 min × 60 s/min). Therefore, substituting and solving for log R give

$$\log R = \frac{(6.22 \times 10^{-4} \, s^{-1})(3600 \, s)}{2.303}$$
$$= 0.972$$

To obtain R, we have to take the antilog. In the log tables we find that 0.972 is the log of 9.38. Therefore,

$$R = 9.38$$

Now that we know the value of the concentration ratio, we can solve for the unknown concentration. Substituting for R,

$$\frac{[N_2O_5]_0}{[N_2O_5]_t} = 9.38$$

If we now solve for the unknown concentration, then

$$[N_2O_5]_t = \frac{[N_2O_5]_0}{9.38}$$
$$= \frac{0.500 \ M}{9.38}$$
$$= 0.0533 \ M$$

One hour after the reaction has begun, the concentration of N_2O_5 will drop to 0.0533 M. If you followed Example 18.6, you see that both methods for solving this problem give essentially the same answer.

PRACTICE EXERCISE 8   Sucrose, $C_{12}H_{22}O_{11}$, reacts slowly with water in the presence of acid to form two other sugars, glucose and fructose, both of which have the same molecular formulas, but different structures (as we noted in Practice Exercise 6).

$$C_{12}H_{22}O_{11} + H_2O \longrightarrow C_6H_{12}O_6 + C_6H_{12}O_6$$
$$\text{sucrose} \qquad\qquad \text{glucose} \quad \text{fructose}$$

The reaction is first order and has a rate constant of $6.2 \times 10^{-5}\ s^{-1}$ at 35 °C when the H^+ concentration is 0.1 M. Suppose that the initial concentration of sucrose in the solution is 0.40 M.
(a) What will its concentration be after 2 hours?
(b) How many minutes will it take for the concentration to drop to 0.30 M?

Second-Order Reactions

For the sake of simplicity, we will only consider the relationship between concentration and time for a second-order reaction of the kind that has a rate law

$$\text{rate} = k[B]^2$$

The equation is rather different from that for a first-order reaction:

$$\frac{1}{[B]_t} - \frac{1}{[B]_0} = kt \qquad\qquad (18.8)$$

Using the same notation as before, $[B]_0$ is the initial concentration of reactant B, and $[B]_t$ is the concentration after a time t. We can use this expression to solve problems similar to those in Examples 18.5 through 18.7.

The expression relating concentration and time for a second-order reaction having a rate law of the type

$$\text{rate} = k[A]^1[B]^1$$

is more complicated than this and won't be discussed in this text.

EXAMPLE 18.8   CALCULATING THE CONCENTRATION OF A REACTANT AFTER A TIME t

Problem: Nitrosyl chloride, NOCl, is a very corrosive, reddish yellow gas that is intensely irritating to the eyes and skin. It decomposes slowly to NO and Cl_2 by a second-order reaction having the rate law

$$\text{rate} = k[NOCl]^2$$

Aqua regia, a mixture of concentrated hydrochloric acid and concentrated nitric acid, takes on a yellow color because of the formation of NOCl.

At a certain temperature, the rate constant for the reaction is $k = 0.020\ L\ mol^{-1}\ s^{-1}$. If the initial concentration of NOCl in a reaction vessel is 0.050 mol/L, what will the concentration be after 30 minutes?

Solution: In this problem we wish to calculate $[NOCl]_t$ by Equation 18.8. Let's begin by tabulating the data.

$$[NOCl]_0 = 0.050\ M$$
$$k = 0.020\ L\ mol^{-1}\ s^{-1}$$
$$t = 30\ min = 1800\ s$$

The equation that we wish to substitute into is

$$\frac{1}{[NOCl]_t} - \frac{1}{[NOCl]_0} = kt$$

which gives

$$\frac{1}{[NOCl]_t} - \frac{1}{0.050\ mol/L} = (0.020\ L\ mol^{-1}\ s^{-1}) \times (1800\ s)$$

Solving for $1/[NOCl]_t$ gives

$$\frac{1}{[NOCl]_t} - 20 \text{ L mol}^{-1} = 36 \text{ L mol}^{-1}$$

$$\frac{1}{[NOCl]_t} = 56 \text{ L mol}^{-1}$$

The value of $[NOCl]_t$ can be obtained by taking the reciprocal.

mol L^{-1} is moles per liter, the units of molarity.

$$[NOCl]_t = 1/(56 \text{ L mol}^{-1})$$
$$= 0.018 \text{ mol L}^{-1}$$
$$= 0.018 \, M$$

PRACTICE EXERCISE 9

Referring to the reaction in Example 18.8, determine how many minutes it would take for the NOCl concentration to drop from 0.040 M to 0.010 M.

18.6 HALF-LIVES

The time required for the concentration of a reactant to be reduced to half of its initial value is called the half-life.

A concept that provides a useful measure of the speed of a reaction, especially for first-order processes, is that of half-life, $t_{1/2}$ — the amount of time required for half of a given reactant to disappear. When the half-life is short, a reaction is rapid because half of the reactant disappears quickly.

Fast reactions are characterized by large values of k and small values of $t_{1/2}$.

For any first-order reaction, the half-life is constant at any given temperature. It is not affected by the initial concentration of the reactant. A typical example is the change that radioactive isotopes undergo during radioactive "decay." (In fact, you have probably heard the term *half-life* used in reference to the lifespan of radioactive substances.)

Iodine-131, an unstable, radioactive isotope of iodine, undergoes a nuclear reaction that causes it to emit a type of radiation and transforms it into a stable isotope of xenon. The intensity of this radiation decreases, or *decays*, with time. Figure 18.3 is a graph of this decay. Notice that the time it takes for the first half of the ^{131}I to disappear is 8 days. During the next 8 days, half of the remaining ^{131}I disappears, and so on. Regardless of the amount begun with, it takes 8 days for half of that amount of ^{131}I to disappear, which means that the half-life of ^{131}I is a constant.

Iodine-131 is used in the diagnosis and treatment of thyroid disorders. The thyroid gland is the only part of the body that uses iodide ions. When a patient is given a small

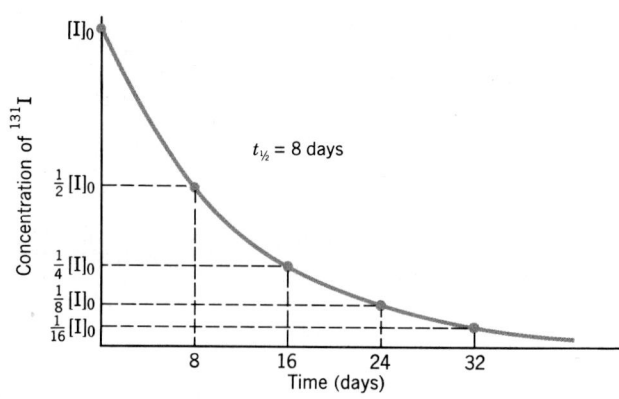

FIGURE 18.3 First-order radioactive decay of iodine-131. Initial concentration is $[I]_0$.

quantity of $^{131}I^-$ mixed with nonradioactive I^-, both are absorbed by the thyroid. This behavior allows for the testing of thyroid activity and is also used to treat certain types of thyroid cancer.

For any reaction, the half-life is related to the rate constant. If the reaction is first order, its $t_{1/2}$ can be obtained from Equation 18.5 by setting $[A]_t$ equal to one-half $[A]_0$.

$$[A]_t = \frac{1}{2}[A]_0$$

Substituting and solving for t, which becomes $t_{1/2}$, give

$$t_{1/2} = \frac{\ln 2}{k} \qquad (18.9)$$

or

$$t_{1/2} = \frac{0.693}{k} \qquad (18.10)$$

$\ln 2 = 0.693$

The half-life of a second-order reaction *does* depend on the initial concentrations of the reactants. We can see this by examining Figure 18.2 (page 741), which shows the decomposition of gaseous HI, a second-order reaction. The reaction begins with a hydrogen iodide concentration of 0.10 M. The concentration drops to half of this value, 0.050 M, in about 125 seconds. If we take 0.050 M as the next "initial" concentration, it drops to half its value, 0.025 M, 250 seconds later (at a total elapsed time of 375 seconds). Thus, halving the initial concentration, from 0.10 M to 0.05 M, causes a doubling of the half-life, from 125 to 250 seconds. For a second-order reaction like the decomposition of HI, the half-life is inversely proportional to the initial concentration of the reactant and is related to the rate constant by the equation

The half-life is not a very meaningful quantity for a second-order reaction *because* its magnitude depends on the concentration.

$$t_{1/2} = \frac{1}{k \times (\text{initial concentration of reactant})} \qquad (18.11)$$

This equation comes from Equation 18.8 by letting

$$[B]_t = \tfrac{1}{2}[B]_0$$

EXAMPLE 18.9 USING HALF-LIVES

Problem: The half-life of iodine-131 is 8 days. What fraction of the initial iodine-131 would be present in a patient after 24 days if none of it were eliminated through natural bodily processes?

Solution: A period of 24 days is exactly three half-lives. Taking the fraction initially present as 1, we can set up a table.

The fraction remaining after n half-lives is $(\tfrac{1}{2})^n$, or simply $1/2^n$.

Half-life	0	1	2	3
Fraction	1	$\frac{1}{2}$	$\frac{1}{4}$	$\frac{1}{8}$

Half of the ^{131}I is lost in the first half-life, half of that disappears in the second half-life, and so on. Therefore, the fraction remaining after three half-lives is $\frac{1}{8}$.

EXAMPLE 18.10 CALCULATING THE HALF-LIFE

Problem: The reaction $2HI(g) \rightarrow H_2(g) + I_2(g)$ has the rate law, rate $= k[HI]^2$, with $k = 0.079$ L mol^{-1} s^{-1} at 508 °C. What is $t_{1/2}$ for this reaction when the initial concentration of HI is 0.050 M?

Solution: We have already found the answer to this problem by inspecting Figure 18.2, but let's calculate $t_{1/2}$ using Equation 18.11, anyway.

$$t_{1/2} = \frac{1}{k \times (\text{initial concentration})}$$

Substituting values given in the problem,

$$t_{1/2} = \frac{1}{(0.079 \text{ L mol}^{-1} \text{ s}^{-1})(0.050 \text{ mol L}^{-1})}$$

$$= 253 \text{ s}$$

Rounding to two significant figures gives $t_{1/2} = 2.5 \times 10^2$ s, which is the same answer we obtained from Figure 18.2.

PRACTICE EXERCISE 10   In Practice Exercise 6, the reaction of sucrose with water was found to be first order with respect to sucrose. The rate constant for the reaction under the conditions of the experiments was $6.17 \times 10^{-4} \text{ s}^{-1}$. Calculate the value of $t_{1/2}$ for this reaction in minutes. How many minutes would it take for three-quarters of the sucrose to react? (*Hint:* What fraction of the sucrose remains?)

PRACTICE EXERCISE 11   Suppose that the value of $t_{1/2}$ for a certain reaction was found to be independent of the initial concentration of the reactants. What can you say about the order of the reaction?

18.7 THE EFFECT OF TEMPERATURE ON REACTION RATE

In any system, increasing the temperature increases the number of reactant molecules having the minimum kinetic energy that they need to react.

In Section 18.2 we mentioned the fact that nearly all reactions proceed faster at higher temperatures. As a rule, the reaction rate increases by a factor of about 2 or 3 for each 10 °C rise in temperature, although the actual amount of increase differs from one reaction to another. To understand why temperature affects reaction rates in this way, we have to examine what actually happens to the molecules in a reaction system.

One of the simplest theories about the way various factors affect reaction rates is the collision theory. The basic postulate of this theory is that the rate of a reaction is proportional to the number of collisions per second among the reactant molecules. Anything that can increase the frequency of collisions should increase the rate. As reasonable as this sounds, for most reactions it is impossible for every one of the collisions between the reactants to actually result in a chemical change. In a gas or a liquid, molecules of the reactants undergo an enormous number of collisions with each other each second. If each were effective, all reactions would be over in an instant. Of all of the collisions that occur, only a *very* small fraction actually results in a chemical change. There are two principal reasons for this.

At the start of the reaction in Figure 18.2, only one out of every billion billion (10^{18}) collisions leads to a net reaction. In each of the other collisions, the reactant molecules just bounce off each other.

In many cases, when two reactants collide, their atoms must be oriented correctly in order for a reaction to occur. For example, later we will see that the reaction

$$2NO_2Cl \longrightarrow 2NO_2 + Cl_2$$

appears to proceed by a mechanism that involves two steps. One of these steps involves the collision of an NO_2Cl molecule with a chlorine atom.

$$NO_2Cl + Cl \longrightarrow NO_2 + Cl_2$$

The orientation of the NO_2Cl molecule is important in this collision, as is shown in Figure 18.4. In Figure 18.4*a*, no Cl_2 can be formed, but in Figure 18.4*b* the orientation is right for the collision to be effective.

The major reason that so few collisions actually lead to chemical change, however, is that prior to collision the reacting molecules — even when correctly oriented — must possess between them a certain minimum kinetic energy (KE), called the activation

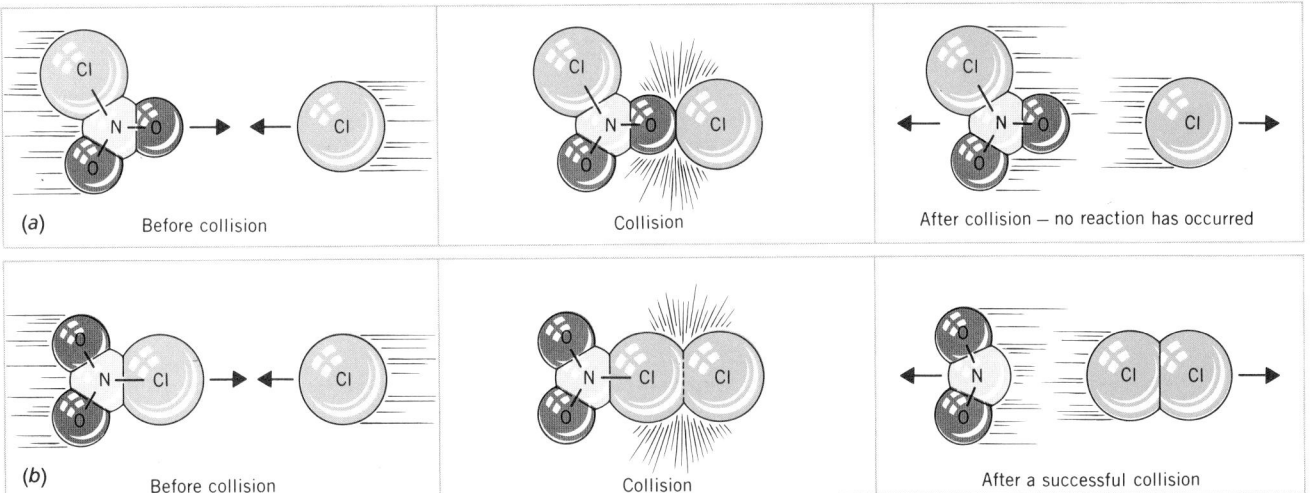

FIGURE 18.4 The importance of molecular orientation during a collision. (a) A collision that cannot produce a Cl_2 molecule. (b) A collision that can produce NO_2 and Cl_2 molecules.

energy, E_a. For nearly all chemical reactions, the activation energy is quite large, and at ordinary temperatures very few molecules are going fast enough to have this minimum energy.

The reason for the activation energy can be understood if we examine in detail what actually takes place during a reaction. Any chemical change involves a reorganization of chemical bonds—old bonds are broken as new ones are formed. For this to happen during a collision, the nuclei of the reacting particles must somehow find themselves in the right locations. This requires that the colliding molecules experience a very energetic collision. For example, if two slow-moving molecules collide, the repulsions between their electron clouds cause the molecules to simply bounce apart. Only fast-moving molecules, which have large kinetic energies, can collide with enough force to enable their nuclei and electrons to overcome repulsions and thereby allow the necessary bond breaking and bond making.

Figure 18.5 shows why the rate of a reaction increases with temperature. Here we have the kinetic energy distributions for a collection of molecules at two different temperatures. Regardless of temperature, the minimum kinetic energy needed for an effective collision—the activation energy—is the same. At the higher temperature, the total fraction of molecules that have this necessary energy is greater than it is at the lower temperature. (This total fraction is indicated by the ratio of the shaded area to the total area under each curve.) In other words, at the higher temperature, a greater fraction of the total number of collisions that occur each second will result in a chemical change. The reactants will therefore disappear faster at the higher temperature.

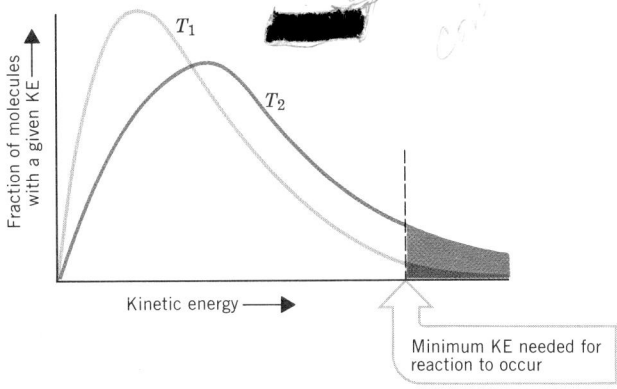

FIGURE 18.5 Kinetic energy distributions for a reaction mixture at two different temperatures. The sizes of the shaded areas under the curves are proportional to the total fractions of the molecules that possess the minimum activation energy.

Transition State Theory

Now let's take a close look at what happens when the reactant molecules come together in a collision. When they collide, the molecules slow down, stop, and then fly apart again. If a reaction occurs during the collision, the particles that separate are chemically different from those that collide. Regardless of what happens to them chemically, however, as the molecules slow down, the total kinetic energy that they possess decreases. Since energy can't disappear, this means that their total potential energy (PE) must increase.

The relationship between the activation energy and the total potential energy of the reactants and products can be expressed graphically by a potential-energy diagram. A typical diagram for a reaction is shown in Figure 18.6. The horizontal axis represents the extent to which the reactants have changed to the products. It follows the path taken by the reaction as reactant molecules come together and change to produce the product molecules that separate after collision. The activation energy appears as a potential energy "hill" between the reactants and products. Only colliding molecules that together have energies at least as large as E_a are able to climb over the hill and produce the products.

FIGURE 18.6 Potential energy diagram for an exothermic reaction.

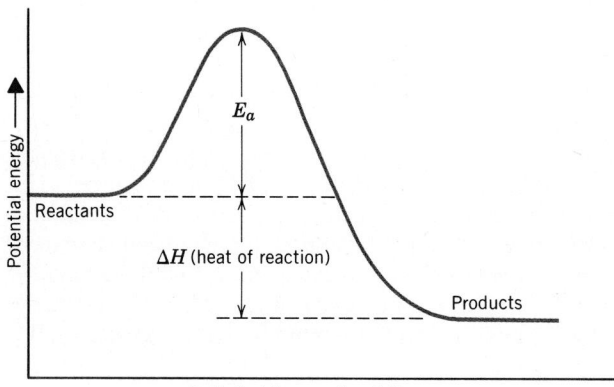

We can use the potential-energy diagram for a reaction to follow the progress of both an unsuccessful and a successful collision. As two reactant molecules collide, they slow down and their kinetic energy is changed to potential energy — they begin to climb the potential-energy barrier toward the products. If their combined initial kinetic energies are less than E_a, they are unable to reach the top of the hill. Instead, they fall back toward the reactants. They come apart chemically unchanged and with their original total kinetic energy; no net reaction has occurred (Figure 18.7a). On the other hand, if their combined kinetic energies are equal to or greater than E_a, and if they are oriented properly, they are able to pass over the activation-energy barrier and form product molecules (Figure 18.7b).

FIGURE 18.7 (a) An unsuccessful collision — molecules separate unchanged. (b) A successful collision — activation-energy barrier is crossed and the products are formed.

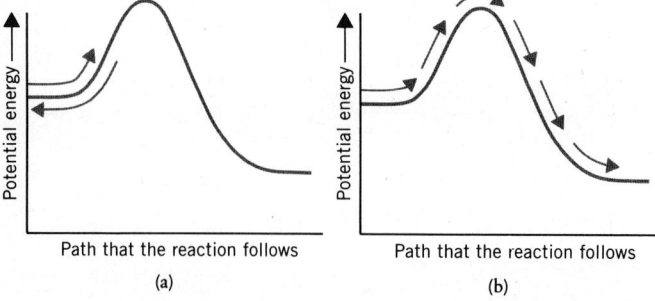

In Figure 18.6 we also find the heat of reaction—the difference between the potential energy of the products and the potential energy of the reactants. The reaction depicted in Figure 18.6 is exothermic because the products have a lower potential energy than the reactants. When a successful collision occurs, the decrease in potential energy appears as an increase in the kinetic energies of the emerging product molecules. The temperature of the system rises during an exothermic reaction because the average kinetic energy of the system increases.

Figure 18.8 is a potential-energy diagram for an endothermic reaction. In this case the products are at a higher potential energy than the reactants and, in terms of the heat of reaction, a net input of energy is needed to form the products. Endothermic reactions produce a cooling effect as they proceed because there is a net conversion of kinetic energy to potential energy. As the total kinetic energy decreases, the average kinetic energy decreases as well and the temperature drops.

In Chapter 5 we saw that when the direction of a reaction is reversed, the sign of its ΔH is changed—a reaction that is exothermic in the forward direction is endothermic in the reverse direction, and vice versa. This suggests that, in general, reactions are reversible. If we look again at the energy diagram for a reaction that is exothermic in the forward direction (Figure 18.6), it is obvious that in the opposite direction the reaction is endothermic. What differs most for the forward and reverse directions is the relative height of the activation-energy barrier (Figure 18.9).

The law of conservation of energy requires that PE + KE = constant during a collision.

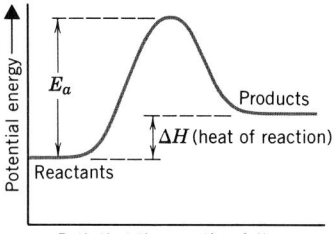

FIGURE 18.8 A potential-energy diagram for an endothermic reaction.

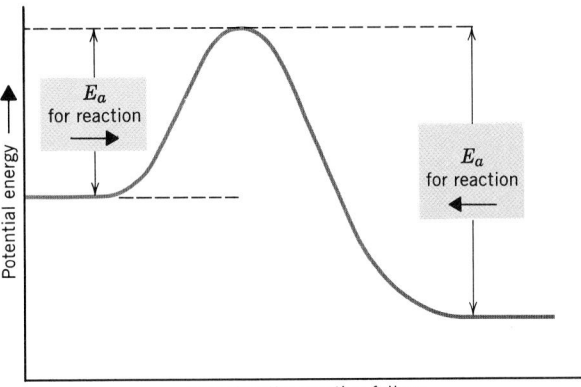

Path that the reaction follows

FIGURE 18.9 Activation-energy barrier for the forward and reverse reactions.

One of the main reasons for studying activation energies is that they provide information about what actually occurs during an effective collision between reactant molecules. For example, suppose that we were studying a reaction between the molecules A_2 and B_2 to form molecules of AB. One way for A_2 and B_2 to react during a collision is depicted in Figure 18.10. The A_2 and B_2 molecules come together and, at the moment of collision, the A—A and B—B bonds are weakened as the new A—B bonds are formed.

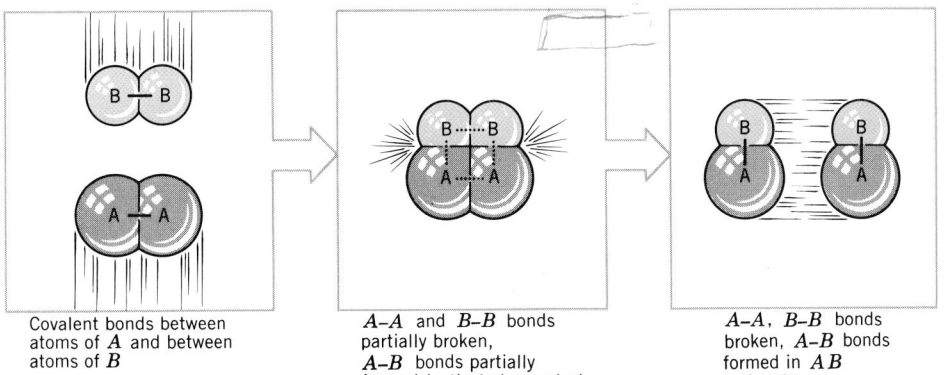

Covalent bonds between atoms of A and between atoms of B

A–A and B–B bonds partially broken, A–B bonds partially formed (activated complex)

A–A, B–B bonds broken, A–B bonds formed in AB molecules

FIGURE 18.10 Reaction between molecules of A_2 and B_2. This illustrates one way that molecules of AB could be formed by collisions between A_2 and B_2.

The location of the transition state on the potential energy diagram for a reaction.

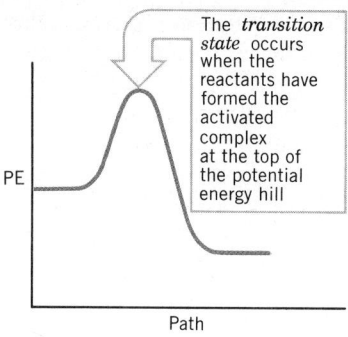

The *transition state* occurs when the reactants have formed the activated complex at the top of the potential energy hill

PE

Path

This brief moment during the reaction is called the transition state. It corresponds to the high point on the potential-energy diagram. The species that exists at that instant, with its partly formed and partly broken bonds, is called the activated complex.

The activation energy provides information about the relative importance of bond breaking and bond making during formation of the activated complex. A very high activation energy, for instance, would suggest that bond breaking contributes very heavily to formation of the activated complex because bond breaking is an energy-absorbing process.

18.8 MEASURING THE ACTIVATION ENERGY

The activation energy can be calculated from rate constants measured at different temperatures.

Changing the temperature alters the rate of a reaction by changing the value of the rate constant, and the amount that k changes depends on the magnitude of the activation energy. When the activation energy is very large, even a small increase in the temperature causes a substantial percentage increase in the number of molecules having sufficient energy to react, so a large change in the rate is observed. On the other hand, if E_a is small, the same temperature rise has a proportionately smaller effect on the number of molecules having enough energy to react. This is because there are already a large number of molecules able to react, and this number doesn't undergo much of a percentage increase when the temperature undergoes this small rise.

Quantitatively, the activation energy is related to the rate constant by a relationship discovered in 1889 by Svante Arrhenius, whose name you may recall from our discussion of electrolytes and acids and bases in Chapter 11. The usual form of the Arrhenius equation is

$$k = A\,e^{-E_a/RT} \tag{18.12}$$

where k is the rate constant, A is a proportionality constant sometimes called the frequency factor, e is the base of the natural logarithm system, R is the gas constant expressed in energy units ($R = 8.314$ J mol^{-1} K^{-1} or $R = 1.987$ cal mol^{-1} K^{-1}), and T is the absolute temperature (expressed in kelvins).

Determining the Activation Energy Graphically

The way Equation 18.12 is normally used is in its logarithmic form. Suppose we take the natural log of both sides. We obtain

$$\ln k = \ln A - E_a/RT$$

Let's rewrite the equation as

$$\ln k = \ln A - (E_a/R)\cdot(1/T) \tag{18.13}$$

We know that the rate constant k varies with the temperature T, which also means that the quantity $\ln k$ varies with the quantity $(1/T)$. These two quantities—$\ln k$ and $1/T$—are our variables, and Equation 18.13 actually is an equation for a straight line. To see this, recall from algebra that a straight line has the general equation

$$y = b + mx$$

where x and y are variables, m is the slope, and b is the intercept of the line with the y axis. In this case, we can make the substitutions

$$y = \ln k$$
$$x = (1/T)$$
$$b = \ln A$$
$$m = -E_a/R$$

Thus,

$$\ln k = \ln A + (-E_a/R) \cdot (1/T)$$
$$\Updownarrow \qquad \Updownarrow \qquad \Updownarrow \qquad \Updownarrow$$
$$y \quad = \quad b \quad + \quad m \quad\quad x$$

How do we make use of this? To determine the activation energy, we make a graph of $\ln k$ (the vertical axis) versus $1/T$ (the horizontal axis), measure the slope of the line, and then use the relationship

Measurement of the slope of a line is reviewed in Appendix A.

$$\text{slope} = -E_a/R$$

to calculate E_a. Example 18.11 illustrates how this is done.

EXAMPLE 18.11   DETERMINING E_a GRAPHICALLY

Problem: The following data were collected for the reaction

$$2NO_2(g) \longrightarrow 2NO(g) + O_2(g)$$

Rate Constant (L mol^{-1} s^{-1})	Temperature (°C)
7.8	400
10	410
14	420
18	430
24	440

Determine the activation energy for the reaction in kilojoules per mole.

Solution: First, let's tabulate the natural logarithm of the rate constant and $1/T$. In doing this, we have to remember to change the temperatures to kelvins before taking the reciprocals.

For example, for the first set of data

$\ln(7.8) = 2.05$
$1/(400 + 273) = 1/673$
$\quad = 1.486 \times 10^{-3}$

We are carrying extra "significant figures" for the purpose of graphing the data.

$\ln k$	$1/T$ (K^{-1})
2.05	1.486×10^{-3}
2.30	1.464×10^{-3}
2.64	1.443×10^{-3}
2.89	1.422×10^{-3}
3.18	1.403×10^{-3}

Then we plot the data as shown in the figure.

We draw the straight line to best fit the data.

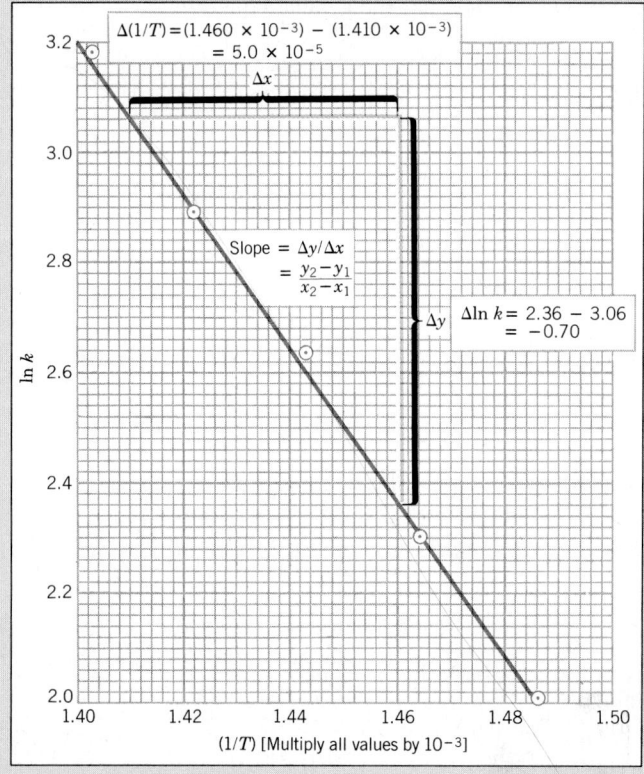

$$\Delta(1/T) = (1.460 \times 10^{-3}) - (1.410 \times 10^{-3})$$
$$= 5.0 \times 10^{-5}$$

$$\text{Slope} = \Delta y/\Delta x$$
$$= \frac{y_2 - y_1}{x_2 - x_1}$$

$$\Delta \ln k = 2.36 - 3.06$$
$$= -0.70$$

$(1/T)$ [Multiply all values by 10^{-3}]

The slope of the graph is obtained as the ratio

$$\text{slope} = \frac{\Delta(\ln k)}{\Delta(1/T)}$$

$$= \frac{-0.70}{5.0 \times 10^{-5} \text{ K}^{-1}}$$

$$= -1.4 \times 10^4 \text{ K}^{-1} = -E_a/R$$

Therefore, after changing signs and solving for E_a we have

$$E_a = R \times (1.4 \times 10^4 \text{ K}^{-1})$$

Using $R = 8.314$ J/mol K gives

$$E_a = 1.2 \times 10^5 \text{ J/mol}$$
$$= 1.2 \times 10^2 \text{ kJ/mol}$$

Calculating the Activation Energy

Sometimes, in obtaining an activation energy, we may not wish to go to the bother of graphing data, or we may have just two rate constants measured at two temperatures. In such cases the activation energy can be obtained algebraically. From Equation 18.12, the following useful relationship can be derived.

$$\ln\left(\frac{k_2}{k_1}\right) = \frac{-E_a}{R}\left(\frac{1}{T_2} - \frac{1}{T_1}\right) \tag{18.14}$$

This equation permits us to calculate the activation energy from two rate constants, k_1 and k_2, measured at the respective absolute temperatures T_1 and T_2. Equation 18.14 can also be written using common logs. In this form it is

$$2.303 \log\left(\frac{k_2}{k_1}\right) = \frac{-E_a}{R}\left(\frac{1}{T_2} - \frac{1}{T_1}\right) \qquad (18.15)$$

As we mentioned in Section 18.5, natural logarithms are easiest to use if you use a scientific calculator to do your computations. Let's look at some examples.

EXAMPLE 18.12   CALCULATING THE ACTIVATION ENERGY

Problem: The decomposition of HI has rate constants $k = 0.079$ mol^{-1} s^{-1} at 508 °C and $k = 0.24$ L mol^{-1} s^{-1} at 540 °C. What is the activation energy of this reaction in kJ/mol?

Solution: The problem, basically, is to substitute numbers into Equation 18.14 or Equation 18.15 and solve for E_a. We will work this problem using natural logs and Equation 18.14. Let's begin by organizing the data — a small table will be helpful. Choose one of the rate constants as k_1 (it doesn't matter which one) and then fill in the table.

	k	T
1	0.079	508 + 273 = 781
2	0.24	540 + 273 = 813

We will also use $R = 8.314$ J mol^{-1} K^{-1} to make it easy to convert the answer to kJ. Substituting into Equation 18.14 gives us

$$\ln\left(\frac{0.24 \text{ L mol}^{-1}\text{s}^{-1}}{0.079 \text{ L mol}^{-1}\text{s}^{-1}}\right) = \frac{-E_a}{8.314 \text{ J mol}^{-1}\text{K}^{-1}}\left(\frac{1}{813 \text{ K}} - \frac{1}{781 \text{ K}}\right)$$

$$\ln(3.0) = \frac{-E_a}{8.314 \text{ J mol}^{-1}\text{K}^{-1}}(0.001230 \text{ K}^{-1} - 0.001280 \text{ K}^{-1})$$

$$1.10 = \frac{-E_a}{8.314 \text{ J mol}^{-1}}(-0.000050)$$

Multiplying both sides by 8.314 J mol^{-1} gives

$$9.15 \text{ J mol}^{-1} = E_a(0.000050)$$

Solving for E_a, we have

$$E_a = \frac{9.15 \text{ J mol}^{-1}}{0.000050} = 1.8 \times 10^5 \text{ J mol}^{-1}$$

Since 1 kJ = 10^3 J,

$$E_a = 1.8 \times 10^2 \text{ kJ mol}^{-1}$$

Before we work another example, it might be worthwhile to mention that the greatest problem students generally have in applying Equations 18.14 and 18.15 is remembering where the 1's and 2's go — which subscripts belong to which k's and T's. Perhaps it is helpful to know that the only effect caused by interchanging a pair of subscripts is that the algebraic sign of E_a becomes negative. The correct sign of E_a is *always* positive, however. (There is always an energy barrier to the reaction, never an energy valley.) Therefore, if you work through a problem and obtain a negative value for E_a, you probably simply switched subscripts. To correct the problem, just make the sign of E_a positive.

EXAMPLE 18.13   USING THE ACTIVATION ENERGY

Problem: The reaction $2NO_2 \rightarrow 2NO + O_2$ has an activation energy of 26.6 kcal/mol. At 400 °C, $k = 7.8$ L mol^{-1} s^{-1}. What is the value of k at 430 °C?

Solution: This problem shows how knowledge of the activation energy and the rate constant at one temperature allows us to calculate the reaction rate at another temperature. To solve the problem, let's use Equation 18.15.

$$2.303 \log\left(\frac{k_2}{k_1}\right) = \frac{-E_a}{R}\left(\frac{1}{T_2} - \frac{1}{T_1}\right)$$

Again, let's organize the data.

	k	T
1	7.8	400 + 273 = 673 K
2	?	430 + 273 = 703 K

This time we must use $R = 1.987$ cal mol^{-1} K^{-1} and express E_a in calories ($E_a = 26,600$ cal/mol).

Next, we substitute values into the right side of the equation and solve for $\log(k_2/k_1)$.

$$\log\left(\frac{k_2}{k_1}\right) = \frac{-26,600 \text{ cal mol}^{-1}}{(2.303)(1.987 \text{ cal mol}^{-1}\text{K}^{-1})}\left(\frac{1}{703 \text{ K}} - \frac{1}{673 \text{ K}}\right)$$

$$= (-5.81 \times 10^3 \text{ K})(-6.3 \times 10^{-5} \text{ K}^{-1})$$

$$\log\left(\frac{k_2}{k_1}\right) = 0.37$$

antilog = 10^x
antiln = e^x

Taking the antilog gives the ratio of k_2 to k_1.

$$\frac{k_2}{k_1} = 2.3$$

As a check on whether you've placed the 1's and 2's in the right places, remember that the k at the higher temperature must be larger than the k at the lower temperature.

Solving for k_2,

$$k_2 = 2.3 \, k_1$$

Substituting the value of k_1 from our data table gives

$$k_2 = 2.3(7.8 \text{ L mol}^{-1} \text{ s}^{-1})$$
$$= 18 \text{ L mol}^{-1} \text{ s}^{-1}$$

PRACTICE EXERCISE 12   The reaction $CH_3I + HI \rightarrow CH_4 + I_2$ was observed to have rate constants $k = 3.2$ L mol^{-1} s^{-1} at 350 °C and $k = 23$ L mol^{-1} s^{-1} at 400 °C. (a) What is the value of E_a for this reaction expressed in kJ/mol? (b) What would the rate constant for this reaction be at 300 °C?

18.9   COLLISION THEORY AND REACTION MECHANISMS

The slow step in a reaction mechanism determines the overall rate of reaction.

At the beginning of this chapter it was stated that one of the benefits to be gained from a study of reaction rates is information about the paths, or *mechanisms,* followed by reactions. For most reactions, the individual chemical steps, called elementary processes, that make up the mechanism cannot actually be observed. The mechanism that a chemist writes is really a theory about what occurs step-by-step as the reactants are converted to the products. Since the individual steps in the mechanism can't be observed, arriving at a chemically reasonable set of elementary processes is not at all a simple task. Making reasonable guesses requires a lot of "scientific intuition" and much more chemical knowledge than has presently been provided to you, so you need not worry about having to predict mechanisms for reactions. Nevertheless, to understand the science of chemistry better, it is worthwhile knowing how the study of reaction rates can provide clues to a reaction's mechanism.

In Section 18.7 we learned that the basic postulate of the collision theory is that the rate of a reaction is proportional to the number of collisions per second between the reactants. We also learned that, for a given set of conditions, the number of effective collisions is only a certain small fraction of the total. If we could somehow double the total number of collisions, the number of effective collisions would be doubled also.

Predicting the Rate Law for an Elementary Process

Let's suppose, now, that we know for a fact that a certain collision process takes place during a particular reaction. For example, suppose that the decomposition of $NOCl$ into NO and Cl_2 actually involves collisions between $NOCl$ molecules and Cl atoms as one step in the mechanism. In other words, the reaction

$$NOCl + Cl \longrightarrow NO + Cl_2 \qquad (18.16)$$

represents an elementary process. What would happen if we were to double the number of Cl atoms in the container? Since there would then be twice as many Cl atoms with which the $NOCl$ molecules could collide, there should be twice as many $NOCl$-to-Cl collisions per second. This should, in turn, double the rate of the reaction given by Reaction 18.16. In other words, *doubling* the Cl concentration would *double* the rate of this elementary process. The rate law for Reaction 18.16 should therefore contain $[Cl]$ raised to the first power.

Similarly, if we were to double the number of $NOCl$ molecules in the container, there would be twice as many $NOCl$ molecules with which Cl atoms could collide. As a result, the collision frequency should *double* and the rate of the reaction should *double*. This means that the $NOCl$ concentration should also be raised to the first power in the rate law for this elementary process. Therefore, the rate law for Reaction 18.16 should be

$$\text{rate} = k[Cl][NOCl]$$

Notice that the exponents in the rate law for this *elementary process* are the same as the coefficients in the chemical equation for the elementary process.

Let's look at another elementary process, one involving collisions between like molecules.

$$2NO_2 \longrightarrow NO_3 + NO \qquad (18.17)$$

If the NO_2 concentration were doubled, there would be *twice* as many individual NO_2 molecules and each would have *twice* as many neighbors with which to collide. The number of NO_2-to-NO_2 collisions per second would therefore be increased by a factor of 4. Increasing the collision frequency by a factor of 4 should also increase the rate by the same factor, so the rate should rise by a factor of 4, which is 2^2. Earlier we saw that when the doubling of concentration leads to a fourfold increase in the rate, the concentration of that reactant is raised to the second power in the rate law. Thus, the rate law for Reaction 18.17 should be

$$\text{rate} = k[NO_2]^2$$

Once again the exponent in the rate law for the *elementary process* is the same as the coefficient in the chemical equation.

The point that these two examples make is that the rate law for an elementary process can be predicted.

You learned in Section 18.4 that when the rate is doubled by doubling the reactant concentration, the exponent on the concentration term is 1.

There are twice as many NO_2 molecules, each of which collides twice as often, so the total number of collisions per second doubly doubles, i.e., it increases by a factor of 4.

> The exponents in the rate law for an elementary process are equal to the coefficients of the reactants in the chemical equation for that elementary process.

It is very important to understand that this rule applies *only* to elementary processes. If all we know is an overall equation, we can only be sure of the exponents in the rate law if we determine them by doing experiments!

Predicting Reaction Mechanisms

How does the ability to predict the rate law of an elementary process help chemists predict reaction mechanisms? To answer this question, let's look at two mechanisms. First, consider the gaseous reaction

$$2NO_2 \longrightarrow 2NO + O_2 \qquad (18.18)$$

It has been experimentally determined that this is a second-order reaction having the rate law

$$\text{rate} = k[NO_2]^2$$

The mechanism that has been proposed for the reaction involves two steps with these elementary processes:

(Notice that if these two reactions are added and NO_3 is canceled from both sides, the net overall reaction, Reaction 18.18, is obtained.)

The NO_3 is called a *reactive intermediate*. We never actually observe the NO_3 because it decomposes so quickly.

When a reaction such as this occurs in a series of steps, one step is very often much slower than the others. In this mechanism, for example, it is believed that the first step is slow and that once the NO_3 is formed, it decomposes quickly to give NO and O_2.

The slow step in a mechanism is called the rate-determining step or rate-limiting step. This is because the final products of the overall reaction cannot appear faster than the products of the slow step. In the mechanism for the decomposition of NO_2, then, the first reaction is the rate-determining step because the final products can't be formed any faster than the rate at which NO_3 is formed.

The rate-determining step is similar to a slow worker on an assembly line. Regardless of how fast the other workers are, the production rate depends on how quickly the slow worker does his or her job. The factors that control the speed of the rate-determining step therefore also control the overall rate of the reaction. This means that the rate law for the rate-determining step is directly related to the rate law for the overall reaction.

The rate law for the rate-determining step is the same as the rate law for the whole reaction.

Because the rate-determining step is an elementary process, we can predict its rate law from the coefficients. The coefficient of NO_2 is 2, so the rate law for the first step would be the same as that found experimentally for the overall reaction. The mechanism, therefore, does not conflict with experimental evidence and could be correct.

Now let's look at the reaction

$$2NO_2Cl \longrightarrow 2NO_2 + Cl_2 \qquad (18.19)$$

which is a first-order reaction that has the experimentally determined rate law

$$\text{rate} = k[NO_2Cl]$$

The rate-determining step for this reaction could not possibly involve collisions of two NO_2Cl molecules because if it did, the reaction would be second order. The actual mechanism here appears to be

$$NO_2Cl \longrightarrow NO_2 + Cl \quad \text{(slow)}$$
$$NO_2Cl + Cl \longrightarrow NO_2 + Cl_2 \quad \text{(fast)}$$

Note that the sum of the elementary processes (after canceling Cl from both sides) gives the equation for the overall reaction (Reaction 18.19) and that the rate-determining step is first order because the coefficient of NO_2Cl is equal to one.

Although chemists may devise other experiments to help prove or disprove the correctness of a mechanism, one of the strongest pieces of evidence is the experimentally measured rate law for the overall reaction. No matter how reasonable a particular mechanism may appear, if its elementary processes cannot yield a predicted rate law that matches the experimental one, the mechanism is wrong and must be discarded.

PRACTICE EXERCISE 13   The reaction of ozone, O_3, with nitric oxide, NO, to form nitrogen dioxide and oxygen

$$NO + O_3 \longrightarrow NO_2 + O_2$$

is one of the reactions involved in the production of smog. It is believed to occur by a one-step mechanism (the reaction above). If this is so, what is the expected rate law for the reaction?

18.10 FREE RADICAL REACTIONS

Reactions that involve reactive species containing unpaired electrons tend to be very rapid and involve chain mechanisms.

A free radical is a very reactive species that contains one or more unpaired electrons. Examples are chlorine atoms that are produced when a Cl_2 molecule absorbs a photon (light) of the appropriate energy:

$$Cl_2 + \text{light energy} \longrightarrow 2Cl \cdot$$

Remember, $E = h\nu$, where ν is the frequency of the light.

(A dot placed next to the symbol of an atom or molecule indicates that it is a free radical.) The high degree of reactivity of free radicals exists because of the tendency of electrons to pair through the formation of either ions or covalent bonds.

Free radicals are important in many gaseous reactions, including those responsible for the production of photochemical smog in urban areas. In biological systems, they appear to be responsible for the aging process as well as many of the harmful effects of radiation. Reactions involving free radicals have useful applications, too. Many plastics are made by polymerization reactions that occur by mechanisms that involve free radicals. In addition, free radicals play a part in one of the most important processes in the petroleum industry, *thermal cracking*. This reaction is used to break C—C and C—H bonds of long-chain hydrocarbons to produce the smaller molecules that give gasoline a higher octane rating. A simple example is the thermal-cracking reaction of butane. When butane is heated to 700 to 800 °C, one of the major reactions that occurs is

$$CH_3-CH_2 \!:\! CH_2-CH_3 \longrightarrow CH_3CH_2 \cdot + CH_3CH_2 \cdot$$

The central C—C bond is shown as a pair of dots, :, rather than the usual dash.

This reaction produces two *ethyl radicals*, $C_2H_5 \cdot$.

Free radical reactions tend to have high initial activation energies because chemical bonds must be broken to form the radicals. This can be accomplished by light energy or

Direct experimental evidence exists for the presence of free radicals in functioning biological systems. These highly reactive species play many roles, but one of the most interesting is their apparent involvement in the aging process. One theory suggests that free radicals attack protein molecules in collagen. Collagen is composed of long strands of fibers of proteins and is found throughout the body, especially in the flexible tissues of the lungs, skin, muscles, and blood vessels. Attack by free radicals seems to lead to cross-linking between these fibers, which stiffens them and makes them less flexible. The most readily observable result of this is the stiffening and hardening of the skin that accompanies aging (or too much sunbathing).

Free radicals also seem to affect fats (lipids) within the body by promoting the oxidation and deactivation of enzymes that are normally involved in the metabolism of lipids. Over a long period, the accumulated damage gradually reduces the efficiency with which the cell carries out its activities. Interestingly, vitamin E appears to be a natural free radical inhibitor. Diets that are deficient in vitamin E produce effects resembling radiation damage and aging.

Still another theory of aging suggests that free radicals attack the DNA in the nuclei of cells. DNA is the substance that is responsible for directing the chemical activities required for the successful functioning of the cell. Reactions with free radicals cause a

People exposed to sunlight over long periods, like this woman from Nepal (a small country between India and Tibet), tend to develop wrinkles because ultraviolet radiation causes changes in their skin. This is particularly evident when compared to the smooth skin of the child that she's holding.

gradual accumulation of errors in the DNA, which reduce the efficiency of the cell and can lead, ultimately, to malfunction and cell death.

Once a free radical is formed, its reactions with other substances tend to have very low activation energies.

heat. Once the free radicals are formed, however, the chemical reactions in which they are involved tend to be very rapid. In many cases, a free radical reacts with a reactant molecule to give a product molecule plus another free radical. Reactions that involve this step are called chain reactions.

Many explosive reactions are chain reactions involving free radical mechanisms. One of the most studied reactions of this type is the formation of water from hydrogen and oxygen. The reactions involved in the mechanism can be described according to their roles in the mechanism. The chain reaction begins with an initiation step that produces free radicals.

$$H_2 + O_2 \xrightarrow{\text{hot surface}} 2OH\cdot \qquad \text{(initiation)}$$

The chain continues with a propagation step, which produces the product plus another free radical.

$$OH\cdot + H_2 \longrightarrow H_2O + H\cdot \qquad \text{(propagation)}$$

The reaction of H_2 and O_2 is explosive because the mechanism also contains branching steps.

Not all chain reactions involve branching steps.

$$\left.\begin{array}{l} H\cdot + O_2 \longrightarrow OH\cdot + O\cdot \\ O\cdot + H_2 \longrightarrow OH\cdot + H\cdot \end{array}\right\} \quad \text{branching}$$

Thus, the reaction of one $H\cdot$ with oxygen leads to the net production of two $OH\cdot$ plus another $H\cdot$. Every time an $H\cdot$ reacts with oxygen, then, there is an increase in the number of free radicals in the system. The free radical concentration grows rapidly, and the reaction rate becomes explosively fast.

Chain mechanisms also contain termination steps. In the reaction of H_2 and O_2, the wall of the reaction vessel serves to remove $H \cdot$, which tends to stop the chain process.

$$2H \cdot + \text{wall} \longrightarrow H_2 \qquad \text{(termination)}$$

18.11 CATALYSTS

Catalysts accelerate chemical reactions by providing an alternative mechanism that has a lower activation energy.

A catalyst, we have learned, is a substance that increases the rate of a chemical reaction without being used up itself. All of the catalyst added at the start of a reaction is present chemically unchanged after the reaction has gone to completion. The catalyst participates by changing the mechanism of the reaction. It provides a path to the products that has a lower activation energy than that of the uncatalyzed reaction, as illustrated in Figure 18.11. Since the activation energy following this new route is lower, a greater fraction of the reactant molecules have the minimum energy needed to react, so the rate of reaction increases.

Catalysts can be divided into two groups—homogeneous catalysts, which are found in the same phase as the reactants, and heterogenous catalysts, which exist as a separate phase. An example of homogeneous *catalysis* is found in the now-outdated lead chamber process used for manufacturing sulfuric acid. To make sulfuric acid, sulfur is first burned to give SO_2, which is then oxidized to SO_3. Dissolving SO_3 in water gives H_2SO_4.

Catalysis is the action or effect produced by a catalyst.

$$S + O_2 \longrightarrow SO_2$$
$$2SO_2 + O_2 \longrightarrow 2SO_3$$
$$SO_3 + H_2O \longrightarrow H_2SO_4$$

Unassisted, the second reaction, oxidation of SO_2 to SO_3, occurs slowly. In the lead chamber process, the SO_2 is mixed with NO, NO_2, air, and steam in large lead-lined reaction chambers. The NO_2 readily oxidizes the SO_2 to give NO and SO_3. The NO is then reoxidized to NO_2.

Because the NO_2 is reformed in the second reaction and is therefore recycled over and over, only small amounts of it are needed in the reaction mixture to do an effective job.

$$NO_2 + SO_2 \longrightarrow SO_3 + NO$$
$$NO + \tfrac{1}{2}O_2 \longrightarrow NO_2$$

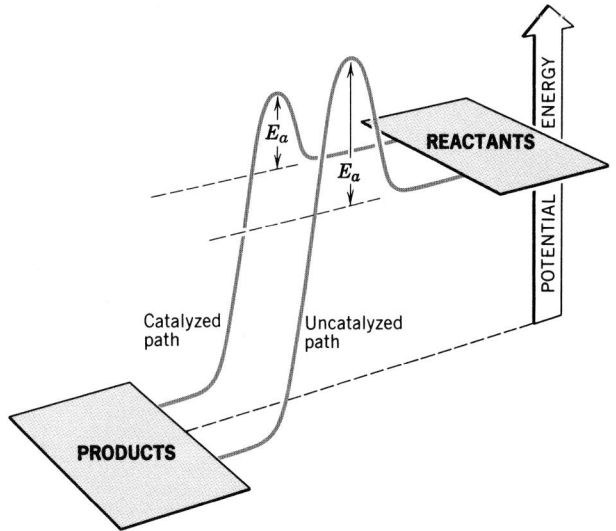

FIGURE 18.11 Effect of a catalyst on a reaction. The catalyst provides an alternative, low-energy path from the reactants to the products.

The photograph on the left shows a catalytic cracking unit at a Standard Oil Company refinery. On the right are beads of a variety of reforming catalysts used in the process.

The NO$_2$ serves as a catalyst by being an oxygen carrier and by providing a low-energy path for the oxidation of SO$_2$ to SO$_3$. The steam in the reaction mixture reacts with the SO$_3$ as it is formed and produces the sulfuric acid.

A heterogeneous catalyst functions by promoting a reaction on its surface. One or more of the reactant molecules are adsorbed on the surface of the catalyst where interaction with the surface increases their reactivity. An example is the synthesis of ammonia from hydrogen and nitrogen by the Haber process.

The synthesis of ammonia is described in detail in *Chemicals in Use II* (page 692).

$$3H_2(g) + N_2(g) \longrightarrow 2NH_3(g)$$

This is one of the most important industrial reactions in the world because ammonia and nitric acid (which is made from ammonia) are necessary for the production of fertilizers. The reaction takes place on an iron catalyst that contains traces of aluminum and potassium oxides. It is thought that hydrogen molecules and nitrogen molecules dissociate while being held on the surface of the catalyst. The hydrogen atoms then combine with the nitrogen atoms to form ammonia. Finally, the completed ammonia molecule breaks away, freeing the surface of the catalyst for further reaction. This sequence of steps is illustrated in Figure 18.12.

FIGURE 18.12 Catalytic formation of ammonia molecules on the surface of a catalyst. Nitrogen and hydrogen are adsorbed and dissociate into atoms that then combine to form ammonia molecules. In the final step, ammonia molecules leave the surface and the whole process can be repeated.

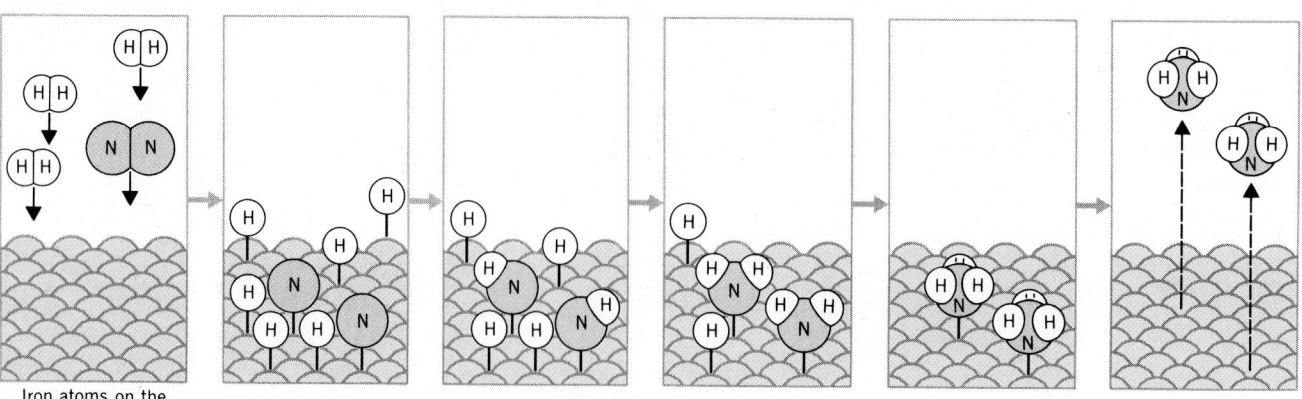

Iron atoms on the catalyst's surface

Heterogeneous catalysts are used in many important commercial processes. The petroleum industry uses catalysts to crack hydrocarbons into smaller fragments and then re-form them into the useful components of gasoline. The availability of catalysts allows refineries to produce gasoline, jet fuel, or heating oil from crude oil in any ratio necessary to meet the demands of the marketplace.

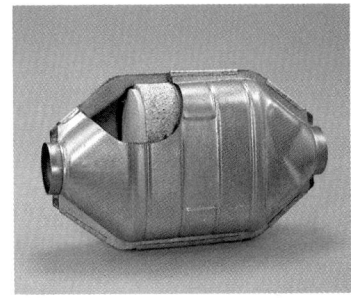

FIGURE 18.13 A modern catalytic converter of the type used in about 80% of new cars. Part of the converter has been cut away to reveal the porous ceramic material that serves as the support for the catalyst.

If you drive a car that requires unleaded gasoline, the car is equipped with a catalytic converter (Figure 18.13) designed to lower the concentrations of pollutants in the exhaust. These pollutants are primarily carbon monoxide, unburned hydrocarbons, and nitrogen oxides. Air is mixed with the exhaust stream and passed over a rhodium catalyst which adsorbs CO, NO, and O_2. The NO dissociates into N and O atoms, and the O_2 also dissociates into atoms. Pairing of nitrogen atoms than produces N_2, and oxidation of CO by oxygen atoms produces CO_2. Unburned hydrocarbons are also oxidized to CO_2 and H_2O.

Unleaded gasoline must be used in cars equipped with catalytic converters because lead — present in gas as tetraethyllead, $Pb(C_2H_5)_4$ — interferes with the active sites on the catalytic surface. The lead "poisons" the surface by destroying its catalytic properties.

The poisoning of catalysts is a major problem in many industrial processes. Methanol (methyl alcohol, CH_3OH), for example, is a promising fuel. It can be made from coal and steam by the reaction

$$C \text{ (from coal)} + H_2O \longrightarrow CO + H_2$$

followed by

$$CO + 2H_2 \longrightarrow CH_3OH$$

A catalyst for this process is copper(I) in solid solution with zinc oxide. However, traces of sulfur, a contaminant in coal, must be avoided because sulfur reacts with the catalyst and destroys its catalytic activity.

The Mobil Oil Company has discovered a catalytic process for converting methanol to gasoline.

SUMMARY

Reaction Rates. The speed or rate of a reaction is controlled by five factors: (1) the nature of the reactants, (2) their ability to meet, (3) their concentrations, (4) the temperature, and (5) the presence of catalysts. The rates of heterogeneous reactions are determined largely by the area of contact between the phases; the rates of homogeneous reactions are determined by the concentrations of the reactants. The rate is measured by monitoring the change in reactant or product concentrations with time.

Rate Laws. The rate of a reaction is proportional to the product of the molar concentrations of the reactants, each raised to an appropriate power. These exponents must be determined by experiments in which the concentrations are varied and the effects of the variations on the rate are measured. The proportionality constant, k, is called the rate constant. Its value depends on temperature but, of course, not on the concentrations of the reactants. The sum of the exponents in the rate law is the order (overall order) of the reaction.

Concentration and Time. Equations relate the concentration of a reactant at a given time t to the initial concentration and the rate constant. The time required for half of a reactant to disap-

pear is the half-life, $t_{1/2}$. For a first-order reaction, the half-life is a constant that depends only on the rate constant for the reaction; it is independent of the initial concentration. The $t_{1/2}$ for a second-order reaction is inversely proportional both to the initial concentration of the reactant and to the rate constant.

Collision Theory. The rate of reaction depends on the number of effective collisions per second, which is only an extremely small fraction of the total number of collisions per second. One reason for the small number of successful collisions is that some collisions require special orientations of the reactant molecules. The major reason, however, is that the molecules must jointly possess a minimum kinetic energy called the activation energy, E_a. As the temperature increases, a larger fraction of reactant molecules have this necessary energy, so more collisions are effective per second and the reaction is faster. The E_a appears as an energy barrier on the potential-energy diagram for the reaction; the heat of reaction is the net potential-energy difference between the reactants and the products. Since reactions are reversible, E_a for forward and reverse reactions can be identified. The species at the high point on the energy diagram is the activated complex and is in the transition state.

Determining the Activation Energy.   The Arrhenius equation relates the activation energy to the variation in the rate constant with temperature. This allows E_a to be obtained either graphically or by calculation using the appropriate form of the Arrhenius equation (Equation 18.14). Calculation requires two rate constants determined at two temperatures, whereas the graphical method requires more rate constants at more temperatures. The activation energy and the rate constant at one temperature can be used to calculate the rate constant at another temperature.

Reaction Mechanisms.   The detailed sequence of elementary processes that lead to the net chemical change is the mechanism of the reaction. Since intermediates usually cannot be detected, the mechanism is a theory. Support for a mechanism comes from matching the predicted rate law for the mechanism with the rate law obtained from experimental data. For the rate-determining step (or any elementary process) the rate law has exponents equal to the coefficients.

Free Radical Reactions.   Species with unpaired electrons tend to be very reactive because of nature's tendency to form electron-pair bonds. Free radicals can be formed from stable species by heat or light of the proper frequency. Reactions involving free radicals are often rapid chain reactions. Some chain reactions are branched—some steps produce more free radicals than they consume. These reactions tend to be explosive.

Catalysts.   Catalysts are substances that affect the rate but are not consumed by the reaction. Homogeneous catalysts are in the same phase as the reactants and provide an alternative mechanism that has a lower activation energy than the uncatalyzed reaction. Heterogeneous catalysts provide a path of lower activation energy by having a surface on which the reactants are adsorbed and react.

REVIEW EXERCISES

Answers to questions whose numbers are printed in color are given in Appendix C.

Factors That Affect Reaction Rate

18.1   What does *rate of reaction* mean in qualitative terms?

18.2   Give an example from everyday experience of (a) a very fast reaction, (b) a moderately fast reaction, and (c) a slow reaction.

18.3   In terms of reaction rates, what is an explosion?

18.4   From an economic point of view, why would industrial corporations want to know about the factors that affect the rate of a reaction?

18.5   Suppose we compared two reactions, one requiring the simultaneous collision of three molecules and the other requiring a collision between two molecules. From the standpoint of statistics, which reaction should be faster? Explain your answer.

18.6   State the five factors that affect the rates of chemical reactions.

18.7   What is a homogeneous reaction? Give an example.

18.8   What is a heterogeneous reaction? Give an example.

18.9   Why are chemical reactions usually carried out in solution?

18.10   What is the major factor that affects the rate of a heterogeneous reaction?

18.11   How does particle size affect the rate of a heterogeneous reaction? Why?

18.12   The rate of hardening of epoxy glue depends on the amount of hardener that is mixed into the glue. What factor affecting reaction rates does this illustrate?

18.13   What effect does temperature have on reaction rate?

18.14   What is a catalyst?

18.15   In cool weather, the number of chirps per minute from crickets diminishes. How can this be explained?

18.16   On the basis of what you learned in Chapter 10, why do foods cook faster in a pressure cooker than in an open pot of boiling water?

18.17   Persons who have been submerged in very cold water and who are believed to have drowned sometimes can be revived. On the other hand, persons who have been submerged in warmer water for the same length of time have died. Explain this in terms of factors that affect the rates of chemical reactions.

Measuring Rates of Reaction

18.18   What units are used to express reaction rate?

18.19   In the reaction $3H_2 + N_2 \rightarrow 2NH_3$, how does the rate of disappearance of hydrogen compare to the rate of disappearance of nitrogen? How does the rate of appearance of NH_3 compare to the rate of disappearance of nitrogen?

18.20   The following data were collected for the reaction $SO_2Cl_2 \rightarrow SO_2 + Cl_2$ at a certain temperature.

$[SO_2Cl_2]$ (mol/L)	Time (s)
0.100	0
0.082	100
0.067	200
0.055	300
0.045	400
0.037	500
0.030	600
0.025	700
0.020	800

Make a graph of concentration versus time and determine the rate of the reaction at $t = 200$ seconds and $t = 600$ seconds.

18.21 The following data were collected for the reaction below at 530 °C.

$$CH_3CHO \longrightarrow CH_4 + CO$$

[CH₃CHO](mol/L)	Time (s)
0.200	0
0.153	20
0.124	40
0.104	60
0.090	80
0.079	100
0.070	120
0.063	140
0.058	160
0.053	180
0.049	200

Make a graph of concentration versus time and determine the reaction rate after 60 seconds and after 120 seconds.

18.22 For the reaction $2A + B \rightarrow 3C$, it was found that the rate of disappearance of B was 0.30 mol L^{-1} s^{-1}. What was the rate of disappearance of A and the rate of appearance of C?

18.23 At a certain temperature, the rate of decomposition of N_2O_5, $2N_2O_5 \longrightarrow 4NO_2 + O_2$, is 2.5×10^{-6} mol L^{-1} s^{-1}. How fast are NO_2 and O_2 being formed?

Concentration and Rate; Rate Laws

18.24 What is a rate law? What is the proportionality constant called?

18.25 What is meant by the order of a reaction?

18.26 What are the units of the rate constant for (a) a first-order reaction, (b) a second-order reaction, and (c) a third-order reaction?

18.27 How must the exponents in a rate law be determined?

18.28 Is there any way of using the coefficients in the balanced equation for a reaction to predict with certainty what the exponents are in the rate law?

18.29 If the concentration of a reactant is doubled and the reaction rate doubles, what must be the order of the reaction with respect to that reactant?

18.30 If the concentration of a reactant is doubled, by what factor will the rate increase if the reaction is second order with respect to that reactant?

18.31 What kind of experiments must be done to determine the exponents in the rate law for a reaction?

18.32 The rate law for the reaction $2NO + O_2 \rightarrow 2NO_2$ is rate $= k[NO]^2[O_2]$. At 25 °C, $k = 7.1 \times 10^9$ L^2 mol^{-2} s^{-1}. What is the rate of reaction when $[NO] = 0.0010$ mol/L and $[O_2] = 0.034$ mol/L?

18.33 The rate law for the decomposition of N_2O_5 is rate $= k[N_2O_5]$. If $k = 1.0 \times 10^{-5}$ s^{-1}, what is the reaction rate when the N_2O_5 concentration is 0.0010 mol/L?

18.34 The rate law for the reaction

$$2HCrO_4^- + 3HSO_3^- + 5H^+ \longrightarrow 2Cr^{3+} + 3SO_4^{2-} + 5H_2O$$

is

$$\text{rate} = k[HCrO_4^-][HSO_3^-]^2[H^+].$$

(a) What is the order of the reaction with respect to each reactant?
(b) What is the overall order of the reaction?

18.35 The rate law for the reaction $NO + O_3 \rightarrow NO_2 + O_2$ is rate $= k[NO][O_3]$. What is the order with respect to each reactant and what is the overall order of the reaction?

18.36 Biological reactions involve interaction of an enzyme with a *substrate*, the substance that actually undergoes the chemical change. In many cases, the rate of reaction depends on the concentration of the enzyme but is independent of the substrate concentration. What is the order of the reaction with respect to the substrate in such instances?

18.37 The following data were collected for the reaction

$$M + N \longrightarrow P + Q$$

Initial Concentrations (mol/L)		Initial Rate of Disappearance of M (mol L⁻¹ s⁻¹)
[M]	[N]	
0.010	0.010	2.5×10^{-3}
0.020	0.010	5.0×10^{-3}
0.020	0.030	4.5×10^{-2}

What is the rate law for the reaction? What is the value of the rate constant?

18.38 Cyclopropane, C_3H_6, is a gas used as a general anesthetic. It undergoes a slow molecular rearrangement to propylene.

cyclopropane   propylene

At a certain temperature, the following data were obtained relating concentration and rate.

Initial Concentrations of C₃H₆(mol/L)	Rate of Formation of Propylene (mol L⁻¹ s⁻¹)
0.050	2.95×10^{-5}
0.100	5.90×10^{-5}
0.150	8.85×10^{-5}

What is the rate law for the reaction?

18.39 The reaction of iodide ion with hypochlorite ion, OCl^- (which is found in liquid bleach), follows the equation

$$OCl^- + I^- \longrightarrow OI^- + Cl^-$$

It is a rapid reaction that gives the following rate data.

Initial Concentrations (mol/L)		Rate of Formation of Cl^-
$[OCl^-]$	$[I^-]$	$(mol\ L^{-1}\ s^{-1})$
1.7×10^{-3}	1.7×10^{-3}	1.75×10^4
3.4×10^{-3}	1.7×10^{-3}	3.50×10^4
1.7×10^{-3}	3.4×10^{-3}	3.50×10^4

What is the rate law for the reaction? Determine the value of the rate constant.

18.40 The formation of small amounts of nitric oxide, NO, in automobile engines is the first step in the formation of smog. Nitric oxide is readily oxidized to nitrogen dioxide by the reaction $2NO(g) + O_2(g) \rightarrow 2NO_2(g)$. The following data were collected in a study of the rate of this reaction.

Initial Concentrations (mol/L)		Rate of Formation of NO_2
$[O_2]$	$[NO]$	$(mol\ L^{-1}\ s^{-1})$
0.0010	0.0010	7.10
0.0040	0.0010	28.4
0.0040	0.0030	255.6

What is the rate law for the reaction? What is the rate constant?

18.41 At a certain temperature the following data were collected for the reaction $2ICl + H_2 \rightarrow I_2 + 2HCl$.

Initial Concentrations (mol/L)		Initial Rate of Formation of I_2
$[ICl]$	$[H_2]$	$(mol\ L^{-1}\ s^{-1})$
0.10	0.10	0.0015
0.20	0.10	0.0030
0.10	0.050	0.00075

Determine the rate law and the rate constant for the reaction.

18.42 The following data were obtained for the reaction of $(CH_3)_3CBr$ with hydroxide ion at 55 °C.

$$(CH_3)_3CBr + OH^- \longrightarrow (CH_3)_3COH + Br^-$$

Initial Concentrations (mol/L)		Initial Rate of Formation of $(CH_3)_3COH$
$[(CH_3)_3CBr]$	$[OH^-]$	$(mol\ L^{-1}\ s^{-1})$
0.10	0.10	1.0×10^{-3}
0.20	0.10	2.0×10^{-3}
0.30	0.10	3.0×10^{-3}
0.10	0.20	1.0×10^{-3}
0.10	0.30	1.0×10^{-3}

What is the rate law for the reaction? What is the value of the rate constant at this temperature?

Concentration and Time

18.43 Give the equations that relate concentration to time for (a) a first-order reaction and (b) a second-order reaction.

18.44 Derive Equation 18.5 from Equation 18.7.

18.45 The compound SO_2Cl_2 decomposes according to the equation

$$SO_2Cl_2(g) \longrightarrow SO_2(g) + Cl_2(g)$$

The reaction is first order with a rate constant $k = 2.2 \times 10^{-5}\ s^{-1}$ at 320 °C. If the initial SO_2Cl_2 concentration in a container is 0.0040 M, what will its concentration be (a) after 1 hour, (b) after 1 day?

18.46 The decomposition of hydrogen iodide follows the equation $2HI(g) \rightarrow H_2(g) + I_2(g)$. The reaction is second order and has a rate constant equal to $1.6 \times 10^{-3}\ L\ mol^{-1}\ s^{-1}$ at 700 °C. The initial concentration of HI in a container was $1.1 \times 10^{-2}\ M$. How many minutes did it take for the concentration to be reduced to $5.0 \times 10^{-4}\ M$?

18.47 The decomposition of HI follows the equation

$$2HI(g) \longrightarrow H_2(g) + I_2(g)$$

The reaction is second order and has a rate constant equal to $1.6 \times 10^{-3}\ L\ mol^{-1}\ s^{-1}$ at 700 °C. At 2.0×10^3 minutes after a particular experiment had begun, the HI concentration was equal to $3.0 \times 10^{-4}\ mol/L$. What was the initial molar concentration of HI in the experiment?

18.48 The concentration of a drug in the body is often expressed in units of milligrams per kilogram of body weight. The initial dose of a drug in an animal was 30.0 mg/kg body weight. After 2.00 hours, this concentration had dropped to 10.0 mg/kg body weight. If the drug is eliminated metabolically by a first-order process, what is the rate constant for the process in units of min^{-1}?

18.49 In the preceding question, what must the initial dose of the drug be in order for the drug concentration 3.00 hours afterward to be 10.0 mg/kg body weight?

18.50 If it takes 45.0 min for the concentration of a reactant to drop to 10% of its initial value in a first-order reaction, what is the rate constant for the reaction in the units min^{-1}?

Half-Lives

18.51 What is meant by the term *half-life*?

18.52 How is the half-life of a first-order reaction affected by the initial concentration of the reactant?

18.53 How is the half-life of a second-order reaction affected by the initial reactant concentration?

18.54 Derive the equations for $t_{1/2}$ for first- and second-order reactions from Equations 18.5 and 18.8, respectively.

18.55 A certain first-order reaction has a rate constant $k = 1.6 \times 10^{-3}\ s^{-1}$. What is the half-life for this reaction?

18.56 The decomposition of NOCl, $2NOCl \rightarrow 2NO + Cl_2$, is a second-order reaction with $k = 6.7 \times 10^{-4}\ L\ mol^{-1}\ s^{-1}$ at 400 K. What is the half-life of this reaction if the initial concentration of NOCl is 0.20 mol/L?

18.57 Using the graph from Review Exercise 18.20, determine the time required for the SO_2Cl_2 concentration to drop from 0.100 mol/L to 0.050 mol/L. How long does it take for the concentration to drop from 0.050 mol/L to 0.025 mol/L? What is the order of this reaction? (*Hint:* How is the half-life related to concentration?)

18.58 Using the graph from Review Exercise 18.21, determine how long it takes for the CH_3CHO concentration to decrease from 0.200 mol/L to 0.100 mol/L. How long does it take the concentration to drop from 0.100 mol/L to 0.050 mol/L? What is the order of this reaction? (*Hint:* How is the half-life related to concentration?)

18.59 The half-life of a certain first-order reaction is 15 minutes. What fraction of the original reactant concentration will remain after 2.0 hours?

18.60 Strontium-90 has a half-life of 28 years. How long will it take for all of the strontium-90 presently on earth to be reduced to $\frac{1}{32}$ of its present amount?

Effect of Temperature on Rate

18.61 In general, to what extent are the rates of most reactions affected by a 10 °C rise in temperature?

18.62 What is the basic postulate of collision theory?

18.63 What two factors affect the effectiveness of molecular collisions in producing chemical change?

18.64 In terms of the kinetic theory, why does an increase in temperature increase the reaction rate?

18.65 Draw the potential-energy diagram for an endothermic reaction. Indicate on the diagram the activation energy for both the forward and reverse reactions. Also indicate the heat of reaction.

18.66 Why does an endothermic reaction lead to a cooling of the reaction mixture if heat cannot enter from outside the system?

18.67 Define *transition state* and *activated complex.*

18.68 Draw a potential-energy diagram for an exothermic reaction and indicate on the diagram the location of the transition state.

18.69 The decomposition of carbon dioxide,

$$CO_2 \longrightarrow CO + O$$

has a very large activation energy of approximately 460 kJ/mol. Explain why this is consistent with a mechanism that involves the breaking of a C=O bond.

Calculations Involving the Activation Energy

18.70 State the Arrhenius equation and define the symbols.

18.71 The following data were collected for a reaction.

Rate Constant (s^{-1})	Temperature (°C)
2.88×10^{-4}	320
4.87×10^{-4}	340
7.96×10^{-4}	360
1.26×10^{-3}	380
1.94×10^{-3}	400

Determine the activation energy for the reaction in kJ/mol both graphically and by calculation using Equation 18.14. For the calculation of E_a, use the first and last sets of data in the table above.

18.72 Rate constants were measured at various temperatures for the reaction

$$HI(g) + CH_3I(g) \longrightarrow CH_4(g) + I_2(g)$$

The following data were obtained.

Rate Constant (L mol^{-1} s^{-1})	Temperature (°C)
1.91×10^{-2}	205
2.74×10^{-2}	210
3.90×10^{-2}	215
5.51×10^{-2}	220
7.73×10^{-2}	225
1.08×10^{-1}	230

Determine the activation energy in kJ/mol both graphically and by calculation using Equation 18.14. For the calculation of E_a, use the first and last sets of data in the table above.

18.73 The decomposition of NOCl, $2NOCl \rightarrow 2NO + Cl_2$, has $k = 9.3 \times 10^{-5}$ L mol^{-1} s^{-1} at 100 °C and $k = 1.0 \times 10^{-3}$ L mol^{-1} s^{-1} at 130 °C. What is E_a for this reaction in kJ/mol? Use the data at 100 °C to calculate the frequency factor.

18.74 The conversion of cyclopropane, an anesthetic, to propylene (see Review Exercise 18.38) has a rate constant $k = 1.3 \times 10^{-6}$ s^{-1} at 400 °C and $k = 1.1 \times 10^{-5}$ s^{-1} at 430 °C.
(a) What is the activation energy in kJ/mol?
(b) What is the value of A for this reaction, calculated using Equation 18.12?
(c) What is the rate constant for the reaction at 350 °C?

18.75 The reaction of CO_2 with water to form carbonic acid,

$$CO_2(aq) + H_2O \longrightarrow H_2CO_3(aq)$$

has $k = 3.75 \times 10^{-2}$ s^{-1} at 25 °C and $k = 2.1 \times 10^{-3}$ s^{-1} at 0 °C. What is the activation energy for this reaction in kJ/mol?

18.76 If a reaction has $k = 3.0 \times 10^{-4}$ s^{-1} at 25 °C and an activation energy of 25.0 kcal/mol, what will the value of k be at 50 °C?

18.77 It was mentioned that the rates of many reactions approximately double for each 10 °C rise in temperature. Assuming a starting temperature of 25 °C, what would the activation energy be, in kJ, if the rate of a reaction were to be twice as large at 35 °C?

18.78 The decomposition of N_2O_5 has an activation energy of 103 kJ/mol and a frequency factor of 4.3×10^{13} s^{-1}. What is the rate constant for this decomposition at (a) 20 °C and (b) 100 °C?

18.79 At 35 °C, the rate constant for the reaction

$$\underset{\text{sucrose}}{C_{12}H_{22}O_{11}} + H_2O \longrightarrow \underset{\text{glucose}}{C_6H_{12}O_6} + \underset{\text{fructose}}{C_6H_{12}O_6}$$

is $k = 6.2 \times 10^{-5}$ s^{-1}. The activation energy for the reaction is 108 kJ/mol. What is the rate constant for the reaction at 45 °C?

Reaction Mechanisms

18.80 What is an elementary process and what is its relationship to a reaction mechanism?

18.81 What is a rate-determining step?

18.82 In what way is the rate law for a reaction related to the rate-determining step?

18.83 A reaction has the following mechanism.

$$2NO \longrightarrow N_2O_2$$
$$N_2O_2 + H_2 \longrightarrow N_2O + H_2O$$
$$N_2O + H_2 \longrightarrow N_2 + H_2O$$

What is the net overall change that occurs in this reaction?

18.84 If the reaction $NO_2 + CO \rightarrow CO_2 + NO$ occurs by a one-step collision process, what would be the expected rate law for the reaction? The actual rate law is rate = $k[NO_2]^2$. Could the reaction actually occur by a one-step collision between NO_2 and CO? Explain.

18.85 Oxidation of NO to NO_2—one of the reactions in the production of smog—appears to involve carbon monoxide. A possible mechanism is

$$CO + \cdot OH \longrightarrow CO_2 + H\cdot$$
$$H\cdot + O_2 \longrightarrow HOO\cdot$$
$$HOO\cdot + NO \longrightarrow \cdot OH + NO_2$$

Write the net chemical equation for the reaction.

18.86 Show that the following two mechanisms give the same net overall reaction.

Mechanism 1

$$OCl^- + H_2O \longrightarrow HOCl + OH^-$$
$$HOCl + I^- \longrightarrow HOI + Cl^-$$
$$HOI + OH^- \longrightarrow H_2O + OI^-$$

Mechanism 2

$$OCl^- + H_2O \longrightarrow HOCl + OH^-$$
$$I^- + HOCl \longrightarrow ICl + OH^-$$
$$ICl + 2OH^- \longrightarrow OI^- + Cl^- + H_2O$$

18.87 The experimental rate law for the reaction $NO_2 + CO \rightarrow CO_2 + NO$ is rate = $k[NO_2]^2$. If the mechanism is

$$2NO_2 \longrightarrow NO_3 + NO \quad \text{(slow)}$$
$$NO_3 + CO \longrightarrow NO_2 + CO_2 \quad \text{(fast)}$$

show that the predicted rate law is the same as the experimental rate law.

Free Radical Reactions

18.88 What is a free radical? How can they be generated from stable molecules?

18.89 What is a chain reaction?

18.90 The reaction of hydrogen and bromine appears to follow the mechanism

$$Br_2 \xrightarrow{h\nu} 2Br\cdot$$
$$Br\cdot + H_2 \longrightarrow HBr + H\cdot$$
$$H\cdot + Br_2 \longrightarrow HBr + Br\cdot$$
$$2Br\cdot \longrightarrow Br_2$$

(a) Identify the initiation step in the mechanism.
(b) Identify any propagation steps.
(c) Identify the termination step.

The mechanism also contains the reaction

$$H\cdot + HBr \longrightarrow H_2 + Br\cdot$$

How does this reaction affect the rate of production of HBr?

18.91 In the upper atmosphere, a layer of ozone, O_3, shields the earth from harmful ultraviolet radiation. The ozone is generated by the reactions

$$O_2 + h\nu \longrightarrow 2O$$
$$O + O_2 \longrightarrow O_3$$

Release of chlorofluorocarbon aerosol propellants and refrigerant fluids such as Freon 12 (CCl_2F_2) into the atmosphere threatens to at least partially destroy the ozone shield by the reactions

$$CCl_2F_2 \longrightarrow CClF_2 + Cl$$
$$Cl + O_3 \longrightarrow ClO + O_2 \qquad (1)$$
$$ClO + O \longrightarrow Cl + O_2 \qquad (2)$$

Explain how reactions 1 and 2 constitute a chain reaction that can destroy huge numbers of O_3 molecules and O atoms that might otherwise react to replenish the ozone.

Catalysts

18.92 How does a catalyst increase the rate of a chemical reaction?

18.93 What is a homogeneous catalyst? How does it function, in general terms?

18.94 What is a heterogeneous catalyst? How does it function?

18.95 What is the difference in meaning between the terms *adsorption* and *absorption*? (If necessary, use a dictionary.) Which one applies to heterogeneous catalysts?

18.96 What does the catalytic converter do in the exhaust system of an automobile? Why can't leaded gasoline be used in cars equipped with catalytic converters?

TEST OF FACTS AND CONCEPTS: CHAPTERS 14–18

The concepts of chemical equilibria (including acid-base equilibria), thermodynamics, electrochemistry, and kinetics are critically important to a thorough understanding of chemistry, because they control what we see happen in the real world. This has been the theme in this group of chapters. These are not easy concepts, and not many students grasp them immediately. Therefore, it would be wise for you to test your knowledge of them now to learn where you might work more to improve your understanding of them and your problem solving skills.

1. Write the appropriate mass-action expression, using molar concentrations, for these reactions.
   (a) $NO_2(g) + N_2O(g) \rightleftharpoons 3NO(g)$
   (b) $CaSO_3(s) \rightleftharpoons CaO(s) + SO_2(g)$
   (c) $NiCO_3(s) \rightleftharpoons Ni^{2+}(aq) + CO_3^{2-}(aq)$

2. At a certain temperature, the reaction $2HF(g) \rightleftharpoons H_2(g) + F_2(g)$ has $K_c = 1 \times 10^{-13}$. Does this reaction proceed far toward completion when equilibrium is reached? If 0.010 mol HF was placed in a 1.00-L container and the system was permitted to come to equilibrium, what would be the concentrations of H_2 and F_2 in the container?

3. At 100 °C, the reaction $2NO_2(g) \rightleftharpoons N_2O_4(g)$ has $K_p = 6.5 \times 10^{-2}$. What is the value of K_c at this temperature?

4. For the reaction $3NO(g) \rightleftharpoons NO_2(g) + N_2O(g)$, calculate K_p at 25 °C, using values of $\Delta G_f°$ from Table 16.2.

5. Consider the reaction
   $$CH_4(g) + Cl_2(g) \rightleftharpoons CH_3Cl(g) + HCl(g).$$
   (a) Calculate $\Delta G'$ for this reaction at 200 °C.
   (b) What is the value of K_p for this reaction at 200 °C?

6. At 1000 °C, the reaction $NO_2(g) + SO_2(g) \rightleftharpoons NO(g) + SO_3(g)$ has $K_c = 3.60$. If 0.100 mol NO_2 and 0.100 mol SO_2 are placed in a 5.00-L container and allowed to react, what will be all the concentrations when equilibrium is reached? What will be the new equilibrium concentrations if 0.010 mol NO and 0.010 mol SO_3 are added to this mixture?

7. For the reaction in the preceding question, $\Delta H° = -9.98$ kcal. How will the equilibrium concentration of NO be affected if
   (a) More NO_2 is added to the container?
   (b) Some SO_3 is removed from the container?
   (c) The temperature of the reaction mixture is raised?
   (d) Some SO_2 is removed from the mixture?
   (e) The pressure of the gases is lowered by expanding the volume to 10.0 L?

8. The molar solubility of silver chromate, Ag_2CrO_4, is 6.7×10^{-5} M. What is K_{sp} for Ag_2CrO_4?

9. What is the pH of a saturated solution of $Mg(OH)_2$?

10. What is the solubility of $Fe(OH)_2$ in grams per liter if the solution is buffered to a pH of 10.00?

11. Suppose 30.0 mL of 0.100 M $Pb(NO_3)_2$ is added to 20.0 mL of 0.500 M KI.
    (a) How many grams of PbI_2 will be formed?

(b) What will be the molar concentrations of all the ions in the mixture after equilibrium has been reached?

12. How many moles of NH_3 must be added to 1.00 L of solution to dissolve 1.00 g of $CuCO_3$? For $CuCO_3$, $K_{sp} = 2.5 \times 10^{-10}$. Ignore hydrolysis of CO_3^{2-}, but consider formation of the complex ion, $Cu(NH_3)_4^{2+}$.

13. At 60 °C, $K_w = 9.5 \times 10^{-14}$. What is the pH of pure water at this temperature? Why can we say that this water is neither acidic nor basic?

14. At 25 °C, the water in a natural pool of water in one of the western states was found to contain hydroxide ions at a concentration of 4.7×10^{-7} g OH^-/L. Calculate the pH of this water and state if it is acidic, basic, or neutral.

15. The first antiseptic to be used in surgical operating rooms was phenol, C_6H_5OH, a weak acid and a potent bactericide. A 0.550 M solution of phenol in water was found to have a pH of 5.07.
    (a) Calculate the K_a and pK_a of phenol.
    (b) Calculate the K_b and pK_b of the phenoxide ion, $C_6H_5O^-$.

16. The pK_a of saccharin, $HC_7H_3SO_3$, a sweetening agent, is 11.68.
    (a) What is the pK_b of the saccharinate ion, $C_7H_3SO_3^-$?
    (b) Judge if sodium saccharinate hydrolyzes in water. If it does, what is the pH of a 0.010 M solution of sodium saccharinate in water?

17. At 25 °C the value of K_b for codeine, a pain-killing drug, is 1.63×10^{-6}. Calculate the pH of a 0.0115 M solution of codeine in water.

18. The pK_b of methylamine, CH_3NH_2, is 3.43. Calculate the pK_a of its conjugate acid, $CH_3NH_3^+$.

19. Ascorbic acid, $H_2C_6H_6O_6$, a diprotic acid is usually known as vitamin C. For this acid, pK_{a_1} is 4.10 and pK_{a_2} is 11.79. When 125 mL of a solution of ascorbic acid was evaporated to dryness, the residue of pure ascorbic acid has a mass of 3.12 g.
    (a) Calculate the molar concentration of ascorbic acid in the solution before it was evaporated.
    (b) Calculate the pH of the solution and the molar concentration of the ascorbate ion, $C_6H_6O_6^{2-}$, before the solution was evaporated.

20. What ratio of molar concentrations of sodium acetate to acetic acid can buffer a solution at a pH of 4.50?

21. Over what pH range must a solution be buffered to achieve a selective separation of the carbonates of barium, $BaCO_3$, ($K_{sp} = 5.0 \times 10^{-9}$), and lead, $PbCO_3$ ($K_{sp} = 7.4 \times 10^{-14}$)? The solution is initially 0.010 M in Ba^{2+} and 0.010 M in Pb^{2+}.

22. A biology experiment requires the use of a nutrient fluid buffered at a pH of 4.85, and 625 mL of this solution is needed. It has to be buffered to be able to hold the pH to within ± 0.10 pH unit of 5.00 even if 5.00×10^{-3} mol of OH^- or 5.00×10^{-3} mol of H^+ ion enter.
    (a) Using tabulated data, pick the best acid and its sodium salt that could be used to prepare this solution.
    (b) Calculate the minimum number of grams of the pure acid and its salt that are needed to prepare this solution.
    (c) What are the molar concentrations of the acid and of its salt in this solution?

23. How would each aqueous solution test, acidic, basic, or neutral? (Assume that each is at least 0.2 M.)
    (a) KNO_3 (b) $CrCl_3$
    (c) NH_4I (d) K_2HPO_4

24. When 50.00 mL of an acid with a concentration of 0.115 M (for which $pK_a = 4.87$) is titrated with 0.100 M NaOH, what is the pH at the equivalence point? What would be a good indicator for this titration?

25. What is meant by the term *spontaneous change*?

26. What is the definition of *internal energy*? State the first law of thermodynamics.

27. What is pressure-volume work? Give the equation that could be used to calculate it.

28. If a gas in a cylinder pushes back a piston against a constant opposing pressure of 3.0×10^5 pascals and undergoes a volume change of 0.50 m^3, how much work will the gas do, expressed in joules?

29. The change in the internal energy, ΔE, is equal to the "heat of reaction at constant volume." Explain the reason for this.

30. Which of these are state functions: E, H, q, S, G, w?

31. Suppose that a gas is produced in an exothermic chemical reaction between two reactants in an aqueous solution. Which quantity will have a larger magnitude, ΔE or ΔH?

32. How is entropy related to statistical probability?

33. What would be the algebraic signs of ΔS for the following reactions?
    (a) $Br_2(\ell) + Cl_2(g) \longrightarrow 2BrCl(g)$
    (b) $CaO(s) + CO_2(g) \longrightarrow CaCO_3(s)$

34. Which of the following states has the greatest entropy: (a) $2H_2O(\ell)$, (b) $2H_2O(s)$, (c) $2H_2(\ell) + O_2(g)$, (d) $2H_2(g) + O_2(g)$, (e) $4H(g) + 2O(g)$?

35. Calculate $\Delta S°$ (in J/K) for the following reactions.
    (a) $H_2O(\ell) + SO_3(g) \longrightarrow H_2SO_4(\ell)$
    (b) $2KCl(s) + H_2SO_4(\ell) \longrightarrow K_2SO_4(s) + 2HCl(g)$
    (c) $C_2H_4(g) + H_2O(g) \longrightarrow C_2H_5OH(\ell)$

36. Calculate $\Delta G°$ in kJ for the following reactions.
    (a) $CaSO_4 \cdot \frac{1}{2}H_2O(s) + \frac{3}{2}H_2O(g) \longrightarrow CaSO_4 \cdot 2H_2O(s)$
    (b) $CH_4(g) + Cl_2(g) \longrightarrow CH_3Cl(g) + HCl(g)$
    (c) $CaSO_4(s) + CO_2(g) \longrightarrow CaCO_3(s) + SO_3(g)$

37. Which of the reactions in the preceding question would appear to be spontaneous?

38. Calculate $\Delta S°$, $\Delta H°$, and $\Delta G'$ (at 400 °C) using energy units of joules or kilojoules for these reactions.
    (a) $CaSO_4 \cdot 2H_2O(s) \longrightarrow CaSO_4 \cdot \frac{1}{2}H_2O(s) + \frac{3}{2}H_2O(g)$
    (b) $NaOH(s) + NH_4Cl(s) \longrightarrow NaCl(s) + NH_3(g) + H_2O(g)$
    (c) $SO_3(g) \longrightarrow SO_2(g) + \frac{1}{2}O_2(g)$

39. Sketch a diagram of an electrolysis cell in which a concentrated solution of NaCl is undergoing electrolysis.
    (a) Label the cathode and anode, including their charges.
    (b) Write half-reactions for the changes taking place at the electrodes.
    (c) Write a balanced equation for the net cell reaction.

40. Suppose that the electrolysis cell described in the previous question contains 250 mL of brine.
    (a) What will be the pH of the solution if the electrolysis is carried out for 20.0 minutes using a current of 1.00 A?
    (b) How many milliliters of H_2, measured at STP, would be evolved if the cell were operated at 5.00 A for 10.0 minutes?

41. What current would be required to deposit 0.100 g of nickel in 20.0 minutes from a solution of $NiSO_4$?

42. In the text, it was mentioned that a large electrolysis cell can produce as much as 900 lb of aluminum in 1 day. What current is required to accomplish this?

43. Sketch a diagram of a galvanic cell consisting of a copper electrode dipping into 1.00 M $CuSO_4$ solution and an iron electrode dipping into 1.00 M $FeSO_4$ solution.
    (a) Identify the cathode and the anode. Indicate the charge carried by each.
    (b) What is the potential of the cell?
    (c) What is the purpose of a salt bridge?
    (d) Write the net cell reaction.

44. Suppose the cell described in the previous question contains 100 mL of each solution and is operated for a period of 50.0 hr at a constant current of 0.10 A. At the end of this time, what will be the concentrations of Cu^{2+} and Fe^{2+}? What will the cell potential be at this point?

45. Use data from Table 17.1 to calculate the value of K_c at 25 °C for the reaction

$$O_2 + 4Br^- + 4H^+ \longrightarrow 2Br_2 + 2H_2O$$

46. A galvanic cell was assembled as follows. In one compartment, a copper electrode was immersed in a 1.00 M solution of $CuSO_4$. In the other compartment, a manganese electrode was immersed in a 1.00 M solution of $MnSO_4$. The potential of the cell was measured to be 1.52 V, with the Mn electrode as the negative electrode. What is the value of $E°$ for the half-reaction, $Mn^{2+}(aq) + 2e^- \longrightarrow Mn(s)$?

47. Calculate $E°_{cell}$ for the reaction

$$3Cu(s) + 2NO_3^-(aq) + 8H^+(aq) \longrightarrow$$
$$2NO(g) + 4H_2O + 3Cu^{2+}(aq)$$

48. List the factors that affect the speed of a chemical reaction.

49. At 25 °C and $[OH^-] = 1.00$ M, the reaction

$$I^-(aq) + OCl^-(aq) \longrightarrow OI^-(aq) + Cl^-(aq)$$

has the rate law: rate = $(0.60 \text{ L mol}^{-1} \text{ s}^{-1})[I^-][OCl^-]$. Calculate the rate of the reaction when
(a) $[I^-] = 0.0100 \ M$ and $[OCl^-] = 0.0200 \ M$
(b) $[I^-] = 0.100 \ M$ and $[OCl^-] = 0.0400 \ M$

50. A reaction has the stoichiometry: $3A + B \rightarrow C + D$. The following data were obtained for the initial rate of formation of C at various concentrations of A and B.

Initial Concentrations (M)		Initial Rate of Formation of C
A	B	(mol L^{-1} s^{-1})
0.010	0.010	2.0×10^{-4}
0.020	0.020	8.0×10^{-4}
0.020	0.010	8.0×10^{-4}

(a) What is the rate law for the reaction?
(b) What is the value of the rate constant?
(c) What is the rate at which C is formed if $[A] = 0.017 \ M$ and $[B] = 0.033 \ M$?

51. If the concentration of a particular reactant is doubled and the rate of the reaction is cut in half, what must be the order of the reaction with respect to that reactant?

52. Organic compounds that contain large proportions of nitrogen and oxygen tend to be unstable and are easily decomposed. Hexanitroethane, $C_2(NO_2)_6$, decomposes according to the equation

$$C_2(NO_2)_6 \longrightarrow 2NO_2 + 4NO + 2CO_2$$

The reaction in CCl_4 as a solvent is first-order with respect to $C_2(NO_2)_6$. At 70 °C, $k = 2.41 \times 10^{-6} \text{ s}^{-1}$ and at 100 °C, $k = 2.22 \times 10^{-4} \text{ s}^{-1}$.
(a) What is the half-life of $C_2(NO_2)_6$ at 70 °C? What is the half-life at 100 °C?
(b) If 0.100 mol of $C_2(NO_2)_6$ is dissolved in CCl_4 at 70 °C to give 1.00 L of solution, what will be the $C_2(NO_2)_6$ concentration after 500 minutes?
(c) What is the value of the activation energy of this reaction, expressed in kilojoules?
(d) What is the reaction's rate constant at 120 °C?

53. Radioactive strontium-90, ^{90}Sr, has a half-life of 28 years.
(a) What fraction of a sample of ^{90}Sr will remain after 168 years?
(b) If the amount of ^{90}Sr remaining in a sample is only one-sixth of the amount originally present, how many years has the sample been undergoing radioactive decay?

54. The reaction $2A + 2B \rightarrow M + N$ has the rate law: rate = $k[A]^2$. At 25 °C, $k = 1.0 \times 10^{-4} \text{ L mol}^{-1} \text{ s}^{-1}$. If the initial concentrations of A and B are 0.250 M and 0.150 M, respectively,
(a) What is the half-life of the reaction?
(b) What will be the concentrations of A and B after 30 minutes?

55. Define *reaction mechanism*, *rate-determining step*, and *elementary process*.

56. The decomposition of ozone, O_3, is believed to occur by the two-step mechanism

$$O_3 \longrightarrow O_2 + O \quad \text{(slow)}$$
$$\underline{O + O_3 \longrightarrow 2O_2 \quad \text{(fast)}}$$
$$2O_3 \longrightarrow 3O_2 \quad \text{(net reaction)}$$

If this is the mechanism, what is the reaction's rate law?

57. Draw a potential energy diagram for an exothermic reaction. Identify the activation energy for both the forward and reverse reactions. Also, identify the heat of reaction.

58. How does a heterogeneous catalyst increase the rate of a chemical reaction?

59. One possible mechanism for the decomposition of ethane, C_2H_6, into ethylene, C_2H_4, and hydrogen,

$$C_2H_6 \longrightarrow C_2H_4 + H_2$$

includes the following steps.
(1) $C_2H_6 \longrightarrow 2CH_3\cdot$
(2) $CH_3\cdot + C_2H_6 \longrightarrow CH_4 + C_2H_5\cdot$
(3) $C_2H_5\cdot \longrightarrow C_2H_4 + H\cdot$
(4) $H\cdot + C_2H_6 \longrightarrow C_2H_5\cdot + H_2$
(5) $H\cdot + C_2H_5\cdot \longrightarrow C_2H_6$
(a) Which steps initiate the reaction?
(b) Which are propagation steps?
(c) Which is a termination step?

Qual Topics I and II

60. What is the difference between qualitative analysis and quantitative analysis?

61. List the metal ions in each of the qualitative analysis cation groups. Describe how each of the groups is separated from a mixture of cations.

62. Suppose no precipitate is formed when HCl is added to a solution that is known to contain one or more metal salts. Which metal ions are *not* present in the solution?

63. Suppose in the analysis of cation group 1 the entire precipitate dissolves in aqueous NH_3. Which ions are not present? Which cation is almost certainly present?

64. Suppose a solution was suspected to contain $(NH_4)_2SO_4$. What test could you perform to determine whether ammonium ion is in the solution?

65. Specify the colors of (a) CdS, (b) NiS, (c) As_2S_3, (d) SnS_2, (e) ZnS, and (f) Sb_2S_3?

66. A solution containing 0.10 M Pb^{2+} and 0.10 M Ni^{2+} is to be saturated with H_2S. What range of pH values could this solution have so that when the procedure is completed one of the ions remains in solution while the other is precipitated as its sulfide?

67. A solution that contains 0.10 M Fe^{2+} and 0.10 M Sn^{2+} is maintained at a pH of 3.00 while H_2S is gradually added to it. What will be the concentration of Sn^{2+} in the solution when FeS just begins to precipitate?

68. A metal sulfide MS has a value of K_{sp} of 4.0×10^{-29}.
(a) What is the value of K_{spa} for this compound?
(b) Calculate the molar solubility of MS in 0.30 M HCl.
(c) Does MS fall into cation group 2 or cation group 3?

The thermal decomposition of N_2O added to the combustion chamber in this drag racer gives an extra boost. This and other oxides of nitrogen are studied in this chapter.

SIMPLE MOLECULES AND IONS OF NONMETALS. PART 1

PREVALENCE OF THE NONMETALLIC ELEMENTS	OXYGEN
THE NOBLE GAS ELEMENTS	NITROGEN
HYDROGEN	OXOACIDS AND OXIDES OF NITROGEN

19.1 PREVALENCE OF THE NONMETALLIC ELEMENTS

Hydrogen, helium, and a few other nonmetals dominate the composition of the universe.

The facts of chemistry — the chemical and physical properties of the elements and their compounds — would be very difficult to comprehend and remember if it were not for the framework of theory that we've described in previous chapters. The principles of atomic structure, thermodynamics, equilibrium, and kinetics have enabled chemists to begin to see the "big picture" and to understand why certain reactions occur and others do not. Now that you've been exposed to this theoretical framework, you too are ready to gain a broader knowledge and understanding of chemical facts. We turn our attention in these next chapters, therefore, to a systematic study of some of the most important elements and their compounds. We begin by examining the origin and abundance of the elements in our universe. Studies of the frequencies and relative intensities of the lines in the spectra of radiation from stars as well as analyses of meteorites have enabled scientists to estimate the elemental composition of the universe. The relative abundances of hydrogen and helium, as estimated by these studies, so dominate this composition that all the other elements seem like impurities, as the data in Table 19.1 show. These two, both nonmetals and the lightest of all elements, also dominate our solar system as seen in Table 19.2. How has this most probably come about?

TABLE 19.1 Elements in the Universe — The Ten Most Abundant

Element	Atomic Number	Abundance[a]
Hydrogen	1	4.0×10^8
Helium	2	3.1×10^7
Oxygen	8	2.2×10^5
Neon	10	8.6×10^4
Nitrogen	7	6.6×10^4
Carbon	6	3.5×10^4
Silicon	14	1.0×10^4
Magnesium	12	9.1×10^3
Iron	26	6.0×10^3
Sulfur	16	3.8×10^3

Data from B. Mason, *Principles of Geochemistry,* 3rd ed. John Wiley & Sons, Inc., New York, 1966.

[a] Number of atoms per 10,000 atoms of silicon.

TABLE 19.2 Elements in the Solar System — The Ten Most Abundant

Element	Atomic Number	Abundance[a]
Hydrogen	1	3.2×10^8
Helium	2	5.0×10^7
Oxygen	8	2.9×10^5
Carbon	6	1.7×10^5
Nitrogen	7	3.0×10^4
Silicon	14	1.0×10^4
Magnesium	12	7.9×10^3
Sulfur	16	6.3×10^3
Iron	26	1.2×10^3
Sodium	11	6.3×10^2

Data from B. Mason, *Principles of Geochemistry,* 3rd ed. John Wiley & Sons, Inc., New York, 1966.

[a] Number of atoms per 10,000 atoms of silicon.

At the center of our sun, $T = 1.5 \times 10^7$ K and the density is 1.5×10^2 g/cm³.

Most solar energy today comes from the fusion of hydrogen and helium nuclei in the sun.

According to the most widely accepted theory, the formation of all stars, suns, planets, and other cosmic bodies can be traced to one original body. To explain the present rate of expansion of the universe, its estimated mass, and the kind of background radiation currently observed, this initial body must have had a density of about 10^{96} g/cm³ and a temperature of about 10^{32} K. It exploded — hence, the theory is dubbed the "big bang" theory — and in just one second its temperature fell to about 10^{10} K, and the universe was on its expanding, evolving way. It consisted of about equal numbers of protons and neutrons as well as electrons, and for about 8 minutes it operated like a huge nuclear fusion system. (We will describe fusion in Chapter 23.) By a number of complex nuclear transformations, all the elements eventually formed from these elementary particles, and in various kinds of stars, these processes continue to occur today.

So much hydrogen and helium remain that hydrogen atoms or hydrogen nuclei constitute 88.6% of all remaining particles in the universe. Helium atoms or nuclei make up another 11.3%, so these two light nonmetals account for 99.9% of all of the remaining atoms in the universe and almost 99% of all of the mass. The rest of the mass is distributed among the other elements, and the common nonmetals — particularly neon, oxygen, nitrogen, carbon, and sulfur — predominate. (See again Tables 19.1 and 19.2.) The metalloid silicon and two metals — magnesium and iron — are also significant.

The Earth's Elements

Oxygen atoms are the most common kind of atom in the earth (Table 19.3), but so much iron occurs in the earth's core that it dominates on a mass basis. Oxygen also dominates

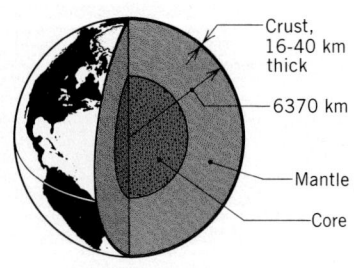

Crust, 16-40 km thick

6370 km

Mantle

Core

The earth's core, mantle and crust

TABLE 19.3 Elements in the Earth — The Ten Most Abundant

Element	Atomic Number	Atom Percent	Weight Percent
Oxygen	8	50	30
Iron	26	17	35
Silicon	14	14	15
Magnesium	12	14	13
Sulfur	16	1.6	1.9
Nickel	28	1.1	2.4
Aluminum	13	1.1	1.1
Calcium	20	0.7	1.1
Sodium	11	0.7	0.6
Chromium	24	0.01	0.3

Data on weight percents from B. Mason, *Principles of Geochemistry,* 3rd ed. John Wiley & Sons, Inc., New York, 1966.

TABLE 19.4 Elements in the Earth's Crust—The Ten Most Abundant

Element	Atomic Number	Atom Percent	Weight Percent	Volume Percent
Oxygen	8	61	47	93.8
Silicon	14	21	28	0.86
Aluminum	13	6.2	8.1	0.47
Hydrogen	1	3.0	0.14	—
Sodium	11	2.6	2.8	1.3
Iron	26	1.9	5.0	0.43
Magnesium	12	1.8	2.1	0.29
Calcium	20	1.6	3.6	1.0
Potassium	19	1.4	2.6	1.8
Titanium	22	0.2	0.44	—

Data from B. Mason, *Principles of Geochemistry*, 3rd ed. John Wiley & Sons, Inc., New York, 1966.

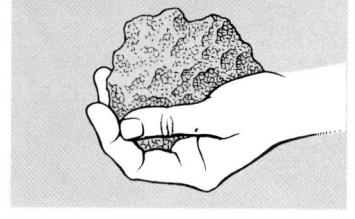

About 90% of the volume of silicate rock is occupied by oxygen atoms.

the composition of the earth's crust (Table 19.4) whether we calculate percentages by atoms, by weight, or by volume.

The earth's atmosphere is 78% nitrogen and 21% oxygen (by volume) with carbon dioxide, argon, and other noble gases supplying nearly all of the rest (Table 19.5). For as long as measurements have been made, the percentage of oxygen has been constant. Although several processes consume oxygen—respiration, decay, combustion, and other oxidations—it is replenished just as rapidly by photosynthesis. As we learned in Chapter 5, this is the process by which plants trap solar energy in molecules of chloro-

Radon isn't in Table 19.5, but wherever rocks contain uranium you'll have radon in the soil and surrounding air. (See Special Topic 23.4.)

TABLE 19.5 Composition of Clean, Dry Air at Sea Level

Component	Concentration
Major components	
Nitrogen, N_2	78.084% (vol/vol)
Oxygen, O_2	20.948% (vol/vol)
Minor components	
Oxides of carbon	
Carbon dioxide, CO_2	320 ppm[a]
Carbon monoxide, CO	0.1 ppm
Oxides of nitrogen	
Dinitrogen oxide, N_2O	0.5 ppm
Nitrogen dioxide, NO_2	0.02 ppm
Oxide of sulfur	
Sulfur dioxide, SO_2	1 ppm
Noble gases	
Helium, He	5.24 ppm
Neon, Ne	18.18 ppm
Argon, Ar	9340 ppm (0.934%, vol/vol)
Krypton, Kr	1.14 ppm
Xenon, Xe	0.087 ppm
Miscellaneous	
Ammonia, NH_3	0 to traces
Hydrogen, H_2	0.05 ppm
Methane, CH_4	2 ppm
Ozone, O_3	0.02–0.07 ppm (seasonal variations)

Data from J. R. Holum, *Topics and Terms in Environmental Problems*, Wiley–Interscience, New York, 1977.

[a] This unit, ppm, means *parts per million*. For example, 320 ppm means 320 liters in 1 million liters of total volume.

TABLE 19.6 Elements in the Human Body

More Than 1% by Atoms or by Weight			0.1–1% by Weight		Less than 0.1% by Weight	
Element	Atom Percent	Weight Percent	Element	Weight Percent	Element	Weight Percent
Hydrogen	61	9.3	Sulfur	0.64	Magnesium	0.04
Oxygen	25.7	62.8	Phosphorus	0.63	Iron	0.005
Carbon	10.6	19.4	Sodium	0.26	Zinc	0.0025
Nitrogen	2.4	5.1	Potassium	0.22	Copper	0.0004
Calcium	0.2	1.4	Chlorine	0.18	Tin	0.0002
					Manganese	0.0001
					Iodine	0.0001
					Molybdenum	0.00002
					Cobalt	0.000004
					Vanadium	0.000003

Percents by weight are from B. Mason, *Principles of Geochemistry,* 3rd ed. John Wiley & Sons, Inc., New York, 1966. Mason's data include other very trace elements that are very likely impurities and not at all essential. His list omits fluorine, selenium, and chromium, which seem to be important to good health but in exceedingly trace quantities.

phyll, a green pigment, and use it to make complex compounds such as carbohydrates from carbon dioxide and water. Oxygen is the other chief product.

In the human body (Table 19.6), four nonmetals — hydrogen, carbon, oxygen, and nitrogen — contribute over 96% of the mass. And three other nonmetals — sulfur, phosphorus, and chlorine — are also important. You can see from all these data that whether we journey to the outer reaches of space or plunge to the molecular level of life, a handful of nonmetallic elements of low atomic weight make up most of the universe. In the rest of this chapter and the next we will study the most familiar chemical and physical properties of these elements.

19.2 THE NOBLE GAS ELEMENTS

The noble gases — helium, neon, argon, krypton, xenon, and radon — are the least reactive of all the elements.

The noble gases — the elements listed in Table 19.7 — are trace components of air, as we saw in Table 19.5, so neon, argon, krypton, and xenon are obtained by the careful, low-temperature distillation of liquid air. Helium, however, is obtained more economically from natural gas. It occurs with natural gas because today's *terrestrial* helium originated from the decay of radioactive elements in the earth's crust, not from the big bang's fusion furnace. In some geological formations helium is mixed with natural gas at

The helium present when the earth first formed could not be held by the earth's gravitational field and was lost into space.

TABLE 19.7 The Noble Gas Elements

Element	Symbol	Atomic Number	Boiling Point (°C)	Melting Point (°C)	Color Emitted from a Discharge Tube
Helium	He	2	−268.934	−272.2[a]	White to pink-violet
Neon	Ne	10	−246.048	−248.67	Red-orange
Argon	Ar	18	−185.7	−189.2	Violet-purple
Krypton	Kr	36	−152.30	−156.6	Pale violet
Xenon	Xe	54	−107.1	−111.9	Blue to blue-green
Radon	Rn	86	− 61.8	− 71	—

[a] Under a pressure of 26 atm.

a concentration as high as 7% (v/v). Radon is obtained from the gaseous mixture above aged radium chloride solutions, where it appears as a product of radioactive decay.

None of the noble gases had been discovered (on earth) when Mendeleev refined his periodic table in 1869, so he did not make provision for them in the table. (Evidence for helium in the sun — helium is named after *helios*, "the sun" — had appeared in late 1868, but only from an otherwise unexplained spectral line in the sun's spectrum.) The discovery of argon (1894), which then urgently needed a place in the periodic table, soon led to the realization that perhaps an entirely new family of elements existed. All eventually were found, and their family name evolved from *rare gases,* which they are not, to *inert gases,* which three are not, to the present name, *noble gases.* Their lack of vigorous activity seemed to parallel the lives of the nobility.

Argon is from a Greek word meaning "lazy" or "idle."

Chemical Properties of the Noble Gases

The central fact about the noble gases is their lack of chemical reactivity. As seen in the margin table, they have high ionization energies suggesting a very low potential for entering into any chemical changes that would transfer electrons away or even share them in bonds. Not until we get to xenon do we have a noble gas with a lower first ionization energy than oxygen. In 1933, Linus Pauling predicted that KrF_6 and XeF_6 should be stable enough to exist, but it was not until 1962 that the first noble gas compound was prepared. It was a compound of xenon, fluorine, and platinum, which formed as follows.

Noble Gas	First IE (kJ mol^{-1})
He	2372
Ne	2080
Ar	1520
Kr	1351
Xe	1169
Rn	1037
Oxygen (O)	1314

$$Xe + 2PtF_6 \xrightarrow{25\ °C} [XeF^+][PtF_6^-] + PtF_5$$

When the mixture of products is warmed to 60 °C, it changes to $[XeF^+][Pt_2F_{11}^-]$.

Thus far, only krypton, xenon, and radon have been found to have chemical reactions, and the only observed stable bonds are to fluorine, oxygen, chlorine, and nitrogen. Krypton evidently stands at a threshold as we descend the noble gases in the periodic table; only one molecular compound, KrF_2, has been found. A few systems with the KrF^+ and $Kr_2F_3^+$ ions are also known.

Xenon, with its larger radius and lower ionization energy, is much more reactive than krypton, and compounds with xenon in oxidation states from $+2$ to $+8$ are known. Most involve fluorine or oxygen or both. In addition, metal ions are present in a few xenon compounds.

Three binary fluorides of xenon are known — XeF_2, XeF_4, and XeF_6. They are prepared by heating xenon and fluorine under pressure in special containers. The exact conditions and starting concentrations determine which fluoride forms. All three binary fluorides are low-melting and volatile, suggesting that they are molecular compounds with covalent bonds rather than ionic compounds. XeF_2 is a linear molecule; XeF_4 has a square planar geometry; and XeF_6 has a distorted, nonrigid, octahedral structure. (We discussed these geometries and the reasons for them in Section 7.9, page 245.) All three fluorides of xenon react with water — XeF_4 and XeF_6 violently so. One product is the oxide XeO_3, which is itself dangerously explosive.

All three binary fluorides are colorless.

The radioactivity of all isotopes of radon has discouraged studies of radon chemistry, but a fluoride is known.

Uses of the Noble Gases

The relatively higher abundances of helium and argon make these the most used of the noble gases, and virtually all of their uses relate to their lack of chemical reactivity. Thus argon, not air, is used as the low-pressure gas inside a light bulb. Either the oxygen or the nitrogen of air would soon react with the hot metal filament and destroy it, but argon is chemically inert to the glowing metal. Fluorescent bulbs also contain argon instead of air.

"Spectacular!" "Impressive is an understatement." "This kind of thing you see once in a lifetime." "For someone of my generation, there is only one comparison for what happened last night; the Woodstock of physics!" Such were some of the emotional outbursts among the over 3000 physicists who attended an early 1987 meeting of the American Physical Society. But why all the excitement? And why should this also interest chemists? Chemistry, of course, is a study of materials, and not long after the physicists met, thousands of chemists gathered at a national meeting of the American Chemical Society and learned about the materials behind the physicists' excitement. They were materials that made possible the resistance-free transmission of electricity at temperatures higher than ever before thought possible. They were *superconductors*.

The Superconducting State Ordinary metals at ordinary temperatures, such as copper at room temperature, have a small but nevertheless significant resistance to the flow of electricity. Such resistance means that a percentage of the electrical energy is converted to heat as the electricity flows in the wires. Although this is okay in a toaster, it is definitely not desirable in the long distance transmission of electricity from power stations to homes and industry. In the latter case it is a waste of energy and energy resources, which can be compensated for only by the use of extremely high (and dangerous) voltages in the overhead power cables. Electrical resistance is also a factor that has limited the capacity of computer chips. Although the individual components on these chips use little power, when thousands of them are crowded together, much heat is generated and the computer chips must be cooled to keep them from burning out. Such potential burnout stands in the way of designing even more densely packed circuits that would be necessary to make computers smaller. With these limitations in mind, it is not at all surprising that scientists and engineers have for years looked wistfully at the phenomenon of superconductivity.

The superconducting material in this photo is the black, saucer-shaped ceramic disk resting in a pool of liquid nitrogen. The cobalt-samarium magnet is the slivery object above it. We can tell that it is levitated above the superconductor because it casts a distinct shadow that nowhere touches the magnet. (Courtesy of AT & T Bell Laboratories, Murray Hill, NJ.)

When a material is in the state in which it offers no resistances to electricity it is called a superconductor. Prior to 1987, only a few materials, nearly all of them metals and metal alloys, could be put into such a state and then only if their temperatures were lowered to just a few degrees above absolute zero. This meant the immersion of the material in a bath of liquid helium, which boils at 4.2 K. To keep the material in a superconducting state, in other words, required refrigeration technology that was too expensive and too cumbersome to permit any important practical applications beyond the bounds of special devices for scientific research.

To make superconductivity practical, materials had to be discovered that could pass over to the superconducting state at temperatures much higher than the

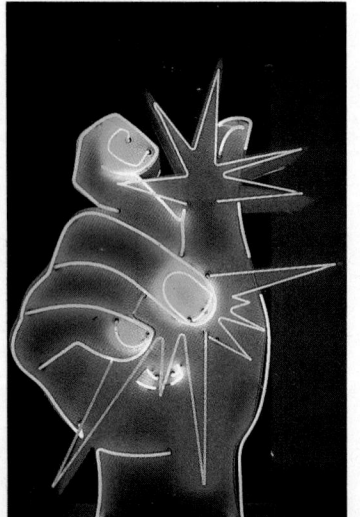

FIGURE 19.1 Different mixtures of noble gases at low pressure in gas discharge tubes make possible the variety of colors of "neon" signs.

The familiar "neon" lights (Figure 19.1) contain mixtures of noble gases whose percentages determine the color seen. Table 19.7 shows the color emitted by each noble gas when used alone in a discharge tube.

When scientists must ensure that no oxygen from the air can reach a mixture of chemicals, they can displace the air with helium or argon. (Nitrogen, quite unreactive itself, can usually be used for this purpose as well.)

Helium liquefies at 4.18 K ($-268.9\ °C$), and liquid helium is used as a coolant where experiments call for very low temperatures. Helium is the only substance known that cannot be changed to a solid by cooling it at atmospheric pressure. It takes a pressure of nearly 26 atm to make helium solidify even at absolute zero. It is quite remarkable that boiling point of liquid helium. Just such a discovery — materials that are superconducting at about 90 K, which is attainable with baths of liquid nitrogen (b. p. 77.4 K) — swept physicists at the 1987 meeting to those emotional heights usually felt only by first night audiences at a major new Broadway hit. The refrigeration technology for producing and handling liquid nitrogen is well known and relatively simple. And liquid nitrogen costs about as much as milk or beer.

One scientist called these developments the most significant happening in science in the latter half of the twentieth century. Many engineering problems have yet to be solved, but the superconducting materials are easy to make and the time between discovery and important applications could well be one or two decades.

Near-Insulators as Superconductors

The new superconducting materials were not metals, but mixed metal oxides doped with rare earth elements and baked as ceramics. One material had a superconducting phase that consisted of a mixed oxide of barium, copper, and yttrium with the formula $Ba_2Cu_3YO_7$. The yttrium can be replaced by samarium, gadolinium, dysprosium, ytterbium, and other rare earths. Conventional wisdom would have regarded such substances as far more likely to be insulators than superconductors.

There appears to be no theoretical upper limit to the temperature at which some substance, yet to be discovered (1987), can be made superconducting. Unverified reports circulated in 1987 of materials made superconducting at temperatures much higher than 90 K. Such possible developments are important because the higher the temperature at which a substance can be made superconducting, the more widespread and the less expensive will be the practical applications. The ultimate goal is to develop a room-temperature superconductor, and teams of scientists around the world are working on it.

Levitation

Besides zero electrical resistance, a substance in its superconducting state permits no magnetic field within itself. A weak magnetic field is actually repelled by a superconductor, and this makes possible the levitating (the floating on air) of magnetized objects. As seen in the figure, a tiny cobalt-samarium magnet ($SmCo_5$) can float above a ceramic bowl fabricated of material made superconducting when cooled by liquid nitrogen. (Too strong a magnetic field can destroy the superconducting state.)

The repulsion that a superconductor has for a magnetized object could make it possible to levitate entire trains and send them at high speed with no more frictional resistance than offered to planes by the surrounding air.

Paired Conduction Electrons and the Resonating Valence Bond State

No universal agreement exists over theories about the mechanism of superconductivity, although the phenomenon has been known since 1911, when it was discovered by Heike Kamerlingh-Onnes (The Netherlands; Nobel prize in physics, 1913). The standard theory in existence prior to 1987 was that of J. Bardeen, L. N. Cooper, and J. R. Schrieffer, U.S. physicists (all sharing the 1972 Nobel prize in physics). This theory had led to the prediction that the maximum temperature at which a superconducting state could exist was 30 K. The discovery of materials that are superconducting well above this limit has forced theorists to ask, "Is there a new mechanism that works for the high-temperature superconductors?"

One of the suggested new mechanisms that appeared in 1987 involves the valence electrons of the material. Electricity is conducted by electrons, and the conduction electrons are not the same as the valence electrons, those that participate in the chemical bonds. The conduction electrons must form pairs to participate in superconductivity, but agitations caused by heat as well as electrostatic repulsions between electrons works against such pairing. According to this theory (P. W. Anderson), the new materials become superconducting when their temperature allows them to enter a *resonating valence-bond* state. In this state, the valence electrons shift back and forth from atom to atom thereby causing distortions that allow conduction electrons to become paired.

there is no combination of temperature and pressure at which dynamic equilibrium exists between all three phases — solid, liquid, and gas — so helium has no triple point (see page 378).

When liquid helium is cooled to 2.17 K, it undergoes startling changes in physical properties. It loses all resistance to flow — its viscosity is lost — and it is now called a *superfluid*. In this state it is able to climb the walls of its container.

The degree of coldness made possible by liquid helium causes an interesting change in the ability of some metals and metal alloys to conduct electricity. When such materials are placed in liquid helium, they entirely lose their electrical resistance and become *superconductors*. The scientific world was startled in early 1987 to learn that ceramic substances could be put into superconducting states at a much higher temperature, at the temperature of liquid nitrogen (bp 77.4 K). Special Topic 19.1 describes some of the implications of this development and why it is seen as a scientific breakthrough of immense significance.

19.3 HYDROGEN

Hydrogen occurs in more compounds than any other element.

Hydrogen is a colorless, odorless, tasteless gas consisting under normal conditions of diatomic molecules, H_2. Bulk quantities are shipped either in its liquid form or through pipelines as a gas. Several factors make its shipment expensive—its flammability, its ability to diffuse through pinholes too small to admit larger molecules such as those in natural gas (chiefly methane or CH_4), and its tendency to make some metals more brittle. Most industrial uses of hydrogen, therefore, occur in *integrated operations,* those that make and use key raw materials on site and thus avoid shipping problems.

Most hydrogen is made industrially by stripping it from molecules of natural gas or oil refinery hydrocarbons. Using propane, C_3H_8, as an example, a mixture of the hydrocarbon and high-temperature steam is passed over a catalyst at about 900 °C where the following reaction occurs irreversibly.

Hydrocarbons are compounds of carbon and hydrogen that occur in natural gas and oil. Propane is just one example.

$$C_3H_8 + 3H_2O \xrightarrow[\text{catalyst}]{900\ °C} 3CO + 7H_2$$

The carbon monoxide and hydrogen are then separated from each other by a combination of chemical and physical methods, which we will not study.

Hydrogen can be made in several ways where only small amounts are needed. Perhaps the oldest method is the reaction of a metal such as zinc with hydrochloric acid (Figure 19.2). It is an oxidation–reduction reaction in which zinc is oxidized and the hydrogen ion is reduced.

$$Zn(s) + 2HCl(aq) \longrightarrow H_2(g) + ZnCl_2(aq)$$

The net ionic equation is

$$Zn(s) + 2H^+(aq) \longrightarrow H_2(g) + Zn^{2+}(aq)$$

Henry Cavendish (1731–1810), an English chemist credited with first recognizing hydrogen as a distinct substance, made it this way.

Calcium metal, which is more easily oxidized than zinc and therefore a stronger reducing agent, can reduce water, and this is another way to make a small quantity of hydrogen in the lab.

$$Ca(s) + 2H_2O \longrightarrow Ca(OH)_2(s) + H_2(g)$$

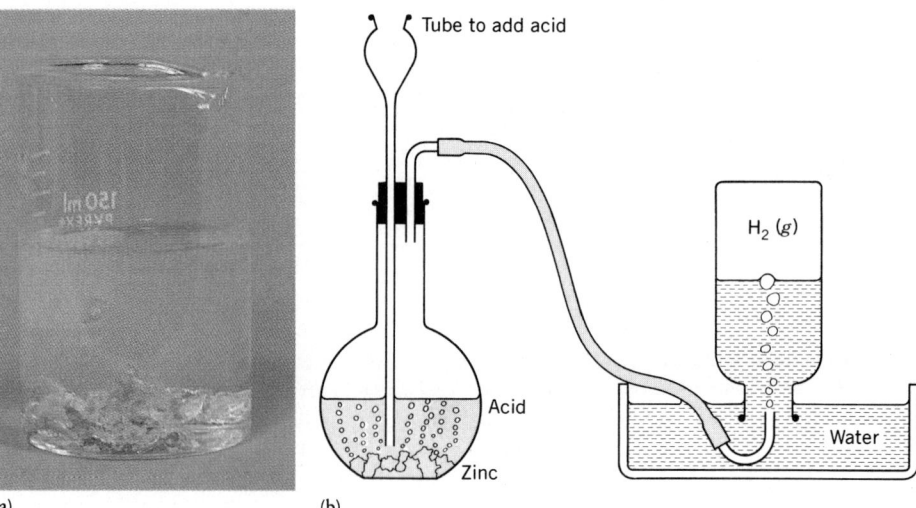

FIGURE 19.2 (a) Hydrogen gas bubbles from a beaker in which zinc metal reacts with hydrochloric acid. (b) In an apparatus such as this, hydrogen gas can be collected in the lab by the reaction of zinc with hydrochloric acid.

(a)   (b)

In labs where dry hydrogen is needed frequently, it is simply obtained from commercially available cylinders of compressed hydrogen.

Isotopes of Hydrogen

Hydrogen is the only element whose isotopes have their own names. The most abundant is commonly called *hydrogen*, but *protium* is used in special situations. Some properties of these isotopes are given in Table 19.8.

For all practical purposes, only protium and deuterium occur naturally, with protium constituting 99.98% of elemental hydrogen. Prior to the development of nuclear weapons and their testing, the tritium content of the atmosphere was estimated to be about 1 tritium atom out of every 10^{17} to 10^{18} hydrogen atoms. This tritium originates from nuclear transformations caused by cosmic rays. Tritium also forms from nuclear fission (Section 23.9), so any leaks of waste gases from a nuclear reactor or a nuclear test put more tritium into the atmosphere. It eventually disappears, because tritium decays radioactively with a half-life of just 12.35 years.

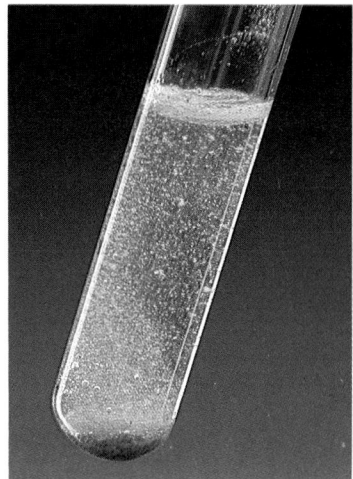

Bubbles of H_2 form when CaH_2 and water react.

TABLE 19.8 Isotopes of Hydrogen

Name	Atomic Symbol	Mass Number	Melting Point (K)	Boiling Point (K)
Protium	H	1	13.957	20.39
Deuterium	D	2	18.73	23.67
Tritium[a]	T	3	20.62	25.04

[a] Radioactive; half-life, 12.35 years; total atmospheric content = about 11 g.

The Deuterium Isotope Effect   Although an element's isotopes generally give identical *kinds* of chemical reactions, the *rates* of these reactions can differ. Both protium and deuterium, for example, react with chlorine; H_2 reacts to give HCl and D_2 reacts to give DCl. At 0 °C, however, the reaction of H_2 with chlorine is 13.4 times as rapid as that of D_2.

A difference in rates shown by different isotopes in the same kind of reaction is called an *isotope effect,* and deuterium's generally slower reactivity than protium is called the deuterium isotope effect. It is the most pronounced isotope effect of all that have been studied. Bonds to D are generally stronger than bonds to H in the sense that the energy of activation of a rate-limiting elementary process is greater for breaking a bond to D than to H. And as we learned in the previous chapter, a larger energy of activation means a slower rate. (We leave to other courses theories about *why* bonds to D are stronger.)

An interesting consequence of the detuerium isotope effect is that if some of the hydrogen atoms in the molecules of the foods you eat, which have numerous H—C bonds, were replaced with deuterium atoms, your rate of body chemistry would become slower, possibly fatally so. If you or your foods and drink contained D_2O, which is called *heavy water,* instead of H_2O, you would also die.

Chemical Properties of Hydrogen

We will continue our study of hydrogen as if it consisted of just the common isotope.

Hydrogen is the only reactive element with no electron below its valence shell, and this has interesting consequences. The $1s^1$ electron configuration of the hydrogen atom uniquely enables it, by the loss of one electron, to change to a *subatomic particle,* the bare proton. This species has an extremely small radius, just 1.5×10^{-3} pm, which we can compare with the radii of alkali metal ions — 60 to 170 pm. Such a small radius gives the bare proton a much more highly localized positive charge than in any other cation. With such a tightly confined positive charge, the bare proton strongly distorts any nearby electron cloud and forces it to set up a chemical bond to the proton. In liquid or solid systems, therefore, H^+ can exist only by being covalently bound to something else, either an ion (as in H_3O^+) or a molecule (as in H_2SO_4). The chemistry of these and other proton-bearing species has been discussed under acids and bases. (In the gaseous state, bare protons are known but only in connection with nuclear phenomena or in special apparatus used in research in certain areas of physics.)

The hydrogen bond is, in part, another consequence of the lack of inner level electrons in H. We first discussed the hydrogen bond in Section 10.3. There we learned that when H is held covalently by strongly electronegative atoms such as F, O, and N, the hydrogen end of the bond has enough $\delta+$ character to set up a hydrogen bond, a force of attraction to the $\delta-$ charge on another F, O, or N.

Forces of attraction, however, do not depend only on the sizes of $\delta+$ and $\delta-$. They also depend on the distance between the charges. Such forces become weaker in proportion to the *square* of the distance between the opposite charges. If the distance were doubled, for example, the attractive force would be reduced by a factor of 2^2 or 4. This distance between $\delta+$ and $\delta-$ can be particularly short in the hydrogen bond because the H atom has no inner level electrons. If it had them, they would require their own room and thus make the distance from $\delta+$ to $\delta-$ greater.

The $1s^1$ configuration of the hydrogen atom also means that hydrogen can obtain a noble gas configuration (helium's $1s^2$) by gaining one electron to become the hydride ion, H^-. The change to H^- resembles the formation of a halide ion from a halogen atom, and the change to H^+ resembles the formation of an alkali metal ion. However, hydrogen is chemically so unlike either the halogens or the alkali metals that many who draw up periodic tables set hydrogen outside any family.

Hydrides

We will continue here with some of the chemistry of hydrides, the binary compounds of hydrogen with other elements. Not all binary hydrides involve the hydride *ion*. They vary in type across the periodic table from the ionic hydrides of the alkali metals, which do involve H^- ions, to the covalent hydrides of the halogens, which can donate H^+ ions.

Ionic Hydrides   The ionic hydrides conduct electricity at or just below their melting points, which indicates the presence of ions, and they give up hydrogen at the *anode* when they are electrolyzed. This indicates that H^- ions are oxidized: $2H^- \rightarrow H_2 + 2e^-$.

H^- can exist at room temperature in the crystals of ionic, saltlike compounds only when the positive ion is from one of the least electronegative elements and thus least able to take an electron from H^-. These are the metals with the lowest ionization energies — generally those in Group IA (the alkali metals) and in Group IIA (the alkaline earth metals). (Beryllium and magnesium in Group IIA have hydrides that are more covalent than ionic.)

The ionic hydrides are made by a direct combination of hydrogen with the metal at high temperatures, for example,

$$Ca(s) + H_2(g) \xrightarrow{\text{300-400 °C}} CaH_2(s)$$

The hydride ion, H^-, in the ionic hydrides is an exceptionally powerful Brønsted base and it is a very strong reducing agent. Thus the ionic hydrides are very sensitive both to oxygen and to water, so they must be stored in environments that rigorously exclude both. Two of them — RbH and CsH — even ignite spontaneously in dry air.

All ionic hydrides react with water as follows, where M represents a metal in either Group IA or IIA.

hydrogen bond (· · ·) in water

Ionic hydrides are electrolyzed from their solutions in molten alkali halides.

Group IA ionic hydrides

$$MH + H_2O \longrightarrow MOH + H_2$$

Group IIA ionic hydrides

$$MH_2 + 2H_2O \longrightarrow M(OH)_2 + 2H_2$$

The reactivities of the ionic hydrides toward water vary widely. The reaction of sodium hydride with water, for example, is even more violent than that of sodium metal (Section 12.6). Calcium hydride, however, reacts so moderately with water that the reaction is used as a small-scale source of hydrogen in the lab.

Some ionic hydrides are used to make other reducing agents. Thus LiH is used to make $LiAlH_4$, lithium aluminum hydride, an important reducing agent in organic chemistry. And NaH is used to make $NaBH_4$, sodium borohydride, another useful but less reactive reducing agent.

Covalent Hydrides   On the right side of the periodic table lie the elements that form the covalent hydrides. Even the *metallic* elements in Groups IIIA, IVA, VA, and VIA form covalent hydrides.

In a covalent hydride an electron pair bond holds H to an adjoining atom. The bond, however, varies widely in polarity from compound to compound because the possible adjoining atoms differ greatly in their relative electronegativities. (The table in the margin gives electronegativities for elements on the right side of the periodic table.) The electronegativities vary enough to put a $\delta+$ on H in some covalent hydrides, a nearly zero charge in others and a $\delta-$ charge in still others.

When there is a $\delta+$ on H, we have acids or H^+ donors like the hydrides of the halogens (HF, HCl, HBr, HI) for example. The Group VIA hydrides (H_2O, H_2S, H_2Se, H_2Te) also tend to be H^+ donors. Their *strengths* as acids vary widely, of course, as we studied in Section 11.8 (where we also learned that the bond polarity is not the only factor affecting the acidities of covalent hydrides). Although water, for example, is an extremely weak acid, when it does give up H it usually gives it up as H^+, not H^-.

When the partial charge on H is zero or nearly so, as in CH_4 (methane), we have a compound not generally considered to be either an acid or a base. Methane, in fact, tends to react as a donor of a hydrogen *atom*. (Such reactions go by free radical chain mechanisms of the type introduced in Section 18.10. We will not study these; they are described in detail in other courses in chemistry.)

In covalent hydrides with a $\delta-$ charge on H, we have compounds that tend to be donors of H^-. Covalent hydrides of some of the representative metallic elements as well as some metalloids are examples, such as the hydrides of boron, aluminum, arsenic, and tin. Although they are not ionic, these hydrides can still supply H^- in chemical reactions. This makes them reducing agents. Some are so reactive that they burst into flame in air. Thus diborane, the simplest hydride of boron, B_2H_6, is a gas that must be handled out of contact with air. It reacts spontaneously with oxygen as follows.

$$B_2H_6(g) + 3O_2(g) \longrightarrow B_2O_3(s) + 3H_2O$$
$$\text{diborane} \qquad\qquad \text{boron oxide}$$

Transition Metal Hydrides   We have seen here a full transition in moving from the ionic hydrides of the representative elements on the far left in the periodic table to the covalent hydrides of the reactive elements on the far right. We go from donors of H^- to donors of H^+. And this change parallels the change in electronegativities across the table. In the center of the chart among the transition metal elements, however, we have a mixed situation.

The transition metal hydrides have an unusually complex chemistry. Some have definite formulas like those of nickel (NiH_2), iron (FeH_2), and uranium (UH_3), but

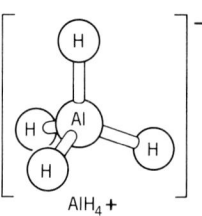

AlH_4^+

Covalent hydrides include some of our most vital compounds, such as water, ammonia, and methane.

Electronegativities (H = 2.1)				
IIIA	IVA	VA	VIA	VIIA
B 2.0	C 2.5	N 3.1	O 3.5	F 4.1
Al 1.5	Si 1.8	P 2.1	S 2.4	Cl 2.9
Ga 1.8	Ge 2.0	As 2.2	Se 2.5	Br 2.8
In 1.5	Sn 1.7	Sb 1.8	Te 2.0	I 2.2

For explaining the unusual structures of the hydrides of boron, William Lipscomb received the 1976 Nobel prize in chemistry.

others do not. Some appear to be more like solutions of hydrogen in the metal, where hydrogen is lodged within interstices (spaces) between the atoms, and the material retains considerable electrical conductivity. They often behave as donors of gaseous H_2, rather than H^+ or H^-.

As an interim summary, we note that hydrides can serve as donors of hydrogen in any form — as H^-, $H\cdot$, H^+, and H_2 — depending on the atom that holds H.

H^- donors   Ionic hydrides and covalent hydrides with $\delta-$ on H (generally metal hydrides)
$H\cdot$ donor CH_4 and similar compounds of carbon and hydrogen
H^+ donors   Covalent hydrides with $\delta+$ on H (generally nonmetal hydrides)
H_2 donors   Some transition metal hydrides

Uses of Hydrogen

In terms of *moles* manufactured, no compound surpasses ammonia.

Most hydrogen goes directly as it is made to the synthesis of ammonia by the Haber–Bosch process. Ammonia, either as itself or in such derivatives as urea, ammonium nitrate, or ammonium sulfate, is the world's most important nitrogen fertilizer. It ranks in the top five of all chemicals made in the United States in terms of total mass — 15 to 18 million tons (800 to 1000 billion moles) in the mid-1980s.

The Haber–Bosch process is the direct combination of hydrogen and nitrogen under heat and pressure in the presence of an iron-based catalyst according to the following overall equation.

$$N_2(g) + 3H_2(g) \xrightarrow[\text{heat, catalyst}]{\text{pressure}} 2NH_3(g)$$

Since 4 volumes of reactant gases are changed into 2 volumes of the product gas, the increased pressure assists the reaction, but the catalyst is crucial. We will discuss this reaction in more detail in Section 19.5. (We explained the sources of N_2 and H_2 for the Haber-Bosch process in Chemicals in Use 11, page 692.)

The makers of oleomargarine and other hydrogenated vegetable oils (e.g., Crisco and Spry) let hydrogen combine with carbon–carbon double bonds and change them to single bonds. Using just parts of molecules to illustrate this reaction, we can write it as follows.

$$-CH{=}CH- + H_2 \xrightarrow[\text{heat, catalyst}]{\text{pressure}} \underset{\underset{H}{|}}{-CH}\underset{\underset{H}{|}}{-CH}- \text{ (or } -CH_2-CH_2- \text{)}$$

double bond
in an edible
oil molecule

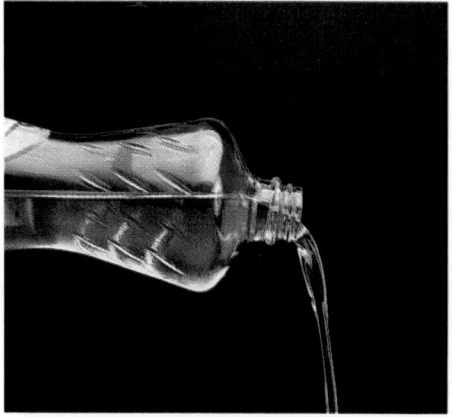

The partial hydrogenation of a liquid vegetable oil gives a solid product similar to an animal fat.

Edible vegetable oils, such as the corn oil in many salad dressings, have molecules with more carbon–carbon double bonds per molecule than in edible solid fats, such as butterfat. These oils and fats are otherwise structurally and chemically alike, but their melting points vary with the number of carbon–carbon double bonds per molecule. By the partial hydrogenation of an abundant (so much-less expensive) vegetable oil, the product loses some of its double bonds and becomes a higher-melting and commercially more valuable material barely distinguishable from an animal fat.

Makers of synthetic detergents, plastics, and other chemicals use hydrogen together with carbon monoxide to convert smaller carbon molecules into larger ones. A simple example of this reaction — called the OXO reaction — is as follows.

$$CH_2{=}CH_2 + H_2 + CO \xrightarrow[\text{heat, catalyst}]{\text{pressure}} CH_3{-}CH_2{-}CH{=}O$$

Another important use of hydrogen is as a rocket fuel. For this purpose it is loaded into fuel tanks as a liquid, and then allowed to mix and react with oxygen in the propulsion chamber.

19.4 OXYGEN

Stable oxides occur for nearly all of the elements, but their properties range from strongly basic to strongly acidic.

The members of the oxygen family are given in Table 19.9. They form hydrides and oxides with similar formulas, as we would expect. Our study will focus on the first two members, oxygen in this chapter and sulfur in the next.

Elemental Oxygen

Oxygen is a colorless, odorless, tasteless gas that in its most stable state under ordinary conditions exists almost entirely as diatomic molecules, O_2. Oxygen has three isotopes,

Liquid hydrogen is used as a rocket fuel.

Isotope	% Abundance
O-16	99.759
O-17	0.0374
O-18	0.2039

TABLE 19.9 The Oxygen Family — Group VIA

Element	Atomic Symbol	Melting Point (°C)	Boiling Point (°C)	Appearance (at Room Temperature and Atmospheric Pressure)
Oxygen	O	−218	−183	Colorless gas
Sulfur	S	113[a]	445	Yellow, brittle solid
Selenium	Se	217	685	Bluish-gray metal
Tellurium	Te	452	1390	Silvery-white metal
Polonium	Po	254	962	An intensely radioactive metal

Some Compounds			
Hydrides		Oxides	
Formula	Boiling Point (°C)	Formula	Boiling Point (°C)
H_2O	100	O_3(ozone)[b]	−112
		SO_2	−10
H_2S	−61	SO_3	−45
		SeO_2	(sublimes) mp > 300
H_2Se	−42	SeO_3	mp 118
H_2Te	−2	TeO_2	Decomposes
—	—	TeO_3	Decomposes
		PoO_2	Decomposes

[a] For orthorhombic sulfur when heated rapidly.

[b] This is an allotrope of oxygen, not a compound, but it can also be thought of as an oxide of oxygen.

with oxygen-16 making up nearly 99.8% of the total. Both liquid and solid oxygen are light blue in color, and both are paramagnetic (as we discussed in Section 8.7).

Oxygen makes up 20.95% (v/v) of dry air and has been at this concentration for an estimated 50 million years. Without oxygen, life is impossible; we could survive only a few minutes. It was first prepared in 1773 by Carl Wilhelm Scheele (1742–1786), a Swedish scientist, and soon thereafter in 1774 by Joseph Priestly (1733–1804), an English scientist. The first to recognize it as a new element, however, was apparently Antoine Laurent Lavoisier (1743–1794), the great French scientist. Priestly made oxygen by focusing sunlight through a "burning lens" (a magnifying glass) onto a mound of mercury(II) oxide, which decomposed as follows.

$$2HgO(s) \xrightarrow{\text{heat}} 2Hg(\ell) + O_2(g)$$

To his great surprise, a candle burned with a far more brilliant flame in the gas that formed than it could in air. When steel wool is heated in a bunsen burner flame, it merely sputters and glows in air and stops doing so when the flame is removed. But when steel wool heated this way is thrust into pure oxygen, it bursts into flame (Figure 19.3). Almost needless to say, anyone who works around pure oxygen or even oxygen enriched air must very carefully exclude all sparks and flames. One scientist has estimated that if the air we breathe were 30% (v/v) oxygen instead of 21%, no forest fire could be put out. (And this is another way of dramatizing the effect of concentration on the rate of a reaction!)

Small amounts of oxygen can be made by heating potassium chlorate in the presence of manganese dioxide (Figure 19.4).

$$2KClO_3(s) \xrightarrow[150-200\ °C]{MnO_2} 3O_2(g) + 2KCl(s)$$

(The manganese dioxide serves as a catalyst; without it a much higher temperature is needed.) For virtually all laboratory needs, however, oxygen in heavy-walled, high-pressure steel cylinders is purchased and dispensed as needed.

Oxygen is obtained in commercial quantities today by the careful distillation of liquid air. Nitrogen boils at 77.4 K, argon at 87.5 K, and oxygen at 90.2 K, so nitrogen and argon distill from liquid air first. To reduce shipping costs, most of the capacity for this process is located near plants that use oxygen. Liquid oxygen, however, can be transported and stored in insulated tanks, and gaseous oxygen can be transmitted in pipelines.

Oxygen generally ranks among the top five chemicals produced in the United States. In the mid-1980s, the annual U.S. production was around 16 million tons (roughly 500 billion moles). Over half was consumed by the production of metals, mostly in the basic oxygen process for making steel (Section 21.2). About a fifth of the oxygen produced is

Lavoisier, often called the father of modern chemistry, was executed by a mob at the guillotine on May 8, 1794, during the French Revolution.

FIGURE 19.3 Steel wool, which has been heated in a bunsen flame, burns brightly when dipped into an atmosphere rich in oxygen.

FIGURE 19.4 Oxygen can be collected in the laboratory in an apparatus such as this in which heat converts potassium chlorate into oxygen and potassium chloride. (The rubber stopper must not be heated nor can $KClO_3$ ever be allowed to come in contact with it or anything else organic. Wastes from this experiment must never be discarded in a container with paper, wood products, or any organic material.)

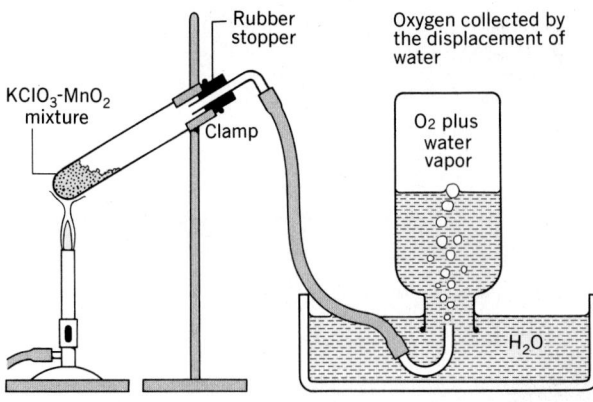

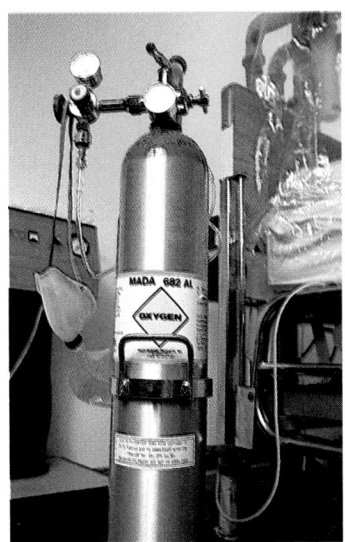

FIGURE 19.5 Liquid oxygen, LOX, was the oxidizing agent that reacted with the fuel to give the energy for propelling American astronauts to the moon and back.

Pure O_2 (or O_2-enriched air) is used when the partial pressure of the oxygen in ordinary air is not high enough for a patient.

used in the chemicals industry to make various oxygen derivatives of hydrocarbons, including plastics. The treatment of sewage, wastewater, and waste paper pulp fluids; the health industry; and rocketry consume most of the small remaining oxygen production. See Figure 19.5. (Oxygen's greatest use, of course, is in nature to support respiration, combustion, and decay.)

Ozone

Ozone is triatomic oxygen, O_3. Different molecular (or crystalline) forms of the same element are called allotropes, so ozone is an allotrope of dioxygen (the name for O_2 when absolute clarity is essential). The O_3 molecule has a bent geometry, and in valence bond (resonance) theory it is viewed as a hybrid of the following resonance structures.

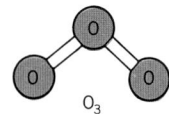

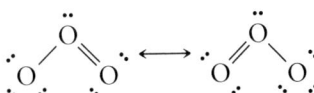

Ozone boils at 161.3 K and melts at 80.7 K. It is a blue, diamagnetic, and unstable gas with a characteristic pungent odor. You may have noticed this odor around sparking motors or in the air during severe electrical storms. Ozone is deep blue as a liquid and violet-black as a solid, but both forms are dangerously explosive, decomposing to oxygen. (Gaseous ozone also decomposes to oxygen, but more slowly.)

Ozone can be synthesized by passing an electrical discharge through dioxygen. Generally oxygen just somewhat enriched in O_3 is sufficient for any need. Some sterilizing lamps work by making a low concentration of ozone from the action of ultraviolet light on the O_2 in air. When absorbed, photons of the right UV frequency and therefore of the right energy can break the O—O bond to make oxygen atoms.

$$O_2(g) \xrightarrow{\text{UV light}} 2O(g)$$

When an oxygen atom collides with an oxygen molecule at the surface of some neutral particle, M, ozone forms.

$$O(g) + O_2(g) + M \longrightarrow O_3(g) + M$$

(M serves to accept some of the energy of collision and thus enables O_3 to survive its formation without instantly breaking apart.) This UV-induced formation of ozone also

Ozone is from the Greek *ozein*, "to smell."

Remember, for light of frequency ν, the energy of a photon is $E = h\nu$, where h is Planck's constant.

occurs in the stratosphere where it is part of a mechanism that screens out most of the sun's dangerous ultraviolet rays and keeps them from reaching us. See Chemicals in Use 13, page 822.

Another source of ozone is a complex series of reactions that occurs in smog exposed to sunshine, as described in Special Topic 19.2.

Ozone has very high standard reduction potentials in both acidic and basic media (which indicates that O$_3$ is a strong oxidizing agent).

$$O_3(g) + 2H^+(aq) + 2e^- \rightleftharpoons O_2(g) + H_2O \qquad E° = +2.08\ V$$
$$O_3(g) + H_2O + 2e^- \rightleftharpoons O_2(g) + 2OH^-(aq) \qquad E° = +1.24\ V$$

In acid, ozone is second only to fluorine (among common chemicals) as an oxidizing agent, and it is used as an oxidant to make compounds (particularly in organic chemistry). It is also used to purify water. Ozone attacks just about anything—trees (Figure 19.6), fabrics, dyes, rubber, lung tissue. This property makes it one of the most serious pollutants in smog, and at a level of only 0.5 µg/m³ (0.5 ppm) the physical activities of

FIGURE 19.6 The effect of ozone on vegetation. Ozone attacks the chlorophyll, a green pigment, in pine needles. Ozone, therefore, is believed to be one factor in the damage to these pines in the Harz Mountains of Lower Saxony, West Germany.

children and the elderly must be curtailed to reduce their inhaling the ozone deeply. Exposure to ozone at a level of 1.0 $\mu g/m^3$ (1.0 ppm) for over 10 minutes is dangerous for all people.

In severe smog episodes, the ozone level can reach 1 ppm.

For water treatment, ozone is just as lethal as chlorine to bacteria and viruses, but the excess ozone decomposes, so no odor or taste is left. (This is also a disadvantage, because the detection of a chlorine odor when chlorine is used is an assurance that enough disinfectant is present all the way to the final water tap.) Many cities in Europe use ozone instead of chlorine for water treatment.

Oxides

We will study the formation of the oxides of many of the elements when we take up these elements. A few general principles and correlations between structure and property can be described here, however.

Except for helium, neon, argon, and possibly krypton—the noble gases—all elements form oxides. Many can be prepared directly from the elements.

Basic and Ionic Oxides   As we learned in Section 2.12, the oxides of some metals are *basic* oxides (basic anhydrides), and most are ionic materials involving the oxide ion, O^{2-}. Since this ion is a powerful proton acceptor, any basic oxide that dissolves in water—generally the Group IA metal oxides and the oxides of calcium, strontium, and barium of Group IIA—does so by a reaction that produces the hydroxide ion. Thus sodium oxide reacts with water as follows to leave a solution of sodium hydroxide.

$$Na_2O(s) + H_2O \longrightarrow 2NaOH(aq)$$

Like all compounds that react vigorously with water, Na$_2$O can remove water from humid air, so it must be stored in tightly capped bottles.

All other Group IA metal oxides react with and dissolve in water by a similar reaction. They must all be stored in a dry atmosphere.

The Group IIA metal oxides all melt above 1900 °C, which indicates that these are also ionic compounds involving the oxide ion. (Beryllium oxide, while having a high melting point, has bonds of mostly a covalent nature.) Magnesium oxide, MgO, is typical of the Group IIA metal oxides (except for that of beryllium). MgO is insoluble in water but does react with water to form the relatively *insoluble* hydroxide. (It should therefore also be stored in a dry atmosphere.) In acids, however, MgO dissolves by the following reaction.

$$MgO(s) + 2H^+(aq) \longrightarrow Mg^{2+}(aq) + H_2O$$

Group IIA Oxide	Melting Point (°C)
BeO	2530
MgO	2800
CaO	2600
SrO	2430
BaO	1923

The oxides of the Group IIA metals melt so high that some are useful for making the bricks that line ovens and furnaces where pottery and china are baked. They conduct heat well, and can withstand exposure to air at the high temperatures needed.

Covalent Metal Oxides   Once we leave the oxides of the Group IA and IIA metals and move among the other metals, we find a mixed situation. Some transition metal oxides have low melting points, suggesting that they consist of relatively nonpolar molecules. Examples are dimanganese heptoxide (Mn_2O_7), melting at 6 °C, ruthenium tetroxide (RuO_4), melting at 25 °C, and osmium tetroxide (OsO_4), melting at 40 °C. (Interestingly, osmium metal has the fourth highest melting point of all the elements—3040 °C.) Let us consider why these metal oxides are covalent in character, not ionic.

OsO$_4$ is very poisonous, and its vapors cause blindness.

The oxidation numbers of the metals in the low-melting covalent metal oxides are particularly high: $+7$ for Mn in Mn_2O_7, $+8$ for Ru in RuO_4 and for Os in OsO_4. If Os existed as a cation in OsO_4, its charge would therefore be $8+$. In other words, we would have a particle with a very high concentration of positive charge. We learned in Section 19.3 how a high concentration of positive charge in the bare proton, H^+, enables it to polarize electron clouds in any chemical species and force the formation of a covalent bond. Because there are inner level electrons in the hypothetical Os^{8+} cation, this ion would not have a radius as small as that of a bare proton. But it would be much smaller than any lesser charged ion because the radii of cations shrink as their positive charges increase. (We discussed this in Section 6.11. See also the table inside the back cover.) Evidently a combination of small size and high charge enables Os^{8+} to force oxide ions, O^{2-}, to provide electrons for covalent bonds to osmium. The very high concentration of charge on the hypothetical Os^{8+} ion, in other words, prevents it from existing as such together with O^{2-} ions, so instead of ionic bonds, there are covalent bonds in OsO_4.

> As a general rule, the higher the oxidation state of the metal atom in its compounds the more covalent and less ionic are the bonds to it.

Of course, as the oxidation number *decreases,* the bonds to the central atom should become less covalent and more ionic. We see this as we move from the +8 state of osmium to its +4 state in its dioxide, OsO_2. The two crystalline forms of osmium dioxide, OsO_2, do not have low melting points, like the tetroxide, suggesting ionic bonds in OsO_2. Between 350 and 400 °C, one oxide changes to the other and at 500 °C, the latter decomposes and does not melt. As we said, this behavior suggests ionic bonds, not covalent bonds, which indicates that Os^{4+} is much less able than Os^{8+} to distort electron clouds and force covalent bonds to form.

The oxidation states of the metals are also +4 in thorium dioxide, ThO_2, and titanium dioxide, TiO_2, and their melting points are very high, 2950 °C for ThO_2 and 1840 °C for TiO_2, suggesting that bonding is more ionic than covalent. (We can only say "suggesting," not "proving." We will see why on page 806.)

TiO_2 is the most commonly used white pigment in paints.

Oxidation State and Oxide Acidities or Basicities   The oxidation state of the metal in a metal oxide also influences the acid–base properties of the compound. Oxides of metals in high oxidation states, like Cr in chromium(VI) oxide, CrO_3, tend to be acidic oxides. CrO_3 melts at 197 °C, which is low enough to suggest that its bonds are covalent in character. And in water it reacts to give an equilibrium mixture in which one species is chromic acid, H_2CrO_4. (The equilibrium will be more fully described in Chapter 21.)

$$
\underset{\text{chromic acid}}{H-O-\overset{\overset{\displaystyle O}{|}}{\underset{\underset{\displaystyle O}{|}}{Cr}}-O-H} + H_2O \;\rightleftharpoons\; \underset{\substack{\text{monohydrogen}\\\text{chromate ion}}}{H-O-\overset{\overset{\displaystyle O}{|}}{\underset{\underset{\displaystyle O}{|}}{Cr}}-O^-} + H_3O^+
$$

In H_2CrO_4, Cr is in a high, +6 oxidation state. It is therefore able to draw electron density from the O—H groups. This polarizes each O—H bond enough to give the H a high $\delta+$ charge, and so H^+ can readily transfer to water to form a hydronium ion.

> As a general rule, the higher the oxidation state of the metal in a metal oxide, the more acidic is the oxide.

Osmium tetroxide, with its high oxidation state for Os, is also an acidic oxide that illustrates this rule. In concentrated potassium hydroxide, for example, it neutralizes hydroxide ion according to the following net ionic equation.

$$
OsO_4(s) + 2OH^-(aq) \longrightarrow \underset{\text{perosmate ion}}{[OsO_4(OH)_2]^{2-}(aq)}
$$

Oxides of metals with lower oxidation numbers are not acidic but are either amphoteric or basic. An *amphoteric oxide* is one that can neutralize either acid or base. Aluminum oxide is a common example. It is a very high-melting oxide in which the bonds are believed to be essentially ionic, not covalent as they are in CrO_3. Al_2O_3 consists of Al^{3+} ions and O^{2-} ions. Because of the oxide ion, O^{2-}, which is a powerful proton-accepting base, Al_2O_3 can react with acids as follows.

$$Al_2O_3(s) + 6H^+(aq) + 9H_2O \longrightarrow 2[Al(H_2O)_6]^{3+}(aq)$$

At the same time that the O^{2-} ion in Al_2O_3 has the properties of a base, the Al^{3+} ion has the properties of an acid, a strong *Lewis acid* or electron-pair acceptor (Section 11.9). The high concentration of positive charge in the small Al^{3+} ion makes this ion able to attract species with electron pairs, like OH^- or H_2O, and accounts for its being a strong Lewis acid. Both the hydroxide ion and the water molecule are Lewis bases. Al^{3+} ions can neutralize such bases, and Al_2O_3 reacts with hydroxide ion as follows and dissolves.

A Lewis acid is an electron pair-accepting species.

$$Al_2O_3(s) + 2OH^-(aq) + 7H_2O \longrightarrow 2[Al(H_2O)_2(OH)_4]^{2-}(aq)$$

Chromium also forms an amphoteric oxide, Cr_2O_3, but another oxide of chromium, CrO, is basic. In Cr_2O_3, Cr is in a $+3$ oxidation state, and in CrO it is in the $+2$ state. Thus the three oxides of chromium that we have mentioned, CrO_3, Cr_2O_3, and CrO, illustrate a general trend. *Among the oxides of a given transition metal, the oxides become more basic and less acidic as the oxidation number of the metal ion decreases.*

CrO is less acidic than Cr_2O_3 because the change of chromium from the $+3$ to the $+2$ state makes the metal ion a weaker Lewis acid. But this does not weaken the ability of the oxygens to function as bases. Quite the contrary; the decrease in oxidation number of the metal makes the oxygens in the metal oxide stronger bases. This is because the electron pairs on the oxygens experience less attraction toward the metal ions when the positive charge on the metal ion is less. The oxygen's electron pairs are more available, therefore, to help the oxygens function as bases. Illustrating this are the powerfully basic oxide ions in the Group IA and Group IIA metal oxides where the metal ions have low oxidation numbers. Similarly, many transition metals in the $+2$ oxidaton state also form basic oxides, like NiO and FeO.

A few metal oxides, for example, MnO_2 and PbO_2, are inert both to acids and to bases (unless the reaction is a redox reaction, not a proton-transfer reaction).

Most of the oxides of the nonmetals are acidic oxides (acidic anhydrides), as was mentioned in Section 2.12. Examples are the oxides of sulfur — SO_3 and SO_2 — as well as carbon dioxide. The relative acidities of SO_3 and SO_2 parallel the oxidation numbers of S in these oxides. When S is in the $+6$ state in SO_3, we have a system analogous to the CrO_3 system, which is also an acidic oxide as we just learned. And SO_3, which reacts with water to give the strong acid, sulfuric acid, is a more acidic oxide than SO_2 where S is in the $+4$ state. SO_2 interacts with water to give a much weaker acidic system (which will be better characterized in the next chapter).

Carbon monoxide is neutral in an acid–base sense, and the common oxide of hydrogen (H_2O) is weakly amphoteric. Their properties are studied elsewhere.

Hydrogen Peroxide

Hydrogen peroxide, H_2O_2, is a colorless liquid that melts at -0.43 °C, boils at 150.2 °C, and has a density of 1.4425 g/mL (25 °C). It must be handled extremely carefully; it is particularly dangerous when pure.

As the figure in the margin shows, the H_2O_2 molecule is not planar. Remember that each O atom has two unshared electron pairs, and their electron clouds as well as those in the O—H bonds repel each other. The twisted structure—called a *skew-chain* structure—accommodates these intramolecular repulsions best.

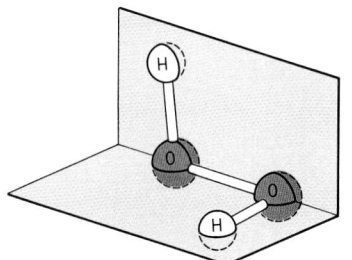

Hydrogen peroxide (solid phase)

Virtually all hydrogen peroxide is made today by complex oxidations of organic compounds which we will not study. It was once made by the hydrolysis of the peroxydisulfate ion, which was synthesized by passing direct-current electricity through sulfuric acid.

$$2H_2O + (O_3S\!-\!O\!-\!O\!-\!SO_3)^{2-}(aq) \longrightarrow 2HSO_4^-(aq) + H\!-\!O\!-\!O\!-\!H(aq)$$
peroxydisulfate ion

Hydrogen peroxide has a pK_a of 11.75, so it is a stronger acid than water, but only slightly stronger an acid than HPO_4^{2-}, for which the pK_a is 12.35. A high pH, therefore, is required if the following equilibrium is to favor the hydroperoxide ion, HO_2^-.

$$OH^-(aq) + H_2O_2(aq) \rightleftharpoons H_2O + HO_2^-(aq)$$

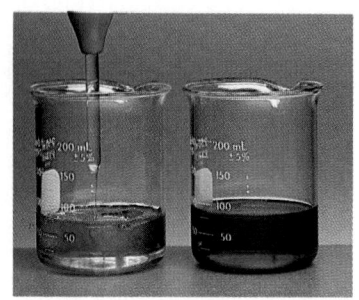

FIGURE 19.7 The permanganate ion is a strong enough oxidizing agent to oxidize hydrogen peroxide, as described in this section. Before hydrogen peroxide is added (right) the color of the solution is deep purple, the characteristic color of $MnO_4^-(aq)$. As H_2O_2 is added (left), the color of the MnO_4^- fades. Oxygen also evolves.

As a proton acceptor, hydrogen peroxide is probably a million times weaker than water, so the following equilibrium strongly favors the reactants.

$$H_3O^+(aq) + H_2O_2(aq) \rightleftharpoons H_2O + H_3O_2^+(aq)$$

The following equations summarize the chief redox properties of the hydrogen peroxide/hydroperoxide ion system in acid and base.

$$H_2O_2(aq) + 2H^+(aq) + 2e^- \rightleftharpoons 2H_2O \qquad E° = +1.77\ V \qquad (19.1)$$

$$O_2(g) + 2H^+(aq) + 2e^- \rightleftharpoons H_2O_2(aq) \qquad E° = +0.69\ V \qquad (19.2)$$

$$HO_2^-(aq) + H_2O + 2e^- \rightleftharpoons 3OH^-(aq) \qquad E° = +0.87\ V \qquad (19.3)$$

Thus in either an acidic solution (Equation 19.1) or a basic solution (Equation 19.3), hydrogen peroxide is a strong oxidizing agent. The *rates* of oxidation are higher in basic media, however. Hydrogen peroxide will operate as a reducing agent — using the *reverse* of equation 19.2 as the half-cell reaction — only toward very strong oxidizing agents, such as MnO_4^- (Figure 19.7).

$$2MnO_4^-(aq) + 5H_2O_2(aq) + 6H^+(aq) \longrightarrow 2Mn^{2+}(aq) + 8H_2O + 5O_2(g)$$

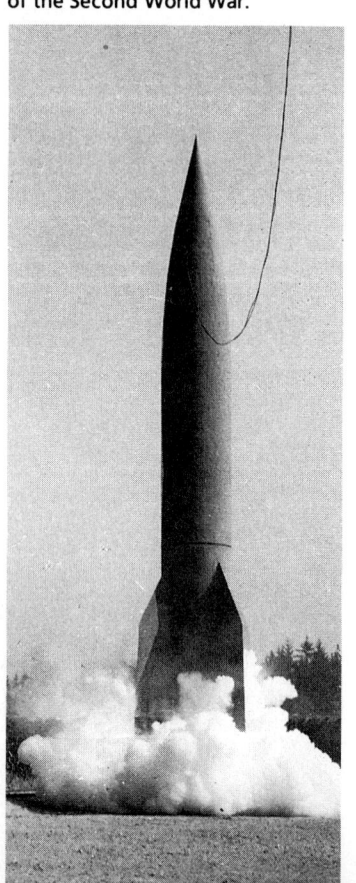

FIGURE 19.8 Hitler sent scores of small explosive-filled rockets, with hydrogen peroxide as the oxidant, to England during the latter stages of the Second World War.

Hydrogen peroxide eventually decomposes to water and oxygen.

$$2H_2O_2(\ell) \longrightarrow 2H_2O + O_2(g)$$

The decomposition is accelerated by almost anything — heavy metal ions, dirt, heat, and even water and oxygen, the products of its decomposition — particularly if traces of base are present. This is why pure hydrogen peroxide is so dangerous. And since oxygen is a product, the decomposition of hydrogen peroxide can start and sustain fires. In fact, 90% hydrogen peroxide was part of the propellant system in the Nazi's V-1 missile during World War II (Figure 19.8).

Even 20% aqueous hydrogen peroxide should be handled only by experienced personnel. Material of this concentration and higher is generally stored in bottles made of special plastic or bottles coated on the inside with wax, because glass surfaces can have alkaline properties that accelerate the decomposition of H_2O_2. The "peroxide" sold in drugstores is 2 to 3% in concentration, and it can be safely handled as a bleach or a disinfectant. If you ever use it to clean a wound, you will see it foam strongly, because substances in blood catalyze its decomposition.

Many hundreds of thousands of tons of hydrogen peroxide are manufactured worldwide each year. It is used chiefly as a bleach for paper pulp, textiles, leather, and edible fats and oils (and, of course, to bleach hair), and the chemicals industry uses it in the manufacture of organic chemicals, polymers, pharmaceuticals, and food products.

Metal Peroxides   Hydrogen peroxide is the "parent" compound of the peroxides, a family of substances with O—O bonds. Metal peroxides such as Na_2O_2 are not, however, made from hydrogen peroxide. Their synthesis will be discussed elsewhere.

As you might expect, the peroxide ion, O_2^{2-}, is a powerful base, so when sodium peroxide is added to water it reacts almost quantitatively as follows to give an alkaline solution of hydrogen peroxide (which then decomposes to water and oxygen as described earlier).

$$Na_2O_2(s) + 2H_2O \longrightarrow 2NaOH(aq) + H_2O_2(aq)$$

Thus solid metal peroxides capable of this kind of reaction provide convenient and safer sources of bleaching and disinfecting power than aqueous hydrogen peroxide itself. Figure 19.9 dramatically shows how the decomposition of sodium peroxide by a trace of water in the presence of an organic material, sugar, can cause a dangerous fire. (And it is not easy to set sugar afire!)

Dilute hydrogen peroxide

(a)

(b)

FIGURE 19.9 (*a*) A drop of water is about to be added to a mixture of sodium peroxide, Na_2O_2, and ordinary sugar. (*b*) The hydrolysis of the peroxide is so exothermic that it ignites the mixture. Sodium peroxide is a powerful oxidizing agent and causes the sugar to burn vigorously.

Superoxides   Of the Group IA metals, only lithium reacts with the oxygen in air to give mostly the simple oxide, Li_2O. (Some Li_2O_2 also forms.)

$$4Li(s) + O_2(g) \longrightarrow 2Li_2O(s)$$
$$\text{lithium oxide}$$

Sodium reacts with the oxygen in air to give mostly the peroxide, Na_2O_2. (Some Na_2O also forms.)

$$2Na(s) + O_2(g) \longrightarrow Na_2O_2$$
$$\text{sodium peroxide}$$

The remaining alkali metals, potassium, rubidium, and cesium, react with oxygen in air to give superoxides with the general formula MO_2. We can illustrate the reaction with potassium.

The simple oxides of Na, K, and Rb are made by indirect methods.

$$K(s) + O_2(g) \longrightarrow KO_2(s)$$
$$\text{potassium superoxide}$$

The superoxides are crystalline, saltlike compounds of the superoxide ion, O_2^-. Since this has one unpaired electron, the superoxides are paramagnetic. They are also very powerful oxidizing agents, being able to oxidize even water in a very vigorous reaction that liberates oxygen.

$$2O_2^-(s) + H_2O \longrightarrow O_2(g) + HO_2^-(aq) + OH^-(aq) \tag{19.4}$$

One product, the hydroperoxide ion, HO_2^-, decomposes to give more oxygen.

$$2HO_2^-(aq) \longrightarrow 2OH^-(aq) + O_2(g)$$

The superoxides also react with carbon dioxide to give carbonates and oxygen. With potassium superoxide, for example, we have

$$4KO_2(s) + CO_2(g) \longrightarrow 2K_2CO_3(s) + 3O_2(g)$$

This reaction can be used to remove CO_2 from air and replace it with O_2. Thus in a closed chamber such as a spacecraft, this reaction can be used to keep the air fresh. Of course, such air also has moisture, so this water would also decompose KO_2, as shown by Equation 19.4. But one product in Equation 19.4 is the OH^- ion, which also combines with CO_2.

$$HO^- + CO_2 \longrightarrow HCO_3^-$$

So CO_2 could be removed from either dry or humid air by KO_2 and be replaced by oxygen.

Because of their powerful oxidizing abilities, the superoxides must be handled carefully by experienced chemists. Any contact of Group IA superoxides with organic materials must be rigorously prevented, because these burn so readily. The ability of potassium, rubidium, and cesium to form superoxide coatings in air means that these metals should be handled only by chemists who know how to minimize the danger of a detonation. It is not likely that you would ever even be shown these elements in any beginning study of chemistry. But we take the trouble to mention a detail such as this to provide a warning that an overreliance on simple periodic relationships to predict properties can lead not just to surprising but even lethal results. "Never do unauthorized experiments in the lab" is a rule that could save your life or someone else's.

19.5 NITROGEN

Ammonia, nitric acid, and their salts are some of the commercially important compounds of nitrogen.

Some properties of the elements of the nitrogen family—nitrogen, phosphorus, arsenic, antimony, and bismuth—are given in Table 19.10. We will study only two members in any detail—nitrogen here and phosphorus in Section 20.2.

Elemental Nitrogen—Its Discovery and Principal Uses

Isotope	% Abundance
N-14	99.63
N-15	0.37

Nitrogen is a colorless, odorless, and tasteless gas that makes up 78% (v/v) of dry air. Daniel Rutherford (1749–1819, England) is generally credited with its discovery (1772).

TABLE 19.10 The Nitrogen Family—Group VA

Name	Atomic Symbol	Melting Point (°C)	Boiling Point (°C)	Important Types of Compounds
Nitrogen	N	−210	−196	Nitrates (fertilizers, explosives) Oxides (air pollutants) Ammonia (fertilizer)
Phosphorus	P	44	281	Phosphates and polyphosphates (detergents, fertilizers)
Arsenic	As	815[a]	613[b]	Arsenates (pesticides)
Antimony	Sb	631	1750	Lead—antimony mixtures (alloys) for storage batteries and type metal
Bismuth	Bi	271	1560	In mixtures with other metals: low-melting alloys for automatic fire alarms and sprinkler systems

[a] Under 28 atm.

[b] Sublimation temperature.

He isolated it by letting a limited supply of air be used for a combustion, which replaced the air's oxygen with carbon dioxide. When he removed the CO_2 by its reaction with potassium hydroxide, the remaining gas would not support life and it suffocated living things. Nitrogen is named after a Greek phrase meaning generator of "nitron," which refers to nitrates. (The French call it *azote*, meaning "nonlife," and the Germans call it *stickstoff*, meaning "suffocating substance.")

Roughly 23 to 25 million tons (700 to 800 billion moles) of nearly oxygen-free nitrogen is produced each year from liquefied air. About a third is sold as liquid nitrogen and the remainder is piped as a gas or shipped in compressed gas cylinders. The largest single use of oxygen-free nitrogen (30%) is in an oilfield operation called enhanced oil recovery. This is a method whereby a gas under pressure — such as nitrogen or carbon dioxide, neither of which reacts with petroleum — helps to force more oil from subterranean deposits. The next largest use — to provide a blanketing atmosphere of an unreactive gas — is in the metals and chemicals processing industry. Companies that make electronic components and semiconductor chips use ultrapure nitrogen (99.9999% pure) for this purpose.

Its low boiling point (77 K) and chemical unreactivity make liquid nitrogen an ideal and relatively inexpensive coolant in research. Manufacturers of frozen seafood, poultry, and meat products use it for fast freezing. An interesting but small-scale use is by commercial sperm banks that store sperm from high-quality cattle, hogs, and sheep. This is sold to farmers and ranchers seeking to upgrade their livestock. We have already described (Special Topic 19.1) how liquid nitrogen is used to put certain mixed metal oxides into superconducting states.

By complex reactions called nitrogen fixation, microorganisms take nitrogen from air and make compounds they need, such as ammonia. Eventually, nitrogen compounds in living things are reconverted to nitrogen by processes of decay, so a huge nitrogen cycle exists in nature. In it, nitrogen nuclei move from the air through the biosphere and back again.

One of the remarkable properties of nitrogen is its wide range of oxidation states (Table 19.11). It occurs in all possible whole-number states, from −3 to +5, as well as in a $-\frac{1}{3}$ state, ten oxidation states in all. Moreover, many nitrogen compounds have nitrogen in more than one state, as in NH_4NO_3, where the −3 and +5 states occur (−3 in NH_4^+ and +5 in NO_3^-). We will take up the *reduced forms* of nitrogen — those in

The reaction is

$$CO_2(g) + KOH(s) \longrightarrow KHCO_3(s)$$

A rubber ball chilled to the temperature of liquid nitrogen will shatter when dropped.

The **biosphere** consists of all those regions of the planet that have living things.

TABLE 19.11 Oxidation States of Nitrogen

Oxidation Number	Examples and Types	
	Formula	Name
−3	NH_3	Ammonia
	NH_4^+	Ammonium ion
	Mg_3N_2, Li_3N	Saltlike nitrides
	BN	Diamondlike nitrides
	$(CN)_2$	Covalent nitrides
−2	N_2H_4	Hydrazine
−1	NH_2OH	Hydroxylamine
$-\frac{1}{3}$	N_3^-	Azide ion
0	N_2	Nitrogen
	$[Ru(NH_3)_5(N_2)]^{2+}$	Complex ions bearing N_2 as a ligand
+1	N_2O	Dinitrogen monoxide (nitrous oxide)
+2	NO	Nitrogen monoxide (nitric oxide)
+3	N_2O_3	Dinitrogen trioxide (nitrous anhydride)
+4	NO_2	Nitrogen dioxide
	N_2O_4	Dinitrogen tetroxide
+5	HNO_3	Nitric acid
	NO_3^-	Nitrate ion
	N_2O_5	Dinitrogen pentoxide (nitric anhydride)

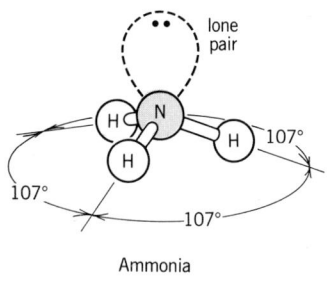

lone pair

107°
107°
107°

Ammonia

Never inhale the vapors coming from a bottle of concentrated aqueous ammonia. Pour this liquid at the hood.

Fritz Haber won the 1918 Nobel prize in chemistry.

which N has negative oxidation numbers and correspond to nitrogen having "gained electrons"—in the remainder of this section. Its *oxidized forms* will be surveyed in the next section.

Ammonia

Probably no other compound can be identified by smell *and correctly named* by as many people as ammonia. This gas, which boils at −33.4 °C and freezes at −77.7 °C, has a very characteristic, pungent odor, and it can be detected at a level in air of only about 50 to 60 ppm. At levels of 100 to 200 ppm, ammonia sharply irritates the eyes and air passages to the lungs. At high concentrations it makes the lungs fill with fluid, which can quickly cause death unless prompt aid is given.

Ammonia is easily liquefied, and it has a relatively high heat of vaporization (1370 J/g). Wide-mouthed vacuum bottles—Dewar flasks—are generally used as its containers, and liquid ammonia is purchased in steel cylinders that keep it under a pressure. It is always handled at an efficient fume hood.

Ammonia is made today by the Haber–Bosch process. Fritz Haber did the theoretical work, while Karl Bosch solved the engineering problems. Theirs was the first major industrial process to benefit from the chemical knowledge accumulated during the nineteenth century on thermodynamics, kinetics, and equilibria. The central reaction in the process is

$$N_2(g) + 3H_2(g) \rightleftharpoons 2NH_3(g) \qquad \Delta H° = -92.38 \text{ kJ}$$

The forward reaction is exothermic, so it would seem that low temperatures should favor it. But the energy of activation is high, and even with catalysts, high temperatures are required. The best catalyst is a mesh made of iron containing small amounts of aluminum oxide and potassium carbonate. Since the forward reaction changes 4 volumes of reactants into 2 volumes of product, high pressure favors it. Most commercial operations today are run in the range of 100 to 350 atm pressure at 450 to 500 °C.

In the mid-1980s, the annual production rate for ammonia was about 16 million tons (860 billion moles). About a quarter went directly for fertilizer (Figure 19.10), and the rest was used to make chemicals such as nitric acid, dyes, pharmaceuticals, and cleaning agents. Some ammonia is used as the heat-exchanger gas in large refrigeration units.

Ammonia's high solubility in water, 51.8 g NH_3/100 g H_2O at 20 °C, arises largely from the ability of its molecules to form hydrogen bonds to water molecules, but some NH_3 reacts with water, as we studied at the end of Section 13.5. Figure 19.11, showing the ammonia "fountain," illustrates how ammonia's solubility in water can be used to create a partial vacuum.

The hydrogen bond ($\cdots$) in aqueous ammonia.

The most familiar reactions of ammonia involve its basicity. Recall that it is not a strong base, when compared with the OH^- ion in water:

$$NH_3(aq) + H_2O \rightleftharpoons NH_4^+(aq) + OH^-(aq) \qquad K_b = 1.8 \times 10^{-5} \ (pK_b = 4.75)$$

But in the presence of a strong proton donor such as the hydronium ion (available in any strong aqueous acid), ammonia is quantitatively converted to the ammonium ion. For example,

$$NH_3(aq) + HCl(aq) \xrightarrow{100\%} NH_4Cl(aq) + H_2O$$

or

$$NH_3(aq) + H_3O^+(aq) \longrightarrow NH_4^+(aq) + H_2O$$

Ammonia is not considered to be flammable, but it can be made to burn in air.

$$4NH_3(g) + 3O_2(g) \rightleftharpoons 2N_2(g) + 6H_2O(g) \qquad K_p = 10^{228} \ (25 \ °C)$$

Despite the exceptionally high equilibrium constant for this reaction, ammonia is stable in air because the rate-limiting step in the multistep process has a high energy of activation. However, the combustion is exothermic enough so that when ammonia is mixed with oxygen in the right proportion, an explosive combustion can be initiated. The heat liberated as the combustion starts is sufficient to activate the combustion of unchanged ammonia.

FIGURE 19.11 The ammonia fountain. Water with a trace of phenolphthalein (which is pink in base and colorless in acid) is being sucked from the beaker into the inverted flask. Initially the flask contained only gaseous ammonia. Then a little water was injected from the squeeze bulb. Because ammonia is so soluble in water, much of the gaseous ammonia dissolved, and that left a partial vacuum in the flask. As the water rushes to fill the vacuum, it turns pink because solutions of ammonia are slightly basic.

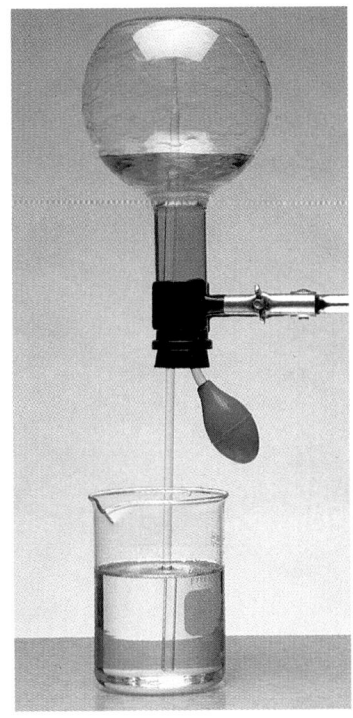

When a platinum-rhodium catalyst is used at a temperature of 750 to 900 °C, the oxidation can be directed toward a species thermodynamically less favored than N_2, namely nitrogen monoxide, NO (Figure 19.12).

$$4NH_3(g) + 5O_2(g) \xrightleftharpoons[750-900\ °C]{Pt/Rh} 4NO(g) + 6H_2O(g) \qquad K_p = 10^{168}\ (25\ °C)$$

This reaction is vital to the Ostwald synthesis of nitric acid from ammonia, as we will study in the next section.

FIGURE 19.12 Ammonia escaping from the aqueous ammonia at the bottom of this beaker reacts with oxygen at the surface of the catalyst, a gauze of platinum. Enough heat is generated to make the gauze glow. Colorless NO, initially produced, soon reacts with oxygen to give NO_2, which is responsible for the reddish-brown color.

Liquid Ammonia as a Solvent   Liquid ammonia dissolves the Group IA and IIA metals, giving solutions with a beautiful blue color. Just as water can hydrate dissolved ions, so molecules of ammonia can *ammoniate* ions in liquid ammonia. To be ammoniated means to be surrounded by molecules of the solvent, liquid ammonia. In very dilute solutions of Group IA and IIA metals in liquid ammonia, both the metal ions *and the electrons* are ammoniated. Ammoniated species are indicated in equations by (*am*) after the formula, as in $M^+(am)$ and $e^-(am)$. The blue color of these solutions is caused by a most unusual species, the ammoniated electron $e^-(am)$. Eventually, the ammoniated materials interact irreversibly with ammonia to make solutions of the metal amides in liquid ammonia. For example, a solution of sodium in liquid ammonia changes as follows, and this is how sodium amide is manufactured.

$$2Na^+(am) + 2e^-(am) + 2NH_3(\ell) \longrightarrow H_2(g) + 2NaNH_2(am)$$
$$\text{sodium amide}$$

Transition metal ions catalyze the reaction.

The amide ion, NH_2^-, in liquid ammonia is analogous to the hydroxide ion in liquid water. Each ion is a powerful base in its respective solvent. Similarly, the ammonium ion in liquid ammonia is analogous to the hydronium ion in liquid water — each is a strong acid in its respective solvent. Thus a whole range of acid–base chemistry exists for liquid ammonia involving NH_4^+ and NH_2^- as the key acidic and basic species. The important acid–base neutralization in liquid ammonia is

$$NH_4^+(am) + NH_2^-(am) \xrightarrow{NH_3(\ell)} 2NH_3(\ell)$$

Thus we can titrate NH_4Cl with KNH_2 in liquid ammonia just as we can titrate HCl with KOH in water, and we can even use phenolphthalein as the indicator for both titrations.

Metal Amides   Sodium and potassium amides are commercially available. As we earlier learned, the amide ion is a powerful base toward water, so sodium amide, for example, reacts quantitatively and rapidly with water as follows.

$$NaNH_2(s) + H_2O \longrightarrow NH_3(aq) + NaOH(aq)$$

But the actual reaction is

$$NH_2^-(s) + H_2O \xrightarrow{100\%} NH_3(aq) + OH^-(aq)$$

Metal amides, therefore, must be stored in a dry atmosphere.

Ammonium Salts   The common inorganic ammonium salts are those with the chloride ion, nitrate ion, sulfate ion, and the various ions from phosphoric acid. Generally, all ammonium salts dissolve readily in water, where they become fully dissociated.

Since the ammonium ion is a weak acid, an aqueous solution of a salt such as ammonium chloride is slightly acidic, as we studied in Section 15.7. If a strong base, such

For NH_4^+, $K_a = 5.5 \times 10^{-10}$. The pH of 1 M NH_4Cl is about 4.7.

as OH⁻, is added to an ammonium salt, it is neutralized and ammonia is released. For example:

$$NH_4Cl(aq) + NaOH(aq) \longrightarrow NH_3(aq) + NaCl(aq) + H_2O$$

or

$$NH_4^+(aq) + OH^-(aq) \longrightarrow NH_3(aq) + H_2O$$

A common test for the ammonium ion is to add some strong base to the solution and test for the evolution of a basic gas. When the concentration of ammonium ion is high, the odor of ammonia also becomes unmistakable during this test.

Several ammonium salts break up when heated strongly, with ammonia being one product. Ammonium chloride, for example, sublimes without melting and decomposes at around 300 °C by the forward reaction in

$$NH_4Cl(s) \overset{heat}{\rightleftharpoons} NH_3(g) + HCl(g) \qquad \Delta H° = +177 \text{ kJ } (K_p = 10^{-16}, 25 °C)$$
$$\Delta S° = +0.29 \text{ kJ}$$
$$\Delta G° = +91 \text{ kJ}$$

Remember that the degree sign in thermochemical symbols data refer to substances in their standard states at 25 °C and 1 atm.

Notice that the enthalpy change, ΔH, is positive, which discourages this reaction from being spontaneous, and essentially no reaction does occur at 25 °C. But the entropy change, ΔS, is also positive, which favors a spontaneous change. (The reaction converts a highly ordered solid into two disordered gases, so the entropy change should be positive.) We learned in Section 16.7 that when both ΔH and ΔS are positive, the temperature is the crucial factor in determining if a reaction is spontaneous. We learned there that the net effect of the enthalpy and entropy factors is found by calculating the free energy change, ΔG, which involves ΔH, ΔS, and T by means of the equation

$$\Delta G = \Delta H - T\Delta S$$

The value of $\Delta G°$, calculated by this equation, is +91 kJ. Since $\Delta G°$ is positive, the reaction at 25 °C cannot be spontaneous. Only by raising T to a high value can we make the $T\Delta S$ term in the equation for ΔG negative enough to change the sign of the free energy from plus to minus. And at 500 °C (773 K), the value of ΔG calculates to be −48 kJ. So the free energy change is now negative, meaning that the decomposition is a spontaneous reaction at the higher temperature.

When the anion of an ammonium salt is an oxidant, it can *oxidize* the ammonium ion when the salt is heated, instead of simply taking H⁺ from it. For example, the N of the ammonium ion is changed from the −3 to the +1 oxidation state when ammonium nitrate is heated, and dinitrogen monoxide forms. The N of the nitrate ion is changed from the +5 to the +1 state. (This reaction is potentially very dangerous and only experienced chemists should attempt to carry it out.)

In an oxidation, the oxidation number becomes more positive.

$$\underset{\substack{\text{ammonium} \\ \text{nitrate}}}{NH_4NO_3(s)} \overset{heat}{\longrightarrow} \underset{\substack{\text{dinitrogen} \\ \text{monoxide}}}{N_2O(g)} + 2H_2O(g)$$

When ammonium dichromate is heated, the N in the NH_4^+ ion is oxidized by the dichromate ion from the −3 to the 0 oxidation state. (This reaction is also potentially dangerous and should be carried out only by experienced chemists.)

$$\underset{\substack{\text{ammonium} \\ \text{dichromate}}}{(NH_4)_2Cr_2O_7(s)} \overset{heat}{\longrightarrow} N_2(g) + \underset{\substack{\text{chromium(III)} \\ \text{oxide}}}{Cr_2O_3(s)} + 4H_2O(g)$$

Nitrides

Nitrides are binary compounds of nitrogen with elements other than hydrogen, but nitrogen is in the −3 oxidation state, the same state as in ammonia. Magnesium nitride, a

When Mg burns in *air*, the chief product is MgO, not Mg_3N_2, but some of this also forms.

typical metal nitride, is made by heating magnesium with nitrogen or with ammonia.

$$3Mg(s) + 2NH_3(g) \xrightarrow{900\ °C} \underset{\text{magnesium nitride}}{Mg_3N_2(s)} + 3H_2(g)$$

The nitrides of Group IA and IIA metals are saltlike compounds with high melting points, suggesting that they are ionic. And they do contain the nitride ion, N^{3-}. They react with water to give a metal hydroxide and ammonia. For example, using lithium nitride,

$$Li_3N(s) + 3H_2O \longrightarrow 3LiOH(aq) + NH_3(aq)$$

The nitrides of nonmetallic elements, for example, BN, $(CN)_2$, S_4N_4, and P_3N_5, involve largely covalent bonds. Their properties differ very widely. Thus boron nitride (BN) in one form melts at about 3000 °C and is very inert. Here the high melting point does *not* mean that the compound consists of ions. Instead, vast sheets of boron and nitrogen atoms, linked covalently, are present and are stacked one on top of the other (Figure 19.13). Thus, BN has a high melting point not because it is ionic but because it is *macromolecular*—its molecules consist of thousands of atoms. (BN is thus an empirical, not a molecular formula.) A high melting point can therefore mean either ionic bonds or a macromolecular system involving covalent bonds.

Macro- signifies "very large."

FIGURE 19.13 Boron nitride consists of stacks of sheets of macromolecules involving covalent bonds between B and N.

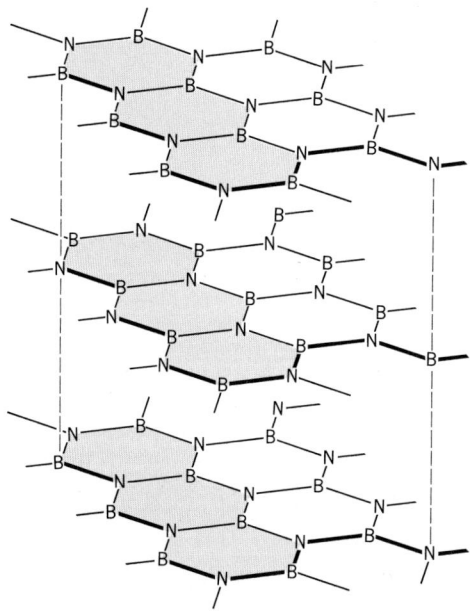

In huge contrast, the nitride of carbon, cyanogen, has the *molecular* formula $(CN)_2$, and it is a gas—a poisonous gas, in fact. The nitride of sulfur, whose molecular formula is S_4N_4, melts at 178 °C, but could detonate when struck or heated too rapidly.

Many transition metals form nitrides, including some that are almost as hard as diamond—for example, vanadium nitride, VN—and some that retain the appearance and electrical conductivity of a metal. We leave the nature of the bonds in these substances to other treatments, but we have learned here to be somewhat cautious in automatically assuming that a high melting point means that a compound must be ionic. It can also mean that it is macromolecular while having covalent bonds.

Hydrazine

Hydrazine, NH_2-NH_2 (often written as N_2H_4), is another reduced form of nitrogen. It is a colorless, toxic liquid that boils at 113.5 °C and melts at 2 °C and is a weaker base than ammonia. Yet, it is a strong enough base to form a family of salts of the hydrazinium ion, $N_2H_5^+$.

Hydrazine is made commercially by the *Raschig* process in which ammonia is oxidized by sodium hypochlorite. The overall reaction is

$$2NH_3(aq) + NaOCl(aq) \longrightarrow \underset{\text{hydrazine}}{N_2H_4(aq)} + NaCl(aq) + H_2O$$

Because "chlorine" bleaches used in home laundries contain the hypochlorite ion, OCl^-, they must never be mixed with household ammonia. The two react, as we just saw, to give toxic hydrazine. Fortunately, traces of Cu^{2+} ion, which are usually present when plumbing is made of copper pipe, catalyze the decomposition of hydrazine. So as laundry products drain way, any hydrazine that may form is decomposed.

Hydrazine is a strong reducing agent, as indicated in the following half-cell reaction, which is just one mode by which hydrazine can function. The large *negative* value of $E°$ indicates that the reverse reaction, which would supply electrons to some acceptor (and thus reduce it), is favored.

$$N_2(g) + 4H_2O + 4e^- \rightleftharpoons N_2H_4(aq) + 4OH^-(aq) \qquad E° = -1.16 \text{ V}$$

Liquid hydrazine and some of its organic derivatives have been used as rocket fuels. When mixed with hydrogen peroxide or with oxygen from liquid oxygen tanks, hydrazine burns violently to produce rapidly expanding gases that give thrust to the rocket. The reaction is very exothermic. For example, the oxidation of N_2H_4 by O_2 is

$$N_2H_4(\ell) + O_2(g) \longrightarrow N_2(g) + 2H_2O(\ell) \qquad \Delta H° = -621.5 \text{ kJ}$$

Hydroxylamine

Hydroxylamine, $HO-NH_2$, is a white solid that melts at 33 °C but must be stored at 0 °C to prevent decomposition. We can think of it as an ammonia molecule in which one H has been replaced by OH, or as a water molecule in which one H has been replaced by NH_2. It is basic; $K_b = 6.6 \times 10^{-9}$, 25 °C. But it is more weakly basic than either ammonia or hydrazine. It is normally purchased as a salt such as the salt with hydrogen chloride, $(NH_3OH)Cl$, in which form it can be stored under ordinary conditions. One of its major uses is in the synthesis of a member of the nylon family of substances.

Hydrazoic Acid

Hydrazoic acid, HN_3, contains nitrogen in the $-\frac{1}{3}$ oxidation state. When pure, it is a colorless liquid and extremely susceptible to explosion. It is a weak acid; $pK_a = 4.75$, 25 °C.

Several salts called *azides* are known. Those of heavy metals, like lead azide, explode when sharply struck, so they have been used to make detonator caps — devices used to set off the explosions of other materials, like gunpowder. The azides of the Group IA metals are not explosive, and sodium azide, NaN_3, is commercially available. The azide ion in aqueous solution behaves something like a halide ion; in fact, it is sometimes called a pseudohalide ion.

The azide ion is linear and symmetrical. In valence bond theory it is described as the hybrid of the following resonance structures.

$$[N\equiv N-N]^- \longleftrightarrow [N=N=N]^- \longleftrightarrow [N-N\equiv N]^-$$

As these structures suggest, the two nitrogen–nitrogen bonds should be, and are, identical in length — 116 pm.

Base	K_b (25 °C)
N_2H_4	8.5×10^{-7}
NH_3	1.8×10^{-5}

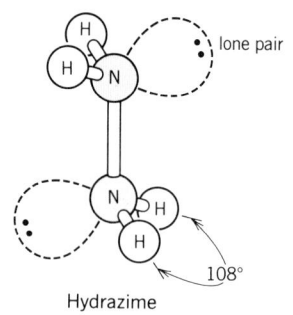

Hydrazine

The combustion of dimethylhydrazine, $(CH_3)_2NNH_2$, and oxygen was used to propel U.S. astronauts to the moon and back.

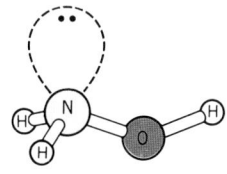

Hydroxylamine

19.6 OXOACIDS AND OXIDES OF NITROGEN

Nitric acid, nitrogen's most important oxoacid, is made from ammonia by the Ostwald process.

Nitric Acid

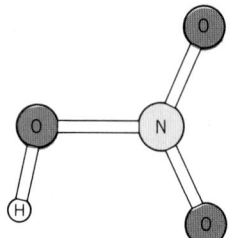

Nitric acid (gas phase)

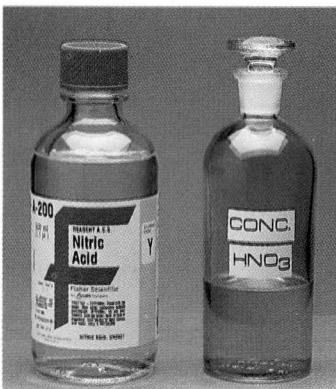

Concentrated HNO_3. On the left, a fresh bottle; on the right, an aged sample.

Commercial "concentrated HNO_3" has a concentration of 16 M and is about 68% (w/w) HNO_3. The pure, anhydrous acid is known only in its solid state. In the liquid or gaseous states, it spontaneously decomposes to nitrogen dioxide and oxygen.

$$4HNO_3 \longrightarrow 4NO_2 + 2H_2O + O_2$$

Sunlight accelerates this reaction, and it also slowly happens in aqueous nitric acid solutions. (Since NO_2 is a reddish-brown gas, aged nitric acid solutions generally have a yellow to reddish color.)

Nitric acid is manufactured from ammonia by the Ostwald process, developed in 1902 by Wilhelm Ostwald (Germany; 1853–1932; Nobel prize, 1909). The availability of synthetic methods to make both ammonia and nitric acid prior to World War I gave Germany an essential freedom from overseas supplies of nitrates and the ability to carry on the war far longer than it would otherwise have been able. Germany had no nitrate deposits, and nitrates are needed to make gunpowder, as we will soon see.

In modern operations of the Ostwald process, a mixture of air with 10% (v/v) NH_3, heated to 850 °C and pressurized to 5 atm, flows through a series of gauzes made of an alloy of platinum and rhodium, which serve as the catalyst. The first reaction is the oxidation of ammonia to nitrogen monoxide (nitric oxide), which we discussed earlier (page 804).

$$4NH_3(g) + 5O_2(g) \xrightarrow[850\ °C,\ 5\ atm]{Pt/Rh\ catalyst} 4NO(g) + 6H_2O(\ell) \qquad \Delta H° = -1170 \text{ kJ}$$

As you can see, this step is very exothermic. The flowing gas stream has to be cooled somewhat, so more air is let in, and the nitrogen monoxide reacts with oxygen to give nitrogen dioxide.

$$2NO(g) + O_2(g) \longrightarrow 2NO_2(g)$$

The flowing gases are now passed into a spray of water in special absorbing towers where newly formed NO_2 reacts with water to give 60% (w/w) nitric acid.

$$3NO_2(g) + H_2O \longrightarrow 2HNO_3(aq) + NO(g)$$

Unchanged components of the initial air plus the NO produced in this last step leave the absorbing towers and are recycled.

The chief problem solved by the Ostwald catalyst is to keep the oxidations from producing either nitrogen, N_2, or dinitrogen monoxide (nitrous oxide), N_2O.

The overall change in the Ostwald process can be written as follows.

$$NH_3(g) + 2O_2(g) \longrightarrow HNO_3(aq) + H_2O$$

The 60% nitric acid produced in the absorbing towers is concentrated to a maximum of 68.5% (w/w) HNO_3 by distillation. At this concentration the solution boils at a constant temperature, and cannot be further separated by distillation. Special methods are needed to obtain a more concentrated nitric acid.

By far the largest use (80%) of nitric acid is in making ammonium nitrate, NH_4NO_3, for fertilizers.

$$NH_3 + HNO_3 \longrightarrow NH_4NO_3$$

Ammonium nitrate is also an important explosive and is an ingredient in many gunpowder "recipes" and in explosives used in mining.

The remaining nitric acid is divided largely between the manufacture of nylon and that of various organic nitro compounds used as explosives, for example, TNT (trinitrotoluene), nitrocellulose, and nitroglycerin. In the production of nylon, nitric acid is used because it is an oxidizing agent, not because it is an acid. And this is usually the reason when nitric acid is picked over hydrochloric or sulfuric acid in many laboratory uses.

Chemical Properties of Nitric Acid

Nitric acid is a strong monoprotic acid, a strong oxidizing agent, and a nitrating agent. We will illustrate each of these.

Nitric acid, like all strong acids, readily neutralizes all proton-accepting bases. The products include nitrate salts, as in the reaction with ammonia, just above, and as the following examples illustrate. (These serve as a review of the typical kinds of reactions of any strong acid.)

With a Metal Hydroxide

$$HNO_3(aq) + KOH(aq) \longrightarrow KNO_3(aq) + H_2O$$

With a Metal Carbonate

$$2HNO_3(aq) + Na_2CO_3(aq) \longrightarrow 2NaNO_3(aq) + H_2O + CO_2(g)$$

With a Metal Bicarbonate

$$HNO_3(aq) + KHCO_3(aq) \longrightarrow KNO_3(aq) + H_2O + CO_2(g)$$

With a Metal Oxide

$$2HNO_3(aq) + MgO(s) \longrightarrow 2Mg(NO_3)_2(aq) + H_2O$$

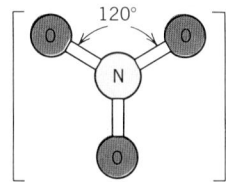

Nitrate ion (all N—O bond lengths are the same)

As we mentioned, nitric acid is used more as an oxidizing agent than as an acid. The following half-reactions and their associated standard reduction potentials—all positive—indicate that in acid the nitrogen in the $+5$ oxidation state of the nitrate ion can be reduced to nitrogen in the $+4$, $+3$, and $+2$ states.

$$NO_3^-(aq) + 4H^+(aq) + 3e^- \rightleftharpoons \underset{\text{(+2 state)}}{NO(g)} + 2H_2O \qquad E^\circ = +0.96 \text{ V}$$

$$NO_3^-(aq) + 3H^+(aq) + 2e^- \rightleftharpoons \underset{\text{(+3 state)}}{HNO_2(aq)} + H_2O \qquad E^\circ = +0.94 \text{ V}$$

$$NO_3^-(aq) + 2H^+(aq) + e^- \rightleftharpoons \underset{\text{(+4 state)}}{NO_2(g)} + H_2O \qquad E^\circ = +0.80 \text{ V}$$

A typical oxidation by nitric acid is its reaction with copper. As we studied on page 462, the products depend on the concentration of the nitric acid. *Concentrated* nitric acid favors the formation of nitrogen dioxide, NO_2.

$$Cu(s) + 4HNO_3(aq) \longrightarrow Cu(NO_3)_2(aq) + 2NO_2(g) + 2H_2O$$

Dilute nitric acid favors the formation of nitrogen monoxide, NO.

$$3Cu(s) + 8HNO_3(aq) \longrightarrow 3Cu(NO_3)_2(aq) + 2NO(g) + 4H_2O$$

In the manufacture of organic nitro compounds, nitric acid functions as a *nitrating agent*. Such an agent creates a covalent bond in an organic molecule between a nitro group (NO_2) and a carbon atom, as in TNT, or from an oxygen atom to the NO_2 unit, as in nitroglycerin.

The yellow spots on skin after a nitric acid spill result from some nitration of organic groups in skin proteins to produce a yellow dye. The spots will wear off, but not wash off.

Dynamite sticks are widely used in mining.

$$
\begin{array}{c}
\text{CH}_3 \\
\text{O}_2\text{N} \quad \text{NO}_2 \\
\\
\text{NO}_2
\end{array}
$$

trinitrotoluene, TNT

$$
\begin{array}{l}
\text{CH}_2-\text{O}-\text{NO}_2 \\
| \\
\text{CH}-\text{O}-\text{NO}_2 \\
| \\
\text{CH}_2-\text{O}-\text{NO}_2
\end{array}
$$

nitroglycerin

An explosive like TNT is a substance that can very suddenly change from a solid or liquid into hot expanding gases. Thus, 2 moles of TNT, a solid, almost instantly change to 15 moles of hot gases plus some powdered carbon, which gives a dark, sooty appearance to the explosion.

$$2C_7H_5N_3O_6(s) \longrightarrow 3N_2(g) + 7CO(g) + 5H_2O(g) + 7C(s)$$
TNT

Nitroglycerin also detonates to give 35 moles of hot gases from 4 moles of explosive, but no carbon forms, so it is used to make "smokeless" powders.

$$4C_3H_5N_3O_9(s) \longrightarrow 6N_2(g) + 12CO(g) + 10H_2O(g) + 7O_2(g)$$
nitroglycerin

To artillery and naval gunners in conventional warfare, smokeless gunpowder means that the field of vision does not become obscured during the action.

To be a *useful* explosive, the substance has to be able to withstand, without detonating, the bumps and jolts of its handling as well as normal ranges of temperatures. And its sudden decomposition should not be catalyzed by materials with which it would have contact during handling. The fortune that Alfred Nobel accumulated, now used to fund the Nobel prizes, came largely from his invention of dynamite, which made nitroglycerin relatively safe to use. The standard dynamite used in the United States is a

Smokeless powder in use.

mixture of nitroglycerin, ammonium nitrate, sodium nitrate, wood pulp, and a trace of calcium carbonate (to neutralize traces of acids that might form during storage).

Nitrate Salts

Nitrates of nearly all metals are known. What is often isolated from the aqueous solution formed when a nitrate is made by the neutralization of a base with nitric acid is a hydrate of the metal nitrate. Thus the Group IIA metals generally crystallize as hydrates that decompose below 100 °C to less hydrated or anhydrous forms.

Nitrates of metals vary widely in thermal stability. Those of Groups IA and IIA are so thermally stable, and have such relatively low melting points that they are used to prepare molten salt baths when heat in a medium range of temperatures is needed to sustain some chemical reaction. At high enough temperatures, however, these nitrates decompose to the corresponding nitrites and oxygen or, when the nitrite is thermally unstable, to the metal oxides and other products. Note, for example, the different ways that sodium and potassium nitrates decompose, one to a nitrite and the other to an oxide.

Nitrate	Melting Point (°C)
$LiNO_3$	255
$NaNO_3$	303
KNO_3	310
$CsNO_3$	414

$$2NaNO_3(s) \xrightarrow{>500\ °C} 2NaNO_2(s) + O_2(g)$$

$$4KNO_3(s) \xrightarrow{>500\ °C} 2K_2O(s) + 2N_2(g) + 5O_2(g)$$

Ammonium nitrate explodes violently at high temperatures, particularly when some easily oxidized impurity is present, to produce nitrogen, oxygen, and water by the following equation.

$$2NH_4NO_3(s) \xrightarrow{>300\ °C} 2N_2(g) + O_2(g) + 4H_2O(g)$$

Detonators can set off the explosion, as well, and ammonium nitrate is commonly a component of explosives, as we noted earlier. A major part of Texas City, Texas, was destroyed in April 1947 from an explosion of ammonium nitrate being loaded into a ship. Nearly 600 lives were lost.

Under milder conditions, ammonium nitrate decomposes to dinitrogen monoxide, and the reaction is a method of preparing this anesthetic gas.

$$NH_4NO_3(s) \xrightarrow{200\text{-}260\ °C} N_2O(g) + 2H_2O(g)$$

With very few exceptions, nitrate salts are soluble in water. The exceptions, for example, the nitrates of mercury(II) and bismuth, appear not to dissolve because they react with water to give an insoluble compound. Thus bismuth nitrate, formed by the addition of Bi_2O_3 to concentrated nitric acid, decomposes when this solution is diluted with water.

$$\underset{\text{bismuth subnitrate}}{Bi(NO_3)_3 + H_2O \longrightarrow BiO(NO_3)(s)} + 2HNO_3(aq)$$

Interestingly, the pentahydrate of bismuth nitrate, $Bi(NO_3)_3 \cdot 5H_2O$, which can be crystallized from the *concentrated* nitric acid solution of bismuth nitrate, gives the same reaction when it is added to water. Efforts to use heat to drive off the water of hydration of hydrated bismuth trinitrate gives bismuth oxide, not the anhydrous nitrate. The Bi^{3+} ion attracts a water molecule, to give $[Bi{-}OH_2]^{3+}$. The central positive charge on Bi^{3+} strongly polarizes the O{-}H bonds in $[Bi{-}OH_2]^{3+}$, and places substantial partial positive charges on the H atoms. So this ion then gives up two protons and becomes BiO^+. This ion can crystallize with a nitrate ion to give bismuth subnitrate, $BiO(NO_3)$. We see here another illustration of the effect of a strong positive charge of a central ion on chemical structure and properties.

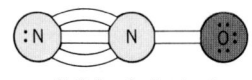

N₂O (Lewis structure)

Oxides of Nitrogen

The oxides of nitrogen (Table 19.12) are compounds with very different reactivities.

Dinitrogen monoxide (nitrous oxide), N_2O, is widely used as a safe, nonexplosive anesthetic ("laughing gas"). At room temperature, N_2O is quite unreactive with most substances, including alkali metals, the halogens, and even ozone. It is therefore widely used as a propellant in aerosol cans. We noted earlier that it can be made by the careful thermal decomposition of ammonium nitrate. When heated sufficiently, N_2O decomposes to nitrogen and oxygen.

$$2N_2O(g) \longrightarrow 2N_2(g) + O_2(g) \qquad \Delta H° = -156.1 \text{ kJ}$$

This exothermic reaction can occur in the cylinder of an auto engine, where 3 moles of gas would form from 2 moles, providing an extra boost and also liberating heat. This heat of decomposition adds to the heat of combustion of the fuel to give extra takeoff power. Drivers of racing cars, therefore, sometimes inject dinitrogen oxide into their fuel lines to give more power to the engine. Since oxygen is produced, it promptly enters into the combustion of the fuel, providing a further boost, so the overall chemistry is more complicated than the equation given above would suggest.

Nitrogen monoxide (nitric oxide), NO, can be made by the reaction of dilute nitric acid with a relatively unreactive metal, like copper, as we studied earlier. NO also forms when fuels are burned in vehicles. It forms in an equilibrium with N_2 and O_2 when these two elements are heated together at very high temperatures as air and fuel react in the cylinders of internal combustion engines.

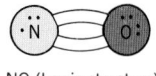

NO (Lewis structure)

$$N_2(g) + O_2(g) \rightleftharpoons 2NO(g) \qquad \Delta H° = +180.4 \text{ kJ } (K_p = 4.1 \times 10^{-31})$$

You can see from the equilibrium constant that at 25 °C the *reactants* are very strongly favored in this equilibrium. Essentially no NO forms at this temperature. At the high temperature in an engine cylinder, however, the equilibrium constant is somewhat higher, so some NO does form. As the exhaust gases cool, the NO does not immediately change back to N_2 and O_2, because the energy of activation for this change is high enough to make this process very slow. Thus, NO is a by-product of the combustion of fuels, and it is a significant air pollutant.

TABLE 19.12 Oxides of Nitrogen

Oxidation State of Nitrogen	Formula	Name	Color	Boiling Point (°C)	Melting Point (°C)
+1	N_2O	Dinitrogen monoxide (nitrous oxide)	None	−89	−91
+2	NO	Nitrogen monoxide (nitric oxide)	None	−152	−164
+3	N_2O_3	Dinitrogen trioxide (nitrous anhydride)	Red-brown	Decomposes	−102
+4	$2NO_2$ ⇅	Nitrogen dioxide[a]	Brown	−11[b]	
+4	N_2O_4	Dinitrogen tetroxide	None		−21
+5	N_2O_5	Dinitrogen pentoxide (nitric anhydride)	None	Decomposes	32

[a] NO_2 and N_2O_4 exist in the presence of each other in both the liquid and the gaseous states. At 25 °C and 760 torr, N_2O_4 is favored in a mixture of these gases; its partial pressure is about 540 torr and that of NO_2 is about 220 torr. In the solid state only N_2O_4 is present.

[b] The temperature at which the *mixture's* vapor pressure equals the atmospheric pressure is −11 °C.

Whenever NO forms and mixes with air at moderate to low temperatures, it combines quickly with oxygen to form nitrogen dioxide, NO_2.

$$2NO(g) + O_2(g) \longrightarrow 2NO_2(g)$$

Thus, when hot auto exhaust gases with traces of NO reach the outside air and are cooled, the NO is oxidized to NO_2. This gas gives the characteristic reddish-brown color to smog (Figure 19.14).

The electron configuration of the NO molecule, in molecular orbital (MO) theory, is similar to that of O_2 (Table 8.4, page 296). However, NO has one less electron than O_2. This electron is unpaired, so NO is paramagnetic. And this odd electron exists in an antibonding $2\pi^*$ MO, which affects the bond order of the N—O bond. It is close to 2.5, not 3 as in N_2, because the odd electron cancels some of the bonding ability of the paired electrons of the π orbitals. The resulting N—O bond distance is 120 pm—roughly intermediate between a triple bond and a double bond.

Dinitrogen trioxide (nitrous anhydride), N_2O_3, is a deeply blue liquid, but is unstable as a gas. As it is warmed above -30 °C, the following equilibrium increasingly shifts in favor of nitrogen monoxide and nitrogen dioxide.

$$N_2O_3(g) \rightleftharpoons NO(g) + NO_2(g)$$

In fact, for chemical studies of dinitrogen trioxide, a mixture of NO and NO_2 gases at -20 °C usually serves the purpose.

N_2O_3 reacts with water to form aqueous nitrous acid.

$$N_2O_3(g) + H_2O \longrightarrow 2HNO_2(aq)$$

N_2O_3 reacts with aqueous base to give aqueous nitrite ion.

$$N_2O_3(g) + 2OH^-(aq) \longrightarrow 2NO_2^-(aq) + H_2O$$

We will look closer at nitrous acid and the nitrite salts later.

Nitrogen dioxide, NO_2, and **dinitrogen tetroxide**, N_2O_4, occur together in an equilibrium.

$$2NO_2(g) \underset{\text{heat}}{\overset{\text{cool}}{\rightleftharpoons}} N_2O_4(g)$$

nitrogen   dinitrogen
dioxide   tetroxide
(reddish brown)   (colorless)

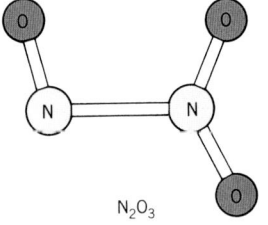

N_2O_3

N₂O₄ has been used as an oxidant in space shuttle engines.

N_2O_4 is called a *dimer* of NO_2 (from *di-*, "two," and the Greek *meros*, "parts"). The formation of N_2O_4 is an example of a *dimerization* reaction. In terms of Lewis structures, we see that this dimerization creates a new N—N bond by the pairing and sharing of the initially unpaired electrons on the N atoms of two NO_2 molecules.

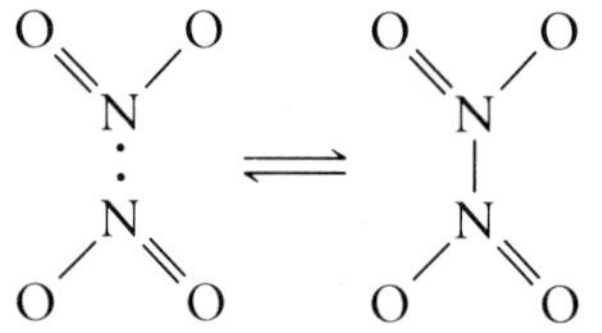

The position of this equilibrium and the color of the system vary with temperature. Pure, solid N_2O_4 is present below -21 °C, the melting point of the system, so there is no color. At -21 °C (the melting point of N_2O_4), there is 0.01% NO_2, enough to make the system pale yellow. The mixture boils at 21 °C, and now it contains 0.1% NO_2, enough to impart a deep reddish-brown color. Above 140 °C, the system is 100% NO_2 and an even darker reddish brown.

Nitrogen dioxide is paramagnetic because each molecule has an unpaired electron. Dinitrogen tetroxide, in which all electrons are paired, is diamagnetic.

When the equilibrium mixture of NO_2 and N_2O_4 is bubbled into aqueous sodium hydroxide, a solution of sodium nitrate and sodium nitrite forms. We can treat the reaction as occurring to NO_2.

$$2NO_2(g) + 2NaOH(aq) \longrightarrow NaNO_3(aq) + NaNO_2(aq) + H_2O$$

As we have already noted, NO_2 forms when NO mixes with oxygen at ordinary temperatures. Thus one pollutant, NO, leads to another, NO_2, which is in equilibrium with N_2O_4. In environmental affairs these nitrogen oxides are collectively referred to as the nitrogen oxide pollutants and given the symbol NO_x. The NO_2 component, of course, is the dangerous member of this pair. Besides contributing the disagreeable color to smog, it is itself poisonous, and to make matters worse, it reacts with rainwater to generate a dilute solution of nitric acid and nitrous acid.

$$2NO_2(g) + H_2O \longrightarrow HNO_3(aq) + HNO_2(aq)$$

This is therefore one source of the acid in *acid rain* (which we will say more about in the next chapter after we've introduced another component).

Dinitrogen pentoxide (nitric anhydride), N_2O_5, is a white solid made by letting

P₄O₁₀ is a powerful dehydrating agent.

P_4O_{10} remove water from concentrated nitric acid by the following reaction. (Nitric anhydride gets its name because it is related to nitric acid by this dehydration.)

$$4HNO_3(aq) + \underset{\substack{\text{tetraphosphorus}\\\text{decaoxide}}}{P_4O_{10}(s)} \xrightarrow{-10\ °C} 2\underset{\substack{\text{dinitrogen}\\\text{pentoxide}}}{N_2O_5(s)} + 4\underset{\substack{\text{metaphosphoric}\\\text{acid}}}{HPO_3(\ell)}$$

As noted in Chapter 18, N₂O₅ decomposes gradually to give NO₂ and O₂.

Solid dinitrogen pentoxide sublimes at 32.4 °C, but it must be handled very cautiously, because it can explode. In the solid state it exists as a combination of the nitronium ion, NO_2^+, and the nitrate ion—as nitronium nitrate ($NO_2^+NO_3^-$). The covalent form is present in the gas phase. It reacts very rapidly with water, even the moisture in humid air, to generate nitric acid.

$$N_2O_5(s) + H_2O \longrightarrow 2HNO_3(aq)$$

Nitrous Acid and the Nitrites

Nitrous acid, HNO_2, is a weak, unstable acid that is known only in solution. It has never been isolated as a pure compound. Its salts, however, are stable. Nitrous acid has a pK_a of 3.35 (at 18 °C), making it a slightly stronger acid than acetic acid (pK_a 4.75, 25 °C). But nitrous acid is still weak enough to make its anion, the nitrite ion (NO_2^-), a good Brønsted base. So when aqueous nitrous acid is needed (and the presence of other ions is unimportant), chemists simply mix cold hydrochloric acid with an equimolar amount of sodium nitrite.

$$NaNO_2(aq) + HCl(aq) \xrightarrow{0\ °C} HNO_2(aq) + NaCl(aq)$$

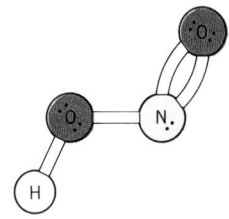

HNO$_2$ (Lewis structure)

If this solution were heated, the nitrous acid would decompose largely as follows.

$$3HNO_2(aq) \xrightarrow{heat} HNO_3(aq) + 2NO(g) + H_2O$$

A gaseous mixture of NO, NO_2, and water vapor exists in equilibrium with HNO_2.

One NO_2^- ion is oxidized to NO_3^-, and two NO_2^- ions are reduced to 2NO.

$$2HNO_2(g) \rightleftharpoons NO(g) + NO_2(g) + H_2O(g)$$

The reverse reaction shows the combination of water with the two oxides—NO and NO_2—that gives N_2O_3, the nitrous anhydride that we studied earlier. The forward reaction, the apparent dehydration of nitrous acid to the equivalent of N_2O_3, explains why N_2O_3 came to be called nitrous anhydride.

Perhaps the most common nitrite salt is sodium nitrite, prepared by bubbling an equimolar mixture of NO and NO_2 (the equivalent of N_2O_3) into aqueous sodium hydroxide.

$$2NaOH(aq) + NO(g) + NO_2(aq) \longrightarrow 2NaNO_2(aq) + H_2O$$

Sodium nitrite can be obtained by evaporating the water; the salt is thermally stable, melting without decomposition at 284 °C. It is very soluble in water and draws moisture from humid air. Sodium nitrite is a food additive. (See Special Topic 19.3.)

SUMMARY

Occurrence of Nonmetals.   Just a few nonmetal elements, notably hydrogen and helium, contribute the greatest number of atoms to the universe. Oxygen is particularly prevalent on earth, although iron contributes the greatest mass, and only a few nonmetals dominate the substances in living systems. In the periodic table, the nonmetals occur in the upper right-hand corner.

Noble Gases.   The noble gas elements — helium, neon, argon, krypton, xenon, and radon — are noted (and used) for their lack of chemical reactivity. Krypton forms fluorides and xenon forms several compounds, but mostly with the most electronegative elements — fluorine and oxygen. The most used noble gases (because they are the most abundant) are helium and argon. They provide inert blanketings where the exclusion of anything reactive, particularly oxygen, is essential, as inside light bulbs. Mixtures of noble gases are used to make the colors of neon lights.

Hydrogen.   Most hydrogen is manufactured from hydrocarbons, but small amounts are made by the reaction of an active metal with acid (or water) or from the reaction of a metal hydride with water. Three isotopes are known — protium (hydrogen), deuterium, and tritium — but the natural abundance of tritium is extremely low. Deuterium gives an important isotope effect.

Most hydrogen is used immediately upon its synthesis in the manufacture of ammonia and related nitrogen fertilizers. Some hydrogen goes to the manufacture of hydrogenated oils, detergents, plastics, and organic chemicals, and some is used as the fuel in rockets.

Binary compounds of hydrogen — the hydrides — exist for nearly all elements except the noble gases. Hydrogen bonds are possible when H is bonded to very electronegative elements — F, O, and N. The ionic hydrides involve the least electronegative elements and can be made by heating the metal with hydrogen. Ionic hydrides are hydride ion donors (and so are reducing agents and bases). The covalent hydrides involve nonmetal elements, but their bonds to hydrogen have partial ionic character. When H is joined to an atom of higher electronegativity, it bears a $\delta+$ charge, so such covalent hydrides are acids. Some are strong acids (e.g., the hydrides of Cl, Br, and I), and others (e.g., H_2O, H_2S) are weak. In a hydride such as CH_4, in which there is virtually no $\delta+$ or $\delta-$ on H, the substance is neither an acid nor a base; if anything, it tends to donate hydrogen atoms. When H is joined to a less electronegative atom, as in many covalent metal hydrides, the compound tends to be a hydride ion donor. Some hydrides of the transition metals are donors of H_2.

Oxygen.   Elements of the oxygen family — oxygen, sulfur, selenium, and tellurium — display a steady transition from a nonmetal to a metalloid. Most oxygen is obtained from liquid air, and over half the amount annually produced is used in the production of metals and alloys. The chemicals industry, water treatment, clinical uses, and rocketry also use oxygen. In nature, oxygen is essential to the respiration of nearly all living things and to the processes of decay and combustion. It is regenerated by photosynthesis.

Ozone (O_3), an allotrope of oxygen, is a powerful oxidizing agent present in smog (where it is harmful to health) and in the stratosphere (where it helps screen out UV radiation that could harm us). Besides laboratory uses, ozone is used by many European cities to purify water.

Binary compounds of oxygen — the oxides — exist for all elements except He, Ne, and Ar. When the oxidation state of the other element in the oxide is particularly high, as in OsO_4 and CrO_3, the oxide is covalent and acidic. As the oxidation number of the other element decreases, the oxide becomes more and more ionic and basic. Thus oxides of Group IA metals and most Group IIA metals as well as several transition metals in their $+2$ oxidation states are ionic and basic oxides. Between the acidic and basic oxides lie some that are amphoteric, but their high melting points suggest that they are ionic. Only a few metal oxides are inert to acids and bases.

Hydrogen peroxide, H_2O_2, is a strong oxidizing agent in both acids and bases, like all peroxides (compounds with O—O bonds). It is unstable — decomposing to water and oxygen — and when pure or even at concentrations above 15–20% it has to be handled very carefully. Most hydrogen peroxide is used as a bleaching agent for paper pulp and fabrics.

Nitrogen.   The elements of the nitrogen family are nitrogen, phosphorus, arsenic, antimony, and bismuth. Nitrogen is the principal component of air and is obtained from it. Its chief uses are in enhanced oil recovery and as an inert blanketing agent. Liquid nitrogen is used to freeze food products and as a coolant. The nitrogen of the Haber–Bosch process for ammonia is taken directly from air. The nitrogen cycle in nature moves nitrogen nuclei from air through microorganisms (by nitrogen fixation) and back to air via decay processes.

In compounds, nitrogen occurs in ten oxidation states ranging from -3 to $+5$ as well as $-\frac{1}{3}$. The Haber–Bosch process converts the hydrogen in methane and the nitrogen in air into ammonia, a fertilizer, and a raw material for nitric acid and other chemicals.

Ammonia is a weak base, but quantitatively neutralizes strong acids. While it is not readily flammable, it can burn (and can even be made to explode). Liquid ammonia dissolves Group IA and IIA metals to give a solvated (ammoniated) electron and a metal ion. These solutions decompose eventually to metal amides (salts of the amide ion, NH_2^-) and H_2. The amide ion is a powerful base that can take a proton from water, leaving the OH^- ion. Liquid ammonia has an acid–base chemistry analogous to that of liquid water. The ammonium ion, NH_4^+, is a proton donor and tends to form slightly acidic aqueous solutions. Ammonium salts are soluble in water.

Hydrazine, N_2H_4, is a toxic liquid, a weaker base than ammonia, and a strong reducing agent. It has been used as a rocket fuel.

Hydroxylamine, NH_2OH, is a white, unstable solid, usually kept as a hydrochloride salt, that is a weaker base than hydrazine and a milder reducing agent.

Hydrazoic acid, HN_3, with N in the $-\frac{1}{3}$ oxidation state, is a liquid susceptible to explosion. It is a weak acid, and its salts are called azides. Heavy metal azides can detonate. The azide ion, N_3^-, is a linear resonance-stabilized ion that behaves much like a chloride ion.

Oxoacids and Oxides of Nitrogen. Nitric acid, the important oxoacid of nitrogen and a strong oxidizing agent, is made from ammonia by the Ostwald process. The oxidation of NH_3 is channeled by the choice of catalyst to NO, which further oxidizes to NO_2, and this reacts with water to give HNO_3 and some NO (which is recycled). Most nitric acid is used to make fertilizers; some is used to make explosives.

Nitric acid gives all the typical reactions of a strong, monoprotic acid, but it is more often used for its oxidizing properties. When concentrated, nitric acid tends to change to NO_2 in its reaction with a metal such as copper. Dilute nitric acid, when an oxidizing agent, tends to change to NO.

Nitrate salts of the Group IA and IIA metals are stable beyond their melting points—which tend to be relatively low for salts—but eventually decompose to nitrites or metal oxides at very high temperatures. (Ammonium nitrate explodes if shocked or heated too rapidly.) Most nitrates are soluble in water, but a few (e.g., bismuth nitrate) react with water. Thus the Bi^{3+} ion of bismuth nitrate changes in water to BiO^+, and $BiO(NO_3)$ forms.

Dinitrogen monoxide (N_2O, nitrous oxide), an anesthetic gas, is relatively unreactive at ordinary temperatures but decomposes when heated to nitrogen and oxygen.

Nitrogen monoxide (NO, nitric oxide) begins to form when nitrogen and oxygen are strongly heated (as in internal combustion engines). It reacts further with oxygen at lower temperatures to give NO_2. With an odd electron in its molecule, NO is paramagnetic.

Dinitrogen trioxide (N_2O_3, nitrous anhydride) exists in the gaseous state only in an equilibrium mixture with NO and NO_2. When the mixture is bubbled into water, nitrous acid forms; if added to aqueous base, then the nitrite ion forms.

Nitrogen dioxide (NO_2) and dinitrogen tetroxide (N_2O_4) exist together in equilibrium. N_2O_4, a colorless substance, is favored at low temperatures, and NO_2, a reddish-brown, paramagnetic substance, is favored at high temperatures. In aqueous base, NO_2 gives both nitrate and nitrite ions. When NO_2 is added to water, both nitric and nitrous acids form, so NO_2 (an air pollutant) contributes to acid rain. In environmental studies, NO and NO_2 are referred to collectively as "nitrogen oxides" and symbolized as NO_x.

Dinitrogen pentoxide (N_2O_5, nitric anhydride), a white unstable solid, reacts with water to give nitric acid. In the solid state, N_2O_5 exists as $(NO_2^+NO_3^-)$, but it is largely covalent in the gas phase.

Nitrous acid (HNO_2) is a weak unstable acid known only in solution. When the solution is heated, HNO_2 decomposes to nitric acid and NO. An equimolar mixture of NO, NO_2, and H_2O vapor exists in the gas phase with molecules of HNO_2. Nitrites have been used as food preservatives.

REVIEW EXERCISES

Answers to questions whose numbers are printed in color are given in Appendix C. Difficult questions are marked with asterisks. Several questions will require you to review concepts from earlier chapters.

Occurrence of Nonmetals

19.1 In the "big bang" theory, the initial body consisted of what particles?

19.2 What two elements are the most abundant in the universe today?

19.3 In very general terms, how did the helium that is now in the universe form?

19.4 Which element contributes the most atoms to planet earth?

19.5 Which two elements make up most of our atmosphere and in what percentages (on a volume/volume basis)?

19.6 Which general processes on earth use oxygen?

19.7 Which general processes on earth regenerate oxygen?

19.8 Which four nonmetals predominate in the human body?

Noble Gases

19.9 What are the names and atomic symbols of the members of the noble gas family?

19.10 Which noble gases can form compounds? Give the formulas of some examples.

19.11 Explain why xenon is more reactive than krypton.

19.12 Describe some uses of the noble gas elements.

19.13 Explain why helium and argon are more widely used than the other noble gases.

19.14 Why is the atmosphere in an electric light bulb a noble gas and not air or nitrogen?

19.15 Describe an unusual property of liquid helium.

Hydrogen

19.16 What kinds of compounds are the chief sources of hydrogen in industry?

19.17 Give the names, atomic symbols, and mass numbers of the isotopes of hydrogen.

19.18 Why have the hydrogen isotopes been given their own names?

19.19 What is the *deuterium isotope effect*?

19.20 Explain why the proton does not exist as a free chemical species but is always associated with some other species (as in H_3O^+).

19.21 What are the three kinds of hydrides? Give the formula and name of an example of each kind.

19.22 Give one physical and one chemical property in which ionic and covalent hydrides differ markedly.

19.23 Complete the following equations. If no reaction occurs, write "no reaction."

(a) $NaH(s) + H_2O \longrightarrow$ (d) $Na(s) + H_2(g) \longrightarrow$

(b) $CaH_2(s) + H_2O \longrightarrow$ (e) $Mg(s) + H_2(g) \longrightarrow$

(c) $HCl(g) + H_2O \longrightarrow$

19.24 How was it determined that a compound like NaH contains the hydride ion?

19.25 What is the largest industrial use of hydrogen?

19.26 Write the overall equation for the Haber–Bosch process.

19.27 Explain why a high pressure helps drive the Haber–Bosch reaction.

19.28 How does hydrogen react with a carbon–carbon double bond? For your answer complete the following equation: $CH_2{=}CH_2 + H_2 \rightarrow$.

19.29 What is the OXO process and why is it important, industrially?

Oxygen

19.30 Give the names and atomic symbols of the members of the oxygen family.

19.31 How did Priestly make oxygen?

19.32 How is the oxygen needed for industrial purposes manufactured?

19.33 Give three major *commercial* uses of oxygen.

19.34 What are the names and formulas of the allotropes of O_2?

19.35 In what specific ways do the terms *isotope* and *allotrope* differ? (In your answer, give examples.)

19.36 How is ozone made in the lab (in general terms)?

19.37 How do ultraviolet light and neutral particles in air work together to convert oxygen to ozone?

19.38 Why is ozone in smog dangerous?

19.39 Give one advantage and one disadvantage of using ozone in water treatment.

Oxides

19.40 The oxides of which representative family of elements in the periodic table (a) are most ionic? (b) Which are most basic? (c) Which are insoluble in water but can still neutralize strong acids? (d) Which are most acidic in water?

19.41 If moist red litmus paper is pressed against powdered MgO, the paper turns blue. Explain.

19.42 The melting point of TiO_2 is 1870 °C; the melting point of OsO_4 is 40 °C. In which solid oxide are the bonds more likely to be ionic? How do the melting points suggest the answer?

19.43 Explain why the bonds in OsO_4 are covalent.

19.44 (a) Write the net ionic equation that describes the conversion of $[Al(H_2O)_6]^{3+}$ into $[Al(H_2O)_2(OH)_4]^-$ by the action of OH^-.

(b) It appears that this conversion consists of proton transfers from four of the six water molecules in the hydrated cation to OH^- ions. What makes these proton transfers possible when they do not occur from the water molecules of a hydrated sodium ion?

19.45 One of the reactions of molybdenum(VI) oxide is

$$MoO_3(s) + 2OH^-(aq) \longrightarrow MoO_4^{2-}(aq) + H_2O$$

From this information, are the bonds in MoO_3 likely to be covalent or ionic? How can you tell?

19.46 Sulfur trioxide is described as an *acidic oxide*. What does this mean?

19.47 $Cr(OH)_3$ is amphoteric. What does this mean? Give equations of possible reactions that illustrate this property.

Peroxides and Superoxides

19.48 What must be true, *structurally*, about any compound to call it a peroxide?

19.49 List some uses of dilute hydrogen peroxide.

19.50 Calculate the oxidation number of O in H_2O_2.

19.51 In what way is pure hydrogen peroxide dangerous?

19.52 (a) Write the K_a expression for the first ionization of hydrogen peroxide.

(b) The pK_a of H_2O_2 is 11.75. What is its K_a?

(c) Calculate the pH of a solution of 1.0 M H_2O_2.

*19.53 Hydrogen peroxide oxidizes H_2SO_3 to SO_4^{2-}. Assume that this reaction were run as a galvanic cell.

(a) Write the net ionic equation for the cell reaction.

(b) Based on data in this chapter and Chapter 17, calculate the standard potential of this cell.

(c) Calculate the equivalent weight of H_2O_2 in this redox reaction.

(d) How many grams of H_2O_2 are required to react with 5.25 g of Na_2SO_3 by this reaction?

19.54 A white solid is either Na_2O or Na_2O_2. When a piece of red litmus paper is dipped into a freshly made aqueous solution of the white solid, its color changes from red to white. Which substance is it? How can you tell? What would happen to the red litmus paper if the solid were the other compound?

19.55 Which member or members of the Group IA metals react(s) with oxygen to give chiefly (a) a peroxide? (b) An oxide? (c) A superoxide?

19.56 Describe by means of equations how potassium superoxide removes carbon dioxide from (a) dry air and (b) humid air.

Reactions Involving Oxygen and Its Compounds

19.57 Complete and balance the following equations.

(a) $NaH(s) + O_2(g) \longrightarrow$ (e) $Na_2O_2(s) + H_2O \longrightarrow$

(b) $H^-(s) + H_2O \longrightarrow$ (f) $Li(s) + O_2(g) \longrightarrow$

(c) $HgO(s) \xrightarrow{\text{high temperature}}$ (g) $H_2O_2(\ell) \xrightarrow{\text{spontaneously}}$

(d) $KClO_3(s) \xrightarrow{\text{high temperature}}$

19.58 Using general descriptions of chemical properties given in this chapter, figure out the products in the following equations and write balanced molecular equations. (Some of these reactions were studied in earlier chapters, particularly Chapter 11.)

(a) $Na_2O(s) + H_2O \longrightarrow$

(b) $Na_2O(s) + HCl(aq) \longrightarrow$
(c) $Na_2O(s) + HNO_3(aq) \longrightarrow$
(d) $Al_2O_3(s) + HCl(aq) \longrightarrow$
(e) $Al_2O_3(s) + NaOH(aq) \longrightarrow$
(f) $K_2O(s) + HCl(aq) \longrightarrow$
(g) $K_2O(s) + H_2O \longrightarrow$
(h) $Na_2O_2(s) + H_2O \longrightarrow$
(i) $Al_2O_3(s) + HBr(aq) \longrightarrow$

19.59 The addition of water to a white solid known to be either Na_2O or Na_2O_2 caused a basic solution to form. A gas evolved. When a burning match was thrust into this gas, the flame flared more brightly. What was the solid and what gas formed? Write the molecular equation.

Nitrogen

19.60 Elemental nitrogen needed for industrial purposes is obtained in what way?

19.61 With respect to enhanced oil recovery using nitrogen, (a) how does it work? (b) Why isn't (cheaper) air used instead of pure nitrogen?

19.62 Why would it be particularly unlikely for nitrogen (a) to occur in a +6 oxidation state? (b) To occur in a −4 oxidation state?

19.63 What is *nitrogen fixation*?

Ammonia

19.64 The value of $\Delta H_{vaporization}$ of liquid ammonia is 1370 J g^{-1}. Suppose its value were 13.7 J g^{-1} instead. Would this make it easier or more difficult to handle liquid ammonia as a solvent? Explain.

19.65 Answer the following questions concerning the Haber–Bosch process.
(a) Why should a low temperature favor the formation of ammonia, thermodynamically?
(b) Why doesn't a low temperature actually favor the reaction?
(c) Why does a high pressure favor the process?

19.66 Write the structures of a water molecule and an ammonia molecule and then add a dotted line to show a hydrogen bond between them.

19.67 Write net ionic equations to explain the following.
(a) How ammonia in water causes an increase in pH.
(b) How aqueous ammonia neutralizes hydrochloric acid.
(c) How gaseous ammonia can burn to give nitrogen and water.

19.68 Write molecular equations for the reaction of $NH_3(aq)$ with each of the following acids (assumed to be in dilute solutions). This constitutes a review of the chemistry of ammonia regardless of the chapter in which the reaction was first described.
(a) HCl(aq)   (d) $H_2SO_4(aq)$ (as a diprotic acid)
(b) HBr(aq)   (e) $HNO_3(aq)$
(c) HI(aq)

19.69 When sodium dissolves in liquid ammonia, ammoniated forms of the sodium ion and the electron form. What does *ammoniated* mean?

19.70 In the liquid ammonia system of acid-base reactions, what ion is the chief proton donor and what ion is the chief proton acceptor?

19.71 How is sodium amide prepared? (Write the equation.)

19.72 What reaction will the amide ion give with the ammonium ion in a liquid ammonia solution? Write the net ionic equation.

19.73 Does the behavior of the amide ion in water tell us that this ion is a stronger or a weaker base than the hydroxide ion? Explain.

19.74 If you add $KNH_2(s)$ to water, what will happen, chemically? (Write a molecular equation and a net ionic equation.)

19.75 At a sufficiently high temperature, $NH_4Cl(s)$, decomposes. Write the molecular equation for this decomposition.

19.76 Write the net ionic equation that explains how a solution of NH_4Cl in water has a pH less than 7.

Nitrides

19.77 When a compound is called a *nitride,* what do we know about it, structurally?

19.78 When a compound is known as a nitride of a Group IIA metal, what do we know (a) about its formula? (b) About the oxidation number of N in it? (c) About its behavior toward water?

19.79 When magnesium is ignited in air it burns with an extremely bright flame, and a white residue remains. When water is sprinkled on this residue, the odor of NH_3 can be detected.
(a) How did NH_3 form? (Write the molecular equation.)
(b) What other compound formed? (Write the molecular equation for its formation.)

Hydrazine

19.80 How is hydrazine prepared? (Write the molecular equation.)

19.81 What is the oxidation number of N in hydrazine?

19.82 Why is it particularly dangerous to let household bleach (of the "chlorine" type) mix with household ammonia?

19.83 When hydrazine is used as a rocket fuel, what provides the thrust? As part of the answer, write a molecular equation.

*19.84 (a) Write the thermochemical equation for the formation of one mole of liquid hydrazine from its elements at 25 °C and 1 atm. The value of ΔH_f° for hydrazine is +50.6 kJ mol^{-1}.
(b) Toward its own elements, is hydrazine *thermodynamically* stable or unstable? How can you tell?
(c) Pure hydrazine is described as being *kinetically* stable with respect to its decomposition to its elements. What does this mean?
(d) The value of ΔG_f° for hydrazine is +149.2 kJ mol^{-1}. Calculate ΔS_f° for hydrazine, and comment on the meaning of its algebraic sign in relationship to the thermochemical equation for its formation.

19.85 The dissolved oxygen present in any highly pressurized, high-temperature steam used in steam boilers can be extremely corrosive at the temperatures used. (The boilers are usually made of an iron alloy.) For the past several years, large steam-boiler installations have used hydrazine to remove this oxygen.

(a) Write the molecular equation for the reaction by which hydrazine does this.

(b) Considering the products of this reaction (part a), are they themselves harmful to the metals of the equipment? Explain.

19.86 Hydrazine in strong acid forms both the $N_2H_5^+$ ion and the $N_2H_6^{2+}$ ion. What are the likely Lewis structures of these ions? According to VSEPR theory, what would be the likely geometry around each N atom in $N_2H_6^{2+}$?

19.87 In a basic solution, hydrazine can be changed to N_2 and H_2O, with $E° = -1.16$ V (25 °C).

(a) Write the half-reaction *in the accepted manner* for this change. (In other words, show it as a reduction.)

(b) On the basis of the information given here and in Chapter 17, could hydrazine reduce $Ag^+(aq)$ to $Ag(s)$? Write the cell equation and calculate the cell potential.

Hydroxylamine and Hydrazoic Acid

19.88 What is the equilibrium equation that shows hydroxylamine, water, and the products of their interaction? Write the expression for K_b for hydroxylamine based on this equilibrium equation.

19.89 Using the values of K_b for ammonia and hydroxylamine given in this text, which is the stronger base?

19.90 Based on the relative basicities of water and ammonia, which site in hydroxylamine, O or N, is likely to be the better proton acceptor? What is the Lewis structure of the protonated form of hydroxylamine?

19.91 What is the equilibrium equation that shows hydrazoic acid, water, and the products of their interaction? Write the expression for K_a for hydrazoic acid based on this equilibrium equation.

19.92 Based on the pK_a of hydrazoic acid (4.75), will a solution of lithium azide in water test slightly acidic, slightly basic, or neutral? Explain with a net ionic equation.

19.93 What property of azides of heavy metal ions should make you careful about ever handling them?

19.94 How does valence bond theory explain the one N—N bond distance observed in the azide ion? (Draw the resonance structures.)

Nitric Acid and Nitrates

19.95 Write the molecular equations for the steps in the Ostwald process.

19.96 What is the chief commercial use of nitric acid?

19.97 After concentrated nitric acid has remained in a bottle exposed to sunlight for some time, the reagent turns from colorless to reddish brown. Write the equation for the reaction responsible for this change in color. What chemical causes the color?

19.98 Write net ionic equations for the reaction of copper with nitric acid (a) when dilute nitric acid is used and is reduced to $NO(g)$, and (b) when concentrated nitric acid is used and is reduced to $NO_2(g)$.

*19.99 The fact that concentrated nitric acid, acting on copper, is changed to NO_2, not NO (which forms when *dilute* nitric acid is used), suggests that excess concentrated nitric acid can *oxidize* NO to NO_2. Answer the following questions using information and equations in this chapter.

(a) What would be the half-reactions for the reaction of HNO_3 with NO to give NO_2?

(b) What would be the resulting cell reaction?

(c) What would be the resulting value of $E°_{cell}$? Does this value suggest that the reaction of HNO_3 with NO would be spontaneous? Explain.

(d) Thinking in terms of the Nernst equation, how might a high concentration of HNO_3 give a favorable value to $E°$?

19.100 (a) Stoichiometrically, what ratio must be particularly high if a liquid or a solid is to be a good explosive?

(b) To be useful and practical, what else must be true about an explosive?

(c) How did Alfred Nobel solve the problem of safely handling nitroglycerin?

19.101 From the ways in which sodium nitrate and potassium nitrate respond to heat above 500 °C, what can we learn about the relative thermal stabilities of sodium nitrite and potassium nitrite?

19.102 What characterizes the nitrates of Group IIA metals with respect to their thermal stability?

19.103 Using equations, describe the behavior of bismuth nitrate toward water.

19.104 The decomposition of ammonium nitrate takes different directions according to the temperature at which it is heated.

(a) Write an equation for this mode at 200–260 °C

(b) Write an equation for this mode at >300 °C.

(c) What do these facts alone possibly tell us about the thermal stability of N_2O and the identities of the products of the thermal decomposition of N_2O?

Oxides of Nitrogen and Nitrous Acid

19.105 The bond order of the N—N bond in N_2O is about 2.5 instead of the value of 3 in N_2. How does molecular orbital theory explain this lower bond order?

19.106 One reason why aerosol cans carry a "do not incinerate" warning is found in the pressure–temperature law of gases. When the propellant is N_2O, there is an additional reason. What is it?

19.107 Explain why N_2O_4 is diamagnetic but NO_2 is paramagnetic.

19.108 How does nitrous acid decompose in water? (Write the equation.)

19.109 The pK_a of HNO_2 (at 18 °C) is 3.35.
  (a) Calculate K_a.
  (b) Write the equilibrium equation on which K_a is based.

19.110 What does the addition of $NaNO_2$ to water do to the pH—raise it, lower it, or leave it unchanged? Write a net ionic equation that explains your answer.

19.111 How is dinitrogen pentoxide prepared? (Write the equation.)

19.112 Draw the Lewis structure of HNO_3.

19.113 Deduce the Lewis structure of the nitronium ion.

19.114 Give the name and formula of the oxide of nitrogen that does the following.
  (a) Gives nitric acid when dissolved in water.
  (b) Forms N_2O_4 when cooled.
  (c) Can be made by heating ammonium nitrate.
  (d) Is unstable in air, being oxidized to NO_2.
  (e) Forms in an automobile cylinder by a reaction between nitrogen and oxygen.
  (f) Is unstable and readily breaks up into two other oxides of nitrogen.
  (g) Is a reddish-brown, poisonous gas.
  (h) Gives the same reactions as a mixture of NO and NO_2.
  (i) Is used by some auto racers to get more power out of the combustion of the fuel.
  (j) Is paramagnetic.
  (k) Is a solid at room temperature.
  (l) Is recycled in the Ostwald process.
  (m) Forms from the decomposition of nitrous acid in water.
  (n) Reacts with aqueous sodium hydroxide to give a mixture of sodium nitrite and sodium nitrate.
  (o) Is responsible for the characteristic color of heavy smog.

Formulas, Names, and Reactions of Substances

19.115 Write the formula of each. (Remember that the formula of an ion must include the kind and amount of charge.)
  (a) Hydride ion
  (b) Potassium amide
  (c) Ammonium sulfate
  (d) Ozone
  (e) Amide ion
  (f) Sodium oxide
  (g) Sodium peroxide
  (h) Sodium hydride
  (i) Ammonium nitrate
  (j) Lithium amide
  (k) Sodium nitride
  (l) Nitric acid
  (m) Sodium amide
  (n) Nitrous acid
  (o) Hydrogen peroxide
  (p) Hydrazine
  (q) Nitric oxide
  (r) Sodium nitrite
  (s) Nitrous oxide
  (t) Nitric anhydride
  (u) Hydroxylamine
  (v) Hydrazoic acid

19.116 Complete and balance the following equations. In many parts, you are expected to figure out the formulas of products from general statements about chemical properties made in the chapter. (Write molecular equations.)
  (a) $Zn(s) + HCl(aq) \longrightarrow$
  (b) $Ca(s) + H_2(g) \longrightarrow$
  (c) $NaH(s) + H_2O \longrightarrow$
  (d) $Mg(s) + HCl(aq) \longrightarrow$
  (e) $KH(s) + H_2O \longrightarrow$
  (f) $CH_3-CH{=}CH_2(g) + H_2(g) \xrightarrow[\text{heat, catalyst}]{\text{pressure}}$
  (g) $CH_2{=}CH_2(g) + CO(g) + H_2(g) \xrightarrow[\text{heat, catalyst}]{\text{pressure}}$
  (h) $HgO(s) \xrightarrow{\text{heat}}$
  (i) $Na_2O(s) + H_2O \longrightarrow$
  (j) $MgO(s) + HBr(aq) \longrightarrow$
  (k) $KClO_3(s) \xrightarrow{\text{heat}}$
  (l) $Al_2O_3(s) + HBr(aq) \longrightarrow$
  (m) $Al_2O_3(s) + KOH(aq) \longrightarrow$
  (n) $CrO_3(s) + H_2O \longrightarrow$
  (o) $H_2O_2(aq) \xrightarrow{\text{heat}}$
  (p) $K_2O_2(s) + H_2O \longrightarrow$

19.117 Complete and balance the following equations. In many parts, you are expected to figure out the formulas of products from general statements about chemical properties made in the chapter. (Write molecular equations.)
  (a) $NH_3(aq) + HCl(aq) \longrightarrow$
  (b) $NH_4Br(s) \xrightarrow{\text{heat}}$
  (c) $NH_4Cl(aq) + NaOH(aq) \longrightarrow$
  (d) $NH_4Cl(am) + NaNH_2(am) \xrightarrow{NH_3(\ell)}$
  (e) $(NH_4)_2Cr_2O_7(s) \xrightarrow{\text{heat}}$
  (f) $Li_3N(s) + H_2O \longrightarrow$
  (g) $NH_3(aq) + NaOCl(aq) \longrightarrow$
  (h) $N_2H_4(\ell) + O_2(g) \longrightarrow$
  (i) $NH_2OH(aq) + HCl(aq) \longrightarrow$
  (j) $HN_3(\ell) + LiOH(aq) \longrightarrow$

19.118 Complete and balance the following equations. In many parts, you are expected to figure out the formulas of products from general statements about chemical properties made in the chapter. (Write molecular equations.)
  (a) $NO(g) + O_2(g) \xrightarrow{\text{heat}}$
  (b) $KOH(aq) + HNO_3(aq) \longrightarrow$
  (c) $N_2O(g) \xrightarrow{\text{heat}}$
  (d) $NO_2(g) + H_2O \longrightarrow$
  (e) $N_2O_3(g) \xrightarrow{\text{heat}}$
  (f) $N_2O_5(s) + H_2O \longrightarrow$
  (g) $N_2O_4(g) \xrightarrow{\text{heat}}$
  (h) $HNO_3(\ell) + P_4O_{10}(s) \longrightarrow$
  (i) $NaNO_3(s) \xrightarrow{>500\,°C}$
  (j) $Bi(NO_3)_3 + H_2O \longrightarrow$
  (k) $N_2O_3(g) + NaOH(aq) \longrightarrow$
  (l) $NaNO_2(aq) + HCl(aq) \xrightarrow{0\,°C}$
  (m) $NaNO_2(aq) + HCl(aq) \xrightarrow{\text{heat}}$

CHEMICALS IN USE
SOME ATMOSPHERIC CHEMISTRY

By the mid-1980s, atmospheric scientists had identified over 40 different reactive chemical species in the atmosphere with concentrations of less than 1 part per million, some as low as one part per trillion (10^{-4} to 10^{-10} %, v/v). And they had studied nearly 200 different chemical reactions. Several reactions affect the ozone level in the stratosphere, which deeply concerns many scientists, advisory committees, and governments. We will see why in this unit.

The Ozone Cycle in the Stratosphere

The stratosphere is the zone of the atmosphere lying just above the troposphere (where we live) and between 16 and 40 km (10 to 25 miles) in altitude. One of its trace gases is ozone. The synthesis and destruction of ozone in the stratosphere converts the ultraviolet energy in solar radiation into heat. This removes a form of energy that would greatly harm plants and animals. In humans, skin cancer is the chief result of overexposure to UV radiation, particularly UV wavelengths of 315 nm and lower. In plants, photosynthesis is reduced.

High-energy ultraviolet radiation can break the bonds in oxygen molecules and generate oxygen atoms.

$$O_2 \xrightarrow[\text{(for best results, } \lambda = 242 \text{ nm or lower)}]{\text{UV radiation}} 2O$$

This reaction of oxygen, not of ozone, largely removes UV radiation of wavelengths below 280 nm, the most dangerous kind. The oxygen atoms made by this reaction then participate in a two-step cyclic process that removes longer-wavelength UV radiation, which is also dangerous to life on earth.

The oxygen atoms formed by the cracking of O_2 molecules can react with oxygen molecules at the surface of a neutral particle.

$$O + O_2 + M \longrightarrow O_3 + M$$

The neutral particle M absorbs some of the collision energy and thus leaves the new O_3 molecule less able to split apart immediately. Instead, what is desired is that the splitting of O_3 be left to UV radiation in the range of $\lambda = 240-320$ nm. This is the chief means by which such UV radiation is removed from incoming sunlight.

$$O_3 \xrightarrow[(\lambda = 240-320 \text{ nm})]{\text{UV radiation}} O_2 + O^*$$

Then, in a later collision at another M, ozone is remade.

$$O_2 + O^* + M \longrightarrow O_3 + M$$

O^* is an electronically excited oxygen atom, which is able to react with oxygen to remake ozone.

The generation of O^* from O_3 thus launches a chemical chain reaction. A chemical chain reaction is a two-step process in which products of the first step react in the second to regenerate the reactant(s) needed to run the first step again. The cracking of one O_2 molecule to give two O atoms sets off two such chains. In each, atoms of O and molecules of O_2 first combine to give ozone and then are recovered when UV radiation splits the ozone. Each chain, in other words, constitutes a natural cycle in the stratosphere — the ozone cycle (Figure 13a) — and we depend on its continuous operation to shield us from virtually all of the UV radiation of incoming sunlight. Anything that interferes with this cycle and reduces the stratospheric ozone level, therefore, poses a threat to human health and to agriculture.

Several mechanisms can potentially affect the ozone cycle, some natural and some caused by industrial activities. For example, N_2O, which occurs naturally at very trace levels, can react with electronically excited oxygen atoms made by the ozone cycle to give NO.

$$O^* + N_2O \longrightarrow 2NO$$

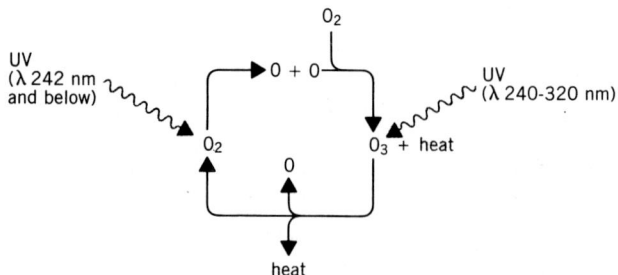

FIGURE 13a The stratospheric ozone cycle.

This removes an oxygen atom needed by the ozone cycle, and the product, NO, can destroy a second ozone molecule, which breaks up another cycle.

$$NO + O_3 \longrightarrow NO_2 + O_2$$
$$NO_2 + O \longrightarrow NO + O_2$$
$$\text{Net:} \quad \overline{O_3 + O \longrightarrow 2O_2}$$

NO is particularly damaging because just one NO molecule sets up a cyclic, chain reaction of its own that removes ozone over and over again — perhaps as many as 100,000 times. There are similar ozone-destroying cycles involving other trace stratospheric gases.

Chlorofluorocarbons and the Ozone Layer

Not much can be done to prevent the migration of N_2O into the stratosphere. It is a natural process, and the net of all such processes is a fairly steady-state global ozone level. (The level, by nature, fluctuates daily, weekly, monthly, and yearly. Determining an *average* annual ozone level is very tricky.) Increases in activities that have reduced the stratospheric ozone level since the 1960s are what have to be seen as grave threats to health and agriculture. The chlorofluorocarbons (CFCs) might be such threats.

The chlorofluorocarbons are a family of volatile, nonflammable, chemically stable, and essentially odorless and tasteless compounds. CFC-11, for example, is $CFCl_3$. It boils at 24 °C and was once the propellant in 50 to 60% of all aerosol cans. CFC-12 is CCl_2F_2, which boils at −30 °C. Their physical properties and chemical inertness make the CFCs ideal as refrigerants for air conditioners, freezers, and refrigerators as well as blowing agents for foam plastics. They have been superior solvents for cleaning computer parts. The CFCs were also the most commonly used aerosol propellants in the United States until their use in nearly all consumer products was banned in 1978. (They are still used for this purpose in other countries, however.)

When released, the CFCs enter the atmosphere where they can persist for a century. Their concentrations in the atmosphere have doubled in the last decade. They become globally distributed and migrate into the stratosphere where, according to mounting evidence, they likely have become an additional stress on the ozone cycle.

Ultraviolet energy can break chlorine atoms out of CFC molecules.

$$CCl_3F \xrightarrow{\text{UV radiation}} CCl_2F + Cl$$
$$CCl_2F_2 \xrightarrow{\text{UV radiation}} CClF_2 + Cl$$

Atomic chlorine destroys ozone and disrupts the ozone cycle by the following chain reaction.

$$Cl + O_3 \longrightarrow ClO + O_2$$
$$ClO + O \longrightarrow Cl + O_2$$
$$\text{Net:} \quad \overline{O_3 + O \longrightarrow 2O_2}$$

The breakup of one CFC molecule can thus initiate the subsequent removal of thousands of ozone molecules.

Getting definitive data directly from the stratosphere is very difficult, but most scientists consider that between 1978 and 1984 the total global ozone declined 3%. If so, roughly 7% more UV radiation now reaches the earth's surface than 10 years ago. The CFCs have almost certainly contributed significantly to this change.

The Antarctic Ozone Hole

Over the Antarctic continent, the ozone level has declined much more than 3% and, since the late 1970s, has followed a particularly intriguing seasonal trend. During the Antarctic winter, the ozone level over most of this continent declines so much that scientists call it the "Antarctic ozone hole." Then the level rebounds in the Antarctic summer. What causes this phenomenon, whether it will continue annually, and whether it is spreading globally are questions that have helped to make several nations take actions at the diplomatic level to try to stabilize the ozone layer. It has been found that concentrations of chlorine in various forms are relatively high over the Antarctic, and the CFCs are suspected of being mostly responsible.

The CFCs constitute perhaps the most easily controlled trace gases affecting the ozone level because they have known sources. In late 1986, negotiations sponsored by the United Nations Environmental Program led to an agreement among member nations to freeze the production of CFC-11 and CFC-12. Further negotiations continued in 1987 when 23 countries endorsed a plan to reduce the world consumption of CFCs 50% by 1999. The potential harm to life on earth from the loss of stratospheric ozone is so great that it would be most imprudent to wait the several years needed to get universally accepted proof that the CFCs are indeed the culprits. By that time, the damage might be too great and incapable of correction.

Questions

1. Write the equation for the stratospheric reaction that removes the shortest-wavelength (and most dangerous) UV radiation.

2. Write the equations that explain how the reaction in Question 1 sets up two ozone cycles. Explain how such cycles convert UV radiation into heat.

3. What particular harm does UV radiation in the range of 240 to 320 nm do to humans? To plants?

4. Using equations, explain how N_2O can interfere with the ozone cycle.

5. Nitrogen and oxygen can be made to combine in the stratosphere to form nitrogen monoxide, NO. Using equations, explain what this species can do to the ozone cycle.

6. What are the CFCs?

7. Using equations, explain how CFC-11 can reduce the stratospheric ozone level.

When acid rain makes a lake too acidic for game fish, heroic measures are needed, as in this use of a helicopter to spread crushed limestone ($CaCO_3$) on Fall Pond in New York's Adirondack

SIMPLE MOLECULES AND
IONS OF NONMETALS. PART 2

ELEMENTAL SULFUR
SULFUR DIOXIDE, SULFUROUS ACID, AND SULFITES
SULFUR TRIOXIDE, SULFURIC ACID, AND SULFATES
OTHER BINARY COMPOUNDS OF SULFUR

PHOSPHORUS
PHOSPHORUS OXIDES, OXOACIDS, AND THEIR SALTS
THE HALOGENS
CARBON

20.1 ELEMENTAL SULFUR

The most stable allotrope of sulfur exists as cyclic S_8 molecules with a crownlike conformation.

Sulfur, the "brimstone" of Biblical times, occurs as a brittle, bright yellow, nonmetallic element (Figure 20.1) in underground deposits, as a component of sulfide and sulfate minerals, and in various forms in natural gas, petroleum, and coal.

FIGURE 20.1 Orthorhombic sulfur.

In the mid-1980s, annual sulfur production in the United States was 12–13 million tons.

Sulfur obtained from gas and oil processing exceeded its production by the Frasch process in the mid-1980s.

Isotope	% Abundance
S-32	95.02
S-33	0.75
S-34	4.21
S-35	0.02

Underground deposits of sulfur—notably in Louisiana, Texas, Mexico, and Poland—were probably formed by the reducing activities of bacteria on sedimentary sulfate minerals. This sulfur is brought to the surface by the Frasch process in which superheated steam is pumped into the deposit (Figure 20.2). The sulfur melts and gathers in pools, and then the pressure of hot air and steam forces molten sulfur to the surface through pipes that deliver it to cooling yards (Figure 20.3).

In most years, the second largest source of sulfur is from the hydrogen sulfide found in natural gas and petroleum. Sulfur dioxide is used to convert hydrogen sulfide to elemental sulfur by the following reaction, which also makes sulfur from the sulfur dioxide.

$$2H_2S(g) + SO_2(g) \longrightarrow 3S(s) + 2H_2O$$

The sulfur dioxide for this process is itself made by the controlled oxidation of some of the hydrogen sulfide in the oil or gas.

$$2H_2S(g) + 3O_2(g) \longrightarrow 2SO_2(g) + 2H_2O$$

The sulfur present in sulfide ores, such as FeS, FeS_2, CuS, and PbS, is used to make sulfuric acid, but the processes generally do not involve first making elemental sulfur. Instead, such ores are roasted (heated strongly) in air to change them to metal oxides and sulfur dioxide. For example,

$$2PbS(s) + 3O_2(g) \xrightarrow{\text{heat}} 2PbO(s) + 2SO_2(g)$$

The sulfur dioxide is fed directly into the manufacture of sulfuric acid. Sulfur dioxide can also be obtained by the reduction of sulfate minerals such as gypsum, $CaSO_4 \cdot 2H_2O$.

There are four isotopes of sulfur (see the table in the margin), but one predominates. Samples of sulfur obtained from different locations and different sources differ more widely in isotopic composition than most common elements. The differences are not great, but they prevent us from giving an atomic weight for sulfur any more precise than 32.06. For example, the sulfur in sulfate ions in ocean water is enriched somewhat in sulfur-34, while the sulfur in terrestrial sulfide ores is slightly depleted in this isotope.

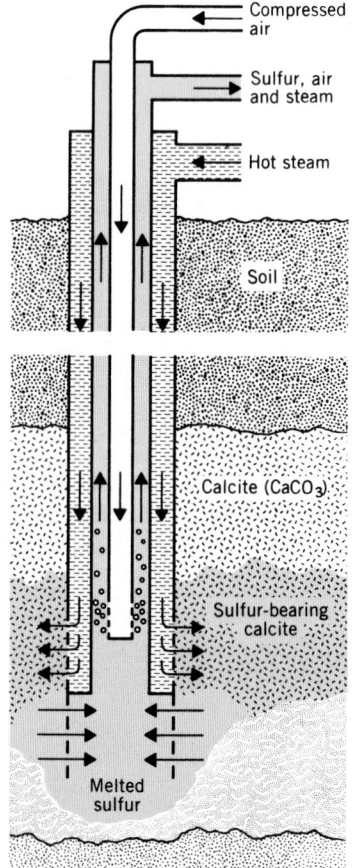

Compressed air

Sulfur, air and steam

Hot steam

Soil

Calcite ($CaCO_3$)

Sulfur-bearing calcite

Melted sulfur

FIGURE 20.2 The Frasch process for extracting sulfur from deep deposits.

FIGURE 20.3 Sulfur emerges in the molten state from the Frasch process and is allowed to cool in huge yards.

The Principal Allotropes of Sulfur

Several allotropes exist for sulfur but we will review only those that are the most common and most easily prepared. Unlike oxygen atoms, sulfur atoms can form chains a few to several thousand sulfur atoms long as well as rings of various sizes.

Two common allotropes are different crystalline forms of cyclooctasulfur, S_8. *Orthorhombic sulfur* (S_α) is thermodynamically the most stable of all sulfur allotropes, and it can be found near certain volcanos as beautiful, large, yellow crystals (Figure 20.1). The eight sulfur atoms in each molecule of S_α form a crownlike ring. This allotrope displays a melting point of 112.8 °C, but only if it is rapidly heated. If S_α is kept unmelted at a temperature of 95.5 °C, the crown S_8 molecules of S_α slowly take up new positions and change into the crystalline form of another common allotrope, S_β, often called *monoclinic sulfur*, which crystallizes in long needles (Figure 20.4). S_β melts at 119 °C, but only if rapidly heated. (Slow heating allows time for some S_8 rings to open. This gives a mixture that can start to melt at a lower temperature.) If left several weeks at ordinary temperature, S_β changes back to the orthorhombic form of S_α.

When either S_α or S_β is melted, some covalent bonds in the S_8 rings break, the rings open, and when the liquid sulfur is cooled, new S—S bonds can form. So when molten sulfur is rapidly cooled, the new S—S bonds reform randomly, some to make rings and others to give chains of varying chain length. With so many kinds of molecules present, neat crystals cannot form, and the rapid cooling of molten sulfur gives another allotrope, *plastic sulfur,* an amorphous solid. It, too, slowly reverts to S_α on long standing.

When sulfur is heated beyond its melting point, some interesting changes occur that illustrate the effect of molecular size on physical properties. See Figure 20.5. Freshly melted sulfur, in which the molecules are mostly S_8 rings, exists first as a straw-yellow, transparent, and mobile fluid. But at 159 °C, a sharp transition occurs, and the density, surface tension, viscosity, and other physical properties dramatically change. Between 160 and 195 °C, the viscosity of molten sulfur increases by a factor of 10^5! In this range, increasing numbers of S_8 rings open to give chains *whose ends join.* Chains may be as long as 200,000 sulfur atoms. Such huge molecules intertangle and do not easily slip by each other during flow, so the fluid is extremely viscous and stays so until the temperature rises well above 200 °C. Now, enough thermal energy is supplied to break the chains into small lengths, and the molten sulfur becomes less and less viscous, until the liquid — by now dark red — boils at 444.60 °C.

Be sure to distinguish between *isotope* and *allotrope*.

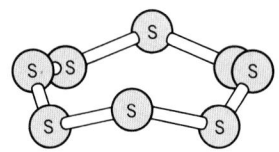

S_8, crown configuration

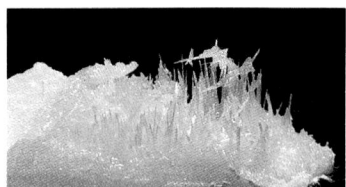

FIGURE 20.4 Monoclinic sulfur, S_β.

Viscosity is resistance to flow. Cold molasses, for example, has a high viscosity.

FIGURE 20.5 Changes in sulfur as it is heated. (*a*) Solid sulfur. (*b*) Molten sulfur just above its melting point. (*c*) Molten sulfur, just below 200 °C, is dark and viscous. (*d*) Boiling sulfur is dark red and no longer viscous.

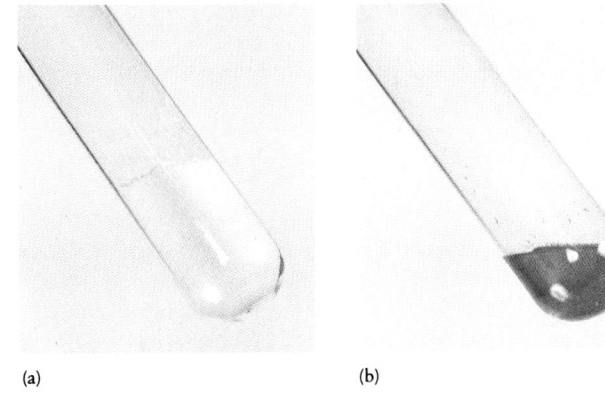

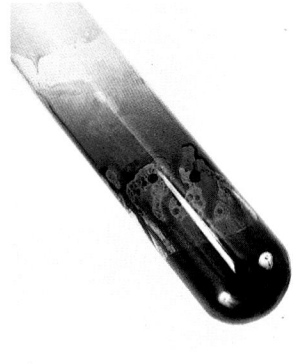

(a)   (b)   (c)   (d)

Chemical Properties of Sulfur — An Overview

When heated (so as to encourage breaking S—S bonds in S_8 sulfur rings), sulfur can be made to combine directly with all other elements except the noble gases and six other elements (gold, platinum, iridium, tellurium, nitrogen, and iodine).

Sulfur's high ionization energy (999.30 kJ mol⁻¹) and relatively high electron affinity (+200 kJ mol⁻¹) mean that it will exist in positive oxidation states largely in combina-

Bonds from S to these six elements can be formed indirectly.

TABLE 20.1 The Oxidation States of Sulfur

Oxidation State	Examples
−2	H_2S, S^{2-}, CH_3SH
−1	S_2^{2-}
0	S
+4	SO_2, SO_3^{2-}
+6	SO_3, HSO_4^{2-}, SO_4^{2-}

tion only with elements of similarly high electron affinities, such as fluorine, oxygen, and chlorine. (Cations such as S_8^{2+} have also been generated, but they are by no means common.)

Table 20.1 summarizes the various oxidation states that sulfur can ordinarily have, together with some examples of typical compounds. Hydrogen sulfide and the metal sulfides are the most common compounds in which sulfur occurs in a negative oxidation state. We will limit our study to the common oxides and oxoacids of sulfur as well as to hydrogen sulfide and a few other types of sulfur compounds.

20.2 SULFUR DIOXIDE, SULFUROUS ACID, AND SULFITES

What is traditionally called sulfurous acid, H_2SO_3, is an aqueous solution of hydrates of sulfur dioxide.

In both environmental and economic terms, the direct combination of sulfur with oxygen, which gives sulfur dioxide, is probably the most important reaction of sulfur. At an ignition temperature of 250 to 260 °C, sulfur burns with a blue flame.

$$S(s) + O_2(g) \longrightarrow SO_2(g)$$

The chief use of sulfur dioxide is to make sulfuric acid, but enormous quantities of SO_2 never make it this far. They enter and pollute the atmosphere. Another important commercial use of SO_2 is the manufacture of sodium sulfite, Na_2SO_3, used principally by the paper industry.

Most sulfur dioxide in the environment forms from the oxidation of the trace amounts of sulfur and sulfur compounds in coal. Because so much coal is burned each year, hundreds of millions of tons of SO_2 are released per year worldwide into the atmosphere. It constitutes a pollutant of major importance, as we will see shortly. Other sources of atmospheric SO_2 are the burning of sulfur-containing oil and the roasting of sulfide ores, as we mentioned earlier.

Sulfur dioxide boils at −10 °C and freezes at −72 °C. It has such a sharp, acrid, and disagreeable odor that if you ever inhale it, you will feel as if you are choking. Because liquid sulfur dioxide boils only a few degrees below 0 °C, gaseous sulfur dioxide is easily liquefied. Like liquid ammonia, liquid sulfur dioxide can be used as a solvent, but the most common use of this liquid is as a refrigerant in heat-exchange systems.

Sulfurous Acid

Sulfur dioxide is very soluble in water, 12% (w/w) at 15 °C. It dissolves by forming hydrates, $SO_2 \cdot nH_2O$, where n varies with concentration, temperature, and pH. From the hydrates, some hydronium ion and hydrogen sulfite ion, HSO_3^-, form, and the presence of these ions has long been explained simply in terms of $H_2SO_3(aq)$, sulfurous acid. Actual molecules of this species — H_2SO_3 — have never been detected in or out of water. So we must regard H_2SO_3 as a convenient simplification, one that we will employ, of the

For simplicity, we will revert again to writing sulfur as S, not as S_8.

The sharp odor of a freshly lit match is caused by SO_2.

real status of aqueous sulfur dioxide. Traditionally, the first ionization of sulfurous acid is therefore represented as follows,

$$H_2SO_3(aq) \rightleftharpoons H^+(aq) + HSO_3^-(aq)$$

Actually, the first ionization should be represented as

$$SO_2 \cdot nH_2O(aq) \rightleftharpoons H_3O^+(aq) + HSO_3^-(aq)$$

The second ionization of sulfurous acid is, of course, the ionization of the hydrogen sulfite ion, HSO_3^-, which does exist in solution.

$$HSO_3^-(aq) \rightleftharpoons H^+(aq) + SO_3^{2-}(aq)$$

For sulfurous acid at 25 °C,
$$K_1 = 1.2 \times 10^{-2}$$
$$K_2 = 6.6 \times 10^{-8}$$

Acid Rain

Sulfurous acid is an acid of moderate strength, so when rain washes gaseous SO_2 from the atmosphere, the rainwater is acidic. Moreover, both oxygen and ozone in smog can convert some SO_2 to SO_3, particularly in sunlight when fine dust is present. SO_3 reacts with water to form sulfuric acid, a strong acid, which also contributes to the acidity of rain when the air contains sulfur oxides.

Nitrogen dioxide, as we learned in Section 19.6, is another major air pollutant that dissolves in water to give acids — both HNO_3, a strong acid, and HNO_2, a weak acid. Thus oxides of nitrogen and sulfur are chiefly responsible for what is commonly called acid rain. Rain as acidic as lemon juice (pH 2.2 – 2.7) has been observed. The record was rain as acidic as vinegar at Pitlochry, Scotland, in April of 1974.

A better term for acid rain is *acid deposition,* because dry dust particles settling on buildings, metals, and soil also carry acidic materials adhering to their surfaces. Acid deposition is a problem particularly in those regions downwind from major users of sulfur-containing fuels. Southern parts of the Scandinavian peninsula receive acid deposition from Germany's Ruhr valley, the English Midlands, and countries of eastern Europe (Figure 20.6). Parts of southern Canada and the northern United States get acid

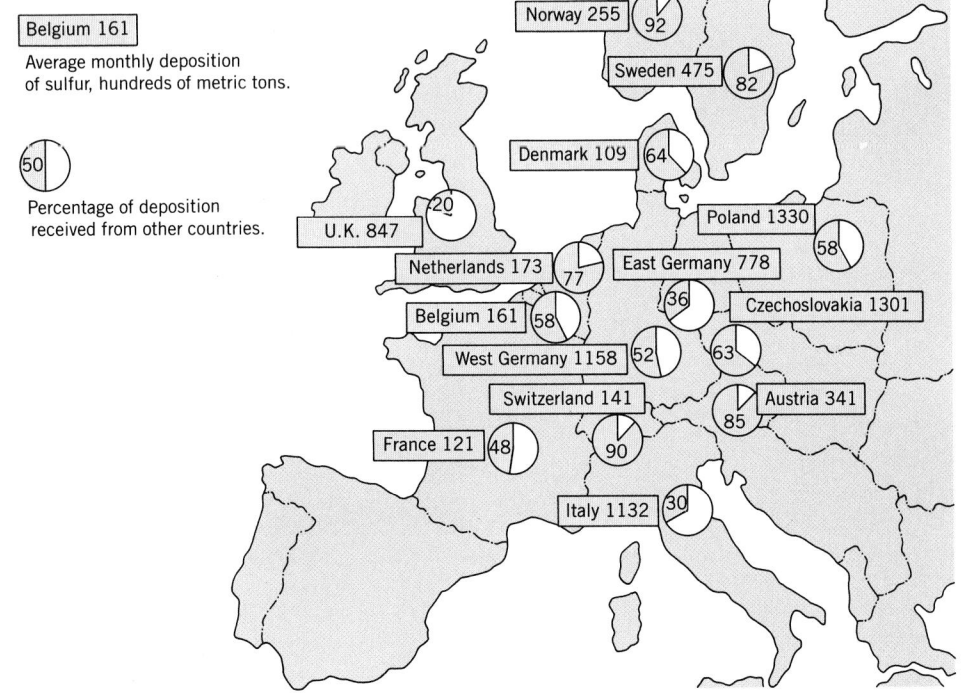

FIGURE 20.6 Deposition of sulfur pollutants in Europe (1983 figures). The pie figures show that most countries receive more of these pollutants than they generate. The numbers in the dark rectangles show that the heaviest depositions occur on Poland and Czechoslovakia. But even the seemingly light depositions on Norway have made many of its southern lakes biological deserts where fish can no longer survive. The metric ton, the unit for the data, is a mass of 1000 kg or about 2200 lb. (*Source:* United Nations Economic Commission for Europe, European Monitoring and Evaluation Program.)

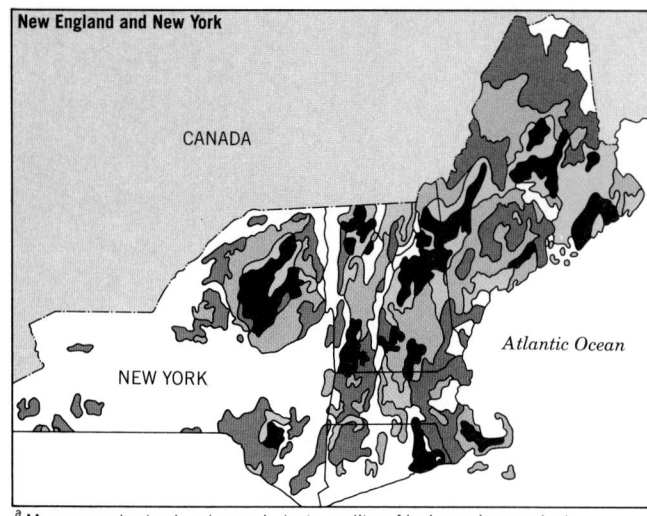

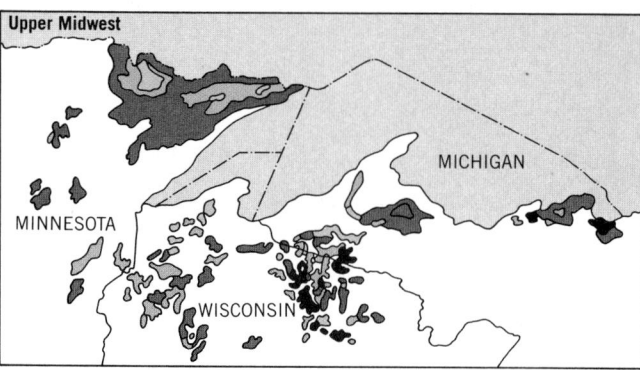

New England and New York

CANADA

NEW YORK

Atlantic Ocean

Upper Midwest

MINNESOTA

MICHIGAN

WISCONSIN

Total alkalinity of surface waters[a]
■ <50 (endangered)
□ 50—99 (crisis area)
▨ 100—199 (moderate to extremely vulnerable)
▨ 200—399 (vulnerable)

[a] Mean annual value in microequivalents per liter of hydrogen ion required for neutralization.

FIGURE 20.7 Areas of significant potential damage from acid deposition are found in New York and the New England states as well as in the northern regions of Michigan, Wisconsin, and Minnesota. The famous Boundary Waters Canoe Area lakes of northeastern Minnesota are vulnerable. Lakes nestled in noncarbonate rock depressions have the least protection. (*Source:* U.S. Environmental Protection Agency, 1985.)

deposition from the great industrial belt curving from Boston to Chicago (Figure 20.7).

Acid deposition makes lakes too acidic for desirable fish to live. Dissolved acids leach calcium and magnesium ions from the soil and so destroy vegetation. Severe damage has occurred to the Black Forest in Germany and in the Erz mountains of Czechoslovakia (Figure 20.8). Acid deposition is corrosive to exposed metals such as railroad rails, vehicles, and machinery, as well as to stone building materials (Figure 20.9). Limestone and marble are particularly sensitive, because they are chiefly calcium car-

FIGURE 20.8 Acid deposition can cause severe damage to trees. Shown in this photo is a tree in West Germany's Black Forest, which is being severely damaged by this form of air pollution.

FIGURE 20.9 Acid deposition can rapidly accelerate the decay of building stone. This statue of Baumberg sandstone is at the Herten Castle in Westphalia, West Germany. On the left, its appearance in 1908 after 206 years of exposure. On the right, the same statue in 1968, only 60 years later, but during exposure to the heavily polluted air of the Rhein-Ruhr region of West Germany.

bonate, and carbonates are dissolved by acids. Several major cathedrals in Europe need constant repair because of the problem.

Less reliance on sulfur-containing coal and oil might be thought to be a solution to the acid deposition problem. No doubt it would, but the alternative is either drastic cutbacks in energy consumption or nuclear power. But nuclear power presently is costlier in every way than power obtained from the burning of coal or oil and it bears its own pollution ills. Meanwhile, as coal and oil are used, the removal of most of the SO_2 from smokestack gases is possible. SO_2, for example, is absorbed by wet calcium hydroxide ("slaked lime") by the following reaction.

$$SO_2(g) + Ca(OH)_2(s) \longrightarrow CaSO_3(s) + 2H_2O$$

Not all SO_2 is removed, however, and given the enormous quantities of coal and oil burned worldwide, large emissions of SO_2 still occur. In a technological sense, the problem is controllable. The questions remaining are mostly political, economic, and diplomatic. According to Swedish scientist Svante Oden, if SO_2 emissions are left uncontrolled, acid deposition will make the earth north of the 45th parallel a chemical desert.

Hydrogen Sulfites, Disulfites, and Sulfites

Although chemical supply houses sell a product called sodium bisulfite (or sodium hydrogen sulfite), $NaHSO_3$, the material in the bottle is mostly sodium disulfite, $Na_2S_2O_5$. In an acidified solution, however, these two compounds are chemically equivalent. Since we are entering a small nomenclatural jungle here, we must proceed carefully.

We can make a *solution* of $NaHSO_3$ by passing SO_2 into aqueous Na_2CO_3.

$$2SO_2(g) + Na_2CO_3(aq) + H_2O \longrightarrow \underset{\substack{\text{sodium hydrogen sulfite}\\\text{(sodium bisulfite)}}}{2NaHSO_3(aq)} + CO_2(g)$$

If we try to isolate $NaHSO_3(s)$ from this solution by boiling off the water, complex events occur that yield some sodium sulfite, Na_2SO_3, some SO_2, and some $Na_2S_2O_5$. We would not obtain solid sodium hydrogen sulfite. If we maintain a continuous presence of SO_2 during the removal of the water, however, the solid isolated is $Na_2S_2O_5$. It contains the disulfite ion, $S_2O_5{}^{2-}$, which forms as follows.

$$2HSO_3{}^-(aq) \rightleftharpoons \underset{\substack{\text{disulfite}\\\text{ion}}}{S_2O_5{}^{2-}(aq)} + H_2O$$

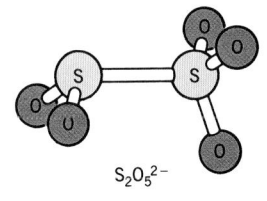

$S_2O_5{}^{2-}$

Sodium disulfite, $Na_2S_2O_5$, is often called *sodium metabisulfite*, particularly on labels of bottles of this chemical. We will return to it shortly.

If, instead of boiling off the water (while SO_2 is present), we add more sodium carbonate to the $NaHSO_3(aq)$, we obtain $Na_2SO_3(aq)$, aqueous sodium sulfite. It is really just an acid–base neutralization:

$$2NaHSO_3(aq) + Na_2CO_3(aq) \longrightarrow \underset{\substack{\text{sodium}\\\text{sulfite}}}{2Na_2SO_3(aq)} + CO_2(g) + H_2O$$

This solution can be evaporated to dryness to give sodium sulfite. Depending on the temperature maintained during this operation, either the heptahydrate ($Na_2SO_3 \cdot 7H_2O$) or the anhydrous form of Na_2SO_3 is made.

Sodium sulfite is the most important sulfite salt. Hundreds of thousands of tons are made per year throughout the world, chiefly for use in the manufacture of paper pulp. A small amount is used as a food additive to prevent bacterial decomposition. Some people

Coal from Montana and Wyoming has less sulfur than coal from Appalachia. Oil or gas with sulfur compounds is called "sour crude" or "sour gas."

$Ca(OH)_2$ is made by thermally decomposing limestone to CaO and then adding water.

People with a severe sulfite allergy should read the ingredient lists on food packages.

are extremely allergic to sulfites, and in the mid-1980s the U.S. government required that all products with sulfites disclose their use and carry a warning.

Although *crystalline* sodium hydrogen sulfite cannot be made, we can still obtain the equivalent of this compound as an *aqueous* solution. (This would be the form used anyway.) Aqueous sodium hydrogen sulfite is available simply by dissolving $Na_2S_2O_5$ in water. The following equilibrium exists in this solution.

$$2HSO_3^-(aq) \rightleftharpoons S_2O_5^{2-}(aq) + H_2O$$

If some other substance were added that reacted with HSO_3^-, the equilibrium would continuously resupply this ion as long as it could, in accordance with Le Châtelier's principle. Of course, at low pH values, any of the following equilibria involving SO_2 or the sulfites shifts to the right, so you can see that the particular species present as the reactant depends on the pH.

$$SO_3^{2-}(aq) + H^+ \rightleftharpoons HSO_3^-(aq)$$
$$HSO_3^-(aq) + H^+ \rightleftharpoons H_2SO_3(aq)$$

In Section 12.9, we studied the sulfites as mild reducing agents. In acid, sulfurous acid participates with the sulfate ion in the following half-reaction.

$$SO_4^{2-}(aq) + 4H^+(aq) + 2e^- \rightleftharpoons H_2SO_3(aq) + H_2O \qquad E° = +0.172 \text{ V}$$

The low value of $E°$ means that the reverse reaction, where H_2SO_3 gives up electrons and functions as a reducing agent, can occur with many compounds—anything that is a stronger oxidizing agent than SO_4^{2-}.

Because sulfurous acid is unstable with respect to sulfur dioxide, the following equilibrium must be remembered whenever a sulfite is used as a reducing agent in an acid solution.

$$H_2SO_3(aq) \rightleftharpoons H_2O + SO_2(g)$$

The escape of SO_2 would mean the loss of reducing ability from the solution (and the need to be working at the hood).

The Sulfite System—A Brief Summary

We can summarize what is of the most practical importance about the "sulfurous acid"–sulfur dioxide system as follows.

The pH of a solution determines whether we have principally SO_2, HSO_3^-, or SO_3^{2-} present.

All species are reducing agents.

A bottle labeled "sodium sulfite" contains Na_2SO_3, but one labeled "sodium hydrogen sulfite" or "sodium bisulfite" contains $Na_2S_2O_5$—called sodium disulfite or sodium metabisulfite, depending on the vendor. But $Na_2S_2O_5$ behaves in solution like $NaHSO_3$ (due attention being paid, of course, to the stoichiometry).

Sulfur dioxide can be expelled from any of these solutions by adding acid and heating the system. Experiments with acidified sulfite solutions should be done at the hood.

20.3 SULFUR TRIOXIDE, SULFURIC ACID, AND SULFATES

The annual worldwide manufacture of over a trillion moles of sulfuric acid makes this acid the most widely used acid in the world.

Sulfur dioxide itself does not burn in air, but it still can be catalytically oxidized by oxygen to sulfur trioxide, the parent compound for sulfuric acid and its salts.

Sulfur Trioxide

Sulfur trioxide exists in the gaseous and liquid states as molecules of SO_3 in equilibrium with molecules of a trimer, S_3O_9. The physical properties of sulfur trioxide, therefore, are for this equilibrium mixture, which is colorless and which boils at 44.6 °C and melts at 16.86 °C. Solid sulfur trioxide, which exists entirely as the trimer when perfectly dry, easily sublimes. For simplicity, we will represent sulfur trioxide as SO_3.

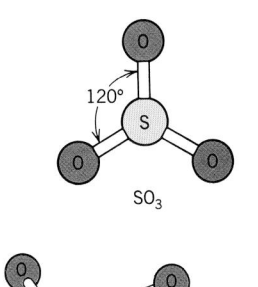

SO_3

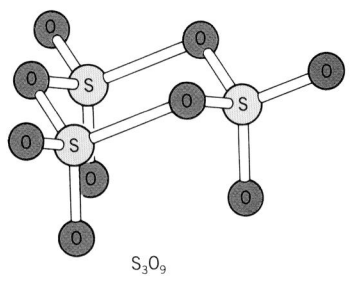

S_3O_9

Sulfur trioxide gas has an even sharper, more choking odor than sulfur dioxide, and its conversion to sulfuric acid in any moist tissue makes it particularly dangerous. It also contributes to the overall problem of acid deposition which we discussed in the previous section.

The oxidation of sulfur dioxide to sulfur trioxide takes place to a slight extent in any sunlit atmosphere polluted with SO_2 and microscopic dust particles.

$$2SO_2 + O_2(g) \longrightarrow 2SO_3(g) \qquad \Delta H° = -191.2 \text{ kJ}$$

Even though this is an exothermic reaction, it occurs very slowly in the absence of a catalyst. Lucky us! If the oxidation of SO_2 occurred spontaneously in air, like the oxidation of NO to NO_2, either the use of sulfur-bearing coal or oil would be banned outright or elaborate and very costly steps would be needed to remove *all* SO_2 from smokestack gases.

Sulfur trioxide, as we have learned, is an acidic oxide, and it is certainly well classified. It reacts violently and very exothermically with water to form sulfuric acid, which we will study in more detail shortly.

$$SO_3(g) + H_2O \longrightarrow H_2SO_4(aq) \qquad \Delta H° = -880 \text{ kJ}$$

Sulfur trioxide neutralizes aqueous solutions of Group IA or IIA hydroxides, forming salts of the hydrogen sulfate ion, HSO_4^-, or the sulfate ion, SO_4^{2-}, depending on the mole proportions taken, for example,

$$SO_3(g) + NaOH(aq) \longrightarrow NaHSO_4(aq)$$

or

$$SO_3(g) + 2KOH(aq) \longrightarrow K_2SO_4(aq)$$

Sulfur trioxide similarly neutralizes bicarbonates and carbonates, for example,

$$SO_3(g) + Na_2CO_3(aq) \longrightarrow Na_2SO_4(aq) + CO_2(g)$$

The sulfur trioxide present in smokestack gases can be removed by passing the gases over solid metal oxides, which react with SO_3 to give sulfates. This is also an acid–base neutralization in which the oxide ion is the Lewis base and SO_3 is the Lewis acid, for example,

$$O^{2-}(s) + SO_3(g) \longrightarrow SO_4^{2-}(s)$$

The mixed iron oxide, Fe_3O_4, which can be regarded as $FeO + Fe_2O_3$, is commonly the metal oxide used.

$$Fe_3O_4(s) + 4SO_3(g) \longrightarrow FeSO_4(s) + Fe_2(SO_4)_3$$

Sulfuric Acid

In the mid-1980s, sulfuric acid was being manufactured in the United States at an annual rate of about 40 million tons (370×10^9 mol), making it the country's largest-volume

In terms of *moles*, ammonia production at 1.2×10^{12} mol/year hugely outdistances sulfuric acid production.

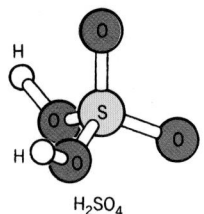

H_2SO_4

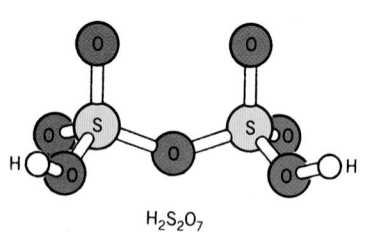

$H_2S_2O_7$

96% H_2SO_4 is 18 M H_2SO_4.

chemical (in terms of tonnage). Sulfuric acid is used in so many ways to make commercial products that a good index for the industrial activity of a nation and its standard of living is the per capita use of this substance.

The Contact Process   Sulfuric acid is manufactured by the contact process, where the name denotes *contact* between SO_2 and a special catalyst during one step. All steps are exothermic, as the data given below show. In the first step of the contact process, sulfur is burned to give sulfur dioxide.

$$S(s) + O_2(g) \longrightarrow SO_2(g) \qquad \Delta H° = -297 \text{ kJ}$$

The sulfur dioxide is passed together with oxygen over heated beds of the catalyst, vanadium pentoxide (V_2O_5), where the SO_2 is oxidized to SO_3.

$$2SO_2(g) + O_2(g) \longrightarrow 2SO_3(g) \qquad \Delta H° = -191.2 \text{ kJ}$$

The SO_3 is now bubbled into concentrated sulfuric acid which reacts with it and traps it as another oxoacid of sulfur, disulfuric acid, $H_2S_2O_7$ (sometimes called pyrosulfuric acid).

$$SO_3(g) + H_2SO_4(\ell) \longrightarrow H_2S_2O_7(\ell)$$

Water is now added, and the disulfuric acid breaks down into sulfuric acid.

$$H_2S_2O_7(\ell) + H_2O \longrightarrow 2H_2SO_4(\ell)$$

The overall value of the enthalpy change for the last two steps is $\Delta H° = -130 \text{ kJ}$. (The reason why the direct addition of SO_3 to water is not used is that it would contact the water vapor first, and produce a fine mist of sulfuric acid that would be hard to trap without its escaping to the atmosphere.)

The addition of water can be controlled to give a final product with a concentration of 95 to 98% (w/w) H_2SO_4, the usual form of commercially available concentrated sulfuric acid. The disadvantage of pure H_2SO_4 is that it melts at 10.4 °C (51 °F) and would therefore be susceptible to solidification at cool temperatures and be harder to handle.

Handling Sulfuric Acid   Some of its properties make concentrated sulfuric acid potentially a very dangerous chemical if used incautiously. Because it is thick and syrupy, concentrated sulfuric acid tends to cling to whatever it is spilled. And all reactions of this acid are very exothermic, including simple dilution with water. An initial sharp rise in temperature when the concentrated acid begins to interact with either water or another material further accelerates chemical changes.

If you spill a drop of concentrated sulfuric acid on the skin, which contains water, expect a chemical burn and a blister regardless of how rapidly you flush the site with cold water. But start flushing the area with water even if you are not sure if the drop is concentrated acid. The only experience most people have with sulfuric acid is its use as the acid in the lead storage batteries of vehicles. So be sure to handle or dispose of such batteries very carefully.

Protective gloves that can be quickly removed as well as safety glasses should be worn when dispensing concentrated sulfuric acid. Give yourself space so that no one will inadvertently bump you as you pour this acid. Never pick up a jug without providing some support at its base, because jug handles have been known to break off. Never carry a jug without setting it inside a special rubber bucket large enough to contain the acid fully if the jug should break. And always remember the first rule about diluting concentrated solutions—always add the concentrated solution slowly, with stirring, to the water. Never add water to the concentrated solution. So much heat can be generated

FIGURE 20.10 Sugar, $C_{12}H_{22}O_{11}$, had been placed in both beakers. Concentrated sulfuric acid was then added to the beaker on the right. The concentrated acid draws the components of water from sugar molecules and leaves behind a spongy mass of elemental carbon.

when the first water hits the concentrated solution that some material boils and spatters out of the container.

Chemical Uses of Sulfuric Acid   As we have indicated, sulfuric acid is the workhorse strong acid in both industry and the laboratory. It has several advantages over other acids. It is the least expensive of the strong acids. It is the only strong acid that can be prepared and shipped in an almost pure form. At or near room temperature it is stable and nonvolatile. It is not a vigorous oxidizing agent unless it is heated. It is a potent dehydrating agent, being powerful enough to gather the pieces of water molecules out of carbohydrate molecules leaving only a spongy mass of carbon (Figure 20.10).

Sulfuric acid, a strong diprotic acid, gives all of the usual reactions of strong acids with active metals, and the oxides, hydroxides, bicarbonates, and carbonates of metals. Whether one or both potential protons from H_2SO_4 are used is largely a matter of the chemist's choice of mole proportions. We need not review these reactions here (but be sure you can write equations illustrating each).

About 70% of the annual sulfuric acid production is consumed to make fertilizers, mostly in a process that converts insoluble phosphate rock into more soluble phosphate forms. Phosphate rock in many mines is principally $Ca_5(PO_4)_3F$, which is very insoluble in water and not of much use as a source of the phosphate ion system needed by growing plants. However, sulfuric acid can be made to react with this rock as follows to give a mixture called *superphosphate,* which is much more soluble in water.

$$2Ca_5(PO_4)_3F + 7H_2SO_4 + 17H_2O \longrightarrow \underbrace{7\,CaSO_4 \cdot 2H_2O + 3Ca(H_2PO_4)_2 \cdot H_2O}_{\text{"superphosphate"}} + 2HF$$

The gypsum ($CaSO_4 \cdot 2H_2O$) in the product is not needed and just dilutes the phosphate portion. The hydrogen fluoride is trapped because otherwise it is a serious air pollutant. At air levels as low as 0.1 ppm it is toxic to certain plants, causing chlorosis and edge and tip burn. In growing animals, HF interferes with bone development, causing bone spurs and oversized bones.

The wheat on the left benefits from phosphate fertilizer, which has been denied to the wheat on the right.

By increasing the proportion of H_2SO_4, H_3PO_4 can be made from phosphate rock.

Chlorosis is the loss of chlorophyll, the green pigment in plants necessary for photosynthesis.

Hydrogen Sulfates and Sulfates

Generally, sulfate salts of metals are more available and more often used than hydrogen sulfate salts. Remember that in an aqueous solution the hydrogen sulfate ion exists in an equilibrium with the sulfate ion in what corresponds to the second ionization of sulfuric

The K_a of a typical weak acid is on the order of 10^{-5}.

In 0.10 M NaHSO$_4$, the HSO$_4^-$ ion is over 30% ionized.

acid. You can see from the acid ionization constant of HSO$_4^-$ that this ion is a moderately strong acid.

$$HSO_4^-(aq) \rightleftharpoons H^+(aq) + SO_4^{2-}(aq) \qquad K_a = 1.0 \times 10^{-2} \ (25 \ °C)$$

Hydrogen sulfate salts exist generally just for the Group IA metals. They have much lower melting points (Table 20.2) than the corresponding sulfates, ranging from 120 to 315 °C (although CsHSO$_4$ decomposes when heated). When an acid in the solid state is needed, sodium hydrogen sulfate is often used.

The Group IA sulfates (Table 20.2) are all high-melting solids—consistent with their ionic nature—and all are quite soluble in water. Their aqueous solutions generally have pH's around 7. Sodium and potassium sulfate are used as chemicals when dissolved sulfate ion is needed at a neutral pH. Some Group IA sulfates are available as hydrates. Chemists use anhydrous sodium sulfate as an inert drying agent for organic solutions because it combines strongly with water to form hydrates.

The Group IIA sulfates (Table 20.2) either decompose or melt at high temperatures and, except for beryllium and magnesium sulfate, are relatively insoluble in water. Most are available as hydrates, and some of these are commercially very important. We have already mentioned in earlier places that the hydrate gypsum, CaSO$_4 \cdot 2H_2O$, is present in alabaster (page 43) and is used in plasterboard. Another hydrate, plaster of paris (CaSO$_4 \cdot \frac{1}{2}H_2O$), can be used in making plaster casts.

The strong ability of MgSO$_4$ to form hydrates makes this an important drying agent, sold under the name Drierite. The thermal stability of magnesium sulfate (below 400 °C) makes it possible to regenerate spent Drierite granules indefinitely.

Sulfur Trioxide, Sulfuric Acid, and the Sulfates— An Overview

We can summarize the most important points about the chemistry studied in this section as follows.

Sulfur trioxide—a powerfully acidic, gaseous, nonmetal oxide—is made from sulfur dioxide and is used to make sulfuric acid.

Sulfuric acid is a very strong, water-soluble, diprotic acid. Even its second ionization constant is high enough to make the hydrogen sulfate ion a moderately strong acid. Sulfuric acid gives all of the usual reactions of strong acids toward metals and bases.

Hydrogen sulfate salts are known just for the Group IA metals, and their aqueous solutions are acidic. The water-soluble Group IA sulfates form neutral aqueous solutions. The Group IIA sulfates, except for those of beryllium and magnesium, are relatively insoluble in water.

TABLE 20.2 Hydrogen Sulfates and Sulfates of Group IA and IIA Metals

| Group IA Salts | | | | Group IIA Salts (Sulfates[a]) | |
| Hydrogen Sulfates | | Sulfates | | | |
Formula	Melting Point (°C)	Formula	Melting Point (°C)	Formula	Melting Point (°C)
LiHSO$_4$	170	Li$_2$SO$_4$	845	BeSO$_4$	Decomposes 550–600
NaHSO$_4$	315	Na$_2$SO$_4$	884	MgSO$_4$	Decomposes 1124
KHSO$_4$	214	K$_2$SO$_4$	1067	CaSO$_4$	1450
RbHSO$_4$	—	Rb$_2$SO$_4$	1060	SrSO$_4$	1605
CsHSO$_4$	Decomposes	Cs$_2$SO$_4$	1019	BaSO$_4$	Decomposes >1600

[a] The hydrogen sulfates do not exist as free solids.

Other Oxoacids of Sulfur and Their Salts

At least 10 oxoacids or oxoacid salts of sulfur are known, but 4 of these acids are unstable and unknown as free, pure compounds. Only salts of these acids have been prepared. We have studied sulfurous acid (unstable) and sulfuric acid and have mentioned disulfuric acid, $H_2S_2O_7$. We will mention just one more.

Thiosulfates The thiosulfate ion, $S_2O_3^{2-}$, can be thought of as the anion of thiosulfuric acid, $H_2S_2O_3$. Although the free acid is very unstable, thiosulfate salts are known.

When sulfur is added to a hot solution of sodium sulfite, the sulfur dissolves by converting the sulfite ion to the thiosulfate ion.

$$S(s) + Na_2SO_3(aq) \xrightarrow{\text{heat}} \underset{\text{sodium thiosulfate}}{Na_2S_2O_3(aq)}$$

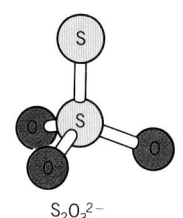

$S_2O_3^{2-}$

If the solution is acidified in an attempt to change the thiosulfate ion to thiosulfuric acid, $H_2S_2O_3$, any such acid that forms decomposes to a complex mixture of products that include sulfur. The sulfur is usually so finely divided that it is colloidal in nature and therefore difficult to separate from the rest of the solution.

The thiosulfate ion is a moderate reducing agent, as the standard reduction potential indicates for its conversion to the tetrathionate ion, $S_4O_6^{2-}$:

$$\underset{\substack{\text{tetrathionate}\\\text{ion}}}{S_4O_6^{2-}(aq)} + 2e^- \rightleftharpoons 2S_2O_3^{2-}(aq) \qquad E^\circ = +0.169 \text{ V}$$

Thus, the thiosulfate ion reduces iodine to the iodide ion as follows.

$$2S_2O_3^{2-}(aq) + I_2(aq) \longrightarrow S_4O_6^{2-}(aq) + 2I^-(aq)$$

As we studied in Section 12.9, this reaction is the basis of an important procedure in quantitative analysis.

Chlorine is a stronger oxidizing agent than iodine, and it oxidizes the thiosulfate ion to the hydrogen sulfate ion.

$$S_2O_3^{2-}(aq) + 4Cl_2(g) + 5H_2O \longrightarrow 2HSO_4^-(aq) + 8H^+(aq) + 8Cl^-(aq)$$

The bleaching industry has used sodium thiosulfate in this reaction to destroy excess chlorine bleaches in fibers.

Tap water from chlorinated water supplies kills fish, so aquarium owners use this reaction to remove chlorine.

Photographer's Hypo The pentahydrate of sodium thiosulfate — $Na_2S_2O_3 \cdot 5H_2O$ — is the familiar *hypo* that photographers use as a fixing agent when they develop film.

Photographic film contains silver salts such as silver bromide and silver iodide in the form of microscopically small crystals. When a picture is taken, these microcrystals are exposed to varying intensities of light. Such exposure initiates chemical changes that make the exposed sites more susceptible than the unexposed sites to the *developer*. This is a chemical that reduces exposed silver ions to silver atoms. The greater the exposure, the greater is the quantity of silver atoms and the darker the "negative" at that site. To fix a negative, the unchanged silver salts have to be removed, and this is the function of aqueous hypo. The thiosulfate ion can displace halide ion from water-insoluble silver halides and form water-soluble complex ions such as $Ag(S_2O_3)^-$ and $Ag(S_2O_3)_2^{3-}$, for example,

$$2S_2O_3^{2-}(aq) + AgBr(s) \longrightarrow Ag(S_2O_3)_2^{3-}(aq) + Br^-(aq)$$

When exposed film is *fixed*, its picture image is irreversibly set and the film can no longer record images.

Nothing happens to the silver atoms. They remain on the negative as the hypo fixer dissolves unchanged silver halides.

20.4 OTHER BINARY COMPOUNDS OF SULFUR

Of the several polysulfanes, H_2S_n, the only thermodynamically stable example is the simplest, H_2S, hydrogen sulfide.

Important binary compounds of sulfur include, besides the oxides, those with hydrogen, the halogens, and nitrogen. We will look at just a few of the dozens of examples.

Hydrogen Sulfide and Related Compounds

Hydrogen sulfide is a gas whose odor of rotten eggs nearly everyone knows. It is a more dangerous poison than hydrogen cyanide, but fortunately its distinctive odor provides warning. You would first notice it at a concentration of only 0.02 ppm. Nonetheless, hydrogen sulfide also desensitizes the nose's sense of small, so it would be dangerous to try to judge the concentration of hydrogen sulfide by the intensity of the odor. At a concentration of only 100 ppm, hydrogen sulfide is lethal, and it acts very rapidly. Its use in the lab, of course, should always be at the hood.

Hydrogen sulfide boils at $-50\ °C$ and freezes at $-83\ °C$. It forms a saturated solution in water at 25 °C with a concentration of about 0.1 M. It can be made by the action of a strong acid on any number of metal sulfides, such as iron(II) sulfide.

$$FeS(s) + 2HCl(aq) \longrightarrow H_2S(g) + FeCl_2(aq)$$

The chief use of hydrogen sulfide in the laboratory is to detect and identify several metal ions that form sulfide precipitates with characteristic colors and solubilities. For this use, the most common preparation of aqueous hydrogen sulfide is by the reaction of water with thioacetamide.

$$
\underset{\text{thioacetamide}}{CH_3\!-\!\overset{\displaystyle \overset{S}{\|}}{C}\!-\!NH_2} + H_2O \longrightarrow \underset{\text{acetamide}}{CH_3\!-\!\overset{\displaystyle \overset{O}{\|}}{C}\!-\!NH_2} + H_2S
$$

For $H_2S(aq)$ at 20 °C, $pK_1 = 6.88$ and $pK_2 = 13.89$.

Hydrogen sulfide is a weak, diprotic acid. Besides numerous, naturally occurring metal sulfide minerals, soluble salts such as NaHS and Na_2S are known and commercially available. The first is an example of a hydrogen sulfide salt and the second, a sulfide salt. Since the anions in both kinds of salts are Brønsted bases, the addition of a strong acid to either salt liberates hydrogen sulfide. These are the major reactions of the sulfide–hydrogen sulfide system, for example,

$$HCl(aq) + NaHS(aq) \longrightarrow H_2S(g) + NaCl(aq)$$
$$2HCl(aq) + Na_2S(aq) \longrightarrow H_2S(g) + 2NaCl(aq)$$

Some sulfide ores:

FeS_2	Iron pyrite
Cu_2S	Chalcocite
ZnS	Sphalerite
HgS	Cinnabar
PbS	Galena
$CuFeS_2$	Chalcopyrite

The same action occurs when a drop of hydrochloric or sulfuric acid is placed on a mineral suspected of being a sulfide. If the odor of rotten eggs soon develops, the field geologist knows that the mineral is a sulfide. (If no odor develops, the opposite conclusion cannot be drawn. Some sulfide minerals, like CuS and HgS, do not dissolve in HCl or H_2SO_4.)

Polysulfanes

When the hydrate of sodium sulfide, $Na_2S \cdot 9H_2O$, is melted with sulfur, a solution forms that gives a yellow oil when it is poured into an excess of dilute, cold hydrochloric acid ($-10\ °C$). It is a mixture of H_2S_x molecules, where x varies from 4 to 7. These are the polysulfanes, and their molecules are chains of sulfur atoms terminating at their ends with S—H bonds. They are not stable, and slowly break up into sulfur and hydrogen sulfide.

Binary Sulfur–Halogen Compounds

Sulfur forms numerous binary compounds with the halogens—six compounds with fluorine alone. Of these, sulfur hexafluoride is quite unique. Alone among the sulfur fluorides, SF_6 is exceptionally stable. It is a colorless, odorless, tasteless, nontoxic, nonflammable, water-insoluble gas at room temperature. Its structure was described in Example 8.2 on page 278. SF_6 is made by burning sulfur in fluorine.

$$S(s) + 3F_2(g) \longrightarrow SF_6(g)$$

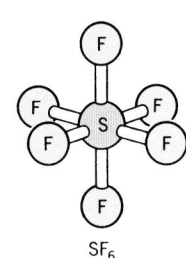

SF_6

SF_6 is so stable and it resists electrical currents so well that it is used as an insulating gas in electrical generators and switches. Few metals attack SF_6, but boiling sodium metal breaks it down to NaF and Na_2S.

Of the many binary sulfur–chlorine compounds, two are commercially important—disulfur dichloride (S_2Cl_2) and sulfur dichloride (SCl_2). The former is used to vulcanize rubber and the latter to make certain organic compounds. Disulfur dichloride is made by the reaction of chlorine with molten sulfur. It is a yellow liquid boiling at 138 °C, and is both extremely foul smelling and toxic. Sulfur dichloride, a red liquid (bp 59 °C), is made from S_2Cl_2 by further action of chlorine in the presence of a catalyst (e.g., $FeCl_3$). Both of these sulfur chlorides react with water to give complex mixtures of products.

Binary compounds of sulfur with bromine and iodine are known, but we will not study them.

Some Properties of Selenium, Tellurium, and Polonium

The oxygen family moves from the nonmetals—oxygen and sulfur—to two metalloids (semimetals)—selenium and tellurium—and ends with a radioactive metal—polonium. Yet there are many similarities among them, and we will look briefly at some.

Selenium, like sulfur, has many allotropes and some involve crownlike rings of eight selenium atoms, Se_8, just like the S_8 rings in S_α and S_β. Ordinary, commercial selenium, however, is more an amorphous, vitreous solid (and black in color). Tellurium has only one crystalline form. (Metallic polonium exists in a simple cubic structure.)

All members of the oxygen family form hydrides of the general formula H_2Z, where we let Z represent the symbol for any member of the family. H_2Se and H_2Te are gases at room temperature and known to be weak acids in water (although stronger than H_2S). H_2Po is a volatile liquid.

Selenium and tellurium form salts with the Group IA metals—M_2Se and M_2Te (where M is the metal ion). They also form salts with Group IIA metals—MSe and MTe. These salts are, of course, similar in formula to the sulfide salts.

Selenium and tellurium, like sulfur, form dihalides. And like sulfur they form dioxides and trioxides. These have behaviors similar to those of SO_2 and SO_3. They react with water, for example, to give acids of similar formulas and acid strengths. (The estimated acid dissociation constants for the first ionization are shown below the names. For comparison, K_1 for H_2SO_3 is 1.2×10^{-2}.)

H_2SeO_3	H_2TeO_3	H_2SeO_4	$Te(OH)_6$
Selenous acid	Tellurous acid	Selenic acid	Telluric acid
3.5×10^{-3}	3×10^{-3}	High	2×10^{-8}

The formula of telluric acid is also represented as H_6TeO_6.

Both selenous and tellurous acids are white solids. Selenic acid resembles sulfuric acid in acidity, but telluric acid differs in formula and is a weak acid.

Chemists have learned to be very cautious about extrapolating from similarities in formulas and physical properties to chemical and biological properties. Experiments have to be done to find out these properties. The compounds we have described here differ in thermal stability and in stability toward water, oxidizing agents (including air), and reducing agents. Almost needless to say, there are differences in physiological properties, too, such as toxicities and odors. Advanced treatments and references go into such matters.

20.5 PHOSPHORUS

The most stable of the many allotropes of phosphorus is red phosphorus.

Phosphorus was discovered in 1669 by a Hamburg alchemist, Hennig Brand, who distilled the residue from boiled-down, well-putrified urine, condensed the vapors under water, and found he had something that glowed in the dark and burst into flame in warm air. All from urine! It must certainly have made Brand's reputation and given hope to those seeking to make gold from baser things. These properties gave the element its name, after the Greek roots *phos,* "light," and *phoros,* "bringing."

Phosphorus is the second member of the nitrogen family, but it contrasts sharply with the first, nitrogen. Phosphorus is a solid. It burns in air. It occurs entirely as one isotope, phosphorus-31. And it exists in several allotropic forms, much like sulfur (which, second in the oxygen family, is also so unlike the first member of its family).

An important structural difference between elemental phosphorus and nitrogen is that nitrogen atoms can find great stability (that is, low energy) by forming *triple* bonds with each other in diatomic N_2 molecules. These moleules are symmetrical and nonpolar and have a low formula mass, so nitrogen is a gas at room temperature. Phosphorus atoms, on the other hand, do not have triple bonds between them. The bonds are single, instead, and phosphorus atoms can form chains and rings. The several allotropes of phosphorus reflect different molecular formulas and geometrical organizations.

Nitrogen, like the other nonmetals of period 2, is a small atom. Because of this, the two nitrogen atoms can approach each other quite closely during bond formation. This allows for effective sideways overlap of $2p$ valence shell orbitals, which is necessary for pi bond formation. As a result, nitrogen can form strong pi bonds and so is able to exist as diatomic molecules with $N{\equiv}N$ triple bonds. Phosphorus, on the other hand, is a much larger atom, and when two phosphorus atoms approach each other to form bonds, the sideways overlap of the phosphorus $3p$ orbitals is poor. Phosphorus, therefore, does not form strong pi bonds with its $3p$ valence orbitals. Each phosphorus, instead, uses three single bonds (sigma bonds), one to each of three neighboring phosphorus atoms.

Phosphorus sometimes uses its $3d$ orbitals in the formation of bonds, for example, to make sp^3d hybrid orbitals in PCl_5.

Two Phosphorus Allotropes

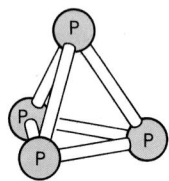

White phosphorus, P_4

White Phosphorus   What Hennig Brand so picturesquely stumbled onto was one of the less stable yet most common allotropes of phosphorus, *white phosphorus,* P_4. See Figure 20.11. It is a waxy, white solid that melts at 44 °C, boils at 280 °C, and dissolves in nonpolar organic solvents such as carbon disulfide or benzene. White phosphorus is extremely toxic, and any kind of direct body contact with it must be avoided.

The four atoms in the P_4 molecule are at the corners of a regular tetrahedron, and the bond angles are only 60°, much smaller than normal for a trivalent atom. The small angles mean poor overlap between atomic orbitals, so the P—P bonds are weak. This probably explains why white phosphorus is so easily attacked by atmospheric oxygen, and it should not be stored except under water and out of contact with air.

In air, white phosphorus spontaneously bursts into flame at about 35 °C. So it has been used by military forces in incendiary devices. In moist air and below 35 °C, the slower reaction of white phosphorus with oxygen causes the glowing that Brand observed. This phenomenon is now called *chemiluminescence* but an earlier name, *phosphorescence,* more closely tied the observed glowing to the element itself.

White phosphorus is produced today on the scale of millions of tons per year by the reduction of calcium phosphate by carbon, with silicon dioxide being used to capture the calcium as a (molten) calcium silicate slag. The reduction requires a very high temperature even though, overall, it is exothermic ($\Delta H° = -3060$ kJ/mol P_4).

The combustion of the CO made here helps provide the energy for the reaction.

$$2Ca_3(PO_4)_2(s) + 10C(s) + 6SiO_2(s) \xrightarrow{1400-1500 \ °C} P_4(g) + 10CO(g) + 6CaSiO_3(\ell)$$

We take matches so much for granted that it's hard to imagine what life was like before safe and inexpensive matches were developed.

The "safety match" was invented in 1855 by J. E. Lundstrom (Sweden). A mixture of mostly potassium chlorate — a strong oxidizing agent — antimony(III) sulfide, and a glue is on the match head. The striking surface on the matchbook cover is a mixture of red phosphorus, antimony(III) sulfide, a little iron(III) oxide, and powdered glass held by glue. During the strike, some heat is produced that converts a tiny trace of red phosphorus to white phosphorus, which instantly ignites. The heat ignites the chemicals in the match-head, and their short blaze ignites the wood or paper of the match stick.

In the "strike-anywhere match," invented in 1898 by H. Sevene and E. Cahan (France), the matchhead contains potassium chlorate, tetraphosphorus trisulfide (P_4S_3), ground glass, and the oxides of zinc and iron, all bound by glue. The scratch gives enough heat to initiate a violent (but small-scale) reaction between $KClO_3$ and P_4S_3, and its heat ignites the matchstick.

Between 80 and 90% of all elemental phosphorus produced is used almost as soon as it is made to manufacture phosphoric acid. Much of the rest is used to make phosphorus trichloride, phosphorus pentachloride, and phosphorus oxychloride for further manufacture of other chemicals.

Red Phosphorus   White phosphorus changes spontaneously to another common allotrope, *red amorphous phosphorus,* when it is carefully heated in the absence of air between 270 and 300 °C. See Figure 20.11. The structure is complex as the term *amorphous* implies.

Red amorphous phosphorus melts at about 600 °C, is essentially nontoxic, can be stored in air, and is altogether far safer than white phosphorus. Both red and white phosphorus are involved in the chemistry of matches, as discussed in Special Topic 20.1. As we noted, many other allotropes are known, but we will not go into them. For simplicity, we will use the monatomic symbol, P, for phosphorus in discussing its reactions.

Binary compounds of phosphorus with all elements except bismuth, antimony, and the noble gases are known. Most can be prepared directly from the elements. We will single out just some of the oxides and oxoacids of phosphorus, some of its halides, and its hydride.

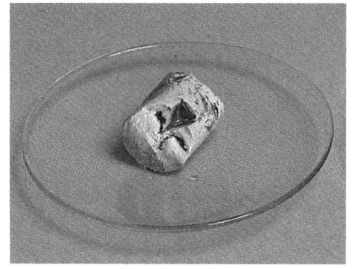

FIGURE 20.11 Phosphorus — its two main allotropes. White phosphorus is above and red amorphous phosphorus is shown below it.

Phosphine

The most stable binary hydride of phosphorus is commonly called *phosphine,* PH_3. It can be made by the hydrolysis of calcium phosphide, which is produced when calcium and phosphorus are heated.

$$\underset{\substack{\text{calcium} \\ \text{phosphide}}}{Ca_3P_2(s)} + 6H_2O \longrightarrow \underset{\text{phosphine}}{2PH_3(g)} + 3Ca(OH)_2(s)$$

Although PH_3 is the structural analog of NH_3, there are few other similarities. Phosphine is an extremely poisonous gas and has a garlic odor. It is over 2.5 thousand times less soluble in water than ammonia, and does not readily react either as a proton acceptor or a proton donor. When ignited, it burns readily in air to give orthophosphoric acid. (Sometimes impurities make it burn spontaneously.)

$$\underset{\text{phosphine}}{PH_3(g)} + 2O_2(g) \longrightarrow \underset{\substack{\text{orthophosphoric} \\ \text{acid}}}{H_3PO_4(\ell)}$$

Its chief use is in the manufacture of certain plastics used to make compounds that make cotton cloth flame resistant.

Phosphorus–Halogen Compounds

Twelve phosphorus halides are known. They occur in three sets according to the general formulas PX_3, P_2X_4, and PX_5 where $X =$ F, Cl, Br, or I. Many mixed halides with different halogen atoms present are also known.

We will look briefly at just two chlorides, PCl_3 and PCl_5, then generalize about some others, and look also at phosphorus oxychloride. We will be interested mainly in just two of their reactions — with water and with oxygen — because these reactants are common (and available in humid air, too).

Both PCl_3 and PCl_5 are made by the direct chlorination of phosphorus.

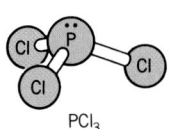

PCl_3

$$2P(s) + 3Cl_2(g) \longrightarrow 2PCl_3(\ell)$$
$$2P(s) + 5Cl_2(g) \longrightarrow (PCl_4)^+(PCl_6)^-(s)$$

Phosphorus trichloride, or phosphorus(III) chloride, is a colorless liquid that boils at 76 °C. It exists as a covalent substance in all phases. It is used to make organic compounds either of phosphorus or of chlorine.

Phosphorus pentachloride, or phosphorus(V) chloride, is a whitish solid that melts at 167 °C. Structurally, it is at the ionic–covalent borderline. See Figure 20.12. In the solid state and in concentrated solutions in polar solvents like nitromethane (CH_3NO_2), it exists as a pair of ions — $(PCl_4)^+(PCl_6)^-$. But in the liquid and gaseous states as well as in nonpolar solvents such as carbon tetrachloride or benzene it occurs as molecules of PCl_5. We will simplify things by using PCl_5 as the formula.

FIGURE 20.12 Phosphorus(V) chloride has the structure on the left in the gaseous state but exists as a pair of ions, on the right, in the solid state.

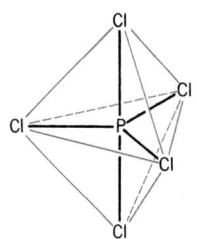
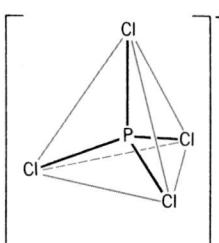
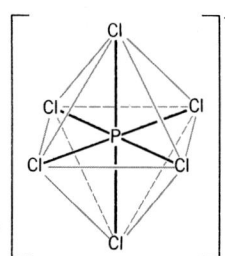

Phosphorus pentachloride is a fairly large-scale industrial chemical (20 to 30 thousand tons per year, worldwide). It is a raw material for making a number of inorganic and organic phosphorus compounds. Some of these are macromolecular and have unusual resistance to flames and solvents and remain flexible at low temperatures. These properties would be useful in fuel hoses, seals, and gaskets in vehicles operating at very low temperatures. (The principal barrier to these applications is the cost of the chemicals involved.)

Both the trichloride and the pentachloride of phosphorus react with water, even the moisture in humid air, to give oxoacids and hydrogen chloride. In humid air, hydrogen chloride forms whitish mists of hydrochloric acid droplets, and so the phosphorus halides are said to *fume* in moist air. Obviously, they must be kept in well-capped bottles.

$$2PCl_3(\ell) + 6H_2O \longrightarrow \underset{\text{phosphorous acid}}{2H_3PO_3(aq)} + 6HCl(aq)$$

$$PCl_5(s) + 4H_2O \longrightarrow \underset{\text{phosphoric acid}}{H_3PO_4(aq)} + 5HCl(aq)$$

This reaction is violent.

Exactly analogous reactions are shown by the tri- and pentabromides.

At room temperature, the phosphorus(III) halides are easily oxidized by pure oxygen to oxohalides, POX_3. The rate of the reaction is slower in air.

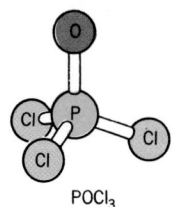

$POCl_3$

$$2PCl_3(\ell) + O_2(g) \longrightarrow \underset{\substack{\text{phosphorus} \\ \text{oxychloride}}}{2POCl_3(\ell)}$$

Since phosphorus is in its highest oxidation state ($+5$) in its pentahalides, these compounds are not oxidized in a similar way.

Phosphorus oxychloride, like the phosphorus chlorides, fumes in moist air, and reacts very readily with water to give phosphoric acid and hydrochloric acid.

$$POCl_3(\ell) + 3H_2O \longrightarrow H_3PO_4(aq) + 3HCl(aq)$$

The same kind of reaction occurs with methyl alcohol, which is structurally "waterlike" because it has an OH group.

$$POCl_3(\ell) + 3CH_3OH \longrightarrow \underset{\text{trimethyl phosphate}}{(CH_3O)_3PO(\ell)} + 3HCl(g)$$

We mention this reaction because it shows how some organophosphate compounds are made. The CH_3 group (the methyl group) in methyl alcohol is just one of literally thousands of hydrocarbon groups that can be joined to OH groups. Hundreds of thousands of tons of various organophosphorus compounds are made annually to serve such widely different uses as flame retardants, insecticides, oil additives, and detergents.

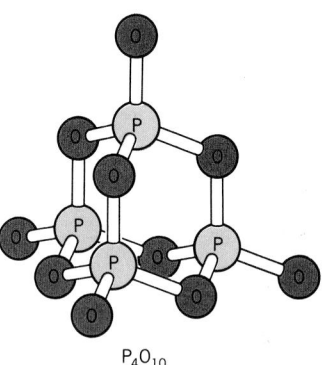

trimethyl phosphate

20.6 PHOSPHORUS OXIDES, OXOACIDS, AND THEIR SALTS

The phosphoric acid system—the acid in its several variations and its several salts—comprises the largest-volume manufactured compounds of phosphorus.

Most people have seen how spectacularly phosphorus burns in air, even if they have not known the chemical details. The product is a cloud of incandescent, white, powdery tetraphosphorus decaoxide, P_4O_{10}, so often seen in fireworks displays (Figure 20.13). (Red amorphous phosphorus is used.)

$$4P(s) + 5O_2(g) \longrightarrow P_4O_{10}(s)$$

In insufficient air, the oxidation goes to the $+3$ oxidation state in tetraphosphorus hexaoxide, P_4O_6.

$$4P(s) + 3O_2(g) \longrightarrow P_4O_6(s)$$

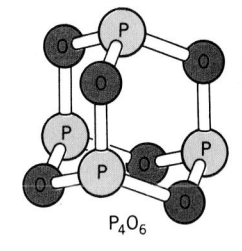

P_4O_{10}

P_4O_6

FIGURE 20.13 Fireworks displays show the combustion of phosphorus in beautiful action. Shown here is a display over the Lincoln Memorial in Washington, DC.

The Phosphoric Acids and Their Salts

The most notable chemical property of P_4O_{10} is its extremely exothermic reaction with water. When it is dusted onto water, the system hisses and spits as the reaction occurs. P_4O_{10} is one of the best desiccating agents (water-removing agents) available to chemists, and it obviously must be stored in well-capped bottles. Its reaction with water is also its major industrial use. As we said, 80 to 90% of all manufactured phosphorus is used to make P_4O_{10}, nearly all of which is promptly used in its reaction with water to make the various phosphoric acids.

The problem with this reaction — ours, not nature's problem — is that we cannot write a simple equation. Any one or a mixture of a family of three phosphoric acids can form depending on the temperature and the degree to which water has been taken up.

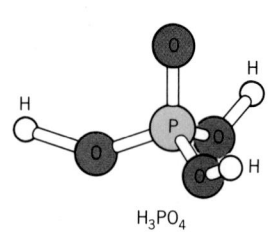

H_3PO_4

85.0% H_3PO_4 is 14.7 M H_3PO_4.

For H_3PO_4,

$$pK_1 = 2.15$$
$$pK_2 = 7.20$$
$$pK_3 = 13.89$$

HPO_3	$H_4P_2O_7$	H_3PO_4
metaphosphoric acid (empirical formula)	diphosphoric acid (pyrophosphoric acid)	orthophosphoric acid (phosphoric acid is the usual name)

If you examine these formulas, you can see that orthophosphoric acid is just metaphosphoric acid to which H_2O has been added, and diphosphoric acid is just two orthophosphoric acids minus a water molecule. The structural formulas of these compounds are complicated by the presence of extended chains or rings, which we will soon study. Each form has its own family of salts.

Phosphoric Acid Orthophosphoric acid is usually called simply *phosphoric acid,* since it is the common type of phosphoric acid in commerce. (We will use this name, too.) Phosphoric acid is sold as an 85% (w/w) solution in water. When pure, it is a solid melting at 42 °C.

Both molten phosphoric acid and the 85% aqueous solution are viscous, syrupy liquids. With three OH groups per molecule plus an O, several hydrogen bonds exist between molecules or to water molecules, and these make it difficult for molecules to slip by each other as they must when the liquid flows.

Phosphoric acid is a moderately strong, triprotic acid. One of its minor uses is to give a tart taste to manufactured soft drinks, but its largest use is to make "triple superphosphate" fertilizers. The reaction is just like the reaction of sulfuric acid with phosphate rock (page 835) — conversion of the phosphate salt to the more soluble dihydrogen phosphate salt — but no calcium sulfate forms to dilute the fertilizer's essential potency as a supplier of "phosphate." Hence the name "triple superphosphate."

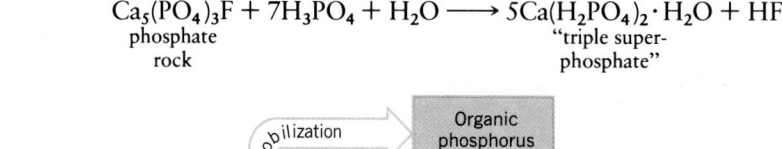

$$Ca_5(PO_4)_3F + 7H_3PO_4 + H_2O \longrightarrow 5Ca(H_2PO_4)_2 \cdot H_2O + HF$$

phosphate rock   "triple super- phosphate"

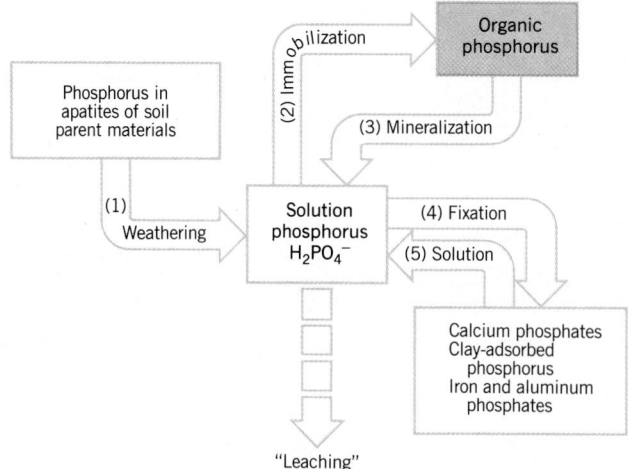

FIGURE 20.14 The phosphorus cycle. Only the principal processes are shown. (From H. D. Foth, *Fundamentals of Soil Science,* 6th ed., New York: John Wiley & Sons, Copyright © 1978. Used by permission.)

Soil without phosphorus (in the form of ions of phosphoric acid) is barren. Plant growth absolutely requires phosphate, because so may key compounds in cells—plants and animals alike—are organophosphate compounds. One of the many cycles in nature is the phosphorus cycle (Figure 20.14) in which phosphate units cycle between living things and minerals.

Salts of Phosphoric Acid   Three kinds of salts of H_3PO_4 are known—those of the dihydrogen phosphate ion, $H_2PO_4^-$, the monohydrogen phosphate ion, HPO_4^{2-}, and the phosphate ion, PO_4^{3-}. The sodium, potassium, calcium, and ammonium salts are the most common, and they are commercially available either as anhydrous salts or as a large variety of different kinds of hydrates. When these salts are used, we have to be aware of their acid–base properties, because they can affect the pH of a solution. We will review these properties briefly.

The three anions of phosphoric acid are all Brønsted bases but they differ widely in strength. Their order of basicity is as follows where the pK_b values are given beneath the formulas.

$$PO_4^{3-} > HPO_4^{2-} > H_2PO_4^-$$
$$pK_b \quad 1.66 \quad\quad 6.80 \quad\quad 11.85$$

The strongest base, PO_4^{3-}, is naturally the conjugate base of the weakest possible acid available to this series, the HPO_4^{2-} ion. Thus a solution of sodium phosphate, Na_3PO_4, in water is basic because OH^- is generated by the following equilibrium with its conjugate acid.

$$PO_4^{3-}(aq) + H_2O \rightleftharpoons HPO_4^{2-}(aq) + OH^-(aq)$$

In a 0.01 M solution of Na_3PO_4 the pH is 11.6, which is roughly equivalent to the basicity of 0.004 M NaOH. (You would not want to expose your skin or eyes to solutions of Na_3PO_4.)

From our study of buffers in Section 15.6, we learned that a buffer is made up of a conjugate acid–base pair. One such pair consists of the $H_2PO_4^-$ and HPO_4^{2-} ions, and this is a buffer system of great importance in living systems. They make up what biochemists call the "phosphate buffer" present inside cells to maintain control over the pH of the cell fluid. It neutralizes H^+ by the reaction:

$$HPO_4^{2-} + H^+ \longrightarrow H_2PO_4^-$$

And it neutralizes OH^- by the following reaction.

$$H_2PO_4^- + OH^- \longrightarrow HPO_4^{2-} + H_2O$$

Salts of $H_2PO_4^-$ are examples of **acid salts,** because they have fairly readily available hydrogen ions. The calcium salt, $Ca(H_2PO_4)_2$, for example, is used in baking powders as an *acidulant*, a substance that can begin to act as an acid as water is added. In water, the anions in calcium dihydrogen phosphate donate protons to bicarbonate ions, which are present in baking powders as solid sodium bicarbonate. This reaction releases carbon dioxide, and the generation of this gas within the batter creates the desired frothy texture of a baked item.

$$Ca(H_2PO_4)_2(aq) + 2NaHCO_3(aq) \longrightarrow 2CO_2(g) + 2H_2O + CaHPO_4(aq) + Na_2HPO_4(aq)$$

The net ionic equation is

$$H_2PO_4^-(aq) + HCO_3^-(aq) \longrightarrow CO_2(g) + H_2O + HPO_4^{2-}(aq)$$

$(NH_4)_2HPO_4$, diammonium phosphate (DAP), is an important agricultural fertilizer.

TRUE

Remember, the *smaller* the pK_b, the stronger is the base.

Polyphosphates

The term *polyphosphates* refers to the anions of the various higher oxoacids of phosphorus, those with two or more phosphorus atoms per molecule and with alternating P—O—P linkages. The chains can extend to as many as ten phosphorus atoms. In each system the geometry is tetrahedral around P.

We can think of the P—O—P network as forming by the removal of a water molecule between molecules of phosphoric acid. When kept above its melting point, phosphoric acid slowly develops an equilibrium with diphosphoric acid.

For $H_4P_2O_7$,

$$pK_1 = 1.0$$
$$pK_2 = 1.8$$
$$pK_3 = 6.57$$
$$pK_4 = 9.62$$

$$\text{H—O—P—O—H + H—O—P—O—H} \rightleftharpoons$$

two molecules of phosphoric acid, H_3PO_4

$$\text{H—O—P—O—P—O—H + H—OH}$$

diphosphoric acid, $H_4P_2O_7$

When heated with excess water, the equilibrium in this system shifts to the left and diphosphoric acid hydrolyzes and changes back to phosphoric acid. All polyphosphate systems can be similarly hydrolyzed to phosphoric acid.

The neutralization of all four available protons of diphosphoric acid gives the diphosphate ion, $P_2O_7^{4-}$, whose structure and geometry are given in the margin. This is the simplest polyphosphate ion. Its calcium salt, $Ca_2P_2O_7$, is used in toothpastes as an inert, insoluble abrasive that does not interfere with fluoride additives.

The sodium salt of a doubly protonated form of $P_2O_7^{4-}$, $Na_2H_2P_2O_7$, is an acid salt, and is also used as an acidulant in baking powders. Its advantage over $Ca_2(H_2PO_4)_2$ (discussed above) is that it does not react with bicarbonate ion except when heated. Hence, bakers can prepare large quantities of batter or dough at room temperature and even store them without the dough-raising action starting prematurely.

Higher members of the polyphosphate family can be thought of as forming by the loss of water between additional molecules of phosphoric acid, much as in the formation of diphosphoric acid. Their structures would share the following features, where the subscript n would indicate the chain length.

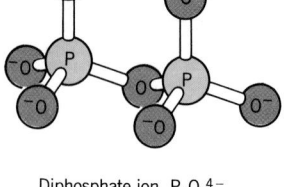

Diphosphate ion, $P_2O_7^{4-}$

$$\text{HO—P—(O—P—)}_n\text{O—P—O—H}$$

polyphosphoric acids

When $n = 3$, the corresponding compound is triphosphoric acid, and like all polyphosphoric acids it can be hydrolyzed back to phosphoric acid.

The tripolyphosphate ion, $P_3O_{10}^{5-}$, as the sodium salt, $Na_5P_3O_{10}$, is the chief "phosphate" in laundry detergents (where they are still permitted). In laundry and cleaning products, tripolyphosphate ions act in many ways, one being as a water-softening agent—a substance that can tie up the "hardness ions" in hard water, Ca^{2+}, Mg^{2+}, and Fe^{3+}, and prevent them from interfering with the surface-active agent, the detergent itself. The regularly spaced negative charges of the tripolyphosphate ion enable it to form water-soluble complex ions with the divalent metal ions of hard water and hold them out of reach of any reaction with the detergent molecules. For example,

$$P_3O_{10}^{5-}(aq) + Ca^{2+}(aq) \longrightarrow CaP_3O_{10}^{3-}(aq)$$

tripolyphosphate ion   complex between Ca^{2+}
                                            and tripolyphosphate
                                            (water-soluble)

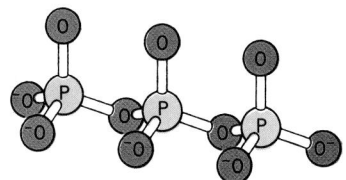

Tripolyphosphate ion, $P_3O_{10}^{5-}$

Sodium tripolyphosphate in a laundry product also boosts the cleansing efficiency of the detergent.

The trouble with phosphates in detergents is that they support the life of algae, so when phosphate-containing waste water gets into lakes, the lakes become unfit for recreation. Many areas, therefore, have banned the sale of phosphate-containing laundry products, since this is deemed much less costly than the removal of phosphates from waste water.

Cyclic types of phosphoric acids and their ions are also known. The simplest is *cyclo*-triphosphoric acid, $H_3P_3O_9$, with three phosphorus atoms alternating with oxygen atoms in six-membered rings. (Its commonly called metaphosphoric acid, but the IUPAC name we have used obviously carries more structural information.) Larger rings with up to ten phosphorus atoms are also known.

Some states around the Great Lakes and some eastern states have banned the sale of phosphate detergents and Na_3PO_4.

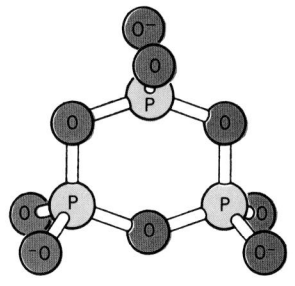

Cyclo–triphosphate ion, $P_3O_9^{3-}$

Other Oxoacids of Phosphorus

Besides phosphoric acid (in its variety of forms), phosphorus forms two other oxoacids in which the phosphorus occurs in lower oxidation states. **Phosphorous acid,** H_3PO_3, is formed when tetraphosphorus hexaoxide (P_4O_6) dissolves in water and also by the reaction of PCl_3 with water.

$$P_4O_6 + 6H_2O \longrightarrow 4H_3PO_3$$
$$PCl_3 + 3H_2O \longrightarrow H_3PO_3 + 3HCl$$

The pure acid is a colorless solid that melts as 70.1 °C. Although the formula of phosphorous acid suggests it is triprotic, it is actually a diprotic acid because one of the hydrogens is bonded directly to the phosphorus atom.

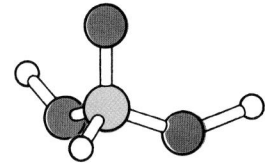

Phosphorous acid

$$\begin{array}{c} H \\ | \\ H-O-P-O-H \\ | \\ O \end{array}$$

phosphorous acid

Phosphorous acid is moderately strong with $pK_{a1} = 1.8$.

Another similar acid is **hypophosphorous acid,** H_3PO_2, the anion of which is formed as one of the phosphorus-containing species when white phosphorus reacts with boiling aqueous base.

$$P_4(s) + 4OH^-(aq) + 4H_2O \longrightarrow 4H_2PO_2^-(aq) + 2H_2(g)$$

hypophosphite
ion

The free acid, a white crystalline solid, can be obtained by acidifying the mixture. It is a moderately strong monoprotic acid ($pK_a = 1.2$), and has two hydrogens covalently bonded to the central phosphorus atom.

$$\begin{array}{c} H \\ | \\ H-P-O-H \\ | \\ O \end{array}$$

hypophosphorous acid

Antimony   Arsenic

Orpiment, As₂S₃

Realgar, As₄S₄

Stibnite, Sb₂S₃

Some Properties of Arsenic, Antimony, and Bismuth

In the nitrogen family, we move from nonmetals that form acidic oxides — nitrogen and phosphorus — to metalloids that form amphoteric oxides — arsenic and antimony — to the last element — bismuth — that is barely a metal and forms a basic oxide. (Bismuth will be discussed in more detail in Chapter 21.)

As we go down the family from arsenic through bismuth, the general reactivity of the elements toward other elements and water decreases.

Arsenic, antimony, and bismuth are all brittle, crystalline, nonductile solids at room temperature, and all poorly conduct electricity. All three form oxides of the same two types we studied for phosphorus — M_4O_6 and M_4O_{10}, where M is again the element in question. Arsenic and antimony form acids, H_3MO_4, similar to phosphoric acid. The hydride of arsenic, AsH_3 (arsine), is much more stable than those of antimony — SbH_3 (stibine) — and of bismuth — BiH_3 (bismuthine), which is extremely unstable. All three hydrides are very toxic.

Trihalides, MX_3, for all four halogens (X) are known. Of the pentahalides, MX_5, only the pentafluorides are known plus the pentachlorides of arsenic and antimony. The oxidation state is $+5$ in the pentahalides, and the pentafluorides are particularly strong oxidizing agents. BiF_5, is noteworthy; it explodes on contact with water, forming ozone and other products!

The sulfides of arsenic, antimony, and bismuth occur in a variety of ratios of elements and colors. The mineral *orpiment*, As_2S_3 has an almost goldlike color. *Realgar*, As_4S_4, is bright orange-red. And *stibnite*, Sb_2S_3, a black compound, was used in the most ancient of times to darken women's eyebrows. The sulfides are important in schemes for the qualitative analysis of these elements.

20.7 THE HALOGENS

The hydrides, metal salts, and a few oxoacids are the most commonly used compounds of the halogen family.

The halogen family consists of fluorine, chlorine, bromine, iodine, and astatine, and some properties are given in Table 20.3. As a name, *halogen* comes from Greek roots meaning "sea-salt producer," after the ability of halogens to combine directly with metals to make salts.

Like the first members of the oxygen and nitrogen families, fluorine is more different from chlorine, the second member, than chlorine differs from the rest of the family. Astatine is intensely radioactive and found only rarely in some uranium deposits. We will not study it.

Fluorine

Fluorine, a pale yellow gas, does not occur in the free state. Its chief minerals are fluorite (CaF_2), cryolite (Na_3AlF_6), and fluoroapatite [$Ca_3(PO_4)_2 \cdot CaF_2$], and only the fluorine-19

TABLE 20.3 The Halogens — Group VIIA

Name	Atomic Symbol	Boiling Point (°C)	Melting Point (°C)	Density (20 °C)
Fluorine	F	−188	−220	1.7 g/L
Chlorine	Cl	−35	−101	3.2 g/L
Bromine	Br	59	−7	3.1 g/mL
Iodine	I	184	114	4.9 g/mL
Astatine	At	No stable isotope occurs		

isotope is present. Henri Moisson (1858–1907), a French chemist, won the 1906 Nobel prize in chemistry for isolating fluorine in its elemental form. Its existence in compounds had long been known, but its unusual chemical reactivity and high reduction potential defeated efforts, launched by the nineteenth century's best chemists, to isolate fluorine itself. Moisson won by electrolyzing a solution of KHF_2 in liquid HF using platinum–iridium alloy electrodes in a platinum vessel.

Today fluorine is manufactured at a rate of thousands of tons per year. Although it can be shipped as a liquid in special casks cooled by liquid nitrogen, most fluorine is used where it is made. Its chief use is to convert uranium minerals into UF_6, a step in the processing of uranium fuel for nuclear power plants. Some fluorine is changed to SF_6, discussed on page 839. There is also a demand for special fluorinating agents to make organofluorine compounds, such as the well-known nonstick coating, Teflon. Such fluorinating agents, made where fluorine is manufactured, include ClF_3 and BrF_3, which are safer to ship and handle than fluorine itself.

Fluorine has the highest reduction potential of all common chemicals, and it reacts violently at room temperature with almost all other elements. Compounds of fluorine can be made with all elements except helium, neon, and argon.

Water *burns* in fluorine to give HF and oxygen.

$$2F_2(g) + 2H_2O \longrightarrow 4HF(g) + O_2(g)$$

Mixed with hydrocarbons, fluorine cracks all carbon–carbon bonds and strips hydrogens, leaving HF and CF_4. You can see why this element must be stored in special metal alloy cylinders, and why more moderate fluorinating agents are useful.

Fluorine owes its extreme reactivity in part to an unusually low F-F bond energy. Because the fluorine atom is so small and the atoms in the F_2 molecule are so close together, there is considerable repulsion between the lone pairs of electrons of the two atoms. This tends to push the atoms apart and thereby destabilizes the molecule, so not very much energy is needed to break the bond.

Hydrogen Fluoride and Hydrofluoric Acid Hydrogen fluoride forms explosively when hydrogen and fluorine are mixed, so it is made instead by the action of concentrated sulfuric acid on fluorospar, calcium fluoride.

$$CaF_2(s) + H_2SO_4(\ell) \longrightarrow 2HF(g) + CaSO_4(s)$$

A rotating alloy steel kiln is used, and gaseous HF evolves as the mixture is heated.

The boiling point of HF is 19.5 °C, near room temperature, so it is easily condensed. It is marketed as a liquid contained in special vessels lined with wax or polyethylene or in cylinders made of alloy steel or polyethylene. Glass containers cannot be used since HF dissolves this material, a reaction we can write using SiO_2 to represent the chief component of glass.

$$4HF(g) + SiO_2(s) \longrightarrow SiF_4(g) + 2H_2O$$

Fluorine is the most electronegative of all elements, and the covalent bond in H—F is quite polarized, much more so than the bond in H—Cl. HF molecules, therefore, much more strongly hydrogen-bond to each other than do those of HCl. Thus HF boils almost 65 degrees higher than HCl despite its lower formula weight. In the solid state, HF molecules join by hydrogen bonds in extended zig-zag chains as follows.

hydrogen bonding ($\cdots$) in HF(s)

The synthesis of F_2 by *chemical* means involving $MnF_6{}^{2-}$ and SbF_5 was reported in 1986—the first nonelectrolytic synthesis of F_2.

The molecular chains in Teflon are thousands of C atoms long:

$$F_2(g) + 2e^- \rightleftharpoons 2F^-(aq)$$
$$E° = +2.87 \text{ V}$$

Hydrogen fluoride dissolves in water to form hydrofluoric acid. Its K_a is 6.8×10^{-4}, so it acts as a weak acid. There is evidence, however, that a high concentration of HF molecules are ionized in solution. But the H$_3$O$^+$ and F$^-$ ions are apparently so strongly associated with each other that they effectively are not independent, and so hydrofluoric acid behaves as if it were a weak acid. Another unusual feature of hydrofluoric acid is the presence of the bifluoride ion, HF$_2^-$, which has no counterpart among the other hydrohalic acids.

$$F^- + \overset{\delta+}{H} - \overset{\delta-}{F} \longrightarrow [F \cdot\cdot H \cdot\cdot F]^-$$
bifluoride ion

Hydrofluoric acid is a very dangerous chemical. For a look at some of the unusual physiological properties of hydrofluoric acid and the fluoride ion, see Special Topic 20.2.

Chlorine

Isotope	% Abundance
Cl-35	75.53
Cl-37	24.47

Chlorine is a poisonous, yellowish-green gas (Figure 20.15). It was first isolated in 1774 by Carl Wilhelm Scheele (1742–1786), a Swedish scientist. What Scheele made was not recognized as an *element*, however, until 1810 by Humphrey Davey (1778–1829), an English chemist. Its name is from the Greek *chloros* meaning "yellow-green," the same association with color as found in the word "chlorophyll."

Chlorine consists of two isotopes—chlorine-35 and chlorine-37—and it is present in nature mostly as the chloride ion in seawater or in deposits of the minerals halite (NaCl) and sylvite (KCl).

Chlorine ranks in the top ten of all chemicals manufactured in the United States. Its production in the mid-1980s amounted to over 10 million tons (over 123 billion moles), most of it made as we discussed in Section 17.4—by the electrolysis of aqueous sodium chloride. Chlorine boils at −10.1 °C, and it is easily liquefied and shipped under pressure.

About 70% of all chlorine production goes to the manufacture of organochlorine products such as the plastic, polyvinyl chloride, used to make clear plastic bottles. Another 20% of the chlorine made is used to disinfect waste water and drinking water supplies and to bleach cotton and paper products. (Hydrogen peroxide has outpaced chlorine as the chief bleaching agent for paper and textiles, however.) The remaining chlorine is used to make a large variety of inorganic compounds needed by the chemicals industry.

FIGURE 20.15 Gaseous chlorine.

Hydrogen Chloride and Hydrochloric Acid When high-purity hydrogen chloride is needed, it is made today by the direct combination of the elements.

$$H_2(g) + Cl_2(g) \longrightarrow 2HCl(g)$$

A 1:1 mole (or volume) mixture of H_2 and Cl_2 is used, and it explodes when exposed to ultraviolet light. Most hydrogen chloride—as much as 90%—however, is obtained as a by-product in the manufacture of organochlorine compounds. It can also be made by heating a mixture of salt (NaCl) and sulfuric acid. Figure 20.16 shows how an apparatus for a laboratory synthesis by this reaction could be set up. The overall reaction can be written as follows.

$$NaCl(s) + H_2SO_4(\ell) \longrightarrow HCl(g) + NaHSO_4(s)$$

Hydrogen chloride boils at $-84.2\ °C$, so it is shipped under pressure in steel containers. Considerable quantities of it are used directly, however, to make its aqueous solution, hydrochloric acid. The concentrated hydrochloric acid of commerce is 12 M and 37% (w/w) in HCl. Its largest industrial use is to dissolve oxide scale from steel and other metals—a process called "pickling."

Hydrochloric acid, of course, is by now a familiar lab chemical to you. Its chemical properties are those of the hydronium ion and the chloride ion, and have been discussed elsewhere in this book. This acid is the acid of "stomach acid," where it is needed to activate a protein-digesting enzyme. And the chloride ion is present in all body fluids.

Oxides and Oxoacids of Chlorine

Three oxides of chlorine are listed in Table 20.4. The most important is ClO_2. We will briefly describe it plus the four oxoacids of chlorine, HClO, $HClO_2$, $HClO_3$, and $HClO_4$.

Hypochlorous Acid and the Hypochlorites When chlorine is bubbled into water, some of what dissolves reacts to give a mixture of hydrochloric acid and hypochlorous acid in the following equilibrium.

$$Cl_2(aq) + H_2O \rightleftharpoons \underset{\substack{\text{hydrochloric} \\ \text{acid}}}{HCl(aq)} + \underset{\substack{\text{hypochlorous} \\ \text{acid}}}{HOCl(aq)}$$

At equilibrium at 25 °C, a saturated solution of chlorine in water has taken up 0.091 mol of Cl_2 per liter. Of this, the molar concentration of unreacted $Cl_2(aq)$ is 0.061 M. And the concentrations of HCl(aq) and HOCl(aq) are each 0.030 M. Thus 67% of the chlorine in saturated chlorine water remains as dissolved Cl_2.

If chlorine is bubbled into a base, the equilibrium shifts to consume all of the added Cl_2 as salts of HCl and HOCl form. Thus when chlorine reacts in an aqueous slurry of

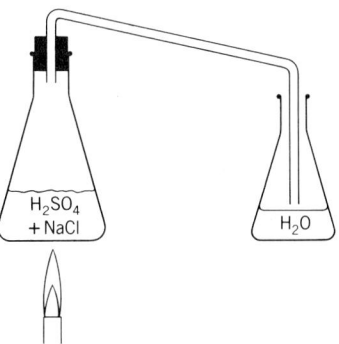

FIGURE 20.16 Small amounts of hydrogen chloride can be made in the lab and converted to hydrochloric acid using the apparatus shown here. Sodium chloride and sulfuric acid are loaded into the flask on the left, which is then heated strongly. Hydrogen chloride is driven into the receiver on the right where it is trapped in water.

Hydrochloric acid is known commercially as *muriatic acid,* and can be purchased under this name in most hardware stores.

The formula for hypochlorous acid is written as HClO or HOCl, but the latter is generally preferred.

Oxoacid of Chlorine	K_a
HOCl	2.9×10^{-8}
$HClO_2$	1.1×10^{-2}
$HClO_3$	1 (approx)
$HClO_4$	Very large

TABLE 20.4 Oxides of Chlorine

Oxidation State of Chlorine	Formula	Name	Color	Boiling Point (°C)	Melting Point (°C)
+1	Cl_2O	Dichlorine monoxide	Yellow-red	2.2	−116
+4	ClO_2	Chlorine dioxide	Gas: yellow Liquid: red	11	−60
+7	Cl_2O_7	Dichlorine heptoxide	Colorless	80[a]	−96

[a] This is a very explosive liquid.

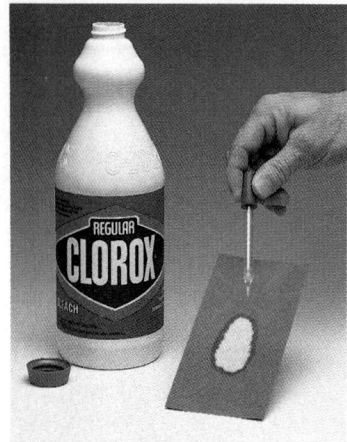

"Chlorine" bleach, NaOCl(aq),
destroys dyes by oxidizing them to
colorless compounds.

The hypochlorite ion can be
written as either ClO⁻ or OCl⁻,
but the latter is preferred.

calcium hydroxide, a mixture of calcium chloride and calcium hypochlorite, called "bleaching powder," can be made.

$$2Cl_2(aq) + 2Ca(OH)_2(s) \longrightarrow \underbrace{Ca(OCl)_2(s) + CaCl_2(s)}_{\text{components of "bleaching powder"}} + 2H_2O$$

Since this mixture is a solid, it is a more convenient way to handle a bleaching agent than gaseous chlorine or a solution of hypochlorous acid. So it is used for general sanitation purposes where special equipment would be inconvenient. But its calcium ion—an ion responsible for water hardness—makes it unattractive for the laundry.

The "liquid bleach" available for households and laundries is a solution of sodium hypochlorite, NaOCl, at a pH of about 11 or higher. The high pH ensures that the following equilibrium favors OCl⁻, not the less table HOCl.

$$HOCl(aq) + OH^-(aq) \rightleftharpoons OCl^-(aq) + H_2O$$

A minor problem with aqueous sodium hypochlorite is that the hypochlorite ion dispro-portionates in base. Disproportionation is the change of a single substance into two or more different substances, usually by an oxidation–reduction process. When ClO⁻ ions disproportionate, some are reduced to Cl⁻ ions and others are oxidized to ClO₃⁻ ions, and the following equilibrium forms.

$$3ClO^-(aq) \rightleftharpoons 2Cl^-(aq) + ClO_3^-(aq) \qquad K_c = 10^{27}$$

Oxidation state of chlorine: +1 ... −1 ... +5

The reaction is slow at room temperature, however, despite how strongly the products are thermodynamically favored at equilibrium. This means that "liquid bleach" has an acceptable shelf life, unless it is heated.

Chlorous Acid and Chlorites   The least stable of all of the oxoacids of chlorine is HClO₂, chlorous acid. It decomposes chiefly by the following reaction when heated, and it cannot be isolated from its aqueous solutions.

$$5HClO_2(aq) \longrightarrow 4ClO_2(aq) + HCl(aq) + 2H_2O$$

With a K_a of 1.1×10^{-2}, chlorous acid is a moderately strong acid.

Salts of chlorous acid are called chlorites. Sodium chlorite, NaClO₂, is manufac-tured as a bleaching agent and as an oxidizing agent particularly suited to destroying organic gases of foul odor in certain manufacturing operations.

Chlorine Dioxide   Chlorine dioxide, ClO₂, is an important bleaching agent used in the manufacture of paper. It is made in the thousands of tons per year by the reduction of NaClO₃. Chloride ion is a strong enough reducing agent for this reaction.

$$2ClO_3^-(aq) + 2Cl^-(aq) + 4H^+(aq) \longrightarrow 2ClO_2(aq) + Cl_2(aq) + 2H_2O$$

Generally, dilute aqueous solutions of ClO₂ are used, because this oxide can explode when pure or even in a concentrated solution.

Chloric Acid and the Chlorates   Chloric acid, HClO₃, is strong acid but known only in aqueous solution. Attempts to isolate it by the evaporation of the solvent lead to a complex disproportionation into chlorine, chlorine dioxide, perchloric acid, and oxygen instead.

$$11HClO_3 \longrightarrow 5HClO_4 + 2ClO_2 + 2Cl_2 + 3O_2 + 3H_2O$$

Salts of chloric acid, such as sodium chlorate (NaClO₃), are stable and are used as oxidizing agents, bleaches, and in the manufacture of explosives and matches.

Perchloric Acid and Perchlorates Perchloric acid, $HClO_4$, is the only oxoacid of chlorine that is reasonably stable, and then only when it is not made anhydrous. The commercial form is a 70 to 72% (w/w) solution. This is a very strong acid, and when it is concentrated it is a *very* strong oxidizing agent, particularly toward almost anything organic — cork, rubber stoppers, filter paper, alcohols — anything that can burn. It can react explosively on contact with these materials.

Salts of perchloric acid, the perchlorates, are likewise dangerous in contact with organic matter. Interestingly, in a *dilute* solution, the perchlorate ion, ClO_4^- is one of the most inert ions, not even serving as an oxidizing agent.

Bromine and Its Major Compounds

Bromine was first isolated in 1826 by a young French chemist, Antone Balard (1802–1876) who found that the addition of aqueous chlorine (called "chlorine water") to water from the Montpellier salt marshes in France turned the water deeply yellow. After isolation and purification, the cause of the color was found to be a liquid element with a terrible smell. The French Academy named it bromine, from *bromos*, which is Greek for "stench."

Balard's basic reaction is still the way that bromine is produced from its most common sources, naturally occurring brines. The reaction in which chlorine oxidizes bromide ion goes by the following equation.

$$Cl_2(aq) + 2Br^-(aq) \longrightarrow 2Cl^-(aq) + Br_2(aq)$$

The bromine is removed by passing either air or steam into the solution, which carries bromine vapors out where they can be condensed and further purified. We can see from the following standard reduction potentials why this reaction goes from left to right, not the reverse.

$$Cl_2(aq) + 2e^- \rightleftharpoons 2Cl^-(aq) \quad E° = +1.36 \text{ V}$$
$$Br_2(aq) + 2e^- \rightleftharpoons 2Br^-(aq) \quad E° = +1.07 \text{ V}$$

Cl_2 has a greater potential for being reduced than Br_2, so its reduction has to be shown as the forward reaction, making the $Br_2/2Br^-$ system run as an oxidation.

The annual consumption of bromine in the United States varied from 175,000 to 200,000 tons per year in the 1980 to 1985 period. Since over half went to make ethylene bromide, a gasoline additive that scavenges lead additives after combustion, the rate of use of bromine is sensitive to the use of leaded gasoline. And this use is declining. (In Special Topic 21.2 we will discuss why leaded gasoline is no longer allowed for new cars.)

Other important uses of bromine are to make organobromine compounds, like methyl bromide (CH_3Br), a pesticide that kills root-destroying worms (nematodes). The manufacture of silver bromide, AgBr, which is needed to make photographic film, also uses some bromine.

Bromine occurs as a nearly 1:1 mixture of two isotopes, bromine-79 and bromine-80. It has a high density, 3.12 g/mL, and a deep red color (Figure 20.17). Bromine boils at 59.5 °C, which makes it quite volatile at room temperature. Not only does bromine have a most irritating odor, its vapors also chemically attack the soft tissue of the nose and throat.

Bromine can cause a severe chemical burn on the skin, so this element must always be handled at the hood. Wear easily discarded protective gloves when you pour it. Do not use a medicine dropper; bromine is too dense, and will dribble out. It is wise to have some photographers' hypo handy when dispensing bromine. By complex reactions, a paste of this compound quickly changes bromine to substances less harmful to the skin, at least in the short term. One reaction that occurs is the following.

$$Na_2S_2O_3 \cdot 5H_2O + 4Br_2 \longrightarrow 2NaHSO_4 + 8HBr$$
hypo (sodium thiosulfate)

Left: aqueous NaBr. Center: $Cl_2(aq)$ has been added and the brown color of Br_2 develops. Right: The newly added globule of CCl_4, on the bottom, dissolves some of the Br_2 and becomes brown.

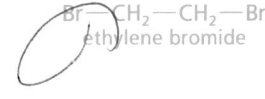

Br—CH₂—CH₂—Br
ethylene bromide

Isotope	% Abundance
Br-79	50.54
Br-81	49.46

FIGURE 20.17 Bromine is one of the only two elements that is a liquid at room temperature. (Mercury is the other.) At room temperature, the vapor pressure of bromine is relatively high, about 220 torr; and it is quite dense, 3.12 g/mL. A combination of vapor pressure and density, therefore, works to make bromine dribble easily from a dropping pipet without any need to squeeze the pipet's rubber bulb, as you can see in this photo.

(The skin should soon be washed with water, of course; the products here are both acids.)

Bromine resembles chlorine in chemical properties, except for reactivity. Bromine combines directly with most elements, but not as many as chlorine. It is more soluble in water than chlorine but, unlike chlorine, bromine does not react with water to any appreciable extent. A saturated solution of bromine in water at 25 °C is 0.21 M in bromine but only 1.2×10^{-3} M in hypobromous acid. This acid is thus a small part of the following equilibrium that is present in "bromine water."

$$Br_2(aq) + H_2O \rightleftharpoons HBr(aq) + HOBr(aq)$$

Hypobromous acid can be prepared by other means, but like hypochlorous acid it is unstable.

The only important acid of bromine is hydrobromic acid, HBr(aq), made from hydrogen bromide, HBr(g). Industrially, this gas is made directly from the elements by heating them in the presence of a platinum catalyst.

$$H_2(g) + Br_2(g) \xrightarrow[\text{Pt}]{200-400 \text{ °C}} 2HBr(g)$$

When hydrogen bromide is bubbled into water, hydrobromic acid results. It is usually sold either at a 50% (w/w) or a 40% (w/w) concentration.

In a small-scale lab preparation of hydrogen bromide, a mixture of sodium bromide and 50% (w/w) sulfuric acid is heated. The gaseous hydrogen bromide that distills from this is passed directly into water to make hydrobromic acid.

$$2NaBr(aq) + H_2SO_4(aq, 50\%) \xrightarrow{\text{heat}} 2HBr(g) + Na_2SO_4(aq)$$

Sulfuric acid of a higher concentration is not used, because it then becomes an oxidizing agent strong enough to oxidize bromide ion to bromine. (This was not a problem in a comparable synthesis of hydrogen chloride, described earlier, because the chloride ion is harder to oxidize.) The oxidation reaction is

$$2Br^-(aq) + 2H_2SO_4(concd) \longrightarrow Br_2(\ell) + SO_2(g) + 2H_2O + SO_4^{2-}(aq)$$

Like hydrochloric acid, hydrobromic acid is a strong acid that gives all of the reactions we have studied for the hydronium ion and the bromide ion. Oxides and oxoacids of bromine similar to some of those of chlorine are known, but they are unstable and seldom used.

Iodine and Its Major Compounds

Iodine was first prepared in 1811 by Bernard Courtois (1777–1838), who observed purple vapors rising from an extract of kelp ashes that he had acidified with sulfuric acid and heated. In his time, kelp was commonly collected, dried, and burned to give ashes from which sodium and potassium salts were obtained. The purple vapor condensed on a cold surface, forming nearly black crystals (Figure 20.18). Others—notably Joseph Gay-Lussac and Humphrey Davy—proved that the crystals were an element, and it was named after the Greek *iodes,* meaning "violet."

Only one isotope of iodine occurs naturally.

FIGURE 20.18 Solid iodine. Note the violet color of the vapor.

Many natural brines contain iodide ion, so when chlorine is bubbled through them, I^- is oxidized to I_2. This is just like the recovery of bromine from brines, only iodide ion is more easily oxidized than bromide ion, as the standard reduction potentials indicate.

The body needs traces of I^- each day to prevent goiter, a disfiguring enlargement of the thyroid gland. Buy and use "iodized" salt, and you'll be safe.

$$Br_2(aq) + 2e^- \rightleftharpoons 2Br^-(aq) \qquad E° = +1.07 \text{ V}$$
$$I_2(s) + 2e^- \rightleftharpoons 2I^-(aq) \qquad E° = +0.54 \text{ V}$$

Bromine, of course, also oxidizes iodide ion, but the calculated cell potential is not as large as with chlorine.

$$2I^-(aq) + Br_2(aq) \longrightarrow I_2(s) + 2Br^-(aq) \qquad E°_{cell} = 0.53 \text{ V}$$
$$2I^-(aq) + Cl_2(aq) \longrightarrow I_2(s) + 2Cl^-(aq) \qquad E°_{cell} = 0.82 \text{ V}$$

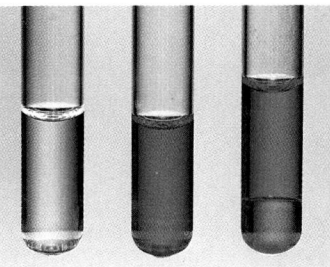

Left: aqueous NaI. Center: Cl_2 has been added, and I_2 starts to form. It colors the solution brown. Right: The newly added globule of CCl_4, on the bottom, dissolves some of the I_2, which has a violet color in this solvent.

$$E^{\circ}_{cell} = \frac{0.0592}{n} \log K_c$$

Although I_2 is insoluble in pure water, it dissolves readily in solutions that contain I^- to give I_3^-. The equilibrium is

$$I_2(s) + I^-(aq) \rightleftharpoons I_3^-(aq)$$

Courtois' isolation of I_2 from the I^- in brines worked because hot H_2SO_4 oxidizes the anion to the element.

These cell potentials actually correspond to equilibrium constants so high that we are fully justified by writing the equations with single, forward arrows. Using the equation given in the margin, as studied in Section 17.8, the equilibrium constant for the first reaction is 8.0×10^{17} and for the second reaction it is 5.0×10^{27}.

Iodine reacts with far fewer elements or compounds than do the other halogens. Although iodine is a poison, it is not as dangerous as the other members of its family. Its vapors irritate the eyes and mucous membranes. It kills bacteria, and backpackers have "recipes" for using iodine to disinfect water. *Tincture of iodine,* a solution of iodine in aqueous ethyl alcohol, is used as an antiseptic.

Worldwide, hundreds of thousands of tons of iodine are produced annually, and roughly half is used to make various organic iodine compounds.

A solution of hydrogen iodide in water, called hydriodic acid, HI(*aq*), is a strong acid, like its counterparts, hydrochloric acid and hydrobromic acid. Unlike them, however, it has little use outside the laboratory. The others are cheaper and are more stable in the presence of oxygen.

HI(*aq*) is made by letting HI(*g*) pass into water. The latter is made by heating sodium iodide with an acid. Only now, the acid must be phosphoric acid, not sulfuric acid, as was used for the exactly analogous synthesis of HBr(*g*). This is because sulfuric acid too easily oxidizes iodide ion back to iodine. Thus the synthesis of HI(*g*) is

$$NaI(s) + H_3PO_4(concd) \longrightarrow HI(g) + NaH_2PO_4(s)$$

Hydriodic acid can be purchased as a concentrated solution, 40 to 55% (w/w) HI. Since oxygen in air, with the aid of light, easily oxidizes the iodide ion to iodine, hydriodic acid is stored in dark brown, well-capped bottles. Even so, the solution slowly darkens as I_2 forms in the following reaction.

$$O_2(g) + 4HI(aq) \longrightarrow 2I_2(s) + 2H_2O$$

Iodine forms a few oxides and oxoacids similar to those of bromine and chlorine, but most are unstable. Iodic acid, HIO_3, a white solid, is a strong acid and a strong oxidizing agent. It is the only one of the HXO_3 types of acids ($X = Cl$, Br, or I) known as a pure compound. Salts of iodic acid, the *iodates,* are commercially available and are used as oxidizing agents in analytical chemistry.

Iodine is in the $+7$ oxidation state in all of these forms of periodic acid.

Periodic acid (pronounced PER-i-o-dic) is another crystalline oxoacid of iodine with enough stability to be made and stored. Depending on the degree of hydration, it exists as HIO_4, $H_4I_2O_9$, and H_5IO_6, with the latter being most common. It is used as an oxidizing agent in organic chemistry.

Some Periodic Correlations among the Halogens

We will draw together here some of the important trends in properties that we have noticed during our study of the halogens and their compounds, both in this chapter and others.

We have seen that as we go down the family of halogens in the periodic table from fluorine to iodine, many properties also diminish in the same order.

IE = Ionization Energy

E° = Standard Reduction Potential

IE for $X(g) \rightarrow X^-(g) + e^-$	F	>	Cl	>	Br	>	I
(kJ mol^{-1})	1680.6		1255.7		1142.7		1008.7
E° for $X_2(aq) + 2e^- \rightleftharpoons 2X^-(aq)$	F_2	>	Cl_2	>	Br_2	>	I_2
(V)	$+2.87$		$+1.36$		$+1.07$		$+0.54$
Reactivity toward other elements	F_2	>	Cl_2	>	Br_2	>	I_2
Basicity of X^-	F^-	>	Cl^-	>	Br^-	>	I^-
Relative electronegativity	F	>	Cl	>	Br	>	I
(Pauling scale values)	4.1		2.9		2.8		2.2

One point that these data show is what we have noted before. The first member of this family differs more from the second than the second and remaining members differ from each other. The values for IE, $E°$, and electronegativity all bear this out.

The order of basicity of X^- ions is just the expected reverse of the order of acidity of HX. ("The stronger the acid, the weaker is its conjugate base.")

In Strengths as Acids: HF < HCl < HBr < HI

Similarly, the order for the values of $E°$, which tells us that F_2 accepts electrons and is far more easily reduced than I_2, means that the order of ease of oxidation of halide ions to halogens, X_2, is the reverse.

In Ease of Oxidation: $I^- > Br^- > Cl^- > F^-$

In these oxidations, the anions are changed to the corresponding elements, and we learned earlier how I^- ions, the most easily oxidized of the halide ions, are oxidized to I_2 molecules by either Br_2 or Cl_2. And we saw how Br^- ions are oxidized to Br_2 molecules by Cl_2. You could not do this oxidation of Br^- using I_2.

We would expect the ionic radii of X^- ions to *increase* as we go down the halogen family, because we go to larger and larger electron clouds.

In Ionic Radii of X^-: $\qquad F^- < Cl^- < Br^- < I^-$
in pm: $\quad$ 133 $\quad$ 184 $\quad$ 196 $\quad$ 220

This means that the smallest ion with the highest concentration of negative charge is the fluoride ion, F^-. For this reason, F^- is strongly hydrated in water, as reflected by the order of the enthalpies of hydration of the halide ions. The hydration of F^- is most exothermic.

In Enthalpies of Hydration: $\qquad F^- > Cl^- > Br^- > I^-$
$kJ\ mol^{-1}$: $-515 \quad -381 \quad -347 \quad -305$

Finally, one of the most obvious changes we would notice if we looked at samples of the halogens is how their colors change. Fluorine is a pale yellow gas. Chlorine is a greenish-yellow gas (or yellowish-green gas — it is somewhat subjective). Bromine is a deeply red-brown liquid. And iodine forms nearly black crystals. No other family of elements displays such large differences in color.

As oxidizing agents:
$F_2 > Cl_2 > Br_2 > I_2$

20.8 CARBON

The important inorganic compounds of carbon are the carbonates, bicarbonates, and cyanides.

The carbon family consists of carbon, silicon, germanium, tin, and lead. Some of their properties are given in Table 20.5. Carbon is the chief constituent of coal, and forms the

TABLE 20.5 The Carbon Family — Group IVA

Element	Atomic Symbol	Melting Point (°C)	Boiling Point (°C)	Ionization Energy (kJ mol⁻¹)	Relative Electro-negativity[a]	Average Bond Dissociation Energy, M—M (kJ mol⁻¹)
Carbon[b]	C	3800[c]	—	1086	2.5	368
Silicon	Si	1420	3280	786	1.8	340
Germanium	Ge	959	2700	761	1.8	188
Tin	Sn	232	2275	708	1.8	151
Lead	Pb	327	1750	715	1.9	—

[a] Pauling scale. [b] Graphite. [c] Under pressure to repress sublimation.

backbones of the hydrocarbon molecules in oil and natural gas. Carbon also occurs widely in carbonate rocks, such as limestone, dolomite, and marble.

Silicon is widely present in the silicate minerals, such as quartz and several others. It is used to make semiconductors for electronic instruments.

Germanium is seldom found in concentrations of minerals. It is isolated from the dusts in the flues of zinc smelters and from coal ash. It is also used to make semiconductors.

The mineral cassiterite, SnO_2, has been the source of tin since antiquity. Bronze is an alloy of about 5 to 10% tin in copper, and when ancient metal workers figured out that bronze made good weapons and tools — better than those of stone — civilization could move out of the stone age. Other tin alloys include pewter (90–95% tin with varying amounts of copper and antimony) and solder (widely varying concentrations of tin and other metals). An alloy of 5 to 10% lead in tin makes organ pipes of the finest tonal properties.

The most important mineral of lead is galena, PbS. Both tin and lead will be studied more in Chapter 21.

Some Similarities and Differences in the Carbon Family

In probably no other family are the differences between the first element and the second and succeeding members more dramatic, and yet several similarities exist in types of compounds. They all form oxides of empirical formula MO_2 (where we let M represent any member of the carbon family). All form tetrahalides, MX_4, except that among those of lead, only PbF_4 is stable at ordinary temperatures. ($PbCl_4$ is stable only below 0 °C; $PbBr_4$ is even less stable; and the existence of PbI_4 is doubtful.) All but lead form hydrides of the formula MH_4. See Table 20.6.

Carbon, alone among all the elements, can form chains of thousands of atoms and rings of three atoms up to very large sizes while retaining the ability to hold other nonmetal atoms. Thus carbon's simplest hydride, CH_4 is just the first member of an enormous family of compounds, the *alkanes*, all answering to the general formula $C_nH_{(2n+2)}$, where n can vary from one to thousands.

In contrast, the value of n in $Si_nH_{(2n+2)}$, the *silanes* with Si–Si chains, goes (so far) only as high as 8. In the *germanes*, $Ge_nH_{(2n+2)}$, n reaches only 5. And in the *stannanes*, $Sn_nH_{(2n+2)}$, only the first two members are known. No corresponding plumbanes have been definitely characterized.

The common name of ethene is ethylene.

Double and triple bonds also can occur between carbon atoms; ethene, $CH_2{=}CH_2$, is only the simplest member of the *alkene* family — a huge family with the general formula C_nH_{2n}. Only in the early 1980s was a compound found with an Si=Si double bond. Carbon atoms can form triple bonds between each other, too, and ethyne,

The common name of ethyne is acetylene.

$H{-}C{\equiv}C{-}H$, is just the first member of a large *alkyne* family, $C_nH_{(2n-2)}$.

No triple bonds between atoms of other members of the carbon family have been observed. We suggested why this might be when we commented on why phosphorus, in the same family as nitrogen, does not form triple bonds, as nitrogen does in N_2 and other compounds. We will look at this again.

TABLE 20.6 Hydrides and Tetrachlorides of Group IVA Elements

Element	Boiling Point (°C)	
	MH_4	MCl_4
Carbon	−161.4	76.7
Silicon	−111.8	59.0
Germanium	−88.1	83.1
Tin	−52.5	114
Lead	Unknown	(a)

[a] Unstable above 0 °C.

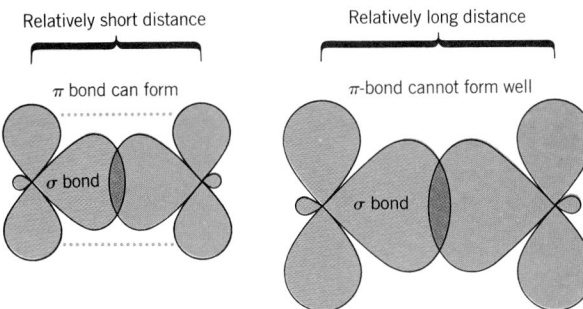

Relatively short distance

π bond can form

σ bond

Relatively long distance

π-bond cannot form well

σ bond

FIGURE 20.19 Better side-to-side overlap of p orbitals occurs when they are closer, as they can be when they are in the second shell, illustrated to the left. When they are in the third shell, illustrated to the right, they are too far away to provide a good overlap.

A double or triple bond requires one or two pi bonds. Those that form between two carbon atoms use $2p$ orbitals. When we drop down to silicon or other members of the carbon family, the p orbitals available are from a higher shell and are therefore larger and on the average are farther away from the atomic nucleus. As Figure 20.19 suggests, the side-to-side overlap of adjacent $2p$ orbitals can be more effective than with $3p$ or $4p$ or $5p$ orbitals. Consequently the formation of double (or triple) bonds between the noncarbon members of this family does not provide the energy lowering that the formation of an equivalent number of single bonds can.

Nowhere do these considerations have a more common yet vivid illustration than in the dramatic difference between the dioxides of carbon and silicon. Carbon dioxide is a gas at room temperature, and silicon dioxide is a solid that makes up quartz (Figure 20.20) and quartz sands. CO_2 exists as discrete molecules, with two carbon–oxygen double bonds, O=C=O. In SiO_2, single covalent bonds interlink alternating Si and O atoms in all directions, with each silicon center being tetrahedral, as seen in the structure in the margin. In effect, then, SiO_2 is a huge macromolecular system and, like all such systems, has a very high melting point.

FIGURE 20.20 Quartz is a particularly beautiful form of silicon dioxide.

As we mentioned, another unique feature of carbon is that its atoms in chains and rings of whatever size retain the ability to form strong covalent bonds to other nonmetals. Hydrogen is bound to carbon in the hydrocarbons, but carbon can also have covalent bonds to oxygen, nitrogen, sulfur, and the halogens. Thus, methyl alcohol, CH_3—OH, is just the simplest member of a huge family of *alcohols*, $C_nH_{(2n+1)}$—OH, where no theoretical limit exists to the value of n. Variations on this theme occur in family after family of compounds of carbon, hydrogen, and oxygen. Further examples are given in Chapter 24.

We will limit our study here to the chief *inorganic* compounds of carbon, particularly its oxides, carbonates, carbonic acid, and a few others. We will not say more in this chapter about silicon, germanium, tin, and lead.

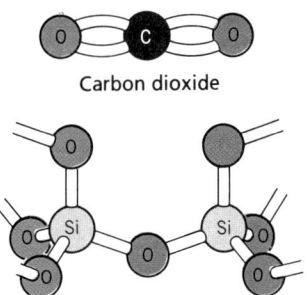

Carbon dioxide

Silicate system

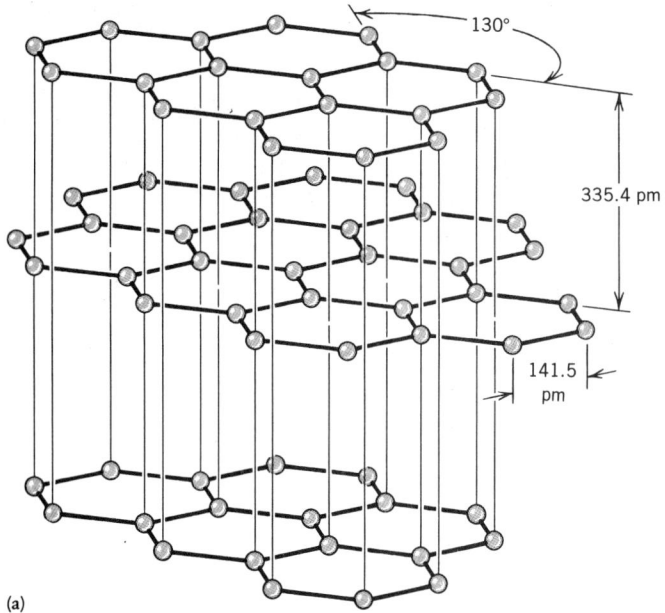

(a)

(b)

FIGURE 20.21 Two allotropes of carbon. (*a*) Graphite structure. (*b*) Diamond structure.

The Chief Allotropes of Carbon

Carbon has two common allotropes—graphite and diamond. The carbon atoms in graphite occur in parallel sheets as sketched in Figure 20.21*a*. Each sheet is made of extended networks of carbon hexagons within which occurs a vast network of delocalized pi electrons. This network permits graphite to conduct electricity, unusual for a nonmetal. The sheets are held to each other by relatively weak forces and it is easy for planes to slide by each other. This is why powdered graphite is a good dry lubricant, and you may have used it to lubricate locks.

The carbon atoms in diamond (Figure 20.21*b*) are joined exclusively by sigma bonds that project tetrahedrally from each carbon. Thermodynamically, diamond is slightly *less* stable than graphite at ordinary pressures. But at ultrahigh pressures and at temperatures high enough to encourage carbon atoms to move about, graphite can be forced into the less stable form, diamond. This is because diamond's density is higher than graphite's so high pressure induces the carbon atoms to rearrange into a more dense packing. Today, chiefly at the General Electric Company, industrial-grade diamonds are made more cheaply from graphite than they can be mined. The opposite is true for gem-quality diamonds.

Diamond is the hardest substance known, and industrial-grade diamonds are used to make abrasives and cutting tools for the toughest of such jobs. Recently, a technology for preparing diamond films has been developed. See Special Topic 20.3. Gem diamonds (Figure 20.22) have, of course, intrigued and attracted all who have ever looked upon them.

For graphite, $\Delta H_{combustion} =$ -393.51 kJ mol^{-1}. For diamond, $\Delta H_{combustion} = -395.41$ kJ mol^{-1}.

FIGURE 20.22 The centerpiece of this diamond necklace is the famous Hope diamond, which has a brilliant blue color, unusual for diamonds.

Carbon Monoxide

Carbon forms several oxides, but only two are particularly stable, CO and CO_2.

Pure carbon monoxide is made on a small scale by heating formic acid with concentrated H_2SO_4, a strong dehydrating agent.

$$\underset{\text{formic acid}}{HCHO_2} \xrightarrow[\text{heat}]{H_2SO_4(concd)} CO(g) + H_2O$$

When a mixture of CH_4 and H_2 at 800 to 1000 °C is passed through a discharge of microwave radiation, molecules crack into atomic carbon and hydrogen. The carbon atoms deposit on a nearby surface, and when the conditions are right they gradually form a film of carbon *with a tetrahedral structure,* not a graphite structure. The film, in other words, is diamond. Smooth films as thick as 1 mm have been prepared. Being diamond films, they are essentially unscratchable.

Some of the potential applications of this technology are in the manufacture of wear-resistant tools and bearings, scientific and medical optical systems, computer equipment, rocket engines, and communications. Sony markets a tweeter loudspeaker with a resonating diamond membrane that is claimed to reproduce sounds of a fidelity never before heard.

A diamond-coated silicon wafer. The fringes of color at the bottom are interference fringes indicating changes in the thickness of the thin film.

CO is always produced when any carbon-containing fuel burns incompletely. But when there is enough oxygen, CO itself burns to CO_2. In fact, CO is used in industry as a fuel whenever some chemical operation makes it as a by-product. Thus the by-product is not wasted, and burning CO to CO_2 solves the problem of disposal of the very poisonous CO.

Because it can be shipped by pipes, CO has advantages over coke, a solid fuel. Coke is essentially carbon made from coal by thermally driving off noncarbon substances. Coke becomes CO by heating it to incandescence and blowing either air or steam over it.

$$2C(s) + O_2(g) \xrightarrow{\text{heat}} 2CO(g)$$

The mixture of gases that leaves this operation contains the nitrogen of the air, of course, but the mixture — called *producer gas* — is still a gaseous fuel.

With steam and coke, *water gas* is made, and this is a mixture of CO and H_2 formed by the following reaction.

$$C(s) + H_2O \xrightarrow{\text{heat}} \underbrace{CO(g) + H_2(g)}_{\text{water gas}}$$

Both components of water gas are combustible, of course.

Carbon monoxide is also a reducing agent, and the metallurgical industry uses it to reduce metal oxides to metals, for example,

$$FeO + CO \xrightarrow{\text{heat}} Fe + CO_2$$
$$CuO + CO \xrightarrow{\text{heat}} Cu + CO_2$$

Carbon monoxide is an important ligand in the chemistry of coordination compounds (compounds that contain complex ions). CO forms neutral compounds called *metal carbonyls* with over a dozen transition metals. Iron pentacarbonyl, $Fe(CO)_5$, is just one example. It is a *pyrophoric* compound, meaning that it spontaneously bursts into flame in air. When it burns, it gives Fe_2O_3 and CO_2. Nickel tetracarbonyl, $Ni(CO)_4$, is another metal carbonyl. It forms during a procedure for purifying nickel metal, as we will study in Chapter 22. Nickel carbonyl is unstable — it explodes at 60 °C — and it is a

The term **greenhouse effect** refers to the insulating effect of the earth's atmosphere caused chiefly by the presence of what are called the *greenhouse gases*— carbon dioxide, stratospheric water vapor, methane, nitrous oxide (N_2O), and chlorofluorocarbons (*Chemicals in Use* 13). Most of the atmospheric N_2O and CH_4 arise from the action of soil bacteria. Although CO_2 makes the largest single contribution to the greenhouse effect, the combined effect of the rest of the greenhouse gases is now slightly greater than that of CO_2 alone.

Heat energy must be radiated from the earth to outer space at exactly the same average rate as it is received from the sun. Otherwise, the earth's average temperature will either increase or decrease. No one knows with certainty what the effect of a small increase in average global temperature would mean. If warmer ocean water seeped under the ice shelves of the Antarctic, for example, could this cause the ice sheets to slip more rapidly into the oceans? If so, the additional water from the melting of all of this ice— both from the Antarctic and from Greenland—would cause the levels of the oceans to rise 200 to 400 feet, inundating some of the world's coastal cities and major population centers. The best estimate of the U.S. Environmental Protection Agency is that the earth will warm by 4 to 5 °C in the twenty-first century, and that this will raise the oceans between 2 and 3 feet. Small changes in average air temperatures could have major changes in world climate patterns and during the last century, the average global air temperature rose by 0.5 °C.

Studies of ocean sediment cores and tree rings indicate that the carbon dioxide level of the atmosphere fluctuated between 200 and 300 ppm until the end of the last century. What concerns scientists and government officials is the steady increase in the concentration of CO_2 in the atmosphere since the time when the burning of coal and oil became a major source of energy. Since about 1900, the average CO_2 level in the atmosphere has risen from 280 to 345 ppm,

and some scientists see this going to 600 ppm in another 50 to 75 years, maybe less. This change alone could increase the average global temperature by 1.5 degrees. The problems of forecasting are huge given our uncertain knowledge about how climate works. But the situation is a "catch-22" case. By the time we know what is really going on, it could be too late to do much about anything that could be a major threat.

The greenhouse gases act as insulators by absorbing electromagnetic energy in the infrared region as this energy speeds from the radiating earth toward outerspace. Then the energy-rich molecules reradiate their extra energy (in all directions), so some of this energy comes back to earth again. In the absence of such insulation, no energy would be returned to the earth, so the higher the level of the greenhouse insulators, the more such energy is reradiated, and the warmer the lower air becomes.

The earth's two main "sinks" for atmospheric CO_2 are the formation of carbonate deposits in the oceans and the process of photosynthesis. Calcium ions and magnesium ions take dissolved CO_2 in ocean water and form insoluble carbonates, but this is a very slow process. Photosynthesis, a much more rapid consumer of CO_2, uses it to make plant materials and oxygen. Photosynthetic work is carried out by microscopic plants (phytoplankton) in the oceans and by trees and other vegetation on land. So we have two reasons for maintaining the great photosynthesizing engines of our planet—the oceans and the great rain forests of equatorial regions. We need to regenerate oxygen; and we need to remove CO_2. But the oceans continue to be polluted, the rain forests are being cut down for timber and to open lands for cultivation, and we continue to burn fossil fuels. Our children and their children will find out how it all comes out. In the meantime large numbers of specialists in atmospheric chemistry, meteorology, and climate are gathering data and testing theories that will permit better forecasting.

dangerous poison and possibly an air pollutant because coal has traces of nickel. When coal burns, some nickel tetracarbonyl forms and escapes.

The ligand property of CO makes it a very poisonous gas. It binds to the Fe^{2+} ions of hemoglobin and blocks the ability of this oxygen carrier in blood to take up and transport oxygen. CO binds about 200 times more strongly than O_2 to hemoglobin. The only antidote for CO poisoning is to let the victim breath oxygen-enriched air. This causes equilibria to shift in favor of releasing CO in accordance with Le Châtelier's principle. Some hospitals place a CO poison victim in a pressurized chamber where the partial pressure of oxygen in the enriched air can be made even higher.

Tobacco smoke at the cigarette tip is about 2 to 5% CO.

These are called *hyperbaric* chambers. Hyperbaric means higher than atmospheric pressure.

Carbon Dioxide and Carbonic Acid

Carbon dioxide is one of the end products of the complete combustion of organic compounds, including the fossil fuels. In the atmosphere, CO_2 contributes to a growing problem called the *greenhouse effect*, discussed in Special Topic 20.4. Industrially, the chief source of CO_2 is as a by-product in the Haber–Bosch process for making ammonia. Hydrogen for this process is now stripped from methane, CH_4, and the carbon emerges eventually as CO_2. Much of the CO_2 made this way is consumed on site to make urea, a solid fertilizer and also a raw material for making urea–formaldehyde polymers.

Urea is made by the following overall reaction.

$$CO_2 + 2NH_3 \xrightarrow[185\ °C]{200\ atm} NH_2\overset{\overset{\displaystyle O}{\|}}{C}NH_2 + H_2O$$
$$\text{urea}$$

Since urea is a solid, it is easily shipped in bags to farm locations where the technology for distributing it on the land is simple. In soil, water hydrolyzes the urea and releases the ammonia.

Carbon dioxide also forms when many metal carbonates and bicarbonates are strongly heated. Limestone — mostly calcium carbonate — is converted to calcium oxide, an important commercial compound, this way.

$$CaCO_3(s) \xrightarrow{\text{heat}} CaO(s) + CO_2(g)$$
$$\text{lime}$$

This reaction is carried out, however, not as much for the CO_2 as for the lime. Between 15 and 20 million tons of lime are made annually in the United States. It is consumed mostly by the metallurgical, chemical, and wastewater treatment industries where it constitutes an inexpensive base. Scrubbing SO_2 from smokestack gases is a growing use of lime.

The thermal decomposition of limestone also occurs in making cement. When finely ground limestone, sand, and clay are strongly heated together in rotating kilns, the product is transformed into *portland cement*. This is a mixture of various calcium silicates, for example, Ca_2SiO_4 and Ca_3SiO_5, and calcium aluminates, for example, $Ca_3Al_2O_6$. When water and sand are added, an unusually strong, stonelike solid — concrete — forms that features large networks of —Si—O—Si—O—Si systems.

The largest use of carbon dioxide — about 50% — is not as a chemical but as a refrigerant. It becomes a solid below $-78.5\ °C$, and solid CO_2 is popularly known as "dry ice." It sublimes at ordinary pressures, but if kept under 5.2 atm, it can be melted to a liquid at $-56\ °C$. CO_2 is thus marketed as a gas or a liquid under pressure in steel cylinders or as "dry ice."

Urea is also the principal form in which waste nitrogen is excreted by animals.

Some common names for CaO are *lime, burnt lime, quick lime,* and *calx.* Ca(OH)$_2$ is called *hydrated lime* or *slaked lime.*

Steel-making uses lime as a fluxing agent to remove impurities in molten iron.

Dry ice — solid CO_2

A familiar use of CO_2 is as a fire extinguisher. A blanket of the more dense CO_2, replacing the less dense air, shuts oxygen away from the burning material, and the fire goes out. It can be used on any kind of fire — wood, paper, gasoline, or electrical — except in those rare situations, probably known only to chemists, where what is burning reacts vigorously not only with oxygen but also with CO_2. Some metal hydrides, like $LiAlH_4$, pose this problem. Magnesium metal fires likewise cannot be put out by CO_2 extinguishers. Mg reduces CO_2 to carbon.

About 25% of CO_2 production goes to the carbonation of soft drinks and "light" beer. A saturated solution of CO_2 in water at 760 torr and 25 °C is 0.033 M in dissolved CO_2 — as in an opened bottle of club soda. Most of the CO_2 molecules exist simply as hydrates. A fraction, however, are present as the weak, unstable acid, carbonic acid, H_2CO_3.

$$CO_2(aq) + H_2O \rightleftharpoons H_2CO_3(aq)$$

Ascarite, a commercial product for removing CO_2 from an air supply, is NaOH deposited on an inert support.

The ratio, $[CO_2(aq)]/[H_2CO_3(aq)]$, in saturated aqueous CO_2 is about 600. However, when it comes to the ability of aqueous CO_2 to neutralize base, all dissolved CO_2, whatever the form, is available, because CO_2 can react directly with aqueous hydroxide ion.

$$CO_2(aq) + OH^-(aq) \longrightarrow HCO_3^-(aq)$$

The bicarbonate ion can also neutralize OH^-.

$$HCO_3^-(aq) + OH^-(aq) \longrightarrow CO_3^{2-}(aq) + H_2O$$

Carbonates and Bicarbonates   The important salts of carbonic acid are the water-soluble bicarbonates and carbonates of the Group IA metals and the water-insoluble carbonates of the Group IIA metals. We have studied their reactions with acids earlier, but let us review the important net ionic equations.

$$CO_3^{2-}(s \text{ or } aq) + H^+(aq) \longrightarrow HCO_3^-(aq)$$
$$HCO_3^-(s \text{ or } aq) + H^+(aq) \longrightarrow H_2CO_3(aq) \rightleftharpoons CO_2(aq) + H_2O$$

In the usual applications of these reactions, so much CO_2 is produced that most fizzes out of the solution.

Carbon Disulfide

When sulfur vapor is passed over hot coke, an interesting sulfur analog of carbon dioxide forms — carbon disulfide, CS_2.

$$C + 2S \xrightarrow{\text{heat}} CS_2$$

The industrial synthesis more frequently used today is the reaction of sulfur with methane in the presence of a catalyst.

$$CH_4 + 4S \xrightarrow[\substack{\text{alumina} \\ (Al_2O_3)}]{600 \text{ °C}} CS_2 + 2H_2S$$

$\ddot{S}=C=\ddot{S}$
carbon disulfide

CS_2 has a low boiling point, 46 °C, and a pleasant odor *when very pure.* Commercial samples have an extremely disagreeable odor owing to impurities. It is very poisonous, and it is also thermally unstable. Vapors of CS_2 have been known to explode on contact with a hot steam pipe or even with boiling water. Above 100 °C, CS_2 vapors spontaneously ignite in air. One important use of carbon disulfide is in the manufacture of rayon and cellophane.

Cyanides and Hydrogen Cyanide

Cyanides and covalent cyano compounds of most elements are known, but we will limit our study just to sodium cyanide, NaCN. The cyanide ion, CN^-, is the conjugate base of a weak acid, hydrogen cyanide, HCN. Thus any soluble cyanide will hydrolyze in water

$:N≡C:^-$
cyanide ion

to give some hydrogen cyanide and hydroxide ion (making the solution basic).

$$CN^-(aq) + H_2O \rightleftharpoons HCN(aq) + OH^-(aq)$$

A quantitative conversion to HCN occurs if a strong acid is present.

$$CN^-(s \text{ or } aq) + H^+(aq) \longrightarrow HCN(aq)$$

Although HCN is very soluble in water, it is easily removed by heating the solution. If it forms at a high enough concentration, some gaseous HCN inevitably materializes. And HCN is an extremely poisonous gas. Survivors say that it has an almond odor. It is the gas traditionally used for executions in gas chambers. Cyanide salts are equally lethal. Whatever the source of the cyanide ion, it kills by forming a complex ion with the copper ion in cells. Copper ions are essential in the operation of an enzyme vital in every cell to the use of oxygen, but this enzyme is deactivated when cyanide ion binds to copper ion.

Carbides

Carbides are binary compounds of carbon with metals. Some are largely saltlike or ionic, some are covalent but macromolecular, and some are called interstitial carbides.

Among the saltlike carbides, some behave as if they contain the C_2^{2-} or acetylide ion. These react with water to give acetylene. Calcium carbide is an example.

In CaC_2, C_2^{2-} is $[:C\equiv C:]^{2-}$, so the triple bond is already formed.

$$\underset{\substack{\text{calcium} \\ \text{carbide}}}{CaC_2(s)} + 2H_2O \longrightarrow \underset{\text{acetylene}}{H-C\equiv C-H(g)} + Ca(OH)_2(s)$$

The carbide lanterns once used by underground miners and even in the headlamps of early autos contained a supply of calcium carbide and a reservoir of water. Illumination resulted from the burning acetylene produced as water dripped on the carbide! Rather ingenious.

Another type of saltlike carbide reacts with water to give methane, so these carbides behave as if they contained the C^{4-} ion. Magnesium carbide is an example.

$$\underset{\substack{\text{magnesium} \\ \text{carbide}}}{Mg_2C(s)} + 4H_2O \longrightarrow \underset{\text{methane}}{CH_4(g)} + 2Mg(OH)_2(s)$$

Silicon carbide, SiC, is a covalent carbide made by heating silicon dioxide with carbon at very high temperatures (2000 to 2600 °C). The common name of its alpha form is *carborundum*. It's almost as hard as diamond. Its crystals fracture to give very sharp edges, so it is an unusually effective abrasive for sandpapers and grinding wheels.

In interstitial carbides, carbon atoms occupy spaces or interstices within the lattices of metal atoms. This leaves the material with many characteristics of a metal, such as conductivity and luster. An industrially important example is tungsten carbide, WC, used to make high-speed cutting tools because it is exceptionally hard and chemically stable even as the tool becomes very hot during use.

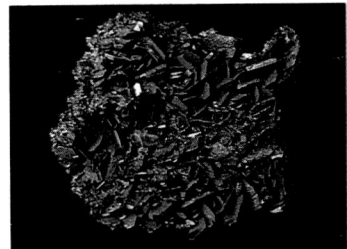
Silicon carbide

Tungsten carbide cutting tool

SUMMARY

Sulfur.   Sulfur occurs as the free element, in sulfide ores, hydrogen sulfide, and sulfate minerals. Of its several allotropes, S_α or orthorhombic sulfur is the most stable, and it consists of crownlike rings of eight sulfur atoms. The same molecules occur in S_β, a form of monoclinic sulfur. Rings and extended chains of sulfur occur in amorphous sulfur. All allotropes eventually revert to S_α. Molten sulfur becomes increasingly viscous until 200 °C, because the sulfur chains become longer. Then, with further heating, the chains break up and liquid sulfur becomes less viscous.

Sulfur forms binary compounds with hydrogen, most metals (sulfides), and the nonmetals (except the noble gases). Its oxides, oxoacids, and their salts are particularly important compounds.

The only stable hydride is hydrogen sulfide, H_2S. Polysulfanes, H_2S_x, are known, but slowly decompose to sulfur and hydrogen sulfide.

Binary sulfur–halogen compounds include the unusually stable SF_6 and both S_2Cl_2 and SCl_2, two liquids.

Selenium, tellurium, and polonium form hydrides like H_2S, as well as salts of these hydrides. They also form halides and oxides similar in formula to those of sulfur.

Oxoacids of Sulfur and Their Salts.   Sulfur burns to give sulfur dioxide, and this forms an aqueous solution commonly considered to be H_2SO_3, and called sulfurous acid. Various hydrated forms of SO_2, and not molecules of H_2SO_3, are present, however. Sulfurous acid, a moderate acid, forms solutions of the hydrogen sulfite ion, HSO_3^-, at slightly basic pH's, and solutions of the sulfite ion, SO_3^{2-}, at high pH's. Just bubbling SO_2 into a solution of a base, like NaOH or $NaHCO_3$, neutralizes the base. Solid samples of a compound such as sodium hydrogen sulfite consist of sodium disulfite, $Na_2S_2O_5$, but this provides hydrogen sulfite ion in solution. Metal sulfites have the sulfite ion in the solid state. The sulfur dioxide/sulfite system is used as a reducing agent. Sulfur trioxide is made by the catalyzed oxidation of sulfur dioxide and is used to make sulfuric acid, a strong, diprotic acid that in water gives all of the reactions of the hydronium ion. Hydrogen sulfate salts of the Group IA metals are known. Sulfate salts of most metals are important chemicals. The sodium salt of the unstable thiosulfuric acid, $H_2S_2O_3$, sodium thiosulfate, is used in photography as a fixer.

Selenium and tellurium form oxoacids similar to those of sulfur.

Acid Rain.   Acid deposition—popularly called acid rain—brings the acidic oxides of sulfur and nitrogen onto lakes, soil, vegetation, building stone, and other surfaces. Sulfur dioxide is the largest contributor, but NO_2 is also important. SO_3 makes up for its being a smaller contributor by forming a strong acid in water. The sulfur oxides originate from the burning of sulfurbearing fossil fuels, so acid deposition is a problem where heavy industries and large power plants operate—in or near populated areas. The acids destroy vegetation and lower the pH of lakes enough to prevent the occurrence of fish.

Phosphorus.   Among the many allotropes of phosphorus, white phosphorus (P_4) and red amorphous phosphorus are commercially important. When white phosphorus is made from the reduction of calcium phosphate by carbon, nearly all is used directly to make phosphorus compounds, like tetraphosphorus decaoxide, P_4O_{10}, phosphorus trihalides, and phosphorus pentahalides. White phosphorus, however, is unstable in air. Its red amorphous allotrope is stable. The binary compound, PH_3 (phosphine), burns in air to give phosphoric acid.

The phosphorus halides react readily with water to give oxoacids—phosphorous acid from phosphorus(III) halides, and phosphoric acid from phosphorus(V) halides. Phosphorus trichloride can be oxidized to phosphorus oxychloride, $POCl_3$, which hydrolyzes to give phosphoric acid. It reacts with alcohols to give organophosphate compounds.

Phosphorus can form extended chains of O—P—O—P—O—P— etc. units. It is harder for phosphorus to participate in triple bonds than nitrogen.

Phosphorus Oxides, Oxoacids, and Salts. P_4O_{10} reacts vigorously with water to give one or a mixture of oxoacids, depending on the relative amount of water. Orthophosphoric acid, H_3PO_4, is the common form referred to when the term "phosphoric acid" is used. Dihydrogen phosphate salts, monohydrogen phosphate salts, and phosphate salts can be made from H_3PO_4.

In its lower oxidation states, phosphorus occurs in phosphorous acid, H_3PO_3, a diprotic acid, and in hypophosphorous acid, H_3PO_2, a monoprotic acid. Reaction of water with P_4O_6 or PCl_3 gives H_3PO_3; reaction of P_4 with boiling base gives the hypophosphite ion, $H_2PO_2^-$.

Salts of diphosphoric acid, $H_4P_2O_7$, are known, but the most widely used polyphosphate system is found in the tripolyphosphate salts of the $P_3O_{10}^{5-}$ ion of detergents, as well as salts of the *cyclo*-triphosphate system, $P_3O_9^{3-}$.

Arsenic, antimony, and bismuth, form the same kinds of oxides (in terms of formulas) as phosphorus, as well as hydrides (arsine, stibine, and bismuthine). Trihalides between these three elements and all four halogens are known, but only the pentafluorides and the pentachlorides of As and Sb are known.

Halogens.   The halogens—fluorine, chlorine, bromine, and iodine (astatine, being radioactive, is excluded here)—display important similarities, but also significant trends in properties as you go down the family in the periodic table. All form metal halide salts. Their oxides and oxoacids, when they form, have similar formulas. Their hydrides are all gases and (except for HF) form strong acids in water. None occurs as the element free in nature but as halide ions in various salts.

In descending the family in the periodic chart, we can say:

1. The hydrohalic acids, $HX(aq)$, become stronger.
2. The standard reduction potentials become smaller.
3. The electronegativity of the element becomes lower.
4. The general reactivity toward anything becomes lower.
5. The basicity of the halide ion becomes weaker.

Fluorine and chlorine are made by electrolysis. Bromine and iodine are made by the oxidation of their halide ions. The hydrogen halides can all be made from metal halides by heating them with sulfuric acid (or phosphoric acid) and driving off the HX gas. (But commercially, most HCl is made as a by-product in the manufacture of organochlorine compounds.) All HX types, except HI, can be made directly from the elements.

Fluorine and water burn. Aqueous chlorine includes hydrochloric acid and hypochlorous acid. The latter is unstable but its salts are known and have bleaching power. Aqueous bromine contains very little HBr and HOBr. Iodine is quite insoluble in water but dissolves in solutions of I^- to give I_3^-. Chlorine dioxide, chlorate salts, perchloric acid, and periodic acid are strong oxidizing agents.

The halogens illustrate a general observation about families of representative elements. The first member differs from the second much more than the second differs from the third or fourth, and so on.

Carbon Family.   The members of the carbon family — carbon, silicon, germanium, tin, and lead — form series of compounds with similar formulas. But carbon is in most respects very different from silicon. The hydrides of carbon (alkanes, alkenes, and alkynes) are extremely numerous. Silanes number only 8, and only one compound with a $Si=Si$ double has been reported. The germanes number 5 and there are only two stannanes and no plumbanes. SiO_2 is markedly different in structure from CO_2, reflecting the difficulty of forming double (or triple) bonds involving p orbitals of the third or higher shell.

Carbon, unlike the other members of the family, can in its extended chains and rings not only hold H atoms but also those of O, N, S, and the halogens.

The chief allotropes of carbon are graphite and diamond. The chief oxides are CO and CO_2. CO is a common fuel in industry. It forms interesting metal carbonyls, and it is very toxic.

Carbon dioxide is an acidic oxide. It's made commercially as a by-product of the Haber process, as a component of producer gas or water gas, and from the thermal decomposition of carbonate rocks. It neutralizes strong bases, and gives some H_2CO_3 in water. This acid, carbonic acid, gives a series of bicarbonate salts (of Group IA metals) and carbonates (of virtually all metals). CO_2 is used as a refrigerant and to make urea, but some is used to carbonate beverages and fill fire extinguishers.

Carbon is present in carbon disulfide, hydrogen cyanide, the cyanide salts, and the carbides. The cyanide ion is a Brønsted base, and both it and HCN are very toxic. The carbides fall roughly into three groups: saltlike carbides like CaC_2, which react with water to give acetylene; those like Mg_2C, which give methane with water, and covalent, high-formula-weight carbides that resist any reaction with water, like silicon carbide (carborundum); and interstitial, metallike carbides, like tungsten carbide.

REVIEW EXERCISES

The more difficult questions are marked with asterisks. Many questions involve a review of material in earlier chapters.

Sulfur

20.1   In what chemical forms does sulfur occur in nature?

20.2   How is the sulfur in hydrogen sulfide, found in natural gas and oil, converted to elemental sulfur? (Write an equation.)

20.3   Give the name and formula of the gaseous compound into which the sulfur in sulfide ores is converted by *roasting* the ore. Write the molecular equation for the roasting of CuS. How is the gaseous product often used?

20.4   What is the name and formula of the most stable sulfur allotrope?

20.5   What is the geometry of the S_8 molecule in orthorhombic sulfur?

20.6   What happens during the conversion of S_α to S_β?

20.7   What is the molecular formula of sulfur in monoclinic sulfur?

20.8   What happens to S_β when it is allowed to stand a long time?

20.9   Why does the viscosity of molten sulfur increase sharply between 160 and 195 °C?

20.10   As the temperature of molten sulfur is increased above 200 °C, the viscosity of the liquid decreases. Explain.

Sulfur Dioxide, Sulfurous Acid, Acid Rain, and the Sulfites

20.11   What human activity increases the level of sulfur dioxide in the atmosphere?

20.12   What name is traditionally given to an aqueous solution of sulfur dioxide? Give the names and formulas of the chief chemical forms of the solute in this solution.

20.13   What is acid rain? Why is the term *acid deposition* more descriptive of the nature of the problem?

20.14   What nonmetal oxides contribute to acid deposition?

20.15   By what chemical means can SO_2 be removed from smokestack gases?

20.16 SO_2 has been described as an acidic oxide. Complete and balance the following reactions that illustrate this.
(a) $NaOH(aq) + SO_2(g) \longrightarrow$
(b) $NaHCO_3(aq) + SO_2(g) \longrightarrow$

20.17   What is the chief sulfur-containing anion in the following?
(a) Solid sodium metabisulfite.
(b) Solid "sodium hydrogen sulfite."
(c) Aqueous sodium hydrogen sulfite.
(d) Solid sodium bisulfite.
(e) Solid sodium sulfite.
(f) Aqueous sodium sulfite.

20.18   Write the equation for the equilibrium involving sulfur-containing species present in aqueous sodium hydrogen sulfite.

*20.19   Using information in this chapter and earlier chapters, write the net ionic equation for the reaction in which aqueous sulfur dioxide, $H_2SO_3(aq)$, reacts with iodine to give sodium sulfate and iodide ion. Using standard reduction potentials, calculate the cell potential for this reaction. Will the reaction be spontaneous?

20.20   The addition of dilute hydrochloric acid to a white powder that was known to be either $NaHSO_3$ or $NaHSO_4$ produced a gas with a sharp, choking odor. What is the name and formula of the gas that formed, and what was the formula of the solid? Write molecular and net ionic equations for the reaction.

20.21   Write both the molecular and net ionic equations for the reaction of hydrochloric acid with (a) $NaHSO_3(aq)$ and (b) $Na_2SO_3(aq)$.

20.22   Hydrobromic acid, $HBr(aq)$, reacts with hydrogen sulfites and sulfites in the same way as hydrochloric acid (Review Exercise 20.21). Write the molecular and net ionic equations for the reaction of $HBr(aq)$ with each of the following.
   (a) $NaHSO_3(aq)$ (c) $K_2SO_3(aq)$
   (b) $Na_2SO_3(aq)$ (d) $KHSO_3(aq)$

Sulfur Trioxide, Sulfuric Acid, and the Sulfates

20.23   Complete and balance the following equations that illustrate how sulfur trioxide is an acidic oxide.
   (a) $SO_3(g) + H_2O \longrightarrow$
   (b) $SO_3(g) + NaHCO_3(aq) \longrightarrow$
   (c) $SO_3(g) + Na_2CO_3(aq) \longrightarrow$
   (d) $SO_3(g) + NaOH(aq) \longrightarrow$

20.24   Write the molecular equations for the individual steps in the *contact process*.

20.25   Why can phosphate rock not be used directly as a source of phosphate for soil enrichment? How does sulfuric acid make this phosphate available?

20.26   Sulfuric acid can just as easily be made as 100% H_2SO_4 as it can as 96% H_2SO_4. Why is sulfuric acid most commonly sold as the 96% form?

20.27   Give five reasons why sulfuric acid is the most frequently used strong acid in industrial applications.

20.28   What properties of 96% sulfuric acid make it a dangerous chemical?

20.29   A solid known to be either $NaHSO_4$ or Na_2SO_4 was dissolved in water and solid Na_2CO_3 was added. The solution fizzed as a gas evolved. What was the gas? What was the original solid? Write a molecular equation for the reaction.

20.30   How would a 1 M solution of $NaHSO_4$ test—acidic, basic, or neutral? (If not neutral, write an equation to explain your answer.)

20.31   How would a 1 M solution of Na_2SO_4 test—acidic, basic, or neutral? (If not neutral, write an equation to explain your answer.)

20.32   What is Drierite? What is it used for? What property allows for its regeneration?

20.33   How does Na_2SO_4 work as a drying agent?

Other Compounds of Sulfur

20.34   How is sodium thiosulfate made? (Write the molecular equation.)

20.35   The "parent" acid of the thiosulfate ion has what name? Is it stable?

20.36   How does the thiosulfate ion react with (a) Cl_2 and (b) I_2? Write net ionic equations.

20.37   Which compound of sulfur—give its formula and name—has a rotten egg odor?

20.38   A mineral found on a geological survey was suspected of containing either FeS or $FeCO_3$. What simple test could the field geologist perform to find out which it was? (Include an equation.)

20.39   Which is the stronger Brønsted base, S^{2-} or O^{2-}? What fact about H_2S given in this chapter enables one to know?

20.40   What formula is used for the polysulfanes, and how are they made?

20.41   Which fluoride of sulfur is exceptionally stable?

20.42   Give the names and formulas of two binary compounds of S and Cl.

20.43   Write the formulas of each.
   (a) Selenous acid   (c) Selenic acid
   (b) Tellurous acid   (d) Telluric acid

20.44   In what structural way does selenium resemble sulfur?

20.45   What are the formulas of the hydrides of selenium and tellurium?

20.46   Complete and balance the following equation, basing your answer on expected similarities to the chemical properties of a corresponding sulfur compound.

$$H_2SeO_4(aq) + 2NaOH(aq) \longrightarrow$$

Phosphorus and Some of Its Binary Compounds

20.47   In terms of the bonding abilities of N and P, what might be a reason nitrogen occurs only as N_2, but phosphorus occurs in several allotropes?

20.48   What allotrope of phosphorus forms directly from the carbon-based reduction of calcium phosphate? Write its formula, and describe the structure of its molecules.

20.49   What makes the allotrope of Review Exercise 20.48 so reactive?

20.50   Why do you suppose red amorphous phosphorus is so much less reactive than white phosphorus?

20.51   What is the name and formula of the binary hydride of phosphorus? How does it compare with NH_3 in (a) solubility in water? (b) Basicity? (c) Ease of oxidation in air? (d) Odor?

20.52   What are the names and formulas of two common binary compounds of phosphorus and chlorine? Write equations for their syntheses.

20.53   Write equations for the reactions of the two binary P—Cl compounds (Review Exercise 20.52) with water, and name the phosphorus-containing products.

20.54   Write equations for the reactions of PBr_3 and PBr_5 with water.

20.55   How is phosphorus oxychloride made? (Write an equation.)

20.56   Write equations that show how phosphorus oxychloride reacts with each of the following.
   (a) Water

(b) CH_3OH (methyl alcohol)
(c) C_2H_5OH (ethyl alcohol)

Oxides of Phosphorus and Oxoacids

20.57 How is P_4O_{10} made? (Write an equation.)

20.58 How is H_3PO_4 made (a) from P_4O_{10}? (b) From phosphate rock?

20.59 What is the formula of diphosphoric acid? How is it related to orthophosphoric acid? (Write the equilibrium equation.)

20.60 How do "superphosphate" and "triple superphosphate" fertilizers differ (a) in manner of preparation (in general terms, not an equation)? (b) In composition?

20.61 Sodium dihydrogen phosphate is an *acid salt*. What does this mean?

20.62 In general terms, what does an *acidulant* do in a batter? Also write a net ionic equation.

20.63 Write the formulas of the following compounds.
(a) Potassium dihydrogen phosphate
(b) Sodium phosphate
(c) Sodium monohydrogen phosphate
(d) Calcium phosphate
(e) Sodium diphosphate
(f) Sodium tripolyphosphate
(g) Sodium *cyclo*-triphosphate
(h) Potassium hypophosphite
(i) Phosphorous acid
(j) Arsine
(k) Stibine

20.64 Write names for the following compounds.
(a) H_3PO_4 (e) $MgHPO_4$
(b) H_3PO_3 (f) Na_2HPO_4
(c) H_3PO_2 (g) NaH_2PO_2
(d) KH_2PO_4 (h) NaH_2PO_3

20.65 Reasoning by analogy to names and formulas of compounds of phosphorus, what would be the most likely names of the following?
(a) H_3AsO_4 (d) NaH_2AsO_4
(b) Na_3AsO_4 (e) $AsCl_3$
(c) H_3SbO_4 (f) $SbCl_5$

The Halogens and Their Compounds

20.66 What are the names and the molecular formulas of the halogens?

20.67 Describe how each of the halogens is made from naturally occurring materials. Write equations.

20.68 Write equations that show how each hydrogen halide is made on an industrial scale.

20.69 Write the equations for the reactions (if any) of water with each of the halogens.

20.70 How does hydrogen fluoride dissolve glass? (Write an equation.)

20.71 Why is hydrogen bonding so much stronger in $HF(\ell)$ than in $HCl(\ell)$?

20.72 What is the principle behind the use of traces of fluoride ion in drinking water as a means of reducing the frequency of tooth decay among children? (Cf. Special Topic 20.2.)

20.73 How is bleaching powder made? (Write an equation.)

20.74 Give the names and formulas of the three oxides of chlorine mentioned in this chapter.

20.75 Compare and contrast the kinds and relative abundances of various solute species in aqueous chlorine and aqueous bromine. (Use both names and formulas.)

20.76 Write the equations for the disproportionations of the following species. (a) ClO^- in base. (b) ClO_3^- in acid.

20.77 I_2 is insoluble in pure water but dissolves in aqueous solutions of NaI or KI. Explain.

20.78 What property makes perchloric acid particularly dangerous when it is pure?

20.79 In a lab-scale preparation of $HI(aq)$, phosphoric acid is used instead of sulfuric acid. What reaction (write the equation) is used, and why is the choice of acid important?

20.80 X and Y are both halogens, and the following reaction takes place:

$$X_2 + 2Y^- \longrightarrow 2X^- + Y_2$$

(a) Which halogen is oxidized, X or Y?
(b) Which halogen is reduced?
(c) If Y is Cl, can X be Br? Explain.
(d) If Y is Br, can X be Cl? Explain.
(e) If Y is I, can X be Br? Explain.

20.81 Arrange the halogens in their correct orders according to the following criteria. Place the symbol of the species that corresponds to the lowest value on the left and the symbol of the species with the highest value on the right in a horizontal row.
(a) The strengths of X_2 as oxidizing agents.
(b) The strengths of $HX(aq)$ as acids.
(c) The electronegativities of X.
(d) The polarities of the bonds in $H-X(g)$.
(e) The basicities of $X^-(aq)$.
(f) The general reactivities toward metals of X_2.
(g) The strengths of X^- as reducing agents.

20.82 What is *tincture of iodine*?

20.83 Why should iodide ion be in the diet?

20.84 What happens chemically to cause the darkening of a solution of concentrated $HI(aq)$? Write the equation.

20.85 What are the names and formulas of two oxoacids of iodine?

Carbon and Its Inorganic Compounds

20.86 Give the names and the atomic symbols of the elements in the carbon family.

20.87 Write structural formulas for the following.
(a) An alkane with 5 carbons per molecule.
(b) A stannane with 2 tin atoms per molecule.
(c) An alkene with 3 carbons per molecule.

(d) A germane with 3 germanium atoms per molecule.
(e) An alkyne with 4 carbon atoms per molecule.

20.88 Describe the structural differences between CO_2 and SiO_2.

20.89 Why is graphite slippery?

20.90 What fact about graphite and diamond makes it possible to convert the more stable graphite into the thermodynamically less stable diamond? How is this transformation done (in general terms)?

20.91 The carbon monoxide used as a fuel in industry is generally made as part of an integrated operation where the CO is burned as it is made. Write equations for the industrial production of CO that gives (a) producer gas and (b) water gas.

20.92 How can small amounts of CO be made in the lab? (Write an equation.)

20.93 How is CO used in the metallurgical industry? (Give an equation of one example.)

20.94 How does CO work as a poison?

20.95 Iron pentacarbonyl is *pyrophoric*. What is the formula of iron pentacarbonyl and what does *pyrophoric* mean? Write an equation.

20.96 What organic compound is the raw material for carbon dioxide when it is to be used to make urea? Write the equation for the synthesis of urea.

20.97 Write the equation for the preparation of lime?

20.98 In general terms, how is portland cement made and what substances are in it?

20.99 What happens chemically when portland cement sets (in general terms)?

20.100 Dry ice sublimes, so how is it possible to make *liquid* CO_2?

20.101 Although relatively little of the carbon dioxide in an aqueous solution of this gas is present as carbonic acid, the solution can be treated as a neutralizer of base as if all of the carbon dioxide were in this form. Explain, using equations.

20.102 All metal carbonates and bicarbonates react alike with strong acids. What are the net ionic equations for these reactions? Write one equation for carbonates and one for bicarbonates (assuming that both are in solution).

20.103 How is carbon disulfide made today? (Write an equation.)

20.104 Describe two dangers associated with handling carbon disulfide.

20.105 A solution of potassium cyanide in water has a pH above 7. Write a net ionic equation that explains this.

20.106 Using a net ionic equation, describe what happens when hydrochloric acid is poured into a concentrated solution of sodium cyanide.

20.107 How does the cyanide ion act as a poison (in general terms)?

20.108 We discussed three types of carbides. Give specific examples—both names and formulas.

20.109 Write the equation for the reaction (if any) of each compound with water. (a) WC, (b) CaC_2, (c) SiC, (d) Mg_2C.

20.110 What is carborundum?

Names, Formulas, and Reactions

20.111 Write the formula of each. (Remember that the formula of an ion is not correct unless it also includes the charge.)
(a) Sulfite ion
(b) Hydrogen sulfate ion
(c) Perchloric acid
(d) Calcium carbide
(e) "Sulfurous acid"
(f) Chloric acid
(g) Sulfate ion
(h) Hydrofluoric acid
(i) Hydrogen sulfite ion
(j) Thiosulfate ion
(k) Bifluoride ion
(l) Chlorous acid
(m) Orthophosphoric acid
(n) Tripolyphosphate ion
(o) Bromate ion
(p) Perchlorate ion
(q) Phosphine
(r) Phosphorous acid
(s) Hypophosphorous acid

20.112 Write the formula of each. (Remember that the formula of an ion is not correct unless it also includes the charge.)
(a) Arsenic acid
(b) Calcium carbide
(c) Stibine
(d) Metaphosphoric acid
(e) Tin(IV) chloride
(f) Magnesium carbide
(g) Diphosphoric acid
(h) Sodium metabisulfite
(i) S_α
(j) Disulfite ion
(k) Gypsum
(l) Selenous acid
(m) White phosphorus
(n) Sodium monohydrogen phosphate
(o) Selenic acid
(p) Fluorite
(q) Cryolite
(r) Halite
(s) Sodium hypobromite
(t) Sodium chlorate
(u) Potassium perchlorate
(v) Calcium hypochlorite
(w) Sodium bromite
(x) Sodium hypophosphite

20.113 Write the name of each.

(a) PCl_3 (k) $HClO_4$
(b) ClO^- (l) PBr_5
(c) H_2SO_4 (m) ClO_3^-
(d) $H_4P_2O_7$ (n) HSO_3^-
(e) $NaClO_4$ (o) HCO_3^-
(f) H_3PO_4 (p) Mg_2C
(g) H_2SO_3 (q) $HClO_3$
(h) PH_3 (r) Cl_2O
(i) HPO_3 (s) $HBrO_3$
(j) Cl_2O_7 (t) H_3PO_2

20.114 Name each of the following.

(a) $Na_2S_2O_5$ (l) $NaHCO_3$
(b) $Na_2S_2O_3$ (m) Na_2CO_3
(c) Na_2SO_3 (n) Na_3AsO_4
(d) $NaHSO_4$ (o) Na_2SeO_3
(e) Na_2SO_4 (p) Na_2SeO_4
(f) Na_3PO_4 (q) Na_2C_2
(g) Na_2HPO_4 (r) $NaClO$
(h) NaH_2PO_4 (s) $NaBrO_3$
(i) $Na_4P_2O_7$ (t) $NaClO_4$
(j) $Na_3P_3O_9$ (u) $NaClO_2$
(k) $Na_5P_3O_{10}$ (v) NaH_2PO_2

20.115 Complete and balance the following equations.

(a) $SO_2(g) + H_2O \longrightarrow$
(b) $SO_3(g) + H_2O \longrightarrow$
(c) $CO_2(g) + H_2O \longrightarrow$
(d) $SO_2(g) + NaOH(aq) \longrightarrow$
(e) $SO_3(g) + NaOH(aq) \longrightarrow$
(f) $CO_2(g) + NaOH(aq) \longrightarrow$
(g) $S(s) + O_2(g) \longrightarrow$
(h) $SO_2(g) + O_2(g) \longrightarrow$
(i) $SO_2(g) + Na_2CO_3(aq) \longrightarrow$
(j) $Na_2SO_3(aq) + HCl(aq) \longrightarrow$
(k) $S(s) + Na_2SO_3(aq) \xrightarrow{\text{heat}}$
(l) $K_2S(aq) + HCl(aq) \longrightarrow$
(m) $KHS(aq) + HCl(aq) \longrightarrow$
(n) $P_4O_{10}(s) + H_2O \xrightarrow[\text{ortho acid)}]{\text{(to the}}$
(o) $P_4O_6(s) + H_2O \longrightarrow$

20.116 Aqueous KOH reacts in exactly the same way with non-metal oxides as aqueous NaOH. Write molecular and net ionic equations for the reaction of aqueous KOH with each of the following substances. (a) $SO_2(g)$, (b) $SO_3(g)$, (c) $CO_2(g)$.

20.117 Write the molecular and net ionic equations for the reaction of hydrochloric acid with each of the following substances.

(a) $NaOH(aq)$ (f) $KHCO_3(aq)$
(b) $NaHCO_3(aq)$ (g) $CaCO_3(s)$
(c) $Na_2CO_3(aq)$ (h) $Ca(OH)_2(s)$
(d) $KOH(aq)$ (i) $Mg(OH)_2(s)$
(e) $K_2CO_3(aq)$ (j) $MgCO_3(s)$

20.118 Hydrobromic acid, $HBr(aq)$, gives reactions just like those of hydrochloric acid (previous exercise). Write molecular and net ionic equations for the reaction of hydrobromic acid with each of the substances listed in Review Exercise 20.117.

20.119 Nitric acid, $HNO_3(aq)$, gives reactions just like those of hydrochloric acid. Write molecular and net ionic equations for the reaction of nitric acid with each of the substances listed in Review Exercise 20.117.

20.120 Sulfuric acid, $H_2SO_4(aq)$, gives reactions just like those of hydrochloric acid (Review Exercise 20.117). Write molecular and net ionic equations for the reaction of sulfuric acid with each of the substances listed in parts a through f of Review Exercise 20.117. Assume that only sulfate compounds, not hydrogen sulfate compounds, form.

20.121 Hydrochloric acid, hydrobromic acid, hydriodic acid, nitric acid, perchloric acid, and sulfuric acid all give very similar reactions with the hydroxides, cyanides, bicarbonates, carbonates, hydrogen sulfites, sulfites, hydrogen sulfides, and sulfides of metal ions. Write the net ionic equation that dilute solutions of any of these strong acids give with each of the following substances at room temperature.

(a) $KOH(aq)$ (e) $K_2SO_3(aq)$
(b) $KHCO_3(aq)$ (f) $KCN(aq)$
(c) $K_2CO_3(aq)$ (g) $KHS(aq)$
(d) $KHSO_3(aq)$ (h) $K_2S(aq)$

20.122 If we define the neutralization of an acid as any reaction that will change hydrogen ions to anything else—for example, H_2O or H_2—then which of the following substances can neutralize hydrochloric acid? For each one that does, write the molecular and net ionic equations.

(a) $Zn(s)$ (e) $Zn(HCO_3)_2(aq)$
(b) $ZnO(s)$ (f) $ZnCO_3(s)$
(c) $Zn(OH)_2(s)$ (g) $ZnSO_3(aq)$
(d) $Zn(NO_3)_2(aq)$ (h) $ZnSO_4(aq)$

CHEMICALS IN USE
INORGANIC POLYMERS

Polymers are large molecules formed by linking together large numbers of small **monomer** units. Polyethylene, $+CH_2-CH_2\frac{}{n}$, made from ethylene, $CH_2=CH_2$, is a typical example. Others include dacron, nylon, and polyvinyl chloride. These are organic polymers whose molecular "backbones" consist mostly of carbon atoms, and typically they tend to become brittle when cold and deteriorate when very hot. They also tend to be flammable, swell in organic solvents, and are usually not compatible with living tissue. (Only a few organic polymers can be used for medical implants.) With inorganic polymers, whose molecular backbones consist of atoms other than carbon, such problems are often minimized.

Silicone Polymers

Among the synthetic inorganic polymers that have received the most widespread uses are the silicones. These polymers contain an alternating silicon–oxygen backbone, and have organic groups attached to the backbone at each silicon atom. Usually, these are methyl groups, $-CH_3$, which we can think of as being derived from methane, CH_4, by the removal of one hydrogen atom. For example, the structure of a typical linear silicone polymer is illustrated below.

$$CH_3-\underset{\underset{CH_3}{|}}{\overset{\overset{CH_3}{|}}{Si}}-O\left(\underset{\underset{CH_3}{|}}{\overset{\overset{CH_3}{|}}{Si}}-O\right)_n\underset{\underset{CH_3}{|}}{\overset{\overset{CH_3}{|}}{Si}}-O-\underset{\underset{CH_3}{|}}{\overset{\overset{CH_3}{|}}{Si}}-CH_3$$

a silicone polymer chain

Silicone polymers are formed by hydrolysis of such compounds as $(CH_3)_2SiCl_2$ and $(CH_3)_3SiCl$. As shown below, this gives $Si-OH$ bonds that subsequently expel the components of water and form $Si-O-Si$ linkages. Thus the first step in the polymerization of $(CH_3)_2SiCl_2$ is

$$Cl-\underset{\underset{CH_3}{|}}{\overset{\overset{CH_3}{|}}{Si}}-Cl + 2H_2O \longrightarrow HO-\underset{\underset{CH_3}{|}}{\overset{\overset{CH_3}{|}}{Si}}-OH + 2HCl$$

which is followed by the expulsion of two hydrogens and an oxygen from between pairs of silicon atoms to give the polymer, as shown in Figure 14a.

Depending on the chain length and the degree to which chains are crosslinked to each other, the silicones are oils, greases, or rubbery solids. Silicone oils repel water and water-based products. For this reason, they have been used in polishes for furniture and cars, and as waterproofing treatments for fabrics and leather (Figure 14b). Silicone greases are used as permanent lubricants in clocks and ball bearings. Silicone resins are used in some paints because of their resistance to peeling and high temperatures. Silicone rubber has excellent electrical insulating properties and is used to coat electrical wire and in electric motors. These rubbery materials also retain their flexibility over wide temperature ranges and can be used in place of natural rubber for both high- and low-temperature applications. Silicones are even found in medical preparations. The polymer with the structure illustrated in Figure 14a is known medicinally as *simethicone* and is a common antigas agent found in antacid products.

Phosphazenes

In **polyphosphazenes,** the polymer backbone consists of alternating phosphorus and nitrogen atoms. They are prepared by polymerizing a ring-shaped molecule, $(PNCl_2)_3$, and then substituting various groups of atoms for the chlorines in the polymer. The general structure of the polyphosphazenes is

$$\cdots -\underset{\underset{X}{|}}{\overset{\overset{X}{|}}{P}}=N-\underset{\underset{X}{|}}{\overset{\overset{X}{|}}{P}}=N-\underset{\underset{X}{|}}{\overset{\overset{X}{|}}{P}}=N-\underset{\underset{X}{|}}{\overset{\overset{X}{|}}{P}}=N-\underset{\underset{X}{|}}{\overset{\overset{X}{|}}{P}}=N \text{ etc.}$$

where X stands for any of a large number of different groups of atoms that can be attached to the phosphorus atoms. The large variety of possible side groups has led to the synthesis of many different polymers with a wide range of properties. The phosphazenes are generally nonflammable, and some are used as flame retardants for other polymers. Some phosphazenes are glasses while others are elastomers that retain their elasticity to low temperatures (in one case, as low as -90 °C). When X contains a nonpolar chain of atoms, as in the group $-O-CH_2-CF_3$, the polymer is water insoluble. In fact, this side group produces flexible polymer films (Figure 14c) that are more strongly water repellant than the silicones. They also do not react with living tissue, which makes this polymer an

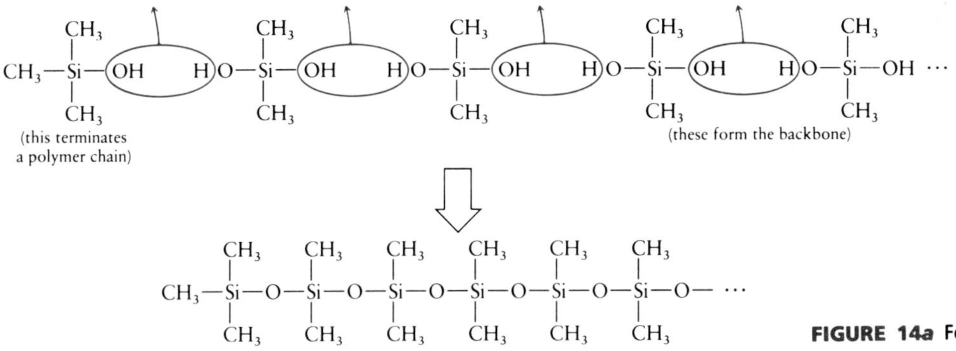

$$CH_3-Si-OH \quad HO-Si-OH \quad HO-Si-OH \quad HO-Si-OH \quad HO-Si-OH \cdots$$

(with CH₃ groups above and below each Si, OH groups circled)

(this terminates a polymer chain) (these form the backbone)

$$CH_3-Si-O-Si-O-Si-O-Si-O-Si-O- \cdots$$

(with CH₃ groups above and below each Si)

FIGURE 14a Formation of a linear silicone polymer.

excellent candidate for tubes used to replace blood vessels. But when a polar group such as —NHCH₃ is attached to the phosphorus, a water-soluble polymer is obtained.

One of the most interesting potential applications of phosphazene polymers is in chemotherapy. Most chemotherapeutic agents are small molecules that can easily diffuse through cell membranes into regions of the body where they aren't wanted and where they can do more harm than good. The results are unwanted side effects. It is hoped that these chemicals can be immobilized by attaching them to phosphazene polymer chains, which are too large to diffuse through cell membranes, and thereby minimize such problems. Some progress in this direction has already been made. It has been demonstrated that certain biologically active molecules can be incorporated into these polymers without losing their activity.

Electrically Conducting Polymers

The earliest and one of the most interesting of these polymers is formed when tetrasulfur tetranitride, S_4N_4, is heated. The ring-shaped molecule initially breaks down into a smaller ring, S_2N_2, which then polymerizes to give crystals of poly(sulfur nitride), $(SN)_x$, a shiny metallic solid. The crystals contain long zig-zag chains of alternating sulfur and nitrogen atoms.

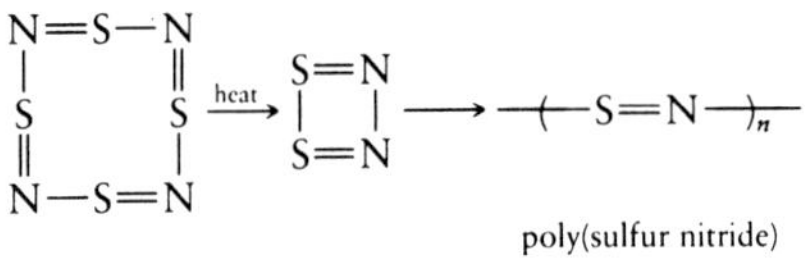

$$\begin{array}{c} N=S-N \\ | \quad\quad \| \\ S \quad\quad S \\ \| \quad\quad | \\ N-S=N \end{array} \xrightarrow{heat} \begin{array}{c} S=N \\ | \quad | \\ S=N \end{array} \longrightarrow (S=N)_n$$

poly(sulfur nitride)

What is amazing about these crystals is that they show metallic conduction along the chains. In fact, they even become superconductors if cooled to near absolute zero (that is, they lose all resistance to electrical conduction). As you might expect, these properties have generated great interest because of their potential practical importance, and research on this and other electrically conducting polymers is actively being pursued.

Questions

1. Identify three problems that organic polymers have, but which important inorganic polymers avoid.
2. Sketch a portion of the molecular backbone of the silicone polymers.
3. What is the function of $(CH_3)_3SiCl$ in the reaction mixture used to prepare a silicone polymer? What would happen if some CH_3SiCl_3 were included?
4. Sketch a portion of the molecular backbone of the phosphazene polymers.
5. What property of poly(sulfur nitride) makes this polymer of special interest?

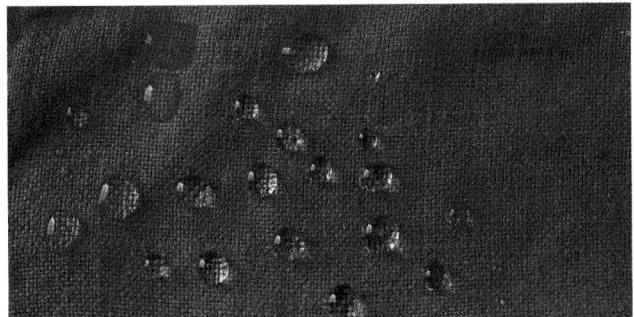

FIGURE 14b Water forms beads on a fabric treated with a silicone polymer.

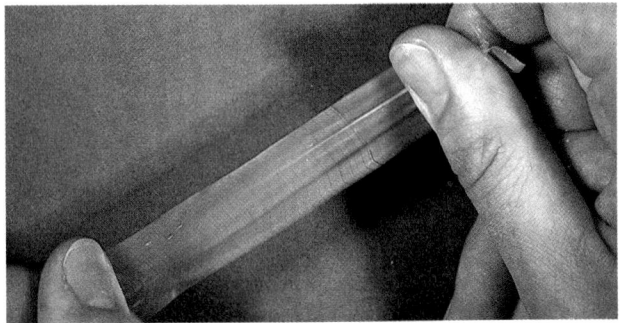

FIGURE 14c A sample of a polyphosphazene polymer film.

The reduction of iron oxide, Fe_2O_3, by aluminum powder is called the thermite reaction. It is so exothermic that the molten metallic iron that's formed is white-hot. Here we see a practical application of the reaction, which is occurring in an apparatus that allows the molten iron to run down into a mold between rail ends, thereby welding the rails together. The thermite reaction is just one of many that we discuss in this chapter about metals.

METALS AND METALLURGY

METALLIC CHARACTER AND THE PERIODIC TABLE

METALLURGY

THE ALKALI METALS

THE ALKALINE EARTH METALS

THE METALS OF GROUPS IIIA, IVA, AND VA

21.1 METALLIC CHARACTER AND THE PERIODIC TABLE

The closer an element is to the lower left corner of the periodic table, the more metallic are its properties.

In Chapter 2 we discussed some of the general physical and chemical properties of metals. Now that you have acquired a background in topics such as electronic structure and bonding, we can look again at metals to gain a better understanding of their similarities and differences.

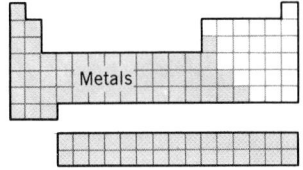

The metals, you recall, make up about 75% of the elements. We encounter some of them every day. Metals such as iron and aluminum are commonly used as structural materials, while other familiar ones such as copper, silver, gold, and platinum are found in coins and jewelry. Some common metals, such as sodium and calcium, are generally seen only in compounds because of the tendency of the free metals to react with air and moisture. There are also metals that are rare and have only limited applications, so we are hardly aware of their existence or the impact that they might have on our lives, even though some of them are necessary components of certain enzymes that keep us alive.

The chemical and physical properties of metals, like those of the nonmetals, are determined by their electron configurations. We have seen that nonmetals react with each other to form molecules in which atoms are attached to each other by covalent bonds. The sharing of electrons in these bonds permits the nonmetal atoms to complete their valence shells and achieve stable electronic structures. Covalent bonding does not occur to any appreciable extent between metal atoms, however. In general, metals have few electrons in their valence shells and therefore they are not able to form enough covalent bonds to achieve an octet. This fact, plus the low ionization energies that metals normally have, is the reason that metals form the "metallic lattice" described in Chapter 10. In a metallic lattice, you may recall, each metal atom loses electrons to the lattice as a whole and all of the metal atoms in the crystal have a share of the pooled electrons. It is the mobility of these electrons that accounts for the unique "metallic luster" and the electrical and thermal conductivity of metals.

We have also seen that metals tend to react with nonmetals to form ionic compounds in which the metal exists as a positive ion. The tendency of any two elements to form an ionic compound is determined by the difference between the amount of energy that is needed to create the ions and the amount of energy that is recovered in the form of the lattice energy—the energy-lowering effect produced by the attraction between the positive and negative ions. Usually, little energy is needed to remove an electron from a metal atom. In addition, metals have few valence electrons, so only a few have to be removed to completely empty the valence shell. As a result, in the reaction between a metal and a nonmetal, the energy input can often be compensated for by the lattice energy, and stable ionic compounds can be formed.

Although metals tend to form positive ions when they react with nonmetals, the completeness of the electron transfer is not the same in every case. In Chapter 7 we saw that the ionic character of a bond—that is, the extent to which the bond is ionic or the extent to which it is polar—depends on the difference in the electronegativities of the two bonded atoms. Because of the way electronegativity varies in the periodic table—increasing from left to right across a period and from bottom to top in a group—the farther away from the lower left corner of the table a metal is located, the more chemically "nonmetallic" it becomes.

Within the periodic table, metallic character decreases from left to right across a period and increases from top to bottom in a group. In period 3, for example, sodium and magnesium are metals with typically basic oxides—their oxides react with acids, but not with bases. Aluminum is also a metal, but its oxide is amphoteric—it reacts with both acids and bases.

Metallic character

Increases

Decreases

AlO_2^- is just one way to represent the aluminum-containing species that exists in basic solution.

$$Al_2O_3(s) + 6H^+(aq) \longrightarrow 2Al^{3+}(aq) + 3H_2O$$
$$Al_2O_3(s) + 2OH^-(aq) \longrightarrow 2AlO_2^-(aq) + H_2O$$

The fact that Al_2O_3 reacts with bases means that it is at least somewhat acidic. Since acidic oxides are normally associated with nonmetals, we can say that aluminum is "less metallic" than sodium or magnesium. Moving farther to the right in period 3 we come to silicon, which is a metalloid. Then we come to the clearly nonmetallic elements—phosphorus, sulfur, chlorine, and argon.

The increase in metallic character going down a group can be seen most clearly in Group IVA. At the top of the group is carbon, a nonmetal whose oxide, CO_2, dissolves in water to give carbonic acid, H_2CO_3. Next are silicon and germanium, both of which are metalloids having the semiconductor properties typical of metalloids. Then come tin and lead, which are metals. Both show some amphoteric properties, but tin more so than lead. Thus tin is less metallic than lead.

21.2 METALLURGY

The science of extracting metals from their naturally occurring compounds and then preparing them for practical use employs many important chemical reactions.

Even before the beginning of recorded history, metals played an important role in the growth and development of society. Archaeological evidence indicates that gold was used in making eating utensils and ornaments as early as 3500 B.C. Silver was discovered at least as early as 2400 B.C., and iron and steel have been used as construction materials since about 1000 B.C. From these earliest times, methods for obtaining metals from their naturally occurring deposits have gradually evolved as more and more knowledge of the behavior of metals was acquired.

Metallurgy is the science and technology of metals. In modern terms, it is primarily concerned with the procedures and chemical reactions that are used to separate metals from their ores and the preparation of metals for practical use.

Gathering Metals and Their Compounds

Metals occur in nature in many different ways. Most of them are found in compounds, either in the earth's crust or in the oceans, although some of the less reactive metals have been able to withstand the ravages of time and are found in the uncombined state. A well-known example is gold.

The distribution of the metals (as well as the nonmetals) throughout the earth is not uniform, which is quite fortunate for us. Localized deposits of particular compounds serve as ores for metals of various kinds. An **ore** is simply a mineral deposit that has a desirable component in a sufficiently high concentration to make extraction of that component economically profitable. The economics of the recovery operation, though, is what distinguishes an ore from just another rock. For example, magnesium occurs in one form of a mineral called olivine, which has the formula Mg_2SiO_4. This compound contains over 30% magnesium, but there is no economical way of extracting the metal from it. Instead, a large amount of magnesium is obtained from seawater, where its concentration is a mere 0.13%. Even though the concentration is much less in the seawater, a profitable method of obtaining the magnesium has been developed, and that makes seawater the principal source of this metal.

Metals from the Sea   The oceans of the world are a storehouse for many different chemicals. These are found not only in the water itself, but also beneath the sea in large deposits. Everyone is aware of the drilling for oil that has taken place offshore in many places in the world. However, perhaps you are less aware of the metals that come from the sea.

The two most abundant cations in seawater are those of sodium and magnesium. Their concentrations are about 0.6 M for Na^+ and about 0.06 M for Mg^{2+}. Sodium chloride, from which sodium is extracted by electrolysis (Section 17.4), can be obtained in a fairly pure form by the evaporation of seawater in large ponds (see Figure 21.1). About 13% of the salt produced in the United States comes from this source. The rest is mined from vast underground deposits, some of which lie quite deep below the earth's surface. (See also *Chemicals in Use* 2, page 86.)

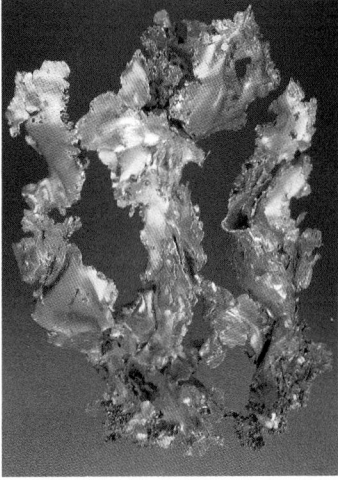
Gold is an element that is found free in nature. Prospectors dream of finding samples like the one shown here, which came from California.

FIGURE 21.1 Salt crystals being harvested in Bonaire, Netherlands Antilles. The salt crystals are formed by evaporation of seawater.

Separation of magnesium from seawater takes advantage of the low solubility of magnesium hydroxide. A batch of seawater is made basic by dissolving lime in it. Lime is calcium oxide, which is prepared by decomposing calcium carbonate (limestone), as we discussed in Section 20.8 on page 863. Near the ocean, sea shells, which are composed chiefly of $CaCO_3$, can also provide a cheap source of this raw material.

When the lime is dissolved in seawater, it reacts to form calcium hydroxide.

$$CaO(s) + H_2O \longrightarrow Ca^{2+}(aq) + 2OH^-(aq)$$

The hydroxide ion precipitates magnesium hydroxide, $Mg(OH)_2$, which is insoluble.

$$Mg^{2+}(aq) + 2OH^-(aq) \longrightarrow Mg(OH)_2(s)$$

This precipitate is then filtered from the seawater and dissolved in hydrochloric acid.

$$Mg(OH)_2(s) + 2H^+(aq) + 2Cl^-(aq) \longrightarrow Mg^{2+}(aq) + 2Cl^-(aq) + 2H_2O$$

Evaporation of the resulting solution yields magnesium chloride, which is dried, melted, and finally electrolyzed to give free magnesium metal.

$$MgCl_2(\ell) \xrightarrow{\text{electrolysis}} Mg(\ell) + Cl_2(g)$$

Another potential source of metals from the sea is the mining of manganese nodules from the ocean floor. Manganese nodules are lumps about the size of an orange that contain significant amounts of manganese (almost 25%) and iron (about 15%). In some places, they occur in large numbers, as Figure 21.2 shows. Although the procedures and equipment used to recover these nodules from the ocean floor are very expensive, the huge number of nodules that is believed to exist makes the concept of deep-ocean mining attractive, more so as land-based sources of these metals become depleted.

Metals from the Earth   It has only been in recent times that we have turned to the oceans as a source of metals. Land-based ores were what provided civilizations with their earliest access to metals, and to this day, most of our metals come from ores that are mined from the earth.

When the source of a metal is an ore that is dug from the ground, considerable amounts of sand and dirt are usually unavoidably mixed in with the ore. Therefore, the ore is normally concentrated before the metal is separated from it to reduce the volume of material that must be processed. How this is done depends on the physical and chemical properties of the ore itself, as well as those of the impurities.

In some cases, the unwanted rock and sand, called gangue, can be removed simply by washing the material with a stream of water. This flushes away the waste and leaves the enriched ore behind. Some iron ores, for example, are treated in this way. This procedure also forms the basis for the well-known technique called "panning for gold" that you've probably seen in movies. A sample of sand that might also contain gold is placed into a shallow pan filled with water. As the water is swirled around, it washes the less dense sand over the rim of the pan, but leaves any of the more dense bits of gold behind.

Flotation is a method commonly used to enrich the sulfide ores of copper and lead. First, the ore is crushed, mixed with water, and ground into a souplike slurry. The slurry is then transferred to flotation tanks (Figure 21.3) where it is mixed with detergents and oil. The oil adheres to the particles of sulfide ore, but not to the particles of sand and dirt. Next, air is blown through the mixture. The rising air bubbles become attached to the oil-coated ore particles and bring them to the surface where they are held in a froth. The detergents in the mixture stabilize the bubbles long enough for the froth and its load of ore particles to be skimmed off. Meanwhile, the sand and dirt settle to the bottom of the tanks, from which they are removed and disposed of.

Many ores must undergo a second round of pretreatment before the metal can be

FIGURE 21.2 A large pile of manganese nodules on the ocean floor.

Panning for gold is an activity many tourists practice when they visit Alaska.

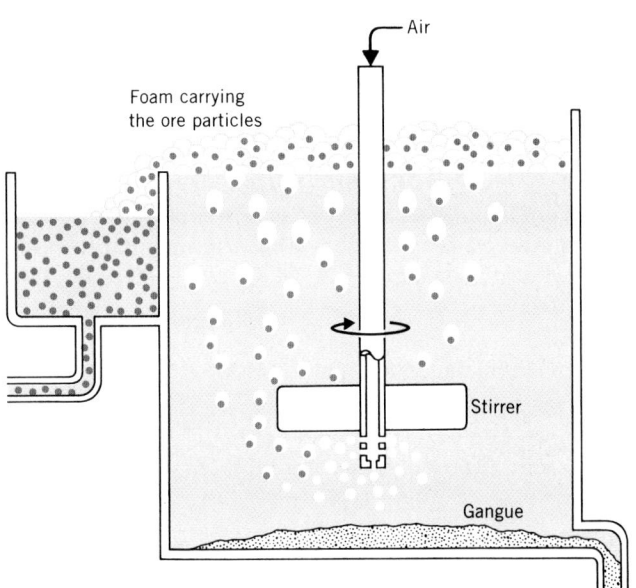

FIGURE 21.3 A flotation apparatus.

recovered from them. For example, after enrichment, sulfide ores are usually heated in air. This procedure, called roasting, converts the sulfides to oxides. This is done because oxides are more conveniently reduced than sulfides. Typical reactions that occur during roasting are

$$2Cu_2S + 3O_2 \longrightarrow 2Cu_2O + 2SO_2$$
$$2PbS + 3O_2 \longrightarrow 2PbO + 2SO_2$$

As described in Chapter 20, a by-product of roasting is sulfur dioxide. In years past, this was simply released to the atmosphere and was a major source of air pollution in the vicinity of an ore treatment plant. Figure 21.4, for example, shows the effect on the environment of the processing of copper ore at Copper Hill, Tennessee. Today we realize that the SO_2 cannot be allowed to escape into the air. One way of removing it from the exhaust gases is to allow it to react with calcium carbonate ($CaCO_3$).

$$CaCO_3(s) + SO_2(g) \longrightarrow CaSO_3(s) + CO_2(g)$$

This method creats another problem, however—the disposal of the solid calcium sulfite. Another way of disposing of the SO_2 is to oxidize it to SO_3. The SO_3 can be converted to sulfuric acid and then sold.

FIGURE 21.4 Heating metal sulfides in air to convert them to their oxides is called roasting and is a common metallurgical process, but the SO_2 that is released must not be allowed to escape into the atmosphere. Here we see the effects of a longtime roasting operation at Copper Hill, Tennessee, where SO_2 was released into the atmosphere.

Aluminum oxide, Al_2O_3, is also called *alumina*.

Aluminum ore, called *bauxite,* must also be pretreated before it can be processed. Bauxite contains aluminum oxide, Al_2O_3, but a number of impurities are also present. To remove these impurities, use is made of aluminum oxide's amphoteric behavior. The ore is mixed with a concentrated sodium hydroxide solution, which dissolves the Al_2O_3.

$$Al_2O_3(s) + 2OH^-(aq) \longrightarrow 2AlO_2^-(aq) + H_2O$$

The major impurities, however, are insoluble in base, so when the mixture is filtered, the impurities remain on the filter while the aluminum-containing solution passes through. The solution is then acidified, which precipitates aluminum hydroxide.

$$AlO_2^-(aq) + H^+(aq) + H_2O \longrightarrow Al(OH)_3(s)$$

When the precipitate is heated, water is driven off and the oxide is formed.

$$2Al(OH)_3(s) \xrightarrow{\text{heat}} Al_2O_3(s) + 3H_2O(g)$$

This purified Al_2O_3 then becomes the raw material for the Hall process discussed in Section 17.4.

Separating Metals from Their Compounds

In general, producing a free metal from one of its compounds involves reduction. This is because, with very few exceptions, metals in compounds have positive oxidation states. They must therefore "gain electrons" to become free elements. The nature of the reducing agent that is needed to provide these electrons depends on how difficult the reduction process is. If the free metal itself is very reactive, only electrolysis can provide enough energy to cause the decomposition of its compounds. This is the reason active metals such as sodium, magnesium, and aluminum are produced electrolytically. Less active metals such as lead or copper require less active reducing agents to displace them from their compounds, and chemical reducing agents can do the job.

A plentiful, and therefore inexpensive, reducing agent used to produce a number of metals is carbon. Coal is usually the source of this carbon. When coal is heated, volatile components are driven off and coke is formed. Coke is composed almost completely of carbon. Carbon is an effective reducing agent for metal oxides because it combines with

Coke is made in a battery of side-by-side coke ovens where coal is heated to drive off volatile materials. Here we see a fresh batch of white-hot coke being pushed from one of the ovens. It will be delivered to a blast furnace where it will be used to reduce iron ore to metallic iron.

the oxygen to form carbon dioxide. For example, after it is roasted, lead oxide is mixed with coke and heated.

$$2PbO(s) + C(s) \xrightarrow{heat} 2Pb(\ell) + CO_2(g)$$

The high thermodynamic stability of CO_2 serves as one driving force for this reaction. The loss of CO_2 is another.

Copper oxide ores can also be reduced with carbon.

$$2CuO + C \xrightarrow{heat} 2Cu + CO_2$$

This step is unnecessary for some copper sulfide ores if the conditions under which the ore is roasted are properly controlled. For example, heating an ore that contains Cu_2S in air can convert some of the Cu_2S to Cu_2O.

$$2Cu_2S + 3O_2 \xrightarrow{heat} 2Cu_2O + 2SO_2$$

At the appropriate point, the supply of oxygen is cut off, and the mixture of Cu_2S and Cu_2O reacts further to give metallic copper.

$$Cu_2S + 2Cu_2O \xrightarrow{heat} 6Cu + SO_2$$

Reduction of Iron Ore Without question, the most important use of carbon as a reducing agent is in the production of iron and steel. The chemical reactions take place in a huge tower called a blast furnace (see Figure 21.5). Some blast furnaces are as tall as a

Heating an ore with a reducing agent is called *smelting*.

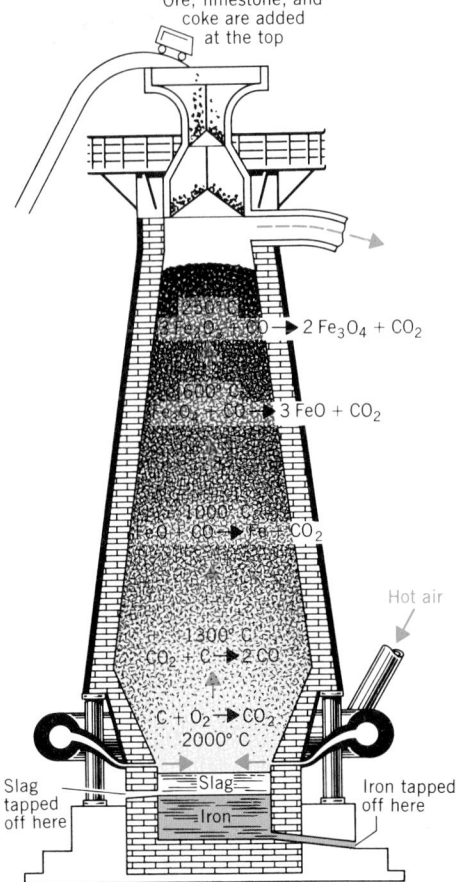

Ore, limestone, and coke are added at the top

$$\text{... + CO} \rightarrow 2\,Fe_3O_4 + CO_2$$
$$\text{... + CO} \rightarrow 3\,FeO + CO_2$$
$$FeO + CO \rightarrow Fe + CO_2$$
$$CO_2 + C \rightarrow 2\,CO$$
$$C + O_2 \rightarrow CO_2$$

Hot air

Slag tapped off here

Slag

Iron

Iron tapped off here

FIGURE 21.5 A typical blast furnace for the reduction of iron ore. The photograph on the right is of a blast furnace at Bethlehem Steel's Sparrows Point plant in Maryland. Blast furnaces like this one operate 24 hours per day.

15-story building and produce up to 2400 tons of iron a day. They are designed for continuous operation, so that the raw materials can be added at the top and molten iron can be tapped off at the bottom. Once started, a typical blast furnace may run continuously for two years or longer before it is worn out and must be rebuilt.

The material put into the top of the blast furnace is called the *charge*. It consists of a mixture of iron ore, limestone, and coke. A typical iron ore consists of an iron oxide—Fe_2O_3, for example—plus impurities of sand and rock. The coke is added to reduce the iron oxide to the free metal. The limestone is added to react with the high-melting impurities to form a slag, which has a lower melting point. The slag can then be drained off as a liquid at the base of the furnace.

To understand what happens in the furnace, it is best to begin with the reactions that take place near the bottom. Here, heated air is blown into the furnace where carbon (from the coke) reacts with oxygen to form carbon dioxide.

$$C + O_2 \longrightarrow CO_2$$

The reaction is very exothermic, and the temperature in this part of the furnace rises to nearly 2000 °C. It is the hottest region of the furnace. The hot CO_2 rises and reacts with additional carbon to form carbon monoxide.

$$CO_2 + C \longrightarrow 2CO$$

This reaction is endothermic, which causes the temperature higher up in the furnace to drop to about 1300 °C. As the carbon monoxide rises through the charge, it reacts with the iron oxides and reduces them to the free metal. The reactions are

$$3Fe_2O_3 + CO \longrightarrow 2Fe_3O_4 + CO_2$$
$$Fe_3O_4 + CO \longrightarrow 3FeO + CO_2$$
$$FeO + CO \longrightarrow Fe + CO_2$$

As the charge settles toward the bottom, molten iron trickles down and collects in a well at the base of the furnace.

The high temperature in the furnace causes the limestone in the reaction mixture to decompose to give calcium oxide.

$$\underset{\text{limestone}}{CaCO_3} \longrightarrow CaO + CO_2$$

The calcium oxide reacts with impurities such as silica (SiO_2) in the sand to form the slag.

$$CaO + SiO_2 \longrightarrow \underset{\substack{\text{calcium}\\\text{silicate}\\\text{(slag)}}}{CaSiO_3}$$

The molten slag also trickles down through the charge. It collects as a liquid layer on top of the more dense molten iron. Periodically, the furnace is tapped, and the iron and slag are drawn off. The iron, which still contains some impurities, is called pig iron.[1] It is usually treated further to produce steel. The slag itself is a valuable by-product. It is used to make insulating materials and is one of the chief ingredients in the manufacture of portland cement.

[1] The name *pig iron* comes from an early method of casting the molten iron into bars for shipment. The molten metal was run through a central channel that fed into sand molds. The arrangement looked a little like a litter of pigs feeding from their mother.

Preparing Metals for Use

Before metals can be used, most of them must be purified, or refined, after they are reduced to the metallic state. For example, the metallic copper that comes from the smelting process is about 99% pure, but before it can be used in electrical wiring, it must be purified further. The electrolytic refining of copper, you recall, was described in Chapter 17.

The conversion of pig iron to steel is the most important commercial refining process. When it comes from the blast furnace, pig iron consists of about 95% iron, 3 to 4% carbon, and smaller amounts of phosphorus, sulfur, manganese, and other elements. Steel contains much less carbon as well as certain other ingredients in very definite proportions. Converting pig iron to steel, therefore, involves removing the impurities and much of the carbon, and adding other metals in precisely controlled amounts.

The production of steel on a large scale began with the introduction of the Bessemer converter in England in 1856. In this method, a batch of molten pig iron weighing about 25 tons is placed in a large tapered cylindrical vessel that is lined with a refractory (very high melting) material. Because the impurities in the iron are usually silicon, sulfur, or phosphorus, whose oxides are acidic, a basic lining of CaO or MgO is normally used. A blast of air is then forced through the molten metal from a set of small holes in the bottom of the container. The oxygen in the air oxidizes the silicon, sulfur, and phosphorus, and the oxides that form react with the CaO or MgO to form a slag. At the same time, the carbon content of the steel is reduced by reaction with oxygen to form CO or CO_2. The conversion of pig iron to steel by this method is rapid, giving an impressive display of fire and sparks, but the exact composition of the finished product is difficult to control.

Because of the variable quality of the steel produced by the Bessemer process, it was largely replaced by a more easily controlled method that makes use of an open hearth furnace, so called because it contains a large shallow hearth (floor) into which the charge is placed. The floor is usually lined with a basic refractory such as CaO or MgO, and the charge consists of a mixture of molten pig iron, Fe_2O_3, scrap iron, and limestone. This mixture is heated to keep it in a molten state by a blast of hot air and burning fuel that is played across its surface (Figure 21.6). The Fe_2O_3 gradually reacts with the carbon in the mixture, forming bubbles of CO_2 that rise to the surface and keep the mixture stirred. The limestone decomposes and the resulting CaO reacts with impurities to form a slag. Other metals, such as manganese or chromium, are also sometimes added to give the steel special properties such as hardness or resistance to corrosion. Finally, after eight to ten hours, the steel is ready and the furnace is tapped. The metal is then fabricated into the various forms required by steel users.

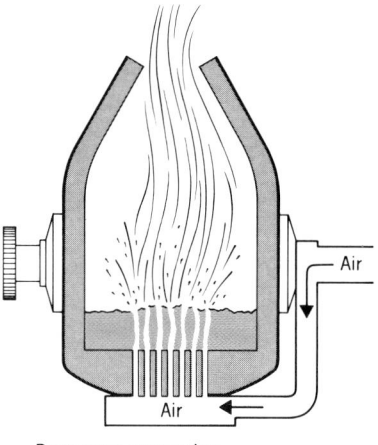

Bessemer converter.

Recall that metal oxides are basic anhydrides.

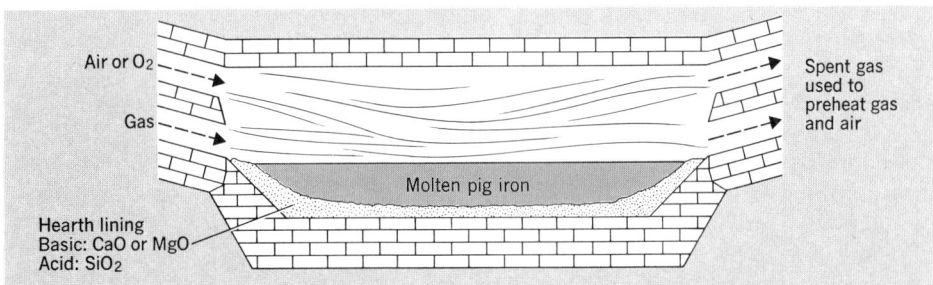

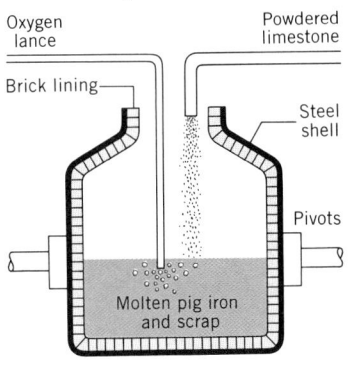

FIGURE 21.6 Cross section of an open hearth furnace.

FIGURE 21.7 Basic oxygen furnace used for the production of steel.

Although the open hearth furnace produces steel of reproducibly high quality, it is slow (and therefore expensive). Fast computers teamed with modern, high-speed methods of analysis have permitted the steel industry to return to a modification of the older Bessemer process. This modern method is called the basic oxygen process. It uses a large pear-shaped reaction vessel that is mounted on pivots and is lined with an insulating layer of special bricks, as shown in Figure 21.7. The charge consists of about 30% scrap iron and scrap steel and about 70% molten pig iron, which melts the scrap. A tube called an *oxygen lance* is then dipped into the charge and pure oxygen is blown through the molten metal. The oxygen rapidly burns off the excess carbon and oxidizes impurities to their oxides. These form a slag with calcium oxide that comes from powdered limestone, which is also added. Finally, other metals are introduced in the proper proportions to give a product with the desired properties. After the steel is ready, the reaction vessel can be tipped to pour out its contents. This method of making steel is very fast. A batch of steel weighing 300 tons can be made in less than an hour.

21.3 THE ALKALI METALS

The metals of Group IA are very reactive and possess very similar chemical and physical properties.

In the previous section we discussed how metals are obtained from their ores, and one of our goals was to understand how the methods used to do this depend on the specific chemical properties of the metals, as well as on the properties of the ores in which the metals are found. Now we will refocus our thinking around the periodic table and consider the chemical and physical properties of the metals in relationship to the groups in which they occur. When we do this, we find the strongest group similarities among the representative metals. We will discuss them first in the remainder of this chapter, and then in Chapter 22 we will complete our study of the metals by considering the properties of the transition elements.

The properties of many metal compounds have already been discussed in our study of the chemistry of the nonmetals in Chapters 19 and 20.

The Metals of Group IA

The elements of Group IA are hydrogen, lithium, sodium, potassium, rubidium, cesium, and francium. Hydrogen, at the top of the group, is a nonmetal, and its chemistry was discussed in Chapter 19. As we saw, hydrogen doesn't really fit well in any group. It owes its location in Group IA to its electron configuration, $1s^1$, rather than to its chemical properties. All the rest of the elements in Group IA are metals, however, and they are generally known by their group name — the alkali metals.

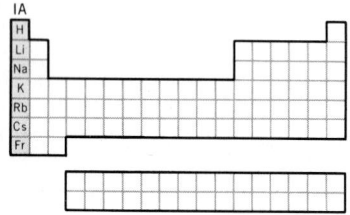

Occurrence

All the alkali metals are so easily oxidized, both by oxygen and by water, that they never are found as free elements in nature; they always occur in compounds. And because nearly all their salts are soluble in water, they are often present as ions in aqueous solution, primarily in seawater, in deep brine wells, and in certain lakes such as the Great Salt Lake in Utah and the Dead Sea on the border between Israel and Jordan. Compounds of the alkali metals are also found in mineral deposits on and below the earth's surface. For example, large underground deposits of solid salt (NaCl) are common, and there are also surface deposits in some dry salt lakes such as the Bonneville Salt Flats in Utah. Large deposits of sodium sulfate are found in North Dakota, Canada, and in the southwestern United States. In the desert regions near the border between Chile and Peru there are vast deposits of sodium nitrate, a compound often referred to in commerce as *Chile saltpeter*. In addition, some clays contain alkali metal ions along with aluminum, silicon, and oxygen.

Saltpeter is KNO_3; Chile saltpeter is $NaNO_3$.

By far, the most abundant of the alkali metals are sodium and potassium. By mass, they rank sixth and seventh among the elements in the earth's crust. It is perhaps no surprise, therefore, that they are also the most biologically important elements in Group IA. Both sodium and potassium ions are important in animals, but potassium ions are far more important than sodium ions in plants. In fact, with few exceptions (seaweed, for example), sodium ions in high concentrations are normally toxic to plants.

Lithium, which is relatively rare, has shown promise in the treatment of mental disorders. Lithium carbonate, for example, is used in the treatment of manic depression.

Sodium chloride occurs in large deposits both below and above ground. (Left) The interior of a salt mine located in Texas. (Right) The Bonneville Salt Flats in Utah.

Both rubidium and cesium are rare and have little commercial importance. Francium has only a fleeting existence because all its isotopes are radioactive and have very short half-lives.

Francium's longest-lived isotope is ^{223}Fr, which has a half-life of only 22 minutes.

Physical Properties

Some of the physical properties of the alkali metals are summarized in Table 21.1. All the elements are typically metallic with a bright luster and high thermal and electrical conductivity. The metals are also soft and have low melting points. This is because their atoms each have only one valence electron; in their metallic lattices they exist as cations with a charge of 1+, and they are only weakly attracted to the "sea" of electrons that surrounds them. These weak attractions are easily overcome, so their lattices are easily deformed (hence, soft) and not too much thermal energy is needed to overcome the attractions and cause the solid to melt.

The photograph on page 5 shows sodium being cut with a knife, which exposes the bright metallic surface of the metal.

TABLE 21.1 Some Properties of the Alkali Metals

Element	Electron Configuration	Ionization Energy (kJ/mol)	Melting Point (°C)	Reduction Potential (V)[a]
Lithium	[He] $2s^1$	520.1	180.5	−3.05
Sodium	[Ne] $3s^1$	495.1	97.8	−2.71
Potassium	[Ar] $4s^1$	418.7	63.7	−2.92
Rubidium	[Kr] $5s^1$	402.9	39.0	−2.93
Cesium	[Xe] $6s^1$	375.6	28.6	−2.92
Francium	[Rn] $7s^1$	—	—	—

[a] For $M^+(aq) + e^- \longrightarrow M(s)$.

Sodium-cooled reactors are described in Chapter 23.

You've already learned that the alkali metals have low ionization energies; their atoms are large and their single electron outside a noble gas core is only weakly held by the atom. This makes it easy to remove the outer electron, so the metals are easily oxidized. (We can also say they are good reducing agents.) Because they are so easily oxidized, the free elements have few practical uses as metals, and none that allows them to be exposed to air or moisture. Recently, however, sodium has shown promise as a coolant in certain kinds of nuclear reactors. Sodium has a low melting point and a reasonably high boiling point, and it conducts heat well. It can be pumped through the core of the reactor, where it readily picks up heat, and then through tubes in a heat exchanger, where the heat is removed.

The ease of oxidation of the alkali metals is reflected in their very negative reduction potentials. But here we observe a curious trend. As you know, within a group the ionization energy decreases from top to bottom. We might well expect, therefore, that it should become progressively easier to oxidize the metal, and progressively more difficult to reduce the M^+ cation going from top to bottom. We do, in fact, observe this from Na to Rb, but notice how far out of line lithium is. It has the most negative reduction potential of all. The reason is related to the fact that reduction potentials deal with the reduction of the metal ion in *aqueous solution*.

$$M^+(aq) + e^- \longrightarrow M(s)$$

Anything that helps to stabilize the metal ion should make it more difficult to reduce, and it is here that lithium ion has a special advantage. The Li^+ ion is very small, so its positive charge is highly concentrated. Also, it is able to get very close to the water molecules that surround it, so it attracts them very strongly. Quantitatively, this is expressed in terms of the ion's hydration energy—the energy that would be released when the cation is placed into its surrounding cage of water molecules. Because of its small size, the lithium ion has a very large hydration energy that must be overcome when the cation is reduced, so Li^+ is very difficult to reduce. The other cations—Na^+, K^+, etc.—are considerably larger, and they have smaller hydration energies that don't differ very much from one to another. They are easier to reduce than Li^+, and the group trend in ionization energy, which is more significant than any differences in the hydration energies of the ions, gives the expected variation in $E°$ as we descend the rest of the group.

Preparation of the Alkali Metals

Because the alkali metals are so easily oxidized, their compounds are very difficult to reduce and there are very few chemical reducing agents able to do the job. For this reason, the electrons for the reduction are usually supplied by electrolysis, especially for lithium and sodium. In carrying out the electrolysis, a molten salt must be used rather than an aqueous solution because water is more easily reduced than the metal ions. For instance, sodium and lithium are usually prepared from their molten chlorides, NaCl and LiCl. Chlorides are used because their melting points tend to be lower than on those of other compounds such as the oxides.

Interestingly, potassium can be made by the reaction of potassium chloride with metallic sodium, even though potassium ions are more difficult to reduce than sodium ions. This is done by passing sodium vapor over molten potassium chloride. The equilibrium

$$KCl(\ell) + Na(g) \rightleftharpoons NaCl(\ell) + K(g)$$

is shifted to the right because potassium is more volatile than sodium. The sodium condenses while the potassium is vaporized and swept away as it's formed. Metallic rubidium and cesium are also prepared in the same way.

Compounds and Reactions

The Group IA metals are the most reactive of all the metals. They are such powerful reducing agents that they readily reduce water. When placed in water, they react vigorously, releasing hydrogen and forming the metal hydroxide. (The reaction of sodium with water is shown on page 464.)

Reduction of H_2O produces H_2; oxidation of H_2O gives O_2.

$$2M(s) + 2H_2O \longrightarrow 2M^+(aq) + 2OH^-(aq) + H_2(g)$$

The alkali metals react with nearly all the elemental nonmetals. Lithium, however, is the only one that reacts directly with N_2 to form a nitride.

$$6Li(s) + N_2(g) \longrightarrow 2Li_3N(s)$$
$$\text{lithium nitride}$$

Jamie

Some of the most interesting reactions of the alkali metals are those that occur with oxygen, O_2. All these metals react rapidly with O_2 when exposed to air, but the ion formed by oxygen varies for the different metals, as we studied in Section 19.4. Lithium reacts with oxygen to form the "normal oxide," Li_2O, which contains the O^{2-} ion.

The normal oxide of sodium can also be made, but the supply of oxygen must be limited. In an abundant supply of oxygen, sodium forms the yellowish white peroxide, Na_2O_2, that contains the peroxide ion, O_2^{2-}. (Sodium peroxide forms even when Na_2O is exposed to air.)

When the remaining alkali metals, potassium through cesium, are exposed to air, they form superoxides. These are yellow solids that contain the paramagnetic superoxide ion, O_2^-, and have the general formula, MO_2. An example is potassium superoxide, KO_2. The normal oxides of K, Rb, and Cs can also be made, but by indirect methods.

When lithium oxide or any of the other normal oxides react with water, the metal hydroxide is formed.

Peroxides and superoxides of the alkali metals are powerful oxidizing agents. Na_2O_2 is used as a bleach because it oxidizes colored substances to give colorless products.

$$Li_2O(s) + H_2O \longrightarrow 2Li^+(aq) + 2OH^-(aq)$$

The reaction is really the hydrolysis of the oxide ion.

$$O^{2-}(s) + H_2O \longrightarrow 2OH^-(aq)$$

Sodium peroxide reacts with water to give NaOH and the anion of hydrogen peroxide, HO_2^-.

$$Na_2O_2(s) + H_2O \longrightarrow 2Na^+(aq) + OH^-(aq) + HO_2^-(aq)$$

In Chapter 19 it was pointed out that H_2O_2 is a weak acid. In very basic solution, the dominant species is the anion HO_2^-.

This is also a hydrolysis reaction, but of the peroxide ion.

$$O_2^{2-}(s) + H_2O \longrightarrow HO_2^-(aq) + OH^-(aq)$$

When superoxides react with water, they decompose to give oxygen, for example,

$$2KO_2(s) + H_2O \longrightarrow 2K^+(aq) + OH^-(aq) + HO_2^-(aq) + O_2(g)$$

Potassium superoxide is employed in gas masks that recirculate the air breathed by the user. As moisture in the exhaled air comes in contact with the KO_2, oxygen is generated, and the KOH formed in the reaction removes carbon dioxide.

A similar use of KO_2 to freshen the air supply in spacecraft was described in Chapter 19.

$$KOH + CO_2 \longrightarrow KHCO_3$$

Chlorides, Hydroxides, and Carbonates   Although the oxides of the alkali metals are interesting compounds, of much greater practical importance are the chlorides, hydroxides, and carbonates. And among these, the most important are those of sodium and potassium. This is simply because sodium and potassium are the most abundant of the Group IA metals.

As we have already mentioned, sodium chloride is a very common chemical, being present in relatively high concentration in seawater and in vast underground deposits. This makes NaCl an inexpensive raw material from which other compounds of sodium can be made. Potassium chloride is much less common. Its principal sources are the minerals *sylvite,* composed primarily of KCl, and *carnalite,* which has a composition corresponding to $KCl \cdot MgCl_2 \cdot 6H_2O$. These minerals are important fertilizers because of the need of plants for both potassium and magnesium.

Because sodium chloride is so plentiful, sodium compounds tend to be far less expensive to produce and therefore far more commercially important than similar compounds of the other alkali metals. A typical example is sodium hydroxide — industry's most important base. (In 1986, NaOH ranked eighth in total industrial production; over 10 million tons were produced.) This compound, known as *lye* or *caustic soda* in commerce, is manufactured from NaCl by electrolysis, as described in Chapter 17. The overall reaction, you may recall, is

$$2NaCl(aq) + 2H_2O \xrightarrow{\text{electrolysis}} 2NaOH(aq) + Cl_2(g) + H_2(g)$$

Potassium hydroxide, KOH, can be made in the same way from KCl, but the amount produced annually is far less. One of its principal uses is as an electrolyte in batteries (for example, the alkaline Zn/MnO_2 battery and the alkaline Zn/HgO battery).

The carbonates represent another important group of compounds. For example, among sodium compounds, sodium carbonate (known as *soda ash* in commerce) is second behind NaOH in annual production. Approximately 8.5 million tons of Na_2CO_3 were produced in 1986. About half of this was used to make glass and the rest was used to make other chemicals, paper, and detergents and in water softening. Almost 80% of the Na_2CO_3 produced each year comes from a mineral called *trona,* $Na_2CO_3 \cdot NaHCO_3 \cdot 2H_2O$. The rest is manufactured from $NaHCO_3$, which is made from NaCl by the **Solvay process.**

The raw materials for the Solvay process are sodium chloride, ammonia, and calcium carbonate (limestone). The limestone is heated which causes it to decompose to CaO plus CO_2 (page 863). The CO_2 from this reaction is bubbled into a cold solution of sodium chloride and ammonia. The dissolved CO_2 reacts with water and ammonia to form ammonium bicarbonate.

$$CO_2(aq) + H_2O + NH_3(aq) \longrightarrow NH_4^+(aq) + HCO_3^-(aq)$$

At this point, the solution contains the ions Na^+, Cl^-, NH_4^+, and HCO_3^-. At low temperatures, sodium bicarbonate is less soluble than any of the other possible salts that could be formed (NH_4Cl, NH_4HCO_3, or NaCl), so $NaHCO_3$ precipitates, leaving ammonium ion and chloride ion in the solution. The net overall reaction, therefore, is

$$NaCl(aq) + CO_2(aq) + H_2O + NH_3(aq) \longrightarrow NaHCO_3(s) + NH_4Cl(aq)$$

Ammonium chloride, which is recovered from the reaction mixture after the sodium bicarbonate is removed by filtration, is then allowed to react with the calcium oxide produced earlier by the decomposition of the limestone.

$$2NH_4Cl + CaO \longrightarrow CaCl_2 + 2NH_3 + H_2O$$

The ammonia formed in this reaction is recycled.

Most of the sodium bicarbonate from the Solvay process is converted to sodium

Sodium hydroxide was discussed in *Chemicals in Use 3,* on page 110.

Solutions of Na_2CO_3 are basic because of hydrolysis of the carbonate ion.

$$CO_3^{2-} + H_2O \rightleftharpoons HCO_3^- + OH^-$$

Calcium chloride, $CaCl_2$, is one of the by-products of the Solvay process.

carbonate by thermal decomposition according to the equation

$$2NaHCO_3(s) \xrightarrow{heat} Na_2CO_3(s) + H_2O(g) + CO_2(g) \qquad (21.1)$$

Dry-chemical fire extinguishers often contain $NaHCO_3$, which decomposes by this reaction when sprayed on a fire. The products of the decomposition smother the flames.

Ordinarily, sodium carbonate is sold in two forms. One, called *washing soda*, is the hydrate $Na_2CO_3 \cdot 10H_2O$, which is formed when Na_2CO_3 is crystallized from water below 35.2 °C. The other, $Na_2CO_3 \cdot H_2O$, is formed when sodium carbonate is crystalized above this temperature.

Sodium bicarbonate, the primary product of the Solvay process, is itself an important chemical. Solutions of it are mildly basic and serve as a buffer because of the ability of the HCO_3^- ion to neutralize both acids and bases.

$$HCO_3^- + H^+ \longrightarrow H_2CO_3$$
$$HCO_3^- + OH^- \longrightarrow CO_3^{2-}$$

For this reason, $NaHCO_3$ is recommended as an additive for swimming pools where it helps control the pH of the water when other chemicals are added to destroy bacteria.

Sodium bicarbonate is commonly called *baking soda* or *bicarbonate of soda*. When added to dough it produces a light and appealing product because it decomposes during baking according to Equation 21.1 above. The tiny bubbles of CO_2 cause the dough to rise as it is baked.

Sodium bicarbonate is also used in various analgesic products such as Alka Seltzer® and Brioschi®. In these products use is made of the ability of the bicarbonate ion to neutralize acids (See Special Topic 18.1 on page 739).

Potassium carbonate, K_2CO_3, is known by the common name *potash* and is found in the ashes of burned plants and wood. It is used in making certain soft soaps, pottery, glass, and other compounds of potassium.

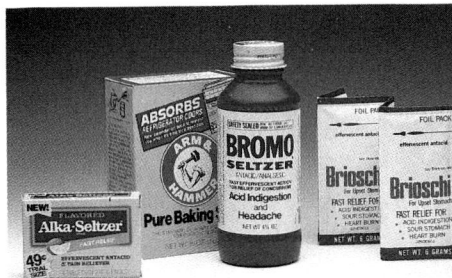

Some consumer products that contain sodium bicarbonate.

Solubilities of Salts

In the laboratory, sodium and potassium salts are common reactants because nearly all of them are soluble in water. Therefore, if a particular anion is needed for a reaction in solution, a chemist almost always reaches for a bottle of its sodium or potassium salt. However, salts of these metals with a given anion are not equally soluble. An interesting and useful generalization is that the sodium salt of a *strong acid* is often more soluble than the potassium salt, whereas the potassium salt of a *weak acid* is often more soluble than the sodium salt. Table 21.2 contains some data that illustrate this.

If you need NO_3^- for a reaction, you can use either $NaNO_3$ or KNO_3.

Spectra

When ions of an alkali metal are introduced into a flame, brilliant colors are produced that are characteristic of the element's atomic spectrum. Sodium salts, for example, give a

TABLE 21.2 Molar Solubilities of Some Sodium and Potassium Salts[a]

		Sodium Salt	Solubility (mol/100 g H_2O)	Potassium Salt	Solubility (mol/100 g H_2O)
Strong Acids					
Hydrochloric	(HCl)	NaCl	0.61	KCl	0.46
Perchloric	($HClO_4$)	$NaClO_4$	1.49	$KClO_4$	0.01
Nitric	(HNO_3)	$NaNO_3$	0.86	KNO_3	0.12
Weak Acids					
Acetic	($HC_2H_3O_2$)	$NaC_2H_3O_2$	1.45	$KC_2H_3O_2$	2.58
Tartaric	($H_2C_4H_4O_6$)	$Na_2C_4H_4O_6$	0.035	$K_2C_4H_4O_6$	0.64
Citric	($H_3C_6H_5O_7$)	$Na_3C_6H_5O_7$	0.25	$K_3C_6H_5O_7$	0.56

[a] Compared at the same temperature.

FIGURE 21.8 The colors imparted to a Bunsen burner flame by (*a*) sodium, (*b*) potassium, and (*c*) lithium.

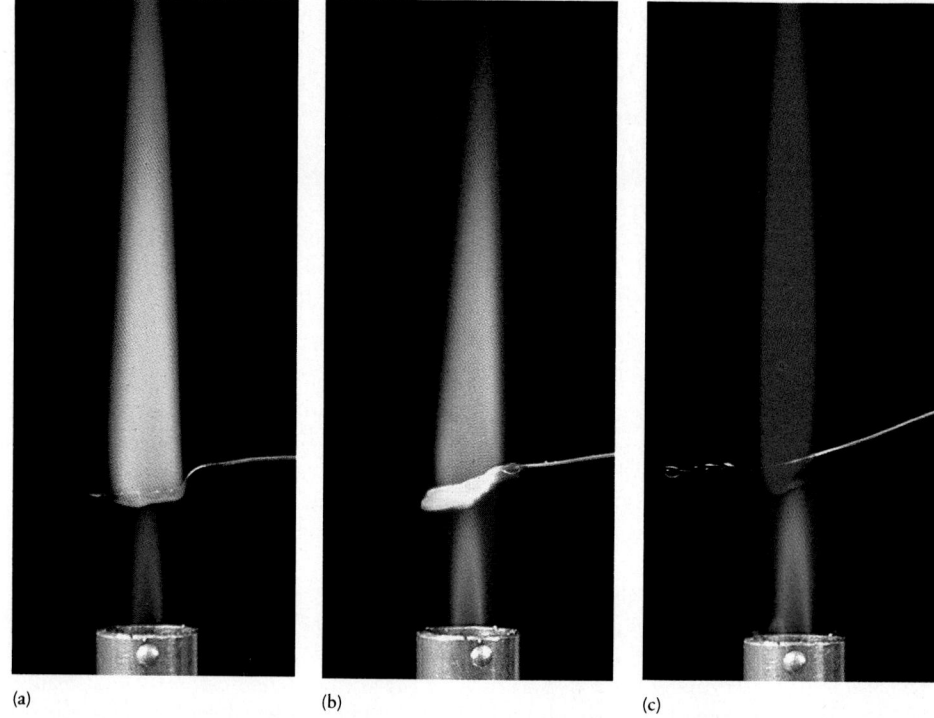

(a)   (b)   (c)

Na_2CO_3 is used in making glass.

bright yellow flame (Figure 21.8). If you've ever softened a glass rod or glass tubing in a Bunsen burner flame, you have seen this yellow color. It is produced by sodium ions that are vaporized from the hot glass.

The yellow flame produced by sodium-containing compounds is so easily recognized that it forms the basis of a flame test for this element. If a sample is suspected to contain sodium, a clean wire is dipped into a solution of the sample and then held in the flame of a Bunsen burner. A bright yellow color in the flame confirms the presence of sodium; the absence of the bright yellow flame means the sample contains no sodium.

Some fireplace logs are soaked in solutions of metal salts to produce brightly colored flames when they are burned.

The other alkali metals for which a flame test is normally used are lithium and potassium. Potassium salts impart a pale violet color to a flame, and lithium salts give a beautiful, deep red color. The pale violet of potassium is sometimes difficult to see, particularly if sodium is present. The yellow from the sodium is so intense that it masks the violet produced by the potassium. Viewing the flame through a special blue glass called *cobalt glass* filters out the yellow and allows the pale violet of the potassium flame to be seen.

21.4 THE ALKALINE EARTH METALS

The metals of Group IIA tend to be less reactive than those alongside in Group IA.

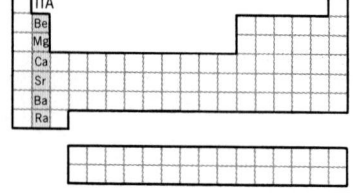

The elements of Group IIA are beryllium, magnesium, calcium, strontium, barium, and radium. They are known as the alkaline earth metals and, as with their neighbors in Group IA, there are strong similarities in chemical and physical properties among the members of the group. However, the similarities are not as striking as among the alkali metals. For example, lithium is much more like cesium than beryllium is like barium. This is the beginning of a general phenomenon that is observed as we move from either the left or right toward the representative elements in the center of the periodic table — there is a progressively larger difference between the properties of the element at the top of the group and the element at the bottom.

Occurrence

The alkaline earth metals are almost as reactive as the metals of Group IA, and therefore they are never found in the elemental state in nature; they always occur in compounds, either in aqueous solution or in mineral deposits in the earth.

The most abundant of the alkaline earth metals are calcium and magnesium. As we learned earlier, magnesium ions are found in relatively large concentration in seawater. Calcium ions are also present in seawater, and marine organisms take the Ca^{2+} from the water to make their calcium carbonate shells. Calcium and magnesium ions are also found in rather large concentrations in certain underground brines. In the United States, particularly rich sources of these ions are brines that are pumped from wells in Michigan.

On land, the alkaline earth metals are found in various mineral deposits. Largest among them are limestone (calcium carbonate, $CaCO_3$) and *dolomite* (a mixed calcium–magnesium carbonate, generally written as $CaCO_3 \cdot MgCO_3$). Many of these limestone deposits occur below the Earth's surface, and groundwater trickling through them often creates spectacular caverns. Another important calcium mineral is gypsum, $CaSO_4 \cdot 2H_2O$, from which plaster is made (see *Chemicals in Use* 1, page 42). The element beryllium is found in the mineral *beryl*, $Be_3Al_2(SiO_3)_6$. You might have seen beryl crystals such as *emerald* and *aquamarine* (Figure 21.9). Strontium and barium occur chiefly as their sulfates and carbonates, and radium is isolated from uranium ores because it is a product of the radioactive decay of uranium.

Calcium is the third most abundant *metal* in the earth's crust.

In seawater, the concentration of Mg^{2+} is 0.056 *M* and the concentration of Ca^{2+} is 0.011 *M*.

For more on limestone caverns, see *Chemicals in Use* 9 on page 440.

Radium was discovered by Marie and Pierre Curie.

FIGURE 21.9 Naturally occurring emerald (left) and aquamarine (right). When cut and polished, they give beautiful gems. Both are forms of the mineral beryl, $Be_3Al_2(SiO_3)_6$.

Physical Properties

Some properties of the alkaline earth metals are given in Table 21.3. Each atom of these elements has a pair of rather loosely held outer electrons that is lost to the metallic lattice. The resulting doubly charged cations are bound more tightly to the surrounding sea of

TABLE 21.3 Some Properties of the Alkaline Earth Metals

Element	Electron Configuration	Ionization Energy (kJ/mol)		Melting Point (°C)	Reduction Potential (V)[a]
		First	Second		
Beryllium	[He] $2s^2$	899	1757	1278	−1.85
Magnesium	[Ne] $3s^2$	737	1450	651	−2.37
Calcium	[Ar] $4s^2$	590	1145	843	−2.76
Strontium	[Kr] $5s^2$	549	1059	769	−2.89
Barium	[Xe] $6s^2$	503	960	725	−2.90
Radium	[Rn] $7s^2$	509	975	700	−2.82

[a] For $M^{2+}(aq) + 2e^- \longrightarrow M(s)$.

electrons than are the singly charged cations in crystals of the alkali metals. Therefore, the Group IIA metals have higher melting points and are harder than their neighbors in Group IA.

The ionization energies of the alkaline earth metals are relatively low, but higher than those of the alkali metals. As a result, even though the Group IIA metals are easily oxidized, they are not quite as effective as reducing agents as are the alkali metals. This is reflected in the reduction potentials of the alkaline earth metals, which tend not to be quite as negative as the alkali metal alongside. Near the bottom of the group, however, the differences become negligible.

Preparation

The only alkaline earth metals to be produced in significant amounts are beryllium and magnesium. They are the only two that do not react readily with air and moisture at room temperature, so they are the only ones that can survive prolonged contact with our environment.

Beryllium is made by electrolysis of molten beryllium chloride. However, $BeCl_2$ is covalent and a poor conductor of electricity, so NaCl must be added to the melt to serve as an electrolyte. Metallic beryllium is used to make windows of X-ray tubes, because of all usable metals it is the most transparent to X rays. The metal is also used in alloys. One is a copper–beryllium alloy from which springs, electrical contacts, and tools are made. Hammers and wrenches made from Be/Cu alloys do not produce sparks when struck against steel and can therefore be used in explosive atmospheres. Beryllium compounds are quite toxic, and tough safety standards must be followed in fabricating beryllium-containing products.

Magnesium is recovered both from seawater and from ores obtained from the earth — chiefly carnalite, $MgCl_2 \cdot KCl \cdot 6H_2O$, and dolomite, $CaCO_3 \cdot MgCO_3$. Recovery from dolomite involves first heating the mixed carbonate to give a mixed oxide.

$$CaCO_3 \cdot MgCO_3 \xrightarrow{\text{heat}} CaO \cdot MgO + 2CO_2$$

This oxide mixture is added to seawater. Reaction of the oxides gives the hydroxides.

$$CaO + H_2O \longrightarrow Ca(OH)_2$$
$$MgO + H_2O \longrightarrow Mg(OH)_2$$

Calcium hydroxide is considerably more soluble than magnesium hydroxide, and the hydroxide ions from the $Ca(OH)_2$ cause Mg^{2+} ions in the seawater to be precipitated as $Mg(OH)_2$. The combined precipitate of $Mg(OH)_2$ is then allowed to settle in large ponds from which it is later recovered. As discussed earlier, the $Mg(OH)_2$ is dissolved in hydrochloric acid, and the resulting solution is evaporated to give solid $MgCl_2$, which is melted and electrolyzed to give metallic magnesium and chlorine.

Beryllium is used in alloys with copper and bronze to give them hardness.

Finely divided BeO is very toxic if inhaled.

Heating the carbonate in air like this is called *calcining*.

Magnesium hydroxide settling ponds at Dow Chemical Company's plant in Freeport, Texas, where magnesium is recovered from seawater.

Because magnesium has such a low density and a moderate strength, it is useful as a structural metal when alloyed with aluminum. Perhaps you have an aluminum-magnesium alloy ladder in your home. Another use of magnesium is in flashbulbs, fireworks, and incendiary bombs. Magnesium burns in air with the evolution of much heat and a very bright light. In a flashbulb, a thin magnesium wire is heated electrically by a battery and this ignites the metal, which burns very quickly in the pure oxygen atmosphere that surrounds it.

Calcium, strontium, and barium have few commercial uses other than as reducing agents in specialized metallurgical operations. They are produced in only limited quantities, usually by electrolysis of their molten chlorides.

Chemical Properties and Group Trends

As you can see from their reduction potentials, the alkaline earth metals are easily oxidized. In fact, all are oxidized easily enough to be attacked by water, at least in principle. However, beryllium and magnesium metals form oxides that are relatively insoluble in water, and these oxides protect the metal from attack. The oxides of the elements below magnesium, however, are reasonably soluble in water. This permits water to come in contact with the metals, and they react to liberate hydrogen. For example,

$$Ca(s) + 2H_2O \longrightarrow Ca(OH)_2(aq) + H_2(g)$$

However, the reactions of these metals with water are not as violent as the reactions of the alkali metals with water. Although magnesium is not attacked by cold water, it does react slowly with hot water and fairly rapidly with steam to form H_2 and $Mg(OH)_2$. Beryllium, on the other hand, does not react at all with water.

Metallic magnesium burns in air. The reaction produces a large amount of heat and light. The product of the reaction, magnesium oxide, can be seen dangling from the burning magnesium ribbon.

A piece of calcium metal reacts with water to liberate H_2 gas, which rises to the surface as bubbles. The reaction also produces $Ca(OH)_2$. Notice that the reaction of calcium with water is much less violent than the similar reaction of sodium with water (page 464).

In their compounds, the alkaline earth metals always exist in the $+2$ oxidation state, corresponding to the "loss" of the outer pair of s valence electrons from each of their atoms. In general, compounds of calcium, strontium, and barium are distinctly ionic, that is, their compounds exhibit properties that are typical of ionic substances. Most magnesium compounds are also ionic, although magnesium does form some covalently bonded *organomagnesium* compounds in which the Mg atom is bonded to carbon atoms in groups derived from hydrocarbons. An example is $Mg(C_2H_5)_2$ in which two "ethyl groups," C_2H_5, are covalently bonded through carbon to the magnesium atom. (Because Mg is less electronegative than C, it still is assigned an oxidation number of $+2$.)

In the case of beryllium, there is no evidence that Be^{2+} ions actually exist. Beryllium compounds such as $BeCl_2$ are covalent. In fact, solid $BeCl_2$ really contains long strands of $BeCl_2$ units linked together by covalent bonds to give a structure that has the general formula $(BeCl_2)_x$, where x is a large number. The structure of these $BeCl_2$ chains is illustrated in Figure 21.10. We can view the structure as being formed by the donation of

$$
\begin{array}{ccc}
 & H & H \\
 & | & | \\
H\!-\!\!\!&C&\!\!\!-\!\!\!C\!\!\!-\!H \\
 & | & | \\
 & H & H
\end{array}
$$
an ethyl group

FIGURE 21.10 The formation of $(BeCl_2)_x$ from $BeCl_2$. (a) Formation of coordinate covalent bonds between the Cl atoms and Be atoms permits Be to complete its octet. The result is long chains of $BeCl_2$ units. (b) The arrangement of the chlorine atoms around each beryllium is approximately tetrahedral.

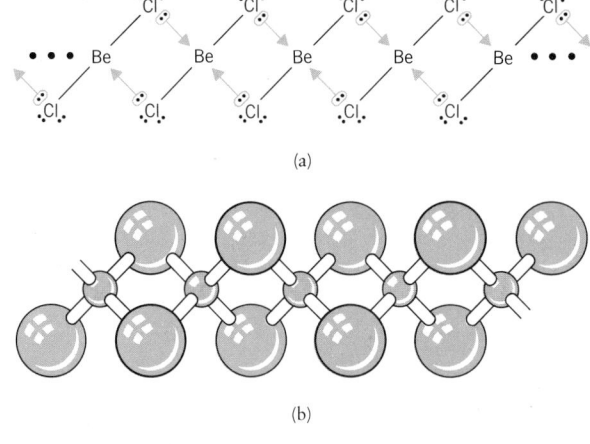

(a)

(b)

When one atom supplies both electrons in the formation of a covalent bond, the bond is called a coordinate covalent bond. Of course, once it is formed, the coordinate covalent bond is no different than any other covalent bond.

The effect of small cation size and high cation charge on the covalent character and acid-base properties of metal oxides was discussed in Chapter 19.

FIGURE 21.11 A small Be^{2+} ion would have such a high concentration of positive charge that it would distort the electron cloud of an anion such as Cl^-. The electron density drawn between the nuclei gives a great deal of covalent character to the bond.

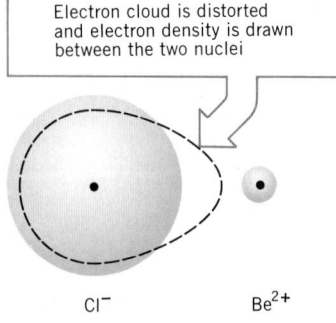

Electron cloud is distorted and electron density is drawn between the two nuclei

Cl^- Be^{2+}

a pair of electrons from a chlorine of one $BeCl_2$ to the beryllium of another. These covalent bonds (we could call them coordinate covalent bonds) permit the Be atoms to complete their octets, and in the process, each Be atom becomes surrounded by four Cl atoms that are arranged approximately at the corners of a tetrahedron.

The large degree of covalent character in the bonds that beryllium atoms form with nonmetals is attributed to the small size and highly concentrated positive charge that a true Be^{2+} ion would have. As illustrated in Figure 21.11, if a Be^{2+} ion were placed next to an anion, its high concentration of positive charge would distort the anion's electron cloud and draw the electron density into the region between the Be^{2+} and the anion. Because this would concentrate the electron density between the two nuclei, it would produce, in effect, a covalent bond. Larger positive ions — Mg^{2+}, for example — are considerably less effective in distorting the electron cloud of an anion because the positive charge is more spread out. As a result, the bonds formed between Mg^{2+} and anions such as Cl^- are considerably less covalent, and therefore much more ionic.

As with the alkali metals, the alkaline earth metals combine directly with most of the nonmetals. Magnesium, however, is the only one that reacts with elemental nitrogen to form a nitride.

$$3Mg(s) + N_2(g) \longrightarrow Mg_3N_2(s)$$

You may recall that lithium is alone among the alkali metals in its reaction with N_2, and we see here a phenomenon repeated again for Be and Al; there are certain similarities between the chemistries of lithium and magnesium, and there are also certain similarities that we will note later between the chemistries of beryllium and aluminum. For lithium and magnesium, these similarities include the formation of certain covalent compounds such as LiC_2H_5 and the $Mg(C_2H_5)_2$ mentioned above, as well as similarities in the solubilities of some of their salts. This parallel behavior of Li and Mg is often referred to as a *diagonal similarity* because of the relative locations of these elements in the periodic table, and it appears to be due to their similar ratios of ionic charge to ionic radius. This causes Li^+ and Mg^{2+} to have about the same effects on adjacent anions and therefore about the same degree of covalent character to their metal–nonmetal bonds.

One of the observations we can make about the properties of the elements in the periodic table is that as we descend a group, the elements become progressively more metallic. We mentioned earlier, for example, that in Group IVA the element at the top is a nonmetal (carbon) and the element at the bottom is a metal (lead). To a much lesser extent, we also observe this trend in Group IIA. All the alkaline earth metals are typically metallic in their physical properties; they are good conductors of heat and electricity, for example. But in some of their chemical properties, we find that beryllium is considerably less metallic, and magnesium slightly less metallic than the heavier members of the group. For example, we find that beryllium compounds are predominantly covalent — a property that we usually associate with the compounds formed by the nonmetallic elements.

And we also find that magnesium forms some covalent compounds, too, but not nearly as many as beryllium.

Another example is the behavior of the metals and their oxides toward hydrogen ions. Earlier you learned that the oxides of metals are basic; either they dissolve in water to form basic solutions or they react with acids. On the other hand, typical oxides of nonmetals, such as SO_2, are acidic and either give acids in water or react with bases.

When we examine the beavior of the alkaline earth metals, we find that they all react with nonoxidizing acids such as HCl to liberate H_2, and their oxides react with HCl to give the corresponding salt and water. Beryllium and its oxide, however, are amphoteric and react with strong base as well.

$Mg + 2HCl \longrightarrow MgCl_2 + H_2$

$$Be(s) + 2H_2O + 2OH^-(aq) \longrightarrow Be(OH)_4^{2-}(aq) + H_2(g)$$
$$BeO(s) + H_2O + 2OH^-(aq) \longrightarrow Be(OH)_4^{2-}(aq)$$

Because BeO is somewhat acidic in its properties, we can say that beryllium is less metallic than the rest of the members of Group IIA, whose oxides do not react with base.

Important Compounds of the Alkaline Earth Metals

The most important compounds of the alkaline earth metals are their chlorides, oxides, hydroxides, carbonates, and sulfates. All the chlorides are water soluble and, except for $BeCl_2$, they are all ionic. Calcium chloride is unusual because of its high affinity for moisture. If calcium chloride is left exposed to moist air, it absorbs so much water that the crystals actually dissolve to form a concentrated solution of $CaCl_2$. The ability of a salt to absorb enough water to dissolve in it is called deliquescence, and calcium chloride is said to be *deliquescent*. Calcium chloride can be purchased in hardware stores (although not always under its chemical name) for use in removing moisture from places with high humidity such as damp basements.

In Chapter 2 you learned that $CaCl_2$ is sometimes spread on dirt roads because the moisture it absorbs keeps the dust down.

The oxides of the alkaline earth metals can be formed by direct combination of the elements.

$$2M(s) + O_2(g) \longrightarrow 2MO(s)$$

However, the usual method of preparation is by thermal decomposition of the carbonate. We have already seen this several times in the decomposition of limestone, which is calcium carbonate. Calcium oxide, known commercially as *lime* or sometimes *quick lime,* is produced in enormous amounts each year. In 1986, it ranked fourth among all chemicals, with almost 16 million tons being made. The reason is that it is inexpensive and a reasonably strong base. When treated with water, a process called *slaking,* it forms calcium hydroxide.

The *lime* that is spread on lawns or gardens is really pulverized limestone or dolomite.

$$CaO(s) + H_2O \longrightarrow Ca(OH)_2(s)$$

Calcium oxide is an important ingredient in portland cement (Section 20.8), and this is one of the first reactions to occur when water is added to the cement.

$Ca(OH)_2$ is called *slaked lime.*

Magnesium oxide can also be made from its carbonate by thermal decomposition. It is much less reactive toward water than CaO, however, especially if the solid is heated first to high temperatures. In fact, as noted earlier, magnesium metal is protected from attack by water by a thin film of MgO that forms on its surface. One of the principal uses of magnesium oxide is in special bricks that withstand high temperatures in furnaces; they're called *refractory* bricks. Magnesium oxide is also used in making paper and in certain antacid preparations.

The hydroxides increase in solubility from $Mg(OH)_2$ to $Ba(OH)_2$.

MgO melts at about 2800 °C.

Magnesium hydroxide is generally made by adding base to a solution containing Mg^{2+}. This precipitates $Mg(OH)_2$.

$$Mg^{2+}(aq) + 2OH^-(aq) \longrightarrow Mg(OH)_2(s)$$

The antacid and laxative known as milk of magnesia is a suspension of $Mg(OH)_2$ in water.

BaSO$_4$ $K_{sp} = 1.1 \times 10^{-10}$
CaSO$_4$ $K_{sp} = 2.4 \times 10^{-5}$

The most important sulfates of the alkaline earth metals are those of magnesium, calcium, and barium. The solubilities of the sulfates decrease going down the group — MgSO$_4$ is quite soluble, whereas BaSO$_4$ is very insoluble. Magnesium sulfate as its hydrate MgSO$_4 \cdot 7H_2O$ is known as *Epsom salts*, and is used in the tanning of leather, to treat fabrics so they readily accept dyes, to fireproof fabrics, as a fertilizer, and medicinally.

The dihydrate of calcium sulfate, CaSO$_4 \cdot 2H_2O$, is called *gypsum*. Partial dehydration of gypsum gives plaster of paris, as mentioned in *Chemicals in Use* 1.

$$2CaSO_4 \cdot 2H_2O(s) \xrightarrow{heat} (CaSO_4)_2 \cdot H_2O(s) + 3H_2O(g)$$
$$\text{gypsum} \qquad\qquad \text{plaster of paris}$$

Usually, the formula for plaster of paris is written as CaSO$_4 \cdot \frac{1}{2}H_2O$, which expresses the same mole ratio of CaSO$_4$ to H$_2$O as (CaSO$_4$)$_2 \cdot$ H$_2$O.

One of the important uses of barium sulfate is in obtaining medical X-ray photographs of the digestive tract. A patient drinks a suspension of BaSO$_4$ in water and then an X-ray photograph is taken. The path of the patient's digestive tract is clearly visible on the film because BaSO$_4$ is opaque to X rays (see Figure 21.12). Even though the barium ion, like most heavy metal ions, is very toxic to the human body, barium sulfate is safe to drink because its solubility is so low. Hardly any Ba^{2+} is absorbed by the body as the BaSO$_4$ passes through the digestive system. Other uses of BaSO$_4$ are based on its whiteness; it is used as a whitener in photographic papers and as a filler in papers and polymeric fibers.

Spectra

Like the alkali metals, certain of the alkaline earth metals give characteristic colors to flames. Calcium salts give an orange-red color, strontium salts produce a bright red (crimson) flame, and barium salts give a yellow-green color. These are shown in Figure 21.13, and are intense enough to serve as flame tests. Like salts of the alkali metals, salts of these alkaline earth metals are used in coloring fireworks displays.

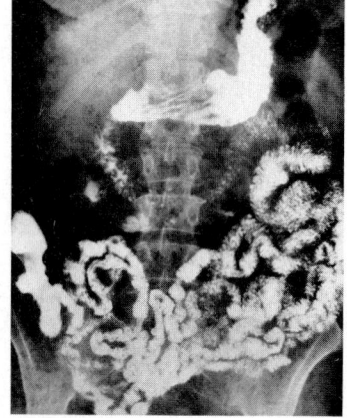

FIGURE 21.12 Barium sulfate, which is opaque to X rays, defines the path of the small intestine in a patient who swallowed a suspension of this solid in water.

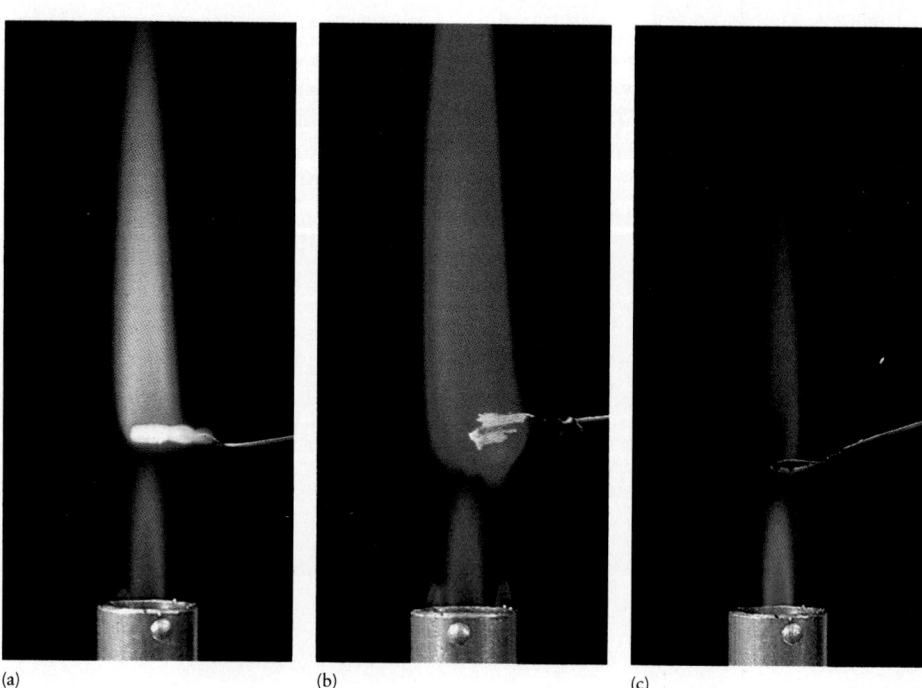

FIGURE 21.13 The colors given to a Bunsen burner flame by (a) calcium, (b) strontium, and (c) barium.

(a)   (b)   (c)

21.5 THE METALS OF GROUPS IIIA, IVA, AND VA

The metals in groups to the right of the transition elements are less reactive than the metals of Groups IA and IIA, and the heavier of them have two oxidation states.

Except for aluminum, the metals in Groups IIIA, IVA, and VA are called post-transition metals because they follow the row of transition elements in their respective periods. The post-transition metals are, in general, considerably less reactive than the metals in Groups IA and IIA. They each have a completed *d* subshell just below the valence shell. Because these *d* electrons are less than 100% effective in screening the nuclear charge, the outer electrons of these elements are held much more tightly than the outer electrons of the alkali or alkaline earth metals.

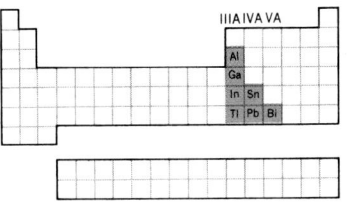

The Metals of Group IIIA

The elements of Group IIIA are boron, aluminum, gallium, indium, and thallium. They have no general group name, as do the elements of Groups IA and IIA. Except for boron, which is a metalloid, all are metals. Some of their properties are given in Table 21.4. Notice that gallium has a melting point of only 29.8 °C; it will melt in the palm of your hand. Gallium also has a very high boiling point, and this large liquid range has made gallium useful for certain types of thermometers.

TABLE 21.4 Some Properties of the Group IIIA Elements

Element	Electron Configuration	Ionization Energy (kJ/mol)			Melting Point (°C)	Reduction Potential (V)[a]
		First	Second	Third		
Boron	[He] $2s^2 2p^1$	801	2427	3660	2200	—
Aluminum	[Ne] $3s^2 3p^1$	577	1816	2744	600	−1.66
Gallium	[Ar] $3d^{10} 4s^2 4p^1$	578	1971	2950	29.8	−0.53
Indium	[Kr] $4d^{10} 5s^2 5p^1$	559	1813	2690	157	−0.33
Thallium	[Xe] $4f^{14} 5d^{10} 6s^2 6p^1$	589	1961	2860	303	+0.72

[a] For $M^{3+}(aq) + 3e^- \longrightarrow M(s)$.

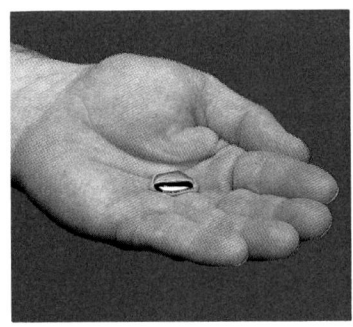

The melting point of gallium is only 29.8 °C. Body temperature is 37 °C, which is high enough to cause the metal to melt in the palm of your hand.

The only really important metal in Group IIIA, as far as we are concerned, is aluminum. On a weight basis, it is the third most abundant *element* in the earth's crust, exceeded only by oxygen and silicon. This also means that it is the most abundant *metal* in the earth's crust. It is found abundantly combined with silicon and oxygen in *aluminosilicates,* which occur in various kinds of rock, such as granite, and in clays. Unfortunately, no one has found an economical way of separating the aluminum from these minerals, so they are not considered aluminum ores. Aluminum is also found in the mineral *cryolite,* which has the formula Na_3AlF_6 and contains the complex ion AlF_6^{3-}. The chief ore of aluminum is *bauxite,* which contains aluminum in the form of a hydrated oxide, $Al_2O_3 \cdot xH_2O$. The purification of bauxite was described on page 880, and the production of aluminum by the Hall process was discussed in Chapter 17. As you know, aluminum is used widely in our modern world for all sorts of purposes, from aluminum foil to airplane construction. Many of its structural uses are based on its low density ("light weight") and high strength.

The earth's crust consists of about 8.8% aluminum by weight.

In its compounds, aluminum always exists in the +3 oxidation state, which corresponds to the "loss" of its three valence electrons. However, an Al^{3+} ion is small and highly charged, and as we saw in the case of beryllium, this can cause considerable distortion of the electron cloud of a neighboring anion and impart a large degree of covalent character to the metal–nonmetal bond. For this reason, many of the compounds of aluminum show substantial covalent character in much the same way as compounds of beryllium.

In capping a tooth, dentists often use a temporary aluminum cover to protect the tooth while the dental laboratory fabricates the permanent crown. If the tooth being treated has also received a fresh silver filling, however, the aluminum temporary crown can't be employed.

A silver filling is a mixture of silver powder and mercury. Mercury dissolves the silver to form an amalgam — a solution of a metal in mercury. This cements the silver particles together. When the silver filling is fresh, however, the silver particles still have mercury on their surface. It hasn't yet been absorbed by the silver. If an aluminum crown is placed over the tooth, the mercury can contact the aluminum and form an amalgam with it. This prevents the adhesion of the protective Al_2O_3 coating and allows aluminum metal in the amalgam to react with oxygen and water in the patient's mouth. The resulting exothermic reaction can make the aluminum crown painfully hot.

Aluminum metal is easily oxidized; the reduction potential of $Al^{3+}(aq)$ is -1.66 V. Fortunately, the reaction between aluminum and oxygen produces a tough oxide coating that adheres tightly to the aluminum surface and protects the metal beneath from further attack. In fact, *anodized aluminum* has an oxide coating that is made deliberately thick by electrolysis. Because of the way it is formed, this coating is porous enough to accept and hold printing inks that would not otherwise stick to the aluminum.

The strong affinity of aluminum for oxygen can be shown experimentally by first dipping the aluminum into acid, which dissolves the oxide coating, and then into a solution of $HgCl_2$. In this second solution, some aluminum dissolves and metallic mercury deposits.

$$3Hg^{2+}(aq) + 2Al(s) \longrightarrow 3Hg(\ell) + 2Al^{3+}(aq)$$

The liquid mercury forms a solution (called an amalgam) with the aluminum at the surface. Aluminum oxide doesn't adhere to the amalgam, and this allows aluminum and oxygen to continue to react — a reaction that generates a great deal of heat and rapidly warms the aluminum to the point where it is too hot to hold. (See Special Topic 21.1)

Aluminum oxide is an extremely stable compound, and has a very exothermic heat of formation ($\Delta H_f^\circ = -1670$ kJ/mol). This fact accounts for some interesting uses of aluminum. For example, one of the most spectacular reactions of aluminum is called the thermite reaction in which iron oxide is reduced to metallic iron by aluminum.

The thermite reaction has $\Delta H^\circ = -847.6$ kJ.

$$2Al(s) + Fe_2O_3(s) \longrightarrow Al_2O_3(\ell) + 2Fe(\ell)$$

The large exothermic heat of formation of Al_2O_3 causes this reaction to be very exothermic, and enough heat is released to raise the temperature to nearly 3000 °C, a temperature at which the products are molten. The hot molten iron formed in the thermite reaction is often used to weld iron and steel parts together, as we see in the opening photograph in this chapter.

Another relatively recent use of the large amount of energy released by the formation of aluminum oxide is in the solid booster rockets used in launching the space shuttle. These boosters stand about 125 feet high and are 12 feet in diameter. The solid propellant in them is prepared by mixing powdered aluminum (the fuel), ammonium perchlorate (NH_4ClO_4, the oxidizer), a small amount of iron oxide powder (a catalyst), and a plastic epoxy resin binder. When first mixed, the propellant has the consistency of peanut butter and a gray color. It is poured into large casting molds and then cured slowly over a period of about 4 days to give a brick-red material with the consistency of a hard rubber eraser. When this rubbery propellant is ignited, it burns fiercely as the ammonium perchlorate oxidizes the aluminum to Al_2O_3. The large amount of energy that's released heats the gases formed by the reduction of the NH_4ClO_4 to very high temperatures. These hot gases expand with great force and help lift the space shuttle from its launch pad.

White clouds of aluminum oxide produced by burning aluminum powder billow from the solid booster rockets that lift the space shuttle *Discovery* from its launch pad at the Kennedy Space Center in Florida. The very exothermic heat of formation of Al_2O_3 is the source of the energy that provides the boosters' thrust.

$(H_2O)_5Al^{3+}$ ⟵ ... ⟹ $\left[(H_2O)_5Al \longleftarrow \ddot{O}: \right]^{2+}$ + $\left[\begin{array}{c} H \\ | \\ O - H \\ | \\ H \end{array}\right]^{+}$

FIGURE 21.14 The aluminum ion draws electrons to itself from the oxygen atoms of the neighboring water molecules. This further polarizes the O—H bonds. This polarization, shown here for one of the six water molecules, makes it easier for an H^+ to be transferred to a molecule of water in the surrounding solvent, thereby producing H_3O^+ in the solution.

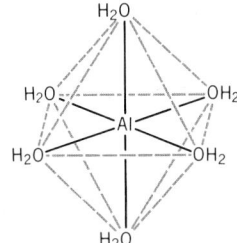

The $Al(H_2O)_6^{3+}$ ion.

In this close-up photo of the drain cleaner Drano®, we see tiny bits of aluminum metal surrounded by crystals of sodium hydroxide that are colored slightly blue with a dye. In action, the aluminum bits react with the dissolved NaOH, and the bubbles of H_2 that are formed cause a stirring effect in the clogged drain.

Aluminum is amphoteric. It dissolves in both acids and bases with the liberation of hydrogen. The reactions are often written

$$2Al(s) + 6H^+(aq) \longrightarrow 2Al^{3+}(aq) + 3H_2(g) \quad \text{(acidic solution)}$$
$$2Al(s) + 2OH^-(aq) + 2H_2O \longrightarrow 2AlO_2^-(aq) + 3H_2(g) \quad \text{(basic solution)}$$

The second reaction illustrates why oven cleaners that contain lye (NaOH) should not be used on aluminum pots and pans—the sodium hydroxide attacks the aluminum and dissolves it. This reaction is exploited, however, by the drain cleaner Drano®, which consists mostly of NaOH along with small bits of aluminum metal. When sprinkled into a clogged drain, the bubbles caused by the release of hydrogen stir the mixture and help it dissolve grease or hair that might be causing the stoppage.

The nature of the aluminum species in an aqueous solution is more complex than indicated by the equations above. For example, in acidic or neutral solutions, the aluminum ion exists as the complex ion $Al(H_2O)_6^{3+}$, in which six water molecules are bound rather tightly to the Al^{3+} ion in an octahedral arrangement. When aluminum salts crystallize from water, they usually contain this complex ion within their crystals.

Aqueous solutions that contain aluminum salts such as $AlCl_3$ or $Al_2(SO_4)_3$ are slightly acidic because of reaction of the $Al(H_2O)_6^{3+}$ ion with water. The attraction of the Al^{3+} ion for the electrons of the water molecules that surround it in the $Al(H_2O)_6^{3+}$ ion pulls some of the electron density away from the O—H bonds, as illustrated in Figure 21.14. This tends to further polarize the already polar O—H bonds of the water molecules, which increases the amount of positive charge on the hydrogen atoms. In turn, this makes it easier to remove the hydrogens as H^+ ions, and these become attached to other water molecules in the solvent to give H_3O^+ ions. The reaction is an equilibrium that can be represented by the equation

$$Al(H_2O)_6^{3+}(aq) + H_2O \rightleftharpoons Al(H_2O)_5OH^{2+}(aq) + H_3O^+(aq)$$

The addition of a base to a solution of an aluminum salt precipitates a *gelatinous* (gelatinlike) aluminum hydroxide (see Figure 21.15). This can be formulated as a neutralization reaction in which hydroxide ion strips protons from three of the water molecules that are attached to the aluminum ion, turning them into hydroxide ions.

$$Al(H_2O)_6^{3+}(aq) + 3OH^-(aq) \longrightarrow Al(H_2O)_3(OH)_3(s) + 3H_2O$$

Aluminum hydroxide is amphoteric because it dissolves in either acid or base. When it dissolves in base, the additional hydroxide ion removes still another proton from a water molecule, and when it dissolves in acid, the hydronium ion adds a proton to an

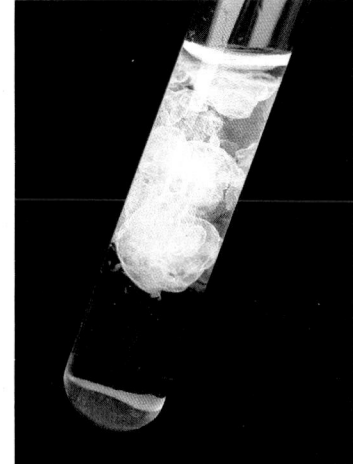

FIGURE 21.15 Addition of aqueous ammonia to a solution that contains $Al(H_2O)_6^{3+}$ makes the solution basic and causes the formation of a gelatinous precipitate of $Al(H_2O)_3(OH)_3$. The solid is usually called aluminum hydroxide and is often written without the water as $Al(OH)_3$.

OH⁻ that's attached to the Al^{3+}. This continues until the $Al(H_2O)_6^{3+}$ ion is formed.

Reaction with Base

$$Al(H_2O)_3(OH)_3(s) + OH^-(aq) \longrightarrow Al(H_2O)_2(OH)_4^-(aq) + H_2O$$

Reaction with Acid

$$Al(H_2O)_3(OH)_3(s) + H_3O^+(aq) \longrightarrow Al(H_2O)_4(OH)_2^+(aq) + H_2O$$
$$Al(H_2O)_4(OH)_2^+(aq) + H_3O^+(aq) \longrightarrow Al(H_2O)_5OH^{2+}(aq) + H_2O$$
$$Al(H_2O)_5OH^{2+}(aq) + H_3O^+(aq) \longrightarrow Al(H_2O)_6^{3+}(aq) + H_2O$$

Notice that the ion formed in basic solution, $Al(H_2O)_2(OH)_4^-$, is equivalent in terms of total atoms to $AlO_2^- + 4H_2O$. Writing the formula AlO_2^- for the aluminum-containing species in basic solution, as we did earlier, is really shorthand for the more complicated substances that actually exist in the solution.

Compounds of Aluminum   Among the important compounds of aluminum are the oxide, the halides, and the sulfates. Aluminum oxide has the formula Al_2O_3, but it occurs in two different crystalline forms that differ greatly in their chemical reactivity. If the gelatinous hydroxide is heated mildly, it loses water and gives an oxide referred to as γ-Al_2O_3.

$$2Al(H_2O)_3(OH)_3 \xrightarrow{\text{heat}} \gamma\text{-}Al_2O_3 + 9H_2O$$

This form of the oxide readily dissolves in both acidic and basic solutions. If γ-Al_2O_3 is heated to temperatures above 1000 °C, its crystal structure changes to that of α-Al_2O_3. This form of the oxide is very resistant to chemical attack. Naturally occurring α-Al_2O_3 is called *corundum*. Its crystals are very hard, and it is commonly used as an abrasive in sandpaper. Aluminum oxide has a very high melting point (2045 °C), and it is also used to make special bricks for the interiors of furnaces.

Large crystals of corundum that contain traces of certain other metals are valued as gems. For example, sapphire is Al_2O_3 that contains small amounts of iron and titanium; ruby is Al_2O_3 containing trace amounts of Cr^{3+} (Figure 21.16). Artificial sapphires and rubies are made by melting very pure aluminum oxide with carefully measured amounts of the appropriate metal oxides and then cooling the mixture in such a way that large crystals are formed. Such artificial gems are difficult to distinguish from the naturally occurring variety, and they find uses as bearings in watches and other mechanical devices, as well as in jewelry.

A compound of aluminum produced in very large quantities, much of which ultimately is converted to the hydroxide, is aluminum sulfate. It is usually made by dissolving bauxite in sulfuric acid.

The hydroxide ions attached to the aluminum in the $Al(H_2O)_3(OH)_3$ originally were water molecules.

Two other gems composed of Al_2O_3 with trace impurities giving them their colors are oriental topaz and oriental amethyst. The oriental topaz contains traces of Fe^{3+} and the oriental amethyst contains traces of Mn^{3+}.

In 1986, $Al_2(SO_4)_3$ ranked 37th in total production among industrial chemicals.

FIGURE 21.16 The gems sapphire (left) and ruby (right). Both are composed of nearly pure aluminum oxide, Al_2O_3. In the sapphire, the Al_2O_3 is contaminated by traces of iron and titanium, which give it its color. The ruby is Al_2O_3 that contains traces of chromium.

$$Al_2O_3(s) + 3H_2SO_4(aq) \longrightarrow Al_2(SO_4)_3(aq) + 3H_2O$$

When aluminum sulfate is crystallized, it forms hydrates with as many as 18 water molecules per formula unit, $Al_2(SO_4)_3 \cdot 18H_2O$.

A major consumer of aluminum sulfate is the paper industry where it is used to adjust acidity — recall that solutions of $Al_2(SO_4)_3$ are acidic — and to make the paper water resistant. Aluminum sulfate is also used in municipal water treatment where it is added to the water along with lime (CaO). The calcium oxide reacts with water to make the solution slightly basic, and this causes gelatinous aluminum hydroxide, $Al(H_2O)_3(OH)_3$, to precipitate. As the precipitate settles to the bottom of the treatment tank, it carries with it finely suspended solids and bacteria.

When a solution that contains equal numbers of moles of aluminum sulfate and sodium sulfate is evaporated, crystals are formed that have the composition $NaAl(SO_4)_2 \cdot 12H_2O$. Similar crystals are formed from solutions that contain both aluminum sulfate and either ammonium sulfate or potassium sulfate. Their formulas are $NH_4Al(SO_4)_2 \cdot 12H_2O$ and $KAl(SO_4)_2 \cdot 12H_2O$. These are examples of double salts — compounds that contain the components of two salts in a definite ratio. A double salt that has the general formula $M^+M^{3+}(SO_4)_2 \cdot 12H_2O$ is called an alum, and salts of this type include double salts formed not only by aluminum, but also by Cr^{3+} and Fe^{3+}. Under carefully controlled conditions, large well-formed crystals of the alums can be grown that are quite beautiful (see Figure 21.17).

$M^+ = Na^+, K^+, NH_4^+$
$M^{3+} = Al^{3+}, Cr^{3+}, Fe^{3+}$

Potassium alum, $KAl(SO_4)_2 \cdot 12H_2O$, is used to treat cotton fibers to make them absorb dyes more easily. Sodium alum, $NaAl(SO_4)_2 \cdot 12H_2O$, is another compound that is used along with sodium bicarbonate to make baking powders. When added to moist dough, the acidity of the aluminum ion in water causes carbon dioxide to be released by reaction of hydronium ion with the bicarbonate ion.

$$H_3O^+(aq) + HCO_3^-(aq) \longrightarrow H_2O + CO_2(g)$$

As noted in Section 20.6, the small bubbles of CO_2 cause the dough to rise and give the finished baked product a light and airy texture.

The anhydrous halides of aluminum are interesting because of similarities with the halides of beryllium; there is a substantial degree of covalent bonding between aluminum and the halide in the solid and particularly in the vapor. For example, aluminum bromide exists in the solid and vapor as Al_2Br_6 — a dimer formed from two $AlBr_3$ units. Similarly, in the vapor aluminum chloride exists in the molecular form Al_2Cl_6. The structures of these molecules are illustrated in Figure 21.18. Their formation from AlX_3 ($X = Cl$ or Br) can be viewed as an attempt by aluminum to complete its octet.

FIGURE 21.17 A crystal of potassium alum, $KAl(SO_4)_2 \cdot 12H_2O$, shown in actual size.

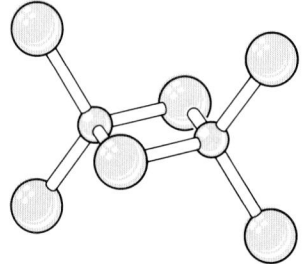

FIGURE 21.18 The molecular structure of Al_2X_6.

In the dimer, aluminum is surrounded by four pairs of electrons, and the halogens are arranged approximately at the corners of a tetrahedron with the aluminum at the center. This is, of course, what is expected based on the VSEPR theory. Note the similarity between the structure of the dimeric Al_2X_6 and the structure of the polymeric $(BeCl_2)_x$ illustrated in Figure 21.10.

When anhydrous aluminum bromide or chloride is dissolved in water, their molecules dissociate and the hydrated aluminum ion, $Al(H_2O)_6^{3+}$ is formed along with the corresponding halide ions. These reactions are very exothermic.

Gallium, Indium, and Thallium   These elements each have an electron configuration that can be represented as ns^2np^1. Each forms compounds in both the $+1$ and $+3$ oxidation states. The lower one corresponds to the loss of the outer p electron, and the higher one to the loss of the pair of s electrons as well. The most noteworthy aspect of the chemistry of these metals is the increasing tendency, going down the group, to form an ion with a $1+$ charge. The reason appears to be that as the atoms become larger, the additional energy needed to remove the pair of s electrons is more than can be easily recovered by bond formation, so as the atoms become larger going down the group there is a lesser tendency to form the higher oxidation state, and an increasing tendency to form the lower oxidation state. Thus, for Ga the $+3$ state is common and the $+1$ state is rare, whereas for Tl both the $+1$ and $+3$ states are well known.

The Metals of Groups IVA and VA

The metals of Group IVA are tin and lead. Bismuth is the only metallic element in Group V. Some of the properties of these elements are given in Table 21.5.

TABLE 21.5 **Some Properties of Tin, Lead, and Bismuth**

Element	Electron Configuration	Melting Point (°C)	Oxidation States (Most Stable in Bold Type)
Tin	[Kr] $4d^{10}5s^25p^2$	232	$+2$, **$+4$**
Lead	[Xe] $4f^{14}5d^{10}6s^26p^2$	328	**$+2$**, $+4$
Bismuth	[Xe] $4f^{14}5d^{10}6s^26p^3$	271	**$+3$**, $+5$

Tin is generally found in nature as its oxide, SnO_2, and can be reduced to the free metal by reaction with carbon (coke).

$$SnO_2 + C \longrightarrow Sn + CO_2$$

Often it is purified further electrolytically in the same manner as copper. The impure tin is made to be the anode of an electrolytic cell, and pure tin is made to be the cathode. As the cell operates, the impure anode gradually dissolves and pure tin is deposited on the cathode.

Elemental tin occurs in three allotropic forms. The most common allotrope is the one you're most familiar with. It is called *white tin* and is the shiny tin coating you find over steel on a "tin can." If tin is kept for long periods below a temperature of 13.2 °C, the white tin gradually changes to a powdery, nonmetallic form called *gray tin*. Tin objects kept in cold climates for long periods sometimes develop lumps as this change begins to take place, and at one time it was thought that some organism was attacking the tin. The phenomenon was called "tin disease." A third allotrope called *brittle tin* is obtained when white tin is heated; its properties are obvious from its name.

Lead occurs as a sulfide, PbS, in the ore *galena*. Roasting in air gives the oxide, which is then reduced by carbon to give the free metal.

$$2PbS + 3O_2 \longrightarrow 2PbO + 2SO_2$$
$$2PbO + C \longrightarrow 2Pb + CO_2$$

The lead from this process often contains silver, gold, and other metals as impurities, and it too can be purified by electrolysis. As you would expect, the silver and gold are recovered to help offset the cost of the electricity.

The Latin name for lead is *plumbum,* and the terms *plumbing* and *plumber* come from the early use of lead for pipes and pipe joints. The metal is also used to make batteries and solder and to manufacture tetraethyllead, $Pb(C_2H_5)_4$, a gasoline octane booster. As you may know, the use of lead-containing additives in gasoline is being phased out because of the environmental dangers. Special Topic 21.2 describes the problems of lead poisoning.

Tetraethyllead, $Pb(C_2H_5)_4$, and similar lead compounds have long been used to improve the performance of gasoline in vehicle engines. Although only about 3 grams of this additive are present per gallon of "leaded" gasoline, over a long enough time lead deposits would build up in engine cylinders. To prevent this, another additive—ethylene bromide—is used.

In scavanging lead from burning gasoline, ethylene bromide converts the lead into compounds volatile enough to be carried out in the exhaust. Of course, lead compounds then enter the atmosphere, from which they soon precipitate. And that's the rub. Lead compounds are poisons whether inhaled on dust or taken in food or drink. In the early 1970s, street sweepings in New York City contained over 2500 ppm of lead (as compared with the average crustal abundance of 16 ppm). Rainfall along the southern California coast was once found to have 40 μg Pb/L. (The "grounds-for-rejection" limit on drinking water is 50 μg Pb/L.)

Alarmed government health officials working with elected people brought about a law requiring new cars to burn unleaded gasoline only. Such cars also are equipped with catalytic mufflers to reduce emissions of carbon monoxide, unburned hydrocarbons, and nitrogen monoxide; and lead compounds poison the catalyst in these mufflers. So it was required that gas tanks be made that would not be able to accept the larger gas nozzles for leaded gasoline. This gasoline is still available for some vehicles.

The lead ion, like most heavy metal ions, can permanently inactivate enzymes. Some consequences in children are lower birthweights, slower development of the neuromotor system, and diminished height. Symptoms of chronic lead poisoning appear when the blood carries only 25 μg/100 mL. They include headaches, loss of appetite, stomach aches, fatigue, and irritability, which in little children are not unusual complaints and normally do not arouse much alarm. In time, however, lead in circulation leads to incurable damage of the brain, liver, and kidneys.

The U.S. Center for Disease Control estimates that 25% of all U.S. children have an excessive body burden of lead, defined as a level of over 25 μg/100 mL of a child's blood. Some specialists in heavy metal pollution at the U.S. Environmental Protection Agency, alarmed particularly by the effect of lead on nerve development, have recommended that this level be reduced to 10 μg/100 mL. Under this definition, over 80% of all U.S. children and 77% of adults would carry an excessive burden.

Among little children growing up in tenements, the eating of flakes of lead-based paints was a bizarre cause of lead-based brain disease in the 1960s, until lead compounds were banned from interior paints.

One study on adult men found that blood pressure increases with increases in the levels of lead in the blood between 10 and 20 μg/100 mL.

Bismuth is sometimes found in nature as the free metal, but usually it occurs as either the oxide or sulfide. Roasting followed by reduction with carbon gives free bismuth.

Bismuth is often an impurity in lead ores.

$$2Bi_2S_3 + 9O_2 \longrightarrow 2Bi_2O_3 + 6SO_2$$
$$2Bi_2O_3 + 3C \longrightarrow 4Bi + 3CO_2$$

It is also a by-product of the production of lead.

Bismuth is one of only a few substances that expand slightly when they freeze. This property makes bismuth ideal for making accurate castings because it expands to fill all the details of the mold. The other principal use of bismuth is in making alloys with unusually low melting points. An example is *Wood's metal*, an alloy consisting of 50% bismuth, 25% lead, 12.5% tin, and 12.5% cadmium. This alloy has a melting point of only 70 °C; it melts when dipped into boiling water! Wood's metal is used to seal the heads of overhead sprinkler systems. A fire will trigger the system automatically by melting the alloy before the temperature has risen very high.

One of the major features of the chemistries of tin, lead, and bismuth is the occurrence of two oxidation states. Tin and lead have valence shell electron configurations corresponding to ns^2np^2 and form compounds in the $+2$ and $+4$ oxidation states. These correspond to the loss of first the pair of outer p electrons and then the pair of outer s electrons. Bismuth has the valence shell configuration $6s^26p^3$; its two oxidation states are $+3$ and $+5$.

Sn $[Kr]4d^{10}5s^25p^2$
Sn^{2+} $[Kr]4d^{10}5s^2$
Sn^{4+} $[Kr]4d^{10}$

As with the elements in Group IIIA, the lower oxidation state becomes increasingly more stable relative to the higher one as we go down the group. Thus, for tin we find many compounds for both the $+2$ and $+4$ states, but the $+2$ state is easily oxidized to the $+4$ state. (Perhaps you remember from Chapter 12 that Sn^{2+} is a useful mild reducing agent for laboratory reactions.) On the other hand, lead in the $+4$ oxidation state is a powerful oxidizing agent, indicating the tendency of lead(IV) to be reduced to the more stable lead(II).

The relative stabilities of the oxidation states of tin and lead can also be seen in their reactions with acids. With a nonoxidizing acid, tin forms the $+2$ ion.

$$Sn(s) + 2HCl(aq) \longrightarrow SnCl_2(aq) + H_2(g)$$

But with nitric acid, an oxidizing acid, the product is tin(IV) oxide, SnO_2.

$$Sn(s) + 4HNO_3(aq) \longrightarrow SnO_2(s) + 4NO_2(g) + 2H_2O$$

Lead forms only the $+2$ state, even when strong oxidizing acids attack it.

$$3Pb(s) + 8HNO_3(aq) \longrightarrow 3Pb(NO_3)_2(aq) + 2NO(g) + 4H_2O$$

The behavior of these two metals toward HNO_3 reflects the greater stability of lead(II) compared with tin(II).

Besides dissolving in acids, tin and lead also dissolve in base. The reactions are

$$Sn(s) + 2OH^-(aq) + 2H_2O \longrightarrow Sn(OH)_4{}^{2-}(aq) + H_2(g)$$
$$\text{stannite ion}$$

$$Pb(s) + 2OH^-(aq) + 2H_2O \longrightarrow Pb(OH)_4{}^{2-}(aq) + H_2(g)$$
$$\text{plumbite ion}$$

SnO and PbO are amphoteric; they dissolve in acids (e.g., HNO_3) and bases (e.g., NaOH).

The stannite ion is very easily oxidized and is therefore a strong reducing agent. The plumbite ion, on the other hand, is much more difficult to oxidize, which once more illustrates the relative stabilities of the oxidation states of these two metals.

For bismuth, the $+3$ state is much more stable than the $+5$ state. Oxidation using O_2 or Cl_2 gives only Bi_2O_3 and $BiCl_3$, for example. Oxidation of Bi_2O_3 to Bi_2O_5 is very difficult and requires a very strong oxidizing agent and special conditions. The compound $BiCl_5$ doesn't exist because bismuth(V) is such a strong oxidizing agent it would oxidize Cl^- to Cl_2.

Compounds of Tin, Lead, and Bismuth

The properties of $SnCl_4$ are those of a molecular substance, not an ionic one.

Tin forms halides in both its $+2$ and $+4$ oxidation states; their general formulas are SnX_2 and SnX_4, where X stands for a halogen. One of the most important of these is tin(II) fluoride (stannous fluoride), SnF_2, an ingredient in many fluoride toothpastes. The tin(IV) halides are all covalent. An example is $SnCl_4$, a colorless liquid that freezes at $-33\ °C$ and boils at $114\ °C$. The covalent nature of the Sn—Cl bonds in $SnCl_4$ can be explained in the same way as the covalent Be—Cl bonds in $BeCl_2$: The small size and high charge of a true Sn^{4+} ion would distort the electron cloud of the chloride ion to such an extent that the electron density drawn between the two nuclei would constitute a covalent bond. Of course, we can also view the bonds in $SnCl_4$ as ordinary covalent bonds formed in the usual way.

$$\cdot Sn \cdot + 4 \cdot \overset{\cdot\cdot}{\underset{\cdot\cdot}{Cl}} \colon \longrightarrow \colon \overset{\cdot\cdot}{\underset{\cdot\cdot}{Cl}} - \overset{\overset{\displaystyle :\overset{\cdot\cdot}{Cl}:}{|}}{\underset{\underset{\displaystyle :\overset{\cdot\cdot}{Cl}:}{|}}{Sn}} - \overset{\cdot\cdot}{\underset{\cdot\cdot}{Cl}} \colon$$

In either case, the result is the same.

The reaction of tin with oxygen gives tin(IV) oxide, SnO_2, rather than SnO. Tin(IV) oxide is used to give glass a transparent electrically conducting surface. Such conducting glass is used in electronic displays. Another oxygen-containing compound of tin is called bis(tributyltin) oxide. Its formula is $[(C_4H_9)_3Sn]_2O$ and it is used in wood-treatment products that prevent rot and in antifouling paints that are applied to boat hulls to prevent the growth of marine organisms such as barnacles.

Lead forms halides in the +2 state, but in the +4 state only the chloride is relatively stable. Lead(IV) is such a powerful oxidizing agent that Br^- and I^- are oxidized by it, so $PbBr_4$ is very unstable and the existence of PbI_4 is doubtful. In fact, even $PbCl_4$ decomposes easily by an internal redox reaction.

$$\overset{\displaystyle\overset{\text{(reduction)}}{\overbrace{\qquad\qquad}}}{\underset{\underset{PbCl_4 \xrightarrow{\hspace{3cm}} Cl_2 + PbCl_2}{}}{\underset{+4\ \ -1\xrightarrow{\text{(oxidation)}}0\qquad +2}{}}}$$

Among the more useful compounds of lead are the oxides. When heated in air, lead forms PbO—a yellow powder called *litharge* that is used in pottery glazes and in making fine lead crystal. If PbO is heated in air it can be oxidized further to give *red lead*—a mixed lead(II)/lead(IV) oxide with the formula Pb_3O_4. This oxide is used in corrosion-inhibiting coatings applied to structural steel. Treatment of solutions of plumbite ion, $Pb(OH)_4^{2-}$, with hypochlorite ion produces yet another oxide, PbO_2. This compound is a strong oxidizing agent, and its most common use is as the cathode material in the lead storage battery.

For many years, a common pigment in paints was a white basic carbonate of lead, $Pb_3(OH)_2(CO_3)_2$, known generally as "white lead." Its use is now severely restricted because of the high toxicity of lead (as mentioned in Special Topic 21.2). As you probably know, children living in old buildings where there are remnants of former coats of lead-based paints have often shown signs of lead poisoning, presumably caused by eating chips of the lead-containing paint. As a pigment, white lead has excellent covering power, but is gradually darkened by exposure to air. This is caused by small amounts of hydrogen sulfide in the air formed by decomposition of organic matter. The H_2S reacts with the lead-based pigment forming black lead sulfide, PbS. Another lead-containing pigment used in artists' oil paints is lead chromate, $PbCrO_4$, which has a bright yellow color.

Bismuth(III) fluoride is ionic, but its other halides, such as $BiCl_3$, are covalent. In water, $BiCl_3$ hydrolyzes—it reacts with water—to give a precipitate having the formula BiOCl (named *bismuthyl chloride*).

$$BiCl_3 + H_2O \longrightarrow BiOCl(s) + 2H^+ + 2Cl^-$$

Other bismuth salts, such as its nitrate, $Bi(NO_3)_3$, and sulfate, $Bi_2(SO_4)_3$, also react with water to form the bismuthyl ion, BiO^+. In fact, a compound formed this way from the nitrate and having the approximate formula $BiO(NO_3)$ (the exact formula depends on how it is prepared) is called bismuth subnitrate and is used medicinally as an antacid in the treatment of gastric ulcers. Other bismuth compounds are also used in both the cosmetic and pharmaceutical industries and, combined, they account for approximately 30% of the bismuth produced each year.

When bismuth is heated in air, its oxide Bi_2O_3 is formed. As noted earlier, this can be further oxidized under extreme conditions to give bismuth in the +5 oxidation state. Compounds such as sodium bismuthate, $NaBiO_3$, which contain bismuth in the +5 oxidation state, are extremely powerful oxidizing agents. For example, the usual analytical test for manganese ion involves treating a solution made acidic with HNO_3 with $NaBiO_3$. Any Mn^{2+} in the solution is oxidized by permanganate ion, MnO_4^-, whose presence is easily detected because of its intense purple color (see Figure 21.19). The reaction is

$$14H^+ + 5BiO_3^- + 2Mn^{2+} \longrightarrow 2MnO_4^- + 5Bi^{3+} + 7H_2O$$

Many states have recently banned the use of bis(tributyltin) oxide in antifouling bottom paints for boats because of its apparent toxic effects on shellfish.

The halides $PbCl_2$, $PbBr_2$, and PbI_2 are relatively insoluble in water, and become less soluble going from the chloride to the iodide.

Pb_3O_4 can be formulated as $(PbO)_2 \cdot (PbO_2)$.

Red lead, Pb_3O_4.

FIGURE 21.19 In acid, yellow $NaBiO_3$ oxidizes the nearly colorless Mn^{2+} ion to the violet colored MnO_4^- ion.

SUMMARY

Metals.   Metals form a metallic lattice; they have too few electrons to form enough covalent bonds to achieve an octet. Metals tend to form ionic compounds with nonmetals. Within the periodic table, elements become more metallic moving from right to left in a period and from top to bottom in a group.

Metallurgy.   Some unreactive metals, such as gold, are found free in nature, but most are combined with other elements. Sodium and magnesium are extracted from seawater. Manganese nodules on the ocean floor are a potential source of manganese and iron. Usually, when an ore is dug from the ground, the metal-bearing component must be enriched by a pretreatment step that removes much of the gangue. Flotation is often used with lead and copper sulfide ores. Sulfide ores are usually roasted to convert them to oxides, which are more easily reduced. Aluminum's amphoteric character is exploited in purifying bauxite.

Metals are obtained from their compounds by reduction. Active metals such as sodium, magnesium, and aluminum must generally be prepared by electrolysis. Carbon is a common chemical reducing agent because it is plentiful and inexpensive. Metallic iron forms in a blast furnace where a charge of iron ore, limestone, and coke (which is made by heating coal strongly in the absence of air) reacts in a stream of heated air. Molten iron and slag flow to the bottom of the furnace and are periodically tapped. Steel is made from this impure iron by removing impurities and adding special metals. The open hearth furnace has been replaced by the basic oxygen process as the most frequently used method for making steel.

The Alkali Metals.   Sodium and potassium are the most common of the alkali metals, all of which are soft, low melting, and very reactive. Lithium has an exceptionally negative reduction potential because of its small size, which causes it to have a very large hydration energy. Sodium is prepared by electrolysis; the heavier alkali metals are displaced from their chlorides by sodium vapor.

All the alkali metals react with water to form H_2 and the metal hydroxide. Only lithium reacts with N_2 to form lithium nitride, and only lithium forms the normal oxide on reaction with O_2. Sodium forms the peroxide, Na_2O_2, and the rest form superoxides, MO_2. Hydrolysis of the peroxide gives H_2O_2; hydrolysis of the superoxide gives H_2O_2 and O_2.

The most important compounds are the chlorides, hydroxides, and carbonates. NaCl is the most plentiful, so sodium compounds are much less expensive than potassium compounds. Electrolysis of aqueous NaCl gives NaOH. KOH is made the same way. Sodium carbonate is made from trona, $Na_2CO_3 \cdot NaHCO_3 \cdot 2H_2O$ and by the Solvay process. (Study the chemistry of the Solvay process on page 888.) $NaHCO_3$ (baking soda) decomposes on heating to give Na_2CO_3.

Sodium salts of strong acids tend to be more soluble in water than the potassium salts; potassium salts of weak acids tend to be more water soluble than the sodium salts. Sodium salts impart a bright yellow color to a flame, lithium salts give a brilliant red, and potassium salts give a violet color.

Alkaline Earth Metals.   These metals are harder, higher melting, and less reactive than their neighbors in Group IA. Only magnesium and beryllium are made in commercial quantities, because these are the only two that don't react with air and moisture. Magnesium is obtained from dolomite and seawater; the metal is formed in the electrolysis of $MgCl_2$. Beryllium is obtained from beryl (which forms the gem stones emerald and aquamarine). Electrolysis of $BeCl_2$ with NaCl as an electrolyte gives Be. Limestone ($CaCO_3$), dolomite ($CaCO_3 \cdot MgCO_3$), and gypsum ($CaSO_4 \cdot 2H_2O$) are rich sources of calcium.

Beryllium does not react with water because of its tough insoluble oxide coating. Magnesium reacts with hot water and steam, but not with cold water. Its surface is also protected by its oxide. Calcium, strontium, and barium react with water to give H_2 and the metal hydroxide. All beryllium compounds are covalent because of the small size of the Be^{2+} ion. Beryllium and its oxide are amphoteric. Magnesium, like lithium, is the only member of its group that reacts with N_2 to give the nitride.

All the chlorides are water soluble and except for $BeCl_2$, which is covalent and exists in the solid state as polymeric $(BeCl_2)_x$, all are ionic. $CaCl_2$ is deliquescent. The oxides are usually made by decomposition of the carbonate. Calcium oxide is called lime and reacts with water to form the hydroxide (slaked lime). Magnesium hydroxide is in milk of magnesia. The hydroxides increase in solubility from $Mg(OH)_2$ to $Ba(OH)_2$. The sulfates decrease in solubility from $MgSO_4$ to $BaSO_4$. Gypsum can be partially dehydrated to give plaster of paris. Barium sulfate is used to obtain medical X-ray photographs of the intestinal tract.

Calcium salts give a red-orange color to a flame, strontium salts give a crimson color, and barium salts give a yellow-green color.

Aluminum.   Aluminum is a very common element in the earth's crust, but can only be obtained easily from a few minerals. Its chief ore is bauxite, Al_2O_3. Aluminum is easily oxidized to the $+3$ state, and in air it forms an oxide coating that protects the metal from further attack. In the thermite reaction, the large heat of formation of Al_2O_3 is used to reduce iron to the molten state. Aluminum is amphoteric. In acidic solution or neutral solution it exists as the complex ion $Al(H_2O)_6^{3+}$. This ion is itself a weak acid. When made basic, a gelatinous precipitate with the formula $Al(H_2O)_3(OH)_3$ is formed that can be dissolved in either acid or base.

Aluminum oxide occurs in two crystalline forms, γ-Al_2O_3, which is soluble in acids and bases, and α-Al_2O_3 (known also as corundum), which is quite inert. Gems such as ruby and sapphire are composed of α-Al_2O_3. Aluminum sulfate is made in large quantities by dissolving bauxite in sulfuric acid, and forms double salts called alums, for example, $KAl(SO_4)_2 \cdot 12H_2O$.

Aluminum is similar in many ways to beryllium. Its anhydrous chloride and bromide are dimeric with the general formula Al_2X_6.

Post-transition Metals.   These elements are characterized by two oxidation states. The lower one comes from losing the outer p electron(s) and the higher comes from the further loss of the outer pair of s electrons. Going down a group, the lower oxidation state becomes progressively more stable relative to the higher one. In Group IIIA, the oxidation states observed are +1 and +3. In Group IVA they are +2 and +4, and in Group VA, bismuth has +3 and +5 oxidation states.

Tin is made from its oxide SnO_2 by reduction with carbon and is purified electrolytically. It forms three allotropes. White tin is metallic; gray tin has the diamond structure and has nonmetallic properties. When heated, the metallic allotrope changes to brittle tin. Tin reacts with oxygen and oxidizing acids such as HNO_3 to form SnO_2. Two kinds of halides are formed, SnX_2 and SnX_4. The tetrahalides such as $SnCl_4$ are covalent. For tin, the +4 oxidation state is easily formed from the +2 state.

Lead is made from its sulfide PbS which is first roasted to give the oxide PbO and then reduced with carbon. It is also purified electrolytically. Lead only occurs in a metallic form. Three oxides are known: PbO (litharge), PbO_2, and a mixed oxide Pb_3O_4 (called red lead). The +2 state is much more stable than the +4 state for lead, and lead(IV) compounds are very strong oxidizing agents. The halides PbX_2 can be made and are relatively insoluble in water. $PbBr_4$ and PbI_4 are very unstable because lead(IV) is so strong an oxidizing agent; $PbCl_4$ decomposes slowly.

Bismuth is obtained from its oxide or sulfide. The sulfide is roasted to give the oxide and then reduced with carbon. Bismuth is also a by-product of the production of lead. The metal is used to make low-melting alloys such as Wood's metal. Bismuth reacts with oxygen and chlorine to form Bi_2O_3 and $BiCl_3$, respectively. The chloride, $BiCl_3$, hydrolyzes in water to give a precipitate of bismuthyl chloride, BiOCl. The +5 state can only be made under severely oxidizing conditions, and bismuth(V) is a very powerful oxidizing agent. Sodium bismuthate, $NaBiO_3$, is used in the qualitative test for manganese where it oxidizes Mn^{2+} to MnO_4^-.

REVIEW EXERCISES

Answers to questions whose numbers are printed in color are given in Appendix C.

Metallic Character and the Periodic Table

21.1 What is a metallic lattice? Why do metals form this kind of solid?

21.2 How does metallic character vary (a) going from left to right across a period? (b) going from top to bottom in a group?

21.3 Choose the more metallic element in each pair: (a) Ga or Tl, (b) Ba or Tl, (c) Mg or K.

21.4 Which compound is most ionic — $MgCl_2$, $AlCl_3$, or RbCl?

21.5 Which oxide is more basic — BeO or MgO?

Metallurgy

21.6 What is *metallurgy*?

21.7 What is an ore? What distinguishes an ore from some other potential source of a metal?

21.8 Why are sodium and magnesium extracted from seawater?

21.9 Give the chemical reactions for the separation of metallic magnesium from seawater.

21.10 What is lime? How is it made from $CaCO_3$?

21.11 What are manganese nodules? Why are they a potential source of metals? Why aren't they being mined now?

21.12 What is *gangue*?

21.13 Why can gold be separated from impurities of rock and sand by *panning*?

21.14 Describe the *flotation process*.

21.15 Write chemical equations for the reactions that occur when Cu_2S and PbS are roasted in air. Write a chemical equation to show how SO_2 from the roasting can be kept from being released to the environment. Why might a sulfuric acid plant be located near a plant that roasts sulfide ores?

21.16 Write chemical equations that show how bauxite is purified.

21.17 Why is reduction, rather than oxidation, necessary to extract metals from their compounds?

21.18 Sodium, magnesium, and aluminum are produced by electrolysis instead of by reduction with chemical reducing agents. Why?

21.19 Why is carbon such a useful industrial reducing agent?

21.20 Write chemical equations for the reduction of PbO and CuO with carbon.

21.21 Copper(I) sulfide can be converted to metallic copper without adding a reducing agent. Explain this using appropriate chemical equations.

21.22 Why is a blast furnace called a *blast* furnace?

21.23 What is the composition of the charge that's added to a blast furnace?

21.24 Describe the chemical reactions involved in the reduction of Fe_2O_3 that take place in a blast furnace. What is the active reducing agent in the blast furnace?

21.25 What is *slag*? Write a chemical equation for its formation in a blast furnace. What are some of its uses?

21.26 What does *refining* mean in metallurgy?

21.27 What is the difference between pig iron and steel?

21.28 What is a Bessemer converter? Why was it replaced by the open hearth furnace?

21.29 Describe the operation of an open hearth furnace. What reactions take place during the production of steel?

21.30 Describe the basic oxygen process. Why has it replaced the open hearth furnace as the chief steel-making method?

The Alkali Metals

21.31   Write the electron configurations of (a) Li, (b) K, and (c) Fr.

21.32   What are the sources of the alkali metals? Why aren't these elements found free in nature?

21.33   What is saltpeter? What is Chile saltpeter?

21.34   Why is francium such a rare metal?

21.35   Which of the alkali metals are most important biologically? Which is most important in plants?

21.36   Why are the alkali metals soft and low melting?

21.37   Which alkali metal has been used as a coolant in a nuclear reactor? Why?

21.38   Explain why lithium has an unusually negative standard reduction potential. In the absence of water, which is expected to be the better reducing agent, lithium or sodium? Why?

21.39   How is metallic sodium prepared? Why can molten KCl be reduced by sodium vapor?

21.40   Complete and balance the following chemical equations.
(a) $Li + H_2O \longrightarrow$
(b) $K + H_2O \longrightarrow$
(c) $Rb + H_2O \longrightarrow$

21.41   Complete and balance the following chemical equations.
(a) $Li + O_2 \longrightarrow$
(b) $Na + O_2 \longrightarrow$
(c) $Rb + O_2 \longrightarrow$

21.42   Complete and balance the following chemical equations.
(a) $Na_2O_2 + H_2O \longrightarrow$
(b) $Na_2O + H_2O \longrightarrow$
(c) $KO_2 + H_2O \longrightarrow$
(d) $NaOH + CO_2 \longrightarrow$

21.43   Why is KO_2 used in a recirculating breathing apparatus?

21.44   Which alkali metal reacts with N_2? Write the chemical equation for the reaction.

21.45   What are the formulas for (a) caustic soda and (b) lye?

21.46   What reactions take place at the anode and cathode in the mercury cell during the electrolysis of aqueous NaCl? Write a chemical equation that shows how NaOH is formed in this apparatus. (If necessary, refer back to Chapter 17.)

21.47   What is the principal use of Na_2CO_3? Give the name and formula for the ore used to obtain Na_2CO_3. Use chemical equations to describe how Na_2CO_3 is made by the Solvay process.

21.48   Write a chemical equation that shows how $NaHCO_3$ functions during the baking of bread.

21.49   Write chemical equations that show how $NaHCO_3$ functions as a buffer.

21.50   Use a chemical equation to show why solutions of Na_2CO_3 are basic. Calculate the pH of a 1.0 M solution of Na_2CO_3.

21.51   Which salt would you expect to be more water-soluble in each of the following pairs? (a) $NaClO_4$ or $KClO_4$, (b) $NaC_2H_3O_2$ or $KC_2H_3O_2$, (c) $KC_2H_3O_2$ or $KClO_4$.

21.52   What color flame is produced by (a) Na^+, (b) K^+, and (c) Li^+?

21.53   In April 1983, 20,000 gallons of nitric acid spilled from a ruptured railroad car at a rail siding near downtown Denver, Colorado. Giant snowblowers were used to blow soda ash on the spill to neutralize the acid. Write a chemical equation for this neutralization reaction.

21.54   Given that concentrated nitric acid has a concentration of 16 M, how many pounds of soda ash were required to completely react with all the HNO_3 spilled by the tank car described in Review Exercise 21.53?

Alkaline Earth Metals

21.55   Write the electron configurations of (a) Mg and (b) Sr.

21.56   Are the alkaline earth metals ever found free in nature? Explain your answer.

21.57   Give the formulas for (a) limestone, (b) dolomite, (c) gypsum, and (d) beryl.

21.58   What is the source of radium?

21.59   Which gems are composed of beryl?

21.60   Why are the alkaline earth metals harder and higher melting than the alkali metals alongside? How does the ease of oxidation of the Group IIA metals compare with that of the Group IA metals?

21.61   How are the alkaline earth metals prepared?

21.62   Which alkaline earth metals react with cold water? Which react only with hot water? Which do not react with water at all?

21.63   Which are the only alkaline earth metals to be prepared in significant quantities? Why these and not the others?

21.64   How is beryllium metal prepared?

21.65   Why is beryllium added to copper or bronze? What is a typical use for one of these alloys?

21.66   Show by means of chemical equations how magnesium is extracted from dolomite.

21.67   Which alkaline earth metal is used in flashbulbs and flares? Why? Explain using a chemical equation.

21.68   How are the properties of magnesium and lithium alike? What is the apparent reason? How do the properties of magnesium and lithium differ?

21.69   Why is $BeCl_2$ covalent, but $MgCl_2$ is ionic? How does the tendency to form covalent bonds vary among the Group IIA elements?

21.70   Describe the structure of $BeCl_2$ in the solid state.

21.71   What compound of a Group IIA element is found in milk of magnesia? Write a chemical equation to show how this compound neutralizes stomach acid, HCl.

21.72   How do the solubilities of the Group IIA sulfates and hydroxides vary going from top to bottom in the group?

21.73   Define *deliquescence*.

21.74 Give two reactions for the formation of MgO.

21.75 How can $Ca(OH)_2$ and $Mg(OH)_2$ be prepared?

21.76 Give uses for the following:
(a) $CaSO_4 \cdot 2H_2O$ (d) $MgSO_4 \cdot 7H_2O$ (g) Be
(b) $CaCO_3$ (e) $CaCl_2$ (h) Mg
(c) $BaSO_4$ (f) $Mg(OH)_2$

21.77 How is plaster of paris made? Write its formula and write a chemical equation for the reaction of plaster of paris with H_2O.

21.78 What color flame is produced by (a) Ca^{2+}, (b) Sr^{2+}, and (c) Ba^{2+}?

Aluminum

21.79 Write the electron configurations for (a) Al and (b) Al^{3+}.

21.80 How does the abundance of aluminum in the earth's crust compare with (a) that of other elements and (b) that of other metals?

21.81 How is aluminum extracted from its ores?

21.82 What is cryolite? What ions are present in it?

21.83 Aluminum is a very reactive metal. Why doesn't it experience rapid and extensive corrosion in air?

21.84 What is *anodized aluminum*?

21.85 Write the chemical equation for the thermite reaction.

21.86 Why is aluminum used as a fuel in the rocket propellant in the booster rockets of the space shuttle?

21.87 Write net ionic equations for the reactions of aluminum with aqueous solutions of (a) HCl and (b) NaOH.

21.88 What is the aluminum-containing species that exists in acidic or neutral solutions? Why do many aluminum salts crystallize from aqueous solutions as hydrates?

21.89 Why is the $Al(H_2O)_6^{3+}$ ion acidic? Write chemical equations that show what happens to the $Al(H_2O)_6^{3+}$ ion when a base is added to a solution that contains it.

21.90 Write chemical equations that show what happens when aluminum hydroxide is dissolved in (a) a base and (b) an acid.

21.91 Why is aluminum hydroxide used in water purification?

21.92 In a strongly basic solution, aluminum exists as an ion that can be written as AlO_2^-. What is a better way of writing the formula for this aluminum-containing ion?

21.93 What differences are there in the properties of γ-Al_2O_3 and α-Al_2O_3? What is the difference between ruby and sapphire?

21.94 What is an *alum*? Why are alums called double salts? Write the chemical formulas for sodium alum and potassium alum. What are some of their uses?

21.95 Sketch the structure of the dimer of $AlCl_3$. How are the chlorines arranged geometrically around the aluminum atoms?

21.96 How is the structure of Al_2Cl_6 similar to the structure of $BeCl_2$ in the solid state? Specify one other way the chemistries of beryllium and aluminum are similar.

Other Metals in Groups IIA, IVA, and VA

21.97 Which are the post-transition metals? Write their electron configurations.

21.98 What oxidation states are observed for (a) Tl, (b) Sn, (c) Pb, and (d) Bi? Account for them in terms of their electron configurations.

21.99 How do the relative stabilities of the oxidation states of Ga, In, and Tl vary? How is this accounted for?

21.100 In general, why are the reactivities of the post-transition metals lower than the reactivities of the metals in Groups IA and IIA?

21.101 Write chemical equations for the preparation of tin, lead, and bismuth from their ores.

21.102 What are the allotropes of tin? What are their properties?

21.103 What property of bismuth makes it useful for making castings? What is Wood's metal? What property is special about this alloy?

21.104 Why is $SnCl_4$ covalent, but $SnCl_2$ is ionic?

21.105 Write chemical equations for the reactions of tin and lead with HNO_3.

21.106 Which is a better oxidizing agent, SnO_2 or PbO_2?

21.107 Write the chemical formulas for the stannite ion and the plumbite ion. Which is the better reducing agent?

21.108 Write net ionic equations for the reactions of a strong base with (a) Sn, (b) SnO, (c) Pb, and (d) PbO.

21.109 Write chemical equations for the reactions of bismuth with (a) O_2 and (b) Cl_2.

21.110 Use the data in Table 14.2 to calculate the molar solubilities in water of (a) $PbCl_2$, (b) $PbBr_2$, and (c) PbI_2.

21.111 Why doesn't $BiCl_5$ exist?

21.112 Why are $PbBr_4$ and PbI_4 so unstable? Write an equation for the decomposition of $PbBr_4$.

21.113 What is the formula for "white lead"? Why does it gradually darken in air when it is used as a paint pigment?

21.114 What lead-containing pigment used by artists is yellow?

21.115 What are the oxides of lead? Give one use of each.

21.116 Write the chemical equation for the hydrolysis of $BiCl_3$. Explain why adding concentrated HCl to a solution that contains this hydrolysis reaction causes BiOCl to dissolve.

21.117 What chemical reaction is involved in the qualitative test for Mn^{2+}? Describe what is observed during the test.

CHEMICALS IN USE
IRON AND STEEL

When ancient peoples discovered iron and how to free it from red-colored "earths," they stumbled onto a material much stronger and better able to form a sharp edge than bronze or brass. They found a metal that was also more abundant than copper, tin, or zinc. These developments created opportunities for tools and weapons of such value that historians have honored their impact on history, as they have recognized earlier periods dominated by stone or bronze, by calling the third age of civilization the Iron Age.

Over 90% of all the metal refined in the world is iron, and nearly all goes into the manufacture of various kinds of steel. The principal ores of iron are various oxides such as hematite (Fe_2O_3), magnetite (Fe_3O_4), and limonite (mostly $2Fe_2O_3 \cdot 3H_2O$), so iron refining requires that these be reduced. As described in Chapter 21, the reducing agent is carbon monoxide, which is generated right within the blast furnace by the incomplete oxidation of carbon in coke. The net chemical equation for the reduction of hematite, for example, is

$$Fe_2O_3 + 3CO \longrightarrow 2Fe + 3CO_2$$

Steel

Iron emerges from the blast furnace in the molten state and it contains a variety of impurities such as carbon, manganese, phosphorus, silicon, and sulfur, some of which render the resulting pig iron too brittle for most uses. This is why the molten metal, before being allowed to cool, is further treated. The chief impurity is carbon (3–4%), and nearly all of it is removed by oxidation. One of the principal industrial uses of oxygen is for this purpose, as discussed in Sections 19.4 and 21.2. The product is steel, but "steel" is a broad name, like "trees." There are many different steels, but all are alloys of iron.

Steel Alloys

Even ancient people realized how much naturally occurring metals could be improved for various purposes by melting them and mixing them. Copper and tin, for example, are used to make bronze, an alloy that is harder and tougher than copper. Copper and zinc are used to prepare brass, which is also stronger than copper. These experiences in improving metals led to extensive studies in im-

proving iron. The result has been three broad classes of steel alloys.

The simplest steel is iron alloyed with small amounts of carbon — generally ranging from about 0.03 to 1.5%. These are called *carbon steels* and may also contain small amounts of other metals such as aluminum and copper. They are relatively inexpensive to make directly from the hot metal that comes from the blast furnace and find uses in many industrial and consumer products such as auto bodies, appliances, knives, various kinds of machinery, ships, containers, and structural steel in buildings.

Somewhat more expensive to make are *low-alloy steels,* which contain from 1 to 5% of metals such as nickel, chromium, manganese, vanadium, molybdenum, titanium, and tungsten. These metals each tend to produce certain special properties when incorporated into the steel. For example, if the end use requires great strength in resisting pulling stresses, then nickel is alloyed with iron. Nickel steel is used, for example, in building long-span bridges, in making armor plate, and in manufacturing bicycle chains.

Where the metal must function at high temperatures, tungsten or vanadium steel is often used. Tool steels containing only carbon as the alloying element lose their ability to hold an edge when used at high speeds, where operating temperatures can cause the tool to glow a dull red. Tungsten and vanadium alloys, however, hold an edge even at these high temperatures, and high-speed metal-cutting tools and drills (Figure 15a) are often made of such alloys.

If the final product must withstand constant high-energy impact battering — the dipper teeth of power shovels used in rock- and earth-moving machinery, for example — then manganese steel is a good choice. Manganese steel is also used to make rifle barrels.

Steel with a trace of niobium has great strength that holds over long periods at high temperatures, and it is used in fabricating atomic reactors.

High-alloy steels are the most expensive to make and their applications depend on their unique properties. Most familiar are the stainless steels, which are used for cooking utensils, chemical manufacturing equipment, and a host of applications where resistance to corrosion is critical. These alloys owe their special properties to chromium present in concentrations that range from about 12 to 18%. Nickel is

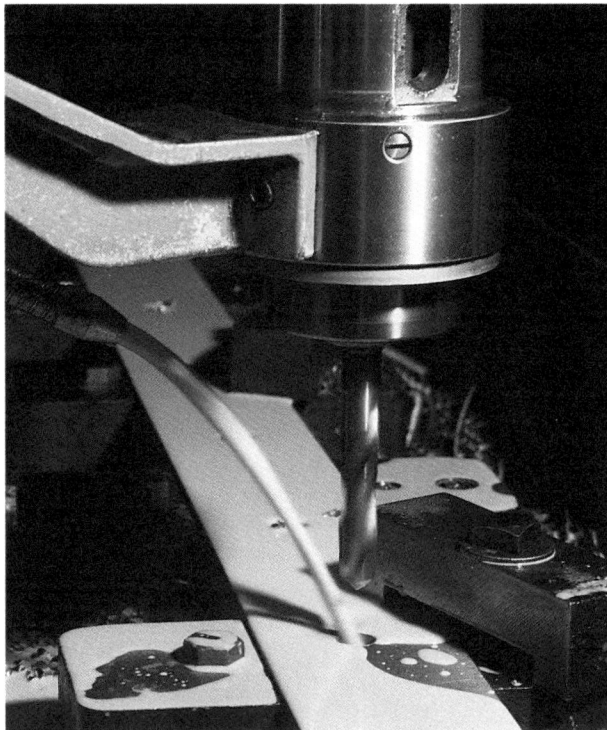

FIGURE 15a A tungsten steel drill operating at high speed is withdrawn from a freshly drilled hole in a steel plate.

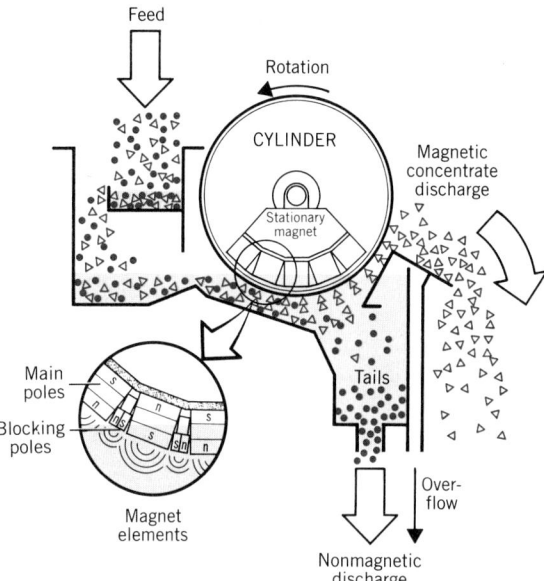

FIGURE 15b An important step in converting taconite rock into taconite ore is the magnetic separation of the desired mineral, magnetite, from the nonmagnetic substances. The final product of the overall operation consists of small spheres of high-grade iron "ore."

also used in these alloys to increase their strength and may be present in concentrations up to about 10%.

If you think about all of the mechanical and electrical devices that depend on various steel alloys and about the many metallic elements that are essential to these alloys — and our discussion by no means covers them all — you can see how much our civilization relies on these substances. Not every industrialized nation has adequate supplies of all these alloying metals for its own needs, so all nations, developed and undeveloped alike, depend on each other.

Taconite

To reduce dependence on scarce resources, a nation can turn to substitutes for as many uses as possible. Today, many plastics have taken the place of metals, for example. Another strategy is to develop new supplies of resources from materials at hand. For example, as the deposits of high-percentage iron ores in the great iron ranges of northern Minnesota dwindled, scientists and engineers at the University of Minnesota developed ways to use a lower-grade mineral, taconite. This very hard rock occurs in great abundance, but it contains just 23% iron, compared with the high-grade ores with over 50% iron. The chief iron mineral in taconite is magnetite, Fe_3O_4, so named because it is attracted to a magnet. This property is exploited in changing taconite into something that existing blast furnaces can use.

Taconite rock is crushed, pulverized, and then carried past a turning cylinder that holds powerful magnets, as seen in Figure 15b. These pull the magnetic ore particles from the nonmagnetic materials (mostly silicate minerals). Next the magnetic powder is mixed with a clay binder, some hard coal, and moisture, and then baked in a rotating drum. The heat changes the magnetite to nonmagnetic hematite, Fe_2O_3, and the rotating action creates golfball-sized spheres. The iron content is now 60 to 65%. Because of its uniform quality in both iron content and physical shape, this modified form of taconite works exceptionally well in blast furnaces. Until scientists and engineers developed this process, however, taconite was nothing more than rock.

Questions

1. Write chemical formulas for (a) hematite, (b) magnetite, and (c) limonite.

2. Write a chemical equation for the reduction of magnetite by carbon monoxide. What is the source of carbon monoxide in the blast furnace?

3. Each mole of carbon in coke could lead to the production of how many moles of iron in the reduction of hematite? Assuming 100% efficiency, how many tons of iron could be made from Fe_2O_3 and one ton of coke?

4. What are the three classes of steel alloys? What special properties are imparted to steel by (a) tungsten, (b) manganese, (c) nickel, and (d) chromium? What elements are present in stainless steel?

5. If a 100.0-g sample of refined taconite is 65.00% iron, how many grams of Fe_2O_3 are present in the sample?

The familiar face of the Statue of Liberty is colored green by a compound of copper, $Cu_2(OH)_2CO_3$, formed by the gradual corrosion of the statue's copper skin. Copper is one of many transition metals that find a myriad of uses, both decorative and structural, in our society. The chemistries of copper and other important transition metals are discussed in this chapter.

TRANSITION METALS
AND THEIR COMPLEXES

THE TRANSITION METALS: GENERAL
    CHARACTERISTICS AND PERIODIC TRENDS
PROPERTIES OF SOME IMPORTANT TRANSITION METALS

COMPLEXES OF THE TRANSITION METALS
ISOMERS OF COORDINATION COMPOUNDS
BONDING IN COMPLEXES

22.1 THE TRANSITION METALS: GENERAL CHARACTERISTICS AND PERIODIC TRENDS

Metals with partially filled *d* subshells often exhibit more than one oxidation state, have compounds that are colored, and form complex ions easily.

In the last chapter, we discussed the properties of the metals that belong to the A-groups in the periodic table. You learned that many of these metals are very reactive, and usually are encountered only in compounds. Only a handful of them — beryllium, magnesium, aluminum, tin, and lead — find many practical uses as free metals. Most of the metals that we come in contact with on a day-to-day basis belong to groups in the center of the periodic table.

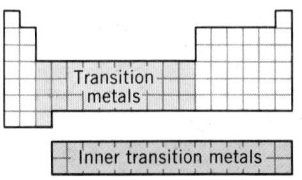

All of the elements between Group IIA and Group IIIA of the periodic table are metals. They are usually divided into two main categories. The transition elements or transition metals are the block of elements generally placed in the body of the table; they consist of the elements in the B-groups plus Group VIII. The inner transition elements are those in the two long rows normally placed just below the main body of the table. The elements in the first of the long rows are called the lanthanides because they follow lanthanum, atomic number 57. Those in the second long row are called the actinides because they follow actinium, atomic number 89. The lanthanides and antinides are rare elements and their chemistry is not particularly important to us in our present studies, so

The lanthanides are also called the *rare earth metals*.

we will have little more to say about them. However, since many of the transition elements are common structural metals, and some of their ions play major roles in certain biochemical reactions, it is worthwhile to learn about them.

As is true with all of the other elements that we have studied, the chemical and physical properties of the transition metals depend on their electron configurations. In Chapter 6, we saw that as we cross a period from left to right, a d subshell is gradually filled as we pass through a row of these elements. Except for the metals in Group IIB, atoms of the transition elements all have partially filled d subshells. This is the main feature that distinguishes them from the representative elements.[1] In fact, the transition elements are often called the *d-block* elements.

There are several properties that many of the transition elements have in common. For example, they generally tend to be hard and to have high melting points. This is particularly true of the elements near the center of each row, as shown in Figure 22.1. The elements with the highest melting points also have the maximum number of unpaired d electrons, which suggests that the d electrons are probably involved to some extent in covalent bonding within the metallic lattice.

Another general characteristic of the transition metals is the occurrence of multiple oxidation states. The $+2$ state is common to many of these elements because many of their atoms have a pair of electrons in their outermost s subshell. Some examples are manganese, iron, and cobalt.

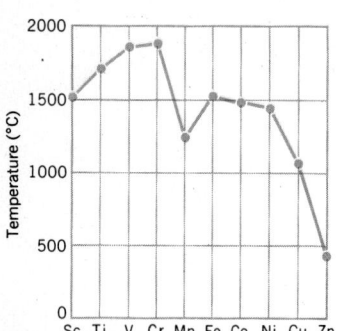

FIGURE 22.1 Melting points in °C of the period 4 transition elements.

$$\text{Mn} \quad [\text{Ar}]3d^5 4s^2$$
$$\text{Fe} \quad [\text{Ar}]3d^6 4s^2$$
$$\text{Co} \quad [\text{Ar}]3d^7 4s^2$$

Each forms an ion by the loss of both $4s$ electrons (Mn^{2+}, Fe^{2+}, Co^{2+}). The underlying $3d$ subshell is fairly close in energy to the $4s$ subshell, however, and not very much energy is needed to remove still another electron to give an ion with a $3+$ charge. We will take a closer look at oxidation states later in this section.

Still another property of the transition metals is the tendency of their ions to combine with neutral molecules or anions to form complex ions. One example is the reaction of the copper(II) ion with ammonia to form the complex ion $Cu(NH_3)_4{}^{2+}$.

Recall that complex ions are often simply called complexes.

The charge on a complex ion is just the sum of the charges of the molecules and ions from which the complex is formed.

$$\underset{\text{pale blue}}{Cu^{2+}(aq)} + 4NH_3(aq) \longrightarrow \underset{\text{deep blue}}{Cu(NH_3)_4{}^{2+}(aq)}$$

You may recall that we discussed this reaction in Chapter 11. The number of complex ions formed by the transition elements is enormous, and their study is a major specialty of chemistry. We will say more about them later in this chapter.

One of the most interesting properties of transition metal compounds is that they are often colored. For instance, all of the compounds of chromium are colored. In fact, chromium gets its name from the Greek word *chroma*, which means "color." Many complex ions also have beautiful colors.

Variations in Atomic Size

In Section 6.11 we learned the general rules that among the *representative elements* the atomic radius increases from top to bottom within a group and decreases from left to right within a period. Among the *transition elements*, however, there are variations to these rules, but they are also understandable in terms of locations within the periodic table. As with the representative elements, there is a decrease in size among the transition elements as we move from left to right across a period. However, the size changes more

---

[1] The Group IIB elements have electron configurations corresponding to $(n-1)d^{10}ns^2$ outside of a noble gas core, and some chemists prefer not to consider them to be true transition elements because their d subshells are complete.

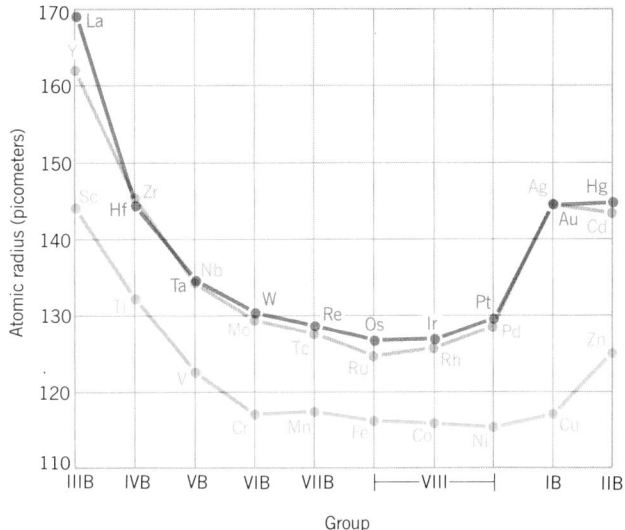

FIGURE 22.2 Size variations among the transition elements.

slowly for the transition elements, and there is a size minimum near the center of each row, as can be seen in Figure 22.2.

The slowness of the size change is because the electrons that are being added as we go from one atom to the next enter a *d* subshell that is actually below the outer shell. Thus, for the transition elements in period 4, the outer electrons are in the 4*s* subshell, but it is the 3*d* subshell that is filled as we cross from left to right. These *d* electrons shield the outer electrons pretty well from the increasing nuclear charge, so the *effective nuclear charge* felt by the outer electrons increases very slowly, and that leads to a slow decrease in size. The size minimum seems to occur because as the *d* subshell becomes more than half-filled, electron–electron repulsions force the *d* subshell to gradually expand in size, and this causes the sizes of the atoms to swell.

The Lanthanide Contraction   When we turn our attention from periods to groups, we find the most unusual variations in atomic size. As we said, we normally expect atomic size to increase in going down a group, and indeed we find this for the elements in Group IIIB. However, for the other groups something odd happens. We do find the expected increase as we go from the first row of transition elements (period 4) to the second row (period 5), as Figure 22.2 shows. The lower plot is for the first-row transition elements and the one just above it is for the second-row transition elements. Thus Ti and Zr are in the same group — Group IVB — and Zr in period 5 has a larger radius than Ti in period 4, which is just what we would expect. Hafnium (Hf), however, is in the third row of transition elements (period 6), and it is essentially identical in size to Zr — a clear "violation" of the expected trend. As you can see in Figure 22.2, all the elements in period 6, except for La, have the same close match in size to their family members in period 5. It's as if something is forcing a contraction of the expected radii for the period-6 elements. This shrinkage in radii of the period-6 elements from their expected values is called the lanthanide contraction. Let's see what causes it.

Lanthanum (La), in Group IIIB, is larger than yttrium (Y), as we would expect based on the normal way size varies down a group among the representative elements. But now, let's consider what happens as we move from left to right across periods 5 and 6. As we go from Y to Zr there is an increase of 1 in the nuclear charge and, as usual, there is a slight decrease in atomic size. As we go from La to Hf in period 6, however, we must pass through the entire series of lanthanide elements. This increases the nuclear charge by 15 as the 4*f* subshell is gradually filled and one more electron is added to the 5*d*. The 4*f* electrons, being two shells beneath the outer shell, are quite effective at shielding the

Zirconium and hafnium occur together in nature because their ions have the same charge and the same size. They are also difficult to separate from each other.

outer electrons from the increasing nuclear charge, *but they are not 100% effective.* Thus, as the lanthanide series is crossed, there is something of a buildup of effective nuclear charge felt by the outer electrons. As a result, the overall increase in effective nuclear charge experienced by the outer electrons on going from La to Hf is considerably more than the increase that occurs on going from Y to Zr. In other words, the outer electrons in Hf feel an extra large effective nuclear charge compared to the outer electrons in Zr. This draws the outer electrons in Hf closer to the nucleus, and causes Hf to be smaller than expected. The net result is that the increase in size that we might have otherwise expected on going from Zr to Hf is canceled. The extra shrinkage experienced by each of the period-6 transition elements that follow La is the lanthanide contraction. Because of it, Zr and Hf are nearly identical in size, and as we noted above, there are also close matches in size for the second and third members of the other groups that are to the right of Group IVB.

Two major consequences of the lanthanide contraction are that the transition elements in period 6 are very dense, and they are very resistant to oxidation. They have unusually high densities because their atoms have nearly twice as much mass as do the atoms of the elements above them, but their sizes are virtually the same. Twice the mass occupying about the same volume leads to about twice the density.

The resistance of the period-6 transition metals to oxidation (loss of electrons) occurs because their outer electrons feel a very large effective nuclear charge. This means that the electrons are tightly held and are difficult to remove. The resistance to oxidation of metals like platinum and gold makes them commercially useful and adds to their overall value. Gold, for example, is used to plate electrical contacts in low-voltage circuits where even small amounts of corrosion would block the flow of current.

Compounds and Oxidation States

If we compare the *formulas* of compounds formed by the B-group elements—the transition elements—with those formed by the elements in the corresponding A-groups, we find certain similarities. For example, the maximum oxidation state for elements in Groups IIIB through VIIB is equal to the group number. This is also true for the elements in Groups IIIA through VIIA. Similar relationships hold true for compounds of the elements in Groups I and II, although both copper and gold (Group IB) are also commonly found in oxidation states higher than $+1$. As a result, elements having the same group number form compounds having the same general formula, when the elements are in their highest oxidation states. Some examples are shown in Table 22.1.

Although similarities exist among the *formulas* of some of the compounds in the A- and B-groups, most of the chemical and physical properties of the A- and B-group elements are quite different. Therefore, it is not wise to attempt to extend the analogy between the groups any farther.

The nine elements that are collectively labeled Group VIII bear no similarities to the other elements in the periodic table in terms of the general formulas of their compounds.

The similarities among formulas of compounds and the maximum oxidation states of the elements form the basis of the A- and B-group designations.

TABLE 22.1 Similarities in Formulas of Compounds of the A- and B-Group Elements

Group Number	Maximum Oxidation State	A-Group Compound	B-Group Compound
I	$+1$	NaCl	CuCl
II	$+2$	$CaCl_2$	$ZnCl_2$
III	$+3$	Al_2O_3	Sc_2O_3
IV	$+4$	CO_2	TiO_2
V	$+5$	Na_3PO_4	Na_3VO_4
VI	$+6$	Na_2SO_4	Na_2CrO_4
VII	$+7$	$KClO_4$	$KMnO_4$

In addition, greater horizontal similarities than vertical similarities are found within this group. For instance, the similarities among iron, cobalt, and nickel are greater than those among iron, ruthenium, and osmium. These horizontal sets of elements are generally spoken of as *triads*. Iron, cobalt, and nickel constitute the *iron triad*, for example.

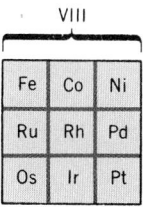

Ferromagnetism

One of the properties we often associate with iron is its strong attraction to magnets. In fact, if you were wondering whether an object were made of iron, you might test it with a magnet. Iron actually is one of three elements that exhibit this strong magnetism, called ferromagnetism; the other two are cobalt and nickel. The origin of ferromagnetism is the same as for paramagnetism — the presence of unpaired electrons. However, in a ferromagnetic material, the effects are about one million times larger than in a substance that is only paramagnetic. Most of the transition elements have atoms with unpaired *d* electrons. Iron atoms, for example, each have four unpaired electrons and manganese atoms each have five. Yet iron is ferromagnetic but manganese is only paramagnetic. Why?

Ferromagnetism is a property of the solid state. In iron, cobalt, or nickel, the individual paramagnetic atoms are spaced *just right* in their crystals to interact with each other very strongly. Huge numbers of individual atomic magnets lock onto each other and line up in regions called domains where their individual magnetic forces combine to produce very strong magnetic effects within each domain. As illustrated in Figure 22.3, in an ordinary piece of any of these metals, these domains are randomly oriented, so the object does not appear to be a magnet. But in the magnetic field produced by another magnet, the domains shift and turn, and each time a domain turns a tremendous number of tiny magnets turn at once. Because so many "atomic magnets" line up with the field, a very strong attraction is felt. In addition, when the external magnetic field is removed, the domains don't immediately become random again. They stay aligned and the metal retains its magnetism; it has become a "permanent magnet." It isn't really permanent, however. Heating it or pounding it with a hammer can provide the jostling needed to randomly orient the domains, and the "permanent" magnetism is lost. Melting the iron, cobalt, or nickel object also destroys the ferromagnetism because the domains of the solid state are lost when the metal is liquefied.

Why aren't manganese or other paramagnetic substances attracted to a magnet as strongly as solid iron, cobalt, or nickel? In any solid substance, the atoms are constantly

Adding the proper amount of copper to manganese adjusts the spacings between manganese atoms and produces a ferromagnetic alloy.

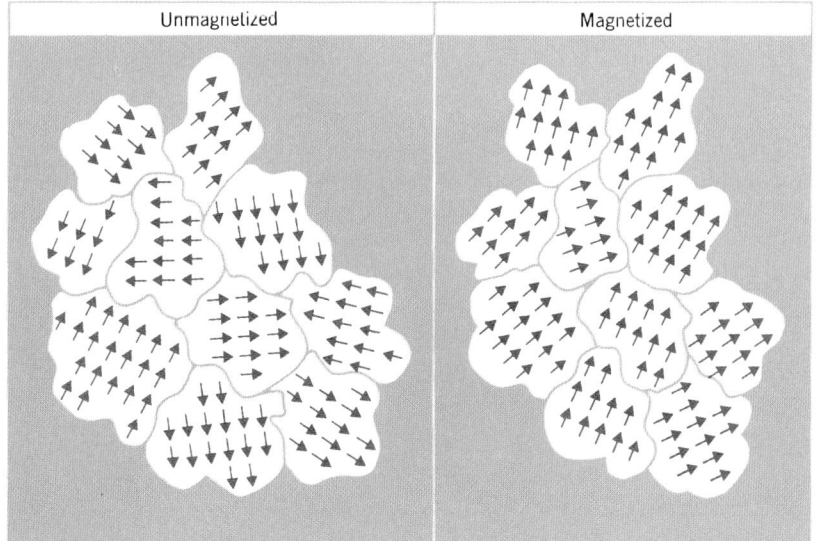

Unmagnetized	Magnetized

FIGURE 22.3 In a ferromagnetic solid, the individual paramagnetic atoms are aligned within domains. In the unmagnetized state, the domains are randomly oriented. When magnetized, the domains shift and become aligned. (From J. E. Brady and G. E. Humiston, *General Chemistry: Principles and Structure*, 4th ed., John Wiley and Sons, NY, 1986. Used by permission.)

jostling each other. In a paramagnetic substance, the interactions between the individual atomic magnets are not strong enough to prevent the collisions from knocking them about and upsetting their attempts to line up. As a result, the individual atomic magnets point in random directions. Even in an external magnetic field, the motions of the atoms prevent large-scale alignment of the tiny atomic magnets, and only a weak attraction is felt. Contrast this to the strong attractions felt by the millions of atomic magnets when a domain turns to be aligned with an external field.

22.2 PROPERTIES OF SOME IMPORTANT TRANSITION METALS

The functioning of our technological society as well as our bodies depends on the variety of chemical and physical properties exhibited by the transition elements.

If you pause for a moment and look around, you will see a number of important transition metals or their compounds. The paint on the wall probably contains titanium (as its dioxide), the coins in your pocket or purse are copper or a copper–nickel alloy, and iron and steel alloys are surely not far from you. Iron and many other transition metals are present in various quantities — some of them in only trace amounts — in your body. Certain transition elements are clearly crucial for our comfort and even our existence. In this section we focus our attention on the specific properties of some of these important metals.

Titanium

Ti $[Ar]3d^2 4s^2$

Titanium is only about 60% as dense as iron, it is very strong, and it is resistant to corrosion. This combination of properties makes it a very useful metal. It is used in place of steel and aluminum in aircraft, and it is used in jet engines because it doesn't lose its strength at high temperatures.

Titanium is found in several minerals, one of the most important of which is rutile, TiO_2. Separation of the metal from its ores is not easy, however, because titanium reacts directly with oxygen, nitrogen, and carbon at the temperatures required for its reduction by ordinary metallurgical methods. Instead, the oxide is heated with carbon and chlorine, which converts it to the chloride, $TiCl_4$.

$$TiO_2 + C + 2Cl_2 \longrightarrow TiCl_4 + CO_2$$

Titanium tetrachloride is then reduced with magnesium under a blanket of argon.

$$TiCl_4 + 2Mg \longrightarrow Ti + 2MgCl_2$$

FIGURE 22.4 Titanium tetrachloride, $TiCl_4$, is a volatile liquid whose vapors react with moisture in the air to give a dense smoke of TiO_2. The reaction was once used by the U.S. Navy to make smoke screens during naval battles.

Besides being involved in the production of titanium metal, $TiCl_4$ is one of the most important compounds of titanium because it is used to prepare most other compounds of this element. Titanium tetrachloride is a clear colorless liquid with a boiling point of only 136 °C and it is composed of covalently bonded molecules similar to $SnCl_4$. The U.S. Navy has used $TiCl_4$ to make smoke screens because it reacts almost instantly with moist air to form a dense fog of TiO_2 and HCl. (See Figure 22.4.)

$$TiCl_4(g) + 2H_2O(g) \longrightarrow TiO_2(s) + 4HCl(g)$$

The high degree of covalence of $TiCl_4$ can be explained in the way we explained the covalence in $SnCl_4$. The small, highly charged Ti^{4+} ion placed next to a Cl^- ion would distort the anion's electron cloud so much that the electron density drawn between the two would constitute a covalent bond.

Although titanium occurs in several oxidation states, just like most of the other transition elements, by far the most stable is the +4 state. $TiCl_4$ is one example of a titanium(IV) compound. Another, and the most commercially important titanium compound, is titanium(IV) oxide, TiO_2. Usually it is called titanium dioxide, and is a brilliant white substance that is the most common white pigment in paint today. It is also used as a brightener in paper.

Vanadium

Vanadium, like titanium, is also difficult to extract from its compounds. Its chief use is in making special alloy steels and cast iron. Adding vanadium makes the alloys more ductile and shock resistant. Vanadium is also used in nuclear reactors because it is highly "transparent" to neutrons.

Vanadium also exhibits multiple oxidation states. Its highest is +5 and corresponds to the "loss" of its $4s$ and $3d$ electrons. The oxide V_2O_5 is the catalyst in the contact process for the manufacture of sulfuric acid. V_2O_5 is a reasonably good oxidizing agent, and in Section 20.3 we saw that it promotes the oxidation of SO_2 to SO_3 by O_2. Vanadium(V) oxide dissolves slightly in water to give acidic solutions, in acids to give the VO_2^+ ion, and in base to give the VO_4^{3-} ion.

The formation of oxocations such as VO_2^+ by metals in high oxidation states is not unusual. It can be viewed simply as a case of extreme hydrolysis of the cation. For this cation, VO_2^+, we think of two water molecules that have been so extensively polarized by the positive charge of the V^{5+} cation that the hydrogens have come off as H^+ ions and become attached to water molecules as H_3O^+. (That explains the acidity of V_2O_5 solutions in water.) The reaction is diagrammed in Figure 22.5.

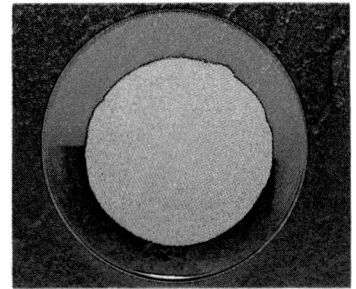

Vanadium(V) oxide, V_2O_5. The compound is also known as vanadium pentoxide.

V $[Ar]3d^34s^2$

Recall from Chapter 19 that the higher the oxidation state of the metal, the more acidic is its oxide.

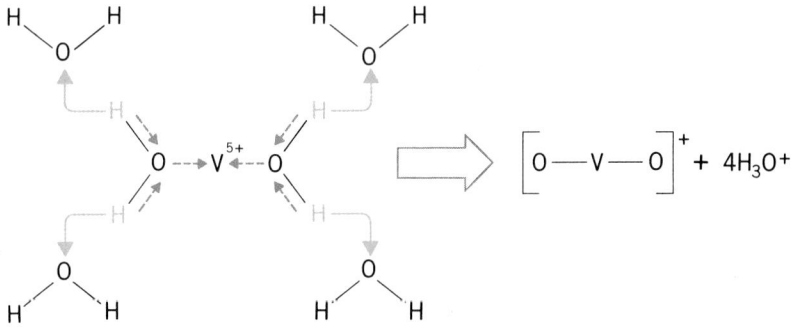

FIGURE 22.5 The formation of the VO_2^+ ion by extensive hydrolysis of the V^{5+} cation.

Vanadium also exists in compounds in the +4, +3, and +2 oxidation states, but they're of little practical importance.

Chromium

This metal is white, lustrous, hard, brittle, and very resistant to corrosion. These properties make it excellent as both a decorative and a protective coating over other metals such as brass, bronze, and steel. Almost everyone is familiar with the *chrome plate* that is deposited electrolytically on automobile parts such as bumpers. Chromium is also used in large amounts to produce alloys. The best known is *stainless steel*—a type of steel alloy that is resistant to corrosion. A typical stainless steel contains about 18% chromium, 10% nickel, plus small amounts of manganese, carbon, phosphorus, sulfur, and silicon, all combined with iron. *Nichrome,* an alloy of chromium and nickel, is often used as the wire heating element in various heating devices.

In compounds, chromium can exist in a number of different oxidation states, the most common of which are +2, +3, and +6. If elemental chromium is dissolved in dilute nonoxidizing acids that have been purged of oxygen, the pale blue chromium(II) ion or chromous ion, Cr^{2+}, is formed.

$$Cr(s) + 2H^+(aq) \longrightarrow Cr^{2+}(aq) + H_2(g)$$

This is the least stable oxidation state of chromium and it is very easily oxidized to the +3 state, even by oxygen. (That's why the dilute acid must be oxygen free to give Cr^{2+}.)

The +3 state is the most stable for chromium. In water, Cr^{3+} actually exists as the violet complex ion $Cr(H_2O)_6^{3+}$, and many chromium(III) salts owe their color to the presence of this ion in their crystals. An example is violet chrome alum, $KCr(SO_4)_2 \cdot 12H_2O$, which is formed when solutions that contain both K_2SO_4 and $Cr_2(SO_4)_3$ are gradually evaporated.

Traces of O_2 are removed from gases such as N_2 by bubbling them through a solution that contains Cr^{2+} ion.

Chrome alum, $KCr(SO_4)_2 \cdot 12H_2O$. The solid and the solution contain the violet complex ion $Cr(H_2O)_6^{3+}$.

In some respects, chromium(III) and aluminum(III) are alike. Solutions that contain the $Cr(H_2O)_6^{3+}$ ion are slightly acidic, just as solutions that contain the $Al(H_2O)_6^{3+}$ ion.

$$Cr(H_2O)_6^{3+} + H_2O \rightleftharpoons Cr(H_2O)_5OH^{2+} + H_3O^+$$

In the +3 state, chromium is also amphoteric. Addition of a base to the violet solution containing $Cr(H_2O)_6^{3+}$ ion precipitates the pale blue-violet gelatinous hydroxide. Making the solution more basic causes the precipitate to dissolve, yielding a deep green solution.

Cr^{3+} is known to form thousands of complexes.

The amphoteric behavior of Cr_2O_3 was discussed in Chapter 19.

$$\underset{\text{violet}}{Cr(H_2O)_6^{3+}(aq)} + 3OH^-(aq) \longrightarrow \underset{\text{blue-violet}}{Cr(H_2O)_3(OH)_3(s)} + 3H_2O$$

$$\underset{\text{blue-violet}}{Cr(H_2O)_3(OH)_3(s)} + OH^-(aq) \longrightarrow \underset{\text{green}}{Cr(H_2O)_2(OH)_4^-(aq)} + H_2O$$

On the left, a solution containing $Cr(H_2O)_6^{3+}$. In the center, the precipitate of $Cr(H_2O)_3(OH)_3$ that is formed when the first solution is made slightly basic. When more base is added (right), the precipitate dissolves and yields a green solution of $Cr(H_2O)_2(OH)_4^-$.

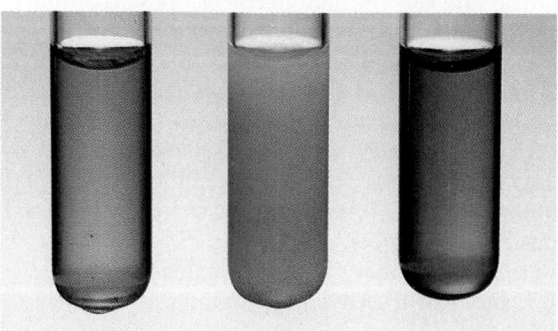

The third important oxidation state of chromium is $+6$. The highly poisonous red-orange oxide, CrO_3, is a powerful oxidizing agent and the acid anhydride of chromic acid, H_2CrO_4. Chromic acid is the primary species in aqueous solutions at very low pH (i.e., in very acidic solutions). At higher pH, two other species predominate, the yellow *chromate ion*, CrO_4^{2-}, and the red-orange *dichromate ion*, $Cr_2O_7^{2-}$. (See Figure 22.6.) There is an equilibrium between CrO_4^{2-} and $Cr_2O_7^{2-}$ that can be written

$$2CrO_4^{2-}(aq) + 2H^+(aq) \rightleftharpoons Cr_2O_7^{2-}(aq) + H_2O$$

In solutions that are acidic, this equilibrium is shifted to the right and dichromate predominates. The principal ion present in basic solutions is CrO_4^{2-} because the reaction is shifted to the left as the H^+ concentration decreases.

The formation of $Cr_2O_7^{2-}$ from CrO_4^{2-} is easier to understand if the equilibrium is written

$$2HCrO_4^-(aq) \rightleftharpoons H_2O + Cr_2O_7^{2-}(aq)$$

As the pH is lowered and the hydrogen ion concentration increases, more and more $HCrO_4^-$ is formed in the solution by adding protons to CrO_4^{2-} ions. Using Lewis structures, we see that removal of the components of water from a pair of $HCrO_4^-$ ions joins the chromium ions with an oxygen bridge.

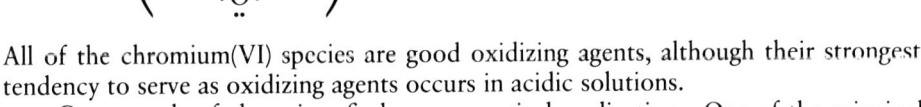

Chromium(VI) oxide, CrO_3.

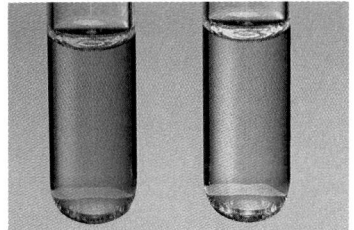

FIGURE 22.6 Dichromate ion, $Cr_2O_7^{2-}$, in the solution at the left has a reddish-orange color. Chromate ion, CrO_4^{2-}, in the other solution is yellow.

All of the chromium(VI) species are good oxidizing agents, although their strongest tendency to serve as oxidizing agents occurs in acidic solutions.

Compounds of chromium find many practical applications. One of the principal uses is in pigments. The oxide Cr_2O_3, for example, is the most stable green pigment known, and it is used for coloring paints, roofing granules, cements, and plaster. Another pigment is called *zinc yellow*, which is actually zinc chromate, $ZnCrO_4$. It is used as a corrosion-inhibiting primer on aluminum and magnesium aircraft parts. Also, as mentioned in Chapter 21, lead chromate is a yellow pigment used in artists' oil colors. Barium chromate, also insoluble in water, is another yellow pigment.

Pigments account for about 35% of the chromium chemicals that are produced annually. Another 25%, principally in the form of $Cr_2(SO_4)_3$, are used in tanning leather, and much of the rest of the chromium-containing chemicals that are made are used in various ways in treating the surfaces of metals for corrosion control.

Chromium(III) oxide, Cr_2O_3.

Manganese

Manganese has many properties that are similar to those of iron. It corrodes in moist air, for example, and it dissolves in dilute acids with the evolution of hydrogen.

$$Mn(s) + 2H^+(aq) \longrightarrow Mn^{2+}(aq) + H_2(g)$$

Crystals and solutions of $MnCl_2 \cdot 6H_2O$ are pink because they contain the complex ion, $Mn(H_2O)_6{}^{2+}$.

A photograph of this reaction appears on page 475.

MnO_2 is the cathode reactant in the dry cell.

The metal's chief uses are as an additive to steel and in the preparation of other alloys such as *manganese bronze,* a copper–manganese alloy, and *manganin,* an alloy of copper, manganese, and nickel whose electrical resistance changes only very slightly with temperature.

The most important oxidation states of manganese are $+2, +3, +4, +6,$ and $+7$. The most stable is the $+2$ state, which is formed by the removal of the outer $4s$ electrons from the manganese atom.

$$Mn([Ar]3d^5 4s^2) \longrightarrow Mn^{2+}([Ar]3d^5) + 2e^-$$

The Mn^{2+} ion thus has a half-filled d subshell—a configuration you may recall that is particularly stable. In water, the ion exists as the pale pink complex $Mn(H_2O)_6{}^{2+}$, and that is the color of most maganese(II) salts which usually crystallize from solutions as hydrates that contain this ion.

When made basic, Mn^{2+} precipitates as pale pink $Mn(OH)_2$, which is easily oxidized in air to the compound $MnO(OH)$, which contains manganese in the $+3$ oxidation state. Again we have a compound in which H^+ has been lost from H_2O or OH^- to give an oxygen firmly attached to a metal ion, a phenomenon that we noted for vanadium(V) in water, where it exists as $VO_2{}^+$.

The least stable oxidation state of manganese is $+7$, which has a strong tendency to be reduced and is therefore a powerful oxidizing agent. The most common compound of manganese(VII) is *potassium permanganate,* $KMnO_4$, which dissolves in water to give deep purple solutions that contain the $MnO_4{}^-$ ion. In an acidic solution, $MnO_4{}^-$ is usually reduced to Mn^{2+}, which has a very pale pink color that is practically invisible if the solution is dilute.

$$\underset{\substack{\text{deep}\\\text{purple}}}{MnO_4{}^-(aq)} + 8H^+(aq) + 5e^- \longrightarrow \underset{\substack{\text{almost}\\\text{colorless}}}{Mn^{2+}(aq)} + 4H_2O$$

As we've noted earlier, many analytical procedures use $KMnO_4$ as an oxidizing agent in titrations. As the $MnO_4{}^-$ ion is added to a reaction mixture from a buret, the Mn^{2+} ion produced by the reduction of $MnO_4{}^-$ has hardly any effect on the color of the solution. When the reaction is complete, however, the next drop of $KMnO_4$ solution that's added produces a noticeable pink color. This signals the endpoint. In effect, then, $KMnO_4$ serves as its own indicator in a titration.

When permanganate ion is reduced in neutral or basic solutions, reduction ceases at the $+4$ state with the formation of *manganese dioxide,* MnO_2.

$$MnO_4{}^-(aq) + 2H_2O + 3e^- \longrightarrow MnO_2(s) + 4OH^-(aq)$$

Manganese dioxide is a common compound from which many other manganese compounds are made. It tends to be nonstoichiometric—the ratio of oxygen to manganese is somewhat variable and is not exactly 2 to 1 as indicated by its formula. Manganese dioxide is found in the manganese nodules scattered on the ocean floor, which were described in Chapter 21. According to one theory, microorganisms may have extracted manganese from the seawater and deposited it as MnO_2 in the nodules.

When MnO_2 is added to molten potassium hydroxide and oxidized with air or potassium nitrate, the green *manganate ion,* $MnO_4{}^{2-}$ is formed. This ion is stable only in very basic solutions. When acidified, a portion of the $MnO_4{}^{2-}$ is oxidized while the rest is reduced. The products are $MnO_4{}^-$ and MnO_2.

$$3MnO_4{}^{2-}(aq) + 4H^+(aq) \longrightarrow 2MnO_4{}^-(aq) + MnO_2(s) + 2H_2O$$

Recall that a reaction such as this, in which one portion of a chemical is oxidized by the rest, is called disproportionation. We say that the $MnO_4{}^{2-}$ disproportionates.

When manganese metal is oxidized in air, the final product is the oxide Mn_2O_3, which contains manganese in the $+3$ oxidation state. In water, the ion slowly oxidizes the solvent, liberating oxygen, which shows that it is a fairly strong oxidizing agent. The Mn^{3+} ion can be stabilized, however, by forming complex ions, and many complexes of manganese(III) are known.

Oxidation of water produces O_2:

$$2H_2O \longrightarrow O_2 + 4H^+ + 4e^-$$

Iron

You are probably more familiar with this metal than with any other. It is relatively inexpensive, and iron and its alloys have such useful physical properties that they have been put to more uses than any other metal.

Iron is the second most abundant metal, next to aluminum, and it is the fourth most abundant element in the earth's crust. The molten core of the earth is thought to be composed mostly of iron and nickel. In the pure state, iron is white and lustrous, but it is not especially hard. It is also quite reactive. It is attacked by nonoxidizing acids such as HCl or H_2SO_4 to generate hydrogen.

$$Fe(s) + 2H^+(aq) \longrightarrow Fe^{2+}(aq) + H_2(g)$$

Oddly enough, iron doesn't react with concentrated nitric acid. Instead, its surface is made quite unreactive; the iron is said to have been made *passive*. Presumably, this is caused by the formation of an adhering oxide coating produced by the strongly oxidizing HNO_3.

The production and metallurgy of iron were discussed in Chapter 21. Its chief ores have traditionally been the red-orange oxide *hematite*, Fe_2O_3, and black *magnetite*, Fe_3O_4. In the United States, rich sources of these ores have been largely depleted and a rock called *taconite*, consisting of fine crystals of magnetite bound together by silicate minerals, is now the focus of attention. (This is discussed in detail in *Chemicals in Use 15*, which precedes this chapter, along with a more thorough discussion of steel alloys.)

The name *hematite* suggests the almost blood-red color of some of these iron ores.

An open-pit iron mine in the Mesabi Range in Minnesota. The hematite (Fe_2O_3) ore in this region has been nearly depleted. Instead, mining operations today focus on processing a rock-hard ore called taconite, which contains magnetic Fe_3O_4.

As a child, you may have learned the hard way—by leaving your bicycle out in the rain—that iron or steel objects rust if they are exposed for any prolonged period to a damp or wet environment. The corrosion or rusting of iron and steel is a severe economic problem in the modern world, and each year huge sums of money are spent to cover iron and steel surfaces with protective films of various kinds to prevent this damaging phenomenon.

The rusting of iron is a complex chemical reaction that involves both oxygen and moisture. Iron won't rust in pure water that is oxygen free, and it won't rust in pure oxygen in the absence of moisture. As noted in Special Topic 17.1, the corrosion process is apparently electrochemical in nature. This is shown in the accompanying diagram. At one place on the surface, iron becomes oxidized in the presence of water and enters solution as Fe^{2+}.

$$Fe(s) \longrightarrow Fe^{2+}(aq) + 2e^-$$

At this location on the metal's surface the iron is acting as an anode.

The electrons that are released when the iron is oxidized travel through the metal to some other location where the iron is exposed to oxygen. This is where reduction takes place (it's a cathodic region on the metal surface), and oxygen is reduced to give hydroxide ion.

$$\tfrac{1}{2}O_2(aq) + H_2O + 2e^- \longrightarrow 2OH^-(aq)$$

The iron(II) ions that are formed at the anodic regions diffuse through the water and eventually contact the hydroxide ions. When this happens, a precipitate of $Fe(OH)_2$ is formed. Oxygen is able to oxidize this compound very easily to give $Fe(OH)_3$, which readily loses water. In fact, complete dehydration gives the oxide,

$$2Fe(OH)_3 \longrightarrow Fe_2O_3 + 3H_2O$$

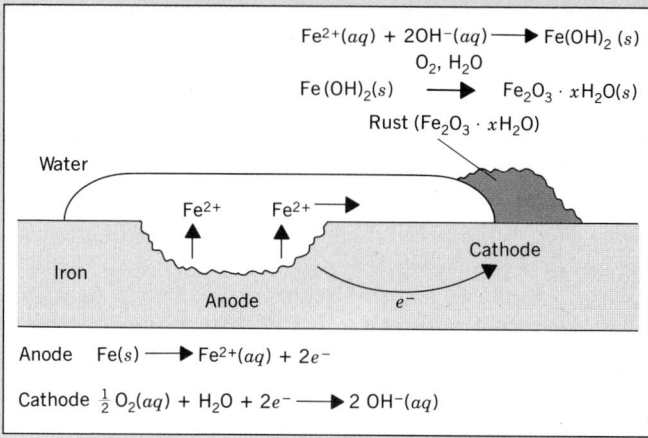

$$Fe^{2+}(aq) + 2OH^-(aq) \longrightarrow Fe(OH)_2(s)$$
$$\phantom{Fe^{2+}(aq) + 2OH^-(aq) \longrightarrow} O_2, H_2O$$
$$Fe(OH)_2(s) \longrightarrow Fe_2O_3 \cdot xH_2O(s)$$
$$Rust\ (Fe_2O_3 \cdot xH_2O)$$

Anode $Fe(s) \longrightarrow Fe^{2+}(aq) + 2e^-$

Cathode $\tfrac{1}{2}O_2(aq) + H_2O + 2e^- \longrightarrow 2OH^-(aq)$

The corrosion of iron. Iron dissolves in anodic regions to give Fe^{2+}. Electrons travel through the metal to cathodic sites where oxygen is reduced, forming OH^-. The combination of the Fe^{2+} and OH^-, followed by air oxidation, gives rust. (From J. E. Brady and G. E. Humiston, *General Chemistry: Principles and Structures,* 3rd ed., John Wiley and Sons, NY, 1982. Used by permission.)

When partial dehydration of the $Fe(OH)_3$ occurs, the resulting material has a composition that lies between that of the hydroxide and that of the oxide, Fe_2O_3. Usually, this solid—which we call rust—is referred to as a *hydrated oxide,* and its formula is normally represented as $Fe_2O_3 \cdot xH_2O$.

This mechanism for the rusting of iron explains one of the more interesting aspects of this damaging process. Perhaps you have noticed that when rusting occurs on the body of a car, the rust appears at and around a break (or a scratch) in the surface of the paint, but the damage extends under the painted surface for some distance. Apparently, the Fe^{2+} ions that are formed at the anode sites are able to diffuse rather long distances to the hole in the paint, where they finally react with air to form the rust.

A problem that has plagued civilization ever since iron was discovered is its reaction with moisture and air. The corrosion product—rust, a hydrated oxide whose formula is usually given as $Fe_2O_3 \cdot xH_2O$—doesn't adhere to the metal. Instead, it falls away, exposing fresh metal to attack. The way that the corrosion of iron occurs and how it can be prevented are treated in Special Topic 22.1 and Special Topic 17.1 (page 712).

Fe^{2+} = ferrous ion
Fe^{3+} = ferric ion

Iron forms compounds in two principal oxidation states, $+2$ and $+3$. They are both relatively stable, both in the solid state in compounds and in aqueous solutions. Unlike the transition metals that appear earlier in period 4, iron forms no compounds in an oxidation state that would, in principle, involve all eight of its $3d$ and $4s$ electrons. The highest oxidation state observed for iron is $+6$, and that state is rare (and, therefore, unimportant to us here). This behavior of iron illustrates a general trend:

The lower oxidation states of the transition metals become progressively more stable relative to the higher ones as we move from left to right across a period.

Iron forms compounds with most of the nonmetals. For example, it reacts with such nonmetals as oxygen, chlorine, sulfur, phosphorus, and carbon when heated. The oxides of iron are FeO, Fe_2O_3, and Fe_3O_4. Crystalline FeO is difficult to prepare. When heated, it undergoes disproportionation to give both Fe and Fe_2O_3. The most common oxide is Fe_2O_3, which is the one found in rust and the ore hematite. As mentioned above, Fe_3O_4 occurs as the mineral magnetite, and as its name suggests, it is magnetic. It contains iron in both the $+2$ and $+3$ oxidation states.

In water, the chemistry of the Fe^{2+} and Fe^{3+} ions is largely the chemistry of their complex ions. For example, in water the iron(II) ion has a pale bluish-green color, which is the color of the $Fe(H_2O)_6^{2+}$ complex ion. Solid salts that contain Fe^{2+} are often pale green because this complex persists in the crystals. In water, iron(III) salts tend to be somewhat acidic and often appear slightly yellow because of hydrolysis of the $Fe(H_2O)_6^{3+}$ ion. The equilibria here are similar to those we described earlier in this chapter to explain the acidity of solutions of Cr^{3+}, and in the last chapter to explain the acidity of solutions that contain Al^{3+}.

$$Fe(H_2O)_6^{3+}(aq) + H_2O \rightleftharpoons \underset{\text{yellow}}{Fe(H_2O)_5OH^{2+}(aq)} + H_3O^+(aq)$$

The reaction has an equilibrium constant, $K_a = 9 \times 10^{-4}$, which makes the $Fe(H_2O)_6^{3+}$ ion a stronger acid than acetic acid!

Addition of base to a solution that contains iron(II) ion precipitates a nearly color-less $Fe(OH)_2$, but if exposed to air the precipitate rapidly turns brown as it is oxidized by oxygen to give brown $Fe_2O_3 \cdot xH_2O$. Addition of base to a solution that contains iron(III) ion produces a reddish-brown gelatinous hydroxide, but unlike those of aluminum and chromium, it is not amphoteric — it does not dissolve in excess base.

$$Fe(H_2O)_6^{3+}(aq) + 3OH^-(aq) \longrightarrow Fe(H_2O)_3(OH)_3(s) + 3H_2O$$

Actually, the composition of the precipitate isn't as clearly defined as the equation above might suggest, and its formula is usually just given as the hydrated oxide we have seen earlier, $Fe_2O_3 \cdot xH_2O$.

As hinted above, iron ions form many complex ions. One that can easily be used to detect the presence of iron(III) in a solution is formed when thiocyanate ion, SCN^-, is added. This anion forms a blood-red complex with Fe^{3+} that is intensely colored; its formula can be represented as $Fe(SCN)_6^{3-}$.

Among the more interesting compounds of iron are those involving its complexes with cyanide ion. Their common names and formulas are ferrocyanide ion, $Fe(CN)_6^{4-}$, and ferricyanide ion, $Fe(CN)_6^{3-}$. Taking the charge of the cyanide ion to be $1-$, you can see that the ferrocyanide ion contains iron(II) (ferrous) and the ferricyanide ion contains iron(III) (ferric). If a solution that contains Fe^{3+} is added to a solution containing $Fe(CN)_6^{4-}$, a blue precipitate known as *Prussian blue* is formed, as shown in Figure 22.7.

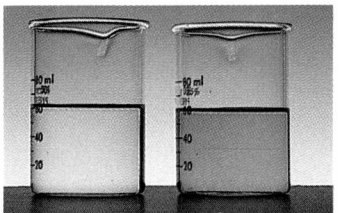

Solutions of iron(II) salts (left) have a pale greenish-blue color caused by the ion $Fe(H_2O)_6^{2+}$. Solutions that contain iron(III), like that on the right, have a yellow color from the hydrolysis of the $Fe(H_2O)_6^{3+}$ ion.

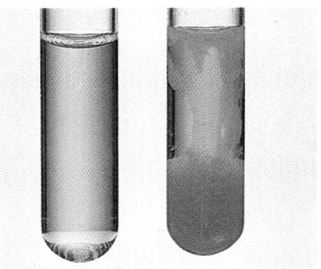

On the left, a solution that contains Fe^{3+}. When base is added, a reddish-brown gelatinous precipitate of $Fe(H_2O)_3(OH)_3$ is formed.

In writing their formulas, gelatinous metal hydroxides such as this are often given in a simplified form: $Fe(OH)_3$.

The formation of the $Fe(SCN)_6^{3-}$ ion is shown in Qual Topic III following this chapter.

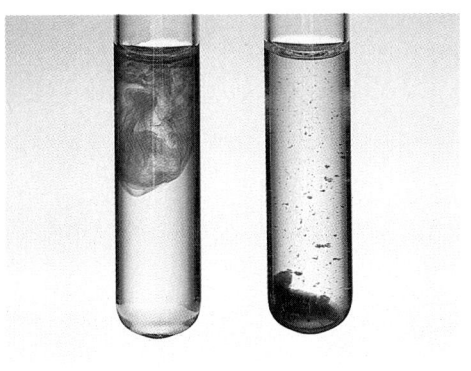

FIGURE 22.7 At the left we see the deep blue color of *Prussian blue* formed when a drop of a dilute solution containing Fe^{3+} is added to a dilute solution of ferrocyanide ion, $Fe(CN)_6^{4-}$. After a few moments, the blue precipitate, $Fe_4[Fe(CN)_6]_3 \cdot 16H_2O$, settles to the bottom of the test tube. On the right an architect studies blueprints of plans for the alteration of a building. The blue dye in the paper is $Fe_4[Fe(CN)_6]_3 \cdot 16H_2O$.

Fe$_4$[Fe(CN)$_6$]$_3 \cdot$16H$_2$O contains both iron(II) and iron(III).

Evidence suggests that its formula is Fe$_4$[Fe(CN)$_6$]$_3 \cdot$16H$_2$O. Interestingly, the exact same compound is formed if Fe^{2+} ion is added to a solution of Fe(CN)$_6$$^{3-}$. The intense blue color of this precipitate is the basis of the blueprint process, which has been used by engineers and architects to copy their plans and drawings. The blueprint paper is made by treating paper with solutions of ammonium ferricyanide, (NH$_4$)$_3$Fe(CN)$_6$, and iron(III) citrate. When the paper is exposed to light, the Fe^{3+} ion of the citrate salt is reduced by the citrate ion to Fe^{2+}, and then the Fe^{2+} reacts with the Fe(CN)$_6$$^{3-}$ to give the deep blue precipitate. Finally, the unreacted ammonium ferricyanide and iron(III) citrate are washed from the paper so no further reaction can occur when the blueprints are used out in the field.

Cobalt

Cobalt is a hard, bluish-white metal that is used mostly in catalysts and alloys. For example, it is combined with chromium and tungsten in an alloy called *stellite,* which retains its hardness even when hot. This property makes stellite useful for high-speed cutting tools (e.g., drill bits) used to machine steel. Cobalt is also used to prepare an alloy called *alnico,* which forms powerful magnets. A typical alnico alloy is composed of 46% Fe, 23% Co, 19% Ni, 8% Al, and 4% Cu.

Chemically, cobalt is somewhat less active than iron, although it dissolves slowly in nonoxidizing acids such as HCl to liberate hydrogen.

$$\text{Co}(s) + 2\text{H}^+(aq) \longrightarrow \text{Co}^{2+}(aq) + \text{H}_2(g)$$

A photograph showing the color of the Co(H$_2$O)$_6$$^{2+}$ ion and a solid containing this complex is on page 432.

There are two important oxidation states of cobalt, +2 and +3. In water, the most stable species is Co^{2+}, which exists as the pink Co(H$_2$O)$_6$$^{2+}$ complex ion. This ion persists in hydrated cobalt(II) salts such as pink crystals of cobalt chloride hexahydrate. On the labels of bottles of this chemical purchased from chemical supply firms, the formula is written CoCl$_2 \cdot$6H$_2$O, but the chemical is really [Co(H$_2$O)$_6$]Cl$_2$. This particular salt is interesting because if it is heated, it loses water and turns blue, as shown in Figure 22.8. It is believed that the nature of the complex changes as indicated by the equilibrium equation

$$\underset{\text{pink}}{[\text{Co(H}_2\text{O})_6]\text{Cl}_2(s)} \rightleftharpoons \underset{\text{blue}}{[\text{Co(H}_2\text{O})_4]\text{Cl}_2(s)} + 2\text{H}_2\text{O}(g)$$

The change is reversible; exposing the blue crystals to moisture allows water to be recaptured and the solid becomes pink again. This color change can, in fact, be used to detect moisture.

Cobalt(III) ion is a strong enough oxidizing agent to oxidize water unless the ion is bound in a complex ion. An enormous number of these complexes have been studied, and we will see some of them later in this chapter.

FIGURE 22.8 *(a)* Pink hydrated CoCl$_2 \cdot$6H$_2$O, which actually contains [Co(H$_2$O)$_6$]Cl$_2$. *(b)* After heating and driving off some of the water, blue [Co(H$_2$O)$_4$]Cl$_2$ remains.

(a)   (b)

Nickel

Nickel is one of modern technology's most useful metals. In the pure state, it resists corrosion, and metals such as iron and steel are frequently given thin protective coatings of nickel by electrolysis. When added to iron, nickel makes the metal more ductile and resistant to corrosion. We saw earlier that nickel and chromium are the chief additives to iron in making stainless steel. Nickel also makes steel resistant to impact, a property that is particularly desirable in armor plate. Combined with copper, nickel produces an alloy called *monel* that is very resistant to corrosion and is very hard and strong. Propeller shafts made of monel are very desirable in boats that operate in the corrosive environments of seawater. Nickel is also used as a catalyst for hydrogenation of organic compounds that contain double bonds.

When recovered from its ores, nickel is only about 96% pure and must be refined. This can be done electrochemically much as the refining of copper described in Chapter 17. Another interesting method of purification is called the *Mond process*. When nickel is warmed in the presence of carbon monoxide, it forms a compound called *nickel carbonyl,* which has the formula $Ni(CO)_4$. It is an interesting compound because it contains nickel in the zero oxidation state. One of the properties of nickel carbonyl is that it is a liquid with a high vapor pressure, so that as it is formed, its vapor is carried away by the stream of carbon monoxide. When nickel carbonyl vapor is passed over a very hot surface, it decomposes into pure metallic nickel and carbon monoxide. The CO is then recycled. The Mond process produces nickel that's about 99.95% pure.

Nickel is a moderately active metal, and it dissolves in nonoxidizing acids such as HCl to give Ni^{2+} and H_2.

$$Ni(s) + 2H^+(aq) \longrightarrow Ni^{2+}(aq) + H_2(g)$$

As with many of the other transition metals we've studied, the nickel(II) ion exists in water as a complex ion; its formula is $Ni(H_2O)_6^{2+}$ and it has an emerald green color. Many crystalline nickel salts are "hydrates" and are green because the crystals contain this complex. Nickel forms many complexes in its +2 oxidation state. Another is the blue $Ni(NH_3)_6^{2+}$ ion that is formed when ammonia is added to a solution that contains Ni^{2+} (see Qual Topic III following this chapter).

The +2 oxidation state is the most stable one for nickel, and this is how we find it in nearly all its compounds. Higher oxidation states are powerful oxidizing agents because they have a strong tendency to be reduced to the +2 state. You may recall that the nickel–cadmium cell described in Section 17.10 uses nickel(IV) oxide as the cathode. The cell reaction during discharge is

$$Cd(s) + NiO_2(s) + 2H_2O \longrightarrow Cd(OH)_2(s) + Ni(OH)_2(s)$$

Note once again that as we move from left to right across the period, the lower oxidation state (in this case, +2) becomes increasingly more stable and the higher oxidation states become less stable.

Copper, Silver, and Gold — The Coinage Metals

Copper, silver, and gold are often called the coinage metals because they have been used for that purpose since ancient times. Each has an outer electron configuration of $(n-1)d^{10}ns^1$. The loss of the single *s* electron gives the +1 oxidation state, which is the reason for the *IB* designation for the group. In the case of copper, a second electron can be lost easily, and most copper compounds contain Cu^{2+}. Gold loses another two electrons rather easily, so it tends to form Au^{3+} rather than Au^+.

All the Group IB metals are commercially valuable. Copper is a metal that every U.S. citizen would recognize because it is used to make the penny (although, today, new

The "nickel" coin is a copper–nickel alloy.

Metal carbonyls were discussed in Section 20.8.

$Ni(CO)_4$ is very toxic.

The complex ion $Ni(H_2O)_6^{2+}$ gives its green color to both crystals and solutions of $NiCl_2 \cdot 6H_2O$.

Nickel compounds are added to glass to give it a green color.

Cu $3d^{10}4s^1$
Ag $4d^{10}5s^1$
Au $5d^{10}6s^1$

pennies are made of zinc with just a thin copper coating). Copper has a very high electrical and thermal conductivity and, for this reason, it is used in electrical wiring. In fact, the largest concentration of copper in the world is said to be the copper wire that lies beneath the streets of New York City. Copper is also fairly resistant to corrosion and is widely used as pipe to carry hot and cold water in homes and office buildings.

Silver has the highest thermal and electrical conductivity of any metal. Its value as a coinage metal, however, makes it too expensive to be used often as an electrical conductor. Silver has a very high luster and, when polished, reflects light very well. This has made it valuable for jewelry and for coating the backs of mirrors. Silver is a soft metal and is usually alloyed with copper. *Sterling silver,* for example, contains 7.5% copper, and silver used in jewelry often contains up to 20% copper. One of silver's most important applications is in photography. Silver compounds tend to be unstable and sensitive to light. (See *Chemicals in Use 10.*)

Everyone knows of gold's value as bullion and as a decorative metal for jewelry. As mentioned earlier, gold is also used occasionally to plate electrical contacts because of its low chemical reactivity. Pure gold is very soft and is particularly ductile and malleable. Gold leaf, used for decorative lettering in signs, is made by pounding gold into very thin sheets. It is so thin that some light passes through it. A stack of 11,000 gold leaflets is only 1 mm thick.

Chemical Properties of Copper

In general, the elements of Group IB are not as reactive as other metals that we've discussed so far. Copper, for example, isn't attacked by nonoxidizing acids such as HCl. As we've noted on several occasions, however, it does dissolve in both dilute and concentrated nitric acid. (See page 461.) Copper also is attacked by hot concentrated sulfuric acid, which is a fairly potent oxidizing agent.

$$Cu(s) + 2H_2SO_4(concd) \xrightarrow{heat} CuSO_4(s) + SO_2(g) + 2H_2O$$

Copper sulfate, the product of this reaction, is one of the more important compounds of copper. We usually encounter it as the blue hydrate $CuSO_4 \cdot 5H_2O$. The dehydration of this compound, you may recall, was discussed in Chapter 1 (page 13).

When heated strongly in air, copper acquires a coating of its black oxide, CuO. Although fairly resistant to corrosion, copper does oxidize slowly in air and, when CO_2 is also present, its surface becomes coated with a green film of $Cu_2(OH)_2CO_3$. The outer surface of the Statue of Liberty is made of copper, and this compound is what gives it its green color. (See the photograph at the beginning of this chapter.)

As mentioned above, copper forms compounds in two oxidation states, +1 and +2. The Cu^+ ion is unstable in water, and disproportionates.

$$2Cu^+(aq) \longrightarrow Cu(s) + Cu^{2+}(aq)$$

However, it can be isolated in compounds such as CuCl, which is insoluble. For example, heating a solution that contains $CuCl_2$ and metallic copper acidified by HCl produces CuCl.

$$Cu(s) + Cu^{2+}(aq) + 2Cl^-(aq) \longrightarrow 2CuCl(s)$$

The most stable oxidation state of copper is +2, and most copper compounds contain the Cu^{2+} ion, often in the form of a complex ion such as the blue $Cu(H_2O)_4^{2+}$ complex. This is what gives the color to crystals of $CuSO_4 \cdot 5H_2O$, and it is the ion present in dilute solutions of most copper salts. Copper(II) ion forms many complexes, and we've already discussed the deep blue $Cu(NH_3)_4^{2+}$ ion formed when aqueous ammonia is added to a solution that contains the $Cu(H_2O)_4^{2+}$. Concentrated solutions of the salt $CuCl_2$ are green because they contain a mixture of yellow $CuCl_4^{2-}$ and blue $Cu(H_2O)_4^{2+}$. And adding base to a solution of a soluble copper(II) salt first precipitates $Cu(OH)_2$ and then redissolves the precipitate to give the complex $Cu(OH)_4^{2-}$. (Notice that copper

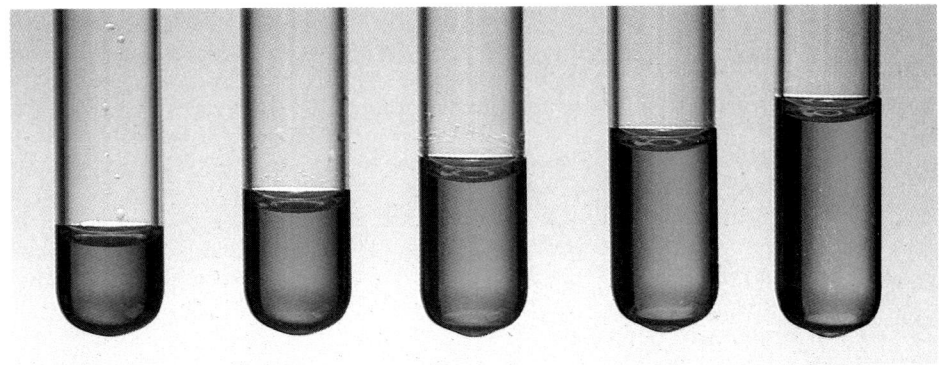

complexes tend to have only four ligands attached to the metal ion, rather than six as in most of the other complexes discussed so far.)

Chemical Properties of Silver   Silver is even more difficult to oxidize than copper, so it isn't attacked by nonoxidizing acids either. But, like copper, silver does dissolve in concentrated and dilute nitric acid, yielding NO_2 and NO as reduction products, respectively, and one of silver's most important water-soluble compounds, silver nitrate ($AgNO_3$). Metallic silver isn't attacked by oxygen of the air, but anyone who has had to polish silverware knows that it does tarnish. The reaction is between silver, traces of H_2S in the air, plus oxygen, and the black tarnish deposit is silver sulfide.

Silver is often found free in nature. It also occurs as AgCl and Ag_2S.

$$4Ag(s) + 2H_2S(g) + O_2(g) \longrightarrow 2Ag_2S(s) + 2H_2O$$

Similar reactions occur if silver utensils are left in contact with sulfur-containing foods such as eggs or mustard.

The only important oxidation state of silver, as far as we are concerned, is $+1$. Many compounds of Ag^+ are water insoluble. Addition of base to Ag^+ yields a brown precipitate of silver oxide.

$$2Ag^+(aq) + 2OH^-(aq) \longrightarrow Ag_2O(s) + H_2O$$

Although silver fluoride is very soluble in water, AgCl, AgBr, and AgI are quite insoluble, and their solubilities decrease from the chloride to the iodide. The precipitates all darken on exposure to light because of *photochemical* decomposition — decomposition promoted by the absorption of photons of appropriate energy. An example is

Silver iodide is used to "seed" clouds to bring on rain.

$$2AgI \xrightarrow{h\nu} 2Ag + I_2$$

The symbol *h*ν over an arrow in a chemical equation means the change is caused by a photon with energy $E = h\nu$.

Silver ion forms many complex ions. In the qualitative analysis for silver, for example, Ag^+ is precipitated as AgCl by the addition of HCl to the "unknown" solution. Ammonia is then added, which causes any AgCl in the precipitate to dissolve. The equilibria involved are

$$AgCl(s) \rightleftharpoons Ag^+(aq) + Cl^-(aq)$$
$$Ag^+(aq) + 2NH_3(aq) \rightleftharpoons Ag(NH_3)_2^+(aq)$$

As NH_3 is added, the second equilibrium is shifted to the right, which decreases the concentration of Ag^+. This causes the first equilibrium also to be shifted to the right, and the AgCl dissolves.

In the final step in the test for Ag^+, HNO_3 is added to the mixture thought to contain the $Ag(NH_3)_2^+$. Hydrogen ion from the acid reacts with NH_3 to form NH_4^+. Since this removes molecular NH_3 from the solution, the second reaction above shifts to the left, which increases the Ag^+ concentration, and eventually the Ag^+ and Cl^- recombine to form a precipitate of AgCl.

Aqua regia is one part concentrated HNO_3 and three parts concentrated HCl, by volume.

Gold is found as the free element in nature.

Zn $3d^{10}4s^2$
Cd $4d^{10}5s^2$
Hg $5d^{10}6s^2$

Chemical Properties of Gold   Gold is so unreactive that even concentrated nitric acid fails to attack it. A mixture of concentrated HNO_3 and HCl, called aqua regia by the alchemists, will dissolve gold slowly. The chloride ion helps the nitrate ion to oxidize the gold by stabilizing the Au^{3+} ion in the form of a complex ion with the formula $AuCl_4^-$.

$$Au(s) + 6H^+(aq) + 3NO_3^-(aq) + 4Cl^-(aq) \longrightarrow AuCl_4^-(aq) + 3NO_2(g) + 3H_2O$$

Although gold forms compounds in both the $+1$ and $+3$ oxidation states, none of them are particularly important. In general, gold compounds have little stability and many are easily decomposed by heat.

Zinc, Cadmium, and Mercury

Zinc, as well as the other members of Group IIB, has a completed d subshell just beneath its outer pair of s electrons, and it is only these outer electrons that are involved when zinc reacts. Both electrons are lost, and zinc only occurs in compounds in the $+2$ oxidation state.

Zinc is found in the ore *zinc blende*, which is zinc sulfide, ZnS. The zinc is recovered by first roasting the ore to give zinc oxide, which is then reduced with carbon. It is a very reactive metal, and its surface quickly acquires a film of a basic carbonate, $Zn_2(OH)_2CO_3$. This coating protects the metal below from further oxidation.

The metal finds many applications in industry. For example, it is often used to provide a protective coating on steel—a process called *galvanizing*. The zinc is applied either in a thin coat electrolytically or in a much thicker coat by dipping the steel object in molten zinc, and it protects the steel in two ways. While the zinc coat is unbroken, of course, the steel is shielded from air and moisture. If the zinc is scratched deeply or worn away, the steel is still kept from corroding by cathodic protection, a phenomenon we discussed in Chapter 17. Zinc is more easily oxidized than iron, and a galvanic cell is formed in which iron is the cathode and zinc is the anode. Zinc is also used in various alloys. Examples are brass—an alloy of copper and zinc—and bronze—an alloy of copper, tin, and zinc. As noted earlier, pennies today are made of zinc and have just a thin coating of copper. Zinc is also used to make castings and it is important in the manufacture of zinc–carbon dry cells and other batteries.

Zinc is easily oxidized and reacts quickly with nonoxidizing acids such as HCl and H_2SO_4 to form Zn^{2+} and H_2, and with strong base to give the complex ion $Zn(OH)_4^{2-}$ and H_2.

$$Zn(s) + 2H^+(aq) \longrightarrow Zn^{2+}(aq) + H_2(g)$$
$$Zn(s) + 2OH^-(aq) + 2H_2O \longrightarrow Zn(OH)_4^{2-}(aq) + H_2(g)$$

Zinc hydroxide is also amphoteric. If base is added to a solution containing Zn^{2+} a precipitate of $Zn(OH)_2$ forms, but dissolves on the addition of more strong base.

$$Zn(OH)_2(s) + 2OH^-(aq) \longrightarrow Zn(OH)_4^{2-}(aq)$$

Zinc forms many complex ions similar to its complex with hydroxide ion. For example, if aqueous ammonia is added to $Zn(OH)_2$, the precipitate dissolves with the formation of an ammonia complex.

$$Zn(OH)_2(s) + 4NH_3(aq) \rightleftharpoons Zn(NH_3)_4^{2+}(aq) + 2OH^-(aq)$$

Zinc compounds are employed for many purposes. Zinc chloride, which is exceptionally soluble in water (38.1 mol/L at 25 °C), has a wide range of uses that extend from embalming, to fireproofing lumber, to the refining of petroleum. Zinc oxide, a white powder, is used in various creams such as sun screens and to make quick-setting dental cements. Zinc sulfide is interesting because it can be used to prepare *phosphors*—

In recent years, pennies have been made of zinc, with only a thin coating of copper. Here we see a new penny that has been dipped in nitric acid, which has dissolved this copper coating to expose the gray zinc beneath.

substances that glow when bathed in ultraviolet light or the high-energy electrons of cathode rays. Such phosphors are used on the inner surfaces of TV picture tubes and the CRT displays of computer monitors, and in devices for detecting atomic radiation (see Section 23.6).

Cadmium is less abundant than zinc and is usually found as an impurity in zinc ores. The free metal is soft and moderately active. Its chief use is as a protective coating on other metals and for making nickel–cadmium batteries.

Chemically, cadmium is like zinc in many respects. For example, it dissolves in nonoxidizing acids to form a $+2$ ion, which corresponds to cadmium's only oxidation state in compounds. Unlike zinc, however, cadmium is not affected by base. For this reason, cadmium can be used as a protective coating on metals that must be exposed to an alkaline environment. Cadmium compounds are quite toxic; if absorbed by the body, they can cause high blood pressure, heart disease, and even painful death.

Mercury is a metal known since ancient times. Its chief ore is the sulfide, HgS, known as *cinnabar*. Roasting the ore in air gives elemental mercury.

$$HgS(s) + O_2(g) \longrightarrow Hg(\ell) + SO_2(g)$$

As you know, mercury is a liquid at room temperature. Its freezing point is $-38.9\ °C$ and it boils at $357\ °C$, and this large convenient liquid temperature range accounts for one of mercury's most familiar uses—as the fluid in thermometers.

One of the useful properties of mercury is its ability to dissolve many other metals to form solutions called amalgams. Dentists use a silver amalgam containing an excess of silver to fill teeth. Gold also forms amalgams easily, and mercury is used to separate gold from its ores. The ore is mixed with mercury which dissolves the gold. Because mercury is so dense, rocks and other debris float on its surface and can easily be removed. The mercury is then distilled away, leaving the gold behind, and condensed to be used again.

Mercury is a much less active metal than either zinc or cadmium. It is unaffected by nonoxidizing acids, but it will dissolve in concentrated nitric acid. The product, $Hg(NO_3)_2$, was once used in the manufacture of felt for hats. Workers often developed severe mercury poisoning, an affliction that leads to nervous disorders, loss of hair and teeth, loss of memory, and even insanity—hence the term "mad as a hatter." Mercury compounds are relatively unstable. The oxide HgO, which forms when mercury is heated at moderate temperatures, decomposes when heated strongly. This thermal instability, in fact, was what allowed Joseph Priestly to discover oxygen in 1774, as discussed in Section 19.4.

In compounds, mercury occurs in two oxidation states, $+1$ and $+2$. In the $+1$ state, corresponding to the loss of only one of its outer s electrons, it exists as the Hg_2^{2+} ion (mercurous ion), which contains a mercury–mercury covalent bond.

$$[Hg\!-\!Hg]^{2+}$$

Addition of chloride ion to a solution of Hg_2^{2+} gives a white precipitate of Hg_2Cl_2, and this reaction is part of the qualitative analytic test for mercury in this oxidation state. Since Ag^+ and Pb^{2+} also give white chloride precipitates, confirmation of mercury(I) is obtained when aqueous ammonia is added, which causes Hg_2Cl_2 to disproportionate.

$$Hg_2Cl_2(s) + 2NH_3(aq) \longrightarrow Hg(\ell) + Hg(NH_2)Cl(s) + NH_4^+(aq) + Cl^-(aq)$$

The compound $Hg(NH_2)Cl$ is called *mercury amido chloride* and contains mercury in the $+2$ oxidation state combined with the amide ion, NH_2^-, and Cl^-. The mixture of black finely divided mercury and white $Hg(NH_2)Cl$ usually appears dark gray or black.

Mercury(I) chloride, which is also known as *calomel*, is very insoluble in water. Its low solubility has permitted therapeutic uses as an antiseptic and as a treatment for syphilis before the discovery of penicillin. Very little mercury is retained by the body

Cadmium and mercury are both toxic, but zinc is needed by the body.

A photograph of liquid mercury is on page 62.

The ease of thermal decomposition of mercury compounds is the reason mercury was discovered so long ago.

Hg_2Cl_2 has been used in agriculture to control root maggots on onions and cabbage.

because so little Hg_2Cl_2 is able to dissolve. It is actually molecular and exists as linear molecules with a Hg—Hg bond.

$$:\overset{\cdot\cdot}{\underset{\cdot\cdot}{Cl}}—Hg—Hg—\overset{\cdot\cdot}{\underset{\cdot\cdot}{Cl}}:$$

Mercury(II) chloride is unusual because even though water soluble it hardly dissociates at all in solution. It exists as molecules of $HgCl_2$ in which covalent bonds exist between Hg and Cl.

$$:\overset{\cdot\cdot}{\underset{\cdot\cdot}{Cl}}—Hg—\overset{\cdot\cdot}{\underset{\cdot\cdot}{Cl}}:$$

However, because of its water solubility, $HgCl_2$ is very poisonous.

Addition of H_2S to a solution containing mercury(II) gives a black precipitate of very insoluble HgS. This is an interesting compound because when it is heated its crystal structure changes and it becomes a brilliant red substance called *vermilion*.

22.3 COMPLEXES OF THE TRANSITION METALS

The tendency of the transition metals to form complexes has made the study of their composition, structure, and bonding a major area of research in chemistry.

In Chapter 11 we introduced you to complex ions, and you learned about the kinds of molecules and ions that serve as ligands and how to write the formulas of complexes. Since then, we've discussed how complex ions influence the solubilities of salts, and in this chapter we've noted often the tendency of the various transition metal ions to form complex ions. The number of such complexes, especially of the transition elements, is enormous and chemists have done a tremendous amount of work studying their properties and chemical reactions. In the space we have available here, we certainly cannot delve very deeply into this subject, so our goal will simply be to give you a taste of it. If you have any difficulty following the discussion because you've forgotten what some of the terms mean, we suggest you review Section 11.10, which begins on page 428.

Coordination Number and Structure

One of the most interesting aspects of the study of complexes is the kinds of structures that are formed. In many ways, this is related to the number of donor atoms attached to the metal ion, and we call this number the coordination number. For example, in the complex $Cu(H_2O)_4{}^{2+}$, the copper is surrounded by the four oxygen atoms that belong to the water molecules, so the coordination number of Cu^{2+} in this complex is 4. Similarly, the coordination number of Cr^{3+} in the $Cr(H_2O)_6{}^{3+}$ ion is 6, and the coordination number of Ag^+ in $Ag(NH_3)_2{}^+$ is 2.

Sometimes the coordination number isn't immediately obvious from the formula of the complex. For example, in Chapter 11 you learned that there are many ligands that contain more than one donor atom that can be used to bind a metal ion. These are called *polydentate ligands* in general. More specifically, they are *bidentate ligands* if they contain two donor atoms, *tridentate ligands* if they contain three, and so forth. Two typical examples of bidentate ligands are ethylenediamine (normally abbreviated *en* in a chemical formula) and oxalate ion, $C_2O_4{}^{2-}$, whose structures are

The bright red mineral cinnabar, also called vermilion, is composed of mercury(II) sulfide, HgS.

This is one exception to the rule that soluble metal salts are fully dissociated into ions in aqueous solution.

$$H_2N—CH_2—CH_2—NH_2$$
ethylenediamine (en)

$$\left[\overset{\cdot\cdot}{\underset{\cdot\cdot}{O}} \diagdown \overset{\cdot\cdot}{\underset{\cdot\cdot}{O}} \atop C—C \diagup \overset{\cdot\cdot}{\underset{\cdot\cdot}{O}} \right]^{2-}$$
oxalate ion

You may recall that these form complexes with an ion of a metal *M* by donating electron pairs from two of their atoms. The complexes that result have ring systems and are called chelates. For ethylenediamine and oxalate ion, complexes are formed as follows:

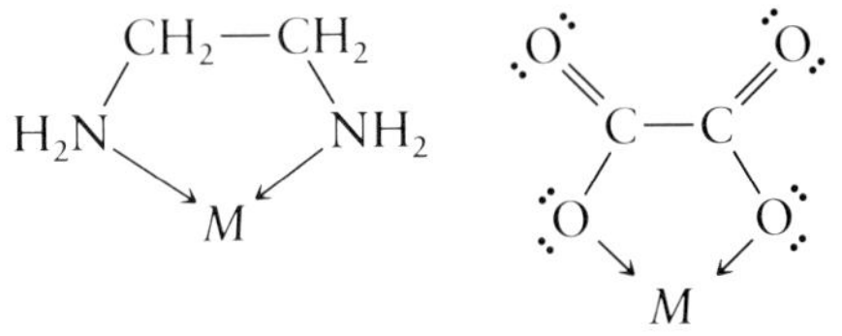

Often, two or more polydentate ligands are able to bind to a metal ion to give complexes with formulas such as $[Cr(en)_3]^{3+}$ and $[Cr(en)_2(H_2O)_2]^{3+}$. In each of these examples, the coordination number of the Cr^{3+} is 6. In the $[Cr(en)_3]^{3+}$ ion, there are three ethylenediamine ligands that each supply two donor atoms, for a total of 6, and in $[Cr(en)_2(H_2O)_2]^{3+}$ the two ethylenediamine ligands supply a total of 4 donor atoms and the two H_2O molecules supply another 2; again the total is 6.

The Chelate Effect   One of the interesting aspects of the complexes formed by ligands such as ethylenediamine and oxalate ion is their stabilities compared to similar complexes formed by monodentate ligands. For example, the complex $Ni(en)_3^{2+}$ is considerably more stable than $Ni(NH_3)_6^{2+}$, even though both complexes have six nitrogen atoms bound to a Ni^{2+} ion. This phenomenon is called the chelate effect, so named because it occurs in compounds that have these *chelate ring* structures.

> Monodentate ligands can supply only one donor atom to a metal.

The reason for the chelate effect appears to be related to the probability of the ligand being removed from the vicinity of the metal ion when a donor atom becomes detached. If one end of a bidentate ligand comes loose from the metal ion, the donor atom cannot wander very far because the other end of the ligand is still attached to the metal. There is a high probability that the loose end will become reattached to the metal ion before the other end can let go, so the ligand appears to be bound tightly. With a monodentate ligand, however, there is nothing to hold the ligand near the metal ion if it becomes detached. The ligand can easily wander off into the surrounding solution and be lost. As a result, a monodentate ligand doesn't behave as if it is as firmly attached to the metal ion as a polydentate ligand.

Geometries of Complexes   In our previous discussions of molecular geometry, we saw that VSEPR theory was quite effective at predicting the structures of molecules and ions in which the central atom was a representative element. Although the underlying principles of VSEPR theory—repulsions between electron pairs that surround the central ion—still influence the structures of complex ions, they don't always give the correct geometries when the central metal ion has electrons in a partially filled *d* subshell. For that reason, we don't attempt to predict the shapes of complexes by the VSEPR approach. Nevertheless, there are certain geometries that are usually associated with particular coordination numbers. We will examine these for the most common coordination numbers: 2, 4, and 6.

> Ordinarily, VSEPR theory isn't used to predict the structures of transition metal complexes because it can't be relied on to always give correct results.

Coordination Number 2   This includes complexes such as $Ag(NH_3)_2^+$ and $Ag(CN)_2^-$. Usually, these complexes have a linear structure such as

$$[H_3N—Ag—NH_3]^+ \quad \text{and} \quad [NC—Ag—CN]^-$$

(Since the Ag^+ ion has a filled *d* subshell, it behaves as any of the representative elements as far as predicting geometry by VSEPR theory, so these structures are exactly what we would expect based on that theoretical model.)

Coordination Number 4   Two common geometries occur when four ligand atoms are bonded to a metal ion — tetrahedral and square planar. These are illustrated in Figure

FIGURE 22.9 Tetrahedral and square planar geometries that occur for complexes in which the metal ion has a coordination number of four. For the copper complex, we are viewing the square planar structure tilted back into the plane of the paper.

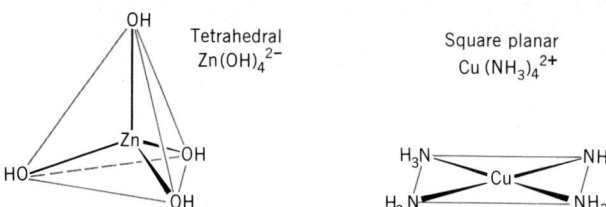

22.9. The tetrahedral geometry is usually found with metal ions that have completed d subshells, such as Zn^{2+}. The complexes $Zn(NH_3)_4^{2+}$ and $Zn(OH)_4^{2-}$ are examples.

Square planar geometries are observed for complexes of Cu^{2+}, Ni^{2+}, and especially Pt^{2+}. Examples are $Cu(NH_3)_4^{2+}$, $Ni(CN)_4^{2-}$, and $PtCl_4^{2-}$. Because they are considerably more stable than the others, the most well-studied square planar complexes are those of Pt^{2+}.

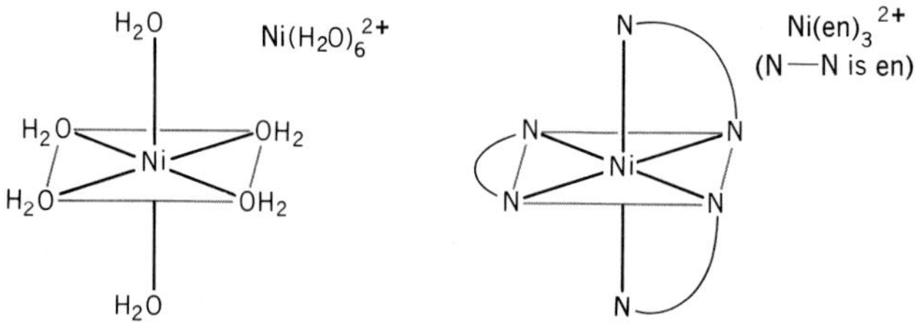

FIGURE 22.10 Octahedral complexes can be formed with either monodentate ligands such as water or with polydentate ligands such as ethylenediamine (en). To simplify the drawing of the ethylenediamine complex, the atoms joining the donor nitrogen atoms in the ligand, $-CH_2-CH_2-$, are represented as the curved line between the N atoms. Also notice that the nitrogen atoms of the bidentate ligand span adjacent positions within the octahedron. This is the case for all polydentate ligands that you will encounter in this book.

Coordination Number 6   The most common coordination number for complex ions is 6. Examples are $[Al(H_2O)_6]^{3+}$, $[Co(C_2O_4)_3]^{3-}$, $[Ni(en)_3]^{2+}$, and $[Co(EDTA)]^-$. With few exceptions, all complexes with a coordination number of 6 are octahedral. This holds true for those formed from both monodentate and bidentate ligands, as illustrated in Figure 22.10. Octahedral coordination is also observed for metal ions in a variety of biologically important complexes. One example is the structure of cyanocobalamin, the form of vitamin B_{12} found in vitamin pills, as shown in Figure 22.11. The principal ligand structure in this and several other biologically active complexes is the square planar arrangement of nitrogen atoms in a polydentate ligand structure called the *porphyrin structure*.

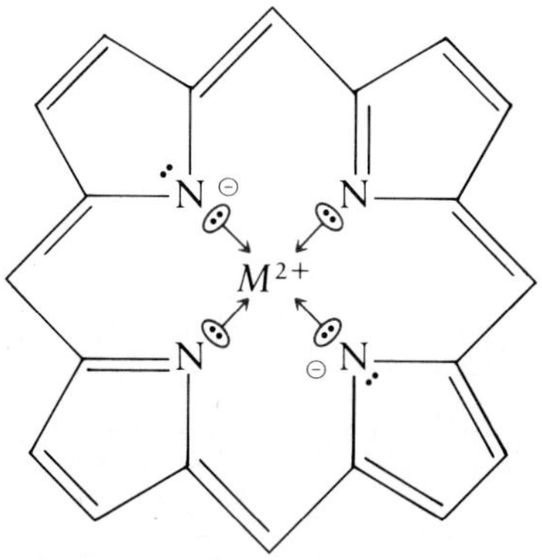

porphyrin structure

This same basic ligand structure is found not only in vitamin B_{12}, but also in hemoglobin and myoglobin, where Fe^{2+} is held (see Figure 24.5), and in chlorophyll, where Mg^{2+} is held.

In describing the shapes of octahedral complexes, most chemists use the simplified drawing of the octahedron shown in Figure 22.12.

PRACTICE EXERCISE 1

What is the coordination number of the metal ion in (a) $[Cr(C_2O_4)_3]^{3-}$, (b) $[Co(C_2O_4)_2Cl_2]^{3-}$, (c) $[Cr(en)(C_2O_4)_2]^-$, and (d) $[Co(EDTA)]^-$?

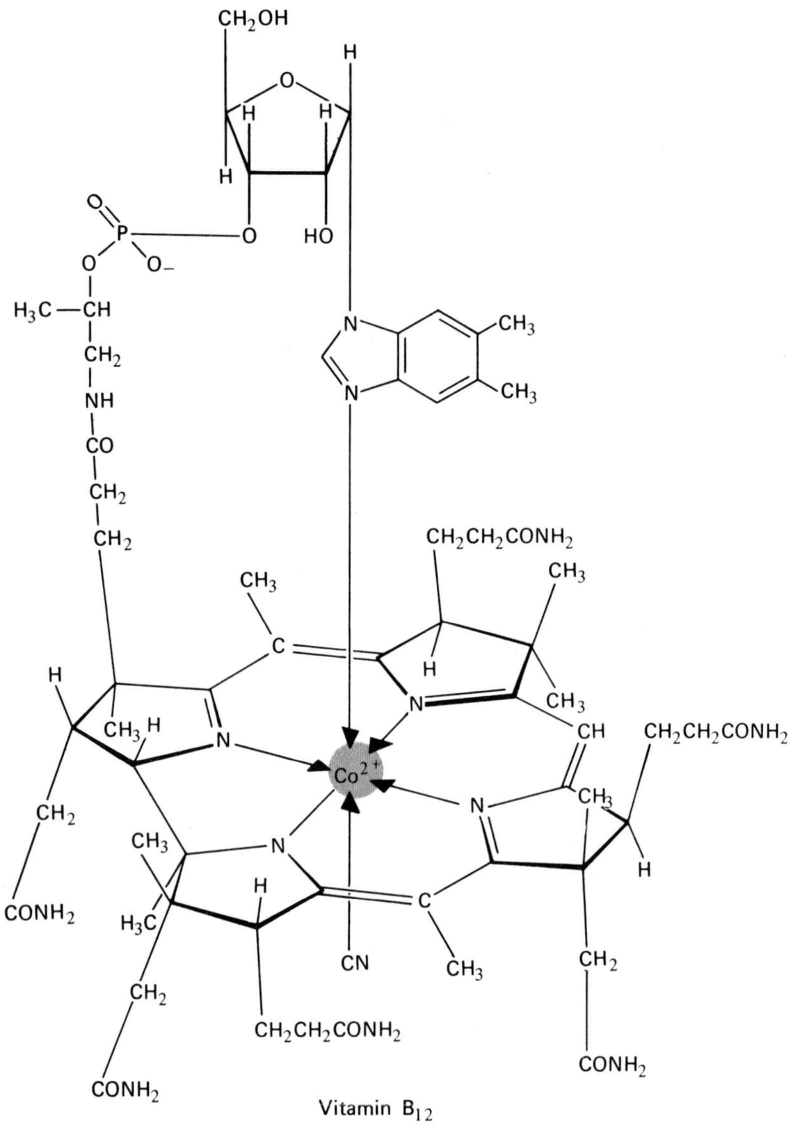

FIGURE 22.11 The structure of cyanocobalamin. Notice the cobalt ion in the center of the square planar arrangement of nitrogen atoms that are part of the porphyrin structure. Overall, the cobalt is surrounded octahedrally by donor atoms.

Vitamin B_{12}

FIGURE 22.12 Drawing the structure of an octahedral complex.

Step 1. Draw a parallelogram to represent a square viewed in perspective.	Step 2. Draw a vertical line upward from the center.	Step 3. Draw a vertical line of the same length downward. Erase the portion that would be hidden by the plane.	Step 4. Draw four lines to the center of the "square"	Step 5. Fill in the symbols for the atoms. For $CrCl_6^{3-}$ we would have the following structure.

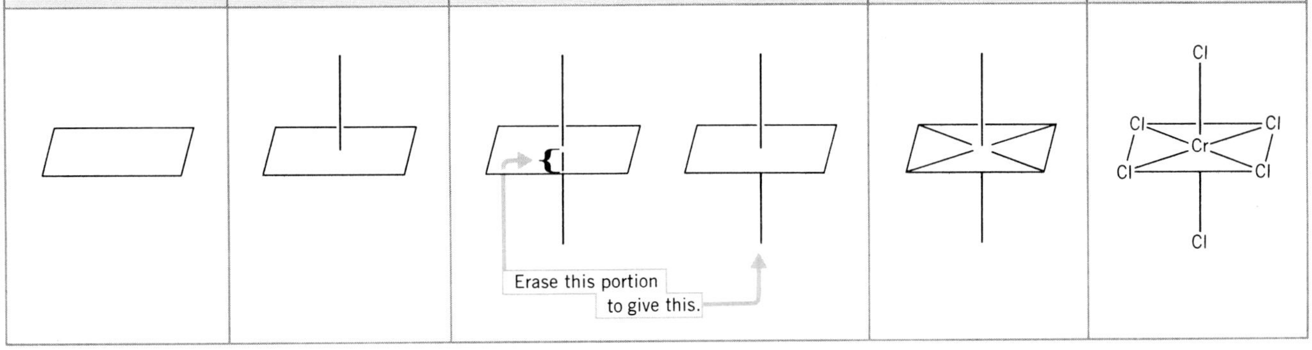

Erase this portion to give this.

22.4 ISOMERS OF COORDINATION COMPOUNDS

It is possible to have two or more different compounds with the same chemical formula because the atoms in the formula can be arranged in different ways.

When you write the chemical formula for a particular compound, you might be tempted to think that you should also be able to predict exactly what the structure of the molecule or ion is. As you may realize by now, this just isn't possible in many cases. Sometimes we can use simple rules to make reasonable structural guesses, as in the discussion of the drawing of Lewis structures in Chapter 7, but these rules apply only to simple species. (And you have been asked to apply them only to those cases where they work!) For more complex substances, there usually is no way of knowing for sure what the structure of the molecule or ion is without performing the necessary experiments to determine the structure.

One of the reasons that structures can't be predicted with certainty from chemical formulas alone is that there are usually many different ways that the atoms in the formula can be arranged. In fact, it is frequently possible to isolate two or more compounds that actually have the same chemical formula. For example, three different solids, each with its own characteristic color and other properties, can be isolated from a solution of chromium(III) chloride. All three have the same overall composition. In the absence of other data, their formulas can be written $CrCl_3 \cdot 6H_2O$. However, experiments have shown that these solids are actually the salts of three different complex ions. Their actual formulas (and colors) are

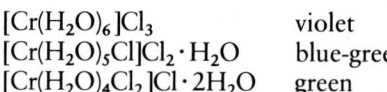

$[Cr(H_2O)_6]Cl_3$	violet
$[Cr(H_2O)_5Cl]Cl_2 \cdot H_2O$	blue-green
$[Cr(H_2O)_4Cl_2]Cl \cdot 2H_2O$	green

Even though their overall compositions are the same, each of these substances is a distinct chemical compound with its own characteristic set of properties.

The existence of two or more compounds, each having the same chemical formula, is known as *isomerism*. In the example above, each salt is said to be an *isomer* of $CrCl_3 \cdot 6H_2O$. For coordination compounds, there are a variety of ways for isomerism to occur. In the example above, isomers exist because of the different possible ways that the water molecules and chloride ions can be held in the crystals. In one instance, all the water molecules serve as ligands, while in the other two, part of the water is present as water of hydration and some of the chloride is bonded to the metal ion. Another example, which is similar in some respects, is $Cr(NH_3)_5SO_4Br$. This "substance" can be isolated as two isomers. One reacts with Ag^+ to give a precipitate of AgBr, and the other reacts with Ba^{2+} to give a precipitate of $BaSO_4$. The first isomer is

$$[Cr(NH_3)_5SO_4]Br$$

which dissolves in water to give the ions $[Cr(NH_3)_5SO_4]^+$ and Br^-. The second isomer is

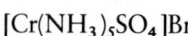

$$[Cr(NH_3)_5Br]SO_4$$

which dissolves in water to give the ions $[Cr(NH_3)_5Br]^{2+}$ and SO_4^{2-}. In the first isomer, the chromium is bonded to five ammonia molecules and a sulfate ion. In the second, the chromium is bonded to five ammonia molecules and a bromide ion. Although both isomers have the same overall composition, they have distinctly different chemical properties, and they are therefore different compounds.

Stereoisomerism

The most interesting kind of isomerism found among coordination compounds is called *stereoisomerism* — differences among compounds that arise as a result of the various

Chromium(III) chloride purchased from chemical supply companies is actually $[Cr(H_2O)_4Cl_2]Cl \cdot 2H_2O$. Its green color in both the solid state and in solution is due to the complex ion, $[Cr(H_2O)_4Cl_2]^+$.

This salt reacts with Ag^+ but not with Ba^{2+}.

This salt reacts with Ba^{2+} but not with Ag^+.

possible orientations of their atoms in space. When stereoisomerism exists, we have compounds in which the same atoms are attached to each other, but they differ in the way those atoms are arranged in space relative to one another.

One form of stereoisomerism is called geometric isomerism, and it is best understood by considering an example. Consider the square planar complexes having the formula $Pt(NH_3)_2Cl_2$. There are two ways to arrange the ligands around the platinum. In one isomer, called the *cis* isomer, the chloride ions are next to each other and the ammonia molecules are also next to each other. In the other isomer, called the *trans* isomer, identical ligands are opposite each other. In identifying (and naming) isomers, *cis* means "on the same side," and *trans* means "on opposite sides."

Cl   NH₃   H₃N   Cl
  \   /   \   /
   Pt   and   Pt
  /   \   /   \
Cl   NH₃   Cl   NH₃
     cis   *trans*

The isomer on the left, *cis*-$Pt(NH_3)_2Cl_2$, is the anticancer drug known as *cisplatin*. It is interesting that only the *cis* isomer is clinically active against tumors. The *trans* isomer is totally ineffective.

Geometric isomerism also occurs for octahedral complexes. For example, consider the ions $Cr(H_2O)_4Cl_2{}^+$ and $Cr(en)_2Cl_2{}^+$. Both can be isolated as *cis* and *trans* isomers.

$Cr(H_2O)_4Cl_2{}^+$

cis   *trans*

$Cr(en)_2Cl_2{}^+$
N—N is en
(ethylenediamine)

cis   *trans*

In the *cis* isomer, the chloride ligands are both on the same side of the metal ion; in the *trans* isomer, the chloride ligands are on opposite ends of a line that passes through the center of the metal ion.

Chirality

There is a second kind of stereoisomerism that is much more subtle than geometric isomerism. This occurs when molecules are exactly the same except for one small difference — they differ from each other in the same way that your left hand differs from your right hand. Before we see how this applies to molecules, however, let's examine how a left hand and a right hand are alike, and how they are different. They certainly are similar in that they each have four fingers and a thumb, but they are also different, as you can tell if you try to fit your right hand into a left-hand glove (Figure 22.13).

FIGURE 22.13 A left-hand glove won't fit the right hand.

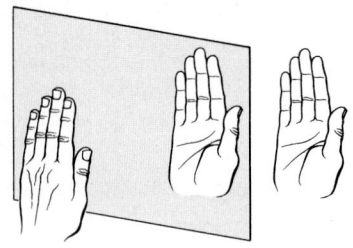

FIGURE 22.14 The image of the left hand reflected in the mirror appears the same as the right hand.

The special relationship between the two hands can be seen if you stand in front of a mirror. If you hold your left hand so that it is facing the mirror, and then look at its reflection as illustrated in Figure 22.14, you will see that it looks exactly like your right hand. (Try it.) Notice that the thumb on the hand in the mirror points in the same direction as the thumb of your right hand, and that both palms are facing you. If it were possible to reach "through the looking glass," your right-hand glove would fit the reflection of your left hand perfectly.

In this analysis, we say that our left and right hands are mirror images of each other. Hands are not unique in this way—in many other pairs of objects, one is the mirror image of the other. For example, if you have two thumbtacks and look at the reflection of one of them in a mirror, it will look exactly like the other. Yet there is a difference between thumbtacks and hands. This difference is seen in a test for "handedness" that we call superimposability.

To see how the test works, bring your left and right hands together so that they are facng each other, palm to palm and thumb to thumb. What you are doing is "imposing them." Superimposing goes beyond that, but it can only be done in the imagination. Pretend that you were able to blend your left hand *into* your right hand, so that it is entirely inside. Will all the parts match perfectly? The answer is no. The fingers may seem to be blending all right, but notice that the fingernails on one hand come out where the fingerprints are on the other hand. The palms do not superimpose either. One palm comes out on the backside of the other hand. What if you turn one hand over to get the palms to superimpose? Then the thumbs point in opposite directions (Figure 22.15). The left and right hands fail the test of superimposability because we can't get all the parts to match at the same time. In the final analysis, *only if two objects pass the test of superimposability may we call them identical.* If we try the same test with the two thumbtacks, we can see that they would fit one inside the other, with no mismatch of parts. One thumbtack is exactly like another, and we can say this because they are superimposable.

A thumbtack and its mirror image are identical; they are superimposable.

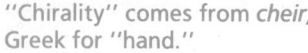

"Chirality" comes from *cheir*, Greek for "hand."

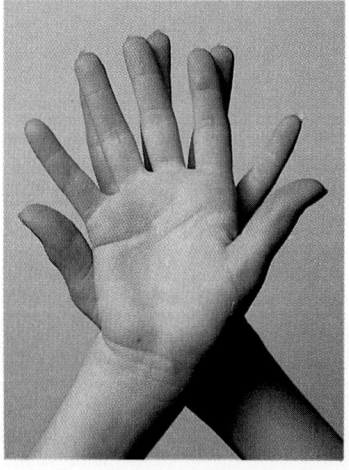

FIGURE 22.15 Your left hand and your right hand are related as object-to-mirror image, but they do not superimpose.

The technical term for handedness is chirality. Two plain thumbtacks do not have chirality, but your left and right hands do. Two gloves in a pair also have chirality. The test for chirality is the test for superimposability. *If one object is the mirror image of the other, and if they are not superimposable, then the objects are chiral.*

Let's move now from the world of hands, gloves, and tacks back to the realm of molecules and ions. Here we can apply the same test for handedness as before to determine whether two structures are actually the same (and therefore not really two structures, but one), or whether they are different.

The most common examples of chirality among coordination compounds occur with octahedral complexes that contain two or three bidentate ligands—for instance,

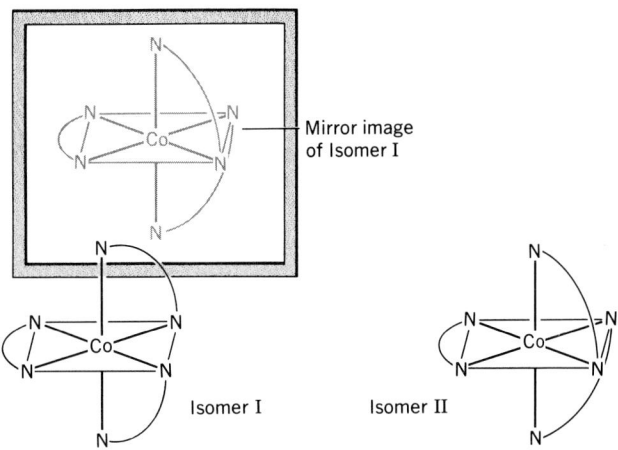

FIGURE 22.16 The two isomers of Co(en)$_3$$^{3+}$. Isomer II is constructed as the mirror image of isomer I. No matter how Isomer II is turned about, it is not superimposable on Isomer I.

Co(en)$_2$Cl$_2$$^+$ and Co(en)$_3$$^{3+}$. For the complex Co(en)$_3$$^{3+}$, the two nonsuperimposable isomers, called **enantiomers,** are shown in Figure 22.16. For the complex Co(en)$_2$Cl$_2$$^+$, only the *cis* isomer is chiral, as described in Figure 22.17.

As you can see, chiral isomers differ in only a very minor way from each other. This difference is so subtle that most of their properties are identical. They have identical melting points and boiling points, and in nearly all of their reactions, their behavior is exactly alike. The only way that the difference between chiral molecules or ions manifests itself is in the way that they interact with physical or chemical "probes" that also have a handedness about them. For example, if two reactants are both chiral, then a given

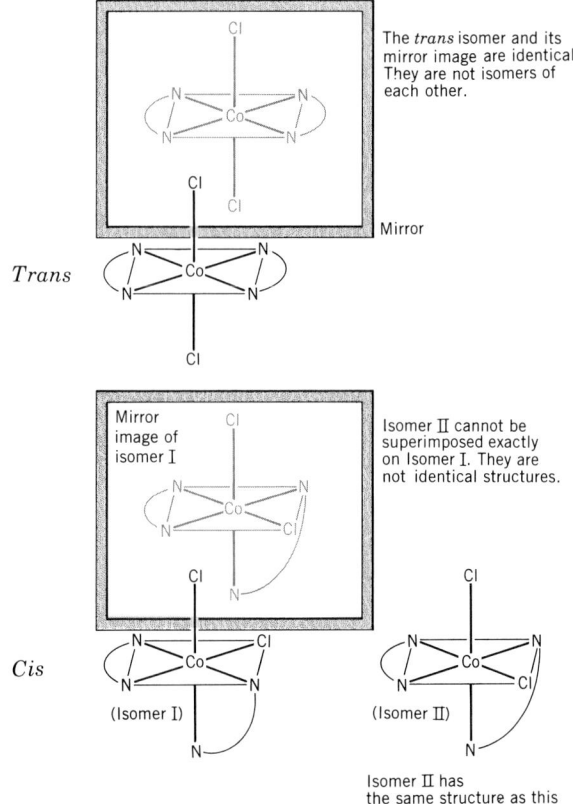

FIGURE 22.17 Isomers of the Co(en)$_2$Cl$_2$$^+$ ion. The mirror image of the *trans* isomer can be superimposed exactly on the original, so the *trans* isomer is not chiral. The *cis* isomer is chiral, however, because its mirror image cannot be superimposed on the original.

Since the molecules of enantiomers have identical internal geometries—bond angles and lengths—and the same atoms or groups are joined to the same central atoms, enantiomers must also have identical polarities. Therefore, enantiomers must and do have identical boiling points, melting points, densities, and other physical properties, except for one difference: *Solutions of enantiomers behave oppositely toward polarized light.*

If you and a friend each have a pair of Polaroid® sunglasses, get them out and try these experiments. The light that gets through the lens of one of these glasses is plane polarized light, which means that its vibrations are only in one plane, not in every plane. Put a second Polaroid® lens over one lens in the first pair of glasses, and then slowly turn one with respect to the other, holding them up to a light. The accompanying figure shows this being done to two pieces of Polaroid film. Find a position of the two lenses where the light that can get through *both* is the brightest. Now rotate one lens 90° with respect to the other. Notice how the light getting through both first gradually grows dimmer and then goes out. Turning one lens is like pinching the vibrations out of a guitar string until there is no sound. Turn one lens another 90°; the brightness comes back, doesn't it? Turn it still another 90°, and it goes out again; go still another 90° and back the brightness comes.

Now imagine the two films positioned to let no light through. Here's a part you will have to imagine because it takes special equipment to show it. If you placed a solution of an enantiomer between the two films it would restore at least some of the light's brightness. You would have to rotate one film a certain measurable number of degrees to shut out the light again, say, 10° *clockwise*. If you repeated this experiment with the other enantiomer, the same thing would

No light gets through where two Polaroid® lenses overlap and are oriented properly with respect to each other.

happen—except you would have to rotate one film 10° *counterclockwise*. This is the only physical difference between enantiomers. They affect polarized light the same number of degrees but in opposite directions. Brightness is restored by the enantiomer because the electron clouds of its molecules force the plane polarized light to experience a twist or rotation in the plane of its vibration. One enantiomer rotates the plane in one direction; the other, an identical number of degrees in the opposite direction. This ability of an enantiomer to rotate the plane of plane polarized light is called *optical activity*. Each enantiomer is said to be optically active, and one enantiomer is sometimes called an *optical isomer* of the other.

isomer of one of them will usually behave slightly differently toward each of the two isomers of the other reactant. As you will see in Chapter 24, this has very profound effects in biochemical reactions, in which nearly all of the molecules involved are chiral.

Chiral isomers like those described in Figures 22.16 and 22.17 also have a peculiar effect on polarized light, which is described in Special Topic 22.2. Because of this phenomenon, chiral isomers are said to be optical isomers.

22.5 BONDING IN COMPLEXES

The colors and magnetic properties of complexes are explained by the way the ligands affect the energies of the *d* orbitals.

Complexes of the transition metals are set apart from the complexes of other metals in two special ways: (1) they are usually colored, whereas complexes of the representative metals are usually (but not always) colorless, and (2) their magnetic properties are often

FIGURE 22.18 Each of these
brightly colored compounds
contains a complex ion of Co³⁺. The
variety of colors arises because of
the different ligands (molecules or
anions) that are bonded to the
cobalt ion in the complexes.

affected by the ligands attached to the metal ion. For example, it is not unusual for a given metal ion to form complexes with different ligands to give a rainbow of colors, as illustrated in Figure 22.18 for a series of complexes of cobalt. Also, because the transition metal ions often have incompletely filled d subshells, we expect to find many of them with unpaired d electrons, and compounds that contain them should be paramagnetic. But for a given metal ion, the number of unpaired electrons is not always the same from one complex to another. For example, Fe^{2+} has four of its six $3d$ electrons unpaired in the $Fe(H_2O)_6^{2+}$ ion, but all of its electrons are paired in the $Fe(CN)_6^{4-}$ ion. As a result, the $Fe(H_2O)_6^{2+}$ ion is paramagnetic and the $Fe(CN)_6^{4-}$ ion is diamagnetic.

To be successful, any theory that attempts to explain the bonding in complex ions must also explain their colors and magnetic properties. One of the simplest theories that does this is the crystal field theory. The theory gets its name from its original use in explaining the behavior of transition metal ions in crystals. It was discovered later that the theory works well for transition metal complexes, too.

In its simplest form, crystal field theory ignores covalent bonding in complexes. It assumes that the basic stability comes just from the electrostatic attractions between the positively charged metal ion and the negative charges of the ligand anions or dipoles. The crystal field theory's unique approach, though, is the way it examines how the negative charges on the ligands affect the energy of the complex by influencing the energies of the d orbitals of the metal ion, and it is this that we focus our attention on here. To understand the theory, therefore, it is very important that you know how the d orbitals are shaped and especially how they are oriented in space relative to each other. The d orbitals were described in Chapter 6, and they are illustrated again in Figure 22.19.

Notice that four of the d orbitals — the ones labeled $d_{x^2-y^2}$, d_{xy}, d_{xz}, and d_{yz} — have four lobes of electron density, and the fifth one — the one labeled d_{z^2} — has two lobes

More complete theories consider the covalent nature of metal–ligand bonding, but crystal field theory nevertheless provides a useful model for explaining the colors and magnetic properties of complexes.

FIGURE 22.19 The shapes and directional properties of the five d orbitals of a d subshell.

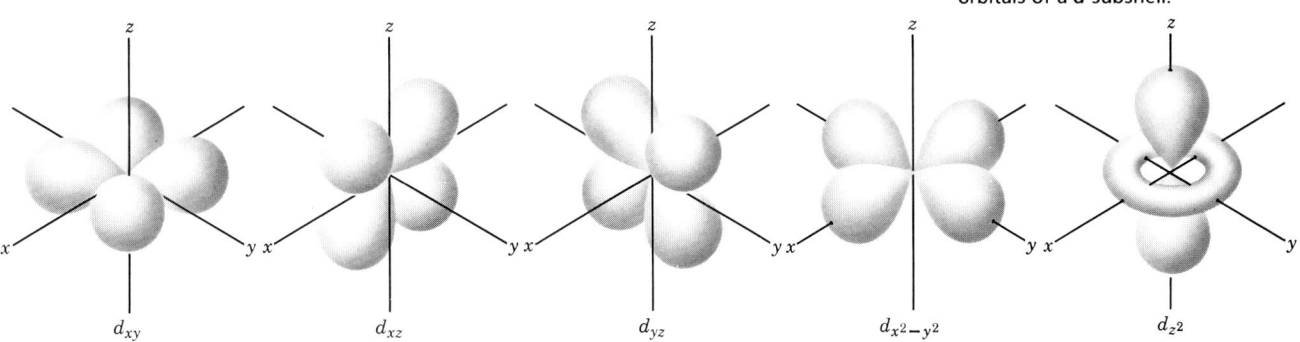

The labels for the *d* orbitals come from the mathematics of quantum mechanics.

that point in opposite directions along the *z* axis plus a small donut-shaped ring of electron density around the center that is concentrated in the *xy* plane.

Of prime importance to us are the *directions* in which the lobes of the *d* orbitals point. Notice that three of them — d_{xy}, d_{xz}, and d_{yz} — point *between* the *x*, *y*, and *z* axes. The other two — the d_{z^2} and $d_{x^2-y^2}$ orbitals — have their maximum electron densities along the *x*, *y*, and *z* axes.

Now let's consider constructing an octahedral complex within this coordinate system. We can do this by bringing ligands in along each of the axes as shown in the figure in the margin. The question we want to answer is "How do these ligands affect the energies of the *d* orbitals?"

As you know, in an isolated atom or ion, all the *d* orbitals of a given *d* subshell have the same energy. This means that an electron will have the same energy regardless of the *d* orbital it occupies. In an octahedral complex, however, this is no longer true. If the electron is in the $d_{x^2-y^2}$ or d_{z^2} orbital, it is forced to be nearer the negative charge of the ligands than if it is in a d_{xy}, d_{xz}, or d_{yz} orbital. Since the electron itself is negatively charged and is repelled by the charges of the ligands, the electron's potential energy will be higher in the $d_{x^2-y^2}$ and d_{z^2} orbitals than it will be if it occupies a d_{xy}, d_{xz}, or d_{yz} orbital. So as the complex is formed, the *d* subshell actually splits into two energy levels, as shown in Figure 22.20. Here we see that regardless of which orbital the electron occupies, its energy increases because it is repelled by the negative charges of the approaching ligands, but it is repelled *more* (and has a higher energy) if it is in an orbital that points directly at the ligands than if it occupies an orbital that points between them.

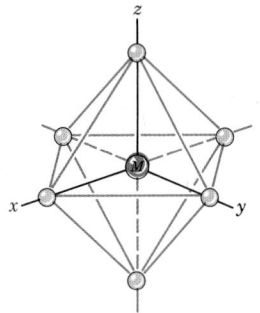

An octahedral complex ion with ligands along the *x y*, and *z* axes.

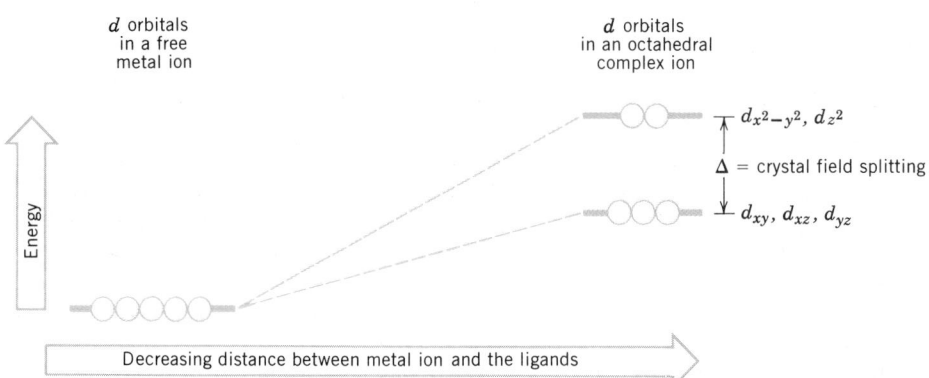

FIGURE 22.20 The changes in the energies of the *d* orbitals of a metal ion as an octahedral complex is formed.

In the octahedral complex, the energy difference between the two sets of *d*-orbital energy levels is called the crystal field splitting. It is usually given the symbol Δ (delta). The magnitude of Δ is very important in determining the colors and magnetic properties of complexes. It depends on a number of factors, one of which is the nature of the ligands attached to the metal. We will have more to say about this later. Another factor that affects the size of Δ is the oxidation state of the metal. As electrons are removed from a metal and the charge on the ion becomes more positive, the ion becomes smaller. This means that the ligands are attracted to the metal more strongly and they can approach the center of the complex more closely. As a result, they also approach the *d* orbitals along the *x*, *y*, and *z* axes more closely, and thereby cause a greater repulsion. This causes a greater splitting of the two *d*-orbital energy levels and a larger value of Δ. Therefore, for a given metal and ligand, the higher the oxidation state of the metal, the larger will be the value of Δ.

Now that we've seen how the ligands affect the energy level diagram for the *d* orbitals, let's examine how it helps us explain some properties. We will consider three of these by way of examples: the relative stabilities of oxidation states, the colors of complexes, and magnetic properties.

Relative Stabilities of Oxidation States

In our discussion of the chemistry of chromium, we mentioned that the Cr^{2+} ion is very easily oxidized to Cr^{3+}. This is really pretty simple to understand based on crystal field theory. In water, we expect each of these ions to actually exist as a complex— $Cr(H_2O)_6^{2+}$ and $Cr(H_2O)_6^{3+}$, respectively. Let's examine the energies and electron populations of the d-orbital energy levels in each of them (Figure 22.21).

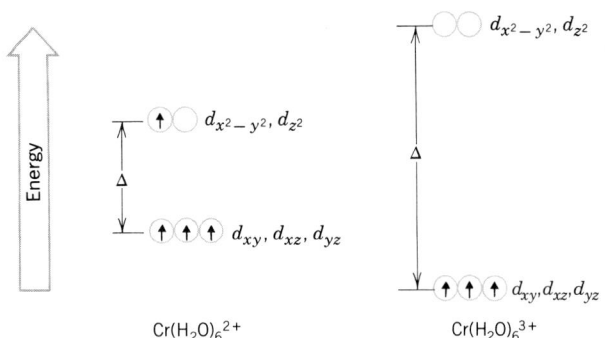

FIGURE 22.21 Energy level diagrams for the $Cr(H_2O)_6^{2+}$ and $Cr(H_2O)_6^{3+}$ ions.

The element chromium has the electron configuration

$$Cr \qquad [Ar]3d^5 4s^1$$

Removing two electrons gives the Cr^{2+} ion, and removing three gives Cr^{3+}.

$$Cr^{2+} \qquad [Ar]3d^4$$
$$Cr^{3+} \qquad [Ar]3d^3$$

Following our usual rules, we distribute these among the various orbitals, but for Cr^{2+} we have to make a choice. Should they all be forced into the lower of the two energy levels, or should they be spread out? From the diagram we see that the fourth electron does not pair with one of the others in the lower d-orbital level. Instead, three electrons go into the lower level and the fourth is in the upper level. We will discuss why this happens later, but for now, let's use the two energy diagrams to explain why Cr^{2+} is so easy to oxidize.

There are actually two factors that favor the oxidation of Cr(II) to Cr(III). First, notice that the electron that is removed comes from the higher energy level, so oxidizing the chromium(II) *removes* a high-energy electron. Second, we note the effects caused by raising the chromium to a higher oxidation state. As we've pointed out, this increases the magnitude of Δ, and as you can see, the energy of the electrons that remain is lowered. Both the removal of a high-energy electron and the lowering of the energy of the electrons that are left behind help make the oxidation occur, and the $Cr(H_2O)_6^{2+}$ ion is very easily oxidized to $Cr(H_2O)_6^{3+}$.

Colors of Complex Ions

When light is absorbed by something, the energy of the photon raises an electron from one energy level to another. In many substances, such as sodium chloride, the energy difference between the highest-energy populated level and the lowest-energy unpopulated level is quite large, so the frequency of a photon that carries the necessary energy lies outside the visible region of the spectrum. The substance appears white because visible light is unaffected; it is reflected unchanged.

For complex ions of the transition metals, the energy difference between the d-orbital energy levels is not very large, and photons with frequencies in the visible region

Remember, $E = h\nu$.

FIGURE 22.22 (a) The electron
distribution in the ground state of
the $Cr(H_2O)_6^{3+}$ ion. (b) Light energy
raises an electron from the lower
energy set of d orbitals to the
higher energy set.

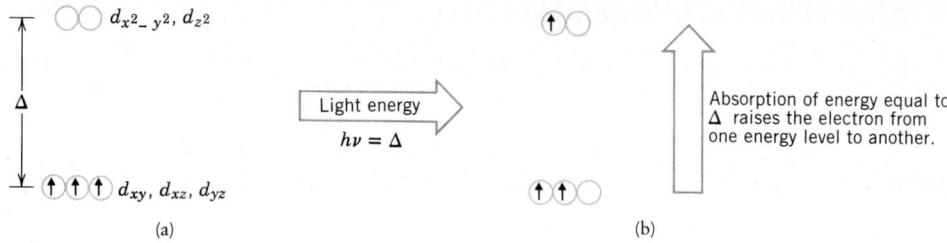

of the spectrum are able to raise an electron from the lower-energy set of d orbitals to the higher-energy set. This is shown in Figure 22.22 for the $Cr(H_2O)_6^{3+}$ ion.

As you probably know, white light contains photons of all the frequencies and colors in the visible spectrum. If we shine white light through a solution of a complex, the light that passes through has all the colors except those that have been absorbed. It is not difficult to determine what will be seen if we know what colors are being absorbed. All we need is a color wheel like the one shown in Figure 22.23. Across from each other on the color wheel are *complementary colors*. Green is the complementary color to red, and yellow is the complementary color to violet. If a substance absorbs a particular color when bathed in white light, the color of the reflected or transmitted light is the complementary color. In the case of the $Cr(H_2O)_6^{3+}$ ion, the light absorbed when the electron is raised from one set of d orbitals to the other has a frequency of 5.22×10^{14} Hz, which is the color of yellow light. This is why a solution of this ion appears violet.[2]

Because of the relationship between the energy and frequency of light, it is easy to see that the color of the light absorbed by a complex depends on the magnitude of Δ; the larger the size of Δ, the more energy the photon must have, and the higher will be the frequency of the absorbed light. For a given metal in a given oxidation state, the size of Δ depends on the ligand. Some ligands give a large crystal field splitting, while others give a small splitting. For example, ammonia produces a larger splitting than water, so the complex $Cr(NH_3)_6^{3+}$ absorbs light of higher energy and higher frequency than

FIGURE 22.23 A color wheel. Colors that are across from each other are called complementary colors. When a substance absorbs a particular color, light that is reflected or transmitted has the color of its complement. Thus something that absorbs red light appears green-blue, and vice versa.

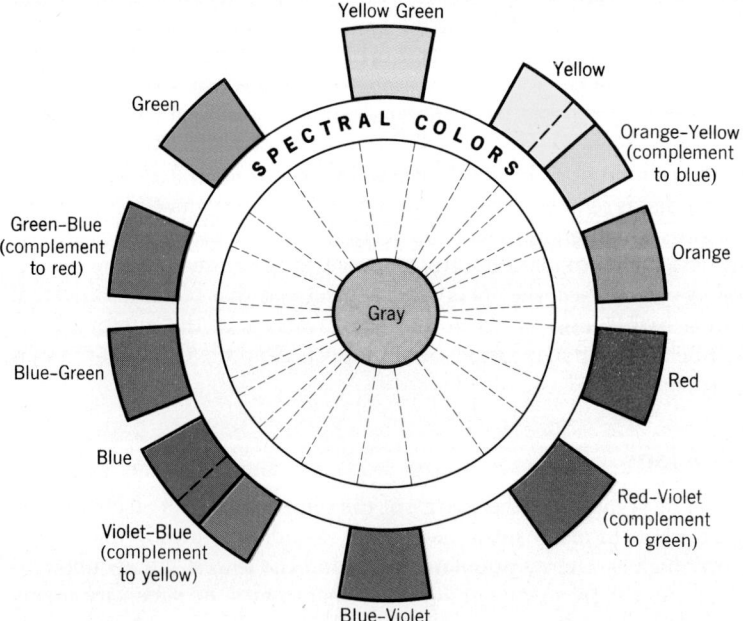

[2] The perception of color is actually somewhat more complex than this because of the varying sensitivity of the human eye to various wavelengths. For example, the eye is much more sensitive to green than to red. If a compound reflects both of these colors with equal intensity, it will appear greenish simply because the eye sees green better than it sees red.

$Cr(H_2O)_6^{3+}$; it absorbs blue light and appears yellow. Because changing the ligand changes Δ, the same metal ion is able to form a variety of complexes with a large range of colors.

A ligand that produces a large crystal field splitting with one metal ion also produces a large Δ in complexes with other metals. For example, cyanide ion is a very effective ligand and always gives a very large Δ, regardless of the metal to which it is bound. Ammonia is less effective than cyanide ion, but more effective than water. Thus, ligands can be arranged in order of their effectiveness at producing a large crystal field splitting. This sequence is called the spectrochemical series. Such a series containing some common ligands arranged in order of their decreasing strength is

$$CN^- > en > NH_3 > H_2O > C_2O_4^{2-} > OH^- > F^- > Cl^- > Br^- > I^-$$

For a given metal ion, cyanide ion produces the largest Δ and iodide produces the smallest.

The order of the ligands can be determined by measuring the frequencies of the light absorbed by complexes.

Magnetic Properties

Let's return to the question of the electron distribution among the d orbitals in Cr^{2+} complexes. As you saw above, this ion has four d electrons, and we noted that in placing these electrons in the d orbitals we had to make a decision about where to place the fourth electron. There's no question about the fate of the first three, of course. They just spread out across the three d orbitals in the lower level with their spins unpaired. In other words, we just follow Hund's rule, which we learned to apply in Chapter 6. But when we come to the fourth electron we have to decide whether to pair it with one of the electrons already in a d orbital of the lower set or to place it in one of the d orbitals of the higher set.[3] If we place it in the lower energy level, we give it extra stability (lower energy), but some of this stability is lost because it takes energy, called the pairing energy, P, to force the electron into an orbital that's already occupied by an electron. On ther other hand, if we place it in the higher level, we are relieved of the burden of pairing the electron, but it also tends to give the electron a higher energy. Thus, for the fourth electron, "pairing" and "placement" work in opposite directions in the way they affect the energy of the complex.

The critical factor in determining whether the fourth electron enters the lower level and becomes paired or whether it enters the higher level with the same spin as the other d electrons is the magnitude of Δ. If Δ is larger than the pairing energy P, then greater stability is achieved if the fourth electron is paired with one in the lower level. If Δ is small compared to P, then greater stability is obtained by spreading the electrons out as much as possible. The complexes $Cr(H_2O)_6^{2+}$ and $Cr(CN)_6^{4-}$ illustrate this well. Water is a ligand that does not produce a large Δ, so $P > \Delta$, and minimum pairing of electrons takes place. That explains the energy level diagram for the $Cr(H_2O)_6^{2+}$ complex in Figure 22.24a. When cyanide is the ligand, however, a very large Δ is obtained, and that leads to

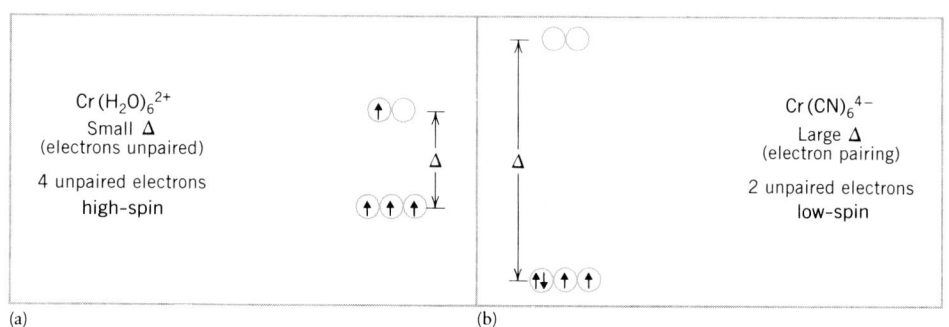

(a)   (b)

FIGURE 22.24 The effect of Δ on the electron distribution in a complex with four d electrons. *(a)* When Δ is small, the electrons remain unpaired. *(b)* When Δ is large, the lower energy level accepts all four electrons and two electrons become paired.

[3] We've never had to make this kind of decision before because the energy levels in atoms were always widely spaced. In complex ions, however, the two d-orbital energy levels are fairly close in energy.

pairing of the fourth electron with one in the lower set of d orbitals. This is shown in Figure 22.24b. It is interesting to note that experimentally it can be demonstrated that $Cr(H_2O)_6{}^{2+}$ has 4 unpaired electrons and the $Cr(CN)_6{}^{4-}$ ion has 2 unpaired electrons.

For octahedral chromium(II) complexes, there are two possibilities in terms of the number of unpaired electrons. They contain either four or two, depending on the magnitude of Δ. When there is the maximum number of unpaired electrons, the complex is described as being high-spin; when there is the minimum number of unpaired electrons it is described as being low-spin. High- and low-spin octahedral complexes are possible when the metal has a d^4, d^5, d^6, or d^7 electron configuration. Let's look at another example — one containing the Fe^{2+} ion, which has the electron configuration

$$Fe^{2+} \quad [Ar]3d^6$$

At the beginning of this section we mentioned that the $Fe(H_2O)_6{}^{2+}$ ion is paramagnetic and has four unpaired electrons, while the $Fe(CN)_6{}^{4-}$ ion is diamagnetic, meaning it has no unpaired electrons. Now we can see why, by referring to Figure 22.25. Water produces a weak splitting and a minimum amount of pairing of electrons. When the six d electrons in the Fe^{2+} ion are distributed, one must be paired in the lower level after filling the upper level. The result is four unpaired d electrons, and a high-spin complex. Cyanide ion, however, produces a large splitting, so $\Delta > P$. This means that maximum pairing of electrons in the lower level takes place, and a low-spin complex is formed. Six electrons are just the right amount to completely fill all three of these d orbitals, and since all the electrons are paired, the complex is diamagnetic.

The crystal field theory that we have developed can be extended to other geometries as well. Although the relative energies of the d orbitals are different, the applications of the theory are basically the same.

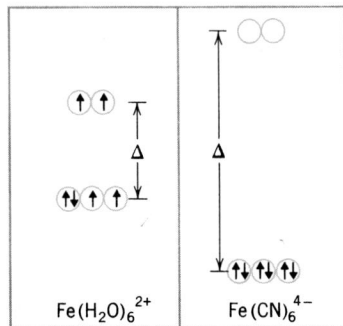

FIGURE 22.25 The distribution of d electrons in (a) $Fe(H_2O)_6{}^{2+}$ and (b) $Fe(CN)_6{}^{4-}$. The magnitude of Δ for the cyanide complex is much larger than for the water complex. This produces a maximum pairing of electrons in the lower energy set of d orbitals in $Fe(CN)_6{}^{4-}$.

SUMMARY

General Characteristics and Periodic Trends. The elements in the center of the periodic table are the transition elements. They have partially filled d subshells (except for Zn, Cd, and Hg); they are generally hard and have high melting points; they exhibit multiple oxidation states; their ions form many complex ions; and many of their compounds are colored. The sizes of the transition metal atoms decrease gradually from left to right. The transition elements in period 6 are nearly the same size as those above them in period 5 because of the lanthanide contraction. As a result, the transition elements in period 6 that occur after the lanthanide series are very dense and are unreactive because of the large effective nuclear charge felt by their outer electrons.

Most of the columns are labeled with a B, and parallels exist between the formulas of A-group and B-group compounds when the elements are in their highest oxidation states. There are few other similarities, however. Group VIII consists of three horizontal triads of elements.

The ferromagnetic elements are iron, cobalt, and nickel. In their crystals, the paramagnetic atoms have their magnetic poles aligned within regions called domains. In an unmagnetized state, the domains point in random directions, but when "permanently" magnetized the domains are aligned.

Properties of Important Transition Metals. Many common substances are composed of transition metal compounds or are the transition metals themselves. This section discusses the properties of some familiar and important ones.

Titanium. This metal is lightweight and resists corrosion. Its chloride, $TiCl_4$, is molecular and reacts readily with water to give TiO_2. Titanium dioxide is an important paint pigment.

Vanadium. Alloy steels contain vanadium to give them ductility and shock resistance. Vanadium pentoxide, V_2O_5, is the catalyst in the contact process for the manufacture of sulfuric acid. Solutions of V_2O_5 are acidic and contain the $VO_2{}^+$ ion, which can be viewed as forming by extensive hydrolysis of the V^{5+} cation.

Chromium. Chromium has many properties that make it useful as a protective coating over other metals. It is used in stainless steel alloys and to make nichrome heating elements. The metal is moderately active and dissolves in nonoxidizing acids, yielding Cr^{2+} and H_2 in the absence of oxygen. All of chromium's compounds are colored. Its principal oxidation states are $+2$, $+3$, and $+6$. Chromium(II) is easily oxidized to Cr^{3+}, which is amphoteric. $Cr(H_2O)_6{}^{3+}$ precipitates a gelatinous hydroxide in base, which redissolves in excess base. Chromium(III) forms many complexes; the $Cr(H_2O)_6{}^{3+}$ ion is violet and is found often in chromium(III) salts. CrO_3 is the acid anhydride of chromic acid, H_2CrO_4, which exists at very low pH. At higher pH there is an equilibrium between $Cr_2O_7{}^{2-}$ and $CrO_4{}^{2-}$. The oxide Cr_2O_3 is green and is used extensively as a pigment.

Manganese.   Manganese corrodes in moist air and dissolves readily in nonoxidizing acids to give H_2 and Mn^{2+}. It is used mostly in alloys. The most important oxidation states of manganese are $+2, +3, +4, +6,$ and $+7$. The $+2$ state is the most stable because the ion has a half-filled d subshell. When precipitated as $Mn(OH)_2$, oxidation to $MnO(OH)$ occurs readily. Mn^{3+} is unstable in water, but can be stabilized by the formation of complex ions. The least stable oxidation state is $+7$ in the MnO_4^- ion. In acidic solution, this permanganate ion is reduced to Mn^{2+} and in basic or neutral solution it is reduced to MnO_2. In acidic solution, MnO_4^- often serves as its own indicator. MnO_2 is a nonstoichiometric compound. Oxidation of MnO_2 by air or KNO_3 in molten KOH gives the MnO_4^{2-} ion. The MnO_4^{2-} ion is stable only in basic solution; it disproportionates in acidic solution.

Iron.   Iron is a common metal and the one most familiar to people. It is moderately active and dissolves in nonoxidizing acids to give H_2 and Fe^{2+}. Its chief ores contain hematite, $Fe_2O_3 \cdot xH_2O$, and magnetite, Fe_3O_4, the oxide in the more recently used taconite. The corrosion of iron is discussed in Special Topic 22.1. The principal oxidation states of iron are $+2$ and $+3$. Oxides include FeO (difficult to prepare), Fe_2O_3, and Fe_3O_4. The ion $Fe(H_2O)_6^{2+}$ is pale green and occurs in various iron(II) salts. Solutions of $Fe(H_2O)_6^{3+}$ are acidic and slightly yellow because of hydrolysis. Addition of base precipitates a gelatinous hydroxide. Iron forms many complex ions. Examples given are $Fe(SCN)_6^{3-}$, $Fe(CN)_6^{4-}$, and $Fe(CN)_6^{3-}$. The reaction of Fe^{2+} with $Fe(CN)_6^{3-}$ gives a blue precipitate of Prussian blue, a reaction that is used to make blueprints.

Cobalt.   Cobalt is less reactive than iron, but does dissolve slowly in nonoxidizing acids. Its chief use is in alloys. Two oxidation states are important, $+2$ and $+3$. In water, Co^{2+} exists as the pink ion $Co(H_2O)_6^{2+}$. Heating pink $[Co(H_2O)_6]Cl_2$ changes the color to blue as dehydration gives $[Co(H_2O)_4]Cl_2$. Cobalt(III) ion oxidizes water, but forms many stable complex ions.

Nickel.   The metal is resistant to corrosion and forms alloys with iron that are impact resistant. Purification of nickel by the Mond process relies on the ease of formation of nickel carbonyl, $Ni(CO)_4$. The metal dissolves in nonoxidizing acids with the evolution of H_2 and the formation of Ni^{2+}. In solution the $Ni(H_2O)_6^{2+}$ ion is green. The only stable oxidation state of nickel is $+2$. The compound NiO_2 is used as the cathode in nickel–cadmium batteries.

Copper, Silver, and Gold.   These are the coinage metals and they are less reactive than the other metals discussed. They do not dissolve in nonoxidizing acids, but copper and silver do dissolve in nitric acid. Gold dissolves in aqua regia. Copper has a high electrical and thermal conductivity and is used to make electrical wire and pipe. Silver is an even better conductor, but it's too expensive to use in these applications. Gold is used mainly as a decorative metal and in specialized electrical applications. Gold leaf is extremely thin gold that's used for decorative purposes.

Copper has two oxidation states, $+1$ and $+2$. Extended exposure of the metal to air gives it a green coating of $Cu_2(OH)_2CO_3$. Heating in air gives the black oxide CuO. The Cu^+ ion disproportionates in water, but can be isolated in solids such as $CuCl$. The most stable state is $+2$, which forms many complex ions.

Silver tarnishes in air and forms a thin film of Ag_2S. The only important oxidation state is $+1$. Addition of base to Ag^+ gives Ag_2O. AgF is soluble, but the other halides are insoluble and light sensitive. The ion forms many complex ions.

Gold has oxidation states of $+1$ and $+3$, but none of its compounds are particularly important. Gold compounds are easily decomposed by heat.

Zinc, Cadmium, and Mercury.   Zinc occurs in the ore zinc blende, ZnS, which is roasted to the oxide and reduced with carbon. It is very reactive metal and forms a protective coating of $Zn_2(OH)_2CO_3$. One of its chief uses is in galvanizing steel where it protects the steel against corrosion by cathodic protection. It is also used in alloys such as brass and bronze, and in batteries. Zinc and its hydroxide dissolve in acids and bases. The only oxidation state is $+2$, and the Zn^{2+} ion forms many complexes.

Cadmium is less common than zinc and occurs as an impurity in zinc ores. It is moderately active and dissolves in nonoxidizing acids, but it does not dissolve in base. Cadmium compounds are very toxic.

Mercury is a liquid over a broad range of temperatures and is used in thermometers. It also forms amalgams, which are solutions of metals in mercury. Mercury does not react with nonoxidizing acids but does dissolve in concentrated HNO_3 to give $Hg(NO_3)_2$. There are two oxidation states, $+1$ (Hg_2^{2+} ion) and $+2$ (Hg^{2+} ion). The Hg_2^{2+} ion has a mercury–mercury covalent bond. Hg_2Cl_2 is insoluble and disproportionates when aqueous ammonia is added; the products are Hg and $Hg(NH_2)Cl$. The very poisonous $HgCl_2$ is soluble, but only a very weak electrolyte. Both Hg_2Cl_2 and $HgCl_2$ are linear covalent molecules.

Complexes.   The coordination number of a metal ion in a complex is the number of donor atoms attached to the metal ion. Polydentate ligands supply two or more donor atoms, which must be taken into account when determining the coordination number from the formula of the complex. Complexes of polydentate ligands are more stable than similar complexes formed with monodentate ligands, because a polydentate ligand is less likely to be lost completely if one of its donor atoms becomes detached from the metal ion. This is the chelate effect. Geometries associated with common coordination numbers are, for coordination number 2, linear; for coordination number 4, tetrahedral and square planar (especially for Pt^{2+} complexes); and for coordination number 6, octahedral. The porphyrin structure is found in many biologically important metal complexes. Be sure you can draw octahedral complexes in the simplified way described in Figure 22.12.

Isomers of Coordination Compounds.   When two or more distinct compounds have the same chemical formula, they are isomers of each other. Stereoisomers have the same atoms attached to each other but the atoms are arranged differently in space. In a *cis* isomer, attached groups of atoms are on the same side of some reference plane through the molecule. In a *trans* isomer, they are on opposite sides. *Cis* and *trans* isomerism is a form of geometrical isomerism. Chiral isomers are exactly the same in every way but one — they are not superimposable on their mirror images. These kinds of isomers exist for complexes of the type $M(AA)_3$, where M is a metal ion and AA is a bidentate ligand, and also for complexes of the type cis-$M(AA)_2a_2$, where a is a monodentate ligand. Chiral isomers that are related as object-to-mirror images are said to be enantiomers and are optical isomers.

Crystal Field Theory.   In an octahedral complex, the ligands influence the energies of the d orbitals by splitting the d subshell into two energy sublevels. The lower one consists of the d_{xy}, d_{xz}, and d_{yz} orbitals; the higher-energy level consists of the d_{z^2} and $d_{x^2-y^2}$ orbitals. The energy difference between the two new d sublevels is Δ, and for a given ligand it increases with an increase in the oxidation state of the metal. For a given metal ion, Δ depends on the ligand.

In the spectrochemical series, ligands are arranged in order of their ability to cause a large Δ. Cyanide ion produces the largest crystal field splitting; iodide ion the smallest. Low-spin complexes result when Δ is larger than the pairing energy — the energy needed to cause two electrons to become paired in the same orbital. Light of energy equal to Δ is absorbed when an electron is raised from the lower-energy set of d orbitals to the higher set, and the color of the complex is determined by the colors that remain in the transmitted light. Crystal field theory can also explain the relative stabilities of oxidation states, in many cases.

REVIEW EXERCISES

Answers to questions whose numbers are printed in color are given in Appendix C.

General Properties of Transition Metals

22.1   Make a sketch of the periodic table. Mark off those regions where you would find (a) the transition elements, (b) the inner transition elements, (c) the lanthanides, (d) the actinides, and (e) the representative elements.

22.2   Give three properties of the transition elements that distinguish them from the representative elements.

22.3   Why is a +2 oxidation state common among the transition elements?

22.4   What is a complex ion? What would be the formula and the charge on a complex ion formed from Cu^{2+} and four OH^- ions?

22.5   Why are many of the transition elements able to exist in more than one oxidation state?

22.6   How do the sizes of the transition metal atoms vary from left to right across a period?

22.7   What is the *lanthanide contraction?* How does it affect the properties of the transition elements in period 6?

22.8   What similarities exist between elements in the A and B groups in the periodic table? Give three examples.

22.9   The density of niobium (Nb) is 8.57 g/cm³. Both Nb and Ta atoms are the same size. Estimate the density of tantalum.

22.10   In the periodic table, what is a triad? Where do we find them?

22.11   Which elements in their pure states are ferromagnetic?

22.12   Why do paramagnetic substances feel such a weak attraction to a magnet compared with ferromagnetic substances?

22.13   Why does iron lose its ferromagnetic properties and become paramagnetic when melted?

22.14   If an iron bar is lined up with its long axis pointing north/south and then given repeated sharp blows with a hammer, it becomes a weak permanent magnet. Explain what happens within the iron bar when this occurs.

Chemistry of Some Important Transition Elements: Titanium, Vanadium, and Chromium

22.15   Why is titanium a commercially useful metal? Why can't its oxide be reduced with carbon by ordinary metallurgical methods? How is metallic titanium made?

22.16   What titanium compound was used by the U.S. Navy to make smokescreens? Write the chemical equation for the reaction that produces the smoke.

22.17   Explain why $TiCl_4$ is covalent.

22.18   What is the formula for the most commercially important compound of titanium? What is its principal use?

22.19   What are two principal uses of elemental vanadium?

22.20   What is V_2O_5 used for?

22.21   In Chapter 21 we described the hydrolysis of $BiCl_3$ to give $BiOCl$. Explain how the BiO^+ ion can be accounted for in terms of the hydrolysis of the Bi^{3+} ion.

22.22   What vanadium-containing species is formed in aqueous solutions of V_2O_5?

22.23   Why is elemental chromium used to plate automobile bumpers?

22.24   What is nichrome? What is it used for?

22.25   What are the common oxidation states of chromium? Write their electron configurations.

22.26   What is chrome alum? How can it be made? What color is it?

22.27   Write a balanced net ionic equation for the oxidation of Cr^{2+} by O_2 in an acidic solution.

22.28   Which is a better reducing agent, Cr^{2+} or Cr^{3+}?

22.29   Write a chemical equation that explains why solutions of Cr^{3+} are slightly acidic.

22.30   Write chemical equations to show what happens when concentrated NaOH solution is gradually added to a solution containing chromium(III) ion. Identify the colors of the different species.

22.31   What relationships exist among CrO_3, H_2CrO_4, CrO_4^{2-}, and $Cr_2O_7^{2-}$?

22.32 What are the colors of the CrO_4^{2-} and $Cr_2O_7^{2-}$ ions?

22.33 Write the equilibria for the ionization of chromic acid.

22.34 The oxide Cr_2O_3 is amphoteric, meaning it is both acidic and basic, but CrO_3 is only acidic. Why is this?

22.35 Explain, using appropriate Lewis structures, how CrO_4^{2-} is changed to $Cr_2O_7^{2-}$ as the pH is lowered.

22.36 What color is Cr_2O_3? What is it used for?

22.37 What are the names of (a) H_2CrO_4, (b) $Cr_2O_7^{2-}$, (c) CrO_4^{2-}, and (d) Cr_2O_3?

22.38 Write the formulas of two yellow chromium-containing pigments.

Chemistry of Some Important Transition Elements: Manganese, Iron, Cobalt, and Nickel

22.39 What are the common oxidation states of manganese? Which is the most stable? Which is most easily reduced to a lower oxidation state?

22.40 How does the reactivity of manganese compare with that of chromium?

22.41 Write molecular equations for the reactions of (a) manganese with hydrochloric acid and (b) manganese(II) chloride with aqueous NaOH.

22.42 What product is formed when freshly precipitated $Mn(OH)_2$ is exposed to air?

22.43 What are the colors of (a) Mn^{2+}, (b) MnO_4^-, and (c) MnO_4^{2-}?

22.44 What are the names of (a) MnO_4^-, (b) MnO_4^{2-}, and (c) MnO_2?

22.45 Under what conditions is the manganate ion stable?

22.46 What properties of MnO_4^- make it particularly useful in redox titrations?

22.47 What is a nonstoichiometric compound? Give an example of one.

22.48 Write half-reactions for the reduction of MnO_4^- in (a) an acidic solution and (b) a basic solution.

22.49 What is a disproportionation reaction? Write a chemical equation for the disproportionation of MnO_4^{2-} in acidic solution.

22.50 Which compound of manganese is used in the common dry cell? What is its function?

22.51 What compounds are found in ores of iron?

22.52 What happens when iron is treated with (a) concentrated hydrochloric acid and (b) concentrated nitric acid?

22.53 What are the common oxidation states of iron?

22.54 What are the formulas of the oxides of iron?

22.55 Write a net ionic equation for the reaction that occurs when base is added to aqueous solutions of (a) ferrous ion and (b) ferric ion.

22.56 What happens to freshly precipitated $Fe(OH)_2$ if it is exposed to air?

22.57 Write formulas for (a) rust, (b) the ion that gives a green color to an aqueous solution of iron(II) salts, (c) Prussian blue, and (d) the gelatinous precipitate that forms when NaOH is added to a solution of iron(III) sulfate.

22.58 Why are solutions of Fe^{3+} slightly acidic?

22.59 Why are solutions of iron(III) salts often slightly yellow?

22.60 Explain why a solution of Fe^{3+} is more acidic than a solution of Fe^{2+}.

22.61 What are the common names for $K_4Fe(CN)_6$ and $K_3Fe(CN)_6$?

22.62 What reactions are involved in the rusting of iron?

22.63 What chemicals are involved in making a blueprint? Describe the chemical reactions that are involved.

22.64 What are the properties and uses of cobalt?

22.65 What is stellite? What is alnico?

22.66 What color is cobalt(II) ion in aqueous solution? What ion is responsible for the color?

22.67 Write a chemical equation that shows what happens when $CoCl_2 \cdot 6H_2O$ is heated. What color change takes place?

22.68 Write a balanced net ionic equation for the oxidation of water by Co^{3+} ion in acidic solution.

22.69 Why is nickel an important metal? What metals are in the "nickel" used to make coins? What is monel?

22.70 Write chemical equations for the reactions involved in the Mond process for purifying nickel.

22.71 What is the oxidation number of nickel in $Ni(CO)_4$? What is the most important oxidation state of nickel?

22.72 Write a chemical equation for the reaction of nickel with hydrochloric acid.

22.73 What is the formula for the ion that gives many nickel compounds a green color in aqueous solution?

22.74 What metals are usually found in stainless steel?

Chemistry of Some Important Transition Elements: The Coinage Metals, Zinc, Cadmium, and Mercury

22.75 What are some typical uses for copper?

22.76 What copper compound is found in copper ores? How is it treated to recover the metal? Describe how impure copper is purified. (If necessary, refer back to Chapters 17 and 21.)

22.77 How does the electrical conductivity of silver compare to that of other metals?

22.78 Which metals are the coinage metals?

22.79 What metal is alloyed with silver in sterling silver and jewelry silver?

22.80 What is gold leaf?

22.81 What oxidation states are observed for copper, silver, and gold?

22.82 Complete and balance the following equations. If no reaction occurs, write N.R.
(a) $Cu + H_2SO_4(dilute) \longrightarrow$
(b) $Cu + H_2SO_4(hot, concd) \longrightarrow$
(c) $Cu + HNO_3(dilute) \longrightarrow$
(d) $Cu + HNO_3(concd) \longrightarrow$

22.83   What is the formula of the copper compound that gives old copper objects a green color?

22.84   How can CuCl be prepared? Write a chemical equation.

22.85   Why are copper sulfate crystals blue?

22.86   Why are concentrated solutions of $CuCl_2$ green? What happens to the color when the solution is diluted? Write chemical equilibria for the formation of the complexes involved and explain why the color change occurs based on Le Châtelier's principle.

22.87   Write chemical equations that show what happens when concentrated NaOH is added to a solution of copper sulfate.

22.88   Based on what you have learned in this section, predict the formula of the complex that is formed by Cu^{2+} and CN^-.

22.89   What test can be used to detect the presence of copper ion in a solution? What are the formulas of the complex ions of copper involved in this test?

22.90   Write chemical equations for the reaction of silver with concentrated and dilute nitric acid.

22.91   What compound is formed when NaOH is added to a solution of silver nitrate?

22.92   How do the solubilities of the silver halides vary?

22.93   Silver bromide can be made by stirring a suspension of solid AgCl in a NaBr solution. Write the equilibria involved and explain how this reaction occurs.

22.94   Write the chemical equation involved in the analytical test for Ag^+.

22.95   Gold(III) is a fairly potent oxidizing agent. Why?

22.96   Explain, in terms of the lanthanide contraction, why gold is a much less reactive element than either copper or silver?

22.97   What are the ores of zinc, cadmium, and mercury?

22.98   Even though zinc is an active metal, it corrodes very slowly. What is the formula of the compound responsible for this?

22.99   Write net ionic equations that show what happens when zinc dissolves in HCl and NaOH.

22.100 Write net ionic equations that show what happens when concentrated base is added gradually to a solution of $ZnCl_2$.

22.101 How does a coating of zinc protect iron from corrosion, even if the coating is partially worn away? Why doesn't tin afford the same protection? (Hint: If necessary, refer to Chapter 17.)

22.102 Why is cadmium sometimes used as a protective coating over steel instead of zinc?

22.103 Zinc is such an active metal that when it reacts with dilute nitric acid, the reduction product is NH_4^+. Write a balanced net ionic equation for this reaction.

22.104 Write net ionic equations for the reaction, if any, of hydrochloric acid with (a) zinc, (b) cadmium, and (c) mercury.

22.105 Write the net ionic equation for the reaction of mercury with concentrated nitric acid.

22.106 What is an amalgam? How is mercury used in the separation of gold from its ores?

22.107 Hg_2Cl_2 is much less toxic than $HgCl_2$ if taken internally. Why?

22.108 What are the structures of Hg_2Cl_2 and $HgCl_2$?

22.109 What reaction occurs when aqueous ammonia is added to Hg_2Cl_2? What is the color of the solid produced in this reaction?

22.110 What is vermilion?

Complexes of the Transition Metals

22.111 How do ethylenediamine and oxalate ion form chelate rings?

22.112 What is a monodentate ligand?

22.113 In what ways is the porphyrin structure important in biological systems?

22.114 What is *coordination number*? What structures are generally observed for complexes in which the central metal ion has a coordination number of 4?

22.115 What is the coordination number of Fe in $[Fe(en)(H_2O)_2Cl_2]$? What is the oxidation number of iron in this complex?

22.116 Explain the chelate effect.

22.117 Sketch the structure of an octahedral complex that contains only monodentate ligands.

22.118 Sketch the structure of the octahedral $[Co(EDTA)]^-$ ion. Remember that donor atoms in a polydentate ligand span adjacent positions in the octahedron.

22.119 NTA is the abbreviation for *nitrilotriacetic acid,* a substance that was used at one time in detergents. Its structure is

The four donor atoms of this ligand are shown in red. Sketch the structure of an octahedral complex containing this ligand. Assume that two water molecules are also attached to the metal ion and that each oxygen donor atom in the NTA is bonded to a position in the octahedron that is adjacent to the nitrogen of the NTA.

Isomers of Coordination Compounds

22.120 What are isomers?

22.121 Define *stereoisomerism, geometric isomerism, chiral isomers,* and *enantiomers.*

22.122 What are *cis* and *trans* isomers? What condition must be fulfilled in order for a molecule or ion to be chiral?

22.123 What is *cisplatin*? Draw its structure.

22.124 Sketch the isomers of the square planar complex $Pt(NH_3)_2ClBr$.

22.125 The complex $Pt(NH_3)_2Cl_2$ can be obtained as two distinct isomeric forms. Make a model of a tetrahedron and show that if the complex were tetrahedral, two isomers would be impossible.

22.126 The complex $Co(NH_3)_3Cl_3$ can exist in *two* isomeric forms. Sketch them.

22.127 Sketch the chiral isomers of $Cr(en)_2Cl_2^+$. Is there a non-chiral isomer?

22.128 Sketch the chiral isomers of $Co(C_2O_4)_3^{3-}$.

22.129 What are optical isomers?

Bonding in Complexes

22.130 On appropriate coordinate axes, sketch and label the five *d* orbitals.

22.131 Explain why an electron in a $d_{x^2-y^2}$ or d_{z^2} orbital in an octahedral complex will experience greater repulsions because of the presence of the ligands than an electron in a d_{xy}, d_{xz}, or d_{yz} orbital.

22.132 Sketch the *d*-orbital energy level diagram for a typical octahedral complex.

22.133 Explain why octahedral cobalt(II) complexes are easily oxidized to cobalt(III) complexes. Sketch the *d*-orbital energy diagram and assume a large value of Δ when placing electrons in the *d* orbitals.

22.134 In which complex do we expect to find the larger Δ?
(a) $Cr(H_2O)_6^{2+}$ or $Cr(H_2O)_6^{3+}$
(b) $Cr(en)_3^{3+}$ or $CrCl_6^{3-}$

22.135 Explain how the same metal in the same oxidation state is able to form complexes of different colors.

22.136 What does the term *spectrochemical series* mean? How can the order of the ligands in the series be determined?

22.137 Which complex should absorb light at the longer wavelength?
(a) $Fe(H_2O)_6^{2+}$ or $Fe(CN)_6^{4-}$
(b) $Mn(CN)_6^{3-}$ or $Mn(CN)_6^{4-}$

22.138 Which complex should be expected to absorb light of the highest frequency, $Cr(H_2O)_6^{3+}$, $Cr(en)_3^{3+}$, or $Cr(CN)_6^{3-}$?

22.139 What do the terms low-spin complex and high-spin complex mean?

22.140 For which *d*-orbital electron configurations are both high-spin and low-spin complexes possible?

22.141 Sketch the *d*-orbital energy level diagrams for $Fe(H_2O)_6^{3+}$ and $Fe(CN)_6^{3-}$ and predict the number of unpaired electrons in each.

22.142 Indicate what happens to the *d*-orbital electron configuration of the $Fe(CN)_6^{4-}$ ion when it absorbs a photon of visible light.

22.143 The complex $Co(C_2O_4)_3^{3-}$ is diamagnetic. Sketch the *d*-orbital energy level diagram for this complex and indicate the electron population of the orbitals.

COMPLEX IONS
IN QUALITATIVE ANALYSIS

As you may have come to realize from our discussions of the metals in Chapters 21 and 22, complex ions are formed by nearly all metal ions. It should come as no surprise, therefore, that we encounter complex ions frequently in qualitative analysis. In general, we come across them in three principal ways: (1) in the form of the "free" metal ions in aqueous solutions, where most of them actually exist as complex ions with water as the ligand, (2) during some of the procedures used to separate cations from each other, and (3) in tests that are used to confirm the presence of specific metal ions.

Metal Salts in Aqueous Solution

Water is an effective electron donor (i.e., Lewis base), and forms complex ions readily with metal ions. Therefore, when a salt dissolves in water, the cation seldom is able to exist as just the simple ion. Instead, the metal ion forms a complex with H_2O. We mentioned many of these complexes previously when we described the chemical properties of specific metals. A few examples are aluminum ion, which exists in aqueous solution as the hydrate $[Al(H_2O)_6]^{3+}$; copper ion, which exists as $[Cu(H_2O)_4]^{2+}$; and nickel ion, which exists as $[Ni(H_2O)_6]^{2+}$. Because of the tendency for metal ions to exist as such complexes, reactions that we think of as involving a simple "free" metal ion actually involve the reaction of its water-containing complex ion. Although we realize this, for simplicity we usually ignore the waters that surround the metal ion in aqueous solution. Thus in writing equations involving the copper ion, we normally use just the symbol Cu^{2+} or perhaps $Cu^{2+}(aq)$. For example, the precipitation of copper as copper sulfide is usually given by the equation

$$Cu^{2+}(aq) + H_2S(aq) \longrightarrow CuS(s) + 2H^+(aq)$$

even though it is more accurately represented by

$$[Cu(H_2O)_4]^{2+}(aq) + H_2S(aq) \longrightarrow CuS(s) + 2H_3O^+(aq) + 2H_2O$$

In qualitative analysis, the characteristic colors of some of the hydrated metal ions can sometimes serve as a clue to the presence or absence of the ions. For example, you learned that the $[Cu(H_2O)_4]^{2+}$ ion has a pale blue color. If a solution to be analyzed has a blue color, one can suspect that it contains copper ion and therefore be especially careful in testing for that ion during the analytical procedures. Similarly, if the solution has a pale green color, nickel can be suspected and carefully tested for, because the $[Ni(H_2O)_6]^{2+}$ ion is green. The presence or absence of color, however, should only raise suspicions. For example, if the solution contains copper ion at low concentration, its color may not be intense enough to be easily seen, but the apparent lack of color doesn't prove that copper ion is absent. On the other hand, if the solution to be analyzed contains two or more colored metal ions, the observed color may not suggest any particular ion. If the unknown solution is colored, however, it does suggest that at least one colored ion is probably present. So before starting an analysis it is useful to note any color observed in the solution to be analyzed.

Complexes Used in the Separation of Ions

In previous Qual Topics, we described how metal ions are separated into cation groups by precipitation reactions involving the formation of their chlorides, sulfides, and hydroxides. Dividing each of these cation groups into the individual metal ions makes use of a variety of reactions, including redox reactions and other precipitation reactions. Also included are reactions involving complex ions, where the *relative tendencies* of various metals to form complex ions with particular ligands are used to selectively dissolve precipitates or to prevent precipitates from forming. We will not be exhaustive in our examination of this topic here, but rather only illustrate the principle with examples.

We already described in Qual Topic I how the metal ions in cation group 1 are separated from each other and tested for. Recall that after $PbCl_2$ is dissolved using hot water, the next step is the addition of aqueous ammonia to the precipitate that could contain AgCl and Hg_2Cl_2. Here, use is made of the strong tendency of silver ion to form the soluble complex $[Ag(NH_3)_2]^+$. As we described earlier, this causes the AgCl to dissolve. Equally important to this step of the analysis is the relatively low tendency of mercury ion (either Hg_2^{2+} or Hg^{2+}) to form a similar complex with ammonia. As a result, only the silver chloride dissolves, which separates silver from mercury.

In cation group 2, the tendency of tin to form a complex with sulfide ion is used to separate this metal from the others in its group. After separating the group 2 sulfide precipitate (SnS_2, PbS, CuS, and Bi_2S_3) from the solution that might contain other cations, the precipitate is treated with a basic solution to which H_2S is added. At high pH, the availability of sulfide ion increases, and the SnS_2 dissolves by forming the complex $[SnS_3]^{2-}$. The other precipitates do not dissolve because their cations have relatively little tendency to form complexes with sulfide ion.

Another example is the separation of zinc ion from aluminum ion, both of which occur in cation group 3. In this group, the ions are precipitated from a slightly basic solution to which H_2S is added. The solid that forms contains NiS, FeS, MnS, ZnS, and $Al(OH)_3$. The initial separation of the cations here is accomplished by making use of the amphoteric behavior of aluminum and zinc. The mixed precipitate is dissolved in acid and the resulting solution is then made strongly basic with sodium hydroxide. Iron, nickel, and manganese precipitate as hydroxides, but the aluminum and zinc remain in solution as complexes of hydroxide ion.

The equations for these reactions are best given as

$$[Al(H_2O)_6]^{3+}(aq) + 4OH^-(aq) \longrightarrow [Al(H_2O)_2(OH)_4]^-(aq) + 4H_2O$$
$$[Zn(H_2O)_4]^{2+}(aq) + 4OH^-(aq) \longrightarrow [Zn(OH)_4]^{2-}(aq) + 4H_2O$$

although they are often simplified by omitting the waters bound to the free metal ion in solution.

$$Al^{3+}(aq) + 4OH^-(aq) \longrightarrow [Al(OH)_4]^-(aq)$$
$$Zn^{2+}(aq) + 4OH^-(aq) \longrightarrow [Zn(OH)_4]^{2-}(aq)$$

In the reactions above, the *similar* behavior of aluminum and zinc is used to separate these two metals from the others in cation group 3. A *dissimilarity* is used next to separate aluminum and zinc from each other. The solution that contains these complexes is acidified to give the "free" metal ions, and then aqueous ammonia is added. Zinc ion forms an ammonia complex readily, but the ammonia complex of aluminum is not sufficiently stable to prevent precipitation of gelatinous aluminum hydroxide in the slightly alkaline ammonia solution. Thus, aluminum hydroxide precipitates, but the zinc remains in solution as its soluble ammonia complex.

$$[Al(H_2O)_6]^{3+}(aq) + 3NH_3(aq) \longrightarrow [Al(H_2O)_3(OH)_3](s) + 3NH_4^+(aq)$$
$$[Zn(H_2O)_4]^{2+}(aq) + 4NH_3(aq) \longrightarrow [Zn(NH_3)_4]^{2+}(aq) + 4H_2O$$

These equations also can be simplified by omitting the waters attached to the metal ions. The equations become

$$Al^{3+}(aq) + 3H_2O + 3NH_3(aq) \longrightarrow Al(OH)_3(s) + 3NH_4^+(aq)$$
$$Zn^{2+}(aq) + 4NH_3(aq) \longrightarrow [Zn(NH_3)_4]^{2+}(aq)$$

Complexes Used to Confirm the Presence of Metal Ions

Many complexes of metal ions have characteristic colors, and sometimes we can use these as a means of confirming the presence of a metal ion in a solution. For example, we know that if ammonia is added to a solution that contains $[Cu(H_2O)_4]^{2+}$, the color deepens dramatically because the water molecules surrounding the copper ion are replaced by NH_3 molecules. This gives the ion $[Cu(NH_3)_4]^{2+}$, which has an even deeper blue color than $[Cu(H_2O)_4]^{2+}$. In fact, this color is so intense that it is used to test for the presence of Cu^{2+} in a solution, even if the copper ion is present in small amounts. After reactions are carried out that would separate the copper from other metal ions, an aqueous ammonia solution is added. If a blue color appears, or if an already blue color intensifies, the presence of copper ion is confirmed. See Figure 1.

FIGURE 1 A dilute solution of copper(II), which contains the $Cu(H_2O)_4^{2+}$ ion, is a very pale blue. Addition of ammonia forms the complex ion $Cu(NH_3)_4^{2+}$, which is a deep blue color.

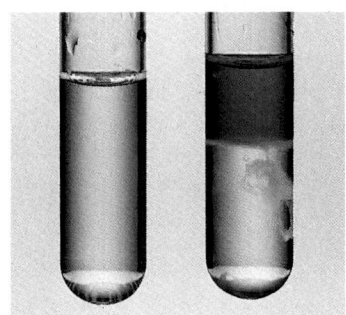

Two other metal ions whose complexes are used for the purpose of detection are Fe^{3+} and Ni^{2+}. If a solution is suspected to contain Fe^{3+}, we can add thiocyanate ion, SCN^-, which forms an intensely colored blood-red complex if the solution contains Fe^{3+} (see Figure 2).

$$Fe^{3+}(aq) + 6SCN^-(aq) \longrightarrow \underset{\text{red}}{[Fe(SCN)_6]^{3-}(aq)}$$

(Actually, depending on the SCN^- concentration, the formula of the red complex can range from $[Fe(SCN)]^{2+}$ to $[Fe(SCN)_6^{3-}]$.)

FIGURE 2 When added to a solution that contains Fe^{3+}, a drop of a solution containing SCN^- ion produces the deep red-colored complex ion, $Fe(SCN)_6^{3-}$.

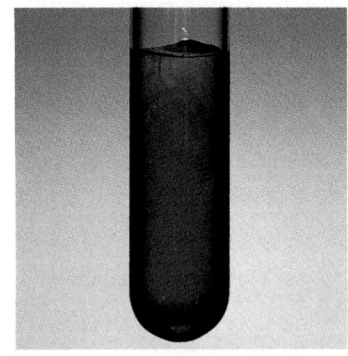

For Ni^{2+}, addition of ammonia produces a blue-colored complex (Figure 3).

$$Ni^{2+}(aq) + 6NH_3(aq) \longrightarrow \underset{\text{blue}}{[Ni(NH_3)_6]^{2+}(aq)}$$

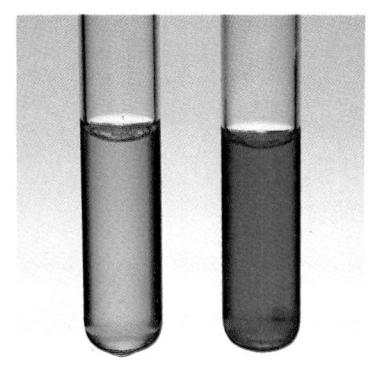

FIGURE 3 A solution of nickel(II) is green because it contains the complex $Ni(H_2O)_6^{2+}$. The addition of aqueous ammonia to this solution gives the deep blue complex $Ni(NH_3)_6^{2+}$.

To differentiate this from the copper–ammonia complex, a chemical called dimethyl-glyoxime (abbreviated H_2DMG) is added, which forms a brick-red complex with the nickel(II) ion that is insoluble in water (see Figure 4).

$$[Ni(NH_3)_6]^{2+}(aq) + 2H_2DMG(aq) \longrightarrow \underset{\text{brick red}}{[Ni(HDMG)_2](s)} + 2NH_4^+(aq) + 4NH_3(aq)$$

Dimethylglyoxime is an organic compound with the structure

$$CH_3—C—C—CH_3$$
$$\qquad \| \quad \|$$
$$HO—N \quad N—OH$$

FIGURE 4 On the left, the complex $Ni(NH_3)_6^{2+}$ in a solution that contains excess ammonia. Addition of a solution containing dimethyl-glyoxime (a weak diprotic acid abbreviated H_2DMG) produces the blood-red insoluble complex $Ni(HDMG)_2$.

In qualitative analysis, a complex ion is also used to confirm the presence of Zn^{2+}, but in this case it is a complex of Fe^{2+} with cyanide ion. This complex is $[Fe(CN)_6]^{4-}$ and has the common name *ferrocyanide ion*. When the potassium salt of this complex is added to a slightly acidic solution that contains Zn^{2+}, it forms a light green (sometimes almost white) precipitate (Figure 5).

$$2K^+(aq) + 3Zn^{2+}(aq) + 2[Fe(CN)_6]^{4-}(aq) \longrightarrow \underset{\text{light green}}{K_2Zn_3[Fe(CN)_6]_2(s)}$$

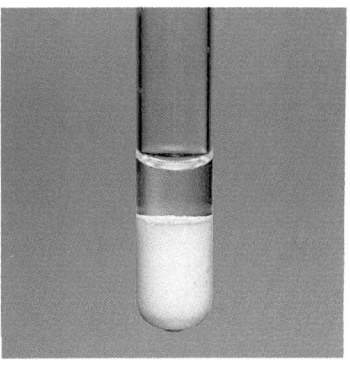

FIGURE 5 The precipitate $K_3Zn_3[Fe(CN)_6]_2$.

Several lasers are focused on a tiny pellet of nuclear fuel during research on nuclear fusion by inertial confinement, one of the topics in this chapter.

NUCLEAR REACTIONS AND THEIR ROLE IN CHEMISTRY

CONSERVATION OF MASS – ENERGY

NUCLEAR BINDING ENERGIES

RADIOACTIVITY

THE BAND OF STABILITY

TRANSMUTATION

DETECTING AND MEASURING RADIATIONS

APPLICATIONS OF RADIOACTIVITY

NUCLEAR FUSION

NUCLEAR FISSION

THE BREEDER REACTOR

23.1 CONSERVATION OF MASS – ENERGY

Whenever an insulated system gains energy by some change, it has lost an equivalent amount of mass, a quantity that can be calculated by the Einstein equation.

Until now, we have said very little about the role of the nuclei of atoms in the study of chemistry. All that we have had to know about a nucleus is that it is there at the center of the atom; that it is responsible for the mass of the atom; and that it has a certain number of protons — the atomic number of the element. This number, of course, determines the number of electrons in a neutral atom.

In this chapter, we will turn our attention inward to the nucleus and examine its structure and properties in more detail. We will study why the nuclei of some isotopes are stable and why others are unstable. We will learn that unstable nuclei emit streams of particles or of radiation and so undergo a kind of reaction new to our study, nuclear reactions, which cause the various kinds of radioactivity you may have read about in the press. We will see how scientists exploit nuclear reactions to study chemical changes, to conduct many kinds of analyses (including the dating of rocks and ancient objects), to treat certain diseases, and to obtain energy. Two particularly important nuclear reactions—fission and fusion—are of special interest because they could be used to destroy us or to supply us with an almost unlimited amount of peaceful energy.

We begin by reexamining two physical laws that we have so far assumed to be separate and independent laws, the laws of conservation of energy and conservation of mass. For *chemical* reactions they are independent, but such is not the case for *nuclear* reactions. Albert Einstein, reflecting on the natures of mass, length, and time on his way to the theory of relativity, saw that these two laws are just different aspects of a deeper, more general law.

With respect to mass, Einstein realized that if other laws of physics were to be "saved"—that is, if they were to hold true in the emerging atomic and nuclear physics—the mass of a particle, m, had to be redefined. Einstein said that the mass of a particle is not actually constant, but is related by the following equation to its velocity, v—its velocity *with respect to the observer*—and the velocity of light, c.

$c = 3.00 \times 10^8$ m/s, the speed of light.

$$m = \frac{m_0}{\sqrt{1 - (v/c)^2}} \qquad (23.1)$$

Notice what happens if the velocity, v, is zero, relative to the observer. The ratio, v/c, is now zero, the whole denominator is 1, and Equation 23.1 reduces to

$$m = m_0$$

The symbol m_0 therefore stands for the particle's *rest mass*. It is the mass that we observe in all measurements when an object is either at rest (from our viewpoint) or is not moving extraordinarily rapidly. Only as the particle's velocity approaches the speed of light, c, does the v/c term in Equation 23.1 measurably affect the denominator and so affect the mass. As v approaches c, the ratio v/c approaches 1, and the whole denominator approaches 0. If it actually reached 0, then the mass, m, of the particle would have become infinitely large (because any value of m_0 divided by zero is infinity)—a physical impossibility. Thus the speed of light is an absolute upper limit on the speed that any particle can have.

Even at $v = 1000$ m/s (about 2250 mph), the denominator (unrounded) is 0.99999333, or within 7×10^{-4}% of 1.

At the ordinary velocities of everyday experience, the mass calculated by Equation 23.1 for any object equals the rest mass to several significant figures. The difference cannot be detected by any devices that measure mass. Thus in all of our normal experiences, including all chemical reactions, the velocities of things are not so large as to appreciably affect their masses, so mass appears to be conserved; and the law of conservation of mass works as an independent law in chemistry.

When Einstein turned his attention from mass to energy, he realized that to save the idea of energy conservation in a universe where mass changes with velocity, he had to postulate that mass and energy are interconvertible just as are potential and kinetic energy. With respect to energy, what is conserved is *all* forms of energy including a system's mass expressed as an equivalent amount of energy. Or, saying this alternatively, what is conserved about mass is all forms of mass including a system's energy expressed as an equivalent amount of mass. The single, deeper law that summarizes these conclusions is now called the law of conservation of mass–energy.

Albert Einstein (1879–1955); Nobel Prize, 1921

Law of Conservation of Mass–Energy   The sum of all the energy in the universe and of all the mass (expressed as an equivalent in energy) is a constant.

The Einstein Equation

Einstein was able to show that when mass converts to energy, the amount of such energy, ΔE, is related to the change in rest mass, Δm_0, by the following equation, now called the Einstein equation.

The Einstein equation is given as $E = mc^2$ in the popular press.

$$\Delta E = \Delta m_0 \times c^2$$

(where c is the velocity of light).

Because the velocity of light is very large (and its square, 9×10^{16} m^2 s^{-2}, is almost beyond imagining), even an enormous change in energy would translate into only a very small change in mass. We can demonstrate this by a calculation involving the energy released when methane burns according to the following thermochemical equation.

$$CH_4(g) + 2O_2(g) \longrightarrow CO_2(g) + 2H_2O(\ell) \qquad \Delta H° = -890 \text{ kJ}$$

The source of this energy is a *loss* of mass that we can calculate by the Einstein equation:

$$\Delta m_0 = \frac{\Delta E}{c^2} = \frac{(890 \text{ kJ})}{(3.00 \times 10^8 \text{ m s}^{-1})^2} \times \frac{1000 \text{ J}}{1 \text{ kJ}} \times \frac{1 \frac{\text{kg m}^2}{\text{s}^2}}{1 \text{ J}}$$

This conversion factor is from the definition of the joule (page 149).

The units of the joule are kg m^2/s^2.

$$= 9.89 \times 10^{-12} \text{ kg} = 9.89 \times 10^{-9} \text{ g} = 9.89 \text{ ng}$$

The 890 kJ released by the combustion of one mole of methane thus originates in the conversion of 9.89 ng of mass into energy. Such a small change in mass — roughly 10 nanograms out of a total mass of 80 grams — cannot be detected by weighing balances. It amounts to a loss of about $1 \times 10^{-7}\%$ of the mass. You can see why we ignore the Einstein equation in working with the stoichiometry of chemical changes.

The importance of the Einstein equation lies in its applications to nuclear physics and related topics in chemistry. The truth of this equation became abundantly clear when atomic fission was first observed in 1939, and since then the equation has been a routine tool among nuclear physicists. We will see how the Einstein equation can help us understand the energy available by nuclear changes in the next section.

23.2 NUCLEAR BINDING ENERGIES

The sum of the rest masses of the nucleons of an atom does not equal the measured mass of its nucleus.

The actual mass of an atomic nucleus is always a trifle smaller than the sum of the rest masses of all of its nucleons — its protons and neutrons. The mass differences changed into energy as the nucleus formed, and was emitted as high-energy electromagnetic radiation. It would cost this much energy to break the nucleus apart into its nucleons again, so the energy is called the binding energy of the nucleus.

We can calculate the binding energy for helium-4 to illustrate. Helium has an atomic number of 2, so in the nucleus of one helium-4 atom there are 2 protons and 2 neutrons. The rest mass of one helium-4 nucleus is known to be 4.001506 amu. However, if we calculate the mass of a helium-4 nucleus just by adding the masses of its four nucleons, we obtain a slightly higher value. The rest mass of a proton is 1.007277 amu and that of a neutron is 1.008665 amu, so we have

1 amu = 1.6606×10^{-24} g.

$$
\begin{array}{lll}
\text{for 2 protons:} & 2 \times 1.007277 \text{ amu} = 2.014554 \text{ amu} \\
+ \text{ for 2 neutrons:} & 2 \times 1.008665 \text{ amu} = \underline{2.017330 \text{ amu}} \\
\text{Calculated mass of the nucleons in helium-4} = 4.031884 \text{ amu}
\end{array}
$$

The difference between the calculated and measured rest masses for the helium-4 nucleus is (4.031884 amu − 4.001506 amu) = 0.030378 amu. The energy equivalent of this difference in mass is calculated using Einstein's equation, where we have to convert amu to kilograms by the relationships 1 amu = 1.6606×10^{-24} g and 1 kg = 1000 g.

$$\Delta E = \Delta m c^2 = \underbrace{(0.030378 \text{ amu}) \frac{(1.6606 \times 10^{-24} \text{ g})}{1 \text{ amu}} \frac{(1 \text{ kg})}{1000 \text{ g}}}_{\Delta m \text{ (in kg)}} \underbrace{\left(3.00 \times 10^8 \frac{m}{s}\right)^2}_{\text{velocity of light}}$$

$$= 4.54 \times 10^{-12} \text{ kg m}^2/\text{s}^2$$
$$= 4.54 \times 10^{-12} \text{ J}$$

There are four nucleons in the helium-4 nucleus, so the binding energy per nucleon is $(4.54 \times 10^{-12}$ J$)/4$ nucleons or 1.14×10^{-12} J/nucleon.

The formation of just one nucleus of helium-4 releases 4.54×10^{-12} J. If we could make Avogadro's number of helium-4 nuclei — the total mass would be only 4 grams — the net release of energy would be

$$6.02 \times 10^{23} \text{ nuclei} \times 4.54 \times 10^{-12} \text{ J/nucleus} = 2.73 \times 10^{12} \text{ J}$$

This amount of energy could keep a 100-watt light bulb lit for nearly 900 years! The formation of a nucleus from its nucleons is called *nuclear fusion* — which we will study further in Section 23.8 — but you can see by our calculation that the energy potentially available from nuclear fusion is enormous.

The larger the binding energy per nucleon, the more stable is the nucleus. Figure 23.1 shows a plot of such energy values versus mass numbers, and you can see that the curve passes through a maximum at iron-56. It is not a sharp maximum, so among the isotopes of intermediate mass numbers we find the most stable isotopes.

As we follow the plot to the highest mass numbers, the nuclei become less stable. Among the heaviest atoms, therefore, we would expect to find the isotopes that could undergo nuclear *fission* — a spontaneous breaking apart into more stable isotopes of intermediate mass number. (We will study fission in more detail in Section 23.9.)

On the other hand, it would be the *fusion* of nuclei of the lowest mass number that would give a gain in nuclear stability as nuclei of higher mass numbers formed. So research on nuclear fusion is concentrated on fusing the nuclei of isotopes of hydrogen.

Quite often you'll see the terms *atomic fusion* and *atomic fission* used for *nuclear fusion* and *nuclear fission.*

FIGURE 23.1 Binding energies per nucleon. The energy unit here is the megaelectron-volt or MeV; 1 MeV = 10^6 eV = 1.602×10^{-13} J. (From D. Halliday and R. Resnick, *Fundamentals of Physics,* 2nd Ed., Revised, 1986. Used by permission.)

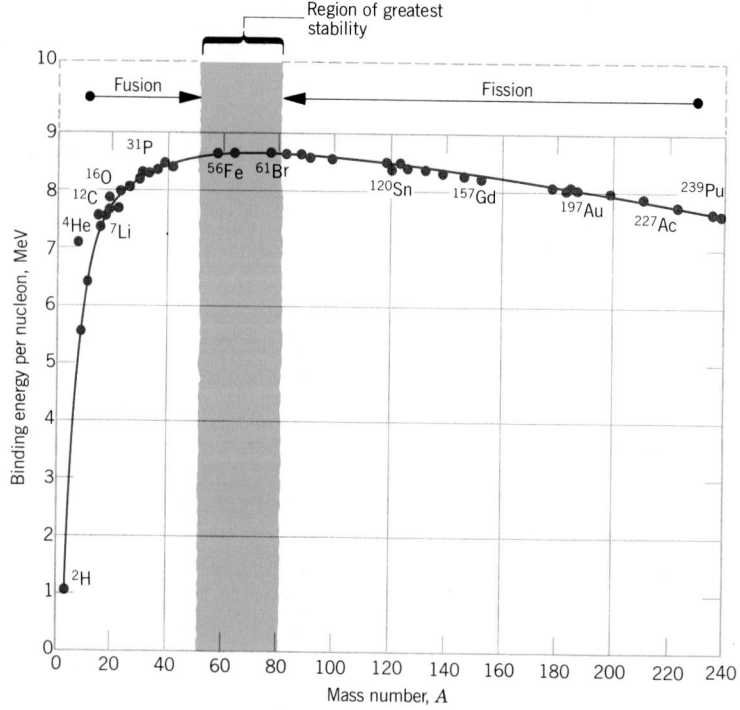

23.3 RADIOACTIVITY

The nuclei of several naturally occurring isotopes spontaneously decay by emitting alpha particles, beta particles, or gamma rays.

Except for hydrogen, all atomic nuclei have more than one proton. Each proton bears a positive charge, and we have learned that like charges repel. So we might well ask how can any nucleus be stable? What keeps the protons from repelling each other? And what holds neutrons within the nuclear package? Without an electrical charge, why don't they just drift away? As physicists addressed such questions, they came to realize that electrostatic forces of attraction and repulsion—the kind present in, say, a crystal of sodium chloride—are not the only forces at work in the atomic regions. Protons do, indeed, repel each other, but another force is also at work in an atomic nucleus, a force that physicists simply call the *strong force*. This force is able to overcome the electrostatic force within a nucleus, and it binds nucleons into a package.

The strong force differs in a particularly important way from the electrostatic force. It decreases far more rapidly with distance. The strong force exerted by one nucleon on any of the others falls to zero even *within* the nucleus. The strong force between two adjacent nucleons, therefore, does not contribute anything to the binding of nucleons on the other side of a nucleus. They have their own internucleon strong forces at work. The electrostatic force of repulsion between protons in a nucleus, however, does not fall off to zero. Protons in one nuclear region repel protons in all other regions. Such repulsions are, of course, toned down by intervening neutrons, because they help to separate the protons.

> The strong force has the same character between any two nucleons: proton–proton, proton–neutron, neutron–neutron.

One consequence of the difference between the strong force and the electrostatic force occurs when nuclei carry large numbers of protons without enough intermingled neutrons to dilute the electrostatic repulsions between the protons. Now the nucleus can be unstable. Fission, however, is a possible consequence of such instability only in one naturally occurring isotope, uranium-235. Other mechanisms are used by all isotopes with unstable nuclei, including also uranium-235, to achieve more stability. These mechanisms involve the ejection of small nuclear fragments and often, simultaneously, the emission of high-energy electromagnetic radiation.

> Adjacent neutrons experience no electrostatic repulsion between each other, only the strong force (of attraction).

Isotopes whose nuclei emit particles or radiation are called radionuclides. The emission itself is called radioactivity or radioactive decay, and materials with this ability are said to be radioactive. About 50 of the approximately 350 naturally occurring isotopes are radioactive. Their radioactivity involves just three kinds of naturally occurring radiation—alpha, beta, and gamma radiation.

Alpha Radiation

Alpha radiation consists of streams of the nuclei of helium atoms, called alpha particles, symbolized as ^4_2He, where 4 is the mass number (sum of protons and neutrons) and 2 is the atomic number (number of protons). The electrical charge on each alpha particle is 2+, but the charge is usually omitted from the symbol.

> Alpha-emitting radionuclides are common among the elements near the end of the periodic table.

Alpha particles are the most massive of any that radionuclides can emit. When ejected (Figure 23.2), they break out and shoot through the atom's electron orbits reaching emerging speeds of up to one-tenth the speed of light. Their size prevents alpha

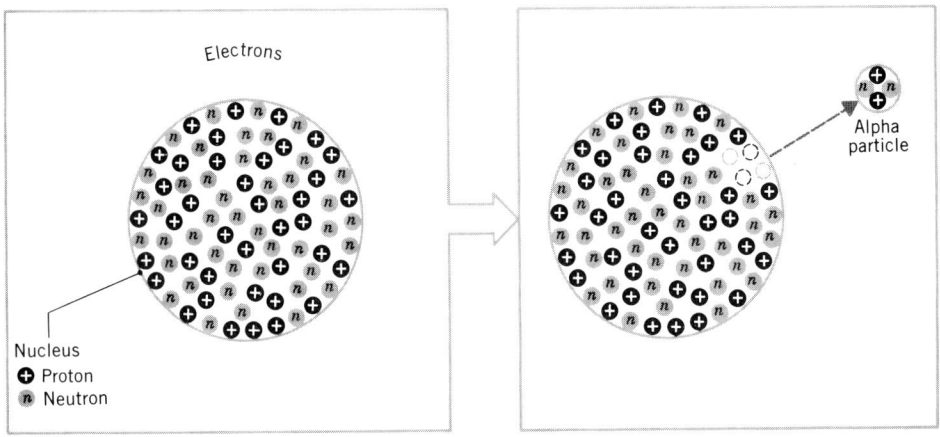

Electrons

Nucleus
⊕ Proton
ⓝ Neutron

Alpha particle

FIGURE 23.2 Emission of an alpha particle.

particles from going far, however. Within a few centimeters of traveling in air, they collide with air molecules, lose kinetic energy, pick up electrons, and become neutral helium atoms. Alpha particles cannot penetrate the skin, although enough exposure causes a severe skin burn. If carried in air or on food into the soft tissues of the lungs or the intestinal tract, emitters of alpha particles can cause serious harm, including cancer.

Nuclear Equations

To symbolize nuclear decay, we construct a nuclear equation, which we can illustrate by the alpha decay of uranium-238 to thorium-234.

$$^{238}_{92}U \longrightarrow {}^{234}_{90}Th + {}^{4}_{2}He$$

Unlike changes that occur in chemical reactions, nuclear reactions produce new isotopes, so we need special principles for balancing nuclear equations. A nuclear equation is balanced when

1. The sums of the mass numbers on each side are equal.
2. The sums of the atomic numbers on each side are equal.

In the above equation you can see that the atomic numbers balance because $(90 + 2) = 92$. And the mass numbers balance, because $(234 + 4) = 238$. Notice that electrical charges are not included, even though we know they are there (initially). The thorium particle initially has two electrons too many, because it formed by the loss of two protons from the uranium-238 atom. And, as we have said, the alpha particle has a charge of $2+$. But these particles eventually pick up or lose electrons as they travel through matter (including air) and become neutral atoms.

Beta Radiation

Naturally occurring beta radiation consists of streams of electrons, which often are called beta particles in this context. In a nuclear equation the beta particle has the symbol $^{0}_{-1}e$, because the electron's mass number is 0 and its charge is -1. Tritium is a beta emitter.

$$\underset{\text{tritium}}{{}^{3}_{1}H} \longrightarrow \underset{\text{helium-3}}{{}^{3}_{2}He} + \underset{\substack{\text{beta particle} \\ \text{(electron)}}}{{}^{0}_{-1}e}$$

Tritium is a synthetic radionuclide.

A beta particle comes from the atom's nucleus, not from an electron orbital. It does not have a prior existence in the nucleus, any more than a photon exists prior to its emission from an excited atom. Instead, the beta particle is created during the decay process, one in which a neutron is transformed into a proton. See Figure 23.3.

Some references symbolize the proton in a nuclear equation as $^{1}_{1}H$.

$$\underset{\substack{\text{neutron} \\ \text{(in nucleus)}}}{{}^{1}_{0}n} \longrightarrow \underset{\substack{\text{beta particle} \\ \text{(emitted)}}}{{}^{0}_{-1}e} + \underset{\substack{\text{proton} \\ \text{(remains in} \\ \text{nucleus)}}}{{}^{1}_{1}p}$$

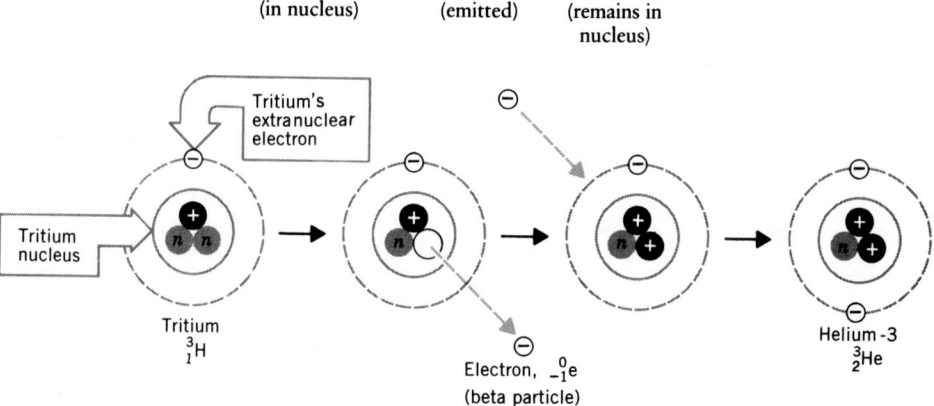

FIGURE 23.3 Emission of a beta particle.

Unlike the alpha particles from a given radionuclide, which are all emitted with the same discrete energy, the beta particles of a given beta emitter emerge with a continuous spectrum of energy, varying from zero up to some characteristic fixed upper limit for the isotope. This fact gave nuclear physicists considerable trouble, because it was an apparent violation of energy conservation (among other difficulties which we will not go into). To solve this problem, Wolfgang Pauli proposed in 1927 that beta emission is accompanied by yet another decay particle. This particle would have to be electrically neutral and virtually, if not exactly, massless. Enrico Fermi named it the *neutrino* ("little neutral one"), and evidence for it eventually was discovered.

Since an electron has only 1/7000th the mass of an alpha particle, a beta particle is less likely to collide with the molecules of the substance through which it travels (such as air). Depending, of course, on the initial kinetic energy of a beta particle, it can travel up to 300 cm in dry air, much farther than alpha particles. Only the highest-energy beta particles can penetrate the skin, however. (Neutrinos, on the other hand, can travel as far as an estimated 3000 light years in water before striking a proton in a water molecule. No wonder they were hard to find.)

1 light year = 9.460×10^{12} km, the distance light travels in 1 year.

Gamma Radiation

Gamma radiation consists of high-energy photons emitted from certain radionuclides. Often gamma radiation accompanies either alpha or beta radiation, and in nuclear equations the symbol for gamma radiation is $^0_0\gamma$, or often simply γ.

Nuclei can have energy levels of their own much as an atom has orbital energy levels. The emission of an alpha or beta particle can leave the nucleus in an excited state, and the emission of a gamma ray photon relaxes the nucleus to a more stable state.

Gamma radiation is an extremely penetrating radiation that is effectively blocked near a source only by shields made of very dense materials, such as lead. Table 23.1

In medicine and pharmacy, the symbol γ (gamma) is also used for *microgram*.

TABLE 23.1 Penetrating Abilities of Some Common Radiations[a]

Type of Radiation	Common Sources	Approximate Energy When from These Sources	Approximate Depth of Penetration of Radiation into		
			Dry Air	Tissue	Lead
Alpha rays	Radium-226 Radon-222 Polonium-210	5 MeV	4 cm	0.05 mm[c]	0
Beta rays	Tritium	0.05 – 1 MeV	6 to 300 cm[b]	0.06 to 4 mm[c]	0.005 to 0.3 mm
	Strontium-90 Iodine-131 Carbon-14				
			Thickness to Reduce Initial Intensity by 10%		
Gamma rays	Cobalt-60 Cesium-137 Decay products of radium-226	1 MeV	400 m	50 cm	30 mm
X rays					
Diagnostic	—	Up to 90 keV	120 m	15 cm	0.3 mm
Therapeutic	—	Up to 250 keV	240 m	30 cm	1.5 mm

[a] Data from J. B. Little, *The New England Journal of Medicine*, vol. 275, pages 929–938, 1966.

[b] The range of beta particles in air is about 30 cm/MeV. Thus a 2-MeV beta particle has a range of about 60 cm in air.

[c] The protective layer of skin is about 0.07 mm thick. To penetrate it, alpha particles need about 7.5 MeV of energy and beta particles, about 0.07 MeV.

describes the penetrating abilities of alpha, beta, and gamma radiations from various sources.

The energy unit in Table 23.1 and the unit most widely used to describe the energies of radiations is the electron-volt. One electron-volt, 1 eV, is the energy an electron receives when accelerated under the influence of 1 volt. It is related to the joule as follows.

$$1 \text{ eV} = 1.602 \times 10^{-19} \text{ J}$$

The electron-volt is obviously an extremely small amount of energy, so multiples are commonly used, such as the kilo-, mega-, and gigaelectron volt.

$$1 \text{ keV} = 10^3 \text{ eV}$$
$$1 \text{ MeV} = 10^6 \text{ eV}$$
$$1 \text{ GeV} = 10^9 \text{ eV}$$

The gamma radiation from cobalt-60, a radionuclide used in cancer therapy, consists of photons with energies of 1.173 MeV and 1.332 MeV. X rays used in diagnosis typically have energies of 100 keV or less.

X Rays

X rays, like gamma rays, are high-energy electromagnetic radiation, but their energies are usually less than those of gamma radiation. X rays are emitted by some synthetic radionuclides, but normally they are generated by special machines.

X rays can be made by directing a high-energy electron beam at a metal target. The beam knocks electrons out of atomic orbitals in the target and creates *holes* — empty or half-filled orbitals. If a hole is created at a low-orbital energy level, then an orbital electron at a higher energy level drops down to fill the hole. This creates a hole higher up, so an electron at a still higher level drops into it, making another hole even higher up. Thus a cascade of electron transitions occurs, and those of the highest energy cause the emission of photons in the X-ray region of the spectrum. The chief difference, therefore, between X rays and gamma rays is that X rays come from transitions involving electron energy levels and gamma rays are from transitions between nuclear energy levels.

In the interconversion of mass and energy,

$$1 \text{ eV} = 1.783 \times 10^{-36} \text{ kg.}$$

EXAMPLE 23.1 BALANCING NUCLEAR EQUATIONS

Problem: Cesium-137, $^{137}_{55}$Cs, is one of the radioactive wastes from a nuclear power plant or an atomic bomb explosion. It emits beta and gamma radiation. Write the nuclear equation for the decay of cesium-137.

Solution: Set up the nuclear equation using the information given and leaving blanks for any information we have to figure out.

$$^{137}_{55}\text{Cs} \longrightarrow {}^{0}_{-1}\text{e} + {}^{0}_{0}\gamma + \underline{\quad}\underline{\quad}$$

space for mass number

space for atomic symbol

space for atomic number

Beta particle   Gamma radiation

Ions of cesium, which is in the same family as sodium, travel in the body to many of the same sites where sodium ions go.

The first thing to do is to figure out the atomic symbol for the isotope that forms. For this we need the atomic number, because then we can look up the symbol in a table. We will figure out the atomic number using a step-by-step process, but with just a little experience, you will be able to do this by inspection. We find the atomic number by using the fact that the sum of the atomic numbers on one side of the nuclear equation must equal the sum on the other side. Therefore, letting Z be the atomic number of the unknown isotope:

$$55 = -1 + 0 + Z$$
$$Z = 56$$

Now we can use the periodic table to find that the atomic symbol that goes with element 56 is Ba (barium). Next, to find out which particular isotope of barium forms, we use the fact that the sum of the mass numbers on one side of the nuclear equation must equal the sum on the other side. Letting A equal the mass number of the barium isotope,

$$137 = 0 + 0 + A$$
$$A = 137$$

The balanced nuclear equation, therefore, is

$$^{137}_{55}\text{Cs} \longrightarrow {}^{0}_{-1}\text{e} + {}^{0}_{0}\gamma + {}^{137}_{56}\text{Ba}$$

PRACTICE EXERCISE 1   Marie Curie (1867–1934) earned one of her two Nobel prizes for isolating the element radium, which soon became widely used to treat cancer. Radium-226, $^{226}_{88}\text{Ra}$, is an alpha and gamma emitter. Write a balanced nuclear equation for the decay of radium-226.

PRACTICE EXERCISE 2   Strontium-90 is one of the many radionuclides present in the radioactive wastes of operating nuclear power plants. Strontium-90 is a beta emitter. Write the balanced nuclear equation for its decay.

Radioactive Disintegration Series

Sometimes one radionuclide decays not to a stable isotope but to another radioactive isotope. This might, in turn, decay to still another radionuclide, and the process can continue through a series of nuclear reactions until a stable isotope forms. Such a sequence is called a radioactive disintegration series. Four such naturally occurring series are known. Uranium-238 is at the head of one, shown in Figure 23.4. Notice that some of the half-lives are extremely short, but that others are very long.

FIGURE 23.4 The uranium-238 radioactive disintegration series. The time given beneath each arrow is the half-life period of the preceding isotope (y = year; m = month; d = day; mi = minute; and s = second).

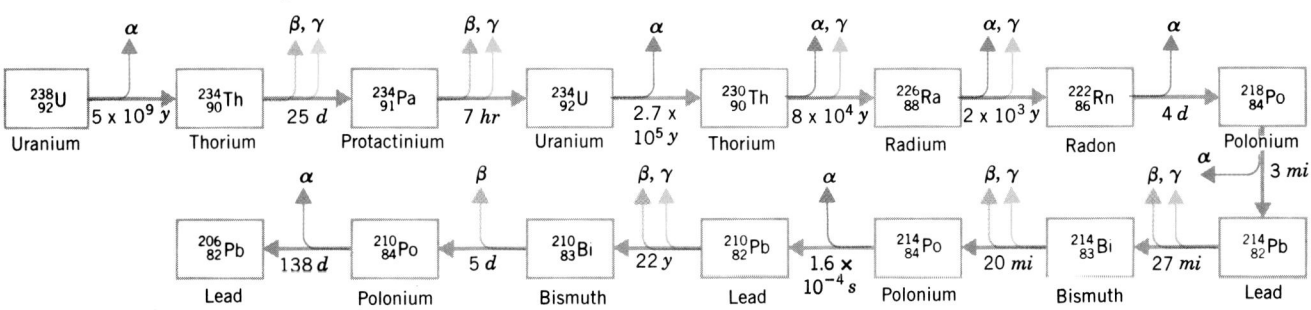

We studied half-lives in Section 18.6. One half-life period in nuclear science means the time it takes for a given sample of a radionuclide to decay to one-half of its initial amount. Since radioactive decay is a first-order process, the period of time taken by one half-life period is independent of the initial number of nuclei. Table 23.2 (page 966) gives the half-lives of several natural and synthetic radionuclides.

Other Radiations

A number of synthetic isotopes emit positrons, particles with the mass of an electron but a positive instead of a negative charge. Its symbol is ${}^{0}_{1}\text{e}$. A positron forms in the nucleus by the conversion of a proton to a neutron. See Figure 23.5. Cobalt-54, for example, is a

The symbol of the positron is sometimes β^+.

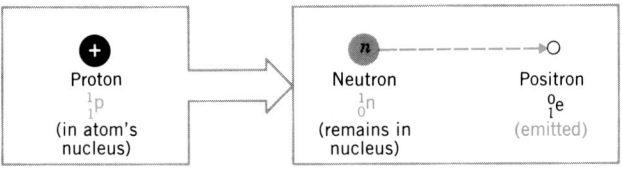

FIGURE 23.5 The emission of a positron replaces a proton by a neutron.

TABLE 23.2 Typical Half-Life Periods

Element	Isotope	Half-Life	Radiations or Mode of Decay
Naturally Occurring Radionuclides			
Potassium	$^{40}_{19}K$	1.3×10^9 yr	beta, gamma
Tellurium	$^{123}_{52}Te$	1.2×10^{13} yr	electron capture
Neodymium	$^{144}_{60}Nd$	5×10^{15} yr	alpha
Samarium	$^{149}_{62}Sm$	4×10^{14} yr	alpha
Rhenium	$^{187}_{75}Re$	7×10^{10} yr	beta
Radon	$^{222}_{86}Rn$	3.82 days	alpha
Radium	$^{226}_{88}Ra$	1590 yr	alpha, gamma
Thorium	$^{230}_{90}Th$	8×10^4 yr	alpha, gamma
Uranium	$^{238}_{92}U$	4.51×10^9 yr	alpha
Synthetic Radionuclides			
Hydrogen	$^{3}_{1}H$ (tritium)	12.26 yr	beta
Oxygen	$^{15}_{8}O$	124 s	positron
Phosphorus	$^{32}_{15}P$	14.3 days	beta
Technetium	$^{99m}_{43}Tc$	6.02 hr	gamma
Iodine	$^{131}_{53}I$	8.07 days	beta
Cesium	$^{137}_{55}Cs$	30 yr	beta
Strontium	$^{90}_{38}Sr$	28.1 yr	beta
Americium	$^{243}_{95}Am$	7.37×10^3 yr	alpha

positron emitter and changes to a stable isotope of iron.

$$^{54}_{27}Co \longrightarrow {}^{54}_{26}Fe + {}^{0}_{1}e$$

The positron is a positive beta particle — a positive electron. As it moves out through the orbital electrons, it soon hits one. (They are attracted to each other, of course.) When such a collision occurs, the positron and the electron annihilate each other. Their masses change into the energy of two photons of gamma radiation called *annihilation radiation photons*, each with an energy of 511 keV. See Figure 23.6.

$$\underset{\substack{\text{electron} \\ \text{(orbital)}}}{{}^{0}_{-1}e} + \underset{\text{positron}}{{}^{0}_{1}e} \longrightarrow \underset{\substack{\text{gamma} \\ \text{radiation}}}{2\,{}^{0}_{0}\gamma}$$

Because a positron destroys a particle of ordinary matter (an electron), it is called a particle of antimatter. To be called antimatter, a particle must have a counterpart among

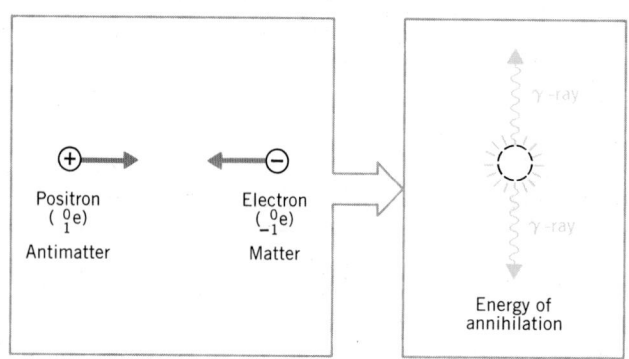

FIGURE 23.6 Gamma radiation is produced when a positron and an electron collide.

particles of ordinary matter, and the two must annihilate each other when they collide. For example a neutron destroys an antineutron.

Positron emission, like beta emission, is accompanied by a chargeless and virtually massless particle. We said that a neutrino accompanies beta emission, but actually it is an antineutrino. The neutrino is what emerges along with positron emission.

Neutron emission is another kind of nuclear reaction. Krypton-87, for example, decays as follows.

$$\ce{^{87}_{36}Kr} \longrightarrow \ce{^{86}_{36}Kr} + \ce{^{1}_{0}n}$$

This is one decay process that does not lead to an isotope of a different element.

Electron capture or K-capture is still another kind of nuclear reaction. It is very rare among natural isotopes, but common among synthetic radionuclides. Vanadium-50 nuclei, for example, can capture orbital K-shell electrons and change to nuclei of stable atoms of titanium.

The K-shell is the shell with the principal quantum number $n = 1$.

$$\underset{\substack{\text{orbital} \\ K\text{-electron}}}{\ce{^{50}_{23}V} + \ce{^{0}_{-1}e}} \xrightarrow{\text{electron capture}} \ce{^{50}_{22}Ti} + \text{X rays}$$

The net effect of electron capture is the conversion of a proton into a neutron. See Figure 23.7.

$$\underset{\substack{\text{proton} \\ \text{in the} \\ \text{nucleus}}}{\ce{^{1}_{1}p}} + \underset{\substack{\text{electron} \\ \text{captured} \\ \text{from the} \\ K \text{ shell}}}{\ce{^{0}_{-1}e}} \longrightarrow \underset{\substack{\text{neutron} \\ \text{in the} \\ \text{nucleus}}}{\ce{^{1}_{0}n}}$$

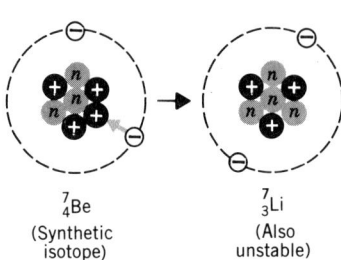

$$\underset{\substack{\text{(Synthetic} \\ \text{isotope)}}}{\ce{^{7}_{4}Be}} \qquad \underset{\substack{\text{(Also} \\ \text{unstable)}}}{\ce{^{7}_{3}Li}}$$

FIGURE 23.7 Electron capture is the collapse of an orbital electron into the nucleus, and this changes a proton into a neutron.

Electron capture does not change an atom's mass number, only its atomic number. It also leaves a hole in the K shell, and the atom emits photons of X rays as other orbital electrons drop down to fill the hole. Moreover, the nucleus that has just captured an orbital electron can emit a gamma ray photon.

It is commonly thought that the rate of radioactive decay of a particular isotope is independent of its oxidation state or any other aspect of its chemical combination or its physical state (pressure and temperature, for example). When decay is by alpha or beta emission, this appears to be true — at least modern instruments have not yet detected any variations. When decay is by electron capture, however, some very small differences in rates of decay have been observed. For example, beryllium-7 decays about 0.3% faster as the metal than as the oxide, BeO. There is more electron density next to the beryllium nucleus in the metal than in the oxide (where the oxidation state of Be is +2), and thus the ability of the nucleus to capture an electron is greater in the metal.

23.4 THE BAND OF STABILITY

The more stable isotopes of the elements fall within a band of stability when all isotopes are placed into an array according to their numbers of neutrons and protons.

No mathematical equation has been found to predict if an isotope will be stable. We might think that because iron-56 has the highest binding energy per nucleon, other iron isotopes should be stable. One isotope, however — iron-53 — has a half-life of only 8.5 minutes, while another — iron-60 — has a half-life of 300,000 years.

In the absence of an equation, scientists have developed some "rules of thumb" to guide the selection of new isotopes that could be candidates for synthesis. Such candidates, of course, should only be those with half-lives long enough to permit detection — greater than 10^{-8} second. One rule of thumb involves the array of all known isotopes,

FIGURE 23.8 The band of stability.

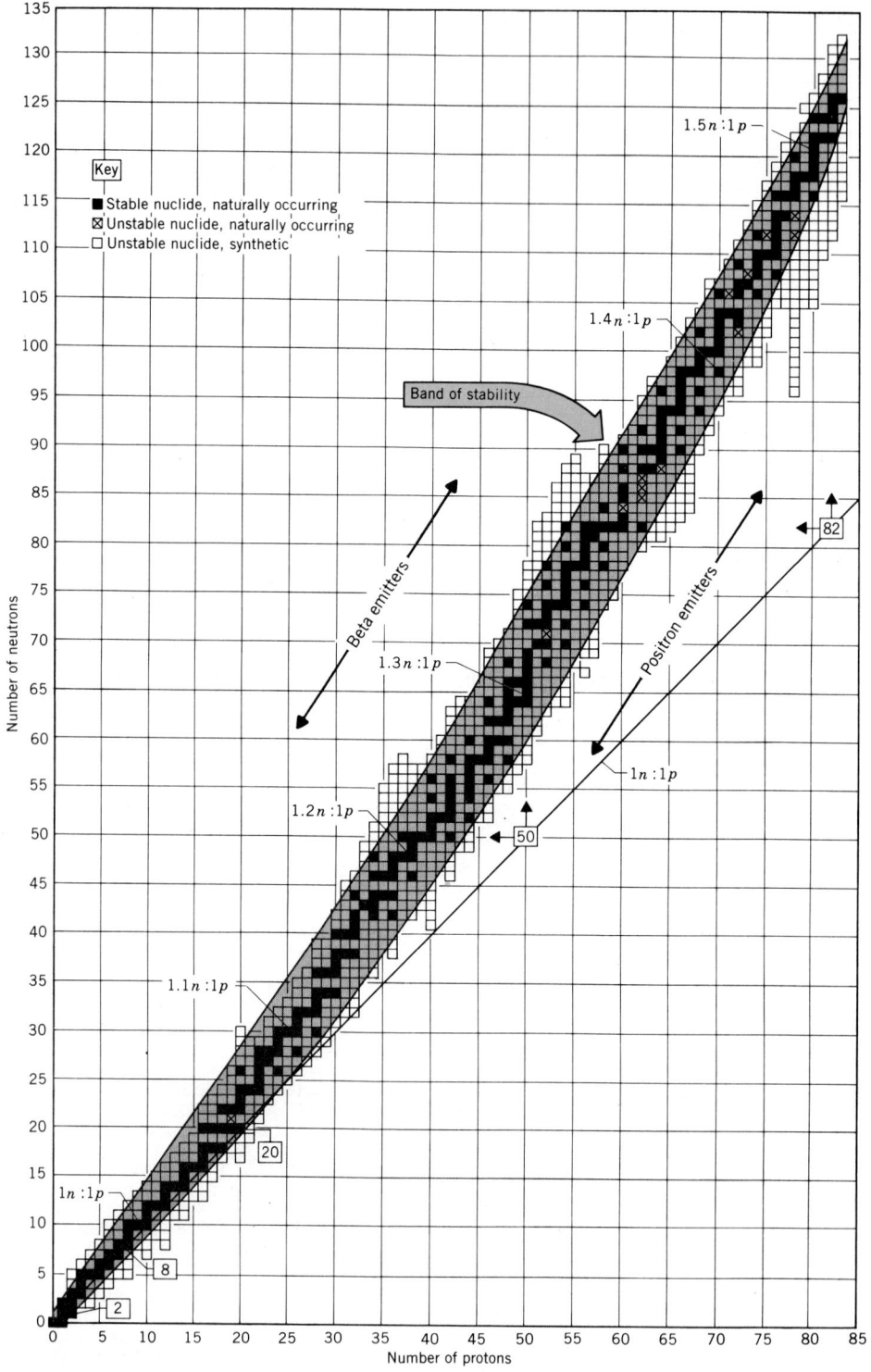

both stable and unstable, shown in Figure 23.8. Each isotope has a location defined by its numbers of protons and neutrons. (No isotope above element 83, bismuth, is plotted because bismuth is the last one to have at least one stable isotope.)

Two curved lines have been drawn on this array to enclose all stable nuclei. Between these two lines lies the band of stability. It includes some radionuclides, because smooth lines could not be drawn to exclude them. Any isotope not represented anywhere on the array, inside or outside the band of stability, probably has a half-life too short to permit its detection (if it indeed exists). Therefore, as a rule of thumb, the search for new isotopes lying at some distance from the band of stability would likely be unsuccessful. For example, an isotope with 50 neutrons and 60 protons would be too unstable to justify the time and money for an attempt to make it.

One useful observation concerning the array in Figure 23.8 involves the ratio of neutrons to protons. With increasing numbers of protons, this ratio gradually moves farther and farther away from a simple 1:1 ratio (indicated in the figure by the straight line). It seems that as more and more protons become packed into a nucleus—and remember that protons repel each other electrostatically—still larger and larger numbers of neutrons are needed to provide compensating strong forces to "dilute" proton–proton electrostatic repulsions.

As indicated in Figure 23.8, isotopes occurring above and to the left of the band of stability tend to be beta emitters and those lying below and to the right are positron emitters. The isotopes not plotted, those with atomic numbers above 83, tend to be alpha emitters. Are there any reasons for these tendencies?

The beta emitters evidently have too high a neutron/proton ratio (which is why they are *above* the band of stability). Remember that beta decay has the overall effect of losing a neutron and gaining a proton:

$$\ce{^1_0n \longrightarrow ^1_1p + ^0_{-1}e}$$

This change would decrease the neutron/proton ratio. The beta decay of fluorine-20, for example, lets the neutron-to-proton ratio in its nucleus decrease from 11/9 to 10/10. This moves the surviving nucleus, that of neon-20, closer to the center of the band of stability.

$$\ce{^{20}_9F \longrightarrow ^{20}_{10}Ne + ^0_{-1}e}$$

$$\frac{\text{neutron}}{\text{proton}} = \frac{11}{9} \qquad \frac{10}{10}$$

Figure 23.9 is a "blowup" of this part of Figure 23.8 showing how this beta emission

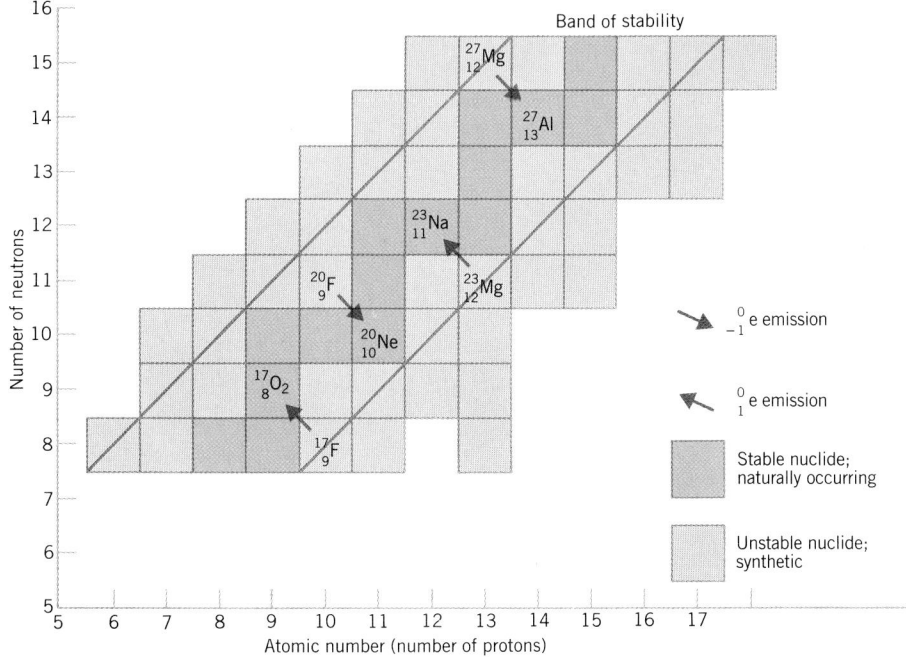

FIGURE 23.9 Beta decay from magnesium-27 and fluorine-20 reduces their neutron-proton ratio and moves them closer to the band of stability. Positron decay from magnesium-23 and fluorine-17 increases this ratio and moves these nuclides closer to the band of stability, too. Shown here is an enlarged section of Figure 23.8.

moves the nucleus over. The beta decay of magnesium-27 to aluminum-27, also shown in Figure 23.9, is another illustration of how the loss of an electron causes a favorable lowering of the neutron/proton ratio and a shift toward the center of the band of stability.

Positron emission has the opposite effect on the neutron/proton ratio as electron emission; it raises this ratio. Evidently, the positron emitters — and they are found only among artificial radionuclides — have too few neutrons for stability, because the net effect in the nucleus of ejecting a positron is to gain a neutron and lose a proton. Thus the nucleus of another isotope of fluorine, fluorine-17, improves its stability by emitting a positron,

$$^{17}_{9}F \longrightarrow {}^{17}_{8}O + {}^{0}_{1}e$$

$$\frac{\text{neutron}}{\text{proton}} = \frac{8}{9} \qquad \frac{9}{8}$$

Figure 23.9 shows how this change moves the nucleus nearer the center of the band of stability. Positron decay by magnesium-23 to sodium-23, also seen in Figure 23.9, produces a similarly favorable shift.

The heavy nuclei above atomic number 83 are alpha emitters, evidently because their nuclei have too many protons for stability. The loss of an alpha particle is the quickest way to get rid of protons.

Odd–Even Rule

Another rule of thumb about stable isotopes is the odd–even rule.

> Odd–Even Rule   When the numbers of neutrons and protons in a nucleus are both even, the isotope is far more likely to be stable than when both numbers are odd.

$^{2}_{1}H$, $^{6}_{3}Li$, $^{10}_{5}B$, $^{14}_{7}N$, and $^{138}_{57}La$ have odd numbers of both protons and neutrons.

Out of the 264 stable isotopes, only five have odd numbers of *both* protons and neutrons, whereas 157 have even numbers of both. The rest have an odd number of one nucleon and an even number of the other. To see this, notice in Figure 23.8 how the horizontal lines with the largest numbers of dark squares (stable isotopes) most commonly correspond to even numbers of neutrons. Similarly, the vertical lines with the most dark squares most often correspond to even numbers of protons.

The odd–even rule is related to the spins of nucleons. Both protons and neutrons have spin, like orbital electrons. When two like particles have paired spins — when their spins are opposite — their combined energy is less than when their spins are not paired. Thus when there are even numbers of protons and neutrons, all spins can be paired, and the system has less energy (and is more stable) than when an odd proton or odd neutron is present. The least stable nuclei tend to be those with both an odd proton and an odd neutron.

Magic Numbers

Another rule of thumb about nuclear stability is based on the magic numbers of nucleons. Isotopes with specific numbers of protons or neutrons — the magic numbers — are more stable than the rest (and, according to a theory whose details we cannot discuss, *ought to be more stable*). The magic numbers are 2, 8, 20, 28, 50, 82, and 126. Figure 23.8 shows where these numbers fall (except for 126).

Notice in Figure 23.8 that an atom with 82 protons and 82 neutrons lies far outside the band of stability even though 82 is a magic number.

When both the number of protons and the number of neutrons are the *same* magic number, the isotope can be quite stable. These are $^{4}_{2}He$, $^{16}_{8}O$, and $^{40}_{20}Ca$. One stable isotope of lead, $^{208}_{82}Pb$, has 82 protons and 126 neutrons, both magic numbers.

The existence of magic numbers supports the hypothesis that there are nuclear energy levels just as there are electron energy levels. Electron orbitals have a set of magic numbers of their own—the maximum number of electrons that can be in a principal energy level. These are 2, 8, 18, 32, 50, 72, and 98 for principal levels 1, 2, 3, 4, 5, 6, and 7, respectively. *Chemically* speaking, there are magic numbers for the total numbers of electrons in atoms. The most chemically stable elements—the noble gases—have 2, 10, 18, 36, 54, and 86 electrons.

23.5   TRANSMUTATION

When a high-speed particle is captured by a nucleus, the nucleus can be permanently changed to that of another element.

The change of one isotope into another is called transmutation, and radioactive decay is just one way that it can happen. Another is by bombarding the atoms of an isotope with high-energy particles. These can be alpha particles from natural alpha emitters, neutrons from atomic reactors, and protons made by the removal of electrons from hydrogen. The electrically charged nature of alpha particles and protons makes it possible for them to be accelerated to ultrahigh energies in special accelerators. With sufficient energy, bombarding particles can shoot through the electron orbitals and bury themselves in nuclei. Beta particles (electrons), however, are strongly repelled by the orbital electrons. Therefore they are seldom able to get through to a nucleus, and they generally are not used to cause transmutations.

Formation and Decay of Compound Nuclei

When a nucleus captures a bombarding particle it becomes a compound nucleus. At the moment of impact, it has all of the energy of the bombarding particle and is therefore unstable. Its additional energy is soon distributed among all of its nuclear particles, but eventually the nucleus has to get rid of the excess energy. It can do this by ejecting a high-energy particle such as a neutron, a proton, or an electron, or by emitting gamma radiation. The overall effect is a transmutation.

Compound here refers only to the idea of *combination*, not to a chemical.

Ernest Rutherford, the British scientist whose research team discovered the atomic nucleus, observed the first transmutation to occur under laboratory conditions. He allowed alpha particles from a natural radionuclide to pass through a chamber containing nitrogen atoms. He discovered that another radiation was produced, one that could be detected at far greater distances than alpha particles could reach. He was able to show that this new radiation consisted of high-energy protons, $_1^1\text{p}$, and Rutherford concluded that they must have resulted from the conversion of nitrogen nuclei into oxygen nuclei. He reasoned that the capture of an alpha particle by a nitrogen nucleus produced a compound nucleus, one of fluorine-18, that soon emitted a proton to form an atom of oxygen-17 (Figure 23.10).

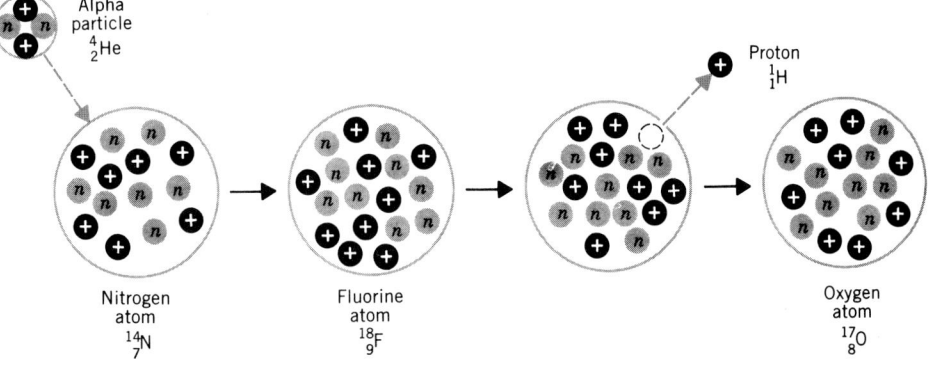

Alpha particle $_2^4\text{He}$

Proton $_1^1\text{H}$

Nitrogen atom $_7^{14}\text{N}$

Fluorine atom $_9^{18}\text{F}$

Oxygen atom $_8^{17}\text{O}$

FIGURE 23.10 Transmutation of nitrogen into oxygen. When the nucleus of nitrogen-14 captures an alpha particle it becomes a compound nucleus of fluorine-18. This then expels a proton and becomes the nucleus of oxygen-17.

The asterisk, *, symbolizes a high-energy nucleus — a compound nucleus.

$$\underset{\substack{\text{alpha}\\\text{particle}}}{{}^{4}_{2}\text{He}} + \underset{\substack{\text{nitrogen}\\\text{nucleus}}}{{}^{14}_{7}\text{N}} \longrightarrow \underset{\substack{\text{fluorine}\\\text{(compound}\\\text{nucleus)}}}{{}^{18}_{9}\text{F}^{*}} \longrightarrow \underset{\substack{\text{oxygen}\\\text{(a rare but}\\\text{stable}\\\text{isotope)}}}{{}^{17}_{8}\text{O}} + \underset{\substack{\text{proton}\\\text{(high-energy)}}}{{}^{1}_{1}\text{p}}$$

Alpha particles and protons pick up speed and energy when they are attracted toward a negatively charged object. The awareness of this fact led to the development of a number of particle accelerators — gigantic machines such as cyclotrons. The first transmutation caused by experimentally accelerated particles was the conversion of lithium-7 to helium-4. The bombarding particles were protons made by using energy to strip electrons from ordinary hydrogen. The protons were accelerated and then focused onto a lithium target. The capture of a proton by a lithium nucleus creates a compound nucleus of beryllium, which promptly breaks in half to give two alpha particles. (These eventually pick up electrons and become helium, as we have noted.)

$$\underset{\substack{\text{proton}}}{{}^{1}_{1}\text{p}} + \underset{\substack{\text{lithium}}}{{}^{7}_{3}\text{Li}} \longrightarrow \underset{\substack{\text{beryllium}}}{{}^{8}_{4}\text{Be}^{*}} \longrightarrow \underset{\substack{\text{alpha particles}}}{2\,{}^{4}_{2}\text{He}}$$

The manner in which a given compound nucleus decays depends on how much energy it has, not on the kind of bombarding particle used to make it. For example, a high-energy, compound nucleus of an atom of aluminum-27 could be made by using alpha particles, protons, or deuterons according to the following reactions.

$$\underset{\substack{\text{alpha}\\\text{particle}}}{{}^{4}_{2}\text{He}} + \underset{\substack{\text{sodium-23}\\\text{atom}}}{{}^{23}_{11}\text{Na}}$$

$$\underset{\substack{\text{proton}}}{{}^{1}_{1}\text{p}} + \underset{\substack{\text{magnesium-26}\\\text{atom}}}{{}^{26}_{12}\text{Mg}} \longrightarrow \underset{\substack{\text{aluminum-27}\\\text{atom with a}\\\text{compound nucleus}}}{{}^{27}_{13}\text{Al}^{*}}$$

A deuteron is the nucleus of a deuterium atom just as a proton is the nucleus of a protium (hydrogen) atom.

$$\underset{\substack{\text{deuteron}}}{{}^{2}_{1}\text{D}} + \underset{\substack{\text{magnesium-25}\\\text{atom}}}{{}^{25}_{12}\text{Mg}}$$

Once formed, the high-energy aluminum-27 nucleus "forgets" how it was made. It only knows how much *energy* it holds. Any of the following decay modes are open to it, depending on this energy. They illustrate how the synthesis of such a large number of synthetic radionuclides has been possible.

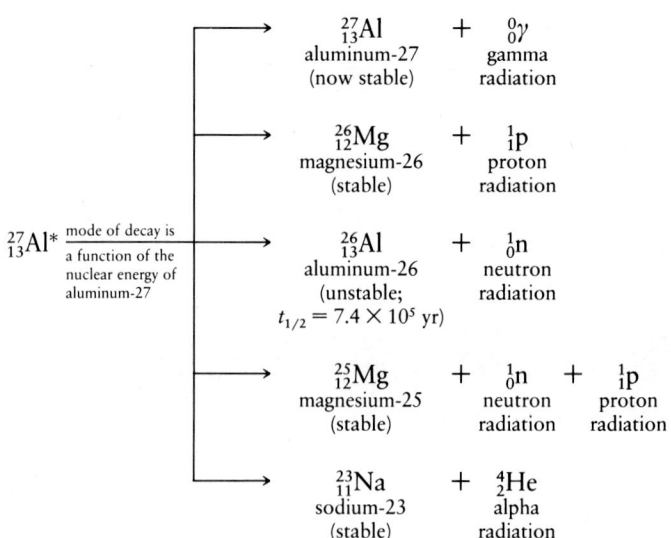

$${}^{27}_{13}\text{Al}^{*} \xrightarrow{\substack{\text{mode of decay is}\\\text{a function of the}\\\text{nuclear energy of}\\\text{aluminum-27}}}$$

$$\underset{\substack{\text{aluminum-27}\\\text{(now stable)}}}{{}^{27}_{13}\text{Al}} + \underset{\substack{\text{gamma}\\\text{radiation}}}{{}^{0}_{0}\gamma}$$

$$\underset{\substack{\text{magnesium-26}\\\text{(stable)}}}{{}^{26}_{12}\text{Mg}} + \underset{\substack{\text{proton}\\\text{radiation}}}{{}^{1}_{1}\text{p}}$$

$$\underset{\substack{\text{aluminum-26}\\\text{(unstable;}\\t_{1/2} = 7.4 \times 10^{5}\text{ yr)}}}{{}^{26}_{13}\text{Al}} + \underset{\substack{\text{neutron}\\\text{radiation}}}{{}^{1}_{0}\text{n}}$$

$$\underset{\substack{\text{magnesium-25}\\\text{(stable)}}}{{}^{25}_{12}\text{Mg}} + \underset{\substack{\text{neutron}\\\text{radiation}}}{{}^{1}_{0}\text{n}} + \underset{\substack{\text{proton}\\\text{radiation}}}{{}^{1}_{1}\text{p}}$$

$$\underset{\substack{\text{sodium-23}\\\text{(stable)}}}{{}^{23}_{11}\text{Na}} + \underset{\substack{\text{alpha}\\\text{radiation}}}{{}^{4}_{2}\text{He}}$$

Radionuclides above Number 83

From the standpoint of practical applications, some of the most useful transmutations have been those that produce radionuclides with sufficiently long half-lives to be isolated and used in research, medicine, and technology. Over a thousand radionuclides have been made by transmutations, and many are isotopes that do not occur in nature. (There are nearly 900 open squares in Figure 23.8 designating known synthetic radionuclides for elements up through atomic number 83.) Many naturally occurring isotopes exist above 83, but all have long half-lives. Many of the other possible isotopes above 83 may well have existed at one time in nature, but had half-lives too short in relation to the age of the planet to have survived into the twentieth century. For example, neptunium-237, which has a half-life of over 2 million years, undoubtedly existed naturally at one time, but the estimated age of the earth is over 4 billion years, long enough for essentially all of the neptunium to have decayed. We know about neptunium-237 today only because it can be made in accelerators.

None of the known elements from 93 (neptunium) and above, the transuranium elements, occurs naturally. All have been made using high-energy bombardment techniques in nuclear reactors and accelerators (Figure 23.11). The synthetic elements 93 to

FIGURE 23.11 In this linear accelerator at Brookhaven National Laboratory on Long Island, New York, protons can be accelerated to just under the speed of light and given up to 33 GeV of energy before they strike their target. (1 GeV = 10^9 eV)

103 complete the actinide series of the periodic table (which begins with element 90). Elements 104, 105, 106, 107, and 109 have also been made. Only trace quantities are ever initially made prior to the announcement of a new element, and this has led to controversies and has complicated the processes of giving acceptable names and symbols. See Special Topic 23.1.

To make the still heavier elements, bombarding particles that are larger than neutrons have been used. These include alpha particles, and highly accelerated cations from such isotopes as carbon-12, nitrogen-15, oxygen-18, and neon-22.

One model of nuclear structure predicts that isotopes of element 114 should have long enough half-lives to be prepared and detected. In fact, there is some evidence from the analysis of meteorites that this element once existed in the solar system.

From a study of certain ancient rocks on this planet — monazite rocks — some scientists think that element 126 existed at one time in the earth's past. Something that was once a part of these rocks caused the formation of circular patterns of color called "halos," patterns that can be duplicated only by using ultrahigh-energy alpha particles. However, no radionuclides known today naturally emit alpha radiation of the necessary energy to cause these halos. Element 126 would have 126 protons in its nucleus — one of the magic numbers. Research designed to make such heavy isotopes is in progress.

Condensed Nuclear Equations

The equations that describe some of the transmutations used to synthesize the new, heavy elements are complicated by multiple processes, so to make the descriptions simpler, scientists often use condensed forms. For example, actinium-227 has been made by bombarding radium-226 with neutrons. Neutron capture by radium-226 is accompanied by simultaneous emission of gamma radiation and the formation of radium-227. This then decays by beta emission to Ac-227. All of this can be symbolized simply as

> The numbers beneath the arrows in this and the remaining equations in this Section are half-lives.

$$^{226}_{88}\text{Ra (n, } \gamma) \, ^{227}_{88}\text{Ra} \xrightarrow[\text{41 min}]{} \, ^{227}_{89}\text{Ac} + \, ^{0}_{-1}\text{e}$$

On the left in this expression we see a term (n, γ). In such a term, the first symbol is of the bombarding particle and the second is of the kind of radiation that is emitted simultaneously with capture of the first. Then follows the product. If this product has a finite half-life and we want to show its own decay, then it can be followed in the conventional way with the decay equation. The expression for the synthesis of neptunium-237 from uranium-238 is

> To change into $^{237}_{92}\text{U}$, $^{238}_{92}\text{U}$ captures *one* neutron and emits two. This is what (n, 2n) signifies in $^{238}_{92}\text{U}(n, 2n)^{237}_{92}\text{U}$.

$$^{238}_{92}\text{U (n, 2n)} \, ^{237}_{92}\text{U} \xrightarrow[\text{6.75 days}]{} \, ^{237}_{93}\text{Np} + \, ^{0}_{-1}\text{e}$$

Often, multiple neutron capture occurs to transmute one element into another. Another synthesis of neptunium-237, for example, involves two successive neutron captures starting with uranium-235 and symbolized as follows.

$$^{235}_{92}\text{U (n, } \gamma) \, ^{236}_{92}\text{U (n, } \gamma) \, ^{237}_{92}\text{U} \xrightarrow[\text{6.75 days}]{} \, ^{237}_{93}\text{Np} + \, ^{0}_{-1}\text{e}$$

Americium, element 95, is made by successive neutron captures starting with plutonium-239.

$$^{239}_{94}\text{Pu (n, } \gamma) \, ^{240}_{94}\text{Pu (n, } \gamma) \, ^{241}_{94}\text{Pu} \xrightarrow[\text{13.2 yr}]{} \, ^{241}_{95}\text{Am} + \, ^{0}_{-1}\text{e}$$

23.6 DETECTING AND MEASURING RADIATIONS

The ability of radiations to generate ions in matter makes their detection and measurement possible.

Physicists have developed many ways to detect, record, and measure radiations from radionuclides. Several are based on one of the most important properties of radiations —their ability to generate ions in the matter they penetrate.

Detection Devices

All the radiations we have thus far described are ionizing radiations. When ions are produced in a gas, even momentarily, the gas becomes a conductor of electricity. A Geiger–Müller tube is one radiation detector based on this fact (Figure 23.12). It is particularly useful in detecting beta and gamma radiation. When a pulse of radiation enters the thin-walled window of the tube, it creates ions in the gas inside. This allows a pulse of electricity to flow from the cathode, the metallized surface of the tube, to the anode, a wire electrode. This flow of current activates an instrument that amplifies the current and records the pulse. The tube and the associated equipment is called a *Geiger counter*.

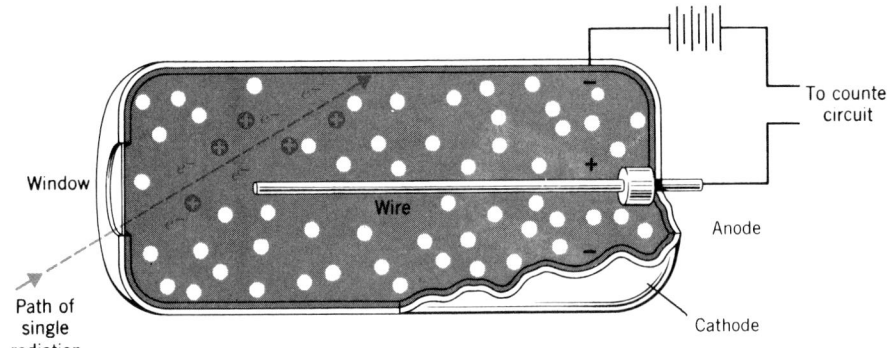

FIGURE 23.12 The Geiger-Müller tube. The white dots within the tube represent atoms of argon at very low pressure.

For visualizing the actual tracks of particles of radiation, several kinds of *cloud chambers* have been invented. When an enclosed space is briefly made supersaturated in the vapor of some fluid—water or alcohol, for example—while it is exposed to radiation, microdroplets of condensed vapor form along any path taken by a particle, as seen in Figure 23.13. The existence of some of the fundamental atomic particles was initially inferred by analyses of photographed tracks produced by cloud chambers.

Atmospheric clouds form in air supersaturated with water vapor and in which condensation is started by dust particles.

FIGURE 23.13 Cloud chamber photo. Where the heavy tracks intersect there were collisions between a neutron and an oxygen atom and a neutron and a carbon atom.

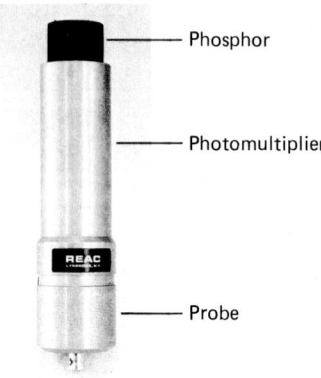

FIGURE 23.14 Scintillation probe. Energy received from radiations striking the phosphor at the top of this probe is processed by the photomultiplier unit and sent to an instrument (not shown) for recording.

Scintillation counters (Figure 23.14) are devices that permit an investigator to see when a collision occurs between a particle and a special surface on the counter. This surface is coated with a substance that gives off a tiny light flash when it is hit by a particle of radiation. For example, if the coating contains a zinc sulfide phosphor, then alpha particles cause visible scintillations. The scintillations can be magnified with electronic equipment and automatically counted.

A *dosimeter* is used to measure the total quantity of radiation received by a surface during a specified interval of time. Some dosimeters (Figure 23.15) use photographic plates that are kept completely shielded from ordinary light but that are sensitive to atomic radiations, including gamma and X rays. The amount of darkening in the developed film is proportional to the dose of radiation received and can be measured. Photographic film is also used to record X rays in medicine. Dense tissues, such as bones, reduce the intensity of X rays or gamma rays. Therefore, if either radiation passes through the body toward a photographic plate, shadows of bones will be cast on the plate and will show up as a familiar "X-ray" picture. For a description of two significant advances in X-ray technology, see Special Topics 23.2 and 23.3.

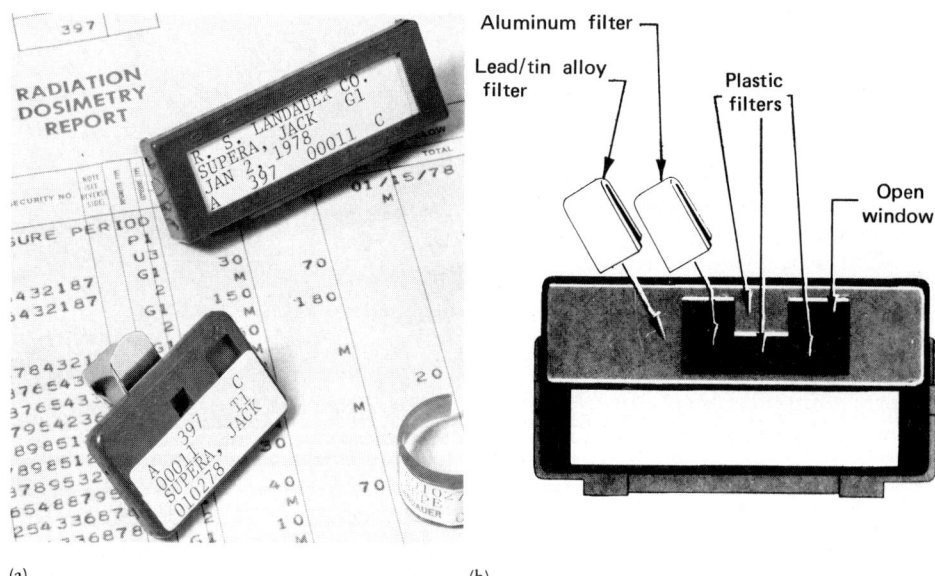

FIGURE 23.15 Dosimeter. *(a)* Badge dosimeters. The doses of various kinds of radiations for predetermined periods of exposure are carefully logged to help in preventing overexposures to people whose jobs require them to use radioactive materials or X rays. *(b)* The heart of the dosimeter (as seen from the back) is this piece that can be slipped out for analysis. By using different filters, different parts of the badge register the exposures to radiations of various energies and penetrating abilities — X rays, beta rays, and gamma rays.

The Curie and the Becquerel

When the radiation department of a hospital or clinic buys radioactive material, it is particularly interested in the activity of the sample. *Activity* refers to the number of nuclear disintegrations per second occurring in the sample. The SI unit of activity is called the becquerel(Bq), which is equal to one disintegration or other kind of nuclear transformation per second.

An older, pre-SI unit of activity that is still widely used was named after Marie Curie, the discoverer of radium (see Practice Exercise 1). One curie, Ci, equals the rate of

The becquerel is named after Henri Becquerel (1852–1908), the discoverer of radioactivity, who won a Nobel prize in 1903.

The CT scan (for computerized tomography) uses extremely brief pulses of X rays, simultaneously focused from many angles, to obtain a cross-sectional picture of some region of a patient. Figure 23.16 shows a picture of one such instrument. The X-ray pulses received from the several angles are processed by a computer and converted into an X-ray image, like that shown in Figure 23.17. Someone has likened the technique to getting a full, cross-sectional view of a cherry pit without cutting open the cherry.

The CT scan, initially dubbed the CAT scan, has been called the greatest development in X-ray techniques since the discovery of X rays, and its principal developers, A. M. Cormack (United States) and G. N. Hounsfield (England), received the 1979 Nobel prize in medicine.

The CT scan is an example of an invasive diagnostic technique, meaning that its use invades the patient with something—here, X rays—that can cause harm. It is used, of course, only when the expected benefits are firmly believed to outweigh such potential harm.

The MRI Scan

Since the CT scan was developed, scientists have given medicine another technique, the MRI scan (for magnetic resonance imaging), initially called the NMR scan (for nuclear magnetic resonance imaging). The MRI scan takes advantage of the weakly magnetic properties of certain isotopes in the molecules of the body—notably protons in water molecules and all organic compounds. These interact with low-energy radiofrequency radiation to generate photos just like X-ray images. There is no evidence that such low-energy radiation poses invasive harm to the patient. Interestingly, the medical community switched from the term NMR scan to MRI scan because of the public's fear of anything associated with the word "nuclear." The nuclear property utilized in the MRI scan, however, has nothing to do with unstable, radioactive matters, only with the abiding presence of the weak magnetism of protons and a few other nuclei of interest. The actual phenomenon used, nuclear magnetic resonance, is the basis of an important analytical method widely used by chemists.

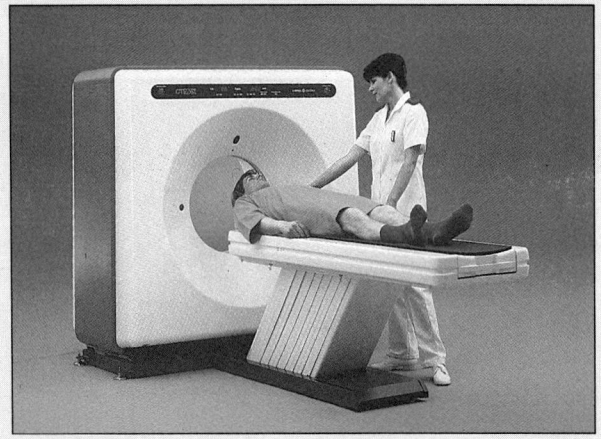

FIGURE 23.16 Instrument for computerized axial tomography, the CT scan. Shown here is General Electric's CT/T Total Body Scanner that can complete an entire scan of the head or the body in as little as 4.8 seconds.

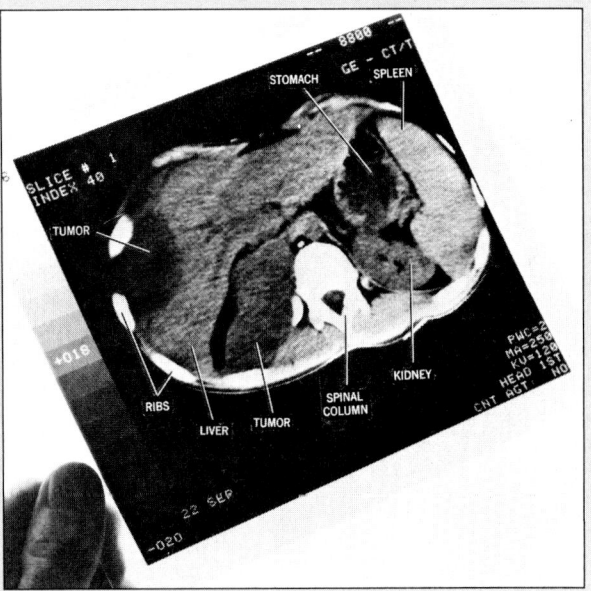

FIGURE 23.17 This cross-sectional view through a patient showing two tumors was obtained with a CT scanner.

disintegrations that occur in a sample of 1.0 g of radium, which is 37 billion disintegrations per second.

$$1 \text{ Ci} = 3.7 \times 10^{10} \text{ disintegrations/s}$$

Thus,

$$1 \text{ Ci} = 3.7 \times 10^{10} \text{ Bq}$$

If a hospital owns a radioactive source rated as 1.5 Ci, the source delivers $1.5 \times 3.7 \times 10^{10} = 5.6 \times 10^{10}$ Bq.

In a PET scan (for positron emission tomography) the gamma radiation from positron–electron collisions is generated within a targeted tissue and then leaves the body to make an image like an X-ray image. Figure 23.18 pictures a PET scan instrument.

To get the positron emitters into the patient, ordinary biochemicals such as glucose or some amino acid are synthesized with at least one atom in each molecule a positron-emitting isotope of a stable atom. For example, oxygen-15 can be used in place of oxygen-16, or carbon-11 for carbon-12, or nitrogen-13 for nitrogen-14.

The positron emitters have short half-lives, so the work has to be done very quickly. Positron-emitting glucose ($t_{1/2} = 124$ s), for example, is made by letting photosynthesizing swiss chard use carbon dioxide labeled with oxygen-15 or carbon-11 to make glucose. The swiss chard is then quickly sacrificed; its glucose is removed and fed intravenously to the patient. Glucose molecules are able to cross the blood–brain barrier and enter brain cells, and they can be tracked by external detectors of gamma radiation. Glucose is the principal source of chemical energy for the brain's work. Abnormalities in glucose metabolism can therefore be pinpointed by identifying which regions

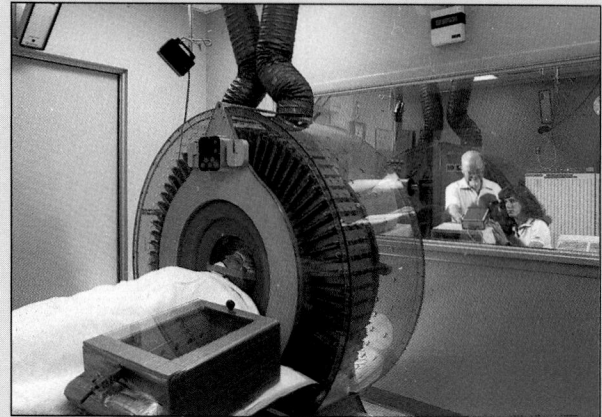

FIGURE 23.18 The PET scan. The patient is undergoing a brain scan using the PET technique.

have become poor glucose users or have accelerated their glucose consumption. The PET scan offers unusually high accuracy in locating trouble spots, thus enabling a neurosurgeon, for example, to find a small tumor and remove it with a reduced likelihood of damaging healthy cells.

The Rad and the Gray

Rad stands for *r*adiation *a*bsorbed *d*ose.

To describe the quantity or dose of radiation absorbed by some material, units of *absorbed dose* (or simply *dose*) have been devised. The most common unit of dose in the United States is the rad, symbolized by rd. One rad corresponds to the absorption of 10^{-5} J per gram of tissue.

The *gray* is named after Harold Gray, a British radiologist.

The SI unit of absorbed dose is called the gray, Gy, and 1 gray corresponds to 1 joule of energy absorbed per kilogram of absorbing material.

$$1 \text{ rd} = 10^{-5} \text{ J/g}$$
$$1 \text{ Gy} = 1 \text{ J/kg}$$
$$1 \text{ Gy} = 100 \text{ rd}$$

According to U.S. government estimates, if a large group of people were to receive 450 rads each, half the group would die in 60 days. Some specialists think that this would happen at lower doses — 350 rads or even as low as 220 rads — unless antibiotics, blood plasma, and bone marrow transplant therapy were available.

The Rem

Rem stands for *r*adiation *e*quivalent for *m*an.

One of the drawbacks of the rad is that radiation damage is a function not just of the absorbed energy but also of the kind of radiation. Neutrons, for example, are more dangerous than beta radiation of the same energy and intensity. To take such differences into account, scientists devised a unit called the rem. To find the dose in rems, the dose in rads is multiplied by a conversion factor that reflects the effectiveness of the *kind* of radiation in causing damage. When alpha radiation is involved, the rem dose has a value of about 20 times the rad dose. With neutron radiation, the rem dose is between 2 and 10 times the rad dose. However, the rem dose is numerically about the same as the rad dose for beta and gamma radiation and for X rays.

One advantage with using rems to describe doses is that, unlike rads, doses in rems are additive. Thus we cannot say that 10 rads to the chest from gamma radiation and 5 rads to the head from neutron radiation add up to 15 rads for the whole body. But after these rads are converted to rems, the figures can be added to find the whole-body dose in rems.

When the accident occurred at the Chernobyl nuclear power plant in Russia on April 26, 1986, some workers nearest the disaster probably received 400 rems almost immediately, according to an estimate released by the Soviet government. In the ensuing four months, 31 people died from radiations. The day after the disaster, the radiation level was 1 rem/hr in the streets of the nearby town of Pripyat. More than 135,000 people near Chernobyl were evacuated.

According to U.S. government guidelines, workers who are exposed to radiations during their work must not receive more than 0.3 rem per week. (A typical chest X ray involves a dose of about 7 mrem (7×10^{-3} rem). A whole-body dose of 25 rems induces noticeable changes in human blood. A set of symptoms called *radiation sickness*— nausea, vomiting, a drop in the white cell count, diarrhea, dehydration, prostration, hemorrhaging, and loss of hair — sets in at 100 rems and is severe at 200 rems. At roughly 400 rems, half of a population each receiving such a dose would die in 60 days. And a 600-rem dose would kill everyone in a week.

Considering how dangerous a few rems are, it is somewhat astonishing that such damage involves extremely small quantities of energy. Obviously, the dangers of radiation cannot be associated with the *heat* equivalent of a rem. Instead, the danger comes from the ability of radiation to create unstable ions or neutral species with unpaired electrons (free radicals). For example, water can interact as follows with radiations.

$$H\text{—}\ddot{\underset{..}{O}}\text{—}H + \begin{matrix}\text{radioactive}\\\text{particle or}\\\text{gamma ray}\end{matrix} \longrightarrow (H\text{—}\dot{\underset{..}{O}}\text{—}H)^+ + \underset{\text{electron}}{_{-1}^{0}e}$$

The new cation, $(H\text{—}\dot{\underset{..}{O}}\text{—}H)^+$, in unstable and one way it can break down is as follows.

$$(H\text{—}\dot{\underset{..}{O}}\text{—}H)^+ \longrightarrow \underset{\text{proton}}{H^+} + \underset{\begin{matrix}\text{hydroxyl}\\\text{radical}\end{matrix}}{:\dot{\underset{..}{O}}\text{—}H}$$

The H^+ might pick up a stray electron and become $H\cdot$, a hydrogen atom. Both $H\cdot$ and $:\overset{..}{\underset{..}{O}}\text{—}H$ are examples of free radicals, and both are decidedly unstable. What they might do depends on what other molecules are in the vicinity. The formation of free radicals and unstable ions can set off a series of unwanted chemical changes inside a body cell that can lead to cancer, tumors, or other equally grave results. For this chemical reason, absorbed radiations can inflict injury all out of proportion to their actual energies. The actual energy in a dose of 600 rems causes the ionization of only one water molecule for every 36 million present, yet 600 rems is a lethal dose.

An average dose of 43 rem was received by those living 3 to 15 km from the Chernobyl plant. For some perspective on this, more cancer, heart disease, and genetic defects are caused by a lifetime of cigarette smoking than are caused by such an exposure, according to risk-analysis experts.

Background Radiation

Scientists, medical personnel, and technicians who work with radionuclides or X-ray machines must have as much protection as practical from repeated exposures to radiations. However, no one can completely escape low-level exposures, because sources of radioactivity are everywhere. Together they make up what is called the natural background radiation, and this gives an average of about 100 mrem of exposure per year per person in the United States. Background radiation includes cosmic rays from above and radiations from radionuclides in the earth. The top 40 cm of soil holds an average of 1

A few years ago an engineer unintentionally but repeatedly set off radiation alarms at his workplace, a nuclear power plant in Pennsylvania. Such alarms are triggered when radiation levels surpass a predetermined value. The problem eventually was traced to his home where the radiation level was 2,700 picocuries per liter of air, compared with the level in the average house of 1 picocurie per liter. (The picocurie is 10^{-12} curie.) The chief culprit was radon-222, a natural intermediate in the uranium-238 disintegration series (Figure 23.4). (One picocurie per liter is equivalent to the decay of two radon atoms per minute.) The engineer had carried radon-222 and its decay products on his clothing from his home to the power plant.

One result of this episode was to send geologists scurrying to map the general area for concentrations of uranium-238. They found that the bedrock of the Reading Prong, a geological formation that cuts across Pennsylvania, New Jersey, New York, and up into the New England states, is relatively rich in uranium.

Radon-222 is a noble gas, and is as *chemically* unreactive as argon, krypton, and the other noble gases. But this decay product of uranium-238 has an unstable nucleus. Its half-life is 4 days, and it emits both alpha and gamma radiation. Being a gas, it diffuses through rocks and soil and enters buildings. In fact, buildings can act much like chimneys and draw radon as a chimney draws smoke. Radon-222 in the air is breathed into the lungs. Because of its short half-life, some of it decays in the lungs before it is exhaled. The decay products are not gases, and several have short half-lives—polonium-218 ($t_{1/2}$ 3 min, alpha), lead-214 ($t_{1/2}$ 27 min, beta, gamma), bismuth-214 ($t_{1/2}$ 20 min, beta, gamma), and polonium-214 ($t_{1/2}$ 1.6×10^{-4} s, alpha and gamma). When these are left in the lungs, their radiation can cause lung cancer, and U.S. government officials believe that about 10,000 of the country's 130,000 annual deaths from lung cancer are caused by indoor radon concentrations.

The Environmental Protection Agency (EPA) has recommended an upper limit of 4 picocuries per liter in homes. Estimates released in 1986 by the EPA were that roughly 1% of the homes in the United States have radon levels that generate radiation of more than 10 picocuries per liter. Those who estimate risks present in day-to-day activities say that such a level poses less a risk to life than the risk of driving a car. The inhabitants and owners of the small percent of homes with high radon levels, however, are faced with hard and costly choices.

Coal usually contains very minute concentrations of radium and thorium, both radioactive, which can escape into the air when coal burns.

gram of radium per square kilometer. Carbon-14 is present in all the food we eat. Inside the body of an adult human, about 5×10^5 nuclear disintegrations occur every minute. Additional exposures come from X rays used for diagnosis or therapy and from pollutants released by nuclear power plants or present in the smoke from coal-burning power

TABLE 23.3 Average Radiation Doses Received Annually by the U.S. Population

Source	Dose (mrem)
Natural background (whole-body dose equivalent)	100
Medical sources	
Diagnostic X rays	20–50
Therapeutic X rays	3–5
Radioisotopes	0.2
Other sources	
Fallout (from atmospheric testing, 1954–1962)	1.5 (5 mrem maximum in early 1960s)
Radioactive pollutants introduced into the environment from nuclear power plants (1970), United States	0.85
Miscellaneous	2.0[a]
Maximum annual limit proposed for the average, general population (by the International Commission on Radiological Protection, 1970), from all sources *exclusive of medical and background*[b]	170

[a] The one-pack-a-day smoker's lungs get at least 40 mrem per year more dose equivalent. Naturally occurring radon, a radioactive, gaseous decay product of naturally occuring radium, decays to radioactive lead that deposits on earth and foliage (including broadleaf tobacco plants).

[b] The Advisory Committee on the Biological Effects of Ionizing Radiation of the National Academy of Sciences–National Research Council sharply criticized this recommendation in its report in late 1972. Americans on the average, receive 100 mrem from natural background according to the committee. To allow the average exposure to climb to 170 mrem would cause a 2% increase in the spontaneous cancer rate.

plants. Table 23.3 summarizes the amounts of radiation in various parts of the background. Special Topic 23.4 describes an unusual contribution to the background caused by radon-222.

Radiation Protection

Although we cannot escape background radiation we ought to minimize exposures from any other sources. Those who routinely work with radioactive materials should use radiation shielding materials, and they should stay as far as practical from the source of the radiation. As the data in Table 23.1 show, gamma radiation and X rays are effectively shielded only by very dense materials such as lead. Low-density materials such as cardboard, plastic, or aluminum are poor shielders, although concrete, if thick enough, is a good and relatively inexpensive material for shielding.

Keeping one's distance from a radioactive source is effective in providing radiation protection because the intensity of radiation diminishes with the *square* of the distance from the source. Thus, if a worker doubles the distance from the source, the intensity of the exposure will be reduced by one fourth. The relationship between distance and intensity is given by the inverse square law,

$$\text{radiation intensity} \propto \frac{1}{d^2}$$

where d is the distance from the source. If the intensity I_1 is known at distance d_1, then the intensity I_2 at distance d_2 can be calculated by the following equation.

$$\frac{I_1}{I_2} = \frac{d_2{}^2}{d_1{}^2}$$

This law applies only to a small source that radiates equally in all directions and with no intervening shields.

Modern, high-speed X-ray film permits shorter exposure times and so reduces radiation hazards to patients and medical people.

EXAMPLE 23.2   USING THE INVERSE SQUARE LAW

Problem: At 1.5 m from a small radioactive source the radiation intensity was 40 units. If the operator moves to a distance 4.5 m from the source, what will be the radiation intensity?

Solution: We substitute into the equation just given.

$$\frac{40 \text{ units}}{I_2} = \frac{(4.5 \text{ m})^2}{(1.5 \text{ m})^2}$$

$$I_2 = \frac{(40 \text{ units})(1.5)^2 \text{m}^2}{(4.5)^2 \text{m}^2}$$

$$= 4.4 \text{ units}$$

PRACTICE EXERCISE 3   If an operator 10 m from a small source is exposed to 1.4 units of radiation, what will be the intensity of the radiation if he moves to 1.2 m from the source?

23.7 APPLICATIONS OF RADIOACTIVITY

Isotope dilution analysis, neutron activation analysis, tracer studies, and archaeological dating are some of the applications of radionuclides.

The chemical properties of radionuclides are the same as those of the stable isotopes of the same element, as we have noted before, and this fact is exploited in several practical applications of radioactivity. The chemical properties, as well as such ordinary physical

The deuterium isotope effect (Section 19.3) is an exception, but it involves *rate*, not kind of reaction.

Technetium-99*m* is an almost ideal radionuclide for diagnostic work. The half-life (6.02 hr) of this gamma emitter is not so short that it decays before it delivers diagnostic benefits. Yet its half-life is short enough so that most of its decay occurs *during* diagnostic work. This enables the maximum use of its gamma radiation for potential benefits, because continued decay afterward can cause only harm. All of the radiation of technetium-99*m* is gamma radiation, so all of it (unlike alpha or beta particles) can leave the body, reach external detectors, and serve some good. The short half-life of this radionuclide means that very small samples will have a sufficiently high activity while leaving behind a very low concentration of its decay products. Technetium-99*m* decays to technetium-99 (a beta emitter, $t_{1/2}$ 2.12×10^5 yr) whose half-life is so long that the activity introduced into the patient this way is very low. Moreover, technetium in any diagnostic form is eliminated by the body. All of these properties mean that the patient is minimally exposed to the dangers of radiation while enjoying the benefits.

Technetium-99*m* in the form of the pertechnitate ion, TcO_4^-, is administered intravenously as one solute in a dilute saline solution. The solution is prepared by letting 0.9% NaCl trickle through a bed of (inert) alumina that also holds radiodecaying molybdenum-99 (as MoO_4^{2-}). See the accompanying figure of what is called the molybdenum-99 cow. Molybdenum-99 decays by beta emission ($t_{1/2}$ 67 hr) to technetium-99*m*.

$$^{99}_{42}Mo \longrightarrow {}^{99m}_{43}Tc + {}^{0}_{-1}e$$

As the saline solution passes through the device, it leaches newly formed pertechnitate ions, which are similar in solubility to the chloride ion. The radiologist, after "milking the cow," uses the solution for diagnostic work.

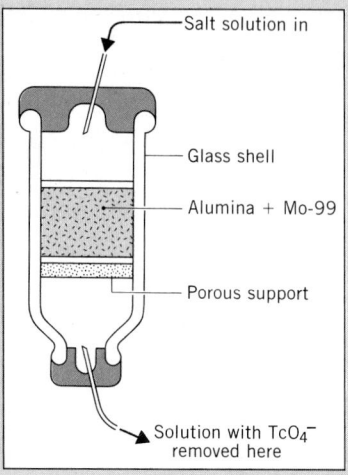

- Salt solution in
- Glass shell
- Alumina + Mo-99
- Porous support
- Solution with TcO_4^- removed here

The molybdenum-99 is made at a facility with an atomic reactor, a source of neutrons. Neutron capture by molybdenum-98 (also as MoO_4^{2-}) generates molybdenum-99.

$$^{98}_{42}Mo + {}^{1}_{0}n \longrightarrow {}^{99}_{42}Mo + {}^{0}_{0}\gamma$$

Once the apparatus is charged, it is sheathed in lead and sent to the clinic or hospital. It lasts about a week before it has to be sent back to the reactor for recharging.

The CT scan (Special Topic 23.2) and the MRI scan (also described in Special Topic 23.2) have taken over much of the diagnostic work of locating brain tumors and other anatomical features. For assessing the *functions* of organs such as the liver, spleen, kidneys, and lungs, however, radionuclides like technetium-99*m* continue to be useful.

properties as solubilities and dialysis, get a radionuclide where the scientist wishes. Then the radiations are exploited for some medical or analytical purpose. Tracer analysis is an example.

Tracer Analysis

In tracer analysis the location of a radionuclide is signaled by its radiations, so if its other chemical and physical properties take it somewhere, it can easily be found. We have already studied an example in Section 18.6 where we learned how iodine-131—a beta emitter, $t_{1/2}$ 8.07 days—is used to test thyroid function.

Tracer analyses are also used to pinpoint the locations of brain tumors, which are uniquely able to concentrate the pertechnitate ion, TcO_4^-, made from technetium-99*m*.[1] This strong gamma emitter is one of the most widely used radionuclides in medicine, as further discussed in Special Topic 23.5.

The tumor appears as a "hot spot" under radiologial analysis.

[1] The *m* in technetium-99*m* means that the isotope is in a metastable form. Its nucleus is at a higher energy level than the nucleus in technetium-99, to which technetium-99*m* decays.

Biochemists have used tracer analysis to find out how plants or animals use simple molecules to make those more complex. A plant's roots, for example, might be allowed to dip for some time into a solution containing bicarbonate ion made with carbon-14 ($t_{1/2}$ 5730 years, beta emitter). Whether the plant uses this ion to make organic compounds can be determined when the plant is harvested and separated into the compounds of interest. Not only is it easy to tell if any compound contains extra carbon-14, but the compound can be further analyzed to find out which of its molecular locations that normally have carbon atoms are made of carbon-14, instead of carbon-12. With this information biochemists are able to construct theories of the biochemical pathways used to make the complex molecule.

Neutron Activation Analysis

Neutron activation analysis is a technique for analyzing the concentration of some element in a substance. It is based on the fact that a number of stable nuclei can be changed into gamma emitters by capturing neutrons.

$$\underset{\substack{\text{isotope} \\ \text{of element } X \\ \text{being analyzed}}}{^M_A X} + \underset{\text{neutron}}{^1_0 n} \longrightarrow \underset{\substack{\text{compound} \\ \text{nucleus}}}{^{(M+1)}_A X^*} \longrightarrow \underset{\substack{\text{stable form} \\ \text{of a new} \\ \text{isotope of } X}}{^{(M+1)}_A X} + \underset{\substack{\text{gamma} \\ \text{radiation}}}{^0_0 \gamma}$$

M = mass number
X = isotope symbol
A = atomic number

But each kind of neutron-enriched nucleus emits gamma radiation at its own unique frequencies, and these are known. By measuring the *frequencies* emitted, therefore, the element can be identified. By measuring the *intensity* of the gamma radiation, the concentration of the element can be determined.

The technique is so sensitive that concentrations as low as $10^{-9}\%$ can be determined. A museum, for example, might have a lock of hair of some famous person long dead. A historian might suspect he was slowly murdered by arsenic poisoning. If so, some arsenic would be in the hair, and neutron activation analysis can find it—without destroying the lock.

Isotope Dilution Analysis

The concentration of some component in a mixture, or the volume of a fluid otherwise hard to measure, can be determined by isotope dilution analysis. The manufacturer of vitamin B-12, for example, using a fermentation system, might need to know how well some batch is going. But such a mixture is typically very complex. Moreover, the concentration of the vitamin is low in the best of circumstances. Here is how the problem is solved by isotope dilution analysis.

Vitamin B-12 is a coordination compound in which the central ion is cobalt, so it is possible to prepare the vitamin with cobalt-60, a gamma emitter. We will symbolize the vitamin with cobalt-60 as B-12*. A known mass of previously prepared B-12* with a known specific activity—the ratio of curies or becquerels per unit of mass—is stirred into a small sample of the fermentation batch. The molecules of the B-12* then fully intermingle with those of the ordinary B-12 already synthesized by the medium. Then a measured but small amount of the batch is withdrawn and its vitamin B-12 (B-12 + B-12*) is extracted and purified until it has a constant specific activity. The extent to which the activity introduced by the B-12* has been diluted can be used to calculate the original amount of B-12. We will put these ideas into the forms of equations and then work an example. We have added a mass m of B-12*:

$$m = \text{mass of B-12* added (in mg)}$$

And its specific activity is

$$S_0 = \text{specific activity of } m \text{ (in } \mu\text{Ci/mg)}$$

Pernicious anemia is a disease caused by a lack of vitamin B-12.

1 Bq = 1 disintegration/s, so *specific activity* means the number of disintegrations per second per unit quantity of material.

$$a = (mg)\left(\frac{\mu Ci}{mg}\right) = \mu Ci$$

So the total activity of the B-12* added, a, is the product

$$a = m \times S_0 \text{ (in } \mu Ci)} \tag{23.2}$$

What we seek is the mass, n, of B-12 initially in the sample:

$$n = \text{mass of B-12 originally present (in mg)}$$

After adding m, the total amount of B-12 of both kinds present is

$$(m + n) = (\text{B-12} + \text{B-12*}) \text{ in the sample (in mg)}$$

Now we experimentally isolate enough of $(m + n)$ to determine the new specific activity, S (in μCi). (We purify the material until the B-12 mixture, B-12 + B-12*, has a constant specific activity.) The values of $(m + n)$ and S give us another expression for the total activity, a:

$$a = (m + n) \times S \tag{23.3}$$

Since the *total* activity, a, does not change, but is simply diluted, the two expressions for a—given by Equations 23.2 and 23.3—must equal each other:

$$m \times S_0 = (m + n) \times S$$

We can solve this equation for n, the amount of B-12 originally present.

$$n = m\left(\frac{S_0}{S} - 1\right) \tag{23.4}$$

We will study an example to see how this works.

EXAMPLE 23.3   ISOTOPE DILUTION ANALYSIS

Problem: A vitamin B-complex tablet with a mass of 4.77 mg was dissolved in water. To this solution was added 0.407 mg of radioactive vitamin B-12 with a specific activity of 2.460 $\mu Ci/mg$. After thorough mixing, a sample of B-12 (including the radioactive form) was isolated and found to have a specific activity of 1.490 $\mu Ci/mg$. Calculate the concentration of vitamin B-12 in the tablet in units of micrograms of B-12 per milligram of B-complex.

Solution: The data are in a form that permits the direct use of Equation 23.4. Letting n equal the mass in milligrams of B-12 in the original sample,

$$n = 0.407 \text{ mg} \times \left(\frac{2.460 \ \mu Ci/mg}{1.490 \ \mu Ci/mg} - 1\right)$$

$$= 0.407 \text{ mg} \times 0.651$$

$$n = 0.265 \text{ mg}$$

This quantity, 0.265 mg, is the same as 265 μg, so the concentration of B-12 in the units specified in the original sample of 4.77 mg is found by

$$\frac{265 \ \mu g}{4.77 \text{ mg}} = 55.6 \ \mu g/mg$$

This B-complex capsule contained 55.6 μg of B-12 per milligram of capsule contents.

PRACTICE EXERCISE 4   A sample of 0.348 mg of radioactive vitamin B-12 with a specific activity of 4.15 $\mu Ci/mg$ was added to 495 mL of a fermentation broth in which B-12 was being formed. Subsequently, 9.70 mL of the broth with a total activity of $9.00 \times 10^{-3} \ \mu Ci$ was isolated. Calculate the concentration of B-12 (in $\mu g/mL$) in the original broth.

Cosmic rays consist of streams of particles that pour into our atmosphere from the sun and outer space. Those that first encounter our outer atmosphere are chiefly high-speed protons, along with some alpha and beta particles and some atomic nuclei up to number 26, iron. These are the *primary cosmic rays*.

As they collide with molecules in the atmosphere, primary cosmic rays create *secondary cosmic rays*, which include all of the subatomic particles, including neutrons. Cosmic rays are thus one source of our background radiation, and one's exposure to them is a function of altitude. Frequent flying in jet airliners or living in high-altitude cities such as Denver, Colorado, provide more exposure to this part of the background than being mostly at low altitudes.

Another application of isotope dilution analysis is in measuring volumes. Suppose, for example, a measured volume of a sample of blood from some animal—say, 1.0 mL —is mixed with a trace of a radionuclide (which must be chemically compatible with the blood). This blood sample now has a specific activity, say 1000 Bq/mL. The sample is then returned to the animal, and it becomes thoroughly mixed as it circulates. Now another 1-mL sample is withdrawn and found to have a specific activity of, say, 10 Bq/mL. The ratio of 10 to 1000 is the same as 1 to 100, so the 1.0-mL blood sample must have been diluted by a factor of 100. Since the original 1.0 mL became part of 100 mL of blood, the original blood volume of the animal was 99 mL.

Radiological Dating

The determination of the age of a geological deposit or an archaeological find through the use of radionuclides in the sample is called radiological dating. It takes advantage of the known half-lives of the radionuclides, and is based partly on the premise that these half-lives have been constant throughout the entire period in question. The premise is strongly supported by the finding that half-lives are insensitive to all external forces such as heat, pressure, magnetic, or electrical stresses.

In geological dating, a pair or isotopes is sought that are related as a "parent" to a "daughter" in a radioactive disintegration series such as the uranium-238 series given in Figure 23.4. Uranium-238 (as "parent") and lead-206 (as "daughter"), for example, have been used. The half-life of uranium-238 is both well known and very long—a necessary feature. A rock specimen whose age is sought is crushed, and the concentrations of uranium-238 and lead-206 are determined. The *ratio* of these concentrations together with the half-life of uranium-238 can then be used to calculate the age of the rock. We will not go into the details of such a calculation.

For dating organic remains such as wooden objects from ancient tombs, scientists can use carbon-14 analysis. This analysis measures the ratio of carbon-14 to stable carbon in the ancient sample and compares it to the same ratio in organic remains that have recently been produced. There are two approaches to the measurements. The older method introduced by its discoverer, J. Willard Libby (Nobel prize in chemistry, 1960), measures the radioactivity of a sample taken from the specimen whose age is sought. It is still widely used because the newer method relies on a very costly apparatus, an accelerator mass spectrometer. This method separates the atoms of carbon-14 from the other isotopes of carbon and counts all of them, not just those that decay. It thus can utilize far smaller samples, it works at much higher efficiencies, and it gives more precise dates. Since it does not depend on the radioactivity of carbon-14, only on its existence as we will next discuss, the method does not illustrate how radioactivity itself can be used for the purpose. We will therefore confine the rest of our discussion largely to Libby's original method.

Throughout history, carbon-14, a beta emitter, has been produced in the atmosphere by the action of cosmic radiation on atmospheric nitrogen. (Special Topic 23.6

FIGURE 23.19 Carbon-14 forms following neutron capture by nitrogen-14.

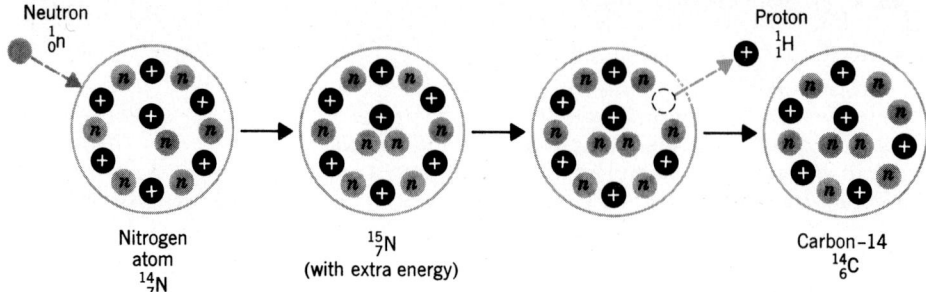

describes cosmic rays.) High-energy neutrons in cosmic rays can transmute nitrogen atoms into carbon-14 atoms according to the following equation. (See also Figure 23.19.)

$$\underset{\substack{\text{neutron}\\ \text{(in cos-}\\ \text{mic ray)}}}{{}^{1}_{0}n} + \underset{\substack{\text{nitrogen}\\ \text{atom in}\\ \text{the at-}\\ \text{mosphere}}}{{}^{14}_{7}N} \longrightarrow \underset{\substack{\text{compound}\\ \text{nucleus}}}{{}^{15}_{7}N^{*}} \longrightarrow \underset{\substack{\text{carbon-14}\\ \text{atom (a}\\ \text{beta}\\ \text{emitter)}}}{{}^{14}_{6}C} + \underset{\substack{\text{high-energy}\\ \text{proton}}}{{}^{1}_{1}p}$$

The newly formed carbon-14 migrates into the lower atmosphere where, together with the ordinary carbon isotopes, it becomes oxidized to carbon dioxide. In this form, carbon-14 enters the earth's biosphere by means of photosynthesis in plants. The plants might be eaten by animals and so become incorporated into bones. Bones are also commonly found in ancient, archaeological sites. Once plants or animals die, the processes of decay eventually return much of their carbon, including any remaining carbon-14, to the atmosphere or the ground in the form of carbon dioxide. Of course, during these changes, cosmic rays continue to generate more carbon-14, and carbon-14 continues to undergo radioactive decay at a rate that does not depend on where it is or how it is *chemically* combined. The net effect is a grand, somewhat steady-state equilibrium in which all living plants and animals have a known and reasonably constant ratio of carbon-14 to carbon-12. As long as the plant or animal is alive, this ratio is maintained. At death, no new carbon-14 can become part of the organism. Its remains have as much of this isotope as they can ever have, and they now slowly lose their carbon-14 by radioactive decay. Because this decay is a first-order process occurring at a constant half-life that is independent of the number of original carbon atoms, the ratio of carbon-14 to carbon-12 can be related to the years that have elapsed between the time of death and the time of the measurement.

The specific activity in all living plants and animals today as well as in the atmosphere is 918 Bq per gram of carbon. The half-life of carbon-14 is 5730 years. Thus, if a bone specimen from an ancient tomb is found to have a specific activity of 459 Bq/g C, the bone has been dead for one half-life period, or for 5730 years. This assumes — and this is a critical assumption — that the steady-state availability of carbon-14 from the atmosphere has remained unchanged since life emerged on earth.

Objects of up to 30,000 years can be dated by this method, but it is most reliable for objects no more than about 7000 years old. The precision in the measurement of specific activity becomes increasingly unacceptable beyond this age. Another problem is that the key assumption that we mentioned in the previous paragraph is not strictly true. The available pool of carbon-14 atoms fluctuates with the intensities of the cosmic ray showers that enter our atmosphere. These intensities vary with the solar activities that produce the cosmic rays. They also vary with slow, very long-term changes in the earth's magnetic field, because this magnetism influences the earth's ability to capture cosmic rays. Still another influence on the ratio of carbon-14 atoms to carbon-12 atoms in the earth's biosphere is the huge injection of carbon-12 atoms into the atmosphere in this century by the large-scale burning of fossil fuels — fuels so ancient that virtually all of their carbon-14 atoms have radioactively decayed. Countering this to some extent is the increased production of carbon-14 caused indirectly by the testing of nuclear weapons.

The net effect of all these influences is that archaeologists have had to work closely with physicists and chemists to calibrate the carbon-14 dating method to other reliable dating methods, such as tree-ring counting. (This method is called dendrochronology.) Each ring that you can see when you look at the end of a log represents exactly one year of growth. The *widths* of the rings reflect fluctuations in the climate, particularly those changes that affect the water supply and the growing season. By studying living, dead, and fossilized trees with overlapping ages, specialists have learned how to use tree rings to date wooden artifacts from as long ago as 7000 years. They have also compared the results to those of carbon-14 dating, and have used these comparisons to prepare correction factors that make carbon-14 dating more precise. For example, an uncorrected carbon-14 dating of a Viking site at L'anse aux Meadows, Newfoundland, gave a date of 895 ± 30 A.D., but when corrected, the date of the settlement became 997 A.D. This is almost exactly the time indicated in Icelandic sagas for Leif Eriksson's landing at "Vinland," which is now believed to be the L'anse aux Meadows site.

In Chapter 18, we studied the equation governing first-order processes.

$$\ln \frac{[A]_0}{[A]_t} = kt \qquad (23.5)$$

We also showed in Chapter 18 that the half-life for a first-order process is given by the equation

$$t_{1/2} = \frac{\ln 2}{k} = \frac{0.693}{k}$$

We can use this equation to find a value for the rate constant, k, given the half-life of carbon-14, 5730 yr.

$$5730 \text{ yr} = \frac{0.693}{k}$$

or

$$k = 1.21 \times 10^{-4} \text{ yr}^{-1}$$

Since the concentration of carbon-14 in a sample is directly proportional to the specific activity of the sample, we can substitute specific activities in the ratio $[A]_0/[A]_t$. Assuming that the specific activity of the carbon-14 in the ancient artifact was the same as today, 918 Bq/g C, we can use this figure for $[A]_0$. We will let s stand for $[A]_t$, the specific activity t years since the time when the remains were fresh. Thus, Equation 23.5 becomes the following equation for use in radiocarbon dating (but not including any correction factors).

When the ratio is taken, the proportionality constant cancels, so we don't have to know what its value is.

$$t = \frac{1}{1.21 \times 10^{-4} \text{ yr}^{-1}} \times \ln\left(\frac{918}{s}\right)$$

$$t = 8.26 \times 10^3 \text{ yr} \times \ln\left(\frac{918}{s}\right) \qquad (23.6)$$

EXAMPLE 23.4 CARBON-14 DATING

Problem: In the late 1940s, a number of manuscripts were found in caves near the northwestern corner of the Dead Sea region of the Middle East, and they came to be called the Dead Sea scrolls. Some were made of copper but others were of parchment. When one parchment scroll was analyzed by the carbon-14 dating method, its specific activity was found to be 708 Bq/g. Calculate the age of this scroll (to two significant figures).

Solution: The data are in the form needed by Equation 23.6.

$$t = 8.26 \times 10^3 \text{ yr} \times \ln\left(\frac{918}{708}\right)$$

$$= 8.26 \times 10^3 \text{ yr} \times \ln 1.30$$
$$= 8.26 \times 10^3 \text{ yr} \times 0.262$$
$$= 2.2 \times 10^3 \text{ yr}$$

The scroll is about 22 centuries old.

PRACTICE EXERCISE 5

In 1926, near the town of Folsom, New Mexico, the remains of an early American Indian culture were unearthed. Carbon-14 dating methods were applied to a bone sample, which was found to have a specific activity of 265 Bq/g. How old was this specimen (to one significant figure)?

23.8 NUCLEAR FUSION

At sufficiently high temperatures and densities, atomic nuclei of deuterium and tritium will fuse to give helium-4, a neutron, and energy.

In the first two sections of this chapter we learned that enormous yields of energy are potentially available from nuclear fusion. There are immensely difficult scientific and engineering problems, however, that have to be solved before a working, producing fusion power station can be realized. Yet the benefits to people worldwide from success seem to be so great that major research efforts are being made in several countries.

The deuterium nucleus has one proton and one neutron.

Deuterium, an isotope of hydrogen, is a key fuel in all approaches to fusion. It is naturally present as 0.015% of all hydrogen atoms, including those of the molecules of the earth's waters. The earth is rich in water, of course, so despite this low percentage, geologist M. King Hubbert has estimated that just 1% of the deuterium in the world's oceans could supply fusion energy equivalent to 500,000 times the combined total of all the energy in the original fossil fuel supply of the earth. The deuterium in just 0.005 cubic kilometers of ocean water would supply the energy needs of the United States for one year. You can see why energy planners consider fusion research to be important.

A volume of 0.005 km³ roughly equals the capacity of 45 supertankers (700,000 gallons each).

The Lawson Criterion

By the mid-1980s, specialists in fusion research became cautiously optimistic that the scientific problems of fusion would be solved by the end of the twentieth century. The central scientific problem is to get the fusing nuclei close enough for a long enough time so that the strong force, which makes nucleons — both protons and neutrons — attract each other, can overcome the electrostatic force, which makes protons repel each other.

As we learned in Section 23.3, the strong force acts over a shorter range than the electrostatic force. Two nuclei on a collision course, therefore, repel each other virtually until they touch. The combined kinetic energies of two approaching nuclei must therefore be very substantial if they are to overcome this electrostatic barrier. And achieving such energies must be accomplished on large numbers of nuclei all at once in batch after batch if there is to be any practical generation of electrical power by nuclear fusion. Relatively isolated fusion events achieved in huge accelerators will not do. The only practical way to give batch quantities of nuclei enough energy is to transfer *thermal* energy to them, and so the overall process is called *thermonuclear fusion*.

If the batches are too large, however, you have a hydrogen bomb.

The atoms whose nuclei we want to fuse must first be stripped of their electrons. Thus a high-energy cost is exacted from the start. The product of this change is an electrically neutral gaseous mixture of nuclei and unattached electrons called *plasma*. Then the plasma must be made so dense that like-charged nuclei are forced to within a distance of 2×10^{-15} m of each other, which means that the density must become about 200 g/cm^3 (as compared with 200 mg/cm^3 under ordinary conditions). Another way to describe this change is that the gaseous plasma must be confined at a pressure of several billion atmospheres long enough for the separate nuclei to fuse.

Physicist J. D. Lawson has calculated that thermonuclear fusion will provide a net production of energy only if the product of the particle density, n, and the confinement time, τ, equals or exceeds a certain value, called the *Lawson number*. When n is in particles per cubic centimeter and τ is in seconds, the Lawson number is thought to be approximately

$$n\tau = 3 \times 10^{14} \text{ s/cm}^3$$

This equation tells us that if a long confinement time were unworkable, then a compensating high particle density would be needed. Or, conversely, if a high particle density were too difficult, then the confinement time would have to be extra long. A workable balance between n and τ is thus an important goal in all approaches to fusion.

The Lawson number and the temperature are the two major scientific parameters involved in the research on successful thermonuclear fusion. The temperature needed is about one million degrees Celsius, which is several times the temperature at the center of our sun.

A high temperature means high kinetic energies of collision.

Two broad approaches are currently being pursued — inertial confinement fusion and magnetic confinement fusion. Both would bring about the following nuclear reaction.

$$\underset{\substack{\text{tritium} \\ \text{nuclei}}}{^{3}_{1}\text{H}} + \underset{\substack{\text{deuterium} \\ \text{nuclei}}}{^{2}_{1}\text{H}} \longrightarrow \underset{\substack{\text{helium} \\ \text{nuclei}}}{^{4}_{2}\text{He}} + \underset{\text{neutron}}{^{1}_{0}\text{n}}$$

The fusion of two deuterium nuclei is another option being studied.

Using isotopes of hydrogen keeps the electrostatic repulsions between fusing nuclei at a minimum, because the nuclei of both tritium and deuterium have single positive charges. Isotopes of any higher atomic number would pose nearly impossible problems, because their nuclear charges are twice as large or larger. In addition, the highest gain in binding energy per nucleon occurs in the fusion of the lightest nuclei.

Inertial Confinement Fusion

Inertial confinement means enclosing the fuel — tritium and deuterium — within a pellet made of glass, plastic, or metal with a diameter of about 1 mm. The pellet is then given so much energy so rapidly from all angles that the material on its outer surface suddenly vaporizes and itself becomes a plasma. This loss of pellet material is similar to the loss of some of the surfaces of the heat shields of reentering space vehicles. A portion of the thermal energy generated by this step is transmitted to the interior of the shell of the pellet as well as into the fuel region. The conversion of the fuel atoms into their own plasma begins as the shell of the pellet explodes outward in all directions.

A batch this small poses no danger whatsoever of a hydrogen bomb explosion.

To every action, such as this explosion, there is an equal but opposite reaction — Newton's third law. The reaction to the explosion of the pellet shell is an *implosion* of the fuel inside — now a plasma — in upon itself. If the explosive action occurs in a spherically symmetrical way, the implosive reaction will also occur this way. The fuel plasma thus is compressed into a much denser, very hot ball of much smaller radius than the fuel had initially. Now fusion can occur. Now more energy is released than was needed to launch these steps, and there is a net gain in energy.

Isaac Newton (1642–1727), an English physicist, is considered to be one of the most important figures in all history.

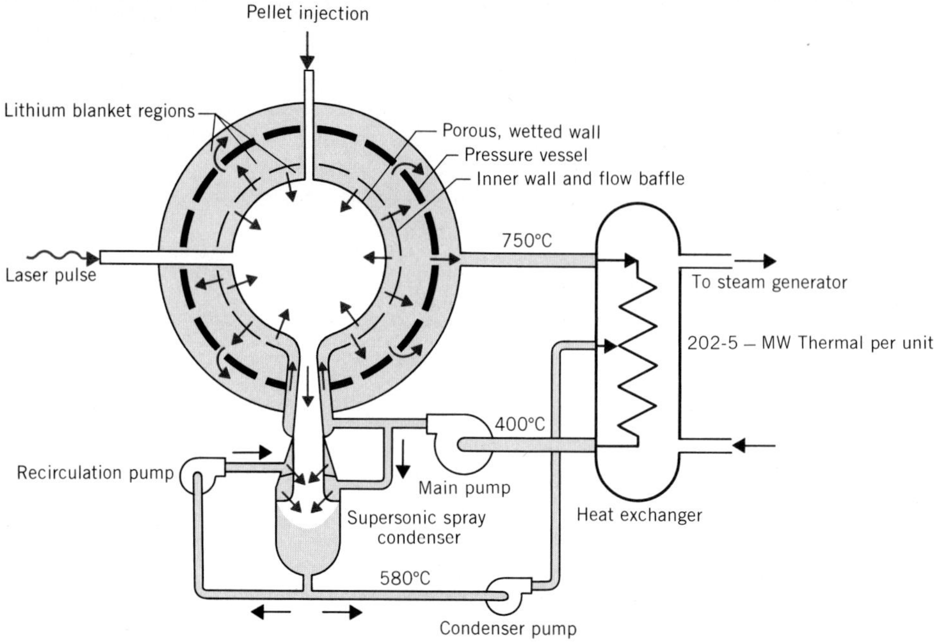

FIGURE 23.20 Fusion by laser implosion. The pellets bled in at the top are hollow beads of thin glass filled with deuterium and tritium. When such a pellet is caught in the intersecting pulses of a laser beam, its contents are heated enough to lose their electrons. Glass from the pellet also flakes off much as a heat shield on a reentering space vehicle. This action collapses the capsule inward creating the intense pressure needed to hold the plasma together long enough to fuse. Projects of this and similar types are being studied. (Drawing from WASH-1239, U.S. Atomic Energy Commission, 1973.)

One experimental device is called the SHIVA after the many-armed Hindu god.

Lithium melts at 180.5 °C and boils at 1336 °C.

The other fuel needed, deuterium, is to be extracted from water, as we mentioned earlier.

The initial delivery of energy can be achieved in more than one way, but the most intensely studied system uses lasers. Figure 23.20 gives a schematic of a laser fusion reactor. Only one porthole for laser pulses is shown here to keep the clutter to a minimum, but pulses from several directions have to be focused simultaneously onto the center of the device. The OMEGA laser system shown in the chapter opening photograph has 24 focused laser beams.

Surrounding the interior of the system is a blanket of molten lithium metal, a good neutron absorber, retained by a wall porous to neutrons. The neutrons will deliver their energy to the lithium by the following nuclear reaction.

$$^{6}_{3}Li + ^{1}_{0}n \longrightarrow ^{4}_{2}He + ^{3}_{1}H$$

The thermal energy in the products of this reaction heats the bulk of the lithium. To keep the lithium from vaporizing as it receives thermal energy, it is pumped through a heat exchanger where its heat vaporizes circulating water and changes it to steam under pressure. This steam then drives a conventional steam-powered electrical generator.

The tritium produced by the reaction of neutrons with lithium is recycled into the fusion reaction, so the system makes one of the two fuel elements needed.

One of the recent developments that has made fusion physicists somewhat optimistic about laser fusion has been the discovery of a way to convert common infrared laser beams into beams of higher frequencies with energies in the ultraviolet region. (Infrared laser energy causes a premature heating of the fuel before it can be compressed to the requisite density.)

Two products form in fusion, alpha particles and neutrons. About 80% of the thermal energy is carried away by the neutrons, which heat the lithium. About 20% of the thermal energy is possessed by the alpha particles. Most of these collide with and transfer their energy to the surrounding fuel, so this energy is not directly delivered to the circulating lithium. What must therefore happen if fusion is to be economically feasible is that the thermal energy of the alpha particles be absorbed by unchanged fuel, help to compress it, and so boost it into fusion. As of the mid-1980s, this self-ignition had not been observed, so the ultimate scientific feasibility of laser fusion had yet to be demonstrated.

At the center of the sun, according to present estimates, the temperature is about 1.5×10^7 K, the pressure is about 2×10^{11} atm, and the density is about 1.5×10^5 kg/m³ (roughly 13 times the density of lead). The matter there, all in the gaseous state, consists mostly of a plasma of protons (35% by mass), alpha particles (65% by mass), electrons, and photons. The conditions are suitable for thermonuclear fusion, and the chief process is called the proton–proton cycle. (We represent protons as ^1_1H and deuterons as ^2_1H; ^0_1e is a positron, and ν is a neutrino.)

The Proton–Proton Cycle

$$2[^1_1H + ^1_1H] \longrightarrow 2\,^2_1H + 2\,^0_1e + 2\nu$$
$$2[^1_1H + ^2_1H] \longrightarrow 2\,^3_2He$$
$$\underline{^3_2He + ^3_2He \longrightarrow ^4_2He + ^1_1H + ^1_1H}$$
$$\text{Net:} \qquad 4\,^1_1H \longrightarrow ^4_2He + 2\,^0_1e + 2\nu$$

The positrons produced combine with electrons in the plasma, annihilate each other, and generate gamma radiation. Virtually all the neutrinos escape the sun and move into the solar system (and beyond), carrying with them a little less than 2% of the energy generated by the cycle. Not counting the energy of the neutrinos, each operation of one cycle generates 26.2 MeV or 4.20×10^{-13} J of energy, which is equivalent to 2.53×10^{12} J per *mole* of alpha particles produced. This is the source of the solar energy radiated throughout our system. It can continue in this way for another 5 billion years.

Magnetic Confinement Fusion

The other approach to the confinement of hot, high-density plasma long enough for fusion to happen uses a magnetic "bottle." No known material has the strength to withstand the required temperatures and pressures of the plasma. So scientists studying magnetic confinement have devised ways to shape and hold magnetic forces in configurations that prevent entrapped plasma from leaking out. Figure 23.21 illustrates a device, called a *tokamak*, which gives the magnetic bottle a donut shape. The plasma can be heated well above the requisite temperature by being bombarded with intense beams of electrically neutral deuterium atoms, using special injectors. By the mid-1980s, however, the Lawson number had yet to be equaled in any Tokamak reactor. The nearest approach was a Lawson number of 1.5×10^{14} s/cm³, too low by a factor of 2.

How thermonuclear fusion is believed to occur in the sun is described in Special Topic 23.7.

The paths of moving, electrically charged particles are changed by magnetic forces.

Tokamak is a Russian term for a magnetic confinement device.

FIGURE 23.21 Fusion by magnetic confinement. *(a)* Cutaway view of a full-scale Ormak reactor nearing the end of its assembly. Two partially assembled sectors are in the foreground. *(b)* Cross-sectional view through the reactor. (The lithium is used in the nuclear cycle that synthesizes the tritium needed for fusion.)

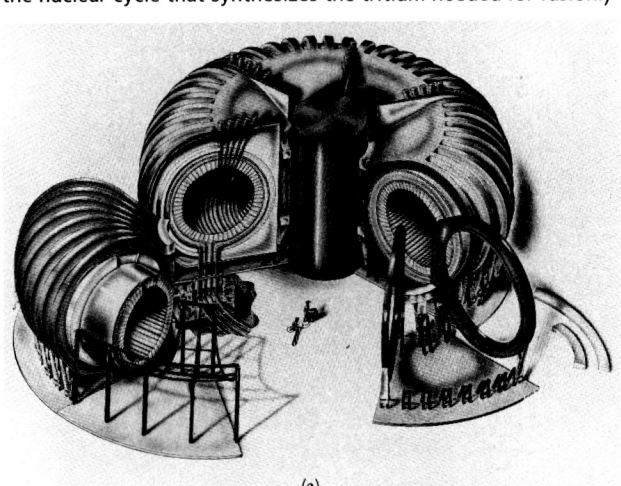

(a)

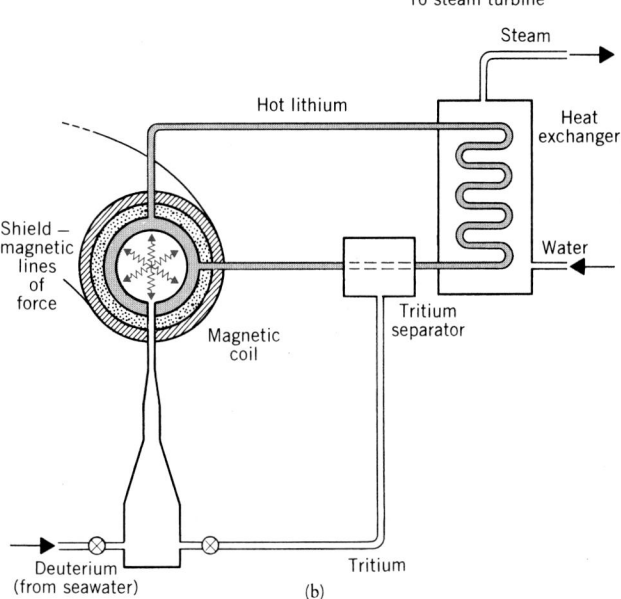

(b)

Thermonuclear fusion is the source of energy released in the explosion of a *hydrogen bomb*. The energy needed to trigger the fusion is provided by the explosion of a fission bomb based either on uranium or on plutonium.

23.9 NUCLEAR FISSION

In fission, a heavy nucleus absorbs a neutron and breaks apart to give much lighter nuclei plus two or more neutrons, and much energy.

The electrical neutrality of the neutron means that no electrostatic force works to hinder its approach to an atomic nucleus. Neutron capture by a nucleus is therefore energetically much easier than the capture of a charged particle, a fact utilized in neutron activation analysis (Section 23.7). Even neutrons whose average kinetic energy puts them in thermal equilibrium with their surroundings at room temperature — these are called *thermal neutrons* — can be captured by nuclei.

Two German chemists, Otto Hahn and Fritz Strassman, discovered in 1939 that when a uranium sample was bombarded with thermal neutrons, several isotopes much lighter than uranium remained. One, for example, was barium-140. Shortly thereafter, two German physicists, Lise Meitner and Otto Frisch, were able to show that one of the rare isotopes in the uranium — uranium-235 — was able to split into roughly two equal parts following the capture of a slow neutron. They named this breakup nuclear fission. The general reaction can be represented as follows.

$$^{235}_{92}U + n \longrightarrow \underset{\substack{\text{a compound}\\\text{nucleus}}}{^{236}_{92}U^*} \longrightarrow \underset{\substack{\text{two smaller}\\\text{nuclides}}}{X + Y} + bn$$

X and *Y* can be large numbers of nuclei with midrange mass numbers. The factor *b* in this equation has a value of 2.47, the average number of neutrons produced by fission events. A typical specific fission is

$$^{235}_{92}U + ^1_0n \longrightarrow ^{236}_{92}U^* \longrightarrow ^{94}_{36}Kr + ^{139}_{56}Ba + 3\,^1_0n$$

What actually fissions is the compound nucleus, that of uranium-236, which has 144 neutrons and 92 protons for a neutron/proton ratio of roughly 1.6. Initially, the emerging krypton and barium isotopes have the same ratio, and this is too high for them. As you can see in Figure 23.8, the neutron/proton ratio for isotopes with 36 to 56 protons is closer to 1.2 to 1.3. Hence, the krypton and barium nuclides that initially form are neutron rich, so they promptly eject neutrons. They can do this with no change in atomic number, of course, just in mass number. Neutrons of much higher energy than thermal neutrons are produced by such events.

As the graph in Figure 23.2 shows, the binding energy per nucleon in uranium-235 (about 7.6 MeV) is lower than the binding energies of the new nuclides (about 8.5 MeV). The net change for a single fission event can be calculated as follows (where we intend only two significant figures in each result).

(8.5 MeV/nucleon) × 94 nucleons =   800 MeV (binding energy in Kr-94)
+ (8.5 MeV/nucleon) × 139 nucleons = 1200 MeV (binding energy in Ba-139)
         Total binding energy of products = 2000 MeV

The binding energy of uranium-235 is

(7.6 MeV/nucleon) × 235 nucleons = 1800 MeV (binding energy in U-235)

The difference in total binding energy is (2000 MeV − 1800 MeV) or 200 MeV. (This has to be taken as just a rough calculation.) This is the energy released by each fission

Otto Hahn received the 1944 Nobel prize in chemistry.

1 MeV = 1.602 × 10⁻¹³ J.

event that goes by the equation given. The energy released by the fission of 1 kg of uranium-235 calculates to be roughly 8×10^{13} J — enough to keep a 100-watt light bulb in energy for 3000 years. The energy available from 1 kg of uranium-235 is equivalent to the energy of 3000 tons of soft coal or 13,200 barrels of oil.

An isotope capable of undergoing fission following neutron capture is called a fissile isotope. The only naturally occurring fissile isotope is uranium-235, whose natural abundance among the uranium isotopes is only 0.72%. Two other fissile isotopes, uranium-233 and plutonium-239, can be made in nuclear reactors.

The neutrons released by fission, if slowed down by collisions with surrounding materials, can be captured by unchanged uranium-235 atoms. Since each fission event produces on the average more than two new neutrons, the potential for a nuclear chain reaction exists, as illustrated in Figure 23.22. A chain reaction is a self-sustaining process whereby products from one event cause one or more new events. Of course, if the mass of uranium-235 is small enough, the loss of neutrons to the surroundings is rapid enough to prevent a chain reaction. However, at the critical mass of uranium-235, this loss is insufficient to prevent a sustained reaction, and the result is the virtually instantaneous fission of the sample — an atomic bomb explosion. To trigger the bomb, therefore, two subcritical masses of uranium-235 (or plutonium-239) are allowed to come together to form a critical mass.

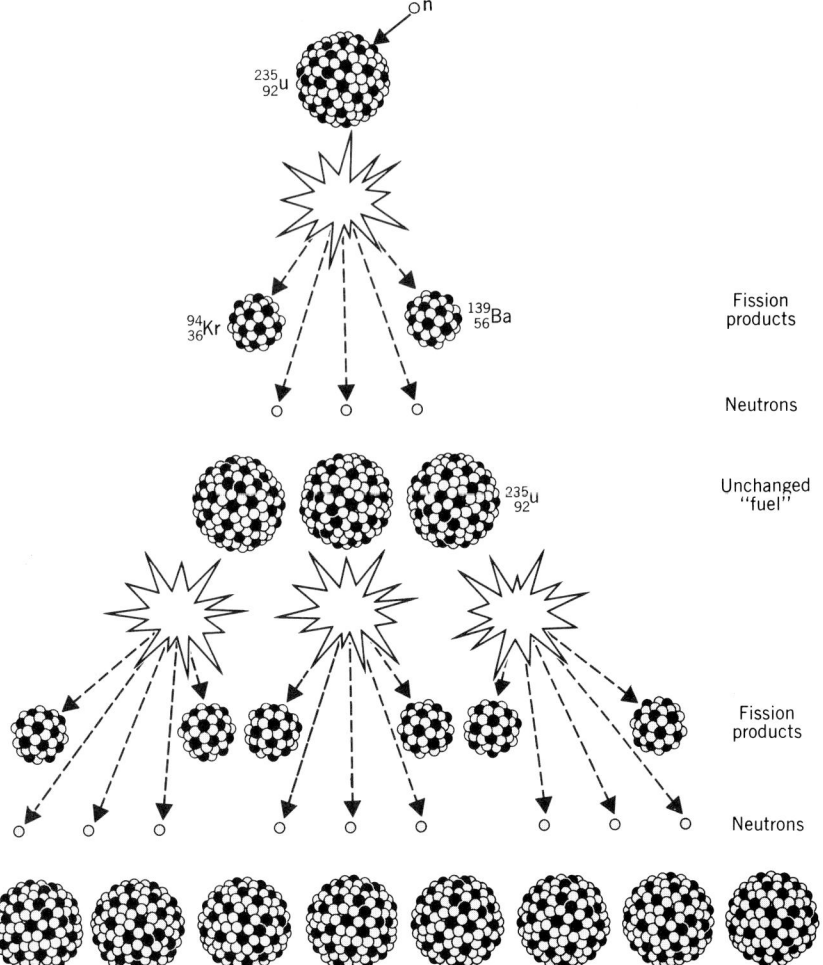

Fission products

Neutrons

Unchanged "fuel"

Fission products

Neutrons

FIGURE 23.22 Nuclear chain reaction. Whenever the concentration of a fissile isotope is high enough (at or above the critical mass), the neutrons released by one fission can be captured by enough unchanged nuclei to cause more than one additional fission events. In civilian nuclear reactors, the fissile isotope is too dilute for this to get out of control. In addition, control rods of nonfissile materials that are able to capture excess neutrons can be inserted or withdrawn from the reactor core to make sure that the heat generated can be removed as fast as it forms.

Electrical Energy from Fission

Virtually all civilian nuclear power plants throughout the world operate on the same very general, overall principles. The heat of fission either directly or indirectly is used to increase the pressure of some gas, which then drives an electrical generator just as high-pressure steam is used in a power plant fueled by coal or oil. In fact, worldwide (as of January 1, 1986), 320 of the 374 operational reactors and 246 of 273 reactors being built or planned also used steam as the working gas. Two main types of reactors eventually send high-pressure steam to turbines—the boiling-water reactor and the pressurized-water reactor. Of the 100 operating reactors in the United States (1986), 62 were the pressurized-water type, so to illustrate principles we will continue the discussion with just this type.

Figure 23.23 is a schematic diagram of a typical pressurized-water reactor. There are two loops, and water moves in both. The primary loop circulates water through the reactor itself, where it picks up thermal energy from fission. Most of this water is kept in the liquid state by being under high pressure (hence the name, pressurized-water reactor).

The hot water in the primary loop transfers thermal energy to the secondary loop at the steam generator (Figure 23.23). This makes steam at high temperature and pressure, which is piped to the turbine. As it drives the turbine, the steam pressure drops. The condenser at the end of the turbine, cooled by water circulating from a river or lake, or from huge cooling towers, forces a maximum pressure drop within the turbine by condensing the steam to liquid water. The returned water is then recycled to high-pressure steam. (In the boiling-water reactor, there is only one coolant loop. The water heated in the reactor itself is changed to the steam that drives the turbine.)

In some reactors, the primary coolant is helium gas. Heavy water (deuterium oxide) is used in some. In others, a liquid metal (e.g., sodium) is used. Each has particular advantages, and we will come back to the liquid sodium reactor later.

Scientists and engineers have tried to design as many safety features as they can think of. First, there is no danger whatsoever of a reactor undergoing an atomic bomb kind of explosion. The bomb requires pure uranium-235 or plutonium-239. The concentration of fissile isotopes in a reactor, however, is in the range of 2 to 4%—with much of the remainder being the common, nonfissile uranium-238. It is not possible, therefore, for a critical mass to form, not even if all the fuel melts into one pool. But this is not the only safety problem, as you probably well know. Since safety and controlled operation of fission are closely related, we will look further at two of the scientific problems.

To have controlled fission take place in a reactor, the leakage of neutrons away from the core must be controlled, and the fast neutrons produced by fission must be

FIGURE 23.23 Nuclear reactor. Shown here is a pressurized water reactor, the type used in most of the nuclear power plants in the United States. Water in the primary coolant loop is pumped around and through the fuel elements in the core, and it carries away the heat of the nuclear chain reactions. It delivers this heat to the water in the secondary coolant loop where steam is generated to drive the turbines. (Drawing from WASH 1261, U.S. Atomic Energy Commission, 1973.)

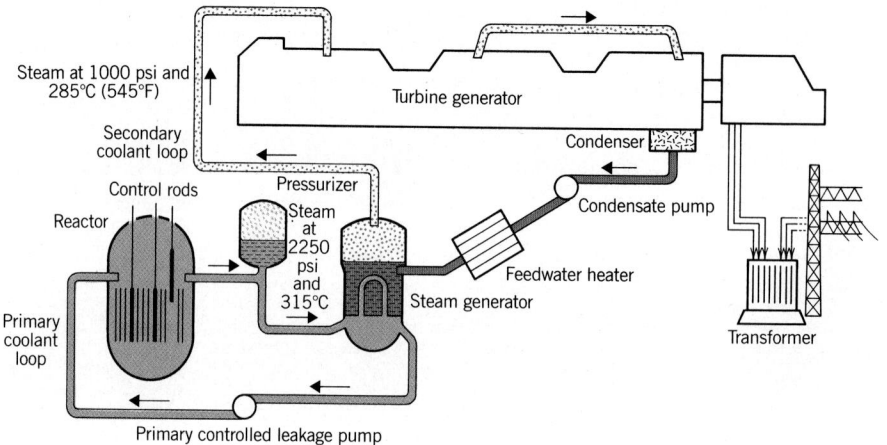

slowed. Leakage occurs at the surfaces of the fuel, whereas neutrons are generated throughout the entire volume. By making the fuel core large enough, the ratio of surface to volume can be reduced to where enough neutrons are retained to cause other fission events.

To slow neutrons down, the fuel core is provided with a *moderator*. Water itself is a good moderator, so the pressurized-water reactor provides it in the primary coolant loop. Heavy water is the moderator in some Canadian power plants. Graphite is also a good moderator, and it was in use in the ill-fated power plant at Chernobyl, Russia.

The safe operation of a reactor requires that the multiplication of slow neutrons be controlled throughout a fission cycle. This cycle begins with the production of fast neutrons (with some leakage), their moderation (also with some leakage), the use of the survivors to cause additional fission events, and a new production of fast neutrons by these events to launch another round of the cycle. The ratio of neutrons at the end of the cycle to those that started it is called the *multiplication factor, k*. For smooth, safe, continuous operation, the value of k in the reactor should be exactly 1. When it is, the reactor is said to be *critical*. When $k > 1$, fission is accelerating and the reactor is supercritical. When $k < 1$, fission is slowing down, and the reactor is subcritical. To be able to start up, shut down, and safely operate the reactor in the meantime, the reactor has to be designed with the built-in ability to go supercritical. Therefore, to bring the system back into the critical mode, the reactor core incorporates control rods typically made of cadmium. When fully inserted, these rods capture enough neutrons to keep the reactor subcritical. If removed long enough, the reactor goes supercritical. By moving the control rods in and out, as you can see, the reactor itself can be controlled.

The nuclear fuel is typically in the form of sintered pellets of uranium or plutonium oxide. These are packed in tubes called *cladding* of small diameter, and bunches of such tubes are bound together and spaced somewhat apart within the reactor with spaces left for control rods. The primary coolant circulates around these tubes. The tubes are typically made of a zirconium alloy, and they must have no pinhole leaks so that gaseous radioactive wastes are contained. As fission continues, wastes build up in the cladding, and they act to dilute unchanged atoms of uranium-235. This slows the rate of fission. In time, the fuel rods are removed and replaced. They still contain uranium-235 as well as some newly synthesized plutonium-239 — made from uranium-238, as we will see in the next section. Both are fissile, so the spent fuel is processed to recover them for recycling. A major issue with nuclear power is the availability of plutonium-239, formed in civilian reactors, for military weapons. Some countries no doubt have acquired nuclear bombs this way.

Two other major issues of nuclear power are the handling of radioactive wastes and the prevention of a loss of the primary coolant. We will discuss the latter first.

Loss of Coolant Emergencies

No single aspect of a reactor's operation is more important than the removal of the heat generated by fission as fast as it is produced. The primary coolant must do this or the temperature of the core will increase very rapidly. The conversion of any water remaining in the primary loop into steam could quickly develop enough pressure to blow the reactor apart. The first of two explosions at the Chernobyl power plant in Russia was almost certainly such a steam explosion. It blew off a cover plate weighing 1000 tons and tore the tops from all 1661 cladding tubes. Each tube became a large-bore shotgun that sent a fireworks display of eight tons of incandescent, radioactive fuel elements and wastes into the night sky visible miles away. A second explosion also occurred, and was also probably a steam explosion. The combined explosions destroyed a large part of the building, and sent a radioactive plume 3200 feet into the atmosphere where it drifted on the prevailing winds. The fire ignited the charcoal used as the moderator, which burned like any barbecue charcoal while some extremely brave firefighters struggled nearly two and a half hours to put it out. According to one estimate, the human exposure in the

A commercial reactor at Fort St. Vrain, Colorado, uses graphite as its moderator.

A metal oxide powder is *sintered* by baking it below its melting point until it becomes hard and ceramiclike.

There may have been other explosions. Only a quarter of the total discharge of radioactive materials occurred the first day.

The reactor that blew held 3×10^9 Ci at the time of the accident. About 3 to 4% of this was released.

northern hemisphere to the radioactive cesium released at Chernobyl alone came to nearly 60% of the human exposure to this isotope from all prior atmospheric tests of nuclear weapons.

Even if insufficient steam is present from which to generate enough pressure for a steam explosion, the temperature in the core can make the cladding melt and its now superhot metal could then react with steam to produce hydrogen. The production of enough hydrogen during a runaway incident can lead to a chemical explosion caused by combustion of the hydrogen. Eventually, the fuel elements melt and gather in a molten mass in which fission energy continues to be generated. The temperature of this mass might rise enough so that it melts its way through the containment structure and on into the ground.

To prevent a core meltdown and a breach of the containment vessel, engineers have installed emergency core cooling systems designed to flood the reactor with water at the first sign of impending danger, before the failure of cladding. While this emergency water temporarily halts a steep increase in temperature, other action is taken to shut the reactor down. The loss of primary coolant occurred in the Three Mile Island disaster (1979, Pennsylvania), and the backup systems worked. Hydrogen was produced, but there was insufficient oxygen to permit an explosion. The mess was still enormous, and the disaster understandably caused widespread alarm, but the emergency systems did work. It took nearly seven years of cleanup work, testing, and litigation to put the *undamaged* Unit 1 at Three Mile Island back in operation. It should be noted that the containment structures used at U.S. nuclear power plants are much more substantial than what was in place at Chernobyl.

The problem at Chernobyl, according to the Soviet government's report, was that plant operators carried out unauthorized tests during which they violated six important rules of operation. Experts outside Russia believe that faulty reactor design contributed greatly as well. Within 5 seconds of the loss of the primary coolant, the reactor's power increased by a factor of 1400 from a low power to 100 times its rated full capacity. The control rods could not be dropped rapidly enough, and nothing could save the situation.

Radioactive Wastes

Cesium-137 has a half-life of 30 years; that of strontium-90 is 28.1 years. Both are beta emitters.

Radioactive wastes from nuclear power plants occur as gases, liquids, and solids. The gases are mostly radionuclides of krypton and xenon, but with the exception of xenon-85 ($t_{1/2}$ 10.4 years), these have short half-lives and decay quickly. Until they do, of course, they must be contained, and this is an important function of the cladding. If radioactive gases escape explosively from reactors, they also include iodine-131, strontium-90, cesium-137, and many other dangerous radionuclides.

Since the human thyroid gland concentrates iodine-131 (a beta emitter), one of the first fears following such an incident is contamination of the water supply as well as the food supply (through dairy products produced by livestock eating contaminated feed). An effective countermeasure to iodine-131 poisoning is to take doses of ordinary iodine (as sodium iodide) to increase statistically the likelihood that the thyroid will take up stable iodine rather than the radionuclide.

One of the many very sad results of Chernobyl was the poisoning of the reindeer herds upon which the Lapps depend so much. They are nomadic people of northern Finland, Sweden, and Norway, and their areas were under the radioactive plume that went from Chernobyl.

The entire reactor of a power plant becomes a solid waste problem when its working life is over.

Some waste products are soluble in water, and they become mixed with the primary coolant. When this is changed, its radioactive solutes must be isolated and stored safely. Solid wastes from nuclear power plants include spent fuel rods and precipitates from liquid wastes. Cesium-137 and strontium-90 pose particular problems to humans. Cesium and sodium are in Group IA; strontium and calcium are in Group IIA. So cesium-137 cations can go where sodium ions go in the body—just about anywhere—and strontium-90 cations can replace calcium ions in bone tissue. But the half-lives of cesium-137 and strontium-90 are short compared with many of the nuclear wastes. Some are so long that solid wastes in general have to be kept away from all human contact for hundreds of centuries—longer than any nation has ever yet endured. Critics of nuclear power argue that until the problem of safely storing radioactive wastes is solved, no new power plants should be authorized. Proponents believe that technology will find a solution. In the meanwhile, no new nuclear power plants were being planned in the United States in the mid 1980s—almost entirely for economic reasons. However, research on safer reactors has continued, as we will survey next.

Liquid Sodium-Cooled Reactors

A reactor that uses liquid sodium as the primary coolant and a metal alloy as the fuel instead of a metal oxide cannot go out of control, not because of emergency features but because of the laws of physics. At least the designers of such a reactor have strong evidence from actual tests to support this claim. The secondary coolant is still water, and if the sodium and water were to mix following a pipe rupture, there could be a chemical explosion as hydrogen forms and burns. But the engineering technology for handling and pumping hot liquid sodium is well known.

The upper limit on the temperature to which water can be heated under pressure and remain a liquid is 374.1 °C, the critical temperature of water (page 380). The materials used to make pipes and turbines begin to suffer unacceptable deterioration from the effects of water vapor at temperatures above 500 °C. But sodium boils at 881 °C, so with liquid sodium as the primary coolant, the reactor can operate at temperatures above water's critical temperature while still being at atmospheric pressure or nearly so. In addition to using liquid sodium as the primary coolant, the entire reactor can be immersed in a huge pool of thousands of liters of liquid sodium. This pool, then, is actually the emergency core coolant, because if the reactor heats up too much, the pool can continue to absorb this heat for long periods of time. During this period, control rods or other strategies can be put into operation. Moreover, as discussed next, when the fuel is in a *metal* form, a shutdown will automatically occur.

The critical pressure of water is 217.7 atm.

Sodium melts at 98 °C. Its specific heat as a liquid at this temperature is 1.385 J/g °C.

The advantage of a metal alloy form of the fuel over a metal oxide form is that heat moves through and away from a metal much faster than through a ceramic oxide. This means that the reactor can operate at a lower and safer temperature. Should the fuel's temperature increase at a dangerous rate, not only is the heat rapidly conducted to the coolant, but the metal form also expands enough to separate individual fissile atoms to greater distances. This slows down the chain reaction without any human or mechanical action. In one test where the circulation of the primary coolant was deliberately stopped, the coolant's temperature did quickly increase by 200 °C, but the sodium did not boil, the pressure did not significantly increase, and the chain reaction soon stopped. Within 5 minutes the temperature of the coolant had decreased essentially to its normal operating value. No commercial, operating nuclear power plant currently employs the features described here, but several groups are engaged in the necessary research and development.

The test was done on a 20-megawatt research reactor, substantially smaller than the 1000-megawatt reactors of many nuclear power plants.

23.10 THE BREEDER REACTOR

In a breeder reactor nonfissile uranium-238 is converted to fissile plutonium-239.

Naturally occurring uranium is 99.275% uranium-238, 0.720% uranium-235, and 0.005% uranium-234. When a reactor uses uranium fuel, naturally occurring uranium is processed to increase the concentration of uranium-235 to 2 to 4%, which means that 96 to 98% of the uranium atoms in the fuel are the nonfissile uranium-238. Therefore, in any operating nuclear reactor fueled with uranium, the uranium-238 atoms are mostly "neu-

FIGURE 23.24 The breeding cycle whereby uranium-238, which is nonfissile, is changed into plutonium-239, which is fissile. (From J. R. Holum, *Topics and Terms in Environmental Problems.* 1978, Wiley-Interscience, New York. Used by permission.)

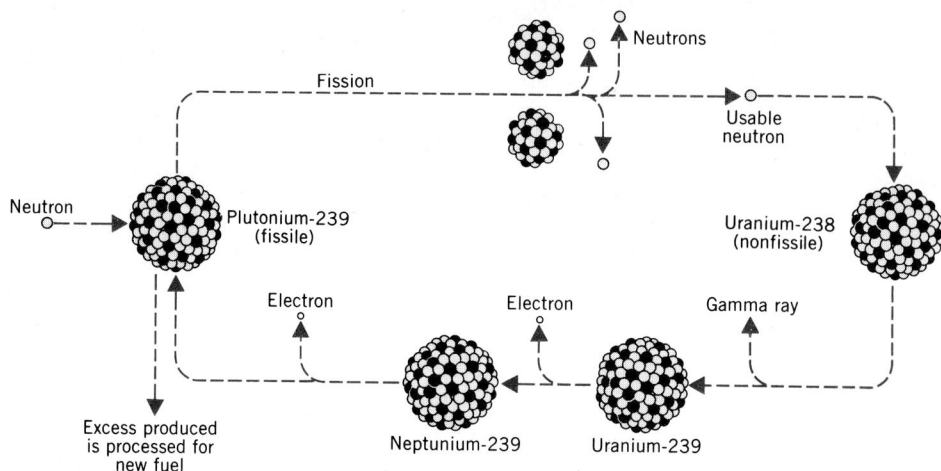

tral stuff." We say "mostly" because the following nuclear reaction of uranium-238 occurs to a small extent, and it produces a fissile isotope, plutonium-239.

$$\,_0^1n + \,_{92}^{238}U \longrightarrow \,_{92}^{239}U \longrightarrow \,_{94}^{239}Pu + 2\,_{-1}^{0}e$$

A nonfissile isotope that can be changed to a fissile isotope is called a **fertile isotope**.

Uranium-fueled nuclear reactors thus manufacture plutonium fuel out of nonfissile uranium-238. The plutonium-239 can be extracted, and used as fuel, too.

If plutonium-239 were mixed with uranium-238 in a reactor, in principle it would be possible to exploit all of the uranium-238 as a source of fuel. The plutonium-239 would serve something of a catalyst role, because it would not be consumed. As seen in Figure 23.24, for every atom of plutonium-239 that fissions and generates energy, another atom of plutonium-239 is made from an atom of uranium-238. The system thus breeds its own fuel from what initially was not a fuel, uranium-238, and a reactor capable of this is called a breeder reactor.

The breeder reactor is sometimes called the *catalytic burner reactor.*

When the costs of all other fuels for power plants increase enough, interest in the breeder will return.

At current prices, the economics of building and using breeder reactors are so decisively unfavorable that such reactors are simply not in the near energy future. Even funds for research on the breeder reactor have not been available in the United States in the 1980s. What continues to be thought provoking about this reactor, however, is the fact that enough uranium-238 is on hand today, stockpiled from the processing of spent nuclear fuel elements, to supply the entire energy needs of the United States for centuries. But the apparent cost per megawatt of power from breeders is simply far too high compared to any other source of power.

SUMMARY

The Einstein Equation.   Einstein's work on the theory of relativity led him to an equation that relates a moving particle's mass to its rest mass and its velocity. Only at velocities approaching the speed of light, however, does the mass differ detectably from the rest mass.

Einstein also realized that mass and energy are interconvertible and that the equation, $\Delta E = \Delta mc^2$ (where c is the speed of light), lets us calculate one from the other. In exothermic events, some mass converts to energy, and in endothermic events, some energy changes to mass. Overall, the total of the energy in the

universe and all the mass calculated as an equivalent of energy is a constant — the law of conservation of mass-energy.

Nuclear Binding Energies.   The sum of the rest masses of the nucleons in a given nucleus exceeds the measured mass of the nucleus, so when a nucleus forms from its nucleons the "lost" mass is converted to energy. This same amount of energy would have to be supplied to break up the nucleus again, so the energy is called the nuclear binding energy. (It is energy the nucleus does not have.) It's usually expressed for each isotope in joules per nucleon. The higher the binding energy per nucleon, the more stable is the nucleus.

A mass number of 56 corresponds to the most stable nuclei, and it would be energetically favorable for nuclei with other mass numbers to undergo changes to achieve it. Thus among the nuclei of higher mass numbers, nuclear fission is a possibility. Among those of lower mass numbers, nuclear fusion would be the route. For all but a few isotopes, the energy barriers to these changes are too high. For those with barriers that can be surmounted, a change by fusion or by fission produces a net yield of energy of enormous amounts.

Radioactivity. Only the nuclear strong force holds protons and neutrons within nuclei. The electrostatic force between protons, which acts to break up nuclei, is overcome by the strong force at short ranges. The ratio of neutrons to protons is thus a parameter in nuclear stability. The naturally occurring radionuclides adjust their neutron/proton ratios and lower their energies by emitting alpha or beta radiation. They can also lose energy by emitting gamma radiation.

The loss of an alpha particle leaves a nucleus with four fewer units of mass number and two fewer of atomic number. Loss of a beta particle leaves a nucleus with the same mass number and an atomic number one unit higher. Gamma radiation, which often accompanies alpha or beta radiation, lets a nucleus shed some energy without a change in mass or atomic number.

Synthetic radionuclides emit these radiations, too, but many are emitters of other things. Some emit positrons—positive electrons—which soon collide with electrons and produce gamma radiation. Other synthetic radionuclides decay by electron capture (K-capture), and as their electron energy levels adjust they emit X rays. Some radionuclides emit neutrons; some, just gamma rays.

Nuclear equations are balanced when the mass numbers and atomic numbers on either side of the arrow respectively balance. The energies of emission are usually described in electron-volts or multiples. $1 \text{ eV} = 1.602 \times 10^{-19}$ J. The mass equivalent to 1 eV is 1.783×10^{-36} kg.

A few very long-lived radionuclides in nature are at the heads of radioactive disintegration series, which represent the successive decays of "daughter" radionuclides until a stable isotope forms. Uranium-238, for example, is the "parent" of a series that ends with lead-206, and in which the gas radon-222 is an intermediate and so becomes an environmental pollutant.

Nuclear Stability. When all known nuclides, stable or not, are located on a plot organized according to numbers of neutrons and protons, the stable nuclides are within a curving band called the band of stability. The curvature reflects the need for increasing neutron/proton ratios at higher and higher atomic numbers so that the extra neutrons can "dilute" the protons and reduce proton–proton repulsions.

Radionuclides can have too high neutron/proton ratios, too, and by ejecting beta particles adjust their ratios downward. Those with neutron/proton ratios too low generally emit positrons to change their ratios upward. Both actions move nuclei nearer the band of stability.

Isotopes whose nuclei consist of even numbers of both neutrons and protons are generally much more stable than those with odd numbers of both—the odd–even rule. Evidently, having all neutrons paired and all protons paired is energetically better (more stable) than having any nucleon unpaired.

Isotopes with specific numbers of protons or neutrons—the magic numbers of 2, 8, 20, 28, 50, 82, and 126—are generally more stable than others. When both protons and neutrons in the nucleus correspond to magic numbers (and the nucleus is not well outside the band of stability), the isotope is particularly stable.

Transmutation. Various kinds of particles—protons, deuterons, alpha particles, and neutrons, for example—are used to bombard atomic nuclei and transmute them to nuclei of different isotopes. When a bombarding particle is captured, the resulting nucleus contains the energy of the captured particle as well as its nucleons. This compound nucleus then promptly decays to a more stable, but generally still radioactive, isotope. Which mode of decay is used by the compound nucleus is a function of its extra energy, not its extra mass. These nuclear reactions have been used to synthesize many radionuclides with medical and industrial applications, including all of the elements from 93 and higher—the transuranium elements.

Detecting and Measuring Radiations. Instruments to detect and measure radiations—Geiger counters, scintillation counters, and dosimeters—take advantage of the ability of radiations to generate ions in air or other matter. These ions can lead to chemical events in the body that result in cancer, tumors, mutations, or birth defects.

Units to describe radiations have been devised, each to help answer a specific question. For example, the electron-volt is used in answering the question, "How much energy does a radiation carry?"

"How *active* is a source?" The curie (Ci) and the becquerel (Bq) are available as units, where $1 \text{ Ci} = 3.7 \times 10^{10}$ disintegrations/s, and $1 \text{ Bq} = 1$ disintegration/s.

"How much energy is absorbed by a unit mass of absorber?" The SI unit of absorbed dose is the gray (Gy), where $1 \text{ Gy} = 1 \text{ J/kg}$. An older unit, the rad, is equal to 0.01 Gy.

"How can we compare doses absorbed by different tissues and caused by different kinds of radiation?" The unit of the rem is used. A 600-rem whole-body dose is lethal, and if all the people of a large population each receive 220–450 rem (depending on the available medical facilities and medications), half will die within 60 days.

The normal background radiation causes millirem exposures per year to most people. Cosmic rays, radionuclides in soil and stone building materials, medical X rays, and releases from nuclear tests or from nuclear power plants all contribute to this background.

Protection against radiations can be achieved by using dense shields (e.g., lead or thick concrete), by avoiding overuse of radionuclides or X rays, and by taking advantage of the inverse square law. This tells us that the intensity of radiation decreases with the square of the distance from its source.

Applications of Radionuclides. Because radiations are easily detected and measured, and because radionuclides have the same chemistry as their stable isotopes, radionuclides are used in medicine, in tracer analysis, neutron activation analysis, isotope dilution analysis, and the dating of geological samples or archaeological finds.

Nuclear Fusion.   The fusion of the nuclei of isotopes of hydrogen, whose nuclear charge is just $1+$, is being tried as a source of commercial energy. The atoms must be heated to a high enough temperature to change them to plasma—a neutral, gaseous mixture of nuclei and electrons. The pressure on the plasma must create a density so high that the bare nuclei, now carrying high thermal energies, can get so near each other that the nuclear strong force overcomes the repulsive electrostatic force (between protons). When the product of the particle density and the confinement time equals or exceeds the Lawson number, fusion can occur.

To create the necessary temperature, particle density, and confinement time, some research teams are using inertial confinement of the fuel (tritium and deuterium) in small hollow pellets. The pellet is subjected simultaneously to several pulses of high-energy lasers symmetrically arrayed about it. The fuel's temperature rises and as the pellet surface explodes outward, the rest implodes inward to cause sharp increases in density and temperature within the fuel.

Another approach is by magnetic confinement whereby tokamaks create magnetic "bottles" for the plasma. Pulses of high-energy, neutral particles then provide the energy for the necessary changes in both the temperature and the density.

Fission and Breeder Reactors.   Uranium-235, which occurs naturally, and plutonium-239, which can be made from uranium-238, are fissile isotopes that serve as the fuel in present-day reactors. When either isotope captures a thermal neutron, it splits apart in one of several ways to give two isotopes having midrange mass numbers, energy, and a net gain in neutrons. The neutrons, in a properly designed system, generate additional fission events, and a nuclear chain reaction becomes possible. If a critical mass of a fissile isotope is allowed to form, the chain reaction proceeds out of control, and it detonates as an atomic bomb explosion.

To control the chain reaction and keep it from becoming supercritical, various reactors have been designed. Pressurized-water reactors are most commonly used, and these have two loops of circulating fluids. The liquid in the primary loop (water) circulates around the reactor core, which consists of cladding tubes that hold fissile isotopes (much diluted by nonfissile material), a moderator (the circulating water, usually), and control rods. The heat of fission transfers to the circulating fluid, which carries it to the secondary loop where it heats water and makes hot steam under high pressure. This steam drives the generator.

If the circulation of fluid in the primary loop fails, the reactor must be shut down immediately, and several safety measures are available. These worked at Three Mile Island, but not at Chernobyl.

Reactors that use liquid sodium as the coolant in the primary loop as well as a metal form for the fuel are being studied as a safer alternative to conventional pressurized-water systems. It appears that such reactors are inherently unable to run out of control, and are not just safe because of backup measures.

The major problems with nuclear energy are the storage of radioactive wastes, the potential for loss-of-coolant accidents, the diversion of plutonium-239 to weapons (nuclear proliferation), and current high economic costs of building safe power plants. The radiations released during normal operations of power plants are very low in relation to the background, but the long-term storage of wastes is still an unsolved problem.

A fuel fabricated from a catalytic amount of plutonium-239 in uranium-238 should, in principle, be able to convert uranium-238 atoms, which are not fissile, to plutonium-239 atoms, which are. This is the idea of the breeder reactor, one that breeds fissile material as it operates and so would utilize an abundant supply of what is now not a fuel, uranium-238.

REVIEW EXERCISES

Answers to questions whose numbers are printed in color are given in Appendix C. Difficult qestions are marked with asterisks.

Conservation of Mass Energy

23.1   In chemical calculations involving chemical reactions we can regard the law of conservation of mass as a law independent of the law of conservation of energy despite Einstein's union of the two. What fact(s) about our everyday world make this possible?

23.2   In working with certain kinds of problems we make a distinction between an object's *mass* and its *rest mass*. What is this distinction, and in what kind of problem do we make it?

23.3   According to Einstein, how can we know that the absolute upper limit on the speed of any object is the speed of light?

23.4   Calculate the mass in kilograms of a 1.00-kg object when its velocity, relative to us, is (a) 3.00×10^7 m/s, (b) 2.90×10^8 m/s, and (c) 2.99×10^8 m/s.

23.5   What is the law of conservation of mass–energy?

23.6   What is the Einstein equation?

23.7   Calculate the mass equivalent in grams of 1.00 kJ.

23.8   In thermochemical calculations we use the older law of conservation of energy as if it were entirely independent of the law of conservation of mass. What makes this possible?

23.9   Calculate the quantity of mass in nanograms that is changed into energy when one mole of liquid water forms by the combustion of hydrogen, all measurements being made at 1 atm and 25 °C.

Nuclear Binding Energies

23.10   Why isn't the sum of the masses of all nucleons in one nucleus equal to the mass of the actual nucleus?

23.11   Calculate the binding energy in joules per nucleon of the deuterium nucleus whose mass is 2.0135 amu.

23.12   Calculate the binding energy in joules per nucleon of the tritium nucleus whose mass is 3.01550 amu.

23.13 Calculate the binding energy in joules per nucleon of the nucleus of an atom of iron-56. The observed mass of one *atom* is 55.9349 amu. What information in this chapter lets us know that no isotope has a *larger* binding energy per nucleon?

23.14 Calculate the binding energy in joules per nucleon of uranium-235. The observed mass of one *atom* is 235.0439 amu.

Radioactivity

23.15 What are the names of the two forces at work within an atomic nucleus? Which provides the net force of attraction? How do they differ in their abilities to act over a distance?

23.16 If the nucleons in a nucleus above mass number 56 could find more stable groupings following the fission of this nucleus, what (in just a general term) keeps any nucleus above this mass number intact?

23.17 What is a radionuclide?

23.18 What three kinds of radiation have been observed from naturally occurring radionuclides?

23.19 Give the composition of each of the following.
(a) Alpha particle   (c) Positron
(b) Beta particle   (d) Deuteron

23.20 Why is the penetrating ability of alpha radiation less than that of beta or gamma radiation?

23.21 Complete these nuclear equations by writing the symbols of the missing particles.
(a) $^{211}_{82}Pb \longrightarrow \,^{0}_{-1}e + \rule{1cm}{0.4pt}$
(b) $^{177}_{73}Ta \xrightarrow{\text{electron capture}} \rule{1cm}{0.4pt}$
(c) $^{220}_{86}Rn \longrightarrow \,^{4}_{2}He + \rule{1cm}{0.4pt}$
(d) $^{19}_{10}Ne \longrightarrow \,^{0}_{1}e + \rule{1cm}{0.4pt}$

23.22 Write the symbols of the missing particles to complete these nuclear equations.
(a) $^{245}_{96}Cm \longrightarrow \,^{4}_{2}He + \rule{1cm}{0.4pt}$
(b) $^{140}_{56}Ba \longrightarrow \,^{0}_{-1}e + \rule{1cm}{0.4pt}$
(c) $^{58}_{29}Cu \longrightarrow \,^{0}_{1}e + \rule{1cm}{0.4pt}$
(d) $^{68}_{32}Ge \xrightarrow{\text{electron capture}} \rule{1cm}{0.4pt}$

23.23 Write a balanced nuclear equation for each of these changes.
(a) Alpha emission from plutonium-242.
(b) Beta emission from magnesium-28.
(c) Positron emission from silicon-26.
(d) Electron capture by argon-37.

23.24 Write the balanced nuclear equation for each of these nuclear reactions.
(a) Electron capture by iron-55.
(b) Beta emission by potassium-42.
(c) Positron emission by ruthenium-93.
(d) Alpha emission by californium-251.

23.25 What is the nuclear equation for each of these nuclear reactions?
(a) Beta emission from aluminum-30.
(b) Alpha emission from einsteinium-252.

(c) Electron capture by molybdenum-93.
(d) Positron emission by phosphorus-28.

23.26 Give the nuclear equation for each of these changes.
(a) Positron emission by carbon-10.
(b) Alpha emission by curium-243.
(c) Electron capture by vanadium-49.
(d) Beta emission by oxygen-20.

23.27 Write the symbols, including the atomic number and mass number, for the radionuclides that would give each of these products.
(a) Fermium-257 by alpha emission.
(b) Bismuth-211 by beta emission.
(c) Neodymium-141 by positron emission.
(d) Tantalum-179 by electron capture.

23.28 Each of these nuclides forms by the decay mode specified. Write the symbols of the radionuclide parents, giving both the atomic number and the mass number.
(a) Rubidium-80 formed by electron capture.
(b) Antimony-121 formed by beta emission.
(c) Chromium-50 formed by positron emission.
(d) Californium-253 formed by alpha emission.

23.29 Iodine-123 is better than iodine-131 for diagnostic work in assessing thyroid function. It has a half-life of 13.3 hours (compared to 8 days for iodine-131), and it decays by electron capture, not beta emission.
(a) Why do these properties make iodine-123 better than iodine-131 for diagnostic work? (Consult Special Topic 23.5, also.)
(b) Write the equation for the decay of iodine-123.

23.30 Write the symbol of the nuclide that forms from cobalt-58 when it decays by electron capture.

23.31 With respect to their modes of formation, how do gamma rays and X rays differ?

23.32 What happens in electron capture to generate X rays?

Nuclear Stability

23.33 What is the band of stability?

23.34 Since the majority of unstable nuclides falls inside the band of stability, what is this band used for?

23.35 Both barium-123 and barium-140 are radioactive, but which is likelier to have the longer half-life? Explain your answer.

23.36 Tin-112 is a stable nuclide but indium-112 is radioactive and has a very short half-life (14 min). What does tin-112 have that indium-112 does not to account for this difference in stability?

23.37 Lanthanum-139 is a stable nuclide but lanthanum-140 is unstable ($t_{1/2}$ 40 hr). What "rule of thumb" concerning nuclear stability is involved?

23.38 As the atomic number increases, the ratio of neutrons to protons increases. What does this fact suggest is a factor in nuclear stability?

23.39 Radionuclides of high atomic number are far more likely to be alpha emitters than are those of low atomic number. Offer an explanation for this.

23.40   Although lead-164 has two magic numbers, 82 protons and 82 neutrons, it is unknown, whereas lead-208 is known and stable. What problem accounts for the fact that lead-164 cannot exist?

23.41   What decay particle is emitted from a nucleus of low to intermediate atomic number but a relatively high neutron/proton ratio? Explain why this particular kind of emission is likeliest.

23.42   What decay particle is emitted from a nucleus of low to intermediate atomic number but with a relatively low neutron/proton ratio? How does the emission of this particle benefit the nucleus?

23.43   What does electron capture do to the neutron/proton ratio of a nucleus? Which kinds of radionuclides are more likely to undergo this change, those above or those below the band of stability?

23.44   If an atom of potassium-38 had the option of decaying by positron emission or beta emission, which route would it likely take and why? Write the nuclear equation.

23.45   Suppose that an atom of argon-37 could decay by either beta emission or electron capture. Which route would it likely take and why? Write the nuclear equation.

*23.46   If a positron is to be emitted spontaneously, how much more *mass* (as a minimum) must an *atom* of the parent have than an *atom* of the daughter nuclide? Explain.

23.47   If we begin with 3.00 mg of iodine-131 ($t_{1/2}$ 8.07 hr), how much remains after six half-life periods?

23.48   A 9.0-ng sample of technetium-99m will still have how much of this radionuclide left after four half-life periods (about 1 day)?

Transmutations

23.49   When vanadium-51 captures a deuteron ($^2_1$D), what compound nucleus forms? Write the symbol. This compound nucleus expels a proton ($^1_1$p). Write the nuclear equation for the overall change caused by bombarding vanadium-51 with deuterons.

23.50   The bombardment of fluorine-19 by alpha particles generates sodium-22 and neutrons. Write the nuclear equation for this transmutation, including the intermediate compound nucleus.

23.51   Gamma ray bombardment of bromine-81 causes a transmutation of which neutrons are one product. Write the symbol of the other product.

23.52   Neutron bombardment of cadmium-115 results in neutron capture and the release of gamma radiation. Write the nuclear equation.

23.53   When manganese-55 is bombarded by protons, neutrons are released. What else forms? Write the nuclear equation.

23.54   Fill in the blanks in the following expression of successive nuclear reactions in the synthesis of a heavy element.

$$^{238}_{92}U\ (\underline{\quad},\ \gamma)\ ^{239}_{92}U \longrightarrow \underline{\qquad} + \ ^{0}_{-1}e$$

23.55   Complete the following expression that describes the synthesis of a heavy element.

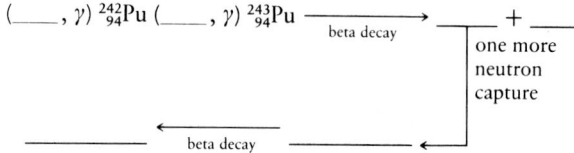

23.56   Curium-244 is made by irradiating plutonium-241. Complete the following sequence for this synthesis by filling in the blanks. (The original sample of plutonium-241 is subjected to a heavy neutron flux throughout this synthesis.)

Detecting and Measuring Radiations

23.57   What specific property of radiations is used by the Geiger counter?

23.58   Dangerous doses of radiations can actually involve very small quantities of energy. Explain.

23.59   What radiation units (common and SI) are used to describe the activity of a sample of radioactive material?

23.60   What energy unit is commonly used to describe the energy of a particle or a photon involved in radiation?

23.61   What is the name of the radiation unit that equals 1 disintegration/s?

23.62   What radiation units (common and SI) are used to describe the quantity of energy absorbed by a mass of tissue?

23.63   A sample giving 3.7×10^{10} disintegration/s has what activity in curies and in becquerels?

23.64   In general terms only, how is the rem related to the rad? (What consideration made the development of the rem useful?)

Applications of Radionuclides

23.65   Why should a radionuclide used in medical diagnostic work have a short half-life? If the half-life is too short, what problem arises?

23.66   Why isn't an alpha emitter used in medical diagnostic work?

23.67   In general terms, explain how neutron activation analysis is used and how it works.

23.68   How do cosmic rays produce carbon-14 in the atmosphere?

23.69   The ratio of carbon-14 to carbon-12 is (or is assumed to be) constant in a living organism but changes after the organism dies. Why is it constant during life?

23.70   What is one assumption in the use of the uranium-to-lead ratio for dating ancient geologic formations?

23.71   List some of the sources of exposure to radiations experienced by everybody.

23.72   A sample of a radioactive waste contains strontium-90, a beta emitter with a half-life of 28.1 yr. The activity of this sample was 0.245 Ci/g. How many years will it take for the activity of strontium-90 in this sample to drop to 1.00×10^{-6} Ci/g?

23.73 A maintenance worker in a laboratory unknowingly was exposed to a sample of radio-labeled sodium iodide made from iodine-131 (beta emitter, $t_{1/2} = 8.07$ days). The mistake was realized 28.0 days after the accidental exposure. The activity of the sample was then measured and found to be 25.6×10^{-5} Ci/g. The lab's safety officer had to know how active the sample was (in Ci/g) at the time of the exposure. Do the calculations that give this information.

23.74 Technetium-99*m* ($t_{1/2} = 6.02$ hr) is used in many kinds of organ scans in medicine. A sample prepared in the early morning for use throughout the day had an activity of 4.52×10^{-6} Ci. What will be the activity of this sample at the end of the day, after 8.00 hr?

23.75 Americium-243 ($t_{1/2} = 7.37 \times 10^3$ yr) is one radionuclide in radioactive waste. If such waste is to be kept out of reach of people and their food and water supplies for 20.0 half-life periods, how many years of safekeeping are involved for this isotope? What will be the activity (in Bq/g) of this isotope after this period if the initial activity is 120 Bq/g?

23.76 Potassium-40, which has a half-life of 1.3×10^9 yr, decays by electron capture to argon-40, a stable isotope. If this decay occurs in dense rocks, the argon cannot escape. A 500-mg sample of such rock was analyzed and found to have 2.45×10^{-6} mol of argon-40 and 2.45×10^{-6} mole of potassium-40. How old was the rock? (And what assumption has to be made about the origin of the argon-40?)

23.77 If a sample of dense rock were found to contain 1.16×10^{-7} mol of argon-40, then how much potassium-40 would also have to be present in the same sample for the rock to be 1.3×10^9 years old? (The argon-40 comes from potassium-40 by a decay having a half-life of 1.3×10^9 years.)

23.78 If exposure from a distance of 1.60 m gave a worker a dose of 8.4 rem, then how far should the worker move away from the source to reduce the dose to 0.50 rem for the same period?

23.79 During work with a radioactive source, a worker was told that she would receive a dose of 50 mrem at a distance of 4.0 m during 30 min of work. What would happen to the dose received if the worker moved closer, to 0.50 m, for the same period?

23.80 To determine the vitamin B-12 content of a multivitamin supplement, the contents of 50 capsules were dissolved in 500 mL of water. To this solution was added 1.58 mg of radioactive vitamin B-12 with a specific activity of 9.67 Ci/mg. After mixing and processing, a 14.6-mL sample was found to have an activity of 0.385 Ci. Calculate the number of micrograms of vitamin B-12 in each capsule, rounding to one significant figure.

23.81 A liquid concentrate used as a food additive contains some vitamin B-12. To 25.0 mL of this concentrate there was added 1.12 mg of radioactive B-12 having a specific activity of 3.08 Ci/mg. After thorough mixing and processing, a sample with a volume of 10.5 mL was found to have an activity of 1.20 Ci. Calculate the concentration of B-12 in the liquid concentrate in units of micrograms per milliliter.

23.82 Crater Lake in southern Oregon formed after the volcanic cone of ancient Mount Mazama collapsed. During long periods of volcanic activity, pumice blown out during one eruption covered thousands of square miles, burying extensive forests. A sample of buried wood with a mass of 152 mg was found to have an activity of 61.2 Bq. Calculate the age of this specimen (to two significant figures).

23.83 A specimen of wood with a mass of 458 mg taken from a support beam in an ancient tomb had an activity of 271 Bq. Calculate the age of this specimen (to two significant figures).

Nuclear Fusion

23.84 There is a gain in binding energy per nucleon when light nuclei fuse to form heavier nuclei. Why, therefore, don't a tritium atom and a deuterium atom, in a mixture of these isotopes, just spontaneously fuse to give helium (and energy)?

23.85 The word *plasma* is used in discussions of nuclear fusion. What does this word mean?

23.86 What is the Lawson criterion and how does it serve in research on nuclear fusion?

23.87 When inertial confinement of tritium and deuterium within a pellet is used, how does laser energy interact with the pellet and initiate nuclear fusion?

23.88 Write the nuclear equation for the fusion of tritium and deuterium.

23.89 What purpose is served by a molten lithium "blanket" around the chamber where lasers and fuel pellets interact?

23.90 When scientists work on fusion using lasers, what will have been achieved when *self-ignition* is observed? What does this term mean and why is self-ignition important?

23.91 What is a tokamak?

Nuclear Fission

23.92 Why is it easier for a nucleus to capture a neutron than a proton?

23.93 What does the term *thermal neutron* mean?

23.94 What is meant by *nuclear fission*?

23.95 What is meant by a *fissile isotope*?

23.96 Which fissile isotope occurs in nature?

23.97 Which fissile isotope is commonly a by-product in the operation of a nuclear reactor using uranium-235 fuel (mixed with uranium-238)?

23.98 What fact about the fission of uranium-235 makes it possible for a *chain reaction* to occur?

23.99 Explain in general terms why fission generates more neutrons than needed to initiate it.

23.100 Complete the following nuclear equation by supplying the symbol for the other product of the fission.

$$^{235}_{92}U + ^{1}_{0}n \longrightarrow ^{94}_{38}Sr + \underline{\hspace{1.5cm}} + 2\,^{1}_{0}n$$

23.101 Both products of the fission in Review Exercise 23.100 are unstable. Using Figures 23.8 and 23.9, what is the most likely way for each of them to decay — by alpha emission, beta emission, or by positron emission? Explain.

23.102 What are some of the possible fates of the extra neutron produced by the fission shown in Review Exercise 23.100?

23.103 Why would there be a *subcritical* mass of a fissile isotope? (Why isn't *any* mass of uranium-235 critical?)

23.104 What purpose is served by a *moderator* in a nuclear reactor?

23.105 What does the term *multiplication factor* mean in connection with a nuclear reactor?

23.106 How would the multiplication factors compare between two reactors when one is operating continuously at full power and the other is operating continuously at half power?

23.107 When a reactor is said to be *critical,* what is true about the multiplication factor within the reactor?

23.108 What do power plant engineers do to keep a reactor from going supercritical?

23.109 If in a pressurized-water reactor a break occurs in a primary coolant line, what is one of the first things to happen within the reactor?

23.110 How might hydrogen gas be generated following a loss-of-coolant accident in a reactor?

23.111 Besides retaining the actual fuel in a reactor, what important function does a cladding tube have?

23.112 Why is there no possibility of an atomic bomb explosion from a nuclear power plant?

23.113 Which system would produce more radioactive wastes, a fusion reactor (assuming one becomes operable) or a fission reactor — each operating at the same power for the same time? Explain.

23.114 What advantages does molten sodium offer as a primary coolant in a nuclear power plant?

23.115 What property does the metallic form of the nuclear fuel have that helps to make a reactor based on it, a reactor also using liquid sodium as the primary coolant, inherently safe in a failure-of-coolant accident?

23.116 What are the major issues raised by opponents of nuclear power, and what do proponents of such power say in response?

Breeder Reactor

23.117 What does it mean when uranium-238 is described as a *fertile isotope?*

23.118 Write the nuclear equation for the breeding of plutonium-239 from uranium-238.

23.119 Why is the breeder reactor sometimes called a *catalytic burner reactor?*

23.120 Why is there any interest at all in the development of a commercial breeder reactor?

TEST OF FACTS AND CONCEPTS: CHAPTERS 19–23

These chapters have concentrated less on theoretical principles that apply to all chemical systems and more on the plain, ordinary chemical properties of important elements and compounds. The "world out there" tends to expect that students of a beginning college chemistry course have a general knowledge of the major chemical properties of the major chemical elements. Perhaps of greater urgency is the expectations of your professors in succeeding chemistry courses. You could be a little embarrassed, for example, if your teacher of organic chemistry discovers that you know little or nothing of, say, the hydride ion, or the behavior of carbon dioxide in aqueous base, or the existence of the amide ion, or the formula for sodium bisulfite, or the fact that nitric acid is an oxidizing agent. Many of the following questions are meant to test your *chemical knowledge*. Don't be discouraged if you can't answer as many as you'd like without going back into the preceding chapters. Chemical knowledge isn't eveyday, ordinary, common knowledge. It takes learning and reviewing . . . and then some more reviewing. But it is important that you be knowledgeable in these things, particularly if you are going to take additional courses in chemistry.

Also included in this Test of Facts and Concepts are questions from Chapter 23 on radioactivity and nuclear chemistry. These are also areas in which a chemistry student should be knowledgeable.

1. Name the two elements whose atoms occur in the greatest numbers in each system.
   (a) Clean, dry air   (d) The earth as a whole
   (b) The human body   (e) The solar system
   (c) The earth's crust   (f) The universe as a whole

2. What factors involving the nucleus of helium might be responsible for the large abundance of helium in the universe?

3. The atoms of which element contribute the most bulk volume to the earth's crust and surface waters?

4. Which four nonmetallic elements make up over 95% by mass of the human body?

5. Give the names and symbols of the elements in Groups IA, IIA, IIIA, IVA, VA, VIA, VIIA, and 0.

6. Which group number corresponds to each family?
   (a) The noble gases   (e) The carbon family
   (b) The halogens   (f) The alkali metals
   (c) The nitrogen family   (g) The alkaline earth metals
   (d) The oxygen family

7. Give the names and formulas of the hydrides of the following.
   (a) Each nonradioactive halogen   (d) Oxygen
   (b) Nitrogen   (e) Sodium
   (c) Phosphorus   (f) Magnesium

8. Give an electron dot structure of each particle.
   (a) A hydrogen atom   (c) A hydride ion
   (b) A hydrogen molecule   (d) The hydrogen ion

9. Which of the particles named in the previous question is the strongest Brønsted base? What is the name and the formula of its conjugate acid?

10. Write balanced molecular equations for a synthesis of each element by the method or by the source specified.
    (a) Hydrogen—from methane
                —from a metal and an acid
                —from a metal hydride
    (b) Oxygen—from potassium chlorate
              —from mercury(II) oxide
              —from hydrogen peroxide
    (c) Chlorine—from sodium chloride
    (d) Bromine—from sodium bromide and chlorine
               —from hydrobromic acid
    (e) Iodine—from sodium iodide and manganese oxide
            —from sodium iodide and chlorine

11. What are the chief industrial sources of each of the following elements? Write molecular equations where chemical reactions are used to prepare the element.
    (a) Hydrogen   (e) Chlorine
    (b) Oxygen   (f) Aluminum
    (c) Nitrogen   (g) Lead
    (d) Iron   (h) Copper

12. Write the molecular equation for the reaction of water with the following.
    (a) Any Group IA hydride
    (b) Any Group IIA hydride
    (c) Any Group VIIA hydride
    (d) Solid sodium oxide
    (e) Solid calcium oxide
    (f) Any Group IA amide

13. Write the molecular equation for the reaction of water with each of the following substances.
    (a) Sulfur trioxide

(b) Sulfur dioxide
(c) Carbon dioxide
(d) Fluorine
(e) Phosphorus trichloride
(f) Tetraphosphorus decaoxide
(g) Sodium
(h) Chromium trioxide
(i) Dinitrogen pentoxide
(j) Thioacetamide
(k) Phosphorus pentabromide
(l) Chlorine

14. Write a molecular equation for the reaction that hydrogen gas can be made to undergo with each substance.
(a) Calcium   (d) Oxygen
(b) Sodium   (e) Iron(III) oxide
(c) Nitrogen   (f) Chlorine

15. Write a molecular equation for the reaction that oxygen gas can be made to undergo with each substance.
(a) Sulfur   (f) Sodium
(b) Sulfur dioxide   (g) Calcium
(c) Carbon   (h) Nitrogen
(d) Phosphorus   (i) Nitric oxide
(e) Phosphine   (j) Hydrazine

16. Element Z forms the following series of oxides: Z_2O_7, Z_2O_3, and ZO.
(a) Which oxide would be the most acidic oxide? Explain.
(b) Which oxide would be the most basic oxide? Explain.

17. Element Z forms a hydride, HZ, that dissolves readily in water to form a solution that conducts electricity and leaves phenolphthalein indicator colorless. In which family is Z likely to be, Group IA or Group VIIA? Explain.

18. In which family of representative elements of the periodic table would one most likely find that all hydrides function neither as acids nor as bases?

19. For which family of the representative elements of the periodic table would all oxides be relatively high-melting solids that dissolve readily and exothermically in water?

20. The oxides of which family of representative elements would include the greatest number of relatively strong oxidizing agents?

21. Write the formulas (including charges) of the superoxide ion, the peroxide ion, and the oxide ion.

22. If the following pairs of substances were mixed together in proper proportions and a method of ignition (if needed) were carried out (or if the mixture were heated), what would you *see*? Observations might include changes in color or odor, or the bubbling out of a gas, or the separation of a solid, or the generation of light. Quite often, more than one of these changes occurs during the same reaction. Write a balanced molecular equation for the reaction, but first describe what you would actually see.
(a) Phosphorus and oxygen
(b) Solid silver bromide and aqueous sodium thiosulfate
(c) Solid sulfur and the oxygen in air
(d) Fluorine gas is bubbled into water

(e) Phosphine and oxygen (from air)
(f) Hydrofluoric acid and glass
(g) Aqueous sodium carbonate and hydrochloric acid
(h) Chlorine gas bubbled into aqueous sodium iodide
(i) A mixture of concentrated hydrobromic acid and concentrated sulfuric acid is heated
(j) A direct electrical current is passed through unstirred concentrated aqueous sodium chloride
(k) A direct electrical current is passed through stirred, concentrated aqueous sodium chloride.
(l) A vial containing dinitrogen tetroxide is warmed slightly.
(m) A colorless solution of concentrated hydriodic acid is left exposed to the air.

23. Write the molecular and net ionic equations for the reaction of hydrochloric acid with each of the following.
(a) Sodium carbonate
(b) Sodium cyanide
(c) Potassium bicarbonate
(d) Sodium hydrogen sulfite
(e) Sodium sulfite
(f) Sodium oxide
(g) Calcium oxide (solid)
(h) Zinc metal
(i) Aqueous ammonia
(j) Calcium carbonate (solid)
(k) Iron(II) sulfide (solid)
(l) Aluminum oxide (solid)

24. Write the molecular and net ionic equations for the reaction of aqueous sodium hydroxide with each of the following.
(a) Aluminum oxide   (e) Dinitrogen trioxide
(b) Nitric acid   (f) Sulfur dioxide
(c) Sulfur trioxide   (g) Nitrogen dioxide
(d) Carbon dioxide   (h) Chlorine

25. Write the name of each compound.
(a) $HOCl$ (n) HNO_2
(b) PBr_5 (o) N_2O_5
(c) H_2SO_3 (p) $NaNH_2$
(d) $HClO_4$ (q) N_2H_4
(e) $LiHCO_3$ (r) HPO_3
(f) $NaHSO_3$ (s) KNO_2
(g) H_3PO_4 (t) H_2O_2
(h) CaC_2 (u) Na_2O_2
(i) $NaOBr$ (v) $Na[Al(H_2O)_2(OH)_4]$
(j) $Na_2S_2O_3$ (w) O_3
(k) PH_3 (x) NaH
(l) N_2O (y) KNH_2
(m) P_4 (z) $HClO_3$

26. Write the formula of each of the following.
(a) Nitric oxide
(b) Sodium hypochlorite
(c) Lithium sulfite
(d) Nitric acid
(e) Magnesium hydride
(f) Sodium amide
(g) Chlorous acid
(h) Calcium carbonate

(i) Potassium nitrite
(j) Sodium hydrogen sulfite
(k) Hydrazine
(l) Ozone
(m) Metaphosphoric acid
(n) Sodium thiosulfate
(o) Phosphine
(p) Pyrophosphoric acid
(q) Sodium tripolyphosphate
(r) Cryolite
(s) Hydroxyapatite
(t) Sodium iodate

27. What are the sulfanes?

28. Sulfur does not dissolve in water but does dissolve in aqueous sodium sulfide. Explain.

29. Write the equations for the steps in the Ostwald process.

30. Write the equations for the steps in the contact process.

31. What trend in each property is observed — an increase or a decrease — as a family of representative nonmetallic elements is descended?
(a) Tendency to become metal-like.
(b) Hydration energy of most common anion.
(c) Electronegativity.
(d) Ionization energy.
(e) Acidity of binary hydride.

32. Give two major industrial uses of each of the following.
(a) Oxygen   (d) Ammonia
(b) Nitrogen   (e) Sulfuric acid
(c) Hydrogen   (f) Nitric acid

33. Why don't metals tend to form covalent bonds to each other?

34. Which of these metals is most metallic in its chemical properties: magnesium, barium, aluminum, tin?

35. Which metals are recovered from seawater? Why?

36. Aluminum occurs in many rocks in the form of minerals called aluminosilicates, yet the only common ore of aluminum is bauxite, which contains a hydrated form of aluminum oxide. Based on what you learned in Chapter 21, give a probable reason why bauxite is the principal ore of aluminum.

37. Why haven't manganese nodules been extensively mined yet?

38. What is the purpose of the flotation method? On what kinds of ores is it normally used?

39. What chemical substances are put into the blast furnace as the *charge?* Give the molecular equations that are involved in the reduction of the iron ore. Give a typical molecular equation for the formation of slag in the blast furnace.

40. Why must iron be refined after it comes from the blast furnace? Why is the basic oxygen process the most used method for making steel?

41. What are the electron configurations of the outer shells of (a) aluminum, (b) lead, (c) barium, (d) potassium, and (e) magnesium?

42. What oxidation states are observed for the following: (a) Na, (b) Ca, (c) Al, (d) Pb, and (e) Cs?

43. Which of the alkali metals are able to react with water? Give a molecular equation to illustrate the reaction.

44. Which of the alkaline earth metals react with water? Write a molecular equation to illustrate the reaction.

45. Complete and balance the following chemical equations. If no reaction occurs, write "N.R."
(a) $Na(s) + O_2(g) \longrightarrow$
(b) $Cs(s) + O_2(g) \longrightarrow$
(c) $K(s) + O_2(g) \longrightarrow$
(d) $NaHCO_3(s) \xrightarrow{heat}$
(e) $Al(s) + Fe_2O_3(s) \xrightarrow{heat}$
(f) $CaCO_3(s) \xrightarrow{heat}$
(g) $Al(s) + OH^-(aq) + H_2O \longrightarrow$
(h) $Mg(s) + O_2(g) \xrightarrow{heat}$
(i) $Ba(s) + H_2O \longrightarrow$
(j) $Al(s) + H^+(aq) \longrightarrow$

46. Which salt is likely to be more soluble in water, (on a mole basis), $KClO_4$ or $NaClO_4$?

47. Give the molecular equations for the formation of $NaHCO_3$ in the Solvay process.

48. What is the composition of dolomite?

49. What is plaster of paris? Give a molecular equation for the reaction that takes place when plaster of paris is moistened.

50. What flame colors are caused by salts of (a) sodium, (b) barium, (c) calcium, (d) potassium, (e) lithium, and (f) strontium?

51. What is the principal chemical component of ruby? What element gives it its red color?

52. Aluminum ion in water is somewhat acidic. Give equations that account for this. Show by means of equations how the gradual addition of base produces a precipitate of aluminum hydroxide, and then causes it to dissolve.

53. What is an alum? Give the chemical formula for potassium alum.

54. What are the oxidation states of (a) lead, (b) tin, and (c) bismuth?

55. Which is easier to oxidize, SnO or PbO?

56. Balance the following equation by the ion-electron method.

$$BiO_3^- + Mn^{2+} \longrightarrow Bi^{3+} + MnO_4^-$$

Explain why this reaction can serve as a sensitive qualitative test for manganese(II) ion in a solution.

57. $PbCl_4$ easily decomposes into $PbCl_2$ and Cl_2, while PbI_4 cannot be made. What is the reason for this?

58. Match the chemical formulas on the left with their common names on the right.

NaOH	Corundum
$NaHCO_3$	Lye
Al_2O_3	Limestone
$CaCO_3$	Gypsum
CaO	Epsom salts
$CaSO_4 \cdot 2H_2O$	Lime
$MgSO_4 \cdot 7H_2O$	Baking soda

59. $BeCl_2$ and $SnCl_4$ are both covalently bonded substances. Why do their metal-halogen bonds have so little ionic character?

60. List four properties that are usually associated with the transition metals.

61. What is the lanthanide contraction? What effect does it have on the densities of the metals that follow lanthanum in the periodic table?

62. Give an important commercial use for each of the following compounds.
    (a) TiO_2 (f) $(NH_4)_3Fe(CN)_6$
    (b) V_2O_5 (g) NiO_2
    (c) $ZnCrO_4$ (h) $AgBr$
    (d) MnO_2 (i) ZnO
    (e) Cr_2O_3 (j) ZnS

63. Explain by means of a chemical equation how we can convert chromate ion into dichromate ion and vice versa by properly adjusting the pH of the solution.

64. Calculate $\Delta H°$, in kilojoules, for the reaction

$$2Al(s) + Fe_2O_3(s) \longrightarrow Al_2O_3(s) + 2Fe(s)$$

65. Give the color of the following.
    (a) CrO_4^{2-} (g) $Co(H_2O)_6^{3+}$
    (b) MnO_4^- (h) $Ni(H_2O)_6^{3+}$
    (c) MnO_4^{2-} (i) $Cu(H_2O)_4^{2+}$
    (d) $Cu(NH_3)_4^{2+}$ (j) $Fe(H_2O)_6^{2+}$
    (e) Fe_2O_3 (k) $Mn(H_2O)_6^{2+}$
    (f) $Cr(H_2O)_6^{3+}$

66. Why is cadmium sometimes used in place of zinc as a protective coating over steel?

67. Which metals are the coinage metals? Which of them dissolve in hydrochloric acid? Which dissolve in nitric acid?

68. Give the electron configurations of the transition metals in period 4 of the periodic table.

69. Give the electron configurations of the following transition metal ions.
    (a) Fe^{3+} (b) Cr^{2+} (c) Ag^+ (d) Zn^{2+}

70. Write chemical equations to describe what happens when concentrated NaOH solution is gradually added to solutions that contain the following cations: (a) Fe^{3+}, (b) Zn^{2+}, (c) Cr^{3+}, (d) Cu^{2+}, (e) Cd^{2+}.

71. Sketch the stereoisomers of (a) $[Co(en)_2Cl_2]^+$ and (b) $[Cr(en)_3]^{3+}$. Label *cis* and *trans* isomers. Which of the structures are chiral?

72. What test is applied to a molecular or ionic structure to determine whether it is chiral?

73. The octahedral complex $[Cr(NH_3)_3Cl_3]$ has two stereoisomers. Sketch their structures. Be sure that the two structures that you draw are indeed different.

74. Sketch an energy level diagram showing the energies of the *d* orbitals in an octahedral crystal field. Label the diagram by identifying the *d* orbitals in each energy level and the crystal field splitting, Δ.

75. The complex $Pt(NH_3)_2Cl_2$ can be isolated as two isomers (*cis*

and *trans*) because it has a square planar structure. Show that if it were tetrahedral, only one isomer (i.e., only one structure) would be possible. Do this by trying to draw another tetrahedral isomer and then show that it can be rotated to give the first one.

76. The nitrite ion gives a very strong crystal field splitting in the complex ion $[Co(NO_2)_6]^{3+}$. How many unpaired electrons would be present in this complex ion?

77. How does crystal field theory explain the colors of complex ions?

78. Suppose that the total mass of the reactants in a chemical reaction was 100.00000 g. How many kilojoules of energy would have to evolve from this reaction if the total mass of the products could be no greater than 99.99900 g?

79. Calculate the binding energy in J/nucleon for the nucleus of the deuterium atom, $_1^2H$. The mass of an atom of deuterium, including its electron, is 2.014102 amu.

80. Which nuclear force acts over the longer distance, the electrostatic force or the nuclear strong force?

81. Why is an isotope of hydrogen a better candidate for nuclear fusion than an isotope of helium?

82. When uranium-238 is bombarded by a particular bombarding particle, it can be transmuted into plutonium-239 and neutrons. Which bombarding particle accomplishes this? How many electrons are produced for each plutonium-239 atom? What is the nuclear equation for this change?

83. If an atom of beryllium-7 decays by the capture of a K-electron, into which isotope does it change? Write the equation.

84. An atom of neodymium-144 decays by alpha emission. Write the nuclear equation.

85. Phosphorus-32 decays by beta emission. Write the nuclear equation.

86. The half-life of samarium-149, an alpha emitter, is 4×10^{14} years. The half-life of oxygen-15 is 124 seconds. Assuming that equimolar samples are compared, which is the more intensely radioactive?

87. The half-life of cesium-137 is 30 years. Of an initial 100 g sample, how much cesium-137 will remain after 300 years?

88. What are some "rules of thumb" that can be used to judge if a particular radionuclide might have a long enough half-life to warrant the effort to make it?

89. The ratio of neutrons to protons in atomic nuclei grows larger with increasing atomic number. What is an explanation for this?

90. Which is likelier to be the more stable isotope of calcium, calcium-40 or calcium-42? (What "rule of thumb" about nuclear stability is involved here?)

91. A particular compound nucleus can form from a variety of combinations of targets and bombarding particles, as in the example of aluminum. What determines how a compound nucleus breaks up?

92. What is the *rad*, and what difficulty with this unit makes the *rem* more useful when assessing possible radiation damage to people?

93. The number of nuclear disintegrations per unit of time that is occurring in a sample of a radionuclide is described by what two units (one of them being in the SI)?

94. A person who hasn't yet studied radioactivity and its applications comes to you and asks you to give a brief description of how each of the following work and what they can be used for. What would be good responses?
    (a) Neutron activation analysis
    (b) Tracer analysis
    (c) Isotope dilution analysis
    (d) Radiocarbon dating
    (e) Atomic fusion
    (f) Atomic fission
    (g) The breeder reactor
    (h) A dosimeter

Qual Topic III

95. Give the colors of the following complexes: (a) $Fe(SCN)_6^{3-}$, (b) $Cu(NH_3)_4^{2+}$, (c) $Ni(NH_3)_6^{2+}$, (d) $Ni(HDMG)_2$, (e) $Ag(NH_3)_2^+$.

96. Suppose a solution contained a mixture of Al^{3+} and Zn^{2+}. What single chemical reagent could be used to separate these metal ions from each other? Write chemical equations to illustrate your answer.

97. Suppose concentrated aqueous NH_3 were added to a solution that contains Fe^{3+}, Zn^{2+}, Ni^{2+}, and Al^{3+}. Which metals would be found in the precipitate that forms? Give the formulas of the compounds that are precipitated. Give the formulas of the metal-containing ions in the solution after the addition of the NH_3.

Sky divers trust their lives to the strengths of synthetic fibers in the fabrics and lines of their equipment. Some important synthetics are described in this chapter.

ORGANIC COMPOUNDS, POLYMERS, AND BIOCHEMICALS

HOW ORGANIC COMPOUNDS ARE CLASSIFIED	ORGANIC AND BIOCHEMICAL COMPOUNDS OF OXYGEN
SATURATED HYDROCARBONS	ORGANIC AND BIOCHEMICAL COMPOUNDS OF NITROGEN
UNSATURATED HYDROCARBONS	NUCLEIC ACIDS AND HEREDITY

24.1 HOW ORGANIC COMPOUNDS ARE CLASSIFIED

Functional groups enable the classification of millions of organic compounds into a small number of families whose members have similar chemical properties.

Organic chemistry is the study of the preparation, properties, identifications, and reactions of those compounds of carbon not classified as inorganic. Inorganic carbon compounds include the oxides of carbon, the bicarbonates and carbonates of metal ions, the cyanides, and a handful of other compounds. All the other compounds of carbon are organic compounds, those whose molecules nearly always have chains of two or more carbon atoms or rings of three or more. These chains and rings serve as the skeletons to which other nonmetal atoms, particularly hydrogen but also O, N, S, or any of the halogens, are covalently attached. Many organometallic compounds are also known in which metal atoms are attached by electron pair bonds.

The study of organometallic compounds is a large field of chemistry.

Virtually all plastics, synthetic and natural fibers, dyes and drugs, insecticides and herbicides, ingredients in perfumes and flavoring agents, and all petroleum products are organic compounds. All the foods you eat consist chiefly of such organic nutrients as carbohydrates, fats and oils, proteins, and vitamins. The substances that make up furs and feathers, hides and skins, and all cell membranes are organic.

Bio- is from a Greek root meaning "life."

The special study of the substances of living cells is biochemistry. A closely related and overlapping field is molecular biology, the study of how living systems make the chemicals of heredity—nucleic acids—and use them to make proteins.

Functional Groups

In Section 20.8 we described how carbon is unique among all the elements in its ability to form chains and rings of carbon atoms that can still hold other nonmetals by strong covalent bonds. We learned that double and triple bonds are possible. And we introduced the one major fact about organic compounds that makes the study of so many a manageable effort, the existence of families defined by functional groups—parts of molecules where most reactions occur. Table 24.1 identifies the more important families, describes the structures of their functional groups, and gives examples of specific family members.

The structural formulas in this table are "condensed." Many C—H bonds, instead of being each symbolized separately by a line, are "understood." We have been doing this all along. We write NH_3 for ammonia, H_2O for water, and CH_4 for methane with the understanding that the hydrogens are individually joined to the central atom by single

TABLE 24.1 Some Important Families of Organic Compounds

Family	Characteristic Structural Features of Molecules[a]	Examples
Hydrocarbons	Only C and H	
	Alkanes: only single bonds	CH_3CH_3, ethane
	Alkenes: C=C	$CH_3CH=CH_2$, propene
	Alkynes: C≡C	HC≡CH, acetylene
	Aromatic: benzene ring	CH_3—⟨benzene ring⟩, toluene
Alcohols	R—OH	CH_3CH_2OH, ethyl alcohol
Aldehydes	(H)R—C(=O)—H	CH_3—C(=O)—H, acetaldehyde
Ketones	R—C(=O)—R′	CH_3—C(=O)—CH_3, acetone
Carboxylic acids	(H)R—C(=O)—OH	H—C(=O)—OH, formic acid
Esters	(H)R—C(=O)—O—R′	CH_3—C(=O)—O—CH_3, methyl acetate
Amines	$R—NH_2$, R—NH—R′, R—N(R″)—R′	CH_3NH_2, methylamine; CH_3NHCH_3, dimethylamine; $CH_3N(CH_3)CH_2CH_3$, dimethylethylamine
Amides	R—C(=O)—N(R″(H))—R′(H)	CH_3C(=O)—NH_2, acetamide

[a] The feature (H)R— means that the group can be either H or a hydrocarbon group. A prime (′) or a double prime (″) means that the hydrocarbon groups may be alike or different.

Cycloalkanes and other cyclic compounds occur so widely that chemists have simplified the structural representations of rings by using polygons. Thus a hexagon stands for cyclohexane. The simplest cycloalkane, cyclopropane, which was once an important anesthetic, is represented by a triangle.

Cyclic compounds can have double bonds as in cyclohexene. Carbon–carbon double bonds, whether in open chains or in rings, are always shown by two lines. They are never "understood."

Rings, of course, can carry substituents, as in ethylcyclohexane, and not all of the ring atoms have to be carbon atoms. They can be oxygen, nitrogen, or sulfur, too, and cyclic compounds with ring atoms other than carbon are called **heterocyclic compounds**. A simple example is tetrahydropyran, which has the basic ring system widely present among some of the forms that molecules of carbohydrates can assume.

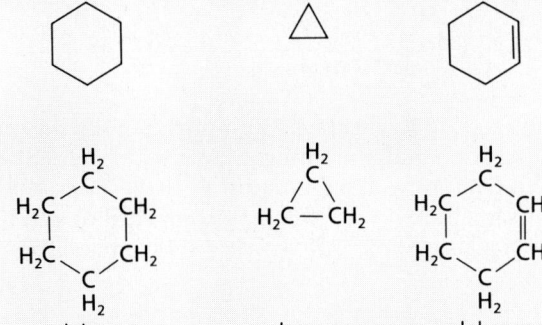

cyclohexane   cyclopropane   cyclohexene   ethylcyclohexane   tetrahydropyran

bonds. We have written CH_3CH_2OH for ethyl alcohol with the understanding that the C—C and the C—O single bonds are understood, that the "skeleton" is C—C—O, and that the hydrogen atoms are attached to this skeleton so as to satisfy the requirement that each carbon have four bonds and each oxygen two. For a review of these matters and how cyclic structures are condensed, see Special Topic 24.1.

The significance of functional groups will be clear when we contrast the alkanes with other families of organic compounds. The alkanes and their cyclic relatives are members of a multifamily "clan" of compounds, the hydrocarbons, that were introduced briefly in Section 20.8. Molecules of alkanes have no functional group—just C—C and C—H single bonds. Because C and H are so alike in electronegativity, these bonds are the least polar single bonds in organic molecules. The alkanes, therefore, are the least polar of all organic compounds. Hence their molecules are least able of all organic molecules to interact with polar or ionic reactants, which include just about all possible reactants. (Important exceptions are some nonmetal elements, like oxygen, the halogens, and sulfur.)

When a polar group of atoms, like the OH group, the C=O group, the NH_2 group, or even a halogen atom, like Cl, is attached to one carbon atom of an otherwise alkane chain or ring, the molecule then has a polar site. Such molecules react with several polar or ionic reactants. And the reactions occur generally just at the polar group—which is why it is called a *functional* group—or with a nearby unit of the hydrocarbon-like remainder of the molecule, which is now more polar because of the group. This is why all compounds with a given functional group display similar reactions. And this is what greatly simplifies the study of organic compounds. We do not have to treat each of the several thousand members of, say, the amine family as individual compounds whose reactions bear no similarity. The reactions of amines are similar from compound to compound, so we can learn just the handful of reactions exhibited by all amines and then adapt this knowledge to a specific example, as needed.

We can illustrate this by the reaction of amines with a strong acid, like HCl(*aq*). Amines have the —NH_2 group and are Brønsted bases, like ammonia. Thus the simplest amine, methylamine, reacts with HCl as follows. (We will show the parallel reaction of ammonia first.)

Amines of low formula weights are responsible for the odor from overripe fish.

$$NH_3(aq) + HCl(aq) \longrightarrow NH_4^+(aq) \qquad + Cl^-(aq)$$
$$\underset{\text{methylamine}}{CH_3NH_2(aq)} + HCl(aq) \longrightarrow \underset{\substack{\text{methylammonium}\\\text{ion}}}{CH_3NH_3^+(aq)} + Cl^-(aq)$$

But octylamine, an amine with 8 carbons per molecule, gives the same kind of reaction.

$$\underset{\text{octylamine}}{CH_3CH_2CH_2CH_2CH_2CH_2CH_2CH_2NH_2(\ell)} + HCl(aq) \longrightarrow$$

$$\underset{\text{octylammonium ion}}{CH_3CH_2CH_2CH_2CH_2CH_2CH_2CH_2NH_3^+(aq)} + Cl^-(aq)$$

Just the NH_2 group changes, and it changes in the identical way in both methylamine and octylamine. We could write equations for hundreds of other amines with hydrochloric acid, but they would all be alike. Instead, we will just summarize the reactions of all compounds with the NH_2 group—actually, with the NH_2 group attached to a *hydrocarbon* group—in just one simple equation, where R represents any and all purely hydrocarbon, alkanelike groups.

$$R-NH_2 + HCl(aq) \longrightarrow R-NH_3^+(aq) + Cl^-$$

The study of organic chemistry is thus not the study of individual compounds, taking them one at a time. It is the study of the common properties of functional groups, how to change one into another, and how each such group affects physical properties. Thus chemists treat molecular structures like complex "maps" that can be "read" with the knowledge of just a handful of "map signs," the functional groups.

24.2 SATURATED HYDROCARBONS

The alkanes and cycloalkanes—the saturated hydrocarbons—have the fewest chemical properties of all organic compounds.

Saturated implies here, as elsewhere, that there is no more room for something, such as no room for holding more atoms or groups.

Four families of hydrocarbons were mentioned in Table 24.1—alkanes, alkenes, alkynes, and aromatic hydrocarbons. Their chief similarities are that they are made exclusively of carbon and hydrogen and they are insoluble in water. The alkane hydrocarbons are examples of saturated organic compounds, compounds with only single bonds regardless of family. Alkenes and alkynes are unsaturated organic compounds, because their molecules have double or triple bonds. The aromatic hydrocarbons are also unsaturated because the carbon atoms of their rings, when represented by simple Lewis structures, also have double bonds. (We discussed the problem of Lewis structures for a compound like benzene in Section 8.8 where we noted that the pi electrons of the double bonds implied by Lewis structures are actually delocalized.)

Another structural feature that we want to stress, even though conventional structural formulas leave it largely to the imagination, is that the atoms in organic molecules, with few exceptions, do not all lie in the same plane, such as the plane of the paper. Any carbon that is bonded to four other atoms has a tetrahedral geometry—as we studied in Section 7.9 (page 247). Its four bonds point to the corners of a tetrahedron.

When a molecule contains a double or a triple bond, the atoms immediately attached to those of the multiple bond do lie in the same plane, but usually the molecule has tetrahedral carbons, too. Propylene, for example—the raw material for polypropylene used to make indoor–outdoor carpeting—has one sp^3 hybridized carbon and two sp^2 hybridized carbons. The geometry at the former is tetrahedral (page 247) and that at the latter is planar (page 246).

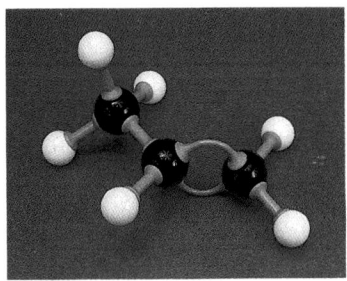

sp^3 hybridized carbon

sp^2 hybridized carbons

propylene

or $CH_3CH=CH_2$

Propylene

Another feature of hydrocarbons and of all other open-chain organic compounds in which single bonds occur is that groups joined by a single bond can rotate with respect to each other about the bond. There is *free rotation* about single bonds in open-chain systems. This is allowed because the overlapping of atomic orbitals, hybrid orbitals or otherwise, in single bonds is not significantly affected by such a rotation. Thus little if any energy barrier exists to such internal motions.

Sources of Alkanes

Virtually all of the usable supplies of hydrocarbons are obtained from the fossil fuels — coal, petroleum, and natural gas. One of the operations in petroleum refining is to boil the crude oil (petroleum freed of natural gas) and then selectively condense the vapors that pass over between prechosen ranges of temperatures. The liquid collected at each range is called a *fraction,* and the operation is fractional distillation. Each fraction is used for a particular purpose, and each consists of a rather complex mixture of compounds — almost entirely hydrocarbons and mostly alkanes. Gasoline, for example, is a fraction boiling roughly between 40 and 200 °C. The kerosene and jet fuel fraction overlaps this range, going from 175 to 325 °C. The molecules of the alkanes in gasoline generally have from 5 to 10 carbon atoms; those in kerosene, from 12 to 18. Paraffin wax is part of the nonvolatile residue of petroleum refining, and it consists of alkanes with over 20 carbons per molecule. Low-boiling fractions of crude oil are used as nonpolar solvents; their molecules have 4 to 8 carbons.

Diesel fuel is another fraction of petroleum.

Straight and Branched Chains and Rings in Alkanes

In a straight-chain alkane, one carbon follows another so that each of the two end-carbons holds 3 H-atoms and each of all of the carbons in between holds 2 H atoms. The ten simplest examples are listed in Table 24.2. They are often described as *open-chain* systems, because they have no rings of atoms and all fit the formula $C_nH_{(2n+2)}$. In branched-chain alkanes, which fit the same general formula, there are hydrocarbon groups attached by C—C single bonds to carbons between the ends of a straight chain alkane. In carbon ring alkanes, or *cycloalkanes,* three or more carbons are linked in a cyclic system, and they have the general formula C_nH_{2n}.

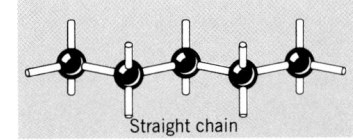

Straight chain

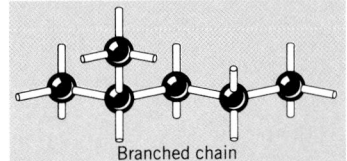

Branched chain

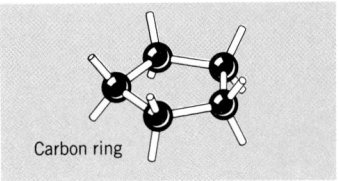

Carbon ring

Naming the Alkanes

Chemists realized almost 100 years ago that no one could remember unsystematic names for all organic compounds. So chemists from several countries established what is now called the Commission on Nomenclature of Organic Chemistry, a committee of the International Union of Pure and Applied Chemistry, or IUPAC, made up of representatives of the chemical societies of all nations. Its various commissions on nomenclature have worked out what we now call the IUPAC rules of nomenclature for several kinds of compounds.

Unsystematic or common names are still widely used for many compounds. The common name of 2-methylpropane is isobutane, for example.

TABLE 24.2 Straight-Chain Alkanes

IUPAC Name	Number of Carbon Atoms	Molecular Formula[a]	Structure	Boiling Point (°C)	Melting Point (°C)	Density (g/mL, 20 °C)
Methane	1	CH_4	CH_4	−161.5	−182.5	
Ethane	2	C_2H_6	CH_3CH_3	−88.6	−183.3	
Propane	3	C_3H_8	$CH_3CH_2CH_3$	−42.1	−189.7	
Butane	4	C_4H_{10}	$CH_3CH_2CH_2CH_3$	−0.5	−138.4	
Pentane	5	C_5H_{12}	$CH_3CH_2CH_2CH_2CH_3$	36.1	−129.7	0.626
Hexane	6	C_6H_{14}	$CH_3CH_2CH_2CH_2CH_2CH_3$	68.7	−95.3	0.659
Heptane	7	C_7H_{16}	$CH_3CH_2CH_2CH_2CH_2CH_2CH_3$	98.4	−90.6	0.684
Octane	8	C_8H_{18}	$CH_3CH_2CH_2CH_2CH_2CH_2CH_2CH_3$	125.7	−56.8	0.703
Nonane	9	C_9H_{20}	$CH_3CH_2CH_2CH_2CH_2CH_2CH_2CH_2CH_3$	150.8	−53.5	0.718
Decane	10	$C_{10}H_{22}$	$CH_3CH_2CH_2CH_2CH_2CH_2CH_2CH_2CH_2CH_3$	174.1	−29.7	0.730

[a] The molecular formulas of the open-chain alkanes fit the general formula C_nH_{2n+2}, where n = the number of carbon atoms per molecule.

IUPAC Rules for Naming Alkanes   In the IUPAC rules, the last syllable in the name designates the family of the compound. The names of all saturated hydrocarbons, for example, end in *-ane*. Hydrocarbons with double bonds have names that end in *-ene*, and those with triple bonds have names that end in *-yne*. Here are the rules for the alkanes.

1. Use the ending *-ane* for all alkanes (and cycloalkanes).

2. Determine what is the longest continuous chain of carbons in the structure, and let this be the *parent chain* for naming purposes.
   For example, view the branched-chain alkane

$$CH_3-CH_2-\overset{\overset{\displaystyle CH_3}{|}}{CH}-CH_2-CH_2-CH_3$$

   as coming from

$$CH_3-CH_2-CH_2-CH_2-CH_2-CH_3$$

   by replacing a hydrogen atom on the third carbon from the left with a CH_3 group.

$$CH_3-CH_2-CH-CH_2-CH_2-CH_3 \longrightarrow CH_3-CH_2-\overset{\overset{\displaystyle CH_3}{|}}{CH}-CH_2-CH_2-CH_3$$

3. Attach a prefix to *-ane* that specifies the number of carbon atoms *in the parent chain*. The prefixes through C-10 are as follows and should be learned. The names in Table 24.2 show their use.

meth-	1 C	hex-	6 C
eth-	2 C	hept-	7 C
prop-	3 C	oct-	8 C
but-	4 C	non-	9 C
pent-	5 C	dec-	10 C

Thus the parent chain of our example has six carbons, so the corresponding alkane is called hexane—*hex* for six carbons and *ane* for being in the alkane family. The branched-chain compound whose name we are devising is regarded as a derivative of this parent, hexane.

4. Assign numbers to each carbon of the parent chain starting from whichever of its ends gives to the location of the first branch the lower of two possible numbers. Thus the correct direction for numbering our example is from left to right.

$$\underset{1}{CH_3}-\underset{2}{CH_2}-\underset{3}{\overset{\overset{\displaystyle CH_3}{|}}{CH}}-\underset{4}{CH_2}-\underset{5}{CH_2}-\underset{6}{CH_3}$$

(correct direction of numbering)

Had we numbered from right to left, the carbon holding the branch would have had a higher number.

$$\underset{6}{CH_3}-\underset{5}{CH_2}-\underset{4}{\overset{\overset{\displaystyle CH_3}{|}}{CH}}-\underset{3}{CH_2}-\underset{2}{CH_2}-\underset{1}{CH_3}$$

(incorrect direction of numbering)

5. Determine the correct name for each branch (or for any other atom or group). We must now pause and learn the names of a few such groups.

The Alkyl Groups   Any branch that consists only of carbon and hydrogen and has only single bonds is called an alkyl group, and the names of all alkyl groups end in -*yl*. Think of an alkyl group as an alkane minus one of its hydrogen atoms. For example,

$$H-\overset{\overset{\displaystyle H}{|}}{\underset{\underset{\displaystyle H}{|}}{C}}-H \xrightarrow{\text{remove one H}} H-\overset{\overset{\displaystyle H}{|}}{\underset{\underset{\displaystyle H}{|}}{C}}- \quad \text{or} \quad CH_3-$$

methane → methyl

$$H-\overset{\overset{\displaystyle H}{|}}{\underset{\underset{\displaystyle H}{|}}{C}}-\overset{\overset{\displaystyle H}{|}}{\underset{\underset{\displaystyle H}{|}}{C}}-H \xrightarrow{\text{remove one H}} H-\overset{\overset{\displaystyle H}{|}}{\underset{\underset{\displaystyle H}{|}}{C}}-\overset{\overset{\displaystyle H}{|}}{\underset{\underset{\displaystyle H}{|}}{C}}- \quad \text{or} \quad CH_3CH_2-$$

ethane → ethyl

Two alkyl groups can be obtained from propane because the middle position is not equivalent to either of the end positions.

$$H-\overset{\overset{\displaystyle H}{|}}{\underset{\underset{\displaystyle H}{|}}{C}}-\overset{\overset{\displaystyle H}{|}}{\underset{\underset{\displaystyle H}{|}}{C}}-\overset{\overset{\displaystyle H}{|}}{\underset{\underset{\displaystyle H}{|}}{C}}-H \xrightarrow[\text{from either end}]{\text{remove one H}} H-\overset{\overset{\displaystyle H}{|}}{\underset{\underset{\displaystyle H}{|}}{C}}-\overset{\overset{\displaystyle H}{|}}{\underset{\underset{\displaystyle H}{|}}{C}}-\overset{\overset{\displaystyle H}{|}}{\underset{\underset{\displaystyle H}{|}}{C}}- \quad \text{or} \quad CH_3CH_2CH_2-$$

propane → propyl

$$H-\overset{\overset{\displaystyle H}{|}}{\underset{\underset{\displaystyle H}{|}}{C}}-\overset{\overset{\displaystyle H}{|}}{\underset{\underset{\displaystyle H}{|}}{C}}-\overset{\overset{\displaystyle H}{|}}{\underset{\underset{\displaystyle H}{|}}{C}}-H \xrightarrow[\text{from the middle}]{\text{remove one H}} H-\overset{\overset{\displaystyle H}{|}}{\underset{\underset{\displaystyle H}{|}}{C}}-\overset{\overset{\displaystyle H}{|}}{C}-\overset{\overset{\displaystyle H}{|}}{\underset{\underset{\displaystyle H}{|}}{C}}-H \quad \text{or} \quad CH_3\overset{\overset{\displaystyle |}{}}{CH}CH_3$$

propane → isopropyl

Since we will not need the names for higher alkyl groups, we will continue with the IUPAC rules for alkanes.

Methyl group

Ethyl group

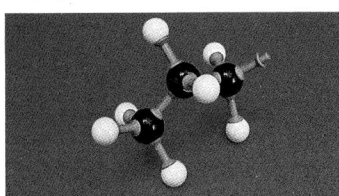

Propyl group

Isopropyl group

6. Attach the name of the alkyl group or other substituent to the name of the parent as a prefix. Place the location number of the group in front of the resulting name and separate the number from the name by a hyphen.

Returning to our original example, its name is 3-methylhexane.

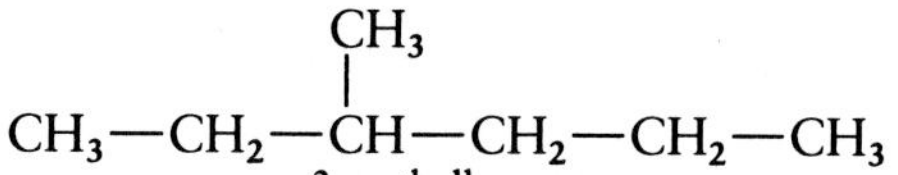

$$CH_3-CH_2-\underset{\underset{\displaystyle CH_3}{|}}{CH}-CH_2-CH_2-CH_3$$

3-methylhexane

7. When two or more groups are attached to the parent, name each and locate each with a number. Always use *hyphens* to separate numbers from words. Here is an example of applying this rule.

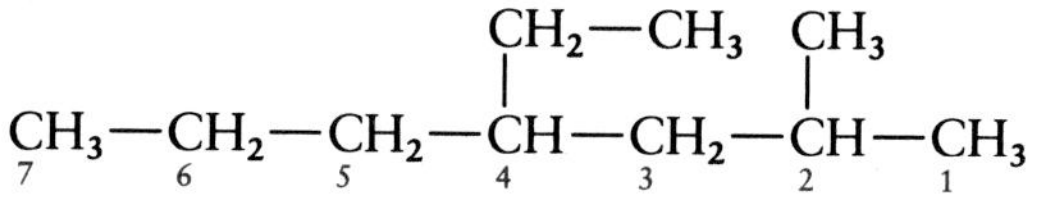

$$\underset{7}{CH_3}-\underset{6}{CH_2}-\underset{5}{CH_2}-\underset{4}{\underset{\underset{\displaystyle CH_3}{|}}{CH}}-\underset{3}{CH_2}-\underset{2}{\underset{\underset{\displaystyle CH_3}{|}}{CH}}-\underset{1}{CH_3}$$

4-ethyl-2-methylheptane

Since 1971, the IUPAC has specified that the names of alkyl substituents be assembled in their alphabetical order.

8. When two or more substituents are identical, use such prefixes as di- (for 2), tri- (for 3), tetra- (for 4), and so forth; and specify the location number of every group. Always separate a number from another number in a name by a *comma*. For example,

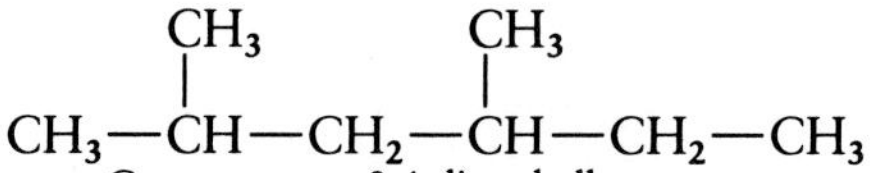

$$CH_3-\underset{\underset{\displaystyle CH_3}{|}}{CH}-CH_2-\underset{\underset{\displaystyle CH_3}{|}}{CH}-CH_2-CH_3$$

Correct name: 2,4-dimethylhexane
Incorrect names: 2,4-methylhexane
3,5-dimethylhexane
2-methyl-4-methylhexane

9. When identical groups are on the *same* carbon, repeat the number of this carbon in the name. For example,

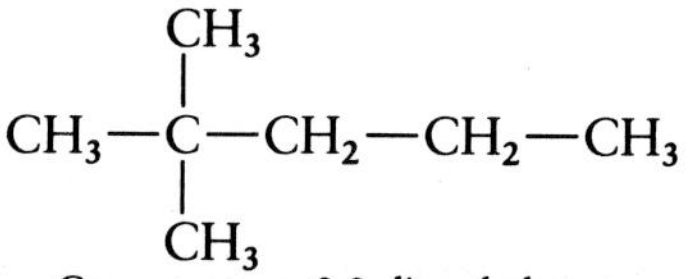

$$CH_3-\underset{\underset{\displaystyle CH_3}{|}}{\overset{\overset{\displaystyle CH_3}{|}}{C}}-CH_2-CH_2-CH_3$$

Correct name: 2,2-dimethylpentane
Incorrect names: 2-dimethylpentane
2,2-methylpentane
4,4-dimethylpentane

These are not all of the IUPAC rules for alkanes, but they will handle all of our needs. Study the following examples of correctly named compounds. Be sure to notice that in choosing the parent chain we sometimes have to go around a corner as the chain zigzags on the page.

2,2-dimethylbutane
not 2-ethyl-2-methylpropane

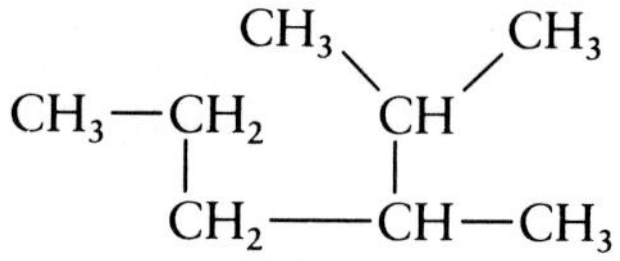

2,3-dimethylhexane
not 2-isopropylpentane

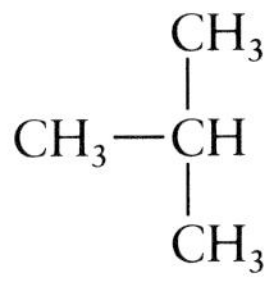

CH₃
|
CH₃—CH
|
CH₃
2-methylpropane
not 1,1-dimethylethane

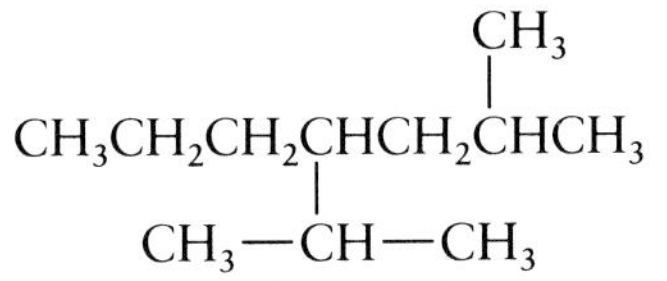

CH₃
|
CH₃CH₂CH₂CHCH₂CHCH₃
|
CH₃—CH—CH₃
4-isopropyl-2-methylheptane
not 4-isopropyl-6-methylheptane

EXAMPLE 24.1   USING THE IUPAC RULES TO NAME AN ALKANE

Problem: What is the IUPAC name for the following compound?

CH₂CH₂CH₂CH₃
|
CH₃
|
CH₃CHCHCHCHCH₂CH₂CH₃
| |
CH₃ CH
/   \
CH₃   CH₃

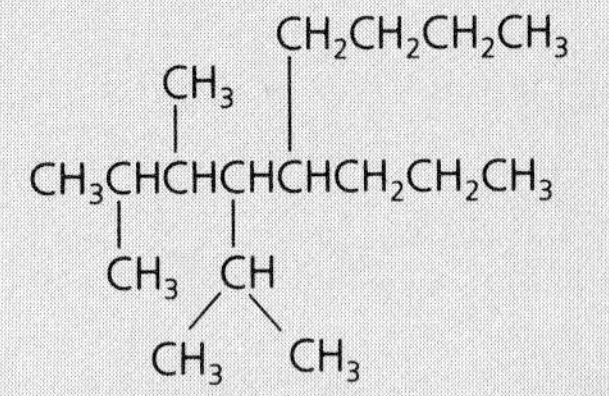

Solution: The compound is an alkane since it is a hydrocarbon with only single bonds, so the ending to the name is *-ane*. The next step is to find the longest chain even if we have to go around corners. This chain is nine carbons long, so the name of the parent alkane is *nonane*. We have to number the chain from left to right, as follows, to reach the first branch with the lower number.

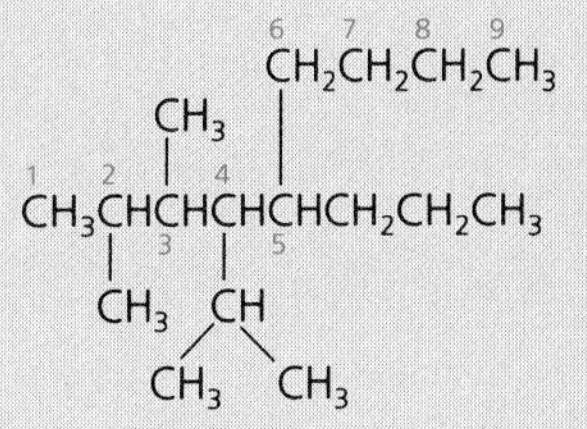

6 7 8 9
CH₂CH₂CH₂CH₃
|
CH₃
|
1 2 4
CH₃CHCHCHCHCH₂CH₂CH₃
3 5
| |
CH₃ CH
/   \
CH₃   CH₃

At carbons 2 and 3 there are one-carbon methyl groups. At carbon 4, there is a three-carbon isopropyl group (not the propyl group, because the bonding site is at the middle carbon of the three-carbon chain). At carbon 5, there is a three-carbon propyl group. (It has to be this particular propyl group because the bonding site is the end of the three carbon chain in the group.) We will assemble these names as follows to make the final name.

2,3-Dimethyl-4-isopropyl-5-propylnonane

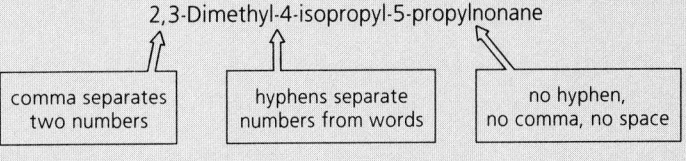

comma separates two numbers | hyphens separate numbers from words | no hyphen, no comma, no space

PRACTICE EXERCISE 1   Write the IUPAC names of the following compounds.

(a) CH₃—CH₂
              \
               CH—CH₃
              /
     CH₂—CH₂
     |
     CH₃

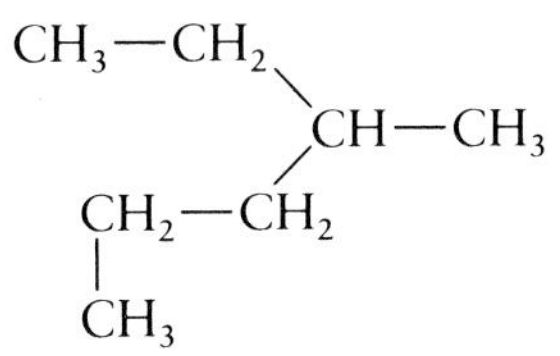

(b) H₃C   CH₂—CH₂—CH₃
        \ |
         CH—CH—CH₂—CH₃
        /
  CH₃—CH
        |
        CH₃

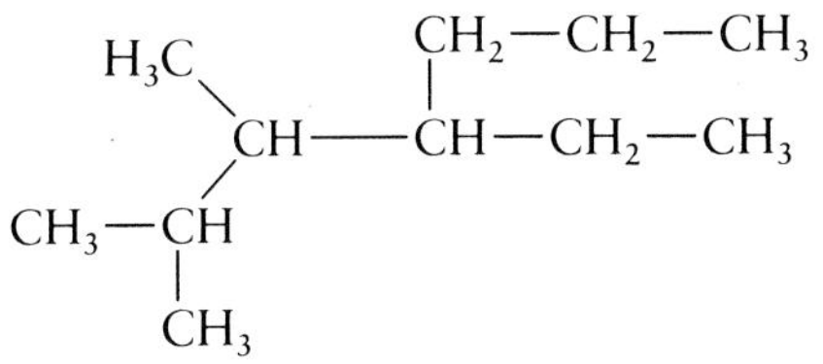

(c)

$$CH_3-CH_2-\underset{\underset{\displaystyle CH_2-CH_3}{|}}{CH}-\underset{\underset{\displaystyle }{|}{CH_3}}{CH}-\underset{\underset{\displaystyle }{|}{CH_3}}{CH}-CH_2-\underset{\underset{\displaystyle }{|}{CH_3}}{CH}-CH_3$$

Isomerism among the Alkanes

Isomerism is the existence of two or more compounds with the same molecular formula. There are two alkanes with the formula C_4H_{10}, for example. They are butane and 2-methylpropane, and you can see that, like any pair of different compounds, they have different properties.

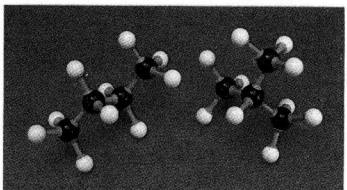

FIGURE 24.1 The isomers of C_4H_{10}. Butane is on the left and 2-methylpropane (isobutane) is on the right.

$$\underset{\substack{\text{butane}\\ \text{bp} -0.5\ °C\\ \text{mp} -138.4\ °C\\ \text{density } 0.622\ \text{g/mL}}}{CH_3CH_2CH_2CH_3} \qquad \underset{\substack{\text{2-methylpropane}\\ \text{bp} -11.7\ °C\\ \text{mp} -159.6\ °C\\ \text{density } 0.604\ \text{g/mL}}}{\overset{\overset{\displaystyle CH_3}{|}}{CH_3CHCH_3}}$$

These compounds are isomers of each other; they share the same molecular formula but have different structures. See also Figure 24.1.

Isomerism is one of the major reasons for the existence of so many organic compounds. As the number of carbons per molecule increases, the number of possible isomers for any given formula becomes astronomic. The alkane C_8H_{18}, for example, has 18 isomers; $C_{10}H_{22}$ has 75 isomers; and $C_{20}H_{42}$ has 366,319 (but just a few have actually been made). For $C_{40}H_{82}$, an estimated 6.25×10^{13} isomers are believed to be possible.

The two isomers of C_4H_{10} are both alkanes, so they have very similar chemical properties. Quite often, isomers belong to different families with dramatically different properties. There are two isomers of C_2H_6O, for example, ethyl alcohol and dimethyl ether. The alcohol is a liquid (bp 78.5 °C) and the ether is a gas (bp −24 °C) at room temperature. The alcohol reacts vigorously with sodium metal, but the ether is inert (at room temperature) toward virtually all common acids and bases or redox reactants.

$$\underset{\text{dimethyl ether}}{CH_3-O-CH_3} \quad + Na \longrightarrow \text{no reaction}$$

$$\underset{\text{ethyl alcohol}}{2CH_3CH_2-O-H} + 2Na \longrightarrow 2CH_3CH_2-O^- + 2Na^+ + H_2$$

Chemical Properties of Alkanes and Cycloalkanes

Alkanes and cycloalkanes are generally stable at room temperature toward such different reactants as concentrated sulfuric acid (or any other common acid), concentrated aqueous bases (like NaOH), and even the most reactive metals. Fluorine attacks virtually all organic compounds, including the alkanes, to give mixtures of products. Like all hydrocarbons, the alkanes burn in air to give carbon dioxide and water. Hot nitric acid, chlorine, and bromine also react with alkanes to give useful products (which we will not study further). Our point here is to show how generally stable alkanes are toward just about anything. When heated at high temperatures in the absence of air, alkanes "crack," which means that they break up into smaller molecules. The cracking of methane yields finely powdered carbon and hydrogen.

$$CH_4 \xrightarrow{\text{high temperatures}} C + 2H_2$$

The carbon is used as a filler in auto tires, and the hydrogen is used either as a fuel or as a chemical. Thus the source of hydrogen for the modern operations of the Haber–Bosch synthesis of ammonia is methane. (See Chemicals in Use 11, page 692.)

The controlled cracking of ethane gives ethylene, one of the most important raw materials in the organic chemicals industry.

$$CH_3—CH_3 \xrightarrow{\text{high temperatures}} CH_2{=}CH_2 + H_2$$

Ethylene is used to make polyethylene plastic items, a number of other polymers, as well as industrial-purpose ethyl alcohol, ethylene glycol (an antifreeze), and compounds made from these.

The IUPAC accepts both ethene and ethylene as the name of $CH_2{=}CH_2$.

24.3 UNSATURATED HYDROCARBONS

Alkenes readily react by addition reactions but aromatic hydrocarbons react mainly by substitution reactions.

The transition from alkanes to the unsaturated hydrocarbons is a large jump from few chemical properties to many. Yet the reactions that alkenes (and alkynes) give are quite different from those given by another unsaturated family of hydrocarbons, the aromatic hydrocarbons.

Alkenes

Hydrocarbons with one or more double bonds are alkenes. The double bond consists of one stronger σ bond and one weaker π bond (as we described in Section 8.5). Some alkenes can be isolated from petroleum and coal, and some occur as plant products. What we must emphasize here, however, is that the alkene double bond occurs widely among compounds that have other groups and that are classified in other ways. Thus the chemical reactions of the alkene double bond occur wherever this bond is found.

Naming the Alkenes   The IUPAC rules for alkene nomenclature are adaptations of those for alkanes. The *parent* chain, however, must include the double bond—so that the compound can have the *-ene* ending of an alkene—even if this makes the parent chain shorter than another that could be found. The parent alkene chain must be numbered from whichever end gives the first carbon of the double bond the lower of two possible numbers. The locations of branches are not a factor. Otherwise, alkyl groups are named and located as before. Some examples of correctly named alkenes are as follows.

No number is needed to name $CH_2{=}CH_2$ ethene or $CH_3CH{=}CH_2$ propene.

$$\underset{\text{1-butene}}{CH_3CH_2CH{=}CH_2} \qquad \underset{\text{2-butene}}{CH_3CH{=}CHCH_3} \qquad \underset{\text{2,5-dimethyl-2-heptene}}{CH_3CH_2\overset{\overset{\displaystyle CH_3}{|}}{C}HCH_2CH{=}\overset{\overset{\displaystyle CH_3}{|}}{C}CH_3} \qquad \underset{\text{cyclohexene}}{\hexagon}$$

Notice that only one number is needed to locate the double bond, the number of the first carbon of the double bond to be reached as the chain is numbered.

Some alkenes have two double bonds and are called *dienes*, some have three double bonds and are *trienes*, and so forth. (Each double bond has to be located by a number.)

The common name of propene is propylene, and other simple alkenes have common names, too.

$$\underset{\text{1,3-pentadiene}}{CH_2{=}CHCH{=}CHCH_3} \qquad \underset{\text{1,4-pentadiene}}{CH_2{=}CHCH_2CH{=}CH_2} \qquad \underset{\text{1,3,5-hexatriene}}{CH_2{=}CHCH{=}CHCH{=}CH_2}$$

Geometric Isomerism among the Alkenes   There is no free rotation at a double bond, because such a motion would break one bond, and this costs too much energy. Alkenes, therefore, have many examples of geometric isomerism, the same kind de-

Figure 8.16, page 285, explains why free rotaton at a double bond is difficult.

scribed in Section 22.5 for coordination compounds. Thus *cis*-2-butene and *trans*-2-butene are geometric isomers of each other. They have not only the same molecular formula, C_4H_8, but also the same organization of atoms and bonds; they differ in the directions taken by groups attached at the double bond.

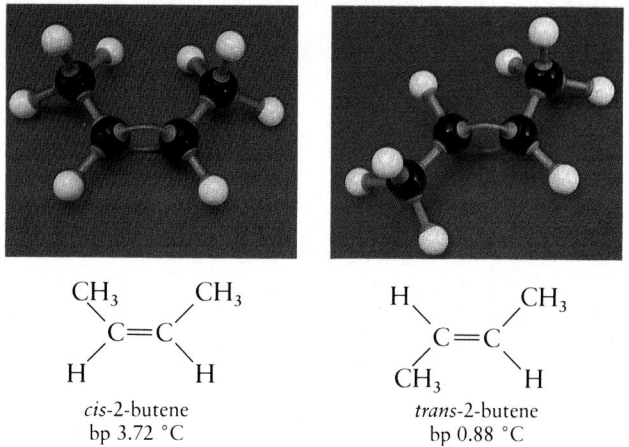

cis-2-butene
bp 3.72 °C

trans-2-butene
bp 0.88 °C

Cis means "on the same side"; *trans* means "on opposite sides." This structural difference gives the two geometric isomers of 2-butene measurable differences in physical properties (as the boiling points show). Since they each have the double bond, however, the chemical properties of *cis*- and *trans*-2-butene are very similar.

Some Chemical Properties of Carbon–Carbon Double Bonds   The carbon–carbon double bond has a rich chemistry because of the π bond. It is a weaker bond than most σ bonds, and its electron density projects outward where electron-seeking reactants are better able to get at it. Thus alkenes react very readily with concentrated sulfuric acid because the π-electron pair confers on the double bond the properties of a Brønsted base. In the presence of H_2SO_4, this pair enables one carbon of the double bond to accept a proton. The other carbon momentarily becomes electron poor. In fact, it becomes a Lewis acid, and the hydrogen sulfate ion, a Lewis base, quickly attaches to it. The product is an alkyl derivative of sulfuric acid. We can illustrate these steps with the reaction between ethene and sulfuric acid.

Because ring structures also lock out free rotation, geometric isomers of ring compounds are also possible. These two isomers of 1,2-dimethylcyclopropane are examples.

H_3C ⟋⟍ CH_3
bp 28 °C
trans

H_3C ⟍⟋ CH_3
bp 37 °C
cis

$CH_3CH_2^+$ is called the ethyl carbocation. Carbocations are generally *very* unstable and exist only fleetingly.

$$CH_2{=}CH_2 + H{-}O{-}\underset{\underset{O}{|}}{\overset{\overset{O}{||}}{S}}{-}O{-}H \longrightarrow CH_2{-}\overset{+}{CH_2} \longrightarrow CH_3{-}CH_2{-}O{-}\underset{\underset{O}{|}}{\overset{\overset{O}{||}}{S}}{-}O{-}H$$

ethene   sulfuric
                  acid

$$\overset{|}{H}$$

ethyl hydrogen
sulfate

$$+\ {}^{-}O{-}\underset{\underset{O}{|}}{\overset{\overset{O}{||}}{S}}{-}O{-}H$$

If water is present with the strong acid, another Lewis base beside the acid anion is available—the water molecule. Now the second step takes a different turn, and an alcohol eventually forms instead of an alkyl hydrogen sulfate. We will just show the overall result.

$$CH_2{=}CH_2 + H{-}O{-}H \xrightarrow{\text{acid catalyst}} CH_2{-}CH_2 \quad \text{or} \quad CH_3CH_2OH$$

ethene $\underset{H}{|} \quad \underset{OH}{|}$ ethanol

2-Butene (either *cis* or *trans*) gives a similar reaction. Note that in the overall result, all that happens is that one of the two bonds at the alkene double bond disappears and the pieces of the water molecule, H and OH, become attached at the different ends of the double bond.

$$CH_3CH{=}CHCH_3 + H{-}OH \xrightarrow{\text{acid catalyst}} CH_3CH{-}CHCH_3 \quad \text{or} \quad CH_3CH_2CHCH_3$$

2-butene · · H OH (2-butanol) · OH

These two kinds of reactions of the double bond—reaction with sulfuric acid and reaction with water—are examples of addition reactions. Other chemicals that add to the double bond are any of the gaseous hydrogen halides, as well as chlorine, bromine, and hydrogen. The hydrogen halides and the two halogens react rapidly at room temperature. Ethene, for example, reacts with bromine to give 1,2-dibromoethane (the DBE used as a lead scavanger in gasoline mentioned in Special Topic 20.3). You can see why this is also an addition reaction.

Alkynes gives similar addition reactions.

$$CH_2{=}CH_2 + Br{-}Br \xrightarrow{\text{room temperature}} CH_2{-}CH_2 \quad \text{or} \quad BrCH_2CH_2Br$$

Br Br (1,2-dibromoethane)

Compared with the mild conditions for the addition of Cl_2 or Br_2, hydrogen requires a higher temperature, a higher pressure, and the presence of a catalyst—powdered platinum, for example. When hydrogen adds to the double bond of an alkene, the product is an alkane and the reaction is called *hydrogenation*. We will illustrate by the reaction of 2-butene (*cis* or *trans*), which adds hydrogen to give butane.

On page 790 we discussed the hydrogenation of vegetable oils to give edible lipids that are then solids at room temperature.

$$CH_3CH{=}CHCH_3 + H{-}H \xrightarrow[\text{catalyst}]{\text{heat, pressure}} CH_3CH{-}CHCH_3 \quad \text{or} \quad CH_3CH_2CH_2CH_3$$

H H

Ozone reacts with anything with carbon–carbon double bonds and breaks the molecules into fragments at each such site. Oxidizing agents in general attack the electron-rich double bond. A variety of types of products form depending on the exact conditions, and we will not pursue this further. Our point is simply to show that the double bond makes any substance that contains it vulnerable to oxidations.

Polymers from Alkenes In the presence of suitable *promoters*, alkenes give a special kind of addition reaction called addition polymerization. Thousands of molecules join together, one after the other, with pieces of the promoter molecules taking end places of very long chains. We will illustrate with ethylene, $CH_2{=}CH_2$.

$$nCH_2{=}CH_2 \xrightarrow{\text{polymerization}} {-}CH_2{-}CH_2{-}CH_2{-}CH_2{-}CH_2{-}CH_2{-}\text{etc.}$$

ethylene · polyethylene

The product is an example of a polymer, a large molecule with repeating structural units. The starting material is called the monomer. To simplify the structure of a polymer without sacrificing any important information, its structure is often condensed to show just the repeating unit. Thus polyethylene has the structure $-(CH_2{-}CH_2)_n-$, where n is several thousand.

poly- many
mono- one
meros- parts

Propylene also polymerizes. Its polymer, polypropylene, is used to make indoor–outdoor carpets, rope, and containers for foods and fluids.

Teflon-coated frying pan

TABLE 24.3 Some Polymers of Substituted Alkenes

Polymer	Monomer	Uses
Polyvinyl chloride (PVC)	$CH_2=CHCl$	Credit cards, soap bottles, insulation, plastic pipe
Saran	$CH_2=CCl_2$ and $CH_2=CHCl$	Packaging film, fibers, tubing
Teflon	$F_2C=CF_2$	Nonstick surfaces for pots and pans
Orlon	$CH_2=CH—C\equiv N$	Fabrics
Polystyrene	$C_6H_5—CH=CH_2$	Foam, plastics, and molded items
Lucite	$CH_2=\underset{\underset{CH_3}{\vert}}{C}—CO_2CH_3$	Coatings, windows, molded items
Natural Polymer		
Natural rubber	$CH_2=\underset{\underset{CH_3}{\vert}}{C}—CH=CH_2$ (Isoprene)	Tires, hoses, boots

$$n\,CH_3—CH=CH_2 \xrightarrow{\text{polymerization}} —\underset{\underset{CH_3}{\vert}}{CH}—CH_2—\underset{\underset{CH_3}{\vert}}{CH}—CH_2—\underset{\underset{CH_3}{\vert}}{CH}—CH_2—\text{etc.}$$

propylene   polypropylene

Many other alkenes as well as dienes polymerize to give useful polymers. One of the polydienes is naturally occurring rubber. Some are mentioned in Table 24.3.

The polyalkenes—often called the *polyolefins* after an older name for alkenes, olefins—are really saturated hydrocarbons once they have formed. So they have all the chemical inertness of alkanes, which is why they are so highly prized for making containers for chemicals, foods, and beverages. But we see in the polyalkenes how sheer molecular size affects *physical* properties. Neither polyethylene nor polypropylene, despite their consisting of nonpolar molecules, dissolves readily in nonpolar solvents, like gasoline, for example. The sheer sizes of polyalkene molecules also makes London forces between them strong enough to give strength to articles made from them. And because polyalkene molecules are extremely long, it is possible to get them organized into overlapping networks that constitute monofilament fibers.

Aromatic Hydrocarbons

Hydrocarbons or their oxygen or nitrogen derivatives that are not aromatic are called *aliphatic* compounds.

The simplest aromatic hydrocarbon, the parent of all the rest, is benzene. Benzene, C_6H_6, is quite unsaturated compared to its nearest saturated relative, cyclohexane, C_6H_{12}. Despite this, benzene does not undergo addition reactions. Let us see why.

As we discussed in Sections 8.6 to 8.8, the benzene molecule has a most unique structure. Its six π electrons are in *delocalized* molecular orbitals. This greatly lowers the energy of the system when compared to the localized picture of the three isolated π bonds implied by either of the two resonance structures for benzene.

resonance structures for benzene   common symbol for benzene

Benzene ring

The energy benzene does *not* have because of this delocalization is its *resonance energy* (or, as it is often called, its *delocalization energy*). Except under the most strenuous reaction conditions of high pressure and temperature, benzene will not give any reaction that disrupts its delocalized π-electron network. To do so, benzene would have to sacrifice the stabilization the network affords. Benzene, instead, undergoes **substitution reactions**, those in which one atom or group in the molecule is replaced by another. For example, benzene reacts with concentrated sulfuric acid (if heated) as follows. (To make our point, we have to use a resonance structure for benzene.)

The pi electron structure of the benzene ring was discussed in Section 8.8.

If benzene were to *add* sulfuric acid, like an alkene, then the following would happen.

product of *adding* sulfuric acid to one double bond

You can see that this addition reaction would break up the three alternating double bond–single bond units that encircle the ring in the resonance structure of benzene. If this happened, the ring-encircling π-electron network of benzene, which gives it its unusual stability, would be broken. So benzene does not add H_2SO_4. It reacts by substitution, instead.

Provided an acid catalyst—usually a Lewis acid catalyst—is present, benzene reacts by substitution with chlorine, bromine, and nitric acid as well as with sulfuric acid. (Recall that Cl_2 and Br_2 readily *add* to alkene double bonds.)

$$C_6H_6 + Cl_2 \longrightarrow \underset{\text{chlorobenzene}}{C_6H_5-Cl} + HCl \qquad \text{(FeCl}_3 \text{ catalyst)}$$

$$C_6H_6 + Br_2 \longrightarrow \underset{\text{bromobenzene}}{C_6H_5-Br} + HBr \qquad \text{(FeBr}_3 \text{ catalyst)}$$

$$C_6H_6 + HNO_3 \longrightarrow \underset{\text{nitrobenzene}}{C_6H_5-NO_2} + H_2O \qquad \text{(H}_2SO_4 \text{ catalyst)}$$

The benzene ring strongly resists attack by any oxidizing agent, because such reactions would also break up the π-electron delocalization that stabilizes the system. Benzene gives no reactions with bases or active metals.

Other Aromatic Compounds   The presence of the benzene ring puts any substance into a very broad organic family known as the **aromatic compounds**. Aromatic compounds overlap all families organized around functional groups. They need not be just hydrocarbons. The name *aromatic* survives from the time when all known examples of aromatic compounds had pleasant aromas, like oil of wintergreen and vanillin. But odor is no longer a criterion. Aspirin, with no odor, is also an aromatic compound, because it has the benzene ring.

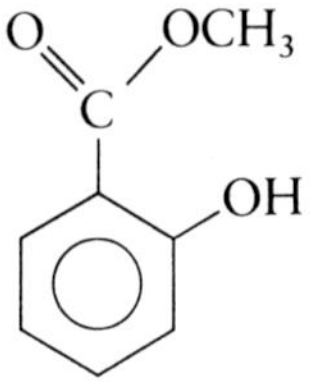

oil of wintergreen 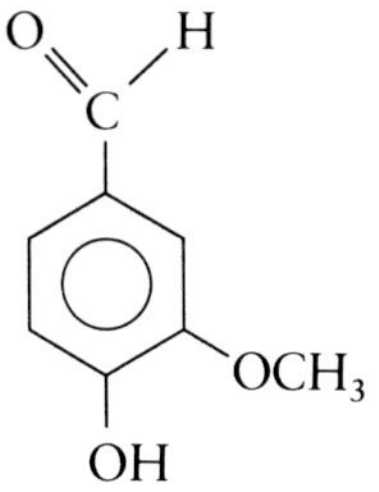   vanillin 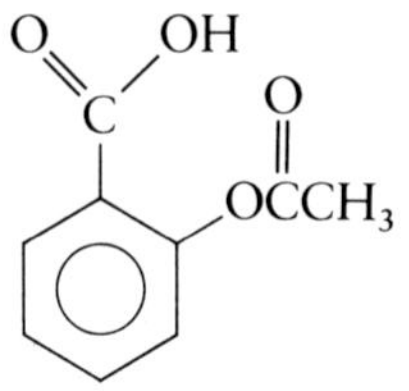   aspirin

Note the variety of functional groups in these substances.

24.4 ORGANIC AND BIOCHEMICAL COMPOUNDS OF OXYGEN

Important organic compounds of oxygen contain either the OH group or the $C=O$ group or both.

The functional groups we will briefly survey in this section are those in alcohols, ethers, aldehydes, ketones, carboxylic acids, and esters. The discussion will involve both specific examples and general family structures in which we exploit the symbol R for a hydrocarbon group. Since carbohydrates and lipids (which include edible fats and oils) are organic compounds of oxygen, we will briefly describe them here as well.

Alcohols

An alcohol is any compound with an OH group attached to a carbon with three other groups also attached by *single bonds*.

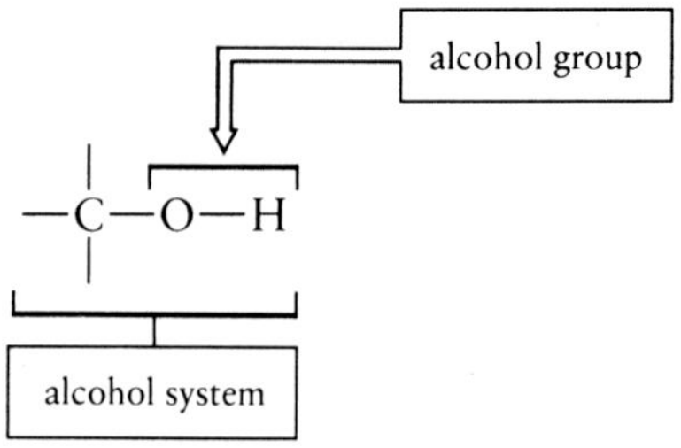

alcohol group

$$-\overset{|}{\underset{|}{C}}-O-H$$

alcohol system

$$R-O-H$$
alcohols

The four simplest (and most common) alcohols are the following. (We will place the IUPAC names in parentheses, but we will not go into the IUPAC rules. Perhaps you can figure out what they most likely are.)

Isopropyl alcohol is sometimes sold as an additive to put in a vehicle's gas tank to prevent gasline freeze. It helps any water to dissolve in the gasoline.

$$CH_3OH \qquad CH_3CH_2OH \qquad CH_3CH_2CH_2OH \qquad CH_3\overset{|}{\underset{\underset{OH}{|}}{C}HCH_3}$$

methyl alcohol   ethyl alcohol   propyl alcohol   isopropyl alcohol
(methanol)   (ethanol)   (1-propanol)   (2-propanol)
bp 65 °C   bp 78.5 °C   bp 97 °C   bp 82 °C

Ethyl alcohol is the alcohol in beverages. It is also the alcohol added to gasoline to make "gasohol."

In the body, methyl alcohol and ethyl alcohol have radically different effects. Methyl alcohol causes blindness and often death. Ethyl alcohol causes loss of coordination and inhibitions and, when abused, ruins the liver and the brain. Isopropyl alcohol is often the "alcohol" present (with some water) in rubbing alcohol.

The OH group is a "waterlike" group, and it participates in hydrogen bonding. Alcohol molecules, therefore, attract each other much more strongly than can alkane molecules. So alcohols have boiling points considerably higher than those of alkanes of comparable formula weights. When an alcohol group is present, a substance is also more soluble in water than hydrocarbons because water and alcohol molecules hydrogen-bond to each other. But when the hydrocarbon part of an alcohol carries five or more carbons, the solubility of the alcohol in water is quite low. Because of their hydrocarbon groups, alcohols are generally soluble in nonpolar solvents.

We have gone into this much detail to illustrate how chemists learn to "read" the most likely physical properties of a compound by assessing the contributions to these properties made by different functional and nonfunctional groups. The chemical properties of organic compounds are similarly "read." For example, the oxygen atom in the OH group makes alcohols Brønsted bases, and so they react with concentrated, strong acids in a proton-transfer reaction. Thus, ethyl alcohol dissolves in sulfuric acid, and some of its molecules enter the following equilibrium.

$$CH_3CH_2OH + H_2SO_4 \rightleftharpoons CH_3CH_2OH_2^+ + HSO_4^-$$

The organic cation is nothing more than an ethyl derivative of a hydronium ion. And remember that all three bonds in H_3O^+ are weak. In the same way, the extra proton on O in $CH_3CH_2OH_2^+$ creates three weak bonds from O, *including the bond to the ethyl group*. This is why alcohols can be made to lose their OH groups in the presence of an acid catalyst. The OH group becomes part of a water molecule as the alcohol also loses an H from a carbon atom. The result is the formation of a carbon–carbon double bond. The overall reaction is an elimination reaction, another kind of organic reaction. We will illustrate this kind of reaction as follows.

$$\underset{\substack{|\;\;\;\;|\\ \text{H} \;\; \text{OH}\\ \text{ethyl alcohol}}}{CH_2CH_2} \xrightarrow[\text{heat}]{\text{acid catalyst}} \underset{\text{ethene}}{CH_2{=}CH_2} + H_2O$$

$$\underset{\substack{\;\;\;\;\;\;\;|\;\;\;\;|\\ \;\;\;\;\;\;\;\text{H} \;\; \text{OH}\\ \text{propyl alcohol}}}{CH_3CH{-}CH_2} \xrightarrow[\text{heat}]{\text{acid catalyst}} \underset{\text{propene}}{CH_3CH{=}CH_2} + H_2O$$

Ethers

Molecules with two alkyl groups bonded to one oxygen are ethers.

$$\underset{\substack{\text{dimethyl ether}\\ \text{bp}\,-23\,°C}}{CH_3{-}O{-}CH_3} \quad \underset{\substack{\text{diethyl ether}\\ \text{bp}\,34.5\,°C}}{CH_3CH_2{-}O{-}CH_2CH_3} \quad \underset{\substack{\text{methyl ethyl ether}\\ \text{bp}\,11\,°C}}{CH_3{-}O{-}CH_2CH_3} \quad \underset{\text{ethers}}{R{-}O{-}R'}$$

Lacking OH groups to set up hydrogen-bonding forces between their molecules, ethers have lower boiling points than alcohols of the same formula weights. As we noted earlier, ethyl alcohol (bp 78.5 °C) and dimethyl ether (bp −23 °C) are isomers, but boil about 100 °C apart.

Aldehydes and Ketones

The carbon–oxygen double bond, C=O, or the carbonyl group, is a polar group present in several families, and it uniquely defines none. What is attached to the carbon

This is just like the reaction of H_2SO_4 with water:

$$H_2O + H_2SO_4 \rightleftharpoons H_3O^+ + HSO_4^-$$

The organic cation collapses by releasing H^+ as the electron pair of the C—H bond swings in to become the π bond.

$$\underset{\substack{|\\ \text{H}}}{CH_2{-}CH_2^+} \longrightarrow CH_2{=}CH_2 + H^+$$

Catalysts and special conditions are often organized around the arrow of an equation.

Diethyl ether was once the chief anesthetic used in surgery.

atom in $C{=}O$ determines the specific family. To be an aldehyde, the carbonyl group must hold an H atom plus another hydrocarbon group (or a second H). To be a ketone, the $C{=}O$ must hold two hydrocarbon groups.

$$
\underset{\substack{\text{carbonyl} \\ \text{group}}}{\diagdown C \diagup^{\displaystyle \overset{O}{\|}}} \qquad
\underset{\substack{\text{aldehyde} \\ \text{group}}}{-\overset{\displaystyle \overset{O}{\|}}{C}-H} \qquad
\underset{\text{aldehydes}}{(H)R-\overset{\displaystyle \overset{O}{\|}}{C}-H} \qquad
\underset{\substack{\text{ketone} \\ \text{system}}}{-\overset{|}{\underset{|}{C}}-\overset{\displaystyle \overset{O}{\|}}{C}-\overset{|}{\underset{|}{C}}-} \qquad
\underset{\text{ketones}}{R-\overset{\displaystyle \overset{O}{\|}}{C}-R'}
$$

Examples of aldehydes and ketones include formaldehyde, the simplest aldehyde — we will continue to use common names and place IUPAC names in parentheses — and acetone, the simplest ketone and the solvent in fingernail polish remover.

$$
\underset{\substack{\text{formaldehyde} \\ \text{(methanal)} \\ \text{bp} -21\,°C}}{H-\overset{\displaystyle \overset{O}{\|}}{C}-H} \qquad
\underset{\substack{\text{acetaldehyde} \\ \text{(ethanal)} \\ \text{bp} \ 21\,°C}}{CH_3-\overset{\displaystyle \overset{O}{\|}}{C}-H} \qquad
\underset{\substack{\text{butyraldehyde} \\ \text{(butanal)} \\ \text{bp} \ 76\,°C}}{CH_3CH_2CH_2-\overset{\displaystyle \overset{O}{\|}}{C}-H} \qquad
\underset{\substack{\text{acetone} \\ \text{(propanone)} \\ \text{bp} \ 56\,°C}}{CH_3-\overset{\displaystyle \overset{O}{\|}}{C}-CH_3}
$$

Lacking OH groups, aldehydes and ketones boil lower than alcohols of comparable formula weights. Yet, the carbonyl group is a polar group, and it helps to make these compounds much more soluble in water than hydrocarbons.

Since the carbonyl group has a double bond, it adds hydrogen (but it does not give stable addition products with other reactants that add to the alkene double bond). We can illustrate the hydrogenations of acetaldehyde and acetone as follows.

$$
\underset{\text{acetaldehyde}}{CH_3-\overset{\displaystyle \overset{O}{\|}}{C}-H} \ + H{-}H \xrightarrow[\text{catalyst}]{\text{heat, pressure}} \underset{\text{ethyl alcohol}}{CH_3-\overset{\displaystyle \overset{O-H}{|}}{\underset{\displaystyle \underset{H}{|}}{C}}-H} \qquad \text{or} \qquad CH_3CH_2OH
$$

$$
\underset{\text{acetone}}{CH_3-\overset{\displaystyle \overset{O}{\|}}{C}-CH_3} + H{-}H \xrightarrow[\text{catalyst}]{\text{heat, pressure}} \underset{\text{isopropyl alcohol}}{CH_3-\overset{\displaystyle \overset{O-H}{|}}{\underset{\displaystyle \underset{H}{|}}{C}}-CH_3} \qquad \text{or} \qquad \overset{\displaystyle \overset{OH}{|}}{CH_3CHCH_3}
$$

Aldehydes and ketones are in separate families because of their remarkably different behavior toward oxidizing agents. Aldehydes are about the most easily oxidized of all organic compounds, but ketones strongly resist oxidation. Potassium dichromate, for example, easily oxidizes acetaldehyde to acetic acid as follows.

$$
\underset{\text{acetaldehyde}}{3CH_3\overset{\displaystyle \overset{O}{\|}}{C}H} + K_2Cr_2O_7 + 4H_2SO_4 \longrightarrow \underset{\text{acetic acid}}{3CH_3\overset{\displaystyle \overset{O}{\|}}{C}OH} + Cr_2(SO_4)_3 + K_2SO_4 + 4H_2O
$$

As you can see, the oxidation affects the H on the carbonyl group of the aldehyde and changes it to the OH group of another family, the carboxylic acids. Ketones lack this H, so they cannot be oxidized this way.

Carbohydrates

Carbohydrates are naturally occurring polyhydroxyaldehydes or polyhydroxyketones — collectively called monosaccharides — or they are compounds that react with water to give monosaccharides. The carbohydrates include common sugar — sucrose — as well as starch and cellulose. The most common monosaccharide is glucose, a pentahydroxyaldehyde, and probably the most widely occurring structural unit in the entire living world. It is the chief carbohydrate in blood. Fructose, a pentahydroxyketone, is present with glucose in honey.

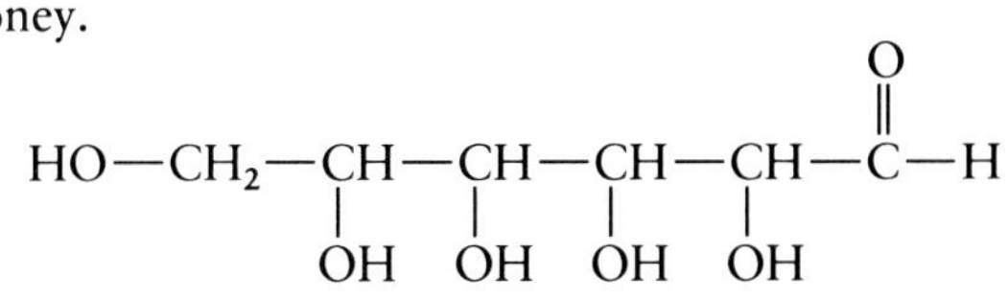

$$HO-CH_2-\underset{\underset{OH}{|}}{CH}-\underset{\underset{OH}{|}}{CH}-\underset{\underset{OH}{|}}{CH}-\underset{\underset{OH}{|}}{CH}-\overset{\overset{O}{\|}}{C}-H$$

glucose (open-chain form)
a polyhydroxyaldehyde

$$HO-CH_2-\underset{\underset{OH}{|}}{CH}-\underset{\underset{OH}{|}}{CH}-\underset{\underset{OH}{|}}{CH}-\overset{\overset{O}{\|}}{C}-CH_2-OH$$

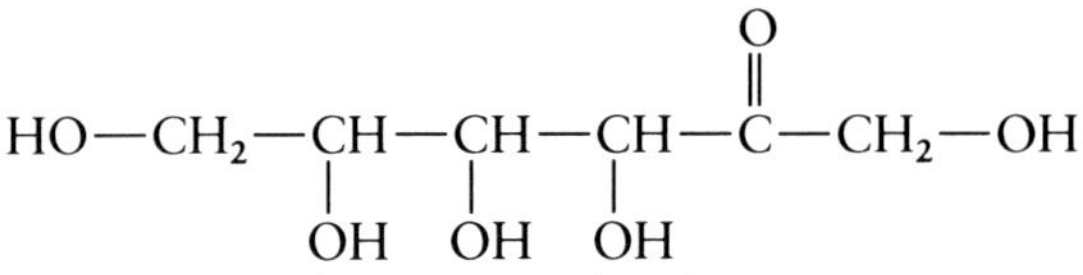

fructose (open-chain form)
a polyhydroxyketone

The structures of carbohydrates are complicated by the ability of their molecules to exist in more than one form. Glucose, for example, can be isolated as two cyclic forms and one open-chain form. See Figure 24.2. In water, all three forms coexist in equilibrium, with the open-chain form making up less than 0.1% of the glucose units. But we know from Le Châtelier's principle that the equilibrium can shift to supply more of any compound in an equilibrium, so glucose gives the chemical reactions of a polyhydroxyaldehyde. With an aldehyde group available, glucose is easily oxidized, for example. And both animals and plants use the chemical energy from the oxidation of glucose to run themselves.

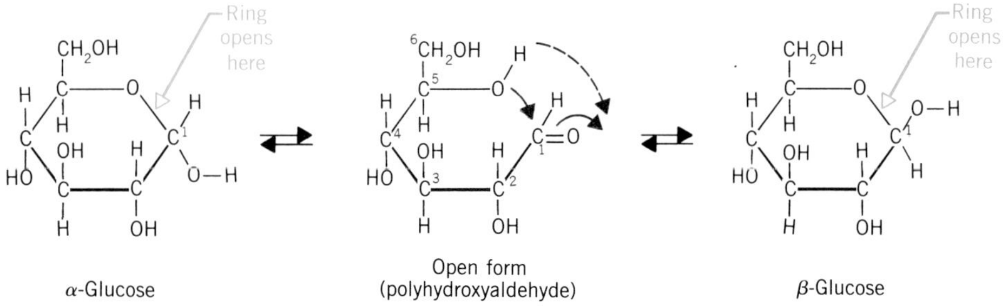

α-Glucose — Open form (polyhydroxyaldehyde) — β-Glucose

Disaccharides are carbohydrates that hydrolyze to give two monosaccharides. Sucrose (cane sugar or beet sugar) is an example, and its hydrolysis gives glucose and fructose. Lactose, or milk sugar, hydrolyzes to glucose and galactose, an isomer of glucose. The monosaccharide units occur in cyclic forms in disaccharides, and they are linked by etherlike bridges. When we digest disaccharides, specific enzymes catalyze the hydrolysis of these bridges.

Polysaccharides are carbohydrates that involve thousands of monosaccharide units linked to each other by oxygen bridges. They include starch, glycogen, and cellulose. The complete hydrolyses of all three yield only glucose, so their structural differences involve details of the oxygen bridges (which we will not go into). Starch, found often in seeds and tubers (e.g., potatoes), is the way plants store glucose units for energy needs. Cellulose is a major constituent of the walls of plant cells, and it makes up nearly 100% of cotton. Glycogen, structurally like starch, is the way that animals store glucose units for energy.

With so many OH groups in glucose and fructose, you can see why they are high-melting solids and very soluble in water.

FIGURE 24.2 The three forms of glucose. The curved arrows in the open form show how bonds become reoriented as the open form undergoes ring closure. The ring can close with the OH at position 1 either down (as in α-glucose) or up (as in β-glucose); it depends on which "face" of the C=O group happens to be exposed to the O at position 5 when closure starts. Free rotation about the C–C bond between positions 1 and 2 creates these options.

To hydrolyze here means to react with water so as to break bonds in the organic molecule and create new functional groups.

Humans have no digestive enzyme to handle the hydrolysis of cellulose, but do have the enzyme for starch.

Carboxylic Acids

The word *carboxyl* is coined from "*carb*onyl + hydr*oxyl*," and the carboxyl group is present in carboxylic acids.

$$\begin{array}{c} O \\ \parallel \\ -C-O-H \end{array} \quad \text{(often written as } -CO_2H \text{ or as } -COOH)$$

carboxyl group

We have often used acetic acid as a typical weak acid. And we have mentioned formic acid as being the agent that causes the pain from an ant sting. Butyric acid causes the particularly vile odor from rancid butter.

$$H-\overset{\displaystyle O}{\overset{\parallel}{C}}-OH \qquad CH_3-\overset{\displaystyle O}{\overset{\parallel}{C}}-OH \qquad CH_3CH_2CH_2-\overset{\displaystyle O}{\overset{\parallel}{C}}-OH \qquad R-\overset{\displaystyle O}{\overset{\parallel}{C}}-OH$$

formic acid (methanoic acid) bp 101 °C   acetic acid (ethanoic acid) bp 118 °C   butyric acid (butanoic acid) bp 164 °C   carboxylic acids

The 5-, 6-, and 8-carbon carboxylic acids provide much of the pungency in stale locker rooms.

Since molecules of carboxylic acids have two polar groups, C=O and OH, which can engage in hydrogen bonding, these acids have relatively high boiling points. They are *associated* compounds, as we illustrated in Section 13.11.

The OH group of carboxylic acids, unlike the OH group in either water or alcohols, is a proton donor. The polar carbonyl group polarizes the O—H bond and puts a larger $\delta+$ charge on H. This makes it easier for the H to transfer away as H^+ than it can without the C=O group. The carboxylic acids are not strong acids, as we know, but all molecules with the carboxyl group neutralize strong base.

$$R-\overset{\displaystyle O}{\overset{\parallel}{C}}-OH + NaOH \longrightarrow R-\overset{\displaystyle O}{\overset{\parallel}{C}}-O^- + Na^+ + H_2O$$

The anions of all carboxylic acids are good Brønsted bases, because their conjugate acid forms are weak acids. Carboxylic acids can be converted to members of the ester family by reacting with alcohols, which we will study next.

Esters

When a carboxylic acid has its H of the O—H group replaced by a hydrocarbon group, like the ethyl group, the compound no longer is acidic. It is now an ester, a compound with the ester group.

$$-\overset{\displaystyle O}{\overset{\parallel}{C}}-O-\overset{\displaystyle |}{\underset{|}{C}}- \qquad (H)R-\overset{\displaystyle O}{\overset{\parallel}{C}}-O-R' \qquad CH_3-\overset{\displaystyle O}{\overset{\parallel}{C}}-O-CH_2CH_2\overset{\displaystyle CH_3}{\overset{|}{C}}HCH_3$$

ester group   ester family   a typical ester—oil of banana (isopentyl acetate)

Esters of low formula weights have some of the most pleasant fragrances of all compounds. The delicate flavors and odors of oranges, raspberries, bananas, apricots, and other fruits are caused by various esters.

Esters can be made in the laboratory by heating a carboxylic acid with an alcohol (in the presence of an acid catalyst). For example, the following equilibrium is established when butyric acid and ethyl alcohol are boiled together with a trace of acid catalyst.

$$CH_3CH_2CH_2\overset{\displaystyle O}{\overset{\displaystyle \|}{C}}-O-H + H-O-CH_2CH_3 \underset{heat}{\overset{H^+}{\rightleftharpoons}} CH_3CH_2CH_2\overset{\displaystyle O}{\overset{\displaystyle \|}{C}}-O-CH_2CH_3 + H_2O$$

butyric acid   ethyl alcohol   ethyl butyrate
                                                                    (pineapple fragrance)

To favor the formation of the ester, an excess of ethyl alcohol is generally used. The stress that this places on the equilibrium causes a shift in favor of the products. We see in this, of course, a way to hydrolyze an ester — to make it react with water and make the ester deliver back its carboxylic acid and alcohol components. When an ester is heated with a large molar excess of water (with an acid catalyst), the parent acid and alcohol are recovered. For example,

$$CH_3\overset{\displaystyle O}{\overset{\displaystyle \|}{C}}-O-CH_2CH_3 + H-OH \overset{heat}{\longrightarrow} CH_3\overset{\displaystyle O}{\overset{\displaystyle \|}{C}}-OH + H-O-CH_2CH_3$$

ethyl acetate   (large excess   acetic   ethyl alcohol
                         used)   acid

The reaction goes essentially to completion because of the large excess of water, so we need not describe this as an equilibrium.

Ester Groups in Polymers—Polyesters   When a *dicarboxylic acid,* one with two carboxyl groups per molecule, is heated with an alcohol, both carboxyl groups can be changed to esters. But suppose that we use not just an ordinary alcohol with one OH group, but an alcohol like ethylene glycol with two OH groups together with a dicarboxylic acid. This sets the stage for making a polymer, a *polyester.* Dacron is an example. It is a polymer of terephthalic acid and ethylene glycol, and we can visualize its formation (at least on paper) along the following lines.

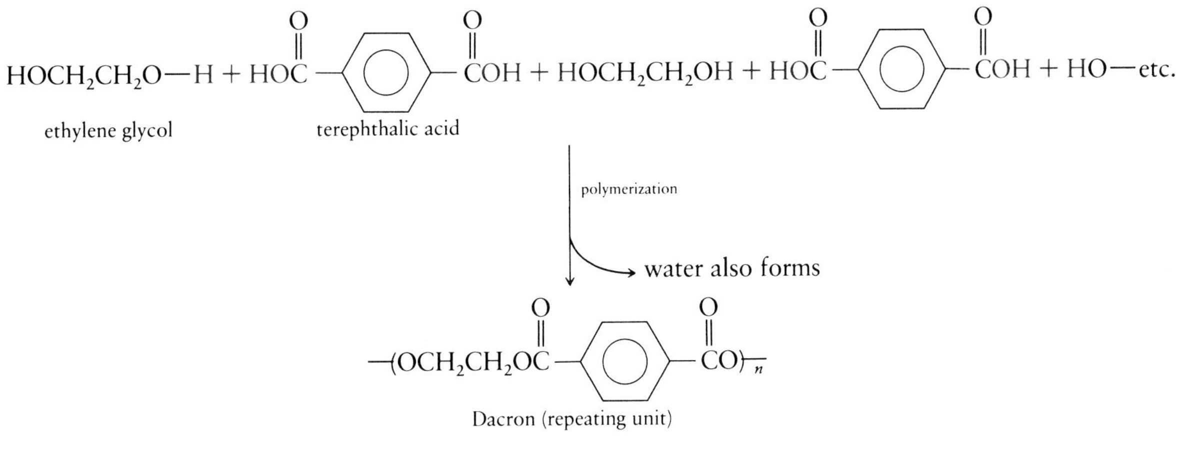

$$HOCH_2CH_2O-H + HO\overset{\displaystyle O}{\overset{\displaystyle \|}{C}}-\hexagon-\overset{\displaystyle O}{\overset{\displaystyle \|}{C}}OH + HOCH_2CH_2OH + HO\overset{\displaystyle O}{\overset{\displaystyle \|}{C}}-\hexagon-\overset{\displaystyle O}{\overset{\displaystyle \|}{C}}OH + HO-\text{etc.}$$

ethylene glycol   terephthalic acid

polymerization

water also forms

$$-(OCH_2CH_2O\overset{\displaystyle O}{\overset{\displaystyle \|}{C}}-\hexagon-\overset{\displaystyle O}{\overset{\displaystyle \|}{C}}O)_{\overline{n}}$$

Dacron (repeating unit)

The actual manufacture of Dacron uses an ester of terephthalic acid, not the free acid.

Dacron molecules are very long, are very symmetrical about the long axis, and have regularly spaced polar groups that help neighboring molecules stick together. It makes an excellent fiber for fabrics. When cast as a thin film, Dacron — now called Mylar — is an exceptionally strong material much prized for use in recording tapes.

Esters in Living Systems — The Lipid Family   Lipids are natural products that tend to dissolve in such nonpolar solvents as diethyl ether or benzene. Lipids, therefore, have molecules that are relatively hydrocarbon-like and nonpolar, and they are all insoluble in water.

The lipid family is huge and diverse. It includes cholesterol and sex hormones, like testosterone. You can see from their structures how largely hydrocarbon-like they are and why the blood (mostly an aqueous fluid) would not be able to hold much cholesterol in solution. The precipitation of excess cholesterol in blood capillaries is associated with heart disease.

Testosterone is a male sex hormone.

cholesterol

testosterone

Lard is hog fat. Tallow is beef fat.

The lipids also include the edible fats and oils in our diets — substances such as olive oil, corn oil, peanut oil, butterfat, lard, and tallow. These are triacylglycerols — esters between glycerol, an alcohol with three OH groups, and any three of several long-chain carboxylic acids called fatty acids. Several important fatty acids have alkene groups. Vegetable oils have more alkene double bonds per molecule than animal fats, and so are said to be *polyunsaturated* materials. The long hydrocarbon chains in the triacylglycerols are what make these substances insoluble in water.

Triacylglycerols are also known as *triglycerides*.

Figure 24.3 describes the structural features of triacylglycerols further. It also shows the reaction — hydrolysis of ester groups — by which we digest the triacylglycerols. Powerful detergents in the bile, a fluid that empties into the digestive tract from the gall bladder, helps to dissolve the triacylglycerols and aid in their digestion.

24.5 ORGANIC AND BIOCHEMICAL COMPOUNDS OF NITROGEN

Amines, amides, amino acids, and proteins are important organic nitrogen compounds.

Nitrogen, like oxygen, introduces polarity into an organic molecule, so organic nitrogen compounds have a rich and varied chemistry.

Amines

The amines are organic derivatives of ammonia in which one, two, or three hydrocarbon groups have replaced hydrogens. For example,

Low-formula-weight amines have "fishy" odors.

$$H-\underset{\underset{H}{|}}{\overset{\overset{H}{|}}{N}}-H$$
ammonia
bp −33.4 °C

$$CH_3-\underset{\underset{H}{|}}{\overset{\overset{H}{|}}{N}}-H$$
methylamine
bp −8 °C

$$CH_3-\underset{\underset{H}{|}}{\overset{\overset{H}{|}}{N}}-CH_3$$
dimethylamine
bp 8 °C

$$CH_3-\underset{\underset{CH_3}{|}}{\overset{\overset{CH_3}{|}}{N}}-CH_3$$
trimethylamine
bp 3 °C

(The hydrocarbon groups need not be alike.)

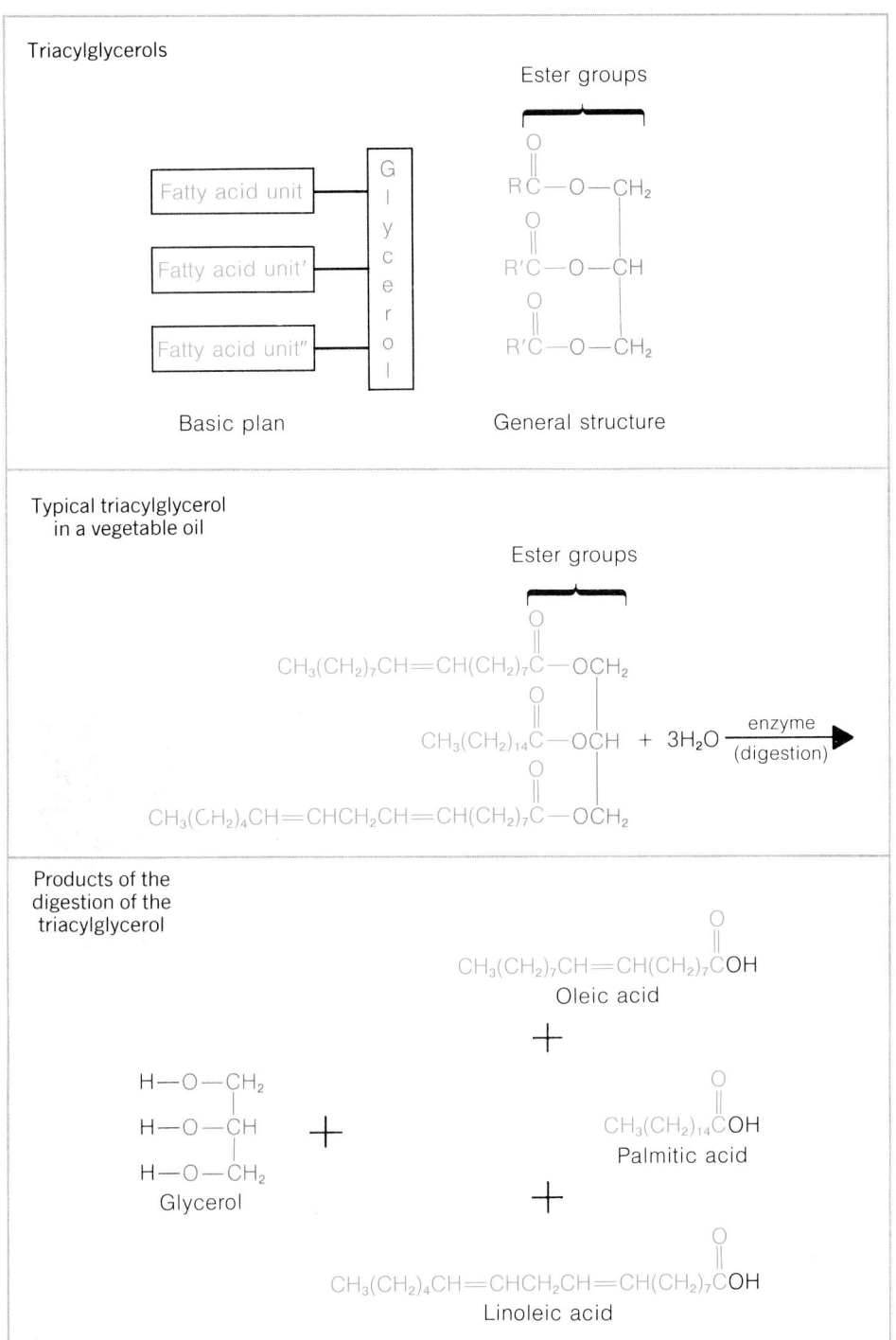

FIGURE 24.3 Triacylglycerols. The structural features of all triacylglycerols are indicated at the top. A specific example is given in the middle, and the products of its hydrolysis (digestion) are shown at the bottom.

As long as the N atom in an amine has one H, the molecule can hydrogen-bond either to like neighbors or to water. But the N—H bond is not as polar as the O—H bond, so amines boil at lower temperatures than alcohols of comparable formula weights. In trialkylamines, like trimethylamine, no H is on N, so intermolecular hydrogen bonding is not possible, and the compound's boiling point does not keep pace with its larger formula weight. Compare, for example, dimethylamine (bp 8 °C) with trimeth-

CH_3CH_2OH, bp 78.5 °C
$CH_3CH_2NH_2$, bp 17 °C
$CH_3CH_2CH_3$, bp −42 °C

ylamine (bp 3 °C), a larger molecule. (Thus we see again how we can learn to "read" properties from structures.)

Amines of low formula weight are soluble in water. Hydrogen bonding facilitates this, of course, but when the total carbon content of an amine molecule reaches six to seven carbons, the substance has become too hydrocarbon-like and the amine becomes relatively insoluble in water.

Like ammonia, the amines are Brønsted bases and accept protons from acids, as we noted in Section 24.1. To give another illustration,

Salts of amines with acids tend to be water-soluble because they are ionic.

$$CH_3CH_2-\underset{\underset{\text{methylethylamine}}{}}{\overset{\overset{H}{|}}{N}}-CH_3(aq) + HCl(aq) \longrightarrow CH_3CH_2-\underset{\underset{\text{methylethylammonium ion}}{}}{\overset{\overset{H}{|}}{\underset{\underset{H}{|}}{N^+}}}-CH_3(aq) + Cl^-(aq)$$

Substituted ammonium ions, like NH_4^+ itself, are weak Brønsted acids. They can all neutralize strong base. For example,

$$\underset{\underset{\text{ion}}{\text{methylammonium}}}{CH_3NH_3^+(aq)} + OH^-(aq) \longrightarrow \underset{\text{methylamine}}{CH_3NH_2(aq)} + H_2O$$

Amino Acids — Building Blocks of Polypeptides   The amino group, NH_2, and the carboxyl group are joined to the same carbon atom in the molecules of several substances called *α-amino acids,* or just amino acids for short.

Just as monosaccharides are the building blocks for polysaccharides, so the amino acids are the building units for polypeptides, huge molecules that make up 99 to 100% of the molecules of all proteins.

Roughly 20 amino acids are known to occur naturally, each differing in the identity of the group *G*, also attached to the α position. (Since not all such groups are purely hydrocarbon, we use *G*, instead of R in the general formula of α-amino acids.)

The general formula of amino acids that we gave is incorrect in one important respect. When a molecule has both a proton-accepting site and a proton-donating group, self-neutralization occurs. The proton goes from the carboxyl group to the amino group, and in solution amino acids exist largely in dipolar ionic forms.

Because their molecules are saltlike, amino acids have high melting points and are relatively soluble in water.

$$^+NH_3-\underset{\underset{G}{|}}{CH}-\overset{\overset{O}{\|}}{C}-O^-$$

dipolar ionic form of α-amino acids

What joins amino acid units in polypeptides are amide linkages, which we will study next, before returning to polypeptides and proteins.

Amides

When a carbonyl group is directly attached to nitrogen, the result is an *amide.*

$$\underset{\substack{\text{amide} \\ \text{system}}}{-\overset{\displaystyle O}{\overset{\|}{C}}-\overset{|}{N}-} \qquad \underset{\substack{\text{formamide} \\ \text{(methanamide)}}}{H-\overset{\displaystyle O}{\overset{\|}{C}}-NH_2} \qquad \underset{\substack{\text{acetamide} \\ \text{(ethanamide)}}}{CH_3-\overset{\displaystyle O}{\overset{\|}{C}}-NH_2} \qquad \underset{\text{urea}}{NH_2-\overset{\displaystyle O}{\overset{\|}{C}}-NH_2}$$

The remaining H atoms on nitrogen of these examples can be replaced by hydrocarbon groups. Such compounds are also amides. For example,

$$\underset{\text{N-methylacetamide}}{CH_3-\overset{\displaystyle O}{\overset{\|}{C}}-NH-CH_3} \qquad \underset{\text{N,N-dimethylacetamide}}{CH_3-\overset{\displaystyle O}{\overset{\|}{C}}-\overset{\overset{\displaystyle CH_3}{|}}{N}-CH_3}$$

The big change in chemical properties in moving from amines to amides is the loss of basicity. Amides are neutral in an acid/base sense. The carbonyl group, being polar and electron withdrawing, drains too much electron density from N to allow it to accept a proton anymore.

Amides can be thought of as forming as follows, where water splits out between a carboxyl group and an amino group. This is not energetically easy to accomplish, so amides are generally made indirectly, but the overall result is the same. (And some amides can be made directly as follows, that is, by heating an acid with ammonia.)

$$\underset{\text{acetic acid}}{CH_3-\overset{\displaystyle O}{\overset{\|}{C}}-OH} + \underset{\text{ammonia}}{H-NH_2} \xrightarrow{\text{heat strongly}} \underset{\text{acetamide}}{CH_3-\overset{\displaystyle O}{\overset{\|}{C}}-NH_2} + H-OH$$

Amides, like esters, can be hydrolyzed. The reaction is the reverse of the one just written, except that an excess of water is used. The hydrolysis of the amide linkage is the chemistry of the digestion of proteins.

Polypeptides and Proteins

The amide bond is called the *peptide bond* in protein chemistry. The steps by which a living organism makes peptide bonds to link amino acid units together are several and complex. (A significant part of the chemistry of heredity deals with these steps, which we will survey later in this chapter.) We will just give the overall result, showing how the two simplest α-amino acids, glycine and alanine, are related to the simplest kind of polypeptide, called a *dipeptide* because it has two amino acid units. The dashed-line arrow signifies several steps.

$$\underset{\text{glycine}}{{}^+NH_3CH_2C\overset{\displaystyle O}{\overset{\|}{}}-O^-} + \underset{\text{alanine}}{H-\overset{\overset{\displaystyle H}{\diagdown}\,{}^+}{\underset{\diagup}{N}}\overset{\displaystyle O}{\underset{\underset{\displaystyle CH_3}{|}}{\overset{\|}{C}}HCO^-} \dashrightarrow \underset{\text{a dipeptide}}{{}^+NH_3CH_2C\overset{\displaystyle O}{\overset{\|}{}}-NH\underset{\underset{\displaystyle CH_3}{|}}{C}H\overset{\displaystyle O}{\overset{\|}{C}}O^-} + H_2O$$

Of course, there is no reason why we could not picture the roles reversed so that alanine acts at its carboxyl end and glycine at its amino end. This results in a different dipeptide (but an isomer of the first).

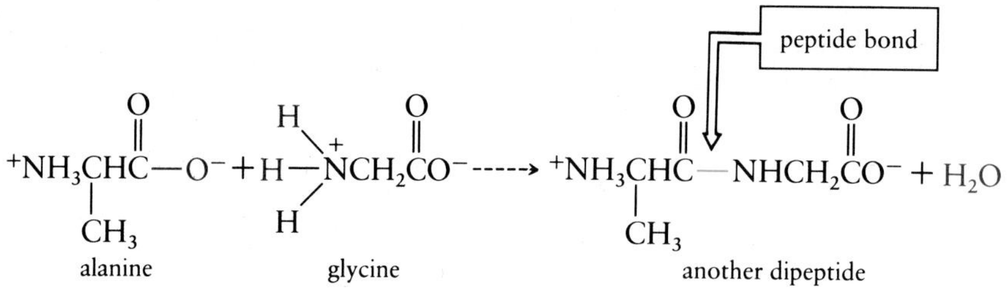

$$\underset{\text{alanine}}{\overset{O}{\underset{CH_3}{^+NH_3CHC}}-O^-} + \underset{\text{glycine}}{\overset{H}{\underset{H}{^+H-NCH_2CO^-}}} \longrightarrow \underset{\text{another dipeptide}}{\overset{O}{\underset{CH_3}{^+NH_3CHC}}-\overset{O}{NHCH_2CO^-}} + H_2O$$

peptide bond

If we think of the two amino acids as being two letters, like W and A, of a polypeptide "alphabet," then we can liken the formation of the two isomeric dipeptides to the formation of two isomeric words, WA and AW.

Notice that each dipeptide has, like any amino acid, an NH_3^+ group at one end and a CO_2^- group at the other. Thus either dipeptide could participate in the formation of another peptide bond *at either end* with a third amino acid such as phenylalanine, another α-amino acid present in proteins. And we will show how three-letter symbols, like Gly for glycine, Ala for alanine, and Phe for phenylalanine, can be used to represent amino acid units.

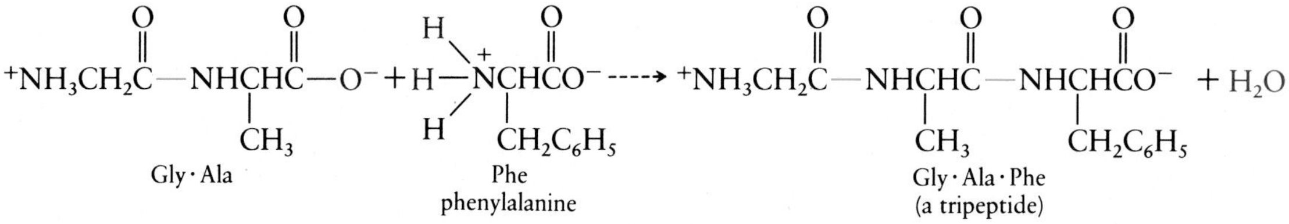

$$\underset{\text{Gly·Ala}}{\overset{O}{^+NH_3CH_2C}-\overset{O}{\underset{CH_3}{NHCHC}}-O^-} + \underset{\substack{\text{Phe}\\\text{phenylalanine}}}{\overset{H}{\underset{H}{^+H-N}}\overset{O}{\underset{CH_2C_6H_5}{CHCO^-}}} \longrightarrow \underset{\substack{\text{Gly·Ala·Phe}\\\text{(a tripeptide)}}}{\overset{O}{^+NH_3CH_2C}-\overset{O}{\underset{CH_3}{NHCHC}}-\overset{O}{\underset{CH_2C_6H_5}{NHCHCO^-}}} + H_2O$$

The product is a tripeptide, Gly·Ala·Phe, because it consists of three amino acid units. The one shown is only one of six possible tripeptides that involves these three different amino acids. The set of all possible sequences for a tripeptide made from glycine, alanine, and phenylalanine is as follows:

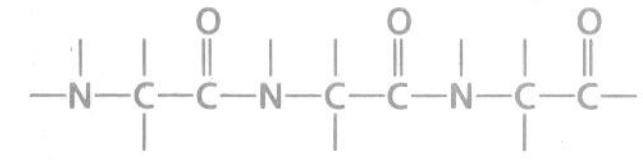

Gly·Ala·Phe   Ala·Gly·Phe   Phe·Gly·Ala
Gly·Phe·Ala   Ala·Phe·Gly   Phe·Ala·Gly

Each of these tripeptides still has groups at each end, $^+NH_3-$ and $-CO_2^-$, that can interact with still another amino acid to make a tetrapeptide. And this product would still have the end groups from which the chain could be extended still further. You can see how a repetition of this many hundreds of times can produce a long polymer, a polypeptide, in which the backbone would have the following structure.

All six tripeptides have the same "backbone":

$$-\overset{|}{\underset{|}{N}}-\overset{|}{\underset{|}{C}}-\overset{O}{C}-\overset{|}{\underset{|}{N}}-\overset{|}{\underset{|}{C}}-\overset{O}{C}-\overset{|}{\underset{|}{N}}-\overset{|}{\underset{|}{C}}-\overset{O}{C}-$$

The open sites at the α positions hold side chains in specific sequences in each polypeptide.

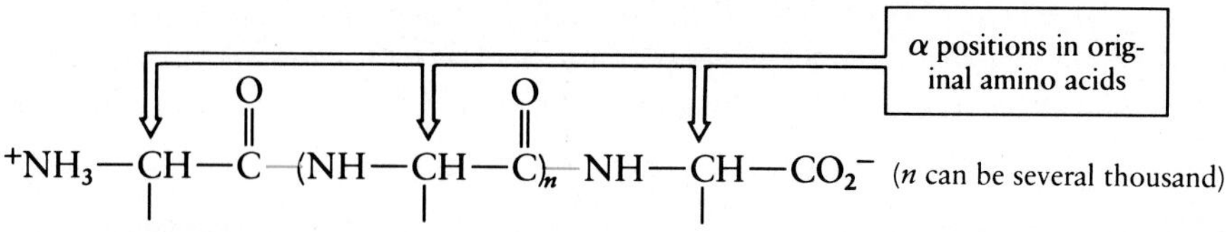

$$^+NH_3-\overset{|}{\underset{|}{CH}}-\overset{O}{C}-(NH-\overset{|}{\underset{|}{CH}}-\overset{O}{C})_n-NH-\overset{|}{\underset{|}{CH}}-CO_2^- \quad (n \text{ can be several thousand})$$

α positions in original amino acids

Each α position bears a specific group, such as the CH_3 group of alanine or the $C_6H_5CH_2$ group of phenylalanine, or any of the other 20 amino acids. All polypeptides are alike in having the nitrogen–carbon–carbonyl repeating system. Polypeptides differ in the lengths of their chains and in the identities and the orders in which the side-chain G groups are strung out on the chains.

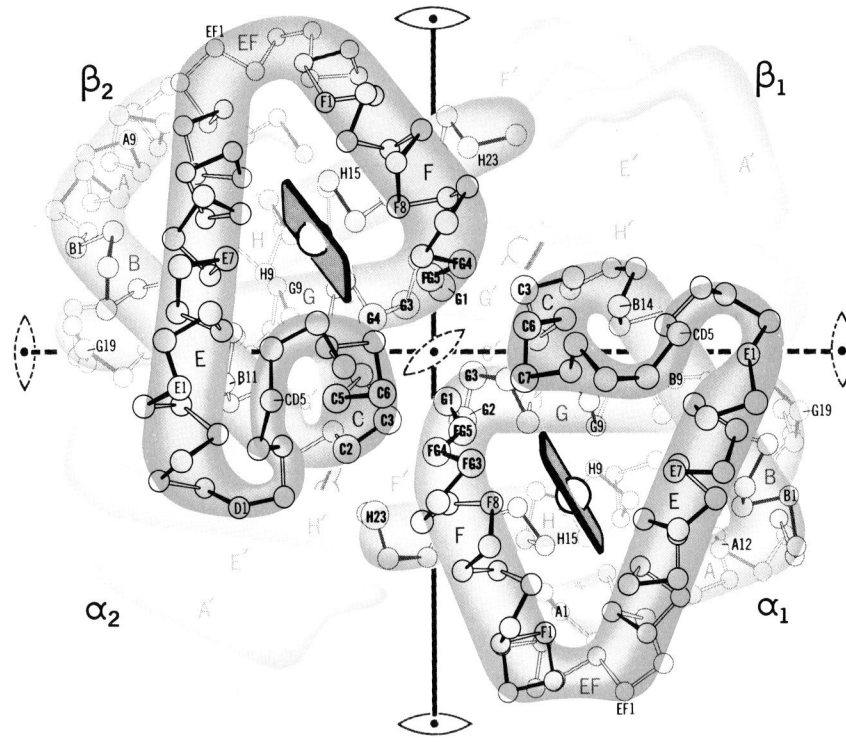

Many proteins consist of single polypeptides. But most proteins involve not just two or more polypeptides associated together but also an organic group and frequently a metal ion. Thus one molecule of hemoglobin, the carrier of oxygen in the blood, is made of four polypeptides—two similar pairs—and one molecule of the organic compound that causes the red color, heme. (See Figure 24.4.) Heme, in turn, holds an iron(II) ion. (See Figure 24.5.) The entire package is a protein, so you can see that "protein" and "polypeptide" are not synonyms.

FIGURE 24.5 Heme.

The Importance of Protein Shape Notice in Figure 24.4 that each polypeptide unit occurs with its strands coiled and that these coils are kinked and twisted. These final shapes are as critical to the ability of heme to function as anything about its molecular

The coiling and twisting reflect water avoidance and water attraction by hydrophobic and hydrophilic side chains.

architecture. Just the exchange of one side-chain G group by another changes the hemoglobin shape and causes a debilitating condition known as sickle cell anemia.

As we have noted before, all enzymes, the catalysts in cells, are proteins. Should an alien chemical distort the shape of an enzyme molecule, the enzyme loses its catalytic ability. Many of our most dangerous poisons act by rendering enzymes unfit. "Dangerous" means capable of causing death or permanent harm even in very small quantities. And enzymes, being catalysts, are never in large concentrations, so whatever attacks them must be avoided. Even changes in the pH of a cell fluid can alter the shape of a protein, so the body must buffer all its fluids.

Nylon — A Synthetic Polyamide

A number of polymers are made from diamines and dicarboxylic acids. One example is nylon. We can visualize its relationship to its monomers by the following equation, which is intended to convey only the overall result of the formation of this polyamide.

$$\text{etc.} - \overset{\overset{\text{O}}{\|}}{\text{C}} - \text{OH} + \text{H} - \underset{\text{1,6-diaminohexane}}{\text{NH(CH}_2)_6\text{NH}} - \text{H} + \underset{\text{hexanedicarboxylic acid}}{\text{HO} - \overset{\overset{\text{O}}{\|}}{\text{C}}(\text{CH}_2)_4\overset{\overset{\text{O}}{\|}}{\text{C}} - \text{OH}} + \text{H} - \text{NH(CH}_2)_6\text{NH} - \text{H} + \text{HO} - \overset{\overset{\text{O}}{\|}}{\text{C}} - \text{etc.}$$

(water also forms)

$$-[\text{NH(CH}_2)_6\text{NH} - \overset{\overset{\text{O}}{\|}}{\text{C}}(\text{CH}_2)_4\overset{\overset{\text{O}}{\|}}{\text{C}}]_n$$
nylon

A raw polymer is called a resin. When made into a finished article it's referred to as a plastic.

Nylon molecules are very long, are very symmetrical around the long axis and have regularly spaced polar sites — in fact, sites that can engage in chain-to-chain hydrogen bonding. These molecules, in other words, are ideally suited to forming filaments and fibers and nylon, as you well know, is widely used to make fabrics. No known moth or microorganism can digest nylon, so it resists such attacks exceptionally well.

24.6 NUCLEIC ACIDS AND HEREDITY

The sequence of side chains on molecules of DNA controls the sequence of side chains on molecules of the polypeptides in enzymes.

The enzymes of an organism are made under the chemical direction of a family of compounds called the nucleic acids. Both the similarities and the uniqueness of every species as well as every individual member of a species depend on structural features of these compounds.

DNA and RNA

The nucleic acids occur as two broad types: RNA, or ribonucleic acids, and DNA, or deoxyribonucleic acids. DNA is the actual chemical of a gene, the individual unit of heredity, and the chemical basis through which we inherit all our physical and biological characteristics. Our purpose in this section is to study the general structural features of the nucleic acids, how DNA molecules direct their own duplications, and how they guide the assembling of amino acid units into a unique sequence in a polypeptide.

Something like 10,000 to 100,000 genes are needed to account for all of the kinds of proteins in a human body.

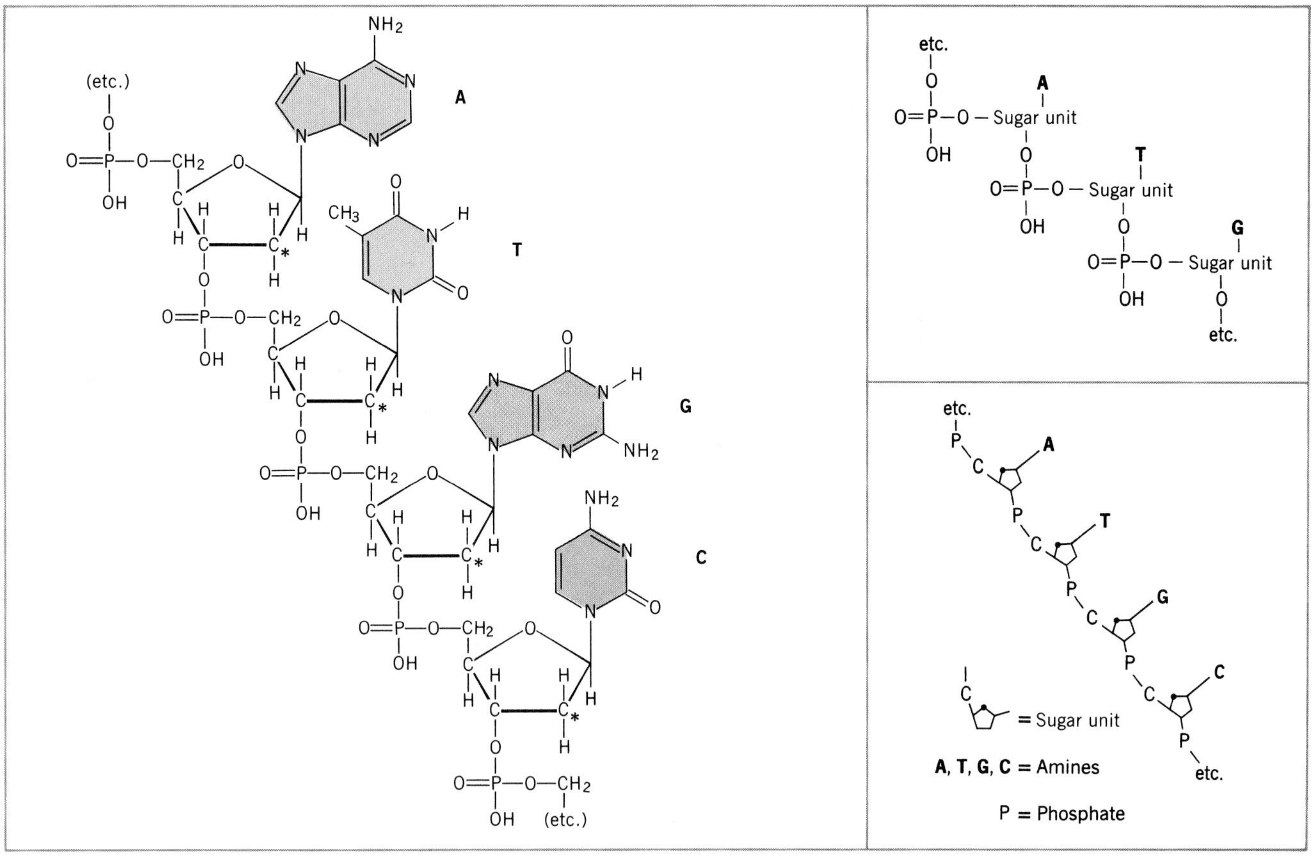

FIGURE 24.6 Nucleic acids. A segment of a DNA chain featuring each of the four DNA bases. When the sites marked by asterisks each carry an OH, the main "backbone" would be that of RNA. In RNA U would replace T. The insets show simplified versions of a DNA strand. (From J. R. Holum, *Elements of General and Biological Chemistry*, 7th ed. © 1986, John Wiley and Sons.)

Figure 24.6 gives a short segment of a DNA molecule together with some simpler representations. The main chains or "backbones" of DNA molecules consist of alternating units contributed by phosphoric acid and a simple monosaccharide. One difference between DNA and RNA is in the monosaccharide. It is ribose in RNA (hence the R in the symbol RNA) and it is deoxyribose in DNA. These two sugars differ by just one OH group (and *deoxy* means "lacking an oxygen unit"). Thus, both DNA and RNA have the following systems, where the G units represent the unique side chains of nucleic acids.

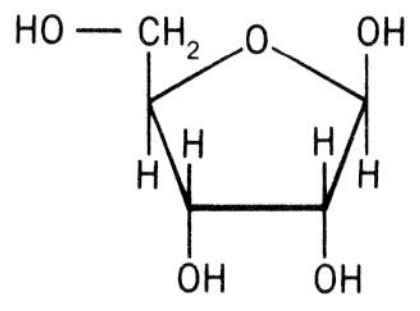

$$G^1 \qquad G^2 \qquad G^3$$
$$\text{phosphate—sugar—phosphate—sugar—phosphate—sugar—etc.}$$

Backbone system in all nucleic acids—many thousands of repeating units long.
In DNA, the sugar is deoxyribose. In RNA, the sugar is ribose.

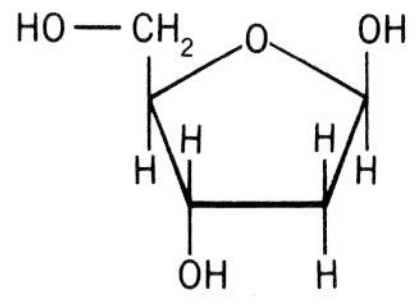

The side chains are all cyclic amines whose molecular shapes have as much to do with their function as anything else. Being amines, these side chains are often referred to as the *bases* of the nucleic acids. The bases are usually represented by single letters — A for adenine, T for thymine, U for uracil, G for guanine, and C for cytosine.

Ribose

Deoxyribose

adenine
A

thymine
T

uracil
U

guanine
G

cytosine
C

A, T, G, and C occur in DNA. A, U, G, and C are in RNA. These few are the "letters" of the genetic alphabet.

The uniqueness of an individual gene lies in two features: the length of the backbone and the sequence in which the bases occur. All of an individual's hereditary information resides in these structural aspects of its DNA molecules.

The DNA Double Helix

In 1953, F. H. C. Crick of England and J. D. Watson, an American, deduced that DNA occurs in cells as two oppositely running strands of molecules intertwined like a spiral staircase and called the DNA double helix. See Figure 24.7. Hydrogen bonds hold the two strands side by side.

Crick and Watson shared the 1962 Nobel prize in physiology and medicine with Maurice Wilkins. Rosalind Franklin contributed an important X-ray diffraction picture of DNA to this work.

FIGURE 24.7 The DNA double helix. In this schematic drawing, the hydrogen bonds that hold the two strands together (given in more detail in Figure 24.8) are indicated by dotted lines.

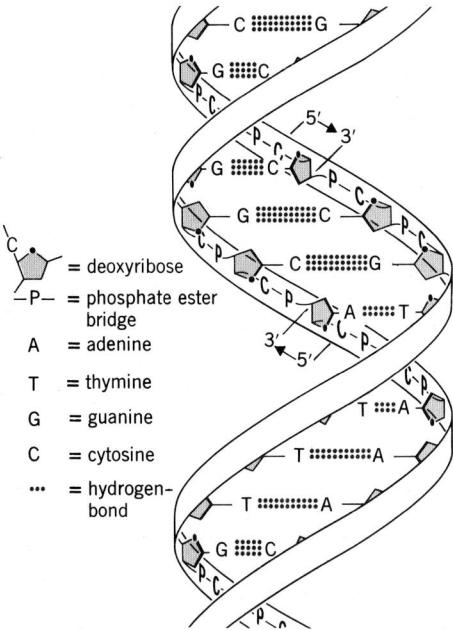

= deoxyribose
—P— = phosphate ester bridge
A = adenine
T = thymine
G = guanine
C = cytosine
··· = hydrogen-bond

The side-chain bases have both N—H groups and O=C units between which hydrogen bonds can exist.

$$-\overset{|}{N}-\overset{\delta+}{H}\cdots\cdots\overset{\delta-}{O}=C$$

hydrogen bonds between bases

However, the best conditions for maximum hydrogen bonding occur in just particular choices of *pairs* of bases. The members of each pair have complementary structures; their functional groups are in exactly the right locations to allow hydrogen bonds between pair members. Figure 24.8 shows how adenine, A, pairs with thymine, T, and how cytosine, C, pairs with guanine, G. A pairs only with T in DNA, never with G or C.

The DNA in one human cell nucleus has an estimated 3 to 5 million base pairs.

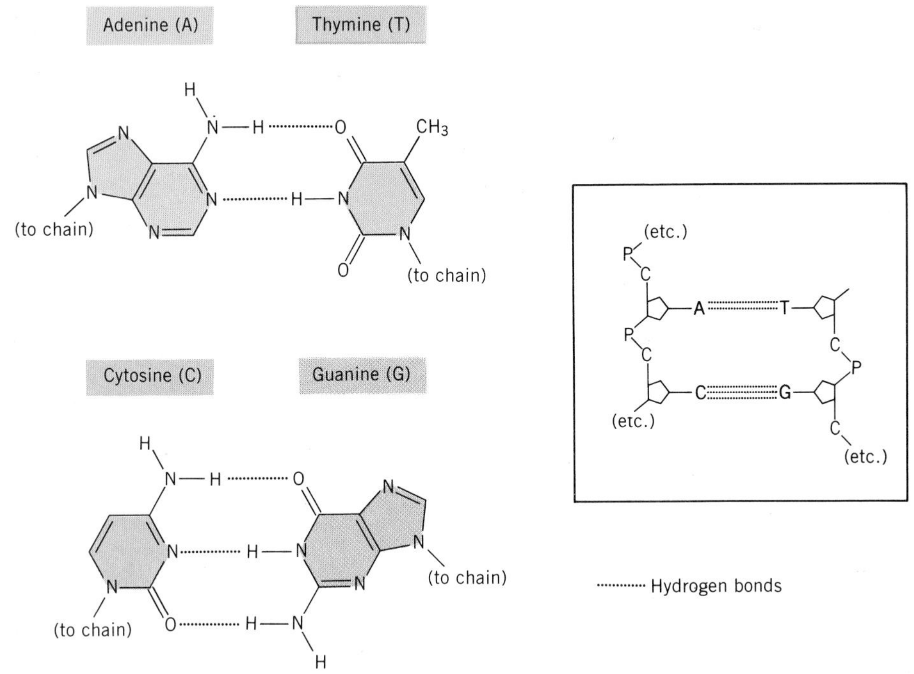

Adenine (A) Thymine (T)

Cytosine (C) Guanine (G)

FIGURE 24.8 Base pairing in DNA. The hydrogen bonds are indicated by dotted lines.

·········· Hydrogen bonds

Adenine, A, can also pair with uracil, U, but U occurs in RNA and the A-to-U pairing is an important factor in the work of RNA, as we will see.

No other pairs are possible; the shapes do not allow others. In DNA, only G can pair with C; only A can pair with T. Thus opposite every G on one strand in a DNA double helix, a C occurs on the other strand. Opposite every A on one strand in DNA, a T is found on the other. And, as we said, the only force holding the strands together is the hydrogen bond multiplied many times by the pairings of thousands of bases. Someone has likened the hydrogen bonds to parts of a zipper—individually weak, but very strong when in numbers.

The actual mole ratios of G to C and of A to T in all DNA samples had previously been found to be 1 to 1, as the Crick–Watson structure of DNA requires.

The Replication of DNA   If a cell's DNA stores all of its genetic information, then prior to cell division, this DNA must have a way to reproduce itself *in duplicate* so that each daughter cell has a complete set. Such reproductive duplication is called DNA replication.

The extraordinary accuracy of replication is the result of the limitations of base pairing as just described. Figure 24.9 gives a broad outline of the way this works. The

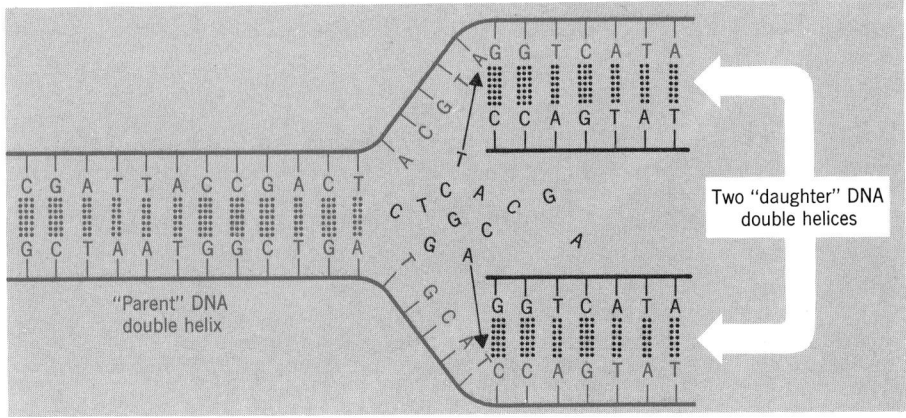

Two "daughter" DNA double helices

"Parent" DNA double helix

FIGURE 24.9 Base pairing and the replication of DNA. (From K. R. Atkins, J. R. Holum, and A. N. Strahler, *Essentials of Physical Science,* © 1978, John Wiley and Sons.)

Structural features of a DNA nucleotide

unattached letters A, T, G, and C in this figure represent the monomers of DNA — rather large molecules made of a phosphate – sugar unit holding one particular base. These monomers, called *nucleotides,* are present in the cellular "soup."

Several enzymes catalyze each step of replication. As it occurs, the two strands of the parent DNA double helix separate, and the monomers for new strands assemble along the exposed single strands. Their order of assembling is determined altogether by the specificity of base pairing. A base such as T on a parent strand can only accept a nucleotide with the base A. Two daughter double helixes result that are identical to the parent, and each carries one strand from the parent. One new double helix goes to one daughter cell and the second goes to the other.

Genes Occur as Segments of DNA Molecules   A single human gene has between 1000 and 3000 bases, but they do not occur continuously on a DNA molecule. In multicelled organisms, a gene is neither a whole DNA molecule nor a continuous sequence in one. Virtually all single genes are divided or split, so a gene actually is made up of *segments* of a DNA chain that, taken together, carry the necessary genetic instruction for a particular enzyme. These separated sections are called exons — a unit that helps to *ex*press a message. The sections of a DNA strand between exons are called introns — units that *in*terrupt the gene. The exact significance of the introns is not yet fully understood, and we will not discuss the speculations.

From Gene to Enzyme — Transcribing and Translating the Genetic Message   Each enzyme in a cell is made under the direction of its own gene. In broad outline, the steps between a gene and an enzyme occur as follows. (We momentarily ignore both the introns and also how the cell ignores them.)

$$\underset{\text{gene}}{\text{DNA}} \xrightarrow{\text{transcription}} \text{RNA} \xrightarrow{\text{translation}} \text{Enzyme}$$

(genetic message is read off in the cell nucleus and transferred to RNA)   (genetic message, now on RNA outside the nucleus, is used to direct poly-peptide synthesis)

The step labeled transcription accomplishes just that — the genetic message, represented by a sequence of bases on DNA, is transcribed into a *complementary* sequence of bases on RNA. (But U, not T, is on the RNA.)

Translation means the conversion of the base sequence on RNA into a side-chain sequence on a new polypeptide. It is like translating from one language (the DNA/RNA base sequences) to another language (the polypeptide side-chain sequence).

The cell tells which amino acid unit must come next in a growing polypeptide by means of a code that relates unique groups of nucleic acid bases to specific amino acids. The code, in other words, enables the cell to translate from the four-letter DNA alphabet (A, T, G, and C) to the 20-letter alphabet of amino acids (the amino acid side chains). To see how the cell does this, we have to learn more about RNA.

The RNA of a Cell

Four types of RNA are involved in the gene – enzyme connection. One is called *ribosomal RNA,* or *r*RNA, and it is packaged together with enzymes in small granules called *ribosomes.* Ribosomes are manufacturing stations for polypeptides.

Another RNA is called *messenger* RNA, or *m*RNA, and it brings the blueprints for a specific polypeptide from the cell nucleus to the manufacturing station (a ribosome). It is the real carrier of the genetic message.

*m*RNA is made from another kind of RNA, *primary transcript* RNA, or *pt*RNA. This is the RNA first made (hence *primary* transcript) at the direction of a DNA unit.

The last type of RNA is called *transfer* RNA, or *t*RNA. It carries the prefabricated parts — the amino acids — from the cell "soup" to the ribosome.

Combinations of a minimum of 3 letters from a 4-letter alphabet are needed to have a unique code for each of 20 amino acids.

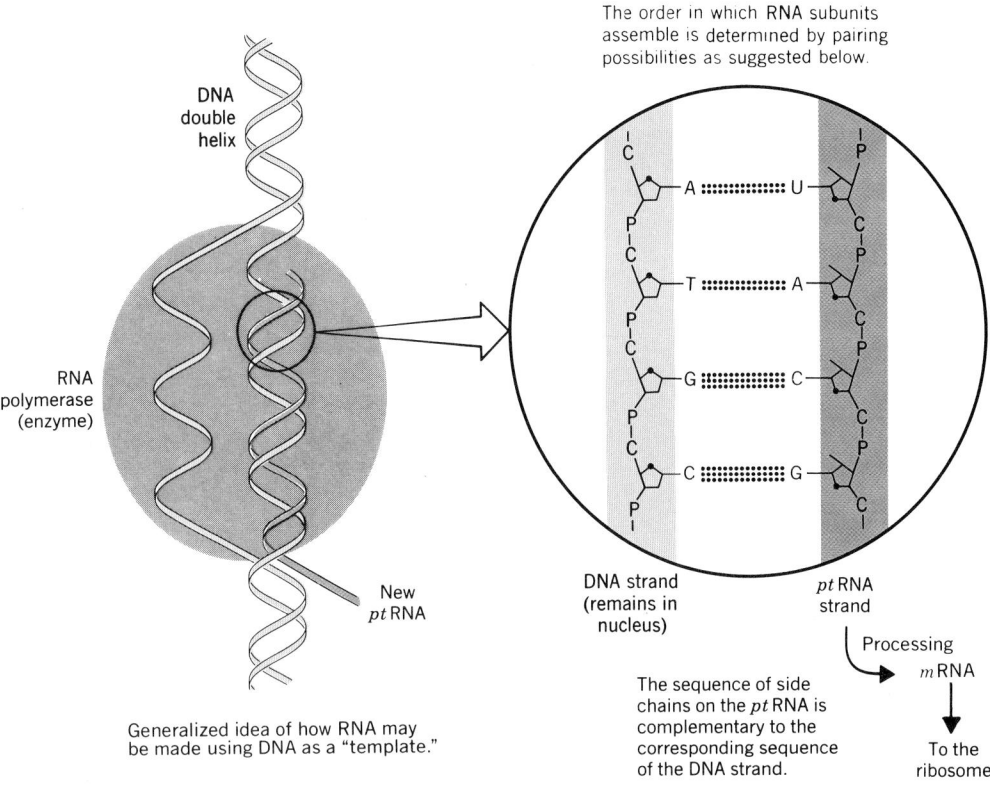

The order in which RNA subunits assemble is determined by pairing possibilities as suggested below.

DNA double helix

RNA polymerase (enzyme)

New *pt* RNA

Generalized idea of how RNA may be made using DNA as a "template."

DNA strand (remains in nucleus)

pt RNA strand

The sequence of side chains on the *pt* RNA is complementary to the corresponding sequence of the DNA strand.

Processing

m RNA

To the ribosome

FIGURE 24.10 The DNA-directed synthesis of *primary transcript* RNA (*pt*RNA). Once made, *pt*RNA is processed to give *m*RNA, which carries the genetic message from the cell nucleus to a protein assembly site. (From J. R. Holum, *Elements of General and Biological Chemistry*, 7th ed. © 1986, John Wiley and Sons.)

*m*RNA-Directed Polypeptide Synthesis

When some chemical signal tells a cell to make more of a particular enzyme, the cell nucleus makes the *pt*RNA that corresponds to the gene. Figure 24.10 outlines the steps, and you can see that the base sequence in *pt*RNA is an exact complement to a base sequence in the DNA being transcribed—both the exon sections of the DNA and the introns. Reactions and enzymes in the cell nucleus then remove those parts of the new *pt*RNA that were caused by introns. It is as if the message, now on *pt*RNA, is edited to remove nonsense sections. The result of this editing is *m*RNA.

The *m*RNA now moves outside the cell nucleus and becomes joined to a ribosome where the enzymes for polypeptide synthesis are found. The polypeptide manufacturing site now awaits the arrival of *t*RNA molecules bearing amino acid units. All is in readiness for polypeptide synthesis—for translation. Figure 24.11 shows how a *t*RNA molecule carries an amino acid unit in one place and at another site has three base pairs with an important role in what follows.

The Genetic Code To use the language of coding, the genetic message is written in code words made of three letters—three bases. Each triplet of bases as it occurs on *m*RNA is called a codon, and each codon corresponds to a specific amino acid. What amino acid goes with which codon constitutes the genetic code. One most remarkable feature of the genetic code is that it is essentially universal. The codon that specifies alanine, for example, in humans also specifies alanine in the genetic machinery of aardvarks, camels, rabbits, stinkbugs, and bacteria. Our chemical kinship with the entire living world is profound.

A triplet of bases complementary to a codon occurs on *t*RNA and is called an anticodon. In Figure 24.11, the triplet AGC was arbitrarily selected to illustrate an anticodon.

FIGURE 24.11 Transfer RNA or *t*RNA. Each *t*RNA consists of a short strand of RNA folded and twisted as indicated. The triplet identified as AGC is an anticodon.

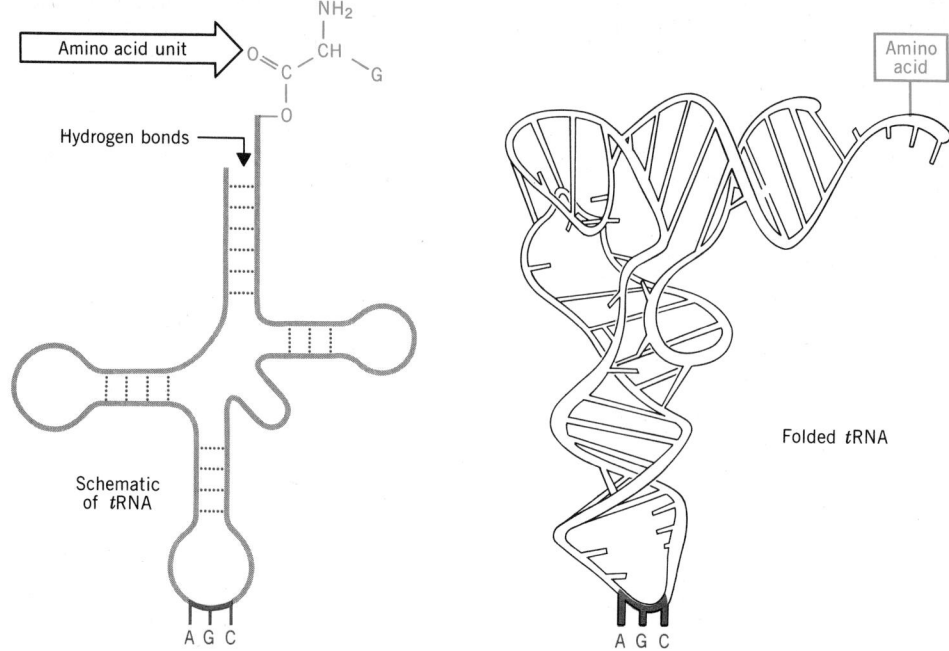

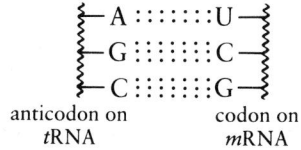

A *t*RNA molecule, bearing the amino acid corresponding to its anticodon, can line up at a strand of *m*RNA only where the anticodon "fits" by hydrogen bonding to the matching codon. If the anticodon is AGC, then the A on the anticodon must find a U on the *m*RNA while its neighboring G and C bases find equally neighboring C and G bases on the *m*RNA. Thus the anticodon and codon line up as follows.

$$
\begin{array}{c}
\text{A} :::::::: \text{U} \\
\text{G} :::::::: \text{C} \\
\text{C} :::::::: \text{G}
\end{array}
$$

anticodon on   codon on
*t*RNA   *m*RNA

*m*RNA codons determine the sequence in which *t*RNA units can line up, and this sets the sequence in which amino acid units become joined in a polypeptide. Figure 24.12 carries this description further, where we enter the process soon after special steps (not shown) have launched it.

In 24.12*a*, a *t*RNA–glycine unit arrives at *m*RNA where it finds two situations—a dipeptide already started and a codon to which the *t*RNA–glycine can fit. In 24.12*b*, the *t*RNA–glycine unit is in place. It is the only such unit that could be in this place. Base pairing between the two triplets—codon to anticodon—ensures this.

In 24.12*c*, the whole dipeptide unit moves over and is joined to the glycine unit by a new peptide bond. This leaves a tripeptide attached via a *t*RNA unit to the *m*RNA.

Since the next codon, GCU, is coded for the amino acid alanine, only a *t*RNA–alanine unit can come into the next site. So alanine, with its methyl group as the side chain, can be the only amino acid that enters the polypeptide chain at this point. These steps repeat over and over until a special codon is reached that is designed to stop the process. The polypeptide is now complete. And it has a unique sequence of side chains that were initially specified by the molecular structure of a gene.

Genetic Defects   With so many steps from gene to enzyme, we should expect many opportunities for things to go wrong. Suppose, for example, that just one base in a gene

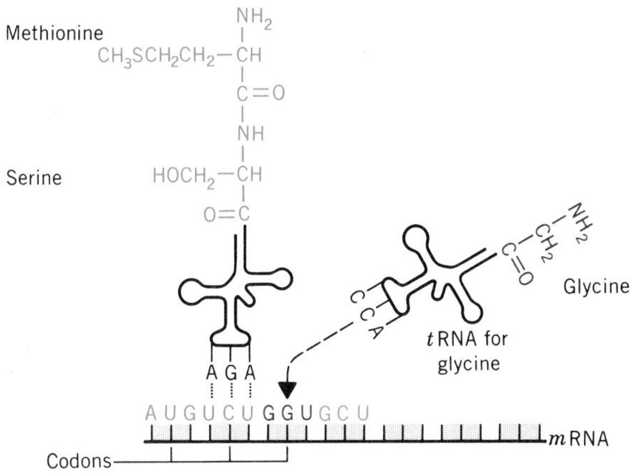

(a) A growing polypeptide is shown here just two amino acid units long. A third amino acid, glycine, carried by its own tRNA is moving into place at the codon GGU. (Behind the assemblage shown here is a large particle called a ribosome, not shown, that stabilizes the assemblage and catalyzes the reactions).

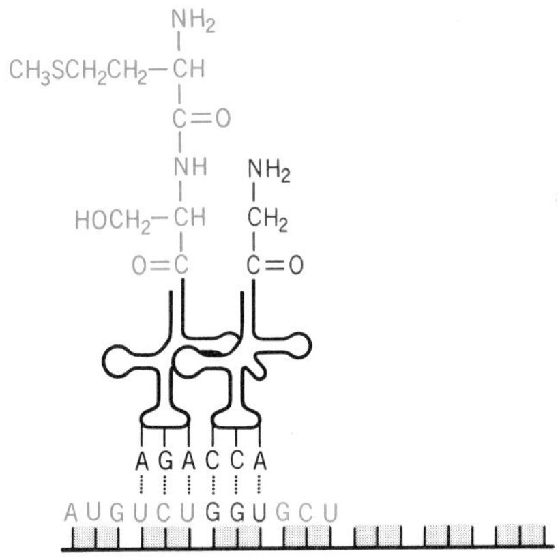

(b) The tRNA- glycine unit is in place, and now the dipeptide will transfer to the glycine, form another peptide bond, and make a tripeptide unit.

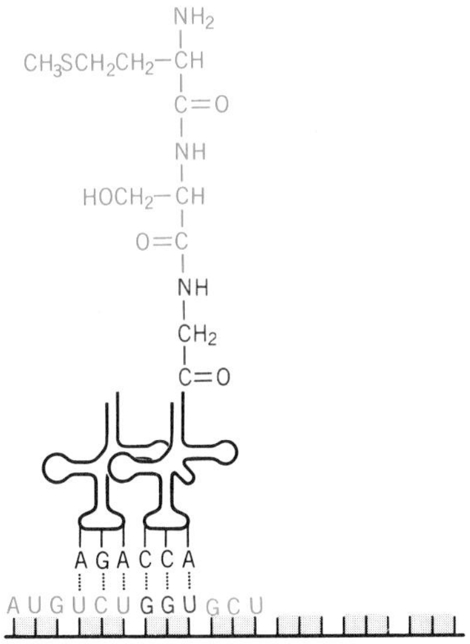

(c) The tripeptide unit has formed, and the next series of reactions that will add another amino acid unit can now start.

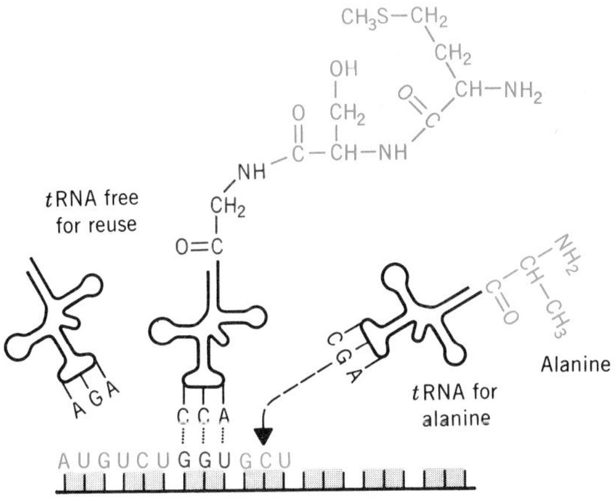

(d) One tRNA leaves to be reused. Another tRNA carrying a fourth amino acid unit, alanine, is moving into place along the mRNA strand to give an assemblage like that shown in part (b). Then the tripeptide unit will transfer to alanine as another peptide bond is made and a tetrapeptide unit forms. A cycle of events thus occurs as the polypeptide forms.

FIGURE 24.12 Polypeptide synthesis at an mRNA strand.

were wrong. Then one base on an *m*RNA strand would be the wrong base—wrong meaning with respect to getting the enzyme we want. This would fundamentally change every remaining codon. If the next three codons on *m*RNA were UCU–GGU–GCU–etc., for example, and the first G were deleted, then the sequence would become UCU–GUG–CUU–etc. All remaining triplets change! You can imagine how this could lead to an entirely different polypeptide than what we want.

Or suppose that the GGU codon were replaced by, say, a GCU codon so that the UCU–GGU–GCU–etc. becomes UCU–GCU–GCU–etc. One amino acid in the resulting polypeptide would not be what we want. Something like this makes the difference, for example, between normal hemoglobin and sickle cell hemoglobin and the difference between health and a tragic genetic disease. About 2000 diseases are attributed to various kinds of defects in the genetic machinery of cells.

Atomic radiations, like gamma rays, beta rays, or even X rays, or chemicals that mimic radiations can cause genetic defects, too. If a stray radiation hits a gene and causes two opposite bases to fuse together chemically, the cell might be reproductively dead. When it came time for it to divide, the cell could not replicate its genes, could not divide, and that would be the end of it. If this happens to enough cells in a tissue, the organism might be fatally affected.

Viruses   Viruses are packages of chemicals usually consisting of nucleic acid and protein. Their nucleic acids are able to take over the genetic machinery of the cells of particular host tissues, manufacture more virus particles, and multiply enough to burst the host cell. Since cancer cells divide irregularly and usually more rapidly than normal cells, some viruses are thought to be among the agents that can cause cancer.

Genetic Engineering   Human insulin and human growth hormone are now being manufactured by a process that involves the production of *recombinant DNA*. It represents one of the important advances in scientific technology of this century. Figure 24.13 shows how it works.

Bacteria make polypeptides using the same genetic code as humans. There are some differences in the machinery, however. For example, bacteria have DNA not only as parts of their chromosomes but also in the form of large, circular, supercoiled DNA molecules called *plasmids*. Each plasmid carries just a few genes, but several copies of a plasmid can exist in one bacterial cell. Each plasmid can replicate independently of the chromosome.

Plasmids can be isolated from bacteria, and they can be given new DNA material—a new gene—that has the base triplets for directing the synthesis of a particular polypeptide. It might even be a polypeptide that is completely alien to the bacteria, such as the subunits of human insulin or human growth hormone. The plasmids are snipped open by special enzymes, the new DNA combines with the open ends, and then the plasmid loops are reclosed. The DNA in these altered plasmids is called recombinant DNA. The technology associated with this is often referred to as *genetic engineering*.

The remarkable feature of bacteria with altered plasmids made of recombinant DNA is that when the bacteria multiply, the new bacteria have within them the *altered* plasmids. And when these multiply, still more altered plasmids are made. Between their cell divisions, the bacteria manufacture the proteins for which they are genetically programmed, including the proteins specified by the recombinant DNA. In this way, bacteria can be tricked into making the *human* proteins we have mentioned. The technology is not limited to bacteria; yeast cells work, too.

People who rely on the insulin of animals—cows, pigs, and sheep, for example—sometimes experience allergic responses. They are also vulnerable to the availability of a continuing high supply of a particular organ in all higher animals, the pancreas, because insulin is isolated from this organ. Therefore the ability to make *human* insulin from easily available bacteria has been a welcome development.

Recombinant DNA technology has interacted with archaeology in an interesting

The protein of a virus is a protective overcoat for its nucleic acid, and it provides an enzyme to catalyze a breakthrough by the virus into a host cell.

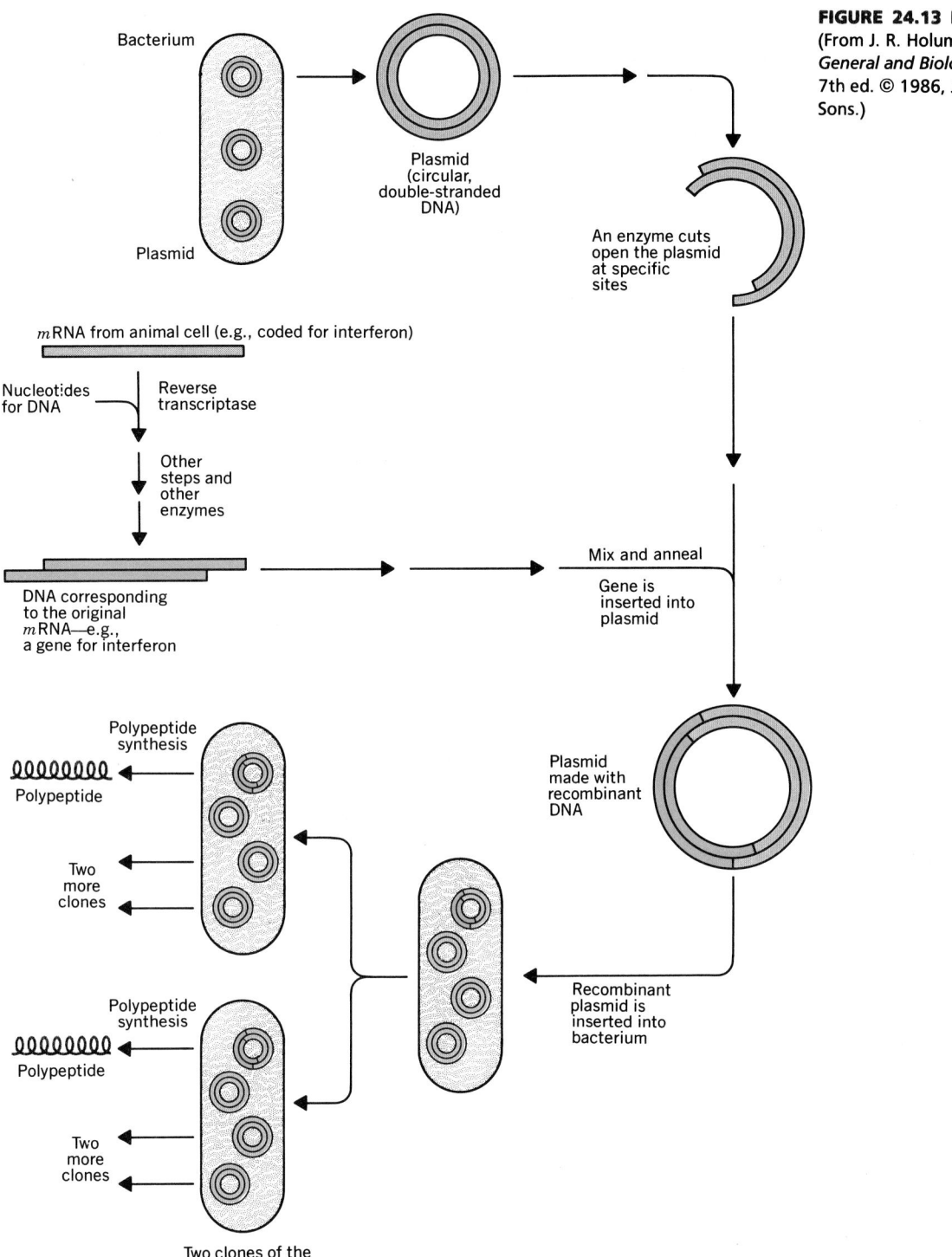

FIGURE 24.13 Recombinant DNA. (From J. R. Holum, *Elements of General and Biological Chemistry*, 7th ed. © 1986, John Wiley and Sons.)

way. In the early 1980s scientists were able to remove segments of DNA molecules from an Egyptian mummy and from an extinct horselike animal (a quagga) and reproduce these segments using bacteria. No segment was long enough to include a complete gene, but the technique no doubt will be developed as another tool for the study of evolutionary history.

SUMMARY

Organic Compounds.   Organic chemistry is the study of the covalent compounds of carbon, except for its oxides, carbonates, bicarbonates, and a few others. Biochemistry is the study of the organic compounds present among living systems. Molecular biology is the study of how the chemicals of heredity, the nucleic acids, control the synthesis of themselves and polypeptides.

The occurrence of functional groups attached to nonfunctional and hydrocarbon-like groups is the basis for the classification of the millions of organic compounds. Members of a family have the same kinds of chemical properties. Their physical properties are a function of the relative proportions of functional and nonfunctional groups.

Saturated Hydrocarbons.   The alkanes and cycloalkanes—saturated compounds found chiefly in fossil fuels—are the least polar and the least reactive of the organic families. Carbon chains are straight or branched and, when open chain, enjoy free rotation about single bonds.

The IUPAC rules for naming alkanes start with the common name ending, -ane, for the family. The prefix denotes the carbon content of the longest, continuous chain, which is numbered from whichever end is nearest a branch. The alkyl groups are alkanes minus a hydrogen.

Alkanes and cycloalkanes in general do not react with strong acids or bases or strong redox reactants. At high temperatures, their molecules crack to give H_2 and useful organic compounds.

Unsaturated Hydrocarbons.   The π bonds of alkenes prevent free rotation at carbon–carbon double bonds and so give rise to geometric (*cis–trans*) isomerism. These bonds also enable alkenes to act like Brønsted bases and undergo addition reactions with hydrogen halides, sulfuric acid, and water (in the presence of an acid catalyst). The alkenes also add hydrogen (under pressure, at an elevated temperature, and in the presence of a metal catalyst). Bromine and chlorine add to alkenes under mild, uncatalyzed conditions. Alkenes also can be monomers that can be polymerized to give polyolefins.

Despite their formal unsaturation, aromatic hydrocarbons, like benzene, do not give the addition reactions shown by alkenes. Instead, they undergo substitution reactions that leave the energy-lowering, π-electron network intact. In the presence of suitable catalysts, benzene reacts with chlorine and bromine to give chlorobenzene and bromobenzene. Nitrobenzene is the product when benzene and nitric acid react. And concentrated sulfuric acid, which readily adds to alkenes, reacts with benzene by substitution.

Organic Compounds of Oxygen.   The main organic families with oxygen are

ROH   R—O—R′   R—C(=O)—H
alcohols   ethers   aldehydes

R—C(=O)—R′   R—C(=O)—OH   R—C(=O)—O—R′
ketones   carboxylic   esters
                 acids

Those with waterlike OH groups (alcohols and carboxylic acids) have much higher boiling points than alkanes of comparable formula weights, and they are much more soluble in water. Hydrogen bonding is behind both properties. The other families are also more soluble in water than the alkanes.

The alcohols undergo an elimination reaction that changes them to alkenes. The ethers have almost as few chemical properties as alkanes. The aldehydes are among the most easily oxidized organic compounds—being oxidized to carboxylic acids. Aldehydes and ketones both add hydrogen (under special conditions) and are thereby changed to alcohols.

The carboxylic acids are weak acids but neutralize strong bases. Their anions are good Brønsted bases. These acids can also be changed to esters by heating them with alcohols. Esters react with water to give back their parent acids and alcohols.

If the acid is dicarboxylic and the alcohol is dihydroxy, the product of ester formation is a polyester, such as Dacron.

Organic Nitrogen Compounds.   The principal organic nitrogen compounds are the amines and the amides. The amines are organic derivatives of ammonia and so are weak bases that can neutralize strong acids. The conjugate acids of amines are good proton donors and can neutralize strong bases. The amides have a carbonyl-to-nitrogen system that is neutral (in an acid–base sense) and that can be hydrolyzed to the parent acid and amine of the amide.

If the acid is dicarboxylic and the amine is diamino, the product of amide formation is a polyamide, like nylon. The polymers most useful for making filaments and fibers are those with very long, symmetrical molecules. If regularly spaced polar groups occur that can attract each other, either dipole to dipole or by hydrogen bonds, the fibers are particularly strong.

Carbohydrates.   The glucose unit is present in starch, glycogen, cellulose, sucrose, lactose, and simply as a monosaccharide in honey. It's the chief "sugar" in blood. As a monosaccharide, glucose is a pentahydroxyaldehyde in one form but it also exists in two cyclic forms. The latter are joined by oxygen bridges in the disaccharides and polysaccharides. The digestion of these substances hydrolyzes the oxygen bridges and releases the monosaccharides.

Lipids.   Lipids are the natural products in living systems that are relatively soluble in nonpolar solvents, not water. They include a large variety, ranging from sex hormones and cholesterol to the triacylglycerols of nutritional importance. The latter are tri-esters of glycerol with three of a number of long-chain, often unsaturated fatty acids. The digestion of the triacylglycerols is the hydrolysis of the ester groups.

Amino Acids, Polypeptides, and Proteins.   The amino acids in nature are mostly compounds with NH_3^+ and CO_2^- groups joined to a carbon atom called the alpha position. About 20 are important to the structures of polypeptides and proteins.

When two amino acids are linked by a peptide bond (the amide bond), the compound is a dipeptide. In polypeptides, several (sometimes thousands of) regularly spaced peptide bonds occur involving as many amino acid units. The uniqueness of a polypeptide lies in its chain length and in the sequence of the side-chain groups.

Some proteins consist of just a polypeptide, but most proteins involve two or more (sometimes different) polypeptides and often a nonpolypeptide, an organic molecule, and a metal ion. The final shape of a protein is as important to its ability to function in a cell as its molecular architecture. The latter, of course, largely determines the shape, but other conditions (like pH) are factors as well.

The proteins serving as enzymes are particularly important, and their vulnerability to changes in shape is a matter every cell must guard.

Nucleic Acids and Polypeptide Synthesis.   Nucleic acids direct the formation of a polypeptide with the correct amino acid sequence needed for an enzyme. All of the genetic messages implied by the term *heredity* are encoded on molecules of DNA by a sequence of side-chain bases. These bases are attached to deoxyribose units of the DNA main chain, a repeating phosphate–sugar system. In DNA, they are adenine (A), thymine (T), guanine (G), and cytosine (C). In RNA, the sugar unit is ribose and uracil (U) replaces (T) in the list of four bases.

Except when replicating or when transcribing its message to primary transcript RNA, DNA exists as a double-stranded molecule twisted into a helical shape. The two strands, which run in opposite directions, are held side by side by interchain hydrogen bonds. Units of A on one chain "fit" by hydrogen bonds to units of T (and only T) on the other chain. Similarly, G and C form a pair. The faithful replication of DNA involves the base pairing of only A with T and only C with G. (In RNA systems, A and U can pair.) An individual gene (at least in higher organisms) consists of sections of a DNA chain, called exons, separated by other sections called introns.

In directing the synthesis of a polypeptide, the DNA that carries the gene for this polypeptide is used to make a molecule of primary transcript RNA. Its bases are an exact complement of all the exons and introns of the DNA being used (except that U appears in RNA where opposite it on DNA a unit of A occurred). The new *pt*RNA is then "edited" to remove those segments arising from the introns of the DNA. This leaves messenger RNA, the carrier of the message of the gene to the ribosomes outside the cell nucleus. The overall change from gene to *m*RNA is called translation. *m*RNA is like an assembly line at a factory (ribosomes) awaiting the arrival of spare parts — the amino acids. These are borne on transfer RNA molecules.

A unit of three bases on *m*RNA constitutes a codon, and it is responsible for getting one and only one amino acid unit into place in a growing polypeptide. A *t*RNA–amino acid unit has, on its *t*RNA part, a set of three bases called an anticodon that can pair to a codon. In this way a *t*RNA–amino acid unit can "recognize" just one place where it can deliver the amino acid to the polypeptide developing at the *m*RNA assembly line. The key that relates codons to specific amino acids is the genetic code and it is essentially the same code for all organisms in nature. The involvement of *m*RNA in directing the synthesis of a polypeptide is called translation.

Even small changes or deletions in the sequences of bases on DNA can have large consequences in the forms of genetic diseases. Viruses are able to use their own nucleic acids to take over the genetic apparatus of the host tissue. In genetic engineering, the genetic machinery of microorganisms is taken over and made to manufacture polypeptides not otherwise synthesized by the organism.

REVIEW EXERCISES

Answers to questions whose numbers are printed in color are given in Appendix C. The more difficult questions are marked with asterisks.

Classifications

24.1   Why are the compounds of carbon generally called *organic* compounds?

24.2   What aspects of chemistry are studied in biochemistry?

24.3   Molecular biology is a branch of biochemistry that concentrates on the study of what processes in living systems?

24.4   In $CH_3CH_2NH_2$, the NH_2 is called the *functional group*. In general terms, why is it called this?

24.5   Which of the following compounds are inorganic?
(a) CH_3OH (c) CCl_4 (e) K_2CO_3
(b) CO (d) $NaHCO_3$

Structural Formulas

24.6   One can write the structure of butane, lighter fluid, as follows.

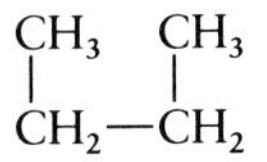

$$CH_3 \quad CH_3$$
$$| \qquad |$$
$$CH_2 - CH_2$$
butane

Are butane molecules properly described as straight chain or as branched chain, in the sense in which we use these terms? Explain.

24.7   Which of the following structures are possible, given the numbers of bonds that various atoms can form?
(a) $CH_2CH_2CH_3$
(b) $CH_3=CHCH_2CH_3$
(c) $CH_3CH=CH_2CH_2CH_3$

24.8   Write full (expanded) structures for each of the following *molecular* formulas. Remember how many bonds the various kinds of atoms must have. In some you will have to use double or triple bonds. (Hint: A trial-and-error approach will have to be used.)
(a) CH_5N (i) N_2H_4
(b) CH_2Br_2 (j) HCN
(c) $CHCl_3$ (k) C_2H_3N
(d) C_2H_6 (l) CH_4O
(e) CH_2O_2
(f) CH_2O
(g) NH_3O
(h) C_2H_2

24.9 Write, neat condensed structures of the following.

(a)

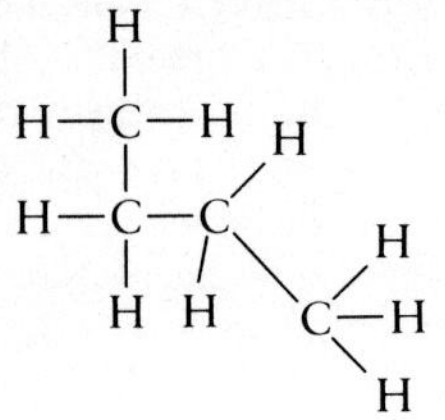

(b)

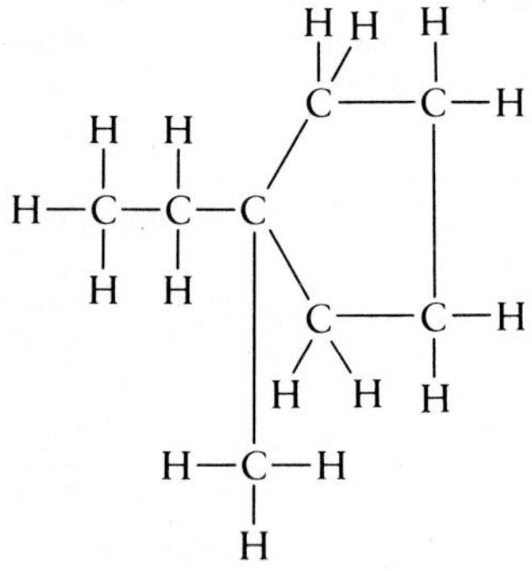

(c)

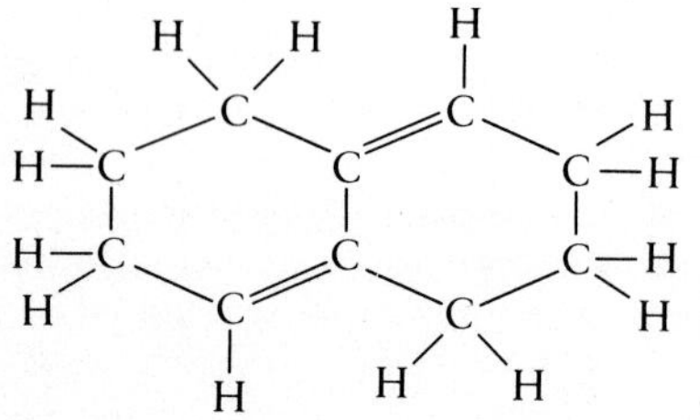

Families of Organic Compounds

24.10 In general terms, why do functional groups impart more chemical reactivity to families that have them? Why don't the alkanes display as many reactions as, say, the amines?

24.11 Name the family to which each compound belongs.
  (a) $CH_3CH=CH_2$
  (b) CH_3CH_2OH
  (c) $CH_3CH_2\overset{O}{\overset{\|}{C}}-O-CH_3$
  (d) $CH_3CH_2CH_2COH$
  (e) $CH_3CH_2CH_2NH_2$
  (f) $HOCH_2CH_2CH_3$
  (g) $CH_3C\equiv CH$
  (h) $CH_3CH_2\overset{O}{\overset{\|}{C}}H$
  (i) $CH_3\overset{O}{\overset{\|}{C}}CH_2CH_2CH_3$
  (j) $CH_3-O-CH_2CH_3$

24.12 Identify, by letter, the compounds in Review Exercise 24.11 that are saturated compounds.

Isomers

24.13 Decide whether the members of each pair are identical, are isomers, or are unrelated.
  (a) $\underset{CH_3}{CH_3}$ and CH_3-CH_3
  (b) 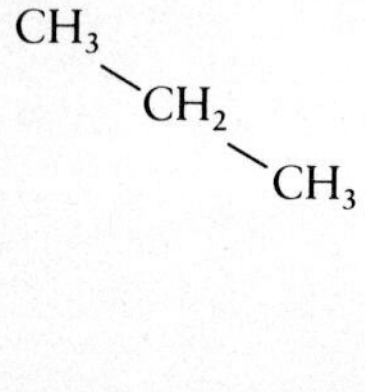 $CH_3-CH_2-CH_3$ and $CH_3-CH_2-CH_2-CH_3$

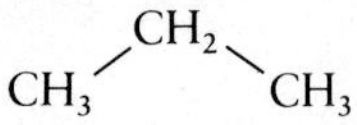

(c) CH_3CH_2-OH and $CH_3CH_2CH_2-OH$
(d) $CH_3CH=CH_2$ and cyclopropane

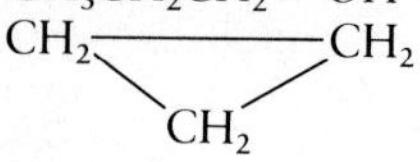

(e) $H-\overset{O}{\overset{\|}{C}}-\underset{CH_3}{CH_3}$ and $CH_3-\overset{O}{\overset{\|}{C}}-\underset{CH_3}{H}$

(f) CH_3CHCH_3 and CH_3CH with CH_3

(g) $CH_3CH_2CH_2-NH_2$ and $CH_3CH_2-NH-CH_3$

(h) $CH_3CH_2\overset{O}{\overset{\|}{C}}-O-H$ and $H-O-\overset{O}{\overset{\|}{C}}CH_2CH_3$

(i) $H-\overset{O}{\overset{\|}{C}}-O-CH_2CH_3$ and $CH_3CH_2-\overset{O}{\overset{\|}{C}}-OH$

(j) $H-\overset{O}{\overset{\|}{C}}-O-CH_2CH_2OH$ and $HOCH_2CH_2-\overset{O}{\overset{\|}{C}}-OH$

(k) $CH_3\overset{O}{\overset{\|}{C}}CH_2CH_3$ and $CH_3CH_2\overset{O}{\overset{\|}{C}}CH_3$

(l)

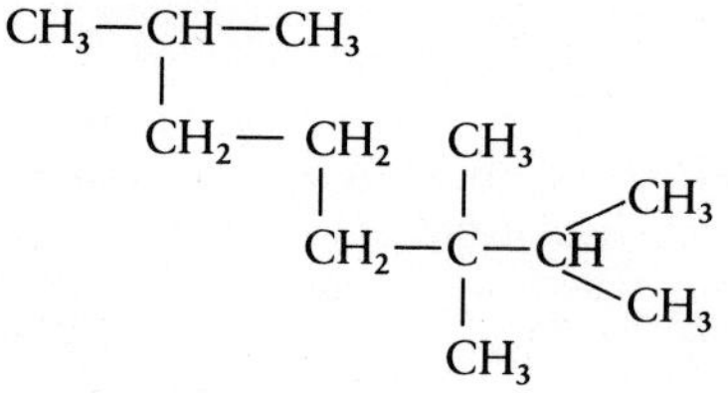

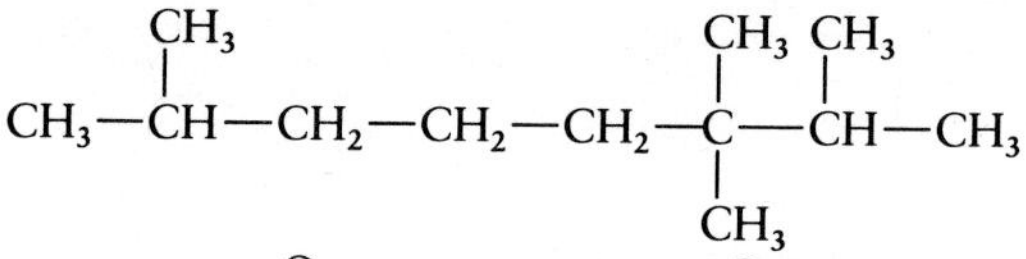

(m) $CH_3-NH-\overset{O}{\overset{\|}{C}}-CH_3$ and $CH_3CH_2\overset{O}{\overset{\|}{C}}NH_2$
(n) $H-O-O-H$ and $H-O-H$

Names of Hydrocarbons

24.14 Write the IUPAC names of the following hydrocarbons.
  (a) $CH_3CH_2CH_2CH_2CH_3$
  (b) $CH_3CH_2CH_2\underset{CH_3}{CH}CH_3$
  (c) $CH_3\underset{CH_3}{CH}CH_2\underset{CH_3}{CH}CH_2CH_3$
  (d) $CH_3CH_2\underset{CH_3}{CH}CH_2\underset{CH_3}{CH}CH_3$
  (e) $CH_3CH_2CH=CHCH_2CH_3$

$$CH_3$$
$$|$$
(f) $CH_3CHCH=CHCH_3$

24.15 Write the condensed structures of the following compounds.
  (a) 2,2-dimethyloctane
  (b) 1,3-dimethylcyclopentane
  (c) 1,1-diethylcyclohexane
  (d) 6-ethyl-5-isopropyl-7-methyl-l-octene
  (e) *cis*-2-pentene

Properties of Unsaturated Hydrocarbons

24.16 What prevents free rotation about a double bond?

24.17 If at either *end* of a carbon–carbon double bond the two atoms or groups are identical, can the alkene exist as a pair of *cis* and *trans* isomers? Write the structures of the *cis* and *trans* isomers, if any, for the following.

  (a) $CH_2=CH_2$

  (b) $CH_3CH=CH_2$

$$CH_3$$
$$|$$
  (c) $CH_3C=CH_2$

$$CH_3$$
$$|$$
  (d) $CH_3C=CHCH_3$

$$CH_3$$
$$|$$
  (e) $CH_3CH=CCH_2CH_3$

$$Cl \quad Cl$$
$$| \quad |$$
  (f) $CH_3CH_2C=CCH_2CH_3$

24.18 Propene is known to react with concentrated sulfuric acid as follows. The reaction occurs in two steps. Write the equation for the first step.

$$CH_3CH=CH_2 + H_2SO_4 \longrightarrow CH_3CHCH_3$$
$$|$$
$$OSO_3H$$

24.19 Write the structures of the products that form when ethylene reacts with each of the following substances by an addition reaction. (Assume that any needed catalysts or other conditions are provided.)
  (a) H_2 (e) HBr(*g*)
  (b) Cl_2 (f) H_2O (in acid)
  (c) Br_2 (g) H_2SO_4
  (d) HCl(*g*)

24.20 Repeat Review Exercise 24.19 using 2-butene.

24.21 Repeat Review Exercise 24.19 using cyclohexene.

24.22 If 1-butene polymerized in the manner of the polymerization of propene, what would be the structure of the polymer? (Write two kinds of structure—one showing three repeating units, and one condensed structure that indicates the repeating unit.)

24.23 Repeat Review Exercise 24.22 using 2-methylpropene.

24.24 In general terms, why doesn't benzene give the same kinds of addition reactions shown by, say, cyclohexene?

24.25 If benzene were to *add* one mole of Cl_2, what would form? What forms, instead, when Cl_2, benzene, and an $FeCl_3$ catalyst are heated together? (Write the structures.)

24.26 Can a compound have just any functional group and be classified as *aromatic*? Being in this class requires what?

Organic Compounds of Oxygen

24.27 Write condensed structures, IUPAC names, and common names for all saturated alcohols with three carbon atoms or fewer.

24.28 Deducing how the IUPAC names for alcohols are fashioned from the examples given, write structures for (a) 1-butanol and (b) 2-butanol.

24.29 "3-Butanol" is not a proper name, but a structure could still be written for it. What is that structure, and what IUPAC name should be used?

24.30 Briefly explain why ethyl alcohol is more soluble in water than ethane.

24.31 Briefly explain how the C—O bond in isopropyl alcohol is weakened when a strong acid is present.

24.32 When ethyl alcohol is heated in the presence of an acid catalyst, ethene and water form; an elimination reaction occurs. When the following alcohol, 2-butanol, is heated under similar conditions, two alkenes form (although not in equal quantities). Write the structures of these two alkenes.

$$CH_3CHCH_2CH_3$$
$$|$$
$$OH$$
2-butanol

*24.33 If the formation of the two alkenes in Review Exercise 24.32 were determined purely as a statistical matter, then the mole ratio of 1-butene to 2-butene should be in the ratio of what whole numbers? Why?

*24.34 The major product of the elimination reaction described in Review Exercise 24.32 is, by far, 2-butene. In the light of Review Exercise 24.33, is the ratio of alkenes that form determined simply by statistics? Suggest a possible thermodynamic reason for the observed major product.

24.35 Which of the following two compounds could be easily oxidized? Write the structure of the product of the oxidation.

$$\quad\quad\quad O \quad\quad\quad\quad\quad\quad O$$
$$\quad\quad\quad \| \quad\quad\quad\quad\quad\quad \|$$
$$CH_3CH_2CH \quad\quad CH_3CH_2CCHCH_3$$
$$\quad\quad\quad\quad\quad\quad\quad\quad\quad\quad\quad |$$
$$\quad\quad\quad\quad\quad\quad\quad\quad\quad\quad\quad CH_3$$
$$\quad\quad A \quad\quad\quad\quad\quad\quad\quad\quad B$$

24.36 Arrange the following compounds in their order of increasing boiling points. Place the identifying letter of the highest boiling compound on the right and arrange the other letters before it in the correct order. Explain the order.

$$CH_3CH_2OCH_3 \quad CH_3CH_2CH_2OH \quad CH_3CH_2CH_2CH_3$$
$$\quad\quad A \quad\quad\quad\quad\quad\quad B \quad\quad\quad\quad\quad\quad C$$

24.37 Repeat Review Exercise 24.36 with respect to increasing solubility in water, and explain the order.

24.38 Arrange the following compounds in their order of increasing boiling points. Place the identifying letter of the highest-boiling compound on the right and arrange the other letters before it in the correct order. Explain the order.

$$CH_3CH_2CH_2OH \qquad CH_3CH_2\overset{\displaystyle O}{\overset{\|}{C}}OH \qquad CH_3CH_2\overset{\displaystyle O}{\overset{\|}{C}}H$$
$$\quad\ \ A \qquad\qquad\qquad\quad B \qquad\qquad\qquad\quad C$$

24.39 Which of the following two compounds has the chemical property given below? Write the structure of the product of the reaction.

$$CH_3CH_2CH_2\overset{\displaystyle O}{\overset{\|}{C}}H \qquad CH_3CH_2CH_2\overset{\displaystyle O}{\overset{\|}{C}}OH$$
$$\qquad\qquad A \qquad\qquad\qquad\qquad\qquad B$$

   (a) Is easily oxidized.
   (b) Neutralizes NaOH.
   (c) Forms an ester with methyl alcohol.

*24.40 If the following two compounds polymerized in the same way that Dacron forms, what would be the repeating unit of their polymer?

$$HO\overset{\displaystyle O}{\overset{\|}{C}}CH_2CH_2\overset{\displaystyle O}{\overset{\|}{C}}OH \quad \text{and} \quad HOCH_2CH_2OH$$

Carbohydrates

24.41 How are carbohydrates defined?

24.42 What monosaccharide forms when the following polysaccharides are completely hydrolyzed? (a) Starch, (b) glycogen, (c) cellulose.

24.43 Glucose exists in three forms only one of which is a polyhydroxyaldehyde. How then can an aqueous solution of any form of glucose give the reactions of an aldehyde?

24.44 What compounds form when sucrose is digested?

24.45 The digestion of lactose gives what compounds?

Lipids

24.46 How are lipids defined?

24.47 Cholesterol is not an ester, yet it is defined as a lipid. Why?

24.48 A product such as corn oil is advertised as "polyunsaturated." But all nutritionally important lipids have unsaturated C=O groups. So what does "polyunsaturated" refer to?

*24.49 Write the structures of the products of the complete digestion of the following triacylglycerol.

$$CH_2-O-\overset{\displaystyle O}{\overset{\|}{C}}-(CH_2)_7CH=CH(CH_2)_7CH_3$$
$$CH-O-\overset{\displaystyle O}{\overset{\|}{C}}-(CH_2)_{16}CH_3$$
$$CH_2-O-\overset{\displaystyle O}{\overset{\|}{C}}-(CH_2)_7CH=CHCH_2CH=CH(CH_2)_4CH_3$$

Organic Compounds of Nitrogen

24.50 Classify the following compounds as amines or amides. If another functional group occurs, name it too.

   (a) $CH_3CH_2NH_2$

   (d) $CH_3\overset{\displaystyle O}{\overset{\|}{C}}CH_2NH_2$

   (b) $CH_3CH_2NHCH_3$

   (e) $CH_3\overset{\displaystyle CH_3}{\overset{|}{N}}CH_2CH_2CH_3$

   (c) $CH_3CH_2\overset{\displaystyle O}{\overset{\|}{C}}NH_2$

   (f) $CH_3CH_2\overset{\displaystyle O}{\underset{\displaystyle CH_3}{\overset{\|}{\underset{|}{N}}}}CCH_3$

24.51 Which of the following species give the reaction specified? Write the structures of the products that form.

$$CH_3CH_2\overset{\displaystyle O}{\overset{\|}{C}}NH_2 \qquad CH_3CH_2CH_2NH_2 \qquad CH_3CH_2NH_3^+$$

   (a) Neutralizes dilute hydrochloric acid.
   (b) Hydrolyzes.
   (c) Neutralizes dilute sodium hydroxide.

24.52 Amines, RNH_2, do not have as high boiling points as alcohols of comparable formula weights. Why?

24.53 Acetamide is a solid at room temperature but butylamine, which has about the same formula weight, is a liquid. Explain.

$$CH_3\overset{\displaystyle O}{\overset{\|}{C}}NH_2 \qquad CH_3CH_2CH_2CH_2NH_2$$
$$\text{acetamide} \qquad\qquad \text{butylamine}$$

*24.54 If the following two compounds polymerized in the same way that nylon forms, what would be the repeating unit in the polymer?

$$NH_2-CH_2CH_2CH_2CH_2-NH_2$$

$$HO-\overset{\displaystyle O}{\overset{\|}{C}}-\bigcirc-\overset{\displaystyle O}{\overset{\|}{C}}-OH$$

24.55 Describe the specific ways in which the monomers for all polypeptides are (a) alike and (b) different.

24.56 What is the peptide bond?

24.57 Write the structure of the dipeptide that could be hydrolyzed to give two molecules of glycine.

24.58 What is the structure of the tripeptide that could be hydrolyzed to give three molecules of alanine?

24.59 What are the structures of the two dipeptides that could be hydrolyzed to give glycine and serine?

24.60 Describe the structural way in which two *isomeric* polypeptides would be different.

24.61 Describe the structural ways in which two different polypeptides can differ.

24.62 Why is a distinction made between the terms *polypeptide* and *protein?*

Nucleic Acids and Heredity

24.63 In general terms only, how does the body solve the problem of getting particular amino acid sequences in a polypeptide rather than a mixture of randomly organized sequences?

24.64 In what specific structural way does DNA carry genetic messages?

24.65 What holds the two DNA strands together in a double helix?

24.66 How are the two DNA strands in a double helix structurally related?

24.67 In what ways do DNA and RNA differ structurally?

24.68 Which base pairs with (a) A in DNA? (b) A in RNA? (c) C in DNA or RNA?

24.69 A specific gene in the DNA of a higher organism is said to be *segmented.* Explain.

24.70 On what kind of RNA are codons found?

24.71 On what kind of RNA are anticodons found?

24.72 What function do each of the following have in polypeptide synthesis? (a) a ribosome, (b) *m*RNA, (c) *t*RNA.

24.73 What kind of RNA forms directly when a cell uses DNA at the start of the synthesis of a particular enzyme? What kind of RNA is made from it?

24.74 The process of *transcription* begins with which nucleic acid and ends with which one?

24.75 The process of *translation* begins with which nucleic acid? What is the end result of translation?

24.76 What is a virus and (in general terms only) how does it infect a tissue?

24.77 Genetic engineering is able (in general terms) to accomplish what?

24.78 What is *recombinant DNA?*

CHEMICALS IN USE

ETHYL ALCOHOL

In the Arabic language *kohol* or *koh'l* once meant a "very fine powder," one used to darken the eyelids and thus enhance the beauty of its user. Perhaps because admirers thought that the use of this fine powder evoked the real essence of its wearer—no one can be sure—the term slowly evolved into meaning "the most subtle part" or the "essence" of a thing. By the sixteenth century, one writer wrote of *alcool vini,* the essence of wine. In time *vini* was dropped and *ethyl* was added so that since about the middle of the last century the essence of wine is now called ethyl alcohol. (*Eth-* came from *ether,* which chemists knew could be made from the essence of wine, plus *-yl* from the Greek *hyle,* signifying "material," that is, material that can be used to make . . . .)

Fermentation

Fermentation is the breaking down of large organic molecules into simpler substances through the action of the enzymes (the catalysts) in yeasts. Thus, ethyl alcohol is the end product of the fermentation of glucose, a sugar present in grape juice and other fruit juices. The overall change when glucose ferments is as follows.

Alcohol Fermentation

$$C_6H_{12}O_6 \xrightarrow{\text{fermentation}} 2CH_3CH_2OH + 2CO_2$$

glucose   ethyl   carbon
                   alcohol   dioxide

No oxygen is needed, and CO_2 is another product (Figure 16a). The set of enzymes that causes this fermentation is inactivated when the alcohol content reaches about 14% (v/v). Therefore, this concentration is roughly the maximum attainable without additional work, such as distillation.

Mead, Beer, and Wine

More than one kind of yeast catalyzes the change of sugar to alcohol, so modern makers of wine and beer try to control this aspect of the process by using carefully chosen and cultured species. Aside from this, the source of the sugary solution largely determines the nature of the product as well as what it is called. *Mead,* which is probably what, in an old expression, was the "nectar of the Gods," is

FIGURE 16a The frothing seen in this fermentation kettle is caused by the evolution of carbon dioxide.

the product of the fermentation of honey in water. *Beer* results from the fermentation of the sugars that form from starches when certain cereal grains sprout. *Wine* is from the fermentation of the sugar (mostly glucose) in grape juice, but the term *wine* is also used for the product of the fermentation of almost any fruit juice. In fact, judging from paintings on the walls of a temple at Thebes in Egypt (dating prior to 2000 B.C.), the first wine to be made in any systematic way was probably prepared from dates.

Although wine was well known to the ruling classes of all ancient peoples, the Greeks are credited with introducing its use to all classes. (The Greeks considered beer to be a beverage only for barbarians.) The Greeks elevated the cultivation of grape vineyards to a high level. Their Roman conquerors were quick to learn from them, and they brought this knowledge to Spain, France, southern Germany and Austria. Perhaps they tried their grape-growing skills in England, too, but the climate was not suitable.

In areas inhospitable to vineyards the skills of making beer and mead developed. The important step in making beer (and, much later, whisky) is in preparing the *malt.* When grains such as wheat and rye germinate, enzymes convert some of the starch into maltose (malt sugar) the same sugar that is quite abundant in corn syrup.

Distilled "Essences"

During the Middle Ages, probably in the twelfth century, someone applied the process of distillation to beers and wines. Ethyl alcohol boils at 78.5 °C, so when wine or beer is distilled, the first fraction is rich in the alcohol, rich enough to burn, in fact. This is why the distilled "essences" came to be known as *aqua ardens* (from the Latin for water that burns — *fire water* in more recent history). The most powerful alcohol that could be then obtained by distillation came to be called *aqua vitae* (from the Latin for water of life, probably owing to its use as a medicine). Descendants of the Vikings still call it akavit.

By the end of the thirteenth century one Raymond Lully added lime, CaO, to aqua vitae and distilled it again. He repeated the process twice more and thus prepared essentially pure alcohol — *absolute alcohol* as it is now called. What he succeeded in doing was removing the water largely by a chemical reaction:

$$CaO + H_2O \longrightarrow Ca(OH)_2$$

Absolute alcohol has a powerful burning taste, and it is a very dangerous substance; nobody tries to drink it. The strongest alcoholic beverages marketed today are between 50 and 60% alcohol.

At one time, as proof to a skeptical customer that a batch of alcohol hadn't been watered down too much, a little of the batch was poured over a small mound of gunpowder. The mixture had to ignite as "proof" that the liquid was really strong. When the concentration of alcohol is 50% (v/v), it passes this test, and this is why "100 proof" alcohol is actually 50% in ethyl alcohol. The percentage is always one-half of the proof rating.

Once the arts of distilling wines and beers spread, countless innkeepers, apothecaries (ancient pharmacists), and monks experimented with various herbs and spices and other concoctions. Wine that is distilled without added herbs or spices is called *brandy* after a Dutch word, *brandewijn,* meaning "burnt (distilled) wine." When special flavors have been added, the beverage is called a liqueur. For example, in 1510, Dom Bernardo Vincelli of the Benedictine monastery at Fècamp (in France), seeking a better medicinal beverage, used a still secret combination of herbs from the monastery garden to prepare a liqueur known today as Benedictine.

As we said earlier, the British Isles aren't particularly hospitable to vineyards. The arts of distilling beer reached their highest state no doubt in this area for this reason. In Scotland, the freshly sprouted malting grains are dried with peat fires, and substances in peat smoke are absorbed to add flavor and color to the final product, whisky. In Ireland, the peat smoke is not allowed to mingle with the drying malt, and the final product (now spelled whiskey, as

FIGURE 16b "Gin Lane", an engraving from 1751 by William Hogarth (1697–1764), the first great English engraver and painter. The sign over the gin shop in the lower left foreground reads "Drunk for a penny; dead drunk for twopence."

in America) has a different flavor. The name "whisky" came from a Gaelic expression, *uisge beatha* or "water of life," a term that became *uisquebaugh* before eventually evolving to its present form.

Abuse of alcohol has been one of the major drug-related health problems since alcoholic beverages were discovered. An angry engraving by Hogarth (Figure 16b) portrayed with little exaggeration the consequences in eighteenth-century England of the very widespread abuse of gin. Research has now confirmed what was long suspected — the compulsive, long-term overuse of alcohol, called alcoholism, causes permanent liver and brain damage.

Questions

1. Give the names and formulas of the products of the alcohol fermentation of glucose.
2. What prevents the complete fermentation of glucose?
3. What is the concentration (% v/v) of ethyl alcohol in 110-proof alcohol?
4. Fermentation occurs to 1.00 kg of 20.0% (w/w) glucose. Assume that all of the glucose ferments and that all of the CO_2 leaves the solution.
   (a) How many grams of ethyl alcohol form?
   (b) What is the alcohol concentration (% w/w) of the final solution?
   (c) How many milliliters of ethyl alcohol form? (The density of ethyl alcohol is 0.789 g/mL.)
   (d) The density of the *solution* is 0.982 g/mL. Using the answer to part c, what is the ethyl alcohol concentration (% v/v)? What is the proof of this solution?

QUALITATIVE ANALYSIS OF IONS

GENERAL APPROACH

SOME GENERAL EXPERIMENTAL TECHNIQUES USED IN
   QUAL ANALYSIS

CATION GROUP-1 ANALYSIS

CATION GROUP-2 ANALYSIS

CATION GROUP-3 ANALYSIS

CATION GROUP-4 ANALYSIS

ANION ANALYSIS

25.1 GENERAL APPROACH

Cations are separated into groups by the use of group reagents.

Qualitative analysis and how it works were described in Qual Topic I where the concept of cation groups was introduced together with a survey of some of the fundamental principles involved. Qual Topics II and III, strategically placed immediately following chapters on basic principles, continued these surveys. In this and succeeding sections we will briefly review these topics and get down to the details of how to carry out a qualitative analysis of a mixture of ions.

   Over the last several decades standard procedures have been developed that use reagents and small test tubes to separate, detect, and confirm the presence of over two dozen cations and a dozen anions. We will limit our study, however, to 15 cations and 8 anions, enough to illustrate the chemical principles. They will amply provide opportunities to practice those skills needed by all scientists—to observe properties, to use scientific judgment, to draw conclusions, and to deal with the unexpected.

Qual Procedures Apply to Samples in an Aqueous Solution   In the real world of analytical chemistry, a sample to be analyzed might come in any state, as a solid metal or metal alloy, for example, or as a granulated or powdered solid instead of the more convenient form of a solution. If the sample isn't in an aqueous solution, it must be

Qualitative analysis is often nicknamed *qual analysis* or just *qual.*

Enough detailed information and directions for qual analysis are given in this chapter to let it serve as a lab manual, too.

Before you start the lab work, learn the principles and reactions involved so that you know what's going on. Those who use the directions blindly *frequently* start over.

converted to one before the qual procedures we are studying can be used. The preparation of such a solution is a separate experimental problem, which we will not discuss here. For our purposes, we assume that the sample being analyzed is an aqueous solution of two or more ions. We also assume that the concentration of each ion is large enough so that a test for it should give unmistakable results.

Qual Procedures are Semimicro in Scale   The volumes of reagents to be used in the procedures range from 1 drop to 5 mL, so the scale of the work is *semimicro*. A smaller or micro scale is possible, and it is studied in more advanced courses. Tests on a larger scale — the macro — are seldom done because they usually take longer, require larger amounts of expensive chemicals, and create greater waste disposal problems.

Separation, Detection, and Confirmation Are the Three Main Steps   A given sample to be analyzed might contain several cations and anions. Different portions of the sample are separately analyzed for these two types of ions.

It is seldom possible to confirm the occurrence of one cation or anion in the presence of the others because any of the others might also react with the test reagent to produce confusing changes. The ions in the sample, therefore, must first be separated enough to allow the tests that detect and confirm any one ion to be free of such interference.

Sometimes detection and confirmation steps roll into one operation using one reagent.

Once the ions are separated, there are tests that individually detect their presence and, usually, additional tests that confirm that the ion detected is unmistakably there. Thus there are three broad steps in the qual analysis of the ions in a given solution — separation, detection, and confirmation. Carelessness in the first step is a guarantee of a sad result — starting over (sometimes with a new and *different* sample). So let us review what we have studied about separations in the earlier Qual Topics.

We will study a shortened version of the full standard set of cations listed in Qual Topic I, page 594.

Group Reagents Sort the Cations into a Few Groups   An overview of the scheme of cation analysis that we will follow is given in Figure 25.1. It indicates the three reagents, called group reagents, that are used to separate the cations into groups. More comprehensive schemes employ a fourth group reagent. (Reagents for detection and confirmation are discussed later.)

Cations Whose Chlorides Are Insoluble Make Up Group 1   The first group reagent is 6 M HCl. Its chloride ion reacts to form a precipitate with only 3 of the 15 possible cations — Ag^+, Hg_2^{2+}, and Pb^{2+}. These three form cation group 1, variously called the *chloride group* or the *silver group*. When 6 M HCl is added to the sample solution containing all group-1 cations, AgCl, Hg_2Cl_2, and $PbCl_2$ precipitate. The precipitate is separated and further analyzed for the individual cations in group 1. The next group reagent is applied to the remaining solution, now quite acidic, which could have as many as 12 cations (13, if some Pb^{2+} escapes separation in group 1, as it usually does).

The full list of cations in group 1 includes no other cations but these three.

Cations That Form Acid-Insoluble Sulfides Are in Group 2   The second group reagent is a saturated solution of hydrogen sulfide — about 0.1 M H_2S. Eight of our possible remaining cations can form insoluble sulfides but, as we learned in Qual Topic II (page 652), some metal sulfides are soluble in acid and others are not. The second group reagent, therefore, is first used on the test solution whose acidity has been adjusted to a pH of about 0.5, roughly the equivalent of 0.3 M HCl. If the pH is allowed to be higher and so more basic, group 3 cations will also precipitate as sulfides. But under the careful control of pH, just three of the cations that have survived the group-1 precipitation — Bi^{3+}, Cu^{2+}, and Sn^{4+} (or Sn^{2+}) — form insoluble sulfides.

As we mentioned (and for reasons we need not discuss here), Pb^{2+} usually escapes separation with the first group reagent, so its sulfide might now also precipitate. It's wise to assume the worst with the lead ion separation, to assume that enough escapes precipitation as $PbCl_2$ to be involved in group 2. If lead is found in group 1, we must be

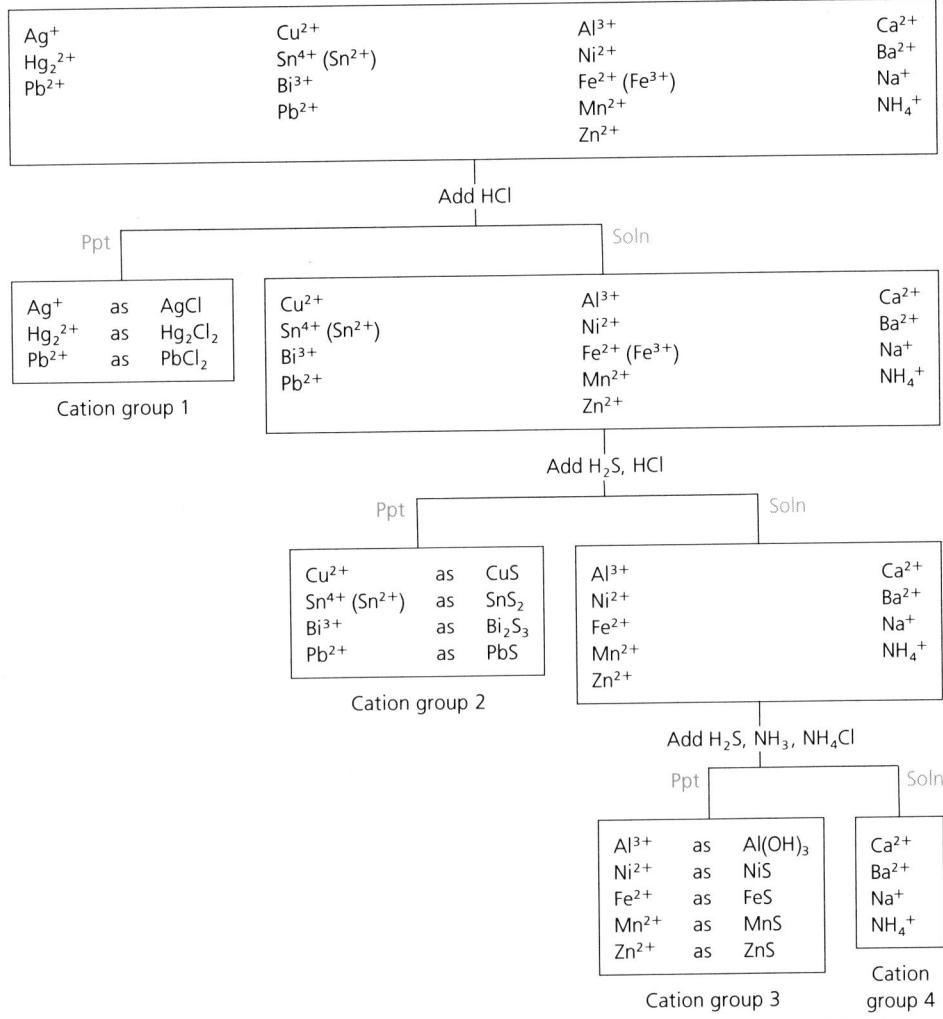

FIGURE 25.1 The cation groups and the group reagents by which they are separated. (Ppt is the abbreviation for precipitate.)

particularly alert to this possibility, and the standard procedures take account of it. Thus we consider that there are four, not three, acid-insoluble sulfides in our scheme, PbS, Bi_2S_3, CuS, and SnS_2. The metal ions in these sulfides — Pb^{2+}, Bi^{3+}, Cu^{2+}, and Sn^{4+} (or Sn^{2+}) — are the group-2 cations, called the *hydrogen sulfide group* or the *copper–arsenic group*. (The arsenic ion, As^{3+}, is one of the cations in more comprehensive schemes, but not in ours.)

Mercury(II), cadmium, arsenic, and antimony are also among the group-2 cations in longer versions of the qual scheme.

Cations That Form Acid-Soluble Sulfides Are in Group 3

The separation of the sulfide precipitate at this stage leaves a relatively strongly acidic solution holding the remaining cations, possibly Mn^{2+}, Ni^{2+}, Fe^{2+}, Al^{3+}, Zn^{2+}, Na^+, NH_4^+, Ca^{2+}, and Ba^{2+}. If Fe^{3+} ion had been present at the time when the sulfides of the group-2 cations had been precipitated, it would now be present as Fe^{2+}. This is because any Fe^{3+} is reduced by hydrogen sulfide (or thioacetamide) to Fe^{2+} under acidic conditions.

$$2Fe^{3+}(aq) + H_2S(aq) \longrightarrow 2Fe^{2+}(aq) + S(s) + 2H^+(aq)$$

But in acid, the sulfide of neither Fe^{3+} nor Fe^{2+} precipitates. Of course, if your analysis begins with a group-3 known or unknown, then either or both of the iron ions could be present.

Cobalt and chromium are present in longer versions of the scheme for the group-3 cations.

The third group reagent is also 0.1 M H_2S, but it is added only after the pH of the remaining test solution has been adjusted to be slightly basic so that the acid-soluble sulfides can precipitate. Of the possible nine remaining cations, five — Mn^{2+}, Ni^{2+}, Fe^{2+} (or Fe^{3+}), Al^{3+}, and Zn^{2+} — will form precipitates under these conditions provided that the solution is not acidic. These five are the group-3 cations, called the *ammonium sulfide group* or the *aluminum–nickel* group. Four of our group-3 cations — Mn^{2+}, Ni^{2+}, Fe^{2+}, and Zn^{2+} — precipitate as sulfides, and the fifth, Al^{3+}, precipitates as its less soluble hydroxide instead.

The hydroxides of some of the other remaining nine cations are also relatively insoluble in water, but two factors keep them from precipitating when the group-3 reagent is used. One factor is experimental; the pH of the solution is adjusted by an ammonia–ammonium chloride buffer to keep the concentration of hydroxide ion low. The other factor is that the sulfide-forming group-3 cations have sulfides that are far less soluble than their hydroxides.

The Remaining Cations Are in Group 4   Of the original mixture of cations in our scheme, there are possibly four remaining to be detected and confirmed — Na^+, NH_4^+, Ca^{2+}, and Ba^{2+}. The solution that remains after the group-3 sulfides have been separated, however, certainly contains one of these, NH_4^+, because NH_4Cl and NH_3 were used in the separation of these sulfides. Moreover, some reagents carry traces of sodium ion impurity, so it could now be present even though the original solution had none. Tests for these two ions, NH_4^+ and Na^+, must therefore be made on a portion of the original sample.

In procedures designed to handle more remaining cations than those in our scheme, a fourth group reagent is used to cause some of the cations to precipitate as insoluble carbonates. This creates the *carbonate group* (Ca^{2+}, Mg^{2+}, Ba^{2+}, and Sr^{2+}), and it leaves three cations in a last group called the *soluble group* — Na^+, K^+, and NH_4^+. We will not need to make this separation because there are tests for detecting and confirming any of the four of our group-4 cations — Na^+, NH_4^+, Ca^{2+}, and Ba^{2+} — in the presence of any of the others.

Anion Analysis Also Involves Groups   Figure 25.2 broadly outlines the scheme that we will use to analyze a solution for its anions. The group-1 anions consist of the anions that form insoluble barium salts in a basic medium — SO_4^{2-}, CO_3^{2-}, and PO_4^{3-}.

The group-2 anions are those anions that form silver salts that are insoluble in dilute nitric acid — Cl^-, Br^-, I^-, and HS^-.

In our scheme, there is only one group-3 anion, NO_3^-. Since all nitrates are soluble, the nitrate ion has to be detected by a special test done on the initial solution after it has been cleared of interfering ions by special methods. The nitrate ion does not itself interfere with the tests used to detect and confirm any other ions, so its separation from the others is not necessary.

25.2   SOME GENERAL EXPERIMENTAL TECHNIQUES USED IN QUAL ANALYSIS

Study this entire section before you begin the laboratory study of qualitative analysis.

Several laboratory techniques are used in qualitative analysis, and we discuss them in detail here. They aren't hard to master.

Tap water in almost all locations is chlorinated, so it contains Cl^- ion.

Distilled Water Must Be Used in Qual Procedures Requiring Water   Nearly all tap water contains ions, including some of the ions being tested for by our qual procedures. "Hard" water has Ca^{2+}, Mg^{2+}, and Fe^{2+} (or Fe^{3+}) — at least one of these — as well as some anions, like Cl^-, SO_4^{2-}, or HCO_3^-. Sodium ion is nearly always present, too. To avoid false tests caused by these ions in tap water, always use distilled water to prepare qual reagents and to give glassware its final rinse.

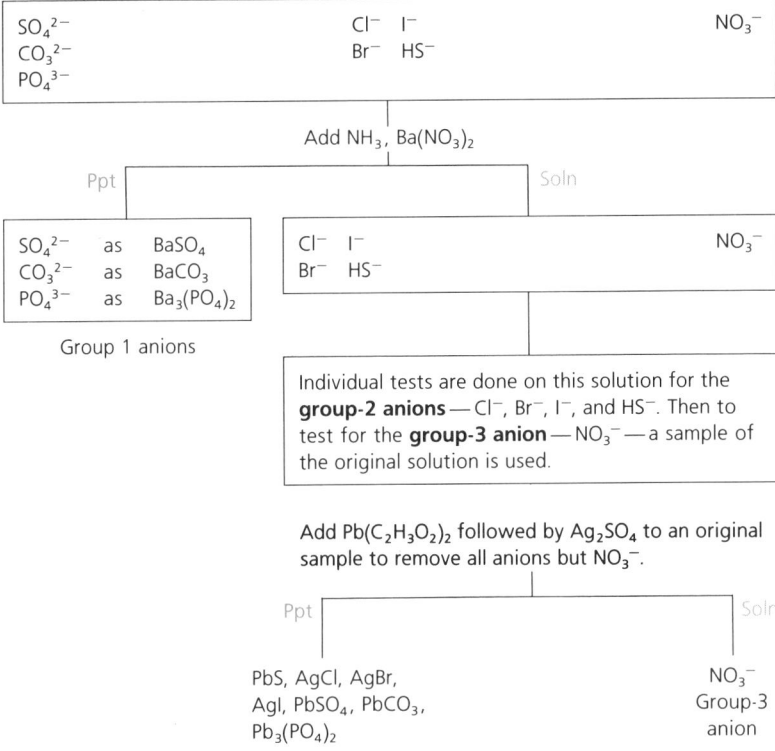

FIGURE 25.2 The anion groups.

Many labs use deionizers to make their "distilled" water, and such water is often called deionized water. For purposes of qual analysis, the two are equivalent, and when we say "distilled water" we mean one form or the other.

Keep your own supply of distilled water in a plastic squeeze bottle (Figure 25.3) supplied at your lab desk. (If you are in a lab where glass wash bottles are used, you will be given instructions on how to prepare one.) The wash bottle has a flexible delivery tube that lets you direct a stream of distilled water where you want it to go. Be sure to use such water when necessary, but only sparingly; it's quite expensive.

A good rule is "Don't assume any glassware is clean enough for your needs unless you personally have washed and rinsed it." Scrub glassware with detergent and brush (supplied near the water taps and sinks). Then rinse the pieces repeatedly with tap water to remove all traces of the detergent. Finally, rinse each piece twice with small amounts of distilled water. Do not use toweling to dry freshly rinsed items; let them drain dry. Invert rinsed test tubes into a clean beaker for air-drying, not onto the (often contaminated) pegs of a test tube rack. (Most of the time, no drying is necessary.) To dry beakers, lay a stirring rod on a few layers of paper toweling on your desk top and invert the beaker so that an edge is propped up slightly by the rod. This allows air to circulate under the beaker and promote drying.

When in doubt, rinse glassware with distilled water.

Volumes of Qual Reagents Are Seldom Measured with Graduated Cylinders

Since qual tests are *qualitative* tests, exact volumes of reagents are almost never needed when such tests are carried out. Instead, volumes are estimated either by counting drops or by judging from the height of a liquid in a test tube. With *aqueous* solutions or water, 20 drops from a medicine dropper and 40 drops from a capillary pipet roughly equal 1 mL. To estimate milliliter amounts, calibrate one of your centrifuge tubes as follows. Trace its outline somewhere in your lab notebook or on a card that you can keep in your

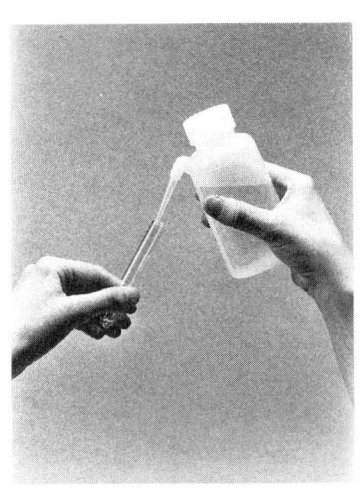

FIGURE 25.3 Plastic wash bottle.

lab desk. Then fill it with 1-mL, 2- mL, up to 5-mL amounts of water (using a graduated cylinder or graduated pipet). At each volume mark your drawing. You will soon be able to estimate volumes using these marks as guides.

You Will Have Your Own Supply of Some Qual Reagents   Several of the qual reagents are frequently used, and it is handy to have your own supply. (This also minimizes the chances of these reagents becoming contaminated through their careless use by others.) If this arrangement is made possible in your lab, your desk will have several dropping bottles. Your first job with these is to clean them and affix your own labels. Remove the droppers and take off their bulbs. Wash and rinse the bottles, the droppers, and the bulbs. (If a bulb is cracked or is otherwise in bad shape, get a replacement at the stock supply.) Then label the bottles. The following list is typical, but your instructor will advise you in detail.

> Don't let a drop of any of these reagents touch your skin without rinsing it off as soon as possible.

0.1 M HCl	1.0 M NH$_3$	5% Thioacetamide
6.0 M HCl	3.0 M NH$_3$	4.0 M NaOH
6.0 M H$_2$SO$_4$	6.0 M NH$_3$	6.0 M HNO$_3$
	Distilled water	

Find the stock solutions of the reagents and fill your dropper bottles about half full.

Less frequently used reagents are available in dropper bottles at the designated reagent shelves together with solutions of "knowns." A *known* in qual analysis is a solution whose solute ions are disclosed to you. There are individual knowns for each cation and anion as well as knowns for the various groups and combinations of groups.

WARNING   When obtaining knowns, get the dropper back into the right bottle.

This rule is quite obvious, but you should mutter it to yourself each time you are at the reagent shelf.

Masses of Solid Reactants Can Also Be Estimated   The precision delivered by a laboratory balance is seldom needed in qual work. Most of the solids you will use have about the same density, so it is possible to estimate masses by the size (meaning volume) of the sample. If the sample is about the size of the head of a paper match, its mass is about 50 mg. A completely loaded semimicro metal spatula holds about 150 mg when heaped and 100 mg when rounded. Just a "tip-load" is about 75 mg. Some time in the lab, when there is no crowding at the balance, check these with sodium chloride using the semimicro spatula you have been issued.

> Remember, of course, never weigh a chemical directly onto a balance pan.

WARNING   A spatula must be cleaned and wiped completely dry before it is used to carry a solid reagent.

Be Sure That Mixtures Are Thoroughly Stirred   Failure to stir mixtures and solutions thoroughly is a common cause of an incorrect qual analysis. Clean glass rods

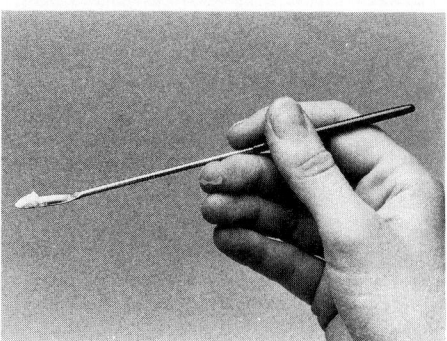

50 mg of solid

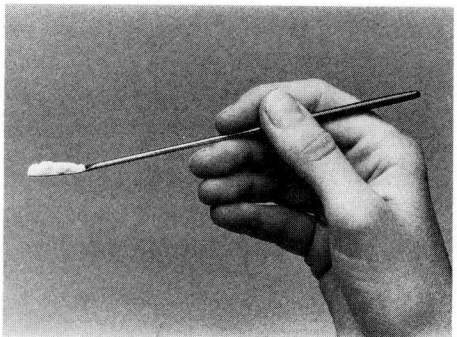

100 mg of solid

These operations begin with a relatively long length of glass rod or tubing, and there is only one way to carry one.

WARNING Hold glass tubing or rod **vertically** and move cautiously when carrying it from one place to another.

Obtain a piece of glass rod or tube and a triangular file, have a towel ready for grasping the glass, and read the entire next paragraph before proceeding.

To make a clean cut, place the rod or tube flat on the bench. Bearing down with some pressure, sharply pull the edge of the triangular file across the glass to make a single, deep scratch crosswise where you want the break to be. Quickly moisten it with a drop of water (or saliva). Grasp the glass firmly in the towel, as shown in Figure 25.4(a), with both thumbs close to but opposite the scratch. Use your fingers to apply pressure to the tubing to snap the glass as you simultaneously exert a pull away from the scratch. It works best if the snap comes soon after you make the scratch because for a moment following the scratch there is some strain in the glass at that place. The moisture seems to increase the strain. Be sure that your fingers are protected by the toweling if the glass happens to break where they are instead of at the scratch mark.

The new ends are very sharp and may even be jagged. If jagged, you can usually break off most of the sharp projection as follows. First, be sure you are wearing eye protection. Then, take the glass, a towel, and a wire mesh square (the kind used to support a beaker on an iron ring at a ring stand) to the nearest receptacle for solid wastes. Remember, whenever you carry glass tubing or rod anywhere, always hold it vertically, and move cautiously. Grasp the glass with a towel close to its jagged end and, holding it over the receptacle, gently strike the jagged projection with the wire mesh, using a glancing motion. This should break off the any sharp spur, and the fragment should fall into the receptacle.

The end is still very sharp, and it must be fire polished. To do this, hold the end to be fire polished in the hottest part of the flame—just above the flame's inner blue cone as illustrated in Figure 25.4(b)—and rotate the glass until you notice its sharp edges softening. You can see this happening as you watch the flame and the hot glass take on a richer and richer yellowish color.

WARNING Hot glass looks exactly like cold glass. Set hot glass aside where you will not pick it up until it is cool.

Hot tubing must never be placed on a painted surface. It can be laid on a stone surface, on a square of wire gauze, or across the top of a beaker. (Make sure it cannot roll anywhere.)

Tapered Stirring Rods To make a tapered stirring rod, before you cut a glass rod to the desired length, you have to soften it in the flame and draw it out until you have given it as narrow a constriction as you want. Obtain a piece of glass rod long enough so that you can comfortably hold it at its ends with your hands as you rotate its center over the hot part of a flame. (If the ends of the rod are sharp, be careful they don't cut you while you're rotating the rod.) By rotating it in the flame, you ensure that it will be evenly heated. You will soon notice that the rod softens in the flame and that if you don't continue to rotate it, it will sag. Now is the moment to remove it from the flame, hold it firmly in a *vertical* position, and gently but firmly pull on its opposite ends. If nothing happens, you haven't softened the center enough, so heat it some more as described just above.

If you pull the softened rod too strongly, you can draw it out to the diameter of a capillary or even pull it fully apart. But you don't want this; you want the end diameter to be about a millimeter after you break the rod and fire polish the ends.

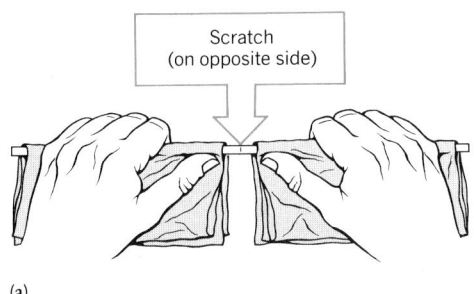

Scratch
(on opposite side)

(a)

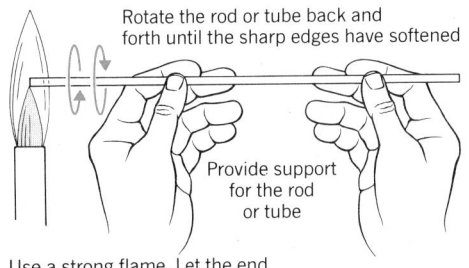

Rotate the rod or tube back and forth until the sharp edges have softened

Provide support
for the rod
or tube

Use a strong flame. Let the end
of the rod be over the tip
of the inner blue cone

Hold in outer edge of flame for tubing

(b)

FIGURE 25.4 Working with soft glass. *(a)* How to hold a freshly scratched glass tube or rod when breaking it. (Imagine the glass is a pencil to be broken in this manner.) *(b)* Fire polishing a glass tube or rod. Remove the glass from the flame as soon as you see its edges become rounded.

It's not a good idea to stopper a test tube with a cork and then shake it. The cork might be contaminated. And never ever stopper a test tube with your thumb!

are usually employed to stir mixtures. Never use a metal spatula; enough of the metal (or its oxide coating) may dissolve to give a positive test for its cation. And never use a pencil or a pen to stir anything. (It's been done!) If your lab desk is equipped with special glass test tubes with cone-shaped bottoms — centrifuge tubes — you will want to have tapered stirring rods so the rod can fully reach the bottom.

Make as many stirring rods as your instructor suggests, including tapered rods if necessary. Use 3-mm glass rod. Be sure to fire polish the ends, as described in Special Topic 25.1.

Stirring rods should be rinsed with distilled water before each use. If you set one aside, rinse it first and then be sure it touches no surface where it can pick up a contaminant.

A test tube normally should not be filled more than half full, particularly if you are going to heat its contents.

Sometimes it is possible to mix the contents of a tube by holding it firmly at the top between the thumb and index finger of one hand and then flicking it several times with a finger of the other hand. Do this carefully so that the contents of the tube don't splash out. Be sure that you hold the tube firmly enough not to drop it but not so tightly that its contents cannot be agitated.

For smooth boiling, place a "boiling chip," for example, a small piece of unglazed pottery, in the hot water bath.

Set Up a Hot Water Bath at the Start of Each Lab Period   A test tube is seldom if ever heated directly with a Bunsen burner in the qual lab. This can produce very uneven heating. And the risk of losing some or all of the contents by sudden boiling is too great. The hot liquid might also hit your hand or face. To avoid this, use a hot water bath (Figure 25.5) into which tubes can safely be set and heated while you do something else. Since it takes some time to get the bath water heated, it's a good idea to set up the bath at the start of each lab period (using tap water). Then it is ready at all times.

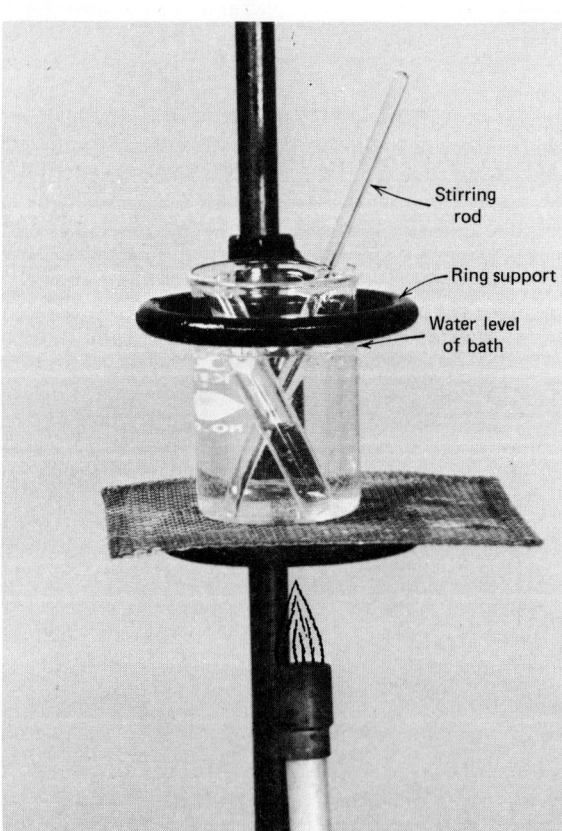

FIGURE 25.5 Hot water bath. Be sure there is a boiling chip in the water and that you have labeled the tubes. From Jo A. Beran, *Laboratory Manual for Fundamentals of Chemistry*, 2nd edition, 1984. John Wiley & Sons, New York. Used by permission.

The flame beneath the hot water bath should be kept no hotter than what is just barely necessary to keep the water very gently boiling. Place this system at the back of your lab bench where your activities or those of other students cannot accidentally upset it.

In the rare situations when a test tube is heated over an open flame, it must be held by a test tube holder and it must never be more than half full. The flame is allowed to play over the surface of the tube *just above the surface of the liquid,* never at the bottom. The tube is agitated from time to time to permit its contents to contact the heated glass surface as well as to mix, and it's a good idea to move the tube in and out of the flame. There is great danger that the tube's hot (and maybe chemically dangerous) contents will erupt or "bump" out of the tube and hit your hands, clothing, or face (not all of which is protected by safety glasses). If you ever have to heat a test tube over a flame, practice it first with half a tube of water. Never let the tube point at anyone.

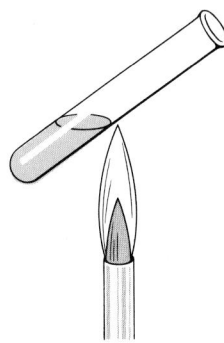

Heat the tube just above the liquid's surface, and slosh the liquid onto the hot surface.

Add Precipitating Reagents Carefully When a qual procedure is expected to produce a precipitate, you should add the precipitating reagent dropwise with stirring to avoid adding too much. Sometimes an excess causes the precipitate to redissolve! Moreover, when precipitates form slowly, their particles tend to be larger and therefore they more readily separate.

Sometimes the expected precipitate doesn't seem to form although the solution becomes cloudy. What you may have is the solid in such a finely divided state that it is colloidal. Then it helps to digest the material. Digestion is done by warming the mixture with the hot water bath and then cooling it slowly. Solid that has reentered the dissolved state when the mixture was warmed can now come back out of solution in the form of larger particles that more readily separate.

Occasionally, specks of solid persist in floating on the surface of the liquid that's been centrifuged. This is a surface tension effect. Try pushing them down with a glass rod. Otherwise, let them stick to the rod and discard them. The specks probably consist of sulfur, which can always be discarded. (You will see how sulfur can form when we go into the details of forming insoluble sulfides and redissolving them.)

The precipitation of a solid must usually be as complete as possible so that its ions cannot interfere with tests on the remaining solution. Such interference is a common cause of mistakes in qual analysis. Before we can describe how to test for the completeness of precipitation, we have to discuss how the precipitate is separated, which we'll do next.

Semimicro Work Uses the Centrifuge The centrifuge (Figure 13.1, page 494) is the only special mechanical device required for the manipulation of chemicals on a semimicro scale. It is used for the rapid and quantitative separation of a precipitate from its supernatant liquid. This is the liquid above the solid and, when produced with a centrifuge, is called the centrifugate. The centrifuge technique lets you separate and manipulate a small mass of precipitate that would be lost on the relatively vast expanse of even a small folded piece of filter paper. This technique is also much faster and more quantitative than gravity filtration, and no solution is lost from being blotted up by a piece of filter paper.

Centrifuges are equipped with removable *inserts.* These are metal sleeves that are attached to the centrifuge's rotor; when you want to centrifuge something you slip the test tube into one of these inserts. Sometimes a test tube breaks as it is being spun in the centrifuge, and then the insert keeps the glass pieces and your solution from the mechanical parts of the machine. Since an insert can be removed for cleaning, someone may forget to put it back. **Always be sure that the inserts are all in place when you use the centrifuge.** Otherwise, your test tube will disappear into the hole where the insert is supposed to be, your solution will spill out and be lost, and the centrifuge will have to be disassembled to retrieve the tube and then cleaned. Not a way to win friends and influence instructors!

As indicated earlier, your lab desk might be equipped with special, tapered centrifuge tubes. If so, the centrifuge you will use will have the appropriate inserts for accepting such tubes. Some centrifuges accept slightly larger, 4-in. (100 mm) standard round-bottomed test tubes. If you have both kinds of tubes, be sure to check the centrifuge to see which inserts are provided, and then make sure that you are using the proper glass tubes for the inserts provided.

All the precipitates encountered in our qual procedures are more dense than the liquid phase, so when a liquid mixture with a precipitate is whirled at high speed in a centrifuge, the solid quickly and firmly settles and packs at the bottom of the centrifuge tube. As we said, sometimes precipitates form in such a finely divided state that they are actually colloidal in size. Their separation at the centrifuge takes longer.

Nearly all labs are equipped with centrifuges with high enough rpms (revolutions per minute) to require two special precautions. The first is a protective cap, which was part of the picture of the centrifuge in Figure 13.1 (page 494). This cap must be in place over the top opening of the centrifuge housing to keep objects (maybe fingers) out of the way of the whirling rotor. The second precaution concerns the need to balance the masses being spun by the centrifuge. When a centrifuge is run with a serious imbalance of masses, it will shudder, shake, and sometimes walk off the bench top. To balance the masses, place a second tube in the centrifuge insert that is opposite the tube holding your solution. The second tube should contain about the same volume of water as the volume of mixture in the other tube. This provides a close enough balance and you don't have to weigh the two tubes.

Let the centrifuge run at least 30 seconds (sometimes longer when colloidal solids are involved), then shut it off, *wait for the spinning to stop completely before lifting the cap,* and remove the tubes. Be sure you have made provision for telling which tube holds your solution and which holds water! This is particularly important in labs where two students share a centrifuge, and one student's solution provides the mass balance for the other student's solution. If you can't tell which tube is which, you probably will have to start over.

It Is Easy to Forget What's in a Tube or Flask

When you set an unlabeled flask on the bench top or an unlabeled tube into a test tube rack, it won't take long for you to become unsure about what it contains. If you are supplied with glassware having small roughened white areas, you can write on these with a soft pencil and later easily erase the marks. Do not use ink or crayons on such areas; they are too hard to clean then. If your glassware lacks these features, then use a wax crayon to write labels.

Decide on a marking scheme that you can carry over to your lab notebook. A combination of letters and numbers works well, but avoid a complicated system. Thus, you might label a tube that holds the first precipitate in a series of tests with the symbol P-1. When you transfer its centrifugate to another tube, this tube might be labeled C-1, and so on. Your instructor might suggest something else.

Centrifugates Are Removed by Decantation or with Capillary Pipets

When the precipitate is dense and very well-packed at the bottom of the centrifuge tube, the centrifugate can be removed simply by decantation (Figure 25.6). But when this procedure might disturb the solid, you can withdraw the centrifugate with a capillary pipet. If you are not supplied with capillary pipets, they are easily made. See Special Topic 25.2.

To remove centrifugate with a capillary pipet, tilt the test tube to about a 30° angle. Squeeze air from the pipet first and then lower its tip to just below the surface of the centrifugate. As you carefully release the bulb, slowly lower the pipet as liquid is withdrawn. Be particularly careful not to stir up the solid. Continue until it is apparent that to take more material would withdraw some of the solid.

If you place the smaller centrifuge tube into an insert meant for a 4-in. test tube, you may have trouble recovering it.

About 30 seconds of spinning time is ample in nearly all circumstances.

Never leave the centrifuge when you're using it. Be prepared to shut if off if you should hear anything crack or if it vibrates too much.

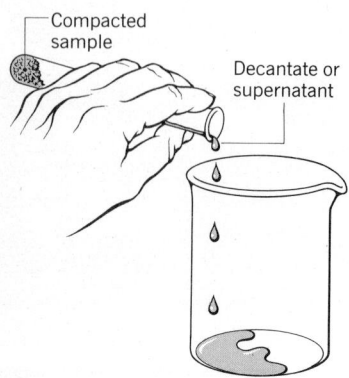

Compacted sample

Decantate or supernatant

FIGURE 25.6 Decantation.

Wax markings do not take very easily to an unclean or wet glass surface.

We will suggest a marking code in later sections.

If you insert the pipet into the centrifugate *before* you squeeze the air from it, the bubbles might disturb the precipitate. If this happens, simply recentrifuge.

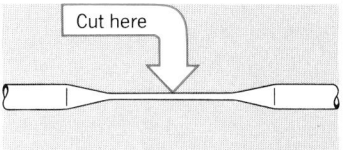
The Centrifugate Can Be Tested for Completeness of Precipitation   To find out if you used enough precipitating reagent when you made the precipitate form, add a drop of the reagent to the centrifugate *before you transfer it*. See Figure 5, Qual Topic I (page 595). If no cloudiness appears, precipitation was complete. Otherwise, centrifuge the mixture to remove the cloudiness, and repeat these steps until the test gives a clear supernatant liquid. Then withdraw it.

Precipitates Are Usually Washed Twice   The precipitate, of course, is wet with the remains of the centrifugate. This liquid, therefore, contains solutes that might interfere with subsequent tests on the precipitate. So the remaining centrifugate that still wets the solid has to be regarded as a contaminant and has to be replaced by water. This is done simply by washing the precipitate with distilled water, usually at least twice.

To wash a precipitate, add the distilled water, thoroughly mix it with the precipitate using a (freshly rinsed) glass stirring rod, centrifuge this mixture, and then withdraw the centrifugate. Do not throw this second centrifugate away until you know whether it should be combined with the first.

The solid, of course, is still wet, but wetted by a solution much more dilute in other possible ions than the previous solution. A second wash generally leaves the residual liquid so dilute in other ions that they can cause no interferences in later tests on the solid. (You can discard this third centrifugate.)

WARNING   Carelessness in the washing of a precipitate is perhaps the single most common source of error in a qual analysis.

Solutions Can Be Concentrated by Evaporation   When an initial centrifugate is combined with the centrifugate obtained by the first washing of a precipitate, the resulting solution might be too dilute for subsequent work. To make this solution more

Sometimes a precipitate is washed with a solution instead of distilled water. When to do this will be made clear in the detailed directions given later.

Use a spurt of distilled water to rinse any precipitate from the stirring rod as you withdraw it.

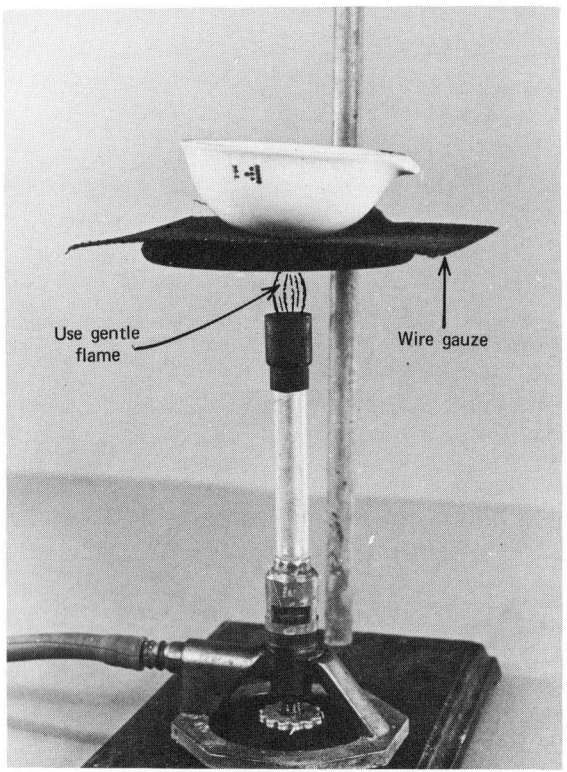
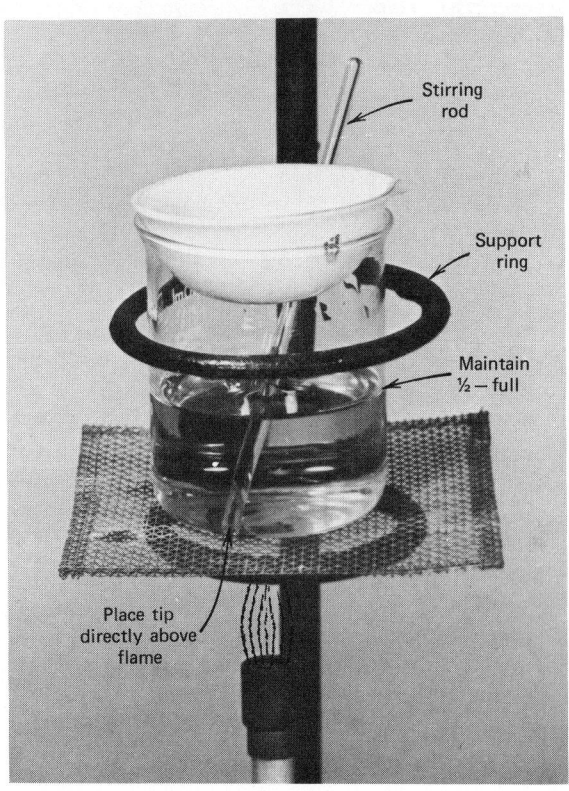

(a)   (b)

FIGURE 25.8 Using an evaporation dish. *(a)* Evaporation of a nonflammable liquid using a direct, low flame. *(b)* Evaporation of a nonflammable liquid over a boiling water bath. From Jo A. Beran, *Laboratory Manual for Fundamentals of Chemistry,* 2nd edition, 1984. John Wiley & Sons, New York. Used by permission.

If you suspect that a drop has spattered onto your skin, first turn off the burner and then rinse the spot with water. (Never leave a burner unattended.)

concentrated, some of its water must be removed by evaporation. To speed up this process, the solution can be placed in an evaporating dish (Figure 25.8). Take the dish with its sample to the hood — it should never be filled to over half its capacity — and set it on a wire gauze supported by an iron ring on a ring stand. Carefully heat the dish with a low Bunsen burner flame. The wire gauze distributes the heat and helps to prevent hot spots. The danger is that the liquid, if boiled too rapidly, will spatter. It's a good idea to practice this technique first with an evaporating dish that holds a third or less of its volume of water.

If a procedure calls for evaporating a solution to dryness, be particularly careful about hot spots near the end. Remove the flame when you judge that only a few drops of water remain to be evaporated. There will be enough heat in the container to finish the process.

A beaker can also be used for evaporation, but spattering can be more of a problem. It helps to stir the liquid from time to time, and you can leave a stirring rod in place for this.

A Reagent, Rather Than Water, Is Sometimes Used to Transfer Solids   Sometimes the solid is to be treated with a reagent and transferred to another tube. Use the reagent itself to aid in this transfer. Add some of the reagent to the centrifuge tube that contains the (washed) solid, stir the mixture well, and pour it into the second container.

When the reagent is one that *dissolves* the solid, allow time for this process before you pour the mixture. To help the solution form, warm the tube in the hot water bath. Use the remainder of the reagent to rinse out any material still clinging to the walls of the centrifuge tube.

One Solid Can Be Separated From Another by Extraction   In many situations, a new precipitate is actually a *mixture* of solid compounds. But one or more of these might be soluble in a special reagent or in hot water while the others remain undissolved. The removal of the soluble solids from the insoluble ones by such a special solvent is called extraction.

The extraction solvent (or solution) is added to the washed precipitate in the tube, and the mixture is stirred with a glass rod or is otherwise agitated. The tube might be warmed in the water bath during this operation. Then the tube is centrifuged, and the centrifugate, now with additional solute, is removed in the usual way. The procedure can be repeated to make the extraction more complete.

Try not to use too much time in centrifuging a hot solution because it might cool enough to cause some solute to precipitate.

pH Paper Is Never Dipped into a Solution   When you have to find out if a solution is acidic or basic, dip the end of a glass rod into it and then touch the rod to litmus paper (or other pH paper, if supplied). This prevents a possible contamination of the solution.

Sometimes you want to see if an acidic gas (perhaps SO_2) or a basic gas (ammonia) is evolving from a solution. To find out, moisten a piece of indicator paper with distilled water and carefully suspend it inside the mouth of the tube or flask. *Do not let it touch the wall anywhere.* The wall may be wet with the solution, which could itself be acidic or basic.

Do Not Put Your Nose Over the Opening of a Tube or Flask to Check for Odor   A noseful of hydrogen sulfide or ammonia can really hurt. So to check the odor of a solution, keep your nose a few centimeters back from the rim of the container and use your hand to waft vapors from the tube toward the nose.

Test Your Skills and the Reagents with Knowns   A known in qual analysis is a solution whose solutes are identified for you. A group-1 known for cations, for example, would include all three group-1 cations in water—Ag^+, Hg_2^{2+}, and Pb^{2+}. Similar knowns are available for individual ions and for the other groups (or instructions are provided to permit you to prepare them). To study the separations of groups, some knowns include ions from more than one group.

No unknown should be analyzed without first applying the procedure and all its tests to a known.

It's easy to understand the reason for this rule. First, analyzing a known gives you the necessary experience in carrying out the procedures. It also lets you see what positive tests look like. It lets you find out if the reagents are satisfactory. Sometimes they are contaminated and give misleading results. And it gives you confidence.

Some Cations Are Detected by Flame Tests   At the ends of Sections 21.3 and 21.4 we described in general how several cations can be detected by the color they produce in a flame test. To perform such a test you need a short length of platinum wire coiled into one very small loop at one end and sealed into a piece of glass at the other end. To remove any interfering impurities from this wire, dip its loop into 6 *M* HCl and bring it slowly up to the outer edge of a Bunsen burner flame (Figure 25.9). Don't put it into the hottest part of the flame, and never thrust it into the central, reducing part (the inner cone) of the flame. Repeat this treatment with acid and heat until the wire causes no color in the flame. Now dip the loop into the solution you want to test, and bring the wet loop slowly up to the flame edge in the usual way. Notice the color, if any. It may not last long. Always do the same test on a single known so that you can get a quick color comparison. Always clean the wire when switching from one solution to another.

See pages 889 and 896.

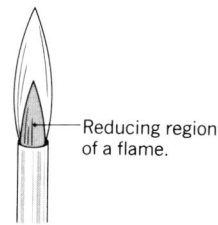

Reducing region of a flame.

Keep a Lab Notebook   Some instructors specify the lab notebook you must use. (If so, it will probably be a sewn bound, not a spiral bound, notebook with cross-hatch

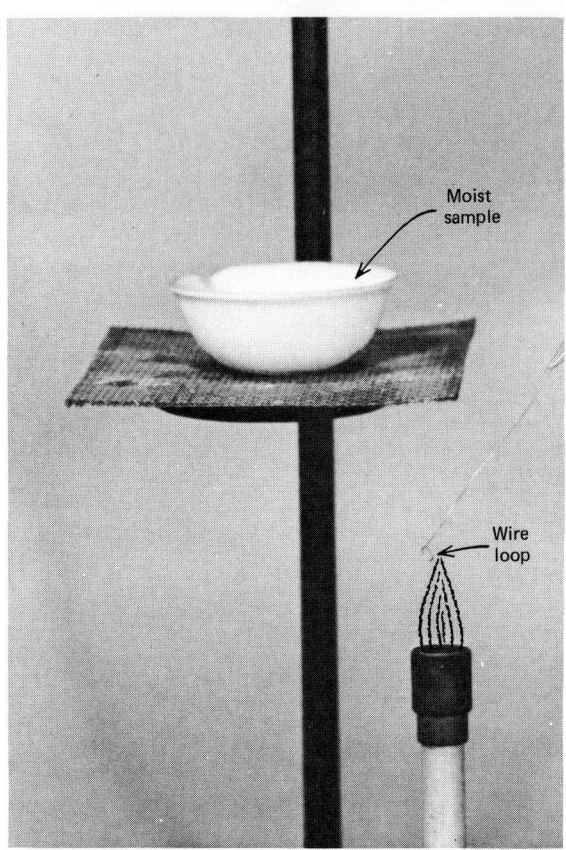

Moist
sample

Wire
loop

FIGURE 25.9 Flame test. From Jo
A. Beran, *Laboratory Manual for
Fundamentals of Chemistry,* 2nd
edition, 1984. John Wiley & Sons,
New York. Used by permission.

Be thorough in identifying
solutions, centrifugates, and
precipitates. Write equations in
your notebook for the reactions
that you are doing in the lab.

rulings.) One of the important purposes of the lab is to get experience in doing experimental work in the way done by scientists. And scientists know from painful experiences (sometimes in beginning labs, sometimes much later when they want to apply for a patent on a new discovery) that observations and data must be recorded promptly and accurately. They must also be recorded in ink. The pages must be numbered. Pages are never torn out; data or other entries are never erased. Entries can be crossed out, however. The first page or two of the notebook are left blank for the development of a table of contents. (In research situations, the chemist and a witness sign and date each page. Then a duplicate of each page is filed in a separate facility. Research notebooks can be purchased that are provided with removable, duplicate pages.)

Data and observations go directly into the lab notebook, never first onto separate pieces of paper, like towel paper or filter paper. (Some instructors are known to confiscate such things, sometimes with a hint of glee.) But observations sometimes come rapidly, so advance planning can be very important.

Your instructor will advise you on the format you should use, but expect to be required to summarize all procedures used, report the results of tests on both knowns and unknowns, and give balanced equations for all reactions observed on the unknown. Figure 25.10 gives an example of how a notebook might look after some tests on a group-1 known.

Be Safety Conscious at All Times Accidents with serious results are never totally prevented, but their frequency can be reduced greatly by common sense and a habit of safety awareness. We've stressed safety throughout the foregoing discussion. The most important safety rule is to think safety. Be safety conscious. Plan ahead each experiment and know the chemistry (and potential hazards) of what you expect to happen.

Experiment: Analysis of Cation group–1 Unknown   date: 4/16

Procedure	Substance	Reagent	Observations	Possibles formulas ppt.	Possibles formulas Solu	Inferences
1A	group 1 unknown	6M HCl	ppt. formed	$AgCl, Hg_2Cl_2$ $PbCl_2$	any in group 2–4	Cation grp 1 ions are present
1B	1P1	hot water	some ppt. seems to dissolve	$AgCl$ Hg_2Cl_2	Pb^{2+}	Pb^{2+} may be present
1C	1C2	K_2CrO_4	yellow ppt	$PbCrO_4$	—	Pb^{2+} is confirmed
1D	1P2	6M NH_3	all ppt dissolves	—	$Ag(NH_3)_2^+$	No Hg_2^{2+} present; Ag^+ probable
1E	1C3	4M HNO_3	white ppt	$AgCl$	—	Ag^+ is confirmed

Conclusion: Unknown 27 contains Pb^{2+} and Ag^+.

FIGURE 25.10 Sample page from a lab notebook.

Here are the other rules.

1. Use protective eye wear. If a chemical still enters the eye, wash the eye in running water at an eyewash fountain for 15 minutes. Get the instructor to help, and seek medical attention.

2. Touch no chemical. Rinse a spill off at once with copious amounts of water. Ask your instructor for help.

3. Clean up chemical spills as soon as possible. Your instructor will help you decide how to do this. The procedure varies with the nature of the chemical.

4. No food or beverage is permitted in the lab. These can absorb and concentrate poisonous fumes. Poisonous fumes can also be made unexpectedly more dangerous when inhaled with tobacco smoke, so there is a No Smoking rule in the lab, too. No radios or similar equipment, either. There are enough other distractions that can affect safety.

5. If you cut or burn yourself, see your instructor at once for help. Except for very minor injuries, you must get medical attention. Always inspect glassware for cracks, particularly "star" cracks. Such glassware MUST be discarded into a designated receptacle.

6. Use the hood when there is the possibility that poisonous vapors will develop, particularly when using thioacetamide to make hydrogen sulfide. (This gas is more poisonous than hydrogen cyanide.) Wash your hands thoroughly when leaving the lab.

7. Never taste a chemical. Check odors cautiously as we described earlier.

8. Work at the hood when you are evaporating a solution.

9. Never let a heated tube point at yourself or a neighbor.

10. Do not dispose of any chemical or solution except where designated by your instructor. (Consider all solutions used in qual analysis to be potentially harm-

ful to the environment. Not all are, but we want to be very careful about what goes down the drain.)

25.3 CATION GROUP-1 ANALYSIS

The group-1 cations form chlorides that are insoluble in cold water.

The three group-1 cations have no color in water.

Of all the cations that chemists regard as "common," only three form sparingly soluble binary chlorides—Ag^+, Hg_2^{2+}, and Pb^{2+}. All three of their chlorides are white. AgCl, however, slowly darkens as it remains exposed to sunlight, because sunlight causes the conversion of some Ag^+ ion into very finely divided metallic silver, which is black.

Throughout all our discussions in this chapter, group will always mean a certain cation group or anion group, not a column in the periodic table.

The group reagent used to isolate the group-1 cations is 6 *M* HCl. Figure 25.11 shows the flow diagram for the analysis, and we will refer to it in our discussion. A **flow diagram** is a display that outlines how the components of a complex mixture are separated and how individual chemical species are finally identified. Special Topic 25.3 describes the meanings of the vertical and horizontal lines, the brackets, and the boxes.

As the data in Table 25.1 show, the chloride of Pb^{2+} is considerably more soluble than the chlorides of Ag^+ and Hg_2^{2+}. Enough lead ion, therefore, might escape the procedure to show up in group 2, also. As we have said before, the procedures for the group-2 cations take this into account, and Pb^{2+} is also a member of group 2. In fact, when a multigroup known or unknown is analyzed, there are advantages in carrying the lead(II) ion entirely into group 2. This can be done by heating the mixture from the addition of 6 *M* HCl before it is centrifuged to bring $PbCl_2$ into solution. We mention this here and refer to it later, but we also assume in the following discussion of group-1 analysis that enough Pb^{2+} will, if present, remain a part of this group to warrant testing for it.

The three group-1 cations also form insoluble sulfides, so the complete removal of at least two of them (Ag^+ and Hg_2^{2+}), if not entirely the third (Pb^{2+}), as insoluble chlorides simplifies the analysis of group 2.

The solubility equilibria for the group-1 metal chlorides are

$$AgCl(s) \rightleftharpoons Ag^+(aq) + Cl^-(aq) \qquad K_{sp} = 1.8 \times 10^{-10} \text{ (25 °C)}$$
$$Hg_2Cl_2(s) \rightleftharpoons Hg_2^{2+}(aq) + 2Cl^-(aq) \qquad K_{sp} = 1.1 \times 10^{-18} \text{ (25 °C)}$$
$$PbCl_2(s) \rightleftharpoons Pb^{2+}(aq) + 2Cl^-(aq) \qquad K_{sp} = 1.7 \times 10^{-5} \text{ (25 °C)}$$

The common ion effect was explained on page 580.

When the 6 *M* HCl is added to the test solution—see Procedure 1A—a small excess is used. This does two things. It provides a small excess of the chloride ion that by the common ion effect, shifts these equilibria to the left and thus even more in favor of the insoluble salts. It also guards against the formation of bismuth(III) oxychloride, if Bi^{3+} ion happens to be present. This group-2 cation hydrolyzes strongly and forms a white, insoluble precipitate of BiOCl according to the forward reaction in the following equilibrium.

The hydrolysis of Bi^{3+} also occurs, but BiO^+ is soluble.

$$Bi^{3+} + H_2O \rightleftharpoons BiO^+ + 2H^+$$

$$Bi^{3+}(aq) + H_2O + Cl^-(aq) \rightleftharpoons BiOCl(s) + 2H^+(aq)$$

As you can see, an excess of H^+ (from the excess HCl) would help keep this equilibrium shifted to the left in favor of the soluble Bi^{3+} cation. (The group-1 cations also hydrolyze,

TABLE 25.1 Solubilities of Cation Group-1 Chlorides

Formula	Solubility (25 °C)		
	mol/L	mmol/mL	mg/mL
AgCl	1.3×10^{-5}	1.3×10^{-5}	1.9×10^{-3}
Hg_2Cl_2	6.5×10^{-7}	6.5×10^{-7}	3.1×10^{-4}
$PbCl_2$	1.6×10^{-2}	1.6×10^{-2}	4.5^a

^{a} At 100 °C, 32 mg/mL.

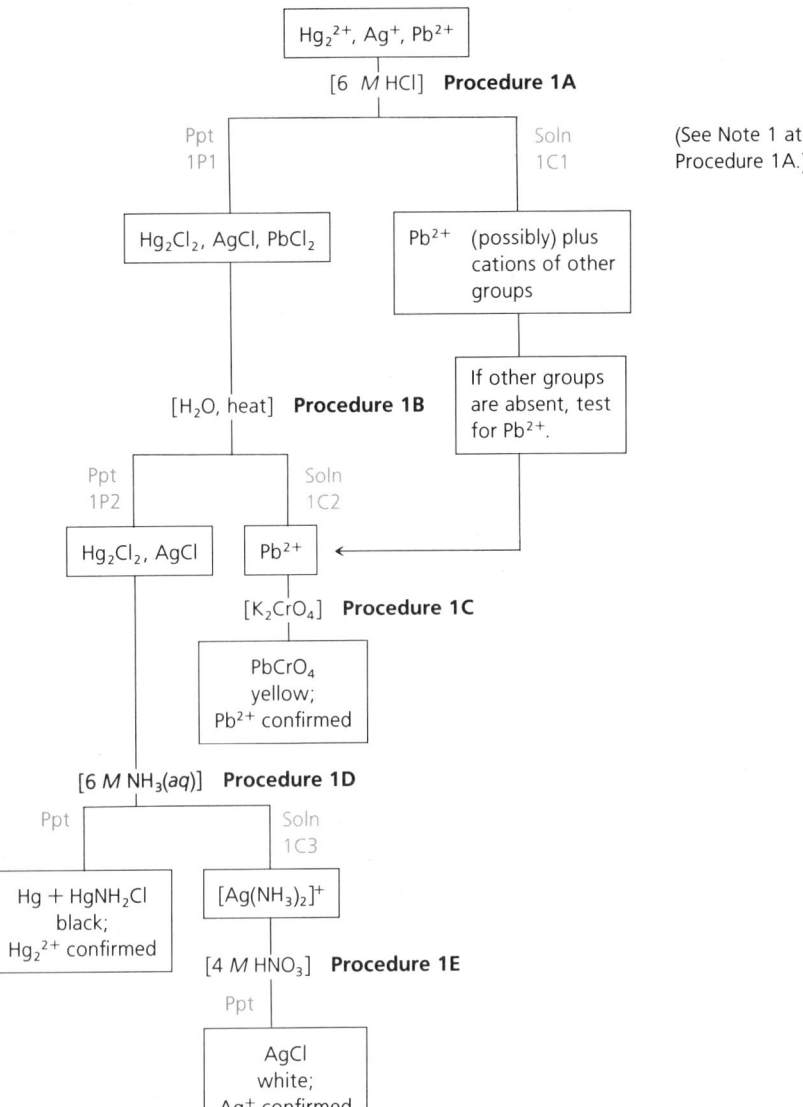

Hg_2^{2+}, Ag^+, Pb^{2+}

[6 *M* HCl]   **Procedure 1A**

Ppt
1P1

Soln
1C1

(See Note 1 at
Procedure 1A.)

Hg_2Cl_2, AgCl, $PbCl_2$

Pb^{2+} (possibly) plus
cations of other
groups

If other groups
are absent, test
for Pb^{2+}.

[H_2O, heat]   **Procedure 1B**

Ppt
1P2

Soln
1C2

Hg_2Cl_2, AgCl

Pb^{2+}

[K_2CrO_4]   **Procedure 1C**

$PbCrO_4$
yellow;
Pb^{2+} confirmed

[6 *M* $NH_3(aq)$]   **Procedure 1D**

Ppt

Soln
1C3

Hg + $HgNH_2Cl$
black;
Hg_2^{2+} confirmed

$[Ag(NH_3)_2]^+$

[4 *M* HNO_3]   **Procedure 1E**

Ppt

AgCl
white;
Ag^+ confirmed

FIGURE 25.11 The flow diagram
for the group-1 cations.

but not as strongly. The knowns and unknowns of the group-1 cations are prepared by
dissolving their nitrates in very dilute nitric acid to inhibit this hydrolysis. The nitrate ion
causes no interferences.)

Too great an excess of the group reagent, HCl, must be avoided when you carry out
Procedure 1A. Two of the group-1 chlorides can redissolve. Both AgCl and $PbCl_2$ form
water-soluble complex ions with chloride ion according to the following equilibria.

$$AgCl(s) + Cl^-(aq) \rightleftharpoons \underset{\text{dichloroargentate(I) ion}}{[AgCl_2]^-(aq)}$$

$$PbCl_2(s) + 2Cl^-(aq) \rightleftharpoons \underset{\text{tetrachloroplumbate(II) ion}}{[PbCl_4]^{2-}(aq)}$$

The Latin names of some metal
ions are used in naming their
complex ions. Thus, *argentate* is
from the Latin *argentum*, silver.
Plumbate is from *plumbum*, lead.

To the extent this happens, some or all of both the cations in these chloro complexes
could be carried forward into the solution holding the cations of higher groups. This
must not be allowed to happen, because both Ag^+ and Hg_2^{2+} would form insoluble
compounds with the group-2 reagent.

Figure 25.11 displays the kind of **flow diagram** that is commonly used in descriptions of schemes for qualitative analysis. These diagrams use features of design in the same way—vertical and horizontal lines, brackets, boxes, and even left–right positions—so once you learn the meanings of these features, you can read a flow diagram like a map.

Flow diagrams are read from top to bottom to indicate how the analysis of a test sample logically flows from beginning to end until the presence of the last constituent has been confirmed. The formulas of the ions for which tests will be outlined are at the top. The first vertical line is interrupted by the formula of a reactant placed within brackets, by [6 *M* HCl] in Figure 25.11, for example. Then the vertical line continues either to a branch, signified by a horizontal line, or directly to a box. Alongside the vertical line, usually by the bracketed formula of the reagent, there is a reference to a procedure. Thus, in Figure 25.11, "Procedure 1A" by [6 *M* HCl] refers to the procedure number where experimental directions are given in this chapter for the use of this reagent.

The first horizontal line in Figure 25.11 shows that the first reagent produces a precipitate and a solution. So the analysis has a branch here, because the precipi-

tate and solution are separately analyzed. Branches are written so that the scheme for the precipitate will continue on the left and that of the solution, on the right.

This first branch in Figure 25.11 leads to two boxes, each holding the formulas of the compounds that can be present and that are made by the reagent just above the branch. The flow of the analysis continues with vertical lines from these boxes, with the lines again interrupted by formulas in brackets. Below these, the vertical lines may lead to other branches, or to boxes. A box may have the formula of a compound whose formation you observe but which is subject to further procedures, or it could describe what is seen in a final confirmatory test for an ion.

We have used codes to designate particular precipitates, like 1P1, or specific solutions (centrifugates), like 1C1. The first digit refers to the ion group number. The letter P or C specifies precipitate or centrifugate, and the last digit specifies the particular precipitate or centrifugate in the scheme for the given ion group. (While C is derived from "centrifugate," it means a solution regardless of how it was separated—at the centrifuge or by decantation.) Thus, 1C2 would mean the second centrifugate (solution) in cation group 1.

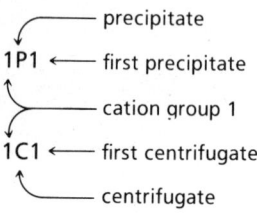

Procedure 1A gives a precipitate, which we might label 1P1, and a centrifugate, which we could label 1C1.

Detection and Confirmation of the Lead(II) Ion   Precipitate 1P1 includes up to three solids, $AgCl$, Hg_2Cl_2, and $PbCl_2$. Of these, $PbCl_2$ is the most soluble in water, and it can be extracted from the other two by hot water—Procedure 1B. The hot mixture is centrifuged, and the still hot centrifugate 1C2 contains the now dissolved $PbCl_2$. The addition of a few drops of K_2CrO_4 solution to 1C2—Procedure 1C—confirms lead(II) ion if a yellow precipitate forms. (It may be necessary to centrifuge the solution to see the yellow solid.) Pb^{2+} forms a very insoluble yellow salt, $PbCrO_4$, with the chromate ion according to the following equilibrium.

$$Pb^{2+}(aq) + CrO_4^{2-}(aq) \rightleftharpoons \underset{\substack{\text{lead(II) chromate}\\ \text{(bright yellow)}}}{PbCrO_4(s)} \qquad K_{sp} = 1.8 \times 10^{-14} \ (14 \ °C)$$

See Figure 6, page 596, in Qual Topic I for a color photo of $PbCrO_4$.

Detection and Confirmation of Mercury(I)   The precipitate 1P2 left by the extraction of $PbCl_2$ can consist of $AgCl$ and Hg_2Cl_2. These two react very differently with ammonia. $AgCl$ dissolves as the diamminesilver(I) complex ion, $[Ag(NH_3)_2]^+$. (This was described in detail on page 596.)

$$AgCl(s) + 2NH_3(aq) \rightleftharpoons [Ag(NH_3)_2]^+(aq) + Cl^-(aq)$$

The other chloride, Hg_2Cl_2, reacts with NH_3 by disproportionation to give a grey or black precipitate consisting of finely divided mercury and mercury(I) amidochloride, $HgNH_2Cl$, which is white. (The black color of the free mercury overpowers this white color to make the precipitate appear black.)

$$Hg_2Cl_2(s) + 2NH_3(aq) \longrightarrow Hg(\ell) + HgNH_2Cl(s) + NH_4^+(aq) + Cl^-(aq)$$

black   white

As with most metals, when mercury metal is finely divided it is black, not shiny.

Thus if precipitate 1P2 turns black following the addition of aqueous ammonia — Procedure 1D — the presence of Hg_2^{2+} ion is confirmed.

This disproportionation reaction occurs because ammonia is able to react with the Hg^{2+} ion *that is already present* in the second of the following equilibria involving the sparingly soluble Hg_2Cl_2. First, there is the usual heterogeneous solubility equilibrium that we wrote before.

$$Hg_2Cl_2(s) \rightleftharpoons Hg_2^{2+}(aq) + 2Cl^-(aq)$$

Then the ultratrace concentration of Hg_2^{2+} present becomes involved in the following (also heterogeneous) equilibrium.

$$Hg_2^{2+}(aq) \rightleftharpoons Hg(\ell) + Hg^{2+} \qquad K_{eq} = 0.0061$$

So little $Hg(\ell)$ is present that you can't notice it. But any chemical species that can lower the concentration of Hg^{2+} will shift this equilibrium to the right to produce enough very finely divided $Hg(\ell)$ to see it as part of a black residue. To cause this to happen, we need a species that could, for example, form a compound of Hg^{2+} that is even less soluble than Hg_2Cl_2. And ammonia is just such a reactant. It reacts with Hg^{2+} to give the very insoluble (and white) mercury(II) amidochloride.

$$Hg^{2+}(aq) + 2NH_3(aq) + Cl^-(aq) \longrightarrow HgNH_2Cl(s) + NH_4^+(aq)$$

Thus this irreversible reaction would draw all equilibria to the right and, if Hg_2Cl_2 is present, the addition of aqueous ammonia shows it by the appearance of a black material.

The Detection and Confirmation of the Silver Ion   The supernatant liquid (1C3) that forms when aqueous ammonia is added to precipitate 1P2 is isolated at the centrifuge. It will contain silver ion as $[Ag(NH_3)_2]^+$ as well as the Cl^- ion. When the NH_3 molecules in $[Ag(NH_3)_2]^+$ are neutralized by dilute nitric acid — Procedure 1E — free Ag^+ is reexposed to Cl^- and the two must again combine and give white, insoluble $AgCl$. Thus a white precipitate or cloudiness at this point confirms the presence of Ag^+.

Procedures for Group-1 Cations

The instructor will show you how to use the centrifuge. Reread Section 25.2 with particular attention to the techniques for decanting, extracting, using glass stirring rods, centrifuging, testing for pH, using a water bath, and keeping a lab notebook.

Carry out the following procedures first on a known solution. Then ask your instructor for your cation group-1 unknown.

Read through a procedure completely before starting it in the lab.

These procedures are applied in the same order when you analyze a multigroup known or unknown.

Procedure 1A: Separation of Group-1 Cations   Add 1 mL of the solution to be tested to a 75-mm test tube (or whatever other tube is supplied for use with your centrifuge). Add 2 drops of 6 *M* HCl. Allow any precipitate to form, and centrifuge. But before you separate the centrifugate, add 2 drops of 6 *M* HCl to the tube and watch for any additional cloudiness to appear. If it does, centrifuge again. (See Note 1.) The result of testing for complete precipitation is pictured in Figure 5, page 595, in Qual Topic I.

Decant the liquid, which is centrifugate 1C1, from the precipitate, 1P1. If you are working with a multigroup known or unknown, this centrifugate is saved for further analysis.

NOTE 1   If the solution being tested included group-2 cations, then the mixture formed by the addition of the group test reagent, HCl, can be warmed *before* it is centrifuged. The intent would be to pass Pb^{2+} ions deliberately into the cation group-2 system.

To do this, place the tube with the uncentrifuged mixture produced by HCl (Procedure 1A) in a boiling water bath and stir its contents frequently with a glass rod. Centrifuge the hot solution to remove any solid AgCl and Hg_2Cl_2—they make up precipitate 1P2. The centrifugate is 1C1. Wash the precipitate (1P2) with 10 drops of very hot (distilled) water to extract out any remaining $PbCl_2$. Centrifuge and add the centrifugate to 1C1.

Procedure 1B: The Extraction of $PbCl_2$ from Precipitate 1P1   Add 10 drops of distilled water to precipitate 1P1 from Procedure 1A. Stir the mixture often with a glass rod as you heat the mixture in a boiling water bath. The intent is to dissolve any $PbCl_2$. Centrifuge the *hot* mixture quickly, and decant the centrifugate (1C2) into a small test tube. Wash (extract) the precipitate (1P2) with 4 drops more of *hot* water to bring out any remaining $PbCl_2$. Centrifuge, and combine the centrifugate with the previous one, 1C2. This should now contain any Pb^{2+} ion. And the precipitate, 1P2, should have any AgCl and Hg_2Cl_2.

Procedure 1C: Detection and Confirmation of Pb^{2+}   Test solution 1C2 for Pb^{2+} by adding 2 drops of 1 M K_2CrO_4. A yellow precipitate confirms the presence of Pb^{2+} (and it might be necessary to centrifuge the mixture to see the solid).

Procedure 1D: Detection and Confirmation of Hg_2^{2+} Ion   To precipitate 1P2, which consists of any AgCl and Hg_2Cl_2, add 5 drops of 6 M NH_3. If a black precipitate appears, Hg_2^{2+} is confirmed.

> NOTE 2   If the entire precipitate dissolves when 6 M NH_3 is added, then Hg_2^{2+} must have been absent, and it is not necessary to centrifuge the mixture. Proceed directly to Procedure 1E. Otherwise, continue with the next paragraph.

Centrifuge the mixture and decant its supernatant liquid. Save the solution, which is 1C3. Add 5 more drops of 6 M NH_3 to the precipitate and then centrifuge again and decant. Combine the two centrifugates, and reserve them for Procedure 1E. Discard any precipitate into a receptacle designated for this waste.

Procedure 1E: Detection and Confirmation of Ag^+ Ion   Add 4 M HNO_3 drop by drop to centrifugate 1C3 from Procedure 1D It holds any $[Ag(NH_3)_2]^+$ ion. Add enough nitric acid to make the centrifugate 1C3 acid to litmus. (See page 1069 for the correct technique.) If a white precipitate or cloudiness appears, it confirms Ag^+.

25.4 CATION GROUP-2 ANALYSIS

The sulfides of cation group-2 will not dissolve in aqueous acid no matter how low the pH.

Of the group-2 cations, Pb^{2+}, Bi^{3+} (or BiO^+), and Sn^{4+} (or Sn^{2+}) are colorless. Hydrated Cu^{2+} is light blue.

The cations of both groups 2 and 3 form insoluble sulfides, but the sulfides of group 2 — the acid-insoluble sulfides, PbS, Bi_2S_3, CuS, and SnS_4 (or SnS) — cannot be made to dissolve in aqueous acid regardless of how low we make the pH. By adjusting the pH of a mixture of group-2 (or higher) cations to a value of 0.5, as a safe cutoff point, we can add H_2S to the mixture and be confident that only those of group 2 can precipitate. Thus, the group reagent for cation group 2 is hydrogen sulfide added after the pH of the test solution has been adjusted to 0.5 (the equivalent of 0.3 M HCl).

In Qual Topic II (page 650) we discussed the theory of the selective precipitation of sulfides (and other compounds). We learned that the sulfide ion, S^{2-}, has never been detected even in very strongly basic solutions, even as strong as 8 *M* NaOH. In acid, therefore, the acid-insoluble sulfides form by reactions between the metal ions and H_2S, not with S^{2-} (or even HS^-). For example, PbS, H_2S, Pb^{2+}, and H^+ coexist in the following equilibrium.

$$PbS(s) + 2H^+(aq) \rightleftharpoons Pb^{2+}(aq) + H_2S(aq)$$

The equilibrium constant for this heterogeneous equilibrium is called the *acid solubility product*, K_{spa}, not just the solubility product (K_{sp}), because it is for an insoluble compound that dissolves not simply by the dissociation of its ions but rather by a chemical reaction with the acid. For PbS, for example, K_{spa} is given by

$$K_{spa} = \frac{[Pb^{2+}][H_2S]}{[H^+]^2} = 3 \times 10^{-7}$$

With the conditions under which the group reagent is used, $[H^+] = 0.3$ mol/L and $[H_2S] = 0.1$ mol/L; the solubility of Pb^{2+}, calculated using this K_{spa} value, is about 3×10^{-7} mol/L, which is roughly 6×10^{-5} mg/mL (or 0.06 μg/mL). Despite the low precision of the K_{spa} value and the resulting uncertainty in this calculated solubility, we can still say with confidence that PbS is extremely insoluble in acid. And PbS is the most soluble of our group-2 sulfides under the experimental conditions used. Next in solubility is CuS ($K_{spa} = 6 \times 10^{-16}$) and, under the same conditions, the calculated solubility of Cu^{2+}, using this K_{spa} value, is about 5×10^{-16} mol/L, which is roughly 3×10^{-14} mg/mL. The sulfides of the remaining group-2 cations are even less soluble.

The least soluble of the sulfides produced by the group-3 analysis, ZnS, has a K_{spa} value of 3×10^{-2}. Using this value, which is arguable, the solubility of ZnS in 0.3 *M* HCl and 0.1 *M* H_2S calculates to be 1.8 mg/mL. We say that this is arguable because ZnS occurs in two crystalline forms, alpha and beta, and we have used here the K_{spa} value for the more soluble beta form. (Table 1 in Qual Topic II, page 652, gives the K_{spa} values for both forms.) Whether either the alpha or beta variety actually does form *initially* is not known. It is certain, however, that freshly precipitated ZnS *readily* dissolves in acid with the evolution of H_2S, which means that it isn't possible to make ZnS precipitate in 0.3 *M* HCl. The next most soluble of our group-3 sulfides, MnS ($K_{spa} = 40$), is extremely soluble in dilute acid. Thus you can see that the group-3 cations cannot precipitate as sulfides under the acidic conditions of the reagent for the group-2 cations.

The flow diagram for the analysis of the group-2 cations is given in Figure 25.12, and you should follow it as you study the following discussion.

Hydrogen Sulfide Is Generated Slowly from Thioacetamide in the Solution We must never forget that hydrogen sulfide is an extremely toxic gas. At a concentration of only 100 ppm in air, it can cause rapid paralysis and death. Fortunately, your nose detects its distinctive rotten egg odor at an air concentration of only about 0.02 ppm. H_2S, however, soon desensitizes your sense of smell, so operations involving this compound are carried out at the fume hood.

There is another safety factor for the use of H_2S that is a property of the way it is prepared for these qual schemes — from thioacetamide. Thioacetamide is a stable, crystalline compound. In the presence of either acid or base, it reacts with water to give acetamide and hydrogen sulfide. Thus hydrogen sulfide is released right within the solution being tested. There is no need to manipulate gaseous H_2S.

$$CH_3-\underset{\text{thioacetamide}}{\overset{\overset{\displaystyle S}{\|}}{C}}-NH_2(aq) + H_2O \longrightarrow CH_3-\underset{\text{acetamide}}{\overset{\overset{\displaystyle O}{\|}}{C}}-NH_2(aq) + H_2S(aq)$$

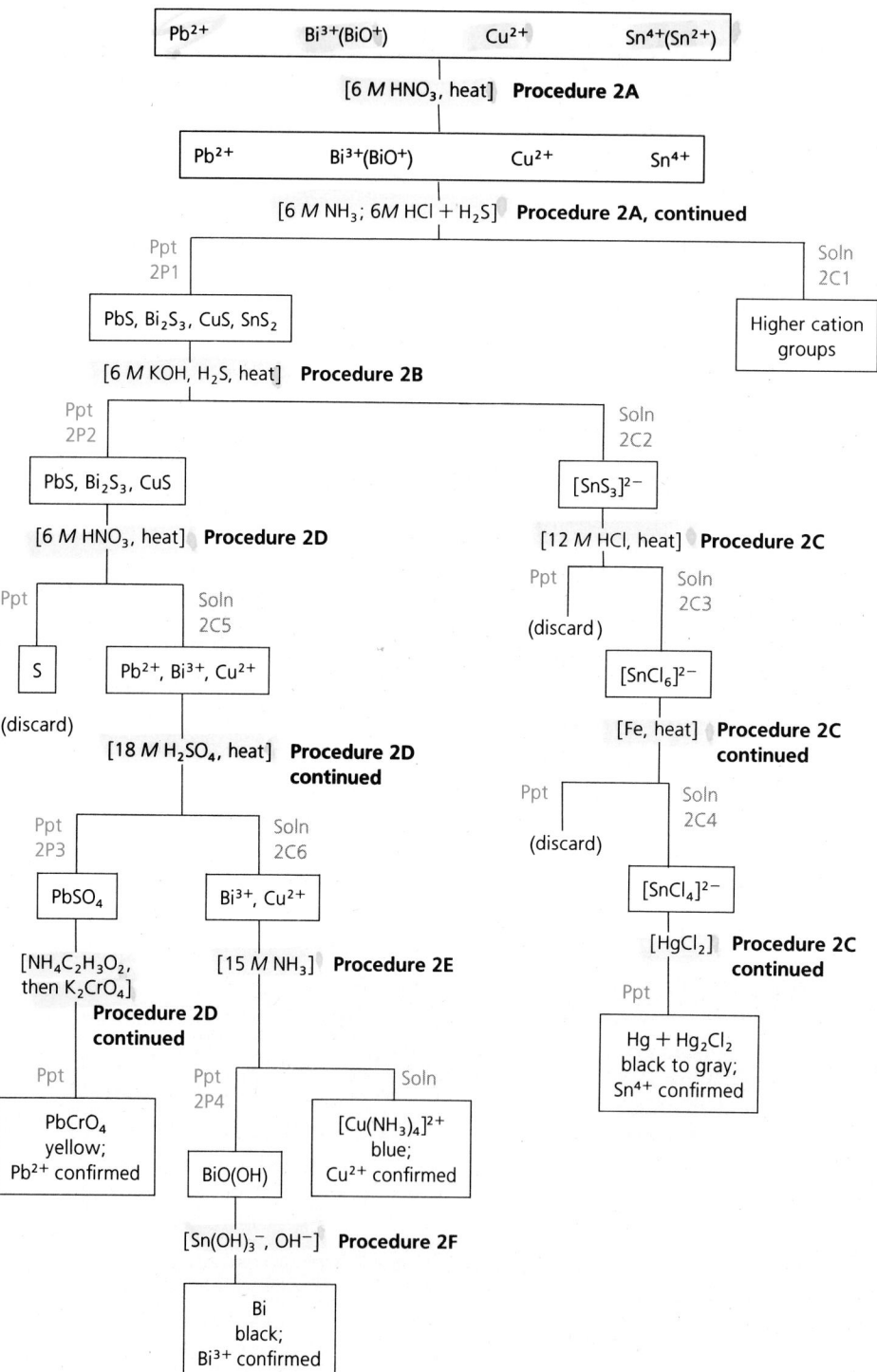

FIGURE 25.12 The flow diagram for group-2 cations.

Acetamide also hydrolyzes. In acid, the reaction is

$$\underset{\text{acetamide}}{CH_3-\overset{\overset{\displaystyle O}{\|}}{C}-NH_2(aq)} + H^+(aq) + H_2O \longrightarrow \underset{\text{acetic acid}}{CH_3-\overset{\overset{\displaystyle O}{\|}}{C}-OH(aq)} + NH_4^+(aq)$$

Unreacted thioacetamide, acetamide, and ammonium ion do not interfere with the subsequent analyses of the group-2 cations. The formation of the ammonium ion, however, consumes some of the hydrochloric acid used to adjust the pH of the test solution, but not enough to make a difference (unless, of course, you take a huge excess of the thioacetamide solution).

In chemistry lab procedures, the "more is better" rule is almost always wrong.

The newly formed hydrogen sulfide *remains almost entirely in the aqueous solution* in which it forms, which is a safety factor, as we said. Another advantage of using thioacetamide is that its release of H₂S is not instantaneous. H₂S forms relatively slowly, and this gives insoluble sulfides the opportunity to form slowly and with better formed crystals. These are easier to separate at the centrifuge than sulfides that suddenly form, as occurs when hydrogen sulfide gas is bubbled into the test solution. Insoluble sulfides formed by this latter method tend to be colloidal. The colloidal particles adsorb negative ions and so acquire negative charges and resist any growth into larger, more dense particles. Moreover, these negatively charged colloidal particles can attract cations from higher groups, which causes incomplete separations of the groups. Always bear in mind that the major reason for separating cation groups from each other is to make it possible to carry out a confirmatory test for one ion without interferences from others.

The Scheme Does Not Distinguish Between Sn⁴⁺ and Sn²⁺ We have to comment here on the appearance of tin in the flow diagram as $Sn^{4+}(Sn^{2+})$. This means that the initial test solution might have tin in either form. If Sn^{2+} is the one present, however, it will very soon be oxidized to Sn^{4+} to make the experimental work easier. Moreover, Sn^{2+} is readily oxidized by the oxygen that enters the solution from the air.

Recall from Chapter 12 that Sn^{2+} is used in the lab as a mild reducing agent, and so becomes oxidized to Sn^{4+}.

$$2Sn^{2+}(aq) + O_2(aq) + 4H^+(aq) \longrightarrow 2Sn^{4+}(aq) + 2H_2O$$

As a result, a solution with Sn^{2+} ion cannot be stored for a long period unless air is excluded. In any case, no attempt is made in the scheme to distinguish the two ions of tin. The sulfides of both are extremely insoluble, but another reason for first converting all Sn^{2+} to Sn^{4+} is that Sn^{4+} forms a better (more crystalline) sulfide that Sn^{2+}. SnS tends to form as a gelatinous solid, and such solids are not easy to centrifuge and work with. Still another reason for converting all Sn^{2+} to Sn^{4+} involves Procedure 2B (Figure 25.12), as we will soon learn.

We avoid the Sn^{2+} problem by first heating the solution of the group-2 cations with nitric acid—Procedure 2A. All Sn^{2+} is thereby oxidized to Sn^{4+}. The mixture in which this oxidation occurs is heated in an evaporating dish until only a moist paste remains. Thus, much of the unused nitric acid decomposes to gases that leave the mixture. This step is important because if nitric acid remained in the mixture it would later oxidize hydrogen sulfide to give colloidal (and pale yellowish) sulfur. This is hard to separate and it could be mistaken for tin(IV) sulfide.

Be very careful to prevent spattering while heating a solution to evaporate it.

The moist paste that forms in this part of Procedure 2A is now neutralized (NH_3) and then reacidified, this time using a nonoxidizing acid, hydrochloric acid, not nitric acid. We need a *nonoxidizing* acid because we do not want to oxidize H₂S to free sulfur. Yet, as we have said several times, we need an *acidic* medium to keep the sulfides from group 3 from precipitating. We also need an acidic medium because we want the group-2 cations to be in solution, and in a basic medium they would precipitate as hydroxides or hydrated oxides. Even in a neutral solution, Sn^{4+} and Bi^{3+} strongly hydrolyze to insoluble oxides. The last reason for using hydrochloric acid is its chloride ion. The Sn^{4+} ion is also able to oxidize H_2S to free sulfur, which we do not want.

$$Sn^{4+}(aq) + H_2S(aq) \longrightarrow Sn^{2+}(aq) + S(s) + 2H^+(aq)$$

The reaction would also give Sn^{2+}, which we are trying to avoid as we said. But Sn^{4+} forms a complex ion with 6 Cl^- ions, $[SnCl_6]^{2-}$, and in this form, tin(IV) cannot oxidize H_2S to S.

The Group-2 Cations Are Now Separated

Now, to complete Procedure 2A, the thioacetamide solution — the equivalent of H_2S — is added. The precipitate (2P1) consists of the sulfides of group 2. The supernatant liquid (2C1) is reserved for group-3 analysis. (If you are working only with a group-2 known or unknown, this liquid can be discarded, of course.)

The sulfides of cation group 2 form as follows.

$$Pb^{2+}(aq) + H_2S(aq) \longrightarrow PbS(s) + 2H^+(aq)$$
$$Cu^{2+}(aq) + H_2S(aq) \longrightarrow CuS(s) + 2H^+(aq)$$
$$Sn^{4+}(aq) + 2H_2S(aq) \longrightarrow SnS_2(s) + 4H^+(aq)$$
$$2BiO^+(aq) + 3H_2S(aq) \longrightarrow Bi_2S_3(s) + 2H_2O + 2H^+(aq)$$

We show the bismuth(III) ion in the form of the bismuthyl ion, BiO^+, because Bi^{3+} strongly hydrolyzes in aqueous acid (at the pH of our solution) according to the forward reaction of the following equilibrium.

$$Bi^{3+}(aq) + H_2O \rightleftharpoons BiO^+(aq) + 2H^+$$

But Bi_2S_3 also forms from the BiO^+ ion as the earlier equation showed.

Two of the possible sulfides, PbS and CuS, are black. Bi_2S_3 is dark brown (almost black), and SnS_2 is yellow. Color photos of some of them can be seen in Figure 2, page 594, in Qual Topic I. When tin(IV) is present, its yellow disulfide, SnS_2, usually precipitates first, so watch for this hint — and it's no more than a hint — of the presence of Sn^{4+}.

Tin Is Next Separated From the Other Group-2 Cations

The next operation — Procedure 2B — extracts Sn^{4+} from the precipitate. In base and aqueous H_2S, SnS_2 reacts with H_2S (or HS^-) to form still another complex ion, SnS_3^{2-}.

$$SnS_2(s) + H_2S(aq) + 2OH^-(aq) \rightleftharpoons [SnS_3]^{2-}(aq) + 2H_2O$$

This brings Sn^{4+} back into solution as $[SnS_3]^{2-}$. This solution, centrifugate 2C2, is separated from the insoluble group-2 sulfides, which now constitute precipitate 2P2.

SnS, if it were present, would not dissolve well in this procedure, so we have still another reason for making sure that any Sn^{2+} in the initial known or unknown is first oxidized to Sn^{4+}.

Detection and Confirmation of Tin

Centrifugate 2C2 is now tested — Procedure 2C — for the presence of tin(IV). It is first heated strongly with HCl, which causes the hexachlorostannate(IV) ion to form once again.

$$[SnS_3]^{2-}(aq) + 6H^+(aq) + 6Cl^-(aq) \longrightarrow [SnCl_6]^{2-}(aq) + 3H_2S(aq)$$

The hot HCl also finishes the hydrolysis of any remaining thioacetamide.

The overall purpose of this part of Procedure 2C is to remove all traces of hydrogen sulfide from the solution. Unless this is done, the final test for tin will be a false-positive test. This test involves the addition of Hg^{2+} ion as $(HgCl_2)$. If H_2S is present, a black precipitate will form even if tin is absent because Hg^{2+} forms a highly insoluble sulfide, HgS. HgS has the same black color that can be expected in a positive test for tin.

To continue with Procedure 2C, we now have any tin ion as $[SnCl_6]^{2-}$ in centrifugate 2C3, which is strongly acidic in HCl. No H_2S is in the solution. (Heat has driven out what might remain.) We next add iron (as freshly polished pieces of iron wire) and heat the system. The purpose is to reduce Sn^{4+} to Sn^{2+}, because the final test for tin requires a tin ion that is a reducing agent, which Sn^{2+} is, but not Sn^{4+}. Iron reduces tin from the $+4$ to the $+2$ state as follows.

$$[SnCl_6]^{2-}(aq) + Fe(s) \longrightarrow [SnCl_4]^{2-}(aq) + Fe^{2+}(aq) + 2Cl^-(aq)$$

After removing any unreacted iron and other solids, a solution of $HgCl_2$ is added to the test solution. Sn^{2+} in $[SnCl_4]^{2-}$, acting now as the mild reducing agent we mentioned, reduces Hg^{2+} (in $HgCl_2$) from an oxidation number of $+2$ either to $+1$, or to 0, or to a mixture of these. When the reduction goes to mercury(I), the white, insoluble Hg_2Cl_2 forms.

$$[SnCl_4]^{2-}(aq) + 2HgCl_2(aq) \longrightarrow [SnCl_6]^{2-}(aq) + \underset{\text{white}}{Hg_2Cl_2(s)}$$

When the concentration of $[SnCl_4]^{2-}$ is high enough, then at least some of the Hg_2Cl_2 formed in this reaction is further reduced to Hg.

$$[SnCl_4]^{2-}(aq) + Hg_2Cl_2(s) \longrightarrow [SnCl_6]^{2-}(aq) + \underset{\text{black}}{2Hg(\ell)}$$

What is often observed when knowns or unknowns containing group-2 cations are examined is the formation of a blackish mixture of $Hg(\ell)$ and $Hg_2Cl_2(s)$, because only some of the Hg_2Cl_2 made in the first of these two reductions is further reduced. We could write the following equation to represent this overall result.

$$2[SnCl_4]^{2-}(aq) + 3HgCl_2(aq) \longrightarrow \underset{\text{black}}{Hg(\ell)} + \underset{\text{white}}{Hg_2Cl_2(s)} + 2[SnCl_6]^{2-}(aq)$$

Thus the formation of a white or grayish black or black precipitate at the end of Procedure 2C confirms the presence of one of the two tin ions in the original solution.

Detection and Confirmation of Lead Now we go back and pick up the analysis of the remaining group-2 sulfides—PbS, Bi_2S_3, and CuS—which constitute precipitate 2P2 (Figure 25.12). These sulfides are now dissolved in hot nitric acid (Procedure 2D). The ability of the nitrate ion in acid to oxidize S^{2-} to S is the driving force for these reactions.

Note: If any Hg_2^{2+} from group 1 was allowed to pass on to group 2, black HgS has formed here, and it will *not* dissolve in nitric acid.

$$3PbS(s) + 8H^+(aq) + 2NO_3^-(aq) \longrightarrow 3Pb^{2+}(aq) + 3S(s) + 2NO(g) + 4H_2O$$
$$Bi_2S_3(s) + 8H^+(aq) + 2NO_3^-(aq) \longrightarrow 2Bi^{3+}(aq) + 3S(s) + 2NO(g) + 4H_2O$$
$$3CuS(s) + 8H^+(aq) + 2NO_3^-(aq) \longrightarrow 3Cu^{2+}(aq) + 3S(s) + 2NO(g) + 4H_2O$$

The sulfur that forms is removed at the centrifuge and discarded. The centrifugate, 2C5, now contains Pb^{2+}, Bi^{3+}, and Cu^{2+}.

Procedure 2D continues with this centrifugate. Concentrated sulfuric acid is added to make Pb^{2+} precipitate as $PbSO_4$ and leave Bi^{3+} and Cu^{2+} in solution.

The finely divided sulfur has only a very pale yellow color, not the bright yellow usually associated with sulfur.

$$Pb^{2+}(aq) + SO_4^{2-}(aq) \rightleftharpoons PbSO_4(s) \qquad K_{sp} = 6.3 \times 10^{-7}$$

It is important to proceed very carefully in using Procedure 2D because the concentration of Pb^{2+} will be low if most was removed in group 1. One experimental problem is that $PbSO_4$ is somewhat soluble in acid, because H^+ and SO_4^{2-} can combine to give HSO_4^-.

$$PbSO_4(s) + H^+(aq) \rightleftharpoons Pb^{2+}(aq) + HSO_4^-(aq) \qquad K_{eq} = 1.2 \times 10^{-6}$$

To suppress the forward reaction of this equilibrium, which represents the dissolving of $PbSO_4$, the mixture is heated strongly. The nitric acid decomposes to oxygen, water, and nitrogen dioxide:

$$4HNO_3(aq) \xrightarrow{heat} O_2(g) + 4NO_2(g) + 2H_2O(g).$$

This leaves excess sulfuric acid as the solvent. But at its boiling point (338 °C), sulfuric acid decomposes as follows.

$$H_2SO_4(\ell) \xrightarrow{heat} SO_3(g) + H_2O(g)$$

The gaseous mixture of SO_3 and H_2O vapor that forms produces a whitish mist in air. These fumes signal that all water and nitric acid have been expelled and that some of the excess sulfuric acid has been decomposed. The supply of $H^+(aq)$ is now low enough so that the $PbSO_4$ does not dissolve when water is next added. (Needless to say, this operation must be done at the hood. SO_3 has an unusually sharp, irritating odor and, in contact with moist tissue in the nose or lungs, SO_3 changes back to sulfuric acid.)

Continuing with Procedure 2D, water is now added to bring Bi^{3+} and Cu^{2+} into solution—centrifugate 2C6—and to leave $PbSO_4$ as the precipitate, 2P3. The very presence of this precipitate strongly indicates Pb^{2+} ion, but it is also possible that the white precipitate consists instead of bismuth in the form of $(BiO)_2SO_4$. As we mentioned earlier in this section, the Bi^{3+} ion strongly hydrolyzes in water to give BiO^+. This ion also forms a white insoluble sulfate, $(BiO)_2SO_4$, which could be mistaken for $PbSO_4$. To prevent such a mistake, precipitate 2P3 is mixed with a solution of ammonium acetate. The acetate ion forms a water soluble complex with Pb^{2+} and thus draws the following equilibrium to the right, which means that $PbSO_4$ dissolves.

$$PbSO_4(s) + 4C_2H_3O_2^-(aq) \rightleftharpoons [Pb(C_2H_3O_2)_4]^{2-}(aq) + SO_4^{2-}(aq)$$

Very little $(BiO)_2SO_4$ dissolves at this point. Thus, ammonium acetate solution dissolves any $PbSO_4$ in precipitate 2P3. But now potassium chromate is added, and the chromate ion, CrO_4^{2-}, is able to take Pb^{2+} from $[Pb(C_2H_3O_2)_4]^{2-}$ and form a yellow precipitate of $PbCrO_4$.

A yellow *precipitate*, not a yellow solution, must form.

$$[Pb(C_2H_3O_2)_4]^{2-}(aq) + CrO_4^{2-}(aq) \longrightarrow \underset{\substack{\text{yellow} \\ \text{(confirms } Pb^{2+})}}{PbCrO_4(s)} + 4C_2H_3O_2^-(aq)$$

Detection and Confirmation of Copper   We next process centrifugate 2C6 (Figure 25.12) according to Procedure 2E. The solution should now contain Bi^{3+} (mostly as BiO^+) and Cu^{2+}. If the concentration of Cu^{2+} is high enough, the solution will have the pale blue color of $[Cu(H_2O)_4]^{2+}$. But to confirm the presence of Cu^{2+}, aqueous ammonia is added. This converts $[Cu(H_2O)_4]^{2+}$ to the much more deeply blue $[Cu(NH_3)_4]^{2+}$ ion.

$$\underset{\text{light blue}}{[Cu(H_2O)_4]^{2+}(aq)} + 4NH_3(aq) \longrightarrow \underset{\text{dark blue}}{[Cu(NH_3)_4]^{2+}(aq)} + 4H_2O$$

So if the color becomes more deeply bluish when aqueous ammonia is added, the presence of Cu^{2+} is confirmed. See Figure 1, page 954, in Qual Topic III.

Confirmation of Bismuth   Aqueous ammonia also contains OH^- ions. So as NH_3 reacts with $[Cu(H_2O)_4]^{2+}$ in centrifugate 2C6, OH^- reacts with Bi^{3+} (or BiO^+) to give a white, insoluble hydroxide, $BiO(OH)$.

$$BiO^+(aq) + OH^-(aq) \longrightarrow BiO(OH)(s)$$

The mixture produced when aqueous ammonia is added to solution 2C6 is therefore centrifuged to separate any white solid. To make sure that any such solid actually contains Bi^{3+}, a final confirmatory test for bismuth is performed. The Bi^{3+} in $BiO(OH)$ is reduced to bismuth metal by freshly prepared stannite ion, $[Sn(OH)_3]^-$, and the metal appears as a black, finely divided solid. The appearance of such a solid confirms bismuth.

The stannite ion is prepared by the reaction of OH^- with Sn^{2+}. When $NaOH(aq)$ is added to $SnCl_2(aq)$, $Sn(OH)_2$ first precipitates. But it is amphoteric, and further OH^- ion converts it to the soluble $[Sn(OH)_3]^-$ ion. These two reactions are

$$Sn^{2+}(aq) + 2OH^-(aq) \longrightarrow Sn(OH)_2(s)$$
$$Sn(OH)_2(s) + OH^-(aq) \longrightarrow [Sn(OH)_3]^-(aq)$$

Then the $[Sn(OH)_3]^-$ ion reacts with $BiO(OH)$ as follows to give bismuth.

$$2BiO(OH)(s) + 3[Sn(OH)_3]^-(aq) + 3OH^-(aq) + 2H_2O \longrightarrow$$
$$\underset{\substack{\text{black} \\ \text{(confirms Bi}^{3+})}}{2Bi(s)} + 3[Sn(OH)_6]^{2-}(aq)$$

The stannite ion must be freshly prepared because it is oxidized rather quickly by the oxygen that dissolves in the solution from the surrounding air.

$$2[Sn(OH)_3]^-(aq) + O_2(aq) + 2OH^-(aq) + 2H_2O \longrightarrow 2[Sn(OH)_6]^{2-}$$

And if the stannite solution is allowed to remain at room temperature long enough, the stannite ion in a basic solution can also disproportionate.

$$2[Sn(OH)_3]^-(aq) \longrightarrow Sn(s) + [Sn(OH)_6]^{2-}(aq)$$

We therefore have two reasons for the need to use freshly prepared stannite solution in the confirmatory test for bismuth.

Procedures for Group-2 Cations

In some of the procedures given in this section you will be directed to wash a precipitate with a dilute solution of ammonium nitrate instead of with water. This calls for an explanation. As we mentioned earlier, some precipitates can form in a very finely divided colloidal state. The particles adsorb like-charged ions on their surfaces and so are able to repel each other and thus cannot grow to larger sizes. The separation of such colloidal materials by the centrifuge is thereby made very difficult.

If the surrounding solution, however, has a high enough concentration of electrolytes, these charges tend to be removed and any colloidal particles can coalesce into particles large enough to permit easy separation at the centrifuge. Thus when insoluble sulfides are first formed, the electrolyte level in the solution is normally high enough to prevent a colloidal dispersion from forming. But if a precipitate of such sulfides is washed with pure water, the electrolyte level might, at one stage, drop low enough to cause at least some of the precipitate to break up into colloidal particles. We have to prevent this from happening, so when it is a danger, we wash the precipitates with a dilute electrolyte solution. We pick an electrolyte whose ions can cause no interference with any operation or test, and ammonium nitrate fits this requirement nicely.

The test solution you will use will either be the centrifugate 1C1 left when the group-1 reagent, HCl, precipitated the group-1 chlorides. Or it will be a special group-2 known or unknown, which you are told has only cations of this group. In any case, these procedures should be carried out on a group-2 known first.

Procedure 2A: Precipitation of the Group-2 Sulfides   Transfer 2 mL of the test solution to an evaporating dish and add 10 to 15 drops of 6 M HNO_3. Heat the solution over a low, direct flame at the hood to drive off enough liquid so that only a moist paste of a residue remains. To minimize spattering, stir the mixture with a glass rod as it is heated. Let the solution cool enough so that it no longer gives off sharp fumes.

If you remove the glass rod from the evaporating dish, don't lay it on the lab bench where it might pick up contaminants.

Using tongs to carry the evaporating dish; return to your desk and add 1 mL of distilled water to the residue. Use some of the water to rinse off the glass rod. Stir the mixture with this rod if necessary to make everything dissolve. By the directions of the next paragraph, you will adjust the pH of this solution to about 0.5 by using a combination of aqueous ammonia and hydrochloric acid. If necessary, go back to page 1069 and review how to test a solution for acidity.

Add 10 to 15 drops of 6 M NH_3 until the solution is basic to litmus. Now add 6 M HCl drop by drop until the solution just turns acidic, and then add two more drops of 6 M HCl. The pH is now close enough to 0.5 to ensure that the reaction with hydrogen sulfide, next, will cause only the group-2 cations to precipitate.

If you drop some thioacetamide solution on your skin it won't cause a burn, but wash the area thoroughly as soon as possible anyway.

Carefully transfer the solution to a clean centrifuge tube, and add 10 to 15 drops of 1 M thioacetamide. Heat the mixture in gently boiling water in a hot water bath for 5 to 6 minutes, then cool the system and centrifuge it. Before removing the supernatant liquid, test it for completeness of precipitation by adding a few drops of the thioacetamide solution. Do not stir the mixture because you do not want to disturb the existing precipitate. But heat the system in the hot water bath a few minutes as you watch it for the appearance of more precipitate in the supernatant liquid. Recentrifuge the mixture if necessary.

The precipitate, 2P1, consists of the sulfides of any group-2 cations present. The centrifugate, 2C1, is saved if you began with a test solution containing group-3 or -4 cations. Otherwise, 2C1 can be discarded.

Procedure 2B: Separation of Tin   To wash the precipitate, 2P1, add a few drops of 2 M NH_4NO_3, stir the mixture, centrifuge, and discard the supernatant liquid. Repeat this washing operation once.

6 M KOH, if left on the skin, will cause a chemical burn.

To the washed precipitate add 2 mL of 6 M KOH and 3 drops of 1 M thioacetamide. Heat the mixture at least 5 minutes at 95°C using the hot water bath. Centrifuge the *hot* mixture and decant the supernatant liquid into a clean centrifuge tube. This solution is centrifugate 2C2, and it contains any tin(IV) ion as $[SnS_3]^{2-}$. The precipitate is 2P2, and it includes any sulfides of copper, bismuth, and lead. Save 2P2 for Procedure 2D.

Procedure 2C: Detection and Confirmation of Tin   To centrifugate 2C2 add 12 M HCl drop by drop until the solution is just acidic to litmus. Then add 12 more drops.

HCl vapors are very irritating to the lungs.

WARNING   Keep 12 M HCl from the skin. Do not inhale its vapors.

Heat this solution in a boiling water bath for 5 to 6 minutes, stirring it with a glass rod from time to time. Then cool the solution, and centrifuge it. This operation transforms $[SnS_2]^{2-}$ into $[SnCl_6]^{2-}$. Moreover, all remaining thioacetamide hydrolyzes to release H_2S into the solution. There isn't much of it, and the heating expels the H_2S from the solution. (Any that might remain will be expelled by the additional heating prescribed in the next paragraph.)

Decant the supernatant liquid, now designated 2C3 (Figure 25.12), into a centrifuge tube and add two clean pieces of iron wire. (First clean any rust from the wire pieces by heating them in enough 6 M HCl to cover them. Then rise them thoroughly with distilled water.) The mixture of 2C3 and iron wire should be heated in a boiling water bath for 3 to 5 minutes. After this time, decant the liquid from the unreacted iron pieces into a clean test tube.

Add 5 drops of 0.2 M $HgCl_2$. If a white, gray, or black precipitate forms, the presence of tin is confirmed.

Procedure 2D: Separation, Detection, and Confirmation of Lead

Precipitate 2P2 is washed twice with 2 to 3 drops of 2 M NH_4NO_3 as described earlier for the washing of 2P1. The washings are discarded. In the next paragraph, you will be directed to use an open flame to heat a mixture in a centrifuge tube to boiling.

WARNING This is potentially dangerous, so be sure to review how to do this as described on page 1065. Be sure your safety glasses are in place and that you do not let the tube point toward anyone.

Add 15 drops of 6 M HNO_3 to 2P2, and heat the mixture over a low flame until it just comes to the boil. Cool the mixture, centrifuge it, and discard any precipitate (free sulfur). Any sulfides of lead, bismuth, and copper are now redissolved, and the centrifugate, 2C5, contains any of these as Pb^{2+}, Bi^{3+} (mostly as BiO^+), and Cu^{2+}.

Transfer 2C5 to an evaporating dish and add 3 to 5 drops of concentrated sulfuric acid. (CAUTION: It can cause very severe burns.) At the hood, heat the mixture with an open flame *very cautiously* until you notice dense, white fumes evolving. (They are a mixture of SO_3 and water vapor.) Let the mixture cool and then gently stir it as you add 10 drops of distilled water. Allow any precipitate — 2P3 ($PbSO_4$) — to settle, and decant the supernatant liquid — 2C6 — into a centrifuge tube for Procedure 2E.

Add 3 drops of 1 M ammonium acetate ($NH_4C_2H_3O_2$) to precipitate 2P3 in the evaporating dish. Stir the mixture until the solid dissolves, and transfer the solution to a small test tube. Add 1 drop of 1 M K_2CrO_4. This will give the *solution* a yellow color, because this is the color of $CrO_4^{2-}(aq)$. What you must look for, however, is a yellow *precipitate* to confirm the presence of lead. You might have to scratch the inside of the tube with a clean glass rod to encourage any precipitate of $PbCrO_4$ to form, but only this yellow *solid* confirms the presence of lead.

Procedure 2E: Separation, Detection, and Confirmation of Copper

To the centrifugate 2C6 add concentrated ammonia (15 M NH_3) drop by drop until the solution is just basic to litmus. (CAUTION: Do not inhale the vapor of concentrated ammonia or let it contact the skin. Work at the hood.) If the solution becomes deep blue in color — the color of $[Cu(NH_3)_4]^{2+}(aq)$ — the presence of copper is confirmed. If a precipitate also forms, centrifuge the solution and continue with Procedure 2F on this precipitate, 2P4.

Procedure 2F: Detection and Confirmation of Bismuth

As discussed earlier, you will need a small volume of freshly prepared stannite ion, $[Sn(OH)_3]^-$ for this procedure. To make it, put 2 drops of 1 M $SnCl_2$ in a small test tube and add 6 M NaOH drop by drop as you agitate the tube until the initial precipitate, $Sn(OH)_2$, just dissolves.

Add several drops of this stannite solution to precipitate 2P4. If a black precipitate forms immediately, the presence of bismuth is confirmed. Be sure to watch for this right away. If you let the $Sn(OH)_3^-$ solution stand by itself long enough, even without any

Once you start Procedure 2F, it must be carried through to its completion during the same lab period.

BiO(OH), a black precipitate will eventually appear because of the disproportionation described earlier, which gives some finely divided tin.

25.5 CATION GROUP-3 ANALYSIS

Alkaline hydrogen sulfide separates the group-3 cations—Mn^{2+}, $Fe^{2+}(Fe^{3+})$, Ni^{2+}, Al^{3+}, and Zn^{2+}—as solids from the cations of the last group of our qual scheme— Ca^{2+}, Ba^{2+}, Na^+, and NH_4^+.

The flow diagram for the analysis of the group-3 cations is given in Figure 25.13. When the discussion mentions specific procedures, refer to this figure.

The group-3 cations are separated from those of group 4 by the formation of precipitates when the test solution is made basic with ammonia and treated with thio-acetamide. We discussed the fundamental principles of separations by selective precipi-tations in Qual Topic II and in Section 25.4. Now that our qual scheme has reached group 3, we must note those additional chemical properties of group-3 and -4 cations that make the scheme possible. Among such properties are the abilities of complexing agents to alter the properties of the metal ions in the complexes that they form, which we studied in Section 14.11 and in Qual Topic III. (You should review these units now, because most of the special terms involving complexes will be used here.) Another important property concerns the unequal tendencies of various metal ions to form complexes with the same ligand.

> Some terms you will need to know are complex or complex ion, ligand, and bidendate ligand.

When Two Insoluble Compounds Are Possible, the Least Soluble Forms   You will recall that the test solution must now be made basic to make it possible for the group-3 cations but not those of group 4 to form precipitates in the presence of hydrogen sulfide. Making a solution basic, of course, also introduces a much higher concentration of hydroxide ion than has been present in test solutions thus far. And most of the cations of groups 3 and 4 can form quite insoluble hydroxides. In our groups 3 and 4, only the hydroxide of the sodium ion in group 4 is very soluble in water. Table 25.2 gives the K_{sp} values for the hydroxides of the remaining group-3 and -4 cations. You can see that all are very insoluble in water, although the solubilities vary greatly. However, at the end of Procedure 3A, which is the action of H_2S on the group-3 and -4 cations in a medium made basic by NH_3, all group-4 cations are still in solution and only the hydroxide of Al^{3+} has formed. The rest of the group-3 cations have formed insoluble sulfides.

> NH_4^+, also in our group 4, does not form an undissociated hydroxide.

Two factors, a chemical property and experimental design, make this result possi-ble. It is a chemical property that when two insoluble compounds are possible, like a sulfide and a hydroxide, the less soluble compound forms first. And the sulfides of

TABLE 25.2 Solubility Product Constants of the Hydroxides of Qual Groups 3 and 4

Qual Group	Cation	Hydroxide	K_{sp} (25 °C)[a]
3	Mn^{2+}	$Mn(OH)_2$	1.6×10^{-13}
	Fe^{2+}	$Fe(OH)_2$	7.9×10^{-16}
	Fe^{3+}	$Fe(OH)_3$	1.6×10^{-39}
	Ni^{2+}	$Ni(OH)_2$	6×10^{-16}
	Al^{3+}	$Al(OH)_3$	3×10^{-34}
	Zn^{2+}	$Zn(OH)_2$	3×10^{-16}
4	Ca^{2+}	$Ca(OH)_2$	6.5×10^{-6}
	Ba^{2+}	$Ba(OH)_2$	3.0×10^{-4}

[a] From A. E. Martell and R. M. Smith, *Critical Stability Constants*, Vol. 4. Plenum Press, New York, 1976.

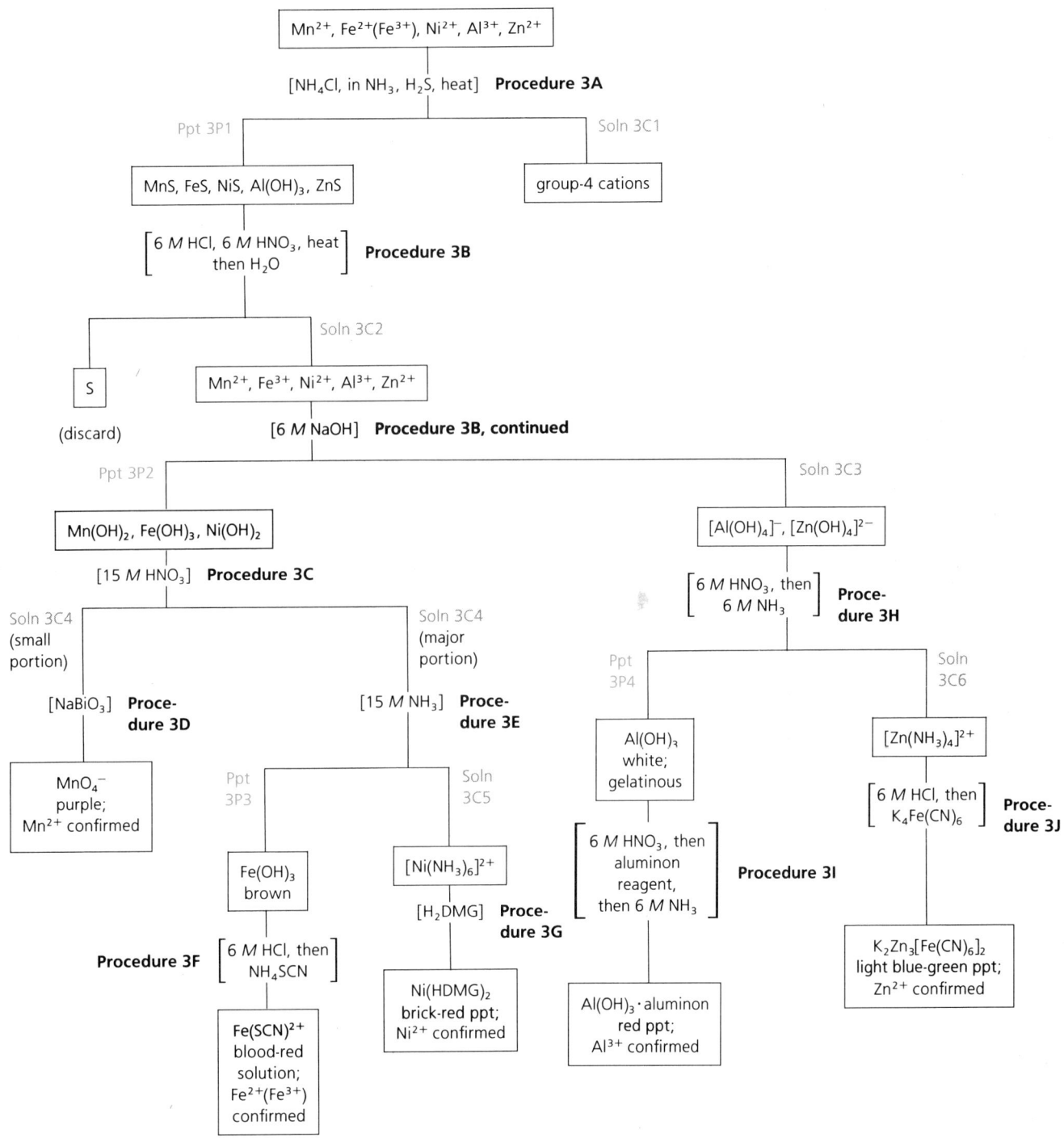

FIGURE 25.13 The flow diagram for group-3 cations.

Mn^{2+}, Fe^{2+}, Ni^{2+}, and Zn^{2+} are all less soluble than their hydroxides under the conditions of Procedure 3A. Although Al_2S_3 is a known compound—it can be prepared directly from the elements—it hydrolyzes completely in water to give $Al(OH)_3$ and H_2S. So H_2S is unable to convert either $Al(OH)_3$ or Al^{3+} (aq) into an insoluble sulfide. The sulfides of all our group-4 cations are soluble in water.

An experimental factor prevents the hydroxides of Ca^{2+} and Ba^{2+} from precipitating. The test solution is made basic, to be sure, but not by using a strong base like NaOH. An ammonia/ammonium chloride buffer is used instead. First, some 2 M NH_4Cl is added and then more than enough 6 M NH_3 is added to make the solution basic. This addition of NH_3 to an already acidic solution creates even more NH_4^+ ion. The final pH after these operations is between 8 and 9. Any OH^- ion is present because of the following equilibrium.

The NH_3/NH_4Cl buffer was discussed on page 628.

$$NH_3(aq) + H_2O \rightleftharpoons NH_4^+(aq) + OH^-(aq)$$

The concentration of OH^- ion, however, is not high enough to make the hydroxides of any group-4 cations precipitate. Yet the concentration of OH^- is high enough to cause $Al(OH)_3$ to form without being so high that $Al(OH)_3$ redissolves as $[Al(OH)_4]^-$.

When H_2S forms from thioacetamide in a group-3 test solution made alkaline by an ammonia/ammonium chloride buffer, the following reactions occur. In this alkaline buffer, the H_2S that initially forms from thioacetamide is assumed to exist almost entirely as HS^-. (We will assume that any iron ion present exists as Fe^{2+}, but following these equations we will return to the Fe^{2+}/Fe^{3+} "problem.")

$$Mn^{2+}(aq) + HS^-(aq) + NH_3(aq) \longrightarrow \underset{\text{peach}}{MnS(s)} + NH_4^+(aq)$$
$$Fe^{2+}(aq) + HS^-(aq) + NH_3(aq) \longrightarrow \underset{\text{black}}{FeS(s)} + NH_4^+(aq)$$
$$Ni^{2+}(aq) + HS^-(aq) + NH_3(aq) \longrightarrow \underset{\text{black}}{NiS(s)} + NH_4^+(aq)$$
$$Al^{3+}(aq) + 3OH^-(aq) \longrightarrow \underset{\text{white}}{Al(OH)_3(s)}$$
$$Zn^{2+}(aq) + HS^-(aq) + NH_3(aq) \longrightarrow \underset{\text{white}}{ZnS(s)} + NH_4^+(aq)$$

Aluminum hydroxide is a gelatinous precipitate that can also be formulated as $Al(H_2O)_3(OH)_3$.

The precipitates have different colors, as indicated, and Figure 4, page 595, of Qual Topic III, has photos of the sulfides. If either iron or nickel is present, regardless of what other ions are also present, any precipitate will appear gray or black.

A black precipitate will hide pink or white precipitates.

The Qual Scheme Does Not Distinguish Between Fe^{2+} and Fe^{3+}   If your test solution is centrifugate 2C1, any iron ions present would be in the +2 oxidation state. This is because hydrogen sulfide (used in the last part of Procedure 2A, Figure 25.12) reduces Fe^{3+}(aq) to Fe^{2+}(aq).

$$2Fe^{3+}(aq) + H_2S(aq) \longrightarrow 2Fe^{2+}(aq) + S(s) + 2H^+(aq)$$

(Remember, Procedure 2A involves a rather strongly acidic solution when H_2S is present, so the Fe^{2+} made in this reaction cannot precipitate as FeS.)

One of the complications, however, of having Fe^{2+}(aq) either form by this reduction or be deliberately made a solute of a special group-3 known (or unknown) is its oxidation to iron(III) by the oxygen that dissolves from the surrounding air.

$$4Fe^{2+}(aq) + O_2(aq) + 4H^+(aq) \longrightarrow 4Fe^{3+}(aq) + 2H_2O$$

Fortunately, this oxidation is slow when the medium is acidic. Moreover, if any Fe^{3+} does form or is otherwise put into the solution, it is reduced in Procedure 3A to the $+2$ oxidation state by hydrogen sulfide. And now that the medium is basic enough, the Fe^{2+} that forms promptly reacts with more H_2S to give a precipitate of FeS. The overall change from $Fe^{3+}(aq)$ in the NH_3/NH_4^+ buffer of Procedure 3A is

$$2Fe^{3+}(aq) + 3HS^-(aq) + 3NH_3(aq) \longrightarrow 2FeS(s) + S(s) + 3NH_4^+(aq)$$

This reaction makes it impossible to tell whether or not the initial test solution was prepared with Fe^{3+}.

The Colors of Ions of Group 3 Are Unreliable Indications of What Ions Are Present

In working with solutions of single salts of group-3 cations, you may notice that some have distinctive colors. See Table 25.3. We cannot, however, reliably use them to tell if a particular ion is present. Some colors are similar. Some are pale and, therefore, show up only in relatively concentrated solutions. Moreover, a mixture of ions of different colors would have a net color that would be difficult to interpret. Finally, some of the hydrated ions—particularly $[Al(H_2O)_6]^{3+}$ and $[Fe(H_2O)_6]^{3+}$—are moderately strong acids, and the color of their aqueous solution (if any) becomes a function of the pH of the medium.

None of the ionization products of $[Al(H_2O)_6]^{3+}$ is colored, but those of $[Fe(H_2O)_6]^{3+}$ are colored. Thus $[Fe(H_2O)_6]^{3+}$ has a pale violet color, but it is seen only at a pH of about 0. This is sufficiently low to suppress the ionization of $[Fe(H_2O)_6]^{3+}$ to $[Fe(H_2O)_5(OH)]^{2+}$, which has a yellowish color. At higher pH values, still more iron(III)-containing species form, and the composite color caused by all such ions (which we will not further describe) makes the solution amber colored. If the pH rises above 3, reddish, colloidal iron-containing gels appear until eventually a reddish-brown precipitate of hydrated iron(III) oxide forms. In great contrast, $Fe^{2+}(aq)$ ion in water does not hydrolyze appreciably.

Nitric Acid Redissolves Group-3 Precipitate 3P1

After separation of the group 3-cations from those of group 4, you will have a precipitate, 3P1, and a centrifugate, 3C1 (which is saved, if warranted, for analysis of the group-4 cations). The precipitate, 3P1, is next dissolved—Procedure 3B—by the action of dilute hydrochloric acid and nitric acid. The nitric acid at this stage aids in dissolving NiS, but it also oxidizes Fe^{2+} to Fe^{3+} and S^{2-} to S. We can represent the reactions with acid as follows.

$$MnS(s) + 2H^+(aq) \longrightarrow Mn^{2+}(aq) + H_2S(aq)$$
$$FeS(s) + 2H^+(aq) \longrightarrow Fe^{2+}(aq) + H_2S(aq)$$

Then

$$3Fe^{2+}(aq) + 4H^+(aq) + NO_3^-(aq) \longrightarrow 3Fe^{3+}(aq) + NO(g) + 2H_2O$$
$$3NiS(s) + 8H^+(aq) + 2NO_3^-(aq) \longrightarrow 3Ni^{2+}(aq) + 2NO(g) + 3S(s) + 4H_2O$$
$$ZnS(s) + 2H^+(aq) \longrightarrow Zn^{2+}(aq) + H_2S(aq)$$
$$Al(OH)_3(s) + 3H^+(aq) \longrightarrow Al^{3+}(aq) + 3H_2O$$
$$3H_2S(aq) + 2H^+(aq) + 2NO_3^-(aq) \longrightarrow 3S(s) + 2NO(g) + 4H_2O$$

Sulfur and Chloride Ion Must Be Removed

The remaining steps of Procedure 3B are meant to remove the sulfur as well as any chloride ion. This ion interferes with the test for Mn^{2+}—Procedure 3D. The sulfur is removed at the centrifuge and the remaining solution is evaporated to a moist residue which is then heated strongly with concentrated

---

TABLE 25.3 Colors of Group 3 Cations in Water

$[Mn(H_2O)_6]^{2+}$	Pale pink
$[Ni(H_2O)_6]^{2+}$	Green
$[Fe(H_2O)_6]^{2+}$	Pale green
$[Fe(H_2O)_6]^{3+}$	Pale violet[a]
$[Fe(H_2O)_5(OH)]^{2+}$	Amber
$[Al(H_2O)_6]^{3+}$	Colorless
$[Zn(H_2O)_6]^{2+}$	Colorless

[a] Seen only when the solution has a pH of 0 or lower.

If your group-3 unknown is colored, however, it is wise to suspect the presence of *at least one* colored cation.

In $[Fe(H_2O)_5(OH)]^{2+}$, 5 H_2O molecules and 1 OH^- ion are coordinated with the central ion.

NO_3^- ion from $HNO_3(aq)$ is an oxidizing agent in three of these reactions, but the other two are simply acid–base neutralizations.

nitric acid. This oxidizes chloride ion to chlorine, which leaves during the evaporation of water from the mixture.

$$6Cl^-(aq) + 2NO_3^-(aq) + 8H^+(aq) \longrightarrow 3Cl_2(g) + 2NO(g) + 4H_2O$$

The result of these steps is the production of solution 3C2.

The Amphoteric Properties of Aluminum and Zinc Hydroxides Enable Their Separation   Procedure 3B closes with the separation of Al^{3+} and Zn^{2+} from the other cations of group 3. These two cations, like the others, form insoluble hydroxides, but these hydroxides are amphoteric and they redissolve in excess OH^-, as we discussed in Qual Topic III. So the addition of strong base to solution 3C2 results in the following reactions. (The insoluble hydroxides are gelatinous. A photo of typical gelatinous solids is given in Figure 3, page 594, of Qual Topic I.)

Leading to 3P2:

$$Mn^{2+}(aq) + 2OH^-(aq) \longrightarrow Mn(OH)_2(s)$$
$$Fe^{3+}(aq) + 3OH^-(aq) \longrightarrow Fe(OH)_3(s)$$
$$Ni^{2+}(aq) + 2OH^-(aq) \longrightarrow Ni(OH)_2(s)$$

Leading to 3C3:

$$Al^{3+}(aq) + 3OH^-(aq) \longrightarrow Al(OH)_3(s)$$

Then the precipitate reacts with more OH^- ion and redissolves:

$$Al(OH)_3(s) + OH^-(aq) \longrightarrow Al(OH)_4^-(aq)$$

For Zn^{2+}:

$$Zn^{2+}(aq) + 2OH^-(aq) \longrightarrow Zn(OH)_2(s)$$

Then additional OH^- ion redissolves the precipitate:

$$Zn(OH)_2(s) + 2OH^-(aq) \longrightarrow Zn(OH)_4^{2-}(aq)$$

As we described in Qual Topic III, these equations are simplified treatments. The equation that represents, for example, the redissolving of the insoluble (and gelatinous) hydroxide of aluminum would better be written as follows.

$$[Al(H_2O)_3(OH)_3](s) + OH^-(aq) \longrightarrow [Al(H_2O)_2(OH)_4]^-(aq) + H_2O$$

The OH^- ion, in other words, takes H^+ from one of the three H_2O molecules coordinated as ligands in $[Al(H_2O)_3(OH)_3]$, leaving another OH^- ion as a ligand. The reaction also leaves the whole complex with an electrical charge, and it is now soluble in the solution.

The Insoluble Hydroxides, 3P2, Are Dissolved and Tested for Mn²⁺, Fe³⁺, and Ni²⁺   Nitric acid is used to dissolve the hydroxides of precipitate 3P2 — $Mn(OH)_2$, $Fe(OH)_3$, and $Ni(OH)_2$ — because we must avoid chloride ion (from HCl), as we said, and we know that all nitrate salts will be soluble. Using an oxidizing acid, moreover, keeps iron in its higher oxidized state, as Fe^{3+} instead of Fe^{2+}. This operation produces solution 3C4.

Neither Fe^{3+} nor Ni^{2+} interferes with the test for Mn^{2+}, so a small portion of solution 3C4 can be tested for Mn^{2+}—Procedure 3D. A powerful oxidizing agent, $NaBiO_3$ (sodium bismuthate), is added. It oxidizes $Mn^{2+}(aq)$ to the purple permanganate ion, and if this dramatic change in color occurs, the presence of Mn^{2+} is confirmed. See Figure 21.19, page 905.

$$2Mn^{2+}(aq) + 5BiO_3^-(aq) + 14H^+(aq) \longrightarrow \underset{\text{purple}}{2MnO_4^-(aq)} + 5Bi^{3+}(aq) + 7H_2O$$

When aqueous ammonia is added to the larger portion of 3C4 with its $Mn^{2+}(aq)$, $Fe^{3+}(aq)$, and $Ni^{2+}(aq)$ ions, Fe^{3+} is precipitated as its extremely insoluble hydroxide but insoluble hydroxides of neither Mn^{2+} nor Ni^{2+} can form. The $Ni^{2+}(aq)$ ion is changed, instead, to the soluble $[Ni(NH_3)_6]^{2+}$ ion, and $Mn^{2+}(aq)$ remains largely as $[Mn(H_2O)_6]^{2+}$.

$$Fe^{3+}(aq) + 3NH_3(aq) + 6H_2O \longrightarrow Fe(H_2O)_3(OH)_3(s) + 3NH_4^+(aq)$$
$$[Ni(H_2O)_6]^{2+}(aq) + 6NH_3(aq) \longrightarrow [Ni(NH_3)_6]^{2+}(aq) + 6H_2O$$

The iron precipitate can also be represented as hydrated iron(III) oxide, $Fe_2O_3 \cdot xH_2O$.

The gelatinous, brown precipitate (3P3) of $Fe(OH)_3$ is separated from the solution (3C5) which contains $[Ni(NH_3)_6]^{2+}(aq)$ (and any manganese species).

To make sure that precipitate 3P3 actually includes the Fe^{3+} ion, it is redissolved with hydrochloric acid and treated with ammonium thiocyanate, NH_4SCN, in Procedure 3F. The appearance of a blood red color confirms the presence of iron. See Figure 2, page 954, of Qual Topic III. The reaction that dissolves $Fe(OH)_3(s)$ is

$$Fe(OH)_3(s) + 3HCl(aq) + 3H_2O \longrightarrow [Fe(H_2O)_6]^{3+}(aq) + 3Cl^-(aq)$$

Then SCN^- displaces a molecule of H_2O from $[Fe(H_2O)_6]^{3+}$ to give the intense color of $[Fe(H_2O)_5(SCN)]^{2+}$

$$[Fe(H_2O)_6]^{3+}(aq) + SCN^-(aq) \longrightarrow \underset{\text{blood-red}}{[Fe(H_2O)_5(SCN)]^{2+}(aq)} + H_2O$$

At higher concentrations of SCN^-, the formulas of the complexes can range from $[Fe(H_2O)_5(SCN)]^{2+}$ to $[Fe(SCN)_6]^{3-}$, but the blood-red appearance is still the same.

If solution 3C5 contains $[Ni(NH_3)_6]^{2+}$, it will have the deep blue color of this complex. (See Figure 3, page 955, of Qualitative Topic III.) To confirm the presence of nickel—Procedure 3G—a special complexing agent, dimethylglyoxime (H_2DMG), is used. This is a bidentate ligand with nitrogens as the donor atoms, and two units form an insoluble, brick-red chelate complex with Ni^{2+}. (See Figure 4, page 955, of Qual Topic III.) No other cations that could be present at this stage give this kind of response to H_2DMG.

$$[Ni(NH_3)_6]^{2+}(aq) + 2H_2DMG(aq) \longrightarrow \underset{\text{brick-red}}{Ni(HDMG)_2(s)} + 4NH_3(aq) + 2NH_4^+(aq)$$

dimethylglyoxime
H_2DMG

The structure of the complex is

Ni(HDMG)$_2$

Hydrogen bonds ($\cdots$) help to stabilize this complex.

Zinc Ion but Not Aluminum Ion Forms a Water-Soluble Complex in Aqueous Ammonia   We now go back and pick up the analysis of solution 3C3, which contains the aluminate and zincate ions, $Al(OH)_4^-$ and $Zn(OH)_4^{2-}$. In Procedure 3H, the solution is neutralized and then made basic again with aqueous ammonia. This brings down the white, gelatinous $Al(OH)_3$ again—precipitate 3P4—but leaves zinc in solution as $[Zn(NH_3)_4]^{2+}$—solution 3C6. The reactions are best represented as follows.

$$[Al(H_2O)_6]^{3+}(aq) + 3NH_3(aq) \longrightarrow [Al(H_2O)_3(OH)_3](s) + 3NH_4^+(aq)$$
$$\text{white (in ppt 3P4)}$$

$$[Zn(H_2O)_4]^{2+}(aq) + 4NH_3(aq) \longrightarrow [Zn(NH_3)_4]^{2+}(aq) + 4H_2O$$
$$\text{colorless (in soln 3C6)}$$

Often, however, you will see these written with the water molecules omitted.

The presence of Al^{3+} in precipitate 3P4 is confirmed in Procedure 3I by the use of a special reagent, aluminon. It is specific for Al^{3+} when ions of chromium, iron, and beryllium are absent. Aluminon is the ammonium salt of an organic acid, aurintricarboxylic acid, whose complex structure need not concern us. The aluminon reagent is added and followed by ammonia. The solution is now again basic, so Al^{3+} precipitates as $Al(OH)_3$ once again, only this time with aluminon present. A red solid complex forms that is flocculant—meaning that it has a low enough density to float on water. In the dye industry, colored flocculant solids in water are called lakes. The appearance of the red lake confirms the presence of aluminum. A red *solid* must form. It was once thought that this solid was the result merely of the adsorption by $Al(OH)_3$ of the aluminon molecules. It is now believed, however, that these molecules form actual chelate complexes with structures that depend sensitively on the conditions in the solution.

The Ferrocyanide Ion Gives a Precipitate with Zn^{2+} and K^+   The test for zinc—Procedure 3J—is straightforward. The $[Zn(NH_3)_4]^{2+}$ ion in solution 3C6 reacts with hydrochloric acid to liberate the Zn^{2+} ion.

$$[Zn(NH_3)_4]^{2+}(aq) + 4H^+(aq) \longrightarrow Zn^{2+}(aq) + 4NH_4^+(aq)$$

Then $K_4Fe(CN)_6$ is added to supply the $[Fe(CN)_6]^{4-}$ and K^+ ions. These form a light green precipitate with Zn^{2+}. (See Figure 5, page 955, of Qual Topic III.)

$$2K^+(aq) + 3Zn^{2+}(aq) + 2[Fe(CN)_6]^{4-}(aq) \longrightarrow K_2Zn_3[Fe(CN)_6]_2(s)$$
$$\text{pale blue-green}$$

The appearance of the pale blue-green precipitate confirms the presence of zinc. (The color is so pale that many see the precipitate as white.)

Procedures for Group-3 Cations

Your first experience with the qual scheme for the group-3 cations should be with a known. You should then work an unknown specific for this group. Then you will be ready for a known or an unknown that includes several groups.

Procedure 3A: Precipitation of the Group-3 Cations   Place 2 mL of the test solution in a centrifuge tube and add 1 mL of 2 M NH_4Cl. Next, add 6 M NH_3 until the solution is basic to litmus and then add 5 more drops. Now add 6 drops of 1 M thioacetamide, and heat the solution for several minutes in a hot water bath at about 95 °C. Centrifuge the mixture and save the centrifugate, 3C1, for group-4 analysis if warranted. The precipitate, 3P1, is processed for the group-3 cations by the next procedure.

Procedure 3B: Separation of Al³⁺ and Zn²⁺ from Mn²⁺, Fe²⁺, and Ni²⁺ Wash precipitate 3P1 with 2 mL of water two times and discard the washings. To the precipitate, add 5 drops of 6 M HCl and 5 drops of 6 M HNO₃ and heat the mixture in a hot water bath at about 95 °C until the black precipitate dissolves. The cloudiness or the pale yellowish precipitate of free sulfur does not dissolve, so centrifuge the solution. Discard the sulfur and continue with the solution, 3C2. Transfer this solution to an evaporating dish, and take it to the hood. With a soft flame carefully heat the solution until there is a moist residue. Do not heat this to dryness. Next add 1 to 2 mL of 15 M HNO₃.

WARNING Concentrated nitric acid must not contact the skin. Avoid its vapors.

Continuing to work at the hood, evaporate this solution over a soft flame until the residue is again moist, not dry. Once again add 1 to 2 mL of 15 M HNO₃ and with a soft flame evaporate the mixture to a moist residue.

Now dissolve the residue in 1 to 2 mL of distilled water, and transfer the solution to a centrifuge tube. Add 15 drops of 6 M NaOH to the tube, and centrifuge the mixture. Decant the centrifugate, 3C3, into a clean centrifuge tube and reserve this solution for later tests that begin with Procedure 3H.

Continue with the precipitate, 3P2, which might have Mn(OH)₂, Fe(OH)₃, and Ni(OH)₂.

Procedure 3C: Dissolving Precipitate 3P2 Add 1 to 2 mL of 15 M HNO₃ to precipitate 3P2, and heat the mixture for several minutes in a hot water bath. Withdraw 10 drops (about 0.5 mL) of the resulting solution and place it in a clean centrifuge tube. Reserve the rest of the solution for Procedure 3E.

Procedure 3D: Detection and Confirmation of Manganese To the 0.5-mL portion of solution 3C4 add a small spatulaful of solid sodium bismuthate, NaBiO₃, and stir the mixture with a glass rod. Centrifuge the mixture. The appearance of a purple color, MnO₄⁻, in the solution confirms the presence of manganese. (If the purple color soon fades, it means that you were not successful in removing all Cl⁻ as Cl₂ in Procedure 3B. MnO₄⁻ oxidizes Cl⁻ to Cl₂, and the color of MnO₄⁻ therefore fades. Just add a bit more of NaBiO₃.)

Procedure 3E: Separation and Detection of Iron Add 10 drops of 2 M NH₄Cl to the larger portion of solution 3C4. Next add 15 M NH₃ drop by drop until the solution is basic to litmus and then add one more drop.

WARNING Avoid the vapors of ammonia. Work at the hood.

A brown precipitate, 3P3, at this stage indicates Fe³⁺. Centrifuge the mixture and reserve the centrifugate 3C5 for the test for nickel in Procedure 3G.

Procedure 3F: Confirmation of Iron Add 6 M HCl, drop by drop, to precipitate 3P3 until it dissolves, and then add 2 drops of 0.1 M NH₄SCN. The appearance of a blood-red color confirms the presence of iron.

Procedure 3G: Detection and Confirmation of Nickel To centrifugate 3C5 add 3 drops of H₂DMG. The appearance of a brick-red precipitate confirms the presence of nickel.

Procedure 3H: The Separation of Aluminum and Zinc We return now to solution 3C3, which should have any aluminum and zinc as [Al(OH)₄]⁻ and [Zn(OH)₄]²⁻.

Add 6 M HNO₃ drop by drop to solution 3C3, which should be in a centrifuge tube, until the solution is acidic. Then add 6 M NH₃ drop by drop until the solution is

As HNO₃ is added, you may see a precipitate of Al(OH)₃ and Zn(OH)₂ form, which then redissolves as more HNO₃ is added.

once again just basic. Add 5 more drops of 6 M NH$_3$, and heat the mixture a few minutes in a hot water bath. If a white gelatinous precipitate — 3P4 — is present, aluminum is indicated.

Centrifuge the mixture, and decant the supernatant liquid — 3C6 — into a centrifuge tube and save it for Procedure 3J.

Procedure 3I: Detection and Confirmation and Aluminum   To confirm that aluminum is present in precipitate 3P4, first wash the solid twice with 2 to 3 mL of *hot* water. Centrifuge the mixture after each washing as needed and discard the washings. To the precipitate, add 2 or 3 drops of 6 M HNO$_3$ — or enough to dissolve the solid. Now add 2 drops of aluminon reagent, which will impart a color to the solution. Next, as you stir the solution, add 6 M NH$_3$ drop by drop until the solution is basic to litmus. Aluminum is confirmed if a red *precipitate* forms. (It may be necessary to centrifuge the system to see the precipitate.)

Procedure 3J: Detection and Confirmation of Zinc   Acidify solution 3C6 by adding 6 M HCl dropwise. Next, add 3 drops of 0.2 M K$_4$Fe(CN)$_6$ and stir the mixture. The appearance of a pale light green or pale blue-green precipitate confirms the presence of zinc. (It might be necessary to centrifuge the mixture to see the solid.)

25.6 CATION GROUP-4 ANALYSIS

A fresh sample of the original, unanalyzed test solution must be used to test for Na$^+$ and NH$_4$$^+$.

The flow diagram for our group-4 cations is given in Figure 25.14. This group is very simple. Its analysis, however, requires three flame tests, so be sure to review the subsection on such tests on page 1069. It is also a good idea to perform flame tests on single-solute knowns so that you can see how the expected colors actually appear.

Tests for Na$^+$ and NH$_4$$^+$ Are Performed on Original Test Solutions   If you are analyzing a multigroup known or unknown, by the time you reach this stage and have a sample of centrifugate 3C1 it will certainly contain ammonium ion. Recall that an ammonia/ammonium ion buffer was part of the group-3 reagent system. Some of the reagents used earlier also were sodium salts, so centrifugate 3C1 also contains sodium ion.

The tests for NH$_4$$^+$ and Na$^+$, therefore, must be carried out on a portion of the original, unanalyzed solution. It is possible to do this because no other cation in the qual scheme interferes with the tests for NH$_4$$^+$ and Na$^+$. No other cation, for example, gives the bright yellow flame test that the sodium ion gives in Procedure 4D. And no other cation in our qual scheme besides NH$_4$$^+$ changes into a *volatile* molecular species capable of turning red litmus blue in Procedure 4E. When a solution with ammonium ion is made more alkaline by a strong base, like NaOH, and heated, ammonia gas escapes.

The pale violet color that K$^+$, if present, would impart to a flame could not be seen in the strong yellow color caused by Na$^+$.

$$NH_4^+(aq) + OH^-(aq) \xrightarrow{\text{heat}} NH_3(g) + H_2O$$

It's easy to put a piece of moist litmus paper over the beaker in which this reaction is occurring and see if it turns blue.

The Chromate Ion Separates Ba^{2+} from Ca^{2+}   The barium and calcium ions of group 4 are separated from the test solution in Procedure 4A by one of the very few

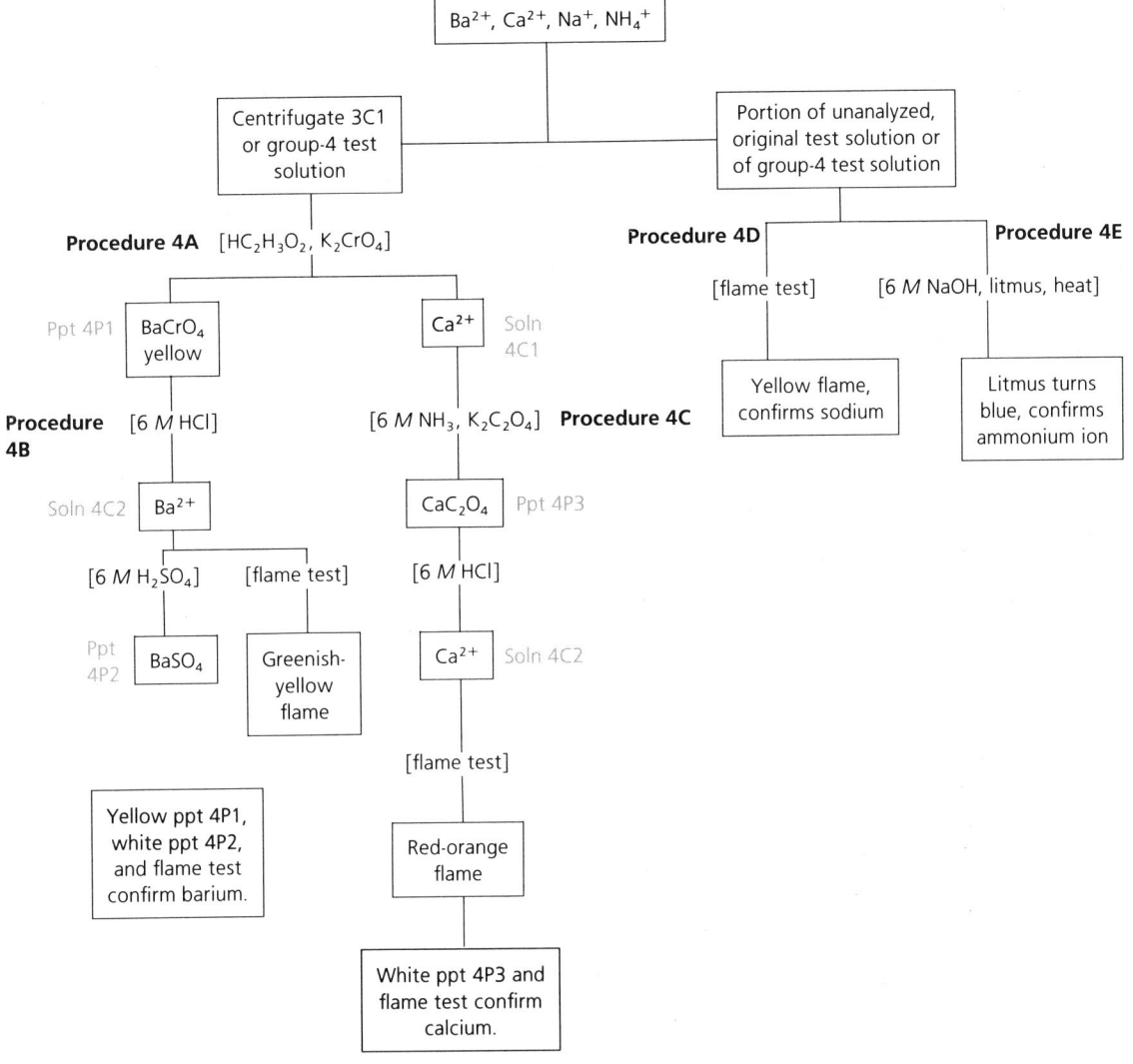

FIGURE 25.14 The flow diagram for the group-4 cations.

anions, CrO_4^{2-}, that produces salts of these cations with significantly different solubilities.

$$Ba^{2+}(aq) + CrO_4^{2-}(aq) \longrightarrow \underset{\text{yellow}}{BaCrO_4(s)} \qquad K_{sp} = 2.1 \times 10^{-10}$$

$BaCrO_4$ is insoluble — precipitate 4P1 — and Ca^{2+} remains in solution 4C1.

$BaCrO_4$ is a yellow compound, so its formation is strong evidence for barium. But it is quick and easy to carry out two confirmatory tests. So in Procedure 4B, the $BaCrO_4$ is redissolved (using hydrochloric acid), and the resulting solution — 4C2 — is tested with sulfuric acid. Ba^{2+} forms the insoluble $BaSO_4$ (white).

$$Ba^{2+}(aq) + SO_4^{2-}(aq) \longrightarrow \underset{\text{white}}{BaSO_4(s)} \qquad K_{sp} = 1.1 \times 10^{-10}$$

Solution 4C2 is also subjected to a flame test, and a greenish-yellow flame is caused by the barium ion. Thus, three pieces of evidence confirm the presence of Ba^{2+}: a yellow precipitate of $BaCrO_4$, a white precipitate of $BaSO_4$, and a greenish-yellow flame test.

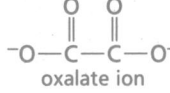
oxalate ion

The Oxalate Ion Precipitates Ca^{2+} If Ca^{2+} is in solution 4C1, it is first detected by Procedure 4C which uses the oxalate ion to form a white, insoluble compound, CaC_2O_4 — precipitate 4P3.

$$Ca^{2+}(aq) + C_2O_4{}^{2-}(aq) \longrightarrow \underset{white}{CaC_2O_4(s)} \qquad K_{sp} = 1 \times 10^{-9}$$

As additional evidence for Ca^{2+}, this precipitate is redissolved in acid and the solution is subjected to a flame test. A red-orange flame is further proof that Ca^{2+} is present.

Procedures for Group-4 Cations

You will need a total of 2 mL of the test solution for Procedures 4A, 4B, and 4C, so set this aside in a centrifuge tube. You will need 5 mL of the *original* test solution for the tests for Na^+ and $NH_4{}^+$. By "original test solution" we mean either the special group-4 known or unknown, which contains the members of no other cation groups, or a sample of the unanalyzed multigroup known or unknown solution.

Procedure 4A: Separation of Ba^{2+} and Ca^{2+} To 2 mL of the test solution in a centrifuge tube, add 10 drops of 6 M $HC_2H_3O_2$ and 5 to 10 drops of 1 M K_2CrO_4. (The potassium salt, not the sodium salt of $CrO_4{}^{2-}$, is used because we don't want sodium ion present when we do a flame test for Ba^{2+} or Ca^{2+}. The K^+ ion does have a flame test, giving a violet color, but it does not obscure the flame tests for either Ba^{2+} or Ca^{2+}. By comparison, the strong, flame-engulfing yellow color given when Na^+ is present in the flame masks most other flame tests.)

If a yellow precipitate forms, the presence of barium is strongly suggested. Centrifuge the mixture, reserve the centrifugate (4C1) for Procedure 4C, and continue with the yellow precipitate (4P1).

Procedure 4B: Confirmation of Ba^{2+} To precipitate 4P1 add 6 M HCl drop by drop until the solid dissolves to form solution 4C2. Perform a flame test on this solution. When Ba^{2+} is present, you will see a greenish-yellow flame.

To the remainder of solution 4C2 add 10 drops of 6 M H_2SO_4. A white precipitate is further evidence for the presence of barium.

Procedure 4C: Detection and Confirmation of Ca^{2+} Add 6 M NH_3 drop by drop to solution 4C1 until it is basic to litmus. Add 5 to 10 drops of 1 M $K_2C_2O_4$ (potassium oxalate), and stir the mixture. The appearance of a white precipitate — 4P3 — strongly indicates the presence of Ca^{2+}, but if no precipitate forms immediately, warm the mixture in a hot water bath for 2 to 3 minutes.

Centrifuge the mixture, discard the centrifugate, and redissolve precipitate 4P3 by adding 6 M HCl drop by drop until the solid has dissolved. Carry out a flame test on this solution. A red-orange flame further confirms the presence of Ca^{2+}.

Procedure 4D: Detection and Confirmation of Na^+ Place a 5-mL portion of the original unanalyzed test solution in an evaporating dish. Using a soft flame, boil off enough of the water to leave a moist residue. Carry out a flame test on this residue. A bright yellow light confirms the presence of Na^+.

Procedure 4E: Detection and Confirmation of NH$_4^+$ Using 1 to 2 mL of distilled water, moisten the residue from Procedure 4D and transfer it to a 50-mL beaker. Moisten a piece of red litmus paper and (with the aid of the moisture) make it adhere to the underside (the convex side) of a small watch glass. You will soon invert this watch glass over the beaker so that the litmus paper will be directly exposed to any fumes coming from the solution in the beaker.

Add 1 to 2 mL of 6 *M* NaOH to the beaker (without letting any of this solution touch the rim of the beaker or the litmus paper). Invert the watch glass over the beaker as described just above. With a soft flame, very carefully warm the contents of the beaker. A change in color of the litmus paper from red to blue confirms the presence of the ammonium ion. (You might also want to check the odor coming from the beaker. Do this cautiously, of course.)

General Unknown   Having completed the analysis of the group-4 known, you may now be assigned a multigroup known or unknown. Be sure to reserve 5 mL of this solution for Procedures 4D and 4E. If you will also be expected to analyze for anions by procedures to be described in the next section, make sure that you reserve enough of the original test solution for that work, too.

25.7 ANION ANALYSIS

Three of the anions in the qual scheme are separated as their insoluble barium salts, three others — all halide ions — are tested as a group, and the detection of the nitrate ion is done on the original solution.

The flow diagram for the analysis of the anions by our qual scheme is given in Figure 25.15. Eight anions are included: SO_4^{2-}, CO_3^{2-}, PO_4^{3-}, Cl^-, Br^-, I^-, HS^-, and NO_3^-. Some, however, are Brønsted bases that change to different anions (or to molecules) as the pH of the solution is made lower. Depending on how low the pH is made, PO_4^{3-} exists in the form of $H_2PO_4^-$ or HPO_4^{2-} or as a mixture of mostly these two anions. CO_3^{2-} is changed by acid first to HCO_3^- and then H_2CO_3 (some of which would break down to CO_2 and H_2O with excess CO_2 bubbling out of the solution). HS^- changes to H_2S as the solution becomes acidic, but some H_2S stays in solution. If the concentration of H_2S were high enough, the solution would have the rotten egg odor of this compound. At no pH, regardless of how highly basic we might make the solution in any usual laboratory situation, will we have S^{2-} ions present at any appreciable concentration, as we have noted earlier. This is why we have designated HS^-, not S^{2-}, as the supplier of sulfide ion.

Thus, the qual scheme for anions does not distinguish between $HS^-(aq)$ and $H_2S(aq)$; among $CO_3^{2-}(aq)$, $HCO_3^-(aq)$, and $H_2CO_3(aq)$; or among $PO_4^{3-}(aq)$, $HPO_4^{2-}(aq)$, $H_2PO_4^-(aq)$, and $H_3PO_4(aq)$. The scheme also cannot distinguish between $SO_4^{2-}(aq)$ and $HSO_4^-(aq)$. But, to have a solution with HCO_3^- or CO_3^{2-} and with HS^- but not H_2S (whose presence would be betrayed by odor), the anion knowns and unknowns — if prepared as aqueous solutions — must be in solutions that are at least slightly basic. In such a medium, HSO_4^- does not exist to any appreciable extent.

A basic solution, of course, has implications for possible cations. Most of the cations in our qual schemes are insoluble in basic solutions, and several cations cannot exist in the presence of HS^- at any pH. Many cations form insoluble carbonates, phosphates, and halides. So if you have a general unknown which is to be analyzed for both cations and anions, use your knowledge of the solubilities of metal hydroxides, oxides, carbonates, halides, and sulfides to help you decide against impossible combinations.

HS^-, of course, is not the sulfide ion, but it *supplies* it to any species that can form an insoluble sulfide.

The eight anions of our scheme and their protonated variations are all colorless.

This paragraph should suggest the importance of "using your head" as you do the tests and interpret your observations.

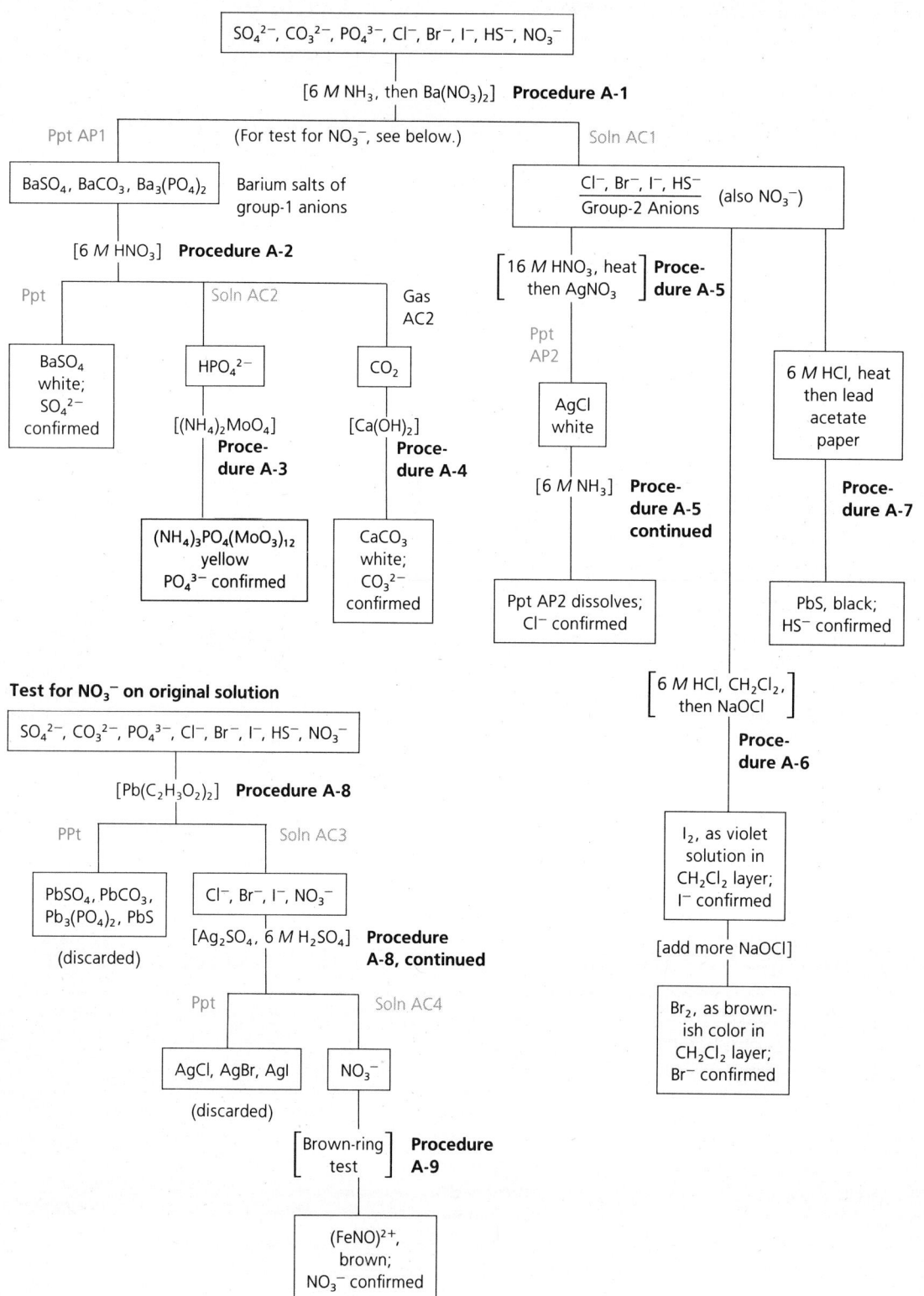

FIGURE 25.15 The flow diagram for anion analysis.

Generally speaking, only the nitrate salts of the cations of our qual scheme are *all* soluble in water. And even then, Bi^{3+} would occur as BiO^+. A student who reported that a general unknown (as an aqueous solution) contained both $Ag^+(aq)$ and $I^-(aq)$ would be asked to think again (and maybe start over). The only cations of our qual scheme that can be present at appreciable concentrations together with all of the anions are Na^+ and NH_4^+.

Let us now see how we can analyze for the anions given at the top of Figure 25.15. Refer to this figure often as you study this section. Keep in mind that when we discuss the detection of the "phosphate ion," or when we use its formula PO_4^{3-}, we use this term or formula for whatever mixture of PO_4^{3-}, HPO_4^{2-}, and $H_2PO_4^-$ may be present at the pH of the particular solution. Similarly, "carbonate ion" or CO_3^{2-} might also stand for HCO_3^-, depending on the pH of the solution.

The Group-1 Anions, SO_4^{2-}, CO_3^{2-}, and PO_4^{3-}, Form Insoluble Barium Salts

Procedure A-1 separates the three anions that make up anion group 1, SO_4^{2-}, CO_3^{2-}, and PO_4^{3-}, as barium salts. This leaves a solution, AC1, that is subjected to individual tests for the remaining anions. We'll look first at the procedures for the anions that form insoluble barium salts, those in precipitate AP1.

Action of Nitric Acid on Precipitate AP1 Might Leave an Undissolved Solid, a Solution, and a Gas

The possible compounds in precipitate AP1 are $BaSO_4$, $BaCO_3$, and $Ba_3(PO_4)_2$. If all three are present when nitric acid is added to it, $BaSO_4$ will remain unaffected. Under the conditions of Procedure A-2, $BaSO_4$ does not dissolve. The persistence of this white precipitate confirms the presence of SO_4^{2-}.

In HNO_3, $BaCO_3$ will dissolve and CO_2 gas will evolve.

$$BaCO_3(s) + 2H^+(aq) \longrightarrow Ba^{2+}(aq) + H_2O + CO_2(g)$$

The detection of gaseous CO_2 (Procedure A-4) involves passing any gases that might be evolving into *limewater*, a solution that is saturated in $Ca(OH)_2$. CO_2 reacts with the Ca^{2+} ion in limewater as follows to give $CaCO_3(s)$. This is a white, water-insoluble solid whose appearance turns the solution milky. This result confirms CO_3^{2-}.

Procedure A-4 requires a special but simple piece of apparatus described later.

If you blew through a straw into limewater, the CO_2 in your exhaled air would turn the solution milky.

$$H_2O + CO_2(g) + Ca^{2+}(aq) \longrightarrow CaCO_3(s) + 2H^+(aq)$$

The third possible solid in precipitate AP1 is $Ba_3(PO_4)_2$, and it will dissolve in nitric acid to liberate $Ba^{2+}(aq)$ and $H_2PO_4^-(aq)$.

$$Ba_3(PO_4)_2(s) + 4H^+(aq) \longrightarrow 3Ba^{2+}(aq) + 2H_2PO_4^-(aq)$$

Thus, the sulfate ion, the carbonate ion, and the phosphate ion are differentiated by the behavior of their barium salts toward acid. Barium sulfate does not dissolve, but barium carbonate and barium phosphate do. CO_2 evolves when the carbonate is present and is detected with limewater. The phosphate ion (or its various protonated forms) is found in the solution (AC2) by the formation of a yellow precipitate with ammonium molybdate, $(NH_4)_2MoO_4$. The reaction is complex and is usually represented as follows.

$$HPO_4^{2-}(aq) + 12MoO_4^{2-}(aq) + 23H^+(aq) + 3NH_4^+(aq) \longrightarrow$$
$$(NH_4)_3PO_4(MoO_3)_{12}(s) + 12H_2O$$
ammonium
phosphomolybdate
(yellow)

The appearance of a yellow precipitate at this point (Procedure A-3) confirms the presence of phosphate ion.

The NO_3^- ion is alone in our anion group 3.

The Slow Relative Rate of Oxidation of Cl^- Is Exploited to Isolate It

Of the five anions in solution AC1, Cl^-, Br^-, I^-, HS^-, and NO_3^-, all but NO_3^- make up anion group 2. All in this group form very insoluble precipitates with silver ion. Although each silver salt has a characteristic color, these colors are not reliable guides to the confirmation of the presence of any given anion, especially if more than one anion is present.

AgCl	AgBr	AgI	Ag_2S
white	pale yellow	yellow	black

If a black precipitate formed on the addition of silver nitrate to a solution of the sodium salts of the anions in anion group 2, there would be no way to tell directly if the blackness of Ag_2S were not masking the colors of one or more of the other silver salts. If the precipitate were not black, then you would have to make a judgment between yellow and pale yellow or between pale yellow and white. In sunlight, moreover, all of these nonblack silver halides slowly darken as electromagnetic energy initiates the conversion of silver ions to silver atoms. What all of this discussion says is that the identifications of the halide ions have to be done carefully, and there has to be some separation of them from each other. Let's see how this is done.

To find the chloride ion, the solution AC1 is *heated briefly* with concentrated nitric acid. This oxidizes all anions that would interfere with the confirmatory test for Cl^- to species that do not interfere. Only because the chloride ion is the most slowly oxidized of the anions present in AC1 can this approach, Procedure A-5, work. So be sure to follow the directions carefully. It is vital that you try the procedures on knowns first, including a known that does not contain Cl^- but does contain Br^-, I^-, or HS^- (or, better, a combination of these).

Do a known for Cl^- and then the same test on tap water to see the difference between a real precipitate of AgCl and cloudiness.

Following the action of hot nitric acid on solution AC1, silver nitrate solution is added, and if Cl^- is present then a white precipitate of AgCl appears. Don't be misled by a faint cloudiness at this stage. Chloride ion is such a common impurity that a trace of AgCl, enough to cloud the solution, is almost inevitable.

If more than a cloudy trace of AgCl appears to form at this time, aqueous ammonia is added. AgCl, but not AgBr or AgI, dissolves fully in the presence of dilute NH_3 because the diamminesilver complex, $Ag(NH_3)_2^+$, readily and completely forms from the more soluble AgCl but not as completely from the extremely insoluble AgBr or AgI. (We studied this in Section 14.1. "Complex Ion Equilibria." Thus, the test that confirms Cl^- is the dissolving of a white precipitate (AgCl) when NH_3 is added.

The Relative Redox Properties of Br^- And I^- as Well as the Colors of I_2 and Br_2 in CH_2Cl_2 Are Used to Identify Br^- and I^-

CCl_4 has commonly been used instead of CH_2Cl_2, but it poses greater health and environmental problems.

Dichloromethane, CH_2Cl_2, is insoluble in water and much more dense than water. It is also a much better solvent than water for Cl_2, Br_2, and I_2. Moreover, Br_2 and I_2 have very characteristic colors in CH_2Cl_2. To bring out these colors in a way that is useful in qual analysis — Procedure A-6 — we first add a small amount of CH_2Cl_2 to solution AC1. Then we make the solution acidic with hydrochloric acid. Next, a dilute solution of sodium hypochlorite, NaOCl, is added dropwise. What you see and the order in which you see it depend on whether both I^- and Br^- or only one of these is present. But let's back up a bit and review the chemistry.

The hypochlorite ion is a moderately strong Brønsted base, so in the presence of a strong acid, like hydrochloric acid, OCl^- changes to HOCl (hypochlorous acid), a weak acid. A strong acid, in other words, shifts the following acid–base equilibrium to the right.

$$OCl^-(aq) + H^+(aq) \rightleftharpoons HOCl(aq)$$

HOCl is able to react further with hydrochloric acid to give an aqueous solution of chlorine; the following equilibrium forms.

$$HOCl(aq) + Cl^-(aq) + H^+(aq) \rightleftharpoons Cl_2(aq) + H_2O$$

What these two equilibria mean is that if we add NaOCl solution dropwise to a solution already acidified with hydrochloric acid we generate Cl_2 in the solution. A mixture of hydrochloric acid and sodium hypochlorite, in other words, is a source of aqueous chlorine, $Cl_2(aq)$.[1]

When aqueous chlorine forms in the presence of a globule of CH_2Cl_2, the chlorine becomes distributed between the CH_2Cl_2 layer and the aqueous layer (with most in the CH_2Cl_2 layer). The following equilibrium, however, makes Cl_2 available in either the water or the CH_2Cl_2 layer.

$$Cl_2(aq) \rightleftharpoons Cl_2(CH_2Cl_2)$$

$Cl_2(CH_2Cl_2)$ means a solution of Cl_2 in CH_2Cl_2.

If a chlorine-consuming reaction occurs in one layer, the equilibrium can shift in accordance with Le Châtelier's principle to supply more Cl_2 to that layer. Now let's consider what will happen if solution AC1 contains both Br^- and I^- ions when Cl_2 becomes available in it. Remember that both ions can be oxidized by Cl_2.

If both $Br^-(aq)$ and $I^-(aq)$ are present in solution AC1, then the aqueous chlorine generated by the first drops of NaOCl will oxidize I^-, the more easily oxidized ion of the two, to I_2.

$$Cl_2(aq) + 2I^-(aq) \xrightarrow{CH_2Cl_2} 2Cl^-(aq) + I_2(CH_2Cl_2) \qquad (25.3)$$
$$\text{violet}$$

The colors of Br_2 and I_2 in water and in a chlorinated solvent are shown on pages 853 and 855 of Section 20.7.

As indicated, the newly formed I_2 enters the CH_2Cl_2 layer, which becomes violet in color. So if I^- is present in AC1, the CH_2Cl_2 layer turns violet after two or three drops of NaOCl solution have been added.

More NaOCl solution is now added, and it oxidizes I_2 to a colorless ion, IO_3^-, according to the following equation.

$$5Cl_2(aq) + I_2(aq) + 6H_2O \longrightarrow 2IO_3^-(aq) + 10Cl^-(aq) + 12H^+(aq)$$

The reaction occurs in the *aqueous* phase, but traces of I_2 are already present in this layer because of the following equilibrium.

Br_2 isn't oxidized by Cl_2 to BrO_3^- (or anything else).

$$I_2(aq) \rightleftharpoons I_2(CH_2Cl_2)$$

As the oxidation of $I_2(aq)$ to $IO_3^-(aq)$ occurs in the aqueous phase, this equilibrium shifts to the left, in accordance with Le Châtelier's principle, until all the I_2 is oxidized. Thus, this oxidation causes the violet color in the CH_2Cl_2 layer to fade, and all iodine is changed to a form that cannot be affected when more aqueous NaOCl is added. We next ask: "What will happen if Br^- is also present as more NaOCl is added?"

When Br^- is also present in solution AC1, the addition of more NaOCl solution at this stage will cause Br^- to be oxidized to Br_2. As it forms, it dissolves in the CH_2Cl_2 layer, *which turns brownish.*

$$Cl_2(aq) + 2Br^-(aq) \xrightarrow{CH_2Cl_2} 2Cl^-(aq) + Br_2(CH_2Cl_2)$$
$$\text{brown}$$

This reaction would have occurred with the *first* drops of NaOCl if Br^- ion but not I^- ion had been present in solution AC1. Cl_2 can oxidize both ions, as we said. Only when

---

[1] Aqueous chlorine can also be prepared simply by bubbling Cl_2 (from a pressurized cylinder of this gas) through distilled water, and some labs will provide freshly made *chlorine water* in this way. Instead of adding NaOCl solution, this chlorine water is added. We'll assume that you will use the sodium hypochlorite/hydrochloric acid approach.

both are present does it attack the more easily oxidized ion, I⁻, first, and we see the violet color first, then the brownish color in the CH_2Cl_2 layer.

If no Br⁻ is present, only I⁻, then the first drops of NaOCl solution would produce the violet $I_2(CH_2Cl_2)$ layer, the color would fade with additional NaOCl, *and it would not to be replaced by a brownish color.*

We can summarize these changes in the color of the CH_2Cl_2 layer as aqueous NaOCl is added dropwise to AC1 (which has been acidified with hydrochloric acid and which has the globule of CH_2Cl_2):

1. If both Br⁻ and I⁻ are present, a violet color appears which then fades (as more NaOCl is added), and a brownish color appears.

2. If I⁻ but not Br⁻ is present, a violet color appears which fades and leaves the CH_2Cl_2 layer colorless.

3. If Br⁻ but not I⁻ is present, the CH_2Cl_2 layer becomes brownish with the first drops of NaOCl solution, and the color does not fade as more NaOCl is added.

4. If neither Br⁻ nor I⁻ is present, the CH_2Cl_2 layer remains colorless as NaOCl is added.

Needless to say, it is very important that you work with a known first.

HS⁻ Is Converted to H₂S and Detected by Odor and by Lead Acetate Paper

If HS⁻ ion is present in a fresh sample of solution AC1 when it is acidified (Procedure A-7), you might be able to detect the rotten egg odor of H_2S. To confirm it, a strip of moistened filter paper containing $Pb(C_2H_3O_2)_2$— lead acetate paper — is held above the solution. Any evolving H_2S will tend to dissolve in the moisture of the paper and be trapped by the following reaction. The strip will turn to the black color of PbS, which confirms that HS⁻ was present.

$$H_2S(aq) + Pb^{2+}(aq) \longrightarrow \underset{\text{black}}{PbS(s)} + 2H^+(aq)$$

The Brown-Ring Test for NO₃⁻ Involves the Formation of [FeNO]²⁺

No precipitate can be used to detect the nitrate ion since all nitrates are soluble; however, the iron(II) ion in acid can reduce the nitrate ion to nitrogen monoxide, NO.

$$NO_3^-(aq) + 4H^+(aq) + 3Fe^{2+}(aq) \longrightarrow NO(aq) + 3Fe^{3+}(aq) + 2H_2O$$

Nitrogen monoxide forms a complex with unchanged iron(II) ion that has a characteristic brown color under the conditions of the procedure—Procedure A-9.

$$Fe^{2+}(aq) + NO(aq) \longrightarrow \underset{\text{brown}}{[FeNO]^{2+}(aq)}$$

Many anions interfere with the brown-ring test. Hot sulfuric acid is one of the reagents in the test, and it can oxidize Br⁻ to Br_2 and I⁻ to I_2, both of which can give a brown color to water, too. Other anions also interfere. So to remove all possible interfering anions, a fresh solution of the known or unknown is treated with precipitating agents that carry all such interfering anions out of solution as insoluble solids. Thus, in Procedure A-8, lead acetate is first added to the test solution to remove HS⁻ (as insoluble PbS), which would form a precipitate with Fe^{2+}. (The Pb^{2+} ion in lead acetate also takes out CO_3^{2-}, PO_4^{3-}, and SO_4^{2-}, but this is incidental.) The solution, AC3, might now possibly have one or more halide ions, so silver sulfate is next added in a continuation of Procedure A-8. (We can't use the more soluble silver nitrate, of course, because this would add the anion we're testing for.) The silver ion from silver sulfate takes out all halide ions, leaving solution AC4. (The sulfate ion, which is now present, does not interfere with the brown-ring test; this test, in fact, uses sulfuric acid.) Now Procedure A-9, the brown-ring test, can be performed on solution AC4.

Procedures for Anion Analysis

To obtain enough of solution AC1 for the analyses for halide ions and HS⁻, you will have to perform Procedure A-1 more than once.

One procedure, A-4, requires a simple gas delivery tube that you make by bending a piece of glass tube and affixing a one-hole stopper to it. Special Topic 25.4 explains how to do this.

Procedure A-1: The Separation of SO_4^{2-}, CO_3^{2-}, and PO_4^{3-}   Transfer 2 mL of the test solution to a centrifuge tube. Add 6 M NH_3 drop by drop until the solution is just basic, and then add 3 or 4 more drops. Now add 0.5 mL of 0.2 M $Ba(NO_3)_2$ and stir the mixture. The appearance of a white precipitate, AP1, indicates one or more of the ions SO_4^{2-}, CO_3^{2-}, or PO_4^{3-}. Centrifuge this mixture and save the centrifugate, AC1, for further tests.

Of course, the absence of a precipitate means that SO_4^{2-}, PO_4^{3-}, and CO_3^{2-} are absent!

Procedure A-2: Test for SO_4^{2-}   You will be adding acid to precipitate AP1. Watch to see if this causes a fizzing, because this will indicate that the carbonate, $BaCO_3$, is present.

Add 2 mL of 6 M HNO_3 to precipitate AP1. Stir the mixture and centrifuge it. The persistence of an undissolved, white precipitate confirms the presence of SO_4^{2-}. Save the centrifugate, AC2, for Procedure A-3.

Procedure A-3: Test for PO_4^{3-}   Add 1 mL of 1 M $(NH_4)_2MoO_4$ to centrifugate AC2, now in a clean centrifuge tube. Stir the mixture and then warm it in a hot water bath for 2 to 3 minutes. A yellow precipitate confirms the presence of PO_4^{3-}.

Procedure A-4: Test for CO_3^{2-}   If no fizzing was noticed in Procedure A-2, this test for the carbonate ion will probably give negative results. If you noticed fizzing (or are doubtful about it), then proceed.

Procedure A-1 has to be repeated for this test for CO_3^{2-}. This time, however, use a 200-mm test tube. When you have carried out Procedure A-1 using this tube, decant the solution (AC1) from the precipitate (AP1) and continue with the precipitate. Have the gas delivery tube ready to attach to the 200-mm test tube which now contains AP1.

Place 3 to 5 mL of a saturated solution of $Ca(OH)_2$, limewater, in a separate 150- or 200-mm test tube and clamp it to a ring stand as pictured in Figure 25.16. What you will next do is add acid to precipitate AP1, quickly attach the gas delivery tube, and insert the exposed end of this tube below the surface of the limewater as seen in Figure 25.16.

Add 5 mL of 1 M HNO_3 to the precipitate, AP1, and quickly attach the gas delivery tube as described just above. The appearance of a milky precipitate in the limewater within a minute or two confirms the presence of CO_3^{2-}.

Procedure A-5: Test for Cl^-   Redo Procedure A-1 to obtain 2 mL of solution AC1. You will next add 15 drops of concentrated (16 M) HNO_3, but read this entire procedure first. Be very careful that none of this acid drops onto your skin or clothing. Should any touch your skin, wash it off in cold running water at once. (Expect a spot of yellow on the skin. It will go away in about a week.) If any nitric acid touches your clothing, do not expect to prevent an acid hole, and do not touch the area with your fingers. (A wet paste of sodium bicarbonate or a towel dampened with dilute aqueous ammonia can be used to neutralize such minor acid spills.) Contact your instructor for help in neutralizing

Obtain a piece of 8-mm soft glass tubing about 20 cm long. You can fire polish its ends either before or after you make the desired bend, but always remember that cool glass and hot glass look alike.

To make the bend, put a flame spreader on your Bunsen burner, light the burner, and adjust the flame to be as uniform and as hot as possible. Grasp the glass tube at each end, and rotate it in the flame as illustrated in the figure on the top below. Let the heated area of the tube be as near one hand as the heat permits. When you feel the glass will bend, make a bend of about 45° or 50° as indicated in Figure 25.16. Set the hot tube aside across the mouth of a beaker to cool.

Do not proceed unless the ends of the bent glass tube are fire polished and cool.

Obtain a one-hole rubber stopper that fits the test tube to be used for the carbonate ion test. (The size of the stopper should be such that about half of its length, but no more, will stick into the mouth of the test tube.) Apply a little glycerol (or saliva) around the edges of the hole at the *wider* end of the rubber stopper. You will be inserting the end of the glass tube *nearer* the bend into the hole at this wider end of the stopper. Wrap the bent tube in a towel or a handkerchief in such a way that if the glass should snap during the next operation the jagged ends will not cut your hand. See the figure on the bottom below. First, apply a small film of glycerol (or saliva) around the outer edge of the end of the glass tube that is to be inserted. Slowly *and simultaneously* twist and push this end of the tube into the hole at the wider end of the stopper until the tube has emerged about 1 cm from the other side. You can now rinse off the lubricant.

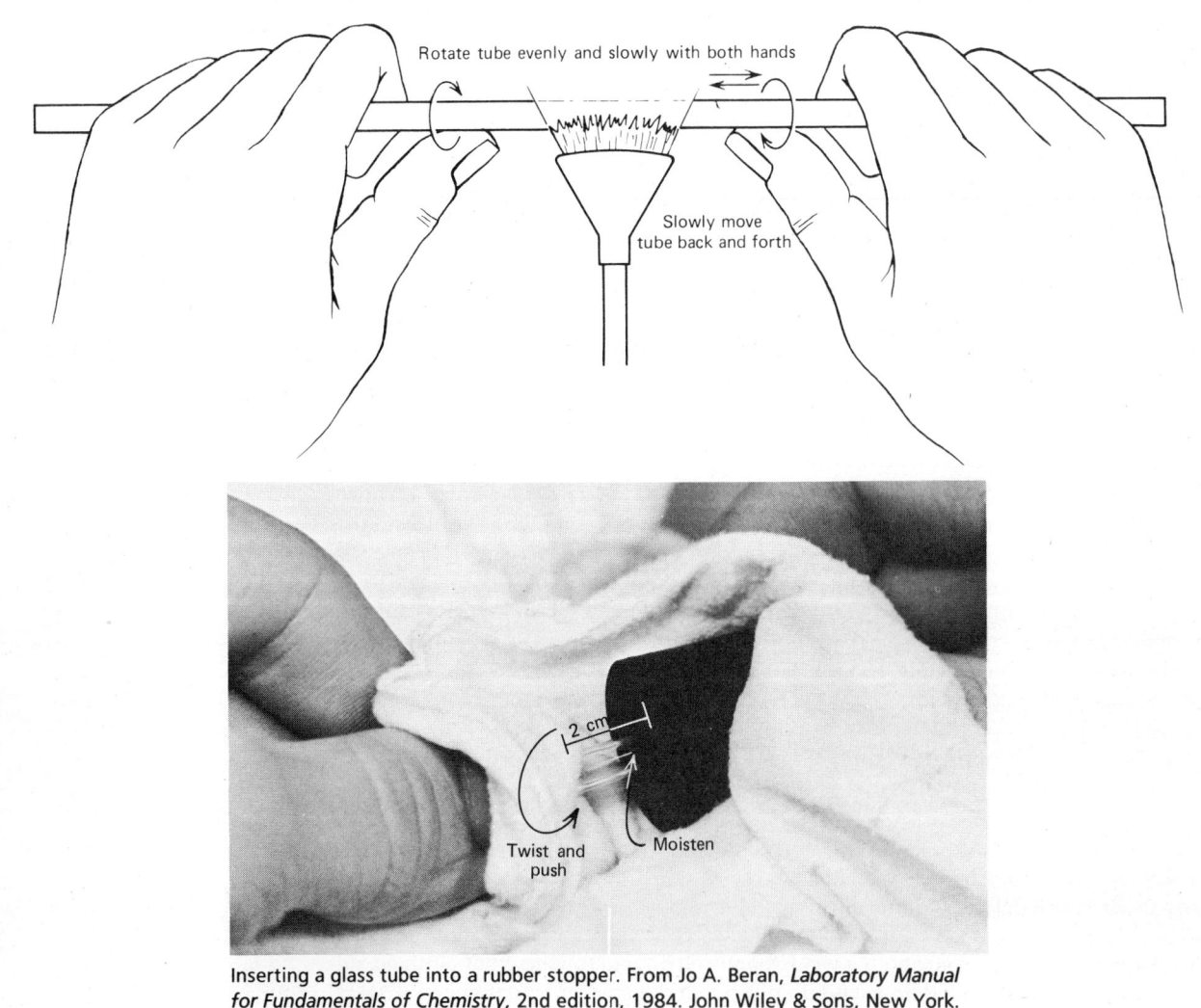

Inserting a glass tube into a rubber stopper. From Jo A. Beran, *Laboratory Manual for Fundamentals of Chemistry,* 2nd edition, 1984. John Wiley & Sons, New York. Used by permission.

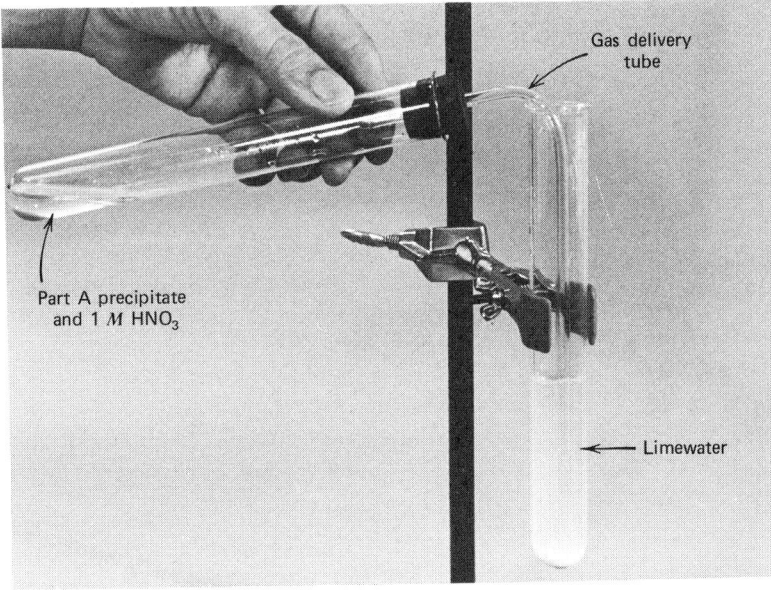

FIGURE 25.16 Test for carbonate ion. Inside the tube on the left, CO_2 is generated by the action of acid on the carbonate ion. The vertical tube contains limewater, a saturated solution of $Ca(OH)_2$. The gas delivery tube must project well below the surface of the limewater, but it should extend only a little way beyond the stopper inside the other tube. From Jo A. Beran, *Laboratory Manual for Fundamentals of Chemistry,* 2nd edition, 1984. John Wiley & Sons, New York. Used by permission.

the acid. If you have had to pause to deal with this kind of situation, discard the solution and start this procedure over.

As soon as you have added the concentrated nitric acid, place the tube in a hot water bath (kept at about 95 °C) for about 30 seconds. Now hold the tube, tilted, in a stream of cold water to cool its contents as quickly as possible. (Stir the mixture with a *glass* rod as you cool it.) Do not let any of the running water enter the tube. If any solid forms, remove it using the centrifuge and discard it.

To the resulting solution, add 10 drops of 0.1 M $AgNO_3$. A white precipitate indicates the presence of Cl^-. To confirm this, add 6 M NH_3 until the precipitate dissolves or until the solution is basic. If the white precipitate dissolves, the presence of Cl^- is confirmed, but you should double-check this by repeating this procedure.

Procedure A-6: Tests for I^- and Br^-　You will be using CH_2Cl_2 in this procedure. Do not inhale its vapors. It is toxic. CH_2Cl_2 is a dense ($d = 1.326$ g/mL), volatile liquid and it tends to dribble out of dropper pipets. Work at the hood when you transfer it so that any drops that escape will not enter the lab. Keep CH_2Cl_2 off the skin. It won't cause a burn, but like all organic liquids it is absorbed through the skin. At the end, discard all wastes in a special waste disposal bottle set aside solely for CH_2Cl_2 wastes.

Obtain a fresh supply of solution AC1 by redoing Procedure A-1. Place 2 mL of this solution in a centrifuge tube and add 6 M HCl dropwise until the solution is acidic to litmus. Then add 6 more drops. Add 5 drops of CH_2Cl_2 (see previous paragraph). Slowly add NaOCl solution, drop by drop. Stir and agitate the mixture thoroughly after each drop. This is a two-phase system, and you want to maximize opportunities for equilibria to be established between the phases. Agitating the mixture helps this process. If the CH_2Cl_2 layer becomes violet in color, the presence of I^- is confirmed. (See page 855 for a color photograph of I_2 in a chlorinated solvent.)

If no violet color appears, but the CH_2Cl_2 layer turns brownish, then I^- is not present but the presence of Br^- is confirmed. (Again, see page 853 for a color photo of Br_2 in a chlorinated solvent.) The procedure can now be ended; you have found Br^- but not I^- in your sample.

If no color appears in the CH_2Cl_2 layer, neither I^- nor Br^- is present, and the procedure can be abandoned.

If the violet color did appear, then continue to add NaOCl. If the violet color fades as you do this, and the CH_2Cl_2 layer takes on a brownish color, then the presence of Br^- is confirmed. Your sample had both Br^- and I^-.

However, if the violet color fades and is *not* replaced by a brownish color in the CH_2Cl_2 layer, your sample had I^- but not Br^-.

Procedure A-7: Test for HS⁻

Prepare a fresh 2-mL sample of solution AC1 by redoing Procedure A-1. Also prepare a strip of moist lead acetate paper as follows. Cut a strip of filter paper roughly 1 cm wide and 4 to 6 cm long, and spread over approximately half its surface a few drops of 2 M $Pb(C_2H_3O_2)_2$ solution.

To solution AC1, add 3 to 5 drops of 6 M HCl. Place the tube in a warm water bath and quickly hold the moist lead acetate paper in the mouth of the test tube. If the paper turns black, the presence of HS^- is confirmed.

Procedure A-8: Isolation of the Nitrate Ion

Place 2.5 mL of the *original* test solution in a centrifuge tube and add 5 to 6 drops of 1 M $Pb(C_2H_3O_2)_2$. Stir the mixture and centrifuge it. Transfer the centrifugate, solution AC3, to another centrifuge tube (and discard the precipitate). To solution AC3, add 10 drops of a saturated solution of Ag_2SO_4 (0.04 M) plus 5 drops of 6 M H_2SO_4. If a precipitate forms, centrifuge and decant the supernatant liquid, which is solution AC4. Its only anions should now be the sulfate ion (from the reagents just used) and the nitrate ion, if present in the original test solution.

Procedure A-9: The Brown Ring Test for NO₃⁻

You will use concentrated sulfuric acid in this procedure, and be very careful in handling this substance. It can cause severe skin burns, and if you suspect even for a moment that any has gotten onto your skin, flush the affected area at once with a copious supply of cold, running water.

You will be pouring concentrated H_2SO_4 from a small bottle, and often the outer surface of such a bottle becomes contaminated by sulfuric acid. To rinse it off, first be sure that the stopper is well seated in the opening of the bottle. Then protect your hand with a towel as you grasp the bottle at its neck in such a way that you keep pressure on the stopper to keep it from easing loose. Hold the bottle under a *very slow* stream of running water, letting the water rinse the outside of the bottle below its neck, the area where you ordinarily would grasp the bottle with your hand. It is, of course, imperative that no water enter the bottle, because an extremely exothermic reaction will occur.

Follow the directions of your instructor about the disposal of the wastes from this experiment. Read through the rest of this procedure before you start it.

Place 1 mL of solution AC4 in a small (75 mm) test tube, add 1 mL of a saturated solution of iron(II) sulfate, and mix the contents of the tube. Place the test tube inside the largest beaker you have that will still permit the mouth of the test tube to extend above

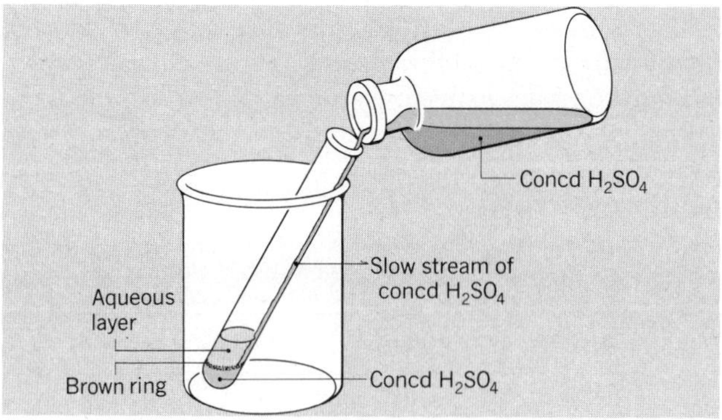

Don't let lead wastes go down the drain.

If you use a rapid stream of water, chances are that some of the acid that might be on the outside of the bottle will splash onto your clothing.

FIGURE 25.17 The brown-ring test for NO_3^-. The concentrated sulfuric acid is slowly poured down the inside of the test tube. Be particularly careful to set this arrangement where nothing can bump it. Discard the wastes only where designated by your instructor.

the beaker's rim. See Figure 25.17. Check to see that the beaker is protected from being tipped. Slowly and carefully, without any agitation of the test tube, pour about 0.5 mL of concentrated H_2SO_4 down the inside of the tube. The intent is to exploit the higher density of the acid, which causes it to sink to the bottom of the tube as a separate layer without becoming mixed with the solution already in the tube. Of course, at the interface between the acid and the solution, where the two liquids meet, some mixing will begin, and this is desired. The reduction of NO_3^- by Fe^{2+} is favored by a highly acidic medium, but we want to keep the key product, NO, from spreading. We want it to form $[FeNO]^{2+}$ in as concentrated a form as possible so that its brown color is plainly seen. Let the system stand for several minutes. If a brown ring develops at the interface, the presence of NO_3^- is confirmed.

SUMMARY

Qualitative Inorganic Analysis.   Experiments for the isolation, detection, and confirmation of 15 cations and 8 anions are described and explained. The test samples, both knowns and unknowns, consist of aqueous solutions in which any particular ion has a concentration high enough to be detected by our procedures. All procedures are semimicro.

A test to detect or confirm one ion might employ a chemical that also reacts with other ions present, so much of the work of qual analysis is to separate the ions from each other before the final tests are made. Specific group reagents are used for this separation.

Cation group 1, the silver group, consists of Ag^+, Hg_2^{2+}, and Pb^{2+}. These are separated from the other cations with 6 M HCl, the cation group-1 reagent, which precipitates these three cations as AgCl, Hg_2Cl_2, and $PbCl_2$. $PbCl_2$ is the most soluble of these, so some Pb^{2+} can escape separation and will be found in the next cation group.

Cation group 2, the hydrogen sulfide group, consists of Bi^{3+} (usually as BiO^+), Cu^{2+}, Sn^{4+} (or Sn^{2+}), and Pb^{2+}. These are separated by using saturated H_2S at a pH of about 0.5, the cation group-2 reagent. The sulfides of these ions, PbS, CuS, SnS_2, and Bi_2S_3, form, and all are insoluble in acid. The H_2S is made by the hydrolysis of thioacetamide, so an aqueous solution, not gaseous H_2S, is actually used. In larger schemes, arsenic is also in group 2, which is why this group is often called the copper–arsenic group.

Cation group 3, the ammonium sulfide group or the aluminum–nickel group, consists of Mn^{2+}, Ni^{2+}, Fe^{2+} (or Fe^{3+}), Al^{3+}, and Zn^{2+}. They form insoluble compounds in 0.1 M H_2S only when the solution is slightly basic, so the cation group-3 reagent is a solution of H_2S in a basic buffer made of NH_3 and NH_4^+. Four of the cations form base-insoluble sulfides, MnS, NiS, FeS, and ZnS. Al^{3+} under these conditions forms $Al(OH)_3$, which is also insoluble.

Cation group 4 in our scheme, the soluble group, consists of Na^+, NH_4^+, Ca^{2+}, and Ba^{2+}. The tests for these can be run without further separations.

The anions fall into three groups. Anion group 1 consists of SO_4^{2-}, PO_4^{3-}, and CO_3^{2-}, which form insoluble barium salts. Anion group 2 consists of Cl^-, Br^-, I^-, and HS^- all of which form insoluble silver salts that are insoluble in dilute nitric acid. Anion group 3 has just one anion, NO_3^-.

Cation Group-1 Analysis.   Since two of the insoluble chlorides of this group can form soluble complex ions with Cl^-, the group reagent (HCl) must not be used in too much excess. Since $PbCl_2$ dissolves in hot water, it is separated from AgCl and Hg_2Cl_2 by such extraction and then detected and confirmed by its yellow precipitate, $PbCrO_4$, with chromate ion. AgCl dissolves in aqueous ammonia, but Hg_2Cl_2 reacts with NH_3 to form a black mixture of Hg and $HgNH_2Cl$. The NH_3 molecules in the $Ag(NH_3)_2^+$ complex are neutralized by nitric acid, and the freed Ag^+ ions precipitate as AgCl using Cl^- already present.

The flow diagram, Figure 25.11, for cation group 1 and the net ionic equations for each reaction should be learned together with the colors of precipitates or solutions that are important to the detection and confirmation of any given ion.

Cation Group-2 Analysis.   The test solution is first heated with an oxidizing acid, HNO_3, to make sure that any Sn^{2+} present will be changed to Sn^{4+}. Then, after the adjustment of the pH of the solution to about 0.5, aqueous thioacetamide is added and H_2S forms. Four cations are able to obtain S^{2-} from this compound and form the acid-insoluble sulfides.

From this precipitated mixture of sulfides, tin is extracted by the action of H_2S in base, which provides S^{2-} to form a water-soluble complex ion, $[SnS_3]^{2-}$, from SnS_2.

The three possible remaining sulfides, Bi_2S_3, CuS, and PbS, are dissolved by nitric acid. It functions again as an oxidizing agent because it dissolves these metal sulfides by oxidizing their sulfide ions to sulfur. Then tests are made for the released metal ions.

The flow diagram for the cation group-2 ion in Figure 25.12 together with the associated net ionic equations and colors of key ions or precipitates should be learned. It is important to learn the reasons for each step, because blindly doing the lab work frequently leads to starting over.

Cation Group-3 Analysis.   The metal sulfides of this group form at a slightly basic pH, but not so basic that the hydroxides of Ca^{2+} and Ba^{2+} would precipitate. Since the hydroxide of Al^{3+} is more insoluble than its sulfide, aluminum ion precipitates as $Al(OH)_3$. But the sulfides, not the hydroxides, of the other cations in this group are the more insoluble. If iron were present as Fe^{3+} instead of Fe^{2+}, it becomes reduced by HS^- to Fe^{2+} anyway, and FeS, not Fe_2S_3, forms.

All the precipitated ions of this group, once separated from group 4, are eventually changed to hydroxides. But sufficient OH^- is able to convert the hydroxides of aluminum and zinc to soluble forms, $Al(OH)_4^-$ and $Zn(OH)_4^{2-}$, and this further splits up the group-3 cations.

The flow diagram for group 3 in Figure 25.13 together with the associated net ionic equations, key colors, and reasons should be learned.

Cation Group-4 Analysis.   Tests for the cations of this group are made with little further separations. Tests for Na^+ and NH_4^+ are run on the initial, unanalyzed test solution. The flow diagram for this group, Figure 25.14, the associated net ionic equations, colors, and reasons should be learned.

Anion Analysis.   The flow diagram for these, which must be learned, is in Figure 25.15. Some breaking down into groups is done. Thus, in anion group 1, we have SO_4^{2-}, PO_4^{3-}, and CO_3^{2-}, which are separated as insoluble barium salts. Anion group 2 has Cl^-, Br^-, I^-, and HS^-. And the NO_3^- ion is all by itself in anion group 3 and is tested for using a portion of the original test solution. The tests for the halide ions depend heavily on their relative ease of oxidation.

REVIEW EXERCISES

Unless specified otherwise, always write net ionic equations for reactions and include physical states— (s), (ℓ), or (g). In exercises where you are asked to write an equation, one or two sentences of additional explanation are also expected. Answers to questions with numbers printed in color are found in an Appendix. A few more difficult questions are identified with asterisks.

*When a question asks what one **does** in a test, the answer should describe what reagent is used or what general procedure (e.g., extraction) is employed. But specific quantities are not requested. When the question concerns what one **sees** in a test, the answer should describe only an observation (color, odor, precipitate, etc.), not an equation.*

General Concepts and Procedures

25.1   In our study, which cations make up the following groups? (Write their formulas.) Wherever the scheme cannot distinguish between cations of the same metal in two oxidation states, give the formulas of both cations.
(a) Cation group 1   (c) Cation group 3
(b) Cation group 2   (d) Cation group 4

25.2   What other names do the following groups have?
(a) Cation group 1   (c) Cation group 3
(b) Cation group 2   (d) Cation group 4

25.3   In broad terms only, what is the purpose of a group reagent and why must the scheme involve them?

25.4   Describe the composition of the group reagents for each of the following groups. (Specify concentrations and relative acidities where appropriate.)
(a) Cation group 1   (c) Cation group 3
(b) Cation group 2

25.5   Write the chemical formulas traditionally used to identify the anions in our scheme. (Ignore here the fact that the exact forms of some depend on the pH.)

Procedures

25.6   Why is the scheme called *semimicro?* (Refer to volumes or masses of reagents and to minimum concentrations of ions.)

25.7   What ions might be present in tap water in most places that would complicate qualitative analysis according to our scheme?

25.8   The particles in a colloidal solid can sometimes be made larger and so more easily centrifuged by *digestion*. What is this procedure and how does it accomplish its goal?

25.9   What do the following terms mean?
(a) Supernatant liquid   (c) Extraction
(b) Centrifugate

25.10   In general terms, what is the reason that a precipitate is washed?

25.11   What are two reasons for carrying out procedures on knowns before doing an unknown?

Cation Group 1

25.12   Draw the complete flow diagram for cation group 1. (You should be able to do this from memory.)

25.13   Write the net ionic equations for the reactions of the group-1 reagent with each of the possible cations in this group.

25.14   What are the colors of the three possible solids that can form when the group-1 reagent is used?

25.15   A slight excess of the group-1 reagent is used. Give two reasons for this, using equations.

25.16   A large excess of the group-1 reagent must be avoided. Why? (Use equations as parts of your answer.)

25.17   How can Pb^{2+} be transferred to cation group 2?

25.18   Write equations for the reactions that confirm each of the following cations. If more than one reaction is used, one after the other in the test, then write both equations. (a) Pb^{2+}, (b) Hg_2^{2+}, (c) Ag^+.

Cation Group 2

25.19 Draw the complete flow diagram for cation group 2. (You should be able to do this from memory.)

25.20 Write the net ionic equations for the reactions of the group-2 reagent with each of the possible cations in this group.

25.21 At about what pH are the cation sulfides of group 2 precipitated?

25.22 What is the molarity of H_2S produced in the scheme by the hydrolysis of thioacetamide?

25.23 Why are some precipitates washed with dilute NH_4NO_3 and not with distilled water?

25.24 Using their formulas, arrange the following sulfides in their order of decreasing solubility in dilute acid, placing the most soluble on the left and the least soluble on the right: CuS, PbS, SnS_2, Bi_2S_3.

25.25 What are four reasons for converting any Sn^{2+} into Sn^{4+} before the group reagent for cation group 2 is used?

25.26 A nonoxidizing acid, HCl, is used to neutralize and then re-acidify the solution after Sn^{2+} is oxidized by HNO_3 to Sn^{4+} but before the group-2 reagent system is added.
(a) Why is a *nonoxidizing* acid used?
(b) Why must an *acid* be used at all? Give two reasons.
(c) Why is HCl, rather than, say, H_2SO_4 used?

25.27 In the net ionic equation for the precipitation of Bi_2S_3, why is it better to write BiO^+ rather than Bi^{3+} as a reactant?

25.28 Write the formulas and the colors of the four possible precipitates that the group-2 reagent can produce.

25.29 Refer to the extraction of tin(IV) sulfide from the other solids that form when the group-2 reagent is used:
(a) Write the equation.
(b) Could tin(II) sulfide be extracted this way?

25.30 Refer to the confirmatory test for tin:
(a) What is the formula and the name of the reactant supplied by the test reagent?
(b) Write the equation for the confirmatory test if just a white product forms. (Identify this product by formula.)
(c) Write the equation for the confirmatory test if a black product forms. (Identify this product by formula.)
(d) Traces of H_2S must have been removed prior to the addition of the reagent in this test. Why? (What else might form and how would it confuse the test?)
(e) How is H_2S removed, and what does this operation do to Sn(IV)? (Write an equation.)
(f) Why must Sn(IV) be changed to Sn(II) prior to the confirmatory test?
(g) Write the equation for the conversion of Sn(IV) to Sn(II).

25.31 After SnS_2 is removed from the mixture of group-2 sulfides, the remaining sulfides are dissolved by chemical reactions. (Write the equations for these reactions.)

25.32 In the confirmatory work for Pb^{2+}, sulfuric acid is added and then the mixture is very strongly heated. Explain why. What reactions occur? Write their equations.

25.33 Refer to the *second* confirmatory test for Pb^{2+}:
(a) Why is this done?
(b) How is $PbSO_4$ dissolved before this second test is done? (Write the equation.)
(c) What does this (referring to part b) accomplish in the overall procedure for this second test?
(d) What does one do and see at the end of this second test for Pb^{2+}? (Write the equation.)

25.34 What does one do and see to confirm Cu^{2+}? Write the equation.

25.35 The confirmatory test for Cu^{2+} also affects BiO^+. In what way? (Write an equation.)

25.36 What does one do and see in the final reaction to confirm bismuth? (Write the equation.)

25.37 Why must the stannite ion be freshly prepared? Write the two equations for what can happen to it when its aqueous solution stands exposed to air.

25.38 Refer to the preparation of the stannite ion:
(a) What chemicals are used? (Write their formulas.)
(b) What reactions occur? (Write the equations.)

Cation Group 3

25.39 Draw the complete flow diagram for cation group 3. (You should be able to do this from memory.)

25.40 Write the net ionic equations for the reactions of the group-3 reagent with each of the possible cations in this group.

25.41 To make the solution basic enough for the group-3 cations to precipitate when H_2S is present, but not so basic as to cause anything in cation group 4 to precipitate, how is the pH of the test solution adjusted?

25.42 In what form does Al^{3+} precipitate when the group-3 reagent is used? (Write its formula.) Why doesn't the sulfide form instead?

25.43 Write the formulas and the colors of the possible insoluble compounds that can form from the use of the group-3 reagent.

25.44 When the unknown is prepared using, say, $Fe(NO_3)_3$ instead of $Fe(NO_3)_2$, why does FeS, not Fe_2S_3, form anyway? (Write an equation.)

25.45 Unless a solution containing Fe^{3+} is kept very acidic (pH about 0), the color is not the pale violet of $[Fe(H_2O)_6]^{3+}$ but a yellow-to-amber color. Explain in general terms.

25.46 What happens to $[Fe(H_2O)_6]^{3+}$ in aqueous solution with a pH above 3 (in general terms)?

25.47 A mixture of HCl(*aq*) and HNO_3(*aq*) is used to dissolve the precipitates formed from the group-3 cations and the group reagent. Write the equations for the dissolving reactions.

25.48 Following the redissolving of the solids (referring to Review Exercise 25.47), Cl^- must be entirely removed.

(a) Why? (In what way can it interfere with a later test?)

(b) By what reaction is Cl^- removed? (Write the equation.)

25.49 What does *amphoteric* mean, and (in general terms) how is this property used in the analysis of the group-3 cations?

25.50 Once the group-3 cations are separated from the rest of the possible cations, they are further separated from each other.

(a) Into what two groups are they separated? (Write the formulas of the group-3 cations in each of these two subgroups.)

(b) What reagent achieves this separation?

(c) Write the equations for the reactions this reagent produces.

25.51 Is an oxidizing acid *needed* to dissolve the nonamphoteric hydroxides of the group-3 cations? Wouldn't HCl also dissolve them? Why isn't HCl used? What other function does the oxidizing acid serve?

25.52 What does one do and see in the confirmatory test for Mn^{2+}? (Write the equation as well.)

25.53 If the color for the positive test for manganese fades, what probably causes this?

25.54 Why do you suppose that neither Fe^{3+} nor Ni^{2+} interferes with the test for Mn^{2+}?

25.55 Why doesn't aqueous NH_3 cause insoluble hydroxides of Mn^{2+} and Ni^{2+} to form? Write the formulas of the products that these two cations do form in the presence of NH_3.

25.56 When aqueous NH_3 is added to a solution with Fe^{3+}, what happens? (Write the equation.)

25.57 Two reactions are used to confirm that the precipitate that forms when $NH_3(aq)$ is mixed with Fe^{3+} actually contains Fe^{3+}.

(a) Write the equations for these two reactions.

(b) What does one *see* in the final positive test for Fe^{3+}?

25.58 The presence of Ni^{2+} is indicated by the *color* of a solution just before the confirmatory test for Ni^{2+} is done.

(a) What procedure produces this color? (Briefly describe what is done.)

(b) If Ni^{2+} is actually present, what color is the solution?

(c) What is the formula of the species responsible for this color?

(d) One cation in group 2 can form a similar species with the same color. What is the formula of this cation?

25.59 Refer to the confirmatory test for Ni^{2+}:

(a) What does one do and see?

(b) What is the equation for the reaction?

(c) What is meant by a *bidendate ligand*?

25.60 A test solution is believed to contain Al^{3+} and Zn^{2+} as their hydrated ions. What reactant can be used to separate one cation from the other? (Write the equations.)

25.61 Describe what one does and sees in the confirmatory test for Al^{3+}.

25.62 A solid is described as being *flocculant*. What does this mean? Give an example.

25.63 What does the term *lake* mean in dye chemistry?

25.64 Two reactions are used in the final test for zinc.

(a) Write their equations.

(b) What does one see in the positive test?

Cation Group 4

25.65 Draw the complete flow diagram for cation group 4. (You should be able to do this from memory.)

25.66 Describe what one does and sees in the test for NH_4^+. (Write the equation.)

25.67 Why must the test for NH_4^+ be done on the original, unanalyzed test solution of a known or unknown?

25.68 How are Ba^{2+} and Ca^{2+} separated from the other group-4 cations? (Write equations.)

25.69 How is Ba^{2+} confirmed? Give two tests, writing relevant equations for each.

25.70 How is the presence of Ca^{2+} first indicated? (Write the equation.)

25.71 How is Ca^{2+} confirmed?

Anion Analysis

25.72 How does a change in the pH of the solution affect the chemical form of each of the following anions?

(a) Phosphate ion

(b) Carbonate ion

(c) Hydrogen sulfide ion

25.73 Refer to anion group 1:

(a) What anions are in it? (Give their formulas as they usually are given in lists of the anions.)

(b) How are they separated from the rest of the anions? (Write equations.)

25.74 Concerning the precipitated forms of the group-1 anions, what chemical reaction occurs when nitric acid is added, and the following anions are present? What do you see and what reaction occurs? (Write equations for reactions.) (a) CO_3^{2-} (b) PO_4^{3-} (c) SO_4^{2-}

25.75 How is phosphate confirmed? What does one do and see? (Write the equation.)

25.76 How is CO_3^{2-} confirmed? What does one do and see? (Write the equation.)

25.77 How is SO_4^{2-} confirmed?

25.78 Describe how one tests for each of the following anions in the presence of the other three: (a) Cl^-, (b) Br^-, (c) I^-, (d) HS^-.

25.79 Refer to the brown-ring test:

(a) The occurrence of which ion (formula) is tested in this procedure?

(b) What is the formula of the species responsible for the brown ring?

(c) Write the equation for the formation of this brown ring.

(d) Before doing this test, how are all interfering anions removed? (Write equations.)

25.80 A solution is known to be either 1 M Na_2SO_4 or 1 M $NaHSO_4$. What simple test would enable one to tell which solute is present?

25.81 An unknown solution was found to have a pH of 1.5 and no odor. What anions (or anion systems) are therefore known to be absent? Explain.

25.82 A solution is known to contain Ag^+, Cu^{2+}, and Ba^{2+}. Which of our eight anions are almost certainly *not* also present? Explain.

Additional Questions

25.83 Several ions for which our qual scheme analyzes have unique colors in aqueous solution. Write the formulas of such colored (hydrated) ions and give their colors.

25.84 What is the color of each of the following species? (Assume that each ion is hydrated and in an aqueous solution and that the color of the solid state is to be given for compounds.) If a species has no color, write "none."
- (a) $AgCl$
- (b) $PbSO_4$
- (c) PbS
- (d) $PbCl_2$
- (e) CuS
- (f) $BiO(OH)$
- (g) CrO_4^{2-}
- (h) ZnS
- (i) $Ag(NH_3)_2^+$
- (j) Hg_2Cl_2
- (k) Bi_2S_3
- (l) Ag_2S
- (m) $Cu(NH_3)_4^{2+}$
- (n) NO_3^-
- (o) SnS_2
- (p) I^-
- (q) SO_4^{2-}
- (r) HgS
- (s) MnO_4^-
- (t) $(Hg + Hg_2Cl_2)$
- (u) FeS
- (v) $Al(OH)_3$
- (w) Cl^-
- (x) NiS
- (y) $AgBr$
- (z) AgI

25.85 Apply the directions for Review Exercise 25.84 to the following species.
- (a) CaC_2O_4
- (b) $[Fe(H_2O)_6]^{3+}$
- (c) $[Ni(NH_3)_6]^{2+}$
- (d) $BaCrO_4$
- (e) $[FeNO]^{2+}$
- (f) $I_2(aq)$
- (g) $I_2(CH_2Cl_2)$
- (h) $K_2Zn_3[Fe(CN)_6]_2$
- (i) $Ni(HDMG)_2$
- (j) $[Fe(H_2O)_5OH]^{2+}$
- (k) MnS
- (l) $CaCO_3$
- (m) $Fe(OH)_3$
- (n) Na_2SO_4
- (o) $[Zn(NH_3)_4]^{2+}$
- (p) $[Fe(H_2O)_5(SCN)]^{2+}$
- (q) $(NH_4)_3PO_4(MoO_3)_{12}$
- (r) $Br_2(aq)$
- (s) $Br_2(CH_2Cl_2)$
- (t) PO_4^{3-}
- (u) HCO_3^-

25.86 Flame tests for the following ions display what colors? (a) Ba^{2+}, (b) Na^+, (c) Ca^{2+}, (d) K^+.

25.87 Imagine for each part that follows that you have been handed an aqueous solution and told that its solute is one of the two given compounds (i.e., no other solutes are present). Describe a simple, semimicro qual test that could be used to decide between the two solutes. (Odor tests, flame tests, litmus tests, and colors can be used.) Assume the availability of any qual reagent, but figure out the *easiest* and *quickest* way to tell which is the solute. (More than one test might be possible, so pick a test that is the easiest to carry out.)

- (a) $NaCl$ and KCl
- (b) $NaCl$ and $NaNO_3$
- (c) $NaNO_3$ and Na_2SO_4
- (d) $Cu(NO_3)_2$ and $Zn(NO_3)_2$
- (e) $Cu(NO_3)_2$ and $Ni(NO_3)_2$
- (f) $Zn(NO_3)_2$ and $Al(NO_3)_3$
- (g) $Fe(NO_3)_3$ and $Ni(NO_3)_2$
- (h) $NaCl$ and Na_2SO_4
- (i) $NaCl$ and $CaCl_2$
- (j) $NaBr$ and NaI
- (k) $AgNO_3$ and KNO_3
- (l) Na_2CO_3 and Na_2HPO_4
- (m) $NaHS$ and $NaHCO_3$
- (n) K_2CrO_4 and K_2SO_4
- (o) $(NH_4)_2SO_4$ and Na_2SO_4 (without a flame test)

25.88 Complete and balance the following equations. If no reaction occurs, write "no reaction." (Many of these reactions actually establish equilibria, but they are not indicated here by the usual double arrows.)
- (a) $Pb^{2+}(aq) + CrO_4^{2-}(aq) \longrightarrow$
- (b) $Hg_2Cl_2(s) + NH_3(aq) \longrightarrow$
- (c) $BiO^+(aq) + H_2S(aq) \longrightarrow$
- (d) $Sn^{4+}(aq) + H_2S(aq) \longrightarrow$
- (e) $[SnCl_4]^{2-}(aq) + HgCl_2(aq) \longrightarrow Hg_2Cl_2(s) + Hg(\ell) +$
- (f) $PbS(s) + H^+(aq) + NO_3^-(aq) \longrightarrow$
- (g) $H_2SO_4 \xrightarrow[\text{strongly}]{\text{heat}}$
- (h) $Sn^{2+}(aq) + OH^-(aq) \text{ [excess]} \longrightarrow$
- (i) $AgCl(s) + NH_3(aq) \longrightarrow$
- (j) $Sn^{2+}(aq) + O_2(aq) + H^+(aq) \longrightarrow$
- (k) $[SnCl_6]^{2-}(aq) + Fe(s) \longrightarrow$
- (l) $Bi_2S_3(s) + H^+(aq) + NO_3^-(aq) \longrightarrow$
- (m) $PbSO_4(s) + C_2H_3O_2^-(aq) \longrightarrow$
- (n) $BiO^+(aq) + OH^-(aq) \longrightarrow$
- (o) $Bi_2S_3(s) + H^+(aq) \longrightarrow$
- (p) $AgCl(s) + Cl^-(aq) \longrightarrow$
- (q) $PbS(s) + H^+(aq) \longrightarrow$
- (r) $Sn^{4+}(aq) + H_2S(aq) \longrightarrow$
- (s) $Bi^{3+}(aq) + H_2O \longrightarrow$

25.89 Complete and balance the following equations. If no reaction occurs, write "no reaction." (Many of these reactions actually establish equilibria, but they are not indicated here by the usual double arrows.)
- (a) $[SnS_3]^{2-}(aq) + H^+(aq) + Cl^-(aq) \longrightarrow$
- (b) $[SnCl_4]^{2-}(aq) + Hg_2Cl_2(s) \longrightarrow Hg(\ell) +$
- (c) $CuS(s) + H^+(aq) + NO_3^-(aq) \longrightarrow$
- (d) $[Pb(C_2H_3O_2)_4]^{2-}(aq) + CrO_4^{2-}(aq) \longrightarrow$
- (e) $PbCl_2(s) + Cl^-(aq) \longrightarrow$
- (f) $Cu^{2+}(aq) + H_2S(aq) \longrightarrow$
- (g) $SnS_2(s) + H_2S(aq) + OH^-(aq) \longrightarrow$
- (h) $[SnCl_4]^{2-}(aq) + HgCl_2(aq)$
  $\longrightarrow Hg_2Cl_2(s) + \text{one more product}$
- (i) $BiO(OH)(s) + [Sn(OH)_3]^-(aq)$
  $+ OH^-(aq) + H_2O \longrightarrow$
- (j) $Al^{3+}(aq) + OH^-(aq) \longrightarrow$
- (k) $CuS(s) + H^+(aq) \longrightarrow$
- (l) $Zn^{2+}(aq) + OH^-(aq) \text{ [excess]} \longrightarrow$

(m) $Fe^{3+}(aq) + NH_3(aq) + H_2O \longrightarrow$
(n) $NH_4^+(aq) + OH^-(aq) \longrightarrow$
(o) $Ba_3(PO_4)_2(s) + H^+(aq) \longrightarrow$
(p) $I_2(aq) + Cl^-(aq) \longrightarrow$
(q) $Mn^{2+}(aq) + HS^-(aq) + NH_3(aq) \longrightarrow$
(r) $FeS(s) + H^+(aq) \longrightarrow$
(s) $Zn^{2+}(aq) + HS^-(aq) + NH_3(aq) \longrightarrow$

25.90 Complete and balance the following equations. If no reaction occurs, write "no reaction." (Many of these reactions actually establish equilibria, but they are not indicated here by the usual double arrows.)

(a) $Al^{3+}(aq) + OH^-(aq)$ [excess] $\longrightarrow$
(b) $Ca^{2+}(aq) + CrO_4^{2-}(aq) \longrightarrow$
(c) $Ba^{2+}(aq) + CrO_4^{2-}(aq) \longrightarrow$
(d) $HOCl(aq) + Cl^-(aq) + H^+(aq) \longrightarrow$

(e) $Cl_2(aq) + I^-(aq) \longrightarrow$
(f) $Fe^{2+}(aq) + HS^-(aq) + NH_3(aq) \longrightarrow$
(g) $Fe^{3+}(aq) + HS^-(aq) + NH_3(aq) \longrightarrow$
(h) $Al(OH)_3(s) + H^+(aq) \longrightarrow$
(i) $Mn^{2+}(aq) + BiO_3^-(aq) + H^+(aq) \longrightarrow$
(j) $Ca^{2+}(aq) + C_2O_4^{2-}(aq) \longrightarrow$
(k) $HCO_3^-(aq) + H^+(aq) \longrightarrow$
(l) $Cl_2(aq) + Br^-(aq) \longrightarrow$
(m) $Ni^{2+}(aq) + HS^-(aq) + NH_3(aq) \longrightarrow$
(n) $Fe^{3+}(aq) + H_2S(aq) \longrightarrow$
(o) $MnS(s) + H^+(aq) \longrightarrow$
(p) $Cl^-(aq) + NO_3^-(aq) + H^+(aq) \longrightarrow NO(g) +$
(q) $BaCO_3(s) + H^+(aq) \longrightarrow$
(r) $Ca^{2+}(aq) + CO_2(aq) + H_2O \longrightarrow$
(s) $Br_2(aq) + Cl^-(aq) \longrightarrow$

Appendix

A

REVIEW OF MATHEMATICS

EXPONENTIALS   LOGARITHMS   GRAPHING   ELECTRONIC CALCULATORS

A.1 EXPONENTIALS

Very large and very small numbers are often expressed as powers of ten. This is called exponential notation or scien-

TABLE A.1

Number	Exponential Form
1	1×10^0
10	1×10^1
100	1×10^2
1000	1×10^3
10,000	1×10^4
100,000	1×10^5
1,000,000	1×10^6
0.1	1×10^{-1}
0.01	1×10^{-2}
0.001	1×10^{-3}
0.0001	1×10^{-4}
0.00001	1×10^{-5}
0.000001	1×10^{-6}
0.0000001	1×10^{-7}

tific notation. Several examples are given in Table A.1, and in the last column the small number up and to the right of each 10 is called the exponent. In the number 10^4, 4 is the exponent, and we say that the number is 10 raised to the 4th power.

What a Positive Exponent Means   Suppose you come across a number such as 6.4×10^4. This is just an alternative way of writing 64,000. In other words, the number 6.4 is multiplied by 10 *four* times.

$$6.4 \times 10^4 = 6.4 \times 10 \times 10 \times 10 \times 10 = 64,000$$

Instead of actually writing out all of these 10s, notice that the decimal point is simply moved four places to the right in going from 6.4×10^4 to 64,000. Here are a few other examples. Study them to see how the decimal point changes.

$$53.476 \times 10^2 = 5,347.6$$
$$0.0016 \times 10^5 = 160$$
$$0.000056 \times 10^3 = 0.056$$

Here is another example that more forcefully shows the value of using an exponential form of a number.

602,000,000,000,000,000,000,000
$$= 6.02 \times 10 \times 10 \times 10 \times 10 \times 10 \times 10 \times 10 \times 10$$
$$\times 10 \times 10 \times 10 \times 10 \times 10 \times 10 \times 10 \times 10 \times 10$$
$$\times 10 \times 10 \times 10 \times 10 \times 10 \times 10$$
$$= 6.02 \times 10^{23}$$

(This number is called Avogadro's number, which is the number of atoms in 12.0 g of carbon.)

In some calculations it is helpful to reexpress large numbers in exponential form. To do this, count the number of places you would have to move the decimal point to the left to put it just after the first digit in the number. For example, you would have to move the decimal point four places left in the following number.

$$6\,0\,5\,3\,0$$
$$\scriptstyle 4\ \ 3\ \ 2\ \ 1$$

Therefore, we can rewrite 60530 as 6.0530×10^4. Thus the number of moves equals the exponent. Now try these problems.

EXERCISE 1

Write each number in expanded form.
(a) 1.2378×10^{10}
(b) 0.000588×10^6
(c) 45.6×10^5

EXERCISE 2

Write each number in exponential form.
(a) 13,000,000,000,000
(b) 15.68
(c) 6,000,456,568

Answers to Exercises 1 and 2
1. (a) 12,378,000,000
    (b) 588
    (c) 4,560,000
2. (a) 1.3×10^{13}
    (b) 1.568×10^1
    (c) 6.000456568×10^9

What a Negative Exponent Means   Imagine you have just encountered a number such as 1.4×10^{-4}. This is an alternative way of writing

$$\frac{1.4}{10 \times 10 \times 10 \times 10}$$

Thus a negative exponent tells us how many times to *divide* by ten. Dividing by ten, of course, can be done by simply moving the decimal point. Thus, 1.4×10^{-4} can be reexpressed as 0.00014. Notice that the decimal point is four places to the *left* of the 1. Here are some examples of

numbers expressed as negative exponentials and their equivalents.

$$3567.9 \times 10^{-3} = 3.5679$$
$$0.01456 \times 10^{-2} = 0.0001456$$
$$45691 \times 10^{-3} = 45.691$$

To change a number smaller than 1 into a negative exponential, we count the number of places the decimal has to be moved to put it to the right of the first digit. The number that we count is the negative exponent. For example, you would have to move the decimal point five places to the right to put it to the right of the first digit in

$$0.000068901$$

This number can be reexpressed as 6.8901×10^{-5}. Try these problems and check your answers.

EXERCISE 3

Write each number in expanded form.
(a) 1.45×10^{-4}
(b) 0.00568×10^{-3}
(c) 45623.34×10^{-4}

EXERCISE 4

Write each number in exponential form.
(a) 0.1004
(b) 0.0000000000000428
(c) 0.00400056

Answers to Exercises 3 and 4
1. (a) 0.000145
    (b) 0.00000568
    (c) 4.562334
2. (a) 1.004×10^{-1}
    (b) 4.28×10^{-14}
    (c) 4.00056×10^{-3}

Multiplying Numbers Written in Exponential Form   The real value of exponential forms of numbers comes in carrying out multiplications and divisions of large and small numbers. Suppose we want to multiply 1000 by 10000. Doing this by a long process fills the page with zeros (and some 1s). But notice the following relationships.

Numbers	1000	$\times$	10000		
	$[10 \times 10 \times 10]$	$\times$	$[10 \times 10 \times 10 \times 10]$	$=$	10,000,000
Exponential form	10^3	$\times$	10^4	$=$	10^7
Exponents	3	$+$	4	$=$	7

If we *add* 3 and 4, the exponents of the exponential forms of 1000 and 10000, we get the exponent for the answer. Suppose we want to multiply 4160 by 20000. If we reexpress each number in exponential form we have

$$[4.16 \times 10^3] \times [2 \times 10^4]$$

This now simplifies to

$$(4.16) \times (2) \times [10^3 \times 10^4]$$

The answer can be easily worked while, at the same time, the decimal point isn't lost. The *exponents are added* to give the exponent we want in the answer, and the numbers 4.16 and 2 are *multiplied*. The product is 8.32×10^7. Thus there are three steps in multiplying numbers that are more conveniently written in exponential form.

1. First write the numbers in exponential form (if they aren't already).
2. Next *multiply* the numbers before the 10s.
3. Finally *add* the exponents of the 10s *algebraically*.

Here are some examples.

EXAMPLE 1
USING EXPONENTIALS

Problem: What is 3650×0.00000568?
Solution: Step 1. $3650 = 3.650 \times 10^3$
$0.00000568 = 5.68 \times 10^{-6}$

Step 2. 3650×0.00000568
$= [3.650 \times 10^3] \times [5.68 \times 10^{-6}]$
$= (3.650 \times 5.68) \times (10^3 \times 10^{-6})$
$= 20.732 \times 10^{[3+(-6)]}$
$= 20.7 \times 10^{-3}$ (rounded)

EXAMPLE 2
USING EXPONENTIALS

Problem: What is 10045×0.025058?
Solution: This works out as follows:

$[1.0045 \times 10^4] \times [2.5058 \times 10^{-2}]$
$= (1.0045) \times (2.5058) \times 10^{(4-2)}$
$= 2.5171 \times 10^2$

For practice, try these problems.

EXERCISE 5

Calculate the products of the following.
(a) $6.20 \times 10^{23} \times 1.50 \times 10^{-4}$
(b) $0.003 \times 0.002 \times 0.004$
(c) $0.00045 \times 45,000,000$

Answers
(a) 9.30×10^{19}
(b) 24×10^{-9} or 2.4×10^{-8}
(c) 20.25×10^3 or 2.025×10^4

Dividing Numbers Written in Exponential Form
Suppose we want to divide 10,000,000 by 1000. Doing this by long division would be primitive to say the least. Notice the following relationships.

$$\frac{10,000,000}{1000} = \frac{10^7}{10^3} = 10^4 = 10000$$

The exponent in the result is $(7 - 3)$. So, to divide numbers in exponential form, we *subtract* exponents. Before continuing, work these problems.

EXERCISE 6

Carry out the following divisions and express the answers in exponential forms.
(a) $1000 \div 10000$
(b) $0.001 \div 0.0001$
(c) $100 \div 0.001$

Answers
(a) $10^3 \div 10^4 = 10^{(3-4)} = 10^{-1}$
(b) $10^{-3} \div 10^{-4} = 10^{[(-3)-(-4)]} = 10^{(-3+4)} = 10^1$
(c) $10^2 \div 10^{-3} = 10^{[(2)-(-3)]} = 10^{(2+3)} = 10^5$

To extend this operation one step, we consider how to divide 0.00468 by 0.0000400. This may be expressed by using exponentials as follows.

$$\frac{4.68 \times 10^{-3}}{4.00 \times 10^{-5}} = \frac{4.68}{4.00} \times 10^{[(-3)-(-5)]}$$
$$= 1.17 \times 10^{[-3+5]}$$
$$= 1.17 \times 10^2$$

This calculation illustrates the three steps for dividing numbers.

1. Write the numbers in exponential form.
2. Carry out the division of the numbers standing before the 10s.
3. Algebraically subtract the exponents, the exponent in the denominator from the exponent in the numerator.

Now try these exercises.

EXERCISE 7

Carry out the following divisions.
(a) $9593 \div 362,000$
(b) $0.035 \div 0.70$
(c) $450 \div 0.090$

Answers
(a) 2.65×10^{-2} $(9.593 \times 10^3 \div 3.62 \times 10^5)$
(b) 5.0×10^{-2}
(c) 5.0×10^3

Suppose you carry out the steps on the following:

$$\frac{1908 \times 0.00456 \times 45,200,000}{0.1664 \times 2,340,000,000 \times 456}$$

In exponential form, it is easiest to collect all of the 10s with their exponents in one place and the other numbers together.

$$\frac{(1.908) \times (4.56) \times (4.52) \times (10^3) \times (10^{-3}) \times (10^7)}{(1.664) \times (2.34) \times (4.56) \times (10^{-1}) \times (10^9) \times (10^2)}$$

$$= 2.21 \times 10^{[(3-3+7)-(-1+9+2)]}$$
$$= 2.21 \times 10^{[7-(10)]}$$
$$= 2.21 \times 10^{-3} \quad \text{(rounded)}.$$

EXERCISE 8

In applying the ideal gas equation studied in Chapter 9, you might have to carry out the following calculation.

$$\text{Pressure} = \frac{0.150 \times 0.0821 \times 373}{0.750} \text{ atm}$$

What is the pressure?

Answer 6.12 atm, rounded

The Pocket Calculator and Exponentials   In scientific work the concept of an exponential is very important because it is much easier to handle either very large or very small numbers in exponential form. The review we've provided in this appendix is intended to help you refresh your memory from secondary school about these kinds of numbers. No doubt you know that almost any scientific calculator gives you the ability to multiply and divide large or small numbers when they are reexpressed as exponentials. We encourage you to study the *Instruction Manual* for your pocket calculator and read Section A.4 of this appendix so that you can do these kinds of calculations easily. However, we caution you that doing this without a basic understanding of exponentials will leave many traps in your path to skill in performing chemistry calculations.

A.2   LOGARITHMS

Common Logarithms   The common logarithm or log of a number N to the base 10 is the exponent to which 10 must be raised to equal the number N. If

$$N = 10^x$$

then the common logarithm of N, or log N, is simply x.

$$\log N = x$$

There is another kind of logarithm called a natural logarithm that we'll take up later. However, when we use the terms "logarithm" or "log N" we always mean *common* logarithms related to the base 10.

Some examples of logs are as follows. When

$N = 10$	$\log N = 1$	because	$10 = 10^1$
$= 100$	$= 2$		$100 = 10^2$
$= 1000$	$= 3$		$1000 = 10^3$
$= 0.1$	$= -1$		$0.1 = 10^{-1}$
$= 0.00001$	$= -5$		$0.00001 = 10^{-5}$

The logs of these numbers were easy to figure out by just using the definition of logarithms, but usually a log isn't this simple. For example, what is the value of x in the following equations?

$$4.23 = 10^x$$
$$\log 4.23 = x$$

To determine the value of x we have to use a table of logarithms—such as Table A.2—or a hand-held calculator with the capability of using logs. Unless you understand how logs work, however, the calculator will get you into considerable trouble. This is why learning how to use logs and log tables is a necessary development.

To learn how to use a log table, we will work an example. However, all log tables give only one part of a log, and a log consists of two parts. The value of log 423,000, for example, is 5.6263, meaning that

$$423,000 = 10^{5.6263}$$

$$\log 423,000 = 5.\overbrace{6263}^{\text{mantissa (0.6263)}}$$

characteristic (5)

The characteristic of a log consists of the digits in front of the decimal point; the mantissa of a log is made up of all the digits after the decimal. *Log tables give mantissas only* (and you have to put in the decimal point yourself). The characteristic is something you figure out, but this is easy because it's the whole-number exponent of 10 when the number N is written in exponential form. Here is one place where being able to write numbers in exponential notation is essential. We have to write (or at least be able to visualize) 423,000 as 4.23×10^5, and the characteristic of the log of this number is simply 5, the exponent. We use the log table to find the mantissa, and *this is always the log of a number between 1 and 10* (after inserting the decimal point). The exponent (characteristic) and the mantissa are then added *algebraically* to give the log.

The reason for this procedure flows out of the properties of exponents when tied to the definition of a logarithm. We know, for example, that the exponent of $10^x \times 10^y$ is $(x + y)$, meaning that

$$10^x \times 10^y = 10^{(x+y)}$$

We also know that we can express 423,000 as

$$423,000 = 10^x \times 10^5 = 10^{(x+5)}$$

Now x has to be some number between 0 and 1 because 423,000 is a number between 100,000 ($10^0 \times 10^5 = 10^5$) and 1,000,000 ($10^1 \times 10^5 = 10^6$); in fact 423,000 = $4.23 \times 100,000$. Now we can use our definition of a log. If

$$423,000 = 4.23 \times 100,000 = 10^{(x+5)}$$

then

$$\log(4.23 \times 100{,}000) = \log 10^{(x+5)}$$
$$= x + 5$$

The "5" in $(x + 5)$ goes with the 100,000 (10^5) and the value of x goes with 4.23. We find this value in the log table:

$$x = 0.6263$$

and

$$\log 423{,}000 = 0.6263 + 5$$
$$= 5.6263$$

In other words,

$$423{,}000 = 10^{0.6263} \times 10^5$$
$$= 10^{5.6263}$$

To summarize, when

$$N = M \times 10^A \text{ (where } M \text{ is between 1 and 10)}$$

then

$$\log N = A + \log M$$

where A is the characteristic of log N and log M (preceded by a decimal point) is the mantissa.

Let's work some examples to see how to use a log table.

EXAMPLE 3
USING A LOG TABLE TO FIND A COMMON LOGARITHM

Problem: What is the value of log 40?
Solution: The first step with any number not between 1 and 10 is to express it in exponential form:

$$40 = 4 \times 10^1$$

Therefore, log 40 = 1 + log 4. (The characteristic of the log is 1.) To find the value of log 4, let your finger move down the left-hand column of the log table to "40." Interpret this to mean "4.0." Now move your finger over so that it is in the column headed by "0" at the top of the table. This brings you to 4.00, and your finger is now over 6021. Interpret this to be 0.6021. (Remember: the table omits all of the decimal points in front of mantissas.) Finally, add the "1" — the characteristic — to 0.6021, the mantissa.

$$\log 40 = 1.6021$$

EXERCISE 9

What are the values of (a) log 50? (b) log 800?

Answers
(a) 1.6990   (b) 2.9031

EXAMPLE 4
USING A LOG TABLE TO FIND A COMMON LOGARITHM

Problem: What is the value of log 423,000?
Solution: In exponential form, $423{,}000 = 4.23 \times 10^5$. Therefore

$$\log 423{,}000 = 5 + \log 4.23$$

To find the log of 4.23, go to the log table and move down the left-hand column to 42 (meaning 4.2), then over to the column headed by a "3". Now we are at 4.23. This gives the mantissa of log 4.23, 0.6263. Therefore, the answer is

$$\log 423{,}000 = 5 + 0.6263$$
$$= 5.6263$$

EXERCISE 10

Find the logs of each of the following numbers.
(a) 625   (b) 12,900

Answers
(a) 2.7959   (b) 4.1106

EXAMPLE 5
USING A LOG TABLE TO FIND A COMMON LOGARITHM OF A NUMBER LESS THAN 1

Problem: What is the value of log 0.000524?
Solution: As usual, we express the number in exponential form: 5.24×10^{-4}. Also as usual, we find the log of 5.24; this is 0.7193. What we now must remember is that we have to add the characteristic, which is -4 (and is treated as an exact number), to the mantissa algebraically:

$$-4 + 0.7193 = -3.2807$$

Remembering this *algebraic* addition is the only extra thing in working with the logs of numbers less than 1.

EXERCISE 11

Find the logs of each of the following numbers.
(a) 0.00000294   (b) 0.872

Answers
(a) −5.5317   (b) −0.0595

In the examples used thus far, the number standing before the 10 in the exponential forms always had no more than three digits. What do we do if there are more? We might want to look up, for example, the value of log 4.238 because we're manipulating some numbers such as 4238 (4.238×10^3). The mantissa we want will be between 0.6263 (for 4.23) and 0.6274 (for 4.24). It is not hard to estimate where the mantissa for 4.238 will be, but instead we'll adopt a simpler rule that will take care of all of the needs in this book. We will round a number like 4.238 or

TABLE A.2 Logarithms

	0	1	2	3	4	5	6	7	8	9
10	0000	0043	0086	0128	0170	0212	0253	0294	0334	0374
11	0414	0453	0492	0531	0569	0607	0645	0682	0719	0755
12	0792	0828	0864	0899	0934	0969	1004	1038	1072	1106
13	1139	1173	1206	1239	1271	1303	1335	1367	1399	1430
14	1461	1492	1523	1553	1584	1614	1644	1673	1703	1732
15	1761	1790	1818	1847	1875	1903	1931	1959	1987	2014
16	2041	2068	2095	2122	2148	2175	2201	2227	2253	2279
17	2304	2330	2355	2380	2405	2430	2455	2480	2504	2529
18	2553	2577	2601	2625	2648	2672	2695	2718	2742	2765
19	2788	2810	2833	2856	2878	2900	2923	2945	2967	2989
20	3010	3032	3054	3075	3096	3118	3139	3160	3181	3201
21	3222	3243	3263	3284	3304	3324	3345	3365	3385	3404
22	3424	3444	3464	3483	3502	3522	3541	3560	3579	3598
23	3617	3636	3655	3674	3692	3711	3729	3747	3766	3784
24	3802	3820	3838	3856	3874	3892	3909	3927	3945	3962
25	3979	3997	4014	4031	4048	4065	4082	4099	4116	4133
26	4150	4166	4183	4200	4216	4232	4249	4265	4281	4298
27	4314	4330	4346	4362	4378	4393	4409	4425	4440	4456
28	4472	4487	4502	4518	4533	4548	4564	4579	4594	4609
29	4624	4639	4654	4669	4683	4698	4713	4728	4742	4757
30	4771	4786	4800	4814	4829	4843	4857	4871	4886	4900
31	4914	4928	4942	4955	4969	4983	4997	5011	5024	5038
32	5051	5065	5079	5092	5105	5119	5132	5145	5159	5172
33	5185	5198	5211	5224	5237	5250	5263	5276	5289	5302
34	5315	5328	5340	5353	5366	5378	5391	5403	5416	5428
35	5441	5453	5465	5478	5490	5502	5514	5527	5539	5551
36	5563	5575	5587	5599	5611	5623	5635	5647	5658	5670
37	5682	5694	5705	5717	5729	5740	5752	5763	5775	5786
38	5798	5809	5821	5832	5843	5855	5866	5877	5888	5899
39	5911	5922	5933	5944	5955	5966	5977	5988	5999	6010
40	6021	6031	6042	6053	6064	6075	6085	6096	6107	6117
41	6128	6138	6149	6160	6170	6180	6191	6201	6212	6222
42	6232	6243	6253	6263	6274	6284	6294	6304	6314	6325
43	6335	6345	6355	6365	6375	6385	6395	6405	6415	6425
44	6435	6444	6454	6464	6474	6484	6493	6503	6513	6522
45	6532	6542	6551	6561	6571	6580	6590	6599	6609	6618
46	6628	6637	6646	6656	6665	6675	6684	6693	6702	6712
47	6721	6730	6739	6749	6758	6767	6776	6785	6794	6803
48	6812	6821	6830	6839	6848	6857	6866	6875	6884	6893
49	6902	6911	6920	6928	6937	6946	6955	6964	6972	6981
50	6990	6998	7007	7016	7024	7033	7042	7050	7059	7067
51	7076	7084	7093	7101	7110	7118	7126	7135	7143	7162
52	7160	7168	7177	7185	7193	7202	7210	7218	7226	7235
53	7243	7251	7259	7267	7275	7284	7292	7300	7308	7316
54	7324	7332	7340	7348	7356	7364	7372	7380	7388	7396

TABLE A.2 Logarithms *(Continued)*

	0	1	2	3	4	5	6	7	8	9
55	7404	7412	7419	7427	7435	7443	7451	7459	7466	7474
56	7482	7490	7497	7505	7513	7520	7528	7536	7543	7551
57	7559	7566	7574	7582	7589	7597	7604	7612	7619	7627
58	7634	7642	7649	7657	7664	7672	7679	7686	7694	7701
59	7709	7716	7723	7731	7738	7745	7752	7760	7767	7774
60	7782	7789	7796	7803	7810	7818	7825	7832	7839	7846
61	7853	7860	7868	7875	7882	7889	7896	7903	7910	7917
62	7924	7931	7938	7945	7952	7959	7966	7973	7980	7987
63	7993	8000	8007	8014	8021	8028	8035	8041	8048	8055
64	8062	8069	8075	8082	8089	8096	8102	8109	8116	8122
65	8129	8136	8142	8149	8156	8162	8169	8176	8182	8189
66	8195	8202	8209	8215	8222	8228	8235	8241	8248	8254
67	8261	8267	8274	8280	8287	8293	8299	8306	8312	8319
68	8325	8331	8338	8344	8351	8357	8363	8370	8376	8382
69	8388	8395	8401	8407	8414	8420	8426	8432	8439	8445
70	8451	8457	8463	8470	8476	8482	8488	8494	8500	8506
71	8513	8519	8525	8531	8537	8543	8549	8555	8561	8567
72	8573	8579	8585	8591	8597	8603	8609	8615	8621	8627
73	8633	8639	8645	8651	8657	8663	8669	8675	8681	8686
74	8692	8698	8704	8710	8716	8722	8727	8733	8739	8745
75	8751	8756	8762	8768	8774	8779	8785	8791	8797	8802
76	8808	8814	8820	8825	8831	8837	8842	8848	8854	8859
77	8865	8871	8876	8882	8887	8893	8899	8904	8910	8915
78	8921	8927	8932	8938	8943	8949	8954	8960	8965	8971
79	8976	8982	8987	8993	8998	9004	9009	9015	9020	9025
80	9031	9036	9042	9047	9053	9058	9063	9069	9074	9079
81	9085	9090	9096	9101	9106	9112	9117	9122	9128	9133
82	9138	9143	9149	9154	9159	9165	9170	9175	9180	9186
83	9191	9196	9201	9206	9212	9217	9222	9227	9232	9238
84	9243	9248	9253	9258	9263	9269	9274	9279	9284	9289
85	9294	9299	9304	9309	9315	9320	9325	9330	9335	9340
86	9345	9350	9355	9360	9365	9370	9375	9380	9385	9390
87	9395	9400	9405	9410	9415	9420	9425	9430	9435	9440
88	9445	9450	9455	9460	9465	9469	9474	9479	9484	9489
89	9494	9499	9504	9509	9513	9518	9523	9528	9533	9538
90	9542	9547	9552	9557	9562	9566	9571	9576	9581	9586
91	9590	9595	9600	9605	9609	9614	9619	9624	9628	9633
92	9638	9643	9647	9652	9657	9661	9666	9671	9675	9680
93	9685	9689	9694	9699	9703	9708	9713	9717	9722	9727
94	9731	9736	9741	9745	9750	9754	9759	9763	9768	9773
95	9777	9782	9786	9791	9795	9800	9805	9809	9814	9818
96	9832	9827	9832	9836	9841	9845	9850	9854	9859	9863
97	9868	9872	9877	9881	9886	9890	9894	9899	9903	9908
98	9912	9917	9921	9926	9930	9934	9939	9943	9948	9952
99	9956	9961	9965	9969	9974	9978	9983	9987	9991	9996

4.244 to 4.24 *before* looking up the mantissa. (Our needs, in other words, will never exceed three significant figures.)

Logs are powerful tools for handling complex multiplications and divisions, and for these uses we have the following rules:

For Multiplications

$$\text{if } \quad N = A \times B$$
$$\log N = \log A + \log B$$

For Divisions

$$\text{if } \quad N = \frac{A}{B}$$
$$\log N = \log A - \log B$$

For Exponentials

$$\text{if } \quad N = A^b$$
$$\log N = b \times \log A$$

EXERCISE 12

By logarithms, find the values of the logarithms of each of the following numbers, N.
(a) $N = 525 \times 0.00346$ (c) $N = 2.68^3$
(b) $N = 422 \div 0.0451$ (d) $N = 81^{1/2}$

Answers
(a) $2.7202 + (-2.4609) = 0.2593$
(b) $2.6253 - (-1.3458) = 3.9711$
(c) $3 \times 0.4281 = 1.2843$
(d) $\frac{1}{2} \times 1.9085 = 0.9543$

Doing multiplications and divisions by means of logarithms works only if we can use a log found by any of the above operations to figure out the actual number that it is a log of. This number is called the **antilogarithm** or the **antilog**. Thus if

$$N = 10^x$$

and

$$\log N = x$$

then

$$N = \text{antilog of } x$$

Finding an antilog is easy. We just use the log table "in reverse." Suppose we found that the log of some number, say, $81^{1/2}$ is 0.9543—as we did in part d of the previous exercise. What number has 0.9543 as its log? This is what we ask when we look for an antilog. Go to the log table and find 9543 among the columns of mantissas. The nearest mantissa to it is 9542, the mantissa for the number 9.0. Thus 9.0 is the antilog of 0.9543 (and we know that 9 is the square root of 81, and should be written as 9.0 to have as many significant figures as there are in 81.)

EXAMPLE 6
FINDING ANTILOGS

Problem: What is the antilog of 3.9711?
Solution: The characteristic is 3; therefore, 10^3 must be part of the antilog. A mantissa of 9711 comes closest to 9713 in the log table, and this corresponds to 9.36. (Don't forget the decimal point; mantissas are always for numbers between 1 and 10.) Therefore,

$$\text{antilog } 3.9711 = 9.36 \times 10^3$$

EXAMPLE .7
FINDING ANTILOGS FOR NEGATIVE LOGARITHMS

Problem: What is the antilog of −4.7899?
Solution: It is now vitally important to remember how this log, −4.7899, was obtained. We would have looked up a mantissa that when *algebraically added* to −5 gives −4.7899. In other words,

$$[-5 + (\text{some number})] = -4.7899$$

This "some number" is the mantissa we seek, and its value is (5 − 4.7899) = 0.2101. The antilog of −4.7899 is really the antilog of [−5 + 0.2101] so the antilog must have 10^{-5} as part of it. In the log tables the number whose log is 0.2101 is (or is closest to) 1.62. Therefore,

$$\text{antilog } (-4.7899) = 1.62 \times 10^{-5}$$

EXERCISE 13

Find the antilog of each of the following numbers.
(a) −8.4422
(b) −0.2468
(c) −1.2233

Answers
(a) 3.61×10^{-9} [Since $-8.4422 = -9 + 0.5578$, and the antilog of 0.5578 = 3.61.]
(b) 5.67×10^{-1} [Since $-0.2468 = -1 + 0.7532$; and the antilog of 0.7532 is 5.67.]
(c) 5.98×10^{-2} [Since $-1.2233 = -2 + 0.7767$, and the antilog of 0.7767 is 5.98.]

EXAMPLE 8
SOLVING A PROBLEM IN DIVISION BY USING LOGS

Problem: What is 0.0456 divided by 0.00382?
Solution: Remember that

$$\log \frac{A}{B} = \log A - \log B$$

$$\log 0.0456 = \log (4.56 \times 10^{-2})$$
$$= -2 + 0.6590$$
$$= -1.3410$$
$$\log 0.00382 = \log (3.82 \times 10^{-3})$$
$$= -3 + 0.5821$$
$$= -2.4179$$

Therefore,

$$\log \frac{0.0456}{0.00382} = -1.3410 - (-2.4179)$$
$$= 1.0769$$

Next, we find the antilog of 1.0769 to get the answer. We can express it this way.

$$\text{antilog } 1.0769 = 10^1 \times 10^{0.0769}$$
$$= (10^1) \times (1.19)$$
$$= 1.19 \times 10^1, \text{ our answer}$$

EXAMPLE 9
SOLVING A PROBLEM IN MULTIPLICATION AND ROOTS USING LOGS

Problem: What is the value of $423 \times 0.456^{1/4}$?
Solution: We solve this by first finding the log of this expression, remembering that we add logs when we want to multiply and we find logs of numbers involving exponentials by multiplying the log of the number by its exponent.

$$\log 423 = \log (4.23 \times 10^2) = 2.6263$$
$$\log 0.456^{1/4} = \frac{1}{4} \times \log 0.456$$
$$= \frac{1}{4} \times \log (4.56 \times 10^{-1})$$
$$= \frac{1}{4} \times (-1 + 0.6590)$$
$$= \frac{1}{4} \times (-0.3410)$$
$$= -0.08525$$

Therefore,

$$\log (423 \times 0.456^{1/4}) = 2.6263 + (-0.08525)$$
$$= 2.5410 \text{ (rounded)}$$

Next we take the antilog.

$$\text{antilog } 2.5410 = 10^2 \times 10^{0.5410}$$
$$= 10^2 \times 3.48$$

Our answer is 3.48×10^2. Since taking roots of numbers, particularly roots higher than square roots, is very difficult without a scientific calculator, you can see one reason why before calculators were invented mathematicians and scientists welcomed the invention of logarithms.

Long strings of multiplications and divisions can be handled through logarithms simply by adding or subtracting logs and then taking the antilog of the result.

EXERCISE 14

Evaluate the following expressions using logarithms to carry out the necessary operations.

(a) $\dfrac{4520 \times 0.000362}{89.2}$

(b) $\dfrac{92^{1/8} \times 0.759 \times 298,000}{56.8 \times 0.0467}$

Answers
(a) 1.83×10^{-2} from $[3.6551 + (-3.4413)] - 1.9504 = -1.7366 = -2 + 0.2634$.
(b) 1.50×10^5 from $[\frac{1}{8} \times 1.9638 + (-0.1198) + 5.4742] - [1.7543 + (-1.3307)] = 0.2455 - 0.1198 + 5.4742 - 1.7543 + 1.3307 = 5.1763$

$$\text{antilog } 5.1763 = 10^5 \times 10^{0.1763}$$
$$= 1.50 \times 10^5$$

Natural Logarithms We have just seen how common logarithms are found and used. They are a special case—made special by our need to use the decimal system—for a general idea in mathematics:

when $\qquad N = B^x$
then $\qquad \log_B N = x \qquad$ (read: "log N to the base B equals x")

where B is called the *base*. Therefore, common logs are base 10 logs.

In many applications in science another base is needed, the base e, where

$$e = 2.7182818 \ldots \qquad \text{(to show just 8 significant figures)}$$

To appreciate the need for this base, you must have a good background in calculus. We can't assume that you have yet acquired such a background, but we can still use the results. Logarithms to the base e are called natural logarithms, and their symbol is ln. Thus if N is our number, then ln N is the natural logarithm of the number.[1]

Just as the logarithm of 10 to the base 10 is 1, so the logarithm of e to the base e is also 1. It can also be shown that

$$\log_e 10 = \ln 10 = 2.303$$

or

$$e^{2.303} = 10$$

Tables of natural logarithms are available, and their values can easily be "called up" with an inexpensive hand-

[1] If you have had calculus, then if ln $x = y$, and we take the derivative of x with respect to y, the result is

$$\frac{d(\ln x)}{dy} = \frac{1}{x}$$

Without the base e we would have no function in mathematics whose derivative is $1/x$, and such a function is often needed both in mathematics and in its uses in all branches of science.

held scientific calculator. They are used in the same way as common logarithms. For most needs, tables of natural logs aren't necessary because of a simple relationship between natural and common logs:

$$\ln N = 2.303 \log N$$

In other words, to find $\ln N$, simply find the common log of N and multiply it by 2.303.

A.3 GRAPHING

One of the most useful ways of describing the relationship between two quantities is by means of a graph. It allows us to obtain an overall view of how one quantity changes when the other changes. Constructing graphs from experimental data, or from data calculated from an equation, is therefore a common operation in the sciences, so it is important that you understand how to present data graphically. Let's look at an example.

Suppose that we wanted to know how the volume of a given quantity of gas varies as we change its pressure. In Chapter 9 we find an equation that gives the pressure–volume relationship for a fixed amount of gas at a constant temperature,

$$PV = \text{constant} \qquad (A.1)$$

where P = pressure and V = volume. For calculations, this equation can be used in two ways. If we solve for the volume, we get

$$V = \frac{\text{constant}}{P} \qquad (A.2)$$

Knowing the value of the constant for a given gas sample at a given temperature allows us to calculate the volume that the gas occupies at any particular pressure. Simply substitute in the values for the constant and the pressure and compute the volume.

Equation A.1 can also be solved for the pressure.

$$P = \frac{\text{constant}}{V} \qquad (A.3)$$

From this equation the pressure corresponding to any particular volume can be calculated.

Suppose, now, that we want to see graphically how the pressure and volume of a gas are related as the pressure is raised from 1.0 to 10.0 atm in steps of 1.0 atm. Let's also suppose that for this sample of gas the value of the constant in our equation is 0.25 L atm. To construct the graph we first must decide how to label and number the axes.

Usually, the horizontal axis, called the *abscissa,* is chosen to correspond to the *independent variable*—the variable whose values are chosen first and from which the values of the *dependent variable* are determined. In our

TABLE A.3 Pressure–Volume Data

Pressure (atm)	Volume (L)
1.0	0.25
2.0	0.125
3.0	0.083
4.0	0.062
5.0	0.050
6.0	0.042
7.0	0.036
8.0	0.031
9.0	0.028
10.0	0.025

example, we are choosing values of pressure (1.0 atm, 2.0 atm, etc.), so we will label the abscissa "pressure." We will also mark off the axis evenly from 1.0 to 10.0 atm, as shown in Figure A.1. Next we put the label "volume" on the vertical axis (the *ordinate*). Before we can number this axis, however, we need to know over what range the volume will vary. Using Equation A.2 we can calculate the data in Table A.3, and we see that the volume ranges from a low of 0.025 L to a high of 0.25 L. We can therefore mark off the ordinate evenly in increments of 0.025 L, starting at the bottom with 0.025 L. Notice that in labeling the axes we have indicated the units of P and V in parentheses.

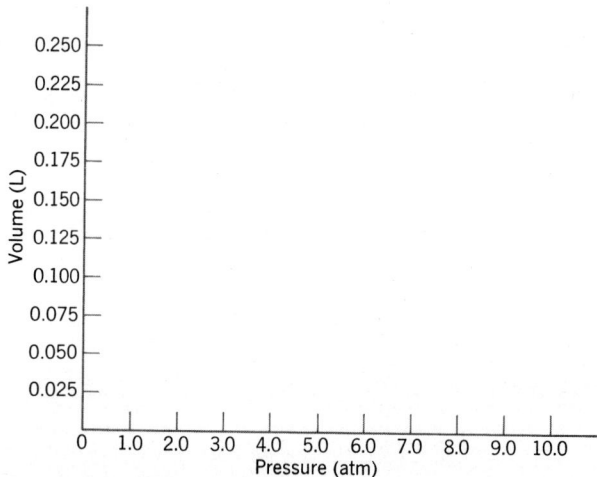

FIGURE A.1 Choosing and labeling the axes of a graph.

Next we plot the data as shown in Figure A.2. To clearly show each plotted point, we use a small circle that has its center located at the coordinates of the point. Then we draw a *smooth* curve through the points.

Sometimes, when plotting experimentally measured data, the points do not fall exactly on a smooth line (Figure A.3). Nature generally is not irregular even though that's what the data appear to suggest. The fluctuations are usually due to experimental error of some sort. Therefore,

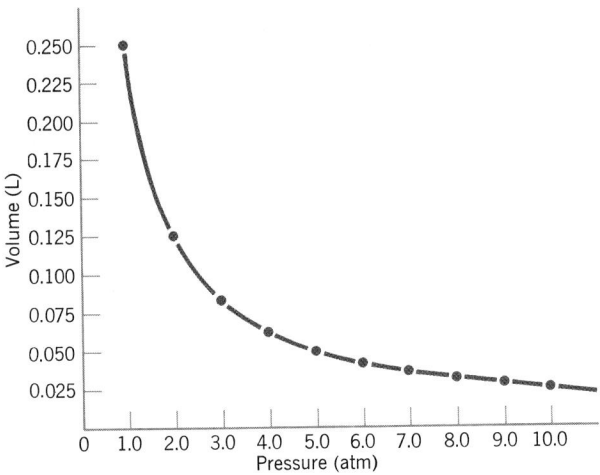

FIGURE A.2 Plotting points and drawing the curve.

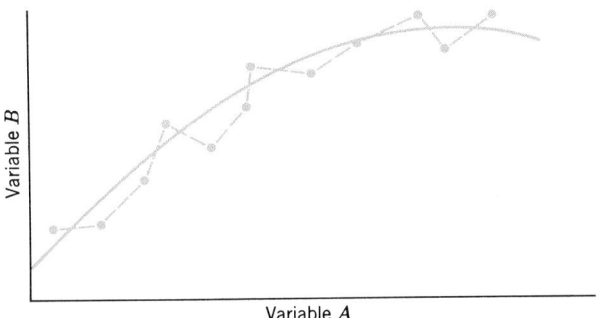

FIGURE A.3 Experimental data do not all fall on a smooth curve when they are plotted. A smooth curve is drawn as close to all of the data points as possible, rather than the jerky dashed line.

rather than draw an irregular line connecting all the data points (the dashed line), we draw a smooth curve that passes as close as possible to all of the points, even though it may not actually pass through any of them.

Slope   One of the properties of curves and lines on a graph is their *slope* or steepness. Consider Figure A.4,

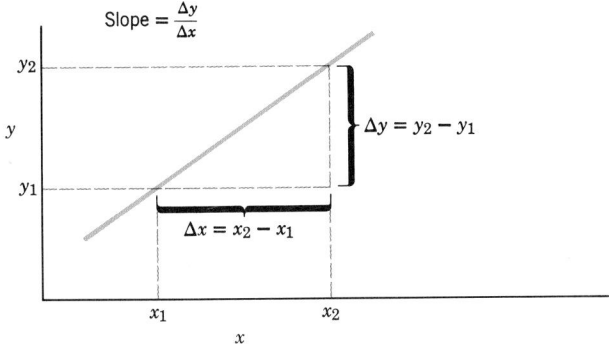

FIGURE A.4 Determining the slope of a straight line.

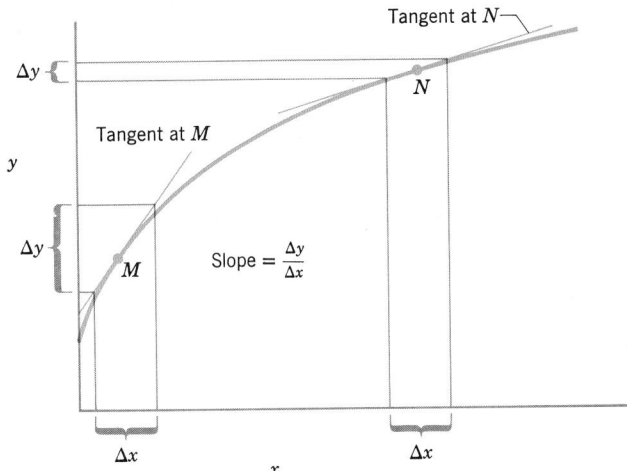

FIGURE A.5 For a given Δx, Δy is much larger at *M* than it is at *N*. Therefore the slope at *M* is larger than at *N*.

which is a straight line drawn on a set of xy coordinate axes. The slope of the line is defined as the change in y, Δy, divided by the change in x, Δx.

$$\text{slope} = \frac{\Delta y}{\Delta x} = \frac{y_2 - y_1}{x_2 - x_1}$$

Curved lines have slopes too, but the slope changes from point to point on the curve. To obtain the slope graphically, we can draw a line that is tangent to the curve at the point where we want to know the slope. The slope of this tangent is the same as the slope of the curve at this point. In Figure A.5 we see a curve for which the slope is determined at two different points. At point *M* the curve is rising steeply and $\Delta y/\Delta x$ for the tangent is large. At point *N* the curve is rising less steeply, and $\Delta y/\Delta x$ for the tangent at *N* is small. The slope at *M* is larger than the slope at *N*.

A.4   ELECTRONIC CALCULATORS

Performing arithmetic calculations is commonplace in all of the sciences, and electronic calculators have made these operations simple and routine. So-called scientific calculators are inexpensive (many cost less than this book) and are able to perform complex functions at the touch of a button. But these electronic wizards are only as versatile as the person who presses the keys.

This section is not intended to teach you how to use your particular calculator — each calculator comes with its own instruction booklet. Instead, our goal is to point out common errors that students make with some kinds of calculators and to explain how to speed up certain types of calculations.

Most scientific calculators use an algebraic entry system in which keys are pressed in the order that the quantities and mathematical operations appear in the computation. Thus

$$2 \times 3 = 6$$

is entered by pressing keys in this order: 2, ×, 3, =. We will indicate this by using boxes to represent the keys.

$$\boxed{2}\;\boxed{\times}\;\boxed{3}\;\boxed{=}$$

When the keys are pressed in this sequence, the answer, 6, appears on the display. In explaining how to use the calculator for certain calculations, we will assume an algebraic entry system.

Scientific Notation

Earlier in this appendix and in Section 1.12 we saw that numbers expressed as

$$2.3 \times 10^5$$
$$6.7 \times 10^{-12}$$

are said to be in *exponential notation* or *scientific notation*. Scientific calculators are able to do arithmetic with numbers such as these, provided they are entered correctly.

Many calculators have a key labeled *EE* or *EXP* that is used to enter the exponential portion of numbers in scientific notation. If we read the number 2.3×10^5 as "two point three times ten to the fifth," the *EE* (or *EXP*) key should be read as ". . . times ten to the. . . ." Thus the sequence of keys that enter this number is

$$\boxed{2},\boxed{.},\boxed{3},\boxed{EE},\boxed{5}$$

Try this on your own calculator and check to be sure that the number is correctly displayed after you've entered it. If not, check the instruction booklet that came with the calculator or ask your instructor for help.

When entering a number such as 6.7×10^{-12}, be sure to press the "change sign" or "+/−" key after pressing the *EE* key. Most calculators will not give the correct entry if the "minus" key is pressed instead of the "+/−" key. For example, try the following sequence of keystrokes on your calculator.

$$\boxed{6}\;\boxed{.}\;\boxed{7}\;\boxed{EE}\;\boxed{+/-}\;\boxed{1}\;\boxed{2}$$

Clear the display and then try this sequence.

$$\boxed{6}\;\boxed{.}\;\boxed{7}\;\boxed{EE}\;\boxed{-}\;\boxed{1}\;\boxed{2}$$

Press the "equals" key and see what happens.

Chain Calculations

It is not uncommon to have to calculate the value of a fraction in which there are a string of numbers that must be multiplied together in the numerator and another string of numbers to be multiplied together in the denominator. For example,

$$\frac{5.0 \times 7.3 \times 8.5}{6.2 \times 2.5 \times 3.9} = ?$$

Many students will compute the value of the numerator and write down the answer, then compute the value of the denominator and write it down, and finally divide the value of the numerator by the value of the denominator (whew!). Although this gives the correct answer, there is a simpler way to do the arithmetic that doesn't require writing down any intermediate values. The procedure is as follows:

1. Enter the first value from the numerator (5.0).
2. Any of the other values from the numerator are entered by multiplication and any of the values from the denominator are entered by division.

The following sequence gives the answer.

$$5.0 \times 7.3 \times 8.5 \div 6.2 \div 2.5 \div 3.9 = 5.13234 \ldots$$

Notice that *each* value from the denominator is entered by division. It also doesn't matter in what sequence the numbers are entered. The following gives the same answer.

$$5.0 \div 6.2 \times 7.3 \div 2.5 \div 3.9 \times 8.5 = 5.13234 \ldots$$

EXERCISE 15

Compute the value of the following fractions:

(a) $\dfrac{3 \times 4 \times 8}{6 \times 2}$

(b) $\dfrac{2.54 \times 1.47 \times 8.11}{9.23 \times 41.3}$

(c) $\dfrac{1}{3.0 \times 2.5 \times 7.3}$

(d) $\dfrac{(2.3 \times 10^{-5}) \times (3.1 \times 10^8)}{(4.3 \times 10^4) \times (8.4 \times 10^{-2}) \times (5.9 \times 10^6)}$

Answers. (a) 8   (b) 0.0794   (c) 0.018
(d) 3.3×10^{-7}
Note that we have followed the rules for significant figures given in Section 1.11.

THE ELECTRON CONFIGURATIONS OF THE ELEMENTS

Atomic Number			Atomic Number			Atomic Number		
1	H	$1s^1$	36	Kr	[Ar] $4s^2 3d^{10} 4p^6$	71	Lu	[Xe] $6s^2 4f^{14} 5d^1$
2	He	$1s^2$	37	Rb	[Kr] $5s^1$	72	Hf	[Xe] $6s^2 4f^{14} 5d^2$
3	Li	[He] $2s^1$	38	Sr	[Kr] $5s^2$	73	Ta	[Xe] $6s^2 4f^{14} 5d^3$
4	Be	[He] $2s^2$	39	Y	[Kr] $5s^2 4d^1$	74	W	[Xe] $6s^2 4f^{14} 5d^4$
5	B	[He] $2s^2 2p^1$	40	Zr	[Kr] $5s^2 4d^2$	75	Re	[Xe] $6s^2 4f^{14} 5d^5$
6	C	[He] $2s^2 2p^2$	41	Nb	[Kr] $5s^1 4d^4$	76	Os	[Xe] $6s^2 4f^{14} 5d^6$
7	N	[He] $2s^2 2p^3$	42	Mo	[Kr] $5s^1 4d^5$	77	Ir	[Xe] $6s^2 4f^{14} 5d^7$
8	O	[He] $2s^2 2p^4$	43	Tc	[Kr] $5s^2 4d^5$	78	Pt	[Xe] $6s^1 4f^{14} 5d^9$
9	F	[He] $2s^2 2p^5$	44	Ru	[Kr] $5s^1 4d^7$	79	Au	[Xe] $6s^1 4f^{14} 5d^{10}$
10	Ne	[He] $2s^2 2p^6$	45	Rh	[Kr] $5s^1 4d^8$	80	Hg	[Xe] $6s^2 4f^{14} 5d^{10}$
11	Na	[Ne] $3s^1$	46	Pd	[Kr] $4d^{10}$	81	Tl	[Xe] $6s^2 4f^{14} 5d^{10} 6p^1$
12	Mg	[Ne] $3s^2$	47	Ag	[Kr] $5s^1 4d^{10}$	82	Pb	[Xe] $6s^2 4f^{14} 5d^{10} 6p^2$
13	Al	[Ne] $3s^2 3p^1$	48	Cd	[Kr] $5s^2 4d^{10}$	83	Bi	[Xe] $6s^2 4f^{14} 5d^{10} 6p^3$
14	Si	[Ne] $3s^2 3p^2$	49	In	[Kr] $5s^2 4d^{10} 5p^1$	84	Po	[Xe] $6s^2 4f^{14} 5d^{10} 6p^4$
15	P	[Ne] $3s^2 3p^3$	50	Sn	[Kr] $5s^2 4d^{10} 5p^2$	85	At	[Xe] $6s^2 4f^{14} 5d^{10} 6p^5$
16	S	[Ne] $3s^2 3p^4$	51	Sb	[Kr] $5s^2 4d^{10} 5p^3$	86	Rn	[Xe] $6s^2 4f^{14} 5d^{10} 6p^6$
17	Cl	[Ne] $3s^2 3p^5$	52	Te	[Kr] $5s^2 4d^{10} 5p^4$	87	Fr	[Rn] $7s^1$
18	Ar	[Ne] $3s^2 3p^6$	53	I	[Kr] $5s^2 4d^{10} 5p^5$	88	Ra	[Rn] $7s^2$
19	K	[Ar] $4s^1$	54	Xe	[Kr] $5s^2 4d^{10} 5p^6$	89	Ac	[Rn] $7s^2 6d^1$
20	Ca	[Ar] $4s^2$	55	Cs	[Xe] $6s^1$	90	Th	[Rn] $7s^2 6d^2$
21	Sc	[Ar] $4s^2 3d^1$	56	Ba	[Xe] $6s^2$	91	Pa	[Rn] $7s^2 5f^2 6d^1$
22	Ti	[Ar] $4s^2 3d^2$	57	La	[Xe] $6s^2 5d^1$	92	U	[Rn] $7s^2 5f^3 6d^1$
23	V	[Ar] $4s^2 3d^3$	58	Ce	[Xe] $6s^2 4f^1 5d^1$	93	Np	[Rn] $7s^2 5f^4 6d^1$
24	Cr	[Ar] $4s^1 3d^5$	59	Pr	[Xe] $6s^2 4f^3$	94	Pu	[Rn] $7s^2 5f^6$
25	Mn	[Ar] $4s^2 3d^5$	60	Nd	[Xe] $6s^2 4f^4$	95	Am	[Rn] $7s^2 5f^7$
26	Fe	[Ar] $4s^2 3d^6$	61	Pm	[Xe] $6s^2 4f^5$	96	Cm	[Rn] $7s^2 5f^7 6d^1$
27	Co	[Ar] $4s^2 3d^7$	62	Sm	[Xe] $6s^2 4f^6$	97	Bk	[Rn] $7s^2 5f^9$
28	Ni	[Ar] $4s^2 3d^8$	63	Eu	[Xe] $6s^2 4f^7$	98	Cf	[Rn] $7s^2 5f^{10}$
29	Cu	[Ar] $4s^1 3d^{10}$	64	Gd	[Xe] $6s^2 4f^7 5d^1$	99	Es	[Rn] $7s^2 5f^{11}$
30	Zn	[Ar] $4s^2 3d^{10}$	65	Tb	[Xe] $6s^2 4f^9$	100	Fm	[Rn] $7s^2 5f^{12}$
31	Ga	[Ar] $4s^2 3d^{10} 4p^1$	66	Dy	[Xe] $6s^2 4f^{10}$	101	Md	[Rn] $7s^2 5f^{13}$
32	Ge	[Ar] $4s^2 3d^{10} 4p^2$	67	Ho	[Xe] $6s^2 4f^{11}$	102	No	[Rn] $7s^2 5f^{14}$
33	As	[Ar] $4s^2 3d^{10} 4p^3$	68	Er	[Xe] $6s^2 4f^{12}$	103	Lr	[Rn] $7s^2 5f^{14} 6d^1$
34	Se	[Ar] $4s^2 3d^{10} 4p^4$	69	Tm	[Xe] $6s^2 4f^{13}$			
35	Br	[Ar] $4s^2 3d^{10} 4p^5$	70	Yb	[Xe] $6s^2 4f^{14}$			

ANSWERS TO PRACTICE EXERCISES AND SELECTED REVIEW EXERCISES

CHAPTER 1

PRACTICE EXERCISES

1. (a) 1 Ni and 2 Cl (c) 3 Ca, 2 P, and 8 O
   (b) 1 Fe, 1 S, and 4 O (d) 1 Cu, 1 S, 9 O, and 10 H

2. 1 Mg, 2 O, 4 H, and 2 Cl

3. $Mg(OH)_2(s) + 2HCl(aq) \rightarrow MgCl_2(aq) + H_2O(l)$

4. (a) m^3 (b) m/s

5. (a) nanometer (b) centimeter (c) kilometer (d) picometer
   (e) millimeter

6. 86 °F = 30 °C, −17.8 °C = −0.0400 °F

7. 90 °F = 32 °C, 85 °F = 29 °C (rounded)

8. 15 °C

9. (a) 1900 (b) 0.08920 (c) 1.82 (d) 14.518

10. (a) 1.785 yd (b) 0.001014 mi

11. (a) 2.3×10^4 (b) 2.17×10^7 (c) 1.5×10^{-3} (d) 2.7×10^{-5}

12. (a) 2700 (b) 35,000,000,000,000,000,000,000,000,000
    (c) 0.000000000002

13. (a) 108 in. (b) 1.25×10^5 cm (c) 0.0107 ft (d) 0.116 L/km

14. $t_F = \dfrac{9\ °F}{5\ °C}\,(30\ °C) + 32\ °F = 86\ °F$

15. 2.70 g/mL

16. 0.272 cm³, 171 g

17. 2.70, 169 lb/ft³

18. 0.902 g/mL, 7.52 lb/gal

REVIEW EXERCISES — Chapter 1

1.26 (a) 1 (b) 2 (c) 8 (d) 4 (e) 8 (f) 10

1.28 3 Ca, 5 Mg, 8 Si, 24 O, 2 H

1.33 (a) 6 (b) 3 (c) 27

1.35 $2C_8H_{18}(l) + 25\ O_2(g) \rightarrow 16CO_2(g) + 18H_2O(l)$

1.49 (a) 0.01 m   (c) 1×10^{12} pm   (e) 1×10^{-3} kg
     (b) 1000 m   (d) 0.1 m   (f) 1×10^{-3} g

1.52 (a) 75 °F (rounded)   (c) 5 °C   (e) 303 K
     (b) 50 °F   (d) 10 °C   (f) 263 K

1.54 98.83 °F

1.56 −269 °C, −452 °F

1.61 (a) 3 (b) 5 (c) 1 (d) 4 (e) 6 (f) 4

1.63 (a) 2.06   (c) 8×10^4 (e) 6×10^{-4}
     (b) 4.02   (d) 0.01216

1.64 (a) 2.45×10^2 (c) 2.87×10^{-3} (e) 4.00×10^{-8}
     (b) 3.10×10^4 (d) 4.50×10^7 (f) 3.24×10^5

1.66 (a) 2100   (c) 3,800,000   (e) 0.346
     (b) 0.000335   (d) 0.00000000046   (f) 85,000

1.68 (a) 2.0×10^4 (c) 1.0×10^3 (e) 2.0×10^{18}
     (b) 8.0×10^7 (d) 2.4×10^5

1.72 4 sig. fig. The number 4.165 has four significant figures.

The numbers in the conversion factor are exact numbers and can be considered to have as many significant figures as we want.

1.73 (a) 1×10^{-4} km   (d) 3.75×10^{-2} L
(b) 5.3×10^3 mg   (e) 125 mL
(c) 5.3×10^{-6} kg   (f) 3.42×10^{-4} mm

1.75 (a) 91 cm   (c) 2.8×10^3 mL   (e) 88 km/hr
(b) 2.3 kg   (d) 237 mL   (f) 80.5 km

1.77 3.6×10^2 mL

1.79 2205 lb

1.81 1.7×10^2 cm

1.85 204 francs

1.87 5.9×10^{12} mi

1.89 1.2 mi/hr

1.91 (a) 62 kg (b) 1.4×10^2 lb

1.93 12.6 mL

1.95 11 g/mL

1.97 0.715

1.99 1.47 g/mL

CHAPTER 2

PRACTICE EXERCISES

1. $^{240}_{94}Pu$

2. Two

3. 26.9814

4. 5.288

5. H = 1.6798 amu, Fe = 93.078 amu, S = 53.43 amu

6. (a) K, Ar, Al (b) Cl (c) Ba (d) Ne (e) Li (f) Ce

7. (a) NaF (b) Na_2O (c) MgF_2 (d) Al_2S_3

8. (a) $CrCl_2$, $CrCl_3$, CrO, Cr_2O_3 (b) CuCl, $CuCl_2$, Cu_2O, CuO

9. (a) Na_2CO_3 (c) $KC_2H_3O_2$ (e) $Fe(C_2H_3O_2)_2$
(b) $(NH_4)_2SO_4$ (d) $Sr(NO_3)_2$

10. potassium sulfide, magnesium phosphide, nickel(II) chloride, iron(III) oxide

11. (a) Al_2S_3 (b) SrF_2 (c) TiO_2 (d) $CrBr_3$

12. phosphorus trichloride, sulfur dioxide, dichlorine heptaoxide

13. hydrofluoric acid, hydrobromic acid

14. sodium arsenate

15. $NaHSO_3$

REVIEW EXERCISES—Chapter 2

2.8   Answer = (c): 8.84 g N/5.05 g O = 1.75 g N/1.00 g O

2.10   (a) 1-to-2 (b) 0.597 g Cl $\times$ 2 = 1.19 g Cl

2.21
	neutrons	protons	electrons
(a)	138	88	88
(b)	8	6	6
(c)	124	82	82
(d)	12	11	11

2.24 $1.9926798 \times 10^{-28}$ g

2.29
	electrons	protons	neutrons
(a)	18	17	18
(b)	23	26	30
(c)	27	29	35
(d)	18	19	20

2.35
atomic mass	79.8
melting point	281 °C
boiling point	717 °C
formula of oxide	XO_2
melting point of oxide	330 °C
density	4.16 g/cm³

2.71   (a) NaBr (b) KI (c) BaO (d) $MgBr_2$ (e) BaF_2

2.73   (a) $Ca + Cl_2 \rightarrow CaCl_2$
(b) $2Mg + O_2 \rightarrow 2MgO$
(c) $4Al + 3O_2 \rightarrow 2Al_2O_3$
(d) $2Na + S \rightarrow Na_2S$

2.77   (a) KNO_3 (c) NH_4Cl (e) $Mg_3(PO_4)_2$
(b) $Ca(C_2H_3O_2)_2$ (d) $Fe_2(CO_3)_3$

2.81   (a) $CdCl_2$ (b) AgCl (c) $ZnCl_2$ (d) $NiCl_2$

2.89 Bi_2O_3, Bi_2O_5

2.101 $Mg(OH)_2 + 2HCl \rightarrow MgCl_2 + 2H_2O$

2.103   (a) $HNO_2 + KOH \rightarrow KNO_2 + H_2O$
(b) $2HCl + Ca(OH)_2 \rightarrow CaCl_2 + 2H_2O$
(c) $H_2SO_4 + 2NaOH \rightarrow Na_2SO_4 + 2H_2O$
(d) $3HClO_4 + Al(OH)_3 \rightarrow Al(ClO_4)_3 + 3H_2O$
(e) $2H_3PO_4 + 3Ba(OH)_2 \rightarrow Ba_3(PO_4)_2 + 6H_2O$

2.107   X is a nonmetal in Group VA of the periodic table.

2.108   (a) calcium sulfide   (d) magnesium carbide
(b) sodium fluoride   (e) sodium phosphide
(c) aluminum bromide   (f) lithium nitride

2.110   (a) silicon dioxide
(b) chloride trifluoride
(c) xenon tetrafluoride
(d) disulfur dichloride
(e) tetraphosphorus decaoxide
(f) dinitrogen pentaoxide (or dinitrogen pentoxide)

2.112   (a) sodium nitrite   (d) ammonium acetate
(b) potassium phosphate   (e) barium sulfate
(c) potassium permanganate   (f) iron(III) carbonate

2.114   (a) Na_2HPO_4 (e) $Ni(CN)_2$ (i) Al_2Cl_6
(b) Li_2Se (f) Fe_2O_3 (j) As_4O_{10}
(c) NaH   (g) SnS_2 (k) $Mg(OH)_2$
(d) $Cr(C_2H_3O_2)_3$ (h) SbF_5 (l) $Cu(HSO_4)_2$

2.116   (a) hypochlorous acid, sodium hypochlorite
(b) iodous acid, sodium iodite
(c) bromic acid, sodium bromate
(d) perchloric acid, sodium perchlorate

CHAPTER 3

PRACTICE EXERCISES

1. 1.534 g of Na

2. (a) 58.5 (b) 324.0 (c) 186.2 (d) 284.1

3. 8.36×10^{24} molecules of H_2O

4. 0.15 mol of Fe

5. 2.0 mol of P and 5.0 mol of O

6. (a) 10.5 g H_2O (c) 32.7 g Fe
   (b) 105 g $C_6H_{12}O_6$ (d) 9.38 g CH_4

7. (a) 5.88 mol NH_3 (c) 0.508 mol Au
   (b) 0.259 mol $C_{27}H_{46}O$ (d) 2.17 mol C_2H_6O

8. Individual NH_3 molecules weigh much less than those of cholesterol, so it takes more to have 100.0 g.

9. 20.0 g

10. 42.10% C, 6.44% H, 51.45% O

11. 82.66% C, 17.43% H

12. $Cr(CO)_6$ (The calculated values for this compound are 23.63% Cr, 32.74% C, and 43.62% O.)

13. $Fe_3P_2O_8$

14. HgCl

15. $BaSO_4$

16. Hg_2Cl_2

17. N_2H_4

REVIEW EXERCISES — Chapter 3

3.2 50.53 g of chlorine

3.4 (a) 106 (b) 191.2 (c) 397

3.10 6.02×10^{23} K atoms. Avogadro's number

3.12 (a) 10.60 g Na_2CO_3 (b) 19.12 g $(NH_4)_2B_4O_7$ (c) 39.65 g $Ca(C_6H_{12}NSO_3)_2$

3.15 2.00×10^{-2} mol $NaNO_3$

3.17 171 g $C_{12}H_{22}O_{11}$

3.19 1.98 mol Tl_2SO_4

3.21 6.02×10^{23} atoms of Ca, 40.1 g of Ca

3.23 9.03×10^{23} N atoms, 72.1 g $(NH_4)_2CO_3$

3.25 826 g $(NH_4)_2HPO_4$

3.27 344 mol $(NH_4)_2HPO_4$

3.29 380 g $NaBO_3$

3.31 3.0 mol NaOH/1.0 mol tallow

3.33 8.5×10^{11} mol NH_3/year

3.35 82.4% N in NH_3, 46.7% N in $CO(NH_2)_2$

3.37 Theoretical percentages are 83.89% C, 10.35% H, and 5.756% N, and these calculated values agree well enough with the experimental values for phencyclidine.

3.45 $NaTcO_4$

3.47 $C_{10}H_{16}N_5P_3O_{13}$ for both the empirical and the molecular formulas.

3.49 empirical formula, CH_2; molecular formula, C_4H_8

3.51 empirical formula, H_2PO_3; molecular formula, $H_4P_2O_6$

3.53 empirical formula, $HgC_2H_3O_2$; molecular formula, $Hg_2C_4H_6O_4$

3.56 (a) NaO (b) 1.437 g Na

(c) Some possible answers are
2.874 g Na (from 2×1.437 g Na)
4.311 g Na (from 3×1.437 g Na)
5.748 g Na (from 4×1.437 g Na), etc., or
0.7185 g Na (from $\frac{1}{2} \times 1.437$ g Na)
0.4790 g Na (from $\frac{1}{3} \times 1.437$ g Na)
0.3593 g Na (from $\frac{1}{4} \times 1.437$ g Na), etc.

(d) 2.875 g Na

(e) $\dfrac{1.437 \text{ g Na}/1.000 \text{ g O in sodium peroxide}}{2.875 \text{ g Na}/1.000 \text{ g O in sodium oxide}} = \dfrac{1}{2}$

Yes, the two compounds do illustrate the law of multiple proportions.

(f) Na_2O

3.58 $C_{18}H_{22}O_2$

CHAPTER 4

PRACTICE EXERCISES

1. (a) $4P + 5O_2 \rightarrow P_4O_{10}$
   (b) $N_2 + 3H_2 \rightarrow 2NH_3$

2. (a) $2Mg + O_2 \rightarrow 2MgO$
   (b) $CH_4 + 2Cl_2 \rightarrow CCl_4 + 4HCl$
   (c) $2NO + O_2 \rightarrow 2NO_2$
   (d) $2NaOH + H_2SO_4 \rightarrow Na_2SO_4 + 2H_2O$
   (e) $CH_4 + 2O_2 \rightarrow CO_2 + 2H_2O$
   (f) $2C_2H_6 + 7O_2 \rightarrow 4CO_2 + 6H_2O$
   (g) $2Al(OH)_3 + 3H_2SO_4 \rightarrow Al_2(SO_4)_3 + 6H_2O$

3. 12 mol Fe

4. 8 mol H_2O

5. (a) First pair $\dfrac{3 \text{ mol } Cl_2}{2 \text{ mol Fe}}$ $\dfrac{2 \text{ mol Fe}}{3 \text{ mol } Cl_2}$

   Second pair $\dfrac{2 \text{ mol Fe}}{2 \text{ mol } FeCl_3}$ $\dfrac{2 \text{ mol } FeCl_3}{2 \text{ mol Fe}}$

   Third pair $\dfrac{3 \text{ mol } Cl_2}{2 \text{ mol } FeCl_3}$ $\dfrac{2 \text{ mol } FeCl_3}{3 \text{ mol } Cl_2}$

   (b) 16 mol $FeCl_3$ (c) 16 mol Fe
   (d) 0.750 mol Cl_2, 0.500 mol $FeCl_3$

6. 11.5 g Fe

7. 13.8 g $K_2Cr_2O_7$

8. 18.0 g I_2

9. 0.0536 mol $KMnO_4$, 8.47 g $KMnO_4$

10. 0.793 mol SiF_4, 82.5 g SiF_4

11. 16.6 g HI, 12.4 g CuI, 6.38 g H_2SO_4, 8.26 g I_2
    The sums of the masses on each side of the arrow are each 27.0 g.

12. 8.47 g $KMnO_4$

13. 82.4 g SiF_4 (Notice the small discrepancy, caused by rounding, between the answer to this exercise and that of Exercise 10.)

14. 11.5 g Fe, 13.8 g $K_2Cr_2O_7$

15. 17.9 g I_2 (Notice the small discrepancy, caused by rounding, between the answer to this exercise and that of Exercise 8.)

16. Limiting reactant is Fe_2O_3. Theoretical yield of
    $Fe = 0.600$ mol.

17. Na is the limiting reactant.

18. 78.5%

19. 5.00 mL solution

20. Dissolve 4.20 g $NaHCO_3$ in a total volume of 250 mL of
    solution.

21. Dissolve 2.25 g of glucose in sufficient water to give
    250 mL of solution.

22. 17.2 mL KOH

23. 61.4 mL HCl

24. Place 2.0×10^2 mL of the 0.50 M NaOH into a graduated
    cylinder or volumetric flask and add sufficient water to
    make the final volume equal to 5.0×10^2 mL.

25. 450 mL

26. 0.17 M

REVIEW EXERCISES — Chapter 4

4.5 (a) $Ca(OH)_2 + 2HCl \rightarrow CaCl_2 + 2H_2O$
    (b) $2AgNO_3 + CaCl_2 \rightarrow Ca(NO_3)_2 + 2AgCl$
    (c) $2Fe_2O_3 + 3C \rightarrow 4Fe + 3CO_2$
    (d) $2NaHCO_3 + H_2SO_4 \rightarrow Na_2SO_4 + 2H_2O + 2CO_2$
    (e) $2C_4H_{10} + 13O_2 \rightarrow 8CO_2 + 10H_2O$

4.7 (a) $Mg(OH)_2 + 2HBr \rightarrow MgBr_2 + 2H_2O$
    (b) $2HCl + Ca(OH)_2 \rightarrow CaCl_2 + 2H_2O$
    (c) $Al_2O_3 + 3H_2SO_4 \rightarrow Al_2(SO_4)_3 + 3H_2O$
    (d) $2KHCO_3 + H_3PO_4 \rightarrow K_2HPO_4 + 2H_2O + 2CO_2$
    (e) $C_9H_{20} + 14O_2 \rightarrow 9CO_2 + 10H_2O$

4.12 (a) The coefficients in the equation, 1:1:1, disclose the
         stoichiometry.
     (b) The specified quantities, 0.1 mol of Mg and 0.1 mol of
         Cl_2, give the scale.

4.14 (a) 50.0 mol O_2
     (b) 8.00 mol CO_2
     (c) 54.0 mol H_2O
     (d) 12.5 mol O_2 needed, 1.00 mol octane needed

4.16 (a) 1.37×10^{-1} mol H_2O form
     (b) 2.74×10^{-2} mol C_4H_{10} burned
     (c) 1.59 g C_4H_{10} burned
     (d) 1.78×10^{-1} mol O_2 or 5.70 g O_2 were used up

4.18 1.45 mol or 241 g of $C_8H_6O_4$

4.20 (a) 50.0 mol Na_2CO_3
     (b) 4.67 mol or 495 g of Na_2CO_3

4.22 (a) 4.05 mol or 226 g of Fe
     (b) 87.6%

4.24 (a) 1.25 mol or 123 g of H_3PO_4
     (b) 2.50 M H_3PO_4
     (c) 2.14×10^{10} mol or 2.99×10^6 metric tons of P_4O_{10}

4.26 CuS

4.28 Empirical formula, C_4H_9; molecular formula, C_8H_{18}

4.31 (a) Zn (b) 37.3 g ZnS (c) 17.7 g S remain

4.33 (a) 253 g $CaSO_3$ (b) 78.3%

4.51 (a) 1.46 g NaCl (b) 7.92 g $C_6H_{12}O_6$ (c) 24.5 g H_2SO_4

4.53 55.6 M H_2O

4.55 15 M NH_3

4.58 (a) 160 mL K_2CO_3 solution and 240 mL $CaCl_2$ solution
     (b) 174 mL K_2CO_3 solution and 261 mL $CaCl_2$ solution

4.60 2.70×10^{-1} g Na_2CO_3, 84.1% pure

4.62 0.106 M Na_2SO_4 and 0.115 M $Pb(NO_3)_2$

4.64 Zn

4.67 3.1 mL of concd HNO_3 needed

4.70 250 mL

CHAPTER 5

PRACTICE EXERCISES

1. 5.2×10^3 J, 1.3×10^3 cal

2. 2.7 kJ, 0.64 kcal

3. $q = -3.7$ kJ, $\Delta H = -74$ kJ/mol H_2SO_4

4. 1.13×10^4 kJ/mol

5. $H_2(g) + (\frac{1}{2})O_2(g) \rightarrow H_2O(l)$, $\Delta H° = -258.9$ kJ

6. $\Delta H° = -44.0$ kJ

7. (a) $\Delta H° = -113.1$ kJ (b) $\Delta H° = -134$ kJ

8. $\Delta H_f° = -2225.6$ kJ/mol or -531.92 kcal/mol

REVIEW EXERCISES — Chapter 5

5.9 1.92×10^6 J, 1.92×10^3 kJ

5.11 1 cal = 3.97×10^{-3} BTU

5.13 KE = 4.5×10^{10} kJ

5.19 3.8 kcal (rounded from 3.75)

5.21 (a) 3.5×10^3 kcal (b) 57 mi

5.23 25.12 J/mol °C

5.38 14.7 kcal or 61.5 kJ

5.39 $q = -9.1 \times 10^2$ cal, -3.8×10^3 J, -13 kcal/mol

5.41 25.7 °C

5.48 $4Al(s) + 2Fe_2O_3(s) \rightarrow 2Al_2O_3(s) + 4Fe(s)$,
      $\Delta H° = -1708$ kcal

5.51 For the single-step process

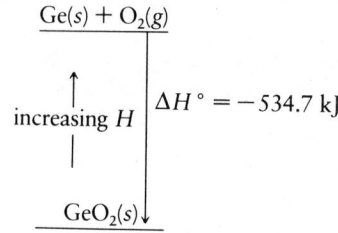

For the two-step process

$$Ge(s) + (\tfrac{1}{2})O_2(g) + (\tfrac{1}{2})O_2(g)$$
$$\Delta H° = -255 \text{ kJ}$$

$$GeO(s) + (\tfrac{1}{2})O_2(g)$$
$$\Delta H° = -280 \text{ kJ (answer)} \qquad \Delta H° = -534.7 \text{ kJ}$$
$$\text{sought)}$$

$$GeO_2(s)$$
$$\Delta H° = -280 \text{ kJ}$$

5.55 $\Delta H° = -171.07$ kJ

5.57 $\Delta H° = 64.8$ kJ

5.59 $\Delta H° = -135$ kJ

5.61 $\Delta H° = -127.8$ kJ

5.63 $\Delta H° = -54$ kJ

5.66 (a) $2C(graphite) + 2H_2(g) + O_2(g) \rightarrow HC_2H_3O_2(l)$,
$\Delta H_f° = -487.0$ kJ/mol^{-1}
(b) $Na(s) + (\tfrac{1}{2})H_2(g) + C(graphite) + (\tfrac{3}{2})O_2(g) \rightarrow$
$NaHCO_3(s), \Delta H_f° = -947.7$ kJ/mol^{-1}
(c) $Ca(s) + S(s) + 3O_2(g) + 2H_2(g) \rightarrow CaSO_4 \cdot 2H_2O(s)$,
$\Delta H_f° = -2021.1$ kJ/mol^{-1}
(d) $Ca(s) + S(s) + (\tfrac{5}{2})O_2(g) + (\tfrac{1}{2})H_2(g) \rightarrow CaSO_4 \cdot \tfrac{1}{2}H_2O(s)$,
$\Delta H_f° = -1575.2$ kJ/mol^{-1}
(e) $C(graphite) + 2H_2(g) + (\tfrac{1}{2})O_2(g) \rightarrow CH_3OH(l)$,
$\Delta H_f° = -238.6$ kJ/mol^{-1}

5.68 $\Delta H_{reaction}° = -130.2$ kJ

5.70 $\Delta H_{reaction}° = 490.7$ kJ

5.72 (a) $C_2H_2(g) + (\tfrac{5}{2})O_2(g) \rightarrow 2CO_2(g) + H_2O(l), \Delta H° = -1299.6$ kJ
(b) $CH_3OH(l) + (\tfrac{3}{2})O_2(g) \rightarrow CO_2(g) + 2H_2O(l), \Delta H° = -726.51$ kJ
(c) $C_4H_{10}O(l) + 6O_2(g) \rightarrow 4CO_2(g) + 5H_2O(l), \Delta H° = -2751.1$ kJ
(d) $C_7H_8(l) + 9O_2(g) \rightarrow 7CO_2(g) + 4H_2O(l), \Delta H° = -3909$ kJ

5.74 (a) $C_{16}H_{32}O_2(s) + 24O_2(g) \rightarrow 16CO_2(g) + 16H_2O(l)$,
$\Delta H_{combustion}° = -2380$ kcal/mol
(b) $\Delta H_f° = -218$ kcal/mol

CHAPTER 6

PRACTICE EXERCISES

1. 5.45×10^{14} Hz

2. 3.21 m

3. For $n = 3$: $3s, 3p, 3d$
For $n = 4$: $4s, 4p, 4d, 4f$

4. nine

5. (a) Mg $1s^2\ 2s^2\ 2p^6\ 3s^2$
(b) Ge $1s^2\ 2s^2\ 2p^6\ 3s^2\ 3p^6\ 3d^{10}\ 4s^2\ 4p^2$
(c) Cd $1s^2\ 2s^2\ 2p^6\ 3s^2\ 3p^6\ 3d^{10}\ 4s^2\ 4p^6\ 4d^{10}\ 5s^2$
(d) Gd $1s^2\ 2s^2\ 2p^6\ 3s^2\ 3p^6\ 3d^{10}\ 4s^2\ 4p^6\ 4d^{10}\ 4f^7\ 5s^2\ 5p^6$
$5d^1\ 6s^2$

6. (a) Na
(b) S
(c) V

7. (a) P [Ne] $3s^23p^3$ (b) Sn [Kr] $4d^{10}\ 5s^2\ 5p^2$

8. (a) N $2s^2\ 2p^3$ (b) Si $3s^2\ 3p^2$ (c) Sr $5s^2$

9. (a) Ni [Ar] $3d^8\ 4s^2$ (b) Ru [Kr] $4d^6\ 5s^2$

10. (a) Se $4s^2\ 4p^4$ (b) Sn $5s^2\ 5p^2$ (c) I $5s^2\ 5p^5$

11. (a) Sn (b) Ga (c) Fe (d) S^{2-}

12. (a) Be (b) C

REVIEW EXERCISES — Chapter 6

6.4 6.64×10^{-24} g

6.14 4.62×10^{13} Hz

6.16 9.40×10^9 Hz

6.18 589 nm

6.25 3.62×10^{-19} J

6.26 6.31×10^{22} photons

6.40 (a) 1 (b) 3

6.42 (a) $n = 2, \ell = 0$ (b) $n = 3, \ell = 2$ (c) $n = 5, \ell = 3$

6.45 (a) $m_\ell = -1, 0, 1$ (b) $m_\ell = -3, -2, -1, 0, 1, 2, 3$

6.47 eleven

6.62 (a) Mg
(b) Ti

6.64 (a) zero (b) three (c) three

6.65 (a) Ni [Ar] $3d^8\ 4s^2$ (c) Ge [Ar] $3d^{10}\ 4s^2\ 4p^2$
(b) Cs [Xe] $6s^1$ (d) Br [Ar] $3d^{10}\ 4s^2\ 4p^5$

6.69 (a) Na $3s^1$ (b) Al $3s^2\ 3p^1$ (c) Ge $4s^2\ 4p^2$ (d) P $3s^2\ 3p^3$

6.72 (a) and (c)

6.79 (a) Na (b) Sb

6.81 Sn

6.83 $Mg^{2+} < Na^+ < Ne < F^- < O^{2-} < N^{3-}$

6.85 (a) Na (b) Co^{2+} (c) Cl^-

6.87 (a) C (b) O (c) Cl

6.92 (a) Cl (b) Br (c) Si

CHAPTER 7

PRACTICE EXERCISES

1. $S(3s^2\,3p^4) + 2e^- \rightarrow S^{2-}(3s^2\,3p^6)$
   $Mg([Ne]3s^2) \rightarrow Mg^{2+}([Ne]) + 2e^-$

2. (a) $\cdot\ddot{Se}\!:$ (b) $\cdot\ddot{I}\!:$ (c) $\cdot Ca\cdot$

3. $\overset{\frown}{Mg}\!\cdot + \cdot\ddot{O}\!: \longrightarrow Mg^{2+} + \left[\,:\ddot{O}\!:\,\right]^{2-}$

4. (a) $H\!:\!\overset{\cdot\cdot}{\underset{\overset{\cdot\cdot}{H}}{P}}\!:\!H$ (b) $:\ddot{S}\!:\!\ddot{F}\!:$
   $\qquad\qquad\qquad\quad :\!\ddot{F}\!:$

5.
   O S O, $\underset{O\ \ O}{N}$,   H O Cl O,

   $\qquad\qquad\qquad\qquad\underset{\overset{\displaystyle O}{H}}{H\ O\ P\ O\ H}$

   (with O and H below)

6. SO_2, 18; PO_4^{3-}, 32; NO^+, 10

7. $:\ddot{O}\!-\!\ddot{F}\!:$, $\left[H\!-\!\overset{\overset{\displaystyle H}{|}}{\underset{\underset{\displaystyle H}{|}}{N}}\!-\!H\right]^+$, $\ddot{O}\!=\!\overset{\cdot\cdot}{S}\!-\!\ddot{O}\!:$, $\underset{O\ \ \ \ \ O}{\overset{\overset{\displaystyle :O:}{\|}}{N}}$,

   $:\ddot{F}\!-\!\overset{\overset{\displaystyle :\ddot{F}:}{|}}{Cl}\!-\!\ddot{F}\!:$, $H\!-\!\overset{\overset{\displaystyle :\ddot{O}:}{|}}{\underset{\underset{\displaystyle :O:}{|}}{Cl}}\!-\!\ddot{O}\!:$

8. $\left[\underset{O\quad O}{\overset{\overset{\displaystyle :O:}{|}}{N}}\right]^- \longleftrightarrow \left[\underset{O\quad O}{\overset{\overset{\displaystyle :O:}{\|}}{N}}\right]^- \longrightarrow \left[\underset{O\quad O}{\overset{\overset{\displaystyle :O:}{|}}{N}}\right]^-$

9. Trigonal bipyramidal

10. ClO_3^-, trigonal pyramidal; XeO_4, tetrahedral; OF_2, nonlinear

11. CO_3^{2-}, planar triangular

12. (a) Br (b) Cl (c) Cl

13. SO_2, BrCl, AsH_3, CF_2Cl_2

REVIEW EXERCISES—Chapter 7

7.11 (a) $\overset{\frown}{Ca} + \overset{\cdot\ddot{Br}:}{\underset{\cdot\ddot{Br}:}{}} \longrightarrow Ca^{2+} + 2\left[\,:\ddot{Br}\!:\,\right]^-$

(b) $\underset{Al}{\overset{Al}{}} \overset{\cdot\ddot{O}:}{\underset{\cdot\ddot{O}:}{\cdot\ddot{O}:}} \longrightarrow 2Al^{3+} + 3\left[\,:\ddot{O}\!:\,\right]^{2-}$

(c) $\underset{K}{\overset{K}{}} + \cdot\ddot{S}\!: \longrightarrow 2K^+ + \left[\,:\ddot{S}\!:\,\right]^{2-}$

7.20 1390 g H_2O

7.21 (a) $:\ddot{Br}\cdot + \cdot\ddot{Br}: \rightarrow :\ddot{Br}\!:\!\ddot{Br}:$

(b) $H\cdot\ \ \cdot\ddot{O}\cdot\ \ \cdot H \rightarrow H\!:\!\ddot{O}\!:\!H$

(c) $H\cdot\ \ \cdot\overset{\cdot\cdot}{\underset{\cdot}{N}}\cdot\ \ \cdot H \rightarrow H\!:\!\overset{\cdot\cdot}{\underset{\overset{\cdot\cdot}{H}}{N}}\!:\!H$
   $\qquad\qquad H$

7.24 (a) PCl_3 (b) CF_4 (c) BrCl

7.32 (a) $\quad$ Cl $\quad$ (b) F P F (c) H P H
   Cl Si Cl $\qquad$ F $\qquad\qquad$ H
   $\qquad$ Cl
   (d) Cl S Cl

7.34 $:\ddot{Cl}:$ $\qquad$ $:\ddot{F}\!-\!\overset{}{P}\!-\!\ddot{F}:$ $\quad$ $H\!-\!\overset{}{P}\!-\!H$ $\quad$ $:\ddot{Cl}\!-\!\ddot{S}\!-\!\ddot{Cl}:$
   $:\ddot{Cl}\!-\!\overset{\overset{:\ddot{Cl}:}{|}}{\underset{\underset{:\ddot{Cl}:}{|}}{Si}}\!-\!\ddot{Cl}:$ $\qquad$ $:\ddot{F}:$ $\qquad\quad$ H

   $\qquad\qquad\qquad\qquad\qquad\qquad\qquad\qquad\qquad :\ddot{Cl}\!-\!\ddot{S}\!-\!\ddot{Cl}:$

7.36 (a) $H\!-\!\ddot{O}\!-\!N\!=\!\ddot{O}:$ (b) $\qquad$ $:\ddot{O}:$
   $\qquad\qquad\qquad\qquad\qquad\qquad H\!-\!\ddot{O}\!-\!\overset{\overset{}{}}{Cl}\!-\!\ddot{O}:$

(c) $H\!-\!\ddot{O}\!-\!\overset{\overset{}{}}{\underset{\underset{:O:}{}}{Se}}\!-\!\ddot{O}\!-\!H$
   $\qquad\qquad\ :\ddot{O}:$

7.38 (a) $\underset{F\qquad F}{\overset{F\qquad F}{Te}}$ (c) $:\ddot{F}\!-\!Xe\!-\!\ddot{F}:$

(b) $\underset{F\quad :F:}{\overset{:F:\ :F:\ :F:}{Cl}}$ (d) $\underset{F\qquad F}{\overset{F\qquad F}{Xe}}$

7.41 (a) $:\ddot{Cl}:$ $\qquad\qquad$ (c) $\left[\underset{:O:}{\overset{\overset{:\ddot{O}:}{|}}{\underset{|}{P}}}\right]^{3-}$
   $:\ddot{Cl}\!-\!\overset{\overset{}{}}{\underset{\underset{:\ddot{Cl}:}{}}{Ge}}\!-\!\ddot{Cl}:$ $\qquad\qquad\ddot{O}\!-\!P\!-\!\ddot{O}$

(b) $\left[\underset{:O:\quad O:}{\overset{\overset{:O:}{\|}}{C}}\right]^{2-}$ (d) $\left[\,:\ddot{O}\!-\!\ddot{O}\!:\,\right]^{2-}$

7.46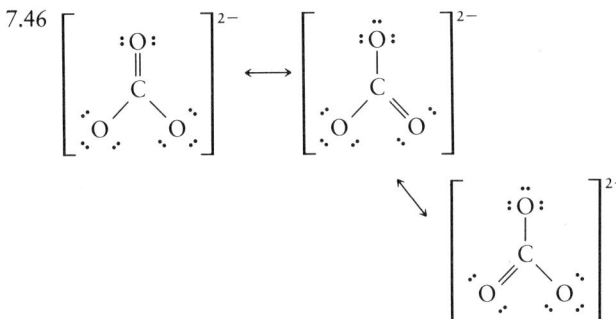

7.47 $CO < CO_2 < CO_3^{2-}$

7.49 $:O\equiv C-\ddot{O}:$ and $:\ddot{O}-C\equiv O:$

7.51 In the nitrate ion, there are three equivalent oxygens, and therefore three resonance structures. In nitric acid, there are two equivalent oxygens, so there are two resonance structures.

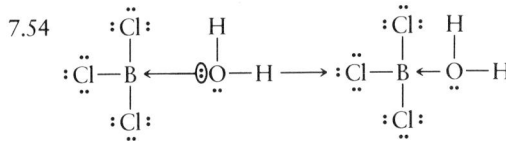

7.54

7.60 (a) nonlinear   (c) trigonal pyramidal
     (b) trigonal bipyramidal   (d) nonlinear

7.63 (a) linear   (c) T-shaped
     (b) square planar   (d) planar triangular

7.72 (a) N (b) I (c) N

7.74 Most polar is Si—F. Partial negative charge is carried by (a) I (b) Cl (c) F (d) N.

7.77 (a) P (b) P (c) F

7.83 $HBr, POCl_3, CH_2O$

CHAPTER 8

PRACTICE EXERCISES

1. The H—Cl bond is formed by the overlap of the half-filled $1s$ orbital of H and the half-filled $3p$ orbital of Cl;

Cl (in HCl) (colored arrow is H electron)

2. The half-filled $1s$ orbitals of the H atoms overlap with the half-filled $3p$ orbitals of the P atom. The expected bond angle is 90°.

P (in PH_3) (colored arrows are H electrons)

3. (a) sp^3 (b) sp^3d

4. (a) sp^3 (b) sp^3d

5. (a) sp^3d^2

(b) P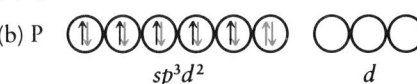
sp^3d^2 d
(colored arrows are Cl electrons)
(c) octahedral

6. $\Delta H_{atom} = 1.94 \times 10^3$ kJ/mol

7. $\Delta H_f^\circ = -77$ kJ/mol

8. NO has 11 valence electrons

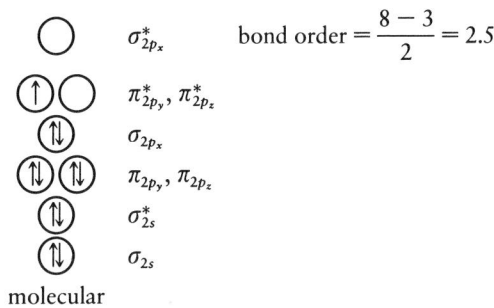 bond order $= \dfrac{8-3}{2} = 2.5$

molecular orbitals

REVIEW EXERCISES—Chapter 8

8.8 The $1s$ orbitals of hydrogen overlap with mutually perpendicular p orbitals of the selenium.

Se (in H_2Se) 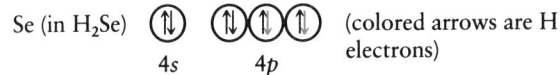 (colored arrows are H electrons)
$4s$ $4p$

8.15 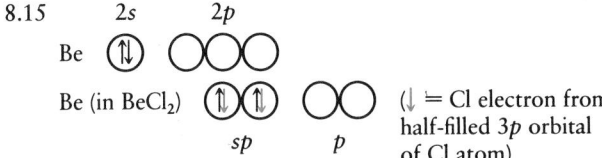 ($\downarrow$ = Cl electron from half-filled $3p$ orbital of Cl atom)
sp p

8.17 sp^3d^2 octahedral

Sn (in $SnCl_6^{2-}$)
sp^3d^2 $5d$

($\uparrow$ = Cl electrons)

8.22 90°, because that is the angle between p atomic orbitals

8.34 The C—C≡C—C arrangement of bonds must be linear.

8.38 Each carbon in C_2Cl_4 is sp^2 hybridized. The double bond consists of a σ bond formed by overlap of an sp^2 orbital from each carbon, plus a π bond formed by the overlap of the unhybridized p orbital on each carbon. Each of the C—Cl bonds is a σ bond formed by overlap of a carbon sp^2 orbital with a half-filled p orbital of chlorine. The molecule is planar, and the expected bond angles are 120°.

8.44 1.17×10^3 kJ/mol

8.46 598 kJ/mol

8.49 307 kJ/mol

8.51 Calculated $\Delta H_f^\circ = -638$ kJ/mol. Percentage difference between calculated and measured $\Delta H_f^\circ = (80/718) \times 100 = 11\%$.

8.53 $+228$ kJ/mol

8.55 CF_4

8.62 For Li_2, the net bond order is 1, but for Be_2, the net bond order is 0.

8.64 (a) 2.5 (b) 1.5 (c) 1.5

CHAPTER 9

PRACTICE EXERCISES

1. With largest side down, pressure = 0.0424 lb/in.². With narrow side down, pressure = 0.164 lb/in.².

2. 0.974 atm, 98.7 kPa, 740 mm Hg

3. 818 torr

4. 719 torr = 0.946 atm, 818 torr = 1.08 atm

5. 1.48×10^3 torr

6. 190 mL

7. 159 torr

8. 732 torr, 299 mL

9. The sample should be cooled to $-29\ ^\circ$C

10. Since $P_1 = P_2$, cancel them from Equation 9.20, which leaves Equation 9.15.

11. When $V_1 = V_2$, they can be canceled from Equation 9.20, which leaves the equation $P_1/T_1 = P_2/T_2$. Since each side of this equation can equal the same constant, say, C^*, we see that, in general, $P/T = C^*$. This is the same as writing $P \propto T$ (providing that the mass and the volume of the gas are kept constant).

12. 2.74×10^3 torr or 3.61 atm

13. (a) $P_2 = P_1 \times \dfrac{V_1}{V_2} \times \dfrac{T_2}{T_1}$

    (b) $V_2 = V_1 \times \dfrac{P_1}{P_2} \times \dfrac{T_2}{T_1}$

    (c) $T_2 = T_1 \times \dfrac{V_2}{V_1} \times \dfrac{P_2}{P_1}$

14. 688 torr

15. 15.7 L

16. 2.09 g/mol. The gas most likely is H_2.

17. 3.8×10^2 mL

18. 57.8 mol or 983 g or (at STP) 1.29×10^3 L of NH_3

19. Hydrogen effuses 2.83 times faster than methane (meaning that it might readily leak through valves and tiny cracks that methane only slowly would leak through).

20. Hydrogen effuses 3.98 times faster than oxygen.

REVIEW EXERCISES—Chapter 9

9.8 1.39 lb/in.²

9.10 744 torr

9.12 0.987 atm

9.14 838 torr

9.16 79.287 kPa for N_2 and 21.332 kPa for O_2

9.25 1.14×10^3 torr

9.27 (a) 200 torr (b) 197 torr

9.29 The data to be plotted are as follows.

Pressure (torr)	Volume (mL)	Pressure (torr)	Volume (mL)
780	97.4	900	84.5
800	95.0	920	82.7
820	92.7	940	80.9
840	90.5	960	79.2
860	88.4	980	77.6
880	86.4	1000	76.0

9.31 706 torr

9.33 265 mL dry CO

9.35 246 mL

9.37 The effusion rate of N_2 is 1.25 times faster than that of CO_2.

9.43 807 torr

9.45 2.79 L

9.47 $-54\ ^\circ$C

9.49 727 torr

9.51 555 K or 282 °C

9.53 24.5 L

9.55 4.09 atm

9.57 204 atm

9.59 154 mol

9.61 4.15 mol or 133 g of O_2 consumed

9.63 (a) 0.714 g/L (b) 1.43 g/L (c) 9.02×10^{-2} g/L

9.65 1.28 g/L

9.67 27.8 g/mol

9.69 88.2 g/mol

9.71 5.00 L

9.73 60 L

9.75 (a) $Zn(s) + 2HCl(aq) \rightarrow H_2(g) + ZnCl_2(aq)$
     (b) 0.409 mol H_2, 26.7 g Zn
     (c) 0.818 mol HCl
     (d) 136 mL of 6.00 M HCl

9.77 (a) 7.64×10^4 L NH_3 (at STP)
     (b) 1.71×10^3 mol H_2SO_4
     (c) 285 L 6 M H_2SO_4

9.79 (a) 6.47×10^4 L NH_3 (at STP)
     (b) 2.89×10^3 mol NH_3
     (c) 49.1 kg NH_3

9.81 (a) 40.00% S and 60.00% O
     (b) empirical formula, SO_3
     (c) molecular formula, SO_3

9.83 (a) 8.65% N
(b) 74.02% C, 7.46% H; empirical formula, $C_{10}H_{12}NO$
(c) molecular formula, $C_{20}H_{24}N_2O_2$

CHAPTER 10

PRACTICE EXERCISES

1. The number of molecules in the vapor will increase and the number in the liquid will decrease.

2. Adding heat will shift the equilibrium to the right, which produces more vapor. More vapor will exert a higher pressure, so the vapor pressure increases with increasing temperature.

3. approximately 75 °C

4. covalent solid

5. molecular solid

6. Sublimation will occur at −10 °C.

7. liquid

REVIEW EXERCISES — Chapter 10

10.9 C_8H_{18}, because the longer molecule has more places along its chain to be attracted to other molecules by London forces

10.13 London forces are present in all of them. Dipole–dipole attractions occur in (a), (c), and (e).

10.22 H_2O, because of its stronger intermolecular attractions (hydrogen bonds)

10.24 H_2O molecules attract each other much more strongly than they are attracted to the nonpolar polyethylene surface.

10.38 2.44×10^3 kJ

10.44 The energy released by condensation warms the air. Warm air rises and this movement of air creates wind currents.

10.45 (a) 0 °C (b) 50.0 g

10.51 ether < acetone < benzene < water < acetic acid

10.64 about 73 °C

10.66 diethyl ether < ethanol < water < ethylene glycol

10.71 The boiling point of H_2S is lower than that of H_2Se because the H_2S molecule is smaller and is not as easily polarized, so in H_2S the London forces are weaker. In H_2O there are strong hydrogen bonds, so it has a higher boiling point than H_2S, in which there are weaker dipole–dipole and London forces.

10.82 one

10.84 Unit cell edge length equals 407 pm.

10.88 molecular

10.90 covalent

10.93 metallic

10.106 (a) solid (b) gas (c) liquid (d) solid, liquid, gas

10.111 The critical temperature of H_2 is below room temperature, which is why at room temperature it cannot be liquefied by the application of pressure. The critical temperature of butane is above room temperature.

CHAPTER 11

PRACTICE EXERCISES

1. (a) $MgCl_2(s) \rightarrow Mg^{2+}(aq) + 2Cl^-(aq)$
(b) $Al(NO_3)_3(s) \rightarrow Al^{3+}(aq) + 3NO_3^-(aq)$
(c) $Na_2CO_3(s) \rightarrow 2Na^+(aq) + CO_3^{2-}(aq)$
(d) $(NH_4)_2SO_4(s) \rightarrow 2NH_4^+(aq) + SO_4^{2-}(aq)$

2. $HCHO_2(aq) + H_2O \rightleftharpoons H_3O^+(aq) + CHO_2^-(aq)$

3. $NH_3(aq) + H_2O \rightleftharpoons NH_4^+(aq) + OH^-(aq)$

4. *Molecular equation*

$$2NaOH(aq) + MgCl_2(aq) \longrightarrow Mg(OH)_2(s) + 2NaCl(aq)$$

Ionic equation

$$2Na^+(aq) + 2OH^-(aq) + Mg^{2+}(aq) + 2Cl^-(aq) \longrightarrow$$
$$Mg(OH)_2(s) + 2Na^+(aq) + 2Cl^-(aq)$$

Net ionic equation

$$Mg^{2+}(aq) + 2OH^-(aq) \longrightarrow Mg(OH)_2(s)$$

5. (a) $Ag^+(aq) + Cl^-(aq) \rightarrow AgCl(s)$
(b) $S^{2-}(aq) + Pb^{2+}(aq) \rightarrow PbS(s)$
(c) no reaction

6. (a) $2HClO_4(aq) + Ni(OH)_2(s) \rightarrow Ni(ClO_4)_2(aq) + 2H_2O$
$2H^+(aq) + 2ClO_4^-(aq) + Ni(OH)_2(s) \rightarrow Ni^{2+}(aq) + 2ClO_4^-(aq) + 2H_2O$
$2H^+(aq) + Ni(OH)_2(s) \rightarrow Ni^{2+}(aq) + 2H_2O$
(b) $Al_2O_3(s) + 3H_2SO_4(aq) \rightarrow Al_2(SO_4)_3(aq) + 3H_2O$
$Al_2O_3(s) + 6H^+(aq) + 3SO_4^{2-}(aq) \rightarrow 2Al^{3+}(aq) + 3SO_4^{2-}(aq) + 3H_2O$
$Al_2O_3(s) + 6H^+(aq) \rightarrow 2Al^{3+}(aq) + 3H_2O$
(c) $H_2CO_3(aq) + 2KOH(aq) \rightarrow K_2CO_3(aq) + 2H_2O$
$H_2CO_3(aq) + 2K^+(aq) + 2OH^-(aq) \rightarrow 2K^+(aq) + CO_3^{2-}(aq) + 2H_2O$
$H_2CO_3(aq) + 2OH^-(aq) \rightarrow CO_3^{2-}(aq) + 2H_2O$

7. $NaF(aq) + HCl(aq) \rightarrow HF(aq) + NaCl(aq)$
$Na^+(aq) + F^-(aq) + H^+(aq) + Cl^-(aq) \rightarrow HF(aq) + Na^+(aq) + Cl^-(aq)$
$F^-(aq) + H^+(aq) \rightarrow HF(aq)$

8. $Ba(OH)_2(aq) + H_2SO_4(aq) \rightarrow BaSO_4(s) + 2H_2O$
The ionic and net ionic equations are the same:

$$Ba^{2+}(aq) + 2OH^-(aq) + 2H^+(aq) + SO_4^{2-}(aq) \longrightarrow$$
$$BaSO_4(s) + 2H_2O$$

9. (a) $(NH_4)_2SO_4(aq) + 2NaOH(aq) \rightarrow 2NH_3(aq) + Na_2SO_4(aq) + 2H_2O$
$2NH_4^+(aq) + SO_4^{2-}(aq) + 2Na^+(aq) + 2OH^-(aq) \rightarrow 2NH_3(aq) + SO_4^{2-}(aq) + 2Na^+(aq) + 2H_2O$
$NH_4^+(aq) + OH^-(aq) \rightarrow NH_3(aq) + H_2O$
(b) $(NH_4)_2SO_4(aq) + Ba(OH)_2(aq) \rightarrow 2NH_3(aq) + BaSO_4(s) + 2H_2O$
The ionic and net ionic equations are the same:

$$2NH_4^+(aq) + SO_4^{2-}(aq) + Ba^{2+}(aq) + 2OH^-(aq) \longrightarrow$$
$$2NH_3(aq) + BaSO_4(s) + 2H_2O$$

10. (a) 3.00×10^{-2} mol HCl (b) 66.7 mL

11. 60.0 mL

12. 3.45×10^{-2} mol Fe^{3+} and 5.18×10^{-2} mol SO_4^{2-}

13. 1.20×10^{-2} mol of $BaSO_4$ will form. Final concentrations: 6.02×10^{-2} M SO_4^{2-}, 0.300 M Mg^{2+}, and 0.480 M Cl^-.

14. (a) 5.41×10^{-3} mol Ca^{2+} (d) 0.601 g $CaCl_2$
    (b) 5.41×10^{-3} mol Ca^{2+} (e) 30.1% $CaCl_2$
    (c) 5.41×10^{-3} mol $CaCl_2$

15. 0.178 M

16. 0.0220 M HCl, 8.03×10^{-2}% HCl

17. for complete neutralization, 2 eq/mol; for partial neutralization (neutralization of one proton), 1 eq/mol

18. 3.00 eq/mol

19. (a) 32.7 g (b) 49.0 g

20. 12.2 g H_2SO_4

21. (a) 0.20 N $Ba(OH)_2$ (b) 0.30 M H_2SO_4

22. 33.8 mL

23. 2.10×10^{-2} eq HNO_3

24. (a) 6.32×10^{-3} eq (b) 6.32×10^{-3} eq (c) 158 g/eq

25. (a) OH^- (c) NO_2^- (e) HPO_4^{2-} (g) H^-
    (b) I^- (d) $H_2PO_4^-$ (f) PO_4^{3-} (h) NH_3

26. (a) H_2O_2 (d) $HC_2H_3O_2$ (g) H_3PO_4
    (b) HSO_4^- (e) NH_3 (h) $H_2PO_4^-$
    (c) HPO_4^{2-} (f) NH_4^+

27.   base   conjugate pair   acid

    $HCO_3^-(aq) + H_2PO_4^-(aq) \rightarrow H_2CO_3(aq) + HPO_4^{2-}(aq)$

                    acid   conjugate pair   base

28.   base   conjugate pair   acid

    $PO_4^{3-}(aq) + HC_2H_3O_2(aq) \rightarrow HPO_4^{2-}(aq) + C_2H_3O_2^-(aq)$

                    acid   conjugate pair   base

29. HPO_4^{2-} neutralizes acid:

    $HPO_4^{2-}(aq) + H_3O^+(aq) \longrightarrow H_2PO_4^-(aq) + H_2O$

    HPO_4^{2-} neutralizes base:

    $HPO_4^{2-}(aq) + OH^-(aq) \longrightarrow PO_4^{3-}(aq) + H_2O$

30. (a) $HBrO_4$ (b) H_2SeO_4

31. $HClO_4 > HClO_3 > HClO_2 > HClO$

32. (a) HBr (b) H_2Te

33. CO_2 is the Lewis acid and OH^- is the Lewis base.

34. $[Ag(S_2O_3)_2]^{3-}$, $(NH_4)_3[Ag(S_2O_3)_3]$

35. $AlCl_3 \cdot 6H_2O$, $[Al(H_2O)_6]^{3+}$

REVIEW EXERCISES—Chapter 11

11.19   (a) $2NH_4^+(aq) + CO_3^{2-}(aq) + Mg^{2+}(aq) + 2Cl^-(aq) \rightarrow$
            $2NH_4^+(aq) + 2Cl^-(aq) + MgCO_3(s)$
            $CO_3^{2-}(aq) + Mg^{2+}(aq) \rightarrow MgCO_3(s)$

(b) $Cu^{2+}(aq) + 2Cl^-(aq) + 2Na^+(aq) + 2OH^-(aq) \rightarrow$
    $Cu(OH)_2(s) + 2Na^+(aq) + 2Cl^-(aq)$
    $Cu^{2+}(aq) + 2OH^-(aq) \rightarrow Cu(OH)_2(s)$

(c) $3Fe^{2+}(aq) + 3SO_4^{2-}(aq) + 6Na^+(aq) +$
    $2PO_4^{3-}(aq) \rightarrow Fe_3(PO_4)_2(s) + 6Na^+(aq) +$
    $3SO_4^{2-}(aq)$
    $3Fe^{2+}(aq) + 2PO_4^{3-}(aq) \rightarrow Fe_3(PO_4)_2(s)$

(d) $2Ag^+(aq) + 2C_2H_3O_2^-(aq) + Ni^{2+}(aq) +$
    $2Cl^-(aq) \rightarrow 2AgCl(s) + Ni^{2+}(aq) + 2C_2H_3O_2^-(aq)$
    $Ag^+(aq) + Cl^-(aq) \rightarrow AgCl(s)$

11.26   a, b, and d are soluble.

11.28   (a) $Na_2SO_3(aq) + Ba(NO_3)_2(aq) \rightarrow$
            $BaSO_3(s) + 2NaNO_3(aq)$
            $2Na^+(aq) + SO_3^{2-}(aq) + Ba^{2+}(aq) + 2NO_3^-(aq) \rightarrow$
            $BaSO_3(s) + 2Na^+(aq) + 2NO_3^-(aq)$
            $SO_3^{2-}(aq) + Ba^{2+}(aq) \rightarrow BaSO_3(s)$

(b) $K_2S(aq) + ZnCl_2(aq) \rightarrow ZnS(s) + 2KCl(aq)$
    $2K^+(aq) + S^{2-}(aq) + Zn^{2+}(aq) + 2Cl^-(aq) \rightarrow$
    $2K^+(aq) + 2Cl^-(aq) + ZnS(s)$
    $S^{2-}(aq) + Zn^{2+}(aq) \rightarrow ZnS(s)$

(c) $2NH_4Br(aq) + Pb(C_2H_3O_2)_2(aq) \rightarrow$
    $2NH_4C_2H_3O_2(aq) + PbBr_2(s)$
    $2NH_4^+(aq) + 2Br^-(aq) + Pb^{2+}(aq) +$
    $2C_2H_3O_2^-(aq) \rightarrow 2NH_4^+(aq) + 2C_2H_3O_2^-(aq) +$
    $PbBr_2(s)$
    $2Br^-(aq) + Pb^{2+}(aq) \rightarrow PbBr_2(s)$

(d) no reaction (all ions cancel)

11.30   (a) $3HNO_3(aq) + Cr(OH)_3(s) \rightarrow Cr(NO_3)_3(aq) + 3H_2O$
            $3H^+(aq) + 3NO_3^-(aq) + Cr(OH)_3(s) \rightarrow Cr^{3+}(aq) +$
            $3NO_3^-(aq) + 3H_2O$
            $3H^+(aq) + Cr(OH)_3(s) \rightarrow Cr^{3+}(aq) + 3H_2O$

(b) $HClO_4(aq) + NaOH(aq) \rightarrow NaClO_4(aq) + H_2O$
    $H^+(aq) + ClO_4^-(aq) + Na^+(aq) + OH^-(aq) \rightarrow$
    $Na^+(aq) + ClO_4^-(aq) + H_2O$
    $H^+(aq) + OH^-(aq) \rightarrow H_2O$

(c) $Cu(OH)_2(s) + 2HC_2H_3O_2(aq) \rightarrow$
    $Cu(C_2H_3H_2)_2(aq) + 2H_2O$
    $Cu(OH)_2$ is insoluble and so is undissociated.
    $HC_2H_3O_2$, acetic acid, is a weak acid and so is
    un-ionized. $Cu(C_2H_3O_2)_2$ is a soluble salt and so is
    dissociated into its ions. H_2O, of course, is a
    nonelectrolyte. Thus, both the ionic and net ionic
    equations are the same and as follows.
    $Cu(OH)_2(s) + 2HC_2H_3O_2(aq) \longrightarrow$
    $\qquad\qquad Cu^{2+}(aq) + 2C_2H_3O_2^-(aq) + 2H_2O$

(d) $ZnO(s) + 2HBr(aq) \rightarrow ZnBr_2(aq) + H_2O$
    $ZnO(s) + 2H^+(aq) + 2Br^-(aq) \rightarrow$
    $Zn^{2+}(aq) + 2Br^-(aq) + H_2O$
    $ZnO(s) + 2H^+(aq) \rightarrow Zn^{2+}(aq) + H_2O$

11.35   (a) Needed: a soluble bicarbonate and a strong acid.
            For example:
            $NaHCO_3(aq) + HCl(aq) \longrightarrow$
            $\qquad\qquad NaCl(aq) + H_2O + CO_2(g)$

(b) Needed: a soluble Fe^{2+} salt and a strong base. For
    example:
    $Fe(NO_3)_2(aq) + 2NaOH(aq) \longrightarrow$
    $\qquad\qquad Fe(OH)_2(s) + 2NaNO_3(aq)$

(c) Needed: a soluble Ba^{2+} salt and a soluble sulfite. For example:

$$Ba(NO_3)_2(aq) + Na_2SO_3(aq) \longrightarrow$$
$$BaSO_3(s) + 2NaNO_3(aq)$$

(d) Needed: a soluble Ag^+ salt and a soluble sulfide. For example:

$$2AgNO_3(aq) + Na_2S(aq) \longrightarrow$$
$$Ag_2S(s) + 2NaNO_3(aq)$$

(e) Needed: a strong acid whose anion forms a soluble salt with Zn^{2+}. For example:

$$ZnO(s) + 2HCl(aq) \longrightarrow ZnCl_2(aq) + H_2O$$

11.36 (a) Not a good choice. The Cu^{2+} would not readily migrate out of its insoluble reactant, $Cu(OH)_2$, into another insoluble material.
(b) A good choice. Both reactants are soluble in water so they are dissociated and their ions are free to move around.
(c) Not a good choice. The CO_3^{2-} would not readily migrate out of its insoluble reactant, $CaCO_3$, into another insoluble material.
(d) A good choice. Both reactants are soluble in water so they are dissociated and their ions are free to move around.
(e) Not a good choice. Both reactants are insoluble in water.

11.38 (a) 3.25×10^{-2} mol K^+ and 3.25×10^{-2} mol OH^-
(b) 1.9×10^{-2} mol Ca^{2+} and 3.8×10^{-2} mol Cl^-
(c) 2.0×10^{-2} mol NH_4^+ and 1.0×10^{-2} mol CO_3^{2-}
(d) 2.8×10^{-2} mol Al^{3+} and 4.2×10^{-2} mol SO_4^{2-}

11.40 2.46 g

11.42 0.63 g

11.43 1.9×10^2 mL CO_2

11.45 100 mL

11.47 (a) 9.00×10^{-3} mol AgCl
(b) 0.150 M NO_3^-, 0.200 M Na^+, and 5.0×10^{-2} M Cl^-

11.49 8.06 mL

11.52 0.120 M

11.54 2.94×10^{-3} mol

11.56 (a) 0.168 g Cl (b) $FeCl_2$

11.58 63.2%

11.60 (a) 1 (b) 2 (c) 3 (d) 1

11.62 (a) 0.250 eq (b) 0.150 eq (c) 0.440 eq (d) 0.350 eq

11.64 (a) 0.375 eq (b) 0.317 eq (c) 0.880 eq (d) 0.405 eq

11.65 17.2 g

11.67 1.00 N H_2SO_4

11.69 0.214 g

11.71 (a) 0.123 N (b) 6.15×10^{-2} M H_2SO_4

11.73 (a) 63.9 g (b) three

11.77 (a)

conjugate
acid   *pair*   *base*

$$HNO_3 + N_2H_4 \rightleftharpoons NO_3^- + N_2H_5^+$$

base conjugate *acid*
pair

(b) *base*

conjugate
pair   *acid*

$$NH_3 + N_2H_5^+ \rightleftharpoons NH_4^+ + N_2H_5$$

acid conjugate *base*
pair

(c) *acid*

conjugate
pair   *base*

$$H_2PO_4^- + CO_3^{2-} \rightleftharpoons HPO_4^{2-} + HCO_3^-$$

base conjugate *acid*
pair

(d) *acid*

conjugate
pair   *base*

$$HIO_3 + HC_2O_4^- \rightleftharpoons IO_3^- + H_2C_2O_4$$

base conjugate *acid*
pair

11.93

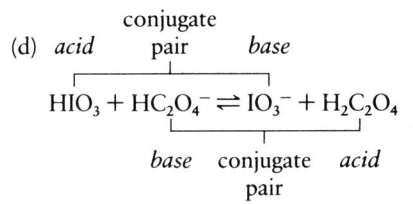

11.97 Cu^{2+} accepts the electron pairs of O atoms of the H_2O molecules as the new bonds in the complex ion form.
(a) Lewis acid: Cu^{2+}, Lewis base: H_2O
(b) ligand: H_2O
(c) water
(d) O is the donor atom because it supplies the electron pair.
(e) copper(II) ion

11.107 Net charge is $1-$. Formula is $Co(EDTA)^-$.

11.110 The bonds are stronger in $[Cu(NH_3)_4]^{2+}$. NH_3 displaces H_2O as the ligand. A solution containing the hydrated copper(II) ion changes to a deeper blue when aqueous ammonia is added and the ammoniated copper(II) ion forms.

CHAPTER 12

PRACTICE EXERCISES

1. Ca is oxidized; Cl_2 is reduced.

2. $Ni = +2$; $Cl = -1$
$Mg = +2$; $Ti = +4$; $O = -2$
$K = +1$; $Cr = +6$; $O = -2$
$S = +6$; $O = -2$

3. Average oxidation number $= -2$

4. (a) $2KCl + MnO_2 + 2H_2SO_4 \rightarrow K_2SO_4 + MnSO_4 + Cl_2 + 2H_2O$

   (b) $2KMnO_4 + 10FeSO_4 + 8H_2SO_4 \rightarrow K_2SO_4 + 2MnSO_4 + 5Fe_2(SO_4)_3 + 8H_2O$

   (c) $4Zn + 10HNO_3 \rightarrow 4Zn(NO_3)_2 + NH_4NO_3 + 3H_2O$

5. $3Cu + 8H^+ + 2NO_3^- \rightarrow 3Cu^{2+} + 2NO + 4H_2O$

6. $4OH^- + 2MnO_4^- + 3C_2O_4^{2-} \rightarrow 2MnO_2 + 2H_2O + 6CO_3^{2-}$

7. (a) *Molecular equation*

$$Mg(s) + 2HCl(aq) \longrightarrow MgCl_2(aq) + H_2(g)$$

   Ionic equation

$$Mg(s) + 2H^+(aq) + 2Cl^-(aq) \longrightarrow Mg^{2+}(aq) + 2Cl^-(aq) + H_2(g)$$

   Net ionic equation

$$Mg(s) + 2H^+(aq) \longrightarrow Mg^{2+}(aq) + H_2(g)$$

   (b) *Molecular equation*

$$2Al(s) + 6HCl(aq) \longrightarrow 2AlCl_3(aq) + 3H_2(g)$$

   Ionic equation

$$2Al(s) + 6H^+(aq) + 6Cl^-(aq) \longrightarrow 2Al^{3+}(aq) + 6Cl^-(aq) + 3H_2(g)$$

   Net ionic equation

$$2Al(s) + 6H^+(aq) \longrightarrow 2Al^{3+}(aq) + 3H_2(g)$$

8. (a) $Mg(s) + Cu^{2+}(aq) \rightarrow Cu(s) + Mg^{2+}(aq)$
   (b) no reaction

9. $2C_4H_{10} + 13O_2 \rightarrow 8CO_2 + 10H_2O$

10. $C_2H_5OH + 3O_2 \rightarrow 2CO_2 + 3H_2O$

11. $4Fe + 3O_2 \rightarrow 2Fe_2O_3$

12. $P_4 + 5O_2 \rightarrow P_4O_{10}$

13. (a) $5Fe^{2+}(aq) + MnO_4^-(aq) + 8H^+(aq) \rightarrow 5Fe^{3+}(aq) + Mn^{2+}(aq) + 4H_2O$
    (b) $6Fe^{2+}(aq) + Cr_2O_7^{2-}(aq) + 14H^+(aq) \rightarrow 6Fe^{3+}(aq) + 2Cr^{3+}(aq) + 7H_2O$

14. (a) $H_2O + HSO_3^-(aq) \rightarrow SO_4^{2-}(aq) + 3H^+(aq) + 2e^-$
    (b) $H_2O + H_2SO_3(aq) \rightarrow SO_4^{2-}(aq) + 4H^+(aq) + 2e^-$

15. $3S_2O_8^{2-}(aq) + 7H_2O + 2Cr^{3+}(aq) \rightarrow Cr_2O_7^{2-}(aq) + 6SO_4^{2-}(aq) + 14H^+(aq)$, 57.9 g $K_2S_2O_8$

16. 6.00×10^2 mL

17. (a) $5Sn^{2+}(aq) + 2MnO_4^-(aq) + 16H^+(aq) \rightarrow 5Sn^{4+}(aq) + 2Mn^{2+}(aq) + 8H_2O$
    (b) 0.120 g Sn
    (c) 40.0% Sn
    (d) 50.7% SnO_2

18. (a) 7.05×10^{-4} mol I_3^-
    (b) 7.05×10^{-4} mol OCl^-
    (c) 0.705 M OCl^-

19. 2 eq/mol

20. 54.0 g/eq

21. 14.0 g Na_2SO_3

22. (a) 5 eq/mol (b) 1.00 N

23. 0.287 N

24. 2.50×10^{-3} mol Sn

25. (a) 5.90×10^{-3} eq $KMnO_4$
    (b) 5.90×10^{-3} eq Na_2SO_3
    (c) 2.00×10^{-2} eq Na_2SO_3
    (d) 1.41×10^{-2} eq OCl^-
    (e) 7.05×10^{-3} mol OCl^-
    (f) 5.25% NaOCl

REVIEW EXERCISES—Chapter 12

12.9   (a) -2 (b) $+4$ (c) zero (d) -3

12.12 (a) $N = +5$, $O = -2$
       (b) $S = +4$, $O = -2$
       (c) $N = +3$, $O = -2$
       (d) $Cr = +6$, $O = -2$

12.14 No, there are no changes in oxidation numbers.

12.16 (a) $Ca = +2$, $V = +5$, $O = -2$
       (b) $Sn = +4$, $Cl = -1$
       (c) $Mn = +6$, $O = -2$
       (d) $Mn = +4$, $O = -2$

12.19 (a) $Br = +1$, $Cl = -1$ (b) $S = +2$, $Cl = -1$

12.21 (a) $2HNO_3 + 3H_3AsO_3 \rightarrow 2NO + 3H_3AsO_4 + H_2O$
       (b) $NaI + 3HOCl \rightarrow NaIO_3 + 3HCl$
       (c) $2KMnO_4 + 5H_2C_2O_4 + 3H_2SO_4 \rightarrow 10CO_2 + K_2SO_4 + 2MnSO_4 + 8H_2O$
       (d) $6H_2SO_4 + 2Al \rightarrow Al_2(SO_4)_3 + 3SO_2 + 6H_2O$

12.23 (a) $Cu + 2H_2SO_4 \rightarrow CuSO_4 + SO_2 + 2H_2O$
       (b) $3SO_2 + 2HNO_3 + 2H_2O \rightarrow 3H_2SO_4 + 2NO$
       (c) $5H_2SO_4 + 4Zn \rightarrow 4ZnSO_4 + H_2S + 4H_2O$
       (d) $I_2 + 10HNO_3 \rightarrow 2HIO_3 + 10NO_2 + 4H_2O$

12.25 (a) $6H^+ + BiO_3^- + 2e^- \rightarrow Bi^{3+} + 3H_2O$ (reduction)
       (b) $10H^+ + NO_3^- + 8e^- \rightarrow NH_4^+ + 3H_2O$ (reduction)
       (c) $2H_2O + Pb^{2+} \rightarrow PbO_2 + 4H^+ + 2e^-$ (oxidation)
       (d) $6H_2O + Cl_2 \rightarrow 2ClO_3^- + 12H^+ + 10e^-$ (oxidation)

12:27 (a) $2H^+ + OCl^- + 2S_2O_3^{2-} \rightarrow S_4O_6^{2-} + Cl^- + H_2O$
       (b) $4H^+ + 3NO_3^- + Cu \rightarrow 2NO_2 + Cu^{2+} + 2H_2O$
       (c) $3AsO_3^{3-} + IO_3^- \rightarrow I^- + 3AsO_4^{3-}$
       (d) $4H^+ + Zn + SO_4^{2-} \rightarrow Zn^{2+} + SO_2 + 2H_2O$
       (e) $10H^+ + NO_3^- + 4Zn \rightarrow 4Zn^{2+} + NH_4^+ + 3H_2O$
       (f) $2Cr^{3+} + 4H^+ + 3BiO_3^- \rightarrow Cr_2O_7^{2-} + 3Bi^{3+} + 2H_2O$
       (g) $I_2 + H_2O + 5OCl^- \rightarrow 2IO_3^- + 2H^+ + 5Cl^-$
       (h) $2Mn^{2+} + 5BiO_3^- + 14H^+ \rightarrow 2MnO_4^- + 5Bi^{3+} + 7H_2O$
       (i) $3H_3AsO_3 + 8H^+ + Cr_2O_7^{2-} \rightarrow 3H_3AsO_4 + 2Cr^{3+} + 4H_2O$
       (j) $2I^- + 3H^+ + HSO_4^- \rightarrow SO_2 + 2H_2O + I_2$

12.30 (a) $6OH^- + 4NBr_3 \rightarrow 2N_2 + 6HOBr + 6Br^-$
       (b) $2OH^- + ClNO_2 \rightarrow NO_3^- + Cl^- + H_2O$
       (c) $3Cl_2 + 6OH^- \rightarrow ClO_3^- + 5Cl^- + 3H_2O$
       (d) $2H_2S + H_2SeO_3 \rightarrow 2S + Se + 3H_2O$
       (e) $2H_2O + 2SO_3^{2-} + MnO_2 \rightarrow Mn^{2+} + S_2O_6^{2-} + 4OH^-$
       (f) $2OH^- + BrO_3F \rightarrow BrO_4^- + F^- + H_2O$
       (g) $3H_2O + 6I^- + XeO_3 \rightarrow 3I_2 + Xe + 6OH^-$
       (h) $2HXeO_4^- + 2OH^- \rightarrow XeO_6^{4-} + Xe + O_2 + 2H_2O$
       (i) $(CN)_2 + 2OH^- \rightarrow CN^- + OCN^- + H_2O$

12.31 (a) *Ionic equation*

$$Mn(s) + 2H^+(aq) + 2Cl^-(aq) \longrightarrow$$
$$Mn^{2+}(aq) + H_2(g) + 2Cl^-(aq)$$

Net ionic equation

$$Mn(s) + 2H^+(aq) \longrightarrow Mn^{2+}(aq) + H_2(g)$$

(b) *Ionic equation*

$$Cd(s) + 2H^+(aq) + 2Cl^-(aq) \longrightarrow$$
$$Cd^{2+}(aq) + H_2(g) + 2Cl^-(aq)$$

Net ionic equation

$$Cd(s) + 2H^+(aq) \longrightarrow Cd^{2+}(aq) + H_2(g)$$

(c) *Ionic equation*

$$Sn(s) + 2H^+(aq) + 2Cl^-(aq) \longrightarrow$$
$$Sn^{2+}(aq) + H_2(g) + 2Cl^-(aq)$$

Net ionic equation

$$Sn(s) + 2H^+(aq) \longrightarrow Sn^{2+}(aq) + H_2(g)$$

(d) *Ionic equation*

$$Ni(s) + 2H^+(aq) + 2Cl^-(aq) \longrightarrow$$
$$Ni^{2+}(aq) + H_2(g) + 2Cl^-(aq)$$

Net ionic equation

$$Ni(s) + 2H^+(aq) \longrightarrow Ni^{2+}(aq) + H_2(g)$$

(e) *Ionic equation*

$$2Cr(s) + 6H^+(aq) + 6Cl^-(aq) \longrightarrow$$
$$2Cr^{3+}(aq) + 3H_2(g) + 6Cl^-(aq)$$

Net ionic equation

$$2Cr(s) + 6H^+(aq) \longrightarrow 2Cr^{3+}(aq) + 3H_2(g)$$

12.34 (a) $3Ag + 4HNO_3 \rightarrow 3AgNO_3 + NO + 2H_2O$
(b) $Ag + 2HNO_3 \rightarrow AgNO_3 + NO_2 + H_2O$

12.36 $Cu + 2H^+ + H_2SO_4 \rightarrow SO_2 + 2H_2O + Cu^{2+}$

12.37 $Cu < Pb < Fe < Al$

12.39 (c) and (d)

12.41 (a) no reaction
(b) $2Cr(s) + 3Pb^{2+}(aq) \rightarrow 2Cr^{3+}(aq) + 3Pb(s)$
(c) $2Ag^+(aq) + Fe(s) \rightarrow 2Ag(s) + Fe^{2+}(aq)$
(d) $3Ag(s) + Au^{3+}(aq) \rightarrow Au(s) + 3Ag^+(aq)$

12.53 (a) $2C_6H_6 + 15O_2 \rightarrow 12CO_2 + 6H_2O$
(b) $C_3H_8 + 5O_2 \rightarrow 3CO_2 + 4H_2O$
(c) $C_{21}H_{44} + 32O_2 \rightarrow 21CO_2 + 22H_2O$
(d) $2C_{12}H_{26} + 37O_2 \rightarrow 24CO_2 + 26H_2O$
(e) $2C_{18}H_{38} + 55O_2 \rightarrow 36CO_2 + 38H_2O$

12.55 $C_6H_{12}O_6 + 6O_2 \rightarrow 6CO_2 + 6H_2O$

12.58 1.95×10^5 kJ released

12.60 CO_2, H_2O, and SO_2
$2C_2H_6S + 9O_2 \rightarrow 4CO_2 + 2SO_2 + 6H_2O$

12.62 (a) $2Zn + O_2 \rightarrow 2ZnO$
(b) $4Al + 3O_2 \rightarrow 2Al_2O_3$
(c) $2Mg + O_2 \rightarrow 2MgO$
(d) $4Fe + 3O_2 \rightarrow 2Fe_2O_3$
(e) $2Ca + O_2 \rightarrow 2CaO$

12.68 $2MnO_4^- + H^+ + 5HSO_3^- \rightarrow 5SO_4^{2-} + Mn^{2+} + 3H_2O$

12.70 $3H_2C_2O_4 + 8H^+ + Cr_2O_7^{2-} \rightarrow 6CO_2 + 2Cr^{3+} + 7H_2O$

12.73 4.45 g Cl_2 formed

12.75 (a) $2Mn^{2+} + 14H^+ + 5BiO_3^- \rightarrow 5Bi^{3+} + 7H_2O + 2MnO_4^-$
(b) 58.7 g $NaBiO_3$

12.77 (a) $5H_2O + 3SO_3^{2-} + 2CrO_4^{2-} \rightarrow 3SO_4^{2-} + 2Cr(OH)_3 + 4OH^-$
(b) 1.63×10^{-2} mol CrO_4^{2-}
(c) 0.848 g Cr
(d) 28.3% Cr by weight

12.79 14.0 mL $KMnO_4$ soln.

12.81 (a) 0.2994 g H_2O_2 (b) 29.94% H_2O_2

12.83 (a) 5.970×10^{-3} mol $C_2O_4^{2-}$ (b) 0.6627 g $CaCl_2$
(c) 25.00% $CaCl_2$

12.85 (a) 5 eq/mol $KClO_3$ (c) 2 eq/mol NiO_2
(b) 6 eq/mol $KClO_3$ (d) 6 eq/mol Cr_2O_3

12.87 54.0 g Na_2CrO_4

12.89 8.01 g SO_2

12.91 35.5 g SO_2

12.93 2.8 g $Na_2S_2O_3$

12.94 1.00 N $NaIO_3$

12.96 8.55×10^{-2} g $Ba(OH)_2$

12.98 2.50×10^{-3} eq K_2CrO_4

12.100 0.1114 N $KMnO_4$

12.102 (a) 5.760×10^{-3} eq MnO_4^-
(b) 1.152×10^{-3} mol MnO_4^-
(c) 6.329×10^{-2} g Mn
(d) 12.66% Mn

CHAPTER 13

PRACTICE EXERCISES

1. 8.83×10^{-4} g O_2 and 1.47×10^{-3} g N_2

2. For CH_3OH, $X = 0.359$ or 35.9 mol%. For H_2O, $X = 0.641$ or 64.1 mol%.

3. Notice that a precision of 4 significant figures is allowed by the data, so compute formula weights accordingly. For NaCl, $X = 0.01334$ or 1.334 mol%. For H_2O, $X = 0.9867$ or 98.67 mol%.

4. 15.3 mol% oxygen

5. 16.0 g CH_3OH

6. 0.400 m NaOH

7. 20 g of HCl solution

8. Add 75 g NaOH to (750 g − 75 g) or 675 g water to make the final mass equal to 750 g.

9. Dilute 7.0 qt of methyl alcohol with sufficient water to make the final volume 20 qt.

10. (a) 16.7 m (b) 22.4 mol% HCl, 77.6 mol% H_2O

11. 29.3% (w/w) benzene

12. For $CaCl_2$, $X = 0.0760$ or 7.60 mol%. For H_2, $X = 0.924$ or 92.4 mol%.

13. 6.82 M HBr

14. 47.0% (w/w) HI

15. 9.02 torr

16. $P_{cyclohexane} = 33.4$ torr, $P_{toluene} = 10.6$ torr, $P_{total} = 44.0$ torr

17. 108.2 °C

18. 171

19. -30 °C

20. 157

21. 168.6 °C. Camphor offers the higher precision.

22. 1.4×10^{-3} mol/L

23. If 100% dissociated, fp $= -0.882$ °C. If 0% dissociated, fp $= -0.441$ °C.

REVIEW EXERCISES — Chapter 13

13.29 (a) $KCl(s) \rightarrow K^+(g) + Cl^-(g)$, $\Delta H = +690$ kJ/mol
$K^+(g) + Cl^-(g) \rightarrow K^+(aq) + Cl^-(aq)$,
$\Delta H = -686$ kJ/mol
(b) $KCl(s) \rightarrow K^+(aq) + Cl^-(aq)$, $\Delta H = +4$ kJ/mol

13.40 0.035 g/L

13.46 $X_{toluene} = 0.55$, $X_{chlorobenzene} = 0.45$

13.48 42 mol% water

13.49 O_2, 14 mol%
CO_2, 5.3 mol%
H_2O, 6.2 mol%

13.51 (a) 0.100 m (b) $X_{glucose} = 1.80 \times 10^{-3}$ (c) $X_{H_2O} = 0.998$

13.55 (a) 1.36 m H_2SO_4 (b) 11.8% H_2SO_4 (c) 1.30 M H_2SO_4

13.56 (a) 1.89 m KNO_3 (b) 16.1% (w/w) KNO_3
(c) 1.75 M KNO_3

13.57 (a) 4.82 m NaCl (b) $X_{NaCl} = 7.98 \times 10^{-2}$, $X_{H_2O} = 0.919$

13.60 (a) 0.277 m HCl (or 0.7% larger than M)
(b) 0.997% (w/w) HCl
(c) For HCl, $X = 0.00496$ or 0.496 mol% HCl. For H_2O, $X = 0.995$ or 99.5 mol% H_2O.

13.64 Mix 37.5 g of the concentrated solution with enough water to bring the final mass to 250 g (i.e., mix with 212.5 g water).

13.66 Mix 3 qt water with 5 qt of the more concentrated antifreeze solution to give a final volume of 8 qt.

13.74 23.8 torr

13.76 151 torr

13.82 (a) 22 mol $C_2H_6O_2$/1000 g H_2O
(b) 1.2×10^3 mL $C_2H_6O_2$/1000 g H_2O
(c) 1.2 qt $C_2H_6O_2$/1.0 qt H_2O

13.84 (a) 101.0 °C (b) -3.72 °C (c) 23.0 torr

13.85 152

13.87 (a) $C_3H_2NO_2$ (b) 168, $C_6H_4N_2O_4$

13.96 1.8×10^2 torr

13.100 17.4 torr

13.102 0.50 m Na_2CO_3

13.107 (a) bp $= 100.51$ °C (b) bp $= 102.04$ °C (c) 2.3

13.108 3% ionized

CHAPTER 14

PRACTICE EXERCISES

1. $K_c = \dfrac{[H_2O]^2}{[H_2]^2[O_2]}$, $K_c = \dfrac{[NH_4^+][OH^-]}{[NH_3][H_2O]}$

2. $K_p = \dfrac{p_{HI}^2}{p_{H_2}p_{I_2}}$

3. $K_c = 57$

4. $K_P = 1.8 \times 10^{36}$

5. Reaction 2.

6. (a) Decrease the amount Cl_2.
(b) Increase the amount of Cl_2.
(c) Increase the amount of Cl_2.
(d) Decrease the amount of Cl_2.
Only the temperature change will alter K_p.
An increase in temperature will lower K_p.

7. [CO] decreased by 0.060 mol/L; [CO_2] increased by 0.060 mol/L

8. $K_c = 4.06$

9. (a) initially, [PCl_3] $= 0.20$ mol/L, [Cl_2] $= 0.10$ mol/L, [PCl_5] $= 0.0$ mol/L
(b) [PCl_3] decreases by 0.08 mol/L, [Cl_2] decreases by 0.08 mol/L, [PCl_5] increases by 0.08 mol/L
(c) equilibrium concentrations: [PCl_3] $= 0.12$ M, [Cl_2] $= 0.02$ M, [PCl_5] $= 0.08$ M
(d) $K_c = 3 \times 10^1$

10. 8.98×10^{-3} M

11. [HI] $= 0.312$ M, [H_2] $=$ [I_2] $= 0.044$ M

12. 1.1×10^{-17} M

13. (a) $K_c = \dfrac{1}{[Cl_2]}$ (b) $K_c = \dfrac{1}{[NH_3][HCl]}$

14. (a) $K_{sp} = [Ba^{2+}][CrO_4^{2-}]$ (b) $K_{sp} = [Ag^+]^3[PO_4^{3-}]$

15. 3.98×10^{-8}

16. 1.0×10^{-10}

17. 3.8×10^{-8}

18. 7.1×10^{-7} M

19. 1.3×10^{-4} M

20. 4.2×10^{-16} M

21. 1.3×10^{-35} M

22. Ion product $= 7.5 \times 10^{-5}$; a precipitate should form.

23. Ion product $= 7.8 \times 10^{-13}$; no precipitate will form.

24. Ion product $= 5.0 \times 10^{-7}$; no precipitate will form.

25. Ion product $= 8.6 \times 10^{-6}$; no precipitate will form.

26. 4.9×10^{-3} mol AgCl per liter of 0.10 M $NH_3(aq)$. AgCl is 380 times more soluble in 0.10 M aqueous NH_3 as in pure water.

27. 4.1 mol NH_3

REVIEW EXERCISES — Chapter 14

14.20 $K_c = 11$ (rounded)

14.22 $K_P = 5.4 \times 10^{-5}$

14.26 (a) increase (c) increase (e) decrease
(b) decrease (d) no change

14.28 (a) increase (c) increase
(b) increase (d) decrease

14.29 (a) and (d)

14.33 (a) no (b) from right to left.

14.34 (a) 0.205 atm (b) 0.505 atm

14.36 $[CH_3OH] = 1.05 \times 10^{-3}$ mol/L

14.38 $K_c = 0.293$

14.40 $[NO_2] = [SO_2] = 0.0621\ M$, $[NO] = [SO_3] = 0.088\ M$

14.42 $[H_2] = [Cl_2] = 3.6 \times 10^{-17}\ M$

14.44 $[NO_2] = [SO_2] = 0.028\ M$

14.46 (a) $\dfrac{[CO]^2}{[O_2]} = K_c$ (c) $\dfrac{[CH_4][CO_2]}{[H_2O]^2} = K_c$

(b) $[SO_2][H_2O] = K_c$ (d) $\dfrac{[H_2O][CO_2]}{[HF]^2} = K_c$

14.48 $[HI] = 6.8 \times 10^{-12}\ M$, $[Cl_2] = 3.4 \times 10^{-12}\ M$, $[HCl] = 1.00\ M$

14.53 (a) $K_{sp} = [Ca^{2+}][F^-]^2$ (d) $K_{sp} = [Fe^{3+}][OH^-]^3$
(b) $K_{sp} = [Ag^+]^2[CO_3^{2-}]$ (e) $K_{sp} = [Pb^{2+}][I^-]^2$
(c) $K_{sp} = [Pb^{2+}][SO_4^{2-}]$ (f) $K_{sp} = [Cu^{2+}][OH^-]^2$

14.57 (a) 1.05×10^{-5} mol $BaSO_4$/1.00 L
(b) $1.05 \times 10^{-5}\ M\ SO_4^{2-}$, $1.05 \times 10^{-5}\ M\ Ba^{2+}$
(c) 1.10×10^{-10}

14.59 2.8×10^{-18}

14.60 5.8×10^{-38}

14.61 8.0×10^{-7}

14.64 (a) $3.3 \times 10^{-7}\ M$ (b) 4.0×10^{-20} (c) 8.4×10^{-9}

14.66 Solubilities per 100 mL: 2.0×10^{-7} g AgCN and 5×10^{-5} g $Zn(CN)_2$. Thus $Zn(CN)_2$ has a larger solubility when expressed in g/100 mL.

14.67 1.0×10^{-18}

14.70 Molar solubility $= 6.7 \times 10^{-5}$ mol/L. In 100 mL there is 6.7×10^{-4} g $CaCO_3$.

14.72 1.3×10^{-4} mol/L

14.74 7.1×10^{-10} mol/L

14.76 Ion product $= 6.0 \times 10^{-3} > K_{sp}$, so a precipitate should form.

14.78 Ion product $= 5.3 \times 10^{-4} < K_{sp}$, so no precipitate will form.

14.80 (a) 0.36 g AgCl
(b) $[Ag^+] = 2.5 \times 10^{-2}\ M$, $[Cl^-] = 7.2 \times 10^{-9}\ M$, $[NO_3^-] = 5.0 \times 10^{-2}\ M$, $[Na^+] = 2.5 \times 10^{-2}\ M$

14.82 $[Sr^{2+}] = 1.3 \times 10^{-3}\ M$

14.88 3.9×10^2 g NaCN

CHAPTER 15

PRACTICE EXERCISES

1. $[H_3O^+] = 1.3 \times 10^{-9}$ mol/L. The solution is basic.

2. (a) pH = 1.70, pOH = 12.30

(b) pH = 11.70, pOH = 2.30
(c) pH = 7.14, pOH = 6.86. Blood is slightly basic.
(d) pH = 10.85, pOH = 3.15

3. (a) $[H^+] = 1.3 \times 10^{-3}$ mol/L
(b) $[H^+] = 1.4 \times 10^{-4}$ mol/L
(c) $[H^+] = 1.5 \times 10^{-11}$ mol/L
(d) $[H^+] = 7.8 \times 10^{-5}$ mol/L
(e) $[H^+] = 2.5 \times 10^{-12}$ mol/L

4. (a) acidic (b) acidic (c) basic (d) acidic (e) basic

5. $K_a = \dfrac{[H^+][NO_2^-]}{[HNO_2]}$

6. $K_a = \dfrac{[H^+][PO_4^{3-}]}{[HPO_4^{2-}]}$

7. For H_2O, $K_a = \dfrac{[H^+][OH^-]}{[H_2O]}$.
For H_3O^+, $K_a = \dfrac{[H^+][H_2O]}{[H_3O^+]}$.

8. $K_a = 1.7 \times 10^{-5}$

9. $K_a = 1.1 \times 10^{-4}$

10. $[H^+] = 3.7 \times 10^{-4}$ mol/L, pH = 3.43

11. (a) 4.2% (b) 13%

12. (a) $K_b = \dfrac{[HCN][OH^-]}{[CN^-]}$

(b) $K_b = \dfrac{[HC_2H_3O_2][OH^-]}{[C_2H_3O_2^-]}$

(c) $K_b = \dfrac{[C_6H_5NH_3^+][OH^-]}{[C_6H_5NH_2]}$

(d) $K_b = \dfrac{[H_3O^+][OH^-]}{[H_2O]}$

13. $K_b = 1.7 \times 10^{-6}$

14. pH = 10.78, pOH = 3.22, $[OH^-] = 6.0 \times 10^{-4}$ mol/L

15. $pK_a = 3.15$

16. acetic acid

17. $pK_b = 3.19$

18. $CN^- + H_2O \rightleftharpoons HCN + OH^-$
$K_b = \dfrac{[HCN][OH^-]}{[CN^-]}$, $pK_b = 4.8$

19. pH = 2.89

20. (a) pH $= pK_a + 1$ (b) pH $= pK_a - 1$

21. pH = 3.92

22. [anion]/[acid] = 1.4

23. pH = 9.07

24. $[H^+] = 7.9 \times 10^{-5}$ mol/L
pH = 4.10
$[HPO_4^{2-}] = 4.5 \times 10^{-13}$ mol/L

25. (a)–(f) can hydrolyze

26. It makes the solution slightly basic.

27. It makes the solution slightly acidic.

28. The pH will be lowered.

29. pH = 8.52

30. pH = 10.23

31. pH = 4.92

REVIEW EXERCISES — Chapter 15

15.4 $[H^+] = [OH^-] = 1.55 \times 10^{-7}$
     pH = pOH = 6.81, pH + pOH = 13.62

15.6 pH = 2.00

15.8 pOH = 0.82, pH = 13.18

15.10 pH = 4.72

15.12 $K_a = \dfrac{[H^+][C_6H_5CO_2^-]}{[C_6H_5CO_2H]}$

15.14 CN^- is the stronger base.

15.16 percentage ionization = 0.42%, pH = 2.38.

15.18 $K_a = 2.3 \times 10^{-2}$, $pK_a = 1.64$

15.20 pH = 3.22, $[H^+] = 6.0 \times 10^{-4}$ mol/L

15.22 $K_b = 5.6 \times 10^{-4}$, $pK_b = 3.25$, $pK_a = 10.75$

15.24 pH = 10.26

15.26 $pK_a = 8.21$, pH = 4.46

15.28 (a) to handle extra OH^-
          $HCO_3^-(aq) + OH^-(aq) \rightarrow H_2O + CO_3^{2-}(aq)$
          to handle extra H^+
          $CO_3^{2-}(aq) + H^+(aq) \rightarrow HCO_3^-(aq)$
       (b) to handle extra OH^-
          $H_2PO_4^-(aq) + OH^-(aq) \rightarrow H_2O + HPO_4^{2-}(aq)$
          to handle extra H^+
          $HPO_4^{2-}(aq) + H^+(aq) \rightarrow H_2PO_4^-(qq)$
       (c) to handle extra OH^-
          $NH_4^+(aq) + OH^-(aq) \rightarrow H_2O + NH_3(aq)$
          to handle extra H^+
          $NH_3(aq) + H^+(aq) \rightarrow NH_4^+(aq)$

15.32 22 g $NaC_2H_3O_2(s)$

15.34 $[NH_3]/[NH_4Cl] = 1.0$

15.36 (a) 37.1 g $HC_2H_3O_2$ and 175 g $NaC_2H_3O_2 \cdot 2H_2O$
       (b) 2.47 M $HC_2H_3O_2$ and 5.92 M $NaC_2H_3O_2$
       (c) pH = 12.60
       (d) pH = 1.40

15.38 (a) moderate acid

       (b) $K_{a_1} = \dfrac{[H^+][HSO_3^-]}{[H_2SO_3]}$ $K_{a_2} = \dfrac{[H^+][SO_3^{2-}]}{[HSO_3^-]}$

       (c) $H_2CO_3(aq) \rightleftharpoons H^+(aq) + HCO_3^-(aq)$

          $K_{a_1} = \dfrac{[H^+][HCO_3^-]}{[H_2CO_3]}$

          $HCO_3^-(aq) \rightleftharpoons H^+(aq) + CO_3^{2-}(aq)$,

          $K_{a_2} = \dfrac{[H^+][CO_3^{2-}]}{[HCO_3^-]}$

       (d) Sulfurous acid

15.40 $[H^+] = 1.0 \times 10^{-1}$ mol/L
       pH = 1.00
       $[HPO_3^{2-}] = 2.6 \times 10^{-7}$ mol/L

15.42 basic

15.44 (a) $(NH_4)_2SO_4$ (b) KF, KCN, $KC_2H_3O_2$ (c) NaI, $CsNO_3$, KBr

15.46 pH = 8.08

15.48 pH = 11.66

15.50 8.6 g NH_4Br per liter

15.56 pH = 8.23, cresol red

15.58 (a) pH = 1.00   (f) pH = 9.3
       (b) pH = 1.37   (g) pH = 10.30
       (c) pH = 3.70   (h) pH = 11.29
       (d) pH = 4.7   (i) pH = 12.52
       (e) pH = 7.00

15.60 (a) pH = 11.11   (d) pH = 5.86   (g) pH = 3.70
       (b) pH = 9.44   (e) pH = 5.28   (h) pH = 2.71
       (c) pH = 6.86   (f) pH = 4.70   (i) pH = 1.48

CHAPTER 16

PRACTICE EXERCISES

1. (a) negative (b) positive

2. (a) negative (b) positive

3. (a) -229 J/K (b) -120.9 J/K

4. $\Delta G^\circ = -1482$ kJ

5. (a) -69.7 kJ (b) -120.1 kJ

6. 788 kJ

7. 614 K or 341 °C

8. $\Delta G^\circ = -70.0$ kJ; the reaction should occur spontaneously.

9. $\Delta G^\circ = +53.9$ kJ, so we should not expect to see any $CaCO_3$ formed by this reaction.

10. $\Delta G' = -29$ kJ, $\Delta G^\circ = +33$ kJ. ΔG becomes more negative as the temperature increases, so the position of equilibrium shifts toward the products.

11. $\Delta G^\circ = -7.96$ kcal

12. $K_P = 0.3$ (rounded from 0.27)

13. $K_P = 3.8 \times 10^4$

REVIEW EXERCISES — Chapter 16

16.12 1 L atm = 101.325 J

16.13 121 J

16.22 (a) $\Delta H^\circ = -17.8$ kJ, favored
       (b) $\Delta H^\circ = -311.42$ kJ, favored
       (c) $\Delta H^\circ = +1084$ kJ, not favored
       (d) $\Delta H^\circ = +65.3$ kJ, not favored
       (e) $\Delta H^\circ = +64.22$ kJ, not favored

16.26 (a) negative   (c) positive   (e) negative
       (b) negative   (d) negative   (f) positive

16.31 Probability of all heads = $\frac{1}{16}$. Probability of 2 heads and 2 tails = $\frac{3}{8}$.

16.33 (a) negative (b) negative (c) negative (d) positive

16.38 (a) $\Delta S^\circ = -198.3$ J/K, not favored
       (b) $\Delta S^\circ = -332.1$ J/K, not favored
       (c) $\Delta S^\circ = +92.6$ J/K, favored
       (d) $\Delta S^\circ = +14$ J/K, favored
       (e) $\Delta S^\circ = +159.6$ J/K, favored.

16.40 (a) -52.8 J/mol K   (d) -868.9 J/mol K
      (b) -74.0 J/mol K   (e) -318 J/mol K
      (c) -90.1 J/mol K

16.42 -269.7 J/K

16.47 $\Delta G_f^\circ = -210$ kJ/mol

16.50 (a) -184.1 kJ   (d) -530.9 kJ
      (b) $+0.6$ kJ   (e) -91.96 kJ
      (c) $+8.6$ kJ

16.52 -146 kJ

16.54 $+1.51$ kcal

16.60 -1299.8 kJ

16.62 -8.42×10^4 kJ/gal C_2H_5OH and -1.23×10^5 kJ/gal C_8H_{18}. On a volume basis, gasoline is a better fuel.

16.68 333 K (60 °C)

16.70 24.2 cal/mol K

16.73 (a) -84.5 kJ, spontaneous
      (b) -144 kJ, spontaneous
      (c) $+835.4$ kJ, not spontaneous
      (d) $+68.7$ kJ, not spontaneous

16.78 (a) -91.9 kJ (b) -171 kJ (c) $+91.8$ kJ

16.82 $K_P = 8.0 \times 10^8$. Yes, because the reaction should go essentially to completion.

16.84 $K = 1$

16.86 $\Delta G' = +21.2$ kJ

16.88 $K_P = 1 \times 10^{121}$

CHAPTER 17

PRACTICE EXERCISES

1. $2H_2O + 2Br^-(aq) \rightarrow Br_2(aq) + H_2(g) + 2OH^-(aq)$

2. 8.29×10^3 mol OH^-

3. 7.35 min

4. 3.67 A

5.

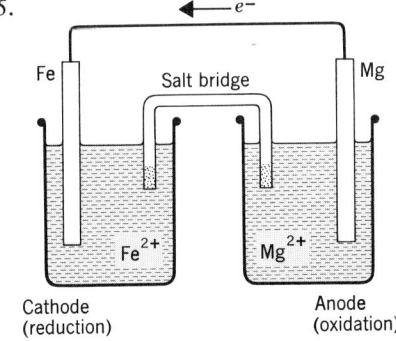

Cathode (reduction)   Anode (oxidation)

6. -2.37 V

7. 1.37 V

8. 2.23 V

9. $Br_2(aq) + H_2SO_3(aq) + H_2O \rightarrow 2Br^-(aq) + SO_4^{2-}(aq) + 4H^+(aq)$

10. (a) -0.56 V, nonspontaneous (b) $+0.02$ V, spontaneous

11. -3.23×10^3 kJ

12. $K_c = 3 \times 10^{-16}$. The small value of K_c means that very small amounts of products are expected to be formed by the time equilibrium is reached.

13. 1.04 V

14. $[Cu^{2+}] = 2 \times 10^{-4}$ M in the first sample and 6×10^{-13} M in the second sample.

15. $[Cr^{3+}] = 1 \times 10^{-3}$ M

REVIEW EXERCISES — Chapter 17

17.18 cathode: copper, anode: bromine (Br_2)

17.20 (a) 2.40×10^3 C (b) 1.2×10^4 C (c) 3.24×10^4 C

17.22 (a) 0.40 $\mathscr{F}$ (b) 0.70 $\mathscr{F}$ (c) 4.50 $\mathscr{F}$ (d) 0.050 $\mathscr{F}$

17.24 3.09×10^{-2} mol Cr^{3+}

17.26 52.2 mL of dry H_2 at STP

17.28 30.9 hr

17.30 106 A

17.32 2.20 g Cl_2

17.33 (a) 2.80×10^{-2} eq V (b) 9.33×10^{-3} mol V (c) 3.00 eq/mol, therefore V^{3+}

17.47 $Al \rightarrow Al^{3+} + 3e^-$ (anode = aluminum)
      $Sn^{2+} + 2e^- \rightarrow Sn$ (cathode = tin)

17.57 -0.34 V

17.62 0.19 V

17.64 cell reaction: $BrO_3^- + 6I^- + 6H^+ \rightarrow 3I_2 + Br^- + 3H_2O$, $E_{cell}^\circ = 0.90$ V

17.66 $4HOCl + 2H^+ + S_2O_3^{2-} \rightarrow 2Cl_2 + H_2O + 2H_2SO_3$

17.68 The reaction will not occur spontaneously.

17.72 1.25×10^3 J

17.73 7×10^2 hr

17.75 -1.63×10^3 kJ

17.77 $K_c = 10$

17.79 $K_c = 6.0 \times 10^8$

17.82 1.86 V

17.83 0.105 V

17.85 7.2×10^{-11} M

17.88 (a) pH $= (E_{cell} - E_{cell}^\circ)/0.0592$ (b) $E_{cell} = 0.589$ V
      (c) pH $= 5.46$

CHAPTER 18

PRACTICE EXERCISES

1. rate $= 9.4 \times 10^{-5}$ mol L^{-1} s^{-1}

2. rate $= 5.0 \times 10^{-3}$ mol L^{-1} s^{-1}

3. (a) 8.0×10^{-2} (b) L mol^{-1} s^{-1}

4. first order with respect to each reactant, second order overall

5. second set, $k = 2.0 \times 10^2$ L^2 mol^{-2} s^{-1}
   third set, $k = 2.0 \times 10^2$ L^2 mol^{-2} s^{-1}
   fourth set, $k = 2.0 \times 10^2$ L^2 mol^{-2} s^{-1}
   fifth set, $k = 2.0 \times 10^2$ L^2 mol^{-2} s^{-1}
   The values of k are the same; k is a constant.

6. First order

7. rate = $k[A]^2[B]^2$, $k = 6.9 \times 10^{-3}$ mol^{-3} L^3 s^{-1}

8. (a) 0.26 (b) 78 min

9. 63 min

10. $t_{1/2} = 18.7$ min; 37.4 min (2 half-lives) are required for three fourths of the sucrose to react.

11. The reaction is first order.

12. (a) $E_a = 1.4 \times 10^2$ kJ/mol (b) 0.30 L mol^{-1} s^{-1}

13. rate = $K[NO][O_3]$

REVIEW EXERCISES—Chapter 18

18.17 Cold water slows down metabolic reactions, including those that are involved in the death of an individual.

18.19 H_2 disappears three times faster than N_2. NH_3 is formed twice as fast as N_2 disappears.

18.21

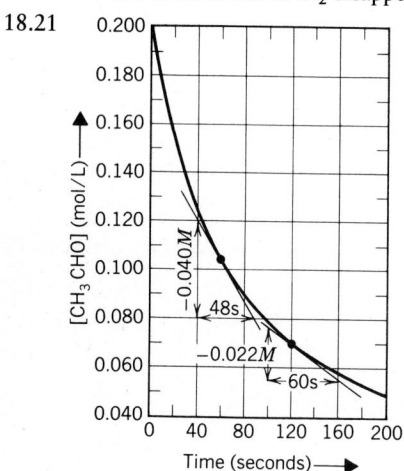

At 60 s,
rate = -8.3×10^{-4} mol L^{-1} s^{-1}.
At 120 s,
rate = -3.7×10^{-4} mol L^{-1} s^{-1}.

18.23 rate of formation of $NO_2 = 5.0 \times 10^{-6}$ mol L^{-1} s^{-1}
rate of formation of $O_2 = 1.3 \times 10^{-6}$ mol L^{-1} s^{-1}

18.26 (a) s^{-1} (b) L mol^{-1} s^{-1} (c) L^2 mol^{-2} s^{-1}

18.32 rate = 1.0×10^{-8} mol L^{-1} s^{-1}

18.35 NO, order = 1; O_3, order = 1; overall order = 2

18.37 rate = $k[M][N]^2$, $k = 2.5 \times 10^3$ L^2 mol^{-2} s^{-1}

18.39 rate = $k[OCl^-][I^-]$, $k = 6.1 \times 10^9$ L mol^{-1} s^{-1}

18.41 rate = $k[ICl][H_2]$, $k = 1.5 \times 10^{-1}$ L mol^{-1} s^{-1}

18.45 (a) 3.7×10^{-3} M (b) 6.0×10^{-4} M

18.47 3.2×10^{-4} M

18.48 $k = 9.17 \times 10^{-3}$ min^{-1}

18.50 $k = 5.1 \times 10^{-2}$ min^{-1}

18.55 4.3×10^2 s

18.58 To go from 0.200 M to 0.100 M requires approximately 66 s. To go from 0.100 M to 0.050 M requires approximately 128 s. The reaction is second order.

18.71 $E_a = 79.1$ kJ/mol

18.73 $E_a = 99$ kJ/mol, $A = 6.8 \times 10^9$ L mol^{-1} s^{-1}

18.77 52.9 kJ/mol

18.79 2.3×10^{-4} s^{-1}

18.84 Rate = $k[NO_2][CO]$. It can't be a one-step mechanism, otherwise the predicted rate law and the actual rate law would be the same.

18.85 $CO + O_2 + NO \rightarrow CO_2 + NO_2$

18.87 For the first step (the rate-determining step), rate = $k[NO_2]^2$.

18.90 (a) first step (b) second and third steps (c) third step This other reaction removes HBr and thereby slows down the rate of formation of HBr.

CHAPTER 19

REVIEW EXERCISES

19.28 $CH_2{=}CH_2 + H_2 \rightarrow CH_3{-}CH_3$ (High pressure, high temperature, and a metal catalyst are required.)

19.44 (a) $[Al(H_2O)_6]^{3+}(aq) + 4OH^-(aq) \rightarrow$
$[Al(H_2O)_2(OH)_4]^-(aq) + 4H_2O$
(b) The small, highly charged Al^{3+} central ion is far more able than Na^+ (with a smaller charge) to attract electron density from O—H bonds in the water molecules. This makes the O—H bonds more polar in the hydrated Al^{3+} ion than in the hydrated Na^+ ion, and the water molecules in the hydrated Al^{3+} ion therefore are more acidic.

19.47 An amphoteric compound is able to function as a base (and so neutralize acids) or as an acid (and so neutralize bases).
as a base $Cr(OH)_3 + 3HCl \rightarrow CrCl_3 + 3H_2O$
as an acid $Cr(OH)_3 + 3NaOH \rightarrow Na_3CrO_3 + 3H_2O$

19.50 -1

19.52 (a) $K_a = \dfrac{[H^+][HO_2^-]}{[H_2O_2]}$
(b) 1.8×10^{-12} (c) 5.8

19.54 The substance is Na_2O_2. Its solution, which contains hydrogen peroxide from the reaction $Na_2O_2 + 2H_2O \rightarrow 2NaOH + H_2O_2$, *bleached* the litmus paper. Had the white solid been Na_2O, the litmus paper would have turned blue, because now the reaction would have generated NaOH without the bleaching agent: $Na_2O + H_2O \rightarrow 2NaOH$.

19.58 (a) $Na_2O(s) + H_2O \rightarrow 2NaOH(aq)$
(b) $Na_2O(s) + 2HCl(aq) \rightarrow 2NaCl(aq) + H_2O$
(c) $Na_2O(s) + 2HNO_3(aq) \rightarrow 2NaNO_3(aq) + H_2O$
(d) $Al_2O_3(s) + 6HCl(aq) \rightarrow 2AlCl_3(aq) + 3H_2O$
(e) $Al_2O_3(s) + 2NaOH(aq) + 7H_2O \rightarrow$
$2Na[Al(H_2O)_2(OH)_4](aq)$
The "(aq)" after NaOH lets us use H_2O as a reactant, too.
(f) $K_2O(s) + 2HCl(aq) \rightarrow 2KCl(aq) + H_2O$
(g) $K_2O(s) + H_2O \rightarrow 2KOH(aq)$

(h) $Na_2O_2(s) + 2H_2O \rightarrow 2NaOH(aq) + H_2O_2(aq)$
(i) $Al_2O_3(s) + 6HBr(aq) \rightarrow 2AlBr_3(aq) + 3H_2O$

19.64 More difficult. With a much lower heat of vaporization, it would evaporate at a much faster rate in an open system at atmospheric pressure.

19.68 (a) $NH_3(aq) + HCl(aq) \rightarrow NH_4Cl(aq)$
(b) $NH_3(aq) + HBr(aq) \rightarrow NH_4Br(aq)$
(c) $NH_3(aq) + HI(aq) \rightarrow NH_4I(aq)$
(d) $2NH_3(aq) + H_2SO_4(aq) \rightarrow (NH_4)_2SO_4(aq)$
(e) $NH_3(aq) + HNO_3(aq) \rightarrow NH_4NO_3(aq)$

19.78 (a) The general formula of Group IIA nitrides is M_3N_2, where M is any metal in Group IIA.
(b) -3
(c) $M_3N_2(s) + 6H_2O \rightarrow 3M(OH)_2(aq) + 2NH_3(aq)$

19.84 (a) $N_2(g) + 2H_2(g) \rightarrow N_2H_4(l)$, $\Delta H° = +50.6$ kJ
(b) Unstable. It forms endothermically.
(c) The energies of activation for all elementary processes in the mechanism of its decomposition are too high to allow its decomposition to proceed at any noticeable rate at room temperature.
(d) $\Delta S_f° = -0.331$ kJ mol^{-1} K^{-1} (-331 J mol^{-1} K^{-1}) The *negative* entropy change, which does not favor a spontaneous change, reflects the shift toward *more* order as 3 mol of gaseous reactants change to 1 mol of liquid product.

19.85 (a) $N_2H_4(aq) + O_2(aq) \rightarrow N_2(g) + 2H_2O$
(b) No. N_2 is much less reactive toward metals than O_2, and the H_2O mingles with the rest of the water.

19.87 (a) $N_2(g) + 4H_2O + 4e^- \rightleftharpoons N_2H_4(aq) + 4OH^-(aq)$, $E° = -1.16$ V
(b) Yes. $E° = +1.96$ V in the following reaction.
$$4Ag^+(aq) + N_2H_4(aq) + 4OH^-(aq) \longrightarrow 4Ag(s) + N_2(g) + 4H_2O$$

19.90 The N is more basic.

$$\begin{array}{c} H \\ | \, + \\ H-N-O-H \\ | \\ H \end{array}$$

19.99 (a) $NO(g) + 2H_2O \rightarrow NO_3^-(aq) + 4H^+(aq) + 3e^-$, $E° = 0.96$ V
$NO_3^-(aq) + 2H^+(aq) + e^- \rightarrow NO_2(g) + H_2O$, $E° = +0.80$ V
(b) $2NO_3^-(aq) + NO(g) + 2H^+(aq) \rightarrow 3NO_2(g) + H_2O$
(c) $E°_{cell} = -0.04$ V, which indicates that the reaction is not spontaneous under the conditions implied by the ° in $E°$ (unit molarities and unit pressures).
(d) The last term in the Nernst equation (page 722) can be made positive enough by having the denominator in the reaction quotient, Q, become larger than the numerator. (This would make the log be the log of a fraction and so would have a negative value.) Since the $E°_{cell}$ term in the Nernst equation is only a very small negative number, it would not take much to tip the whole value of the resulting $E°$ to the positive side and make the

reaction spontaneous. At a high concentration of HNO_3, we would easily make the denominator in Q larger than the numerator and so make the value of $E°$ positive and the oxidation of NO by HNO_3 a spontaneous reaction.

19.101 KNO_2 must be thermally unstable (unlike $NaNO_2$), because KNO_3 thermally decomposes to K_2O, not KNO_2, under conditions ($t > 500$ °C) that decompose $NaNO_3$ to $NaNO_2$, and not to Na_2O.

19.106 N_2O decomposes, two volumes of it breaking up into three volumes of gaseous products: $2N_2O(g) \rightarrow 2N_2(g) + O_2(g)$. This would further increase the pressure in the can.

19.116 (a) $Zn(s) + 2HCl(aq) \rightarrow ZnCl_2(aq) + H_2(g)$
(b) $Ca(s) + H_2(g) \rightarrow CaH_2(s)$
(c) $NaH(s) + H_2O \rightarrow NaOH(aq) + H_2(g)$
(d) $Mg(s) + 2HCl(aq) \rightarrow MgCl_2(aq) + H_2(g)$
(e) $KH(s) + H_2O \rightarrow KOH(aq) + H_2(g)$
(f) $CH_3-CH=CH_2(g) + H_2(g) \rightarrow CH_3-CH_2-CH_3(g)$
(g) $CH_2=CH_2(g) + CO(g) + H_2(g) \rightarrow CH_3-CH_2-CH=O(l)$
(h) $2HgO(s) \rightarrow 2Hg(l) + O_2(g)$
(i) $Na_2O(s) + H_2O \rightarrow 2NaOH(aq)$
(j) $MgO(s) + 2HBr(aq) \rightarrow MgBr_2(aq) + H_2O$
(k) $2KClO_3(s) \rightarrow 2KCl(s) + 3O_2(g)$
(l) $Al_2O_3(s) + 6HBr(aq) \rightarrow 2AlBr_3(aq) + 3H_2O$
(m) $Al_2O_3(s) + 2KOH \rightarrow 2KAlO_2(aq) + H_2O$ or $Al_2O_3(s) + 2KOH(aq) + 7H_2O \rightarrow 2K[Al(H_2O)_2(OH)_4](aq)$
(n) $2CrO_3(s) + H_2O \rightarrow H_2Cr_2O_7(aq)$
(o) $2H_2O_2(aq) \rightarrow 2H_2O + O_2(g)$
(p) $K_2O_2(s) + 2H_2O \rightarrow 2KOH(aq) + H_2O_2(aq)$

19.118 (a) $2NO(g) + O_2(g) \rightarrow 2NO_2(g)$
(b) $KOH(aq) + HNO_3(aq) \rightarrow KNO_3(aq) + H_2O$
(c) $2N_2O(g) \rightarrow 2N_2(g) + O_2(g)$
(d) $2NO_2(g) + H_2O \rightarrow HNO_2(aq) + HNO_3(aq)$
(e) $N_2O_3(g) \rightarrow NO(g) + NO_2(g)$
(f) $N_2O_5(s) + H_2O \rightarrow 2HNO_3(aq)$
(g) $N_2O_4(g) \rightarrow 2NO_2(g)$
(h) $4HNO_3(l) + P_4O_{10}(s) \rightarrow 2N_2O_5(s) + 4HPO_3(l)$
(i) $2NaNO_3(s) \rightarrow 2NaNO_2(s) + O_2(g)$
(j) $Bi(NO_3)_3(aq) + H_2O \rightarrow BiO(NO_3)(s) + 2HNO_3(aq)$
(k) $N_2O_3(g) + 2NaOH(aq) \rightarrow 2NaNO_2(aq) + H_2O$
(l) $NaNO_2(aq) + HCl(aq) \rightarrow HNO_2(aq) + H_2O$
(m) $3NaNO_2(aq) + 3HCl(aq) \rightarrow HNO_3(aq) + 2NO(g) + 3NaCl(aq) + H_2O$

CHAPTER 20

REVIEW EXERCISES

20.16 (a) $NaOH(aq) + SO_2(g) \rightarrow NaHSO_3(aq)$
(b) $NaHCO_3(aq) + SO_2(g) \rightarrow NaHSO_3(aq) + CO_2(g)$

20.19 $I_2(s) + H_2SO_3(aq) + H_2O \rightarrow 2I^-(aq) + SO_4^{2-}(aq) + 4H^+(aq)$
$E° = +0.37$ V, so the reaction is spontaneous as written.

20.22 (a) $NaHSO_3(aq) + HBr(aq) \rightarrow H_2O + SO_2(g) + NaBr(aq)$
$HSO_3^-(aq) + H^+(aq) \rightarrow H_2O + SO_2(g)$

(b) $Na_2SO_3(aq) + 2HBr(aq) \rightarrow H_2O + SO_2(g) + 2NaBr(aq)$
$SO_3^{2-}(aq) + 2H^+(aq) \rightarrow H_2O + SO_2(g)$

(c) $K_2SO_3(aq) + 2HBr(aq) \rightarrow H_2O + SO_2(g) + 2KBr(aq)$
$SO_3^{2-}(aq) + 2H^+(aq) \rightarrow H_2O + SO_2(g)$

(d) $KHSO_3(aq) + HBr(aq) \rightarrow H_2O + SO_2(g) + KBr(aq)$
$HSO_3^-(aq) + H^+(aq) \rightarrow H_2O + SO_2(g)$

20.23 (a) $SO_3(g) + H_2O \rightarrow H_2SO_4(aq)$

(b) $SO_3(g) + 2NaHCO_3(aq) \rightarrow Na_2SO_4(aq) + H_2O + 2CO_2(g)$

(c) $SO_3(g) + Na_2CO_3(aq) \rightarrow Na_2SO_4(aq) + CO_2(g)$

(d) $SO_3(g) + 2NaOH(aq) \rightarrow Na_2SO_4(aq) + H_2O$

20.29 CO_2 evolved, because $NaHSO_4$ is an acid salt that reacts with sodium carbonate as follows.

$2NaHSO_4(aq) + Na_2CO_3(s) \longrightarrow$
$2Na_2SO_4(aq) + H_2O + CO_2(g)$

20.38 Drop some hydrochloric acid on the mineral. An odor of rotten eggs (from H_2S) would signify FeS. The H_2S forms by the equation

$FeS(s) + 2HCl(aq) \longrightarrow FeCl_2(aq) + H_2S(g)$

Fizzing without the rotten-egg odor would indicate $FeCO_3$.

$FeCO_3(s) + 2HCl(aq) \longrightarrow FeCl_2(aq) + CO_2(g) + H_2O$

20.46 $H_2SeO_4(aq) + 2NaOH(aq) \rightarrow Na_2SeO_4(aq) + 2H_2O$

20.54 $2PBr_3(l) + 6H_2O \rightarrow 2H_3PO_3(aq) + 6HBr(aq)$
$PBr_5(s) + 4H_2O \rightarrow H_3PO_4(aq) + 5HBr(aq)$

20.56 (a) $POCl_3(l) + 3H_2O \rightarrow H_3PO_4(aq) + 3HCl(aq)$

(b) $POCl_3(l) + 3CH_3OH \rightarrow (CH_3O)_3PO_4(l) + 3HCl(g)$

(c) $POCl_3(l) + 3CH_3CH_2OH \rightarrow (CH_3CH_2O)_3PO_4(l) + 3HCl(g)$

20.64 (a) orthophosphoric acid ("phosphoric acid")
(b) phosphorous acid
(c) hypophosphorous acid
(d) potassium dihydrogen phosphate
(e) magnesium monohydrogen phosphate
(f) sodium monohydrogen phosphate
(g) sodium hypophosphite
(h) sodium dihydrogen phosphite

20.65 (a) arsenic acid
(b) sodium arsenate
(c) antimonic acid
(d) sodium dihydrogen arsenate
(e) arsenic trichloride
(f) antimony pentachloride

20.80 (a) Y^- is oxidized. (b) X_2 is reduced. (c) No (see next).
(d) Yes, Cl_2 is a stronger oxidizing agent than Br_2.
(e) Yes, Br_2 is a stronger oxidizing agent than I_2.

20.87 (a)

(which can also be written as $CH_3CH_2CH_2CH_2CH_3$)

(b)

(c) $CH_3-CH=CH_2$

(d)

(e) $CH_3-C\equiv C-CH_3$ or $CH_3-CH_2-C\equiv C-H$

20.105 The cyanide ion, a Brønsted base, hydrolyzes to generate some OH^- ion, which raises the pH above 7.

$CN^-(aq) + H_2O \longrightarrow HCN(aq) + OH^-(aq)$

20.109 (a) no reaction
(b) $CaC_2(s) + 2H_2O \rightarrow HC\equiv CH(g) + Ca(OH)_2(s)$
(c) no reaction
(d) $Mg_2C(s) + 4H_2O \rightarrow CH_4(g) + 2Mg(OH)_2(s)$

20.117 (a) $HCl(aq) + NaOH(aq) \rightarrow H_2O + NaCl(aq)$
$H^+(aq) + OH^-(aq) \rightarrow H_2O$

(b) $HCl(aq) + NaHCO_3(aq) \rightarrow H_2O + CO_2(g) + NaCl(aq)$
$H^+(aq) + HCO_3^-(aq) \rightarrow H_2O + CO_2(g)$

(c) $2HCl(aq) + Na_2CO_3(aq) \rightarrow H_2O + CO_2(g) + 2NaCl(aq)$
$2H^+(aq) + CO_3^{2-}(aq) \rightarrow H_2O + CO_2(g)$

(d) $HCl(aq) + KOH(aq) \rightarrow H_2O + KCl(aq)$
$H^+(aq) + OH^-(aq) \rightarrow H_2O$

(e) $2HCl(aq) + K_2CO_3(aq) \rightarrow H_2O + CO_2(g) + 2KCl(aq)$
$2H^+(aq) + CO_3^{2-}(aq) \rightarrow H_2O + CO_2(g)$

(f) $HCl(aq) + KHCO_3(aq) \rightarrow H_2O + CO_2(g) + KCl(aq)$
$H^+(aq) + HCO_3^-(aq) \rightarrow H_2O + CO_2(g)$

(g) $2HCl(aq) + CaCO_3(s) \rightarrow H_2O + CO_2(g) + CaCl_2(aq)$
$2H^+(aq) + CaCO_3(s) \rightarrow H_2O + CO_2(g) + Ca^{2+}(aq)$

(h) $2HCl(aq) + Ca(OH)_2(s) \rightarrow 2H_2O + CaCl_2(aq)$
$2H^+(aq) + Ca(OH)_2(s) \rightarrow 2H_2O + Ca^{2+}(aq)$

(i) $2HCl(aq) + Mg(OH)_2(s) \rightarrow 2H_2O + MgCl_2(aq)$
$2H^+(aq) + Mg(OH)_2(s) \rightarrow 2H_2O + Mg^{2+}(aq)$

(j) $2HCl(aq) + MgCO_3(s) \rightarrow H_2O + CO_2(g) + MgCl_2(aq)$
$2H^+(aq) + MgCO_3(s) \rightarrow H_2O + CO_2(g) + Mg^{2+}(aq)$

20.121 (a) $H^+(aq) + OH^-(aq) \rightarrow H_2O$
(b) $H^+(aq) + HCO_3^-(aq) \rightarrow H_2O + CO_2(g)$
(c) $2H^+(aq) + CO_3^{2-}(aq) \rightarrow H_2O + CO_2(g)$
(d) $H^+(aq) + HSO_3^-(aq) \rightarrow H_2O + SO_2(g)$
(e) $2H^+(aq) + SO_3^{2-}(aq) \rightarrow H_2O + SO_2(g)$
(f) $H^+(aq) + CN^-(aq) \rightarrow HCN(g)$
(g) $H^+(aq) + HS^-(aq) \rightarrow H_2S(g)$
(h) $2H^+(aq) + S^{2-}(aq) \rightarrow H_2S(g)$

CHAPTER 21

REVIEW EXERCISES

21.3 (a) Tl (b) Ba (c) K

21.5 MgO

21.38 Li^+ has a much larger hydration energy than Na^+ and so is stabilized in water to a much larger extent. That makes Li^+ more difficult to reduce than Na^+, and in the presence of water Li is the more easily oxidized. In the absence of water, Na should be the better reducing agent.

21.40 (a) $2Li + 2H_2O \rightarrow 2LiOH + H_2(g)$
(b) $2K + 2H_2O \rightarrow 2KOH + H_2(g)$
(c) $2Rb + 2H_2O \rightarrow 2RbOH + H_2(g)$

21.42 (a) $Na_2O_2 + H_2O \rightarrow 2Na^+ + OH^- + HO_2^-$
(b) $Na_2O + H_2O \rightarrow 2Na^+ + 2OH^-$
(c) $2KO_2 + H_2O \rightarrow 2K^+ + OH^- + HO_2^- + O_2$
(d) $NaOH + CO_2 \rightarrow NaHCO_3$

21.50 $CO_3^{2-}(aq) + H_2O \rightleftharpoons HCO_3^-(aq) + OH^-(aq)$
For 1.0 M Na_2CO_3, the calculated pH is 12.16.

21.51 (a) $NaClO_4$ (b) $KC_2H_3O_2$ (c) $KC_2H_3O_2$

21.54 1.4×10^5 lb

21.79 (a) Al $1s^22s^22p^63s^23p^1$ (b) Al^{3+} $1s^22s^22p^6$

21.110 (a) 1.6×10^{-2} M $PbCl_2$ (b) 8.1×10^{-3} M $PbBr_2$
(c) 1.3×10^{-3} M PbI_2

21.112 Pb^{4+} is such a powerful oxidizing agent that it tends to oxidize the halide ion to the free halogen:

$$PbBr_4 \rightarrow PbBr_2 + Br_2$$

CHAPTER 22

REVIEW EXERCISES

22.4 A complex ion is formed when one or more neutral molecules or anions become bonded to a metal ion. The formula for the complex is $[Cu(OH)_4]^{2-}$.

22.9 $d = 16.7$ g/cm^3

22.14 Hammering the bar jars the domains and allows them to become aligned with the earth's magnetic field.

22.21 $Bi^{3+} + H_2O \longrightarrow [Bi-O\!\!\begin{smallmatrix}H\\ \\H\end{smallmatrix}]^{3+} \xrightarrow{H_2O} BiO^+ + 2H^+(aq)$

Extensive hydrolysis of the O—H bonds of the H_2O bound to the Bi^{3+} allows the transfer of H^+ ions to H_2O molecules of the solvent.

22.27 $4H^+(aq) + 4Cr^{2+}(aq) + O_2(g) \rightarrow 4Cr^{3+}(aq) + 2H_2O$

22.34 The high oxidation number of the Cr in CrO_3 causes the Cr—O bonds in this oxide to be largely covalent. In Chapter 19 you learned that covalent oxides tend to be acidic. Cr_2O_3 is more ionic and, therefore, more basic, so it shows both acid and base properties.

22.41 (a) $Mn(s) + 2HCl(aq) \rightarrow MnCl_2(aq) + H_2(g)$
(b) $MnCl_2(aq) + 2NaOH(aq) \rightarrow Mn(OH)_2(s) +$
2NaCl(aq)

22.55 (a) $Fe(H_2O)_6^{2+}(aq) + 2OH^-(aq) \rightarrow Fe(OH)_2(s) + 6H_2O$
(b) $Fe(H_2O)_6^{3+}(aq) + 3OH^-(aq) \rightarrow Fe(H_2O)_3(OH)_3(s) +$
$3H_2O$

22.60 Fe^{3+}, because of its smaller size and larger positive charge, causes a greater polarization of the H_2O molecules that surround it, and this makes it easier for the hydrogens to be removed as H^+.

22.68 $4Co^{3+}(aq) + 2H_2O \rightarrow 4Co^{2+}(aq) + O_2(g) + 4H^+(aq)$

22.82 (a) no reaction
(b) $Cu + 2H_2SO_4 \rightarrow CuSO_4 + SO_2 + 2H_2O$
(c) $3Cu + 8HNO_3 \rightarrow 3Cu(NO_3)_2 + 2NO + 4H_2O$
(d) $Cu + 4HNO_3 \rightarrow Cu(NO_3)_2 + 2NO_2 + 2H_2O$

22.88 $Cu(CN)_4^{2-}$

22.93 $AgCl(s) + Br^-(aq) \rightleftharpoons AgBr(s) + Cl^-(aq)$
AgBr is less soluble than AgCl, so addition of Br^- drives the equilibrium to the right as AgCl dissolves and AgBr precipitates.

22.101 Zn is more easily oxidized than Fe. When the two are in contact, the Fe becomes the cathode and Zn the anode, so Zn is oxidized while the Fe is protected. However, Fe is more easily oxidized than Sn. When Fe and Sn are in contact, it is the Fe that is oxidized while the Sn is protected.

22.103 $4Zn(s) + NO_3^-(aq) + 10H^+(aq) \rightarrow 4Zn^{2+}(aq) +$
$NH_4^+(aq) + 6H_2O$

22.115 coordination number = 6

22.119

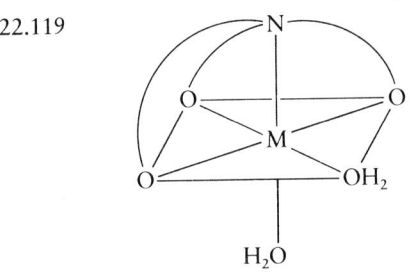

22.124

22.126

22.133

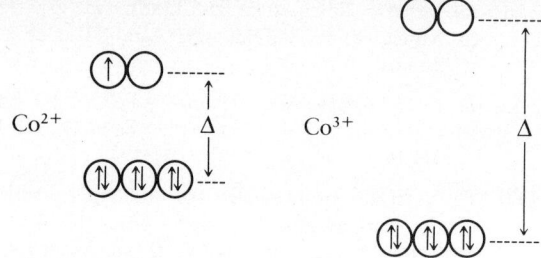

Oxidizing Co^{2+} to Co^{3+} removes a high-energy electron and also produces a larger value of Δ. Both factors favor this oxidation by lowering the energy of (i.e., stabilizing) Co^{3+}.

22.134 (a) $Cr(H_2O)_6{}^{3+}$ (b) $Cr(en)_3{}^{3+}$

22.137 (a) $Fe(H_2O)_6{}^{2+}$ (b) $Mn(CN)_6{}^{4-}$

22.141 $Fe(H_2O)_6{}^{3+}$ $Fe(CN)_6{}^{3-}$

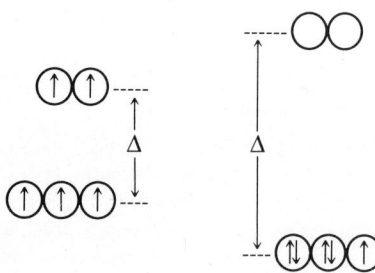

CHAPTER 23

PRACTICE EXERCISES

1. $^{226}_{88}Ra \rightarrow {}^{0}_{0}\gamma + {}^{4}_{2}He + {}^{222}_{86}Rn$

2. $^{90}_{38}Sr \rightarrow {}^{0}_{-1}e + {}^{90}_{39}Y$

3. 1.0×10^2 units

4. 1.51 g/mL

5. 1×10^4 yr

REVIEW EXERCISES—Chapter 23

23.4   (a) 1.02 kg (b) 3.91 kg (c) 12.3 kg

23.9   3.18 ng

23.11 1.8×10^{-13} J/nucleon

23.13 1.37×10^{-12} J/nucleon. In Figure 23.1, iron-56 is shown to have the highest binding energy per nucleon.

23.21   (a) $^{211}_{83}Bi$ (b) $^{177}_{72}Hf$ (c) $^{216}_{84}Po$ (d) $^{19}_{9}F$

23.23   (a) $^{242}_{94}Pu \rightarrow {}^{4}_{2}He + {}^{238}_{92}U$ (c) $^{26}_{14}Si \rightarrow {}^{0}_{1}e + {}^{26}_{13}Al$
(b) $^{28}_{12}Mg \rightarrow {}^{0}_{-1}e + {}^{28}_{13}Al$ (d) $^{37}_{18}Ar + {}^{0}_{-1}e \rightarrow {}^{37}_{17}Cl$

23.25   (a) $^{30}_{13}Al \rightarrow {}^{0}_{-1}e + {}^{30}_{14}Si$ (c) $^{93}_{42}Mo + {}^{0}_{-1}e \rightarrow {}^{93}_{41}Nb$
(b) $^{252}_{99}Es \rightarrow {}^{4}_{2}He + {}^{248}_{97}Bk$ (d) $^{28}_{15}P \rightarrow {}^{0}_{1}e + {}^{28}_{14}Si$

23.27   (a) $^{261}_{102}No$ (b) $^{211}_{82}Pb$ (c) $^{141}_{61}Pm$ (d) $^{179}_{74}W$

23.35   Barium-140, because both its mass number and atomic number are even.

23.37   Lanthanum-139, because it has 82 neutrons, a magic number.

23.44   The more likely process is beta emission because it

produces a product with a lower neutron/proton ratio: $^{38}_{19}K \rightarrow {}^{0}_{-1}e + {}^{38}_{20}Ca$

23.47 4.69×10^{-2} mg left

23.49 $^{53}_{24}Cr; {}^{51}_{23}V + {}^{2}_{1}D \rightarrow {}^{53}_{24}Cr \rightarrow {}^{1}_{1}H + {}^{52}_{23}V$

23.51 $^{80}_{35}Br$

23.53 $^{55}_{26}Fe; {}^{55}_{25}Mn + {}^{1}_{1}H \rightarrow {}^{1}_{0}n + {}^{55}_{26}Fe$

23.54 $^{1}_{0}n, {}^{239}_{93}Np$

23.56 $^{241}_{94}Pu({}^{1}_{0}n, \gamma){}^{242}_{94}Pu({}^{1}_{0}n, \gamma){}^{243}_{94}Pu \longrightarrow {}^{243}_{95}Am + {}^{0}_{-1}e$

$^{0}_{-1}e + {}^{244}_{96}Cm \longleftarrow {}^{244}_{95}Am \longleftarrow ({}^{1}_{0}n, \gamma)$

23.72   502 yr

23.74 1.80×10^{-6} Ci

23.76 1.3×10^9 yr. This assumes that all of the argon-40 came from potassium-40.

23.78   6.6 m

23.80   5 µg

23.82 6.8×10^3 years old

23.86   that the product of the particles/cm³ (n) and the confinement time (τ) be $n\tau = 3 \times 10^{14}$ s/cm³

23.88 $^{3}_{1}H + {}^{2}_{1}H \rightarrow {}^{4}_{2}He + {}^{1}_{0}n$

23.90   Self-ignition means that the thermal energy of the alpha particles formed in fusion be able to heat unchanged fuel and help to compress it prior to continued fusion. Without self-ignition, fusion using lasers probably cannot succeed.

23.100 $^{235}_{92}U + {}^{1}_{0}n \rightarrow {}^{94}_{38}Sr + {}^{139}_{54}Xe + 2 {}^{1}_{0}n$

23.101 Both will be beta emitters. Both are above the band of stability, and by emitting beta particles they can move closer to it.

23.102 Escape to the surroundings. Capture by unfissioned nuclei and so initiate more fission.

23.106 The same multiplication factor for both.

23.112 No critical mass can form because the atoms of the fissionable isotopes are so much diluted by those of the unfissionable isotopes.

CHAPTER 24

PRACTICE EXERCISE

1. (a) 2-ethylpentane
(b) 4-ethyl-2,3-dimethylheptane
(c) 5-ethyl-2,4,6-trimethyloctane

REVIEW EXERCISES—Chapter 24

24.5   b, d, and e are considered inorganic

24.7   (a) impossible (too few H atoms on carbon 1)
(b) impossible (too many H atoms on carbon 1)
(c) impossible (too many H atoms on carbon 3)

24.8   (a)

```
    H   H
    |   |
H —  C — N — H
    |
    H
```

(b)

```
    H
    |
Br — C — Br
    |
    H
```

(c)

$$H-\underset{\underset{Cl}{\overset{Cl}{|}}}{\overset{Cl}{\overset{|}{C}}}-Cl$$

(d)

$$H-\underset{\underset{H}{\overset{H}{|}}}{\overset{H}{\overset{|}{C}}}-\underset{\underset{H}{\overset{H}{|}}}{\overset{H}{\overset{|}{C}}}-H$$

(e)

$$H-\overset{\overset{\displaystyle O}{\|}}{C}-O-H$$

(f)

$$H-\overset{\overset{\displaystyle O}{\|}}{C}-H$$

(g)

$$H-\underset{\overset{|}{H}}{\overset{|}{N}}-O-H$$

(h) $H-C\equiv C-H$

(i)

$$H-\underset{\overset{|}{H}}{\overset{|}{N}}-\underset{\overset{|}{H}}{\overset{|}{N}}-H$$

(j) $H-C\equiv N$

(k)

$$H-\underset{\overset{|}{H}}{\overset{|}{C}}-C\equiv N$$

or $H-\overset{\overset{\displaystyle H}{|}}{C}=C=N-H$

(l)

$$H-\underset{\overset{|}{H}}{\overset{|}{C}}-O-H$$

24.11 (a) alkene
(b) alcohol
(c) ester
(d) carboxylic acid
(e) amine
(f) alcohol
(g) alkyne
(h) aldehyde
(i) ketone
(j) ether

24.12 b, e, f, and j

24.13 (a) identical
(b) identical
(c) unrelated
(d) isomers
(e) identical
(f) identical
(g) isomers
(h) identical
(i) isomers
(j) isomers
(k) identical
(l) identical
(m) isomers
(n) unrelated

24.14 (a) pentane
(b) 2-methylpentane
(c) 2,4-dimethylhexane
(d) 2,4-dimethylhexane
(e) 3-hexene
(f) 4-methyl-2-pentene

24.17 no
(a)–(d) no *cis*–*trans* isomers for these

(e)

$$\underset{H}{\overset{CH_3}{\diagdown}}C=C\underset{CH_2CH_3}{\overset{CH_3}{\diagup}}$$ and $$\underset{H}{\overset{CH_3}{\diagdown}}C=C\underset{CH_3}{\overset{CH_2CH_3}{\diagup}}$$

(f)

$$\underset{Cl}{\overset{CH_3CH_2}{\diagdown}}C=C\underset{Cl}{\overset{CH_2CH_3}{\diagup}}$$ and $$\underset{Cl}{\overset{CH_3CH_2}{\diagdown}}C=C\underset{CH_2CH_3}{\overset{Cl}{\diagup}}$$

24.19 (a) CH_3CH_3
(b) $Cl-CH_2CH_2-Cl$
(c) $Br-CH_2CH_2-Br$
(d) CH_3CH_2-Cl
(e) CH_3CH_2-Br
(f) CH_3CH_2-OH
(g) $CH_3CH_2-O-SO_3H$

24.28 (a) $CH_3CH_2CH_2CH_2OH$ (b) $CH_3CH_2\underset{\overset{|}{OH}}{\overset{}{C}}HCH_3$

24.32 $CH_2{=}CHCH_2CH_3$ and $CH_3CH{=}CHCH_3$

24.33 3:2 of $CH_2{=}CHCH_2CH_3$ to $CH_3CH{=}CHCH_3$, because there are 3 H atoms on carbon 1 that are available as candidates for loss with the OH group, but only 2 H atoms on carbon 3 available to eliminate with the OH group. (The major product, incidentally, is 2-butene, so statistical factors do not control the course of the reaction.)

24.36 C < A < B. This is the order of increasing molecular polarity, and compound B is able to participate in hydrogen bonding.

24.40

$$\pm OCH_2CH_2O\overset{\overset{\displaystyle O}{\|}}{C}CH_2CH_2\overset{\overset{\displaystyle O}{\|}}{C}\mp_n$$

24.49

$$HOCH_2\underset{\overset{|}{OH}}{\overset{}{C}}HCH_2OH + CH_3(CH_2)_7CH{=}CH(CH_2)_7\overset{\overset{\displaystyle O}{\|}}{C}OH$$

$$+ CH_3(CH_2)_{16}\overset{\overset{\displaystyle O}{\|}}{C}OH$$
$$+ CH_3(CH_2)_4CH{=}CHCH_2CH{=}CH(CH_2)_7\overset{\overset{\displaystyle O}{\|}}{C}OH$$

24.50 (a) amine
(b) amine
(c) amide
(d) keto group + amine
(e) amine
(f) amide

24.51 (a) $CH_3CH_2CH_2NH_2 + HCl \rightarrow$
$CH_3CH_2CH_2NH_3^+ + Cl^-$

(b) $CH_3CH_2\overset{\overset{\displaystyle O}{\|}}{C}NH_2 + H_2O \rightarrow CH_3CH_2\overset{\overset{\displaystyle O}{\|}}{C}OH + NH_3$

(c) $CH_3CH_2NH_3^+ + OH^- \rightarrow CH_3CH_2NH_2 + H_2O$

24.54

$$\pm NHCH_2CH_2CH_2CH_2NH\overset{\overset{\displaystyle O}{\|}}{C}-\!\!\bigcirc\!\!-\overset{\overset{\displaystyle O}{\|}}{C}\mp_n$$

24.57

$$^+NH_3CH_2\overset{\overset{\displaystyle O}{\|}}{C}-NHCH_2\overset{\overset{\displaystyle O}{\|}}{C}O^-$$

24.59

$$^+NH_3\underset{\overset{|}{CH_3}}{\overset{}{C}}H\overset{\overset{\displaystyle O}{\|}}{C}-NHCH_2\overset{\overset{\displaystyle O}{\|}}{C}O^- \text{ and}$$

$$^+NH_3CH_2\overset{\overset{\displaystyle O}{\|}}{C}-NH\underset{\overset{|}{CH_3}}{\overset{}{C}}H\overset{\overset{\displaystyle O}{\|}}{C}O^-$$

CHAPTER 25[1]

REVIEW EXERCISES

25.11 To obtain experience with the procedures. To see what positive tests look like. To check on the reagents.

25.15 (1) To exploit the common ion effect (the common ion being Cl^-) to force the following kind of equilibrium to the right and ensure complete precipitation. (M^{2+} represents a group 1 cation. Even though one of

[1] This chapter, "Qualitative Analysis of Ions," appears only in the *Qual Version* of this book.

them, Ag^+, is singly charged, the idea is the same.)

$$M^{2+}(aq) + 2Cl^-(aq) \rightleftharpoons MCl_2(s)$$

(2) To use the hydrogen ion in the group reagent to keep any possible Bi^{3+} or (BiO^+) from precipitating as BiOCl with the group 1 cations. The extra H^+ ion helps shift the following equilibrium to the left.

$$Bi^{3+}(aq) + H_2O + Cl^-(aq) \rightleftharpoons BiOCl(s) + 2H^+(aq)$$

25.16 The otherwise insoluble chlorides of two group 1 cations can form soluble complexes with excess chloride ion:

$$AgCl(s) + Cl^-(aq) \rightleftharpoons [AgCl_2]^-(aq)$$
$$PbCl_2(s) + 2Cl^-(aq) \rightleftharpoons [PbCl_4]^{2-}(aq)$$

25.24 $PbS > CuS > SnS_2 > Bi_2S_3$.

25.25 (1) Sn^{2+} is slowly oxidized by oxygen dissolved in the solution to Sn^{4+} anyway. (2) SnS tends to precipitate as a gelatinous solid that is more idfficult to handle than the more crystalline SnS_4. (3) SnS does not form a soluble complex with S^{2-} as well as does SnS_2.

25.32 A high concentration of H^+ ion tends to dissolve $PbSO_4(s)$ by shifting the following equilibrium to the right.

$$PbSO_4(s) + H^+(aq) \rightleftharpoons Pb^{2+}(aq) + HSO_4^-(aq)$$

The extra H^+ ion comes from the nitric acid present as well as sulfuric acid. The heating decomposes the nitric acid:

$$4HNO_3(aq) \xrightarrow{heat} O_2(g) + 4NO_2(g) + 2H_2O(g)$$

The heating also decomposes some of the sulfuric acid:

$$H_2SO_4(l) \xrightarrow{heat} H_2O(g) + SO_3(g)$$

Now when water is added, the concentration of H^+ will be too low to make $PbSO_4$ dissolve.

25.35 As NH_3 molecules form the deep blue complex with Cu^{2+} (the confirmatory test for Cu^{2+}), they also make the solution basic enough to bring down bismuth as an insoluble hydroxide of the BiO^+ ion:

$$BiO^+(aq) + OH^-(aq) \rightarrow BiO(OH)(s)$$

25.37 Oxidation by O_2 (dissolved from air into the solution) as well as disproportionation happen to the stannite ion when its solution is allowed to stand.
Oxidation:

$$2[Sn(OH)_3]^-(aq) + O_2(aq) + 2OH^-(aq) + 2H_2O \longrightarrow$$
$$2[Sn(OH)_6]^{2-}(aq)$$

Disproportionation:

$$2[Sn(OH)_3]^-(aq) \longrightarrow Sn(s) + [Sn(OH)_6]^{2-}(aq)$$

25.44 Fe^{3+} is reduced to Fe^{2+} by H_2S:

$$2Fe^{3+}(aq) + H_2S(aq) \longrightarrow 2Fe^{2+}(aq) + S(s) + 2H^+(aq)$$

(The Fe^{2+} then forms a precipitate of FeS with more H_2S.)

25.48 (a) $Cl^-(aq)$ interferes with the final test for manganese.
(b) $6Cl^-(aq) + 2NO_3^-(aq) + 8H^+(aq) \rightarrow 3Cl_2(g) + 2NO(g) + 4H_2O$.

25.51 An oxidizing acid is needed to keep iron in the Fe^{3+} state. HCl would dissolve the precipitate, but then it would also furnish Cl^-, which interferes with the test for manganese.

25.54 Neither Fe^{3+} nor Ni^{2+} can be further oxidized by the powerful oxidizing agent, BiO_3^- (and neither forms a precipitate with this anion). Thus the BiO_3^- is left to oxidize Mn^{2+} to MnO_4^-.

25.60 Aqueous ammonia is the reactant. Zn^{2+} forms a soluble complex, but Al^{3+} precipitates as its hydroxide.

$$[Zn(H_2O)_4]^{2+}(aq) + 4NH_3(aq) \longrightarrow$$
$$[Zn(NH_3)_4]^{2+}(aq) + 4H_2O$$
$$[Al(H_2O)_6]^{3+} + 3NH_3(aq) \longrightarrow$$
$$[Al(H_2O)_3(OH)_3](s) + 3NH_4^+(aq)$$

25.67 Aqueous ammonia has already been added to the test solution as part of various procedures.

25.74 The barium salts of these anions constitute the precipitate. Thus:
(a) If CO_3^{2-} (as $BaCO_3$) is present, CO_2 evolves when HNO_3 is added. (A slight fizzing will be seen.)

$$BaCO_3(s) + 2H^+(aq) \longrightarrow Ba^{2+}(aq) + CO_2(g) + H_2O$$

(b) If PO_4^{3-} [as $Ba_3(PO_4)_2$] is present, it will dissolve, but unless no $BaSO_4$ is present, you really won't be able to see this. It's hard to tell, just by looking, if just one solid goes into solution when another remains as a solid.

$$Ba_3(PO_4)_2(s) + 4H^+(aq) \longrightarrow 3Ba^{2+}(aq) + 2H_2PO_4^-(aq)$$

(c) If SO_4^{2-} (as $BaSO_4$) is present, you won't see any change to it. $BaSO_4$ does not dissolve.

25.78 Refer to Figure 25.15, procedures A-5, A-6, and A-7.

25.80 1 M $NaHSO_4$, an acid salt, turns blue litmus red.

25.81 Both CO_3^{2-} and HS^- ions are absent. CO_3^{2-} (or HCO_3^-) decomposes to CO_2 at this acidity. HS^- gives H_2S (with the rotten-egg odor) at this pH.

25.82 Certainly *not* present are HS^-, CO_3^{2-}, PO_4^{3-}, SO_4^{2-}, or any of the halide ions. These form very insoluble compounds with one or more of the given cations. Only the NO_3^- from our list can be present.

25.87 (a) Flame test — yellow if NaCl, violet if KCl.
(b) $AgNO_3$ solution — a precipitate if NaCl, otherwise, no reaction.
(c) $BaNO_3$ solution — a precipitate if Na_2SO_4, otherwise, no reaction.
(d) Aqueous NH_3 — a deep blue solution if $Cu(NO_3)_2$; otherwise, no change.
(e) H_2DMG solution — a brick-red precipitate if $Ni(NO_3)_2$, but not with $Cu(NO_3)_2$.
(f) Aqueous NH_3 — a precipitate if $Al(NO_3)_3$; otherwise, no change.
(g) Aqueous NH_3 — a precipitate if $Fe(NO_3)_3$ (its hydrated oxide); a deep blue solution if $Ni(NO_3)_2$.
(h) $AgNO_3(aq)$ — a precipitate if NaCl; otherwise, no change. [One could also use $Ba(NO_3)_2$, which gives a precipitate with SO_4^{2-} in Na_2SO_4 but not with Cl^-.]

(i) Flame test. Yellow if NaCl; red-orange if Ca^{2+}. One could also use $K_2C_2O_4$: a precipitate with $CaCl_2$ but not with NaCl.

(j) Chlorine water ($NaOCl$ + HCl) with a globule of CH_2Cl_2 present. The I^- in NaI, changing to I_2, would give a violet color to the globule. Otherwise, the globule becomes brownish as the Br^- ion in NaBr is oxidized to Br_2.

(k) Add any solution that contains Cl^-, such as HCl, or NaCl; a precipitate if $AgNO_3$ otherwise, no change.

(l) Any dilute acid. Fizzing (CO_2) if Na_2CO_3; otherwise, no evident change.

(m) Any dilute acid. If NaHS, a rotten-egg odor, otherwise an odorless fizzing.

(n) Just color alone. K_2CrO_4 is yellow but K_2SO_4 is colorless.

(o) Add dilute NaOH. An ammonia odor if $(NH_4)_2SO_4$; otherwise, no apparent change.

25.89 (a) $[SnS_3]^{2-}(aq) + 6H^+(aq) + 6Cl^-(aq) \rightarrow$ $[SnCl_6]^{2-}(aq) + 3H_2S(aq)$

(b) $[SnCl_4]^{2-}(aq) + Hg_2Cl_2(s) \rightarrow 2Hg(l) + [SnCl_6]^{2-}(aq)$

(c) $3CuS(s) + 8H^+(aq) + 2NO_3^-(aq) \rightarrow 3Cu^{2+}(aq) + 3S(s) + 2NO(g) + 4H_2O$

(d) $[Pb(C_2H_3O_2)_4]^{2-}(aq) + CrO_4^{2-}(aq) \rightarrow PbCrO_4(s) + 4C_2H_3O_2^-(aq)$

(e) $PbCl_2(s) + 2Cl^-(aq) \rightarrow PbCl_4^{2-}(aq)$

(f) $Cu^{2+}(aq) + H_2S(aq) \rightarrow CuS(s) + 2H^+(aq)$

(g) $SnS_2(s) + H_2S(aq) + 2OH^-(aq) \rightarrow [SnS_3]^{2-}(aq) + 2H_2O$

(h) $[SnCl_4]^{2+}(aq) + HgCl_2(aq) \rightarrow Hg_2Cl_2(s) + [SnCl_6]^{2-}(aq)$

(i) $2BiO(OH)(s) + 3[Sn(OH)_3]^-(aq) + 3OH^-(aq) + 2H_2O \rightarrow 2Bi(s) + 3[Sn(OH)_6]^{2-}(aq)$

(j) $Al^{3+}(aq) + 3OH^-(aq) \rightarrow Al(OH)_3(s)$

(k) $CuS(s) + H^+(aq) \rightarrow$ No reaction

(l) $Zn^{2+}(aq) + 4OH^-(aq)$ [excess] $\rightarrow Zn(OH)_4^{2-}(aq)$

(m) $Fe^{3+}(aq) + 3NH_3(aq) + 6H_2O \rightarrow Fe(H_2O)_3(OH)_3(s) + 3NH_4^+(aq)$

(n) $NH_4^+(aq) + OH^-(aq) \rightarrow NH_3(aq) + H_2O$

(o) $Ba_3(PO_4)_2(s) + 4H^+(aq) \rightarrow 3Ba^{2+}(aq) + 2H_2PO_4^-(aq)$

(p) $I_2(aq) + Cl^-(aq) \rightarrow$ No reaction

(q) $Mn^{2+}(aq) + HS^-(aq) + NH_3(aq) \rightarrow MnS(s) + NH_4^+(aq)$

(r) $FeS(s) + 2H^+(aq) \rightarrow Fe^{2+}(aq) + H_2S(aq)$

(s) $Zn^{2+}(aq) + HS^-(aq) + NH_3(aq) \rightarrow ZnS(s) + NH_4^+(aq)$

Appendix

D

TABLES OF SELECTED DATA

The following pages contain tables of selected data, compiled from a variety of sources, that you may find helpful in this and other chemistry courses that you may take. Another useful source of chemical and physical data is the *Handbook of Chemistry and Physics* edited by Robert C. Weast and published by CRC Press. A copy of a recent edition is no doubt available in your school library.

TABLE D.1 Thermodynamic Data for Selected Elements, Compounds, and Ions (25 °C)

Substance	ΔH_f° (kJ/mol)	S° (J/mol K)	ΔG_f° (kJ/mol)
Aluminum			
Al(s)	0	28.3	0
Al³⁺(aq)	−524.7	—	−481.2
AlCl₃(s)	−704	110.7	−629
Al₂O₃(s)	−1676	51.0	−1576.4
Al₂(SO₄)₃(s)	−3441	239	−3100
Arsenic			
As(s)	0	35.1	0
AsH₃(g)	+66.4	223	+68.9
As₄O₆(s)	−1314	214	−1153
As₂O₅(s)	−925	105	−782
H₃AsO₃(aq)	−742.2	—	—
H₃AsO₄(aq)	−902.5	—	—

TABLE D.1 Thermodynamic Data for Selected Elements, Compounds, and Ions (25 °C) *(Continued)*

Substance	ΔH_f° (kJ/mol)	S° (J/mol K)	ΔG_f° (kJ/mol)
Barium			
Ba(s)	0	66.9	0
Ba²⁺(aq)	−537.6	9.6	−560.8
BaCO₃(s)	−1219	112	−1139
BaCrO₄(s)	−1428.0	—	—
BaCl₂(s)	−860.2	125	−810.8
BaO(s)	−553.5	70.4	−525.1
Ba(OH)₂	−998.22	−8	−875.3
Ba(NO₃)₂(s)	−992	214	−795
BaSO₄(s)	−1465	132	−1353
Beryllium			
Be(s)	0	9.50	0
BeCl₂(s)	−468.6	89.9	−426.3
BeO(s)	−611	14	−582
Bismuth			
Bi(s)	0	56.9	0
BiCl₃(s)	−379	177	−315
Bi₂O₃(s)	−576	151	−497
Boron			
B(s)	0	5.87	0
BCl₃(g)	−404	290	−389
B₂H₆(g)	+36	232	+87
B₂O₃(s)	−1273	53.8	−1194
B(OH)₃(s)	−1094	88.8	−969
Bromine			
Br₂(l)	0	152.2	0
Br₂(g)	+30.9	245.4	3.11
HBr(g)	−36	198.5	53.1
Br⁻(aq)	−121.55	82.4	−103.96
Cadmium			
Cd(s)	0	51.8	0
Cd²⁺(aq)	−75.90	−73.2	−77.61
CdCl₂(s)	−392	115	−344
CdO(s)	−258.2	54.8	−228.4
CdS(s)	−162	64.9	−156
CdSO₄(s)	−933.5	123	−822.6
Calcium			
Ca(s)	0	41.4	0
Ca²⁺(aq)	−542.83	−53.1	−553.58
CaCO₃(s)	−1207	92.9	−1128.8
CaF₂(s)	−741	80.3	−1166
CaCl₂(s)	−795.8	114	−750.2
CaBr₂(s)	−682.8	130	−663.6
CaI₂(s)	−535.9	143	—
CaO(s)	−635.5	40	−604.2
Ca(OH)₂(s)	−986.6	76.1	896.76
Ca₃(PO₄)₂(s)	−4119	241	−3852
CaSO₃(s)	−1156	—	—
CaSO₄(s)	−1433	107	−1320.3

TABLE D.1 Thermodynamic Data for Selected Elements, Compounds, and Ions (25 °C) *(Continued)*

Substance	ΔH_f° (kJ/mol)	S° (J/mol K)	ΔG_f° (kJ/mol)
Calcium *(Continued)*			
$CaSO_4 \cdot \frac{1}{2}H_2O(s)$	-1573	131	-1435.2
$CaSO_4 \cdot 2H_2O(s)$	-2020	194.0	-1795.7
Carbon			
$C(s, \text{graphite})$	0	5.69	0
$C(s, \text{diamond})$	$+1.88$	2.4	$+2.9$
$CCl_4(l)$	-134	214.4	-65.3
$CO(g)$	-110	197.9	-137.3
$CO_2(g)$	-394	213.6	-394.4
$CO_2(aq)$	-413.8	117.6	-385.98
$H_2CO_3(aq)$	-699.65	187.4	-623.08
$HCO_3^-(aq)$	-691.99	91.2	-586.77
$CO_3^{2-}(aq)$	-677.14	-56.9	-527.81
$CS_2(l)$	$+89.5$	151.3	$+65.3$
$CS_2(g)$	$+117$	237.7	$+67.2$
$HCN(g)$	$+135.1$	201.7	$+124.7$
$CN^-(aq)$	$+150.6$	94.1	$+172.4$
$CH_4(g)$	-74.9	186.2	-50.79
$C_2H_2(g)$	$+227$	200.8	$+209$
$C_2H_4(g)$	$+51.9$	219.8	$+68.12$
$C_2H_6(g)$	-84.5	229.5	-32.9
$C_3H_8(g)$	-104	269.9	-23
$C_4H_{10}(g)$	-126	310.2	-17.0
$C_6H_6(l)$	$+49.0$	173.3	$+124.3$
$CH_3OH(l)$	-238	126.8	-166.2
$C_2H_5OH(l)$	-278	161	-174.8
$HCHO_2(g)$ (formic acid)	-363	251	$+335$
$HC_2H_3O_2(l)$ (acetic acid)	-487.0	160	-392.5
$CH_2O(g)$ (formaldehyde)	-108.6	218.8	-102.5
$CH_3CHO(g)$ (acetaldehyde)	-167	250	-129
$(CH_3)_2CO(l)$ (acetone)	-248.1	200.4	-155.4
$C_6H_5CO_2H(s)$ (benzoic acid)	-385.1	167.6	-245.3
$CO(NH_2)_2(s)$ (urea)	-333.5	104.6	-197.2
$CO(NH_2)_2(aq)$	-319.2	173.8	-203.8
$CH_2(NH_2)CO_2H(s)$ (glycine)	-532.9	103.5	-373.4
Chlorine			
$Cl_2(g)$	0	223.0	0
$Cl^-(aq)$	-167.2	56.5	-131.2
$HCl(g)$	-92.5	186.7	-95.27
$HCl(aq)$	-167.2	56.5	-131.2
$HClO(aq)$	-131.3	106.8	-80.21

TABLE D.1 Thermodynamic Data for Selected Elements, Compounds, and Ions (25 °C) (Continued)

Substance	ΔH_f° (kJ/mol)	S° (J/mol K)	ΔG_f° (kJ/mol)
Chromium			
Cr(s)	0	23.8	0
$Cr^{3+}(aq)$	−232	—	—
$CrCl_2(s)$	−326	115	−282
$CrCl_3(s)$	−563.2	126	−493.7
$Cr_2O_3(s)$	−1141	81.2	−1059
$CrO_3(s)$	−585.8	72.0	−506.2
$(NH_4)_2Cr_2O_7(s)$	−1807	—	—
$K_2Cr_2O_7(s)$	−2033.01	—	—
Cobalt			
Co(s)	0	30.0	0
$Co^{2+}(aq)$	−59.4	−110	−53.6
$CoCl_2(s)$	−325.5	106	−282.4
$Co(NO_3)_2(s)$	−422.2	192	−230.5
CoO(s)	−237.9	53.0	−214.2
CoS(s)	−80.8	67.4	−82.8
Copper			
Cu(s)	0	33.15	0
$Cu^{2+}(aq)$	+64.77	−99.6	+65.49
CuCl(s)	−137.2	86.2	−119.87
$CuCl_2(s)$	−172	119	−131
$Cu_2O(s)$	−168.6	93.1	−146.0
CuO(s)	−155	42.6	−127
$Cu_2S(s)$	−79.5	121	−86.2
CuS(s)	−53.1	66.5	−53.6
$CuSO_4(s)$	−771.4	109	−661.8
$CuSO_4 \cdot 5H_2O(s)$	−2279.7	300.4	−1879.7
Fluorine			
$F_2(g)$	0	202.7	0
$F^-(aq)$	−332.6	−13.8	−278.8
HF(g)	−271	173.5	−273
Gold			
Au(s)	0	47.7	0
$Au_2O_3(s)$	+80.8	125	+163
$AuCl_3(s)$	−118	148	−48.5
Hydrogen			
$H_2(g)$	0	130.6	0
$H_2O(l)$	−286	69.96	−237.2
$H_2O(g)$	−242	188.7	−228.6
$H_2O_2(l)$	−187.8	109.6	−120.3
$H_2Se(g)$	+76	219	+62.3
$H_2Te(g)$	+154	234	+138
Iodine			
$I_2(s)$	0	116.1	0
$I_2(g)$	+62.4	260.7	+19.3
HI(g)	+26	206	+1.30
Iron			
Fe(s)	0	27	0
$Fe^{2+}(aq)$	−89.1	−137.7	−78.9

TABLE D.1 Thermodynamic Data for Selected Elements, Compounds, and Ions (25 °C) *(Continued)*

Substance	ΔH_f° (kJ/mol)	S° (J/mol K)	ΔG_f° (kJ/mol)
Iron *(Continued)*			
$Fe^{3+}(aq)$	−48.5	−315.9	−4.7
$Fe_2O_3(s)$	−822.2	90.0	−741.0
$Fe_3O_4(s)$	−1118.4	146.4	−1015.4
$FeS(s)$	−100.0	60.3	−100.4
$FeS_2(s)$	−178.2	52.9	−166.9
Lead			
$Pb(s)$	0	64.8	0
$Pb^{2+}(aq)$	−1.7	10.5	−24.4
$PbCl_2(s)$	−359.4	136	−314.1
$PbO(s)$	−217.3	68.7	−187.9
$PbO_2(s)$	−277	68.6	−219
$Pb(OH)_2(s)$	−515.9	88	−420.9
$PbS(s)$	−100	91.2	−98.7
$PbSO_4(s)$	−920.1	149	−811.3
Lithium			
$Li(s)$	0	28.4	0
$Li^+(aq)$	−278.6	10.3	—
$LiF(s)$	−611.7	35.7	−583.3
$LiCl(s)$	−408	59.29	−383.7
$LiBr(s)$	−350.3	66.9	−338.87
$Li_2O(s)$	−596.5	37.9	−560.5
$Li_3N(s)$	−199	37.7	−155.4
Magnesium			
$Mg(s)$	0	32.5	0
$Mg^{2+}(aq)$	−466.9	−138.1	−454.8
$MgCO_3(s)$	−1113	65.7	−1029
$MgF_2(s)$	−1113	79.9	−1056
$MgCl_2(s)$	−641.8	89.5	−592.5
$MgCl_2 \cdot 2H_2O(s)$	−1280	180	−1118
$Mg_3N_2(s)$	−463.2	87.9	−411
$MgO(s)$	−601.7	26.9	−569.4
$Mg(OH)_2(s)$	−924.7	63.1	−833.9
Manganese			
$Mn(s)$	0	32.0	0
$Mn^{2+}(aq)$	−223	−74.9	−228
$MnO_4^-(aq)$	−542.7	191	−449.4
$KMnO_4(s)$	−813.4	171.71	−713.8
$MnO(s)$	−385	60.2	−363
$Mn_2O_3(s)$	−959.8	110	−882.0
$MnO_2(s)$	−520.9	53.1	−466.1
$Mn_3O_4(s)$	−1387	149	−1280
$MnSO_4(s)$	−1064	112	−956
Mercury			
$Hg(l)$	0	76.1	0
$Hg(g)$	+61.32	175	+31.8
$Hg_2Cl_2(s)$	−265.2	192.5	−210.8
$HgCl_2(s)$	−224.3	146.0	−178.6

TABLE D.1 Thermodynamic Data for Selected Elements, Compounds, and Ions (25 °C) (Continued)

Substance	ΔH_i° (kJ/mol)	S° (J/mol K)	ΔG_i° (kJ/mol)
Mercury (Continued)			
HgO(s)	−90.83	70.3	−58.54
HgS(s,red)	−58.2	82.4	−50.6
Nickel			
Ni(s)	0	30	0
NiCl$_2$(s)	−305	97.5	−259
NiO(s)	−244	38	−216
NiO$_2$(s)	—	—	−199
NiSO$_4$(s)	−891.2	77.8	−773.6
NiCO$_3$(s)	−664.0	91.6	−615.0
Ni(CO)$_4$(g)	−220	399	−567.4
Nitrogen			
N$_2$(g)	0	191.5	0
NH$_3$(g)	−46.0	192.5	−16.7
NH$_4^+$(aq)	−132.5	113	−79.37
N$_2$H$_4$(l)	+50.6	121.2	+149.4
NH$_4$Cl(s)	−314.4	94.6	−203.9
NO(g)	+90.4	210.6	+86.69
NO$_2$(g)	+34	240.5	+51.84
N$_2$O(g)	+81.5	220.0	+103.6
N$_2$O$_4$(g)	+9.16	304	+98.28
N$_2$O$_5$(g)	+11	356	+115
HNO$_3$(l)	−174.1	155.6	−79.9
NO$_3^-$(aq)	−205.0	146.4	−108.74
Oxygen			
O$_2$(g)	0	205.0	0
O$_3$(g)	+143	238.8	+163
OH$^-$(aq)	−230.0	−10.75	−157.24
Phosphorus			
P(s,white)	0	41.09	0
P$_4$(g)	+314.6	163.2	+278.3
PCl$_3$(g)	−287.0	311.8	−267.8
PCl$_5$(g)	−374.9	364.6	−305.0
PH$_3$(g)	+5.4	210.2	+12.9
P$_4$O$_6$(s)	−1640	—	—
P$_4$O$_{10}$(s)	−2984	228.9	−2698
H$_3$PO$_4$(s)	−1279	110.5	−1119
Potassium			
K(s)	0	64.18	0
K$^+$(aq)	−252.4	102.5	−283.3
KF(s)	−567.3	66.6	−537.8
KCl(s)	−436.8	82.59	−408.3
KBr(s)	−393.8	95.9	−380.7
KI(s)	−327.9	106.3	−324.9
KOH(s)	−424.8	78.9	−379.1
K$_2$O(s)	−361	98.3	−322
K$_2$SO$_4$(s)	−1433.7	176	−1316.4

TABLE D.1 Thermodynamic Data for Selected Elements, Compounds, and Ions (25 °C) (Continued)

Substance	ΔH_f° (kJ/mol)	S° (J/mol K)	ΔG_f° (kJ/mol)
Silicon			
Si(s)	0	19	0
$SiH_4(g)$	+33	205	+52.3
$SiO_2(s,alpha)$	−910.0	41.8	−856
Silver			
Ag(s)	0	42.55	0
$Ag^+(aq)$	+105.58	72.68	+77.11
AgCl(s)	−127.1	96.2	−109.8
AgBr(s)	−100.4	107.1	−96.9
$AgNO_3(s)$	−124	141	−32
$Ag_2O(s)$	−31.1	121.3	−11.2
Sodium			
Na(s)	0	51.0	0
$Na^+(aq)$	−240.12	59.0	−261.91
NaF(s)	−571	51.5	−545
NaCl(s)	−413	72.38	−384.0
NaBr(s)	−360	83.7	−349
NaI(s)	−288	91.2	−286
$NaHCO_3(s)$	−947.7	102	−851.9
$Na_2CO_3(s)$	−1131	136	−1048
$Na_2O_2(s)$	−510.9	94.6	−447.7
$Na_2O(s)$	−510	72.8	−376
NaOH(s)	−426.8	64.18	−382
$Na_2SO_4(s)$	−1384.49	149.49	−1266.83
Sulfur			
S(s,rhombic)	0	31.8	0
$SO_2(g)$	−297	248	−300
$SO_3(g)$	−396	256	−370
$H_2S(g)$	−20.6	206	−33.6
$H_2SO_4(l)$	−813.8	157	−689.9
$H_2SO_4(aq)$	−909.3	20.1	−744.5
$SF_6(g)$	−1209	292	−1105
Tin			
Sn(s,white)	0	51.6	0
$Sn^{2+}(aq)$	−8.8	−17	−27.2
$SnCl_4(l)$	−511.3	258.6	−440.2
SnO(s)	−285.8	56.5	−256.9
$SnO_2(s)$	−580.7	52.3	−519.6
Zinc			
Zn(s)	0	41.6	0
$Zn^{2+}(aq)$	−153.9	−112.1	−147.06
$ZnCl_2(s)$	−415.1	111	−369.4
ZnO(s)	−348.3	43.6	−318.3
ZnS(s)	−205.6	57.7	−201.3
$ZnSO_4(s)$	−982.8	120	−874.5

TABLE D.2 Heats of Formation of Gaseous Atoms From Elements in Their Standard States

Element	ΔH_f° (kJ mol^{-1})[a]	Element	ΔH_f° (kJ mol^{-1})
Group IA		Group IVA	
H	217.89	C	716.67
Li	161.5	Si	450
Na	108.2	Group VA	
K	89.62	N	472.68
Rb	82.0	P	332.2
Cs	78.2	Group VIA	
Group IIA		O	249.17
Be	324.3	S	276.98
Mg	146.4	Group VIIA	
Ca	178.2	F	79.14
Sr	163.6	Cl	121.47
Ba	177.8	Br	112.38
Group IIIA		I	107.48
B	560		
Al	329.7		

[a] All values in this table are positive because formation of the gaseous atoms from the elements is endothermic: it involves bond breaking.

TABLE D.3 Average Bond Energies

Bond	Bond Energy (kJ mol^{-1})	Bond	Bond Energy (kJ mol^{-1})
C—C	348	C—Br	276
C=C	612	C—I	238
C≡C	960	H—H	436
C—H	412	H—F	565
C—N	305	H—Cl	431
C=N	613	H—Br	366
C≡N	890	H—I	299
C—O	360	H—N	388
C=O	743	H—O	463
C—F	484	H—S	338
C—Cl	338	H—Si	376

TABLE D.4 Vapor Pressure of Water as a Function of Temperature

Temp. (°C)	Vapor Pressure (torr)	Temp. (°C)	Vapor Pressure (torr)	Temp. (°C)	Vapor Pressure (torr)	Temp. (°C)	Vapor Pressure (torr)
0	4.58	26	25.2	51	97.2	76	301.4
1	4.93	27	26.7	52	102.1	77	314.1
3	5.68	28	28.3	53	107.2	78	327.3
4	6.10	29	30.0	54	112.5	79	341.0
5	6.54	30	31.8	55	118.0	80	355.1
6	7.01	31	33.7	56	123.8	81	369.7
7	7.51	32	35.7	57	129.8	82	384.9
8	8.04	33	37.7	58	136.1	83	400.6
9	8.61	34	39.9	59	142.6	84	416.8
10	9.21	35	41.2	60	149.4	85	433.6
11	9.84	36	44.6	61	156.4	86	450.9
12	10.5	37	47.1	62	163.8	87	468.7
13	11.2	38	49.7	63	171.4	88	487.1
14	12.0	39	52.4	64	179.3	89	506.1
15	12.8	40	55.3	65	187.5	90	525.8
16	13.6	41	58.3	66	196.1	91	546.0
17	14.5	42	61.5	67	205.0	92	567.0
18	15.5	43	64.8	68	214.2	93	588.6
19	16.5	44	68.3	69	223.7	94	610.9
20	17.5	45	71.9	70	233.7	95	633.9
21	18.6	46	75.6	71	243.9	96	657.6
22	19.8	47	79.6	72	254.6	97	682.1
23	21.1	48	83.7	73	265.7	98	707.3
24	22.4	49	88.0	74	277.2	99	733.2
25	23.8	50	92.5	75	289.1	100	760.0
						101	787.6

TABLE D.5 Solubility Product Constants

Salt	Solubility Equilibrium	K_{sp}
Fluorides		
MgF_2	$MgF_2(s) \rightleftharpoons Mg^{2+}(aq) + 2F^-(aq)$	6.6×10^{-9}
CaF_2	$CaF_2(s) \rightleftharpoons Ca^{2+}(aq) + 2F^-(aq)$	3.9×10^{-11}
SrF_2	$SrF_2(s) \rightleftharpoons Sr^{2+}(aq) + 2F^-(aq)$	2.9×10^{-9}
BaF_2	$BaF_2(s) \rightleftharpoons Ba^{2+}(aq) + 2F^-(aq)$	1.7×10^{-6}
LiF	$LiF(s) \rightleftharpoons Li^+(aq) + F^-(aq)$	1.7×10^{-3}
PbF_2	$PbF_2(s) \rightleftharpoons Pb^{2+}(aq) + 2F^-(aq)$	3.6×10^{-8}
Chlorides		
$CuCl$	$CuCl(s) \rightleftharpoons Cu^+(aq) + Cl^-(aq)$	1.9×10^{-7}
$AgCl$	$AgCl(s) \rightleftharpoons Ag^+(aq) + Cl^-(aq)$	1.8×10^{-10}
Hg_2Cl_2	$Hg_2Cl_2(s) \rightleftharpoons Hg_2^{2+}(aq) + 2Cl^-(aq)$	1.2×10^{-18}
$TlCl$	$TlCl(s) \rightleftharpoons Tl^+(aq) + Cl^-(aq)$	1.8×10^{-4}
$PbCl_2$	$PbCl_2(s) \rightleftharpoons Pb^{2+}(aq) + 2Cl^-(aq)$	1.7×10^{-5}
$AuCl_3$	$AuCl_3(s) \rightleftharpoons Au^{3+}(aq) + 3Cl^-(aq)$	3.2×10^{-25}
Bromides		
$CuBr$	$CuBr(s) \rightleftharpoons Cu^+(aq) + Br^-(aq)$	5×10^{-9}
$AgBr$	$AgBr(s) \rightleftharpoons Ag^+(aq) + Br^-(aq)$	5.0×10^{-13}
Hg_2Br_2	$Hg_2^{2+}(s) \rightleftharpoons Hg_2^{2+}(aq) + 2Br^-(aq)$	5.6×10^{-23}
$HgBr_2$	$HgBr_2(s) \rightleftharpoons Hg^{2+}(aq) + 2Br^-(aq)$	1.3×10^{-19}
$PbBr_2$	$PbBr_2(s) \rightleftharpoons Pb^{2+}(aq) + 2Br^-(aq)$	2.1×10^{-6}
Iodides		
CuI	$CuI(s) \rightleftharpoons Cu^+(aq) + I^-(aq)$	1×10^{-12}
AgI	$AgI(s) \rightleftharpoons Ag^+(aq) + I^-(aq)$	8.3×10^{-17}
Hg_2I_2	$Hg_2I_2(s) \rightleftharpoons Hg_2^{2+}(aq) + 2I^-(aq)$	4.7×10^{-29}
HgI_2	$HgI_2(s) \rightleftharpoons Hg^{2+}(aq) + 2I^-(aq)$	1.1×10^{-28}
PbI_2	$PbI_2(s) \rightleftharpoons Pb^{2+}(aq) + 2I^-(aq)$	7.9×10^{-9}
Hydroxides		
$Mg(OH)_2$	$Mg(OH)_2(s) \rightleftharpoons Mg^{2+}(aq) + 2OH^-(aq)$	7.1×10^{-12}
$Ca(OH)_2$	$Ca(OH)_2(s) \rightleftharpoons Ca^{2+}(aq) + 2OH^-(aq)$	6.5×10^{-6}
$Mn(OH)_2$	$Mn(OH)_2(s) \rightleftharpoons Mn^{2+}(aq) + 2OH^-(aq)$	1.6×10^{-13}
$Fe(OH)_2$	$Fe(OH)_2(s) \rightleftharpoons Fe^{2+}(aq) + 2OH^-(aq)$	7.9×10^{-16}
$Fe(OH)_3$	$Fe(OH)_3(s) \rightleftharpoons Fe^{3+}(aq) + 3OH^-(aq)$	1.6×10^{-39}
$Co(OH)_2$	$Co(OH)_2(s) \rightleftharpoons Co^{2+}(aq) + 2OH^-(aq)$	1×10^{-15}
$Co(OH)_3$	$Co(OH)_3(s) \rightleftharpoons Co^{3+}(aq) + 3OH^-(aq)$	3×10^{-45}
$Ni(OH)_2$	$Ni(OH)_2(s) \rightleftharpoons Ni^{2+}(aq) + 2OH^-(aq)$	6×10^{-16}
$Cu(OH)_2$	$Cu(OH)_2(s) \rightleftharpoons Cu^{2+}(aq) + 2OH^-(aq)$	4.8×10^{-20}
$V(OH)_3$	$V(OH)_3(s) \rightleftharpoons V^{3+}(aq) + 3OH^-(aq)$	4×10^{-35}
$Cr(OH)_3$	$Cr(OH)_3(s) \rightleftharpoons Cr^{3+}(aq) + 3OH^-(aq)$	2×10^{-30}
Ag_2O	$Ag_2O(s) + H_2O \rightleftharpoons 2Ag^+(aq) + 2OH^-(aq)$	1.9×10^{-8}
$Zn(OH)_2$	$Zn(OH)_2(s) \rightleftharpoons Zn^{2+}(aq) + 2OH^-(aq)$	3.0×10^{-16}
$Cd(OH)_2$	$Cd(OH)_2(s) \rightleftharpoons Cd^{2+}(aq) + 2OH^-(aq)$	5.0×10^{-15}
$Al(OH)_3$ (alpha form)	$Al(OH)_3(s) \rightleftharpoons Al^{3+}(aq) + 3OH^-(aq)$	3×10^{-34}
Cyanides		
$AgCN$	$AgCN(s) \rightleftharpoons Ag^+(aq) + CN^-(aq)$	2.2×10^{-16}
$Zn(CN)_2$	$Zn(CN)_2(s) \rightleftharpoons Zn^{2+}(aq) + 2CN^-(aq)$	3×10^{-16}
Sulfites		
$CaSO_3$	$CaSO_3(s) \rightleftharpoons Ca^{2+}(aq) + SO_3^{2-}(aq)$	3×10^{-7}
Ag_2SO_3	$Ag_2SO_3(s) \rightleftharpoons 2Ag^+(aq) + SO_3^{2-}(aq)$	1.5×10^{-14}
$BaSO_3$	$BaSO_3(s) \rightleftharpoons Ba^{2+}(aq) + SO_3^{2-}(aq)$	8×10^{-7}

TABLE D.5 Solubility Product Constants *(Continued)*

Salt	Solubility Equilibrium	K_{sp}
Sulfates		
$CaSO_4$	$CaSO_4(s) \rightleftharpoons Ca^{2+}(aq) + SO_4^{2-}(aq)$	2.4×10^{-5}
$SrSO_4$	$SrSO_4(s) \rightleftharpoons Sr^{2+}(aq) + SO_4^{2-}(aq)$	3.2×10^{-7}
$BaSO_4$	$BaSO_4(s) \rightleftharpoons Ba^{2+}(aq) + SO_4^{2-}(aq)$	1.1×10^{-10}
$RaSO_4$	$RaSO_4(s) \rightleftharpoons Ra^{2+}(aq) + SO_4^{2-}(aq)$	4.3×10^{-11}
Ag_2SO_4	$Ag_2SO_4(s) \rightleftharpoons 2Ag^+(aq) + SO_4^{2-}(aq)$	1.5×10^{-5}
Hg_2SO_4	$Hg_2SO_4(s) \rightleftharpoons Hg_2^{2+}(aq) + SO_4^{2-}(aq)$	7.4×10^{-7}
$PbSO_4$	$PbSO_4(s) \rightleftharpoons Pb^{2+}(aq) + SO_4^{2-}(aq)$	6.3×10^{-7}
Chromates		
$BaCrO_4$	$BaCrO_4(s) \rightleftharpoons Ba^{2+}(aq) + CrO_4^{2-}(aq)$	2.1×10^{-10}
$CuCrO_4$	$CuCrO_4(s) \rightleftharpoons Cu^{2+}(aq) + CrO_4^{2-}(aq)$	3.6×10^{-6}
Ag_2CrO_4	$Ag_2CrO_4(s) \rightleftharpoons 2Ag^+(aq) + CrO_4^{2-}(aq)$	1.2×10^{-12}
Hg_2CrO_4	$Hg_2CrO_4(s) \rightleftharpoons Hg_2^{2+}(aq) + CrO_4^{2-}(aq)$	2.0×10^{-9}
$CaCrO_4$	$CaCrO_4(s) \rightleftharpoons Ca^{2+}(aq) + CrO_4^{2-}(aq)$	7.1×10^{-4}
$PbCrO_4$	$PbCrO_4(s) \rightleftharpoons Pb^{2+}(aq) + CrO_4^{2-}(aq)$	1.8×10^{-14}
Carbonates		
$MgCO_3$	$MgCO_3(s) \rightleftharpoons Mg^{2+}(aq) + CO_3^{2-}(aq)$	3.5×10^{-8}
$CaCO_3$	$CaCO_3(s) \rightleftharpoons Ca^{2+}(aq) + CO_3^{2-}(aq)$	4.5×10^{-9}
$SrCO_3$	$SrCO_3(s) \rightleftharpoons Sr^{2+}(aq) + CO_3^{2-}(aq)$	9.3×10^{-10}
$BaCO_3$	$BaCO_3(s) \rightleftharpoons Ba^{2+}(aq) + CO_3^{2-}(aq)$	5.0×10^{-9}
$MnCO_3$	$MnCO_3(s) \rightleftharpoons Mn^{2+}(aq) + CO_3^{2-}(aq)$	5.0×10^{-10}
$FeCO_3$	$FeCO_3(s) \rightleftharpoons Fe^{2+}(aq) + CO_3^{2-}(aq)$	2.1×10^{-11}
$CoCO_3$	$CoCO_3(s) \rightleftharpoons Co^{2+}(aq) + CO_3^{2-}(aq)$	1.0×10^{-10}
$NiCO_3$	$NiCO_3(s) \rightleftharpoons Ni^{2+}(aq) + CO_3^{2-}(aq)$	1.3×10^{-7}
$CuCO_3$	$CuCO_3(s) \rightleftharpoons Cu^{2+}(aq) + CO_3^{2-}(aq)$	2.3×10^{-10}
Ag_2CO_3	$Ag_2CO_3(s) \rightleftharpoons 2Ag^+(aq) + CO_3^{2-}(aq)$	8.1×10^{-12}
Hg_2CO_3	$Hg_2CO_3(s) \rightleftharpoons Hg_2^{2+}(aq) + CO_3^{2-}(aq)$	8.9×10^{-17}
$ZnCO_3$	$ZnCO_3(s) \rightleftharpoons Zn^{2+}(aq) + CO_3^{2-}(aq)$	1.0×10^{-10}
$CdCO_3$	$CdCO_3(s) \rightleftharpoons Cd^{2+}(aq) + CO_3^{2-}(aq)$	1.8×10^{-14}
$PbCO_3$	$PbCO_3(s) \rightleftharpoons Pb^{2+}(aq) + CO_3^{2-}(aq)$	7.4×10^{-14}
Phosphates		
$Mg_3(PO_4)_2$	$Mg_3(PO_4)_2(s) \rightleftharpoons 3Mg^{2+}(aq) + 2PO_4^{3-}(aq)$	6.3×10^{-26}
$SrHPO_4$	$SrHPO_4(s) \rightleftharpoons Sr^{2+}(aq) + HPO_4^{2-}(aq)$	1.2×10^{-7}
$BaHPO_4$	$BaHPO_4(s) \rightleftharpoons Ba^{2+}(aq) + HPO_4^{2-}(aq)$	4.0×10^{-8}
$LaPO_4$	$LaPO_4(s) \rightleftharpoons La^{3+}(aq) + PO_4^{3-}(aq)$	3.7×10^{-23}
$Fe_3(PO_4)_2$	$Fe_3(PO_4)_2(s) \rightleftharpoons 3Fe^{2+}(aq) + 2PO_4^{3-}(aq)$	1×10^{-36}
Ag_3PO_4	$Ag_3PO_4(s) \rightleftharpoons 3Ag^+(aq) + PO_4^{3-}(aq)$	2.8×10^{-18}
$FePO_4$	$FePO_4(s) \rightleftharpoons Fe^{3+}(aq) + PO_4^{3-}(aq)$	4.0×10^{-27}
$Zn_3(PO_4)_2$	$Zn_3(PO_4)_2(s) \rightleftharpoons 3Zn^{2+}(aq) + 2PO_4^{3-}(aq)$	5×10^{-36}
$Pb_3(PO_4)_2$	$Pb_3(PO_4)_2(s) \rightleftharpoons 3Pb^{2+}(aq) + 2PO_4^{3-}(aq)$	3.0×10^{-44}
$Ba_3(PO_4)_2$	$Ba_3(PO_4)_2(s) \rightleftharpoons 3Ba^{2+}(aq) + 2PO_4^{3-}(aq)$	5.8×10^{-38}
Ferrocyanides		
$Zn_2[Fe(CN)_6]$	$Zn_2[Fe(CN)_6](s) \rightleftharpoons 2Zn^{2+}(aq) + Fe(CN)_6^{4-}(aq)$	2.1×10^{-16}
$Cd_2[Fe(CN)_6]$	$Cd_2[Fe(CN)_6](s) \rightleftharpoons 2Cd^{2+}(aq) + Fe(CN)_6^{4-}(aq)$	4.2×10^{-18}
$Pb_2[Fe(CN)_6]$	$Pb_2[Fe(CN)_6](s) \rightleftharpoons 2Pb^{2+}(aq) + Fe(CN)_6^{4-}(aq)$	9.5×10^{-19}

TABLE D.6 Formation Constants of Complexes (25 °C)

Complex Ion Equilibrium	K_{form}
Halide complexes	
$Al^{3+} + 6F^- \rightleftharpoons [AlF_6]^{3-}$	2.5×10^4
$Al^{3+} + 4F^- \rightleftharpoons [AlF_4]^-$	2.0×10^8
$Be^{2+} + 4F^- \rightleftharpoons [BeF_4]^{2-}$	1.3×10^{13}
$Sn^{4+} + 6F^- \rightleftharpoons [SnF_6]^{2-}$	1×10^{25}
$Cu^+ + 2Cl^- \rightleftharpoons [CuCl_2]^-$	3×10^5
$Ag^+ + 2Cl^- \rightleftharpoons [AgCl_2]^-$	1.8×10^5
$Pb^{2+} + 4Cl^- \rightleftharpoons [PbCl_4]^{2-}$	2.5×10^{15}
$Zn^{2+} + 4Cl^- \rightleftharpoons [ZnCl_4]^{2-}$	1.6
$Hg^{2+} + 4Cl^- \rightleftharpoons [HgCl_4]^{2-}$	5.0×10^{15}
$Cu^+ + 2Br^- \rightleftharpoons [CuBr_2]^-$	8×10^5
$Ag^+ + 2Br^- \rightleftharpoons [AgBr_2]^-$	1.7×10^7
$Hg^{2+} + 4Br^- \rightleftharpoons [HgBr_4]^{2-}$	1×10^{21}
$Cu^+ + 2I^- \rightleftharpoons [CuI_2]^-$	8×10^8
$Ag^+ + 2I^- \rightleftharpoons [AgI_2]^-$	1×10^{11}
$Pb^{2+} + 4I^- \rightleftharpoons [PbI_4]^{2-}$	3×10^4
$Hg^{2+} + 4I^- \rightleftharpoons [HgI_4]^{2-}$	1.9×10^{30}
Ammonia complexes	
$Ag^+ + 2NH_3 \rightleftharpoons [Ag(NH_3)_2]^+$	1.6×10^7
$Zn^{2+} + 4NH_3 \rightleftharpoons [Zn(NH_3)_4]^{2+}$	7.8×10^8
$Cu^{2+} + 4NH_3 \rightleftharpoons [Cu(NH_3)_4]^{2+}$	1.1×10^{13}
$Hg^{2+} + 4NH_3 \rightleftharpoons [Hg(NH_3)_4]^{2+}$	1.8×10^{19}
$Co^{2+} + 6NH_3 \rightleftharpoons [Co(NH_3)_6]^{2+}$	5.0×10^4
$Co^{3+} + 6NH_3 \rightleftharpoons [Co(NH_3)_6]^{3+}$	4.6×10^{33}
$Cd^{2+} + 6NH_3 \rightleftharpoons [Cd(NH_3)_6]^{2+}$	2.6×10^5
$Ni^{2+} + 6NH_3 \rightleftharpoons [Ni(NH_3)_6]^{2+}$	2.0×10^8
Cyanide complexes	
$Fe^{2+} + 6CN^- \rightleftharpoons [Fe(CN)_6]^{4-}$	1.0×10^{24}
$Fe^{3+} + 6CN^- \rightleftharpoons [Fe(CN)_6]^{3-}$	1.0×10^{31}
$Ag^+ + 2CN^- \rightleftharpoons [Ag(CN)_2]^-$	5.3×10^{18}
$Cu^+ + 2CN^- \rightleftharpoons [Cu(CN)_2]^-$	1.0×10^{16}
$Cd^{2+} + 4CN^- \rightleftharpoons [Cd(CN)_4]^{2-}$	7.7×10^{16}
$Au^+ + 2CN^- \rightleftharpoons [Au(CN)_2]^-$	2×10^{38}
Complexes with other monodentate ligands	
$Ag^+ + 2CH_3NH_2 \rightleftharpoons [Ag(CH_3NH_2)_2]^+$ methylamine	7.8×10^6
$Cd^{2+} + 4SCN^- \rightleftharpoons [Cd(SCN)_4]^{2-}$ thiocyanate ion	1×10^3
$Cu^{2+} + 2SCN^- \rightleftharpoons [Cu(SCN)_2]$	5.6×10^3
$Fe^{3+} + 3SCN^- \rightleftharpoons [Fe(SCN)_3]$	2×10^6
$Hg^{2+} + 4SCN^- \rightleftharpoons [Hg(SCN)_4]^{2-}$	5.0×10^{21}
$Cu^{2+} + 4OH^- \rightleftharpoons [Cu(OH)_4]^{2-}$	1.3×10^{16}
$Zn^{2+} + 4OH^- \rightleftharpoons [Zn(OH)_4]^{2-}$	2×10^{20}
Complexes with bidentate ligands[a]	
$Mn^{2+} + 3\,en \rightleftharpoons [Mn(en)_3]^{2+}$	6.5×10^5
$Fe^{2+} + 3\,en \rightleftharpoons [Fe(en)_3]^{2+}$	5.2×10^9
$Co^{2+} + 3\,en \rightleftharpoons [Co(en)_3]^{2+}$	1.3×10^{14}
$Co^{3+} + 3\,en \rightleftharpoons [Co(en)_3]^{3+}$	4.8×10^{48}
$Ni^{2+} + 3\,en \rightleftharpoons [Ni(en)_3]^{2+}$	4.1×10^{17}

TABLE D.6 Formation Constants of Complexes (25 °C) *(Continued)*

Complex Ion Equilibrium	K_{form}
Complexes with bidentate ligands[a]	
$Cu^{2+} + 2\ en \rightleftharpoons [Cu(en)_2]^{2+}$	3.5×10^{19}
$Mn^{2+} + 3\ bipy \rightleftharpoons [Mn(bipy)_3]^{2+}$	1×10^{6}
$Fe^{2+} + 3\ bipy \rightleftharpoons [Fe(bipy)_3]^{2+}$	1.6×10^{17}
$Ni^{2+} + 3\ bipy \rightleftharpoons [Ni(bipy)_3]^{2+}$	3.0×10^{20}
$Co^{2+} + 3\ bipy \rightleftharpoons [Co(bipy)_3]^{2+}$	8×10^{15}
$Mn^{2+} + 3\ phen \rightleftharpoons [Mn(phen)_3]^{2+}$	2×10^{10}
$Fe^{2+} + 3\ phen \rightleftharpoons [Fe(phen)_3]^{2+}$	1×10^{21}
$Co^{2+} + 3\ phen \rightleftharpoons [Co(phen)_3]^{2+}$	6×10^{19}
$Ni^{2+} + 3\ phen \rightleftharpoons [Ni(phen)_3]^{2+}$	2×10^{24}
$Co^{2+} + 3C_2O_4^{2-} \rightleftharpoons [Co(C_2O_4)_3]^{4-}$	4.5×10^{6}
$Fe^{3+} + 3C_2O_4^{2-} \rightleftharpoons [Fe(C_2O_4)_3]^{3-}$	3.3×10^{20}
Complexes of other ligands	
$Zn^{2+} + EDTA^{4-} \rightleftharpoons [Zn(EDTA)]^{2-}$	3.8×10^{16}
$Mg^{2+} + 2NTA^{3-} \rightleftharpoons [Mg(NTA)_2]^{4-}$	1.6×10^{10}
$Ca^{2+} + 2NTA^{3-} \rightleftharpoons [Ca(NTA)_2]^{4-}$	3.2×10^{11}

[a] en = ethylenediamine

bipy =

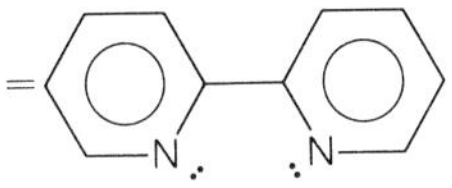

bipyridyl

phen =

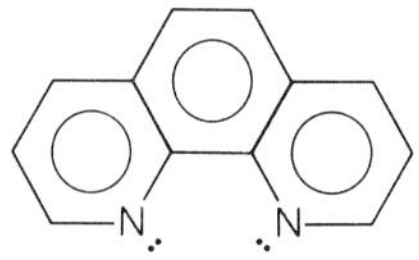

1,10-phenanthroline

$EDTA^{4-}$ = ethylendiaminetetraacetate ion

NTA^{3-} = nitrilotriacetate ion

TABLE D.7 Ionization Constants of Weak Acids and Bases

Monoprotic Acids		K_a
$HC_2O_2Cl_3$	Trichloroacetic acid (Cl_3CCO_2H)	2.2×10^{-1}
HIO_3	Iodic acid	1.69×10^{-1}
$HC_2HO_2Cl_2$	Dichloroacetic acid (Cl_2CHCO_2H)	5.0×10^{-2}
$HC_2H_2O_2Cl$	Chloroacetic acid (ClH_2CCO_2H)	1.36×10^{-3}
HNO_2	Nitrous acid	7.1×10^{-4}
HF	Hydrofluoric acid	6.8×10^{-4}
$HOCN$	Cyanic acid	3.5×10^{-4}
$HCHO_2$	Formic acid (HCO_2H)	1.8×10^{-4}
$HC_3H_5O_3$	Lactic acid ($CH_3CH(OH)CO_2H$)	1.38×10^{-4}
$HC_4H_3N_2O_3$	Barbituric acid	9.8×10^{-5}
$HC_7H_5O_2$	Benzoic acid ($C_6H_5CO_2H$)	6.28×10^{-5}
$HC_4H_7O_2$	Butanoic acid ($CH_3CH_2CH_2CO_2H$)	1.52×10^{-5}
HN_3	Hydrazoic acid	1.8×10^{-5}
$HC_2H_3O_2$	Acetic acid (CH_3CO_2H)	1.8×10^{-5}
$HC_3H_5O_2$	Propanoic acid ($CH_3CH_2CO_2H$)	1.34×10^{-5}
$HOCl$	Hypochlorous acid	3.0×10^{-8}
$HOBr$	Hypobromous acid	2.1×10^{-9}
HCN	Hydrocyanic acid	6.2×10^{-10}
HC_6H_5O	Phenol	1.3×10^{-10}
HOI	Hypoiodous acid	2.3×10^{-11}
H_2O_2	Hydrogen peroxide	1.8×10^{-12}

Polyprotic Acids		K_{a1}	K_{a2}	K_{a3}
H_2SO_4	Sulfuric acid	Large	1.0×10^{-2}	
H_2CrO_4	Chromic acid	5.0	1.5×10^{-6}	
$H_2C_2O_4$	Oxalic acid	5.6×10^{-2}	5.4×10^{-5}	
H_3PO_3	Phosphorous acid	3×10^{-2}	1.6×10^{-7}	
H_2SO_3	Sulfurous acid	1.2×10^{-2}	6.6×10^{-8}	
H_2SeO_3	Selenous acid	4.5×10^{-3}	1.1×10^{-8}	
H_2TeO_3	Tellurous acid	3.3×10^{-3}	2.0×10^{-8}	
$H_2C_3H_2O_4$	Malonic acid ($HO_2CCH_2CO_2H$)	1.4×10^{-3}	2.0×10^{-6}	
$H_2C_8H_4O_4$	Phthalic acid	1.1×10^{-3}	3.9×10^{-6}	
$H_2C_4H_4O_6$	Tartaric acid	9.2×10^{-4}	4.3×10^{-5}	
$H_2C_6H_6O_6$	Ascorbic acid	7.9×10^{-5}	1.6×10^{-12}	
H_2CO_3	Carbonic acid	4.5×10^{-7}	4.7×10^{-11}	
H_3PO_4	Phosphoric acid	7.1×10^{-3}	6.3×10^{-8}	4.5×10^{-13}
H_3AsO_4	Arsenic acid	5.6×10^{-3}	1.7×10^{-7}	4.0×10^{-12}
$H_3C_6H_5O_7$	Citric acid	7.1×10^{-4}	1.7×10^{-5}	6.3×10^{-6}

TABLE D.7 Ionization Constants of Weak Acids and Bases *(Continued)*

Weak Bases		K_b
$(CH_3)_2NH$	Dimethylamine	9.6×10^{-4}
CH_3NH_2	Methylamine	4.4×10^{-4}
$CH_3CH_2NH_2$	Ethylamine	4.3×10^{-4}
$(CH_3)_3N$	Trimethylamine	7.4×10^{-5}
NH_3	Ammonia	1.8×10^{-5}
N_2H_4	Hydrazine	9.6×10^{-7}
NH_2OH	Hydroxylamine	6.6×10^{-9}
C_5H_5N	Pyridine	1.5×10^{-9}
$C_6H_5NH_2$	Aniline	4.1×10^{-10}
PH_3	Phosphine	10^{-28}

TABLE D.8 Standard Reduction Potentials (25 °C)

$E°$ (volts)	Half-Cell Reaction
+2.87	$F_2(g) + 2e^- \rightleftharpoons 2F^-(aq)$
+2.08	$O_3(g) + 2H^+(aq) + 2e^- \rightleftharpoons O_2(g) + H_2O$
+2.05	$S_2O_8^{2-}(aq) + 2e^- \rightleftharpoons 2SO_4^{2-}(aq)$
+1.82	$Co^{3+}(aq) + e^- \rightleftharpoons Co^{2+}(aq)$
+1.77	$H_2O_2(aq) + 2H^+(aq) + 2e^- \rightleftharpoons 2H_2O$
+1.695	$MnO_4^-(aq) + 4H^+(aq) + 3e^- \rightleftharpoons MnO_2(s) + 2H_2O$
+1.69	$PbO_2(s) + SO_4^{2-}(aq) + 4H^+(aq) + 2e^- \rightleftharpoons PbSO_4(s) + 2H_2O$
+1.63	$2HOCl(aq) + 2H^+(aq) + 2e^- \rightleftharpoons Cl_2(g) + 2H_2O$
+1.51	$Mn^{3+}(aq) + e^- \rightleftharpoons Mn^{2+}(aq)$
+1.49	$MnO_4^-(aq) + 8H^+(aq) + 5e^- \rightleftharpoons Mn^{2+}(aq) + 4H_2O$
+1.46	$PbO_2(s) + 4H^+(aq) + 2e^- \rightleftharpoons Pb^{2+}(aq) + 2H_2O$
+1.44	$BrO_3^-(aq) + 6H^+(aq) + 6e^- \rightleftharpoons Br^-(aq) + 3H_2O$
+1.42	$Au^{3+}(aq) + 3e^- \rightleftharpoons Au(s)$
+1.36	$Cl_2(g) + 2e^- \rightleftharpoons 2Cl^-(aq)$
+1.33	$Cr_2O_7^{2-}(aq) + 14H^+(aq) + 6e^- \rightleftharpoons 2Cr^{3+}(aq) + 7H_2O$
+1.24	$O_3(g) + H_2O + 2e^- \rightleftharpoons O_2(g) + 2OH^-(aq)$
+1.23	$MnO_2(s) + 4H^+(aq) + 2e^- \rightleftharpoons Mn^{2+}(aq) + 2H_2O$
+1.23	$O_2(g) + 4H^+(aq) + 4e^- \rightleftharpoons 2H_2O$
+1.20	$Pt^{2+}(aq) + 2e^- \rightleftharpoons Pt(s)$
+1.07	$Br_2(aq) + 2e^- \rightleftharpoons 2Br^-(aq)$
+0.96	$NO_3^-(aq) + 4H^+(aq) + 3e^- \rightleftharpoons NO(g) + 2H_2O$
+0.94	$NO_3^-(aq) + 3H^+(aq) + 2e^- \rightleftharpoons HNO_2(aq) + H_2O$
+0.91	$2Hg^{2+}(aq) + 2e^- \rightleftharpoons Hg_2^{2+}(aq)$
+0.87	$HO_2^-(aq) + H_2O + 2e^- \rightleftharpoons 3OH^-(aq)$
+0.80	$NO_3^-(aq) + 4H^+(aq) + 2e^- \rightleftharpoons 2NO_2(g) + 2H_2O$
+0.80	$Ag^+(aq) + e^- \rightleftharpoons Ag(s)$
+0.77	$Fe^{3+}(aq) + e^- \rightleftharpoons Fe^{2+}(aq)$
+0.69	$O_2(g) + 2H^+(aq) + 2e^- \rightleftharpoons H_2O_2(aq)$
+0.54	$I_2(s) + 2e^- \rightleftharpoons 2I^-(aq)$
+0.49	$NiO_2(s) + 2H_2O + 2e^- \rightleftharpoons Ni(OH)_2(s) + 2OH^-(aq)$
+0.45	$SO_2(aq) + 4H^+(aq) + 4e^- \rightleftharpoons S(s) + 2H_2O$
+0.401	$O_2(g) + 2H_2O + 4e^- \rightleftharpoons 4OH^-(aq)$
+0.34	$Cu^{2+}(aq) + 2e^- \rightleftharpoons Cu(s)$
+0.27	$Hg_2Cl_2(s) + 2e^- \rightleftharpoons 2Hg(l) + 2Cl^-(aq)$
+0.25	$PbO_2(s) + H_2O + 2e^- \rightleftharpoons PbO(s) + 2OH^-(aq)$
+0.2223	$AgCl(s) + e^- \rightleftharpoons Ag(s) + Cl^-(aq)$
+0.172	$SO_4^{2-}(aq) + 4H^+(aq) + 2e^- \rightleftharpoons H_2SO_3(aq) + H_2O$
+0.169	$S_4O_6^{2-}(aq) + 2e^- \rightleftharpoons 2S_2O_3^{2-}(aq)$
+0.16	$Cu^{2+}(aq) + e^- \rightleftharpoons Cu^+(aq)$
+0.15	$Sn^{4+}(aq) + 2e^- \rightleftharpoons Sn^{2+}(aq)$
+0.14	$S(s) + 2H^+(aq) + 2e^- \rightleftharpoons H_2S(g)$
+0.07	$AgBr(s) + e^- \rightleftharpoons Ag(s) + Br^-(aq)$
0.00	$2H^+(aq) + 2e^- \rightleftharpoons H_2(g)$
−0.13	$Pb^{2+}(aq) + 2e^- \rightleftharpoons Pb(s)$
−0.14	$Sn^{2+}(aq) + 2e^- \rightleftharpoons Sn(s)$
−0.15	$AgI(s) + e^- \rightleftharpoons Ag(s) + I^-(aq)$
−0.25	$Ni^{2+}(aq) + 2e^- \rightleftharpoons Ni(s)$
−0.28	$Co^{2+}(aq) + 2e^- \rightleftharpoons Co(s)$
−0.34	$In^{3+}(aq) + 3e^- \rightleftharpoons In(s)$

TABLE D.8 Standard Reduction Potentials (25 °C) (Continued)

$E°$ (volts)	Half-Cell Reaction
−0.34	$Tl^+(aq) + e^- \rightleftharpoons Tl(s)$
−0.36	$PbSO_4(s) + 2e^- \rightleftharpoons Pb(s) + SO_4^{2-}(aq)$
−0.40	$Cd^{2+}(aq) + 2e^- \rightleftharpoons Cd(s)$
−0.44	$Fe^{2+}(aq) + 2e^- \rightleftharpoons Fe(s)$
−0.56	$Ga^{3+}(aq) + 3e^- \rightleftharpoons Ga(s)$
−0.58	$PbO(s) + H_2O + 2e^- \rightleftharpoons Pb(s) + 2OH^-(aq)$
−0.74	$Cr^{3+}(aq) + 3e^- \rightleftharpoons Cr(s)$
−0.76	$Zn^{2+}(aq) + 2e^- \rightleftharpoons Zn(s)$
−0.81	$Cd(OH)_2(s) + 2e^- \rightleftharpoons Cd(s) + 2OH^-(aq)$
−0.83	$2H_2O + 2e^- \rightleftharpoons H_2(g) + 2OH^-(aq)$
−0.88	$Fe(OH)_2(s) + 2e^- \rightleftharpoons Fe(s) + 2OH^-(aq)$
−0.91	$Cr^{2+}(aq) + e^- \rightleftharpoons Cr(s)$
−1.16	$N_2(g) + 4H_2O + 4e^- \rightleftharpoons N_2O_4(aq) + 4OH^-(aq)$
−1.18	$V^{2+}(aq) + 2e^- \rightleftharpoons V(s)$
−1.216	$ZnO_2^-(aq) + 2H_2O + 2e^- \rightleftharpoons Zn(s) + 4OH^-(aq)$
−1.63	$Ti^{2+}(aq) + 2e^- \rightleftharpoons Ti(s)$
−1.66	$Al^{3+}(aq) + 3e^- \rightleftharpoons Al(s)$
−1.79	$U^{3+}(aq) + 3e^- \rightleftharpoons U(s)$
−2.02	$Sc^{3+}(aq) + 3e^- \rightleftharpoons Sc(s)$
−2.36	$La^{3+}(aq) + 3e^- \rightleftharpoons La(s)$
−2.37	$Y^{3+}(aq) + 3e^- \rightleftharpoons Y(s)$
−2.37	$Mg^{2+}(aq) + 2e^- \rightleftharpoons Mg(s)$
−2.71	$Na^+(aq) + e^- \rightleftharpoons Na(s)$
−2.76	$Ca^{2+}(aq) + 2e^- \rightleftharpoons Ca(s)$
−2.89	$Sr^{2+}(aq) + 2e^- \rightleftharpoons Sr(s)$
−2.90	$Ba^{2+}(aq) + 2e^- \rightleftharpoons Ba(s)$
−2.92	$Cs^+(aq) + e^- \rightleftharpoons Cs(s)$
−2.92	$K^+(aq) + e^- \rightleftharpoons K(s)$
−2.93	$Rb^+(aq) + e^- \rightleftharpoons Rb(s)$
−3.05	$Li^+(aq) + e^- \rightleftharpoons Li(s)$

Glossary

This glossary has the definitions of all of the Key Terms that were marked in boldface and in color throughout the chapters plus a few additional terms. The numbers in parentheses that follow the definitions are the numbers of the sections in which the glossary entries received their principal discussions.

NOTE: References to Chapter 25 are for terms that appear in the qualitative analysis version of this text.

Absolute Zero   0 K. The lowest temperature attainable. (1.10)

Acceptor Ion   The central metal ion in a complex ion. (11.10)

Accuracy   Freedom from error. The closeness of a measurement to the true value. (1.11)

Acid   Arrhenius theory — a substance that produces hydronium ions (hydrogen ions) in water. (2.12, 11.2)
Brønsted theory — a proton donor. (11.7, 15.3)
Lewis theory — an electron pair acceptor. (11.9)

Acid–Base Indicator   A dye with one color in acid and another color in base. (11.5, 15.2, 15.10)

Acidic Anhydride   An oxide that reacts with water to make the solution acidic. (2.12)

Acidic Solution   An aqueous solution in which $[H^+] > [OH^-]$. (15.1, 15.2)

Acid Ionization Constant (K_a) $K_a = ([H^+][A^-])/[HA]$ for the equilibrium $HA \rightleftharpoons H^+ + A^-$ (15.3)

Acid Rain   Rain made acidic by dissolved sulfur and nitrogen oxides. (20.2)

Acid Salt   A salt of a partially neutralized polyprotic acid, for example, $NaHSO_4$ or $NaHCO_3$. (2.12, 20.6)

Acid Solubility Product   The special solubility product expression for metal sulfides in dilute acid that is related to the equation for their dissolving. For a divalent metal sulfide, MS,

$$MS(s) + 2H^+(aq) \longrightarrow M^{2+}(aq) + H_2S(aq)$$

$$K_{spa} = \frac{[M^{2+}][H_2S]}{[H^+]^2} \qquad \text{(Qual Topic II)}$$

Actinide Elements (Actinide Series)   Elements 90–103. (2.12, 22.1)

Activated Complex   The chemical species that exists with partly broken and partly formed bonds in the transition state. (18.7)

Activation Energy   The minimum kinetic energy that must be possessed by the reactants to give an effective collision (one that produces products). (18.7)

Activity Series   A list of metals in their order of reactivity as reducing agents. (12.6)

Addition Reaction   The addition of a molecule to a double or triple bond. (24.3)

Alcohol   An organic compound whose molecules have the OH group attached to tetrahedral carbon. (24.4)

Aldehyde   An organic compound whose molecules have the group —CH=O. (24.4)

Alkali Metals The Group IA elements (except hydrogen)—lithium, sodium, potassium, rubidium, and cesium. (2.7, 21.3)

Alkaline Dry Cell A zinc–manganese dioxide galvanic cell of 1.54 V used commonly in flashlight batteries. (17.10)

Alkaline Earth Metals The Group IIA elements—beryllium, magnesium, calcium, strontium, barium, and radium. (2.7, 21.4)

Alkalis (a) The alkali metals. (2.7) (b) Hydroxides of the alkali metals; strong bases. (11.2)

Alkane A hydrocarbon whose molecules have only single bonds. (20.6, 24.2)

Alkene A hydrocarbon whose molecules have one or more double bonds. (20.6, 24.3)

Alkyl Group An organic group of carbon and hydrogen atoms related to an alkane but with one less hydrogen atom (e.g., CH_3-, methyl; CH_3CH_2-, ethyl). (24.2)

Alkyne A hydrocarbon whose molecules have one or more triple bonds. (20.6, 24.3)

Allotrope One of two or more forms of an element. (19.4)

Alpha Particle (^4_2He) The nucleus of a helium atom. (23.3)

Alpha Radiation A high-velocity stream of alpha particles produced by radioactive decay. (23.3)

Alum A double salt with the general formula:

$$M^+M^{3+}(SO_4)_2 \cdot 12H_2O,$$

such as potassium alum, $KAl(SO_4)_2 \cdot 12H_2O$. (21.4)

Amalgam A solution of a metal in mercury. (22.2)

Amide An organic compound whose molecules have the group
$$-\overset{O}{\overset{\|}{C}}NH_2, \quad -\overset{O}{\overset{\|}{C}}NHR, \quad or \quad -\overset{O}{\overset{\|}{C}}NR_2. \quad (24.5)$$

Amine An organic compound whose molecules contain the group $-NH_2$, $-NHR$, or $-NR_2$. (24.5)

α-Amino Acid One of 20 monomers of polypeptides. (24.5)

Amorphous Solid A noncrystalline solid. A glass. (10.13)

Ampere (A) The SI unit for electric current; one coulomb per second. (17.3)

Amphiprotic Compound A compound that can act either as a proton donor or a proton acceptor; an amphoteric compound. (11.7)

Amphoteric Compound A compound that can react as either an acid or a base. (11.7)

Angstrom (Å) $1 \text{ Å} = 10^{-10}$ m = 100 pm = 0.1 nm. (6.11)

Anion A negatively charged ion. (2.9)

Anion Group 1 Anions that form insoluble salts with Ba^{2+}: SO_4^{2-}, CO_3^{2-}, and PO_4^{3-}. (25.7)

Anion Group 2 Cl^-, Br^-, I^-, and HS^-. (25.7)

Anion Group 3 NO_3^- (25.7)

Anode The positive electrode in a gas discharge tube. The electrode at which oxidation occurs during an electrochemical change. (6.1, 17.2)

Antibonding Molecular Orbital A molecular orbital that denies electron density to the space between nuclei and destabilizes a molecule when occupied by electrons. (8.7)

Anticodon A triplet of bases on a *t*RNA molecule that pairs to a matching triplet—a codon—on an *m*RNA molecule during *m*RNA-directed polypeptide synthesis. (24.6)

Antimatter Any particle annihilated by a particle of ordinary matter. (23.3)

Aqua Regia One part concentrated nitric acid and three parts concentrated hydrochloric acid (by volume). (22.2)

Aqueous Solution A solution that has water as the solvent. (4.5)

Aromatic Compound An organic compound whose molecules have the benzene ring system. (24.3)

Arrhenius Acid See *Acid.*

Arrhenius Base See *Base.*

Arrhenius Equation An equation that relates the rate constant of a reaction to the reaction's activation energy. (18.8)

Association The joining together of molecules by hydrogen bonds. (13.11)

Atmosphere, Standard (Atm) 101,325 Pa. The pressure that supports a column of mercury 760 mm high at 0 °C; 760 torr. (5.5, 9.2)

Atom A neutral particle having one nucleus; the smallest representative sample of an element. (1.6, 2.3)

Atomic Mass Unit (amu) 1.6606×10^{-24} g; $\frac{1}{12}$th the mass of one atom of carbon-12. (2.3)

Atomic Number The number of protons in a nucleus. (2.3)

Atomic Spectrum The line spectrum produced when energized or excited atoms emit electromagnetic radiation. (6.2)

Atomic Weight The average mass (in amu) of the atoms of the isotopes of a given element as they occur naturally. (2.4)

Atomization Energy ($\Delta H_{atom.}$) The energy needed to rupture all of the bonds in one mole of a substance in the gas state and produce its atoms, also in the gas state. (8.6)

Avogadro's Number 6.02×10^{23}; the number of particles or formula units in one mole. (3.3)

Avogadro's Principle Equal volumes of gases contain equal numbers of molecules when they are at identical temperatures and pressures. (9.7)

Background Radiation The atomic radiation from the natural radionuclides in the environment and from cosmic radiation. (23.6)

Balance An apparatus for measuring mass. (1.10)

Balanced Equation A chemical equation that has on opposite sides of the arrow the same number of each atom and the same net charge. (1.6, 4.2, 11.3)

Band of Stability The envelope that encloses just the stable nuclides in a plot of all nuclides constructed according to their numbers of neutrons versus their numbers of protons. (23.4)

Barometer An apparatus for measuring atmospheric pressure. (9.2)

Base Arrhenius theory—a substance that releases OH^- ions in water. (2.12)
Brønsted theory—a proton acceptor. (11.7, 15.3)
Lewis theory—an electron pair acceptor. (11.9)

Base Ionization Constant, K_b $K_b = ([BH^+][OH^-])/[B]$ for the equilibrium $B + H_2O \rightleftharpoons BH^+ + OH^-$. (15.4)

Base Units   The units of the fundamental measurements of the SI. (1.9)

Basic Anhydride   An oxide that can neutralize acid or that reacts with water to give OH^-. (2.12)

Basic Oxygen Process   A method to convert pig iron into steel. (21.2)

Basic Solution   An aqueous solution in which $[H^+] < [OH^-]$. (15.1, 15.2)

Battery   One or more galvanic cells arranged to serve as a practical source of electricity. (17.10)

Becquerel (Bq)   1 disintegration/s. The SI unit for the activity of a radioactive source. (23.6)

Bent Molecule   A molecule that is nonlinear. (7.9)

Bessemer Converter   A now obsolete device for changing pig iron to steel. (21.2)

Beta Particle ($_{-1}^{0}e$)   An electron emitted by radioactive decay. (23.6)

Beta Radiation   A stream of electrons produced by radioactive decay. (23.6)

Bidentate Ligand   A ligand that has two atoms that can become simultaneously attached to the same metal ion. (11.10, 22.3)

Binary Acid   An acid with the general formula H_nX, where X is a nonmetal. (2.12, 11.8)

Binary Compound   A compound composed of two different elements. (2.9)

Binding Energy, Nuclear   The energy equivalent of the difference in mass between an atomic nucleus and the sum of the masses of its nucleons. (23.2)

Biochemistry   The study of the organic substances in organisms. (24.1)

Blast Furnace   A structure in which iron ore is reduced to iron. (21.2)

Body-Centered Cubic (bcc) Unit Cell   A unit cell having identical atoms, molecules, or ions at the corners of a cube plus one more particle in the center of the cube. (10.10)

Boiling Point   The temperature at which the vapor pressure of the liquid equals the atmospheric pressure. (10.9)

Bond Angle   The angle formed by two bonds that extend from the same atom. (8.2)

Bond Dipole   A dipole within a molecule associated with a specific bond. (7.11)

Bond Dissociation Energy   See *Bond Energy*.

Bond Distance   See *Bond Length*.

Bond Energy   The energy needed to break one mole of a particular bond to give electrically neutral fragments. (7.3, 8.6)

Bonding Molecular Orbital   A molecular orbital that introduces a buildup of electron density between nuclei and stabilizes a molecule when occupied by electrons. (8.7)

Bond Length   The distance between two nuclei that are held by a chemical bond. (7.3)

Bond Order   The *net* number of pairs of bonding electrons. bond order = $\frac{1}{2} \times$ (No. of bonding e^- − No. of antibonding e^-). (8.7)

Boundary   That which divides a system from its surroundings. (5.3)

Boyle's Law   See *Pressure–Volume Law*.

Bragg Equation $n\lambda = 2d \sin \theta$. The equation used to convert X-ray diffraction data into crystal structure. (10.11)

Branched-Chain Compound   An organic compound in whose molecules the carbon atoms do not all occur one after another in a continuous sequence. (24.2)

Branching Step   A step in a chain reaction that produces more chain-propagating species than it consumes. (18.10)

Breeder Reactor   A device in which a fissile isotope is continuously made from a fertile isotope. (23.10)

Brine   An aqueous solution of sodium chloride, often with other salts. (17.4)

Brønsted Acid   See *Acid*.

Brønsted Base   See *Base*.

Brownian Movement   The random, erratic motions of colloidally dispersed particles in a fluid. (13.1)

Buffer   (a) A pair of solutes that can keep the pH of a solution almost constant if either acid or base is added. (b) A solution containing such a pair of solutes. (15.6)

Buret   A long tube of glass usually marked in mL and 0.1-mL units and equipped with a stopcock for the controlled addition of a liquid to a receiving flask. (11.5)

By-product   The substances formed by side reactions. (4.4)

Calibration   A procedure to check the accuracy and precision of any apparatus used to make a measurement. (4.6)

Calorie (cal)   4.184 J. The energy that will raise the temperature of 1.00 g of water from 14.5 to 15.5 °C. (In popular books on foods, the term *Calorie*, with a capital C, means 1000 cal or 1 kcal.) (5.1)

Calorimeter   An apparatus used in the determination of the heat of a reaction. (5.4)

Calorimetry   The science of measuring the quantities of heat that are involved in a chemical or physical change. (5.4)

Carbohydrates   Polyhydroxyaldehydes or polyhydroxyketones or substances that yield these by hydrolysis and that are obtained from plants or animals. (24.4)

Carbon Family   Group IVA in the periodic table — carbon, silicon, germanium, tin, and lead. (20.8)

Carbonyl Group   The carbon–oxygen double bond, $C=O$. (24.4)

Carboxylic Acid   An organic compound whose molecules have the carboxyl group, $-CO_2H$. (24.4)

Catalysis   Rate enhancement caused by a catalyst. (18.2, 18.11)

Catalyst   A substance that in relatively small proportion accelerates the rate of a reaction without being permanently chemically changed. (18.2, 18.11)

Cathode   The negative electrode in a gas-discharge tube. The electrode at which reduction occurs during an electrochemical change. (6.1, 17.2)

Cathode Ray   A stream of electrons ejected from a hot metal and accelerated toward a positively charged site in a vacuum tube. (6.1)

Cation   A positively charged ion. (2.9)

Cation Group 1 Ag^+, Hg_2^{2+}, and Pb^{2+}. Also called the chloride group or the silver group. (25.1, 25.3)

Cation Group 2 Pb^{2+}, $Bi^{3+}(BiO^+)$, Cu^{2+}, $Sn^{4+}(Sn^{2+})$. Also called the hydrogen sulfide group or the copper-arsenic group. (25.1, 25.4)

Cation Group 3 Mn^{2+}, Ni^{2+}, $Fe^{2+}(Fe^{3+})$, Al^{3+}, and Zn^{2+}. Also called the ammonium sulfide group or the aluminum-nickel group. (25.1, 25.5)

Cation Group 4 Na^+, NH_4^+, Ca^{2+}, and Ba^{2+}. (25.1, 25.6)

Cell Potential, E_{cell}   The emf of a galvanic cell when no current is drawn from the cell. (17.6)

Cell Reaction   The overall chemical change that takes place in an electrolytic cell or a galvanic cell. (17.2)

Celsius Scale   A temperature scale on which water freezes at 0 °C and boils at 100 °C (at 1 atm) and that has 100 divisions called Celsius degrees between these two points. (1.10)

Centimeter (cm)   0.01 m. (1.10)

Centrifugate   The liquid that is obtained by centrifuging a solid-liquid mixture; the supernatant liquid from this operation. (25.2)

Chain Reaction   A self-sustaining change in which the products of one event cause one or more new events. (18.10, 23.9)

Change of State   Transformation of matter from one physical state to another. In thermochemistry, any change in a variable used to define the state of a particular system—a change in composition, pressure, volume, or temperature. (5.3, 10.5)

Charge-to-Mass Ratio   The ratio of a particle's electrical charge to its mass. (6.1)

Charles' Law   See *Temperature–Volume Law.*

Chelate   A complex ion containing rings formed by polydentate ligands. (22.3)

Chelate Effect   The extra stability found in complexes that contain chelate rings. (22.3)

Chemical Bond   The force of electrical attraction that holds atoms together in compounds. (2.5, 7.1)

Chemical Change   A change that converts substances into other substances; a chemical reaction. (1.2)

Chemical Energy   The potential energy of chemicals that is transferred during chemical reactions. (1.3)

Chemical Equation   A before-and-after description that uses formulas and coefficients to represent a chemical reaction. (1.6, 4.2, 11.3)

Chemical Property   The ability of a substance, either by itself or with other substances, to undergo a change into new substances. (1.4)

Chemical Reaction   A change in which new substances (products) form from starting materials (reactants). (1.2)

Chemistry   The study of the compositions of substances and the ways by which their properties are related to their compositions. (1.1)

Chirality   The "handedness" of an object; the property of an object (such as a molecule) that makes it unable to be superimposed onto a model of its own mirror image. (22.4)

cis Isomer   A stereoisomer whose uniqueness is in having two groups on the same side of some reference plane. (22.4, 24.3)

Codon   An individual unit of hereditary instruction that consists of three, side-by-side, side chains on a molecule of *m*RNA. (24.6)

Coefficients   Numbers in front of formulas in chemical equations. (1.6)

Coinage Metals   Copper, silver, and gold. (22.2)

Coke   Coal that has been strongly heated to drive off its volatile components and that is mostly carbon. (20.8, 21.2)

Colligative Property   A property such as vapor pressure lowering, boiling point elevation, freezing point depression, and osmotic pressure whose physical value depends only on the ratio of the numbers of moles of solute and solvent particles and not on their chemical identities. (13.7)

Collision Theory   The rate of a reaction is proportional to the number of collisions that occur each second between the reactants. (18.7)

Colloidal Dispersion (Colloid)   A homogeneous mixture in which the particles of one of more components have at least one dimension in the range of 1 to 1000 nm—larger than those in a solution but smaller than those in a suspension. (13.1)

Combustion   A rapid reaction with oxygen accompanied by a flame and the evolution of heat and light. (12.8)

Common Ion   The ion in a mixture of ionic substances that is common to the formulas of at least two. (14.10)

Common Ion Effect   The solubility of one salt is reduced by the presence of another having a common ion. (14.11)

Complex Ion (or simply a *Complex*)   The combination of one or more anions or neutral molecules (ligands) with a metal ion. (11.10, 14.11, 22.1)

Compound   A substance consisting of chemically combined atoms from two or more elements and present in a definite ratio. (1.5, 2.1)

Compound Nucleus   An atomic nucleus carrying excess energy following its capture of some bombarding particle. (23.5)

Concentrated Solution   A solution that has a large ratio of the amounts of solute to solvent. (4.5)

Concentration   The ratio of the quantity of solute to the quantity of solution (or the quantity of solvent). (See *Molal Concentration, Molar Concentration, Mole Fraction, Normality, Percent Concentration.*) (4.6)

Condensation   The change of a vapor to its liquid state. (10.5)

Conformation   A particular relative orientation or geometric form of a flexible molecule. (8.3)

Conjugate Acid   The species in a conjugate acid–base pair that has the greater number of H^+ units. (11.7, 15.5)

Conjugate Acid–Base Pair   Two substances (ions or molecules) whose formulas differ by only one H^+ unit. (11.7, 15.5)

Conjugate Base   The species in a conjugate acid–base pair that has the fewer number of H^+ units. (11.7, 15.5)

Conservation of Energy, Law of   See *Law of Conservation of Energy.*

Conservation of Mass–Energy, Law of   See *Law of Conservation of Mass–Energy.*

Contact Process   An industrial synthesis of sulfuric acid from sulfur in which the oxidation of sulfur dioxide to sulfur trioxide is catalyzed by contact with vanadium pentoxide. (20.3)

Continuous Spectrum   The electromagnetic spectrum corresponding to the mixture of frequencies present in white light. (6.2)

Contributing Structure   One of a set of two or more Lewis structures used in applying the theory of resonance to the structure of a compound. A resonance structure. (7.7)

Conversion Factor   A ratio constructed from the relationship between two units such as 2.54 cm/1 in., from 1 in. = 2.54 cm. (1.13)

Coordinate Covalent Bond   A covalent bond in which both electrons originated from one of the joined atoms, but otherwise like a covalent bond in all respects. (7.8)

Coordination Compound   A complex or its salt. (11.10, 22.3)

Coordination Number   The number of donor atoms that surround a metal ion. (22.3)

Copolymer   A polymer made from two or more different monomers. (24.6)

Core Electrons   The inner electrons of an atom that are not exposed to the electrons of other atoms when chemical bonds form. (6.7)

Coulomb (C)   The SI unit of electrical charge; the charge on 6.25×10^{18} electrons; the amount of charge that passes a fixed point of a wire conductor when a current of 1 A flows for 1 s. (6.1, 17.3)

Covalent Bond   A chemical bond that results when atoms share electron pairs. (7.3)

Covalent Crystal   A crystal in which the lattice positions are occupied by atoms that are covalently bonded to the atoms at adjacent lattice sites. (10.12)

Critical Mass   The mass of a fissile isotope above which a self-sustaining chain reaction occurs. (23.9)

Critical Point   The point at the end of a vapor pressure versus temperature curve for a liquid and that corresponds to the critical pressure and the critical temperature. (10.14)

Critical Pressure (P_c) The vapor pressure of a substance at its critical temperature. (10.14)

Critical Temperature (T_c) The temperature above which a substance cannot exist as a liquid regardless of the pressure. (10.14)

Crystal Field Splitting (Δ) The difference in energy between sets of d orbitals in a complex ion. (22.5)

Crystal Field Theory   A theory that considers the effects of the polarities or the charges of the ligands in a complex ion on the energies of the d orbitals of the central metal ion. (22.5)

Crystal Lattice   The repeating symmetrical pattern of atoms, molecules, or ions that occurs in a crystal. (10.10)

Curie (Ci)   A unit of activity for radioactive samples, equal to 3.7×10^{10} disintegrations/s. (23.6)

Dalton's Atomic Theory   Matter consists of tiny, indestructible particles called atoms. All atoms of one element are identical. The atoms of different elements have different masses. Atoms combine in definite ratios by atoms when they form compounds. (2.2)

Dalton's Law of Partial Pressures   See *Partial Pressures, Law of.*

Data   The information (often in the form of physical quantities) obtained in an experiment or other experience or from references. (1.7)

Decimal Multipliers   Factors—exponentials of 10 or decimals—that are used to define larger or smaller SI units. (1.9)

Decomposition   A chemical reaction that changes one substance into two or more other substances. (1.5)

Deliquescent Compound   A compound able to absorb enough water from humid air to form a concentrated solution. (3.1, 21.4)

Delocalization Energy   The difference between the energy a substance would have if its molecules had no delocalized molecular orbitals and the energy it has because of such orbitals. (8.8)

Delocalized Molecular Orbital   A molecular orbital that spreads over more than two nuclei. (8.8)

Density   The ratio of an object's mass to its volume. (1.14)

Derived Unit   Any unit defined solely in terms of base units. (1.9)

Deuterium   The isotope of hydrogen with a mass number of 2; ^2_1H. (19.3)

Deuterium Isotope Effect   The reduction in a reaction rate when a deuterium atom instead of a hydrogen atom is involved in the rate-limiting step. (19.3)

Dialysis   The passage of small molecules and ions, but not species of a colloidal size, through a semipermeable membrane. (13.10)

Diamagnetism   The condition of not capable of being attracted to a magnet. (6.6)

Diaphragm Cell   An electrolytic cell used to manufacture sodium hydroxide by the electrolysis of aqueous sodium chloride. (17.4)

Diffraction   Constructive and destructive interference by waves. (6.4)

Diffraction Pattern   The image formed on a screen or a photographic film caused by the diffraction of electromagnetic radiation such as visible light or X rays. (6.4)

Diffusion   The spontaneous intermingling of one substance with another. (9.9)

Digestion   In qualitative analysis, the heating of a suspension of a precipitate in water to encourage the growth of larger crystals. (25.2)

Dilute Solution   A solution in which the ratio of the quantities of solute to solvent is small. (3.5)

Dipole   Partial positive and partial negative charges separated by a distance. (7.10)

Dipole–Dipole Attractions   Attractions between molecules that are dipoles. (10.2)

Dipole Moment   The product of the sizes of the partial charges in a dipole multiplied by the distance between them; a measure of the polarity of a molecule. (7.10)

Diprotic Acid   An acid that can furnish two H^+ per molecule. (2.12)

Disaccharide   A carbohydrate whose molecules can be hydrolyzed to two monosaccharides. (24.4)

Disproportionation   A redox reaction in which a portion of a substance is oxidized at the expense of the rest, which is reduced. (20.7)

Dissociation   The separation of preexisting ions when an ionic compound dissolves or melts. (11.1)

DNA   Deoxyribonucleic acid; a nucleic acid that hydrolyzes to deoxyribose, phosphate ion, adenine, thymine, guanine, and cytosine, and that is the carrier of genes. (24.6)

DNA Double Helix   Two oppositely running strands of DNA held in a helical configuration by interstrand hydrogen bonds. (24.6)

Donor Atom   The atom on a ligand that makes an electron pair available in the formation of a complex. (11.10)

Double Bond   A covalent bond consisting of one sigma bond and one pi bond. (7.4)

Double-Replacement Reaction (Metathesis Reaction)   A reaction of two salts in which cations and anions exchange partners (e.g., $AgNO_3 + NaCl \rightarrow AgCl + NaNO_3$). (11.3)

Double Salt   A salt consisting of two different salts combined in a definite ratio. (21.5)

Ductility   A metal's ability to be drawn (or stretched) into wire. (2.8)

Dynamic Equilibrium   A condition in which two opposing processes are occurring at equal rates. (10.5)

Effective Nuclear Charge   The net positive charge an outer electron experiences as a result of the partial screening of the full nuclear charge by core electrons. (6.11)

Effusion   The movement of a gas through a very tiny opening into a region of lower pressure. (9.9)

Effusion, Law of (Graham's Law)   The rate of effusion of a gas is inversely proportional to the square root of its molecular weight (M) when the pressure and temperature of the gas are constant.

$$\text{effusion rate} \propto \frac{1}{\sqrt{M}} \quad (\text{constant } P \text{ and } T) \qquad (9.9)$$

Einstein Equation $\Delta E = \Delta m_0 c^2$ where ΔE is the energy obtained when a quantity of rest mass, Δm_0, is destroyed, or the energy lost when this quantity of mass is created. (23.1)

Electrochemical Change   A chemical change that is caused by or that produces electricity. (17.1)

Electrochemistry   The study of electrochemical changes. (17.1)

Electrolysis   The production of a chemical change by the passage of electricity through a solution that contains ions or through a molten ionic compound. (17.2)

Electrolysis Cell   An apparatus for electrolysis. (17.2, 17.4)

Electrolyte   A compound that conducts electricity either in solution or in the molten state. (11.1)

Electrolytic Cell   See *Electrolysis Cell*.

Electrolytic Conduction   The transport of electrical charge by ions. (17.2)

Electrolyze   To pass electricity through an electrolyte and cause a chemical change. (17.2)

Electromagnetic Energy   Energy transmitted by wavelike oscillations in the strengths of electrical and magnetic fields; light energy. (6.2)

Electromagnetic Radiation   The successive series of oscillations in the strengths of electrical and magnetic fields associated with light, microwaves, gamma rays, ultraviolet rays, infrared rays, and the like. (6.2)

Electromagnetic Spectrum   The distribution of frequencies of electromagnetic radiation among various types of such radiation—microwave, infrared, visible, ultraviolet, X, and gamma rays. (6.2)

Electromotive Force (emf)   The voltage that is produced by a galvanic cell and can make electrons move in a conductor. (17.6)

Electron (e^- or $_{-1}^{0}e$)   A subatomic particle with a charge of 1− and mass of 0.0005486 amu (9.10939×10^{-28} g) that occurs outside an atomic nucleus. The particle that moves when an electric current flows. (2.3)

Electron Affinity (EA)   The energy change (usually expressed in kJ/mol) that occurs when an electron adds to an isolated gaseous atom or ion. (6.13)

Electron Capture   The capture by a nucleus of an orbital electron and that changes a proton into a neutron in the nucleus. (23.3)

Electron Cloud   Because of its wave properties, an electron's influence spreads out like a cloud around the nucleus. (6.10)

Electron Configuration   The distribution of electrons in an atom's orbitals. (6.7)

Electron Density   The concentration of the electron's charge within a given volume. (6.10)

Electronegativity   The relative ability of an atom to attract electron density toward itself when joined to another atom by a covalent bond. (7.10)

Electronic Structure   The distribution of electrons in an atom's orbitals. (6.2, 6.7)

Electron Pair Bond   A covalent bond. (7.3)

Electron Spin   The spinning of an electron about its axis that is believed to occur because the electron behaves as a tiny magnet. (6.6)

Electron-Volt (eV)   The energy an electron receives when it is accelerated under the influence of 1 V; equal to 1.6×10^{-19} J. (23.3)

Electroplating   Depositing a thin metallic coating on an object by electrolysis. (17.4)

Element   A substance in which all of the atoms have the same atomic number. A substance that cannot be broken down by chemical reactions into anything that is both stable and simpler. (1.5, 2.6)

Elementary Process   One of the individual steps in the mechanism of a reaction. (18.9)

Elimination Reaction   The loss of a small molecule from a larger molecule as in the elimination of water from an alcohol. (24.4)

Empirical Facts   Facts discovered by performing experiments. (1.7)

Empirical Formula   A chemical formula that uses the smallest whole-number subscripts to give the proportions by atoms of the different elements present. (3.5)

Emulsifying Agent   A substance that stabilizes an emulsion. (13.1)

Emulsion   A colloidal dispersion of one liquid in another. (13.1)

Enantiomers   Stereoisomers whose molecular structures are related as an object to its mirror image but that cannot be superimposed. (22.4)

Endergonic   Descriptive of a change in which a system's free energy increases. (16.7)

Endothermic   Descriptive of a change in which a system's internal energy increases. (5.3)

Endpoint   The moment in a titration when the indicator changes color and the titration is ended. (11.5, 15.10)

Energy   Something that matter possesses by virtue of an ability to do work. (1.3, 5.1)

Energy Level   A particular energy an electron can have in an atom or a molecule.

Enthalpy (H)   The heat content of a system. (5.3, 16.2)

Enthalpy Change (ΔH)   The difference in enthalpy between the initial state and the final state for some change. (5.3, 16.2)

Enthalpy of Solution   The enthalpy change that accompanies the formation of a solution from the solute and the solvent. (13.3)

Entropy   The thermodynamic quantity that describes the degree of randomness of a system. The greater the disorder or randomness, the higher is the statistical probability of the state and the higher is the entropy. (16.5)

Enzyme   A catalyst in a living system and that consists of a protein. (24.5)

Equilibrium Constant   The value that the mass action expression has when the system is at equilibrium. (14.3)

Equilibrium Law   The mathematical equation for a particular equilibrium system that sets the mass action expression equal to the equilibrium constant. (14.3)

Equilibrium Vapor Pressure of a Liquid   The pressure exerted by a vapor in equilibrium with its liquid state. (10.7)

Equilibrium Vapor Pressure of a Solid   The pressure exerted by a vapor in equilibrium with its solid state. (10.7)

Equivalence Point   The moment in a titration when the number of equivalents of the reactant added from a buret equals the number of equivalents of another reactant in the receiving flask. (15.10)

Equivalent (eq)   (a) The amount of an acid (or base) that provides 1 mol of H^+ (or 1 mol of OH^-). (b) The amount of an oxidizing agent (or a reducing agent) that can accept (or provide) 1 mol of electrons in a particular redox reaction. (11.6, 12.10)

Equivalent Weight   The mass in grams of one equivalent. (11.6, 12.10)

Ester   An organic compound whose molecules have the group
$-C(=O)-O-C$. (24.4)

Ether   An organic compound in whose molecules two hydrocarbon groups are joined to an oxygen. (24.4)

Ethyl Group CH_3CH_2-. (24.2)

Evaporate   To change from a liquid to a vapor. (10.4)

Exact Number   A number obtained by a direct count or that results by a definition and that is considered to have an infinite number of significant figures. (1.11)

Exergonic   Descriptive of a change in a system in which there is a free energy decrease. (16.7)

Exon   One of a set of sections of a DNA molecule (separated by introns) that, taken together, constitute a gene. (24.6)

Exothermic   Descriptive of a change in which energy leaves a system and enters the surroundings. (5.3)

Exponential Notation   See *Scientific Notation.*

Extensive Property   A property of an object that is described by a physical quantity whose magnitude is proportional to the size or amount of the object (e.g., mass or volume). (1.4)

Face-Centered Cubic (fcc) Unit Cell   A unit cell having identical atoms, molecules, or ions at the corners of a cube and also in the center of each face of the cube. (10.10)

Factor-Label Method   A problem-solving technique that uses the correct cancellation of the units of physical quantities as a guide for the correct setting-up of the solution to the problem. (1.13)

Fahrenheit Scale   A temperature scale on which water freezes at 32 °F and boils at 212 °F (at 1 atm) and between which points there are 180-degree divisions called Fahrenheit degrees. (1.10)

Family of Elements   See *Group.*

Faraday ($\mathscr{F}$)   One mole of electrons; 96,500 coulombs. (17.3)

Faraday Constant ($\mathscr{F}$)   96,500 coulomb/mol e^-. (17.3)

Fatty Acid   One of several long-chain carboxylic acids produced by the hydrolysis (digestion) of a lipid. (24.4)

Ferromagnetism   The ability possessed by iron, cobalt, and nickel to have strong magnetism, much stronger than that of paramagnetic substances. (22.1)

Fertile Isotope   An isotope that can be converted by nuclear reactions into a fissile isotope. (23.10)

First Law of Thermodynamics   A formal state of the law of conservation of energy. See *Law of Conservation of Energy*. (16.2)

First-Order Reaction   A reaction with a rate law in which rate $= k[A]^1$, where A is a reactant. (18.4, 18.5)

Fissile Isotope   An isotope capable of undergoing fission following neutron capture. (23.9)

Fission   The breaking apart of atomic nuclei into smaller nuclei accompanied by the release of energy; the source of energy in nuclear reactors. (23.9)

Flame Test   The use of the color produced when a trace of a substance is thrust into a flame to identify the ion(s) present. (21.3, 25.2, 25.6)

Flocculent   A cotton-tuft-like appearance and texture of a precipitate suspended in water. (25.6)

Flotation   A method for concentrating sulfide ores of copper and lead by bubbling air through a slurry of oil-coated ore particles. The sulfides, but not soil or other rock particles, stick to the rising air bubbles and collect in the foam at the surface. (21.2)

Flow Diagram   A display of a scheme of qualitative analysis. (25.3)

Force   Anything that can cause an object to change its motion or direction. (9.2)

Formation Constant (K_{form})   The equilibrium constant for an equilibrium involving the formation of a complex ion. Also called the *stability constant*. (14.11)

Formula   A representation of the composition of a substance that uses the chemical symbols of the elements present and subscripts that describe the proportions by atoms. (1.6, 2.9)

Formula Unit   A particle that has the composition given by the chemical formula. (2.9)

Formula Weight   The sum of the atomic weights of all of the atoms represented in the chemical formula. (3.2)

Fossil Fuels   Coal, oil, and natural gas. (5.1)

Fractional Crystallization   The precipitation of one component of a mixture in solution; a technique of purification. (13.3)

Free Energy   See *Gibbs Free Energy*.

Free Radical   An atom, molecule, or ion that has one or more unpaired electrons. (18.10)

Frequency (v)   The number of cycles per second of electromagnetic radiation. (6.2)

Frequency Factor   The proportionality constant, A, in the Arrhenius equation. (18.8)

Fuel Cell   An electrochemical cell in which electricity is generated from the reactions that accompany the burning of a fuel. (17.10)

Functional Group   The group of atoms of an organic molecule that enters into a characteristic set of reactions that are independent of the rest of the molecule. (24.1)

Fusion   (a) Melting. (10.6) (b) The formation of atomic nuclei by the joining together of the nuclei of lighter atoms. (23.9)

G   See *Gibbs Free Energy*.

ΔG   See *Gibbs Free Energy Change*.

$\Delta G°$ See *Standard Free Energy Change*.

Galvanic Cell   An electrochemical cell in which a spontaneous redox reaction produces electricity. (17.5)

Gamma Radiation   Electromagnetic radiation with wavelengths in the range of 1 Å or less (the shortest wavelengths of the spectrum). (23.3)

Gangue   The unwanted rock and sand that are separated from an ore. (21.2)

Gas Constant, Universal (R) $R = 0.0821 \dfrac{\text{liter atm}}{\text{mol K}}$ (9.7)

Gas Law, Combined   For a given mass of gas, the product of its pressure and volume divided by its Kelvin temperature is a constant. $PV/T = $ a constant. (9.7)

Gas Law, Ideal $PV = nRT$ (9.7)

Gay-Lussac's Law of Combining Volumes   When measured at the same pressure and temperature, the volumes of gaseous reactants and products are in ratios of simple, whole numbers. (9.7)

Generalization   A summary statement that describes experimentally observed properties for a class of things. (1.7)

Genetic Code   The correlation of codons with amino acids. (24.6)

Geometric Isomer   One of a set of isomers that differ only in geometry. (22.4, 24.3)

Geometric Isomerism   The existence of isomers whose molecules have identical atomic organizations but different geometries; *cis–trans* isomers. (22.4, 24.3)

Gibbs Free Energy (G)   A thermodynamic quantity that relates enthalpy, (H), entropy (S), and temperature (T) by the equation $G = H - TS$. (16.7)

Gibbs Free Energy Change (ΔG)   The difference given by $(\Delta H - T\Delta S)$. (16.7)

Glass Electrode   A hollow glass tube containing the chemicals needed to complete an electrode and fitted with a thin-walled membrane at its bottom across which a potential difference can be created whose magnitude depends on the concentration of some ion in solution (e.g., H^+) and the kind of reference electrode used. (17.9)

Gradient   The presence of a change in value of some physical quantity with change in location within a system. (9.9)

Graham's Law   See *Effusion, Law of*.

Gram (g)   0.001 kg. (1.10)

Gray (Gy)   The SI unit of radiation-absorbed dose. 1 Gy = 1 J/kg. (23.6)

Ground State The lowest energy state of an atom or molecule. (6.3, 6.5)

Group A vertical column of elements in the periodic table. (2.6, 2.7)

Group Reagent In qualitative analysis, a chemical or reagent used to separate the ions of a particular group from the ions of other groups. *See also* Anion Group, Cation Group. (25.1)

ΔH See *Enthalpy Change.*

ΔH_{atom} See *Atomization Energy.*

$\Delta H°$ See *Standard Heat of Reaction.*

$\Delta H°_{combustion}$ See *Standard Heat of Combustion.*

$\Delta H°_f$ See *Standard Heat of Formation.*

ΔH_{fusion} See *Molar Heat of Fusion.*

$\Delta H_{sublimation}$ See *Molar Heat of Sublimation.*

$\Delta H_{vaporization}$ See *Molar Heat of Vaporization.*

Haber–Bosch Process An industrial synthesis of ammonia from nitrogen and hydrogen under pressure and heat in the presence of a catalyst. (19.3, 19.5)

Half-Cell That part of a galvanic cell in which either oxidation or reduction takes place. (17.5)

Half-Life ($t_{1/2}$) The time required for a reactant concentration or the mass of a radionuclide to be reduced by half. (18.6, 23.3)

Half-Reaction A hypothetical reaction that constitutes exclusively either the oxidation or the reduction half of a redox reaction and in whose equation the correct formulas for all species taking part in the change are given together with enough electrons to give the correct electrical balance. (12.4)

Halogen Family Group VIIA in the periodic table—fluorine, chlorine, bromine, iodine, and astatine. (2.7, 20.7)

Hard Water Water with dissolved Mg^{2+}, Ca^{2+}, Fe^{2+}, or Fe^{3+} ions at a concentration high enough (above 25 mg/L) to interfere with the use of soap. (11.10)

Heat Capacity The quantity of heat needed to raise the temperature of an object by 1 °C. (5.2)

Heat of Combustion The heat evolved in the combustion of a substance. (5.4)

Heat of Combustion, Standard See *Standard Heat of Combustion.*

Heat of Formation, Standard See *Standard Heat of Formation.*

Heat of Reaction The heat exchanged between a system and its surroundings when a chemical change occurs in the system. (5.3)

Heat of Reaction, Standard See *Standard Heat of Reaction.*

Henderson–Hasselbalch Equation An equation that relates the pH of a buffer solution to the molar concentrations of the members of the buffer and the pK_a of the Brønsted acid in the buffer.

$$\text{pH} = pK_a + \log \frac{[\text{anion}]}{[\text{acid}]} \qquad (15.6)$$

Henry's Law See *Pressure–Solubility Law.*

Hertz (Hz) 1 cycle/s; the SI unit of frequency. (6.2)

Hess's Law For any reaction that can be written in steps, the standard heat of reaction is the same as the sum of the standard heats of reaction for the steps. (5.6)

Hess's Law Equation For the change $aA + bB + \cdots \rightarrow nN + mM + \cdots$,

$$\Delta H° = [n\Delta H°_f(N) + m\Delta H°_f(M) + \cdots] - [a\Delta H°_f(A) + b\Delta H°_f(B) + \cdots]$$
$$= [\text{sum of } \Delta H° \text{ of products}] - [\text{sum of } \Delta H° \text{ of reactants}] \quad (5.7)$$

Heterogeneous Catalyst A catalyst that is in a different phase than the reactants and onto whose surface the reactant molecules are adsorbed and where they react. (18.11)

Heterogeneous Equilibrium An equilibrium involving more than one phase. (14.9)

Heterogeneous Mixture A mixture that has two or more phases with different properties. (1.5)

Heterogeneous Reaction A reaction in which not all of the chemical species are in the same phase. (14.9, 18.2)

High-Spin Complex A complex ion or coordination compound in which there is the maximum number of unpaired electrons. (22.5)

Homogeneous Catalyst A catalyst that is in the same phase as the reactants. (18.11)

Homogeneous Equilibrium An equilibrium system in which all components are in the same phase. (14.9)

Homogeneous Mixture A mixture that has only one phase and that has uniform properties throughout. (1.5)

Homogeneous Reaction A reaction in which all of the chemical species are in the same phase. (14.9, 18.2)

Hund's Rule Electrons that occupy orbitals of equal energy are distributed with unpaired spins as much as possible among all such orbitals. (6.7)

Hybrid Atomic Orbitals Orbitals formed by mixing two or more of the basic atomic orbitals of an atom and that make possible more effective overlaps with the orbitals of adjacent atoms than do ordinary atomic orbitals. (8.3)

Hydrate A compound that contains molecules of water in a definite ratio to other components. (1.6)

Hydrated Ion An ion surrounded by a cage of water molecules that are attracted by the charge on the ion. (11.1)

Hydration The development in an aqueous solution of a cage of water molecules about ions or polar molecules of the solute. (11.1, 13.2)

Hydration Energy The enthalpy change associated with the hydration of gaseous ions or molecules as they dissolve in water. (13.3)

Hydride (a) A binary compound of hydrogen. (b) A compound containing the hydride ion (H^-). (19.3)

Hydrocarbon An organic compound whose molecules consist entirely of carbon and hydrogen atoms. (12.8, 24.2, 24.3)

Hydrogen Bond An extra strong dipole–dipole attraction between a hydrogen bound covalently to nitrogen, oxygen, or fluorine and another nitrogen, oxygen, or fluorine atom. (10.3)

Hydrogen Electrode   The standard of comparison for reduction potentials and for which $E°_{H^+}$ has a value of 0.00 V (25 °C, 1 atm) when $[H^+] = 1\ M$ in the reversible half-cell reaction: $2H^+(aq) + 2e^- \rightleftharpoons H_2(g)$. (17.6)

Hydrolysis of Ions   The reactions of ions with water to affect the pH of the solution. (15.8, 15.9)

Hydronium Ion H_3O^+. (2.12)

Hydrophilic Group   A polar molecular unit capable of having dipole–dipole attractions or hydrogen bonds with water molecules. (13.2)

Hydrophobic Group   A nonpolar molecular unit with no affinity for the molecules of a polar solvent, like water. (13.2)

Hypothesis   A tentative explanation of the results of experiments. (1.7)

Ideal Gas   A hypothetical gas that obeys the gas laws exactly. (9.3)

Ideal Gas Law $PV = nRT$. (9.7)

Ideal Solution   A hypothetical solution that would obey the vapor pressure–concentration law (Raoult's law) exactly. (13.3, 13.7)

Incompressible   Incapable of losing volume under increasing pressure. (10.4)

Indicator   A chemical put in a solution being titrated and whose change in color signals the endpoint. (11.5, 15.2, 15.10)

Induced Dipole   A dipole created when the electron cloud of an atom or a molecule is distorted by a neighboring dipole or by an ion. (10.3)

Inert Gas   Any of the noble gases—Group 0 of the periodic table. (2.7, 19.2)

Initiation Step   The step in a chain reaction that produces reactive species that can start chain propagation steps. (18.10)

Inner Transition Elements   Members of the two long rows of elements below the main body of the periodic table—elements 58–71 and elements 90–103. (2.7, 22.1)

Inorganic Compound   A compound made from any elements except those compounds of carbon classified as organic compounds. (2.7)

Instability Constant (K_{inst})   The reciprocal of the formation constant for an equilibrium in which a complex ion forms. (14.11)

Instantaneous Dipole   A momentary dipole in an atom, ion, or molecule caused by the erratic movement of electrons. (10.3)

Intensive Property   A property whose physical quantity is independent of the size of the sample, such as density or temperature. (1.4)

Intermolecular Attractions   Attractions *between* neighboring molecules. (10.2)

Internal Energy (E)   The sum of all of the kinetic energies and potential energies of the particles within a system. (16.2)

International System of Units (SI)   The successor to the metric system of measurements that retains most of the units of the metric system and their decimal relationships but employs new reference standards. (1.9)

Intron   One of a set of sections of a DNA molecule that separate the exon sections of a gene from each other. (24.6)

Inverse Square Law   The intensity of a radiation is inversely proportional to the square of the distance from its source. (23.6)

Ion   An electrically charged particle on the atomic or molecular scale of size. (2.5)

Ion-Electron Method   A method for balancing redox reactions that uses half-reactions. (12.4)

Ionic Bond   The attractions between ions that hold them together in ionic compounds. (7.1)

Ionic Character   The extent to which a covalent bond has a dipole moment and is polarized. (7.10)

Ionic Compound   A compound consisting of positive and negative ions. (2.5)

Ionic Crystal   A crystal that has ions located at the lattice points. (10.12)

Ionic Equation   A chemical equation in which soluble, strong electrolytes are written in dissociated or ionized form. (11.3)

Ionic Reaction   A chemical reaction in which ions are involved. (11.3)

Ionization Energy (IE)   The energy needed to remove an electron from an isolated, gaseous atom, ion, or molecule (usually given in units of kJ/mol). (6.12)

Ionization Reaction   A reaction of chemical particles that produces ions. (11.2)

Ionizing Radiation   Any high-energy radiation—X rays, gamma rays, or radiations from radionuclides—that generates ions as it passes through matter. (23.6)

Ion Product   The mass action expression for the solubility equilibrium involving the ions of a salt and equal to the product of the molar concentrations of the ions, each concentration raised to a power that equals the number of ions obtained from one formula unit of the salt. (14.10)

Ion Product Constant of Water (K_w) $K_w = [H^+][OH^-]$. (15.1)

Isomer   One of a set of compounds that have identical molecular formulas but different structures. (22.4, 24.2, 24.3)

Isomerism   The existence of sets of isomers. (22.4, 24.2)

Isotope Dilution Analysis   An analytical method for determining volumes or concentrations by measuring the reduction in specific activity of radioactive material when it is diluted into a larger volume. (23.7)

Isotopes   Constituents of the same element but whose atoms have different mass numbers. (2.2)

IUPAC Rules   The formal rules for naming substances as developed by the International Union of Pure and Applied Chemistry. (24.2)

Joule (J)   The SI unit of energy. $1\ J = 1\ kg\ m^2/s^2$ and $1\ J = 4.184$ cal (exactly). (5.1)

K   See *Kelvin*.

K_a See *Acid Ionization Constant*.

K_b See *Base Ionization Constant*.

K_{sp} See *Solubility Product Constant*.

K_{spa} See *Acid Solubility Product Constant*

K_w See *Ion Product Constant of Water*.

K-Capture   See *Electron Capture*.

Kelvin (K)   One degree on the Kelvin scale of temperature identical in size to the Celsius degree. (1.10)

Kelvin Scale   The temperature scale on which water freezes at 273.15 K and boils at 373.15 K and that has 100-degree divisions called kelvins between these points. K = °C + 273.15. (1.10)

Ketone   An organic compound whose molecules have the carbonyl group (C=O) flanked by hydrocarbon groups. (24.4)

Kilocalorie (kcal)   1000 cal. (5.1)

Kilogram (kg)   The base unit for mass in the SI and equal to the mass of a cylinder of platinum–iridium alloy kept by the International Bureau of Weights and Measures at Sèvres, France. 1 kg = 1000 g. (1.10)

Kilojoule (kJ)   1000 J. 1 kJ = 4.184 kcal. (5.1)

Kinetic Energy (KE)   Energy of motion. KE = $(\frac{1}{2})mv^2$. (1.3)

Kinetic Theory of Gases   A set of postulates used to explain the gas laws. A gas consists of an extremely large number of very tiny, very hard particles in constant, random motion. They have negligible volume and, between collisions, experience no forces between themselves. (9.10)

Lake   A colored, flocculent solid suspended in water. (25.6)

Lanthanide Contraction   The decrease in the sizes of the atoms between elements 58 and 71 that causes the elements that follow the lanthanides to have unusually small radii. (22.1)

Lanthanide Elements   Elements 58–71. (2.7, 22.1)

Lattice Energy   Energy released by the imaginary process in which isolated ions come together to form a crystal of an ionic compound. (7.1)

Law   A description of behavior (and not an *explanation* of behavior) based on the results of many experiments. (1.7)

Law of Combining Volumes   See *Gay-Lussac's Law of Combining Volumes*.

Law of Conservation of Energy   The energy of the universe is constant; it can be neither created nor destroyed but only transferred and transformed. (5.3, 16.2)

Law of Conservation of Mass   No detectable gain or loss in mass occurs in chemical reactions. Mass is conserved. (2.1)

Law of Conservation of Mass–Energy   The sum of all the mass in the universe and of all of the energy, expressed as an equivalent in mass (calculated by the Einstein equation), is a constant. (23.1)

Law of Definite Proportions   In a given chemical compound, the elements are always combined in the same proportion by mass. (2.1)

Law of Gas Effusion   See *Effusion, Law of*.

Law of Heat Summation (Hess's Law)   For any reaction that can be written in steps, the standard heat of reaction is the sum of the standard heats of reaction for the steps. (5.6)

Law of Multiple Proportions   Whenever two elements form more than one compound, the different masses of one element that combine with the same mass of the other are in a ratio of small whole numbers. (2.2)

Law of Partial Pressures   See *Partial Pressures, Dalton's Law of*.

Lead Storage Battery   A galvanic cell of about 2 V involving lead and lead(IV) oxide in sulfuric acid. (17.10)

Le Châtelier's Principle   When a system that is in dynamic equilibrium is subjected to a disturbance that upsets the equilibrium, the system undergoes a change that counteracts the disturbance and, if possible, restores the equilibrium. (10.8, 14.7)

Lewis Acid   An electron pair acceptor. (11.9)

Lewis Base   An electron pair donor. (11.9)

Lewis Structure (Lewis Formula)   A structural formula drawn with Lewis symbols that uses dots and dashes to show the valence electrons and shared pairs of electrons. (7.2)

Lewis Symbol   The symbol of an element that includes dots to represent the valence electrons of an atom of the element. (7.2)

Ligand   A molecule or an anion that can bind to a metal ion to form a complex. (11.10, 14.11, 22.3)

Like-Dissolves-Like Rule   Strongly polar and ionic solutes tend to dissolve in polar solvents and nonpolar solutes tend to dissolve in nonpolar solvents. (13.2)

Limiting Reactant   The reactant that determines how much product can form when nonstoichiometric amounts of reactants are used. (4.4)

Lipid   Any substance found in plants or animals that can be dissolved in nonpolar solvents. (24.4)

Liter (L)   1 dm³. 1 L = 1000 mL = 1000 cm³. (1.10)

London Forces   Weak attractive forces caused by instantaneous dipole-induced dipole attractions. (10.3)

Lone Pair   A pair of electrons in the valence shell of an atom that is not shared with another atom. An unshared pair of electrons. (7.9)

Low-Spin Complex   A coordination compound or a complex ion with electrons paired as much as possible in the lower energy set of *d*-orbitals. (22.5)

Magic Numbers   The numbers 2, 8, 20, 28, 50, 82, and 126, numbers whose significance in nuclear science is that a nuclide in which the number of protons or neutrons equals a magic number has nuclei that are relatively more stable than those of other nuclides nearby in the band of stability. (23.4)

Magnetic Quantum Number (m_ℓ)   A quantum number that can have values from $-\ell$ to $+\ell$. (6.5)

Malleability   A metal's ability to be hammered or rolled into thin sheets. (2.8)

Manganese Nodules   Lumps the size of an orange found on the ocean floor that contain large concentrations of manganese and iron. (21.2)

Manometer   A device for measuring the pressure within a closed system. The two types—closed end and open end—differ according to whether the operating fluid (e.g., mercury) is exposed at one end to the atmosphere or not. (9.2)

Mass   A measure of the amount of matter that there is in a given sample. (1.3)

Mass Action Expression   A fraction in which the numerator is the product of the molar concentrations of the products, each raised to a power equal to its coefficient in the equilibrium equation, and the denominator is the product of the molar concentrations of the reactants, each also raised to the power that equals its coefficient in the equation. (For gaseous reactions, partial pressures can be used in place of molar concentrations.) (14.3)

Mass Number   The numerical sum of the protons and neutrons in an atom of a given isotope. (2.3)

Matter   Anything that has mass and occupies space. (1.3)

Maximum Work   The maximum amount of energy that can be harnessed as work (instead of released as heat) and equal to the standard Gibbs free energy change for the reaction. (16.9)

Mechanism of a Reaction   The series of individual steps (called elementary processes) in a chemical reaction that gives the net, overall change. (18.1, 18.9)

Melting Point   The temperature at which a substance melts; the temperature at which a solid is in equilibrium with its liquid state. (10.5)

Mercury Battery   A galvanic cell of about 1.35 V involving zinc and mercury(II) oxide. (17.10)

Metal   An element or an alloy that is a good conductor of electricity, that has a shiny surface, and that is malleable and ductile; an element that normally forms positive ions and has an oxide that is basic. (2.8)

Metallic Crystal   A solid having positive ions at the lattice positions that are attracted to a "sea of electrons" that extends throughout the entire crystal. (10.12)

Metalloids   Elements with properties that lie between those of metals and nonmetals, and that are found in the periodic table around the diagonal line running from boron (B) to astatine (At). (2.8)

Metallurgy   The science and technology of metals, the procedures and reactions that separate metals from their ores, and the operations that create practical uses for metals. (21.2)

Metathesis Reaction   See *Double-Replacement Reaction.*

Meter (m)   The SI base unit for length. (1.10)

Methyl Group CH_3—. (24.2)

Metric System   A decimal system of units for physical quantities. (1.8)

Micelle   A colloidal-sized group of ions, such as the anions of a detergent, that have clustered to maximize the contact between the ions' hydrophilic heads and water and to minimize the contact between the ions' hydrophobic parts and water. (13.2)

Milliliter (mL)   0.001 L. 1000 mL = 1 L. (1.10)

Millimeter (mm)   0.001 m. 1000 mm = 1 m. (1.10)

Millimeter of Mercury (mm Hg)   A unit of pressure equal to 1/760 atm. 760 mm Hg = 1 atm. 1 mm Hg = 1 torr. (9.2)

Miscible   Mutually soluble. (13.2)

Mixture   Any matter consisting of two or more substances physically combined in no particular proportion by mass. (1.5)

Model, Scientific   A picture or a mental construction derived from a set of ideas and assumptions that are imagined to be true because they can be used to explain certain observations and measurements (e.g., the model of an ideal gas). (9.10)

Molal Concentration (m)   The number of moles of solute in 1000 g of solvent. (13.6)

Molality   The molal concentration. (13.6)

Molar Concentration (M)   The number of moles of solute per liter of solution. The molarity of a solution. (4.6)

Molar Heat Capacity   The heat that can raise the temperature of 1 mol of a substance by 1 °C; the heat capacity per mole. (5.2)

Molar Heat of Fusion   The heat absorbed when 1 mol of a solid melts to give 1 mol of its liquid at constant temperature and pressure. (10.6)

Molar Heat of Sublimation   The heat absorbed when 1 mol of a solid sublimes to give 1 mol of its vapor at constant temperature and pressure. (10.6)

Molar Heat of Vaporization   The heat absorbed when 1 mol of a liquid changes to 1 mol of its vapor at constant temperature and pressure. (10.6)

Molarity   See *Molar Concentration.*

Molar Mass   The number of grams per mole of a compound or an element. (3.3)

Molar Solubility   The number of moles of solute required to give 1 L of a saturated solution of the solute. (14.10)

Molar Volume, Standard   The volume of 1 mol of a gas at STP, 22.4 L/mol. (9.7)

Mole (mol)   The SI unit for amount of substance; the formula weight in grams of an element or compound; a quantity of chemical substance that contains 6.02×10^{23} formula units.

Molecular Biology   The chemistry of nucleic acids and how these substances direct the synthesis of polypeptides. (24.1)

Molecular Compound   A compound consisting of neutral (but often polar) molecules. (2.5)

Molecular Crystal   A crystal that has molecules or individual atoms at the lattice points. (10.12)

Molecular Equation   A chemical equation that gives the full formulas of all of the reactants and products and that is used to plan an actual experiment. (11.3)

Molecular Formula   A chemical formula that gives the actual composition of one molecule. (2.5, 3.6)

Molecular Orbital (MO)   An orbital that extends over two or more atomic nuclei. (8.7)

Molecular Orbital Theory (MO Theory)   A theory about covalent bonds that views a molecule as a collection of positive nuclei surrounded by electrons distributed among a set of bonding and antibonding orbitals of different energies. (8.1)

Molecular Weight   See *Formula Weight.*

Molecule   A neutral particle composed of two or more atoms combined in a definite ratio of whole numbers. (1.6, 2.5, 7.3)

Mole Fraction   The ratio of the number of moles of one component of a mixture to the total number of moles of all components. (13.6)

Mole Percent   The mole fraction of a component expressed as a percent. (13.6)

Monodentate Ligand   A ligand that can attach itself to a metal ion by only one atom. (11.10, 22.3)

Monomer   A substance of relatively low formula weight that is used to make a polymer. (24.3)

Monoprotic Acid   An acid that can furnish one H^+ per molecule. (2.12)

Monosaccharide   A carbohydrate that cannot be hydrolyzed. (24.4)

Nernst Equation $E_{cell} = E_{cell}^{\circ} - \dfrac{0.0592}{n} \log Q.$ (17.9)

Net Ionic Equation   An ionic equation from which spectator ions have been omitted. It is balanced when both atoms and electrical charges balance. (11.3)

Neutralization, Acid–Base   The destruction of an acid by a base or of a base by an acid. (2.12, 15.10)

Neutral Solution   A solution in which $[H^+] = [OH^-]$. (15.1, 15.2)

Neutron (n, $^1_0 n$)   A subatomic particle that has a charge of zero and a mass of 1.008665 amu (1.674954×10^{-24} g) and that exists in all atomic nuclei except those of the hydrogen-1 isotope. (2.3, 6.1)

Neutron Activation Analysis   A technique to analyze for trace impurities in a sample by studying the frequencies and intensities of the gamma radiations they emit after they have been rendered radioactive by neutron bombardment of the sample. (23.7)

Neutron Emission   A nuclear reaction in which a neutron is ejected. (23.3)

Nickel–Cadmium Battery   A galvanic cell of about 1.4 V involving cadmium and nickel(IV) oxide. (17.10)

Nitrating Agent   A chemical that places a nitro group, NO_2, into a molecule. (19.6)

Nitrogen Family   Group VA in the periodic table—nitrogen, phosphorus, arsenic, antimony, and bismuth. (19.5)

Nitrogen Fixation   The conversion of atmospheric nitrogen into chemical forms used by plants that is accomplished by soil microorganisms with the assistance of plants. (19.5)

Noble Gases   Group 0 in the periodic table—helium, neon, argon, krypton, xenon, and radon. (2.7, 19.2)

Node   A place where the amplitude or intensity of a wave is zero. (6.5)

Nomenclature   The names of substances and the rules for devising names. (2.12, 24.2)

Nonelectrolyte   A compound that in its molten state or in solution cannot conduct electricity. (11.1)

Nonlinear Molecule   A molecule in which the atoms do not lie in a straight line. (7.9)

Nonmetal   A nonductile, nonmalleable, nonconducting element that tends to form negative ions (if it forms them at all) far more readily than positive ions and whose oxide is likely to show acidic properties. (2.8)

Nonmetallic Element   An element without metallic properties; an element with poor electrical conductivity. (2.8)

Nonoxidizing acid   An acid in which the anion is a poorer oxidizing agent than the hydrogen ion (e.g., HCl, H_2SO_4, H_3PO_4). (12.5)

Nonstoichiometric Compound   A substance in which the elements are combined in a not-quite-whole-number ratio by moles. (22.2)

Nonvolatile   Descriptive of a substance that has a high boiling point and a low vapor pressure and that does not evaporate. (13.7)

Normal Boiling Point   The temperature at which the vapor pressure of a liquid equals 1 atm. (10.9)

Normality (N)   The number of equivalents of solute per liter of solution. (11.6, 12.10)

Nuclear Equation   A description of a nuclear reaction that uses the special symbols of isotopes, that describes some kind of nuclear transformation or disintegration, and that is balanced when the sums of the atomic numbers on either side of the arrow are equal and the sums of the mass numbers are also equal. (23.3)

Nuclear Reaction   A change in the composition or energy of the nuclei of isotopes accompanied by one or more events such as the radiation of nuclear particles or electromagnetic energy, transmutation, fission, and fusion. (23.1)

Nucleic Acids   Polymers in living cells that store and translate genetic information and whose molecules hydrolyze to give a sugar unit (ribose from ribonucleic acid, RNA, or deoxyribose from deoxyribonucleic acid, DNA), a phosphate, and a set of four or five nitrogen-containing, heterocyclic bases (adenine, thymine, guanine, cytosine, and uracil). (24.6)

Nucleon   A proton or a neutron. (2.3)

Nucleus   The hard, dense core of an atom that holds the atom's protons and neutrons. (2.3, 6.1, 23.1)

Octahedral Molecule   A molecule in which the planar surfaces of an octahedron are created by connecting adjacent nuclei with imaginary lines. (7.9)

Octet (of electrons)   Eight electrons in the valence shell of an atom. (7.4)

Octet Rule   An atom tends to gain or lose electrons until its outer shell has eight electrons. (7.4)

Odd–Even rule   When the numbers of protons and neutrons in an atomic nucleus are both even, the isotope is more likely to be stable than when both numbers are odd. (23.4)

Open Hearth Furnace   A furnace used to convert pig iron into steel. (21.2)

Optical Isomers   Stereoisomers other than geometric *(cis-trans)* isomers that include substances that can rotate the plane of plane polarized light. (22.4)

Orbital   An electron waveform with a particular energy and a unique set of values for the quantum numbers n, ℓ, and m_ℓ. (6.5)

Orbital Diagram   A diagram showing an atom's orbitals in which the electrons are represented by arrows to indicate paired and unpaired spins. (6.7)

Order (of a Reaction)   The sum of the exponents in the rate law is the *overall* order. Each exponent gives the order of the reaction with respect to a specific reactant. (18.4)

Ore   A substance in the earth's crust from which an element or compound can be extracted at a profit. (21.2)

Organic Chemistry   The study of the compounds of carbon that are not classified as inorganic. (24.1)

Organic Compound   Any compound of carbon other than a carbonate, bicarbonate, cyanide, cyanate, carbide, or gaseous oxide. (2.13, 12.8, 24.1)

Osmosis   The passage of solvent molecules, but not those of solutes, through a semipermeable membrane; the limiting case of dialysis. (13.10)

Osmotic Pressure   The back pressure that would have to be applied to prevent osmosis; one of the colligative properties. (13.10)

Ostwald Process   An industrial synthesis of nitric acid from ammonia. (19.6)

Overlap   A portion of two orbitals from different atoms that share the same space in a molecule. (8.2)

Oxidation   A change in which an oxidation number increases (becomes more positive). A loss of electrons. (12.1)

Oxidation Number   The charge an atom in a molecule or ion would have if all of the electrons in its bonds belonged entirely to the more electronegative atoms; the oxidation state of an atom. (12.2)

Oxidation Number-Change Method   A method for balancing redox equations by exploiting changes in oxidation numbers. (12.3)

Oxidation–Reduction Reaction   A chemical reaction in which changes in oxidation numbers occur. (12.1)

Oxidation State   See *Oxidation Number.*

Oxidizing Acid   An acid in which the anion is a stronger oxidizing agent than H^+ (e.g., $HClO_4$, HNO_3). (12.5)

Oxidizing Agent   The substance that causes oxidation and that is itself reduced. (12.1)

Oxoacid   An acid that contains oxygen besides hydrogen and another element (e.g., HNO_3, H_3PO_4, H_2SO_4). (2.13, 11.8)

Oxygen Family   Group VIA in the periodic table — oxygen, sulfur, selenium, tellurium, and polonium. (19.4)

Pairing Energy   The energy required to force two electrons to become paired and occupy the same orbital. (22.5)

Paramagnetism   The weak magnetism of a substance whose atoms, molecules, or ions have unpaired electrons. (6.6)

Partial Charge   Charges at opposite ends of a dipole that are fractions of full $1+$ or $1-$ charges. (7.10)

Partial Pressure   The pressure contributed by an individual gas to the total pressure of a gas mixture. (9.4)

Partial Pressure, Law of (Dalton's Law of Partial Pressures)   The total pressure of a mixture of gases equals the sum of their partial pressures. (9.4)

Pascal (Pa)   The SI unit of pressure equal to 1 newton/m²; 133.3224 Pa $= 1$ torr. (9.2)

Pauli Exclusion Principle   No two electrons in an atom can have the same values for all four of their quantum numbers. (6.6)

Peptide Bond   The amide linkage in molecules of polypeptides. (24.5)

Percentage Composition   A list of the percentages by weight of the elements in a compound. (3.4)

Percentage Concentration   A ratio of the amount of solute to the amount of solution expressed as a percent. (13.6) Weight/weight (w/w): The grams of solute in 100 g of solution. Volume/volume (v/v): The volumes of solute in 100 volumes of solution.

Percentage Ionization   [(Amount of species ionized)/(initial amount of species)] $\times$ 100. (15.3)

Percentage by Weight   (a) The number of grams of an element combined in 100 g of a compound. (3.4) (b) The number of grams of a substance in 100 g of a mixture or solution. (13.6)

Percentage Yield   The ratio (taken as a percent) of the mass of product obtained to the mass calculated from the reaction's stoichiometry. (4.4)

Period   A horizontal row of elements in the periodic table. (2.6, 2.7)

Periodic Law   The properties of the elements are a periodic function of their atomic numbers. (2.7)

Periodic Table   A display of the symbols of all of the elements, in order of increasing atomic number, in rows (periods) and columns (groups) that organize the families of the elements into columns. (2.6, inside front cover)

Peroxide   A compound whose molecules have two oxygen atoms joined by a single bond. (19.4)

pH $-\log[H^+]$. (15.2)

Phase   A homogeneous region within a sample. (1.5)

Phase Diagram   A pressure–temperature graph on which are plotted the temperatures and the pressures at which equilibrium exists between the states of a substance. It defines regions of T and P in which the solid, liquid, and gaseous states of the substance can exist. (10.14)

Photon   A unit of energy in electromagnetic radiation equal to $h\nu$, where ν is the frequency of the radiation and h is Planck's constant. (6.2)

Photosynthesis   The use of solar energy by a plant to make high-energy molecules from carbon dioxide, water, and minerals. (5.1)

Physical Change   An event unaccompanied by a chemical reaction. (1.5)

Physical Law   A relationship between two or more physical properties of a system, usually expressed as a mathematical equation, that describes how a change in one property affects the others.

Physical Property   A property that can be specified without reference to another substance and that can be measured without causing a chemical change. (1.4)

Physical State   The condition of aggregation of a substance's formula units, whether as a solid, a liquid, or a gas. (1.6)

Pi Bond (π Bond)   A bond formed by the sideways overlap of a pair of p orbitals that concentrates electron density into two separate regions that lie on opposite sides of a plane that contains an imaginary line joining the nuclei. (8.5)

Pig Iron   The impure iron made by a blast furnace. (21.2)

pK_a $-\log K_a$. (15.4)

pK_b $-\log K_b$. (15.4)

pK_w $-\log K_w$. (15.4)

Planar Triangular Molecule   A molecule in which a central atom holds three other atoms located at the corners of an equilateral triangle and that includes the central atom at its center. (7.9)

pOH $-\log[OH^-]$. (15.2)

Polar Covalent Bond (Polar Bond)   A covalent bond in which more than half of the bond's negative charge is concentrated around one of the two atoms. (7.10)

Polar Molecule   A molecule in which individual bond polarities do not cancel and in which, therefore, the centers of density of negative and positive charges do not coincide. (7.10)

Polyatomic Ion   An ion composed of two or more atoms. (2.9)

Polydentate Ligand   A ligand that has two or more atoms that can become simultaneously attached to a metal ion. (11.10, 22.3)

Polymer   A substance consisting of macromolecules that have repeating structural units. (24.3)

Polymerization   A chemical reaction that converts a monomer into a polymer. (24.3)

Polypeptide   A polymer of α-amino acids that makes up all or most of a protein. (24.5)

Polyprotic Acid   An acid that can furnish more than one H^+ per molecule. (2.12, 15.7)

Polysaccharide   A carbohydrate whose molecules can be hydrolyzed to hundreds of monosaccharide molecules. (24.4)

Position of Equilibrium   The relative amounts of the substances on both sides of the double arrows in the equation for an equilibrium. (10.8)

Positron (^0_1p)   A positively charged particle with the mass of an electron. (23.3)

Post-transition Metal   A metal that occurs in the periodic table to the right of a row of transition elements. (2.9)

Potential Energy   Stored energy. (1.3)

Precipitate   A solid that separates from a solution usually as the result of a chemical reaction. (4.5)

Precipitation   In chemistry, the formation of a precipitate. (4.5)

Precision   How reproducible measurements are; the fineness of a measurement as indicated by the number of significant figures reported in the physical quantity. (1.11)

Pressure   Force per unit area. (9.2)

Pressure–Concentration Law   See *Vapor Pressure–Concentration Law.*

Pressure–Solubility Law (Henry's Law)   The concentration of a gas dissolved in a liquid at any given temperature is directly proportional to the partial pressure of this gas above the solution. (13.5)

Pressure–Temperature Law   The pressure of a given mass of gas is directly proportional to its Kelvin temperature if the volume is kept constant. $P \propto T$. (9.6)

Pressure–Volume Law (Boyle's Law)   The volume of a given mass of a gas is inversely proportional to its pressure if the temperature is kept constant. $V \propto 1/P$. (9.4)

Pressure–Volume Work (P–V Work)   The energy transferred as work when a system expands or contracts against the pressure exerted by the surroundings. At constant pressure, work $= P\Delta V$. (16.2)

Principal Quantum Number (n)   The quantum number that defines the principal energy levels and that can have values of 1, 2, 3, . . . , ∞. (6.5)

Products   The substances produced by a chemical reaction and whose formulas follow the arrows in chemical equations. (1.6)

Propagation Step   A step in a chain reaction for which one product must serve in a succeeding propagation step as a reactant and for which another (final) product accumulates with each repetition of the step. (18.10)

Property   A characteristic of matter. (1.4)

Propyl Group $CH_3CH_2CH_2-$. (24.2)

Protein   A macromolecular substance found in cells that consists wholly or mostly of one or more polypeptides that often are combined with an organic molecule or a metal ion. (24.5)

Proton   (a) A subatomic particle, ^1_1p or ^1_1H, that has a charge of 1+ and a mass of 1.007276 amu (1.672623×10^{-24} g) and that is found in atomic nuclei. (6.1) (b) The name often used for the hydrogen ion and symbolized as H^+. (2.3)

Proton Acceptor   A Brønsted base. (11.7)

Proton Donor   A Brønsted acid. (11.7)

Pure Substance   An element or a compound. (1.5)

Qualitative Analysis   The use of experimental procedures to determine what elements are present in a substance. (3.4)

Qualitative Observation   Observations that do not involve numerical information. (1.8)

Quantitative Analysis   The use of experimental procedures to determine the percentage composition of a compound or the percentage of a component of a mixture. (3.4)

Quantitative Observation   An observation involving a measurement and numerical information. (1.8)

Quantized   Descriptive of a discrete, definite amount as of *quantized energy.* (6.2)

Quantum   The energy of one photon. (6.2)

Quantum Mechanics   See *Wave Mechanics.* (6.5)

Quantum Number   A number related to the energy, shape, or orientation of an orbital, or to the spin of an electron. (6.3)

R   see *Gas Constant, Universal.*

Rad (rd)   A unit of radiation-absorbed dose equal to 10^{-5} J/g or 10^{-2} Gy. (23.6)

Radiation   The emission of electromagnetic energy or nuclear particles. (23.3)

Radioactive   The ability to emit various atomic radiations or gamma rays. (23.3)

Radioactive Decay   The change of a nucleus into another nucleus (or into a more stable form of the same nucleus) by the loss of a small particle or a gamma ray photon. (23.3)

Radioactive Disintegration Series   A sequence of nuclear reactions beginning with a very long-lived radionuclide and ending with a stable isotope of lower atomic number. (23.3)

Radioactivity   The emission of one or more kinds of radiation from an isotope with unstable nuclei. (23.3)

Radiological Dating   A technique for measuring the age of a geologic formation or an ancient artifact by determining the ratio of the concentrations of two isotopes, one radioactive and the other a stable decay product. (23.7)

Radionuclide   A radioactive isotope. (23.3)

Raoult's Law   See *Vapor Pressure–Concentration Law.*

Rare Earth Metals   The lanthanides. (2.7, 22.1)

Rate   A ratio in which a unit of time appears in the denominator, for example, 40 mile hr^{-1} or 3.0 mol $L^{-1}s^{-1}$. (18.3)

Rate Constant   The proportionality constant in the rate law; the rate of reaction when all reactant concentrations are 1 *M*. (18.4)

Rate-Determining Step (Rate-Limiting Step)   The slowest step in a reaction mechanism. (18.9)

Rate Law   An equation that relates the rate of a reaction to the molar concentrations of the reactants raised to powers. (18.4)

Rate of Reaction   How quickly the reactants disappear and the products form; usually expressed in units of mol $L^{-1} s^{-1}$. (18.1)

Reactant, Limiting   See *Limiting Reactant.*

Reactants   The substances brought together to react and whose formulas appear before the arrow in a chemical equation. (1.6)

Reaction Quotient (*Q*)   The numerical value of the mass action expression. See *Mass Action Expression.* (14.3)

Recombinant DNA   DNA in a bacterial plasmid altered by the insertion of DNA from another organism. (24.6)

Redox Reaction   An oxidation–reduction reaction. (12.1)

Reducing Agent   A substance that causes reduction and is itself oxidized. (12.1)

Reduction   A change in which an oxidation number decreases (becomes less positive and more negative). A gain of electrons. (12.1)

Reduction Potential   A measure of the tendency of a given half-reaction to occur as a reduction. (17.6)

Rem   A dose in rads multiplied by a factor that takes into account the variations that different radiations have in their damage-causing abilities in tissue. (23.6)

Replication   In nucleic acid chemistry, the reproductive duplication of DNA double helixes prior to cell division. (24.6)

Representative Element   An element in one of the A-groups in the periodic table. (2.7)

Resin   A raw, unfabricated polymer used to make all or most of a plastic article. (24.5)

Resonance   A concept in which the actual structure of a molecule or polyatomic ion is represented as a composite or average of two or more Lewis structures, which are called the resonance or contributing structures (and none of which has real existence). (7.7)

Resonance Energy   The difference in energy between a substance and its principal resonance (contributing) structure. (8.6)

Resonance Hybrid   The actual structure of a molecule or polyatomic ion taken as a composite or average of the resonance or contributing structures. (7.7)

Resonance Structure   A Lewis structure that contributes to the hybrid structure in resonance-stabilized systems; a contributing structure. (7.7)

Reversible Process   A process that occurs by an infinite number of steps during which the driving force for the change is just barely greater than the force that resists the change. (16.9)

Ring, Carbon   A closed-chain sequence of carbon atoms. (24.2)

RNA   Ribonucleic acid; a nucleic acid that gives ribose, phosphate ion, adenine, uracil, guanine, and cytosine when hydrolyzed. It occurs in several varieties.
*m*RNA: Messenger RNA, the director of polypeptide assembly.
*pt*RNA: Primary transcript RNA, the RNA made directly when DNA is used to specify the base sequences — both introns and exons.
*r*RNA: Ribosomal RNA, the RNA that, together with proteins, provides the polypeptide assembly sites.
*t*RNA: Transfer RNA, the carrier of amino acid units to polypeptide assembly sites. (24.6)

Roasting   Heating a sulfide ore in air to convert it to an oxide. (21.2)

Salt   An ionic compound in which the anion is not OH^- or O^{2-} and the cation is not H^+. (2.11)

Salt Bridge   A tube that contains an electrolyte that connects the two half-cells of a galvanic cell. (17.5)

Saturated Organic Compound   A compound whose molecules have only single bonds. (24.2)

Saturated Solution   A solution in which there is an equilibrium between the dissolved and the undissolved states of the solute.

A solution that holds as much solute as it can at a given temperature. (4.5)

Scientific Method   The observation, explanation, and testing of an explanation by additional experiments. (1.7)

Scientific Notation   The representation of a quantity as a decimal number between 1 and 10 multiplied by 10 raised to a power (e.g., 6.02×10^{23}). (1.12)

Secondary Quantum Number (ℓ)   The quantum number whose values can be 0, 1, 2, . . . , $n-1$, where n is the principal quantum number. (6.5)

Second Law of Thermodynamics   Whenever a spontaneous event takes place, it is accompanied by an increase in the entropy of the universe. (16.5)

Second-Order Reaction   A reaction with a rate law of the type rate $= k[A]^2$ or rate $= k[A][B]$, where A and B are reactants. (18.4, 18.5)

Semiconductor   A substance that conducts electricity weakly. (2.8)

Shell   All of the orbitals associated with a given value of n (the principal quantum number). (6.5)

SI (International System of Units)   The modified metric system adopted in 1960 by the General Conference on Weights and Measures. (1.9)

Side Reaction   A reaction that occurs simultaneously with another reaction (the main reaction) in the same mixture to produce by-products. (4.4)

Sigma Bond (σ Bond)   A bond formed by the head-to-head overlap of two atomic orbitals and in which electron density becomes concentrated along and around the imaginary line joining the two nuclei. (8.5)

Significant Figures   The number of digits in a physical quantity that are known to be certain plus one more. (1.11)

Silver Oxide Battery   A galvanic cell of about 1.5 V involving zinc and silver oxide and used when miniature batteries are needed. (17.10)

Simple Cubic Unit Cell   A cell or unit of crystal structure that has atoms, molecules, or ions only at the corners of a cube. (10.10)

Single Bond   A covalent bond in which a single pair of electrons is shared. (7.4)

Skeleton Equation   An unbalanced equation showing only the formulas of reactants and products. (12.4)

Slag   A relatively low-melting mixture of impurities that forms in the blast furnace or other furnaces used to refine metals. (21.2)

Soap   A salt of a fatty acid whose anions form micelles and which is able to hold oil and grease particles in suspension in water. (13.3)

Sol   The colloidal dispersion of a solid in a fluid. (13.1)

Solubility   The ratio of the quantity of solute to the quantity of solvent in a saturated solution that is usually expressed in units of g solute/100 g solvent at a specified temperature. (4.5)

Solubility Product Constant (K_{sp})   The equilibrium constant for the solubility of a salt that, for a saturated solution, is equal to the product of the molar concentrations of the ions, each raised to a power equal to the number of its ions in one formula unit of the salt. (14.10) See also *Acid Solubility Product*.

Solute   Something dissolved in a solvent to make a solution. (4.5)

Solution   A homogeneous mixture in which all particles are of the size of atoms, small molecules, or small ions. (1.5, 4.5, 13.1)

Solvation   The development of a cagelike network of a solution's solvent molecules about a molecule or ion of the solute. (13.2)

Solvation Energy   The enthalpy of the interaction of gaseous molecules or ions of solute with solvent molecules during the formation of a solution. (13.2)

Solvay Process   An industrial synthesis of sodium carbonate from sodium chloride, ammonia, and carbon dioxide. (21.3)

Solvent   A medium, usually a liquid, into which something (a solute) is dissolved to make a solution. (4.5)

sp Hybrid Orbital   A hybrid orbital formed by mixing one s and one p atomic orbital. The angle between a pair of sp hybrid orbitals is 180°. (8.3)

sp^2 Hybrid Orbital   A hybrid orbital formed by mixing one s and two p atomic orbitals. The angle between two sp^2 hybrid orbitals is 120°. (8.3)

sp^3 Hybrid Orbital   A hybrid orbital formed by mixing one s and three p atomic orbitals. The angle between two sp^3 hybrid orbitals is 109.5°. (8.3)

sp^3d Hybrid Orbital   A hybrid orbital formed by mixing one s, three p, and one d atomic orbital. sp^3d hybrids point to the corners of a trigonal bipyramid. (8.3)

sp^3d^2 Hybrid Orbital   A hybrid orbital formed by mixing one s, three p, and two d atomic orbitals. sp^3d^2 hybrids point to the corners of an octahedron. (8.3)

Specific Gravity   The ratio of the density of a substance to the density of water. (1.14)

Specific Heat   The quantity of heat that will raise the temperature of 1 g of a substance by 1 °C, usually in units of cal g^{-1} °C^{-1} or J g^{-1} °C^{-1}. (5.2)

Spectator Particle   An ion or molecule whose formula appears in an ionic equation identically on both sides of the arrow, that does not participate in the reaction, and that is excluded from the net ionic equation. (11.3)

Spectrochemical Series   A listing of ligands in order of their ability to produce a large crystal field splitting. (22.5)

Spin Quantum Number (m_s)   The quantum number associated with the spin of a subatomic particle and for the electron can have a value of $+\frac{1}{2}$ or $-\frac{1}{2}$. (6.6)

Spontaneous Change   A change that occurs by itself without outside assistance. (16.3)

Stability Constant   See *Formation Constant*.

Stabilization Energy   See *Resonance Energy*.

Standard Cell Potential ($E°_{cell}$)   The potential of a galvanic cell at 25 °C when all ionic concentrations are exactly 1 M and the partial pressures of all gases are 1 atm. (17.6)

Standard Conditions of Temperature and Pressure (STP)   273 K (0 °C) and 1 atm (760 torr). (9.7)

Standard Enthalpy Change ($\Delta H°$)   See *Standard Heat of Reaction.*

Standard Entropy ($S°$)   The entropy of 1 mol of a substance at 25 °C and 1 atm. (16.6)

Standard Entropy Change ($\Delta S°$)   The entropy change of a reaction when determined with reactants and products at 25 °C and 1 atm and on the scale of the mole quantities given by the coefficients of the balanced equation. (16.6)

Standard Free Energy Change ($\Delta G°$) $\Delta H° - T\Delta S°$. (16.8)

Standard Free Energy of Formation ($\Delta G_f°$)   The value of $\Delta G°$ for the formation of *one* mole of a compound from its elements in their standard states. (16.8)

Standard Heat of Combustion ($\Delta H°_{combustion}$)   The enthalpy change for the combustion of one mole of a compound under standard conditions. (5.7)

Standard Heat of Formation ($\Delta H_f°$)   The amount of heat absorbed or evolved when one mole of the compound is formed from its elements in their standard states. (5.7)

Standard Heat of Reaction ($\Delta H°$)   The enthalpy change of a reaction when determined with reactants and products at 25 °C and 1 atm and on the scale of the mole quantities given by the coefficients of the balanced equation. (5.5)

Standard Reduction Potential   The reduction potential of a half-reaction at 25 °C and 1 atm when all ion concentrations are 1 M and the partial pressures of all gases are 1 atm. (17.6)

Standard Solution   Any solution whose concentration is accurately known. (11.5)

Standard State   The condition in which a substance is in its most stable form at 25 °C and 1 atm. (5.7)

Standing Wave   A wave whose peaks and nodes do not change position. (6.5)

State Function   A function or variable whose value depends only on the initial and final states of the system and not on the path taken by the system to get from the initial to the final state. (P, V, T, H, S, and G are all state functions.) (5.3)

State of Matter   Solid, liquid, or gas. (See also *Standard State.*) (1.6)

State of a System   The set of specific values of the physical properties of a system — its composition, physical form, concentration, temperature, pressure, and volume. (5.3)

Stereoisomerism   The existence of isomers whose structures differ only in spatial orientations (e.g., geometric isomers and optical isomers). (22.4)

Stock System   A system of nomenclature that uses Roman numerals to specify oxidation states. (2.13)

Stoichiometry   A description of the relative quantities by moles of the reactants and products in a reaction as given by the coefficients in the balanced equation. (4.1)

Stopcock   A device on a buret that is used to control the flow of titrant. (11.5)

STP   See *Standard Conditions of Temperature and Pressure.*

Straight-Chain Compound   An organic compound in whose molecules the carbon atoms are joined in one continuous open-chain sequence. (24.2)

Strong Acid   An acid that is essentially 100% ionized in water. A good proton donor. An acid with a large value of K_a. (11.2)

Strong Base   Any powerful proton acceptor. A base with a large value of K_b. A metal hydroxide that dissociates essentially 100% in water. (11.2)

Strong Electrolyte   Any substance that ionizes or dissociates in water to essentially 100%. (11.2)

Structural Formula   A chemical formula that shows how the atoms of a molecule or polyatomic ion are arranged, to which other atoms they are bonded, and the kinds of bonds (single, double, or triple). (7.3)

Sublimation   The conversion of a solid directly into a gas without passing through the liquid state. (10.4)

Subshell   All of the orbitals of a given shell that have the same value of their secondary quantum number, ℓ. (6.3)

Substitution Reaction   The replacement of an atom or group on a molecule by another atom or group. (24.3)

Supercooled   The condition of a substance in its liquid state below its freezing point. (10.13)

Supercooled Liquid   A liquid at a temperature below its freezing point. An amorphous solid. (10.13)

Supercritical Fluid   A substance at a temperature above its critical temperature (10.14)

Superimposability   A test of structural chirality in which a model of one structure and a model of its mirror image are compared to see if the two could be made to blend perfectly, with every part of one coinciding simultaneously with the parts of the other. (22.4)

Supernatant Liquid   In a solid-liquid mixture, the liquid phase that normally rests on top of the solid. (25.2)

Superoxide   A compound of the O_2^- ion. (19.4)

Supersaturated Solution   Any solution that contains more solute than a saturated solution would ordinarily hold at the given temperature. (4.5)

Surface Tension   A measure of the amount of energy needed to expand the surface area of a liquid. (10.4)

Surfactant   A substance that lowers the surface tension of a liquid and promotes wetting. (10.4)

Surroundings   That part of the universe other than the system being studied and separated from the system by a real or an imaginary boundary. (5.3)

Suspension   A homogeneous mixture in which the particles of at least one component are larger than colloidal particles (>1000 nm in at least one dimension). (13.1)

System   That part of the universe under study and separated from the surroundings by a real or an imaginary boundary. (5.3)

$t_{1/2}$ See *Half-Life.*

Temperature   A measure of the hotness or coldness of something. (1.3)

Temperature–Volume Law (Charles' Law)   The volume of a given mass of a gas is directly proportional to its Kelvin temperature if the pressure is kept constant. $V \propto T$. (9.5)

Termination Step   A step in a chain reaction in which a reactive species needed for a chain propagation step disappears without helping to generate more of this species. (18.10)

Tetrahedral Molecule   A molecule with a central atom bonded to four other atoms located at the corners of an imaginary tetrahedron. (7.9)

Theoretical Yield   The yield of a product calculated from the reaction's stoichiometry. (4.4)

Theory   A tested explanation of the results of many experiments. (1.7)

Thermal Property   A physical property, like heat capacity or heat of fusion, that concerns a substance's ability to absorb heat without changing chemically. (5.2)

Thermite Reaction   The very exothermic reaction $2Al + Fe_2O_3 \rightarrow 2Fe + Al_2O_3$. (21.5)

Thermochemical Equation   A balanced chemical equation accompanied by the value of $\Delta H°$ that corresponds to the mole quantities specified by the coefficients. (5.5)

Thermochemistry   The study of the energy changes of chemical reactions. (5.3)

Thermodynamic Equilibrium Constant (K)   The equilibrium constant that is calculated from $\Delta G°$ (the standard free energy change) for a reaction at T K by the equation $\Delta G° = RT \ln K$. (16.12)

Thermodynamics   The study of the laws that govern the energy and entropy changes of physical and chemical events. (16.1)

Thermometer   A device for the measurement of temperatures. (1.10)

Third Law of Thermodynamics   For a pure crystalline substance at 0 K, $S = 0$. (16.6)

Titrant   The solution added from a buret during a titration. (11.5)

Titration   An analytical procedure in which a solution of unknown concentration is combined slowly and carefully with a standard solution until a color change of some indicator or some other signal shows that equivalent quantities have reacted. Either solution can be the titrant in a buret with the other solution being in a receiving flask. (11.5)

Torr   A unit of pressure equal to 1/760 atm. 1 mm Hg. (9.2)

Tracer Analysis   The use of radioactivity to trace the movement of something or to locate the site of the radioactivity. (23.7)

Transcription   The synthesis of mRNA at the direction of DNA. (24.6)

trans **Isomer**   A stereoisomer whose uniqueness lies in having two groups that project on opposite sides of a reference plane. (22.4, 24.3)

Transition Elements   The elements located between Groups IIA and IIIA in the periodic table. (2.7, 22.1)

Transition Metals   The transition elements. (2.7, 22.1)

Transition State   The brief moment during an elementary process in a reaction mechanism when the species involved have acquired the minimum amount of potential energy needed for a successful reaction, an amount of energy that corresponds to the high point on a potential energy diagram of the reaction. (18.7)

Transition State Theory   A theory about the formation and breakup of activated complexes. (18.7)

Translation   The synthesis of a polypeptide at the direction of a molecule of mRNA. (24.6)

Transmutation   The conversion of one isotope into another. (23.5)

Transuranium Elements   Elements 93 and higher. (2.7)

Traveling Wave   A wave whose peaks and nodes move. (6.5)

Triacylglycerol   An ester of glycerol and three fatty acids. (24.4)

Trigonal Bipyramidal Molecule   A molecule with a central atom holding five other atoms that are located at the corners of a trigonal bipyramid (two three-sided pyramids that share a common face. (7.9)

Triple Bond   A covalent bond in which three pairs of electrons are shared. (7.4)

Triple Point   The temperature and pressure at which the liquid, solid, and vapor states of a substance can coexist in equilibrium. (10.14)

Triprotic Acid   An acid that can furnish three H^+ ions per molecule. (2.12)

Tritium (^3_1H)   The isotope of hydrogen with a mass number of 3. (19.3)

Tyndall Effect   The scattering of light by colloidally dispersed particles that gives a milky appearance to the mixture. (13.1)

Uncertainty Principle   There is a limit to our ability to measure a particle's speed and position simultaneously. (6.10)

Unit Cell   The smallest portion of a crystal than can be repeated over and over in all directions to give the crystal lattice. (10.12)

Universal Gas Constant (R)   The ratio of PV to nT for gases, 0.08205 L atm/K mol. (9.7)

Unsaturated Compound   A compound whose molecules have one or more double or triple bonds. (24.2)

Unsaturated Solution   Any solution with a concentration less than that of a saturated solution of the same solute and solvent. (4.5)

Vacuum   An enclosed space containing no matter whatsoever. A *partial vacuum* is an enclosed space containing a gas at a very low pressure. (9.2)

Valence Bond Theory   A theory of covalent bonding that views a bond as being formed by the sharing of one pair of electrons between two overlapping atomic or hybrid orbitals. (8.1)

Valence Electrons   The electrons of an atom in its valence shell that participate in the formation of chemical bonds. (6.9)

Valence Shell   The electron shell with the highest principal quantum number, n, that is occupied by electrons. (6.9)

Valence Shell Electron Pair Repulsion Theory (VSEPR Theory)   The geometry of a molecule at any given central atom is determined by the repulsions of both bonding and nonbonding (lone pair) electrons in the valence shell. (7.9)

Van der Waals' Equation   An equation of state for a real gas that corrects the ideal gas law for the excluded volume of the gas and for intermolecular attractions. (9.12)

Van't Hoff Factor   The ratio of the observed freezing point depression to the value calculated on the assumption that the solute dissolves as un-ionized molecules. (13.11)

Vapor Pressure   The pressure exerted by the vapor above a liquid (usually referring to the *equilibrium* vapor pressure when the vapor and liquid are in equilibrium with each other). (9.4, 10.7)

Vapor Pressure–Concentration Law. (Raoult's Law)   The vapor pressure of one component above a mixture of molecular compounds equals the product of its vapor pressure when pure and its mole fraction. (13.7)

Visible Spectrum   That region of the electromagnetic spectrum whose frequencies can be detected by the human eye. (6.2)

Vital Force Theory   The belief (prior to 1828) that the synthesis of an organic compound could take place with the aid of a vital force present only in living systems. (24.1)

Volatile   Descriptive of a liquid that has a low boiling point, a high vapor pressure at room temperature, and therefore evaporates easily. (13.7)

Volt (V)   The SI unit of electromotive force or emf in joules per coulomb. 1 V = 1 J/C. (17.6)

Voltaic Cell   See *Galvanic Cell.*

VSEPR Theory   See *Valence Shell Electron Pair Repulsion Theory.*

Wavelength (λ)   The distance between crests in the wavelike oscillations of electromagnetic radiations. (6.1)

Wave Mechanics   A theory of atomic structure based on the wave properties of matter. (6.5)

Weak Acid   An acid with a low percentage ionization in solution; a poor proton donor; an acid with a low value of K_a. (11.2)

Weak Base   A base with a low percentage ionization in solution; a poor proton acceptor; a base with a low value of K_b. (11.2)

Weak Electrolyte   A substance that has a low percentage ionization or dissociation in solution. (11.2)

Weighing   The operation of measuring the mass of something. (1.10)

Weight   The force with which something is attracted to the earth by gravity. (1.3)

Wetting   The spreading of a liquid across a solid surface. (10.4)

X Ray   A stream of very high-energy photons emitted by substances when they are bombarded by high-energy beams of electrons or are emitted by radionuclides that have undergone K-electron capture. (23.3)

Yield, Percentage   The ratio, given as a percent, of the quantity of product actually obtained in a reaction to the theoretical yield. (4.4)

Yield, Theoretical   The quantity of a product calculated by the stoichiometry of the reaction. (4.4)

Zero-Order Reaction   A reaction whose rate is independent of the concentration of any reactant. (15.4)

Zinc–Carbon Dry Cell   A galvanic cell of about 1.5 V involving zinc and manganese dioxide under mildly acidic conditions. (17.10)

Photo Credits

Chemicals in Use 4
Fig. 4a: OPC, Inc.   Fig. 4b: General Chemical Company.

Chapter 6
Opener: Courtesy Eaton Corporation.   Page 183: Courtesy General Cable Corporation.   Page 186: *Nature*, Courtesy AIP Niels Bohr Library.   Page 189: Courtesy AIP Niels Bohr Library, W. F. Meggers Collection.   Courtesy Perkin Elmer Corporation.   Page 190: From "The Gift of Color," Eastman Kodak Company.   Fig. 6.14: Bruce Coleman.   Page 198: Courtesy Perkin Elmer Corporation.   Page 203: E. R. Degginger/Bruce Coleman.

Chemicals in Use 5
Fig. 5a: (left) William K. Sacco/Yale Peabody Museum; (right) Courtesy TRW.

Chapter 7
Opener: Smithsonian Institution.

Chapter 8
Opener: E. R. Degginger/Bruce Coleman.

Chapter 9
Opener: Swanson.   Page 308 and Fig. 9.1: OPC, Inc. Page 310: B. B. C. Hulton Picture Library.   Page 313: Courtesy of the New York Public Library, Picture Collection. Fig. 9.7: Joe Viesti.   Fig. 9.10: Joseph P. Kakareka.   Page 332: OPC, Inc.

Chapter 10
Opener: Dan Guravich/Photo Researchers.   Page 355: Jen and Des Bartlett.   Page 356 (top): Courtesy of Melroe Company: Fig. 10.7: Robert J. Capece.   Page 357: Adrian Davies/Bruce Coleman.   Page 358: OPC, Inc.   Page 362: Robert J. Capece.   Page 366: Courtesy Mountain House. Page 368: Courtesy Corning.   Fig. 10.17: Robert J. Capece. Page 370: Eric Grave/Science Source/Photo Researchers. Page 373: (left) Courtesy Professor M. H. F. Wilkins, Biophysics Department, Kings College, London; (right) Courtesy John Mach Jr., photo by Irving Geis.   Page 376: Robert J. Capece.

Chemicals in Use 8
Fig. 8a: P. Belzeaux/Rapho-Photo Researchers.   Fig. 8b: Kimble Glass Company.

Chapter 11
Opener: Steven Groer.   Page 391: The Bettmann Archive. Page 395: (top) Courtesy Permutit Company, Division of Sybron Corporation.   Figs. 11.1 a–c and 11.2: OPC, Inc. Page 401: Robert J. Capece.   Fig. 11.4: OPC, Inc.   Page 404: The Denver Post.   Fig. 11.5: Peter Lerman.   Fig. 11.6: OPC, Inc.   Fig. 11.7 a and b: Peter Lerman.   Fig. 11.8: OPC, Inc.

Chemicals in Use 9
Fig. 9b: Fred E. Mang Jr./National Park Service.

Chapter 12
Opener: Bureau of Engraving and Printing.   Page 459 and Fig. 12.1: OPC, Inc.   Figs. 12.2, 12.3 a–c, and 12.4: Peter Lerman.   Page 464 and Fig. 12.5 a–c: OPC, Inc.   Page 469: (top) Robert J. Capece.   Fig. 12.7: Courtesy Bethlehem Steel.   Fig. 12.8 and pages 473 and 475: Peter Lerman.

Chemicals in Use 10
Fig. 10a: H. W. Kitchen/National Audubon Society.   Fig. 10b: USDA.

Chapter 13
Opener: M. Timothy O'Keefe/Bruce Coleman.   Fig. 13.1c: Courtesy International Equipment Company.   Fig. 13.5: OPC, Inc.   Page 497: USDA.   Page 510: Scott Slobodian.

Chapter 14
Opener: Jessica Anne Ehlers/Bruce Coleman.

Qualitative Topic 1
Figs. 1–6: OPC, Inc.

Chapter 15
Opener: Mark Sherman/Bruce Coleman.   Fig. 15.2: Peter Lerman.   Fig. 15.3: Courtesy Fisher Scientific.   Figs. 15.8 a–d: OPC, Inc.

Qualitative Topic 2
Fig. 1: OPC, Inc.

Chapter 16
Opener: USDA.   Fig. 16.3a: Courtesy Burton Corporation. Fig. 16.3b: Courtesy U. S. Fish and Wildlife Service.   Photo by Bruce Halstead.   Fig. 16.3c: Lowell J. Georgia/Photo Researchers.   Page 666: United Press International.   Fig. 16.6: Susan Leavines/Photo Researchers.

Chapter 17
Opener: Peter C. Poulides.   Fig. 17.3: Jim Brady and Kathy Bendo.   Page 699: OPC, Inc.   Page 700: The Royal Institution.   Page 704: Kaiser Aluminum.   Page 705: Courtesy Kennecott.   Fig. 17.11: Jim Brady and Kathy Bendo.   Page 712: Steve McCutcheon.   Page 725: Fisher Scientific.   Fig. 17.18: OPC, Inc.

Chapter 18
Opener: Bruce Coleman.   Page 739: (top) Courtesy Miles Laboratories.   Fig. 18.1: USDA.   Page 766: Leonard Lee Rue III/Photo Researchers.   Page 769: (left) Courtesy American Petroleum Institute, (right) Courtesy Englehard Corporation.   Fig. 18.13: Courtesy AC Spark Plug.

Chapter 19
Opener: Courtesy BBDO.   Fig. 19.1: Ginger Chih/Peter Arnold.   Page 784 (top): Courtesy AT & T Bell Labs. Fig. 19.2a: Peter Lerman.   Page 787: OPC, Inc.   Page 790 (left): USDA; (right) Robert J. Capece.   Page 791: Courtesy NASA.   Fig. 19.3 and page 793 (right): OPC, Inc.

Fig. 19.5: Courtesy NASA.   Fig. 19.6: Uta Hoffman/German Information Center.   Fig. 19.7: OPC, Inc.   Fig. 19.8: Bettmann Newsphotos.   Page 799: (top) Robert J. Capece.   Fig. 19.9: OPC, Inc.   Fig. 19.10: Thomas Hovland/Grant Heilman.   Fig. 19.11: OPC, Inc.   Fig. 19.12: Peter Lerman.   Page 808: OPC, Inc.   Page 810: (top) IRECO; (bottom) Courtesy US Navy.   Photo by PHC. Jeff Hilton.   Fig. 19.14: Peter Arnold.

Chapter 20

Opener: New York State Department of Environmental Conservation.   Fig. 20.1: Yale Peabody Museum.   Fig. 20.3: Leo Touchet.   Fig. 20.4: OPC, Inc.   Fig. 20.5: Peter Lerman.   Fig. 20.8: Uta Hoffman.   Fig. 20.9: Schmidt-Thomsen, Landesdenkmalamt, Westfallen-Lippe, Munster, West Germany.   Page 832: Safra Nimrod.   Fig. 20.10: OPC, Inc.   Page 835: (right) Courtesy of Dr. Noble R. Usherwood, Potash and Phosphate Institute.   Fig. 20.11: Peter Lerman.   Fig. 20.13: Courtesy USDA.   Page 848: (top) Peter Lerman, (top and bottom center) Yale Peabody Museum, (bottom) Brian J. Skinner.   Figs. 20.15 and 20.18, and pages 852–855: OPC, Inc.   Fig. 20.20: Brian J. Skinner.   Fig. 20.22: Smithsonian Institution.   Page 863: OPC, Inc.   Page 865: (top) Yale Peabody Museum, (bottom) Mula and Haramaty/Phototake.

Chemicals in Use 14

Fig. 14*b*: Safra Nimrod.   Fig. 14*c*: Courtesy of H. R. Allcock, Pennsylvania State University.

Chapter 21

Opener: Courtesy Orgo-Thermit, Inc.   Fig. 21.1: Courtesy International Salt Company.   Page 877: (top) Smithsonian Institution.   Fig. 21.2: Courtesy NOAA.   Page 878: (bottom) Peter B. Kaplan/Photo Researchers.   Fig. 21.4: Courtesy TVA.   Photo by Kollar.   Page 880 and Fig. 21.5*b*: Courtesy Bethlehem Steel.   Page 885: (left) Dan McCoy/Rainbow; (right) Courtesy International Salt Company.   Page 889: OPC, Inc.   Fig. 21.8*a–c*: Yoav/Phototake.   Fig. 21.9: H. Stern/Photo Researchers.   Page 892: Courtesy Dow Chemical.   Page 893: OPC, Inc.   Fig. 21.12: Lester V. Bergman & Assoc.   Fig. 21.13*a–c*: Yoav/Phototake.   Page 897: OPC, Inc.   Page 898: Courtesy NASA.   Page 899: (top) Robert J. Capece.   Fig. 21.15: (bottom) OPC, Inc.   Fig. 21.16: (left) R. Rowan/Photo Researchers; (right) Runk-Schoenberger/Grant Heilman.   Fig. 21.17: Peter Lerman.   Page 905 (top) and Fig. 21.19: OPC, Inc.

Chemicals in Use 15

Fig. 15*a*: Courtesy Bethlehem Steel Corporation.

Chapter 22

Opener: Wesley Bocxe/Photo Researchers.   Figs. 22.4, 22.6, and pages 919–922: OPC, Inc.   Page 923: Grant Heilman/Grant Heilman Photography.   Page 925 and Fig. 22.7*a*: OPC, Inc.   Fig. 22.7*b*: Safra Nimrod.   Fig. 22.8: Peter Lerman.   Pages 927, 929, and 930: OPC, Inc.   Page 932: William K. Sacco/Peabody Museum, Yale.   Page 936: OPC, Inc.   Figs. 22.13, 22.15, and page 938: Robert J. Capece.   Page 940: Courtesy Polaroid Corp.   Fig. 22.18: Jim Brady and Kathy Bendo.

Qualitative Topic 3

Figs. 1–5: OPC, Inc.

Chapter 23

Opener: Dr. R. L. McCrory, Laboratory for Laser Energetics. Page 958: Philippe Halsman.   Fig. 23.11: Brookhaven National Laboratory.   Fig. 23.13: Courtesy Lawrence Radiation Laboratory.   Fig. 23.14: Courtesy Nuclear Equipment Chemical.   Fig. 23.15*a*: Courtesy Landaur.   Figs. 23.16 and 23.17: Courtesy General Electric—Medica Systems Group.   Fig. 23.18: Brookhaven National Laboratory. Fig. 23.21: Courtesy Oak Ridge National Laboratory.

Chapter 24

Opener: Didier Givois/Photo Researchers.   Pages 1015–1022 and Fig. 24.1: Robert J. Capece.   Page 1024: Courtesy Regal Ware.

Chemical in Use 16

Fig. 16*a*: Courtesy Wine Institute.   Fig. 16*b*: The Metropolitan Museum of Art, Harris Brisbane Dick Fund, 1932.

Chapter 25

Fig. 25.3 and page 1062: OPC, Inc.   Figs. 25.5, 25.8, 25.9, page 1104, and Fig. 25.16: Courtesy J. A. Beran, *LABORATORY MANUAL FOR FUNDAMENTALS OF CHEMISTRY*, third edition, John Wiley & Sons, 1988.

Index

RELATIONSHIPS AMONG UNITS
(Values in boldface are exact.)

Length
1 in. = **2.54** cm
1 ft = **30.48** cm
1 yd = **91.44** cm
1 mi = **5280** ft
1 ft = **12** in.
1 yd = **36** in.

Volume
1 liq. oz = **29.57353** mL
1 qt = **946.352946** mL
1 gallon = **3.785411784** L
1 gallon = **4** qt = **8** pt
1 qt = **2** pt = **32** liq. oz

Mass
1 oz = **28.349523125** g
1 lb = **453.59237** g
1 lb = **16** oz

Pressure
1 atm = **760** torr
1 atm = **101,325** Pa
1 atm = 14.696 psi (lb/in.^{2})
1 atm = 29.921 in. Hg

Energy
1 cal = **4.184** J
1 eV = 1.6022×10^{-19} J
1 eV/molecule = 96.49 kJ/mol
1 eV/molecule = 23.06 kcal/mol
1 J = 1 kg m^2 s^{-2} = 10^7 erg

PHYSICAL CONSTANTS

Rest mass of electron m_e = 0.0005485712 amu ($9.1093897 \times 10^{-28}$ g)
Rest mass of proton m_p = 1.00727605 amu ($1.67262305 \times 10^{-24}$ g)
Rest mass of neutron m_n = 1.008665 amu (1.674954×10^{-24} g)
Electronic charge e = $1.60217733 \times 10^{-19}$ C
Atomic mass unit
(amu or u) = $1.6605665 \times 10^{-24}$ g
Gas constant R = 0.082057 L atm mol^{-1} K^{-1}
= 8.31441 J mol^{-1} K^{-1}
= 1.9872 cal mol^{-1} K^{-1}
Molar volume of an
ideal gas = 22.41383 L (at STP)
Avogadro's number = 6.022137×10^{23} things/mol
Speed of light in a
vacuum c = 2.99792458×10^8 m s^{-1}
Planck's constant h = $6.6260755 \times 10^{-34}$ J s
Faraday constant $\mathscr{F}$ = 96,485.309 C mol^{-1}

LABORATORY REAGENTS

Values are for the average concentrated reagents available commercially.

Reagent	Percent (w/w)	Mole solute / Liter solution	Gram solute / 100 mL solution
NH_3	29	15	26
$HC_2H_3O_2$	99.7	17	105
HCl	37	12	44
HNO_3	71	16	101
H_3PO_4	85	15	144
H_2SO_4	96	18	177